GENERAL CHAIR'S WELCOME

Welcome to the vibrant city of New Orleans and the 39th Design Automation Conference (DAC). DAC yearly attracts thousands of engineers who are eager to understand the latest developments in electronic design automation (EDA) as well as to hear about the most exciting chip and system designs. DAC's primary mission is to facilitate technical interchanges among design automation researchers and developers, the engineers who use DA systems and the vendors who provide both DA systems and silicon. This is primarily enabled by bringing together a wide variety of outstanding individuals in each of these areas, representing a true diversity of backgrounds - ranging from students to chief executive officers.

DAC is the premier conference for the presentation of research and development work on tools for designing chips and systems, and new methodologies for circuit and embedded system designs. It is the preferred conference for presenting next generation electronic circuits and system design experiences. DAC continues to be the forum where the EDA industry exhibits leading edge products and makes new product announcements. This year the conference continues to highlight the convergence of embedded systems software design tools and EDA hardware design tools in support of large complex systems-on-a-chip. Besides a strong presence in the technical program (through a special track, keynote speakers, panels, and invited presentations), the "Embedded System Showcase", on the exhibit floor, highlights companies with offerings in this area.

More than 225 companies will be exhibiting the latest commercially available products at DAC this year. Many of these companies are hosting demo suites to provide a more complete demonstration of their product offerings. For the first time this year, DAC is featuring "hands-on" tutorials in which vendors instruct attendees in the capabilities of their tools. Another first this year is the DAC Pavilion, a DAC sponsored booth which brings highlights of the technical program, interviews and lively discussion topics to those on the exhibit floor.

These Proceedings record an outstanding technical program which was assembled under the leadership of this year's Co-Program Chairs, Luciano Lavagno and David Blaauw, and Panel Chair, Sharad Malik. This year, 491 papers from around the world were submitted and reviewed by over 600 EDA and system design professionals. These reviews, along with a detailed examination by the Technical Program Committee, yielded the 147 papers included in these Proceedings and presented at the conference. These papers are complemented by eight panel sessions. A number of special sessions featuring renowned invited speakers embrace emerging and hot topics in the world of electronic design and its automation.

I want to express my sincere appreciation to all those who contributed to making DAC the premier conference for electronic design technology: the Authors, Speakers, Reviewers, Session Organizers and Session Chairs. A particular thanks to the DAC Executive Committee, the Technical Program Committee, and the Exhibitor Liaison Committee for their dedication. Also to the outstanding efforts of MP Associates, Inc. who consistently ensure that the conference and the exhibits run like a well-oiled machine.

DAC is sponsored by ACM/SIGDA, IEEE Circuits and Systems Society, and the EDA Consortium. Their members represent the breadth and depth of the EDA community, and we are thankful for their continued and active support. Also this year, for the first time, the conference is being held in technical co-operation with the IEEE Solid State Circuits Society, further representing the designer community.

We wish you a very productive week in this vibrant and eclectic city. We trust that you will find these Proceedings to be a valuable resource and reference for many years to come.

Bryan Ackland
General Chair
39th Design Automation Conference

PROCEEDING OF THE 39TH DESIGN AUTOMATION CONFERENCE®

The Association for Computing Machinery
1515 Broadway
New York, New York 10036

Notice to Past Authors of ACM-Published Articles

ACM intends to create a complete electronic archive of all articles and/or other material previously published by ACM. If you have written a work that was previously published by ACM in any journal or conference proceedings prior to 1978, or any SIGNewsletter at any time, and you do NOT want this work to appear in the ACM Digital Library, please inform permissions@acm.org, stating the title of the work, the author(s), and where and when published.

ACM Order Number 477020
ISBN 1-58113-461-4

IEEE Catalog Number 02CH37324

Library of Congress Number 85-644924

ISSN 0738-100X

Additional copies may be ordered prepaid from:

ACM Order Department
P.O. Box 11405
New York, New York, 10286-1405

(U.S.A. and Canada)
Fax: 1-800-342-6626

(all other countries)
Fax: +1-212-944-1318
E-mail: orders@acm.org

Additional copies of this publication are available from:

IEEE Service Center
P.O. Box 1331
445 Hoes Lane
Piscataway, NJ 08855-1331

1-800-678-IEEE
1-732-981-1393
1-732-981-9667 (Fax)
833-233 (Telex)
customer.service@ieee.org (email)

Printed in the U.S.A.

EXECUTIVE COMMITTEE

GENERAL CHAIR

Bryan Ackland
Agere Systems, Inc.
101 Crawfords Corner Rd., Rm. 4E-508
Holmdel, NJ 07733-1900
(732) 949-7248
bda@agere.com

VICE/FINANCE CHAIR

Ian Getreu
3450 SW Sherwood Pl.
Portland, OR 97201
(503) 222-7330
getreu@aol.com

TECHNICAL PROGRAM CO-CHAIR (METHODS)
David Blaauw
Univ. of Michigan
Dept. of EECS
1301 Beal Ave.
Ann Arbor, MI 48109-2122
(734) 763-4526
blaauw@umich.edu

TECHNICAL PROGRAM CO-CHAIR (TOOLS)
Luciano Lavagno
Cadence Berkeley Labs.
2001 Addison St., 3rd Fl.
Berkeley, CA 94704
(510) 647-2810
luciano@cadence.com

NEW INITIATIVES CHAIR
Ivo Bolsens
Xilinx, Inc.
2100 Logic Dr.
San Jose, CA 95124
(408) 626-5931
ivo.bolsens@xilinx.com

PAST CHAIR
Jan Rabaey
Univ. of California
EECS Dept.-Cory Hall
Berkeley, CA 94720
(510) 666-3111
jan@eecs.berkeley.edu

TUTORIAL CHAIR
Ellen M. Sentovich
Cadence Berkeley Labs.
2001 Addison St., 3rd Fl.
Berkeley, CA 94704
(510) 647-2807
ellens@cadence.com

PANEL CHAIR
Sharad Malik
Princeton Univ.
Dept. of EE
Princeton, NJ 08544
(609) 258-4625
sharad@ee.princeton.edu

ELECTRONIC MEDIA CHAIR
Steven P. Levitan
Univ. of Pittsburgh
EE Dept., 348 Benedum Eng. Hall
Pittsburgh, PA 15261
(412) 648-9663
steve@ee.pitt.edu

EDA INDUSTRY CHAIR
William H. Joyner, Jr.
Semiconductor Research Corp.
PO Box 12053
Research Triangle Park, NC 27709
(919) 941-9472
joyner@src.org

EXECUTIVE COMMITTEE (cont.)

PUBLICITY CHAIR
Nanette Collins
37 Symphony Rd., Unit A
Boston, MA 02115
(617) 437-1822
nanette@nvc.com

EUROPE/MIDDLE EAST REPRESENTATIVE
Ahmed A. Jerraya
TIMA Lab.
46, Avenue Felix Viallet
Grenoble Cedex, France F-38031
(33) 476-574759
ahmed.jerraya@imag.fr

ASIA/S. PACIFIC REPRESENTATIVE
Hidetoshi Onodera
Kyoto Univ.
Communications/Computer
Engineering Dept.
Sakyo-ku Kyoto 606-8501 Japan
(81) 75-753-5314
onodera@i.kyoto-u.ac.jp

ACM/SIGDA REPRESENTATIVE
Robert Walker
Kent State Univ.
Dept. of Computer Science
351 Math & CS Bldg.
Kent, OH 44242
(330) 672-9105
walker@cs.kent.edu

IEEE CAS REPRESENTATIVE
Giovanni De Micheli
Stanford Univ.
Gates Computer Science Bldg.
Rm. 333
Stanford, CA 94305-9030
(650) 725-3632
nanni@galileo.stanford.edu

EDA CONSORTIUM REPRESENTATIVE
Larry Eberle
Synopsys, Inc.
700 E. Middlefield Rd.
Mountain View, CA 94043-4033
(650) 548-4418
leberle@synopsys.com

CONFERENCE MANAGER
Kevin Lepine
MP Associates, Inc.
5305 Spine Rd., Ste. A
Boulder, CO 80301
(303) 530-4562
kevin@dac.com

EXHIBITS MANAGER
Lee Wood
MP Associates, Inc.
5305 Spine Rd., Ste. A
Boulder, CO 80301
(303) 530-4562
lee@dac.com

TECHNICAL PROGRAM COMMITTEE

David Blaauw
Technical Program Co-Chair
Univ. of Michigan, Dept. of EECS
1301 Beal Ave.
Ann Arbor, MI 48109-2122
(734) 763-4526
blaauw@umich.edu

Luciano Lavagno
Technical Program Co-Chair
Cadence Berkeley Labs.
2001 Addison St., 3rd Fl.
Berkeley, CA 94704
(510) 647-2810
luciano@cadence.com

Miron Abramovici
Agere Systems, Inc.
600 Mountain Ave., Rm. 2C-225
Murray Hill, NJ 07974-0636
(908) 582-3933
miron@agere.com

Charles J. Alpert
IBM Corp.
11400 Burnet Rd., MS 904-6G016
Austin, TX 78758
(512) 838-1045
alpert@austin.ibm.com

Luca Benini
DEIS Univ. di Bologna
viale risorgimento 2
Bologna, Italy 40136
(39) 051-202-3782
lbenini@deis.unibo.it

Jo Dale Carothers
The Univ. of Arizona, Dept. of ECE
Tucson, AZ 85721
(520) 621-8733
carothers@ece.arizona.edu

Mustafa Celik
Monterey Design Systems
894 Ross Dr., Ste. 100
Sunnyvale, CA 94089
(408) 548-4663
celik@mondes.com

Chaitali Chakrabarti
Arizona State Univ., Dept. of EE
Tempe, AZ 85287-5706
(480) 965-9516
chaitali@asu.edu

Kwang-Ting (Tim) Cheng
Univ. of California, Dept. of ECE
Santa Barbara, CA 93106
(805) 893-7294
timcheng@ece.ucsb.edu

Abhijit Dharchoudhury
Intel Corp.
1501 S. Mopac Expy., Ste. 400
Austin, TX 78746
(512) 314-0093
abhijit.dharchoudhury@intel.com

Marco Di Natale
Univ. of Pisa, Dept. of CE
Scuola Superiore S. Anna via Carducci, 40
Pisa, Italy 56100
(39) 050-883-453
marco@ssup.it

Limor Fix
Intel Semiconductors Ltd.
MTM Ind. Ctr., PO Box 498
Haifa, Israel 31000
(972) 4-865-5256
limor.fix@intel.com

Masahiro Fujita
Univ. of Tokyo, Dept. of EE
7-3-1 Hongo, Bunkyo-Ku
Tokyo, Japan 113-8656
(81) 3-5841-6673
fujita@ee.t.u-tokyo.ac.jp

Georges G. Gielen
Katholieke Univ., ESAT-MICAS
Kasteelpark Arenberg 10
Leuven, Belgium B-3001
(32) 16-321-047
georges.gielen@esat.kuleuven.ac.be

Soha M. Hassoun
Tufts Univ., Dept. of EECS
161 College Ave.
Medford, MA 02421
(617) 627-5177
soha@eecs.tufts.edu

Xiaobo (Sharon) Hu
Univ. of Notre Dame, Dept. of CSE
Notre Dame, IN 46556
(219) 631-6015
shu@cse.nd.edu

Ahmed A. Jerraya
TIMA Lab.
46, avenue Felix Viallet
Grenoble Cedex, France F-38031
(33) 476-574759
Ahmed.Jerraya@imag.fr

Seiji Kajihara
Kyushu Institute of Tech., Dept. of CSE
680-4 Kawazu
Lizuka, Japan 820-8502
(81) 948-29-7665
kajihara@cse.kyutech.ac.jp

TECHNICAL PROGRAM COMMITTEE (cont.)

Timothy Kam
Intel Corp.
Strategic CAD Labs., JF4-211
5200 NE Elam Young Pkwy.
Hillsboro, OR 97124-6497
(503) 264-7536
tkam@ichips.intel.com

Kurt Keutzer
Univ. of California, Dept. of EECS
566 Cory Hall, #1770
Berkeley, CA 94720-1770
(510) 642-9267
keutzer@eecs.berkeley.edu

Krzysztof Kuchcinski
Lund Institute of Tech.
Dept. of CS, Box 118
Lund, Sweden 221 00
(46) 46-222-3414
krysztof.kuchcinski@cs.lth.se

Tadahiro Kuroda
Keio Univ., Dept. of EE
Hiyoshi 3-14-1, Kohoku-ku
Yokohama, Japan 223-8522
(81) 45-566-1534
kuroda@elec.keio.ac.jp

Enrico Macii
Politecnico di Torino
Dip. Di Elettronica e Info.
Corso Duca degli Abruzzi 24
Torino, Italy 10129
(39) 011-564-7074
enrico@athena.polito.it

T.M. Mak
Intel Corp.
2200 Mission College Blvd., SC12-604
Santa Clara, CA 95052
(408) 765-4543
t.m.mak@intel.com

Malgorzata Marek-Sadowska
Univ. of California, Dept. of ECE
Santa Barbara, CA 93106
(805) 893-2721
mms@ece.ucsb.edu

Grant E. Martin
Cadence Design Systems, Inc.
555 River Oaks Pkwy.
San Jose, CA 95134
(510) 647-2804
gmartin@cadence.com

Yusuke Matsunaga
Kyushu Univ., Dept. of CS and CE
6-1 Kasuga Koen, Kasuga
Fujuoka, Japan 816-8580
(81) 92-583-7621
matsunaga@c.csce.kyushu-u.ac.jp

Kartikeya Mayaram
Oregon State Univ.,
Dept. of ECE, Owen 220
Corvallis, OR 97331
(541) 737-2972
karti@ece.orst.edu

Renu Mehra
Synopsys, Inc.
700 E. Middlefield Rd.
Mountain View, CA 94043
(650) 584-1516
renu@synopsys.com

Noel Menezes
Intel Corp.
2111 NE 25th Ave., MS JFT-104
Hillsboro, OR 97124
(503) 264-8259
noel.menezes@intel.com

Shin-ichi Minato
NTT Network Innovation Labs.
1-1 Hikarinooka
Yokosuka-shi, Japan 239-0847
(81) 468-59-3021
minato@ieee.org

Steven M. Nowick
Columbia Univ., Dept. of CS
New York, NY 10027
(212) 939-7000
nowick@cs.columbia.edu

Ralph Otten
Delft Univ. of Tech.
ITS/ENS, Mekelweg 4
Delft, The Netherlands CD2628
(31) 15-278-1600
otten@akebono.ics.ele.tue.nl

Joel R. Phillips
Cadence Berkeley Labs.
555 River Oaks Pkwy., MS 1A1
San Jose, CA 95134
(408) 944-7983
jrp@cadence.com

Carl Pixley
Synopsys, Inc.
19500 NW Gibbs Dr.
Beaverton, OR 97006
(503) 748-2405
cpixley@synopsys.com

Miodrag Potkonjak
Univ. of California, Dept. of CS
3532G Boelter Hall
Los Angeles, CA 90095-1596
(310) 825-0790
miodrag@cs.ucla.edu

TECHNICAL PROGRAM COMMITTEE (cont.)

PANEL SUB-COMMITTEE

Dennis Brophy
Model Technology, Inc.
10450 SW Nimbus Ave., Bldg. R-B
Portland, OR 97223-4347
(503) 526-1694
dennisb@model.com

Deirdre Hanford
Synopsys, Inc.
700 E. Middlefield Rd.
Mountain View, CA 94043-4033
(650) 584-4201
ryan@synopsys.com

Andrew B. Kahng
Univ. of California at San Diego
Dept. of CSE, MC-C-014
La Jolla, CA 92093-0114
(858) 822-4884
abk@cs.ucsd.edu

Tokinori Kozawa
STARC
Onariman BN Bldg.
6 chome 16-10, Shimbashi
Minatoku, 105-0004, Japan
(81) 33-436-1250
kozawa@starc.or.jp

Philippe Magarshack
STMicroelectronics, CRD Grp. VP
850 rue Jean Monnet
Crolles Cedex, France 38926
(33) 4-769-26498
philippe.magarshack@st.com

Shishpal S. Rawat
Intel Corp.
MS FM5-108
Folsom, CA 95630
(916) 356-6639
shishpal.s.rawat@intel.com

Rob A. Rutenbar
Carnegie Mellon Univ.
2118 Hammerschlag Hall
5000 Forbes Ave.
Pittsburgh, PA 15213-3890
(412) 268-3334
rutenbar@ece.cmu.edu

STUDENT DESIGN CONTEST JUDGES

Robert C. Frye
Agere Systems, Inc.
600 Mountain Ave.,
Rm: 2T-204
Murray Hill, NJ 07974
(908) 582-5353
rcf@agere.com

Shane Bell
Intel Corp.
334 South St.
SHR1-1/S30
Shrewsbury, MA 01545
(508) 841-1613
shane.bell@intel.com

Jim Goodman
Lumic Electronics Inc.
18 Antares Dr., Ste. 200
Ottawa, ON, Canada K2E 1A9
(613) 224-9926
jimg@lumictech.com

Ivan Kourtev
Univ. of Pittsburgh
Dept. of EE
336 Benedum Hall
Pittsburgh, PA 15261
(412) 624-9088
ivan@ee.pitt.edu

Steve Maciejewski
Advanced Micro Devices
One AMD Pl.
PO Box 3453, MS 363
Sunnyvale, CA 94088-3453
(408) 774-7763
stevenj.maciejewski@amd.com

H. Alan Mantooth
Univ. of Arkansas
Dept. of EE
Bell Engr. Ctr. 3217
Fayetteville, AR 72701
(479) 575-4838
mantooth@engr.uark.edu

Grant Martin
Cadence Design Systems, Inc.
2001 Addison St., 3rd Fl.
Berkeley, CA 94704
(510) 647-2804
gmartin@cadence.com

EXHIBITOR LIAISON COMMITTEE

Bryan Ackland
General Chair, 39th DAC
Agere Systems, Inc.
101 Crawfords Corner Rd.
Rm. 4E-508
Holmdel, NJ 07733-1900
(732) 949-7248
bda@agere.com

Nanette Collins
Publicity Chair, 39th DAC
37 Symphony Rd., Unit A
Boston, MA 02115
(617) 437-1822
nanette@nvc.com

Sue Drouin
Mentor Graphics Corp.
8005 SW Boeckman Rd.
Wilsonville, OR 97070-7777
(503) 685-1183
sue_drouin@mentorg.com

Larry Eberle
EDA Consortium Representative, 39th DAC
Synopsys, Inc.
700 E. Middlefield Rd.
Mountain View, CA 94043-4033
(650) 584-4418
leberle@synopsys.com

William H. Joyner, Jr.
EDA Industry Chair, 39th DAC
Semiconductor Research Corp.
PO Box 12053
Research Triangle Park,
NC 27709
(919) 941-9472
joyner@src.org

Peter Kalenak
Numerical Technologies, Inc.
70 W. Plumeria Dr.
San Jose, CA 95110
(408) 273-4316
pkalenak@numeritech.com

Darby Mason-Merchant
Xilinx, Inc.
2100 Logic Dr.
San Jose, CA 95124
(408) 879-5228
darby@xilinx.com

Mark Miller
Tera Systems, Inc.
2105 S. Bascom Ave.,
Ste. 135
Campbell, CA 95008
(408) 879-1990
mmiller@terasystems.com

Steve Pollock
Real Intent
3910 Freedom Cir.,
Ste. 102A
Santa Clara, CA 95054
(408) 982-5412
pollock@realintent.com

Richard Porter
Innoveda, Inc.
293 Boston Post Rd.
W. Marlboro, MA 01753
(508) 303-5434
rporter@innoveda.com

Gary Smith
Dataquest
251 River Oak Pkwy.
San Jose, CA 95134
(408) 468-8271
gary.smith@dataquest.com

Lee Wood
Exhibit Manager, 39th DAC
MP Associates, Inc.
5305 Spine Rd., Ste. A
Boulder, C0 80301
(303) 530-4562
lee@mpassociates.com

OPENING KEYNOTE ADDRESS

Hajime Sasaki
Chairman of the Board
NEC Corp., Tokyo, Japan

Paradigm Shift in Semiconductor Design: Challenges on the CAD System

Abstract: Last year, the Semiconductor industry suffered the severest recession in its history, withering by 32%. There had been no case where the final market of electronic products itself shrank. Concurrently, there have been some symbolic events in the industry. First, Moore's Law has become uncertain. In order to maintain the law, we must come up with novel processes for each new generation. Secondly, vendors have begun to shift from mass-production oriented business schemes to the system LSI business. This shift to application specific devices requires a drastic review of the value chain.

Semiconductor design in the new era can be divided into three areas. Area 1 deals with the traditional design hierarchy, such as function design, logic design and layout design. In this area, a revolution in function design is especially required that is represented by C language-based design. Area 2 deals with the new challenges of pursuing shrinking technology. We should be prepared for problems resulting from the application of new materials, represented by electric current leakage through the gate insulator. Also, interconnect structure is another challenging problem. The merger of design and process is the key. Area 3 deals with comprehensive issues such as IP management and mass production know-how.

The industry will become much more dependent on design. The challenge is how we should cultivate competitive designers both in quantity and quality. The "design project leader" must cover all three areas above. The corporate structure of IDM does not carry a competitive advantage, as long as design and process is divided into further sub-segments. When the industry can provide a solution that best meets its customer's needs and niceties from every angle, the industry will take off again and reach a new prosperous stage.

Biography: After receiving a master's degree of engineering in the field of electrical engineering from the graduate school of the University of Tokyo in 1961, he joined NEC Corporation. He served as General Manager of the VLSI Development Division and General Manager of the Microcomputer Products Division before being elected to the Board of Directors in 1988. He was appointed Senior Executive Vice President in 1996 in charge of semiconductors operations, and in 1999 elected to his current position of Chairman of the Board. Mr. Sasaki received the commendation from the Minister of State for Science and Technology, as a person of scientific and technological merits in 1995. In 2000, he received the Third Millennium Medal from IEEE. Also he was elected as a member of NAE (National Academy of Engineering). In 2001, he received the Robert N. Noyce Medal from IEEE and was elected as an IEEE Life Fellow.

THURSDAY KEYNOTE ADDRESS

Jerry Fiddler
Chairman and Co-Founder
Wind River, Alameda, CA

Software and Silicon - Where's the Equilibrium?

Abstract: Silicon, EDA, and embedded software companies have largely lived in different universes. Despite repeated predictions that they would move closer together, and some real attempts to do so, those predictions have mostly proved false. Now, however, technology and economic trends are making it necessary and inevitable that suppliers find new ways to solve our customers problems. The complexity of the devices being built require solutions that cross the comfortable boundaries of the past. This talk will examine the trends and problems driving this.

Biography: Fiddler co-founded Wind River in 1981 with David Wilner, a colleague from Fiddler's tenure at University of California's Lawrence Berkeley Laboratory. Since establishing Wind River, Fiddler has taken on whatever role was necessary, from chief and only programmer to president and chairman of the board. Today Wind River, with operations worldwide, is the market share leader in embedded technology and continues to innovate and define the market.

Beginning with his tenure at the University of California's Lawrence Berkeley Laboratory, Fiddler has earned a reputation as an expert in the design and implementation of real-time systems. His work at Berkeley Lab gave him the ability to approach problems from a broader perspective and also taught him how to determine the pieces needed to build a system to solve a problem, skills that are key in developing embedded technology. He has published and presented numerous papers and articles on embedded technology and the rise of hidden, or embedded, computers in everyday life.

Fiddler graduated from the University of Illinois with a double baccalaureate in music and photography, and he received an M.S. in computer science from the same institution. Fiddler continues to follow his love of music with a home music studio and his band, XAZ, for which he writes songs and plays acoustic, electric, and classical guitars, as well as guitar synthesizer.

2002 Best Paper Award

This year, awards are made for the best papers in three categories. Winners were determined from detailed reviews of the accepted papers in the technical sessions. Each award is accompanied by a plaque. The awards are given by ACM/SIGDA (Special Interest Group on Design Automation), IEEE/CAS (Institute of Electrical and Electronics Engineers/Circuits and Systems Society) and EDA Consortium (Electronic Design Automation Consortium).

DESIGN TOOLS

Paper 4.3: *Guaranteed Passive Balancing Transformations for Model Order Reduction*

Author: Joel R. Phillips
Affiliation: Cadence Berkeley Labs., San Jose, CA

Author: Luca Daniel
Affiliation: Univ. of California, Berkeley, CA

Author: Miguel Silveira
Affiliation: INESC, Lisboa, Portugal

DESIGN METHODOLOGY

Paper 29.2: *Remembrance of Circuits Past: Macromodeling by Data Mining in Large Analog Design Spaces*

Authors: Hongzhou Liu, Amit Singhee, Rob A. Rutenbar, L. Richard Carley
Affiliation: Carnegie Mellon Univ., Pittsburgh, PA

EMBEDDED SYSTEMS

Paper 3.1: *A Universal Technique for Fast and Flexible Instruction-Set Architecture Simulation*

Authors: Achim Nohl, Gunnar Braun, Andreas Hoffmann, Oliver Schliebusch, Rainer Leupers, Heinrich Meyr
Affiliation: Aachen Univ. of Technology, Aachen, Germany

Marie R. Pistilli Women in EDA Achievement Award

Ann Rincon - Senior Engineer, IBM Corp., Microelectronic Division, Burlington, VT
For her significant contributions in helping women advance in the field of DA technology.

ACM/SIGDA Outstanding New Faculty Award

Charlie Chung-Ping Chen - Univ. of Wisconsin, Madison, WI
For a junior faculty member early in his academic career who demonstrates outstanding potential as an educator and researcher in the field of electronic design automation.

2002 IEEE CASS Fellows

Georges G. E. Gielen - Katholieke Univ., Leuven, Belgium
For contributions to computer-aided design and design automation of analog and mixed-signal integrated circuits and systems.

Sharad Malik - Princeton Univ., Princeton, NJ
For contributions to electronic design automation techniques in logic and embedded software synthesis.

Lawrence T. Pileggi - Carnegie Mellon Univ., Pittsburgh, PA
For contributions to simulation and modeling of integrated circuits.

Kaushik Roy - Purdue Univ., West Lafayette, IN
For contributions to the power-aware design of digital circuits.

Ellen June Yoffa - IBM Research, Yorktown Heights, NY
For technical, professional, and business leadership in electronic design automation.

CAD Transactions Best Paper Award

Floorplanning Using a Tree Representation, IEEE Transactions on Computer Aided Design of Integrated Circuits and Systems, vol. 20, no. 2, pp. 281-289, February 2001.

Pei-Ning Guo - Mentor Graphics Corp., San Jose, CA
Toshihiko Takahashi - Niigata Univ., Niigata, Japan
Chung-Kuan Cheng - Univ. of California at San Diego, La Jolla, CA
Takeshi Yoshimura - NEC Corp., Kawasaki, Japan

VLSI Transactions Best Paper Award

A Framework for Reconfigurable Computing: Task Scheduling and Context Management, IEEE Transactions on Very Large Scale Integration (VLSI) Systems, vol. 9, no. 6, pp. 858 - 873, December 2001.

Rafael Maestre - Morpho Technologies, Irvine, CA
Fadi J. Kurdahi - Univ. of California, Irvine, CA
Milagros Fernandez - Univ. Complutense, Madrid, Spain
Roman Hermida - Univ. Complutense, Madrid, Spain
Nader Bagherzadeh - Univ of California, Irvine, CA
Hartej Singh - University of California, Irvine, CA

Outstanding Young Author Award

Yehea I. Ismail - Northwestern Univ., Evanston, IL
Effects of Inductance on the Propagation Delay and Repeater Insertion in VLSI Circuits
IEEE Transactions on Very Large Scale Integration (VLSI) Systems, vol. 8, no. 2, pp. 195-206, April 2000 (with E.G. Friedman).

The Association for Computing Machinery/Special Interest Group on Design Automation (ACM/SIGDA) presents its Distinguished Service Awards

Cheng-Kok Koh - Purdue Univ., West Lafayette, IN
For exemplary service to ACM/SIGDA and the EDA industry -- as Co-director of SIGDA's CDROM Project, as SIGDA's Travel Grants Coordinator, and as Editor of the SIGDA Newsletter.

Steven P. Levitan - Univ. of Pittsburgh, Pittsburgh, PA
For over a decade of service to ACM/SIGDA and the EDA industry -- as DAC University Booth Coordinator, Student Design Contest organizer, founder and promoter of SIGDA's web server, and most recently, Chair of ACM/SIGDA from 1997 to 2001,

The P. O. Pistilli Undergraduate Scholarships for Advancement in Computer Science and Electrical Engineering

The objective of the P. O. Pistilli Scholarship program is to increase the pool of professionals in Electrical Engineering, Computer Engineering and Computer Science from under-represented groups (women, African American, Hispanic, Native American, and physically challenged). In 1989, ACM Special Interest Group on Design Automation (SIGDA) began providing the program. Beginning in 1993, the Design Automation Conference provided the funds for the scholarship and SIGDA continues to administer the program for DAC. DAC normally funds two or more $4000 scholarships, renewable up to 5 years, to graduating high school seniors. In 1999 the IEEE Circuits and Systems Society also began to sponsor these scholarships.

The 2002 winners will be announced at the Conference. The 2001 winners were:

2001 DAC P. O. Pistilli Undergraduate Scholarships

Tenisha Evonne Austin - attending the University of Michigan
Sylvia Blythe Glassco - attending Yale University
Joelle B. Arnold - attending the Franklin W. Olin College of Engineering
Jieyu Hao - attending Massachusetts Institute of Technology
Mariya Sergeyevna Sadakova - attending Macalester College
Todd R. Roman - attending the University of Arizona
Erika Lin - attending Massachusetts Institute of Technology

For more information about the P. O. Pistilli scholarship, please contact Dr. Cherrice Traver, EE/CS Dept., Union College, Schenectady, NY 12308. email: traverc@union.edu

Design Automation Conference Graduate Scholarships

Each year the Design Automation Conference sponsors several $24,000 scholarships to support graduate research and study in Design Automation (DA), with emphasis in "design and test automation of electronic and computer systems". Each scholarship is awarded directly to a university for the Faculty Investigator to expend in direct support of one or more DA graduate students.

The criteria for granting such a scholarship expanded in 1996 to include financial need. The criteria are: the academic credentials of the student(s); the quality and applicability of the proposed research; the impact of the award on the DA program at the institution; and financial need. Preference is given to institutions that are trying to establish new DA research programs.

Prof. **Florin Balasa**, Univ. of Illinois, Chicago, IL
Students: **Sarat Chandra Maruvada, Karthik Krishnamoorthy**
Novel Exploration Techniques in Device-Level Placement for Analog and Radio Frequency Blocks

Prof. **Dinesh K. Bhatia**, Univ. of Texas at Dallas, Richardson, TX
Students: **Shankar Balachandaran, Parivallal Kannan**
Congestion and Routability Estimation for Large ASICs

Prof. **Krishnendu Chakrabarty**, Duke Univ., Durham, NC
Student: **Lei Li**
Algorithms and Tools for Plug-and-Play System-on-a-Chip Test Automation

Prof. **Ingrid Verbauwhede**, Univ. of California, Los Angeles, CA
Student: **Patrick Schaumont**
Domain Specific Tools and Methods for Application in Security Processor Design

Information on next year's DAC scholarship award program will be available on the DAC web page: http://www.dac.com.

DAC 2002 Student Design Contest Winners

Operational Category:

1st Place *Design and Performance Testing of a 2.29 Gb/s Rijndael Processor (session 41.1)*
Patrick R. Schaumoont, Ingrid M. Verbauwhede, Henry Kuo - Univ. of California, Los Angeles, CA

2nd Place *System Design of iBadge for Smart Kindergarten*
Ivo Locher, Sung I. Park, Andreas Savvides - Univ. of California, Los Angeles, CA

3rd Place *A Low Noise Switched-Capacitor Interface Electronics for Sub-micro Gravity Resolution Micromachined Accelerometers*
Haluk Kulah - Univ. of Michigan, Ann Arbor, MI

Conceptual Category:

1st Place *A Microsystem for Near-Patient Accelerated Clotting Time Blood Tests*
Steven M. Martin, Roy H. Olsson, Richard B. Brown - Univ. of Michigan, Ann Arbor, MI

2nd Place *Design of a Crossbar Switch Chip for Use in a Demonstration System of an Optoelectronic Multi-Chip Module*
Jason D. Bakos, Donald M. Chiarulli - Univ. of Pittsburgh, Pittsburgh, PA

3rd Place (tie) *Highly Parallel DNA Sequence Matching and Alignment Processor*
A. T. Patzer - Duke Univ., Durham, NC

3rd Place (tie) *Low-Jitter Power-Aware Non-PLL Clock Generator for GHz Microprocessors*
Chulwoo Kim, Inchul Hwang - Univ. of Illinois, Urbana, IL
Sung Mo Kang - Univ. of California, Santa Cruz, CA

Honorable Mention:

A Low-Power Line Driver Using Resonant Charging With Reduced High-Order Frequency Components
Clemens Schlachta, Burkart Voss, Manfred Glesner - Darmstadt Univ., Darmstadt, Germany

Systematic Design of a 200 Ms/s 8-bit Interpolating/Averaging A/D Converter (session 29.4)
Jan Vandenbussche - Katholieke Univ., Leuven, Belgium

Power Minimization for Digital Optical Interconnects
Xiaoqing Wang, Fouad Kiamilev, Jeremy Ekman - Univ. of Delaware, Newark, Delaware

REVIEWERS

A total of 491 manuscripts were submitted to the 2002 DAC. The Conference Executive and Technical Program Committees wish to acknowledge the time and effort spent by the following people who reviewed these manuscripts and returned the review forms completed. Our thanks to all of those who participated and contributed to the success of the Conference.

Nizar Abdallah
Bryan Ackland
Andrea Acquaviva
Will Adams
Allon Adir
Vishwani Agrawal
Antoine Aiello
Kathir Aingaran
Hussain Al Asaad
Nada Aleksic
Mohamed Ali
Dhamin Al Khalili
Eli Almog
Fadi Aloul
Charles Alpert
Said AlSarawi
Gustavo Alves
Abbes Amira
Jason Anderson
Kishore Andra
Mohab Anis
Kurt Antreich
Yuri Apanovich
Rafael Aquino
Harm Arts
Sigal Asaf
Veronique Aubert
Michel Auguin
Nadine Azemard
Mustafa Badaroglu
Amer Baghdadi
Xiaoliang Bai
Brian Bailey
M Balakrishnan
Florin Balasa
Kaustav Banerjee
Francisco Barat
Robert Batten
Shabbir Batterywala
Kia Bazargan
Michael Beattie
Derek Beatty
Murat Becer
Bernd Becker
Bill Bell
Imed BenDhaou
Maarten Berkens
Valeria Bertacco
Jayanta Bhadra
Sanjukta Bhanja
Sarvesh Bhardwaj
Swarup Bhunia
David Binkley
Alexandre Birguer
Srinivas Bodapati
Alessandro Bogliolo
Cristiana Bolchini
Surendra K. Bommu
Bruno Bougard
Youcef Bourai
Ahmed Bouridane
Elaheh Bozorgzadeh
Eric Bracken
Gunnar Braun
Marnik Brunfaut
Thomas Buechner
Jerry Burch
Jeff Burns
Gianpiero Cabodi
Kerin Cao
Juan-Antonio Carballo
Joao Cardoso
Luca Carloni
Andy Carpenter
Francky Catthoor
Mustafa Celik
Krishnendu Chakrabarty
Sricharan Challa
Francis Chan
ChihWei Chang
Henry Chang
Naehyuck Chang
Shih-Chieh Chang
YiCheng Chang
YiShing Chang
Karam Chatha
Alexander Chatzigeorgiou
Tiberiu Chelcea
Charlie Chen
Chunhong Chen
Dahe Chen
Deming Chen
Howard Chen
Liang Chen
Qiushi Chen
Sao Jie Chen
Tsung Hao Chen
Wei Chen
Zhanping Chen
Kwang-Ting (Tim) Cheng
Charles Chiang
David Chinnery
Eli Chiprout
Hoon Choi
Kiyoung Choi
Seung-Hoon Choi
Philip Chong
Umakanta Choudhury
Chris Chu
Robert Chun
Leah Clark
Bobby Compton
George Constantinides
Martin Coors
Roberto Corgnati
Radu Cornea
Henk Corporal
Bernard Courtois
Andrew Crews
Alfredo Cruz
Eduardo da Costa
Walter Daems
John A. Darringer
Florentin Dartu
Ali Dasdan
Fernando de Bernardinis
Eduardo de la Torre
Hugo de Man
Werner de Rammelaere

Saumya Debray
Stanislaw Deniziak
Dirk Desmet
Lewis Dewey
JeanPhilippe Diguet
Thomas Dillinger
Eugene Ding
Li Ding
Pedro Diniz
Charutosh Dixit
Stephane Donnay
Udeerna Doppalapudi
Frederic Doucet
Bill Dougherty
Steven Drager
Rolf Drechsler
Johan Driesen
Milenko Drinic
Anthony Drumm
Ashvin Dsouza
Nikil Dutt
Shantanu Dutt
Satyajit Dutta
Martin Dzbor
Kerstin Eder
Martyn Edwards
Stephen Edwards
Petru Eles
Abe El-Fadel
Norman Elias
Peeter Ellervee
Chris Ellingham
Gary Ellis
Farzan Fallah
Ray Farbarik
Morteza Fayyazi
Dominique Federici
Jessica Feng
Fabrizio Ferrandi
Craig Files
William Fornaciari
Laurent Fournier
Kenneth Francken
Jean Arnaud Francois
Bruno Franzini
Mark Fredrickson
Ana Freitas
Martin Frerichs
Eby Friedman
Hisanori Fujisawa
Franco Fummi
Paolo Gai
Kaushik Gala
Vijay Gangaram
Youxin Gao
Alberto Garcia
Olivier Gautherot
Daniel Geist
Joachim Gerlach
Reinhard Gerndt
Vassilios Gerousis
Andreas Gerstlauer
Soheil Ghiasi
Alon Gluska
Sandeep Goel
Victor Mauro Goulart Ferreira
Helmut Graeb
Thorsten Groetker
John Grout
Flavius Gruian
Xinli Gu
Kiran Gullapalli
Sumit Gupta
Matthew Guthaus
Malay Haldar
Bill Halpin
Scott Hamm
Omar Hammami
Manish Handa
Ziyad Hanna
Pavan Hanumolu
Kevin Harer
Alan Hartman
Zafar Hasan
Masanori Hashimoto
David J. Hathaway
Karsten Haug
Terumine Hayashi
Jim Heaton
Ulrich Heinkel
Krupnova Helena
Domenik Helms
Joerg Henkel
Dwight Hill
Mokhtar Hirech
James Hoe
Inki Hong
Xianlong Hong
Michael Hsiao
Changchao Hsieh
Harry Hsieh
Alan Hu
Bo Hu
Jiang Hu
Xiaobo (Sharon) Hu
Yutao Hu
Jing Reng Huang
Christophe Hui Bon Hoa
William Hung
Syed Hussain
Sunyoung Hwang
Paolo Ienne
Yukihiro Iguchi
Yonghee Im
Leandro Indrusiak
Akihiko Inoue
Tomoo Inoue
Adrian Ionescu
Norris Ip
Ali Iranli
Yehea Ismail
Lubomir Ivanov
Balakrishnan Iyer
Madhu Iyer
Jae Young Jang
Abhijit Jas
Murali Jayapala
Goeran Jerke
Ahmed Jerraya
Pradip Jha
Michael Ji
Yanbin Jiang
YiMin Jiang
Yunjian Jiang
Yujia Jin
Bruce Johnson
Kevin Jones
Lech Jozwiak
Knut Just
Hilary Kahn
Andrew B. Kahng
Daher Kaiss
Kerim Kalafala
Sudhakar Kale
Praveen Kalla
Gila Kamhi
Mattan Kamon

Robert Kanzelman
William Kao
Bhaskar Karmakar
Tanay Karnik
Ryan Kastner
Yoshiyuki Kawakami
Masamichi Kawarabayashi
Udo Kebschull
Brion Keller
Mahesh Ketkar
Kamal Khouri
Chris Kim
Chulwoo Kim
Daehong Kim
DaeHwan Kim
Dong Wook Kim
Euiseok Kim
Jae-Joon Kim
Taewhan Kim
Yong Kim
Darko Kirovski
Andreas Kirschbaum
Michael Kishinevsky
Polen Kission
Akira Kitajima
Robert Klenke
Alfred Koelbl
Cheng-Kok Koh
Reiner Kolla
Gangadhar Konduri
Farinaz Koushanfar
Andrzej Krasniewski
Volodymyr Kratyuk
Bharat Krishna
Harsha Krishnamurthy
Dilip Krishnaswamy
Florian Krohm
Angela Krstic
Lars Kruse
Prabhakar N. Kudva
Yuji Kukimoto
James Kukula
Chidamber Kulkarni
Shashi Kumar
Eren Kursun
Timothy Kurzweg
Hee Hwan Kwak
Feipei Lai
Minghorng Lai

Marcello Lajolo
William Lam
Li Lao
Rudy Lauwereins
Mihai Lazarescu
JenqKuen Lee
Kyoung Keun Lee
Miriam Leeser
Gunther Lehmann
Kwok Shing Leung
Rainer Leupers
Regis Leveugle
Moshe Levinger
Jackel Lew
Ping Chung Li
Wencheng Li
Stan Liao
Sung Lim
Chieh Lin
Tao Lin
Alvin Ling
Jing-Jia Liou
Giuseppe Lipari
Jinfeng Liu
Marisa Lopez Vallejo
Yufeng Luo
Tony Ma
Luca Macchiarulo
Alberto Macii
Jean Christophe Madre
Jan Madsen
Philippe Magarshack
Ali Mahdoum
Babu Mailachalam
Yossi Malka
Charles Malley
Mahesh Mamidipaka
Ion Mandoiu
Rick Mankin
Maq Mannan
Nazanin Mansouri
Ali Manzak
Henri Maramis
Pol Marchal
Diana Marculescu
Malgorazata Marek-Sadowska
Tiziana Margaria
Erik Marinissen
Igor Markov

Ewout Martens
Grant Martin
Thomas Martin
Peter Marwedel
Guido Masera
Hiroo Masuda
Karti Mayaram
Matthias Meerwein
Seapahn Meguerdichian
Amit Mehrotra
Bingfeng Mei
Noel Menezes
Shin-ichi Minato
Miguel Miranda
Prabhat Mishra
Maher N. Mneimneh
Firas Mohamed
Sundararajarao Mohan
Robert Montoye
InHo Moon
Fernando Moraes
Francisco Assis Moreira do Nascimento
Daniel Moritz
Vasily Moshnyaga
Matthew Moskewicz
Daniel Mosse
Chandra Moturu
Mark Moulin
Vahram Mouradian
Zhen Mu
Michael Muench
Arindam Mukherjee
Brian Mulvaney
Murali Mohan Nagaraj
Ravi Naiknaware
Walid Najjar
Alan Nakamoto
Tsuneo Nakata
Gi-Joon Nam
Alberto Nannarelli
Sanjiv Narayan
Sridhar Narayanan
Sani Nassif
Ognen Nastov
Suriyaprakash Natarajan
Lasse Natvig
Danial Neebel
Kyle Nelson

Leor Nevo
Huy Nguyen
Tuyen Nguyen
Dan Nicolaescu
Jan Niehof
Achim Nohl
Thomas Nora
Adrian Nunez
Jhonny Oberg
Altan Odabasioglu
Takanori Okuma
Arlindo Oliveira
Mauro Olivieri
C.K. Ong
Frank Oppenheimer
Robert Palermo
David Pan
Davide Pandini
Debashis Panigrahi
Chris Papachristou
Subramaniam Parasuraman
Alice Parker
Ganapathy Parthasarathy
Claudio Passerone
Roberto Passerone
Nelson Passos
Rakesh Patel
Srinivas Patil
Bipul Paul
Adam Pawlak
Michael Payer
Helvio Peixoto
Rajesh Pendurkar
Yang Peng
Zebo Peng
Sakthivel Periyasamy
Armita Peymandoust
Ian Phillips
Karen Pieper
Lawrence T. Pileggi
Peter Pirsch
Vijay Pitchumani
Massimo Poncino
Ralf Popp
Adam Postula
Nachiketh Potlapally
Mukul Prasad
Anurag Pulincherry
Carla Purdy
Kiran Puttaswamy
Iksoo Pyo
Gang Qu
Gang Quan
Stefano Quer
Rajesh Radhakrishnan
Vijay Raghunathan
Anand Ragunathan
Soumyendu Raha
Sreeja Raj
Daler Rakhmatov
Logie Ramachandran
Hema Ramamurthy
Yajun Ran
Abhishek Ranjan
Rajeev Ranjan
C.P. Ravikumar
Alain Raynaud
Philippe Raynaud
Marco Re
Sudhakar Reddy
Andre Reis
Ricardo Reis
Peyman Rezvani
Lluis Ribas Xirgo
Kai Richter
Teresa Riesgo
Luca Rijnders
Jeffrey Roehr
Oriol Roig
Marly Roncken
Frederic Rousseau
Bruno Rouzeyre
Philippe Royannez
Shanqjang Ruan
Daryl Rudusky
Per Runeson
Hans Sahm
Biranchinath Sahu
Fabio Salice
Kewal Saluja
Debasis Samanta
Pablo Sanchez
Sachin Sapatnekar
Steve Sapiro
Majid Sarrafzdeh
Ryuichi Satomura
Christian Sauer
Nicolae Savoiu
Prashant Saxena
Patrick Schaumont
Louis Scheffer
Heinz Schlebusch
Ulf Schlichtmann
Oliver Schliebusch
Eike Schmidt
Klaus Schneider
Thorsten Schubert
Philip Schumaker
Robert Schwencker
Donatello Sciuto
Luc Semeria
Ellen Sentovich
Kenshu Seto
Marco Sgroi
Niraj Shah
Mahmoud Shahram
Saghir Shaikh
Nikhil Sharma
Rohit Sharma
Rupesh Shelar
C.-J. Richard Shi
Toshiyuki Shibuya
Eung Shin
Youngsoo Shin
Tom Shiple
Wen Tsong Shiue
Aviral Shrivastava
Sandeep Shukla
Gil Shurek
Cristina Silvano
Mathias Silvant
Miguel Silveira
Tajana Simunic
Amit Singh
Montek Singh
Sanjay Singh
Mukund Sivaraman
Sasa Slijepcevic
Gerald Sobelman
Lambert Spaanenburg
Sorin Andrei Spanoche
Raghukiran Sreeramaneni
Vamsi Srikantam
Srikanth Srinivasan
Ankur Srivastava
Ted Stanion
Balsha Stanisic

Martin Stark
Stergios Stergiou
Ken Stevens
Leon Stok
Neal Stollon
Dirk Stroobandt
Haihua Su
Wenquan Sui
Weikai Sun
Peter Sutton
Goro Suzuki
Shivakumar Swaminathan
William Swartz
Frank Szorc
Abdallah Tabbara
Bassam Tabbara
Andiyono Tan
Sheldon Tan
Xiaoping Tang
Tuna Tarim
Alexander Taubin
Clark Taylor
John Taylor
Steven Teig
Volker Thamm
Michael Theobald
Meera Thiagarajan
Mitch Thornton
Ruiqi Tian
Hiroyuki Tomiyama
Lionel Torres
Masahiko Toyonaga
Steve Trimberger
Yatin Trivedi
Anne Trullemans
Chung-Kuan Tsai
Mel Tsai
Emre Tuncer
Raimund Ubar
Durgam Vahia
Werner van Almsick
Geert Van der Plas
Pieter van der Wolf
Arjan van Genderen
Piet Vanassche
Arnout Vandecappelle
Jan Vandenbussche
Gerd Vandersteen
Harsh Vardhan
Milan Vasilko
Antonio Velasco
Miroslav Velev
Giacomino Veltri
Stefano Vercelli
Ward Vercruysse
Carl Verdonck
Wim Verhaegen
Frederik Vermeulen
Massimo Violante
Praveen Vishakantaiah
Ashok Vittal
Jefferry Vo
Martin Vogels
Harald Vranken
Huu Nghia Vuong
Arno Wagner
Duncan Walker
Robert Walker
Piet Wambacq
Albert Wang
Haibo Wang
Jyh Wang
Li.-C. Wang
Pinghann Wang
Sying Jyan Wang
Ting Chi Wang
Zhong Wang
Scott Weber
Norbert Wehn
Liqiong Wei
Yen Cheng Wen
Andreas Wieferink
Brian Wilson
Greg Wilson
Steven Wilton
Michael Wirthlin
Jens Wittenburg
Francis Wolff
Yaron Wolfsthal
Christophe Wolinski
Jennifer L. Wong
Franz Wotawa
Chengwen Wu
Lifeng Wu
Wen Jer Wu
Yu Liang Wu
Jie Xu
Masaaki Yamada
Koichi Yamashita
Shigeru Yamashita
Fred Yang
Xiaojian Yang
Martelloni Yannick
Hiroto Yasuura
Lun Ye
Gin Yee
C.Y. Steve Yeh
Chingwei Yeh
Chantal Ykman
Sungjoo Yoo
Tak Young
Tiejun Yu
Jiann Yuan
Xin Yuan
Vittorio Zaccaria
Roberto Zafalon
Maurizio Zamboni
Arek Zawada
Jindrich Zejda
Alexander Zemliak
Xuan Zeng
Peter Zepter
Nacer Eddine Zergainoh
Tianhao Zhang
Yumin Zhang
Min Zhao
Shuqing Zhao
Lyudmila Zinchenko
Yann Zinzius
Avi Ziv

Table of Contents

Session 3
Design Innovations for Embedded Processors

Chair: Gang Qu
Organizers: Grant E. Martin, Majid Sarrafzadeh

Session 4
Passive Model Order Reduction

Chair: Jacob White
Organizers: Jaijeet Roychowdhury, Mustafa Celik

Session 5
New Perspectives in Physical Design

Chair: Steven Teig
Organizers: Ralph Otten, Timothy Kam

Session 6
PANEL: Tools or Users: Which is the Bigger Bottleneck?

Chair: Andrew B. Kahng
Organizer: Bob Dahlberg

Session 7
Special Session: Life after CMOS: Imminent or Irrelevant?

Chairs: Dennis Sylvester, Kaustav Banerjee
Organizers: Dennis Sylvester, Kaustav Banerjee

Session 8
Formal Verification

Chair: Yaron Wolfsthal
Organizers: Carl Pixley, Karem Sakallah

Session 9
High Level Specification and Design

Chair: Andreas Kanstein
Organizers: Limor Fix, Shin-ichi Minato

Session 10
Timing Abstraction

Chair: Mark Hahn
Organizers: Chandu Visweswariah, Narendra V. Shenoy

Session 11
Special Session: E-Textiles

Chair & Organizer: Majid Sarrafzadeh

Session 12
PANEL: Analog Intellectual Property: Now? Or Never?

Chair: Stephen Ohr
Organizers: Linda Marchant, Philippe Magarshack

Session 13
Low-Power System Design

Chair: Giovanni De Micheli
Organizers: Renu Mehra, Enrico Macii

Session 14
Fabric-Driven Logic Synthesis

Chair: Maciej Ciesielski
Organizers: Malgorzata Merek-Sadowska, Steven Nowick

Session 15
Memory Management and Address Optimization in Embedded Systems

Chair: Nikil Dutt
Organizers: Diederik Verkest, Luca Benini

Session 16
Special Session: Optics: Lighting the Way to EDA Riches?

Chair: Jaijeet Roychowdhury
Organizers: Jaijeet Roychowdhury, Joel R. Phillips

Session 17
PANEL: Nanometer Design: What Hurts Next...?

Chair: Lawrence T. Pileggi
Organizers: Rob A. Rutenbar, Andrew B. Kahng

Session 18
Novel DFT, BIST and Diagnosis Techniques

Chair: Rathish Jayabharathi
Organizers: Kwang-Ting (Tim) Cheng, T. M. Mak

Session 19
Case Studies in Embedded System Design

Chair: Wayne Wolf
Organizers: Anand Raghunathan, Xiaobo (Sharon) Hu

Session 20
Theoretical Foundations of Embedded System Design

Chair: Rajesh Gupta
Organizers: Annette Reutter, Donatella Sciuto

Session 21
Equivalence Verification

Chair: Ziyad Hanna
Organizer: Shin-ichi Minato

Session 22
PANEL: Whither (or Wither?) ASIC Handoff

Chair: Michael Santarini
Organizers: Sudhakar Jilla, Mark Miller

Session 23
Embedded Software Automation: From Specification to Binary

Chair: Joerg Henkel
Organizers: Marco Di Natale, Xiaobo (Sharon) Hu

Session 24
Applications of Reconfigurable Computing

Chair: Ivo Bolsens
Organizers: Grant E. Martin, Kurt Keutzer

Session 25
New Test Methods Targeting Non-Classical Faults

Chair: Rob Aitken
Organizers: Miron Abramovici, T. M. Mak

Session 26
Special Session: How Do You Design a 10M Gate ASIC?

Chair: Ahmed A. Jerraya
Organizers: Ahmed A. Jerraya, Kurt Keutzer

Session 27
Power Distribution Issues

Chair: Sachin Sapatnekar
Organizers: Abhijit Dharchoudhury, Tadahiro Kuroda

Session 28
Advances in Synthesis

Chair: Marek Perkowski
Organizers: Soha M. Hassoun, Yusuke Matsunaga

Session 33
Advanced Simulation Techniques

Chair: L. Miguel Silveira
Organizers: Georges Gielen, Kartikeya Mayaram

Session 34
Design Methodologies Meet Network Applications

Chair: Anand Raghunathan
Organizers: Anand Raghunathan, Marco Di Natale

Session 35
Advances in Analog Modeling

Chair: Alan Mantooth
Organizers: Joel R. Phillips, Kartikeya Mayaram

Session 36
Advances in Timing and Simulation

Chair: David J. Hathaway
Organizers: Louis Scheffer, Narendra V. Shenoy

Session 37
PANEL: Formal Verification Methods: Getting around the Brick Wall

Session 38
Routing and Buffering

Session 39
System on Chip Design

Session 40
Timing Analysis and Memory Optimization for Embedded Systems

Session 41
Processors and Accelerators for Embedded Applications

Chair: Chris Rowen
Organizers: Kurt Keutzer, Majid Sarrafzadeh

Session 42
PANEL: What's the Next EDA Driver?

Chair: Jan Rabaey
Organizers: Joachim Kunkel, Dennis Brophy

Session 43
Cross-Talk Noise Analysis and Management

Chair: Cheng-Kok Koh
Organizers: Kaushik Roy, Noel Menezes

Session 44
Test Cost Reduction for SOCS

Chair: Yervant Zorian
Organizers: Seiji Kajihara, Kwang-Ting (Tim) Cheng

Session 45
Scheduling Techniques for Embedded Systems

Chair: Rolf Ernst
Organizers: Diederik Verkest, Donatella Sciuto

Session 46
Special Session: Designing SoCs for Yield Improvement

Chair: Srivaths Ravi
Organizers: Anand Raghunathan, Alfred E. Dunlop

Session 47
Advances in SAT

Chair: Joao Marques-Silva
Organizers: Malgorzata Merek-Sadowska, Soha M. Hassoun

Session 48
Inductance and Substrate Analysis

Chair: Noel Menezes
Organizers: Jaijeet Roychowdhury, Mustafa Celik

Session 49

Development of Processors and Communication Networks for Embedded Systems

Chair: Jan Rabaey

Organizers: Grant E. Martin, Majid Sarrafzadeh

Session 50

Moving Towards More Effective Validation

Chair: Magdy Abadir

Organizers: Carl Pixley, Masahiro Fujita

Session 51

Special Session: Energy Efficient Mobile Computing

Chair: Enrico Macii

Organizer: Mani Srivastava

Session 52

Floorplanning and Placement

Chair: Ralph Otten

Organizers: Ralph Otten, Steven Teig

Session 53
Circuit Effects in Static Timing

Chair: Jamil Kawa
Organizers: Chandu Visweswariah, Louis Scheffer

Session 54
Design Space Exploration for Embedded Systems

Chair: Henk Corporaal
Organizers: Jo Dale Carothers, Luca Benini

Session 55
Behavioral Synthesis

Chair: Petru Eles
Organizers: Ahmed A. Jerraya, Krzysztof Kuchcinski

PANEL: Wall Street Evaluates EDA

Chair: Aart de Geus, Synopsys

Organizers: Sharon Turnoy, Synopsys and Deirdre Hanford, Synopsys

Abstract

The EDA sector is capturing unprecedented attention on Wall Street. With seven IPOs in 2001 alone and strong performance by the EDA "blue chips," the industry has gained new prominence with the capital markets. In this panel, Aart de Geus will moderate a discussion between representatives of the various constituencies who play a role in shaping Wall Street's opinion of EDA: financial analysts, portfolio managers, venture capitalists, CEOs, and the press.

Questions discussed will include: How do investors and analysts currently view EDA? What contributes to that perception? What factors drive EDA's current favor with Wall Street, and why is the sector "hot" compared to two to three years ago? How do we sustain that favor? Given the highly complex nature of our industry, how do investors decipher the strength of a particular EDA firm? Is EDA tied to the semiconductor industry's performance? What role does the press play in shaping the view?

Panelists

Moshe Gavrielov
Verisity

Richard Goering
EE Times

Lucio Lanza,
Lanza techVentures

Vishal Saluja
Seligman Technology Group

Jay Vleeschhouwer
Merrill Lynch & Company Inc.

IP Delivery for FPGAs Using Applets and JHDL

Michael J. Wirthlin and Brian McMurtrey
Brigham Young University, 459 CB
Provo, UT 84664
wirthlin@ee.byu.edu,mcmcurtre@ee.byu.edu

ABSTRACT

This paper introduces an FPGA IP evaluation and delivery system that operates within Java applets. The use of such applets allows designers to create, evaluate, test, and obtain FPGA circuits directly within a web browser. Based on the JHDL design tool, these applets allow structural viewing, circuit simulation, and netlist generation of application-specific circuits. Applets can be customized to provide varying levels of IP visibility and functionality as needed by both customer and vendor.

Categories and Subject Descriptors

B.6.3 [**Logic Design**]: Design Aids—*Simulation, Hardware description languages*

General Terms

Design

Keywords

Intellectual Property, JHDL, Applet, FPGA

1. INTRODUCTION

The rapid growth in the density of Field Programmable Gate Arrays (FPGAs) has prompted the use of third-party intellectual property (IP) in large FPGA designs. Many vendors now offer a variety of design cores targeted specifically for one of many FPGA technologies. The advantages of IP-based design are well known – using third-party IP can reduce the design time, improve the design quality, and facilitate the design of systems too large to design without pre-designed cores.

While the use of third-party cores may improve design productivity, the process of delivering the IP from a vendor to a customer is not straightforward. The IP delivery process must balance the conflicting demands of IP visibility for the customer and IP protection for the vendor. For the customer, visibility is necessary to properly evaluate and validate the IP. Such visibility may require simulation of the IP, structural understanding of the IP, or some form of detailed IP characterization. The more visibility available to the customer, the more confidence he or she has that the IP operates as specified. Further, greater visibility eases the ability to integrate the IP within the customer's design. For the vendor, such customer needs conflict with the need to protect the IP from unauthorized use or duplication. If too much of the design is exposed to the customer, unauthorized use or reproduction may occur and significantly reduce the value of the IP. Successful IP delivery systems must balance the conflicting needs of user visibility and vendor IP protection.

The approach for IP evaluation and delivery described in this paper involves the use of Java applets executing within a user's web browser. A potential user may evaluate a given FPGA circuit by accessing a web page and interacting with the applet that executes within their local browser. IP evaluation applets can be written to provide various levels of both IP visibility and protection.

1.1 IP Evaluation and Delivery Using Applets

Java applets are executable binaries that operate *within* a web browser when accessed by the user. When the user accesses a web page containing an applet, both the `html` code used to render the web page and the Java binaries are downloaded by the browser. The browser then executes the Java binaries within a local Java Virtual Machine and displays the applet within the browser window. Java applets have been written for a wide variety of applications and web interfaces.

Custom applets can be used as a convenient and powerful way to evaluate and deliver electronic intellectual property on-line. Based on the user's license, a custom applet is presented that offers the appropriate IP evaluation and delivery functionality. Such applets may provide IP characterization, simulation, viewing, and other features as necessary.

Providing IP delivery executables in the form of applets has several advantages over the delivery of user-installed software. First, users do not have to install the software used for IP delivery or evaluation. Providing the executables online simplifies the process for the user and allows the potential customer to evaluate the IP with less effort. Second, providing the executable on the web server insures that the customer accesses the most recent code every time the associated web page is accessed. The online executables can be updated to reflect improvements and bug fixes by the vendor. Rather than redistributing the IP executables every time problems are resolved, vendors need only update executables found on their online server. Third, it is relatively easy to control the executables accessed by the user. Based on user profiles, the web server can provide an executable applet customized to the needs or license of the user. Such user-specific executables will vary the level of observability and control provided to the module generator applet.

The applet delivery mechanism described by this paper uses the JHDL design environment – a design tool written in Java for creating FPGA circuits[4]. This tool allows designers to create, simulate, and netlist FPGA circuits by writing Java programs that instance appropriate FPGA library cells and wires. While JHDL has been used primarily as a stand-alone design tool, it offers a number of tools and aids for describing, evaluating, and delivering FPGA IP over the internet.

1.2 Related Work

A variety of techniques for web-based IP evaluation and delivery have been proposed and demonstrated. Similar approaches to the JHDL applet approach described by this paper will be introduced and contrasted below.

1.2.1 Web-CAD

The Web-CAD methodology provides the infrastructure for simulating IP over the internet using a client-server architecture[2]. Using the internet to communicate simulation information, this methodology allows users to evaluate IP *without* actually acquiring simulation models. In this system, protected IP is simulated at a remote server where the details of the simulation model are hidden from the user. Users can simulate protected IP within their simulation environment by transferring simulation information with the remote server using a publicly available socket library.

Unlike the Web-CAD methodology, the JHDL applet approach does not transfer simulation events across the network. Instead, a protected IP executable is sent over the internet in the form of a Java applet. Simulating the IP directly on the user's machine will result in increased simulation speed by avoiding the relatively long latency associated with a network.

1.2.2 JavaCAD

Like the Web-CAD methodology, the JavaCAD project provides an infrastructure for simulating and evaluating circuits over the internet using remote method invocation (RMI) [1]. This system also provides the capability of delivering detailed area, power, and delay estimates of the IP in question. Like the JavaCAD project, the IP presented in JHDL Applets is described in Java and simulated using a custom simulator. Unlike JavaCAD, the JHDL approach uses the network to deliver IP applets rather than transfer simulation data over the internet.

1.2.3 JBits

The JBits tool developed at Xilinx is another Java-based tool used for delivering FPGA IP cores [5]. This tool delivers pre-placed IP cores by *modifying* the configuration bitstream of the user. Because the IP is delivered in the form of changes to a proprietary configuration bitstream, the structure of the IP is hidden from the user. While this tool currently does not use the internet to deliver IP cores, JBits could easily be incorporated into online Java applets.

1.3 Overview

This paper will begin by providing an overview of the JHDL design environment and discussing its ability to describe FPGA intellectual property. Next, the notion of IP "executables" will be introduced along with an example of such an executable using JHDL. The use of JHDL within Java applets will be discussed and a detailed example of a JHDL applet for IP evaluation and delivery will be provided. Next, various scenarios for IP delivery applets will be suggested and the paper will conclude with directions for future work.

2. JHDL

JHDL is a open-source design environment developed at BYU used for creating high-performance FPGA designs[3, 4]. Based on the Java programming language, users create FPGA designs by writing Java programs that instance FPGA components and wires found within the supported JHDL libraries. These Java files are compiled using conventional Java compilation tools and simulated within the Java Virtual Machine (JVM). The circuit is described using conventional Java programming constructs and class objects are "constructed" by instancing the object within a JHDL hardware system object.

The Java code shown below demonstrates the construction of a simple full-adder within JHDL using the `and2`, `or3`, and `xor3` library elements:

```
public FullAdder(Node parent, Wire a, Wire b,
                 Wire ci, Wire s, Wire co) {
    ...
    Wire t1 = new Xwire(this,1);
    Wire t2 = new Xwire(this,1);
    Wire t3 = new Xwire(this,1);
    new and2(this,a,b,t1);
    new and2(this,a,ci,t2);
    new and2(this,b,ci,t3);
    new or3(this,t1,t2,t3,co);/* co is carry out */
    new xor3(this,a,b,ci,s);  /* s is output */
  }
```

This example begins by constructing three intermediate one-bit wires: `t1`, `t2`, and `t3`. The carry out logic is created using three `and2` primitives and the `or3` primitive (i.e. `co=a&b | a&ci | b&ci`). Finally, the sum logic is created using a three-input exclusive or gate.

A variety of design aids are available for circuits constructed within JHDL. These tools can be used for manipulating, viewing, and simulating JHDL circuit objects. With an open API to the circuit structure, application-specific tools can be created to manipulate or interact with JHDL circuit objects.

2.1 Schematic Viewer

Within the design tool is a schematic viewer for displaying the structure of the JHDL circuit. This viewer allows the designer to quickly view the structure and hierarchy of a circuit described by the Java code. In addition, the designer can view the relative layout of FPGA circuits that include performance enhancing placement attributes. Other tools are available for viewing memory contents, signal waveforms, and circuit hierarchy.

2.2 Circuit Netlister

To interface with other tools and design flows, JHDL provides an API for converting a JHDL circuit object into a user-defined data interchange format. Through this API, the structure, interconnect, hierarchy and properties of a circuit described in JHDL is exposed and can be regenerated in one of many possible formats. JHDL currently creates EDIF and VHDL netlists and effort is being made to support other netlist formats such as Verilog. If necessary, user-defined textual or binary interchange formats can be created by exploiting this API.

2.3 Circuit Simulator

An important component of the JHDL design suite is a built-in circuit simulator. This simulator allows designers to quickly test their JHDL circuit within an interactive simulation environment running in the Java Virtual Machine. Simulation models for several FPGA technologies are supported and behavioral models of non-FPGA circuitry can easily be created within Java class objects. This simulator provides an open API for interacting with the simulation environment in an application-specific manner. User defined viewers, functions, testbenchs, or third-party simulators can be linked to the JHDL simulator through this API.

JHDL has been used by several organizations to generate large, high-performance FPGA designs. These designs range in complexity from small, single-chip controllers to large, multi-FPGA designs for automatic target recognition and sonar beamforming. Most designs target FPGA coprocessor boards that provide multiple FPGAs, memory, programmable interconnect and a high-speed host interface. The ability to describe the structure of an FPGA circuit in a general-purpose programming language and the set of design aids provided by JHDL has aided the development of many general-purpose FPGA designs and reconfigurable computing applications.

3. JHDL FOR IP DELIVERY

Although JHDL has been used effectively as a design tool, it is especially useful for creating parameterizable module generators. The programming constructs and data structures within Java allow module generator executables to create complex circuit structures based on user parameters and application-specific optimizations. Most module generators exploit technology-specific mapping optimizations and use relative placement to improve timing. Many FPGA module generators have been created in JHDL including a variety of arithmetic, signal processing, logic, and memory modules.

An instance of a JHDL module generator can be created by calling the Java class constructor of the module with the appropriate parameters and options. Once the circuit has been created within JHDL, the user may export the circuit by generating a netlist in one of the supported interchange formats (i.e. EDIF, VHDL, or Verilog). This netlist can then be imported into a users conventional design environment. The following example demonstrates how an instance-specific module is created within JHDL.

3.1 Constant Coefficient Module Generator

A module generator that builds optimized constant coefficient multipliers was created for the Xilinx Virtex FPGA[9]. This module generator creates optimized, preplaced constant coefficient multipliers using partial-product look-up tables. To minimize the area and latency of this circuit, the generated circuit is customized to the specific constant, signal widths, and parameters specified by the user.

An application-specific instance of this constant coefficient multiplier is created by *constructing* the module generator object within Java. The class constructor is called and given all of the appropriate user-specific parameters needed to create the optimized, user-specific instance of the IP. The constructor for this multiplier is shown below:

```
public VirtexKCMMultiplier(Node parent,
                           Wire multiplicand,
                           Wire product,
                           boolean signed_mode,
                           boolean pipelined_mode,
                           int constant);
```

To construct this multiplier, a user would first create `Wire` objects for the multiplicand input and product output signals. The size of these wires is specified by the user when the wires are created. The resulting circuit structure and size is determine by the widths of these signal inputs. For example, if the user specifies an 8-bit multiplicand input, 8-bit constant multiplier, and a 12-bit product, the module generator will create an optimized 8x8 multiplier that provides only the top 12-bits of the product.

Once the wires are created, the user creates variables to represent the various parameters needed by the module generator. In this case, `boolean` variables are used to indicate whether the multiplier is signed and pipelined and an `int` variable is used to represent the actual constant used by the constant multiplier. Once the `Wire` objects and parameter variables are available, the circuit object can be created by instancing the module generator with the Java `new` operator. The following code fragment demonstrates the construction of an 8x8 constant multiplier with 12-bit output using the constant -56.

```
Wire m = new Wire(this,8);  //8-bit input
Wire p = new Wire(this,12); //12-bit output
boolean signed = true;
boolean pipelined = true;
int c = -56;                //constant

new VirtexKCMMultiplier(this,m,p,signed,
                        pipelined, c);
```

After the constant coefficient multiplier has been created, standard JHDL tools may be used to manipulate, view, simulate, or output the circuit into an interchangeable format as needed by the designer.

3.2 Module Generator Executables

While most users of JHDL directly instance module generators within a top-level JHDL design, custom executable programs can be written to deliver a circuit outside of the JHDL design environment. Rather than requiring users to write Java code, an executable can be provided that offers a simple user interface for creating, evaluating, manipulating, and delivering circuits to the user. Although these executables are based on the JHDL design infrastructure, the details of Java and JHDL are hidden from the user.

For example, a simple graphical user interface for delivering the constant coefficient multiplier circuit is shown in

Figure 1. This executable program allows a user to create an application-specific instance of the multiplier by selecting the appropriate module parameters (i.e. bitwidths, constant value, etc.). After selecting parameters, the user may evaluate the multiplier by obtaining area and timing estimates or even simulating the design.

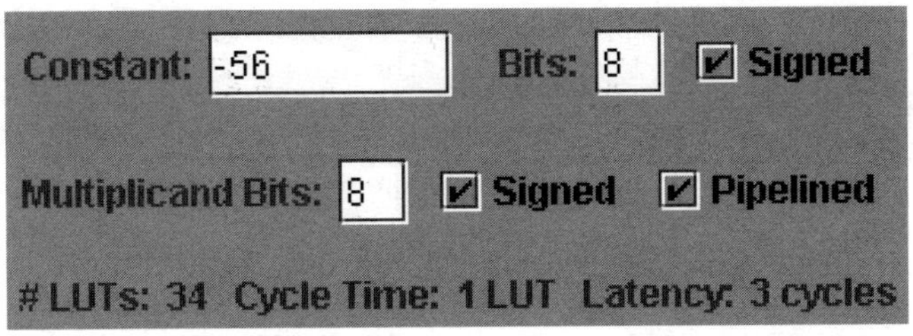

Figure 1: Graphical User Interface for Constant Coefficient Multiplier.

This executable is relatively straightforward – it contains the JHDL code for generating a multiplier circuit, the FPGA technology libraries, and a circuit estimator. Other more customized executables can be provided that offer different levels of IP visibility. If more visibility is necessary, a structural viewer and netlist generator can be added. If less visibility is desired, the vendor can remove the simulation capability of the executable.

Using an executable to deliver IP allows the vendor to control the content, functionality, and opacity of the IP on an individual basis. A custom Java executable can be created and delivered that is customized to the needs of both the customer and vendor. By controlling the content and opacity of the IP executable, vendors may determine the features available for evaluation as well as the visibility into the delivered IP. Specifically, the IP executable may include any of the following JHDL tools:

Structural circuit viewer The circuit viewer allows to user to browse the hierarchy and structure of a generated design. A structural circuit viewer may be included within the executable to provide the user the ability to examine the circuit structure and interface (see Figure 3).

Executable simulation model The JHDL simulator may be included within the executable to allow simulation of the generated circuit. If necessary, the self-contained functional simulator can be used in conjunction with a user's own simulation tools to evaluate the functionality of the IP within a larger design.

Programatic circuit generator interface Most reusable circuit modules for FPGAs are parameterizable and require user intervention to determine the bit-widths, style, pipelining and other parameters of the circuit in question. IP executables may provide an interface that exposes the parameters and options available to the user of the IP. Using this interface, customers may experiment with various parameters to estimate the speed, size and cost of the IP.

Layout view For many FPGA circuit modules, the relative placement or layout of the module has a significant impact on the overall timing and placement of the final FPGA design. A view of the layout for pre-placed FPGA macros provides the user with feedback on the size, shape, and layout of a circuit module under review. Users may explore various placement and layout options of a macro without seeing the underlying circuit structure or netlist.

Circuit netlisting The executable must provide a way for delivering the IP in a form acceptable by the user's conventional tool chain. For licensed users, the circuit netlisting libraries can be included to generate netlists of the IP in the appropriate netlist format. Structural VHDL, Verilog, or EDIF may be generated with the appropriate technology specific mapping constraints for the circuit module of interest.

These and other application-specific IP evaluation and delivery tools may be organized into a single executable on a customer by customer basis. Two such executable configurations are depicted in Figure 2. For passive customers that wish to browse a given circuit and its corresponding characteristics, limited visibility into the IP is provided. As suggested by the left configuration, an executable can be created that contains an interface to the IP and a basic circuit estimator. For more active or licensed customers, an executable can be created that provides greater visibility. As shown by the configuration on the right, the IP delivery executable may contain circuit layout viewers, high-level structural views, and a simulation model. A wide variety of such IP delivery executables may be created by combining the various JHDL design aids.

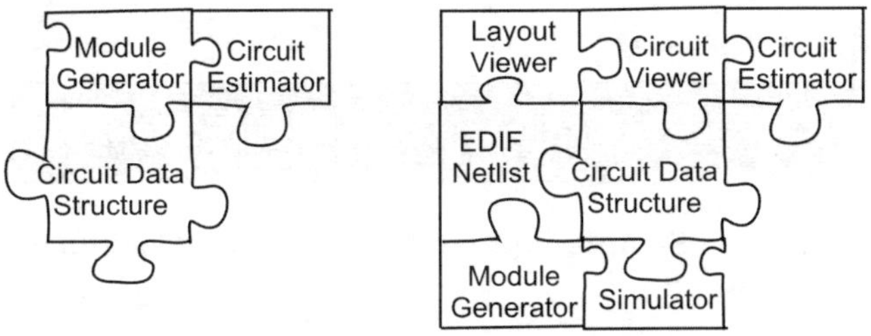

Figure 2: Two configurations of an IP delivery executable.

4. IP DELIVERY APPLETS

A simple way of delivering such JHDL circuit generator executables to a customer is through the internet in the form of Java applets. Rather than delivering circuit generators directly to the customer via mail or email, customers can directly access and execute these generator applets within their standard web browser. Delivering circuit generators through the internet offers a number of advantages over traditional methods: users do not need to install software on their machine, the IP delivery applets can be maintained at a central server, and access to such applets can be controlled through a variety of web-based security measures. By accessing applets over the internet, users are able to evaluate, test, and obtain intellectual property from an IP provider directly from a web browser.

There are a variety of ways in which the IP delivery concepts using JHDL applets can be used. This section will outline two specific scenarios in which the JHDL infrastructure has been used to test the IP-delivery capability within our laboratory. Sample JHDL IP delivery applets can be found and tested online at the following web URL: *http://www.jhdl.org/applets*.

4.1 Constant Coefficient Multiplier Applet

Figure 3 demonstrates a simple applet used for evaluating and delivering the constant coefficient multiplier described earlier. This applet uses a JHDL module generator to create an optimized constant multiplier based on the user's application-specific needs. Rather than exposing the software API to the user, this applet provides a GUI interface to the user for specifying the circuit parameters and evaluating the suitability and quality of the resulting design.

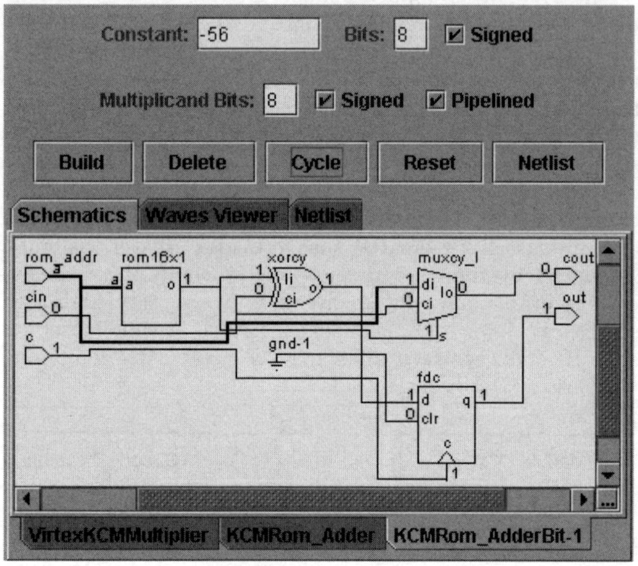

Figure 3: Sample Applet for Constant Coefficient Multiplier.

A user evaluates a specific instance of the IP by selecting the desired parameters and pressing the "build" button. When this event occurs, the applet creates the constant multiplier by calling the module generator constructor. At this point, the circuit is created within the JHDL data structures and any of the JHDL tools can be used to manipulate, netlist and view the circuit structure. In the applet example of Figure 3, a schematic viewer is drawn and a user can interactively explore the structure and hierarchy of the created circuit.

If the JHDL simulator is included with an applet, the user is able to simulate the behavior of the generated circuit directly within the web browser. Using the **Cycle** and **Reset** buttons, a user may cycle the simulator and browse the state of the circuit throughout the circuit hierarchy. The history of the circuit state can be recorded and viewed using the JHDL waveform viewer.

Once a circuit module and its application-specific parameters are selected by a user, the applet may provide a way of delivering the circuit to the user in a reusable form. Since a data structure representing the circuit is resident within the applet, any one of the JHDL circuit netlisting tools can be used to generate a reusable representation of the circuit. For this specific applet, a **Netlist** button is given that generates an EDIF netlist of the application-specific circuit. The generated netlist is displayed in a scrollable text window for the user to browse or copy.

4.2 Black Box Simulation Applet

The previous applet example demonstrates a relatively transparent IP delivery applet – the users of such applet can browse the hierarchy of a circuit, simulate the internals of the circuit, and generate a reusable netlist from the instance-specific circuit. In many cases, the applet executables will need to limit access to the internals of the generated circuits.

One approach for limiting internal access is to provide a "black box" simulation model of the generated circuit. Like the previous "transparent" example, a user would interact with the applet to choose instance-specific circuit parameters and options to build a circuit. Unlike the previous example, the user does not have the ability to browse the hierarchy of the circuit or obtain a netlist. Instead, the applet includes a self-contained simulation model of the intellectual property. This simulation model may be operated interactively by the user or integrated into the user's conventional design flow.

Figure 4 demonstrates a black box simulation model of two IP delivery applets operating within a user's web browser. In this scenario, the user wishes to simulate two external circuits with other components in a complete system simulation. To create this simulation environment, the user visits the web page associated with each IP and constructs the IP within the applet. Next, the user establishes a connection between each applet and the system simulator using an application-specific communication protocol such as RMI or socket communication[1]. Once an appropriate communication link established between these simulators, the entire system can be simulated together without exposing the internals of the applet-based IP.

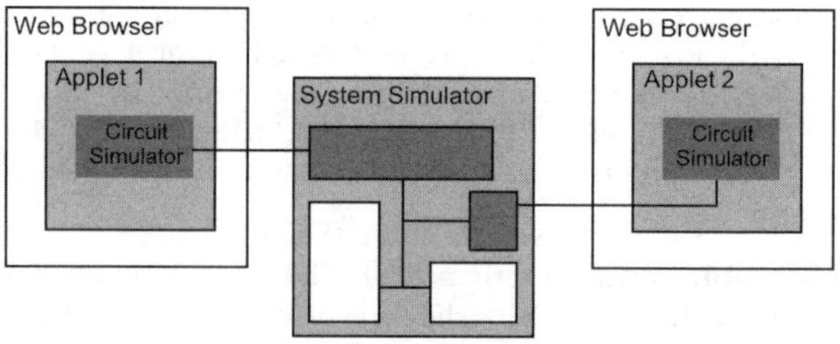

Figure 4: Integrating Black Box Simulation Model Into Simulation Environment.

Several JHDL designs were tested using this black-box simulation methodology. A simulation wrapper was created to interface the JHDL black-box simulator with a Verilog simulation using PLI[8]. Simulation events are exchanged over network sockets and a custom communication protocol. By exposing the JHDL black-box simulator, a user can evaluate intellectual property within their design environment without exposing any proprietary information.

[1]Establishing network connections between applets operating on a client browser and other client applications violates the default applet security model and requires explicit permission from the user.

This approach of black-box simulation is similar to the client-server simulation architectures proposed by [1] and [2]. In these architectures, IP can be simulated on a customer client machine through a pre-arranged communication protocol with a vendor supported server. The IP is protected by keeping the simulation models at the vendor site and only providing the appropriate simulation state. Unlike these methods, the IP in this approach is simulated on the *client* machine. This offers the the potential for IP simulation with much lower network latency than possible with off-site simulation models.

4.3 IP Applet Security

An important concern associated with IP delivery using Java Applets is preserving the security of the intellectual property contained within the applet. Since the IP applet executable is accessible by the user, a sophisticated user may attempt to uncover the IP by reverse engineering the applet. IP applets may need to provide additional measures of security to properly protect the IP. Techniques such as Java class file obfuscation and class encryption may be added to increase the security of the IP. Other more standard approaches for IP protection may be used such as watermarking[7] and hardware metering[6].

4.4 Applet Download Time

An important issue that faces users of Java applets is the size of the program binary downloaded by the browser. Since the binaries are loaded by the browser the first time the web page is accessed, large binaries may require an unreasonable amount of time and network bandwidth. To address this issue, the binaries associated with the JHDL design tool are partitioned into a number of smaller, more specific Jar archive files[2]. This allows a given applet to require only those Jar files required by the applet code and avoid wasting network bandwidth with unused class files. As shown in Table 1, the executables required by the applet of Figure 3 involve four such jar files for a total of 795 kB. Other jar files may be provided for applets requiring additional functionality or technology libraries.

File	Size	Description
JHDLBase.jar	346 kB	JHDL Classes & Simulator
Virtex.jar	293 kB	Xilinx Virtex Library
Viewer.jar	140 kB	Schematic Viewers
Applet.jar	16 kB	Module Generator & Applet
Total	795 kB	

Table 1: JAR Files Used By Constant Multiplier Applet

5. CONCLUSIONS

Exploiting the functionality of the JHDL design environment within Java applets provides a powerful and flexible approach for web-based IP evaluation and delivery. By using the JHDL infrastructure, applets can be created with a variety of IP evaluation and delivery features. Schematic viewers, simulators, netlist generators and other related functionality can be combined into an applet that provides appropriate levels of visibility along with IP protection. Since these applets run on a browser and are accessed from the web, IP evaluation is made available to customers with little or no external software. Further, customers will always access the latest revisions of the IP applet software when the applet binaries are downloaded.

Future efforts to improve JHDL applets include creating applets for more complicated IP, investigating more secure delivery techniques, and developing applets that delivering more than one IP module. As additional IP evaluation features are developed and interfaces with more tools are provided, JHDL based applets will provide a powerful alternative for web-based delivery of FPGA IP.

[2]Java Jar files are compressed archive files used to collect a number binary class files and other program resources.

6. REFERENCES

[1] M. Dalpasso, A. Bogliolo, and L. Benini. Virtual simulation of distributed ip-based design. In *Proceedings of the 36th Design Automation Conference, DAC 1999*, pages 50–55, 1999.

[2] A. Fin and F. Fummi. A Web-CAD methodology for IP-core anaylsis and simulation. In *Proceedings of the 37th Design Automation Conference, DAC 2000*, pages 597–600, 2000.

[3] B. Hutchings, P. Bellows, J. Hawkins, S. Hemmert, B. Nelson, and M. Rytting. A CAD suite for high-performance FPGA design. In *Proceedings of the IEEE Workshop on FPGAs for Custom Computing Machines*, pages 12–24. IEEE Computer Society, IEEE, April 1999.

[4] B. L. Hutchings and B. E. Nelson. Using general-purpose programming languages for FPGA design. In *37rd Design Automation Conference (DAC)*, pages 561–566, Los Angeles, CA, June 2000.

[5] P. James-Roxby and S. A. Guccione. Automated extraction of run-time parameterisable cores from programmable device configurations. In *Proceedings of the IEEE Symposium on FPGAs for Custom Computing Machines*, pages 153–161, April 2000.

[6] F. Koushanfar and G. Qu. Hardware metering. In *Proceedings of the 38th Design Automation Conference, DAC 2001*, pages 490–493, 2001.

[7] J. Lach, W. H. Mangione-Smith, and M. Potkonjak. Robust FPGA intellectual property protection through multiple small watermarks. In *Proceedings of the 36th Design Automation Conference, DAC 1999*, pages 831–836, 1999.

[8] S. Mittra. *Principles of Verilog PLI.* Kluwer Academic Publishers, 1999.

[9] M. J. Wirthlin and B. McMurtrey. Efficient constant coefficient multiplication using advanced FPGA architectures. In *Field-Programmable Logic and Applications. Proceedings of the 11th International Workshop, FPL 2001*, pages 555–564, August 2001.

Watermarking Integer Linear Programming Solutions

Seapahn Megerian
Computer Science Department
University of California Los Angeles
seapahn@cs.ucla.edu

Milenko Drinić
Computer Science Department
University of California Los Angeles
milenko@cs.ucla.edu

Miodrag Potkonjak
Computer Science Department
University of California Los Angeles
miodrag@cs.ucla.edu

ABSTRACT
Linear programming (LP) in its many forms has proven to be an indispensable tool for expressing and solving optimization problems in numerous domains. We propose the first set of generic watermarking techniques for integer-LP (ILP). The proof of authorship by watermarking is achieved by introducing additional constraints to limit the solution space and can be used as effective means of intellectual property protection (IPP) and authentication. We classify and analyze the types of constraints in the ILP watermarking domain and show how ILP formulations provide more degrees of freedom for embedding signatures than other existing approaches. To demonstrate the effectiveness of the proposed ILP watermarking techniques, the generic discussion is further concretized using two examples, namely Satisfiability and Scheduling.

Categories and Subject Descriptors
K.6.5 [**Management of Computing and Information Systems**]: Security and Protection -- *Digital Watermarking*

General Terms
Algorithms, Economics, Theory, Legal Aspects.

Keywords
Digital Watermarking, Intellectual Property Protection

1. INTRODUCTION
Recently, the advent of modularized hardware and software libraries and core-based system-on-chip (SOC) design methodologies fueled a flurry of research focused on intellectual property (IP) protection (IPP). One of the techniques that has been shown to be extremely effective for IPP is watermarking (WM) due to its potential simplicity, strong authorship proofs, flexibility, and low overhead. Watermarking techniques utilize the fact that generally, tasks used in IP design rely on computationally intractable optimization problems having numerous solutions of similar quality. By introducing additional constraints corresponding to the author's signature, they limit the solution space and hence make a convincing authorship proof if a particular solution can be shown to belong to the constrained solution space. Numerous Watermarking methods have been proposed for different stages of the design process, including system synthesis, logic synthesis, technology mapping, and physical design [Kah98, Hon95, Meg00].

Linear programming is an efficient method for minimizing a linear objective function under linear inequality constraints. Integer linear programming (ILP) is a special form of linear programming with the additional restriction that variables are integers. While any linear programming instance can be solved in polynomial time, solving ILP is NP-hard. Still, numerous effective techniques and packages have been developed that solve fairly large ILP instances in reasonable run times. The popularity of ILP as a backbone to many optimization algorithms has prompted the need for effective ILP watermarking techniques. In this paper we classify and discuss a generic set of techniques that can be used to watermark the solutions of ILP formulations. In addition to the generic-level discussion, we illustrate the methods using two specific problems, namely Satisfiability (SAT) and Scheduling. In addition to their wide range of applications, these two problems are well suited to complement our discussions since SAT is decision oriented while Scheduling is an optimization problem. Although problem-specific watermarking techniques have already been proposed for both SAT and Scheduling, we show the effectiveness and power of ILP watermarking techniques and discuss why they can be superior.

1.1. An ILP Watermarking Example
To motivate the ILP watermarking discussion, we use a very simple SAT example. We watermark the solution of the simple instance by embedding signature constraints in its corresponding ILP formulation in two ways. Consider the five-variable formula:

$$F = (x_1 + x_2 + x_3) \cdot (x_4 + x_5)$$

There are a total of 21 possible satisfying truth assignments out of the 32 total assignments for F. We assume that any of the valid solutions are equally likely to occur so the probability that a specific solution is selected as the final answer is 21/32. The ILP formulation corresponding to the SAT instance represented by F can be stated as:

$$\begin{array}{lll} \text{Minimize:} & \varnothing & \\ \text{Where:} & x_1 + x_2 + x_3 & \geq 1 \\ & x_4 + x_5 & \geq 1 \end{array}$$

For the sake of brevity, we ignore the complemented variables and assume that all variables are defined as 0-1 integers. A watermark is essentially a set of characteristics imposed in the final ILP solution by limiting the solution space. The following formulation demonstrates one such constraint using the ILP objective function:

$$\begin{array}{lll} \text{Maximize:} & x_1 - x_2 - x_3 + x_4 - x_5 & \\ \text{Where:} & x_1 + x_2 + x_3 & \geq 1 \\ & x_4 + x_5 & \geq 1 \end{array}$$

This new ILP formulation still represents the original SAT instance corresponding to F. The new objective function signature here is obtained using the pseudo-random bit string "10010" (see 5.2). The new solution space has been reduced to 1 out of the original 21 valid assignments, an indication of the WM strength. Note that for this example, this is the strongest possible watermarking solution.

Another method for imposing watermarking constraints in ILP formulations is the use of augmented variables. Consider the addition of a new integer variable x_a in the original formulation:

DAC 2002, June 10-14, 2002, New Orleans, Louisiana, USA.

Maximize: x_a

Where: $x_1 + x_2 + x_3 - x_a \geq 1$

$x_4 + x_5 - x_a \geq 1$

$x_a \geq 0$

It is easy to verify that the new formulation has 3 solutions of equal quality. Such use of augmented variables in order to impose new and complex relationships among unrelated constraints is one of the main advantages of using ILP WM. Previously proposed WM methods do not provide the flexibility and power of ILP techniques to impose complex relationships among groups of constraints and variables. Generally, practical solution instances are much larger and provide a rich environment for embedding long signatures by simultaneously applying several different WM techniques.

1.2. Paper Organization

The remainder of this paper is organized as follows: Section 2 contains the related work. Section 3 contains the preliminary discussion including definitions, notations, and assumptions. Section 4 presents the generic ILP watermarking techniques and constraint classifications. Sections 5 and 6 present the details of ILP watermarking specific to SAT and Scheduling with supporting experimental results and analysis, followed by the conclusion.

2. RELATED WORK

Imperfections of the human visual and auditory system have been used as the basis for hiding data in still images, audio, and video [Bor96]. However, watermarking IP where no errors are allowed, such as designs and programs, require different techniques. Protocols for watermarking active IP have been developed at the physical level [Kah98], technology mapping and template matching [Meg00], and behavioral synthesis [Hon95]. Qu et al. address the questions of the quality of the solution subspace and the difficulty of finding a solution from such subspace in [Qu00].

Boolean Satisfiability and Scheduling have numerous applications in CAD and other domains. Many design problems such as test pattern generation, circuit delay computation, logic optimization, and combinational equivalence checking can be formulated as SAT problems [Mar00]. An example of a heuristics for solving SAT problem is presented in [Bay96]. Summery of important classical scheduling theory results for real-time computing can be found in [Sta95] and [Pau89]. An example of an ILP-based formulation for Scheduling is presented in [Geb91].

3. PRELIMINARIES

3.1. SAT

Let $U=\{u_1, u_2, \ldots, u_m\}$ be a set of Boolean variables. A *clause* over U is a set of literals that represents their disjunction. A clause is *satisfied* by a truth assignment if and only if at least one of its members is true under that assignment. The Satisfiability problem is hence defined as [Gar79]:

Problem: ***SATISFIABILITY (SAT)***
Instance: *A set U of variables; A collection C of clauses over U.*
Question: *Is there a satisfying truth assignment of C ?*

Usually SAT problems are stated in the from of a formula. Section 1.1 contains an example of a SAT formula and its corresponding ILP formulation. In our notation, x_i' indicates the logical complement of the variable x_i. For simplicity, we use the same names to represent the Boolean SAT variables and the ILP variables. We assume that all SAT ILP variables are defined as 0-1 integers unless otherwise specified.

3.2. Scheduling

We use the synchronous data flow (SDF) as our primary computational model [Lee87]. SDF is a special case of general data flow, in which the number of data samples produced or consumed by each node is specified a priori. The syntax of a targeted computation is defined as a hierarchical control-data flow graph (CDFG) [Rab91]. The CDFG represents the computation as a flow graph, with nodes, data edges, and control edges. The semantics underlying the syntax of the CDFG format is that of the SDF flow model. Operation scheduling is the process of partitioning the set of operations in a CDFG into groups such that the operations in the same group are executed concurrently in the same control step. The task of operation scheduling is to determine the amount of resources necessary for execution of operations in a CDFG.

The two general approaches for scheduling are heuristic-based methods [Pau89] and ILP [Geb91]. In this paper, we restrict our attention to ILP for time-constrained scheduling, where for a given number of control steps, the set of arithmetic and logical operations in a CDFG is partitioned such that each partition corresponds to a single control step. The goal of scheduling here is to minimize the number of functional units necessary for the execution of all operations in the CDFG. The problem can formally be stated as:

Problem: ***Time-Constrained SCHEDULING of a CDFG.***
Instance : *A CDFG C, a time constraint t, parameter n.*
Question: *Does there exist a Scheduling of C into t control steps, such that the number of operations from C in each control step is less than n and all precedence constraints in C are satisfied?*

The problem of Time-Constrained Scheduling of a CDFG is computationally intractable as it is essentially the same problem as precedence constrained scheduling, a well known NP-complete problem [Gar79]. To present the ILP formulation of the time-constrained scheduling problem [Geb91], we define variables x_{ij} as follows:

$$x_{i,j} = \begin{cases} 1 & \text{if operation } j \text{ is scheduled in control step } i. \\ 0 & \text{Otherwise.} \end{cases}$$

$1 \leq i \leq n$, $1 \leq j \leq m$, where n is the number of the given clock cycles and m is the number of operations in the CDFG. The condition that each operation has to be scheduled in exactly one control step is expressed through the equation:

$$(1) \qquad \forall j \quad \sum_{i=1}^{n} x_{i,j} = 1$$

This condition is usually replaced with a tighter condition taking the earliest and the latest possible times when an operation can be scheduled into consideration.

$$(2) \qquad \forall j \quad \sum_{i=k}^{l} x_{i,j} = 1$$

where $k = ASAP(x_{i,j})\ \forall i$, and $l = ALAP(x_{i,j})\ \forall i$. $ASAP(x_{i,j})$ and $ALAP(x_{i,j})$ refer to *as soon as possible* and *as late as possible* heuristic algorithms for scheduling. The data flow in the CDFG is restricted with precedence constraints, formulated as:

$$(3) \qquad \sum_{i=1}^{n} (n-(i+1))x_{i,j_1} \geq 1 + \sum_{i=1}^{n} (n-(i+1))x_{i,j_2}$$

where j_1 is the operation in the CDFG whose output is used as an input by the operation j_2. We also define variables x_i as:

$$(4) \qquad \forall i \quad x_i = \sum_{j=1}^{m} x_{i,j}$$

The variable x_i represents the actual number of functional units for each control step i. The problem of minimizing the number of functional units necessary for execution of the CDFG under the above constraints can now be formulated as:

Minimize: y
Where: $x_i < y \quad \forall i$

This definition can trivially be extended to the allocation problem, separating the conditions for number of functional units to different types of units.

4. GENERAL ILP WATERMARKING

The global flow for ILP watermarking is shown in Figure 1. The main strategy is to add a number of additional constraints and try to satisfy as many of them as possible. The watermarking constraint selection is made according to a pseudorandom number generated in correspondence with a signature. The constraint encoding procedure yields sufficiently randomized constraints to hinder the detection and forging of signatures. Figure 2 shows the encoding process: SHA is a one-way hash function, RSA is a public key encryption system, and RC4 is a stream cipher (SHA and RSA are used in the PGP software package). RC4 outputs a cryptographically strong pseudorandom bit-stream that the constraint encoding step uses to generate the WM constraints. Verification of a signature involves both testing for signature presence and matching the signature to the text file of the alleged owner. Demonstrating that the signature is in the solution is done by showing an unusual presence of enough WM constraints. A signature can be shown to correspond to the text file and the owner's public key using PGP.

The types of methods used to embed watermarks in ILP formulations have three main classifications: (i) *augmented variable definition*, (ii) *objective function*, and (iii) *ILP constraints*. In the following subsections, we provide an overview of each of these methods.

4.1. Variable Definitions

The variables used in ILP formulation of optimization and decision problems fall into two categories: (i) *natural* and (ii) *augmented*. *Natural* variables are variables that directly represent aspects of the original problem. *Augmented* variables on the other hand, have no apparent relationship to the problem and often do not have any physical meaning. The ability to define and use augmented variables in ILP formulations is a powerful method of imposing additional constraints in order to improve watermarking strength, which is often not possible with other methods.

Augmented variables enable the imposition of complex relationships among the natural variables and constraints. These new dependencies are key factors that make the ILP based watermarking methods more powerful than previously proposed algorithm-specific techniques. Section 5 and 6 provide several examples to demonstrate how augmented variables can be used to impose specific watermarking relationships among independent constraints.

4.2. Objective Function

Generally, each ILP formulation is composed of an objective function F_0 and a set of constraints. Any number of additional objective functions can be used for signature encoding. The general objective function F_{WM} used in WM ILP has the following structure:

$$F_{WM} = \alpha_0 F_0 + \alpha_1 F_1 + \alpha_2 F_2 + \alpha_3 F_3 + \ldots$$

where F_i (i>0) denote augmented watermarking objective functions and coefficients α_i are real valued constants.

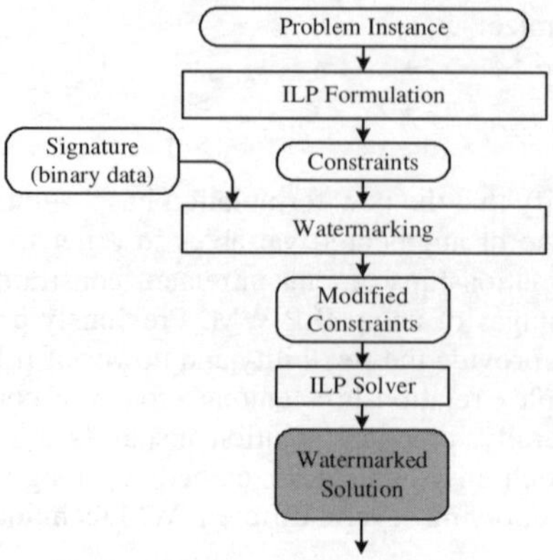

Figure 1. Global Flow For ILP Watermarking

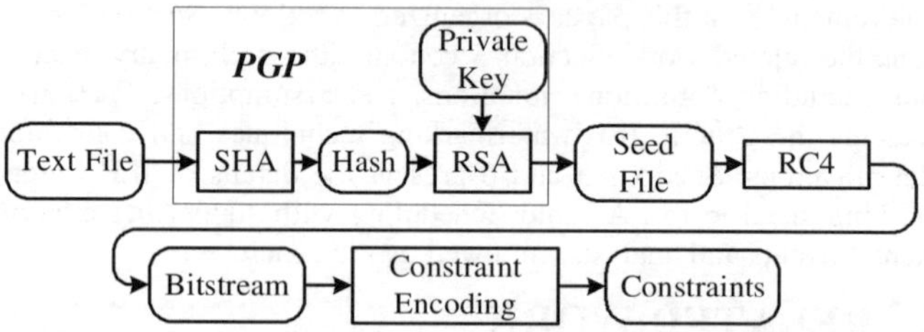

Figure 2. ILP Watermarking - Constraint Encoding

Note that in optimization intensive applications, $\alpha_0 >> \alpha_i$ for i>0 since the quality of the final solution must not be allowed to degrade. The augmented objective functions F_i can be any arbitrary linear function of the natural and augmented ILP variables. Problem specific attributes can guide the construction of such functions in order to improve WM strength and preserve solution quality. Certain decision problems such as SAT, do not have any inherent objective functions in their corresponding ILP formulation and hence provide the most flexibility for embedding such signatures.

4.3. ILP Constraints

Modifying the original ILP constraints is the third and most broad method of embedding WM constraints. To ensure solution correctness, almost all such techniques manipulate the original set of ILP constraints such that the new resulting constraints are more restrictive. As the case with ILP variables, ILP constraints can be either *natural* or *augmented*. Natural constraints arise from the problem specification and care must be taken to ensure that they are not violated as a result of the watermarking process. Augmented constraints do not generally have any obvious relation to the problem at hand and can provide great freedom in embedding signatures. Furthermore, each type of constraint can be classified as *positive* or *negative*. A *positive* constraint specifies the value or range of values that each variable should be assigned to, while a *negative* constraint specifies undesirable assignments.

5. ILP WATERMARKING – SAT

5.1. SAT WM: Variable Definitions

Normally, there is little freedom in redefining the natural set of variables in SAT. However, augmented variables allow the imposition of complex relationships among the existing and/or augmented constraints. Consider the following constraints:

$$x_1 + x_2 + x_5 + x_9 + x_{21} \geq 3$$
$$x_3 + x_7 + x_{11} + x_{17} \geq 2$$

- Define augmented integer variable x_a and modify constraints:

$$\begin{aligned} x_1 + x_2 + x_5 + x_9 + x_{21} - x_a &= 3 \\ x_3 + x_7 + x_{11} + x_{17} - x_a &= 2 \\ x_a &\geq 0 \end{aligned}$$

In this simple example, the new variable x_a defines a new relationship between the two original constraints. So, although previously each constraint could be satisfied independently, here, there is a new underlying dependency that relates seemingly unrelated constraints. This notion can further be used to include multiple sets of constraints and multiple augmented variables. However, care must be taken to ensure that the additional constraints do not cause a satisfiable formula to become unsatisfiable and vice versa.

5.2. SAT WM: Objective Functions

As stated earlier, SAT is purely a decision problem and hence does not pose any inherent objective functions. This additional degree of freedom provides powerful means of encoding SAT watermarking constraints. For instance, consider a SAT formulation with 10 variables and a 10-bit pseudo-random bit sequence (signature) S:

$$S = \text{“1011001011”}$$

- Define a new SAT ILP objective function F_1 by mapping each variables x_i to a bit $s_i \in S$. We add the variable x_i if the corresponding bit b_i=1 and subtract x_i if b_i=0:

$$F_1 = x_1 - x_2 + x_3 + x_4 - x_5 - x_6 + x_7 - x_8 + x_9 + x_{10}$$

Maximizing F_1 in the ILP in effect is an attempt to preserve the signature S in the form of variable assignments. Here, if all variables are assigned according to the bits in S, then the WM strength is maximized. However, such an assignment may not be a valid solution. The scheme can be enhanced using the first- and second-order heuristic objective functions presented in [Qu99]. The functions essentially calculate the likelihood of particular assignments for each variable in a valid solution. This likelihood information can be used to improve the mapping process of the signature string to the WM objective function, thus increasing the WM strength. Specifically, the heuristics can be used to decide if each bit $b_i \in S$ should be mapped to the variable x_i or its complement x_i' (+ or – in the objective function) in order to increase the likelihood that assigning variables according to S would result in a valid solution.

5.3. SAT WM: ILP Constraints

Let us now consider ILP constraint manipulation techniques for watermarking. Each *natural* constraints in the ILP corresponds to a clause in the SAT formula. Consider the constraint:

$$x_1 + x_2 + x_5 + x_9 + x_{21} \geq 1$$

- We can reduce the solution space by *removing variables* from such constraints. This has the same effect as *deleting literals* from corresponding clauses that is discussed in [Qu99]. For example, we can replace the constraint above with the new constraint assuming that the new constraint does not make the formula unsatisfiable:

$$x_2 + x_5 + x_9 \geq 1$$

- We can further limit the solution space by adding new *augmented ILP constraints.* In addition to adding new constraints similar to the ones above (see *adding clauses* in [Qu99]) we may opt to impose special characteristics on groups of variables such as:

$$\begin{aligned} x_3 + x_4' + x_9 + x_{11}' + x_{17} &\geq 3 \\ x_1' + x_5 + x_6 + x_{12} + x_{20} &\geq 4 \\ x_1 + x_2 + x_5 + x_9 &\leq 2 \end{aligned}$$

Augmented constraints can be combined with augmented variables to produce even more signature embedding options.

5.4. SAT Experimental Results

We compared the performance of the ILP WM approach to the optimization intensive SAT-solver results presented in [Qu99]. We use the same set of DIMACS benchmark instances. Table 1 lists the DIMACS benchmarks and the number of variables and clauses for each instance. The “Opt. Bits” column shows the number of watermarked bits as reported in [Qu99]. For each instance, we embedded a comparable size random signature in each instance and used LP_SOLVE, a freely available (GNU) mixed integer linear program solver. In all cases, the ILP solver produced the watermarked results with the run-times reported in minutes in column six, on a 333 MHz Sun Ultra 5 Workstation with 128 MB RAM. It is expected that larger instances can be watermarked equally effectively using more sophisticated commercial ILP solvers.

Table 1. SAT WM: ILP vs. Optimization Intensive Technique

Instance	Vars	Clauses	Opt. Bits	ILP Bits	ILP Run time
ii8a1.cnf	66	186	1400	1500	1
ii8a2.cnf	180	800	3100	3400	5
ii8a3.cnf	264	1552	6700	6500	3
ii8a4.cnf	396	2798	3900	4600	7
ii8b1.cnf	336	2068	6800	7200	11
ii8b2.cnf	576	4088	8900	8800	35
ii8c1.cnf	510	3065	7500	7900	22

6. ILP WATERMARKING – SCHEDULING

Formulating scheduling as an ILP problem opens new dimensions of imposing watermarking constraints that are impossible or difficult to achieve using other methods. In this section we present a few examples to demonstrate how generic ILP watermarking can apply to Scheduling.

6.1. Scheduling WM: Variable Definitions

- Define augmented variables $x_{i,j}$ that satisfy the equation $\sum_{i=k}^{l} x_{i,j} = 1$ and impose a number of precedence constraints expressed using equations of type (3). The new variables are not added to equations for variable x_i as they are not affecting the quality of the final solution. Here, we exploit a degree of freedoms specific to ILP formulations of scheduling that does not have a physical interpretation.

- Define an augmented variable x_i corresponding to the addition of a new control step. The direct consequence is the quality of the final solution. As shown in Table 2, this technique does not reduce the number of solutions and it should be used in conjunction with other techniques. However, the feasible solutions are altered as a consequence of the constraints and which can serve as a proof of authorship. The only exception is when this technique enables the reduction of number of functional units necessary for execution of operations in a CDFG.

- Define an augmented variable z_i,

$$z_i = x_{i_1,j_1} + x_{i_2,j_2} + \ldots + x_{i_g,j_g} + x_{i_1} + x_{i_2} + \ldots + x_{i_h}$$

under the constraint that $a_i \leq z_i \leq b_i$, , where a_i and b_i are specified positive integers. Note that this type of constraint does not have any real interpretations in the Scheduling domain, yet provides another degree of freedom for watermarking solutions. Many more constraints of this nature with varying structures can be imposed.

6.2. Scheduling WM: Objective Functions

- Introduce correlation constraints for variables x_i:

$$x_{i_1} + x_{i_2} + \ldots + x_{i_g} = x_{i_{g+1}} + x_{i_{g+2}} + \ldots + x_{i_{g+h}}$$

This technique directly affects the objective function. The introduced correlation does not have any direct association to the scheduling problem.

- Maximize α, the weighted total number of satisfied constraints, $\sum_{i=1}^{g} w_j c_i = \alpha \quad j \in C$, where C is the set of constraints.

Various watermarking constraints introduced in this section can be applied alone or in conjunction to one or more of the other techniques. The ILP solver may be unable to solve the problem under all posed constraints. Here, we introduce another objective function which will allow the ILP solver to satisfy as many WM constraints as possible, while still obtaining a good quality solution.

6.3. Scheduling WM: ILP Constraints

- Fix variable $x_{i,j}$ to 1: $x_{i,j} = 1$, and $x_{g,j} = 0$ for $g \neq i$.

For solution feasibility, the range of i is in [*ASAP, ALAP*]. This technique corresponds to fixing an operation from CDFG to a particular control step.

- Fix variable $x_{i,j}$ to 0: $x_{i_1,j} = 0, x_{i_2,j} = 0, \ldots, x_{i_h,j} = 0, k \le i_i \le l$.

This constraint in effect forbids an operation from being scheduled in one or more control steps and is thus a *negative* constraint.

- Enforce minimum distance between two variables with existing precedence constraint:

$$\sum_{i=1}^{n} (n-(i+1)) x_{i,j_1} \ge d_{\min} + \sum_{i=1}^{n} (n-(i+1)) x_{i,j_2}$$

Minimum distance $d_{\min}$, where $d_{\min}$ is a positive integer, represents the minimum number of clock cycles that j_1 is scheduled before j_2.

- Enforcing maximum distance between two variables with existing precedence constraint: The constraints in this case have essentially the same form as the previous case with the exception that the positive integer variable $d_{\max}$ is used instead of d_{min} and the inequality is '$\le$' instead of '$\ge$'. Variable $d_{\max}$ represents the maximum number of clock cycles that j_1 is scheduled before j_2.

- Enforcing minimum distance between two variables without existing precedence constraint:

$$\sum_{i=1}^{n} (n-(i+1)) x_{i,j_1} \ge d_{\min} + \sum_{i=1}^{n} (n-(i+1)) x_{i,j_2}$$

Variables that do not have explicit precedence constraints but have implicit precedence constraints as a result of transitivity of preceding relations are constrained by this technique in such a way that the implicit order is preserved.

- Enforcing maximum distance between two variables without existing precedence constraint: Implicit precedence constraint restricts this technique to cases where $d_{\max}$ is greater than the critical path between j_1 and j_2.

6.4. Impacts of WM on Solution Space

Encoding the signature into constraints is the final step in embedding the signature in the ILP formulation. This step relies on the selection of WM techniques and their associated parameters. We use the first block of bits of a signature bit-stream for technique-selection and the rest for parameter selection. For each of the presented techniques, we attempt to embed the WM such that after satisfying all constraints, there is still a non-zero probability that a valid solution exists. For example, if a variable $x_{i,j}$ is being fixed to 1, the applied procedure must choose i in the range [*ASAP, ALAP*]. Similarly, if the critical path in the CDFG between two variables is more than $d_{\max}$ while enforcing a maximum distance constraint, then the solution to the problem becomes infeasible. Due to the intractability of Scheduling, the proposed procedure cannot guarantee that the solution will exist. Thus, such decisions almost always rely on problem specific heuristics. One attempt at such a heuristic is the α maximization technique presented in Section 6.2, that relaxes the constraints and enables easier solution search for the ILP solver. The weights for each constraint are inversely proportional to the ratio of their resulting average solution space reduction.

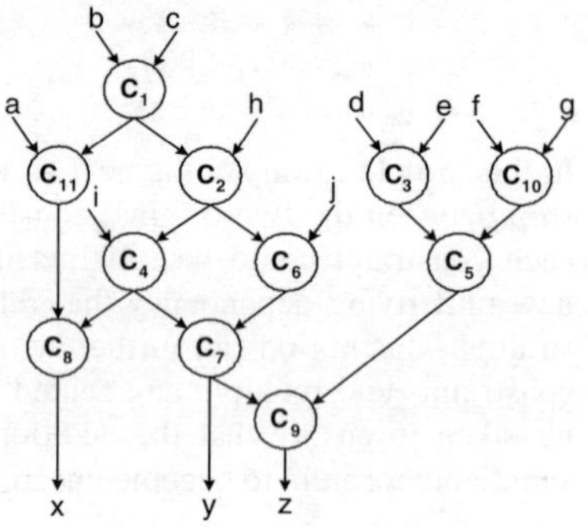

Figure 3. A CDFG Example

Consider the CDFG shown in Figure 3, where variable $x_{i,j}$ corresponds to operation C_j and variable x_i corresponds to the control step i. Suppose that the required number of control steps for the execution of this CDFG is 8. Here, the minimum number of functional units required for execution is 3. Table 2 lists the corresponding *ASAP* and *ALAP* assignment for each operation.

Table 2. ASAP/ALAP Scheduling for the CDFG in Figure 3.

OP.	C_1	C_2	C_3	C_4	C_5	C_6	C_7	C_8	C_9	C_{10}	C_{11}
ASAP	1	2	1	3	2	3	4	4	5	1	2
ALAP	4	5	6	6	7	6	7	8	8	6	7

There are 44,715 valid operation-to-control-step (scheduling) assignments of the CDFG in Figure 3. Let us consider the 36-bit sequence 39D6B4C86$_{\text{hex}}$ as our signature. If we only use the "fix variable $x_{i,j}$ to 1" technique then our encoding will produce:

$x_{7,8} = 1$; $x_{i,8} = 0,\ i \neq 7$
$x_{6,11} = 1$; $x_{i,11} = 0,\ i \neq 6$
$x_{2,5} = 1$; $x_{i,5} = 0,\ i \neq 2$
$x_{7,9} = 1$; $x_{i,9} = 0,\ i \neq 7$

We use the first 5 bits of the sequence for selection of the particular variable $x_{i,j}$. Index j is obtained by the formula $j = (code)_{\text{mod}11} + 1$, where *code* refers to the decimal number corresponding to the first 5 bits of the sequence. The reason for using 5 bits instead of 4, the minimal number, is to make the selection as fair as possible [Qu00]. Usage of only 4 bits would make the likelihood of selecting the first 6 variables twice as large as the rest. Index i is selected similarly, using 4 bits, and normalizing the number to the range of feasible solutions (in the [ASAP, ALAP] range). If the sequence decoding results in the selection of a variable that is already scheduled, that selection is discarded. After applying the first constraint, the possible number of solutions is reduced to 12,515, after the second constraint the number of solutions drops to 2,793, and after the next two constraints, to 29 and 7 respectively.

If we select the "fix variable $x_{i,j}$ to 0" technique from Section 6.3, using the same encoding as in the previous case ($x_{7,8} = 0$; $x_{6,11} = 0$; $x_{2,5} = 0$; and $x_{7,9} = 0$), the number of possible solutions is reduced at a much slower rate. After applying the first constraint, the number of solutions drops to 32,200, after the second constraint, to 28,437, and after the last two constraints to 28,251 and 24,137 respectively.

Clearly, different methods of converting the signature to WM constraints are possible. We can expect that the "fix variable $x_{i,j}$ to 1" reduces the solution space more than the "fix variable $x_{i,j}$ to 0". Due to the nature of Scheduling ILP formulations, the *positive* constraint of setting a variable to 1 forces several more to 0. The solution space reduction caused by other techniques can be seen in Table 4. The number of possible solutions and minimal number of functional units needed for execution of all operations of the CDFG in Figure 3 is shown in Table 3, as the allowed maximum number of control cycles is changed. For some techniques, the length of the signature in our example is not sufficient to show enough correlation between the technique and the change in solution space size. The impact of the remaining techniques highly depends on the selected variables and the actual enforced constraints.

Table 3. Impact of Increasing Cycles on Number of Solutions

	Total number of cycles available					
	6	7	8	9	10	11
F. Units	2	2	2	2	2	1
Solutions	155	4,175	44,715	257,506	1,435,227	6,944

Table 4. Impact of Watermarking on Solution Space

Technique	WM Sequence [bits]			
	9	**18**	**27**	**36**
Define new variable $x_{i,j}$	-	44715	-	39012
Fix variable $x_{i,j} = 1$	12515	2793	29	7
Fix variable $x_{i,j} = 0$	32200	28437	28251	24237
4	16030	16030	4354	184
5	36438	36438	30685	30305
6	-	44715	-	823
7	-	44715	-	44647

4 – Min distance between variables with existing precedence.
5 – Max distance between variables with existing precedence.
6 – Min distance between variables without existing precedence.
7 – Max distance between variables without existing precedence.

Table 5. Scheduling WM Experimental Results

Application		4K bits signature		16K bits signature	
Name	**# of Ops**	**P_c**	**Perf. Overhead**	**P_c**	**Perf. Overhead**
G721	758	10^{-20}	0.2%	10^{-49}	0.7%
Epic	872	10^{-29}	0.1%	10^{-85}	0.9%
PEGWIT	658	10^{-23}	0.8%	10^{-46}	1.6%
PGP	1755	10^{-76}	0.3%	10^{-178}	1.2%
GSM	802	10^{-44}	0.9%	10^{-125}	1.3%
JPEG.c	1422	10^{-68}	0.3%	10^{-289}	2.5%
MPEG2.d	1372	10^{-47}	0.2%	N/A	N/A

6.5. Scheduling Experimental Results

To quantify the performance of the ILP-based watermarking approach for Scheduling, we compared the performance to the localized WM methodology presented in [Kir99]. Table 5 lists the instances used and the number of operations in each CDFG. The next set of results are for embedding 4K bit- and 16K-bit random signatures. The P_c column shows the calculated likelihood of coincidence as defined in [Kir99]. It estimates the ratio of the number of high quality solutions in the constrained watermarked space with respect to the non-constrained solution space, which can be used as an approximation for WM strength. Note that the exact calculation of this ratio is intractable due to the exponentially-increasing size of the solution space. The performance overhead column lists the run-time overhead in solving the watermarked ILP formulation vs. the original formulation. In most cases, the ILP-based method was able to produce the specified WM solutions fairly rapidly. However, the limitation of our ILP solver (LP_SOLVE) became apparent on larger instances such as MPEG2.d.

7. CONCLUSION

We presented a set of generic watermarking techniques for ILP solutions. We classified the general types of constraints in the ILP watermarking domain and discussed how each can be used to embed a signature. We also provided specific examples and detailed discussion from the SAT and Scheduling domains and showed several tradeoffs and problem specific attributes in each case. Experimental results indicate that ILP watermarking methods have tremendous potentials for embedding strong watermarks with little overhead.

8. REFERENCES

[Bay96] R.J. Bayardo and R. Schrag. "*Using CSP Look-Back Techniques To Solve Exceptionally Hard SAT Instances.*" Principles and Practice of Constraint Programming, pp.46-60, 1996.

[Bor96] A.G. Bors and I. Pitas. "Image Watermarking Using DCT Domain Constraints." *International Conference on Image Processing*, Vol.3, pp.231-244, 1996.

[Geb91] C.H. Gebotys, and M.I. Elmasry. "*Simultaneous Scheduling And Allocation For Cost Constrained Optimal Architectural Synthesis.*" Proceedings of IEEE Design Automation Conferecnce, pp. 2-7, 1991.

[Gar79] M. R. Garey and D. S. Johnson. *Computers And Intractability: A Guide To The Theory Of NP-Completeness*. W.H. Freeman, 1979.

[Hon95] I. Hong, M. Potkonjak. "*Behavioral Synthesis Techniques for Intellectual Property Protection.*" DAC, pp. 849-854, 1999.

[Kah98] A. Kahng, et al. "*Watermarking Techniques For Intellectual Property Protection.*" DAC, pp. 776-781, 1998.

[Kir99] D. Kirovski, and M. Potkonjak. "*Localized Watermarking: Methodology And Application To Operation Scheduling.*" IEEE International Conference on Computer-Aided Design, pp.596-9, 1999.

[Lee87] E.A. Lee, D.G. Messerschmitt. "*Synchronous Dataflow.*" Procs. of the IEEE, Vol.75, (no.9), pp.1235-45, 1987.

[Mar00] J.P. Marques-Silva and K.A. Sakallah. "*Boolean Satisfiability in Electronic Design Automation.*" DAC, pp. 675–680. 2000.

[Meg00] S. Meguerdichian and M. Potkonjak. "*Watermarking While Preserving The Critical Path.*" DAC, pp. 108-111, June 2000.

[Pau89] P.G. Paulin and J.P. Knight. "*Force-Directed Scheduling For The Behavioral Synthesis Of ASICs.*" IEEE Trans. on Computer-Aided Design of Integrated Circuits and Systems, Vol.8, pp.661-679, 1989.

[Qu99] G. Qu, J.L. Wong, and M. Potkonjak. "*Optimization-Intensive Watermarking Techniques For Decision Problems.*" 36th ACM/IEEE Design Automation Conference Proceedings, pp. 33-36, June 1999.

[Qu00] G. Qu, J.L. Wong, M. Potkonjak. "*Fair Watermarking Techniques.*" Procs of Design Automation Conference, pp.55-60, 2000

[Sta95] J. Stankovic, et al. "*Implications Of Classical Scheduling Results For Real-Time Systems.*" IEEE Computer, Vol.28, pp. 16-25, 1995.

[Rab91] J. Rabaey, et. al. "*Fast Prototyping Of Data Path Intensive Architectures.*" IEEE Design & Test of Computers, Vol.8, pp.40-51, 1991.

This material is based upon work supported by the National Science Foundation Career award under Grant No. ANI-9734166.

Model Design Using Hierarchical Web-Based Libraries

Fabrice Bernardi
University of Corsica
UMR CNRS 6134
Corte, France
bernardi@univ-corse.fr

Jean-François Santucci
University of Corsica
UMR CNRS 6134
Corte, France
santucci@univ-corse.fr

ABSTRACT

Design tools can be profitably associated with libraries of reusable modeling components that will make the description and also the validation of the models much easier. Furthermore, applications of today and tomorrow will be increasingly based on three fundamental technologies: Object Orientation, Client/Server and Internet. We propose in this article an object-oriented architecture for the definition of Web-based hierarchical models libraries. The originality of our approach lies in the facts that it is based on : (i) a notion of genericity of use, (ii) notions like inheritance and abstraction links between the stored models and (iii) Web-based storing and consulting libraries procedures.

Categories and Subject Descriptors

J.6 [**Computer Applications**]: Computer-Aided Design; D.2.11 [**Software**]: Software Engineering—*Software Architectures*; D.2.13 [**Software**]: Software Engineering—*Reusable Software*

General Terms

Design, Management

Keywords

models reuse, models libraries, Web-based access, abstraction hierarchy

1. INTRODUCTION

The design of complex manufactured systems is a task requiring a lot of time to be achieved. One way to speed up this task is to develop methodologies for deriving and for using reusable design components. Recently a set of research work has been oriented towards this direction. We can highlight for example : (i) that components represent over 70% of product cost in the EDA industry[15] ; (ii) a multi-company project called ECIX (Electronic Component Information Exchange) and presented in [5]. In the same time, computer simulations have been becoming increasingly complex and require efficient system simulation framework [14]. Furthermore, increasing size and geographical separation of design data and teams has created a need for network-based design environments [6]. Usually, Design tools are profitably associated with libraries of reusable modeling components that will make the description of the models and also their validation much easier [4]. Storing models in a common generic library has several benefits. First, the genericity of this storage service can be offered to various modeling and simulation environments. Second, a common library allows environments to share information so they can interact each other, and third, modeling components can be shared by several users. This last point is the most important since it allows a design team enabling an efficient collaborative work.

Applications of today and tomorrow will be increasingly based on three fundamental technologies [13]: Object orientation allowing applications to be viewed in terms of natural objects; Client/Server allowing application components to behave as service consumers (client) and service providers (server); The Web allowing to access resources located all around the world. A Web-based access is a very interesting perspective for a generic modeling components library, since it allows an user-friendly remote access using a simple Web browser. We describe in this article an object-oriented architecture for the definition of Web-based hierarchical models libraries. The libraries are based on a hierarchical organization of concepts relating to the class of applications. The leaves of the hierarchical structure represent the elements which can be used in model descriptions. Information concerning the elements can be placed at the appropriate level in the hierarchy, thus, eliminating needless repetition when the elements are being defined. The originality of our approach lies in the facts that it is based on a strong notion of genericity of use, and notions like the inheritance and abstraction links between the stored models. We describe how managing inheritance between stored models can improve their classification and their accuracy, and how the abstraction hierarchy allows to describe a same model at various detail levels.

The paper is organized as follows. Section 2 presents the foundation of our work: the component-oriented hierarchical modeling and simulation process. Section3 introduces the notion of models library with all its inherent specificities. Section 4 discusses our approach for the construction of a Web-based access to a models library. Finally, section 5 concludes this paper and provides some perspectives of work.

2. COMPONENT ORIENTED MODELING AND SIMULATION PROCESS

We introduce in this section the basic concepts of the hierarchical components-oriented modeling and simulation process : (i) the abstraction hierarchy, (ii) the notion of modeling component and (iii) a complete modeling and simulation process.

2.1 Abstraction Hierarchy

One of the most difficult tasks in the field of modeling and simulation of complex systems is to choose a good level of detail. In all Domains, models are built at a precise abstraction level. The abstraction level of a model determines the amount of information which is contained in the model. The quantity of information in a model decreases with the abstraction levels: a model described at a low abstraction level will contain more information than a model described at a higher abstraction level [2].

Determining the correct abstraction level refers to selecting the quantum of information that must be included in the model to help address the modeling goals. So, well defining the abstraction level is an important step in modeling, and is often done very early in the modeling process described in section 2. A model described according to several abstraction levels is said a "hierarchical model".

2.2 Notion of Modeling Component

Since many years, reusable modeling components appear to be the ideal paradigm for implementing simulation models [15, 11]. We define a modeling component as a model that can be described following various abstraction levels, presenting well-defined communication interfaces (called ports) and that can be reused in various contexts.

2.3 Component-Oriented Process

We introduce in this part a theoretical modeling process that can be applied to all component-oriented modeling processes. This process can be divided in eight steps given in Figure 1. The first step (box 1) is to formalize the problem: presentation, objectives to be reached, validation criterions,... The second step (box 2) is to identify the components to be used. Starting from this identification, components are : (i) selected in a models library, or (ii) built from scratch (box 3). Once all the components are well defined, the newly created ones are stored in a models library for further retrieval (box 4). The next step is to build the global model with the previously selected or defined components (box 5). The main advantage of this modeling process lies in this step, since it consists only in interconnecting components. Once the global model is well-defined and seems acceptable, the simulator is build and the simulation is performed (box 6). The next step (box 7) concerns the validation of the model defined in box 5. If the validation step presents good results, the model can be used (box 8), otherwise it is necessary to go back and perform changes either in the components identification (box 2), in the components selection (box 3) or in the global model building (box 5).

3. DESCRIPTION OF A MODELS LIBRARY

When we deal with CAD (Computer Aided Design), a modeling component is most of the time a software component written in a given language and presenting some input and output ports. This component is used in a software framework (called modeling and simulation environment) able to handle it. A persistent model is a model that can exists before being used in the context of the environment, and that can exists even after its use. We define a models library as an object-oriented architecture, allowing a model designer to store and retrieve persistent models, directly reusable in their own modeling and simulation environment.

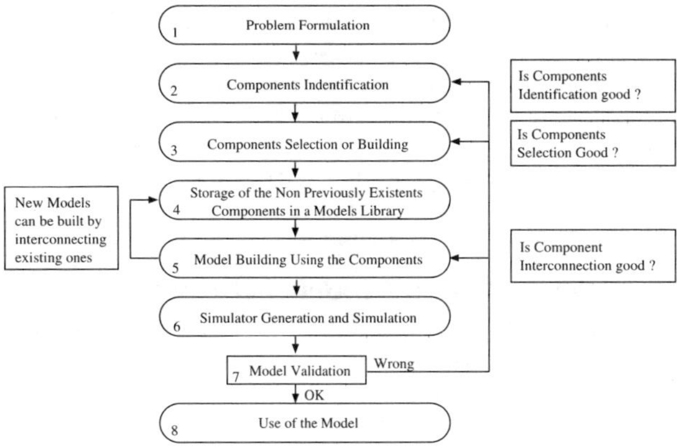

Figure 1: Components Oriented Process

3.1 Basic Concepts

A models library is structured according to two paradigms: the application domains and the abstraction levels. Furthermore the three following requirements should be taken into account : a models library must allow the storage independence from the considered application domain, must manage an inheritance hierarchy between the stored models, and must manage the abstraction links between the stored models. The notion of library we introduce here is different from the one commonly used in the Computer Science. We can compare it to the notion of Object Database [9], since we want to store directly reusable models rather than software functions or objects used for well-defined tasks. One of the most important objectives of a models library, as defined in this paper, is to be independent from the storage mode. That means that a library must be able to store fundamentally different models coming from fundamentally different modelling and simulation environments. This independence implies that we must dissociate the contents from the format of the stored models, and also that the communication interfaces of a library with the external applications must be identical for all kinds of models.

We define, in order to be as precise as possible, two notions: "context-in" and "context-out" models. A context-out model is an abstraction of a model. It presents a structure allowing it to be stored in a models library. A context-in model is a context-out model extracted from a Library and formatted so as to be directly reusable in its environment.

In order to avoid a repetition of properties inside same kinds of models, a models library must deal with an inheritance between the stored models. This inheritance between a parent model and its children allows to store these shared properties inside special models (called parent models), dramatically simplifies the children models, and facilitates the maintenance of the whole Library. Furthermore, this inheritance hierarchy inside a library provides all the classical benefits of object inheritance: automatic properties transmission and methods overloading if components are described using algorithmic functions. The importance of a good models library is that the model designer can be supplied with reasonable alternatives [12]. The choice between these alternatives can be strongly facilitated if the inheritance hierarchy is structured in a smart way.

3.2 Elements of a Models Library

Building a Models Library is as creating a high level representation of the models and their relations. In order to meet the requirements stated above, we define five kinds of objects to be stored in a Library object.

Domain elements allow performing a classification between theoretical domains of the stored models (Ex: DEVS Simulation [16], High Level Synthesis, VHDL Test).

Application Domain elements allow performing a classification between the application domain inside a given domain (Ex: Microelectronic, Energetic,...).

Classification Intermediate Model elements, belonging to an Application Domain, allow the creation of a classification hierarchy between the storage objects. This kind of model is not a storage model.

Inheritance Intermediate Model elements, which are storage models, allow the share of characteristics with its children through an inheritance mechanism.

Finally, Model File elements represent context-out models. It is the basic storage element of the Library.

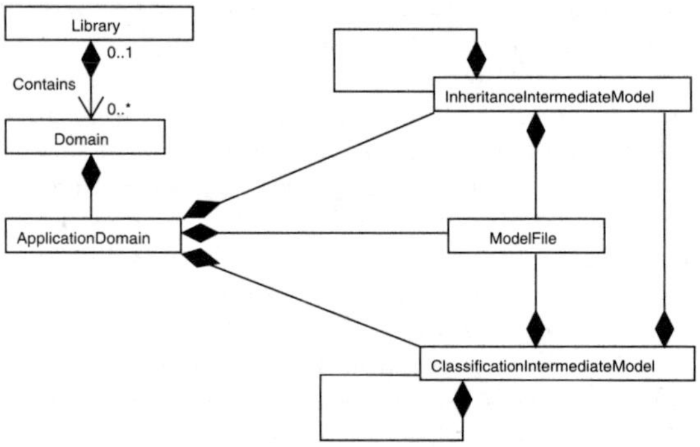

Figure 2: Links Between Elements

Figure 2 presents how are organized the links between these elements.

4. WEB-BASED ARCHITECTURE

We present in this section our approach for the definition of an architecture allowing to access the storage engine through a network or the Web. In the first part we introduce the basics of the Web-based architecture while the second part exposes the Remote Access package, the key of Web-based approach we have defined .

4.1 Basics of the Approach

The Web can serve as an operating system, and as a distribution channel for applications [7]. The main characteristics of the Web are: ease of navigation and use, ease of publishing content, new distribution models and enabling of a network-centric computing paradigm. Our Libraries Architecture is built on a core storage engine, and provides a set of interfaces enabling a remote access. This is one of the main features of our approach since it allows a design team to work on the same models stored on a storage server.

The basic idea is that the storage engine is running on a server and that the client access the models using a Web browser or directly from the environment upon a local network or even over the Internet, following the classical 3-tiers architecture. Clients do not need to know how the models are stored on the server, they should only access them for consulting, adding or removing them. We wanted to provide to a final user the easiest way to access the models libraries remotely. We found that the best approach was to use a Web browser since people are more and more aware of this kind of software even if they are not computer specialists.

This approach but appears viable only in the cases where the user wants to consult or manage a library. This thought led us to develop another complementary approach: if the final user wants to use the models library inside a modeling and simulation environment, we provide the computer specialist in charge with the environment the capacity to connect it with the storage engine using a set of simple APIs. The reader can note that we built these APIs in order to hide all the complexity of the storage engine.

4.2 The Remote Access Package

One of the main advantages of our libraries architecture is that it is independent from the format of the model. The storage engine included in the package Models Library contains two servers, a server acting as a file transfer server, and another one able to process instantiated and compiled objects. The Remote Access package uses a combination of these two servers in order to allow a remote access to the storage engine through interfaces defined in the Storage Engine package. These two servers use very simple communication protocols and two different connectors between the storage engine and the modeling and simulation environment.

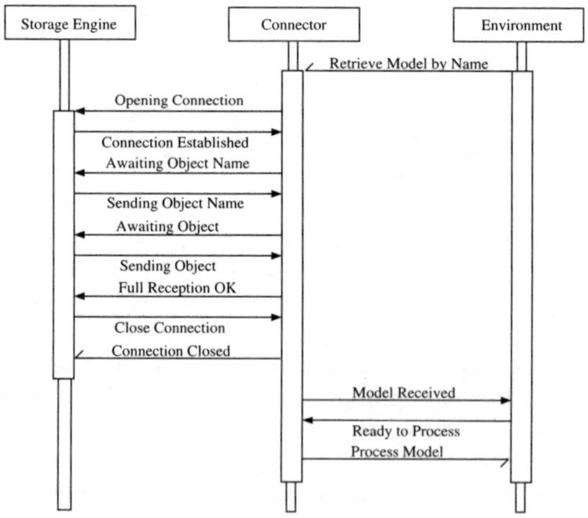

Figure 3: The Simple Model Object Transfer Protocol, SMOTP

The Simple Model Object Transfer Protocol (SMOTP, Figure 3) is used by the server in charge with the processing of previously instantiated and compiled models. Since these kinds of objects already contain all the characteristics of the associated modeling component, the transfer protocol is very simple. When the connection is opened, the dialog between the connector and the storage engine consists only in transferring the name of the model and the associated object.

The Simple Model File Transfer Protocol (SMFTP) is used by the server in charge with models provided as description or sources files. The main difference with the SMOTP lies in the fact that the abstraction level must be specified, since it usually does not appear in the model file. However, these protocols are very similar and very simple. That was our objectives since we had in mind to keep the efficiency of the storage engine.

4.3 Software Realization

The implementation of the whole software packages has been performed using the Java language, and we used the servlet/Java Web Start approach for the remote part of our work. There are many advantages in adopting such an approach [8]: first, servlets are persistent since they are loaded once by the Web server and can maintain services between requests; Secondly, servlets are fast since they are loaded once, they offer much better performance than over CGI approaches; Lastly, servlets are platform-independent and extensible since they are written in Java and since they can take advantage of all the Java benefits.

We use a set of interfaces contained in the Storage Access package as a kind of "connector" between them. The main advantage of this approach is that modifications on the servlets or on the storage engine can be performed independently, since the Storage Access will not change.

In our current implementation, context-out models can be of three types: Compiled Java objects, Java sources files or C++ sources files. The Library Storage engine can deal with these three types and their own specificities. If the modeller wants to store compiled Java objects, the engine creates a Proxy object for each model to be stored. This Proxy is built from the ".class" file using a de-serialization mechanism. In order to avoid replication, the engine uses a single-instance architecture: there is only one instance of a model that lives in the server (this was inspired by the J2EE EJB model [1]). A modeller can also store source files in the Library. Our engine is able to deal with Java and C++ source files using some mechanisms based on the XML language, a text-based language allowing to dissociate the contents from the format of data [10]. The basic idea is to transform the source code of a model in XML following predefined DTDs (Document type Definition), and to instantiate and to associate an object in the engine with this file. This approach has many advantages: we can produce DTDs for any kind of programming languages in order to extend the possibilities, we can work independently on the file without performing tasks on the associated object, and we can modify directly from the Proxy object some parameters of the model.

5. CONCLUSION

We introduced in this paper our approach for the definition of models libraries enabling the reusability of modeling components. These models libraries are built on the concepts of modeling components, abstraction hierarchy and genericity of use. They allow the model designer to perform its model design very quickly, once models have been introduced in such a library. We provided also this user with a set of APIs allowing him to process his models through some Web-based APIs and graphical interfaces accessible using a simple Java-enabled Web browser. Our approach has been validated in two different domains: VHDL test and DEVS-based environmental modeling and simulation [3]. We have three main perspectives of work. We saw that we are currently able to store instantiated Java objects. The next step will be to develop the storage engine in order to enabling it to store C++ instantiated objects. This will be done with a massive use of XML-based techniques. The second perspective is to define a meta-language in order to help the models design. Finally, we want to study how to manage and maintain with the most efficiency a distribution of storage engines over a network and over multiple hosts.

6. REFERENCES

[1] K. Ahmed and C. Umrysh. *Developing Enterprise Java Applications with J2EE(TM) and UML.* Addison-Wesley Pub Co, 2001.

[2] P. Benjamin, J. Erraguntla, D. Delen, and R. Mayer. Simulation modeling at multiple levels of abstraction. In *Proceedings of the 1998 Winter Simulation Conference*, 1998.

[3] F. Bernardi, E. de Gentili, and J. Santucci. Reusable models integration in a devs-based modelling and simulation environment. In *Proceedings of ESS2001*, 2001. Marseille, France.

[4] A. Breneuse, J. Top, J. Broenink, and J. Akkermans. Libraries of reusable models: Theory and application. *Simulation*, (71), 1998.

[5] D. Cottrell. Electronic component information exchange (ecix). In *Proceedings of the DAC 1997*, 1997. session 35.2.

[6] G. Konduri and A. Chandrakasabn. A framework for collaborative and distributed web-design. In *Proceedings of the DAC 1999*, 1999.

[7] J. Kuljis and R. Paul. A review of web-based simulation: Whither we wander ? In *Proceedings of the 2000 Winter Simulation Conference*, 2000.

[8] K. Moss. *Java Servlets, second edition.* McGraw-Hill, 1999.

[9] ODMG. *ODMG 3.0 Specifications.* Morgan Kaufmann Publishers, 2000.

[10] W. Pardi. *XML in Action.* Microsoft Press, 1999.

[11] H. Praehofer, J. Sametinger, and A. Stritzinger. Building reusable simulation components. In *Proceedings of WEBSIM2000, Web-Based Modelling & Simulation*, 2000. San Diego, CA, USA.

[12] R. Rosenberg. The bond graph as an unified database for engineering system design. *Journal of Engineering for Industry*, 97, 1975.

[13] A. Umar. *Cient/Server Internet Environments.* Prentice Hall, 1997.

[14] P. van den Hamer, W. van der Linden, P. Bingley, and N. Schellingerhout. A system simulation framework. In *Proceedings of the DAC 2000*, pages 699–705, 2000.

[15] R. Wadhwani. Component and library management tutorial. In *Proceedings of the DAC 1991*, 1991. session 1.1.

[16] B. Zeigler. *Theory of Modeling and Simulation.* Academic Press, 1976.

Behavioral Synthesis via Engineering Change

Milenko Drinić
Computer Science Department
University of California
Los Angeles, CA 90095
milenko@cs.ucla.edu

Darko Kirovski
Microsoft Research
One Microsoft Way
Redmond, WA 98052
darkok@microsoft.com

ABSTRACT

Engineering change (EC) is a technique that enables a designer to rapidly perform minor specification alternations while minimally resynthesizing only small portions of the specification throughout several levels of design abstraction. In this paper, we introduce the first EC-based synthesis technique for coordinated design optimization in multiple steps. The technique has four phases: optimization region identification, feedback formulation, resynthesis in first step, and finally resynthesis in the second design step. To demonstrate the technique, we focus on behavioral synthesis and transformation, scheduling, and register assignment steps. We developed a generic EC-based approach for design optimization during multiple consecutive synthesis steps. Next, we show how one can use EC to enhance coordinated application of transformations and scheduling, and scheduling and register assignment.

Categories and Subject Descriptors

B.5.2 [**Register-Transfer-Level Implementation**]: Design AidsOptimization

General Terms

Design

Keywords

Engineering change, transformations, scheduling, register assignment

1. INTRODUCTION

The modern design process is long and complex with numerous subtasks that interact in convoluted ways. The conceptual complexity of applications and quantitative exponential growth of silicon platforms will make the process additionally cumbersome. Even if we restrict our attention to behavioral synthesis only, we see a number of complex tasks (transformation, partitioning, resource allocation, scheduling, assignment, register and interconnect allocation and assignment, module and clock selection, and memory allocation) interacting.

While each of these steps is associated with solving NP-hard problems, the real difficulty is in the layered nature of subtasks and their mutual impact. For example, the standard behavioral synthesis flow is transformations - scheduling - register assignment. During scheduling the primary goal is to minimize the number of execution units and interconnect. However, decisions made during scheduling greatly impact the register requirements. Since the area of a register is almost comparable to the area of an adder or a barrel-shifter and often registers dominate power consumption, it would be advantageous to consider the impact on all components on the final cost during each of steps as well as their interaction. However, the fundamental principle of separation of concerns with implication and emphasis on modularity, precludes this direction.

Nevertheless, there is an elegant and powerful need to address this issue without violating the principle of separation of concerns. Interestingly, the basis for solving this problem has unexpected roots: EC. EC is a technique that facilitates minor specification changes while inducing the need for minimal resynthesis of only small parts of the specification throughout other levels of the design abstraction. Until now, EC has been widely and solely used for enabling localized functional and timing changes, most often during debugging or after initial specification alternations. We introduce the first EC-based synthesis technique for coordinated design optimization in multiple steps. The technique has four phases: optimization region identification, feedback formulation, resynthesis in the first step, and finally resynthesis in the second design step. We focus our effort on the use of EC as a tool for coordinated optimization between multiple tasks in the design flow in the behavioral synthesis steps: transformation, scheduling, and register assignment steps. The starting point is a generic EC-based approach for design optimization during multiple consecutive synthesis steps. We demonstrate how a designer can use EC to enhance the coordinated application of transformations and scheduling, scheduling and register assignment.

1.1 Motivational Example

Consider an example of a control-data flow graph (CDFG) with ten multiplications and seven additions shown in Figure 1(a). The CDFG has been scheduled in seven control steps. For the given schedule, the computation requires at least three multipliers and two adders. Note that because of dependencies among operations, the operations scheduling alone cannot further reduce the hardware cost.

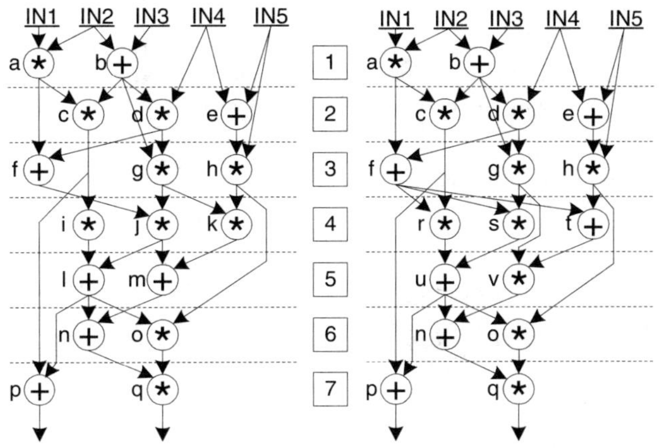

(a) An example of a scheduled CDFG

(b) The resulting CDFG after applied full cycle of the transformations-scheduling EC optimization procedure

Figure 1: An example of application of the transformations-scheduling EC iterative optimization procedure.

The scheduling bottleneck occurs in control step 4 where three multiplications need to be executed concurrently. Instead of proceeding with the solution as is into subsequent design steps, the scheduling algorithm is modified such that it indicates its bottleneck to the previous design step. In the CDFG from Figure 1(a), operations i, j, and k are selected for the transformations procedure with the highest priority. In order to enable flexibility in transformations, operations l and m (as operations whose inputs are outputs of bottleneck operations i, j, and k), and operations c, f, g, and h (as operations whose outputs are inputs to bottleneck operations) are also marked for changes with the transformations procedure. These operations are ranked by two different criteria. First, operations scheduled closer to the bottleneck control step have a higher priority in the operation selection process for transformations (g and h) then the operations further from it (c). Second, operations of the same type (g and h) as the bottleneck operations are assigned a higher rank than the selected operations of a different type (f).

Transformations are selected from the library of transformations, and applied to the operations indicated by the scheduling. After applying replication on operation j, inverse distributivity on i, j, and l, and j' (replicated node), k, and m scheduler identifies another bottleneck, again in control step 4. This time the bottleneck and the highest priority operations are additions. The resulting CDFG and its schedule after applied distributivity on one addition from control step 4, and one multiplication from control step 5 nodes r is shown in Figure 1(b). This CDFG and its scheduling are balanced so the global procedure exits the iterative loop and continues synthesis subsequent lower design levels. More details for transformations-scheduling EC iterative optimization procedure is presented in Subsection 5.1.

2. RELATED WORK

Engineering change recently attracted a great deal of attention. However, EC efforts have longer history. In 1986, Shinsha et al. [9] proposed an incremental technique for supporting changes during physical design. Kirovski et al [4] defined the engineering change problem as constraint manipulation where the instance of the problem is restructured in such a way that it supports EC.

An inclusive survey of behavioral synthesis can be found in [2]. Scheduling and Register assignment (regularly abstracted as graph coloring) are standard tasks in both compilers [7] and behavioral synthesis [2]. The difficulty and suitability of optimization of the scheduling problem is function of both computation complexity and underlying computational model [5], targeted architecture and optimized metrics. We focus on the synchronous data-flow model [5] of computation that provides an exceptional trade-off in terms of the number and importance of application domains, and suitability for optimization.

Kernighan and Lin introduced the first iterative improvement heuristic and applied it to graph partitioning [3]. Numerous improvements on the basic strategies have been proposed over the years [1]. Iterative improvement has been also applied to many other optimization problems [6].

3. PRELIMINARIES

We have chosen as a computational model the synchronous data flow (SDF) model [5]. We restrict our attention to homogeneous SDF (HSDF), where each atomic operational unit consumes and produces exactly one sample per each input and output in every execution cycle. The syntax of a targeted computation is defined as a control-data flow graph (CDFG) [8]. The CDFG represents the computation as a flow graph, with nodes, data edges, and control edges. The semantics underlying the syntax of the CDFG format is that of the SDF model.

Behavioral synthesis transforms a given behavioral specifications into a RTL description from the input behavioral descriptions (or algorithms). This high-level abstraction provides a more powerful and complete way to explore the design space.

Scheduling is the process of partitioning a CDFG into groups of operations such that the operations from the same group are executed in a single control step. The objective of the scheduling procedure is the minimization of the total number of control steps and the total hardware cost. These objectives are, in general, in contradiction, therefore one must consider possible trade-offs when the objective function of the scheduling algorithm is constructed.

Register assignment is a process of binding each variable to a storage unit. The goal of register assignment is minimization of the total number of used registers. Many register allocation algorithms for CDFGs without loops focus either on unconditional register sharing, or conditional register sharing. Register assignment can be performed by coloring the interval graph.

4. GENERIC DESIGN FLOW

Consider a pair of synthesis/optimization tasks A and B, where task B uses the output of task A as its input. The quality of the solution provided by task B is highly dependent on the quality of the solution provided by task A. The solution of task A can be of high quality based on criteria ensuing from the level of abstraction of task A. At the same time, task B could be prevented, upfront, from obtaining a good quality solution because its input is such that there is little or no room for task B to perform good synthesis measured on effectiveness by the criteria of task B.

We propose a new EC-based synthesis technique that in-

```
Run task A
Run task B
While the solution of B unbalanced
  Identify bottleneck region BR of B
  Identify ε-neighborhood of BR
  Rank operations/nodes
  Run task A on BR and ε-neighborood
  Run task B on modified BR and ε-neighborood
end while
```

Figure 2: Pseudo-code of iterative segment of the generic design flow for two consecutive synthesis/optimization tasks A and B.

tegrates two subsequent abstraction level tools through iterative collaboration. The key idea behind this approach is the passing information from the lower level to the higher level of abstraction, and the resynthesis of only small portions of the design specifications. The pseudo-code of the generic design flow is shown in Figure 2. It depicts the part of the full design flow which includes collaboration of two consecutive tasks.

Initially, both tasks, A and B are run without any changes. When task B finishes synthesis, it is necessary to analyze the solution, and formulate feedback for task A. This analysis includes identification of portions of the design that represent bottlenecks of the design causing the design to be unbalanced. Task B ranks variables, operations, or nodes in such a way that highest ranked items are those entailing the highest cost in the bottleneck. Immediate, ε-neighborhood of the bottleneck region is also selected, and ranked accordingly. The last step of the formulation of the feedback for task A is to make guidelines for its objective function. These guidelines indicate in which the direction optimization should proceed, and they are given in form of parameters included in the objective function of the task A.

After the feedback for the task A has been formulated, the task A performs a resynthesis procedure only on the portion of the design specified by the feedback. The resynthesis procedure is guided with the feedback also through a set of objectives. For example, if task B is an operation scheduling procedure, besides the selection and ranking of operations that need to be transformed, it indicates which type of operations needs to be minimized. Upon completion of task A, task B is run again, only on the bottleneck region. Since both tasks are performed only on small portions of the design, the total running time is not significantly increased.

5. APPLICATIONS OF EC SYNTHESIS

5.1 Transformations - Scheduling Procedure

The conceptual complexity of transformations makes design synthesis difficult to be algorithmically defined through them. A localization of a targeted area for the optimization can open a clear path for a set of transformations toward the specified goal. We have built a library of transformations from a restricted set which includes algebraic laws and redundancy manipulation, and sub-operational level transformations. The latter set has a powerful property of enabling a wide range of transformations from different sets.

The goal of transformations is to minimize the weighted total number of operations in the computation trough transformations. We define the objective function which needs to be minimized during an optimization in the following way:

$$OF = \sum_{\mathcal{T}_i=\mathcal{T}_1}^{\mathcal{T}_n} R_{\mathcal{T}_i} \cdot \alpha_{\mathcal{T}_i} \cdot |op_{\mathcal{T}_i}| - \omega$$

where $R_{\mathcal{T}_i}$ is the sum of ranks, $\alpha_{\mathcal{T}_i}$ represents the hardware cost, $|op_{\mathcal{T}_i}|$ is the cardinality of operations of type $\mathcal{T}_i$, and ω is proportional to the total number of available transformations.

Optimize CDFG with transformations
Schedule CDFG
While $|op_{\mathcal{T}}^{cs_i}| - AVG(|op_{\mathcal{T}}^{cs}|) > C1$
Select:
$op_{\mathcal{T}}^{cs_i}$; $op_j | out(op_i) = in(op_{\mathcal{T}}^{cs_i})$; $op_k | in(op_k) = out(op_{\mathcal{T}}^{cs_i})$
Rank:
$R(op_{\mathcal{T}}^{cs_i}) = HR$; $R(op^{cs_l}) = LR > 0, |cs_l - cs_i| = max$
$LR < R(op_k), R(op_j) < HR, op_k \neq op^{cs_l}, op_k \neq op^{cs_l}$
Transform selected op
Reschedule selected op
end while

Figure 3: Pseudo-code of computational transformations - scheduling EC iterative optimization procedure.

The pseudo-code of transformation - scheduling EC iterative optimization procedure is depicted in Figure 3. In the initial run of transformations, the rank of each operation is a constant, and equal for all operations. In a scheduled CDFG, the bottleneck region is identified as a set of operations entailing the highest cost. These operations (labeled as $op_{\mathcal{T}}^{cs_i}$) are all of the same type $\mathcal{T}$, belong to a single control step cs_i, and are assigned the highest rank HR. In order to enable a reasonably wide spectrum of transformations, the operations whose inputs are outputs of $op_{\mathcal{T}}^{cs_i}$, and those whose outputs are inputs to of $op_{\mathcal{T}}^{cs_i}$ are also selected and assigned a non-zero rank. Higher rank is assigned to these operations if: (i) they are scheduled closer to the bottleneck control step, and (ii) they are of type $\mathcal{T}$.

The iterations between transformations and scheduling are continued while there is a control step entailing significantly more resources than the remainder of control steps on the average. Once when the peaks of hardware requirements are eliminated, it is not likely that further iterations which are performed locally will bring a substantial benefit. An example of transformation - scheduling EC iterative optimization procedure is introduced as the motivational example in Subsection 1.1.

5.2 Scheduling - Register Assignment Procedure

The pseudo-code of scheduling - register assignment EC iterative optimization procedure is depicted in Figure 4. The initial scheduling of CDFG implies that the complete EC iterative optimization procedure described in Subsection 5.1 has been processed. The cut in the variable lifetime graph over a single control step cs_i, $Cut(cs_i)$ is defined as a number of live variables in that control step. The condition for entering the iterative collaboration of two design stages is that the largest $Cut(cs_i)$ is larger for more than empirical constant $C2$ than the average cut over a single control step, $AVG(Cut(cs))$.

The information where each variable is consumed is lost on the interval graph level. The register assignment (graph coloring) algorithm passes only the information about variables are in $Cut(cs_i)$ and the number of the critical step cs_i. This cut is causing the minimal required number of registers to be also nine. On the abstraction level of CDFG, operations belonging to the bottleneck region are selected by three criteria.

Design name	Number of operations	Active area [mm^2]								Improvement [%]		
		Original		Trans.-Sche.		Sche.-Reg.Asg.		Trans.-Sche.-Reg.Asg.				
		EU	Regs	EU	Regs	EU	Regs	EU	Regs	EU	Regs	Total
volterra	30	0.96	1.17	0.95	1.1	0.96	1.09	0.95	1.06	1.0	9.4	5.6
conv5	45	1.19	2.41	1.09	2.14	1.19	2.05	1.09	1.92	8.4	20.3	16.4
LinearCntrl3	56	3.92	9.11	3.71	8.21	3.92	6.56	3.71	6.14	5.4	32.6	24.4
Differentiator	80	1.16	3.40	1.04	3.16	1.16	2.68	1.04	2.60	10.3	23.5	20.2
Winogradfft11	104	2.56	6.29	2.13	5.41	2.56	4.34	2.13	4.02	16.8	36.1	30.5
DSkais55	137	3.00	10.12	2.52	8.50	3.00	6.78	2.52	6.21	16.0	38.6	33.4
fir100	302	4.18	16.41	3.39	13.33	4.18	10.83	3.39	9.93	18.9	39.5	35.3

Table 1: Impact of EC based behavioral synthesis on active area of the design. Active area for four different design flows are shown: (*i*) original, (*ii*) enhanced with only transformation-scheduling iterative procedure, (*iii*) enhanced with only scheduling-register assignment iterative procedure, (*iv*) enhanced with both iterative procedures. Active area shown separately for execution units (EU) and registers.

Schedule CDFG
Assign registers
While $Cut(cs_i) - AVG(Cut(cs)) > C2$
Select:
$op_j \in cs_i$; $op_l | v_k \in Cut(cs_i) \wedge v_k \rhd cs_i \wedge$
$LT(v_k) \geq AVG(LT(v)) \wedge in(op_l) = v_k$;
$op_m | LT(v_m) \ll AVG(LT(v)) \wedge out(op_m) = in(op_j)$
$\vee out(op_m) = in(op_l)$
Rank:
$R(op_j) = HR$; $R(op_m) = LR > 0$, $LR < R(op_l) < HR$
Reschedule selected *op*
Reassign registers for affected variables
end while

Figure 4: Pseudo-code for scheduling-register assignment EC iterative optimization procedure.

First, operations op_j in the critical step cs_i are selected and assigned the highest priority. Second, consider a variable v_k with a long lifetime LT which is consumed (indicated by $\rhd$ in the pseudo-code in Figure 4) both in cs_i by an operation already selected into the bottleneck region, and by one or more operations outside of cs_i. This variable is highly likely causing the bottleneck in register assignment, so operations op_l consuming v_k are selected for rescheduling. The third criterion has a purpose to make the bottleneck region more amenable to rescheduling. By this criterion, operations op_m, whose outputs are inputs of already selected operations by the first two criteria, are also chosen to be added in the bottleneck region. The restriction is established by the lifetime of the produced variables by op_m which needs to be short compared to average lifetime of variables.

As a scheduling and register assignment algorithm we have used the standard HYPER synthesis script [8]. The objective functions of both optimization procedures were augmented to accommodate the ranking of operations, variables, or nodes, and limited scope determined by the bottleneck region and its ε-neighborhood.

6. EXPERIMENTAL RESULTS

The selected benchmarks for evaluating the effectiveness of our approach include designs of the volterra filter, a convolution, a linear controller, Winogradfft11 transformation, a differentiator, and FIR filters (DSkais55, and fir100).

The experimental results are presented in Table 1. We have conducted four different types of experiments, and mapped solutions on $.25\mu$ technology. The initial results (shown in the third and the forth column) have been obtained using the standard HYPER synthesis script [8]. In the next six columns, active area is shown for execution units and register of the following design flow enhancements: only transformation-scheduling EC iterative optimization, only scheduling-register assignment, and these two optimizations combined. The area improvement for execution units, registers and the overall improvement are shown in the last three columns of Table 1. During our experiments, we have encountered only minimal run time overheads not exceeding 5% of the original run time for each benchmark.

7. CONCLUSION

In this work, we have introduced the first behavioral synthesis technique based on EC for coordinate design optimization. The collaboration is established between of two subsequent design stages where the latter one formulates and presents feedback information based on its bottleneck regions, to the first design stage. The developed technique assumes multiple resynthesis steps of two design stages which are performed in alternating order. The generic EC-based design approach has been introduced with examples of application of the technique to transformations and scheduling, and scheduling and register assignment. We have demonstrated the effectiveness of the new technique on a number of benchmarks while entailing minimal run-time overhead.

8. REFERENCES

[1] C.J. Alpert and A.B. Kahng. Recent Directions in Netlist Partitioning: A Survey. Integration: The VLSI Journal, Vol.19, (no.1-2), pp.1–81, 1995.

[2] G. De Micheli. Synthesis and optimization of digital circuits. McGraw-Hill, 1994.

[3] B.W. Kernighan and S. Lin. An Efficient Heuristic Procedure for Partitioning Graphs. Bell System Technical Journal, Vol.49, (no.2), pp.291–308, 1970.

[4] D. Kirovski and M. Potkonjak. Engineering change: methodology and applications to behavioral and system synthesis. Proc. of Design Automation Conference, pp.604–609, 1999.

[5] E.A. Lee and D.G. Messerschmitt. Synchronous dataflow. Proceedings of the IEEE, Vol.75, (no.9), pp.1235–45, 1987.

[6] S. Lin and B.W. Kernighan. An effective heuristic algorithm for the traveling-salesman problem. Operations Research, Vol.21, (no.2), pp.498–516, 1973.

[7] S. S. Muchnik. Advanced Compiler Design & Implementation. Morgan Kaufmann Publishers, Inc., 1997.

[8] J. Rabaey, C. Chu, P. Hoang, and M. Potkonjak. Fast prototyping of data path intensive architectures. Design & Test of Computers, Vol.8, (no.2), pp.40–51, 1991.

[9] T. Shinsha, T. Kubo, Y. Sakataya, J. Koshishita, and K. Ishihara. Incremental logic synthesis through gate logic structure identification. Proc. of Design Automation Conference, pp.391–397, 1986.

A Universal Technique for Fast and Flexible Instruction-Set Architecture Simulation

Achim Nohl, Gunnar Braun,
Oliver Schliebusch, Rainer Leupers,
Heinrich Meyr
Integrated Signal Processing Systems
Templergraben 55, 52056 Aachen, Germany
nohl@iss.rwth-aachen.de

Andreas Hoffmann
LISATek Inc.
190 Sand Hill Circle
Menlo Park, CA 94025
andreas.hoffmann@lisatek.com

ABSTRACT

In the last decade, instruction-set simulators have become an essential development tool for the design of new programmable architectures. Consequently, the simulator performance is a key factor for the overall design efficiency. Based on the extremely poor performance of commonly used *interpretive* simulators, research work on fast *compiled* instruction-set simulation was started ten years ago. However, due to the restrictiveness of the compiled technique, it has not been able to push through in commercial products. This paper presents a new retargetable simulation technique which combines the performance of traditional compiled simulators with the flexibility of interpretive simulation. This technique is not limited to any class of architectures or applications and can be utilized from architecture exploration up to end-user software development. The work-flow and the applicability of the so-called *just-in-time cache compiled simulation* (JIT-CCS) technique will be demonstrated by means of state of the art real world architectures.

Categories and Subject Descriptors

I.6.3 [**Simulation and Modeling**]: Simulation Support Systems; I.6.3 [**Simulation and Modeling**]: Model Validation and Analysis; D.3.2 [**Programming Languages**]: Design Languages—*LISA*; C.0 [**General**]: Modeling of Computer Architecture

General Terms

Design, Languages, Performance

Keywords

Retargetable simulation, compiled simulation, instruction set architectures

DAC 2002, June 10-14, 2002, New Orleans, Louisiana, USA.

1. INTRODUCTION

Instruction-set simulators are an integral part of a today's processor and software design process. Their important role within the architecture exploration, early system verification and pre-silicon software development phase is undisputable. However, the growing complexity of new programmable architectures has a negative impact on the simulation performance. Taking into account Moore's law, research work on a high performance simulation technique for instruction-set simulators was started in 1991 [10]. The so-called compiled simulation technique is rooted in the domain of RTL level hardware simulation and was applied successfully to instruction-set simulators. Nevertheless, the fast compiled simulation technique has not been able to succeed in commercial tools. This is due to the fact that the speed-up is expensively paid with restrictions and assumptions contradicting with the requirement of today's applications and architectures.

The major restriction for the utilization of compiled simulators is the requirement for static program code. This limits the compiled technique to a small class of digital signal processors (DSP). In contrast to typical DSP applications which are signal processing algorithms, micro-controller (μC) architectures usually run an operating system (OS). The significant characteristic of operating systems, run-time dynamic program code, conflicts with the limitation of compiled simulators. However, even for DSP architectures real-time operating systems are increasingly gaining importance. This paper presents a simulation technique which meets the requirements for both, high simulation speed and maximum flexibility. The so-called *just-in-time cache compiled simulation* (JIT-CCS) technique can be utilized within the architecture design as well as for end-user software development. The presented technique is integrated in the retargetable LISA processor design platform [2]. A generator back-end for the LISA 2.0 processor compiler has been developed, which automatically constructs a JIT-CCS simulator from a LISA machine description.

The rest of this paper is organized as follows: Section 2 presents related work concerning compiled simulation and result caching techniques. The restrictions of existing simulation techniques and their consequences on the applicability are worked out in section 3. Subsequently, section 4 describes the work-flow of the just-in-time cache compiled simulation technique. Finally, benchmark results for various state of the art architectures are presented in section 5.

2. RELATED WORK

Research work on instruction-set architecture simulation has been an active research topic since the early days of programmable architecture design. Within the scope of the EMBRA project [5] a high performance simulator for the MIPS R3000/R4000 processor has been developed. The objective is similar to the one presented in this paper – providing highest flexibility with maximum performance. Similar to FastSim [3] based simulators, the performance gain is achieved by dynamic binary translation and result caching. However, EMBRA is a non-retargetable simulator and restricted to the simulation of the MIPS R3000/R4000 architecture.

Work on *retargetable* fast simulators using a machine description language was published within the scope of the FACILE project [4]. The simulator generated from a FACILE description utilizes the *Fast Forwarding* technique to achieve reasonably high performance. Fast forwarding is similar to compiled simulation and uses result caching of processor actions, indexed by a processor configuration code. Previously cached actions can be replayed directly in a repeated occurrence of a configuration. Due to the assumption that program-code is run-time static, dynamic program-code cannot be simulated. Furthermore, no results on the applicability of FACILE for VLIW or irregular DSP architectures have been published.

A retargetable tool suite which allows *cycle- and bit-true modelling* of pipelined processors is based on the EXPRESSION [1] language. Previous publications have shown its suitability for modelling real-world architectures like the Motorla DSP 56k or Texas Instruments TMS320C6000, however, no results are available on the performance of the generated simulators.

Retargetable *compiled* simulators based on an architecture description languages have been developed within the Sim-nML (FSim) [9], ISDL (XSSIM) [8] and MIMOLA [11] projects. Due to the simplicity of the underlying instruction sequencer, it is not possible to realize processor models with more complex pipeline control mechanisms like Texas Instruments TMS3206000 at a cycle accurate level. A further retargetable approach, which is based on machine descriptions in ANSI C, has been published by Engel and Fettweis [6]. However, only results for a single proprietary DSP architecture are available so far. Moreover, all of the presented compiled simulation approaches are qualified by the limitations that result from the compiled principle.

In summary, none of the above approaches combines retargetability, flexibility, and high simulation performance at the same time. The LISA language has proven to be capable of retargeting fast compiled instruction-set simulators for various real-world DSP architectures [2]. To finally overcome the flexibility restrictions of compiled simulators a novel simulation technique is presented in the following sections.

3. PROCESSOR SIMULATION

The broad spectrum of today's instruction-set simulation techniques starts with the most flexible[1] but slowest interpretive technique. Compared to the interpretive technique, much higher simulation speed is achieved by compiled simulation, however, the gain is expensively paid with a loss of flexibility. This section critically compares the existing simulation techniques. Especially,the impact of the tradeoff between flexibility and performance on the suitability for different application domains is discussed.

[1]Here flexibility refers to the ability to address all kinds of architectures and applications.

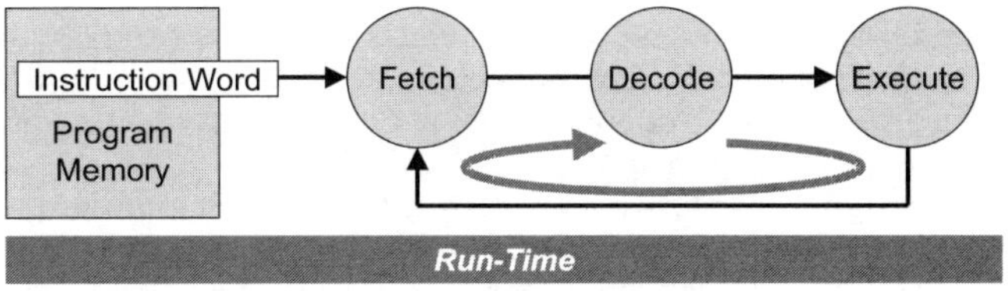

Figure 1: Interpretive Simulation Workflow

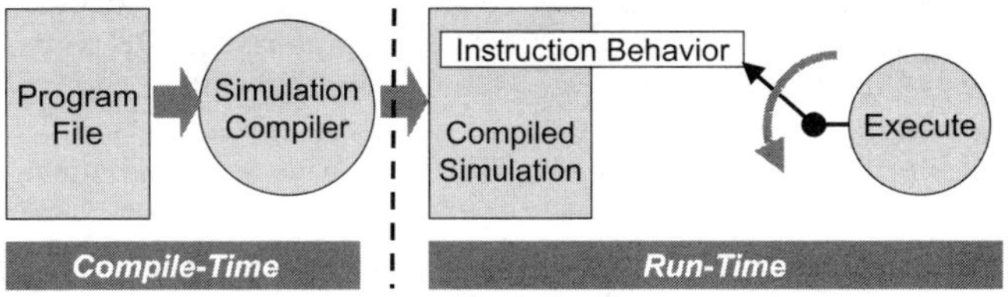

Figure 2: Compiled Simulation Workflow

3.1 Interpretive Simulation

An interpretive simulator is basically a virtual machine implemented in software, which interprets the loaded object code to perform appropriate actions on the host as shown in figure 1. Similar to the operation of the hardware, an instruction word is fetched, decoded, and executed at runtime (simulation loop), which enables the highest degree of simulation accuracy and flexibility. However, the straightforward mapping of the hardware behavior to a software simulator has major disadvantages. Unlike in real hardware, instruction decoding is a very time consuming process in a software simulator. Compared to the operation of the functional units, which can be easily transformed into equivalent instructions on the host, the decoder is characterized by control flow. Especially for today's VLIW architectures the decoding overhead dominates.

3.2 Compiled Simulation

The objective of compiled simulation is to improve the simulation performance. Considering instruction-set simulation, efficient run-time reduction can be achieved by shifting time-consuming operations from the simulator run-time into an additional step before the simulation (compile-time). This step is performed by a tool called *simulation compiler* (see figure 2). Depending on architectural and application characteristics, the degree of compilation varies. All compiled simulators have in common that a given application is decoded at compile-time. Based on the results of the decoding phase the simulation compiler subsequently selects and sequences the appropriate host operations that are required to simulate the application. Since the time-consuming instruction scheduling is still performed at run-time *(dynamically scheduled)*, *statically scheduled* [7] compiled simulators also move the instruction scheduling into the compilation phase. However, all compiled simulators rely on the assumption that the complete program code is known before the simulation starts and is furthermore run-time static.

3.3 Importance of Flexibility

Many application domains are excluded from the utilization of compiled simulators. An overview of these domains is given in figure 3 and the details are discussed in the next paragraphs.

The integration of compiled simulators into embedded system environments is not possible, since the prime requirement, predictable program code, is not fulfilled when using external program memories.
Furthermore, applications with run-time dynamic program code, as provided by operating systems (OS), cannot be addressed by compiled simulators. Run-time changes of program code cannot be considered by a simulation compiler. Today's embedded systems consists of multiple processor cores and peripherals which make an underlying OS indispensable. However, compiled simulators only allow the isolated simulation of applications which is not sufficient for the verification of a complete HW/SW system. Considering novel architectural features, especially in the domain of low power architectures, multiple instruction-sets are widely used to reduce power and memory consumption. These architectures can switch to a compressed instruction-set at run-time. For instance the ARM core family provides a so-called "Thumb" instruction-set. This dynamic instruction-set switching cannot be considered by a simulation compiler, since the selection depends on run-time values and is not predictable.

Application Domain	Compiled Simulation Restriction
- System Integration - External Memories	Program code is not predictable before runtime.
- Operating Systems - Boot Loading	Dynamic program code is not allowed
- Multiple Instruction Sets	Run-Time switch between different instruction sets cannot be considered before simulation.
- Large Applications	Compiled simulation is qualified by an enormous memory usage.

Figure 3: Compiled Simulation Restrictions

Compiled simulation of large applications requires an enormous amount of memory (approx. 1000x, depending on the architecture) compared to interpretive techniques. As long as the host memory is big enough, the high memory consumption does not noteworthy decrease the performance. This is due to the fact that program execution concentrates on small local parts of the code, and thus cache effects are not noticeable. Of course for multi-processor simulation of embedded systems or processor arrays, the memory efficiency of the simulator becomes increasingly important.

Summarizing the above arguments the enormous performance gain of compiled simulators succumbs to their restrictiveness. This implies that the presented application areas are still dominated by the slow interpretive technique. But driven by the ever increasing complexity of applications, architectures, and systems the urgent requirement for a high performance simulation technique arises. This technique has to provide the maximum degree of flexibility together with a high performance.

4. JIT CACHE COMPILED SIMULATION

The *Just-In-Time Cache Compiled Simulation (JIT-CCS)* technique presented in this paper has been developed with the intention to combine the full flexibility of interpretive simulators with the speed of the compiled principle. The basic idea is to integrate the simulation compiler into the simulator. The compilation[2] of an instruction takes place at simulator run-time, *just-in-time* before the instruction is going to be executed. Subsequently, the extracted information is stored in a simulation cache for the direct reuse in a repeated execution of the program address. The simulator recognizes if the program code of a previously executed address has changed and initiates a re-compilation. First of all, this method offers the full flexibility, as the interpretive technique does. However, it has to be investigated whether the performance of the compiled simulation can be achieved. The next paragraph discusses the theoretically achievable performance and gives a comparison to the compiled technique.

[2]In this context compilation does not refer to invoking a C compiler at run-time. The compilation process is discussed in section 4.2.

4.1 Performance

The total compiled simulation time $t_{app,cs}$ is equivilant to the total instruction execution time $t_{ex,total}$, which is the product of the instruction execution count n_{exec} and the average execution time per instruction $\tilde{t}_{ex,insn}$. For the just-in-time compiled simulation the total simulation time $t_{app,js}$ is additionally made up by the total compilation time $t_{comp,total}$. Under the assumption that the program code is constant the total compilation time only depends on the instruction count of the application n_{insn} and the average time for compiling a single instruction $\tilde{t}_{comp,insn}$.

$$t_{app,cs} = t_{ex,total} \quad (1)$$
$$t_{app,js} = t_{ex,total} + t_{comp,total} \quad (2)$$
$$t_{ex,total} = \tilde{t}_{ex,insn} * n_{exec}$$
$$t_{comp,total} = \tilde{t}_{comp,insn} * n_{insn}$$

Based on the equations (1) and (2) the instruction throughput p_{cs} (3) and p_{js} (4) of the two simulation techniques can be deduced.

$$p_{cs} = \frac{1}{\tilde{t}_{ex,insn}} \quad (3)$$
$$p_{js} = \frac{1}{\tilde{t}_{ex,insn} + \tilde{t}_{comp,insn} * \frac{n_{insn}}{n_{exec}}} \quad (4)$$

In (5) it is shown that for a growing number of instruction executions the performance of the just-in-time compiled simulator converges to the performance of a compiled simulator.

$$\lim_{n_{exec} \to \infty} p_{js} = p_{cs} \quad (5)$$

The number of repeatedly executed instructions needed for a good convergence of the compiled simulation speed very well corresponds to the conditions provided by real-world applications. This is due to the fact that most programs behave according to the 90/10 rule: 90% of execution time is spent in 10% of the code. For instance the proportion $n_{exec}/n_{insn} = 256$ corresponds to a loop that is iterated 256 times. Together with the valid assumption that $\tilde{t}_{comp,insn} = 4 * \tilde{t}_{exec,insn}$, 98.5% of the compiled simulation performance is achieved. The validity of the assumptions made here is proved in section 5.

4.2 Just-in-Time Simulation Compiler

The complete design of the JIT-CCS is based on the integration of the simulation compiler into the simulator. In fact, the run-time compilation of instructions requires a new concept for the simulation compiler. Based on the separation of the simulation compiler and the simulator the following conditions are fulfilled for traditional compiled simulators: the instruction decode time is not critical since it does not influence the simulation performance. Furthermore, the simulation compiler is allowed to generate C-code which is subsequently compiled by a C-compiler to build a compiled simulator. Obviously, this is not true for an integrated simulation compiler. Especially, the run-time call of a C-compiler is not practical.

In order to explain the work-flow of the just-in-time simulation compiler, some important characteristics of a LISA processor model have to be introduced. A LISA model is a mixed structural/behavioral description of a processor. The structural part keeps the definition of processor resources like registers, memories and pipelines. The processor's instruction-set including instruction-coding, assembly syntax, functional behavior, and timing is contained in so-called LISA *operations*. A single processor instruction can be composed by multiple LISA operations. The following example shows an excerpt of a LISA processor description for a simple `ADD` instruction. Beside the definition of the program memory and the register resources, two examples for LISA operations are given. The operation *ADD* implements the binary coding, assembly syntax, and the functional behavior of the processor instruction `ADD`. The instruction-set information of the operand registers *src1*, *src2*, and *dst* are referenced from the inferior LISA operation *Register*. The operation *Register* describes the binary coding, assembly syntax of a single register within the processor's register file. A reference to the respective resource is returned for the use in the functional behavior of operation *ADD*.

```
RESOURCE {
    PROGRAM_MEMORY byte8  prog_mem[0x0..0x1000];
    REGISTER       word32 R[1..15];
    }
OPERATION ADD {
    DECLARE { GROUP dst,src1,src2 = {Register}}
    CODING  { 0b01011 0b0000 src1 src2 dst}
    SYNTAX  { "ADD" dst "," src1 "," src2 }
    BEHAVIOR{ dst = src1 + src2}
    }
OPERATION Register {
    DECLARE { LABEL index; }
    CODING  { index=0bx[4]}
    SYNTAX  { "R" index }
    EXPRESSION { R[index]}}
```

The presented structure of LISA processor models enables the following procedure. The behavioral C code of all LISA operations is pre-compiled into C-functions which are part of the simulator. The JIT simulation compiler selects the appropriate operations, which are required to simulate an instruction, on the basis of the coding information. References to the selected C-functions are subsequently stored in the simulation cache. These references are utilized by the simulator to execute the instructions' behavior.

4.3 Simulation Cache

The just-in-time simulation compiler stores the compiled data within a so-called simulation cache. The simulation cache is indexed by the non-ambiguous instruction address. For simplification purpose the cache size is considered to be unlimited first. Every time an instruction is going to be executed, the just-in-time simulation compiler looks up the cache entry that corresponds to the current program address. Before the previously cached data is used to accelerate the simulation, the validity is verified. Therefore, each cache entry contains a key, the instruction word, which is compared to the simulator's instruction register contents. If an instruction word change is detected, re-compilation is initiated at run-time. Often, processors with multi-word instructions, or parallel instructions within a VLIW bundle, require that the key is made up by more than one instruction register. Figure 4 illustrates the workflow of the JIT-CCS. The simulator is going to execute the instruction at address 4. Therefore the compiled data for the instruction at address 4 **(1)** is requested. The simulation cache management looks up the address in the simulation cache **(2)** and indicates that this address has not been compiled before. Therefore the JIT compiler is invoked **(3)** which stores the data together with the instruction word `ea10` in the cache. Eventually a reference to the table entry is returned to the simulator **(4)**.

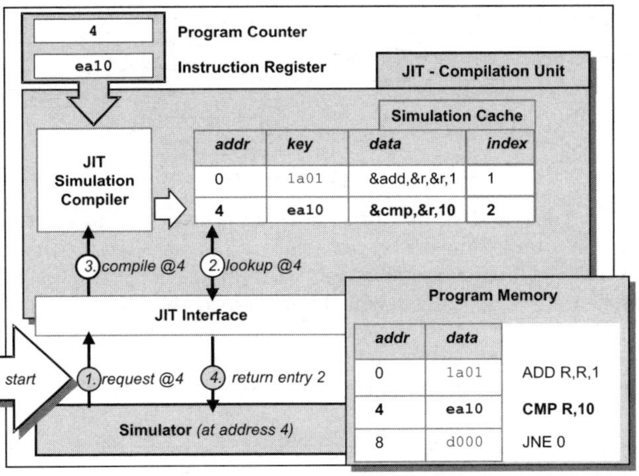

Figure 4: JIT-CCS Workflow

The presented workflow of the JIT compiled simulator shows that the major disadvantage of the compiled simulation technique, the requirement for fixed and predictable program code, does not apply for the JIT compiled simulator. However, another important parameter has to be considered – the application size. The size of the program code must be assumed to be run-time dynamic. Since the cache has a run-time static size, which is usually smaller than the application size, a replacement strategy is needed. Furthermore, a unique but not biunique mapping function has to be defined which describes the assignment of program memory addresses to cache addresses (hash-function). The cache key which has been initially used to identify changed program code, is now additionally employed to detect cache collisions. These collisions result from the fact that multiple program memory addresses share a single entry in the cache.

An important concern is the minimization of the cache administration overhead. Since the instruction compile time $\tilde{t}_{comp,insn}$ defines the penalty for a cache miss, the overhead for the cache administration has to be significantly smaller. To fulfill this constraint, a simple one-level cache with a direct address mapping was chosen (see figure 5). Each program memory address corresponds to exactly one entry in the cache. When a cache collision occurs the respective entry is directly overwritten by the new instruction. The direct address mapping assures a minimum cache miss rate ϵ_{miss} for spatially coherent code. Loops, the time critical parts of a program, directly benefit from this characteristic if the loop kernel fits into the cache.

The convergence of the JIT-CCS *(just-in-time cache compiled simulator)* instruction throughput to the performance

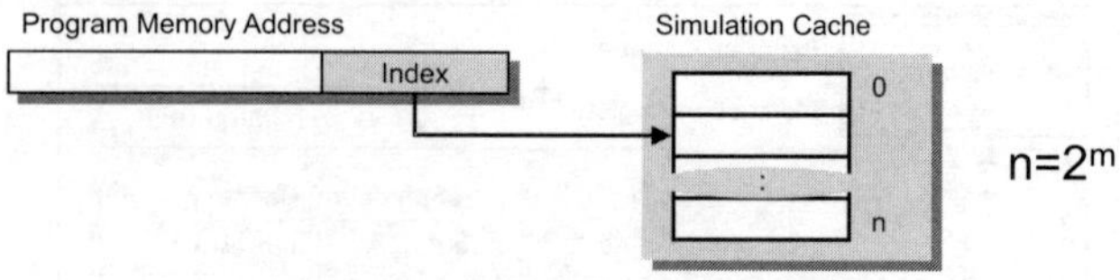

Figure 5: Direct Mapping

of the compiled simulator is now significantly characterized by the simulation cache miss rate ϵ_{miss} as shown in equation (6) (disregarding the cache administration). The results presented in section 5 will demonstrate that already small caches, which are sufficient to store the most time-critical loop of an application, achieve a very good performance approximation of the traditional compiled technique.

$$p_{js_{CS}} \approx \frac{1}{\tilde{t}_{ex,insn} + \tilde{t}_{comp,insn} * \epsilon_{miss}} \quad (6)$$

$$\epsilon_{miss} = f(cachesize) \quad (7)$$

4.4 Pipelined Processor Models

Previous explanations of the JIT-CCS have assumed that each cache address can be overwritten directly in a following simulator control step. While this is perfectly true for instruction accurate simulators where a complete instruction is processed within each control step, this assumption is not maintainable in case of cycle bases simulation of pipelined architectures. The lifetime of an instruction within the pipeline is no longer predictable. Therefore cache entries of instructions currently processed in the pipeline have to be protected against being overwritten. This is achieved by maintaining a look-up table that keeps the instruction addresses which are currently in use. Since the number of (VLIW) instructions present in the pipeline at the same time cannot exceed the number of pipeline stages, the size of the look-up table is defined by the number of stages. Concerning the cache access, three cases have to be considered :

1. cache hit,
2. cache miss, cache address not in use, and
3. cache miss, cache address in use.

In case of a cache hit an instruction can be taken directly from the compiled simulation cache. In the second case the instruction is not present in the cache, but the corresponding cache entry may be overwritten with the recently compiled data. Furthermore, the instruction address is registered in the look-up table. Table entries are written in a circular manner to displace instructions which have already left the pipeline. In the third case, a cache miss occurs and it is determined that the cached instruction which is to be replaced still remains in the pipeline. One solution would be a set-associative cache, however, due to the very unlikely occurrence of this case the simulator switches to interpretive mode for the particular instruction.
Figure 6 illustrates this procedure. The instruction `MOV Y,1` at address `8000` is going to be decoded. Therefore the cache address is generated (1) and the corresponding entry is looked up in the cache (2). A cache collision is detected because the cache entry keeps the instruction `JMP 8000` from address `0`. Before the entry can be overwritten the pipeline-protection table has to be checked (3) whether the `JMP` instruction is still in use or not. Since this is the case, the instruction `MOV Y,1` is not cache compiled but interpreted instead, without updating the cache.

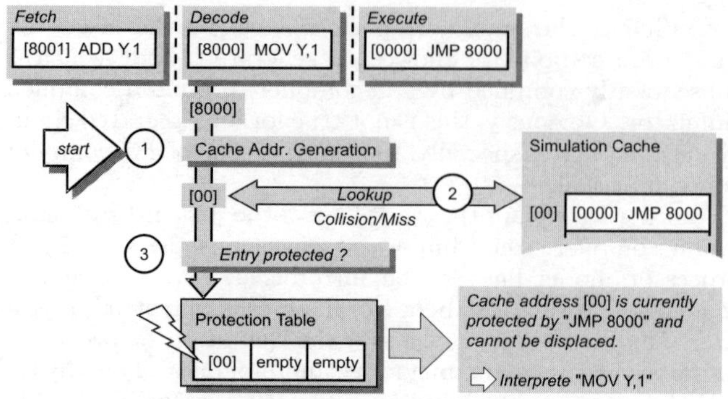

Figure 6: Cache Collision Scenario

5. RESULTS

The applicability and the efficiency of the retargetable just-in-time cache compiled simulation technique has been evaluated on the basis of various LISA processor models. The simulation results of the Advanced Risc Machines ARM7 and ST Microelectronics ST200 LISA models are presented in this section. The processor models have been developed within previous work and the generated tools were verified accurately against the vendor tools. Within this section the instruction throughput of the JIT-CCS will be compared to the performance of the interpretive and traditional compiled technique (both generated from LISA 2.0 descriptions). Therefore, the applications that have been selected for the benchmark fulfill the requirements of the traditional compiled simulation technique: constant program code. Furthermore, it is investigated how the performance of the JIT simulator is influenced by the cache size.
Performance results of the different generated simulators were obtained using an 1200 MHz Athlon PC, 768 MB RAM running the Microsoft Windows 2000 operating system. The generated simulator code has been compiled using the Microsoft Visual C++ 6.0 compiler with optimizations turned on (/O2).

Figure 7 shows the benchmark results of instruction accurate simulators for the ARM7 running a jpeg2000 codec. The dark-grey bars visualize the simulation performance for different simulators in MIPS (million instructions per second). The two leftmost bars show the performance of the compiled and interpretive simulation techniques. The remaining dark-grey bars correspond to the JIT-CCS performance for different cache sizes. The light-grey bars visualize the relative cache-miss rate for the examined cache configurations.

It is clearly recognizable that the JIT-CCS simulation performance is improving with a growing cache size. This effect becomes clear when looking at the continuously decreasing cache-miss rate. Examining the results it can be seen that more than 95% of the compiled simulation performance is achieved with a 4096 entries size cache. This cache allocates less than 2MB of host memory. Compared to the compiled simulator which requires approx. 23MB for all 47354 instructions of the application[3] the JIT-CCS is very memory efficient. Considering big applications such as operating systems it is obvious that also the memory consumption of compiled simulators can become a serious problem. In contrast to compiled simulators the JIT-CCS is characterized by a user defined performance vs. memory trade-off. Due to the execution locality of programs the JIT-CCS perfor-

[3]The memory consumption of the compiled simulation is proportional to the application size.

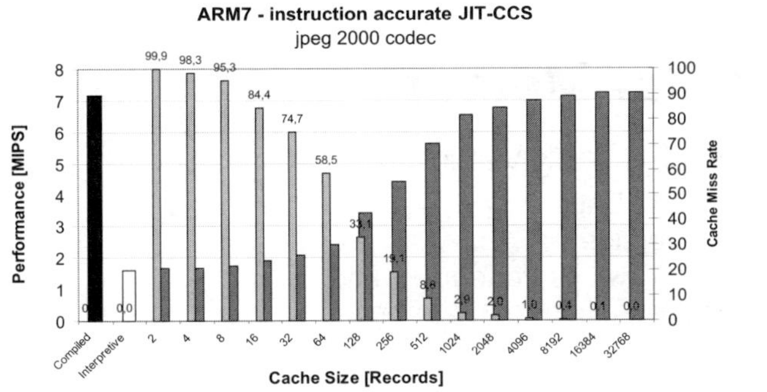

Figure 7: Simulator Performance ARM7 – jpg2000

mance saturation is achieved rapidly, resulting in manageable small simulation caches. Unlike the voluminous jpeg

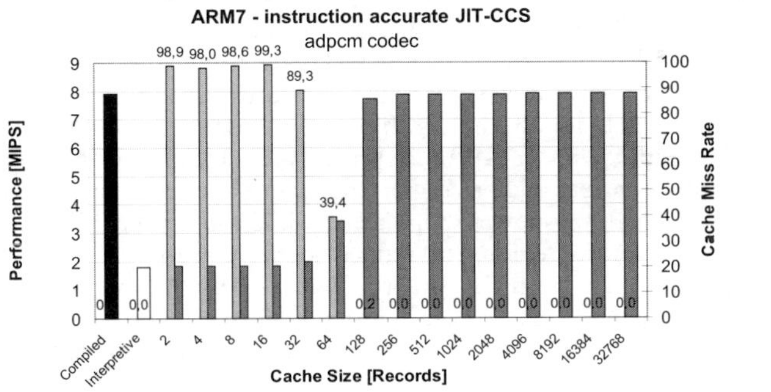

Figure 8: Simulator Performance ARM7 – adpcm

2000 codec with large (>1024 instructions) loop kernels, typical DSP applications like the adpcm codec require much smaller caches. The biggest loop of the adpcm codec compiled for the ARM7 spans 122 instructions. The benchmark results for the adpcm codec presented in figure 8 reflect this characteristic. Here a cache size of 128 entries is sufficient to achieve 95% of the compiled simulation performance. The investigation of further applications has shown, that a reasonable big cache size (approx. 4096-16384) is absolutely sufficient for a >95% approximation of the compiled performance. Almost the same applies to the instruction accurate JIT-CCS of the ST200 architecture. Figure 9 shows the simulation results of an adpcm codec on this machine.

Compared to the ARM7 benchmark results, the extremely poor performance of the interpretive simulator is conspicuous. This is due to the complex instruction decoder which consequentially also influences the performance of the JIT-CCS for small caches. The ST200 processor is a VLIW architecture with many possible parallel instruction combinations. The 128bit instruction word (VLIW) allows multiple combinations of parallel instructions with and without extensions for immediate values. Together with distributed

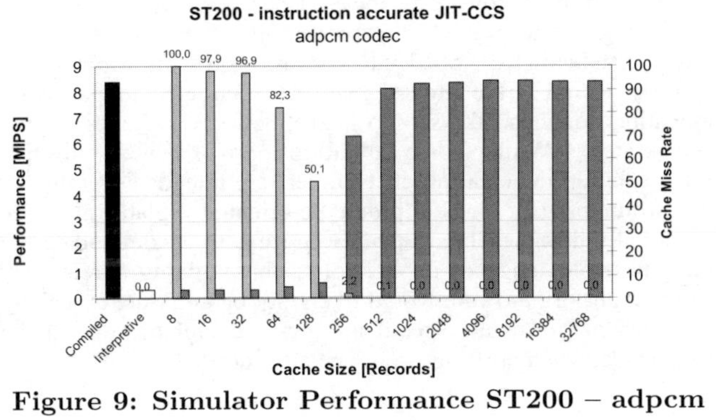

Figure 9: Simulator Performance ST200 – adpcm

op-codes the instruction decoding process of the software simulator dominates the actual behavioral execution. Here, a cache size of at least four entries is required to store the parallel instructions of the VLIW instruction word.

6. CONCLUSION AND FUTURE WORK

In this paper, we presented a novel instruction-set simulation technique, addressing retargetability, flexibility, and finally high performance at the same time. Results for various real world architectures have proven the applicability and the high simulation speed of this technique. Future work will concentrate on modelling further real world architectures and verifying the presented simulation methodology. Beyond simulation, current research aims at the retargeting of HLL compilers and the generation of synthesizeable HDL descriptions from a LISA 2.0 processor model.

7. REFERENCES

[1] A. Halambi and P. Grun and V. Ganesh and A. Khare and N. Dutt and A. Nicolau. EXPRESSION: A Language for Architecture Exploration through Compiler/Simulator Retargetability. In *Proc. of the Conference on Design, Automation & Test in Europe*, 1999.

[2] A. Hoffmann and T. Kogel and A. Nohl and G. Braun and O. Schliebusch and A. Wieferink and H. Meyr. A Novel Methodology for the Design of Application Specific Instruction Set Processors (ASIP) Using a Machine Description Language. *IEEE Transactions on Computer-Aided Design*, 20(11):1338–1354, 2001.

[3] E. Schnarr and J.R. Larus. Fast Out-Of-Order Processor Simulation Using Memoization. In *Proc. 8th Int. Conf. on Architectural Support for Programming Languages and Operating Systems*,1998.

[4] E. Schnarr and M.D. Hill and J.R. Larus. Facile: A Language and Compiler For High-Performance Processor Simulators. In *Proc. of the Int. Conf. on Programming Language Design and Implementation*, 1998.

[5] E. Witchel and M. Rosenblum. Embra: Fast and Flexible Machine Simulation. In *Proc. of the Conf. on Measurement and Modeling of Computer Systems*, 1996.

[6] F. Engel and J. Nuhrenberg and G.P. Fettweis. A Generic Tool Set for Application Specific Processor Architectures. In *Proc. of the Int. Workshop on HW/SW Codesign*, 1999.

[7] G. Braun and A. Hoffmann and A. Nohl and H. Meyr. Using Static Scheduling Techniques for the Retargeting of High Speed, Compiled Simulators for Embedded Processors from an Abstract Machine Description. In *Proc. of the Int. Symposium on System Synthesis*, 2001.

[8] G. Hadjiyiannis and S. Hanono and S. Devadas. ISDL: An Instruction Set Description Language for Retargetability. In *Proc. of the Design Automation Conference*, 1997.

[9] M. Hartoog J.A. Rowson and P.D. Reddy and S. Desai and D.D. Dunlop and E.A. Harcourt and N. Khullar. Generation of Software Tools from Processor Descriptions for Hardware/Software Codesign. In *Proc. of the Design Automation Conference*, 1997.

[10] C. Mills, S. Ahalt, and J. Fowler. Compiled instruction set simulation. *Software - Practice and Experience*, 21(8):877–889, 1991.

[11] R. Leupers, J. Elste, and B. Landwehr. Generation of interpretive and compiled instruction set simulators. In *Proc. of the Asia South Pacific Design Automation Conference* , 1999.

A Fast On-Chip Profiler Memory

Roman Lysecky, Susan Cotterell, Frank Vahid*
Department of Computer Science and Engineering
University of California, Riverside
{rlysecky, susanc, vahid}@cs.ucr.edu, http://www.cs.ucr.edu/~vahid
*Also with the Center for Embedded Computer Systems at UC Irvine

ABSTRACT

Profiling an application executing on a microprocessor is part of the solution to numerous software and hardware optimization and design automation problems. Most current profiling techniques suffer from runtime overhead, inaccuracy, or slowness, and the traditional non-intrusive method of using a logic analyzer doesn't work for today's system-on-a-chip having embedded cores. We introduce a novel on-chip memory architecture that overcomes these limitations. The architecture, which we call ProMem, is based on a pipelined binary tree structure. It achieves single-cycle throughput, so it can keep up with today's fastest pipelined processors. It can also be laid out efficiently and scales very well, becoming more efficient the larger it gets. The memory can be used in a wide-variety of common profiling situations, such as instruction profiling, value profiling, and network traffic profiling, which in turn can be used to guide numerous design automation tasks.

Categories and Subject Descriptors

B.3.0 [**Memory Structures**]: General.

General Terms: Performance, Design.

Keywords: Profiling, system-on-a-chip, platform tuning, adaptive architectures, low power, embedded CAD, binary tree, memory design, embedded systems.

1. INTRODUCTION

Profiling an application executing on a microprocessor is a technique needed to solve a wide variety of optimization problems. In the domain of computing, *profiling* generally means to determine the relative frequency of code regions of interest as a program executes, ranging from fine-grained items like individual statements or basic blocks, to coarser-grained items such as loops or subroutines. The term has also been used to refer to determining the relative frequencies of values that a variable takes on during program execution.

Profiling appears as part of the solution for a tremendous variety of program and hardware optimization and design automation problems. For example, profiling has long been used to find the most frequently executed subroutines of an application, so that a programmer might focus on optimizing those subroutines [10]. Profiling has been used in compilers to map frequently executed code and data to non-interfering cache regions [15] to improve performance. The approach in [8] proposes using profiling to generate alternate subroutine versions for common cases, with the program then using run-time profiling to pick the best version. Likewise, the approach in [14] uses profiling information to synthesize hardware optimized to the most common situations. Dynamic binary translation methods profile in order to store the translation results of frequent code regions, for improved performance as well as power [13], while dynamic optimization methods search for the hottest blocks for runtime recompilation [3]. The approaches in [4][9] use profiling to detect frequent loops to map to a special address region that an architecture would then map to a small low-power loop cache, while the approach in [12] compresses those regions to reduce memory traffic and hence power. The approach in [6] profiles values of variables or subroutine parameters to detect pseudo-constants that can aid a compiler in optimizing for performance, or even for reduced energy [7].

Most previous profiling approaches, being intended for desktop computing systems, introduce runtime overhead. In particular, either they insert additional code into the application binary, or they interrupt the processor at particular intervals to sample the processor's registers. However, for embedded systems, runtime overhead is often not acceptable, since very tight real-time constraints must be met. Thus, embedded system designers in the past relied on logic analyzers to non-intrusively profile an executing application – though even this was cumbersome and hence not a common feature in design automation techniques.

The trend of increasing chip transistor capacity has led to systems-on-a-chip (SOCs). While providing tremendous advantages in terms of cost, size, performance and power, SOCs have the drawback of low accessibility to the internal components. Thus, logic analyzer probes cannot be connected to arbitrary buses inside the SOC to achieve profiling. Although SOCs typically come with means for accessing internal registers through external pins (e.g., using the JTAG standard [11]), such access is accomplished by stopping normal application execution and then serially scanning the register contents in or out. This access approach incurs large runtime overhead, being intended for test and debug purposes rather than profiling.

Fortunately, the same transistor capacity trend that has led to SOCs has enabled hardware-based approaches to profiling. While on-chip profiling hardware in the past has been limited to high-volume high-performance microprocessors, such hardware can today be added to embedded system prototyping platforms. *Platforms* [16][18] are predesigned SOCs targeted to particular application domains, like set-top boxes, network switches, digital cameras, etc. While some platforms are oriented towards implementation in actual products, others are intended specifically for prototyping. These prototype-oriented platforms are intentionally designed larger than necessary, to accommodate the widest possible range of applications. Thus, adding a relatively small amount of hardware would likely not be an issue, especially since such platforms are specifically designed for use during the design stage, when profiling would be most needed.

DAC 2002, June 10-14, 2002, New Orleans, Louisiana, USA.

Designing the additional hardware that profiles the executing application non-invasively yet accurately is non-trivial, since updating profile counts fast enough is hard. Existing fast memory designs, even advanced content-addressable memories, have several limitations (which we will describe). Thus, prior profiling hardware typically only counted easy to detect events (like a cache miss) or could be configured to detect only one particular pattern on a bus (like a specific address).

We therefore have developed a memory architecture specifically intended for profiling. The profiler memory can be used to keep counts of hundreds or thousands of bus patterns simultaneously, in contrast to previous profiler hardware. Yet, the memory has a simple interface to the bus it is monitoring, is very size efficient, and can be composed from a collection of standard register files or memories surrounded by a small amount of additional logic. By using a novel pipelined binary tree architecture, our profiler memory achieves single-cycle throughput, meaning complete accuracy even when monitoring very fast sequences of patterns, such as the address bus of modern processors that fetch one instruction per cycle. The binary tree is implemented with a separate module for each tree level, and thus scales well, becoming even more efficient the larger it gets.

In this paper, we define the on-chip profiling problem and discuss related work. We introduce our profiler memory architecture, and describe its use. We show how the memory's modular structure results in timing and area efficient layouts for any size memory. We provide data on the memory's size and performance, and illustrate its use on several benchmarks.

2. PROBLEM DEFINITION

Code profiling requires that we count the frequency of target addresses appearing on a microprocessor's address bus, during some given period. We assume that the set of target addresses has been given to us. They could correspond to statements, blocks, loops, subroutines, variables, arrays, or any combination thereof. Our approach is independent of how these addresses are selected, what they represent, or how they will be used.

We assume the address bus being monitored contains the actual addresses of interest, meaning we are monitoring the bus before any translation to cache addresses (if a cache exists). We also assume that virtual memory is not being used (the common case in embedded systems), so that the physical addresses on the address bus need not be translated further to virtual addresses. We assume target addresses could appear every clock cycle. We assume the profiling circuit operates at the same (or lower) clock frequency than the bus being monitored. Specifically, we do not assume that the profiling circuit has the luxury of running at a higher clock frequency.

We state the problem generally as follows. We must monitor a given bus B, having a width w, for a period of M clock cycles. During each cycle t, a pattern p_t appears on B, thus forming an input pattern set $P=\{p_1, p_2, \ldots p_M\}$. A pattern is simply a combination of w 0s and 1s on the bus. We are given a target pattern set $TP=\{tp_1, tp_2, \ldots tp_n\}$; each target pattern is a combination of w 0s and 1s. Our goal is to maintain a set of target pattern counts $C=\{ctp_1, ctp_2, \ldots ctp_n\}$, such that ctp_i equals the numbers of times pattern tp_i was seen on B during the period M. Specifically:

$$ctp_i = \sum_{t=1}^{M} \begin{Bmatrix} 1, & if\ p_t = tp_i \\ 0, & otherwise \end{Bmatrix}$$

This more general problem formulation can also represent problems other than code profiling. For example, the bus B might consist of both an address bus and data bus, so that we might perform value profiling of a set of variables. The bus could correspond to an internal bus of a router chip, for which we might want to profile the traffic to determine the most common destinations, to optimize their routing table lookup speed [20].

We assume our method of monitoring and counting must be non-intrusive. In other words, we cannot introduce extra clock cycles or change the patterns occurring over bus B. Non-intrusiveness is important in embedded systems, since hard real-time constraints often must be met, and precise timing of input/output operations is often imperative. Increasing runtime by even the smallest amount can lead to constraint violations, radically different system behavior, and/or system failure.

3. PROFILING APPROACHES

We can view the general solution to the profiling problem as involving three parts: detecting patterns, checking if detected patterns match a target pattern, and updating the target pattern counts when a match is found. We categorize solution approaches into two categories: software-based and hardware-based.

3.1 Software-Based Profiling

Software-based profiling approaches use the instruction-execution resources of the processor itself to solve the profiling problem.

A common software-based approach involves "instrumenting" the application by adding code to count frequencies of the desired code regions [10]. For example, if we wish to count the frequency of execution of a subroutine, we can add code to the beginning of the subroutine that increments a counter variable declared for that subroutine. This approach is straightforward and flexible, yet incurs significant runtime overhead, especially if the granularity of the code regions being profiled is fine. There are additional problems with this approach in an embedded system. For example, the additional code increases code size, which may not be allowable if the non-instrumented program already used all the program memory (a common situation). The additional code may also cause register spills, thus further modifying program behavior. Furthermore, instrumenting requires a special compiler or a post-compilation binary instrumentation tool.

To reduce runtime overhead, other profiling approaches use sampling techniques [1][8]. Such methods interrupt the microprocessor at certain intervals, and then read the program counter and other internal registers. These methods tradeoff accuracy for reduced overhead. However, in embedded systems, even the reduced overhead may not be acceptable. Furthermore, while the overhead may be reduced, the disruption of runtime behavior during the interrupt itself can be very significant in a real-time system. A related method assumes the existence of a multi-tasking operating system. These methods add an additional task to perform profiling, in place of an interrupt [22]. However, the disadvantages are the same as the interrupt approach.

Others use a simulation-based approach to profiling [5]. The application is run on an instruction-set simulator, with the simulator keeping track of profile information. While accurate, this approach is extremely slow, especially when simulating a system-on-a-chip. Simulating an application for several hours may cover only a few seconds of real time, so this may not exercise realistic code execution. Furthermore, setting up such simulations can be difficult if not impossible for embedded systems, due to the complex external environments that must also be modeled.

3.2 Hardware-Based Profiling

Due to the above limitations, embedded system designers have often resorted to using a logic analyzer, where the analyzer's probes are placed directly on the bus to be monitored. The logic analyzer can store a pattern trace, which can be later processed to derive profile information. To avoid the problem of enormous

Figure 1: Conceptual view of ProMem: (a) sorted binary tree for target pattern set, (b) storing each level (stage) in target pattern memories (TPM) with corresponding memory addresses *A* shown, and (c) example showing how address is formed searching for pattern *e* in each stage.

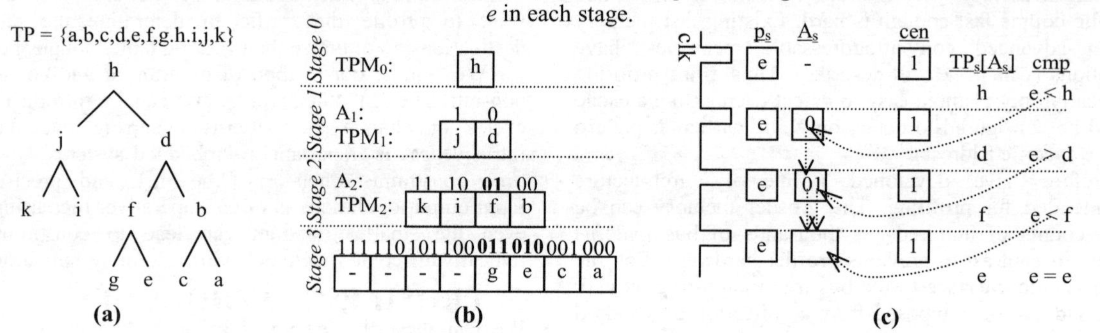

traces filling up available trace memory, certain patterns can be programmed into the analyzer, which the analyzer uses to trigger trace storage for a given period. To perform profiling as we defined earlier, one could simply input the target patterns as triggering patterns. However, a drawback of logic analyzers is their high cost and their ineffectiveness in the era of SOCs.

Some microprocessors include on-chip hardware to assist software developers to profile an executing program [19][21]. Such hardware consists primarily of event counters. Monitored events typically include cache misses, pipeline stalls, branch mis-predictions, and so on. Some counters may be configured to detect a particular bus address, but there are typically only a few of these since having many of them would be costly in terms of size. Thus, the programmer must dynamically reconfigure those counters during program execution to obtain a more complete profile. Not only can this lead to inaccuracy, but also reconfiguring those counters requires additional software instructions, thus bringing us back to some of the problems with software-based profiling.

Recently, a programmable coprocessor for profiling has been proposed, primarily intended to detect performance-related events in high-performance processors [23]. The memory we propose could be used in conjunction with such a coprocessor to be able to detect large pattern sets on a bus.

A hardware technology that might seem to solve the problem of detecting patterns in large pattern sets is content-addressable memories (CAM). CAMs provide fast searches for a key in a large data set. Given the data, a CAM returns the address at which the key (data) resides in a memory. There are several CAM implementation types. One type uses a fully associative memory, which simultaneously compares every location with the key. Fully associative memories do not scale well to larger memories – they become very large and difficult to place and route due to all the comparison logic, and their access time increases due to the high capacitive load involved in sending the key to all memory addresses simultaneously. Thus, newer versions of CAMs use a regular static or dynamic RAM coupled with a smart controller. The controller implements an efficient lookup data structure in the memory, such as a binary tree, or a Patricia Trie [20]. Lookups of binary trees are of the order of log of the number of data items, while lookups for a Patricia Trie are linear with respect to the number of characters (or bits) in the key.

CAMs thus do not satisfy our goals for a profiler memory. While CAMs built using fully associative memories may have single-cycle access time, they do not scale well to larger target pattern sets. On the other hand, CAMs built from regular RAM require multiple cycles for lookup, which does not support the situation where target patterns appear on the bus on every cycle.

Likewise, hash tables do not satisfy the goal either, since they require multiple cycles to compute the hashing function, and may require additional cycles in case of a collision. Finally, approaches that queue the input patterns only help if such input does not appear on every cycle.

4. A SELF-PROFILER MEMORY ARCHITECTURE

The key to building an efficient self-profiler memory is to recognize that we do not need single cycle lookup, or even single-cycle write, but rather just single-cycle update throughput. In other words, the memory must be able to accept a new pattern every cycle. The memory need not actually update the count field of a matching target pattern until many cycles later.

4.1 Pipelined Binary Tree

To achieve a throughput of one new pattern every cycle, we implement our ProMem memory structure using a pipelined binary tree structure to find the location of the given input pattern. Figure 1(a) shows an inorder binary tree for a target pattern set *TP*={a, b, c, d, e, f, g, h, i, j, k}. We will implement the tree by using a separate module for each tree level, with each module implementing a pipeline stage. Figure 1(b) illustrates the target patterns that would appear in a memory (target pattern memory, or TPM) in each module, assuming a four stage ProMem design. We place the target patterns in memory such that the following property holds:

> *A node's children have the same high-order address bits as the parent's address, with the low order address bit of each child indicating left (1) or right (0) child.*

This relationship can be observed, for example, in the bolded bits of Figure 1(b): *f* has address 01 in TPM_2, and has left child *g* and right child *e* with addresses 011 and 010 in TPM_3.

Before presenting the architecture of each module, we will describe conceptually the manner in which each module searches for matches. Figure 1(c) illustrates the search for a single input pattern *e* in the tree of Figure 1(b). We compare *e* with the root node *h*. Since $e < h$, the address A_1 passed to stage 1 is 0. At stage 1, we compare *e* with $TPM_1[0]$, or *d*. Since $e > d$, the address A_2 passed to stage 2 is 01. At stage 2, we compare *e* with $TPM_2[01]$, or *f*. Since $e < f$, the address A_3 passed to stage 3 is 010. Finally, at stage 3, we compare *e* with $TPM_3[010]$, or *e*. Since equal, we found a match, and we then increment a count value associated with $TPM_3[010]$. Those count values are actually stored in a separate memory in a stage's module, as we will see later.

The above example illustrated how one item would be searched for in the tree. Figure 2 extends the example by continuing to feed input patterns to the root node on each clock cycle, illustrating the pipelined and hence single-cycle throughput of ProMem. We wish to search for input patterns *e*, *d*, and *w*, which appear in that order on a bus during consecutive clock

Figure 2: Example search for input patterns P = {e, d, w}, showing pipelined behavior of ProMem.

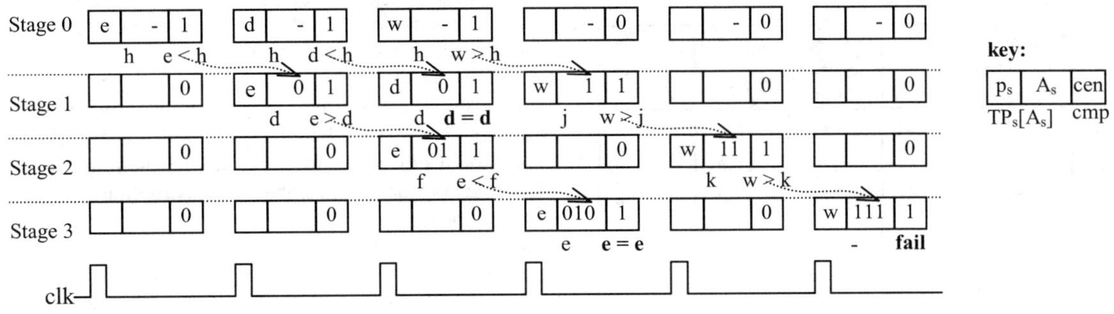

cycles. We see the same sequence as the above example for *e* as it progresses through the pipeline, though in this figure we can also see how the enable signal *cen* is updated. On the second clock cycle, we see *d* enter the pipeline at Stage 0, from where it is passed to Stage 1 as the right child. In Stage 1, a match is found. Notice that this match disables the continued search during subsequent cycles in Stages 2 and 3, by setting *cen* to 0. *w* progresses through the tree following left children, failing when it hits a node with no target pattern stored.

4.2 Memory Architecture

Figure 3(a) shows the design of a single module within ProMem. Each level's module consists of registers to latch the input pattern, address, and enable signal (collectively referred to as the Pipeline Register), a memory containing the target patterns (TPM), a memory containing target patterns' corresponding counts (CM), a comparator, an incrementer, and some required logic.

From the pipeline register, the input pattern will be compared against the contents of the TPM_s at address A_{s}_i. The comparator used in the comparison has two outputs: a greater than comparison and an equal to comparison. If the current stage is enabled by the *cen_i* signal and the input pattern is equal to the target pattern, the *cen_o* signal will be set to 0, indicating the pattern was found and

Figure 3: ProMem Module Design: (a) design of a module for stage *s* (> 0), (b) structure of ProMem using modules (each stage is actually twice the size of the previous stage).

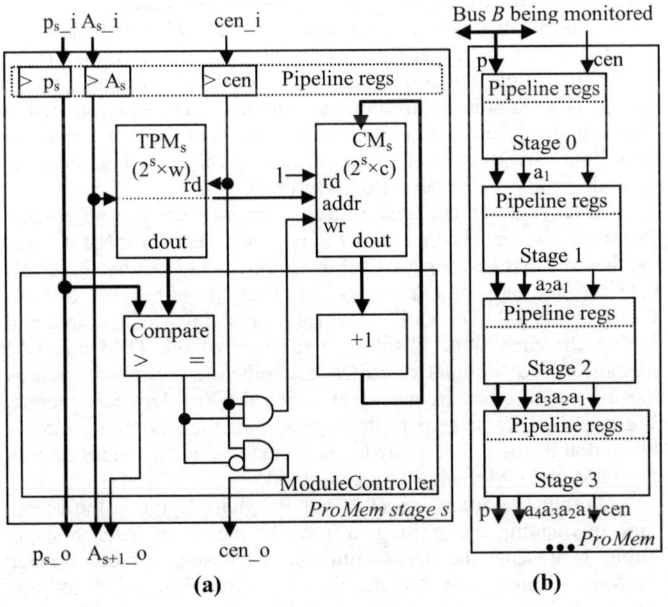

no further searching is required. In addition, the write port of the CM will be enabled to increment the associated count value. However, because this requires a read and write of the same memory location, we will need to use a memory that has an independent read port and an independent write port. Such a memory will be slightly larger than a single port memory but not as large as a true dual port memory. Alternatively, if the input pattern does not match the current target pattern, the result from the greater than comparison will be concatenated with the input address, A_{s}_i, to create the address for the next stage, A_{s+1}_o, and *cen_o* will be set to 1, indicating we have not found the target pattern and need to continue searching.

To address the issue of locations within our binary search that do not contain valid addresses, we implement a simple scheme that extends the width of the TPM by one bit. When a valid target pattern is written to the TPM, it is concatenated with single bit with the value of 1. Then, to compare the target pattern to the input pattern, we also concatenate the input pattern with a 1 bit before it reaches the comparator. If the two values when compared are equal, we have found the target pattern. Otherwise, to determine if the current target pattern is valid, we check the most significant bit of the current target pattern. If the bit is 0, this indicates that we do not have a valid target pattern, in which case we stop searching the binary tree as no valid patterns will be found within that sub-tree.

Although the above design can be used for each stage of our binary search tree structure, using this design for the first stage would not be efficient. With only a single entry within both the TPM and CM, implementing the TPM and CM as actual memories would be inefficient. Instead, the TPM and CM of the first stage are implemented using two registers. In addition, the input to the first stage will only consist of the enable signal, *cen_i*, and the target pattern. All other computation and logic will be implemented in the same manner as the remaining stages.

To construct the entire ProMem structure, we simply connect a module to the module for the next stage, as shown in Figure 3(b). Thus, to achieve a ProMem design that can handle a target pattern set with 1023 entries, we will need to create a binary search tree structure with 10 stages. However, using the modular design for each level, extending the number of target patterns is as simple as adding another level to the design.

5. ADDITIONAL CONSIDERATIONS

We thus far have introduced the main feature of ProMem, namely its pipelined structure enabling single-cycle throughput when monitoring a bus. We now discuss how to write the target patterns into ProMem, how to read out the counts after the monitoring period is complete, extensions to support ranges, and layout considerations.

We initially considered memory mapping every internal register. The memory driver software could then store the target patterns, and read the counts, in any manner desired. This option

Table 1: ProMem Area and Timing Results (8 stages, 32-bit target patterns).

Stage	TPMs + CMs (Gates)	Pipeline Register (Gates)	Module Controller (Gates)	Module Controller (% total)	Total (Gates)	Access Time (ns)
0	774	268	640	38%	1682	4
1	1618	274	713	27%	2606	4
2	3156	284	753	18%	4193	4
3	6468	298	812	11%	7577	4
4	12925	307	823	6%	14054	4
5	25870	313	934	3%	27117	4
6	53254	322	995	2%	54572	4
7	115477	327	1287	1%	117090	4
Total	**219541**	**2393**	**6957**	**3%**	**228891**	**4**

has the undesirable requirement that all the modules be connected to a single bus, whether a processor's peripheral bus, or an internal bus connected through a bridge to that peripheral bus. In either case, running a bus to all modules is costly in terms of area. Furthermore, each module would then require additional logic to interface to that bus to detect and process write and read requests.

A second option was to connect all registers in a scan chain. Connecting the registers in sorted order would require excessive wires between stages. Connecting in breadth first order eliminates that problem. Another potential issue is the compatibility of this scan chain with that used for testing, and the potential complexity of the driver software requiring access to JTAG ports. We leave this potentially viable option for future work.

We currently implement a scheme that makes use of the existing pipeline structure, by adding a few additional control lines and an up-counting register *ptr* added to each module. *ptr* is one bit wider than the module's input address. We initialize *ptr* to zero upon memory reset. We add a write enable signal *wen_i* to each module, and a 1-bit register *wen* as part of the module's pipeline register. If a module at stage *s* detects *wen*=1, it writes the input pattern *p* to *TPM[ptr]* and increments *ptr*, but only if the high-order bit of *ptr* was 0. If the high-order bit was 1, this means all the words in TPM have already been written. In this case, the module instead passes the pattern *p* and the *wen* signals to the next stage. Such functionality results in the entire ProMem target pattern contents being written in breadth first order (right to left for each level). Thus, the driver software must provide the target patterns in breadth-first order – a straightforward task. We assume the target patterns can be written to the bus being monitored. Otherwise, a second input for a target pattern would be required *on Stage 0 only*.

In addition to the *wen_i* signal, we also add the *val_i* signal to the pipeline register. This signal will be used to indicate that the target pattern being written is valid. To do so, the *val_i* input and the target pattern will be concatenated together and written to the TPM. It is important to note that the driver software must initialize all locations when loading a new target pattern set to ensure all entries within the TPM contain the correct valid bit.

Reads of the counts are done similarly. However, we must first reset all the *ptr* counters to 0 – we include an additional input signal *rstptr_i* and 1-bit register *rstptr* to each module, which asserts the clear signal on *ptr* and also gets passed on to the next stage. We must therefore hold this signal high and then low for a certain number of cycles (two times the number of stages) to clear all the *ptr* counters. We also add a read enable signal *ren_i* and a 1-bit register *ren*. Reading then results in a breadth first outputting of the target memory contents onto the pattern output of the last stage. Thus, the driver software must reorder this output into sorted order – again, a straightforward task.

ProMem can be extended to support ranges. For example, we may want to count how many times patterns appear with a value between 1 and 10, 11 and 20, 21 and 30, etc. Such functionality could be useful to count, for example, the number of accesses to a large array, or the number of instructions executed within a subroutine or loop. We simply use two TPMs to store the target pattern ranges. The first TPM will contain the lower bounds of the ranges, and the second TPM will contain the upper bounds of the ranges. As before, the target ranges will be stored in sorted order within the binary tree structure (the ranges must be non-overlapping). When searching to detect if an input pattern is part of the target range, we will use two comparators to simultaneously compare the input pattern with both the lower and upper bound of the target range. If an input pattern is not within the target range, the output from greater than comparison with the lower bound will determine the low order address bit for the next stage.

Finally, another important concern in the design of ProMem is the ability to create an efficient layout. We would like to create a layout that is as tight as possible. In addition, we must ensure that the number of wires and length of wires connecting each module is at a minimum. Assuming each module has a rectangular layout, we can tightly place each module of the ProMem in a spiral fashion, with the output put of each stage abutted with the input of the nest stage. However, in addition, we must ensure that the module corresponding to level 0 is placed on the boundary of the entire structure. Furthermore, the register files or memories within each module can come from a standard register file or memory implementation, which already have efficient layouts.

6. RESULTS

We implemented ProMem using VHDL. We designed ProMem as a combination of stages that are connected together in a top-level entity. Because stage 0 of our ProMem has a different structure then the remaining stages, it was implemented as its own entity. The remaining stages, however, use a single entity that contains a generic *STAGE*, which specifies the stage of the instantiated entity. Furthermore, all stages contain a generic *WIDTH*, which specifies the *WIDTH* of the target patterns and the associated count values. Using this implementation approach, adding another stage to the design is as simple as instantiating another stage and connecting the outputs of the previous stage to the inputs of the new stage. We tested our VHDL code at structural RTL level as well as the gate-level description generated from synthesis.

Table 1 shows the area results from synthesizing an 8-stage ProMem using 32-bit target patterns. The ProMem was synthesized with Synopsys Design Compiler [17] using the UMC 0.18 technology and memory libraries provided by Artisan Components [1]. The table provides a breakdown of the area and timing for each stage, displaying the size of the TPM and CM memories, the ModuleController, and pipeline register as well as the maximum clock frequency at which the ProMem can operate. We see that the size of each stage within the design is roughly twice that of the previous stage, as expected since the sizes of both the TPM and CM memories are doubled.

Another important aspect of our ProMem design is the access time for reading and writing to/from the memories in each stage. Table 1 presents the access time for each stage of our 8-stage ProMem design. For all stages, an access time of 4 ns was achieved. More importantly, as each stage within the ProMem

Table 2: Comparison of ProMem with fully-associative CAM.

Entries	ProMem (Gates)	CAM (Gates)	% Reduction
1	1682	1006	-67%
3	4288	4081	-5%
7	8481	8319	-2%
15	16058	16627	3%
31	30113	33276	9%
63	57230	68066	16%
127	111802	145100	23%
255	228891	309831	26%

grew in size, the access time did not increase, in contrast to a CAM design. The timing results were obtained with Design Compiler using the synthesized design along with the timing information in the UMC technology library.

In addition to the total size of our design, we would like to see how much overhead the supporting logic with our ProMem adds to the overall design. Thus, Table 1 also presents the percentage of the total design corresponding to the ModuleController. We see that the overhead per stage decreases for deeper stages. We also see that for a ProMem with 255 entries, the ModuleController only consists of 3% percent of the total design size.

Table 2 compares our ProMem design to a fully-associative CAM for various numbers of entries. For a small number of entries, the overhead of the ModuleController within the ProMem structure results in a larger size than a design using a CAM. However, for as little as 15 entries, our ProMem design results in smaller overall size than the CAM approach. Furthermore, for a design with 255 entries our ProMem design is 26% smaller than the alternative design. Furthermore, as mentioned earlier, fully-associative CAMs will have slower access time as the size grows since each pattern must be sent to every entry.

7. CONCLUSIONS

Profiling is a key to numerous design automation problems that optimize a program and/or architecture. A prototype-oriented SOC platform that includes on-chip profiling hardware, along with a software interface to that hardware, can enable profiling in real-time embedded environments. We introduced a new memory architecture, based on a binary tree, that can monitor a bus for patterns on every clock cycle. The memory has a simple interface to the monitored bus, and scales very well as it gets bigger.

8. ACKNOWLEDGEMENTS

This work was supported in part by the National Science Foundation (CCR-9876006) and the UC MICRO program.

9. REFERENCES

[1] Anderson, J., et al. Continuous Profiling: Where Have All the Cycles Gone? 16th ACM Symp. of Operating Systems Design, 1997.

[2] Artisan Components, Inc. UMC .18 Technology Library, http://www.artisan.com, 2001.

[3] Bala, V., E. Duesterwald, and S. Banerjia. Dynamo: A Transparent Dynamic Optimization System. ACM SIGPLAN Conference on Programming Language Design and Implementation (PLDI), June 2000.

[4] Bellas, N., et al. Energy and Performance Improvements in Microprocessor Design Using a Loop Cache. ICCD, pp. 378-383, 1999.

[5] Burger, D. and T. M. Austin. The SimpleScalar tool set, version 2.0. Tech. Rep. CS-1342, University of Wisconsin-Madison, June 1997.

[6] Calder, B., P. Feller and A. Eustace. Value Profiling. MICRO, pp. 259-269, 1997.

[7] Chung, E.Y., L. Benini and G. De Micheli. Automatic Source Code Specialization for Energy Reduction. ISLPED, 2001.

[8] Dean, J., et al. ProfileMe: Hardware Support for Instruction-Level Profiling on Out-of-Order Processors. MICRO, 1997.

[9] Gordon-Ross, A., S. Cotterell and F. Vahid. Exploiting Fixed Programs in Embedded Systems: A Loop Cache Example. IEEE Computer Architecture Letters, Jan. 2002.

[10] Graham, S.L., P.B. Kessler and M.K. McKusick. gprof: a Call Graph Execution Profiler. SIGPLAN Symp. on Compiler Construction, pp. 120-126, 1982.

[11] IEEE, IEEE 1149.1 Standard Test Access Port and Boundary-Scan Architecture, http://standards.ieee.org, 2001.

[12] Ishihara, T., H. Yasuura. A Power Reduction Technique with Object Code Merging for Application Specific Embedded Processors. DATE, March 2000.

[13] Klaiber, A. The Technology Behind Crusoe Processors. Transmeta Corporation, http://www.transmeta.com, 2000.

[14] Lakshminarayana, G., et al. Common-Case Computation: A High-Level Technique for Power and Performance Optimization. DAC, pp. 1-5, 1999.

[15] Pettis, K. and R.C. Hansen. Profile Guided Code Positioning. ACM SIGPLAN Conference on Programming Language Design and Implementation (PLDI), June 1990.

[16] Semiconductor Industry Association. International Technology Roadmap for Semiconductors: 1999 edition. Austin, TX: International SEMATECH, 1999.

[17] Synopsys, Inc. Design Compiler, http://www.synopsys.com, 2001.

[18] Vahid, F., T. Givargis. Platform Tuning for Embedded Systems Design. IEEE Computer, Vol 34, No. 3, pp. 112-114, March 2001.

[19] Vtune Environment, Intel Corp., http://developer.intel.com/vtune.

[20] Waldvogel, M., et al. Scalable High Speed IP Routing Lookups, SIGCOMM 97, 1997.

[21] Zagha, M., B. Larson, S. Turner, and M. Itzkowitz. Performance Analysis Using the MIPS R10000 Performance Counters. Supercomputing, Nov. 1996,

[22] Zhang, X., et al. System Support for automatic Profiling and Optimization. Proceedings of the 16th Symp. on Operating Systems Principles, 1997.

[23] Zilles, C.B. and G.S. Sohi. A Programmable Co-processor for Profiling. International Symp. on High-Performance Computer Architectures, 2001.

Design of an One-cycle Decompression Hardware for Performance Increase in Embedded Systems

Haris Lekatsas
NEC USA
4 Independence Way
Princeton, New Jersey 08540
lekatsas@nec-lab.com

Jörg Henkel
NEC USA
4 Independence Way
Princeton, New Jersey 08540
henkel@nec-lab.com

Venkata Jakkula
NEC USA
4 Independence Way
Princeton, New Jersey 08540
jakkula@nec-lab.com

Abstract

Code compression is known as an effective technique to reduce instruction memory size on an embedded system. However, code compression can also be very effective in increasing processor-to-memory bandwidth and hence provide increased system performance. In this paper we describe our design and design methodology of the first running prototype of a one-cycle code decompression unit that decompresses compressed instructions on-the-fly. We describe in detail the architecture that enables decompression of multiple instructions in one cycle and we present the design methodologies and tools used. The stand-alone decompression unit does not require any modifications on the processor core. We observed up to 63% performance increase with 25% in average over a wide variety of applications running on the hardware prototype under various system configurations.

Categories and Subject Descriptors

B.3 [**Hardware**]: Memory Structures; C.3 [**Computer Systems Organization**]: Special-purpose and Application-based Systems-*Real-time and embedded systems*

General Terms

Algorithms, Design, Performance

1. INTRODUCTION

Designers in the high volume and cost-sensitive embedded system's market face many design challenges as the market demands single-chip-solutions to meet constraints like lower system cost, higher functionality, higher level of performance and lower power-consumption requirements for increasingly complex applications.

Code compression/decompression has emerged as an important field to address parts of these problems in embedded system designs. Typically, a system that runs compressed code is decompressing the code as it is needed by the CPU. The decompression can be performed outside the CPU (e.g. between L2 cache and main memory or between L2 cache and CPU) or inside the CPU as part of the instruction fetch unit or decoding unit. In any case only parts of the whole code will be decompressed at a time (on an instruction-level or block-level, for example) in order to minimize the memory requirements. The code is being compressed as a step following code compilation and then it is burned into Flash or ROM from where it is loaded during system reset into the main memory or L2 cache. That implies that the software cannot be changed and must be fixed.[1] If, however, the software application does change then the code has to be compressed again and the Flash or ROM has to be burned again. This is typically not a restriction for embedded systems that do not allow the user to change the software.

The primary goal in code compression[2] has traditionally been to reduce the instruction memory size although newer approaches have shown that it can even lead to performance increases of an embedded system.

Application areas for code compression/decompression are embedded systems in the areas of personal wireless communication (e.g. cell phones), personal computing (e.g. PDAs), personal entertainment (e.g. MPEGIII players) and other embedded systems where memory size and performance is a constraint like automotive control etc.

We introduce a code compression/decompression technology that is based on a commercial parameterizable processor core. We have designed a 1-cycle decompression unit that enhances the performance of the core system by an average of 25% and facilitating a code size reduction of 35% on average. Note, that the core has an already densely designed instruction set architecture as it's instruction size varies (2 instruction types exist: 16-bit instructions and 24-bit instructions). Our technology is generic as it can be applied to any RISC processor. The only application-specific parts are the processor and memory interfaces as far as the hardware is concerned as well as some modules of the software design flow. We have recently finished a hardware prototype that demonstrates the advantages of our technology in terms of performance increase and instruction memory size reduction.

This paper is structured as follows: Section 2 describes related work in code compression. Section 3 presents the compression and decompression algorithms and provides insight in the decompression process. Section 4 focuses on the hardware implementation and the design methodology we followed. Section 5 provides experimental results, while Section 6 concludes the paper.

[1]Note that this characteristic prohibits the system from using self-modifying code.

[2]Note that we use in the following the term "compression" or "decompression" when we denote a technology that runs code that has been compressed before the system runs and where code is being decompressed when the system runs i.e. on-the-fly.

2. RELATED WORK

In the following we report some of the most relevant work for code compression approaches.

Wolfe and Chanin developed the Compressed Code RISC Processor (CCRP), which was the first system to use cache misses to trigger decompression [9]. Their decompression engine is designed as part of the cache refill hardware. The instructions in each L1 cache block are separately Huffman-encoded so that each block can be individually decompressed without requiring decompression of other blocks in advance. The authors report a 73% compression ratio on the MIPS architecture.

CodePack is used in IBM's embedded PowerPC systems [3]. Their scheme resembles CCRP in that it is part of the memory system. The CPU is unaware of compression, and a table maps between the native and compressed address spaces. The decompression engine accepts L1-cache miss addresses, retrieves the corresponding compressed bytes from main memory, decompresses them, and returns native PowerPC instructions to the L1-cache. CodePack achieves 60% compression ratio on PowerPC.

Software decompression is also possible, simplifying the hardware design and allowing the decompression to be selected at runtime. Lefurgy et al. [6] proposed two hardware mechanisms to support software decompression. Another technique that can be carried out purely in software is a dictionary method proposed by Liao et al. [7] where mini-subroutines are introduced replacing frequently appearing code fragments.

Ishiura and Yamaguchi [4] proposed a compression scheme for VLIW processors based on automated field partitioning. They keep the size of the decompression tables small by producing codes for sub-fields of instructions. Benini et al. [2] limit the dictionary size by selectively compressing instructions. Okuma et al. [8] proposed an encoding technique that takes into account fields within instructions. Yoshida et al. [10] proposed a logarithmic-based compression scheme which can result in power reduction as well.

Our work differs from previous work in the following ways: 1) our decompression engine is capable of decompressing 1 to 2 instructions per cycle to meet the CPU's demand at no time penalty 2) it is completely transparent 3) it has been fully implemented and is running on a commercial FPGA board.

3. ALGORITHMS FOR THE ARCHITECTURAL DESIGN

In this section we describe techniques that enable the decompression engine to decompress in just one cycle. Though table-based dictionary encoding is a well known compression technique [1], advanced architectural design strategies are necessary to meet the one-cycle design constraint. Here is a summary of the most important constraints:

- Decompressing in one cycle
- Meeting the clock cycle delay budget
- No modification to the CPU should be necessary. That means the decompression engine has to interface to the CPU and to the memory hierarchy such that both CPU and memory hierarchy are not aware of the existence of the decompression engine.
- The decompression engine should be considerably smaller than the CPU.

Most of the architectural design strategies are independent of the CPU but some of them are especially designed for usage with our platform. A particular characteristc of the CPU we used is that the instruction set contains two types of instructions: 16-bit and 24-bit.

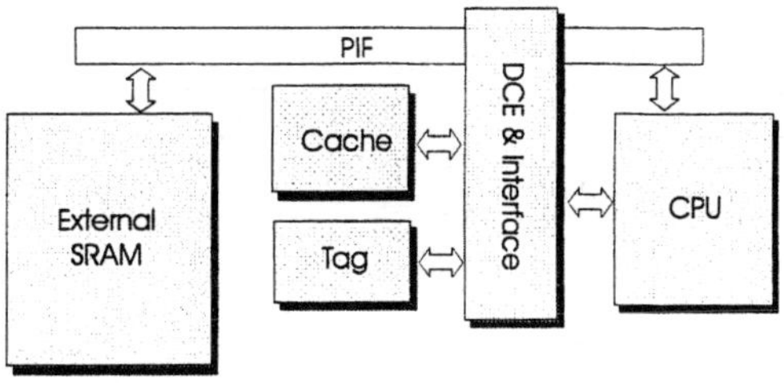

Figure 1: Basic system architecture

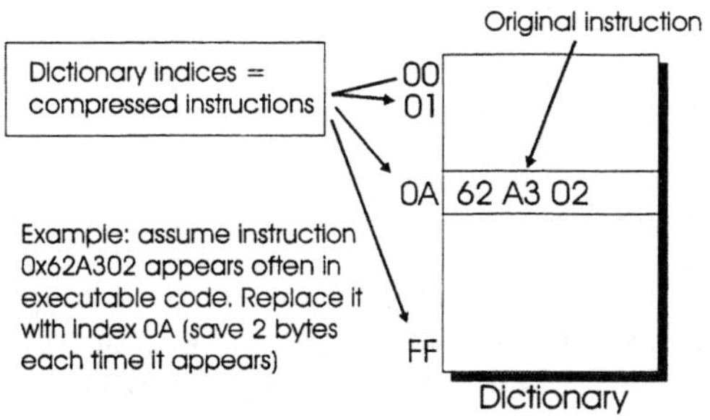

Figure 2: Principle of dictionary based compression technique

3.1 Dictionary-based decompression table

We deploy a dictionary technique to compress frequently appearing instructions (according to program statistics). Dictionary coding refers to a class of compression methods that replace sequences of symbols with indices to a table. This table is called the "dictionary". The indices are the codewords in the compressed program. An entry in the dictionary holds either an original symbol or a sequence of symbols. This encoding reduces the data if the indices are shorter than the replaced symbols and if the dictionary is compact. The main advantage is that the indices are usually of fixed-length and thus simplify the decompression logic to access the dictionary and also reduce the decompression latency. Fig. 2 illustrates the dictionary compression algorithm. We use a dictionary table of 256 entries as a good compromise between compression ratio and table size. The table is duplicated in order to enable simultaneous decoding of two compressed instructions per cycle [3]. Due to the limited size of the table some instructions of the code are left uncompressed. We use two different encodings for 24-bit instructions, namely 8-bit and 16-bit codes. When we use 8-bit codes the whole codeword is used to index the decompression table. When we use 16-bit codes the 8 MSB bits are used to access the decompression table, while the 8 LSB bits are used to signal the type of instruction (i.e. whether it is compressed and what is its size). Our encoding allows us to use the 4 LSB bits of any instruction to determine its size.

The codes guarantee that differentiation between compressed and uncompressed instructions can be accomplished and decompression is done only when necessary. Fig. 2 shows the principle of dictionary encoding while the high-level system architecture is given in Fig. 1.

3.2 Decoding Tree Algorithm

The most significant part of the decompression algorithm is com-

[3] We found this to be more efficient than using a dual-port register file.

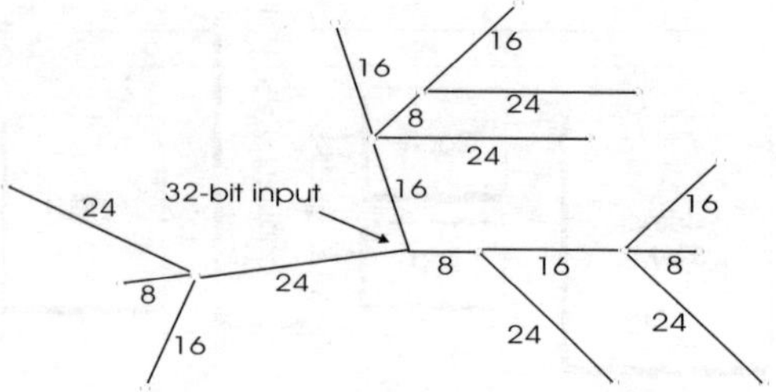

Figure 3: The structure of the 32 bit wide decompression tree

prised by the tree logic. The tree receives 32-bit instruction data from the cache (compressed) and produces decompressed instructions for the CPU. Since compressed data will expand to more than 32 bits, the tree also stores any extra decompressed or compressed bits for the next cycle.

The tree structure is shown in Fig. 3. It illustrates all the possible sequences of compressed instructions that can be subject to decompression by the decompression engine within any one cycle.

At each node we show all instruction sizes that can possibly occur. Since our technology compresses 24-bit instructions into 8-bit or 16-bit instructions, and since the commercial CPU we use has a mixture of 16-bit and 24-bit instructions, there are only three different sizes appearing in an actual compressed program. Hence, our tree starts with three branches corresponding to these sizes (8, 16 and 24). The root node is pointed to by the "32-bit input" arrow. It then continues with all cases until 32 bits have been received (in certain cases some instruction combinations will not end on a 32-bit word boundary). After decompression any instructions/instruction-parts that do not fit in a 32-bit word boundary are buffered and sent to the CPU in the next cycle. The tree will reach terminal nodes when 32 bits have been collected to ensure that the CPU will always receive 32 bits per cycle. The basic registers and elements that we will use in the following to describe the tree logic are:

- We denote cw as a compressed 32-bit word coming from the cache, and $cw(i)$ is the i-th bit of cw.
- The register that holds data decompressed in the previous cycle is $prev\_decomp$. This is needed when, for example, the decompression engine decompresses two 16-bit instructions into their original 24-bit sizes and thus exceeds the 32 bit limit the CPU can handle. The remaining 16 bits of the second decompressed instruction are stored in the $prev\_decomp$ register for use in the next cycle.
- The register that holds data compressed in the previous cycle is $prev\_comp$. This is used when it is not possible to decompress the whole instruction in the current cycle because part of it belongs to the next word. It is therefore necessary to store the part of the compressed instruction that is available in the current cycle in the $prev\_comp$ register.
- Bytes to be sent directly from the next compressed data stream are held in register $bytes\_next$. Consider the case where there are two 24-bit uncompressed instructions. Since the decompression engine can only receive and produce 32 bits, it will only receive 8 bits of the second instruction from the cache. The bits needed to decode this instruction are always held in the first four bits. Therefore it is possible to know the type of the instruction but it is impossible to send more than 8 bits. The $bytes\_next$ register is used to inform the decompression engine that it needs to forward the next two bytes directly to the CPU in the next cycle without any decoding.
- We denote $output$ as the 32-bits that are sent to the CPU (this is not a register; the output pins are written to directly).

We differentiate between types of instructions by denoting their lengths as follows: 24 (a 24-bit uncompressed instruction), $24 \rightarrow 8$ (a 24-bit compressed to 8 bits instruction), $24 \rightarrow 16$ (a 24-bit compressed to 16 bits instruction) and $16(n)$ (a narrow instruction).

Tree algorithm

```
If (cw(3) == 0 & cw(27) == 0)
    // 24 bits + 24 bits
    output = cw(31 - 0)
    bytes_next = 2
    prev_decomp = empty
    prev_comp = empty
Elsif (cw(3) == 0 & cw(27) == 1 & cw(26 - 24) == 110)
    // 24 bits + 8 bits compressed
    output = Table(cw(31 - 24)) + cw(23 - 0)
    bytes_next = 0
    prev_decomp = 2 bytes
    prev_comp = empty
Elsif (cw(3) == 0 & cw(27) == 1 & cw(26 - 24) == 111)
    // 24 bits + 16 bits compressed
    output = Table(cw(31 - 24)) + cw(23 - 0)
    bytes_next = 0
    prev_decomp = 1 byte (MSB bits of NOP)
    prev_comp = 1 byte (from 16 bits comp. data)
Elsif (cw(3) == 0 & cw(27) == 1 & cw(26 - 24) == xxx)
    // 24 bits + 16 bits narrow
    bytes_next = 1
    prev_decomp = empty
    prev_comp = empty
...
```

Figure 4: Tree algorithm

We now present a fragment of the tree's algorithm structure. Listing all possible cases would take too much space (there are 25 different cases). In figure 4 we show four cases that should be enough for the reader to re-construct the whole logic. Consider an example with the following pattern of instructions coming from the cache: a 24, $24 \rightarrow 8$, $24 \rightarrow 16$, 16(n) (see Fig. 5). The above instructions would arrive at the decompression engine in two cycles of 32 bit streams.

1) Cycle 1: In the first cycle the decompression engine receives a 24-bit uncompressed and a 8-bit compressed instruction. 24 bits of the first instruction form the LSB 24 bits of the outgoing bit stream (to the CPU). The 8-bit compressed data is used to obtain the 24-bit instruction from the decompression table. The first byte of the resulting 24 bits of the decompression table is used to form the MSB 8 bits of the 32 bits outgoing bit stream to the CPU core, and the remaining 2 bytes are stored in the $prev\_decomp$ register. The dce_pc (i.e. decompression engine program counter; more details follow later) is incremented by 4.
2) Cycle 2: In the second cycle the decompression engine receives a 16-bit compressed instruction and a 16-bit uncompressed instruction. The decompression engine assembles the $prev\_decomp$ register's 16 bits as the LSB 16 bits of the outgoing 32 bit-stream. The LSB 16 bits of the incoming compressed instruction are used to obtain a 24-bit instruction from the decompression table. The resulting LSB 16 bits of the decompressed instruction's

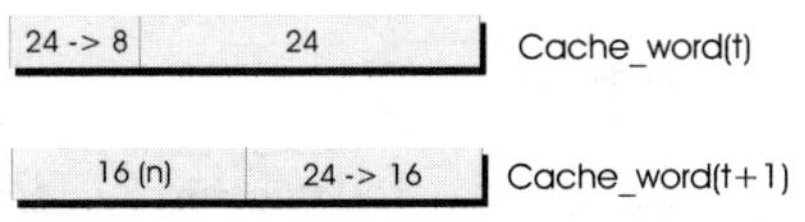

Figure 5: Example for decoding using the tree

24 bits are used to form the MSB 16 bits of the outgoing 32 bit-stream to the CPU core and the remaining MSB 1 byte is stored in the *prev_decomp* register. The MSB 16-bit regular narrow instruction from the cache memory is stored in the *prev_comp* register. The *dce_pc* is incremented by 4.

An important problem associated with code compression is how to locate branch/jump/call targets in compressed code since the original offsets will not point to "correct" locations after compression [4].

Our approach is to change branch offsets to point to the new compressed space addresses first proposed by Lefurgy et al. [5]. Branch offsets are patched after the compression phase. To simplify our design, we forced our targets to be aligned at word boundaries, as this simplifies the decoding tree. Patching the branch offsets involves a) compressing the original program and creating a map between native instruction addresses and compressed instruction addresses, and b) using the map to patch the offset-fields in the original branch instructions so they use offsets that correspond to the compressed address space.

An implication of using this technique is that the decompression engine must maintain its own program counter (*dce_pc*). When the application starts execution its value equals the CPU's program counter. As the execution progresses it diverges from the CPU's program counter. Using a separate program counter for the decompression engine also means that cache hits or misses must be triggered according to the *dce_pc*'s value and not the *cpu_pc*'s value. Our interface described in the next section takes care of this problem.

4. DESIGN ISSUES

Fig. 7 shows a block diagram of the decompression engine. The engine consists of the decoding logic and a 24-bit wide by 256 entries SRAM (decompression table). The Decompression engine's decoding logic is triggered by the arrival of incoming compressed or uncompressed 32-bit instruction words from the cache. The engine examines the first four bits (LSB bits), which form the select lines for the multiplexer in Fig. 7 and determine the type of the instruction.

The decompression table is accessed and the corresponding address contents return the decompressed 24-bit instruction. In some cases, there might be instances where the 32-bit compressed instruction stream may contain a part of an instruction and the remaining part might be in the next 32-bit compressed instruction stream. In that case the decoding logic provides two 16-bit registers (*prev_decomp* and *prev_comp*) to store the previous portion of the instruction. These registers are shown as the *pre_part_instr* register in Fig. 7. Furthermore, two bits (*pre_part_set* signal in Fig. 7) are used to indicate that there is a portion of the previous instruction stored. In addition, a 2-bit register (*bytes_next*) is used to indicate whether 8 or 16 bits of the *pre_part_instr* registers comprise the previous instruction. The dotted lines in Fig. 7 denote alternative paths for MUX selection or for copying data from the input stream to the output register directly.

[4]In the following we will use the term *branch* loosely to mean any instruction that alters the program counter such as a call, a jump, etc.

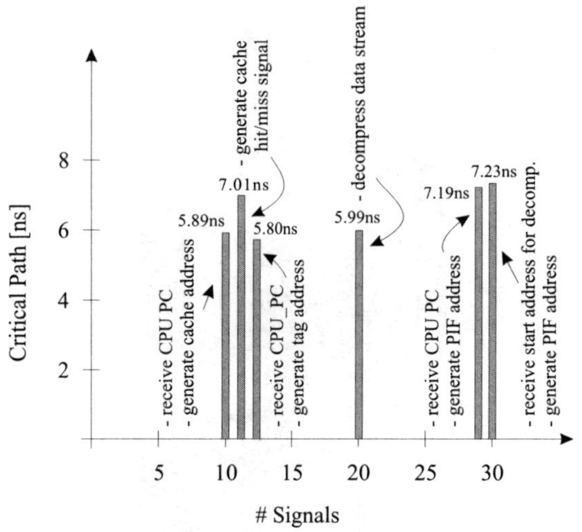

Figure 6: Critical path

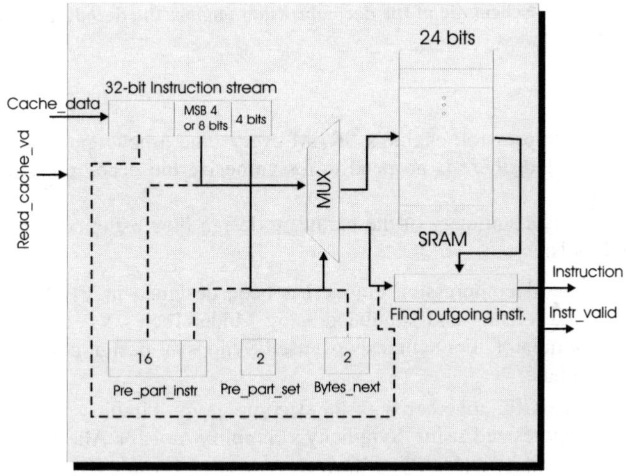

Figure 7: Decompression block diagram

Fig. 8 shows the schematic of the decompression engine. The top part consists of the SRAM-based decoding table. The bottom part shows the circuits for the decompression engine program counter (*dce_pc*) the PIF (Processor InterFace) address generator and the cache/tag control logic.

Fig. 9 illustrates our design flow for building the decompression engine and for generation of the compressed executable. On the left side our proprietary tools (gray area) are shown which take a compiled object file and generate a compressed object file. This procedure involves one pass which gathers statistics, i.e. determines which are the most frequent instructions for inclusion in the dictionary. The second pass performs compression and also stores the most frequent instructions in the dictionary. The third pass handles the branch/call/jump targets. Linking and generation of an compressed executable code follows. Finally, it is loaded into the system SRAM.

Note that although a new decompression table must be loaded in

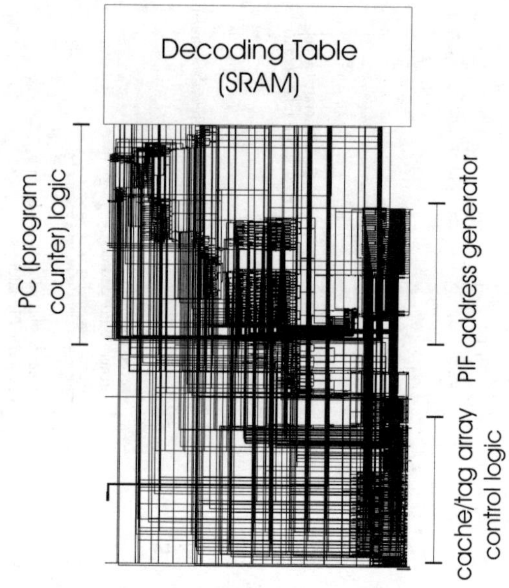

Figure 8: **Schematic of the decompression engine; the decoding table is SRAM based**

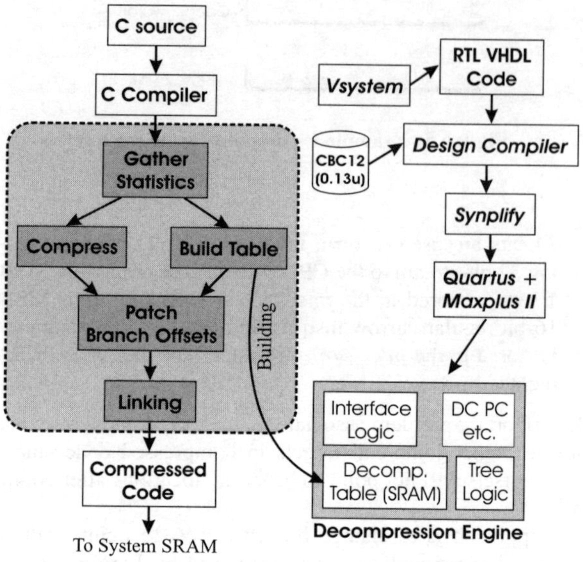

Figure 9: **Design Flow**

the decompression engine's SRAM every time a new application is compiled, there is no need to re-synthesize the decompression engine.

Here is a summary of the hardware design flow using commercial tools:

1) The decompression engine has been designed in VHDL and Verilog and simulated using Model-Tech's Vsystem simulator. For synthesis we used Synopsys' design compiler.
2) The CPU together with the decompression circuitry was synthesized using Synplicity's Synplify tool for Altera's APEX20KE family of devices.
3) The database format of Synplify was used by the Quartus and Maxplus II tools to place and route. The Quartus and Maxplus II tools generated the output format POF (Programmable logic Object File) and SOF (SRAM Object File) files of the design.
4) The generated POF files were then used to program Altera's APEX20KE device on the board.

5. RESULTS

The aim of the design of the code decompression unit was to guarantee that under no circumstances a delay of an additional clock cycle would occur since this would potentially prohibit its usage in a hard real-time system. Furthermore, the architecture should lead to a significantly increased system performance. Since we designed the decompression unit for a location between L2 cache and CPU both, instruction cache and main memory would profit from the compressed code. As a major effect, the instruction cache holds effectively more instructions and thus causing a higher cache hit ratio.

In order to increase the performance of the whole system we not only had to guarantee that the number of clock cycles is reduced compared to the non-compressed case, but we also had to avoid exceeding the cycle time budget. The results of this effort are shown in Fig. 6: it shows the six most critical paths in the design with the total maximum delay time for each path and the number of signals to be handled. It also gives the major functionality comprised by a certain path. An interesting observation is that the path *"decompress data stream"* (fourth bar) is not the most critical path as it accounts for only 5.99ns as opposed to 7.01ns for generating cache signals. We had a budget of 8.5ns for the whole clock cycle that would guarantee no additional delay. We managed to reduce this constraint by around 1.5ns (7.01ns to 8.5ns). The key in achieving this goal was to distribute the necessary computation equally among the most critical paths. The result can be seen in Fig. 6 where the six most critical paths are all within a time window of 1.12ns. Since the cycle time is guaranteed we can in the following concentrate to discuss the performance by solely comparing the number of clock cycles it takes to execute an application with and without code compression applied.

The applications have been chosen to demonstrate various effects and to show the advantages of our decompression architecture for diverse characteristics of applications like application area (video, animation, algorithmic), application domain (data dominated vs. control dominated) and size of the applications The applications are the following: the commonly available compress program (*"cpr"*) from SPEC95, a real–time Diesel engine control algorithm (*"diesel"*), an algorithm for computing 3D vectors for a motion picture (*"i3d"*), a real–time HDTV Chromakey algorithm (*"key"*), a complete MPEGII encoder (*"mpeg"*), a smoothing algorithm for digital images (*"smo"*) and a trick animation algorithm (*"trick"*).

Table 1 shows the performance results given as the number of clock cycles it takes to execute. Compared are the compressed cases to the un-compressed cases. Furthermore, each application's results are shown for various instruction cache sizes. The cache sizes have been chosen around the saturation point of the instruction cache i.e. the cache sizes below and slightly above the point where a further increase would not lead to a noticeable drop in cache misses. This point is typically the best choice for a designer as it represents a good compromise between costs (i.e. cache size) and obtained result (performance). As we can see from the table, the performance increases are significant as there is an average per-

Appl.	Inst. Cache	Cmp	Exec Time [cycles]	Performance improvement [%]
cpr	1k	no	187,777,618	n/a
	1k	yes	165,037,748	12.11
	2k	no	159,948,542	n/a
	2k	yes	155,885,849	02.54
	4k	no	154,460,263	n/a
	4k	yes	154,321,248	0.09
diesel	128	no	242,332	n/a
	128	yes	153,274	36.75
	512	no	233,441	n/a
	512	yes	85,136	63.53
	1k	no	84,724	n/a
	1k	yes	41,523	50.99
i3d	128	no	117,697	n/a
	128	yes	62,156	47.19
	512	no	58,475	n/a
	512	yes	42,418	27.46
	1k	no	44,158	n/a
	1k	yes	42,771	03.14
key	256	no	181,519,564	n/a
	256	yes	174,077,262	04.10
	512	no	178,888,128	n/a
	512	yes	174,755,812	02.31
	1k	no	174,318,687	n/a
	1k	yes	173,516,821	0.46
mpeg	2k	no	9,236,642	n/a
	2k	yes	5,024,733	45.60
	4k	no	5,142,418	n/a
	4k	yes	4,313,460	16.12
	8k	no	4,310,808	n/a
	8k	yes	4,271,149	0.92
smo	128	no	13,269,732	n/a
	128	yes	5,582,576	57.93
	512	no	3,658,805	n/a
	512	yes	3,657,890	0.25
	1k	no	3,716,882	n/a
	1k	yes	3,715,693	0.32
trick	256	no	5,416,167	n/a
	256	yes	2,347,908	56.65
	512	no	3,473,197	n/a
	512	yes	1,626,314	53.16
	1k	no	2,848,322	n/a
	1k	yes	1,551,766	45.52

Table 1: Performance results of all applications with instruction cache size as a parameter

formance boost of 25%.

A summary of the best results achieved for each application is given in Fig. 10 as a percentage improvement.

Note that the performance increase comes solely from the decompression hardware as any other system configurations in the comparisons are identical. In fact, in both cases (compressed and un-compressed) the hardware circuitry used is the same as the decompression hardware is simply bypassed in the non-compressed case.

6. CONCLUSIONS

In this paper we presented the design, architecture and prototyping hardware of a decompression unit that allows to run an embedded system using compressed code. It is the first running prototype that decompresses instructions in just one cycle and without any additional cycle time delay. The whole system is running as a hardware prototype comprising our decompression unit, a CPU, caches, memories and peripherals. Throughout a wide range of applications we observed a performance increase of up to 63% with 25% in average. Our technology is not limited to a certain CPU

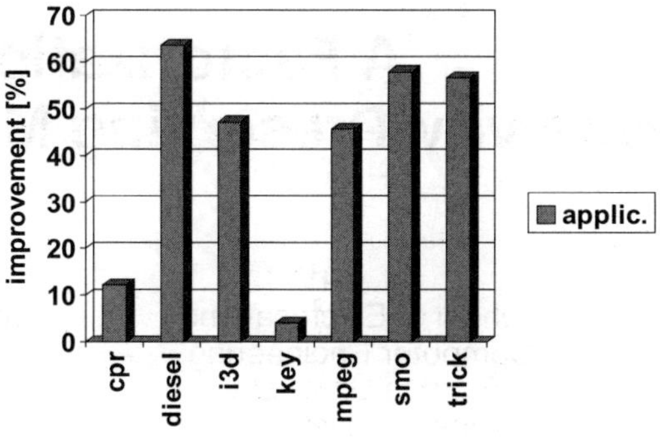

Figure 10: Best performance improvements for each application

and might achieve similar results when used with other CPUs. The limitation of our technology is that it can only be applied to systems that do not modify their instruction code during execution. This is the case for most embedded systems.

7. REFERENCES

[1] T.C. Bell, J.G. Cleary, and I.H. Witten. *Text Compression*. Prentice Hall, New Jersey, 1990.

[2] L. Benini, A. Macii, E. Macii, and M. Poncino. Selective Instruction Compression for Memory Energy Reduction in Embedded Systems. *IEEE/ACM Proc. of International Symposium on Low Power Electronics and Design (ISLPED'99)*, pages 206–211, 1999.

[3] IBM. CodePack PowerPC Code Compression Utility User's Manual. *Version 3.0*, 1998.

[4] N. Ishiura and M. Yamaguchi. Instruction Code Compression for Application Specific VLIW Processors Based on Automatic Field Partitioning. *Proceedings of the Workshop on Synthesis and System Integration of Mixed Technologies*, pages 105–109, 1998.

[5] C. Lefurgy and T. Mudge. Code Compression for DSP. *CSE-TR-380-98, University of Michigan*, November 1998.

[6] C. Lefurgy, E. Piccininni, and T. Mudge. Reducing Code Size with Run-time Decompression. *Proceedings of the International Symposium of High-Performance Computer Architecture*, January 2000.

[7] S.Y. Liao, S. Devadas, and K. Keutzer. Code Density Optimization for Embedded DSP Processors Using Data Compression Techniques. *Proceedings of the Chapel Hill Conference on Advanced Research in VLSI*, pages 393–399, 1995.

[8] T.Okuma, H.Tomiyama, A.Inoue, E.Fajar, and H.Yasuura. Instruction Encoding Techniques for Area Minimization of Instruction ROM. *International Symposium on System Synthesis*, pages 125–130, December 1998.

[9] A. Wolfe and A. Chanin. Executing Compressed Programs on an Embedded RISC Architecture. *Proceedings of the International Symposium on Microarchitecture*, pages 81–91, December 1992.

[10] Y. Yoshida, B.-Y. Song, H. Okuhata, and T. Onoye. An Object Code Compression Approach to Embedded Processors. *Proceedings of the International Symposium on Low Power Electronics and Design (ISLPED)*, ACM:265–268, August 1997.

A Factorization-Based Framework for Passivity-Preserving Model Reduction of RLC Systems*

Q. Su
School of Electrical and Computer Engineering
Purdue University
West Lafayette, IN
qsu@ecn.purdue.edu

V. Balakrishnan
School of Electrical and Computer Engineering
Purdue University
West Lafayette, IN
ragu@ecn.purdue.edu

C.-K. Koh
School of Electrical and Computer Engineering
Purdue University
West Lafayette, IN
chengkok@ecn.purdue.edu

ABSTRACT

We present a framework for passivity-preserving model reduction for RLC systems that includes, as a special case, the well-known PRIMA model reduction algorithm. This framework provides a new interpretation for PRIMA, and offers a qualitative explanation as to why PRIMA performs remarkably well in practice. In addition, the framework enables the derivation of new error bounds for PRIMA-like methods. We also show how the framework offers a systematic approach to computing reduced-order models that better approximate the original system than PRIMA, while still preserving passivity.

Categories and Subject Descriptors

G.1.3 [**NUMERICAL ANALYSIS**]: Numerical Linear Algebra—*linear systems*; F.2.1 [**ANALYSIS OF ALGORITHMS AND PROBLEM COMPLEXITY**]: Numerical Algorithms and Problems—*computations on matrices*

General Terms

Algorithms

Keywords

Model Reduction, Large Scale Systems, RLC interconnect, Passivity Preserving, Factorization

1. INTRODUCTION

As VLSI technology advances, integrated circuits are designed with ever-decreasing sizes, and ever-increasing speeds of operation, with a consequence that RLC interconnect effects have an increasing impact on many critical design criteria [5]. Therefore, accurate modeling of RLC interconnect has become very important. A typical interconnect model usually involves thousands or even millions of tightly coupled RLC components, whose direct simulation can stretch the limit of computing resources. A standard practice that addresses such issues is that of model reduction, i.e., finding an approximate model of the original system with far fewer variables. It is highly desirable that this approximate model inherit many of the properties of the original system such as stability and passivity.

Model reduction of linear time-invariant systems is a well-studied topic. Some well-known techniques are moment-matching and projection methods. While these methods are well understood in theory, the challenges in the context of model reduction of VLSI interconnect have arisen from the large-scale nature of these models. Some work towards addressing these challenges are the asymptotic waveform evaluation (AWE) technique due to Huang, Pillage, and Rohrer [14, 20], and Padé approximations through moment matching [6]. For state-space models, Krylov subspace computation techniques such as the Arnoldi and the Lanczos iterations have proved to be very attractive; see for example Padé approximation via Lanczos iterations in [7, 10], and via Arnoldi iterations in [21]. Krylov methods have also been used to efficiently compute a basis for the principal controllable and observable subspaces; see [9]. The reduced-order model then can be obtained by projecting the state-space on these subspaces [12, 13]. Another model reduction technique whose implementation for VLSI interconnect has been recently studied is the balance-and-truncate method (see for example, [18, 11]). The underlying numerical problem with this techniques is the solution of two large-scale Lyapunov equations. Efficient implementation of the balance-and-truncate method is the subject of [17, 2].

The drawback with some (but not all) of the methods described above is that the reduced-order models obtained through them do not inherit the important property of *passivity* of the original model. It has been recognized that when orthogonal projections are employed prior to model reduction, passivity is indeed inherited by the reduced-order model, and methods based on this observations have found favor with the VLSI community. Perhaps the best known method in this class is PRIMA [19], which is based on orthogonal projection of the state vector onto a lower dimensional subspace. It has been recognized that in practice, PRIMA offers excellent time and frequency domain approximation properties. However, there exist no error bounds for PRIMA in the spirit of the system-theoretic error bounds such as those for the balance-and-truncate method.

The main contributions of the this paper are a framework for passivity-preserving model reduction of RLC systems that includes

*This work was supported in part by NSF under contract number CCR-9984553, SRC under contract number 99-TJ-689, and Purdue Research Foundation.

PRIMA as a special case. The underlying idea is that in order to preserve passivity, the order reduction must be performed not on the original model, but on a "factor", which is analogous to a square root of the real part of the transfer function. This framework offers a number of advantages: It provides a new interpretation for PRIMA, and offers a qualitative explanation as to why PRIMA performs remarkably well in practice. In addition, the framework enables the derivation of new error bounds for PRIMA-like methods. Perhaps most important, the framework offers a systematic approach to computing reduced-order models that better approximate the original system than PRIMA, while still preserving passivity. This improvement comes at the expense of additional computational cost.

2. A FACTORIZATION-BASED MODEL REDUCTION FRAMEWORK FOR PASSIVE SYSTEMS

Given an RLC circuit comprising only passive linear elements, we can extract the time-domain equations as follows:

$$\begin{aligned} E\dot{x} &= Ax + Bu, \\ y &= Cx + Du, \end{aligned} \tag{1}$$

where $x(t) \in \mathbb{R}^N$, $u(t) \in \mathbb{R}$ and $y(t) \in \mathbb{R}$. Hence $E \in \mathbb{R}^{N\times N}$, $A \in \mathbb{R}^{N\times N}$, $B \in \mathbb{R}^{N\times 1}$, $C \in \mathbb{R}^{1\times N}$ and $D \in \mathbb{R}$. We will consider only single-input single-output systems in this paper; the extension of the results presented herein to multi-input multi-output systems is straightforward. We also assume in the sequel that E is invertible; this assumption is made purely for ease of exposition, in [23] it was shown that a general RLC system with noninvertible E in (1) can be transformed to an equivalent state space realization where this assumption holds. Since E is assumed to be invertible, without loss of generality, we assume $E = I$.

As (1) describe a stable linear time-invariant system, all the eigenvalues of A have negative real parts. Moreover, as system (1) is passive, we must have

$$H(j\omega) + H^*(j\omega) \geq 0 \text{ for all } \omega,$$

where $H(s) = C(sI - A)^{-1}B$ is the transfer function. It turns out that from system theory [1, 3], stability and passivity are equivalent to the existence of a symmetric, positive-definite matrix P such that

$$M \triangleq \begin{bmatrix} A^TP + PA & PB - C^T \\ B^TP - C & -(D + D^T) \end{bmatrix} \leq 0. \tag{2}$$

The matrix P can be interpreted as representing the energy stored in the system: It is readily verified that as a consequence of (2), we have

$$\begin{aligned} & \frac{d}{dt}(x(t)^T Px(t)) \\ = \ & x(t)^T(A^TP + PA)x(t) + 2x(t)^TPBu(t) \\ = \ & \begin{bmatrix} x(t) \\ u(t) \end{bmatrix}^T M \begin{bmatrix} x(t) \\ u(t) \end{bmatrix} + 2u(t)y(t) \\ \leq \ & 2u(t)y(t), \end{aligned}$$

so that for every $T \geq 0$, we have

$$x(T)^TPx(T) \leq x(0)^TPx(0) + 2\int_0^T u(t)y(t)dt. \tag{3}$$

For VLSI systems, $u(t)y(t)$ has the interpretation of the instantaneous power input to the system, so that (3) means that the energy stored in the system never exceeds the initial energy plus the energy input to the system, i.e., it is passive.

Another consequence of (2) is that we have a factorization of $H(s) + H(-s)$. Let

$$M = -\begin{bmatrix} L^T \\ D_v^T \end{bmatrix} \begin{bmatrix} L & D_v \end{bmatrix},$$

where $[L \ \ D_v]$ has r linearly independent rows ($r \leq N$). Then, it is readily verified that

$$H(s) + H(-s) = V(-s)^TV(s),$$

where $V(s) = L(sI - A)^{-1}B + D_v$. We will refer to $V(s)$ as a generalized factor of $H(s)$. Note that although $H(s)$ is a scalar transfer function, the generalized factor $V(s)$ is a column vector with r entries.

The generalized factor V suggests a model reduction scheme that preserves passivity and stability for the system (1).

Factorization-based model-reduction algorithm (FMR)

Step 1. Find any P satisfying (2).

Step 2. Form $V(s)$, and find a stable reduced-order approximant V_r.

Step 3. Form H_r so that it is stable and satisfies $H_r(s) + H_r(-s) = V_r(-s)^TV_r(s)$.

Note that $H_r(s)$ is guaranteed to be stable and passive. In Step 2 above, the reduced-order model can be obtained by any model reduction technique. However, the balanced-truncation method appears to be the natural one for a number of reasons:

1. Although $V(s)$ is a single-input multi-output transfer function, its model reduction via balanced-truncation is particularly simple. Let W_c be the controllability Gramian of the realization (A, B, L, D_v) of V, given by

   $$AW_c + W_cA^T + BB^T = 0.$$

   The observability Gramian of the realization (A, B, L, D_v) of V is simply the matrix P. Then, with $W_c = XX^T$ and $P = YY^T$ being square root decompositions, let the singular value decomposition (SVD) of X^TY satisfy

   $$X^TY = U\Sigma V^T \approx U_k\Sigma_kV_k^T,$$

   where the diagonal entries of Σ_k contain the significant Hankel singular values of $V(s)$. Then, it can be shown that a realization of H_r derived from a balanced truncation of V to V_r is

   $$(T_lAT_r, T_lB, CT_r, D)$$

   where

   $$T_l = \Sigma_k^{-\frac{1}{2}}V_k^TY_k^T, \quad T_r = X_kU_k\Sigma_k^{-\frac{1}{2}}.$$

   We make the important observation that the matrix L and D_v need not be explicitly computed in order to obtain H_r. All we need are square root decompositions of W_c and P.

2. When the balanced-truncation method is used, we have a bound on the approximation error [25, 4]:

   $$\begin{aligned} & \|H - H_r\|_{L_\infty} \\ \leq \ & \tfrac{1}{2}(4N+1)2\Sigma_{k+1}^N\sigma_i(4\Sigma_1^k\sigma_i + 2\Sigma_{k+1}^N\sigma_i), \end{aligned} \tag{4}$$

   where σ_i are the Hankel singular values of $V(s)$.

3. PRIMA IN THE CONTEXT OF FMR

In order to implement FMR, we first need to find one generalized factor. This corresponds to finding some P satisfying (2). While this is a convex feasibility problem [3], solving it for a large-scale scenario as that encountered in VLSI CAD models is a daunting task. However, in the case when the system matrices in (1), are obtained via a modified nodal analysis (MNA), such a P is readily available. In such cases, the state space equations are as follows [19]:

$$\begin{aligned} \hat{E}\dot{x} &= \hat{A}x+\hat{B}u, \\ y &= \hat{C}x, \end{aligned}$$

where the state-space matrices have the following structure.

$$\hat{E}=\begin{bmatrix} F & 0 \\ 0 & L \end{bmatrix} \geq 0, \quad \hat{B}=\hat{C}^T,$$
$$\hat{A}=\begin{bmatrix} G & N \\ -N^T & 0 \end{bmatrix}, \text{ with } \hat{A}+\hat{A}^T \leq 0, \tag{5}$$

As shown in [23], we can always find an equivalent state space realization (E,A,B,C,D), where E is invertible. Further, without loss of generality, we may take $E=I$, and

$$\begin{bmatrix} A^T+A & B-C^T \\ B^T-C & -(D+D^T) \end{bmatrix} \leq 0. \tag{6}$$

*The most important observation with the inequality (6) is that $P=I$ satisfies inequality (2), that is, from the form of the MNA equations, we have "**for free**" a generalized factor.*

Once again, with

$$\begin{bmatrix} A^T+A & B-C^T \\ B^T-C & -(D+D^T) \end{bmatrix} = -\begin{bmatrix} L^T \\ D_v^T \end{bmatrix}\begin{bmatrix} L & D_v \end{bmatrix},$$

we have

$$H(s)+H(-s)=V(-s)^TV(s),$$

where $V(s)$ has a state-space realization (A,B,L,D_v). Note that the observability Gramian of this realization is simply the identity (this follows from $A+A^T+L^TL=0$), so that the balanced truncation method corresponds to simply projecting the state vector on the principal eigenspace of the controllability Gramian, an orthogonal projection. As the principal eigenspace of the controllability Gramian coincides with the principal subspace of the matrix

$$C=\begin{bmatrix} A^{-1}B & A^{-2}B & \cdots \end{bmatrix}, \tag{7}$$

this method is, in principle, the same as PRIMA. In other words:

We can interpret PRIMA as an instance of our FMR algorithm, when the generalized factor corresponding to $P=I$ is used.

Indeed, we can demonstrate further connections. Suppose the Smith method with shift $p<0$ is used to calculate the Gramian W_c approximately [22, 2]:

$$W_c=\Sigma_{i=0}^{k}A_p^iB_pB_p^T(A_p^i)^T,$$

where $A_p=(pI+A)^{-1}(pI-A)$, $B_p=\sqrt{-2p}(pI+A)^{-1}B$. The first k eigenvectors of W_c in V span the same column space as Krylov(A_p,B_p,k). Thus, the PRIMA method applied around the expansion point p can be interpreted as a special case of the FMR algorithm, with the generalized factor corresponding to $P=I$, and with the modified Smith method used to approximately compute W_c.

While these comments are true in principle, computational issues will lead to minor numerical differences. One issue is how an approximation to the principal eigenspace of the controllability Gramian of the realization (A,B,L,D_v) is computed: Practical implementations of PRIMA obtain this approximation by computing a basis for the first few columns of the matrix C in (7), while the FMR algorithm with $P=I$ corresponds to computing the principal eigenspace by directly computing an approximation to the Gramian.

In addition to a new interpretation for PRIMA, the above development offers an error bound, in principle, for PRIMA:

$$\|H-H_r\|_{L_\infty} \leq \frac{1}{2}(4N+1)2\Sigma_{k+1}^{N}\sigma_i(4\Sigma_1^k\sigma_i+2\Sigma_{k+1}^{N}\sigma_i), \tag{8}$$

where σ_i are the square roots of the eigenvalues of the controllability Gramian W_c.

4. AN IMPROVED IMPLEMENTATION OF FMR

Motivated by the observation that the error bound in (4) depends crucially on the matrix P that satisfies (2) (specifically, they depend on the square roots of the eigenvalues of W_cP, where W_c is the controllability Gramian of (A,B)), with "smaller" values of P being better, we present a systematic procedure for obtaining $P<I$, thereby improving on the performance of PRIMA. We assume that the system is modeled by equations (5). Without loss of generality, we have the same assumption as in last section: $E=I$ and the inequality (6) holds.

We first explore the set of P that satisfy (2) for such systems. Applying the Schur complement lemma [3], (2) is equivalent to

$$A^TP+PA+(PB-C^T)(D+D^T)^{-1}(B^TP-C)\leq 0, \tag{9}$$

if D is nonzero. In the case where $D=0$, (2) is equivalent to

$$A^TP+PA\leq 0, \quad PB=C^T, \tag{10}$$

which can be transformed to an equivalent Riccati inequality of the form

$$\tilde{A}Q+Q\tilde{A}^T+QRQ+S\leq 0 \tag{11}$$

with some lower order $(N-1)$ matrices $\tilde{A}$, R and S. For simplicity of the presentation, we assume $D\neq 0$. The procedure for obtaining a smaller solution P for (10), or equivalently a smaller solution Q for (11), for the case $D=0$ is the same as that for the case $D\neq 0$.

Note that $P=I$ is one feasible solution to (9); this is the "natural" solution obtained from the form of the MNA equations, and the one corresponding to PRIMA, when it is interpreted in the context of our FMR algorithm.

We next turn to the problem of finding the "smallest" P satisfying (9). It turns out that there is indeed such a minimal element [24, 15]: There exists $P_{\min}>0$ such that any P satisfying (9) also satisfies $P\geq P_{\min}$. This matrix $P_{\min}$ satisfies (9) with equality, i.e.,

$$\begin{aligned} &A^TP_{\min}+P_{\min}A \\ &+(P_{\min}B-C^T)(D+D^T)^{-1}(B^TP_{\min}-C)=0, \end{aligned} \tag{12}$$

It also turns out that the quantity $A^TP_{\min}+P_{\min}A$ is rank one, so that the corresponding generalized factor $V_{\min}(s)$ is *scalar*. This is referred to as the *spectral factor* in the system theory literature [8]. Thus, computing $P_{\min}$ exactly requires the solution of the Riccati equation (12). The corresponding computation is prohibitive even for moderate-sized problems. Indeed, for moderate-sized problems, the preferred solution method for solving equation (9) is iterative, with each iteration requiring the exact solution of a Lyapunov equation. Therefore, computing $P_{\min}$ directly is impractical for VLSI systems.

In summary, the case $P = I$ corresponds to PRIMA, and the optimal $P_{\min}$, while it exists, is prohibitively expensive to compute. We therefore propose a scheme that finds a "sub-optimal" P, one that satisfies (9), and also has the property that $P_{\min} \le P \le I$, thereby guaranteeing a better error bound than with PRIMA.

With $\mathcal{P}$ denoting the set of $P \ge 0$, we define a function $f : \mathcal{P} \to R$ by

$$\begin{aligned} f(P) &= \lambda_{\max}(A^T P + PA \\ &\quad + (PB - C^T)(D + D^T)^{-1}(B^T P - C)), \end{aligned}$$

where $\lambda_{\max}(\cdot)$ denotes the maximum eigenvalue of a symmetric matrix. Thus, $P > 0$ is a solution of (9) if and only if $f(P) \le 0$. It is easily established that f is a convex function, that is, for all $P_1, P_2 \in \mathcal{P}$ and for all $\mu \in [0,1]$,

$$f(\mu P_1 + (1-\mu)P_2) \le \mu f(P_1) + (1-\mu) f(P_2). \quad (13)$$

Note that from the definition of f, we have that $f(I) \le 0$, and that $f(P_{\min}) = 0$. Suppose that we can find some $0 \le P_1 < I$ with $f(P_1) > 0$. As shown in Figure reffig:algfig, by convexity of the function f, and the fact that $f(I) \le 0$, we can find some $P = \mu P_1 + (1-\mu)I$ between P_1 and I, such that $f(P) \le \mu f(P_1) + (1-\mu) f(I) = 0$. Such P is guaranteed to be a solution to (9), because $f(P) \le 0$; and yield a better error bound than PRIMA, because P is between P_1 and I, hence $P \le I$.

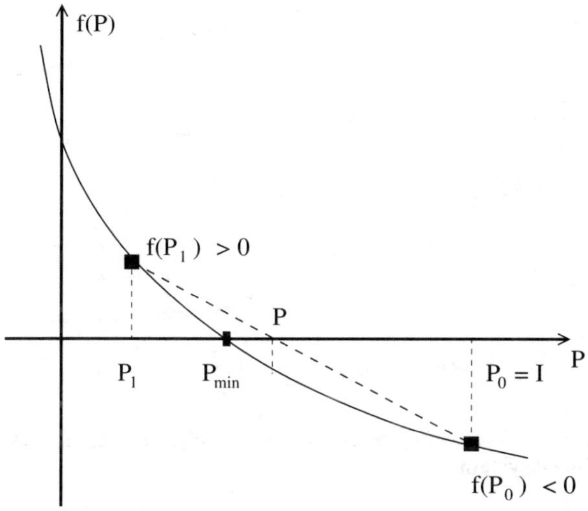

Figure 1: A systematic procedure for obtaining P with $0 < P \le I$ and $f(P) \le 0$.

It now remains to find P_1 such that $0 \le P_1 < I$ with $f(P_1) > 0$. One possible choice is $P_1 = 0$; this follows from (9). There are some other choices for P_1, for example, take P_1 as the solution of the Lyapunov equation

$$\begin{aligned} &(A - B(D+D^T)^{-1}C)^T P_1 + P_1(A - B(D+D^T)^{-1}C) \\ &+ C^T(D+D^T)^{-1}C = 0. \end{aligned}$$

It is easily shown that $0 < P_1 < I$ and $f(P_1) > 0$, as desired.

We can run several iterations of the above procedure to obtain better (i.e., "smaller") and better P. Indeed, this iterative procedure can be adapted to compute $P_{\min}$. However, each iteration requires the solution of one Lyapunov equation. As we will demonstrate through numerical examples, it suffices to run one iteration to obtain significant improvement over PRIMA.

Computational cost over PRIMA

We now briefly discuss the amount of additional computation required by the improved FMR scheme over PRIMA. As PRIMA in principle corresponds to using $P = I$, the additional computational costs comes mainly from the evaluation of $f(P)$ which involves some matrix multiplications during the search of a smaller P. This is approximately $4N^3$.

5. NUMERICAL RESULTS

We present two numerical examples that are representative of the performance of the factorization-based passivity-preserving model reduction framework presented in this paper.

Example 1

We consider a randomly generated passive system with state space realization (A, B, C), where $A + A^T < 0$, and $B = C^T$. There are 50 state components. We apply the passivity preserving model-reduction algorithm (FMR) to this system to obtain reduced systems of order 13 using three different symmetric, positive-definite matrices: (i) $P = I$, (ii) $0 < P_{\min} < P < I$, and (iii) $P = P_{\min}$. We compute P in (ii) using one step of the technique outlined in Section 4. $P_{\min}$ is obtained by solving the Riccati equation (12). With these three matrices, we form three different generalized factors for the original system, and apply the approximate balanced truncation method on them to obtain three different reduced systems. For purposes of comparison, we also apply the PRIMA model-reduction algorithm from [19] on the original system. Figures 2, 3 and 4 show the relative errors and the frequency responses of the four reduced systems. From these plots, it is evident that the reduced order system from FMR with $P = P_{\min}$ gives us the most accurate approximation. The reduced order system from FMR with $0 < P_{\min} < P < I$ approximates the original system better the one with $P = I$. All the reduced systems from FMR have lower peak errors than the reduced system from PRIMA.

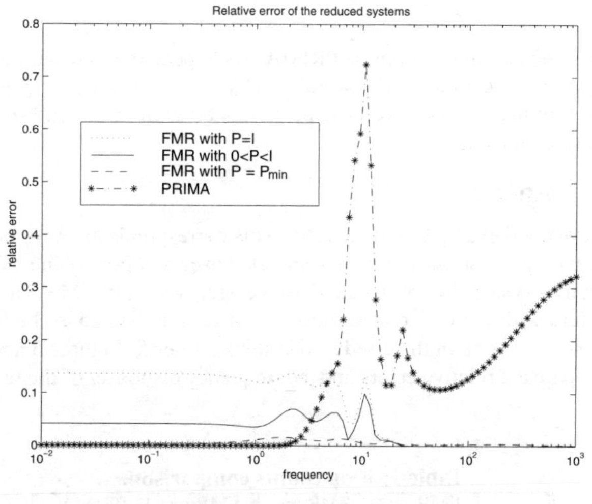

Figure 2: Relative error of the reduced systems.

The flop counts of these four model reduction methods are summarized in Table 1. We can see that FMR with $P = I$ uses roughly

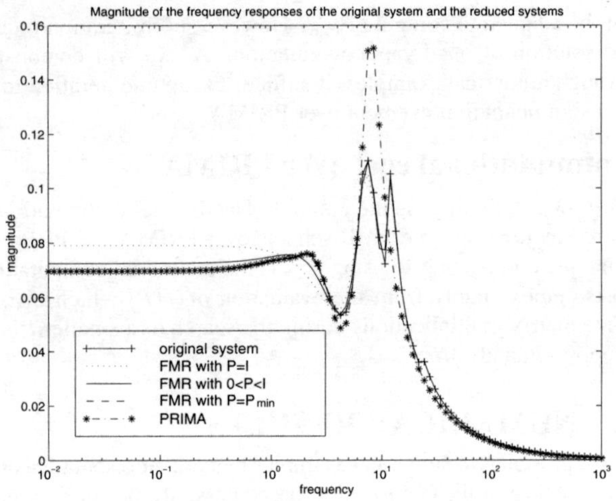

Figure 3: Magnitude of the frequency responses of the original system and the reduced systems.

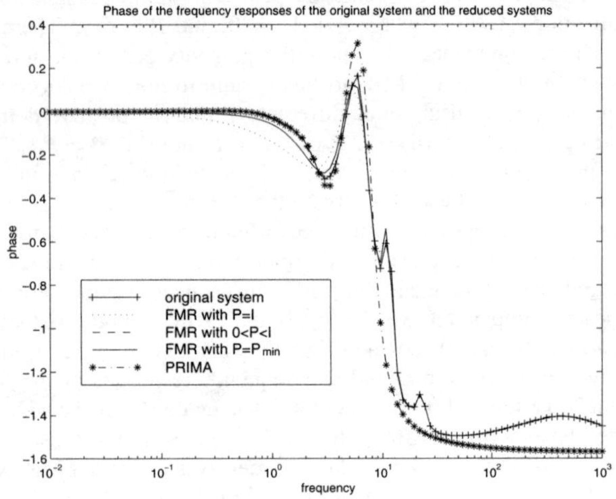

Figure 4: Phase of the frequency responses of the original system and the reduced systems.

the same amount of flops as PRIMA. As expected, FMR with $P = P_{\min}$ costs the most. FMR with $0 < P_{\min} < P < I$ provides a good compromise; it improves the performance with some additional computational cost.

Example 2

The second example is from [16]. This corresponds to an on-chip planar square spiral inductor suspended over a copper plane. The original system has 500 states. As before, we apply FMR on the system with the different solutions P to (2) (as defined in the first example) to obtain three reduced systems, order 5. Figures 5 and 6 compare the relative errors and the frequency responses of the three

Table 1: Flop counts comparison.

	FMR on $P = P_{\min}$	FMR on $0 < P < I$	FMR on $P = I$	PRIMA
flops	41×10^6	2.4×10^6	0.7×10^6	0.5×10^6
ratio	79	4.7	1.3	1

reduced systems with those of the reduced system generated by PRIMA. The performance of the three FMR reduced systems follows the same trends exhibited in the preceding example. Moreover, the FMR reduced systems have lower peak errors than the PRIMA reduced system. The flop counts of these four model reduction methods are summarized in Table 2.

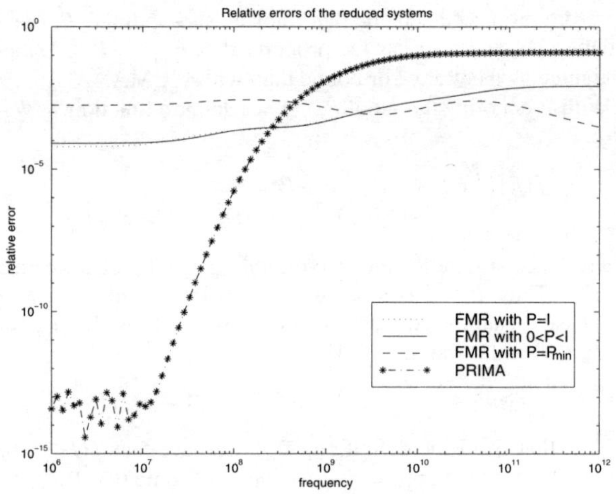

Figure 5: Relative error of the reduced systems.

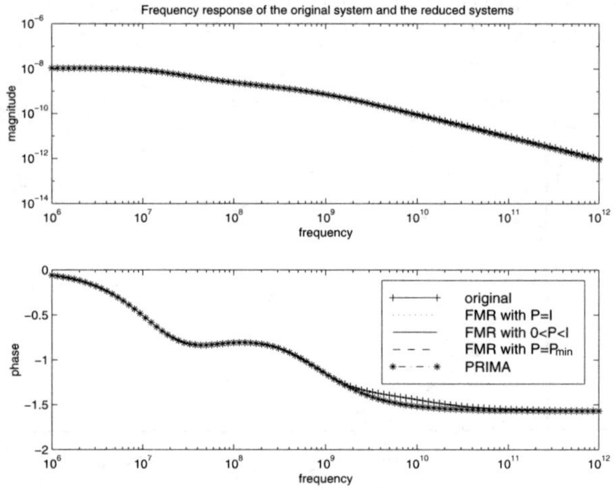

Figure 6: Frequency responses of the original system and the reduced systems.

6. CONCLUSIONS

We have presented a factorization-based framework for passivity-preserving model reduction that includes PRIMA as a special case. We have shown that this framework provides a new interpretation for PRIMA, as well as enabling the derivation of an error bound for PRIMA in principle. In addition, we have a systematic method for improving the performance of PRIMA-like methods.

7. REFERENCES

[1] B. Anderson and S. Vongpanitlerd. *Network analysis and synthesis: a modern systems theory approach.* Prentice Hall, 1973.

Table 2: Flop counts comparison for the RLC circuit.

	FMR on $P = P_{\min}$	FMR on $0 < P < I$	FMR on $P = I$	PRIMA
flops	334×10^8	24×10^8	4.7×10^8	3.4×10^8
ratio	98	7.1	1.4	1

[2] V. Balakrishnan, Q. Su, and C.-K. Koh. Efficient balance-and-truncate model reduction for large scale systems. In *Proc. American Control Conf.*, pages 4746–4751, Arlington, VA, June 2001.

[3] S. Boyd, L. El Ghaoui, E. Feron, and V. Balakrishnan. *Linear Matrix Inequalities in System and Control Theory*, volume 15 of *Studies in Applied Mathematics*. SIAM, Philadelphia, PA, June 1994.

[4] X. Chen and T. Wen. Positive realness preserving model reduction with h_{infty} norm error bounds. *IEEE Trans. on Circuits and Systems - I: Fundamental Theory and Applications*, 42(1):23–29, 1995.

[5] C.-K. Cheng, J. Lillis, S. Lin, and N. Chang. *Interconnect Analysis and Synthesis*. John Wiley and Sons, Inc., New YorK, 2000.

[6] E. Chiprout and M. Nakhla. Generalized moment-matching methods for transient analysis of interconnect networks. In *Proc. 29th ACM/IEEE Design Automation Conference*, pages 201–206, Anaheim, CA, June 1992.

[7] P. Feldmann and R. W. Freund. Efficient linear circuit analysis by Padé approximation via the Lanczos process. *IEEE Trans. Computer-Aided Design*, 14:639–649, May 1995.

[8] B. A. Francis. *A course in* $\mathbf{H}_\infty$ *Control Theory*, volume 88 of *Lecture Notes in Control and Information Sciences*. Springer-Verlag, 1987.

[9] R. W. Freund. Reduced-order modeling techniques based on Krylov subspaces and their use in circuit simulation. Technical Report 98-3-02, Numerical Analysis Manuscript, Bell Laboratories, Murray Hill, New Jersey, Feb. 1998.

[10] K. Gallivan, E. Grimme, and P. V. Dooren. Padé approximation of large-scale dynamic systems with Lanczos methods. In *Proc. IEEE Conf. on Decision and Control*, pages 443–448, Dec. 1994.

[11] M. Green and D. J. N. Limebeer. *Linear Robust Control*. Information and System sciences. Prentice Hall, Englewood Cliffs, NJ, 1995.

[12] E. Grimme, K. Gallivan, V. de Clippel, and P. V. Dooren. Approximate solves in Krylov-based modeling methods. In *Proc. IEEE Conf. on Decision and Control*, pages 3849–3854, Dec. 1997.

[13] P. K. Gunupudi and M. S. Nakhla. Model-reduction of nonlinear circuits using Krylov-space techniques. In *Proc. 29th ACM/IEEE Design Automation Conference*, pages 13–16, 1999.

[14] X. Huang. *Padé Approximation of Linear(ized) Circuit Responses*. PhD thesis, Carnegie Mellon University, Pittsburgh, PA, 1990.

[15] P. Lancaster and L. Rodman. *Algebraic Riccati Equations*. Oxford University Press, 1995.

[16] J. Li, F. Wang, and J. White. An efficient Lyapunov equation-based approach for generating reduced-order models of interconnect. In *Proc. Design Automation Conference*, pages 1–6, 1999.

[17] J. Li and J. White. Efficient model reduction of interconnnect via approximate system Gramians. In *Proc. Int. Conf. on Computer-Aided Design*, pages 380–383, 1999.

[18] B. C. Moore. Principal component analysis in linear systems: Controllability, observability, and model reduction. *IEEE Trans. Aut. Control*, AC-26(1):17–32, Feb. 1981.

[19] A. Odabasioglu, M. Celik, and L. T. Pileggi. PRIMA: Passive reduced-order interconnect macromodeling algorithm. *IEEE Trans. Computer-Aided Design of Integrated Circuits and Systems*, 17(8):645–654, 1998.

[20] L. T. Pillage and R. A. Rohrer. Asymptotic waveform evaluation for timing analysis. *IEEE Trans. Computer-Aided Design*, 9:352–366, 1990.

[21] L. M. Silveira, M. Kamon, and J. White. Efficient reduced-order modeling of frequency-dependent coupling inductances associated with 3D interconnect structures. In *Proc. 32nd ACM/IEEE Design Automation Conference*, pages 376–380, June 1995.

[22] R. A. Smith. Matrix equation $XA + BX = C$. *SIAM Journal on Applied Mathematics*, 16(1):198–201, 1968.

[23] Q. Su, V. Balakrishnan, and C.-K. Koh. Efficient approximate balanced truncation of general large-scale RLC systems via Krylov methods. In *Proc. ASPDAC/Int. Conf. VLSI Design*, Bangalore, India, Jan. 2002.

[24] J. C. Willems. Least squares stationary optimal control and the algebraic Riccati equation. *IEEE Trans. Aut. Control*, AC-16(6):621–634, Dec. 1971.

[25] K. Zhou. Remarks on stochastic model reduction. *IEEE Trans. Aut. Control*, 32(4):356–358, 1987.

Model Order Reduction for Strictly Passive and Causal Distributed Systems

Luca Daniel
University of California, Berkeley
dluca@eecs.berkeley.edu

Joel Phillips
Cadence Berkeley Labs
jrp@cadence.com

ABSTRACT

This paper presents a class of algorithms suitable for model reduction of distributed systems. Distributed systems are not suitable for treatment by standard model-reduction algorithms such as PRIMA, PVL, and the Arnoldi schemes because they generate matrices that are dependent on frequency (or other parameters) and cannot be put in a lumped or state-space form. Our algorithms build on well-known projection-based reduction techniques, and so require only matrix-vector product operations and are thus suitable for operation in conjunction with electromagnetic analysis codes that use iterative solution methods and fast-multipole acceleration techniques. Under the condition that the starting systems satisfy system-theoretic properties required of physical systems, the reduced systems can be guaranteed to be passive. For distributed systems, we argue that *causality* of the underlying representation is as important a consideration as *passivity* has become.

Categories & Subject Descriptors: B.7.2 Simulation, B.8.2 Performance Analysis & Design Aids, G.1.1 Interpolation G.1.2, Approximations, I.6 Simulation & Modeling.
General Terms: Algorithms, Performance, Design,
Keywords: Passive reduced order modeling, Distributed systems.

1. INTRODUCTION

Design of modern high-performance electronic systems such as RF circuits, optical transceiver ICs, and global digital signal interconnect requires careful attention to physical modeling so that the intrinsic physical limitations of implementation processes can be accounted for, and undesirable effects such as substrate coupling, signal integrity, and electromagnetic interference (EMI) can be minimized. A high degree of physical fidelity is necessary and can only be achieved by detailed analysis, such as with electromagnetic field solvers. At the same time, effective design of complicated systems requires simple models. Hence, model reduction is now a standard procedure for obtaining simple models of complicated physical systems. Much research has been performed in the model reduction field over the past decade [1, 2, 3], intended to address three primary issues: (a) Model accuracy. (b) Numerically stable and computationally practical generation of models of arbitrary order. (c) Generation of models that are "well-behaved" when embedded into a simulation tool with models of other physical elements.

A reasonable procedure to satisfy the last issue is to require that the models themselves do not possess non-physical properties. For example, components such as interconnect do not generate energy: they are *passive*.

"Lumped" RLC circuits, can be typically represented by matrices that are independent of frequency. For such lumped systems, positive-realness preserving procedures such as those based on congruence transforms [2, 3] are sufficient to guarantee that the reduced models of passive full systems are passive as well. *However, when accounting for high frequency effects, "distributed" systems represented by frequency dependent matrices are typically encountered.* For example, frequency dependent matrices are generated by integral-equation based field solvers that employ full-wave kernels, special Green functions for lossy dielectrics/substrates, or frequency dependent basis functions [4].

There are several approaches to distributed model reduction that essentially convert the model reduction problem to an interpolation or data-fitting problem [5, 6] where it is irrelevant whether the original systems is distributed or lumped. In our experience all of the data-fitting like approaches are limited in some aspect, and to the best of our knowledge, there is no approach that can simultaneously guarantee all the above three requirements. Krylov-based model reduction schemes for lumped systems [1, 3], on the other hand, routinely satisfy all these conditions, so we desire to extend their capabilities to distributed systems.

As input, our algorithm takes a time-invariant state-space-like frequency-domain model whose matrix descriptors may be a function of frequency. As output, it produces a time-invariant state-space model with frequency independent matrix descriptors and whose transfer function is a rational approximant of the original (infinite-order, possibly irrational) transfer function. The algorithm requires only matrix-implicit operations such as matrix-vector products, hence it is suitable for incorporation into modern fast integral equation solvers.

2. PASSIVITY AND CAUSALITY

In this section we will be concerned with properties of an abstract system $\mathcal{H} : \mathcal{X} \to \mathcal{X}$, transforming vector input signals $\mathbf{u}$ into vector output signals $\mathbf{y} = \mathcal{H}\mathbf{u}$ within a space of signals $\mathcal{X}$. For simplicity of exposition we will assume that the system inputs, $\mathbf{u} : \mathbb{R}^+ \to \mathbb{R}^p$ represent port voltages, and that the outputs $\mathbf{y} : \mathbb{R}^+ \to \mathbb{R}^p$ represent port currents, or the converse (the inputs are currents and the outputs voltages). The Laplace-domain representation of the system $\mathcal{H}$ is then a matrix $\mathbf{H}(s)$, s.t. $\mathbf{y}(s) = \mathbf{H}(s)\mathbf{u}(s)$, where $\mathbf{u}(s)$ and $\mathbf{y}(s)$ are the Laplace-domain representations of inputs $\mathbf{u}(t)$ and outputs $\mathbf{y}(t)$. Hence, $\mathbf{H}(s)$ is an immittance function: either an admittance matrix $\mathbf{Y}(s)$, or an impedance matrix $\mathbf{Z}(s)$. Let us introduce two inner products in $\mathcal{X}$, the standard inner product $\langle \mathbf{u}, \mathbf{y} \rangle = \int_{-\infty}^{\infty} \mathbf{y}(t)^T \mathbf{u}(t)dt$, and a product which acts on truncated signals $\langle \mathbf{u}, \mathbf{y} \rangle_\tau = \langle \mathbf{u}_\tau, \mathbf{y} \rangle = \langle \mathbf{u}, \mathbf{y}_\tau \rangle = \int_{-\infty}^{\tau} \mathbf{y}(t)^T \mathbf{u}(t)dt$, where $\mathbf{u}_\tau(t) \equiv \{\mathbf{u}(t)$ if $t \leq \tau, 0$ if $t > \tau\}$.

If $\mathbf{u}$ and $\mathbf{y}$ are port current/voltage pairs, $\langle \mathbf{u}, \mathbf{y} \rangle_\tau$ is the total energy dissipated by the system up to time τ. We will generally work in the space of signals $x \in \mathcal{X} = \mathcal{L}_2$ that have finite norm $||x||$ for any τ, where $||x||^2 = \langle x, x \rangle_\tau$.

Passive Systems. A *passive* system is a system that cannot produce energy. For the systems of interest here we may define:

DEFINITION 1 (PASSIVITY). *A system $\mathcal{H} : \mathcal{X} \to \mathcal{X}$ is passive if $\langle \mathbf{u}, \mathcal{H}\mathbf{u} \rangle_\tau \geq 0$, $\forall\, \tau \in \mathbb{R}^+$, $\forall\, \mathbf{u} \in \mathcal{X}$, $\mathbf{u} : [0, \tau] \to \mathbb{R}^p$.*

In practice, almost all systems of interest for model reduction are non-ideal and contain some loss. That is, they internally consume energy. If a system consumes energy, it is said to be *strictly passive*.

DEFINITION 2 (STRICT PASSIVITY). *A system $\mathcal{H}$ is strictly passive if there is a $\delta \in \mathbb{R}^+$ s.t. $\langle \mathbf{u}, \mathcal{H}\mathbf{u} \rangle_\tau \geq \delta ||u_\tau||^2$, $\forall\, \tau \in \mathbb{R}^+$, $\forall\, \mathbf{u} : [0, \tau] \to \mathbb{R}^p$.*

Positive Real Systems. A related concept in network theory is that of positive realness.

DEFINITION 3 (POSITIVE REALNESS). *A matrix valued function $\mathbf{H}(s)$ is positive-real [7] if*

$$\overline{\mathbf{H}(\bar{s})} = \mathbf{H}(s), \tag{1}$$

$$\mathbf{H}(s) \text{ is analytic in } Re(s) > 0, \tag{2}$$

$$\Pi_{\mathbf{H}}(s) \equiv \mathbf{H}(s) + \mathbf{H}(s)^* \geq 0 \text{ in } Re(s) > 0. \tag{3}$$

DEFINITION 4 (STRICT POSITIVE REALNESS). *A matrix valued function $\mathbf{H}(s)$ is strictly-positive-real [8] if there exists an $\varepsilon \in \mathbb{R}^+$ s.t. $\mathbf{H}(s - \varepsilon)$ is positive-real.*

Positive realness is of interest because of its relation to passivity for lumped networks:

THEOREM 1. *A system $\mathcal{H}$ with rational system transfer function $\mathbf{H}(s)$ is passive and stable if and only if $\mathbf{H}(s)$ is positive-real [7].*

In the context of model reduction, the implication for state-space systems is that if a reduction algorithm for lumped RLC networks produces models with positive-real transfer functions $\mathbf{H}(s)$, then it generates guaranteed passive models. Often the positivity condition (3) can be restricted to the imaginary axis, because of the following result [7]:

THEOREM 2. *A rational $\mathbf{H}(s)$ is positive-real if and only if (1) and (2) hold, and*

$$\Pi_{\mathbf{H}}(i\omega) \geq 0, \ \forall\, \omega \in \mathbb{R} \tag{4}$$

except for simple poles $i\omega_0$ of $\mathbf{H}(s)$, where the residue matrix must be nonnegative definite [7]. $\mathbf{H}(s)$ is strictly positive real if the inequality is strict [7, 8].

THEOREM 3. *If a rational matrix valued function $\mathbf{H}(s)$ is a representation of a passive system, then $\mathbf{H}^{-1}(s)$ is positive-real.*

Causal Systems. A *causal* system is a system whose output depends only on past inputs, not future inputs.

DEFINITION 5 (CAUSALITY). *A system $\mathcal{H}$ is causal if $\mathcal{H}\mathbf{u} = \mathcal{H}\mathbf{u}_\tau$, $\forall\, \tau \in \mathbb{R}^+$, $\forall\, \mathbf{u} : [0, \tau] \to \mathbb{R}^p$.*

All physical systems are causal. Hence, causality is a necessary property of all models intended to be used in any simulator that has a concept of time. However, it is often neglected when modeling distributed systems. When constructing model reduction algorithms for distributed systems, we must keep in mind that the condition in Equation (4) is not sufficient by itself to insure passivity. In particular, it is known that passive systems are required to be causal [9, 10]. If a system is not causal, it cannot be passive.

EXAMPLE 1. *Consider the one port ($p = 1$) network function $\mathbf{Z}_{skin}(i\omega) = \mathbf{R}_0 + \mathbf{R}_{ac}\sqrt{|\omega|}$ that is commonly used as a model for the "resistance" of interconnect in the skin-effect regime. This function satisfies the condition $\Pi_Z(i\omega) > 0$, $\forall\, \omega \in \mathbb{R}$. However, it is* not *a representation of any passive system, because it is not a causal function. In fact, it can be shown that any physical, passive network function that is purely real must be constant with respect to the frequency ω. In Section 6.3 we further illustrate the noncausality of this model.*

Algorithms that generate successively better rational approximations of non-causal systems (for example, $Z_{\text{skin}}(i\omega)$) must in some limit fail to be passive, for example by generating unstable approximants. Therefore we will require that all the systems we manipulate be causal. The algorithms we shall present shortly will produce strictly positive-real (and thus strictly passive) approximations to causal and strictly positive-real (and thus strictly passive) system functions $\mathbf{H}(s)$.

3. TECHNICAL BACKGROUND

Distributed Systems in Descriptor Form. Let us assume the original distributed system (e.g. an interconnect network) has been described, for instance by the discretization step of an integral equation method, in terms of a frequency dependent matrix $\mathbf{Z}(s)$. $\mathbf{Z}(s)$ describes the couplings between all the discretization basis functions and may be very large in the applications of interest. Many integral equation methods, when applied to distributed systems, produce $\mathbf{Z}(s)$ as a linear combination of matrices. One example is $\mathbf{Z}(s) = \mathbf{R}(s) + s\mathbf{L}(s)$, where $\mathbf{R}(s)$ and $\mathbf{L}(s)$ can still be in general frequency dependent. Incidentally this particular form for $\mathbf{Z}(s)$ may be advantageous for our approach, although in general not necessary. We assume input and output information is of interest at some "ports" of the network for which the model is to be generated. The frequency-domain description of the system can be written as

$$[\mathbf{R}(s) + s\mathbf{L}(s)]\mathbf{i}_m(s) = \mathbf{B}\mathbf{v}_p(s), \quad \mathbf{i}_p(s) = \mathbf{B}^T\mathbf{i}_m(s), \tag{5}$$

where $\mathbf{v}_p(s) \in \mathbb{C}^p$ and $\mathbf{i}_p(s) \in \mathbb{C}^p$ are Laplace-domain representations of voltages and currents at the p defined ports of interest, $\mathbf{i}_m(s) \in \mathbb{C}^n$ are the internal currents, and $\mathbf{B} \in \mathbb{R}^{n \times p}$ is a matrix relating ports to internal currents. In this case $\mathbf{u} = \mathbf{v}_p$ represents the system inputs (voltages), $\mathbf{y} = \mathbf{i}_p$ the system outputs (currents), and $\mathbf{x} = \mathbf{i}_m$ the internal states (also in this case currents). The transfer function from inputs to outputs is $\mathbf{H}(s) = \mathbf{B}^T[\mathbf{R}(s) + s\mathbf{L}(s)]^{-1}\mathbf{B}$, $\mathbf{i}_p(s) = \mathbf{H}(s)\mathbf{v}_p(s)$, and one view of model reduction is that it seeks an approximation to the transfer function $\mathbf{H}(s)$.

Projection Methods. The Krylov-subspace based model reduction algorithms [1, 3, 2] are projection methods. One of the key insights in reducing distributed systems is that **projection methods, and their approximation properties, apply to lumped systems *as well as to distributed ones.***

Given a matrix $\mathbf{Q} \in \mathbb{R}^{n \times q}$, an orthogonal projection method obtains a rational approximant (of dimension q), or, equivalently, a reduced model, via the congruence transformations

$$\hat{\mathbf{R}}(s) = \mathbf{Q}^*\mathbf{R}(s)\mathbf{Q},\ \hat{\mathbf{L}}(s) = \mathbf{Q}^*\mathbf{L}(s)\mathbf{Q},\ \hat{\mathbf{B}} = \mathbf{Q}^*\mathbf{B}. \tag{6}$$

The post-multiplication by matrix $\mathbf{Q}$ corresponds to projecting the original state vector of variables $\mathbf{i}_m$ into the reduced space of basis functions represented by the q columns of matrix $\mathbf{Q}$: $\hat{\mathbf{i}}_m = \mathbf{Q}\mathbf{i}_m$. The pre-multiplication by matrix $\mathbf{Q}^*$ in (6) reduces the number of equations to the new size of the state q.

After the congruence transformation step the linear system is

$$[\hat{\mathbf{R}}(s)+s\hat{\mathbf{L}}(s)]\hat{\mathbf{i}}_m = \hat{\mathbf{B}}\mathbf{v}_p(s), \quad \mathbf{i}_p(s) = \hat{\mathbf{B}}^*\hat{\mathbf{i}}_m(s) \tag{7}$$

where $\hat{\mathbf{R}}, \hat{\mathbf{L}} \in \mathbb{C}^{q\times q}; \hat{\mathbf{i}}_m(s) \in \mathbb{C}^q, \hat{\mathbf{B}} \in \mathbb{R}^{q\times p}$. At this point, the projection operation is conceptual; the system of (7) contains a frequency-parameter and is thus infinite-dimensional. We will discuss finite-dimensional realizations later in the paper.

For lumped systems, the most popular approaches to model reduction generate the columns of the matrix $\mathbf{Q}$ by matching moments of the frequency domain response $\mathbf{H}(s)$, or, equivalently, derivatives of the time domain response. An exhaustive discussion on this step is not among the purposes of this work. However, we mention here that one possible way to obtain a very simple projection matrix $\mathbf{Q}$ for the congruence transformation in (6) is to construct each of the q columns of $\mathbf{Q}$ by solving the original full system (5) at q different frequency points [2]. When the underlying system matrices $\mathbf{R}(s)$ and $\mathbf{L}(s)$ are defined implicitly, as in large-scale electromagnetic codes, this is not an inconvenience, and is in fact preferred as derivative information can be awkward to obtain.

The importance of congruence transformations in generating well-behaved reduced models stems from the following theorem.

THEOREM 4. *If $\mathbf{Z}(s)$ is positive-real, then $\hat{Z}(s) = \mathbf{Q}^*\mathbf{Z}(s)\mathbf{Q}$ is also positive-real.*

PROOF. Conditions (1) and (2) in Definition 3 are easily verified. Condition (3) follows from observing that $\Pi_{\hat{\mathbf{Z}}}(s) = \mathbf{Q}^*\Pi_{\mathbf{Z}}(s)\mathbf{Q}$, so $\text{range}(\Pi_{\hat{\mathbf{Z}}}) \subset \text{range}(\Pi_{\mathbf{Z}})$. □

The theorem implies that if the given infinite-dimensional system to be reduced is passive and can be described by a positive-real system matrix $\mathbf{Z}(s)$, then the model with smaller descriptor matrices obtained from the projection operation will represent a *passive* system as well, although still infinite-dimensional.

LEMMA 1. *Given a system of the form in (5), if $\hat{\mathbf{Z}}(s) = \hat{\mathbf{R}}(s) + s\hat{\mathbf{L}}(s)$ is [strictly] positive-real, $\hat{\mathbf{H}}(s)$ is [strictly] positive-real [3].*

The proof follows from Theorems 3 and 4. From this Lemma, one can observe that one way to obtain a positive-real $\hat{\mathbf{H}}(s)$ is to find a positive-real $\hat{\mathbf{Z}}(s)$. Note particularly that positive-realness of $\hat{\mathbf{Z}}(s)$ implies that all the poles and zeros of $\hat{\mathbf{H}}(s)$ are in the left half-plane.

4. THE OPTIMAL GLOBAL INTERPOLATION APPROACH

To illustrate some of the problems encountered in model reduction for distributed systems, consider the algorithm presented in [11]. The central approach of the algorithm is a Taylor expansion of the system matrix descriptor $\mathbf{Z}(s) \approx \sum_{k=0}^{N} \mathbf{Z}_k s^k$, using polynomials as interpolants. A standard Krylov method is then applied to a system constructed from the Taylor expansion. This approach does not generate well-behaved models because the Taylor approximation is not *globally* well-behaved. In fact, all polynomials diverge in the $s \to \infty$ limit. Hence, although good accuracy can be achieve in a given frequency band of interest, *global* properties such as positive-realness cannot be guaranteed. Empirically, the resulting reduced models are often found to have unstable poles, the models are not passive, and thus the algorithm is of little practical value.

In our approach, we will seek to combine approximation of the $\mathbf{Z}(s)$ internal matrix descriptors with a Krylov method as in [11]. But our method differs in the following fundamental aspect.

4.1 The key idea

Almost all systems for which one would wish to extract reduced models are non-ideal (non-ideality is why they must be modeled in detail) and so contain a small amount of loss. These systems are strictly passive and can typically be described by strictly positive real system matrices.

The key idea of our algorithm is based on the observation that **if a system descriptor is strictly positive-real to begin with, a *globally* and *uniformly* convergent interpolant will eventually (for a large enough order of the interpolant) be positive-real as well.** (This will be seen in the proof of Theorem 5.) Furthermore, a well-chosen global interpolant will be positive-real for low enough orders to be practical. Local approximations based on Taylor and Padé typically do not have these properties.

There is one more point, subtle yet of great importance, that we wish to underline before proceeding with the main algorithm. In our algorithm, as in others [3], we require that, not only the transfer function $\mathbf{H}(s)$ of the given large system be strictly positive-real, but also that its internal system matrix descriptor $\mathbf{Z}(s)$ be strictly positive real (i.e. that the state-space description be internally positive-real). However, as discussed in Section 2 in a physical system, $\mathbf{H}(s)$ must also be causal. Hence, as for the the positive realness property, we shall require that not only the transfer function $\mathbf{H}(s)$ of the given large system be strictly positive-real *and causal* but also that its internal system matrix descriptor $\mathbf{Z}(s)$ be strictly positive real *and causal* (i.e. that the state-space description be internally positive-real *and causal*). In this case we can restrict our search for approximations of $\mathbf{Z}(s)$ to the set of stable, positive-real interpolants. For non-causal $\mathbf{Z}(s)$, either accuracy or stability/passivity would have to be eventually sacrificed.

4.2 Proposed Algorithm

We propose an eight step procedure, which we term **G**lobal **R**ational **I**nterpolation, **P**assive (GRIP):

1. Obtain/estimate/given a set of q points at which the transfer function at the network ports $\mathbf{H}(s) \in \mathbb{C}^{p\times p}$ is to be matched,
2. Compute the basis $Q \in \mathbb{C}^{n\times q}$ for the projection operation (see Section 3).
3. Project the internal system matrices $\mathbf{R}(s), \mathbf{L}(s) \in \mathbb{C}^{n\times n}$ to obtain smaller $\hat{\mathbf{R}}(s), \hat{\mathbf{L}}(s) \in \mathbb{C}^{q\times q}$ as in (6). Note that this is a conceptual operation; the reduced matrices are still frequency-dependent, so the system is still of potentially infinite order.
4. Perform a *global* and *uniform* interpolation of the (projected) internal system matrices $\hat{\mathbf{R}}(s), \hat{\mathbf{L}}(s)$ (see Section 5).
5. Check the passivity (see [12]), and accuracy of the matrix interpolants. If not passive, or if matrix interpolants are not accurate, go to Step 4 and increase the order N of the global interpolant.
6. Check the accuracy of the reduced model transfer function $\hat{\mathbf{H}}(s)$. If not accurate, go to Step 1 and add additional matching points q.
7. Realize as state-space system.
8. Perform a second-stage guaranteed-passive optimal reduction step, if desired [13].

Steps 1,2,3, and 6 are standard in lumped-system model reduction. Various approaches are possible, and many are described in the literature. As they are not the main focus here, they will not be discussed further. Step 5 can be performed solving the Lur'e equation in the Positive-Real Lemma [12], for which computational procedures are available in the literature [12]. Step 7 is dependent on how Step 4 is performed, but is always possible if Steps 4-6 are feasible.

It can be noticed that our algorithm is posed in such a way that if it terminates, accuracy, stability, and passivity are guaranteed. However, we have not yet shown that it is possible to construct specific instantiations that *will* terminate. Such task is equivalent to finding for the key Step 4 a suitable interpolant that is guaranteed to converge globally and uniformly. To this purpose, in the next Section, one possible choice will be described.

5. A LAGUERRE-BASIS IMPLEMENTATION

5.1 Choosing the global uniform interpolant

Several approaches are possible to the global interpolation problem. One possibility is to use algorithms developed for general-purpose interpolation or data-fitting [6] that guarantee passivity by construction. These algorithms are very computationally demanding. For many applications of interest, a simpler alternative is available. First, it is advantageous (although not necessary) to find some decomposition of matrix $\mathbf{Z}(s)$ into for instance some matrices $\mathbf{R}(s)$ and $\mathbf{L}(s)$ for which the individual matrix entries do not have sharply discontinuous behavior in the frequency parameter s. Many integral-equation-based electromagnetic field solvers for distributed systems already produce such a decomposition. In order to use our method, particular attention will need to be dedicated to making sure that such solvers generate strictly positive-real and causal system matrices.

Second, the frequency dependency of projected matrices $\hat{\mathbf{R}}(s)$ and $\hat{\mathbf{L}}(s)$ in (6) can then be captured for instance using the set of basis functions [14], $E_k(s) = \left(\frac{\lambda-s}{\lambda+s}\right)^k$, where λ is a positive real number. In this way we can write:

$$\hat{\mathbf{R}}(s) = \sum_{k=0}^{\infty} \hat{\mathbf{R}}_k E_k(s), \quad \hat{\mathbf{L}}(s) = \sum_{k=0}^{\infty} \hat{\mathbf{L}}_k E_k(s). \tag{8}$$

The basis created by the functions $E_k(s)$, sometimes called the Laguerre basis, is a member of a larger family [15] of bases, all of which consist of sets of stable rational functions orthonormal over the imaginary axis $s = i\omega$. An interesting contrast with the Taylor series approach is that the $E_k(s)$ are, in a sense, band-limited. For $|\omega| > \lambda$, the $E_k(s)$ have monotonic magnitude, and for $|\omega| < \lambda$, they are nearly equi-ripple, much like Chebyshev polynomials. This implies that with suitable choice of λ, the approximations to $\mathbf{R}(s), \mathbf{L}(s)$ will be well behaved outside the approximation interval, and convergence will be fast within it.

The Laguerre basis is particularly interesting because, under the bilinear transformation, $s = \lambda(1-z)/(1+z)$, the series expansion in terms of the basis functions $E_k(s)$ is mapped to a Fourier series of complex exponentials, since $E_k(s) = z^k$, where $z = e^{i\phi}, \phi \in [0, 2\pi)$. The problem of rationally approximating the matrix functions $\hat{\mathbf{R}}(s), \hat{\mathbf{L}}(s)$ is reduced to the problem of approximating a function on the circle via a Fourier series, or equivalently computing a Discrete Fourier Transform (DFT), as the entries of $\hat{\mathbf{R}}(s), \hat{\mathbf{L}}(s)$ may be approximated term-wise.

5.2 Global Uniform Convergence

If the entries of $\hat{\mathbf{R}}(s), \hat{\mathbf{L}}(s)$ are smooth when mapped to the circle, then the partial sums $\hat{\mathbf{R}}^{(N)}(s), \hat{\mathbf{L}}^{(N)}(s)$ converge uniformly to $\hat{\mathbf{R}}(s), \hat{\mathbf{L}}(s)$

$$\hat{\mathbf{R}}^{(N)}(s) = \frac{1}{N}\sum_{k=0}^{N-1} \hat{\mathbf{R}}_k z^k, \quad \hat{\mathbf{L}}^{(N)}(s) = \frac{1}{N}\sum_{k=0}^{N-1} \hat{\mathbf{L}}_k z^k. \tag{9}$$

If $\hat{\mathbf{R}}(s), \hat{\mathbf{L}}(s)$ are not smooth, but are continuous, it is still possible to obtain uniformly convergent approximates by summing the Fourier series in the sense of Cesaro [16]. Practically speaking, this means replacing the summations (9) by

$$\tilde{\mathbf{R}}^{(N,C)}(s) = \frac{1}{N}\sum_{k=0}^{N-1} \hat{\mathbf{R}}_k \left[1 - \frac{k}{N}\right] z^k, \quad \tilde{\mathbf{L}}^{(N,C)}(s) = \frac{1}{N}\sum_{k=0}^{N-1} \hat{\mathbf{L}}_k \left[1 - \frac{k}{N}\right] z^k. \tag{10}$$

Equivalently, we may say that as successive approximates, we take the arithmetic means of the partial sums $\hat{\mathbf{R}}^{(N)}(s), \hat{\mathbf{L}}^{(N)}(s)$, rather than the partial sums themselves. Summation in this manner has the property of suppressing the Gibbs effect, and also insuring uniform convergence on a broader class of functions.

LEMMA 2. *If the entries of $\hat{\mathbf{R}}^{(N)}(s), \hat{\mathbf{L}}^{(N)}(s)$ are continuous when mapped to the circle, $\tilde{\mathbf{R}}^{(N,C)}(s), \tilde{\mathbf{L}}^{(N,C)}(s)$ converge uniformly to $\hat{\mathbf{R}}^{(N)}(s), \hat{\mathbf{L}}^{(N)}(s)$ as $N \to \infty$ [16].*

Shortly we will need the following definitions:

DEFINITION 6 (STRONG-η CONDITION). *Satisfied for $\hat{\mathbf{Z}}(s)$ if $\Pi_{\hat{\mathbf{Z}}}(s) - \eta I \geq 0$ for any $\eta > 0$.*

where $\Pi_{\hat{\mathbf{Z}}}(s)$ was defined in (3).

DEFINITION 7 (WEAK-η CONDITION). *Satisfied for $\hat{\mathbf{Z}}(s)$ if for any $\varepsilon \geq 0$, there is an $\eta > 0, \eta < \varepsilon$ s.t. $\Pi_{\hat{\mathbf{Z}}}(s) + \eta I > 0$.*

From Lemma 2 we obtain a major result of the paper:

THEOREM 5. *Given a system description $\hat{\mathbf{Z}}(s) = \hat{\mathbf{R}}(s) + s\hat{\mathbf{L}}(s)$ where matrices $\hat{\mathbf{R}}(s)$ and $\hat{\mathbf{L}}(s)$ are causal, strictly positive real, and continuous on the imaginary axis, there exists an integer N and coefficients $\tilde{\mathbf{R}}_k^{(N,C)}, \tilde{\mathbf{L}}_k^{(N,C)}$ for the partial sums in (10) such that the matrix rational function $\tilde{\mathbf{Z}}(s) = \tilde{\mathbf{R}}^{(N,C)}(s) + s\tilde{\mathbf{L}}^{(N,C)}(s)$ is a positive-real rational interpolant of $\hat{\mathbf{Z}}(s)$ whose error can be bounded from above by any chosen positive constant.*

PROOF. Property (1) follows by construction as the E_k satisfy (1). Property (2) also follows by construction, since by inspection the E_k have poles only in the left half-plane. Due to Theorem 2, proof of (4) is now sufficient to complete the proof. Case 1: $\Pi_{\mathbf{Z}}$ is Strongly-η. From Lemma 2, if $\hat{\mathbf{R}}(s)$ and $\hat{\mathbf{L}}(s)$ are continuous when mapped to the circle, $\tilde{\mathbf{R}}^{(N,C)}(s), \tilde{\mathbf{L}}^{(N,C)}(s)$ converge uniformly and so does $\tilde{\mathbf{Z}}(s) = \tilde{\mathbf{R}}^{(N,C)}(s) + s\tilde{\mathbf{L}}^{(N,C)}(s)$. Thus $\forall \eta > 0$, $\exists N$ s.t. $||\tilde{\mathbf{Z}}(i\omega) - \hat{\mathbf{Z}}(i\omega)||_2 < \eta/4$, $\forall \omega \in \mathbb{R}$. Hence $||\Pi_{\tilde{\mathbf{Z}}}(i\omega) - \Pi_{\hat{\mathbf{Z}}}(i\omega)||_2 \leq 2||\tilde{\mathbf{Z}}(i\omega) - \hat{\mathbf{Z}}(i\omega)||_2 \leq \eta/2$, $\forall \omega \in \mathbb{R}$. $\Pi_{\tilde{\mathbf{Z}}}(i\omega) > \Pi_{\hat{\mathbf{Z}}}(i\omega) - ||\Pi_{\tilde{\mathbf{Z}}}(i\omega) - \Pi_{\hat{\mathbf{Z}}}(i\omega)||_2 > \eta - \eta/2$, so[1] $\Pi_{\tilde{\mathbf{Z}}} > \eta/2$ and $\Pi_{\tilde{\mathbf{Z}}}$ is Strongly-η, which implies (4) (see [8]) and that $\tilde{\mathbf{Z}}$ is strictly positive-real. Case 2: $\Pi_{\hat{\mathbf{Z}}}$ is [strictly] positive-real but not Strongly-η. Choose any $\eta > 0$ and Map $\hat{\mathbf{Z}} \to \hat{\mathbf{Z}} + \eta I$. $\hat{\mathbf{Z}}$ is now Strongly-η. Go to Case 1. [2] □

Theorem 5 proves that an order of interpolation N large enough *does exist* and therefore that the algorithm in Section (4.2 terminates. A practical algorithm would require a small N. The order of the interpolant is related to the smoothness of the function being approximated. Hence, although we could use this algorithm to approximate $\mathbf{H}(s)$ or $\mathbf{Z}(s)$ directly, that would require evaluation of an awful lot of matching points around resonances, and most likely a very large order of the interpolant. A small N is instead needed when the algorithm is used on some internal decomposed matrices $\mathbf{R}(s)$ and $\mathbf{L}(s)$ which are almost always continuous within a given band of interest. Out-of-band smoothness (for example, for delay functions, which create essential singularities at ∞) can be insured by filtering operations which must be designed to preserve passivity and causality of the original matrices.

5.3 Computing the DFT Coefficients

The DFT coefficients in the sums (9) can be efficiently calculated for instance using a Fast Fourier Transform (FFT) algorithm.

[1] In this we needed to use the result that perturbations of a Hermitian matrix ($\Pi_{\hat{\mathbf{Z}}}$) result in perturbations of the eigenvalues bounded by the 2-norm of the perturbation.

[2] The upshot of all this is that the Strongly-η condition is slightly stronger than strict positive-realness and may not be satisfied for all strictly passive systems. But by introducing an additional error of $O(\eta)$ (i.e., roughly doubling the interpolation error bound) through the η-shifting procedure we may guarantee strict positive-realness of the final model. An alternative is to not perform the η-shifting, in which case we may prove that $\tilde{\mathbf{Z}}$ is Weakly-η, which allows $\tilde{\mathbf{Z}}$ to have an excess energy gain of $O(\eta)$. Since we may drive $\eta \to 0$, neither deviation is of practical consequence in systems with loss modeled over a finite bandwidth.

That is equivalent to evaluating the DFT coefficients via numerical integration using the trapezoidal rule. For functions on a circle, the trapezoidal rule is near-optimal, and achieves rapid (spectral) convergence for smooth functions. Hence the steps for one possible global approximation procedure are

1. For a desired interpolation order N, choose the size M of the FFT as some power of two: $M = 2^n > N$.
2. Calculate the frequency points s_k on the imaginary axis corresponding to the M equally spaced FFT points $z_k = exp(j2\pi k/M)$, $k = 1,\ldots,M$ on the unit circle using the bilinear transform: $s_k = \lambda(1-z_k)(1+z_k)$, where $\lambda = 2\pi f_0$ is a parameter to be chosen around the center of the frequency band of interest for the system response.
3. Use (6) to evaluate each individual projected matrix $\hat{\mathbf{R}}(s_k)$ and $\hat{\mathbf{L}}(s_k)$ at the selected frequency points s_k, $k = 1,\ldots,M$.
4. Use an FFT algorithm to calculate the M coefficients $\hat{\mathbf{R}}_k$ and $\hat{\mathbf{L}}_k$ in (9) from the sequences $\hat{\mathbf{R}}(s_k)$ and $\hat{\mathbf{L}}(s_k)$, $k = 1,\ldots,M$.
5. Apply to each of the M FFT coefficients the Cesaro's transformation in (10) and obtain the coefficients $\tilde{\mathbf{R}}_k$ and $\tilde{\mathbf{L}}_k$.

Note that, since the $\mathbf{R}(s)$ and $\mathbf{L}(s)$ matrices usually satisfy conjugate symmetry relations, $\mathbf{R}(s), \mathbf{L}(s)$ need to be evaluated at only half the points on the circle. Also, once the M Cesaro's FFT coefficients are available one can construct at no additional cost several interpolants of increasing order $N < M/2$ simply truncating the sums in (10) to the first N coefficients.

5.4 Realization

In this section we describe how to perform Step 7 in the general algorithm 4.2, realization as a state-space model. Having performed the global rational approximation on the projected matrix functions $\hat{\mathbf{R}}(s)$ and $\hat{\mathbf{L}}(s)$, the system (5) is now:

$$\left[\frac{1}{M}\sum_{k=0}^{N-1}\tilde{\mathbf{R}}_k z^k + \lambda\left(\frac{1-z}{1+z}\right)\frac{1}{M}\sum_{k=0}^{N-1}\tilde{\mathbf{L}}_k z^k\right]\hat{\mathbf{i}}_m = \hat{\mathbf{B}}\mathbf{v}_p \qquad (11)$$

where $\tilde{\mathbf{R}}_k$ and $\tilde{\mathbf{L}}_k$ contain already the Cesaro's correction (10). We then collect the terms corresponding to the same powers of z, define an augment state

$$\mathbf{x} = [\ z^{N-1}\mathbf{i}_m^T \quad z^{N-2}\mathbf{i}_m^T \quad \ldots \quad z^2\mathbf{i}_m^T \quad z\mathbf{i}_m^T \quad \mathbf{i}_m^T\ ]^T. \qquad (12)$$

and produce a finite dimension discrete linear system, Substituting $z = (\lambda - s)/(\lambda + s)$ we obtain the continuous and final system realization

$$\begin{aligned} s\mathbf{E}_c\mathbf{x}_c &= \mathbf{A}_c\mathbf{x}_c + \mathbf{B}_c\mathbf{v}_p \\ \mathbf{i}_p &= \mathbf{C}_c\mathbf{x}_c \end{aligned} \qquad (13)$$

where

$$\mathbf{E}_c = \begin{bmatrix} I & I & \ldots & 0 & 0 \\ 0 & I & \ldots & 0 & 0 \\ \ldots & \ldots & \ldots & \ldots & \ldots \\ 0 & 0 & \ldots & I & I \\ 2\mathbf{F}_N & 2(\mathbf{F}_N+\mathbf{F}_{N-1}) & \ldots & 2(\mathbf{F}_N+\ldots+\mathbf{F}_2) & \mathbf{F}_N+\ldots+\mathbf{F}_1-\mathbf{F}_0 \end{bmatrix}$$

$$\mathbf{A}_c = \lambda\begin{bmatrix} I & -I & \ldots & 0 & 0 \\ 0 & I & \ldots & 0 & 0 \\ \ldots & \ldots & \ldots & \ldots & \ldots \\ 0 & 0 & \ldots & I & -I \\ 0 & 0 & \ldots & 0 & -(\mathbf{F}_N+\ldots+\mathbf{F}_1+\mathbf{F}_0) \end{bmatrix},$$

$$\mathbf{B}_c = -2\lambda[\ 0 \ \ 0 \ \ 0 \ \ 0 \ \ \ldots \ \ 0 \ \ \hat{\mathbf{B}}\ ]^T,$$

$$\mathbf{C}_c = \hat{\mathbf{B}}^*[\ 0 \ \ 0 \ \ 0 \ \ 0 \ \ \ldots \ \ 0 \ \ I\ ],$$

$$\mathbf{F}_k = \begin{cases} \frac{1}{M}[\tilde{\mathbf{R}}_0 + \lambda\tilde{\mathbf{L}}_0] & k = 0, \\ \frac{1}{M}[\tilde{\mathbf{R}}_k + \tilde{\mathbf{R}}_{k-1} + \lambda\tilde{\mathbf{L}}_k - \lambda\tilde{\mathbf{L}}_{k-1}] & 1 \le k \le N-1, \\ \frac{1}{M}[\tilde{\mathbf{R}}_{N-1} - \lambda\tilde{\mathbf{L}}_{N-1}] & k = N. \end{cases}$$

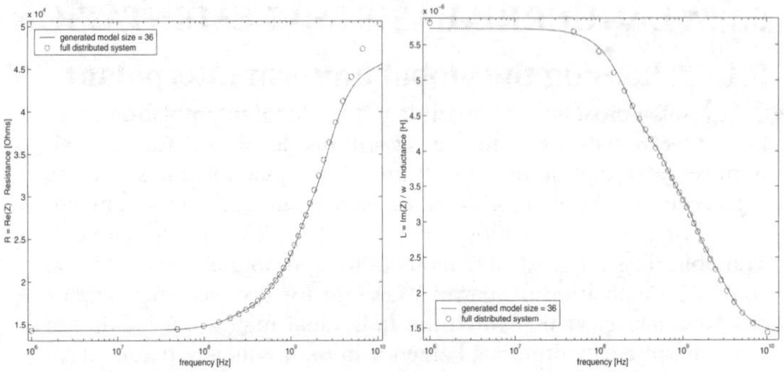

Figure 1: Real part and imaginary part divided by ω of the frequency response for the lossy substrate example.

6. EXAMPLES

6.1 Effect of Lossy Substrate on Line Impedance

The geometry in this example consists of two wires over a lossy substrate. The two wire volumes are discretized into short and thin filaments using a set of piece-wise constant basis functions. A standard Galerkin technique is used to calculate the resistance and partial inductance matrices $\mathbf{R}(s)$ and $\mathbf{L}(s)$. *A frequency dependent Green function is used in the kernel of the Galerkin integration to account for the effects of the lossy substrate. Hence, the resulting matrices are frequency dependent.* The system, before model reduction, appears as in (5). The descriptor matrices have been projected to a reduced space of size $q = 4$ obtained by solving the original full system at frequencies $f = 0$, $0.4GHz$, $1GHz$, $2.4GHz$. which correspond to the points on the unit circle: $z = 1$, $exp(-j\pi/4)$, $-j$, $exp(-j3\pi/4)$. As center frequency for our band of interest we have chosen $\lambda = 2\pi 10^9$. In this example we have chosen $M = 64$ points for the FFT size. Fig. 4.b shows the real part of the DFT coefficients for the inductance matrices. Coefficients $32,\ldots,63$ are very small indicating the original system matrix is causal. We have then truncated the DFT series to N=8 coefficients, producing a final model of order 36. Fig. 1 compares the frequency response of the original full system with the frequency response of the final realized linear state space model. All the poles of the realized model are in the half-plane $Re(s) < 0$, hence stable. Using the Positive-Real Lemma [12], we confirmed that the generated system is passive.

6.2 Full-Wave PEEC kernel

In this example we consider two parallel wires 4um wide, 1um thick and 750 um long. The two wires are separated by 3cm. The wire volumes are discretized into short and thin filaments using a set of piece-wise constant basis functions. A Galerkin technique is used to calculate the resistance and partial inductance matrices $\mathbf{R}(s)$ and $\mathbf{L}(s)$. *Since the separation between the two wires is not small compared to the minimum wavelength of interest, a frequency dependent full-wave Green Function needs to be employed in the kernel of the Galerkin integration. Hence, the resistance and inductance matrices are frequency dependent.* The system, before model reduction, appears as in (5). The descriptor matrices have been projected to a reduced space of size $q = 4$ obtained by solving the full system at frequencies $f = 0$, $1.3GHz$, $3.3GHz$, $8GHz$. which correspond to the points on the unit circle: $z = 1$, $exp(-j\pi/4)$, $-j$, $exp(-j3\pi/4)$. As center frequency we have chosen $\lambda = 2\pi \times 3.3GHz$. In this example we have chosen $M = 128$ points for the FFT size. We have then truncated the DFT series to N=23 coefficients, producing a final model of order 96. Fig. 2 compares the frequency response of the original full system with the frequency response of the final realized linear state space model. Fig. 3 shows that all the poles of the realized model are in the half-plane $Re(s) < 0$, hence the system is stable. Using the Positive-Real Lemma [7], we confirmed that the generated system is passive.

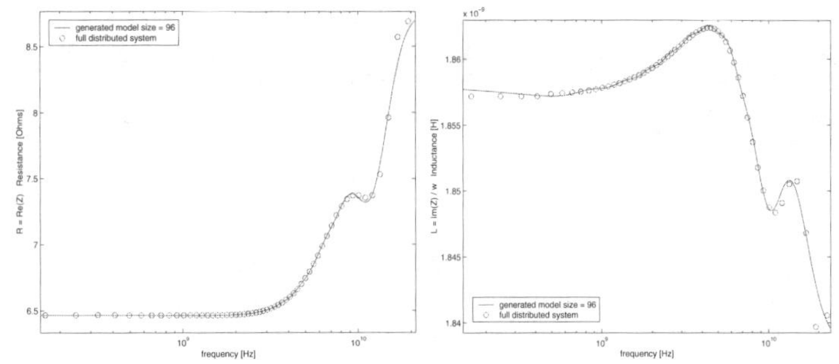

Figure 2: Real part and imaginary part divided by ω of the frequency response for the full-wave example.

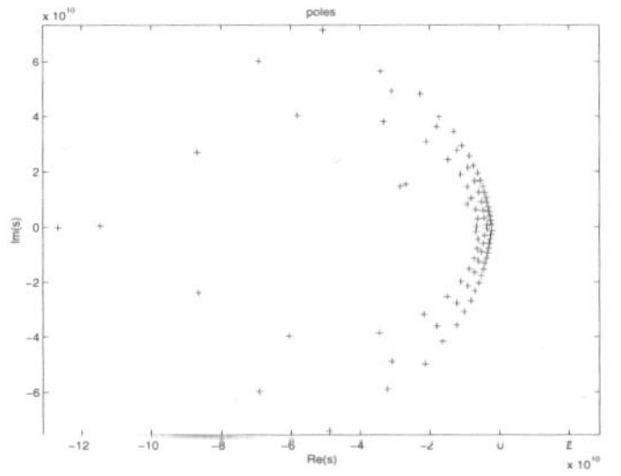

Figure 3: All poles of the reduced fullwave model are stable.

6.3 A non-causal example

Since $\mathbf{R}(s)$ and $\mathbf{L}(s)$ are frequency domain representations, when we calculate their FFT interpolants we obtain something related to their time domain impulse response (actually the impulse response of a discrete-time system obtained by sampling the continuous-time system at rate λ). Since the FFT produces M coefficients that repeat periodically, the $k = M/2, \ldots, M-1$ coefficients are related to negative-time part of the time domain impulse response of $\mathbf{R}(s)$ and $\mathbf{L}(s)$. In Fig. 4.a we show the FFT coefficients of a common *non-causal* example mentioned in Section 2: $\mathbf{Z}_{\mathbf{skin}}(i\omega) = \mathbf{R}_0 + \mathbf{R}_{ac}\sqrt{|\omega|}$ used as a model for the "resistance" of interconnect in the skin-effect regime. We can easily notice in such Figure that the coefficients $k = M/2, \ldots, M-1$ related to non-causal coefficients of the time domain response are non-zero. If a model order reduction is attempted on such an originally non-causal system, one will obtain non-stable models. Alternatively one could deliberately ignore the non-causal coefficients and set them to zero before beginning the reduction. However in this case stable but highly inaccurate models will be produced.

In our approach we require therefore that the original system descriptor matrices $\mathbf{R}(s)$ and $\mathbf{L}(s)$ be causal. This means checking that the non-causal $k = M/2, \ldots, M-1$ coefficients of the FFT be zero except for some aliasing phenomena. Fig. 4.b shows the real part of the DFT coefficients for the $\mathbf{L}(s)$ in example 6.1. One can easily verify that such original system matrix descriptor is actually a causal one.

7. CONCLUSIONS

In this paper we have presented a class of algorithms for guaranteed passive model order reduction of strictly passive and causal linear systems with frequency dependent matrices (distributed systems). Our approach is based on the key idea that if a system is strictly positive-real to begin with, a *globally* and *uniformly* convergent interpolant will eventually (for a large enough order of the interpolant) be positive-real as well.Laguerre basis are a set of well-behaved uniformly convergent interpolation functions whose coefficients can be conveniently calculated using the FFT algorithm. An implementation using a Laguerre basis as interpolant is given and examples are presented. While the Laguerre basis reduces the infinite order of the original distributed system to a finite order, a standard Krylov subspace congruence transformation can still be employed to reduce the size of the matrices. The algorithm is also perfectly compatible with fast matrix-vector product algorithms.

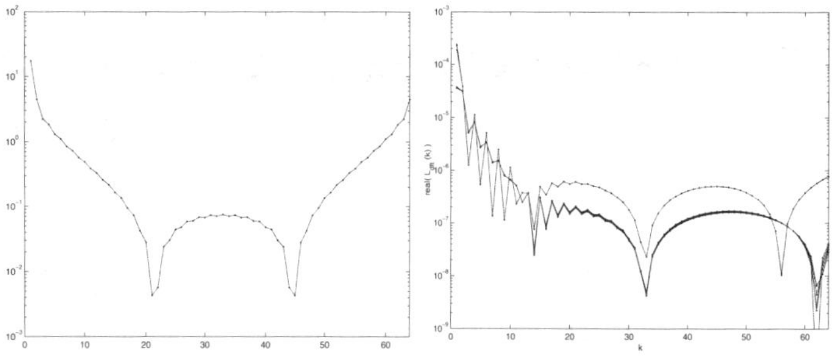

a) b)

Figure 4: a) FFT coefficients for the non-causal example $\mathbf{Z}_{\mathbf{skin}}(i\omega) = \mathbf{R}_0 + \mathbf{R}_{ac}\sqrt{|\omega|}$ used as a model for the skin-effect "resistance". b) FFT coefficients of the inductance matrix for the lossy substrate example. Note that non-causal coefficients $32-63$ are in a) very large and in b) very small.

8. REFERENCES

[1] P. Feldmann and R. W. Freund. Reduced-order modeling of large linear subcircuits via a block Lanczos algorithm. In *Design Automation Conference*, pages 474–479, June 1995.

[2] Eric Grimme. *Krylov Projection Methods for Model Reduction*. PhD thesis, Coordinated-Science Laboratory, University of Illinois at Urbana-Champaign, Urbana-Champaign, IL, 1997.

[3] A. Odabasioglu, M. Celik, and L. T. Pileggi. Prima: Passive reduced-order interconnect macromodeling algorithm. *IEEE Trans. on Computer-Aided Design of Integrated Circuits and Systems*, 17(8):645–54, August 1998.

[4] L. Daniel, A. Sangiovanni-Vincentelli, and J. White. Using conduction modes basis functions for efficient electromagnetic analysis of on-chip and off-chip interconnect. In *Design Automation Conference*, June 2001.

[5] Eli Chiprout and Michael S. Nakhla. Analysis of interconnect networks using complex frequency hopping (CFH). *IEEE Trans. Computer-Aided Design*, 14(2):186–200, February 1995.

[6] Carlos P. Coelho, Joel R. Phillips, and L. Miguel Silveira. A convex programming approach to positive real rational approximation. In *International Conference on Computer Aided-Design*, pages 245–251, San Jose, CA, November 2001.

[7] B. D. Anderson and S. Vongpanitlerd. *Network Analysis and Synthesis*. Prentice-Hall, Englewood Cliffs, New Jersey, 1973.

[8] J. T. Wen. Time domain and frequency domain conditions for strict positive realness. *IEEE Transactions on Automatic Control*, 33:988–992, 1988.

[9] M. R. Wohlers. *Lumped and distributed passive networks*. Academic Press, New York, 1969.

[10] H. M. Nussenzveig. *Causality and dispersion relations*. Academic Press, New York, 1972.

[11] J. R. Phillips, E. Chiprout, and D. D. Ling. Efficient full-wave electromagnetic analysis via model order reduction of fast integral transforms. In *Design Automation Conference*, June 1996.

[12] W. F. Arnold and A. J. Laub. Generalized eigenproblem algorithms and software for algebraic Riccati equations. *Proc. IEEE*, 72:1746–1754, 1984.

[13] J. Phillips, L. Daniel, and M. Silveira. Guaranteed passive balancing transformations for model order reduction. In *Design Automation Conference*, June 2002.

[14] N. E. Wu and G. Gu. Discrete Fourier transform and h_∞ approximation. *IEEE Transactions on Automatic Control*, 35(9):1044–1046, September 1990.

[15] B. Ninness and F. Gustafsson. A unifying construction of orthonormal bases for system identification. *IEEE Transactions on Automatic Control*, 42:515–521, 1997.

[16] G. Bachmann, L. Narici, and E. Beckenstein. *Fourier and wavelet analysis*. Springer-Verlag, 2000.

Guaranteed Passive Balancing Transformations for Model Order Reduction

Joel Phillips
Cadence Berkeley Labs
Cadence Design Systems
San Jose, CA, U.S.A.
jrp@cadence.com

Luca Daniel
EECS Department
University of California Berkeley
Berkeley, CA, U.S.A.
dluca@eecs.berkeley.edu

L. Miguel Silveira
INESC ID - Cadence Euro Labs
IST - Tech. U. Lisbon, Portugal
lms@inesc-id.pt

ABSTRACT

The major concerns in state-of-the-art model reduction algorithms are: achieving accurate models of sufficiently small size, numerically stable and efficient generation of the models, and preservation of system properties such as passivity. Algorithms such as PRIMA generate guaranteed-passive models, for systems with special internal structure, using numerically stable and efficient Krylov-subspace iterations. Truncated Balanced Realization (TBR) algorithms, as used to date in the design automation community, can achieve smaller models with better error control, but do not necessarily preserve passivity. In this paper we show how to construct TBR-like methods that guarantee passive reduced models and in addition are applicable to state-space systems with arbitrary internal structure.

Categories & Subject Descriptors: B.7.2 Simulation, B.8.2 Performance Analysis & Design Aids, I.6 Simulation & Modeling.
General Terms: Algorithms, Performance, Design,
Keywords: Passive reduced order modeling, Truncated balanced realization.

1. INTRODUCTION

Model reduction has been an active research field in design automation over the past decade. In an integrated circuits context, initial interest in model reduction techniques stemmed from efforts to accelerate analysis of circuit interconnect. More recently, model reduction has come to be viewed as a method for generating compact models from all sorts of physical system modeling tools. Because of the need to obtain accurate high-order models at reasonable computational cost, the Krylov-subspace model reduction methods [1] have occupied the forefront of research over the past five years.

Recently it has become apparent that, while very suitable for analysis of large-scale systems, the models generated from the Krylov techniques such as PRIMA and PVL are not necessarily as compact (that is, small in order) as is desired [2, 3]. Therefore, another approach, already well-developed in the control literature has been receiving renewed attention in the electronic design automation community: that of Truncated Balanced Realization (TBR) [4].

Truncated balanced realization algorithms (and their close relatives that generate optimal norm approximants [5]) are of importance in their own right. For small systems, a few hundred states or so, they generate superior reduced models, as well as computable bounds on the reduction error. For large systems, direct application of the techniques used to balance and truncate the systems is computationally infeasible since the computations required have $O(n^3)$ complexity when performed directly (n is the order of the system to be reduced). Therefore the TBR methods are of more interest when combined with iterative Krylov-subspace procedures. One formulation of this method is to directly solve the large Lyapunov equations via a Krylov subspace method [6, 7]. The reduced models are obtained directly from the reduced Lyapunov equation. Another viewpoint is to obtain an initial reduced model via some initial reduction or approximation technique and then further compress it using a TBR method. This second viewpoint is somewhat more general since the initial approximation can be generated by any desired method, for example rational fitting [8] or a now standard Krylov-subspace technique [2, 9].

An issue with the TBR type methods that has not been addressed in most of the above mentioned works is that they cannot be relied on to preserve passivity. The technique in [2] uses a passivity-preserving initial reduction, but follows this reduction with a standard TBR method. There is no guarantee that the second TBR step will not destroy the passivity of the initial model. More problematic, no means is given in [2] to determine if the final model is passive – or not.

Less widely appreciated is another dilemma: Krylov methods such as PRIMA have practical issues that prevent their wide application to systems outside the class of RLC circuits. These methods rely on congruence transformations to preserve positive-realness of the matrices that are internal to the state-space representation. However, whether or not a state-space model represents a passive system is a property of the input-output transfer function, not a property of the internal representation. Many passive systems are not conveniently put into a form for which algorithms such as PRIMA are applicable: they may have asymmetric or non-positive semidefinite system and descriptor matrices. It may not be possible to perform a change of basis to a convenient form without destroying sparse structure that may be present in the system, meaning that for large-scale systems such an approach is infeasible.

Further, positive realness is not necessarily the right property to seek. If the state-space model represents scattering (S) parameters of a passive system, the system is passive if the norm of the S-parameter matrix is bounded by unity, and so even the transfer function has no relation to positive-realness. Such systems cannot be reduced by congruence with any passivity guarantees. To the best of our knowledge, no effective truly general-purpose passivity-preserving algorithms are now widely available.

In this paper we discuss TBR-like model reduction algorithms that can preserve system passivity, have computable error bounds, and, unlike other algorithms such as PRIMA, pose no constraints on the internal structure of the state-space model. We describe variants that preserve both positive-realness (useful for systems that represent Y or Z parameters) and bounded-realness (useful for systems that represent S parameters). These algorithms can be applied directly to a given state-space description, or can be used as the second stage of a Krylov-subspace based procedure [6, 7, 2].

2. BACKGROUND

2.1 State-Space Models

Given a state-space model in descriptor form,

$$E\frac{dx}{dt} = Ax(t) + Bu(t), \quad y(t) = Cx(t) + Du(t) \tag{1}$$

where $E, A \in \mathbb{R}^{n\times n}, B \in \mathbb{R}^{n\times p}, C \in \mathbb{R}^{p\times n}, D \in \mathbb{R}^{p\times p}, u(t) \in \mathbb{R}^{p}$, model reduction algorithms seek to produce a similar system with reduced $\tilde{E}, \tilde{A} \in \mathbb{R}^{q\times q}, \tilde{B} \in \mathbb{R}^{q\times p}, C \in \mathbb{R}^{p\times q}$, of order q much smaller than the original order n, but for which the outputs $y(t)$ and $\tilde{y}(t)$ are approximately equal for inputs $u(t)$ of interest. Often the transfer functions $H(s) = D + C(sE - A)^{-1}B$ and its reduced counterpart $\tilde{H}(s)$ is used as a metric for approximation : if $||\tilde{H}(s) - H(s)|| < \varepsilon$ for some given allowable error ε and allowed domain of the complex frequency variable s, the reduced model is accepted as accurate.

2.2 Passivity

When modeling passive systems – those that cannot produce energy internally – non-passive reduced models may cause non-physical behavior later in circuit simulators, such as by generating energy at high frequencies that causes erratic or unstable time-domain behavior. If $H(s)$ represents the Y (admittance) or Z (impedance) parameters of a system, positive-realness of $H(s)$ implies that the underlying state-space description is a representation of a passive system [10]. The function $H(s)$ is positive-real (PR) if

$$H^*(s) = H(s^*), \tag{2}$$

$$H(s) \text{ is analytic in } \mathrm{Re}(s) > 0. \tag{3}$$

$$H(s) + H(s)^H \geq 0 \text{ in } \mathrm{Re}(s) > 0. \tag{4}$$

If $H(s)$ represents the S (scattering) parameter matrix, then to represent a passive system, it is necessary that $H(s)$ be bounded-real [10]. A function $H(s)$ is bounded-real(BR) if (2) and (3) hold and in addition

$$I - H^H(s)H(s) \geq 0 \text{ in } \mathrm{Re}(s) > 0. \tag{5}$$

2.3 Krylov Methods

Recently developed model reduction methods suitable for application to large systems are based on Krylov-subspace techniques. Mathematically, the reduced models are obtained via a projection operation

$$\tilde{E} = W^T EV \quad \tilde{A} = W^T AV \quad \tilde{B} = W^T B \quad \tilde{C} = CV. \tag{6}$$

For example, PRIMA [1] constructs $V = W$ by using the Arnoldi algorithm, thereby spanning a Krylov subspace of $A^{-1}E$. Because of the moment-matching properties of Krylov-subspaces, the reduced transfer function $\tilde{H}(s)$ will agree with the original $H(s)$ up to the first q derivatives.

The PRIMA algorithm has another interesting property. Given a starting passive model, if the original state-space model can be formulated with positive-semidefinite A and E and $B = C^T$, then the transfer function of the final reduced model will be positive-real, meaning the reduced system is also passive. This is essentially because the projection operation in (6) becomes a congruence transform for $W = V$, and since congruence transforms preserve positive-semi-definiteness, the reduced $\tilde{E}, \tilde{A}$ inherit the numerical range properties of their parents, implying that the reduced function $\tilde{H}(s)$ is positive-real. Note however that it is entirely possible to have systems with positive-real $H(s)$, and thus underlying passive models, for which the conditions necessary for using PRIMA do *not* hold. Such systems cannot be reduced in a guaranteed positive-real manner via congruence transformations. Likewise, such techniques cannot guarantee bounded-real reduced models from bounded-real starting systems.

3. TRUNCATED BALANCED REALIZATIONS

Complementary model reduction techniques are based on truncated balanced realization. We are mostly interested in applying TBR procedures as the second stage of a composite model reduction procedure [2], the first stage being reduction by a Krylov-based projection method. Note that the most of algorithms in [6, 7], are essentially equivalent to a first-stage Krylov projection followed by a second-stage TBR procedure. We first discuss the most commonly used approach before presenting passivity-preserving variants. It is common in this literature to assume $E = I$. When E is non-singular, the mapping $E \to I, A \to E^{-1}A, B \to E^{-1}B$ will put a descriptor system into this form. It is very common in electrical engineering applications to have situations where E is in fact singular and this procedure cannot be performed, but in the situations of interest to us, where an initial projection step has taken place, usually E is non-singular, and so to facilitate comparisons with the literature and somewhat simplify the computational procedure, we will assume the system can be manipulated (possibly implicitly) into a form where $E = I$. We do emphasize that it *is* possible to formulate the computational procedure to work with E directly, and this is necessary when E is singular [1] but we have chosen not to do this as a matter of convenience.

3.1 Standard Approach

The TBR procedure as first presented in [4] is centered around information obtained from the controllability Grammian W_c, which can be obtained from solving the Lyapunov equation

$$AW_c + W_cA^T = -BB^T \tag{7}$$

for W_c, and the observability Grammian W_o, which can be obtained from the dual Lyapunov equation

$$A^T W_o + W_oA = -C^T C. \tag{8}$$

Under a similarity transformation of the state-space model,

$$A \to T^{-1}AT, \quad B \to T^{-1}B, \quad C \to CT \tag{9}$$

the state-space model, and the transfer function, are invariant (only the internal variables are changed). The grammians, however, vary under the rules

$$W_c \to T^{-1}W_cT^{-T}, \quad W_o \to T^T W_oT \tag{10}$$

and so are not invariant. The TBR procedure is based on two observations about W_o and W_c. First, the eigenvalues of the product

[1] For example, Eqn. (7), becomes $AW_cE^T + EW_cA^T = -EBB^TE^T$.

Algorithm 1. **Truncated Balanced Realization (TBR)**

1. *Solve $AW_c + W_cA^T = -BB^T$ for W_c*
2. *Solve $A^TW_o + W_oA = -C^TC$ for W_o*
3. *Compute Cholesky factors $W_c = L_cL_c^T$, $W_o = L_oL_o^T$,*
4. *Compute SVD of Cholesky product $U\Sigma V = L_o^TL_c$ where Σ is diagonal positive and U,V have orthonormal columns*
5. *Compute the balancing transformations $T = L_cV\Sigma^{-1/2}$, $T^{-1} = \Sigma^{-1/2}U^TL_o^T$*
6. *Form the balanced realization as $\hat{A} = T^{-1}AT$, $\hat{B} = T^{-1}B$, $\hat{C} = CT$*
7. *Select reduced model order and partition $\hat{A},\hat{B},\hat{C}$ conformally*
8. *Truncate $\hat{A},\hat{B},\hat{C}$ to form the reduced realization $\tilde{A},\tilde{B},\tilde{C}$*

W_cW_o are invariant. These eigenvalues, the Hankel singular values, contain useful information about the input-output behavior of the system. In particular, "small" eigenvalues of W_cW_o correspond to internal sub-systems that have a weak effect on the input-output behavior of the system and are therefore close to non-observable or non-controllable or both.

Second, since the Grammians transform under congruence, and as any two symmetric matrices can be simultaneously diagonalized by an appropriate congruence transformation, it is possible to find a similarity transformation T that leaves the state-space system dynamics unchanged, but makes W_o and W_c equal and diagonal. In these coordinates, with $W_c = W_o = \Sigma$, we may partition Σ into

$$\Sigma = \begin{bmatrix} \Sigma_1 & 0 \\ 0 & \Sigma_2 \end{bmatrix} \tag{11}$$

where Σ_1 describes the "strong" sub-systems to be retained and Σ_2 the "weak" sub-systems to be deleted. Conformally partitioning the matrices as

$$A = \begin{bmatrix} A_{11} & A_{12} \\ A_{21} & A_{22} \end{bmatrix}, \quad B = \begin{bmatrix} B_1 \\ B_2 \end{bmatrix}, \quad C = \begin{bmatrix} C_1 & C_2 \end{bmatrix}, \tag{12}$$

and truncating the model, retaining $\tilde{A} = A_{11}, \tilde{B} = B_1, \tilde{C} = C_1$ as the reduced system, therefore has the effect of deleting the "weak" internal subsystems. A complete TBR algorithm [11] is shown as Algorithm 1.

We now turn to the question of when TBR procedures produce passive reduced models.

3.2 Symmetrizable Systems

It turns out that there is a special system case, of relevance to integrated circuits applications, for which the standard TBR procedure (Algorithm 1) always produces positive-real reduced models. Suppose that the state-space model is *symmetric*, that is $A = A^T, B = C^T$, and furthermore A is negative-semidefinite. Since $\mathrm{Re}\{sI - A\} = -\frac{1}{2}(A + A^T) = -A \geq 0$, the system is positive-real. From Eqns. (7) and (8) it follows that $W_o = W_c$. From inspecting step 5 in Algorithm 1, we find that $T^{-1} = T^T$. *Thus the similarity transformation is a congruence transformation.* The reduced $\tilde{A}$ must be negative-semidefinite, and we will likewise have $\tilde{B} = \tilde{C}^T$. Therefore the reduced system is positive-real. This would seem to be a similar situation as we have in PRIMA, but it is actually far more general. The reason is that the balancing transformation is essentially unique as explained in [4], meaning that we have the following broader result:

THEOREM 1. *Suppose a state-space system is transformable under similarity to a system of the form in Eqn. (1), with $E = E^T, A = A^T, A \leq 0$. A reduced model generated via Algorithm 1 is positive-real.*

In contrast, the positive-realness preserving properties of congruence transformations depend on the coordinate system used and are *not* preserved under similarity transformations.

Some systems that fall into the symmetrizable class are RL and RC circuits in MNA form, and reductions of such forms via congruence. All systems, however, do not fall into this class, and more powerful techniques are needed to preserve passivity in TBR methods.

Algorithm 2. **Positive-Real TBR (PR-TBR)**

1. *Solve Eqns. (13)-(15) for X_c and their duals for X_o.*
2. *Proceed with steps 3-8 in Algorithm 1, with X_c for W_c and X_o for W_o.*

3.3 Positive Real Conditions

We will show in Section 5 that the TBR procedure of Algorithm 1 does not necessarily produce passive models. In making assessments about passivity, we require a tool that can assess the positive-realness of a state-space model in a global manner. One such tool is the positive-real lemma [10], which states that $H(s)$ is positive-real if and only if there exist matrices $X_c = X_c^T, J_c, K_c$ such that the Lur'e equations:

$$AX_c + X_cA^T = -K_cK_c^T \tag{13}$$

$$X_cC^T - B = -K_cJ_c^T, \tag{14}$$

$$J_cJ_c^T = D + D^T \tag{15}$$

are satisfied, and $X_c \geq 0$ (X_c is positive-semidefinite). X_c is analogous to the controllability Grammian. In fact, it is the controllability Grammian for a system with the input-to-state mapping given by the matrix K_c. It should not be surprising that there are a dual set of Lur'e equations for $X_o = X_o^T > 0, J_o, K_o$ that are obtained from Equations (13)-(15) by the substitutions $A \to A^T, B \to C^T, C^T \to B$. The dual equations have a corresponding observability quantity $X_o \geq 0$ for a positive-real $H(s)$. It is easy to verify that X_c, X_o transform under similarity transformation just as W_c, W_o (Eqn. 10), that their eigenvalues are invariant, and in fact in most respects they behave as the Grammians W_c, W_o.

3.4 Guaranteed Passive Balanced Truncations

It should be clear than the Lur'e equations can be solved for the quantities X_c, X_o which may then be used as the basis for a TBR procedure. We present this as Algorithm 2 and call it PR-TBR, as it preserves positive-realness of the transfer function.

THEOREM 2. *Algorithm 2 applied to systems with positive-real transfer functions produces reduced models with positive-real transfer functions.*

PROOF. From the form of the partitioning, (11) and (12), likewise partitioning either K_c or K_o, it is clear that the reduced system,

Algorithm 3. **Bounded-Real TBR (BR-TBR)**

1. *Solve Eqns. (19)-(21) and their duals for X_c, X_o.*
2. *Proceed with steps 3-8 in Algorithm 1, with X_c for W_c and X_o for W_o.*

in the PR-balanced coordinates, satisfies

$$A_{11}\Sigma_1 + \Sigma_1 A_{11}^T = -K_1 K_1^T \quad (16)$$

$$\Sigma_1 C_1^T - B_1 = -K_1 J_c^T, \quad (17)$$

$$J_c J_c^T = D + D^T. \quad (18)$$

Therefore the reduced system satisfies the Lur'e equations with positive semi-definite Σ_1 ($\Sigma_1 \geq 0$ as $\Sigma \geq 0$). By the positive-real lemma, the reduced system is positive-real. □

We emphasize that Theorem 2 holds regardless of the internal form of the state-space system. Again, this is not true for congruence based procedures. Finally, we note that, just as for TBR, error bounds are available for PR-TBR (see [12] where a similar technique was proposed). A discussion of those bounds is beyond the scope of this paper, but we will state that in most cases we tried, the bounds obtained are competitive with those of TBR.

3.5 Bounded-Real Conditions

To obtain equivalent TBR procedures that guarantee a final transfer function that is bounded-real, useful when working with transfer functions representing S-parameters, we need the bounded real equations

$$AY_c + Y_c A^T = -BB^T - K_o K_o^T \quad (19)$$

$$Y_c C^T + BD = -K_o^T J_o \quad (20)$$

$$J_o J_o^T = I - D^T D \quad (21)$$

and corresponding dual equations that are satisfied with $Y_c \geq 0, Y_o \geq 0$ if the system transfer function is bounded-real. Algorithm 3 performs truncated balanced realization while guaranteeing the boundedness of the reduced transfer function [2].

3.6 A hybrid approach

In many cases, while not guaranteed by construction, it is often the case that the TBR approximants produced by Algorithm 1 turn out to be positive-real. Therefore we propose Algorithm 4, which performs the TBR procedure, solves the positive-real (or bounded-real) equations on the reduced model to check its passivity, and if it turns out not to be passive, discard it and proceeds to Algorithm 2 (or Algorithm 3). There is an advantage in this procedure as often the TBR approximates are more accurate for a given order than PR-TBR. Because of the cubic scaling of cost, it is relatively cheap, compared to the cost of the TBR reduction, to check a reduced model for passivity since the reduced system is presumably of lower order. As often the TBR models are passive, the net effect of the composite algorithm is to approximately double the cost in the worst case, versus usually getting better models at smaller cost (PR-TBR "costs" more than TBR) in the more-common average case.

Algorithm 4, which appropriately combines all of the previously presented algorithms, can be used as generic flow for generating accurate guaranteed passive reduced-order models of systems with arbitrary structure.

[2]The bound does not have to be unity; it can be any positive constant.

Algorithm 4. **Hybrid TBR**

1. *Perform Algorithm 1*
2. *Using the reduced model matrices $\tilde{A}, \tilde{B}, \tilde{C}$, solve Eqns. (13)-(15) for $\tilde{X}_c$ (or Eqns. (19)-(21)).*
3. **if** *Eqns. (13)-(15) (or Eqns. (19)-(21)) are solvable and $X_c > 0$, then terminate and return $\tilde{A}, \tilde{B}, \tilde{C}$.* **else** *discard TBR-reduced model and proceed with Algorithm 2 (or 3).*

4. COMPUTATIONAL CONSIDERATIONS

Solution of the Lur'e equations and solution of algebraic Riccati equations (AREs) are closely related. An overview of basic numerically robust computational procedures is given in [13]. Extensions of the positive-real lemma are available for models in the descriptor form where E is singular such that the transfer function cannot be put into the standard form [14]. To obtain a simpler procedure, consider the matrix pencil $\lambda\mathcal{E} - \mathcal{A}$,

$$\mathcal{E} = \begin{bmatrix} I & 0 & 0 \\ 0 & I & 0 \\ 0 & 0 & 0 \end{bmatrix}, \quad \mathcal{A} = \begin{bmatrix} A & 0 & B \\ 0 & -A^T & C^T \\ C & -B^T & D+D^T \end{bmatrix}. \quad (22)$$

Suppose that via some means we have computed an invariant subspace $Z \in \mathbb{R}^{(2n+p)\times n}$ that satisfies $\mathcal{E}Z\Lambda = \mathcal{A}Z$, $\Lambda \in \mathbb{R}^{n\times n}$, of the special form

$$Z = \begin{bmatrix} I \\ X \\ \tilde{X} \end{bmatrix} \quad (23)$$

where $I, X \in \mathbb{R}^{n\times n}, \tilde{X} \in \mathbb{R}^{p\times n}$. Then from the invariance condition $\mathcal{E}Z\Lambda = \mathcal{A}Z$, it can be verified that X is indeed the solution to Eqns. (13)-(15). To compute an invariant subspace of such a special form, the rank-p singularity of the pencil is first compressed [13] using a QR factorization of $\mathcal{A}$ to reduce the dimension of the pencil to $2n$. Then, we find an invariant subspace $Z_r \in \mathbb{R}^{2n\times n}$ (for example via the QZ method) of the form

$$Z_r = \begin{bmatrix} X_1 \\ X_2 \end{bmatrix}. \quad (24)$$

X can be computed as $X = X_2 X_1^{-1}$. For a positive-definite X, we need the subspace that corresponds to stable eigenvalues of the pencil.

Two more issues deserve our attention even though, for space limitations, we cannot discuss them in detail. The TBR technique in Algorithm 1 often leads to reduced models that present a mismatch in DC gains when compared to the original unreduced model, as the algorithm tends to perform better at high frequencies than at DC. Transformation to the *reciprocal system* [15], which maps $s = 0$ into $s = \infty$, can help produce better approximations at low frequencies, as can frequency weighting.

The second issue is related to computational cost. While the cost of all balancing procedures presented is in principle cubic (due to the need for solving Lyapunov-type equations), such cost is not overwhelming if the algorithm is being applied to a system that results from a prior reduction. Furthermore, even if that is not the case, one can still directly solve the large Lyapunov equations via a Krylov subspace method [6, 7], and obtain the reduced models directly from the reduced Lyapunov equation.

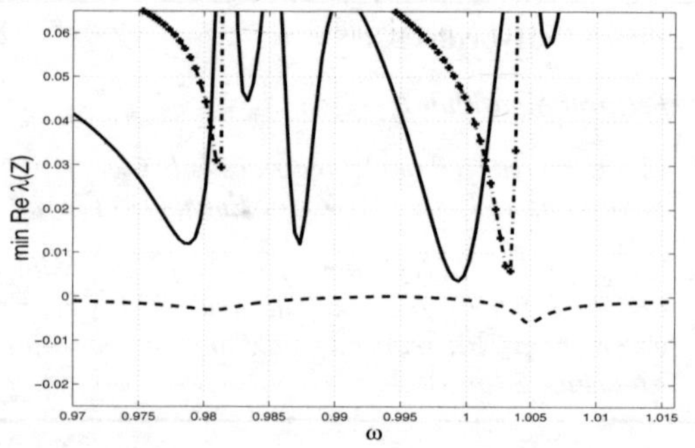

Figure 1: Minimum eigenvalue of transfer functions used to illustrate non-positive-real reduced model generated by standard TBR procedure. Solid line shows original (positive-real) order-26 transfer function, dashed line shows TBR result of order 7, dashed-dot line shows PR-TBR result of order 7. Note negative sign of TBR results indicating non-positive-realness.

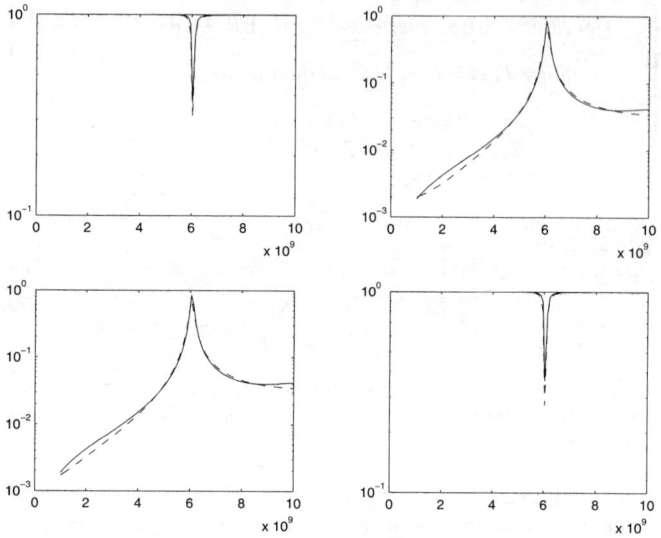

Figure 2: Magnitude of rational function fit and reduced model for S-parameters tabulated by full-wave field solver. Solid line shows initial data and order-42 rational fit (complete overlap). Dashed line shows order-6 reduced model obtained via BR-TBR.

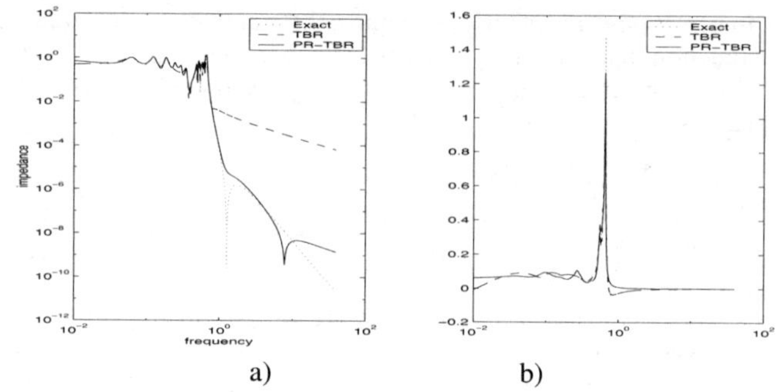

Figure 3: ***Left: Magnitude of Y_{12} for LC line. Right: Minimum eigenvalue of symmetric part of reduced model transfer function. Note that the minimum eigenvalue the TBR model drops below zero for some frequencies, indicating non-passivity.***

5. RESULTS

In this section we show examples that illustrate the relevance of the various algorithms presented in this paper.

5.1 A non-passive ROM generated by TBR

First we demonstrate empirically that standard TBR (Algorithm 1) can generate models that are not passive by examining a 26-state lumped circuit model of a crystal filter. We generated all the possible TBR models of orders 1-26, and used the positive real lemma to inspect them for positive-realness (equivalent to passivity in this case). Several of the models were found to be non-passive (see Figure 1). We then generated all the possible PR-TBR models. All were found to be positive-real as expected.

5.2 A symmetrizable system

Our first example is a spiral inductor modeled with the magnetoquasistatic electromagnetic tool FASTHENRY. This example appeared in [2]. The initial system of around 1500 states is reduced to an initial 60-state positive-real model using PRIMA. Since this order is still considered excessive, the model is then further reduced using TBR. In [2], it was commented that the reduced models after the TBR procedure appeared to be passive, but no explanation was given. Here we have rigorously checked, using the positive-real lemma, that the models were indeed passive, and gave a proof as to why that should be the case.

5.3 A bounded-real example from rational function fitting

In the next example we consider the bounded-real variant of the TBR procedure (BR-TBR). First, a rational fitting method was used to fit a high-order model to tabulated 2-port S-parameter data originating from a full-wave EM field solver. The fitting algorithm, which has provision for automatic estimation of model order, was tuned to a conservative setting, and generated an order-42 initial model that was nearly an exact fit to the data in the given frequency range. The resulting 42-state model was much larger than desired for final simulation, so the BR-TBR procedure was used to reduce the model to six states. The results are shown in Figure 2. The reduced model had norm bounded by unity, indicating that the reduced model represented a passive element. Several models of orders six to eight were also generated by both TBR and congruence transform strategies, but all had H_∞ norms ranging from 1.05 to 1.9, i.e. they were not passive.

5.4 A PEEC Connector

This example features a connector structure from Teradyne Inc. composed of 18 pins with a ground shield around and between the conductors. This structure was previously used [3] to illustrate a PEEC formulation based on PRIMA that generates passive reduced-order models. While the resulting model was indeed provably passive, disappoint reductions were reported, due mostly to the inability of the PRIMA algorithm to zero-in on the relevant modes of the system. In fact volume discretization of the interior of the conductors in order to properly model skin-effect leads to the appearance of various internal subsystems that have negligible effect in the structure impedance but which can fool the PRIMA algorithm. In [9] the same example was used to illustrate a two-step algorithm for RLC order reduction based on PRIMA followed by TBR, in an apparent attempt to solve the above problem. Significant order reductions were reported after the 2nd step of reduction as TBR is able to determine that those modes are not observable nor controllable. While this clearly shows that further reduction after the PRIMA stage is possible and indeed desirable, passivity was no longer guaranteed in the final, smaller models.

Here we have used the same example and checked the passivity of reduced-order models of various orders. We believe that the modes that are being discarded by TBR are related to the internal subsystems resulting from skin-effect modeling. As such the character of the problem after the initial PRIMA reduction is predominantly RL, a type of system for which we know that TBR is passive(see Section 3.2). Once more we generated all the possible TBR models for the system obtained after the PRIMA reduction and used the positive real lemma to inspect them for positive-realness (again equivalent to passivity in this case). Due to the *almost* symmetric nature of the systems, almost all the models we obtained were found to be passive. However, models of order 19 and 29 were found to be non-passive, a problem if the model is to be used in time-domain simulations. This example shows once more that TBR can indeed lead to large reductions in model-order but can produce non-physical models. The example also presents a strong case for using the generic flow presented earlier (see Section 3.6 and Algorithm 4). Since the majority of the TBR-produced models are likely to be passive it is advantageous to obtain such a model, check it for passivity and only compute the PR-TBR algorithm if that check fails (alternatively compute another model of slightly different order and check it, which is cheap since TBR essentially produces models of all orders simultaneously).

5.5 An RLC Line

For our next example we use a 40-segment uniform RLC line that is L-dominated. The values of the line were chosen to be $R = 25$, $C = L = 0.39894$. For the purpose of comparison we computed 25th order models using both TBR and PR-TBR. Figure 3-a) shows the low-frequency behavior of the exact line impedance as well as that obtained using the two models. For this particular case it turns out that PR-TBR performs much better than regular TBR in terms of the model error. More important, however is the result shown in Figure 3-b) where we plot the minimal eigenvalue of the symmetric part of the transfer function as a function of frequency. As can be seen from the plot, the minimal eigenvalue for the TBR model can go below zero at some frequencies which implies that the model is non-passive and may produce non-physical responses when used in time-domain simulations. In fact, on this example, *almost none* of the models produced by TBR were passive. Only very high order models exhibiting an almost exact match to the transfer function over the entire frequency axis were passive. In contrast, all the models produced by the PR-TBR method were found to be passive, as expected.

6. CONCLUSION

In this paper we presented a family of algorithms that can be used to compute guaranteed passive, reduced-order models of controllable accuracy for state-space systems with arbitrary internal structure.

The algorithms presented are similar to the well-known truncated balanced realization (TBR) techniques and share some of their advantages, such as computable error bounds. However, unlike standard TBR techniques, the algorithms presented have been shown to produce provably passive reduced-order models. In addition, unlike other techniques known to also produce passive reduction, the algorithms presented pose no constraints on the internal structure of the state-space. They are thus equally well applicable to systems that represent for instance Y or Z parameters as well as systems that represent S parameters. An hybrid algorithm was also presented where a TBR model is first computed, then checked for passivity and the passive-TBR algorithm is only used if that check fails.

We have experimented with our techniques in a large number of settings and have shown that they can be used as standalone procedures or as part of second step reductions for systems with a large number of unknowns, perhaps replacing the usual TBR procedure. We have thus applied our method to obtain reduced models of various structures, namely the two-port impedance of a crystal filter, a spiral inductors, a large connector and an RLC line. All models were found to be accurate and passive. All previously known techniques failed to produce acceptable models in some of the examples used.

7. REFERENCES

[1] Altan Odabasioglu, Mustafa Celik, and Lawrence Pileggi. Prima: Passive reduced-order interconnect macromodeling algorithm. In *International Conference on Computer Aided-Design*, pages 58–65, San Jose, California, November 1997.

[2] Mattan Kamon, Frank Wang, and Jacob White. Generating nearly optimally compact models from krylov-subspace based reduced-order models. *IEEE Transactions on Circuits and Systems II: Analog and Digital Signal Processing*, 47(4):239–248, April 2000.

[3] Nuno Marques, Mattan Kamon, Jacob White, and L. Miguel Silveira. A mixed nodal-mesh formulation for efficient extraction and passive reduced-order modeling of 3d interconnects. In 35th *ACM/IEEE Design Automation Conference*, pages 297–302, San Francisco, CA, USA, June 1998.

[4] Bruce Moore. Principal Component Analysis in Linear Systems: Controllability, Observability, and Model Reduction. *IEEE Transactions on Automatic Control*, AC-26(1):17–32, February 1981.

[5] Keith Glover. All optimal Hankel-norm approximations of linear multivariable systems and their $\mathbf{L}^{\infty}$-error bounds. *International Journal on Control*, 39(6):1115–1193, June 1984.

[6] Jing-Rebecca Li, F. Wang, and J. White. Efficient model reduction of interconnect via approximate system grammians. In *International Conference on Computer Aided-Design*, pages 380–383, SAN JOSE, CA, November 1999.

[7] I. M. Jaimoukha and E. M. Kasenally. Krylov subspsace methods for solving large Lyapunov equations. *SIAM Journal on Numerical Analysis*, 31:227–251, 1994.

[8] Carlos P. Coelho, Joel R. Phillips, and L. Miguel Silveira. A convex programming approach to positive real rational approximation. In *International Conference on Computer Aided-Design*, pages 245–251, San Jose, CA, November 2001.

[9] L. Miguel Silveira, Nuno Marques, Mattan Kamon, and Jacob White. Improving the efficiency of parasitic extraction and simulation of 3d interconnect models. In *ICECS'99 - IEEE International Conference on Electronics, Circuits and Systems*, pages 1729–1736, Pafos, Cyprus, September 1999.

[10] B. D. Anderson and S. Vongpanitlerd. *Network Analysis and Synthesis*. Prentice-Hall, Englewood Cliffs, New Jersey, 1973.

[11] A. J. Laub, M. T. Heath, C. C. Paige, and R. C. Ward. Computation of system balancing transformations and other applications of simultaneous diagonalization algorithms. *IEEE Transactions on Automatic Control*, 32:115–122, 1987.

[12] X. Chen and J. T. Wen. Positive realness preserving model reduction with h_∞ norm error bounds. *IEEE Transactions on Circuits and Systems I*, 42:23–29, 1995.

[13] W. F. Arnold and A. J. Laub. Generalized eigenproblem algorithms and software for algebraic Riccati equations. *Proc. IEEE*, 72:1746–1754, 1984.

[14] C. Oara and A. Varga. Computation of general inner-outer and spectral factorizations. *IEEE Transactions on Automatic Control*, 45:2307–2325, 2000.

[15] Y. Liu and B.D.O Anderson. Singular perturbation approximation of balanced systems. In *Proceedings of the 28th IEEE Conference on Decision and Control*, pages 1355–1360, 1989.

Uncertainty-Aware Circuit Optimization

Xiaoliang Bai
ECE Dept. UC San Diego
La Jolla, CA 92037
xibai@ece.ucsd.edu

Chandu Visweswariah
Philip N. Strenski
IBM T. J. Watson Research Center
Yorktown Heights, NY 10598
{chandu, strensk}@watson.ibm.com

David J. Hathaway
IBM Microelectronics
Essex Junction, VT 05452
dhathawa@us.ibm.com

ABSTRACT

Almost by definition, well-tuned digital circuits have a large number of equally critical paths, which form a so-called "wall" in the slack histogram. However, by the time the design has been through manufacturing, many uncertainties cause these carefully aligned delays to spread out. Inaccuracies in parasitic predictions, clock slew, model-to-hardware correlation, static timing assumptions and manufacturing variations all cause the performance to vary from prediction. Simple statistical principles tell us that the variation of the limiting slack is larger when the height of the wall is greater. Although the wall may be the optimum solution if the static timing predictions were perfect, in the presence of uncertainty in timing and manufacturing, it may no longer be the best choice. The application of formal mathematical optimization in transistor sizing increases the height of the wall, thus exacerbating the problem. There is also a practical matter that schematic restructuring downstream in the design methodology is easier to conceive when there are fewer equally critical paths. This paper describes a method that gives formal mathematical optimizers the incentive to avoid the wall of equally critical paths, while giving up as little as possible in nominal performance. Surprisingly, such a formulation *reduces* the degeneracy of the optimization problem and can render the optimizer more effective. This "uncertainty-aware" mode has been implemented and applied to several high-performance microprocessor macros. Numerical results are included.

Categories and Subject Descriptors

G.1.6 [**Numerical Analysis**]: Optimization—Constrained optimization

General Terms

Design, Algorithms, Reliability, Performance

Keywords

Uncertainty, circuit tuning, transistor sizing, nonlinear optimization, process variation, performance optimization

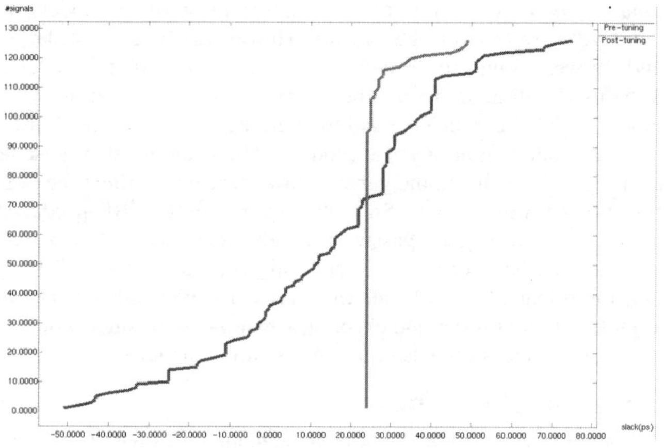

Figure 1: Pre- and post-tuning slack histogram.

1. INTRODUCTION AND MOTIVATION

In the quest for high performance, much research effort has gone into using optimization methods to solve the transistor sizing problem [2, 1, 10, 12, 4, 3]. It has long been speculated that aggressive optimization of a circuit drives the design into a corner of the process space, causing its yield to suffer. One approach to solve this problem is to model the random and systematic effects that impact yield and use them to estimate and/or maximize yield. This task is obviously an important but daunting one. Another approach is to change the design so that it is as insensitive as possible to variations and uncertainties. Instead of attacking the slack of just the most critical path or paths to the exclusion of all other considerations, what is required is an optimization strategy that targets the entire *distribution* of slacks.

A well-tuned circuit has a large number of equally critical paths, which manifests itself as a *wall* in the slack histogram. Fig. 1 shows a slack histogram of a S/390 microprocessor execution-unit macro before and after tuning. A point (x, y) in the histogram implies that there are y signals with a slack of x or worse. The process of optimizing the circuit typically involves stealing transistor width from the off-critical paths and speeding up the critical path. The obvious outcome is a large number of equally critical paths, as shown in Fig. 1. The slack of the circuit has improved from -51 ps to +24 ps in this example, but 95 signals are now equally critical.

Unfortunately, the basis upon which the circuit is optimized has its limitations. Static timing makes a number of simplifying assumptions. The prediction of parasitics may be less than perfect. Model-to-hardware correlation problems often arise. The design may undergo further downstream tweaking and restructuring. Clock skew may not be correctly taken into account. Unpredictable process variations may cause delays to change from their nominal values. In the face of all this uncertainty, having a wall of critical paths is undesirable. In [14], the authors list all the sources of uncertainty and advocate defining a "window of uncertainty" in the slack histogram. As long as the slack of any signal is in the window, there is no guarantee that the circuit will operate at the required frequency. Hashimoto et al. [11] go further and show that it is possible for careful optimization of a circuit to actually *degrade* the performance due to manufacturing variations. The impact of the height of the wall is demonstrated in a dramatic fashion in [11].

Consider an example of a mathematical optimizer tuning a circuit. To begin with, assume two paths are equally critical. As these paths are speeded up, their slacks become equal to the third critical path; now the optimizer must simultaneously work on *three* equally critical paths. As this process continues and optimality is approached, there are diminishing returns. Nonetheless the optimizer will not hesitate to make the number of equally critical paths larger even to gain a fraction of a pico-second. In fact, the less critical paths which are being down sized to provide device width (and perhaps reduced loading) for critical paths are *more* exposed to process variations because of small devices and large slews. Thus the quest for optimality in the limiting slack while excluding other considerations can be dangerous.

Without attempting to model the actual uncertainties, this paper proposes a method to avoid the wall of equally critical paths during optimization. A penalty is added to the objective function to give the optimizer an incentive to avoid a tall wall. The chosen form of the penalty has several good properties that are crucial to implementing a working solution to this problem. The resulting circuits have more appealing slack histograms, while paying a negligible price for the better slack distribution. In the face of uncertainty, the circuit with the better slack distribution offers both better performance and better insensitivity to variations. After optimization, if the required slack is still not met, restructuring is much easier when the number of equally critical paths is manageable. An interesting by-product of the new formulation is that the inherent degeneracy in the problem formulation is reduced, causing the optimizer to be more effective.

The outline of the rest of the paper is as follows. Section 2 presents the penalty function to implement uncertainty-aware tuning. Properties of the penalty function and choice of parameters are discussed. Section 3 extends this formulation to accommodate arrival time pruning. Numerical results and slack histograms are shown in Section 4. Alternative formulations of the optimization problem are discussed in Section 5 before the paper concludes with Section 6.

2. UNCERTAINTY-AWARE TUNING

2.1 Background

Assuming delay minimization, the optimization problem is formulated as

$$
\begin{array}{lrcll}
 & \text{minimize} & & z & \\
\text{s.t.} & z & \geq & AT_i + RAT_i & i \in PO \\
\text{s.t.} & AT_j & \geq & AT_i + d_{ij}(x, s_i) & j \in (IN \cup PO), \\
 & & & & i \in fanin(j) \\
\text{s.t.} & s_j & \geq & s_{ij}(x, s_i) & j \in (IN \cup PO), \\
 & & & & i \in fanin(j) \\
\text{s.t.} & \sum_{i \in T} A_i(x_i) & \leq & A_{target} & \\
\text{s.t.} & \sum_{j \in T} L_{ij}(x_j) & \leq & L_{target_i} & i \in PI \\
\text{s.t.} & \beta_{lower_i} & \leq & \beta_{smallest}(x) & i \in G \\
\text{s.t.} & \beta_{largest}(x) & \leq & \beta_{upper_i} & i \in G \\
\text{s.t.} & X_{lower_i} & \leq & x_i & i \in T \\
\text{s.t.} & x_i & \leq & X_{upper_i} & i \in T \\
\text{s.t.} & s_i & \leq & S_{upper_i} & i \in (IN \cup PO)
\end{array}
\tag{1}
$$

where the following notation is employed.

$$
\begin{array}{rcl}
z & = & \text{auxiliary variable that represents cycle time} \\
AT & = & \text{arrival time variables } \textit{after} \text{ launching} \\
 & & \text{clock edge} \\
RAT & = & \text{required arrival times at POs } \textit{before} \\
 & & \text{capturing clock edge}^1 \\
s & = & \text{slew variables} \\
x & = & \text{transistor widths} \\
d_{ij} & = & \text{nonlinear delays of timing graph edges} \\
s_{ij} & = & \text{nonlinear slews of timing graph edges} \\
A_i & = & \text{contribution of transistor } i \text{ to circuit area} \\
L_{ij} & = & \text{contribution of transistor } j \text{ to loading of PI } i \\
PO & = & \text{set of primary outputs} \\
PI & = & \text{set of primary inputs} \\
IN & = & \text{set of internal nodes} \\
T & = & \text{set of tunable transistors} \\
G & = & \text{set of tunable gates} \\
\beta_{smallest} & = & \text{smallest P to N path}-\text{strength ratio} \\
\beta_{largest} & = & \text{largest P to N path}-\text{strength ratio} \\
A_{target} & = & \text{area target} \\
L_{target} & = & \text{input loading limits on PIs} \\
\beta_{lower} & = & \text{lower bounds on } \beta \text{ ratio} \\
\beta_{upper} & = & \text{upper bounds on } \beta \text{ ratio} \\
X_{lower} & = & \text{lower bounds on } x \\
X_{upper} & = & \text{upper bounds on } x \\
S_{upper} & = & \text{upper bounds on slews.}
\end{array}
\tag{2}
$$

The above formulation has some interesting properties [4]. Note that the arrival times and slews are variables of the problem. The delays and slews of each arc, the β ratios, the input loading and the contribution of each transistor's width to the total area are the nonlinear elements of the formulation. At the solution, the arrival times and slews on the critical path will have the "timing-correct" value, all others may not (they can take one of several equally correct values). Thus off-critical delay and slew constraints and variables are often redundant or degenerate. The formulation as shown above assumes downstream propagation of the largest slew of any input edge incident on a timing point, i.e., what is commonly referred to as "max-slew propagation."

The above formulation typically creates the situation where a tuned circuit has a large number of equally critical signals, as shown in the post-tuning slack histogram of Fig. 1. The reason is that the optimizer will strive to decrease the effective cycle time z until no further improvement is possible, thus causing a large number of paths to be equally critical.

2.2 Penalty function

To prevent this situation, we propose addition of a penalty term corresponding to each primary output. As the arrival time of the signal approaches criticality, we would like this penalty to grow in size. In delay minimization, we unfortunately cannot judge criticality in an absolute sense, since the effective cycle time (z) of the circuit changes as the optimization progresses. Hence, we formulate the penalty for each primary output arrival time in relation to the effective cycle time z. Define a *separation*

$$d_i = z - AT_i - RAT_i \; i \in 1, 2, \cdots, N \tag{3}$$

where N is the cardinality of the set PO. This separation represents the criticality of each primary output as the optimization evolves, and is relative to the most critical path at any stage of the optimization. We would like this separation to be as large as possible for all primary outputs. Of course, at least one of the primary outputs will define the effective cycle time, and the separation for that one will be zero.

[1] By convention, required arrival times are either expressed *before* a capturing clock edge or *after* a launching clock edge.

The basic idea is to change the objective function from z to

$$z + k\sum_{i=1}^{N} P(d_i) \tag{4}$$

where k is a weighting constant and P a suitably chosen penalty function such that the optimizer strives to balance reduction in z with increase in desirable separation.

2.3 Choice of penalty function

There are several requirements on the penalty function. First, since it will be used in a numerical optimizer, it is required to be continuous and smooth. Second, the penalty should be a monotonically decreasing function of d_i, reducing to a negligibly small value after a "required" separation is achieved. Third, its value should be well defined at both zero and negative values of separation (this rules out, for example $P = 1/d_i$). Finally, we would like to control the rate at which the penalty function drops off as the separation increases, since this will provide a knob to adjust for expected amounts of uncertainty.

A simple choice that satisfies all these criteria is

$$P(d_i) = e^{-d_i/\sigma} \tag{5}$$

leading to an objective function which is

$$z + k\sum_{i=1}^{N} e^{-(z-AT_i-RAT_i)/\sigma}. \tag{6}$$

Of course, several other choices of the penalty function are possible, but this simple choice works well in practice. The penalty starts at k for the most critical primary output and reduces by a factor e for each σ of separation that a primary output obtains relative to the most critical path. Beyond about 3σ or 4σ, relatively speaking, further separation is no longer valuable.

2.4 Choice of k and σ

The purpose of the penalty function is to introduce *downward* pressure on all primary output arrival times relative to z. Unfortunately, this results in *upward* pressure on z. Thus the parameters of the penalty function must be chosen in such a way as to make sure that there is overall downward pressure on z. In other words, the optimizer should not artificially increase z to be higher than the actual cycle time of the circuit just to obtain a reduction in the separation penalty terms. We term this situation "lift-off" of z.

One (conservative) way to ensure this criterion is to make sure the gradient of the penalty function with respect to z is always positive. Differentiating (6) with respect to z, we obtain

$$1 - \frac{k}{\sigma}\sum_{i=1}^{N} e^{-d_i/\sigma} > 0. \tag{7}$$

The biggest possible value of the summation is N (*all* primary outputs equally critical) and the smallest value is 1 (one primary output critical, all others separated by a lot). Thus we will be safe provided

$$1 - \frac{kN}{\sigma} > 0 \text{ or } k < \frac{\sigma}{N}. \tag{8}$$

Therefore if k is less than σ/N, lift-off is impossible. If it is more than σ, lift-off is guaranteed. In between these two extremal values, the behavior depends on the relative arrival times in a particular test case.

For a circuit with 50 primary outputs, considering both rising and falling arrival times, we have 100 timing points. Thus, if we choose $\sigma = 10$ ps, k should be less than 0.1. In practice, k can be chosen slightly larger without deleterious effects. However, satisfying (8) guarantees us that z will not lift off.

It turns out that lift-off is not really harmful since as soon as z artificially increases, the separation term for the most critical signal is smaller than k and all the other separation terms also decrese, so the gradient of the objective function with respect to z quickly increases till z will not increase further. Suppose the amount of lift-off in z at optimality is l, then the objective function can be rewritten as

$$l + z' + k'\sum_{i=1}^{N} e^{-(z'-AT_i-RAT_i)/\sigma} \tag{9}$$

where $z' = z - l$ and $k' = ke^{-l/\sigma}$ and therefore we obtain a solution to a different problem with a smaller k but no lift-off.

2.5 Area minimization and tradeoff modes

The formulation in (1) can easily be modified to create a tradeoff mode which minimizes a weighted sum of area and effective cycle time or an area mode which minimizes area subject to timing constraints. In the tradeoff mode, the area constraint is just moved into the objective function with an appropriate weight factor. In area mode, area is the objective function. In this case, z is replaced by the actual desired cycle time of the circuit being optimized (optionally offset by a desired positive slack), and therefore is a constant. Nonetheless, all of the above analysis and the choice of the penalty function for these two additional modes are unchanged.

3. ARRIVAL TIME PRUNING

As was mentioned earlier, the formulation in (1) suffers from a high degree of redundancy and degeneracy. Off-critical arrival time and slew constraints can be tight, yet have zero Lagrange multipliers since they do not influence the objective function. One way in which this problem is addressed is by arrival time pruning [15]. Arrival time pruning results in a mixed path/block formulation. In fact, for small circuits and "non-bushy" circuits, *every* arrival time variable is often pruned, leading to a completely path-oriented formulation. From an uncertainty-awareness point of view, this pruning has benefits, but raises new problems.

The benefit is that the "separation" is not just forced at the primary outputs, but at the source of all sub-paths from unpruned internal arrival times to the primary outputs. The disadvantage is that the separation variable d_i of (3) can no longer be defined in a straightforward manner since the arrival time at the primary output might have been eliminated due to pruning. Indeed, a typical primary output arrival time constraint will be of the form

$$z \geq AT_1 + d_{1,2} + d_{2,3} + \cdots + d_{i-1,i} + RAT_i \tag{10}$$

where the d values are the delays along the sub-path from the internal node 1 to the primary output i. Hence the separation variable for this minimax constraint is

$$d = z - AT_1 - d_{1,2} - d_{2,3} - \cdots - d_{i-1,i} - RAT_i \tag{11}$$

which is not only nonlinear unlike (3), but a function of a large number of variables. This does not lend itself to a penalty function as described in the previous section.

We make the observation that most optimizers convert inequalities to equalities by the addition of a slack variable. Although this is not to be confused with slack as defined by a static timing analysis tool, the conceptual meaning of the two slacks is similar. Thus (10) would become

$$\begin{aligned} &z - AT_1 - d_{1,2} - d_{2,3} - \cdots - d_{i-1,i} - RAT_i - u = 0 \\ &u \geq 0 \end{aligned} \tag{12}$$

where u is an auxiliary slack variable introduced by the optimizer. In this case, we can define the penalty function in terms of this slack variable as $P(u)$. The procedure would be simply to keep track of the slack variables of all the minimax constraints (of the form $z \geq$ something) and then add a penalty to the objective function which is $k\sum_{i=1}^{N} e^{-u_i/\sigma}$ where N is the number of minimax constraints. Since

u_i is always positive, each penalty term is at most k when $u_i = 0$ and then decays quickly as the slack variable increases in value. The key difference is that the number of penalty terms in this method can be much larger than the method of the previous section, so the same care must be taken in the choice of k and σ to make sure that there is eventually sufficient downward pressure on z. Also note that *two-step updating* [8] which takes advantage of the fact that slack variables appear in the merit function in a particular analytic form is no longer applicable to the slack variables that serve this second purpose as separation variables.

With this new formulation, there is downward pressure on the arrival times in *every* subpath feeding a primary output, since the critical ones need to reduce to allow reduction of z, and the others need to decrease to allow increased separation. The downward pressure is transitively passed along the worst path feeding each of these subpaths, thus reducing the arrival time and slew degeneracy of the formulation.

4. RESULTS

The methods described in both the previous sections have been implemented in a circuit optimization package called EinsTuner, which has been applied to several generations of PowerPC and S/390 microprocessors. The main components of EinsTuner are a static, transistor-level timer EinsTLT [13], a fast time-domain simulator and gradient calculator SPECS [16, 9] and a general-purpose nonlinear optimization package LANCELOT [7, 5, 6]. The problem is formulated as in (1), including arrival time pruning [15]. The delays and slews of individual logic blocks are computed by running a time-domain simulation under the covers; gradients with respect to transistor widths, input slews and output loads are obtained by the adjoint method. The resulting optimization problems can be very large and the resulting run time quite long. Recently, a circuit with 47,000 transistors was tuned. The optimization problem had 140,000 variables and 110,000 constraints and took over 4 days of CPU time to solve.

The straightforward implementation of uncertainty-aware tuning in EinsTuner was achieved by modifying the pruning algorithm to leave the primary outputs unpruned and adding a penalty function as in (4). The *group function* facility of LANCELOT [7] was exploited to express the penalty function. This method will be referred to as method 1.

In the second method, all arrival times were allowed to be pruned. The slack variables of the resulting minimax constraints were used as separation variables to form the penalty function, again expressed as a group function. Minimax constraints were submitted to the optimizer as equality constraints and the slack variables and their bounds were explicitly managed by EinsTuner. Thus LANCELOT did not know that these were indeed slack variables, and did not attempt to apply two-step updating to them. This method will be referred to as method 2.

This section presents three different types of results. First, slack histograms are shown to demonstrate the avoidance of the "wall," and the effect of varying k and σ are shown. Second, the performance of a circuit with and without uncertainty-aware tuning is compared in the face of variations. Third, numerical results are presented which show that uncertainty-aware tuning can be accomplished with little to no loss of circuit performance.

4.1 Slack histograms

A 38-bit PowerPC incrementer was tuned with and without uncertainty-aware tuning. This is a difficult circuit on which to achieve separation because of the symmetry and regularity of the critical paths. The resulting slack histograms are shown in Fig. 2. The slack histograms in this section were produced using method 1 of uncertainty awareness; method 2 results look qualitatively similar. Although the uncertainty-aware tuning has a limiting slack that is 8 ps worse than the nominally tuned circuit, the height of the wall has been reduced in more than half.

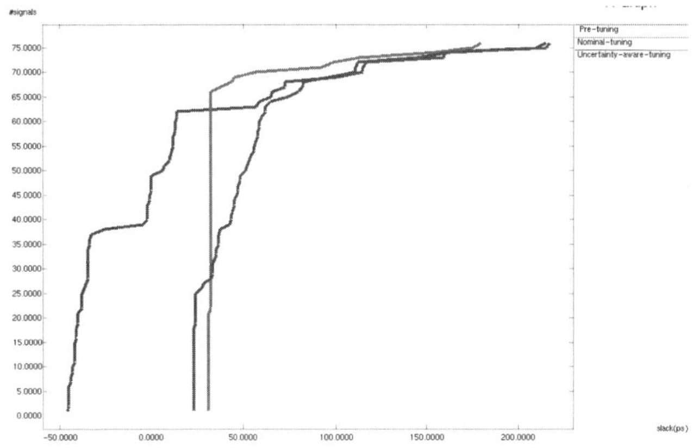

Figure 2: Slack histogram for nominal and uncertainty-aware tuning.

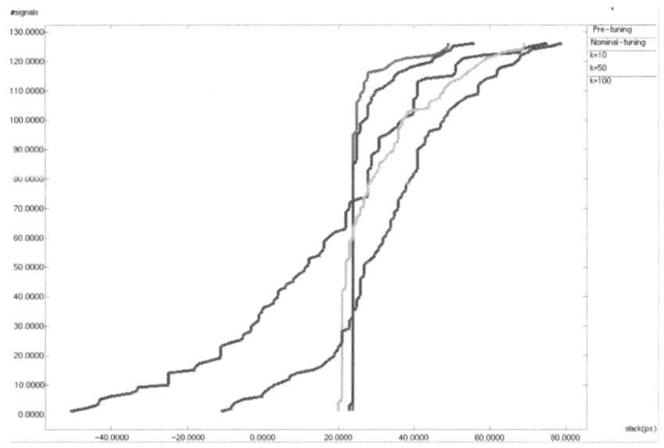

Figure 3: Effect of varying k.

As k and σ are changed, the amount of separation can be controlled, but at progressively higher cost in the limiting slack. Fig. 3 shows the effect of varying k at a fixed σ and slack histograms obtained by varying σ while holding k fixed are shown in Fig. 4. These experiments were performed on the same S/390 execution-unit macro as Fig. 1.

4.2 Degradation due to variations

It is clear that "separation" can be obtained at a relatively small cost in the limiting slack. In many cases, there is almost no price to pay in terms of the limiting slack, as will be shown in the subsequent section. But is the uncertainty-aware circuit really better? A simple Monte Carlo experiment was fashioned to gain insight into this question. Ideally, the nominally tuned circuit and uncertainty-aware circuit would be manufactured and the parametric yields of the two chips compared. Obviously, such an experiment is impractical.

Instead, each point in the slack histograms of the 38-bit PowerPC incrementer (Fig. 2) was *randomly perturbed* according to a 6σ-truncated Gaussian distribution of the form

$$sample = \mu + \sigma\sqrt{2}g(K(2r-1)) \quad (13)$$

where μ is the mean of the distribution (the mean arrival time was left unchanged), σ^2 is the variance, g is the inverse *erf* function, K is $erf(3/\sqrt{2})$ and r is a random number between 0 and 1. In other words, the arrival time of each primary output was assumed to have a truncated Gaussian distribution with no correlation to other primary outputs. While this is clearly untrue in real life, the crude experiment serves to capture the intent of uncertainty-aware tuning. For the nominally tuned circuit, 10,000 circuits were "manufac-

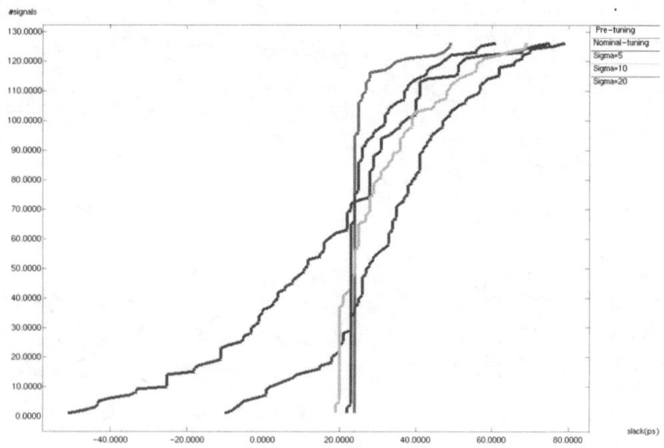

Figure 4: Effect of varying σ.

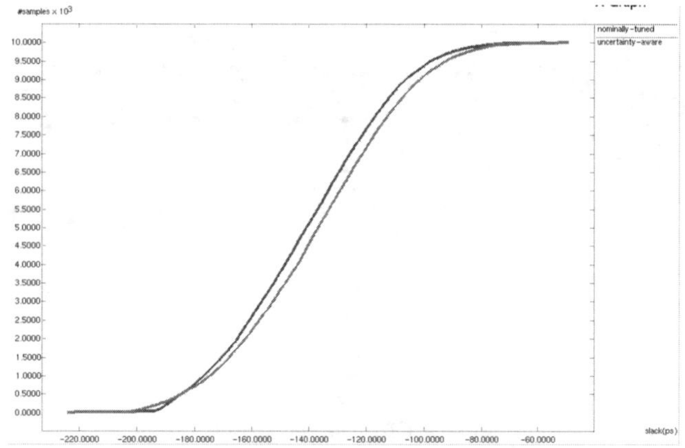

Figure 5: Estimated slack histograms after random perturbation for 38-bit PowerPC incrementer.

tured" by means of a Monte Carlo analysis whereby each primary output arrival time was randomly perturbed. The resulting circuits were "sorted" in order of increasing performance and the resulting distribution in the performance of the chips is shown in Fig. 5. The same analysis was repeated for the uncertainty-aware circuit, with the same random number seeds for each new "chip." The results are superposed in Fig. 5. The perturbation caused both results to perform worse than the nominal circuit, which is the well-known behavior of the max function in the face of variations [11]. Nonetheless, the reduction of the steepness of the wall in the uncertainty-aware circuit made it the better circuit in the face of variations. It can be seen that for a given performance, the uncertainty-aware circuit has a higher parametric yield, and for a given parametric yield, the uncertainty-aware circuit allows a higher ship frequency. Even when the magnitude of the yield improvement is small, it can make a huge difference in reducing costs and enhancing profitability.

The same experiment was repeated on the execution unit circuit of Fig. 1 and the resulting performance distributions are shown in Fig. 6. Thus even for these regular datapath circuits, benefits can be realized from separation.

4.3 Numerical results

A set of 39 test cases ranging in size from 6 to 13,046 transistors were tuned in three different ways: nominal tuning in delay minimization mode, uncertainty-aware tuning method 1 and uncertainty-aware tuning method 2. All other constraints such as the area limit, β ratio constraints, slew limits and input loading

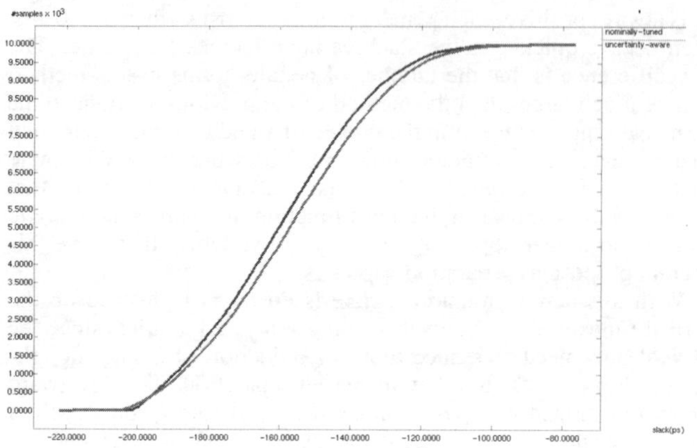

Figure 6: Estimated slack histograms after random perturbation for S/390 execution unit macro.

Table 1: Comparison of nominal and uncertainty-aware optimization.

Test case	# of FETs	Slack (ps)			
		Pre	Nominal	Mth.1	Mth.2
ioaareg_3t2n	22	-2.5	7.3	6.9	6.4
inv3	6	720.0	758.0	759.0	759.0
ibmTechGates	22	83.3	132.0	132.0	131.5
c17	28	683.0	738.0	738.0	738.0
iiiff	26	-161.0	-62.4	-61.1	-60.1
a3_3	34	634.0	718.0	718.0	718.0
andy_graph	34	634.0	651.0	654.0	650.0
full_adder	46	632.0	687.0	687.0	687.0
iqia	72	645.0	668.0	668.0	668.0
tgmux	62	-106.5	-65.9	-66.1	-66.4
alu1b	102	583.0	618.0	617.0	618.0
gain_experiment	162	-118.5	-41.4	-41.4	-41.6
rtt_incr321	716	-129.1	-41.4	-46.5	-48.0
ibbx_compare	824	514.0	554.0	554.0	554.0
exdmsk	882	609.0	683.0	680.0	680.0
ioaareg_dl	772	-22.1	39.1	38.0	36.1
c8	584	432.0	590.0	588.0	587.0
exdcgn_mac	1270	-16.5	30.3	31.1	30.8
epdx3p_mac	802	-54.5	-0.8	0.3	1.1
ifti_inc38	880	454.0	530.0	530.0	529.0
vadd_byte	1428	-77.7	6.2	7.3	7.7
epdlzd_mac	1346	-30.6	101.6	100.9	99.7
rtincrp	1725	-221.9	-115.9	-118.5	-117.5
exda1mx_mac	1792	-0.5	40.3	40.2	40.2
incrementer	1554	447.0	511.0	507.0	509.0
ififbreg_ge56	1400	5.0	424.0	422.0	417.0
exdcdr_mac	1856	-407.9	-325.7	-329.2	-320.1
rtbg58cmp	1500	-42.1	29.9	30.3	29.4
ioperdf_cmpr_al64p	2256	-12.5	164.9	165.6	164.8
idopcmp_mac	3448	10.4	50.7	52.5	51.0
idagiqu_smry_gr	2332	-0.5	33.7	33.2	33.0
exdadd_mac2	4258	-388.3	56.3	58.1	57.0
exdadd_mac	4258	-265.8	-85.9	-86.6	-87.4
iooacrd_mac	5184	-200.7	-200.1	-200.1	-200.1
epdinc_mac	7199	-46.2	83.9	84.6	84.4
rtoutmx_mac	7805	-95.1	-10.9	-5.3	0.7
exdblu_mac	8054	-619.6	-504.5	-500.5	-504.2
idfast_mac	8836	-480.7	-473.7	-473.5	-479.7
ibbhtdf_rlm	13046	11.7	62.6	60.7	64.8

constraints were maintained the same in all three cases. Table 1 shows the results. In most cases, the loss of performance due to uncertainty-aware tuning is very small. As discussed previously, the new method eliminates some redundancy and degeneracy in the formulation, causing the optimizer to be more effective as can be seen in some test cases, particularly the large ones. Thus added slack separation can be obtained at little or no performance cost.

5. ALTERNATIVE FORMULATIONS

This section proposes two different formulations to make the uncertainty awareness more effective. It can be argued that the objective of providing variation tolerance is only important at the primary outputs and latch points of a circuit. But we can only achieve this if we have some separation, or positive timing slack, on all internal points in the fan-in cone of each primary output or latch point. And if we assume that all delay edges in the design are equally subject to variation, we can achieve maximum variation tolerance by ensuring that we have separation at the maximum number of internal delay edges, where separation on a delay edge from i to j means that $AT_i + d_{ij} < RAT_j$. Here, RAT_j is defined relative to the clock launching edge, unlike the primary output required arrival time assertions which are defined as times before the capturing clock edge. In method 1 described above, we require separation only on the worst path feeding each primary output. By reusing the slack variables of the minimax constraints in method 2, we are able to extend this separation pressure to all sub-paths terminating on a primary output. This section discusses methods to give the optimizer incentive to encourage separation at *all* timing points of a circuit.

One way to achieve this goal is to introduce a required arrival time variable (*RAT*) at each timing point. These required arrival times are defined relative to the launching clock edge, as above. The *RAT* variables form a parallel set of constraints that are otherwise identical to those of the arrival time variables. However, these constraints are "anchored" by the z variable rather than by the arrival time assertions at the primary inputs (no constraints are introduced relating the *RAT* variables to the *AT* assertions at the primary inputs). Thus the *AT* variables are constrained to grow "up" from the primary input *AT*s, while the *RAT* variables are constrained to grow "down" from the z variable. The difference between the *RAT* and *AT* variables at a timing point is the timing slack, and by treating this as a separation variable and adding a corresponding penalty term to the objective function, we can ensure that downward pressure is applied to every *AT* variable and upward pressure is applied to every *RAT* variable. Along the critical path, the separation will be zero, and on every off-critical path, the separation will correspond to the positive slack that a static timer would predict, relative to the most critical path. The only drawback is that this method would lead to a substantial increase in the number of variables and constraints.

Another way of achieving the same goal is to define a separation for every unpruned arrival time in the circuit relative either to z or to some constant arrival time that is chosen to be larger than z. Then a penalty term can be introduced to ensure downward pressure on each arrival time such that wherever it is possible to achieve separation, the optimizer will have incentive to do so.

An advantage of the former (*RAT*-variable) approach using an exponential or similarly decaying penalty term is that the maximum pressure is applied to the timing points with the smallest timing slack, and thus the optimizer is given more incentive to increase the slack of the worst paths. In contrast, if we apply a separation penalty relative to a single common value, a decaying penalty function is inappropriate, since the natural separation of each *AT* variable from this common value will vary significantly, leading to uneven separation pressures. Thus a linear penalty term is used (e.g., a simple sum of all AT variables in the problem), and the optimizer may increase the separation at a number of points with already large positive timing slacks at the cost of more valuable separation at a smaller number of nearly-timing-critical points. An advantage for the optimizer of both of these approaches is that they completely eliminate degeneracy in the timing and slew constraints.

6. CONCLUSIONS AND FUTURE WORK

Mathematical optimization of digital circuits causes a "wall" of equally critical paths, which is unsuitable to downstream restructuring in the design methodology and leaves a circuit extremely vulnerable to manufacturing variations. This paper proposed and demonstrated a simple yet effective method of avoiding the wall at little or no cost in circuit performance. In fact, the new method makes the optimizer more effective by eliminating degeneracy in the problem. The method was extended to the situations when arrival times are pruned by making the slack variables of the optimizer play a dual role as separation variables. Detailed numerical results on high-performance microprocessor macros were presented to show the effectiveness of the method.

7. ACKNOWLEDGMENTS

The authors gratefully acknowledge discussions with Sani Nassif, Jim Warnock and Jochen Jess, and the help of members of the EinsTuner team, particularly Jun Zhou, Mei-Ting Hsu, Cindy Washburn and Andy Conn.

8. REFERENCES

[1] R. K. Brayton, G. D. Hachtel, and A. L. Sangiovanni-Vincentelli. A survey of optimization techniques for integrated-circuit design. *Proceedings of the IEEE*, 69(10):1334–1362, October 1981.

[2] R. K. Brayton and R. Spence. *Sensitivity and optimization*, volume 2 of *CAD of Electronic Circuits*. Elsevier Scientific Publishing Co., Amsterdam, The Netherlands, 1980.

[3] C.-P. Chen, C. N. Chu, and D. F. Wong. Fast and exact simultaneous gate and wire sizing by Lagrangian Relaxation. *IEEE Transactions on Computer-Aided Design of ICs and Systems*, 18(7), July 1999.

[4] A. R. Conn, I. M. Elfadel, W. W. Molzen, Jr., P. R. O'Brien, P. N. Strenski, C. Visweswariah, and C. B. Whan. Gradient-based optimization of custom circuits using a static-timing formulation. *Proc. 1999 Design Automation Conference*, pages 452–459, June 1999.

[5] A. R. Conn, N. I. M. Gould, and P. L. Toint. Global convergence of a class of trust region algorithms for optimization with simple bounds. *SIAM Journal on Numerical Analysis*, 25:433–460, 1988. See also same journal, pp. 764–767, volume 26, 1989.

[6] A. R. Conn, N. I. M. Gould, and P. L. Toint. A globally convergent augmented Lagrangian algorithm for optimization with general constraints and simple bounds. *SIAM Journal on Numerical Analysis*, 28(2):545–572, 1991.

[7] A. R. Conn, N. I. M. Gould, and P. L. Toint. *LANCELOT: A Fortran Package for Large-Scale Nonlinear Optimization (Release A)*. Springer Verlag, 1992.

[8] A. R. Conn, L. N. Vicente, and C. Visweswariah. Two-step algorithms for nonlinear optimization with structured applications. *SIAM Journal on Optimization*, 9(4):924–947, September 1999.

[9] P. Feldmann, T. V. Nguyen, S. W. Director, and R. A. Rohrer. Sensitivity computation in piecewise approximate circuit simulation. *IEEE Transactions on Computer-Aided Design of ICs and Systems*, 10(2):171–183, February 1991.

[10] J. P. Fishburn and A. E. Dunlop. TILOS: A posynomial programming approach to transistor sizing. *IEEE International Conference on Computer-Aided Design*, pages 326–328, November 1985.

[11] M. Hashimoto and H. Onodera. Increase in delay uncertainty by performance optimization. *Proc. IEEE International Symposium on Circuits and Systems (ISCAS)*, pages 379–382, 2001.

[12] W. Nye, D. C. Riley, A. Sangiovanni-Vincentelli, and A. L. Tits. DELIGHT.SPICE: An optimization-based system for the design of integrated circuits. *IEEE Transactions on Computer-Aided Design of ICs and Systems*, CAD-7(4):501–519, April 1988.

[13] V. B. Rao, J. P. Soreff, T. B. Brodnax, and R. E. Mains. EinsTLT: transistor-level timing with EinsTimer. *Proc. TAU*, December 1999.

[14] S. E. Rich, M. J. Parker, and J. Schwartz. Reducing the frequency gap between ASIC and custom designs: a custom perspective. *Proc. 2001 Design Automation Conference*, pages 432–437, June 2001.

[15] C. Visweswariah and A. R. Conn. Formulation of static circuit optimization with reduced size, degeneracy and redundancy by timing graph manipulation. *IEEE International Conference on Computer-Aided Design*, pages 244–251, November 1999.

[16] C. Visweswariah and R. A. Rohrer. Piecewise approximate circuit simulation. *IEEE Transactions on Computer-Aided Design of ICs and Systems*, 10(7):861–870, July 1991.

Congestion-driven Codesign of Power and Signal Networks *

Haihua Su † Jiang Hu † Sachin S. Sapatnekar * Sani R. Nassif †

† IBM Corp.
11501 Burnet Rd.
Austin, TX 78758
{haihua,jianghu,nassif}@us.ibm.com

* ECE Dept, Univ. of Minnesota
200 Union St. SE
Minneapolis, MN 55455
sachin@ece.umn.edu

ABSTRACT

We present a global wire design methodology that simultaneously considers the performance needs for both signal lines and power grids under congestion considerations. An iterative procedure is employed in which the global routing is performed according to a congestion map that includes the resource utilization of the power grid, followed by a step in which the power grid is adjusted to relax the congestion in crowded regions. This adjustment is in the form of wire removal in noncritical regions, followed by a wire sizing step that overcomes the effects of wire removal. Experimental results show that the overall routability can be significantly improved while the power grid noise is maintained within the voltage droop constraint.

Categories and Subject Descriptors

B.8.2 [**Performance and Reliability**]: Performance Analysis and Design Aids

General Terms

Algorithms

Keywords

wire congestion, codesign, signal routing, power grid noise

1. INTRODUCTION

The role of interconnect has become increasingly critical in nanometer design and the need to meet stringent performance constraints has resulted in strong contention for scant routing resources. A major consumer of these resources is the power distribution network, which contains dense grids. On the other hand, global wires also compete for the same routing resources, as they often require shortest-path routes to meet their own performance requirements. Traditionally, these two have been designed independently, with the routing needs for a regular power grid being determined first, after which the remaining resources are calculated to provide routing resource budgets for the signal nets.

As the number and criticality of global signal wires becomes more dominant, such a methodology becomes unsustainable as the initial budgets may often be entirely unreasonable. Therefore, in nanometer design, there is a strong need for a unified approach to the design of signal wires and power grids, with an integrated approach to routing resource management.

While it is convenient to build a regular power grid with a constant pitch (defined as the distance between adjacent wires in the grid), some degrees of freedom exist and it is desirable that they be exploited. For instance, in regions where the demand for routing resources from signal nets is high, a sparser power grid may be used as long as the performance constraints on the supply and ground lines can be met; likewise, signal nets are well advised to avoid the hot spots of the chip if possible, since these may need a locally dense power grid.

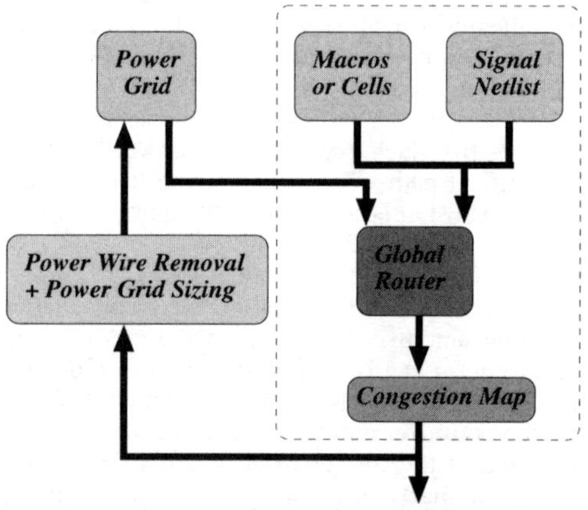

Figure 1: Congestion-driven power grid design and global routing.

The idea of managing wire congestion in signal routing has long been a significant objective in global routing. Various congestion-driven techniques include sequential routing (e.g., [6]), rip-up-and-reroute (e.g., [20]), and multicommod-

*This work was supported in part by the NSF under award CCR-0098117 and by the SRC under contract 99-TJ-714.

ity flow based methods (e.g., [16]) have been proposed. Most of these techniques aim at solving the problem solely at the routing stage, assuming that the total routing resources are fixed. Recent publications [1, 11] have presented techniques for simultaneous global routing and resource allocation under performance constraints.

Power grid optimization techniques have also been studied in [17, 19, 21]. All of these aim at minimizing total power wire area subject to the voltage droop and/or electromigration constraints, formulating the problem as a nonlinear program. The work in [12] presents a technique for shield insertion in a predesigned power grid to control inductive effects.

To the best of our knowledge, no published work performs a concurrent optimization of the power grid along with signal wires under routing congestion constraints, and this is the subject of the work presented in this paper. Our proposed congestion-driven flow is illustrated in Fig. 1. The dashed rectangle corresponds to a more conventional global routing flow. In essence, our approach presents a new flow that adds a feedback loop that permits the readjustment of the signal routing budgets by altering the power grid appropriately. Our approach incorporates a tight coupling between power grid adjustments and the routing of signal wires to exploit the altered congestions that result from these adjustments, and aims to solve problems with severe congestion constraints where conventional techniques are inadequate.

2. POWER GRID-AWARE SIGNAL ROUTING

As in global routing, we tessellate the entire chip into an array of grid cells, as shown in Fig. 2(a), and use the wiring information across the boundaries between neighboring grid cells to estimate the signal wire congestion distributions. We denote the width, in μm, of a boundary b between two neighboring grid cells as $W(b)$. This width represents the limited resources that must be shared on each layer by the supply lines and the signal lines that traverse the boundary, as shown in Fig. 2(b). In other words, the number of wires crossing b is inherently limited by the width $W(b)$, and $W(b)$ may partly or wholly be occupied by the crossing wires.

We represent the total width occupied by power grid wires on boundary b as $P(b)$. If a power grid wire p_i has a track width of $w(p_i)$, which includes its wire width and the required spacing from an adjacent wire, and there are a set of such wires, $p_1, p_2, ...p_m$, that cross b, then $P(b) = \sum_{i=1}^{m} w(p_i)$ represents the space that is unavailable for signal wires to cross the boundary. Therefore, we subtract this quantity from the boundary width to obtain the space available for signal wires as $W(b) - P(b)$. Typically, a uniform track width $\bar{w}$ is applied to all the signal wires in the congestion estimation or global routing stage. Hence, signal wire congestion is often expressed in terms of the number of wiring tracks and the number of tracks available for signal wires is $T(b) = \lfloor \frac{W(b)-P(b)}{\bar{w}} \rfloor$. If there are $S(b)$ signal wires that cross a boundary b, then the overflow on b is $max(0, S(b) - T(b))$. All tile boundaries with positive overflow values form a congestion map for a chip. The wire density at b is represented as $S(b)/T(b)$, measuring the congestion of b. A common objective for global routing is to ensure that there is no boundary with the wire density greater than one, i.e., $S(b) \leq T(b)$.

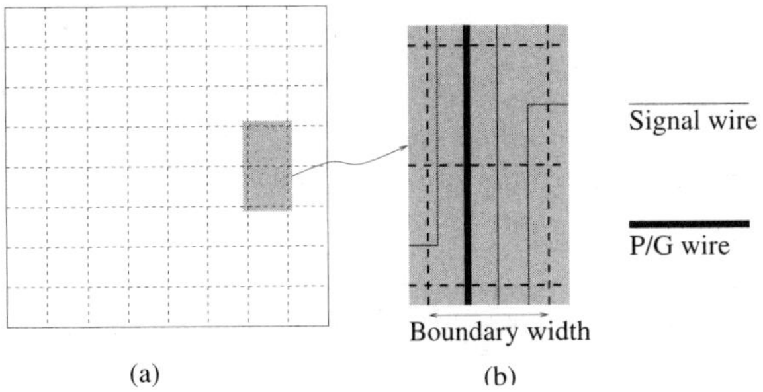

Figure 2: Wire congestion estimation based on tessellation on a chip.

In this work, the topology selection for the power grid is tightly coupled with the requirements of global routing for signal wires, and it is important to estimate the congestion contribution of the signal wires by implementing a fast global routing procedure. The technique used for this purpose is described in the remainder of this section, but it may be noted that this procedure may be replaced by any other computationally efficient method that shares a similar goal.

At the beginning of the algorithm, we perform a coarse global routing of all signal nets to obtain an estimate of the distribution of signal wire congestion at various boundaries, and feed this information to power/ground optimizer. The global routing technique used here is similar to [1], where a Steiner tree is initially constructed for each signal net using the AHHK algorithm [2] without considering congestion. After the initial Steiner trees have been constructed, an iterative rip-up-and-reroute procedure is applied to further reduce the wire congestion. The outer loop, consisting of signal routing followed by power grid optimization, is repeated iteratively until the constraints are satisfied or no further improvement is possible. In each iteration, the global routing solution from the previous iteration is used as a starting point for the rip-up-and-reroute procedure, with updated congestion values being used to direct the routing.

The rip-up-and-reroute procedure processes each net sequentially using the fixed net ordering heuristic proposed in [13], continuing until the maximum wire density is no greater than one, or after two complete iterations, since the congestion reduction after the second iteration is limited. Each net undergoing rip-up-and-reroute is entirely deleted and then rerouted using an algorithm similar to the min-max tree algorithm in [6].

3. POWER SUPPLY NOISE ANALYSIS

As is usually done in power grid analysis work [4, 5], we use the following linear circuit model to analyze the voltage droop noise of the power distribution network:

- The power grid is modeled as a resistive mesh.
- The cells/blocks are modeled as time-varying current sources connected between the power and ground plane.
- Decoupling capacitors are modeled as single lumped capacitors connected between power and ground.
- The top-level metal is connected to a package modeled as an inductance connected to an ideal constant voltage source.

This model leads to a large-scale linear circuit for power grid analysis, which can be efficiently analyzed using the technique described in [14, 18]. The wire width optimization procedure for the supply network, described in Section 4.2, also requires the computation of gradients, and this is performed using the simulation framework. Specifically, the transient adjoint sensitivity analysis technique is applied to calculate the sensitivity of the noise metric with respect to every tuning parameter, which, in our case, is the width of every power wire.

The noise and sensitivity analysis techniques used here are substantially similar to our work on decoupling capacitor (decap) placement [18] and have therefore been described only very briefly here. The only difference is that the tuning parameters in decap placement are the decap values, while in this work, the tuning parameters correspond to the width of each power grid wire. Consequently, current waveforms instead of voltage waveforms must be stored for adjoint sensitivity computations in this work.

4. POWER GRID DESIGN SCHEME

Starting with a dense grid that is guaranteed to meet the constraints on the supply network, our scheme iteratively sparsifies the grid to ease the wire congestion. In each iteration of the loop in Fig. 1, the power grid is adjusted using a two-step technique:

- In the first step, a wire removal heuristic is used to make the grid less dense in some regions, with due consideration paid to both the congestion information and the power grid noise.
- This is followed by a wire sizing step that readjusts the sizes of wires in the grid to compensate for the loss of this wire.

Therefore, our procedure ensures satisfactory performance of the power grid while utilizing just enough routing resources, so that the resources available to signal routing are maximized.

4.1 Power grid wire removal heuristic

As stated in Section 2, the congestion is measured in proportion to the overflow value on each tile edge. All power wires that lie in congested regions are potential candidates for removal, except those that lie within hot spots where the voltage droop is significant. The rationale behind this is that whenever a power wire is removed, the performance of the overall grid is compromised, and this is all the more noticeable if this wire lies within a hot spot. Therefore, we define a power wire as *critical* if the worst-case voltage droop on it is beyond some specified threshold VDP_{th}. Critical wires are not candidates for removal even if they lie in congested regions.

In addition, we define the *criticality* (Crt) of each non-critical power wire as the reciprocal of the RMS distance of that wire to critical nodes with nonzero noise within its neighborhood. Since the *criticality* of a power wire is only defined for non-critical wires, all these distances have non-zero values. For example, when a vertical non-critical power wire crosses a horizontal tile boundary located between $[x_L,\ x_R]$, the criticality of this wire to its closest noisy region is

$$Crt_p = \frac{K_c}{\sqrt{\sum_{i \text{ critical and } |x_p - x_i| \leq \Delta} (x_p - x_i)^2}} \quad (1)$$

where x_p is the x-coordinate of the vertical power wire, x_i is the average of the two boundary terminals x_L and x_R, K_c is the total number of critical nodes within some distance Δ to the tile center and x_i is the x-coordinate of the critical node i. In our experiments, we take Δ as 1.5 to 2 tile edge lengths. It is clear to see that the larger the criticality of a power wire is, the closer this wire is to one of the hot spot region. Given several candidate power wires across one tile boundary with its overflow value larger than OV_{th}, the criticality of each wire determines an optimal order of removal for these wires.

The power wire removal heuristic proceeds as follows. First, the tile boundaries are sorted in decreasing order of their overflow values and all non-critical power wires are sorted according to their criticality values. Next, non-critical power wires are removed in the order of the sorted tile boundaries, provided the overflow value at the boundary is larger than OV_{th}. Several practical considerations must be taken into account. Firstly, the power wire removal process should be accompanied by dynamically updating the overflow value on every tile edge, so that subsequent wire removal is based upon the updated congestion information. Secondly, a reasonable number of power wires must be removed in each iteration, and this depends on the values of OV_{th} and VDP_{th}. Since the choice of these numbers is necessarily empirical and cannot be entirely relied on, we assert an upper bound for the number of power wires to be removed in any iteration to conserve the amount of computation required in the wire sizing step discussed in Section 4.2. In our experiments, this upper bound is chosen to be 6% of the total number of power wires. If a set of non-critical power wires are across one tile boundary, remove them in their increasing order of criticality. Through this rule, non-critical power wires with less criticality are removed first.

4.2 Power grid sizing

The second step in the adjustment of the power grid is related to sizing the wires in the grid to compensate for the increased voltage droop after the wire removal step. The problem is directly formulated as a nonlinearly constrained nonlinear programming problem as follows:

$$\begin{aligned} &\text{minimize} && Area(w_j) = \sum_{j=1}^{N_{wire}} l_j \times w_j \\ &\text{subject to} && w_{min} \leq w_j \leq w_{max},\ j = 1 \cdots N_{wire} \\ &\text{and} && Z(w_j) < \epsilon \end{aligned} \quad (2)$$

where ϵ is a very small number, and N_{wire} is the total number of wires in the grid.

The objective function that minimizes the total power wire area is consistent with the goal of congestion reduction. The first constraint restricts every power wire width to lie within a realistic range that is technology dependent. The second constraint requires the definition of the parameter Z, which is an effective metric for noise measurement that was proposed in [7]. This idea for one node is illustrated in Fig. 3, which shows the voltage waveform of one node on the V_{dd} grid. Z is the sum of each individual nonzero noise z_j. This metric finds the integral of the noise violation and is zero if all constraints are satisfied. It punishes larger violations more severely than minor violations, and since it incorporates both the magnitude and time axes together, it is observed to be more practical than one that considers only the worst-case noise violation.

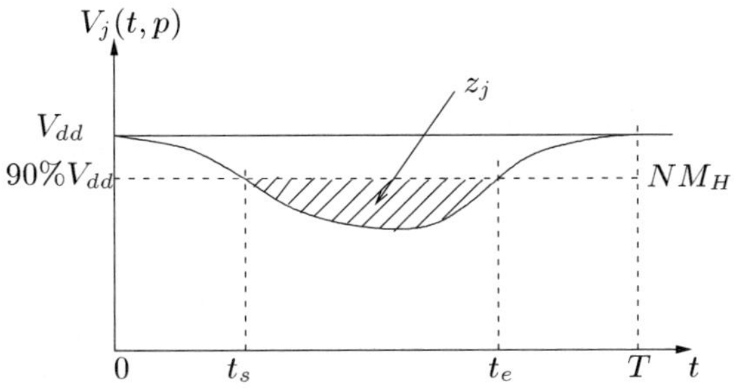

Figure 3: Illustration of the voltage droop at a given node in the V_{dd} power grid. The area of the shaded region corresponds to the integral z at that node.

The nonlinear constraint function Z can be obtained by transient analysis of the power grid circuit, and its sensitivity with respect to all the variables w_j can be calculated using the adjoint method discussed in [7, 18].

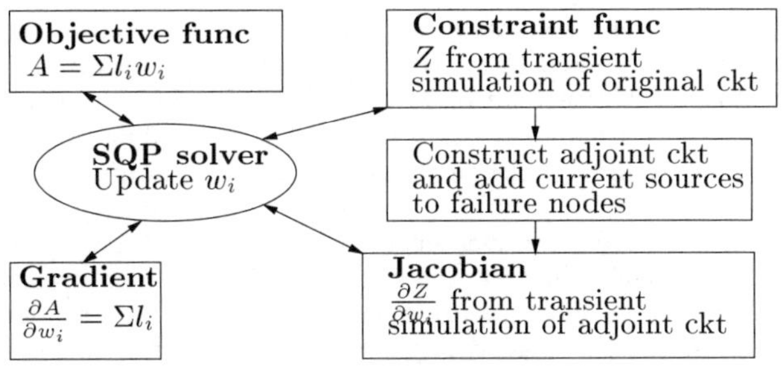

Figure 4: Power grid sizing procedure.

We use a standard Sequential Quadratic Programming (SQP) solver [3, 10] to solve the optimization problem. This solver requires users to provide subroutines to evaluate the objective and constraint functions and their derivatives with respect to each decision variable. The evaluations that are required for the SQP solver are illustrated in Fig. 4.

Practically, it is observed that the SQP solver converges slowly near the optimum. However, an approximate convergence is sufficiently good for our purposes, so that we require Z to be near-zero rather than exactly zero by choosing a sufficiently small ϵ.

5. OVERALL FLOW OF THE ALGORITHM

Having described all of the individual pieces of the algorithm, we now outline the overall flow of the procedure. The starting point is a given tessellation of a chip and a given initial power grid construction, and the sequence of steps can be summarized as follows:

1) An initial power-grid aware global routing step is carried out to route the signal nets. Each net is first routed without considering congestion, followed by an iterative rip-up-and-reroute, described in Section 2, using the updated congestion information.

2) Based on the routes used by the power grid and the signal routes, a congestion map is generated and the overflow on each tile boundary is calculated. All tile boundaries whose overflow value exceeds the threshold OV_{th} are identified and are sorted in decreasing order of this value.

3) A transient simulation of the power grid is performed to identify critical power wires on which the worst-case voltage droop is above the threshold, VDP_{th}.

4) Sort the non-critical power wires in increasing order of the criticality.

5) Based on the congestion map generated in Step 2, non-critical power wires are removed according to the heuristic described in Section 4.1.

6) To compensate for the removal of these wires, the remaining power grid wires are resized using the nonlinear optimization procedure described in Section 4.2.

7) The congestion maps are updated, and the global routing is updated by performing rip-up-and-reroute based on the new congestions. At this point, the iterative loop is invoked so that the updated power grid and congestion map are fed back to step 3. The stopping criterion for the iteration is that the maximum overflow should be no greater than 1, or that no further improvement is possible. The latter is easily identified if it is detected that the changes in the congestion map after rip-up-and-reroute are insignificant, and that no further deletion of the power wires is possible.

The initial routing takes $O(MN^3)$ time, where N is the number of pins for each net and M is the total number of nets. The rip-up-and-reroute has a complexity of $E \log \log E$, where E in our case is the total number of tile edges. The complexity of power wire removal is $O(KlogK + PlogP + KPE)$, where K is the total number of tile edges with negative overflow values, P is total number of power wires and E is total number of tile edges. The first term stands for the complexity of sorting these tile edges, the second term is the worst-case complexity of sorting the criticality of non-critical power wires, and the third term comes from the power wire removal and dynamic tile edge overflow update process. The power grid analysis and sensitivity analysis has the complexity of one LU decomposition and two forward/backward substitutions. In practice, the cost of these computations is just over $O(n)$ for a sparse positive definite matrix, where n is the total number of nodes and inductance branches. For a nonlinear optimization problem with w decision variables, advanced implementations of the SQP solver have $O(w^2)$ cost. Therefore the worst-case complexity for our wire sizing procedure ends up to be $O(I(n + w^2))$, where I is the total number of iterations. The efficiency of the gradient-based SQP solver relies largely on the initial solution and its distance away from an optimum. Practically, it is seen that the optimal solution can be reached in a limited number of iterations.

6. EXPERIMENTAL RESULTS

The power grid analyzer and the global router have both been implemented in C++. The power grid removal and sizing scheme, and the overall congestion-driven power grid design and global routing flow has been written using Tcl. The wire size optimization is performed using an off-the-shelf SQP solver [10]. All experiments are performed on an Intel Pentium-IV 1.8GHz machine with 256M memory running Redhat Linux 7.0. The entire procedure is encapsulated in a flow called PSiCo (**P**ower-**Si**gnal **Co**design).

Circuit	B	N	Tile size
apte	9	77	31x36
ami33	33	112	30x28
ami49	49	368	33x34
playout	62	1294	30x26
ac3	27	200	38x36
hc7	77	430	34x39
a9c3	147	1148	33x31

Table 1: Test circuit parameters.

We have tested our flow on seven benchmark circuits obtained from the authors of [1]. All designs correspond to a $0.18\mu m$ technology and a supply voltage of 1.8V. The time-varying current sources modeling the behavior of each functional block was not originally available in these benchmark circuits. These waveforms are constructed by modifying current waveforms from several industrial circuits by adjusting their magnitude according to the area of each block. However, this is not a critical assumption since our method is applicable to the exact waveforms where available.

Circuit	Method	D_{max}	Overflow
apte	Traditional	2.50	27
	PSiCo	1.00	1
ami33	Traditional	3.00	122
	PSiCo	1.00	0
ami49	Traditional	2.14	192
	PSiCo	1.00	0
playout	Traditional	1.94	158
	PSiCo	1.05	2
ac3	Traditional	2.50	316
	PSiCo	1.00	0
hc7	Traditional	2.13	518
	PSiCo	1.00	0
a9c3	Traditional	1.33	229
	PSiCo	1.00	0

Table 2: A comparison of the congestion improvement after global routing using the conventional approach and our new method.

Initially, six layers of regularly distributed power grids with fixed wire widths are generated for all of these examples, with each layer containing only horizontal or vertical wires. Since a majority of the global routes are typically seen on the third and fourth metal layers, M3 and M4, we perform the global routing and power grid sizing on these two layers, assuming that the other layers are processed separately. The wire widths on these layers are assumed to be constrained within the range of $0.8\mu m$ and $4\mu m$ in our experiments. The initial power grid is constructed with a constant pitch in M3 and M4 such that under the given set of time-varying current sources that represent each block, the voltage droop on the entire power grid is within a threshold, which in our experiments, is chosen to be 0.18V. Table 1 lists the characteristics of each circuit, in terms of the total number of blocks B and nets N and the tile sizes.

The performance of PSiCo can be described in terms of two components: the global routing solution, and the performance of the final power grid, and these results are shown in Tables 2 and 3, respectively. The wire congestion results of PSiCo are compared in Table 2 against the results of the traditional rip-up-and-reroute method, which corresponds to the result at the end of Step 1 in Section 5. For each circuit, the maximum density D_{max}, calculated as the maximum ratio of the utilization to the capacity across any tile boundary. Also shown is the overflow, i.e., the total amount by which the tile boundary capacities are violated, summed over all boundaries. It can be seen that for all the cases PSiCo gives much better congestion results than the conventional method.

Performance metrics for the initial power grid and the power grid obtained by PSiCo are listed in Table 3, in specific, the power wire area as a percentage of the area available on the two layers, the total number of nodes, the number of wires (W) in the power grid and the worst-case voltage droop. These numbers after optimization are shown in the table and may be compared with the corresponding numbers before optimization. It is seen that a well-designed power grid can save up to 12% of the total chip area while providing the optimal amount of resources needed for signal routing.

In each of the test cases, all grids whose worst-case voltage droop exceeds 0.14V (stricter than $10\% V_{dd}$ because layers M3 and M4 instead of M1 and M2 are considered) are marked as critical wires not to be removed. The voltage droop constraint (noise margin) for noise computation is chosen to be 0.17V, which is slightly stricter than $10\% V_{dd}$. To aid the rate of convergence, we heuristically terminate the SQP solver when we detect that the worst-case voltage droop in the circuit is below 0.18V, and it can be seen that the worst-case voltage droop always satisfies this requirement. The optimized noise integral, Z, for each circuit can be seen to be very small, and is nonzero due to the fact that voltage droops between 0.17V and 0.18V are tagged as "violations" by the SQP solver.

Lastly, Table 4 lists the total number of iterations and the CPU time, in minutes, required for each circuit. It can be seen that the number of iterations is typically small, and that the CPU times for these circuits are very reasonable.

7. CONCLUSION

We have proposed a new design flow for the codesign of signal routes and power grids. The technique is guided by congestion maps, and proceeds by removing non-critical power wires greedily in congested areas, and rerouting the signal wires according to the updated congestions. The effects of removing power wires are compensated for by a gradient-based wire sizing scheme. Experimental results for several benchmark circuits are presented in this paper. Future work includes applying this method to designs from industry to show the effectiveness of this method on larger industrial circuits.

8. REFERENCES

[1] C. J. Alpert, J. Hu, S. S. Sapatneker, and P. G. Villarrubia. A Practical Methodology for Early Buffer

Circuit	Before Optimization				After Optimizatioin				
	Area	Nodes	W	Droop	Area	Nodes	W	Droop	Z (V $\times$ ns)
apte	16.8%	3411	133	0.178V	10.7%	3383	123	0.176V	2.84e-3
ami33	16.1%	11226	171	0.178V	8.90%	10245	156	0.179V	0.40e-3
ami49	19.3%	14986	198	0.178V	14.1%	13921	183	0.179V	3.65e-3
playout	19.3%	18981	223	0.170V	8.44%	16341	193	0.174V	0.10e-3
ac3	21.0%	4189	147	0.178V	9.38%	3985	128	0.180V	0.76e-3
hc7	22.4%	8882	215	0.180V	16.0%	8632	201	0.178V	7.05e-3
a9c3	19.3%	24168	251	0.175V	11.4%	22539	233	0.180V	0.85e-3

Table 3: Optimal power grid results.

	apte	ami33	ami49	playout	ac3	hc7	a9c3
CPU(min)	3.6	7.6	9.8	18.1	6.0	7.3	28.8
# Iter.	6	2	1	5	2	3	2

Table 4: CPU time statistics and number of iterations for the benchmarks.

and Wire Resource Allocation. In *Proc. Design Automation Conference*, pages 189–194, Las Vegas, NV, June 2001.

[2] C. J. Alpert, T. C. Hu, J. H. Huang, A. B. Kahng, and D. Karger. Prim-Dijkstra Tradeoffs for Improved Performance-Driven Routing Tree Design. *IEEE Transactions on Computer-Aided Design of ICs and Systems*, 14(7):890–896, July 1995.

[3] P. T. Boggs and J. W. Tolle. *Sequential Quadratic Programming.* Cambridge Univ. Press, Cambridge, 1995.

[4] H. H. Chen and D. D. Ling. Power Supply Noise Analysis Methodology for Deep-Submicron VLSI Chip Design. In *Proc. Design Automation Conference*, pages 638–643, Anaheim, CA, June 1997.

[5] H. H. Chen and J. S. Neely. Interconnect and Circuit Modeling Techniques for Full-Chip Power Supply Noise Analysis. *IEEE Transactions on Components, Packaging, and Manufacturing Technology, Part B*, 21(3):209–215, August 1998.

[6] C. Chiang and M. Sarrafzadeh. Global Routing Based on Steiner Min-max Tree. *IEEE Transactions on Computer-Aided Design*, 9(12):1318–1325, December 1990.

[7] A. R. Conn, R. A. Haring, and C. Visweswariah. Noise Considerations in Circuit Optimization. In *Proc. International Conference on Computer-Aided Design*, pages 220–227, San Jose, CA, November 1998.

[8] S. W. Director and R. A. Rohrer. The Generalized Adjoint Network and Network Sensitivities. *IEEE Transactions on Circuit Theory*, 16(3):318–323, August 1969.

[9] P. Feldmann, T. V. Nguyen, S. W. Director, and R. A. Rohrer. Sensitivity Computation in Piecewise Approximation Circuit Simulation. *IEEE Transactions on Computer-Aided Design of ICs and Systems*, 10(2):171–183, February 1991.

[10] P. E. Gill, W. Murray, M. A. Saunders, and M. H. Wright. *User's Guide for SOL/QPSOL: A Fortran Package for Quadratic Programming.* Department of Operations Research, Stanford University, July, 1983.

[11] J. Hu and S. S. Sapatnekar. A Timing-constrained Algorithm for Simultaneous Routing of Multiple Nets. In *Proc. International Conference on Computer-Aided Design*, pages 99–103, San Jose, CA, November 2000.

[12] J. D. Z. Ma and L. He. Simultaneous Signal and Power Routing under Keff Model. In *Proc. International Workshop on System-Level Interconnect Prediction*, pages 175–182, Sonoma, CA, April 2001.

[13] R. Nair. A Simple Yet Effective Technique for Global Wiring. *IEEE Transactions on Computer-Aided Design*, 6(2):165–172, March 1987.

[14] S. R. Nassif and J. N. Kozhaya. Fast Power Grid Simulation. In *Proc. Design Automation Conference*, pages 156–161, Los Angeles, CA, June 2000.

[15] L. T. Pillage, R. A. Rohrer, and C. Visweswariah. *Electronic and System Simulation Methods.* McGraw-Hill, New York, NY, 1995.

[16] E. Shragowitz and S. Keel. A Global Router Based on a Multicommodity Flow Model. *Intergration: the VLSI Journal*, 5(1):3–16, March 1987.

[17] H. Su, K. H. Gala, and S. S. Sapatnekar. Fast Analysis and Optimization of Power/Ground Networks. In *Proc. International Conference on Computer-Aided Design*, pages 477–480, San Jose, CA, November 2000.

[18] H. Su, S. S. Sapatnekar, and S. R. Nassif. An Algorithm for Optimal Decoupling Capacitor Sizing and Placement for Standard Cell Layouts. To appear in *Proc. International Symposium on Physical Design*, San Diego, CA, April 2002.

[19] X. Tan, C. J. R. Shi, D. Lungeanu, J. Lee, and L. Yuan. Reliability-Constrained Area Optimization of VLSI Power/Ground Networks Via Sequence of Linear Programmings. In *Proc. Design Automation Conference*, pages 156–161, New Orleans, LA, June 1999.

[20] B. S. Ting and B. N. Tien. Routing and Techniques for Gate Array. *IEEE Transactions on Computer-Aided Design*, 2(4):301–312, October 1983.

[21] X. Wu, X. Hong, Y. Cai, C. K. Cheng, J. Gu, and W. Dai. Area Minimization of Power Distribution Network using Efficient Nonlinear Programming Techniques. In *Proc. International Conference on Computer-Aided Design*, pages 153–157, San Jose, CA 2001.

[22] M. Zhao, R. V. Panda, S. S. Sapatnekar, T. Edwards, R. Chaudhry, and D. Blaauw. Hierarchical Analysis of Power Distribution Networks. In *Proc. Design Automation Conference*, pages 481–486, Los Angeles, CA, June 2000.

On Metrics for Comparing Routability Estimation Methods for FPGAs

Parivallal Kannan, Shankar Balachandran, Dinesh Bhatia
Center for Integrated Circuits and Systems
Erik Jonsson School of Engineering and Computer Science
University of Texas at Dallas
PO Box 830688, Richardson, TX 75083, USA
{parik, shankars, dinesh}@utdallas.edu

ABSTRACT

Interconnect management is a critical design issue for large FPGA based designs. One of the most important issues for planning interconnection is the ability to accurately and efficiently predict the routability of a given design on a given FPGA architecture. The recently proposed routability estimation procedure, fGREP [6], produced estimates within 3 to 4% of an actual detailed router. Other known routability estimation methods include RISA [5], Lou's [7] method and Rent's rule based methods [1] [11] [9]. Comparing these methods has been difficult because of the different reporting methods used by the authors. We propose a uniform reporting metric based on comparing the estimates produced with the results of an actual detailed router on both local and global levels. We compare all the above methods using our reporting metric on a large number of benchmark circuits and show that the enhanced fGREP method produces tight estimates that outperform most other techniques.

1. INTRODUCTION

Interconnect prediction is the process of estimating the routing resource requirement and/or utilization, before actually performing the routing process. For the FPGA design flows, the term routability estimation is more appropriate, as the routing resources are fixed for a given device. Given a particular FPGA device and a design to be mapped onto the device, routability estimation is the process of identifying the number of routing elements needed to perform a complete routing. If the device has enough routing elements to satisfy the requirement, then the design is routable. The estimation process attempts to produce routing demand values on every programmable routing element on the device. Useful parameters derived from the estimation process are the *peak routing demand* and the *routing demand distribution*. For a successful routing, the device must satisfy both the peak routing demand and the routing demand distribution requirements. The problem of interconnect prediction has been studied extensively for the ASIC design flows. Most of the recent work in interconnect prediction and routability estimation is based on the Rent's rule [1]. Recently, some efforts have been made to address the problem for FPGAs [9].

2. ESTIMATION METHODS

In general, a good routability estimation method should be :

Fast - Routability estimation is typically used inside other physical design tools and hence speed is critical

Generic - The method should be independent of the FPGA device architecture

Usable - The estimation should produce usable results for a wide variety of applications. Typically we are interested in peak routing demand, channel utilization and routing demand distribution

Accurate - Undoubtedly, the estimation should be reflective of actual numbers obtained from a detailed router. Either the numbers should directly correlate with the detailed routing results or at least the ratios should match over a large set of benchmarks.

Some of the routability estimation methods that satisfy these requirements and currently under wide usage are fGREP [6], RISA [5], Lou's method [7] and Rent's rule derivatives [11] [9]. In this paper, we analyze and compare all the above methods. fGREP is a theoretical method that uses the concept of routing flexibility to model routability. RISA [5] is an empirical method based on the wiring distribution map, derived from a large number of randomly generated optimal Steiner trees. Lou's method [7] is based on the ratio of the number of paths that use a specific routing region to the total number of paths possible. Yang et. al. [11] proposed the use of Rent's rule for congestion estimation.

All the methods except fGREP were originally proposed for ASIC design flows. However, they are generic enough in their formulation that a direct translation to an FPGA design flow is trivially possible. We have implemented all these methods for a generic island style FPGA architecture as described in [3]. To compare the quality of the estimates, we use the VPR [3] FPGA physical design suite's detailed router. In the following sections, we explain these methods in detail.

3. FGREP

fGREP is a theoretical routability estimation method in which the estimation model is based on the concept of routing flexibility over the routing elements. fGREP models the routing fabric of the FPGA as a graph $G(V, E)$, where V represents the channels in the FPGA and E represents the switchboxes. For each $v_i, v_j \in V$, there exists an edge $< v_i, v_j > \in E$ *iff* channels v_i and v_j share a switch box between them. Since it operates on the routing graph and makes no assumptions of the routing architecture, fGREP can be applied to any FPGA architecture without any modifications.

A net $n_k \in N$, where N is the netlist, is made up of a set of terminals $T_k \in V$. Every terminal $t_k^i \in T_k$, exacts a certain routing demand called the *terminal-demand*, on all the routing elements inside the net bounding box. According to fGREP, this terminal-demand on a routing element at a distance of $l = q$ from the terminal is proportional to the total number of elements at the same distance l from the terminal. The distance is measured on a breadth-first search tree on the routing graph, with the terminal as the root. The set of equidistant ($l = q$) routing elements from a terminal t_k^i, is defined as the level set,

$$LS_k^{iq} = \{v_j \in V | l_{ij} = q\} \quad (1)$$

The terminal-demand on the routing element v_j is then,

$$TD_k^{ij} = \frac{1}{|LS_k^{iq}|}; where, l_{ij} = q \quad (2)$$

fGREP then derives a quantity called the *net-demand* for all routing elements inside the net bounding box. The net-demand of v_j is defined as the terminal-demand due to the terminal with the lowest value for the distance metric l_{ij}.

$$ND_k^j = TD_k^{ij} | l_{ij} = min(\forall_i l_{ij}) \quad (3)$$

The final *routing-demand* on the routing element v_j due to all the nets in the netlist is then,

$$D^j = \sum_{k=1}^{\#nets} ND_k^j \quad (4)$$

The terminal-demands due to all the terminals of all the nets are calculated by performing a breadth first traversal of the routing graph with the terminal v_i as the root vertex. At each step of the traversal, the level set is enumerated and the demands are assigned. The net-demand is then calculated from the terminal demands as per Equation 3. The total routing element demand due to all the nets is then calculated as per Equation 4.

4. ENHANCEMENTS TO FGREP

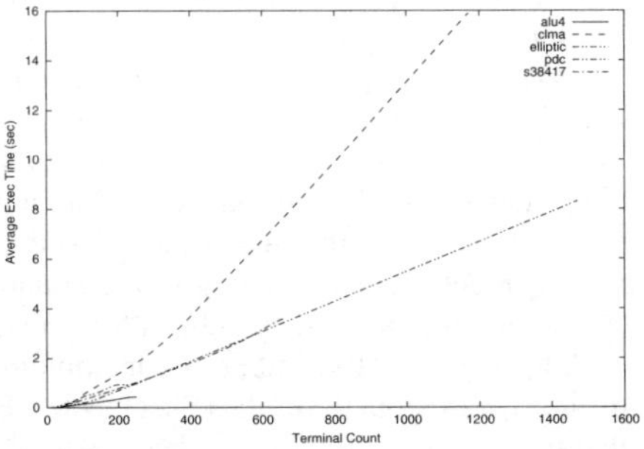

Figure 1: Terminal Count vs Avg. Execution Time

The runtimes of fGREP are high for large circuits with many high-fanout nets. If E_k is the set of routing elements in the bounding box of a net n_k and T_k the set of terminals, then the runtime is proportional to $|E_k| \times |T_k|$. Typically high-fanout nets span the entire device and hence add a severe penalty to the fGREP runtimes. This effect is clearly illustrated in Figure 1, which plots the average fGREP execution time per net against the number of terminals in the net, for a few standard benchmark circuits.

4.1 Zone Limited Search

Equation 2 produces the routing demands due to one net on all the routing elements inside the net bounding box. It can be observed that the operation produces *zones of influence* around each terminal. All the routing elements inside a zone have their net-demands produced by a single common terminal. Conversely, for all the routing elements outside a terminal's zone of influence, the net-demands are produced by other terminals. Hence, theoretically, the terminal-demands due to a terminal v_i need be assigned for only those elements in its zone of influence. The runtime complexity of the operation in Equation 3, is then only of the order of $|E_k|$, instead of $|E_k| \times |T_k|$. We propose to use this technique of limiting the terminal-demand assignments to the zone elements only to reduce the runtimes of fGREP.

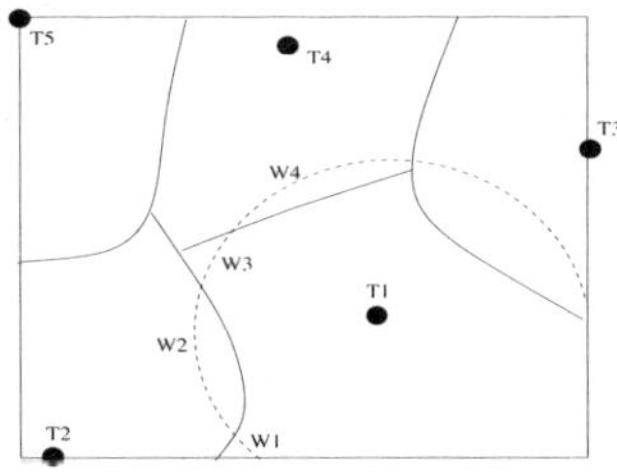

Figure 2: Zones, Wavefront Clipping and Fragmentation

The zone limited search technique cannot be directly applied to the fGREP method. fGREP relies on the enumeration of the level set as per Equation 1. The level sets are identified by starting a breadth-first search from the terminal. At each step of the search, the current elements form a wavefront, and expands outwards. By limiting the search to within the terminal's zone alone, fragmenting and clipping of the wavefront can occur. This has the effect of reducing the level set size and produce artificially high terminal-demand values. Fig 2 illustrates the clipping and fragmenting of the search wavefront. The wavefront for the terminal T_1 is shown by the dotted line. The zones are marked by solid lines. The wavefront W of T_1 is divided into 4 arcs, two of which (W_1 and W_3) are inside T_1's zone while the other two (W_2 and W_4) are outside the zone. According to fGREP, the cardinality of the level set at this distance will be $|W_1| + |W_3|$, which is far less than that of the complete wavefront. This will produce very high terminal-demands for the routing elements on this wavefront, which is clearly erroneous.

4.2 Zone limited Parallel Search

To overcome the wavefront clipping and fragmenting effects described above, the complete wavefront has to be maintained at all stages. But to gain speedup, the search has to be limited to the terminal's zone of influence alone. A simple yet effective solution to the problem is to maintain the complete wavefront for the terminal as long as at least one routing element on the wavefront is still contained in its zone. We propagate breadth first traversals from all terminals of a net simultaneously, every time maintaining the complete wavefront. If at any point, all the elements on the wavefront of a terminal are outside its zone (that is no element on the wavefront got its terminal-demand from this terminal) then we have completely discovered its zone and stop the search for that terminal alone. This process terminates when all the terminals' zones are discovered.

We observed that this enhancement to fGREP is very effective and resulted in up to 30X speedup in execution times while producing the same estimates. The estimation results and the run-times of fGREPcan be obtained from [6].In further discussion we will refer to the enhanced fGREP method only, which we shall refer to as fGREP2.

5. RISA

RISA [5] is a very fast empirical estimation technique. It was originally proposed for ASIC design flows, but can be readily adapted to FPGA design flows. The method involves multiplying the net bounding box of a net by a pin-count dependent net-weight q, to produce the routing demand due to the net. The net-weights for various pin-counts are generated from a one time operation involving the production of a *wiring distribution map, (WDM)*. A WDM for a given pin-count M is produced by adding up and normalizing demands from optimal steiner trees for K sets of M random points (K is very large, $\approx$ 10000). The mean value of the WDM thus obtained is called the net-weight for a pin-count of M. Chang [5] provides experimental values for the net-weights for $1 \leq M \leq 50$ and net-weights for higher values of M, can be obtained by a linear regression process.

Having obtained the net-weights, the original expressions for estimating the routing demand on a routing region R_i due to a net n_k is,

$$D_k^{i,horizontal} = q \times \frac{w \times l}{Y \times L}; D_k^{i,vertical} = q \times \frac{w \times l}{X \times W} \quad (5)$$

where, (W, L) is the dimension of the routing region and (X, Y) is the dimension of the net bounding box of the net n_k and (w, l) is the overlap between the routing region and the net bounding box. For FPGAs, the routing region is a channel and hence $W = L = w = l = 1$. So, Equation 5 becomes,

$$D_k^{i,horizontal} = q \times \frac{1}{Y}; D_k^{i,vertical} = q \times \frac{1}{X} \quad (6)$$

The total routing demand on a routing element on the FPGA, as per RISA is then the sum of the demands due to all the nets. It can be observed that the routing demand due to a net is the same over all the routing elements in its net bounding box and is calculated once for every net.

6. LOU'S METHOD

Lou et. al.'s estimation method [7] is a recent routing estimation method developed for placed circuits in ASIC design flow. The chip area is divided into rectangular grids called *regions*. This method produces routing demands on different regions of the chip. Any given net in a circuit poses routing demands only on the regions that are in the bounding box of the net. For a two-terminal net, the demand assigned to a routing region is the number of tracks that the net needs in that particular region. This is dependent on the ratio of the number of shortest paths that use that particular region to the total number of shortest paths possible in routing the net.

Assume a net n_k with two terminals, one placed in the lower left corner and the other in the top right corner of a $M \times N$ grid in the chip. Let the total number of shortest paths possible to route this net be $F(M, N)$. This value can be calculated as

$$F(M,N) = \begin{cases} F(M-1,N)+F(M,N-1) & M,N \geq 2 \\ 1 & M,N = 1 \end{cases}$$

The demand on a particular region (i, j) is clearly dependent on the number of paths $F(i, j)$ that pass through the region.

In addition to the number of paths possible through a region, the usage is also dependent on whether the path bends in a particular location or not. If a particular path does not bend in (i, j), it places a demand of one full track in the direction of path and zero demand in the other direction for that location. However, if a path bends in (i, j), that path will place a demand of only half track in each of the directions. Considering both the number of paths through a particular region and the nature of the bends that these paths make, the number of tracks needed by the net n_k in region (i, j) is calculated as $P_x^k(i, j)$ and $P_y^k(i, j)$, the horizontal and vertical usages for the net n_k respectively. The usage matrix for the bounding box gives the *probability usage matrix* $P^k(M, N)$ where every matrix entry is a pair $(P_x^k(i, j)), (P_y^k(i, j))$. The regions outside the bounding box are assigned zero demands. For the formulae that are used to calculate the exact usages, refer to [7] .

In [7], Lou. et. al. suggest that the multi-terminal nets be decomposed as a Minimum Spanning Tree (MST) or a Rectilinear Steiner Tree(RST) and then use the two-terminal model for each such pair. The usage matrices of all these pairs are calculated and summed up to get demand due to the net. The final demand due to all the nets on a region is the sum of the individual demands due to each net similar to Equation 4.

In [2], we adapted the Lou's model to FPGA design flows. A rectangular grid was overlaid on an island-style FPGA, where a single grid element corresponds to a switchbox and its four incident *half-channels*. Every channel in the FPGA will be on two grid elements in this overlaid grid. Lou's method is used on this grid for estimation. The horizontal and vertical usages estimated for a region are then assigned to the half-channels. Thus demands are obtained on a channel by channel basis. Experiments on 20 large FPGA benchmarks are reported in [2].

7. ENHANCEMENTS TO LOU'S MODEL

Lou et. al. [7] note that MST decomposition may result in inaccuracies because of over-counting on the overlapping net segments. They do not provide any solution to alleviate the problem. In [2], we established that the problem with segment overlaps applies to any two-terminal decomposition, including the RST based decomposition. Further, we claimed that even if a clever net decomposition may remove large net overlaps, the bounding boxes of the pairs may still overlap thus resulting in inaccuracies. Here we propose a simple yet effective enhancement to the Lou's model that addresses the overlapping-segment problem in this section.

We decompose the two-terminal nets using MST. For all the two-terminal pairs of a single net, we calculate the probability usages for the two-terminal bounding region. For the regions that are on overlapping bounding boxes, we assign the *maximum* of the demands due to different pairs, similar to fGREP's scheme in Equation 3. We call this scheme Lou(Max). This is in contrast to [7], where the authors add up the demands on the overlapping segments. The enhancement is based on the observation that the regions in the overlapping segments actually belong to the bounding box of one single net and adding up the demands will just result in over-estimation on those regions. By finding the maximum of the demands, we do not over-estimate on these segments. Further, we also ensure that the demand assigned is at least as much as the demand posed by any of the two-terminal pairs. This simple enhancement improves the quality of the estimation by as much as 103% compared to the FPGA model proposed in [2]. This is clearly illustrated in the Table 1, where the peak demands estimated by both the methods are compared against the actual peak demands obtained by performing detailed routing using VPR.

8. RENT'S RULE BASED METHOD

Rent's rule is an empirical observation that states the relationship between the number of blocks B in a circuit and the number of

Table 1: Comparison of Lou's Method with Lou(Max)

Circuit	W_{lou}	$W_{lou(max)}$	W_{vpr}	$\%better$
alu4	17.724	13.866	11	35.07
clma	25.760	18.849	13	53.16
dsip	18.260	11.032	7	103.26
s298	19.638	12.423	8	90.18
spla	23.063	19.197	15	25.77

external connections P of the circuit. Specifically,

$$P = T_b B^r \quad (7)$$

where T_b is the average number of interconnections per block and r is the Rent exponent. Yang et. al. [11] list reports of Rent's rule being successfully used to estimate wirelength, and state that since wirelength is a measure of routing demand, Rent's rule can be used for estimating congestion. They also propose methods to estimate the peak routing demand and regional routing demand using Rent's rule. The methods are proposed for ASIC design flows and only the peak routing demand formulation can be directly ported to FPGAs. The regional routing demand method uses expressions for total interconnect lengths that may not be valid for FPGAs. There is no work that uses Rent's rule to estimate routing demand on local regions for FPGAs. For our comparison purpose, we use the peak routing demand estimation method alone from [11]. Yang et. al. [11] perform a rough min-cut based placement to estimate the Rent's parameter. We perform the same operations on a placed circuit, but do area partitioning instead.

The basic idea behind the method is to recursively partition a self-similar circuit and place it hierarchically on a layout divided into equal rectangular tiles. The number of edges crossing these tile boundaries become the routing demand of that region. The cut-set sizes $C_{i,j}$ that result from such partitioning at every level i follow the relation,

$$C_{i,1} = C_{i,2} = \cdots = C_i, \; i = 1, 2, \cdots, 2H \quad (8)$$

where H is the number of hierarchical levels. Also, the ratio between the cut sizes of two successive levels i and $i+1$ is given by $\alpha = \frac{C_{i+1}}{C_i} = 2^{-r}$, where r is the Rent's exponent. Using this expression the upper bound of the maximum routing demand of a tile boundary, C_{max}, is derived. By extending this method for a uniform distribution of cut nets over the partitioning area, the authors have arrived at a modified upper bound of the number of crossings to be

$$C_{max} < \frac{C_1}{\sqrt{N_c}}\left(\frac{1}{2} + 2\alpha\right)\frac{\sqrt{N_c}\,\alpha^{2H} - 1}{2\alpha^2 - 1} \quad (9)$$

where N_c is the total number of gates in the circuit. For any design, r and C_1 can be found by recursive bipartition. The authors also note C_1 corresponds to the Region II of Rent's curve where there is usually a small dip in the number of wires that come out of blocks. A correction is applied to Eqn. 9 where C_h/α^{h-1} is used instead of C_1. In our experiments we have used $h = H/2$ which corresponds to the middle hierarchies. In our experimentation we take a placed FPGA circuit and identify the cuts across different partitioning levels by dividing the areas recursively. Rent's exponent is calculated as the slope of the log-log plot of the number of cells and the number of nets plotted over different partitioning levels. We use the FPGA dimension instead of $\sqrt{N_c}$ in Equation 9. We consider the resulting C_{max} for the circuit as the peak routing demand estimated by this method.

9. OTHER METHODS

Brown et. al. [10] developed a stochastic model for routability based on the probabilities of making individual C_T two-point connections. The probability that a connection C_i can be routed, $P(R_{C_i})$, is modeled as a sequence of conditional events on the switches in the path. Routability of the circuit is then given as Routability $= \frac{1}{C_T}\sum_{i=1}^{C_T} P(R_{C_i})$. Routability is the percentage of nets that can be successfully routed in the FPGA's routing architecture. Predictions are done for island-style FPGAs by varying the switch box and connection-box flexibilities (F_S and F_C respectively). For every circuit a single routability number is produced. The results are validated with the number of unrouted nets produced by an actual router. The method does not produce global or local routing demands.

In [9], Parthasarathy et. al. perform placement by modifying the cost function in VPR to account for interconnection. Rent's exponent for the routing architecture(p_a) and the design(p_d) is calculated and the minimum device size is scaled by a factor of $C = N_{gates}^{\frac{p_d}{p_a}-1}$ to accommodate the design's complexity. For every change in the placement, the changes in local Rent exponent of the design p_d^l is calculated. The modified cost function is $\sum_{n=1}^{N_{nets}}(1 + |p_d^n - p_a|)(wire\_length_n)$. The authors compare the track usage of VPR's placement and the modified placement. However, the scaling of device(C) that was performed is not mentioned. VPR is known to produce very tight placements and hence the device scaling factor is of critical importance.

Chan et. al. [4] predict the routability of unplaced circuits using Rent's exponent based wirelength calculations. They derive expressions for the routing requirements inside and outside all routing regions and then assign confidence of routability on a scale of 3 - Unroutable, Easily routable and Marginal.

10. ESTIMATION QUALITY METRIC

In this section we propose a uniform reporting metric for routability estimation methods. The driving need for such a metric is the requirement to be able to quantitatively compare the various estimation methods for specific applications. Most reports on estimation methods do not try to compare the estimation quality with known results, which can be independently asserted. fGREP is a notable exception, in that it compares its results with a well known detailed router. RISA incorporates its estimation technique inside a placement loop and compares the placement quality with that of the solution produced by the same placement loop without routability estimation. The two placement solutions are routed with a global router and two indices *overCon* and *dsCost* are used for comparison. The first index represents the number of global grids with demand higher than supply and the latter represents the linear summation of the demand overshoot. RISA does not report which global router was used. In [7], there is no quantitative comparison with detailed or global router results. Only two congestion maps, one estimated and another produced by a commercial router are reported. As in RISA, the details about the router are not reported. In [11], the peak routing demand estimates are compared with a *L-Shaped* global router. The global router approximates all routes as L shapes along the net perimeter, and is a crude approximation to regular global and detailed routers. From the above discussion, it is evident that the estimation community needs a common metric by which existing and new estimation methods can be compared and independently verified.

The metric should be based on well known, real world routers and should capture both regional and global estimation quality. For FPGA research, a very commonly used place and route tool suite is the VPR [3]. VPR is available as a free download in source form and can be freely modified for research purposes. VPR's detailed router is based on the PathFinder [8] negotiated congestion algorithm and is generally considered to produce very tight results among the FPGA research community. VPR and other PathFinder based routers derived from it currently hold the FPGA place and route challenge [3]. So, using the VPR's detailed router as the common denominator for comparison among various estimation methods is a valid choice.

We propose four parameters as adequate and required quality metrics. They are the *Peak Demand error*(W), *Mean of regional errors*(μ), the *standard deviation of the errors*(σ) and the *runtimes*(t). Global estimation quality can be stated by comparing the peak routing demands with the maximum channel width needed by VPR to route a placed design. For regional estimation quality, a more detailed metric has to be designed. We are interested in the correlation between the estimates produced on local routing regions and the actual channel utilization reported by VPR. A good measure of the correlation which is very easy to calculate is the mean of the errors between the estimated and routed channel utilization. The standard deviation of the errors adequately represents the distribution of the errors.

11. EXPERIMENTATION AND RESULTS

We have implemented fGREP2, enhanced Lou's method, Rent's rule based estimation method as per [11] and RISA for a generic island style FPGA architecture. We compare all these methods using the four parameters that we derived in the previous section. We used the largest benchmarks available from the ISCAS-89 benchmark set. All experiments were conducted on a standard 800MHz Pentium PC running Linux.

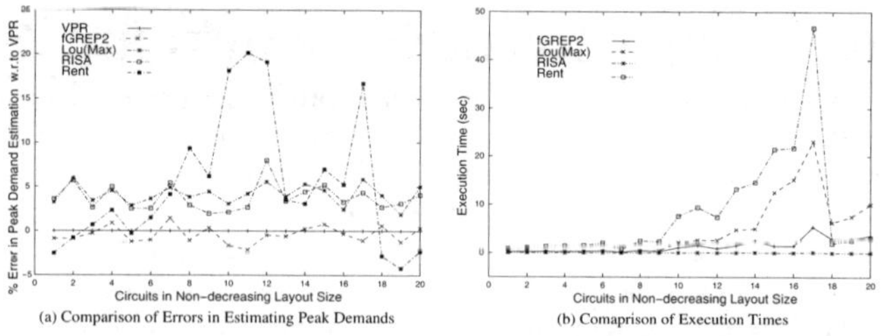

Figure 3: Comparison of Peak Routing Demands W and Execution Times

The results are tabulated in Tables 2 and 3. The Table 2 tabulates the peak routing demands and execution times for VPR and all the estimation methods. The first 5 columns in the Table 2 show the peak routing demands as per VPR, fGREP2, RISA, Lou's method and the Rent's rule based method respectively. The column labelled T_{VPR} is the execution time for the VPR's detailed router run with a fixed channel width of $W = W_{VPR}$. The last four columns in the Table 2 show the execution times for the estimation methods. All times are shown in seconds. The row labelled *Totals* in the Table 2 gives the sum of the tracks and runtimes over all the benchmark circuits. The row labelled $\frac{W_{error}}{W_{vpr}}$ gives the percentage error of the peak routing demand estimates, with respect to VPR. The row labelled $\frac{T_{vpr}}{T_{est}}$ gives the ratio of the execution times of the estimation methods with respect to VPR's detailed router. Table 3 tabulates the mean and standard deviation of the errors between the estimation methods and VPR's detailed router.

Fig 3(a) plots the peak routing demands as estimated by all the methods with respect to VPR. It can be seen that fGREP2 is the

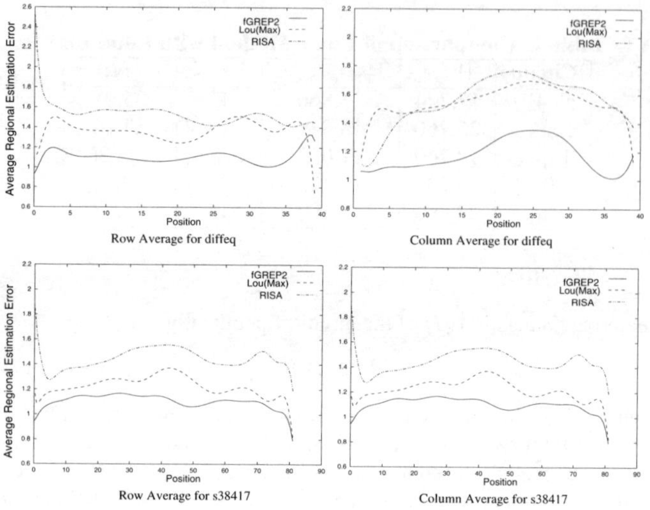

Figure 4: Average of the regional estimation errors

closest to VPR. The Rent's rule based method is very much away from VPR. The estimates are also randomly distributed and do not show any distinct trend. Similar results have been reported by the authors themselves in [11]. RISA and Lou's method are better but they too lack distinct trends and have random swings of about 3 to 5 tracks. To gain sufficient confidence levels, an estimation method should either closely match real world results or at least show comparable estimates for different nets. Random swings indicate that there are a number of circuits for which the estimation method is completely wrong, while there are other circuits for which the estimates are better.

Fig 3(b) plots the execution times for all the estimation methods. It can be seen that RISA has the lowest execution times, closely followed by fGREP2.

The accuracy of the regional estimation is best captured by the mean and standard deviation as explained before. To better illustrate the effect, we plot the row and column averages of the local estimation errors. Fig 4 shows the row and column averages of the estimation errors for the benchmark circuit *diffeq* and *s38417*. In these plots, the smaller the numbers, the better is the estimate. It can be seen that fGREP2 produces the best regional estimates followed by RISA and Lou's method.

12. CONCLUSION

We analyzed all the routability estimation methods available till date and proposed enhancements to two of the methods fGREP and Lou's method. We discussed the need for consistent reporting of routability estimation results and proposed a minimum set of comparison metrics. We implemented four routability estimation methods for FPGAs and compared them using our new metric. We conclude that the method fGREP2 is the most accurate while RISA is the fastest.

13. REFERENCES

[1] H. B. Bakoglu. *Circuits, Interconnections, and Packaging for VLSI*. Addison Wesley, Reading, MA, 1990.

[2] S. Balachandran, P. Kannan, and D. Bhatia. On Routing Demand and Congestion Estimation for FPGAs. In *Proc. 15th Intl. Conf. on VLSI Design and 7th ASP-DAC*, January 2002.

Table 2: *Peak* Routing Demands and Execution Times (seconds)

Circuit	VPR	fGREP2	RISA	Lou(Max)	Rent	T_{VPR}	T_{fGREP2}	T_{RISA}	$T_{Lou(Max)}$	T_{Rent}
alu4	11	9.802	13.506	13.866	10.717	38	0.2654	0.0125	0.2927	1.541
apex2	12	10.92	14.911	15.815	21.322	59	0.4884	0.0225	0.4148	2.494
apex4	13	12.12	18.716	18.931	12.195	43	0.2533	0.0132	0.2230	1.091
bigkey	9	7.760	12.105	10.866	4.7610	101	3.0146	0.0269	7.5189	2.602
clma	13	11.92	17.327	18.848	29.713	549	4.8111	0.1359	23.047	46.68
des	8	8.308	12.075	13.021	5.6691	76	3.6362	0.0631	10.030	3.015
diffeq	8	8.881	12.995	12.535	10.324	23	0.2035	0.0078	0.2679	1.433
dsip	7	7.606	9.699	11.032	4.1760	181	2.8295	0.0196	6.2362	1.977
elliptic	11	10.53	18.979	16.557	30.111	237	0.9493	0.0317	2.6474	7.408
ex1010	12	11.39	15.376	15.894	15.499	131	1.4727	0.0479	4.7849	13.20
ex5p	14	13.12	17.571	17.231	11.461	35	0.2031	0.0115	0.1862	0.904
frisc	14	12.37	16.117	17.063	32.125	150	1.0845	0.0463	2.1932	7.600
misex3	11	10.72	13.649	14.430	11.682	45	0.2824	0.0145	0.2600	1.377
pdc	16	16.24	20.418	21.287	19.067	460	2.4975	0.1031	4.9885	14.61
s298	8	8.340	9.963	12.423	14.150	60	0.2792	0.0104	0.4135	2.332
s38417	8	8.764	13.192	12.597	14.963	203	1.3543	0.0352	12.530	21.43
s38584.1	9	8.676	12.283	11.437	14.238	248	1.6993	0.0346	15.158	21.69
seq	12	10.99	14.530	15.629	13.468	53	0.3905	0.0194	0.3782	2.072
spla	15	12.87	17.663	19.197	35.149	132	1.4808	0.0597	2.6312	9.422
tseng	6	7.430	11.431	10.890	10.118	28	0.1486	0.0058	0.2133	0.717
Totals	217	208.785	292.446	299.549	320.908	2852	27.344	0.7216	94.416	163.595
$\frac{W_{error}}{W_{vpr}}\%$	-	3.79	34	38	47.9	-	-	-	-	-
$\frac{T_{vpr}}{T_{est}}$	-	-	-	-	-	-	104X	3952X	30X	17.5X

[3] V. Betz and J. Rose. VPR: A New Packing, Placement and Routing Tool for FPGA research. In *Field-Programmable Logic and Applications*, pages 213–222. Springer-Verlag, Berlin, Sep 1997.

[4] P. K. Chan, M. D. Schlag, and J. Y. Zien. On Routability Prediction for Field Programmable Gate Arrays. In *Proc. 30th DAC*, 1993June 1993.

[5] C.-L. E. Chang. RISA: Accurate and efficient Placement Routability Modeling. In *Proc. of ICCAD*, 1994.

[6] P. Kannan, S. Balachandran, and D. Bhatia. fGREP - Fast Generic Routing Demand Estimation for Placed FPGA Circuits. In *11th International Workshop on Field-Programmable Logic and Applications, FPL*. Springer-Verlag, Berlin, August 2001.

[7] J. Lou, S. Krishnamoorthy, and H. Sheng. Estimating Routing Congestion using Probabilistic Analysis. In *Proceedings of the 2001 International Symposium on Physical Design (ISPD)*, 2001.

[8] L. McMurchie and C. Ebeling. PathFinder: A Negotiation-Based Performance-Driven Router for FPGAs. In *ACM Symp. on FPGAs, FPGA95*, pages 111–117. ACM, 1995.

[9] G. Parthasarathy, M. Marek-Sadaowska, A. Mukherjee, and A. Singh. Interconnect Complexity-aware FPGA Placement Using Rent's Rule. In *Proc. Intl. Workshop on System Level Interconnect Prediction (SLIP)*, April 2001.

[10] S.Brown, J.Rose, and Z.G.Vranesic. A Stochastic Model to Predict the Routability of Field Programmable Gate Arrays. *IEEE Transactions on CAD*, pages 1827–1838, Dec 1993.

[11] X. Yang, R. Kastner, and M. Sarrafzadeh. Congestion Estimation During Top-down Placement. *IEEE Transactions on CAD*, 21, Jan 2002.

Table 3: μ and σ w.r.to VPR's Detailed Router

Circuit	fGREP2		RISA		Lou(Max)	
	μ	σ	μ	σ	μ	σ
alu4	1.471	1.107	1.726	1.307	1.601	1.257
apex2	1.690	1.301	1.960	1.584	1.773	1.416
apex4	1.927	1.425	2.246	1.776	1.773	1.454
bigkey	0.577	0.733	0.881	0.794	0.623	0.879
clma	1.735	1.266	1.932	1.495	1.791	1.405
des	0.589	0.645	0.809	0.777	0.629	0.811
diffeq	1.125	0.821	1.527	1.174	1.439	1.116
dsip	0.425	0.681	0.901	0.670	0.459	0.838
elliptic	1.415	1.076	1.621	1.285	1.482	1.186
ex1010	1.654	1.223	1.908	1.436	1.763	1.346
ex5p	1.868	1.470	2.272	1.832	1.799	1.424
frisc	1.720	1.313	1.784	1.387	1.782	1.358
misex3	1.555	1.180	1.849	1.462	1.712	1.299
pdc	2.163	1.553	2.287	1.767	2.251	1.752
s298	1.180	0.850	1.317	0.992	1.593	1.234
s38417	1.104	0.823	1.628	1.246	1.402	1.095
s38584.1	1.120	0.854	1.544	1.172	1.316	1.026
seq	1.720	1.286	1.970	1.547	1.790	1.414
spla	2.021	1.476	2.126	1.667	1.897	1.488
tseng	0.937	0.731	1.173	0.938	1.139	0.931

PANEL: Tools or Users: Which is the Bigger Bottleneck?

Chair: Andrew B. Kahng, Univ. of California at San Diego, La Jolla, CA

Organizers: Bob Dahlberg, ReShape, Inc., Mountain View, CA

Abstract

As chip design becomes ever more complex, fewer design teams are succeeding. Who's to blame? On one hand, tools are hard to use, buggy, *not interoperable*, and have missing functionality. On the other hand, there is a wide range of engineering skills within the user population, and tools can be abused within flawed methodologies. This panel will quantify and prioritize the key gaps, including interoperability, that must be addressed on both sides.

Panelist Statements

Ron Collett
Numetrics Management Systems, Santa Clara, CA

Tools, users and interoperability are all bottlenecks. This is well known. Each impacts design productivity and therefore cycle time and product development cost. Thus, it is not a question of which one of these is "the" bottleneck, but rather what is going to be the bottleneck on a particular project. The impact of each varies from project to project. The dominant bottleneck will vary depending on design complexity and time-to-market constraints. Let us assume that there will always be bottlenecks and thus a range of uncertainty in the product development process. The most important issue, then, is determining the impact of a particular set of tools, users and interoperability levels on cycle time for a particular chip design project. This must be done at the outset of the project and updated throughout the project, as design specifications and time-to-market constraints invariably change. Thus, the most important question is how to best manage inevitable bottlenecks and therefore the ever-present risk inherent in each development project.

Patrick Groeneveld
Eindhoven University of Technology / Magma

We must continue the drive toward complete IC design automation, to the point where most actions become almost push-button. The key is a carefully tuned RTL to GDSII design flow, in which hundreds of smaller steps encode a wealth of design methodology know-how and experience. Treating 'logic synthesis', 'placement' and 'routing' as big monolithic design steps is no longer compatible with achieving timing closure, managing on-chip parasitics and dealing with the many DSM peculiarities. Rather, harmonic orchestration of the many tooling steps in the flow that determines overall implementation success, and is more relevant than quality of individual point algorithms. A robust IC implementation flow simultaneously addresses many design constraints and DSM issues: timing closure, electromigration, crosstalk noise, multiple power supply regions with voltage drop constraints, etc. are just a few. The key is to find the proper equilibrium between the conflicting design goals and constraints. Building, debugging and tuning such a flow is the primary added value of a next generation EDA company.

An integrated data model is the enabling technology of a successfully automated RTL-to-GDSII design flow. The data model supplies reliable data on parasitic coupling capacitances, wire delays, system timing, congestion/DRC etc., even in the earlier steps of the flow when little physical data is known. Running the steps directly on the data model ensures fast run times and consistency of design data. For instance, the effect of inserting an additional buffer on a wire results in a change of the net list and wiring that in turn triggers a fast incremental update of the timing and congestion data. Combined with Gain Based Synthesis technology and proper integration of design constraints, this can ensure timing closure without iteration. The ultimate result is a single integrated tool that maps RTL to a working chip.

Lavi Lev
Cadence Design Systems, Inc., San Jose, CA

The main cause of failure and bottlenecks in developing and deploying complex SOC chips into the market is the inability of tools and people to interoperate. If one looks at the development and deployment philosophy of the EDA industry, one can come to the conclusion that complex SOCs are conceived, designed, and fabricated by a single individual using a single EDA company's tool to develop the SOC. In reality, developing a SOC requires hundreds of people who are superb technical problem solvers, but who leave a lot to be desired in the areas of communication and teaming. As a result, a huge number of bugs and schedule delays have nothing to do with the technology; they are the direct result of poor coordination and synchronization between dispersed team and pieces of data.

The EDA industry does not focus on the areas of coordination and synchronization that beg for automation. This is because most EDA engineers have never experienced the joy of building a chip. Instead, they almost exclusively concentrate on what they know how to do best - solving technology problems. To make things even worse, no single EDA provider alone can provide all the tools needed to develop and deploy a SOC. As a result, these large teams are spending millions of dollars and endless frustrating hours trying to make all the tools work with each other. Two components are required to streamline and simplify the complex task of building large chips. First, we require an Open, Industry Standard database (e.g., OpenAccess) that will allow tools from various EDA providers to interoperate. Second, we require involvement of EDA companies in providing tools to assist with the task of managing and synchronizing all the elements that participate in large IC design projects. This requires the bringing in and blending of chip design talent into the EDA industry to enable creation of practical solutions for the market.

Nancy Nettleton
Sun Microsystems, Inc., Palo Alto, CA

Asking the question "Tools or Users (or Interoperability): Which is the Bottleneck?" points the finger at the most readily available culprits but leaves the larger questions unasked. How about project structures set up to handle small designs with less than 20 engineers that collapse under the weight of 15M VDSM gates? How about design managers who haven't seen a transistor since they put down Weste & Eshraghian suddenly trying to handle both cell placement and signal integrity in disparate, uncommunicative tools that interpret timing and noise differently? How about business teams used to ASIC service who think that parts can be fabbed at UMC by design contracting services ?

Recent dramatic increases in chip design complexity have coincided rather unfortunately with fundamental shifts in supporting business models. This collision of increased complexity with changing business models has spread and exacerbated today's design bottlenecks. Three examples are as follows. (1) Wide availability of VDSM silicon process technologies has put microprocessor-like complexity within reach of more design teams who do not have the microprocessor margins that support extensive CAD organizations. Chip integration and design size issues that would have affected a handful of microprocessor design teams seven years ago are impacting hundreds of design teams today. (2) Tight-pitched VDSM metal enables us to pack millions more transistors onto a die, but engineers are waking up to the bad dream of post-route signal integrity iterations every bit as painful as the old timing closure loop. (3) System signaling strategies cause major upticks in volume and complexity of analog design, just as workhorses of system analog design (ASIC suppliers) are sapped by increasing cost of VDSM fabs. System teams who push performance envelopes with tricky signaling strategies based on, complex analog and package design must deal with multiple players, instead of one party who understands all of the technologies involved (device, package, test). Instead of focusing on system logic, these design teams must shop for or design their own analog IP, then try to mate it with separate fab and package houses. This increases risks of missed specs or market windows.

So, tools, users, interoperability: bottlenecks, or symptoms of industry changes that have not yet been absorbed? If we focus on addressing the fundamental project and technology changes within the emerging business models, then solutions to these three "bottlenecks" will naturally fall out.

Paul Rodman
ReShape, Inc., Mountain View, CA

" 'Tis a poor craftsman that blames his tools". This old saw was coined in a world where craftsmen of the pre-industrial age had great understanding of the materials and the tools with which they practiced their trade. Unlike the craftsmen of old, today's chip designers rarely have an opportunity to gain in-depth familiarity with a particular set of tools or materials. Thanks to Moore's Law, IC design tools and materials are not static; each becomes more and more complicated to use. The tools used on any chip design are new (even mature tools have releases every six to 12 months), which means a new UI to learn, and new bugs and instabilities to contend with. And vendors always sing their siren's song of new features to fix last year's problems. It is no wonder that people blame their tools at the end of projects.

Will tool suppliers change? Not likely! And who knows more about how to build chips: Tool developers or Tool users? Clearly, it is users who build chips, and when EDA vendors have tried to tell users how to do their job, they've always failed. Will we see a Consumer Reports for tools and processes? www.deepchip.com has tried, but there is just so much time a passionate moderator and kindhearted users can devote to sharing what they know.

Ultimately, today's chip-designer craftsman will likely fall victim to the industrial age. Just as standard part innovation enabled the industrial age, a standard part foundation is now being laid in our industry to make automation possible. We see the rise of standard silicon intellectual property building elements. EDA tools have undergone de facto standardization in RTL simulation, logic synthesis, place and route, and DRC/LVS. Process technology may be going this way too, e.g., standardizing on a single, open 0.10-micron semiconductor manufacturing process. With such standardized tools and design elements in place, best in class methods can be codified to automate the building of chips. Yes, automation comes at a cost: craftsmen's flexibility, and the ability to optimize at the leaf level, are lost. But automation of methods will enable fast throughput, and allow engineers to do many "what if" experiments to globally optimize the design, predictably yielding smaller and faster designs.

Lambert van den Hoven
Philips, Eindhoven, The Netherlands

Tools, Users and Interoperability are all causing productivity problems. That designs are successfully made is a combination of experience (do's and don'ts), adoption of less-ideal design methodologies (e.g. combination of flat and hierarchical methodologies), guardbanding of portions of a design (e.g., for timing), trade-offs between available time and risk (e.g., test coverage), a strong drive for re-use at higher levels, a patchwork for tool deficiencies, and sometimes luck. It is no surprise that almost every design requires a redesign to meet original specs (cf. Numetrics). Though the trend line is for fewer redesigns per gate count, rapid increase in complexity, impact of advanced process technologies, and prototyping costs represent growing concerns.

Tools have become more complex due to increased functionality, demand for higher capacity of individual tools, tighter connections/interfaces to the tool environment, and the need for improved user interfaces. A big challenge lies in the area of test and QA scenarios that are often in direct conflict with market pressure to release. The focus is still too much on single-tool testing, while the end user is only helped by a combination of tools suited for specific subflows. Our own experience is that most problems found in subflow release (qualification) are due to weak interfaces or lack of appropriate standards.

Interoperability has been a discussion topic for many years. Standadization far beyond the level of direct interfaces has been attempted, but without much success. Recently, new initiatives have again started to tackle this problem. New elements (e.g. open source based standards) may lead to some successes, but expectations are sometimes optimistic. A minimum requirement is to develop interoperability rules and standards for technology parameters, libraries, and IP blocks (both in hardware and software, based on various re-use levels (e.g. hard/soft macros)).

Users can be grouped into various categories, ranging from individual point-tool end user to design manager/supervisor. While the end user is more focused on functionality and stability, overall management attention is on throughput time and cost. Since truly intuitive solutions are often not available, proper (= efficient) use of point tools requires extensive training, augmented by on-the-job experience (i.e., to comprehend do's and don'ts).

Life Is CMOS: Why Chase the Life After?

George Sery, Shekhar Borkar, Vivek De
Intel Corporation
RN3-21, 2200 Mission College Boulevard
Santa Clara, CA 95052-8119
001-408-765-9367
george.sery@intel.com

ABSTRACT

This paper discusses potential solutions to the CMOS device technology scaling at gate lengths approaching 10nm. Promising circuit and design techniques to control leakage power are described. Energy-efficient microarchitecture trends for general-purpose microprocessors are elucidated.

Categories and Subject Descriptors

B.7 INTEGRATED CIRCUITS

B.7.1 Types and Design Styles – *Microprocessors and microcomputers, VLSI.*

General Terms

Performance, Design

Keywords

Technology scaling, Leakage control, Microarchitecture

1. INTRODUCTION

We are encountering several challenges in maintaining historical rates of performance improvement and energy reduction with CMOS technology scaling as we enter the sub-100nm technology generation. Some of the key bottlenecks are related to reducing device parasitics such as source/drain resistances and gate overlap capacitances. Excessive subthreshold and gate oxide leakage are also emerging as serious problems. In addition, energy efficiency of the microarchitecture of general-purpose microprocessors is starting to play a more critical role in the performance vs. power and area trade-offs. Potential solutions to the device technology scaling challenges at gate lengths approaching 10nm are discussed in Section 2. Section 3 describes some promising circuit and design techniques to control leakage power. Energy-efficient microarchitecture trends are elucidated in Section 4.

2. DEVICE TECHNOLOGY CHALLENGES

2.1 Performance and Energy Scaling

In the sub-200nm technology generation, it is difficult to maintain traditional constant-field supply voltage scaling (Figs. 1 & 2) due to the non-scalability of threshold voltage from excessive leakage current considerations. Essentially, the electric field across the gate dielectric has been increasing by 10% per generation. This has been made possible by the superior long-term reliability offered by physically thinner gate oxides. In order to improve the delay of driving constant capacitance loads such as those posed by interconnects in high performance microprocessor designs, device saturation current per unit width must remain constant or increase from one technology generation to the next. This has been accomplished by reducing the rate of supply voltage scaling, clearly at the expense of escalating switching power density.

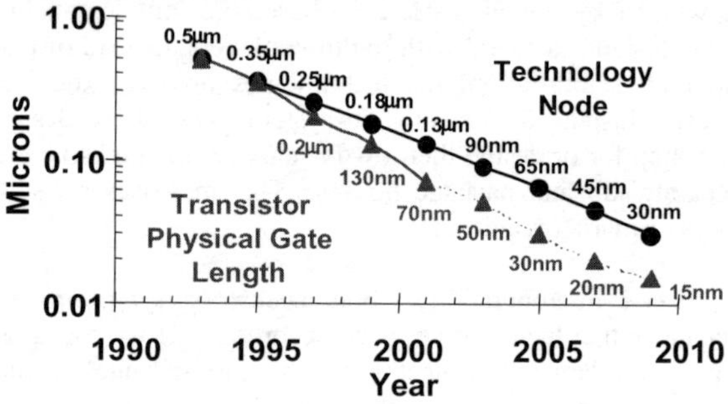

Fig. 1: Physical gate length scaling trend

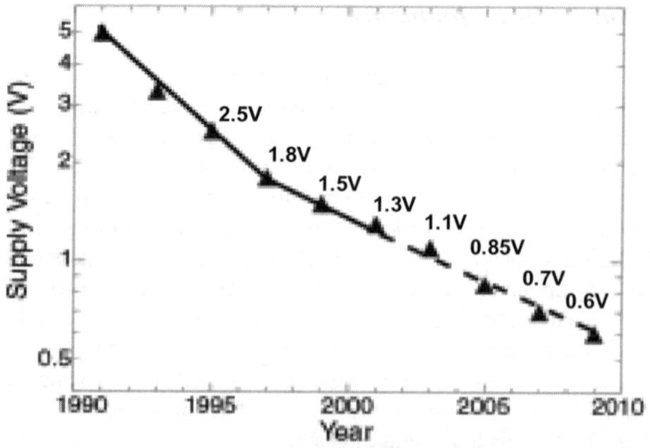

Fig. 2: Supply voltage scaling trend

Extrinsic source/drain resistance, gate overlap capacitance and junction capacitance do not scale in a desired fashion. This limits the circuit delay improvements achievable from large intrinsic device saturation currents. Reducing the depth of the source/drain junction extension or tip improves short-channel effects and thus allows a shorter gate length. However, tip depths less than 50nm causes the tip resistance to become so large that drive current becomes smaller (Fig. 3). Increasing the tip doping concentration beyond the solid solubility limit can help alleviate this problem to some extent. Furthermore, abruptness of the doping profiles, both in vertical and lateral directions, must be increased.

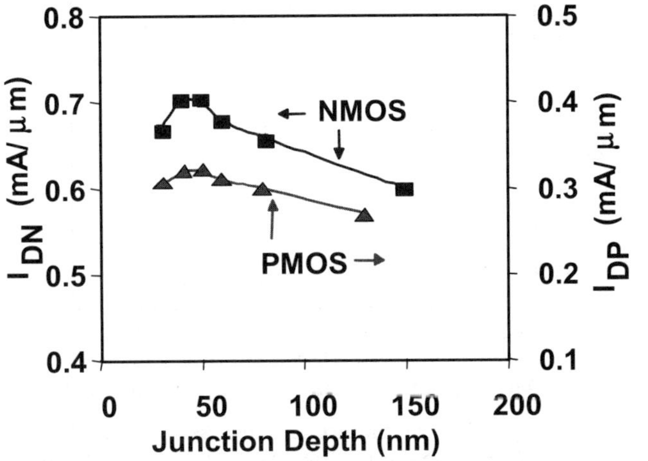

Fig. 3: Optimal source/drain extension depth

Spreading resistance from the inversion layer to the source/drain extension region also limits drive current. This makes it difficult to scale the gate overlap length below 10nm. In addition, increasing channel doping, demanded by sub-surface punch through control and short-channel effect reduction, causes capacitance of the gate-edge junction sidewall to increase with technology scaling and degrades delays of wide-OR circuits such as bitlines in the cache. In spite of all these limitations, device CV/I delays well beyond a terahertz is achievable at sub-1V supply voltages using a traditional planar bulk CMOS device structure in the 15-30nm gate length regime (Figs. 4 & 5).

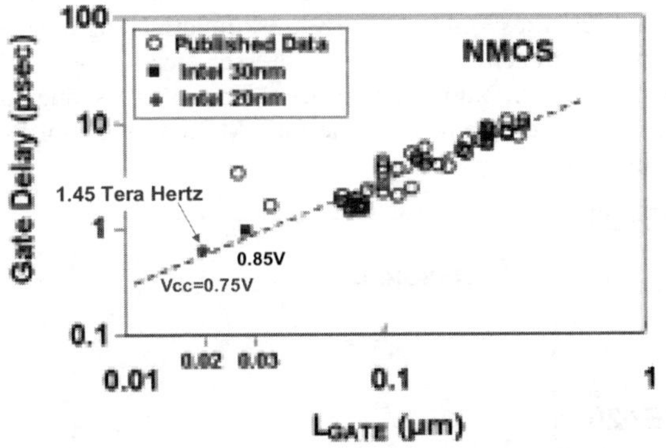

Fig. 4: CV/I delay scaling trends for bulk CMOS

A fully-depleted SOI device structure, referred to as the Depleted Substrate Transistor (DST), is promising for alleviating many of the challenges discussed before (Fig. 6). The subthreshold swing is much steeper compared to either bulk or partially depleted SOI devices when the silicon film thickness is below 30nm, resulting in a fully depleted channel. This allows Vt to be reduced for a specific leakage target and boosts drive current. Furthermore, the oxide layer below the silicon channel completely eliminates sub-surface punch through and junction leakage currents. Therefore, channel doping can be reduced. This reduces the gate-edge junction sidewall capacitance dramatically.

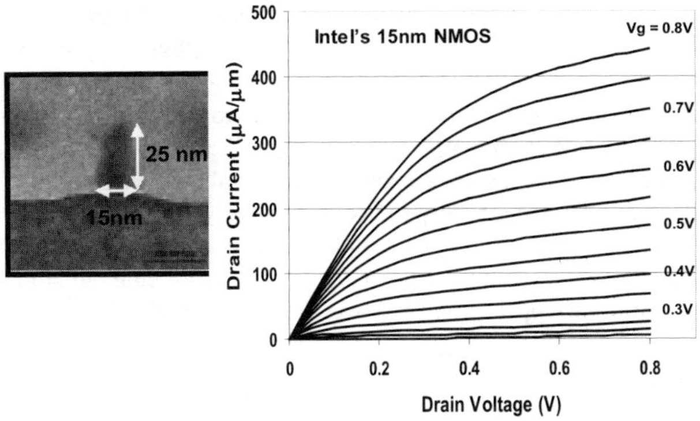

Fig. 5: Intel's 15nm bulk NMOS transistor

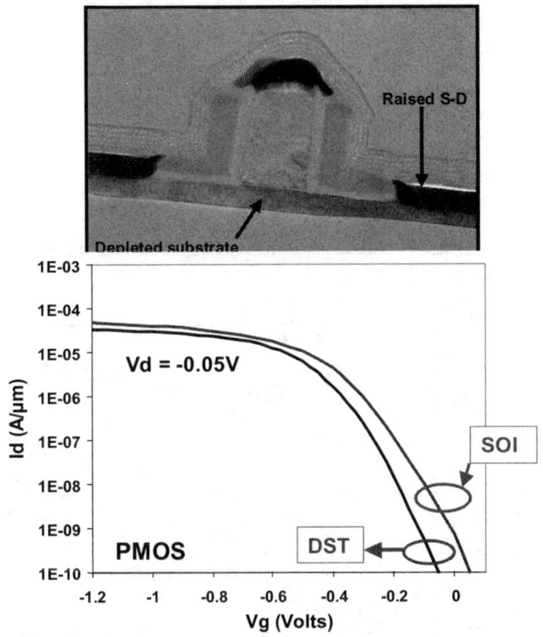

Fig. 6: Depleted substrate transistor (DST) with raised S/D

The source/drain extension depth in DST can be scaled by simply scaling the silicon thickness to improve short-channel effects. The buried oxide layer also serves as a diffusion stopper and creates more abrupt vertical doping profiles in the source/drain region. When combined with a raised source/drain structure, the drive current improvement due to lower parasitic resistance is as much as 30%. The main challenge associated with further development of DST with conventional polysilicon gates is achieving a sufficiently tight control of the silicon film thickness since the threshold voltage is quite sensitive to the film thickness. This problem can be alleviated to some extent by migration to a metal gate electrode whose work function is chosen appropriately to provide the appropriate Vt. Two different metals may be needed for NMOS and PMOS. When the gate length is pushed to the DIBL limit, threshold voltage sensitivity to variations in silicon film thickness will still need to be dealt with. In any case, DST provides a promising scaling path to sub-20nm technology generation.

2.2 Subthreshold and Gate Oxide Leakages

Subthreshold leakage current of a transistor is increasing by ~5X per generation. At high temperature, it exceeds 1000nA/um in sub-100nm technology nodes (Fig. 7). As the physical gate oxide thickness approaches sub-10Å regime, gate oxide leakage becomes larger than 100A/cm^2 (Fig. 8) due to direct band-to-band tunneling. Although gate oxide leakage increases weakly with temperature, it accelerates exponentially with increase in supply voltage at a rate of 2X larger leakage for every 100mV increase in voltage. Junction leakage is an additional component of concern since it is increasingly dictated by tunneling as channel doping concentrations approach $5X10^{18}$ cm^{-3} in the channel.

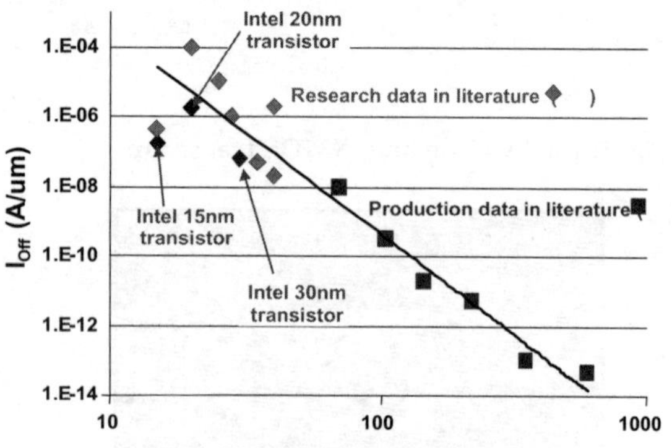

Fig. 7: Subthreshold leakage scaling trend

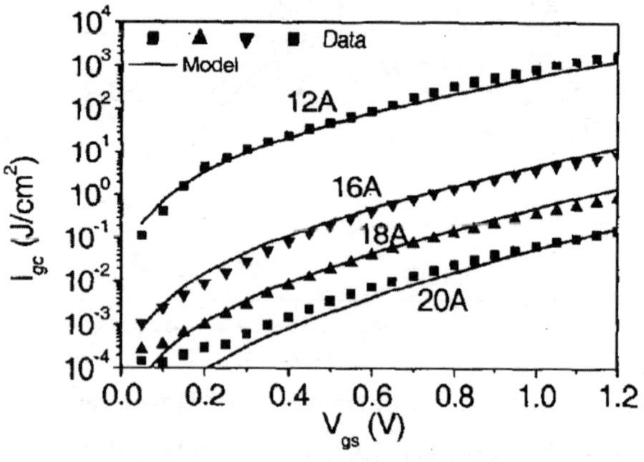

Fig. 8: Gate oxide leakage scaling with thickness and voltage

There is a compelling need to use high-K dielectrics as the gate dielectric material to replace silicon dioxide or oxynitride. High-K dielectric materials would allow electrical thickness of the gate dielectric to be scaled to provide large capacitance per unit area, while keeping the tunneling leakage per unit area within acceptable limits due to larger physical thickness. Scaling of electrical oxide thickness is essential to provide sharp subthreshold swing, large drive current and control short-channel effects. Characteristics of several candidate high-K dielectrics are compared in Fig. 9. HfO_2 and ZrO_2 provide the smallest gate leakage, 2-3 orders of magnitude smaller than oxide, for a target electrical thickness in the sub-10Å regime. Since the bandgap reduces with increasing permittivitty, gate leakage due to thermal emission dominates for materials with very high K values. Thus, Ta_2O_5 is not as attractive for this application. Of course, many process integration challenges need to be resolved and silicon-dielectric interface quality needs to be improved for these new dielectric materials to provide improvements to CMOS circuit delay.

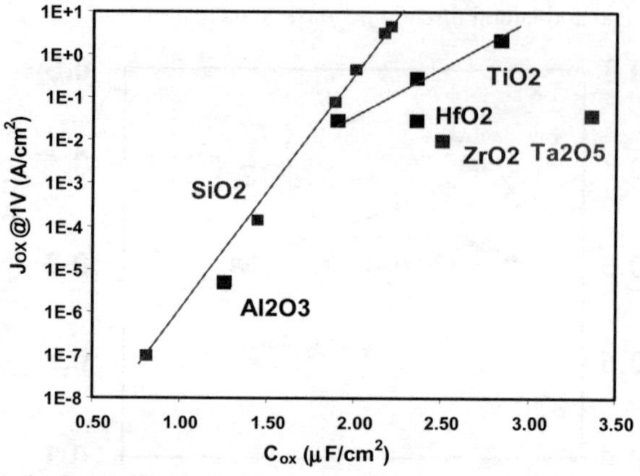

Fig. 9: Capacitance and leakage of high-K dielectric materials

The scaling of electrical gate oxide thickness is limited by poly depletion and separation of inversion layer charge from the oxide-silicon interface at high vertical fields due to quantum-mechanical (QM) effects (Fig. 10). Each of these effects adds 5Å to the effective electrical oxide thickness at the highest gate voltage. These effects become more significant as gate voltage increases. Thus, when averaged over the entire gate voltage range from Vt to the maximum supply voltage, their impacts on drive current are less severe. Nevertheless, in order to maximize the benefit of migrating to a high-K gate dielectric, poly depletion should be reduced or eliminated. Increasing poly doping beyond the solid solubility limit is desirable. Transition to a metal gate fully eliminates poly depletion. But metal gates with appropriate work functions for NMOS and PMOS must be identified and process integration issues must be resolved. Combining metal gates with DST to set the threshold voltage by work function engineering is a promising approach that also addresses the poly depletion problem. However, additional process complexities due to two different gate electrode metals, one for NMOS and one for PMOS, will be incurred.

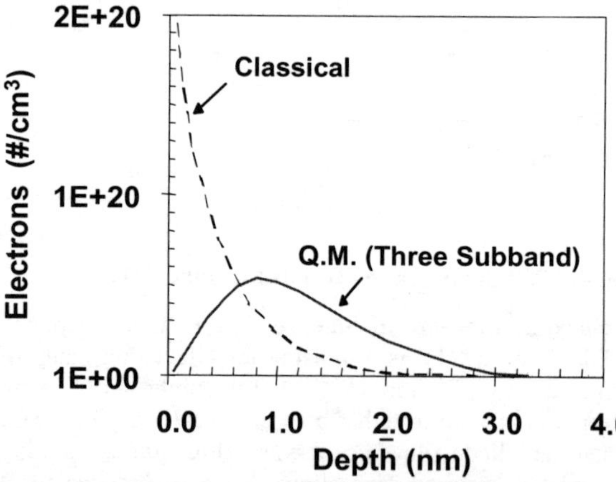

Fig. 10: Increase in electrical oxide thickness by QM effects

3. LEAKAGE CONTROL TECHNIQUES

Leakage power is becoming a larger fraction of the total active power of microprocessors (Fig. 11). This poses serious challenges for heat removal and power delivery in high performance processors. Excessive leakage power can also cause thermal runaway during burn-in, and impact burn-in cost. Subthreshold leakage dominates at high temperature and gate oxide leakage is a significant contributor to the burn-in leakage power due to the higher voltage used. Of course, standby leakage power at room temperature also needs to be kept sufficiently small for battery-operated systems.

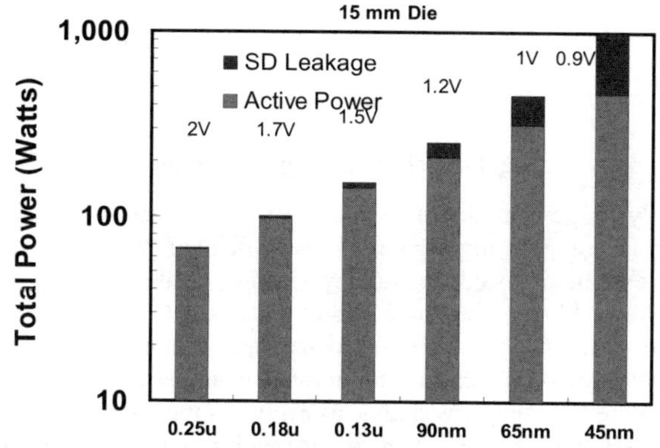

Fig. 11: Switching and leakage power scaling trends

3.1 Dual-Vt and Body Bias

Dual-Vt designs can reduce leakage power during active operation, burn-in and standby. Two Vt's are provided by the process technology for each transistor. Performance-critical transistors are made low-Vt to provide the target chip performance. Rest of the transistors are made high-Vt to minimize leakage power. Since the full-chip frequency is dictated by only a fraction of transistors in the critical paths, this selective Vt assignment is possible without degrading overall chip performance achievable by using a single low-Vt transistor everywhere. Fig 12 shows an example circuit block, where all low Vt design provides 24% delay improvement over all high Vt design. Notice that as you start inserting low Vt devices (Y axis), the delay improves (X axis). Only 34% of the total transistor width needs to be low-Vt in this example, to get the same frequency as using low-Vt everywhere. Typically, low-Vt device leakage is 10X higher than high-Vt. Thus, by carefully employing low-Vt up to 34% of the total width, 24% delay improvement is possible with ~3X increase in leakage, compared to all high-Vt design.

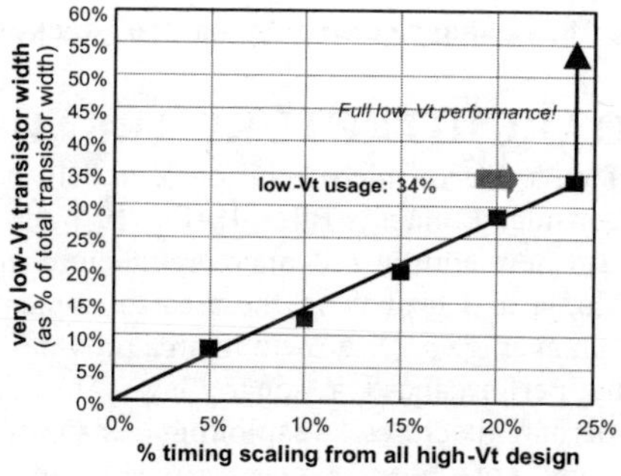

Fig. 12: Performance vs. leakage in dual-Vt designs

Another technique to reduce leakage power during burn-in and standby is to apply reverse body bias (RBB) to the transistors to increase Vt since high performance is not required during these modes. There is an optimal reverse body bias value that minimizes leakage power. Using reverse body bias values larger than this value causes the junction leakage current to increase and overall leakage power to go up. In sub-100nm technology generation, approximately 500mV RBB is optimal. 2-3X reduction in leakage current is achievable. However, effectiveness of RBB reduces as channel lengths become smaller or Vt values are lowered (Fig. 13). Essentially, the Vt-modulation capability by RBB weakens as short-channel effects become worse or body effect diminishes due to lower channel doping. Therefore, RBB becomes less effective with technology scaling and as leakage currents are pushed higher by shorter L or lower Vt.

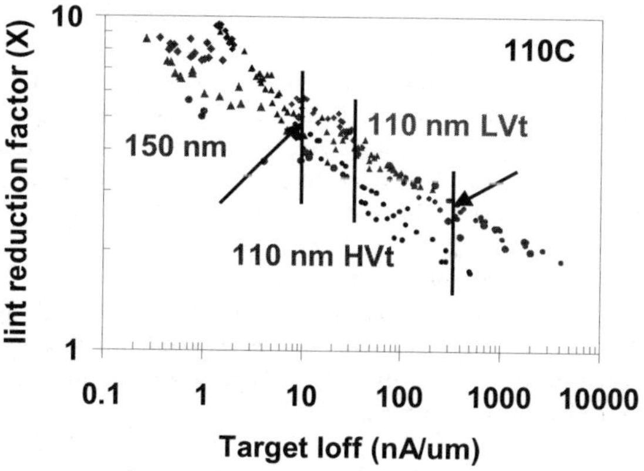

Fig. 13: Subthreshold leakage reduction by reverse body bias

3.2 Stack Effect

Leakage current through series-connected transistors or transistor "stacks", with more than one device "off", is at least an order of magnitude smaller than that through a single device (Fig. 14). This so-called "stack effect" can be exploited for leakage reduction in circuits. The stack effect factor, defined as the ratio of single device leakage to stack leakage, increases as the DIBL factor becomes larger and supply voltage increases. As the rate of supply voltage scaling diminishes and DIBL effects become stronger with technology scaling, the effectiveness of leakage reduction by stacks becomes higher (Fig. 15).

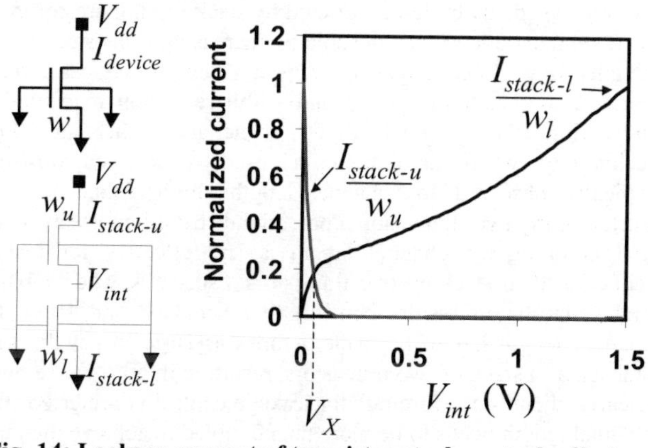

Fig. 14: Leakage current of transistor stacks – stack effect

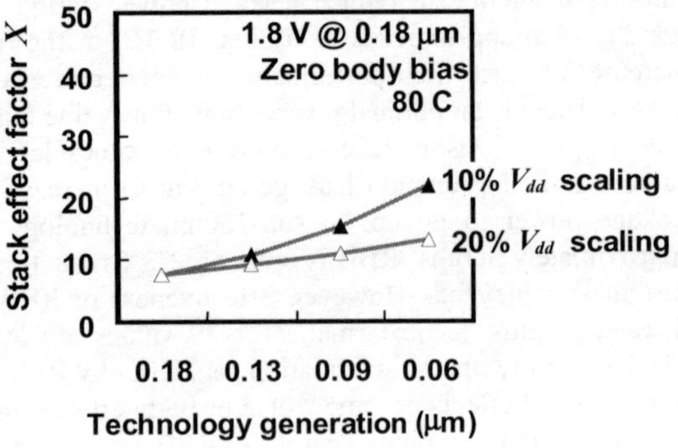

Fig. 15: Scaling of stack effect factor

Leakage reduction by stack effect can be exploited by converting a single transistor to a two-transistor stack in a logic circuit. The widths of these transistors can be half of the original size or other combinations can be chosen to preserve same input capacitance load as the original single device. Leakage vs. delay trade-off provided by this "stack forcing" technique applied to both high-Vt and low-Vt devices is illustrated in Fig. 16. Clearly, stack forcing can be used to emulate additional higher Vt devices without increasing process complexity. Stack forcing can be applied to transistors in non-critical paths in single-Vt or dual-Vt designs to reduce overall chip leakage power without impacting chip performance. Also, robustness of leakage-sensitive circuits can be improved by this technique.

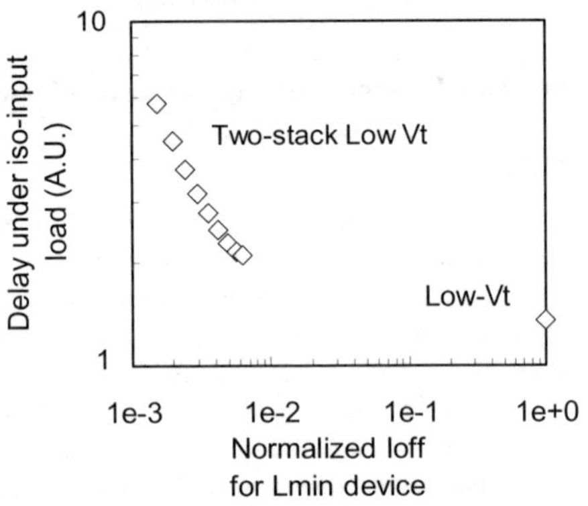

Fig. 16: Leakage vs. delay trade-offs by stack forcing

Leakage vs. delay trade-offs offered by stack forcing are compared with similar trade-offs achievable by increasing transistor channel lengths (Fig. 17). Increasing transistor length reduces leakage because of threshold roll-off and width reduction mandated by preserving the original input capacitance. In sub-100nm technology, where halo doping is used, reverse Vt roll-off is typically observed for channel lengths higher than nominal. Furthermore, two-dimensional potential distribution effects dictate that doubling the channel length is less effective for leakage reduction than stacking two transistors, especially when DIBL is high. Simulation results confirm this behavior and show that channel length has to be made 3 times as large to get the same leakage as a stack of two transistors, resulting in 60% worse delay. Clearly, then "stack forcing" for leakage control is preferred if the channel length needs to be more than doubled to achieve the target low leakage.

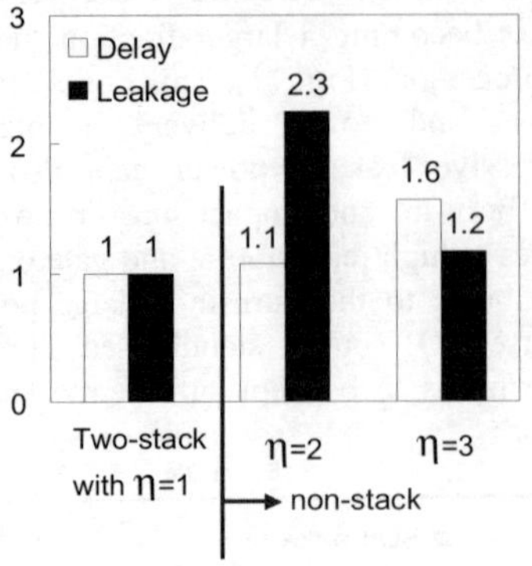

Fig. 17: Stack forcing vs. longer channel length

Typically large circuit blocks contain some series-connected devices in complex logic gates. These so-called "natural stacks" can be exploited to reduce standby leakage. Leakage power of a large circuit block, such as a 32-bit static CMOS Kogge-Stone adder, depends strongly on the primary input vector (Fig. 18). The total "off" device width and the number of transistor stacks with two or more "off" devices change as primary input vectors change. This causes the leakage power to vary with input vector. When a circuit block is "idle", one can store the input vector that provides least amount of leakage at the primary input flops. This can reduce standby leakage power by 2X. There is no performance overhead since this pre-determined input vector can be encoded in the feedback path of the input flip-flop. The minimum time required in standby mode, so that the energy overhead for entry and exit into this mode is less than 10% of the leakage energy saved, is 10's of uS. This time reduces further with technology scaling as leakage levels increase, making this technique more attractive. Of course, EDA tools will be needed to identify this "lowest leakage" input vector efficiently during design phase for each circuit block.

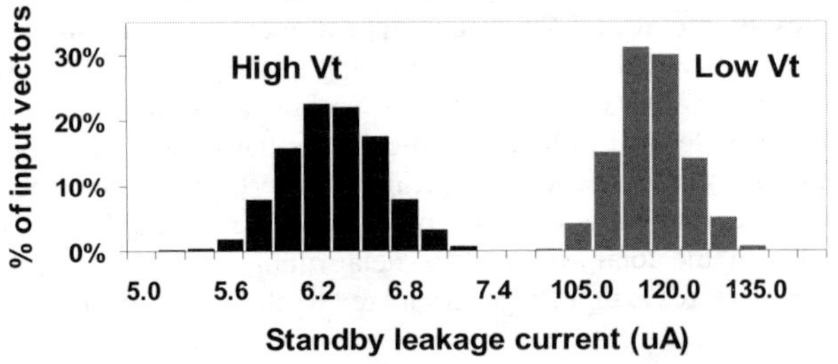

Fig. 18: Leakage control by natural stacks

4. MICROARCHITECTURE TRENDS

To evaluate effectiveness of microarchitecture in delivering higher performance, consider Pollack's rule [4]. Fig 19 plots growth in performance of a new and an old microarchitecture in the same process technology, and growth in the area to implement them. Notice that on an average a 2X growth in area provides only 1.4X increase in the performance—a square law. This shows that traditional microarchitectures, exploiting instruction level parallelism, have not been power efficient in delivering performance.

This is further elaborated in Fig 20, which shows estimated increase in die area, performance, and power due to microarchitecture advances such as super-scalar, dynamic, and netburst. The growth in area and power reflects growth in the number of transistors, and power hungry circuit styles employed for implementation. Notice that each advance has consumed about 2X power delivering 40% more performance. Therefore, we must find alternate energy efficient microarchitectures to continue to deliver higher performance.

Applications will have to lend themselves to incorporate thread-level parallelism, followed by multi-processing to deliver near-linear performance with power. Furthermore, certain application tasks could be easily served by special purpose hardware on the die tailored for the applications, and thus power efficient.

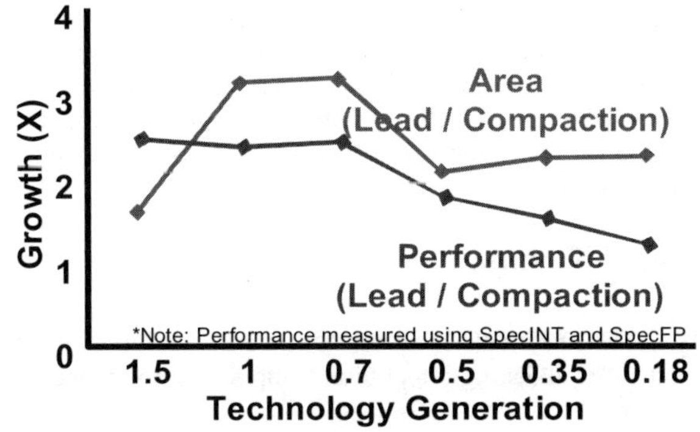

Fig. 19: Microarchitecture efficiency trends

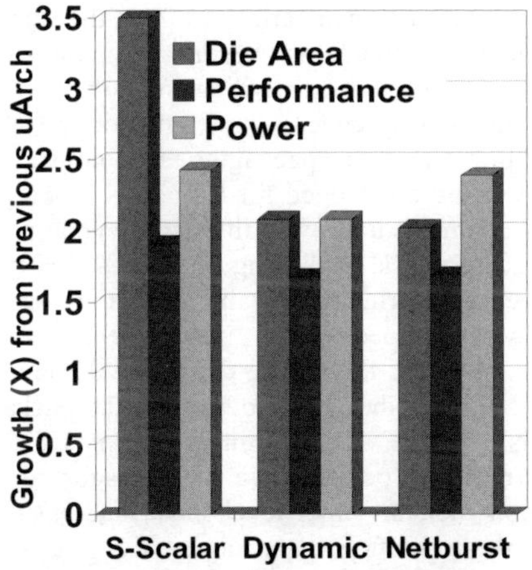

Fig. 20: Performance, power and area trade-offs of general purpose microarchitectures

Fig 21 compares estimated active power density of logic and static memory in a given process technology. Memory power density tends to be an order of magnitude lower than that of logic. This is because only a part of the memory is accessed at any given time. Also, memory transistors can withstand relatively higher threshold voltages, reducing the leakage power compared to logic. To make up for the loss of transistor performance, memory operations can be pipelined, with modest increase in latency.

Therefore future microarchitectures could exploit lower power density of memory to stay on the performance trend, and yet lower active and leakage power. The trend is already evident as shown in Fig 22, which plots cache memory transistors in microprocessors in several technology generations. Future microarchitectures will use even bigger caches to continue to deliver higher performance.

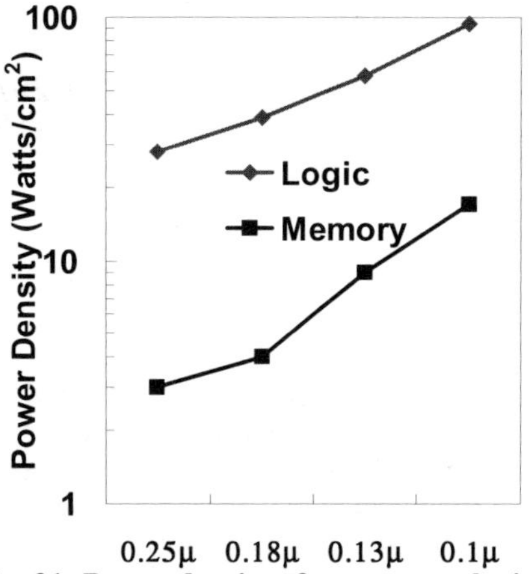

Fig. 21: Power density of memory vs. logic

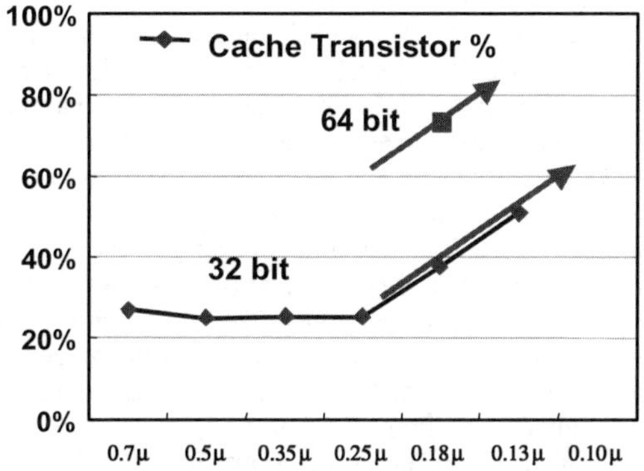

Fig. 22: On-chip memory integration trends

5. CONCLUSIONS

This paper has described CMOS scaling challenges for gate lengths approaching 10nm, and potential solutions in circuits and microarchitecture. These solutions may appear difficult, but are more mature and less risky than other proposed alternatives for CMOS. That is why CMOS is it, for now, and for the foreseeable future.

6. REFERENCES

[1] R. Chau et al., "30 nm physical gate length CMOS transistors with 1.0 ps n-MOS and 1.7 ps p-MOS gate delays", *IEDM 2000*, pp. 45-48.

[2] R. Chau, "30nm and 20nm Physical Gate Length CMOS Transistors", *Silicon Nanotechnology Workshop*, 2001.

[3] R. Chau et al., "A 50nm depleted-substrate CMOS transistor (DST)", *IEDM 2001*, pp. 621-624.

[4] Fred Pollack; New Microarchitecture Challenges in the Coming Generations of CMOS Process Technologies; *Micro32*, 1999.

The Next Chip Challenge: Effective Methods for Viable Mixed Technology SoCs

Dr. H. Bernhard Pogge
IBM Microelectronics
2070 Route 52, Internal Zip: 47B
Hopewell Junction, NY 12533
845-894-1858
pogge@us.ibm.com

Abstract

The next generation of computer chips will continue the trend for more complexity than their predecessors. Many of them will contain different chip technologies and are termed SoCs (System on a Chip). They present to the process community, the system and circuit communities, as well as to the design and test communities major new challenges. On the other hand they also offer at the same time also new opportunities!. For one, the desire to bring more functionality onto a single chip tends to require additional processing, which in turn results in various degrees of device compromises. The chips will also tend to become larger due to the added device content, and this generally will impact the yieldability of the final chip. And such chips will require potentially new approaches to validate the intended design performances. Chip sector reuse must also be brought into the discussion and wherever possible into practice. The net effect implies higher chip costs. Much of the industry's efforts are therefore focused in addressing these challenges; however, so far, not yet very successfully. The alternative has been to continue in the placement of chips onto substrate modules. Yet, this solution creates practical limits on achievable wiring densities and bandwidth, due to the spacing requirements of the C4 interconnection. Furthermore, every C4 joint is associated with a signal delay of about 50 psec. All of these handicaps would potentially benefit greatly from new SoC methods, starting with the fabrication methodology and extending it into the chip design and test areas.

Such a direction has been set in motion. The opportunity for a uniquely new chip fabrication method has emerged by combining a set of somewhat diverse processes. It is based on a judicious selection of process elements from the traditional chip area and combined with those of a somewhat more recent chip packaging process methodology. This approach results in overcoming simultaneously all of the key current process limitations as experienced with today's SoC chip designs, as well as eliminates certain chip packaging technology handicaps. Yet, it does not require the need for new process tooling. It relies on currently existing process tooling and process methodologies.

This new process direction has been found to be quite applicable to a number of desirable SoC device designs, and offers new opportunities for yet another expansion of the current semiconductor technology base over the next few years. However, effective SoC designs and fabrications require a much closer and earlier collaboration between the process, design and test communities.

DAC 2002, June 10-14, 2002, New Orleans, Louisiana, USA.

General Terms: Design

Key Words: SoCs (System on a Chip); Chip Fabrication methods; Chip/Packing integration; Chip Subsector concepts

Introduction

Electronic devices are continuing to become more and more complex with each new generation, while at the same time their respective device elements are becoming smaller and smaller. This environment of device crowding and complexity continues to generate major challenges – not only for the chip process community, but also for the packaging technologist. The latter is continuously being challenged for his ability to keep up with a pace that is commensurate with the rapidly changing advances of the chip sectors. The packaging sector has clearly been for a number of years the primary restraint for further maximizing the potential system speeds of many semiconductor chip technologies. Whereas the wiring density at the chip level is in the "micron" regime, the wiring density in the package remains in the "mils" regime. To minimize the impact as much as possible, current high performance package substrates consist of a combination of many inter-level metal and insulator layers. Yet, even these complex packages are not sufficient to effectively accommodate the needs for today's high performance chips, much less so future chips.

However, new challenges are also being confronted by the device technologist, due to the fact that there is a general desire by the system designers to have more and more device capabilities on a single chip. This implies at least two major directions: larger chips containing more content, while at the same time intermixing more diverse chip technologies. Both have a strong influence on the overall yields of such chips, and in turn, on the ultimate fabrication costs. Large chips do not yield very well, and complex multi-technology chips require much more processing, and thereby also influencing the potential chip yields. And in fact, some multi-technology chips are not possible to be fabricated with today's traditional chip process capabilities because of general incompatible material properties. Furthermore, the need to test these types of mixed technology chips require new characterization methods and test procedures.

This general environment has therefore spawned a great deal of new thinking in terms of finding process alternatives to overcome some of these limitations. But the process remains somewhat sequential. That is, only after the process community has verified the ability to fabricate some of these new chips, the chip and system designers begin to consider them into their application space. And yet later, the test organizations scurry to find means for effective test methodologies.

Traditionally, the multi-layer ceramic (MLC) package has been used to package different type of technology chips on a common substrate. This allows for individual chip testing prior to packaging those chips. But the wiring density limitations of the C4 technology in such packages do create major barriers for advanced high performance devices. This environment would greatly benefit from new and innovative chip fabrication directions. One such recent direction has been a widening interest in stacked chips and 3D type chips. The latter is based on vertical interconnections through one chip in order to reach another. This work has been going on in several research laboratories for the past 5 - 7 years (1-3); yet, to date, no one has published definitive device results actually verifying the theoretical benefits. Such concepts have a number of major handicaps for which solutions have not been brought forward. One major hurdle is the inefficient removal of heat generated by any high performing set of chips stacked on top of each other. An alternative proposal is the concept of Si interposers serving as intermediates between the chip and the package. However, such interposer is not cheap to fabricate, since it requires extensive deep RIE etching for forming high aspect ratio vias, as well as subsequent filling of these deep vias. Furthermore, do they really serve as an effective solution, which could alternatively be achieved with simpler structures, such as thin film wiring? Again, the concept of interposers has been around for a number of years (4,5), but the actual use of them in any product seems to remain absent.

For the past few years, there has also been an increased interest in the creation of more innovative new packages to achieve closer chip-to-chip interconnection opportunities. GE's HDI (6-8) is one such example. The original concept is based on placing chips into substrate cavities, filling in the spaces surrounding the chips, and subsequently locating the positions of the chips and interconnecting them with some wiring technique. Quite a few process refinements have been made by GE's researchers over the years, but this initiative has not yet enjoyed any industry wide acceptance. More recently, slightly similar work was published by the research group at the Frauenhofer Institute in Berlin (9). Other approaches for more closely interconnecting chips to each other tend to be based on some version of C4 flip chip technology serving as the interconnect elements. However, the available C4 dimensions tend to place a practical limit on the possible wiring densities. Furthermore, each such C4 represents a signal delay of as much as 50 psec, so that every chip-to-chip interconnect will incur a total delay of about 100 psec. Since many advanced chip designs depend on about 30% of the total C4s for interconnections, the total delay budget for many chip designs, due to interconnects, can become significant.

As we therefore attempt to reach out into the future, what will the challenges and the potential solutions be which would further the impressive successes of the past? The evolutionary progress has been mapped out with some degree of certainty (i.e., the SIA roadmap). And it points to a number of roadblocks! However, looking back into the past, it becomes somewhat obvious that many of the technology breakthroughs are not the result of such evolutionary thinking, but rather relate to the introduction of more revolutionary approaches creating new process opportunities. The evolutionary forces are always active, and some level of progress will be made. On the other hand, revolutionary progress is less clearly defined in terms of a timetable or impact, but it will more than likely be elements of such innovations, which offer new directions and opportunities for further technology advances.

In the future, mobile systems are anticipated to predominate the semiconductor chip sector. This will bring additional pressures on improved chip packages; that is, lighter and smaller . Will entirely new packages be required, and can they be realized with simply evolutionary thinking? Or would this environment benefit for the chip sector to partner with the packaging sector by creating some level of technology marriage?

The T&J chip/package concept to create effective SoCs

The net chip content of many future chips seems to lie in the realm of SoCs, although there is some debate whether it may be SoPs (System-on-a-package). Viewing either version from the viewpoint of their respective traditional definitions, each of them has specific advantages and disadvantages. But by taking a new process approach - the concept of the T&J (Transfer & Joining) chip/package methodology – these respective handicaps are minimized, while maintaining and capitalizing on their respective collective advantages. Hence, it is possible to realize an effective SoC product, which does retain its inherent advantages; yet, incorporates the advantageous elements of an SoP structures. In essence, the T&J concepts migrates elements, traditionally done in the Thin Film sector of the Packaging house into the S/C house (Fig. 1).

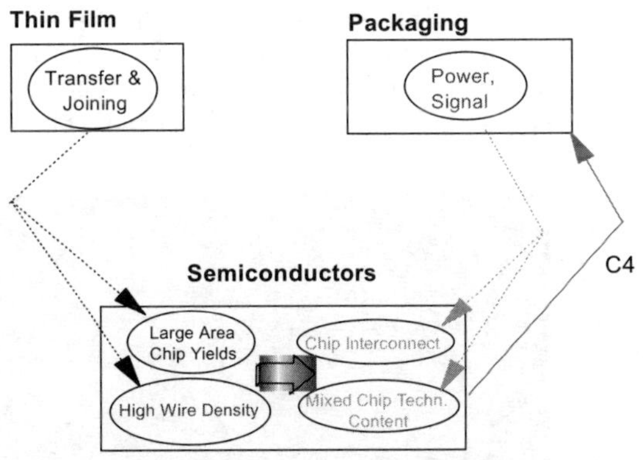

Figure 1. Technology Merging/Bridging

The T&J methodology relies on the initial fabrication of an interconnect wiring/insulator film structure on a glass wafer or plate; placing, attaching and bonding chips and/or full wafers to that metal interconnect structure; and subsequently removing the glass wafer or plate. This process sequence leaves behind the interconnect wiring tying together the attached and bonded chips and/or wafer as is shown in Fig 2.

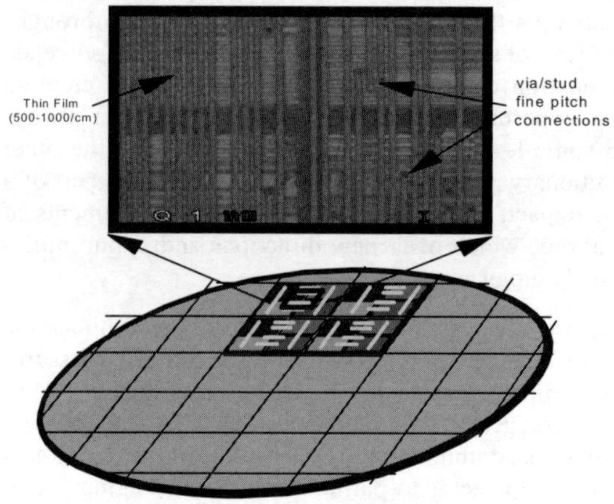

Figure 2. The S&T Generated SoC

In essence, the chip-to-chip interconnections, which previously relied on C4s, have been replaced by thin film wiring elements. The methodology allows for much higher wiring densities (up to a factor of 100) in comparison to C4s. This can have a major positive impact on the net device bandwidth. It also allows for chip-to-chip placements as close as about 25 - 60 um, as opposed to the typical 3 mm spacing of C4 based chip placements. Furthermore, the chip/wafer placement accuracies are well within1um. An additional major advantage of this fabrication methodology is the ability for parallel processing of all the parts and a minimal heat budget, the latter of which consists of only the bonding process itself. This alleviates the reliance on round wafers. In fact, it is preferable and desirable to utilize rectangular plate formats to minimize edge wafer losses.

The T&J concept also results in the ability to form large SoC chips from sub-sector (macros) parts, which (because of their respective processing) can also optimized. These will inherently be smaller chips and therefore offer much higher yields than a comparable monolith large chip. And of course the methodology allows for very diverse mixed chip content within an SoC chip.

Experimental

The T&J fabrication methodology can be done in existing manufacturing tools. Silicon chips were derived from 200 mm wafers. Chip placements were done on either 125 mm square glass plates or round 200 mm glass wafers. Most of the experimental data are based on the use of a single metal film on a silicon wafer added to the top of the normal BEOL (back end of the line) film structure, and a single metal film on the glass support plate and/or wafer, although this is not a limiting factor. Appropriate via and stud structures were generated to create the vertical interconnect bonding between the two metal film layers. A laser method was utilized for the removal of the glass plate and/or wafer from the metal interconnect structure. Different test vehicles were used to assess the various process elements.

Discussion

The T&J process is based on the attachment of chips to existing interconnect wiring separately fabricated on a supporting glass wafer (plate), which itself is only a temporary support for the interconnect wiring. The device chips are terminated with a appropriate insulating adhesive and patterns with via openings. In parallel, the interconnect wiring is built up on the glass wafer (plate), the number of layers being dictated by the specific device and SoC design requirements. The upper level of this interconnection wiring is terminated with an appropriate pattern of metal studs which mirror the via pattern on the device chips. Since the glass wafer serves as the receiver of the chips and/or full wafer, a single mask creating the stud pattern dictates all of the placement positioning for the vias on the chips. This allows for very precise chip positioning and chip-to-chip interconnections. After chip placement, the combined structure is heated to facilitate the full bonding of the chips to the interconnection studs. Chip positioning can be monitored through the glass wafer. See Fig. 3.

After proper bonding, the back sides of the attached chips are planarized and given a backside support with the use of a simple silicon wafer. After this, the glass wafer is removed with a laser ablation technique to leave behind the metal interconnect layers tightly bonded to the chips. The resulting metal interconnects and interconnected chip set structure is now ready to be appropriately diced and further processed for subsequent packaging needs However, it should be pointed out that in this instance the chip-to-chip interconnection already exist, leaving

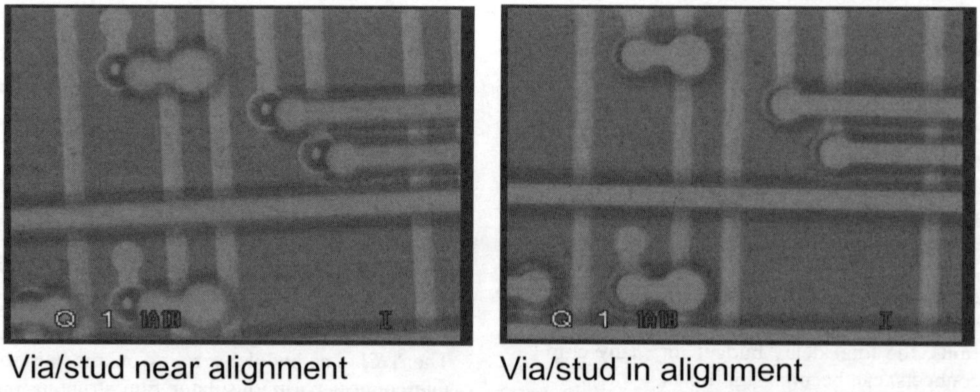

Figure 3. Via/Stud Alignment of Interconnect Wiring layers

only the need for power and signal from the package., and thus much fewer total C4s.

It should also be pointed out that with the T&J concept of fabricating the various chip macro sectors in parallel, the device and circuit design community must realize that the interconnect hierarchy needs to be carefully laid out. In today's large area SoCs, some of the interconnect wiring may be placed at some lower metal level because of either convenience or available space. However, for effective multi-technology chip content SoCs, this interconnect wiring should be brought up to a common set of metal layers. This allows for better testing and diagnostic capabilities, more efficient yield learning of the individual chip sub-sectors, and easier reuse of chip sectors for different designs.

Electrical data was generated for these types of via/stud interconnections to verify not only the ease of fabrication, but also the quality of the electrical interconnect structures. A key process element in the T&J process concept relates to the transferred wiring structure and its integrity when crossing the open gaps between chips which will exist between the attached chips. This is shown in Fig. 4 for different wiring dimensions and compared to similar control structures placed simultaneously on adjacent flat surfaces. No differences are evident. A spacing of 60 um does therefore not create any concerns and can be easily crossed.

Conclusion

The T&J process methodology is a very viable means for achieving true SoC chips without their traditional respective handicaps. That is, true SoCs allow for an effective mixturing of diverse technology chips into a single chip format. By appropriate and selective use of the available building blocks, which exist in both the Package (in this instance, within their Thin Film sector) and the S/C houses, it is possible to create multi-functional chips into a single SoC format. Alternatively, large SoC chips can be fabricated with this technique, and these need not be shackled by the typical yield handicaps associated with large area chips. However, it is important to point out, that it is not entirely feasible to simply take a large chip, section it into smaller ones, and then rebuild the chip back up into a large chip via the T&J method. Physically, this is possible; however, practically the breakage of already existing circuit design needs to be considered from the beginning, not as an afterthought or as a possible remedy for an existing chip yield crisis. In fact, as has often been pointed out in the recent literature, large area chips need the cooperative collaboration between the system, the circuit, the test, and the process engineer from the start of the device desig cycle. This collaboration should also be fostered for any mixed chip designs, since the overall result may be significantly improved, or may even be made much more viable, through such efforts.

References

1. G. Ruhl et al., Applied Surface Science 91, 382 (1995).
2. P. Ramm et al., Microelectronic Engineering 37/38, 39 (1997).
3. J. Burns et al., 2001 IEEE International Solid State Circuits Conference 17, 268 (2001).
4. F. Kroener, German Patent No. 10004647 (Feb. 3, 2000).
5. M. Matsuo et al., 2000 Electronic Components and Technology Conference, 1455 (2000).
6. R. Fillion, Symposium Proceedings, Dupont Symp. High Density Interconnect,, 5 (1990).
7. R. Fillion et al., Symposium Proceedings, 1992 ISHM Int. Symp. Microelectron., 391 (1992).
8. R. Fillion et al., Circuit World 21, 28 (1995).
9. J. Wolf et al., Intern. J. Microcircuits and Electronic Packaging 21, 119 (1998).

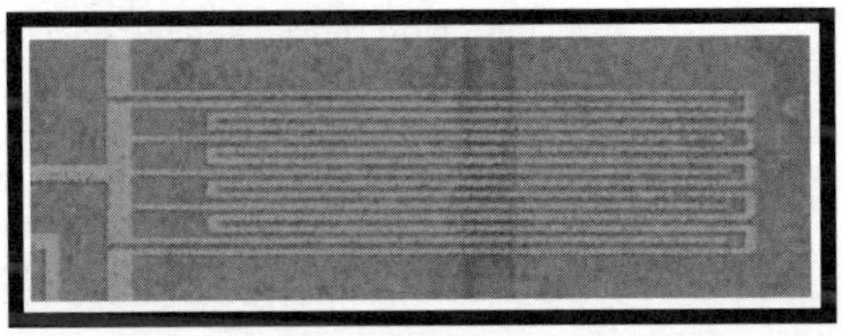

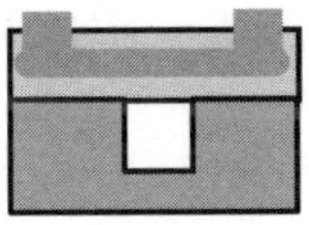

Gap crossing	25um	20um	15um	10um	5um
# of crossings	480	480	720	720	720
Yield	100%	100%	100%	100%	100%
Resistance (Ω)	4.254	3.725	3.977	5.328	10.47

Surface crossing	25um	20um	15um	10um	5um
# of crossings	240	240	360	360	360
yield	100%	100%	100%	100%	100%
Resistance (Ω)	4.314	3.323	4.566	5.580	10.37

Figure 4. Electrical Data for T&J Interconnect Wiring

Few Electron Devices: Towards Hybrid CMOS-SET Integrated Circuits

Adrian M. Ionescu, Michel J. Declercq
Santanu Mahapatra
Swiss Federal Institute of Technology Lausanne
CH-1015 Lausanne, Switzerland
+41 21 693 3978 / 3974 / 4609
{adrian.ionescu, michel.declercq, santanu.mahapatra}@epfl.ch

Kaustav Banerjee
Center for Integrated Systems
Stanford University
Stanford, CA 94305, USA
001 650 724 2909
kaustav@cis.stanford.edu

Jacques Gautier
CEA-DRT – LETI/DTS – CEA/GRE
17 avenue des Martyrs
38054 Grenoble cedex 9, France
33 4 38 78 48 56
jgautier@cea.fr

ABSTRACT

In this paper, CMOS evolution and their fundamental and practical limitations are briefly reviewed, and the working principles, performance, and fabrication of *single-electron transistors* (SETs) are addressed in detail. Some of the unique characteristics and functionality of SETs, like unrivalled integration and low power, which are complementary to the sub-20 nm CMOS[1], are demonstrated. Characteristics of two novel SET architectures, namely, *C-SET* and *R-SET*, aimed at logic applications are compared. Finally, it is shown that combination of CMOS and SET in hybrid ICs appears to be attractive in terms of new functionality and performance, together with better integrability for ULSI, especially because of their complementary characteristics. It is envisioned that efforts in terms of compatible fabrication processes, packaging, modeling, electrical characterization, co-design and co-simulation will be needed in the near future to achieve substantial advances in both memory and logic circuit applications based on CMOS-SET hybrid circuits.

Categories and Subject Descriptors

B.7 INTEGRATED CIRCUITS
B.7.1 Types and Design Styles – *Advanced Technologies*

General Terms

Design, Experimentation, Measurement, Performance

Keywords

Nanoelectronics, Single-Electron Transistors, Ultimate CMOS, Hybrid CMOS-SET Circuits, Low power, Inverter, Quantizer.

1. INTRODUCTION

The modern low power electronics originated with the invention of the bipolar transistor in 1947 as an extension of the vacuum tube: the requirements for several watts of tube power and several hundred volts for its anode were replaced by unrivalled tens of mW [1]. Next breakthroughs, related to the first MOS transistor and then, CMOS circuit in 1963, [2], have pushed these limits even further. Is this story going to repeat? The Single-Electron-Transistor (SET), [3, 4], is finally nothing more than somewhat similar evolution or natural extension of the three-terminal MOSFET. Will SET replace it at the nanometer scale? Or will SET have to co-exist with CMOS in the near foreseeable future? This paper attempts to propose some answers and to trace some perspectives *without the ambition of any prophecy*.

1. PUSHING CMOS LIMITS: *The 10 nm WALL*

During the last several years the International Technology Roadmap for Semiconductors (ITRS) has experienced continuous accelerations and revisions, bringing forward the advent of new generation of devices in industrial production (Fig. 1) In its recent 2001 version [5], ITRS projects the introduction of 45 nm-node CMOS devices into production by 2010. This requires transistor gate lengths in the sub-25 nm range operating at extremely low supply voltages (~0.7-0.8V). Scaling MOSFET below 50 nm exacerbates multiple challenges like the control of *short channel effects* (SCE) and the increase of I_{on} (>700μA/μm, at room temperature) while keeping I_{off} acceptably low (1000pA/μm, at room temperature, for low operating power logic). Conventional MOSFET architecture fails to provide such performance for feature sizes below 20 nm due to the simple fact that control of SCE results in severe limitations in the current drive capability.

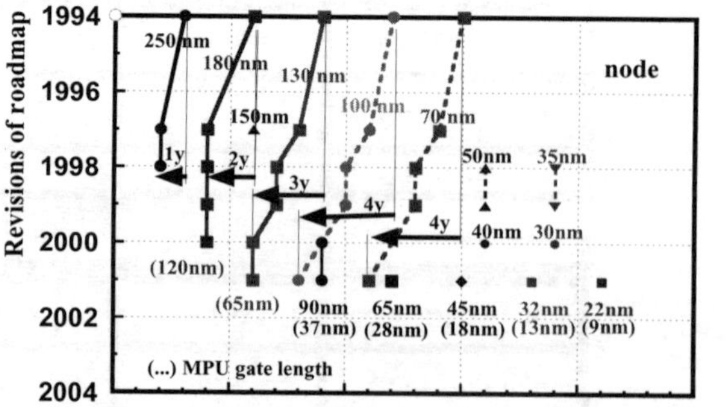

Figure 1. Frequent revisions of the semiconductor roadmap related to fast advancement have resulted in a more-than-4-year shift in terms of projected gate length.

---

[1] It is worth noting that these comments uniquely refer to present status of reported device principles and architectures.

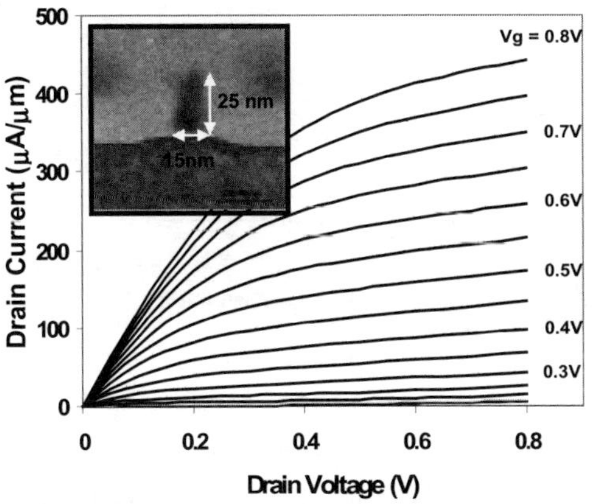

Figure 2. Intel's 15 nm channel length MOSFET. Courtesy of George Sery, Intel Corporation, Santa Clara, USA.

Until now, the microelectronics industry has almost continuously identified innovative methods for aggressive MOSFET scaling to increase performance. Recently, Intel has proposed both a bulk THz MOSFET [6] with gate length of 15 nm (Fig. 2) and a "depleted substrate" architecture that combines some key modifications of the traditional CMOS architecture: *ultra-thin SOI film* instead of bulk silicon, and, *high-k dielectric* instead of the traditional SiO_2 gate dielectric. It is worth noting that even with these solutions, fundamental limitations, expected for the nm dimensions, are not really addressed. Some of the new key fundamental problems [3, 5] of MOSFET-inspired devices for sub-10nm channel length are expected to be: (i) *electrostatic limits*, (ii*) source-to-drain tunnelling*, (iii) *carrier mobility* (iv) *process variations, and* v) *static leakage*. Simultaneously, and perhaps even more critical is the *power scaling concern*. It appears that *emerging device architectures* (outlined in the next Section) can offer CMOS more lifetime and provide solutions to continue scaling into the nanometer range, or at least until the *10 nm wall* is reached.

2. EMERGING DEVICE ARCHITECTURES

2.1 Double/Multi-Gate and FinFETs

Double/multi-gate MOSFETs [7] or FinFET [8] (Fig. 3), with ultrathin Si-film, are expected to constitute mainstream nanometer CMOS technology. Their key advantages, [9], are *excellent scalability* due to superior immunity against SCE effects, near-ideal subthreshold slope, high near-ballistic drive-current and transconductance, and low subthreshold intrinsic capacitance. The design of optimal multi-gate MOSFET requires new insights into the underlying physics and especially the quantum mechanics of the carriers confined in sub-10 nm films, [9].

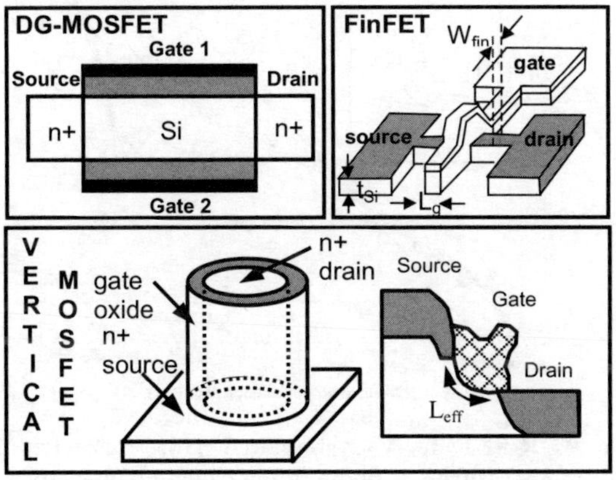

Figure 3. Emerging MOSFET device architectures.

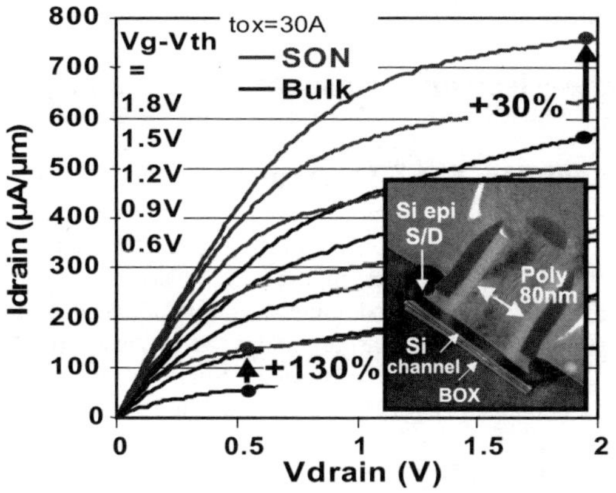

Figure 4. ST's 80 nm Silicon-On-Nothing (SON) transistor. Courtesy of Th. Skotnicki, ST Microelectronics, Crolles, France.

FinFET (Fig. 3) is a successful alternative inspired by DG-MOSFET that uses e-beam lithography for fin patterning. Recent realizations [8] with 20 nm gate length, less than 10 nm Si-film and 2.1 nm gate oxide, have shown (for W_{fin}=40 nm) NMOS and PMOS drive currents of 500 µA/µm and 380 µA/µm, respectively, which with *conventional* definition of double-gate means 1000 µA/µm and 760 µA/µm, respectively. The associated FinFET I_{off} is less than 1nA/mm.

2.2 Silicon-On-Nothing (SON)

Silicon-on-nothing (SON) [10, 11] is an innovative process that *provides "super-SOI" using bulk-silicon*, with a quasi-total suppression of SCE and DIBL and excellent electrical performances due to an extremely thin silicon (5-20 nm) and buried dielectric (10-30 nm). In sharp contrast to any SOI technology (SIMOX, BESOI, Smart Cut, ELTRAN, ELO, etc.), SON's nano-scale silicon film and buried insulator are defined by epitaxy on a bulk substrate. The buried oxide is not continuous (contrary to conventional SOI) and is located only under the gate and the spacers (Fig. 4). Compared to SOI, the advantages of such solutions are reduced series resistance and easier silicidation. 80 nm channel length SON NMOS (Fig. 4), with significant gain compared to similar devices made on pure bulk silicon (drain current of 750 µm/µm and 25 nA/µm I_{off} with 150 mV DIBL), have been recently demonstrated, [11].

2.3 Vertical MOSFET

The vertical MOSFET configuration [12, 13] provides other new device options for higher levels of functional integration. Its vertical channel is *lithography quasi-independent*, using vertically formed pillars (Fig. 3). Arbitrary doping profiles, heterojunctions and multiple devices (epitaxially grown) can also be incorporated into the vertical channel structure, [12].

2.4 Ballistic MOSFET

For dimensions less than 10 nm, CMOS-inspired architectures enter ballistic transport, which, hopefully, is advantageous. *Channel length less than carrier mean free path* defines the basis of ballistic transport, where, with no collisions, carrier mobility does not make any sense and significant current increase is experienced. Low doped thin device bodies are suitable for ballistic MOSFET (also eliminating parameter fluctuations related to discrete/random doping). New conduction mechanisms, [14, 15]: *thermo-ionic emission* and *quantum tunneling* have to be considered. However, the notion that reduction of L allows the voltage/power to be scaled does no longer holds for ballistic devices because the current is limited by electron supply at source.

With all these new emerging devices, there are, of course, many exciting opportunities and developments; however, they do not include any disruptive modification of MOSFET basic principle for which real fundamental limitations arrive below 10 nm gate length.

3. Single-Electron Transistors: *A Not So Different World?*

3.1 Principles: C-SET versus R-SET

The history of *few-electron electronics* [4] probably started in 1951 when C. Gorter explained for the first time the Coulomb blockade phenomenon. Only much later, in 1985, D. Averin and K. Likharev [16, 17] formulated the 'orthodox theory' of single-electron tunneling that describes the charge transport under the influence of Coulomb blockade and allows the exploitation of Single-Electron Transistor (SET). The *capacitive* C-SET architecture (Fig. 5) is similar to the MOSFET; it has a source (S), a drain (D) and a gate (G), the main difference being that the MOS channel is replaced by an ultra-small *conductive island* separated by *two tunnel barriers* from source and drain. The operation of SET exploits the discreteness of the number of electrons in their conductive island, which is in contrast with the MOSFET where, because of the highly transparent boundaries between S/D and the inversion channel, single electron charging is not experienced.

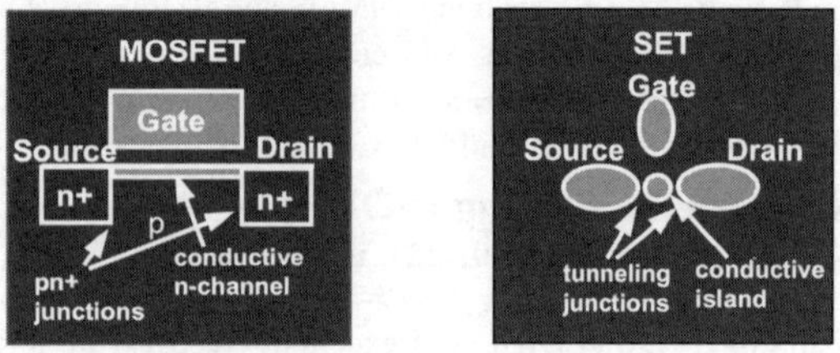

(a) (b)

Figure 5. Schematic of: (a) MOSFET and (b) C-SET.

Thus, proper SET operation needs *opaque barriers* in order to localize electrons in the quantum dot (or island). Application of the energy uncertainty principle results in a tunnel resistance R_T that should satisfy:

$$R_T > h/e^2 = R_Q \cong 26k\Omega \tag{1}$$

With condition (1) satisfied, the quantum mechanical uncertainty of electron location is covered up. In order to have full functionality, the charging energy of the island must be larger than the thermal fluctuations:

$$E_C = e^2/2C_\Sigma > k_B T \tag{2}$$

This condition can be fulfilled either by working with low temperature, *T*, and/or with very small island to ground capacitance, C_Σ. SET operation can now be briefly explained: at low V_{DS} there is no drain current (the current suppression being called *Coulomb blockade*, CB) since any tunneling would lead to an increase in the total energy and at low enough temperatures tunneling is rather low (Fig. 6). There are two ways to overcome CB: one is to increase the V_{DS} voltage up to a threshold voltage, V_T, where the current starts to rise with V_{DS}. The second is, of course, to increase the temperature (Figs. 6, 7). A key property of SET is that V_T is a periodic function of V_{GS} (Figs. 7, 8) providing the specific SET signature called Coulomb oscillations. They relate to the fact that at some external biasing:

$$Q_0 = C_G V_G = e(n + 1/2) \tag{3}$$

one electron can tunnel from source to the island and then to the drain even at negligible V_{DS}. It appears that the SET transconductance, $g_m = dI_D/dV_{GS}$, can have *both positive and negative values* (Fig. 8), depending uniquely on the gate voltage, which is a key difference with respect to MOSFETs. This is a real advantage if one aims to mirror CMOS circuit architectures in SET, because the *equivalent SET circuits would require the use of a unique type of device*. On the other hand, the drawback of SET lies on its low voltage gain that is limited by the capacitance ratio C_G/C_Σ (lower than few units at room temperature, [3, 4].

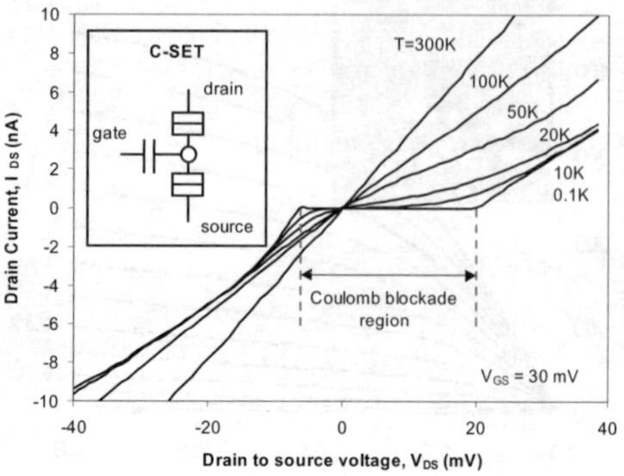

Figure 6. I_{DS}-V_{DS} characteristics of C-SET for various temperatures, T, highlighting Coulomb blockade.

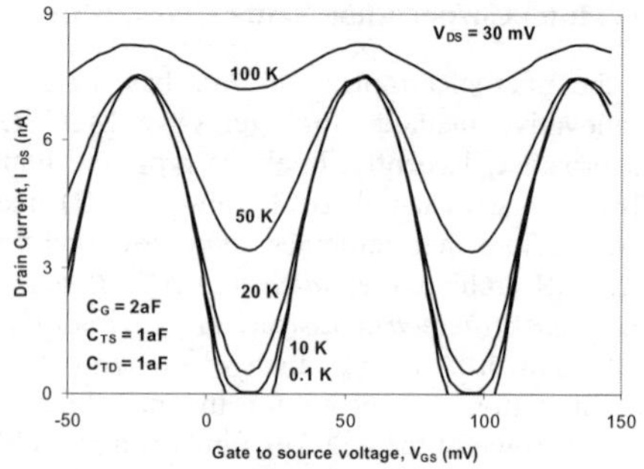

Figure 7. SET typical oscillations of I_{DS} versus V_{GS}, (periodic peaks and valleys) at various operating temperatures, T.

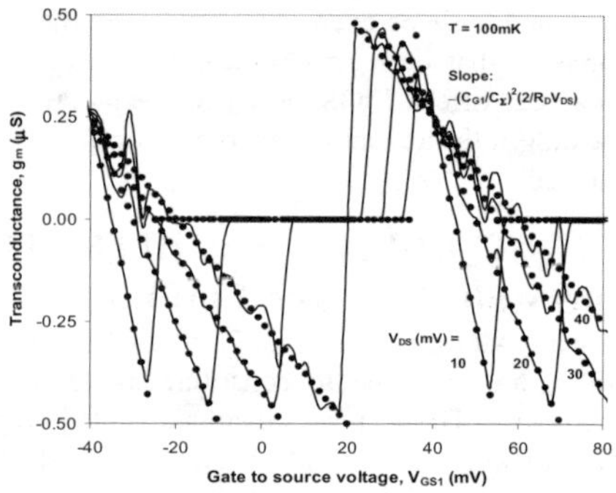

Figure 8. C-SET transconductance, g_m, vs. V_{GS} with V_{DS} as parameter at T=100mK: simulated with SIMON (solid line) and with our recent analytical model, MIB [24] (symbols).

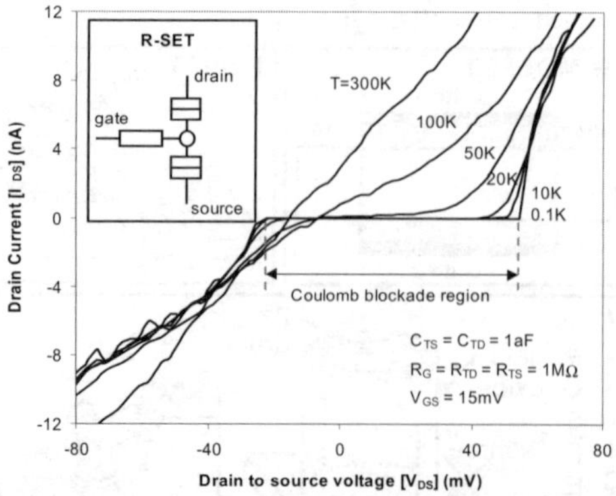

Figure 9. R-SET I_{DS}-V_{DS} simulated [18] characteristics for various temperatures, T, highlighting Coulomb blockade.

Note that SET principle can apply not only to well known C-SET (with capacitive gate), but also to R-SET (Fig. 9) for which the gate architecture is resistive. As shown by Korotkov [18] and demonstrated by Fig. 10, Coulomb blockade can be successfully experienced by R-SET, and even if its I_D-V_{GS} characteristics are no longer periodic, its usefulness for logic applications exists.

3.1.1 Dimensions for Room Temperature Operation

SET operation at room temperature is conditioned by the possibility to provide a very small C_Σ capacitance associated with the quantum dot. The plot reported in Fig. 10 exploits Kirihara's criteria, [19]: $T_{max}<e^2/(40kC_\Sigma)$, to deduce the maximum temperature at which SET can properly operate under Coulomb blockade. It is clearly shown that, for room temperature operation (300 K), SET needs total capacitance of less than around 0.2aF and an island radius less than 0.5nm. These are real challenges for SET device and circuit technology that are briefly described under §3.1.4.

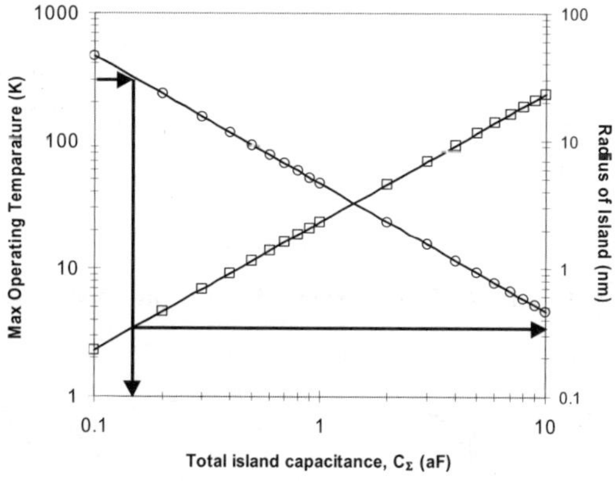

Figure 10. Maximum C-SET operating temperature, T_{max}, (with Kirihara's criteria [19]) vs. total island capacitance, C_Σ.

3.1.2 Background Charge Sensitivity

One of the drawbacks usually mentioned for C-SET is its high sensitivity to background charge that could jeopardize its proper functionality in digital circuits [3]. Materials quasi-free of impurities, R-SET-inspired device architectures (insensitive to background charge because of the resistive-gate coupled to the SET island), variable C-SET or compensation by refresh adapted cycles [4], *at device level*, and neural network architectures that cope with intrinsic errors, *at circuit level*, are some possible answers to this question.

3.1.3 SET Simulation and Modeling

It is clear that a deeper comprehension of SET underlying physics and its high potential for new functionality could help the take off of this new device. Simulation of SET devices and circuits has been proposed via (I) *Monte Carlo* (MC) simulators like SIMON [20], MOSES [21] and CAMSET [19], (II) *macro-modeling* [22] and (III) *analytical models* [23, 24]. More advanced analytical models and related electrical characterization are eagerly needed. Recently, the MIB has been proposed [24], a quasi-analytical model that enables SET co-simulation with CMOS up to a temperature $T=e^2/(40kC_\Sigma)$. MIB is founded on the "orthodox" theory of single electron tunneling and is able to model both symmetric and asymmetric SETs, with single- or double-gate. In MIB both SET drain current and (trans)conductance are analytically modeled via two separated current components: the *harmonically-connected* S/D tunneling currents (considered independent of temperature) and the *thermal* current. For accurate simulation, MIB model requires that interconnect capacitance be much larger than C_Σ and $|V_{DS}|<e/C_\Sigma$.

3.1.4 SET co-Fabrication with CMOS

First experimental SETs were fabricated by Fulton and Dolan [4], and, Kuzmin and Likharev, [17]. Various solutions have since been developed and reported [4], such as: shadow evaporation (Al/Al_2O_3 process), nano-imprint, use of STM/AFT tips for nano-oxidation, deposition of sub-10 nm thick films with nano-grains and self-assembly techniques. One key successful technology that *enables SET and CMOS electronics to co-exist* with encouraging performance is PADOX [25] (or its evolution into V-PADOX [25]). It exploits the pattern dependent (because of gradients of mechanical stress) oxidation of silicon in order to provide sub-lithographic dimensions (<10 nm) of SET islands in simple or twin architectures. Fabrication of SETs for both logic and memory with undulated ultra-thin (<5 nm) polysilicon film on SOI is another successful technique [26] (Fig. 11). Recently, LETI [27] has shown a MOS-inspired alternative to fabricate SETs, called MOS-SET. It exploits Coulomb oscillations (Fig. 12) due to dopant fluctuations in ultra-short MOSFET channel and/or S/D potential barriers (provided by adapted doping in source and drain regions) that act as opaque tunneling junctions.

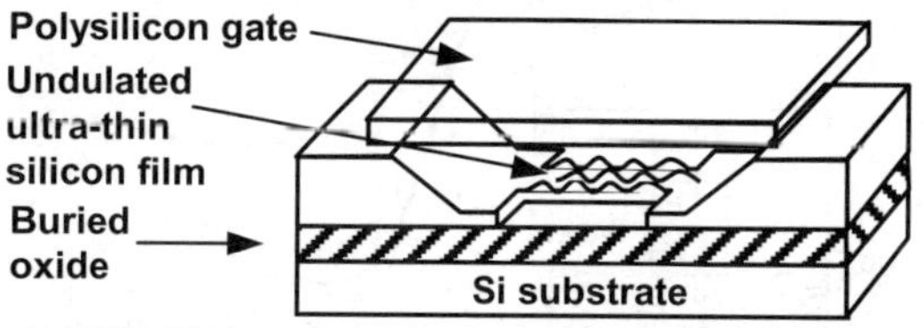

Figure 11. SOI-SET with undulated ultra-thin SOI film [26].

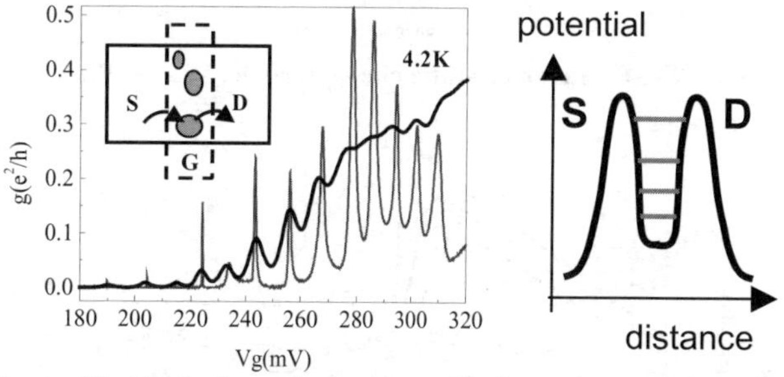

Figure 12. Typical conductance oscillations observed in LETI's [27] MOS-SET, revealing Coulomb blockade.

3.2 Digital SET: *To Mimic or Not to Mimic CMOS?*

One key question for SET logic circuit applications is about the efficiency of a CMOS-like circuit approach. Many successful logic applications have been reported by mimicking CMOS, but real competing performance with CMOS still remains to be demonstrated. A different, eccentric yet exciting, alternative is the *wireless* SET logic [28] where *external* electrical fields are used for both operation and power supply (thus, *no interconnect wires are required*); however, such a realization has not yet been experimentally verified.

3.2.1 C-SET and R-SET Few-Electron Inverters

In the following, performances of the C-SET inverter and, *for the first time*, R-SET inverter are briefly analyzed. Fig. 13 depicts typical V_{out}-V_{in} of realistic C-SET inverter as a function of temperature, *T*. With similar parameters and high-resistive gate (R_G=10 MΩ), we report a fully functional R-SET inverter (Fig. 14). We demonstrate here that C-and R-SET inverter performances are rather different. Figs. 15a and 15b depict the inverter gain, dV_{out}/dV_{in}, which highlights the first key difference: R-SET inverter has a much higher gain than C-SET, but degrades much faster with temperature. Noise margins (defined using CMOS criteria, $|dV_{out}/dV_{in}|=1$) are calculated and plotted together with thermal voltage, kT/q, in order to find out the domain for which real functionality can be expected (negligible thermal fluctuations). A remarkable feature of C-SET inverter is that the point at which $dV_{out}/dV_{in}=-1$, is insensitive to the temperature (Fig. 15). R-SET has better noise margins at low temperature, but their degradation with T is much faster than for C-SET. Fig. 16 demonstrates that with C_{Σ} in the range of aF, inverter functionality is expected up to few tens of K. The crucial advantage of SET is its low power consumption (on the orders of 10^{-10}-10^{-8}W/gate) supported by manipulation of single electrons with low voltages/currents (mV/nA); C-SET consumes even less power than R-SET. SET logic circuits can be source of surprising behavior: Fig. 14 demonstrates that SET inverter currents behave totally different than for CMOS: their "transition" region has negligible current/consumption (degrading with T). It follows that dynamic power of SETs is quasi-negligible and their ultra-low power consumption is mainly static, in contrast with ideal CMOS.

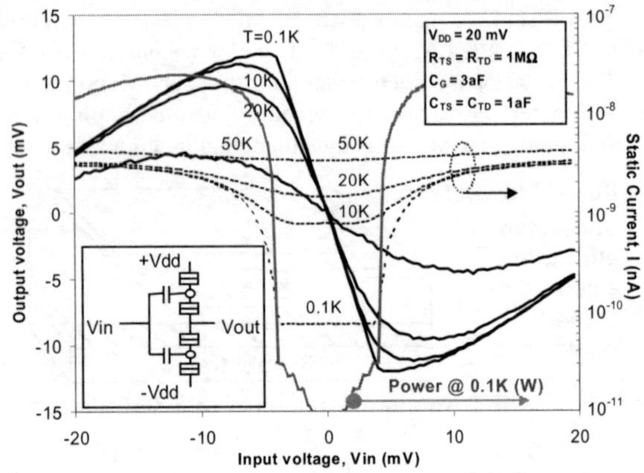

Figure 13. C-SET inverter static characteristic, V_{out} vs. V_{in}.

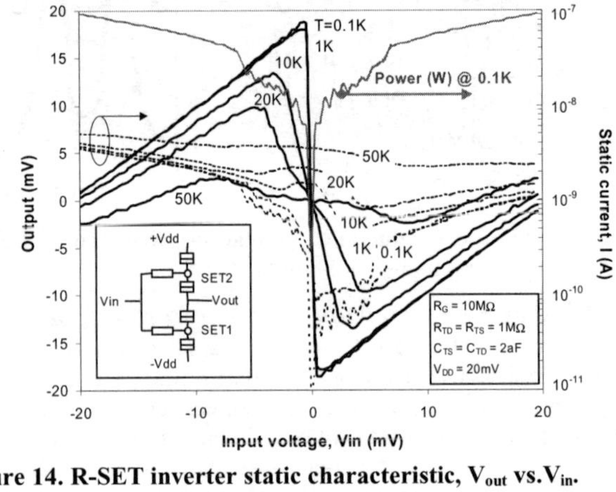

Figure 14. R-SET inverter static characteristic, V_{out} vs. V_{in}.

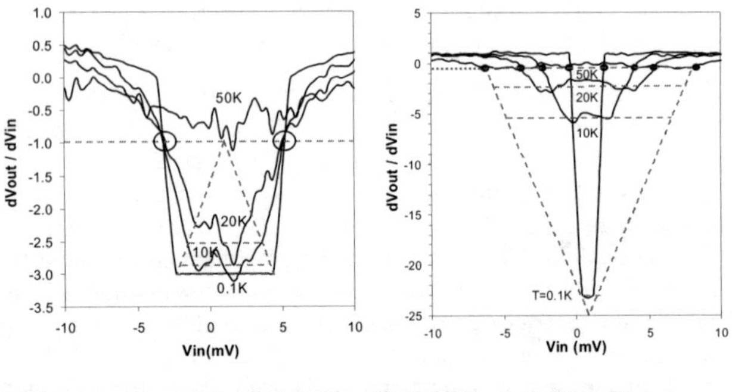

Figure 15. dV_{out}/dV_{in} of C-SET and R-SET at various T.

In Fig. 17a the transient characteristics of a C-SET inverter are reported. SET performance at high frequency is not limited by their intrinsic speed (electron tunneling is a fast process, with time constant of the orders of 10^{-15}s), but by the limited ability to drive high capacitance loads, C_L (Fig. 17b). Based on rather surprising observation that, in terms of current density, SET is not inferior to CMOS (with a size of 1 nm, the SET current density can be as high as a few of 100 µA/µm), we demonstrate in Fig. 17b that with (only three) parallel connected SET-inverters to drive C_L, the delay can be improved by a factor of 2 to 5 (depending on C_L). This parallel driving architecture requires practically ideal clock-skew that can be obtained with a symmetrical design of the interconnect-to-input gates. Its integration with a VPADOX-like process could result in quasi-identical parallel-connected inverters with smaller size than CMOS.

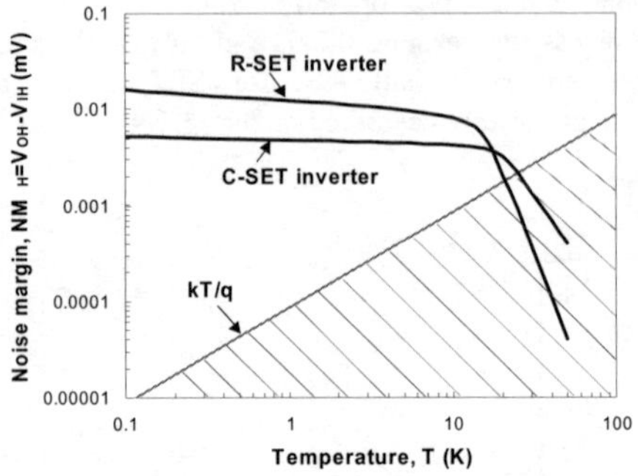

Figure 16. C- and R-SET inverter noise margins, NM_H, vs. T.

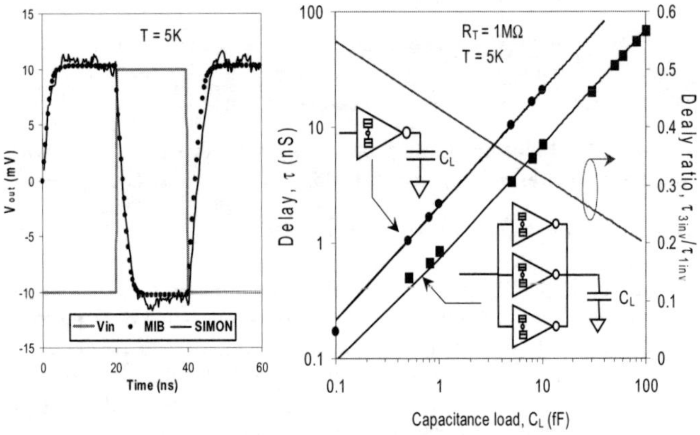

(a) (b)

Figure 17. (a) C-SET inverter transient and (b) Delay, τ, vs. capacitance load, C_L, with 1 and 3-parallel driving inverters.

3.2.2 Single Electron Devices for Ultimate Memory

Ultimate memory with single/few electron devices reflects the possibility *to store one bit of information by the trapping/de-trapping of one single electron*. Their advantages are clear: better than 10^{11}-10^{12} bits/cm^2 storage density with ultra-low power (less than 10^{-9} W/gate). With the few reported successful realizations, it turns out that single electron memories are very close to a major breakthrough in this field. Yano's [29] memory with nanogranular thin polysilicon film, Tiwari's [30] memory with nanocrystals included in MOSFET's gate oxide, and some other SET/FET hybrid few-electron memory architectures have shown reliable operation at room temperature with outstanding integration/power. For the mid-term, prior to ultimate single-electron memory, some other new memory concepts have been proposed, like the *non-volatile random access memory* (NOVORAM) [31] that exploits crested barriers and, with the use of SOI technology, has excellent potential for nanometer scaling.

4. **Hybrid CMOS-SET Architectures:** *Illusion or Useful Compromise?*

We think it is now time for the "old" extraordinary, high-performance CMOS and the "novel", still-promising SET, to start exploring a common, joint evolution. Recent reports demonstrate some new, unique functionality and high performance that cannot be achieved by any other alternative than combining CMOS and SET in hybrid CMOS/SET IC architectures (Fig. 18). Uchida et al., recently reported a future foreseeable hybrid SET/CMOS ULSI (Fig. 19) [26], which exploits the particular advantages of SETs and CMOS and, when possible uses CMOS to compensate for SET drawbacks. Nonvolatile single-electron memory (NVM) circuits for *programmable SET/FET logic* [26, 32], *multiple-valued* logic/SRAM [33, 34], new *quantizer* circuits [34, 35] and other applications that exploit the negative-resistance SET blocks [4] have also been proposed. It appears that a *triple effort* is required for the success of hybrid CMOS/SET: (I) first, is on the development of common technological platform, (II) second, is on enabling advanced SET/CMOS co-simulation and design, and, (III) third, on innovative development/demonstration of new functionality of hybrid IC architectures tolerant to background charge effects (and - why not? - with new types of logic).

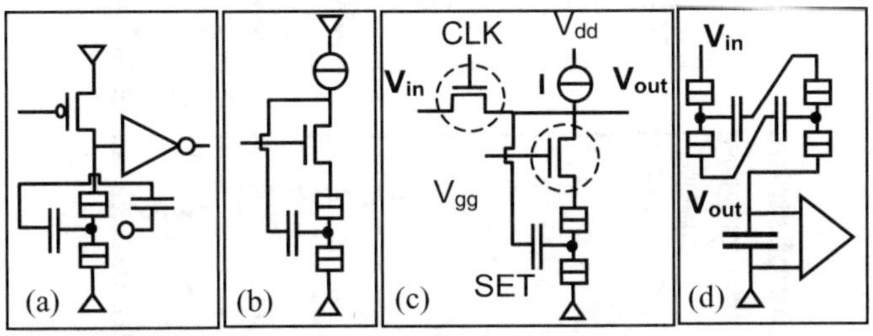

Figure 18. Hybrid SET/CMOS ICs: (a) NVM circuit for programmable logic (undulated-film SET) [26], (b) multiple-valued SRAM cell (PADOX) [32], and (c, d) quantizers [34, 35].

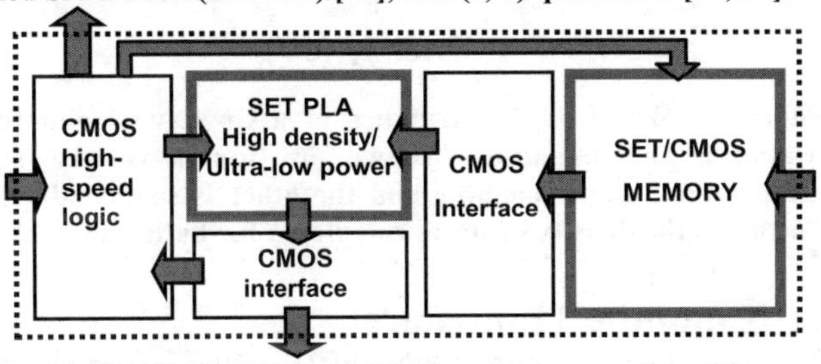

Figure 19. Hybrid SET/CMOS ULSI architecture, after [26].

5. CONCLUSIONS

Tremendous progress in microelectronics has pushed the MOSFET dimension towards the 10 nm limit, which is expected to impact basic working principles of MOSFETs. In the near future it is then probable that CMOS will need to share its domination on modern ICs with fundamentally new devices that use a few electrons, like SETs. It appears that CMOS and SETs are rather complementary: SET is the *champion of low-power consumption* and of new functionality while CMOS advantages like high-speed, driving, voltage gain and input impedance can compensate exactly for SET's intrinsic drawbacks. Moreover, *unrivalled integration with high performance and new functionality* are expected for *hybrid CMOS-SET architectures*.

6. REFERENCES

[1] J. D. Meindl, "A history of low power electronics: how it began and where it's headed", *Proc. of Int. Symp. Low Power El. and Design*, pp. 149 -151, 1997.

[2] J. D. Meindl, "Low power microelectronics: Retrospect and prospect", *Proceedings of IEEE*, vol. 83, no. 4, pp. 619–635, 1995.

[3] K. Likharev, "Sub-20nm Electron Devices", Advanced Semiconductor and Organic Nano-Techniques", Ed. H. Morkoc, *Academic Press*, 2002.

[4] C. Wasshuber, "Computational Single-Electronics", *Springer Ver.*, Wien, 2001.

[5] The International Technology Roadmap for Semiconductors (ITRS), 2001 Eds.

[6] Private communications, George Sery, Intel Corporation.

[7] Y. Taur et al., "CMOS scaling into the nanometer regime", *Proceedings of IEEE*, vol. 85, no. 4, pp. 486-504, 1997.

[8] Y. Choi et al., "Sub-20nm FinFET Technologies', *Technical Digest IEDM*, pp. 421-424, 2001.

[9] L. Ge and J.G. Fossum, "Analytical modeling of quantization and volume inversion in thin Si-film DG MOSFETs", *IEEE Trans. on El. Dev.*, vol. 49, no. 2, pp. 287-294, 2002.

[10] M. Jurczak et al., "Silicon-On-Nothing (SON) - an innovative process for adv. CMOS", *IEEE Trans. on El. Dev.*, vol. 47, no. 11, pp. 2179-2187, 2000.

[11] S. Monfray et al., "First SON (Silicon-On-Nothing) MOSFETs with perfect morphology and high electrical performance", *IEDM Technical Digest*, pp. 645-648, 2001.

[12] J.D. Plummer, "Silicon MOSFETs (conventional and non-traditional) at the scaling limit", *Proc. of Device Research Conference 2000*, pp. 3-7, 2000.

[13] T. Schultz et al., "Short-channel vertical sidewall MOSFETs", *IEEE Trans. on El. Dev.*, vol. 48, no. 8, pp. 1783-1788, 2001.

[14] M. Lundstrom and Z. Ren, "Essential Physics of Carrier Transport in Nanoscale MOSFETs", *IEEE Trans. on El. Dev.*, vol. 49, no. 1, pp. 133-141, 2002.

[15] D. Munteanu, J.L. Autran, "Two-dimensional Modeling of Quantum Ballistic Transport in Ultimate Double-Gate SOI Devices". *Proc. of ULIS 2002*, Munich, Germany, pp. 119-122, 2002.

[16] D. Averin and K. Likharev, "Single electronics: a correlated transfer of single electrons and Copper pairs in systems of small tunnel junctions", in *Mesoscopic phenomena in solids*, pp. 173-271, 1991.

[17] K.K. Likharev, "Single-electron devices and their applications", *Proceedings of IEEE*, vol. 87, Issue 4, pp. 606 -632, 1999.

[18] A.N. Korotkov, "Single-electron transistor controlled with a RC circuit", *Phys. Rev. B*, vol. 49, pp. 16518-16522, 1994.

[19] M. Kirihara, N. Kuwamura, K. Taniguchi, and C. Hamaguchi, "Monte Carlo study of single-electronic devices," in *Ext. Abst. Int. Conf. on Solid State Devices and Materials*, Yokohama, Japan, 1994, pp. 328–330.

[20] C. Wasshuber, H. Kosina, and S. Selberherr, "SIMON - A Simulator for Single-Electron Tunnel Devices and Circuits", *IEEE Transactions on Computer Aided Design of IC and Systems*, vol. 16, no. 9, pp. 937-944, 1997.

[21] R. H. Chen, A. N. Korotkov, and K. K. Likharev, "Single-electron transistor logic", *Appl. Phys. Lett.*, vol. 68, pp. 1954–1956, 1996.

[22] Y. S. Yu, S. W. Hwang, and D. Ahn, "Macromodeling of single electron transistors for efficient circuit simulation", *IEEE Trans. on El. Dev.*, vol. 46, No. 8, pp 1667-1671, 1999.

[23] K. Uchida et al.,"Analytical single-electron transistor (SET) model for design and analysis of realistic SET circuits", *Jpn. J. Appl. Phys.*, vol. 39, Part 1, No. 4B, pp. 2321-2324, 2000.

[24] S. Mahapatra, A.M. Ionescu and K. Banerjee, "A Quasi-Analytical SET Model for Few Electron Circuit Simulation", *IEEE El. Dev. Lett.*, to appear, 2002.

[25] Y. Takahashi et al., "Silicon single-electron devices and their applications", *Proc. of IEEE International 2000 Symposium on ISMVL*, pp. 411 -420, 2000.

[26] K. Uchida et al., "Programmable single-electron transistor logic for low-power intelligent Si LSI", *Digest of ISSCC 2002*, pp. 206-207, 2002.

[27] M. Sanquer et al., "Coulomb blockade in low-mobility nanometer size Si MOSFET's", *Phys.Rev.*, B 61, pp.7249-7252, 2000.

[28] A.N. Korotkov, "Wireless single-electron logic by alternating electric field", *Appl. Phys. Lett.*, vol. 67, pp. 2412-2414, 1995.

[29] K. Yano et al., "Single-electron memory for giga-to-terra bit storage", *Proceedings IEEE*, vol. 87, pp. 633-651, 1999.

[30] S. Tiwari et al., "A silicon nanocrystals based memory", *Appl. Phys. Lett.*, vol. 68, pp. 1377-1379, 1996.

[31] K.K. Likharev, "Riding the crest of a new wave in memory [NOVORAM]", *IEEE Circuits and Devices Magazine*, vol. 16, pp. 16-21, 2000.

[32] K. Uchida et al., "Room operation of multifunctional single-electron transistor logic", *Tech. Digest IEDM*, pp.863-865, 200.

[33] H. Inokawa, A. Fujiwara and Y. Takahashi, "A multiple-valued single-electron SRAM by the PADOX process", *Proc. 6th Int. Conf. on Solid-State and IC Technology*, pp.205 -208, 2001.

[34] H. Inokawa, A. Fujiwara and Y. Takahashi, "A multiple-valued logic with merged Single-Electron and MOS Transistors", *Tech. Digest IEDM*, pp. 147-150, 2001

[35] S. Mahapatra, A. M. Ionescu, K. Banerjee and M. Declercq, "A SET Quantizer Circuit Aiming at Digital Comm. System", *ISCAS 2002,* (to appear), 2002.

Carbon Nanotube Field-Effect Transistors and Logic Circuits

R. Martel*, V. Derycke, J. Appenzeller, S. Wind, and Ph. Avouris
IBM T. J. Watson Research Center, Route 134
Yorktown Heights, NY 10598
* phone : 914-945-2164 and e-mail : Martelr@us.ibm.com

ABSTRACT

In this paper, we present recent advances in the understanding of the properties of semiconducting single wall carbon nanotube and in the exploration of their use as field-effect transistors (FETs). Both electrons and holes can be injected in a nanotube transistor by either controlling the metal-nanotube Schottky barriers present at the contacts or simply by doping the bulk of the nanotube. These methods give complementary nanotube FETs that can be integrated together to make inter- and intra-nanotube logic circuits. The device performance and their general characteristics suggest that they can compete with silicon MOSFETs. While this is true when considering simple prototype devices, several issues remain to be explored before a nanotube-based technology is possible. They are also discussed.

Categories and Subject Descriptors

B.6.0 [**Logic Design**]: General – *novel logic devices.*

General Terms

Measurement, Performance, Design, Experimentation.

Keywords

Nanoelectronics, Carbon Nanotube, Semiconductor, Field-Effect Transistor, FET, Schottky Barrier, Circuits, Inverter, Logic Gate, SWNT.

1. INTRODUCTION

The single-wall carbon nanotube (SWNT) is a one-dimensional conductor that can be either metallic or semiconducting depending on the arrangement of the carbon atoms (see Figure 1). [1] The band-gap, Eg, of a semiconducting nanotube depends strongly on its diameter, d, as $Eg \propto 1/d$. Fig.1 presents two examples of possible structures for a 1.4nm diameter SWNT with their corresponding electronic structures calculated using a tight-binding model. [1] Both metallic and semiconducting nanotubes are attractive for applications. On one hand, the conductivity of metallic nanotubes and their robustness are attracting interest for future interconnects. [2] On the other hand, the semiconducting carbon nanotubes exhibit the desired properties for making field-effect transistors and logic switches [3-9]. In principle, the combined properties of both kinds of nanotubes can form a unique set of building blocks for future electronics.

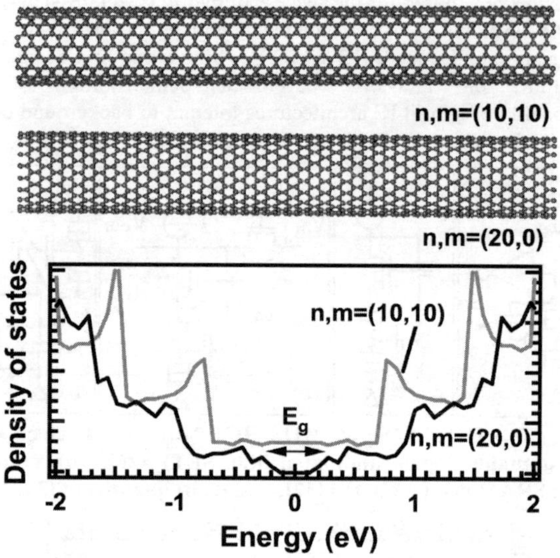

Figure 1. Model of the structure of a single-wall nanotube. Although the nanotubes shown on top have the same composition, one is metallic and the other is semiconducting. Bottom is the density of states calculated for both tubes.

2. Nanotube Transistors

Early experiments [3,4] on SWNT FETs (CNFETs) demonstrated that they behave much like conventional semiconductor transistors. There are, however, some important differences: i) the carbon nanotube is one-dimensional which greatly reduces the scattering probability. [10] As a result, the devices may operate in the ballistic regime. [10,11] ii) The nanotube conducts essentially on its surface where all the chemical bonds are saturated and stable. Therefore, there is no need for careful passivation of the interface between the nanotube channel and the gate dielectric, i.e. there is no equivalent of the silicon/silicon dioxide interface. iii) The Schottky barrier at the metal-nanotube contact is the active switching element in an intrinsic nanotube device. [12] These properties make the one-dimensional transistor action in nanotube unique and interesting. Here we review their characteristics, the properties of the barriers, and discuss the integration of nanotube devices into logic circuits.

Early nanotube FETs used a non-local back-gate with the nanotube side-bonded to noble metal electrodes. This arrangement gave large contact resistances and poor characteristics [3-5].

Since then, significant improvements in the performance of CNFETs have been achieved. New structures are now built with top-gate geometry and they resemble, in many respects, conventional silicon CMOS devices. An example of the characteristics of a top-gated CNFET is shown in Fig. 2. This CNFET is contacted with titanium carbide electrodes and the gate electrode is made of Al separated by 15 nm of gate oxide (SiO_2). [13] Its relevant parameters are presented in Table 1. They are also compared with an earlier back-gated CNFET and to a conventional Si MOSFET of 100nm gate length. First, we observe a very large improvement compared to the early nanotube devices. This improvement is mainly due to the scaling of the dimensions (mostly the gate oxide thickness) and the adoption of a better device geometry. In addition, the comparison of key parameters suggests that the performance of CNFETs, once the device is scaled properly, would be comparable to, if not better than, conventional Si MOSFET.

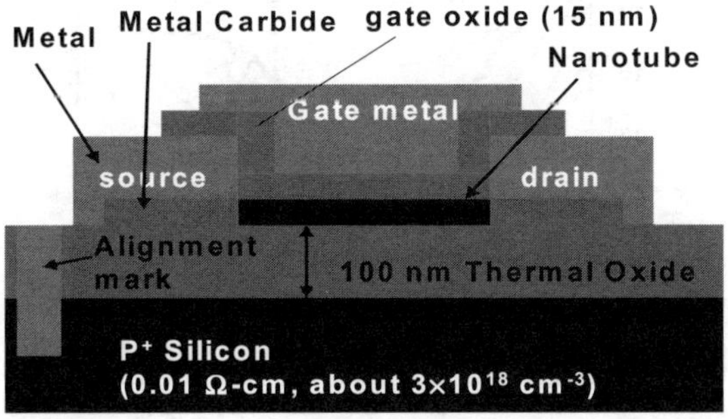

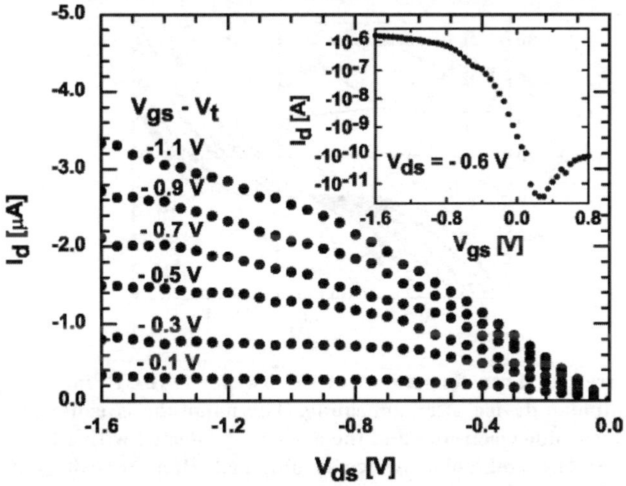

Figure 2. Characteristics of a single-wall nanotube field-effect transistor. The parameters and structure are shown in the schematics above.

While CNFETs show improvement with scaling, we note that their scaling properties are not conventional. [12] They rather follow the behavior of a Schottky barrier MOSFET, i.e. the switching behavior involves mostly the contacts, as oppose to the bulk of the tube. [12] In this respect, the extrapolation of the measured performance for those nanotube transistors to shorter channel devices, i.e. in the limits of scaling, is still unclear. The actual performance, however, suggests that the nanotube transistors, in the structure of an array, will provide enough gain and fan-out for applications. Moreover, it is possible that the CNFET may outperform silicon MOSFET in the limits of scaling.

Table 1 Comparison of the characteristics of CNFETs and Si MOSFETs

p-type FETs	Back-gated CNFET [6]	Top gated CNFET [13]	Si MOSFET [19]
Gate length (nm)	1030	260	100
Gate oxide thickness (nm)	150	15	0.8
Trans-conductance	~0.3 μS/tube (244 μS/μm)*	~3μS/tube (2100 μS/μm)*	460 μS/μm
Subthreshold slope (mV/dec)	730	130	80
ON/OFF ratio	10^5	10^6	10^6

* The transconductance is normalized by the width of the nanotube, i.e. 1.4 nm. A more realistic normalization and comparison with Si MOSFET would take into account an array of nanotubes as discussed in Ref. 6.

3. Nanotube-Metal Schottky Barrier

The CNFETs measured in air, as in Fig. 2, are p-type, i.e. the tubes conduct holes upon applying a negative gate voltage and they show no evidence of electron conduction even at very large positive gate voltage. In addition, the CNFETs in air also show ohmic IVs at small source-drain voltage in the ON state of the device (i.e. at negative gate bias). This situation changes when the devices are annealed in vacuum. This annealing step removes the adsorbed oxygen and yields a reversible transformation of the CNFET from p- to n-type. An example of such a transformation is presented in Fig. 3. Initially, a p-type CNFET is annealed in vacuum and transformed into a n-type device. Then, oxygen is slowly introduced into the chamber and IVs are acquired at intermediate stages of the transformation. This simple annealing technique is described elsewhere. [7,14,15] The intermediate stages (blue curves) are ambipolar, i.e. the tube can conduct both electrons and holes. This transformation can be easily rationalized by considering the presence of tunneling barriers at the contact. [14, 15] Each situation is presented in Figure 4. In Fig. 4a, the Fermi level at the metal-nanotube junction in air is closer to the valence band of the nanotube. This leads to hole conduction and p-type behavior. The annealing step leads to a change in the line-up of the bands at the junction (Fig. 4c) and to lowering of the barrier for electron injection. This modification at the contacts also introduces an increase of the barrier height for hole injection, leading to an n-type CNFET. [15] The intermediate stage, however, simply occurs when the contact Fermi level is around mid-gap. This is the situation illustrated in Fig. 4b. This special situation gives similar barriers for electron and hole injection and the device is ambipolar. [14]

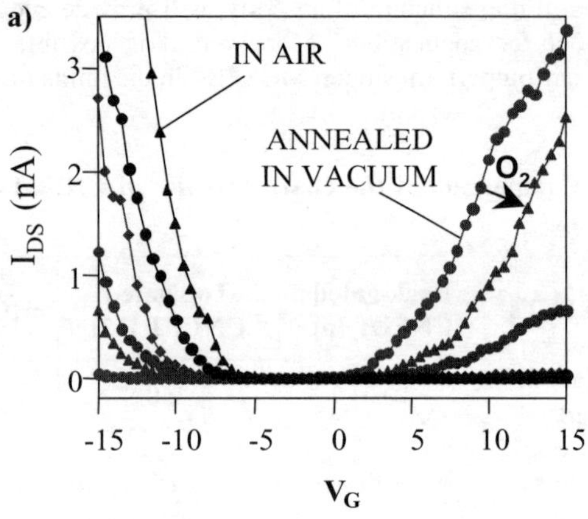

Figure 3 Transformation from n- to p-type of a CNFET as a function of oxygen dose. This conversion is reversible.

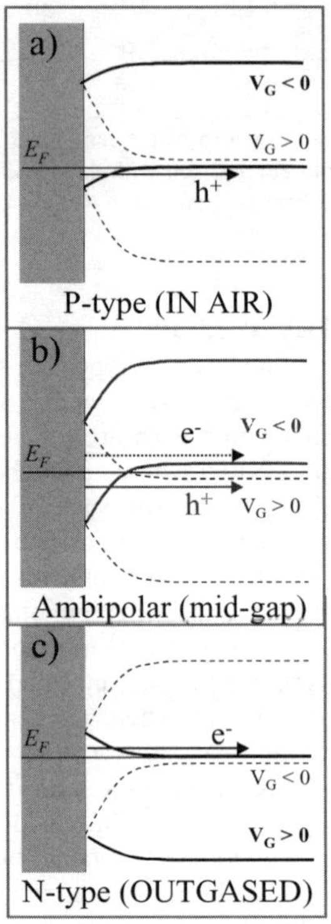

Figure 4.

Schematic of the bands in the vicinity of the contacts as a function of the electrostatic gate field for p-type (a), ambipolar (b), and n-type (c) CNFETs. The effect of annealing the device in vaccum is to change the situation at the contacts from a) to c). Introducing air (oxygen) reverses the process and converts back the device to the situation illustrated in b) first (partial conversion) and finally in a).

3.1 Ambipolar Nanotube Transistors

The ambipolar CNFET is particularly interesting in that it allows us to explore in depth the properties of a one-dimensional (1D) metal-nanotube junction. Although the situation for the ambipolar device is consistent with a mid-gap injection, which implies that the barriers for electron and hole injection are high (Eg/2~300mV), the IVs at room temperature are ohmic (see Figure 5b). [14] This behavior is not expected in presence of a conventional Schottky barrier. This peculiar observation led us to investigate the temperature dependence of the IV curves for an ambipolar device. The results are presented in Figure 5 for the case of hole transport. The results are essentially the same in the region of electron accumulation. In fact, the measured barrier heights for both electrons and holes are very small, at least 20 times lower than expected (~15meV instead of ~300meV). Therefore, the Schottky barrier in the strong accumulation of either holes or electrons is so thin that the junction is quasi-transparent for carrier injection, i.e. there is a very efficient tunneling through the barrier. However, the shape of the barrier of this 1D junction presents also a tail that extends deeper into the nanotube channel. This extension of the barrier is too thick for tunneling to occur unless the source-drain field is larger than the height of the effective barrier introduced by the tail (i.e. V > 15mV). This tail is very particular to nanotube-metal 1D junction and essentially induce the effective barrier measured. [14,16] The shape and the thickness of this 1D Schottky barrier are also non-uniform and strongly depend on the gate field. [17] As a result, the injection of carriers may become very asymmetrical depending on the conditions at the contact. [15]

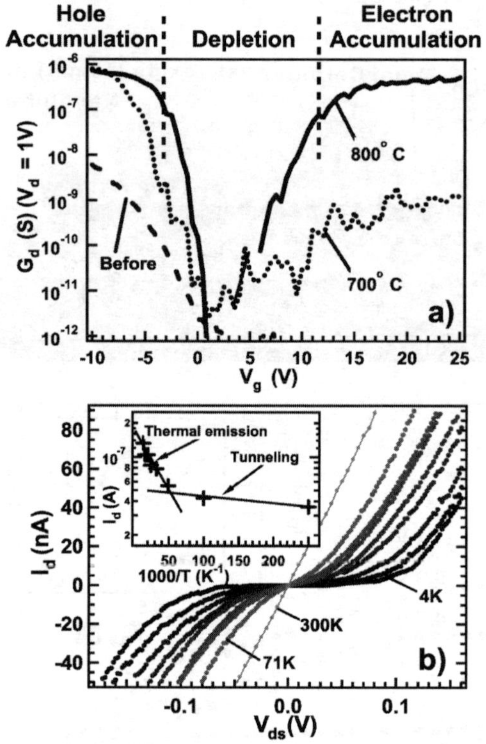

Figure 5. Transformation of a p-type CNFET into an ambipolar device after annealing. The nanotube is contacted with carbide electrodes and the device is protected with a CVD oxide. The ambipolar device is obtained after annealing at 800°C in argon and it is stable in air. The barriers of this 1D junction between the nanotube and the carbide contacts are characterized in the lower figure as a function of temperature.

4 Doping and Contact Injection

As mentioned earlier, the CNFET in air is p-type. The p-character was initially assumed to be due to a charge transfer by a high workfunction metal contact with this transfer accounting for some hole doping. The impact of oxygen was reported earlier and the conclusion was that oxygen also dopes the nanotube with holes, thereby making it p-type. [18] We tested these assumptions using selective doping of the nanotube with an electron donor (potassium). The idea was to reverse the doping caused by oxygen. The results are presented in Fig. 6. They are very different when compared with the results of annealing followed by exposure to oxygen (Fig. 3). Therefore, they support a very different conclusion as discussed below.

As expected, increasing the doping level in the nanotube with potassium leads to the conversion from p- to n-type (curves 1-12, Fig. 6a). This transformation is similar to the previous one obtained by annealing in vacuum. However, the details of the transformation are very different. i) Potassium is known to effectively dope the nanotube with the main effect being a shift of the threshold voltage toward more negative gate voltage (see the trend in curves 1-6). This shift is not observed in the annealing experiment (Fig. 3). ii) The intermediate stages do not involve an ambipolar device. It is therefore a different process: The doping with potassium is consistent with the doping of the bulk of the tube, as illustrated in Fig. 6 b and c. The conversion to n-type only occurs when the level of electron doping is high and significantly affects the barrier thickness at the contacts. Moreover, the absence of the ambipolar stage implies that the doping of the bulk of the nanotube does not shift the Fermi-level at the contacts. The barriers at the contact are therefore pinned and do not change as a function of bulk doping. In contrast, the main effect of oxygen on the n-type CNFET in Fig. 3 is not to dope the nanotube but rather to shift the Fermi-level at the contacts and modify the contact barriers.

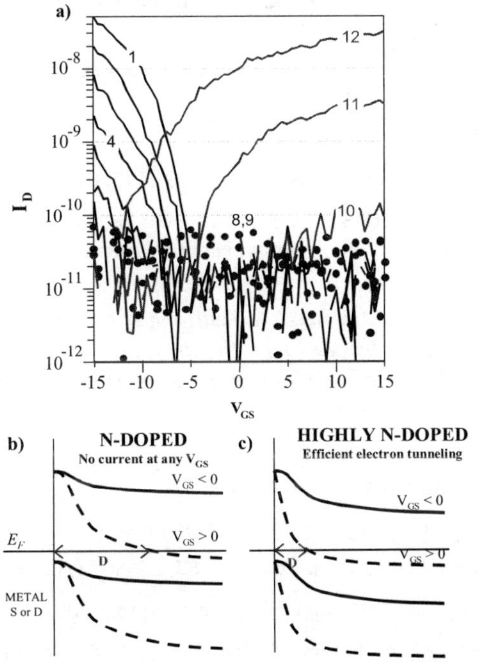

Figure 6. a) Effect of potassium doping on a CNFET. The FET is initially p-type. Doping is done by step from 1 to 12. For higher doping level, the device becomes n-type (red curves 10-12). b) and c) Schematic of the bands in the region of the metal-nanotube contact. b) The case of an n-doped CNFET that was not previously outgassed (curves 8-9), and c) the same but for a higher doping level. D corresponds to the barrier thickness.

5. Nanotube Logic Gates

Having developed ways to fabricate p- and n-type CNFETs with local gating, the next step is to produce integrated circuits out of them. Of particular interest are logic gates that form the basis of today's computer logic. Our approach is to use complementary CNFETs placed in circuit to operate simple logic functions. This kind of nanotube-based circuit is the analogue to the conventional CMOS based logic gates and has the same advantages. That is, logic circuits based on complementary devices aim to consume low power, favor higher gain, be stable, and allow easy implementation in integrated circuits. In this last section, we present our recent advances in making logic functions out of nanotube devices.

A simple example is presented in Fig. 7. By bonding together two ambipolar CNFETs stable in air, we fabricated a logic gate. An offset voltage (V_{shift}) between the isolated transistor back-gates allows adjustment of the threshold of the p- and n-CNFET characteristics so that the inversion function is optimal. The response of this circuit is abrupt and the operating voltage is well below 5V.

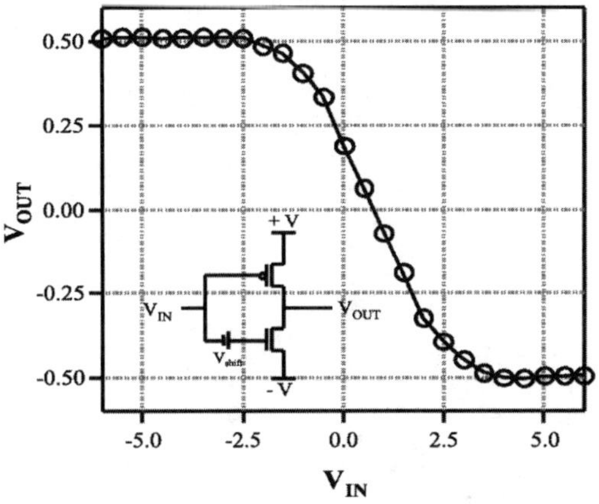

Figure 7 Logic device ("NOT" gate) made with two ambipolar CNFETs. The circuit consists of an inverter device shown in the inset (|V|=0.5V). An oxide layer protects the devices so that the inverter circuit functions in air. [19]

However, the ultimate integration in nanotube electronics would be to build a circuit along the length of the same nanotube, i.e. produce an *intra-nanotube* logic circuit. We were able to implement this integration as shown in Figure 8. The top of Fig. 8 shows an atomic force microscope (AFM) image of a small single-wall nanotube bundle deposited on top of gold electrodes. The entire device is covered by a polymer film (PMMA). Then, a window is opened in the PMMA using electron beam lithography. Last, potassium is used to dope half of the bundle through this window to produce an n-CNFET, while the other CNFET, protected by the PMMA film, remains p-type. The doping is adjusted so that the thresholds for switching of the p- and n-CNFETs overlap. This design indeed leads to an intra-molecular NOT gate. The relative position of the two thresholds voltages was adjusted by choosing the appropriate doping level.

Although the intra-nanotube inverter shown in Fig. 8 is not optimized in terms of structure, it has already a gain of about 1.6. Therefore, its output can be used to drive another gate or a more complicated logic circuit. However, the gain of the gate depends solely on the characteristics of the CNFETs. As shown before, the improvements in the CNFETs and their device geometry (in particular the gate oxide) will further increase the performance of the nanotube-based inverter.

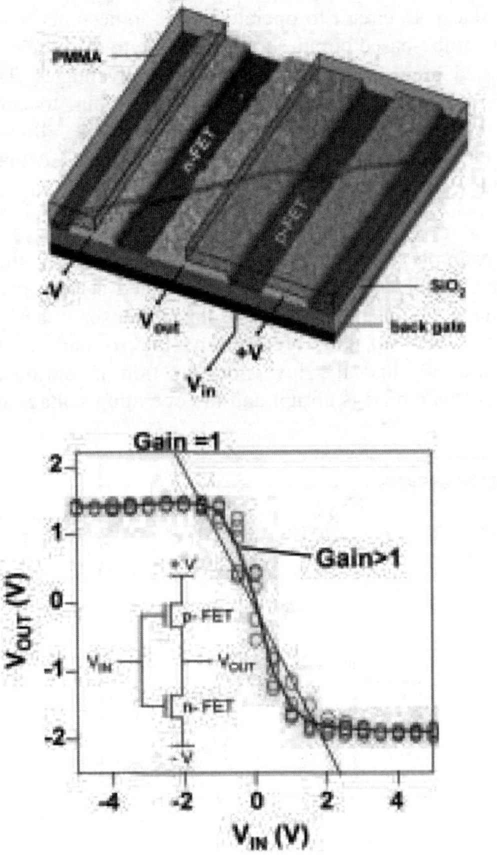

Figure 8. Intra-nanotube inverter. Top: AFM image of the device. A single-wall nanotube is placed to bridge several gold electrodes and produces two p-type CNFETs along the same tube. Then, the CNFETs are patterned with resist (PMMA) and doped selectively with potassium. The window in the PMMA allows the conversion of only one device into n-CNFET. The other is protected and remains p-type. Bottom: Characteristics of the intra-molecular inverter (V= ±2V). The straight line corresponds to an output/input gain of one.

6. Conclusion and Perspectives

The prospect for nanotube-based electronics is attractive but it is not realistic to expect new products soon. Several issues need to be resolved. For example, as-grown nanotubes come in a mixture of semiconducting and metallic types. There is currently no method to selectively separate them. Moreover, there is still no effective ways to make controlled arrays of nanotubes. In addition, as much as we have learned to date, the understanding of the physics of the nanotube devices is also at an early stage. Furthermore, significant advances in processing the material and in controlling the chemistry of nanotubes are needed. Nevertheless, the nanotube transistors show very high performance and integration capabilities. The use of independent gates, such as the top-gate device in Fig. 2 or local bottom-gate structures as in Ref [8], and the capability to make complementary devices allow for the construction of more complex circuits. Overall, the nanotube presents all the desirable properties needed for future electronic applications.

7. ACKNOWLEDGMENTS

The authors thank J. Tersoff, S. Heinze, F. Stern, T. Theis, and P. H. Wong for discussions. Assistance by K. Chan, J. Bucchinano, and B. Ek was also essential for this work.

8. REFERENCES

[1] Carbon Nanotubes: Synthesis, Structure Properties and Applications, M. Dresselhaus, G. Dresselhaus, Ph. Avouris, Editors (Springer-Verlag, Berlin, 2001).

[2] S. Frank, P. Poncharal, Z.L. Wamg and W.A. DeHeer, Science **280**, 1744 (1998); P. J. de Pablo et al., Appl. Phys. Lett. **74**, 323 (1999); Philip G. Collins et al., Phys. Rev. Lett. **86**, 3128 (2001); B. Q. Wei, R. Vajtai, and P. M. Ajayan, Appl. Phys. Lett. **79**, 1172 (2001).

[3] S. Tans, A. Verschueren, C. Dekker, Nature (London) **393**, 49 (1998).

[4] R. Martel et al., Appl. Phys. Lett. **73**, 2447 (1998).

[5] H. T. Soh, et al., Appl. Phys. Lett., 75 (1999) 627.

[6] R. Martel, H.-S. P. Wong, K. Chan, and Ph. Avouris, Proceedings *IEDM*, 152, 2001.

[7] V. Derycke, R. Martel, J. Appenzeller and Ph. Avouris, Nanoletters 1, 453 (2001).

[8] A. Bachtold et al., Science 294, 1317 (2001).

[9] X. Liu et al. Appl. Phys. Lett. 79, 3329 (2001).

[10] J. Appenzeller, et al. Appl. Phys. Lett. **78**, 3313 (2001); C. T. White and T. N. Todorov, Nature (London) **393**, 240 (1998).

[11] M. S. Fuhrer, M. Forero, A. Zettl, and P. L. McEuen , *AIP Conference Proceedings, Electronic Properties of Novel Materials – Molecular Nanostructures*, (Editors: H. Kuzmany, J. Fink, M. Mehring and S. Roth) 2001, p. 401.

[12] J. Appenzeller et al. submitted.

[13] S. Wind et al., Appl. Phys. Lett., in press.

[14] R. Martel et al. Phys. Rev. Lett. **87**, 256807 (2002).

[15] V. Derycke et al., Appl. Phys. Lett. in press 2002.

[16] F. Leonard and J. Tersoff, Phys. Rev. Lett. **84**, 4693 (2000).

[17] S. Heinze et al. submitted.

[18] P. G. Collins et al., Science 287 (2000) 1801; G. U. Sumanasekera et al., Phys. Rev. Lett. 85 (2000) 1096; K. Bradley et al. Phys. Rev. Lett. 85 (2000) 4361.

[19] T. Ghani et al. Proceedings of IEDM p.215 (1999).

[20] R. Martel, V. Derycke, and Ph. Avouris, *AIP Conference Proceedings, Electronic Properties of Novel Materials – Molecular Nanostructures*, (Editors: H. Kuzmany, J. Fink, M. Mehring and S. Roth) 2001, p. 543.

Efficient State Representation for Symbolic Simulation

Valeria Bertacco
Computer Systems Lab
Stanford University
Stanford, CA 94305
valeria@stanford.edu

Kunle Olukotun
Computer Systems Lab
Stanford University
Stanford, CA 94305
kunle@ogun.stanford.edu

ABSTRACT

Symbolic simulation is attracting increasing interest for the validation of digital circuits. It allows the verification engineer to explore all, or a major portion of the circuit's state space without having to design specific and time-consuming test stimuli. However, the complexity and unpredictable run-time behavior of symbolic simulation have limited its scope to small-to-medium circuits.

In this paper, we propose a novel approach to symbolic simulation that reduces the size of the BDDs of the state vector while maintaining an exact representation of the set of states visited. The method exploits the decomposition properties of Boolean functions. By restructuring the next-state functions in their disjoint support components, we gain a better insight in the role of each input variable. Consequently, we can simplify the next-state functions without significantly sacrificing the simulation accuracy. Our experimental results shows that this approach can be used in effectively reducing the memory requirements of symbolic simulation while surrendering only a small portion of the design's state space.

Categories and Subject Descriptors

B.6.3 [**Logic Design**]: Design Aids—*Verification, Simulation*; B.8 [**Hardware**]: Performance and Reliability

General Terms

Design, Verification, Performance, Theory

Keywords

Formal Verification, Symbolic Simulation, BDDs

1. INTRODUCTION

Validating the functionality of digital circuits and systems is an increasingly difficult task. This is due to the growing complexity of the designs that has not been accompanied by improvements in functional verification techniques.

Logic simulation is still the mainstream approach for the validation of large synchronous systems ([1, 2]) because of its scalability : CPU time is proportional to the design size and test length. Simulation is also flexible: Practical cycle-based simulators allow for circuits with multiple clocks and the ability to mix cycle-based and event-based simulation.

Unfortunately, the fraction of the design space which can be explored by simulation is miniscule, especially for large designs. Only one state and one input combination of the design under test are visited during each simulation cycle. Moreover the test stimuli must be hand crafted by the designer to cover those areas of the design that she wishes to validate.

On the other side of the verification spectrum, are symbolic exploration techniques. These methods have significantly increased the reach of formal verification by making it possible to analyze systems with many states. However practical designs are still out of the grasp of these techniques.

Symbolic methods attempt to validate a system by exploring and checking its behavior under all the possible input stimuli. The advantage of this approach is that a property can be verified for all the possible executions of a circuit and without the need for the designer to create the stimuli. However, these methods have not become mainstream solutions so far, mainly because their computational complexity is very high. As a consequence, they are only able to validate successfully designs only up to a few hundreds latches.

One symbolic exploration approach that has been used is *symbolic model checking*. The basic idea underlying this method is to use BDDs ([3]) to represent all the functions involved in the validation and the set of states that have been visited during the exploration. The primary limitation of this approach is that the BDDs that need to be constructed can grow extremely large, exhausting the memory resources of the simulation host machine and/or causing severe performance degration. In order to overcome this limitation, various solutions have been proposed that try to contain the size of the BDDs involved, for instance: [4, 5, 6].

An alternative approach is *symbolic simulation*. This method verifies a set of scalar tests with a single symbolic vector. Symbolic functions (represented by BDD) are assigned to the inputs and propagated through the circuit to the outputs. (see Figure 1. below). This method is used in [7] and has the advantage that large input spaces can be covered *in parallel* with a single symbolic sweep of the circuit. Again, the bottleneck of the approach lies in the explosion of the BDD representations. Various techniques have been suggested to approximate the functions represented in order to contain the BDDs within reasonable limits: [8, 9].

DAC 2002, June 10-14, 2002, New Orleans, Louisiana, USA

In this paper, we present a novel technique for symbolic simulation that uses a new, parametric representation for the functions at the sequential elements of the circuit. This representation produces BDDs that are more compact (*i.e.*, have fewer nodes) than the original ones, while at the same time constitute an *exact* representation of the state of the circuit. It is generated dynamically during the simulation exploiting the properties of disjoint support decomposition of a Boolean function.

In the remainder of the paper we review the ideas of symbolic simulation and of disjoint support decomposition. We then present our new techniques for containing the size of the BDDs in symbolic simulation. We conclude by presenting experimental results and directions for future work.

2. BACKGROUND

Let $\mathcal{B}$ denote the set $\{0,1\}$. A symbolic variable is a variable defined over $\mathcal{B}$. A logic function F is a mapping $F : \mathcal{B}^m \rightarrow \mathcal{B}^n$. The **range** of a function F is the set of n-tuples that can be asserted by F. It will be denoted by $\mathcal{R}(F)$. The ith component of F will be denoted by $f_i : \mathcal{B}^m \rightarrow \mathcal{B}$. We use lower case for single output functions and upper case for multiple output functions.

The *1-cofactor* of a function f w.r.t. a variable v is the function $f_{v=1}$ obtained by substituting 1 for v in f. Similarly, the *0-cofactor* is obtained by substituting 0 for v in f.

We say that a function $f : \mathcal{B}^m \rightarrow \mathcal{B}$ **depends** on a variable x_i iff the Boolean difference $\partial f / \partial x_i = f_{x_i} \oplus f_{\overline{x_i}}$ is not the constant function 0. In the most general case when F is a multiple output function, we say that $F : \mathcal{B}^m \rightarrow \mathcal{B}^n$ depends on a variable x_i, if at least one of its components f_i depends on it.

The **support** of a logic function is the set of variables f depends on and it is indicated by $\mathcal{S}(f)$.Two functions f, g are termed **disjoint-support** if they share no support variables, *i.e.*, $\mathcal{S}(f) \cap \mathcal{S}(g) = \emptyset$. The size of a support set is indicated by $|\mathcal{S}(f)|$.

We assume functions to be represented by their Binary Decision Diagrams. We refer the reader to [3, 10] for a tutorial introduction to BDDs.

A **synchronous logic circuit** is defined by a 6-tuple:

- an ordered set $(i_1, \cdots, i_m)$ and $(o_1, \ldots, o_p)$ of Boolean input and output symbols,
- an ordered set $(s_1, \ldots, s_n)$ of Boolean state symbols,
- next-state function $\delta : S \times I : \mathcal{B}^{n+m} \rightarrow S : \mathcal{B}^n$,
- output function $\lambda : S \times I : \mathcal{B}^{n+m} \rightarrow O : \mathcal{B}^p$,
- and an *initial assignment* S_0 of the state symbols.

The next section reviews the symbolic simulation algorithm, which applies to a synchronous logic circuit. Section 2.2 provides an overview of disjoint support decompositions as we will use them in presenting our approach.

2.1 Symbolic Simulation

Symbolic simulation refers to the iterative symbolic exploration of the state space of a synchronous circuit. The circuit is initialized at time step 0 with the initial assignment S_0 to the state symbols and with the set of variables $IN_{@0} = \{i_{1@0}, \cdots, i_{m@0}\}$ for the input symbols. At each time step k, the expression of the primary outputs and state variables is computed, in terms of the variables in $\{in_{@0}, \cdots, in_{@k}\}$. At the end of the time step, the vector of Boolean functions for the state symbols $S_{@k} : \mathcal{B}^{mk} \rightarrow \mathcal{B}^n$ represents all the states that can be visited by the circuit at step k . Figure 1 shows the flow just described using a time-unrolled version of the circuit.

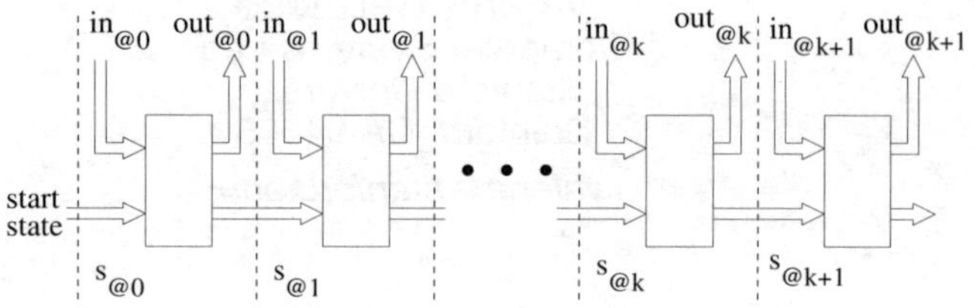

Figure 1: Iterative model of symbolic simulation

Bugs are found by checking at every step that the function $OUT_{@k} : \{in_{@0}, \cdots, in_{@k}\} \rightarrow \mathcal{B}^p$ represents a set of legal values for the outputs of the circuit. When an illegal output combination is found, $out_{@k}$ reports all the possible input combinations that expose it. Theoretically the iteration can be repeated indefinitely, although typically the BDDs for $S_{@k}$ and $OUT_{@k}$ outgrow the memory resources available.

2.2 Disjoint Support Decompositions

The disjoint support decomposition of a scalar function $F : \mathcal{B}^m \rightarrow \mathcal{B}$, consists of finding other, simpler functions L and A_i such that:

$$F(x_1, \cdots, x_m) = L(A_1(x_1, \cdots, x_{A_1}), A_2(x_{A_1+1}, \cdots, x_{A_2}), ..)$$

with $\mathcal{S}(A_i) \cap \mathcal{S}(A_j) = \emptyset$, $\forall i, j$.

An exact solution to this problem has been proposed by Ashenhurst [11] and Curtis [12] using *decomposition charts*. This decomposition algorithm has exponential complexity on the number of variables in $\mathcal{S}(f)$. Ashenhurst also proved in [11] that there is a unique decomposition for a function once we pick a canonical representation for associative operations.

More recently, methods based on the BDD representation of a function have been suggested in [13, 14]. In particular, the method in [13] has complexity quadratic in the size of the BDD of the function to decompose, in the worst case.

The disjoint support decomposition can be applied recursively to each of the A_i components leading to a **block tree** representation of a Boolean function as in Figure 2, where each block represents a Boolean function with a single output and inputs that have pair wise disjoint support. The leaves of the tree are the input variables of the function.

Based on the naming convention in [13], each block can represent either an **associative operator** (AND/OR/XOR) or a **prime function** , that is, a complex function that cannot be decomposed any further with disjoint support inputs.

3. STATE FUNCTION RE-ENCODING

The method we propose in this paper is based on the observation that at each symbolic simulation step k, it is possible to substitute the state function $S_{@k} : \mathcal{B}^{mk} \rightarrow \mathcal{B}^n$ with a new function $DS_{@k}$ such that $\mathcal{R}(S_{@k}) = \mathcal{R}(DS_{@k})$ without affecting the results of the simulations; namely: 1) The set of outputs that can be generated by the circuit and 2) the set of states the circuit can reach at each cycle. If we can find a suitable function $DS_{@k}$ that also has a smaller

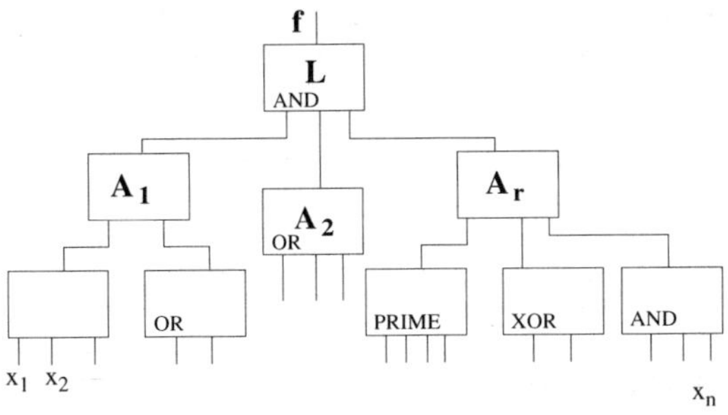

Figure 2. Block tree representation of a Boolean function

BDD representation (*i.e.*, fewer BDD nodes), then we can control the size of the Boolean expression and improve on the performance of symbolic simulation.

We now present various transformations that accomplish this objective. The first technique applies to the state function as a whole, the other two are specific to the decomposition type of a block: either prime function or associative operator. For reasons of readability, the proofs of the theorems presented are reported in the Appendix.

3.1 Reduction at Free Points

By producing the disjoint support decomposition of each component of $S_{@k}$, we obtain a vector of block trees. While each element of the vector has a tree decomposition with no reconvergence, it is now possible that two or more elements intersect at some intermediate node of their trees. An example situation is produced in Figure 3. We call this structure a **decomposition graph**.

From now on, we use the terms *decomposition graph F* and *function F* interchangeably to refer to a multiple output function F. We also drop the subscript $_{@k}$ whenever referring to the state vector S at step k.

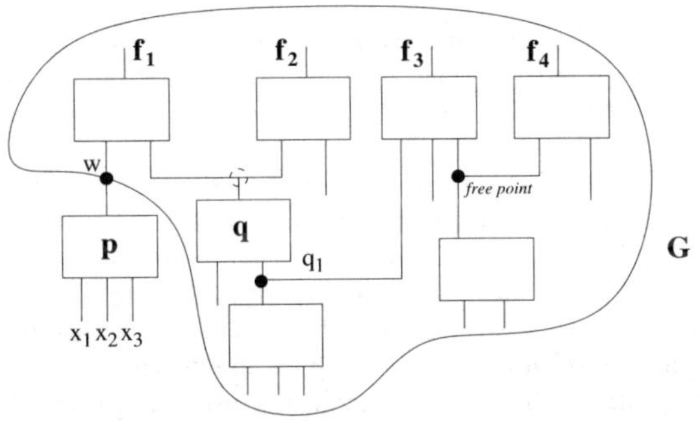

Figure 3. The decomposition of a vector of functions

Definition 1. A **free point** p in a decomposition graph of F is a function corresponding to an output of a block in the graph. It has the property that, if we substitute the sub-graph rooted at the point with a new input variable w, the new function G is disjoint support with the function rooted at p:

$$F(x_1, \cdots, x_m) = G(p, x_{p+1}, \cdots, x_m) \circ p(x_1, \cdots, x_p) \quad (1)$$

and $\mathcal{S}(G) \cap \mathcal{S}(p) = \emptyset$.

Figure 3 shows three free points with darkened circles. Note that the output of p is a free point since none of the variables in the support of p appears in the support of other parts of the graph. On the other hand, the dashed circle at q is not a free point since, if we split the graph at that node, the functions G and q obtained would still share the input q_1. The following theorem shows that we can use free points to simplify the decomposition graph.

THEOREM 1. *Given a decomposition graph for a multiple output Boolean function* $F(x_1, \cdots, x_m) : \mathcal{B}^m \rightarrow \mathcal{B}^n$, *a free point* $p(x_1, \cdots, x_p) : \mathcal{B}^p \rightarrow \mathcal{B}$ *in it, and the function* $G(p, x_{p+1}, \cdots, x_m) : \mathcal{B}^{m-p+1} \rightarrow \mathcal{B}^n$, *obtained by substituting the function* $p()$ *with the new input variable* p *in the graph of* F, $\mathcal{R}(F) = \mathcal{R}(G)$.

Thus, we can substitute all the free points with new variables and generate a new state function G with a smaller representation. A simple traversal of the graph is sufficient to discover all the free points with maximal support, that is, all the free points whose support is not contained in any other free point of the decomposition graph. The transformation of free sub-graphs with new variables produces a new function G, with $|\mathcal{S}(G)| \leq |\mathcal{S}(F)|$.

Example 1. Consider the decomposition graph of Figure 4. Figure 4(a) shows all the free points of the graph. Only the circled free points are maximal. Figure 4(b) shows the new, reduced, function obtained. Note that we can re-use any input variable of a free sub-graph as the new variable at the free point. □

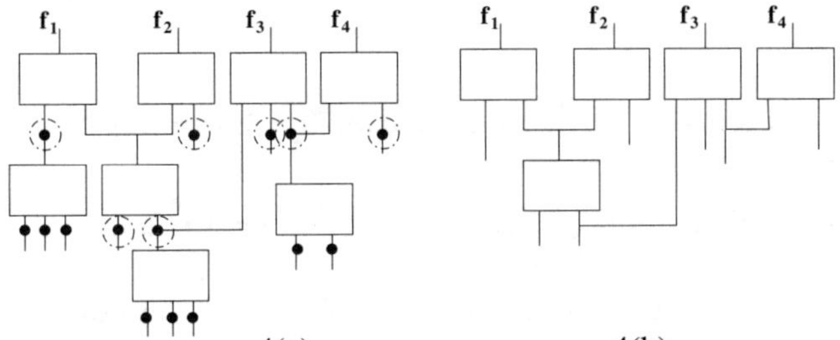

Figure 4: Free points reduction

3.2 Elimination of Prime functions

As mentioned in Section 2.2, each block of a decomposition is either termed a prime function or it is an associative operator. We found that, if a prime function satisfies certain conditions, we can remove it from the decomposition graph, along with all of its sub-graph and substitute it with a fresh input variable.

THEOREM 2. *Given a prime function* $r(r_1, \cdots, r_r)$ *in a decomposition graph* F, *if all of its inputs, except at most one, are free points, than the decomposition graph* G *obtained by substituting the new variable* r *for function* $r()$, $F(x_1, \cdots, x_m) = G(r, \cdots, x_m) \circ r(r_1, \cdots, r_r)$ *is such that* $\mathcal{R}(F) = \mathcal{R}(G)$.

A possible structure of the graph F is represented in Figure 5(a): All the input to block r are free points, except for r_1. We can then remove the block r and substitute it with a new input variable obtaining the graph in Figure 5(b) without affecting the range of the function. Note that input variables r_2 and r_3 are not needed anymore.

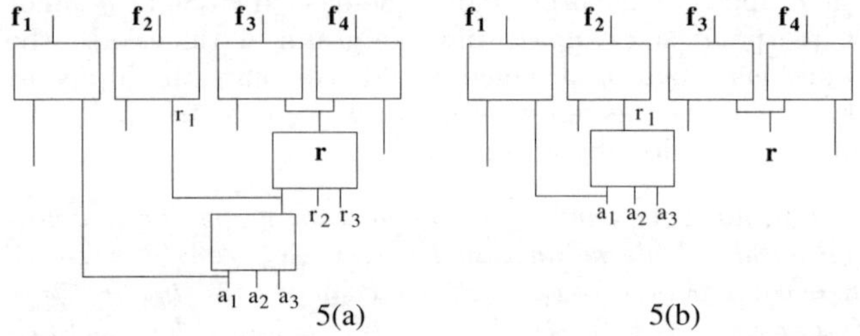

Figure 5. General case for prime function elimination: (a) before and (b) after

Example 2. The testbench s1196 from the IWLS suite contains the blocks reported in Figure 6 in its next state function at step 10 of symbolic simulation. The variables names are just indices corresponding to the variables in the support of the state function. Since the prime function has two inputs that are free points and only one input that has multiple fanout, we can completely eliminate this portion of the graph and just substitute it with an input variable. □

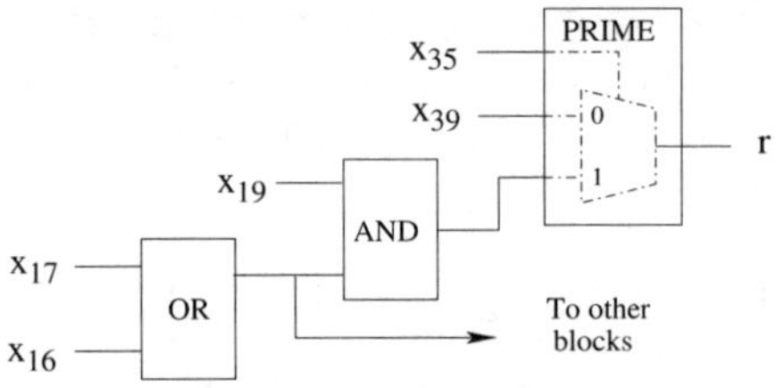

Figure 6: Prime elimination for Example 2.

3.3 Removal of non-dominant variables

Under certain conditions, an input variable can be removed from the decomposition graph without affecting its range.

Example 3. Consider the following 3-outputs function: $f_1 = AND(b, e)$, $f_2 = AND(e, OR(a, b, d))$, $f_3 = XOR(a, c)$.

The range of this function is $\mathcal{B}^3 \setminus \{101, 100\}$. Notice that we can remove the variable a from the function, by assuming it has value 0, without changing the range spanned by F : $f_1 = AND(b, e)$, $f_2 = AND(e, OR(b, d))$, $f_3 = c$ has still range $\mathcal{B}^3 \setminus \{101, 100\}$.

We could do this simplification because the range of the function for $a = 1$ is a subset of the range for $a = 0$. □

Definition 2. An input variable of a decomposition graph has a **non-dominant value 0** iff it fans out only to blocks that are decomposed through OR or XOR associative operators. It has a **non-dominant value 1** iff it fans out only to blocks that are AND or XOR decompositions. Otherwise it does not have a non-dominant value.

Note in particular that a variable may have a non-dominant value 0 and a non-dominant value 1 simultaneously if it fans out only to XOR decompositions. The theorem below shows that in the most general case, a variable that fans out only to associative operators can be removed from the decomposition graph if it has a unique non-dominant value for the whole graph.

THEOREM 3. *If a decomposition graph F has an input variable v with non-dominant value $k \in \{0, 1\}$, and each of the blocks (i.e., intermediate single-output functions) that have v in their fanin have at least one* free point *in their fanin, then:* $\mathcal{R}(F) = \mathcal{R}(F_{v=k})$

3.4 Approximating

Even when all of the previous techniques fail, we still want to maintain a compact representation of the state function $S_{@k}$, so that we can keep progressing in the simulation. When the state function exceeds a configurable threshold value, we choose a variable to set to a constant value. The variable with fanout to the maximum number of blocks is selected to maximize our ability to discover a reduced exact parametric representation at the next attempt. In choosing the variable, we only consider decomposition blocks that have input variables in their fanin, the intuition being that these blocks are closer to become free points.

We found experimentally that often, after eliminating a variable by setting it to constant as described, we discovered additional free points or variables with non-dominant values.

4. EXPERIMENTAL RESULTS

We built a symbolic simulator to experiment with our solution and linked it to the CUDD package [15]. At the end of each simulation step we performed the decomposition of the next state symbolic vector and the set of reductions described in Section 3. We set an upper limit of 2500 nodes for the state vector size. We run our simulation on the sequential circuits of the IWLS '89/'93 benchmark suite using a PC based on a 1.7Ghz Pentium processor equipped with 2GB of memory. The results are presented in Table 1.

For each circuit we report the number of primary inputs, primary outputs, sequential elements, combinational logic gates and the number of cycles we run the the simulation.

The next five columns report how many times we were able to apply our reductions of Section 3: **FP** is the number of free point substitutions, **PR** is the number of prime function eliminations, **VD** the number of non-dominant variables removals, **NL** counts the cases where no exact transformation could be applied, but the state vector was within the limit size and **SV** counts the times when we cofactored out a maximally shared variable as described in Section 3.4. Note that during a single simulation step we may apply more than one technique until we reduce the state vector within limits or until there is no exact parametrization possible. The values reported in Table 1 indicate that the conditions that allow condensing the state vector are frequently met in all the circuits.

We then estimated the number of simulation vectors that are run in parallel, as follows: At each simulation step, we produce a fresh set of input symbols at the primary inputs of the circuit. At the sequential inputs, the number of stimuli presented in parallel is equal to the cardinality of the $\mathcal{R}(NS)$. Thus the number of vectors that we simulate in parallel is: $\#vectors = 2^{\#In} \cdot \mathcal{R}(S_k)$. Column **Vectors** reports the log_2 $\#vectors$, which corresponds to the equivalent number of input wires that are maintained symbolic at each cycle. The number of traces which we carry forward to the

Circuit	In	Out	FF	Gates	steps	FP	PR	VD	NL	SV	Vectors	Traces	plain	Mem	Sec
s1196	14	14	18	790	100	99	5	0	1	0	25.33	25.33	0	22.7	16.60
s1423	17	5	74	830	100	9	0	1	21	144	37.60	35.92	710	35.5	108.95
s298	3	6	14	197	100	16	0	0	20	117	10.35	9.31	45	16.7	67.95
s641	35	24	19	436	100	52	0	21	6	548	43.94	40.71	629	254.0	740.0
s713	35	23	19	480	100	40	0	14	10	531	43.98	41.06	651	154.0	470.4
s953	16	23	29	658	100	59	0	1	3	527	22.95	20.09	745	45.1	106.4
clma	382	82	33	24482	60	38	0	0	21	1	390.56	390.53	1	50.2	28.51
mm4a	7	4	12	310	100	0	0	0	82	40	16.10	15.80	88	30.7	31.11
s38417	38	304	1426	20281	2	2	0	0	0	0	inf	inf	15	33.4	0.05
s38584	38	304	1426	20281	2	1	0	0	1	0	inf	inf	29	42.2	0.06
s5378	35	49	163	3232	2	2	0	0	0	0	54.0	54.0	0	11.3	164.4
s9234	36	39	135	3019	7	1	0	5	1	0	63.49	63.49	0	18.7	350.5
sbc	40	56	27	1143	100	184	0	214	4	932	51.05	46.61	877	66.1	1123.5

Table 1: Experimental results

next cycle is given by: $\#traces = 2^{\#In - \#SV} \cdot \mathcal{R}(S_k)$. Again, we report the bit-equivalent value of the number of traces. In order to compute these values, we needed to compute the size of the reached set at every cycle: All the cases where **steps** is less than a 100 have been limited by this computation.

It is clear from the table that we can achieve a high level of parallelism, in the order of tens of bits per cycle, while keeping a low memory profile.

We compared our results with a plain symbolic simulator that limits the size of the state functions by simply cofactoring out as many variables as needed at the end of each cycle: column **plain** reports the number of variables evaluated to constant by this simulator and should be compared to column **SV**. Most often our approach allows us to evaluate to constant fewer variables than a straightforward approach. It's worth noting that there are cases like *s1196* where both simulators do not need to perform any approximation, but the usage of our tecnique allowed us to mantain simpler BDDs and complete the simulation faster.

For the last two columns, we re-run the tests without calculating the size of the state space reached at every cycle and we report the memory profile in MBytes and the execution time needed for running the specified simulations.

5. CONCLUSIONS AND FUTURE WORK

We have presented a new approach for constructing a parametric representation of the state vector during symbolic simulation. The construction exploits the property of disjoint support decomposition of the symbolic state vector. It applies transformations that preserve the range of the vector function while reducing the size of its BDD representation.

Our experiments indicate that the conditions required to perform these transformations often exist. The results show that these transformations are a valuable tool for improving the performance of symbolic simulation while maintaining the quality of its results. We are currently working on finding additional configurations that lead to exact transformations and on improving the quality of the fall-back action (measured as ratio between number of BDD nodes eliminated to number of elements of the state set lost) when no exact transformation can be found.

Acknowledgments

The authors would like to thank Maurizio Damiani and David Dill for their valuable suggestions during the development of this work. This research has been supported by Georgia Tech Marco contract #B-12-D00-S5.

6. REFERENCES

[1] Z. Barzilai, J. Carter, B. Rosen, and J. Rutledge. Hss-a high-speed simulator. *IEEE Trans. on CAD/ICAS*, pages 601–617, July 1987.

[2] C. Hansen. Hardware logic simulation by compilation. In *DAC, Proceedings of the Design Automation Conference*, pages 712–715, June 1987.

[3] R. Bryant. Graph-based algorithms for boolean function manipulation. *IEEE Trans. on Computers*, 35(8):677–691, August 1986.

[4] I. Moon, J. Kukula, K. Ravi, and F. Somenzi. To split or to conjoin: The question in image computation. In *DAC, Proceedings of the Design Automation Conference*, pages 23–28, June 2000.

[5] K. Ravi and F. Somenzi. High density reachability analysis. In *Proc. ICCAD*, pages 154–158, November 1995.

[6] G. Cabodi, P. Camurati, L. Lavagno, and S.Quer. Disjunctive partitioning and partial iterative squaring: an effective approach for symbolic traversal of large circuits. In *DAC, Proceedings of the Design Automation Conference*, pages 728–733, June 1997.

[7] R. Bryant, D. Beatty, K. Brace, K. Cho, and T. Sheffler. Cosmos: a compiled simulator for mos circuits. In *DAC, Proceedings of the Design Automation Conference*, pages 9–16, June 1987.

[8] J. Bergmann and M. Horowitz. Improving coverage analysis and test generation for large designs. In *Proc. ICCAD*, pages 580–583, November 1999.

[9] V. Bertacco, M. Damiani, and S. Quer. Cycle-based symbolic simulation of gate-level synchronous circuits. In *DAC, Proceedings of the Design Automation Conference*, pages 391–396, June 1999.

[10] R. Bryant. Symbolic boolean manipulation with ordered binary–decision diagrams. *ACM Computing Surveys*, 24(3):293–318, September 1992.

[11] R. Ashenhurst. The decomposition of switching functions. In *Proceedings of the International Symposium on the Theory of Switching*, Part I 29, pages 74–116, 1957.

[12] H. A. Curtis. *A New Approach to the Design of Switching Circuits.* Van Nostrand, Princeton, N.J., 1962.

[13] V. Bertacco and M. Damiani. The disjunctive decomposition of logic functions. In *Proc. ICCAD*, pages 78–82, November 1997.

[14] Fpga design by generalized functional decomposition. In T. Sasao, editor, *Logic Synthesis and Optimization*,

chapter 11. Kluwer Academic, 1993.

[15] CUDD-2.3.1. *http://vlsi.Colorado.edu/~fabio.*

[16] O. Coudert, C. Berthet, and J.C. Madre. Verification of synchronous sequential machines using symbolic execution. In *Workshop on Automatic Verification Methods for FSM*, volume 407 of *LNCS*, pages 365–373, June 1989.

APPENDIX

Proof of Theorem 1. Consider the function $F(x_1, \cdots x_m)$ and compute its range by splitting on the input variables [16]:

$$\mathcal{R}(F) = \mathcal{R}(F_{x_1=0}) \cup \mathcal{R}(F_{x_1=1})$$

By applying this equation recursively over all the variables $(x_1, \cdots x_p)$ in the support of p, we obtain:

$$\mathcal{R}(F) = \bigcup_{(i_1,\cdots,i_p)\in\mathcal{B}^p} \mathcal{R}(F_{x_1=i_1,x_2=i_2,\cdots,x_p=i_p}) \quad (2)$$

Using Equation (1): $F_{x_1=i_1,x_2=i_2,\cdots,x_p=i_p} = G_{p=i_w}$ where $i_w = p(i_1, \cdots, i_p) \in \{0,1\}$ since p evaluates to a constant. Substituting in Eq.(2) we finally obtain: $\mathcal{R}(F) = \bigcup_{i_w\in\{0,1\}} \mathcal{R}(G_{p=i_w}) = \mathcal{R}(G)$ □

Proof of Theorem 2. We distinguish two cases:

1. All the inputs of the prime block are free points. Then the output of the free block is also a free point and the theorem reduces to the hypothesis of Theorem 1.

2. The prime block r has one input that it is not a free-point, say r_1, without loss of generality. All the other inputs to the prime function: $(r_2, \cdots, r_r)$ are still free points and we can assume that have been reduced to input variables by Theorem 1. In the most general case, r_1 is a single output function of other input variables that are in the support of both G and r: $\mathcal{S}(r_1) = (a_1, \cdots, a_p)$. The function F has then the form:

$$\begin{aligned} F(a_1 \cdots a_p, \cdots r_r, \cdots x_m) = G(r, a_1 \cdots a_p, r_1, \cdots x_m) \\ \circ\, r(r_1, \cdots, r_r) \circ r_1(a_1, \cdots, a_p) \end{aligned} \quad (3)$$

Let's proceed again by computing the $\mathcal{R}(F)$ by recursively splitting on the input variables:

$$\mathcal{R}(F) = \bigcup_{(i_1,\cdots,i_p)\in\mathcal{B}^p} \mathcal{R}(F_{a_1=i_1,a_2=i_2,\cdots,a_p=i_p}) \quad (4)$$

For each different assignment $(i_1, \cdots, i_p)$, r_1 evaluates to a constant value: $i_r = r_1(i_1, \cdots, i_p) \in \{0,1\}$. Substituting the expansion of F as in Eq.(3):

$$F_{a_1=i_1,\cdots,a_p=i_p} = G_{a_1=i_1,\cdots,a_p=i_p,r_1=i_{r_1}} \circ r_{r_1=i_{r_1}} \quad (5)$$

Note that we cannot drop the cofactors w.r.t. the a_i in G because r_1 is not a free point and thus its inputs fan out to other nodes of the graph.

Now, the function $r_{r_1=i_{r_1}}(r_2, \cdots, r_r)$ is a free point and as such it can be substituted by a new free variable r. We show now that it is not possible that $r_{r_1=i_{r_1}}(r_2, \cdots, r_r)$ reduces to a constant for any value of i_{r_1}. Infact, if that was the case r could be expressed as $r = r_1 \otimes r_{res}(r_2, \cdots, r_r)$, where $\otimes$ is either AND or OR and $\mathcal{S}(r_1) \cap \mathcal{S}(r_{res}) = \emptyset$, and r would then have a disjoint support decomposition through an associative operator and would not be a prime function.

By carrying on the substitution $r = r_{r_1=i_{r_1}}(r_2, \cdots, r_r)$, Eq. (5) reduces to: $F_{a_1=i_1,a_2=i_2,\cdots,a_p=i_p} = G_{a_1=i_1,a_2=i_2,\cdots,a_p=i_p}$ which substituted into Eq. (4) proves the theorem. □

Proof of Theorem 3. For a generic function F, we have:

$$\mathcal{R}(F) = \mathcal{R}(F_{v=k}) \cup \mathcal{R}(F_{v=\bar{k}}) \quad (6)$$

We now show that under the conditions specified:

$$\mathcal{R}(F_{v=\bar{k}}) \subseteq \mathcal{R}(F_{v=k}) \quad (7)$$

and Eq.(6) reduces to $\mathcal{R}(F) = \mathcal{R}(F_{v=k})$.

Let's consider first the case where $k = 0$ and let's label each of the functions that have v in their fanin $x(v, p, x_1 \cdots)$, $y(v, q, y_1 \cdots)$, $w(v, r, w_1 \cdots)$ ... where p, q, r ... are the free points in each of them and x_i, y_i, w_i ... are other variables the functions depend on. The x,y,w,... functions can only be OR or XOR decompositions by hypothesis.

We can then express F using the composition of these functions: $F = G(x, y, w \cdots, x_1 \cdots y_1 \cdots) \circ x(v, p, x_1..) \circ y(v, q, y_1..) \cdots$

Note that, in general, x_i, y_i, w_i ... are also in the fanin of G. Let's now compute the two cofactor of F w.r.t. v:

$$F_{v=0} = G(x, y, w \cdots, x_1 \cdots w_w \cdots) \circ x(0, p, x_1..) \circ \cdots$$
$$F_{v=1} = G(x, y, w \cdots, x_1 \cdots w_w \cdots) \circ x(1, p, x_1..) \circ \cdots$$

In order to show the inclusion of the ranges of Eq. (7), we are going to represent each range as a union of ranges by cofactoring the variables in the support of x, y, w, ... one function at a time starting with $x()$:

$$\mathcal{R}(F_{v=0}) = \bigcup_{(x_1\cdots x_x)\in\mathcal{B}^x} \mathcal{R}(G(x \cdots x_1 \cdots) \circ x(0 \cdots)_{x_1=i_{x_1},\cdots})$$

$$\mathcal{R}(F_{v=1}) = \bigcup_{(x_1\cdots x_x)\in\mathcal{B}^x} \mathcal{R}(G(x \cdots x_1 \cdots) \circ x(1, \cdots)_{x_1=i_{x_1},\cdots})$$

We distinguish two cases for each x,y,w,... function:

1) **x is a OR decomposition**. When all the $(x_1, \cdots, x_x)$ are zero, for $F_{v=1}$, x evaluates to the constant value 1. For $F_{v=0}$, $x = p$. In all the other cases x evaluates to 1. By grouping all the component ranges so that to distinguish the special case from all the others , we can simplify the expressions:

$$\mathcal{R}(F_{v=0}) = \mathcal{R}(G(p, \cdots 0 \cdots)) \cup \bigcup_{(x_1,\cdots,x_x)\neq \mathbf{0}} \mathcal{R}(G(1, y, \cdots x_1 \cdots)_{x_1=i_{x_1},\cdots}$$

$$\mathcal{R}(F_{v=1}) = \mathcal{R}(G(1, \cdots 0 \cdots)) \cup \bigcup_{(x_1,\cdots,x_x)\neq \mathbf{0}} \mathcal{R}(G(1, y, \cdots x_1 \cdots)_{x_1=i_{x_1},\cdots}$$

It can be easily seen that the first reange for $F_{v=1}$ is a subset of the corresponding range for $F_{v=0}$, while the rest of the expression is identical.

2) **x is an XOR decomposition**. In $F_{v=1}$, $x = XNOR(p, x_1, \cdots, x_x)$. For $F_{v=0}$, $x = XOR(p, x_1, \cdots, x_x)$. We can again group all the component ranges so that to distinguish the cases where $XOR(x_1, \cdots, x_x) = 0$ from the ones where $XOR(x_1, \cdots, x_x) = 1$:

$$\mathcal{R}(F_{v=0}) = \bigcup_{XOR(x_1,\cdots,x_x)=0} \mathcal{R}(G(p, y, \cdots x_1 \cdots)_{x_1=i_{x_1},\cdots} \cup \bigcup_{XOR(x_1,\cdots,x_x)=1} \mathcal{R}(G(\bar{p}, y, \cdots x_1 \cdots)_{x_1=i_{x_1},\cdots}$$

$$\mathcal{R}(F_{v=1}) = \bigcup_{XOR(x_1,\cdots,x_x)=0} \mathcal{R}(G(\bar{p}, y, \cdots x_1 \cdots)_{x_1=i_{x_1},\cdots} \cup \bigcup_{XOR(x_1,\cdots,x_x)=1} \mathcal{R}(G(p, y, \cdots x_1 \cdots)_{x_1=i_{x_1},\cdots}$$

And it can be observed that the two components of each expression match. It follows: $\mathcal{R}(F_{v=0}) = \mathcal{R}(F_{v=1})$.

This procedure can be applied recursively for each of the other functions y, ..., by computing and grouping all the cofactors for the sets of input variables $(y_1 \cdots y_y)$, ...

For the case where $k = 1$, the functions x,y,w, ... can now only be AND or XOR decompositions. The proof can be obtained by substituting AND for OR and 1 for 0 in the proof just discussed. Finally for the case where the input variable v has both a non-dominant value 0 and 1, we can just use any of the two value-specific proofs. □

Handling Special Constructs in Symbolic Simulation

Alfred Kölbl[1] James Kukula[2] Kurt Antreich[1] Robert Damiano[2]

[1]Institute for Electronic Design Automation, Technical University of Munich, 80290 Munich, Germany

[2]Advanced Technology Group, Synopsys Inc., Beaverton, OR-97006, USA

ABSTRACT

Symbolic simulation is a formal verification technique which combines the flexibility of conventional simulation with powerful symbolic methods. Some constructs, however, which are easy to handle in conventional simulation need special consideration in symbolic simulation. This paper discusses some special constructs that require unique treatment in symbolic simulation such as the symbolic representation of arrays, an efficient symbolic method for storing arrayed instances and the handling of symbolic data-dependent delays. We present results which demonstrate the effectiveness of our symbolic array model in the simulation of highly regular structures like FPGAs, memories or cellular automata.

Categories and Subject Descriptors

B.5.2 [**Hardware**]: Register-Transfer-Level Implementation—*Design Aids, Verification*; B.6.3 [**Hardware**]: Logic Design—*Design Aids, Verification*; B.7.2 [**Hardware**]: Integrated Circuits—*Design Aids, Verification*

General Terms

Verification

Keywords

Symbolic Simulation, Formal Verification

1. INTRODUCTION

With the increasing complexity and tight time-to-market schedules of today's digital systems, it is becoming more and more difficult to design correct circuits for these systems. The importance of error-free design is stressed by the fact that design teams spend 60% to 80% of the overall design time for verification. This high effort is necessary because the traditional simulation-based verification approaches cover only a small percentage of the design's state space, limiting their ability to detect subtle design errors. One way to alleviate this problem is the use of formal methods which prove the correctness of the design on a mathematical model. Unfortunately, today's formal verification tools are still limited to small or medium sized designs. Recently, some techniques evolved that pursue a slightly different way. These techniques combine the traditional simulation-based approach with formal symbolic manipulations. They provide higher state space coverage while still retaining all the advantages of simulation. One of these techniques is symbolic simulation.

The objective of symbolic simulation is to broaden the single trace of conventional simulation to a large number of traces that are simulated concurrently. In conventional simulation, a pattern of explicit values is applied to the circuit inputs and the simulator computes the explicit values at the circuit outputs. As only one pattern is simulated per simulation run, 2^n (where n is the number of inputs) runs are necessary to prove the correctness of the design. In symbolic simulation, for each input a symbolic variable is introduced which represents all values this input may take (e.g. 0 and 1). The symbolic simulator then computes symbolic expressions for each output in terms of the input variables. These expressions implicitly represent the values of the outputs for all possible input assignments. Thus, if n symbolic variables are applied to the circuit inputs, one symbolic simulation run concurrently simulates 2^n patterns in parallel.

While simple gates and many RTL-constructs are relatively easy to handle in symbolic simulation, some constructs need special consideration. One, e.g., is the symbolic representation of arrays. Arrays have several applications in digital systems design. They serve as abstraction for memories, queues, stacks or are used to store arrayed instances in hierarchical circuit descriptions. In this paper, we present a generic symbolic model for large arrays. Section 2 introduces the model and discusses the application of this model as representation of memories in symbolic simulation. Section 3 demonstrates the application of this model for the efficient storage of arrayed instances in designs with a high degree of regularity.

Another special construct in symbolic simulation is the delay statement with a data-dependent delay expression. In event-driven symbolic simulation, putting an event on the queue for each time step where the delay expression is satisfied would eventually blow up the queue. We propose a solution to this problem in Section 4. Finally, in Section 5 we present some experimental results.

2. SYMBOLIC ARRAYS

Arrays are frequently used in Hardware Description Languages as abstract model for memories. The example on the left hand side of Fig. 1 shows how such a memory can be accessed in a Verilog symbolic simulator.

Verilog specification	Symbolic execution
reg mem[3:0];	**array** mem[3:0]
reg [1:0] index;	**reg** [1:0] index
reg data;	**reg** data
initial begin	**initial** {
index = $symbol(2);	index := (i_1, i_0)
data = $symbol(1);	data := d
mem[index] = data;	write(mem, index, data)
data = mem[index];	data := read(mem, index, data)
end	}

Figure 1: Symbolic memory example.

The system task *$symbol(number)* is used to introduce the given number of new symbolic variables in the simulation run. In this example, *index* is a 2-bit wide register, therefore two symbolic variables are generated here. The right hand side of Fig. 1 shows the sequence of statements as they are executed by the symbolic simulator. For *index*, two symbolic variables i_1 (denoting the most significant bit of vector *index*) and i_0 (denoting the least significant bit) are created. For the scalar register *data*, just one symbolic variable d is generated. The

DAC 2002, June 10-14, 2002, New Orleans, Louisiana, USA.

read- and write accesses to the memory array are translated to calls to the simulator procedures *write* and *read* with the name of the array, the index and the data expression as arguments. The implementation of these procedures is described in more detail below. In conventional simulation, *index* and *data* always have explicit values and thus the accessed array location is unique. In the symbolic case, however, the difficulty is that not just *data*, but also *index* may have symbolic values. If we explicitly represent each bit of the array, each array location stores a symbolic function denoting the current state of the bit. Writing to the memory with symbolic *data* and *index* can then be achieved by the following computation [6], where ite denotes the if-then-else operator ite(f, g, h) $= f \cdot g + \overline{f} \cdot h$:

$$\begin{array}{lcllcl} \text{mem[0]} & := & \text{ite(index == 0, data, mem[0])} & = & \overline{i_1} \cdot \overline{i_0} \cdot d \\ \text{mem[1]} & := & \text{ite(index == 1, data, mem[1])} & = & \overline{i_1} \cdot i_0 \cdot d \\ \text{mem[2]} & := & \text{ite(index == 2, data, mem[2])} & = & i_1 \cdot \overline{i_0} \cdot d \\ \text{mem[3]} & := & \text{ite(index == 3, data, mem[3])} & = & i_1 \cdot i_0 \cdot d \end{array}$$

Figure 2: Write access to explicit bit array.

So for array location mem[0], only if $index == 0$ ($\overline{i_1} \cdot \overline{i_0} = 1$) then $data$ is stored in this location, otherwise mem[0] retains its old value. If we assume that the array is initialized to zero, the array locations store the symbolic expressions on the right hand side of Fig. 2 after the write access. Note that although all array locations are accessed, for a particular assignment to the symbolic variables i_0, i_1, only one location is different from the initial state and thus, for each assignment only one location is written.

Explicitly storing each bit of the array only works for relatively small arrays. For large arrays, conventional simulators employ sparse techniques which only allocate space for actually accessed array locations. As usually only a very small portion of the array is accessed, arrays of virtually any size can be realized.

The idea of a sparse representation of large arrays in symbolic simulation is also subject of [12, 14]. In that work, Velev and Bryant propose to store symbolic accesses to a memory in a list containing entries of the form (c, i, d), where c is a Boolean expression denoting the set of conditions for which this entry is defined, i is the index expression denoting the memory location, and d is the data expression denoting the contents. While each write access to the memory appends a new entry to the list, each read access scans the list and collects the necessary information from the entries. This approach has a few disadvantages. First, scanning this list in every read operation can be a computationally expensive task if there is a large number of list entries. As their primary application is symbolic trajectory evaluation [11, 3, 1] where usually only few cycles are simulated, it is unlikely that they will run into problems. In a general symbolic simulator, however, the number of memory writes and thus the length of the list can grow very large. A second disadvantage is that their list representation is not canonical, i.e., comparing two memory contents for equality or containment [13] becomes a nontrivial task because the lists of both memories must be scanned and matched.

The array model proposed in this paper has the same benefits as the approach of Velev and Bryant but removes the mentioned disadvantages, and the operations on this model are even simpler.

Our goal is to capture the complete content of the array implicitly in just one function, while keeping access to each individual array location simple. We achieve this by constructing the function f for an array of size n from the expansion

$$f = \sum_{i=0}^{n-1} mem[i] \cdot \phi_i,$$

where $mem[i]$ denotes the symbolic expression that is stored in the array at location i. In order to keep accessing the memory locations simple, the functions ϕ_i must form an orthonormal set $\{\phi_0, \dots, \phi_{n-1}\}$, i.e.,

$$\sum_{i=0}^{n-1} \phi_i = 1 \quad \text{and} \quad \forall_{i \neq j} \phi_i \cdot \phi_j = 0$$

holds. Then, each array location j can easily be accessed by conjoining f with the corresponding ϕ_j, because

$$f \cdot \phi_j = \sum_{i=0}^{n-1} mem[i] \cdot \phi_i \cdot \phi_j = mem[j] \cdot \phi_j$$

To construct the orthonormal functions ϕ_i, we introduce $k = \lceil \log_2 n \rceil$ location variables l which are used to encode all the array locations. Then, $\phi_0 = \overline{l_k} \cdot \ldots \cdot \overline{l_1} \cdot \overline{l_0}$, $\phi_1 = \overline{l_k} \cdot \ldots \cdot \overline{l_1} \cdot l_0$, $\phi_2 = \overline{l_k} \cdot \ldots \cdot l_1 \cdot \overline{l_0}$, $\phi_3 = \overline{l_k} \cdot \ldots \cdot l_1 \cdot l_0$, $\ldots$ $\phi_{(2^k-1)} = l_k \cdot \ldots \cdot l_1 \cdot l_0$.

This encoding is not just orthonormal, it further simplifies accessing one particular array location, because now $mem[i]$ is simply the cofactor of f with respect to ϕ_i: $mem[i] = f|_{\phi_i}$.

For example, an array of size four is now described by the function

$$\begin{aligned} f = & (\overline{l_1} \cdot \overline{l_0}) \cdot mem[0] + (\overline{l_1} \cdot l_0) \cdot mem[1] + \\ & (l_1 \cdot \overline{l_0}) \cdot mem[2] + (l_1 \cdot l_0) \cdot mem[3]. \end{aligned}$$

Each assignment to the location variables references one particular location in the array.

Further on, we will explain our model using BDDs [4] because BDDs are the underlying representation for symbolic expressions in our simulator. A BDD for this function is given in Fig. 3. We assume that all four array locations are initialized to zero. For reasons of clarity, BDD reduction rules have not been applied to the left hand side BDD.

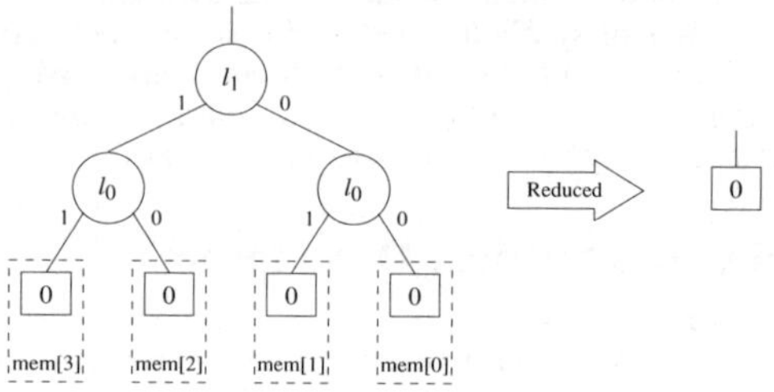

Figure 3: BDD for array of size four initialized to all zero.

The sparse character of our approach comes into play when the standard BDD reduction rules are applied because then, memory locations with the same content are shared. The reduced BDD in our example is simply a constant zero, actually representing four memory locations.

A write operation to an array with a given size at the location specified by a vectored index, given data and an optional enable expression is performed by the following procedure:

procedure write(BDD array, int size, BDD[] index, BDD data, BDD enable)
{

charbdd := $\prod_{i=0}^{\lceil \log_2 size \rceil - 1} (l_i \oplus index_i)$

charbdd := charbdd · enable

array := ite(charbdd, data, array)

}

Figure 4: Array write procedure.

First, a characteristic function is built which yields one if the location variables are equal to the corresponding index bits. This characteristic function determines the array locations that are accessed by the write. Then, the array is updated such that whenever *charbdd* is one, data is written into the memory, otherwise the memory remains unchanged. Some examples are discussed in more detail:

- If *index* is a constant expression, the write access directly addresses a single location.
- If *index* is the two bit symbolic expression from the example in Fig. 1, i.e., $index = (i_1, i_0)$, $data = d$ and the memory is initialized with zero, we obtain the BDD on the left hand side of Fig. 5. It can be seen that the memory locations store exactly the same expressions as in Fig. 2 where the bits were stored explicitly. Note that location variables are only arranged on the top of the BDD for clarity. Generally, they can be interleaved with the other variables in order to achieve a compact representation. Here, e.g., it is a good idea to choose the order $i_1 \prec l_1 \prec i_0 \prec l_0$ resulting in a much simpler BDD.
- If there are two write operations to two different array locations with the same data, e.g., one with $index = (0,0)$, $data = d$ and

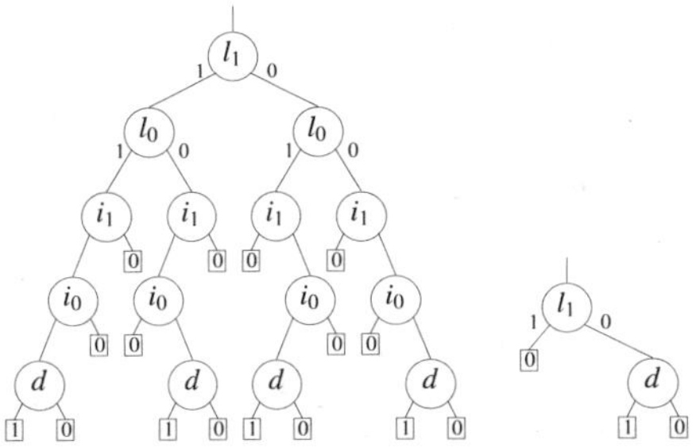

Figure 5: Example BDDs for symbolic write access.

the other with $index = (0,1)$, $data = d$, both memory locations will be automatically shared in the BDD. Assuming that the array was initialized to zero, we obtain the BDD on the right hand side of Fig. 5 for this particular example.

A read access on an array is performed by substituting the location variables by the expressions in the index vector:

procedure read(BDD array, int size, BDD[] index)
{
  data := array($l_0 := index_0, l_1 := index_1, l_2 := index_2, \ldots$)
  return data
}

Figure 6: Array read procedure.

Example: Reading with symbolic $index = (j_1, j_0)$ from the array on the left hand side of Fig. 5 returns the same BDD just that l_1 is replaced by j_1 and l_0 is replaced by j_0. The resulting expression can be interpreted as follows: For each assignment where $(\hat{j}_1, \hat{j}_0) = (\hat{i}_1, \hat{i}_0)$, the data expression d is returned, otherwise zero. So only if the read-index (j_1, j_0) is equal to the index (i_1, i_0) of the previous write operation, data d is returned.

One feature of the approach of Velev and Bryant is the initialization of the memory with symbolic variables, i.e., whenever a memory location is read which was not previously written, a new symbolic variable is returned. This is used to model memories where the initial state is undefined. A special trick is that the number of introduced symbolic variables is maximum the number of read accesses to uninitialized locations, because new symbolic variables are not assigned to memory locations but to memory accesses.

In our model, we also support this feature. We achieve this by initializing each location of the memory with the same dummy variable. This dummy variable marks the locations as unaccessed. As all locations are initialized with the same variable, the reduced BDD for the array is simply the dummy variable itself.

procedure read(BDD array, int size, BDD[] index)
{
  data := array($l_0 := index_0, l_1 := index_1, l_2 := index_2, \ldots$)
  if ($dummy \in$ support(data)) then
  {
    createNewSymbolicVariable(*newvar*)
    data := data(*dummy* := *newvar*)
    write(array, size, index, data, 1)
  }
  return data
}

Figure 7: Modified read procedure.

Now, whenever the dummy variable is in the support of the result of a read operation, the read access happened on an uninitialized location. Then, we replace the dummy variable in the data expression by a newly generated symbolic variable and write this expression back into the memory. The modified read procedure is described in Fig. 7. Note that although a read operation may touch several locations, for each particular assignment of symbolic variables, only one location is read. This is the reason why one symbolic variable is enough because each read operation can only access exactly one location.

Summarizing, the proposed model implements a sparse array for symbolic simulation providing a very compact representation because identical array locations are automatically shared in the BDD. The complete content of the memory is stored in one BDD, removing the need to scan a list when gathering information about the array content. Additionally, our representation is canonical making it easy to compare two memories for equality or containment. Equality, e.g., can be checked in constant time by just comparing the two BDDs for equality. And, last but not least, read- and write-operations on the memory are much simpler than in [14].

3. SYMBOLIC INSTANCE ARRAYS

Modern Hardware Description Languages provide constructs for storing multiple instances of the same type in an arrayed instance. In Verilog, e.g., the code in Fig. 8 creates an instance array of three AND-Gates.

module andgate(c, a, b);
  output c;
  input a, b;

  assign c = a & b;
endmodule

module top;
  reg [2:0] o, i1, i2;

  andgate a[2:0] (o, i1, i2);
endmodule

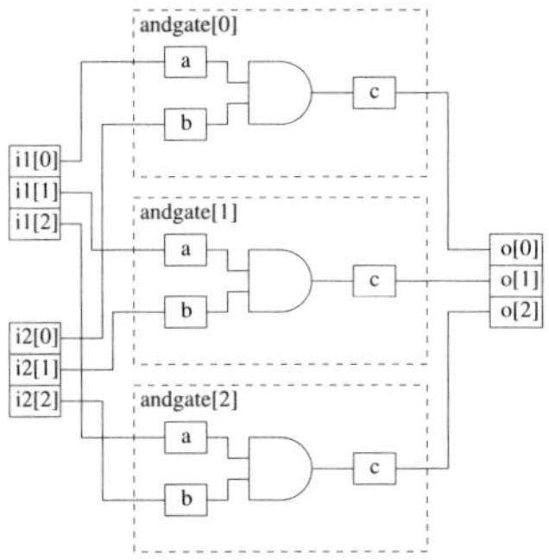

Figure 8: Explicitly expanded instance array

Conventional simulators build an internal netlist from this description where the instance array is expanded and each instance is generated separately. The expanded array and the connections from the signals in module top to the signals internal to the instance array are shown on the right hand side of Fig. 8.

In a symbolic simulator, a natural advancement of our array model is to keep the signals in the arrayed instances in symbolic arrays instead of explicitly expanding them. This is illustrated in Fig. 9. Signals a, b and c are now stored in symbolic arrays, where the index in the array is given by the number of the instance. Note that this is different from bitparallel simulation because register values are not stored explicitly but in form of a BDD which in most cases is much more compact due to the sharing of equal values. This theoretically enables the compression of an unlimited number of instances in such an instance array.

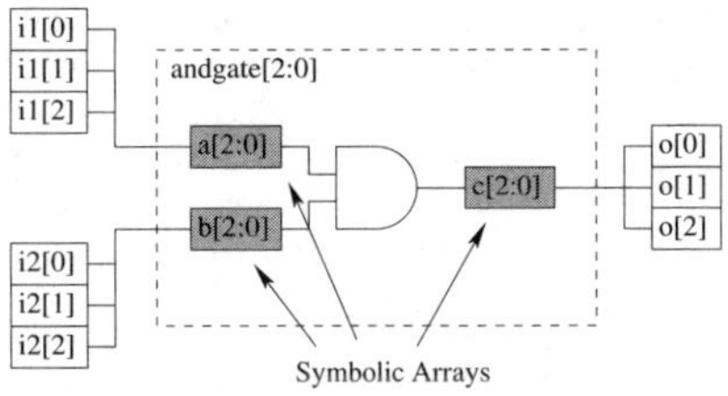

Figure 9: Symbolic instance array.

Whenever the value of the input register *i1* (*i2*) changes, the symbolic value of *a* (*b*) is updated by a call to procedure *write*. For example, if $i1[2]$ changes, *a* is updated by calling $write(a, size, 2, i1[2])$, where *size* denotes the number of instances. On the other hand, if the symbolic output register *c* changes, the values of register *o* can be determined by calling $o[0]$ = *read*(*c*, *size*, 0), $o[1]$ = *read*(*c*, *size*, 1), $o[2]$ = *read*(*c*, *size*, 2).

Using the symbolic array model has not only the effect that signal values in such an instance array are stored efficiently. Because all arrays derived from the same instance array use the same location variables, it also enables the concurrent simulation of all the instances. Unlike in conventional simulators where each instance must be computed separately, with symbolic simulation, we can simulate all instances at once. Although the symbolic computation is more complex, it can be faster for a large number of instances.

Instead of computing the locations of array *c* from arrays *a* and *b* separately for each instance, we can directly compute the symbolic function for *c* from the symbolic functions of *a* and *b*. The validity of this statement is demonstrated by the following theorem.

THEOREM 1. *Let a, b be Boolean functions expressed by expansions with respect to a common orthonormal set* $\{\phi_0, \phi_1, \ldots, \phi_{n-1}\}$, *i.e.,*

$$a = \sum_{i=0}^{n-1} a_i \cdot \phi_i \quad and \quad b = \sum_{i=0}^{n-1} b_i \cdot \phi_i,$$

where a_i (b_i) *denotes the content of the register a (b) for instance i. Let* $\odot$ *be an arbitrary binary operator. Then*

$$\sum_{i=0}^{n-1} (a_i \odot b_i) \cdot \phi_i = a \odot b,$$

i.e., computing the operator $\odot$ *for all instances separately has the same effect as computing* $\odot$ *for the symbolic array functions.*

PROOF. A proof for this theorem can be found in [2]. □

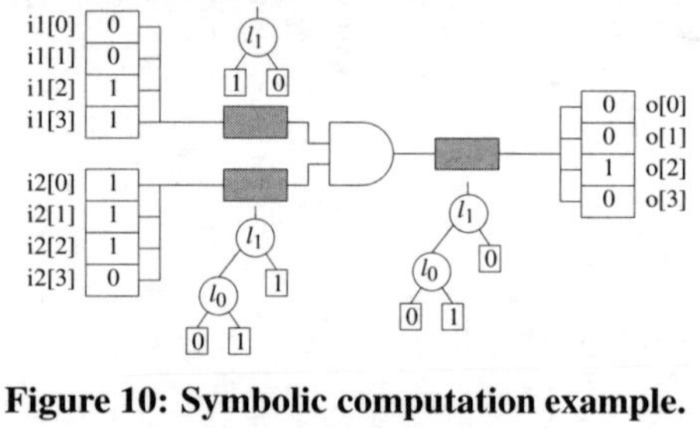

Figure 10: Symbolic computation example.

The complete procedure is demonstrated in the example in Fig. 10. Consider an instance array containing four AND-Gates. First, the explicit values external to the instances are converted into symbolic arrays resulting in the given BDDs. Then, the content of the output array is computed by a symbolic simulation of the instance. For an AND-Gate, the output is determined by conjoining the two input BDDs. Finally, the output array is converted back to an explicit bit-vector.

Summarizing, the used array model is the same as in Section 2 except that the location variables are used to reference different instances instead of memory locations. Because our array model uses one function for each array, which additionally is based on an orthonormal expansion, all instances in such an array can be simulated concurrently.

Although we only allow compression of instance arrays in our implementation, in the extreme case, it is also possible to compress all instances in a circuit of the same type into an symbolic instance array. From our experiments which are discussed in Section 5, however, we doubt that this would make practical sense for general circuits. Furthermore, our approach is not limited to simple gates. Arbitrary complex modules or RTL descriptions can be handled exactly the same way.

We couldn't find any published article about a similar technique, except that recently, Innologic [7] advertised a method called "Hierarchical compression" which sounds similar to our approach.

4. SYMBOLIC DELAYS

Another challenge for an event-driven symbolic simulator is the special case of a symbolic data-dependent delay. The example in Fig. 11 demonstrates the use of a data-dependent delay statement in Verilog.

```
Verilog specification            Symbolic execution

reg [2:0] data;                  reg [2:0] data
reg a;                           reg a

initial begin                    initial {
                                   control := 1
  data = $symbol(3);               data := (d2, d1, d0)
  #(data | 3'b001);                delay := data | 3'b001 = (d2, d1, 1)
                                   schedule(label, delay, control)
                                   returnToSimulator()
                                 label:
  a = 1;                           a = ite(control, 1, a)
end                              }
```

Figure 11: Symbolic data-dependent delay example.

Here, execution of the code is delayed by a number of simulation time steps which is determined from the expression $(data|3'b001)$, i.e., the 3-bit vector data is disjoined with binary value 001. In conventional simulation, this expression is evaluated at runtime resulting in a constant value which is used for the delay. When processing the delay statement, the simulator will schedule an event on the event-queue which causes the execution of the code following the delay statement after the specified time period.

In symbolic simulation, we face the problem that if the expression depends on symbolic variables, evaluation of the expression at runtime yields a function vector instead of a constant value. Thus, we have to schedule several events, one for each possible delay value that is described by the symbolic function vector. Such a delay statement actually splits the current control flow into several disjoint execution paths, each one with a different delay.

This procedure is described in more detail on the right hand side of Fig. 11. As described in [8], we introduce an internal variable *control* to maintain the symbolic condition for the execution of each statement as it is simulated. Initially, *control* is set to one. Then, the bitvector *data* is initialized with three symbolic variables (d_2, d_1, d_0) where d_2 denotes the most significant bit and d_0 denotes the least significant bit of the vector. As one symbolic variable represents both values 0 and 1, vector *data* represents (000), (001), (010), (011), (100), (101), (110) and (111). Next, the delay statement is simulated. First, we evaluate the delay expression which yields the function vector (d_2, d_1, 1). For each assignment to the symbolic variables in *data*, this vector describes a delay value which is given in the following table:

data	(data | 3'b001)	delay	data	(data | 3'b001)	delay
000	001	1	100	101	5
001	001	1	101	101	5
010	011	3	110	111	7
011	011	3	111	111	7

So, for the assignments *data* = (000) and *data* = (001), i.e., $\overline{d_2} \cdot \overline{d_1} = 1$, we have to schedule an event with a delay of 1 time unit, for *data* = (010) and *data* = (011), we have to schedule an event with a delay of 3 units and so on. Thus, the current execution path which is described by *control* is split into four disjoint execution paths, each with a different delay and *control* expression:

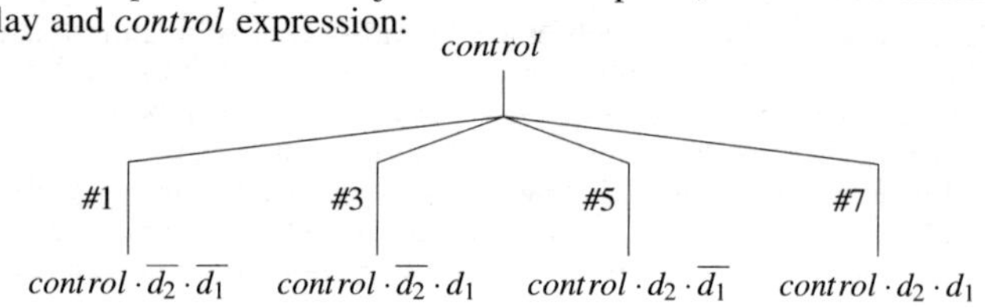

For the first execution path, *control* is conjoined with $\overline{d_2} \cdot \overline{d_1}$ because this path is only executed under the condition that the delay is one, which is only true if $d_2 = 0$ and $d_1 = 0$. Similar considerations apply to the other execution paths. The *control* conditions are saved in the events and will be restored when the event is processed by the simulator. Thus, the assignment statement at *label* will be executed at four different times with four different *control* expressions, namely after one time unit with $control = \overline{d_2} \cdot \overline{d_1}$, after three time units with $control = \overline{d_2} \cdot d_1$, after five time units with $control = d_2 \cdot \overline{d_1}$ and after seven time units with $control = d_2 \cdot d_1$. To ensure that different execution paths don't influence each other, assignments in symbolic simulation are always performed conditionally. This is reflected by the assignment a = ite(*control*, 1, a), because only if the *control* expression is true, one is assigned to *a*, otherwise *a* remains unchanged. So, after one time unit, only for the first execution path where $control = \overline{d_2} \cdot \overline{d_1}$, *a* becomes 1. After three time units, *a* becomes 1 also for the second execution path and so on.

The task is now to determine from the delay function vector which delay values are possible and which assignment to the symbolic variables leads to this particular delay value. In the following we will describe our approach.

Let (f_{n-1}, ..., f_1, f_0) be the delay function vector of size *n* (in our example, (f_2, f_1, f_0) = (d_2, d_1, 1)). First, we compress the information in the delay function vector into one function which implicitly contains all the information. Similar to the symbolic array approach, this is achieved by constructing the characteristic function of the vector. We introduce $n-1$ delay variables $e_0, e_1, \ldots, e_{n-1}$:

$$\chi = \prod_{i=0}^{n-1} (e_i \overline{\oplus} f_i)$$

This characteristic function is one for a combination of delay variables e and symbolic variables in the functions f_i, if the combination of symbolic variables results in a delay described by the e variables. We will demonstrate this in more detail on the BDD for the characteristic function χ in Fig. 12.

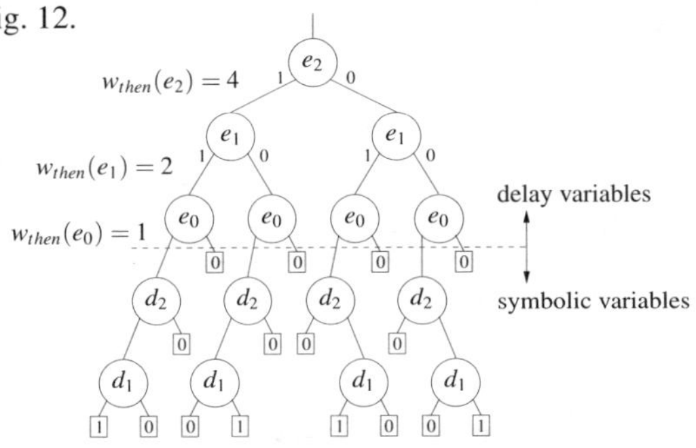

Figure 12: Characteristic delay function.

Each minterm of this characteristic function describes a valid delay value and the corresponding assignment to symbolic variables. The minterm $(e_2, e_1, e_0, d_2, d_1, d_0) = (0,0,1,0,1,0)$, e.g., denotes a delay of 1 ($(e_2, e_1, e_0) = (0,0,1)$) for the assignment $(d_2, d_1, d_0) = (0, 1, 0)$.

Thus, a naive approach would be to iterate over all minterms in χ and schedule the delays with the corresponding conditions. Unfortunately, there is a potential problem with this scheme. Consider a 32-bit symbolic register b. This register may take 2^{32} different values. If such a register is used in a delay statement #(b), there are actually 2^{32} different delays, and for each of them we must schedule an event. This would result in a memory blow-up of the event queue.

Therefore, we take a different approach. We do not schedule all events at once but one after the other. The next event is not scheduled until the previous one was processed. This does not influence the simulation effort, because we will still put the 2^{32} events on the queue. Simulating all the 2^{32} time steps may take too long, however, because we avoid a memory blow-up of the event-queue, we can at least simulate as long as we can tolerate. To reduce the simulation effort, our approach can be combined, e.g., with the method described in [10].

In our case, we only schedule an event for the smallest possible value of the characteristic delay function. The function is stored in the event. Now, whenever this event is removed from the queue, a new event is generated for the next smallest delay value. This does not affect the current simulation step, because it is guaranteed that the new event will be generated for a later time step.

Now, our goal is to find out, which assignment of e_n, ..., e_0 is the smallest one that is possible. If delay variables e are ordered on top of the other symbolic variables, we can find the solution by simply traversing the BDD depth first, always choosing the else-branch on each e node first, i.e., we check if there is a path to the one terminal for $e_2 = 0, e_1 = 0, e_0 = 0$, then for $e_2 = 0, e_1 = 0, e_0 = 1$ and so on. In the general case, where e variables and symbolic variables are intermixed, this problems transforms to a shortest path problem in the BDD where the weights of the then-edges of the e variables correspond to their position in the vector. So $w_{then}(e_0) = 2^0$, $w_{then}(e_1) = 2^1$, $w_{then}(e_2) = 2^2, \ldots, w_{then}(e_{n-1}) = 2^{n-1}$. The weights of the else edges are all zero. So the shortest path in our sense is the path with the smallest weight. Additionally, with this modeling, the weight of the shortest path is also the delay for that path. So, when computing the shortest path, we get the delay and the corresponding assignment to the e variables for that path. Computing the shortest path on a BDD is a standard approach and is described in detail in [5, 9].

The complete schedule procedure is outlined here.

Given the delay function vector $(f_{n-1}, \ldots, f_1, f_0)$ of size n, the *control* condition and *label*.

1. Compute the characteristic function $\chi = \prod_{i=0}^{n-1} (e_i \overline{\oplus} f_i)$.
2. Compute the shortest path with weights.
   → Smallest delay is the weight of the shortest path.
3. Determine cube function c from the delay variables e in the shortest path.

   The shortest path computation returns an assignment to the variables in the path. In our example, this is $e_2 = 0$, $e_1 = 0$, $e_0 = 1$, $d_2 = 0$ and $d_1 = 0$. From this assignment, we construct a cube function c comprising only delay variables: $c = \overline{e_2} \cdot \overline{e_1} \cdot e_0$. Delay variables that are not in the path are set to zero.
4. Schedule an event for *label* with the smallest delay and control expression $control \cdot \chi|_c$. Store the characteristic function $\chi \cdot \overline{c}$ for the next-smallest delay and the current time in the event.

   The *control* condition for this particular delay is given by the cofactor of χ with respect to the cube function c. We eliminate the current delay from χ by computing $\chi \cdot \overline{c}$, such that the next shortest path computation will return the next smallest delay value.

When the event is processed by the simulator, the following steps are performed:

Given an event with a *label*, a *control* expression, the characteristic function and the time of the first event.

1. If there is a non null characteristic function stored in the event
   → Compute the shortest path again and schedule a new event for the next-smallest delay. The event must be scheduled relative to the time when the schedule procedure was invoked for the first time. That's why this time was also stored in the event in the schedule procedure.
2. Restore *control* from the event.
3. Continue simulation at *label* with *control*.

5. RESULTS

The symbolic array model, the arrayed instances approach and the support for symbolic delays were implemented in our symbolic Verilog simulator. We applied symbolic simulation to the low-level verification of a DES encryption circuit synthesized for a Xilinx Virtex XCV600 FPGA. The FPGA is modelled as a generic array of configurable slices, each containing two lookup-tables, two registers and some additional logic [15]. The configurable interconnect is implemented by a memory which stores the values of all wires in the FPGA. The slices can access these values according to their configuration. So, whenever a wire is read or written, the memory is accessed. At simulation startup, the configuration for the DES circuit is read into the FPGA model. Then, the generic model which now implements the DES circuit is simulated.

All slices were stored in one symbolic instance array. Without using the symbolic arrays, simulation was not possible because the simulator ran out of memory when expanding the array of slices. Symbolic arrays were also used to model the configuration memory and the interconnect. A typical symbolic simulation run resulted in about 7000 write accesses and 15000 read accesses with both symbolic and non-symbolic addresses.

Due to BDD size explosion, we couldn't simulate the complete design with all inputs set to symbolic variables. We could, however, verify the behavior of the individual building blocks of the circuit and were also able to simulate the complete DES design with some inputs set to constant values. For a constant key, e.g., the first 4 of the 16 stages of the DES algorithm could be simulated with 64 bit of symbolic input data, the next 5 stages with 32 bit and the remaining ones with 20 bit, which is still much more than can be achieved by conventional simulation.

Then, we conducted some experiments demonstrating the characteristics of symbolic instance arrays. As a symbolic simulator may also be used to perform binary simulations, we were interested in comparing against a leading-edge commercial conventional simulator. Because our symbolic technique mainly targets highly regular structures, we wanted to see if our approach can outperform such a highly optimized commercial simulator for these applications.

To examine the characteristics of our approach, we must be able to control the data values that are stored in the instance array. An FPGA configured for a particular design is not well suited for this task. A better model is a cellular automaton. Therefore, we performed a simulation of the 2-dimensional cellular automaton *Game of Life* introduced by John Conway in 1970. This automaton consists of a large number of identical cells ordered in a regular grid. Each cell is directly connected to all of its eight neighbor cells as shown in Fig. 13.

Grid size	# cells	Conv. Build		Conv. Run		Symb. Build		Symb. Run	
		Mem(MB)	Time(s)	Mem(MB)	Time(s)	Mem(MB)	Time(s)	Mem(MB)	Time(s)
256x256	65536	154	34.5	51	21.9	8	2.5	12	8.7
512x512	262144	602	156.5	206	84.1	8	2.5	12	10.1
1024x1024	1048576	memout	-	-	-	8	2.5	13	11.3
2048x2048	4194304	memout	-	-	-	8	2.5	13	12.0
4096x4096	16777216	memout	-	-	-	8	2.5	13	14.3

Table 1: Game of Life with different grid sizes.

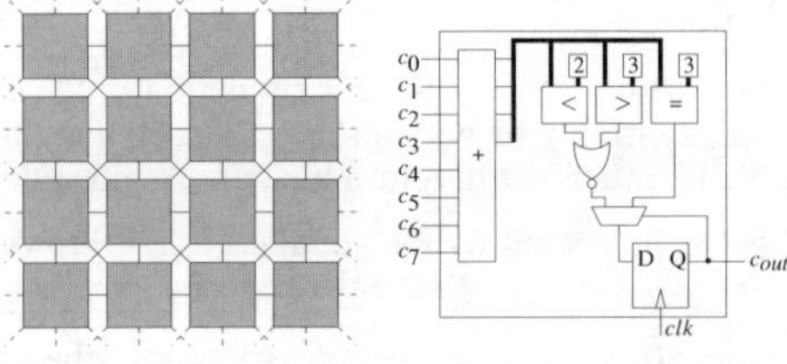

Figure 13: Game of Life grid and cell.

The state of a cell can either be dead or alive. The next state of each cell is computed from the states of its neighbor cells by a specific rule. If a dead cell has exactly three living neighbors, it becomes alive. If a living cell has less than two or more than 3 living neighbors, it dies due to loneliness or overpopulation. A block diagram of a cell is shown on the right hand side of Fig. 13. We implemented a gate-level model of such a cell and instructed our simulator to use the symbolic instance array technique for them.

We consider *Game of Life* as a good example because each cell is reasonable complex ($\sim$ 50 Gates), all the simulation effort is spent in the cells, typical applications require large grids, and the number of active cells is usually very small compared to the grid size.

In our first experiment, we simulated 100 generations of the cellular automaton with different grid sizes. The automaton was initialized with an oscillating pattern with about 600 living cells on average in each generation. Such a sparse distribution of living cells is typical for most *Game of Life* patterns. Usually, the ratio of alive to dead cells is below 1%. In Table 1, we compare build and run times of a commercial conventional simulator with our symbolic simulator. All simulations were performed on a 2GB Linux machine running at 900MHz. The conventional simulator couldn't simulate the grid sizes bigger than 512x512 because it obviously failed in expanding the instance arrays in the build process. The symbolic simulator on the other hand only instantiated one single cell which implicitly stores the data of all the cells in the grid. Thus, build time and memory were constant for all grid sizes. Furthermore, run time of the symbolic simulator is much less sensitive to the grid size. Because of the sparse distribution of living cells, it also outperformed the highly optimized conventional simulator for the small grid sizes.

Finally, we are interested in the behavior of the symbolic simulator for patterns which are non-sparse and randomly distributed, which is assumed to be the worst case for our approach. The initial state of each cell was determined randomly with a given probability. In the diagram in Fig. 14, we simulated such a random pattern for 10 generations on a grid of 512x512 with different probabilities and compared the runtimes of the conventional simulator with our symbolic simulator. Probabilities varied from 1% to 99% in steps of 10%. As expected, for sparse

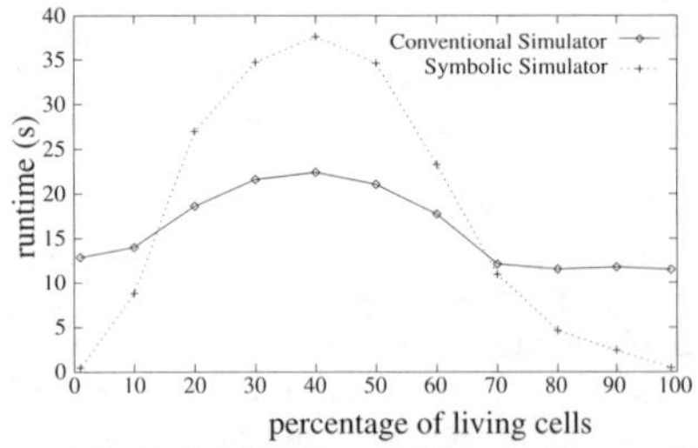

Figure 14: Initialization with random pattern.

distributions, i.e., either very few living cells or very many living cells, the symbolic approach outperformed the conventional simulator. For patterns where dead and living cells are distributed almost equally, the conventional simulator was about twice as fast as our simulator. The asymmetric curve is caused by the asymmetric rule for dead and alive cells. Actually, this speed penalty is still not bad considering that our prototype competes against a highly optimized commercial product. However, it can be seen that our simulator is much more sensitive to the distribution than the conventional simulator (which is also not constant because to some extent, it also simulates only those cells whose values have changed). Nevertheless, when it comes to memory consumption, the symbolic approach with 26MB requires clearly less memory than the conventional simulation with 206MB.

Summarizing, the symbolic approach for instance arrays is very efficient for large, regular structures like memories, cellular automata or FPGAs where the number of active cells is sparse. Even in non-sparse applications, the use of the symbolic technique can make sense if the memory requirements are tight. On the other hand, in applications with only small instance arrays or in circuits with predominantly sequential computations, non-symbolic simulation is the better choice.

6. CONCLUSION

We have proposed an array model for symbolic simulation which effectively implements sparse memories by sharing memory locations with equal values. The complete array content is stored in one function which enables the use of the array model for instance arrays. Besides the advantage that a large number of instances are stored with low memory consumption, it can also be used to speedup simulation because all instances in the array can be simulated at once. We have demonstrated the benefits of our approach for the simulation of highly regular structures. Notably, this approach enables a symbolic simulator to outperform a conventional simulator even for binary simulation. Eventually, we have proposed a way of handling symbolic delays. We have introduced an approach which avoids a potential memory blowup of the event-queue for data-dependent delays containing symbolic expressions.

7. REFERENCES

[1] D. Beatty and R. Bryant. Formally verifying a microprocessor using a simulation methodology. In *ACM/IEEE Design Automation Conference (DAC)*, pages 596–602, 1994.

[2] F. M. Brown. *Boolean Reasoning — The Logic of Boolean Equations*. Kluwer Academic Publishers, 1990.

[3] R. Bryant, D. Beatty, and C.-J. Seger. Formal hardware verification by symbolic ternary trajectory evaluation. In *ACM/IEEE Design Automation Conference (DAC)*, pages 397–402, 1991.

[4] R. E. Bryant. Graph-based algorithms for boolean function manipulation. *IEEE Transactions on Computers*, 35(8):677–691, Aug. 1986.

[5] G. De Micheli. *Synthesis and Optimization of Digital Circuits*. McGraw–Hill, Inc., New York, 1994.

[6] A. J. Hu, D. L. Dill, A. J. Drexler, and C. H. Yang. Higher-level specification and verification with BDDs. In *Computer Aided Verification (CAV 93), Lecture Notes in Comp. Sci., Springer-Verlag 1993*, 1993.

[7] Innologic Systems: www.innologic-systems.com, 2001.

[8] A. Klbl, J. Kukula, and R. Damiano. Symbolic RTL simulation. In *ACM/IEEE Design Automation Conference (DAC)*, pages 47 – 52, 2001.

[9] B. Lin and F. Somenzi. Minimization of symbolic relations. In *IEEE/ACM International Conference on Computer-Aided Design (ICCAD)*, pages 88–91, Nov. 1990.

[10] C. B. McDonald and R. E. Bryant. Symbolic timing simulation using cluster scheduling. In *ACM/IEEE Design Automation Conference (DAC)*, 2000.

[11] C.-J. Seger and R. Bryant. Formal verification by symbolic evaluation of partially-ordered trajectories. In *Formal Methods in System Design*, volume 6(2), pages 147–190, 1995.

[12] M. N. Velev and R. E. Bryant. Efficient modeling of memory arrays in symbolic ternary simulation. In *First International Conference on Tools and Algorithms for the Construction and Analysis of Systems (TACAS '98)*, 1998.

[13] M. N. Velev and R. E. Bryant. Verification of pipelined microprocessors by correspondence checking in symbolic ternary simulation. In *International Conference on Application of Concurrency to System Design (CSD'98)*, pages pp 200–212, 1998.

[14] M. N. Velev, R. E. Bryant, and A. Jain. Efficient modeling of memory arrays in symbolic simulation. In *Computer Aided Verification (CAV 97), Lecture Notes in Comp. Sci., Springer-Verlag 1997*, pages pp. 388–399, 1997.

[15] Xilinx Product Data Sheets, Virtex 2.5V FPGAs: www.xilinx.com.

A Hybrid Verification Approach : Getting Deep into the Design

Scott Hazelhurst
School of Computer Science
University of the Witwatersrand
Johannesburg, South Africa
scott@cs.wits.ac.za

Osnat Weissberg
Logic and Validation Technology
Intel Corporation
Haifa, Israel
osnat.weissberg@intel.com

Gila Kamhi
Logic and Validation Technology
Intel Corporation, Haifa, Israel
gila.kamhi@intel.com

Limor Fix
Logic and Validation Technology
Intel Corporation, Haifa, Israel
limor.fix@intel.com

ABSTRACT

One method of handling the computational complexity of the verification process is to combine the strengths of different approaches. We propose a hybrid verification technology combining symbolic trajectory evaluation with either symbolic model checking or SAT-based model checking. This reduces significantly the cost (both human and computing) of verifying circuits with complex initialisation, as well as simplifying proof development by enhancing verification productivity. The approach has been tested on current Intel designs.

Categories and Subject Descriptors

B.6.3 [**Logic Design**]: Design Aids—*verification*; F.3.1 [**Specifying and Verifying and Reasoning about Programs**]: mechanical verification

General Terms

Verification, Theory

Keywords

symbolic model checking, symbolic trajectory evaluation, hybrid verification

1. INTRODUCTION

The need for and benefits of formal verification (FV) have been accepted for some time. However, *symbolic model checking* (SMC) [11], one of the more automated and therefore more popular FV techniques, while a very valuable method for verifying commercial sequential designs, is still limited with respect to the size of the verifiable designs. This capacity problem also affects productivity: not only is effort needed to decompose proofs into simpler proof obligations on the modules of the design, the abstractions and assumptions needed increase the chance of getting false failure reports. Another reason for false failure reports is the difficulty of initialising an abstracted design. This results in spurious failures, which in turn causes the verifier to get into a loop of model checking and specification modification, which consequently hinders productivity.

1.1 MIST: The Hybrid Approach

This paper introduces a hybrid model checking approach (which we call MIST) that makes use of a complementary FV technique, *Symbolic Trajectory Evaluation* (STE) [7] to empower symbolic model checking-based verification.

In the hybrid approach, MIST, the user provides the design under test and the specification just as if using a classic model checker. Additionally, s/he explicitly specifies the initialising behaviour of the design under test or the desired initial state/states for the model checker. The verification flow consists of two phases:

1. STE performs initialising computation and calculates the state (or set of states) which the design would be in at the end of this initialisation process.
2. Using the set of states in the previous step as the starting point, a SAT/BDD-based model checker completes the verification.

1.2 Contributions of research

MIST provides more control over verification and simplifies specification. The resulting benefits are not only improvement in performance, but also productivity advantages due to finding errors more quickly and reducing the chances of spurious counter-examples. The different aspects of this are detailed below:–

Improving the specification process: SMC, though an automatic method, requires significant human intervention in modelling. We need to abstract or prune to bring the design within the capacity range of model checkers, and part of this abstraction requires building abstract models of our pruned model's environment. MIST uses STE's ability to treat large state spaces directly to provide a more accurate and cheaper process of abstraction.

Finding bugs quickly: Generating counter-examples is faster with MIST since it does not require the reconstruction of the initialisation sequence. By making use of previous counter-example traces, the verifier can find multiple failures fast and consequently back-trace to the root cause more efficiently. Experience with real-life designs shows that the use of STE to initialise the design and bring it to error-prone areas by making use of previous counter-examples can reduce the overall verification time and help in debugging failures efficiently. Precise initialisation of the design by STE also reduces the causes of spurious failures and enables the verifier to move quickly from the specification debugging to the model debugging stage.

Improving capacity and performance of automatic verification: Since the expensive image computations in initialisation can be handled by STE, especially for designs that require a long initialisation sequence, the overall cost of verification can be reduced and we may be able to verify larger systems.

Enhancing SAT-based model checking: A unique contribution is the application of MIST to SAT-based bounded model checking (BMC). SAT-based BMC methods (e.g. [2]) have recently been introduced as a complementary technique to BDD-based symbolic model checking. The cost of BMC is very dependent on the bound used. By allowing fast penetration into the design by STE, MIST reduces the bound needed for the problem.

1.3 Structure of paper

Section 2 provides background and summarises related work. In Section 3, we provide an overview of our hybrid approach to verification. Section 4 presents experimental results illustrating the benefits of the hybrid model checking over classic model checking using current Intel designs. Section 5 concludes.

2. RELATED WORK

2.1 Symbolic Model Checking

Symbolic model checking [11] takes a state-transition graph of the model under test and a property specified in an expressive temporal logic such as CTL or LTL and determines which states satisfy the property. Verification can be performed by computing all states reachable from the initial states and checking that these states satisfy the property. This is done by traversing the state transition graph, representing sets of states symbolically as BDDs, which provide a *characteristic* representation of a state space. SMC's primary limitation is that the BDDs encountered at each iteration can grow very large leading to a blow-up in memory or to a verification time-out. It may be impossible to perform image computation because of the BDDs involved in the intermediate computations. In our experiments we have used Forecast [5], Intel's model checker, a state-of-the art tool that has many optimisations and features that allow it to be used on industrial problems. Nevertheless, a severe capacity problem is witnessed when applied on verification sessions that require many image computation steps.

An important contribution of MIST is its application to bounded model checking. Thunder, our SAT-based model checker [5], turns a symbolic model checking problem to a bounded model checking problem. BMC finds counterexamples of limited length k, and thus it targets falsification and partial verification. To fully verify a property we increase the bound until there are no counterexamples less than the diameter of the FSM. The diameter is very large in some examples, and there's no easy way to compute it in advance. In practice the choice of the bound is critical: too small may fail to find a counter-example, while too large may make the verification problem intractable. Performance is very dependent on bound length. MIST helps because STE can be used for a number of cycles, before switching to BMC, thereby reducing the bound needed for BMC. Consequently, harder BMC problems can be solved.

2.2 Symbolic Trajectory Evaluation

While STE [7] also uses a temporal logic for specification and an FSM for the representation of the circuit, there are important distinctions in relation to SMC. The heart of the STE algorithm is symbolic simulation. Specification is given by assertions of the form $\langle A \Longrightarrow C \rangle$ where A and C are two temporal logic formulas. A, the antecedent, describes (symbolically) input or stimulus to the circuit while C, the consequent, gives the desired resultant behaviour.

The STE algorithm uses the antecedent to initialise the circuit. STE computes a representative, symbolic sequence of states which satisfies the antecedent. Using the values of this sequence as initialising values, STE symbolically executes the circuit and then checks to see that the consequent is also satisfied.

What makes STE efficient is its novel state representation: the use of X or don't care values induces a lattice structure on the state space. STE also uses a *parametric* representation of the state space rather than the characteristic representation (see [8] for a discussion of the difference between these methods). Together, this enables STE to deal with much larger circuits than SMC.

STE complements SMC. STE is relatively insensitive to circuit size. So, directly representing a system with tens of thousands of latches and simulating for thousands of steps can often be done with relatively low cost. However, the temporal logic which STE supports is much weaker. Recent work on generalised symbolic trajectory evaluation (GSTE) [14] has significantly extended the power of STE-based algorithms to support richer specifications (to all ω-regular properties). We have taken a different approach, where we empower SMC capacity and performance by making use of STE technology and make use of this hybrid verification flow to address shortcomings of debugging and initialisation specification in an SMC-based FV environment.

2.3 Hybrid approaches

Since verification is at least NP-hard, there is no one way that can solve all verification problems efficiently. Our approach, MIST, tackles the verification problem by combining different approaches so as to complement strengths and weaknesses. There has recently been some work in this direction.

Ho *et al.* [9] combined the use of scalar and symbolic simulation together with BMC for improving automatic test generation and unreachability analysis. Later work [13] uses SMC to verify abstractions of their circuits, construct hard counter-examples using ATPG and then use simulation-based techniques to check whether counter-examples are traces of the underlying models.

MIST shares some ideas in common with the idea of using *hints* [3]. Hints, typically constraints on the model's primary inputs, are used to improve the performance of CTL model checking by guiding state space search. MIST differs in purpose (we use the primary inputs to model environment behaviour) and technology.

Technically, the use of both characteristic and parametric representations and the conversion between them is critical for MIST (see [8] for a discussion). [10] developed a general method for creating parametric vectors for parametric representation, but focussed on methods for common predicates used in verification. More recently, Aagaard *et al.* [1] used a method based on Shannon's expansion to partition a verification problem into separate parts. Each part was characterised by a predicate (in characteristic representation) before being converted to parametric form for STE.

MIST converts the other way – from parametric to characteristic – to facilitate STE and SMC hand-shake. Conversion in this direction is easier (but can still be very expensive in general). Because of the nature of computation that STE is doing here, the conversion is cheap in MIST.

Yuan *et al.* [15] propose hybrid approaches that they call saturated simulation that performs a partial traversal of the state space while covering controller interactions. At each step they use symbolic model checking to compute the full set of states reachable in one step from the current step. Although this work has a resemblance to ours, the aim there is falsification and partial traversal of the state space where our approach encompasses the full traversal and addresses the initialisation problem.

Comparison with MIST: MIST introduces a novel integration of STE and SMC that resolves the initialisation problem in SMC based formal verification and simplifies proof development. To our knowledge, there has not been significant work that tackles the initialisation problem in formal property verification, though this topic has been focus for research for the field of formal equivalence verification [12]. Moreover, MIST enhances both the capacity and performance of SAT and BDD-based SMC and helps the debugging process by letting the verifier focus on critical, error-prone areas.

3. MIST: A HYBRID SCHEME

This section presents MIST, the hybrid approach that combines SMC and STE to exploit their respective strengths. STE is used for computations where the focus of the work is computing sets of states, whereas SMC is used for that part of the computation where we are focussed on proving properties.

The scenario envisaged for the use of the hybrid approach is as follows. The user has a model, M, and some property ϕ to be proved. STE cannot be used because its logic is not expressive enough for ϕ. SMC cannot be used because of capacity constraints; dealing with these capacity constraints has not only the obvious computational costs, but more importantly the human costs of modelling abstractions are expensive and reduce productivity.

3.1 Existing flow

To verify M using SMC alone we do the following:

- Build a pruned model M' of M (a relatively automated step)
- Build environment E' (a model of M''s environment and that part of M not modelled in M'). Commonly, initialising behaviour is modelled as part of E', and E' contains relatively detailed knowledge of the circuit behaviour.
- Use SMC to verify the property ϕ of the pruned model M' composed with the environment E'.

The disadvantages of the existing flow are the capacity and modelling constraints discussed in previous sections.

3.2 Proposed flow

The proposed flow works as follows

- Build M', a pruned model of M (the same step as above). The initialising behaviour and inputs of M are given by an STE antecedent A.
- Use STE to exercise the *unpruned* model M under the influence of A. STE's ability to deal with the large unpruned model easily provides the key benefits of enhancements of performance and simplification of modelling.
- The run computes a symbolic set of states which gives, S, the set of states of the machine after initialisation.
- Prove ϕ of M' using SMC/BMC starting from the state set S.

The MIST flow addresses the problems discussed in Section 3.1: construction of initialising behaviour is likely to be more efficient and less error-prone; and STE's computation being significantly faster, we are able to make large performance gains.

MIST Correctness: In principle, STE's computation to find the set of states after initialisation completes is the same computation that SMC would do. There are two provisos. (1) The property ϕ should say nothing about the initialisation phase (since STE cannot capture the path semantics of SMC's temporal logic in this phase of the computation). Provided we just need to know the set of states after initialisation completes (and not the paths required to get there), there is no difference between the two flows in what is being computed, just in how it is being computed. (2) Correctness relies on the correctness of the antecedent A. Of course, in the traditional flow correctness relies on the correct modelling of environmental constraints, and we argue that it is easier to do this in MIST than in the traditional flow.

3.3 Generating the initialising behaviour

MIST requires the initialising sequence of the model M. This sequence may be the circuit's reset behaviour, or any other behaviour that the user wishes (so 'initialises' may be from the circuit's or the user's perspective). An example of the former is the external stimulus to the circuit on physical resetting. An example of the latter is a sequence of actions to bring the circuit into a state or set of states from which the verifier is particularly interested in exploring.

Specifying external stimulus for initialising: This mode is very useful since we can reduce the cost of modelling the environment, which is a lengthy and error-prone process. In our approach, computation can be done by STE on the (large) original model as STE is able to cope with large state spaces directly. In the standard flow, the verifier has to build a model that describes how the pruned-off parts of the model behave. Therefore the hybrid approach reduces the work required by the user. The user can also have much higher confidence in the result as the validity of the result will not be affected by the modelling of the environment.

Providing external stimulus is particularly useful when the circuit has a relatively long reset behaviour. Here significant reduction in computation times will be seen too, since the computation of the reset behaviour by STE is extremely efficient compared to SMC. The longer the reset sequence, the greater the savings.

While we anticipate that describing the reset sequence of the circuit will be the most common way in which external stimulus will be used, MIST is not restricted to this use. Any sequence of external stimuli can be used (and we emphasise that the stimuli can be symbolic). This enables a verifier to drive deep into the state space during the more focused, earlier debugging phases of verification.

Example: For an example circuit, the reset sequence takes 50 cycles. The *reset* line is high for the first 50 cycles and low thereafter. A line called *netInit* toggles up and down a few times, and the mode lines of the circuit (collectively called *configFlag*) is given a set of values. This could be expressed more formally by a temporal logic formula such as (note the use of symbolic values for input):

```
clock_ticks &
always [0-49] reset = 1 &
always [49-99] reset = 0 &
always [0-9,20-29,40-49] netInit=0&
always [10-19,30-39,50-99] netInit=1&
always [0-99] configFlags=[a,b,c,d]
```

Providing an initialising sequence: An alternative approach, useful in the debugging phase where it is important to find counter-examples, is to use a prefix of a trace of states for initialising. Here we have as a starting point a known sequence of states that the circuit goes through: this sequence of states may be symbolic or partial (we may only have a partial state description for each instant). This can be used to drive our machine M into an interesting starting point.

This is very useful in specification debugging (an unfortunate fact of life) where we can use the same computation several times to find specification errors.

A typical use of providing initialisation sequences is finding multiple counter-examples (MCEs). When a failed verification produces a counter-example, it is often useful to have many counter-examples, but finding MCEs is expensive [4, 6]. The set of MCEs often forms a tree-like structure: i.e., they share a long common prefix (because you need to get to an 'interesting' stage before any failure happens). So we can skip the first part of the counter-example by replaying it with STE to get to the interesting part, before switching to SMC/BMC-based approaches to find MCEs. In BMC, the counter-example found depends on the bound chosen; so by choosing different bounds in the second phase, we can find MCEs. SMC always finds the shortest counter-example, so replaying the prefix will always lead to the same counter-example. However, we can introduce symbolic values in the prefix or use MCE-finding techniques (e.g. [6]) from the end point of the prefix to find MCEs. This improves the performance of these techniques which are very sensitive to counter-example length.

We can also re-use the result of one STE run in many SMC verifications — useful in the debugging phase, and when a number of properties share the same common initialisation.

3.4 Prototype tool

We have built a prototype using this hybrid approach based on Forecast and Thunder, state-of-the-art industrial-strength tools that support both STE and SMC (both BDD-based and SAT-based model checking algorithms). The user provides

- the model description in RTL format
- pruning directives which indicate which part of the RTL model should be pruned for SMC
- the initialisation information
- the properties to be proved.

When verification starts, STE runs on the original model. When it terminates, we have the initial states for SMC. This is converted from the parametric to the characteristic representation (this conversion takes into about the Xs). The SMC tool is then invoked: first the large model is pruned automatically using the pruning directives. The resultant model is then model-checked taking into account the starting state. This is shown in Figure 1. Note, that in conventional model checking, the STE component is not there. Although in MIST we must provide some additional information, the benefit is the reduced cost in modelling the environment and the performance improvements.

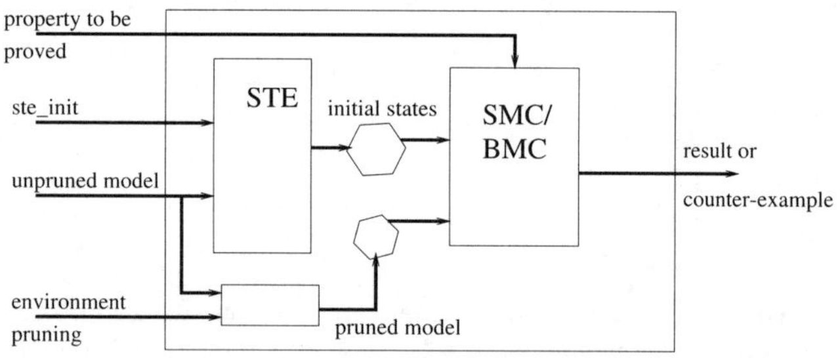

Figure 1: Overview of MIST Prototype System

4. EXPERIMENTS

We evaluated MIST prototype flow against classic SMC and SAT-based BMC flows using Forecast and Thunder, Intel's symbolic and bounded model checkers. The potential benefits of MIST (enhanced specification, capacity and performance boost over classic SMC and BMC, and enhanced debugging) drove our experiments. We assessed each benefit by making use of real verification and falsification sessions on two different current Intel designs.

Properties proved include liveness and safety properties, and focus on proving the correctness of the control-path. The typical format of a property is

```
assuming x, y, z: always (a ==> b).
```

where x, y, and z are LTL liveness or safety properties and a and b are LTL safety properties. (These are examples that occur in real designs — our method doesn't rely on formulas of this structure.)

Section 4.1 compares the MIST and SMC/BMC flows. 64 verification sessions on a current Intel design were used for evaluation. The key benefit demonstrated is improvement in the specification process. Initialisation is given using the full, unpruned model in MIST, but in the classical model checking flow, additional assumptions are needed on the pruned model. We also measure a number of performance parameters (CPU time and number of variables) to show the improvement of capacity and performance.

Section 4.2 assesses the benefit of the MIST flow in improving the debugging phase of circuit verification. We show, by running a set of experiments comparing MIST to both classic SMC and BMC flows, how MIST helps in this phase (when verification sessions mainly fail), and how MIST can be used to improve the finding of counter-examples.

4.1 Verification of a FIFO queue

The first case study was an example from the Intel Pentium III microprocessor. This RTL describes the control of a pseudo-FIFO queue (a FIFO queue with several tail pointers). Verification deals with the correctness of relationships between these pointers. Large parts of the circuit share similar initialisation sequences and thus this case study can be easily extended to new specifications. The circuit had been verified using standard techniques. It contains 15500 variables in the full model, and the verification has been broken into 64 separate model checking tasks. Different verification sessions are used to hierarchically compare different subsets of the reference model and implementation. Properties are in the style: *if the RTL and the reference model agree so far, then they will agree in the next clock cycle.*

Specifying the initialisation sequence: In the original 'standard' proof, reset behaviour is modelled by assumptions, FSMs that observe and restrict the behaviour of the RTL model. Some operations are especially expensive in this working model, particularly the counting of time. For example, the specification 'reset should be high during the first 5 cycles' is modelled by a 3 bit counter that counts the time since startup and makes sure reset is true. In STE, the assignments are coded using an antecedent, similar to test vectors in simulation. Such specification is straight forward. You simply drive reset appropriately during the first 5 cycles.

In general, initialisation sequences lend themselves well to STE-style specification, which can be done in a straight forward way. In this case the initialisation sequence can be broken into the following lemmas:

- clocks are active, and clock enables are high, and reset is active for 5 cycles and then inactive forever;
- No outside events are allowed in the second part of reset, and for some time after reset becomes inactive;
- Testability features are disabled;

- Configuration flags are stable (do not change over a sampling period)–achieved via symbolic values.

The full initialisation sequence costs 2 variables in STE versus 20 in SMC specification (needed for the building the abstraction of the environment, in particular the counters needed to describe the reset input behaviour).

A benefit of the STE style of initialisation is that it is done on the full RTL model (before pruning and abstraction) and thus is easier to specify in terms of RTL primary inputs. On the other hand, the SMC specification must take into account the abstractions/pruning done and thus is lower-level and deals with the model's internal signals. Usually initialisation sequences are well-defined and documented over the primary inputs, but not as well-defined or justifiable over the internal signals.

Debugging the initialisation sequence: Once the initialisation sequence specification is written, one needs to debug to ensure its correctness (as with any formal specification). The main problem in initialisation sequence specification is that it may be partial, i.e. some relevant signals are not being initialised. In STE, this is debugged by running the sequence and checking for Xs (undefined values). Once all relevant Xs are eliminated, and the key signals have the correct value, the initialisation is ready for use as input to the model checker.

Problems in the initialisation sequence in SMC manifest as spurious counter-examples, usually because some of the initialisation in the RTL has been abstracted out. These are hard to reconstruct and solve. Debugging the SMC initialisation took weeks. Switching to STE style took a few days. Of course, we had some benefit of experience with the model, but on the other hand we had no experience with the MIST methodology before this example.

Experimental results of performance improvement: We ran 64 real test cases using Forecast in the traditional flow and in the MIST flow. Figures 2 shows how the two methods compared by showing for each experiment, how long the verification took. Table 1 summarises this data and shows the relative performance of the MIST flow to the standard flow, by showing how many experiments showed what improvement (or degradation) of performance. 84% of the test cases led to improvement – in 61% of cases MIST was at least twice as fast.

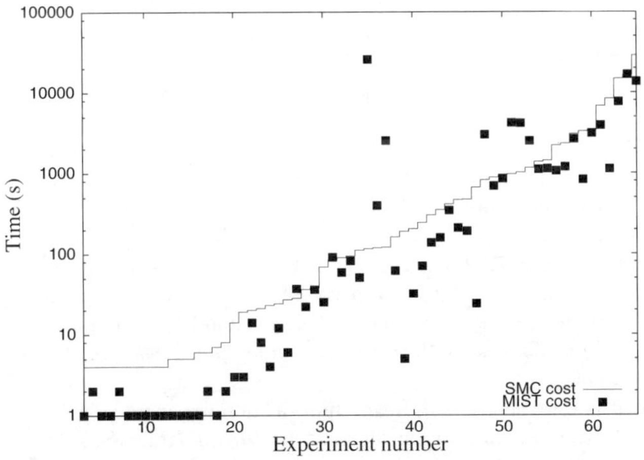

Figure 2: ***MIST versus SMC performance: Time is measured in seconds and shown on a log scale.***

Other metrics also show improvement under MIST. For example, the number of variables needed in verification is reduced on average by 22% (the average number of variables used in the standard flow is 120, with the smallest case being 69 and the largest case being 185). There are some cases where MIST leads to longer results. These are likely caused by the different initialisations which sometimes has a negative effect on performance.

r	< 0.5	$[0.5, 1)$	$[1, 2)$	$[2, 3)$	$3 \leq$
	7	3	15	13	26

Table 1: ***Relative performance of MIST to standard flow: r gives the MIST speedup; entries in the table show number of cases with speed-up in that range.***

We have not done a comparison using SAT-based technologies: in all these cases, the verifications are true and we have no reasonable way of choosing bounds for BMC.

4.2 Internal communication circuit

In this experiment we looked at the verification of a bus protocol from an Intel Pentium III mobile microprocessor. The circuit had been verified using standard techniques. However, the process was extremely expensive, especially in the debugging phase. Some verification runs took days to complete. The circuit has extensive initialisation (124 steps). Taking this circuit, we verified using both the standard and MIST flow five properties, which we call A, B, C, D and E. These are large problems with between 330 and 540 variables needed for model checking.

For all properties we used Forecast. The table below shows the results, the time taken for verification. The cost of the STE component was very small, in the order of 10s. There is a small reduction (about 4%) of the number of variables needed in the MIST flow.

Case	MIST Flow Time (s)	Forecast Flow Time (s)
A	75294	129230
B	7588	24989
C	54045	54725
D	642	66654
E	2643	68779

We also examined test case E using Thunder, comparing the standard flow to MIST. Figure 3 shows the results. We did the verification using MIST and the standard flow with different bounds in order to understand the effect of the choice of bound. A verifier using MIST would adjust the bound set downwards: if they would have used a bound of length b in the standard flow, they would use a bound of length $b - i$ in the MIST flow, where i is the length of the initialising sequence. In this experiment $i = 125$. Thus we compare the case where the standard flow has bound 225 and the MIST flow has bound 100, the case where the MIST flow has bound 120 and the standard flow has bound 245 and so on. In the figure, the x-axis is labelled by the bounds (MIST/Standard) and the y-axis by time.

These results also show that using MIST we can explore deeper into the state space. For example, in less than the time required by the standard flow to get to step 225, using MIST we can go to step 325. This shows that the verifier can guide the system into an interesting state and continue exploration from there.

Finding multiple counter-examples: We also assessed the effectiveness of the MIST flow on finding multiple counter examples. We used multiple counter example finding techniques [6] on a test case using both MIST and standard flows. Using MIST, the time required to generate multiple counter examples dropped from 16409s to 1368s.

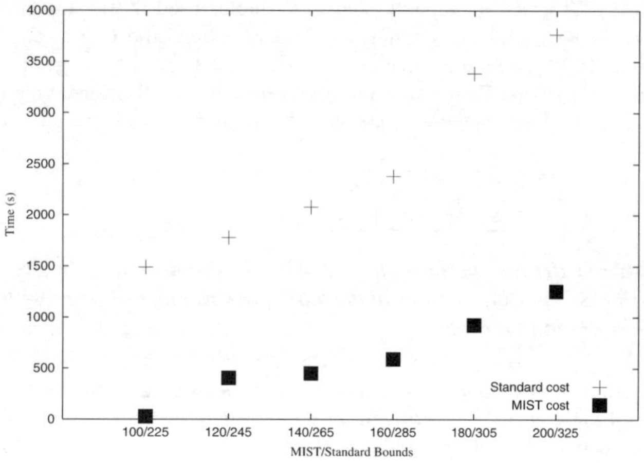

Figure 3: ***MIST versus BMC performance***

5. CONCLUSION AND FUTURE WORK

MIST is a hybrid method of verification using STE, BMC, and SMC. We are able to use the power of STE to deal with large circuits directly to improve ease of specification, performance, and productivity.

We tested this method with current Intel tools and designs. The experiments show that the modelling of initialisation can be simplified and made more precise, thereby increasing productivity and the confidence in the meaningfulness of the verification result, as well as the chances of getting spurious counter-examples. We can use the unpruned model efficiently to extract out useful information in the initialisation phase that can be used in the later phase of verification.

The application of our hybrid approach on real-life industrial test cases shows that MIST can significantly boost the performance and capacity of SAT/BDD-based symbolic model checking. Furthermore, our methodology enables the verification engineer to have much more control over the verification process, facilitating a better debugging environment.

Future work: First, we would like to look at how the cooperation between STE and SMC can be generalised further (not just during the initialisation phase). Can we use the complementary strengths of the two techniques together to extend performance and simplify the modelling phase? Second, the experimentation showed that the performance is very sensitive to the initial states. We believe there is more scope in leveraging even better performance here, and more work needs to be done to understand the best way of automating the flow. Finally, the insight that initialisation is a very important part of the verification of many systems may be helpful in other flows and verification methodologies. This is a point which we wish to take further.

Acknowledgements: We warmly thank the following for their generous help: Evguenia Belfer for her help in applying MIST on real-life Intel designs; Ranan Fraer, Moshe Y. Vardi for their review, useful comments and suggestions; Shlomit Ozer helped in the development of MIST; and Robert Beers and Tom Schubert originally suggested the use of STE for initialising symbolic model checking.

6. REFERENCES

[1] M.D. Aagaard, R.B. Jones, and C.-J.H. Seger. Formal verification using parametric representations of boolean constraints. In *Proc. 36th ACM/IEEE Design Automation Conf.*, pp. 402–407, 1999.

[2] A. Biere, A. Cimatti, E. M. Clarke, M. Fujita, and Y. Zhu. Symbolic model checking using SAT procedures instead of BDDs. In *Proc. 36th ACM/IEEE Design Automation Conf.*, pp. 317–320, 1999.

[3] R. Bloem, K. Ravi, and F. Somenzi. Symbolic Guided Search for CTL Model Checking. In *Proc. 37th. ACM/IEEE Design Automation Conference*, pp. 29–34, 2000.

[4] E. Clarke, O. Grumberg, K. McMillan, and X. Zhao. Efficient generation of counterexamples and witnesses in symbolic model checking. In *Proc. of 32nd ACM/IEEE Design Automation Conf.*, 1995.

[5] F. Copty, L. Fix, R. Fraer, E. Giunchiglia, G. Kamhi, A. Tacchella, and M.Y. Vardi. Benefits of bounded model checking at an industrial setting. In *Proc. 13th Int. Conf. on Computer-Aided Verification*, Lecture Notes in Computer Science 2102, pp. 436–453. Springer-Verlag, 2001.

[6] F. Copty, A. Irron, O. Weissberg, N. Kropp, and G. Kamhi. Efficient Debugging in a Formal Verification Environment. In *Conf. on Correct Hardware Design and Verification Methods (CHARME 2001)*, pp. 275–292, September 2001.

[7] S. Hazelhurst and C.-J.H. Seger. Symbolic Trajectory Evaluation. In Kropf, T (Ed). *Formal Hardware Verification: Methods and Systems in Comparison*. LNCS 1287. Springer-Verlag, Berlin, 1997, pp. 3–79.

[8] S. Hazelhurst. On Parametric and Characteristic Representations of State Spaces. *Technical Report 2002-1*, School of Computer Science, University of the Witwatersrand, March 2002. Available as ftp://ftp.cs.wits.ac.za/pub/research/reports/TR-Wits-CS-2002-1.ps.gz.

[9] P.-H. Ho, T. Shiple, K. Harer, J. Kukula, R. Damiano, V. Bertacco, J. Taylor, and J. Long. Smart Simulation using Collaborative Formal and Simulation Engines. In *Proc. IEEE/ACM International Conf. on Computer Aided Design*, pp. 120–126, November 2000.

[10] P. Jain and G. Gopalakrishnan. Efficient symbolic simulation-based verification using the parametric form of boolean expressions. *IEEE Trans. Computer-Aided Design of Integrated Circuits and Systems*, 13(8):1005–1015, August 1994.

[11] K.L. McMillan. *Symbolic Model Checking*. Kluwer Academics, 1993.

[12] C. Pixley, S. W. Jeong, and G. D. Hachtel. Exact Calculation of Synchronization Sequences Based on Binary Decision Diagrams. In *Proc. ACM/IEEE Design Automation Conf.*, pp. 620–623, June 1992.

[13] D. Wang, P.-H. Ho, J. Long, J. Kukula, Y. Zhu, T. Ma, and R. Damiano. Formal Property Verification by Abstraction Refinement with Formal, Simulation and Hybrid Engines. In *Proc. 38th ACM/IEEE Design Automation Conf.*, pp. 35–40, 2001.

[14] J. Yang and C.-J.H. Seger. Introduction to generalized symbolic trajectory evaluation. In *Proc. of ICCD*, September 2001.

[15] J. Yuan, J. Shen, J. Abraham, and A. Aziz. On Combining Formal and Informal Verification. In *Proc. CAV '97*, pp. 376–387, 1997.

Can BDDs compete with SAT solvers on Bounded Model Checking?

Gianpiero Cabodi Paolo Camurati Stefano Quer

Politecnico di Torino
Dip. di Automatica e Informatica
Turin, ITALY

ABSTRACT

The usefulness of Bounded Model Checking (BMC) based on propositional satisfiability (SAT) methods has recently proven its efficacy for bug hunting. BDD based tools are able to verify broader sets of properties (e.g. CTL formulas) but recent experimental comparisons between SAT and BDDs in formal verification lead to the conclusion that SAT approaches are more robust and scalable than BDD techniques.

In this work we extend BDD-based verification to larger circuit and problem sizes, so that it can indeed compete with SAT-based tools. The approach we propose solves Bounded Model Checking problems using BDDs. In order to cope with larger models it exploits approximate traversals, yet it is exact, i.e. it does not produce false negatives or positives. It reaps relevant performance enhancements from mixed forward and backward, approximate and exact traversals, guided search, conjunctive decompositions and generalized cofactor based BDD simplifications.

We experimentally compare our tool with BMC in NuSMV (using mchaff as SAT engine), and we show that BDDs are able to accomplish large verification tasks, and they can better cope with increasing sequential depths.

Categories and Subject Descriptors

B.6.3 [**Logic Design**]: Design Aids — verification

General Terms

Algorithms, verification

Keywords

Formal verification, model checking, BDDs, SAT

1. INTRODUCTION

BDD-based symbolic manipulation has been one of the most widely used core technologies in the verification domain, over the last decade. Recently, however, this methodology has been challenged by approaches using propositional satisfiability (SAT) solvers for Bounded Model Checking (BMC) problems. The basic goal of BMC is to disprove properties by searching counter-examples of bounded length.

This has shown great efficacy in bug hunting whenever productivity is a key issue (e.g. in industrial settings). Several recent papers have compared BDD and SAT based model checking and worked at mixed approaches. Even though no definite conclusion can be drawn, researchers and engineers agree that SAT tools are complementary to BDD based ones, and they are able to easily solve many model checking hard problems for BDDs. Quite often SAT-based BMC performs well compared to BDD-based model checking, even though it shows its limits when facing sequentially deep bugs.

Our main target in this work is not to confute or debate all good qualities of SAT based BMC. We rather explore possible ways to complement it with BDD based symbolic verification, and we use SAT based BMC as a term of comparison for experimental valuation.

In order to achieve the proposed goal, we follow some of the trends proposed in the past years to face the BDD explosion problem. We exploit *approximate traversals* to simplify exact verification, we follow the *guided search* paradigm when combining *forward and backward* traversals to mutually focus and narrow their search area, we finally exploit conjunctive partitioning and cofactor based simplifications to keep BDD sizes under control.

Some major contributions of our approach are:

- A new BDD based verification strategy for Bounded Model Checking. Bounding the sequential depth of the verification problem enables us to specifically target optimizations for individual traversal iterations
- A technique for exact verification (without false negatives or positives), exploiting approximate traversals as a way to drive and drastically simplify bug hunting efforts in exact symbolic traversals.
- A set of optimizations within image and pre-image procedures, sharing the common goal of keeping BDDs as much partitioned and simplified as possible, in order to avoid memory blow-ups.

Our experimental results show that we are able to check properties on circuits outside the scope of state-of-the-art BDD based verification tools. A further minor contribution of our work is to introduce an automatic property generation strategy that allows us to compare invariant verification tools on sequential models or circuits of unknown functionality and/or with no available property.

2. PRELIMINARIES

Our verification framework handles Finite State Machines described by their transition relation TR with initial state set S. We check an invariant property P by attempting to prove (or disprove) the reachability of its complement T (target state set) from S.

Standard BDD-based *exact forward verification* (Figure 1) is a breadth-first least fix-point visit of the state space that starts from S and tries to find a path to T.

On the fly tests for intersection with target are done at each iteration, thus avoiding full computation of reachable states whenever the property is violated (T is reached) before the fix-point. Conterexamples are computed working on the array of state sets R.

```
FwdVer (TR, S, T)
    k = 0
    R_k = New = S
    while (New ≠ ∅)
        if ((T · New) ≠ ∅)
            return (CounterEx (R))
        k = k + 1
        Next = Img (TR, New)
        New = Next · ¬R_{k-1}
        R_k = R_{k-1} + New
    return (PASS)
```

Figure 1: Breadth-First Exact Forward Verification.

The basic idea in SAT based BMC is to consider only paths of bounded length k and to construct a propositional formula that is satisfiable iff there is a counter-example (a path from S to T) of the same length. For that reason it targets falsification and partial verification rather than full verification. In order to fully verify a property one needs to look for longer and longer counter-examples.

The propositional formula describing a path from s_0 to s_k requires that there is a sequence of transitions from s_i to s_{i+1} for all $0 \leq i < k$:

$$path(s_0, \ldots, s_k) = \mathsf{TR}(s_0, s_1) \cdot \ldots \cdot \mathsf{TR}(s_{k-1}, s_k)$$

BMC tools implement different checks, characterized by decreasing complexity and verification capability.

The first one, referred to as bound k, looks for counter-examples of any length from 0 to k:

$$(s_0 = \mathsf{S}) \cdot path(s_0, \ldots, s_k) \cdot (\mathsf{T} \cdot s_0 + \ldots + \mathsf{T} \cdot s_k)$$

The second one, referred to as exact k, looks for paths of length exactly equal to k:

$$(s_0 = \mathsf{S}) \cdot path(s_0, \ldots, s_k) \cdot \mathsf{T} \cdot s_k$$

The third one, referred to as exact-assume k, looks for paths of length K strictly reaching T only at the last cycle:

$$(s_0 = \mathsf{S}) \cdot path(s_0, \ldots, s_k) \cdot (\overline{\mathsf{T} \cdot s_0}) \cdot \ldots \cdot (\overline{\mathsf{T} \cdot s_{k-1}}) \cdot (\mathsf{T} \cdot s_k)$$

Both BDD and SAT based verification tasks are usually preceded by model decompositions, reductions (e.g., Cone of Influence - COI), and abstractions, often able to bring down even very large problems to manageable sizes.

Let us finally note that in this work we use $\downarrow$ to denote a generalized cofactor function, defined as:

$$(f \downarrow g)(x) = \begin{cases} f(x) & if \;\; g(x) = 1 \\ - & if \;\; g(x) = 0 \end{cases}$$

In other words, g can be viewed as care set for f, that can be arbitrarily simplified in the domain subspace where $g = 0$.

```
BoundedTraversal (TR, S, T, k)
    Frontier_0 = S
    for (i = 0; i < k; i++)
        if (Frontier_i · T ≠ ∅)
            return (FAILURE)
        Frontier_{i+1} = Img (TR, Frontier_i)
    return (PASS)
```

Figure 2: Bounded Traversal.

Due to this degree of freedom, several cofactor operators have been proposed in the literature. "Constrain" cofactor was proposed as image restrictor for symbolic manipulations and canonical decompositions, whereas "restrict", and other function simplification oriented operators, were mainly conceived for logic synthesis optimizations. Since our main purpose is function simplification and BDD reduction given a care set, we use the restrict cofactor in our implementation.

3. RELATED WORKS

In [1] we combined approximate forward and exact backward reachability analysis. The proposed methodology was applied to sequential equivalence verification, and it was used for Automatic Test Pattern Generation of stuck-at faults. The original implementation used a static variable ordering framework, with image and pre-image computations based on transition functions. Our work follows [1] in its idea of combining approximate forward and exact backward traversals, using frontier sets computed in the forward direction to guide and simplify backward traversals. We extend the methodology to bounded invariant checking. Moreover, we exploit optimized image and pre-image procedures based on transition relations, conjunctive partitioning and cofactor based simplifications inspired by [2].

In [3] authors use a combination of approximate forward and backward reachability analysis. The proposed algorithm attempts to prove the mutual reachability between initial and failure states by iteratively performing over-approximate forward and backward traversals. Each new traversal increases the accuracy of the approximation, and the property is proved whenever a forward (backward) traversal reaches a fix-point outside its target. False negatives are possible because counter-examples are generated with approximate state sets. A major contribution of their work is the use of overlapping projections to obtain tighter approximations than with disjoint partitions. Our work shares with them the idea of focusing and guiding more accurate traversals with previously done approximate ones. But our method ends with an exact traversal (we have no false negatives), and we use individual frontier sets instead of fix-point reached states as tighter constraints for symbolic search.

4. BDD BASED BOUNDED MODEL CHECK

A straightforward implementation of BDD based BMC is presented in [4]. The proposed algorithm is an adaptation of standard breadth-first search. TR is the transition relation, S is the start (init) state set, T is the target set (the set of bad states).

The adoption the "bounded" search paradigm removes the fix-point check and the necessity to store the reachable state set (see Figure 2). Memory efficiency is thus increased. Moreover, the authors fully integrated the implementation with a threshold based partitioning strategy called "prioritized-traversal",

Our approach is similar to [4] in its main loop, but we address more aggressive strategies to reduce the BDD size of frontier sets. Some of the optimizations work at the level of traversal iterations, and they are shown in this section. Other ones are discussed in the next section, as they are related to inner steps of image/preimage computations.

Our starting idea is to combine exact and approximate forward and backward search. One (or more) initial approximate traversal(s) is (are) used to guide and simplify an exact verification task. In order to represent (and differentiate) the various state sets involved in different traversal strategies, we adopt the following notations:

- F_i: Generic frontier set in exact forward traversal starting from initial state set S ($F_0 = S$). Index i increases from S to T. We do not compute F in our approach, but we use it as a term of comparison while describing the combined forward-backward traversals.
- B_i: Frontier sets in exact $k-$bounded backward traversal starting from target state set T ($B_k = T$). Index i decreases from T to S. We do not explicitly compute B, but we use it the same way as F.
- $Step_i$: Set of states that can be found at the $i-$th step of paths connecting S to T. This set can be formally represented as the intersection between forward and backward frontiers ($Step_i = F_i \bigcap B_i$).
- F_i^+: Frontier sets in over-approximate forward traversal starting from initial state set S ($F_0 = S$).
- FB_i: Frontier sets in exact backward traversal over approximate forward frontiers (F_i^+). Computation of FB sets is guided and simplified using previously computed F^+ sets.

In the sequel we first introduce a BMC procedure (EXACTFWDBWDBMC, Figure 3) which corresponds to the bound k check of [4] (see section 2). We then introduce other bounded verification procedures as variations of the first one. The EXACTFWDBWDBMC procedure is able to make checks on paths of length k from S to T, so it might miss shorter mismatch sequences. As in SAT based BMC this is may be a good trade-off accuracy for performance. We describe it in detail, with proofs of correctness and a discussion on performance issues.

```
EXACTFWDBWDBMC (TR, S, T, k)
    F0+ = S
    for (i = 1; i ≤ k; i++)
        Fi+ = IMG+ (TR, F(i-1)+)
    FBk = T ↓ Fk+
    for (i = k; i > 0; i−−)
        FB(i-1) = PREIMG (TR, FBi) ↓ F(i-1)+
    if (F0+ · FB0 = 0)
        return (PASS)
    else
        return (COUNTEREX (TR, F+, FB ))
```

Figure 3: Forward Backward Exact BMC.

The procedure first computes over-approximate forward frontier sets. Each entry in the F^+ array overestimates the frontier set of an exact traversal:

$$F_i^+ \supseteq F_i$$

Entries in the FB array represent the frontier sets of an exact bounded backward traversal (B_i), each one simplified using the corresponding forward entry as care set.

$$FB_i = B_i \downarrow F_i^+$$

Indexes in backward traversal are used from k down to 0 so that corresponding frontiers in forward and backward directions have the same index. The generic $FB_i \cdot F_i^+$ set includes all states in $Step_i$:

$$FB_i \cdot F_i^+ \supseteq Step_i$$

So the combined FB and F^+ arrays implicitly include all paths connecting S to T. Furthermore, all states in $FB_i \cdot F_i^+$ are guaranteed to be backward reachable from T:

$$FB_i \cdot F_i^+ \subseteq B_i$$

The forward-backward algorithm guarantees exact bounded verification, as expressed by the following theorem:

THEOREM 1. *Let* FB *be the array of frontier sets produced by the* EXACTFWDBWDBMC *procedure, then*

$$FB_0 \cdot S = 0 \Leftrightarrow \mathsf{InvariantTrue}$$

A key issue for performance is keeping BDD sizes under control by means of generalized cofactor simplifications and threshold based control over conjunctions.

- Frontier sets in approximate forward traversals (F_i^+) are "clustered" on a threshold basis. Each set is a conjunction of terms with disjoint [1] support: $F_i^+ = \prod_{j \in Groups} f_i^j$ (see Section 5). As in transition relation clustering, we partially compute products as far as the size of their result is under a chosen threshold.
- Accuracy of the over-approximate forward traversal is not as important as in approximate model check [3], where the goodness of a verification task heavily relies on the ability of an approximate model to represent the exact behavior. The main purpose of approximate traversal in our solution is to generate good care sets for effective BDD simplifications throughout an exact traversal. Since accuracy often implies larger BDDs, even in approximate traversals, we achieve better results with intermediate solutions (trading off accuracy for BDD size).
- Frontier sets in exact backward traversal (FB_i) are always computed and manipulated in their cofactored form, i.e. the PREIMG procedure directly works with the F_i^+ set in order to simplify $FB_i = B_i \downarrow F_i^+$ while computing it (see next paragraph). Furthermore, FB_i sets are generated in conjunctively decomposed forms, using a technique derived from [2].

A second procedure we adopt is an extension of EXACTFWDBWDBMC, able to find existing minimum length counterexamples within the chosen k bound (see exact k, section 2). The procedure (FWDBWDBMC) is shown in Figure 4.

Forward frontier sets are here over-approximations of reachable state sets: $F_i^+ \supseteq R_i$. This is due to the fact that at each iteration we take the union of the approximate image with the initial state set S, so that the i-th frontier set includes all states reachable from S in at most i steps.

[1] Overlapping projections are also possible, as adopted in [1, 3], but our present implementation is limited to an approximate image procedure with non-overlapping components.

```
FwdBwdBMC (TR, S, T, k)
    F_0^+ = S
    for (i = 1; i ≤ k; i++)
        F_i^+ = Img^+ (TR, F_{i-1}^+)+S
    FB_k = T ↓ F_k^+
    for (i = k; i > 0; i−−)
        FB_{i-1} = PreImg (TR, FB_i) ↓ F_{i-1}^+
        if (S · FB_i ≠ 0)
            return (Trace (TR, F_{i..h}^+, FB_{i..k})
    return (PASS)
```

Figure 4: Forward Backward BMC.

Moreover, the check for failure is done (on the fly) at each backward iteration, and it guarantees minimal length counter-examples. It is worth noticing that FwdBwdBMC has the same cost of ExactFwdBwdBMC whenever the initial state set is included in the image of the initial state set itself, so the F^+ arrays in the two procedures are exactly the same. The only extra effort paid by FwdBwdBMC is testing intersection with the initial set at each backward iteration.

A third verification procedure (FwdBwdFwdBMC, see Figure 5) is based on a couple of forward and backward approximate traversals, followed by a final exact forward one. Here we can exploit the combined contribution of approximate forward (F^+) and backward (FB^+) frontiers to guide exact search through a narrower subspace. Moreover, the initial part of this procedure has an effect similar to the approximate forward-backward verification of [3] in the case of successful checks (property is proved correct right after the approximate traversals).

```
FwdBwdFwdBMC (TR, S, T, k)
    F_0^+ = S
    for (i = 1; i ≤ k; i++)
        F_i^+ = Img^+ (TR, F_{i-1}^+ +S)
    FB_k^+ = T ↓ F_k^+
    for (i = k; i > 0; i−−)
        FB_{i-1}^+ = PreImg^+ (TR, FB_i^+) ↓ F_{i-1}^+
    FBF_0 = F_0^+ · FB_0^+
    for (i = 1; i ≤ k; i++)
        FBF_i = Img (TR, FBF_{i-1} · F_i^+) ↓ FB_i^+
        if (S · FBF_i · FB_i^+ ≠ 0)
            return (Trace (TR, FBF_{0..i}, FB_{0..i}^+))
    return (PASS)
```

Figure 5: Forward Backward Forward BMC.

5. PERFORMANCE ISSUES IN IMAGE AND PRE-IMAGE COMPUTATIONS

Let us concentrate now on the inner steps of image and pre-image computations, involving some of the most effective optimizations for the overall performance.

Approximate image is computed as

$$F_i^+ = \text{Img}^+(\mathsf{TR}, F_{i-1}^+) = \prod_{j \in Groups} \text{Img}(\mathsf{TR}_j, F_{i-1}^+)$$

where

$$\mathsf{TR} = \prod_{j \in Groups} \mathsf{TR}_j$$

is the clustered transition relation (each group is in turn a statically generated product of clusters). Partial images (of individual group) are conjoined under threshold control, so the result F_i^+ is a conjunctively partitioned BDD.

Exact pre-image exploits two levels of partitioning and cofactor simplifications. First of all, the cofactoring term F_{i-1}^+ (see Figure 3) is given as a parameter to the PreImgWithCare function, in order to be used as a care set for inner operations:

$$\begin{aligned} FB_{i-1} &= \text{PreImg}(\mathsf{TR}, FB_i)\downarrow F_{i-1}^+ \\ &= \text{PreImgWithCare}(\mathsf{TR}, FB_i, F_{i-1}^+)\downarrow F_{i-1}^+ \end{aligned}$$

The latter function is implemented as the classical linear "and-exist" (or "relational product") with early quantification, where the generic step (computing a new intermediate product $P_j = \exists_{x_j}(P_{j-1} \cdot \mathsf{TR}_j)$) is optimized as follows

$$\begin{aligned} P_j &= \exists_{x_j}(P_{j-1} \cdot \mathsf{TR}_j)\downarrow F_{i-1}^+ \\ P_j &= \exists_x P_j \cdot (P_j \downarrow \exists_x P_j) \end{aligned}$$

Each partial product is first cofactored with the care set F_i^+, then the basic decomposition step of [2] is applied, exploiting the sets of early quantification variables as variable layers driving the decomposition (instead of the variable ordering). The latter step produces a conjunctively decomposed pre-image, that is finally clustered under threshold control.

Approximate pre-images in FwdBwdFwdBMC (see Figure 5) are computed like approximate images, but we adopt for them a fully dynamic group selection strategy, driven by BDD size (as for clustering), since static pre-computation of groups gave too rough over-approximations in the backward mode: Preimage is exact as far as the size of its intermediate result is below a threshold, otherwise it switches to approximate by dynamically restarting a new group.

6. EXPERIMENTAL RESULTS

The presented technique is implemented in a program called FBV (Forward-Backward Verifier), running on top of the Colorado University Decision Diagram (CUDD) package. In this section we present an experimental comparison between this tool and the SAT-based BMC tool NuSMV [5] (version 2.0.2), using both the internal and a state-of-the-art SAT solver, mchaff [6], as slave engines. We also experimented with the BDD-based verification procedure of NuSMV and VIS, but we do not report data on these runs since they could only verify the smaller presented models.

We show data for a few ISCAS'89 benchmarks and some other circuits and models taken from [7] and from the NuSMV distribution [5]. They have different sizes, most of them outside the range of problems manageable by state-of-the-art BDD based verifiers. The models are originally available in different formats (blif, Verilog, SMV), so to cope with them we implemented or modified a set of translators working in different directions.

Properties (invariants) are either available with the original description or automatically generated (e.g., for the ISCAS'89 circuits) as described in the following paragraph. Albeit artificially generated properties may not be as meaningful as designer given ones, that was the only way we had to cope with the lack of publicly available and suitable benchmarks. For any given property we performed a COI reduction before starting the verification phase.

6.1 Automatic invariant property generation

Automatic invariant property generation was adopted for two reasons:

1. We had no property and no specific knowledge about the functionality of the ISCAS benchmarks,
2. We aimed at measuring the ability of a verifier to explore arbitrary properties in the state space.

We thus generated target state sets T as sets with increasing Hamming distance from the initial state set S. The invariant property requires the T states to be unreachable.

As an example, let us suppose to have a model with 5 state variables $(d_0, d_1, d_2, d_3, d_4)$. and initial state $S = (0,0,0,0,0)$. We have just one state with Hamming distance 5 from S, i.e, $(1,1,1,1,1)$, we have 6 states at Hamming distance ≥ 4, i.e, $(1,1,1,1,1)$, $(0,1,1,1,1)$, $(1,0,1,1,1)$, $\ldots$, $(1,1,1,1,0)$, and so on so forth. In the first case our invariant property is $\overline{(d1 \cdot d2 \cdot d3 \cdot d4 \cdot d5)}$. Our experience is that the shorter the Hamming distance from the initial state, the easier is to falsify the property, whereas most properties generated using high Hamming distances are proved correct (i.e. state space regions distant from initial state are unreachable).

6.2 Experimental data

Our experiments ran on a 600 MHz Alpha IV5 Workstation with 512 MByte of main memory, running RedHat Linux. We compare the results attained by our BDD based tool (FBV) with NuSMV – mchaff.

Figure 6 shows a time comparison on a couple of selected test cases. We plot execution time for the same property checked with increasing values of the bound.

SAT-based BMC are extremely efficient with small values of the bound, but FBV is more performing with larger bounds. We show data for both mchaff and the default/internal SAT solver in NuSMV (NuSmv-SAT). In both the cases NuSMV is used as a front-end to generate the CNF formulation of the problem. The overall behavior seems to confirm the expected degradation in terms of performance of SAT solvers with increasing values of the bound. This is intuitively explained by the linear increment of the number of variables and clauses, directly affecting the depth-first decision procedure. In the case of BDDs no additional variable is required for new traversal iterations, and performance is strictly related to the total amount of BDD nodes. As a consequence our technique is more scalable as far as our optimizations are able to avoid BDD blow-up.

The above observations are confirmed by all data in Table 1, where we report a performance comparison (over a broader set of circuits) in terms of both memory and CPU time. Models are sorted by the original (before the COI extraction) number of state variables (column # SV). Properties proved correct within the chosen bound are denoted by *pass*, properties falsified by *fail*. We avoid reporting data for the NuSmv-SAT SAT solver, since mchaff always outperformed it. Columns # Var. and # Clauses indicate the size of the generated SAT problems, in terms of CNF variables and clauses. The CPU time is reported in different columns. The Setup column indicates the time spend to generate the CNF clauses for the SAT solver or the BDDs and the transition relation for the FBV tool. SAT reports the running time for mchaff. Trav shows the time spent by FBV on symbolic traversals.

NuSMV – mchaff were unable to deal with some of the larger instances FBV solved, because of time or memory constraints. The above failures are usually related to large bound values.

7. CONCLUSIONS

We explore possible ways to face larger problems with BDD based symbolic verification and to make BDD-based tools efficient enough to compete with SAT-based tools. We propose to mix forward and backward approximate and exact traversals, guided search, conjunctive decompositions and generalized cofactor based BDD simplifications, to obtain relevant performance enhancements.

We experimentally compare our tool with state of the art BDD and SAT based model checkers. Our experience leads to the following two conclusions. First of all, we are able to deal with larger problems than other BDD-based tools. Secondly, our methodology seems to be more scalable with increasing bounds compared to SAT-based BMC tools.

To sum up, on the one hand we deem our BDD based methodology is complementary to SAT based BMC, especially when looking for sequentially deep bugs. On the other hand, SAT tools might be able to attach bugs on problems even larger than the one analysed in this paper.

8. REFERENCES

[1] G. Cabodi, P. Camurati, and S. Quer. Efficient State Space Pruning in Symbolic Backward Traversal. In *Proc. Int'l Conf. on Computer Design*, pages 230–235, Cambridge, Massachussetts, October 1994.

[2] G. Cabodi. Meta-BDDs: A Decomposed Representation for Layered Symbolic Manipulation of Boolean Functions. In Gérard Berry, Hubert Comon, and Alan Finkel, editors, *Proc. Computer Aided Verification*, volume 2102 of *LNCS*, pages 118–130, Paris, France, July 2001. Springer-Verlag.

[3] S. G. Govindaraju and D. L. Dill. Verification by Approximate Forward and Backward Reachability. In *Proc. Int'l Conf. on Computer-Aided Design*, pages 366–370, San Jose, California, November 1998.

[4] F. Copty, L. Fix, R. Fraer, E. Giunchiglia, G. Kamhi, A. Tacchella, and M. Y. Vardi. Benefits of Bounded Model Checking at an Industrial Setting. In Gérard Berry, Hubert Comon, and Alan Finkel, editors, *Proc. Computer Aided Verification*, volume 2102 of *LNCS*, pages 435–453, Paris, France, July 2001. Springer-Verlag.

[5] A. Cimatti, E. M. Clarke, F. Giunchiglia, and M. Roveri. NuSMV: a new Symbolic Model Verifyer. In *Proc. Computer Aided Verification*, volume 1633 of *LNCS*, pages 495–499. Springer-Verlag, July 1999.

[6] M. Moskewicz, C. Madigan, Y. Zhao, L. Zhang, and S. Malik. Chaff: Engineering an Efficient SAT Solver. In *Proc. 38th Design Automat. Conf.*, Las Vegas, Nevada, June 2001.

[7] K. Ravi and F. Somenzi. Hints to Accelerate Symbolic Traversal. In *Correct Hardware Design and Verification Methods (CHARME'99)*, pages 250–264, Berlin, September 1999. Springer-Verlag. LNCS 1703.

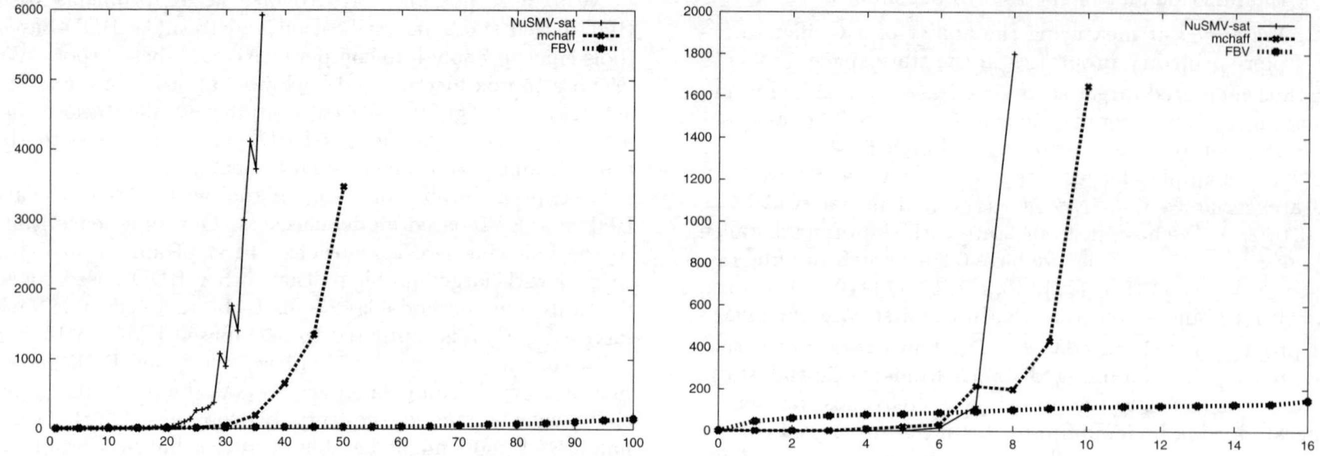

(Circuit s9234) (Circuit am2901)

Figure 6: BMC execution time (sec) versus bound value.

Model	# SV	Property	Bound	NuSMV – mchaff # Var.	# Clauses	Mem. [MByte]	Time [sec] Setup	SAT	FBV Mem. [MByte]	Time [sec] Setup	Trav
vsaR	66	P_1 (fail)	20	46294	134479	29	1	46	60	1	24
am2901	68	P_1 (pass)	10	23488	67443	42	2	1648	11	3	111
		P_1 (fail)	16	37360	107757	–	4	*ovf*	15	3	145
tcas	146	P_1 (fail)	11	258875	770491	27	27	578	35	2	100
philo$_{100}$	200	P_1 (pass)	100	432672	1227634	*ovf*	38	–	57	7	1740
s9234	211	P_1 (pass)	80	213608	597565	141	34	1023	19	10	70
		P_1 (fail)	81	216266	605026	141	35	1998	38	10	151
		P_2 (pass)	80	213608	597565	109	35	853	35	10	100
		P_2 (fail)	81	216266	605026	109	36	535	70	10	189
		P_3 (pass)	80	213608	597565	142	37	1699	37	10	143
		P_3 (fail)	81	216266	605026	141	38	1497	64	10	227
		P_4 (pass)	240	639370	1792771	*ovf*	50	–	37	10	290
		P_4 (fail)	241	641546	1798786	*ovf*	30	–	66	10	387
		P_5 (pass)	240	639370	1792771	*ovf*	48	–	39	10	401
		P_5 (fail)	241	641546	1798786	*ovf*	31	–	70	10	1917
s15850.1	534	P_1 (pass)	19	99703	268876	47	5627	5	33	14	56
		P_1 (fail)	20	104826	282943	43	5653	28	40	14	167
		P_2 (pass)	75	386667	1056856	209	5775	358	35	14	239
		P_2 (fail)	76	391791	1070926	174	5650	1534	80	14	602
s13207.1	638	P_1 (pass)	29	136108	359630	64	43	23	53	15	105
		P_1 (fail)	30	140707	371965	50	44	78	64	15	302
		P_2 (pass)	55	255682	680340	117	74	128	53	15	203
		P_2 (fail)	56	260281	692675	92	74	542	71	15	506
		P_3 (pass)	109	504028	1346430	294	56	5625	59	15	413
		P_3 (fail)	110	508627	1358765	195	55	5330	108	15	1021
		P_4 (pass)	215	991522	2653940	*ovf*	94	–	63	15	834
		P_4 (fail)	216	996121	2666275	*ovf*	99	–	137	15	2086
		P_5 (pass)	429	1975708	5293630	*ovf*	215	–	113	15	1823
		P_5 (fail)	430	1980307	5305965	*ovf*	217	–	166	15	4507
s38584.1	1426	P_1 (pass)	10	169198	464752	70	52	13	114	161	451
		P_1 (fail)	11	201874	556846	63	65	42	177	161	2842
s35932	1728	P_1 (pass)	31	540248	1494490	243	63	7	125	37	293
		P_1 (fail)	32	557448	1542529	205	65	2054	177	37	3758

Table 1: Comparison between BDD-based and SAT-based Bounded Model Checking. *ovf* means overflow on memory or time (memory limit 500 MBytes, time limit 10000 sec). – means data not available.

RTL C-Based Methodology for Designing and Verifying a Multi-Threaded Processor

Luc Séméria[1], Andrew Seawright[2], Renu Mehra[1], Daniel Ng[3], Arjuna Ekanayake[4], Barry Pangrle[1]

[1]Synopsys, Inc. [2]O-In Design Automation, Inc. [3]Broadcom, Inc.
{lucs,renu,pangrle}@synopsys.com andrew@0-in.com dng@broadcom.com [4]arjuna@stanfordalumni.org

Abstract – A RTL C-based design and verification methodology is presented which enabled the successful high speed validation of a 7 million gate simultaneous multi-threaded (SMT) network processor. The methodology is centered on statically scheduled C-based coding style, C to HDL translation, and a novel RTL-C to RTL-Verilog equivalence checking flow. It leverages improved simulation performance combined with static techniques to reduce the amount of RTL-Verilog and gate-level verification required during development.

Categories – B.5.2 [**Register-Transfer-Level Implementation**] Design Aids: *Automatic synthesis, Hardware description languages, Optimization, Simulation, Verification.*

General Terms -- Design, Verification, Performance, Languages.

Keywords – C/C++, RTL, design, verification, formal equivalence checking.

1. INTRODUCTION

With increasing design complexity, different techniques have been developed to bridge the gap between the amount of logic that can be put on a chip and the design and verification effort necessary to build such a chip. Design reuse and moving to higher levels of abstraction are two promising techniques to boost productivity. For the development of high-performance processors, the benefits of standard behavioral synthesis techniques are not obvious. Nevertheless, the need for an efficient and fast design entry language becomes even more crucial. In this paper, we present how a register transfer level (RTL) C/C++ design methodology is used to accelerate the design and verification of a complex processor.

C and C++ can be efficiently compiled enabling fast simulation and a flexible way to scale the number of concurrent simulations. In the case of a processor, a fast synthesizable model also represents a perfect instruction set simulator, which facilitates the early development of software as well as system verification early in the design process.

Using a programming language for hardware implementation has several challenges. The first challenge consists of defining a fast and accurate representation of the hardware. Speed is a must to reduce simulation time compared to RTL Verilog. RTL C also has to be accurate so that the design can be debugged early on, at the source level. In this paper, we present our RTL C coding style. To increase simulation speed, our C model is statically scheduled and can use ANSI C types. For higher accuracy, a C++ object library is used to represent bit-accurate data types. Random initialization of variables is also performed to check reset behavior.

The second challenge consists of integrating a C-based synthesis flow into the overall ASIC design flow. This can be done by automatically translating the RTL C description into RTL hardware-description language (HDL), or directly into gates. This translation step represents an extra step in the ASIC design flow, and can add additional risks. C to HDL translation tools are fairly recent and have not yet been validated on many large production designs. Moreover, evolving and competing C/C++ standards result in various interpretations making it difficult to develop and use such tools.

A C lint tool is used at the design entry to catch the most common coding style errors early on in the design process before C to HDL translation. Further, formal equivalence checking is commonly used to verify the different synthesis steps. Our methodology provides a novel formal equivalence checking flow to compare the RTL C and the HDL generated. It uses a separate and additional C to HDL translation step, followed by equivalence checking between the two translated RTL HDL models.

A final challenge is to provide a method to integrate a C-based design core with IP blocks coded using traditional RTL HDL descriptions. Co-simulation is used to integrate RTL C models with other IP Verilog models. Co-simulation is also used to validate the RTL Verilog and netlists within our C testbench environment.

The contributions of this paper are the following. We present a verification flow for C-based design. The design entry is bit-accurate RTL C model that is automatically translated into HDL for synthesis. Equivalence checking is used to formally verify the RTL C and HDL models are equivalent. Simulation of the design is also performed at each step of the design process.

The rest of this paper is organized as follows. Related work is presented in Section 2. The C/C++-based design methodology is then introduced in Section 3. Our RTL C coding style is described in Section 4. In Section 5, we present our framework for equivalence checking between HDL and RTL C. Section 6 is our result section: the simulation performance of our RTL C model is presented along with examples of bugs found in our design using equivalence checking.

2. RELATED WORK

Several C/C++ coding styles have been used in the past two decades in the industry [17,24] as well as in the research community to describe hardware both for modeling and synthesis [7,20]. In general, the C/C++ language is both extended and restricted [15]. It is extended to support hardware data-types such as bit-vector, 3-state logic, etc. and, sometimes, to support parallelism using communicating processes and reactivity. On the other hand the C/C++ lan-

DAC 2002, June 10-14, 2002, New Orleans, Louisiana, USA.

guage is also restricted to prevent the usage of non-synthesizable constructs.

C/C++ programs are sequential whereas hardware is parallel by nature. Several techniques are usually used to represent hardware. The main difference is on the implementation of parallelism and reactivity. Both parallelism and reactivity may be implicit using a run-time simulation kernel as in SystemC [12,18], Cynlib [5], Spec C [9] or Esterel C [21]. The semantics may then be synchronous/cycle-based, asynchronous/event-driven, or a mix of both. This run-time behavior of the system is well suited for higher levels of abstractions, at the behavioral or system level or to model mixed hardware/software systems at the transaction level. However, it is not optimal at the RT level. There is an overhead for scheduling the different tasks, with the risk of running each task more than once at each clock cycle, which is not very efficient. This is addressed by static scheduling techniques. Different static coding styles have been used in the industry and in commercial simulation and synthesis environments. Such environments include RTL-C from Cynergy [4] and CycleC from C Level Design (acquired by Synopsys) [2]. They are a good fit for processor design.

A synthesizable subset of the language is also defined. Some research has been done on synthesizing floating point variables [19], pointers, and dynamically allocated memory [23]. Using these techniques, recursions could also be synthesized. These features mainly make sense at the behavioral or system level. In the case of C++, the synthesis of user-defined classes (i.e. classes that are not part of the hardware C++ library) and methods remains an open issue.

The synthesis path itself can be implemented in two different ways. One method consists of translating the RTL C description into an RTL HDL description [2,3,6,7]. This RTL description is then synthesized using existing synthesis tools. Synopsys CoCentric SystemC Compiler [11] may also synthesize hardware directly without the intermediate step of generating HDL code.

3. OVERALL METHODOLOGY

In this section, we present an overview of our C-based design and verification methodology. This methodology has been used in the development of a simultaneous multi-threaded (SMT) network service processor. The processor executes MIPS compatible code and is capable of forwarding 25 million packets per second – enough performance for 10 gigabit applications such as web switches, edge routers, traffic shapers and network storage servers. The design has about 7M gates. The core of the processor which accounts for 60%-70% of the overall design, is implemented in C at the RT level. The processor uses TSMC 0.15μ process and is targeted for 300MHz.

An overview of our C-based methodology is shown in Figure 1. Our design entry is a statically scheduled RTL C model. It is integrated with our C testbench environment for simulation. Both ANSI C fundamental types and bit-accurate types are supported. ANSI C fundamental types provide faster simulation whereas bit-accurate types model the hardware types more precisely.

The C code is translated into Verilog for implementation using a commercial tool. An internal lint tool has been developed to check the coding style before C to HDL translation.

Equivalence checking is performed between RTL C and Verilog HDL. We use our own translation tool to convert the RTL C model into a verification model in HDL. Then a commercial RTL to

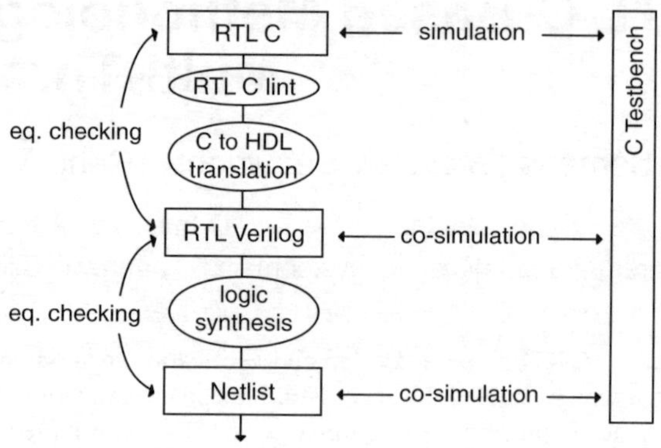

Figure 1: Verification Methodology

RTL equivalence checking tool [1,8,11,13] is used to verify the synthesized HDL against the verification HDL model. This step validates the HDL generated model against the C golden model. It validates the translation tool itself and also its interpretation of the C/C++ model: arithmetic operations, register mapping, and component instantiations.

Once the RTL verilog is generated and checked, our methodology follows the standard ASIC flow. Equivalence checking is used to formally validate the transformations performed on the design during synthesis (and scan insertion). The RTL verilog and the synthesized netlist are also verified using co-simulation with the C testbench environment.

4. RTL C MODELING

4.1 Coding style

The core of our processor is modeled in cycle-accurate C. This C model is statically scheduled. While static scheduling is more efficient, it requires more architecture work early on in the design process. In our static method, the design is partitioned in time into a set of slices. Figure 2 illustrates the use of slices in a simple example of a counter.

The slices represent the logic that is computed with specific ordering in the clock cycle. The design blocks contribute C functions called scheduling functions for execution in each of the slices. We found the static scheduling method to be a natural fit for efficiently modeling processor pipelines.

The design is described in C as a collection of logical blocks. The blocks are coded to define the imported, exported and internal signals and registers for the block, and to define the exported slice functions for the block. There is a methodology for reading signals between slices and between design blocks to ensure the design is proper and synthesizable.

In the RTL C simulation, the blocks are executed and communicate via the global scheduler which calls the slice functions of all the blocks in a simple static scheduling loop. Our methodology supports multiple instantiations of a given logical block.

In the verilog model for the complete design, the wiring of the instantiations of the logical blocks is generated automatically from a common file that describes the interconnect between the different logical blocks. A connectivity checker tool verifies that the connections described in the connectivity file correspond to the connections in the RTL C simulation model.

For synthesis, each logical block is individually translated (in isolation of the other blocks) into a Verilog module. This verilog module may contain hierarchy. Since logic synthesis doesn't perform scheduling of operations, module instantiation is directly inferred from the source code: C functions called inside slice scheduling functions become implemented as sub-modules.

4.2 Synthesizable subset

C and C++ are programming languages. As a result many of their features are of little use for hardware modeling, especially at the RT level.

In our methodology, unbounded loops and recursion are not supported. In order to optimize the synthesis results, pointers may only be used for parameters passed by reference in function calls. Operations on pointers, such as pointer arithmetic and type casting, are not permitted. Non-recursive data structures are allowed and embedded arrays and structures are supported. However, out-of-bound array accesses within structures are forbidden.

Structures are heavily used throughout our design. They are a convenient way of representing the set of input/output ports, registers and combinational signals for a given block. The example shown in Figure 2 uses several `structs`: `rCNT` represents the registers, `CNTimport` represents the imported signals and `CNTExport` the exported signals.

Since logic synthesis doesn't perform register allocation, registers are inferred directly from the source code. A naming convention is used for registers. They are synthesized by having a mandatory assignment from the `d` variable to the `q` variable conditional on the clock in the `RisingEdge` scheduling function (`CNTRisingEdge()` in Figure 2).

A naming convention is also used for ports and combinational logic signals. Ports may either be generated by analyzing the code [2] or by simply analyzing Import and Export signals.

4.3 Data types

Fundamental C/C++ data-types are limited to 8b, 16b, 32b and 64b lengths. The data-types used in hardware on the other hand may be of any length (e.g. 1b, 48b, etc.). In our methodology, two implementations of data types may be used. For the fast simulation speed, the hardware bit vector data types can be mapped to ANSI-C data types. For example 1-8 bits to `char`, 9-16 bits to `short`, etc. A slower but more accurate and elegant way of representing such data-types is to use C++ classes (e.g. the SystemC `sc_uint<>` class) that implement the masking functions.

As shown in the file `common.h` in Figure 2, the SystemC library is used if `BIT_ACC` is defined. Otherwise, ANSI-C data-types (`unsigned char`, `short`, `int` and `long long`) are used.

C++ objects can also be used to implement multi-state logic (e.g. 0, 1, X and Z states). However, as mentioned by Bening and Foster [14], the use of 'X' logic values is not advised at the RT level and should mainly be used at the gate level. Random initialization is a safer and more accurate way of representing the initial state on signals in hardware at the RT level.

Our implementation uses a modified version of SystemC that implements random initialization, in addition to zero initialization,

```
/* common.h */
#ifdef BIT_ACC     // use SystemC data types
#include <systemc.h>
typedef sc_uint<1> U1;
typedef sc_uint<9> U9;
typedef sc_uint<16> U16;
typedef sc_uint<32> U32;
typedef sc_uint<48> U48;
#else              // use ANSI C data types
unsigned char U1;
unsigned short U9;
unsigned short U16;
unsigned int U32;
unsigned long long U48
#endif

typedef struct { U9 d; U9 q; } R9;
```

```
/* counter.local.h */
typedef struct {
  U1 enableS0;
  U1 resetS0;
} CNTImport_s;

typedef struct {
  R9 counter;
} rCNT_s;
```

```
/* counter.export.h */
typedef struct {
  U8 resultS0;
  U1 overflowS0;
} CNTExport_s;

extern CNTExport_s CNTExport;
```

```
/* counter.c */
#include "common.h"
#include "counter.local.h"
#include "counter.export.h"
#include "io.export.h"

CNTImport_s CNTImport;
rCNT_s      rCNT;
CNTExport_s CNTExport;

void CNTRisingEdge() {
  if(system_clock) {
     rCNT.counter.q = rCNT.counter.d;
  }
}

/* slice 0 */
void CNTS0() {
  CNTExport.resultS0 =
              get_slice(0,7,rCNT.counter.d);
  CNTExport.overflowS0 =
              get_bit(8,rCNT.counter.d);
}

/* slice 1 */
void CNTImportS1() {
  CNTImport.enableS0 = IOExport.enableS0;
  CNTImport.resetS0 = IOExport.resetS0;
}

void CNTS1() {
 rCNT.counter.d = rCNT.counter.q; // default
  if(CNTImport.enableS0) {
     rCNT.counter.d = rCNT.counter.q + 1;
  }
  if(CNTImport.resetS0) {
     rCNT.counter.d = 0;
  }
}
```

Figure 2: Example of a counter module in RTL C code

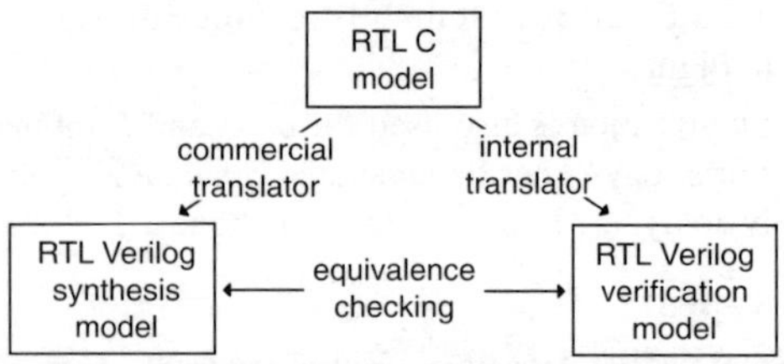

Figure 3: Formal equivalence checking between RTL C and Verilog

using a random seed. This is implemented in C/C++ using random assignments in constructor member functions. The seed can either be selected randomly for regressions or specified by the user to reproduce errors.

5. FORMAL EQUIVALENCE CHECKING

5.1 RTL C vs. Verilog equivalence checking

This section describes how we formally check that the generated verilog description matches the RTL C description. As we have seen in Section 3, current equivalence checkers can formally verify that two HDL models are equivalent. In this work, we do not try to replace existing equivalence checkers. Instead, we developed our own C to verilog translator tailored for our RTL C subset and coding style. Existing equivalence checking tools are then used to check that the verilog model used for synthesis is equivalent to our verification verilog model, as shown on Figure 3.

This translation tool differs from existing C to verilog translation tools not only because it only supports a given coding style, but also because the generated code is targeted for verification rather than synthesis. The emphasis is on matching the bit-accurate C/C++ semantics rather than trying to optimize synthesis results. The implementation of the tool is presented below. Some results and examples of bugs found in the design and in existing translation tool are then presented in Section 6.2.

5.2 C to verilog translation

Our RTL C to HDL translation was implemented using the SUIF compiler framework [10, 25]. The different steps are shown on Figure 4 and are detailed in the rest of this section.

After preprocessing, the SUIF front-end pass generates an intermediate SUIF representation of the code. The front end is slightly modified to keep track of `typedef` constructs and internal data-types for bit-accuracy. The resulting SUIF intermediate representation is then transformed in each pass in order to remove all C constructs that do not make sense in verilog. The next steps represent source-to-source C transformations. They are followed by a C to verilog translation step and by verilog code generation. Each of these steps are defined in the following sub-sections.

- **Function inlining and forward substitution**

The first transformation on the source code consists of inlining functions that are defined (functions that are not defined are assumed to be mapped to components). Pointers may be used to pass parameters by reference. Forward substitution[1] is used to propagate the pointers' values. After propagation, loads (...=`*p`) and stores (`*p`=...) are replaced by direct variable assignments. Forward substitution is used instead of constant propagation [22] because

1. Forward substitution is a pass that replaces a copy operation by a reevaluation of the expression. [22]

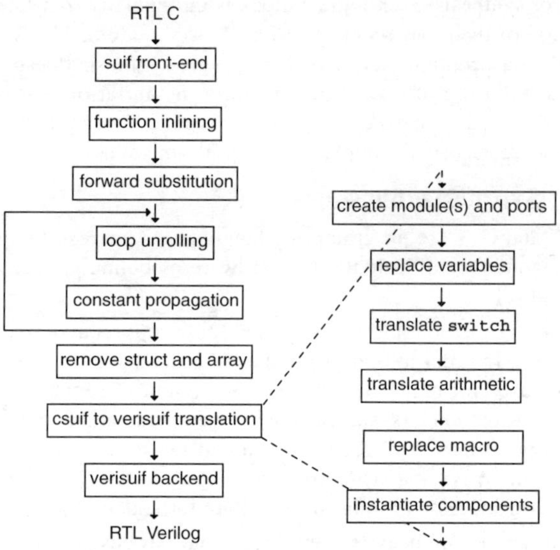

Figure 4: Implementation of xlc2v for the translation of RTL C code into verilog for equivalence checking

pointers to structure fields and array elements are defined using complex arithmetic operations (e.g. `p=&a[ i] .f`).

- **Loop unrolling and constant propagation/folding**

After function inlining, loop unrolling is performed and constant propagation [22] is used to propagate the value of the loop index (and other induction variables). In the case of nested loops where an inner loop depends on the indexes of the outer loops, multiple iterations are necessary.

- **Synthesis of structures and arrays**

At this stage all array accesses and structure accesses can be statically resolved. Structures are replaced by a set of variables that represent their different fields. The dot '`.`' in the `struct` is replaced by an underscore character '`_`'. Array variables are broken into a set of elements by appending the element index to the array variable name. For example, the field `rBLK.packet[ 2] . header.byte[ 2]` corresponds to the variable `rBLK_packet2_header_byte2`. Name clashes are reported as errors.

The case of a structure variable passed by value or by reference in a function call is trickier to implement [2]. Instead of breaking the structure into a set of separate variables, a bit vector is created. Its size equals the sum of the widths of all the fields in the structure.

All of this is possible because pointer arithmetic and out-of-bound array accesses are not part of our synthesizable subset.

- **SUIF to verisuif translation**

This step takes a C intermediate representation and generates a verilog intermediate representation. The verilog IR and back-end used were developed by French et al. [16] and extended for our needs.

First the top-level verilog modules are created. In the case of a logic block, the output ports correspond to the Export signals. The input ports correspond to the Export signals assigned to Import signals. Another technique [2] is to map signals that are assigned but never read to output ports and signals that are read but never assigned to input ports. In the case of internal modules, parameters passed by value are mapped to input ports and parameters passed by reference (i.e. pointers) are mapped to output ports.

All variables are then uniquified and replaced by verilog `regs` with the correct bit-width. The next few passes modify the code itself. First the `switch` statements in C are translated to verilog `case` statements. Then, all arithmetic operations are replaced to match the SystemC bit-accurate semantics in verilog. All intermediate computations are performed on 64bit and their result is truncated at the final assignment (only `sc_uint<>` SystemC datatype is supported).

Several macros are used to represent verilog operations such as concatenation, bit selection and range selection. They are translated into their verilog equivalent. Functions that are not macros and not inlined are then mapped to components. Components are instantiated and temporary `regs` and wires are connected to their ports.

The resulting verisuif intermediate format is then translated into verilog using the verisuif back end.

6. RESULTS

In this section, we first present our RTL C simulation performance results to illustrate the advantage of using RTL C model in term of speed. Results are then presented for formal equivalence checking between RTL C and HDL. Several bugs found using equivalence checking are described.

6.1 Simulation results

The simulation performance results are shown in Table 1. The simulation example used here represents about one third of our chip, equivalent to a 2M gate design. These results are obtained on a Linux server with dual x86 Intel Xeon 730MHz processors. The same C testbench environment is used for all models. Co-simulation is used to validate the RTL Verilog model with the Synopsys VCS simulator [11] and the C testbench running in two separate processes. The Verilog model simulated is the synthesis model. We expect the SUIF-generated verification model would be slower.

Simulation Model	cycle per second	speed-up over HDL Verilog
Pseudo Cycle Accurate C	5864	150.0
RTL-C ANSI-C	1812	46.0
RTL-C Bit Accurate	127	3.2
Verilog RTL	39	1.0

Table 1: Simulation Performance

These results show that the RTL ANSI C model is about 46 times faster than the RTL verilog model and only 3 times slower than our pseudo cycle-accurate behavior model. Using C++ bit-accurate data types leads to a 14x slow down in terms of performance compare to ANSI C. The bit-accurate simulation could however be accelerated using simple compiler optimization techniques (e.g. to perform computations on 32b instead of 64b). The bit-accurate model remains however about 3x faster than verilog simulation. The Verilog results include the overhead of PLI and co-simulation using IPC. A faster co-simulation environment could be implemented using LWT and no PLI (e.g. using Synopsys DirectC [11]).

6.2 Equivalence checking analysis

Our implementation flow uses the tool presented in Section 5 for RTL C to HDL translation for generating a verification model. The C Level Design's System Compiler translation tool [2] is used for synthesis. Equivalence checking is then performed on the two RTL HDL codes using either Verplex Conformal LEC tool [13] or Synopsys Formality [11]. Results for the translation of some of the blocks of our chip are shown on Table 2. The number of lines for both the C and Verilog models are reported after removing comments and empty lines. The number of lines in Verilog RTL is from 2x to 7x more than the equivalent RTL C. This mainly comes from loop unrolling and also from matching the bit-accurate C semantics in Verilog. The translation time is measured on a Linux server with x86 Intel Xeon 730MHz processors. Even though the SUIF compiler framework was not optimized for performance, the translation time scales fairly well on our designs.

Block Name	C Lines	Verilog Lines	Translation time
C	2,597	5,034	1 min 43s
I	969	2,243	32s
L	1,258	7,737	50s
O	440	1,376	14s
P	22,409	150,845	32 min 11s
Q	6,199	35,583	6 min 19s
R	23,386	60,552	17 min 20s

Table 2: C to Verilog translation results for several logic blocks using internal tool xlc2v.

Several problems have been caught early on in our design as well as in the translation tool for synthesis. They are listed in Table 3. The types of bugs are sorted in two categories. The first category marked as *class D* represents the design and coding-style bugs that created incorrect behavior or made our translation tool fail. The second category marked as *class T* represents bugs found in C-Level System Compiler tool [2]. Overall, we found that System Compiler performed well. Using equivalence checking we were able to find the problems early in the project and easily work around them.

A detailed discussion of two of the bugs from this list follows.

Types of Bug	Class
Arithmetic mismatches between C and Verilog semantics leading to design bug	D
Bug in implementation of concat() macros	T
Variable passed by value to a function updated after the function is called	T
Out of bounds array accesses	D+T
Passing a structure field by reference through function call	T
Variable read before initialized	D
Parameter name mismatch between .h and .c (positional vs. name based disagreement during synthesis of module instances from function calls)	D
use of (x = 1) instead of (x == 1)	D
Variable assigned in multiple non-blocking statements	D
Use of m bit output from a function for a n-bit logic (m != n) during synthesis of function call	D
Incorrect syntax/usage of #ifdefs	D

Table 3: Types of Bugs Found

Example 1. *Arithmetic mismatches between C and Verilog are a common problem. With the options we have selected for synthesis, System Compiler generates a verilog code that "looks" very similar to the C code but that may not match its semantics. The typical error is something like the following operation:*

```
o1 = ((b4 + c4) > 0xF);
```

where `o1` *is a 1b variable and* `b4` *and* `c4` *are 4b variables. The carry is lost in the verilog generated for synthesis because the operation is performed on 4b instead of 32b in ANSI C (or 64b for SystemC).*

The correct Verilog code would look like:

```
o1 = (({1'b0,b4}+{1'b0,c4}) > 5'hF);
```

This problem also occurs with right shifts (>>) and other types of comparisons (e.g., <, >, ==, !=). It is often found inside of the condition statement in `if` *or* `(?:)` *constructs, which makes it very hard to debug during simulation.*

Example 2. *Bugs were also found in the System Compiler tool for parameters passed by reference in function calls for functions mapped to components. The following slice calls the function* `foo`*, mapped to a component.*

```
void BLKS1() {
    a.d = a.q;
    foo(a.d,&out);
    if(reset) a.d = 0;
}
```

The verilog code generated for synthesis is the following:

```
foo foo1(a_d, out);
always @(a_q or reset) begin
    a_d = a_q;
    if(reset) a_d = 0;
end
```

The problem in this case is that `foo(a.d,&out)` *reads the value of* `a.d` before *reset in the C code, whereas in the Verilog HDL code, it reads the value of* `a_d` after *reset. A correct Verilog code would have been:*

```
foo foo1(tmp_reg, tmp_wire);
always @(a_q or tmp_wire or reset)
begin
    a_d = a_q;
    tmp_reg = a_d;
    out = tmp_wire;
    if(reset) a_d = 0;
end
```

These types of bugs may be very hard to isolate using simulation. A workaround is to prevent the use of write-read-write sequences. This may sometimes be too restrictive and formal equivalence checking is the best way to make sure the code is correct.

7. CONCLUSION

The RTL-C design and verification methodology presented enabled the successful implementation and rapid validation of an extremely complex SMT network processor. The methodology provides a high speed RTL simulation model from a statically scheduled coding style and provides RTL-C to RTL-Verilog equivalence checking to further leverage the simulation performance advantage by reducing the amount of RTL-Verilog and gate simulation required during development.

ACKNOWLEDGMENT

This work was done at Clearwater Networks. The definition of this design methodology involved the work of several members of the design and verification teams. We would like to thank Forrest Brewer, Steve Hatala, Jeff Handong, Dick Hessel, Jeff Huynh, Ed Jacobs, Phil Lowe, Enric Mussoll, Mario Nemirovsky, James Reilley, Nandu Sampath, Soumya Seshadri, Pierre-Xavier Thomas, Dan Williams, Tom Yeh. We would also like to thank the engineers at C-Level Design.

REFERENCES

[1] Avant! corp, Design Verifyer, http://www.avanticorp.com/

[2] C Level Design, C2HDL, http://www.cleveldesign.com/

[3] CoWare, N2C, http://www.coware.com/

[4] Cynergy System Design, http://www.cynergysd.com

[5] Forte Design Systems, Cynlib, http://www.forteds.com/products/cynlib.html

[6] Frontier Design, A|rt Builder, http://www.frontierd.com/

[7] IMEC OCAPI http://www.imec.be/ocapi/

[8] Mentor Graphics, Formal Pro, http://www.mentorg.com

[9] SpecC Technology Open Consortium http://www.specc.gr.jp

[10] Suif compiler framework http://suif.stanford.edu/

[11] Synopsys tools, http://www.synopsys.com/

[12] SystemC, http://www.systemc.org/

[13] Verplex, http://www.verplex.com

[14] Lionel Bening and Harry Foster, "*Principles of Verifiable RTL Design: a functional coding style supporting verificaion processes in Verilog*," 2nd ed., Kluwer Academic Publishers, 2001

[15] Giovanni De Micheli, "*Hardware Synthesis from C/C++*," proc. Design, Automation and Test in Europe, pp. 382-383, Munich, 1999.

[16] Robert French, Monica Lam, Jeremy Levitt, and Kunle Olukotun, "*A General Method for Compiling Event-Driven Simulations*," proc.. Design Automation Conference, June 95.

[17] Dan Joyce, Robert Stets, Andreas Nowatzyk, "*C++ Design, Verification and Automatic Conversion to Synthesizable Verilog on a Large Processor*," Proc. HDLCON, Santa Clara, Feb. 01.

[18] Stan Liao, Steve Tjang, Rajesh Gupta, "*An efficient Implementastion of Reactivity for Modeling Hardware in the Scenic Design Environment*," proc. Design Automation Conference, pp.70-75, June 97.

[19] H. Keding, M. Willems, M. Coors, H. Meyr, "*FRIDGE: A Fixed-Point Design And Simulation Environment*," proc. Design Automation and Test in Europe, pp. 429-435, 1998.

[20] David Ku and Giovanni De Micheli, "*High-Level Synthesis of ASICs under Timing and Synchronization Constraints*", Kluwer Academic Publishers, Boston, MA 1992.

[21] Luciano Lavagno and Ellen Sentovich, "*ECL: A Specification Environment for System-Level Design*," proc. Design Automation Conference, New Orleans, pp. 511-516, June 99.

[22] Steven Muchnick, "*Advanced Compiler Design and Implementation*," Morgan Kaufman Publishers, San Francisco, California, 1997.

[23] Luc Séméria, Koichi Sato, Giovanni De Micheli, "*Synthesis of Hardware Models in C with Pointers and Complex Data Structures*," IEEE trans. on VLSI, pp. 743-756, vol. 9, n. 6, Dec. 01.

[24] Kazutoshi Wakabayashi and Takumi Okamoto, "*C-Based SoC Design Flow and EDA Tools: An ASIC and System Vendor Perspective*," IEEE trans. on CAD, vol. 19, n. 12, pp. 1507-1522, Dec. 00.

[25] Robert Wilson, Robert French, Christopher Wilson, Saman Amarasinghe, Jennifer Anderson, Steve Tjiang, Shih-Wei Liao, Chau-Wen Tseng, Mary Hall, Monica Lam, and John Hennessy "*Suif: An Infrastructure for Research on Parallelizing and Optimizing Compilers*", ACM SIPLAN Notices 28(9), pp.67-70, Sept. 1994.

High-Level Specification and Automatic Generation of IP Interface Monitors *

Marcio T. Oliveira
Department of Computer Science
University of British Columbia
oliveira@cs.ubc.ca

Alan J. Hu
Department of Computer Science
University of British Columbia
ajh@cs.ubc.ca

ABSTRACT

A central problem in functional verification is to check that a circuit block is producing correct outputs while enforcing that the environment is providing legal inputs. To attack this problem, several researchers have proposed monitor-based methodologies, which offer many benefits. This paper presents a novel, high-level specification style for these monitors, along with a linear-size, linear-time translation algorithm into monitor circuits. The specification style naturally fits the complex, but well-specified interfaces used between IP blocks in systems-on-chip. To demonstrate the advantage of our specification style, we have specified monitors for various versions of the Sonics OCP protocol as well as the AMBA AHB protocol, and have developed a prototype tool that automatically translates specifications into Verilog or VHDL monitor circuits.

Categories and Subject Descriptors

B.5.2 [**Register-Transfer Level Implementation**]: Design Aids; B.6.3 [**Logic Design**]: Design Aids; C.0 [**Computer Systems Organization**]: General—*Systems specification methodology*; J.6 [**Computer-Aided Engineering**]: Computer-aided design (CAD)

General Terms

Documentation, Languages, Verification

Keywords

Formal Verification, Regular Expressions, Pipelining, Alternation

1. INTRODUCTION

Standard design practice is block-based — the design task is carved into small pieces to be tackled by an individual designer or a small team. In the past, block boundaries and interfaces have been casually negotiated face-to-face among the designers. This informal negotiation does not scale with the push for higher productivity and complexity. In addition, we would like to reuse pre-designed and pre-verified IP blocks — either designed previously in-house or purchased from third-party IP suppliers. As a result, the trend is towards designing with large, complex blocks with well-defined functionality and interfaces.

This trend generates two complementary verification problems: how to verify that a block behaves properly *in its intended environment* without having to model and verify the rest of the system, and how to verify that a system behaves properly without having to instantiate all blocks and flatten the design. Current verification practice for the first problem is to create by hand an abstracted environment model for formal verification or a testbench for simulation-based verification. Current practice for the second problem is to create by hand abstract models of the blocks and the system, or else to attempt to verify the whole system and suffer from state explosion in formal verification or slow simulation speeds and poor coverage during system-level simulation. In either case, this practice is labor-intensive, error-prone, and results in time-consuming false error reports if the models are too flexible or missed bugs if the models are too strict.

Several groups have proposed interface-monitor-based methodologies (e.g., [8, 13, 7]) to address this problem. The common theme is to create a monitor circuit that watches the interface between a block and the rest of the system and flags any violations of the interface protocol (Figure 1). The key insight, empirically confirmed in several case studies, is that designing a passive, declarative monitor is easier than designing an active stub to model the environment. Furthermore, because the monitor is symmetric between the block and the rest of the system, the same monitor can be used to verify both the block with the system abstracted as well as the system with the blocks abstracted, thereby supporting a compositional/hierarchical verification style. The monitor also provides a precise documentation of the interface, on which formal sanity checks can be applied. Finally, it is possible to convert a monitor circuit automatically into a testbench (stimulus generator) for simulation-based verification [15]. The advantages of a monitor-based methodology are compelling.

Unfortunately, although impressive monitors have been built [12], creating a monitor for a complex protocol is a challenging task. This paper introduces a high-level specification style designed explicitly to simplify specification of interface monitors. Our goal is to provide an extremely easy way to generate monitors for common interface idioms. With numerous emerging standards for system-on-chip interconnect, the need for a simple, concise, and readable way to specify interface protocols is clear. Being able to translate these high-level specifications automatically into monitor circuits allows tapping the power of monitor-based methodologies. By using our specification style, IP suppliers will be able to formally verify that their cores conform to an interface protocol as well as supply a monitor for that protocol that is both easily human-readable and directly usable by verification tools.

1.1 Related Work

Verifying a block with respect to an environment has been a concern since the beginnings of hardware verification. CTL Model Checking [6], for example, assumes a closed system and implicitly requires writing environment models. Theoretical results show that modifying CTL model checking to explicitly handle environmental inputs (rather than closing the system with an environment model) imposes a large complexity cost [9]. For space reasons, we discuss here only the two lines of research that are the immediate parents

*This work was supported in part by an NSERC research grant and the MITACS Network of Centres of Excellence.

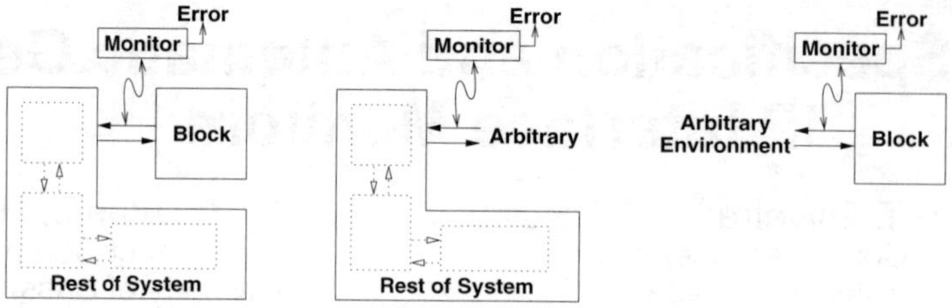

Figure 1: A monitor circuit watches the interface and flags any violations of the protocol. The block and system can be formally verified separately. The monitor can also be converted into a simulation testbench.

of the current work: interface monitors and regular expressions.

Using monitor circuits to encapsulate properties to be checked is an established idea. Indeed, a practical, industrial-strength verification methodology can be built on extensive use of monitors [4]. Our work was directly motivated by the elegance and power of monitor-based approaches to interface specification [8, 13, 7]. The emphasis of those works was mainly on the value of this way of thinking; little emphasis is on the specification language. Our focus is on the specification language; we seek to harness those results by providing a shortcut to specifying monitors.

The other parent of our work is the use of extended regular expressions to generate finite-state machines. For property specification, both Intel's ForSpec[2] and IBM's Sugar[3] include regular expressions and have ardent adherents. The most direct influence on the present work, however, is earlier work from the synthesis community: Production-Based Specification [11]. This work uses an extended regular expression language to specify state machines, which are synthesized in polynomial time into circuits, never explicitly building a deterministic finite-state machine and thereby avoiding a potential blowup. Production-based specification has proven to be particularly well-suited to synthesizing protocol state machines, and hence was a natural starting point for our research.

1.2 Contributions

The most obvious contribution of our research is the demonstration that regular expressions work very well for specifying IP interface monitors. That statement, however, is actually false. Existing specification styles based on regular expressions do rather poorly as soon as the interface protocol becomes as complex as typical system-on-chip interconnect protocols. However, by introducing two novel extensions — storage variables and a pipelining operator — we have created a specification style that does work very well for interface monitors. The new extensions require a new algorithm for translating specifications into monitor circuits. We have implemented this algorithm in a prototype tool that translates specifications into monitor circuits in Verilog or VHDL. Finally, we have demonstrated the usefulness of our specification style by developing monitors for two standards for system-on-chip interconnect: large portions of ARM's AMBA AHB high-performance bus protocol [1] and several versions of OCP (Open Core Protocol) originated by Sonics [14].

2. SPECIFICATION STYLE

We introduce the specification style incrementally, starting with regular expressions, and then introducing productions, storage variables, and pipelining. Examples taken from the AMBA AHB protocol will illustrate the concepts. We will try to provide enough information for readers unfamiliar with AHB to understand the examples. The complete AHB models are too large to include in a paper, but we will present a complete master monitor for a simple variant of OCP in the next section.

Fundamentally, a regular expression specifies a language, which is just a set of strings, which are sequences of characters, which are drawn from some alphabet. Analogously, we start at the very beginning with the alphabet of our specification style. The interface between a block and the rest of the system consists of a bunch of wires: some are inputs to the block; some are outputs. For example, an AHB slave device interfaces to the system via several wires, such as `HADDR[31:0]`, a 32-bit address input; `HRDATA[31:0]`, a 32-bit data output; `HWRITE`, `HTRANS[1:0]`, `HSIZE[2:0]`, `HBURST[2:0]`, which are control signals describing the type and size of a transfer; and `HREADY`, `HREADYOUT`, and `HRESP[1:0]`, which are hand-shaking and response signals. All of these wires are inputs to the monitor, which passively watches their values. Accordingly, the fundamental building-block of our specifications is an assignment of values to the wires on the interface at a given clock cycle.

For convenience, we allow the user to specify any Boolean formula on the interface wires. For example, the AHB protocol defines encodings for the different transfer and response types, so we allow the user to specify:

```
define idle   = !HTRANS[0] & !HTRANS[1];
define busy   =  HTRANS[0] & !HTRANS[1];
define nonseq = !HTRANS[0] &  HTRANS[1];
define seq    =  HTRANS[0] &  HTRANS[1];

define okay  = !HRESP[0] & !HRESP[1];
define error =  HRESP[0] & !HRESP[1];
define retry = !HRESP[0] &  HRESP[1];
define split =  HRESP[0] &  HRESP[1];
```

Any Boolean formula on the interface wires (and `define`d formulas) is a *primitive expression*.

Given primitive expressions, we can define regular expressions recursively in the usual manner. Any primitive expression is a regular expression. A regular expression concatenated to another regular expression is a regular expression. We use a comma as our concatenation operator. For example, the AHB specification defines different response codes for a slave to signal to the master:

```
wait_state -> !HREADY & okay;
okay_resp  ->  HREADY & okay;
error_resp -> (!HREADY & error) , (HREADY & error);
retry_resp -> (!HREADY & retry) , (HREADY & retry);
split_resp -> (!HREADY & split) , (HREADY & split);
```

The error, retry, and split responses take two cycles: the first with `HREADY` low, the second with `HREADY` high. The choice (denoted by `||`) between regular expressions is a regular expression. For example, to specify that a transfer can be one of four different kinds, we can write:

```
transfer -> idle_trans || busy_trans ||
            nonseq_trans || seq_trans;
```

Finally, we have the Kleene closure to denote repetition:

```
resp -> wait_state* , (okay_resp ||
        error_resp || split_resp || retry_resp);
```

The above expression specifies the response phase to be any number of wait states, followed by one of the response types.

For notational convenience, we use *productions* as they were defined in production-based specifications [11]. We have actually been using productions already in the preceding paragraph. The symbol to the left of the -> operator is defined to be an abbreviation for the regular expression on the right-hand side. To guarantee that specifications correspond to finite-state machines, productions cannot be recursive.

The above definitions are the same as earlier regular-expression specification styles and appear to be convenient for describing protocols. The behavior of an AHB slave device, for example, is simply a sequence of transfers or idle periods: `slave -> (slave_idle || transfer)*` where `slave_idle` means `HREADY` is low or the slave is not selected, and `transfer` is defined as above. Describing the full details of typical IP block interface protocols, however, quickly reveals the limitations of a pure regular-expression specification style.

The first major obstacle is persistent storage of information. In AHB, for example, a slave device can reply with a *split* response, indicating that it needs a long time to complete the request. The interface monitor for a split-capable slave should remember the ID numbers of all masters who have splits pending, to ensure that a slave does not signal completion of a split transaction that has not happened. Encoding such information with regular expressions is possible, but painful: for *every possible value of the saved information*, the user must write a slightly modified version of every production that is affected. Instead, we propose *storage variables* as a simple alternative. The user can declare finite-state variables as part of the specification. At any point in a regular expression, values can be assigned to the storage variables. The values of the storage variables are available in any Boolean formula defining a primitive expression. The monitor for an AHB slave could have a 16-bit storage variable, with one bit for each possible master to indicate whether it has had a request that has been split. Whenever the slave issues a split, the corresponding bit is set; whenever the slave issues a split completion, the corresponding bit is checked.

The other major obstacle is pipelining. Almost all high-performance interfaces are pipelined to some degree. Most specifications describe the cycle-by-cycle behavior of an interface, but unfortunately, pipelining is extremely hard to specify (or understand) at the cycle-by-cycle level. Trying to specify pipelining via regular expressions or any other cycle-by-cycle style requires the user to entangle all the possible parallel behaviors by hand, resulting in a difficult, error-prone specification process and an unreadable specification. Instead, pipelining is most naturally understood as an operation that overlaps sequential operations (Figure 2). In the AHB protocol, the arbitration phase, address (request) phase, and data (response) phases are all pipelined. The official AMBA specification document [1] describes these phases sequentially in English, and then presents timing diagrams to attempt to show how they entangle in pipelined operation. Our solution is to provide an explicit pipelining operator, similar to the concatenation operator. For example, for an AHB slave monitor, a transfer has an address phase followed by a response phase:

```
idle_trans -> (idle & HSEL & HREADY) , okay_resp;
busy_trans -> (busy & HSEL & HREADY) , okay_resp;
nonseq_trans -> (nonseq & HSEL & HREADY) , resp;
seq_trans -> (seq & HSEL & HREADY) , resp;
```

(`HSEL` indicates this slave is selected; `HREADY` is the handshake that indicates the address phase is complete.) However, the address and response phases are pipelined, so that the response phase of one transfer occurs at the same time as the address phase of the next transfer. In our specification style, we simply replace the concatenation operator with the pipeline operator `@`:

```
idle_trans -> (idle & HSEL & HREADY) @ okay_resp;
busy_trans -> (busy & HSEL & HREADY) @ okay_resp;
nonseq_trans -> (nonseq & HSEL & HREADY) @ resp;
seq_trans -> (seq & HSEL & HREADY) @ resp;
```

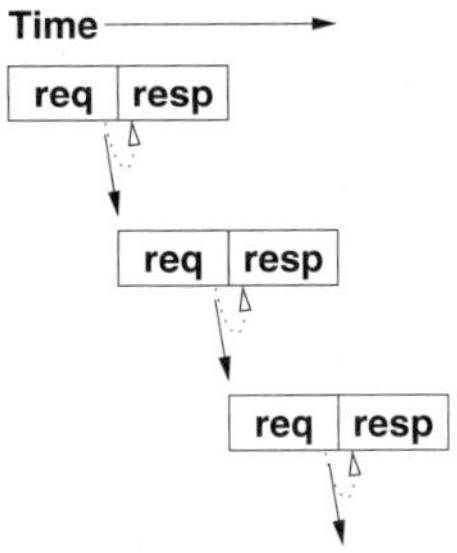

Figure 2: This figure shows multiple pipelined transactions, where each transaction has a request phase and a response phase: `(req@resp)*`. Our pipeline operator marks the point where the next computation overlaps the current one. At that point, we fork a new "thread" to complete the current transaction (dotted arrow), while the current thread continues with the rest of the regular expression, if any (solid arrow).

The semantics of the pipeline operator are that the thread of control forks into two sub-threads when the pipeline operator is encountered: one sub-thread continues with the regular expression as if the right-hand operand of the pipeline operator did not exist, the other sub-thread focuses only on the right-hand operand, ignoring the rest of the regular expression. The thread accepts a string only if both sub-threads accept (Figure 2). Multistage pipelines are easily specified as `(a @ (b @ (c @ ...)))`.

The ability to generate multiple threads where all threads must accept (pipelining) or at least one thread must accept (choice) resembles alternating automata, which in general require a double-exponential state-space blow-up to convert into ordinary automata [5]. To guarantee efficient translation into monitor circuits, we impose a few restrictions. First, we require the expression contained within a Kleene star not to accept the empty string. Known constructions can normalize regular expressions to obey this restriction [10], but our implementation does not currently include this step. Second, we forbid non-deterministic choice: we allow the choice operator, but the choices must be distinguishable within the first clock cycle. In practice, this restriction is not a problem because protocols are typically designed to make it easy to determine immediately what action is occurring. Finally, we allow at most one thread at a time to execute in a pipeline stage. For example, the expression `(a@(b,c))*` generates an error when the second repetition arrives at the `b` while the first repetition's pipeline sub-thread is still at the `c`. This restriction corresponds to allowing only one transaction at a time to use the hardware resources devoted to a pipeline stage.

We close this section with an example of the clarity of our specification. The official AMBA AHB specifications require a two-cycle response for a retry (shown above in the production for `retry_resp`), but also require an immediate OK response to a busy transfer. In August 2001, the question arose in the `comp.lang.verilog` and `comp.sys.arm` newsgroups of what happens if a slave starts to issue a retry (to the previous transfer) in the same cycle that the master issues a busy transfer, e.g.:

Clock Cycle:	1	2	3
Request:	Sequential	Busy	
Response:		Retry1	Retry2 or OK??

The sequential transfer at cycle 1 needs a two-cycle retry response

```
/* Inputs and Outputs of Master */
/* The monitor treats these as inputs. */
input SCmdAccept, SResp[1:0], SData[31:0];
output MAddr[31:0], MCmd[2:0], MData[31:0];

/* Response codes defined in standard. */
/* NULL, Data VAlid, and ERRor */
define null_resp = !SResp[0] & !SResp[1];
define dva_resp  =  SResp[0] & !SResp[1];
define err_resp  =  SResp[0] &  SResp[1];

/* Commands defined in standard. */
define cmd_idle  = !MCmd[0] & !MCmd[1] & !MCmd[2];
define cmd_write =  MCmd[0] & !MCmd[1] & !MCmd[2];
define cmd_read  = !MCmd[0] &  MCmd[1] & !MCmd[2];

master -> (cmd_idle || transfer)*;

transfer -> write_transfer || read_transfer;

/* SCmdAccept indicates that the slave
   has accepted the command. */

write_transfer ->
    (cmd_write & !SCmdAccept)*, (cmd_write & SCmdAccept);

read_transfer ->
    (cmd_read & !SCmdAccept)* ,
    (wait_state_resp || instant_resp);

wait_state_resp ->
    (cmd_read & SCmdAccept & null_resp) ,
    null_resp* , response;

instant_resp -> (cmd_read & SCmdAccept & dva_resp) ||
                (cmd_read & SCmdAccept & err_resp);

response -> dva_resp || err_resp;
```

Figure 3: A complete interface monitor for a Basic OCP master that can only handle one outstanding transaction at a time.

during cycles 2 and 3, but a busy transfer at cycle 2 demands an immediate OK response at cycle 3. Because of the pipelining, the questioner was unsure what would happen. Of the several answers offered in the newsgroup, most were wrong. In our specification, the seq_trans and nonseq_trans productions have a pipelined resp phase, which means that one sub-thread enforces that the response phase behaves correctly while another sub-thread continues to the next transaction. For a retry response, the retry_resp production shows that HREADY must be low in cycle 2, whereas the busy_trans production shows that HREADY must be high in cycle 2. Since the slave controls HREADY, this means that the busy transfer cannot occur at cycle 2, regardless of what the master requests. The only legal behavior is for the request at cycle 2 to be ignored (slave_idle in the productions above), and for the second cycle of the retry response to occur at cycle 3.

3. EXAMPLE: OCP

In the preceding section, we actually saw most of the specification for an AMBA AHB slave monitor. The AHB master monitor is too complex to present here, but for the sake of completeness, we would like to present a complete monitor for the interface between a master device and the rest of a system. Figure 3 shows the entire monitor for a master in a simple version of the Open Core Protocol (OCP) [14]. OCP is actually a very broad, parameterized family of protocols, spanning an enormous range of performance and cost objectives. The interface monitor shown is for a Basic OCP master with only one transaction in-flight at a time.

In Basic OCP, the master presents a command at the same time as the address, as well as the data if the command is a write. These values must be held constant until the slave accepts the command by asserting SCmdAccept, which could happen in the same cycle that the command is presented. Writes are posted (no response required once the slave accepts the command), but read commands have a response phase during which the slave sends data back to the master. The slave can insert zero or more wait states by keeping SResp set to the null response. The slave terminates the response phase by setting SResp to indicate that the data is valid or an error occurred. SResp can go non-null in the same cycle as SCmdAccept is asserted, which can be in the same cycle as the master presents a command.

Our specification in Figure 3 starts by declaring the interface wires, and then defining the command and response encodings. The first production defines the behavior of the monitor — in this case, we declare that the master's interface should exhibit a sequence of idle commands or transfers. A transfer can be a write or a read. A write transfer consists of zero or more states waiting for the slave to accept the command, followed by the slave's accepting the write. The read transfer is a bit more complicated, starting with zero or more states waiting for the slave to accept the command, followed by either an instantaneous response or a response with zero or more wait states.

For simplicity, the specification in Figure 3 does not check that the master holds the address and data values constant if the slave does not accept the command immediately. To enforce this requirement, we need only make a few changes to the specification. First, we would declare some storage variables to remember the values of the address and data:

```
internal hold_addr[31:0] = 0;
internal hold_data[31:0] = 0;
```

Next, we modify the transfers so that if the slave does not accept the command immediately, we remember the address (and data if applicable):

```
write_transfer ->
    (cmd_write & SCmdAccept) /* Same cycle accept */
    ||
    (
     (cmd_write & !SCmdAccept)
        /* Store original address and data. */
        { hold_addr <- MAddr; hold_data <- MData; } ,
     (cmd_write & !SCmdAccept &
        (hold_addr == MAddr) & (hold_data == MData)
     ) * ,
     (cmd_write & SCmdAccept &
        (hold_addr == MAddr) & (hold_data == MData))
    );
```

The read transfer production is modified similarly. To enforce the same constraints using regular expressions without storage variables would require separate read and write productions for each possible value of the address and data.

4. TRANSLATION INTO CIRCUITS

The translation process starts by macro-expanding all productions, since the productions cannot be recursive, resulting in a single (extended) regular expression for the monitor. In theory, this expansion can produce an exponential size blow-up, but in practice, this is often not a problem. The translation from an extended regular expression to circuits can best be understood as recursively building a circuit for each sub-expression, so the structure of the circuit exactly matches the structure of the regular expression. The circuit passes activation signals from sub-circuit to sub-circuit, corresponding to possible parses of the input string by the regular expression. We will elaborate on this construction below.

Our translation is similar to previous work in efficiently converting regular expressions into circuits [11, 10]. The key differences of our algorithm are building a monitor circuit, rather than a recognizer circuit, handling storage variables, and handling pipelining.

The first difference is that we are interested in monitoring the on-going behavior of an interface, rather than recognizing a regular language, which was the focus of previous work. A recognizer asserts its "OK" output only when the input sequence is a string in the language of the regular expression. A monitor, on the other hand, asserts its OK output as long as the sequence seen so far has not done anything not permitted by the regular expression. Accordingly, our logic that tracks the correspondence between the interface and the regular expression (the activation signals) is essentially the same as previous work, but the logic to generate the OK signal is completely different (described below).

Storage variables are simply registers in the circuit, whose output are available to the logic generated for primitive expressions. Assignments to storage variables are enabled by the same activation signals that track the parsing of the input string.

Pipelining is the most difficult difference. Intuitively, we will create a single thread for each pipeline stage, and the circuit for each thread behaves roughly like previous translations of regular expressions into circuits. The monitor is satisfied only if all active threads are satisfied. Additional bookkeeping is required to track the exact status of each thread.

More precisely, take the (macro-expanded) parse tree for the monitor's regular expression and delete the right-hand-operand edges of all pipeline operators, resulting in several, disjoint parse (sub-)trees. Our restrictions on the specifications (deterministic choice and single thread per pipeline stage) guarantee that each sub-tree will support exactly one thread. Each thread i will generate a thread enable output $tenable_i$ and a thread OK output tok_i. The monitor is satisfied as long as for all threads, $tenable_i \Rightarrow tok_i$. (We use $\Rightarrow$ to denote logical implication.)

Each regular (sub-)expression is converted into a circuit that can read all storage variables and interface wires. The circuit also has an activate-in input a_i, an activate-out output a_o, a circuit-enabled output e, an OK output ok, and an "OK-plus" output ok_p:

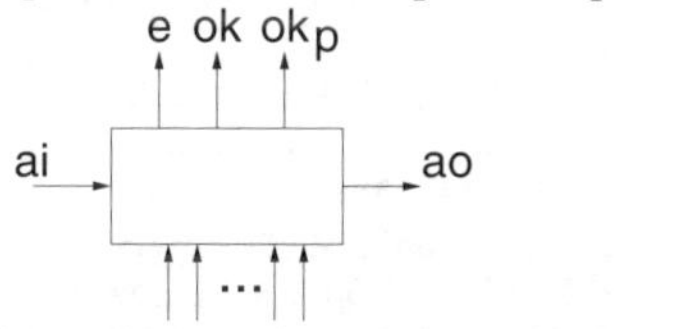

Intuitively, the activate signals indicate where a thread is in the regular expression, the enabled signal e indicates if this sub-circuit is enabled (is trying to match the interface signals), and the OK signal indicates that the sub-circuit is enabled and agrees with the current values on the interface wires. The OK-plus signal is a technical detail needed to handle the possibility of recognizing the empty string with a Kleene star; intuitively, it indicates that the sub-circuit is OK at this point even if all stars (zero or more repetitions) became plusses (one or more repetitions). Inductively, given an extended regular expression:

Base Case: If the expression is a primitive expression, build the combinational logic to evaluate the Boolean formula for the primitive expression. Let f denote the output of this formula. The enable output e is equal to the activate input a_i. Both ok and ok_p are set to $a_i \wedge f$. The activate-out signal is $a_i \wedge f$ delayed by one clock signal (one flip-flop in the circuit).

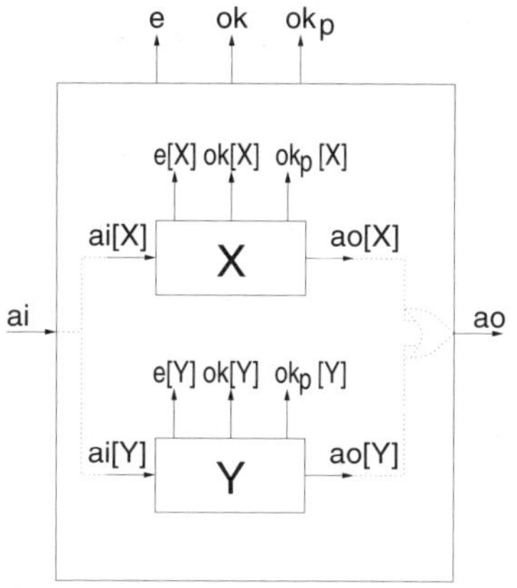

Figure 4: Circuits are built recursively from the circuits for their sub-expressions. The dotted lines show the construction for the activation signals for the choice operator $X||Y$.

Choice: If X and Y are regular expressions with corresponding circuit translations, then build the circuit for $X \mid\mid Y$ from the circuits for X and Y as follows: (Denote the signals for X's circuit with $[X]$, similar for Y. See Figure 4.)

$$a_i[X] = a_i \qquad a_i[Y] = a_i \qquad a_o = a_o[X] \vee a_o[Y]$$

connects that activation signals, and:

$$e = e[X] \vee e[Y] \qquad ok = ok[X] \vee ok[Y] \qquad ok_p = ok_p[X] \vee ok_p[Y]$$

generates the enable and OK signals.

Sequence: Similarly, build the circuit for X , Y as follows:

$$a_i[X] = a_i \qquad a_i[Y] = a_o[X] \qquad a_o = a_o[Y]$$

connect the activation signals so that X goes first, and then activates Y in sequence. The constructions $e = e[X] \vee e[Y]$ and $ok_p = ok_p[X] \vee ok_p[Y]$ are intuitive — the circuit is enabled or "OK-plus" if either sub-circuit is enabled or OK-plus. The OK signal needs extra clauses

$$ok = (e[X] \Rightarrow ok[X]) \wedge (e[Y] \Rightarrow ok[Y]) \wedge (ok[X] \vee ok[Y])$$

because X or Y might consist of a Kleene star, and the construction for the Kleene star is always OK as soon as the circuit is activated, regardless of the values on the interface wires (because the star allows matching zero copies of the repeating expression). The extra clauses prevent these empty-match OK signals from propagating erroneously.

Pipeline: For X @ Y, all of X's signals are connected to the corresponding signals for the circuit for X @ Y, since the current thread ignores Y. At the same time, a new thread for Y gets activated: $a_i[Y] = a_o[X]$.

Kleene Star: For X*, connect the activation signals as $a_i[X] = a_i \vee a_o[X]$ and $a_o = !ok_p[X] \wedge (a_i \vee a_o[X])$. The enable output is $e = e[X] \vee a_o[X]$, the OK output is $ok = ok[X] \vee a_i \vee a_o[X]$, and the OK-plus output is $ok_p = ok_p[X]$. Because of the repetition, the circuit self-activates, so the $a_o[X]$ signal appears in several formulas. The Kleene star accepts the empty string, so the a_i signal appears combinationally in the equations for ok and a_o (as well as indirectly in e for the first cycle of X's activation). Here, we see the use of the ok_p signal: the activate output is disabled if X is truly matching the interface (rather than vacuously matching because of a Kleene star). The deterministic choice restriction prevents the case where a_o should be true at the same time as $ok_p[X]$.

Earlier, we imposed three restrictions to allow efficiently building a monitor: (1) normalization to eliminate empty strings within Kleene stars, (2) deterministic choice, and (3) one thread per pipeline stage. The first two can be handled statically using standard techniques. To enforce the third restriction, we augment our monitor to generate a pipeline-violation error. Intuitively, for each pipeline operator X @ Y, trying to activate Y when it is already running generates a pipeline-violation error. A complication, however, is that the signals from the already running thread and the new activation can interfere. The easiest way around this complication is to generate three versions of every signal in the construction above: the regular version as described already; a primed version, which ignores any new activation; and a double-primed version, which tracks only the first cycle of a new activation. The formulas for the primed signals are identical to the ones above, with primed signal names replacing unprimed signal names, except for the initial thread activate signal $a'_i[Y] =$ false instead of $a_i[Y] = a_o[X]$, and at the base case latches a'_o is driven by the same flip-flop (unprimed) that drives a_o. The formulas for the double-primed signals are also identical to the regular signals, except with double-primed signal names and the base case a''_o is always false rather than driven by the flip-flop. (Considerable redundancy could be eliminated, but this construction is easy to explain and implement.) A pipeline-violation error occurs whenever $a_i[Y] \wedge ok'_p[Y]$. The thread enable and ok signals are defined as $tenable = e[Y]$ and $tok = (e'[Y] \Rightarrow ok'[Y]) \wedge (e''[Y] \Rightarrow ok''[Y])$.

The size of the generated circuit and the runtime of the algorithm are both linear in the size of the (macro-expanded) specification, because the construction does constant work and generates constant-sized circuitry for each operator in the specification. The top-level $tenable_i \Rightarrow tok_i$ circuitry can be accounted for in the cost of the pipeline operators.

We have implemented the above translation into a tool that outputs Verilog or VHDL. Generating any output format that is as expressive as sequential circuits should be straightforward.

5. CONCLUSION AND FUTURE WORK

We have presented a novel, high-level specification style for interface monitors, as well as a linear-size, linear-time translation algorithm into monitor circuits. The specification style naturally fits the interfaces used between IP blocks in systems-on-chip. We hope that these results will facilitate the use and broaden the adoption of monitor-based verification methodologies.

In the short term, we need to improve the translation tool. Our current implementation is rather crude. A revised, more robust tool would be better suited for public distribution. In addition, many optimizations are possible and should be implemented. To access the tool flows of other verification researchers, our tool will have to be able to translate specifications into the lower-level specifications used by other tools. In particular, we do not anticipate difficulties translating from our specifications into the specifications used in the interface-monitor research that inspired this work [8, 13, 7].

The macro-expansion of productions can blow-up the size of the regular expressions. Additional research is needed to see if there are practical ways to avoid this problem, such as by introducing new language features to simplify the expressions or better translations that schedule reuse of circuitry.

On the theoretical side, the precise semantic model and expressive power of our specification style are worth exploring. Is our unrestricted specification style exactly equivalent to alternating automata? How much power do we lose by imposing restrictions? Are there other trade-offs between expressive power and efficiency that would be practically useful?

Exploring the theory and semantics would also clarify obscure points in our work. For example, in our current language, the interaction between storage variables and pipelining is analogous to multiprogramming with shared variables. If a variable is used by only a single thread, the results are intuitive, but if multiple threads access a variable, subtle interactions can result. The language may need to be augmented with different classes of storage variables that behave differently with respect to pipelining. In some cases, we want the current behavior, with threads able to communicate across pipeline stages by accessing shared variables. In other cases, we want each pipeline stage to have its own copy of a variable, with values forwarded from one stage to the next when threads are activated. In this model, the variable holds values that follow a transaction through the pipeline.

An important line of future work is to gain experience with this specification style on additional, real interface protocols. Applying the specification style in practice is the only way to validate the adequacy of the specification style, or determine what additional features are needed.

6. REFERENCES

[1] ARM Limited. *AMBA Specification (Rev 2.0)*. 13 May 1999.

[2] R. Armoni, L. Fix, A. Flaisher, R. Gerth, B. Ginsburg, T. Kanza, A. Landver, S. Mador-Haim, E. Singerman, A. Tiemeyer, M. Y. Vardi, Y. Zbar. The ForSpec temporal logic: A new temporal property-specification language. *Tools and Algorithms for the Construction and Analysis of Systems: 8th Intl Conf*, pp. 296–311. LNCS 2280. Springer, 2002.

[3] I. Beer, S. Ben-David, C. Eisner, D. Fisman, A. Gringauze, Y. Rodeh. The temporal logic sugar. *Computer-Aided Verification: 13th Intl Conf*, pp. 363–367. LNCS 2102. Springer, 2001.

[4] L. Bening, H. Foster. *Principles of Verifiable RTL Design*. 2nd Ed. Kluwer, 2001.

[5] A. K. Chandra, D. C. Kozen, L. J. Stockmeyer. Alternation. *JACM*, 28(1):114–133, 1981.

[6] E. M. Clarke, E. A. Emerson. Design and synthesis of synchronization skeletons using branching time temporal logic. *Workshop on Logics of Programs*, pp. 52–71, 1981. Published as LNCS 131. Springer, 1982.

[7] M. S. Jahanpour, E. Cerny. Compositional verification of an ATM switch module using interface recognizer/suppliers (IRS). *Intl High-Level Design, Validation, and Test Workshop*, pp. 71–76. IEEE, 2000.

[8] M. Kaufmann, A. Martin, C. Pixley. Design constraints in symbolic model checking. *Computer-Aided Verification: 10th Intl Conf*, pp. 477–487. LNCS 1427. Springer, 1998.

[9] O. Kupferman, M. Y. Vardi. Module checking. *Computer-Aided Verification: 8th Intl Conf*, pp. 75–86. LNCS 1102. Springer, 1996.

[10] P. Raymond. Recognizing regular expressions by means of dataflow networks. *23rd Intl Coll on Automata, Languages, and Programming*, pp. 336–347. LNCS 1099. Springer, 1996.

[11] A. Seawright, F. Brewer. High-level symbolic construction techniques for high performance sequential synthesis. *30th Design Automation Conf*, pp. 424–428. ACM/IEEE, 1993.

[12] K. Shimizu, D. L. Dill, C.-T. Chou. A specification methodology by a collection of compact properties as applied to the Intel Itanium Processor Bus protocol. *Correct Hardware Design and Verification Methods: 11th IFIP WG 10.5 Adv Research Working Conf*, pp. 340–354. LNCS 2144. Springer, 2001.

[13] K. Shimizu, D. L. Dill, A. J. Hu. Monitor-based formal specification of PCI. *Formal Methods in Computer-Aided Design*, pp. 335–353. LNCS 1954. Springer, 2000.

[14] Sonics Incorporated. *Open Core Protocol Specification 1.0*. Document Version 1.2.

[15] J. Yuan, K. Shultz, C. Pixley, H. Miller, A. Aziz. Modeling design constraints and biasing in simulation using BDDs. *Intl Conf on Computer-Aided Design*, pp. 584–589. IEEE/ACM, 1999.

Achieving Maximum Performance: A Method for the Verification of Interlocked Pipeline Control Logic

Kerstin Eder*
University of Bristol
Department of Computer Science
Woodland Road, MVB
Bristol BS8 1UB, GB
eder@cs.bris.ac.uk

Geoff Barrett
Broadcom DSL BU
320 Bristol Business Park
Coldharbour Lane
Bristol BS16 1EJ, GB
gbarrett@broadcom.com

ABSTRACT

Getting the interlock logic which controls pipeline flow correct is an important prerequisite for maximising pipeline performance. Unnecessary pipeline stalls can only be eliminated when they can be distinguished from those stalls which are necessary to preserve functional correctness.

We propose a method for deriving a maximum pipeline performance specification from a complete functional specification of the pipeline control logic. The performance specification can be used to generate simulation testbench assertions. On the other hand, the specification can serve as a basis for formal property checking. The most promising aspect of our work is, however, the potential to synthesise the actual control logic from its formal description.

Categories and Subject Descriptors

B.5.2 [**Register-Transfer-Level Implementation**]: Design Aids—*Verification*; B.5.1 [**Register-Transfer-Level Implementation**]: Design—*Control Design, Pipeline*

General Terms

Performance, Verification

Keywords

Pipeline Stall, Interlock Logic, Verification

1. INTRODUCTION

1.1 Motivation

Unnecessary pipeline stalls decrease processor performance. Ensuring that a pipelined microprocessor utilises its maximum pipeline throughput is therefore an important task for processor designers. The complexity of today's microprocessor architectures demands a systematic and rigorous approach; one that can be provided by applying formal methods.

This paper describes the method developed while verifying the performance of the FirePath processor. FirePath [9] is a Broadcom proprietary LIW/SIMD processor targeted at communications and multimedia applications, its first use being embedded within a DSL system chip. The processor is based on a pipelined dual instruction multiple data microarchitecture. Pipeline flow is controlled via an interlock logic that prevents functional hazards. To achieve best performance, the pipeline should only stall when this is required to guarantee functional correctness. For example, competition for a completion bus results in pipeline congestion, thus potentially stalling all final pipeline stages, which did not win the grant for the completion bus. Stalls in a completion pipeline stage can cause preceding stages to stall.

From a performance point of view, the instructions in the pipeline should move into the next pipeline stage whenever possible. Hence, a performance bug is a pipeline stall for which there is no functional justification. Maximum performance can be achieved if there are no unnecessary stalls. With our method we can derive a maximum performance specification from a specification of the control logic which describes all functional constraints that cause a pipeline stall by applying fixed-point iteration. This performance specification can be included in a verification testbench in the form of an assertion. Alternatively, the specification can be used as a basis for formal property checking.

Our method is applicable to any pipelined microprocessor architecture that uses interlock logic to control pipeline flow, provided the functional specification of the flow control logic has the properties described in Section 3.1. These properties are by no means restrictive, in fact they are what one would naturally expect when specifying pipeline control flow.

A very promising aspect of our approach is that the functional specification can serve as the basis for RTL code synthesis. We are confident that the basic HDL code for the interlock logic can be generated fully automatically from its functional specification. This is one topic of our current research.

This paper introduces our method by means of a case study in Section 2. For this purpose and to focus our attention on the approach (rather than the architecture itself) we introduce a simple pipelined processor architecture in Section 2.1. The performance specification of our example architecture will be developed from its functional specification in Section 2.2. Section 3 establishes the theoretical soundness of our approach and outlines assumptions

*This publication is one result of work that was performed during a collaborative summer project between the University of Bristol and Broadcom Corporation. Kerstin Eder is grateful for the support of both organisations during this time.

DAC 2002, June 10-14, 2002, New Orleans, Louisiana, USA.

that need to be satisfied to apply the method. Our results are presented in Section 4, which is followed by some conclusions and an outlook on further work in Section 5.

1.2 Related Work

Approaches to verify the correctness of pipelined microprocessor designs range from traditional simulation-based verification via automatic property or model checking [4, 2] to semi-automatic theorem proving [1]. Combined approaches have also been successfully applied [3].

In practice, simulation-based verification is still widely used to gain confidence in design correctness; this also applies to pipeline flow control. Kohno and Matsumoto describe in [5] a test program generation tool which supports the verification of pipeline design. Their approach requires a pipeline behaviour specification to be provided. Based on this specification their tool automatically generates test cases. The pipeline behaviour specification contains pipeline definitions comprising pipeline stages, corresponding data hazards, bypassing mechanisms and resource usage for each instruction category. Any additional stall conditions are specified in explicit stall definitions. Compared to the specification used in our approach, their pipeline behaviour specification is very similar, but also much more detailed since it is describing the behaviour on instruction category level while ours is, except for instructions which enforce an explicit pipeline stall, independent of the actual instruction set. The assertions obtained from our specification can easily be added to existing test benches which, as we have experienced, facilitates the integration of our method into an existing verification environment.

In [6], Kroening and Paul introduce a tool that supports the design of pipelined microprocessors from prepared sequential machines which are later used as reference models. Prepared in this context means, amongst other things, that the hardware of the original sequential machine has been partitioned into pipeline stages. The design method of pipelining sequential architectures is further explained in [7]. Under the assumption that a prepared sequential machine is given, they present a method that adds a stall engine and also forwarding logic. The correctness of the resulting pipelined design with respect to the original sequential design is then proven with the theorem prover PVS [8]. Designers are expected to provide register mappings for the intermediate results of all pipeline stages. Again, the logic description of the stall engine is very similar to our functional specification. From our point of view the tool proposed by Kroening and Paul is beneficial if the design is developed in a systematic sequential top-down refinement approach. However, from our experience, a combination of top-down and bottom-up design is often necessary to meet the tight time-to-market schedules engineers are faced with in practice. Our method only requires a functional specification of the pipeline flow control logic. Sufficient information to develop a functional specification is often available early in the design cycle *ie* in the microarchitectural description of the pipelined processor design. Like Kroening and Paul we are ultimately aiming to generate HDL code for the pipeline flow control logic from our functional specification.

2. AN EXAMPLE CASE STUDY

2.1 Example Architecture

To demonstrate our method we introduce a simple pipelined microprocessor architecture. Alongside the architectural description we also informally introduce our modelling language by giving a formal description of the features relevant to our analysis.

Our example architecture comprises of two pipes *long* and *short*

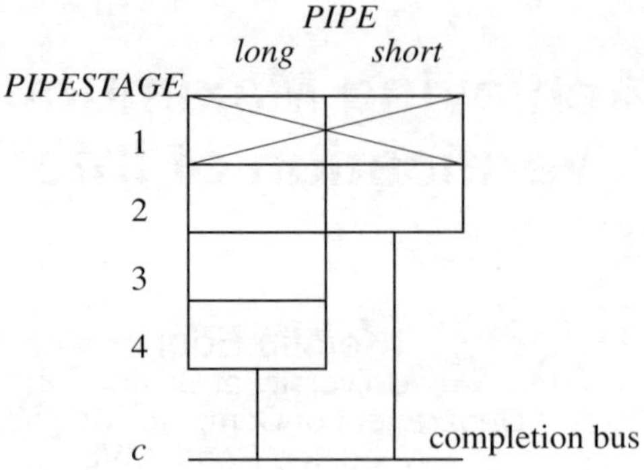

Figure 1: Example pipeline architecture with two pipes and one completion bus.

which share one combined fetch/decode/issue stage. The *long* pipe has two execution stages and one writeback stage, and the *short* pipe has one execution/writeback stage. The first stages of both pipelines operate in lock step, *ie* progress is only made in synchrony. Note that the pipeline stages in Figure 1 have been indexed starting from the fetch/decode/issue stages.

For our formal description we define the sets/types:

$$PIPE = \{long, short\}$$
$$PIPESTAGE = \{1, 2, 3, 4\}$$

For flow control purposes each pipe stage has a moving_or_empty (*moe*) flag, which propagates backwards to its predecessor and indicates whether the stage is moving forward on the next cycle or is currently empty. It is used to determine whether the stage will block the preceding stage. If the *moe* flag is set, the respective stage is ready to take an instruction from the preceding stage on the next cycle. If the *moe* flag is clear, the preceding pipe stage gets blocked, *ie* does not move forward on the next cycle; it stalls.

Further, each pipe stage has a require_to_move (*rtm*) flag which propagates forwards to its successor and indicates whether the contents in that stage intends to move on. So when a stage does not require_to_move, it means either it is not valid (*ie* contains a pipeline bubble) or, if valid it does not need to move into the next pipeline stage (mostly because its processing finishes at that stage and no writeback is required).

BOOLEAN p.s.moe
BOOLEAN p.s.rtm
where $p \in PIPE$
$s \in PIPESTAGE$

A special (instruction-specific) flag, which can only be accessed in the fetch/decode/issue stage of the *long* pipe, indicates whether the machine is in a wait state.

BOOLEAN op_is_WAIT

Our example architecture has eight registers. The register use is recorded on an eight bit scoreboard, where the register address is used to access the scoreboard bit-array. Each instruction has one source and one destination register.

$$REGADDRESS = \{7..0\}$$
$$SDREG = \{src, dst\}$$

BOOLEAN $scb[8]$
REGADDRESS $p.1.r.regaddr$
where $p \in PIPE$
$r \in SDREG$

$$
\begin{array}{ll}
& SPEC_{func} \\
= & \phi(\langle long.4.moe, long.3.moe, long.2.moe, long.1.moe, short.2.moe, short.1.moe\rangle) \\
= & (long.req \wedge \neg long.gnt \rightarrow \neg long.4.moe) \\
\wedge & (long.3.rtm \wedge \neg long.4.moe \rightarrow \neg long.3.moe) \\
\wedge & (long.2.rtm \wedge \neg long.3.moe \rightarrow \neg long.2.moe) \\
\wedge & ((long.1.rtm \wedge \neg long.2.moe \\
& \vee\ op\_is\_WAIT \\
& \vee\ \neg short.1.moe \\
& \vee\ \exists r : SDREG\ .\ \exists a : REGADDRESS\ .\ long.1.r.regaddr = a \wedge scb[a] \wedge c.regaddr \neq a) \\
& \rightarrow \neg long.1.moe) \\
\wedge & (short.req \wedge \neg short.gnt \rightarrow \neg short.2.moe) \\
\wedge & ((short.1.rtm \wedge \neg short.2.moe \\
& \vee\ \neg long.1.moe \\
& \vee\ \exists r : SDREG\ .\ \exists a : REGADDRESS\ .\ short.1.r.regaddr = a \wedge scb[a] \wedge c.regaddr \neq a) \\
& \rightarrow \neg short.1.moe)
\end{array}
$$

Figure 2: A functional specification describing the necessary pipeline stalls.

The final stages of both execution subpipes connect to a completion bus named *c*; the *short* pipe has higher priority for completion than the *long* pipe. Each pipe has a dedicated signal to request the completion bus. There is one grant signal for each pipe. To target a register the completion bus holds the target register's address.

$$
\begin{array}{l}
BOOLEAN\ p.req, p.gnt \\
REGADDRESS\ c.regaddr \\
where\ p \in PIPE
\end{array}
$$

This concludes the example architecture. Note that the example processor is simpler than the FirePath in several ways. These include the fact that FirePath is two-sided and has more and deeper execution pipes than the example. Furthermore FirePath has several pipeline decouple stages (shunts), interrupt logic and several completion buses. FirePath has been successfully verified with our method. To keep the example comprehensible, however, we restrict this paper to the given simple architecture.

2.2 Specification

2.2.1 *Functional Specification*

Our aim is to demonstrate that the pipeline flow control allows maximum performance. By maximum performance we understand that there are no unnecessary pipeline stalls. In the setting described in Section 2.1 this means that there are no situations where the *moe* flag is clear when it should be set.

We can start from a specification, which describes under which conditions it is necessary to stall a pipe stage to avoid hazards. In practice, this information is typically contained in the microarchitectural description of the processor, and hence should be accessible at an early stage during processor design. The following specification describes the individual constraints under which a pipeline stall is necessary in our example architecture.

We start with the completion stages which compete for access to the completion bus. A completion stage does not move on when it does not win the grant to use the completion bus.

$$
\begin{array}{l}
long.4.rtm \wedge \neg long.gnt \rightarrow \neg long.4.moe \\
short.2.rtm \wedge \neg short.gnt \rightarrow \neg short.2.moe
\end{array}
$$

Note that for the completion stages the bus request flag should be set when the *rtm* flag is set, which transforms the above statements to:

$$
\begin{array}{l}
long.req \wedge \neg long.gnt \rightarrow \neg long.4.moe \\
short.req \wedge \neg short.gnt \rightarrow \neg short.2.moe
\end{array}
$$

Note further that the completion logic, *eg* the arbitration scheme of the bus, can also be included in the functional specification. For the sake of simplicity we will concentrate on the *moe* flags only.

To prevent hazards caused by overwriting, the *long* pipe's intermediate stages should stall when their preceding pipe stage stalls.

$$
\begin{array}{l}
long.2.rtm \wedge \neg long.3.moe \rightarrow \neg long.2.moe \\
long.3.rtm \wedge \neg long.4.moe \rightarrow \neg long.3.moe
\end{array}
$$

In the same way, a fetch/decode/issue stage should stall when the respective issue pipe is stalled.

$$
\begin{array}{l}
long.1.rtm \wedge \neg long.2.moe \rightarrow \neg long.1.moe \\
short.1.rtm \wedge \neg short.2.moe \rightarrow \neg short.1.moe
\end{array}
$$

Further, if the machine is in a wait state then the *long* pipe can not issue.

$$
op\_is\_WAIT \rightarrow \neg long.1.moe
$$

In addition, both the *long* and the *short* pipe operate in lock step during fetch, decode and issue.

$$
\begin{array}{l}
\neg long.1.moe \rightarrow \neg short.1.moe \\
\neg short.1.moe \rightarrow \neg long.1.moe
\end{array}
$$

Note that the above two statements describe the logical equivalence of the *moe* flags of the initial stages of both pipelines. We use two implications instead of one equivalence. This makes our specification compositional with respect to individual pipeline stages, which facilitates the maintenance of our specification.

If a source or destination register is outstanding then the instruction cannot be issued. A register is outstanding if it is on the scoreboard, *ie* its scoreboard flag is set, and it is not bypassed, *ie* it will not be written to in the current cycle, which means it is not a completion target register.

$$
\begin{array}{l}
\forall p : PIPE\ .\ \exists r : SDREG\ .\ \exists a : REGADDRESS\ . \\
p.1.r.regaddr = a \wedge scb[a] \wedge c.regaddr \neq a \rightarrow \neg p.1.moe
\end{array}
$$

This statement completes our specification. To get a more compact description we can transform the above specification such that there is exactly one statement for each pipeline stage as given in Figure 2.

$$
\begin{array}{ll}
 & SPEC_{perf} \\
= & (long.req \wedge \neg long.gnt \leftarrow \neg long.4.moe) \\
\wedge & (long.3.rtm \wedge \neg long.4.moe \leftarrow \neg long.3.moe) \\
\wedge & (long.2.rtm \wedge \neg long.3.moe \leftarrow \neg long.2.moe) \\
\wedge & ((long.1.rtm \wedge \neg long.2.moe \\
 & \vee\ op\_is\_WAIT \\
 & \vee \neg short.1.moe \\
 & \vee \exists r : SDREG\ .\ \exists a : REGADDRESS\ .\ long.1.r.regaddr = a \wedge scb[a] \wedge c.regaddr \neq a) \\
 & \leftarrow \neg long.1.moe) \\
\wedge & (short.req \wedge \neg short.gnt \leftarrow \neg short.2.moe) \\
\wedge & ((short.1.rtm \wedge \neg short.2.moe \\
 & \vee \neg long.1.moe \\
 & \vee \exists r : SDREG\ .\ \exists a : REGADDRESS\ .\ short.1.r.regaddr = a \wedge scb[a] \wedge c.regaddr \neq a) \\
 & \leftarrow \neg short.1.moe)
\end{array}
$$

Figure 3: A maximum performance specification.

An assignment to the vector of moving_or_empty flags comprising

$$\langle long.4.moe, long.3.moe, long.2.moe, long.1.moe, short.2.moe, short.1.moe\rangle$$

satisfies this specification iff

$$\vdash \phi(\langle long.4.moe, long.3.moe, long.2.moe, long.1.moe, short.2.moe, short.1.moe\rangle).$$

2.2.2 *Performance Specification*

The functional specification in Figure 2 contains formulas of the form *condition* $\rightarrow \neg moe$, where *condition* represents a disjunction of the individual constraints that should lead to a pipeline stall. Note that a violation of this specification occurs when *condition* is satisfied, but the *moe* flag is set, corresponding to a situation where despite the condition holding, the respective pipeline stage signals to its predecessor that it is either moving or empty, so it could be overwritten on the next cycle. The result is that the pipe is moving even though it should not, which will cause a hazard. This corresponds to a functional bug; we therefore call a specification of the above form a *functional* specification.

However, our aim is to identify cases where the pipeline stalls although it is safe to move on. Formally, this can be expressed by specifying that $\neg moe \wedge \neg condition$ should not occur, where *condition* represents a disjunction of all individual constraints that should lead to a pipeline stall. This is logically equivalent to the formula $\neg moe \rightarrow condition$; stating that if the *moe* flag is not set then we would expect *condition* to hold. If *condition* does not hold, then we have indeed identified an unnecessary pipeline stall. A specification containing formulas of this form will be called a *performance* specification.

The performance specification that corresponds to the functional specification introduced in Section *2.2.1* is given in Figure 3. It can be included into a testbench in the form of an assertion. Alternatively, the performance specification can serve as a basis for property checking; we have left property checking for future work.

2.2.3 *Concluding Remarks on our Case Study*

For obvious reasons, a design should satisfy both its functional specification *and* its performance specification. Hence, the combined specification would contain formulas of the form *condition* $\leftrightarrow \neg moe$, expressing that the pipeline stalls if and only if *condition* is satisfied. The combined pipeline specification thus corresponds to the specification obtained by changing all $\rightarrow$ in Figure 2 into $\leftrightarrow$.

Because the focus of our project was on performance verification, we only intended to verify the performance part of the combined specification. In practice, it is common to use other verification methods to gain confidence in the functional correctness of the design.

To include the assertions into a testbench, what remains to be done is to translate them into the HDL used for RTL design and simulation. In addition, the signals referred to in the assertion need to be connected to the corresponding signals in the RTL code of the pipeline design. Close cooperation with designers is required to ensure that the RTL signals have the intended semantics.

3. DERIVING THE PERFORMANCE SPECIFICATION FROM THE FUNCTIONAL SPECIFICATION

In the previous section we gave an informal and intuitive understanding of the relationship between the functional and the performance specification. The careful reader will have noticed that it is possible to satisfy our functional specification and yet never move at all. There are, in fact, many possible implementations of this specification of varying performance. So we must ask which of these specifications we require. Of course, we want the best one, *ie* the one which stalls least often. In this section, we will show that the best solution exists and prove that it is derived by changing each $\rightarrow$ in Figure 2 into $\leftrightarrow$ as we have indicated to obtain the combined functional and performance specification in the previous section. In general, the best solution may be more complicated than this but only if control flows in both directions along the pipeline.

3.1 Properties of the Functional Specification

We first establish some properties of the functional specification. It is important that the functional specification can be provided in a form that satisfies these properties; they are preconditions for deriving a maximum performance specification. Section 3.2 describes how the derivation is performed.

Clearly, the functional specification given in Section *2.2.1* does not uniquely determine the values that should be assigned to the *moe* flags. Note for instance that our specification is satisfied if all *moe* flags are set to false.

$$\vdash \phi(\langle False, False, False, False, False, False\rangle) \quad (1)$$

Assigning false to all *moe* flags blocks the entire pipeline. But, from a functional point of view, correctness is certainly preserved when the pipeline stalls completely. This is why the specification in Section 2.2.1 is called a functional (correctness) specification.

Note further that our specification can be split into two separate pipeline specifications of the form:

$$\phi(\langle long.4.moe, \ldots, long.1.moe\rangle) = \bigwedge_{i=4}^{1} long.i.F(\neg long.i+1.moe, \neg short.i.moe) \rightarrow \neg long.i.moe$$

and

$$\phi(\langle short.2.moe, short.1.moe\rangle) = \bigwedge_{i=2}^{1} short.i.F(\neg short.i+1.moe, \neg long.i.moe) \rightarrow \neg short.i.moe$$

where the function F, which describes the individual stalling constraints, is constructed using only conjunction and disjunction (on its argument variables), *ie* it is *monotonic*. With suitable indexing the above specification can be further generalised to:

$$\phi(\langle moe\rangle) = \bigwedge(\langle F\rangle(\neg\langle moe\rangle) \rightarrow \neg\langle moe\rangle)$$

such that

$$\langle F\rangle[i](\neg\langle moe\rangle) \rightarrow \neg\langle moe\rangle[i]$$

From the monotonicity of F it follows that ϕ is *disjunctive* in the sense that if two assignments to *moe* flag vectors satisfy ϕ then their bitwise disjunction also satisfies ϕ:

$$\frac{\vdash \phi(\langle moe\rangle_1) \quad \vdash \phi(\langle moe\rangle_2)}{\vdash \phi(\langle moe\rangle_1 \vee \langle moe\rangle_2)} \tag{2}$$

where

$$\langle moe_{n..1}\rangle_1 \vee \langle moe_{n..1}\rangle_2 = \langle\langle moe\rangle_1[n] \vee \langle moe\rangle_2[n], \ldots, \langle moe\rangle_1[1] \vee \langle moe\rangle_2[1]\rangle$$

Now we can show that:

$$F(\neg(\langle moe\rangle_1 \vee \langle moe\rangle_2)) \rightarrow \neg(\langle moe\rangle_1 \vee \langle moe\rangle_2)$$

Assume: $\vdash \phi(\langle moe\rangle_1)$ which means $F(\neg\langle moe\rangle_1) \rightarrow \neg\langle moe\rangle_1$ and $\vdash \phi(\langle moe\rangle_2)$ which means $F(\neg\langle moe\rangle_2) \rightarrow \neg\langle moe\rangle_2$.

Proof:

	$F(\neg(\langle moe\rangle_1 \vee \langle moe\rangle_2))$	
$=$	$F(\neg\langle moe\rangle_1 \wedge \neg\langle moe\rangle_2)$	
$\rightarrow$	$F(\neg\langle moe\rangle_1) \wedge F(\neg\langle moe\rangle_2)$	since $\neg\langle moe\rangle_1 \wedge \neg\langle moe\rangle_2 \rightarrow \neg\langle moe\rangle_i$ where $i \in \{1,2\}$ and F monotonic
$\rightarrow$	$\neg\langle moe\rangle_1 \wedge \neg\langle moe\rangle_2$	by assumption
$=$	$\neg(\langle moe\rangle_1 \vee \langle moe\rangle_2)$	q.e.d.

Properties 1 and 2 of our functional specification form the theoretical foundation for the derivation of the performance specification. The two properties do not restrict the pipeline control flow logic. Establishing the first property is trivial, since our specification does not state anything about when pipeline stages do not stall. To implement pipeline functionality that allows instructions to enter at the fetch/decode/issue stage, progress though the intermediate stages, and exit at a completion stage or possibly earlier, the control flows backwards, starting from the completion stages. For any such pipeline the conditions which cause necessary pipeline stalls can be specified by defining a suitable function F which takes the negated moving_or_empty flags as arguments. Monotonicity should be a natural property of the resulting functional specification.

3.2 Derivation

To get a unique description of the values that should be assigned to the *moe* flags to achieve maximum performance we need to add, to the functional specification, the requirement that the desired set of assignments to *moe* flags is the most liberal, *ie* the assignment which makes more *moe* flags true than any other.

It is important to prove that there is a unique most liberal assignment. To prove this we need to show that there is at least one assignment possible; this is the one that sets all flags in the *moe* vector to false - see property (1), and that or'ing the individual flags of two valid assignments results in another valid assignment; which is established through property (2).

Under the assumption that the functional specification has these two properties, the most liberal assignment to *moe* flags, written $\langle MOE\rangle$, can be obtained by disjunctively combining the individual bits of all valid assignments to *moe* flags.

$$\frac{\langle MOE\rangle = \bigvee\langle moe\rangle \text{ s.t. } \vdash \phi(\langle moe\rangle)}{\vdash \phi(\langle MOE\rangle)} \tag{3}$$

Fixed-point iteration can now be applied to find the least stalling solution. The resulting specification $\phi(\langle MOE\rangle)$ states under which conditions most *moe* flags are set, *ie* how maximum pipeline performance can be achieved. It establishes that, to get maximum performance, each individual *moe* flag should be assigned as follows:

$$\langle MOE\rangle[i] := \neg\langle F\rangle[i](\neg\langle MOE\rangle) \tag{4}$$

This confirms our intuitive approach from Section 2.2.2 and formally establishes the relationship between the functional and the performance parts of our combined specification.

It remains to show that $\langle MOE\rangle$ is indeed the solution that provides the maximum performance. We need to show that any valid assignment to *moe* flags is subsumed by the assignment in *MOE*.

Suppose: $\vdash \phi(\langle moe\rangle')$.

Assume: $\langle moe\rangle'[j] \rightarrow \langle MOE\rangle[j]$ for $j > i$ under the given indexing. This assumption is based on the observation that the *moe* flag of a final stage depends not on *moe* flags but solely on a completion bus grant which is treated as a constant in F.

We can now prove for each *moe* flag i in the vector that:

$$\langle moe\rangle'[i] \rightarrow \langle MOE\rangle[i]$$

Proof:

$\langle moe\rangle'[i]$	$\rightarrow$	$\neg\langle F\rangle[i](\neg\langle moe\rangle')$	by 4
	$\rightarrow$	$\neg\langle F\rangle[i](\neg\langle MOE\rangle)$	F monotonic
	$=$	$\langle MOE\rangle[i]$	by 4

Hence:

$$\langle moe\rangle' \rightarrow \langle MOE\rangle \qquad \text{q.e.d.}$$

Note that the inductive proof above is possible because control flows, starting from the completion stages, to the respective predecessor pipeline stages only. If control flows in both directions along the pipeline the best solution may be more complicated.

4. RESULTS

Our aim was to identify possible performance bugs in the pipeline flow control logic of the FirePath processor. A performance bug, in this context, is defined as an unnecessary pipeline stall. Using the architecture and microarchitecture manuals, and in close collaboration with designers, we investigated all constraints that stall the pipeline to preserve functional correctness, and wrote them down in the form of a functional specification that has the properties discussed in Section 3.1. This specification is already a valuable reference for design engineers. We have shown how a maximum performance specification can be obtained from the functional specification. Ideally the combined specification is used as a basis for formal property checking. Alternatively it can be translated into testbench assertions which are checked during simulation.

The specification of the FirePath pipeline design is now a permanent part of the processor's testbench. It ensures that any modifications of the pipeline flow control logic preserve the initial intent. Even the best simulation is by no means exhaustive, hence the fact that the assertions are not triggered during simulation does not imply that the design satisfies the specification. A more thorough approach is to use a property checking tool instead of simulation. Our investigation has found that the specification can also be translated into statements that can be verified by property checkers. Running a property checker means exhaustive verification and is therefore preferable to simulation.

The project took 10 weeks of one person's time from a cold start on the processor architecture and microarchitecture (and Verilog). In summary, we uncovered inefficiencies in the pipeline control flow, and also some incorrect initialisation values of control signals. The completion logic has been redesigned as a consequence of our analysis, resulting in efficiency increase at the pipeline completion stages.

Another achievement of this project is the gain in design understanding amongst engineers. The entire pipeline flow control was formalised and is now documented in the functional specification which serves as a design reference. It greatly helped to clarify how the pipeline flow is controlled, and bridged gaps when parts were designed by different teams. In several cases, functional equivalence of different implementations needed to be established before a more abstract description was accepted across the design teams. An instance of this, which is not addressed in our example architecture, are shunt stages, where the same effect can be achieved via various implementations, all of which should satisfy the same functional specification on a more abstract level.

5. CONCLUSION AND FURTHER WORK

Interlocked pipeline flow control is designed to prevent hazards by stalling the pipeline. The design of the interlock logic is perceived to be tricky and debugging it can delay the design process considerably. There is a danger of introducing unintended pipeline stalls during debugging and more generally whenever the design is modified.

While the prevention of hazards is very important to achieve functional correctness, the detection and prevention of unnecessary pipeline stalls is crucial to maximise pipeline performance and thus obtain high processor performance. The aim of our project was to develop a specification that detects and thus helps to prevent unnecessary pipeline stalls.

We believe that our method can be applied to any pipelined microprocessor design that uses interlock logic to prevent hazards, provided the functional specification of the flow control logic can be given in a form that satisfies the properties detailed in Section 3.1. These properties are not restrictive, in fact they state what one would naturally expect when specifying pipeline control flow.

The simplicity of our approach is based on the fact that it relies purely on a functional understanding of the pipeline flow control logic, which many designers are familiar with at an early stage of the design process. In addition, the specification is, except for instructions which enforce an explicit pipeline stall, independent of the actual instruction set, allowing the control flow to be developed and verified in separation from the data flow. We have found that our method can already deliver useful results, even when the design is not yet complete.

For this pilot project we have derived the performance specification manually. We are now working on a tool which, given a functional specification that has the properties mentioned in Section 3.1, generates the corresponding performance specification and also Verilog/VHDL assertions. This tool will form the first module of a (semi-)automatic pipeline flow control design environment. Ultimately, we would like to generate the HDL code that implements the pipeline flow control logic from the functional specification. This is much more ambitious. Issues like timing and the introduction of shunt stages to decouple pipelines if signal propagation times cannot meet cycle times need to be addressed. A project that aims to show that this is feasible is currently in progress.

6. REFERENCES

[1] M. Bickford and M. Srivas. Verification of a pipelined microprocessor using CLIO. In *Workshop on Hardware Specification, Verification and Synthesis: Mathematical Aspects*, volume 408 of *Lecture Notes in Computer Science*. Springer, 1989.

[2] A. Biere, E. Clarke, R. Raimi, and Y. Zhu. Verifying safety properties of a PowerPC microprocessor using symbolic model checking without BDDs. In *11th International Conference on Computer-Aided Verification*, Lecture Notes in Computer Science, pages 60–71. Springer, 1999.

[3] J. R. Burch and D. L. Dill. Automatic verification of pipelined microprocessor control. In *6th International Conference on Computer-Aided Verification*, volume 818 of *Lecture Notes in Computer Science*, pages 68–80. Springer, 1994.

[4] E. M. Clarke, O. Grumberg, and D. A. Peled. *Model Checking*. MIT Press, 1999.

[5] K. Kohno and N. Matsumoto. A new verification methodology for complex pipeline behavior. In *DAC 2001 Conference Proceedings*, June 2001.

[6] D. Kroening and W. J. Paul. Automated pipeline design. In *DAC 2001 Conference Proceedings*, June 2001.

[7] S. Müller and W. Paul. *Computer Architecture: Complexity and Correctness*. Springer, 2000.

[8] S. Owre, N. Shankar, J. M. Rushby, and D. W. J. Stringer-Calvert. *PVS System Guide*. SRI International, version 2.3 edition, September 1999. Available at `http://pvs.csl.sri.com`.

[9] S. Wilson. Broadcom's FirePath Processor Architecture. In *Embedded Processor Forum*, June 2001.

Formal Verification of Module Interfaces against Real Time Specifications

Arindam Chakrabarti
Dept. of EECS
UC Berkeley, USA
arindam@cs.berkeley.edu

Pallab Dasgupta P.P. Chakrabarti Ansuman Banerjee
Dept. of Computer Science & Engineering
Indian Institute of Technology Kharagpur, INDIA
pallab,ppchak,ansuman@cse.iitkgp.ernet.in

ABSTRACT

One of the main concerns of the designer of a circuit module is to guarantee that the interface of the module conforms to specific protocols (such as PCI Bus, AMBA bus or Ethernet) by which it interacts with its environment. The computational complexity of verifying such open systems under all possible environments has been shown to be very hard (EXPTIME complete [10]). On the other hand, designers are typically required to guarantee correct behavior only for specific valid behaviors of the environment (such as a valid PCI Bus environment). Designers attempt to model these behaviors through an appropriate test bench for the module. In this paper we present a *module verifier tool* based on a proposed real time temporal logic called Open-RTCTL, which allows combined specification of the correctness properties and the input environments. The tool accepts the design in a subset of Verilog. By making the designer specify the environment constraints, we are able to verify a module in isolation, and thereby avoid the state explosion problem due to composition of modules. We present experimental results on modules from the Texas-97 Benchmark circuits [14] to demonstrate the space/time efficiency of the tool.

Categories and Subject Descriptors

B.7.2 [**Hardware**]: Integrated Circuits—*Verification*

General Terms

Verification

Keywords

Formal Verification, Temporal Logic

1. INTRODUCTION

Temporal logic model checking [4, 5] has emerged as one of the most powerful techniques for automated formal verification of hardware. In this approach, the design is modeled as a finite state non-deterministic transition system. The correctness property that needs to be verified on the design is specified in terms of a temporal logic formula. Computation Tree Logic (CTL) is one of the most popular temporal logics, and is the formal specification language behind many verification tools such as SMV [12], VIS [15] and FormalCheck [6].

The bulk of research in model checking has focussed on the verification of *closed systems*, that is, systems which are self contained and have no external inputs. The truth of properties specified in temporal logics such as CTL, LTL, and their real time extensions TCTL [1] and RTCTL [8] are not constrained by the inputs to the system. In many cases, state transitions enabled by different inputs are treated as non-deterministic transitions, and the semantics of these logics are interpreted over such non-deterministic FSMs.

In the other extreme, there has been research on the verification of *open systems* where a property is true if it holds regardless of the inputs to the system. Verification of open systems (referred to as *module checking*) has been shown to be much more complex than the verification of closed systems [10, 11]. For example, while CTL verification in *closed systems* is polynomial in the size of the system times the length of the formula, CTL verification in modules is EXPTIME-complete.

A natural way to verify a module (or component) lies in appropriate modeling of the system as well as the environment under which it is expected to work. Working under the assumption of a closed system or an adversarial open system are unsuitable in practice.

The *Module Verifier Tool* presented in this paper originates from the observation that the designer of a module is typically required to guarantee that the module works correctly under certain *valid* environments only. Typically, the designer attempts to model these environments through an appropriate test bench for the design. For example, the designer of a PCI compliant device may be required to validate the device under valid PCI Bus environments only.

In this paper, we present a real time logic called Open-RTCTL which allows the designer to express the correctness property and the environment under which the property is expected to hold, in a unified way. This also allows the designer to decompose the correctness specification based on different input scenarios.

The Module Verifier Tool accepts the design in a subset of Verilog and provides a succinct temporal framework for specifying both the environment constraints as well as the correctness properties. Designers may use this tool to elim-

DAC 2002, June 10-14, 2002, New Orleans, Louisiana, USA.

inate as many errors as possible from the modules they design before integrating them with the other modules in the circuit. By making the designer specify the environment constraints, we are able to verify a module in isolation, and thereby avoid the state explosion problem due to composition of modules. The tool is also useful in the *assume-guarantee* style of reasoning [9].

We present experimental results on the Texas-97 Verification Benchmarks [14]. We discuss a scheme to incorporate automatic generation of Open-RTCTL properties from protocols specified as *timing diagrams.*

The paper is organized as follows. Section 2 describes the architecture of the tool. Section 3 presents the logic Open-RTCTL which is used to specify correctness properties, and a symbolic model checking algorithm for Open-RTCTL verification. Section 4 outlines the abstraction mechanism used in our tool. Section 5 presents experimental results. Section 6 discusses translation of timing diagrams to Open-RTCTL properties.

2. THE ARCHITECTURE OF THE TOOL

The architecture of the Module Verifier Tool is shown in Fig 1. The tool accepts the design in a subset of Verilog. In the current version of the tool, the correctness specification is input as a set of properties specified in a language called *Open-RTCTL* which we present in the next section. The correctness specification includes both timed and untimed properties.

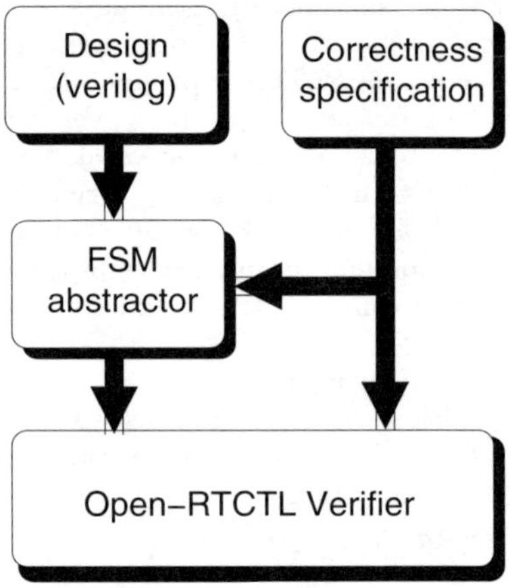

Figure 1: The Module Verifier Tool

From the given set of correctness properties, we extract the relevant state variables of the module which affect the interface protocol. Using a cone-of-influence reduction, we abstract a minimal FSM model of the given design (module). The Open-RTCTL verifier accepts the FSM model and the correctness properties as input and verifies the correctness of the design.

Timing diagrams are a popular form of representation of interface protocols. We present our approach to translate timing diagrams to Open-RTCTL properties. Since timing diagrams are commonly used in the industry, such an interface will alleviate the designer from having to form a complete set of Open-RTCTL properties which covers the correctness specification of the interface protocol.

3. CORRECTNESS SPECIFICATION

In this section, we outline the logic Open-RTCTL which forms the basis of the property specification syntax of the Module Verifier Tool. The untimed version of this logic (which does not express any quantitative timing properties) was presented in [7]. The logic presented here supports the *bounded until* operator, and can express timing properties.

Formally, we define a module as a tuple, $J = \langle \mathcal{AP}, S, \mathbb{I}, \tau, s_0, \mathcal{F} \rangle$, where:

- $\mathcal{AP}$ is a set of atomic propositions,
- S is a finite set of states,
- $\mathbb{I}$ is a finite set of inputs,
- $\tau : S \times 2^{\mathbb{I}} \rightarrow S$ is the next-state function. Given a state $s_i \in S$ and input vector η, $s_j = \tau(s_i, \eta)$ is the next state of s_i under input η,
- $s_0 \in S$ is the initial state,
- $\mathcal{F} : S \rightarrow 2^{\mathcal{AP}}$ is a labeling of states with atomic propositions true in that state.

A *path*, π, in the module is an infinite sequence of state-input pairs, $(s_0, \eta_0), (s_1, \eta_1), \ldots$, such that for all i, $s_i \in S$, $\eta_i \in 2^{\mathbb{I}}$ and $s_{i+1} = \tau(s_i, \eta_i)$. s_0 is called the starting state of π. Since the module has a finite set of states, one or more states will appear multiple number of times on a path. In other words, a path (as defined here) is an infinite *walk* over the state transition graph.

We further define a $\mathcal{I}$-*consistent* path as follows. A boolean constraint $\mathcal{I}$ over the inputs $\mathbb{I}$ is said to be satisfied by a input vector η iff $\mathcal{I}$ evaluates to true on the input η. Given a boolean constraint $\mathcal{I}$ over the inputs $\mathbb{I}$, a path $\pi = (s_0, \eta_0), (s_1, \eta_1), \ldots$ is $\mathcal{I}$-*consistent* if η_i satisfies $\mathcal{I}$ for all i. It may be noted that since the next state of a state is well defined for every input vector, there exists at least one $\mathcal{I}$-consistent path starting from each state whenever $\mathcal{I}$ is satisfiable. Since many different input vectors may satisfy $\mathcal{I}$, there may be multiple $\mathcal{I}$-consistent paths from a state.
The formal syntax of Open-RTCTL is as follows. p denotes an atomic proposition. a and b are positive integers.

- Each $p \in \mathcal{AP}$ is an Open-RTCTL formula,
- If f and g are Open-RTCTL formulas, then so are $\neg f$, $f \vee g$, $f \wedge g$, $EX^{\mathcal{I}} f$, $AX^{\mathcal{I}} f$, $E(f\ U^{\mathcal{I}}_{[a,b]}\ g)$, $A(f\ U^{\mathcal{I}}_{[a,b]}\ g)$, where $\mathcal{I}$ is a boolean formula over inputs in $\mathbb{I}$.

The syntax of Open-RTCTL is obtained by augmenting the X and U operators of RTCTL [8] with an input constraint, $\mathcal{I}$. We use $E(f\ U^{\mathcal{I}}\ g)$ as a shortform for $E(f\ U^{\mathcal{I}}_{[0,\infty]}\ g)$. We also use the usual short forms $EF^{\mathcal{I}} f$ for $E(true\ U^{\mathcal{I}}\ f)$, and $AG^{\mathcal{I}} f$ for $\neg EF^{\mathcal{I}} \neg f$.
The semantics of Open-RTCTL is defined over a module $J = \langle \mathcal{AP}, S, \mathbb{I}, \tau, s_0, \mathcal{F} \rangle$. We use the notation $s \models f$ to indicate that the Open-RTCTL formula f is true at state s of the module. Likewise, we use the notation $\pi \models \psi$ to indicate that a Open-RTCTL path formula, ψ (of the form $f U^{\mathcal{I}}_{[a,b]}\ g$) is true on the path π. The formal semantics of Open-RTCTL is as follows:

- $\forall s \in S, s \models True$ and $s \not\models False$
- $s \models p$ iff $p \in \mathcal{F}(s)$

- $s \models \neg f$ iff $s \not\models f$
- $s \models f \wedge g$ iff $s \models f$ and $s \models g$
- $s \models f \vee g$ iff $s \models f$ or $s \models g$
- $s \models EX^{\mathcal{I}} f$ iff $\exists s' = \tau(s, \eta)$ such that $s' \models f$ and the input vector η satisfies the constraint $\mathcal{I}$
- $s \models AX^{\mathcal{I}} f$ iff $s \models EX^{\mathcal{I}} f$ and $\forall s'$, such that $s' = \tau(s, \eta)$ for some η satisfying $\mathcal{I}$, we have $s' \models f$
- $\pi \models fU^{\mathcal{I}}_{[a,b]}\ g$ iff π is $\mathcal{I}$-consistent, and $\exists (s_i, \eta_i) \in \pi$ such that $a \leq i \leq b$ and $s_i \models g$, and for all (s_j, η_j) preceding (s_i, η_i) in π, $s_i \models f$
- $s \models E(fU^{\mathcal{I}}_{[a,b]}\ g)$ iff there exists a path π starting from s, such that $\pi \models fU^{\mathcal{I}}_{[a,b]}\ g$
- $s \models A(fU^{\mathcal{I}}_{[a,b]}\ g)$ iff $s \models E(fU^{\mathcal{I}}_{[a,b]}\ g)$ and for each $\mathcal{I}$-consistent path π starting from s, we have $\pi \models fU^{\mathcal{I}}_{[a,b]}\ g$

As mentioned earlier, an $\mathcal{I}$-consistent path always exists from a state s provided that $\mathcal{I}$ is satisfiable. In a Open-RTCTL formula, $\mathcal{I}$ is used to specify the possible input patterns under which we are to verify the module, and hence it is natural to assume that $\mathcal{I}$ is non-empty (satisfiable). Otherwise, the formulas are vacuously false at all states. It is for this reason that we have the requirement of $s \models EX^{\mathcal{I}} f$ in the semantics of $s \models AX^{\mathcal{I}} f$ and the requirement of $s \models E(fU^{\mathcal{I}}_{[a,b]}\ g)$ in the semantics of $s \models A(fU^{\mathcal{I}}_{[a,b]}\ g)$.

Though in the formal definition of a module, we have an explicit transition from a state s_i for every input vector, it is often natural to expect that the set of input vectors enabling a transition of the module is succinctly specified in the form of a boolean formula. It is convenient to construct BDDs for such succinct transition relations without actually creating the explicit FSM. The following example illustrates a typical succinct module where each transition is labeled by the boolean formula representing the set of input vectors for which the transition is taken. We also illustrate the syntax and semantics of Open-RTCTL through the following example.

EXAMPLE 1. Figure 2 shows a simple module in succinct form with $S = \{s_1, s_2, s_3, s_4, s_5, s_6\}$, $\mathcal{AP} = \{f, g, h\}$ and $\mathbb{I} = \{i_1, i_2\}$. s_1 is the initial state. The atomic propositions true in a state is shown beside the state. The set of input vectors enabling a given transition is shown as a boolean formula beside the transition. For example, the transition from s_2 to s_1 is enabled by two input vectors, namely $(1,0)$ and $(1,1)$, which are represented by the boolean formula i_1.

Let us consider the following Open-RTCTL formulas:

$$\varphi_1 = E(f\ U^{\mathcal{I}}_{[0,2]}\ g)$$
$$\varphi_2 = A(f\ U^{\mathcal{I}}_{[0,2]}\ g)$$

The formula φ_1 asks whether *there exists* inputs satisfying the boolean constraint $\mathcal{I}$, which drives the system to a state satisfying g through states satisfying f. The formula φ_2 asks whether *all inputs* which satisfy $\mathcal{I}$ drive the system to a state satisfying g through states satisfying f.

Suppose $\mathcal{I} = i_1$. Let us examine the truth of the formulas at the state s_1. The constraint $\mathcal{I} = i_1$ enables the transitions from s_1 to s_3 and s_4. Since g is true at s_4, φ_1 is true at s_1.

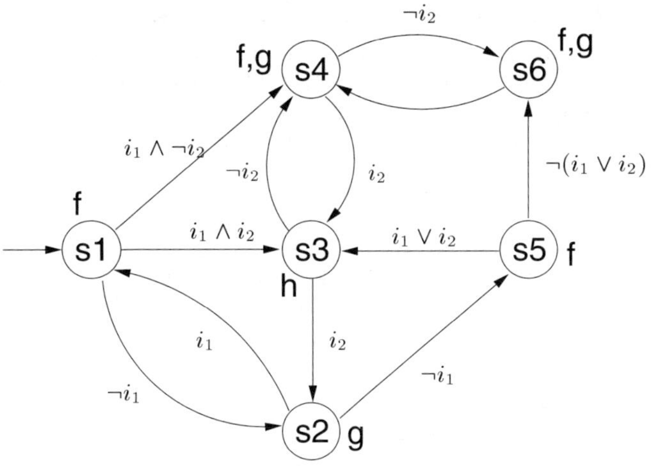

Figure 2: A simple module

On the other hand, since both g and f are false at s_3, φ_2 is false at s_1.

Suppose $\mathcal{I} = \neg i_1$. Now from s_1, only the transition to s_2 is enabled. Since g is true at s_2, both φ_1 and φ_2 are satisfied at s_1. Let us examine the truth of these formulas at s_5. The constraint $\mathcal{I} = \neg i_1$ enables the transition from s_5 to s_6. It also enables the transition from s_5 to s_3. This is because, the transition from s_5 to s_3 is enabled by a collection of three input vectors $\eta_0 = (0,1)$, $\eta_1 = (1,0)$ and $\eta_2 = (1,1)$. Thus the transition shown in the diagram actually consists of three alternative transitions each enabled by a different input vector, but all leading to the same next state s_3. Since η_0 satisfies $\mathcal{I} = \neg i_1$, the transition from s_5 to s_3 is enabled. It is now easy to see that φ_1 is true at s_5, but φ_2 is false at s_5 since g is not true at s_3.

Let us now consider the following Open-RTCTL formulas, where $\mathcal{I} = i_1$:

$$\varphi_3 = E(f\ U^{\mathcal{I}}_{[3,5]}\ h)$$
$$\varphi_4 = E(f\ U^{\mathcal{I}}_{[5,5]}\ h)$$

φ_3 is true at s_1, due to the existence of the path $s_1, s_4, s_6, s_4, s_3, \ldots$, where the state s_3 satisfying h occurs at the 4th cycle, and all previous states satisfy f. On the other hand, φ_4 is false at s_1, since there is no satisfying path from s_1 which reaches s_3 in the 5th cycle.

The syntax of Open-RTCTL allows us to nest formulas with different input constraints. For example, consider the following query:

> *From the state s_1, by applying inputs satisfying the constraint $i_1 \wedge i_2$ are we always able to reach within 4 cycles any state satisfying g where a negative edge of the input i_1 produces a state satisfying f?*

The above query can be expressed in Open-RTCTL as a formula:

$$\varphi = A(true\ U^{\mathcal{I}_1}_{[0,4]}\ g \wedge AX^{\mathcal{I}_2}\ f)$$

where $\mathcal{I}_1 = i_1 \wedge i_2$ and $\mathcal{I}_2 = \neg i_1$. The subformula $g \wedge AX^{\mathcal{I}_2} f$ is true at the states s_2 and s_6. Now, all $\mathcal{I}_1$-consistent paths from s_1 have the prefix $s_1, s_3, s_2, \ldots$, which reaches s_2 in 2 cycles. Hence φ is true at s_1. □

3.1 Symbolic Open-RTCTL verification

In this section we outline a BDD-based symbolic model checking algorithm for Open-RTCTL verification. The next-state function τ is represented as a BDD for the relation, $N(V, I, V')$ where V denotes the present state, I denotes the input vector, and V' denotes the next state.

The verification procedure *Check* takes the Open-RTCTL formula to be checked as its argument and returns an OBDD that represents the states of the system which satisfy the formula. The procedure *Check* recursively handles formulas of the form $EX^{\mathcal{I}}$ f, $AX^{\mathcal{I}}$ f, E[f $U^{\mathcal{I}}_{[a,b]}$ g] and A[f $U^{\mathcal{I}}_{[a,b]}$ g] as follows:

Check($EX^{\mathcal{I}}$ f) = CheckEX($\mathcal{I}$, Check(f))
Check($AX^{\mathcal{I}}$ f) = CheckAX($\mathcal{I}$, Check(f))
Check(E[f $U^{\mathcal{I}}_{[a,b]}$ g]) = CheckEU(a, b, $\mathcal{I}$, Check(f), Check(g))
Check(A[f $U^{\mathcal{I}}_{[a,b]}$ g]) = CheckAU(a, b, $\mathcal{I}$, Check(f), Check(g))

Given the BDD $N(V, I, V')$ for the transition relation, the BDD $C(I)$ for the input constraint $\mathcal{I}$, and the BDDs $f(V')$ and $g(V')$ returned by Check(f) and Check(g) respectively, we have:

CheckEX($\mathcal{I}$, Check(f)) =
  $\exists V' \exists I\, [f(V') \wedge N(V, I, V') \wedge C(I)]$
CheckAX($\mathcal{I}$, Check(f)) =
  $\forall V' \exists I\, [f(V') \wedge N(V, I, V') \wedge C(I)]$
CheckEU(0, 0, $\mathcal{I}$, Check(f), Check(g)) = $g(V')$
CheckEU(0, b, $\mathcal{I}$, Check(f), Check(g)) =
  $g(V') \vee (f(V') \wedge CheckEX(\mathcal{I}, Z(V')))$
  where $Z(V')$ = CheckEU(0, b-1, $\mathcal{I}$, Check(f), Check(g))
CheckEU(a, b, $\mathcal{I}$, Check(f), Check(g)) =
  $g(V') \vee (f(V') \wedge CheckEX(\mathcal{I}, Z(V')))$
  where $Z(V')$ = CheckEU(a-1, b-1, $\mathcal{I}$, Check(f), Check(g))

We compute CheckAU() similarly, except that we use CheckAX() instead of CheckEX(). For the *unbounded until operators*, we use the usual fixpoint computations [3, 5]:

CheckEU(0, ∞, $\mathcal{I}$, Check(f), Check(g)) =
  lfp $Z(V')\, [g(V') \vee (f(V') \wedge CheckEX(\mathcal{I}, Z(V')))]$
CheckAU(0, ∞, $\mathcal{I}$, Check(f), Check(g)) =
  lfp $Z(V')\, [g(V') \vee (f(V') \wedge CheckAX(\mathcal{I}, Z(V')))]$

4. DESIGN EXTRACTION

Extraction of a FSM from a high level language such as Verilog is one of the most difficult tasks in formal verification. In this section we highlight some of the salient features of this part of our tool.

The tool supports a subset of Verilog HDL. It supports single bit register and wire data types with boolean operators defined on them. Also supported are `initial` and event-guarded `always` blocks, procedural assignments, if-then-else conditional statements, and continuous assignments.

Module instantiations and multi-bit data types are not supported in the current implementation. However, it is easy to translate a Verilog description into our supported subset. We have translated many of the modules from the Texas97 benchmark circuits (which are in Verilog) into our supported subset.

The tool performs a static analysis of the given design, during which it identifies the control and data dependencies between inputs, outputs, registers and wire variables. At this stage it detects whether race conditions exist in the circuit.

It is common practice to design modules which are *edge triggered*, that is, actions within the module are triggered by changes in the signal values. These are represented in Verilog by the `@posedge(signal)`, `@negedge(signal)`, and `@(signal)` constructs. In order to detect a change in a variable, the module must remember the value of the variable in the previous *fetch-update* cycle. In our tool, for a variable a which appears within a "@" construct, we create two variables, namely a and a' to represent respectively the current and previous value of the variable.

During the abstraction of a FSM from the given design, we use a cone-of-influence reduction based on the set of control variables that are of interest to us. Currently, this set of variables can be declared by the designer. Alternatively, the tool can extract the set of relevant variables from the set of Open-RTCTL properties that are to be verified.

5. RESULTS

We present experimental results of the tool on modules from the Texas-97 Verification Benchmarks [14]. These benchmarks consist of industrial grade circuit modules specified in Verilog. The verifier was run on a 333 MHz SUN sparc workstation with 128 MB RAM over SunOS 5.8. Our tool uses the CUDD BDD-package [13] for generating the BDDs.

Table 1 shows the results of our tool. The FSM reduction time includes the time required by the design analyzer to parse the circuit, the time required by the abstraction algorithm to prune the circuit, and the time required by CUDD to create the BDDs. The size of the reduced BDDs are shown in the third column. The verification time denotes the time required after extraction by the Open-RTCTL verifier to check the number of properties given in the fourth column. All times refer to CPU time.

It is important to note that since we are able to analyze one module at a time, the BDD sizes are within feasible limits. This supplements the efficiency of the Open-RTCTL verifier in the overall complexity of verification. As a result most of the modules were verified in less than a second, which appears promising considering the size of the original Texas-97 benchmark circuits.

6. TIMING DIAGRAMS

In industrial practice, it is common to represent the timing specification of an interface protocol in the form of a timing diagram. This has prompted researchers in the formal verification community to develop techniques for automated verification of circuits against their timing diagrams [2].

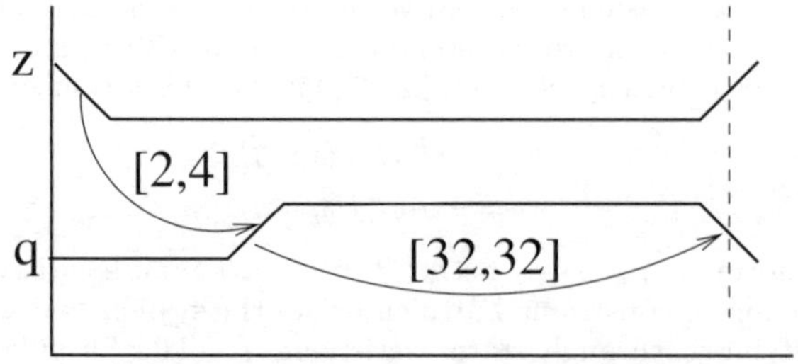

Figure 3: Sample Timing Diagram

We are integrating a technique for translating constraints specified using timing diagrams into Open-RTCTL proper-

Circuit Module	FSM Reduction & BDD Creation time (sec)	BDD size (no of nodes)	No. of Open-RTCTL Queries	Verification time (sec)
PCI Local Bus				
PCI Master	36.50	14786	3	2.30
PCI Target	28.62	34859	3	7.60
MSI Cache Coherence Protocol				
3-proc sys arbiter	0.11	249	3	0.12
2-proc sys arbiter	0.30	170	2	0.60
MPEG System Decoder				
timestamp	0.18	1140	2	0.98
prefixcode	0.05	173	2	0.14
parsepack	0.16	1159	2	0.72
packstart	0.10	324	2	0.37
PI Bus				
slave	1.80	4004	2	1.83

Table 1: Results of Open-RTCTL Verification

ties. Unlike previous work in verification of timing diagrams, we consider open systems. Therefore, some of the waveforms in a timing diagram are inputs to the given module, where as some others are expected outputs of the given module. Our task is to verify the expected output waveform when presented with the given input waveform.

As an example, consider the timing diagram shown in Fig 3. Suppose the given module M has input z and output q as shown in the figure. The protocol starts with `negedge(z)` which is expected to cause a `posedge(q)` between 2 to 4 cycles. Thereafter, exactly after 32 cycles both z and q toggle. Further, by virtue of the start of the protocol, we are given that initially z is high and q is low.

To verify whether a given implementation of M satisfies the timing diagram, we verify the following Open-RTCTL formula at states which serve as possible initial states of the protocol:

$$AX^{\neg z}\ A(\neg q\ U^{\neg z}_{[2,4]}\ (q \wedge A(q\ U^{\neg z}_{[32,32]}\ \neg q)))$$

Note that through this formula, we do not assert that z toggles exactly 32 cycles after `posedge(q)` occurs. For a compositional verification of the whole system, this has to be verified on the module which produces z as its output.

7. CONCLUSIONS

Designers like to eliminate as many errors as possible from the modules they design before integrating them with the other modules in the circuit. The Module Verifier Tool provides a simple way to analyze a single module at a time and allows the designer to express, in a unified way, the property to be verified and the inputs for which it is expected to hold. This can eliminate many bugs prior to simulation or compositional verification. Open-RTCTL is also useful as a specification language for assume guarantee styles of reasoning.

8. ACKNOWLEDGEMENTS

The authors acknowledge the support of Sun Microsystems, USA, for this work. Pallab Dasgupta further acknowledges the support of the Indian National Science Academy. P.P. Chakrabarti further acknowledges the support of Dept. of Sci & Tech, Govt. of India.

9. REFERENCES

[1] Alur, R., and Henzinger, T., Real time logics: Complexity and Expressiveness. *Information and Computation*, **104**, 1, 35-77, 1993.

[2] Amla, N., Emerson, E.A., Kurshan, R.P., and Namjoshi, K., RTDT: A Front-End for Efficient Model Checking of Synchronous Timing Diagrams, In *Proc. of CAV'2001*, 2001.

[3] Burch, J.R., Clarke, E.M., Long, D.E., McMillan, K.L., and Dill, D.L., Symbolic model checking for sequential circuit verification. *IEEE Trans. on CAD*, **13**, 4, 401-424, 1994.

[4] Clarke, E.M., Emerson, E.A., and Sistla, A.P., Automatic verification of finite-state concurrent systems using temporal logic specifications, *ACM Trans. on Prog. Lang. and Sys.*, 8(2):244-263, 1986.

[5] Clarke, E.M., Grumberg, O., and Peled, D.A., *Model Checking*, MIT Press, 2000.

[6] *Cadence Formalcheck Tool*, http://www.cadence.com/datasheets/formalcheck.html

[7] Dasgupta, P., Chakrabarti, P.P., and Chakrabarti, A., Open Computation Tree Logic for Formal Verification of Modules, In *Proc. of ASPDAC/VLSI Design Joint Conference*, Jan 2002, India.

[8] Emerson, E.A., Mok, A.K., Sistla, A.P. and Srinivasan, J., Quantitative temporal reasoning, In *Proc. of CAV'99*, France, 1990.

[9] Grumberg, O., and Long, D.E., Model Checking and Modular Verification, *ACM Trans. on Prog. Lang. and Sys.*, 16:843-872, 1994.

[10] Kupferman, O. and Vardi, M.Y., Module Checking. In *Proc. of CAV'96*, LNCS 1102, 75-86, 1996.

[11] Kupferman, O. and Vardi, M.Y., Module Checking revisited. In *Proc. of CAV'97*, LNCS 1254, 1997.

[12] *The SMV Model Checker*, http://www-cad.eecs.berkeley.edu/~kenmcmil/smv/

[13] Somenzi, F., *CUDD: CU Decision Diagram Package, Release 2.3.0, User's Manual*, Dept. of Electrical and Computer Engg., Univ. of Colorado, Boulder, 1998.

[14] *Texas-97 Verification Benchmarks*, http://www-cad.eecs.berkeley.edu/Respep/Research/vis/texas-97/

[15] VIS: A system for verification and synthesis, The VIS Group, In *Proc. of CAV'96*, LNCS 1102, 1996.

Automated Timing Model Generation

Ajay J. Daga, Loa Mize, Subramanyam Sripada, Chris Wolff, Qiuyang Wu
Synopsys, Inc.
2025 NW Cornelius Pass Rd., Hillsboro, OR 97124
503-547-6900
{ajdaga, loa, ssubram, wolff, qwu}@synopsys.com

ABSTRACT
The automated generation of timing models from gate-level netlists facilitates IP reuse and dramatically improves chip-level STA runtime in a hierarchical design flow. In this paper we discuss two different approaches to model generation, the design flows they lend themselves to and results from the application of these model generation solutions to large customer designs.

Categories and Subject Descriptors
J.6: ComputerApplication.CAD.

General Terms
Design, Performance, Algorithm, Verification

Keywords
Static Timing Analysis, Model Generation. EDA.

1. INTRODUCTION
System-on-a-chip (SoC) design is performed in an inherently hierarchical manner. Intellectual property (IP) is integrated on-chip with new blocks that are often designed by geographically dispersed teams. Timing models are important in such a design flow because they are a compact means of exchanging the interface timing information for blocks. The automatic generation of timing models from the gate-level netlist for a design is key to sustaining a hierarchical SoC design flow for two reasons:

1) To improve capacity and runtime of static timing analysis (STA) in a hierarchical design flow.

   Timing models are compact and accurate representations of the timing characteristics of a block. The use of timing models in place of the full gate-level netlist for a block is key to improving performance and capacity associated with chip-level STA. This is particularly the case for complex SoC designs whose gate counts exceed 5M gates.

2) To disseminate interface timing information for intellectual property (IP) blocks while hiding implementation details.

   From an IP reuse standpoint, timing models help protect IP by limiting the visibility that end users have to the implementation details of IP blocks, while still preserving information required to verify IP integration on a SoC.

DAC 2002, June 10-14, 2002, New Orleans, Louisiana, USA.

Automatic model generation is possible in PrimeTime [1] using two approaches: extracted timing models (ETMs) and interface logic models (ILMs). ETMs abstract the interface behavior of a design and replace the original netlist with a library cell that contains timing arcs between pins. These setup, hold and delay arcs are a function of input transition time and output load - for this reason the model is context independent. ILMs discard the netlist, constraints and back-annotation associated with internal register-to-register paths on a design. Rather than abstract, ILMs preserve only, but in entirety, the portion of a design that impacts and is impacted by the external world. Like ETMs, ILMs are also context independent.

For design styles where the transitive fanout of input ports and the transitive fanin to output ports are registered, ILMs offer significant runtime and capacity improvements without sacrificing accuracy. For this reason, ILMs are well suited for hierarchical timing signoff. ETMs offer significant performance improvements over the original netlist, but because circuit timing is abstracted there is some loss of accuracy. For this reason, ETMs are well suited during the early stages of a design flow, where fast exploration of design alternatives is important. Also, because they hide all implementation details, ETMs lend themselves to IP reuse scenarios. Finally, because the generated model is a library cell containing timing arcs in industry-wide standards such as lib [2] and Stamp [3], ETMs are portable and easy to use across EDA tools.

In the following sections we discuss the ILM and ETM model generation approaches in further detail. We then discuss issues related to the validation of a model against the original gate-level netlist. Finally, we present results on the application of ILMs and ETMs to ten large customer designs.

Prior work in the area of automated timing model generation has focussed on the automatic derivation of block clocking requirements [4] and on the characterization of interface delays without traversing false paths using user-specified mode information [5]. For mainstream IP reuse and hierarchical STA flows, our experience indicates that customers already have a good understanding of the clocking requirements for a block and so the automatic derivation of this information is not crucial. The main contribution of this paper is its comprehensive study, using large customer designs (some exceeding 700K gates), of model generation runtime, model accuracy and model performance for two fundamentally different model generation approaches.

2. INTERFACE LOGIC MODELS
ILMs embody a structural approach to model generation, where the original gate-level netlist is replaced by another gate-level netlist that contains only the interface logic of the original netlist. Interface logic contains all circuitry leading from I/O ports to edge-triggered registers called interface registers. The clock tree

leading to interface registers is preserved in an ILM. Logic that is only contained in register-to-register paths on a block is not in an ILM.

2.1 ILM Generation

ILM generation requires identifying the interface logic on a design and then writing out the netlist, constraints and back-annotations for interface logic in an appropriate format.

The following is identified as belonging to interface logic:

1) All cells contained in timing paths leading from input ports to either an edge-triggered register or output ports at which these paths terminate. If a transparent latch is encountered then it is treated as combinational logic and path tracing continues through the latch until an edge-triggered register is encountered.
2) All cells contained in timing paths leading to output ports from either edge-triggered registers or input ports at which these paths originate. Transparent latches are handled as described previously.
3) The clock tree that drives interface registers is a part of interface logic. This includes any registers in the clock tree. Clock-gating circuitry that is driven by external ports is a part of interface logic, while clock-gating circuitry that is driven by registers on the block is not a part of interface logic.

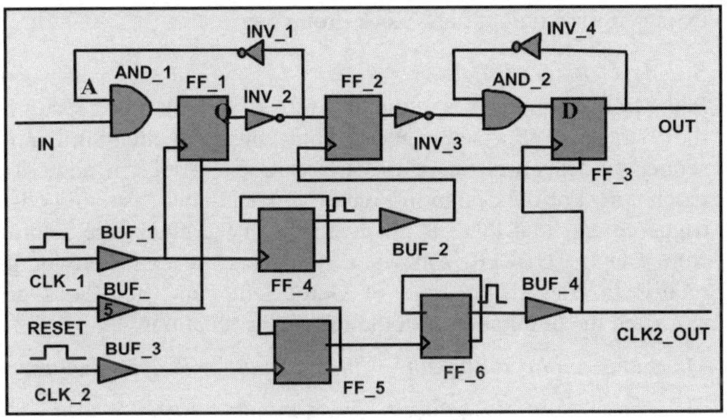

Figure1: Example gate-level netlist.

Figure 1 shows an example gate-level netlist for a block (ports and pins on which clocks are defined are marked ⎍) and Figure 2 its resulting interface logic. Note how cells `INV_*`, `FF_2`, and `AND_2` are not a part of the ILM. These cells are only instantiated in register-to-register-paths. Note how the cell `AND_1` is a part of the ILM though it does have an input pin, `A`, that belongs to a register-to-register path. The input pin `A` on `AND_1` is unconnected in the ILM. Similarly, the output pin `Q` on `FF_1` and the input pin `D` on `FF_3` are also unconnected in the ILM. `FF_1` and `FF_3` are interface registers. Notice how the clock tree for these interface registers is a part of the ILM. Registers `FF_5` and `FF_6` that belong to this clock tree are also a part of the ILM, but notice that the feedback paths on these registers are not a part of the ILM. Also notice how cells `FF_4` and `BUF_2` that belong to the clock tree leading to an internal register are not a part of the ILM.

When identifying interface logic it is necessary to ignore the fanout of input ports connected to global chip-level signals that feed internal registers on a block. Examples of such ports are clocks, reset and scan_enable. In Figure 1, the fanout of ports `CLK_1`, `CLK_2` and `RESET` are ignored because otherwise all registers on a design will be instantiated in the ILM. The clock tree leading to interface registers is identified as belonging to interface logic by processing the fanin of interface register clock pins.

For transparent latch based designs, it is necessary to specify the

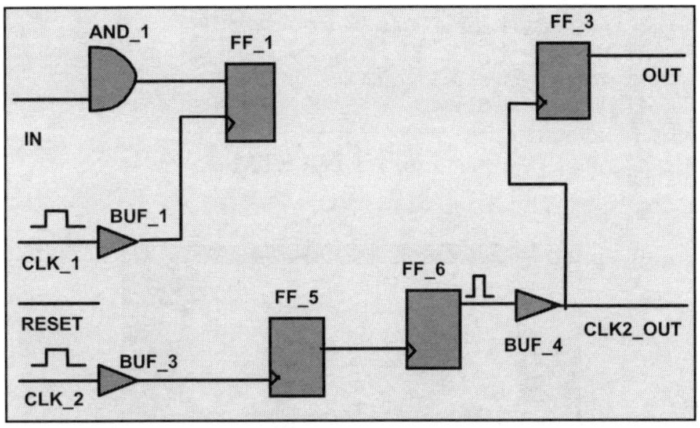

Figure2: ILM for example gate-level netlist.

maximum number of levels of latch borrowing in I/O paths. By default transparent latches are viewed much like combinational logic cells and fanout or fanin traversal continues through a latch until an edge-triggered flip-flop is encountered. When the number of levels of latch borrowing is specified, fanout or fanin traversal stops at a latch whose level is greater than the specified borrowing level. Specifying one level of latch borrowing, for example, results in the first two latches encountered in an I/O path being a part of the ILM. The first latch can borrow, while the second functions as an edge-triggered register that does not borrow.

When delay calculation is performed using the ILM it is necessary to maintain all pins connected to interface logic nets. This ensures that the pin and wire capacitance for interface logic nets is accurate. The inclusion of all pins on interface logic nets for the ILM in Figure 2 results in the addition of pins `FF_4/CP` and `INV_4/A`.

In SDF flows where cell and net delays are annotated to an ILM it is not necessary to maintain accurate pin capacitance information for a net. Also, if clock tree synthesis has not been performed then non-interface-logic pins are not maintained in the ILM for ideal clock nets.

3. EXTRACTED TIMING MODELS

Extracted timing models differ from ILMs in that the interface logic for a block is replaced by context-independent timing relationships between pins on a library cell. In this section, we discuss the different types of ETMs and issues related to their automatic generation.

3.1 Extracted Timing Model Types

There are two types of extracted models, library cell and wrapper with core cell, illustrated in Figure 3. The library cell ETM replaces an entire design with a single library cell. The extracted library cell contains timing arcs between external pins. Internal pins are introduced only when there are clocks defined on internal pins of the design. The library cell is generated in industry standard formats such as lib and Stamp. In addition to the extracted library cell a constraint file is generated that defines exceptions (false paths and multi-cycle paths) that need to be applied to the library cell when it is instantiated at a higher level of abstraction.

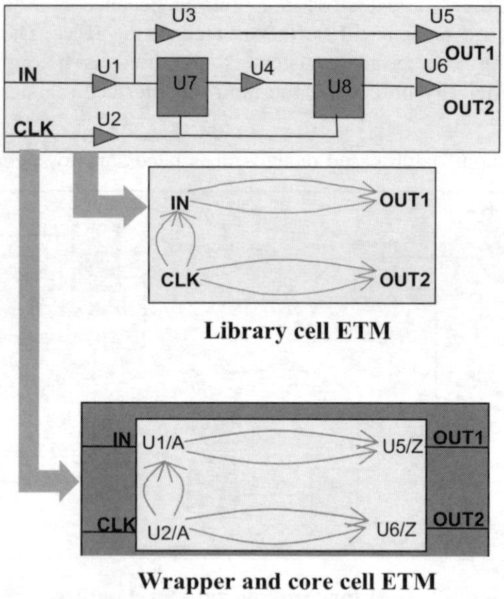

Figure3: Two types of ETMs.

With wrapper and core cell, the extracted model preserves boundary nets and parasitics on boundary nets. This model is more accurate than the library cell because it allows boundary parasitics on blocks to be stitched together with chip-level parasitics when the model is instantiated at a higher level of abstraction. This, in turn, allows inter-block net delays to be calculated with high accuracy by taking into account the complete parasitics for a chip-level net. The wrapper and core cell correspond to a view of the original design in terms of a wrapper design that instantiates a core block, as shown in Figure 3. The extracted core cell is a pin-to-pin replacement of the core block and is instantiated in the extracted wrapper design. Parasitics information is written out in industry standard SPEF [6] format. A constraint file that maintains exceptions that need to be applied to the extracted model is written out just as for a library cell model.

A library cell model is preferred in IP reuse scenarios and for import into third-party simulation and place & route tools. This is primarily because the library cell model is a port-to-port replacement for the original design and is written out in industry standard formats. The wrapper and core cell model is useful for import into tools that can handle the additional level of hierarchy introduced by the wrapper design. The benefit of being able to handle such a model is the additional accuracy it provides.

3.2 Extracted Timing Arcs

The objective of model extraction is to establish context-independent timing relationships between input and output ports on a design. These timing relationships are represented by timing arcs between pins on a library cell. Extracted timing arcs can be classified into three distinct types: combinational delay arcs, sequential delay arcs, constraint arcs.

3.2.1 Combinational Delay Arcs

Combinational paths from an input port to an output port result in the extraction of either two or four combinational delay arcs. These arcs capture the minimum and maximum timing delays for inverting and non-inverting paths from an input port to an output port. Each extracted combinational delay arc has associated with it timing sense information that indicates whether the arc is inverting or non-inverting and whether it is representative of minimum or maximum delay. Minimum and maximum delay arcs represent lower and upper bounds on circuit delay for a single operating condition, and so are indicative of delay variations resulting from circuit topology rather than from process variations. For the design in Figure 3, a pair of non-inverting combinational delay arcs are extracted between IN and OUT1 (there are no inverting paths between IN and OUT1).

3.2.2 Sequential Delay Arcs

Paths to an output port that start at interface registers that are similarly clocked, i.e. by the same clock and by the same clock edge, result in the extraction of a pair of minimum and maximum sequential delay arcs. A set of similarly clocked interface registers is referred to as a clock group. Extracted sequential arcs are relative to a clock and a specific edge of a clock. The clock edge that a sequential delay arc is relative to is a function of whether interface registers in a clock group are rising or falling edge triggered and whether there is inversion in the path leading from the clock definition point to register clock pins. For the design in Figure 3, given that U8 is rising-edge triggered and there is no inversion in the clock tree from CLK to U8/CLK, a pair of sequential delay arcs are extracted from the rising edge of CLK to OUT2. Extracting sequential delay arcs requires summing clock and data path delays for each interface register in a clock group. While minimum and maximum delay arcs are computed for paths from each interface register in a clock group to an output port, only a single worst-case minimum and maximum delay arc is extracted for a output port, clock group pair.

3.2.3 Constraint Arcs

Paths from an input port to the data pins of a clock group result in the extraction of a setup and a hold constraint arc. Similar to sequential delay arcs, these arcs are relative to a specific edge of a clock port. For the design in Figure 3, given that U7 is rising-edge triggered and that there is no inversion in the clock tree leading from CLK to U7/CLK, setup and hold arcs are extracted for IN relative to the rising edge of CLK. Setup and hold arcs are extracted in compliance with the following equations:

1) *Setup = Max data path - Min clock path + Register setup*
2) *Hold = Max clock path - Min data path + Register hold*

As with sequential delay arcs, though setup and hold times are computed for paths from an input port to each interface register in a clock group, only a single worst-case setup and hold arc is extracted for a input port, clock group pair.

3.3 Extracted Timing Tables

Associated with each timing arc are tables that define timing values as a function of input transition time and output load. By extracting timing tables, rather than scalar timing values, the generated model is context independent and applicable for varying transition times and output loads. There could be different critical paths for different transition times and the tables are merged appropriately to handle this.

3.3.1 Delay Tables

Associated with each extracted delay arc (combinational and sequential) are a pair of two-dimensional delay tables and a pair of two-dimensional transition tables. Rise and fall delay tables define timing delays for rising and falling transitions at an output port as a function of input transition time and output load. Rise and fall transition tables define rising and falling transition times at an output port as a function of input transition time and output load. Both the numbers of points in extracted delay and transition tables and the input transition and output load range over which delay is characterized is controlled by a user.

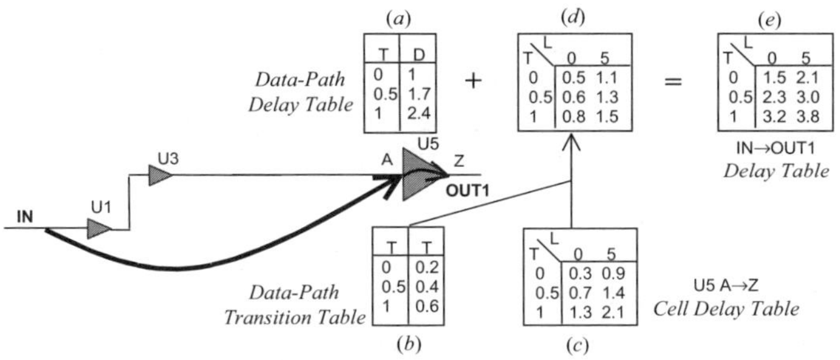

Figure 4: Extracting delay tables.

Consider extraction of the rise delay table for the non-inverting arc from IN to OUT1 in Figure 4 (assume that boundary net delays are zero). A two-dimensional cell delay table (*c*) for U5 defines the rise delay at output pin Z relative to pin A. Model extraction establishes the rise-to-rise (*rr*) transition table (*b*) from IN to U5/A. This table captures transition times at U5/A as a function of transition times on IN. The transition table at U5/A is used to index into the cell rise delay table for U5 to yield an intermediate table (*d*) that describes the delay contributions through U5 as a function of the load on OUT1 and the transition time on IN. This intermediate table is summed with the delay table (*a*) from IN to U5/A that captures the delay at U5/A as a function of transition times on IN. This yields the two-dimensional rise delay table (*e*) for the arc from IN to OUT1.

3.3.2 Constraint Tables

Associated with each extracted constraint arc is a pair of constraint tables. Rise and fall constraint tables define setup or hold times, depending on the type of arc, for rising and falling transitions at an input port. These tables are a function of input transition time and clock transition time.

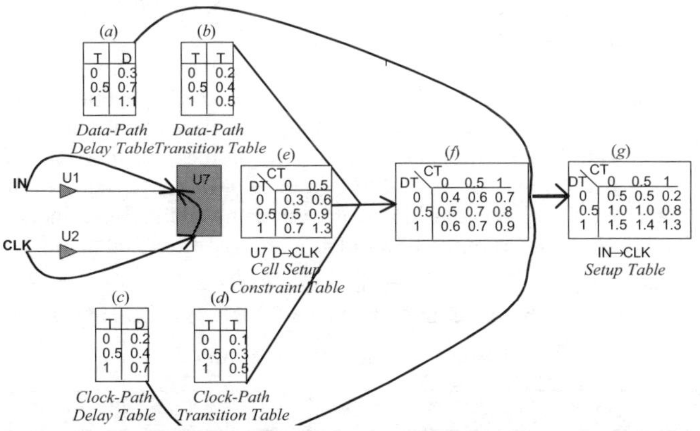

Figure 5. Extracting constraint tables

Consider the computation of the fall constraint table for the setup arc between IN and the rising edge of CLK shown in Figure 5. This computation first requires extracting the transition tables associated with the maximum fall-to-fall (*ff*) path from IN to U7/D (*b*) and the minimum *rr* clock-path from CLK to U7/CLK (*d*). These two tables are used to index into the two-dimensional fall constraint table (*e*) for the setup arc from D to CLK in the library cell for U7. The resulting intermediate table (*f*) defines setup times on register U7 as a function of transition times on IN and CLK. Next, this intermediate table is combined with the delay tables for the maximum *ff* path from IN to U7/D (*a*) and the minimum *rr* path from CLK to U7/CLK (*c*). This is done in accordance with equations in section 3.2.3 to yield the desired fall constraint table for the setup arc from IN to the rising edge of CLK.

3.4 Exception Handling

Exceptions such as false paths and multi-cycle paths (MCPs) are handled during model extraction. Exceptions that only refer to the clocks on a design are simply written out to the generated constraint file and must be reapplied to the extracted model when it is instantiated in a design. False paths that refer to pins or ports on a design are accounted for during model extraction.

Multi-cycle exceptions that are found to apply to all paths leading to or from a port are written out to the constraint file and must be reapplied to the extracted model. The MCP written to the constraint file only refers to pins and clocks on the extracted model and not to internal design pins. For example if an MCP were defined through U1/Z in Figure 6, because it applies to all paths from IN, it would be written out to the constraint file and apply to all paths through port IN that are clocked by CLK.

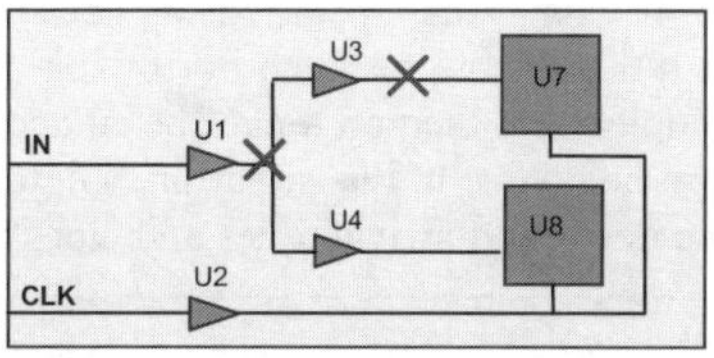

Figure 6: MCP handling during model extraction.

When an MCP only applies to some, but not all, paths to or from a port then the impact of this exception is taken into account when extracting a timing arc. For example, in Figure 6, assume a MCP defined two-cycle clocking through U3/Z. Assume a clock period of 10ns, a data-path delay from IN to U7/D of 17ns and a data-path delay from IN to U8/D of 6ns. Also, assume that the clock-path delay from CLK to the register clock pins is zero and the setup times for both registers is also zero. In this case, a setup arc of 6ns is extracted from IN to CLK as a result of the path from IN to U8/D. A setup arc of 7 ns is extracted from IN to CLK as a result of the path from IN to U7/D after reducing the data-path delay to U7/D (17ns) by the clock cycle shift (10ns). As 7ns is more worst-case than 6ns, a single-cycle setup arc of 7ns is extracted between IN and CLK. So, when a MCP does not apply to all paths to or from a port, then clock periods are taken into account to generate a worst-case arc appropriate for single-cycle clocking of the port.

3.5 Transparent Latch Handling

When the interface logic for a design contains transparent latches, these need to be handled appropriately. Transparent latches allow propagation from the D pin to the Q pin (referred to as a borrowing latch) if the arrival time on the D pin is later than the open edge of its clock. Transparent latches do not permit propagation from the D pin to the Q pin (referred to as a non-borrowing latch) if the arrival time on the D pin occurs before the open edge of its clock. The open edge of a positive level sensitive latch is the rising edge of CLK.

Transparent latches are handled during model extraction by establishing, based on arrival times defined on input ports and specified clock periods, whether a path from an input port to a latch causes borrowing or not. If the path causes borrowing then path traversal continues through the latch until either a non-borrowing latch or edge-triggered flip-flop is encountered. If the path does not case cause borrowing then path tracing stops and a setup arc is extracted relative to the open-edge of the latch. The borrowing behavior of latches does not impact the extracted hold arc. Hold arcs are always extracted relative to the close edge of the first interface register (latch or flip-flop) encountered while tracing paths from input ports.

Latch borrowing can result in paths being traced through latches from an input port to an output port. In this case, a delay arc is extracted from the input port to output port. When tracing paths from interface registers to output ports, transparent latches are handled by tracing through borrowing latches and generating a sequential delay arc from a clock port to an output port. If a non-borrowing latch is encountered then no arc is extracted because this represents an internal register-to-register path.

Any clock cycle shift resulting from borrowing through several levels of latches is not accounted for in the extracted timing arcs and must be accounted for by applying MCPs to the model. Also, the model will match netlist timing as long as the borrowing behavior at the time of model generation continues to hold when the model is used. If input arrival times or clock periods change sufficiently from what was specified at the time of model generation then it is possible for latch borrowing behavior to also change. The model would then no longer be an accurate reflection of original design behavior. In this situation, a new model must be generated for the new input arrival times and clock periods.

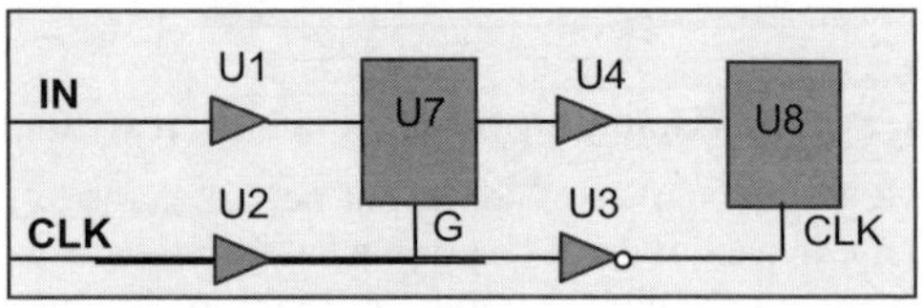

Figure 7: Latch handling during model extraction.

For an illustration of how transparent latches are handled during model extraction consider the design in Figure 7. If the positive-level sensitive latch U7 does not borrow, then a setup time relative to the rising edge of CLK is extracted. The setup time for this arc is influenced by the data-path delay from IN to U7/D and the clock-path delay from CLK to U7/G. If U7 does borrow, then path tracing continues through U7 to U8 (a rising-edge triggered flip-flop), and a setup arc is extracted relative to the falling edge of CLK. This setup arc is impacted by the data-path delay from IN, through U7/D, to U8/D, by the clock-path delay from CLK to U8/CLK and by the setup time between U8/D and U8/CLK.

4. MODEL VALIDATION

Once a model has been generated it is necessary to compare it to the original design and establish that discrepancies between the model and design are within acceptable bounds. This is referred to as model validation and is performed for a given user-specified context (input arrival times and transition times, output loads and clock periods). Model validation takes place by loading the design and model standalone, capturing relevant interface timing parameters for both, and then comparing these parameters. For completeness it is often desirable to validate a model and design at two different contexts; a "min" environment (small loads, fast transition times) and a "max" environment (large loads, slow transition times).

Model validation ensures that timing arc values are within user specified tolerances. This is done by comparing the absolute discrepancy between worst-case timing slacks for all unique input-to-clock, clock-to-output and input-to-output paths on the model and design. This can also be accomplished through an absolute or percentage comparison of the discrepancy between interface timing arc values. Percentage comparison of interface timing slacks is inappropriate because timing slack values are influenced by the specified context (input arrival times, clock periods, etc.). On the other hand, absolute discrepancies between interface timing slacks for the same context applied to model and design can be completely attributed to a discrepancy between model and design. An attractive aspect of comparing interface timing arc values is that such comparisons are independent of some aspects of the user-specified context such as input arrival times and clock periods.

Model validation also ensures that the capacitance values seen at input ports and the transition times at output ports match. This is necessary to ensure that in addition to correctly capturing interface delays and constraints the model presents its external environment with the same input load and output drive characteristics as the original design.

Finally, model validation ensures that design rules such as max transition time on input ports and max load for output ports are the same for both model and design. This ensures that the same design rule violations are reported for model and design.

Once the timing parameters for a model have been established as falling within user-specified tolerances of the original design the model is ready for use in place of the original design.

5. EXPERIMENTAL RESULTS

In this section, we present results from the application of ILMs and ETMs to ten customer designs. The objective was to study model generation runtimes, the performance improvement provided by models in comparison to the original design and the accuracy of the models in comparison to the original design. For all designs a wrapper and core cell ETM was generated. The calculation of delays and critical paths is referred to as "update timing".

Information on the ten designs is shown in Table 1. Six of the designs had RC back-annotation, one had SDF and the remaining three designs had no back-annotation. Some of the designs had transparent latches in the interface logic while others did not. The number of cells on the designs varied from 20K to 260K - this corresponds to a gate count ranging from about 60K gates to 800K gates. The minimum clock periods on these designs ranged from 5ns to 36ns, or about 25MHz to 200MHz. The number of interface timing relationships that were compared between model and design, as part of model validation, ranged from 200 to 6000.

Chart 1 shows model generation runtimes (in seconds) for the ten designs. As a baseline, the time taken to update the timing of the

Table 1: Design information.

	1	2	3	4	5	6	7	8	9	10
# Cells	29277	26540	35254	30449	137557	43850	94925	263490	23441	55499
# Latches	240	0	12	3	3807	2225	0	4565	557	0
Back-annotation	RC	RC	RC	RC	RC	RC	-	SDF	-	-
Clock Period (ns)	8.3	8	16	36	26	6	5	6.9	6.66	6
Timing Relns. Verified	356	204	3489	3206	2006	3544	3084	1646	6114	5392

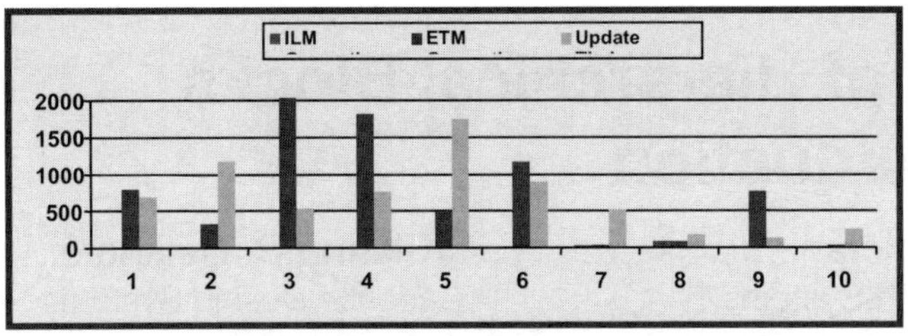

Chart 1: Model generation runtimes.

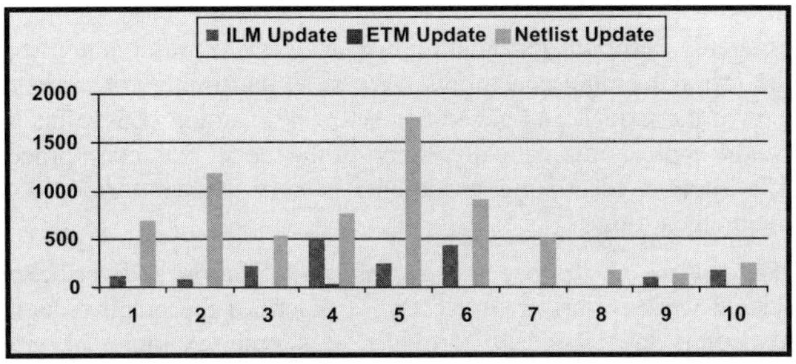

Chart 2. Performance improvements with ETMs and ILMs

original design is shown. As may be observed, ILM generation times are insignificant because of the simplicity of the task - simply discard register-to-register logic, constraints and back-annotation and write out what remains. ETM generation times are significant when a design has RC back-annotation because of the effort involved in characterizing circuit delay for multiple input transition times. In some cases ETM generation times were about 7x update timing times, while in others ETM generation was even faster than update timing. On average ETM generation takes about 3x longer than update timing.

Chart 2 compares update timing runtimes on the original design, ILM and ETM. Across the board, update timing on ETMs only took a few seconds, resulting in performance improvements that ranged from 20x to 1200x when compared to update timing on the original design. ILM performance improvements ranged from about 1.5x to 14x. On average, ETM performance improvements were about 250x relative to the original design, while ILM performance improvements were about 5x. This is not surprising given that ILMs preserve all interface logic while ETMs abstract interface logic into timing arcs. The large model generation times spent characterizing design timing has its payoff when the ETM is used in place of the original design.

Charts 3 and 4 measure ETM and ILM accuracy, respectively. These charts show the percentage of verified interface timing relationships that had discrepancies between netlist and model of less than 10ps and between 10 and 100ps. Across all designs ILMs tend to match the original design within 10ps; 98.4% of all interface timing relationships on an ILM matched the netlist within 10ps. This is not surprising given that ILMs preserve interface logic without modification.

On designs with SDF back-annotation, or no back-annotation, ETMs also had greater than 90% of interface timing relationships with a discrepancy of less than 10ps. For designs with RC back-annotation, however, ETMs tend to be within 100ps of the original design. Overall, across all ten designs, 99.6% of interface timing slacks for ETMs were within 100ps of the original design. For clock speeds of 200MHz, 100ps error is 2% of the clock cycle

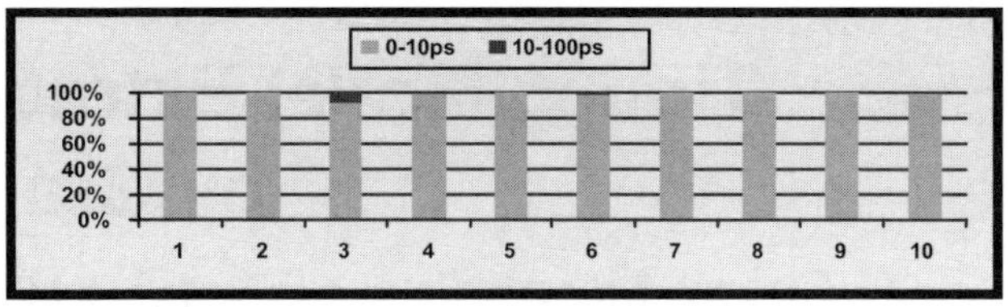

Chart 3: ILM accuracy.

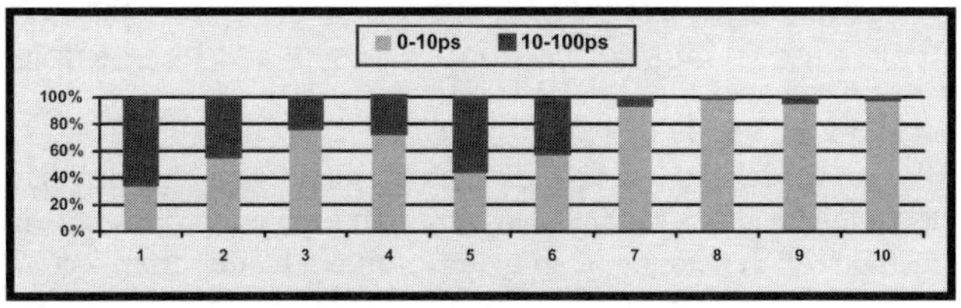

Chart 4: ETM accuracy.

and is insignificant. It is equally likely for model interface relationships to be optimistic or pessimistic when compared to the original design.

6. SUMMARY

In this paper we have presented two approaches to automated timing model generation from gate-level netlists: interface logic models and extracted timing models. Interface logic models discard register-to-register paths on a design but preserve all interface logic without modification. Extracted timing models abstract interface logic into context-independent timing arcs.

Experimental results show that interface logic models provide 5x runtime improvement while preserving original design timing with no accuracy loss. This makes interface logic models well suited for hierarchical timing signoff. Extracted timing models provide greater than 100x runtime improvements with high accuracy. This makes extracted models well suited for IP reuse scenarios where it is important to hide circuit implementation details and early in the design flow where fast design exploration is important. Our experience indicates that different model generation solutions will continue to be appropriate for different design scenarios.

REFERENCES

[1] Synopsys, "PrimeTime: Synopsys Static Timing Solution", Technical White Paper, July 1999.

[2] Synopsys, "Library Compiler Reference Manual", Version 2001.08, August 2001.

[3] Synopsys, "PrimeTime Modeling User Guide", Version 2001.08, August 2001.

[4] S.V. Venkatesh et al., "Timing Abstraction of Intellectual Property Blocks", CICC'97, May 1997.

[5] H. Yalcin et al., "An Advanced Timing Characterization Method Using Mode Dependency", DAC'01, June 2001.

[6] IEEE Std 1481-1999, "IEEE Standard for Integrated Circuit Delay and Power Calculation System", June 1999.

Timing Model Extraction of Hierarchical Blocks by Graph Reduction

Cho W. Moon
Cadence Design Systems
15015 Avenue of Science
San Diego, CA 92128, USA
cmoon@cadence.com

Harish Kriplani
Cadence Design Systems
555 River Oaks Parkway
San Jose, CA 95134, USA
kriplani@cadence.com

Krishna P. Belkhale
Cadence Design Systems
555 River Oaks Parkway
San Jose, CA 95134, USA
belkhale@cadence.com

ABSTRACT

Timing model extractor builds a timing model of a digital circuit for use with a static timing analyzer. This paper proposes a novel method of generating a gray box timing model from gate-level netlist by reducing a timing graph. Previous methods of generating timing models sacrificed accuracy and/or did not scale well with design size. The proposed method is simple, yet it provides model accuracy including arbitrary levels of latch time borrowing and capability to support timing constraints that span multiple blocks. Also, cpu and memory resources required to generate the model scale well with size of the circuit. The generated model can provide a capacity improvement in timing verification by more than two orders of magnitude.

Categories and Subject Descriptor

T1.3 [**Design Tools Track**]: Static timing analysis and timing verification

General Terms: Algorithms

1. INTRODUCTION

Timing extraction or block characterization refers to the process of creating a timing model of a digital circuit for use with a static timing analyzer. Timing extraction plays an important role in hierarchical top-down flow and bottom-up IP authoring flow by reducing the complexity of timing verification and by providing a level of abstraction which hides the implementation details of IP blocks.

Three most desired features in timing extraction are accuracy, efficiency, and usability. The model must preserve the timing behavior of the original circuit and produce accurate results including correct transparent latch behaviors and timing violations. The model also needs to be efficient in terms of the resources needed to generate the model and in terms of the final model size. Finally, the model must be easy to use with existing static timing analyzers. This includes easy model instantiation, easy-to-follow timing reports and capability to allow original constraints to remain valid after a block is replaced by the model.

In this paper, we present a simple method to generate a timing model by reducing the timing graph. The method can generate a model that is accurate within a specified tolerance and is very efficient. Also, our formulation makes it easy to preserve and apply the original timing constraints. We view the timing constraints as part of the model, and generate a set of new timing constraints that can be applied automatically as part of the model extraction process. The support for timing constraints is very critical for top-down hierarchical flows.

The outline of the paper is as follows. Section 2 describes the related work in this area. Section 3 describes the graph reduction algorithm in detail and provides a simple example of graph reduction in action. Model size reduction by using anchor points is described in section 4, followed by some experimental results in section 5. Section 6 concludes with a summary and future work.

2. RELATED WORK

The timing model can be classified into two types: black box model and gray box model. Black box models have no internal visibility into the block: all the timing information relates to the pins at the boundary of the block. Gray box models, on the other hand, have internal pins that can have delay arcs and check arcs associated with them.

2.1 Black Box Models

One of the well known black box model generators comes from Pearl[1]. Users supply a set of input slew values and output load values and the tool performs path tracing to determine all port-to-port path delays and relevant timing checks. Although black box models have been used widely for many years, they suffer from the following drawbacks:

1. The model size can be larger than the original timing graph size. Figure 1 shows that the removal of internal pin can lead to more delay arcs (from 6 to 9) and a larger model.

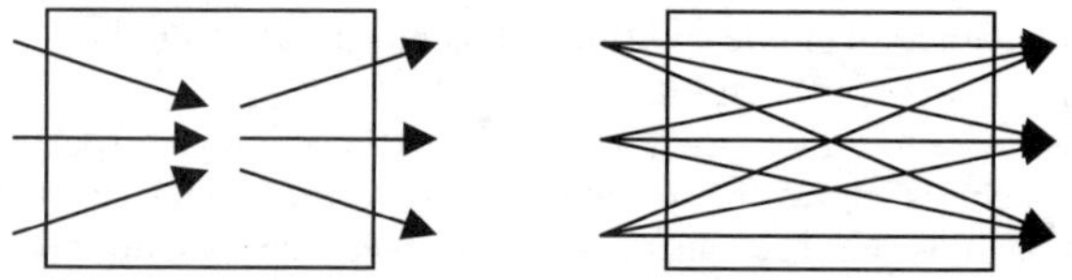

Original timing graph w/ 6 arcs   Black box model w/ 9 arcs

Figure 1: Black box models can be larger than original netlist

2. Only limited latch behavior can be modeled. The model can capture the latch time borrowing behavior of the original netlist for some given clock waveforms. If the clock waveforms change after the model is extracted, the model becomes invalid.
3. Support for assertions is limited, even for those assertions that are fully contained in the block. Only assertions that originate from and terminate at boundary ports can be fully supported. For example, multi cycle paths that do not originate from or

terminate at boundary ports cannot be supported. Also, assertions that span multiple blocks cannot be supported. For example, consider a multi cycle path that originates from block A and terminates in the middle of block B. The black box model for block B cannot support this multi cycle path assertion.

Clock characterization work of [2] addressed some of the drawbacks of the black box model by capturing register/latch to register/latch constraints in a clock feasibility graph. The resulting model consists of a black box for port-to-port path delays plus timing checks and a feasibility region in terms of clock pulse widths and clock edge separation. A separate checker external to the standard timing analysis tool is needed to verify if clock waveforms lie within the feasibility region. This extension allows for limited context insensitivity to clock waveforms in modeling latch behavior. Latch model is limited because increasing the level of transparency incurs some runtime penalty. Allowing arbitrary levels of transparency becomes prohibitively expensive. Also, since the basic underlying model is a black box, there is no guarantee that the model size is smaller than the original timing graph size. The work did not address the support for assertions.

There is a recent work [3] that uses symbolic simulation to generate black box models. This work is targeted mainly for transistor-level circuits where identifying registers or latches can be difficult. The model generation begins from a set of valid input arrival ranges. Simulation is used to grow the given ranges as much as possible in all input directions. The final input arrival ranges are mapped to setup/hold checks relative to some clock signals. This method does not generate port-to-port path delays and does not consider the dependency on input slews or output loads nor support for assertions.

2.2 Gray Box Models

There is not much published work that can deal with arbitrary levels of transparency in latches using gray box model. Latches can be transformed to registers or combinational gates but such transformations lead to models that are too conservative and do not allow for time borrowing. There is a latch path compression work [4] that collapses latch paths instead of individual latches. The extent of compression is controlled by specifying the desired level of latch transparency. Although it has been successfully used for many latch-based designs, the method cannot guarantee a reduction in model size. In some cases, the model size actually increases after latch path compression[4]. The method does not scale well with the number of latch paths or with the level of latch transparency.

3. TIMING GRAPH REDUCTION

We propose a novel method for generating a timing model by reducing the original timing graph. Timing graph is reduced by iteratively removing pins and merging arcs until no further change is possible. This is similar to the manipulation of timing graph in the context of static circuit optimization [5] but has different objectives and cost functions. The main advantages of this approach are

- it is simple
- it guarantees a reduction in model size
- it allows arbitrary levels of latch time borrowing and scales well with design size
- it allows the original assertions to remain valid after a block is replaced by the model

Latch time borrowing and support for the application of original assertions are accomplished by retaining some internal pins. Latch behavior is preserved by retaining all latch input pins and the corresponding latch output pins. Since all the latches are essentially retained, this allows for arbitrary levels of latch time borrowing as in the original netlist. We found that the number of these internal latch pins is not a limiting factor in model generation time or in model size, even for very large industrial latch-based designs. Also, all internal pins associated with assertions are retained. This enables the assertions to remain valid after a block is replaced by the extracted model in the original context, including assertions that are not completely contained in the block.

The outline of this section is as follows. Timing graph will be defined first. Then models for combinational as well as sequential circuits will be described. The last section discusses the reduction algorithm and gives an example of graph reduction on a simple example.

3.1 Timing Graph

Timing graph G is a three-tuple $G = (P, D, C)$, where P is a set of pins, D a set of delay arcs and C a set of check arcs. Delay arcs and check arcs originate from and terminate at pins: $D \subseteq (P \times P)$, $C \subseteq (P \times P)$. Associated with each delay (check) arc is a transition matrix which defines valid transitions (rising or falling) between the source (signal) pin and the sink (reference) pin. Each check arc has a type such as setup, hold, recovery, removal, etc.

A timing graph representation of a simple D-type register is shown in Fig. 2. Delay arcs are represented as solid arrows and check arcs as dotted arrows. The transition matrix for CLK $\rightarrow$ Q delay arc looks like

CLK	Q	Transition
Rising	Rising	True
Rising	Falling	True
Falling	Rising	False
Falling	Falling	False

Delay (check) values are associated with delay (check) arcs. Delay or check values can be linear functions, lookup tables or delay equations.

3.2 Combinational Models

For the sake of exposition, timing models can be divided into two parts:

- Combinational part deals with interaction among delay arcs and are captured in the combinational model.

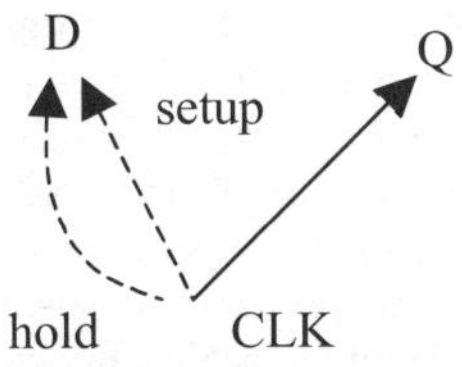

Figure 2: Timing graph for a D-type register

- Sequential part deals with interaction between delay arcs and check arcs. Such interactions are captured in the sequential model.

The two main operations in the combinational model that reduce the timing graph are

- s-merge (serial merge) and
- p-merge (parallel merge).

s-merge takes two delay arcs (*d1* and *d2*) in series and creates a new arc (*d3*) from the source of the first arc (*d1*) to the sink of the second arc (*d2*) to represent the "sum" of *d1* and *d2*. For example, if arc *d1* has a constant delay of 1 and *d2* a delay of 2, then *d3* will have a delay of 3. p-merge takes two arcs (*d1* and *d2*) in parallel (sharing both source and sink pins) and updates *d1* to represent the "worst" of *d1* and *d2*. For example, if *d1* has a constant delay of 3 and *d2* a delay of 4, *d1* will be updated to have a worst delay of 4.

3.2.1 s-merge (serial merge)

s-merge is a fundamental reduction operation that allows removal of pins from a timing graph. Whenever s-merge is performed, arc delays need to be computed. For example, consider two arcs *arc1* and *arc2* which are in series having the following delay values:

arc1		
Input Slew	Output Slew	Arc Delay
is1	os1	d1

arc2		
Input Slew	Output Slew	Arc Delay
os1	os2	d2

The new arc *arc3* resulting from s-merge between *arc1* and *arc2* will have the following delay:

arc3		
Input Slew	Output Slew	Arc Delay
is1	os2	d1 + d2

Note that we use "lazy" slew computation so that output slew and arc delay are computed only for new input slew values. Also, the load-dependent delay computation is confined to the last arc that drives an output port.

Accuracy in s-merge operation is controlled by the number of input slews/output loads that are retained for delay arcs that originate from input ports or terminate at output ports. For table-based libraries, the break points in the table provide the initial range of input slews and output loads. For linear or equation-based libraries, such breakpoints can be characterized. There is no loss in accuracy in s-merge as long as delay calculation is performed using the same input slews and output loads from the initial breakpoints. From the initial set of input slews and output loads, some slew/load values can be removed if such removal does not lead to errors exceeding some desired level. For example, consider a set of three input slews {*s1*, *s2*, *s3*} for a delay arc originating from an input port. Slew value *s2* can be removed from the set if the error between the delay specified for *s2* and the delay computed by interpolating between *s1* and *s3* is within the desired tolerance. Initial errors in slews tend to decrease after several s-merge operations in CMOS designs.

3.2.2 p-merge (parallel merge)

p-merge leads to a substantial reduction in model extraction time as it reduces the number of arcs which need to be processed in the next stage of computation. Given two parallel delay arcs, p-merge chooses the "worst" of the two arcs and allows the other arc to be discarded. Consider two parallel arcs, *arc1* and *arc2,* with the same timing characteristics as in section 3.2.1 except that both have the same input slews *is1* (assume $d1 < d2$).

If a late path is of interest, p-merge picks output slew and arc delay from *arc2* to update *arc1*. The resulting delay for *arc1* becomes

arc1		
Input Slew	Output Slew	Arc Delay
is1	os2	d2

Conversely, if an early path is of interest, *arc1* will remain unchanged.

Note that p-merge only preserves the "worst" (latest and earliest) port-to-port delays.

3.3 Sequential Models

Sequential models involve delay arcs and check arcs. All check arcs can be classified into two main groups. One group is called the "setup" group where the data signal is expected to arrive before the reference or clock signal. Examples of "setup" group are setup, recovery, skew, clock separation, etc. The other group is called the "hold" group where the reference signal is expected to arrive before the data signal. Examples of "hold" group are hold and removal. To preserve latch behavior, all latch input pins and latch output pins are retained but all registers are removed. The operations associated with the removal of sequential elements are similar to those for combinational models. To correctly model the interaction between delay arcs and check arcs, two different s-merge operations are used: forward s-merge and backward s-merge. Sequential p-merge is essentially the same as combinational p-merge. The only difference is that only check arcs of the same type can be merged.

3.3.1 Forward s-merge

Forward s-merge operation is used when register/latch clock pins are removed. When a delay arc and a check arc meet at a clock pin, a new check arc is created to represent a new check value between the old data signal (signal end of the check arc) and the new clock signal (source of the delay arc). Let *d* denote the delay value on the delay arc and *c* the check value on the check arc. For setup (hold) group, the new check value becomes $c - d$ ($c + d$).

3.3.2 Backward s-merge

Backward s-merge operation is used when register input pins are removed. When a delay arc and a check arc meet at a data pin, a new check arc is created to represent a new check value between the source of the delay arc and the old clock signal (reference end of the check arc). For setup (hold) group, the new check value becomes $c + d$ ($c - d$).

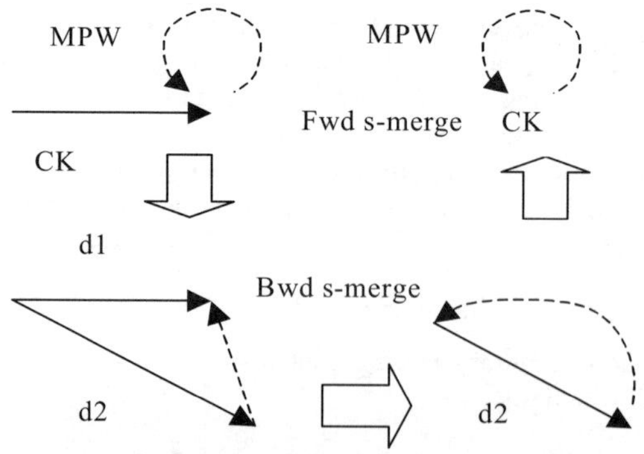

Figure 3: Modeling self-loop check arcs

3.3.3 *Self-loop Check Arcs*

There are some special check arcs where both the signal end and the reference end point to the same pin. Such timing checks include, but are not limited to, minimum pulse width (MPW) and minimum period (MP) checks on clocks. By using forward and backward s-merge operations, such self-loop check arcs can be modeled with correct clock path delays (including asymmetrical rise/fall) and slew propagation.

Self-loop check arcs may be "expanded" as a check arc with no self-loop by conceptually duplicating the incoming delay arc. For example, consider the timing graph fragment for some clock path with a MPW check arc in Fig. 3. The delay arc is duplicated to produce two arcs *d1* and *d2* and now check arc is no longer a self-loop arc. First, backward s-merge is performed on delay arc *d1* and the check arc. Then, forward s-merge is performed on delay arc *d2* and the check arc from the previous step. This leads to another self-loop check arc at port CK that correctly reflects the clock path delay and the slew propagation. Note that no delay arcs or pins need to be duplicated here. The duplication in Fig. 3 is for the sake of exposition only.

3.4 Graph Reduction Algorithm

We start with a timing graph that is acyclic. The timer that builds the timing graph is responsible for breaking combinational cycles. The graph is reduced one pin at a time by visiting each internal pin in breadth first search (BFS) order. The reduction is repeated until no further changes are possible. Fig. 4 shows the core loop of the algorithm. `removePin()` is the main routine which performs all the merge operations. Pseudo code for `removePin()` is given in Fig. 5. `postProcess()` takes care of delay arcs or check arcs which originate from primary input pins (or preserved internal pins) and terminate at primary output pins (or preserved internal pins). For such arcs, delay computation routines need to be called explicitly as no s-merge or p-merge operations are done. A pin cannot be removed if it is associated with assertions or if it is needed as an anchor point. The preservation of internal pins will be discussed in section 4. Self-loop check arcs are processed as check arcs in Fig. 5.

```
reduceGraph(graph){
  changed = 1;
  while (changed) {
    changed = 0;
    for each pin of graph in BFS order {
      if (removePin(pin)) {
        changed = 1;
  } } }
  postProcess(graph);
}
```

Figure 4: reduceGraph() code

BFS traversal guarantees that

- all the incoming delay arcs to a pin have already been processed before the pin is removed, and
- the delay arcs are characterized with respect to a minimum number of input slew values.

However, BFS traversal does not guarantee that both the signal end and the reference end of all check arcs are processed before forward s-merge or backward s-merge operations. For example, consider the

```
removePin(pin){
 if (!canRemovePin(pin)) {
    retainPin(pin);
    return 0;
 }
 for each incoming delay arc d1 to pin {
    for each outgoing delay arc d2 from pin {
      d3 = s-merge(d1, d2);
      for each delay arc d parallel to d3 {
        p-merge(d3, d);
        delete d;
    } }
    for each check arc c1 having pin as sig {
      c3 = backward-s-merge(d1, c1);
      for each check arc c parallel to c3 {
        p-merge(c3, c);
        delete c;
    } }
    for each check arc c2 having pin as ref {
      c4 = forward-s-merge(d1, c2);
      for each check arc c parallel to c4 {
        p-merge(c4, c);
        delete c;
  } } }
  delete pin;
  return 1;
}
```

Figure 5: removePin() code

timing graph in Figure 6. When pin *c1* is removed, the range of slew values coming from the signal end (*d2*) of the check arc is not available. In this case, we characterize the check arc with respect to all possible slew values (usually fewer than 6 slew values are sufficient to characterize a check arc). Later when pin *d2* is processed, we re-characterize the check arc for appropriate slew values.

Fig. 7 gives an example of graph reduction on a circuit with two registers and one latch. The final model includes two internal pins to preserve the latch behavior. Note that the setup/hold checks on *reg1* are modeled as self-loop setup/hold check arcs on clock *C*.

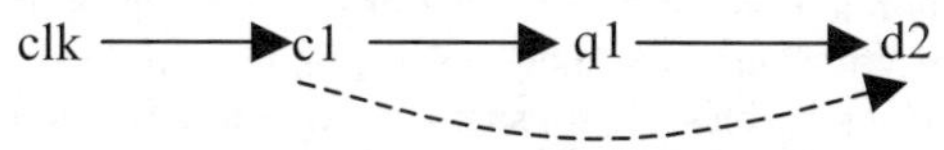

Figure 6: Some check arcs need re-characterization

4. RETAINING INTERNAL PINS

We retain internal pins in the timing graph for several reasons. One reason is to model latch behavior that is correct even in the presence of clock waveform changes. Another reason is to support all original assertions such as multi cycle paths, false paths, generated clocks, etc., including those that span multiple blocks. We retain all internal pins associated with assertions and re-write the original assertions in terms of the retained internal pins. The new assertions become a part of the model and, if the library format allows it, they may even be embedded within the timing model and applied

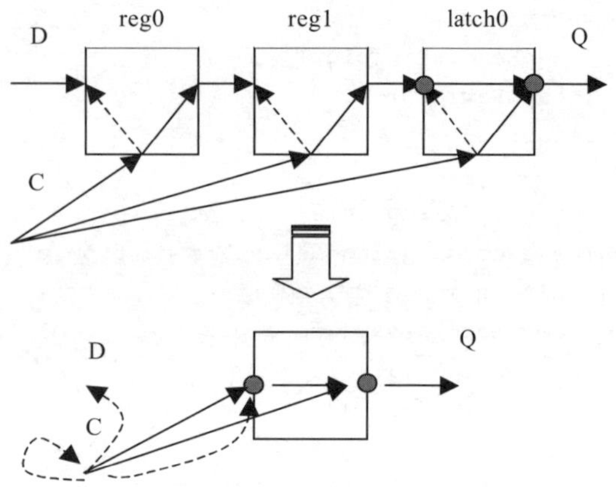

Figure 7: Example of model extraction on a design with two registers and a latch

```
retainPin(pin) {
  for each incoming delay arc d of pin {
    compute_delay (d);
  }
  combine all incoming input slews at pin;
  for each outgoing delay arc d from pin {
    compute_delay(d);
} }
```

Figure 8: retainPin() code

automatically after the model is read. Internal pins retained for these reasons are called *preserved pins*. Another important reason for retaining internal pins is to avoid increasing the model size. Such pins are called *anchor points*. Essential to supporting both preserved pins and anchor points is the `retainPin()` operation whose pseudo code is given in Figure 8. Associated with each retained pin is a range of input slew values which are obtained from the incoming delay arcs.

4.1 Anchor Points

Anchor points are existing internal pins which, if removed, may lead to an increase in model size. To estimate the model size, we use the number of delay arcs. We assume that all the delay arcs make equal contribution to the final model size. Given this assumption, the identification of anchor points is similar to the problem of finding a set of pruning pins in the context of static circuit optimization [5]. The objective here is to minimize the number of delay arcs, not the number of variables. Also, since we want to characterize the delay arcs with respect to a minimum number of input slew values, we do not want to remove internal pins at random without first processing their incoming delay arcs. This constrains the order in which anchor points are identified and retained.

We define *gain* at an internal pin as

$$(\#\text{incoming delay arcs} \times \#\text{outgoing delay arcs}) - \#\text{incoming delay arcs} - \#\text{outgoing delay arcs}.$$

This represents an increase in the number of delay arcs if the pin is removed. For example, in Figure 1, the number of delay arcs is 6 before the removal of the internal pin. If the pin is removed, the number of delay arcs becomes 9. Thus, the gain is 3. Any pin that has a positive gain and is *observable* becomes a candidate for anchor points. A pin is *observable* if there exists a path from the pin to a primary output or to a preserved pin. Observability prevents anchor points from being formed in the transitive fanin cone of pins that will eventually be removed. For example, register input pins are not observable and are eventually removed. Anchor points with a gain value greater than or equal to some threshold are identified by `canRemovePin()` in Figure 5. Once anchor points are identified, they are valid for a particular pass. One pass constitutes a complete sweep of all internal pins. Anchor points must be re-identified in the subsequent passes. Table 1 presents the results of varying the gain threshold on a simple 32-bit combinational multiplier. The cpu time and memory represent the runtime and the memory needed to extract the model on a 750MHz Sun Fire 880 machine running Sun OS 5.8. With a very high value of gain threshold (like 1000), we can remove all anchor points. We provide an option to change the gain threshold but the default is set to 1 and the guarantee in model size reduction compared to the original timing graph only comes with a threshold value of 1. We found this greedy 1-gain heuristic to be very effective.

Table 1: Results of Varying Gain Parameter

Gain	cpu sec	mem MB	#anchor points	#delay arcs	Model size KB	#passes
1	2.7	10.6	359	3312	896	5
2	2.8	10.9	244	3295	1024	4
3	2.9	11.0	178	3411	1095	4
5	2.8	11.0	176	3472	1104	4
50	4.0	12.4	49	5210	2089	3
1000	5.9	14.6	0	3104	1543	2

4.2 Limitation of Anchor Point Heuristic

Although the multi-pass greedy heuristic is fast and effective, this does not guarantee an optimal solution. As a result, the greedy method may produce a slightly larger model than an approach that has a more global view of the graph. Fig. 7 shows an example of such sub-optimality for the 1-gain greedy heuristic. The greedy heuristic fails to delete pins *D* and *E* because the removal of either pin leads to an increase in the number of delay arcs. However, if both pins are removed, the total number of delay arcs would go down. We have not found this to be a big limitation in more than 70 industrial designs that we have tested so far. Nevertheless, we are investigating ways to remedy this limitation.

Table 2: Timing Model Extraction Results

ckt	#gates	#regs& latches	#I/O ports	#int pins intial	#arcs initial	#int pins final	#arcs final	#passes	Model extraction time / mem (sec / MB)	Pre-extraction timing analysis time / mem (sec / MB)	Post-extraction timing analysis time / mem (sec / MB)	model size (MB)
ex1	108K	1855	133	116079	157660	215	1452	3	20.8 / 62.4	11.5 / 43.8	1.5 / < 1.0	1.3
ex2	129K	7	2322	117959	279820	12703	54803	3	41.3 / 126.1	17.5 / 51.0	6.1 / < 1.0	46.4
ex3	307K	7742	777	320697	514033	21	636	3	113.4/162.9	30.3/ 109.1	1.2 / < 1.0	1.0
ex4	456K	17949	180	467049	649922	204	545	4	109.4/463.6	56.1/ 235.1	1.3 / < 1.0	0.3

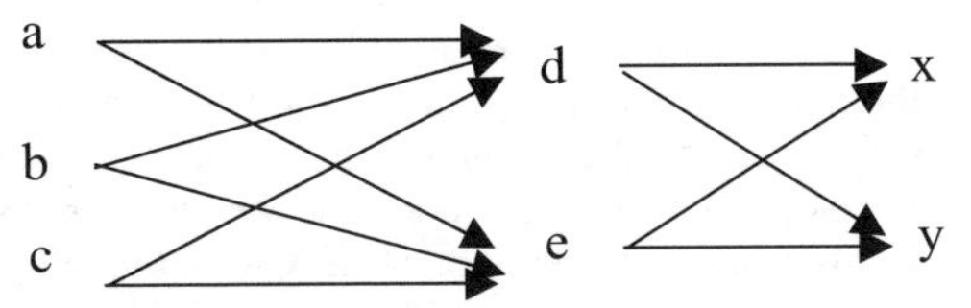

Figure 9: Greedy heuristic fails to remove d & e pins

5. EXPERIMENTAL RESULTS

Table 3: Results showing accuracy of models

ckt	Timing analysis on original netlist		Timing analysis on extracted model	
	Early slack	late slack	early slack	late slack
ex1	-11857.34	-11857.34	-11857.34	-11857.34
ex2	-3.74	-87.37	-3.76	-87.35
ex3	-1.85	-1.85	-1.85	-1.85
ex4	-7.47	-7.47	-7.47	-7.47

Tables 2 and 3 present results of graph reduction on four industry designs. Multi-pass 1-gain greedy heuristic was used on all of them. Reported cpu times and memory usage are on 750MHz Sun Fire 880 machine running Sun OS 5.8. The number of internal pins (final) includes all preserved pins and anchor points. The models were written out in Cadence Timing Library Format (TLF) version 4.4. The runtime includes characterization of clock networks using insertion delays.

The resources needed to generate a timing model are about 2-4x in terms of cpu time and 1.4-2.5x in terms of memory usage compared to performing static timing analysis. This is very efficient since static timing analysis is performed using only a single input slew value and a single output load value, whereas model extraction considers, on the average, about 4 input slew values and 4 output load values. Timing analysis was performed by examining 1000 worst paths in both late mode and early mode (with a maximum of 500 reported paths per endpoint). All the worst late paths and the worst early paths in the model match those in the original circuit within 1% (the specified tolerance value) as can be seen in Table 3.

The design *ex2* led to an out of memory error (> 4 GB) for a black box generator based on [1] and a model size of 5 GB for a gray box generator based on [4]. The design has only 7 latches and the logic is mostly combinational in nature. We identified several anchor points with gain values of 395 and 384. Removal of such anchor points leads to long runtime, large memory usage and large models. We produced a 1% accurate model in 41.3 sec that has a reasonable size (46.4 MB) for this design.

As can be seen from Table 2, extracted timing model leads to significant gains in capacity in terms of both runtime and memory usage. If netlists are replaced by timing models, we gain up to 43x in terms of runtime and up to 235x in terms of memory usage in timing verification.

6. CONCLUSIONS AND FUTURE WORK

We have presented a novel method of generating a timing model by graph reduction. This method is simple and is shown to be quite effective in increasing the capacity of timing verification. The generated model provides a plug-in replacement capability for blocks for static timing analysis and it is well suited for top-down hierarchical flows. It takes 1.4-4x more computational resources to generate an accurate model than to perform timing analysis. The method guarantees a reduction in model size compared to the original timing graph and lends nicely to a formulation where constraints that span multiple blocks can be supported. Also, the model allows for arbitrary levels of latch time borrowing.

As discussed in section 4.2, the greedy heuristic may not always produce an optimal solution. Future research will examine ways to further minimize the model size. Also, work is under way to provide support for bottom-up IP characterization flows and an option to generate black box models for applications that are not capable of reading gray box models.

7. REFERENCES

[1] Cherry, James J. Pearl: A CMOS Timing Analyzer, Design Automation Conference, (1988), 148-159.

[2] Venkatesh, S.V., Palermo, R., Mortazavi, M., Sakallah, K.A. Timing Abstraction of Intellectual Property Blocks, Custom Integrated Circuit Conference (1997), 99 – 102.

[3] McDonald C.B., Bryant, R.E. A Symbolic Simulation-Based Methodology for Generating Black-Box Timing Models of Custom Macrocells, ICCAD (2001), 501 – 506.

[4] Segal, Russell B. Extracting accurate and efficient timing models of latch-based designs, U.S. Patent 5,790,830 (August 4, 1998).

[5] Visweswariah, C., Conn, A.R. Formulation of Static Circuit Optimization with Reduced Size, Degeneracy and Redundancy by Timing Graph Manipulation, ICCAD (2000), 244 – 251.

Efficient Stimulus Independent Timing Abstraction Model Based on a New Concept of Circuit Block Transparency

Martin Foltin
mxf@fc.hp.com
(970) 898 3755

Brian Foutz*

Sean Tyler
sct@fc.hp.com
(970) 898 7286

Hewlett Packard Corporation, 3404 E. Harmony Rd., MS-32
Fort Collins, CO 80528

ABSTRACT

We have developed a new timing abstraction model for digital circuit blocks that is stimulus independent, port based, supports designs with level triggered latches, and can be input into commercial STA (Static Timing Analysis) tools. The model is based on an extension of the concept of latch transparency to circuit block transparency introduced in this paper. It was implemented, tested and is being used in conjunction with transistor level STA for microprocessor designs with tens of millions of transistors. The STA simulation times are significantly shorter than with gray box timing models, which can decrease the overall chip timing verification time. The model can also be used in the intellectual property encapsulation domain.

Categories and Subject Descriptors

B.7.2 [**Integrated Circuits**]: Design Aids–*simulation, verification*

General Terms

Performance, Design, Verification.

Keywords

Timing analysis, timing model, VLSI design, circuit optimization

1. INTRODUCTION AND MOTIVATION

High performance VLSI designs, such as microprocessors or high performance chip-sets, rely on custom designed circuits to achieve maximum performance. To achieve sufficient accuracy, the timing analysis for such designs needs to be performed on the transistor level. Static Timing Analysis (STA) [1] tools are commonly employed to achieve complete path coverage in complex designs. However, a full chip transistor level STA for modern designs with tens of millions of transistors is prohibitively expensive.

Therefore, the design is usually divided into smaller blocks, with transistor level STA performed separately for each block. A timing abstraction model of the block is generated based on the STA results. This model is subsequently instantiated in a parent block. In the parent block, the STA tool will trace paths through timing abstraction models rather than through individual transistors. This process is repeated all the way up to the full chip level. This is the basis of hierarchical design methodologies [2]. Size and quality of timing abstractions determines timing accuracy and the overall turn-around time of the entire design.

The timing abstraction model must reproduce all characteristics of the block needed to time a circuit that contains this block. It should abstract away all unnecessary information from the block. Black box – like timing models assign parameters characterizing the block timing properties to block ports. These models are often used in designs with combinational logic and edge triggered latches (flip-flops). High performance designs, however, mostly employ level triggered latches to take advantage of time borrowing between latches to reach high operating frequencies. Black box models do not adequately model such designs because they do not support transparency, i.e., they cannot model input→output paths that pass level triggered latches.

Several timing models have been developed so far that can model transparent paths. The most relevant models are described in Section 2 below. The disadvantage of these models is that they are either not port based or they are input stimulus dependent. The models that are not port based retain some or all of the internal sequential nodes representing latches, domino logic, etc., from the modeled circuit. Timing analysis of circuits with such models is characterized by a large number of timing paths, long run times and large memory requirements, especially at the top hierarchy levels. The models that are input stimulus dependent are incompatible with hierarchical timing methodologies. In these methodologies, the timing of the design is optimized in an iterative manner. The timing abstraction model is generated based on the context files (i.e., the stimulus, slew rates and loads) rolled down from the previous STA iteration. Between the iterations, the blocks will change to fix timing violations. Hence, the context of the timing model instantiated for the current STA iteration may be different from the context used to create this model. This requires the model to be stimulus independent, i.e., to reproduce faithfully the timing of the block for any input signal arrival times.

Timing abstraction models find application also in the intellectual property encapsulation domain, where there is a need to accurately represent a block's timing behavior without exposing the block's internal implementation. Furthermore, the timing abstraction models may simplify design transfer between different

* Now with IBM Test Design Automation, Endicott, NY.

DAC 2002, June 10-14, 2002, New Orleans, Louisiana, USA.

design teams. Stimulus independence is desired in these applications as well.

In this work we have developed a new timing abstraction model that is lightweight, port based, stimulus independent, supports level triggered latches and can be input into commercial STA tools. The model can be used on any level of hierarchy where it can be mixed with transistors and/or gates. The model was implemented, tested and is used in conjunction with transistor level STA for designs with millions of transistors. We will show that the model performance is significantly better and memory requirements are significantly smaller than for industry standard models.

2. PREVIOUS WORK

A straight-forward, flexible and accurate but not very efficient timing abstraction model for circuit blocks with level triggered latches is the gray box model [3]. This model abstracts away all combinational logic between the sequential elements and replaces it by delay arcs. All sequential elements (latches, domino logic, etc.) are retained and represented by model nodes. Timing checks performed on each model node are the same as on the corresponding sequential element in the modeled circuit. Timing analysis with gray box models will contain many internal paths that are unnecessary for verification of the timing behavior of the block in context of the upper level block. For designs with millions of transistors, the full chip timing analysis with gray boxes has long runtimes and high memory and storage demands. Advantages of the model are that it is clock frequency independent and the model paths are relatively easy to debug.

Striving to improve timing analysis efficiency, Segal et al. [4] developed a timing model that collapses together similar chains of latches in the modeled circuit. Two paths from a given input port are considered similar if the corresponding latches on the first and on the second path are controlled by the same clock signals. The delay arcs between the collapsed latches and from the last latch to output ports are adjusted so that i) the abstraction correctly models the total path delays, and ii) the setup check on each collapsed latch will equal to the most critical setup check from the two (or more) corresponding latches before they were collapsed. Additional latches are needed to model the clock→output paths from each original latch, but the authors conclude that the model will usually have fewer latches than the original design.

In a different approach, Venkatesh et al. [5] developed a simple abstraction model that represents all internal timing constraints of the block (the setup, hold and latch transparency requirements on all sequential nodes) in terms of a reduced set of requirements for clock edge separations and pulse widths. This was achieved by employing the symbolic formulation for signal arrival times, signal departure times and timing constraints on all sequential elements based on the general critical paths model [6]. Although this timing abstraction is stimulus dependent and falls more into the domain of optimal clocking, it is worth mentioning in context of this work because the underlying general critical paths model [6] provides numerical framework that may be used to derive the time constraints for signals at input ports in our new model.

Synopsys recently introduced an enhanced black box timing model that they named "transparent black" model [7]. For signals at input ports, this model measures the worst-case setup/hold time to any latch node through a chain of transparent latches. This is a significant enhancement to the conventional black box model that measures the worst-case setup/hold time only to the first latch node. Unfortunately, the model stops short of representing the input→output transparent paths in a way that would be independent on input stimulus. This prevents the model from being used in hierarchical design methodologies as described in the Introduction section above.

3. NEW TIMING ABSTRACTION MODEL

The new timing abstraction model presented in this work is a simplified circuit or a connectivity network that reproduces the most relevant timing characteristics of the modeled circuit block needed for timing analysis of this block in context of a larger circuit where the block is instantiated. The model is based on extension of the concept of latch transparency to circuit block transparency. Two different data path types are usually analyzed in STA – the longest paths and the shortest paths. In the discussion below we will describe timing abstraction model elements for the longest data paths that pass through sequential elements (latches, domino logic, etc.). Modeling of shortest data paths is simpler because they cannot pass transparently through the block. For shortest paths, the model gets simplified to a black box - like format. All purely combinational input→output paths are simply modeled by delay arcs (or delay nets) connecting the corresponding input and output ports.

3.1 Concept of Circuit Block Transparency

The level triggered latch is transparent when the clock signal connected to its clock pin is active (high or low, depending on the latch design). If the data signal arrives at the latch input before the clock signal becomes active, the signal does not pass to the latch output. In such case, the critical path will be triggered by the clock signal transition rather than by the data signal transition. This concept can be extended from latch to circuit block. Unlike a single latch, the block usually has several output ports. Hence, we need to define transparency separately for each output port. We classify the given output port as transparent if and only if the data signal arrives at this port no earlier than the latest clock signal that can get to this output port. Note that when considering the latest clock signal, this is not limited only to clock signals from last sequential elements in the fan-in of the given output port, but also includes clock signals from any preceding sequential elements as long as they pass transparently through the chain of latches (or domino logic, etc.) all the way to the given output port.

This port-based definition of transparency is more restrictive than considering transparency individually for each path. This is illustrated in Figure 1. The data signal from input **IN_1** arrives at latch **LAT_1** while the clock signal on this latch is active, so if we would consider just the path **IN_1→OUT**, this path would be transparent. However, the clock signal on **CK_2→OUT** path passing latches **LAT_2** and **LAT_3** arrives at output **OUT** later than the data signal on the **IN_1→OUT** path. Therefore, according to our definition of transparency, the block will not be transparent for the **IN_1→OUT** path. Blocking the **IN_1→OUT** path is entirely within the spirit of STA that strives to limit the analyzed paths only to the most critical (i.e., in this case the longest) paths. The designer is not interested in how the **IN_1→OUT** path propagates downstream from our block because

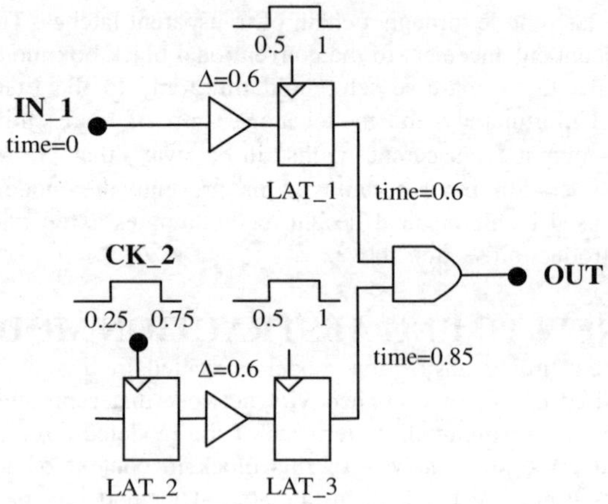

Figure 1. Example circuit with CK_2→OUT path that is more critical than IN_1→OUT data path

it merely trails the more critical path that reaches block output from the **CK_2→OUT** direction. Our definition of transparency therefore limits the number of unnecessary paths leaving the modeled block. Considering that such paths may branch several times in successive blocks, this definition may significantly reduce the number of paths to be traced in the design and therefore improves the STA efficiency and reduces designer's time to analyze STA results.

A timing abstraction model that is consistent with the block transparency definition from previous paragraphs can be designed as follows. Central to this design is that all transparent paths converging to a given output port **OUT** pass through a single latch attached to this port that serves as a transparency "gate" determining whether or not the given path will pass transparently to the output port. This assures the stimulus independence of our model. If, for example, for a different stimulus at input port **IN_1** the **IN_1→OUT** path from Figure 1 becomes longer than the **CK_2→OUT** path, than the signal on this path will arrive at the input of the model latch while the latch is open and the path will be traced to the output port.

The model latch will be controlled by signal from a derived clock node. The opening time of this latch can be easily obtained from results of STA analysis of the modeled block – simply take the output arrival time of the most critical clock→**OUT** path from the path report. Reference this time to a pin clock (i.e., a clock port of the block) that will be connected to the derived clock via delay arc (or delay net). Sections 3.2 and 3.3 will explain which pin clock to use to assure proper cycle advancement across the block boundaries. The model will be valid as long as the phase shift of clock waveforms between the model generation and the model instantiation is the same for all clock ports of the block. The clock frequencies and pulse widths on clock ports may not change. These are reasonable assumptions in most design environments, since the clock distribution tree and clock frequency usually do not change between individual iterations in the timing verification process. For the intellectual property encapsulation application, separate sets of model parameters may be generated for different clock frequencies.

Derivation of the latch closing time is slightly more complicated. Let's denote by **T(OUT)** the set of all input ports in the modeled circuit that may be connected for some input stimulus by a transparent path to the given output port **OUT.** Let's denote by **R(IN_*i*)** the latest time a signal can arrive at input port **IN_*i*∈T(OUT)** before a timing violation would occur on any sequential element that can be reached by any path from that input port **IN_*i***. This time can be obtained, for example, from the general critical paths model [6] where it is referred to as the required time. If **D(IN_*i*)** is the longest path delay between the input port **IN_*i*** and the output port **OUT**, then the closing time **TC** of the model latch attached to the output port **OUT** is obtained from formula

$$TC = \max \forall\, IN\_i \in T(OUT),\ R(IN\_i) + D(IN\_i) \quad (1)$$

This assures that all transparent paths can pass through this latch without timing violation.

To obtain the set of transparent paths **T(OUT)** and the transparent path delays **D(IN_*i*)**, we can, for example, perform a second STA run on the modeled block with stimulus on each input port **IN_*i*** set to the time **R(IN_*i*)**.

3.2 Timing Model of an Example Circuit

Figure 2 shows an example circuit block with two input ports, **IN_1** and **IN_2**, two output ports, **OUT_1** and **OUT_2**, and three transparent paths, **IN_1→OUT_1**, **IN_2→OUT_1** and **IN_2→OUT_2**. Figure 3 shows the timing abstraction model for this block. As discussed above, the timing model has one latch attached to each output port that serves as a transparency "gate" (see Output Latch in Figure 3). The timing model also has one latch attached to each input port that checks for timing violations on the most critical sequential element in the modeled circuit that can be reached from the given input port **IN** (see Input Latch in Figure 3). The closing time of this Input Latch is set to the time **R(IN)** discussed before. In general, the opening time **TO'** of the input model latch can be obtained from formula

$$TO' = \min \forall\, OUT\_j \in T'(IN),\ V(OUT\_j) - D(OUT\_j), \quad (2)$$

where **T'(IN)** is the set of all output ports **OUT_*j*** in the modeled circuit that may be reached for some input stimulus by a transparent path from the input port **IN**, **V(OUT_*j*)** is the opening time of the model output latch attached to output port **OUT_*j*** and **D(OUT_*j*)** is the longest transparent path delay between the input port **IN** and the output port **OUT_*j***. Each Input Latch is connected by delay arcs (or delay nets) with delays **D(OUT_*j*)** to

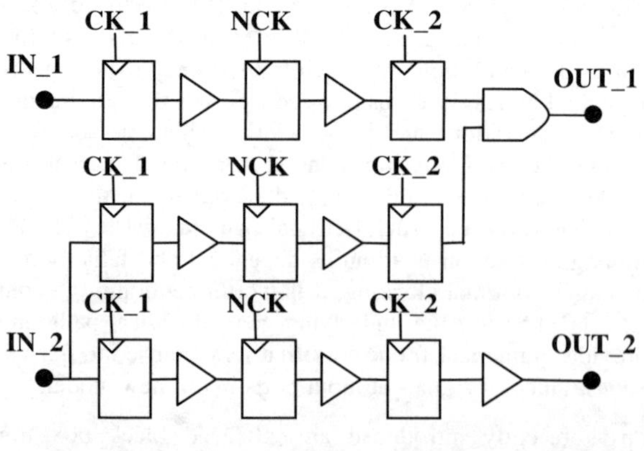

Figure 2. Example circuit with transparent paths IN_1→OUT_1, IN_2→OUT_1, and IN_2→OUT_2

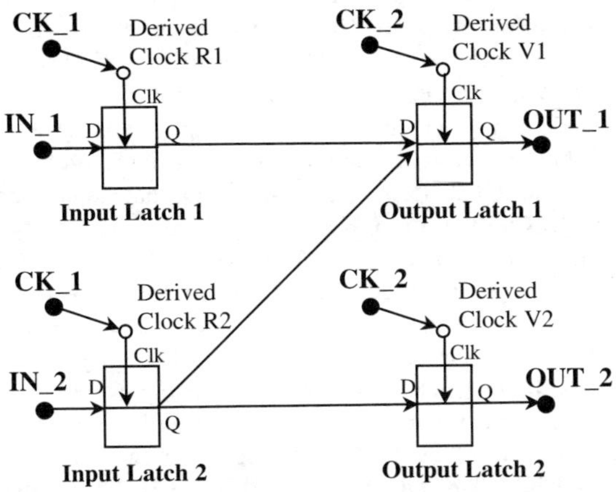

Figure 3. Timing abstraction model for the example circuit from Figure 2

Output Latches *j* attached to output ports **OUT_*j*∈T'(IN)**.

Let's consider STA of a parent block that contains, among other circuit elements, one or more timing abstraction models instantiated in place of children blocks. The timing path entering the timing abstraction model through input port **IN** will first pass through the Input Latch attached to this port. The timing slack on this latch will be exactly equal to the slack that would be observed in full transistor level STA of the modeled block on the most critical sequential element that can be reached from the input port **IN**. The name of this element will be encoded in the Input Latch name, so that if there is a negative slack on the Input Latch, the designer will know which sequential element in the modeled circuit is failing most and by how much. If there is no violation, the timing path will branch and propagate to Output Latch nodes *j* connected to output ports **OUT_*j*∈T'(IN)**. Depending on the actual signal arrival on Output Latch node (that is function of signal arrival at input port **IN**), the Output Latch will determine if the path should go transparently to the associated output port.

Note that the number of internal nodes in this timing abstraction model scales with the number of ports rather than with the number of paths. For most circuit blocks, there will be one internal node per each input port (the Input Latch) and one internal node per each output port (the Output Latch). The number of timing checks to be performed is limited to one check per each input port plus one check per each input port – output port pair that is connected by a transparent path for at least one input stimulus. Both the number of internal nodes, the number of timing checks and the number of paths are significantly smaller than in gray-box timing abstraction models. In spite of its light weight, the model provides all the most important timing information that designers may need – it flags the most critical timing violation for each input port and it models correctly the critical transparent input→output paths.

3.3 Cycle Advancement at Block Boundaries

If the sequential elements connected through combinational logic to output port **OUT** in the modeled circuit are controlled by pin clocks derived from different reference clocks or reference clock edges, we may need to attach more than one Output Latch to output port **OUT** in the timing model. This is to assure proper cycle count adjustment across the block boundaries as the signal passes from the "driver" Output Latch in the timing abstraction model to "receiver" latches in adjacent blocks downstream on the path. For similar reasons we may need to attach more than one Input Latch to input port **IN**.

In Section 3.1 we defined the set **T(OUT)** of all input ports that may be connected for some input stimulus by a transparent path to the given output port **OUT**. In this section we divide this set to subsets **T(OUT, CLK, CLK_edge)** according to the reference clock **CLK** and reference clock edge **CLK_edge** associated with the opening clock edge on the last sequential element along the transparent path **IN_*i*→OUT**, where **IN_*i*∈T(OUT)**. We will assign one Output Latch to each subset **T(OUT, CLK, CLK_edge)**. The opening time for each Output Latch is calculated similarly like in Section 3.1 except that now we consider only the most critical clock→**OUT** path that has the opening clock edge on the last sequential element derived from the reference clock **CLK** and reference clock edge **CLK_edge**. The Output Latch closing time is calculated from formula (1) where we replace **T(OUT)** by **T(OUT, CLK, CLK_edge)**.

To illustrate this, Figure 4 shows an example circuit that is modeled by the timing abstraction model shown in Figure 5. The modeled circuit has three transparent paths leading to output port **OUT**. The last sequential elements on these paths are open when the pin clocks **CK_3**, **CK_6** and **CK_9**, respectively, are high. The opening (rising) edges of **CK_3** and **CK_6** are derived from rising edge of reference clock **CLK**. The opening edge of **CK_9** is derived from falling edge of reference clock **CLK**. Therefore, the first two transparent paths will be in subset **T(OUT, CLK, R)**. The third path will be in subset **T(OUT, CLK, F)**. In the timing abstraction model, the first two transparent paths will pass through Output Latch controlled by a derived clock connected to pin clock **CK_3**. That is because the most critical clock→**OUT** path that determines opening time of the Output Latch for the first subset **T(OUT, CLK, R)** is the **CK_1→OUT**. The last latch on this path is controlled by pin clock **CK_3**. The third transparent path in the timing model will pass through a second Output Latch controlled by a derived clock connected to pin clock **CK_9**.

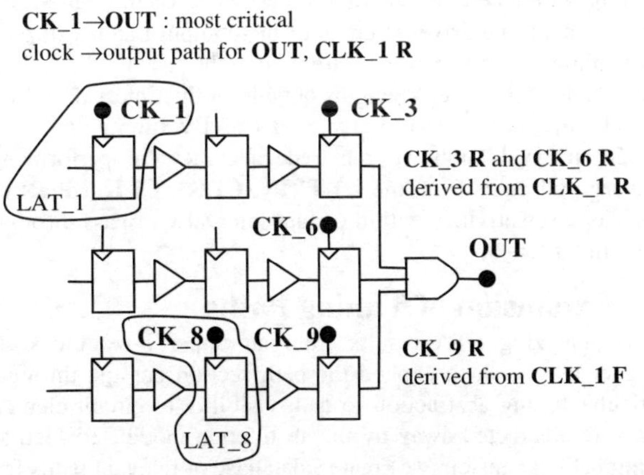

Figure 4. Example circuit with 3 transparent paths to output port OUT.

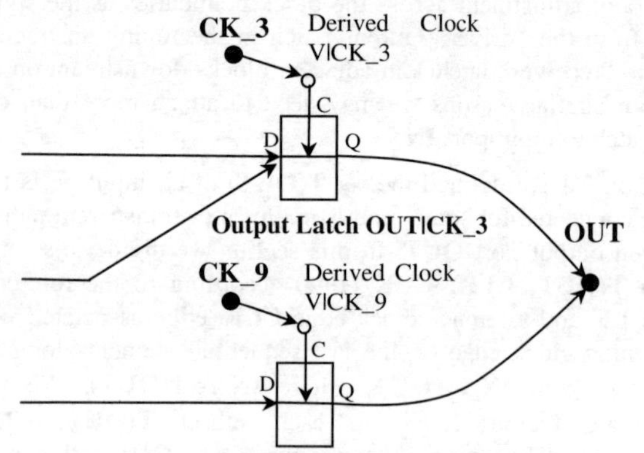

Figure 5. Output section of the timing abstraction model for the example circuit from Figure 4

Similarly, multiple Input Latches will be attached to each input port **IN** if the set **T'(IN)** defined in Section 3.2 needs to be divided to subsets according to the reference clock **CLK'** and reference clock edge **CLK_edge'** associated with the closing clock edge on the first sequential element on the **IN→OUT_*j*** transparent path.

3.4 Modeling Slope and Load Sensitivity

In this timing abstraction model, the model elements that are on block boundaries are the Input and Output Latches. The driver strength parameters of the Output Latch connected to output port **OUT** are derived from the driver strength of the last circuit element connected to that output port in the modeled circuit. Likewise, the slope (slew-rate) sensitivity parameters on Input Latch connected to input port **IN** are derived from the slope sensitivity of the first circuit element connected to that input port in the modeled circuit. In some circuit blocks it may happen that different paths from the set **T(OUT, CLK, CLK_edge)** have different load sensitivities. In the current model implementation, all paths from the set **T(OUT, CLK, CLK_edge)** that have driver strengths within certain interval (200 Ω) will pass through a single Output Latch. The driver strength of this Output Latch will be set to the minimum driver strength from all paths passing through this latch. Paths with driver strengths outside of this interval will pass through different Output Latch node(s). By tuning the driver strength interval knob we can trade accuracy for performance. Similarly, all paths from the set **T'(IN, CLK', CLK_edge')** that have slope sensitivities within certain interval will pass through a single Input Latch.

3.5 Expansion of Timing Paths

When analyzing STA results from an upper level block, the designer may wish to expand the path section passing through a particular timing abstraction to gain visibility to circuit elements that were abstracted away by the abstraction model. To facilitate timing path expansion, we create a database of relevant paths from the modeled circuit that can be plugged into the upper level path in place of path sections from the timing abstraction model. The database will contain the abstracted node names, stage delays and the clock information for the abstracted sequential elements. The Input Latch and Output Latch node names in the abstraction model are designed to serve as pointers to paths in this database. For each subset **T(OUT, CLK, CLK_edge)** there will be one path in the database that will correspond to path Type 4 from Figure 6 (i.e., the most critical clock→**OUT** path). The Output Latch node name must unambiguously identify this subset. This is achieved by including the output port name **OUT**, pin clock name **CK2** and the **CK2** opening edge in the Output Latch name (see Figure 6). Note that the **CK2** opening edge is derived from one and only one reference clock and edge (**CLK**, **CLK_edge**). Similarly, for each subset **T'(IN, CLK', CLK_edge')** there will be one path in the database that will correspond to path Type 1 from Figure 6 (i.e., the most critical **IN**→sequential element path). The Input Latch name is derived analogously to the Output Latch name (see Figure 6). The database will contain up to one transparent path of Type 3 (see Figure 6) for each pair of subsets **T'(IN, CLK', CLK_edge')** and **T(OUT, CLK, CLK_edge)**. The Input Latch and Output Latch names on path Type 3 will unambiguously identify this pair of subsets. The corresponding path from the database will be also used to expand path of Type 2, but a note will be printed for such path that it does not pass transparently through the block for the actual signal arrival time at input port **IN**.

The other reason for including the port and clock names as sub-strings in the Input and Output Latch names is to assure uniqueness for these latch names. For example, if the most critical sequential element **LAT_3** is in the fan-out of two different input ports **IN1** and **IN2**, the timing abstraction will contain two different Input Latch nodes **LAT_3|SL|IN1|CK1|f** and **LAT_3|SL|IN2|CK1|f.**

3.6 Timing Model Verification

For the purpose of model verification, the STA analysis will be performed on the instantiated timing model with the same stimulus that was used in the full transistor level STA run on the modeled block. All paths from the full transistor level STA analysis will be categorized according to **IN**→sequential element, **IN→OUT** and Clock→**OUT** types. Only the most critical **IN**→sequential element path will be considered for each combination **(IN, CLK', CLK_edge')**; only the most critical Clock→**OUT** path will be considered for each combination **(OUT, CLK, CLK_edge)**; and the longest **IN→OUT** transparent path will be considered for each combination **(IN,**

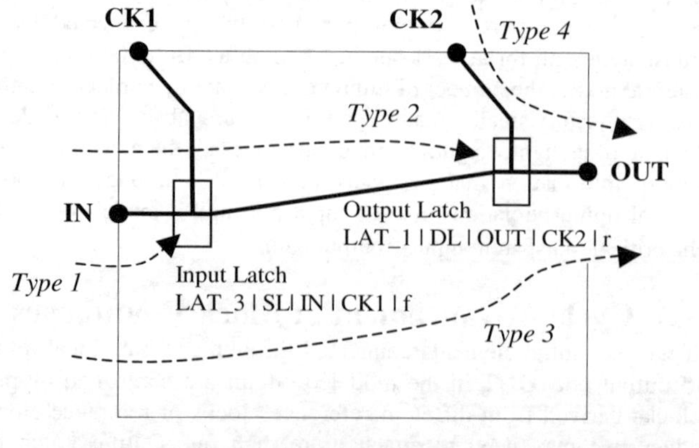

Figure 6. Timing path types passing through the timing abstraction model and the model node naming convention.

CLK',CLK_edge'), (OUT, CLK, CLK_edge). Timing slacks on these paths will be compared with timing slacks on the corresponding paths from the STA of the timing abstraction model. The model will pass verification if the slack difference is smaller than a preset threshold (~ 5ps in our implementation).

4. RESULTS

The timing abstraction model developed in this work can be input into commercial STA tools. For example, in design environments that employ an STA engine supporting gray box - like timing abstractions, the new timing model can be represented by very lightweight gray box. In such case, the model Input Latches and Output Latches are represented by gray box nodes and model delay arcs are represented by gray box time arcs. This representation has far fewer internal nodes than a conventional gray box. For benchmarking, we choose a typical industrial block with about 2 million transistors that contains lower level blocks spanning two levels of hierarchy. The same STA engine is used both for STA with conventional gray boxes and with the new timing models represented by lightweight gray boxes. Because of the large number of lower level blocks with various types of logic (including domino and internal loops), the performance results presented in this section represent a good statistical average. In general, the model generation times will increase with the number of transparent paths in the model. The performance numbers are shown in Table 1 that lists the sums of generation times and runtimes for both representations for each hierarchy level as well as summed over all 3 levels. The performance advantage of the new model gets better as we ascend up the hierarchy level. This is due to the port-based nature of the model. For example, on the Level 3, the new abstraction model generation + runtime is 16-times shorter than for the conventional gray box. The memory requirements on this level are 8-times smaller and the model requires 23-times less disk storage space. Based on extrapolating these trends to the full chip hierarchy level, we anticipate even more performance gain on that level. When summed up over all 3 hierarchy levels, the total STA runtime that includes all model generation times and run times is 3–times shorter for the new timing models than for conventional gray boxes. Again, we anticipate that this factor will increase as we include performance numbers from the full chip level STA.

To test accuracy of the new abstraction model, we employed the verification algorithm from Section 3.6. Timing slacks on timing paths were compared for the full transistor level STA, the STA with the new timing model and the STA with the conventional gray box. We found that the new timing models yield results that agree with the full transistor level STA equally well or slightly better than the results from STA with conventional gray boxes.

5. CONCLUSIONS

We have developed a new timing abstraction model that is stimulus independent, port based, supports transparency and can be input into commercial STA (Static Timing Analysis) tools. Both the number of internal nodes, the number of timing checks and the number of paths in the model are significantly smaller than in the industry standard gray box timing abstraction models. The model size reduction is achieved by eliminating less critical paths, not by giving up accuracy. Performance studies revealed that STA for blocks containing the new timing models is several times faster than with conventional gray box models. The model limitations are that it is clock frequency dependent and for blocks with multiple ported clocks, the phase shift of clock signals between model generation and model instantiation must be the same for all ported clocks. The new timing abstraction model may find application also in the domains of intellectual property encapsulation and design transfer between different design teams.

6. REFERENCES

[1] Hitchcock, R.B., Sr., Smith, G.L., Cheng, D.D., Timing analysis of computer hardware, IBM J. Res. Develop. 26 (1982) 100-105

[2] McDonald, C., Indermaur, T., Buckley, M., Timing analysis for the PA-8000, Integrated System Design (isdmag.com), February 1997, Cover Story

[3] Avidan, J., Circuit analyzer of black, gray and transparent elements, U.S. Patent No. 6158022

[4] Segal, R.B., Extracting accurate and efficient timing models of latch-based designs, U.S. Patent No. 6023568

[5] Venkatesh, S.V., Palermo, R., Mortazavi, M., Sakallah, K.A., Timing abstraction of intellectual property blocks, IEEE 1997 Custom Integrated Circuits Conference, 99-102

[6] Burks, T.M., Sakallah, K.A., Mudge, T.N., Critical paths in circuits with level-sensitive latches, IEEE Trans. on VLSI systems, 3 (1995) 273-291

[7] PathMillPlus, Synopsys Products and Solutions Web page http://www.synopsys.com/products/analysis/pathmillplus_ds.html

Table 1. Performance comparison of the new timing abstraction models with the conventional gray box abstractions for a microprocessor block with 2 million transistors. Model generation + run times are shown in normalized time units.

Hierarchy Level	**Total**	**Level 3** (1 block)	**Level 2** (2 blocks)	**Level 1** (44 blocks)
New Abstraction Generation + Run Time (rel. u.)	12.90	1.00	4.00	7.90
Conventional Gray Box Generation + Run Time (rel. u.)	38.84	16.07	14.02	8.74
New Abstraction Maximum Memory (MB)	503	164	405	503
Conventional Gray Box Maximum Memory (MB)	1369	1369	814	468
New Abstraction File Size (MB)	65.1	4.4	11.2	49.5
Conventional Gray Box File Size (MB)	336.1	103.4	113.5	119.2

An Implication-based Method to Detect Multi-Cycle Paths in Large Sequential Circuits

Hiroyuki Higuchi
Fujitsu Laboratories Ltd.
4-1-1, Kamikodanaka, Nakahara-Ku,
Kawasaki 211-8588, Japan
higuchi@flab.fujitsu.co.jp

ABSTRACT

This paper proposes a fast multi-cycle path analysis method for large sequential circuits. It determines whether or not all the paths between every flip-flop pair are multi-cycle paths. The proposed method is based on ATPG techniques, especially on implication techniques, to utilize circuit structure and multi-cycle path condition directly. The method also checks whether or not the multi-cycle path may be invalidated by static hazards in combinational logic parts. Experimental results show that our method is much faster than conventional ones.

Categories and Subject Descriptors

B.6.3 [**Logic Design**]: Design Aids

General Terms

Algorithms, Designs, Verification

Keywords

multi-cycle path, sequential circuits, implication, ATPG

1. INTRODUCTION

Due to the steadily increasing demand for high performance integrated circuits, timing verification and optimization are becoming more and more important. The key for good timing verification and optimization is to compute circuit delay accurately and quickly.

The most common and easiest approach to compute circuit delay is based on topological delay. Topological delay can be too conservative because it ignores false paths and multi-cycle paths. False paths and multi-cycle paths relax timing constraints, which can be utilized in logic synthesis, layout, ATPG for delay faults, and static timing analysis. A false path in a circuit is a path that is never activated because of the circuit functionality and/or delay values of the circuit components. The problem of false paths in logic circuits have been studied extensively in recent years [1, 2, 3, 7]. Since the number of paths may be exponential in circuit size, path-based analysis may suffer from the combinatorial explosion.

A multi-cycle path in a sequential circuit is a combinational path that does not have to propagate signals in single clock cycle. Multi-cycle paths have also been investigated in literature to some extent [5, 8, 9, 10]. Multi-cycle path analysis methods can be grouped into path-based [5, 10] ones and non-path-based ones [8, 9]. Path-based methods suffer from the combinatorial explosion in the same way as false path analysis. Non-path-based methods determine whether or not all the paths between a given flip-flop(FF) pair are multi-cycle paths. Therefore they do not suffer from the combinational explosion. While it is rare for every path between an input and an output to be false paths, the same is not true of multi cycle paths. Two methods for non-path-based analysis have been proposed; symbolic traversal based one [8] and SAT-based one [9]. Though SAT-based method is faster than symbolic traversal based one, it is still not applicable to large circuits.

In this paper we propose a fast analysis method to detect multi cycle paths in large sequential circuits. Our method belongs to the category of non-path-based methods and is based on automatic test pattern generation(ATPG) techniques, especially on implication techniques, to utilize circuit structure and multi-cycle path conditions directly. We also show that conventional non-path-based methods for multi-cycle path analysis can be optimistic because they do not consider static hazards in combinational logic parts. Our method checks whether or not detected multi-cycle paths are valid even when static hazards are taken into account. It uses static sensitization and co-sensitization conditions.

The rest of this paper is organized as follows. In Section 2, we review definitions and notations. In Section 3, we define the problem of detecting multi-cycle paths. In Section 4, we propose a new multi-cycle path detection method. In Section 5, we also propose a method to check whether or not the multi-cycle paths detected by the method described in Section 4 are really multi-cycle paths under the existence of static hazards. In Section 5, we give experimental results.

2. DEFINITIONS

2.1 Circuits and Paths

We consider synchronous sequential circuits. They con-

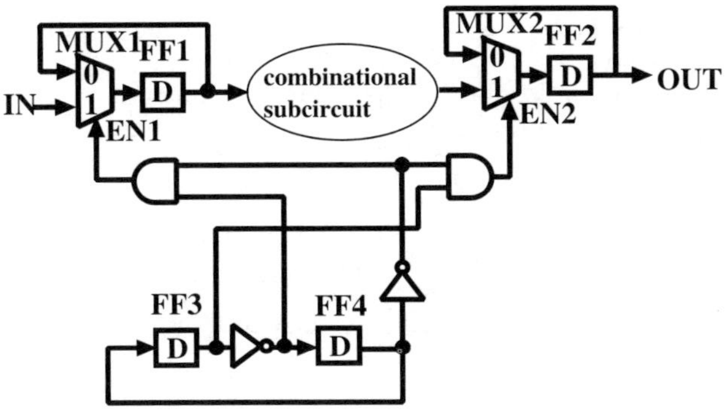

Figure 1: Example of multi-cycle paths

sist of an interconnection of combinational logic gates and clocked FFs. We assume that all the FFs are positive edge triggered D-FF driven by one clock. No direct combinational feedback is allowed.

A **combinational path** P in a sequential circuit is an alternating sequence of gates and edges $\{g_0, e_0, g_1, e_1, \ldots, g_{m-1}, e_{m-1}, g_m\}$, where g_0 is an primary input or an FF and g_m is an primary output or a fanin gate of an FF. Edge $e_i, 0 \leq i \leq m-1$, is an **on-input** of P which connects gate g_j to gate g_{j+1}. An edge is called a **side-input** of P with respect to e_i if it is connected to g_{i+1} but not originated from g_i. The **controlling value** at an input to a gate is the value that determines the output value of the gate independent of the other inputs. The **non-controlling value** to a gate is the complementary value of the controlling value. We say a gate g_i has the **controlled value** if one of its inputs has a controlling value; otherwise, we say g_i has a **non-controlled value**. A combinational path is simply called a path in this paper.

A signal value at the output of gate g_i is simply denoted by g_i. A signal value at the output of gate g_i at time t is denoted by $g_i(t)$.

2.2 Multi-Cycle Paths

A **multi-cycle path** in a sequential circuit is a path that does not have to propagate signals in single clock cycle. A **k-cycle path** is a path that is allowed to use k clock cycles to propagate signals. A **single-cycle path** is a path that is not a multi-cycle path. For example, consider a circuit shown in Fig.1. In the circuit IN and OUT are a primary input and a primary output respectively. FFs FF_3 and FF_4 constitute a 4-cycle gray code counter. The state transitions of the counter is as follows: $(0,0) \to (0,1) \to (1,1) \to (1,0) \to (0,0) \to \cdots$. Multiplexer MUX_1 selects primary input data IN when $(FF_3, FF_4) = (0,0)$. Then FF_1 is set to the value IN when $(FF_3, FF_4) = (0,1)$. On the other hand, MUX_2 selects the output of the combinational subcircuit when $(FF_3, FF_4) = (1,0)$. Then FF_2 is set to the value when $(FF_3, FF_4) = (0,0)$. Since the counter (FF_3, FF_4) requires 3 clocks to go from state (0,1) to state (0,0), the paths from FF_1 to FF_2 are 3-cycle paths. This means that the paths are allowed to use 3 clock cycle to propagate signals. As a consequence, the timing constraint of the paths can be relaxed from 1 clock cycle to 3 clock cycle.

A **multi-cycle FF pair** (FF_i, FF_j) is an ordered pair of FFs such that every path from FF_i to FF_j is a multi-cycle path. In Fig.1, FF pair (FF_1, FF_2) is a multi-cycle FF path.

2.3 Path Sensitization Conditions

A path is **statically sensitizable** if there exists an input vector such that all the side-inputs along the path settle to non-controlling values for that vector. This sensitization condition is delay-independent. In this paper we say that a set of paths is statically sensitizable if at least one of the paths is statically sensitizable.

A path $P = \{g_0, e_0, \ldots, g_{m-1}, e_{m-1}, g_m\}$ is **statically co-sensitizable** to a 1(0) if there exists an input vector v satisfying the following conditions [3]:

1. $g_m(v) = 1(0)$,
2. for each $g_i, 1 \leq i \leq m$, if g_i has a controlled value, then the edge e_{i-1} presents the controlling value of g_i.

In this paper we say that a path is statically co-sensitizable if the path is statically co-sensitizable to at least one logic value. Static co-sensitization condition is also delay-independent. Note that any statically sensitizable path is also statically co-sensitizable. In this paper we say that a set of paths is statically co-sensitizable if at least one of the paths is statically co-sensitizable.

Given a delay assignment for a delay model, a path $P = \{g_0, e_0, \ldots, g_{m-1}, e_{m-1}, g_m\}$ is **sensitizable** (or **true**) if and only if there exists an input vector satisfying the following conditions at each gate $g_i, 1 \leq i \leq m$ [2, 3]:

1. If all inputs of g_i have non-controlling values, then g_{i-1} must be the last input of g_i to present the non-controlling value.
2. If at least one input of g_i has a controlling value, then g_{i-1} must be the earliest input of g_i to present the controlling value.

Note that for any delay assignment any sensitizable path is also statically co-sensitizable and any statically sensitizable path is also sensitizable. This means that static sensitization and static co-sensitization gives lower bound and upper bound to the exact sensitization condition respectively.

3. PROBLEM FORMULATION

3.1 Condition for Multi-Cycle FF Pairs

A pair of FFs (FF_i, FF_j) is a multi-cycle FF pair if and only if whenever a transition occurs at the source FF_i, at least one of the following two conditions is satisfied:

1. The transition is not propagated to the sink FF_j.
2. The transition is propagated to the sink, but
   (a) the transition at the sink is never observed at any primary output, and
   (b) for any FF FF_k, (FF_j, FF_k) is a multi-cycle FF pair under the assumption that a transition is propagated from FF_i to FF_j in the previous clock cycle.

It should be mentioned here that the signal at the source is not necessarily stable for more than one clock cycle before the signal is required at the sink. When the value after a transition at the source is overwritten before it is used at the sink, the transition never arrive at the sink. Then you can say that the path is infinite cycle path, that is, false in the situation.

Condition 2 is difficult to check because the analysis may require traversal of many states. In addition to this, Condition 2 can be viewed as some kind of timing budget borrowing from the subsequent FF pair. Thus we consider only Condition 1 in this paper. We further simplify the condition 1 by using the following relationship:

- $FF_j(t+1) = FF_j(t+2) \Rightarrow$ "A transition at FF_i at time $t+1$ is not propagated to FF_j." [1]

Thus we use the following sufficient condition to detect multi-cycle FF pairs:

- (FF_i, FF_j) is a multi-cycle FF pair if $FF_i(t) \neq FF_i(t+1) \Rightarrow FF_j(t+1) = FF_j(t+2)$.

Since this is a sufficient condition, FF pairs detected by using the condition are always multi-cycle FF pairs when we do not consider delay in combinational logic parts. The condition is the same as in [8, 9]. Note that while [8] take into account reachable states, [9] and our method assume that all the states can be reachable. [8] may detect more multi-cycle paths than [9] and ours.

3.2 Multi-Cycle FF Pair Detection Problem

In the next section we consider the following multi-cycle FF pair detection problem:

Inputs: a synchronous sequential circuit

Outputs: all the multi-cycle FF pairs (FF_i, FF_j) that satisfy the following condition:

$$FF_i(t) \neq FF_i(t+1) \Rightarrow FF_j(t+1) = FF_j(t+2)$$

We refer the condition above as MC condition.

4. IMPLICATION-BASED ANALYSIS FOR DETECTING MULTI-CYCLE PATHS

In this section we propose a fast algorithm for the multi-cycle FF pair detection problem described in the previous section. As you can see, the MC condition is nothing but implication relation. Thus our method utilizes implication procedure as much as possible to detect multi-cycle paths efficiently, while conventional non-path-based methods attempt to check the condition by BDD-based state traversal [8] or SAT solver [9].

4.1 Basic Flow of the Analysis

To check the MC condition, we need to deal with 2 time frames. Therefore we use the expanded combinational logic part in 2 time frames. The checking is done by using implication procedure and backtrack search for ATPG. Implication procedure directly attempts to check the MC condition. The backtrack search tries to prove that there is no input pattern which violates MC condition. Note that the step should also justify the premise of the MC condition, that is, there exists at least one pair of inputs and states which satisfies $FF_i(t) = FF_i(t+1)$. If the justification fails, the step answers that the MC condition holds.

The overall flow of our algorithm is as follows:

1) Check if there exists a path between the FFs in each FF pair. Drop the FF pair if there is no path.

2) Carry out random pattern simulation to drop FF pairs not satisfying the MC condition.

3) Expand the combinational logic part into 2 time frames. Let the first time frame be t.

4) For each remaining FF pair (FF_i, FF_j), do Steps 4.1)-4.2).

   4.1) For each possible assignment $(FF_i(t), FF_j(t+1)) = \{(0,0), (0,1), (1,0), (1,1)\}$, do Steps 4.1.1)-4.1.5).

   4.1.1) $FF_i(t+1) \leftarrow \overline{FF_i(t)}$.

   4.1.2) Carry out implication procedure, that is, assign as many mandatory values as possible.

   4.1.3) If $FF_j(t+2) = FF_j(t+1)$ or contradiction occurs during implication procedure, go to Steps 4.1). If $FF_j(t+2) = \overline{FF_j(t+1)}$, then identify (FF_i, FF_j) as a single-cycle FF pair and go to Step 4).

   4.1.4) Search an input pattern such that $FF_j(t+2) \neq FF_j(t+1)$.

   4.1.5) If an input pattern is found, identify (FF_i, FF_j) as a single-cycle FF pair and go to Step 4).

   4.2) Identify (FF_i, FF_j) as a multi-cycle FF pair. □

Though this algorithm is to detect multi-cycle FF pairs, it can be easily extended to detect k-cycle FF pairs ($k = 3, 4, \ldots$) by increasing the number of time frames in Step 3.

4.2 Example

For example, let us apply the algorithm described in the previous subsection to the circuit in Fig.1.

After Step 1, the following 9 FF pairs remain among 16 FF pairs: (FF_1, FF_1), (FF_1, FF_2), (FF_2, FF_2), (FF_3, FF_1), (FF_3, FF_2), (FF_3, FF_4), (FF_4, FF_1), (FF_4, FF_2), (FF_4, FF_3). After Step 2, the following 5 FF pairs remain: (FF_1, FF_1), (FF_1, FF_2), (FF_2, FF_2), (FF_3, FF_2), (FF_4, FF_1). Here we omit the explanation of the detail of random pattern simulation.

In Step 3 the combinational logic part is expanded into 2 time frames. In Step 4 let us choose (FF_1, FF_2), for example. In Step 4.1 let us choose assignment $(FF_1(t), FF_2(t+1))=(0,0)$. In Step 4.1.1 value $\overline{FF_1(t)}(=1)$ is assigned to $FF_1(t+1)$ to analyze such the case that a rise transition occurs at FF_1 at time $t+1$. Figure 2 shows the value assignment after implication procedure in Step 4.1.2. In the figure, a signal value in a circle represents the value assigned before the implication procedure and the other values represent implied values. Since we obtain $FF_2(t+2) = 0$, we can say that the signal at FF_2 never change at time $t+2$. This implies that the MC condition is satisfied. Thus when $(FF_1(t), FF_1(t+1), FF_2(t+1)) = (0,1,0)$, (FF_1, FF_2) are identified as a multi-cycle FF pair.

Similarly, the analysis is done for the assignments $(FF_1(t), FF_2(t+1)) = (0,1), (1,0), (1,1)$ one by one. If the FF pair cannot be identified as either single-cycle or multi-cycle FF pair, then the analysis is continued in Step 4.1.4 by

[1]The converse is not always true, because even when a transition at FF_i is not propagated to FF_j, a transition from another FF may be propagated.

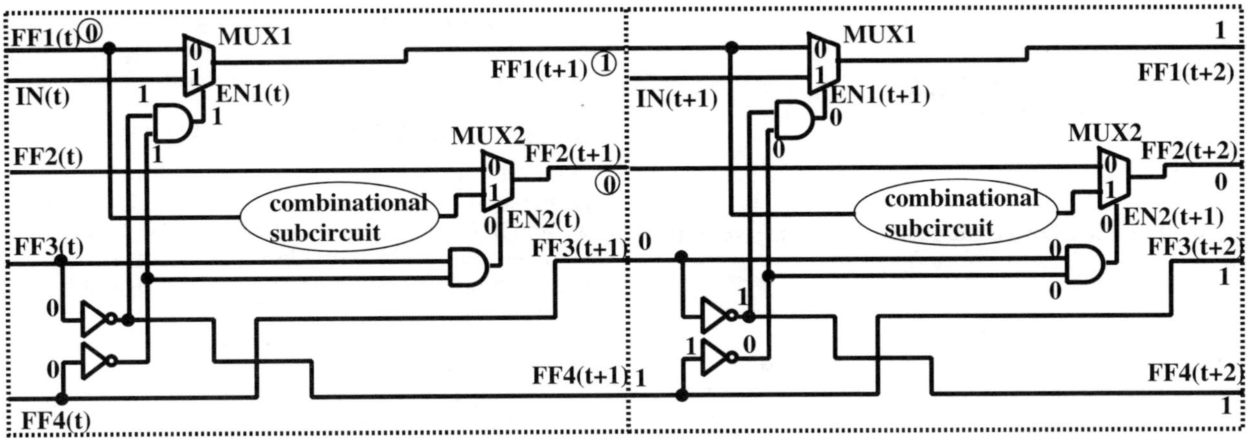

Figure 2: Example of implication procedure

using ATPG. In the circuit shown in Fig.1, all 5 candidates after random pattern simulation are identified as multi-cycle FF pairs.

4.3 Random Pattern Simulation

Random pattern simulation is done by parallel pattern simulation. Specifically, a random pattern of one word length is assigned to each primary input and each output of FFs and simulation is done for two clock cycles. Then it is checked whether or not for each FF pair (FF_i, FF_j) there is a bit position such that the following condition is satisfied:

$$FF_i(t) \neq FF_i(t+1) \wedge FF_j(t+1) \neq FF_j(t+2).$$

If there is such a bit position, the pair is dropped as a single-cycle FF pair. The above logic operation is done by bit-wise operation in parallel. The simulation for 2 clocks are repeated for a given number of patterns.

4.4 Implication

Local implication procedure attempts to assign as many mandatory values as possible by propagating already assigned values. A contradiction during implication procedure indicates that the combination of already assigned values is impossible. In that case there is no need to check whether or not the given FF pair is a multi-cycle FF pair for the assignment.

Global implication like static learning [11] can detect more implication relations than local implication.

4.5 ATPG

Multi-cycle FF pair detection problem has the following properties:

- Several value assignments have already been done before ATPG is applied.
- The target fault is likely to be redundant.

From these point of view, we adopted D-algorithm based method because it assigns values to internal nodes directly and tries to detect contradictions faster than PODEM based method.

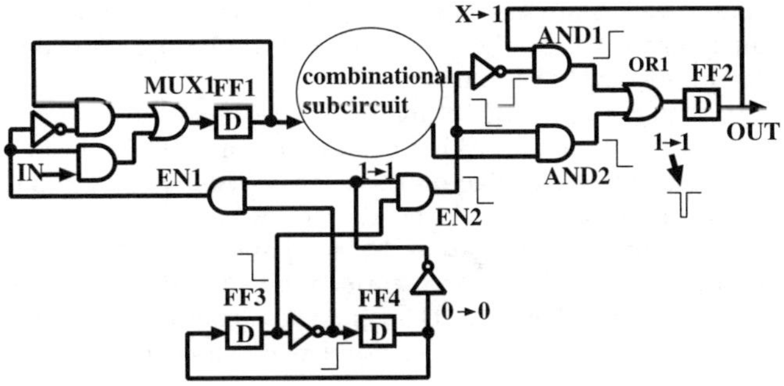

Figure 3: Example of static hazards

5. TAKING INTO ACCOUNT STATIC HAZARDS IN COMBINATIONAL LOGIC

In this section we show that multi-cycle FF pairs detected by our method and the conventional non-path-based methods are not necessarily multi-cycle FF pairs if we take into account static hazards in the combinational logic part. Then we give methods to detect possible static hazards in delay-independent manners.

5.1 Effects of Static Hazards

The following proposition is not always true if we take into account static hazards in the combinational logic part:

- If the MC condition for an FF pair is true, then the FF pair is a multi-cycle FF pair.

For example, let us assume that the circuit shown in Fig.1 is technology-mapped as in Fig.3. Only each multiplexer is replaced with 2 AND gates, 1 OR gates, and 1 NOT gate. Let us consider FF pair (FF_3, FF_2). The pair was identified as a multi-cycle FF pair according to the MC condition. The value assignment in Fig.3 shows the implied assignment when $(FF_3(t), FF_3(t+1), FF_2(t+1)) = (1, 0, 1)$ and $FF_2(t+2) = 1$. Assignment $FF_2(t+2) = 1$ is added because the FF pair satisfies the MC condition. From Fig.3 you can see that the input of FF_2 may have a static hazard when the output of AND_1 is slower than that of AND_2.

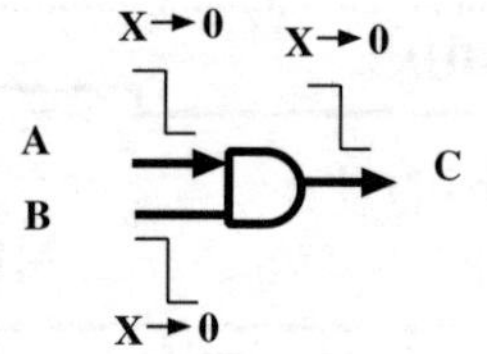

Figure 4: Unsensitizable situation in static sensitization

Since the method shown in Section 4 and conventional non-path-based methods only consider the stable function value, they overlook static hazards. When the FF pair is regarded as a multi-cycle FF pair and its timing constraints are relaxed, problems may occur. For example, assume that the output of AND1 become much slower by either replacing the AND gate with slower one or detouring the routing. Then the signal transition at AND1 exceeds the clock boundary. That may violate the circuit functionality. Therefore it is important to check the existence of static hazards for identified multi-cycle FF pairs.

It requires a lot of computation cost to exactly detect static hazards for a given FF pair, because static hazards can occur at every edge on the paths between the FF pair and in addition it should be checked whether or not each hazards can be propagated. In this paper we utilize static sensitization condition and static co-sensitization condition to detect static hazards. By using these conditions, we can analyze static hazards in delay-independent manners.

5.2 Checking based on Static Sensitization

In case of Fig.3 static sensitization condition can find the static hazard, because the path through AND_1 and OR_1 is statically sensitizable. Thus static sensitization can determine whether or not a static hazard occurs. However checking based on static sensitization may introduce conservative results on propagating the static hazard, because an sensitizable path is not always statically sensitizable. For example, Fig.4 shows a case where a path from A to C is not statically sensitizable, because B has a controlling value and blocks the path from A to C. Note that the values in the first clock cycle is unknown(X) because we should take into account static hazards. Suppose that the path from A to C is detected as a multi-cycle path by our method shown in Section 4 and that checking based on static sensitization find a static hazard and the static hazard disappears at the AND gate in Fig.4. This implies that the static hazard disappears under the condition that the side-input B blocks the on-input at the second clock. Threfore if a path from B to C is also detected by a multi-cycle path and the timing constraint is relaxed, then the static hazard may propagate, which means that the timing constraint of path from A to C cannot be relaxed. Thus multi-cycle FF pairs verified by static sensitization condition may introduce dependency between them.

5.3 Checking based on Static Co-Sensitization

Since a sensitizable path is always statically co-sensitizable, it does not suffer from dependency between multi-cycle paths. That is, multi-cycle FF pairs verified by static co-sensitization are always multi-cycle FF pairs. However there may not be static hazards on the path, because static co-sensitization gives upper bound to the exact sensitization. In Fig.4 the path from A to C is statically co-sensitizable.

6. EXPERIMENTAL RESULTS

We have implemented a multi-cycle FF pair detection program based on the method described so far. The program is written in C++ language and runs on Pentium III(866MHz, main memory 512MByte)+FreeBSD4.5R+gcc version 2.95.3 with -O2 option. ATPG techniques are based on [6]. The experiments were done on ISCAS89 benchmark circuits [4].

In the experiments random pattern simulation was continued until no FF pairs were dropped during 32×10 consecutive patterns. The number of backtracks in ATPG was limited to 50 except for "s9234.1","s9234","s13207.1","prolog". For these circuits the maximum numbers of backtracks were set to 650,150,150,51, and static learning [11] was done before ATPG.

Table 1 shows the experimental results of detecting multi-cycle FF pairs without checking static hazards. We compared our method with conventional SAT-based method [9]. CPU time reported in [9] was measured on Pentium II (500MHz, main memory 512MByte). In table 1, Columns "In", "FF", "FF-pair" show the numbers of the primary inputs, FFs, and the topologically connected FF pairs respectively. Column "ours" shows the results of our methods. Column "[9]" represents the results of conventional SAT-based method [9]. Column "MC-pair" indicates the number of detected multi-cycle FF pairs. Column "CPU(sec)" represents CPU time in seconds. Entry "–" means that the data were not reported in [9].

Table1 indicates that our method can detect multi-cycle FF pairs efficiently for all the benchmark circuits. All the FF pairs were identified as single-cycle FF pairs or multi-cycle FF pairs. For circuits with more than 1,000 FFs, our method can detect all the multi-cycle FF pairs in reasonable time. The table also shows that the number of multi-cycle FF pairs is approximately one-tenth as large as the number of all the topologically connected FF pairs.

The proposed method is much faster than the conventional SAT-based one [9], though machines on which CPU times were measured are different. The number of the multi-cycle FF pairs detected by our method is the same as that by conventional one except for the self-loop FF pairs, which seem to be excluded beforehand in the conventional method.

Table 2 shows the total numbers of identified FF pairs and CPU times in each analysis step. Column "Sim.", "Implication", and "ATPG" represent random pattern simulation, implication procedure, and ATPG respectively. Rows "single cycle" and "multi cycle" indicates the numbers of FF pairs identified as single-cycle paths and multi-cycle paths in each step respectively. Data in the parentheses show the ratio of the number of the detected FF pairs. The table shows that almost all the single-cycle paths can be dropped by random pattern simulation and more than 80% of all the multi-cycle paths can be detected by implication procedure. It is probably the main reason why our method is faster than the SAT-based method.

Table 3 shows the total number of multi-cycle paths after static hazard checking and total CPU times required for the checking. Row "before" indicates data before static hazard checking. Rows "sensitize" and "co-sensitize" representns data after checking based on static sensitization and static

Table 1: Experimental results of detecting multi-cycle FF pairs without checking static hazards

circuit	In	FF	FF-pair	ours MC pair	ours CPU (sec)	[7] CPU (sec)
s27	4	3	7	0	0.3	0.5
s208.1	10	8	36	28	0.3	-
s208	11	8	36	0	0.3	0.5
s298	3	14	70	3	0.3	0.9
s344	9	15	89	1	0.3	1.3
s349	9	15	89	1	0.3	1.3
s382	3	21	146	13	0.3	2.0
s386	7	6	36	4	0.3	0.9
s400	3	21	146	13	0.3	-
s420.1	18	16	136	120	0.4	-
s420	19	16	88	0	0.3	1.1
s444	3	21	146	13	0.3	2.6
s499	1	22	484	379	0.6	-
s510	19	6	36	3	0.3	1.4
s526	3	21	144	7	0.4	2.5
s526n	3	21	144	7	0.3	2.5
s635	2	32	528	0	0.5	-
s641	35	19	115	1	0.3	1.9
s713	35	19	115	1	0.3	2.6
s820	18	5	25	0	0.3	1.7
s832	18	5	25	0	0.3	1.6
s838.1	34	32	528	496	0.9	-
s838	35	32	192	0	0.3	3.0
s938	34	32	538	496	0.9	-
s953	16	29	156	29	0.5	6.2
s967	16	29	156	29	0.5	-
s991	65	19	71	20	0.4	-
s1196	14	18	20	0	0.3	2.2
s1238	14	18	20	0	0.4	2.4
s1423	17	74	1765	47	0.4	60.4
s1488	8	6	36	0	0.3	2.0
s1494	8	6	36	0	0.4	2.0
prolog	36	136	578	29	0.6	-
s1269	18	37	288	3	0.3	-
s1512	29	57	513	1	0.4	-
s3271	26	116	899	0	0.5	-
s3330	40	132	549	0	0.5	-
s3384	43	183	1831	0	0.5	-
s4863	49	104	628	0	1.0	-
s5378	35	179	1200	55	0.9	-
s6669	83	239	2179	0	0.6	-
s9234.1	36	211	2681	37	12.3	-
s9234	19	228	2830	168	11.7	-
s13207.1	62	638	3411	580	18.8	-
s13207	31	669	3716	937	13.2	-
s15850.1	77	534	11873	320	7.6	-
s15850	14	597	15363	3756	9.6	-
s35932	35	1728	4763	0	2.3	-
s38417	28	1636	33852	240	36.1	-
s38584.1	38	1426	16372	17	9.2	-
s38584	12	1452	17978	1622	13.9	-
Total		10908	125504	9476	152.4	(103.5)

Table 2: Results of each analysis step

	Sim.	Implication	ATPG
single cycle	107974 (86%)	240 (0.2%)	7914 (6.3%)
multi cycle	0 (0%)	7712 (6.1%)	1764 (1.4%)
CPU(sec)	31.0	29.0	74.4

Table 3: Results of static hazard checking

	MC-pair	CPU(sec)
before	9,476	152.4
sensitize	8,084	57.1
co-sensitize	5,570	62.5

co-sensitization respectively. It shows that one-tenth of the multi-cycle FF pairs detected without static hazard checking may have static hazards at the input of FFs and three-tenth of them may depend on one another. CPU times required to detect static hazards are practical.

7. CONCLUSIONS

We have proposed a fast multi-cycle path analysis method for large sequential circuits. We have also shown that conventional non-path-based methods can be optimistic because they do not consider static hazards and introduced a method to detect static hazard in delay-independent manners. Experimental results indicates that the method is much faster than conventional methods.

8. REFERENCES

[1] D. Brand and V. S. Iyengar. Timing Analysis using Functional Relationships. In *Proc. of IEEE International Conference on CAD-86*, pages 126–129, Nov. 1986.

[2] H.-C. Chen and D. H.-C. Du. Path Sensitization in Critical Path Problem. *IEEE Trans. Comput.-Aided Design Integrated Circuits*, 12(2):196–207, Feb. 1993.

[3] S. Devadas, K. Keutzer, and S. Malik. Computation of Floating Mode Delay in Combinational Circuits: Theory and Algorithms. *IEEE Trans. Comput.-Aided Design Integrated Circuits*, 12(12):1913–1923, Dec. 1993.

[4] D. B. F. Brglez and k. Kozminski. Combinational Profiles of Sequential Benchmark Circuits. In *Proc. of the International Symposium on Circuits and Systems*, May 1989.

[5] A. P. Gupta and D. P. Siewiorek. Automated Multi-Cycle Symbolic Timing Verification of Microprocessor-based Designs. In *Proc. of the 31th ACM/IEEE Design Automation Conference*, pages 113–119, 1994.

[6] Y. Matsunaga and M. Fujita. Enhanced Unique Sensitization for Efficient Test Generation. *IEICE Trans. on Information and Systems*, E76-D(9):1114–1120, Sept. 1993.

[7] P. C. McGeer, A. Saldanha, R. K. Brayton, and A. Sangiovanni-Vincentelli. Delay Models and Exact Timing Analysis. In T. Sasao, editor, *Logic Synthesis and Optimization*, pages 167–189. Kluwer Academic Publishers, 1993.

[8] K. Nakamura, K. Takagi, S. Kimura, and K. Watanabe. Waiting False Path Analysis of Sequential Logic Circuits for Performance Optimization. In *Proc. of IEEE/ACM International Conference on CAD-98*, pages 388–393, Nov. 1997.

[9] K. Nakamura, S. Maruoka, S. Kimura, and K. Watanabe. Multi-Cycpe Path Detection based on Propositional Satisfiability with CNF Simplification using Adaptive Variable Insertion. *IEICE Trans. on Fundamentals*, E83-A(12):2600–2607, Dec. 2000.

[10] W.-C. Lai, A. Krstic, and K.-T. Cheng. Functionally Testable Path Delay Faults on a Microprocessor. *IEEE Design & Test of Computers*, oct 2000.

[11] M. Schulz, E. Trishler, and T. Sarfert. SOCRATES: A Highly Efficient Automatic Test Pattern Generation System. In *Proc. of International Test Conference*, pages 1016–1026, 1987.

The Wearable Motherboard: A Framework for Personalized Mobile Information Processing (PMIP)

Sungmee Park
School of Textile & Fiber Engineering
Georgia Institute of Technology
Atlanta, Georgia 30332
+1 404-894-6529
sungmee.park@tfe.gatech.edu

Kenneth Mackenzie
College of Computing
Georgia Institute of Technology
Atlanta, Georgia 30332
+1 404-894-1704
kenmac@cc.gatech.edu

Sundaresan Jayaraman[1]
School of Textile & Fiber Engineering
Georgia Institute of Technology
Atlanta, Georgia 30332
+1 404-894-2490
sundaresan.jayaraman@tfe.gatech.edu

ABSTRACT

Textiles and computing share a synergistic relationship, which is being harnessed to create a new paradigm in personalized mobile information processing (PMIP). In this paper, we provide an overview of this "interconnection" between the two fields and present the vision for "E-Textiles," which represents the convergence of the two fields. We discuss the role of the Georgia Tech Wearable Motherboard in pioneering this paradigm of "fabric is the computer" and serving as a framework for PMIP. Finally, recent research in this area resulting in the realization of a "computational *fabric* network" is discussed.

1. TEXTILES & COMPUTING

John Kay's invention of the flying shuttle in 1733 sparked the *first* Industrial Revolution, which led to the transformation of *industry* and subsequently of civilization itself. Yet another invention in the field of textiles – the Jacquard head by Joseph Marie Jacquard (*circa* 1801) – was the first *binary information processor*. At any given point, the thread in a woven fabric can be in one of two *states* or positions: on the face of the fabric or on the back. *Pattern* cards were punched or cut according to the required fabric design. A hole in the card signified that the thread would appear on the face of the fabric, while a blank meant that the end would be left down and appear on the back of the fabric. The Jacquard head was used on the weaving loom or machine for raising and lowering the *warp* threads to form desired patterns based on the *lifting plan* or *program* embedded in the cards. Thus the Jacquard mechanism set the stage for modern day binary information processing.

Ada Lovelace, the benefactor for Charles Babbage who worked on the Analytical Engine (the predecessor to the modern day computer), is said to have remarked, "The Analytical Engine weaves algebraic patterns just as the Jacquard loom weaves flowers and leaves." The Jacquard mechanism that inspired Babbage and spawned the Hollerith punched card has been instrumental in bringing about one of the most profound technological advancements known to humans, viz., the *second* Industrial Revolution also known as the Information Processing Revolution [1]. In fact, when Intel introduced its Pentium class of microprocessors in the 1990s, one of the advertisements had a "fabric of chips" emerging from a weaving machine; this picture eloquently captured the essence of chip making – a true blending of art and science – much like the design and production of textiles!

More recently, the design and realization of the Georgia Tech Wearable Motherboard represents a significant advancement in the integration of textiles and computing paving the way for the paradigm of "fabric is the computer" [2, 3].

2. THE E-TEXTILES VISION: "FABRIC IS THE COMPUTER"

Figure 1 depicts the vision for "E-Textiles" or the paradigm of "fabric is the computer." The major facets illustrate the various "building blocks" of the system that must be seamlessly integrated to realize the vision, starting with the underlying physical fabric or "*Platform*." The design of this platform or infrastructure involves the exploration of materials, structures and manufacturing technologies. The second key facet for realizing this paradigm of a true computational *fabric* is the "*Interconnect Architecture*" in the fabric, which involves the design and incorporation of physical data paths and interconnection technologies, i.e., the realization of "textile electrical circuits." Integration of sensors, microchips and other devices (e.g., for communication and control) is critical for the realization of an "intelligent" E-Textiles for *any* application, say for example, battlefield management, and therefore, "*Hardware Integration*" constitutes the third facet or building block shown in Figure 1. Issues related to information processing such as fault tolerance in light of manufacturing defects and Quality of Service (QoS) *within* the E-Textile and *between* the E-Textile and external agents/devices are critical for the incorporation and optimal utilization of computing resources, and therefore, "*Software*" is the fourth facet of the E-Textile continuum. And finally, as shown in the figure, a set of underlying *performance metrics* ranging

[1] To whom correspondence should be addressed.

DAC 2002, June 10-14, 2002, New Orleans, Louisiana, USA.

from the physical dimensions (of the resulting structure/system) to costs, manufacturability and data flow rates must be utilized to assess the successful realization and performance of the desired E-Textile. Thus this paradigm of "fabric is the computer" represents a fascinating area of research that fosters, nay necessitates, collaboration amongst scientists and engineers from a variety of disciplines including textiles, computing and communications, sensor technologies and application domains (e.g., medicine, space, military).

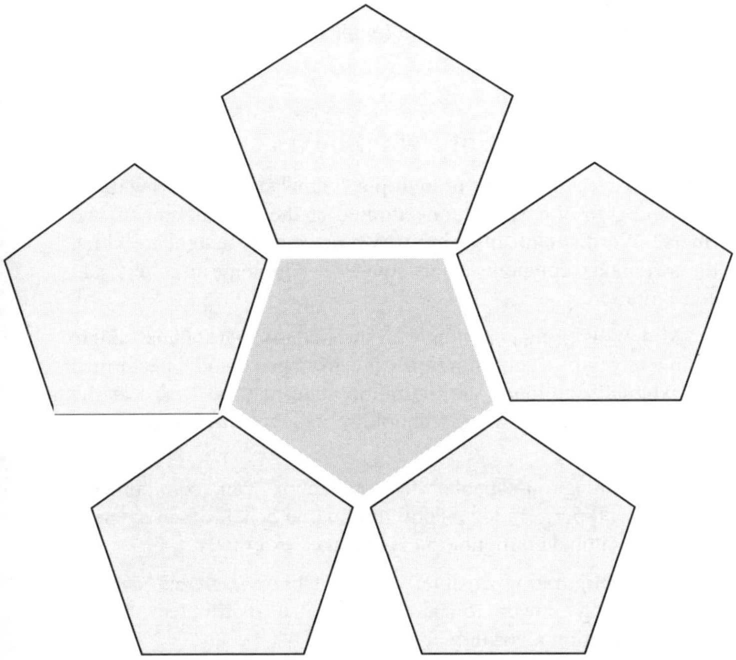

Figure1. The E-Textiles Vision

3. THE WEARABLE MOTHERBOARD

The Georgia Tech Wearable Motherboard (GTWM) provides an extremely versatile framework for the incorporation of sensing, monitoring and information processing devices. It uses optical fibers to detect bullet wounds, and special sensors and interconnects to monitor the body vital signs of individuals. The principal advantage of GTWM is that it provides, for the first time, a very systematic way of monitoring the vital signs of humans in an *unobtrusive* manner (Figure 2). Appropriate sensors have been "plugged" into this motherboard using the developed Interconnection Technology and attached to any part of the individual being monitored, thereby creating a flexible wearable monitoring device. The flexible data *bus* integrated into the structure transmits the vital signs information to monitoring devices such as an EKG Machine, a temperature recorder, a voice recorder, etc. The *bus* also serves to transmit information *to* the sensors (and hence, the wearer) from external sources, thus making GTWM a valuable information infrastructure [3]. GTWM is lightweight and can be worn easily by anyone -- from infants to senior citizens.

Figure 2. Georgia Tech Wearable Motherboard

Although it started off as a "textile engineering" endeavor, the research has led to an even more groundbreaking contribution with enormous implications: the creation of a *wearable integrated information infrastructure* that has opened up entirely new frontiers in personalized information processing, healthcare and telemedicine, and space exploration, to name a few [4]. Until now, it has not been possible to create a personal information processor that was customizable, wearable and comfortable; neither has there been a garment that could be used for unobtrusive monitoring of the vital signs of humans on earth or in space. A Special Issue of LIFE Magazine entitled *Medical Miracles for the Next Millennium* (Fall 1998) featured the Smart Shirt as one of the "21 Breakthroughs that Could Change Your Life in the 21st Century" thus clearly demonstrating the potential impact of the Wearable Motherboard technology on humans in the next century.

4. ADVANCING THE PARADIGM: RECENT RESEARCH

While the interconnection technology in GTWM facilitates the "routing" of information (i.e., signals) from any point to any other point on the wearable motherboard, the data paths are pre-determined. To realize the concept of a "programmable" computing device as an *integral* part of the fabric/garment in which "the fabric is the computer," there is a need to design and develop the *architecture* for the wearable motherboard that would seamlessly integrate the hardware, software and softw*ear* (fabric/garment) components or building blocks of the system. Such an architecture will facilitate the real-time reconfiguration of the system 'building blocks' that would lead to the realization of a truly adaptive and responsive wearable computational fabric system thus resulting in *pervasive/invisible* information processing.

Research is being carried out to explore the feasibility of creating such an E-Textile to demonstrate PMIP using the wearable motherboard. As the first step, the information from one or more sensors (e.g., electrocardiogram or voice) is being routed through the "soft" interconnects in the fabric to the desired output points using FPGAs (field-programmable gate arrays).

4.1 The Approach

A "switchbox" approach has been chosen to combine the conductive fibers of an E-Textile into a programmable network. The switchbox approach is to treat the conductive fibers like the wiring resources in an FPGA to which switching components can be added at strategic intersections. A key problem in the switchbox architecture is to tolerate loose manufacturing tolerances since it is not known in advance as to which wires are connected to which pins on the switchbox. Therefore, we suggest a design for a single-chip, integrated switchbox and, simultaneously, build a demonstration prototype using off-the-shelf components.

4.2 The Architecture

The architecture consists of conductive fibers in the fabric plus switchboxes, which are affixed (like buttons) atop intersections of the fibers. The architecture leads to three decisions [5]:

1. Placement of switchboxes;
2. Complexity of wiring; and
3. Complexity of switchboxes.

4.2.1 Placement of Switchboxes

The manufacturing tolerance for placement of switchboxes can be tight or loose.

a. Exact: switchboxes placement tolerance could be good enough to place switchbox contacts atop particular conductive fibers. This option requires precise manufacturing tolerance or a large fiber-to-fiber pitch.

b. Close: the placement tolerance is a small factor larger than the fiber pitch.

c. Random: switchbox placement is uncontrolled.

Since one of the overall objectives of this research is to produce such E-Textiles in a typical manufacturing environment, option (b) has been chosen since it is similar to fastener placement in apparel manufacturing. We assume that a switchbox covering several fibers can be placed such that it will contact a particular fiber but we do not know which switchbox contact will actually make the contact.

4.2.2 Complexity of Wiring

The conductive fibers in the E-Textiles can be continuous or cut at each switchbox.

a. Continuous fibers are easily manufactured.

b. Cut fibers lead to a richer interconnect with higher local bandwidth between points in the fabric.

From a manufacturing standpoint, option (a) is easier and so fibers (yarns) that are continuous in the fabric have been chosen. We address restricted bandwidth by making the switchboxes more capable.

4.2.3 Complexity of Switchboxes

Switchboxes can be built at multiple "grain" sizes, i.e., how much computing resources are concentrated at the intersection of two fibers. We distinguish grain sizes based on the extent to which the switchbox component participates in topology discovery and configuration.

a. Minimalist, e.g., a single transistor or gate at a single intersection: configuration is managed and performed externally although configuration state may be stored at the intersection, e.g., using technology analogous to floating gates in VLSI.

b. Communication-capable: the switchbox can communicate with, and perform local configuration on behalf of, an external agent. Global configuration is managed externally.

c. Self-configuration-capable: the switchbox contains enough processing power to participate in a distributed, global configuration algorithm.

We chose option (b), switchboxes capable of self-configuring to the point of establishing communication with an external agent which then manages global configuration.

Thus, we chose an architecture for the computational fabric in which the electronic elements are of moderate capability, are placed deliberately, but inexactly, and without making cuts in the fabric.

4.3 Technical Issues and Decisions

We have identified three key issues with our architecture and proposed solutions for them. They are: power distribution, configuration information distribution and automatic discovery of topology.

4.3.1 Power Distribution

Power distribution is difficult with inexact placement of components because power connections are usually distinguished from signal connections. We believe it may be possible to power ordinary integrated circuits via any pin using diode structures similar to existing static discharge protection structures. However, for the first demonstration prototype, we use distinguished power wires in the fabric by providing enough spacing to account for our placement tolerance (unlike signal wires).

4.3.2 Configuration Information Distribution

Configuration information distribution is similarly difficult because FPGAs and microcontrollers typically expect configuration information to be presented on specific pins. This problem is not fundamental, however. We use an FPGA-within-an-FPGA technique to address the problem: we use a statically-defined FPGA that includes configurable elements within it. Thus we can emulate an FPGA that accepts configuration information from multiple (initially any) pins.

4.3.3 Discovery of Topology

Topology discovery can be externally or internally managed and sequential or parallel. We adopt a conservative approach: an external agent sequences the discovery and the discovery proceeds sequentially from the element nearest the external agent.

4.4 Realization of the Architecture

The first step in realizing the architecture through a prototype demonstration has been to define an ideal switchbox element that we believe is buildable using current technology. The next step is to begin prototyping an approximation to the ideal device to serve as a proof-of-concept.

The ideal device is an EEPROM-based FPGA in a custom plastic package that contains insulation-displacement-style connectors instead of pins. The device would be press-fit onto the fabric like a fastener. Once configured, the device would provide digital communications between points on the fabric including sensors, effectors and communications devices that attach to the switchboxes or directly to conductive fibers that cross a switchbox.

The prototype device is a 2.8"x1.8" protoboard containing a small EEPROM-based FPGA (Altera EPM7160S) plus a microcontroller (Motorola HC11) as shown in Figure 3. The board connects to the fabric using two standard 26-pin insulation-displacement connectors (IDCs) ordinarily used for ribbon cable.

The fabric in Figure 3 contains conductive fibers at a density of 10 per inch. The 26-pin IDCs contain contacts at a pitch of 20 to the inch. We find that every fiber makes contact with some contact in the connector but the position of that contact is off by up to two positions in the connector. In the connector, the leftmost and rightmost fibers are dedicated to power buses while the center fibers carry signals, up to seven signals in each direction.

Figure 3. Prototype board, 2.8"x1.8", containing the FPGA and microcontroller. Site-specific sensor/effector/communication devices attach as "daughtercards."

The chosen FPGAs were physically integrated into the fabric (see Figure 4). Software was developed to demonstrate the "in-fabric" network. One of the FPGAs communicated with an external agent (a Linux-based personal computer) that was responsible for managing the global configuration of the FPGAs in the fabric by sequencing the "discovery" in the fabric beginning with that initial FPGA.

Figure 4. The PMIP Network in a Fabric

Two software modules were created; the first was to "demonstrate" the pin-connection discovery algorithm implemented in the system to identify the connections between the various pins on the FPGAs in the fabric and to display the connection paths. This enables discovery of the interconnects on the fly after the manufacturing has been carried out and there is no *a priori* knowledge of the specific connections between the elements in the fabric. The second module discovers the connections and displays the paths on the screen as the discovery process proceeds when the FPGA is powered. To demonstrate the flow of information in the fabric network through the soft interconnects, a potentiometer was attached as a daughterboard to one of the FPGAs and whenever it was "twiddled" (see Figure 4), the resulting change is displayed on the screen (Figures 5 and 6). This recent effort demonstrates the feasibility of realizing a programmable network in a fabric through *soft* interconnects and paves the way for the continued exploration of the "fabric is the computer" paradigm pioneered by the Georgia Tech Wearable Motherboard.

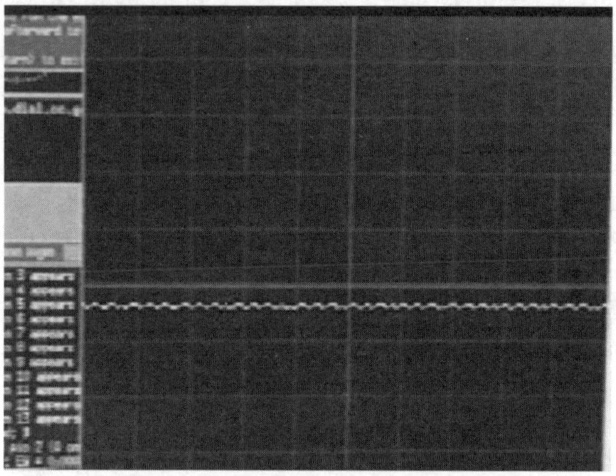

Figure 5. The Steady State Display

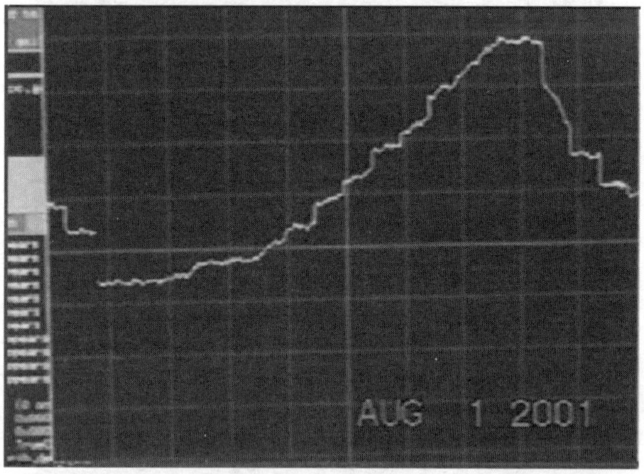

Figure 6. Flow of Information through the PMIP Fabric.

CONCLUDING REMARKS

A well-designed information processing system should facilitate the access of information **A**nytime, **A**nyplace by **A**nyone – the three ***A***s. The 'ultimate' information processing system should not only provide for large bandwidths, but also have the ability to see, feel, think, and act. In other words, the system should be totally 'customizable' and be in tune with the human. Of course, clothing is probably the only element that is 'always there' and in complete harmony with the individual (at least in a civilized society!). And, textiles provide the ultimate flexibility in system design by virtue of the broad range of fibers, yarns, fabrics, and manufacturing techniques that can be deployed to create products for desired end-use applications. A new avenue of research and exploration has opened up for the development of an integrated "textile-computing" system that can serve as a true information processing framework with the ability to sense, feel, think and act based on the end-user stimuli and/or the operational environment. This "fabric is the computer" paradigm exemplified by the Georgia Tech Wearable Motherboard demonstrates the feasibility of realizing personalized mobile information processing (PMIP) and gives new meaning to the term human-machine symbiosis in the context of pervasive/invisible computing.

ACKNOWLEDGEMENTS

The authors would like to thank Dr. Bob Graybill of DARPA for providing funding under Contract # F30602-00-2-0564 for the current research exploring the paradigm of "fabric is the computer." Initial research on the Wearable Motherboard was carried out under Contract # N66001-96-C-8639 from the US Department of Navy.

REFERENCES

[1] Jayaraman, S., "Designing a Textile Curriculum for the '90s: A Rewarding Challenge", *Journal of the Textile Institute*, 1990, 81(2) 185-194.

[2] The Georgia Tech Wearable Motherboard™: The Intelligent Garment for the 21st Century, http://www.gtwm.gatech.edu.

[3] Gopalsamy, C., Park, S., Rajamanickam, R., and Jayaraman, S, "The Wearable Motherboard™: The First Generation Of Adaptive And Responsive Textile Structures (ARTS) For Medical Applications", *Journal of Virtual Reality*, 1999, 4:152-168.

[4] Park, S., Gopalsamy, C., Rajamanickam, R., and Jayaraman, S., "The Wearable Motherboard™: An Information Infrastructure or Sensate Liner for Medical Applications", in Studies in Health Technology and Informatics, IOS Press, Vol. 62, pp. 252-258, 1999.

[5] Mackenzie, K., Hudson, D., Maule, S., Park, S., and Jayaraman, S., "A Prototype Network Embedded in Textile Fabric," Proceedings of CASES 2001, International Conference on Compilers, Architecture and Synthesis for Embedded Systems, November 16-17, Atlanta, Georgia, pp. 188-194, 2001.

Challenges and Opportunities in Electronic Textiles Modeling and Optimization*

Diana Marculescu, Radu Marculescu, and Pradeep K. Khosla
Department of Electrical and Computer Engineering
Carnegie Mellon University
{dianam,radum,pkk}@ece.cmu.edu

ABSTRACT - This paper addresses an emerging new field of research that combines the strengths and capabilities of electronics and textiles in one: electronic textiles, or e-textiles. E-textiles, also called Smart Fabrics, have not only "wearable" capabilities like any other garment, but also local monitoring and computation, as well as wireless communication capabilities. Sensors and simple computational elements are embedded in e-textiles, as well as built into yarns, with the goal of gathering sensitive information, monitoring vital statistics and sending them remotely (possibly over a wireless channel) for further processing. Possible applications include medical (infant or patient) monitoring, personal information processing systems, or remote monitoring of deployed personnel in military or space applications. We illustrate the challenges imposed by the dual textile/electronics technology on their modeling and optimization methodology.

Categories and Subject Descriptors: I.6 [Simulation and Modeling]: Modeling methodologies; B.8.2 [Performance and reliability]: performance analysis and design aids.

General terms: design, performance

1. INTRODUCTION

Electronic textiles or e-textiles are a new emerging interdisciplinary field of research which brings together specialists in information technology, microsystems, materials, and textiles. The focus of this new area is on developing the enabling technologies and fabrication techniques for the economical manufacture of *large-area*, *flexible*, *conformable* information systems that are expected to have unique applications for both consumer electronics and military industry.

E-textiles offer new challenges for designers and CAD tool developers due to their unique requirements, cutting across from the system to the device and technology:

- The need for a new *model of computation* intended to support widely distributed applications, with highly *unreliable* behavior, but with stringent constraints on *longevity* of the system.
- *Reconfigurability and adaptability* with low computational overhead. E-textiles rely on simple computing elements embedded into fabric or directly into active yarns. As operating conditions change (environmental, battery lifetime, etc.), the system has to adapt and reconfigure on-the-fly to achieve a better functionality.
- *Device and technology* challenges imposed by embedding simple computational elements into fabric, by building yarns with computational capabilities, or by the need of unconventional power sources and their manufacturing in filament or yarn form.

There have been a handful of attempts to design and build prototype computational textiles. In one of the earliest accounts [1], the authors demonstrate attaching off-the-shelf electrical components such as microcontrollers, surface mount LEDs, piezoelectric transducers etc., to traditional clothing material, transforming the cloth into a breadboard of sorts. In fabrics which contain conductive strands, these may be used to provide power to the devices, as well as to facilitate communication between devices. In [2], the authors extend the work presented in [1], detailing methods by which items such as user interfaces (keypads) and even chip packages, may be constructed directly by a textile process. The routing of electrical power and communications, through a wearable fabric, was addressed in [3] where the authors provide a detailed account of physical and electrical components for routing electricity through suspenders made of a fabric with embedded conductive strands. A complete apparel with embedded computing elements is described in [4]. The authors describe a jacket designed to be worn in the arctic environment, which augments the capabilities of the wearer with a global positioning system, sensors (accelerometers, conductivity electrode, heart rate monitors, digital thermometers) and heating. However, all these approaches are single design points and do not provide a methodology for the evaluation and validation of e-textile systems.

It is our belief that in support of e-textiles a new design paradigm is needed, as well as a more general model of computation which supports *local computation* and *inexpensive communication* among computational elements. In the classic design cycle (Fig.1(a)), the application is mapped onto a given platform architecture, under specified constraints (performance, area, power consumption). When these constraints are met, the prototype is tested, manufactured and used for running the application. In the case of e-textiles (Fig.1(b)), wide area textile networks are built onto e-textiles, with no prescribed functionality. To achieve high yields (that is, low defect rate), as well as high fault-tolerance later in the lifetime cycle, *regularity* is important. The application is modified so as to expose as much local computation as possible, via partitioning or refinement. At power-up, the application is mapped so as to optimize different metrics of interest (such as, quality of results, power consumption, operational longevity, fault-tolerance) and later re-mapped whenever operating conditions change.

Although e-textiles ostensibly present distributive computing challenges similar to those currently being pursued in adaptive control network and pervasive computing, the specific characteristics and demands of an e-textiles info-computing environment add new dimensions to those challenges. In fact, e-textiles impose specific challenges as opposed to other applications in the general area of networked systems:

*This research has been supported in part by DARPA under contract no. DABT63-96-C-0083 and SRC under grant 2001-HJ-898.

DAC 2002, June 10-14, 2002, New Orleans, Louisiana, , USA.

What e-textiles are *not*

- Classical data networks. While the underlying structure of an e-textile application implies the existence of many processing elements, connected in a Textile-Area Network (TAN), they have limited processing and storage capabilities, as well as very limited power consumption budgets. Hence, classic techniques and inherent methodologies for coping with mapping an application, communication among nodes and dealing with network failures (including congestion), are not appropriate. In addition, having very limited processing capabilities, e-textiles are not the equivalent of "desktops/laptops on a fabric," restricting significantly the range of applications that can be mapped on them.
- Networked embedded systems (including wireless sensor networks). Wireless sensor networks share with e-textile based applications the limited power budgets, and somewhat the limited processing and storage capabilities. However, communication among processing elements is wired, and thus, much less expensive than the wireless communication. In addition, as opposed to ad-hoc networks, the topological location of different processing elements is fixed throughout the lifetime of the application (although the mapping of the application on processing elements may change). Lastly, e-textiles must have low manufacturing costs, and thus, the defect rate of the processing nodes (and physical links) will be much higher and very different than in the case of wireless networks. More limited processing and storage capabilities, in conjunction with higher failure rates imply that existing body of research for sensor networks is not applicable directly to TANs.

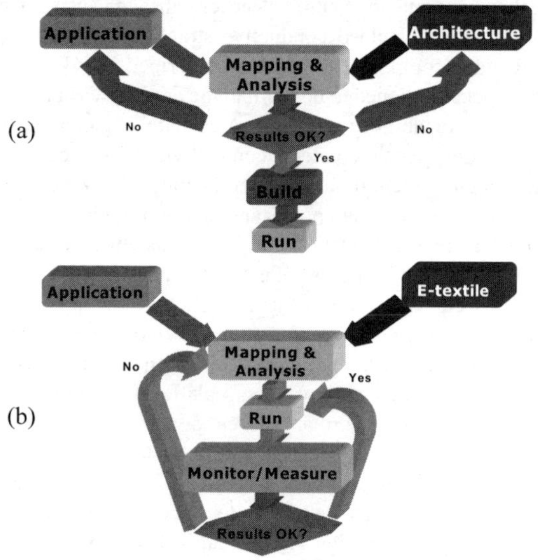

Fig.1 Classic design (a) vs. e-textile lifetime cycle (b)

What e-textiles are

E-textiles are "living designs," consisting of highly unreliable low-cost simple components and interconnect. These characteristics are in fact shared by other non-silicon computing systems (such as those based on nanoelectronics). In many ways, e-textiles systems will be the conformable, fabric equivalent of rigid body, embedded MEMS. However, unlike MEMS, e-textiles require coping with *tears* in the fabric, whether it is due to regular wear of the fabric, or due to unexpected damage. The salient features of e-textiles are:

- limited processing, storage and energy per computational node
- potential failures for *both* nodes and communication links
- highly *spatially and temporally correlated* node and/or link failures due to topological placement or due to external events
- need for scalability and flexibility of resource management, as well as local vs. global management trade-off
- active or smart links which embed simple logic (passive or active components)

In this paper, we present our vision on developing a methodology for *modeling*, *analysis* and *middleware support* for e-textiles that includes:

1. The *Model of Colloidal Computing* (MC^2). Operational longevity and effective usage of e-textiles has to be supported by an entirely new computational and architectural model. We call such a model colloidal computing (MC^2)[5]: simple computation particles are dispersed in a communication medium which is inexpensive, (perhaps) unreliable, yet sufficiently fast. This type of clustering into computation intensive kernels with loose inter-particle communication, but tight intra-particle communication is typical not only for the underlying hardware, but also for the actual software application that runs on it. In the MC^2 model the application is viewed as a set of "nano-particles" loosely communicating so as to exchange information. Likewise, the underlying architecture has local computation, as well as remote communication capabilities, when needed.
2. An *analytical and simulation-based* framework that allows to explore the design space early on for energy/performance/fault-tolerance trade-offs leading to designs that are feasible and easily adaptable to the application profile. Such a framework is able to provide optimal power/performance operating points, increase operational lifetime under prescribed fault-tolerance limits, and ensure application-driven adaptability during the entire e-textile lifetime cycle.
3. A *Port-Based Adaptable Component Architecture* (PBACA) that allows every computational node to reconfigure its computation and communication with other nodes, in real-time. The motivation stems from the ultra-redundant and distributed component networks that comprise an e-textiles computing environment with the ability to sense, act, store, emit. Such capabilities require a lightweight software infrastructure that enables resource management, communication, global and local tasking, and fault-tolerance between components and across the entire system.

The rest of the paper is structured as follows: Section 2 presents the colloidal model MC^2. Section 3 addresses the problem of application modeling, partitioning and mapping. Section 4 presents the architecture and communication modeling, while fault-tolerance management is addressed in Section 5. Finally, some experimental results are presented in Section 6.

2. COLLOIDAL COMPUTING

The model of colloidal computing supports local computation and inexpensive communication among computational elements: simple computation particles are dispersed in a communication medium which is inexpensive, (perhaps) unreliable, yet sufficiently fast (Fig.2). This concept has been borrowed from physical chemistry [6][1] since some of its properties and characteristics resemble the features that an e-textile based computational model would need. Specifically:

- In case of unstable colloidal suspensions, colloidal particles tend to *coalesce* or *aggregate* together due to the Van der Waals and electrostatic forces among them. *Coalescing* reduces surface area, whereas *aggregation* keeps all particles together, without merging.

1.*Colloid* [käl´oid] = a substance consisting of very tiny particles (1 nm and 1000 nm) suspended in a continuous medium, such as a liquid, a solid, or a gaseous substance.

Similarly, the resources of a classic system are *coalesced* together in a compact form, as opposed to the case of e-textile-based computation where useful work can be spread among many, small, (perhaps) unreliable computational elements that are dynamically *aggregated* depending on the needs (Fig.3). *Dynamic* or *adaptive aggregation* is explicitly done whenever operating conditions change (e.g., failure rate is too high or battery level is too low).

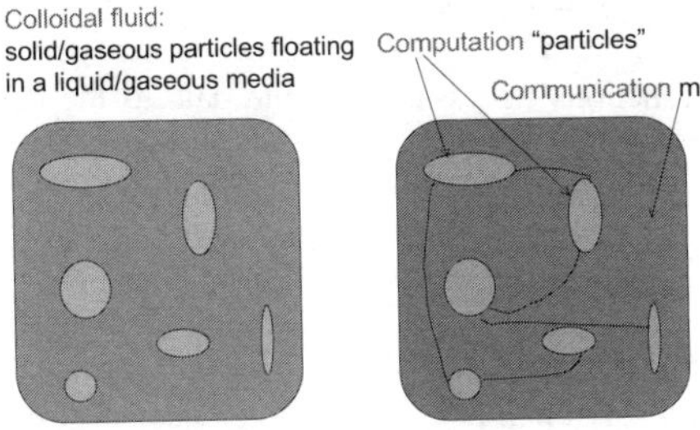

Fig.2 Colloidal computing model

- The MC^2 model supports a wide range of computational granularities, from *computation embedded in the yarn*, up to simple sensors, clusters of sensors, or even specialized computing elements for *local* and *global* resource management, in a tiered, *hierarchical organization*. Perhaps the most striking difference between classical, performance-oriented computing systems and smart fabric-based systems is the possibility of exporting some of the computational power onto the links or active yarns. Since transistor densities of $10{,}000/cm^2$ are possible in the next 3-4 years [7], some functionality can be deployed onto this extra logic in the form of A/D converters or simple computing elements (i.e. adders, multipliers, pseudorandom generators).

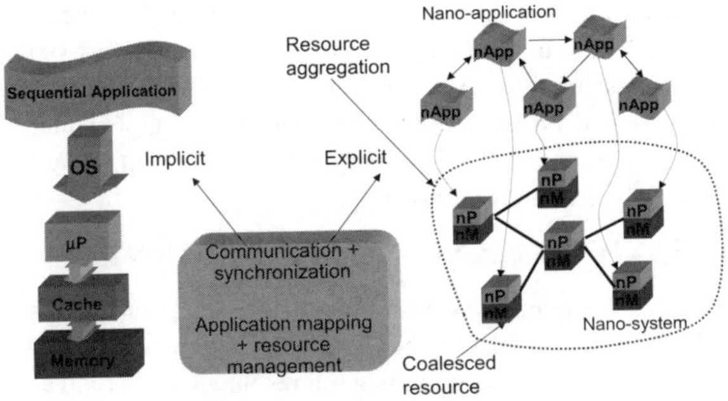

Fig.3 Coalesced vs. aggregated resources

2.1 Explicit vs. Implicit Communication and Management

We propose to apply the MC^2 model to both the application software *and* architecture platform. Specifically, most of applications under consideration consist of a number of computational kernels with high spatial locality, but low degree of communication among them. Such kernels (typically associated with media or signal processing applications) can thus be mapped on separate computational "particles" that communicate infrequently for exchanging results.

The underlying architecture of the e-textile is also complying with the proposed MC^2 model. Both the architecture configuration and the mapping of the application on the computational elements is done dynamically, at power-up and whenever the system becomes "unstable". As an analogy, in case of unstable colloids, a minimal energy configuration is achieved via *coalescing* (e.g., oil in water) or *aggregation* (e.g., polymer colloids).

In e-textiles, a "stable" configuration is one that achieves the required functionality, within prescribed performance, power and probability of failure limits. We propose to employ *aggregation* (Fig.3) or *dynamic connection* of computational "particles" based on their state (i.e. functional or not; idle or active) and their probability of failure so as to achieve a required quality of service (QoS) (characterized by performance, power and fault-tolerance). The mapping and reconfiguration process of the application onto the underlying architecture is achieved via *explicit* mechanisms, as opposed to classic computing systems where mapping and resource management is done via *implicit* mechanisms.

Re-organization and re-mapping need *thin middleware clients*, sufficiently simple to achieve the required goals without prohibitive overhead. In addition, fault and battery modeling and management become intrinsic components for achieving certain levels of quality of results or *operational longevity.*

We describe in the sequel some of the issues that are critical to the e-textile application lifetime, namely application modeling and partitioning, architecture and communication modeling, as well as fault modeling and management.

3. APPLICATION MODELING AND PARTITIONING

The fundamental problem which should be addressed in order to efficiently map complex applications onto e-textiles is the issue of *concurrency.* Given that the e-textiles contain many computational particles distributed on large surfaces, it is of crucial importance to expose the entire concurrency available at application (behavioral) level since this will dictate the ultimate limit of parallelism that may be obtained when using real hardware and software on e-textiles. To this end, we propose *trace-driven application modeling* for e-textiles. This is motivated by our observation that, constraining a given application with various input traces leads to very different *clusterings* of the probability distribution that characterize the application itself. Tuning the target e-textile architecture to this large spectrum of different probability distributions is the most important development in obtaining efficient mappings with respect to certain performance metrics.

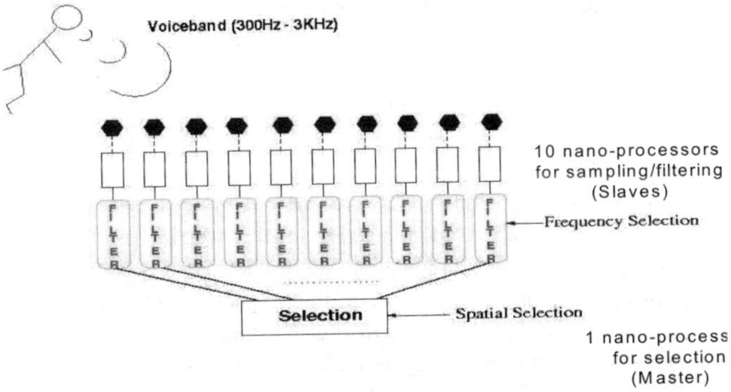

Fig.4 Acoustic beamforming application

A practical example from acoustic beamforming illustrates our point (Fig.4). The acoustic beamforming application uses a processor (master) in conjunction with an array of sensors (slaves) to provide a versatile form of spatial filtering [9]. The entire process graph corresponding to the acoustic beamforming application can be easily modeled using a *Producer-Consumer* paradigm as in [8]. The *Producer* and the *Consumer* are the *computational particles*, while the buffer between them is the *communication medium* in the MC^2 model. For different configurations, we get in a fully analytical manner all the metrics of interest (e.g., throughput, utilization, average response time, mean time to failure or *performability*[1]). Although less accurate than simulation, the analytical approach avoids lengthy off-line profiling simulations for evaluating performance. While for 10 processes we have to deal with a space of 236K states, for a scenario involving 20 processes we may have more than 14G states. This will dramatically increase the time complexity needed in any simulation-based environment for performance evaluation.

4. ARCHITECTURE, COMMUNICATION AND FAULT MODELING

The *architecture* modeling step consists of developing abstract specifications of the e-textile platform and then producing analytic models that reflect the behavior of that particular specification [10]. The model of the *communication architecture* relies on information gathered during and an initial *topology discovery* step which reveals the set of computational and communication resources that are available. Then, based on the performability threshold set for the application at hand, application's computation particles are mapped to available hardware so as to achieve a certain functionality. As operating conditions change (e.g., permanent failures due to fabric tear-off, or intermittent failures due to battery depletion), the application has to be remapped and/or communication links re-routed.

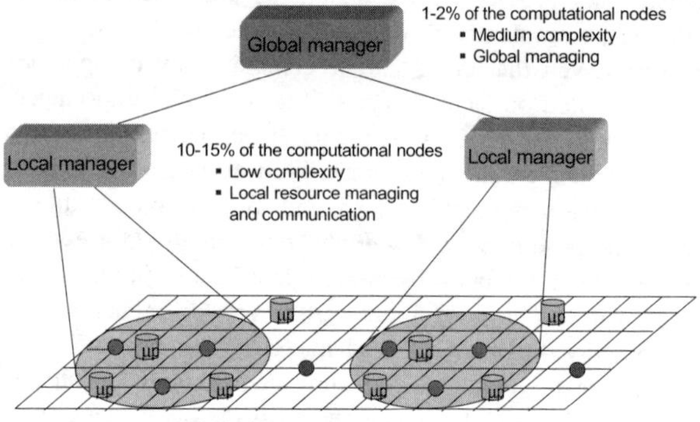

Fig.5 Tiered e-textile architecture

In addition, to better tackle scalability, we propose a tiered architecture (Fig.5) with *local management* of resources (including power consumption, performability, fault tolerance or QoS) and *global coordination* of local managers. Such an architecture offers the advantage of better scalability, as well as fault-tolerance. Specifically, the vast majority of the nodes are very simple computational elements, possibly embedded onto active links. These *nano-nodes* are responsible mostly for computation, and take very simple local decisions (e.g., randomly route data to the neighbors). The next level is comprised of several *micro-nodes* doing local resource management on *clusters* of nano-nodes that cooperate closely to achieve a certain goal (e.g., beamforming). Micro-nodes have more computational capabilities and may run a lightweight operating system kernel that deals with process migration among nano-nodes and overall cluster-level resource management (including energy, quality, fault-tolerance). At the highest level, very few *macro-nodes* are available to realize global resource management and overall coordination of local managers for the achievement of a certain goal (such as, a certain quality under given fault-tolerance constraints, or a certain operational longevity under prescribed battery lifetime limits).

1.Performability is defined as the probability of maintaining a certain level of performance on a finite time horizon.

We also assume that the power distribution system follows the same *clustered* organization, with nodes sharing a power source. Power and battery management is also done locally, and only coordinated globally, whenever an entire region has to be remapped for quality reasons.

Of particular interest is the modeling of the active yarns (*links*) that are used to ensure local and global communication. As opposed to classical distributed systems, for e-textiles the links between nodes may incorporate tens to a few hundreds of transistors and then can perform simple functions necessary to route data to the destination nodes. This is extremely useful if we want to move some of the computational complexity from the nodes onto the yarns and ensure a smoother migration of tasks from one region to another on the e-textile fabric.

4.1 Fault Modeling and Analysis

While the problem of reliability or fault-tolerance of systems with unreliable components has been addressed before, the vast majority of approaches for large networked systems consider only the failure of links or consider that failure probabilities are independent. This is not necessarily true for the case of e-textiles where physical locations of nodes and links are close one to another. In fact, one important characteristic that sets e-textiles apart from classic networking applications is the possibility of *joint* failures for nodes and links. Moreover, the probabilities of failure for nodes and links are *highly correlated*. This comes in sharp contrast with real data networks that assume either node or link failures, but not both.

In addition, the way faults affect the overall behavior is very different, depending on the level in the hierarchy. Graceful quality degradation in a fault-tolerant environment is achieved at *micro-level* by using spare or redundant nodes available in a region or cluster of simple *nano-nodes* that cooperate to achieve a certain goal. Extra redundant nodes are used to run code "migrated" from defective nano-nodes, or nodes running low on battery. In contrast, at macro-level, entire regions can be migrated on demand by the global manager, whenever there is no more redundancy to be exploited at lower levels for achieving a certain level of quality.

5. FAULT-TOLERANCE MANAGEMENT

The model of the communication architecture relies on information gathered during an initial topology discovery step which reveals the set of computational and communication resources that are available. Then, based on the performability threshold set for the application at hand, application's computation particles are mapped to real hardware so as to maximize the overall fault-tolerance. As operating conditions change (e.g., permanent failures due to fabric tear-off, or intermittent failures due to battery depletion), the entire application has to be remapped and/or communication links re-routed (Fig.6). What is unique for e-textiles, is the concept of *cost function of a region* which represents a weighted function of quality of results (QoR), power and fault-tolerance constraints that we use to guide the application mapping and remapping process onto a TAN. Such reconfiguration mechanisms assume that redundancy exists for both nodes and links, such that whenever quality decreases under a certain threshold the virtual communication channel is rerouted or remapped on a different physical channel (Fig.6(a)), or the entire application is remapped onto another set of processing nodes (Fig.6(b)).

Many possibilities exist for implementing redundancy. In a fixed infrastructure, the logical implementation of redundancy is to duplicate resources, with one resource taking over on the failure of the other. This is usually achieved via code migration from a non-functional node to a spare, fully functional one. Code migration is generally a difficult problem, as it could, in the worst case, require the movement of the entire state of an executing application. However, for e-textiles, migration of running applications can be made feasible by restricting the amount of application state that must be preserved. A particular advantage in e-textiles is that, since the environment in which they operate is inherently error-prone, applications can already be constructed to minimize state and facilitate rapid restartability.

In general, process migration [11] makes the fundamental assumption that each node will provide an identical execution environment. At the minimum, the processor architectures of the nodes should be identical, or a hardware abstraction layer (such as a virtual machine) must provide a uniform platform for execution.

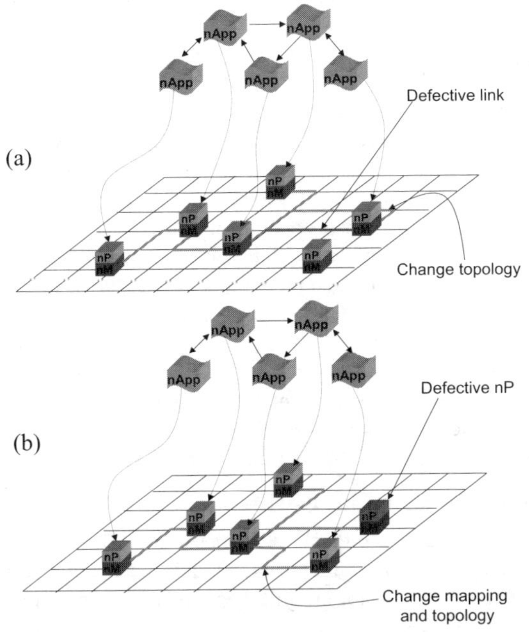

Fig.6 Dynamic reconfiguration in the presence of failures

While migrating among nano-nodes will be very different from migration at micro/macro-node or cluster/region level, a virtual machine may provide a uniform abstraction across varied hardware platforms.

5.1 Middleware Architecture

The ultra redundant and distributed component networks that comprise an e-textiles info-computing environment requires a middleware infrastructure that enables resource management, communication, global and local tasking, and fault tolerance between components and across the entire system. This fabric middleware software should foster locally optimized computation that result in near globally optimal behavior while simultaneously routing around component and communication failures resulting from fabric wear and tear. To this end, we believe that the Port-Based Adaptable Component Architecture (PBACA) approach is well suited to meet these demands [12].

In short, the PBACA middleware suite facilitates the development and deployment of self-adaptive, distributed, e-textile applications. Unlike sequential programming models that require an application to be a single stream of instructions, PBACA can utilize the MC^2 model allowing simultaneous streams of instructions. To exploit the power of PBACA, a solution to a problem must be decomposed into a hierarchy of interconnected tasks. We consider a task to be some flow of execution that takes zero or more inputs, produces zero or more outputs, and may modify some internal state. These input and output points are called *ports*, and the fundamental unit of execution is a Port-Based Module (PBM). For e-textiles platforms, a single PBM is the most fine grain unit of code able to execute on nano-nodes which are comprised of dedicated computational nodes or lighter weight elements embedded in active links. Whereas a PBM represents the most basic unit of execution, a Port Based Process (PBP) represents the most basic unit of self-reconfiguration and self-adaptability. In other words, a PBP is the smallest unit of an application that can measure its performance, quality of service, or operational longevity and take steps to improve those metrics. PBPs are both recursive and hierarchical in that they are composed of sets of interconnected PBMs or other PBPs. In typical e-textiles applications, a PBP will execute on a micro-node's cluster of nano-nodes to achieve logical or spatial sub-goals while satisfying local constraints. Similarly, the cooperating PBPs that form a macro-PBP work together to achieve the overall goal and satisfy the global requirements (such as fault tolerance constraints, or an operational longevity under prescribed battery lifetime limits.)

The overall PBACA middleware approach will foster plug and play integration within an e-textiles system, between e-textiles systems, and between e-textiles systems and existing/future computing systems. Furthermore, once integration is established, higher-level processing may be off-loaded to more powerful computational elements to fit the other information spaces requirements.

6. E-TEXTILE VALIDATION

As a newly emerging field, e-textiles impose new challenges not only on modeling and analysis, but also on simulation tools used for validation. Specifically:

- The simulation environment must be built around a multi-processor architectural instruction set simulator able to provide detailed power/performance numbers for the processing elements.
- Of particular interest is the process of modeling active yarns (*links*) that are used on e-textiles to ensure local and global communication. As opposed to classical distributed systems, in case of e-textiles the links between nodes can perform simple functions to help the data routing process. This is extremely useful in ensuring a smoother migration of tasks from one region to another on the e-textile fabric.
- In addition, the simulation testbed must include detailed battery models for predictive *power* and *fault-tolerance management*, sufficiently accurate network simulation, as well as *failure modeling* and support for *code migration* at nano- and macro-level (region).
- In addition to modeling the nonlinear characteristics of battery storage depletion, the *local resource manager* must store and manage information on how many batteries are instantiated, and on the communication architecture itself.
- Since processors, active or passive links may each fail independently, or may have correlated failures, multiple *failure models* have to be supported. The simulation environment should enable modeling of failure probabilities, mean duration of failures and failure probability distributions, for each of the nodes and communication links. Furthermore, failures between nodes and links may be correlated and a *correlation coefficient* can be specified between any given network segment (active or passive) and any processing element (node). In TANs, correlated failures may appear due to situations such as power and communication being simultaneously severed (as it is likely that all system wiring will be bundled together).

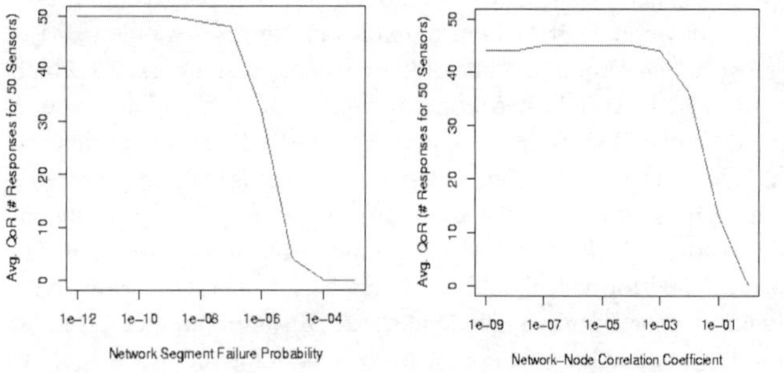

Fig.7 Impact of link failure rates and failure correlation among links and nodes on quality of results (QoR)

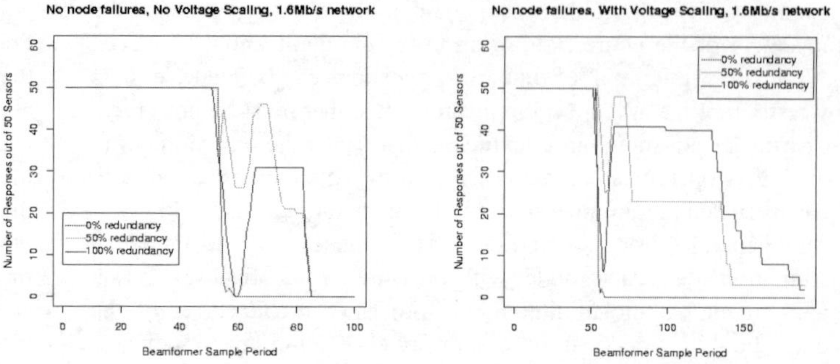

Fig.8 Effects of dynamic voltage scaling during code migration

- Alternatively, failures may appear due to ambient conditions such as high humidity or electromagnetic interference leading to failures in both nodes and links. Lastly, nodes may fail as a side-effect of failures in the network (for example, increased activity can lead to increased peak power consumption, leading to increased failure probability).

To validate results achieved via analytical modeling, we have developed a simulation environment which provides an infrastructure for investigating the many aspects of e-textile based systems. Specifically, we have investigated the impact of link failure rates and correlations among failures on QoR, as well as the feasibility of code migration, with and without voltage scaling.

We show in Fig.7 some results on the effect of *link failure rates*, as well as *correlation among link and node failures* on the operational longevity. The driver application in both cases is the acoustic beamforming application which was configured for 50 sensors that retrieve information and do some local processing before sending data to a central node doing selection of the beam signal. In Fig.7 we show the quality of results (defined as the average number of responses received from slave nodes per time unit) as a function of the network failure rate and the network-node correlation coefficient. As it can be seen, increased link failure rates affect dramatically the quality of the results, while shortening significantly the operational longevity of the system. Correlation among node and link failures have also a negative effect on the overall system quality.

Similar behavior can be observed if a real battery model is used for the local power source (per region or cluster). In such cases, since battery capacity information can be used to *predict* when the node will fail due to insufficient power resources, process migration on a spare/redundant node can be scheduled accordingly. The amount of redundancy, the speed of the communication medium, as well as the communication architecture all have a significant impact on the migration process. The more nodes try to migrate on spare, fully functional nodes, the higher the collision rate on the communication medium is, and thus, the quality in fact decreases. As shown in Fig.8, having more redundancy does not necessarily mean better quality for the same operational lifetime, due to finite speed and capacity of the communication channel. However, since process migration is a non time-critical task, to alleviate this problem, dynamic voltage scaling can be used during the migration process. By slowing the clock speed in combination with voltage supply scaling, collisions due to concurrent process migrations may be avoided and better operational longevity can be achieved with increased redundancy, as shown in Fig.8.

7. CONCLUSION

In this paper, we have addressed a few fundamental issues of an emerging new field of research called *electronic textiles*. Since this new venue of research is likely to introduce *revolutionary* changes in the classic design style of today electronic systems, there will be significant long term impact on both industry (potential e-textile manufacturers, CAD vendors, IP providers) as well as academic research community. Although production of e-textile-based products is limited today, it is very likely to grow sometime in the near future. Thus the DA community has to be ready to deliver tools and techniques for designing, testing and reconfiguring such products. We believe that considering fundamentally new issues in the context of e-textiles can potentially harvest fruitful research and significant results with major impact for the DA community at large.

8. ACKNOWLEDGMENTS

The authors would like to thank the students in CMU's *COATNET* team (especially Phillip Stanley-Marbell and Ted Pham), as well as Bob Graybill of DARPA for insightful comments on our work.

REFERENCES

[1] E. R. Post and M. Orth. Smart Fabric, or Wearable Clothing. In *Proc. Intl. Symp. on Wearable Computers*, pp. 167-168, Oct. 1997.

[2] E. R. Post, M. Orth, P. R. Russo, and N. Gershenfeld. E-broidery: Design and fabrication of textile-based computing. *IBM Systems Journal*, 39(3&4):840-860, 2000.

[3] M. Gorlick. Electric Suspenders: A Fabric Power Bus and Data Network for Wearable Digital Devices. *Proc. Intl. Symp. on Wearable Computers*, Oct. 1999.

[4] J. Rantanen, T. Karinsalo, M. Mkinen, P. Talvenmaa, M. Tasanen, and J. Vanhala. Smart Clothing for the Arctic Environment. *Proc. Intl. Symp. on Wearable Computers*, pp. 15-23, Oct. 2000.

[5] R. Marculescu and D. Marculescu. Does Q=MC^2? (On the relationship between quality in electronic design and model of colloidal computing). *Proc. IEEE/ACM Intl. Symp. on Quality in Electronic Design*, March 2002.

[6] Paul C. Hiemenz, Raj Rajagopalan. Principles of Colloid and Surface Chemistry. Chapters 1 and 13, Marcel Dekker, 3rd Edition, 1992.

[7] Electronic Textiles Concept Session, STA/DARPA, Jan. 2001.

[8] R. Marculescu, A. Nandi. Probabilistic Application Modeling for System-Level Performance Analysis. In *Proc. DATE*, Munich, Germany, March 2001.

[9] B. D. Van Veen, K. M. Buckley. Beamforming: A Versatile Approach to Spatial Filtering. In *IEEE ASSP Magazine*, April 1988.

[10] A. Nandi, R. Marculescu. System-Level Power/Performance Analysis for Embedded Systems Design. In *Proc. ACM/IEEE Design Automation Conference,* Las Vegas, NV, June 2001.

[11] P. Stanley-Marbell and D. Marculescu. Code Migration in Low Power Networked Embedded Systems.CMU-CSSI Technical Report, 2002.

[12] K. R. Dixon, T. Q. Pham, and P. K. Khosla. Port-based adaptable agent architecture. In *Proc. Intl. Workshop on Self-Adaptive Software*, 2000.

PANEL: Analog Intellectual Property: Now? Or Never?

Chair: Stephen Ohr, EETimes

Organizers: Linda Marchant, Cayenne Communication LLC, Chapel Hill, NC and Philippe Magarshack, STMicroelectronics, Grenoble, France

Abstract

There is considerable controversy as to whether or not the trade in analog intellectual properties (IP) will ever represent a viable business opportunity. One school of thought suggests that analog design will always be too specialized to constitute a major market; another school of thought says this will be big business if certain nagging technical problems are solved. While the demand for analog interface components is high, the ability of IP creators to render it in a tradable format is limited. And the ability of digital design teams to successfully utilize analog IPs — without a considerable amount of handholding — is similarly limited. EDA tools here are looked upon as both culpable and offering the best hopes for the future.

This panel of experts — representing analog designers, analog EDA tool providers, silicon foundries and analog IP vendors — bring their own points of view on some of the business and technology issues which need to be resolved to provide the context for analog IP development and trade. Among the open questions:

- Is IP created and validated within the design environment for re-use a more productive approach than imported IP? How many 'traditional' analog designers would admit they are still using kit parts and breadboards — maybe even SPICE and manual IC layout techniques — today in their every-day job?
- What CAD tools are needed to help analog designers? Are newly emerging EDA technologies, designed to enhance analog design productivity, maturing rapidly enough to be accepted by designers?
- What is the future of analog designs at the very low voltage-swings coming with sub-100nm CMOS?
- Though silicon foundries need to process a wide set of external IP offerings in order to allow their users to build complete systems, are foundries seeing enough activity in analog IP designs to justify specialized fab runs or the kind of process tuning that would allow analog and digital IPs to coexist on the same chip?
- How likely – and how soon – can we get to analog IP development and trade?

Technology Issues

Process scaling has enabled unprecedented levels of integration. The most advanced systems-on-chip (SoC) now embodies 200 million transistors on one IC, and that number is expected to increase exponentially over the next decade. Rather than create new designs from a blank slate, design teams are reusing previously-designed blocks and importing the work of other design teams into their efforts. As a consequence, analysts have predicted an exponential growth in the market for IPs — that functional IC blocks will be traded externally, and well as between design departments of a large IC company. Though ARM and MIPS CPU cores now constitute the largest number of blocks traded between companies, analysts project that future trade will include other CPU and DSP cores, memories, communications protocol controllers and multimedia decoders, as well as bus interface components.

With more and more SoC makers hoping to include analog functional blocks as a means of differentiating their designs, it is therefore tempting to believe that analog IP can be created, traded and integrated with the same tools and methodologies with which digital IP are moved about. But analog has a different heritage and calls forth a different set of design issues. The simple act of moving a high-speed digital signal off-chip (the use of an analog line driver) can produce more heat and noise, and utilize more power than the SoC design team is prepared to cope with. The problem becomes more complex with the sophistication of the analog blocks that are incorporated: cable drivers and bus interface components present one level of difficulty; operational amplifiers and phased locked loops present another level; data converters and RF interfaces put most digital design teams into uncharted territory.

Currently, every manufacturer who licenses an analog IP must provide an engineer (or an engineering team) along with the IP just to get it working properly in the customer's design. There are two reasons for this: First, analog designs are typically process-specific and do not scale with shrinking CMOS geometries. Unlike digital circuits that can appear to gain performance in a process shift from (say) 0.18-microns to 0.13-microns, analog circuits can loose performance and, in some case, cease working entirely. Dynamic range is severely curtailed with submicron CMOS, and parasitics (like interconnect resistivity and gate capacitance) play a much larger role. Analog functions must frequently be entirely redesigned in the move from one process environment to another, but there are limitations on the extent to which redesign will help. Participants in panels at this year's International Solid States Circuits Conference (ISSCC), for example, were openly terrified of the leakage currents that would appear at 100- and 90-nm CMOS geometries. Second, analog functional blocks — performing well in isolation — will often be disabled on a digital SoC substrate. Digital switching noise, bubbling up through the substrate, is frequently identified as the culprit.

Newly emerging EDA technologies, designed to enhance analog design productivity, provide some relief. There are two sets of tools needed: One set would enable designers — regardless of whether their primary expertise is analog or digital — to create analog IP cell blocks more rapidly. The expression "analog synthesis," analysts agree, has been abused because it will not (on the near horizon) offer the same kind of movement from high-level language description to gate-level implementations. However, there are many new-generation model-generating, circuit-resizing and optimization tools — many available from silicon foundries — that will speed the process of moving analog functions from one geometry to another. Some new tools provide "design tuning" that allows IPs to verified within the design environment where they are created. The second set of tools —

speedier full chip simulators — would allow analog IPs to verified in the same environment with digital IPs.

But the question remains whether the tools will develop rapidly enough to satisfy analog IP creators. Even among digital design teams, there seems to be trepidation and concern whether the tools will continue to give trustworthy information at 100-nm and below. And these concerns will keep analog designers on their toes — well before the 100-nm mark.

The Positions summarized

Mike Brunoli, Chief Technology Officer, Nurlogic Design, Inc., San Diego, CA
Mike Brunoli retains a steadfast focus on the silicon, particularly new-generation CMOS. Analog IP trade is viable but difficult because of the scaling imposed by each new process generation. The newest challenge, for example, is current leakage from devices within 100-nm circuits. The can be repaired with dual oxides, Brunoli suggests, but the circuits won't offer much head room at 1V. Better tools and methodology are required to provide openness and access to data. Without them, even a minor tweak becomes a major roadblock, he says.

Masao Hotta, Director Advanced Analog Technology Division, Hitachi Ltd., Tokyo, Japan
One of Hitachi's keys to success is in cooperation among its design teams. Hitachi has worked on consumer video devices with hundreds of millions of transistors. The transfer of intellectual property is not only feasible, but also necessary — at least among departments in the same company. This has required the coordinated efforts of large engineering teams, as well as use of specialized signoff tools.

Felicia James, VP, IC Technology, Cadence Design Systems, Inc, San Jose, CA.
Felicia James is on a mission: As a mixed-signal IC designer (analog and communications products) in a large IC company, she would be constantly bumping up against the limitations of the EDA tools. These would entail broken design flows, netlists that would not translate from one operation to the next, simulators that would stall, and IC design tools that did not have the capacity to help large designs with the latest design rules. In her current role, she is working to ensure that EDA tools offer thorough support for IP creation and its integration with larger systems-on-chip.

Rudy Koch, Director R&D, Wireless Products, Infineon Technologies AG, Munich Germany
Does EDA understand RF? Dr. Koch worries. If any designer has had criticism of analog IC design tools, their faults are magnified by their application to radio frequency projects. The traditional challenges of simulating transistor-level behavior are complicated by their application to GHz wavelengths. Dr. Koch has been critical of the EDA industry for its attention to large digital circuits and IP blocks, a focus that typically neglects analog and RF. The industry's new focus on signal integrity and substrate coupling issues do not address the voltage and current concerns of analog designers, nor the waveguide geometry issues of those who work in RF.

Roy McGuffin, President and CEO, Antrim Design Systems, Inc., Scotts Valley, CA
As someone involved in the creation and transfer of analog intellectual property, Roy McGuffin looks at the manpower requirement. He says it is difficult to ship a softcoded IP with considerable engineering handholding and support. Selling analog intellectual property has become synonymous with renting out your design time, he jests. There is some help that can be provided by next-generation EDA tools in terms of capturing the physics of the silicon, the behavior of analog circuits and the engineering know-how to integrate analog and digital function blocks.

Andrew J. Moore, Ph.D Design Services Marketing (EDA) TSMC, North America.
Select from prototypes, Moore recommends. From a foundry perspective, the most important guarantor of success in analog semiconductor design is siliconized components; that is, working analog IPs. To that end, TSMC provides dozens of affordable analog prototyping shuttles per year. There is one per month for users of the TSMC 0.18 um MM/RF process, for example. The majority of IP vendors who use these shuttles are analog IP vendors, who have taped out hundreds of components on more than 20 test chips in the last year. From 3rd party vendors and internal teams, there is a selection of siliconized, "off-the-shelf" IPs for analog, including A/D, D/A, PLL, and LVDS components.

Task Scheduling and Voltage Selection for Energy Minimization

Yumin Zhang
Synopsys, Inc.
700 East Middlefield Road
Mountain View, CA 94043
yumin@synopsys.com

Xiaobo (Sharon) Hu Danny Z. Chen
Department of Computer Science and Engineering
University of Notre Dame
Notre Dame, IN 46556, USA
{shu, chen}@cse.nd.edu

Categories and Subject Descriptors

I.2.8 [**Problem Solving, Control Methods, and Search**]: Scheduling

General Terms

Algorithms, Design

Keywords

voltage selection, task scheduling

ABSTRACT

In this paper, we present a two-phase framework that integrates task assignment, ordering and voltage selection (VS) together to minimize energy consumption of real-time dependent tasks executing on a given number of variable voltage processors. Task assignment and ordering in the first phase strive to maximize the opportunities that can be exploited for lowering voltage levels during the second phase, i.e., voltage selection. In the second phase, we formulate the VS problem as an Integer Programming (IP) problem and solve the IP efficiently. Experimental results demonstrate that our framework is very effective in executing tasks at lower voltage levels under different system configurations.

1. INTRODUCTION

Energy consumption has become a primary concern in today's IC systems. Energy minimization techniques applied at higher design levels tend to be more effective than techniques applied at lower levels [4]. Processors that can operate at variable supply voltages or be switched to sleep mode while idling are becoming available [1, 11]. Two main system level energy saving techniques are: voltage selection (VS) (also called voltage scheduling [4]) which selects processor's supply voltage according to the performance requirement, and power management (PM) [2] which shuts down a processor when it is idle. In general VS is more effective than PM [10]. The switching of voltages values can be done on the fly and the incurred overhead can be ignored if such switching does not happen very frequently [16]. In real-time systems where tasks can take tens or hundreds of thousands of cycles to execute, applying VS judiciously can achieve a large amount of energy saving [17].

In this paper, we consider real-time dependent tasks with deadlines. These tasks are to be executed on variable voltage processors. Our challenge is to find a system implementation that consumes the least amount of energy. A system implementation is determined by the ***task assignment*** (which task runs on which processor), the ***task ordering*** (the execution order of tasks on each processor), and the ***voltage selection*** (which cycles of each task use which voltage level). We refer the combination of task assignment and ordering as task scheduling in the rest of the paper.

Most related work, e.g., [3, 10, 12, 13, 17], concentrates on saving energy of independent tasks or on a single processor. However, tasks in real-world applications usually have control or data dependencies and many systems have multiple processors. Approaches in [8, 15] solve the energy minimization problem for dependent tasks on multiple variable voltage processors. [8] assumes a given task assignment, while [15] assumes a given task scheduling. However, task scheduling without consideration of VS may not present the best energy saving potential. Furthermore, the approach in [8], which evaluates eligible tasks one by one, fails to recognize the joint effect of slowing down certain tasks simultaneously. The joint effect is critical in finding global optimal voltage settings and an example in Section 2 explains the effect in much detail. The heuristics in [15] try to distribute slacks evenly among tasks, which does not necessarily lead to best energy saving because the slowdown of different tasks has different effect on other tasks and contributes differently to the overall energy saving. To achieve the best energy saving, the scheduling should try to present the best potential for VS and the VS decision should be made in a global manner.

In this paper, we present a framework that integrates task scheduling and voltage selection together to minimize energy consumption of dependent tasks on systems with a given number of variable voltage processors. The integration is important because scheduling results determine the potential energy saving that can be achieved through VS. In order to find the scheduling with the best slowing down opportunity in the first phase, we apply an earliest-deadline-first (EDF) scheduling which can be proved to be optimal for a single processor, and a scheduling with priority-based task ordering and a best-fit processor assignment for multiple processors. Tasks after scheduling can be modeled as a Direct Acyclic Graph (DAG). In the second phase, we formulate the VS problem on a DAG as an Integer-Programming (IP) problem. Each task can have cycles running at differ-

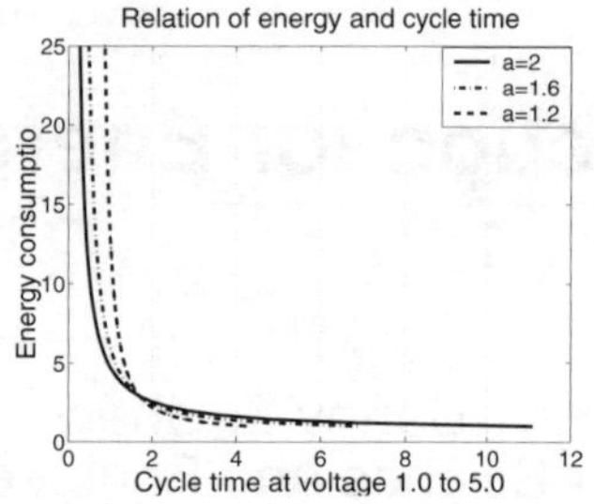

Figure 1: Convex relation of energy and delay

ent voltages. We have implemented the framework and conducted experiments on different systems and tasksets. The results show that our framework is very effective in reducing energy consumption at the system level. Our IP approach for solving the VS problem can slowdown 58% more cycles than a baseline scaling VS approach.

Our main contributions can be summarized as:

First integrated framework: we solve the energy minimization problem of dependent tasks on variable voltage processors in two phases with the first phase task scheduling being guided by the VS approach in the second phase.

Exact algorithm for VS problem: our IP formulation solves the VS problem exactly and we prove the continuous voltage and special discrete voltages case are polynomial-time solvable. The ILP for general discrete voltages is solved with approximation in short runtime and the approximation is tested to be within 97% of the optimal solution.

Substantial energy saving: the framework can slowdown 7-98% of cycles with very short runtime. It can be used in design space exploration to meet energy target and find the best system configuration.

2. PRELIMINARIES AND A MOTIVATIONAL EXAMPLE

The number of cycles, N_u, that task u needs to finish remains a constant during VS, while processor's cycle time CT, u's delay d_u and the the dominant part of total energy consumption, dynamic energy consumption, E_u, change with the supply voltage V_{dd}. CT, d_u and E_u can be computed as

$$CT = \frac{k \cdot V_{dd}}{(V_{dd} - V_{th})^a} \quad (1)$$

$$d_u = N_u \cdot CT = N_u \cdot \frac{k \cdot V_{dd}}{(V_{dd} - V_{th})^a} \quad (2)$$

$$E_u = N_u \cdot C_u \cdot V_{dd}^2 \quad (3)$$

where k is a device related parameter, V_{th} is the threshold voltage, a ranges from 2 to 1.2 and C_u is the effective switching capacitance per cycle. We depict the relation of E_u and d_u for different a in Figure 1. It is clear that E_u is a convex function of d_u [12] and both are functions of V_{dd}. Note that our work allows tasks to have different power characteristics, such as switching capacitance.

We use a DAG to represent a taskset. In the DAG, each node u represents a task u, while an edge $e(u, v)$ indicates that v can only start after u finishes. Deadline dl_u constrains u's finish time and T_{con} on a set restricts all tasks' finish time. A taskset after scheduling can still be represented as a DAG with additional edges, if necessary, to capture processor sharing. The DAG for a 5-task set before and after scheduling to two processors is shown in Figure 2 (a) and (b). **Slack** is the maximum amount of time that a task can be slowed down without violating timing constraints.

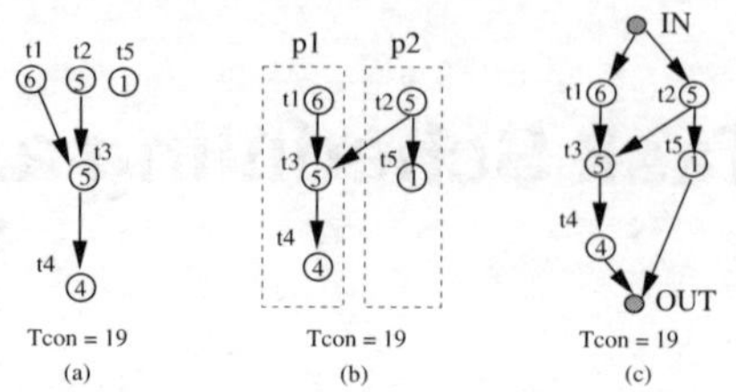

Figure 2: (a) A 5-task set (b) Tasks scheduled on P_1 and P_2 (c) DAG with IN and OUT

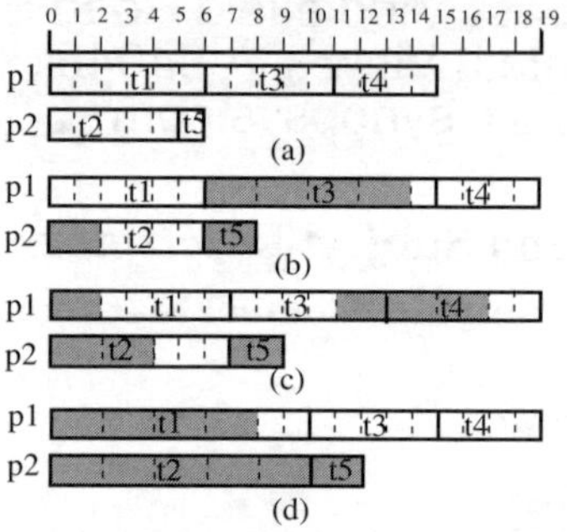

Figure 3: VS results by different approaches. Cycles in shade are executed at V_l.

2.1 A motivational example

We use the example in Figure 2 to show that different VS approaches on multiple processors perform quite differently. This example motivated us to make VS decisions in a global manner. In Figure 2 (b), the number inside each task is N_u, the number of cycles needed to finish the task. Assume P_1 and P_2 can operate on V_h and V_l and $V_h > V_l$. For simplicity, assume CT_{V_h} is 1 time unit, CT_{V_l} is 2, and the average energy saving per slowdown cycle is the same for all tasks on P1 and P2. Let T_{con} be 19.

The VS results by approaches in [8, 15] are shown in Figure 3 (b) and (c). Figure 3 (a) shows the execution of tasks on P_1 and P_2 when the two processors are always on V_h and Figure 3 (d) shows an optimal solution for this problem. It is clear that slowing down t_1 and t_2 simultaneously is the best even though slowing down each will take away the potential of slowing down t_3, t_4 and t_5. Also the even distribution of slack is not optimal in saving energy. The number of cycles at V_l is 10 in Figure 3 (d) and this is an optimal solution, which is is 67% more than the 6 V_l cycles by approach in [8] and 43% more than the 7 V_l cycles by approach in [15]. Our VS approach, which is to be presented in the next section, achieves the optimal solution for this example.

3. VOLTAGE SELECTION

Our approach integrates task scheduling and voltage selection together and strives to achieve the maximum energy saving on variable voltage processors. The framework can be used to optimize energy at the system level in a design

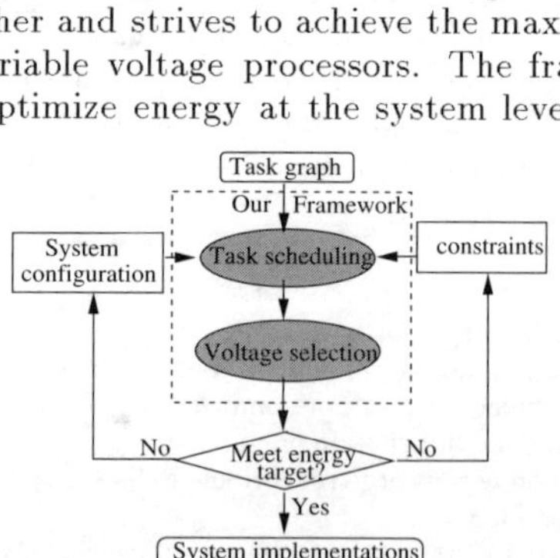

Figure 4: Design flow with our approach

flow like the one shown in Figure 4. In this section, we will present an IP formulation that solves the VS problem on a given scheduling of dependent tasks on multiple processor. A simplified version of the IP formulation can be used to solve the VS problem on a single processor. The IP formulation solves the VS problem exactly and it provides guidance for task scheduling to be discussed in Section 4.

3.1 Unified IP formulation for VS

We need to guarantee that after the VS technique is applied, tasks with deadlines still finish before their deadlines and the last task on each processor finishes no later than T_{con}. That is the delay on each path in the DAG should still not be greater than T_{con}. However, the number of paths in a DAG can be exponential to the number of edges and makes it not practical to do optimization on a path base. This is the reason that VS approaches on single processor systems cannot be readily extended to solve the VS problem on multiple processors. Here we adopt a model proposed in [5] to formulate timing constraints on a DAG not on paths, but on nodes and edges. We add two nodes, IN and OUT, edges from IN to the first task on each processor, and edges from the last task on each processor to OUT to form a new DAG. Figure 2 (c) shows the new DAG, $G(V, E)$, for the taskset in Figure 2 (a). Execution time of task u at the highest supply voltage V_h is a constant T_u, and its deadline dl_u remains a constant. T_{OUT} and T_{IN} are set to be 0. T_{con} is the timing constraint on the DAG. Besides task's delay d_u, we associate another variables D_u with u. Timing constraints on $G(V, E)$ in Figure 2 (c) can be modeled as

$$D_{OUT} - D_{IN} \leq T_{con} \quad (4)$$

$$D_v - D_u - d_u \geq 0 \quad \forall e(u,v) \in E \quad (5)$$

$$D_u + d_u \leq dl_u \quad \forall u \text{ with deadline} \quad (6)$$

$$d_u \geq T_u, \; D_u \geq 0, \text{ int}, \; \forall u \in V \quad (7)$$

Intuitively, D_u represents u's start time. For a feasible scheduling, if D_{IN} is set to be 0, the above constraints guarantee that tasks with deadlines will finish before their deadlines and the finish time of all task is not greater than T_{con}.

Single processor systems are a special case of systems with multiple processors. The timing constraints in (4)-(7) on multiple processors also constrain the timing on a single processor. However, the scheduling on a single processor has special features that we can utilize to simplify the constraints. The main feature on a single processor is that all tasks are on one path. The constraints for single processor can be simplified as follows.

$$\sum_{u \in V} d_u \leq T_{con} \quad (8)$$

$$\sum_{v \in B_u} d_v \leq dl_u \; \forall u \text{ withdeadline} \quad (9)$$

$$d_u \geq T_u, \text{ int}, \; \forall u \in V \quad (10)$$

where B_u is the set that includes u and all tasks scheduled before u, d_u's are variables whose values need to be determined, T_{con}, dl_u and T_u are constants.

The objective of VS is to minimize the sum of each task u's energy consumption, $\sum_{u \in V} E_u$. Combing the objective and the constraints in (4)-(7) for multiple processors and constraints in (8)-(10) for single processor, we have the IP formulation for the VS problem. To trade the increase of delay for energy saving, we need to establish the relationship between d_u and E_u. If the supply voltage can change continuously, d_u can also change continuously. In this case, E_u is a convex function of d_u, as depicted in Figure 1, and we have $E_u = f(d_u)$, where $f(\cdot)$ is a convex function. Substituting E_u with $f(d_u)$, we have the IP formulation for the continuous voltage case.

In the discrete voltage case, only a certain number of voltages are available. For example, the Athlon 4 Processor from AMD [1] can operate at five voltage levels. The discreteness of voltage levels implies that a task's delay can only take discrete values. Let the highest voltages be V_h and other m available voltages be $V_1, V_2, \ldots, V_m$, where $V_i < V_h, 1 \leq i \leq m$. We introduce m new variables, $N_{u,i}$, to represent the number of cycles that task u is executed at V_i. For any given V_i, cycle time CT_i and energy consumption per cycle $E_{u,i}$ are constants as computed in (1) and (3) in Section 2. The total number of cycles u takes, N_u, remains a constant. d_u and E_u are computed as linear functions of $N_{u,i}$.

$$d_u = T_u + \sum_{i=1}^{m} N_{u,i}(CT_i - CT_h) \quad (11)$$

$$E_u = C_u \cdot \left(\sum_{i=1}^{m} N_{u,i} V_i^2 + (N_u - \sum_{i=1}^{m} N_{u,i}) \cdot V_h^2\right) \quad (12)$$

Substituting (11)-(12) as E_u and d_u and adding $\sum_i N_{u,i} \leq N_u$, we have the ILP for the discrete voltage case, where D_u's and $N_{u,i}$'s are variables.

3.2 Solving the IP problem

In general, IP problems are hard to solve. But the IP problem for the continuous voltage case on systems with a given number of processors is polynomial time solvable. A convex programming solver, like *COPL_LC* from [6], can be used to solve the problem in polynomial time.

THEOREM 1. *The VS problem of executing dependent tasks on systems with one or more processors with continuous voltage is polynomial time solvable.*

Proof: The VS problem is formulated as an IP problem with objective $\sum_u f(d_u)$, constraints in (4)-(7) for multiple processors, and constraints in (8)-(10) for a single processor. The objective function is the sum of convex functions, while the coefficients of variables D_u and d_u in (4)-(7), and variable d_u in (8)-(10) are 1, 0 or -1. Thus the IP formulation has a totally unimodular constraint matrix for both single processor and multiple processors. An IP problem is polynomial time solvable if the objective function is a separable convex function and constraint matrix is totally unimodular [9]. The IP formulation for the continuous voltage case has the same form as such a polynomial time solvable IP problem and thus the VS problem on a given number of processors that can operate on continuous voltage is polynomial time solvable. □

The ILP problem for some special discrete values is also polynomial-time solvable on a single processor and multiple processors. If there is only one lower voltage V_l and $CT_{v_h} - CT_{V_l} = 1$, the ILP problem for one processor and multiple processors both have a totally unimodular constraint matrix. Such an ILP problem is polynomial time solvable. An LP solver, such as *LP_SOLVE* from [7], can be used to get the integer solutions in polynomial time. To solve the ILP problem for general discrete voltages in short runtime, we employ an efficient approximation. First the integer constraint on variables is relaxed and the corresponding LP problem is solved. Then we use the floor integer of the non-integer solutions for $N_{u,i}$ in the LP as the integer solutions for $N_{u,i}$ in the ILP to guarantee that the approximation satisfies the timing constraints. This approximation is tested to be very efficient by experiments in Section 5.

4. TASK SCHEDULING

Task scheduling should provide the maximum slowing down potentials for voltage selection to utilize for energy saving. In this section, we present the scheduling that is guided by our VS approach presented in the last section. For a single processor, we apply an EDF scheduling, which can be proved to be optimal in providing slowdown opportunities. For multiple processors, we use a priority-based scheduling in order to provide the best energy saving opportunity.

4.1 Scheduling on a single processor

It is clear that individual task's deadline imposes direct constraint on the design space of delays of tasks that are scheduled before it, as shown in constraints in (9). Energy saving is affected by the design space of tasks' delays because the delay will be traded for energy saving. To maximize the design space of delays, tasks with earlier deadlines should be scheduled earlier in order to avoid imposing constraints on more tasks. We apply an EDF scheduling in the following recursive algorithm $EDF(S, t)$. The input to the algorithm is the taskset S, and t is initialized to be 0. Denote PRE_u as the set of u's predecessors. The output of the algorithm is the same taskset with each task's start time s_u being assigned and the order on the processor being decided.

```
EDF(S, t)
Begin
    If S ≠ φ
        If ∃ u with dl_u
            Find u with dl_u = min (dl_v|∀v ∈ S);
        Else Find a u ∈ S;
        Endif
        S1 = PRE_u; S2 = S − S1 − {u};
        EDF(S1, t);
        s_u = t; t = t + T_u;
        EDF(S2, t);
    Endif
End
```

We have proved that the EDF scheduling provides the biggest solution space for the single processor VS problem.

THEOREM 2. *The EDF scheduling is optimal for the VS process to save energy on a single processor*

Due to space limit, the proof is omitted. The EDF scheduling is feasible if a feasible scheduling exists [14]. Combining the EDF scheduling which is optimal and our IP formulation that solves the VS problem exactly, our framework solves the general energy minimization problem of dependent tasks on a single variable voltage processor exactly.

4.2 Scheduling on multiple processors

The EDF scheduling is optimal on a single processor, but not on multiple processors because tasks will be on multiple paths and affect tasks on these paths differently. In the following example, the scheduling that does not follow the EDF ordering presents more slowdown opportunities. Consider the taskset in Figure 5 (a), an EDF scheduling in Figure 5 (b), and another scheduling that does not follow

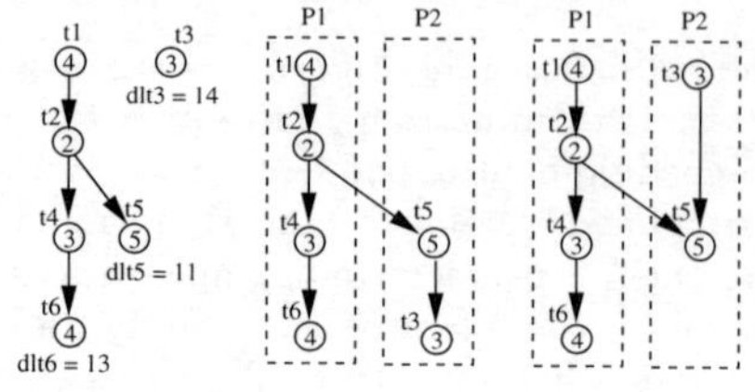

Figure 5: Scheduling by different approaches

the deadline order in Figure 5 (c). Using our IP formulation on the same system used in the example in Figure 3, the optimal number of cycles at V_l is 0 and 3 for the scheduling in Figure 5 (b) and (c), respectively. This example tells us that other information needs to be combined with deadline to find a scheduling that provides the best slowdown opportunity in multiprocessor systems.

To provide more energy saving opportunities in the scheduling for VS to utilize, we should try to avoid putting tasks on unnecessary paths which might pose tighter constraints, making unnecessary long paths and putting unrelated tasks on the same path, and try to let less tasks constrained by smaller deadlines. To achieve these goals, we use a priority to select one task at each scheduling step, find a best-fit processor for the selected task, and repeats the process until all tasks are scheduled.

The priority relates to a task's deadline, dependencies and the usage of processors in the system. Tasks are assigned a latest finish time that they must meet for them or their successors to meet deadlines. Task u's latest finish time lft_u is defined as

$$lft_u = \min(dl_u, lft_v - T_v | \forall v, e(u,v) \in E) \quad (13)$$

Leaf task's latest finish time is its deadline or T_{con} if it does not have a deadline. Start from leaf tasks and traverse the taskset DAG backward, we can assign each task a latest finish time. Denote task u's ready time when all u's predecessors finish as r_u, its earliest start time when it is ready and there is a processor available as es_u, the number of processors in the system as N, and processor P_i's available time as a_{P_i}. r_u, es_u and a_{P_i} are updated during the scheduling.

We build the scheduling step by step. At each scheduling step, eligible tasks's priorities, PRI_u, are evaluated.

$$PRI_u = lft_u + es_u \quad (14)$$

$$es_u = \max(r_u, \min(a_{P_i} | \forall i \le N)) \quad (15)$$

Task u with the smallest PRI_u is assigned to a best-fit processor, which is selected according to the following three steps. s_u, and a_{P_i} are set accordingly:

1. select P_i **if** $a_{P_i} = r_u$. That is to select the processor that is available the same time when u is ready. In this case, $s_u = r_u = a_{P_i}$ and $a_{P_i} = s_u + T_u$. If there is no such P_i,

2. select P_i **if** $a_{P_i} < r_u$, $a_{P_i} > a_{P_j} | \forall j \neq i$, $a_{P_j} < r_u$. That is to select the processor that becomes available the latest among all processors available before u is ready. In this case, $s_u = r_u$ and $a_{P_i} = s_u + T_u$. If there is no such P_i,

3. select P_i **if** $a_{P_i} \le a_{P_j}$, $\forall j \neq i$. That is to select the processor that is available the earliest. In this case, $s_u = a_{P_i}$ and $a_{P_i} = s_u + T_u$.

After the selection and assignment of one task, a new set of eligible tasks are evaluated and another task is selected and assigned to its best-fit processor. The step repeats until all tasks are scheduled. If the priority-based scheduling is not feasible, we can increase the weight of the latest finish time in the priority function and try to get a feasible scheduling.

5. EXPERIMENTAL RESULTS

To evaluate the effectiveness of our framework in saving energy, we implemented it and conducted experiments on various tasksets and systems. 9 tasksets are generated with *TGFF* [18] where each set consists 10 to 500 tasks. Since single processor systems are a special case of multiprocessor systems and we have proved that the EDF scheduling on a single processor is optimal, we concentrate the experiments on multiprocessor systems. We tested systems consisting of 2 to 8 processors that can operate at 2 and 4 different voltages. (Most commercially available variable voltage processors have 2 to 5 voltage levels.) First, we investigate the effectiveness of our IP approach for solving the VS problem by comparing the number of slowdown cycles with a scaling VS approach on the same scheduling. Then we test how our scheduling helps saving energy by comparing the VS results on different scheduling. The energy saving by our framework on different system configurations are also investigated.

The first experimental system consists of 5 processors that can operate at two different voltages, V_h and V_l. For simplicity, we assume that $CT_{V_l} = 4$, $CT_{V_h} = 1$, the energy saving per cycle at V_l is the same for all tasks. (Note there is no such requirement in our IP formulations.) Thus the total number of slowed down cycle SN represents the energy saving that is proportional to $SN(V_h^2 - V_l^2)$.

We use our scheduling in the first phase and alternate different VS approaches in the second phase to test how much improvement our IP approach provides. Since the approach in [8] also deals with task ordering and the approach in [15] has other objectives during their VS process, we feel it is not an apple-to-apple comparison to compare our IP approach with the two directly. We compare the number of cycles at V_l by our IP approach with a scaling VS approach that is also discussed in [8]. It scales down the delay of all tasks by the ratio of timing constraint over critical path length on a given scheduling. The approach in [15] which distributes slack evenly shares the same spirit with the scaling approach. Table 1 summarizes the performance of our IP approach and the scaling approach. Column NT, NC, T_{con} and T_{cri} show the number of tasks, number of task cycles, timing constraints and critical path delay at V_h, respectively. Timing constraint T_{con} is set to be 1.5 times of T_{cri}. Column SC and IP show the number of cycles at V_l by the scaling approach and our IP approach, and Column IMP shows the improvement of number of cycles at V_l by our IP approach (IP) over the scaling (SC), which is computed as $((IP) - (SC))/(SC)$. It is clear that on average our approach can execute 58% more cycles at V_l and save more energy than the scaling approach.

The approximation we use to get the integer solutions for the ILP problem is very effective. The objective of the ILP, total number of slowdown cycles, after the approximation of integer solutions is within 97% of the objective achieved with non-integer solutions from solving the LP. The objective of the LP is the upper limit of the objective that the corresponding ILP can achieve. This means that solutions by the approximation is within at least 97% to the optimal solution of the ILP problem. The whole process including scheduling, solving LP and approximation finishes within seconds for all tasksets.

Table 1: Performance by scaling and our approach

set	NT	NC	T_{cri}	T_{con}	SC	IP	IMP %
s1	9	81	49	74	11	37	236
s2	50	422	111	167	51	104	104
s3	101	922	243	365	116	225	94
s4	151	1501	464	696	188	413	120
s5	213	2988	722	1083	413	644	56
s6	245	2871	652	978	376	557	48
s7	305	2643	572	858	314	495	58
s8	463	4310	910	1365	541	755	40
s9	514	4633	949	1424	556	808	45
ave.	228	2263	519	779	285	449	58

Table 2: Performance on different scheduling

set	NT	NC	T_{con}	EDF	PEDF	IMP %
s2	50	422	167	89	104	17
s3	101	922	365	200	225	13
s7	305	2643	858	432	495	15
ave.	152	1329	463	240	275	14

To test how our scheduling helps providing more slowdown opportunities, we alternate two scheduling approaches in the first phase, one is our priority-based ordering and best-fit processor selection approach (PEDF), and another is a baseline approach that uses an EDF ordering and selects the processor that becomes available the earliest (EDF). Our IP formulation is applied in the second phase to minimize energy. The number of cycles at V_l for several representative tasksets on the same 5-processor experimental system used in the VS comparison are summarized in Table 2. In the table, Column EDF and PEDF show the number of cycles at V_l on the baseline scheduling and on our priority-based scheduling, respectively, and Column IMP shows the improvement of cycles at V_l on our scheduling (PEDF) over the baseline scheduling (EDF). The results show that our scheduling can provide 14% more slowing down opportunities than a baseline scheduling.

It is clear that our first-phase priority-based scheduling and second-phase IP approaches all outperform other respective approaches. Here we want to explore how the framework performs on different system configurations. We first investigate how the number of processors affects the energy saving. With fixed timing constraints, systems with more processors should have shorter critical path length and more tasks with more slacks. The slack can be traded for energy saving. Table 3 shows the number of cycles at V_l on systems with different number of processors. Processors can operate at two values and the cycle time changes from 1 unit to 4 unit when voltage is switched from V_h to V_l, which is the same as in the first two experiments. The energy saving increases dramatically when the number of processors increases from 2 to 5, but not so much when the number increases from 5 to 8. This is because the limited parallelism among tasks has been exploited by the increases of number of processors from 2 to 5. Additional processors do not provide much benefit if there are not many tasks can run in parallel. If tasks are independent, every task can run in parallel and we should expect the increase of number of processors lead to energy saving increase more evenly.

The number of voltages and voltage values also affect the energy saving. With more voltages available, the search space for VS problem is bigger and energy savings should

Table 3: Savings on different number of CPUs

set	NC	T_{con}	2 cpus	5 cpus	8 cpus
s1	81	66	9	31	31
s2	422	266	29	268	397
s3	922	581	67	588	803
s4	1501	989	118	905	1318
s5	2988	1861	208	1939	2910
s6	2871	1760	197	1862	2840
s7	2643	1601	177	1728	2640
s8	4310	2616	287	2832	4310
s9	4633	2789	310	3076	4633
ave.	2263	1392	156	1470	2209
per.	100%		7%	65%	98%

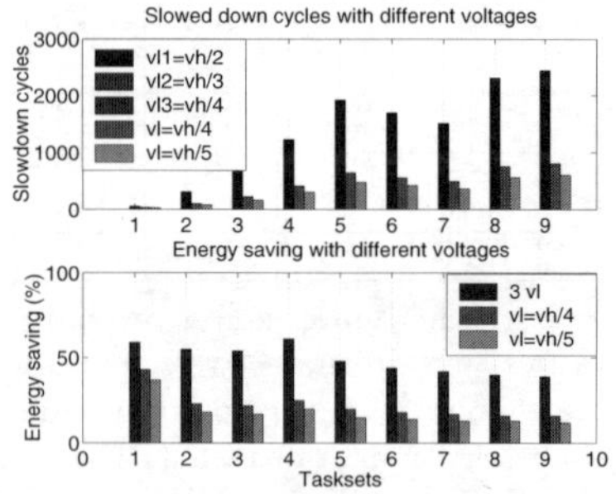

Figure 6: Energy saving of different voltages

be higher. For different voltage values, a lower value increases the cycle time more and may prevent other cycles from slowing down if there is not enough slack. However, the energy saving per cycles is more for the lower voltage value. It is not immediately clear how voltage values affect the overall energy saving. We did experiments on the same 9 tasksets on three 5-processor systems. The configurations for the three systems are Sys1: $V_{l1} = V_h/2$, $V_{l2} = V_h/3$, and $V_{l3} = V_h/4$, Sys2: $V_l = V_h/4$, and Sys3: $V_l = V_h/5$. Assume that cycle time increases from 1 to 2, 3, 4 and 5 when voltage changes from V_h to $V_h/2$, $V_h/3$, $V_h/4$, and $V_h/5$. This is a rough estimation but it should not make any conclusion we draw invalid. The number of slowdown cycles and the energy saving in the three systems are shown in the upper and lower part in Figure 6. In the figure, the first bar corresponds to data for Sys1, the second for Sys2, and the third for Sys3. The results show that our approach is capable of utilizing the bigger design space provided in Sys1 and achieves more energy savings when more voltages are available. In Sys1, where processors can operate at three different levels, most cycles are slowed down to operate at V_{l1}, which is closest to V_h. The number of cycles operating at the other two lower voltages are so few that they are invisible in Figure 6. Sys2 with $V_l = V_h/4$ has more cycles slowed down and saves more energy than Sys3 with $V_l = V_h/5$. This is because that the closer a V_l is to V_h, the more cycles can be slowed down and results in more energy saving even though the saving per cycle is smaller for a V_l that is closer to V_h. Of course, if timing constraint is so loose that every cycle can be executed at the lowest voltage, then the lower voltage, the more energy saving. Continuous voltage is the ideal case and should give the best energy saving.

6. CONCLUSION

In this paper, we present a two-phase framework that integrates task scheduling and voltage selection together to achieve the maximum energy saving of executing dependent tasks on one or multiple variable voltage processors. In the first phase, the EDF scheduling for a single processor is optimal in providing slowdown opportunities and the priority-based scheduling for multiple processors provide more slowing down opportunities than a baseline scheduling. Our IP formulation in the second phase is the first exact algorithm for the VS problem. The IP can be solved optimally in polynomial-time for the continuous voltage and some special discrete voltage cases, or solved efficiently by our approximation for general discrete voltages. Experimental results show that our framework can slow down 7-98% cycles in very short time. Due to its high quality solutions and low computational cost, our framework can be used for design space exploration.

7. ACKNOWLEDGMENTS

This research was supported in part by the National Science Foundation under Grant CCR-9988468 and MIP-9701416.

8. REFERENCES

[1] AMD Athlon 4 Processor, *Data Sheet Reference #24319*, Advanced Micro Devices, Inc., 2001.

[2] L. Benini, A. Bogliolo, and G. De Micheli, "A survey of design techniques for system-level dynamic power management," *IEEE Transactions on VLSI systems*, June 2000, pp. 299-316.

[3] T. Burd, T. Pering, A. Stratakos, and R. Brodersen, "A dynamic voltage scaled microprocessor system," *IEEE Journal of Solid-State Circuits*, vol. 35, 2000, pp. 1571-1580.

[4] A. Chandrakasan and R. Brodersen, *Low Power Digital CMOS Design*, Kluwer Academic Publishers, 1995.

[5] W. Chuang, S. Sapatnekar, and I. Hajj, "Delay and area optimization for discrete gate sizes under double-sided timing constraints," *CICC'93*, pp. 9.4.1-9.4.4.

[6] *http://dollar.biz.uiowa.edu/col.*

[7] *ftp://ftp.es.ele.tue.nl/pub/lp_solve.*

[8] F. Gruian, K. Kuchcinski, "LEneS: task scheduling for low-energy systems using variable supply voltage processors," *ASP-DAC'01*, pp. 449-455.

[9] D. Hochbaum and J. Shanthikumar, "Convex separable optimization is not much harder than linear optimization," *Journal of ACM*, vol. 37, No. 4, 1990, pp. 843-862.

[10] I. Hong, D. Kirovski, G. Qu, M. Potkonjak, and M. Srivastava, "Power optimization of variable voltage core-based systems," *DAC'98*, pp. 176-181.

[11] *http://www.chips.ibm.com:80/products/powerpc/chips.*

[12] T. Ishihara and H. Yasuura, "Voltage scheduling problem for dynamically variable voltage processors," *ISLPED'98*, pp. 197-202.

[13] Y. Lin, C. Hwang and A. Wu, "Scheduling techniques for variable voltage low power designs," *ACM Transaction on Design Automation of Electronic Systems*, vol. 2, no. 2, 1997, pp. 81-97.

[14] C. Liu and J. Layland, "Scheduling algorithms for multiprogramming in a hard-real-time environment," *Journal of the ACM*, vol. 20, no. 1, 1973, pp. 46-61.

[15] J. Luo and N. Jha, "Power-conscious joint scheduling of periodic task graphs and a periodic tasks in distributed real-time embedded systems," *ICCAD'00*, pp. 357-364.

[16] N. Namgoong, M. Yu, and T. Meng, "A high-efficiency variable-voltage CMOS dynamic DC-DC switching regulator," *ISSCC'97*, pp. 380-381.

[17] G. Quan and X. Hu, "Energy efficient fixed-priority scheduling for real-time systems on voltage variable processors," *DAC'01*, pp. 828-833.

[18] *http://helsinki.ee.Princeton.EDU/ dickrp/tgff*

Battery-Conscious Task Sequencing for Portable Devices Including Voltage/Clock Scaling*

Daler Rakhmatov and Sarma Vrudhula
Center for Low Power Electronics
ECE Department, University of Arizona
Tucson, Arizona 85721
daler/sarma@ece.arizona.edu

Chaitali Chakrabarti
Center for Low Power Electronics
EE Department, Arizona State University
Tempe, Arizona 85287
chaitali@asu.edu

ABSTRACT

Operation of battery-powered portable systems can no longer be sustained once a battery becomes discharged. Maximization of the battery lifetime is a difficult task due to nonlinearity of battery behavior that depends on the characteristics of the system load profile. We address the problem of task sequencing without and with voltage/clock scaling that shapes the profile so that the battery lifetime is maximized. We developed an accurate analytical battery model and validated it with measurements taken on a real lithium-ion battery used in a pocket computer. We use the model as a basis for a unique battery-conscious cost function and utilize its properties to develop several novel algorithms, including insertion of recovery periods and voltage/clock scaling for delay slack distribution.

Categories and Subject Descriptors

J.6.2 [**Computer-Aided Engineering**]: Computer-Aided Design

General Terms

Algorithms, Performance, Design

Keywords

Battery, modeling, low-power design, scheduling, voltage scaling

1. INTRODUCTION

In portable battery-powered electronic systems maximizing the battery lifetime is an important problem. Consequently, a designer needs an accurate model of a battery. Previously, we have proposed a high-level analytical model that produces accurate lifetime predictions and permits a trade-off between the accuracy and the amount of computation performed [9]. The present work shows how this model can be utilized for single-processor load scheduling without and with voltage/clock scaling.

*This work was carried out at the National Science Foundation's State/Industry/University Cooperative Research Centers' (NSF-S/IUCRC) Center for Low Power Electronics (CLPE). CLPE is supported by the NSF (Grant EEC-9523338), the State of Arizona, and a consortium of companies from the microelectronics industry (visit the CLPE web site http://clpe.ece.arizona.edu).

DAC 2002, June 10-14, 2002, New Orleans, Louisiana, USA.

Several published papers have considered the battery issues to improve system operation [2, 5, 6]. The authors of [2] used the VHDL-based simulation model from [1] to expose the impact of different dynamic power management policies on the battery lifetime. Both single-battery and dual-battery powered systems were considered. In [6] the authors addressed static scheduling of tasks with real-time constraints. Evaluation of the proposed method was based on the battery model combining Peukert's law [4] and empirical analysis due to [8]. Lifetime improvements reported in [2, 6] must be interpreted with care, since the results are heavily biased by the properties of a model used to represent the battery. Power-aware scheduling under timing constraints was described in [5]. The authors used a NASA/JPL Mars Pathfinder rover as a motivating application with the two power sources: a battery and a solar panel. The objective was to utilize the solar panel (the "free" energy source) as much as possible and minimize the energy drawn from the battery. However, the scheduler was only aware of the presence of an alternative energy source, not battery behavior itself.

In this paper, we propose an approach to battery load profile synthesis based on the high-level battery model from [9]. This model not only accurately predicts the battery lifetime under various discharge conditions, but also provides an analytical expression that can be used as a cost function to guide the battery lifetime optimization process. The specific problems that we address are (1) task sequencing without voltage/clock scaling and (2) task sequencing with voltage/clock scaling. While tackling these problems, we account for and take advantage of charge recovery effects. We also demonstrate that predictions of the model are in close correspondence with *measurements* performed on a real lithium-ion battery used in a pocket computer.

2. MOTIVATING EXAMPLE

In this section we demonstrate, by simple examples, the impact of task sequencing on a battery. Moreover, we also provide justification for the use of the battery model from [9] in our algorithms. All the results are obtained by performing experiments with a real battery.

The experimental setup consisted of a lithium-ion battery used in the ITSY pocket computer [10], the programmable electronic load Agilent 6060B, and the host computer recording experimental data. The open-circuit voltage of the battery was 4.2 V and the cutoff voltage was set to 3.0 V. The electronic load operated in the constant-current mode. Variable-current profiles were generated as a piece-wise constant-current profile (a staircase). The battery voltage was sampled every second, and once the voltage dropped below the cutoff level the load was disconnected from the battery. Recharging was performed in the constant-current mode at 800 mA, until the battery voltage recovered to its open-circuit value.

First, we conducted ten constant-current discharge experiments to estimate parameters for our battery model and to see how well the model predicts battery lifetime under constant loads. The discharge currents ranged from 1011 mA to 123 mA, and the lifetimes ranged from 30 minutes to over 300 minutes. The model fit is shown in Figure 1. The maximum error of lifetime predictions is 4%, with the average of 2%.

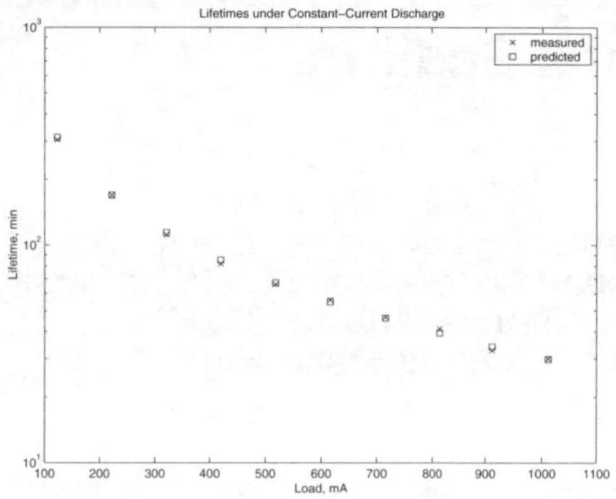

Figure 1: Lifetimes for Various Constant-Current Loads.

Next, we generated five variable-current profiles shown in Figure 2. We selected four currents of certain duration (1011 mA for 10 minutes, 814 mA for 15 minutes, 518 mA for 20 minutes, and 222 mA for 15 minutes) and sequenced them in different order to obtain profiles P1-P4. The length and total charge demand of each of the four profile is 60 minutes and 36010 mA-min, respectively. Note that in P1, after all the four currents were applied, the battery was discharged under the constant rate of 222 mA until the cutoff voltage was reached. The sole purpose of the last load 222 mA is to determine how much residual charge is left.

Profile	Measured L_m min	Measured C_m mA-min	Predicted L_p min	Predicted C_p mA-min	Lifetime Error %	Charge Error %
P1	64.9	37098	66.9	37542	3.1	1.2
P2	54.0	29944	54.4	30348	0.7	1.3
P3	55.8	32591	55.0	31940	1.4	2.0
P4	58.4	35181	57.5	34715	1.5	1.3
P5	67.5	34965	67.0	34706	0.7	0.7

Table 1: ITSY Lifetime and Delivered Charge.

According to these experiments, profile P1 is the best, and profile P2 is the worst sequence for the battery. Note that load currents in P1 (P2) are decreasing (increasing). This important result is accurately predicted by the battery model. In P1 the battery survives all the four loads and remains operational for extra 4.9 minutes under 222 mA (residual 1088 mA-min charge). In P2, however, the battery fails to service the last 6.0 minutes under 1011 mA (undelivered 6066 mA-min charge). For these two profiles, the difference in the total delivered charge is as much as 20% of 36010 mA-min. The other alternatives P3 and P4 are neither better than P1 nor worse than P2, as predicted by the model and demonstrated by the measurements.

Voltage/clock scaling plays a critical role in minimizing energy consumption, thus improving the battery lifetime. To obtain profile P5, for example, we took P2 and changed the failing 10-minute load of 1011 mA to a 20-minute load of 518 mA to reflect a hypothetical change in the system voltage.[1] The profile length increased from 60 minutes to 70 minutes, and the battery failed after 67.5 minutes. The total delivered charge is 34965 mA-min, which is an improvement over P2 with 29944 mA-min.

Table 1 shows the measured/predicted lifetimes (L_m and L_p, respectively) and the measured/predicted delivered charges (C_m and C_p, respectively) for the profiles P1-P5. Note that the charge errors were within 2%, while the maximum lifetime error was 3%.

[1]This is a pessimistic scenario, since the charge consumption is usually reduced after the supply voltage is scaled down.

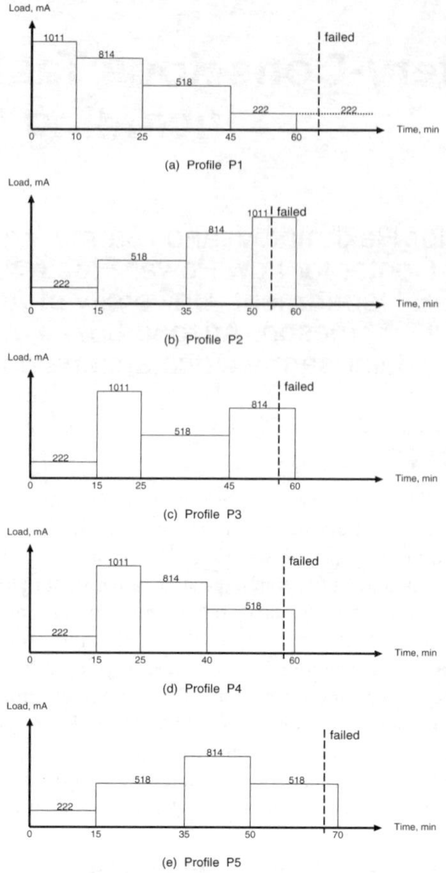

Figure 2: Experimental Load Profiles for ITSY Battery.

One can see that the model adequately captures the trend in battery behavior observed in the experiments, with very small prediction errors.

3. COST FUNCTION

In this section we describe the input and the output of the battery model and define a cost function for the task sequencing procedures. Given n tasks, we associate a 3-tuple with each task k: the current I_k drawn from the battery, the duration Δ_k, and the start time t_k.

Battery Model: The input to the model is the *load profile*, described by the following three sets: the set of currents $S_I = \{I_k \mid k = 0, 1, ..., n-1\}$, the set of durations $S_\Delta = \{\Delta_k \mid k = 0, 1, ..., n-1\}$, and the set of start times $S_t = \{t_k \mid k = 0, 1, ..., n-1\}$. The profile always starts at time 0. Let T denote the profile length, i.e. the latest finish time, $T = max_k\{t_k + \Delta_k\}$.

In addition to S_I, S_Δ, and S_t, the model requires specifications of two battery parameters α and β. These parameters are estimated based on the battery profiling data collected from several constant-current discharge experiments. For instance, the ITSY battery has $\alpha = 39668$ and $\beta = 0.57$. The units of α are mA-min, and the units of β^2 are 1/min. Intuitively, α represents the battery *capacity*, and β captures *nonlinearity* in battery behavior.

Let

$$F(x, y, z, \beta) = z - y + 2\sum_{m=1}^{\infty} \frac{e^{-\beta^2 m^2 (x-z)} - e^{-\beta^2 m^2 (x-y)}}{\beta^2 m^2}. \quad (1)$$

The output of the model is the battery lifetime L – the smallest positive number in the interval $[0, T]$ that satisfies the following equation:

$$\alpha = \sum_{k=0}^{n-1} I_k F(L, min\{L, t_k\}, min\{L, t_k + \Delta_k\}, \beta). \quad (2)$$

It is possible that such a number does not exist, i.e. the battery survives the profile (all tasks are completed). In such cases, $L =$ **NULL**. Procedure $ComputeLifetime(\cdot)$ is used for computing L. Further details are given in [9].

Cost Function: For a given profile of length T, let

$$\sigma = \sum_{k=0}^{n-1} I_k F(T, t_k, t_k + \Delta_k, \beta). \quad (3)$$

Intuitively, σ is the charge that the battery has lost by time T. If $\sigma < \alpha$, then the battery is still operational at time T. We use σ as our cost function to be *minimized*.

For a valid profile, the latency T must not exceed a given delay budget B. In addition the battery may not fail at any time $t \leq T$, i.e.

$$\alpha \geq \sum_{k=0}^{n-1} I_k F(t, min\{t, t_k\}, min\{t, t_k + \Delta_k\}, \beta) \ , \ \forall\, t \leq T. \quad (4)$$

Note that, if for some load k its start time $t_k \geq t$, then it does not contribute to the value of the sum. Condition (4) is necessary to account for relaxation effects that may mask a failure taking place before T. It requires that no sub-profile of length $t \leq T$ is too extreme for the battery to survive.

The following property of the function $F(\cdot)$ is critical to many of the results in this paper.

THEOREM 1. *For $0 \leq \Delta \leq t + \Delta \leq T$, function $F(T, t, t+\Delta, \beta)$ is (a) monotonically increasing in t and (b) monotonically decreasing in T.*

Remark: Note that our battery model does not involve battery voltage. The cutoff voltage (i.e. the voltage at which the battery is declared to be discharged) is abstracted to α, and a user is expected to specify the load profile as a discrete set of current values drawn from the battery during application execution. Sampling the battery current at the frequency of less than 1Hz will suffice [7]. If such information is not available, a user may assume the average battery voltage V_{ave}, and let the load profile $i(t) = P(t)/(\varepsilon V_{ave})$, where ε is the efficiency of a DC-DC converter and $P(t)$ is the power consumption of an application. One can assume that ε is constant and $P(t) \propto V^2(t) f(t)$ [3], where $V(t)$ is the converter output voltage and $f(t)$ is the clock rate. Then $i(t) \propto V^2(t) f(t)$.

4. TASK SEQUENCING (NO SCALING)

The formulation of the battery-aware task sequencing problem without voltage/clock scaling is shown in Figure 3. The input consists of sets S_I and S_Δ, a directed acyclic task graph $\vec{G}$, the delay budget B, and the battery parameters α and β. The output is the set S_t of start times of tasks. The objective is to minimize σ given by Equation (3), subject to the following three constraints:

(1) **dependency constraint:** task dependencies are preserved,

(2) **delay constraint:** the latency does not exceed the delay budget,

(3) **endurance constraint:** at any time within a schedule, the battery is alive.

We tackle the sequencing problem in three steps: (i) greedy sequencing, (ii) incremental recovering, and (iii) local compressing. The first step is *greedy sequencing*, subject to dependencies, with no idling allowed. The length of the resulting profile is $T = \sum_{k=0}^{n-1} \Delta_k$. In this case, the delay constraint is already satisfied, under the assumption that B is never less than the sum of all task durations. However, the endurance constraint may be violated, which calls for the second step – *incremental recovering*. In the second step, we insert idle periods so that the profile no longer fails. As much recovery as necessary is exercised, and the load ordering is not changed. Thus, the endurance constraint and dependencies are not violated; however, the delay budget may be exceeded due to recovery delay penalty. In the third step, *local compressing*, we attempt to place light loads inside idle periods, subject to dependencies, so that load relaxation is still present but recovery delay penalty is reduced. Thus, the purpose of the third step is to reduce the profile length T to the budget B. Details of these three procedures are shown in Figure 3.

Greedy Sequencing: The input is the task graph $\vec{G}$ as well as S_I and S_Δ, and the output is S_t. Dependencies are preserved by construction: no task is scheduled unless its predecessors have already been scheduled. If there are several candidates competing for the current place in a sequence, then the procedure selects the one with the heaviest weight among them. Once a task is scheduled, it is removed from the task graph.

The weight of a node is determined as follows. Let a subgraph induced by some node p be denoted by $\vec{G}_p$, and let the function $Mean(\vec{G}_p)$ return the mean current of the nodes in $\vec{G}_p$. The weight $w(p)$ is defined as the maximum of I_p and $Mean(\vec{G}_p)$. Such weighing is preferred over $w(p) = I_p$ for each task p (e.g. a task with very low current may have a successor with very large current, whose execution will be delayed until its predecessor is scheduled).

If there are no dependencies, then $w(p)$ equals to I_p for each node p. In this case, the procedure sequences tasks in nonincreasing order of their currents. Such a profile has the minimum cost σ (for $T = \sum_{k=0}^{n-1} \Delta_k$), in accordance with Theorem 2.

THEOREM 2. *For a task sequence with no idle periods allowed, $\sum_{k=0}^{n-1} I_k F(T, t_k, t_k + \Delta_k, \beta)$ is minimized, if $I_i \geq I_j \Rightarrow t_i \leq t_j$ for all $0 \leq i, j \leq n-1$.*

Incremental Recovering: The input is an initial profile specified by S_I, S_Δ, and S_t (the output of the greedy sequencing step) as well as the battery parameters α and β. The output is the task sequence with inserted idle periods and a flag indicating whether recovering failed or succeeded. Idle periods are inserted only if the battery fails before time T; otherwise, the load profile is returned as it is.

Let δ denote the length of an idle period. Before computing δ, the procedure checks whether δ will be less than the delay budget B. This check is performed as follows. Let q denote the failing task, i.e. $L \in [t_q, t_q + \Delta_q]$. First, the procedure assumes that $\delta = B$ and creates temporary set $\hat{S}_t$, where start times $\hat{t}_k$ of each task k following and including q are increased by B (e.g. $\hat{t}_q = t_q + B$). Then, it checks whether task q is recovered, i.e. $L \notin [\hat{t}_q, \hat{t}_q + \Delta_q]$. If $\delta \geq B$, then the procedure terminates with **FAILURE**, since it will be impossible to compress the profile so that the delay constraint is met. If δ is guaranteed to be less than B, then it is computed by $MinRecoveryLength(\cdot)$.[2] Finally, $InsertRecovery(\cdot)$ places the idle period of length δ at t_q and adds δ to t_q and to the start times of tasks following q in the sequence. This process is repeated until either all failures are repaired or recovery is no longer feasible.

At the absence of the delay constraint, the proposed approach provides an optimal failure recovering solution for a given load profile, in accordance with Theorem 3.

THEOREM 3. *Let q be the failing task, and let σ_q denote the cost of the subprofile of length $t_q + \Delta_q$. Let δ denote the length of an idle period inserted within the subprofile in question (the length of the resulting subprofile becomes $t_q + \Delta_q + \delta$). (1) The cost of the new subprofile is minimized, if an idle period is inserted at t_q. (2) If an idle period is placed earlier than t_q, its length must be greater than δ to achieve the same cost as in (1).*

Local Compressing: The input is the task graph $\vec{G}$, the recovered load profile (specified by S_I, S_Δ, and S_t), the delay budget B,

[2] Note that we set the upper bound on δ to be B. One (naive) implementation of $MinRecoveryLength(\cdot)$ would be to increase δ from 0 to B with small time increments, say τ, and check whether L is still in $[t_q + \delta, t_q + \Delta_q + \delta]$. (After inserting an idle period the start time of q is shifted by δ). In other words, one needs to increase δ until the cost of the subprofile of length $t_q + \Delta_q + \delta$ becomes less than α.

```
PROBLEM: Battery-Conscious Task Sequencing
    INPUT: S_I, S_Δ, \vec{G}, B, α, β
    OUTPUT: S_t
    OBJECTIVE: min{σ = \sum_{k=0}^{n-1} I_k F(T, t_k, t_k + Δ_k, β)}
    CONSTRAINTS:
    (1) if q depends on p in \vec{G}, then t_q ≥ t_p + Δ_p
    (2) T ≤ B
    (3) ∀ t ≤ T, \sum_{k=0}^{n-1} I_k F(t, min{t, t_k}, min{t, t_k + Δ_k}, β) ≤ α
```

```
TaskSequenceConstruction (\vec{G}, S_I, S_Δ)
    S_t = GreedySequencing(\vec{G}, S_I, S_Δ)
    {S_t, flag} = IncrementalRecovering(S_I, S_Δ, S_t, α, β)
    IF flag = SUCCESS
        {S_t, flag} = LocalCompressing(\vec{G}, S_I, S_Δ, S_t, α, β)
    RETURN {S_t, flag}
```

```
GreedySequencing (\vec{G}, S_I, S_Δ)
    FOR all p ∈ \vec{G}
        w(p) = max{I_p, Mean(\vec{G}_p)}
    now = 0
    WHILE \vec{G} is not empty
        S = {p | Parents(p, \vec{G}) = ∅}
        Select q | w(q) = max{w(p) | p ∈ S}
        t_q = now,  now = t_q + Δ_q
        Remove q from \vec{G}
    RETURN S_t
```

```
IncrementalRecovering (S_I, S_Δ, S_t, α, β)
    L = ComputeLifetime(S_I, S_Δ, S_t, α, β)
    WHILE L ≠ NULL
        Find q | L ∈ [t_q, t_q + Δ_q]
        \hat{S}_t = InsertRecovery(B, t_q, S_t)
        L = ComputeLifetime(S_I, S_Δ, \hat{S}_t, α, β)
        IF L ∈ [\hat{t}_q, \hat{t}_q + Δ_q]
            RETURN {S_t, FAILURE}
        δ = MinRecoveryLength(α, β, q, S_I, S_Δ, S_t)
        S_t = InsertRecovery(δ, t_q, S_t)
        L = ComputeLifetime(S_I, S_Δ, S_t, α, β)
    RETURN {S_t, SUCCESS}
```

```
LocalCompressing (\vec{G}, S_I, S_Δ, S_t, B, α, β)
    T = max_k{t_k + Δ_k}
    IF B ≥ T
        RETURN {S_t, SUCCESS }
    T_new = T,  change = TRUE
    WHILE change = TRUE
        {R_s, R_f} = IdentifyIdlePeriods(S_I, S_Δ, S_t)
        M = |R_s|,  \hat{S}_t = S_t,  T_old = T_new,  change = FALSE
        WHILE M > 0
            \hat{S} = {p | t_p ≥ R_f[M]}
            Find q | I_q = min{I_p | p ∈ \hat{S} AND Parents(p, \vec{G}) ∉ \hat{S}}
            Remove q from \hat{S}
            t_q = R_s[M]
            S_t = UpdateStartTimes(S_t)
            {S_t, flag} = IncrementalRecovery(S_I, S_Δ, S_t, α, β)
            IF flag = FAILURE
                S_t = \hat{S}_t
            ELSE
                T_new = max_k{t_k + Δ_k}
                IF B ≥ T_new
                    RETURN {S_t, SUCCESS }
                IF T_old ≤ T_new
                    S_t = \hat{S}_t
                ELSE
                    change = TRUE
                    BREAK
            M = M − 1
    RETURN {S_t, FAILURE }
```

Figure 3: Task Sequencing without Voltage/Clock Scaling.

and the battery parameters α and β. The output is the new profile specification and the flag indicating whether compression succeeded or failed.

First, the length T of the original profile is compared with the budget B. If the profile length is within the budget, then the profile is not changed. Otherwise, we identify idle periods and construct two sets R_s (idle start times) and R_f (idle finish times). For example, the start time of the k-th idle period (counting from the beginning of the profile) is recorded as $R_s[k]$, and its finish time is recorded as $R_f[k]$. Next, we select the latest unvisited idle period M and collect the tasks scheduled after M into the set $\hat{S}$. In $\hat{S}$, we find task q such that I_q is as low as possible provided that starting q at $R_s[M]$ will not violate dependencies. Once q is placed at $R_s[M]$, the start times of the remaining tasks in $\hat{S}$ are updated by $UpdateStartTimes(\cdot)$, so that their initial ordering is preserved, but initial idle periods among them are omitted. Then, new idle periods are inserted, if necessary. If the length of this new profile is decreased, then it becomes the current solution, and the outer **WHILE**-loop is repeated. Otherwise, profile changes are ignored, and the next latest unvisited idle period is considered (the inner **WHILE**-loop is repeated).

Local compressing is terminated, if (a) changes can no longer reduce the length of the current profile, or (b) the current profile meets the delay budget. Note that modifications are applied starting from the end of the load profile, since the lightest loads are normally scheduled late. Light load may reduce the total recovery time if placed inside an idle period of interest. We consider idle periods starting from the latest so that the minimal portion of the profile is changed. In other words, the procedure places lighter tasks into later idle periods, thus attempting to approach the deadline at a minimal cost penalty.

5. TASK SEQUENCING WITH SCALING

Simultaneous voltage/clock scaling is a very effective way to reduce energy consumption [3]. From the battery lifetime maximization perspective, scaling techniques are critical in shaping the load profile appropriately. In this section we state the problem of task sequencing with voltage/clock scaling and provide two procedures for improving delay-constrained load profiles. Note that the battery model involves only task durations and currents drawn from the energy source by a DC-DC converter. Since voltage/clock scaling is applied to the energy consumer (the output side of a DC-DC converter), changes in voltage levels and clock frequencies must be translated into changes in the task currents and durations before the model can be used.

We assume that the system can operate at any voltage V_i from the ordered set $S_V = (V_0, V_1, ..., V_{K-1})$ at the corresponding maximum frequency ϕ_i from the ordered set $S_\phi = (\phi_0, \phi_1, ..., \phi_{K-1})$. The elements of S_V and S_ϕ are in ascending order. Let I_{ik} and Δ_{ik} respectively denote the current and the duration of task k, when the system is operating at V_i with the clock frequency ϕ_i. Figure 4 shows the formulation of the battery-conscious voltage/clock scaling problem. The objective is to assign each task k to a specific voltage V_i at the frequency ϕ_i (thus determining the task current I_{ik} and the task duration Δ_{ik}) and the start time t_k, so that the resulting profile cost σ is minimized and the following constraints are satisfied: (1) task dependencies are preserved, (2) the profile length is within the delay budget, and (3) the battery survives all the tasks. Unlike the case of task sequencing without voltage/clock scaling, we need to generate not only S_t, but also S_I and S_Δ.

Scaling Based on Lowest-Power Initial Solution: We start with the lowest-power profile, which satisfies the endurance constraint, but may violate the delay constraint. In this profile all the tasks are assigned to V_0 (the lowest supply voltage level), so that the energy consumption is minimized. We use $LookUp(\cdot)$ to look up, in a user-specified table, task currents and durations for a given voltage. For example, for task k at V_0 and ϕ_0, the current and duration are $\{I_k, \Delta_k\} = LookUp(k, V_0, \phi_0) = \{I_{0k}, \Delta_{0k}\}$. Thus, we generate the initial sets S_I and S_Δ. Initial S_t is produced by $GreedySequencing(\cdot)$.

Let L be the computed lifetime of the initial profile $\{S_I, S_\Delta, S_t\}$.

PROBLEM: Battery-Conscious Voltage/Clock Scaling

INPUT: S_V, S_ϕ, $\vec{G}$, B, α, β
OUTPUT: S_I, S_Δ, S_t
OBJECTIVE: $min\{\sigma = \sum_{k=0}^{n-1} I_{ik}F(T,t_k,t_k+\Delta_{ik},\beta)\}$
CONSTRAINTS:
(1) if q depends on p in $\vec{G}$, then $t_q \geq t_p + \Delta_p$
(2) $T \leq B$
(3) $\forall\, t \leq T,\ \sum_{k=0}^{n-1} I_{ik}F(t,min\{t,t_k\},min\{t,t_k+\Delta_{ik}\},\beta) \leq \alpha$

VoltageClockScaling $(S_V,S_\phi,\vec{G},B,\alpha,\beta)$
$K = |S_V|,\quad n = |\vec{G}|$
FOR $k = 0,1,...,n-1$
$\{I_k,\Delta_k\} = LookUp(k,V_0,\phi_0)$
$S_t = GreedySequencing(\vec{G},S_I,S_\Delta)$
$T = \sum_{k=0}^{n-1}\Delta_k$
$L = ComputeLifetime(S_I,S_\Delta,S_t,\alpha,\beta)$
IF $T \leq B$ **AND** $L =$ **NULL**
RETURN $\{S_I,S_\Delta,S_t,$ **SUCCESS**$\}$
IF $T > B$ **AND** $L \neq$ **NULL**
RETURN $\{S_I,S_\Delta,S_t,$ **FAILURE**$\}$
IF $T \leq B$ **AND** $L \neq$ **NULL**
$\{S_t, flag\} = IncrementalRecovering(S_I,S_\Delta,S_t,\alpha,\beta)$
IF $flag =$ **SUCCESS**
$\{S_t, flag\} = LocalCompressing(\vec{G},S_I,S_\Delta,S_t,\alpha,\beta)$
RETURN $\{S_I,S_\Delta,S_t,flag\}$
$\{S_I,S_\Delta,S_t,flag\} = LatencyReduction(S_V,S_\phi,S_I,S_\Delta,S_t,B,\alpha,\beta)$
IF $flag =$ **SUCCESS**
$\{S_I,S_\Delta,S_t\} = SlackUtilization(S_V,S_\phi,S_I,S_\Delta,S_t,B,\alpha,\beta)$
RETURN $\{S_I,S_\Delta,S_t,flag\}$

LatencyReduction $(S_V,S_\phi,S_I,S_\Delta,S_t,B,\alpha,\beta)$
FOR $x = 0,1,...,n-1$
$k =$ the x-th task in the sequence
FOR $i = 1,...,K-1$
$\{\hat{I}_k,\hat{\Delta}_k\} = LookUp(k,V_i,\phi_i)$
$\{\hat{S}_I,\hat{S}_\Delta,\hat{S}_t\} = UpdateLoadProfile(\hat{I}_k,\hat{\Delta}_k,S_I,S_\Delta,S_t)$
$T = max_l\{\hat{t}_l + \hat{\Delta}_l\}$
$L = ComputeLifetime(\hat{S}_I,\hat{S}_\Delta,\hat{S}_t,\alpha,\beta)$
IF $L \neq$ **NULL**
BREAK
$\{S_I,S_\Delta,S_t\} = \{\hat{S}_I,\hat{S}_\Delta,\hat{S}_t\}$
IF $T \leq B$
RETURN $\{S_I,S_\Delta,S_t,$ **SUCCESS**$\}$
RETURN $\{S_I,S_\Delta,S_t,$ **FAILURE**$\}$

SlackUtilization $(S_V,S_\phi,S_I,S_\Delta,S_t,B,\alpha,\beta)$
$n = |S_I|$
$T = max_l\{t_l + \Delta_l\}$
IF $T \geq B$
RETURN $\{S_I,S_\Delta,S_t\}$
FOR $x = n-1,n-2,...,0$
$k =$ the x-th task in the sequence
Find $j \mid \{I_k,\Delta_k\} = LookUp(k,V_j,\phi_j)$
FOR $i = j-1,j-2,...,0$
$\{\hat{I}_k,\hat{\Delta}_k\} = LookUp(k,V_i,\phi_i)$
$\{\hat{S}_I,\hat{S}_\Delta,\hat{S}_t\} = UpdateLoadProfile(\hat{I}_k,\hat{\Delta}_k,S_I,S_\Delta,S_t)$
$T = max_l\{\hat{t}_l + \hat{\Delta}_l\}$
IF $T > B$
BREAK
$\{S_I,S_\Delta,S_t\} = \{\hat{S}_I,\hat{S}_\Delta,\hat{S}_t\}$
RETURN $\{S_I,S_\Delta,S_t\}$

Figure 4: Task Sequencing with Voltage/Clock Scaling.

Note that $T = \sum_{k=0}^{n-1}\Delta_{0k}$. If $T \leq B$ and $L =$ **NULL**, then we have already obtained final solution. On the other hand, if $T > B$ and $L \neq$ **NULL**, then we abort our attempts to generate a feasible profile. There are two non-trivial cases: (a) $T \leq B$ and $L \neq$ **NULL**, and (b) $T > B$ and $L =$ **NULL**.

In case (a), the delay budget is met, but the battery does not survive the profile. Since the voltage level is the lowest, the task currents cannot be reduced. Therefore, only idle period insertion is applicable: we perform *IncrementalRecovering*$(\cdot)$, followed by *LocalCompressing*$(\cdot)$.

In case (b), the battery survives the profile, but the delay budget is exceeded. Therefore, some tasks must be assigned to a higher voltage/clock (resulting in greater currents but shorter durations) to satisfy the delay constraint. We use *LatencyReduction*$(\cdot)$ to reduce the profile length so that the budget is met. Figure 4 provides further details. At each iteration of the outer **FOR**-loop, we consider the earliest unvisited task k (at position x in the task sequence), and its voltage is scaled up as much as possible, provided that the profile remains failure-free. Upon assigning new voltage/clock to k, we call *UpdateLoadProfile*$(\cdot)$ to update the load profile, move to the next task (at position $x+1$ in the task sequence), and perform voltage/clock up-scaling again. Intuitively, this procedure attempts to generate a nonincreasing sequence of currents without causing failures before all the tasks are completed.

Slack Utilization: Given a failure-free profile with positive delay slack (i.e. $L =$ **NULL** and $T < B$), we may be able to improve its quality by scaling voltage/clock down, which decreases task currents but increases task durations. The corresponding procedure, *SlackUtilization*$(\cdot)$, is shown in Figure 4. We consider tasks one by one, starting from the profile's end. The voltage/clock for each considered task is down-scaled as much as possible, provided that the delay constraint is not violated. Thus, this procedure attempts to reduce the battery stress toward the end of the discharge process. Voltage/clock down-scaling does not introduce any failures, in accordance with the following theorem.

THEOREM 4. *Assume that a given task sequence is failure-free. Let voltage and clock be scaled down for some tasks. The resulting profile is still failure-free.*

6. RESULTS

The battery lifetime depends on both the load profile and the battery parameters. In this section we let $\alpha = 40000$ and $\beta = 0.2$. Figure 5 shows the lifetime surface for the simplest case of a constant-current discharge. Note that for sufficiently large β the battery behaves as a practically ideal power source, i.e. the lifetime increases almost linearly as α grows. We use small $\beta = 0.2$ to expose nonlinear effects in battery behavior.

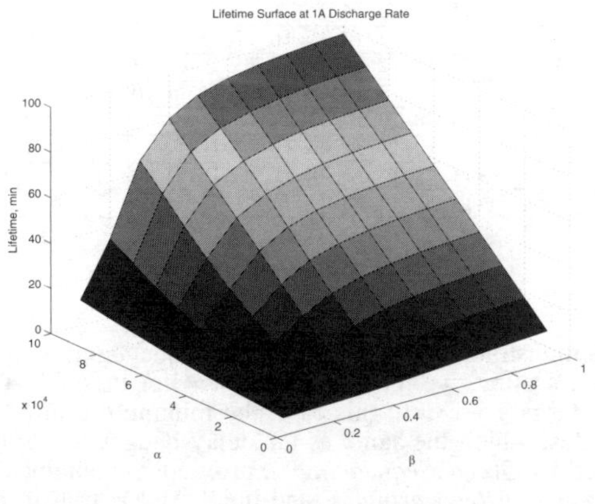

Figure 5: Lifetimes for Various α and β under Constant-Rate Discharge.

Consider a set of tasks described in Table 2. There are two supply voltages V and $V/2$, and each task is associated with the two possible battery current values and the two possible durations. We

Task	Voltage V		Voltage $V/2$		Parents
	I, mA	Δ, min	I, mA	Δ, min	
T1	1000	5	125	10	-
T2	750	5	93	10	-
T3	500	10	62	20	T1
T4	250	10	31	20	-
T5	800	5	100	10	T2, T3
T6	600	5	75	10	T4, T5
T7	400	10	50	20	T1
T8	200	10	25	20	T2, T7

Table 2: Task Currents and Durations.

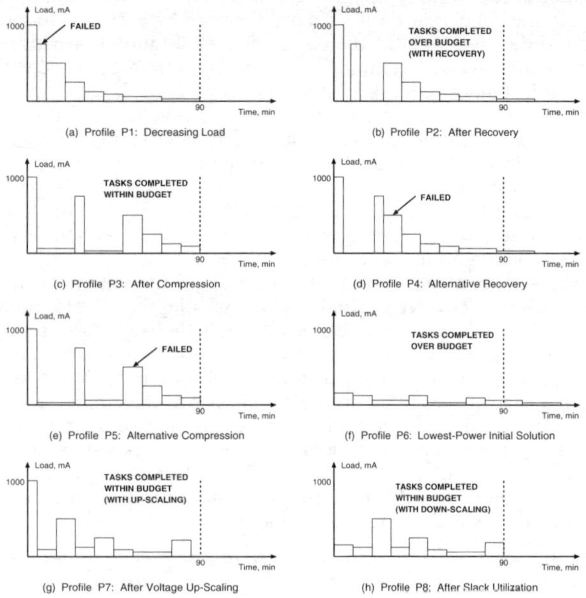

(a) Profile P1: Decreasing Load (b) Profile P2: After Recovery (c) Profile P3: After Compression (d) Profile P4: Alternative Recovery (e) Profile P5: Alternative Compression (f) Profile P6: Lowest-Power Initial Solution (g) Profile P7: After Voltage Up-Scaling (h) Profile P8: After Slack Utilization

Figure 6: Generated Load Profiles.

let the delay budget B equal 90 minutes. To illustrate the proposed methods, we examine eight profiles P1-P8, shown in Figure 6. Table 3 shows the task sequence as well as the corresponding length T and the cost σ, for each profile.

Profile	Task Sequence	T, min	σ @ T, mA-min
P1	(T1, T2, T3, T4, T5, T6, T7, T8)	90	23435
P2	(T1, T2, T3, T4, T5, T6, T7, T8)	106	23180
P3	(T1, T7, T2, T8, T3, T4, T5, T6)	90	29558
P4	(T1, T2, T3, T4, T5, T6, T7, T8)	106	23292
P5	(T1, T8, T2, T7, T3, T4, T5, T6)	90	29646
P6	(T1, T2, T3, T5, T4, T6, T7, T8)	120	9886
P7	(T1, T2, T3, T5, T4, T6, T7, T8)	85	30139
P8	(T1, T2, T3, T5, T4, T6, T7, T8)	90	26103

Table 3: Profile Lengths and Costs.

To demonstrate the impact of task sequencing without voltage/clock scaling, we fix the voltage at V for tasks T1-T4 and at $V/2$ for tasks T5-T8. In this case, the minimum profile length is 90 minutes, which the same as the delay budget B. Profile P1 is generated by $GreedySequencing(\cdot)$, profile P2 is obtained from P1 by $IncrementalRecovering(\cdot)$, and finally, P2 is transformed into profile P3 by $LocalCompressing(\cdot)$.

Dependencies allow the tasks to be sequenced in decreasing order of their currents in P1. The battery, however, fails after 8.6 minutes; therefore, idle period insertion follows. Since there is no delay slack, P2 with two idle periods violates the delay constraint by 16 minutes.[3] However, the profile length of P3 is successfully reduced to 90 minutes by compression. Note that P1 and P3 are two task sequences with no idle periods, and as expected, the cost of P1 is lower than the cost of P3 (23435 vs. 29558 mA-min). Profile P1 demonstrates the need for the endurance constraint. For P1 $\sigma < \alpha$ but $L \neq$ **NULL**, which means that a battery failure has been masked by relaxation effects.

Profile P4 shows an alternative to P2. The total idling time in P2 and P4 is the same, 16 minutes. In P4, however, the two idle periods from P2 are combined into one. Note that the battery can no longer survive P5, and as expected, the cost of P2 is lower than the cost of P4. Profile P5 shows an alternative to P3. P5 differs from P3 in that tasks T7 and T8 are swapped. Even though the costs of P3 and P5 are practically the same, P5 violates both the dependency constraint and the endurance constraint.

Profiles P6, P7, and P8 demonstrate the impact of voltage/clock scaling. P7 is obtained by $LatencyReduction(\cdot)$ from the lowest-power initial solution P6, where all the tasks are assigned to $V/2$. The length of P6 is 120 minutes, and the delay slack is –30 minutes. In profile P7, which is failure-free and within the delay budget, the voltage for tasks T1, T3, T4, and T8 is scaled up to V. Note that a 5-minute delay slack is available in P7. After applying $SlackUtilization(\cdot)$ one obtains profile P8, whose cost is lower than that of P7 by more than 10% (26103 vs. 30139). Note that among the eight considered profiles, only P3, P7, and P8 are feasible (i.e. all the constraints are satisfied), and the cost of P8 is the best.

7. CONCLUSION

Since the battery behavior exhibits a nonlinear dependency on the characteristics of the system load profile, achieving a provably optimal lifetime is not an easy task. We addressed the problem of battery-conscious task sequencing without and with voltage/clock scaling. We employed an accurate analytical battery model as a basis for a unique cost function and developed several efficient heuristics for synthesizing battery-efficient load profiles. Our algorithms include insertion of recovery periods and voltage/clock scaling for delay slack distribution.

8. REFERENCES

[1] L. Benini, G. Castelli, A. Macii, E. Macii, M. Poncino, and R. Scarsi. A discrete-time battery model for high-level power estimation. *Proc. DATE*, 2000.

[2] L. Benini, G. Castelli, A. Macii, and R. Scarsi. Battery-driven dynamic power management. *IEEE Design and Test*, March-April 2001.

[3] T. Burd and R. Brodersen. *Energy Efficient Microprocessor Design*. Kluwer, Boston, 2002.

[4] D. Linden. *Handbook of Batteries*. McGraw-Hill, New York, 1995.

[5] J. Liu, P. Chou, N. Bagherzadeh, and F. Kurdahi. Power-aware scheduling under timing constraints for mission-critical embedded systems. *Proc. DAC*, 2001.

[6] J. Luo and N. Jha. Battery-aware static scheduling for distributed real-time embedded systems. *Proc. DAC*, 2001.

[7] T. Martin. *Balancing Batteries, Power, and Performance: System Issues in CPU Speed-Setting for Mobile Computing*. Ph.D. Dissertation, Carnegie Melon University, 1999.

[8] M. Pedram and Q. Wu. Design considerations for battery-powered electronics. *Proc. DAC*, 1999.

[9] D. Rakhmatov, S. Vrudhula, and D. Wallach. A model for battery lifetime analysis for organizing applications on a pocket computer. *To appear IEEE Trans. VLSI Systems*.

[10] M. Viredaz and D. Wallach. Power evaluation of a handheld computer: a case study. *Compaq WRL Research Report 2001/1*, May 2001.

[3] The first idle period lasts 3 minutes, and the second idle period lasts 13 minutes.

An Energy Saving Strategy Based on Adaptive Loop Parallelization*

I. Kadayif
Microsystems Design Lab
Pennsylvania State University
University Park, PA 16802
kadayif@cse.psu.edu

M. Kandemir
Microsystems Design Lab
Pennsylvania State University
University Park, PA 16802
kandemir@cse.psu.edu

M. Karakoy
Department of Computing
Imperial College
London SW7 2BZ, UK
karakoy@doc.ic.ac.uk

ABSTRACT

In this paper, we evaluate an adaptive loop parallelization strategy (i.e., a strategy that allows each loop nest to execute using different number of processors if doing so is beneficial) and measure the potential energy savings when unused processors during execution of a nested loop in a multi-processor on-a-chip (MPoC) are shut down (i.e., placed into a power-down or sleep state). Our results show that shutting down unused processors can lead to as much as 67% energy savings with up to 17% performance loss in a set of array-intensive applications. We also discuss and evaluate a processor pre-activation strategy based on compile-time analysis of nested loops. Based on our experiments, we conclude that an adaptive loop parallelization strategy combined with idle processor shut-down and pre-activation can be very effective in reducing energy consumption without increasing execution time.

Categories and Subject Descriptors

D.3.4 [**Programming Languages**]: Processors—*Compilers, Optimization*

General Terms

Design, Experimentation, Performance

Keywords

Adaptive Parallelization, Multiprocessing, Energy Consumption

1. INTRODUCTION AND MOTIVATION

The computer industry's remarkable success in squeezing ever-more transistors into an ever-smaller area of silicon is tremendously increasing the computational abilities of electronic devices. As a result, very complex functions can be performed in a single chip. Multi-Processor-on-a-Chip (MPoC) is a single chip architecture that contains multiple processors with some amount of on-chip memory (SRAM), synchronization logic, I/O, and interconnect. Unlike a multi-PU (multi-processing unit) machine where the processing units operate in synchrony, in an MPoC, each processor can operate independently from other processors and synchronization is necessary only when processors share data. In fact, in executing a given application, we may want to use only a subset of the available processors if doing so is beneficial for some purpose. The remaining processors can stay idle or can even be used for executing some other application.

Note that while VLIW/superscalar processors may provide a certain level of parallelism, as indicated by Verbauwhede and Nicol [13], they are not scalable to provide high levels of performance needed by future applications, particularly those in next-generation wireless environments. On top of this, the power consumed by these architectures does not scale linearly as the number of execution units is increased. This is due to complexity of instruction dispatch unit, instruction issue unit, and large register files. Recently, automatic loop parallelization technology developed for array-intensive applications has been shown to be very effective [15]. We believe that array-intensive embedded applications can also take advantage of this technology and derive significant performance benefits from the on-chip parallelism and low-latency synchronization provided by an MPoC.

Energy concerns are becoming increasingly pressing in embedded system design. Particularly, the proliferation of embedded devices raised battery energy consumption to a first-class status in system design [3]. Note that dynamic (switching) energy consumption in an MPoC-based architecture is determined largely by the number of executing processors, whereas the static (leakage) energy depends on the number of powered-on processors. A given MPoC can consume a significant amount of energy and optimizing its energy consumption will be even more important in the future.

Adaptive parallelization is a compiler-directed optimization technique that tunes the number of processors for each part of the code according to its inherent parallelism. For example, intrinsic data dependences in a given nested loop may prevent us from using all processors. In such a case, trying to use more processors (than necessary) can lead to an increase in execution time due to increased inter-processor communication/synchronization costs. Similarly, a small loop bound may also suggest the use of fewer processor (than available) to execute a given loop. Loops in particular present an excellent optimization scope for adaptive parallelization. Since in general each loop might require a different number of processors to achieve the best performance, it might be useful to change the number of processors across the loops. Previous research on large scale parallel machines [4] reports that adaptive loop parallelization (that is, executing each loop with the best number of processors instead of fixing the number of active processors throughout the entire life of application) can be effective in maximizing the utilization of processors.

When adaptive loop parallelization is employed in a given MPoC, the unused (idle) processors can be shut down to conserve energy. Depending on the inherent degree of parallelism in different nests of a given code, such a strategy can lead to significant savings in energy. This is because shutting down a processor reduces its dynamic and leakage energy. However, one has to pay a *re-synchronization penalty* when a processor placed into a power-down (sleep)

*This work was supported in part by NSF Career Award #0093082.

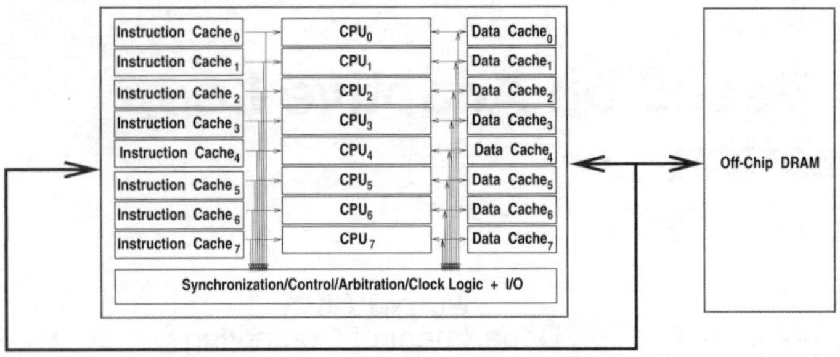

Figure 1: MPoC architecture.

state is requested to participate computation (e.g., in the next nest). The magnitude of this cost depends on the time it takes to bring the processor back from the power-down state. As will be discussed in this paper, in some cases, it might be useful to *pre-activate* a processor before it is actually needed so as to ensure that it is ready when it is required to perform computation. Such a pre-activation strategy can, if successful, eliminate the performance penalty due to re-synchronization and reduce energy consumption further.

This paper makes the following contributions:

•It evaluates an adaptive parallelization strategy and measures the potential energy savings when unused processors in the MPoC are shut down. Our results show that shutting down unused processors can lead to as much as 67% energy savings with up to 17% performance loss in a set of array-intensive applications.

• It presents a processor pre-activation strategy based on compile-time analysis of nested loops. Based on our experiments, we conclude that an adaptive loop parallelization strategy combined with idle processor shut-down and pre-activation can be very effective in reducing energy consumption without increasing execution time. Our experiments with pre-activation indicate a 39% reduction in energy consumption (on average) as compared to a scheme without energy management.

Apart from academic interest ([11, 9]), chip multiprocessor architectures are also finding their ways into commercial products. For example, Sun's MAJC-5200 [10] is a general-purpose multiprocessor system on a chip integrating two processors, a memory controller, a PCI controller, two high bandwidth I/O controllers, a data transfer engine, and a crossbar interfacing all the blocks. It is suitable for multimedia computing and networking applications. Sun Microsystems, IBM, and others are starting to produce systems that take advantage of chip multiprocessing [7]. In fact, early tests are showing that two processors in a single module outperform multiple discrete processors by 50% or more. By putting two processors on a single piece of silicon, engineers are taking advantage of shorter distances and faster bus speeds.

The remainder of this paper is organized as follows. The next section discusses the MPoC architecture assumed in this paper. Section 3 introduces our experimental platform and the array-intensive codes used in this study, and presents our experimental methodology. Section 4 discusses an adaptive parallelization strategy and presents the energy savings and performance penalties due to shutting down unused processors. Section 5 presents a processor pre-activation strategy and gives empirical data showing its effectiveness in saving energy. Section 6 discusses future work, and finally, Section 7 offers our conclusions.

2. MPOC ARCHITECTURE

We assume that a number of processors are integrated on a single die using an on-chip cache memory architecture as shown in Figure 1. In this work, we assume that each processor has a simple pipelined architecture that issues one instruction per cycle. As stated in [11], using simple, identical processors allows the design and verification costs for a single CPU core to be lower, and amortizes those costs over a larger number of processor cores. We assume that each processor has its own data cache and instruction cache. While in this work we focus on a single level on-chip data cache memory, it is straightforward to extend this architecture to designs that contain multiple (on-chip) cache memories and to designs that employ a scratch pad memory (a software-managed on-chip SRAM) instead of conventional caches. It is important to note, however, that if there is a shared level-two (L2) cache in the system, we also need to take into account the energy consumption and performance penalty of its arbitration logic. We assume the existence of a large, off-chip memory (DRAM) whose access time and per access energy consumption are much higher than the corresponding values for on-chip caches. This DRAM space is shared by all processors in the MPoC. We also assume the existence of a simple synchronization mechanism between CPUs to ensure correct execution when they work on the same data in parallel. When a processor is shut-down, we assume that the corresponding caches are also shut-down.

To put a given processor-cache memory pair into the power-down state, different methods can be used [6, 16]. A common characteristic of these techniques is that when the processor/cache is placed into power-down state, the energy consumption is reduced significantly. The magnitude of this energy reduction due to power-down state is expressed as an energy reduction factor (ERF) as will be detailed later in the paper. Another characteristic is that a re-synchronization penalty (RP) is incurred when a processor/cache pair in the power-down state is accessed (attempted to be re-activated). Instead of experimenting with specific values of ERFs and RPs, we chose to vary these parameters to cover a large number of design alternatives. This allows us to evaluate our approach under a variety of scenarios. Note that although we focus here on a single power-down state only, it is also possible to extend our approach to cases where multiple power-down states exist with different ERF and RP values.

This architecture has an important advantage over traditional multi-FU machines such as VLIW/superscalar processors. Since each processor can work independently, we can assign work to each of them at the source code level. For example, we can utilize existing source-level loop parallelization techniques [2, 15] to distribute computation across processors. For array-intensive applications with regular (compile-time predictable) data access patterns, exploiting parallelism opportunities at a large granularity (source level) is expected to return larger benefits than low-level parallelism optimization techniques. In particular, for codes that can be parallelized into multiple threads, the simple MPoC processors working together will perform better than a more complicated wide-issue superscalar machine or a VLIW architecture.

3. PLATFORM AND METHODOLOGY

We used an in-house, cycle-accurate energy simulator [14] to measure the energy consumed in different components of the MPoC such as processors, caches, off-chip memory, clock circuitry, interconnect between processors and caches, and interconnect between caches and off-chip memory. Our simulator takes as input a configuration file and an input code written in C and produces as output the energy distribution across different hardware components and performance data (execution cycles). Each processor is modeled as a simple five-stage pipeline that issues a single instruction at each cycle. Extensive clock gating [5] has been employed to reduce energy consumption.

The energy model used by our simulator for interconnect is transition-sensitive; that is, it captures the switching activity on buses. Since simple analytical energy models for cache memories have proved to be quite reliable, our simulator uses an analytic cache model [12] to capture cache energy. This model takes into account the cache topology, number of hits/misses and write policy, and returns the total amount of energy expended in the cache. We also assume the existence of an off-chip memory and assume a fix per access energy consumption of 4.95 nJ (as in [12]). The simulator uses predefined, transition-sensitive models for each functional unit to estimate the energy consumption of the core. The current implementation does not model the control circuitry in the core. This is not a major problem in this study as the energy consumed by the datapath is much larger than the energy consumed by the control logic due to simple control logic of each MPoC processor (single issue, no speculation). All functional unit energy models are for 0.35 micron technology and have been validated to be accurate (within 10%) [14]. The clock subsystem of the target architecture is implemented using a first level H-tree and a distributed driver approach that supplies clocking to four main units: data cache, instruction cache, register file, and datapath (pipeline registers). The simulated architecture uses static CMOS gates and single-phase clocking for all sequential logic while all memory structures are based on the classic 6T cell. We also model the impact of gating at different levels and for different units. The clock network model was validated to be within 10% error from circuit simulation values. More details can be found in [14].

We also model the energy consumption due to spawning multiple threads to be executed on processors, synchronizing them at the end of each parallel loop, and re-synchronization penalty (waking up time). This is achieved by calculating the energy consumed due to executing the code fragments that perform these spawning/synchronization/re-synchronization tasks. We assume a simple bus-based synchronization mechanism where each processor participates in the synchronization process by setting a bit in its control register. Future work will consider more sophisticated network architectures and more efficient synchronization mechanisms. Unless stated otherwise, all caches in the MPoC are 4KB, 2-way set-associative with a write-back policy and a line (block) size of 32 bytes. The MPoC in consideration has 8 identical processors. The cache access latency is 2 cycles and an off-chip memory access takes 80 cycles.

Figure 2 lists the thirteen benchmark codes used in this study and their important characteristics. `3step-log`, `full- search`, `hier`, and `parallel-hier` are four different motion estimation implementations; `aps`, `bmcm`, `eflux`, and `tsf` are from Perfect Club benchmarks; and `btrix` and `tomcat` are from Spec benchmarks. `adi` is from Livermore kernels and the remaining codes are array-based versions of the DSPstone benchmarks. The second column gives total input size and the third column shows the number of nests in each code. The remaining columns will be explained in the next section.

For each benchmark code, we run three versions: (i) The original version. In this version, we assume that only the processors (and their caches and interconnects) that participate computation consume dynamic energy; but, each processor/cache pair (whether it participates the computation or not) consumes leakage energy. (ii) In this version, which is detailed in Section 4, the processors that do not participate computation are placed into power-down state. Note that, due to the re-synchronization cost, such an approach leads to a performance penalty. (iii) To eliminate this performance penalty, this version (detailed in Section 5) employs a pre-activation strategy using which the processors are pre-activated before they are actually needed. All three versions use clock gating [5] where possible. All necessary code modifications have been implemented using the SUIF compiler infrastructure [1] as a source-to-source translator.

In the rest of this paper, unless otherwise stated, when we mention *energy consumption,* we mean the energies consumed in caches, interconnects, off-chip memory, processor, and the clock circuitry. All optimized energy consumptions are given as values *normalized* with respect to the energy consumed in these components by the original version (version (i)). We do not consider the energy consumed in the control circuitry and I/O modules as the energy impact of our adaptive parallelization strategy on these units is expected to be minimal. It is also possible to extend our power management strategy to selectively shut down these components when the compiler detects that they will not be exercised by a given loop.

4. ADAPTIVE PARALLELIZATION AND IMPACT OF PROCESSOR SHUT-DOWN

Most published work on parallelism [2, 15] is static; that is, the number of processors that execute the code is fixed throughout the execution. In adaptive parallelization, on the other hand, the number of processors is tailored to the specific needs of each code section (e.g., a nested loop in array-intensive applications). For instance, an adaptive parallelization strategy can use 4, 6, 2, and 8 processors to execute the first four nests in a given code. There are two important issues that need to be addressed in designing an effective adaptive parallelization strategy:

•*Mechanism:* How is the number of processors to execute each code section determined? There are two ways of determining the number of processors per nest: dynamic approach and static approach. The first option is adopting a fully-dynamic strategy whereby the number of processors (for each nest) is decided in the course of execution (at run-time). While this approach is expected to generate better results once the number of processors has been decided (as it can take run-time code behavior and resource constraints into account), it may also incur a significant performance overhead during the process of determining the number of processors. This overhead can easily offset the potential benefits of adaptive parallelism. In the second option, the number of processors for each nest is decided at compile-time. This approach has a compile-time overhead but it does not lead to much run-time penalty. In this paper, we adopt the static approach. It should be emphasized, however, that although our approach determines the number of processors statically at compile-time, the activation/deactivation of processors and their caches occurs dynamically at run-time.

•*Policy:* What is the criterion based on which we decide the number of processors? In this study, we used the execution time as the main criterion to decide the number of processors for each nest. This is because our objective is to optimize energy consumption with as little negative impact as possible on performance. In order to determine the number of processors that results in the best execution time for a given nest, we used profile data. That is, using our simulator, we executed the nest with different number of processors and selected the one with the minimum execution time. While this profile-based strategy can increase the compilation time, in many embedded systems, large compilation times can be tolerated as these systems typically run a single (or a small set of) application(s). We plan to make experiments with other criteria (policies) in the future.

It should be mentioned that there are several circumstances that might lead to an increase in overall execution time when the number of processors is increased. As mentioned earlier in the paper, the intrinsic data dependences might lead to excessive synchronization overhead if more processors (than necessary) are used. Also, sometimes (particularly in small loops), the overhead of creating a large number of threads to execute a loop nest in parallel and then synchronizing them after the nest execution can offset the benefits from adaptive parallelism.

Figure 3 shows, for each nest of each benchmark code in our experimental suite, the number of processors that generated the best execution time. The data presented in this figure clearly shows that in many cases, using only a small subset of processors (recall that our MPoC has 8 processors) generates the best result. This is a strong motivation for shutting off unused processors to save energy. The fourth column of Figure 2 gives the average number of

Benchmark	N1	N2	N3	N4	N5	N6	N7	N8	N9
3step-log	1	1	5						
adi	4	5							
aps	1	1	1						
bmcm	1	1	2	4					
btrix	2	1	7	6	1	3	8		
eflux	2	3							
full-search	2	2	6						
hier	1	1	3	3	2	1	5		
lms	2	1	2	2					
n-real-updates	4	4	4						
parallel-hier	3	3	1	1	2				
tomcat	2	1	3	1	2	4	1	8	2
tsf	1	7	2	4					

Figure 3: Number of processors that generates the best execution time for each nest of each benchmark code in our experimental suite.

Benchmark Name	Input Size	Number of Nests	Average Number of Processors	Dynamic Energy	Leakage Energy L=0.1	L=0.5	L=1
3step-log	203.54KB	3	2.33	85168.83	13046.13	65230.66	130461.31
adi	60.65KB	2	4.50	1695.33	228.22	1141.08	2282.16
aps	137.23KB	3	1.00	680.27	322.63	1613.15	3226.30
bmcm	32.84KB	4	2.00	2033.76	350.46	1752.30	3504.61
btrix	5.89MB	7	4.00	54039.68	6321.93	31609.65	63219.30
eflux	86.70KB	2	2.50	7964.98	1055.58	5277.89	10555.78
full-search	203.54KB	3	3.33	485396.25	59391.22	296956.09	593912.19
hier	203.54KB	7	2.29	47842.54	8306.58	41532.89	83065.78
lms	80.00KB	4	1.75	2419.83	608.04	3040.19	6080.37
n-real-updates	20.00KB	3	4.00	1484.88	177.74	888.72	1777.44
parallel-hier	203.54KB	5	2.00	77027.02	24578.78	122893.91	245787.83
tomcat	70.40KB	9	2.66	9136.33	1306.42	6532.08	13064.17
tsf	52.02KB	4	3.50	2941.12	719.35	3596.73	7193.47

Figure 2: Important characteristics of the benchmarks codes used in our experiments. All energy values are in microJoules.

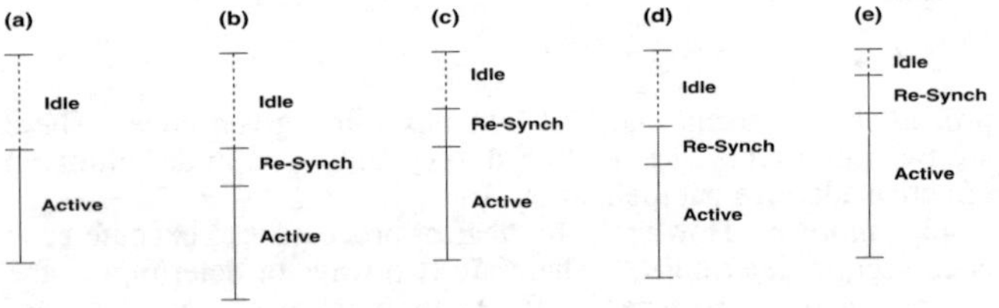

Figure 4: (a) Original idle/active pattern (execution profile). (b) Saving energy using shut-down and incurring performance penalty. (c) Ideal pre-activation strategy. (d-e) Wrong pre-activation timings.

processors per nest for each benchmark.

Having explained our mechanism and policy for determining the number of processors for executing each nest, we next focus on our modifications to the input code. Once the number of processors for each nest has been determined, our strategy inserts (using SUIF) the processor activation/deactivation calls in the code. A processor activation call brings a processor from the power-down state to the active state and takes a specific amount of time to complete (re-synchronization penalty). A deactivation call, on the other hand, places an active processor into the power-down state. We assume that it returns immediately; i.e., it does not incur any additional penalty. Our framework however is general enough to accommodate scenarios where deactivation calls can also have energy and performance costs. Each activation/deactivation call takes the processor id as input parameter and returns a status code indicating whether the action has successfully been performed.

After inserting activation/deactivation calls, our approach performs two optimizations. First, it ensures that in moving from one nest to another, the current active/idle status of processors is maintained as much as possible. The second optimization that we perform targets at reducing the number of activation/deactivation calls inserted in the code when there exists conditional flow of execution. For example, if there is an if-then-else construct with a separate loop in each branch, the compiler hoists the activation/deactivation calls (to be inserted for each loop) above the said construct if both the nests demand the same number of processors.

Note that this energy optimization scheme, while simple to implement, does not pay any attention to reducing the performance overhead due to re-synchronization penalty. Consequently, this approach can result in an increase in execution time. Whether such an increase can be tolerated or not depends largely on the potential energy gains. To illustrate this, let us consider the execution profile of a single processor given in Figure 4(a). The execution profile is broken up into two pieces, each corresponding to a separate loop nest. The processor is idle in the first nest and active (used) in the second. Using our energy-optimization approach gives the modified execution profile shown in Figure 4(b). It is easy to observe from this profile that when the processor is requested to perform computation in the second nest, it first needs to wait some amount of time (re-synchronization penalty, RP). Let t_a, t_i, and t_s denote the active period, idle period, and re-synchronization time, respectively (in cycles). Also, let e_a, e_i, and e_s denote the per cycle energy consumptions for, respectively, the active period, idle period (only when the processor is shut down), and re-synchronization period. Note that, as a result of energy optimization, the length of the original execution profile increases from $t_i + t_a$ to $t_i + t_s + t_a$. The energy consumption of the profile, on the other hand, changes from $(t_i + t_a)e_a$ to $t_i e_i + t_s e_s + t_a e_a$. Assuming conservatively that $e_s = (e_a + e_i)/2$ and $e_i = k e_a$ where k is a coefficient less than one (that is, the energy reduction factor, ERF), this scheme saves energy if and only if:

$$(t_i + t_a)e_a \; > \; t_i e_i + t_s(\frac{e_a + e_i}{2}) + t_a e_a$$

$$> \; (k t_i + t_a + \frac{k+1}{2} t_s) e_a.$$

Consequently, energy savings are possible if:

$$\frac{k+1}{2} t_s < (1-k) t_i.$$

We have evaluated the impact of this energy saving strategy using our benchmark codes. Figure 5 shows the normalized energy consumptions (with respect to version (i)). To cover different scenarios, we performed experiments with different L = (*leakage energy per cycle*)/(*dynamic energy per access*) ratios. This ratio is used for all hardware components of interest. Specifically, we experimented with three different values of L: 0.1, 0.5, and 1. We believe that as leakage is becoming an important part of overall energy budget [6], the experiments with large L values will be more indicative of future trends. In particular, the experiments with L=1 try to capture the anticipated importance of leakage energy in future. Note that leakage is expected to become the dominant part of energy consumption for 0.10 micron (and below) technologies for the typical internal junction temperatures in a chip [6].

The fifth column in Figure 2 gives the overall dynamic energy consumption for version (i). The last three columns show the leakage energy consumption for these three different L values for the same version. In our experiments, we also modified the energy reduction factor (ERF). Specifically, we used ERF=0.1 and ERF=0.2. Note that these values are in reasonable range for several leakage saving techniques such as input vector control and supply gating [6]. The results in Figure 5 include all energy overheads of re-synchronization as well as the energy consumed during spawning threads and synchronizing them following the loop execution. We see that savings for configurations (L=0.1;ERF=0.1) and (L=0.1; ERF=0.2) are 13.9% and 13.3%, respectively. When we increase L to 0.5, the savings for the cases ERF=0.1 and ERF=0.5 move to 43.9% and 41.8%, respectively. Finally, with (L=1.0;ERF=0.1) and (L=1.0;ERF=0.2), the energy savings climb up to 59.7% and

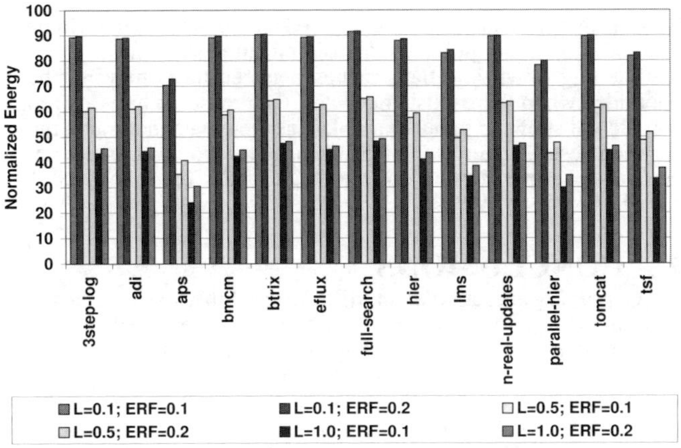

Figure 5: Normalized energy consumption (version (ii)).

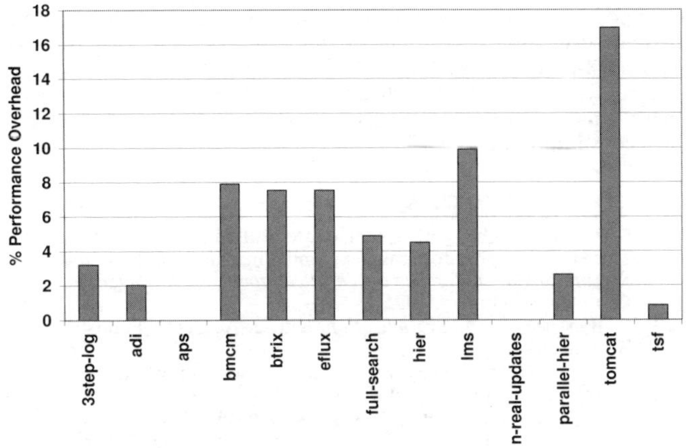

Figure 6: Percentage performance penalty (version (ii)).

57.1%. All these values are obtained assuming that the energy consumed during re-synchronization is the average of active and idle energies and that it takes 20 milliseconds to wake up the processor/cache. Note that this value is very large even for aggressive resource shut-down strategies such as power supply gating [6]. So, in practice, the energy savings that can be obtained from our strategy might be even larger.

While these results indicate large energy savings, delaying waking up a processor until it is actually needed can hurt performance. Figure 6 shows for each benchmark the increase in execution time (over version (i)) when processor shut-down is employed. We observe that the increase in execution time ranges from 0% to 17%, averaging on 5.3%. The reason that two benchmarks (`aps` and `n-real-updates`) do not experience any performance overhead is the fact that each loop in these two codes work with the same number of processors (see Figure 3); consequently, there is no need for processor activation/de-activation between the nests. While this performance penalty may be tolerable for some embedded designs, we also noted that when the re-synchronization penalty was doubled (40 msec), the performance penalties also almost doubled (the detailed results are omitted).

5. PROCESSOR PRE-ACTIVATION

Pre-activation is a strategy that minimizes the performance impact of energy optimizations. It is implemented by activating a resource earlier than the time it will actually be required. The objective here is to eliminate the resynchronization latency that will occur before the resource can start to function normally. Previous work focused on pre-activation of DRAM memory modules [8] and I/O peripherals [3]. In this section, we demonstrate how pre-activation of inactive processors can improve performance and energy behavior of MPoC.

While our processor shut-down strategy explained above can lead to large energy savings, it also increases execution time. For example, employing our approach increases the length of the execution profile shown in Figure 4(a) by t_s cycles (see Figure 4(b)). In this section, we propose a *pre-activation strategy* in which a processor in the power-down state is activated before it is actually needed. The objective here is to ensure that the processor will be ready (i.e., got synchronized) when it is required to participate computation.

The ideal use of this approach is illustrated in Figure 4(c). In this case, the processor remains in the power-down state only for a period of $t_i - t_s$. The last t_s portion of the original idle period is spent in re-synchronization. Consequently, when the processor is requested, it would have just finished the re-synchronization period. An important issue here is to determine the exact point in the code to start re-synchronization. This may not be very trivial; because re-synchronization penalty is given in terms of cycles and it needs to be re-expressed in terms of loop iterations as this (i.e., loop iteration) is the only unit we can use (at source level) to insert activation/deactivation calls. Essentially, in order to pre-activate a processor in the power-down state (for a specific nest), we need to determine an iteration (in the previous nest) before which the processor is activated. Let us assume that each iteration of this previous nest takes C cycles and that the re-synchronization penalty is t_s cycles. Consequently, in the ideal scenario, the processor in question should be activated (i.e., should enter to the resynchronization period) before the last $\lceil \frac{t_s}{C} \rceil$ iterations of the loop. If this is done properly, for our current example, we obtain the execution profile shown in Figure 4(c).

Note that the length of the execution profile in Figure 4(c) is the same as that of Figure 4(a). Its energy consumption, on the other hand, is

$$\begin{aligned} E &= (t_i - t_s)e_i + t_s e_s + t_a e_a \\ &= (kt_i + t_a + \frac{1-k}{2} t_s)e_a \end{aligned}$$

assuming, as before, that $e_s = (e_a + e_i)/2$ and $e_i = ke_a$. Comparing this expression with the original (unoptimized) energy consumption, we can see that this approach is beneficial if

$$\frac{t_s}{2} < t_i.$$

It should be noticed that this ideal pre-activation strategy is better (easier to satisfy) than the power-down scheme discussed in the previous section from both the energy and performance viewpoints.

Figures 4(d) and (e) illustrate, on the other hand, the scenarios where this ideal pre-activation strategy does not happen. In Figure 4(d), the processor activation is delayed. In this case, the length of the original profile is increased (by the amount of the delay in processor activation). In comparison, in Figure 4(e), the processor is activated earlier than necessary. In this case, there is no increase in execution profile length; however, depending on how early the processor is activated, this can cause a significant amount of energy consumption. This is because the re-synchronization period has a fixed length, beyond which the processor is up and starts to consume energy.

The preceding discussion indicates that processor pre-activation can be an effective technique provided that it is done at appropriate moment. There might be several reasons why it may not be possible to achieve ideal pre-activation. First, the point at which the activation call is to be inserted may not be evident from the text of the program. For example, the loop during which a processor needs to be activated (as it is required to participate a computation in the next loop) may have N iterations and our pre-activation strategy can determine that the processor(s) should be activated before the last M iterations are entered. In order to achieve this, we need to split the iteration space of the loop (this is called loop splitting [15]). This, in turn, can increase the code size which may not be tolerated in some embedded environments. Second, the previous

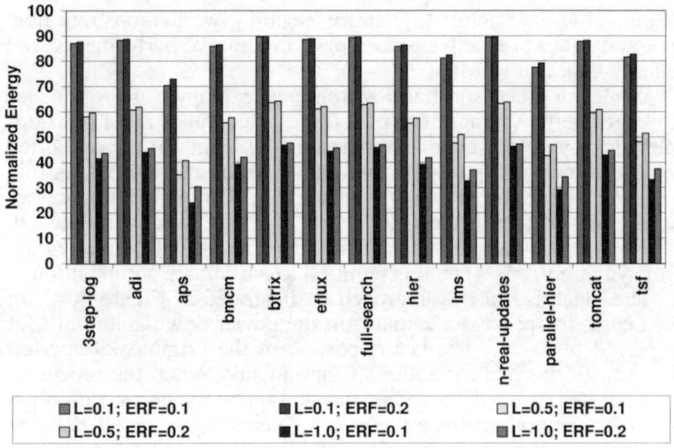

Figure 7: Normalized energy consumption (version (iii)).

loop may not have sufficient number of iterations, in which case we can either consider the next previous nest, or (if it is not possible to do so), we might have to activate the processor(s) later than optimal point, thereby incurring performance penalty.

To evaluate our processor pre-activation strategy (that is, version (iii)), we performed another set of experiments. The results given in Figure 7 (normalized with respect to version (i)) indicate that energy savings due to configurations (L=0.1; ERF=0.1), (L=0.1; ERF=0.2), (L=0.5; ERF=0.1), (L=0.5; ERF=0.2), (L=1.0; ERF=0.1), and (L=1.0; ERF=0.2) are 15.5%, 15.0%, 46.1%, 44.4%, 61.7%, and 59.8%, respectively. The caches (per processor) used in these experiments are 4KB, 2-way set-associative. When we compare these results with the corresponding values given in the previous section, we see that pre-activation is beneficial from an energy perspective too. We also observed that except for one benchmark (`btrix`) the compiler was able to easily insert the pre-activation calls. In `btrix`, in order to insert the pre-activation calls, the compiler needed to split [15] two nests (using SUIF); this led to a 3% increase in the executable size and a 2% degradation in performance. To conclude, processor pre-activation is beneficial from both energy and performance perspectives.

In order to approximate an embedded DRAM, we conducted another group of experiments where the per access off-chip memory energy is reduced to one tenth of its original value. Since this reduces the contribution of DRAM to the overall energy budget (and we do not perform any energy optimization for DRAM), we observed an increase in energy savings due to processor shut-down combined with pre-activation. As an example, with this new value of per-access memory energy, the energy savings for (L=1.0;ERF=0.1) and (L=1.0;ERF=0.2) increased to 68.1% and 63.6%, respectively.

6. FUTURE WORK

This work is a first step in our effort for evaluating and optimizing energy consumption of a complete system based on MPoC. Consequently, it can be extended in several ways:

• We plan to model communication between processors more accurately. Given current trends [6], we can expect that interconnect energy will be more and more pressing in the future. Consequently, to reach an accurate system level energy evaluation, it is critical that interconnect energy should be captured accurately. This issue will be even more important as we attempt to characterize different on-chip interconnection networks from an energy perspective.

•Another way of utilizing multiple processors in an MPoC is to execute (parallelize) code portions *speculatively* when it is not possible to parallelize the application statically. Since speculative parallelization may cause significant wasted energy (in cases where speculation fails), there are important tradeoffs between energy and performance.

•We also plan to augment our current model with energy models of other system components such as I/O devices, software-managed SRAMs, instruction caches, and embedded DRAMs. This will help us to develop more global energy optimization strategies.

•Shutting down idle data caches between nests may not be a good idea when there exists inter-nest data reuse. While this problem did not show up in our examples (as there were not much reuse between nests), a more sophisticated scheme should consider inter-nest reuse before shutting down a data cache. This is an issue that we will visit in the future.

7. CONCLUSIONS

The primary attraction of an MPoC is its ability to shorten execution times for applications that can be parallelized at source level. Due to simple processor architectures (as compared to complex superscalar/VLIW architectures with aggressive branch prediction and speculation) combined with a carefully-designed low-overhead interconnect, these architectures can bring energy benefits as well. Based on the observation that in a given array-intensive code not all the nests require the maximum number of processors in the MPoC, in this paper, we evaluated an adaptive loop parallelization strategy combined with selective shut-down of unused processors. To eliminate potential performance penalty due to energy management, we also proposed a processor pre-activation strategy. Our experiments with an 8-processor MPoC and different parameters (e.g., cache size, re-synchronization overhead, and per-access off-chip memory energy) indicated that our approach is successful in reducing energy consumption.

8. REFERENCES

[1] S. P. Amarasinghe, J. M. Anderson, M. S. Lam, and C. W. Tseng The SUIF compiler for scalable parallel machines. In Proc. *the Seventh SIAM Conference on Parallel Processing for Scientific Computing,* February, 1995.

[2] U. Banerjee. *Loop Parallelization.* Kluwer Academic Publishers, Boston, 1994.

[3] L. Benini, G. De Micheli. System-level power optimization: techniques and tools. *TODAES* 5(2): 115-192 (2000).

[4] N. Carriero, D. Gelernter, D. Kaminsky, and J. Westbrook. Adaptive parallelism with Piranha. *Technical Report 954,* Yale University, February 1993.

[5] A. Chandrakasan and R. Brodersen. *Low Power Digital CMOS Design.* Kluwer Academic Publishers, 1995.

[6] A. Chandrakasan, W. J. Bowhill, and F. Fox. *Design of High-Performance Microprocessor Circuits.* IEEE Press, 2001.

[7] Chip Multiprocessing. *ITWorld.Com,* http://www.itworld.com/Comp/1092/CWSTO54343/

[8] V. Delaluz, M. Kandemir, N. Vijaykrishnan, A. Sivasubramaniam, and M. J. Irwin. DRAM energy management using software and hardware directed power mode control. In Proc. *the 7th International Conference on High Performance Computer Architecture,* Monterrey, Mexico, January 2001.

[9] V. Krishnan and J. Torrellas. A chip multiprocessor architecture with speculative multithreading. *IEEE Transactions on Computers,* September 1999.

[10] MAJC-5200. http://www.sun.com/microelectronics/MAJC/5200wp.html

[11] K. Olukotun, B. A. Nayfeh, L. Hammond, K. Wilson, and K. Chang. The case for a single chip multiprocessor. In Proc. *the 7th Int'l Conference on Architectural Support for Programming Languages and Operating Systems,* ACM Press, New York, 1996, pp. 2–11.

[12] W.-T. Shiue and C. Chakrabarti. Memory exploration for low power embedded systems. In Proc. *the Design Automation Conference,* 1999.

[13] I. Verbauwhede and C. Nicol. Low power DSPs for wireless communications. In Proc. *ISLPED'00,* Rapallo, Italy, 2000.

[14] N. Vijaykrishnan, M. Kandemir, M. J. Irwin, H. Y. Kim, and W. Ye. Energy-driven integrated hardware-software optimizations using SimplePower. In Proc. *International Symposium on Computer Architecture,* June 2000.

[15] M. Wolfe. *High Performance Compilers for Parallel Computing,* Addison-Wesley Publishing Company, 1996.

[16] W. Zhang, N. Vijaykrishnan, M. Kandemir, M. J. Irwin, D. Duarte, and Y. Tsai. Exploiting VLIW schedule slacks for dynamic and leakage energy reduction. In Proc. *the 34th Annual International Symposium on Microarchitecture,* Austin, TX, December 2001.

River PLAs: A Regular Circuit Structure

Fan Mo and Robert K. Brayton
Dept. of EECS, University of California, Berkeley
{fanmo,brayton}@eecs.berkeley.edu

ABSTRACT

A regular circuit structure called a River PLA and its re-configurable version, Glacier PLA, are presented. River PLAs provide greater regularity than circuits implemented with standard-cells. Conventional optimization stages such as technology mapping, placement and routing are eliminated. These two features make the River PLA a highly predictable structure. Glacier PLAs can be an alternative to FPGAs, but with a simpler and more efficient design methodology.

Categories and Subject Descriptors

B.6.3 [**Logic Design**]: Design Aids — Automatic synthesis.

General Terms: Algorithms.

Keywords: Programmable Logic Array, River routing.

1. INTRODUCTION

In modern VLSI design, regular circuit structures are becoming preferable for two reasons. First, from layout to mask, manufacturers transform the geometric patterns to achieve better exposure and imaging quality [8], and for this, regular structures are preferable, because they have fewer layout patterns. The second more important reason is that regular structures make the prediction of circuit characteristics easier. Standard-cells have a somewhat regular structure, however its regularity only lies in the row arrangement. Routing results in numerous layout patterns. The synthesis, technology mapping, placement and routing flow associated with the standard-cell designs has the so-called timing-closure problem, the main cause of which is lack of predictability at early synthesis stages.

Programmable Logic Arrays (PLAs) are a regular structure. The design methodology for PLAs is quite mature [7]; it is a sum-of-products (SOP) minimization, which is a basis for modern logic synthesis. The result of a SOP minimization can be mapped directly to a PLA. Technology mapping, required in standard-cell designs, is not necessary; neither are placement and routing for a single-PLA implementation. Another advantage of the PLA is that its design flow is library free. With the rapid shrinking of feature sizes, design methodologies based on library cells suffer from a scaling problem; changes in design rules may require a complete re-characterization of the library cells, which is very costly.

However, single-PLA structures are relatively weak in expressing logic functions. Thus multi-level logic has become the mainstream synthesis technique [6]. In multi-level logic minimization, the entire circuit is represented as a network of nodes where each node is a SOP logic function (a Boolean network). A common approach is to optimize both the entire network as well as each node, and then transform the circuit to a network of library cells via technology mapping. A natural step is to build a network of PLAs (NPLA) from the minimized network without technology mapping [10]. Thus some desirable features of single-PLAs such as technology-mapping free synthesis are preserved. However, NPLAs require placement and routing and the placement of PLAs is not at the gate but at the block level. So far, block level placement is not as well developed as gate level placement. Also, even though each PLA in the NPLA is regular, global regularity is low because of placement and routing (this can also require additional metal layers [10]). In fact, global regularity of the NPLAs may be even worse than for standard-cells, which have a row structure at least.

To eliminate global irregularity of NPLAs caused by placement and routing, a multiple-PLA structure called River PLA (RPLA) is proposed. Given a Boolean network, the nodes are clustered into a stack of PLAs. In each PLA, the outputs and some of the input signals are ordered and fed to the next PLA via river routing. The advantages of RPLAs are:

1. Since the RPLA is a stacked structure and the river routing is quite regular, the RPLA has both local and global regularity.
2. The output buffer and the input buffer of the next PLA in the stack are merged, reducing area and improving speed.
3. The RPLA needs only two metal layers.
4. The design methodology for RPLAs maintains the advantages of the synthesis for NPLAs, but no placement and routing are necessary.
5. Since RPLAs have dedicated buffers, which only drive a known number of loads, and river routing has fixed shapes and involves short distances, the synthesizer is immune to buffering problems. The computation of the total delay becomes a simple summation of the delays of all the PLAs.

Thus, the RPLA is a good alternative to standard-cells. For future Deep Sub-Micron (DSM) designs in which regularity may be a key issue, the regularity of RPLAs should be an advantage.

A programmable version of RPLAs is called a Glacier PLA (GPLA), targeting re-configurable applications, where a block of hardware is embedded in a system-on-a-chip, and the functionality of the block is re-configured whenever the application changes. GPLAs, unlike Field Programmable Logic Array (FPGA), have fixed wiring; thus they need no routing. Although FPGAs have a regular array structure, the topology of a routed net can be irregular, which is hard to predict during synthesis. The granularity of GPLAs is coarser than FPGAs, so to implement a chain of logic functions, the buffers involved in the GPLAs are less than in FPGAs. One may argue that the finer granularity of FPGAs leads to higher area utilization. Although true, a portion of the logic resources in FPGAs are wasted for either routability or performance considerations. In contrast, GPLAs waste area only because no appropriate logic can fill in. The design methodology for the FPGAs is similar to that for the standard-cells, except the technology mapping targets a LUT structure. Therefore, they also have problems with buffering and routing prediction.

The remainder of this paper is organized as follows. The circuit structure is described in Section 2, and area and delay formulations given in Section 3. The design methodologies are discussed in Section 4. Experimental results are presented and discussed in Section 5, and Section 6 concludes.

2. STRUCTURE Of RPLAs AND GPLAs

As shown in Figure 1, a RPLA consists of a stack of multiple-output PLAs. Since some of the nets feed more than one PLA in the stack, the output signals of a PLA may come from both the OR-plane (defined as OR-signals) and the AND-plane (defined as AND-through-signals). Those nets ending at the current PLA are called AND-only-signals. The AND-through-signals and the OR-signals of each PLA are ordered and connected to the next PLA in the stack via river routing. Metal layer one and two are assigned to the routing of the AND-though and OR signals respectively. The interconnections via river routing are local and hence short. The left sides of the PLAs are aligned. The primary inputs come in at the bottom. The primary outputs can be retrieved from the right side from anyone of the PLAs or the top. When they are derived at the right side, river routing (still on metal levels one and two) is used to bring the outputs of the PLA OR-planes to the right boundary.

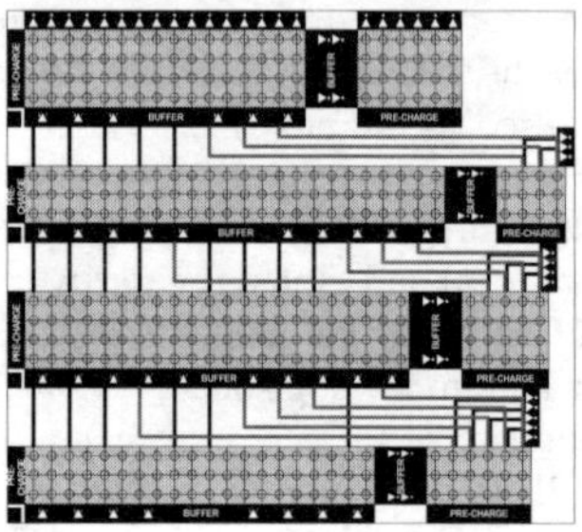

Figure 1. Circuit structure.

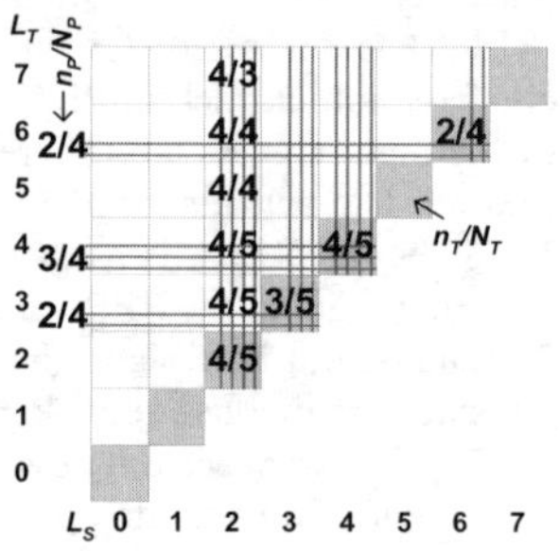

Figure 2. Provide/consume diagram.

A RPLA structure can be characterized by the following parameters: n_L is the number of the levels. $n_T(s,t)$ is a $(n_L+1)\times(n_L+1)$ matrix, the number of the nets starting at level s and reaching level t, where $s\leq t$. The primary input nets have starting level of 0. $n_T(t,t)$ denotes the number of outputs of the PLA on level t. It is conceivable that $n_T(s,t)\geq n_T(s,t')$ for $t\geq t'$, because with increasing level, part of the nets coming from lower levels are no longer used by any nodes, while the nodes on the higher levels create their own outputs, which become the new nets. $n_I(t)$ is the number of inputs to the PLA on level t, which is equal to $\sum_{s=0}^{t-1} n_T(s,t)$. $n_O(t)$ is the number of outputs of the PLA on level t, which is exactly $n_T(t,t)$. $n_P(t)$ is the number of product terms on level t. For the GPLA, limits exist for each of the parameters. We use capital letters to define the limits: N_L is the limit for the number of levels, $N_T(s,t)$ is the limit for the number of the nets starting at level s and reaching level t, and $N_P(t)$ is the limit for the number of product terms on level t. It makes no sense to have a GPLA composed of PLAs with different width. Because of physical design issues, a column in the AND-plane (actually containing two bits, one for the positive signal, one for the complement) and a bit in the OR-plane may have different widths. To achieve a uniform PLA width, it is required that all

$$N_T(t,t)+r_{OA}\cdot\sum_{s=0}^{t-1}N_T(s,t) \qquad t=1,2,..,N_L$$

are nearly the same, where r_{OA} is the width ratio of an OR-plane column to an AND-plane column. Normally this ratio is about 0.5. It is convenient to express the net and product term distribution by a provide/consume diagram as shown in Figure 2. This representation will be frequently used in the discussion of the GPLA synthesis algorithm. A valid GPLA implementation should satisfy $n_L\leq N_L$, $n_P\leq N_P$ and $n_T(s,t)\leq N_T(s,t)$ for all $1\leq t\leq n_L$ and $0\leq s<t$.

3. AREA AND DELAY FORMULATION

The area and delay of a RPLA are both explicitly expressed in terms of the PLA contents, so the objective function in the optimization can be evaluated with good accuracy. This contrasts with standard-cell design where the area and delay are hard to predict during technology independent logic optimization. The width, height and area of the PLA on level t are:

$$s_X(t)=W_{PA}+W_I\cdot n_I(t)+W_{BM}+W_O\cdot[n_O(t)+1]$$
$$s_Y(t)=W_{BI}+W_P\cdot(n_P(t)+1)+W_R\cdot[n_O(t)+1]$$

The number of horizontal wires in the river routing region is exactly $n_O(t)$, as will be shown in Section 4.1.3. The "+1" terms in the formulae account for the tracking lines, if a dynamic PLA structure is adopted. All the constants are defined in Table 1.

Table 1. Definitions of PLA parameters

W_{PA}	Width of pre-charging circuit of the AND plane
W_I	Width of a pair of complementary input columns
W_{BM}	Width of the intermediate buffer
W_O	Width of an OR plane column
W_{BI}	Width of the input switch
W_P	Width of a product line
W_R	Width of a wire in the river routing area

The height of a RPLA is the sum of the heights of the PLAs, and the width is the largest width:

$$S_X=\max_{t=1}^{n_L}s_X(t), \qquad S_Y=\sum_{t=1}^{n_L}s_Y(t),$$

The non-uniformity of the PLA widths results in white space on the right side of those PLAs whose widths are smaller than the RPLA width. The GPLA has the same area formulation, except everything in the formulae is constant. Since buffers isolate the PLAs in the stack, the delay computation is just a summation of the delays of all the PLAs. In delay analysis, accuracy depends only on the delay model, because given a synthesized PLA, there is nothing unknown such as fanout number etc. [1]. We use a simple delay model, assuming the delay of a transistor or a buffer is a constant delay plus a variable part linearly proportional to the number of loads. For one product line in the AND-plane, the delay is the time the first transistor on the line starts to turn on, plus its turn-on transition time. The first transistor being turned on is the one with the minimum number of transistors in the corresponding AND column:

$$dA_p=d_{BI}+\delta_{BI}[n_P C_{poly}+\min_i(n_{TRi}C_{GS})],$$

where d_{BI} and δ_{BI} are the constant and coefficient of the linear delay model for the input switch, $n_P C_{poly}$ forms the load of an AND column, n_{TRi} is the number of the transistors on the i-th AND column (only those AND columns on which there is a transistor connecting to the product line p are considered), and C_{GS} is the transistor gate capacitance. The turn-on transition time depends on the load plus the number of transistors on the product line:

$$dP_p = d_{TR} + \delta_{TR}\left(n_{TRp}C_{DS} + C_{BM}\right),$$

where d_{TR} and δ_{TR} are the constant and coefficient of the linear delay model for the turn-on transistor, n_{TRp} is the number of transistors in the product line p, C_{DS} the transistor load capacitance, and C_{BM} the capacitance provided by the intermediate buffer. The delay of the input switch plus the AND-plane is:

$$d_{AND} = \max_p\left(dA_p + dP\right).$$

Similar results hold for d_{OR}. The delay computation for GPLAs is the same.

4. DESIGN METHODOLOGY

4.1. Synthesis of RPLAs

4.1.1. Multi-level logic minimization.

A multi-level logic minimizer is called to generate an initial Boolean network.

4.1.2. Node level-placement.

Node level-placement is possible because some nodes have flexibility in their levels. Suppose the level of node d is $l(d)$. If $l_{HFI}(d)+1<l(d)$ or $l(d)<l_{LFO}(d)-1$, where $l_{HFI}(d)$ is the highest fanin level and $l_{LFO}(d)$ is the lowest fanout level, then the node is said to be down-movable or up-movable. In this case, the level of a node d can be chosen within the range of $l_{HFI}(d)$ and $l_{LFO}(d)$. Notice that the flexibilities of different nodes may be correlated, because of the fanin/fanout relations. It has the good effect of decreasing the number of possible combinations, and thus shrinking the solution space to be searched. Simulated annealing is suitable for this, since the solution space is reduced due to the level dependencies, and the evaluation of area and delay is straightforward. The objective function is chosen as a weighted sum of the area and delay. In the area and delay computation mentioned above, every variable can be trivially derived from the configuration, except n_P, the number of product terms. The exact value of n_P requires a SOP minimization. One could call the SOP minimizer each time the cost is to be computed to get the exact product number; this might not be slow, because the PLA in a RPLA is usually not big. A faster but more conservative way is to use ns_P, the sum of the product numbers of all the nodes in the cluster, which is an upper bound of the number of product terms of the final multiple-output PLA. A trade-off is that when ns_P exceeds a threshold, we do not call the SOP minimizer but simply use the value ns_P. Otherwise a SOP minimization is performed. After the level of the flexible nodes is determined, the nodes on the same level are assigned to a multiple-output PLA and further optimized with a SOP minimizer.

4.1.3. Net ordering.

Each net is given two variables describing its starting level and ending level, denoted by $s(n)$ and $e(n)$. The starting level is the level of the PLA driving the net, and the ending level of the last PLA using this net as an input. The net ordering algorithm can be explained via the example shown in Figure 3. First, order the nets in the descending order of $e(n)$. For the nets with the same $e(n)$, order them in ascending order of $s(n)$. This results in Figure 3(a). There is no restriction on the nets with the same $s(n)$ and the same $e(n)$, and they will always go side by side. Then, starting from left to right, put each net in the leftmost slot that is vacant. Figures 3(b) to (d) illustrate the process. Applying this algorithm will not generate any twisting in the routing layers, meaning that river routing with two metal layers is feasible. Net ordering done in this example leads to a RPLA layout shown in Figure 1.

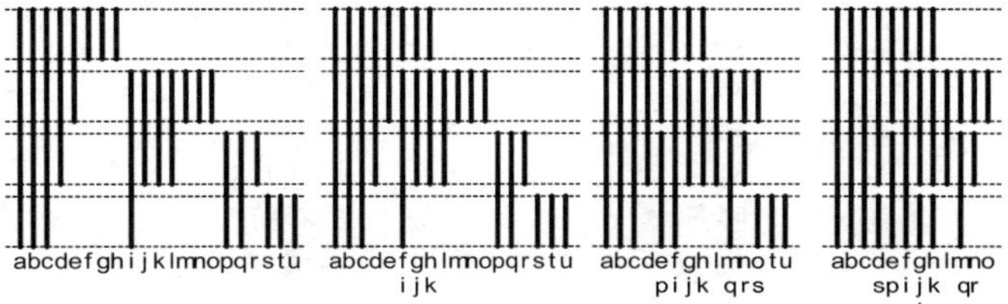

Figure 3. Net ordering example.

This simple scheme causes a little inaccuracy in the computation of PLA widths. Notice that in the bottom PLA in Figure 3(d), there is a vacant slot at the position "l". This makes the real AND column number of the bottom PLA larger than the number of its inputs, which is used to calculate the PLA size as described in Section 3. As long as such PLA is not the widest in the RPLA stack, all the computations remain the same. Experiments have demonstrated that the case where the widest PLA has empty columns is rare.

4.2. Synthesis of GPLAs

The synthesis of GPLAs starts with a multi-level logic minimization with level control. It is followed by an iteration of node level-placement and methods to fix different violations. The iteration either completes with a valid solution found, or quits after a given number of trials and switches to a larger GPLA configuration. If successful, a post optimization is performed to improve speed. In the following context, we use the term "N_P violation" to define the violation of $n_P>N_P$, "N_A violation" to define $n_T(s,t)>N_T(s,t)$ where $s<t$, and "N_O violation" to define $n_T(t,t)>N_T(t,t)$.

4.2.1. Multi-level logic minimization.

This step is similar to that in RPLA synthesis, except the number of levels is controlled to be no greater than N_L during the minimization.

4.2.2. Node level-placement.

As in the RPLA design flow, simulated annealing is used to place the nodes on levels. However, the cost function is the total number of violations. The following Subsections describe the methods for fixing violations remaining after node level-placement.

4.2.3. Net shifting.

Net shifting provides "detours" for nets causing N_A violations. Starting from the bottom left of the provide/consume diagram, we handle N_A violations one by one. The following pseudo code describes how a N_A violation of $e_V=(n_T(s_V,t_V)-N_T(s_V,t_V))$ excessive signals at (s_V,t_V) is handled by net shifting:

```
for (t=tV−1; t>sV, t−−) {
    sn=min(NT(t,t)-nT(t,t), NP(t)-nP(t));  //no. of spare lines.
    if (sn≤0) continue; else dn=min(sn, eV); //no. of detours.
    nT(sV,tt)−=dn for all tt=t+1 to nL;
    nT(t,t)+=dn; nP(t)+=dn; nT(t,tt)+=dn for all tt=t+1 to nL;
    if (eV==0) return success;
}
return failure;
```

An example is shown in Figure 4, where the N_A violations at (s=2,t=5) and (s=2, t=7) are eliminated through net shifting.

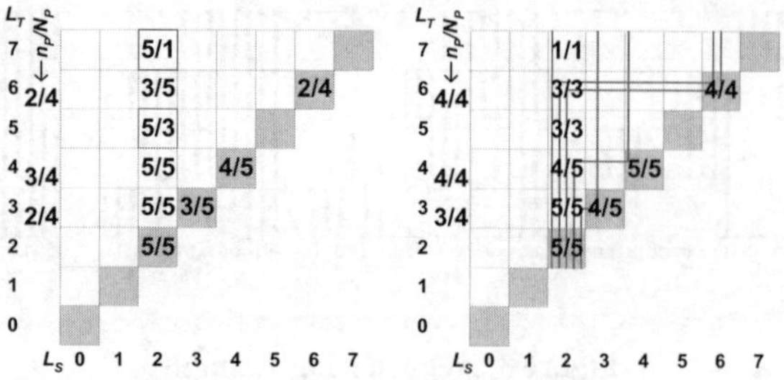

Figure 4. An example of net shifting.

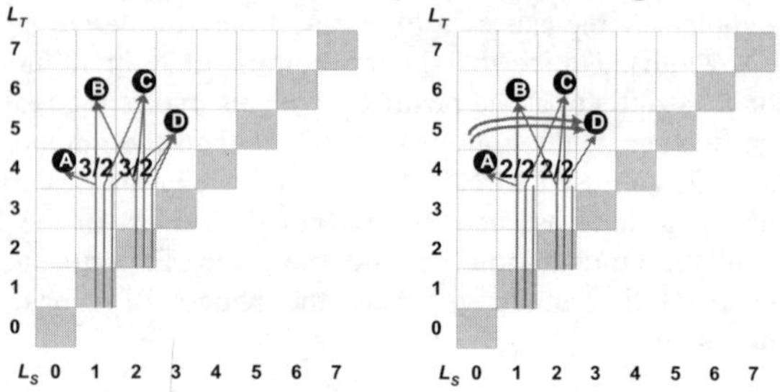

Figure 5. An example of node collapsing.

4.2.4. *Node decomposition.*

The objective of this operation is to reduce the number of product terms on a level with the N_P violation. The decomposition operation factors a node and makes its divisors new nodes in the network. The new node may not have trouble in finding a place in the stack. Area minimized multi-level logic synthesis aims at minimizing the total number of literals, which is a good estimate of the total number of transistors if the circuit is to be implemented with gates. The decomposition does not change the total number of literals, so the initial synthesis step may not generate this decomposition. However, in the synthesis of GPLAs, the first goal is to get a valid implementation for the given configuration. Using more nodes or more literals should not be a problem as long as the mapping is valid. To choose the best node to be decomposed, we simply examine the nodes on the level with the highest N_P violations and select the one having the most product terms. Since decomposition of a node will generate new nodes and the new nodes may affect other levels, the node decomposition is only performed on one node at a time.

4.2.5. *Node collapsing.*

The node collapsing operation deals with the N_A violations. N_A violations mean that the current and higher levels use excessive nets from the lower levels. A remedy is to decrease the need, of the current and higher levels, for nets coming from lower levels. The collapse operation removes intermediate nets from the inputs of a node and replaces them by primary inputs. This operation can be supported by reserving sufficient vertical lines for the primary input nets. We choose the node using the most unique intermediate nets crossing the violation level. The number of unique nets is the saving achieved by collapsing the node. So the algorithm checks all nodes on or above the level with the highest N_A violations to identify a node with the maximum number of unique nets crossing the violation level and collapses it. An example is shown in Figure 5, where node D is collapsed because it used two unique nets passing through level 4.

4.2.6. *Node elimination.*

N_O violations are handled via node elimination. This operation eliminates a node and collapses its function into all its fanout nodes. Eliminating a node immediately reduces $n_T(t,t)$ by one. Node elimination effectively makes a copy of the current node function in all its fanout nodes, so the implementation of the functions in the fanout nodes may cause unexpected growth in the product terms. Thus the node to be eliminated should be chosen carefully. The first criterion is the one with the least fanouts. If a tie occurs, choose the one with the least product terms. Further ties can be resolved by choosing the one with the least input nets.

4.2.7. *Post optimization for performance.*

Post optimization tries to improve the performance, because the previous steps focus on seeking a valid rather than a fast implementation. Post optimization is a greedy approach, accepting any node displacement that preserves validity but reduces total delay. The algorithm terminates when no further improvement can be found.

5. EXPERIMENTAL RESULTS

In the experiment, three structures are compared: standard-cells with three-layer (3 metal layers) over-the-cell-routing (SC), four-layer NPLAs (NPLA) and two-layer RPLAs (RP). Each example starts with a *SIS* optimization on the initial Boolean network and creates a network with n_{L0} levels [2]. Then the number of levels is gradually reduced via partial node collapsing. For each number of levels, n_L, the circuit is technology mapped twice with *SIS*, one with area priority and one with delay priority[1], generating SC-a(n_L) and SC-d(n_L). We assume the SC designs have 100% area utilization and are 100% routable, so no placement and routing is performed. A clustering algorithm, similar to the one proposed in [10], is adopted to build the NPLAs; however there is no area or delay priority. The placement of the NPLAs is done by a force-directed macro-cell placer [4], producing NPLA(n_L). Two RPLAs are synthesized, one with 100% weight on area and one with 50% weight on area and 50% weight on delay, generating RP-a(n_L) and RP-d(n_L). A 0.35-micron technology is assumed for all circuits. The NPLA clustering and the RPLA mapping algorithms are written in JAVA; and all other procedures are within *SIS* package and *ESPRESSO* [3]. The programs ran on a Dec Alpha 8400 5/625 workstation. Sixteen combinational examples from the LGSynth 91 benchmark set [5] were tested. The gate counts of the examples range from 300 to 17K. The results are given in Table 2. To give a better view of the results, we normalize the area and delay data and plot them in Figure 6. For each example, the normalization is with reference to the data of the SC-a(n_{L0}). The layouts of all three types of implementations for alu2(n_{L0}=10) are shown in Figure 7; routing is done by the *Cadence Warp Router*. The experimental results show that in general, fewer levels mean faster speed but larger area. RPLAs (2 layers) needs on average 23% more area than SCs (3 layers), and NPLAs (4 layers) needs on average 31% more area. Note that RPLAs and NPLAs do not use 100% of the area. RPLAs have an average area utilization of 78%, and 76% for NPLAs. If the module is to be embedded in a big chip, other circuits may occupy this white space. In terms of the delay, on average RPLAs are 4% slower than SCs, and NPLAs 9% slower. Figure 6 shows that SCs have cleaner shapes for the area/delay trade-off curves, while NPLAs and RPLAs have more scattered

[1] SIS command "map -m 0 –AF" is used for area-prior tech-mapping, and "map -n 1 –AFG" for delay-prior tech-mapping.

distributions. Synthesis times of SCs are several times faster than for NPLAs and RPLAs. In the RPLA case, the synthesis time is exactly the design time. However, for SCs and NPLAs, additional time is needed to complete the placement and routing. In addition, the placement and routing may change the delay achieved by the synthesis. Thus it is possible that the final layouts do not meet the requirements, which may invoke another run of the design flow.

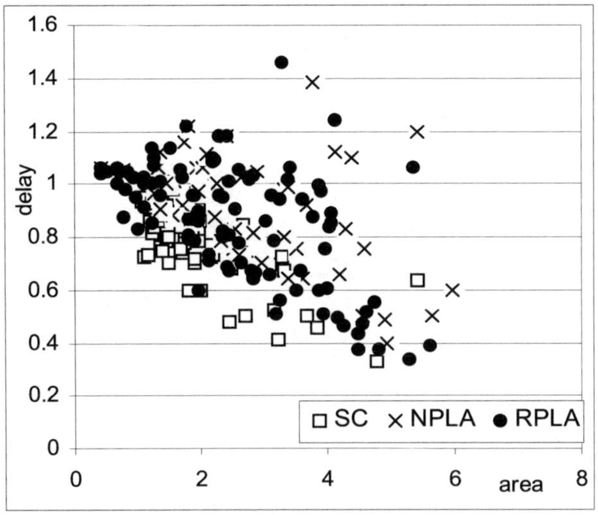

Figure 6. Normalized area/delay.

Table 3. Configuration parameters.

	GPLA			*Xilinx* FPGA		
	G1K	G5K	G10K	4002	4020	4062
max.no.primary input	85	185	384	64	224	384
max.no.primary output	214	540	1132			
max. no. levels	12	18	24			
no. programmable bits	54k	456k	1.41M	61k	522k	1.43M

Table 4. Comparison of GPLA and FPGA

Example	# PI	# PO	G1K	G5K	G10K	4002	4020	4062
apex7	49	37	●	●	●	○	●	●
alu2	10	6	●	●	●		●	●
C1355	41	32	●	●	●	○	●	●
apex6	135	99		●	●		○	●
alu4	14	8		●	●		●	●
x3	135	99		●	●		○	●
C2670	234	139			●		○	●
k2	45	43		●	●		●	●
C3540	50	22		●	●		●	●
dalu	75	16		●	●		●	●
i8	133	81		●	●		●	●
i10	257	224			●		○	○
C5315	178	123		●	●		○	●
C6288	32	32			●		●	●
C7552	206	107			●		○	●
des	256	245			●		○	○

● means a successful mapping; and ○ means that the logic can be fully mapped, but the I/O ports are insufficient.

Three GPLA configurations are considered, named G1K, G5K and G10K respectively. The three FPGAs used for comparison are the *Xilinx* XC4000XL series, 4002, 4020, and 4062, which have similar numbers of programmable bits as G1K, G5K and G10K respectively. Some important parameters are listed in Table 3 [9]. The same LGSynth 91 benchmark examples are tested. The GPLA synthesis algorithm is written in JAVA, except the multi-level logic minimization uses *SIS* [2] and the SOP minimization uses *ESPRESSO* [3]. The FPGAs are synthesized by *Synplicity Synplify Pro* 6.2.4. Due to the lack of a technology library for the programmable devices, we do not have area and delay information available in the experiment. Only a qualitative measurement of "whether the circuit can be implemented in the configuration" is given. The results are shown in Table 4. For the FPGA implementations, some of the failures are caused by insufficient I/Os, not insufficient logic resources. If I/O limit is not an issue, the results show that GPLAs and FPGAs have similar powers for implementing the benchmark circuits.

6. CONCLUSIONS

RPLAs and their design methodology are described. The high regularity of the RPLAs is attractive in DSM integrated circuit design, since it greatly relieves the burden of analyzing huge amount of layout patterns. More important, is that RPLAs have a simple design methodology; conventional design flow steps of technology mapping, placement and routing are eliminated. Thus there is no outer loop in the design for timing closure and hence the design cycle is shortened. Experimental results show that, compared to other structures, RPLAs provide similar delays. The RPLAs implemented with 2 metal layers consume on average 23% more area than standard-cells with 3 layers, however on average 22% of the area overhead is white space, which might be utilized by other circuits. RPLAs offer a good alternative to the standard-cell structure, providing higher circuit regularity and easy design methodology. The GPLA is a re-configurable version of the RPLA. The contents of the PLAs are programmable, but all PLA sizes and interconnections are fixed. Experimental results show that GPLAs can be a viable alternative to the FPGA structure. Extension to sequential circuits is easy for RPLAs and GPLAs. The space under the river routing wires can be used to build flip-flops. One more metal layer (third layer) can be used to route feedback wires from higher level PLAs to lower level PLAs, and these can also be completed via river routing.

7. ACKNOWLEDGEMENTS

This work was supported by GSRC (grant from MARCO/DARPA 98DT-660, MDA972-99-1-0001).

8. REFERENCES

[1] C. Papachristou and A. Pandya, "A design scheme for PLA-based control tables with reduced area and time-delay cost", IEEE Transactions on CAD, vol. 9, no. 5, May 1990, 453-472.

[2] E. Sentovich, K. Singh, L. Lavagno, C. Moon, R. Murgai, A. Saldanha, H. Savoj, P. Stephan, R. Brayton and A. Sangiovanni-Vincentelli, "SIS: A system for sequential circuit synthesis", Tech. Rep., UCB/ERL M92/41, Electronics Research Lab, University of California, Berkeley, May 1992

[3] R. Rudell and A. Sangiovanni-Vincentelli, "Multiple-valued minimization for PLA optimization", IEEE Transactions on CAD, vol. 6, Sep 1987, 727-750

[4] F. Mo, A. Tabbara and R.Brayton, "A Force-Directed Macro-Cell Placer", Proceedings of ICCAD, Nov 2000

[5] http://www.cbl.ncsu.edu/pub/Benchmark_dirs/LGSynth91/

[6] R. Brayton, G. Hachtel and A. Sangiovanni-Vincentelli, "Multi-level logic synthesis', Proc. of IEEE, vol. 78, Feb. 1990

[7] R. Brayton, G. Hachtel, C. McMullen and A. Sangiovanni-Vincentelli, "Logic minimization algorithms for VLSI synthesis", Kluwer Academic Publishers, 1984

[8] "Silicon VLSI Technology", Chapter 5, Lithography, edited by J.D. Plummer, M.D. Deal and P.B. Griffin, Prentice Hall, 2000

[9] Xilinx Datasheet, XC4000E and XC4000X series field programmable gate arrays, version 1.6, May 1999.

[10] S. Khatri, R. Brayton and A. Sangiovanni-Vincentelli, "Cross-talk immune VLSI design using a network of PLAs embedded in a regular layout fabric", Proceedings of ICCAD, Nov 2000, 412-418

Table 2. Comparison of SC-a, SC-d, NPLA, RPLA-a and RPLA-d

		gate count		area (1000 um^2)					delay (ns)					area utilization			synthesis time (minute)				
Circuit	#level	SC-a	SC-d	SC-a	SC-d	NPLA	RP-a	RP-d	SC-a	SC-d	NPLA	RP-a	RP-d	NPLA	RP-a	RP-d	SC-a	SC-d	NPLA	RP-a	RP-d
alu2	10	644	875	34.8	47.25	15	14.1	14.4	8	6.1	8.5	8.5	8.3	80%	79%	76%	0.5	0.5	1.9	2.4	2.6
	6	888	1091	48	58.95	46	29	33	7.4	5.9	8.5	8.4	8.2	73%	77%	73%	0.5	0.5	1.8	2.2	2.1
	5	972	1222	52.5	66	55	38.4	38.4	7.2	5.6	8.1	8.2	8	82%	78%	78%	0.5	0.5	1.6	2.2	2.2
alu4	16	1175	1522	63.45	82.2	49	42.6	46	15.6	15	16.5	16.6	16.3	74%	75%	75%	0.5	0.5	2.3	3.1	3
	10	1494	1716	80.7	92.7	80	50.9	62.3	12.9	12.3	15	15.3	14.8	80%	78%	73%	0.6	0.6	2.5	3	3
	7	1958	2219	105.8	119.9	109	69.5	78.8	12	11.2	14.3	14.2	13.3	80%	83%	74%	0.8	0.8	3	2.9	2.8
	5	3119	3333	168.5	180	162	120	148.8	11.5	11	12.9	12.2	12.8	68%	80%	82%	0.9	1	3	2.8	3
apex6	7	1180	1572	63.75	84.9	93	85.9	86.3	14	13.2	14	14.1	13.4	92%	90%	87%	0.8	0.8	2.3	3	3
	4	1380	1575	74.55	85.05	99.2	122.5	119.1	13.6	12.1	12.3	12.3	12.1	88%	90%	89%	1	1.1	2.4	3.4	3.5
apex7	8	350	430	18.9	23.25	14.8	9.5	12.9	6.4	5.8	6.6	6.7	6.4	90%	97%	82%	0.4	0.4	1.8	2	1.9
	5	425	522	22.95	28.2	25.5	14.7	18.8	5.2	4.5	5.8	5.6	5.3	75%	98%	85%	0.4	0.4	1.8	1.9	1.8
C1355	10	755	900	40.8	48.6	70.8	61.8	61.8	9.4	8.1	10.9	10.7	10.7	72%	61%	61%	0.5	0.5	2	2	2.2
	7	813	947	43.95	51.15	79.4	50.1	50.1	8.8	7.9	10	9.4	9.4	77%	75%	75%	0.6	0.5	2.2	2.8	2.5
	5	1113	1477	60.15	79.8	89.3	77.2	80.2	8.2	7.4	9.4	9	8.2	70%	80%	77%	0.6	0.6	2.3	3	2.9
C2670	12	1119	1516	60.45	81.9	159.4	158.4	205	17.4	16.1	17.8	18.4	17.7	80%	75%	71%	0.8	0.8	4	4.4	3.9
	9	1733	2213	93.6	119.6	200.4	170.8	218.1	17	15.9	17	18	16.4	72%	79%	82%	1	1.1	4.2	4	3.9
	6	2150	2450	116.1	132.3	259.4	182.5	240.3	16.8	14.2	14.5	15	13.2	71%	69%	72%	1.5	1.6	5	4.4	4.2
C3540	23	2044	2780	110.4	150.2	150	136.5	140.4	25.8	22.4	28.9	29.4	28.4	81%	76%	75%	0.8	0.9	4.2	4.8	3.9
	13	3575	5286	193.1	285.5	203	189.3	200.3	20	19.9	22.4	26.4	20.8	73%	77%	73%	1	1.2	3.3	4.9	4.2
	9	4397	6363	237.5	343.7	315	200.8	234.2	18.8	18	21	20.5	18.4	71%	77%	70%	1.8	2	3.8	4.4	4.4
	6	6819	11091	368.3	599	504	310.6	389.2	17.3	16.4	19.4	17.4	15.4	68%	72%	76%	2	2.4	4	4.4	4.3
C5315	15	2633	3133	142.2	169.2	290.4	312.4	236.7	29	27	30.8	31.4	30.5	77%	73%	71%	1	1	4.9	4.8	4.2
	8	3500	4027	189	217.5	420	378.3	439.1	25	26.1	20.3	20.3	19	73%	91%	83%	1.5	1.8	4.8	4.8	5.1
	6	4444	5247	240	283.4	500	450	508.2	22	25	22	22.7	19.6	67%	98%	88%	2.2	2.4	5.6	5.3	5.9
C6288	25	6086	7861	328.7	424.5	694	725.1	732.7	50	44	55.8	55.1	54.6	80%	65%	65%	5	5.3	8.4	9	10.4
	20	7763	8813	419.3	476	1238	1083	1129	47	46.1	69.2	72.9	53	78%	73%	77%	5	5.2	9	12.2	13
	14	11883	16388	641.7	885	1785	1356	1759	45.2	42.2	60	62.2	53.1	79%	82%	79%	5.8	5.8	10.2	12	10.5
C7552	18	3388	5055	183	273	609.9	692.3	730	50.4	40.2	40.2	43.9	30.5	80%	61%	67%	3	3.1	8.2	8.2	9
	14	5283	6119	285.3	330.5	770	710	780	40.2	30.3	33.1	30	23.2	80%	77%	69%	3.4	3.4	12	8	8.9
	9	6805	9180	367.5	495.8	830	760	820	30	25.4	25.4	25	19	75%	72%	77%	4	4.1	11	9.5	9.9
	6	8352	10894	451.1	588.3	902	820	880	24.3	20.6	20.1	22	18.8	72%	80%	82%	4.9	5	11.2	10	10.3
dalu	14	1327	1691	71.7	91.35	130	127.4	164.9	18	16.4	21.9	21.9	21.2	79%	77%	72%	1.2	1.3	4	3.2	3.4
	8	1638	2000	88.5	108	174.3	174	175	16.9	16	21.3	21.2	18.2	77%	77%	78%	1.4	1.4	4	3.3	3.9
	5	2222	2472	120	133.5	207.3	140.1	181.8	15.1	14.9	18.8	16.1	16.3	65%	94%	84%	1.5	1.5	3.9	3.8	3.8
des	8	5038	5569	272.1	300.8	999	887	1093	36	26.2	33	34	30.2	82%	84%	82%	2	2.3	4.5	5.9	5.1
	5	11116	16511	600.3	891.6	1620	1292	1527	26.4	26	21.4	20	14	81%	78%	81%	4.9	4.8	6	8.3	8.9
i8	7	1633	2186	88.2	118.1	204.5	216.3	216.7	14.8	14.2	12.3	12	10	88%	79%	78%	0.9	1	3.1	3.9	3.2
	5	1713	1997	92.55	107.9	220.5	187.7	173.8	14.1	13.7	11.2	10.8	8.9	79%	86%	86%	0.9	1.1	2.9	3.3	4
i10	22	3611	4833	195	261	450	459	470	43.7	43.2	34.2	35	30.2	77%	71%	77%	1.5	1.8	5.2	4.9	5.2
	13	7777	9194	420	496.5	702	555	620	35.6	32.5	28.2	28.2	22.3	80%	80%	82%	2.9	3.3	5.9	5.1	5.5
	9	11388	13250	615	715.5	957	770	890	22.9	21.9	21.1	22.3	20.5	72%	82%	85%	3.3	3.8	5.9	8.2	8.7
	6	13916	17222	751.5	930	1103	902	1029	19.9	14.3	22	22.4	14.8	74%	92%	88%	5	5.5	6	9	9.1
k2	14	1638	2261	88.5	122.1	200.8	203.9	207.1	27.5	20.6	27.5	26.3	26.1	80%	76%	75%	1	1.4	5.9	5	5.7
	8	1913	2783	103.4	150.3	230.2	232.3	254.3	20.2	15	20.1	21.4	18.2	79%	86%	82%	2	2	6.4	6.8	6.4
	4	4050	5675	218.7	306.5	299.4	250.8	288.4	18.8	14.8	17.7	18	15.4	73%	90%	85%	3.7	4.1	6	7.2	6.9
x3	8	1413	1683	76.35	90.9	150	94.9	143	14.8	14.2	14.4	15.8	14.2	71%	89%	76%	0.7	0.8	1.8	2	2.1
	4	1522	1788	82.2	96.6	169	138.3	150	12.4	11.6	13	12.8	12.7	62%	90%	90%	0.8	0.8	1.8	1.9	2

Layout-Aware Synthesis of Arithmetic Circuits

Junhyung Um and Taewhan Kim
Department of Electrical Engineering & Computer Science
and Advanced Information Technology Research Center(AITrc)
Korea Advanced Institute of Science and Technology

ABSTRACT

In deep sub-micron (DSM) technology, wires are equally or more important than logic components since wire-related problems such as crosstalk, noise are much critical in system-on-chip (SoC) design. Recently, a method [12] for generating a partial product reduction tree (PPRT) with optimal-timing using bit-level adders to implement arithmetic circuits, which outperforms the current best designs, is proposed. However, in the conventional approaches including [12], interconnects are not primary components to be optimized in the synthesis of arithmetic circuits, mainly due to its high integration complexity or unpredictable wire effects, thereby resulting in unsatisfactory layout results with long and messed wire connections. To overcome the limitation, we propose a new module generation/synthesis algorithm for arithmetic circuits utilizing carry-save-adder (CSA) modules, which not only optimizes the circuit timing but also generates a much regular interconnect topology of the final circuits. Specifically, we propose a two-step algorithm: (*Phase 1*: CSA module generation) we propose *an optimal-timing CSA module generation algorithm* for an arithmetic expression under *a general CSA timing model*; (*Phase 2*: Bit-level interconnect refinements) we *optimally refine the interconnects between the CSA modules* while retaining the global CSA-tree structure produced by Phase 1. It is shown that the timing of the circuits produced by our approach is equal or almost close to that by [12] in most test-cases (even without including the interconnect delay), and at the same time, the interconnects in layout are significantly short and regular.

Categories and Subject Descriptions
B.2.4. [**Arithmetic and Logic Structures**]: High-Speed Arithmetic – *Algorithms, Cost/Performance*

General Terms: Algorithms, Design and Peformance

Keywords: Carry-save-adder, layout, high performance

1. INTRODUCTION

The optimization of arithmetic expression has been the subject of intensive investigation in the several phases of design process. A lot of techniques for synthesizing arithmetic circuits are proposed [1]-[12]. A carry-save-adder (CSA) module [20] is one of the most frequently used implementations in the operation level since it has an advantage of fast timing possibility with, if any, little area penalty.

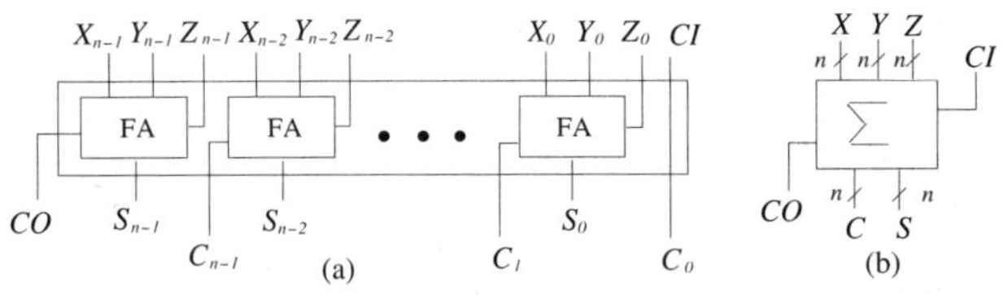

Figure 1: A structure of an n-bit CSA

Figure 1(a) shows the structure of n-bit CSA. (The block symbol in Figure 1(b) represents a CSA module.) An n-bit CSA consists of n disjoint full adders (FAs). Note that unlike normal adders (e.g., RCA (ripple-carry-adder) and CLA (carry-lookahead-adder)), there is no carry propagation between the FAs in the CSA. Thus, for a sufficiently large value of n, the CSA implementation becomes much faster and also relatively smaller in size than the implementations of normal adders.

There are several works related to efficient representation, allocation and transformation of CSAs in VLSI design automation. Kim *et al.* [10] performs the CSA transformation in three steps: (a) an identification of operation tree to be transformed, (b) a translation of the expression of the identified tree into an addition expression, and (c) a conversion of the addition expression into a CSA-tree. Given a non-cyclic dataflow graph of arithmetic computations, the algorithm iteratively performs the three steps until there is no candidate expression of tree to be transformed. A few extensions to the work in [10] were proposed. Mathur and Saluja [13] improved the implementation of the first step in [10] by extracting a characterization of safe grouping of operations to be transformed. Um and Kim [14] improved the procedure of the third step in [10] by providing an optimal-timing CSA-tree allocation algorithm using a constant CSA delay model. Further, Kim and Um [15] extended the scope of the CSA transformation to include operations across multiplexors, across design boundary, and across multiplications. On the other hand, Yu *et al.* [16] extended the scope of the CSA transformation to include operations across registers. Kim and Kim[17] proposed a design exploration algorithm by trading timing and area of the circuits using CSAs. In addition, Yu *et al.* [18] proposed a carry-save representation for DFG (dataflow graph) to make it easy to solve the CSA allocation and retiming together.

Contrary to the CSA module (word-level) based implementations mentioned above, there have been numerous research works on the design of fast arithmetic circuits, especially multipliers (e.g., [19]), using *bit-level FAs* as basic implementation components. (An extensive survey can be found in [20, 21].) The main issue of the bit-level arithmetic optimization is implementing fast Partial Product Reduction Trees (PPRTs) using FAs. Oklobdzija *et al.* [11] proposed an optimal algorithm, called *three-greedy* algorithm, for constructing a minimum-delay PPRT using FAs under a restricted FA timing model. Further, Stelling *et al.* [12] proposed an algo-

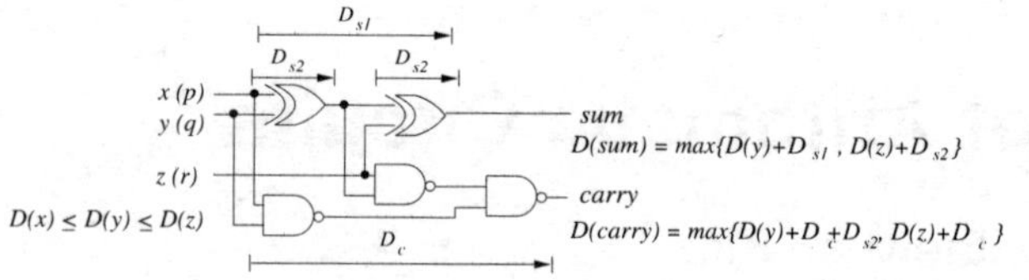

Figure 2: Timing model for FA

rithm called *two-greedy*, which is an improvement of the *three-greedy* algorithm, by developing a number of new techniques for analyzing and designing PPRTs.

Clearly, in terms of timing, the bit-level implementation using FAs is superior to the word-level implementation using CSAs. Conversely, in terms of layout efficiency for interconnects, the CSA implementation is superior to the FA implementation because the wires in the former are regularly grouped into buses. Note that in DSM technology, wires are equally or more important than logic components in layout since wire-related problems such as crosstalk, noise are much critical in system-on-chip (SoC) design. However, most of the previous works have focused on minimizing the circuit timing, and not much seriously taken into account the interconnect optimization for easy layout during the arithmetic circuit implementation.

In this paper, we propose a layout-aware synthesis algorithm for arithmetic circuits utilizing carry-save-adders (CSAs), which not only optimizes the circuit timing but also generates a much reliable interconnect topology of the final circuit. Specifically, we propose a two-step algorithm: (*Phase 1*: CSA module generation) we propose *an optimal-timing CSA module generation algorithm* for an arithmetic expression under *a general CSA timing model*; (*Phase 2*: Bit-level interconnect refinements) we *optimally refine the interconnects between the CSA modules* while retaining the global CSA-tree structure obtained in Phase 1. From experiments, it is shown that the timing of the circuits produced by our approach is equal or almost close to that of the optimal bit-level implementation by the work in [12] in most testcases (even without including the interconnect delay), and at the same time, the interconnects in layout are significantly short and regular.

2. PRELIMINARIES

2.1 CSA Timing Model

Let $D(x)$ denote the arrival time of x if x is a 1-bit signal, and $D(X) = max\{D(x_1), D(x_2), \cdots, D(x_n)\}$ if X is an n-bit vector signal $(x_1, x_2, \cdots, x_n)$.[1] Since n-bit CSA consists of n disjoint FAs, let us check the timing model of FA first.

Let p, q, r be the three input ports of an FA, and *carry* and *sum* be the carry and sum output ports of the FA, respectively. Let x, y and z ($D(x) \leq D(y) \leq D(z)$) be the bit addends, each assigned to p, q, and r. Based on a typical logic implementation for FA (See Figure 2.), the time needed by an FA to generate its *sum* output from its inputs amounts to the 2-level XOR delay. Thus, we use two delay parameters D_{s1} and D_{s2} to model the delay to *sum* output where D_{s1} indicates the delay from q to *sum* port and D_{s2} the delay from r to *sum* port. Similarly, we use D_c (2-level NAND delay) to indicate the *carry* delay of an FA. We refer the FA timing model mentioned above to as the *general timing model of FA*. Consequently, the arrival times for *sum* and *carry* signals from inputs are calculated by $D(sum) = max\{D(y) + D_{s1}, D(z) + D_{s2}\}$ and $D(carry) = max\{D(y) + D_c + D_{s2}, D(z) + D_c\}$. Note that the PPRT reduction algorithm in [12] used the general FA timing model and produces an optimal-timing FA-tree allocation.

We now define *a general CSA timing model* by extending the general timing model of FA as follows. According to the internal structure of CSA, in addition to the three (normal) input ports, a CSA has 1-bit carry input port. Let X, Y, Z ($D(X) \leq D(Y) \leq D(Z)$) be the n-bit vector addends, assigned to the three normal input ports of the CSA, and s be the signal assigned to carry input port. Then, the delays to *sum* and *carry* of the CSA are defined by $D(sum) = max\{D(Y) + D_{s1}, D(Z) + D_{s2}\}$ and $D(carry) = max\{D(y) + D_c + D_{s2}, D(z) + D_c\}$. Note the CSA-tree allocation algorithm in [14] produces an optimal-timing CSA-tree under the constraint of $D_{s1} = D_{s2}$, but is not optimal under the general CSA timing model. We propose a new CSA allocation algorithm (section. 3.1) which produces an optimal-timing CSA-tree under the general CSA timing model.

2.2 Motivating Example

Let us begin with an example of synthesizing an arithmetic circuit to see how our proposed approach is essentially different with the conventional methods. Suppose we want to add seven operands X_1, $\cdots$, X_7. Each operand is 3-bit wide, and its word-level and bit-level arrival times are shown in the table in Figure 3(a). Figure 3(b) shows the optimal-timing CSA-tree structure, which we call *circuit1*, produced by the algorithm in [14] under a restriction on the CSA timing model. On the other hand, Figure 3(c) shows the optimal-timing FA-tree structure, which we call *circuit2*, produced by the algorithm in [12] under the general FA timing model. From the comparisons between *circuit1* and *circuit2*, we can find that the interconnects in *circuit1* look more regular (i.e., grouped into buses) than those in *circuit2* since *circuit1* comes from a coarse-grained (word-level) module generation, but *circuit2* is faster than *circuit1* since *circuit2* comes from a fine-grained (bit-level) cell generation.

The motivation of our approach is to make use of both advantages of [14] (word-level) and [12] (bit-level). Figure 3(d) shows an improved optimal-timing CSA-tree structure, which we call *circuit3*, produced by our algorithm under the general CSA timing model. Consequently, the timing of *circuit3* (=11) is shorter than that of *circut1* (=13), and the interconnects in *circuit3* are still as regular as that in *circuit1*. We further improve the timing of *circuit3* by FA-level interconnect reordering. *circuit4* in Figure 3(e) shows the resultant CSA-tree structure optimized from *circuit3*, in which the FAs marked with dotted circle indicate the reordering of FA input connections. For example, in FA_9 its first and third connections are switched to reduce its output timings. Since the original CSA-level circuit structure is preserved during the (local) FA-level interconnect reordering, the degree of interconnect regularity in *circuit4* is retained in a global prospective. Nevertheless, the timing (=10) of critical paths in *circuit4* is reduced, and much close to the optimal (bit-level) timing (=9) in *circuit2*.

Finally, Figures 3(f) and (g) show the layouts (assumed a simple symmetrical FPGA-style mapping) with optimized routability for *circuit2* and *circuit4*, respectively, where the heavy line in Figure 3(f) indicates the longest net. Note that the layout in Figures 3(f) is optimal in the sense that we were not able to reproduce a layout for *circuit2* that has less number of the longest nets and/or connection blocks with maximum routing density than the layout in Figure 3(g). This means that, in addition to the gate delay, when we take into account the interconnect delay in measuring the performance of the final circuit, the overall timing of *circuit4* can be much shorter than that of *circuit2*. Furthermore, it is clear that this trend will be very likely as the circuit size increases.

In fact, *circuit3* and *circuit4* are produced by Phase 1 and Phase 2 of our proposed two-step approach. The comparison of the layouts clearly indicates that, among the four circuits in Figure 3, *circuit4* is the most efficient structure in terms of timing and layout in synthesizing arithmetic circuits. We now provide the details on the two

[1] $D(X) = 0$ if X is a constant number.

Word	Bit	Delay(Word)	Delay(Bit)
X_1	$x_{1,2}\ x_{1,1}\ x_{1,0}$	2	0, 0, 2
X_2	$x_{2,2}\ x_{2,1}\ x_{2,0}$	2	1, 0, 2
X_3	$x_{3,2}\ x_{3,1}\ x_{3,0}$	4	3, 0, 4
X_4	$x_{4,2}\ x_{4,1}\ x_{4,0}$	7	7, 1, 1
X_5	$x_{5,2}\ x_{5,1}\ x_{5,0}$	5	4, 3, 5
X_6	$x_{6,2}\ x_{6,1}\ x_{6,0}$	6	0, 6, 1
X_7	$x_{7,2}\ x_{7,1}\ x_{7,0}$	9	8, 8, 9

(a) Operands to be added

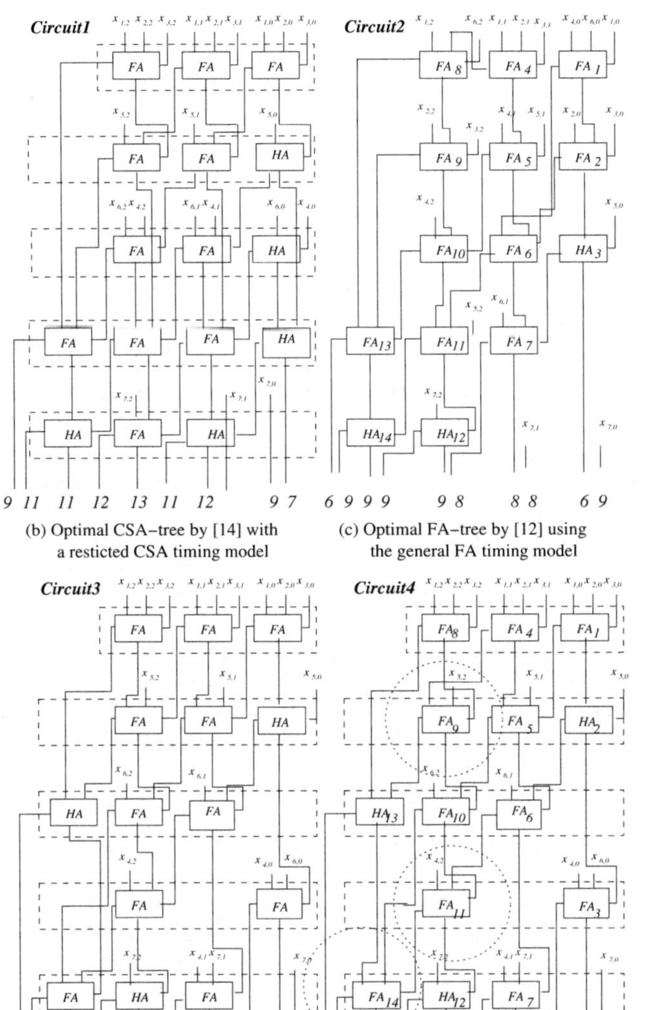

(b) Optimal CSA–tree by [14] with a resticted CSA timing model

(c) Optimal FA–tree by [12] using the general FA timing model

(d) Optimal CSA–tree by ours using thegeneral CSA timiing model (Phase 1)

(e) Optimal CSA–tree by ours using FA–level interconnect reordering (Phase 2)

(f) Layout for the result in (c)

(g) Layout of our result in (e)

Figure 3: An example showing the effectiveness of our approach, in terms of timing and layout

phases of our approach.

3. A NEW SYNTHESIS APPROACH

The proposed layout-conscious synthesis approach is performed in two optimization phases. Given an arithmetic expression to be implemented, in *Phase 1*, we propose a timing-optimal CSA-tree construction algorithm, which we refer to as *partial-greedy-CSA* algorithm, for an arithmetic expression under the general CSA timing model. In *Phase 2*, we propose a layout-aware HA merging technique (called *HA-merge*) to refine the CSA-tree structure (*Phase 2.1*), and an optimal FA input reordering technique (called *FA-ref*) to further improve the timing of the circuit without deteriorating the CSA block-level interconnect structure (*Phase 2.2*).

3.1 Phase 1: Optimal-Timing CSA Allocation

The CSA allocation problem is stated as: *(Problem 1) Given an addition expression*[2] *of*

$$F = X_1 + X_2 + \cdots + X_m + s_1 + s_2 + \cdots + s_n + c \quad (1)$$

where $X_1, \cdots X_m$ are multi-bit addends, $s_1, \cdots s_n$ are single-bit addends and c is a constant, and the arrival times of the addends, for the general delay model for CSA (defined in section 2.1), determine a CSA-tree structure for F with a minimum delay.

Let us first consider a restricted case of Problem 1.

Subproblem 1: Problem 1 with $c = 0$ and $m - 1 \geq n$.

The inequality $m - 1 \geq n$ implies that the number of multi-bit addends is large enough for all single-bit addends to be used as carry inputs to CSAs.

Let $\mathcal{R}$ be the set of all possible CSA-trees for Subproblem 1, in which all single-bit addends are necessarily used as carry inputs to CSAs. Let us consider the following CSA-tree construction strategy, which we call *partial-greedy-CSA* algorithm: We allocate CSAs iteratively to sum up all the addends such that at each iteration, we allocate a new CSA with three multi-bit addends, in which two of them have the first and second earliest arrival times among the multi-bit addends, and one single-bit addend with the earliest arrival time assigned to the CSA inputs. Note that the CSA-trees produced by the *partial-greedy-CSA* algorithm will have different timings and structures depending on the way of selecting the remaining multi-bit addend at each iteration. (See Figure 4.) Let $\mathcal{A}$ be the set of all CSA-trees that can be produced by the *partial-greedy-CSA* algorithm.

We define a number of notations for the ease of our discussion: Let CSA_i denote the CSA created at the ith iteration of *partial-greedy-CSA* method. a_i, b_i and d_i ($D(a_i) \leq D(b_i) \leq D(d_i)$) denote the multi-bit addends connected to the three normal input ports (from left to right, X, Y, Z in Figure 1) of CSA_i, respectively, and s_i denote the single-bit addend to the *carry-in* port of CSA_i. In addition, $carry_i$ and sum_i denote *carry* and *sum* signals of CSA_i, respectively. Let T be a circuit generated by performing n iterations of CSA allocation in *partial-greedy-CSA* method. We use $D(x,T)$ to denote the arrival time of signal x in CSA-tree T, and $D(T)$ is set to $max\{D(x_1,T),\cdots D(x_n,T)), D(y_1,T),\cdots D(y_n,T)\}$ where x_i and y_i, $i = 1,\cdots n$, are *carry* and *sum* signals of CSA_i in T, respectively.[3]

Lemma 1 *There exists a CSA-tree $T' \in \mathcal{A}$ such that $D(T') \leq D(T)$ for any $T \in \mathcal{R}$. (We omit all the proofs due to the space of limitation.)*

Lemma 1 suggests that using *partial-greedy-CSA* method can reduce the search space for finding an optimal CSA-tree from $\mathcal{R}$.

[2]The expression can contain subtraction and/or multiplication operations. As to converting it into an addition expression, refer to [10].

[3]We simply use $D(v)$ to indicate $D(v,T)$ if the meaning of T in $D(v,T)$ is implicitly clear in our discussion.

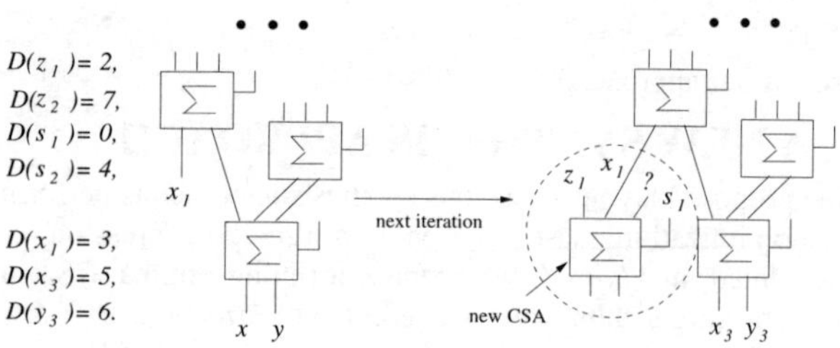

Figure 4: An illustration of *partial-greedy-CSA* algorithm

Note that the procedure of *partial-greedy-CSA* method is quite similar to the *two-greedy FA allocation* method proposed in [12]. The differences are (i) our CSA allocation solution considers the carry-out vector of a CSA as *any* CSA input during the next iterations of CSA allocation, whereas [12]'s FA allocation solution (in bit-weight 2^i) uses the carryout signal of an FA *only* as input to an FA allocated at the next bit-weight (i.e., 2^{i+1}); (ii) In addition to the multi-bit inputs, the CSA allocation problem has another input type, i.e., single-bit inputs, which have a freedom to be used either as normal inputs to CSAs or as carry inputs to CSAs, affecting the value of $D(carry)$ only. We found that some of the properties in [12] regarding the *two-greedy FA allocation* method also hold for our *partial-greedy-CSA* method. Now, let us consider an extended version of Subproblem 1.

Subproblem 2: Problem 1 with $c = 0$ and $m - 1 < n$.

The inequality $m - 1 < n$ implies that some of the single-bit addends must be used as normal inputs to CSAs. Thus, we should decide *which* and *how many* of the single-bit addends are to be used as normal inputs to CSAs. Let k be the number of single-bit addends to be used as normal inputs to CSAs. Then we can easily show $k \geq \lceil \frac{n-m+1}{2} \rceil$, from the fact that converting one single-bit addend to a multi-bit addend to be used as a normal input to a CSA requires one additional CSA and thus, resulting in one more carry input ports available, and at the same time, one less single-bit addends.

First, let us consider which of the single-bit addends are to be used as normal inputs to some of the CSAs when the number k is fixed. We claim the following:

Lemma 2 *To select k single-bit addends to be used as normal inputs to CSAs, the addends with the earliest arrival times will be selected for optimal-timing CSA-tree allocation.*

Moreover, the following lemma answers to the question of how many of the single-bit addends are to be used as normal inputs to some of the CSAs. we show that for optimal-timing CSA-tree allocation, the value of k should be chosen as $k = \lceil \frac{n-m+1}{2} \rceil$.

Lemma 3 *The optimal number of single-bit addends, k, to be used as normal inputs to CSAs for optimal-timing CSA allocation is* $\lceil \frac{n-m+1}{2} \rceil$.

Finally, we complete the solution of Problem 1.

Subproblem 3: Problem 1 with $c \neq 0$.

If c is a negative number, we need to do a sign-extension for the constant. This means that we shall treat it as a multi-bit addend. However, if it is a positive number, there are two choices: (1) breaking c into a form of $1 + 1 + ... + 1 (= c)$ to be used as carry inputs to CSAs, or (2) treating c as one multi-bit addend. (Obviously, using a strategy of mixing options (1) and (2) will produce inferior timings.) If $c > 0$ and $m - 1 \geq c + n$, we can use all single-bit and constant addends as carry inputs to CSAs (i.e., option (1)) because *partial-greedy-CSA* algorithm with the option (1) and the timing of CSA structure with $c = 0$ are basically the same. However, if $c > 0$ and $m - 1 < c + n$, we claim that treating c as a multi-bit addend (i.e., option (2)) produces a better timing.

Theorem 1 *When we use* partial-greedy-CSA *method for CSA allocation, if either $c > 0$ and $m - 1 < c + n$ or $c < 0$ we should treat c as a multi-bit addend, and if $c > 0$ and $m - 1 \geq c + n$, we should break c into into a form of $1 + 1 + ... + 1 (= c)$ and use them as carry inputs to CSAs.*

We use a branch and bound search to find an optimal-timing CSA-tree structure by applying *partial-greedy-CSA* allocation strategy. Since the CSA allocation algorithm in [14] produces an optimal-timing CSA-tree using the CSA timing model constrained by $D_{s1} = D_{s2}$, we can use the timing of the CSA-tree, produced by [14] under the setting of the value of D_{s2} to the value of D_{s1} ($D_{s2} \leq D_{s1}$), as *an upper bound* of the optimal-timing in the branch and bound search. In addition, we devise a number of pruning techniques based on the following properties to speed up the execution of the branch and bound search. Let T be a CSA-tree produced by *partial-greedy-CSA* method, i.e., $T \in \mathcal{A}$.

Property 1 *If there are CSA_i, $CSA_j \in T$ such that $i < j$,[4] $D(d_i) \leq D(b_j)$ and $D(d_i) > D(a_j)$, there is a CSA-tree $T' \in \mathcal{A}$ such that $D(T') \leq D(T)$.*

Property 2 *If there are CSA_i, $CSA_j \in T$ such that $i < j$, $D(b_j) < D(d_i) \leq D(d_j)$ and $D(d_j) \geq D(d_i) + D_{s1} - D_{s2}$, there is a CSA-tree $T' \in \mathcal{A}$ such that $D(T') \leq D(T)$.*

Property 3 *If there are CSA_i, $CSA_j \in T$ such that $i < j$ and $D(d_i) > D(d_j)$, there is a CSA-tree $T' \in \mathcal{A}$ such that $D(T') \leq D(T)$.*

3.2 Phase 2: Optimal Refinement of CSA Timing

The objective of this phase is to reduce the timing of the CSA-tree further by refining the bit-level interconnect structures while maintaining the topology of the word-level CSA-tree structure obtained in Phase 1. Phase 2 is a two-step optimization: (Phase 2.1) Layout-aware *HA (half-adder) merging step* which merges a pair of HAs in a CSA-tree into an FA and (Phase 2.2) *FA refinement step*, which locally refines the FA's inputs.

Phase 2.1 (HA-merging): For the CSA allocation in Phase 1, if the bit-widths of three (normal) inputs to be added using a CSA are not totally equal,[5] or some input has a constant bit-value (e.g., logic-0 or 1) in a certain bit-position, some of the FAs in the CSA will be degenerated into HAs.

For each bit-column, from the rightmost bit column to the leftmost, we perform *HA-merge* from top to bottom, the leaf HAs first and the root HAs last. During the top-down HA-merge process, if the sum signal of an HA is fed to an input to another HA (i.e., tightly connected), we combine the two HAs into one FA, as shown in Case 1 of Figure 5(a). There are two possible ways of connecting the carry signal of the new FA, as described in the example in Figure 5(a) (i.e., trans1 and trans2). We have an important property for the HA-merge for Case 1.

Property 4 *If an HA-merge is performed for Case 1 in Figure 5(a), the timing of* trans2 *is always better than or equal to that of* trans1 *when the general timing model for HA (similar to that of FA) is used. Further, the timing of* trans2 *is always better than or equal to that of* case1_org.

Consequently, according to Property 4, when we encounter a pair of tightly connected HAs (i.e., Case 1) we apply HA-merge to the HAs and produce a merged FA in a form like trans2 in Figure 5(a).

On the other hand, if two HAs are not tightly connected as shown case2_org in Figure 5(b), *we do not directly merge the two HAs*

[4]Note that CSA_i is the CSA created at the i-th iteration of the *partial-greedy-CSA* method.

[5]This mostly happens when a multiplication is decomposed into a set of shift-add operations.

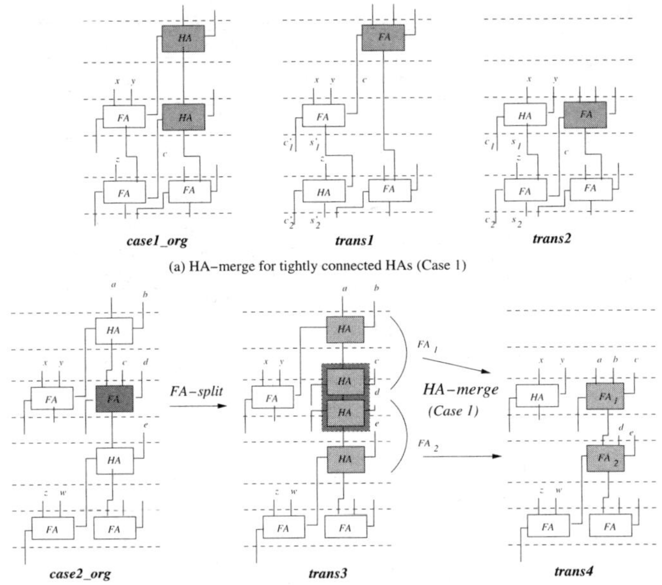

(a) HA–merge for tightly connected HAs (Case 1)

(b) HA–merge for loosely connected HAs (Case 2)

Figure 5: Examples for layout-aware HA-merge for (Case 1) tightly connected HAs and (Case 2) loosely connected HAs.

in the same way as that of Case 1 because the merger causes a (global) distortion of the interconnection topology of the initial CSA-tree. Instead, we split the FA, one of whose inputs is directly fed from the upper HA, into two HAs in a chain and merge the upper HA with the decomposed HA that is directly connected to the upper HA into an FA. Thus, we perform HA-merge process for Case 2 in two steps: (Step 1) *FA-split* to convert Case 2 into Case 1, and (Step 2) *HA-merge* according to the procedure in Case 1. Figure 5(b) shows an example of the two steps. We have the following property, which justifies our HA-merge process.

Property 5 *If an HA-merge is performed for Case 2 in Figure 5(b), the timing of* trans4 *is always better than or equal to that of* case2_org *when the general timing model for HA is used.*

According to Properties 4 and 5, the proposed HA-merge can be performed iteratively, from top (leaf) CSAs to bottom of CSA-tree, which monotonically reduces the timing of the CSA-tree at each iteration.

Phase 2.2 (FA-refinement): In this phase, we perform a bit-level timing refinement while preserving the word-level global interconnect structure obtained in Phase 2.1. Note that Phase 1 used a word-level arrival times of input addends during the CSA-tree allocation. That is, if there is an n-bit addend $X = (x_{n-1}, x_{n-2} \cdots x_0)$, Phase 1 uses $max\{D(x_{n-1}), D(x_{n-2}) \cdots D(x_0)\}$ as the value of $D(X)$.

Since we try to preserve the word-level interconnect structure of the CSA-tree during this phase, we perform a sort of FA-level connection restructuring *inside of* the corresponding CSA block. Consequently, the values of the arrival times of Y_1, Y_2 and Y_3 determine the order of input connections to CSA_i, which in turn determines uniform relative orders of the inputs of all FAs in CSA_i, as shown in Figure 6(b). In Phase 2.2, for each FA we reorder its input connections to reduce the FA's output timings, as shown in Figure 6(c).

We perform the FA's input reordering for a CSA-tree topologically, the FAs in the leaf CSAs first and the FAs in the root CSAs last, which we call *FA-ref* method for FA's input reordering. We claim that *FA-ref* is timing-optimal.

Theorem 2 *When the bit-level rewiring for a CSA-tree is allowed only for the inputs to the same FA, the timing reduced by FA-ref method is optimal.*

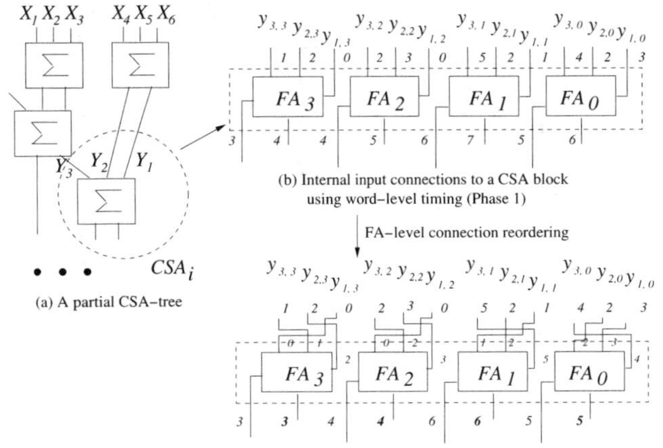

(a) A partial CSA–tree

(b) Internal input connections to a CSA block using word–level timing (Phase 1)

(c) Optimized input connections to FAs in a CSA block using bit–level timing (Phase 2.2)

Figure 6: An example of illustrating FA-refinement.

4. EXPERIMENTAL RESULTS

We conducted a set of experiments to check the effectiveness of the proposed layout-conscious module generation/synthesis algorithm, consisting of *Partial-Greedy-CSA*, *HA-merge* and *FA-ref* techniques. Our algorithm was implemented in C++ and executed on a Pentium-3 500MHz Linux machine. We tested our algorithm on a set of random arithmetic expressions and benchmark examples [22], and then compared our results with that produced by the Stelling's bit-level delay-optimal[6] algorithm [12].

After the synthesis using ours and Stelling's for each testcase, we performed, for each synthesized design, placement[7] using Fiduccia and Mattheyses's partitioning algorithm [23], and routing[8] using Lee's maze routing algorithm [24]. We used the Synopsys Design Compiler [25] with $lcbg10pv(0.35u)$ technology library [26] to set the timing parameters for CSAs and FAs. We used a simple formula $\alpha \cdot L^2$ to approximate the delay of the wire connecting two FA cells where L is the number of blocks on the corresponding routing path. α is a parameter to reflect the relative delay difference between gates and wires, the value of which will depend on the layout style and target library used.

• **Synthesizing arbitrary arithmetic expressions:** We tested our algorithm using a set of arithmetic expressions, and the results are summarized in Table 1. The bit-level arrival times of the operands in the expressions are randomly generated in between 0.0 and 2.0ns. The table shows the critical path and longest net delays for the designed by ours and Stelling's [12]. Figure 7 show the layouts for the designs produced by Stelling's and ours for Exp.3 where we have 61% shorter critical path delay (highlighted in the figure). The comparisons indicate that the circuit structures produced by ours significantly reduce the critical path delay and the longest net delay compared to Stelling's. Note that as the α value increases (i.e., wire delay being more dominant than gate delay), the performance improvement is more significant.

• **Synthesizing filter designs:** We also tested our algorithm using high-level synthesis benchmark designs, and the results are summarized in Table 2. We assumed the arrival times of input operands are 0. The results indicate that our synthesis algorithm reduces the critical path delay 22% on the average compared to Stelling's. Furthermore, the reductions are consistent. In summary, the comparisons of the results strongly suggest that considering layout effects

[6] 'optimal' when considering gate delays only.

[7] The target layout we used is gate array (lattice) style for simplicity, but our approach can be applied to any layout style such as standard cell and FPGA.

[8] We set the channel capacity in the layout to 4.

Exp.	α	Stelling[12] crit. path delay/ long. net delay	Ours crit. path delay/ long. net delay	Improvement crit. path delay/ long. net delay
Exp. 1	0.04	11.27 / 3.24	8.87 / 1.96	21% / 40%
	0.02	6.78 / 1.62	5.59 / 0.98	18% / 40%
	0.01	4.54 / 0.81	3.95 / 0.49	13% / 40%
Exp. 2	0.04	14.69 / 6.76	8.14 / 3.24	45% / 52%
	0.02	7.89 / 3.38	4.86 / 1.62	38% / 52%
	0.01	4.49 / 1.69	3.22 / 0.81	28% / 52%
Exp. 3	0.04	46.43 / 17.64	18.22 / 9.00	61% / 49%
	0.02	23.79 / 8.82	9.50 / 4.50	60% / 49%
	0.01	12.48 / 4.42	5.14 / 2.25	59% / 49%
Exp. 4	0.04	21.92 / 9.00	14.43 / 6.76	34% / 25%
	0.02	12.16 / 4.50	8.49 / 3.38	30% / 25%
	0.01	7.28 / 2.25	5.52 / 1.69	24% / 25%
Exp. 5	0.04	18.42 / 9.00	13.41 / 6.76	27% / 25%
	0.02	10.42 / 4.50	7.96 / 3.38	24% / 25%
	0.01	6.42 / 2.25	5.24 / 1.69	18% / 25%
Exp. 6	0.04	19.27 / 9.00	11.88 / 3.24	38% / 64%
	0.02	10.87 / 4.50	7.48 / 1.62	31% / 64%
	0.01	6.67 / 2.25	5.28 / 0.81	21% / 64%
Avg.	0.04	29.06 / 9.11	12.49 / 5.16	**38% / 43%**
	0.02	11.99 / 4.56	7.31 / 2.58	**34% / 43%**
	0.01	6.98 / 2.28	4.73 / 1.29	**27% / 43%**

Table 1: Results for a set of arithmetic expressions Exp.1 : $X_1 + \cdots + X_6$; Exp.2 : $X_1 \cdot X_2 + X_3 + X_4$; Exp.3 : $X_1 + X_2 + X_3 - X_4 + X_5 - X_6 - X_7$; Exp.4 : $X_1 + \cdots + X_8$; Exp.5 : $X_1 \cdot X_2 + X_3 \cdot X_4 + X_5 \cdot X_6$; Exp.6 : $X_1 \cdot X_2 + X_3 \cdot X_4 - X_5 \cdot X_6 - X_7 \cdot X_8$;

at the early (architectural-level) stage of the synthesis process can considerably improve the performance of the final circuits, and our proposed synthesis approach does the job quite well.

5. CONCLUSIONS

In this paper, we presented a new RT-level synthesis approach for arithmetic circuits, which considers two important design objectives: fast timing and easy layout. As the interaction between architectural/logical (synthesis) and physical (layout) domains becomes much important issue in DSM technology, considering layout effects at an early stage of the synthesis process is necessary. In the respect, we believe that the proposed new two-phase approach, which (Phase 1) *produces an optimal-timing arithmetic circuit using fast CSA (carry-save-adder) modules under general CSA timing model*, and (Phase 2) *refines the circuit structure obtain in Phase 1 by developing layout-aware optimal HA-merging and FA-input-reordering techniques*, can be used effectively in synthesizing (arithmetic) datapaths. From a set of benchmark filter designs, we confirm that the proposed approach produces designs with (gate) timing comparable to the optimal bit-level timing by [12] while greatly reducing the complexity of interconnects, resulting in much shorter critical path delay (22% reduction on average over [12]) and longest net delay (26% reduction) in the final circuits.

Acknowledgment : This work was supported by the Korea Science and Engineering Foundation (KOSEF) through the Advanced Information Technology Research Center(AITrc).

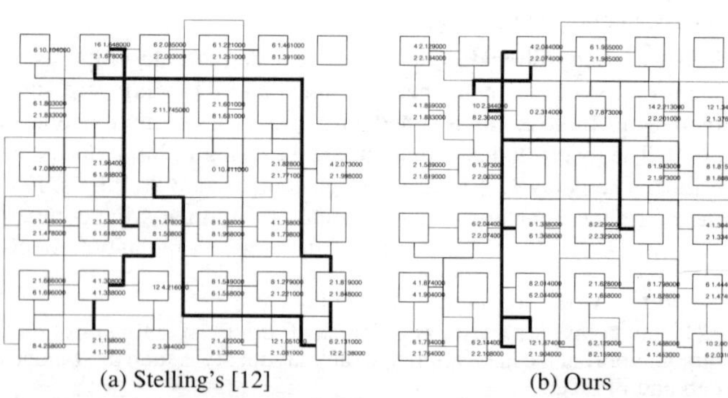

(a) Stelling's [12] (b) Ours

Figure 7: Comparison of layout results for Exp.3. **(The heavy lines indicate the critical paths.)**

Filter Design	α	Stelling's[12] crit. path delay/ long. net delay	Ours crit. path delay/ long. net delay	Improvement crit. path delay long. net delay
Lowpass	0.04	16.11 / 6.76	12.08 / 4.84	25% / 28%
	0.02	8.52 / 3.38	6.48 / 2.42	24% / 28%
	0.01	4.72 / 1.69	3.68 / 1.21	22% / 28%
Laplace	0.04	12.26 / 4.84	12.08 / 4.84	4% / 0%
	0.02	6.74 / 2.42	6.48 / 2.42	4% / 0%
	0.01	3.98 / 1.21	3.78 / 1.21	5% / 0%
Wavelet	0.04	46.43 / 17.64	18.22 / 9.00	61% / 49%
	0.02	23.79 / 8.82	9.59 / 4.50	60% / 49%
	0.01	12.48 / 11.86	5.14 / 2.25	59% / 49%
IIR	0.04	21.38 / 14.44	20.02 / 6.76	6% / 53%
	0.02	11.23 / 7.22	10.50 / 3.38	6% / 53%
	0.01	6.15 / 3.61	5.74 / 1.69	7% / 53%
Complex	0.04	18.68 / 11.56	16.53 / 11.56	12% / 0%
	0.02	9.99 / 5.78	8.69 / 5.78	6% / 0%
	0.01	5.61 / 2.89	4.77 / 2.89	7% / 0%
Avg.	0.04	22.97 / 11.05	15.79 / 7.40	**22% / 26%**
	0.02	10.05 / 6.13	8.26/3.70	**21% / 26%**
	0.01	5.49 / 4.60	5.49 / 4.60	**21% / 26%**

Table 2: Results for a set of benchmark designs.

6. REFERENCES

[1] M. Potkonjak and J. Rabaey, "Optimizing Resource Utilization Using Transformations", *Proc. of ICCAD*, 1991.

[2] M. Potkonjak and J. Rabaey, "Maximally Fast and Arbitrarily Fast Implementation of Liner Computations", *Proc. of ICCAD*, 1992.

[3] D. Lobo and B. Pangrle, "Redundant Operation Creation: A Scheduling Optimization Technique", *Proc. of DAC*, 1991.

[4] A. Nicolau and R. Potasma, "Incremental Tree Height Reduction for High-level Synthesis", *Proc. of DAC*, 1991.

[5] R. Hartley and A. E. Casavant, "Tree-Height Minimization in Pipelined Architectures", *Proc. of ICCAD*, 1989.

[6] R. Hartley and A. E. Casavant, "Optimized Pipelined Networks of Associative and Commutative Operators", *IEEE Trans. on CAD* , Vol. 13, No. 11, 1994.

[7] K. K. Parhi, "High-Level Algorithm and Architecture Transformations for DSP Synthesis", *Journal of VLSI Signal Processing*, No. 9, 1995.

[8] M. Janssens, F. Catthoor, and H. De Man, "A Specification Invariant Technique for Operation Cost Minimization in Flow-graphs", *Proc. of 7th International Symposium on High-Level Synthesis*, 1994.

[9] M. Janssens, F. Catthoor, and H. De Man, "A Specification Invariant Technique for Regularity Improvement between Flow-Graph Clusters", *Proc. of EDAC*, 1996.

[10] T. Kim, W. Jao, and S. Tjiang, "Circuit Optimization using Carry-Save-Adder Cells", *IEEE Trans. on CAD*, Vol.17, No.10,1998.

[11] V. G. Oklobdzija, D.Villeger, and S.S. Liu, "A Method for Speed Optimized Partial Product Reduction and Generation of Fast Parallel Multipliers using an Algorithmic Approach", *IEEE Trans. on Computers*, Vol. 45, No. 3, 1996.

[12] P. Stelling, C.U. Martel, V.G. Oklobdzija, and R. Ravi, "Optimal Circuits for Parallel Multipliers", *IEEE Trans. on Computers*, Vol. 47, No. 3, 1998.

[13] A. Mathur and S. Saluja, "Improved Merging of Datapath Operators using Information Content and Required Precision Analysis", *Proc. of DAC*, 2001.

[14] J. Um and T. Kim, "An Optimal Allocation of Carry-Save-Adders in Arithmetic Circuits", *IEEE Trans. on Computers*, Vol. 50, No. 3, 2001.

[15] T. Kim and J. Um, "A Practical Approach to the Synthesis of Arithmetic Circuits using Carry-Save adders", *IEEE Trans. on CAD*, Vol. 19, No. 5, 2000.

[16] Z. Yu, M.-L. Yu, and A.N. Willson Jr., "Signal Representation Guided Synthesis using Carry-Save Adders for Synchronous Data-path Circuits", *Proc. of DAC*, 2001.

[17] Y. Kim and T. Kim, "Accurate Exploration of Timing and Area Trade-offs in Arithmetic Optimization using Carry-Save-Adders", *Journal of Circuits, Systems and Computers*, Vol. 10, Nos. 5&6, 2000.

[18] Z. Yu, K.-Y. Khoo, and A.N. Willson Jr., "The Use of Carry-Save Representation in Joint Module Selection and Ret iming", *Proc. of DAC*, 2000.

[19] C.S. Wallace, "A Suggestion for a Fast Multiplier", *IEEE Trans. on Computers*, Vol. 13, 1964.

[20] K. Hwang, *Computer Arithmetic: Principles, Architecture and Design*, Jhon Wiley and Sons, 1979.

[21] E. Swartzlander, *Computer Arithmetic*, Vols. 1&2, IEEE CS Press, 1990.

[22] N. D. Dutt, "High-Level Synthesis Design Repositories", *http://www.ics.uci.edu/ dutt*.

[23] C. Fiduccia, R. Matteyses, "A Linear-time Heuristic for Improving Network Partitions", *Proc. of DAC*, 1982.

[24] C. Lee, "An Algorithm for Path Connections and its Applications", *IRE Trans. on Electronic Computers*, VEC-10. pp. 346-365, 1961.

[25] Synopsys Inc., *Design Compiler User Guide*, 2000.

[26] LSI Logic Inc., *G10-p Cell-Based ASIC Products Databook*, 1996.

Automatic Data Migration for Reducing Energy Consumption in Multi-Bank Memory Systems*

V. De La Luz
Microsystems Design Lab
Pennsylvania State University
University Park, PA 16802
delaluzp@cse.psu.edu

M. Kandemir
Microsystems Design Lab
Pennsylvania State University
University Park, PA 16802
kandemir@cse.psu.edu

I. Kolcu
Computation Department
UMIST
Manchester M60 1QD, UK
i.kolcu@co.umist.ac.uk

ABSTRACT

An architectural solution to reducing memory energy consumption is to adopt a multi-bank memory system instead of a monolithic (single-bank) memory system. Some recent multi-bank memory architectures help reduce memory energy by allowing an unused bank to be placed into a low-power operating mode. This paper describes an automatic data migration strategy which dynamically places the arrays with temporal affinity into the same set of banks. This strategy increases the number of banks which can be put into low-power modes and allows the use of more aggressive energy-saving modes. Experiments using several array-dominated applications show the usefulness of data migration and indicate that large energy savings can be achieved with low overhead.

Categories and Subject Descriptors

B.3 [**Hardware**]: Memory Structures

General Terms

Design, Experimentation, Performance

Keywords

Energy Consumption, Multi-Bank Memories, Data Migration

1. INTRODUCTION

The high energy consumption of main memories make them target of many energy-conscious optimization techniques. On current computer systems, main memory is one of the energy bottlenecks of the architecture [2]. This is especially true for mobile applications which are typically memory-intensive (array-dominant such as signal and video processing). In addition, embedded applications are gradually becoming more data-centric with stringent memory requirements (both for storage and speed), causing vendors to incorporate large storage capacities into their offerings. An architectural solution to this problem is to adopt a multi-bank memory system instead of a monolithic (single-bank) memory architecture. In a multi-bank memory architecture, unused memory banks (idle banks) can be disabled, thereby saving energy. However, in many circumstances, disabling a memory bank completely would not be possible because banks usually hold some useful data (even if they are not frequently accessed). Rather, some recent memory systems (e.g., [10]) provide low-power operating modes which enable large energy savings without completely disabling memory banks. More specifically, a bank that is not currently servicing a memory request can be put in a low-power operating mode. A major problem with these energy-saving modes is that they incur additional delay (called resynchronization cost or resynchronization penalty) when a bank in one of these modes needs to service a memory request. The magnitude of this delay (performance penalty) depends on the specific power mode; typically, a more energy-saving mode also incurs more delay. Consequently, it is advantageous to keep memory banks in idle conditions for long durations of time so that the resynchronization penalties can be compensated.

Our goal in this paper is to reduce the overall energy consumption of a given application by reducing the contribution of the component due to a multi-bank main memory with low-power operating modes. One way of achieving this is to cluster data accesses in a small number of memory banks and place the remaining banks (idle banks) into a low-power operating mode. In many cases, the default access pattern may not lead to clustered accesses and intrinsic dependences between data accesses and those due to control flow prevent the most suitable computation transformation from clustering the accesses. An alternative approach is to modify the data layout in memory such that the data elements that are accessed contemporaneously (i.e., those with temporal affinity) are stored in a small set of banks. As in the computation transformation case, such data transformations enable us to place a large number of (idle) memory banks into a low-power operation mode.

Previous compiler-oriented work has focused on implementing static data transformations [6] using which the array layouts in memory are transformed and these transformed layouts remain in effect throughout the execution of the code. Among the data transformations investigated are array interleaving and array co-location. In the operating system (OS) area, Lebeck et al. [7] has studied the OS support for exploiting low-power operating modes. In many applications, the data access patterns are not fixed during the entire execution and vary dynamically as the program executes. In an array-dominated embedded application, different arrays might be used together in different parts of the code. So, there is a need for a dynamic data co-location scheme taking into account temporal affinity (proximity) between data structures.

In this paper, we present a dynamic data migration scheme which co-locates the data elements automatically (as the program executes) to take advantage of the variations in inter-array access patterns. Our approach is based on a runtime library called Data Mi-

*This work is supported in part by NSF Career Award #0093082.

grator (DM) which samples the arrays that are accessed together and keeps temporal access relations between arrays using a table. From time to time, this table is scanned and the arrays with high number of contemporaneous accesses are determined. These arrays are then dynamically migrated across memory banks to ensure that they reside in a small set of banks. The proposed runtime system has been implemented and a simulation framework to evaluate its effectiveness has been built. Our results obtained using applications that manipulate large datasets residing in multi-bank memory configurations show that dynamic data migration improves energy behavior by as much as 40%.

Next section gives a brief review of a multi-bank memory architecture and explains low-power operating modes. Section 3 presents the important characteristics of our data migration scheme. Section 4 gives a description of our implementation (Data Migrator) and illustrates how it interacts with application execution. Section 5 describes our experimental platform, gives information about the codes executed, and reports experimental data showing the usefulness of dynamic data migration. Section 6 presents a summary of the paper.

2. MEMORY ARCHITECTURE

In this work, we assume a multi-bank/multi-module memory system. Each bank can contain multiple modules and is assumed to be a single word wide. Modules/banks can be placed into low-power operating modes independently. One example of this type of memory architecture is RDRAM [10] in which each bank contains a single module and can work with four operating modes: active, standby, nap, and power-down. A bank can service a read/write request only in the active mode. If it is not servicing a memory request, it can be placed into one of low-power operating modes (standby, nap, or power-down). In this paper, we assume a single module per bank; consequently, we use the terms module and bank interchangeably.

Each low-power mode is implemented by disabling specific parts of the DRAM chip and can be described using two metrics: energy consumption and resynchronization cost. The first metric is the per access (or per cycle) energy consumption when the module is in a specific mode whereas the second metric is the time (in cycles) it takes to bring a module back to the active mode (i.e., performance penalty). Typically, lower the energy consumption, higher the resynchronization cost. Figure 1 shows the energy consumptions (per access) and resynchronization costs for the operating modes used in this study. Obviously, there is a tradeoff between energy saving and performance penalty when selecting a low-power mode to use for a given idle bank. The time between successive accesses to a given bank (called inter-access time) is the major determining factor in selecting the most suitable low-power operating mode. The inter-access time can be estimated either by an optimizing compiler and can be predicted by a hardware/runtime mechanism. Both of these strategies have been investigated in a previous paper [6]. In this study, we adopt a hardware-based strategy in which the bank inter-access times are predicted using a prediction logic attached to memory controller. The rationale behind the hardware-based predictor is that if a memory bank has not been accessed in a while, then it is not likely to be needed in the near future (that is, inter-access times are predicted to be long). A threshold is used to determine the idleness of a bank after which it is transitioned to a low-power mode. More specifically, in our current implementation, after 10 cycles of idleness, the corresponding bank is put in standby mode. Subsequently, if the bank is not referenced for another 100 cycles, it is transitioned into the nap mode. Finally, if the bank is not referenced for a further 1,000,000 cycles, it is put into power-down mode. Whenever the bank is referenced, it is brought back into the active mode incurring the corresponding resynchronization costs (based on what mode it was in). Implementation details of this hardware-based scheme are beyond the scope of this paper and can be found in [6].

Operating Mode	Energy Consumption (nJ)	Resynchronization Cost (cycles)
Active	3.570	0
Standby	0.830	2
Nap	0.320	30
Power-Down	0.005	9,000

Figure 1: Energy consumptions (per access) and resynchronizations costs for our operating modes. During transitions from a low-power mode to the active mode, a full active mode energy is assumed to be spent.

3. DATA MIGRATION FOR LOW POWER

Automatic data migration is beneficial when a poor initial placement of the data on memory banks is responsible for large energy consumption and the techniques based on computation restructuring are not able to improve the energy behavior of the application. The idea behind data migration is to bring together the datasets that have originally been placed into different banks but have frequently been accessed together lately. Such datasets (e.g., arrays) are said to exhibit temporal affinity. The expectation is that if two (or more) datasets have been accessed together in the past, they will continue to be accessed together in the near future. By bringing the arrays with temporal affinity together, we increase the chances that fewer number of banks need to be active in a given period of time. In this paper, we focus on array-dominated/loop nest based codes which can be found in domains such as image/video processing and scientific computing. This paper reports experimental results for a single program environment, and does not consider the existence of a virtual memory system (i.e., the data migrator deals directly with physical addresses). There are several important issues that need to be addressed in a data migration scheme for low power. Below we discuss these issues and explain the rationale behind our design and implementation choices.

3.1 Granularity of Migration

The migration granularity might be an entire array or an array region (section). While a fine-granular migration scheme has the potential of localizing the accessed data elements into smaller number of banks better (that is, exploiting a finer-granular temporal affinity), it also entails dividing a given array across banks in a noncontiguous manner. That is, different portions of an array can be stored in different memory banks and sections of different arrays can be interleaved. This, in turn, makes addressing the array in question more difficult (as we work with physical addresses). Consequently, in this work, we chose to migrate entire arrays (when migration is beneficial).

3.2 Detection of Temporal Affinity

To place data elements that are likely to be accessed contemporaneously in the same memory bank, we need to capture which data elements have recently been accessed together. This can be done by sampling the data accesses at runtime (dynamically). Obviously, cycle-by-cycle sampling would affect performance negatively; so, a strategy might be adopting a sampling threshold. More specifically, at regular intervals (sampling threshold), the data accesses are sampled, and the information collected is used to update a table called affinity table. Each entry in the affinity table is a triplet of the form $(U_i, U_j, count_{ij})$ which indicates that arrays U_i and U_j have been accessed together $count_{ij}$ times (up to the current point in the execution). If the current sampling indicates that U_i and U_j have been accessed together (since the last sampling), there are two possibilities. If we do not have any entry (triplet) in the affinity table for this relation between U_i and U_j, we create such an entry, set its count value to 1, and insert it in the affinity table. If, on the other hand, such an entry already exists in the table, we simply increment the corresponding count value. In cases where the sampling indicates that there are $k > 2$ arrays, U_1, U_2, U_3,...,U_k, with temporal affinity, we update all entries $(U_i, U_j, count_{ij})$ where $1 \leq i < j \leq k$ in the table (or create these entries if they are not already there). The size of the affinity table is another issue that needs to be considered. In many data-intensive image/video codes, the number of arrays are limited; this also puts a limit to the table size. Nevertheless, if the size of the table is expected to be large, one might also consider using a small table and managing it as a fully-associative cache. In this case, if, during sampling, we detect a new temporal relation and the table is full, we need to discard a table entry. Possible choices here are to select the entry with the minimum count value or to select the least recently accessed entry.

3.3 Tracing Temporal Affinity

The count values in table entries should be decremented from time to time to adapt the counts to the most recent locality. This is mainly to prevent an old temporal affinity relation from dominating the current (active) affinity relations. For example, imagine two arrays U_i and U_j that have been used together in the past and led to the accumulation of a large $count_{ij}$ value. It might happen that if this count value is never decreased it can still be considered as the dominant count even after a long time this temporal affinity relation has died. One approach would be adopting a threshold, a regular interval, using which the count values are decremented periodically. This, however, involves an extra overhead from the performance viewpoint. A better scheme would be decrementing count values at sampling points. That is, when we increase the count values of the currently accessed arrays (or when we create new entries in the table), we can also decrement the count values of all other (not updated) entries in the table. Such a strategy obviously rewards the most recent temporal affinity relations and punishes the relations that have not been active most recently. In our current implementation, we adopt this strategy.

3.4 Initiating Migration

Another important issue is to decide when the migration should start. Typically, we can select a migration period (migration threshold), a regular interval, at which the affinity table is checked to search for the potential arrays to be migrated. Selecting a suitable migration threshold for a given application and its access pattern is a critical issue. A very small threshold may incur a large overhead, whereas a very large threshold may miss important opportunities for bringing the arrays that have most recently been accessed together. It is not easy to determine a suitable migration threshold without extensive experimentation. Consequently, we experimented with different migration threshold values for each benchmark code we considered. As explained above, the affinity table maintains the temporal affinity relations between array pairs. Independent of how frequently the entries in this table are checked, we need to have a strategy for determining the arrays to be migrated. This is explained next.

3.5 Determining the Arrays to be Migrated

Given an affinity table, we can use different strategies to select the arrays to be migrated. Different strategies have different overhead and performance tradeoffs. In this study, we mainly consider a class of strategies called the first m triplet (mFT) strategies. A 1FT strategy looks at (when the migration period arrives) only the triplet with the highest count value and tries to bring the involved arrays together as much as possible. Similarly, a 2FT strategy looks at the two triplets with the two highest count values, and so on. As we increase the number of triplets to consider, we can expect that it is going to take longer time to decide an appropriate migration scheme (i.e., how the arrays should actually be migrated) as more triplets will usually involve more arrays. A straightforward implementation of an mFT strategy just determines the entries with the highest m count values, picks up the arrays names, and tries to implement a migration strategy as explained below such that the arrays whose

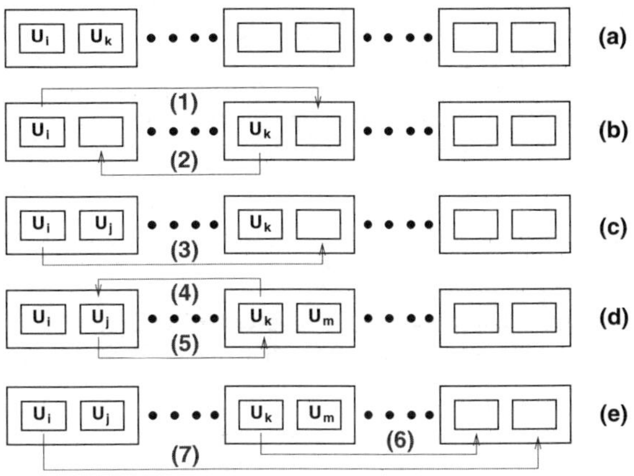

Figure 2: Five different migration scenarios that involve two arrays (U_i and U_k). For clarity, only three banks are explicitly shown.

names appear in each selected triplet are brought together in a single bank (or in the minimum number of banks). In case of conflicts (i.e., different triplets demand different migrations for the same array), we can give priority to the triplet with the higher count value. An alternative class of strategy, which we call the limit-n (nLM) strategy considers a triplet if and only if it's count value is larger than or equal to n. Note that it is also possible to adopt a hybrid strategy by combining the mFT and nLM strategies. For example, a conservative strategy can consider a triplet (that is, the arrays in the triplet for migration) if and only if its count value is larger than n and this value is among the highest m count values.

3.6 Scheduling Movements

Once the arrays to be migrated are selected, a strategy should be decided upon using which the array transfers (migrations) between memory banks are scheduled. The complexity of the schedule depends on the number of arrays involved in the migration. Let us first give a description of the scheduling algorithm when only a single triplet (i.e., two arrays) is involved. Later, we explain how this algorithm can be extended to work when more than two arrays are involved. Figure 2 illustrates five different scenarios assuming that we would like to honor the triplet $(U_i, U_k, count_{ik})$; that is, we want to bring arrays U_i and U_k together. If these two arrays already reside in the same bank as shown in Figure 2(a), nothing needs to be done. In Figure 2(b), on the other hand, these two arrays are in different banks and there is sufficient space in each bank for the other array. In this case, we have two options. We can either migrate U_i to the bank that holds U_k, or vice versa. These two actions are marked using (1) and (2) in Figure 2(b). Figure 2(c) depicts a scenario similar to that in Figure 2(b) except that the bank that holds U_i does not have sufficient space for U_k. Consequently, we have only one option: migrating U_i to the bank that holds U_k. This action is marked (3) in the figure. In Figure 2(d), both the banks that hold the arrays are full. In this case, we need to find an array (or arrays) in the bank that holds U_i and interchange the locations (banks) of this array and U_k; or alternately, we need to find an array (or arrays) in the bank that holds U_k and interchange the locations of this array and U_i. Figure 2(d) depicts the former case and the necessary migration activities are marked using (4) and (5). In cases where we may not be able to find a suitable array to interchange with U_i or U_k (as, for example, such an interchange will distort another important temporal affinity relation), we might want to consider to migrate U_i and U_k, together, to another memory bank as depicted in Figure 2(e). In the figure, these migration activities are marked using (6) and (7). Obviously, it is a good idea to select a bank that already contains some arrays instead of a new bank which would have otherwise been completely disabled (consuming no power). Given two arrays that need to be brought together, our current implementation checks the feasibility of the options in Figure 2 in the order given, and selects the first option that applies to the case at hand. Note that, in general, our strategy for migration tries to minimize the number of migrations. In cases where we have multiple alternatives (to achieve the same result) as in Figure 2(b), we take into account the sizes of the arrays as well; that is, everything else being equal, it is preferable to migrate the smaller array to the bank that holds the larger one.

When more than two arrays are involved in the migration (e.g., as a result of taking into account more than one table entry), scheduling migrations becomes a more challenging problem. Although not shown here due to lack of space, it can be proved that optimal migration problem that involves multiple table entries is NP-hard. Therefore, a polynomial-time solution is unlikely to exist. Instead, we propose here a heuristic solution that works well in practice. In the first step of our heuristic, we (logically) cluster all the involved arrays into groups. Each group contains the arrays that are expected to be in the same (set of) bank(s) for the best memory energy behavior. As an example, consider a scenario where we consider three triplets simultaneously: (U_1,U_2,$count_{12}$), (U_3,U_4,$count_{34}$), and (U_2,U_5,$count_{25}$). Here, the first group contains arrays U_1, U_2, and U_5 whereas the second group contains arrays U_3 and U_4. In the second step, we rank the groups according to their relative importance. This ranking helps us later in resolving the conflicts if different groups demand different (conflicting) migrations. To rank the groups, we use the concept of group weight (denoted gv_k for a given group k) which is simply the sum of the counts of the triplets that supply the arrays in the group. In our current example, gv_1 is $count_{12}$ + $count_{25}$ and gv_2 is $count_{34}$. In the third step, we start with the the group with the highest weight and try to co-locate the arrays in this group as much as possible using minimum number of migrations. Our current implementation achieves this starting with the pair (of arrays) with the highest temporal affinity and tries to bring these arrays together. It then moves to the second pair (of highest affinity) and tries to satisfy this relation, and so on. Note that a given array in a group can be involved in multiple affinity relations. The approach then moves to the second group (i.e., the one with the second highest group weight), and in migrating it, it tries not to re-locate any of the arrays which have been involved in the previous migration (i.e., the one that involves the arrays belonging to the group with the highest weight). The algorithm continues in this fashion until all the groups have been processed.

3.7 Backup Policies

We need backup policies for at least two reasons. First, it might happen that (as a result of our array selection strategy such as nLM) we may not be able to find a triplet in the affinity table to use for migration. In this case, a backup strategy could be just not performing migration at this point (a low-overhead solution). Another backup strategy could be relaxing the strategy that we use to select the triplet and re-probing the affinity table. Although this latter alternative will have a larger overhead (as compared to the former), we can expect that it will perform better. Second, we need a backup policy when the arrays that we want to co-locate already reside in the same bank(s). As in the previous case, we can adopt here at least two alternatives. We can either continue with the execution and drop out migration at this point, or we can re-probe the affinity table to select alternative arrays for migration. Note, however, that in this last case, the migration decisions to be taken should not cause the arrays that have been determined (during the first probe, before re-probing) to be co-located but have been found to be residing on the same bank(s) to be separated.

3.8 Data Freezing and Defrosting

Arrays that exhibit different access patterns in different parts of the computation may be repeatedly migrated from one memory bank to another. Such frequent data migrations can easily offset potential energy gains from optimal data placements in banks. To avoid these costly frequent migrations, it is possible to freeze these arrays in specific banks (also called pinning in this paper) after a certain number of migrations. These frozen arrays can be defrosted after some time so that they become eligible again for migration. To decide whether an array should be pinned or not, we associate a migration counter with each array (initialized to zero). Each time an array is migrated this counter is incremented by one. When it reaches a certain value (migration limit), we can freeze the array on the bank where it currently resides. The migration counters can be decremented at regular intervals (defrosting threshold). To reduce overhead, this interval might be a multiple of the sampling threshold.

3.9 Overheads

There are two major sources of overhead in our migration scheme. The first is that the overall time spent by the program in the migration-related activities. These activities include sampling for temporal affinity, determining the arrays to be migrated, the migration overhead (time) itself, and updating counters (e.g., those keep track of the number of migrations, etc.). This cost is highly implementation dependent and can be made a small proportion of overall execution time in a careful implementation. Second, data migration consumes some amount of energy during copying data from one bank to another. Again, this cost can be decreased by reducing the number of arrays to be migrated, by using fast memory-to-memory copies (e.g., block copies), and by not migrating arrays unless it is really beneficial to do so.

4. DATA MIGRATOR (DM)

The ideas presented in the previous section have been implemented in a software package called Data Migrator (DM). The DM is designed as a runtime library which contains a set of functions and runs as a single thread of control when activated. The programmer initializes the DM by supplying the values for a set of parameters (to a *DM_initialize* routine). After that, the user activates the library by calling a special *DM_activate* routine. From this point on, the DM runs (as a separate thread) parallel with the application. It interrupts the application at regular intervals which are determined by the parameters supplied in the *DM_initialize* routine (more specifically, by the sampling threshold).

For ease of implementation, all thresholds are set to multiples of the sampling threshold which is used to interrupt the execution of the application. At each interrupt, the DM code is executed. A sketch of this code is given in Figure 3. Once activated, the DM thread checks for the occurrence of three events (using the three outermost 'if' statements shown in Figure 3). The first statement is always true as the DM thread is invoked when sampling threshold arrives. The thread then visits the affinity table and updates it (i.e., increments the count values for the triplets whose arrays pairs are accessed and decrements the count values for all other triplets). After that, the DM thread checks whether the migration threshold has arrived. If it has, the thread runs the migration policy (e.g., mFT, nLM, or a hybrid policy). As a result of this policy, we might have arrays that need to be migrated. If such arrays exist, we migrate them as explained in the previous section; and if not, we run the backup policy (and migrate arrays if required). In any case, we need to update the migration counter and freeze the array(s) if the counter overflows. Finally, we need to check whether the defrost threshold has arrived, and if so, make the frozen arrays migratable.

5. EXPERIMENTS

This section explores the overall energy improvement due to array migration and quantifies its energy/performance overheads. Figure 4 gives important characteristics of the array-dominated codes

```
if (sampling threshold arrives) then
 update affinity table;
if (migration threshold arrives) then
 {
  run migration policy;
  if (there are arrays to be migrated) then
   migrate arrays;
  else run backup policy;
  for each array involved in migration do
   update migration counter;
   if (migration counter overflows) then
    freeze the array;
 }
if (defrost threshold arrives) then
 initialize migration counters for all frozen arrays;
```

Figure 3: The DM thread.

Benchmark	Source	Input Size (MB)	Base (mJ)	Overhead	ST	MT
adi	Livermore	19.2	1,492	2.5%	1	100
apsi	Perfect Club	42.4	499	16.2%	1	100
btrix	Spec	48.8	719	7.6%	1	100
eflux	Perfect Club	33.5	98,501	4.8%	1	500
phods	-	32.9	37,785	2.8%	1	1,000
tomcatv	Spec	28.0	19,850	7.7%	1	100
vpenta	Spec	44.0	13,714	6.8%	1	1,000

Figure 4: Benchmark codes used in the experiments and their important characteristics.

that we used in our experiments. We measured the benefits of dynamic data migration using a set of seven array-dominated codes. phods is a motion estimation code whereas other codes are from array-based benchmarks. The third column gives the total input size (in MBs), and the fourth column gives the memory system energy consumption (in millijoules) for the base case (see below) using four 16MB memory banks (i.e., a total of 64 MB off-chip memory). The memory system energy includes the energy consumed in DRAM banks as well as the energy spent in 16KB, 2-way set-associative data cache with a line size of 32 bytes. The fifth column reports the percentage increase in the execution time of applications due to migration (i.e., performance overhead). We see that except for one case, the performance overhead of array migration is bounded by 8%. Finally, the last two columns give the sampling thresholds (denoted ST) and migration thresholds (denoted MT), respectively, used in majority of our experiments (in units of thousand cycles). In all experiments, we set the back-up policy to do nothing as we have found that this policy generated comparable results to more aggressive second choice-based policies without incurring as much overhead. Also, all the energy figures reported here include the energy overhead of data migration. We also used a powerful back-end compiler which performs all major classical low-level optimizations such as instruction scheduling and global register allocation [9].

5.1 Energy Benefits

Figure 5 presents the normalized energy consumptions when dynamic data migration (the 1FT policy) is employed with respect to a base case which does not use migration but tries to determine the best possible static data layout (array mapping) using the technique discussed in [6]. To evaluate the energy impact of array migration, we experimented with two initial data mappings. In the first mapping, which we call sequential mapping, the arrays are stored in memory banks according to their declaration order in the code, the first array starting from the first location in the first bank, the second array starting from the location next to the end of the first array, and so on. In the second mapping, called random mapping, the arrays are assigned to memory locations randomly, making sure that they do not overlap and all elements of a given array are stored consecutively. Experimenting with two initial mappings allows us

Benchmark	Sequential Mapping		Random Mapping	
	8×8MB	4×16MB	8×8MB	4×16MB
adi	92.5	96.0	92.5	96.0
apsi	68.4	61.4	68.2	61.1
btrix	105.0	100.0	105.0	100.0
eflux	92.8	94.3	90.5	91.8
phods	72.7	61.6	70.0	60.3
tomcatv	66.2	60.4	66.0	61.1
vpenta	93.9	83.7	92.8	81.7

Figure 5: Energy consumptions with the mFT policy normalized with respect to the base case.

Benchmark	Migration Threshold		Sampling Threshold	
	MT1	MT2	ST1	ST2
adi	200	400	25	50
apsi	200	400	25	50
btrix	200	400	25	50
eflux	1,000	2,000	100	250
phods	2,000	4,000	250	500
tomcatv	200	400	25	50
vpenta	2,000	4,000	250	500

Figure 6: Migration and sampling thresholds (in units of thousand cycles).

to evaluate the robustness of our data migrator. The base case strategy, on the other hand, statically (at compile time) determines the best array mapping (which is valid throughout the execution of the code); therefore, the initial data mapping does not matter (i.e., it is not taken into account).

In our experiments, we used two memory bank configurations: (i) 8 banks, each 8 MB (denoted 8×8MB), and (ii) 4 banks, each 16 MB (denoted 4×16MB). Note that in both the configurations the total memory capacity is 64 MB. The results in Figure 5 show that array migration reduces the energy consumption by approximately 19%, on the average, over the base case. In only one benchmark, dynamic migration increases the energy consumption (as it increases the number of active banks as compared to the base case). All of these results have been obtained using the sampling and migration thresholds given in Figure 4. We only report the results with the 1FT policy as increasing the number of triplets considered did not make a significant difference for our benchmark codes. More specifically, only in two codes (eflux and vpenta), the 2FT policy brought a small benefit over the 1FT policy (less than 2.5%). It should also be mentioned that the number of times that the affinity table is probed for migration ranged (in our codes) between 226 (in apsi) and 10,468 (in tomcatv).

5.2 Sensitivity to the Sampling and Migration Thresholds

To measure the sensitivity of the energy benefits presented above to the sampling and migration thresholds used, we performed another set of experiments (with the 4×16MB configuration and sequential initial layout), where we modified (increased) the sampling and migration thresholds. The new sampling thresholds (ST1 and ST2) and migration thresholds (MT1 and MT2) used are given in Figure 6 for each benchmark in our suite.

The second and third columns in Figure 7 report the (percentage) variation (increase) in energy benefits (with respect to our base sampling thresholds) when the sampling threshold is ST1 and ST2, respectively, while keeping the migration threshold constant at the base value. The fourth and fifth columns give the variation in energy benefits (with respect to our default migration threshold) when we set the migration threshold to MT1 and MT2, respectively (the sampling threshold is kept at its base value). We observe from these results that varying the sampling threshold does not make too much difference whereas migration threshold variations generally generate larger variations. In other words, even larger sampling thresholds allow DM to capture the temporal affinities between arrays. And, working with a large migration thresholds seems to perform

Benchmark	Sampling Threshold		Migration Threshold	
	ST1	ST2	MT1	MT2
adi	-0.5%	-0.6%	0.0%	0.0%
apsi	-1.0%	-1.0%	18.6%	18.8%
btrix	-2.3%	-2.6%	0.0%	0.0%
eflux	0.0%	0.0%	-8.3%	-8.3%
phods	-0.1%	-0.2%	18.5%	19.0%
tomcatv	-2.4%	-2.8%	19.6%	19.6%
vpenta	-0.8%	-0.8%	7.0%	7.0%

Figure 7: Energy sensitivity due to sampling and migration thresholds (4×16MB and sequential initial layout). A positive value (negative value) indicates a reduction (increase) in energy with respect to the original threshold.

'n' value	phods		tomcatv	
	8 × 8MB	4 × 16MB	8 × 8MB	4 × 16MB
5,000	93.1	88.1	89.4	90.2
10,000	90.7	73.3	76.2	78.6
20,000	82.0	63.2	66.7	**50.0**
40,000	**63.4**	**51.9**	64.0	72.1
80,000	66.1	68.8	**58.3**	83.2
160,000	77.4	76.0	74.7	91.6
320,000	91.2	93.5	89.5	97.4

Figure 8: Energy consumptions with the nLM policy normalized with respect to the base case.

better except for one benchmark (eflux). This is because, in general, a larger migration threshold better distinguishes between false temporal affinities (i.e., affinities with short duration) and real, beneficial temporal affinities.

5.3 Experiments with the nLM Policies

We also conducted experiments with the nLM migration policy. Out of the seven codes we experimented, only two codes (phods and tomcatv) benefited from the nLM policy more than the mFT policy for some 'n' values. The results are given in Figure 8 (the first column gives the 'n' values in 'nLM'). We observe from these results that there exist some 'n' values that generate better results than those obtained through the 1FT policy. For the phods benchmark, this value is 40,000 for both the memory configurations (8×8MB and 4×16MB). For tomcatv, on the other hand, this value is 80,000 for the 8×8MB bank configuration and 20,000 for the 4×16MB bank configuration. Among the values that we experimented with (shown in the first column), these are the optimal ones and are better than the corresponding mFT value. A larger 'n' value delays migration (and sometimes misses the migration opportunity) whereas a smaller 'n' value migrates data too frequently and leads to data freeze (and the associated defrosting cost).

5.4 Energy Overhead of Migration

The cost of energy overhead depends on whether the underlying architecture supports block copy operations that can copy a number of array elements from one portion of the memory to another (in a single instruction). Figure 9 shows the energy overhead (as percentage of the overall memory energy consumption) incurred when different block copy operations are used. Each block copy operation can be defined by the number of array elements that it can copy in a single instruction execution; the larger this number is, the better, as the time (hence, energy) spent during copy is reduced. The x-axis in the figure gives nine alternate block copies (for a given K parameter, the Kth alternative can copy $(64 * K)$ bytes/10 cycles). In the experiments above, we set K to 1. We see from Figure 9 that except for one case, the energy overhead is always below 5%, even if we do not use a very aggressive copy operation. Therefore, for our array-intensive benchmark codes, the runtime energy and

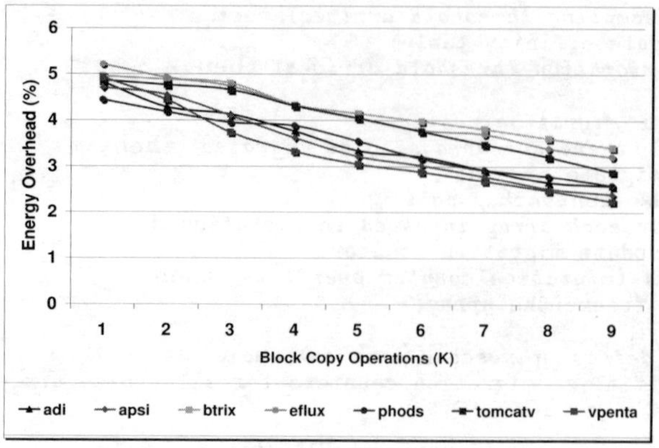

Figure 9: % energy overhead for data migration.

performance benefits due to data migration clearly overwhelms the potential overheads.

6. SUMMARY AND FUTURE WORK

A key design methodology for low-power design is aggressive hardware and software optimizations [2, 1]. Optimizing memory system energy can be crucial in many embedded image and video processing applications. This paper presented an automatic data migration strategy which co-locates arrays with temporal affinity in a small set of memory banks, thereby reducing the energy consumption. Our initial experience with the Data Migrator tool that implements data migration for low power is very promising. The issue of how far the memory system energy can be reduced is interesting to address. Along this direction, we would like to investigate different migration policies and combine these policies with compiler [6] and operating system [7] based techniques to reach a unified solution to the problem.

7. REFERENCES

[1] L. Benini and G. De Micheli. System-level power optimization: techniques and tools. *ACM Transactions on Design Automation of Electronic Systems,* 5(2), pp.115-192, April 2000.

[2] F. Catthoor, S. Wuytack, E. D. Greef, F. Balasa, L. Nachtergaele, and A. Vandecappelle. *Custom Memory Management Methodology – Exploration of Memory Organization for Embedded Multimedia System Design,* Kluwer Academic Publishers, June, 1998.

[3] R. Chandra, S. Devine, B. Verghese, A. Gupta, and M. Rosenblum. Scheduling and page migration for multiprocessor compute servers. In Proc. *the 6th International Conference on Architectural Support for Programming Languages and Operating Systems,* October 1994.

[4] A. Chandrakasan, W. J. Bowhill, and F. Fox. *Design of High-Performance Microprocessor Circuits.* IEEE Press, 2001.

[5] T. M. Chilimbi and J. R. Larus. Using generational garbage collection to implement cache-conscious data placement. In Proc. *the International Symposium on Memory Management,* October 1998.

[6] V. Delaluz, M. Kandemir, N. Vijaykrishnan, A. Sivasubramaniam, and M. J. Irwin. DRAM energy management using software and hardware directed power mode control. In Proc. *the 7th International Conference on High Performance Computer Architecture,* Monterrey, Mexico, January 2001.

[7] A. R. Lebeck, X. Fan, H. Zeng, and C. S. Ellis. Power aware page allocation. In Proc. *the 9th International Conference on Architectural Support for Programming Languages and Operating Systems,* November 2000.

[8] J. R. Lorch and A. J. Smith. Software strategies for portable computer energy management. *IEEE Personal Communications,* pp. 60–73, June 1998.

[9] S. S. Muchnick. *Advanced Compiler Design Implementation.* Morgan Kaufmann Publishers, San Francisco, California, 1997.

[10] 128/144-MBit Direct RDRAM Data Sheet, Rambus Inc., May 1999.

Exploiting Shared Scratch Pad Memory Space in Embedded Multiprocessor Systems

Mahmut Kandemir
Microsystems Design Lab
Pennsylvania State University
University Park, PA 16802
kandemir@cse.psu.edu

J. Ramanujam
Dept. Elec. & Comp. Engr.
Louisiana State University
Baton Rouge, LA 70803
jxr@ece.lsu.edu

A. Choudhary
Dept. Elec. & Comp. Engr.
Northwestern University
Evanston, IL 61801
choudhar@ece.nwu.edu

ABSTRACT

In this paper, we present a compiler strategy to optimize data accesses in regular array-intensive applications running on embedded multiprocessor environments. Specifically, we propose an optimization algorithm that targets the reduction of extra off-chip memory accesses caused by inter-processor communication. This is achieved by increasing the application-wide reuse of data that resides in the scratch-pad memories of processors. Our experimental results obtained on four array-intensive image processing applications indicate that exploiting inter-processor data sharing can reduce the energy-delay product by as much as 33.8% (and 24.3% on average) on a four-processor embedded system. The results also show that the proposed strategy is robust in the sense that it gives consistently good results over a wide range of several architectural parameters.

Categories and Subject Descriptors B.3 [**Hardware**] Memory Structures; D.3.4 [**Software**] Programming Languages: Processors [Compilers]

Terms Algorithms, management, performance.

Keywords Embedded multiprocessors, energy consumption, scratch pad memories, access patterns, compiler optimizations, data tiles.

1. INTRODUCTION

The mobile/embedded computing device market is projected to grow to 16.8 million units in 2004, representing an average annual growth rate of 28% over the five year forecast period [10]. This brings up the issue of optimizations in application design, system software and circuit/architectural levels for resource-constrained devices, where battery energy and limited storage capabilities are brought into spotlight. While there have been significant strides made in optimizing energy and performance using circuit and architectural level optimizations, not many studies have looked at the problem of software level optimizations for mobile/embedded systems. Consequently, any effort in this direction will help us to realize the goal of achieving high performance and low energy consumption in mobile/embedded devices.

As microprocessors grow more and more powerful, designers are building larger and ever more sophisticated systems to solve complex problems. In particular, in the area of digital signal and video processing, specialized microprocessors are getting more powerful, making them attractive for a wide range of embedded applications. As a result of this, embedded system designers are now using multiple processors in a single system (either in the form of SoC or in the form of a multiprocessor board) to address the computational requirements of a wide variety of applications. In future, we can expect that more and more embedded systems will be based on multiprocessor architectures.

Many applications from image and video processing domains have significant data storage requirements in addition to their pressing computational requirements. Previous studies [5, 13] show that high off-chip memory latencies and energy consumptions are likely to be the limiting factor for future embedded systems. Therefore, compiler-based techniques for optimizing memory access patterns are extremely important. An optimizing compiler can be particularly useful in embedded image and video processing applications as many of these applications exhibit regular (compile time analyzable and optimizable) data access patterns. Previous work [2, 9] shows that regular data access patterns found in array-dominated applications can be better captured if we employ a scratch pad memory (a software-managed SRAM), instead of a more traditional data cache.

In this paper, we present a compiler-based strategy for optimizing energy consumption and memory access latency of array dominated applications in a multiprocessor based embedded system. Specifically, we show how a compiler can increase data sharing opportunities when multiple processors (each one is equipped with a scratch pad memory) operate on a set of arrays in parallel. *This is in contrast with previous work on scratch pad memories that exclusively focused on single processor architectures.* In this paper, we make the following major contributions:

- We show that inter-processor communication requirements in a multiprocessor embedded system can lead to extra off-chip memory requests and present a compiler strategy that eliminates these extra off-chip memory requests. Our strategy is implemented using an experimental compiler infrastructure [1] and targets array-dominated embedded applications.
- We report performance and energy consumption data showing the effectiveness of our optimization strategy. The results show that exploiting inter-processor data sharing can reduce energy-delay product by as much as 33.8% (and 24.3% on average) on a four-processor embedded system. The results also show that the proposed strategy is robust in the sense that it gives consistently good results when several architectural parameters are varied over a wide range.

The rest of this paper is organized as follows. In the next section, we discuss our multiprocessor architecture and execution model. In Section 3, we present details of our optimization strategy. In Section 4, we introduce our experimental setup and report experi-

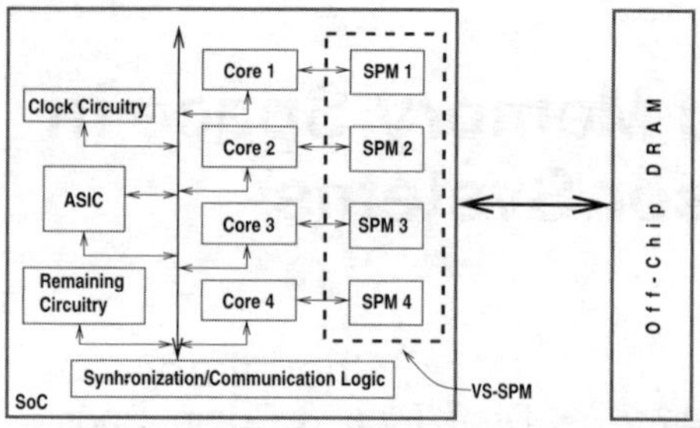

Figure 1: A VS-SPM based system architecture.

mental results. In Section 5, we offer our conclusions and give an outline of the planned future work on this topic.

2. ARCHITECTURE, EXECUTION MODEL

A scratch pad memory (SPM) is a fast SRAM that is managed by software (the application and/or compiler) [11, 9, 2, 14]. It can be used for optimizing both data and instruction accesses. In applications that exhibit regular data access patterns (which are compile-time analyzable and optimizable), an SPM can outperform a conventional data cache memory as software can lead to a better flow of data to/from SPM as compared to the conventional cache, whose management of the flow of data is at a fine-grain level (cache line granularity), controlled by hardware, and mostly independent of the application [9].

A *virtually shared scratch pad memory* (VS-SPM), on the other hand, is a shared SRAM space made up by individual SPMs of multiple processors. In this paper, we focus on a multiprocessor on-a-chip architecture, as shown in Figure 1. In this architecture, we have a system-on-a-chip (SoC) and an off-chip DRAM that can hold data as well as instructions. The SoC holds multiple processor cores (with their local SPMs), inter-processor communication/synchronization mechanism, clock circuitry, and some ASIC. The SPMs of individual processors make up a VS-SPM. Each processor has fast access to its own SPM as well as to SPMs of other processors. With respect to a specific processor, the SPMs of other processors are referred to as remote SPMs. Accessing remote SPMs is possible using fast on-chip communication links between processors. Accessing off-chip DRAM, however, is very costly in terms of both latency and energy. Since per access energy and latency of VS-SPM are much lower than the corresponding values of DRAM, it is important to make sure that as many data requests (made by processors) as possible are satisfied from the VS-SPM. Obviously, this can be achieved by maximizing the reuse of data in the VS-SPM. While an application programmer can achieve this by carefully chorographing data flow to/from VS-SPM, it is clear that this would be overwhelming for her/him. Instead, we show in this paper that automatic compiler support can achieve very good results by optimizing the reuse of data in the VS-SPM. It should be noted that each processor in the SoC can also contain an instruction cache and/or a loop cache. Their management, though, is orthogonal to the VS-SPM management.

An example system that is similar to the architecture considered in this paper is the VME/C6420 from Blue Wave Systems [3]. This system uses a 9-port crossbar on the board. Four of these ports connect to identical processors with their local SRAM modules, four to high-bandwidth I/O connections, and the ninth to an off-board memory. The ports on the crossbar enable high data movement rates (typically, 264 Mbytes/sec). Using crossbar links increases the net bandwidth provided by the system. Consequently, accessing local SRAM module and remote SRAM modules are much faster and much less energy consuming than accessing the off-chip DRAM. The local SRAM modules in Blue Wave correspond to local SPMs in our system; consequently, the compilation techniques for both the systems should be very similar (despite the fact that one of these is board-based and the other is SoC based).

The execution model in a VS-SPM based architecture is as follows. The system takes as input a *loop-level* parallelized application. In this model, each loop nest is parallelized as much as possible. In processing a parallel loop, all processors in the system participate computation and each executes a subset of loop iterations. When the execution of a parallel loop has completed, the processors synchronize using a special construct called *barrier* before starting the next loop. The synchronization and communication between processors is maintained using fast on-chip communication links. Based on the parallelization strategy, each processor works on a portion of each array in the code. Since its local SPM space is typically much smaller than the portion of the array it is currently operating on, it divides its portion into chunks (called *data tiles*) and operates on one chunk at a time. When a data tile has been used, it is either discarded or written back into off-chip memory (if modified).

In order to improve the reuse of data in the VS-SPM, one can consider *intra-processor data reuse* and *inter-processor data reuse.* Intra-processor data reuse corresponds to optimizing data reuse when considering the access pattern of each processor in isolation. Previous work presented in [11] and [9] addresses this problem. It should be noted, however, that exploiting intra-processor reuse only may not be very effective in a VS-SPM based environment. This is because intra-processor data reuse has a local (*processor-centric*) perspective and does not take inter-processor data sharing effects into account. Such effects are particularly important in applications where data regions touched by different processors overlap. This is very common in many array-intensive image processing applications. Inter-processor data reuse, on the other hand, focuses on the problem of optimizing data accesses considering access patterns of all processors in the system. In other words, it has an *application-centric* view of data accesses.

In order to see the difference between application-centric and processor-centric views, let us consider the code fragment in Figure 2 which performs Jacobi iteration over two $N \times N$ square arrays U_1 and U_2. In this code fragment, $f(.)$ is a linear function and $parfor$ indicates a parallel for-loop whose iterations are to be distributed (evenly) across processors available in the system. Note that since this loop does not have any data dependences, both the for-loops are parallel. Since our approach works on an already parallelized program, we do not concern ourselves with the question of how the code has been parallelized (i.e., by application programmer or by an optimizing compiler). Assuming that we have four processors (P_1, P_2, P_3, and P_4) as shown in Figure 1, the portion of array U_2 accessed by each processor is shown in Figure 3(b). Each processor is responsible from updating an $(N/2) \times (N/2)$ sub-array of U_2. The portions accessed by processor 1 from array U_1 are shown in Figure 3(a). This portion is very similar (in shape) to its portion from array U_2 except that it also includes some elements *shared* (accessed) by other processors. In the remainder of this paper, such elements are called *non-local elements* (or border elements). Assuming that the arrays U_1 and U_2 initially reside in off-chip DRAM, each processor brings a data tile of its U_1 sub-array and a data tile of its U_2 sub-array from DRAM to VS-SPM, updates the corresponding elements, and stores the U_2 data tile back in the DRAM. Figure 3(c) shows seven data tiles (from U_2) that belong to processor 4. This processor accesses these tiles starting from tile 1 and ending with tile 7.

Let us first see how each processor can make effective use of its local SPM space (processor-centric optimization). We focus on processor 1, but our discussion applies to any processor. Processor 1 first brings a data tile from U_2 and a data tile from U_1 from off-chip memory to its SPM. After computing the new values for its U_2 data tile, it stores this data tile back in off-chip memory and

```
parfor (i=2; i <= N-1; i++)
 parfor (j=2; j <= N-1; j++)
  U2[i][j] += f(U1[i][j-1]+U1[i-1][j]+U1[i][j+1]+U1[i+1][j])
```

Figure 2: Jacobi iteration.

proceeds by bringing new data tiles from U_1 and U_2. However, to exploit reuse, it keeps the last row of the previous U_1 data tile in SPM. This is because this row is also needed when computing the elements of the new U_2 tile. This optimization is referred to as local buffering or processor-centric data reuse.

While such a tiling strategy makes effective use of the SPM space as far as intra-processor reuse is concerned, it fails to capture inter-processor data reuse. Let us consider how our four processors execute this nest in parallel. Figure 3(d) illustrates the scenario where four processors are working on their first data tiles from array U_1 (assuming row-block data tiles). Let us focus on processor 3; similar discussion applies to other processors as well. This processor, while working on its first data tile, needs data from processors 1, 2, and 4. More specifically, it needs an entire row from processor 1 (that is, the last row of processor 1's last tile), a single element from processor 2, and two elements from processor 4. These non-local elements are also shown in Figure 3(d). It should be noted that processor 4 can supply these elements immediately from its local SPM. This is because these two array elements are part of the data tile it is currently working on. However, processors 1 and 2 need to perform off-chip memory accesses for the data required by processor 3 as these data are not currently in their local SPMs. Obviously, these *extra off-chip memory requests* (that is, memory requests performed due to inter-processor communication requirements only) will be very costly. A similar scenario occurs when we consider a different data tile shape. Figure 3(e) shows processor access patterns (to array U_1) when a square data tile is used. It also shows the non-local elements required by processor 4 when it is operating on its first data tile. We can easily see that none of these elements are in any SPM (as other processors are also working on their first data tiles). Consequently, in order for processor 4 to complete its computation on its first tile, other processors need to perform extra off-chip memory accesses. It should also be noted that it is not a good idea to try to bring the non-local elements to the local SPM and keep them there until they are requested by other processors. This is because such a strategy would lead to keeping data in the SPM without much reuse and decrease overall SPM space utilization.

However, since the code fragment in Figure 2 does not exhibit any data dependence, the processors do not have to stick to the same tile processing order; that is, they do not have to process their data tiles in the same order. In particular, they can bring (and process) their data tiles in such a way that whenever one processor needs a non-local array element (to perform some computation), the needed data can be found in some remote SPM. If this can be achieved for all non-local accesses, we can eliminate all the extra off-chip memory accesses due to inter-processor communication. Figures 3(f) and (g) show how extra off-chip memory requests can be eliminated when row-block and square data tiles are used, respectively. Note that in these scenarios each processor has a different tile access pattern. The following section present an automatic compiler-directed tile processing strategy to achieve this and Section 4 measures potential benefits of doing so.

3. COMPILER SUPPORT

There are at least two sub-problems in compiling array-dominated applications for a VS-SPM based environment:

- *Data Tile Shape/Size Selection:* The first step in compilation is to determine the shape and sizes of data tiles. The important parameters in this process are the available SPM space and data access pattern of the application. While this problem is important, it is beyond the scope of this paper. In this paper, we assume rectilinear tile shapes and that all processors have the same SPM capacity and operate with identical data tiles (whose size is determined by the local SPM size).
- *Tile Access Pattern Detection:* In this step, which we also call *scheduling,* given a data tile shape/size, we want to determine a data tile access pattern (for all processors) such that extra off-chip memory accesses (due to inter-processor communication) are eliminated. In the rest of this section, we present a compiler technique that addresses this problem.

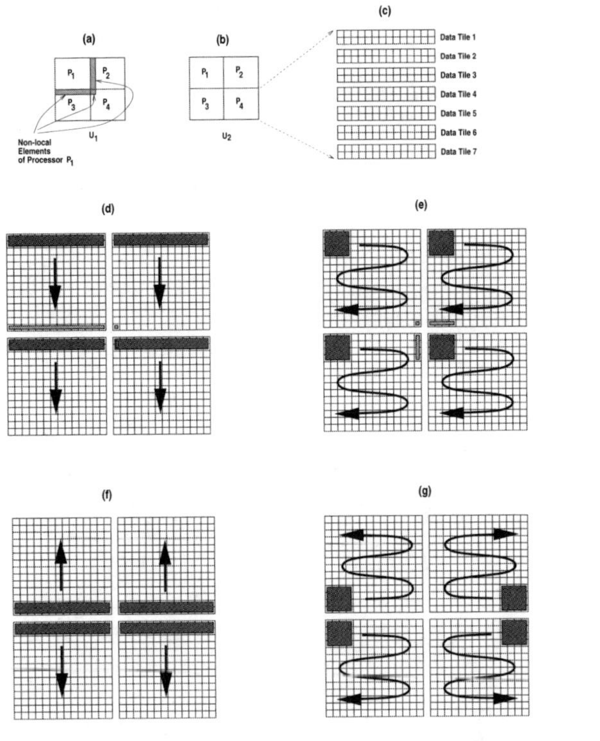

Figure 3: (a-b) Local array portions of processors for the arrays accessed in Figure 2. (c) Data tiles to be processed by a processor. (d-g) Different tile access patterns.

The degree of freedom [4] of a given data tile indicates the its movement capability on data space. For instance, the degree of freedom for the data tile given in Figure 3(d) is one as it can move only in one direction (in vertical direction), whereas the degree of freedom for the data tile shown in Figure 3(e) is two as it can move in both horizontal and vertical directions.

We define a *tile access pattern matrix* (*scheduling matrix*), denoted $\mathcal{H}$, which determines the order in which the data tiles are accessed. The dimensions of a scheduling matrix are decided based on the degree of freedom. Let us focus on two-dimensional arrays; our results extend to higher-dimensional arrays as well. In the two-dimensional case, if the degree of freedom is one, the scheduling matrix is 2×1; if the degree of freedom is two, the scheduling matrix is 2×2. For nested loops that do not contain flow-dependences, the tile access patterns and their corresponding scheduling matrices for a degree of freedom of 1 and 2 are given in Figures 4(a) and (b), respectively. Each column of H corresponds to an axis and the value of the vector in a column denotes the direction of the access along the corresponding axis. In addition to the elements of each column, the ordering of the columns of H is extremely important. The first column denotesAs an example,

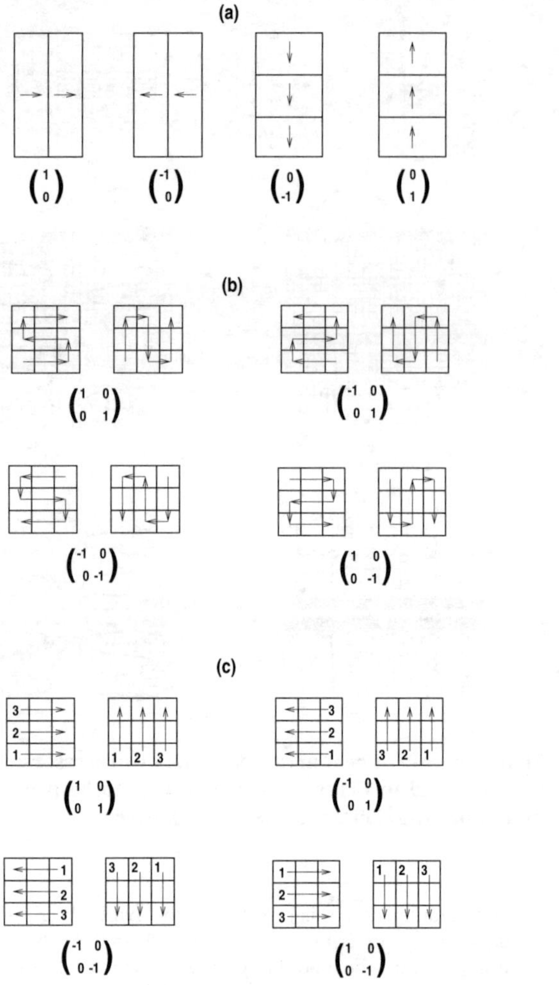

Figure 4: Tile access patterns and scheduling matrices for two-dimensional arrays accessed in a nested loop.

the matrix $\mathcal{H} = \begin{pmatrix} 1 & 0 \\ 0 & -1 \end{pmatrix}$ corresponds to positive direction of movement (at the beginning of the pattern) in the x-axis and negative direction of movement in the y-axis.

We also define *direction vectors* for each processor with respect to its neighbors. For example, the direction vector of processor 1 with respect to processor 4 (see Figure 3(a)) is: $\vec{v}_{1,4} = \begin{pmatrix} 1 \\ -1 \end{pmatrix}$. The direction vector of processor i with respect to processor j is denoted $\vec{v}_{i,j}$. Essentially, this is a vector whose elements indicate the signs (positive denoted by $+1$, negative denoted by -1 and zero by 0) of the vector distance from the co-ordinates of processor i to the co-ordinates of processor j.

The objective of our tile access pattern strategy should be minimizing the extra off-chip memory accesses. This can be achieved by making sure that whenever a processor needs a non-local array element, some other processor can supply it from its SPM. To achieve this, our scheduling strategy uses the following result (proof omitted for lack of space):

Lemma 1. Consider two processors i and j with their scheduling matrices $\mathcal{H}_i$ and $\mathcal{H}_j$, respectively. Schedules denoted by these matrices eliminate extra off-chip memory accesses (when considering only these two processors) if and only if they satisfy the following equality called the *scheduling equality*: $\mathcal{H}_i^T \vec{v}_{i,j} = \mathcal{H}_j^T \vec{v}_{j,i}$.

Our scheduling algorithm consists of three steps:

(1) Assign a *symbolic* scheduling matrix to each processor. The rank of the scheduling matrix will be equal to the degree of freedom of data tiles.

(2) Construct scheduling equalities for each processor pair using direction vectors and scheduling matrices. These equalities collectively represent the constraints that need to be satisfied for eliminating all extra DRAM accesses due to inter-processor communication.

(3) Initialize the scheduling matrix of a processor with an arbitrary schedule (e.g., using one of the matrices in Figures 4(a) or (b)) and compute the corresponding scheduling matrices (tile access patterns) of the remaining processors by solving the scheduling equalities.

As an example application of this algorithm, consider the nest in Figure 2 assuming four processors and a degree of freedom of 2 (Figure 3(e)). In the first step, we assign symbolic scheduling matrices for processors. Let $\mathcal{H}_i = \begin{pmatrix} h^i{}_{1,1} & h^i{}_{1,2} \\ h^i{}_{2,1} & h^i{}_{2,2} \end{pmatrix}$ be the scheduling matrix for processor i. In the second step of our scheduling algorithm, we compute scheduling equalities. For example, scheduling equalities for processor 2 are:

$$\begin{pmatrix} h^2{}_{1,1} & h^2{}_{2,1} \\ h^2{}_{1,2} & h^2{}_{2,2} \end{pmatrix}\begin{pmatrix} -1 \\ -1 \end{pmatrix} = \begin{pmatrix} h^3{}_{1,1} & h^3{}_{2,1} \\ h^3{}_{1,2} & h^3{}_{2,2} \end{pmatrix}\begin{pmatrix} 1 \\ 1 \end{pmatrix}$$

$$\begin{pmatrix} h^2{}_{1,1} & h^2{}_{2,1} \\ h^2{}_{1,2} & h^2{}_{2,2} \end{pmatrix}\begin{pmatrix} 0 \\ -1 \end{pmatrix} = \begin{pmatrix} h^4{}_{1,1} & h^4{}_{2,1} \\ h^4{}_{1,2} & h^4{}_{2,2} \end{pmatrix}\begin{pmatrix} 0 \\ 1 \end{pmatrix}$$

$$\begin{pmatrix} h^2{}_{1,1} & h^2{}_{2,1} \\ h^2{}_{1,2} & h^2{}_{2,2} \end{pmatrix}\begin{pmatrix} -1 \\ 0 \end{pmatrix} = \begin{pmatrix} h^1{}_{1,1} & h^1{}_{2,1} \\ h^1{}_{1,2} & h^1{}_{2,2} \end{pmatrix}\begin{pmatrix} 1 \\ 0 \end{pmatrix}$$

Equalities for other processors are constructed in a similar fashion. In the last step, we determine individual scheduling matrices. To do so, we initialize the scheduling matrix of a processor to a schedule and find the scheduling matrices of the remaining processors. For example, if we set: $\mathcal{H}_1 = \begin{pmatrix} h^1{}_{1,1} & h^1{}_{1,2} \\ h^1{}_{2,1} & h^1{}_{2,2} \end{pmatrix} = \begin{pmatrix} 1 & 0 \\ 0 & 1 \end{pmatrix}$, we find $\mathcal{H}_2 = \begin{pmatrix} -1 & 0 \\ 0 & 1 \end{pmatrix}$;$\mathcal{H}_3 = \begin{pmatrix} 1 & 0 \\ 0 & -1 \end{pmatrix}$;$\mathcal{H}_4 = \begin{pmatrix} -1 & 0 \\ 0 & -1 \end{pmatrix}$. These scheduling matrices give us the tile access patterns shown in Figure 3(g). Note that these access patterns do not incur any extra off-chip memory accesses originating from inter-processor communication.

The scheduling algorithm for a nested loop that may contain flow-dependences is the same as that of a loop that does not contain any flow-dependences except that the scheduling matrix selected should respect all data dependences. Particularly, when flow-dependences [15] are involved, we cannot have snake-like tile access patterns shown in Figure 4(b). Instead, for the cases with a degree of freedom of 2, we can employ the tile access patterns and the associated scheduling matrices shown in Figure 4(c). The numbers attached to arrows in this figure indicate the order of accesses. Our approach has two steps. First, we run only the first two steps of the algorithm given for the non-flow dependence case, and obtain all scheduling equalities. Then, instead of just initializing one of the scheduling matrix to an arbitrary form and determining the others (as in the previous case), we try all possible scheduling matrices for one of the processors and, for each case, we determine the corresponding scheduling matrices for other processors. In other words, instead of just one solution, we come up with multiple solutions. Among these solutions, we select the one (if any) that does not violate any data dependences. In case we have no such a matrix, we employ a default scheduling scheme that does not break any dependences. Obviously, the default strategy will not necessarily eliminate extra DRAM accesses.

As an example, let us consider the successive-over-relaxation (SOR) loop shown in Figure 5. In this code fragment, if we ap-

```
for (i=2; i <= N-1; i++)
 parfor(j=1; j <= N-1; j++)
  U_1[i][j] += g(U_1[i-1][j]+U_1[i][j+1]+U_1[i+1][j])
```

Figure 5: SOR iteration.

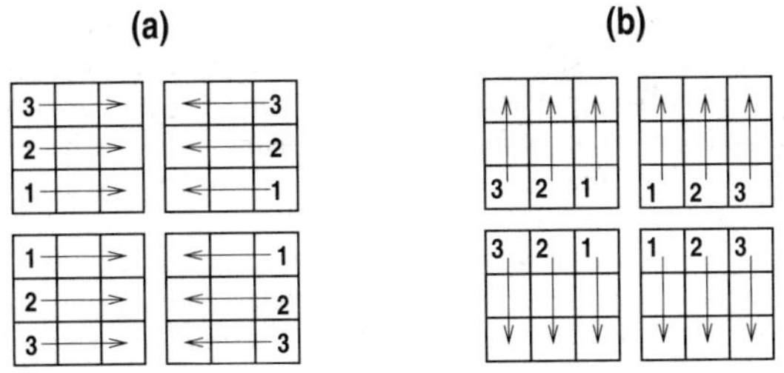

Figure 6: Illegal (a) and legal (b) access patterns for the code fragment in Figure 5.

ply our three-step strategy, one possible schedule would have the following scheduling matrices (assuming four processors): $\mathcal{H}_1 = \begin{pmatrix} 1 & 0 \\ 0 & 1 \end{pmatrix}; \mathcal{H}_2 = \begin{pmatrix} -1 & 0 \\ 0 & 1 \end{pmatrix}; \mathcal{H}_3 = \begin{pmatrix} 1 & 0 \\ 0 & -1 \end{pmatrix}; \mathcal{H}_4 = \begin{pmatrix} -1 & 0 \\ 0 & -1 \end{pmatrix}$. This is shown in Figure 6(a). Since the schedules for processors 2 and 4 violate data dependences, they are not acceptable. Another schedule has these scheduling matrices: $\mathcal{H}_1 = \begin{pmatrix} -1 & 0 \\ 0 & 1 \end{pmatrix}; \mathcal{H}_2 = \begin{pmatrix} 1 & 0 \\ 0 & 1 \end{pmatrix}; \mathcal{H}_3 = \begin{pmatrix} -1 & 0 \\ 0 & -1 \end{pmatrix}; \mathcal{H}_4 = \begin{pmatrix} 1 & 0 \\ 0 & -1 \end{pmatrix}$. The tile access pattern corresponding to this schedule is given in Figure 6(b). We note that this is a legal schedule. In should be noted however that in the case of flow-dependences, it may not always be possible to eliminate all extra off-chip memory accesses.

4. EXPERIMENTS

4.1 Experimental Setup

Our experimental setup consists of a compiler environment and an in-house simulator. Our optimization algorithm is implemented using the SUIF (Stanford University Intermediate Format) experimental compiler infrastructure [1]. SUIF consists of a small, clearly documented kernel and a toolkit of compiler passes built on top of the kernel. The kernel defines the intermediate representation, provides functions to access and manipulate the intermediate representation, and structures the interface between compiler passes. The toolkit currently includes C and Fortran front-ends, a loop-level parallelism and locality optimizer, an optimizing MIPS back-end, a set of compiler development tools, and support for instructional use. The output of SUIF is a C code which can be compiled using the native compiler of the platform in question.

To test the effectiveness of our strategy, we used four array-dominated applications (written in C) from the image processing domain: `3D`, `dfe`, `splat`, and `wave`. `3D` is an image-based modeling application that simplifies the task of building 3D models and scenes. `dfe` is a digital image filtering and enhancement code. `splat` is a volume rendering application which is used in multi-resolution volume visualization through hierarchical wavelet splatting. It is used primarily in the area of morphological image processing. And finally, `wave` is a wavelet compression code that targets specifically medical applications. This code has a characteristic that it can reduce image data to an extremely small fraction of its original size without compromising image quality significantly. These C programs are written so that they can operate on images of different sizes. The total input sizes used in our experiments

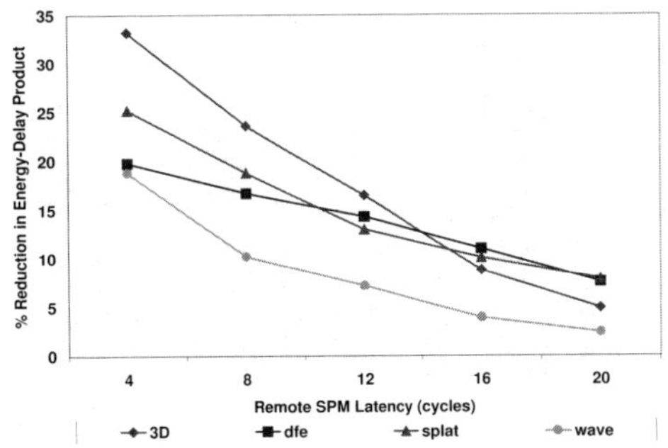

Figure 7: % savings in energy-delay product (base configuration).

were 305KB, 286KB, 635KB and 628KB for `3D`, `dfe`, `splat` and `wave`, respectively.

Our simulator takes a parallel code written in C as input and simulates a multiprocessor architecture. Each simulated processor is a 100MHz MIPS 4Kp core with a five-stage pipeline that supports four execution units (integer, multiply-divide, branch control, and processor control). Each core has thirty-two, 32-bit general-purpose registers. We assume a local SPM (SRAM) attached to each core with an access latency of 2 cycles. We also simulate inter-processor communication. To test the robustness of strategy, we experimented with different remote SPM access latencies (from 4 cycles to 16 cycles). We also assumed an extra 1 cycle latency for each inter-processor synchronization operation. The simulated off-chip DRAM is 4MB with an access latency of 80 cycles. The simulator outputs the number of accesses to each SPM, number off off-chip accesses, number and volume of inter-processor communications, and overall execution time of the application. To parallelize the applications, we adopted an aggressive strategy in which all the loops in a given nest (except the innermost one) are parallelized (provided that data dependences allow that). The energy model used for SPMs is very similar to that of a cache memory [12] except that it assumes full associativity and does not consider a tag array. The energy model used by our simulator for interconnects is transition-sensitive and is very similar to that presented in [16]. Since we focus only on data memory performance and energy, we do not report data on instruction accesses and datapath activity. In computing the executing cycles, however, all the cycles spent in datapath (including the stall cycles) are accounted for.

4.2 Results

Due to space concerns, we present data for only energy-delay product. Energy-delay product is a suitable metric that allows us to evaluate the impact of an optimization on both energy and performance (execution cycles) [6, 8]. In our context, the energy component of the energy-delay product corresponds to the memory system energy due to data accesses. This includes the energy consumed in SPMs, interconnect between processors, off-chip DRAM accesses, and in the interconnect between SoC and off-chip DRAM. The delay is the overall parallel execution time of the application. In our experiments, we compare two different versions of each benchmark. The first version is the code with only local SPM optimizations; it does *not* try to eliminate extra DRAM accesses due to inte-processor communication. The second version is the result of our VS-SPM based optimization strategy. All results presented in this subsection are *percentage improvements* brought by the second version over the first version.

Figure 7 gives the energy-delay product improvements for our base configuration with different values of remote SPM access latency. In the base configuration, the access latencies for local SPM and off-chip DRAM are 2 and 80 cycles, respectively. Also, the

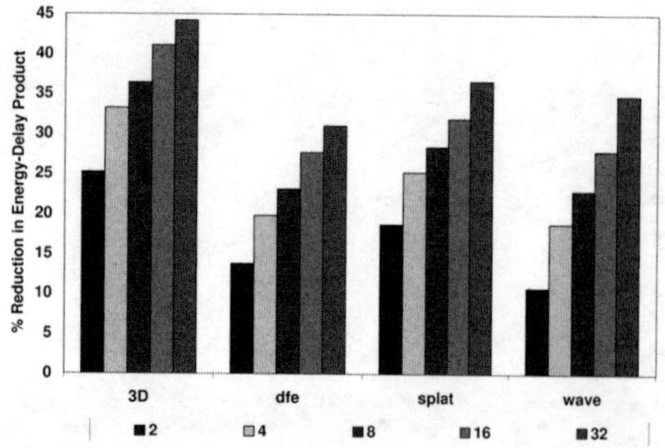

Figure 8: % savings in energy-delay product (sensitivity to the number of processors).

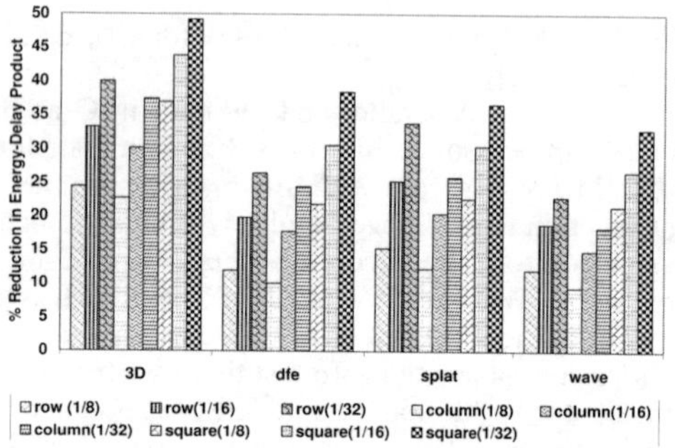

Figure 9: % savings in energy-delay product (sensitivity to the shape of the tile and size of the available SPM space).

number of processors is four, the row-block data tiles are used, and the available local SPM space is 1/8th of the local portion of the processor (this ratio is called the *slab ratio*). Note that the slab ratio can be changed by changing the SPM size or array sizes; in our experiments, we changed the SPM size. It should be mentioned that when the remote SPM latency is increased, the corresponding per access energy consumption (for remote SPM) is also proportionally increased. We observe from these results that our approach generates best results with the smallest remote SPM latencies (as expected). The average improvements when remote SPM latency is 4 and 20 are 24.3% and 5.7%, respectively.

To measure the sensitivity of our approach to the number of processors, we performed another set of experiments. We used processor sizes of 2, 8, 16, and 32. All other parameters are the same as in the base configuration. The results shown in Figure 8 (which also include our default processor size of 4) indicate that the effectivess of our strategy increases when the number of processors is increased. This is because an increase in the number of processors leads to an increase in inter-processor communication volume, and consequently, the version without VS-SPM optimization performs very frequent extra DRAM accesses. Therefore, we observe an increase in percentage improvements.

The last set of experiments gauge the sensitivity of our strategy to the shape of the tile and size of the available SPM space. Each bar in Figure 9 indicates the shape of the tile (row-block, column-block, or square) and the slab ratio (within parantheses). It can be observed that our approach is more effective with smaller slab ratios. Note that a smaller slab ratio represents more pressure on data memory. Considering the fact that applications are processing larger and larger data sets, this result is encouraging. We also observe that we save more when square tiles are used. This is due to the fact that the performance (and energy behavior) of the original codes is extremely poor when square tiles are used.

5. CONCLUSIONS AND FUTURE WORK

Efficient compilation techniques for complex programs are particularly important for array-dominated embedded applications. As microprocessors grow more and more powerful, designers are building more sophisticated embedded/mobile systems. An important trend in such systems is the increased use of multiple processors. This paper presents a compiler-directed optimization strategy for exploiting software-controlled, shared SRAM space in a multiprocessor embedded system. Our approach is oriented towards eliminating extra off-chip DRAM accesses caused by inter-processor communication. The results obtained using four array-intensive applications show that significant reductions in energy-delay product are possible using this approach. This work can be extended in several ways. First, we plan to perform experiments with multiple levels of SPMs and hybrid multiprocessor environments that consist of both SPMs and data caches. Second, we would like to investigate different compiler optimizations for optimizing inter-processor communication and study the impact of these optimizations on off-chip memory accesses. Third, we want to explore compiler optimizations that exploit array sharing between different loop nests in an embedded multiprocessor environment. Finally, we would like to extend this work to heterogeneous multiprocessor architectures.

Acknowledgments We acknowledge the support of the National Science Foundation through grants 0093082, 0073800, and 0121706.

6. REFERENCES

[1] S. Amarasinghe, J. Anderson, M. Lam, and C. Tseng. The SUIF compiler for scalable parallel machines. In Proc. *7th SIAM Conference on Parallel Processing for Scientific Computing,* 1995.

[2] L. Benini, A. Macii, E. Macii, and M. Poncino. Increasing energy efficiency of embedded systems by application-specific memory hierarchy generation. *IEEE Design & Test of Computers,* pages 74–85, April-June, 2000.

[3] Blue Wave Systems. http://www.bluews.com/

[4] R. Bordawekar, A. Choudhary, and J. Ramanujam. Automatic optimization of communication in compiling out-of-core stencil codes. In Proc. *10th ACM International Conference on Supercomputing,* 1996.

[5] F. Catthoor, S. Wuytack, E. D. Greef, F. Balasa, L. Nachtergaele, and A. Vandecappelle. *Custom memory management methodology.* Kluwer, 1998.

[6] A. Chandrakasan, W. Bowhill, and F. Fox. *Design of High-Performance Microprocessor Circuits.* IEEE, 2001.

[7] Dinero IV Trace-Driven Uniprocessor Cache Simulator. URL: http://www.cs.wisc.edu/~markhill/DineroIV/

[8] R. Gonzales and M. Horowitz. Energy dissipation in general purpose microprocessors. *IEEE Journal of Solid-State Circuits,* pages 1277–1284, Sep. 1996

[9] M. Kandemir, J. Ramanujam, M. Irwin, N. Vijaykrishnan, I. Kadayif, and A. Parikh. Dynamic management of scratch-pad memory space. In Proc. *38th Design Automation Conference,* 2001.

[10] *Mobile Computing Devices: A New Era in Personal Computing,* August 2000. Computer Market Dynamics.

[11] P. R. Panda, N. D. Dutt, and A. Nicolau. Efficient utilization of scratch-pad-memory in embedded processor applications. In Proc. *European Design & Test Conference,* 1997.

[12] W-T. Shiue and C. Chakrabarti. Memory exploration for low power, embedded systems. In Proc. *36th Design Automation Conference (DAC'99),* 1999.

[13] N. Vijaykrishnan, M. Kandemir, M. J. Irwin, H. Y. Kim, and W. Ye. Energy-driven integrated hardware-software optimizations using SimplePower. In Proc. *Int. Symp. Computer Architecture,* 2000.

[14] L. Wang, W. Tembe, and S. Pande. Optimizing on-chip memory usage through loop restructuring for embedded processors. In Proc. *9th International Conference on Compiler Construction,* 2000.

[15] M. Wolfe. *High Performance Compilers for Parallel Computing.* Addison-Wesley, 1996.

[16] Y. Zhang, Y. Chen, W. Ye, and M. J. Irwin. System-level interconnect power modeling. In Proc. *the 11th International ASIC Conference,* 1998.

Address Assignment Combined with Scheduling in DSP Code Generation

Yoonseo Choi and Taewhan Kim
Department of Electrical Engineering & Computer Science
and Advanced Information Technology Research Center(AITrc)
Korea Advanced Institute of Science and Technology, KOREA
yschoi@jupiter.kaist.ac.kr tkim@cs.kaist.ac.kr

Abstract – One of the important issues in embedded system design is to optimize program code for the microprocessor to be stored in ROM. In this paper, we propose an integrated approach to the DSP address code generation problem for minimizing the number of addressing instructions. Unlike previous works in which code scheduling and offset assignment are performed sequentially without any interaction between them, our work *tightly couples offset assignment problem with code scheduling* to exploit scheduling on minimizing addressing instructions more effectively. We accomplish this by developing a fast but accurate two-phase procedure which, for a sequence of code schedules, finds a sequence of memory layouts with minimum addressing instructions. Experimental results with benchmark DSP programs show improvements of 13%-33% in the address code size over *Solve-SOA/GOA* [7].

Categories and Subject Descriptors

C.3 [**Special-purpose and application-based systems**]: [Signal processing systems]

General Terms

Algorithms, Performance

Keywords

Offset assignment, Scheduling, Code Generation

1. INTRODUCTION AND RELATED WORK

The increasing complexity of designing embedded VLSI systems has made the traditional DSP compilers unable to meet the very tight constraints of code size for on-chip program and real-time performance. Unfortunately, due to irregular architectures such as non-homogeneous register sets, specialized functional units and registers, even compilers available for commercial DSP processors generate very inefficient code in terms of size and performance.[1, 2].

Storage assignment, i.e., optimization of memory layout for program variables is an important part of code generation. It was shown that, for a set of programs in MediaBench [4] that were compiled for Motorola *DSP56000* family, more than 55% of the instructions involve address registers.[3] Consequently, optimizing address assignment could lead to a significant reduction on code size and program execution time.

Many architectures (e.g., the VAX, TI TMS320C2x DSP family, most embedded controllers) provide indirect addressing modes with auto-increment/decrement address arithmetic. These architectures also provide a set of dedicated address generation units (AGUs) that perform fast address computation in parallel to the central data path and contain a separate adder/subtractor for performing next-address computations. Thus, a high instruction-level parallelism can be achieved by generating many auto-increment/ decrement addressing modes to access variables, which requires a careful placement of variables in memory. Consequently, contrary to the traditional compilers, DSP compilers should carefully determine the relative location of variables in memory.

The storage assignment with a single address register in AGU (the *simple offset assignment* problem, SOA) was first studied by Bartley [5] and Liao *et. al* [1, 6]. They modeled the problem as a graph theoretic optimization problem. Liao *et. al* showed that the SOA problem is equivalent to the *maximum-weighted Hamiltonian path cover* (MWPC) problem and proved that it is NP-complete. They proposed a heuristic to solve SOA based on the Kruskal's maximum spanning tree algorithm. Liao *et. al* [1, 6] also extended the approach to the case with multiple address registers in AGU (the *general offset assignment* problem, GOA). Leupers and Marwedel [7] have extended the work done by Liao *et. al* by proposing a *tie-breaking* heuristic and a variable partitioning method to improve the quality of SOA/GOA solution. Leupers and David [8] have solved GOA problem with arbitrary register file sizes and auto-increment ranges. Sudarsanam *et. al* [9] take into account the utilization of auto-increment with increments varying from $-l$ to $+l$. Gebotys [10] modeled the problem of assigning address registers to every variable access in the code given a fixed memory layout as a network flow problem and solved it optimally.

All of the previous approaches [1, 5, 6, 7, 8, 9] have addressed the SOA/GOA problem as a separate code optimization problem for a given exact access sequence of the program variables. However, since the access sequence of the variables is one of the most critical factors which affect the quality of SOA/GOA solutions, the memory layouts produced by the previous approaches will be SOA/GOA solutions that were locally optimized. In this context, we study the problem of integrating the SOA/GOA with code scheduling to exploit the effect of scheduling on minimizing the size of address code (in assembly) more fully and effectively. Our solution is naturally extended to incorporate reordering of input-operands of commutative operations[1] (e.g., $a = b + c$). To our knowledge, the work done by Rao and Pande [11] has addressed the optimization of access sequence of the variables by a (local) reordering of commuta-

[1] The optimization is performed at the level of an intermediate representation like three-address instruction that has two source operands.

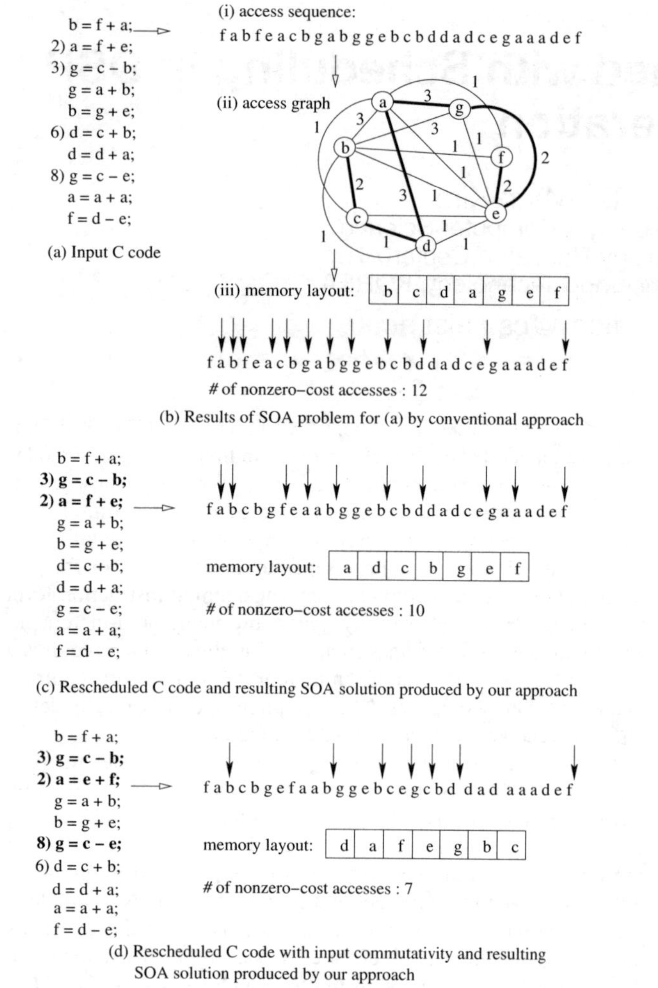

Figure 1: Motivational example illustrating the effectiveness of code scheduling on address assignment.

tive input-operands. However, their solution is confined to the SOA problem only. Lim *et. al* [12] addressed scheduling effect on the SOA problem. Their approach aimed to make graph sparser by an exhaustive search algorithm with pruning techniques. However, it is not true that sparser graphs always lead to cheaper MWPC cost than that of denser graphs.

The key contribution of our work is an effective integration of code scheduling (together with input-operand permutation) which we call Sch-SOA/GOA, into the SOA/GOA heuristics *Solve-SOA/GOA* [1, 6, 7].

2. MOTIVATING EXAMPLE

As an example illustrating the effectiveness of code scheduling on the problem of assigning frame-relative offsets to program variables, consider the C program in Figure 1(a). Suppose we want to solve a simple offset assignment (SOA) problem. That is, we assumed to have a target architecture with a single address register (AR) with only the indirect and auto-increment/decrement address modes. Figure 1(b-i) shows the access sequence of the variables corresponding to the input C code in Figure 1(a).

Liao *et. al* [1] modeled SOA by an undirected weighted graph, called *access graph*, $G(V,E)$ where each node $v \in V$ corresponds to a unique variable and an edge $e \in E$ between nodes x and y exists with weight $w(e)$ if x and y are adjacent to each other $w(e)$ times in the access sequence. The access graph corresponding to the access sequence in Figure 1(b-i) is shown in Figure 1(b-ii).[2] An optimal offset assignment is to find a path cover P in $G(V,E)$ that minimizes the quantity (i.e., total address arithmetic instructions):

$$C(G,P) = \sum_{\forall e \; of \; edges \notin P} w(e) \tag{1}$$

Finding a maximum weighted path cover (MWPC) minimizes the quantity of C. The heavy lines in Figure 1(b-ii) beginning from b and ending at f form a path cover, The path was obtained by applying Liao *et al.*'s SOA solution [1] with Leupers and Marwedel's *tie-breaking* rule [7]. As a result, the corresponding memory layout is shown in Figure 1(b-iii) where $C = w(a,b) + w(a,c) + w(a,e) + w(a,f) + w(b,d) + \cdots + w(e,g) = 12$.

On the other hand, the left side of Figure 1(c) shows a rescheduled C code where the highlighted statements indicate the change of schedule. The access sequence for the rescheduled code and the resulting memory layout are shown on the right side of Figure 1(c). The total number of nonzero-cost addressing instructions is reduced to 10. Finally, Figure 1(d) shows the further optimized access sequence and memory layout obtained when scheduling is considered together with input permutation for commutative operations. The number of nonzero-cost addressing instructions is then reduced to 7, which is 41% improvement over the schedule in Figure 1(a).

The example in Figure 1 clearly reveals that the offset assignment problem is very sensitive to the access sequence of the variables. Since the existing compilers have designed the offset assignment as a post-pass optimization (after code generation) they failed to globally optimize the access sequence. To overcome this limitation, we propose *a new offset assignment algorithm tightly coupled with code scheduling* to exploit the effect of scheduling on minimizing address arithmetic instructions more fully and effectively. We then extend the scheduling problem to include input permutation for commutative operations to enhance the quality of the assignment further.

3. A UNIFICATION OF CODE SCHEDULING AND ADDRESS ASSIGNMENT

Based on the motivating example in section 2, we propose an effective two-phase algorithm for address assignment integrated with code scheduling. First, we describe an overview of the proposed algorithm in section 3.1, followed by the detailed description on our two-phase algorithm: (Phase 1) *a generation of initial schedule and address assignment* in section 3.2, and (Phase 2) *a stepwise refinement of the solution* obtained in Phase 1 in section 3.3.

3.1 An overview

The input C program is composed of basic blocks, which are given as a partially ordered list of code operations. Our current work focuses on basic block-level address code optimization The input to our algorithm is *dependency graph* of operations (i.e., instructions), that is an intermediate representation of a basic block. A node in the dependency graph represents an operation, and an arc from node a to b indicates that a must precede b in execution.

The objective of our algorithm is to schedule operations in the dependency graph and determines a complete sequence of variable accesses, so that the size of resulting address code is minimized.

Figure 2 summarizes the flow of our offset assignment algorithm combined with scheduling/operand-permutation, called Sch-SOA/GOA. The proposed algorithm is performed in two-steps:

[2]The access sequence involved in operation $x = y \; op \; z$ is '*yzx*' where *op* represents an arithmetic operation. The edges with zero weight are not shown in the access graphs for brevity.

```
Sch-SOA/GOA: schedule-driven SOA/GOA (Dependency_graph)
 /* Phase 1 */
 • Generate access sequence S_init and schedule SCH_init by
     calling Sch-SOA/GOA-init(Dependency_graph); /* sec. 3.2 */
 • Generate access graph G from S_init;
 • Apply Solve-SOA/GOA to G and determine C in Eq.(1).;
 /* Phase 2 */
 • Set COST_min = C;
 • Set SCH_best = SCH_init;
 while (COST_min is updated) {
   • Set COST_current = COST_min;
   • Set SCH_current = SCH_best;
   while (there is a 'reschedulable' operation) { /* constrained by
                                        data-flow dependencies */
     foreach 'reschedulable' operation (to execution step j) {
       • Reschedule the operation;
       • Update the previous path cover solution
           to account for the reschedule; /* sec. 3.3 */
       if (operation is commutative) {
           • Update the path cover solution by exploiting
               input commutativity; /* sec. 3.3 */
       }endif
       • Compute C for the path cover of the reschedule;
       • Undo the reschedule;
     }endforeach
     • Reschedule the operation with the smallest C;
     if (C < COST_min) {
       • Update COST_min to current C,
           and update SCH_best accordingly;
     }endif
     • Lock the operation at the execution step;
   }endwhile
   • Unlock all the operations;
 }endwhile
 • return (Solve-SOA/GOA(G of SCH_best));
```

Figure 2: The overall flow of the proposed schedule-driven SOA/GOA algorithm.

(Phase 1) generating an initial schedule and address assignment from the dependency graph of each basic block derived from the input C program, and (Phase 2) iteratively improving the SOA/GOA solution obtained in Phase 1 at the outer *while-loop*. An operation scheduled at an execution step is called "reschedulable" to another execution step, say j, if scheduling the operation at j does not violate the data-flow dependency constraint. For every reschedulable operation, its C value is computed. Among the operations, the operation with the least value of C is selected, and rescheduled it to the corresponding execution step. Once the operation is rescheduled, it will be locked at that clock step for the rest of the execution at the inner *while-loop*. The outer *while-loop* continues until it is not able to generate a schedule and SOA/GOA solution whose C value is less than that of the minimal C found so far during the previous iterations.

The time complexity of Sch-SOA/GOA depends on the time to generate a path cover solution for a trial of rescheduling at the *for-loop*, which will be discussed in detail in section 3.3.

3.2 Initial Schedule and Address Assignment

It is important to produce a good initial schedule and address assignment because the quality of the final address assignment by reschedule is greatly affected by its initial solution. We focus our discussion on generating an initial schedule and its SOA solution for the reason that our extension to GOA solution is rather straightforward with minor modifications.

We construct an access sequence incrementally, one execution step at a time, from the first execution step to the last. An operation in the dependency graph is said to be a *ready operation* for a certain execution step if all of its predecessors have already been scheduled in the previous execution steps. At each iteration, our algorithm selects the most "promising" operation and schedules it at the current execution step, and appends its variable accesses to the end of the current partial access sequence. After the (partial) access sequence is augmented according to the selection of an operation from the ready operations, the algorithm repeats for the next execution step until all the operations in the dependency graph have been scheduled for execution.

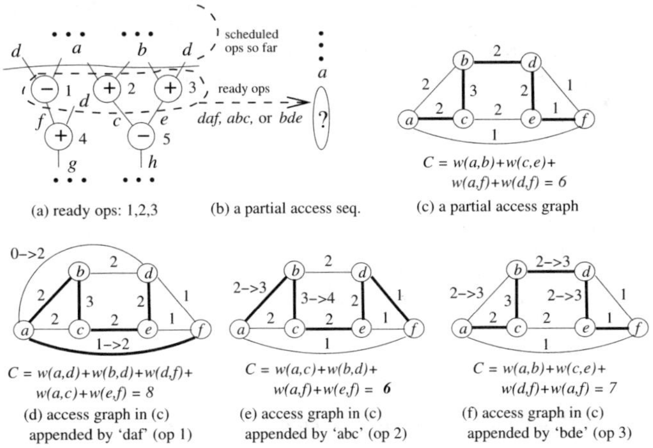

Figure 3: An example of showing the selection of operation, among ready operations, to minimize C in $Eq.(1)$.

The algorithm selects, among ready operations, the operation with minimum additional cost of C in $Eq.(1)$ for the augmented access graph. For example, suppose we have already generated an access sequence (ended by accessing variable a) partially up through execution step $t-1$, as depicted in Figure 3(a). Now, we want to select an operation to be executed in execution step t, and accordingly, append the variable accesses to the current partial access sequence '$\cdots a$' (denoted by S_{curr}), as indicated in Figure 3(b). Suppose we have an access graph (denoted by G_{curr}), shown in Figure 3(c) corresponding to S_{curr}. The path (denoted by P_{curr}), with heavy lines indicates a solution of maximum weighted path cover obtained by applying heuristic *Solve-SOA* [7]. Then, $C(G_{curr}, P_{curr}) = 6$ as shown in Figure 3(c).

Let us now consider all possible schedules of ready operations at the next execution step. Figures 3(d), (e) and (f) show the access graphs with augmented weights and the computation of their C values when operations 1, 2 and 3 in Figures 3(a) are selected for execution at execution step t, respectively. Consequently, we choose operation 2 because it results in the minimum increase of C value. Thus, we attach the access sequence '*abc*' (for operation 2) to S_{curr}, resulting in '$\cdots aabc$'. The ready operations are then updated and the process repeats until there is no more operations in the list.

Figure 4 summarizes the flow of our SOA algorithm combined with code scheduling, called Sch-SOA-init. It is a variant of list scheduling. For each trial of scheduling ready operations, a maximum weighted path cover for the updated access graph is obtained by applying *Solve-SOA* [7], the time complexity of which is $O(m \cdot \log m)$ where m is the number of edges in access graph. Thus, the total time of our algorithm is bounded by $O(n^2 \cdot m \cdot \log m)$ where n is the number of operations since the outer *for-loop* is executed n times and at each time, at most n ready operations are tested using *Solve-SOA*. However, since the number of ready operations is very small in practice, the run time is actually bounded by $O(n \cdot m \cdot \log m)$.

3.3 Iterative Improvement Techniques

The core of our algorithm is to generate, efficiently but accurately,

Sch-SOA-init: schedule-driven initial SOA (*Dependency_graph*)
- Set n = the number of operations in *Dependency_graph*;
- Set R = the operations in *Dependency_graph* with no predecessors;
- Set $S = \phi$;

for t = 1 to n {
  foreach $op_i \in R$ {
  - Schedule op_i to execution step t;
  - Update access graph G by augmenting the weights associated with the variable access transitions in op_i;
  - Find a maximum weighted path cover, P, for G;
  - Compute $C(G,P)$ in *Eq.*(1);
  - Undo the schedule and G;

  } **endforeach**
  - Schedule op_i with the smallest $C(G,P)$ to t;
  - Update access sequence S by appending access sequence for op_i;
  - Update access graph G and ready operation set R accordingly;

} **endfor**
- **return** (S, *Schedule*);

Figure 4: The proposed algorithm for generating an initial SOA solution.

the address assignment solution (i.e., path cover which minimizes the quantity of C in *Eq.*(1) when an operation is rescheduled from execution step i to j. Clearly, the rescheduling changes the the access sequence, and thus, changes weights of some edges in the access graph. A path cover solution can be obtained simply by applying *Solve-SOA/GOA* [7]. However, this definitely requires an excessive run time when the problem size is large. However, from the fact that each reschedule changes the execution order locally, causing a small change of edge weights in the access graph, and a path cover solution with a minimum C value for the previous schedule has already been known, it is natural to ask whether there is *a way to find a path cover with a minimum C value for the current schedule rapidly by exploiting the previous path cover solution.* To do this, we develop a fast and comprehensive path cover computation procedure based on the previous solution.

Figure 5(a) shows an example of code rescheduling. According to the reschedule, as depicted in Figure 5(b), the weights of edges $(x1,x2)$, $(x3,x4)$ and $(x6,x7)$ are decremented by 1, and the weights of $(x1,x4)$, $(x3,x7)$ and $(x2,x6)$ are incremented by 1. Thus, we can easily find that the number of edge weights to update by a reschedule of an operation is bounded by a constant (=6), independently of the size of access graph. Further, the amount of change for each edge is in [-3, +3].[3]

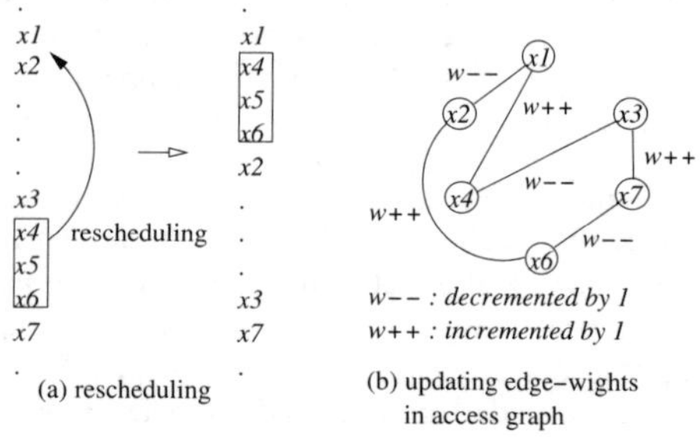

Figure 5: An example of operation rescheduling and update of edge weights.

Schedule-SOA: Let P_{old} denote the path cover (i.e. *solve-SOA* solution) in access graph $G(V,E_{old})$. Suppose some of edge weights in $G(V,E_{old})$ are updated according to a reschedule of an operation. Let $G(V,E_{new})$ denote the updated access graph.

[3]'3' occurs when $x1$=$x2$, $x3$=$x4$, $x6$=$x7$ in Figure 5 whereas '-3' occurs when $x1$=$x4$, $x3$=$x7$, $x2$=$x6$.

The problem is to find a new path cover, P_{new}, for $G(V,E_{new})$ that minimizes the quantity of C in *Eq.*(1). We determine P_{new} according to the following rules:

- **Case 1**: $C(G(V,E_{old}),P_{old}) > C(G(V,E_{new}),P_{old})$.
  Note that $C(G(V,E_{old}),P_{old})$ is the C value for the access graph before the reschedule using path cover P_{old}, and $C(G(V,E_{new}),P_{old})$ is the C value after the reschedule using the same path cover P_{old}. Case 1 belongs to the case that the C value obtained by using P_{old} for the updated (new) G is smaller than that by P_{old} for the old G. In this situation, we simply use the P_{old} as a new path cover for $G(V,E_{new})$ as long as the C value using P_{old} for $G(V,E_{new})$ decreases. Since the number of edges whose weights are updated is at most 6, computing the C value for $G(V,E_{new})$ using P_{old}, can be done in $O(1)$. In Figure 6, we use P_{old} as a path cover, P_{new}, for the new G, since the C value for the new G is reduced even when P_{old} is used.

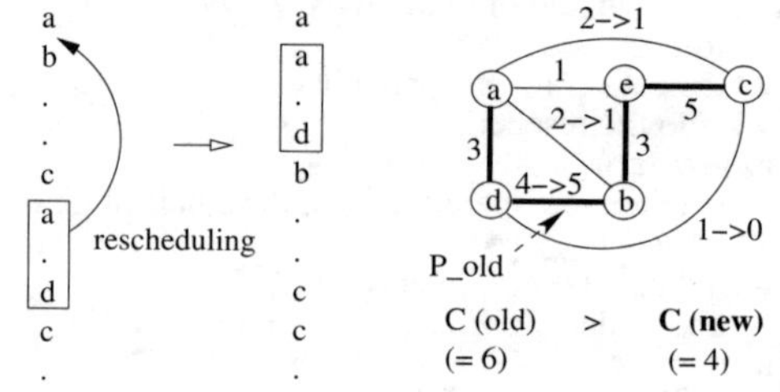

Figure 6: Determination of a path cover for an updated G in Case 1.

The described procedure for determining a path cover for $G(V,E_{new})$ in Case 1 has an optimal property for a special case.

Property 1 Let P_{old} be an optimal path cover for $G(V,E_{old})$, and $w_o(e)$ and $w_n(e)$ denote the weights of edge e in $G(V,E_{old})$ and $G(V,E_{new})$, respectively. We define $\delta(e)$ to be $w_o(e) - w_n(e)$ if $w_n(e) < w_o(e)$, and 0 otherwise. Then, P_{old} is an optimal path cover for $G(V,E_{new})$ if the following conditions are satisfied.

(1) $w_o(e) \geq w_n(e), \forall e \notin P_{old}$ and
(2) $\sum_{e \notin P_{old}} \delta(e) > \sum_{e \in P_{old}} \delta(e)$

proof Let P_i be a path cover for $G(V,E_{new})$ which satisfies $C(G(V,E_{old}),P_i) > C(G(V,E_{old},P_{old})$. We define edge sets

$$S_1 = \{e | e \in E, e \notin P_{old}, e \notin P_i\},$$
$$S_2 = \{e | e \in E, e \in P_{old}, e \notin P_i\},$$
$$S_3 = \{e | e \in E, e \notin P_{old}, e \in P_i\}.$$

We first show the inequality

$$\sum_{e \in S_2} w_n(e) \geq \sum_{e \in S_3} w_n(e). \quad (2)$$

From the assumption that P_{old} is an optimal path cover for $G(V,E_{old})$,

$$\sum_{e \in S_2} w_o(e) > \sum_{e \in S_3} w_o(e).$$

Thus,

$$\sum_{e \in S_2} w_o(e) - 1 \geq \sum_{e \in S_3} w_o(e). \quad (3)$$

According to *condition 2* and the fact that the total sum of the decreases of edge weights (from E_{old} to E_{new}) is at most 3, it is true that $\sum_{e \in P_{old}} \delta(e) \leq 1$, from which

$$\sum_{e \in S_2} w_n(e) \geq \sum_{e \in S_2} w_o(e) - 1. \quad (4)$$

Thus, from Eq.(3) and Eq.(4) and $\sum_{e \in S_3} w_o(e) \geq \sum_{e \in S_3} w_n(e)$ according to *condition 1*, we have

$$\sum_{e \in S_2} w_n(e) \geq \sum_{e \in S_3} w_o(e) \geq \sum_{e \in S_3} w_n(e). \quad (5)$$

To show $C(G(V,E_{new}),P_i) \geq C(G(V,E_{new}),P_{old})$, by the definitions of S_1, S_2 and S_3, $S_1 \cup S_2 = \{e | e \in E, e \notin P_i\}$ and $S_1 \cup S_3 = \{e | e \in E, e \notin P_{old}\}$, and using the inequality in *Eq.*(5), we derive

$$\begin{aligned} C(G(V,E_{new}),P_i) &= \textstyle\sum_{e \in S_1} w_n(e) + \sum_{e \in S_2} w_n(e) \\ &\geq \textstyle\sum_{e \in S_1} w_n(e) + \sum_{e \in S_3} w_n(e) \\ &= C(G(V,E_{new}),P_{old}). \quad \square \end{aligned}$$

- **Case 2**: $C(G(V,E_{old}),P_{old}) \leq C(G(V,E_{new}),P_{old})$.
This is the case where using P_{old} for $G(V,E_{new})$ does not reduce the C value. Thus, we find another path cover by applying *Solve-SOA*. If we maintain a sorted edge-list for $G(V,E_{old})$ according to the edge weights, finding a path cover for $G(V,E_{new})$ using *Solve-SOA*, can be completed in $O(e)$ where e is the number of edges in $G(V,E_{new})$. In Figure 7, since the C value for $G(V,E_{new})$ using P_{old} is not smaller than that for $G(V,E_{old})$ using P_{old}, we compute P_{new} for $G(V,E_{new})$ applying *Solve-SOA*, generating a reduced C value (i.e., from 6 to 5) for $G(V,E_{new})$.

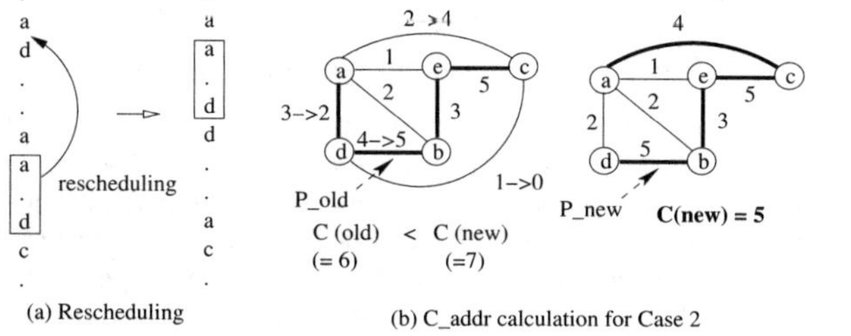

Figure 7: Determination of a path cover for an updated *G* in Case 2.

Schedule-GOA: GOA is the generalization of SOA towards a number k (> 1) of ARs. A reasonable approach proposed by Liao *et al.* [1] and Leupers and Marwedel [7] is to partition the variables into k disjoint subsets $V_1, \cdots, V_k$, for each of which a distinct AR is used. The objective is to minimize

$$C^{tot} = \sum_{\forall V_i,\ i=1,\cdots k} C_i \quad (6)$$

where C_i is the quantity of C in *Eq.*(1) for the access subgraph containing the variables only in V_i. Then, the problem is reduced to k SOA problems.

At each trial of rescheduling, we selectively apply *Solve-GOA* [7] to the updated G. Let G be the access graph for the previous schedule, and $G_1, \cdots, G_k$ be the access subgraph corresponding to the partitioned vertex sets $V_1, \cdots, V_k$, respectively. Also, let $P_1, \cdots, P_k$ be the path covers for $V_1, \cdots, V_k$, respectively. From the change of weights in G by the current reschedule, we calculate C for each of G_i, $i = 1, \cdots, k$ using path cover P_i. According to the change of the value of C, we partition the variables in $V_1, \cdots, V_k$ into two subsets, V_{good} and V_{bad}: If there is a reduction on the value of C, all the variables in the corresponding V_i are assigned to V_{good}. Otherwise, the variables are assigned to V_{bad}. Then, for each access subgraph whose variables are all in V_{good}, we use the corresponding previous path covers as the path covers for the current reschedule. However, for the variables in V_{bad}, we apply *Solve-GOA* to the access subgraph for V_{bad}, and generate new access subgraphs and their path covers.

Figure 8 shows an example of our selective application of *Solve-GOA*. Figure 8(a) shows an access graph that was partitioned into three subgraphs containing variables $V_1 = \{a,b,c\}$, $V_2 = \{d,e,f,g\}$ and $V_3 = \{h,i,j,k\}$ each for the previous schedule. Then, by the changes of weights in the access graph due to a reschedule, it results in $V_{good} = V_1$ and $V_{bad} = V_2 \cup V_3$. Consequently, the path cover for V_1 does not changed, whereas the partitioning and path covers for V_{bad} are updated as shown in Figure 8(b). As a result, the total cost reduction of C^{tot} in *Eq.*(2) by the reschedule is (1+1+2)-(0+1+0) = 3.

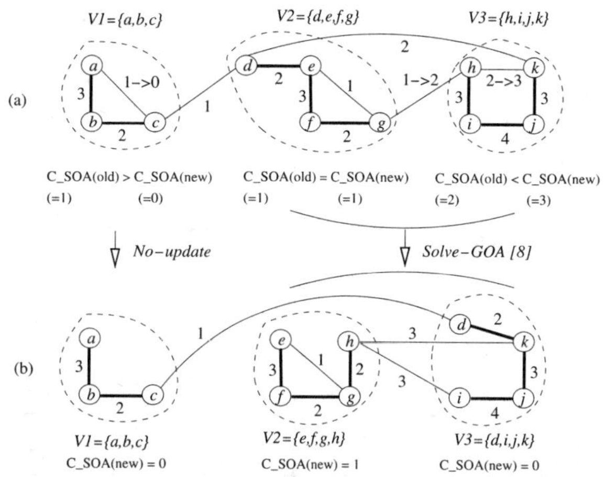

Figure 8: Determination of path covers for Sch-GOA. C_SOA represents *C* value for the access graph corresponding to the partition.

Exploiting operand-input commutativity: For a commutative operation such as $f = a + b$, the access order can be either abf or baf. For a commutative operation to be rescheduled, our algorithm generates the access graph for each alternative of variable access sequence, and compute the new path cover from the access graph according to the procedure described in Sch-SOA/GOA. Then, we choose the access sequence that has the smaller value of C.

4. EXPERIMENTAL RESULTS

The proposed address code generation algorithm was implemented in C++ and executed on an Intel Pentium3 computer. We tested a set of benchmark programs and randomly generated designs. We considered both SOA and GOA. We assumed that the architectures have a sufficiently large number of temporary registers to store all temporary variables in C-code to avoid the memory accesses due to memory spill values. Since our algorithm is performed at the operation code-level, and not directly at the assembly code-level of target machine, the architecture specific pointer operations with increment/decrement (*p++) was assumed to be achieved with a single variable access operation in code, like LAR p, LACL *, +.

Table 1 shows comparisons of the results in terms of address code size produced by OFU (the order of the first use) offset assignment, Solve-SOA with tie-breaking [7], and the proposed Sch-SOA for the randomly generated C programs. $|V|$ and $|S|$ represent the number of variables and the access sequence length, respectively. The last column shows the run time of Sch-SOA. Sch-SOA reduces the address code size up to 74% and 27% over those by OFU and Solve-SOA, respectively.

Table 2 shows the results produced by Solve-GOA [7], and the proposed Sch-GOA for the randomly generated C programs. $|AR|$ represents the number of address registers (ARs). The comparison of the results indicates that Sch-GOA performs well compared to Solve-GOA, reducing the address code size up to 36% more than that of Solve-GOA. Moreover, the reductions are consistent.

Table 3 shows comparisons of results for benchmark C programs when $|AR|$=1 (i.e., SOA). FIR and BIQUAD (biquad_one_section) are taken from DSPstone benchmark suite. COMP (complex multiplier) and ELLIP (elliptical wave filter) are taken from high-level

|V|/|S|	# addr_instr			gain over(%)	time
	OFU	Sol-SOA[7]	Sch-SOA	OFU/[7]	(sec.)
5/10	3.8	1.8	1.4	63.1/22.2	0
5/20	11.1	6.9	5.8	47.7/15.9	0
15/20	13	5.4	4.5	65.3/16.6	0.01
10/50	33.2	25.1	18.9	43.0/24.7	0.02
40/50	35.8	13.3	10.2	71.5/23.3	1.4
10/100	68.2	54.2	42.5	37.6/21.5	0.1
50/100	80.7	48.2	36.9	54.2/23.4	17
80/100	71.3	25.3	18.4	74.1/27.2	45
100/200	167.3	101.4	80.29	52/20.8	1125

Table 1: SOA results for randomly generated codes.

|V|/|S|	|AR|	# addr_instr		gain over	time
		Solve-GOA[7]	Sch-GOA	[7] (%)	(sec.)
10/50	2	15.6	11.2	28.2	0.08
10/50	4	6.8	5.4	20.5	0.03
40/50	4	7.4	4.7	36.4	3.50
40/50	8	8.8	8.1	7.9	1.08
10/100	4	11.3	8.1	28.3	0.04
10/100	8	5	5	0.0	0.02
50/100	8	20.3	14.1	30.5	1.80
80/100	8	10.9	8.6	21.1	6.86
100/200	8	54.9	46	16.2	144

Table 2: GOA results for randomly generated codes.

synthesis benchmark designs. GAULEG (Gauss-Legendre weights and abscissas), GAUHER (Gauss-Laguerre weights and abscissas) and GAUJAC (Gauss-Jacobi weights and abscissas) are taken from [13]. Overall, the improvements by Sch-SOA are 53% and 34% on the average compared to OFU and Solve-SOA [7], respectively.

design ($\|V\|/\|S\|$)	# addr_instr			gain over
	OFU	Sol-SOA[7]	Sch-SOA	OFU/[7] (%)
FIR (5/20)	7	4	2	71.4/50
BIQUAD (10/38)	20	16	11	45/31.2
COMP (10/18)	13	8	6	53.8/25
ELLIP (45/100)	84	49	36	57.1/26.5
GAULEG (23/73)	25	20	12	52/40
GAUHER (11/59)	23	21	13	43.4/38.0
GAUJAC (23/148)	111	76	50	54.9/34.2
avg.				53.9/34.9

Table 3: SOA results for benchmark designs.

Figure 9 graphically summarizes the comparisons of the address code sizes produced by Solve-GOA [7] and Sch-GOA with the changes of the number of ARs. The numbers in X-axis indicates the number of address instructions. As the number of ARs increases the improvement by Sch-GOA is less. This is mainly due to the fact that the number of ARs is relatively large enough to fully utilize the auto-increment/decrement addressing modes without the exploitation of scheduling. However, overall our improvements are 13%-33% of address code reduction over Solve-GOA and it took within 16 seconds when $|AR| = 2$ and 4 seconds when $|AR|=8$.

5. CONCLUSIONS

In this paper, we proposed a new technique for optimizing address instructions for DSP code generation by utilizing code scheduling. Specifically, we solve the SOA/GOA (offset assignment) problem coupled with scheduling in two steps: (1) generating an initial schedule and address assignment and (2) iteratively improving the SOA/GOA solution. To support a large number of iterations, we

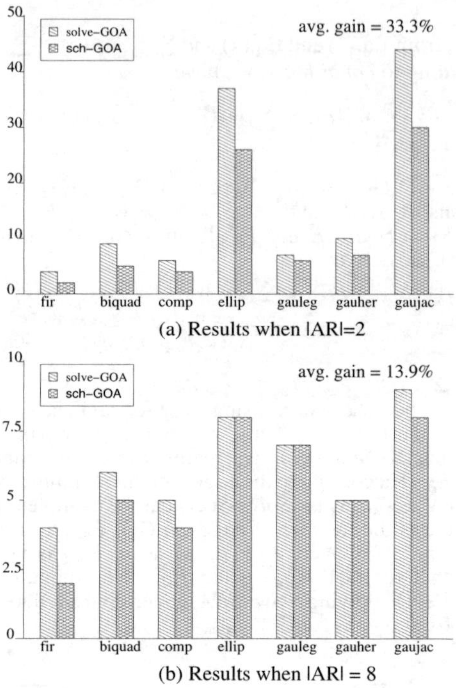

Figure 9: GOA results for benchmark designs.

devised *a set of techniques for fast evaluation of the SOA/GOA solution for a reschedule based on the solution for the previous schedule.* Based on iterative improvement technique, the proposed algorithm may be slow for a large code. However, by limiting the number of iterations or reschedulable operations, it can have moderate speed. Since our approach is oriented towards reducing addressing code size, it may not be suitable for superscalar or VLIW DSPs and DSP codes with ralatively few addressing codes.

Acknowledgment : This work was supported by the Korea Science and Engineering Foundation (KOSEF) through the Advanced Information Technology Research Center(AITrc).

6. REFERENCES

[1] S. Liao, *et. al*, "Storage Assignment to Decrease Code Size," *Proc. SIGPLAN on PLDI*, 1995.

[2] P. Marwedel, G. Goossens (Editors), *Code Generation for Embedded Processors,* Kluwer Academic Publishers, 1995.

[3] S. Udayanarayanan and C. Chakrabarti, "Address Code Generation for Digital Signal Processors," *Proc. DAC*, 2001.

[4] C. Lee, M. Potkonjak, and W. H. Mangione-Smith, "Mediabench: A Tool for Evaluating and Synthesizing Multimedia and Communications Systems," *Proc. Int. Sym. on Microarchitecture*, 1997.

[5] D. H. Bartley, "Optimizing Stack Frame Accesses for Processors with Restricted Addressing Modes," *Software - Practice and Experience*, Vol. 22, No. 2, 1992.

[6] S. Liao, *et. al*, "Storage Assignment to Decrease Code Size," *ACM TOPLAS*, Vol. 18, No. 3, 1996.

[7] R. Leupers and P. Marwedel, "Algorithm for Address Assignment in DSP Code Generation," *Proc. ICCAD*, 1996.

[8] R. Leupers and F. David, "A Uniform Optimization Technique for Offset Assignment Problem," *Proc. ISSS*, 1998.

[9] A. Sudarsanam, S. Liao, and S. Devadas, "Analysis and Evaluation of Address Arithmetic Capabilities in Custom DSP Architectures," *Proc. DAC*, 1997.

[10] C. Gebotys, "DSP Address Optimization using a Minimum Cost Circulation Technique," *Proc. ICCAD*, 1997.

[11] A. Rao and S. Pande, "Storage Assignment using Expression Tree Transformation to Generate Compact and Efficient DSP Code," *Proc. ACM SIGPLAN on PLDI*, 1999.

[12] S. Lim, J. Kim and K. Choi, "Scheduling-based Code Size Reduction in Processors with Indirect Addressing Mode," *Proc. CODES*, 2001.

[13] W. H. Press, *et. al* (Editors), *Numerical Recipes in C: The Art of Scientific Computing*, Cambridge University Press, 1993.

Multifunctional Photonic Integration for the Agile Optical Internet

Edward H. Sargent
Nortel Networks – Canada Research Chair in Emerging Technologies
Department of Electrical and Computer Engineering
10 King's College Road, Toronto, Ontario, Canada
416 946 5051
ted.sargent@utoronto.ca

ABSTRACT
An agile, transparent optical network is emerging. This paper enumerates the functions that will be needed at nodes in order to add transparency and agility to the network while robustly assuring optical fibre channel performance. One key requirement will be scalable integration of multiple functions on a platform. The paper presents recent results along one strategic axis for integration – the use of functionalized self-organized photonic crystals and heterostructures thereof to control the flow and features of light.

Categories and Subject Descriptors
C.2.1 [Network Architecture and Design]: Enabling technologies.

General Terms
Experimentation, Theory.

Keywords
Agile optical networks, reconfigurability, optical performance monitoring, optoelectronic integration, photonic crystals, optical nonlinearity, electro-optics, optical polymers, semiconductor nanocrystals.

1. STATE OF THE OPTICAL NETWORK
Fibre-optics has made available tremendous bandwidth for long-distance, high-speed communications. Increasingly, fibre-optics is making its way toward the network edge, with significant commercial activity in metropolitan-area networks and a reemergence of interest in access and local-area optical networking.

Bandwidth breakthroughs – exponentially-growing time-domain rates and dense wavelength-division multiplexing – engender new bottlenecks. An electronic implementation of a switch so massive as to enable grooming and routing of the information contents of multiple fibres, each conveying terabits of data per second, would be onerous. Some degree of data forwarding may in the future be executed in the optical domain.

2. CHALLENGE: ENABLING THE AGILE OPTICAL NETWORK
In the emerging agile optical network, much of the traffic not originating at or being consumed by a given node will pass through, largely untouched, in the optical domain.

Agility is needed in this network as traffic patterns evolve and new nodes are added. The ability to establish new lightpaths – optically transparent links across multiple physical nodes to an intended destination – is in part enabled by the introduction of optical switches which can route optical signals to different spatial ports. Combined with wavelength-domain multiplexers and demultiplexers, these allow switching of different wavelength-domain channels among fibres. Some degree of wavelength conversion may be used to resolve conflicts in wavelength use in a given fibre. Optical switches are reaching maturity and seeing commercial deployment; reconfigurable wavelength conversion is farther from being a widespread practical reality.

These building-blocks of optical switching – cost-reduced multiplexers and demultiplexers, reliable large-port-count 3-D MEMS switches, and wavelength converters – introduce a new set of challenges. Optical transparency along a diversity of paths of time-varying lengths, losses, dispersions, and accumulated nonlinearities implies a need to groom signals dynamically along multiple performance axes. Some components – optical amplifiers are one example – may introduce interchannel interactions following reconfiguration events.

Switching may not be executed without awareness of and sensitivity to its implications on channels sharing a link.

3. NEED: INTELLIGENT OPTICAL SWITCHING NODES
In addition to raw switching, mux/demux, and wavelength conversion capabilities, the agile optical network requires nodes which combine and interconnect a multiplicity of further functions, among them:

- power monitoring and adjustment
- chromatic dispersion monitoring and compensation

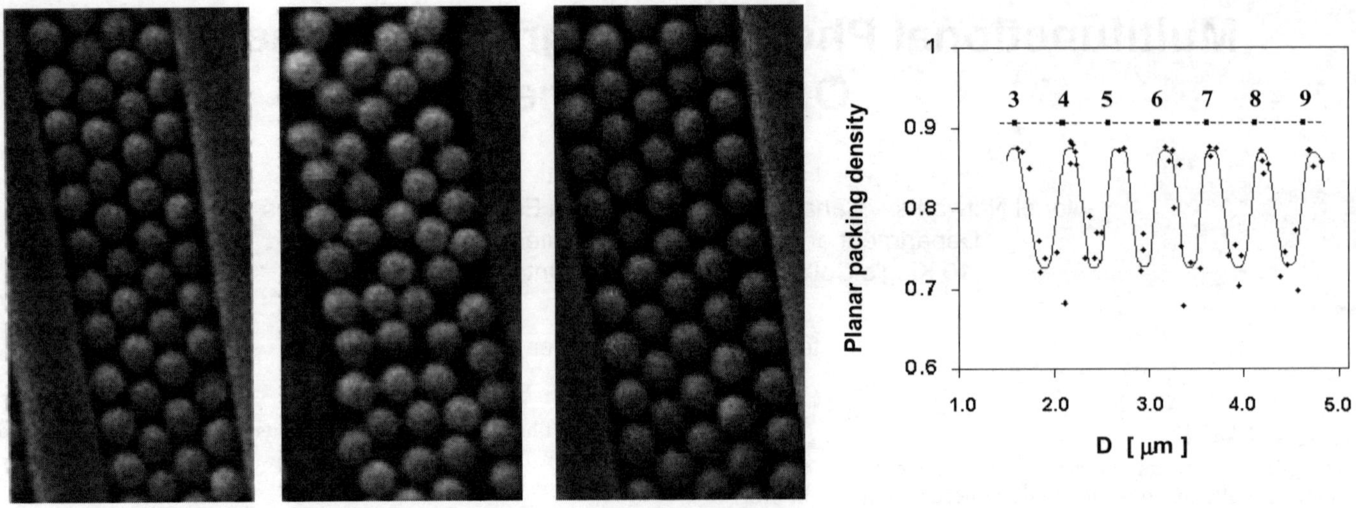

Figure 1. Colloid deposited on mesoscopically patterned substrate. *Left:* From left to right, the width of the groove are 2.22, 2.51 and 2.22 μm The particle diameter is 0.58 μm. *Right:* Planar packing density as a function of the groove width.

- polarization mode dispersion monitoring and compensation
- channel add/drop which is adiabatic on the scale of fibre amplifier response times
- optoelectronic transmit/receive
- control plane interface, logic, and protocol implementation.

At the level of more basic optical and electronic functionalities, these may entail, across a prospectively broad spectral range:

- photodetection and time-domain gating
- light emission
- optical amplification and attenuation
- modulation of refractive index
- low-loss waveguiding
- electronic transistor action
- microwave-frequency guiding.

4. ENGINEERING A SEMICONDUCTOR FOR LIGHT

In some, though admittedly not all, respects, what is needed is a semiconductor for light.

Semiconductors provide a versatile foundation for electronics. Electronic Bloch waves propagate or ar forbidden in a periodic crystal. Energy-momentum dispersion relations for electrons in periodic media give rise to bandgaps, and in turn to the possibility of carrier confinement and anticonfinement through heterostructures. Bandgaps – regions of forbidden energy – give leverage to small levels of impurities in influencing the type and extent of conductivity in donor-rich (n-type) and acceptor-rich (p-type) materials.

Photonic crystals are three-dimensionally periodic materials which may potentially be used to enhance control over the flow of light. Increasingly well-ordered photonic crystals have been demonstrated in recent years. However, a high degree of ordering is a necessary but not a sufficient condition to control the flow of light. Also needed is a means of placing and shaping the photonic crystals to build mesoscopic devices and connect them together.

Colloidal crystals – arrays of spherical particles – provide a self-organizing basis for the realization of photonic crystals, analogous to atomic crystal growth techniques. Polymer spheres are synthesized to have a diameter similar to the period of lightwaves. These materials may be tailored to resonate with certain colours of light, including the wavelengths used in fibre-optic communications systems.

Recently, successful growth of colloidal crystals on textured substrates has been reported to result in long-range ordering in photonic crystals [1]. Holography was used to pattern a photosensitive polymer covering a conductive substrate in order to produce conductive and dielectric features on the surface.

Charged colloid particles were deposited on the substrate by electrophoresis. The particles assembled only on in the conductive areas of the substrate.

The work makes possible precise control over sphere placement and ordering in the crystal through the use of a suitable pattern.

In the absence of any surface pattern, the colloid particles assembled in a random close-packed fashion. However, when grooves were used which were sufficiently large to accommodate several spheres, the presence of a surface pattern induced a hexagonal ordered packing of the particles in the grooves. This occurred only when the groove width corresponded to the width of an integer number of rows of spheres. When this condition was not met, the arrangement of the spheres was not periodic, and the planar packing density decreased below the maximum theoretical value of 0.907, as shown in Fig. 1(b).

The use of narrower grooves designed to accommodate a single sphere in each groove was also demonstrated. Figure 2 shows different packing scheme, square and hexagonal, observed on the same sample where the ratio of the grating period to the sphere diameter is approximately 0.81, between the theoretical values for the square packing arrangement, 0.707, and the hexagonal close-packed arrangement, 0.866

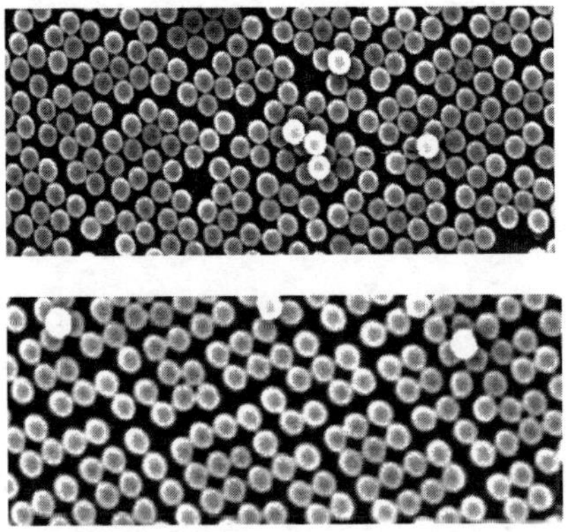

Figure 2 Colloid deposited on microscopically patterned substrate, where the period of the grating is approximately 0.81 sphere diameter. From top to bottom: the grating before colloid deposition, hexagonal packing and square packing.

Thus, photonic crystals can be placed where they are needed and surrounded by a variety of different polymers and glass materials as required. Not only can the degree of order now be controlled, but also the nature of photonic crystal packing – hexagonal versus square, for example. The nature of the photonic crystal lattice determines how waves of light interact with the photonic crystal.

To complement experimental progress/progress in fabrication, a set of tractable design principles and methodologies is needed. In semiconductors, this is available through spatial band diagrams. The energy-momentum bandstructure is captured through a small number of parameters, such as band edge positions, effective masses, the energetic structure seen by charge carriers in devices. In semiconductor heterostructures, band offsets provide an additional degree of freedom in guiding, manipulating, and confining charge carriers.

The analogous picture is rapidly emerging in the photonic case. An envelope approximation formalism was recently developed for 3-D photonic crystal heterostructures [3, 4]. It requires knowledge only of the bulk bandstructure and heterostructure design. The method is mathematically analogous, but distinct from, the envelope formalism widely deployed throughout semiconductor quantum electronic engineering. The method has been applied to photonic crystal waveguides (Figs. 3 and 4), predicting within 1% accuracy [4] the frequencies of guided modes and providing excellent agreement in the shapes of waveguided mode shapes, compared with results from full, computationally-intensive 3D electromagnetic vectorial solution. Guided modes have been shown to be allowed for wavevectors where the curvature of a band in a direction perpendicular to the plane of the waveguide has the same sign as the refractive index contrast between the core and the cladding. Elementary waveguide theory has been shown to provide the basis for the computation of mode shapes and dispersion relations.

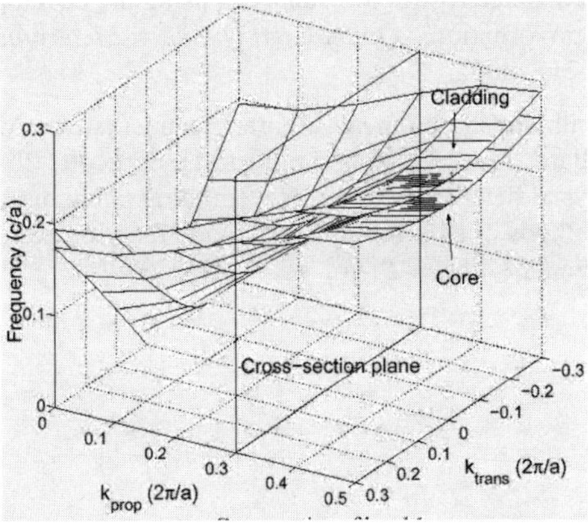

Figure 3. Photonic crystal waveguiding - confinement and modes with the envelope formalism.

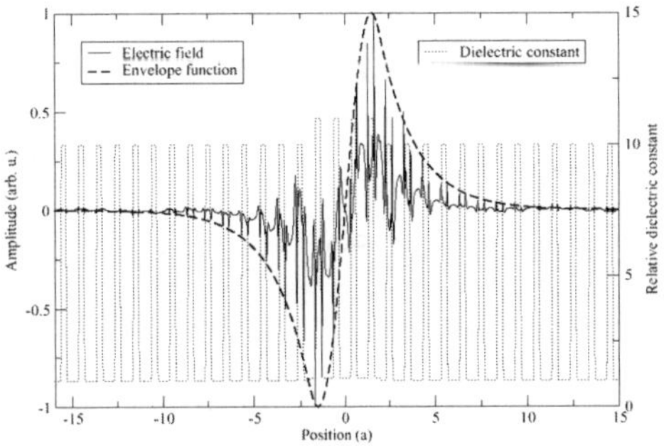

Figure 4. Odd mode of the first band through exact numerical and approximate envelope analyses.

5. CONCLUSIONS

Optoelectronic integration has been discussed and explored since the 1970's. Its time may finally have come, for the agile optical network will rely on scaling optoelectronic functions and densities far beyond what is currently available. Photonic crystals may play a role in increasing control over the flow of light, and in processing the information-bearing signals which it conveys.

6. ACKNOWLEDGMENTS

We thank Nortel Networks and NSERC for their support of this research. The work summarized in the present work is reported in detail in references [1-4] and was carried out by research assistants Mathieu Allard, Emanuel Istrate, Mathieu Charbonneau-Lefort and Joyce Poon and by collaborators Rob Golding and Prof. Eugenia Kumacheva.

7. REFERENCES

[1] E. Kumacheva, R.K. Golding, M. Allard, E.H. Sargent, "Colloid crystal growth on mesoscopically patterned surfaces: Effect of confinement," *Advanced Materials*, vol. 14, pp 221-224 , 2002.

[2] M. Allard, R. Golding, E. Kumacheva, E. H. Sargent, "Control over ordering in colloidal crystals using surface templating," accepted, *2002 Conference on Lasers and Electro-Optics*.

[3] E. Istrate and E. H. Sargent, "The photonic analogue of the graded heterostructure: analysis using the envelope approximation," *Optical and Quantum Electronics*, vol. 34, p. 217, 2002.

[4] Mathieu Charbonneau-Lefort, Emanuel Istrate, Mathieu Allard, Joyce Poon, and Edward H. Sargent, "Photonic crystal heterostructures: Waveguiding phenomena and methods of solution in an envelope function picture," *Physical Review B*, vol. 65, p. 125318, 2002.

Computer Aided Design of Long-Haul Optical Transmission Systems

James G. Maloney
PhotonEx
200 MetroWest Technology Dr
Maynard MA 01754
978-723-2203
jmaloney@photonex.com

Brian E. Brewington
PhotonEx
200 MetroWest Technology Dr
Maynard MA 01754
978-723-2235
brewington@photonex.com

Curtis R. Menyuk
PhotonEx
200 MetroWest Technology Dr
Maynard MA 01754
978-723-2284
menyuk@photonex.com

ABSTRACT

We present a general overview of the role of computer models in the design and optimization of commercial optical transmission systems. Specifically, we discuss (1) the role of modeling in a commercial setting, (2) achieving the proper balance between accuracy and computation speed, (3) model verification against experiment, and (4) case studies demonstrating the benefits of modeling. Ideally, experiments are preferable to models when describing system performance, particularly to support claims of a system's functionality to a customer. However, modeling is often the only choice for many of the problems that a commercial networking company must solve. Because there are design parameter spaces that are either too expensive or too time-consuming to verify experimentally, the main role of modeling in industry is to study what experiments cannot. For example, when an analytical solution of a statistical problem is infeasible, a common modeling solution is to perform Monte Carlo trials to study the statistical behavior. Another typical modeling task involves looking at variations of hardware that would be prohibitively expensive to acquire and test. Implementing modeling in industry involves a balance between three needs: cost-efficiency, time-efficiency, and accuracy. We will discuss the approaches we have taken at PhotonEx to meet these needs: leveraging academic research, developing "reduced models" and utilizing computational clusters. Specifically, we will use case studies to illustrate the application of these approaches to modeling long-haul optical transmission systems.

Categories and Subject Descriptors

J.6 [**Computer Applications**]: Computer-Aided Engineering - *computer-aided design (CAD) and computer-aided manufacturing (CAM).*

General Terms

Algorithms, Measurement, Performance, Design, Experimentation, Theory, Verification

Keywords

Optical Communication, Long-Haul (LH) Transmission, Ultra-Long Haul (ULH) Transmission, Optical Modeling

A Fast Optical Propagation Technique for Modeling Micro-Optical Systems

Timothy P. Kurzweg[*], Steven P. Levitan[*], Jose A. Martinez[*], Mark Kahrs[*], Donald M. Chiarulli[+]

[*]University of Pittsburgh, Department of Electrical Engineering, Pittsburgh, PA, USA,
tim@ee.pitt.edu, steve@ee.pitt.edu, jmarti@ee.pitt.edu, kahrs@ee.pitt.edu

[+]University of Pittsburgh, Department of Computer Science, Pittsburgh, PA, USA, don@cs.pitt.edu

ABSTRACT

As designers become more aggressive in introducing optical components to micro-systems, rigorous optical models are required for system-level simulation tools. Common optical modeling techniques and approximations are not valid for most optical micro-systems, and those techniques that provide accurate simulation are computationally slow. In this paper, we introduce an angular frequency optical propagation technique that greatly reduces computation time while achieving the accuracy of a full scalar formulation. We present simulations of a diffractive optical MEM Grating Light Valve to show the advantages of this optical propagation method and the integration of the technique into a system-level multi-domain CAD tool.

Categories and Subject Descriptors

I.6.5 [**Simulation and Modeling**]: Model Development - *modeling methodologies*

General Terms

Algorithms, Design

Keywords

Optical Propagation, Angular Spectrum, CAD, Optical Micro-systems, Optical MEMS

1. INTRODUCTION

It is well known that optics can provide advantages to micro-systems in terms of speed, bandwidth and reduced power [12]. However, by adding this optical domain to micro-systems, many new challenges are introduced in system design. In these multi-domain systems, optical effects, such as diffraction, interference, and scattering, are critical to the success or failure of the designs. Therefore, optical modeling and simulation are crucial in the design stage of the micro-system.

Currently, multi-domain micro-systems are simulated by point-tools using component level models performed at the physical level. These tools are typically domain specific and computationally complex. In contrast, system-level tools are designed to include multiple domains and allow efficient system simulation, by modeling components by their functionality rather than their physical construction. Established MEM (micro-electrical-mechanical) system-level modeling tools exist, with physical device models being extracted into behavioral-level models. However, optical propagation models are not easily integrated into these existing tools.

When optical wavefronts interact with the small feature sizes of micro-systems, many of the common optical propagation modeling techniques become invalid, and full vector or scalar solutions to Maxwell's equations are required for accurate simulation [7]. However, these accurate solutions are computationally intensive, making interactive design between system designer and CAD tool almost impossible. As more optical components are introduced into micro-systems and the systems become more complex, the demand for computationally efficient simulation tools increases. Therefore, the problem of optical modeling in MOEM (micro-electro-opto-mechanical) systems is two-fold: first, a rigorous model is needed to model optical propagation, and, second, the model must be computationally efficient.

In this paper, we present a solution for both parts of the problem. We introduce an optical propagation model that provides valid results in the micro-system domain, with a computational algorithm that allows for interactive CAD design. We first provide motivation for modeling tools for MOEM systems and components by examining a Grating Light Valve device (GLV). This is followed by a brief background of optical propagation modeling, from which we determine a valid scalar technique for optical MEM system modeling. Next, a description of the angular spectrum technique, which we use to reduce the computational load, is provided. System-level simulations and analyses of a GLV system follow. We conclude with a summary and discussion of future work.

2. MOTIVATION

To provide motivation for our work, we first examine one of the more promising MOEM components, the Grating Light Valve [1]. This device has many display applications, including digital projection, HDTV, and

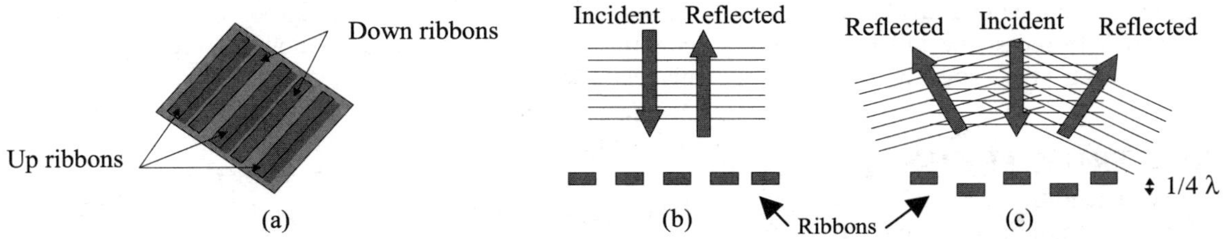

Figure 1: GLV Device (a) Top View and Side View Operation for (b) Up Ribbons and (c) Down Ribbons

vehicle displays. The GLV is simply a MEM (micro-electrical-mechanical) phase grating made from parallel rows of reflective ribbons. When all the ribbons are in the same plane, incident light that strikes normal to the surface reflects 180 degrees off the GLV. However, if alternating ribbons are moved down a quarter of a wavelength, a "square-well" diffraction pattern is created, and the light is reflected at an angle from that of the incident light. The angle of reflection depends on the width of the ribbons and the wavelength of the incident light. Figure 1 shows the ribbons, from both a top and side view, and also the reflection patterns for both positions of the ribbons.

The GLV component is fabricated using standard silicon VLSI technology, with ribbon dimensions approximately 3-5μm wide and 20-100 μm long [1]. Each ribbon moves through electro-static attraction between the ribbon and an electrode fabricated underneath the ribbon. This electro-static attraction moves the ribbons only a few hundred nanometers, resulting in an approximate switching time of 20 nsec. Since the GLV depends on a diffractive phenomenon to direct the light beam, a rigorous modeling technique is required for modeling the GLV system.

3. OPTICAL PROPAGATION

Optical propagation can be modeled completely by the solution of Maxwell's equations for both the electric field vector, $\vec{E}$, and the magnetic field vector, $\vec{H}$ [4]. Solutions for these fields are typically done with a Beam Propagation Method (BPM), using a Fourier Transform, or a finite-difference time-domain (FDTD) algorithm. This vector method is valid for optical propagation in micro-systems, however, the computation time and memory requirements are extremely demanding.

To reduce the computational resources of the vector solutions, a scalar representation is commonly used. Scalar optics are defined by summarizing vectors $\vec{E}$ and $\vec{H}$ by a single complex scalar, *U*. This replacement is valid if the propagation medium is dielectric, isotropic, homogenous, nondispersive, and nonmagnetic. Propagation through free-space meets these requirements. No longer are we solving in 3D, as the scalar function represents a complex 2D wave function.

This complex scalar must satisfy the Helmholtz wave equation, $(\nabla^2 + k^2)U = 0$, where, the wave number, $k=2\pi/\lambda$. With use of Green's theorem, the Rayleigh-Sommerfeld formulation is derived from the wave equation for the propagation of light in free-space [4]:

$$U(x,y,z) = \frac{z}{j\lambda}\iint_{\Sigma} U(\xi,\eta,0)\frac{\exp(jkr)}{r^2}\,\partial\xi\,\partial\eta$$

where, $r = \sqrt{z^2 + (x-\xi)^2 + (y-\eta)^2}$, Σ is the area of the aperture, and *z* is the distance that the light is propagated from an aperture plane ($z = 0$) to an observation plane. It is assumed that the two planes are parallel, with coordinate systems (ξ,η,0) in the aperture plane and (x,y,z) in the observation plane, as seen in Figure 2. The formulation is valid as long as both the propagation distance and the aperture size are greater than the wavelength of light. These restrictions are based on the boundary conditions of the Rayleigh-Sommerfeld formulation, and the fact that the electric and magnetic fields cannot be treated independently at the boundaries of the aperture [4].

To compute the complex wave front at the observation plane, we divide both the aperture plane and the observation plane into gridded meshes, as seen in Figure 2. Both planes are commonly meshed into NxN regions, where N is the number of mesh points along the side of a square. Using a direct integration technique we have successfully shown results using the Rayleigh-Sommerfeld method [7]. However, this direct integration method is computationally intensive, with a computational order of $O(N^4)$.

The far (Franhofer) and near (Fresnel) field approximations of the scalar formulation reduce the computational intensity, using a FFT for optical propagation. However, we have previously shown that these techniques are not valid for typical micro-system dimensions [7]. We have also shown that efficient Gaussian beam propagation

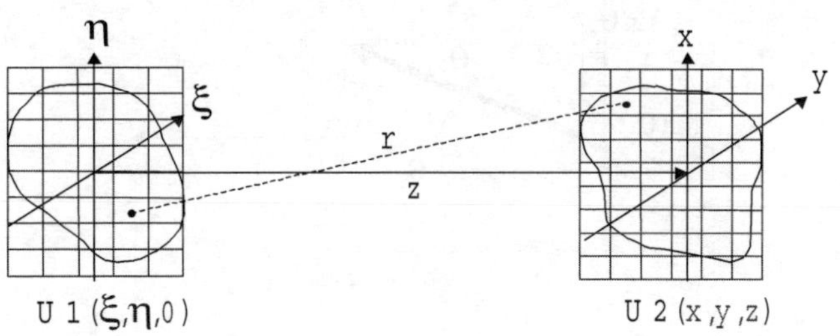

Figure 2: Aperture and Observation Coordinate System in the Rayleigh-Sommerfeld Approximation

techniques have limitations for optical MEM simulation. In the interest of reducing the computational load of using a full scalar technique, we look to recast the full Rayleigh-Sommerfeld formulation using the angular spectrum technique.

4. ANGULAR SPECTRUM TECHNIQUE

As an alternative to direct integration over the surface of the wavefront, the Rayleigh-Sommerfeld formulation can also be solved using a technique that is similar to solving linear, space invariant systems. In this case, the complex wavefront is analyzed across its surface with a Fourier transform. By using the Fourier transform, the complex optical wavefront is reduced into a set of simple linear exponential functions. This transform identifies the components of the angular spectrum, which are plane waves traveling in different directions away from the surface [4].

Examining the angular spectrum, we look at light propagating from an aperture plane at z=0 to a parallel observation plane. The wave function U(x,y,0) has a 2D Fourier transform, $A(v_x,v_y,0)$, in terms of angular frequencies, v_x and v_y.

$$A(v_x, v_y, 0) = \iint U(x, y, 0)\exp[-j2\pi(v_x x + v_y y)]\,\partial x \partial y$$

where, $v_x = \sin\theta_x / \lambda$ and $v_y = \sin\theta_y / \lambda$.

$\sin(\theta_x)$ and $\sin(\theta_y)$ are the directional cosines of the plane wave propagating from the origin of the coordinate system, as seen in Figure 3. *A* is the complex amplitude of the plane wave decomposition defined by the specific angular frequencies.

To propagate the complex wave function to a parallel plane, a propagation phase term is used as a transfer function. The relationship of propagation in the frequency domain between $A(v_x,v_y,0)$ and $A(v_x,v_y,z)$ has been computed by satisfying the Helmhotz equation with the propagated complex wave function, U(*x,y,z*) [4]:

$$A(v_x, v_y, z) = A(v_x, v_y, 0)\exp\left\{jz2\pi\sqrt{\frac{1}{\lambda^2} - v_x^2 - v_y^2}\right\}$$

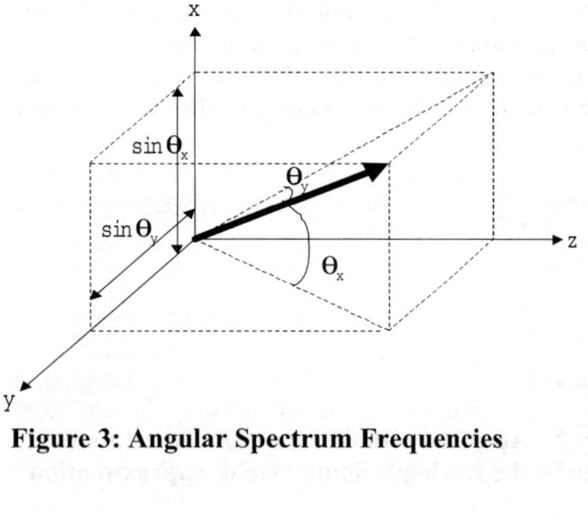

Figure 3: Angular Spectrum Frequencies

This describes the phase difference that each of the plane waves, differentiated by the angular, or spatial, frequencies, experiences due to the propagation between the parallel planes. Therefore, the wave function after propagation can be transformed back into the spatial domain with the following inverse Fourier transform:

$$U(x, y, z) = \iint A(v_x, v_y, 0)\exp\left\{jz2\pi\sqrt{\frac{1}{\lambda^2} - v_x^2 - v_y^2}\right\}\times$$
$$\exp[j2\pi(v_x x + v_y y)]\,\partial v_x \partial v_y$$

It is interesting to note that the output complex wavefunction is simply the convolution of the input wave function and the propagation phase function. Using Fourier transform theory, the convolution in the spatial domain is performed by a multiplication in the frequency domain.

It is appropriate to discuss the physical effect of evanescent waves, which are defined in the case when, $1/\lambda^2 - v_x^2 - v_y^2 < 0$. These waves carry very little power and die out in a couple of wavelengths of propagation [6]. In our simulations, we ignore these waves.

The angular spectrum method is restricted to propagation between parallel planes that share a common center. Removing these restrictions has been the goal of recent research. Tommoasi and Bianco have determined how to propagate to a plane that is tilted with respect to initial plane [10]. Delen and Hooker have determined a way to allow offsets in the observation plane [3]. We summarize these two methods next.

For arbitrary angles between the aperture plane, U(ξ,η,ζ) and the observation plane, U(x,y,z), a mapping of the spatial frequencies in each plane's coordinates system must occur. This mapping is possible due to the fact that the phase accumulation term does not change when the waves propagate to an observation plane that is not normal to the aperture plane. It can be found that the rotational matrix, *M*, relating (ξ,η,ζ) to (x,y,z), can be used to relate spatial frequencies in the two coordinate systems by [10]:

$$(x, y, z)^t = M(\xi, \eta, \zeta)^t \qquad (v_\xi, v_\eta, v_\zeta)^t = M^t(v_x, v_y, v_z)^t$$

In the new tilted coordinate system, the incoming spatial frequencies are perceived as having spatial frequency corresponding to the outgoing coordinate system. For example, the incoming aperture plane wave having spatial frequencies (0,0) correspond to angle (-ϕ,0) in the observation plane with a ϕ tilt in the y-direction. In all cases, even if the spatial frequencies are re-mapped, the amplitude of the plane wave remains constant.

For an observation plane whose center is offset from the propagation axis of the aperture plane, the Fourier shifting theorem can be used to solve for the complex wave function [3]. With this relation between the offset of the coordinate systems, the function for free-space propagation between offset planes is:

$$U(x,y,0)=\iint A'(v_x,v_y,0)\exp\left\{jz2\pi\sqrt{\frac{1}{\lambda^2}-v_x^2-v_y^2}\right\}\times$$
$$\exp[j2\pi(v_x x+v_y y)]\,\partial v_x \partial v_y$$

where,

$$A'(v_X,v_Y,0)=A(v_X,v_Y,0)\exp[j2\pi(v_x(x'-x)-v_y(y'-y)]$$

The advantage of using the angular spectrum to model light propagation is that the method is based on Fourier transforms. In CAD tools, the Fourier transform can be implemented by one of the numerous Fast Fourier Transform (FFT) techniques. The computational order of the FFT for a 2D input, is $O(N^2\log_2N)$, obviously more efficient when compared to the direct integration method. We show this speed increase later through example.

In continuous theory, the angular spectrum method is an exact solution of the Rayleigh-Sommerfeld formulation. However, when solving the algorithm on a digital computer, a Discrete Fourier Transform (DFT) must be used, resulting in the accuracy of the angular spectrum method depending on the resolution, or spacing, of the aperture and observation plane meshing. We call the physical size of the aperture and observation planes the "bounding box", defining the size of the optical wavefront being propagated. Since the complex wave function is only non-zero for a finite space in the bounding box, the signal is not always bandwidth limited, and the Nyquist sampling theory does not always apply. It can be shown, however, that the resolution of the aperture and observation meshing must be $\lambda/2$ or smaller [3]. For many simulation systems without large degrees of tilt and hard diffractive apertures, the resolution can be coarser. In systems with high tilts, the resolution is most sensitive. With a mesh spacing of $\lambda/2$, the angular spectrum decomposition will ensure plane waves propagating from aperture to observation plane in a complete half circle, that is, between –90 and +90 degrees.

Other inaccuracies that can occur when using a discrete Fourier transform are aliasing and window truncation. Aliasing occurs when frequencies exist greater than the critical sampling frequency. In this case, these high frequencies are "folded over" into the sampled frequency range [2]. The effect of this is seen in our simulations as optical power "reflecting" off of the walls of the bounding box. If significant optical power reflects off the wall, interference between the propagating beam and these reflections can occur, resulting in inaccurate optical waveforms. The same effect can be seen when the bounding box truncates the signal. Truncation occurs when the waveform propagates into the edges of the bounding box. The simplest solution toward accurate results is having sufficient zero padding around the optical waveform, reducing the chance the waveform is aliased or truncated off the walls of the bounding box.

This work has been implemented in our system-level, multi-domain CAD tool, Chatoyant [7]. Chatoyant supports modeling and simulation in the optical, electrical, and mechanical domains. The electrical and mechanical simulations are implemented using a piece-wise linear Modified Nodal Analysis technique [8]. The optical simulation supports both Gaussian propagation and the angular spectrum technique presented in this paper.

5. GLV SIMULATIONS

The advantages of the angular spectrum method are best seen through example. We simulate and analyze the GLV system, as presented earlier, in Chatoyant. For the simulation of the GLV, we examine one optical pixel. A projected pixel is diffracted from a GLV composed of 4 ribbons, two stationary and two that are movable [1]. Each ribbon has a length of 20 µm and a width of 5 µm. Ideally, there is no gap between the ribbons, however, in reality, a gap is present and is a function of the feature size of the fabrication. Although this gap can be modeled in our tool, in these simulations, we provide an ideal GLV simulation with no gap.

The GLV is modeled as a phase grating, where the light that strikes the down ribbons propagates a half of a wavelength more than the light that strikes the up ribbons. In our model, light reflecting from the down ribbons is multiplied by a phase term. The phase term is similar to a propagation term through a medium: $U_{down\_ribbon} = U\exp(j2kd)$, where, d is the distance that the ribbon is moved towards the substrate, typically $\lambda/4$ for the GLV.

Far-field diffraction theory states the diffracted angle reflected from the square-well grating is [5]: $\theta = q\lambda / a$, where, q is the diffraction mode (0, ±1, ±2, ±3,...), a is the period of the diffractive grating, and θ is in radians. In the special case of a square well, when light is diffracted by a grating with a displacement of $\lambda/4$ (a $\lambda/2$ optical path difference after reflection), all the optical power is diffracted from the even modes into the odd modes [9].

In the first simulation, the standard operation of the GLV is verified. We assume an incident plane wave of green light (λ=520 nm) striking the grating, with the square-well period defined by the ribbon width, and no gap. We simulate the GLV in both cases, that is, when all the ribbons are on the same plane and when the alternating ribbons are moved downward a distance of $\lambda/4$. In this example, the light is reflected off of the grating and propagated 1000 µm to an observation plane. A bounding box of 400x400 µm is used, with N equal to 2048. Intensity contours of the observation plane are presented in Figure 4(a and b).

When the grating is moved into the down position, all of the optical power is not transferred into the expected odd far-field diffractive modes. This is seen in the center of Figure 4(b), as small intensity clusters are scattered between the $\pm1^{st}$ modes. This scattering is a near field

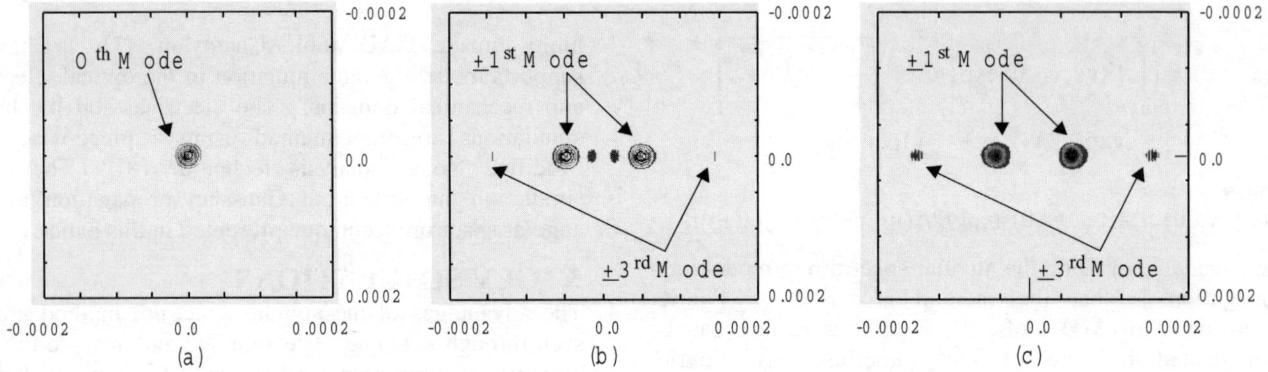

Figure 4: GLV Operation (a) All Ribbons Up (b) Alternating Ribbons Down (c) Fraunhofer Approximation

effect, and demonstrates that in this system, light propagating 1000 µm is not in the far-field. If a designer used a tool propagating with the Fraunhofer far-field approximations, these scattering effects would not be detected. For example, when running the same simulation on LightPipes [11], a CAD tool using the Fraunhofer approximation for optical propagation, only the far-field pattern of light diffracted into the 1st and 3rd modes is seen, as presented in Figure 4(c). When comparing this result to Figure 4(b), it is shown that far field approximation is not valid for this propagation distance. Through this example we have shown that using the angular frequency technique, we achieve the full Rayleigh-Sommerfeld accuracy, while obtaining the same computational speed of using the Fraunhofer approximation.

To show the advantage of the angular spectrum method, we compare the run time of the above simulation with the run time using the direct integration method. With N=2048, the FFT simulation takes about 1.5 minutes. The direct integration technique takes approximately 5.5 days to finish. If N is reduced to 1024, the simulation completes in approximately 25 seconds, whereas the direct integration simulation takes approximately 32 hours. These simulations were run on a 1.7 GHz dual-processor PC running Linux, with 2 GB of main memory.

In the next simulation, we perform a transient sweep of the ribbon movement, from 0 to 150 nm. The rest of the system setup is exactly the same as before. However, this time, we simulate the normalized power efficiency captured in the 1st diffraction mode for different ribbon depths. To simulate this, a circular detector (radius=12.5µm) is placed on the positive 1st mode. Figure 5 is a graph that shows the simulated normalized power efficiency in this first mode. As the ribbons are moved downward, more optical power is diffracted into the non-zero modes. As the ribbons reach the $\lambda/4$ point, almost all the diffractive power is in the $\pm 1^{st}$ mode. Figure 5 also includes intensity contours of selected wavefronts during the transient simulation, along with markings of the system origin and circular detector position. From these wavefronts, interesting diffractive effects can be noted. As expected, when there is little or no ribbon movement, all the light is in the 0th mode. However, with a little ribbon movement, it is interesting to note that the 0th mode is "steered" at a slight angle from the origin. As the ribbons move downward about $\lambda/8$, the energy in the $\pm 1^{st}$ modes are clearly defined. As the gratings move closer to the $\lambda/4$ point, the power is shifted from the 0th mode into the $\pm 1^{st}$ modes, until there is a complete switch. As the ribbons move past the $\lambda/4$ point, optical power shifts back into the 0th mode.

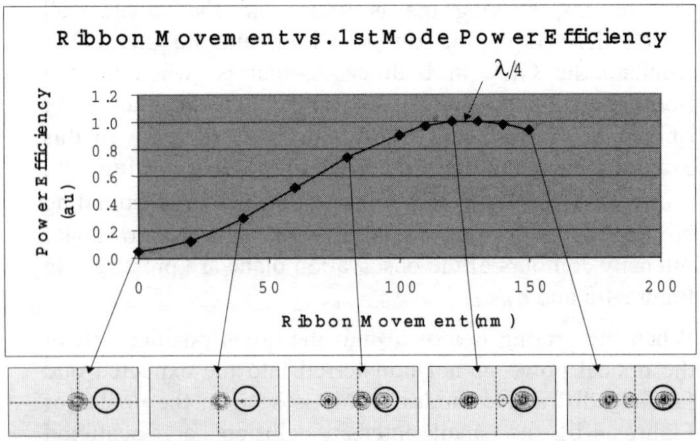

Figure 5: Transient Analysis of Ribbon Movement and Intensity Contours

In the final simulation, we present a full system-level example as we expand the system to show a complete end-to-end link used in a configuration of a color projection system. The system is shown in Figure 6. In this system, we model light, passing through a color wheel, striking a prism, reflecting off the GLV device, past a screen, focused by a lens, and striking a detector [1]. In this system, when the GLV ribbons are all up, the screen blocks the light's 0th mode and the pixel is not displayed. When the alternating

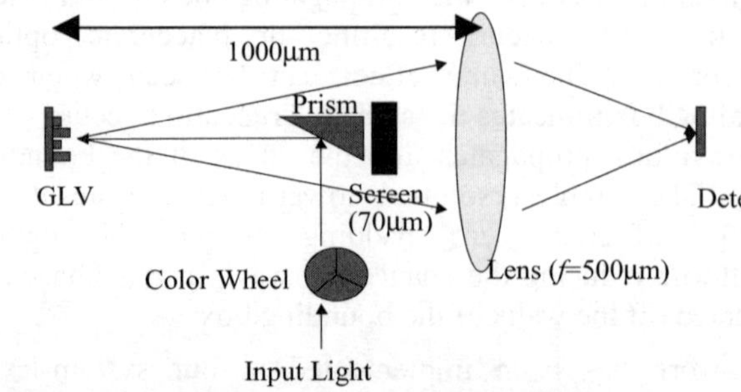

Figure 6: End-to-end GLV Display Link

ribbons are pulled down, the lens focuses the light found in the $\pm1^{st}$ modes and converges them to the center of the system, displaying the pixel. Using a spinning color wheel to change the wavelength of the incident light, a frame-sequential GLV projection system uses red (680 nm), green (530 nm), and blue (470 nm) light on the same grating [1]. Since the same grating is used for all wavelengths of light, the grating movement is tuned for the middle frequency: 130 nm ($\lambda_{green}/4$). During this simulation, we use a hybrid approach for the optical modeling. For the propagation through the color wheel and the prism, we use Gaussian propagation. Since propagating through these components does not diffract the beam, this Gaussian technique is not only efficient, but valid. However, as soon as the light propagates past the prism component, we switch the optical propagation technique to our full scalar method to accurately model the diffraction off of the GLV device. The remainder of the simulation is propagated with the scalar technique.

We analyze the system by looking at the amount of optical power that is being received on a centered circular detector (radius 10μm) for the different wavelengths of light, since we are using the same GLV that is tuned for the green wavelength for all wavelengths. A sweep of the distance between the focusing lens and the detector plane is simulated for 0 to 1500 μm, when the GLV ribbons are pulled down. The graph in Figure 7 shows the normalized power received on the circular detector for each wavelength along with selected intensity contours of the green wavefront as the beam propagates past the lens. For clarity, the detector's size and position is added onto the intensity contours. For distances under 600 μm, the light remains in its two positive and negative 1st modes, as the convergence of the beams has not occurred, resulting in zero power being received on the center detector. As expected, each of the wavelengths focuses at a different rate, as shown by each the wavelength's specific curve in Figure 7. However, it is seen that all wavelengths focus and achieve detected

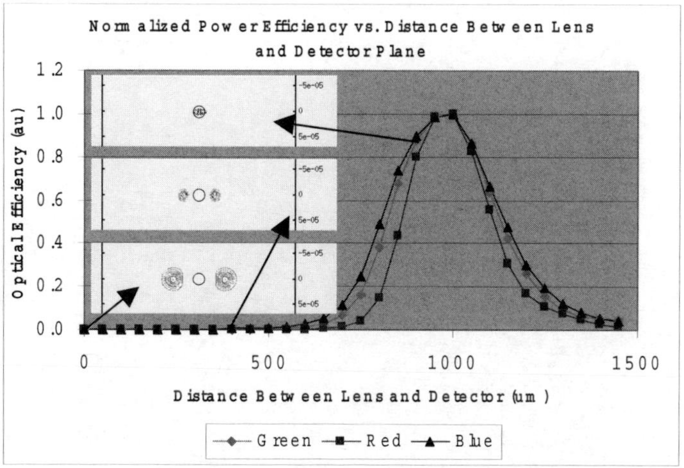

Figure 7: Wavelength Power vs. Distance Propagated

maximum power at a distance past the lens of 1000 μm, or twice the lens' focal length. At this point, all three colors project on top of each other, creating a color pixel in the focal plane. With additional optics, this focal plane can be projected to a screen outside the projector. This simulation has shown that the grating, although tuned for the green wavelength, can be used for all three wavelengths.

6. SUMMARY

In this paper, we have demonstrated a promising optical propagation technique that can greatly speed-up simulation of optical micro-systems. We have shown how to achieve full scalar diffraction with the use of a FFT, by using the angular spectrum of light. This reduces an $O(N^4)$ direct integration problem to $O(N^2 \log N)$. With the angular spectrum method, we can achieve the speed of the Fraunhofer technique, without any of the required approximations of using that technique. We have used the GLV as a typical micro-system to motivate our work, and presented full system level simulation results of this device. In future work, we plan to extend the angular spectrum technique to support roughness and non-planar surfaces.

We would like to acknowledge the support of NSF grant CCR-9988319 and DARPA grant AFOSR F49620-01-1-0536 under the NeoCAD program.

7. REFERENCES

[1] Bloom, D.M., "The Grating Light Valve: Revolutionizing Display Technology," Photonics West, Projection Displays III, 1997.

[2] Briggs, W.L. and Henson, V.E., The DFT: An Owner's Manual for the Discrete Fourier Transform, (SIAM, 1995).

[3] Delen, N., Hooker, B., "Free-space Beam Propagation Between Arbitrarily Oriented Planes Based on Full Diffraction Theory: a Fast Fourier Transform Approach," JOSA, Vol. 15, No. 4, April 1998, pp. 857-867.

[4] Goodman, J.W., Introduction to Fourier Optics, Second Edition (The McGraw-Hill Companies, Inc., 1996).

[5] Hecht, E., Optics, Second Edition (Addison-Wesley Publishing Company, 1987).

[6] Kowarz, M.W., "Diffraction Effects in the Near Field," Ph.D. Thesis, University of Rochester, NY, 1995.

[7] Kurzweg, T.P., Levitan, S.P., Martinez, J.A., Marchand, P.J., Chiarulli, D.M., "Diffractive Optical Propagation Techniques for a Mixed-Signal CAD Tool," Optics in Computing (OC2000), Quebec City, CA, June 18-23, 2000.

[8] Martinez, J.A., Kurzweg, T.P., Levitan, S.P., Marchand, P.J., Chiarulli, D.M., "Mixed-Technology System-Level Simulation", Analog Integrated Circuits and Signal Processing, Vol. 29, pp. 127-149, October/November 2001.

[9] Solgaard, O., "Integrated Semiconductor Light Modulators for Fiber-Optic and Display Applications," Ph.D. Thesis, Stanford University, 1992.

[10] Tommasi, T., Bianco, B., "Frequency Analysis of Light Diffraction Between Rotated Planes," Optics Letters, Vol. 17, No. 8, April 1992, pp.556-558.

[11] Vdovin, G., LightPipes Manual, http://guernsey.et.tudelft.nl

[12] Wu, M.C., "Micromachining for optical and Optoelectronic Systems," Proceedings of the IEEE, Vol. 85, No. 11, November 1997, pp. 1833-1856.

PANEL: Nanometer Design: What Hurts Next...?

Chair: Larry Pileggi, Carnegie Mellon University

Organizers: Rob A. Rutenbar, Carnegie-Mellon University and Andrew Kahng, Univ. of California, San Diego

Abstract

Every year, the design and EDA communities are besieged by dire warnings about the impending doom of "design as we know it." Every year, another unpleasant physical effect from the evil depths of deep submicron physics surfaces, compromising our designs in new and vile ways. Every year, the same story: more nanometer woes. Rather than endorse a new winner in this year's race for the "next worst thing" from the nanometer arena, this panel gathers a set of world-class technology experts to debate what effects are hiding just around the next corner, waiting to pounce on the unwary tool or chip designer. Which among these is really the most important, when will it happen, and why?

Panelists

Bob Brodersen
U.C. Berkeley

Anthony Hill
Texas Instruments

John Kibarian
PDF Solutions

Desmond A. Kirkpatrick
Intel

Mark Lavin
IBM

Mitsumasa Koyanagi
Tohoku University, Japan

Low-Cost Sequential ATPG with Clock-Control DFT

Miron Abramovici
Agere Systems - Murray Hill, NJ
miron@agere.com

Xiaoming Yu and Elizabeth M. Rudnick
University of Illinois - Urbana, IL
{xiaoming,liz}@crhc.uiuc.edu

Abstract: We present a new clock-control DFT technique for sequential circuits, based on clock partitioning and selective clock freezing, and we use it to break the global feedback loops and to generate ***clock waves*** to test the resulting sequential circuit with self-loops. Clock waves allow us to significantly reduce the complexity of sequential ATPG. Unlike scan, our non-intrusive DFT technique does not introduce any delay penalty; the generated tests may be applied at speed, have shorter application time, and dissipate less power.

Categories and Subject Descriptors: B.8.1 [Performance and Reliability]: Reliability, Testing, and Fault-Tolerance

General Terms: Algorithms, Design, Reliability

1. Motivation

Automatic test-pattern generation (ATPG) for sequential circuits is an extremely expensive computational process, so that ATPG algorithms working on complex circuits can spend many hours of CPU time and still obtain poor results in terms of fault coverage. Because of the difficulty of the sequential ATPG problem, the electronics industry has given up the idea that complex circuits can be tested without intrusive design for testability (DFT) techniques, such as scan design. However, scan-type DFT introduces delay penalties resulting in performance degradation, and scan tests require long test application times, increase the power consumption, and are difficult to run at-speed. Because many defects create delay faults, at-speed testing has become essential in achieving good defect coverage.

2. Main Contributions

In this paper, we introduce a **new clock-control DFT technique** based on clock partitioning and selective clock freezing. In a sequential circuit, a feedback loop may be *local* or *global.* A local loop includes only one flip-flop (FF) and is also called a *self-loop*; any loop with two or more FFs is called global. A pipeline is a loop-free (acyclic) sequential circuit. First we use clock control to temporarily freeze a subset of FFs to break all global loops; this creates a *near-acyclic circuit* where every FF is either feedback-free or has a self-loop; we will refer to such a circuit as a ***loopy pipe***, since it has a pipeline structure if we ignore the self-loops. Because loopy pipes do not have global feedback, they are structurally simpler than sequential circuits with both local and global feedback. Many partial-scan design methods, starting with [6], scan FFs to break all global feedback, so that the resulting partial-scan circuit is a loopy pipe. However, sequential ATPG for a loopy pipe can still be difficult. In fact, counters, which are notoriously difficult for sequential ATPG, are often implemented by loopy pipes.

DAC 2002, June 10-14, 2002, New Orleans, Louisiana, USA.

A major contribution of the paper is using clock control to generate **clock waves**. A clock wave is a novel clocking scheme that allows a loopy pipe to be tested as a pipeline. We describe **modeling techniques for loopy pipes** tested with clock waves to allow combinational ATPG techniques to be used. We present **a new sequential ATPG algorithm**, called *WAVEXPRESS*, that **detects most faults in a sequential circuit using combinational techniques**, which are at least one order of magnitude faster than sequential ones.

Our DFT technique, called *CLOCKWAVE,* **does not introduce any delay penalty, has small area overhead, and the generated tests are applied at speed.** Compared with scan tests, *WAVEXPRESS* vectors have **shorter test application time** and dissipate **less power**.

The remainder of this paper is organized as follows. Section 3 reviews prior work related to our new approach. Section 4 shows the use of clock freezing to cut global feedback. Section 5 introduces our new testing paradigm. Section 6 presents the *CLOCKWAVE* DFT. Section 7 describes the modeling techniques used in *WAVEXPRESS*. Section 8 discusses the algorithm, while Section 9 shows the results of our preliminary implementation. Section 10 presents conclusions.

3. Related Prior Work

Our approach uses *clock-freezing* [2], which temporarily suspends the sequential behavior of the circuit by freezing its clock, and applies several vectors without changing the current state. In this way, the current state is fully exploited before it is changed. We couple clock freezing with *clock partitioning* [3][4][8][9][23], a DFT technique that divides the FFs sharing the same clock into several groups, so that in test mode, each group can be independently clocked. Note that clock partitioning implies *selective clock freezing*, as the clocks not activated can be considered temporarily frozen. Clock partitioning increases the testability of the sequential circuit, by reducing dependency among FF values and introducing many more state transitions in the state transition graph, thus making states that are illegal or difficult to reach, easier to reach in test mode. As a result, some faults that were impossible or difficult to detect in the original circuit become easier to test. Clock partitioning has been used for delay-fault testing [10], to cut global feedback [8], to keep global loops in separate partitions [4], and to reduce the detrimental effects of sequential reconvergence on fault detectability [23]. A clock partitioning technique that allows a sequential circuit to be decomposed into a set of overlapping loopy pipes has been used to develop a new at-speed BIST method [26].

An important property of a pipeline is that all or most of its faults can be detected by combinational algorithms [12]. To test a fault in the pipeline of Figure 1a, we apply one input vector, then we clock twice to propagate the input vector throughout the entire circuit. Since the registers serve only to transmit the data, we can make them *transparent* for ATPG, and model the pipeline by a combinational circuit as shown in Figure 1b. Every combinational vector generated on this model is held constant while all registers are clocked; the number of required clocks is equal to the sequential depth of the pipeline. Note that keeping the same vector constant while clocking the circuit a number of times is also very useful for testing general sequential circuits [19].

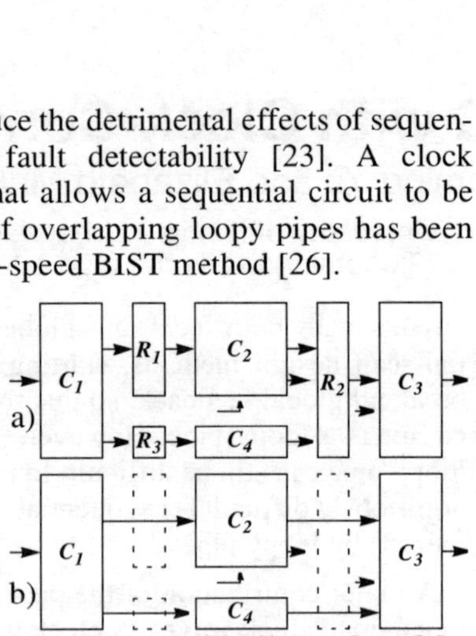

Figure 1. Pipeline and model

4. Breaking Global Feedback

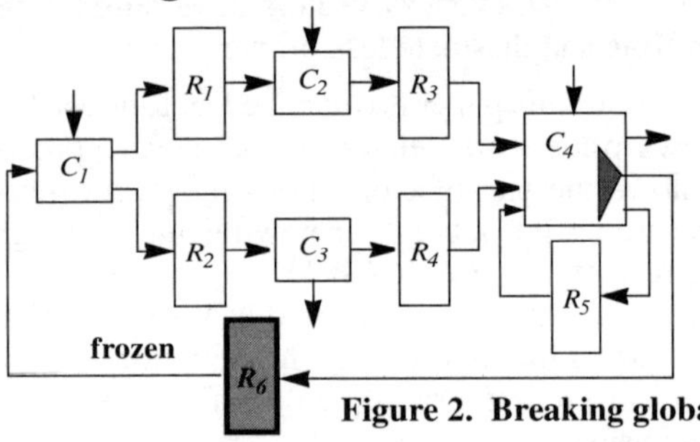

Figure 2. Breaking global loops

In Figure 2 [12], assume that every register R_i has an independent clock (the C_j blocks are combinational). Freezing R_6 is sufficient to cut the two (register-level) global loops of the circuit. The remaining circuit is a loopy pipe because of the self-loop involving R_5. Note that some of the primary inputs (PIs) of the loopy pipe (provided by the outputs of R_6) are frozen, and that a subcircuit of C_4 (indicated by a dashed triangle) remains "invisible" as long as the frozen register is not clocked. The selection of the registers to freeze is heuristic and tries to minimize the invisible regions.

5. The Clock-Wave Testing Paradigm

Figure 3 illustrates the structure of a loopy pipe (the logic driving POs is not shown). For every FF we build its data cone, consisting of all the logic feeding its *D* input. Tracing back to construct a cone, we stop at PIs and FF outputs. Since a loopy pipe has no global feedback, we can levelize the FFs if we ignore the self-loops. We start by assigning level 1 to every FF fed only by PIs; then to every FF fed by FFs whose maximum level is *i* we assign level *i*+1. The highest level *d* is the sequential depth of the loopy pipe. We assume that by clock partitioning, *all FFs at the same level i have an independent clock* $Clock_i$.

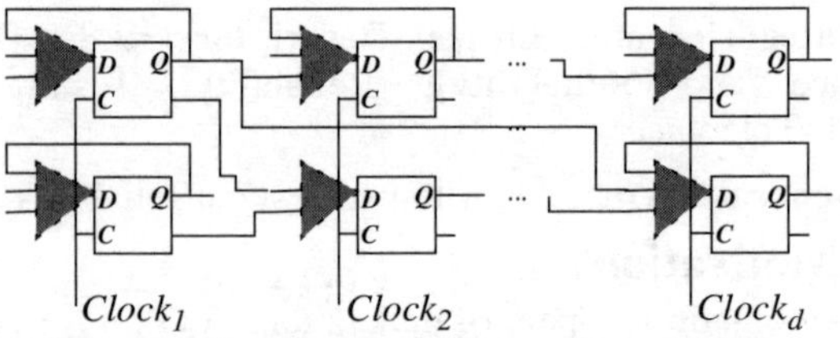

Figure 3. Loopy pipe with partitioned clocks

The basic step in our new paradigm for testing a loopy pipe consists of *keeping a PI vector constant while applying a* ***clock wave***, that is, a sequence that applies a pulse to $Clock_1$, then to $Clock_2$,..., and eventually to $Clock_d$. In this way the applied vector propagates throughout the circuit, with *every FF being clocked only once*. The input vector is kept constant for *d*+1 clock cycles; cycle *d*+1 is needed to allow the changes in the level-*d* FFs to propagate to POs.

6. *CLOCKWAVE* DFT

Figure 4a illustrates the clock control logic for a loopy pipe with *d* levels embedded in a circuit with a frozen register, which is treated as an additional level *d*+1. The role of the additional PIs EG_1 and EG_2 will be explained shortly; now assume that EG_1=EG_2=1. Every FF has a built-in clock-enabling AND gate. FFs at level *i* are enabled by the same EL_i (enable level *i*) signal. In normal mode *N_Mode*=1, so that all EL_i signals are active, and the main clock *MClock* propagates to all FFs. In test mode *N_Mode*=0, and the clock propagation is under the control of *d*+1 enabling signals EN_i, generated by a circular shift register with *d*+1 FFs. The initial

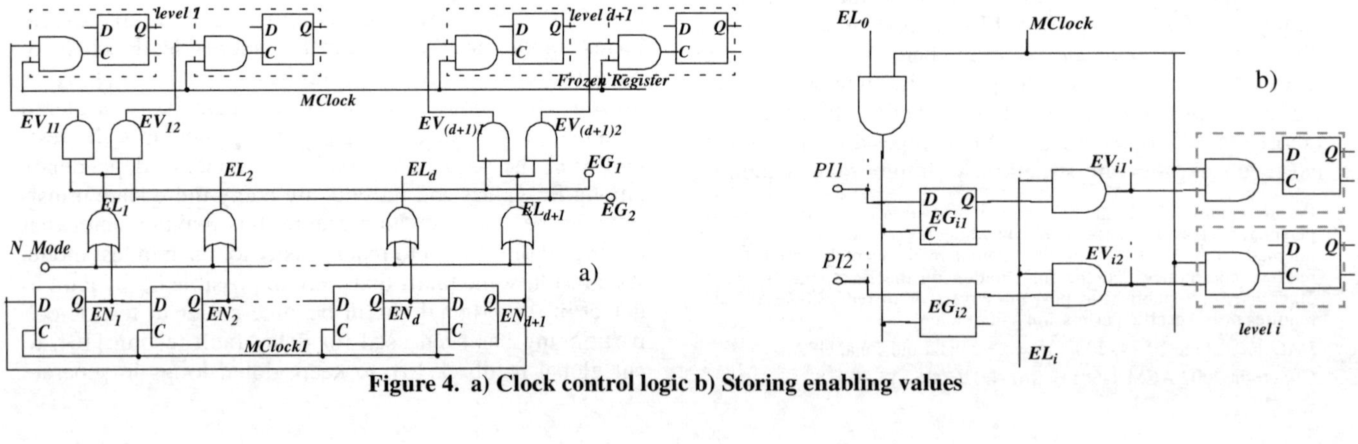

Figure 4. a) Clock control logic b) Storing enabling values

state in the shift register is 10...0, so that one level at a time will be enabled in the proper sequence. The shift register is clocked by a different phase of the main clock (*MClock1*).

The EG_j (enable group j) signals allow further clock partitioning of the FFs at the same level i. The enable value for group j at level i is $EV_{ij} = AND(EL_i, EG_j)$. The timing of *MClock1* is determined so that the *EV* values will be stable before the arrival of the next *MClock* pulse. In Figure 4a we have two groups at every level (only one FF in each group is shown). While the normal PIs are kept constant for d+1 clock cycles, the *EG* signals may change before every clock pulse. They can be shared among all levels, because the levels are clocked one-by-one; hence in every cycle the *EG* signals determine the clock partition for the current level. In normal mode, all *EG* PIs are always set to 1.

The same mechanism is used to hold the state of the frozen register for several clock waves, by setting all the *EG* signals to 0 in the d+1 cycle. Note that even when the frozen state is not changed, cycle d+1 is used to observe POs.

We could save hardware by using only one AND gating *MClock* for all the FFs in the same group, but this is likely to cause clock skewing. Having clock gating within every FF, the clock-tree distribution path to FFs (routing of *MClock*) is not changed in test mode; hence *DFT cannot cause clock skewing*.

The scheme in Figure 4a requires G additional PIs, where G is the maximum number of groups needed at any level. For circuits where G spare PIs are not available, Figure 4b presents an alternative DFT scheme that needs no additional PIs, but requires additional FFs EG_{ij} to store the enable values for every group j at every level i. These FFs must be set before the wave starts, so they are treated as an additional level 0, obtained by adding another FF (EN_0) to the circular shift register. Data values for the *EG* FFs are supplied by "normal" PIs; this is possible because *MClock* does not reach any functional FF when level-0 FFs are enabled. This scheme can provide a large number of enabling signals without additional PIs. In normal mode, the *EG* FFs are always set to 1.

Figure 5 shows the timing diagram for a circuit having a frozen register and a loopy pipe with d=4, using the second DFT scheme. Each vector is held for cycles 1 to 5. *MClock* pulse 0 loads *EG* FFs, and *MClock* pulse 5 is for the frozen register. The POs are observed after *MClock* pulse 4.

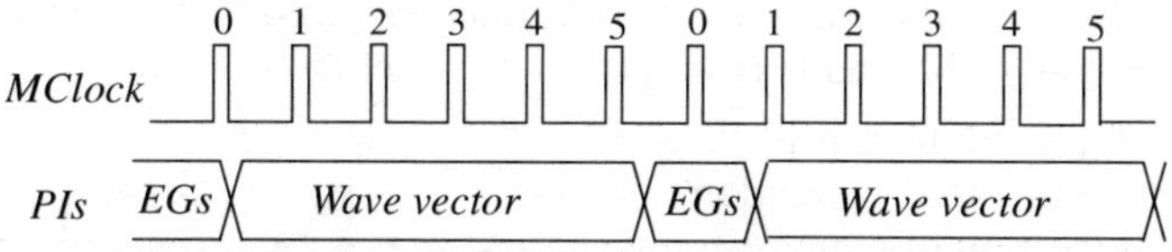

Figure 5. Timing for the second DFT scheme with d=4

For a circuit with n FFs, the main component of the area overhead is the n clock-enabling AND gates, which is less than the area overhead in scan design. In addition, we have d+1 FFs and d+1 OR gates for the generation of the clock wave. Let G_i be the number of clock groups at level i. For the first DFT scheme, we need G=max$\{G_i\}$ PIs, while for the second scheme we need G_i FFs and G_i AND gates at every level i. But usually the total number of FFs for clock control is much smaller than n, so the total area overhead will still be less than that required for full-scan.

An important feature of *CLOCKWAVE* is that the operating frequency of the circuit is not affected by the clock control mechanism, since *no delays are introduced in data paths*. Unlike scan design, *CLOCKWAVE* is *fully compatible with at-speed testing*. Compared to scan-based tests, our tests have *shorter test application time* and dissipate *less power*, because scan operations are not performed, and only a subset of the FFs are concurrently clocked.

7. Modeling Techniques

Figure 6 shows the combinational model of a self-loop FF. The clock is also treated as a regular PI [1]: *CK*=1 denotes the application of a clock pulse that transfers the *D* input into the new state *Q*, while *CK*=0 means that a clock pulse is not applied, so the FF holds its previous state *QF*, represented by an additional PI. Note that *QF* may also influence *Q* through the data cone, reflecting the effect of the feedback loop.

QF is modeled as a *frozen* PI, so that our ATPG algorithm, *WAVEXPRESS*, will not be allowed to change its value. When we start with *QF=X* denoting an unknown initial state, if *WAVEXPRESS* needs *Q* set to a binary value, this model will force it to set *CK*=1 and to assign other inputs of the data cone to values that allow the FF to be initialized.

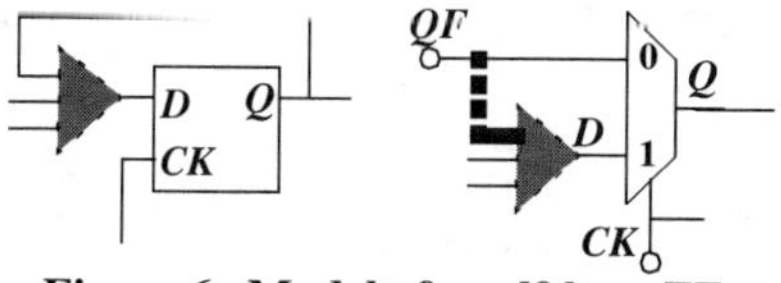

Figure 6. Model of a self-loop FF

The same model is used for all FFs. For frozen FFs we force *CK*=0, and for transparent FFs we set *CK*=1. Then we obtain a combinational circuit, which will be our model for one time-frame (Figure 7). We distinguish between the frozen PIs carrying the previous state of internal FFs (*QFIs*), and those that carry the state of the frozen register (*QFFs*). Clocks are modeled as independent PIs; the figure shows only one clock per level, but, in general, we will have several such clocks at every level. *This combinational model is possible because every internal FF is clocked at most once during a clock wave.* The combinational vector generated in this time-frame is kept constant while the clock wave is applied. Here "one time-frame" corresponds to d+1 conventional time-frames.

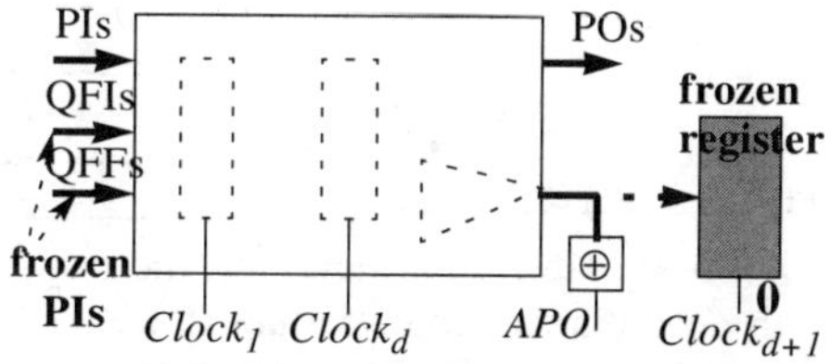

Figure 7. Model for one time-frame

To make the faults from the "invisible" logic also observable within one time-frame (without clocking the frozen register), we add an additional PO (*APO* in Figure 7) that observes all the inputs of the frozen register via an XOR tree. This technique [11][24], used to improve observability, is independent of clock control, has negligible impact on timing, and is compatible with at-speed testing.

8. The *WAVEXPRESS* Algorithm

WAVEXPRESS is an *opportunistic algorithm* that tries to detect all detectable faults with minimal computational effort. Unlike most conventional ATPG algorithms, which work on a selected target until it is detected or proven untestable, *WAVEXPRESS* abandons the current target if it cannot detect it quickly, and moves to other targets that may be easier to detect in the current state. Figure 8 depicts the basic flow of our algorithm, which starts with a single time-frame (Figure 7), and tries to detect as many faults as possible without changing the current frozen state. Before targeting faults, *WAVEXPRESS* analyzes the set of targets and eliminates many faults whose detection is not possible given the currently frozen PI values. After all potentially detectable faults have been targeted within a single time-frame, *WAVEXPRESS* switches to a model with k time-frames, starting with k=2. After the first fault is detected (by a sequence of k vectors), *WAVEXPRESS* reverts to the one-time-frame loop with a new state of the frozen register. The number of time-frames k is increased only when not even one fault could be detected with k time-frames.

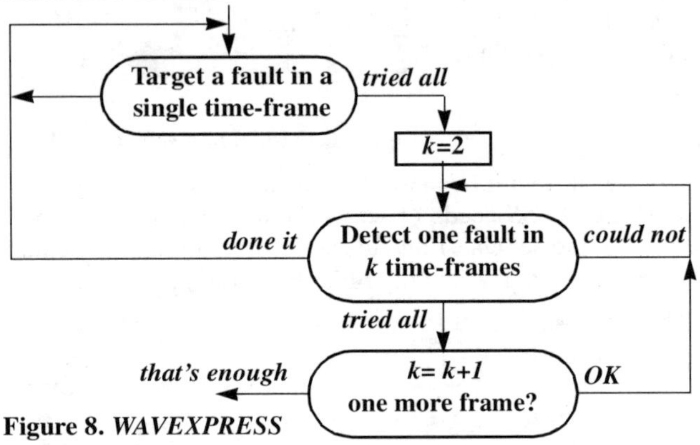

Figure 8. *WAVEXPRESS*

All assignable PIs, including the non-frozen clocks, start with an unknown logic value. If *WAVEXPRESS* detects the selected target fault, every generated vector is expanded into a sequence of d+1 identical vectors and their associated clock wave. The values of the clock PIs are mapped into settings for the corresponding enabling values. Clocks left unassigned are not activated, to save power and (possibly) to preserve fault effects stored in internal FFs. Any unassigned PIs are set to random binary values. The resulting sequence is fault simulated using the PROOFS fault simulator [21]. Note that in the test generation model, the POs are observed only after the d-level FFs have been clocked, but in the fault simulation model they are observed after every clock. This is equivalent to applying several random vectors in between the generated vectors, and these vectors may detect additional faults. The results of PROOFS are ported back into *WAVEXPRESS* to enable fault dropping and reporting fault effects stored in FFs. The two programs communicate via *sockets*.

Figure 9 illustrates the operation of the first loop of *WAVEXPRESS*, showing two consecutive time-frames. After a vector is generated in the first one, the changed values of internal FFs are propagated to the *QFI* inputs of the next time-frame before a new fault is targeted; these values are frozen for ATPG. Except for this value transfer (denoted by the dotted line in the figure), consecutive time-frames are not connected, so *WAVEXPRESS* works separately with each time-frame. The values of frozen register *QFFs* are the same in every time-frame.

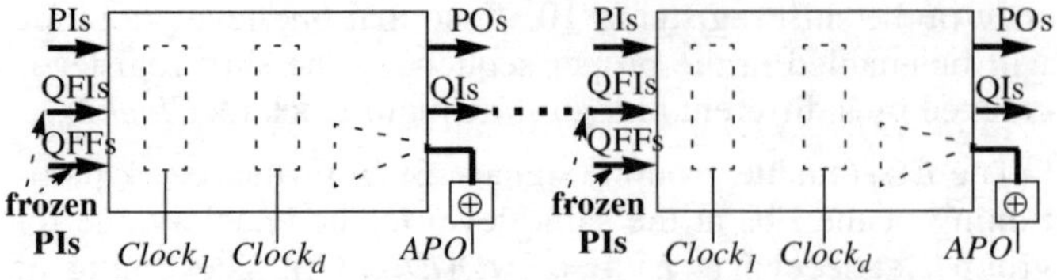

Figure 9. Sequence of separated time-frames

Figure 10 depicts a model with two time-frames, used after all targets have been tried with one time-frame. The previously frozen register is now allowed to change, and its clock ($Clock_{d+1}$) is now an assignable PI in the first time-frame. No inputs are frozen in the second time-frame, as every QF input is connected with its corresponding Q output from the first time-frame. All POs in the first time-frame are ignored, since none of the remaining targets could be detected at these POs. Similarly, in a model with k time-frames, all POs in the first k-1 time-frames are ignored. For m PIs, the size of the search space for k separated time-frames is $k2^m$, while for k connected time-frames is 2^{km}.

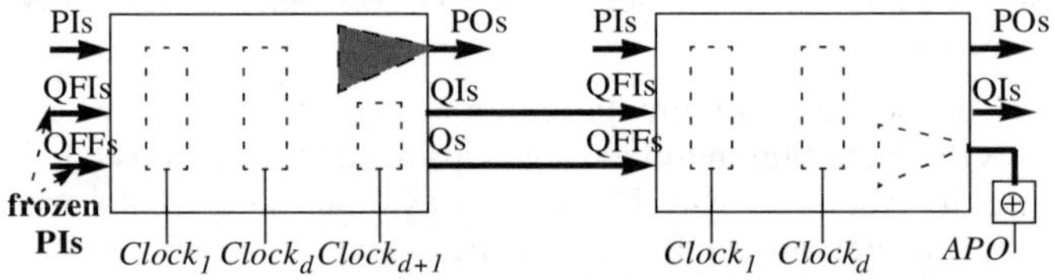

Figure 10. Sequence of connected time-frames

Like FASTEST [15], *WAVEXPRESS* backtraces objectives to PIs, but its logic value system avoids backtracing toward the PIs with frozen values. Backtracing can span multiple time-frames. Since decisions are done only as PI assignments, *WAVEXPRESS does not do explicit state justification, and it never has to justify illegal states.*

Selecting target faults: As preprocessing steps, we use: (1) the algorithms FIRES [14] and FUNI [17], running on the circuit with partitioned clocks, to identify combinationally and sequentially untestable faults; (2) the combinational test generator ATOM [13], running on a full-scan model of the circuit, to identify the rest of the combinationally untestable faults. To reduce the ATPG run-time, we remove the identified untestable faults from the set of target faults.

Since the frozen PI values may preclude the activation or the observation of many still undetected faults, *WAVEXPRESS* removes these faults from the current set of target faults, so that only potentially detectable faults will be targeted. For example, in the circuit in Figure 11, R s-a-1 and P s-a-1 cannot be activated, and no fault in the shaded areas can be observed. Faults whose effects are stored in currently observable frozen or internal FFs are always

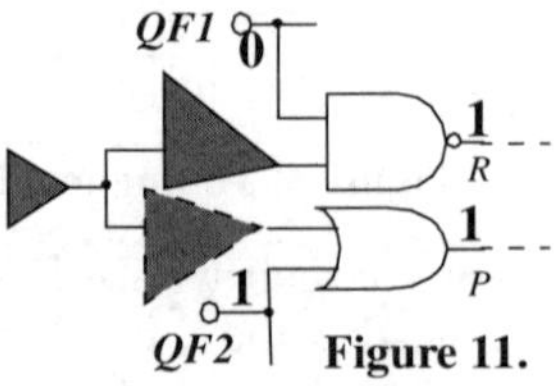

Figure 11.

included in the set of targets. The set of targets is incrementally updated after every clock wave.

WAVEXPRESS relies on additional heuristics based on testability measures to select a target most likely to be detected in the current state. We also enforce a limit on the number of times the same fault is going to be targeted.

9. Results

We compare a preliminary implementation of *WAVEXPRESS* with (1) HITEC - a deterministic test generator [20], (2) GATEST - a test generator based on genetic algorithms [25], (3) a commercially available sequential ATPG (S_ATPG), and (4) ATOM [13]. S_ATPG was run on a SPARC SUNW Ultra-Enterprise computer; the other programs were run on a Pentium III 700 computer. The two machines have comparable performance.

Table 1: Circuits data

Circuit	PIs	POs	FFs	SLs	Faults	Untst.	Frz.	*D*	*G*	*EG*
s208	11	2	8	8	215	0	0	8	1	8
s420	19	2	16	16	430	0	0	16	1	16
s838	35	2	32	32	857	0	0	32	1	32
ctr_12	2	1	12	12	636	13	0	12	1	12
ctr_16	2	1	16	16	940	13	0	16	1	16
div16	33	34	50	48	2147	176	2	19	3	35
pcont2	9	8	24	0	11272	3846	16	3	16	24
piir8o	9	8	56	0	19936	4648	8	5	8	8
write_s	15	13	18	16	446	3	2	11	2	11
tx_s	24	24	23	23	772	27	3	12	3	13
alpha_s	26	25	16	13	759	46	3	10	3	11

Table 1 provides the data for the circuits we have experimented with. The first three are ISCAS89 benchmarks; *ctr_12* and *ctr_16* are 12-bit and 16-bit up-down counters with a single PO off the most significant bit. *piir8o* is an optimized finite-impulse response filter [7], *div16* is a 16-bit divider [7], and *pcont2* is a controller circuit used in DSP applications [7]. *write_s*, *tx_s*, and *alpha_s* are telecommunication circuits from [5]. *SLs* is the number of self-loops, *Untst* is the number of untestable faults identified in preprocessing, *Frz* is the number of temporarily frozen FFs, *D* is the sequential depth of the circuit ($D=d+1$ if a register is frozen, otherwise $D=d$), *G* is the number of additional PIs needed in the first DFT scheme, and *EG* is the number of enable FFs used in the second scheme. The last two parameters were computed to create maximum clock partitioning, where each FF has a separate clock, but this is not needed in practice. Although the first five circuits have no global feedback, they are still highly sequential because the high values of *D*.

Table 2 summarizes the ATPG results. Note that ATOM uses full-scan, *WAVEXPRESS* uses *CLOCKWAVE* DFT, and the other programs handle the original circuit without DFT. S_ATPG was run only on four circuits, and GATEST could not run the counters. All run-times are in seconds. *FC* is the detectable fault coverage, computed after the subtracting *Untst* from the total fault count. *Vec* is the number of vectors. *Tst* is the number of tests generated by *WAVEXPRESS*, where each test corresponds to *D* constant vectors. *Conv* stands for "converted," being the number of faults found untestable by HITEC or GATEST, that were detected by *WAVEXPRESS*. *Msd* is the number of faults that were detected by HITEC or GATEST, but were missed by *WAVEXPRESS*.

Although *CLOCKWAVE* is a non-intrusive DFT, *WAVEXPRESS* obtains much higher fault coverage than the other sequential ATPG programs, and is much faster, often by one or two orders of magnitude. For most circuits, the *WAVEXPRESS* stuck-fault coverage was above 90%. In practice, such a test applied at-speed usually achieves higher defect coverage than a test with higher stuck-fault coverage that cannot run at-speed [18]. As expected, full-scan leads to the highest stuck-fault coverage and the fastest run-time for most circuits. However, there are additional costs for scan: area overhead, test application time, delay penalty, and power dissipation. For the largest circuits (*pcont2* and *piir8o*), the area overhead for scan is about twice the area overhead for *CLOCKWAVE*. For *pcont2*, the test application time for the *WAVEXPRESS* vectors is 300 clock cycles, while the full-scan vectors need 2925 cycles (assuming 2 scan chains). For *piir8o*, the corresponding numbers are 850 and 5115

Table 2: ATPG Results

Circuit	WAVEXPRESS					HITEC			S_ATPG			GATEST			ATOM		
	Time	FC%	Conv.	Msd.	Tst.	Time	FC%	Vec.	Time	FC%	Vec.	Time	FC%	Vec.	Time	FC%	Vec.
s208	1	91.6	78	0	63	1	63.7	147				4	62.3	100	0	100	65
s420	7	90.9	251	3	106	16	41.6	152				14	41.4	113	0	100	108
s838	26	91.2	530	2	227	4411	29.6	161				38	29.3	103	0	100	190
ctr_12	21	96.1	63	4	184	4866	86.7	4865	1403	65.0	1167	-	-	-	0	100	99
ctr_16	22	96.0	173	3	268	9865	77.7	7479	5200	56.0	905	-	-	-	0	100	90
div16	107	90.8	202	94	98	968	85.1	289				155	86.3	425	0.3	100	203
pcont2	169	98.9	4056	8	100	8930	44.3	7	7867	65.6	20	544	91.7	143	38	96.3	225
piir8o	1049	98.6	5135	11	170	11053	65.1	20	13948	54.2	17	2376	98.5	506	27	98.4	341
write_s	81	93.2	8	20	89	370	94.5	5664				20	60.0	180	0	100	93
tx_s	84	89.9	30	4	122	5794	60.8	2305				38	48.5	177	0	100	109
alpha_s	197	62.7	23	23	43	6522	28.0	1412				22	14.0	38	0	100	136

(assuming 4 scan chains). Five circuits have zero or negligible delay penalty, but for the other six the performance degradation caused by scan ranges from 2.5% to 14.3%. Compared to normal operation of the circuit, the average power dissipation during the application of *WAVEXPRESS* vectors is about *D* times smaller, because the clock waves reduce the average activity by a factor of *D* (assuming the logic is uniformly distributed among levels). Since usually the power dissipation during scan increases relative to the normal operation, it follows that the power dissipation during scan is at least *D* times larger than the power dissipated by the *WAVEXPRESS* vectors.

In most circuits, the number of faults missed by *WAVEXPRESS* is low. Since in practice, most faults not detected by a conventional sequential ATPG are untestable, we can conclude that *WAVEXPRESS* usually detects most of the sequentially detectable faults. However, *WAVEXPRESS* cannot determine whether the undetected faults are untestable.

For all but three circuits, *WAVEXPRESS* could detect every fault with no more than four time-frames. Five time-frames were needed only for several faults in *div16*, *tx_s*, and *alpha_s*. In most circuits, most faults are detected with one or two time-frames. This is remarkable since all the circuits are deeply sequential.

In the current *WAVEXPRESS* implementation, we did not use any of the advanced sequential ATPG techniques such as [16]; we expect such techniques to further increase the performance of our program.

10. Conclusions

Our *CLOCKWAVE* DFT technique significantly reduces the complexity of sequential ATPG, allowing most faults in deeply sequential circuits to be detected by a low-cost algorithm (*WAVEXPRESS*). The main factor contributing to the reduction in complexity is handling of *D* conventional time-frames as a single time-frame; for example, *D* is 32 for *s838* and 19 for *div16*. *CLOCKWAVE* is a non-intrusive technique that introduces no performance degradation or clock skews, and its area overhead is smaller than that needed for scan DFT. *WAVEXPRESS* vectors may be applied at-speed, and therefore can be used for delay-fault testing. Compared to scan-based tests, *WAVEXPRESS* vectors have shorter test application time and dissipate about *D* times less power.

For some circuits, additional non-scan DFT techniques, such as partial reset [22], may be used to complement *CLOCKWAVE*. *CLOCKWAVE* is also applicable to multiple clock domains, by separately partitioning each clock.

References

[1] M. Abramovici, J. J. Kulikowski, P. R. Menon, and D. T. Miller, "SMART and FAST: Test Generation for VLSI Scan-Design Circuits," *IEEE Design & Test of Computers*, August 1986.

[2] M. Abramovici, K. Rajan, and D. T. Miller, "FREEZE!: A New Approach for Testing Sequential Circuits," *Proc. 29th Design Automation Conf.*, June 1992.

[3] V. D. Agrawal, S. C. Seth, and J. S. Deogun, "Design for Testability and Test Generation with Two Clocks," *Proc. 4th Intn'l. Symp. on VLSI Design*, pp. 112-117, January 1991.

[4] S. Baeg and W.A. Rogers, "A Cost-Effective Design for Testability: Clock Line Control and Test Generation Using Selective Clocking," *IEEE Trans. on CAD*, vol. 18, no. 6, pp. 850-861, June 1999.

[5] F. Fummi, M. Boschini, X. Yu, and E. M. Rudnick, "Sequential Circuit Test Generation Using a Symbolic/Genetic Hybrid Approach," *Journal of Electronic Testing*, vol. 17, pp. 321-330, June 2001.

[6] K-T. Cheng and V. D. Agrawal, "A Partial Scan Method for Sequential Circuits with Feedback," *IEEE Trans. on Computers*, vol. 39, pp. 544-547, April 1990.

[7] V. Chickermane, J. Lee, and J.H. Patel, "A Comparative Study of Design for Testability Methods Using High-level and Gate-Level Descriptions," *Proc. Intn'l. Conf. on CAD*, pp. 620-624, November 1992.

[8] K. L. Einspahr, S. C. Seth, and V. D. Agrawal, "Clock Partitioning for Testability," *Proc. 3rd IEEE Great Lakes Symp. on VLSI*, pp.42-46, March 1993.

[9] K. L. Einspahr, S. C. Seth, and V. D. Agrawal, "Improving Circuit Testability by Clock Control," *Proc. 6th IEEE Great Lakes Symposium on VLSI*, pp. 288-293, March 1996.

[10] W-C. Fang and S.K. Gupta, "Clock Grouping: A Low Cost DFT Methodology for Delay Testing," *Proc. 31st Design Automation Conf.*, pp. 94-99, 1994.

[11] J. R. Fox, "Test-Point Condensation in the Diagnosis of Digital Circuits," *Proc. of IEE*, vol. 124, no. 2, pp. 89-94, February 1977.

[12] R. Gupta, R. Gupta, and M. A. Breuer, "The Ballast Methodology for Structured Partial Scan Design," *IEEE Trans. on Computers*, vol. 39, no. 4, pp. 538-544, April 1990.

[13] I. Hamzaoglu and J. H. Patel, "New Techniques for Deterministic Test Pattern Generation," *Proc. VLSI Test Symp*, pp. 446-452, April 1998.

[14] M. Iyer, D.E Long, and M. Abramovici, "Identifying Sequential Redundancies Without Search," *Proc. 33rd Design Automation Conf.*, June 1996.

[15] T.P. Kelsey, K.K. Saluja, and S.Y. Lee, "An Efficient Algorithm for Sequential Circuit Test Generation," *IEEE Trans. on Computers*, vol. 42, no 11, pp. 1361-1371, November 1993.

[16] X. Lin, I. Pomeranz, and S. M. Reddy, "Techniques for Improving the Efficiency of Sequential Circuit Test Generation," *Proc. Intn'l. Conf. on CAD*, November 1999.

[17] D.E Long, M. Iyer, and M. Abramovici, "FILL & FUNI: Algorithms To Identify Illegal States and Sequentially Untestable Faults," *ACM Trans. on Design Automation of Electronic Systems*, July 2000.

[18] P. Maxwell, R. Aitken, V. Johansen, and I. Chiang, "The Effect of Different Test Sets on Quality Level Prediction: When is 80% Better than 90%?," *Proc. Intn'l. Test Conf.*, pp. 358-364, 1991.

[19] L. Nachman, K. K. Saluja, S. Upadhyaya, and R. Reuse, "A Novel Approach to Random Pattern Testing of Sequential Circuits," *IEEE Trans. on Computers*, vol. C-47, no.1, pp. 129-134, January 1998.

[20] T. M. Niermann and J. H. Patel, "HITEC: A Test Generation Package for Sequential Circuits," *Proc. European Conf. on Design Automation*, pp. 214-218, 1991.

[21] T. M. Niermann, W. -T. Cheng, and J. H. Patel, "PROOFS: A Fast, Memory-Efficient Sequential Circuit Fault Simulator," *IEEE Trans. CAD*, vol. 11, no. 2, pp. 198-207, February 1992.

[22] P. Parikh, M. Abramovici, B. Mathew, and D. Saab, "On Selecting Flip-Flops for Partial Reset," *Proc. Intn'l. Test Conf.*, October, 1993

[23] K. B. Rajan, D. E. Long, and M. Abramovici, "Increasing Testability by Clock Transformation (Getting Rid of Those Darn States)," *Proc. VLSI Test Symp.*, April 1996.

[24] E. M. Rudnick, V. Chickermane, and J. H. Patel, "An Observability Enhancement Approach for Improved Testability and At-Speed Test," *IEEE. Trans. CAD*, vol. 13, no. 8, pp. 1051-1056, August 1994.

[25] E. M. Rudnick, J.H. Patel, G.S. Greenstein, and T.M. Niermann, "A Genetic Algorithm Framework for Test Generation," *IEEE Trans. CAD*, vol. 16, no. 9, pp. 1034-1044, Sept. 1997.

[26] J. Shin, X. Yu, E. M. Rudnick, and M. Abramovici, "At-Speed Logic BIST Using a Frozen Clock Testing Strategy," *Proc. Intn'l. Test Conf.*, pp. 64-71, October 2001.

Effective Diagnostics through Interval Unloads in a BIST Environment

Peter Wohl
Synopsys Inc.
Williston VT 05495
wohl@synopsys.com

John A. Waicukauski
Synopsys Inc.
Tualatin OR 97062
johnwaic@synopsys.com

Sanjay Patel
Synopsys Inc.
Beaverton OR 97006
sanjayp@synopsys.com

Greg Maston
Synopsys Inc.
Denver CO 80220
gmaston@synopsys.com

ABSTRACT

Logic built-in self test (BIST) is increasingly being adopted to improve test quality and reduce test costs for rapidly growing designs. Compared to deterministic automated test pattern generation (ATPG), BIST presents inherent fault diagnostic challenges. Previous diagnostic techniques have been limited in their diagnosis resolution and/or require significant hardware overhead. This paper proposes an interval-based scan-unload method that ensures diagnosis resolution down to gate-level faults with minimal hardware overhead. Tester fail-data collection is based on a novel construct incorporated into the design-extensions of the standard test-interface language (STIL). The implementation of the proposed method is presented and analyzed.

Categories and Subject Descriptors: B.8.1 **[Performance and Reliability]:** Reliability, Testing and Fault-Tolerance.

General Terms: Algorithms, Design.

Keywords: built-in self-test (BIST), fault diagnosis.

1. INTRODUCTION

Test represents a significant portion of the total cost of digital circuits. Larger and more complex designs lead to greater increases in the cost of automated test equipment, a trend that is projected to continue and accelerate [1]. Design-for-test (DFT) methods to reduce test cost include scan ([2], [3]) and, increasingly, BIST [4], [5].

Figure 1 outlines the logic BIST architecture used in this work, based on the STUMPS architecture [6]. Values from a pseudo-random pattern generator (PRPG) are loaded into the internal scan chains of the design to be tested, and the scan-chain outputs are unloaded into a signature analyzer that performs test-response compaction. After several cycles, the state of the signature analyzer is compared to the known "signature" of the fault-free design; a mismatch indicates that at least one erroneous value was unloaded from the scan chains. The PRPG is commonly implemented as a linear-feedback shift register (LFSR) and the signature analyzer is often implemented as a modified LFSR known as a multiple-input signature register (MISR) [5], [7], [8]. To reduce test application time, most DFT schemes use a multiplicity of parallel scan chains. A combinational phase shifter (Figure 1) converts the unidimensional stream of pseudo-random values generated by the LFSR into a two-dimensional array of values to load parallel scan chains [7], [9], [10]. The values unloaded from scan chains are fed into a MISR that must have at least as many bits as there are scan chains [7]. A combinational [space-] compactor (Figure 1) is inserted between the scan chain outputs and the signature analyzer inputs, making it possible to use a smaller MISR [11].

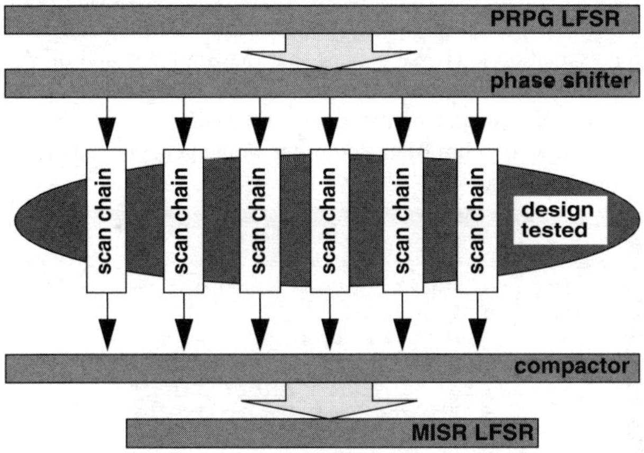

Figure 1. Logic BIST architecture.

When test patterns fail, it is desirable to quickly locate the cause of failure to one or more circuit gates so that possible manufacturing or design errors can be corrected, thereby improving yield. Failures are located through diagnostic at the logical level; failure analysis identifies physical defects. Diagnostic relies on extensive space (scan cell) and time (pattern cycle) failure information to identify logical failures [3]. Unlike ATPG patterns, diagnosing BIST pattern failures is more difficult because only the compressed final signature is available. Analysis of the failing signature can identify the failing test vectors; but, as the number of failing vectors increases, the complexity of this method also increases and its identification abilities diminish [12]. To identify an increased number of failing vectors requires additional hardware as cyclic registers [13], [14]. Another method utilizes the quotient instead of the signature, providing improved diagnosis but requiring substantial additional hardware [15]. Various scan-cell partitioning schemes have also been presented, which require additional hardware and apply the same test multiple times with different partition elements to stepwise refine the failing candidates [16], [17]. In other methods, patterns are re-run multiple times based on a binary search technique used to identify the failing patterns, but a large number of steps may be required and signatures must be computed and analyzed for each search step [18]. In another scheme, the BIST ses-

DAC 2002, June 10-14, 2002, New Orleans, Louisiana, USA.

sion is repeated several times with the MISR programmed to collect multiple signatures for different polynomials [19]. However, the maximum number of scan cells with faulty responses that can be identified depends on how many signatures are collected. Similarly, multiple signatures can be gathered by pseudo-randomly masking out different sets of scan cells each time the BIST session is repeated [20].

This paper presents a fast, very-low hardware overhead diagnostic method that identifies not just the failing scan cells, but the failing gates(s). Section 2 presents the general concept of interval-based self test. Section 3 details the method of unloading all scan data for a selected interval. Section 4 describes the tester flow and section 5 the simulator flow, each based on a single pattern file, without requiring re-generation of BIST data.

2. INTERVAL-BASED LOGIC BIST

The self-test is divided into independent intervals of a fixed number of test patterns; e.g., 256 patterns. Each interval consists of an ordered sequence of events shown in Figure 2, in standard STIL representation [21]. We assume that the BIST test and diagnostic operations are controlled by a tester; therefore, the STIL program in Figure 2 is loaded on the tester. (If the BIST test was fully independent, an on-chip BIST controller would generate the necessary control signals.)

```
// STIL macro definition:
MacroDefs {
"compare_substitute" { /*empty*/ }
"bist_load_unload" {
  ScanStructures bist;
  Loop 100 { // chains are 100 long
      Macro "compare_substitute";
      V { "bist_clk"=P; "clk"=P; }
  }}}
// STIL BIST intervals:
Pattern "pattern" {
"interval 0":
Call "load_unload" {//load PRPG & MISR
  "lfsr_si"=1110101...; }
Macro "bist_setup"; // optional
Loop 256 { // 256 patterns
  Macro "bist_load_unload";
  V { "bist_clk"=P; "clk"=P; }
}
V { "misr_comp" = L; } // MISR compare
"interval 1": /* similar */
"interval 2": /* similar */
....
}
```

Figure 2. STIL description of BIST interval containing 256 patterns; scan chains are 100 cells long.

At the beginning of each interval, the "load_unload" procedure (not shown in Figure 2) is called to initialize, through scan, the PRPG and MISR. Initializing the PRPG and MISR at the beginning of each interval ensures that the intervals are independent of each other. Also, the PRPG is seeded with an initial value targeted to detect certain hard-to-test faults [22]. The MISR is initialized to a value so that the fault-free signature at the end of the interval is 0, simplifying signature compare [7]. Next, an optional "bist_setup" macro is called to set up scanning of the internal scan chains. In boundary-scan based BIST, the "bist_setup" macro sets the JTAG controller to the appropriate state. The 256 test patterns in the interval are run by the Loop statement; each iteration calls the "bist_load_unload" macro (shown above the interval in Figure 2) and then pulses, in a single vector (V), the "bist_clk" and the system "clk". The "bist_clk" clocks all BIST hardware: the PRPG, MISR and BIST controller; the system "clk" is used both for shift and system capture in Figure 2. In general, the shift clock (used in the macro "bist_load_unload") may be different from the capture clock; multiple capture clocks could also be used. The "bist_load_unload" macro (Figure 2) pulses "clk" to shift values into the scan chains defined in "ScanStructures bist" while, at the same time, pulsing "bist_clk" to advance the PRPG and compress unloaded values into the MISR. The macro "compare_substitute" called by "bist_load_unload" can be defined as an empty macro for normal BIST application and has no effect; a non-empty "compare_substitute" will be used during diagnostic (section 3).

During the first pattern, scan cells have unpredictable values that must be prevented from entering the MISR; this is achieved by ANDing off each compactor input (Figure 3) during the first application of "bist_load_unload" [11]. The gating signals c0 and c1 are provided on-chip by a small counter -- part of the BIST controller. During the first unload of each interval c0=c1=0; at all other times c0=c1=1; in diagnosis mode (section 3) c0 and c1 have different values, thus a single gating signal could not be used. At the end of each interval, a logic OR gate compares the MISR content to 0; the interval test has failed if the MISR is non-0.

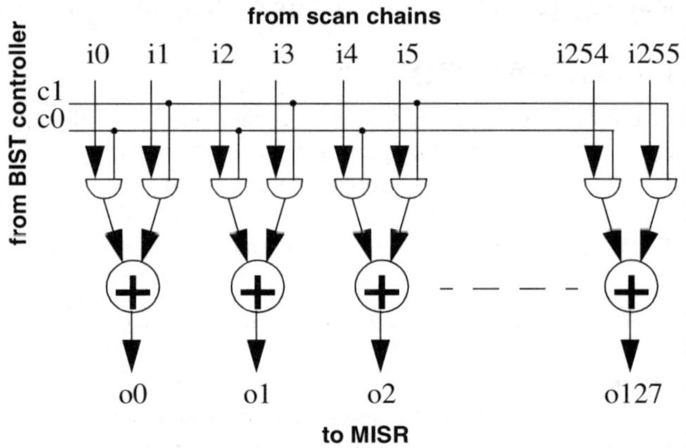

Figure 3. Compactor core showing gating of inputs i0..i255 controlled by signals c0, c1.

Figure 4 shows the timing of the shift and capture operations in non-diagnostic mode. While the shift clocks can be applied at a lower speed (to reduce power consumption), the sequence marked "AC" (from the last shift clock to the capture clock) must be applied fast enough to ensure effective launch-on-last-shift transition-fault test. The timing of the clocks is also stored in the STIL file but not shown in Figure 2. The internal shift clock ("clk") and the BIST clock ("bist_clk") are both pulsed every cycle. Although the bist_clk pulse at the time of the capture cycle (the gray clock pulse) is logically not needed, it simplifies the control circuitry and ensures uniformity of clock-induced noise.

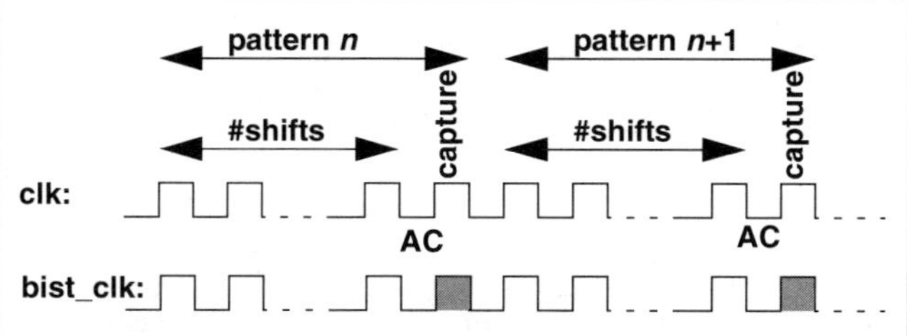

Figure 4. Clocking in non-diagnostic mode.

3. UNLOADING INTERVAL DATA

Unlike previously proposed BIST test windows ([23]), intervals are independent of each other and can be reordered or run individually in much the same way as individual deterministic ATPG patterns. Each interval is fully defined by the PRPG and MISR initial values and the clock(s) pulsed. Test windows define intermediate MISR fault-free signatures used to quickly determine the failing patterns (window) for diagnostic. Intervals achieve the same fail patterns resolution (256 patterns) but do not require the additional overhead of salvaging future windows by adjusting their signatures to fit past observed errors [23].

To make failure diagnosis of the failing gate possible, the exact position of every erroneous bit unloaded from the scan chains must be known. Diagnostic time is less critical than test time therefore, in a logic BIST environment, it is desirable and practical to unload all scan values without using compaction and signature analysis [24]. To minimize hardware overhead, the compactor (Figure 3) supports a "pass-through" mode for diagnostic and the scan values are unloaded from the MISR (Figure 5), eliminating the need for added scan-chain MUXing to bypass the compactor and MISR. The AND gates used in the compactor to gate off scan data during the first unload are re-used during diagnostic to gate off selected scan chains so that data from the remaining chains are passed through unmodified. For an *m*-bit MISR, one bit from each of the first *m* scan chains is first passed unaltered through the compactor, captured in the *m* cells of the MISR, and then scanned out of the MISR. This process is repeated to obtain values from all scan cells of all scan chains [11]. Once all scan unload data has been collected, diagnostic is performed using fault simulation similar to diagnostic of deterministic ATPG, achieving the same resolution down to the failing gates(s) [3].

The compactor (Figure 3) and the MISR (Figure 5) can operate in several modes, controlled by the BIST controller: scan, signature analyzer, or diagnosis mode. The diagnosis operation for a design with 256 scan chains, a 2-to-1 compactor and a 128-bit MISR requires the following steps:

(1) Initialize MISR to 0 through scan: diag=1, c0=c1=0, scen=1; all values shifted into the MISR are 0.

(2) Capture in MISR outputs of group 0 scan chains (chains connected to i0, i2, ..., i254): diag=1, c0=1, c1=0, scen=0; values are captured into MISR cells through XOR gates that have all inputs 0 except the "o" inputs from the compactor.

(3) Scan out MISR, get 1 bit of every group 0 chain through so, re-initialize MISR to 0: diag=1, c0=c1=0, scen=1.

(4) Capture in MISR outputs of group 1 scan chains (chains connected to i1, i3, ..., i255): diag=1, c0=0, c1=1, scen=0; similar to step 2.

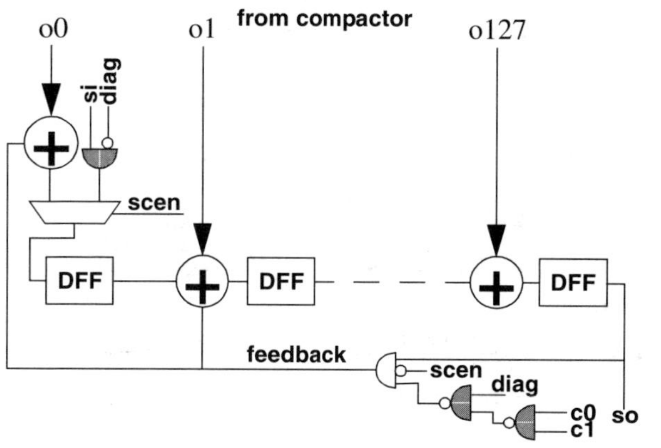

Figure 5. MISR architecture; the gates added for diagnosis support are gray.

(5) Scan out MISR, get 1 bit of every group 1 chain through so, re-initialize the MISR to 0: diag=1, c0=c1=0, scen=1.

(6) If more bits in scan chains, shift scan chains and go to 2.

Note that the hardware added for diagnosis is only 3 gates (Figure 5), 2 signals (diag, one extra "c" gating signal) and a counter in the BIST controller to control c0, c1. The signal diag enables diagnosis operation in the MISR and in the BIST controller; diag is constant 1 during diagnosis and 0 during normal operation. A primary input can be used to directly control diag, or it can be generated internally by the BIST controller when its state machine enters diagnosis mode as controlled by a limited-pin interface, such as boundary scan (JTAG).

Failing intervals are re-run on the tester using diagnosis mode of pattern operation to collect diagnosis data. The pattern block of the STIL file, shown by example under the "STIL BIST intervals" comment in Figure 2 is unchanged, but a different set of procedures and macros are referenced and linked with STIL's PatternBurst construct (Figure 6) [21]. The macro "compare_substitute" is no

```
PatternBurst "normal" {
  // by default uses unnamed MacroDefs
  // which includes empty
  // "compare_substitute"
  Patlist { "pattern" // see Figure 2
}}
PatternBurst "diagnosis" {
  MacroDefs "diagnosis";// see Figure 7
  Start "interval 2"; // first failing
  Stop "interval 4; // last failing
  Patlist { "pattern" // see Figure 2
}}
```

Figure 6. STIL selection of normal vs. diagnosis mode.

longer empty, but defined as in Figure 7. Every shift in "bist_load_unload" is preceded by a call to "compare_substitute" (Figure 2). In the first phase (Figure 7), one value from each of the first 128 scan chains is passed through the compactor and captured into the MISR ("bist_clk" is pulsed). Next, the Loop scans out the

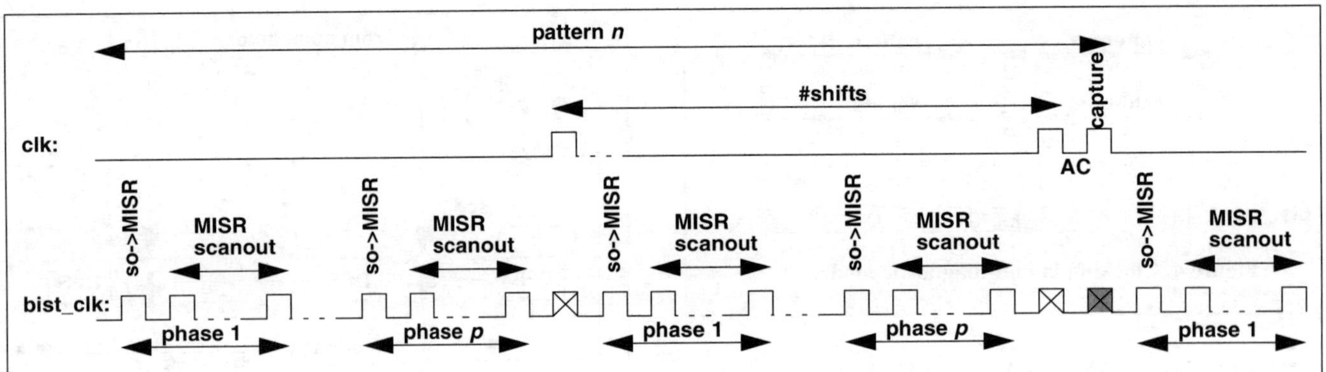

Figure 8. Clocking in diagnostic mode.

```
MacroDefs "diagnosis" {
"compare_substitute" {
  C { "clk"=0; } // Condition clk OFF
  // 1st phase: first 128 scan chains
  V { "bist_clk"=P; } // so->MISR
  Loop 128 { // MISR is 128 bits
      V { "bist_clk"=P; "lfsr_so"=S; }
  }
  // 2nd phase: next 128 scan chains
  V { "bist_clk"=P; } // so->MISR
  Loop 128 {
      V { "bist_clk"=P; "lfsr_so"=S; }
  }
  // more phases if needed
}}
```

Figure 7. STIL definition of diagnostic macro for a MISR of m = 128 bits.

MISR. The "S" character on the scanout pin "lfsr_so" references a Waveform that contains a "CompareSubstitute" event for the output strobe operation, but the value (compare High or compare Low) is not known; instead, the tester substitutes the High or Low value measured from the device-under-test, thus accumulating diagnostic data each time an "S" is encountered [25]. This process is repeated in the second phase (and subsequent phases if needed; e.g., 4 phases for 512 scan chains) to obtain values from all scan cells of all scan chains.

In this diagnostic operation, the MISR is re-used as a shift register to shift out the content of all scan cells (Figure 5). Figure 8 shows clocking in diagnostic mode, focusing on the same events as shown in Figure 4, clocking in normal mode. In diagnostic mode (Figure 8), each shift clock of every pattern is preceded by a complete unload of scan values, which requires p MISR unload phases. Unlike Figure 4, the BIST clock ("bist_clk") is pulsed in Figure 8 while the internal shift clock ("clk") is off. Each MISR unload phase starts by copying scanout values to the MISR (so->MISR), followed by a complete MISR scanout during which the PRPG and the internal scan chains are not pulsed. To preserve AC-testing capability, there is no MISR unload between the last shift and the capture; therefore, the data collected in diagnostic mode does not include the last-shift data, which is part of the data compressed in non-diagnostic mode. As with non-diagnostic timing (Figure 4), the PRPG is advanced by pulsing "bist_clk" every time "clk" is pulsed; for the MISR the "bist_clk" pulses (marked with X in Figure 8) during "clk" pulses are unnecessary because diagnostic data collection starts after the "clk" pulses and completes before next "clk" pulse.

4. TESTER FLOW

In a typical deterministic ATPG test flow, ATPG generates a large deterministic pattern file which is loaded and run on the tester; failing patterns can be selectively re-run in diagnostic mode. It is important that the very same pattern file is re-used for the diagnostic run so that the ATPG tool is not required to produce another large file. The diagnostic run typically requires additional tester settings to only run selected patterns and to log fail data. The fail data file produced by the tester is then returned (possibly after format translation) to the ATPG tool that performs diagnosis by fault simulation [3].

The same flow is maintained with BIST patterns by using the described diagnostic data collection method. BIST ATPG generates a very small BIST pattern file (Figure 2) that contains only the interval data (PRPG and MISR initial values and clocks pulsed in each interval). The BIST pattern file is loaded and run on the tester; failing intervals can be selectively re-run in diagnostic mode. As with deterministic ATPG, the very same pattern file is re-used for the diagnostic run so that additional ATPG runs or other files are not required. The diagnostic run requires additional tester settings to only run selected intervals (using STIL's "start" and "stop" operations in PatternBurst diagnosis) and to select the diagnostic procedures and macros from the STIL file (Figure 6). The diagnostic data file collected by the tester is then returned (possibly after format translation) to the ATPG tool that performs diagnostic by fault simulation. The diagnostic data file is much smaller than a deterministic patterns file because it contains only unload (no load) data of only selected intervals, and unload values are encoded with only 1 bit (2 states -- high and low); unlike deterministic patterns unload, 2 bits are not needed for BIST unloads because X values never propagate to the MISR.

For example, the diagnostic data file for a 10,000 scan cell design requires only 320 KB. For increased diagnosis resolution of multiple faults, multiple diagnostic data files can be collected for multiple intervals and returned to the ATPG tool for diagnostic. The diagnostic flow is enabled by usage of the CompareSubstitute construct in STIL (Figure 7) ([25]) and is not achievable in languages such as WGL ([26]) that lack such construct.

To support deterministic ATPG diagnostic, some testers implement the ability to unwind the hardware pipeline, starting with the first miscompare, to find the cycle corresponding to each failure. Failures occurring within a limited window from the first failure can be logged in one pass; to log additional fail data, the first miscompares are masked out and the test is repeated. If a separate BIST diagnostic pattern file is generated (in a WGL-based flow), the same fail-data logging mechanism can be used for BIST diagnostic. Supporting CompareSubstitute (for a STIL-based flow) requires the tester to efficiently log a larger volume of data, but does not require selectively masking out previously logged bits. Some testers implement this, or a similar mechanism for known-good-die response logging.

5. SIMULATOR FLOW

Before testing a device, it is common that at least a subset of the patterns are first verified though simulation (commonly Verilog [27] or VHDL [28]) for logical and timing accuracy. Verification is particularly important for BIST because the BIST structures (PRPG, phase-shifter, compactor, MISR, BIST controller) are simulated only functionally by the ATPG tool (for increased performance), whereas Verilog or VHDL simulation verifies the structural (gate-level) design. For verification, patterns are written out by the ATPG tool in a testbench format that provides stimuli to a simulator and compares the simulator outputs with the expected ATPG outputs. Both deterministic ATPG and BIST patterns can be verified through this scheme. To verify the diagnostic capability of deterministic ATPG, a fault can be injected in the simulated design; and, in turn, the simulator mirroring a tester, could produce a fail file that the ATPG tool then diagnoses.

Verification of the BIST diagnosis flow is even more important because the ATPG tool simulates all BIST structures functionally, including the mechanisms of unloading scan-chain values through the MISR. For this verification, the testbench is written so that a simulator compile-time definition controls whether the simulation is done in normal or diagnostic mode (using the corresponding STIL procedures and macros translated into Verilog or VHDL); this allows the same BIST pattern file to be used for either normal or diagnostic verification. In diagnostic mode, the simulator mirrors a tester and collects the diagnostic data file by scanning out the MISR. The file is then diagnosed by the ATPG tool in a special fail-only mode in which only the failing scan cell (not the internal gate) is identified to aid in debugging simulator differences. Diagnosing the internal gate is unnecessary because a simulator mismatch in simulating internal gates would have shown up as a MISR miscompare when simulating non-diagnostic mode BIST patterns; diagnostic-mode simulator mismatches can be caused only by functional and structural representation differences of the BIST structures involved in the diagnostic data collection. Evidently, if the simulator run matches the ATPG expected data, then the fail-only diagnostic reports no miscompares.

6. RESULTS

The simulator flow presented in section 5 was used to test the diagnosis method presented on designs ranging from 200K to 1M gates. A typical session, obtained for a 1M gate design, is detailed below.

(1) Generated BIST patterns in 32-pattern intervals.

(2) Saved patterns as Verilog testbench.

(3) Ran Verilog simulation and confirmed that MISR signatures match the expected values.

(4) Edited the Verilog netlist to inject a stuck-at fault.

(5) Re-ran Verilog simulation and confirmed that MISR signatures now miscompare (intervals 0 and 14 miscompared).

(6) Re-ran interval 0 Verilog simulation in diagnosis mode to generate the interval data (section 3).

(7) Read the file produced by the simulator into the ATPG tool and performed fault diagnosis.

The output of the diagnosis run is shown in Figure 9. First, the interval is simulated and the simulated values are compared to the actual values to identify the failing scan cells. Then, fault simulation is performed on these patterns to determine faults that best explain the failing scan cells. The summary shows the number of defects found (1) and the total CPU time (10 seconds on a 1 GHz Pentium®). Next, the fault candidates are listed: there are two faults marked with "DS" (detected by simulation), followed by their equivalent faults marked with "--". The fault injected at step (4) above was a stuck-at 0 (sa0) on output pin S of XOR 179345: the full pin-pathname of this fault is listed on the last line in Figure 9, preceded by two equivalent faults. The first 2 lines in Figure 9 list another defect candidate: the equivalent faults sa0 on output CO of AND 179152 and input B of buffer 179221 could cause the same fault signature. Diagnosis successfully identified the fault simulated. If this were real silicon, failure analysis would follow to analyze the fault candidates physical locations and identify the defect. In many cases, a single fault candidate is identified making failure analysis very simple; the two fault candidates in this case would still be easy to analyze because of their proximity.

7. CONCLUSIONS

The BIST diagnostic method presented allows identification of the failing gates -- not just the failing scan cells -- offering the same diagnosis resolution as deterministic ATPG test. Only minimal hardware is added for diagnostic: a few gates and internal signals controlled by the BIST controller. Determining if BIST will run in normal or diagnosis mode can be done using a "diag" primary input or, in a reduced-pin implementation, by a "start diagnosis" instruction to the BIST controller. Any number of fails can be handled and binary searches or other time-consuming methods are not required. Instead, the BIST test is composed of small, independent intervals. Only failing intervals need to be re-run in diagnostic mode to collect all scan data of the interval. Both the tester and the simulator flows use the same BIST pattern file for normal and diagnostic mode. The device-under-test timing in diagnostic mode is the same as in normal mode, preserving AC-test capabilities. The streamlined tester flow relies on a tester's ability to process the new CompareSubstitute event (added to the standard STIL language). Traditional (WGL-based) tester flows are supported with a separate diagnostic pattern file.

ACKNOWLEDGMENT

We are very grateful to Bill Lloyd and Martin Bell for careful reviews and insightful comments.

```
Failures for interval 0: #patterns=32, clocks= bist_clk clk
---------------------------------------------------
interval  pat#  chain     pos#  gate_id  simv/expv
--------  ----  --------  ----  -------  ---------
0         9     c359      34    389863     1 / 0
0         9     c359      35    389864     0 / 1
0         9     c359      36    389865     0 / 1
Failure summary: #failing_pats=1, #failures=3, #passes=564569
---------------------------------------------------
Diagnosis summary for failure file pat.diag
#failing_patterns=1, #defects=1, #unexplained_fails=0, CPU=10.20
Warning: Internal pattern set is now deleted. (M133)
-----------------------------------------------------------------------------
Fault candidates for defect 1: #failing_patterns_explained=1
-----------------------------------------------------------------------------
val  code  pin_pathname  (module_name)
---  ----  ---------------------------------------------
sa0  DS    core_i/compare_nf/b03_eq_port_fifo_read_60/add_136/U7/B
sa0  --    core_i/compare_nf/b03_eq_port_fifo_read_60/add_136/U6/CO
sa0  DS    core_i/compare_nf/b03_eq_port_fifo_read_60/U26/B2
sa0  --    core_i/compare_nf/b03_eq_port_fifo_read_60/U26/B1
sa0  --    core_i/compare_nf/b03_eq_port_fifo_read_60/add_136/U7/S
```

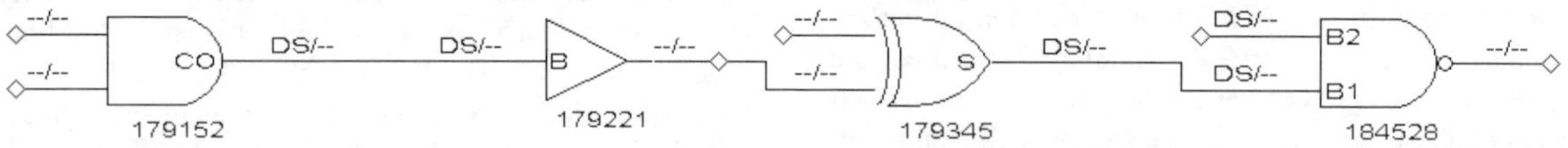

Figure 9. Diagnosis output: text (top) and schematic view of the fault candidates.

REFERENCES

[1] Semiconductor Industry Association (SIA), *International Technology Roadmap for Semiconductors (ITRS)*, 1999.

[2] M. Abramovici, M.A. Breuer, A.D. Friedman, *Digital Systems Testing and Testable Design*, IEEE Press, 1990.

[3] E.B. Eichelberger, E. Lindbloom, J.A. Waicukauski, T.W. Williams, *Structured Logic Testing*, Prentice-Hall, 1991.

[4] H.J. Nadig, "Testing a Microprocessor Product Using a Signature Analysis", *International Test Conference* 1978, pp.159-169.

[5] V.D. Agrawal, C.R. Kime, K.K. Saluja, "A Tutorial on Built-In Self-Test, Part 1: Principles", *IEEE Design & Test* 1993, Vol. 10, No.1, pp. 73-82.

[6] P.H. Bardell, W.H. McAnney, "Self-Testing of Multichip Logic Modules", *International Test Conference* 1982, pp.200-204.

[7] P.H. Bardell, W.H. McAnney, J. Savir, *Built-In Test for VLSI: Pseudorandom Techniques*, John Wiley & Sons, 1987.

[8] V.D. Agrawal, C.R. Kime, K.K. Saluja, "A Tutorial on Built-In Self-Test, Part 2: Applications", *IEEE Design & Test* 1993, Vol. 10, No.2, pp. 69-77.

[9] J. Rajski, N. Tamarapalli, J. Tyszer, "Automated Synthesis of Large Phase Shifters for Built-In Self-Test", *International Test Conference* 1998, pp. 1047-1056.

[10] J. Rajski, N. Tamarapalli, J. Tyszer, "Automated Synthesis of Phase Shifters for Built-In Self-Test Applications", *IEEE Transactions on Computer-Aided Design of Integrated Circuits and Systems* 2000, Vol. 19 No. 10, pp. 1175-1188.

[11] P. Wohl, J.A. Waicukauski, T.W. Williams, "Design of Compactors for Signature-Analyzers in Built-In Self Test", *International Test Conference* 2001, pp.54-63.

[12] W.H. McAnney, J. Savir, "There is Information in Faulty Signatures", *International Test Conference* 1987, pp. 630-636.

[13] J. Savir, W.H. McAnney, "Identification of Failing Tests with Cycling Registers", *International Test Conference* 1988, pp. 322-328.

[14] J. Ghosh-Dastidar, D. Das, N.A. Touba, "Fault Diagnosis in Scan-Based BIST Using Both Time and Space Information", *International Test Conference* 1999, pp. 95-102.

[15] R.C. Aitken, V.K. Agarwal, "A Diagnosis Method Using Pseudo-Random Vectors without Intermediate Signatures", *International Conference on Computer-Aided Design* 1989, pp. 574-580.

[16] I. Bayraktaroglu, A. Orailoglu, "Deterministic Partitioning Techniques for Fault Diagnosis in Scan-Based BIST", *International Test Conference* 2000, pp. 273-282.

[17] J. Ghosh-Dastidar, N.A. Touba, "A Rapid and Scalable Diagnosis Scheme for BIST Environments with a Large Number of Scan Chains", *VLSI Test Symposium* 2000, pp. 79-85.

[18] P. Song, F. Motika, D. Knebel, R. Rizzolo, M. Kusko, J. Lee, M. McManus, "Diagnostic Techniques for the IBM S/390 600 MHz G5 Microprocessor", *International Test Conference* 1999, pp. 1073-1082.

[19] Y. Wu, S. Adham, "BIST Fault Diagnosis in Scan-Based VLSI Environments", *International Test Conference* 1996, pp. 48-57.

[20] J. Rajski, J. Tyszer, "Fault Diagnosis in Scan-Based BIST", *International Test Conference* 1997, pp. 894-902.

[21] IEEE Std 1450-1999, *IEEE Standard Interface Test Language (STIL) for Digital Test Vectors* 1999.

[22] B. Könemann, "LFSR-Coded Test Patterns for Scan Designs", *European Test Conference*, Munich, 1991.

[23] J. Savir, "Salvaging Test Windows in BIST Diagnostics", *VLSI Test Symposium* 1997, pp. 416-425.

[24] J.A. Waicukauski, V.P. Gupta, S.T. Patel, "Diagnosis of BIST Failures by PPSFP Simulation", *International Test Conference* 1987, pp. 480-484.

[25] IEEE P1450.1, *Extensions to STIL for Semiconductor Design Environments,* http://grouper.ieee.org/groups/1450/dot1/index.html

[26] Summit (Fluence) Inc., *Waveform Generation Language (WGL) TDS Release 9.1.*

[27] IEEE Standards Department, "Verilog Hardware Description Language", IEEE-1364, 1994.

[28] IEEE Standards Department, "IEEE Standard VHDL Language Reference Manual", IEEE-1076-1987, IEEE, NY, 1988.

On Output Response Compression in the Presence of Unknown Output Values

Irith Pomeranz[1]
School of ECE
Purdue University
W. Lafayette, IN 47907
pomeranz@ecn.purdue.edu

Sandip Kundu
Intel Corp.
Austin, TX 78746
sandip.kundu@intel.com

Sudhakar M. Reddy[2]
ECE Dept.
University of Iowa
Iowa City, IA 52242
reddy@engineering.uiowa.edu

Abstract

A circuit may produce unknown output values during simulation of an input sequence due to an unknown initial state or due to the existence of tri-state elements. For circuits tested using BIST, unknown output values make it impossible to determine a single unique signature for the fault free circuit. To accommodate unknown output values in a BIST scheme, we describe a procedure for synthesizing a minimal logic block that replaces unknown output values by a known constant. The proposed procedure ensures that the BIST scheme will be able to detect all the faults detectable by the input sequence applied to the circuit while allowing a single unique signature to be obtained.

Categories and Subject Descriptors

Hardware, Arithmetic and Logic Structures, Testing.

General Terms

Algorithms, Reliability.

1. Introduction

Built-in self-test (*BIST*) [1]-[4] is becoming the accepted solution for at-speed testing of high-performance circuits [5]. Typical *BIST* circuitry is shown in Figure 1 (excluding the logic in the dashed box). Test vectors for the circuit-under-test (*CUT*) are produced by the test-pattern generator (*TPG*). The output response analyzer (*ORA*) compresses the output response of the *CUT* and produces a signature which can be used to determine whether the circuit is fault free or faulty.

One of the problems encountered in output response analysis for complex circuits (e.g., circuits containing buses and tri-state elements) is that the output values produced by the circuit under the sequence generated by the *TPG* may be unknown at certain time units and on certain outputs. Even when deterministic test generation is carried out, it is impossible to characterize and avoid all the situations under which, e.g., bus contention will occur and unknown output values will be produced. Under a test sequence produced by a *TPG*, bus contention is even more difficult to avoid. When it occurs, it may result in unpredictable output values. If the output sequence of the fault free circuit contains unknown values, a single predictable signature cannot be computed. One way to resolve this problem is to compute several signatures, one for each specification of the unknown values. However, this becomes impractical if the number of unknown values is large. A more practical solution is to ensure that the *ORA* does not compress the value of an output at a time unit when it is unknown. *BIST* circuitry appropriate for this case is shown in Figure 1. The block called *CMP* (which stands for *compress*) determines which output values at which time units will be compressed by the *ORA*, and which values will be ignored. Ignored values are replaced by the *CMP* block with a constant (0 in Figure 1). Specifically, the *CMP* output determines whether the multiplexer (*MUX*) output that drives the *ORA* will come from the *CUT* or will be replaced by the constant 0. This ensures that the *ORA* design does not have to be changed, and the same *ORA* used without the logic in the dashed box of Figure 1 manipulates only specified, i.e., {0,1} values, in Figure 1. As a result, a single unique fault free signature can be computed for the circuit. We point out that other variations on the *CMP* and *ORA* blocks are possible.

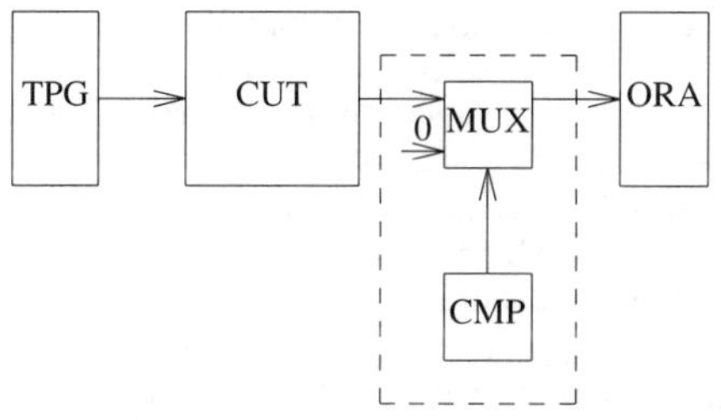

Figure 1: ***BIST*** **circuitry with unknown output values**

When the *CUT* of Figure 1 is a multi-output circuit, there is a multiplexer on every *CUT* output, and the *CMP* block produces a separate control line for every multiplexer.

In this work, we consider the synthesis of a minimal-size *CMP* block in the case of a multi-output synchronous sequential *CUT* whose output sequence contains unknown values. We assume that the *TPG* produces a single input sequence which is used for testing the *CUT*, and that the *CUT* in response produces an output sequence that contains unknown values at specific time units and on specific outputs. The *CMP* block is suitable only for this output sequence. If the *TPG* is changed such that the locations of the unknown values in the output sequence change, the *CMP* block must be redesigned to match the new output sequence produced by the *CUT*. Although we describe the *CMP* synthesis procedure for the case of non-scan circuits, the proposed synthesis procedure can also handle scan circuits with single or multiple scan chains.

1. Research supported in part by NSF Grant No. CCR-0098091 and in part by SRC Grant No. 2001-TJ-950.
2. Research supported in part by NSF Grant No. CCR-0097905 and in part by SRC Grant No. 2001-TJ-949.

DAC 2002, June 10-14, 2002, New Orleans, Louisiana, USA.

A problem similar to the one addressed here is considered in [6]. However, no details are available in [6] to allow a comparison with the method proposed here.

The paper is organized as follows. In Section 2 we formulate the problem of minimizing the size of the *CMP* block. In Section 3 we describe a procedure for synthesizing a *CMP* block based on the formulation of Section 2. We present experimental results in Section 4. Section 5 concludes the paper.

2. Problem formulation

We use the following notation. The output sequence produced by the fault free circuit in response to the *TPG* sequence is denoted by Z. The value of output i at time unit u obtained under the *TPG* sequence is denoted by $Z[u,i]$. For a circuit with a test sequence of length L and O primary outputs, Z is an $L \times O$ array. An output value $Z[u,i]$ can be any one of the three values $\{0,1,U\}$ where U stands for an unknown value.

We define an $L \times O$ array W_0 that indicates which output values at which time units are known and may be compressed by the *ORA*, and which values must not be compressed (i.e., they must be replaced by a constant) since they are unknown. The array W_0 is defined as follows. If $Z[u,i]=U$, $W_0[u,i]=0$ to indicate that the value of output i at time unit u must not be compressed. If $Z[u,i]=0$ or 1, $W_0[u,i]=x$ to indicate that the value of output i at time unit u may be compressed.

We also define an array W_{max} as follows. If $Z[u,i]=U$, $W_{max}[u,i]=0$ to indicate that the value of output i at time unit u must not be compressed. If $Z[u,i]=0$ or 1, $W_{max}[u,i]=1$ to indicate that the value of output i at time unit u is compressed. According to W_{max}, every known output value is compressed.

Our goal is to find an $L \times O$ array W with $\{0,1\}$ entries based on which the *CMP* block shown in Figure 1 will be synthesized. We refer to a value $Z[u,i]$ such that $W[u,i]=1$ (i.e., such that $Z[u,i]$ will be compressed) as an *observed* value, and we refer to a value $Z[u,i]$ such that $W[u,i]=0$ as an *unobserved* value. We thus imply that if the effect of a fault f propagates to output i at time unit u and $Z[u,i]$ is compressed, f will be detected. We point out that detection of f is not guaranteed in this case since aliasing may occur and mask the fault effects. However, since aliasing is rare and methods to avoid it exist, the assumption that f will be detected is justified.

The array W that indicates which *CUT* output values will be compressed and which values will not be compressed must satisfy the following conditions.

(1) Observing the output values corresponding to pairs (u,i) where $W[u,i]=1$ is sufficient to detect all the circuit faults that are detectable by the *TPG* sequence. In other words, the observed values must be such that for every detectable fault f, the effect of f propagates to at least one output i at a time unit u where $W[u,i]=1$.

(2) $W[u,i] \le W_{max}[u,i]$ for every u and i. This condition guarantees that if $W_{max}[u,i]=0$, $W[u,i]$ will be 0 and the corresponding entry will not be compressed. For the remaining time units and outputs, $W_{max}[u,i]=1$ and both 0 and 1 values can be used in W.

(3) The *CMP* logic corresponding to W is minimal. We discuss next a specific implementation of the *CMP* logic based on a given array W, and the requirements for minimizing it.

An example of an array W for a circuit with $O=4$ outputs and a *TPG* sequence of length $L=8$ is shown in Figure 2. In this array, $W[u,i]=0$ if an entry must not be compressed because $Z[u,i]$ is unknown, and $W[u,i]=1$ if the value of output i at time unit u must be observed to detect a target fault. The empty boxes indicate that $W[u,i]=x$. These values are don't-cares and they correspond to output values that may be compressed or not to minimize the size of the *CMP* logic.

We consider W as an incompletely specified truth table with inputs representing the L time units $u=0,1,\cdots,L-1$ and O outputs corresponding to the *CUT* outputs. To keep track of the time unit u the *CMP* block contains a counter, which is reset to 0 when the *TPG* is initialized, and incremented using the *CUT* (and *TPG*) clock.

It is possible to use a general-purpose logic synthesis procedure to find a minimal implementation of the *CMP* block. However, for practical circuits, the *TPG* sequence length L and the number of outputs O are typically large, and general-purpose logic synthesis procedures may be inefficient for such problem sizes. Instead, we use the following approach. We construct the *CMP* block from the basic blocks introduced in [7] and referred to as *comparison units*. The input to a comparison unit is the time unit u. Using the decimal form of u, a comparison unit produces the output value 1 if and only if $LB \le u \le UB$, where LB and UB are constants specific to each comparison unit. The maximum number of two-input gates in a comparison unit with m inputs is $2m-1$. The number of gates can be significantly smaller, depending on LB and UB.

To use comparison units for implementing the *CMP* block, we partition the array W into rectangles of the same logic value. Each rectangle will be implemented by a single comparison unit. For example, consider the rectangle marked with solid lines in Figure 2, which is defined by $2 \le u \le 4$ and $2 \le i \le 3$ and contains the 1's for $(u,i)=(2,3)$ and $(4,2)$. Since this rectangle does not contain any 0's, we can use a comparison unit with $LB=2$ and $UB=4$ to implement it. This comparison unit will be used as part of the implementation of the *CMP* outputs driving the multiplexers on *CUT* outputs 2 and 3. It will ensure that the value of output 3 at time unit 2 and the value of output 2 at time unit 4 are compressed by the *ORA* to detect target faults. Additional values will be compressed, belonging to outputs 2 and 3 at time units 2, 3 and 4. Although these values are not necessary for detecting target faults, observing them will increase the ability to detect unmodeled defects.

In general, our goal is to find a minimal number of rectangles to cover all the entries (u,i) such that $W[u,i]=1$. Each rectangle will be implemented using a single comparison unit. All the comparison units for a given *CUT* output will then be ORed to form the corresponding output of the *CMP*. For the example of Figure 2, we use the rectangles shown in the figure. The *CMP* implementation in this case is shown in Figure 3.

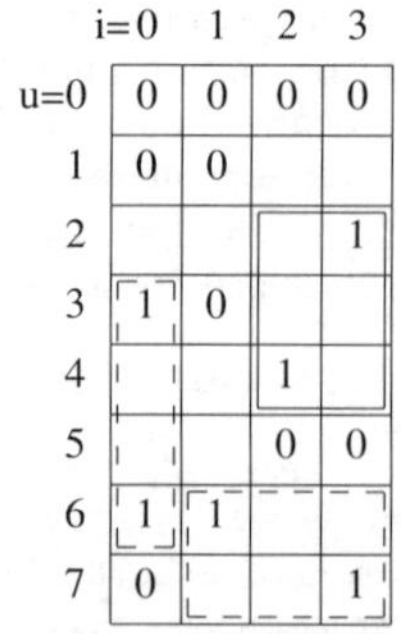

Figure 2: Rectangles for Example 1

Instead of rectangles to cover the 1 values in W, it is possible to find rectangles to cover the 0 values in W. With a *CMP* block implementing a cover of the 0 values, a *CUT* output value should be replaced by a constant value when the corresponding

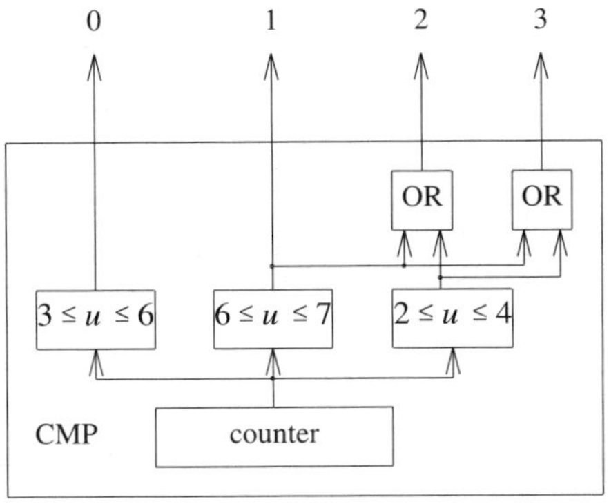

Figure 3: A *CMP* block for Example 1

output of the *CMP* block is 1. We use the cover that results in the smaller number of comparison units. We point out that implementing the *CMP* block based on rectangles that cover W simplies the synthesis procedure but may result in a comparison unit based implementation which is larger than necessary.

3. Synthesis procedure for *CMP* blocks

Synthesis of the *CMP* block consists of two tasks. (1) Determining the time units and outputs where the *CUT* output will be compressed (or observed) to ensure that all the faults can be detected while avoiding unknown output values. This task is equivalent to finding an array W that satisfies the conditions stated in Section 2. (2) Using W to find a minimal number of rectangles for a comparison unit based implementation of the *CMP* block. As part of the proposed synthesis procedure, we carry out these tasks simultaneously to ensure that the smallest possible number of rectangles (and hence comparison units) will be selected. We describe the synthesis procedure in the following subsections.

We already defined in Section 2 the array W_0, which provides information about the necessary 0's in any array W that satisfies the conditions of Section 2. In Subsection 3.1 we define entries that *must* be 1 in W to ensure that all the target faults (the faults detectable by the *TPG* sequence) are detected. These entries are defined based on faults that can only be detected by the *TPG* sequence at a single time unit, on a single output. In Subsection 3.2 we describe a procedure that, given an array W with $\{0,1,x\}$ entries, finds a minimal number of rectangles to cover its 0 (or 1) values. This procedure is used in Subsection 3.3 to guide the selection of additional 1 values in W to ensure that *all* the target faults can be detected by the observed values of W. In Subsection 3.4, this procedure is used for selecting the final set of rectangles that will define the implementation of the *CMP* block.

3.1. Necessary 1's in W

Consider a fault f, which is detected by the *TPG* sequence on a single primary output i at a single time unit u. To ensure that f is detected after output response analysis, we must set $W[u,i]=1$. We find faults such as f by an n-detection ($n > 1$) fault simulation process. This process drops a fault f from simulation only after f has been detected n times, on n different combinations of outputs and time units. We point out that to identify entries that must be 1 in W, it is sufficient to use $n=2$. However, in order to collect information that will be useful in Subsection 3.3, we prefer to use a larger value of n.

For every fault f, we find a set $det(f)$, which is the set of time units and primary outputs (u,i) such that f is detected by the *TPG* sequence at time unit u on output i. We then set $W[u,i]=1$ as follows. Initially, $W=W_0$ to ensure that all the output values that must be ignored are set to 0. For every fault f, if $det(f)$ contains a single entry (u,i), we set $W[u,i]=1$ to ensure that f can be detected after output response analysis. We refer to this procedure as Procedure 1.

3.2. Finding rectangles for a given array W

In this subsection we describe a procedure that, given an array W with $\{0,1,x\}$ entries, finds a minimal number of rectangles to cover its α values, where $\alpha \in \{0,1\}$. When covering the α values with rectangles, no rectangle must contain the value $\bar{\alpha}$. This procedure will be used in Subsections 3.3 and 3.4.

In the proposed procedure, a rectangle R_j is defined by its time unit and output ranges, $R_j=[u_{sj},u_{ej}]\times[i_{sj},i_{ej}]$. The proposed procedure initially includes every pair (u,i) such that $W[u,i]=\alpha$ in a separate rectangle. Thus, if $W[u,i]=\alpha$, we define a rectangle $R=[u,u]\times[i,i]$. Starting from this initial set of rectangles, the procedure *combines* as many pairs of rectangles as possible. The combination of two rectangles $R_1=[u_{s1},u_{e1}]\times[i_{s1},i_{e1}]$ and $R_2=[u_{s2},u_{e2}]\times[i_{s2},i_{e2}]$ results in the rectangle $R_3=[u_{s3},u_{e3}]\times[i_{s3},i_{e3}]$ where $u_{s3}=\min\{u_{s1},u_{s2}\}$, $u_{e3}=\max\{u_{e1},u_{e2}\}$, $i_{s3}=\min\{i_{s1},i_{s2}\}$ and $i_{e3}=\max\{i_{e1},i_{e2}\}$. The combination of R_1 and R_2 is accepted if R_3 does not contain any $\bar{\alpha}$ value of W, i.e., if there does not exist a pair (u,i) such that $W[u,i]=\bar{\alpha}$, $u_{s3}\le u\le u_{e3}$ and $i_{s3}\le i\le i_{e3}$.

If R_1 and R_2 are combined into R_3, we check whether there is any other rectangle R_4 which is contained in R_3. A rectangle R_4 is contained in R_3 if $u_{s4}\ge u_{s3}$, $u_{e4}\le u_{e3}$, $i_{s4}\ge i_{s3}$ and $i_{e4}\le i_{e3}$. If R_4 is contained in R_3, we remove R_4 from the set of rectangles.

In each iteration of the procedure, referred to as Procedure 2, every pair of rectangles R_{j1}, R_{j2} is considered. If the rectangles can be combined, they are replaced by the combined rectangle R_{j3}. Additional rectangles are then removed if they are contained in R_{j3}. The procedure terminates if no pair of rectangles can be combined in a complete iteration.

3.3. Setting additional 1's in W

In this subsection, we return to the array W defined by Procedure 1, and we describe the setting of additional 1 values in W. These values are selected to ensure that all the detectable circuit faults can be detected on the observed values indicated by W. In the selection process, we use the n-detection information collected by Procedure 1, and a cover of the 0's in W by rectangles obtained using Procedure 2.

Consider the array W at an arbitrary stage of its construction and a fault f which is not detected under W. We say that f is not detected under W if, for every (u,i) such that f is detected at time unit u on output i, $W[u,i]=x$. We observe that the array W already contains some 0 and 1 values selected by Procedure 1, and 1 values selected for faults considered before f. These values require a certain number of rectangles. When adding a 1 value to detect an undetected fault f, we would like to ensure as much as possible that the number of rectangles required to cover the 0 or 1 values in W will not go up. We achieve this as follows. We use Procedure 2 to find a cover of the 0 values in W by rectangles. Let the rectangles found be $R_1,R_2,\cdots,R_m$. If possible, we select for f an entry (u,i) such that f is detected at time unit u on output i, and (u,i) is not contained in any one of the rectangles R_j, $1\le j\le m$.

For every fault f, the candidate pairs (u,i) are the ones contained in $det(f)$ found by Procedure 1. Of these pairs, we select one which is not contained in any one of the rectangles R_j,

$1 \le j \le m$, and which appears in the maximum number of sets $det(f')$ of undetected faults f'. If no such pair can be found, we select any pair out of $det(f)$ that appears in the maximum number of sets $det(f')$ of undetected faults f'.

The yet-undetected faults are considered in an order such that faults with smaller sets $det(f)$ are considered earlier. For these faults we have fewer options to set values of W to 1 so as to detect the fault. We prefer to consider such faults first in an attempt to minimize the total number of 1's in W.

The procedure, referred to as Procedure 3, uses a simulation process that allows it to determine which faults are detected under a given array W with values $\{0,1,x\}$. This simulation process is applied every time a 1 value is added to W. This is needed since the size of a set $det(f')$ computed by Procedure 1 is limited to n. Thus, if f' is detected more than n times by the *TPG* sequence, it may be detected under W because $W[u,i] = 1$ even if $(u,i) \notin det(f')$.

3.4. Selecting a set of rectangles for *CMP*

Using W obtained at the end of Procedure 3, we apply Procedure 2 to find two sets of rectangles. The set C_0 is a cover of the 0's in W, and the set C_1 is a cover of the 1's in W. The implementation of the *CMP* block is based on the smaller of C_0 and C_1.

4. Experimental results

We applied the synthesis procedure described in Section 3 to ISCAS-89 and ITC-99 benchmark circuits. To investigate the synthesis procedure in the case where all the detectable circuit faults can be detected by the *TPG* sequence, we used test sequences generated by the test generation and test compaction procedures from [8]-[10].

The benchmark circuits we consider do not have tri-state elements that result in unknown output values. In addition, the circuits are synchronized after a small number of time units following which the output sequence is fully specified. To imitate the situation where unknown values exist throughout the output sequence, we injected unknown values randomly into the output sequence of the *CUT*. For this purpose, we considered every time unit u and every output i. With probability p, we changed the value at time unit u on output i from its actual value to a U. We experimented with $p = 0$, 0.005, 0.01 and 0.05.

In the implementation of Procedure 1 we used $n = 20$. The results are shown in Table 1. For every value of p we show the number of 1's in W at the end of Procedure 3, and the number of rectangles required to cover all the 0's or all the 1's of W (whichever is smaller). It can be seen that the number of rectangles is significantly lower than the number of 1's. Thus, we are able to cover large numbers of 1's (or 0's) using each rectangle.

In practice, the percentage of unknown output values in the output sequence of the circuit should be minimized by designing the *TPG* appropriately. Moreover, it is possible to design the *TPG* and the *CMP* block together so as to minimize the size of the *CMP* block. Thus, it is expected that in practice, the *CMP* block will have to deal with relatively small numbers of unknown output values that will fit well into rectangles. As a result, the number of rectangles, which indicates the number of comparison units in the implementation of the *CMP* block, is expected to be relatively low.

In the last column of Table 1 we show the normalized run time of Procedure 3 for the case where $p = 0.05$. Denoting the run time of Procedure 3 by RT_3 and the run time required to fault simulate the test sequence applied to the circuit by RT_0, the normalized run time reported is RT_3/RT_0. The normalized run time

Table 1: Experimental results

	p=0		p=0.005		p=0.01		p=0.05		n.
circuit	1's	rct	1's	rct	1's	rct	1's	rct	time
s298	37	1	37	3	38	3	39	10	13.60
s344	31	1	31	2	31	2	34	8	6.19
s382	52	1	51	7	52	12	50	24	33.58
s400	51	1	52	8	51	13	52	21	35.26
s444	49	1	49	8	49	11	49	24	33.35
s526	45	1	45	9	45	16	44	29	38.52
s641	93	3	94	7	93	9	96	29	24.56
s820	178	1	177	19	182	25	184	66	59.42
s1196	309	1	310	10	310	16	312	53	34.06
s1423	58	1	58	10	61	14	60	35	8.23
s1488	316	1	320	16	324	29	318	84	113.69
s5378	337	5	338	49	343	67	354	115	252.12
b01	11	1	11	1	11	2	11	4	0.56
b02	8	1	8	1	8	1	8	2	0.00
b03	32	1	32	2	32	4	33	11	9.19
b04	93	1	98	4	99	7	99	23	27.17
b06	32	1	32	2	32	3	31	6	5.86
b09	23	1	23	2	23	3	23	5	7.64
b10	40	1	39	4	40	5	40	14	19.82
b11	42	1	44	7	43	9	45	17	16.77

provides an indication of the simulation effort in Procedure 3 (the other components of the procedure are relatively fast).

5. Concluding remarks

Complex circuits may produce unknown values in response to an input sequence due to the existence of tri-state elements and due to unknown initial states. Unknown values may appear throughout the output sequence of the circuit. Such unknown values make it difficult to determine the signature of the fault free circuit computed by an output response analyzer which is part of a BIST scheme. We described a procedure for synthesizing a minimal logic block that allows unknown values to be replaced by a known constant while ensuring that the BIST scheme can detect all the detectable faults. The implementation of this logic block was based on building blocks called comparison units. Each comparison unit allows us to avoid a different range of time units where unknown values may appear on a circuit output. The use of comparison units simplifies the synthesis procedure.

References

[1] P. H. Bardell, W. H. McAnney and J. Savir, *Built-In Test for VLSI Pseudorandom Techniques*, Wiley, 1987.

[2] E. B. Eichelberger, E. Lindbloom, J. A. Waicukauski and T. W. Williams, *Structured Logic Testing*, Prentice-Hall, 1991.

[3] V. D. Agrawal, C. R. Kime and K. K. Saluja, "A Tutorial on Built-In Self-Test, Part 1: Principles", IEEE Design & Test of Computers, March 1993, pp. 73-82.

[4] V. D. Agrawal, C. R. Kime and K. K. Saluja, "A Tutorial on Built-In Self-Test, Part 2: Applications", IEEE Design & Test of Computers, June 1993, pp. 69-77.

[5] Y. Zorian, E. J. Marinissen and S. Dey, "Testing Embedded-Core Based System Chips", ITC-98, pp. 130-143.

[6] G. Eide, "Embedded Deterministic Test - DFT Technology for Low-Cost IC Manufacturing Test", www.mentor.com/dft.

[7] I. Pomeranz and S. M. Reddy, "On Synthesis-for-Testability of Combinational Logic Circuits", DAC-95, pp. 126-132.

[8] M. S. Hsiao, E. M. Rudnick, and J. H. Patel, "Sequential Circuit Test Generation Using Dynamic State Traversal", ED&TC-97, pp. 22-28.

[9] R. Guo, S. M. Reddy and I. Pomeranz, "PROPTEST: A Property Based Test Pattern Generator for Sequential Circuits Using Test Compaction", DAC-99, pp. 653-659.

[10] I. Pomeranz and S. M. Reddy, "Vector Restoration Based Static Compaction of Test Sequences for Synchronous Sequential Circuits", ICCD-97, pp. 360-365.

Software-Based Diagnosis for Processors

Li Chen and Sujit Dey
Dept. ECE, University of California at San Diego, La Jolla, CA 92093-0407
{lichen, dey}@ece.ucsd.edu

ABSTRACT

Software-based self-test (SBST) is emerging as a promising technology for enabling at-speed test of high-speed microprocessors using low-cost testers. We explore the fault diagnosis capability of SBST, in which functional information can be used to guide and facilitate the generation of diagnostic tests. By using a large number of carefully constructed diagnostic test programs, the fault universe can be divided into fine-grained partitions, each corresponding to a unique pass/fail pattern. We evaluate the quality of diagnosis by constructing diagnostic-tree-based fault dictionaries. We demonstrate the feasibility of the proposed method by applying it to a processor example. Experimental results show its potential as an effective method for diagnosing larger processors.

Categories and Subject Descriptors

B.8.1 [**Performance and Reliability**]: Reliability, Testing, and Fault-Tolerance.

General Terms

Algorithms, Measurement, Reliability, Experimentation.

Keywords

Microprocessor, self-test, instruction, diagnostics.

1. INTRODUCTION

As microprocessor speed continues to rise beyond the gigahertz range, there is a growing need for at-speed testing. Traditionally, functional testing has been used in industry for applying at-speed tests, in which high-speed, high-pin-count functional testers are used to apply billions of architectural validation (AV) vectors to the device-under-test. As the clock speed, I/O pin count, and length of the AV tests continue to increase, functional testers will become prohibitively expensive, making functional testing economically infeasible. An alternative solution is at-speed scan. Unlike functional testing, scan cannot duplicate the exact operational environment in the normal operational mode, such as the cross-die temperature variation. The difference between the test environment and the normal operational environment may lead to over-testing or under-testing in speed-binning, neither of which is desirable.

In attempt to achieve low-cost and accurate at-speed testing, software-based self-test (SBST) [1]-[3] has emerged as a promising technology for testing performance-critical ICs such as microprocessors. Prior to self-test, test programs are loaded to the on-chip memory by a low-cost tester with low-speed and low-pin-count. The processor then performs self-test by executing the test programs. Upon completion of the test, the test responses or test response signatures can be unloaded from the on-chip memory, again using a low-cost tester. The use of low-cost testers makes SBST an economical solution for at-speed testing, while the application of tests in the normal operational mode enforces the test accuracy.

An additional benefit of SBST is that it can be applied in the absence of scan chains, making it a suitable choice for designs that cannot tolerate scan-induced performance overhead on their critical paths (e.g., high-end microprocessors). The absence of scan chains, however, poses a significant challenge for fault diagnosis.

It is a known fact that the complexity of diagnostic test generation is much higher than that of detection test generation. Though deterministic methods for generating diagnostic tests are available for combinational circuits [4], sequential circuits are much too complex to be handled by the same approach. As a result, currently, the existence of scan chains is a must in order to allow diagnosis of large sequential circuits [5].

Recently, there have been several proposals on generating diagnostic tests for sequential circuits by modifying existing detection tests. In [6], genetic algorithm is used to improve the diagnostic capability of detection tests. In [7], diagnostic tests are generated by eliminating certain vectors in the existing detection test sequence in order to un-detect certain faults. A prerequisite for both methods is a high-coverage detection test set for the sequential circuit under test, as neither method is able to distinguish faults that cannot be detected by the given detection tests. Hence, the success of these methods depends on the success of sequential test generation techniques. Currently, many efforts have been invested in making sequential Automatic Test Pattern Generation (ATPG) techniques feasible for large sequential circuits. However, the applicability of sequential ATPG on complex sequential circuits such as processors is yet to be proven.

As an alternative to sequential ATPG, SBST has been shown to have the potential of successfully generating tests for a particular type of sequential circuits – microprocessors. If properly modified, these tests could possibly achieve a high diagnostic capability. In addition, SBST has a high potential in diagnosis, as functional information is an invaluable resource for guiding and facilitating the generation of diagnostic tests. Our goal in this work is to explore the diagnostic potential of SBST by developing methods for software-based diagnosis.

Previous research in fault diagnosis can be grouped under three categories: diagnostic test generation [4][6][7], diagnosis using realistic fault models [5][8], and fault dictionary management [9][10]. We focused on the generation of diagnostic tests and made the following assumptions during the development of our method. First, it is known that by analyzing test response values rather than pass/fail results, the same diagnostic resolution can be achieved with a smaller test set [6]. For simplicity, we choose to analyze only pass/fail results. This assumption can be eliminated with more sophisticated fault simulation tools. Secondly, accurate diagnosis depends on the use of realistic fault models [8]. Thus, the use of stuck-at fault model may lead to inaccurate diagnosis for bridge faults. On the other hand, it has been reported in [6] that there is a strong correlation between the diagnostic measures for stuck-at faults and for bridge faults. Thus, for the purpose of diagnostic test generation, it is acceptable to evaluate the quality of the diagnostic tests using the single stuck-at fault model.

The rest of the paper is organized as follows. In Section 2, we introduce our methodology for software-based diagnosis, which is based on the careful construction of a large number of diagnostic test programs. We evaluate the quality of diagnosis by

This work is supported by MARCO/DARPA Gigascale Silicon Research Center (GSRC).

DAC 2002, June 10-14, 2002, New Orleans, Louisiana, USA.

constructing diagnostic-tree-based fault dictionaries. In Section 3, we demonstrate the proposed method on a processor core. In Section 4, we draw conclusions and discuss future research directions.

2. METHODOLOGY

We describe our methodology for software-based diagnosis by drawing analogy to traditional fault diagnosis methods, which are based on the application of test patterns.

2.1 Generating diagnostic test programs

In traditional fault diagnosis methods, the basic unit of test is a *test pattern*. We refer to them as *test pattern-based diagnosis*. Given a set of test patterns, two faults can be distinguished if they exhibit different behaviors at the primary outputs (POs).

In the SBST method introduced in [1], test patterns are applied to internal modules of a processor using test programs composed of processor instructions. The test response at the POs of the internal module cannot be checked *immediately* after the application of the test pattern. Rather, the test program captures the module's response and stores it to the on-chip memory for later analysis. Thus, in software-based diagnosis, the basic unit of test is a *test program*. We perform diagnosis by analyzing the test responses corresponding to these test programs, which become available in the on-chip memory after the execution of the test program.

Compared with test pattern-based diagnosis, software-based diagnosis faces new challenges in achieving a high diagnostic resolution, as a test program typically detects more faults than a test pattern.

In test pattern-based diagnosis, a solution to combat low diagnostic resolution is to construct new test patterns that can cause two previously undistinguishable faults to behave differently. If the test patterns are to be generated deterministically, structural analysis needs to be performed to find the right patterns. Otherwise, a large number of fault simulations are needed in order to search for the desired test patterns. Both options are expensive for large sequential circuits.

In software-based diagnosis, with the aid of explicit functional information, it is easier to generate test programs to distinguish certain faults without structural analysis or fault simulation. For example, faults on the data bus may be distinguished by loading and storing specific values; faults in the instruction decode unit may be distinguished by executing different instructions. It should be pointed out that due to the complex structure of a processor, it is impossible to foresee all faults detected by a test program. Thus, given an observed test response, the fault candidates should be determined by fault simulation rather than by speculation. Nonetheless, knowledge of functional information can facilitate the search for the right test program.

In order to achieve a high diagnostic resolution, we propose the use of a large number of carefully constructed diagnostic test programs. Each programs is designed to cover as few faults as possible, while the union of all test programs covers as many faults as possible. To achieve these goals, several principles need to be followed in the construction of these test programs:

1. Reduce the *variety* of instructions in each test program. If possible, use one type of instruction only.
2. Reduce the *number* of instructions in each test program. Each test program contains only the essential instructions for (a) applying one specific test pattern to an internal module of the processor, and (b) observing test responses off the target module by storing them to memory.
3. Create multiple copies of the same test program, which are varied to observe test responses on *different outputs* from the target module. This is to differentiate faults that cause errors to propagate to different nodes.

We illustrate the above principles by comparing detection test programs with diagnostic test programs. The detection test program described in [1] includes subroutines targeting at individual sub-modules of a processor. For example, a subroutine may be targeted at testing the adder logic in the ALU. To increase code density, a loop may be used to execute the ADD instruction with an array of operand values. Thus, the subroutine includes not only the ADD instruction, but also instructions needed for constructing the loop, such as branch instructions, SUB (subtraction) instruction, etc. The resulting detection test program is not suitable for diagnosis due to its large coverage caused by a large variety of instructions and a large number of ADD tests performed. To construct a diagnostic test program that detects as few faults as possible, we follow the principle of using only the essential instructions for applying one ADD test to the ALU. An example test program can be found in [12], where only three types of instructions are used, and only one ADD test is applied. Construction of diagnostic test programs for a processor core following the above three principles can be found in Section 3.1.

2.2 Analysis of diagnostic results using diagnostic tree

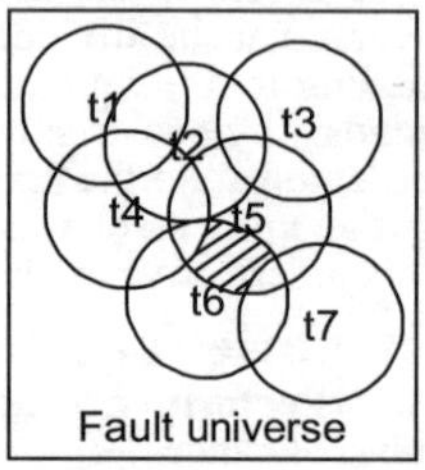

Figure 1. Partitioning the fault universe

Having constructed the diagnostic test programs, the next step is to evaluate the resulting diagnostic resolution, or, how many faults remain indistinguishable from each other.

Groups of indistinguishable faults can be identified with the construction of a fault dictionary. A fault dictionary entry has two fields, with the first being a unique pass/fail pattern and the second being the group of indistinguishable faults that leads to the pass/fail pattern. Figure 1 shows an abstract view of the dictionary, where each circle represents faults detected by a test program (*t1* through *t7*) and each partition in the fault universe corresponds to a group of indistinguishable faults. E.g., the shaded area corresponds to faults causing both *t5* and *t6* to fail but others to pass.

Through fault simulation, we determine the set of faults detected by each test program. By performing set operations, we determine the fault candidates corresponding to any pass/fail pattern. The same method has been used for test pattern-based diagnosis.

Based on the fact that many pass/fail patterns are impossible due to the correlation among test programs, it is efficient to enumerate only the *possible* pass/fail patterns. Thus, we choose to build the fault dictionary using a diagnostic tree-based approach, in which impossible pass/fail patterns can be detected as early as possible.

Diagnostic tree has been developed through numerous works and is regarded as one of the most space-efficient ways for storing a fault dictionary [9][10]. Here we illustrate the organization of the diagnostic tree we have used and several methods for limiting the tree size.

Given the detection list shown in Figure 2(a), Figure 2(b) shows the corresponding diagnostic tree.

Each tree node is associated with two elements: a fault list and the number of faults in the list. The root node contains the list of all faults. Each level in the tree is associated with a test *T*. The left/right child of a node *N* corresponds to faults in *N* causing the *T* to fail/pass. A fault list no long needs to be divided if it (a) contains no faults, or (b) contains only one fault. In case (a), an impossible pass/fail pattern is detected (e.g., (*t0*,*t1*)=(F,F)). In case (b), we found a pass/fail pattern which can uniquely identify a single fault (e.g., (*t0*,*t1*)=(P,F)). Clearly, only collapsed faults should be considered in order to facilitate the timely termination of tree branches. Each non-zero leaf node in the diagnostic tree corresponds to a group of indistinguishable faults. Their pass/fail response can be determined by tracing back to the root node.

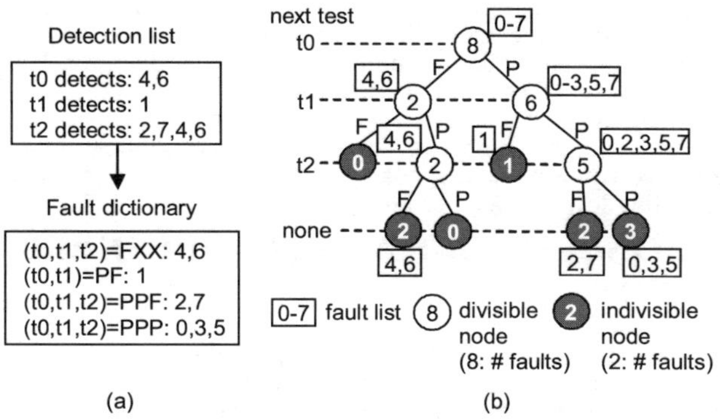

Figure 2. Diagnostic Tree

We built the diagnostic tree from the root up using preorder traversal. The pseudo code can be found in [12].

In a deep diagnostic tree, many impossible pass/fail patterns can be detected near the root. Thus, the tree-size is much smaller than that of a full binary tree. In fact, the tree size is bounded by $F{\cdot}(T{+}1)$ rather than 2^{T+1}-1, where F is the number of collapsed faults (i.e., the maximum number of leaf nodes in the tree), and T is the number of test programs (i.e, T+1 is the depth of the tree).

The tree size can be further reduced by eliminating *redundant test programs*, which we defined as test programs that cannot further divide the existing fault partitions. Consider the diagnostic tree shown in Figure 2. Let *t3* be a test program detecting faults {1,4,6}. Assuming *t0*, *t1*, and *t2* have been applied, *t3* is a redundant test program.

2.3 CAD framework for evaluating diagnostic test programs

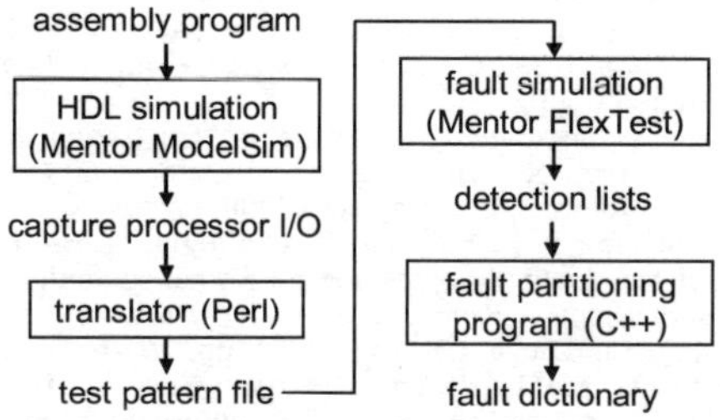

Figure 3. Evaluating diagnostic test programs

To evaluate the quality of a given set of diagnostic test programs, we used the CAD framework shown in Figure 3. To create a detection list for each test program, we first simulate the execution of the test program using an HDL simulation tool. We then capture the I/O signals at the boundary of the processor and translate them into the test pattern format readable by a fault simulator. The detection list of the test program is produced by a full-processor fault simulation. To generate the fault dictionary from the detection lists, we developed a fault-partitioning program based on the diagnostic tree-based method described in Section 2.2. The fault-partitioning program also identifies redundant test programs so as to eliminate them from the final set of diagnostic test programs.

3. EXPERIMENTAL RESULTS

We demonstrate the proposed software-based diagnostic method by constructing diagnostic test programs for a processor core named PARWAN [11]. PARWAN is an 8-bit accumulator-based microprocessor capable of executing 23 different instructions. The synthesized version of PARWAN contains 1785 equivalent NAND gates and 53 flip-flops. The total number of faults is 3940. The number of collapsed faults is 1912. The quality of the diagnostic test programs is evaluated in terms of the size of indistinguishable fault groups in the fault universe.

3.1 Composition of diagnostic test programs

Based on the three principles described in Section 2.1, we constructed 3609 diagnostic test programs for PARWAN. Detailed contents of the test programs can be found in [12]. The composition of the test programs is shown in Table 1. In order to achieve a high diagnostic resolution, each test program aims at applying a specific test pattern to an internal module of the processor using a specific instruction. Thus, we categorize the test programs under module names and instruction names.

Table 1. Diagnostic test programs for PARWAN

Module	Test program types	# test programs
ALU	LDA test	128x5
	ADD test	64x5
	AND test	64x5
	SUB test	64x5
	NOT test	128x5
SHU (Shifter)	ASL test	128x5
	ASR test	128x5
PC	JMP test	64
CTRL	CTRL test	25
	Total	3609

With a three-bit control input, the ALU has six functional modes, corresponding to six instructions: LDA (load from accumulator), STA (store to accumulator), ADD (addition), AND (logic AND), SUB (subtraction), and NOT (inversion). Thus, six different types of test patterns can be applied to the ALU. Note that it is implied that the STA instruction must be used in every test program to store the test response to memory. Thus, we used five types of test programs, corresponding to the other five instructions. In each test program, with control inputs fixed by the instruction type, we apply pseudo random patterns to the data inputs using pseudo random operand values. E.g., 128 operand values are used for LDA tests. We then create 5 copies of each test program, each observing a subset of outputs from the ALU. One test program uses the STA instruction for observing the data output, while the other four combines STA instruction and branch instructions to observe the four status outputs, including *v* (overflow), *c* (carry), *z* (zero), and *n* (negative).

Similarly, two different types of test programs are constructed for the Shifter (SHU), consisting of ASL (arithmetic shift left) tests and ASR (arithmetic shift right) tests.

For the Program Counter (PC), test patterns are applied using the jump instruction (JMP). Each test program jumps from one specific memory location to another. To facilitate the observation of the test response, the test programs are designed in a way such that a particular value is stored to memory *only if* the jump is executed properly.

Finally, 25 test programs are used to partition faults within the Controller (CTRL). For the 19 non-branch instructions, one test program is used to exercise each instruction, with the exception of the STA instruction. For the 4 branch instructions, two test programs are used to exercise each instruction, including a branch-taken case and a branch-not-taken case. By analyzing the state transition diagram, we have concluded that any additional test program will not further partition the faults in the Controller.

3.2 Evaluation of diagnostic test programs

We use the CAD framework shown in Figure 3 to evaluate the quality of the diagnostic test programs. On a Sun workstation with a 300MHz Ultra-SPARC-IIi processor and 128MB of physical memory, the typical CPU time for running HDL simulation and fault simulation for a test program in Table 1 is 1.8 sec and 10 sec, respectively. With 3609 test programs, the total CPU time required is 11.83 hours.

Table 2 shows the quality of the diagnostic test programs. We first divide the test programs into sub groups. We eliminate redundant test programs from each sub groups by running the fault-partitioning program. We then combine all irredundant test programs and used the fault-partitioning program again to derive the final fault dictionary.

In Table 2, Column 1 shows the names of the sub groups, with "ALL" denoting the union of all irredundant tests extracted from all sub groups. For each group of test programs, Column 2 shows the number of test programs. Column 3 shows the number of irredundant test programs. Column 4 shows the number of indistinguishable fault groups in the fault universe. Column 5/6 shows the number of faults that failed/passed all test programs. Column 7 shows the average size of all fault groups, *excluding* the all-F and all-P groups. Column 8 shows the number of non-zero nodes in the diagnostic tree. Column 9 shows the CPU time required for running the fault-partitioning program.

Table 2. Quality of diagnostic test programs

	# tests	# irre.	# groups	all-F	all-P	ave group size	# tree nodes	CPU time [sec]
ADD	320	72	261	560	813	2.081	28633	22.7
AND	320	47	109	544	981	3.617	9487	8.4
SUB	320	55	231	575	819	2.262	29695	23.8
LDA	640	37	68	524	1081	4.652	14258	14.2
NOT	640	41	108	540	1002	3.491	21165	19.9
ASL	640	58	146	464	955	3.424	34075	29.1
ASR	640	59	126	472	979	3.718	28045	24.5
JMP	64	37	122	654	981	2.308	3323	2.7
CTRL	25	25	174	472	621	4.762	1378	1.0
ALL	**431**	**259**	**746**	**339**	**307**	**1.702**	**57830**	**46.1**

Out of 3609 test programs, only 259 are irredundant. Note that this is different from the total number of tests in the ALL group, as there are redundancies among irredundant tests extracted from different sub groups. The small number of tests implies a small diagnostic tree, as well as short test application time. Note that the set of irredundant test programs depends on the order in which test programs are added to the diagnostic tree.

Given the fault dictionary resulting from all irredundant test programs, we extracted the group size distribution shown in Table 3. For a group size *N*, frequency *F* denotes the number of groups that contain *N* faults. Multiplying *N* by *F*, we obtained the fault distribution shown in Figure 4. Note that excluding the all-F and all-P groups, the majority of the faults belong to groups of small sizes. The average size of these groups, as shown in Table 3, is 1.702.

Table 3. Group size distribution

Group size	1	2	3	4	5	6	7	8	9
Frequency	525	125	54	10	5	10	3	2	4

Groups of size 13, 16, 22, 28, 307, 339: one each

Groups of size 26: 2

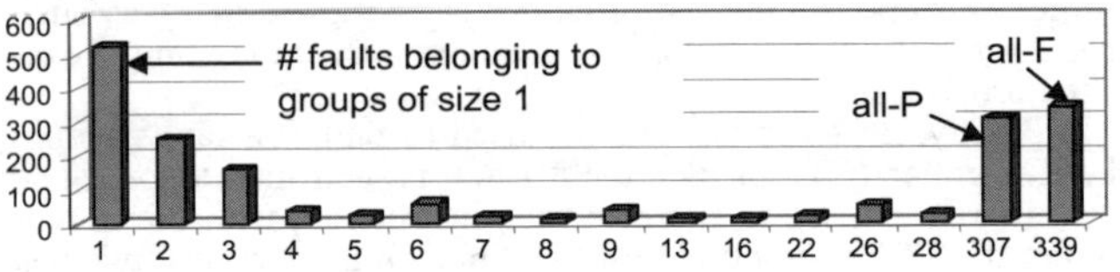

Figure 4. Fault distribution

The all-P group contains faults that cannot be detected by any of the diagnostic test programs. The 307 collapsed faults translate to 382 uncollapsed faults, resulting in a fault coverage of 90.63% out of 3940 faults. This is close to the fault coverage reported in [1] (91.42%), which is that of a detection test program prepared based on the SBST method. As we have discussed in the introduction, the number of diagnosable faults is limited by the number of detectable faults. As reported in [1], on PARWAN, SBST achieved higher fault coverage than sequential ATPG (46.82%) and even full-scan without test points (82.18%).

The all-F group contains faults that are detected by all test programs. A rough inspection of the fault list shows that these faults are located on functionally critical nodes in the processor, such as certain nodes in the instruction decode unit and tri-state buffers controlling the busses. These faults can be further divided by inspecting test response values in addition to pass/fail results. However, it is quite possible that a significant fraction of the all-F faults result in the same empty test response, as their critical location cause the processor's inability to execute *any* instruction. This fundamental limitation indicates that a hybrid approach may be needed for these hard-to-diagnose faults. Nonetheless, an overwhelming majority of faults in a processor (datapath faults and even a larger number of faults in control logic) can be handled by software-based diagnosis, which has the *unique* advantage of enabling low-cost at-speed test.

4. CONCLUSIONS AND FUTURE DIRECTIONS

Though a promising technology for enabling at-speed test with low-cost testers, SBST poses a great challenge for fault diagnosis. To address this problem, we have developed a method for enabling software-based diagnosis, in which a large number of carefully constructed diagnostic test programs are used to divide the fault universe into fine partitions, each corresponding to a unique pass/fail pattern. To demonstrate its feasibility, we applied the proposed method to a processor core. Experimental results show its potential as an effective method for diagnosing larger processors.

The fault simulation time required for deriving the detection lists for all diagnostic test programs is a major concern when scaling the proposed method to large industrial processors. Though the simulation time can be reduced with mixed-level fault simulation and distributed computing, more efficient test generation methods need to be developed in order to reduce the number of diagnostic test programs to be fault-simulated. For example, more deterministic test programs can be used to diagnose programmer-visible faults (e.g., faults on data busses and in register banks) and datapath components with regular structures. In addition, hierarchical approach may be used to reduce the problem size, that is, the number of faults to be considered in one fault simulation.

5. REFERENCES

[1] L. Chen and S. Dey, "Software-based self-testing methodology for processor cores," *IEEE Trans. Computer-Aided Design*, vol.20, no.3, March 2001, pp. 369-380.

[2] W.-C. Lai, A. Krstic, and K.-T. Cheng, "On testing the path delay faults of a microprocessor using its instruction set," *Proc. 18th VLSI Test Symp.*, Montreal, Canada, May 2000, pp. 15-20.

[3] S. Almukhaizim, P. Petrov, and A. Orailoglu, "Low-cost, software-based self-test methodologies for performance faults in processor control subsystems," *Proc. IEEE 2001 Custom Integrated Circuits Conference*, San Diego, CA, May 2001, pp. 263-6.

[4] T. Grüning, U. Mahlstedt, and H. Koopmeiners, "DIATEST: A fast diagnostic test pattern generator for combinational circuits," *Proc. Int. Conf. on Computer-Aided Design*, Santa Clara, CA, Nov. 1991, pp. 194-197.

[5] S. Venkataraman and S.B. Drummonds, "Poirot: applications of a logic fault diagnosis tool," *IEEE Design & Test of Computers*, vol.18, no.1, Jan.-Feb. 2001. pp. 19-30.

[6] X. Yu, J. Wu, and E.M. Rudnick, "Diagnostic test generation for sequential circuits," *Proc. Int. Test Conf. 2000*, Atlantic City, NJ, Oct. 2000, pp. 225-34.

[7] I. Pomeranz and S.M. Reddy, "A diagnostic test generation procedure based on test elimination by vector omission for synchronous sequential circuits," *IEEE Trans. Computer-Aided Design*, vol.19, no.5, May 2000, pp. 589-600.

[8] J. Wu and E.M Rudnick, "Bridge fault diagnosis using stuck-at fault simulation," *IEEE Trans. Computer-Aided Design*, vol.19, no.4, April 2000, pp. 489-495.

[9] V. Boppana, I. Hartanto, and K. Fuchs, "Full fault dictionary storage based on labeled tree encoding," *Proc. 14th VLSI Test Symp.*, Princeton, NJ, April 1996, pp. 174-9.

[10] B. Chess and T. Larrabee, "Creating small fault dictionaries," *IEEE Trans. Computer-Aided Design*, vol.18, no.3, March 1999, pp. 346-356.

[11] Z. Navabi, *VHDL: Analysis and modeling of digital systems*, New York, McGraw-Hill, 1993.

[12] http://esdat.ucsd.edu/~lichen/dac2002, additional information related to this paper.

Design of a High-Throughput Low-Power IS95 Viterbi Decoder

Xun Liu Marios C. Papaefthymiou
Advanced Computer Architecture Laboratory
Department of Electrical Engineering and Computer Science
University of Michigan, Ann Arbor, Michigan 48109

ABSTRACT

The design of high-throughput large-state Viterbi decoders relies on the use of multiple arithmetic units. The global communication channels among these parallel processors often consist of long interconnect wires, resulting in large area and high power consumption. In this paper, we propose a data-transfer oriented design methodology to implement a low-power 256-state rate-1/3 IS95 Viterbi decoder. Our architectural level scheme uses operation partitioning, packing, and scheduling to analyze and optimize interconnect effects in early design stages. In comparison with other published Viterbi decoders, our approach reduces the global data transfers by up to 75% and decreases the amount of global buses by up to 48%, while enabling the use of deeply pipelined datapaths with no data forwarding. In the RTL implementation of the individual processors, we apply precomputation in conjunction with saturation arithmetic to further reduce power dissipation with provably no coding performance degradation. Designed using a 0.25 μm standard cell library, our decoder achieves a throughput of 20 Mbps in simulation and dissipates only 450 mW.

Categories and Subject Descriptors

B.7.1 [**Integrated Circuits**]: Types and Design Styles—*Algorithms implemented in hardware*

General Terms

Design, Performance

Keywords

Communications, Pipelining, Bus reduction

1. INTRODUCTION

Viterbi decoders (VDs) are widely used in digital wireless communication systems due to their powerful error correction capabilities. The quality of a VD design is mainly measured by 3 criteria: coding gain, throughput, and power dissipation. High coding gain results in low data transfer error probability. High throughput is necessary for high-speed applications such as 802.11a wireless LAN. The design of VDs with high coding gain and throughput is made challenging by the need for low power, since VDs are often placed in communication systems running on batteries.

The design of single-chip VDs has been a very active research area for the past 15 years. Figure 1 shows the throughput, power dissipation, and number of states for a large collection of VD designs [1, 3, 8, 27]. In general, these VDs fall within the region between the two dashed lines. While small-state VDs can have throughput over 1 Gbps, throughput decreases with the increase in the number of states. Consequently, the design of high-throughput large-state VDs, though crucial for applications such as satellite communications [2], has remained largely unexplored.

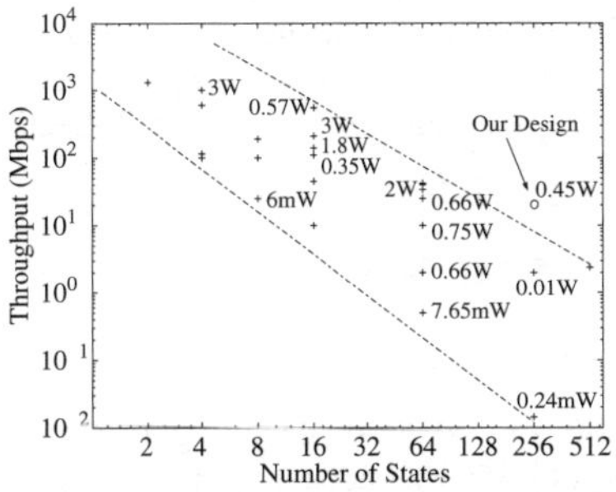

Figure 1: Performance of various VDs.

The main challenge in implementing high-throughput large-state VDs with low power dissipation is the rapid growth in the computational complexity of a VD with the increase in its state count. Although parallel arithmetic processors can be introduced to speed up the computation, they often generate extremely complex interconnect routing problems, degrading both system throughput and power dissipation. Therefore, achieving high speeds using a large-state VD with low power dissipation requires careful consideration of global data transfer and interconnect issues.

In this paper, we present a low-power design methodology for building high-throughput large-state VDs. We also discuss the application of our methodology to the design of a 256-state rate-1/3 IS95 VD. Our VD chip achieves high power efficiency due to our comprehensive design approach that focuses on the reduction of global data transfers, the minimization of global buses, and the maximization of datapath pipeline depths with no data forwarding requirements. The proposed data-transfer oriented optimization consists of 3 steps. Iterative bi-partitioning is first performed on the data flow graph of the VD to cluster all operations with minimal global data exchanges. This step is highly effective and reduces the number of data transfers by up to 75% compared with previously published partitioning approaches. In the second step, operations are packed into slices across different partitions to minimize the volume of global buses. Without stalling any of the processors, this optimization decreases global buses by up to 48% compared with the introduction of buses between every pair of processors. Finally, the execution order of the operation slices is scheduled so that the datapaths of the computation processors are deeply pipelined without requiring any data forwarding circuitry. This scheduling step provides high computation capacity and avoids power dissipation due to data hazard detection and data transfer.

In addition to our inter-processor level optimizations, we describe a novel application of precomputation for reducing power dissipation at the processor level. Specifically, we combine precomputation with saturation arithmetic computation to reduce the power dissipation of datapaths by shutting down datapath circuits that do not contribute to the decoding results. Our approach is provably effective in reducing power dissipation without degrading decoding quality.

Our high-level optimization results in a low-power VD architecture consisting of 8 parallel processors connected by 4 global buses. Our VD was synthesized using a 0.25 μm standard cell library. The processors were placed in an array structure manually and then routed automatically. In simulations using layout, our decoder achieves a throughput of 20 Mbps while dissipating only 450 mW. To our knowledge, this dissipation is the lowest among published VDs with the same number of states and throughput.

The remainder of the paper has 7 sections. In Section 2, we give background on the Viterbi algorithm, previous VD designs, and several techniques for low-power parallel DSPs. Section 3 describes our partitioning approach. Our packing method is presented in Section 4. Our scheduling scheme is presented in Section 5. Section 6 describes the architecture and implementation of our VD. Chip performance is summarized in Section 7. Section 8 concludes our paper.

2. BACKGROUND

2.1 Viterbi decoding

Figure 2 demonstrates Viterbi decoding using a simple example. Figure 2(a) shows a Viterbi encoder with 3 registers and, therefore, 8 possible states. In each clock cycle, one bit I is shifted into the encoder, and 2 bits $\{O_1, O_0\}$ are generated. Figure 2(b) gives the trellis representation of the state diagram in the decoder. The objective is to compute the state sequence in the encoder using the received code $\{C_1, C_0\}$. Each row of vertices represents a single state labeled on the left. Each column represents the states at different times. The edges describe the possible state transitions and are labeled by the corresponding encoder outputs. For example, the edge $u \rightarrow t$ represents the transition from state 0 to state 0. It is labelled with $\{00\}$, the encoder output of this transition. Similarly, the edge $v \rightarrow t$ is labeled $\{11\}$. Both fanout and fanin of every vertex are 2, and the trellis topology is identical regardless of the code $\{C_1, C_0\}$.

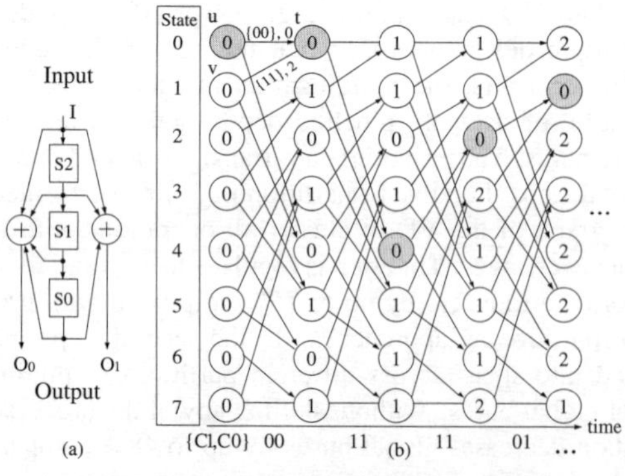

Figure 2: Viterbi decoding example.

Viterbi decoding proceeds in 3 steps. In the first step, each edge e is assigned a weight W_e, which is the difference between the edge label and the code $\{C_1, C_0\}$. For example, $W_e(v \rightarrow t) = 2$, because the Hamming distance between $\{11\}$ and $\{00\}$ is 2. Similarly, $W_e(u \rightarrow t) = 0$.

In the second step, the weight L of each vertex at column $n+1$ is computed recursively by:

$$L = \min\{L_1 + W_1, L_2 + W_2\} , \tag{1}$$

where L_1 and L_2 are vertex weights of the two predecessors in column n, and W_1 and W_2 are the weights of the corresponding edges. This computation is called add-compare-select (ACS) recursion. Figure 2(b) shows all vertex weights, with arrows representing the selection results. (Vertex weights in the first column are initialized to zero, assuming no prior information indicating the initial state of the VD.)

The final step is called *survivor path tracing*. Intuitively, the minimum-weight path through the trellis represents the most likely state transitions. Starting from the minimum-weight vertex in the last trellis column, a path is traced backward based on the decisions in the ACS update steps. The result is marked by the grey vertices in Figure 2(b).

VD architectures can be classified into three groups: serial, parallel, and intermediate. Serial approaches use only one ACS processor to compute all the vertex weights sequentially. They are only applicable to low-throughput large-state VDs due to their lack of processing capacity.

Parallel architectures consist of the same number of ACS processors as that of states. The main challenge in these designs is to route the connections, called shuffle exchange networks, among all processors. Various graph representations such as de Bruijn graphs [15], hypercubic networks, cube-connected cycles [4], and ring structures [21] achieve only limited interconnect reduction and are not effective for VDs with hundreds of states. Locally connected array architectures avoid global interconnects but suffer from low computation efficiency, because processors are used to transfer data as well as compute path weights [9]. Bit-serial computation approaches reduce interconnect routing by transferring data onc bit at a time, but their throughput is severely limited. Due to the data dependency of adjacent trellis columns, pipelining cannot be used in parallel architectures. Interleaving requires changes in communication protocols and is inconvenient to apply [14].

Intermediate solutions, where each ACS processor is shared by more than one trellis state, are promising for large-state VDs [2, 20]. Many such architectures have been proposed, including systolic arrays [5] and 4-level cluster networks [11]. These approaches tend to focus on scalability, overlooking power and performance optimization of the decoder. No operation scheduling has been used to reduce global buses. In [12], operation scheduling was discussed but only in the restricted context of in-place memory update [18]. Pipelining of operations was explored in [6]. However, that work did not present any hardware implementation.

2.2 Low-power parallel DSP system design

In addition to VDs, parallel architectures have been applied to numerous computation-intensive DSP problems [7, 10, 19, 24]. In such systems, multiple computation and memory units operate in parallel to achieve high throughput and low power [13]. The literature on this topic is very extensive, and this section gives a very brief and certainly non-exhaustive discussion of related work.

Parallel hardware synthesis research addresses operation partitioning, scheduling, and binding problems, which are known to be NP-complete and are solved using various heuristics [23]. Design automation research in parallel synthesis is increasingly focusing on data transfer and scheduling over global buses [22], because power dissipation is dominated by data storage and transfer through global interconnect. Such data-transfer oriented scheduling is data-dependent, and problem-specific solutions are often applied [26]. The application of scheduling and binding to reduce interconnect power was investigated in [16]. That work relies on the simulated annealing algorithm and was applied only on small designs such as 2-step FIR filters.

3. REDUCING GLOBAL DATA TRANSFERS

We chose to implement our VD using an intermediate architecture due to its potential for achieving low power. As is the case with all systems consisting of multiple processors, global data transfers among various parallel datapath units could contribute significantly to power dissipation due to long interconnects. This section describes our operation partitioning approach that assigns operations to different processors so that the total communication volume is minimized. As the first step of our optimization methodology, this partitioning lays the ground for the subsequent operation packing for minimal global buses, described in Section 4.

Our analysis focuses on the operations that compute the vertex weights of a single trellis column, since there are no data dependencies among them. In our scheme, the weight of each state is stored in the processor that computes it. Consequently, we build a directed data flow graph (DFG) representing the data transfer pattern of the Viterbi algorithm as follows. For each state in the trellis, a vertex is introduced in the DFG. A directed edge $u \rightarrow v$ is introduced if and only if the weight update of v depends on the weight of u in the previous column. Figure 3 illustrates DFG construction using the simple 8-state VD example from Section 2. Figure 3(a) shows adjacent trellis columns and their connections. The DFG in Figure 3(b) is generated by merging the two vertices of the same state together.

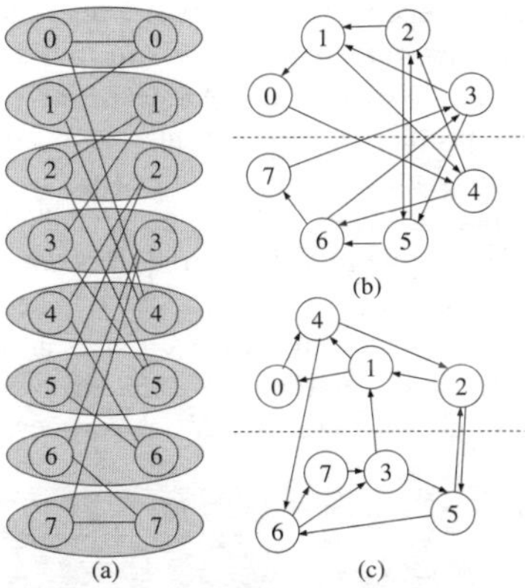

Figure 3: Operation partitioning.

The problem of partitioning all operations to minimize data transfer volume can be cast as a multi-way partitioning problem with minimal cut-size and can be solved using iterative bi-partitioning. Figure 3(c) gives the solution for optimally partitioning the DFG of the 8-state VD into 2 parts. Its global transfer volume is 4, a 50% reduction compared with the simple partitioning approach in Figure 3(b).

Figure 4 shows the effectiveness of our operation partitioning on the IS95 VD. It gives the ratio of global, i.e. inter-partition, data transfers to all data transfers versus the number of partitions. For comparison, we show the results of [25], the only approach we found in previous VD designs that uses partitioning specifically for interconnect reduction. Our partitioning approach reduces the

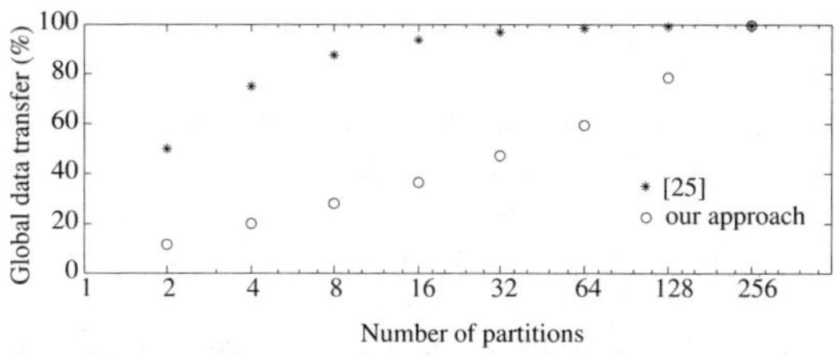

Figure 4: Reduction of global data transfers.

global transfer volume by up to 75%. The largest absolute reduction is achieved in the range of 4 to 32 partitions. Global data transfers increase monotonically with the number of partitions. For 64 or more partitions, more than half of the total data transfers are global, suggesting increased power dissipation in systems with high computation parallelism. In these cases, introducing more parallel processors to increase throughput is inefficient in terms of power dissipation. Pipelining individual processors is a promising alternative for high-throughput and low-power VD implementations. However, pipelining may cause data hazards and often requires power-consuming data forwarding. To that end, Section 5 describes our operation scheduling scheme for non-forwarding pipelining.

4. MINIMIZING GLOBAL BUSES

This section describes our operation packing scheme for global bus minimization. To perform global data transfers, we use a hierarchical bus structure that is derived directly from our partitioning procedure. Each bi-partitioning step generates a unique type of buses. Figure 5(a) shows the bus structure resulting from 8-way partitioning. Blocks $P_i, i = 0, \ldots, 7$, represent the operation clusters. The dotted lines illustrate the partitioning cuts. The triple-line arrow in the middle gives the highest level cut and points to the corresponding bus. Similarly, the double-line and single-line arrows represent the cuts in the following iterative bi-partitioning procedure and indicate their associated buses. Consequently, bus types can be represented using a full binary tree with all operation partitions as the leaves shown in Figure 5(b). A *bus configuration* can be described by a non-negative integer assignment to all non-leaf nodes in the binary tree. For instance, the assignment in Figure 5(b) represents two buses connecting all processors and one bus connecting $p4$ and $p5$. These buses form a partial order, and a bus u can *cover* another bus v if and only if v is at the same position as u or in the subtree rooted at u.

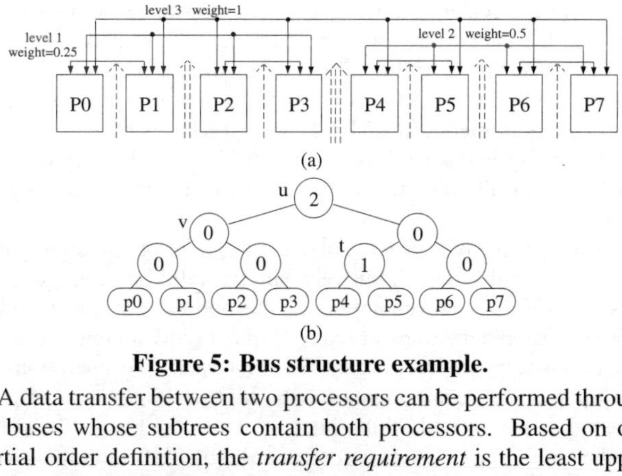

Figure 5: Bus structure example.

A data transfer between two processors can be performed through all buses whose subtrees contain both processors. Based on our partial order definition, the *transfer requirement* is the least upper bound of the two processor leaves in the data transfer. For example, in Figure 5(b), the transfer requirement of a read access between $p4$ and $p5$ is a t-type bus. The weight update operation of a trellis vertex k will require two data transfers if both fanin vertices of k are not in the same partition as k. The *operation requirement* of updating the weight of k is computed by adding the transfer requirements of all data transfers involved. Effectively, this definition ensures that all transfers of an operation can be performed in the same cycle. For example, if an operation requires $p4$ to read $p3$ and $p5$, its operation requirement is one u-type bus and one t-type bus.

In our parallel VD system, operations are executed in *slices* simultaneously by different processors. Each slice contains exactly one operation from each partition. The *slice requirement* is the sum of operation requirements of the operations within. A bus configuration can *satisfy* a slice if and only if each transfer requirement in the slice is covered by a unique bus within the configuration. *Operation packing* is the procedure of forming all slices given a partitioning. A bus configuration is *valid* for a packing solution S, if it satisfies all the slices in S.

Bus resources are quantified by assigning a weight to each bus. The weight of a bus B is proportional to the number of processors connected to B and is normalized by the total number of processors used. The *cost* of a bus configuration is the sum of weights of all buses within. The problem of *operation packing for minimal buses* is to find a packing solution which has a minimum-cost valid bus configuration.

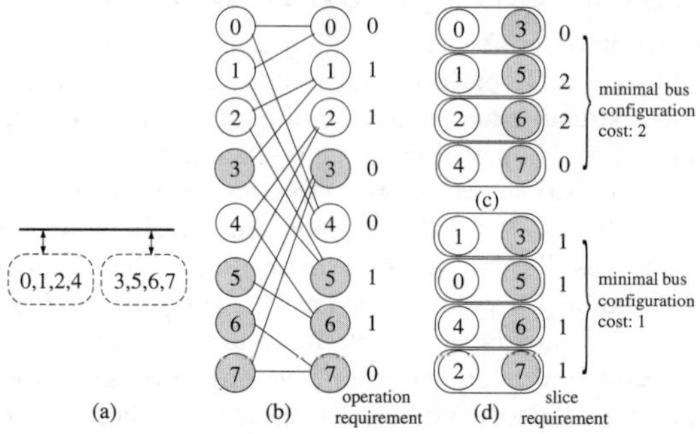

Figure 6: Packing for minimal bus weight.

Figure 6 uses the 8-state VD example to explain the procedure of operation packing for minimal buses. Figure 6(a) shows the VD structure consisting of two processors represented by blocks and connected by buses. Based on the partitioning result from Figure 3(c), operations executed by these two processors are listed within the blocks. Since only one bus type exists, the transfer requirement is a scalar number. Figure 6(b) shows part of the trellis, in which the operation requirement for the weight update of each state is listed on the right. Figure 6(c) gives a packing result in which each pair of operations on the same row form a slice. Since the slice requirement of $\{1,5\}$ is 2, at least 2 buses have to be used. If the packing is changed to the one in Figure 6(d), only one bus can handle all the data transfers, resulting in a 50% reduction of bus resources.

We designed a heuristic to solve the problem of operation packing for minimal buses. Intuitively, since a valid bus configuration must satisfy all slice requirements, the optimal packing solution should distribute the transfer requirements evenly among all slices. Our heuristic procedure, called HS, first packs the operations so that the transfer requirements for the highest level buses of each slice vary by at most 1. Subsequently, it sets the highest level bus number in the bus configuration accordingly. HS then iteratively proceeds to the next lower level and balances the bus requirement by swapping operations between pairs of slices. At each level, the swapping is done in such a way that the current bus configuration is kept valid. The packing procedure terminates when bus numbers at the lowest level of the bus configuration are computed.

Table 1: Bus reduction through operation packing.

Partitions	2	4	8	16	32	64	128
BC_{HS}	2	2	4	8	16	42	111
BC_{simple}	2	3	7	15	31	63	127
Reduction(%)	0	33	43	47	48	33	13

Table 1 shows the results of applying our packing heuristic on the IS95 VD. The numbers of partitions are listed in Row 1. Given any partition number, we applied iterative bi-partitioning followed by our heuristic HS. Row 2 lists the minimal total bus cost BC_{HS}. For comparison, the results BC_{simple} of a simple bus structure where one bus is introduced between any two partitions is given in Row 3. This simple approach still has to introduce stalls, when an operation currently reads two data from another partition, due to lack of bus resources. Procedure HS not only guarantees no-stall execution but also reduces the global buses by 31% on the average. Our approach is most effective in the range of 8 to 32 partitions.

5. NON-FORWARDING SCHEDULING

Data forwarding is often used to resolve data hazards in pipelined datapaths. However, in parallel systems, data hazards occur not only within but also among processors. Data dependencies checking and global data transfers can cause excessive interconnect routing overhead and power dissipation. In this section, we present our operation scheduling scheme for maximizing the depth of non-forwarding pipelines.

In Viterbi decoding, data dependencies only happen between adjacent trellis columns. These limited and fixed data dependencies allow us to use operation scheduling to avoid data hazards without using data forwarding or stalling. In the remainder of this section, we refer to the number of pipeline stages between data fetching and writing as the datapath pipeline depth, because pipelining beyond this range does not cause data hazards.

Since parallel processors execute operations concurrently, we must consider data dependencies between operation slices. The problem of *non-forwarding scheduling* is to find a slice execution order such that the slices in the pipeline never depend on each other's result. The problem can be formulated as an ordering problem on a dependency graph, which is constructed as follows. For each operation slice, a vertex is introduced in the graph. If the weight update of any state in a slice u requires data from a state in another slice v, an edge $u \rightarrow v$ is introduced. The computation is N-stage pipelineable if and only if there exists an order of all the vertices such that, for every k, $0 < k < N$, the kth vertex has no outgoing edges connected to the last $N - k$ vertices.

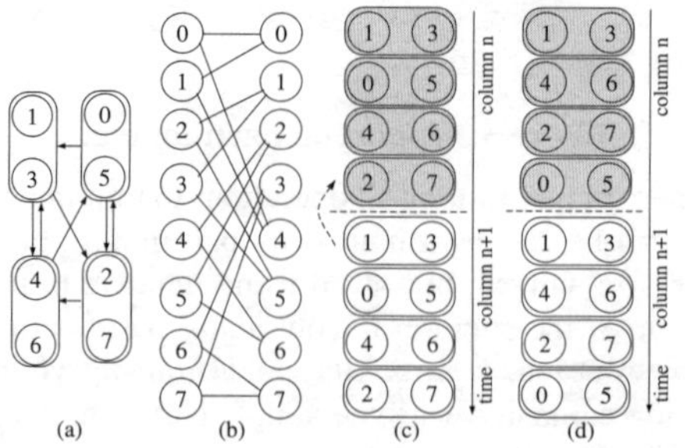

Figure 7: Scheduling for non-forwarding pipelining.

We demonstrate the effectiveness of non-forwarding scheduling using the previous 8-state VD example. Figure 7(a) gives the dependency graph. The vertices are derived based on the partitioning and packing results from Figures 3(c) and 6(d), respectively. The directed edges are added according to the trellis data transfer in

Table 2: Maximum non-forwarding pipeline depth.

Partitions	2	4	8	16	32	64	128
P_{depth}	63	28	9	2	0	0	0

Figure 7(b). Figure 7(c) shows the execution of slices using the order in Figure 6(d). The weight computation of column n is shaded. Since there is data dependency between slices $\{1,3\}$ and $\{2,7\}$, pipelining is not possible without data forwarding or stalling. However, pipelining can be applied without data forwarding or stalling if the slices are scheduled as in Figure 7(d).

We implemented a heuristic scheduler to maximize pipeline depth without data forwarding. Intuitively, since the data dependencies occur only between the first and last several slices in the execution order, our scheduler places the slices with fewer dependencies earlier. It first picks the slice with the least dependencies and then iteratively picks the next slice such that the upper bound of the pipeline depth derived by the chosen slices is maximized.

Table 2 shows the results from the application of our scheduler to the IS95 VD. The non-forwarding pipeline depth is drastically reduced as the number of partitions increases, reflecting the reduction in the number of operations in each partition and the simultaneous execution of increasingly many operations. From this table, it is evident that architectures with 16 or more parallel processors allow very limited or zero non-forwarding pipeline depth.

6. SYSTEM ARCHITECTURE

Using our analysis in Sections 3, 4, and 5, we have designed a system with 8 processors (subcores) that has low global communication volume, a small number of global buses, and a large non-forwarding pipeline depth. The block diagram of our design is shown in Figure 8. The 8 subcores are connected by 4 global address/data (A/D) buses. Each subcore is a 16-bit pipelined microprocessor, whose internal structure is detailed on the right-hand side of Figure 8.

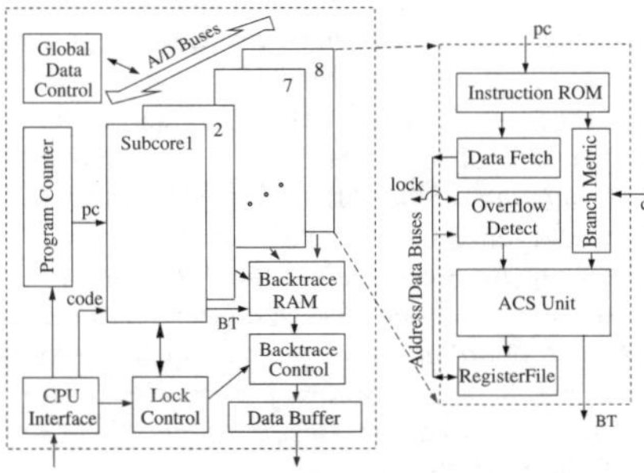

Figure 8: Viterbi decoder architecture.

The **instruction ROM** is derived from our high-level data transfer oriented optimization. The 16-bit instruction format is given in Figure 9. The *Ex* bits indicate the sources of two operands (local register file or global buses). The *Addr/port* fields give the local register file addresses or global port numbers. The *ICodes* and *D* are the label and decoding data of the state transition edge associated with the first operand. The label and decoding data of the state transition edge for the second operand can be derived using bit-inversion of *ICodes* and *D*, respectively.

The **data fetch unit** gets the operands based on the instructions. For each global bus, we use a centralized control. Synchronized with the subcores, global buses deliver the right data at the right time. The data fetch unit just latches the data in and no bus conflict detections are needed, resulting in low power dissipation.

The **branch metric unit** computes the difference between the received codes x_i and the label y_i of each trellis edge $\mathcal{D}$. We used soft-

15	14-10	9	8-4	3-1	0
Ex1	Addr/Port1	Ex2	Addr/Port2	ICodes	D

Figure 9: Instruction format.

decision Viterbi decoding with 8-level quantization, thus avoiding the loss of coding gain associated with hard-decision decoding [17]. Since there are 3 codes for each state transition, $\mathcal{D}$ is calculated as follows:

$$\mathcal{D} = \sum_{i=0,1,2} (x_i - y_i)^2 . \qquad (2)$$

Our **ACS unit** is a 16-bit datapath pipelined into 3 stages. A novel power saving technique applied to the design of our ACS unit is the combination of precomputation and saturation computation that allows overflow. Specifically, once one or two operands are detected as overflows, part of the ACS is shut off to save power. In previous VD designs, datapaths were adjusted to hold the maximal path weight, and no overflow was allowed. Therefore, the trellis had to be frequently backtraced when the *maximal* path weight exceeded the data limit of the registers. This approach may result in unnecessary power consumption. Since in the Viterbi algorithm, the result is the path with minimum weight, the computation is valid as long as the minimum path weight does not exceed the data limit of the register.

In our VD design, we use saturation computation, in which weights exceeding the maximal data value $M = 2^{16} - 1$ are represented as overflows. All path weights are monitored as they are written into the register files. When the minimum path weight is greater than or equal to $M/2$, all finite path weights are decreased by $M/2$. Overflows remain as overflows, however. Our approach degrades the VD quality only when the weight of the correct path overflows. This condition requires the weight of the correct path to exceed that of an incorrect one by $M/2$. The probability of such an event is less than 2^{-500} in our design. (Proof omitted due to page limits.)

The **register file** contains 64 registers. Two 16-bit registers are used for each of the 32 states in the corresponding partition to decouple the write and read accesses. The register file has 1 write port and 4 read ports, 3 of which are global read ports. The global read accesses are pipelined into 3 stages to handle delays due to the long interconnects.

With the data overflow detection, we perform backtracing at the frequency determined by the decoding latency requirement. Our **backtrace memory** consists of two independent parts, each handling 4 subcores. We use a ping-pong architecture to handle synchronized reads and writes with a single port. Because the write addresses follow a round-robin pattern, we divide the backtrace memory into two banks for odd and even addresses so that every two consecutive writes use either bank exactly once. Therefore, we are able to read the memory at a speed twice as slow as the write accesses. This speed is fast enough, because read accesses are performed in the backtracing procedure, which is much slower than the write accesses during the path weight computation.

7. CHIP SUMMARY

Our VD was implemented using *Verilog* and synthesized using *Design Analyzer* from Synopsys and a 0.25 μm standard-cell library. The layout was generated using *Silicon Ensemble* from Cadence in a hierarchical fashion. Each sub-module was laid out as a standard cell array. Manual floorplanning was performed to minimize the length of global buses. Figure 10 shows the floorplan and layout of our VD. The subcores are rotated so that the register files (RF) and the fetch units (FU) can access the global channels easily. The backtrace control (BTC) and memory (BTM) are placed in the middle as well as the output buffer (SB).

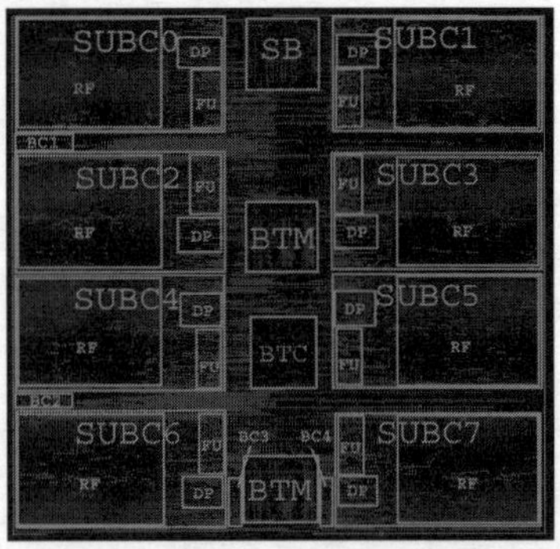

Figure 10: Viterbi decoder layout.

Table 3: Chip Summary.

technology	0.25 μm	metal layers	5
chip size	$2.4 \times 2.4 mm^2$	transistor count	325K
core frequency	640MHz	supply voltage	2.5V
throughput	20 Mpbs	power	450 mW

After layout, the capacitance of all interconnects was extracted, and the timing of each gate was adjusted according to the output load. Our design was then resimulated using *Verilog* with the back-annotated timing and capacitance. The power dissipation was estimated using *PrimePower* from Synopsys. Table 3 summarizes our chip performance. Our decoder has a maximal throughput of 20 Mbps while dissipating only 450 mW.

8. CONCLUSION

In this paper, we present a comprehensive methodology for the design of a high-throughput and low-power 256-state rate-1/3 IS95 Viterbi decoder. Our high-level design optimization focuses on the power reduction of global data communications. We perform a 3-phase partitioning, packing, and scheduling optimization to reduce global data transfer volume, minimize global buses, and implement deep pipelining without data forwarding. Furthermore, we apply precomputation in conjunction with saturation computation to reduce power dissipation within each datapath.

The resulting VD is a parallel system with 8 pipelined processors connected through 4 global buses. Generated using commercial ASIC tools in a 0.25 μm standard cell library, our decoder has a throughput of 20 Mbps and a power dissipation of 450 mW, the lowest among published VDs with 256 states and same throughput.

9. ACKNOWLEDGMENTS

This research was supported in part by NSF under Grant No. CCR-0082876 and by ARO under Grant No. DAAD-19-99-1-0304.

10. REFERENCES

[1] P. J. Black and T. H. Meng. A 1-Gb/s four-state sliding block Viterbi decoder. *IEEE J. of Solid-State Circuits*, 32(6):797–805, June 1997.

[2] M. Bóo, F. Argüello, J. D. Bruguera, R. Doallo, and E. L. Zapata. High-performance VLSI architecture for the Viterbi algorithm. *IEEE Trans. Communications*, 45(2):168–176, Feb. 1997.

[3] Y. Chang, H. Suzuki, and K. K. Parhi. A 2-Mb/s 256-state 10-mW rate-1/3 Viterbi decoder. *IEEE J. of Solid-State Circuits*, 35(6):826–834, June 2000.

[4] P. Chau and K. Stephen. Scaling and folding the Viterbi algorithm trellis. In *Workshop on VLSI Signal Processing*, pages 479–489, 1992.

[5] F. Daneshgaran and K. Yao. The iterative collapse algorithm: A novel approach for the design of long constraint length Viterbi decoders - Part I. *IEEE Trans. Communications*, 43(2):1409–1418, Feb. 1995.

[6] H. Dawid, S. Bitterlich, and H. Meyr. Trellis pipeline-interleaving: a novel method for efficient Viterbi decoder implementation. In *IEEE International Symposium on Circuits and Systems*, May 1992.

[7] H. De Man, F. Catthoor, G. Goossens, J. Vanhoof, J. V. Meerbergen, S. Note, and J. Huisken. Architecture-driven synthesis techniques for VLSI implementation of DSP algorithms. *Proc. of the IEEE*, 78(2):319–335, Feb. 1990.

[8] G. Fettweis and H. Meyr. High-speed parallel Viterbi decoding: Algorithm and VLSI-architecture. *IEEE Communications Magazine*, pages 46–55, May 1991.

[9] P. G. Gulak and T. Kailath. Locally connected VLSI architectures for the Viterbi algorithm. *IEEE J. on Selected Areas in Communications*, 6(3):527–537, Apr. 1988.

[10] R. Hartenstein, J. Becker, M. Herz, R. Kress, and U. Nageldinger. A partitioning programming environment for a novel parallel architecture. In *International Parallel Processing Symposium*, pages 544–548, 1996.

[11] H. Li and C. Chakrabarti. A new architecture for the Viterbi decoder for code rate k/n. *IEEE Trans. Communications*, 44(2):158–164, Feb. 1996.

[12] H. Lin and C. B. Shung. General in-place scheduling for the Viterbi algorithm. In *International Conf. on Acoustics, Speech, and Signal Processing*, pages 1577–1580, 1991.

[13] R. S. Martin and J. P. Knight. Optimizing power in ASIC behavioral synthesis. *IEEE Design and Test of Computers*, 13(2):58–70, 1996.

[14] S. R. Meier. A Viterbi decoder architecture based on parallel processing elements. In *Global Telecommunications Conference*, pages 1323–1327, 1990.

[15] B. K. Min and N. Demassieux. A versatile architecture for VLSI implementation of the Viterbi algorithm. In *International Conf. on Acoustics, Speech, and Signal Processing*, pages 1101–1104, 1991.

[16] P. Prabhakaran, P. Banerjee, J. Crenshaw, and M. Sarrafzadeh. Simultaneous scheduling, binding and floorplanning for interconnect power optimization. In *International Conf. on VLSI Design*, pages 423–427, 1999.

[17] J. G. Proakis. *Digital Communications*. McGraw-Hill Inc., New York, 1995.

[18] C. M. Rader. Memory management in a Viterbi decoder. *IEEE Trans. Communications*, 29(9), Sept. 1981.

[19] L. Raffo, S. P. Sabatini, M. Mantelli, A. D. Gloria, and G. M. Bisio. Design of an ASIP architecture for low-power visual elaborations. *IEEE Trans. VLSI Systems*, 5(1), Mar. 1997.

[20] C. B. Shung, H. Lin, R. Cypher, P. H. Siegel, and H. K. Thapar. Area-efficient architectures for the Viterbi algorithm- part I: Theory. *IEEE Trans. Communications*, 41(4):636–644, Apr. 1993.

[21] J. Sparsø, H. N. Jørgensen, E. Paaske, S. Pedersen, and T. Rübner-Petersen. An area-efficient topology for VLSI implementation of Viterbi decoders and other shuffle-exchange type structures. *IEEE J. of Solid-State Circuits*, 26(2):90–96, Feb. 1991.

[22] A. Sudarsanam and S. Malik. Memory bank and register allocation in software synthesis for ASIPs. In *Design Automation Conference*, pages 388–392, June 1995.

[23] C. Wang and K. K. Parhi. High-level DSP synthesis using concurrent transformations, scheduling, and allocation. *IEEE Trans. Computer-Aided Design of Integrated Circuits and Systems*, 14(3):274–295, Mar. 1995.

[24] J. P. Wittenburg, W. Hinrichs, J. Kneip, M. Ohmacht, M. Bereković, H. Lieske, H. Kloos, and P. Pirsch. Realization of a programmable parallel DSP for high performance image processing applications. In *Design Automation Conference*, pages 56–61, June 1998.

[25] C.-M. Wu, M. Shieh, C.-H. W, and M. Sheu. An efficient approach for in-place scheduling for path metric update in Viterbi decoders. In *International Symposium on Circuits and Systems*, pages 61–64, May 2000.

[26] S. Wuytack, F. Catthoor, G. D. Jong, and H. J. De Man. Minimizing the required memory bandwidth in VLSI system realizations. *IEEE Trans. VLSI Systems*, 7(4):433–441, Dec. 1999.

[27] A. K. Yeung and J. M. Rabaey. A 210Mb/s radix-4 bit-level pipelined Viterbi decoder. In *IEEE International Solid-State Circuits Conference*, pages 88–90, 1995.

A Detailed Cost Model for Concurrent Use With Hardware/Software Co-Design

Daniel Ragan
Elect. Systems Cost Modeling Lab.
University of Maryland
College Park, MD 20742
danragan@wam.umd.edu

Peter Sandborn
Elect. Systems Cost Modeling Lab.
University of Maryland
College Park, MD 20742
sandborn@eng.umd.edu

Paul Stoaks
Foresight-Systems, Inc.
8217 Shoal Creek Blvd, Suite 100
Austin, Texas 78757
paul@foresight-systems.com

ABSTRACT

Hardware/software co-design methodologies generally focus on the prediction of system performance or co-verification of system functionality. This study extends this conventional focus through the development of a methodology and software tool that evaluates system (hardware and software) development, fabrication, and testing costs (dollar costs) concurrent with hardware/software partitioning in a co-design environment.

Based on the determination of key metrics such as gate count and lines of software, a new tool called *Ghost*, evaluates software and hardware development, fabrication, packaging and testing costs. *Ghost* enables optimization of hardware/software partitioning as a function of specific combinations of hardware foundries and software development environments.

Categories and Subject Descriptors

E3 [**HW/SW co-design**]: *specification, model., co-simulation and performance analysis, system-level scheduling and partitioning.*

General Terms
Design, Economics.

Keywords
Cost Modeling, Cost-Performance Trade-off.

1. INTRODUCTION

Hardware/software co-design leads to accelerated system design and performance improvement through dynamic trade-off analysis. Co-design has been a popular topic during the past dec-ade in electronics because designers perceive that performance, reliability, size, and cost can be improved through judicious partitioning of functionality between hardware and software.

System requirements include constraints on functionality, performance, task scheduling, packaging, cost, size, weight, volume, power, reliability, and software and hardware modularity. A variety of alternative architectures may exist that meet both functional and performance standards. Meeting constraints while attempting to optimize the design brings about a number of challenges. One significant gap in co-design methodology is the automation of detailed concurrent cost evaluation. In fact, the majority of system design work categorizing itself as "cost-performance trade-off" is dominated by performance with either very simplistic or purely qualitative cost treatments.

No matter what methodology is used when making system design decisions, dollar cost is often a dominant factor. While cost can be the easiest to measure, it may be the hardest to predict. This is because there is never a simple bottom line; behind the final system cost are a multiplicity of other cost factors, including software and hardware design, hardware manufacturing, software development, software and hardware testing, time-to-market factors, yield, and a host of life cycle costs including reliability, maintenance, and disposal. Ideally, all of these costs should be visible and considered during system design, and especially when design-partitioning decisions are made.

The next section in this paper discusses existing work on the concurrent evaluation of performance and cost during hardware/software co-design. Section 3 summarizes the cost modeling methodology developed in this work. Section 4 presents a case study of the JPEG encoder in a digital camera.

2. EXISTING WORK

The majority of existing work on hardware/software co-design focuses on resource utilization (processor, memory, size, power consumption, timing, etc.). Very few studies have attempted to examine the cost impacts of co-design decisions.

In the DARPA RASSP program, DeBardelaben *et al* [1] developed a design methodology that involved the development of executable specifications, cost-driven architecture selection, task assignment, task scheduling, and performance modeling. For software cost models, they used COCOMO [2] and REVIC [3]. These models allowed the prediction of software development and maintenance effort, and software development schedule. In order to utilize these models, they estimated software size, using a non-linear reuse technique, [4]. For hardware, DeBardelaben *et al* assume only COTS hardware is used and thereby hardware production costs were obtained from vendor quotes. In related work by Egolf [5], a more detailed analysis of ASIC design/redesign, and testing costs was developed.

Also in the DARPA RASSP program, PRICE Systems in collaboration with Ascent Logic built a general design requirements capture and tracking tool called RDD-100. This tool was integrated with PRICE's parametric cost estimating tools. PRICE built software to read the RDD-100 architecture and constructed an integrated hardware/software work breakdown structure of the candidate architecture. Using the work breakdown structure, PRICE returned the cost to RDD-100, where it was assigned to the corresponding architecture blocks.

Other synthesis approaches either factor in only hardware costs, [6]-[9], or only software costs, [10] during the system design process. In [11], both hardware and software costs are determined from libraries consisting of functional units with their corresponding performance, and dollar costs.

The majority of co-design and synthesis approaches that address dollar cost do not derive a cost, but rather only accumulate the costs of predefined functional units (hardware and/or software) ignoring many critical economic effects such as production volume, testing, development, and yield. The few studies that do include detailed cost modeling, e.g, [1], are more focused on software development costs than non-recurring or recurring hardware costs. In this paper we show how the best partitioning choice can be made based on both performance and detailed cost calculations by presenting a method of incorporating comprehensive hardware and software cost modeling into the architecture selection and partitioning process to minimize system cost while maintaining system performance standards.

3. COST MODELING APPROACH

In this section, we describe the composition of the software and hardware cost models that are utilized to estimate chip costs and discuss the implementation of the software and hardware cost models in a software tool called *Ghost*.

3.1 Software Cost Modeling

For costing software development, several tools exist including COCOMO, REVIC and commercial tools. These software cost estimators use nominal effort equations derived from labor effort required for developing software as related to the size of the software system measured in source lines of code or delivered source instructions.

In this study, software size was estimated using a form of function point analysis [12] known as feature point counting [13]. Based on software language, any feature point count can be converted to the number of source lines of code (SLOC) using a conversion factor. One Feature Point is equivalent to 10 to 320 SLOC depending on the language. For this work, the language of interest, Java, is considered a level six language and converts each Feature Point to 53 SLOC, [14]. Once a SLOC count is known, effort and scheduling equations are derived. In particular, the embedded mode of COCOMO and REVIC are the most applicable for the software costing of an embedded digital system, e.g., the digital camera considered in this paper (Section 4). Adjustments made by DeBardelaben aimed at COTS-based embedded DSP systems were also utilized [15]. The software cost analysis process is summarized in Figure 1.

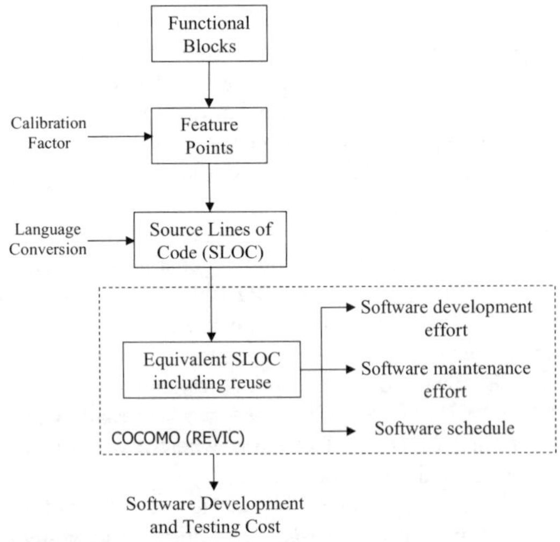

Figure 1: Software cost analysis process.

The converted SLOC counts from the function/feature point counting process are entered into software cost estimating equations based on COCOMO. The methodology developed in this paper computes software size with reuse taken into account, software development effort, software maintenance effort, and the software schedule. The software schedule is based on REVIC development time [4] with the multiplicative effects on time with projects of increasing effort set to 6.2 [15].

The level of accuracy of the estimate provided by this software cost model is proportional to the user's confidence in the software size estimate and the description of the software development environment. Through calibration, the risks associated with estimating software development have been reduced.

3.2 Hardware Cost Modeling

The hardware cost modeling process is shown in Figure 2. The key metric, gate count, allows the calculation of values for I/O count, core area, die size, etc. These factors are coupled with others such as number-up, chip design costs, wafer fabrication, packaging, automatic test pattern generation, and manual test pattern generation, to obtain the final cost.

The number of die I/O is determined using Rent's rule, which relates the chip functionality and number of gates to the number of signal and control I/O on the die. The core area of the die is determined using Donath's formulation of average interconnections length from Rent's rule [16]. Based on the number of gates on the chip, N_g, the average interconnection length in units of gate pitch, $\overline{R}$, can be determined from the interconnection pitch, p_w, number of interconnection layers, n_w, utilization efficiency of interconnections, e_w, and the fan-out of gates, f_g. Using the average interconnection length, the gate dimension d_g, and the die core dimension D_c, can be obtained by setting the supply and demand for wiring equal.

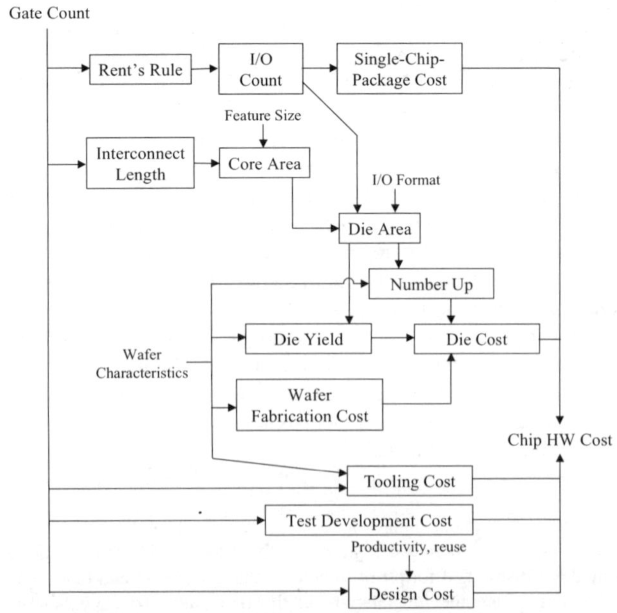

Figure 2. Hardware cost analysis process.

$$d_g = \frac{f_g \overline{R} p_w}{e_w n_w} \text{ and } D_c = \sqrt{N_g} d_g \quad (1)$$

For peripherally bonded die with a single row of bond pads on all sides, the die area is given by the maximum of two limits [17]:

$$\text{peripheral area 1} = \left(2\, \text{length}_{pad} + \text{pitch}_{pad} \left\lfloor N_p / 4 \right\rfloor\right)^2 \quad (2)$$

$$\text{peripheral area 2} = N_p \text{width}_{pad} \text{length}_{pad} + \left(1 + kN_p\right) D_c^2 \quad (3)$$

length_{pad} and width_{pad} = length and width of a die bond pad;

pitch_{pad} = minimum center-to-center pitch of die bond pads;

N_p = total number of die I/O;

k = the fractional increase in the core die area necessary to accommodate redistribution of I/O to the periphery of the die, approximately constant in the range 0.00074-0.00079 [18].

Equation (2) is the I/O limited chip area, and (3) is the peripheral redistribution limited area. For an area array bonded die, the die area is given by the maximum of two limits [17],

$$\text{array area 1} = \left(\text{pitch}_{pad} \left\lceil \sqrt{N_p} \right\rceil\right)^2 \quad (4)$$

$$\text{array area 2} = N_p \text{length}_{pad} \text{width}_{pad} + D_c^2 \quad (5)$$

Equation (4) is the I/O limited chip area, and (5) is the bond pad area limitation, assuming that active circuitry cannot be placed under the bond pads.

Using the die area and the wafer characteristics, the individual un-yielded die cost can be determined. Yield is calculated using Poisson, Murphy, or Seeds models depending on the die area and defect distribution assumed.

For peripheral I/O single chip package costs are calculated using a price versus pin count assumption for plastic or ceramic packages. The forms of these models are shown in (6) and (7).

$$\text{Plastic packages: } \text{cost}_{scp} = 0.01e^{1.16(\log(N_p)) - 2.09} \quad (6)$$

$$\text{Ceramic QFP: } \text{cost}_{scp} = 0.01e^{1.157(\log(N_p)) - 0.912} \quad (7)$$

$$\text{BGA: } \text{cost}_{scp} = 0.13(N_p) \quad (8)$$

The chip yield after packaging and test is computed from the fault coverage (fraction of the total defects that are identified in the test) and the actual yield of the die on the wafer,

$$\text{yield}_{chip} = \text{yield}_{wd}^{(1-\text{fault coverage})} \quad (9)$$

The fraction of chips that are available for board assembly is,

$$\text{pass fraction} = \text{yield}_{wd}^{\text{fault coverage}} \quad (10)$$

The chip cost after the functional test in Figure 3 is given by,

$$\text{cost}_{cumulative} = \frac{\text{cost}_{\text{previous steps}} + \text{cost}_{\text{test step}}}{\text{pass fraction}} \quad (11)$$

where $\text{cost}_{\text{previous steps}}$ is the accumulated cost of all steps up to but not including the test step. Note, all costs prior to sawing the wafer into separate die must be allocated over the die that continue through the process after sawing.

Chip design costs are calculated using the model in [19], which describes a financial model for the design of system-on-chip ICs based on engineering productivity. It examines the impact of design reuse on productivity, and the resulting effect on the cost of design. The model assumes that reusable blocks (intellectual property) are designed and added to a central repository or library. New chips are designed using a combination of custom circuit design and previously designed blocks from the library. Design costs, therefore, are made up of a combination of intellectual property costs and new chip development costs.

Using the model, chip design costs can be explored by varying the level of reuse, the amount of IP developed in a given year, engineering productivity, the reuse cost factor, and the reuse integration factor.

Test pattern generation costs depend on each block in the system, the fault coverage desired, the test pattern generator, and the type of test pattern generation, [20]. Total test development costs are calculated as the combination of the automatic test pattern generation cost, manual test pattern generation costs (when required fault coverage is not achieved through automatic generation), and test application cost.

Tooling (masking) cost at low volumes of production has a large impact on the total per chip cost. The tooling cost is based on the chosen wafer fabrication process. In this study two processes are utilized as examples, a 0.18 μm process and a 0.35 μm process, (Table 3, hardware foundries 1 and 2 respectively). Typical masking costs for the 0.18 μm process run between \$300k-\$340k, independent of die size and production quantity. Masking costs for the 0.35 μm process are less expensive due to the larger feature size, ranging from \$80k-\$100k.

3.3 Implementation and Usage – *Ghost*

The cost modeling discussed in this section is implemented in a tool called *Ghost*. *Ghost* is intended to be used during conceptual design (trade-off analysis). Because of *Ghost's* positioning as a trade-off analysis tool it intentionally operates at a general level. Partitioning options identified by *Ghost* as viable can be subjected to more detailed analysis. The usage model for *Ghost* is shown in Figure 3. *Ghost* implements the cost models described in Sections 3.1 and 3.2 and combines them with the user's choice of software development environment ("software foundry") and the particular fabricator/process used to fabricate the hardware ("hardware foundry"); see Tables 3 and 4 for examples. *Ghost* allows the optimum hardware/software partitioning to be chosen for a specific combination of wafer fabricator and software development environment.

4. DIGITAL CAMERA CASE STUDY

In this section, we present a detailed case study for a JPEG encoder chip used in a special-purpose digital camera focusing on estimating performance and cost of the chip when the chip's functionality is partitioned into different HW/SW combinations.

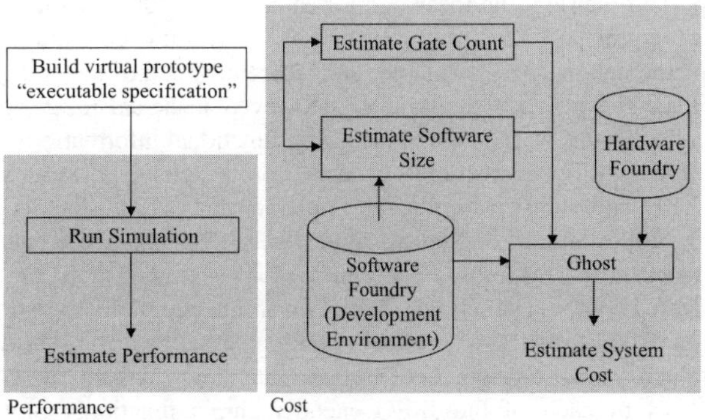

Figure 3. *Ghost* usage model.

The functional requirements list for a digital camera is long, and many requirements are outside the scope of this example, however, the refresh rate is a currently identifiable bottleneck for digital cameras, therefore, when evaluating system performance for the example in this paper, the critical constraint will be that the camera should maintain a 30 frames-per-second (FPS) or higher refresh rate on the viewfinder during JPEG encoding.

Two software tools are utilized to complete the co-design process. The first, Foresight Co-Design, allows us to examine the virtual prototype of the digital camera, choose implementations of the JPEG encoder chip, and assess performance (refresh rate) changes with different implementations, and *Ghost* performs the cost modeling process.

Using Foresight Co-Design's the components of the system (both functional and architectural) are specified using a combination of data flow diagrams, state transition diagrams, and VHDL-based procedural language. Foresight Co-Design's discrete event simulation engine is then used to execute and evaluate the system specification. By mapping the functional components to their implementing resources in the architectural model, system performance can be assessed taking into account all resource constraints. Trade-off analysis can be performed for different possible configurations by changing parameters or mappings and re-evaluating the system.

For the digital camera, high-level functional and architectural models of the image processing system were constructed and three configurations of JPEG encoder implementations were evaluated; all hardware, all software, and mixed hardware/software (Table 1).

Table 1. JPEG encoder implementation options

All software (SW)	• JPEG encoder functionality mapped to threads on the microprocessor core • Data loss occurs due to input buffer overflowing • Refresh rate of 25 FPS • JPEG encoding time = 80 mS
MIXED	• DCT and Quantization functions mapped to the ASIC, other processes mapped to software threads • No input buffer overflow • Refresh rate of 33 FPS • JPEG encoding time of 40 mS
Semi-custom ASIC	• JPEG encoder functionality mapped to HW implemented as semi-custom ASIC blocks • Bus congestion due to loading by JPEG encoder • Refresh rate of 35 FPS • The JPEG encoding time is about 11mS

For the purpose of this study, several key metrics had to be obtained to predict system cost based on each architectural implementation. For a prediction of hardware cost, the most important metric to be measured for each block in the system is the gate count. For this example a synthesized reference implementation was available and therefore could be used to produce the needed gate counts. However, in actual usage, gate count estimates would be obtained from functional information such as VHDL RTL descriptions or data flow graphs. For the SW, MIXED and semi-custom ASIC implementations, gate counts are zero, 8,000, and 16,000 gates, respectively. For the prediction of software costs, the most important metric is number of SLOC or delivered source instructions. Based on architectural blocks and the target characteristics of Feature Point analysis, software size is obtained from Foresight Co-Design's data flow diagrams through analysis of each of the JPEG encoder chip's functional blocks, which was then converted to lines of code.

In order to generate a feature point count that can be converted to source lines of code, each block in the JPEG encoder unit was analyzed for inputs, outputs, inquires, logic files, interfaces, and algorithms. The data flow diagram provides a good picture of inputs and outputs to the block, and the information contained within each block provided answers about the four other factors of interest. The ASIC solution is considered to have no feature points in software. To calibrate the feature point estimates to the existing Java code for the JPEG encoder [21], another data point was desired so that three data points in total were known. An alternative mixed implementation, where the Huffman encoder was assigned to hardware as well, was also evaluated to generate a feature point count ("Mixed #2" in Table 2).

Once these three feature point counts are known, they are multiplied by the conversion factor of 53 [15], corresponding to the conversion from feature points to Java code size, to arrive at a source code count. The results of the feature point analysis on the implementations and the corresponding actual code counts are shown in Table 2. When the source lines derived from feature points are compared to actual lines, each implementation provides a ratio of 2.1 estimated to actual.

Table 2. JPEG encoder lines of code comparison

Implementation	Feature Point Count	Lines of Code (Est.)	Lines of Code (Actual)	Ratio
All software	54	2852	1357	2.102
Mixed #1	41	2149	1026	2.095
Mixed #2	27	1447	690	2.097

Once gate counts and source lines of code are known for the three implementations, the data is entered into the *Ghost* cost model to calculate total system cost. Two sets of foundry data were generated for both hardware and software (Tables 3 and 4).

4.1 Case Study Results

In this section, we present example cost-performance results for various foundry combinations, different production quantities and different levels of reuse. Production quantities ranging from 1000 to 100,000 units were examined. Reuse levels of 0 to 40% were used. Reuse is included in the cost models for chip design (reuse of gate library IP) and software development (code reuse).

The first trend that was observed was in the makeup of the individual cost contributions (design, fabrication, testing, packaging, tooling, software development) of the total cost. Figure 4 shows a plot for the MIXED implementation using hardware foundry 1 (HW1) and software foundry 1 (SW1). At low production quantities (1000 units), design cost and tooling (wafer-masking) cost are the largest factors, dominating the total cost. As production quantity and reuse are increased, both the design and tooling percentages decrease, and fabrication cost becomes significant. This is reasonable because design and tooling costs, are nonrecurring costs, amortized over the entire production quantity. Fabrication and packaging, are recurring costs that remain constant independent of production level. Therefore, the percentage makeup of the total cost for fabrication increases steadily while design, tooling, and testing will decrease. The "Recurring" category on the plot indicates the limit as the production volume approaches infinity.

Similar trends are also observed for hardware foundry 2. Fabrication costs are slightly higher while masking costs are lower due to the change in wafer foundry characteristics (Table 3).

More interesting results were found when the total cost figures were examined for the implementations based on the partitioning of hardware and software.

Table 3. Hardware foundries

Input	HW1	HW2
Wafer diameter	8"	6"
Die-to-die spacing	0.05"	0.05"
Edge scrap	0.4"	0.4"
Wafer format	No centerline cuts	No centerline cuts
Technology	CMOS	CMOS
Min. feature size	0.18 (μm)	0.35 (μm)
Defect density	3 (defects/in2)	3 (defects/in2)
Utilization efficiency	0.5	0.5
Metal layers	6	4
Fab. wafer cost	1600 ($)	1600 ($)
Wafer probing cost	0.003 ($/IO)	0.003 ($/IO)
Probe coverage	0.75	0.75
Sawing cost	0.1 ($)	0.1 ($)
Sawing yield	0.999	0.999
Bonding format	Peripheral	Peripheral
Bond pad width	63.5 (μm)	63.5 (μm)
Bond pad length	76.2 (μm)	76.2 (μm)
Bond pad pitch	0.1016 mm	0.1016 mm

In Figure 5 a medium level of production is assumed (10,000 units). The software foundry remains a smaller factor in total cost than hardware, and combinations utilizing software foundry 2 are slightly less expensive. Software foundry 1 combinations peak between 25% and 50% custom hardware, making the minimum cost the 0% hardware or the software implementation. The software foundry 2 combinations peak similarly, and minimized cost is at 0% hardware. The software implementation does not meet performance standards so the ASIC implementation is the best choice for all foundry combinations. It is important to observe on this plot that the foundry combinations meet at 0% hardware and 100% hardware. This happens because at 100% hardware, the software foundry is irrelevant, causing the combinations HW1/SW1 and HW1/SW2 to end in the same place because only hardware costs are included. The same effect happens for hardware foundry 2. At 0% hardware, this effect is reversed, and the hardware foundry is irrelevant. Therefore, HW1/SW1 and HW2/SW1 will meet and HW1/SW2 and HW2/SW2 will meet. When the production volume is increased

Table 4. Software foundries

Input	SW1	SW2
Assessment/assimilation	0.04	0.04
Software understanding	0.3	0.3
COTS components modified	0	0
COTS code modified	0	0
COTS integration modified	0	0
Annual change traffic	0.15	0.15
Sched. compression/expan.	1	1
All labor rates	80 ($/hr)	80 ($/hr)
Application complexity	1	1
Requirements volatility	1	0.91
Required reliability	1	0.88
Database size	1	0.94
Product complexity	1	0.85
Required reusability	1.1	1
Execution-time constraint	1.11	1
Main-storage constraint	1	1
Virtual-machine volatility	1	0.87
Computer turnaround time	1	0.87
Analyst capability	1	1.19
Application experience	0.91	1
Prog. team capability	0.86	1
Virtual-machine experience	1	1.1
Language experience	0.96	1
Modern prog. pract. (use)	0.91	1
Use of software tools	1	1.1
Classified security appl.	1.1	1
Platform	1.1	1
Req. development schedule	1.04	1
Req. maintenance reliability	1	0.98
Modern maint. prog. pract. (use)	1	1

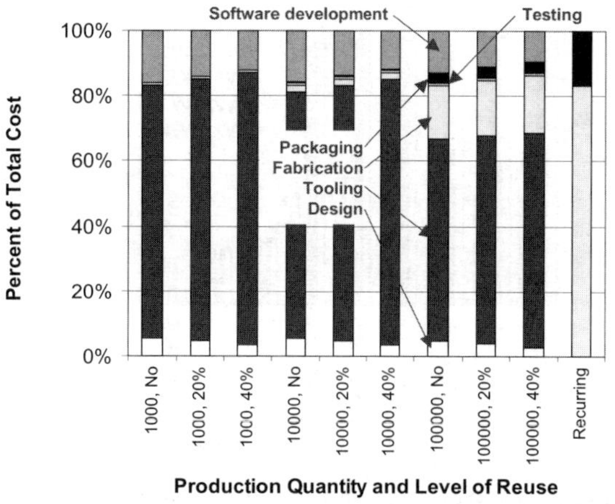

Figure 4. Cost breakdown for MIXED implementation using hardware foundry 1 (HW1) and software foundry 1

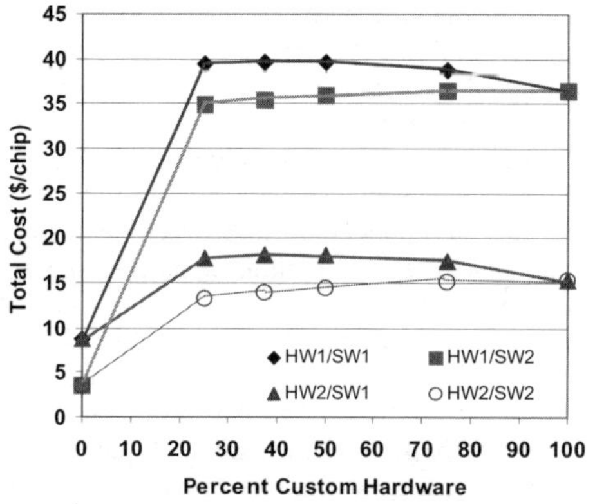

Figure 5. Total cost per chip for 10,000 units based on foundry combination.

there is no longer a peak in cost between 25% and 50% hardware and the optimal costs to fall at 0% hardware. Now, the MIXED implementations must be chosen due to performance constraints. It is interesting to note that although the combination of HW1/SW2 starts out less expensive than that of HW2/SW1, it becomes more expensive at approximately 5% hardware.

Hardware and software reuse was also examined at all levels of production. As the level of reuse increases, cost decreases due to design costs and software development costs decreasing. At low levels of production, reuse affects costs more because design cost and software development cost are both non-recurring and have fewer units to be amortized over. A change in those cost values has a greater affect on total cost values than at larger production rates. The effect of reuse at a medium level of production is shown in Figure 6. We can observe, for instance, that although the MIXED implementation may not be less expensive than the ASIC implementation with no reuse, if the JPEG encoder producer could increase levels of reuse to 40%, the MIXED implementation at 40% reuse could be less expensive than the ASIC implementation with no reuse. However, reuse of IP and software is effectively coupled in this approximation (which would not be the case in reality) so that 20% reuse in the ASIC implementation could allow 20% reuse in the MIXED implementation; this cost improvement is assumed to be present across implementations. Therefore, reuse acts as an

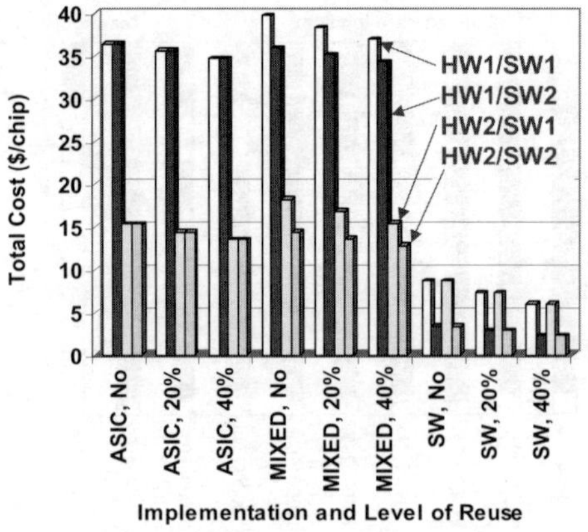

Figure 6. Total cost per chip for 10,000 units based on foundry combinations with reuse.

improvement only within each implementation. Figure 7 examines the expected cost and implementation based on trend-lines drawn through known data points. The plane drawn on the plot is set at the throughput constraint of 30 FPS. The points where the trend-lines for the different foundry combinations intersect the plan are denoted by stars, and are plotted to the corresponding percent custom hardware and cost. Because we have only considered performance ratings for the three implementations (ASIC, MIXED, and SW), we cannot draw conclusions about points in between. In many cases, at large production quantities the optimized partitioning would be between 0% and 50% hardware. Because the MIXED implementation's refresh rate of 33 FPS (Table 1) still exceeds the refresh rate specification of 30 FPS, it may be concluded that a partition can be found that just meets the specification and is less expensive than the MIXED implementation. Figure 7 shows that while we have stated for hardware foundry 1 combinations at 10,000 units that the 100% custom hardware implementation would be optimal, an implementation exactly at the throughput constraint (if one exists) would, in fact, be less expensive. This point falls between 0% and 50% custom hardware, hardware foundry 2 combinations remain less expensive at 100% custom hardware. Despite this demonstration, we cannot conclude as to whether points falling exactly at the 30 FPS throughput constraint could be met by a realizable implementation of hardware and software.

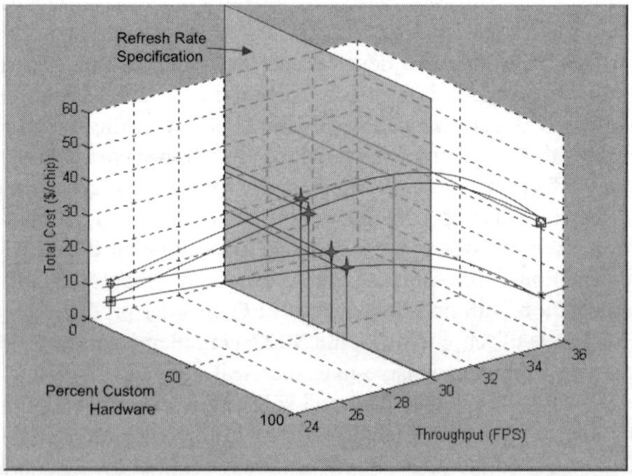

Figure 7. Chip cost and performance based on implementation at 10,000 units.

5. SUMMARY AND CONCLUSIONS

In this paper, we have shown that modern hardware/software co-design methodologies can benefit from the use of comprehensive cost models early in the design process. The ability to view the effects of hardware/software partitioning choices on overall system performance and cost, may prove indispensable in future design methodologies. This paper presented a cost versus performance trade-off-based hardware/software co-design approach that enhances design space exploration early in the system development phase, allowing designers to make educated partitioning choices to meet cost and performance goals and standards. It must, however, be noted that the methodology discussed in this paper ignores the economic impact of several important system attributes, such as flexibility and upgradeability provided by software solutions.

6. REFERENCES

[1] DeBardelaben, J. Madisetti, V. and Gadient, A. J. Incorporating cost modeling in embedded-system design. IEEE Design & Test of Computers 14, 3, (July-Sept 1997), 24-35.

[2] Boehm, B. Software Engineering Economics. Prentice-Hall, Inc., 1981.

[3] REVIC Software Cost Estimating Model User's Manual Version 9.2. U.S. Air Force Analysis Agency, 1994.

[4] Boehm, B., Clark, B.K., Horowitz, E., Madachy, R., Selby, R.W., and Westland, C. Cost models for future software processes: COCOMO 2.0. Annals of Software Eng (1995).

[5] Egolf, T. W. Virtual Prototyping of Embedded Digitial Systems: Hardware/Software Codesign, Integration, and Test. Ph.D. Dissertation, Georgia Inst. of Tech., (Dec. 1997).

[6] Gajski, D., Vahid, F., Narayan, S. and Gong, J. Specification and Design of Embedded Systems. Prentice-Hall, Inc., NJ, 1994.

[7] Ernst, R., Henkel, J. and Benner, T. Hardware-software cosynthesis for micorcontrollers," IEEE Design & Test of Comp. 10, 4, (Dec 1993), 64-75.

[8] Gupta, R. and De Micheli, G. Hardware-software cosynthesis for digital systems. IEEE Design & Test of Computers 10, 3, (Sept 1993), 29-41.

[9] Kalavade, A. and Lee, E. A Hardware-software codesign methodology for DSP applications. IEEE Design & Test of Comp. 10, 3, (Sept 1993), 16-28.

[10] Fornaciari, W., Salice, F, and Bondi, U. Development cost and size estimation starting from high-level specifications. Proc. Int. Symposium on Hardware/Software Codesign (April 2001), 86-91.

[11] Bakshi, S. and Gajski, D. Hardware/software partitioning and pipelining. in Proceedings Design Automation Conference (June 1997), 713-716.

[12] Albrecht, A. Measuring Application Development Productivity. Programming Productivity: Issues for the Eighties. IEEE Comp Society Press. (1981).

[13] Jones, T.C. Programming Productivity. McGraw-Hill, Inc., 1986.

[14] Jones, T.C. Table of Programming Languages and Levels – Version 8.2. Software Productivity Research, Burlington MA, 2001.

[15] DeBardelaben, J. An Optimization-Based Approach for Cost-Effective Embedded DSP System Design. Ph.D. Dissertation, Georgia Institute of Technology (May 1998).

[16] Donath, W. Equivalence of memory to random logic. IBM J. of Res. and Dev. 58, 5, (Sept 1974), 401-407.

[17] Sandborn, P., Abadir, M., and Murphy, C. The tradeoff between peripheral and area array bonding of components in multichip modules. IEEE Trans. CPMT, Part A 17, 2, (June 1994), 249-256.

[18] Dehkordi, P.H. and Bouldin, D.W. Design for packagability: The impact of bonding technology on the size and layout of VLSI dies. Proc. IEEE MCM Conf. (1993), 153-159.

[19] Keating, M. A Financial Model for Design Reuse. Design Reuse Group, Synopsis, Inc., 1998.

[20] Dislis, C., Dick, J.H., Dear, I.D. and Ambler, A.P. Test Economics and Design for Testability, Ellis Horwood, New York NY, 1995.

[21] Weeks, J.R. JPEG Encoder, BioElectroMech and the Independent JPEG Group, www.obrador.com /essentialjpeg/jpeg.htm, 1998.

Efficient Code Synthesis from Extended Dataflow Graphs for Multimedia Applications

Hyunok Oh Soonhoi Ha
The School of Electrical Engineering and Computer Science
Seoul National University
Seoul 151-742, KOREA
TEL : +82-28807292
{oho,sha}@comp.snu.ac.kr

ABSTRACT

This paper presents efficient automatic code synthesis techniques from dataflow graphs for multimedia applications. Since multimedia applications require large size buffers containing composite type data, we aim to reduce the buffer sizes with fractional rate dataflow extension and buffer sharing technique. In an H.263 encoder experiment, the FRDF extension and buffer sharing technique enable us to reduce the buffer size by 67%. The final buffer size is no more than in a manual reference code.

Categories and Subject Descriptors: C.3 [Computer System Organization] : Special-purposed and application- based systems - real time and embedded systems.

General Terms: Algorithms, Design, Experimentation.

Keywords: memory optimization, software synthesis, multimedia, dataflow

1. Introduction

As system complexity increases and fast design turn-around time becomes important, automatic code synthesis from dataflow program graphs is a promising high level design methodology for rapid prototyping of complicated multimedia embedded systems. COSSAP[1], GRAPE[2], and Ptolemy[3] are well-known design environments, especially for digital signal processing applications, with automatic code synthesis facility from graphical dataflow programs.

In a hierarchical dataflow program graph, a node, or a block, represents a function that transforms input data streams into output streams. The functionality of an atomic node is described in a high-level language such as C or VHDL. An arc represents a channel that carries streams of data samples from the source node to the destination node. The number of samples produced (or consumed) per node firing is called the output (or the input) sample rate of the node. In case the number of samples consumed or produced on each arc is statically determined and can be any integer, the graph is called a synchronous dataflow graph (SDF)[4] that is widely adopted in aforementioned design environments.

DAC 2002, June 10-14, 2002, New Orleans, Louisiana, USA.

Memory efficient code synthesis from static dataflow models has been an active research subject to reduce the gap in terms of memory requirements between the automatically synthesized code and the manually optimized code[5]. In the synthesized code, function codes of atomic blocks are stitched together according to the given execution schedule from the compile-time analysis. And each arc is assigned a buffer memory whose size is at least the maximum amount of samples accumulated on the arc at run time. Many works have been done to minimize the buffer requirements by optimal scheduling[4][5] or by buffer sharing[6][7][8]. While most previous works do not differentiate the composite data type from the primitive data type, we can do much better by taking special account for the composite data type, particularly for multimedia applications.

In this paper, we aim to reduce the buffer size requirements in the automatically synthesized code no more than in the manually optimized code especially for the multimedia applications. At first, we introduce a new dataflow extension called fractional rate dataflow(FRDF) in which fractional number of samples can be produced or consumed.

Composite data types, such as video frame or network packet, are used extensively in recent networked multimedia applications, and become the major consumer of scarce memory resource. However, existent dataflow models have inherent difficulty of efficiently expressing the mixture of a composite data type and its constituents: for example, a video frame and macro blocks. A video frame is regarded as a unit of data sample in integer rate dataflow graphs, and should be broken down into multiple macro blocks, explicitly consuming extra memory space. In the proposed FRDF model, a constituent data type is considered as a fraction of the composite data type. Thus no explicit data type conversion is needed.

Next, we propose a new technique to share buffers among different size of data samples. To our best knowledge, no existent technique deals with this problem if dataflow graphs contain delays. When synthesizing software from an SDF/FRDF graph according to the order of block executions or "schedule", a buffer array is allocated to each arc to store the data samples between the source and the destination blocks. Buffer sharing can reduce the allocated buffer memory size. Such sharing decision can be made at compile-time through life-time analysis of data samples, which is a well-known compilation technique.

A key difference between the proposed technique and the previous approaches is that we separate the local pointer buffers from global

This research is supported by National Research Laboratory Program (number M1-0104-00-0015) and Brain Korea 21 program.
The RIACT at Seoul National University provides research facilities for this study.

data buffers explicitly in the synthesized code. This separation provides more memory sharing chances especially when the number of local buffer entries becomes more than one. Let us examine Figure 1(a) which illustrates a simplified version of an H.263 encoder algorithm where 'ME' node indicates a motion estimation block, 'Trans' is transform coding which includes DCT and Quantization, and 'InvTrans' consists of inverse transform coding and image reconstruction. Each sample between nodes is a frame of 176x144 byte size which is large enough to ignore local buffer size. The diamond symbol on the arc between 'ME' and 'InvTrans' denotes an initial data sample, which is the previous frame in this example. If we do not separate local buffers from global buffers then we need three frame buffers as shown in Figure 1(b) since buffers *a* and *c* overlap their life-times at 'ME', *a* and *b* at 'Trans', and *b* and *c* at 'InvTrans'; otherwise we can use only two frames. Figure 1(c) shows the allocation of local buffers and global buffers, and the mapping of local buffers to global buffers. The detailed algorithm and code synthesis techniques will be explained in section 4.

In section 2, we discuss the fractional rate dataflow model with an H.263 encoder example. Section 3 explains the overview of the proposed buffer sharing technique. In section 4, we explain how to determine the minimum local buffer sizes and their mappings to the minimum global buffers assuming that all data samples have the same size. The sharing technique is extended to a graph with data samples of the different sizes and delays in section 5. Finally, we present some experimental results in section 6, and make conclusions in section 7.

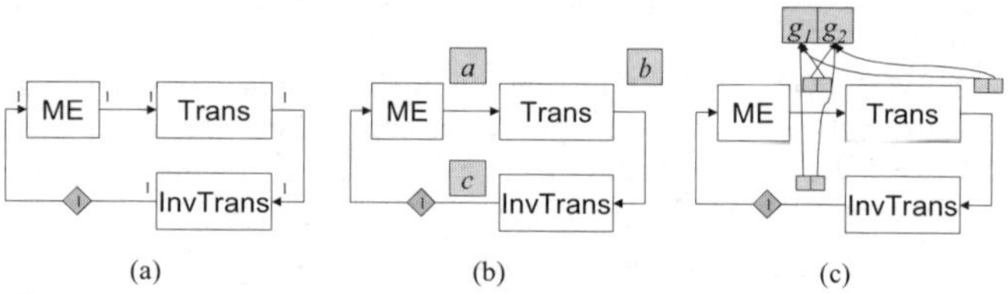

Figure 1. (a) Simplified H.263 Encoder (b) conventional approach (c) separation of global buffer and local buffer

2. Fractional Rate Dataflow

Figure 2 (a) shows an SDF subgraph of an H.263 encoder example. The ME block receives the current and the previous frames to compute the motion vectors and pixel differences, and the D block breaks down the input frame into 99 macro blocks to be encoded in the next EN block. Since the ME block regards a 176x144 video frame as the unit data sample while the EN block a 16x16 macro block, the D block is inserted for explicit data type conversion. This subgraph has the following schedule of block execution: 1(ME,D)99(EN) meaning that after executing the ME and D blocks once, we execute the EN blocks 99 times. Note that the buffer requirements on arcs α and β both have the 176x144 frame size.

Compared with the hand-optimized code with the program graph, we observe that the motion estimation block need not produce the frame-size output at once. Since the ME block is the most time consuming and runs for each macro block, it had better generate output samples at the unit of macro block for shorter latency. The proposed FRDF enables implicit type conversion from the video frame type to the macro block type at the input arc of the ME block (Figure 2(b)). The fraction rate 1/99 indicates that the macro block is 1/99 of the video frame in its size. How to divide the fraction depends the frame type and should be specified by the programmer before run time. For each execution of the ME block, it consumes a macro-block from the input frame, and produces a macro-block output. Thus, we do not need the D block any more since type conversion is already made inside the ME block. A possible schedule becomes 99(ME,EN). Now, the buffer size on arc α becomes the macro-block size. Formal definition and static analysis of FRDF model is omitted due to space limitation[9].

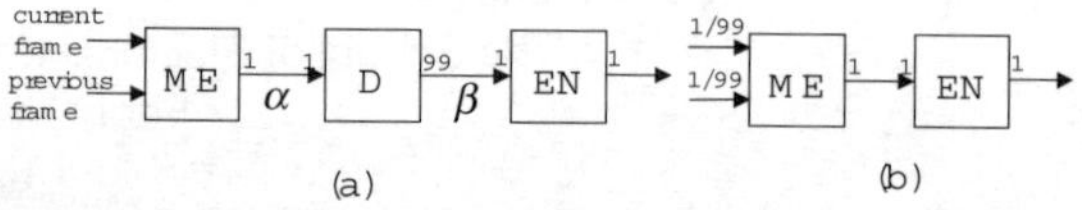

Figure 2. A subgraph of an H.263 encoder example: (a) SDF representation (b) FRDF representation

3. Buffer Sharing Technique

To generate a code from the given FRDF graph, the block execution schedule is determined at compile time. In the buffer sharing technique, global buffers store the live data samples of composite type while the local pointer buffers store the pointers to the global buffer entries. Since multiple data samples can share the buffer space as long as their lifetimes do not overlap, we should examine the lifetimes of data samples. We denote *s(a,k)* as the *k-th* stored sample on arc *a* during an iteration cycle. Consider an example of figure 3(a) with the associated schedule. Within an iteration cycle, two samples are produced and consumed on arc *a*: they are denoted by *s(a,1)* and *s(a,2)* respectively. Arc *b* has an initial sample *s(b,1)* and two more samples, *s(b,2)* and *s(b,3)*, during an iteration cycle.

The lifetimes of data samples are displayed in the *sample lifetime chart* as shown in Figure 3(b) where the horizontal axis indicates the abstract notion of time: each invocation of a node is considered to be one unit of time. The vertical axis indicates the memory size and each rectangle denotes the lifetime interval of a data sample. Note that each sample lifetime defines a single time interval whose start time is the invocation time of the source block and the stop time is after the invocation of the destination block. The lifetime interval of sample *s(b,2)* is [B1, C1]. We take special care of initial samples. The lifetime of sample *s(b,1)* is carried forward from the last iteration cycle while that of sample *s(b,3)* is carried forward to the next iteration cycle. We denote the former-type interval as a *tail lifetime interval*, or shortly a tail interval, and the latter as a *head lifetime interval*, or a head interval. In fact, sample *s(b,3)* at the current iteration cycle becomes *s(b,1)* at the next iteration cycle. To distinguish iteration cycles, we use $s_k(b,2)$ to indicate sample *s(b,2)* at the *k-th* iteration. Then, in Figure 3, $s_1(b,3)$ is equivalent to $s_2(b,1)$. And, the sample lifetime that spans multiple iteration cycles is defined as a *multi-cycle lifetime*.

From the sample lifetime chart, it is obvious that the minimum size of global buffer memory is the maximum of the total memory requirement of live data samples over time. We summarize this fact as the following lemma without proof.

Lemma 1

The minimum size of global buffer memory is equal to the maximum total size of live data samples during an iteration cycle.

We map the sample lifetimes to the global buffers: an example is shown in Figure 3(c) where g(k) indicates the k-th global buffer. In case all data types have the same size, an interval scheduling algorithm can optimally map the sample lifetimes to the minimum size of global buffer memory.

Sample lifetime is distinguished from local buffer lifetime since a local buffer may store multiple samples during an iteration cycle. Consider an example of Figure 3(a) where the local buffer sizes of arcs *a* and *b* are set to be 1 and 2. We denote *B(a,k)* as the k-th local buffer entry on arc *a*. Then, the *local buffer lifetime chart* becomes as drawn in Figure 3(d). Buffer *B(a,1)* stores two samples, *s(a,1)* and *s(a,2)*, to have multiple lifetime intervals during an iteration cycle.

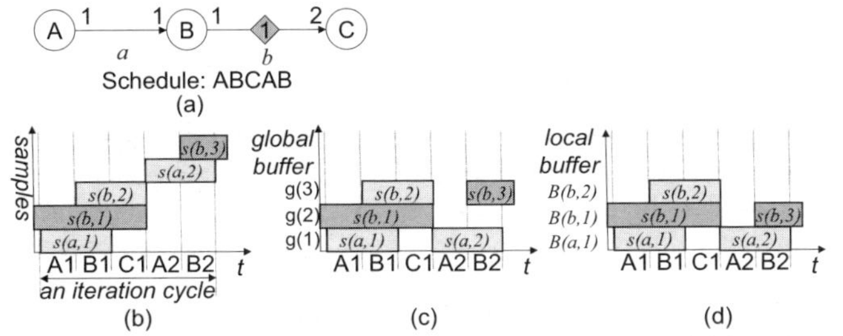

Figure 3. (a) An example SDF graph with an initial delay between B and C illustrated by a diamond (b) the sample lifetime chart (c) the global buffer lifetime chart (d) the local buffer lifetime chart.

Now, we state the problem we have to solve as follows.

Problem:
Determine both the local buffer sizes and the global buffer sizes for the objective of minimizing the sum of them, with the given program graph and the sample lifetime chart that can be obtained from the given schedule.

Since the size of composite data type is usually much larger than that of pointer type in multimedia applications of interest, reducing the global buffer size is more important than reducing the local pointer buffers. Therefore, our sharing heuristic consists of two phases: the first phase is to map the sample lifetimes within an iteration cycle into the minimum size of global buffers ignoring local buffer sizes, and the second phase is to determine the minimum local buffer sizes that store the sample pointers to the given global buffers.

3.1 Global Buffer Minimization

Recall that a sample lifetime has a single interval within an iteration cycle. When all samples have the same data size, the interval scheduling algorithm is known to be an optimal algorithm to find the mapping to the minimum global buffer size. We summarize the interval scheduling algorithm in Figure 4.

```
1:U is a set of sample lifetimes;
2:P is an empty global buffer lifetime chart.
3:While (U is not empty) {
4: Take out a sample lifetime x with the earliest start time from U.
5: Find out a global buffer whose lifetime ends earlier than the start
      time of x:
6: Priority is given to the buffer that stores samples on the same arc
      if exists.
7: If no such global buffer exists in P, create another global buffer.
8: Map x to the selected global buffer
9:}
```

Figure 4. Interval scheduling algorithm

Consider an example of Figure 3(a), whose global buffer lifetime chart is the same as that is displayed in Figure 3(c). After samples *s(a,1)*, *s(b,1)* and *s(b,2)* are mapped into three global buffers, *s(a,2)* can be mapped to all three buffers. Among the candidate global buffers, we select one that already stores *s(a,1)* according to the policy of line 6 of Figure 4. The reason of this priority selection is to minimize the local buffer sizes, which will be discussed in the next section.

3.2 Local Buffer Size Determination

The global buffer minimization algorithm in the previous phase runs for one iteration cycle while the graph will be executed repeatedly. The next phase is to determine the minimum local buffer sizes that are necessary to store the pointers of data samples mapped to the global buffers. Initially we assign a separate local buffer to each live sample during an iteration cycle. Then, the local buffer size on each arc becomes the total number of live samples within an iteration cycle: Each sample occupies a separate local buffer. In Figure 3(a) for instance, two local buffers are allocated on arc *a* while three local buffers on arc *b*.

What is the optimal local buffer size? The answer depends on when we set the pointer values, or when we *bind* the local buffers to the global buffers. If binding is performed statically at compile-time, we call it *static* binding. If binding can be changed at run-time, it is called *dynamic* binding. In general, the dynamic binding can reduce the local buffer size significantly with run-time overhead of setting pointer values.

3.2.1 Dynamic Binding Strategy

Since we can change the pointer values at run-time in dynamic binding strategy, the local buffer size of an arc can be as small as the maximum number of live samples at any time instance during an iteration cycle.

Consider arc b of Figure 3(a) that has an initial sample and needs only two local buffers since there are at most two live samples at the same time. The global buffer lifetime chart does not repeat itself at the next iteration cycle. Lifetime patterns of local buffers *B(b,1)* and *B(b,2)* are interchanged at the next iteration cycle as shown in Figure 5. We can repeat this pointer assignment at every iteration cycle dynamically. The repetition periods of pointer assignment for arcs with initial samples may span multiple iteration cycles. The next section is devoted to computing the repetition period of pointer assignment for the arcs with initial samples.

Suppose an arc *a* has *N* live samples and *M* local buffers within an iteration cycle. Since the local buffers are accessed sequentially, a local buffer has at most $\lceil N/M \rceil$ samples and the pointer to sample *s(a,k)* is stored in *B(a, k* mod *M)*. After the first phase completes, we examine the mapping results of the allocated sample in a local buffer to the global buffers at the code generation stage. If the mapping result of the current sample is changed from the previous one, we have to insert code to alter the pointer value at the current schedule instance. Note that it incurs both memory overhead of code insertion and time overhead of runtime mapping.

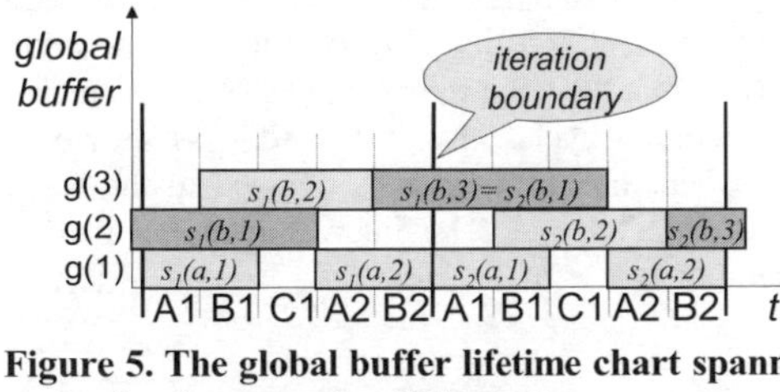

Figure 5. The global buffer lifetime chart spanning two iteration cycles for the example of Figure 3.

3.2.2 Static Binding Strategy

If we use static binding, we may not change the pointer values of local buffers at run time. It means that all allocated samples to a local buffer should be mapped to the same global buffer. For example, arc *a* of Figure 3 needs only one local buffer for static binding since two allocated samples are mapped to the same global buffer. How many buffers do we need for arc *b* of Figure 3 for static binding?

To answer this question, we extend the global buffer lifetime chart over multiple iteration cycles until the sample lifetime patterns on the arc become periodic. We need to extend the lifetime chart over two iteration cycles as displayed in Figure 5. Note that the head lifetime interval of $s_2(b,3)$ is connected to the tail lifetime interval of $s_3(b,1)$ in the next repetition period. Therefore, four live samples are considered during the repetition period that consists of two iteration cycles. The problem is to find the minimum local buffer size *M* such that all allocated samples on each local buffer are mapped to the same global buffer. The minimum number is 4 on arc *b* in this example since $s_3(b,1)$ can be placed at the same local buffer as $s_1(b,1)$.

How many iteration cycles should be extended is an equivalent problem to computing the repetition period of pointer assignment for dynamic binding case. Refer to the next section for detailed discussion.

4. Repetition Period of Sample Lifetime Patterns

Initial samples may make the repetition period of the sample lifetime chart longer than a single iteration cycle since their lifetimes may span to multiple cycles. For simplicity, we assume that all samples have the same size in this section. This assumption will be released in the next section.

First, we compute the iteration length of a sample lifetime. Suppose *d* initial samples stay alive on an arc and *N* samples are newly produced for each iteration cycle. Then, *N* samples on the arc are consumed from the destination node. If *d* is greater than *N*, the newly produced samples all live longer than an iteration cycle. Otherwise, *N-d* newly created samples are consumed during the same iteration cycle while *d* samples live longer. We summarize this fact in the following lemma.

Lemma 2.
If there are *d* initial samples on an arc and *N* samples are newly created per iteration cycle on an arc, the lifetime interval of $d \bmod N$ newly created samples on the arc spans $\lceil d/N \rceil + 1$ iteration cycles and that of $N-(d \bmod N)$ samples spans $\lceil d/N \rceil$ iteration cycles.

Let *p* be the number of iteration cycles in which a sample lifetime interval lies. Figure 6 illustrates two patterns that a sample lifetime interval can have. A sample starts its lifetime at the first iteration cycle with a head interval and ends its lifetime at the *p-th* iteration with a tail interval. Note that the tail interval at the *p-th* iteration also appears at the first iteration cycle. The first pattern, as shown in Figure 6(a), occurs when the tail interval is mapped to the same global buffer as the head interval. The interval mapping pattern repeats every *p-1* iteration cycles in this case.

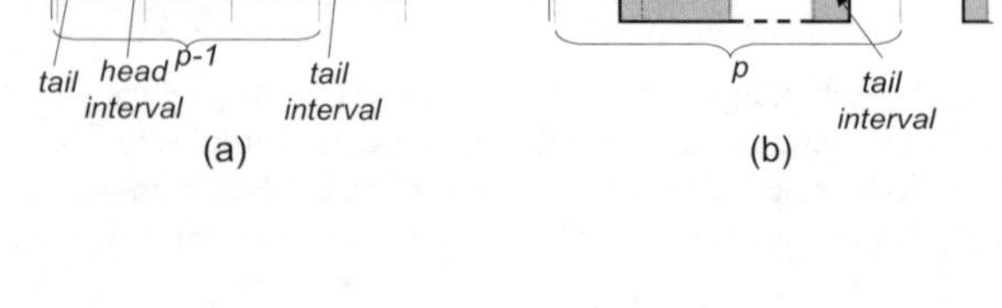

Figure 6. Illustration of a sample lifetime interval: (a) when the tail interval is mapped to the same global buffer as the head lifetime interval, and (b) when the tail interval is mapped to a different global buffer and there is no chained multi-cycle sample lifetime interval.

The second pattern appears when the tail interval is mapped to a different global buffer. To compute the repetition period, we have to examine when a new head interval can be placed at the same global buffer. Figure 6 (b) shows a simple case that a new head interval can be placed at the next iteration cycle. Then, the repetition period of the sample lifetime pattern becomes *p*. More general case occurs when another multi-cycle sample lifetime on a different arc is chained after the tail interval. A multi-cycle lifetime is called *chained* to a tail interval when its head interval is placed at the same global buffer. The next theorem concerns this general case. Due to space limitation, we skip the proof of the theorem.

Theorem 2. Let t_i be the tail interval and h_i the head interval of sample *i* respectively. Assume the lifetime of sample *i* spans p_i+1 and t_i is chained to the lifetime of sample *i+1* for *i = 1* to *n-1*. The interval mapping pattern repeats every $\sum_{i=1}^{n} p_i$ iteration cycles if interval t_n is chained back to the lifetime of sample *1*. Otherwise it repeats every $\sum_{i=1}^{n} p_i + 1$ iteration cycles.

We apply the above theorem to the case of Figure 3 (b) where head interval *s(b,3)* and tail interval *s(b,1)* are mapped to the different global buffers. And, the sample lifetime spans 2 iteration cycles. Therefore, the repetition period becomes 2 and Figure 5 confirms it.

Another example graph is shown in Figure 7(a), which is identical to the simplified H.263 encoder example of Figure 1. There is a delay symbol on arc CA with a number inside which indicates that there is an initial sample, *s(c,1)*. Assume that the execution order is ABC. During an iteration cycle, sample *s(c,1)* is consumed by A and a new sample *s(c,2)* is produced by C as shown in Figure 7(b). By interval scheduling, an optimal mapping is found like Figure 7(c). By theorem 2, the mapping patterns of *s(c,1)* and *s(c,2)* repeat every other iteration cycles since head interval *s(c,2)* is not mapped to the same global buffer as tail interval *s(c,1)*. Initial samples also affect the lifetime patterns of samples on the other arcs if they are mapped to the same global buffers as the initial samples are mapped to. In Figure 7(c), sample *s(b,1)* are mapped to the same global buffer with *s(c,1)* while *s(a,1)* with *s(c,2)*. As a result, their lifetime patterns also repeat themselves every other iteration cycles. The summary of repetition periods is displayed in Figure 7(d).

Recall that the repetition periods determine the period of pointer update in the generated code with dynamic binding strategy, and the size of local buffers in the generated code with static binding strategy. Figure 7(e) and (f) show the code segments that highlight the difference. In this example, using dynamic binding strategy is more advantageous.

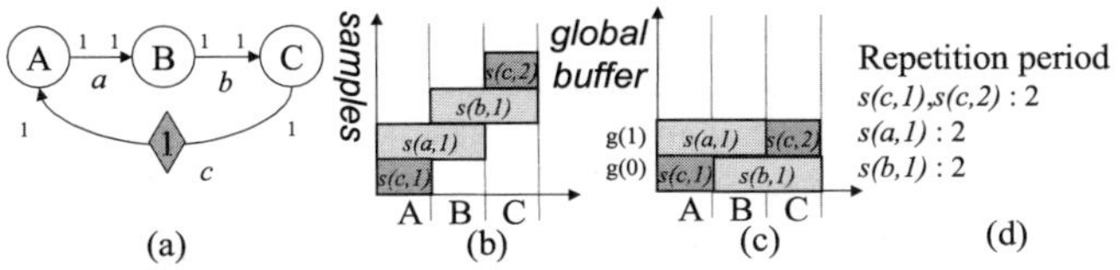


```
struct Frame g[2];
main()
{
  struct G *a, *b, *c[2]={g,g+1};
  int in_A=0, out_C=1;

  for(int i=0;i<max_iteration;i++) {
    { a = c[(i+1)%2];
      // A's codes. Use c[in_A] and a.
      in_A = (in_A+1)%2;
    }
    { b = c[i%2];
      // B's codes. Use a and b.
    }
    { // C's codes. Use b and c[out_C].
      out_C = (out_C+1)%2;
    }
  }
}
```

(e)

```
struct Frame g[2];
main()
{
  struct G *a[2]={g+1,g}, *b[2]={g,g+1}, *c[2]={g,g+1};
  int in_A=0, out_A=0, in_B=0, out_B=0, in_C=0, out_C=1;

  for(int i=0;i<max_iteration;i++) {
    { // A's codes. Use c[in_A] and a[out_A].
      in_A = (in_A+1)%2; out_A = (out_A+1)%2;
    }
    { // B's codes. Use a[in_B] and b[out_B].
      in_B = (in_B+1)%2; out_B = (out_B+1)%2;
    }
    { // C's codes. Use b[in_C] and c[out_C].
      in_C = (in_C+1)%2; out_C = (out_C+1)%2;
    }
  }
}
```

(f)

Figure 7. (a) A graph which is equivalent to Figure 1(a) (b) Life-time intervals for an iteration cycle (c) Placement of intervals (d) Determination of repetition period (e) Generated code with dynamic binding (f) with static binding

Up to now, we assume that all samples have the same size. The next section will discuss the extension of the proposed idea to more general case, where samples of different sizes share the same global buffer space.

5. Different Sample Size

We are given sample lifetime intervals which are determined after the scheduled execution order of blocks. The optimal assignment problem of local buffer pointers to the global buffers is nothing but to pack the sample lifetime intervals into a single box of global buffer space. Since the horizontal position of each interval is fixed, we have to determine the vertical position, which is called the "vertical offset" or simply "offset". The bottom of the box, or the bottom of the global buffer space has offset 0. The objective function is to minimize the global buffer space. Recall that if all samples have the same size, interval scheduling algorithm gives the optimal result. Unfortunately, however, the optimal assignment problem with intervals of different sizes is known to be an NP-hard problem. The lower bound is evident from the sample lifetime chart; it is the maximum of the total sample sizes live at any time instance during an iteration. Therefore, we propose a simple but efficient heuristic algorithm. If the graph has no delays (initial samples), we can repeat the assignment every iteration cycle.

The proposed heuristic is called LOES(lowest offset first and earliest start time first). As the name implies, it assigns intervals in the increasing order of offsets, and in the increasing order of start times as a tie-breaker. At the first step, the algorithm chooses an interval that can be assigned to the smallest offset, among unmapped intervals. If more than one intervals are selected, then an interval is chosen which starts no later than others. The earliest start time first policy allows the placement algorithm to produce an optimal result when all samples have the same size since the algorithm is equivalent to the interval scheduling algorithm.

The detailed algorithm is depicted in Figure 8. In this pseudo code, U indicates a set of unplaced sample lifetime intervals and P a set of placed intervals. At line 5, we compute the feasible offset of each interval in U . Set C contains intervals whose feasible offsets are lowest among unplaced intervals at line 7. We select the interval with the earliest start time in C at line 9 and place it at its feasible offset to remove it from U and add it to P. This process repeats until every interval in U is placed.

```
1: Procedure LOES(U is a set of sample lifetimes) {
2:  P←{}.
3:  While(U is not empty) {
4:    /* compute feasible offsets of every interval in U with P */
5:    computeLowestOffset(U,P);
6:    /* 1st step: choose intervals with the smallest feasible offset from
       U */
7:    C ←findIntervalsWithLowestOffset(U);
8:    /* 2nd step(tie-breaking) : interval scheduling */
9:    select interval x with the earliest arrival time from C;
10:   remove x from U.
11:   P←P ∪ {x}.
12:  }
13:}
```

Figure 8. Pseudo code of LOES (Lowest Offset first and Earliest Start-time first) algorithm

If a graph has initial samples and samples have different sizes this method breaks down as Figure 9 illustrates with a simple example. We assume that "h" and "t" indicate the head interval and the tail interval of the same sample lifetime respectively. At the second iteration, interval h should be placed as shown in Figure 9(b) since it is extended from the first cycle. The head interval h prohibits interval x from lying on contiguous memory space at the second iteration. Such splitting is not allowed in the generated code since the code regards each sample as a unit of assignment. To overcome this difficulty, multi-cycle intervals do not share the global buffer space with other intervals with different sample size. The modified LOES heuristic is omitted due to space limitation.

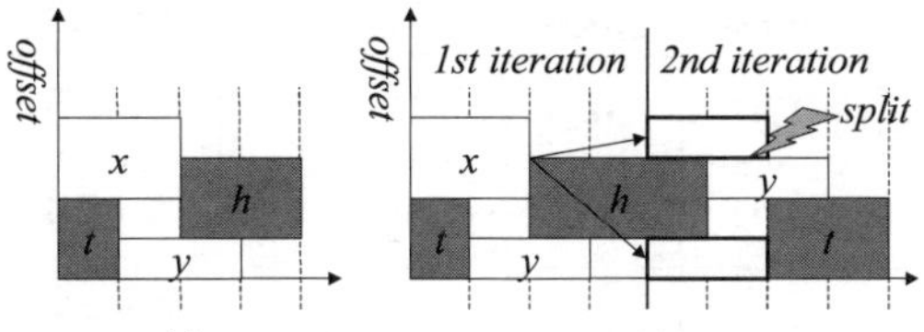

(a) (b)

Figure 9. (a) After interval allocation (b) Interval x is split at the second iteration

6. Experiments

In this section, we demonstrate the performance of the proposed technique with three real-life examples: a jpeg encoder, an MP3 decoder, and an H.263 encoder. We have implemented the proposed technique in a design environment called PeaCE[10].

Table 1 summarizes the performance improvement from the proposed techniques. The third row indicates the unoptimized buffer requirement in the synthesized code from the initial SDF graph. The fourth row denotes the case of buffer sharing of same size samples without separating local and global buffers. It corresponds to the results expected from the previous works. The fifth and the last rows show the results from the proposed technique of buffer sharing of different size samples without and with buffer separation respectively.

A jpeg encoder example represents the first and the simplest class of program graphs in which all non-primitive data samples have the same size and no initial delay exists. In this case, we do not have to separate the local pointer buffer and the global data buffer, which is taken into account in the proposed technique. As a result, we can reduce the memory requirements to one third as the third and the fourth rows of Table 1 illustrate.

An MP3 decode example is composed of three kinds of different size samples. It represents the second class of graphs that have different size samples but no initial delay sample. No separation between local and global buffers is necessary because there is no initial delay in this example. However, the proposed algorithm that shares the buffer space between different size samples reduces the memory requirement by 52% compared with any sharing algorithm that shares the buffer space among only equal size samples. The fourth and the fifth rows show the performance difference.

As for an H.263 encoder example, we make two versions: the simplified version as discussed in the first section and the full version. The simplified version is an example of the third class of graphs in which all non-primitive data samples have the equal size and initial delay samples exist. As discussed earlier, separation of local buffers from global buffers allows us to reduce the buffer space to optimum as the sixth row reveals. The full version of an H.263 encoder example represents the fourth and the most general class of graphs that consist of different size samples and initial samples. The H.263 encoder example include 4 different size sample sizes and 8 initial delay samples on eight different arcs. The proposed technique can reduce the memory requirement by 40% compared with the unshared version. On the other hand, a sharing technique reduces the buffer size only 23% if both buffer separation and sharing between different samples are not considered.

Table 1. Comparison of buffer memory requirements

Example	JPEG	MP3	Simplified H.263	H.263
# of samples	6	336	3	1804
no sharing	1536B	36KB	111KB	659KB
sharing of same size	512B	23KB	111KB	510KB
sharing w/o separation	512B	11KB	111KB	510KB
sharing with separation	-	-	74KB	396KB

Table 2. Comparison of synthesize codes with reference code for the H.263 encoder in FRDF

Example	Ref. Code	FRDF only	FRDF & buffer sharing
H.263	350KB	225KB	219KB

The SDF model has a limitation that it regards a sample of non-primitive type as a unit of data delivery. In an H.263 encoder example, the SDF model needs an additional block that explicitly divides a frame into 99 macro blocks, paying non-negligible data copy overhead and extra frame-size buffer space. In the manually written reference code, such data copy is replaced with pointer operation. Therefore, we apply the proposed technique to an extended SDF model, called fractional rate dataflow (FRDF). With the FRDF model, we could remove such data copy overhead. And, the proposed buffer sharing technique further reduces the memory requirement by 37% more than the reference code.

7. Conclusion

In this paper, we present efficient code synthesis techniques from dataflow graphs for multimedia applications. Since multimedia applications require large size buffers containing composite type data, we propose two techniques to minimize data buffer size: fractional rate dataflow(FRDF) extension and buffer sharing based on local and global buffer separation.

Through the proposed techniques, automatic software synthesis from a dataflow program graph achieves the comparable code quality with the manually optimized code in terms of memory requirement, which is demonstrated by experiments with three real-life examples.

In this paper, we assume that the execution order of blocks is given from the compile-time scheduling. In the future, we will develop an efficient scheduling algorithm which minimizes the memory requirement based on the proposed technique.

8. REFERENCES

[1] Synopsys Inc., 700 E. Middlefield Rd., Mountain View, CA94043, USA, COSSAP User's Manual.

[2] R. Lauwereins, M. Engels, J. A. Peperstraete, E. Steegmans, and J. Van Ginderdeuren, "GRAPE: A CASE Tool for Digital Signal Parallel Processing", IEEE ASSP Magazine, vol. 7, (no.2):32-43, April, 1990

[3] J. T. Buck, S. Ha, E. A. Lee, and D. G. Messerschimitt, "Ptolemy: A Framework for Simulating and Prototyping Heterogeneous Systems", Int. Journal of Computer Simulation, vol. 4, pp. 155-182, April, 1994.

[4] P. K. Murthy, S. S. Bhattacharyya, and Edward A. Lee, "Joint Minimization of Code and Data for Synchronous Dataflow Programs", Journal of Formal Methods in System Design, vol. 11, no. 1, July, 1997.

[5] S. S. Bhattacharyya, P. K. Murthy, and E. A. Lee, "APGAN and RPMC: Complementary Heuristics for Translating DSP Block Diagrams into Efficient Software Implementations", DAES, vol. 2, no. 1, pp. 33-60, January, 1997.

[6] P. K. Murthy, and S. S. Bhattacharyya, "Shared Buffer Implementations of Signal Processing Systems Using Lifetime Analysis Techniques", IEEE TCAD, vol. 20, no. 2, Feb., 2001.

[7] S. Ritz, M. Willems, H. Meyr, "Scheduling for Optimum Data Memory Compaction in Block Diagram Oriented Software Synthesis", Proceedings of ICASSP, p.2651-2653, 1995

[8] E. D. Greef, F. Catthoor, and H. D. Man, "Array Placement for Storage Size Reduction in Embedded Multimedia Systems", Proceedings of the International Conference on Application-Specific Array Processors, pp. 66-75, July 1997.

[9] H. Oh and S. Ha, "Fractional Rate Dataflow Model and Efficient Code Synthesis for Multimedia Applications," accepted by LCTES/SCOPES '02, June, 2002.

[10] http://peace.snu.ac.kr/research/peace, PeaCE : Codesign Environment.

Transformation Based Communication and Clock Domain Refinement for System Design

Ingo Sander
Royal Institute of Technology
Stockholm, Sweden
ingo@imit.kth.se

Axel Jantsch
Royal Institute of Technology
Stockholm, Sweden
axel@imit.kth.se

ABSTRACT

The ForSyDe methodology has been developed for system level design. In this paper we present formal transformation methods for the refinement of an abstract and formal system model into an implementation model. The methodology defines two classes of design transformations: (1) semantic-preserving transformations and (2) design decisions. In particular we present and illustrate communication and clock domain refinement by way of a digital equalizer system.

Categories and Subject Descriptors

B.7.2 [**Integrated Circuits**]: Design-Aids; J.6 [**Computer-Aided Engineering**]: Computer-Aided Design (CAD)

General Terms

Design, Theory

Keywords

System Design, System Modeling, Design Refinement

1. INTRODUCTION

The increasing capacity of integrated circuits makes it possible to integrate more and more functionality on a single chip. A SoC (System-on-a-Chip) architecture can include a variety of components, such as analog parts, micro controller cores, digital signal processor cores, memories, IP blocks and custom hardware. Software, running on a number of different processors, has to be designed to coordinate these components. While such architectures allow for totally new application areas, it is not obvious how to design such applications.

Keutzer et al. discuss system-level design in [7]. They point out, that "to be effective a design methodology that addresses complex systems must start at high levels of abstraction" and underline that an "essential component of a new system design paradigm is the orthogonalization of concerns, i.e. the separation of various aspects of design to allow more effective exploration of alternative solutions". In particular, a design methodology should separate (1) function (what the system is supposed to do) from architecture (how it does it) and (2) communication from computation.

They "promote to use formal models and transformations in system design so that verification and synthesis can be applied to the advantage of the design methodology" and believe that "the most important point for functional specification is the underlying mathematical model of computation".

These arguments strongly support the ForSyDe (Formal System Design) methodology as many of their main requirements on a system design methodology are not only part of our methodology, but establish the foundations of ForSyDe.

The ForSyDe methodology addresses the design of SoC applications. Starting with a formal system model, that captures the functionality of the system at a high abstraction level, it provides formal design transformation methods for a transparent refinement process of the system model into an implementation model, which serves as a starting point for synthesis into hardware and software.

In this paper we present the refinement process in our methodology. We discuss semantic-preserving transformations and design decisions and illustrate them by the refinement of an equalizer system.

2. RELATED WORK

Edwards et al. [5] use the tagged signal model, which is further elaborated by Lee and Sangiovanni-Vincentelli in [8], to classify and analyze several models of computation, in particular discrete event models, communicating finite state machines, synchronous models and data flow process networks. It appears, that different models fundamentally have different strength and weaknesses, and that attempts to find their common features result in models that are very low level and difficult to use.

According to the tagged signal model our system model can be classified as synchronous computational model. The ForSyDe system model is based on the perfect synchrony hypothesis, that also forms the base for the family of the synchronous languages. According to Benveniste and Berry "the synchronous approach is based on a relatively small variety of concepts and methods based on deep, elegant, but simple mathematical principles" [2]. The synchronous assumption implies a total order of events and leads to a clean separation between computation and communication and gives a solid base for formal methods.

Balarin et al. [1] argue, that the synchronous assumption, though very convenient from the analyzing point of view, imposes a too strong restriction on the implementation, as it has to be "fast enough". They advocate a GALS (globally asynchronous locally synchronous) approach and implement it in their methodology as a network of Codesign FSM's communicating via events. Each CFSM is a synchronous FSM, but the communication is done with events that can occur at any time and independently. The CFSM network is inherently nondeterministic. Balarin et al. argue, that this enables them to easily model the unpredictability of the reaction delay of a CFSM both at the specification and at the implementation level, while they admit, that nondeterminism makes the design and verification process more complex.

We argue, that the advantages of a deterministic synchronous system model outweigh the disadvantages. Nondeterminism in the system model implies, that all possible solutions have to be considered, which is a heavy burden put on the designer's shoulders. The task of developing and verifying a system model for a SoC application is so complex, that a system model has to be deterministic, thus avoiding the unnecessary complexity of nondeterminism. The fact, that SoC applications will be implemented on at least partly asynchronous architectures does not justify a nondeterministic approach. We think, that the synthesis of the system model into a partly asynchronous implementation should be part of the synthesis process and not already be decided at the system level.

Approaches like the MASCOT methodology [4] which integrates data and control flow at the system specification level, using the two languages Matlab and SDL, are successful in that they provide an integrated simulation environment. They fail with respect to full integration because they cannot provide an integrated formal analysis and synthesis. For instance, all heterogeneous approaches advocate separate synthesis and design flows for the different domains.

To overcome this difficulty internal representations like FunState [14] have been developed to integrate a heterogeneous system model into one internal representation. However, when compiling high level, abstract and application oriented models into a combined internal representation based on common, low level primitives, most application oriented features are lost and with them optimization opportunities that rely on the meaning of these application oriented concepts.

Functional languages have been used in other research projects in electronic design. The parallel programming community has used functional languages to derive parallel programs from a functional specification [13]. They use skeletons to structure a problem. This formulation is then transformed using cost measures into an efficient implementation for a chosen computer architecture. Reekie [11] has used Haskell to model digital signal processing applications. Similarly to us he modeled streams as infinite lists and used higher-order functions to operate on them. Finally, semantic-preserving methods were applied to transform a model into a more efficient representation. This representation was not synthesized to hardware or software. Lava [3] is a hardware description language based on Haskell. It focuses on the structural representation of hardware and offers a variety of powerful connection patterns. Lava descriptions can be translated into VHDL and there exist interfaces to formal method tools. Hardware ML (HML) [9] is a hardware description language, that is based on the functional programming language Standard ML. Though HML uses some features of Standard ML, such as polymorphic functions and its type system, it is mainly an improvement of VHDL - there is a direct mapping from HML constructs into the corresponding VHDL constructs.

3. THE FORSYDE METHODOLOGY

3.1 The Design Process

The ForSyDe design process starts with the development of a formal abstract functional system model, written in the functional language Haskell. This model is then refined inside the functional domain by a stepwise application of well defined design transformations into an implementation model, which is discussed in Section 4. As the implementation model is a refined version of the system model, the same validation and verification methods can be applied to both models. In the partitioning phase, the implementation model is partitioned into hardware and software blocks, which are mapped on architectural components. Only now, in the code generation phase, we leave the functional domain to produce VHDL or C/C++ for the hardware and software parts. Code generation in our methodology is discussed in [12].

3.2 The System Model

The system model reflects the design principles of the ForSyDe methodology. In order to allow for formal design on a high abstraction level, the system model has the following characteristics:

- It is based on a *synchronous computational model*, which cleanly separates computation from communication.
- It is *purely functional* and *deterministic*.
- It uses *ideal data types* such as lists with infinite size
- It uses the concept of well defined process construction templates, called *skeletons*, which implement the synchronous computational model.
- It is expressed in the executable functional language Haskell, with a well defined *formal semantics*.

The system model abstracts from implementation details, such as buffer sizes and low-level communication mechanisms. This enables the designer to focus on the functional behavior on the system rather than structure and architecture. This abstract nature leaves a wide design space for further design exploration and design refinement, which is supported by our transformational refinement techniques (Section 4).

In order to formally describe our computational model, we follow the denotational framework of Lee and Sangiovanni-Vincentelli [8]. They define signals as a set of events, where each event e has a tag t and a value v, i.e. $e = (t, v) \in T \times V$. As our system model is synchronous, T is the set of natural numbers, and all signals have the same set of tags. In order to model the absence of an event, a data type D can be extended into a data type $D_\perp$ by adding the special value $\perp$, which is used to model the absence of a value. Absent values are used to establish a total order of events when dealing with signals with different or aperiodic event rates.

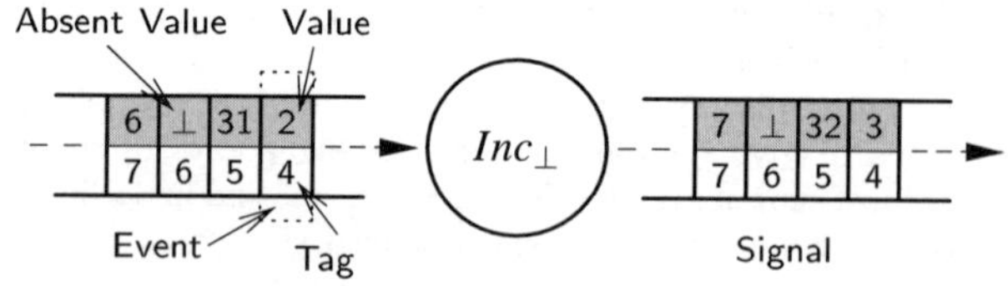

Figure 1: Modeling of Signals and Processes

Figure 1 illustrates the modelling of signals and the behavior of processes. During the event cycle n a process processes the events of each signal with the tag n and outputs the result at the same tag n.

We denote a signal s with $\{x_1, x_2, \ldots\}$, where the tag is given by the position in the signal. To model a process we use the concept of skeletons as *process constructors*. A skeleton is a higher-order function like *mapSY*. *mapSY* takes a combinatorial function as argument and produces a process. In this paper, we denote the processes created by *mapSY* and a function f as the name of the function in capital letters, in this case F. A process F is then created by $mapSY(f)$ and formally defined by

$$\begin{aligned} F &= mapSY(f) \\ F\{x_1, x_2, \ldots\} &= \{f(x_1), f(x_2), \ldots\} \end{aligned}$$

The process $Inc_\perp$ of Figure 1 is also modeled by means of the higher-order function *mapSY*, that maps a function $inc_\perp$ on all values of a signal. The function $inc_\perp$ differs from the increment function *inc* as it is able to process absent values. The function $inc_\perp$ can be generated by the higher-order function Ψ. It takes a function $f : D \rightarrow D$ and generates $f_\perp : D_\perp \rightarrow D_\perp$. Ψ is defined by

$$\begin{aligned} \Psi(f) &= f_\perp \\ \text{where} \quad f_\perp(x) &= \begin{cases} \perp & x = \perp \\ f(x) & \text{otherwise} \end{cases} \\ &\forall f : D \rightarrow D \end{aligned}$$

We use the functional language Haskell as our modeling language. Haskell fits well in our modeling technique since it is a functional language, that supports many of our concepts, such as higher-order functions and laziness, and has a formal semantics. Thus the implementation of higher-order functions like *mapSY* and Ψ is straight forward. We implement the synchronous computational model with skeletons, like *mapSY*. A skeleton is a higher-order function, that takes *combinatorial functions*, i.e. functions that have no internal state, and *values* as input and produces a process as output. Processes consume signals and generate signals as output. Hence, skeletons can be viewed as *process constructors*. The ForSyDe library defines skeletons for more than one computational model, but for the ForSyDe system model only *synchronous skeletons* are used.

The concept of skeletons give the following benefits:

- Due to the construction of processes by skeletons and combinatorial functions, we get a *clean separation between synchronization and computation*. Synchronization is expressed by skeletons and computation by the supplied combinatorial functions.
- There is a family of skeletons that include a *local state*. But there is no global state in the model, which would make it more difficult to reason about the system model.
- Skeletons have a *structural hardware and software interpretation*. This means, that a system model, which is composed of skeletons also has an interpretation in hardware, software or a mix of both.
- As skeletons are higher-order functions, the work on *correctness preserving transformations* [10] can be used to transform a system model inside the functional domain into a more effective implementation model [15].

The process $Inc_\perp$ illustrates the clean separation between synchronization and computation. The skeleton *mapSY* processes an event in each event cycle, while the function $inc_\perp$ computes the output value for each input value.

The hardware interpretation of this process is a combinatorial block implementing the function $inc_\perp$ with one input and one output. The software interpretation is the implementation of $inc_\perp$ as a software function taking one value as input and producing one output value.

The skeleton *mooreSY* is an example for a skeleton with a local state. It models a finite state machine of Moore type. The skeleton takes a function *ns* to calculate the next state as first argument, a function *out* to calculate the output as second argument and a value s_0 for the initial state as last argument. Thus the process

$$Moore = mooreSY(ns, out, init)$$

implements a finite state machine. In hardware the process *Moore* can be implemented as an FSM, where the next state decoder implements the function *ns*, the output decoder the function *out* and the memory elements store data based on the data type for s_0. The general hardware interpretation is visualized in Figure 2.

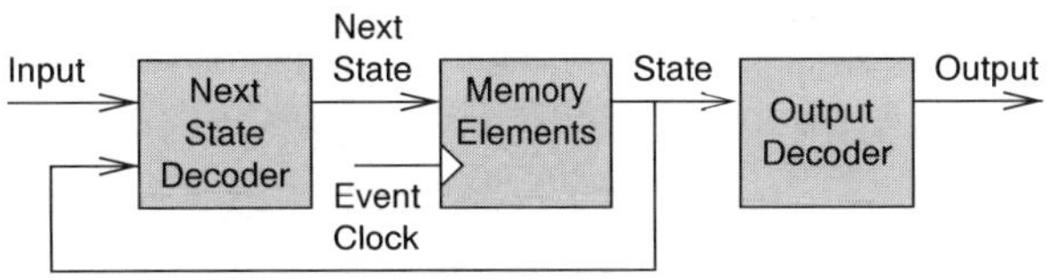

Figure 2: Hardware Interpretation of *mooreSY*

Functions can be glued together by function composition. A new function h is composed of two functions f and g by application of the composition operator $\circ$ according to

$$\begin{aligned} h &= f \circ g \\ h(x) &= f(g(x)) \end{aligned}$$

As processes are functions function composition can also be used for process composition. We can also build a network of processes by expressing it as a set of equations. Using this technique we can express a hierarchical system model.

4. REFINEMENT OF THE SYSTEM MODEL

The system model is stepwise refined through the use of well defined design transformations from an initial specification model S_0 to a final implementation model S_n (Figure 3).

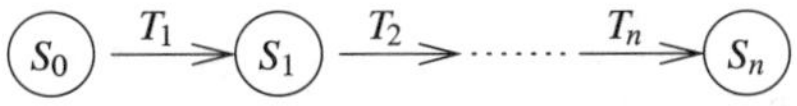

Figure 3: Refinement through Design Transformation

There are two classes of transformation techniques:

Semantic Preserving Transformations Semantic preserving transformations do not change the meaning of the model, i.e. the transformed model behaves in the same way as the original model. Semantic preserving transformations are mainly used to optimize the model for synthesis.

Design Decisions Design Decisions change the meaning of a model. A typical design decision is the refinement of an infinite buffer into a fixed-size buffer with n elements. While such a design decision clearly modifies the semantics, the transformed model may still behave in the same way as the original model. For instance, if it is possible to prove, that a certain buffer will never contain more than n elements, the ideal buffer can be replaced by a finite one of size n.

Before we illustrate the refinement methodology with the system model of the equalizer we want to point out, how powerful simple semantic-preserving transformations can be, as they are performed on a system model, that is a pure composition of functions. Figure 4 shows the power of the formal and functional approach. (1) Processes can easily be move over block borders and (2) two processes

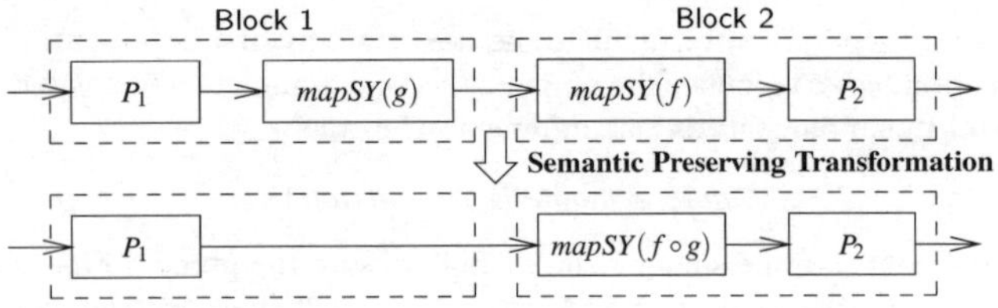

Figure 4: A Semantic Preserving Transformation

can be combined and possibly optimized by use of function composition. In this case we use the semantic-preserving transformation

$$mapSY(f) \circ mapSY(g) = mapSY(f \circ g) \qquad (1)$$

Another useful semantic-preserving transformation is

$$\Psi(f) \circ \Psi(g) = \Psi(f \circ g) \qquad (2)$$

To the best of our knowledge optimizations across process and block borders have not been reported for any other system modeling and design methodology. We believe it is very difficult to achieve in most other approaches because process boundaries tend to be very "hard" and moving a piece of functionality from one process to another would involve a major redesign of the processes and their interfaces. This is particularly true for heterogeneous approaches when the two processes are in different modeling domains. In contrast, in the ForSyDe methodology it is a comparably easy step because processes are semantically just like any other function.

We illustrate our refinement methodology by means of the system model of an equalizer which has also been described in [4]. The main task of the equalizer (Figure 5) is to adjust the audio signal according to the *Button Control*. In addition, the bass level must not exceed a predefined threshold to avoid damage to the speakers. This specification is naturally decomposed into four functions shown in Figure 5.

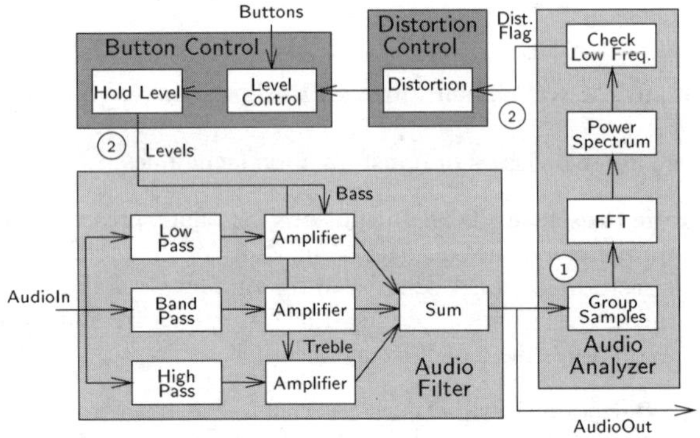

Figure 5: Subsystems of the Equalizer

The *Button Control* subsystem monitors the button inputs and the override signal from the subsystem *Distortion Control* and adjusts the current bass and treble levels. This information is passed to the subsystem *Audio Filter*, which receives the audio input signal, and filters and amplifies it according to the current bass and treble levels. This signal, the output signal of the equalizer, is analyzed by the *Audio Analyzer* subsystem, which determines, whether the bass exceeds a predefined threshold. The result of this analysis is passed to the subsystem *Distortion Control*, which decides, if a minor or major violation is encountered and issues the necessary commands to the *Button Control* subsystem. In the following we use the equalizer model to discuss two refinement techniques as indicated in Figure 5. In addition to the transformation techniques described in the following sections ForSyDe includes data type and memory refinement, which are beyond the scope of this paper.

4.1 Refinement of the Clock Domain

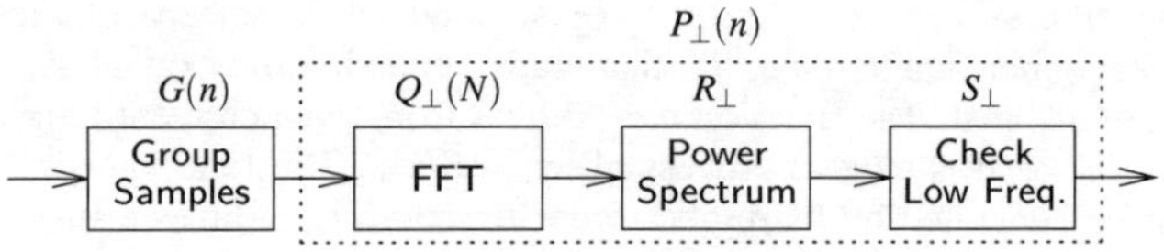

Figure 6: The AudioAnalyzer

Figure 6 shows the *Audio Analyzer* subsystem, which includes a Fast-Fourier Transform (FFT) algorithm. This function *fft* takes a vector of 2^k samples and produces the corresponding FFT result. Vectors are denoted as $< x_1, x_2, \ldots, x_n >$. The FFT algorithm is used to determine the frequency spectrum of a signal. It is implemented in the process *FFT* ($Q_\perp(n)$). The process *Power Spectrum* ($R_\perp$) calculates the power spectrum. The process *Check Low Frequencies* ($S_\perp$) analyzes if the power of the low frequencies exceeds a threshold and issues a warning in this case, which is sent to the *Distortion Control*. The FFT algorithm takes a vector of 2^k samples and calculates a frequency spectrum. The process *Group Samples* ($G(n)$) reads 2^k samples and groups them into a vector of size 2^k. However since we use a synchronous computational model in the system model the grouping process has to produce an output event for each input event. $G(n)$ is defined formally as

$$G(n)(\{x_1, \ldots, x_n, \ldots\}) = \{\underbrace{\perp, \ldots, \perp}_{n-1}, < x1, \ldots, x_n >, \ldots\}$$

As all processes $Q_\perp(n)$, $R_\perp$, $S_\perp$ are constructed with the skeleton *mapSY* we can use the equations 1 and 2 to replace these processes by $P_\perp(n)$.

$$\begin{aligned} p(n) &= s \circ r \circ q(n) \\ P_\perp(n) &= mapSY(\Psi(p(n))) \end{aligned}$$

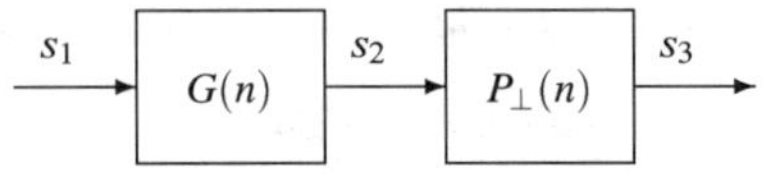

$$s_1 = \{x_1, x_2, \ldots, x_n, \ldots\}$$
$$s_2 = \{\underbrace{\perp, \ldots, \perp}_{n-1}, \underbrace{< x_1, \ldots, x_n >}_{1}, \ldots\}$$
$$s_3 = \{\underbrace{\perp, \ldots, \perp}_{n-1}, \underbrace{p(< x_1, \ldots, x_n >)}_{1}, \ldots\}$$

Figure 7: A Mathematical Model of the Audio Analyzer

We can now model the *Audio Analyzer* as illustrated in Figure 7.

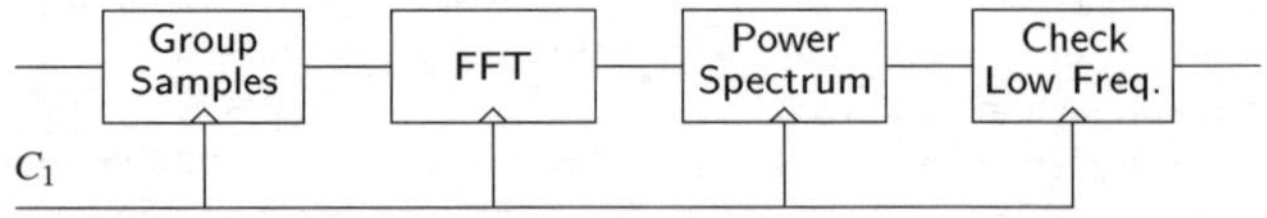

Figure 8: Direct Implementation of the Audio Analyzer

The process $P_\perp(n)$ has to process all absent values. This is not a drawback for the specification phase, but a direct implementation as shown in Figure 8 can make no use of the fact, that the FFT has only to be calculated at each 2^k-th clock cycle.

Such an implementation will be very slow, since the computation of the FFT function is clearly the most time consuming and will determine the overall system performance.

In order to get a more efficient specification the ForSyDe methodology allows to introduce synchronous sub-domains into the system model during the refinement process. These synchronous sub-domains use another set of tags. The introduction of a new sub-domain is done by well defined transformations, which are semantic-preserving though they introduce an additional timing domain into the model.

We present four processes, that can be used to introduce a synchronous sub-domain. There are two processes for *down-sampling* and two processes for *up-sampling*. Each of these processes comes in two versions, either with *head* or *tail synchronization*. The processes are defined formally as

$$U_H(n)(\{x_1, x_2, \ldots\}) = \{x_1, \underbrace{\perp, \ldots, \perp}_{n-1}, x_2, \underbrace{\perp, \ldots, \perp}_{n-1}, \ldots\}$$
$$U_T(n)(\{x_1, x_2, \ldots\}) = \{\underbrace{\perp, \ldots, \perp}_{n-1}, x_1, \underbrace{\perp, \ldots, \perp}_{n-1}, x_2, \ldots\}$$
$$D_H(n)(\{x_1, x_2, \ldots, x_n, x_{n+1}, \ldots\}) = \{x_1, x_{n+1}, \ldots\}$$
$$D_T(n)(\{x_1, x_2, \ldots, x_n, \ldots x_{2n}, \ldots\}) = \{x_n, x_{2n}, \ldots\}$$

ForSyDe allows only transformations, that use an *appropriate* composition of up-sample and down-sample processes. These combinations are well defined, e.g. the identities $D_H(n) \circ U_H(n)$ and $D_T(n) \circ U_T(n)$. An example for a *non-causal* and thus not allowed composition is shown in Figure 9.

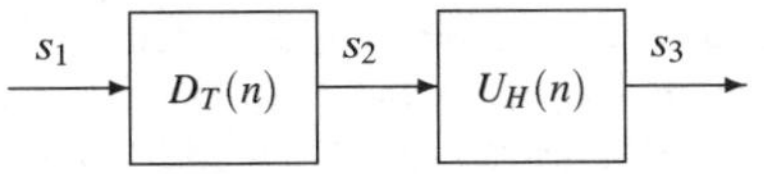

$$s_1 = \{x_1, x_2, \ldots, x_n, \ldots x_{2n}, \ldots\}$$
$$s_2 = \{x_n, x_{2n}, \ldots\}$$
$$s_3 = \{x_n, \underbrace{\perp, \ldots, \perp}_{n-1}, x_{2n}, \underbrace{\perp, \ldots, \perp}_{n-1}, \ldots\}$$

Figure 9: Non-Causal Composition of Sampling Processes

Using the special characteristic of the grouping process $G(n)$ we can derive the identity

$$G(n) = U_T(n) \circ D_T(n) \circ G(n).$$

Using another identity

$$P_\perp \circ U_T(n) = U_T(n) \circ P_\perp$$

it follows that

$$\begin{aligned} P_\perp(n) \circ G(n) &= P_\perp(n) \circ U_T(n) \circ D_T(n) \circ G(n) \\ &= U_T(n) \circ P_\perp(n) \circ D_T(n) \circ G(n) \\ &= U_T(n) \circ P(n) \circ D_T(n) \circ G(n) \end{aligned} \quad (3)$$

In the last step we have replaced $P_\perp(n)$ with $P(n)$ since $D_T(n) \circ G(n)$ does not produce any absent values. Analyzing equation 3 we conclude that the process $P(n)$ processes events only at each n-th tag and thus can be implemented with a slower clock.

Based on these considerations we define a semantic-preserving transformation that introduces a synchronous sub-domain. We implement a synchronous sub-domain with a clock frequency f_{C_2} that is $n = 2^k$ times slower than the clock frequency f_{C_1} of the main synchronous domain. The result of this transformation of the *Audio Analyzer* is illustrated in Figure 10.

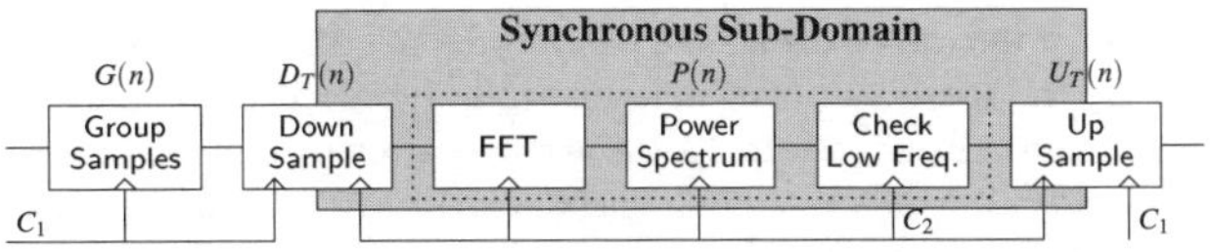

Figure 10: The Audio Analyzer after Refinement

4.2 Communication Refinement

The system model uses the same synchronous communication mechanism between all its subsystems. This is a nice feature for modeling and analyzing, since partitioning issues and special interfaces between subsystems have not to be taken into account in this phase. However, large systems are usually not implemented as one single unit, but are partitioned into hardware and software blocks communicating with each other via a dedicated communication protocol. The ForSyDe methodology offers transformations of a synchronous communication into other protocols. Looking at the equalizer example, we observe, that the aperiodic data rate of the *Button Control* and the *Distortion Control* subsystem is much lower than the data rate of the *Audio Filter* and *Audio Analyzer*. We decide to implement the *Button Control* and *Distribution Control* in software and the *Audio Filter* and *Audio Analyzer* in hardware. For the communication between these parts we implement a handshaking protocol with *Send* and *Receive* processes.

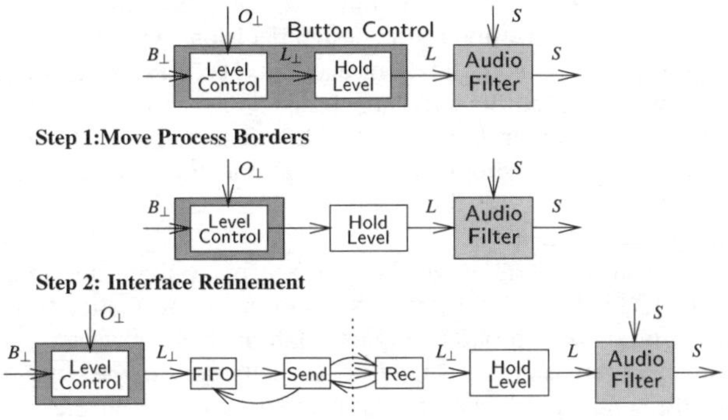

Figure 11: Refinement into a Handshake Protocol

In this paper we focus on the refinement of the synchronous interface between the *Button Control* and the *Audio Filter* subsystems, which is shown in Figure 11. The figure also shows the data types of the signals. Please note, that all the data types $B_\perp$, $O_\perp$, $L_\perp$ are extended data types, containing absent values. The block *Hold Level*, implemented by the synchronous skeleton *holdSY*, outputs the last valid value, when receiving an absent value as shown in Figure 12. Thus, it can also be viewed as a transformation unit between Ψ-extended and non-Ψ-extended data types.

The refinement is done in two steps. First, we move the block *Hold Level* out of the subsystem *Button Control* in order to implement the interface between the block *Level Control* and the block *Hold Level*. The second step is to refine the interface into a handshake protocol.

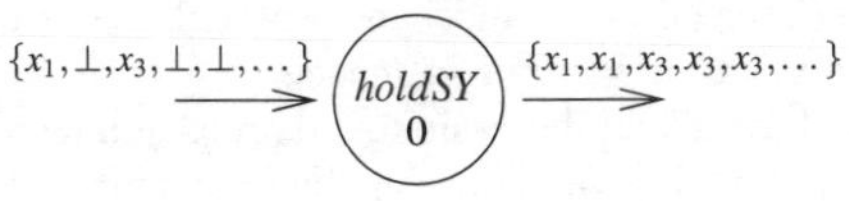

Figure 12: The Process *holdSY*

Here we introduce a FIFO, a *Send* and a *Receive* process. When *Send* is idle, it tries to read data from the FIFO. Then it sends the message `DataReady` to the *Receiver* and after receiving the message `Ready`, it sends the data. The *Receiver* sends a message `Ack`, when the data is received.

The handshake protocol implies a delay of several cycles for each event, as *Send* and *Receive* are synchronous processes. This means, that the timing behavior of the refined interface is different from the original interface. This does also mean, that the *Audio Filter* will not process exactly the same combination of values in each event cycle as in the system model.

These consequences have to be taken into account, when interfaces are refined. In this case, it can be shown that the refined interface still behaves in practice as the system model, if we make two assumptions.

1. The average data rate of the block *Level Control* is lower than the data rate of the *Audio Filter*. If the FIFO is correctly dimensioned there will be no buffer overflow in the FIFO and all values reach the *Audio Control* after a small number of event cycles.

2. The output function of the *Audio Filter* does not change significantly, if the input signals of the *Level Control* are delayed. That is clearly the case, as a small delay of the level signal only delays the change of the amplitude for the same small time, but does not effect the signals shape.

These assumptions point to obligations on other design activities. A further formalization of the design decisions will allow to make all assumptions and obligations explicit. The FIFO buffers have to be dimensioned sufficiently large based on a separate analysis. This will imply a further design decision as illustrated in [12]. Assumptions about the environment and the application, such as the kind of expected input signal, have to be validated to justify the applied design decisions.

We can now synthesize the interface by applying the methods described in [12]. The sole purpose of our transformation is to *prepare* for an asynchronous implementation. Note however, that the model we have derived is not truly asynchronous in the sense that it is still completely *deterministic* without nondeterministic channel delays. Of course, the channel can be modeled more realistically if desired. In the ForSyDe methodology we suggest to avoid a nondeterministic model but to use a stochastic channel model instead, which ForSyDe supports with stochastic skeletons [6].

5. CONCLUSION

This paper presents the refinement of an abstract system model into a more detailed implementation model in the ForSyDe methodology. Our main contribution is the development of a method, that allows the stepwise refinement of a system model into an implementation model by the application of well defined design transformations without leaving the functional domain. There are two classes of transformations, semantic-preserving transformations and design decisions. We have illustrated the use of both types of transformations with a digital equalizer system. By means of well defined transformations we refined by the system model with only one synchronous domain into a model with an additional synchronous clock domain. We also presented the formal refinement of a synchronous interface into a handshake protocol, that is used for communication via asynchronous channels.

We have formulated the basic foundations and techniques of the ForSyDe methodology and have applied them manually for several designs. We will continue our work with the development of tool support in order to automate the design flow.

6. REFERENCES

[1] F. Balarin, M. Chiodo, P. Giusto, H. Hsieh, A. Jurecska, L. Lavagno, A. Sangiovanni-Vincentelli, E. M. Sentovich, and K. Suzuki. Synthesis of software programs for embedded control applications. *IEEE Transactions on Computer-Aided Design of Integrated Circuits and Systems*, 18(6):834–849, June 1999.

[2] A. Benveniste and G. Berry. The synchronous approach to reactive and real-time systems. *Proceedings of the IEEE*, 79(9):1270–1282, September 1991.

[3] P. Bjesse, K. Claessen, M. Sheeran, and S. Singh. Lava: Hardware design in Haskell. In *International Conference on Functional Programming*, 1998.

[4] P. Bjuréus and A. Jantsch. MASCOT: A specification and cosimulation method integrating data and control flow. In *Proceedings of the Design and Test Europe Conference (DATE)*, 2000.

[5] S. Edwards, L. Lavagno, E. A. Lee, and A. Sangiovanni-Vincentelli. Design of embedded systems: Formal models, validation, and synthesis. *Proceedings of the IEEE*, 85(3):366–390, March 1997.

[6] A. Jantsch, I. Sander, and W. Wu. The usage of stochastic processes in embedded system specifications. In *Proceedings of the Ninth International Symposium on Hardware/Software Codesign*, April 2001.

[7] K. Keuzer, S. Malik, A. R. Newton, J. M. Rabaey, and A. Sangiovanni-Vincentelli. System-level design: Orthogonolization of concerns and platform-based design. *IEEE Transaction on Computer-Aided Design of Integrated Circuits and Systems*, 19(12):1523–1543, December 2000.

[8] E. A. Lee and A. Sangiovanni-Vincentelli. A framework for comparing models of computation. *IEEE Transactions on Computer-Aided Design of Integrated Circuits and Systems*, 17(12):1217–1229, December 1998.

[9] Y. Li and M. Leeser. HML, a novel hardware description language and its translation to VHDL. *IEEE Transactions on VLSI*, 8(1):1–8, February 2000.

[10] A. Pettorossi and M. Proietti. Rules and strategies for transforming functional and logic programs. *ACM Computing Surveys*, 28(2):361–414, June 1996.

[11] H. J. Reekie. *Realtime Signal Processing*. PhD thesis, University of Technology at Sydney, Australia, 1995.

[12] I. Sander and A. Jantsch. System synthesis based on a formal computational model and skeletons. In *Proceedings IEEE Workshop on VLSI'99*, pages 32–39, Orlando, Florida, April 1999. IEEE Computer Society.

[13] D. Skillicorn. *Foundations of Parallel Programming*. Cambridge University Press, 1994.

[14] K. Strehl, L. Thiele, M. Gries, D. Ziegenbein, R. Ernst, and J. Teich. FunState - an internal design representation for codesign. *IEEE Transactions on Very Large Scale Integration (VLSI) Systems*, 9(4):524–544, August 2001.

[15] W. Wu, I. Sander, and A. Jantsch. Transformational system design based on a formal computational model and skeletons. In *Forum on Design Languages 2000*, Tübingen, Germany, September 2000.

Model Composition for Scheduling Analysis in Platform Design

Kai Richter, Dirk Ziegenbein, Marek Jersak, Rolf Ernst
Institute of Computer and Communication Network Engineering
Technical University of Braunschweig
D-38106 Braunschweig / Germany
{richter|ziegenbein|jersak|ernst}@ida.ing.tu-bs.de

Categories and Subject Descriptors

C.0 [**General**]: System Architecture; C.3 [**Computer Systems Organization**]: Special-Purpose and Application-Based Systems—*real-time and embedded systems*; C.4 [**Computer Systems Organization**]: Performance of Systems

General Terms

Algorithms, Performance, Verification

Keywords

Platform-Based Design, Performance Analysis, Scheduling, Formal Analysis

Abstract

We present a compositional approach to analyze timing behavior of complex platforms with different scheduling strategies. The approach uses event interfacing in order to couple previously incompatible analysis techniques which provide subsystem and component behavior. Based on these interfaces, event propagation using abstract models is used to derive global system timing properties.

1. INTRODUCTION

Embedded system platforms consist of a combination of different processor types, specialized memories, and weakly programmable or fixed function components with a communication infrastructure consisting of busses, switches or point-to-point connections. Programmability and configurable software architectures are key to adapt platforms to a variety of applications. The verification of such platforms faces two main problems, verification of the system function and verification of the platform implementation, i.e. the adherence to system timing constraints, memory requirements, or power consumption. In this paper, we will focus on timing verification.

In traditional hardware design, system timing can usually be derived by hierarchical composition of individual component timing. This is possible since in most cases, component control is single threaded following a fixed control sequence which only depends on input patterns. Optimization is concerned with an optimal static schedule of operations. Behavioral synthesis closely follows this approach [7].

Embedded software adds process preemption to enable another class of scheduling strategies. Preemptive and time-driven scheduling introduces timing dependencies between functionally independent processes.

Heterogeneous platforms take the next step of target system complexity combining several preemptive and non-preemptive scheduling strategies in one system, e. g. a static schedule on a DSP and a priority-driven schedule on a microcontroller. Communication adds to behavioral complexity by introducing additional resource sharing strategies.

Buffering changes the memory requirements and can lead to internal event bursts.

The analysis of such a platform is a major challenge and is currently limited to simulation approaches, such as in VCC [1], or Seamless [11]. The known limitations of simulation such as incomplete coverage and corner case identification are aggravated since many of the design errors only result from system integration requiring detailed knowledge which is often not available to the integrator. As a result, simulation does not provide conservative system properties, in particular timing. In case certain timing properties have to be guaranteed, static formal analysis is the only option.

Target system timing analysis can be divided in two parts: analysis of a single process executed on a target system component (i. e. the core execution time), and analysis of resource sharing effects on the response time of a process based on given core execution times. For the first part, there are many recent contributions combining implicit or explicit program path analysis and cycle-true processor modeling, such as [9, 3, 17, 5]. In addition, process communication can be determined to analyze communication channel load and timing [17]. For the second part, there is a huge amount of work, mainly in the domain of real-time operating systems [10, 6]. These approaches capture system timing using equations and provide appropriate solution algorithms. Unfortunately, a single coherent scheduling strategy for a given system, whether single or multiprocessor, is assumed in literature. There are very few exceptions which consider special cases of more complex architectures, such as [13] analyzing response times for static-priority process scheduling combined with a TDMA bus protocol. However, it is doubtful that a general approach to a self-contained solution for arbitrarily complex platform architectures can be found, mainly because of the highly complex dependencies in such systems.

The approach presented in this paper is based on analysis coupling rather than finding self-contained solutions. We identify architecture components for which an appropriate analysis exists. Then, we combine these analysis approaches in order to obtain a compositional description of the complex system-level timing behavior. This is not an easy task, mainly because of the incompatibilities of the used event models. Most of the analysis approaches in literature assume certain input event models that lead to process or communication execution. In effect, the input events determine the system workload. In distributed platforms, the input events of one component result from the output events of the connected components. For our compositional approach, it is required that the output event models of one architecture component are compatible with the input event models of the connected components. Otherwise, they can not be reasonably coupled. Interestingly, the importance of output event models has been widely neglected. In this paper, we will first present ways to derive output event models from the existing analysis approaches. We will see that these output event models are usually different from the input event models. To overcome the problem of incompatible models at inputs and outputs, we use transformation functions to adapt event models. Finally, we show the actual composition of analysis approaches. We introduce the concept of analysis domains as the key entities for the composition process. These analysis domains are represented by abstract timing characteristics rather than detailed behavior.

The paper is organized as follows. The next section reviews existing work in the area of real-time scheduling analysis and event models. Section 3 contains some preliminaries: the derivation of output event models and further analysis assumptions. In Section 4, we show how analyzable subsystems (analysis domains) can be identified and represented. The actual composition is explained in Section 5. After an example of a complex analysis problem in Section 6, we conclude the paper with an outlook on future work.

2. STATE OF THE ART

2.1 Analyzing Scheduling Strategies

In the area of real-time operating systems, there are several substantially different resource sharing (scheduling) strategies. Preemptive scheduling based on static priorities (e. g. rate-monotonic), dynamic priorities (e. g. earliest deadline first), and time-slicing (e. g. time division multiple access and round robin) are among the most important strategies. For each of the mentioned strategies, a set of analysis approaches exist. The approaches take the scheduling strategy and the core execution times as input to a self-contained mathematical description in order to calculate conservative bounds on the response times of processes.

In the early 70s, Liu and Layland proposed a preemptive priority-driven scheduling to guarantee deadlines for periodic hard real-time processes [10]. They considered a static (Rate Monotonic) and a dynamic (Earliest Deadline First) priority assignment and provided a formal analysis framework for both. In [6], Kopetz and Gruensteindl proposed TTP (time triggered protocol) for communication scheduling in distributed systems and presented an analysis. TTP implements the TDMA (time division multiple access) scheduling strategy. Both contributions assume a periodic activation of processes. Recent extensions of the mentioned work allow periodic activation with jitter, e. g. [15], and arbitrary deadlines and burst [8] for static-priority scheduling. Sprunt et. al. [16] analyze the influence of sporadic process activation. Unfortunately, all mentioned approaches assume a single coherent scheduling strategy for a given system, whether single or multiprocessor. Very few exceptions consider special cases of more complex architectures, such as [13] analyzing response times for static priority process scheduling combined with a TDMA bus protocol.

So far, there is no general approach to a self-contained solution for arbitrarily complex systems. In contrast, we propose a compositional approach. Currently, this is not possible due to incompatible assumptions on the type of process activation, e. g. periodic, jitter, burst. In the existing analysis approaches, the activation of processes is usually captured using abstract event models.

2.2 Event Models

In the literature on real-time analysis, there are four event models of major importance. A simple and efficient assumption is a stream of *periodic* input events. Here, the arrival of events can be captured by a single parameter, the period T. Often, periodic events are allowed to deviate with respect to their period. This adds another parameter to the periodic model, the *jitter* (J). Other models capture *bursts* of events. A burst is characterized by a number of events (burst length b) within a given time interval (the outer period T). This outer period may also jitter (J). If known, a minimum time distance (the inter-arrival time t) between two successive events within a burst can be specified. *Sporadic* events are captured by the minimum inter-arrival time t, only. Gresser [4] provides a more general model. It introduces vectors of sequential time intervals. However, the model is not applicable to most of the analysis approaches.

In a recent publication [14], we have presented an approach to *interface* between different event models. Table 1 shows simple transformation functions, which we call event model interfaces (EMIFs), between these models. An EMIF transforms the representation of an event stream from one event model into the abstract parameters of another event model. For instance, a periodic event stream with the period T_X can be captured by the burst event model, when we set the burst length to $b_Y = 1$, the outer burst period to $T_Y = T_X$, and the minimum inter-arrival time to $t_Y = T_X$.

Note that such EMIFs do not modify the actual event streams, rather they transform the abstract representation of a single event stream. The transformations are uni-directional, and can not be found for all combinations of event models. For instance, an event stream with burst can not be captured by the parameters of a purely periodic event model. In this case, we need to adapt the event stream itself. This can be done by adding functionality to the system, the so called event adaptation functions (EAF). A periodic event stream with jitter can be re-synchronized by means of a buffer with a periodic output issue rate. The same applies to the re-synchronization of event streams with burst. For detailed information about EMIFs and EAFs and the corresponding formalisms, we refer to [14]. There, we also derive worst-case buffer sizes for the EAFs and additional event delays which result from re-synchronization.

3. PRELIMINARIES

For our proposed composition of existing analysis approaches, the output event models of one architecture component have to be compatible to the assumed input event models of the connected components. However, output event models have been widely neglected in literature, and the analysis approaches do not characterize them by themselves. Instead of going into the details of the analysis approaches, we rather use their abstract characteristics (input event models and response times) to derive the output event models.

3.1 Output Event Models

The basic idea of deriving output event models is simply based on abstract event propagation through architecture components. An input event activates a dedicated function inside the component. A corresponding output event will occur after the function is completed, i. e. after the corresponding response time. When having a constant response time, each event experiences the same propagation delay. Thus, the output model is identical to the input model. However, in complex software systems only upper and lower bounds for the response time will be given, e. g. due to data dependent process execution times or a varying number of preemptions by other processes. For a periodic input model, the output is not purely periodic anymore, but will experience a jitter that equals the difference between the upper (t^+_{resp}) and the lower (t^-_{resp})response time bound. If the input events already arrive with a certain jitter, this jitter is additionally propagated to the output. In other words, both jitters –the internal and the external– are superposed:

$$J_{\text{out}} = J_{\text{in}} + \underbrace{t^+_{\text{resp}} - t^-_{\text{resp}}}_{\text{internal jitter}}$$

A more complicated situation can be found for input event models with burst. An overview of how output event models can be derived from the analysis characteristics (input event model and response time) is given in Table 2.

Input Event Model	Output Event Model
sporadic:	sporadic:
t_{in}	$t_{\text{out}} = \max\left(t^-_{\text{resp}}, t_{\text{in}} - (t^+_{\text{resp}} - t^-_{\text{resp}})\right)$
periodic:	periodic with jitter:
T_{in}	$T_{\text{out}} = T_{\text{in}}, J_{\text{out}} = t^+_{\text{resp}} - t^-_{\text{resp}}$
periodic with jitter:	periodic with jitter:
$T_{\text{in}}, J_{\text{in}}$	$T_{\text{out}} = T_{\text{in}}$
	$J_{\text{out}} = J_{\text{in}} + t^+_{\text{resp}} - t^-_{\text{resp}}$
periodic with burst:	periodic with burst and jitter:
$T_{\text{in}}, b_{\text{in}}, t_{\text{in}}$	$T_{\text{out}} = T_{\text{in}}$
	$b_{\text{out}} = b_{\text{in}}$
	$t_{\text{out}} = \max\left(t^-_{\text{resp}}, t - (t^+_{\text{resp}} - t^-_{\text{resp}})\right)$
	$J_{\text{out}} = t^+_{\text{resp}} - t^-_{\text{resp}}$

Table 2: Deriving Output Event Models

3.2 Input Event Interfaces

Above, we have shown how output event models can be derived. In our compositional approach, we will use the output event models of one component as the input event model of the connected components, possibly incorporating model transformations. In Section 2, we mentioned that the assumed input event models of the existing analysis approaches can be categorized into four classes. Additionally, some of the analysis approaches require a few more properties. For

$EMIF_{X \to Y}$	Y=periodic	Y=jitter	Y=burst	Y=sporadic
X=periodic	$T_Y = T_X$ (identity)	$T_Y = T_X,\ J_Y = 0$	$T_Y = T_X,\ b_Y = 1,\ t_Y = T_X$	$t_Y = T_X$
X=jitter	—	$T_Y = T_X,\ J_Y = J_X$ (identity)	—	$t_Y = T_X - J_X$
X=burst	—	—	$T_Y = T_X,\ b_Y = b_X,\ t_Y = t_X$ (identity)	$t_Y = t_X$
X=sporadic	—	—	—	$t_Y = t_X$ (identity)

Table 1: Event Model Interfaces for the Transformation of Event Models

example, the analysis of Liu and Layland [10] requires that the process response times do not exceed the corresponding activation periods. Lehoczky [8] provides an analysis assuming the same input event model (periodic) but without the additional constraint. In general, assuming a certain model is not enough to analyze scheduling. We will therefore use the term "input event interface" to describe an "input event model" together with additional assumptions. Correspondingly, we use the term "output event interface". We assume these "input event interfaces" to be provided by the analysis techniques employed.

4. ANALYSIS DOMAINS

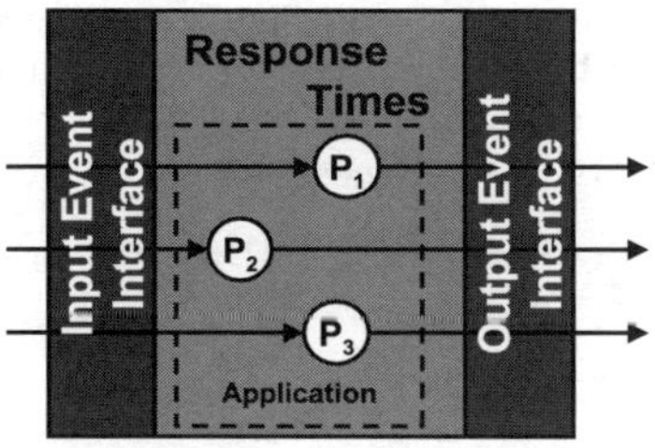

Input Event Interface:	event model purely periodic($T_{P_i,in}$)
	additional assumption from analysis in [10]: $\forall P_i : t^+_{P_i,resp} < T_{P_i,in}$
Response Times:	`GetResponseMax(), GetResponseMin()`
	Example for `GetResponseMax()` for the analysis in [10]:
	`GetResponseMax()` = smallest positive roots of:
	$t^+_{P_i,resp} = t^+_{P_i,core} + \Sigma_{\forall P_j \in HP(P_i)} \left(\left\lceil \frac{t^+_{P_i,resp}}{T_{P_j,in}} \right\rceil \times t^+_{P_j,core} \right)$
Output Event Interface:	periodic ($T_{P_i,out}$) with jitter ($J_{P_i,out}$):
	$T_{P_i,out} = T_{P_i,in}$ $J_{P_i,out} = t^+_{P_i,resp} - t^-_{P_i,resp}$

Figure 1: Representation of Analysis Domain

As described in the introduction, we assume that a platform consists of disjoint parts that use different scheduling strategies. In the following, we will call these parts scheduling domains. In the design, these scheduling domains are coupled via input and output events between domains. We have introduced event interfaces to describe the interfacing between these domains. We assume that for each scheduling domain there is an analysis technique that allows to derive the timing of the domain. As described before, there is a host of solutions in this field which can directly be employed. This disjoint set of scheduling domains defines a corresponding set of analysis domains. More precisely, an analysis domain is an architecture component or subsystem for which an analysis technique is available. Figure 1 shows an analysis domain example.

As previously mentioned, the availability of analysis approaches for an architecture component strongly depends on the scheduling strategy of the component. If several analysis approaches exist, as it is the case for preemptive static-priority scheduling, one candidate has to be selected. The input event interface of the component is constrained by the analysis. If we can find an analysis for the given scheduling strategy, and if the input events are compatible with the assumed interface, the analysis allows us to calculate conservative response times of processes. From these, we can further derive the output event interface, as explained in Section 3.1. The process of analyzing architecture components and deriving analysis domains is depicted in Figure 2.

5. DOMAIN COUPLING

As previously mentioned, the well known scheduling analysis approaches assume certain input event interfaces. But as mentioned in Section 2.2, the output event interfaces do not necessarily directly fit any of the possible input interfaces of the connected analysis domains. In this section, we explain how such seemingly incompatible analysis domains can be coupled. We start by finding reasonable ways to *adapt* event models for comparatively simple systems without any feed-back in the process graph as well as in the analysis itself. Then, we will hierarchically compose compatible domains. Finally, we will investigate several types of feed-back within this analysis process.

5.1 Feed-Forward

We use an example, depicted in Figure 3, to demonstrate the coupling of analysis domains. In this example, the analysis domains have already been selected. The system consists of two analysis domains consisting of one processor each, CPU_1 and CPU_2. Either of the processors executes two processes. The processes P_1 and P_2 are activated by events coming from the systems environment. On completion, they activate the processes P_3 and P_4 which themselves produce output to the environment. From the specification, we know that process P_1 is activated periodically with a fixed period $T_{P_1,in} = 40$ms. Process P_2 is also activated periodically ($T_{P_2,in} = 20$ms) but with a jitter $J_{P_2,in} = 5$ms. The processes on CPU_1 are scheduled according to the rate-monotonic priority assignment [10], while a round robin scheduler alternately assigns CPU_2 to either of its processes. The time slots for P_3 and P_4 are $t_{P_3,slot} = 5$ms and $t_{P_4,slot} = 3$ms, respectively. The core execution time intervals of the processes are $[15, 17]$ms for P_1, and $[8, 11]$ms, $[10, 11]$ms, and $[3, 5]$ms for the processes P_2, P_3, and P_4, respectively. We are looking for conservative (upper and lower) bounds on the response times for each event that is input to the system until a corresponding event is output to the environment.

For simplicity, we would like to use the approach of Liu and Layland [10] to analyze the processes on CPU_1. For the processes on CPU_2, we want to use the analysis from [2]. An interesting point is that we do not even need to know about the details of the mentioned analysis approaches. We only need to know the restrictions on the input event interfaces and the algorithms to calculate the response times. In contrast to finding a dedicated self-contained analysis for the given problem (like e. g. [13] does), the event model interfaces (EMIFs) from [14] will help us to combine both analysis domains with only little additional effort.

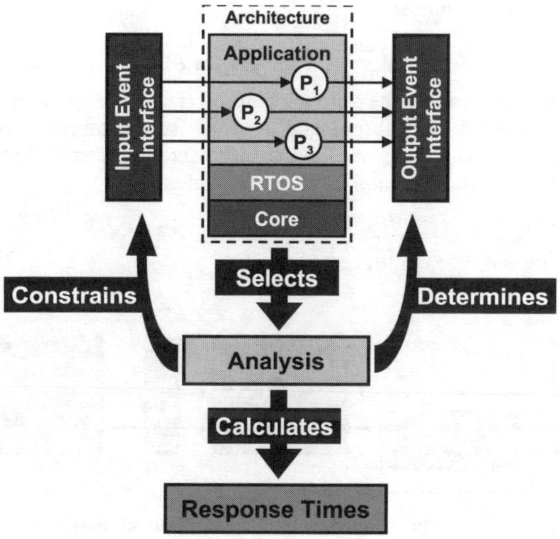

Figure 2: Local Analysis of Component

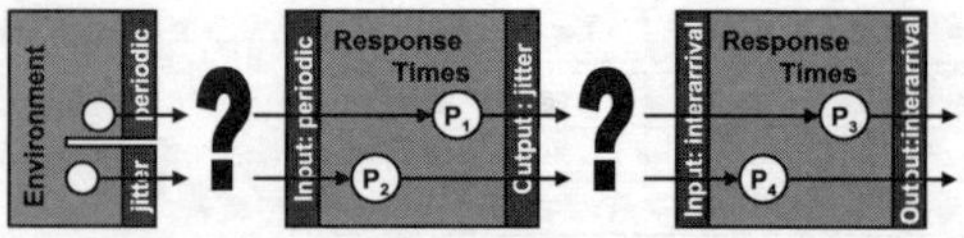

Figure 3: Coupling of Analysis Domains

In the example, the selected analysis approaches assume the input of CPU_1 to be purely periodic (this is not a limitation of the approach, just an example), while only minimum inter-arrival times are required to analyze the timing of CPU_2. Our first problem is the jitter on the input events for P_2. We can not use a simple model transformation, but have to transform the event stream itself by means of an appropriate event adaptation function. As explained in Section 2.2, we need an event buffer of size "one event" with an output issue rate of $T_{P_2,\text{in}} = 20\text{ms}$ to adapt the input event stream to the required input interface of the analysis domain of CPU_1. By this, we re-synchronize the event stream to its period and eliminate the jitter. As can be seen in Figure 4, the EAF is represented the same way as the analysis domains. Although they have no real function, i. e. they just copy input events to output events, they have a dedicated timing behavior in terms of input and output event interfaces, and a response time, which in this case equals the jitter of the input event stream. Now, we can do the selected analysis. We obtain the response times:

$$
\begin{aligned}
t_{P_1,\text{resp}} &= [23, 39]\text{ms} \\
t_{P_2,\text{resp}} &= [8, 11]\text{ms}
\end{aligned}
$$

Obviously, the response times are less than the corresponding periods, and the input interface is met. The output event model can be easily derived. We obtain two periodic event streams with jitter, as described in Section 3.1:

$$
\begin{aligned}
T_{P_1,\text{out}} &= T_{P_1,\text{in}} = 40\text{ms} \\
J_{P_1,\text{out}} &= t^+_{P_1,\text{resp}} - t^-_{P_1,\text{resp}} = 39\text{ms} - 23\text{ms} = 16\text{ms} \\
T_{P_2,\text{out}} &= T_{P_2,\text{in}} = 20\text{ms} \\
J_{P_2,\text{out}} &= t^+_{P_2,\text{resp}} - t^-_{P_2,\text{resp}} = 11\text{ms} - 8\text{ms} = 3\text{ms}
\end{aligned}
$$

Now, we have the output interface of the analysis domain of CPU_1. The next task is to meet the input interface of CPU_2, where minimum inter-arrival times $t_{P_3,\text{int,in}}$ and $t_{P_4,\text{int,in}}$ (known from sporadic process activation) are required. Again, we use the EMIF supplied by [14]. In contrast to the timed buffer for the input of P_2, we find out that the interface of CPU_2 can be met without the need of additional interface components. The event stream can be directly input. However, to be able to do the analysis for CPU_2, we have to transform the representation of the event stream. We have periodic event streams with jitter. We find the appropriate EMIF in Table 1 and obtain:

$$
\begin{aligned}
t_{P_3,\text{int,in}} &= T_{P_1,\text{out}} - J_{P_1,\text{out}} = 24\text{ms} \\
t_{P_4,\text{int,in}} &= T_{P_2,\text{out}} - J_{P_2,\text{out}} = 17\text{ms}
\end{aligned}
$$

In contrast to the re-synchronization of P_2's input events, no additional components are needed to transform the event stream, as depicted in Figure 4. We just add some math to the models. The response times calculated by the analysis in [2] are:

$$
\begin{aligned}
t_{P_3,\text{resp}} &= [13, 20]\text{ms} \\
t_{P_4,\text{resp}} &= [3, 15]\text{ms}
\end{aligned}
$$

Again, we can compute the output event interface. We again propagate the input model to the output and account for the above mentioned "internal" jitter. However, we have to *subtract* the internal jitter, since we are looking for *minimum* inter-arrival times:

$$
\begin{aligned}
t_{P_3,\text{int,out}} &= t_{P_3,\text{int,in}} - (t^+_{P_3,\text{resp}} - t^-_{P_3,\text{resp}}) = 17\text{ms} \\
t_{P_4,\text{int,out}} &= t^-_{P_4,\text{resp}} = 3\text{ms}
\end{aligned}
$$

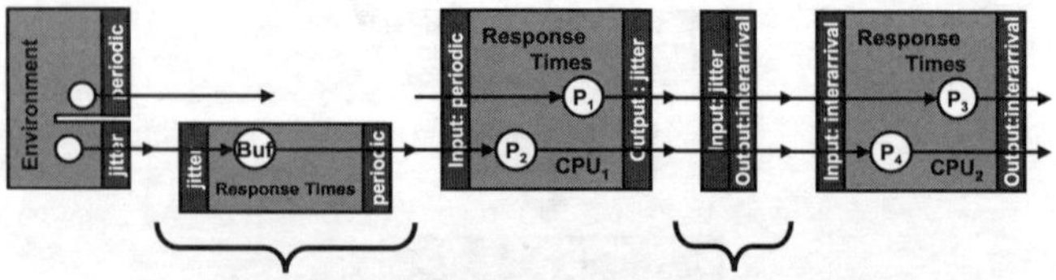

Figure 4: Use of EMIFs and EAFs

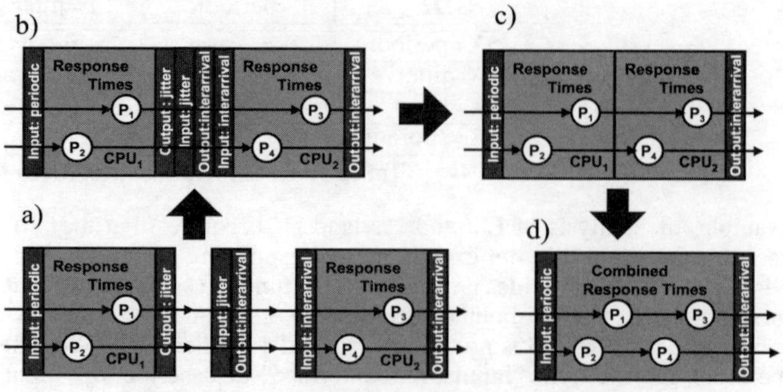

Figure 5: Composition of Analysis Domains

5.2 Composition

Now, after we have successfully added EMIFs and EAFs to the system, all output→input event interfaces match. In other words, no more adaptation is required, and the domains can be directly connected together, as shown in Figures 5(a) and (b). Now, we can eliminate all internal interfaces in our representation (Figure 5(c)) and calculate the accumulated response times along each path:

$$
\begin{aligned}
t_{P_1\to P_3,\text{resp}} &= t_{P_1,\text{resp}} + t_{P_3,\text{resp}} \\
t^+_{P_1\to P_3,\text{resp}} &= 59\text{ms} \\
t^-_{P_1\to P_3,\text{resp}} &= 36\text{ms} \\
t_{P_2\to P_4,\text{resp}} &= t_{P_2,\text{resp}} + t_{P_4,\text{resp}} \\
t^+_{P_2\to P_4,\text{resp}} &= 26\text{ms} \\
t^-_{P_2\to P_4,\text{resp}} &= 11\text{ms}
\end{aligned}
$$

We obtain a system, that is characterized by an input event interface (same as obtained for the original analysis domain CPU_1), an output event interface (obtained for domain CPU_2), and response times for the propagation of events. Following the definitions in 4, we obtain a single analysis domain for our entire sub-system (Figure 5(d)). This single domain can be further hierarchically combined with other domains, and so on. By this, we are now able to apply compositional timing analysis, which is well known from hardware design, to heterogeneous and highly dynamic preemptive software systems.

So far, we have shown, how analysis domains can be coupled and composed into larger analysis domains. We did so by selecting appropriate EMIFs and EAFs to adapt the event models and streams, respectively. We showed that internal EMIFs can eliminated after domain composition, since they do neither add functionality nor architecture to the system. They are just used to couple the analysis approaches. EAFs have to be treated differently when embedded within a newly composed analysis domain. First, they add functionality (buffering) which consumes time. Second, they add architecture (buffers, timers). These influences have to be considered during the composition process.

The pseudo-algorithm for the composition of analysis domains concludes this section. We assume that the scheduling domains (architecture components including RTOS or bus protocols) and the corresponding available analysis techniques have already been identified. Then, we can can define the analysis as an iterative process:

1. select scheduling domain for which all input event models are already determined
2. select analysis approach for scheduling domain (to obtain analysis domain)
3. if necessary, adapt event models (using EMIFs and EAFs)
4. determine process (or communication) response times and output event interfaces from corresponding analysis domain
5. determine output event interfaces
6. repeat from step 1 until all scheduling domains have been analyzed

With wisely selected analysis approaches and event interfaces for the domains, this algorithm will find a solution, if there exists one with respect to the compositional approach. In practice, one usually needs to back-track several times and modify these selections. Finally, the domains can be merged as shown in Figure 5.

5.3 Feed-Back

So far, only a simple feed-forward example was investigated. Selection of analysis methods was straight forward. Often, several analysis approaches are avaliable for a given problem. As previously mentioned, they substantially differ in the restrictions they impose on the input event interface. And they differ in complexity. Usually, the simpler the analysis, the more limited is the supported input event interface, and, as a result, the higher is the event adaptation overhead in terms of buffers, timers, etc. A reasonable design flow should balance between analysis complexity and input interface limitations. However, in the beginning of the analysis, it is often not clear which analysis is most suitable. Thus, similar to the actual design, also the analysis might consist of several cycles of selecting, rejecting and back-tracking of analysis decisions. Even in feed-forward systems, several analysis methods might be applicable for a given problem. However, re-analysis becomes really important when we have more complex systems including *cyclic* dependencies.

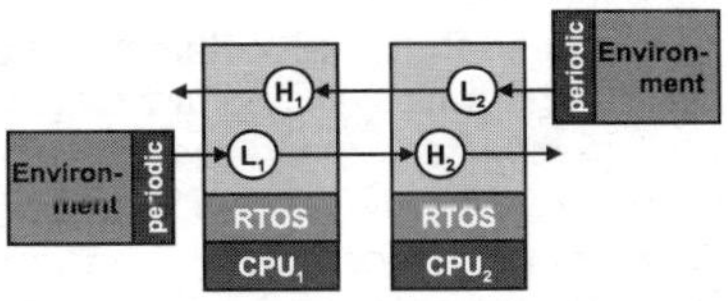

Figure 6: Example System with Cyclic Timing Dependencies

An example for a cyclic dependency of process behavior is given in Figure 6. The output of P_{L_1} (Low priority on component 1) activates P_{H_2} (High priority on component 2) which itself preempts P_{L_2} whose output activates P_{H_1}. And with P_{H_1} preempting P_{L_1}, all processes cyclically depend on each other with respect to timing. Note that this is a rather artificial example. However, similar but less obvious situations are likely to be found in large real-world designs. The challenge of resolving cyclic dependencies will be shown using a very simplified example.

From the specification, we know that the input events of P_{L_1} and P_{L_2} arrive periodically with the periods $T_{P_{L_1},\text{in}}$ and $T_{P_{L_2},\text{in}}$. Furthermore, we assume constant core execution times which are small enough not to violate the input event interface of Liu and Layland's analysis. In the feed-forward example from Sec. 5.1, we started the analysis from system inputs to system outputs by propagating event models. This is not possible here, since there is no dedicated starting point with all parameters given. Hence, we need to start the analysis with making assumptions on the unknown events.

We may start simply by assuming periodic input events for CPU_1. We work from P_{L_1}. The period of P_{H_1} is not yet known. Thus, we just ignore P_{H_1}. As a result, P_{L_1} is not preempted, and its output is periodic, too. Now, we can analyze the timing behavior of CPU_2. We know the two input event models (both periodic), and analyze the behavior. We find out that P_{L_2} can be preempted by P_{H_2}, resulting in a jitter for the output of P_{L_2}, and our initial assumption about periodic input models for P_{H_1} is invalidated. Thus, we reject our initial assumption and try again starting with periodic events with jitter. This procedure has to be continued until the input event interfaces converge. This is not necessarily the case. We carried out two simple experiments. In both experiments, the input periods of the environmental events are $T_{P_{L_1},\text{in}} = 20\text{ms}$ and $T_{P_{L_2},\text{in}} = 30\text{ms}$. We use a very simplistic response time analysis for demonstration. At each activation, we iteratively calculate the worst-case number of process preemptions similar to the approach in [10]. The best-case number of preemptions is set to zero. Experiment one assumes the core execution times of all four processes to be 9ms, while these are 11ms in experiment 2. We start at P_{L_1} and follow the dependencies until we reach a convergence point, i. e. we meet a previously assumed event interface after one iteration. At each step, we calculate new internal jitters. The results of experiment 1 can be seen in Table 3. We see, that the algorithm terminates after three steps, i. e. the input jitters of P_{H_1} and P_{H_2} do not change anymore. Experiment 2 (without table) does not converge. The input jitters of P_{H_1} and P_{H_2} alternatingly increase and can not be bounded. Hence, the input interface assumed by the selected analysis is not met.

Step	$J_{P_{H_2},\text{in}} =$ $J_{P_{L_1},\text{out}}$	$J_{P_{H_1},\text{in}} =$ $J_{P_{L_2},\text{out}}$	comment
INIT	0	—	P_{L_1} not preempted
1	0	**9**	P_{L_2} preempted by P_{H_1}
2	**9**	9	P_{L_1} preempted by P_{H_2}
TERM	9	**9**	P_{L_2} preempted by P_{H_1}

Table 3: Experiment 1: $t_{P_i,\text{core}} = 9\text{ms}$

The two experiments show that, while purely feed-forward event propagation through properly described local analysis domains will always lead to a solution, the feedback situation is more complicated (as it is in practice). There are cases where the feedback approach used in the example does not lead to a solution (i. e. a convergence), even though there exists a valid schedule in practice. This can be proved by analyzing the average processor utilization. With 91,67% ($\frac{55}{60}$), the system is in fact schedulable, i. e. there is a point in time where total system workload will be zero and buffers are empty. However, response time analysis is more complicated than only guaranteeing schedulability.

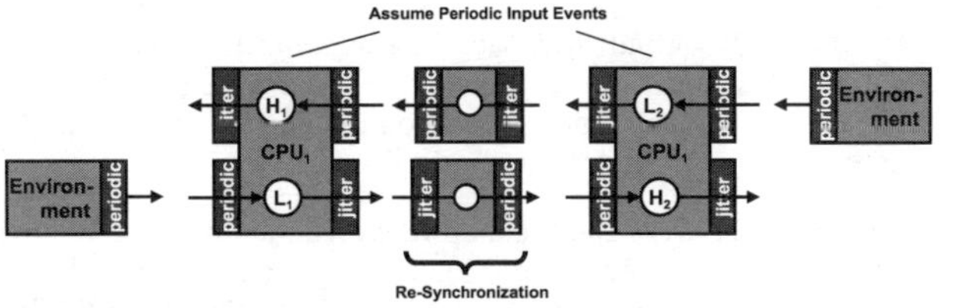

Figure 7: Cyclic Dependencies Resolved by Re-Synchronization

This effect is known from schedulability analysis and is due to the abstraction underlying the event propagation. It is often possible to enforce convergence by inserting buffers for synchronization, as depicted in Figure 7. Here, analysis domains with periodic input event interfaces can be used. Trivially, the iterative propagation of event models terminates after the first cycle. The introduction of buffers to decouple strong interdependencies is what an experienced designer would do in a manual design. As a key advantage of our approach, we can automatically synthesize the required adaptation function for a given analysis domain.

A similar situation can be found when the processes graph itself contains cycles, e. g. feed-back loops in control engineering. The same problems apply. However, cyclic process graphs are much better understood than cyclic non-functional dependencies. Designers are used to insert buffers to store events within the cycle in order to avoid deadlock. Finding a general theory for the feedback case is part of our current research.

6. EXAMPLE

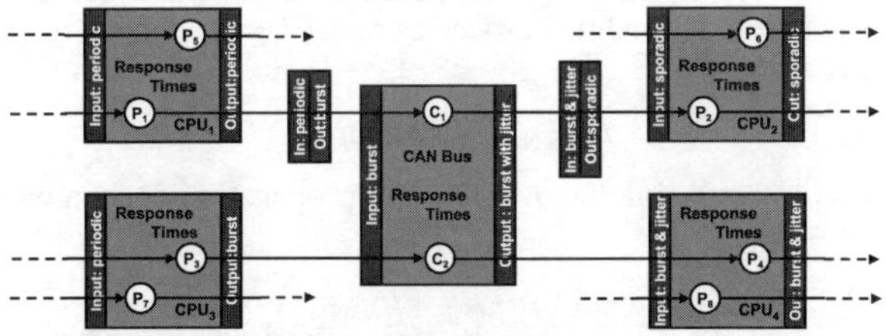

Figure 8: Example of Four Processors and One Bus

Figure 8 shows the analysis domains of a complex heterogeneous platform architecture. It consists of four processors which are connected via a CAN bus [12]. Processor CPU_1 runs a deterministic static scheduler. CPU_2 runs a round robin scheduler. The other processors run static-priority schedulers. The arbitration of packets on the CAN bus is based on static priorities for communicated packets. We have already select appropriate analysis approaches for each of the five analysis domains.

Process P_1, implemented on CPU_1, sends data to process P_2 on CPU_2 over the communication channel C_1. Similarly, processes P_3 and P_4 exchange data via C_2. The other four processes represent the resource sharing influences on the processors. However, they do not use the bus.

From the environment (indicated by the dashed lines) we know that P_1 is activated periodically ($T_{P_1,in} = 10ms$). The timing behavior of CPU_1 is completely determinate, e. g. periodic execution of simple digital signal processing functions with a fixed response time ($t_{P_1,resp} = 4ms$). Thus, the output is purely periodic, too ($T_{P_1,out} = 10ms$). P_3 is also activated periodically ($T_{P_3,in} = 10ms$). However, is has a low priority on CPU_3 and is heavily preempted by P_7 resulting in a bursty output behavior ($T_{P_3,out} = 100ms, t_{P_3,int,out} = 2ms, b_{P_3,in} = 10$). On the bus, the channel C_2 has high priority, while C_1 has low priority. We are interested in the timing behavior of the path $P_1 \rightarrow C_1 \rightarrow P_2$.

In general, two corner-case situations on the bus can be distinguished:

- P_3 outputs a burst of packets to the bus. Since the corresponding channel C_2 has a higher priority, the (lower-priority) packets from P_1 are blocked until the end of the burst. The communication delay of channel C_1, and thus, the overall latency of the path under consideration is high.
- P_3 does not output any packets for some time between two bursts (because it is preempted by P_7). Packets coming from P_1 are not blocked, communication delay on C_1 is low, and the path latency will be low, too.

First, we have to analyze the bus before we can analyze the behavior of P_2. We apply our event model interfacing technique to adapt the event models between P_1 and C_1. From Table 1, we obtain the burst representation of P_1's output: $T_{C_1,in} = 10ms, t_{C_1,int,in} = 10ms, b_{C_1,in} = 1$. The transmission of each packet over the bus constantly requires 5ms. For the analysis, we can directly use Lehoczky's [8] approach. The fact that single bus packets can not be preempted regardless of the priority is considered assuming semaphore locking. As a result, we obtain:

$$\begin{aligned} t_{C_1,resp} &= [5ms, 30ms] \\ t_{C_2,resp} &= [5ms, 28ms] \end{aligned}$$

We see that, although the C_1-packets are to be sent over the bus periodically, the packet response time heavily varies. The best- and worst-case packet response times correspond to the above distinguished situations. From the analysis, we also obtain C_1's output behavior. There are $b_{C_1,out} = 10$ packets within an outer period of $T_{C_1,out} = 100ms$. This results from the influence of C_2-bursts. The minimum inter-arrival time of communicated packets equals the packet transmission delay ($t_{C_1,int} = 5ms$). In other words, the bursty behavior on C_2 turns the periodic send-requests on C_1 into burst communication with similar characteristics.

Now, we can analyze the behavior of P_2. The event model interface (EMIF) between C_1 and P_2 is taken from Table 1, and the analysis of the round robin scheduled processes on CPU_2 results in a response time interval for P_2:

$$t_{P_2,resp} = [7ms, 15ms]$$

It remains to determine the overall event propagation delay along the path $P_1 \rightarrow C_1 \rightarrow P_2$:

$$\begin{aligned} t_{P_1 \rightarrow C_1 \rightarrow P_2,resp} &= t_{P_1,resp} + t_{C_1,resp} + t_{P_2,resp} \\ &= 4ms + [5ms, 30ms] + [7ms, 15ms] \\ &= [16ms, 49ms] \end{aligned}$$

7. CONCLUSION

We have presented a novel compositional approach for scheduling analysis in platform design. We identified analysis domains, i. e. architecture components, scheduling strategies, and appropriate analysis approaches. Using the concept of abstract event adaptation and propagation, we were able to couple initially incompatible scheduling analysis approaches. After coupling, we hierarchically combined the analysis domains. As a result, we were able to formally analyze the timing behavior of complex platforms, combining several preemptive and non preemptive scheduling strategies in one system. The variety of avaliable event model interfaces and adaptation functions shows the generality of our approach.

Our contribution provides a new quality of timing verification in platform design. Conservative bounds on system response times can be provided for highly complex, heterogeneous platforms. We are currently implementing the analysis procedure presented in this paper, including a repository for efficiently storing, identifying, and executing the existing analysis approaches.

Our current research in this area includes but is not limited to finding heuristics to resolve cyclic analysis domain dependencies, and extending the set of event adaptation functions.

8. REFERENCES

[1] Cadence. *Cierto VCC Environment.* http://www.cadence.com/products/vcc.html.

[2] Beatriz Asensio Calvo. Real-time analysis of time-driven scheduling – a quantitative comparison of Round Robin and TDMA. Technical report, Technical University of Braunschweig, September 2001.

[3] Christian Ferdinand and Reinhard Wilhelm. On predicting data cache behavior for real-time systems. In *Proceedings of International Conference on Languages, Compilers, and Tools for Embedded Systems (LCTES)*, pages 16–30, 1998.

[4] K. Gresser. An event model for deadline verification of hard real-time systems. In *Proceedings 5th Euromicro Workshop on Real-Time Systems*, pages 118–123, Oulu, Finland, 1993.

[5] A. Hergenhan and W. Rosenstiel. Static timing analysis of embedded software on advanced processor architectures. In *Proceedings of Design, Automation and Test in Europe (DATE '00)*, pages 552–559, Paris, March 2000.

[6] H. Kopetz and G. Gruensteidl. TTP - a time-triggered protocol for fault-tolerant computing. In *Proceedings 23rd International Symposium on Fault-Tolerant Computing*, pages 524–532, 1993.

[7] David C. Ku and Giovanni de Micheli. *High-Level Synthesis of ASICs Under Timing and Synchronization Constraints.* Kluwer Academic Publishers, Boston, Massachusetts, 1992.

[8] J. Lehoczky. Fixed priority scheduling of periodic task sets with arbitrary deadlines. In *Proceedings Real-Time Systems Symposiom*, pages 201–209, 1990.

[9] Yau-Tsun Steven Li and Sharad Malik. *Performance Analysis of Real-Time Embedded Software.* Kluwer Academic Publishers, 1999.

[10] C. L. Liu and J. W. Layland. Scheduling algorithms for multiprogramming in a hard-real-time environment. *Journal of the ACM*, 20(1):46–61, 1973.

[11] Mentor Graphics. *Seamless Co-Verification Environment.* http://www.mentorg.com/seamless/.

[12] Philips Semiconductors. *Controller Area Network CAN.* http://www.semiconductors.philips.com/can/.

[13] P. Pop, P. Eles, and Z. Peng. Bus access optimization for distributed embedded systems based on schedulability analysis. In *Proc. Design, Automation and Test in Europe (DATE 2000)*, Paris, France, 2000.

[14] K. Richter and R. Ernst. Event model interfaces for heterogeneous system analysis. In *Proc. of Design, Automation and Test in Europe Conference (DATE'02)*, Paris, France, March 2002.

[15] L. Sha, R. Rajkumar, and S. S. Sathaye. Generalized rate-monotonic scheduling theory: A framework for developing real-time systems. *Proceedings of the IEEE*, 82(1):68–82, January 1994.

[16] B. Sprunt, L. Sha, and J. Lehoczky. Aperiodic task scheduling for hard real-time systems. *Journal of Real-Time Systems*, 1(1):27–60, 1989.

[17] F. Wolf and R. Ernst. Execution Cost Interval Refinement in Static Software Analysis. *Journal of Systems Architecture.*

Timed Compiled-Code Simulation of Embedded Software for Performance Analysis of SOC Design

Jong-Yeol Lee
Division of EE, Dept. of EECS, KAIST
373-1, Guseong-dong, Yuseong-gu,
Daejeon, Republic of Korea 305-701
jylee@ics.kaist.ac.kr

In-Cheol Park
Division of EE, Dept. of EECS, KAIST
373-1, Guseong-dong, Yuseong-gu,
Daejeon, Republic of Korea 305-701
icpark@ics.kaist.ac.kr

ABSTRACT
In this paper, a new timing generation method is proposed for the performance analysis of embedded software. The time stamp generation of I/O accesses is crucial to performance estimation and architecture exploration in the timed functional simulation, which simulates the whole design at a functional level with timing. A portable compiler is modified to generate time-deltas, which are the estimated cycle counts between two adjacent I/O accesses, by counting the cycles of the intermediate representation (IR) operations and using a machine description that contains information on a target processor. Since the proposed method is based on the machine-independent IR of a compiler, the method can be applied to various processors by changing the machine description. The experimental results show that the proposed method is effective in that the average estimation error is about 2% and the maximum speed-up over the corresponding instruction-set simulators is about 300 times. The proposed method is also verified in a timed functional simulation environment.

Categories and Subject Descriptors
J.6 [**Computer-aided Engineering**]: Computer-aided design (CAD)

General Terms: Algorithms, Performance, Design

Keywords: Timed functional simulation, compiler, SOC

1. INTRODUCTION
According to advances in design methodology and semiconductor technology, many of today's complex embedded systems are implemented as system-on-a-chips (SOCs). In such an SOC, processor cores and other system components available as intellectual properties (IPs) are integrated on a single chip. In addition to the design complexity, the lifetime of a design becomes shorter than ever before. It is very important, therefore, to reduce the design time as minimal as possible in order to fit the design in a narrow time-to-market window. Since software design is more flexible than hardware design and design errors can be corrected more easily even in the later design stage, an increasing amount of system functionality is being implemented in software running on the processor cores. An example of an SOC is shown in Fig. 1, where a processor, memories and functional blocks are connected with a system bus.

In SOC design, the fast and accurate performance analysis of a target system at a higher level of abstraction is greatly required to reduce the exploring time of the design space of the target system and to obtain guidelines on proceeding to the lower level design. As the processor core and other system components are connected with one or more buses in an SOC, the bus contention is one of the major factors that determine the target system performance. To identify how and when the bus contention occurs, the analysis of the timing at which the system components including the software component accesses the bus is required.

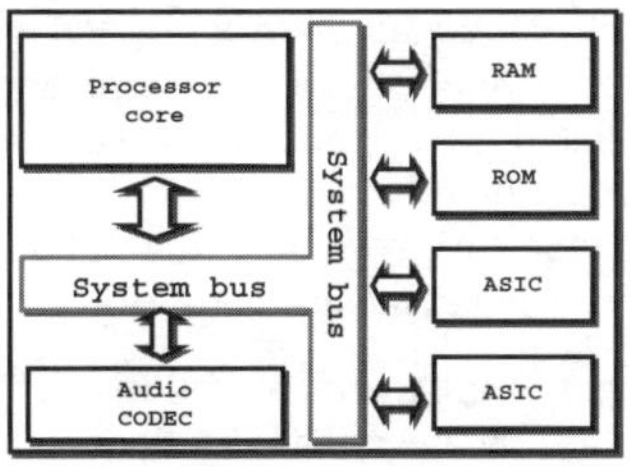

Fig. 1. Example of an SOC.

Various techniques have been proposed for the estimation of software timing in the mixed hardware/software systems [1]-[7]. Some works mandate long execution times since they use instruction-set simulators for cycle-accurate results. The authors of [1] addressed the problem of annotating the original C code with timing estimates guessing the compiler behavior. However, their method shows good results only on the codes generated by POLIS system since the codes always have the same structure. In [2], GNU C compiler is utilized to generate "assembler-level" C code, which can be annotated with timing information and used as a simulation model. This approach needs to run the simulation model and different simulation models for other processors. In [3], an annotated C code is generated from a functional C code or the object code generated by a target compiler and the annotated C code is simulated on a separate simulator. In Seamless [6], a method called "Compiled Code Execution" is supported. In the method, the embedded software is compiled for a host machine instead of a target CPU. As the software executes on the host, the I/O transactions are trapped and selectively translated into bus cycles that excite the surrounding hardware. The results of the method may be inaccurate since there is no way to make up for the discrepancy between the host-machine and target CPU. Innoveda's Virtual-CPU [7] also provides compiled-code execution capability. However, no timing information is available from the compiled-code execution of the embedded software.

A simulation technology called timed functional simulation can make it possible to simulate the whole design at a functional level with timing by estimating the timing of the software components. In timed functional simulation, only the accesses in the software component to other hardware components (peripherals) are considered and the memory-mapped I/O method is assumed. Furthermore, since the memory location mapped to peripherals must remain unchanged throughout the program execution, they must be declared in the source code as specially qualified global variables which are called *memory/IO variables*, and should not participate in optimizations that would increase, decrease, or delay any reference to the variables since the variables can have their values altered in ways not under the control of user code [10].

In this paper, we introduce a new method to generate the timings for memory/IO accesses by using a compiler. The IR codes generated for various processors are virtually all the same regardless of processors for the same source code [8]. The proposed method exploits compiled-code execution and shows decreased execution time compared to instruction-set simulators. A compiler is modified to generate executables that run on a host machine. The executable estimates and reports the times at which memory/IO accesses occur by counting the execution cycles of the IR operations between adjacent accesses to memory/IO variables.

For accurate estimation, the proposed method needs the machine description of a target machine. One section of the machine description is a table called an *operation table*, which specifies the size and execution cycle of each IR operation when the IR operation is translated into machine instructions. The machine description also contains other information such as the type of a processor, the number of registers, the number and kinds of functional units, and the latency of memory.

The rest of the paper is organized as follows. The proposed timed functional simulation is described in Section 2. The proposed time-delta generation method and its two inputs are explained in Section 3 and Section 4, respectively. In Section 5, two enhancements made to increase the accuracy the time-delta generation are described. After showing the experimental results in Section 6, conclusions are addressed in Section 7.

2. TIMED FUNCTIONAL SIMULATION

Fig. 2(a) shows a conceptual block diagram of timed functional simulation. For timed functional simulation, an algorithm described in a high-level language is compiled into an executable, called a wrapper, that runs on a host machine and has a time-delta generation function. A time-delta is an estimated cycle count between two adjacent memory/IO variable access operations. An example of time-deltas, δ_1, δ_2 and δ_3, is shown in Fig. 2(b), where OP_1, OP_2, and OP_3 are memory/IO operations. Let us assume that OP_1 is a transfer to P_1, OP_2 is an access to MEM that is missed in CACHE, and OP_3 is an access to MEM that is hit in CACHE. Only the timing between memory/IO operations is estimated in the time-delta generation since it is determined in software execution. To estimate the time-delta, the wrapper counts the execution cycles of IR operations between two memory/IO operations. Since the IR code is usually generated for scalar processors, the IR operations are scheduled to increase the accuracy of time-deltas for superscalar or VLIW processors. The wrapper also contains routines necessary for interrupt processing. The bus functional model (BFM) translates the memory/IO accesses into real chip signals according to the target CPU specification. The access time of memory/IO operations is determined by cache simulation, memory latency and the BFM that models the interaction between a target CPU and peripherals. For example, the memory latency determines the timing of the operations that load or store data from/to memory. Fig. 2(c) shows the timing in a timed functional simulation in which time-deltas shown in Fig. 2(b) are simulated. The access time of OP_1, T_1 is determined in the BFM by simulating the bus signals of an embedded processor. For OP_2 and OP_3, the accesses are checked in CACHE to decide whether the data is in CACHE. These checking times are denoted as T_2 and T_4. As the OP_2 causes a cache miss, a memory access is invoked through the BFM, which is denoted as T_3. Therefore, the access time of OP_2 is the sum of T_2 and T_3.

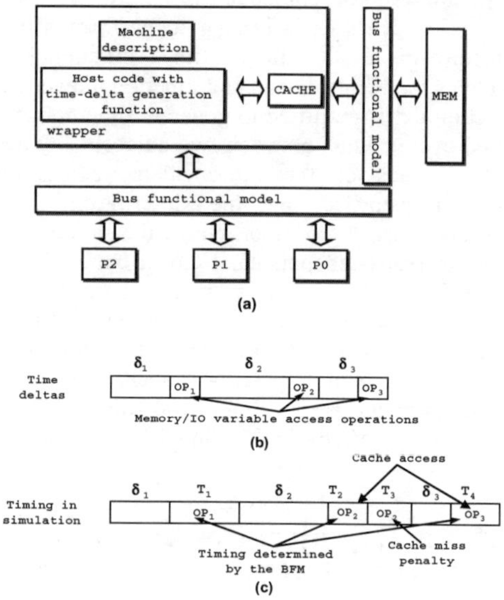

Fig. 2. (a) Conceptual block diagram of timed functional simulation, (b) Time-deltas, and (c) Timing in simulation.

The memory block (MEM) and peripherals (P0, P1, and P2) in Fig. 2(a) can be software models in high-level languages such as C, hardware description language (HDL) models or a mixture of software models and HDL models. By changing the machine description and Verilog tasks called in the BFM, other target CPUs can be easily simulated. One of the advantages of timed functional simulation is that the behavior and timing of the memory/IO signals of a target CPU can be simulated although the simulation is performed at functional level.

3. TIME-DELTA GENERATION

In this paper, a method based on the portable compiler concept is proposed [8]. Since general compilers focus on the functionality and optimization of the compiled code and have no functionalities related to the timing generation of the compiled code, it is necessary to modify compilers to generate the time-deltas. To generate the time-deltas, methods to indicate the memory/IO operations and to estimate the cycles elapsed between the indicated memory/IO operations must be provided. The proposed method is based on the IR of a compiler in order to exploit the portability of compilers. An optional field, called *an indicator field*, is added in the IR data structure to indicate the locations of memory/IO operations in the IR code. The indicator fields are set to the name of the variables that are accessed in the memory/IO operations identified by analyzing and searching the IR code for the memory/IO variables. A variable in the IR code is accessed either by directly referring the variable name or by loading the address of the variable into a register. Therefore, the IR accessing the designated variables can be identified by searching the IR operation related to the variable names or by analyzing the IR code to find registers used for storing the addresses of the designated variables. The cycle counts between memory/IO operations are estimated by accumulating the cycle counts of IR operations. Since the cycle count of an IR operation is determined by a target machine, the cycle count can be obtained from the machine description. To record the obtained cycle counts in the IR operations, another newly defined optional field, called a *counter field*, is added in the IR data structure. An example of the data structure of an IR operation is shown in Fig. 3(a), where counters for IR operations and indicators for memory/IO accesses are represented as two optional fields, `TSGEN_COUNTER` and `TSGEN_INDICATOR`.

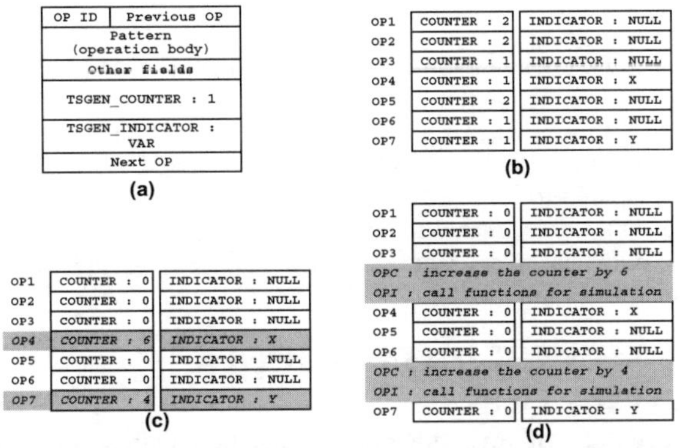

Fig. 3. (a) Data structure of IR operation with counter and indicator fields. The shaded fields represent optional fields. (b) Sequence of IR operations. (c) Sequence of IR operations after merging optional fields in (b). (d) Sequence of IR operations obtained by inserting IR operations, OPC and OPI, for counters and indicators.

After the optional fields are set for each IR operation, the optional fields are merged to save execution time if there is no memory/IO access in a straight-line code such as basic blocks. A sequence of IR operations after merging optional fields is shown in Fig. 3(c), where the values of the counter fields of OP4 and OP7 are replaced by the sum of the counter fields of the preceding operations and the counter fields of the preceding operations are set to zero.

The merged optional fields are translated into IR operations in the final compilation step so that the added IR operations should not affect the optimization results. The counter fields are translated into IR operations that increase the total cycle counts by the cycle count in the counter field. The total cycle counts is stored in a global variable, `tsgen_total_cycles`, defined in the library used in time-delta generation. The indicator fields are translated into IR operations that calculate the time-deltas using the value of `tsgen_total_cycles` at the previous indicator and the current value of `tsgen_total_cycles`. The IR operations translated from the indicator field also contain calls to necessary functions at each memory/IO access to simulate the interaction with the outside of the target CPU. For example, when the target CPU has a data cache, a cache simulation function is called. In Fig. 3(d), a sequence of IR operations obtained by inserting IR operations, OPC and OPI, for counters and indicators is shown. The first OPC increases the total cycle counts by six that is the value of the counter field of OP4. The operations, OPIs, are inserted for indicator fields whose values are not null. After inserting appropriate IR operations, the IR code is translated into machine instructions.

The proposed approach has two advantages. First, since the proposed method is based on the IR code that is independent of target machines, it can be easily applied to different processors with a small modification. The second advantage is that the estimated cycle counts are more accurate than other methods because the IR operations can be modified in such a way that the effects of compiler optimizations are reflected in the time-delta generation.

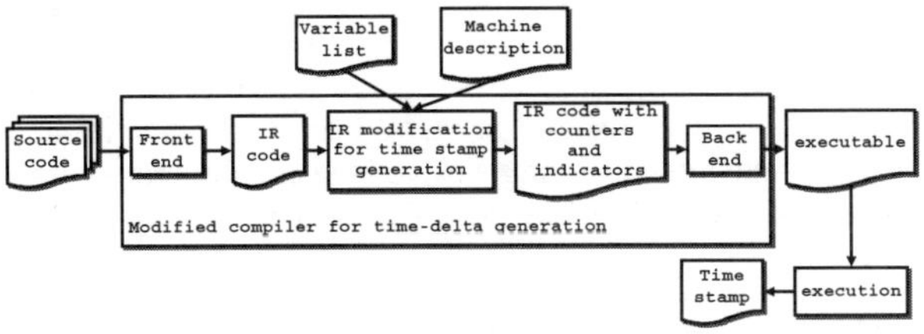

Fig. 4. Overview of time-delta generation.

Fig. 4 shows an overview of the proposed time-delta generation method. The variables to be observed are registered in the variable list. The machine description contains the information on the target processor needed to generate time-deltas.

4. VARIABLE LIST AND MACHINE DESCRIPTION

As shown in Fig. 4, time-delta generation needs two inputs, the variable list and machine description. In this section, the inputs are described.

4.1 Variable list

The variable list contains the list of variable names to be observed. The variables must be global variables or static variables in functions or block boundary. The IR code is searched for the IR operations that access the variables in the variable list.

4.2 Machine description

The machine description contains the information of a target processor, such as the execution cycle of each IR operation, processor architecture, memory parameter, and cache configuration. Since the IR of a compiler is independent of target machines, the IR modification routine in Fig. 4 can be used independent of target machines and only the machine description needs to be changed for other target machines.

- **Operation table**

The machine description contains a section called an operation table that specifies the size and execution cycle of an IR operation when the operation is translated into machine instructions. For example, in the machine description of M-CORE, the IR operation that loads a datum from memory to a register takes two cycles since the load instruction in M-CORE is executed in two cycles. The default value is one for all the IR operations, but the value can be specified by the user. To enhance the accuracy of time-delta generation, the granularity difference between an IR operation and the corresponding machine instruction code is considered in determining the execution cycle of the IR operation. A load operation is an example of granularity difference. A single load operation in an IR code is translated into multiple instructions since an additional instruction that loads an address into a register and hence additional cycles are needed. In case of ARM7, the additional instruction that loads an address into a register takes three cycles, and thus $LOAD_{ARM7}$ which represents the cycles consumed for a load in ARM7 is six cycles. $STORE_{ARM7}$ that represents the cycles consumed for a store in ARM7 is five cycles.

- **Processor architecture**

The machine description contains the type of a target processor. Since the IR code is generated for a scalar processor, for VLIW or superscalar processors such as PowerPC, additional scheduling is needed for generating more accurate time-deltas. In the proposed method, the IR code is scheduled based on the information such as the number of functional units, IR operations executed on each functional unit and the number of registers in order to estimate the real schedule on the target machine. When the average number of instructions executed in a single cycle is available, time-deltas can be generated by scaling the time-deltas obtained from a scalar machine.

The number of registers in the target machine is described in the processor architecture section of the machine description. When the number of general-purpose registers of a target machine is smaller than that of a host machine, restricting the number of general-purpose registers used in simulation enhances the accuracy of time-delta generation. For example, M-CORE has only 16 general-purpose registers and restricting the number of register used in the proposed time-delta generation tool yields a good estimation result. This is because the number of general-purpose registers affects the result of compiler optimizations. Especially, the number of instructions that save and restore registers into and from memory is heavily affected by the number of registers.

A compiler must be modified to use only the registers specified in the machine description so that the time-delta generation tool can be used for simulating various processors. This can be achieved by making the register allocation routine take parameters specifying the number of registers used for register allocation. Since the register allocation routine of a compiler generally deals with only general-purpose registers, restricting the registers is more effective for a target processor with few general-purpose registers.

- **Memory parameter**

The machine description also contains the latency of memory that determines that latency of load instructions. In the timed functional simulation, a load operation must wait until data is ready from the memory, whereas a store operation can go on to the next operation by writing data into a write buffer.

- **Cache configuration**

To generate time-deltas for a target processor with caches, the parameters for cache configuration such as the size of cache, the number of ways, associativity, replacement policy, and write policy are described in the machine description. A cache simulator [11] is provided in the time-delta generation library in order to include the effects of data and instruction caches. Data cache is simulated by calling the cache simulator at every memory access. When the configuration of cache is given in the machine description, an IR operation representing a call to a cache simulation function is inserted at the location of an indicator. In addition to the function call, IR operations that save the address and data of each memory access as arguments of the cache simulation function are inserted. Instead of assembly instructions, the IR operations are used because direct insertion of assembly instructions is possible only after the assembly code is generated and in that case, the inserted assembly instructions may cause some problems in branch operations and require register saving and restoring not to disturb the data flow.

Instruction cache is simulated by assigning a virtual address to each IR operation since the real address is known after code generation and it is very hard to translate the program counter of each machine instruction into the address of IR operation. Since the size of instructions is important in instruction cache simulation, the size of an IR operation is obtained from the operation table. Each IR operation is assigned a virtual address when IR code is generated. Similarly to data cache, IR operations that call a cache simulation function with a virtual address are inserted before each IR operation. Cache miss penalty is modeled as additional clock cycles.

To speed up the simulation with a cache, the hit or miss of a cache access can be estimated by comparing a random number and the desired hit ratio, R_{hit}, of the cache. To estimate the hit or miss of a cache access, a random number between 0.0 and 1.0 is generated and compared to R_{hit}. If the random number is larger than R_{hit}, the cache assess is estimated to cause a miss and a cache function that deals with cache misses is called. Otherwise, the access is estimated as a hit.

5. ENHANCED TIME-DELTA GENERATION

The accuracy of time functional simulation will be enhanced a target compiler is provided and interrupt processing is considered.

5.1 Target compiler

When a compiler for a target CPU is available, the time-delta can be made more accurate by using the target compiler. The cycle counts between accesses to memory/IO variables can be calculated by compiling an application using the target compiler, which is used for generating the IR code to be translated into the assembly code in the final compilation step. The calculated cycle counts are stored in the counter fields of host-machine IR operations instead of the cycle counts in the operation table. Since the variables to be observed are specially qualified not to be optimized, the accesses to the memory/IO variables are not deleted after optimization. Therefore, the access corresponding to an access in target assembly code can be always found in the host-machine executable. To find the accesses to the designated memory/IO variables in the target IR code, the IR code is searched in the same way that is used to set the indicator fields.

5.2 Interrupts

To process interrupts in timed functional simulation, a method to process interrupts must be provided in the time-delta generation. In the proposed method, interrupt service routines (ISRs) are assumed to be provided by the user. Unlike ISSs in which interrupt requests are checked after each instruction, interrupt requests cannot be checked in the execution of compiled-code, and thus they are processed using traps on a host machine to service the interrupts as soon as possible after interrupts are requested.

When an interrupt is requested from a peripheral hardware, the wrapper causes a software trap on the host with the trap number indicating the requested interrupt. In the host-machine ISR, a target ISR corresponding to the trap number is identified and the target ISR provided by the user is called.

6. EXPERIMENTAL RESULTS

We developed a time-delta generation tool by modifying GNU C compiler (GCC) [9]. The code quality of GCC is reported to be better than commercial compilers for some machines. In the modified GCC, two newly defined optional fields, the counter and indicator fields, are inserted in the RTL code that is generated from the front end. The library needed for time-delta generation, such as cache simulators and host-machine ISRs for interrupt processing, was also developed.

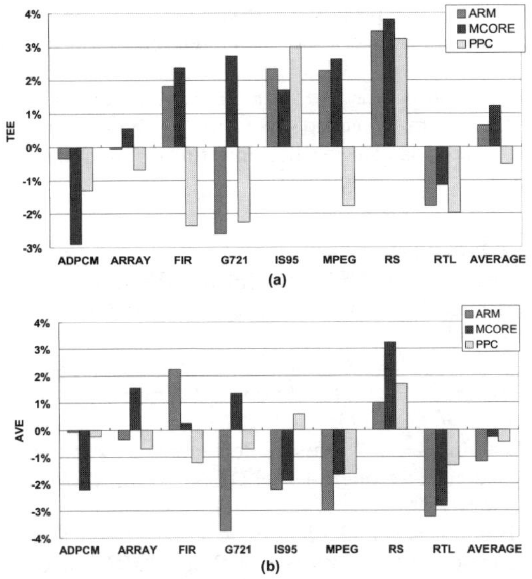

Fig. 5. Estimation errors using unoptimized host executables. (a) TEE. (b) AVE.

To show the accuracy of the time-delta generation tool (*tsgcc*), we compared the results of tsgcc to those of instruction set simulators (ISSs) of target processors for a set of test programs. The target processors used in our experiments are ARM7TDMI [12], M-CORE [14] and PowerPC 601 [13], which are most frequently used in embedded systems. As PowerPC 601 is a superscalar processor capable of issuing and retiring three instructions per clock, a list scheduling is performed in the time-delta generation tool. For cache simulation, it is assumed that PowerPC 601 contains a 32-Kbyte, eight-way set associative, unified cache and the cache line size is 64 bytes. The cache uses write-back write policy and a least-recently-used replacement policy.

Fig. 5 shows the total estimation error (*TEE*) and average error (*AVE*) of time-deltas when the host executables are generated without machine-dependent optimizations. TEE and AVE are calculated as follows.

$$TEE = \frac{C_{target} - C_{tsgcc}}{C_{target}} \text{ and}$$

$$AVE = AVERAGE(\frac{\delta_{target} - \delta_{tsgcc}}{\delta_{target}}),$$

where C_{target} and C_{tsgcc} are the total number of execution cycles measured on the ISS and that estimated by tsgcc, respectively, and δ_{target} and δ_{tsgcc} are time-deltas measured on the ISS and that estimated by tsgcc, respectively. The AVE represents the average errors between two time-deltas generated from the ISS and tsgcc. The average TEE over benchmark programs is about 0.64% for ARM, 1.22% for M-CORE and -0.51% for PowerPC. The TEE of M-CORE is larger than those of ARM and PowerPC, because of the large granularity difference between M-CORE instruction and RTL operations, whereas the difference is small in cases of ARM and PowerPC.

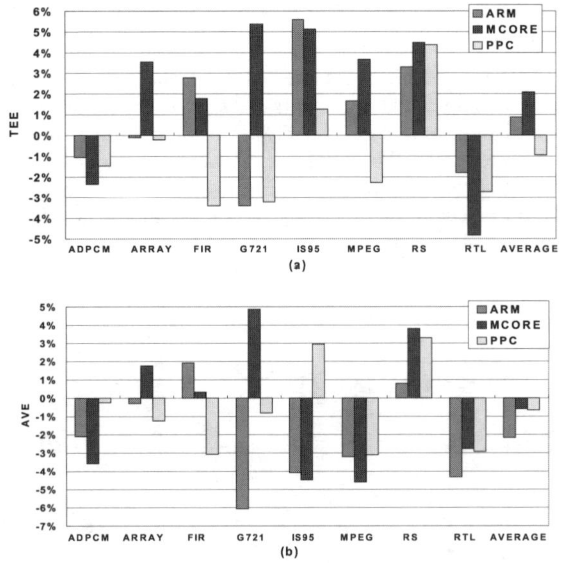

Fig. 6. Estimation errors using optimized host executables. (a) TEE. (b) AVE.

The errors for optimized host executables are shown in Fig. 6. The average TEE increases to 0.87% in ARM, 2.10% in M-CORE, and -0.95% in PowerPC. The average AVE also increases to -2.17% in ARM, -0.58% in M-CORE and -0.65% in PowerPC.

As shown in Fig. 5 and Fig. 6, the proposed method shows good accuracy in both unoptimized and optimized cases because it is based on the IR of a compiler. When an IR operation is deleted, inserted or moved in the back-end optimization, the counter of the deleted IR operation is also deleted, inserted or moved since the counter is associated with the optional field of the IR data structure. This means that the effect of optimization is directly reflected in the proposed time-delta generation.

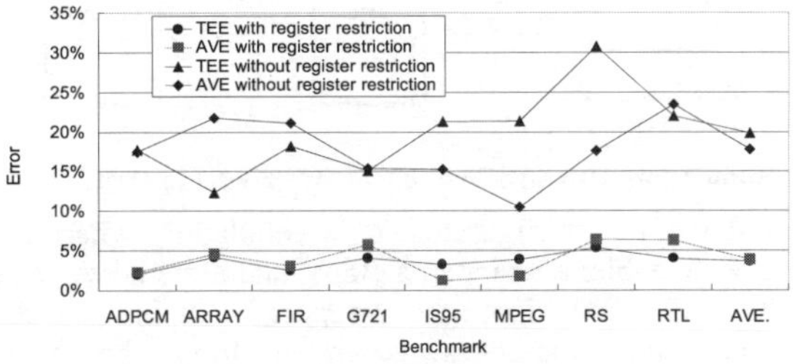

Fig. 7. Estimation errors with and without register restriction in M-CORE.

Fig. 7 shows the estimation errors when the number of registers used in tsgcc is not restricted to the number of registers in M-CORE. The TEE and AVE without register restriction are experimented to show the effect of the number of registers. In this case, the TEE and AVE are increased to 19.84% and 17.82%, respectively, while the TEE and AVE with register restriction are 3.65% and 3.92%, respectively. Comparing the errors in Fig. 7, we can know that tsgcc can underestimate the execution time if more registers are provided than the real CPU has. However, with restricting the number of registers to that of M-CORE, the error is reduced since the smaller number of registers leads to more load and store operations. This result shows that restricting the number of registers is very effective in enhancing the accuracy of the time-delta generation.

The differences in TEEs obtained in the cache simulation and the estimation using random numbers are shown in Fig. 8, where the average difference in TEE is 0.93%. The simulation is speeded up by six times when the estimation is used instead of the cache simulation. From the results in Fig. 8, we can know that the estimation of cache misses using random numbers is effective in reducing simulation times.

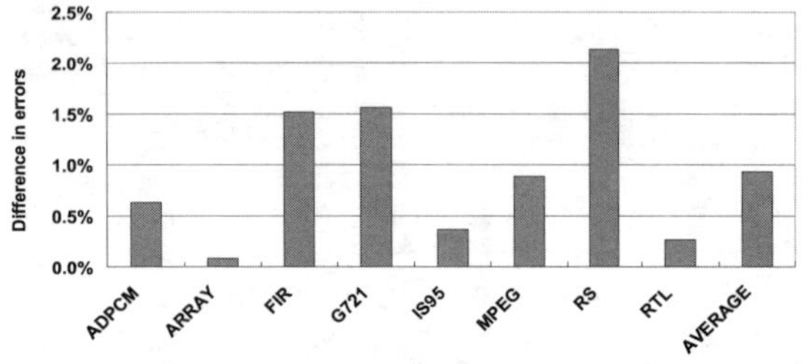

Fig. 8. Differences in TEEs obtained in the cache simulation and the estimation using random numbers and $R_{hit} = 0.95$.

Table 1. Speed-up of host-code execution over ISS.

Program	ARM	MCORE	PPC
ADPCM	4.02	3.86	4.04
G721	18.91	26.44	25.85
MPEG2	342.05	298.25	280.75

The speed-up of compiled-code execution over ISS is shown in Table 1. The maximum speed-up occurs with the MPEG2 encoder program. The time-delta generation tool runs about 300 times faster than ISS. The speed-up becomes larger in programs with longer execution time.

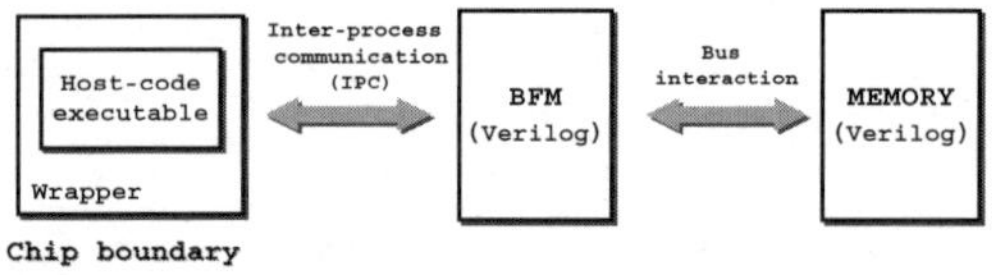

Fig. 9. Simulation with a BFM and memory model in HDL.

Fig. 9 shows the configuration of a simulation, where a host-machine executable, a wrapper, a BFM, and a memory model are used. In the simulation, the wrapper contains codes that compensate the time-deltas generated from the host-code executable and communicates with the BFM using inter-process communication. The wrapper also contains the ISR of the host machine. The BFM and memory is described in Verilog HDL. The BFM generates bus signals to drive the memory model. The shaded area in Fig. 9 represents the boundary of a target CPU, which is ARM7TDMI. The simulation is performed with the selected global variables stored in the memory model. The simulation was successful even for random interrupt requests and we could observe the interaction between the host executable and memory model through the BFM.

7. CONCLUSION

In this paper, we have presented a new time-delta generation method. The generation of accurate time-delta is crucial to performance estimation and architecture exploration. In the proposed method, a compiler is modified to generate time-deltas by counting the cycles of the IR operations and using the machine description that contains information on a target processor such as the number of the cycles of each IR operation and the number of registers. Since the proposed method is based on the IR of a compiler, the method can be easily retargeted by changing the machine description. The experimental results show that the proposed method is effective in that the average error is about 2% and the maximum speed-up over ISS is about 300 times. The proposed method is also verified in a timed functional simulation environment.

8. REFERENCES

[1] K. Suzuki et. al., “Efficient Software Performance Estimation Methods for Hardware/Software Codesign,” In *Proc. 33rd DAC*, 1996, pp. 605-610.

[2] M. Lajolo, et. al., “A Compilation-based Software Estimation Scheme for Hardware/Software Codesign,” *In Proc. CODES*, 1999.

[3] J. Bammi, et. al., “Software Performance Estimation Strategies in a System-Level Design Tool,” *In Proc. CODES*, 2000.

[4] C. Brandolese, et. al, “Source-level Execution Time Estimation of C Prgorams,” *In Proc. CODES*, 2001.

[5] M. Lajolo, et. al, “A case study on modeling shared memory access effects during performance analysis of HW/SW systems,” *In Proc. CODES*, 1998.

[6] B. Bailey, et. al. Hardware/Software Co-Simulation. [Online]. Available:http://www.mentor.org/seamless/papers.

[7] B. Bank. Innoveda’s Virtucal-CPU™ Co-Verification Environment. [Online]. Available:http://www.innoveda.com/products/datasheets_HTML/vcpu.asp.

[8] A. Aho, R. Sethi, and J. Ullman, *Compilers – Principles, Techniques, and Tools*. Reading, MA: Addison-Wesley, 1986.

[9] R. M. Stallman. Using and Porting GNU CC. [Online]. Available: http://gcc.gnu.org/onlinedocs/gcc-2.95.3/gcc.html.

[10] S. P. Harbison and G. L. Steele Jr., *C A REFERNECE MANUAL*. Englewood Cliffs, New Jersey:Prentice Hall, 1995.

[11] M. Hill and A. Smith, "Evaluating Associativity in CPU Caches," *IEEE Trans. on Computers*, C-38, 12, December 1989, p.1612-1630.

[12] ARM Ltd. ARM7TDMI Data Sheet, August 1995.

[13] IBM. PowerPC 601 RISC Microprocessor User’s Manual, 1993.

[14] Motorola. M-CORE-MMC2001 Reference Manual, 1998.

Automated Equivalence Checking of Switch Level Circuits

Simon Jolly
FourSticks Pty. Ltd.
2/259 Glen Osmond Road,
Frewville, South Australia, 5063.
61-8-83385500
sjolly@foursticks.com

Atanas Parashkevov
Motorola Inc.
2 Second Avenue, Mawson Lakes,
South Australia, 5095.
61-8-81683687
aparashk@asc.corp.mot.com

Tim McDougall
Motorola Inc.
2 Second Avenue, Mawson Lakes,
South Australia, 5095.
61-8-81683709
tmcdouga@asc.corp.mot.com

ABSTRACT

A chip that is required to meet strict operating criteria in terms of speed, power, or area is commonly custom designed at the switch level. Traditional techniques for verifying these designs, based on simulation, are expensive in terms of resources and cannot completely guarantee correct operation. Formal verification methods, on the other hand, provide for a complete proof of correctness, and require less effort to setup. This paper presents Motorola's Switch Level Verification (SLV) tool, which employs detailed switch level analysis to model the behavior of MOS transistors and obtain an equivalent RTL model. This tool has been used for equivalence checking at the switch level for several years within Motorola for the PowerPC, M*Core and DSP custom blocks. We focus on the novel techniques employed in SLV, particularly in the areas of pre-charged and sequential logic analysis, and provide details on the automated and integrated equivalence checking flow in which the tool is used.

Categories and Subject Descriptors

J.6 [**Computer-Aided Engineering**]: Computer-Aided Design.

General Terms: Algorithms, Design, Verification.

Keywords: Custom design, switch level analysis, equivalence checking, formal verification, MOS circuits, VLSI design.

1. INTRODUCTION

To keep up with the challenges poised by the constantly increasing digital circuit complexity, several levels of abstraction in circuit representation are typically utilized. Register-Transfer Level (RTL) describes a circuit at a high level of Boolean functions and data flow within the circuit. Gate level representation provides a structural (schematic) description of a circuit as an interconnection of basic blocks having a known and relatively simple Boolean functionality. Switch level representation contains an interconnection of switches (transistors) and gates that implement the desired functionality of a circuit.

Copyright 2002 ACM 1-58113-461-4/02/0006...$5.00. RTL is often the preferred abstraction level for most functional design activities because it provides the highest level of productivity. RTL models serve as a reference for design implementation. However, any RTL design has to be translated into an equivalent switch level design as a necessary step prior to the fabrication of a physical chip. This translation can be performed using synthesis flows and tools such as Synopsys Design Compiler [17]. When a chip has to meet stringent operating requirements (e.g. speed or power) certain parts of the chip may be manually implemented and carefully tuned at the switch level. This practice is commonly referred to as custom design.

When the reference RTL view and the switch level view of the same digital circuit is created independently, there is a clear verification problem: it has to be ascertained that the two views have the same functionality. Traditional verification techniques such as simulation or emulation do not provide a complete guarantee of correctness and can be very expensive in terms of necessary resources. Formal methods such as combinational equivalence checking [12] can be applied to complement traditional approaches and provide a proof of equivalence between two functional descriptions of a circuit. RTL and gate level models are both suitable functional representations and thus equivalence checking tools can operate directly on them. A switch level model, however, is a purely structural representation. Therefore, to enable the use of equivalence checking at the switch level, the actual functionality of the circuit has to be obtained first. This is achieved by applying switch level analysis techniques [2][3].

A novel switch level analysis platform called the SLV (Switch Level Verification) tool has been developed at Motorola over the last four years and used within an automated formal equivalence checking flow. Given a switch level model, SLV can produce an equivalent RTL model in a hardware description language such as Verilog [9] suitable as an input to an equivalence-checking tool. SLV has been successfully applied to other flows, e.g. in the automated generation of test models from switch level designs [16].

The design of the algorithms behind SLV has been driven by a number of design goals and constraints. Firstly, the algorithms of choice had to cover the unique blend of design styles in use at Motorola. Secondly, a switch level model had to be analyzed with the bare minimum of additional design annotation by verification engineers. Thirdly, the analysis had to be as rigorous as possible—deriving a wrong RTL model is, in practice, worse than not completing the analysis at all. Finally, the generated RTL had to be a compact, human-friendly functional representation that preserves the original structure of the circuit. Satisfying these requirements in a practical tool requires a number of novel techniques. Two key contributions to switch level analysis—handling pre-charged logic and structural loops—are the main subjects of this paper.

2. PRELIMINARIES

The choice of semantics for the components in a switch level model affects the range of circuit behaviors that can be accurately modeled. SLV employs the semantics of the Verilog hardware description language (HDL) which was chosen as the primary language for the tool. There are, however, several significant characteristics that can be modeled in the Verilog HDL that are ignored by SLV. In the absence of complete formally specified semantics for Verilog, semantics have been implied from a reference HDL simulator.

At the switch level, a MOS transistor is represented as a switch. Switches usually have two switched terminals and a control terminal that correspond respectively to transistor source, drain and gate terminals. SLV models numerous refinements to this switch model, with the most fundamental distinction being between NMOS and PMOS device models. PMOS switches close when logical 0 is applied to the control terminal, and NMOS switches close when logical 1 is applied to the control terminal. Other refinements include variations on the number of controlling terminals and resistance between the channel terminals. The flow of a signal between the switched terminals is always assumed to be bi-directional and of zero delay.

Gates are devices that model logic relationships between signals. Gates have a set of input and output terminals and define a mapping from the logic value at each input terminal to the logic value at each output terminal. The set of gate types modeled by SLV and the driving strength of the signal output by gates is defined in [9]. In SLV, gates have zero delay.

Nets serve as interconnection points in a circuit. They are assumed to have no resistance, inductance or delay associated with them. Nets are also assumed to have no associated capacitance, except in the analysis of pre-charge logic. Three classes of nets serve special purposes in a circuit:

- Supplies: Circuits are connected to a power supply with two terminals. In the switch level model, these terminals provide a source of constant logical level 0 and 1.
- Ports: Circuits have an interface that is comprised of input, output and inout port connections. It is assumed that the inputs are Boolean and that they are stable through each period of clocks.
- Clocks: Clocks are a special kind of input port periodically alternating their value. There are two types of clocks: pre-charge clocks are used in the analysis of pre-charge logic; master clocks are used in the analysis of sequential behavior.

The channel terminals of switches in a circuit form closely connected partitions called channel-connected components (CCC) [2]. A CCC comprises of a set of switches, a set of gates and a corresponding set of nets that are used to connect these components. The gate terminals of CCC switches and the input nets of CCC gates form the set of input nets that belong to a CCC (called controlling nets in [12]). According to basic laws of electronics, the logic values of all nets in a CCC can be determined from the logic values of the inputs of that CCC.

The input nets of a CCC are further divided into two distinct sub-sets. Internal CCC input nets are driven by the CCC itself and represent feedback (self-dependencies) within a CCC. External CCC input nets are either driven by another CCC or are top-level input ports. CCC outputs are a set of nets belonging to a CCC which serve as external inputs to other CCCs or as circuit output or inout ports.

Each net in the circuit can assume one of the following logic values:

- 0, when there is a path of conducting switches from this net to a source of 0, and no path of conducting switches from this net to a source of 1 that has a greater or equal driving strength than this path. Under these conditions, it is said that this net is *pulled down*.
- 1, when there is a path of conducting switches from this net to a source of 1 and no path of conducting switches from this net to a source of 0 that has a greater or equal driving strength than this path. Under these conditions, it is said that this net is *pulled up*.
- X, when there is a path of conducting switches from this net to both a source of 0 and a source of 1 that have equal driving strengths. Under these conditions, it is said that the net is pulled up and down simultaneously. A net in state X is a part of a short-circuit (DC) path.
- Z, when there is no path of conducting switches to either a source of 0 or a source of 1. We call such nets *floating*.

The primary function of the SLV tool is to determine the logic values of circuit output and inout ports as functions of the logic values applied at the circuit input and inout ports. The evaluation processes result in the logic functions of a net being determined and stored in the following forms:

- The global pull-up and pull-down functions are in terms of the circuit input and inout ports and any state-storing nets in the circuit. These Boolean functions describe the conditions under which paths of switch devices drive the net to a source of logic 1 or 0 respectively. This form is used for many electrical design checks.
- The local pull-up and pull-down functions are in terms of the external inputs to the Channel Connected Component (CCC) containing the net. These Boolean functions describe the conditions under which the net is driven by a path of switch devices to a source of logic 1 or 0 respectively within the CCC. This form is generally used in the production of output.

3. OUR EQUIVALENCE CHECKING FLOW

3.1 Basic Switch Level Analysis

The computation of the pull-up and pull-down functions of a net (referred to as net evaluation) is based on an explicit path enumeration technique [12] that operates within a CCC. It is well known that the evaluation of a net in a CCC using explicit path enumeration can potentially require the exploration of a number of paths exponential with the number of switches in the CCC [2] [12]. SLV avoids this problem by utilizing a series parallel CCC compression algorithm that reduces the circuit graph of the CCC switch network by alternatively applying parallel and then serial compression to the network until no further compression is possible [13]. The explicit path enumeration is also accompanied by the identification of false and self-dependent paths as identified by Brzozowski and Yoeli [6].

The semantics of Verilog discrete signal strengths [9] are incorporated into the evaluation of pull-up and pull-down functions for nets in a two step process:

1. A driving strength is calculated for each path of switches during explicit path enumeration. This driving strength is a function of the driving strength of the net driving the path and the number of resistive switches in the path. Separate pull-up and pull-down functions are maintained for each driving strength during this phase of the analysis.
2. The pull-up and pull-down functions for each driving strength are combined to create single pull-up and pull-down functions for the net.

Supplies are defined to have a driving strength of Supply0/1. Circuit input and inout ports have a driving strength of Strong0/1. The driving strengths associated with a gate output are Strong0/1, unless otherwise specified for a particular gate instance in the input model.

The strength of a signal arriving at a net via a path of switch devices from a supply net, an input port, an inout port, or gate output is calculated by applying the following rules for strength reduction for each of the switches in the path:

1. A non-resistive switch device in the path of switches passes a strength through its channel unchanged, except that a Supply strength is reduced to a Strong strength.
2. A resistive switch device reduces the strength of the signal passing through its channel in accordance with the rules in [9].

For each path that SLV identifies during explicit path enumeration, its driving strength is calculated according to these rules. Separate pull-up and pull-down function pairs are maintained for each of the defined signal strengths. At the completion of explicit path enumeration in the CCC, these functions are combined to form single pull-up and pull-down functions. For simplicity, consider a system with functions of three driving strengths, where a1, b1, and c1 represent the pull-up functions with decreasing strength, and a0, b0, and c0 represent the pull-down functions with decreasing strength. The combined pull-up and pull-down functions of a net `out' are then given by:

```
out.pull-up = a1 | (~a0.~a1).b1 | (~a0.~a1.~b0.~b1).c1
out.pull-down = a0 | (~a0.~a1).b0 | (~a0.~a1.~b0.~b1).c0
```

which simplifies to:

```
out.pull-up = a1 | ~a0.b1 | ~a0.~b0.c1
out.pull-down = a0 | ~a1.b0 | ~a1.~b1.c0
```

3.2 Analysis of Pre-charged Logic

A pre-charge (or dynamic) net has two distinct phases of operation: the pre-charge phase and the evaluate phase. The pre-charge phase is assumed to be the first phase of the dynamic clocking scheme, during which the pre-charge net is unconditionally charged to either logical 1 or logical 0. The evaluate phase is assumed to be the last phase of the dynamic clocking scheme, during which the pre-charge net is conditionally discharged, so that it maintains its pre-charge logic level if the required conditions for discharge are not satisfied. Any dynamic clock phases between the pre-charge and evaluate phases are assumed to be transient, and do not contribute to the required output model.

Figure 1 illustrates a typical example of pre-charge logic in which the net labeled `out' is a pre-charge net. During the pre-charge phase, the dynamic clock input 'clk' is logical 0. The capacitor 'C' is pre-charged to logical 1 since there are no conducting paths to a supply of logical 0 and there is a conducting path to a supply of logical 1. In the evaluate phase, the dynamic clock input is logical 1. If 'a' and 'b' are both logical 1, then there is a path to a supply of logical 0 and 'C' is discharged. Otherwise 'C' maintains a logical 1. This implements a NAND logic gate.

The presence of logic that is dependent on dynamic clocks in the discharging network, and which prevents discharge paths from conducting in the pre-charge phase of the clocking scheme is referred to as 'footed' pre-charge logic. The example shown in Figure 1 illustrates footed pre-charge logic due to the NMOS transistor that prevents the discharge path in the pre-charge phase when 'clk' is logical 0. The absence of footing logic is referred to as 'non-footed' pre-charge logic.

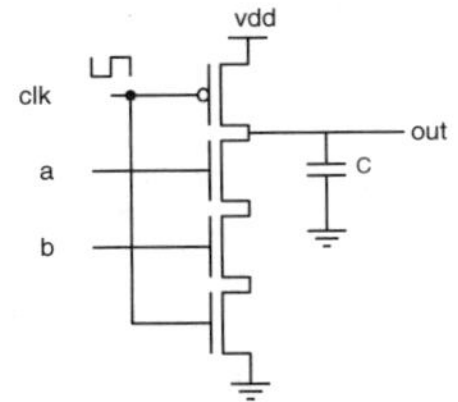

Figure 1 Example of a dynamic nand gate

The major tasks in the analysis of pre-charge logic are to identify nets that are pre-charged and analyzed under a dynamic clocking scheme (termed 'pre-charge nets'), and, for each pre-charge net, to derive the pre-charge functionality and obtain the static behavior.

3.2.1 Identifying Pre-charge Nets

There are two elements of the switch level model that are specific to the analysis of pre-charge logic, being net capacitance and dynamic clock waveforms. All nets that are connected to a transistor gate or the input to a logic gate are assumed to have a capacitance and are capable of holding a logical value. An alternative to capacitors are weak state-holding structural loops that maintain the logical level of a net while it is not driven by a stronger signal.

A dynamic clock waveform defines the clocking scheme for the circuit and is specified by the user. SLV only requires the user to define the dynamic clock inputs of the top-level block in the design hierarchy as it is capable of handling gated clock designs and propagates the clocking scheme throughout the logic in the clock tree during the circuit analysis.

In the initial stages of CCC evaluation, a CCC is marked as being dynamic if it has one or more dynamic clock external inputs. The dependence of this CCC on a dynamic clock input is used to suggest the presence of pre-charge logic. The nets within a dynamic CCC are subsequently examined and can be classified as either a pre-charge net, or a dynamic clock (indicating presence of gated clock logic in this CCC).

A net is labeled as a dynamic clock if either the net is specified as such and associated with a waveform by the user, or if a dynamic CCC drives the net and it is not a pre-charge net and its global functions are dependent on a dynamic clock BDD variable (see Table 1).

A net in a dynamic CCC is labeled as a pre-charge net if the net may float (i.e. it is not driven to a logical 1 or 0) in the evaluate phase of the dynamic clocking scheme and:

1. The net has level restoring logic indicated by a sequential loop; or
2. The local functions of the net are dependent on the capacitance of the net (see Table 1).

The SLV tool has a user selectable ability to ignore net capacitance and only classify pre-charge nets based on condition 1 above, which is useful in the analysis of a particular design style.

In order to model net capacitance in a dynamic CCC, each capacitor C_i in the CCC is associated with two new independent local variables, V_{Ci}^1 and V_{Ci}^0 respectively, representing the pull-up and pull-down drive of the capacitor. The subsequent analysis of each dynamic CCC output derives local pull-up and local pull-down functions that may be dependent on both CCC inputs and any V_{Ci} of the CCC. Immediately after explicit path enumeration the type of a net can be inferred from the dependence of the local functions on V_{Ci}, and the dependence of the global functions on dynamic clock input variables, as shown in Table 1. In this table C_a represents the capacitor variables of the net that is currently being evaluated. Note that bad charge sharing is defined to occur when a net with capacitance is locally dependent on the capacitance of any other net in the same CCC, and the two nets are pre-charged to opposing values.

Table 1 Net Classification in Dynamic CCCs

Local functions depend on V_{Ca}	Local functions depend on V_{Ci} other than V_{Ca}	Net can float in the evaluate phase	Net Classification
Yes	Yes	Yes	This is a pre-charge net. There is the potential for bad charge sharing if capacitors in the local functions are pre-charged to opposite values.
Yes	No	Yes	This is a pre-charge net.
All other combinations			The net is a dynamic clock if its global functions are dependent on a dynamic clock BDD variable.

In order to avoid interfering with subsequent processing steps, once this net classification is complete, logical 0 is substituted for each V_{Ci} in the local pull-up and pull-down functions. Following this, the local pull-up and pull-down functions become what they would be for a non-dynamic CCC. If the net has not been marked as a pre-charge net then no further processing is required, otherwise the local and global functions of the pre-charge net are modified to model the static behavior.

3.2.2 Modifying the Functions of a Pre-charge Net

One of two distinct modifications is made to a pre-charge net's local and global functions, depending on whether the net is pre-charged to logical 1 or logical 0. Therefore, as a first step in modifying the functions of a pre-charge net, its global functions are used to determine if it is pre-charged to logical 1 or logical 0. This is achieved by substituting the actual value of all dynamic clock inputs in the pre-charge phase into the net's global functions. This result indicates either pull-up or pull-down pre-charge as shown in Table 2.

The heuristic presented in Table 2 determines the pre-charge behavior for a wide variety of dynamic design styles including gated clocks, footed or non-footed pre-charge logic, pre-charge paths containing resistive transistors or logic gates with weak driving strengths, and pre-charge level restoring logic.

Table 2 Heuristic for Determining Pre-charge Behavior

Resulting global pull-up	Resulting global pull-down	Additional Conditions	Interpretation
Constant 1	Not constant 1	N/A	Pull-up pre-charge
Not constant 0	Constant 0	N/A	Pull-up pre-charge
Constant 0	Not constant 0	N/A	Pull-down pre-charge
Not Constant 1	Constant 1	N/A	Pull-down pre-charge
Modified	Unchanged	Resulting global pull-up and pull-down are complementary	Resistive pull-up pre-charge
Unchanged	Modified	Resulting global pull-up and pull-down are complementary	Resistive pull-down pre-charge
Modified	Modified	Resulting global pull-up and pull-down are complementary and the strongest driving path is a pull-up path	Resistive pull-down pre-charge
Modified	Modified	Resulting global pull-up and pull-down are complementary and the strongest driving path is a pull-down path	Resistive pull-up pre-charge
All other combinations			The net is not a pre-charge net.

Based on the pre-charge behavior, the local and global functions of a pre-charge net are modified to model the static footed behavior of the net. The first step in this process is the determination of the global pre-charging function, which is the Boolean function representing the global variable condition that results in the pre-charge behavior of the net. The technique for determining the global pre-charging function uses the pre-charge behavior from Table 2 and is shown in Table 3. In Table 3 'gpu' and 'gpd' respectively represent the global pull-up and pull-down functions of the pre-charge net, 'Universal' represents the universal quantification operation, and 'Constrain' represents the BDD constraining function. The local pre-charging function is determined by translation from the global pre-charging function.

Table 3 Determining the Global Pre-charging Function

Pre-charge Value	Resistively Pre-charged	Global Pre-charging Function
1	Yes	gpu.Constrain(~gpd).Universal(<All BDD variables not representing dynamic clocks>)
1	No	gpu.Universal(<All BDD variables not representing dynamic clocks>)
0	Yes	gpd.Constrain(~gpu).Universal(<All BDD variables not representing dynamic clocks>)
0	No	gpd.Universal(<All BDD variables not representing dynamic clocks>)

The functionality of footing logic is added to the local and global functions of the pre-charge net by the following operations:

Pre-charged to logic 1: pull-dn=pull-dn&~(pre-charging-function)
Pre-charged to logic 0: pull-up=pull-up&~(pre-charging-function)

The footing logic is first added to the global functions of the pre-charge net, and non-footed dynamic logic is implemented if this translation modifies the global functions of the pre-charge net. A similar translation is only applied to the local pull-down function of the pre-charge net if non-footed logic is being modeled.

The final step in modeling the footed static behavior of the pre-charge net is to apply a translation that represents the level-restoring behavior of the net when no discharge paths conduct. If the pre-charge net is pre-charged to logical 1, then its local and global functions are modified according to the following scheme:

pull-up = pull-up | ~(pull-down function)
pull-down = ~(pull-up function)

Similarly, if this net is pre-charged to 0, then its local and global functions are modified as follows:

pull-down = pull-down | ~(pull-up function)
pull-up = ~(pull-down function)

SLV provides two modes of modes of output representation for pre-charge nets, the first includes explicit references to the dynamic clocks in the behavior of pre-charge nets, and the second makes no reference to dynamic clocks. This allows users to select the mode that suits their design style.

3.3 Analysis of Structural Loops

Current generation equivalence checking tools rely on a close structural correspondence between the circuits being compared. In practice, a one-to-one mapping between the state points (latches) of the two circuits is required to identify the cones of logic that need to be compared. To apply these tools to a switch level block, one has to determine its sequential behavior by reliably identifying the state points as intended by the logic designer and not introducing any additional ones.

Traditional switch level symbolic analysis derives a gate level model that contains fine-grained sequential behavior introduced by a special simulation clock that is not part of the original design, and associated unit-delay gates [2][4]. Therefore, this gate level model is not suitable for comparison with an RTL model as is; further processing is required to abstract the detailed timing information and obtain an FSM representation suitable for formal verification [10][11]. However, the algorithms performing this abstraction rely on symbolic reachability analysis, which is likely to limit their capacity. Thus, some commercial tool vendors are relying on pattern-matching techniques to identify the latches in a design [7][14], while others require users to pinpoint the state points in the input design [12].

From a practical point of view, sequential behavior in circuits is implemented either by structural loops (latches, keepers and half-keepers), or by storing electrical charge (dynamic logic, dynamic memory arrays). Since the verification of dynamic memory arrays is not an application SLV is currently targeted at, this section focuses on the analysis of structural loops in the switch level circuit.

The primary objective of structural loop analysis is to determine whether a loop is combinational, sequential, or oscillatory. A secondary requirement is the ability to produce an RTL description for the loop in case it is not found to be oscillatory. Our approach to this extends [12][15].

Firstly, we consider structural loops involving two or more CCCs (called external CCC loops). Graph-based algorithms are used to traverse the connectivity of the CCCs in the switch level circuit from the outputs to the inputs. Any closed loops are broken by inserting two new independent boolean variables v_0 and v_1 representing, correspondingly, the pull-down and pull-up functions of the net N closing the loop (referred to as a *shadow* net) [12][15]. Note that selecting, as a shadow net, any of the nets on the boundaries between CCCs that form it will break a loop. To aid in the subsequent equivalence-checking step, SLV employs a set of heuristics guiding the choice of a shadow net.

During the analysis of the logic cone driving a shadow net, the variables v_0 and v_1 are used to resolve the circular dependencies in the logic and the global pull-down function F_0 and pull-up function F_1 are computed. Examining the dependencies of F_0 and F_1 on v_0 and v_1 performs the loop analysis. We introduce a function depend(F, v) that evaluates to 0 when $F|_{v=1}=F|_{v=0}$ and to 1 otherwise. Then, a so-called loop signature LS is computed as follows:

$$LS = depend(F_1, v_1)*8 + depend(F_1, v_0)*4 + depend(F_0, v_1)*2 + depend(F_0, v_0)$$

By definition, LS has an integer value between 0 and 15 characterizing the self-dependencies in the corresponding structural loop. For example, LS=0 implies that the loop is purely combinational (this has been previously noted in [12]). Similarly, LS=9 implies that the loop is likely to be sequential, although a further check is required to ensure that there is no risk of uncontrolled oscillation of that loop (see below).

Table 4 Loop Signature Interpretation

LS	Inferred type of loop / Action
0	Combinational loop; the pair of boolean functions does not depend on either of the temporary boolean variables.
1	Sequential loop implementing a half-latch that can keep a value of 0 but not a value of 1.
8	Sequential loop implementing a half-latch that can keep a value of 1 but not a value of 0.
9	Sequential loop implementing a latch.
other	F_0 and F_1 are simplified under the assumption that the shadow net can only take boolean values; LS is re-computed.

The complete set of LS values and their interpretations are provided in Table 4. In some cases, the value of LS does not uniquely identify the nature of the structural loop. To avoid that, the value of LS is re-computed after F_0 and F_1 are simplified under the assumption that the shadow net can only take Boolean values. This is achieved by substituting any occurrence of v_0 in F_1 by $\sim v_1$, and any occurrence of v_1 in F_0 by $\sim v_0$.

This step is, of course, based on an assumption that is later discharged by verifying that the net cannot have X or Z value. Note that the re-computed loop signature is guaranteed to have a value of 0, 1, 8, or 9.

When the structural loop under analysis is found to be combinational, the only remaining task is to substitute v_0 with F_0 and v_1 with F_1 in the global functions of all nets that depend on the temporary Boolean variables. However, if the loop is inferred to be of sequential nature, the extra steps of functional decomposition and a check for oscillatory behavior are required. The functions F_0 and F_1 are decomposed into sub-functions as follows:

$$F_0 = X_0 \mid Y_0.v_0 \mid Z_0.\sim v_0$$
$$F_1 = X_1 \mid Y_1.v_1 \mid Z_1.\sim v_1$$

where functions X_0, Y_0, Z_0 do not depend on v_0 and do not share common cubes, and, similarly, functions X_1, Y_1, Z_1 do not depend on v_1 and do not share common cubes. These functions have the following interpretations:

- X_0 (X_1) covers the case when F_0 (F_1) is independent of v_0 (v_1). Since the loop is sequential, this is the condition under which the loop's environment "writes" a value into the loop's latch;
- Y_0 (Y_1) covers the case when F_0 (F_1) take their previous value v_0 (v_1). In terms of structural loop behavior this is the condition under which the loop's latch maintains its stored value;

- Z_0 (Z_1) covers the case when F_0 (F_1) take the negation of their previous values $\sim v_0$ ($\sim v_1$). Thus, if $Z_0 \neq 0$ or $Z_1 \neq 0$ the structural loop exhibits oscillatory (unstable) behavior. This clearly points to a design error and prevents the algorithm from obtaining a functionally equivalent RTL description of the input switch level circuit.

The above decomposition of F_0 and F_1 not only allows us to identify unwanted oscillatory loops, but also provides input to the final step of RTL output generation for the loop. One way of describing the loop behavior in the Verilog language (assuming X_0 and X_1 do not share common cubes) is:

```
reg N;
always @(<list of nets that X0 and X1 depend on>)
    If (X0 | X1)
        N = X1;
```

In practice, further analysis is applied to identify asynchronous and synchronous set/reset signals, clocks and data inputs and produce a much more detailed and human-friendly style of RTL description. Loops that implement latches are distinguished from loops that implement keepers based on a functional dependence of X_0 and X_1 on one or more of the master clock inputs defined by the user, and different styles of output are generated. Other types of loop behavior analysis on the decomposed F_0 and F_1 are possible and actually employed in SLV. For example, $X_0=0$ and $X_1=0$ implies the presence of a latch that cannot be written to and, therefore, can be regarded as a source of unknown value (logical X). Such conditions are reported to the user.

A separate technique has been developed for the analysis of structural loops that occur within a single CCC (internal CCC loops). Although some specific but common cases such as Brzozowski-Yoeli loops [6] can be dealt with using rather simple techniques, the design goals of SLV and the requirements of its user community required a more general solution to this problem. In our presentation of the internal CCC loop resolution technique, we consider a CCC with m external inputs with functions $e_1, e_2, \ldots, e_m$, and n>0 internal inputs. Again, we notionally break the loops at the internal inputs by introducing a pair of temporary variables—u_k for the local pull-up function and d_k for the local pull-down function at the k-th input—one for each internal input.

Next, a system of boolean equations B is formed. The actual local pull-up (U_k) and pull-down (D_k) functions for all internal inputs are obtained in terms of e_i, u_i, and d_i. Then, for any given internal input k, the following equations must hold when the signals at the inputs of the CCC are stable:

$$u_k = U_k$$
$$d_k = D_k$$

The 2n equations obtained in this way form the system of equations B. In essence, B captures the conditions under which all nets in the CCC hold a stable value. Solving B with respect to u_k and d_k using a method such as Gaussian elimination can lead to one of the following results:

- If B has no solutions, then the CCC cannot reach a stable state, and it is therefore established that at least one of the internal CCC loops results in oscillation. This is a design error and no RTL output is produced;
- If B has exactly one solution, then u_k and d_k are uniquely determined in terms of e_i. Therefore, all structural loops in the CCC are of combinational nature and the local functions of all nets inside the CCC can be obtained by substituting the temporary boolean variables u_k and d_k with their solutions in B;
- If B has more than one solution, then at least one of the internal CCC loops is sequential. Further analysis is required to handle this case.

Two observations are important in order to understand our approach to handling the latter of the three cases above. Firstly, even if B does not have a unique solution, it is quite possible that u_k and d_k do have a unique solution for a given k. Thus, we can rule out the possibility of a certain internal input being a state storing net without having to solve B. Secondly, u_k and d_k not having a unique solution for a given k implies that either (1) the k-th internal input is a state storing net, or (2) the k-th internal input is combinationally dependent on at least one state storing point inside the CCC.

The analysis of the case when B has more than one solution proceeds in an iterative fashion:

1. The set of pairs of temporary boolean variables u_k and d_k that have unique solutions are identified, and their nets are marked as non-state storing ones. Substituting the identified pairs of boolean variables with their unique solutions modifies the system of equations B. This step reduces the number of the remaining free temporary variables in B.
2. If there are no more free variables in B, all internal inputs have been categorized, and there is no more work to do.
3. One of the remaining unresolved nets is picked up and selected as a state storing one. The system of equations B is modified to make the corresponding temporary boolean variables be parameters rather than unknown entities in B. This step makes B more constrained with respect to the remaining free temporary variables. The three steps above are repeated starting with point 1. above.

3.4 Tool Integration and Flow Automation

The compareS2RTL tool encapsulates an automated flow for performing formal equivalence checks between two designs, each design being at the switch, gate, or RTL abstraction level. The compareS2RTL tool controls the flow of data between each tool that it employs, providing a single generic interface to the user that allows different back-end tools or modules to be plugged-in with minimal impact to the development environment. Typically the compareS2RTL tool is applied to a switch level design and an RTL design; the compareS2RTL flow for this is illustrated in Figure 2.

The compareS2RTL tool employs SLV to create an RTL model from the switch level design. The equivalence check is subsequently performed between the extracted RTL model and the RTL design using a functional equivalence checker such as Tuxedo-LEC [18]. A sequential model checker is applied if the equivalence checker did not determine that the two designs are equivalent, which can often be a result of differences by the tool in mapping sequential elements between the designs.

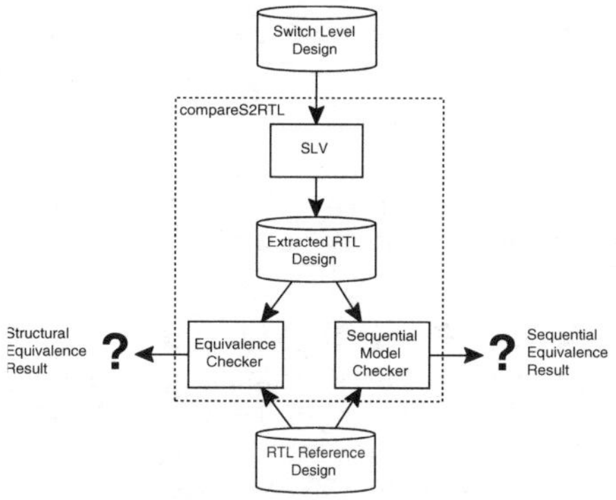

Figure 2 Equivalence Checking using compareS2RTL

The compareS2RTL tool is very versatile, allowing the user full control of which tools and options are used. It provides a single interface for the user to become familiar with and automatically controls the passage of information between tools. Another benefit is that the compareS2RTL tool automatically passes BDD variable ordering information between tools, resulting in fewer reordering operations and improved performance.

4. CONTRIBUTIONS AND RELATED WORK

The main contribution of this paper is the description of two novel switch level analysis algorithms. The first algorithm provides automated identification and analysis of pre-charge logic in the presence of complex logic structures such as gated clocks, non-footed logic, and charge sharing. The other algorithm is concerned with structural loop analysis and classifies loops as combinational, sequential, or oscillatory (unstable).

Many of the fundamental technologies for symbolic analysis of switch level circuits grew out of efforts to increase the speed and capacity of switch level simulation. Bryant pioneered several important techniques in this field, including novel methods for switch level modeling [2], CCC analysis [3][4] and the representation of boolean logic with reduced ordered binary decision diagrams [5]. While this research clearly demonstrated the utility of switch level analysis, the development of software to generate an output model suitable for formal equivalence checking was not its primary objective.

A tool called Anamos reads a switch level model in a Spice-like language and generates an equivalent gate level model [2][4]. The gate level modeling style employs fine-grained sequential behavior introduced by a special simulation clock that is not part of the original design and unit-delay gates to represent sequential behavior. This gate level model is suitable for simulation but not appropriate for comparison with an RTL model as is. In contrast to this, the techniques for dynamic logic and structural loop analysis in SLV ensure the preservation of the original state storing points in the input design and produce a netlist that is suitable as input to equivalence checking tools.

Several commercial and academic tools and tool flows are based on the switch level analysis engine implemented in Anamos. Significantly, experiments have attempted to implement algorithms that process the output of Anamos with the objective of transforming it into a more usable form of HDL [10][11]. These algorithms rely on symbolic reachability analysis, which is likely to limit their capacity. A major advantage of our approach to structural loop analysis is the fact that each loop is analyzed in isolation avoiding such computationally intensive. This has a positive effect on the size of circuits that can be handled. This increase in tool capacity with respect to the background art has been achieved without a compromise in accuracy or rigor of analysis, and these algorithms have exposed subtle custom design problems on a regular basis.

Verity [12] implements efficient algorithms for verifying the behavior of switch level models with static combinational and simple dynamic logic. This has seen the tool used with great success in commercial custom VLSI design settings. However, Verity cannot perform analysis of structural loops and depends on user input to resolve these. Unlike SLV, Verity cannot generate an RTL representation of a switch level design, which limits this tool's utility to formal verification flows only.

Pattern matching techniques have also been used for switch level circuit extraction [7][14]. Tools employing this approach repetitively attempt to match parts of the input design and replace them with RTL components from the library. Unfortunately, there is no guarantee that the component library is both correct and complete. The switch level analysis in SLV, on the other hand, is purely functional and does not rely on pre-defined structures and patterns, which translates in an ability to handle a wide variety of design styles.

5. EXPERIMENTAL RESULTS

SLV has been in production use across Motorola for several years and has been applied to the verification of key PowerPC, M*Core and DSP custom blocks. As a generic switch level analysis platform, SLV has also successfully targeted a number of additional applications—library characterization, legacy design re-engineering, and generation of gate level test views from switch level logic [16].

Table 5 Benchmark Results from Running SLV

Design	Transistors	Loops	Latches	Dyn. Cells	CPU Time (s)	Memory (MB)
1	18	2	2	0	0.1	3.5
2	157	0	0	12	0.1	3.5
3	5615	174	158	16	0.7	9.3
4	5814	395	0	395	11.7	19
5	7329	557	0	557	21.0	47
6	37549	1304	1270	0	11.7	35
7	89667	3771	3456	0	111.0	35

Experimental results for SLV are presented in Table 5. SLV was executed for all benchmarks on a Sun Blade 1000 with 1024 MB of RAM. The benchmarks range from simple flip-flops and adders to complex Motorola proprietary designs. Table 5 demonstrates the ability of SLV to analyze a representative set of switch-level designs including a mixture of structural loops, dynamic, and sequential logic.

6. CONCLUSIONS

This paper presented a complete framework for formal verification of designs at the switch level. The algorithms that make SLV include novel contributions to the state of the art in the areas of pre-charged logic analysis and structural loop and sequential logic analysis. These improvements made possible the satisfaction of all high-level requirements set for SLV from the very beginning of the project—coverage of a wide set of design styles, minimal change and added overhead for established custom design practices, robustness, and user-friendly RTL output comparable to manually written models.

One area for improvement in SLV is increasing the size of the blocks that can be handled in a single run. One approach that looks particularly promising is utilizing a non-canonical representation for boolean functions such as Boolean Expression Diagrams [1], coupled with a joint use of BDD- and SAT-based [8] decision techniques.

7. ACKNOWLEDGMENTS

We would like to acknowledge the importance of the contribution of Vladimir Grudinin, Neil Boroky, and Craig Biggs at various stages of the development of SLV. This work would not be possible without the immense support of the Design Verification group at Motorola by Carl Pixley, Michael Burns, Jun Yuan, and Kurt Shultz.

8. REFERENCES

[1] Andersen, H.R., and Hulgaard, H. Boolean Expression Diagrams. LICS'97, pp. 88-98, 1997.

[2] Bryant, R.E. Boolean analysis of MOS circuits. IEEE Transactions on CAD-ICS, vol. 6, no. 4, pp. 634-649, July 1987.

[3] Bryant, R.E. Algorithmic aspects of symbolic switch network analysis. IEEE Transactions on CAD-ICS, vol. 6, no. 4, pp. 618-633, July 1987.

[4] Bryant, R.E. Extraction of gate level models from transistor circuits by four-valued symbolic analysis. Fujitsu Labs Ltd., April 1991

[5] Bryant, R.E. Graph-based algorithms for Boolean function manipulation. IEEE Transactions on Computers, vol. C-35, pp.677-691, August 1986.

[6] Brzozowski, J.A., and Yoeli, M. Combinational Static CMOS Networks. Integration, the VLSI Journal, 5, pp. 103-122, 1987.

[7] Groupe Bull. Generalized recognition of gates: User Guide. Version 06, April 1996.

[8] Gu, J., Purdom, P.W., Franco, J., and Wah, B.W. Algorithms for the satisfiability (SAT) problem: a survey. DIMACS Series in Discrete Mathematics and Theoretical Computer Science, American Mathematical Society, pp. 19-152, 1997.

[9] IEEE. 1364-1995, Verilog Language Reference Manual. 1995.

[10] Jain, S., Bryant, R.E., and Jain, A. Automatic clock abstraction from sequential circuits. DAC'95, pp. 707-711, 1995.

[11] Kam, T., and Subrahmanyam, P.A. Comparing layouts with HDL models: a formal verification technique. IEEE Transactions on CAD-ICS, vol. 14, no. 4, pp. 503-509, April 1995.

[12] Kuehlmann, A., Srinivasan, A., and LaPotin, D.P. Verity—a formal verification program for custom CMOS circuits. IBM Journal of Research and Development, vol. 39, no 1,2, January/March 1995.

[13] Kundu, S. GateMaker: A Transistor to Gate Level Model Extractor for Simulation, Automatic Test Pattern Generation and Verification. IEEE International Test Conference, pp. 372-381, 1998.

[14] Laurentin, M., Greiner, A., and Marbot, R. DESB, a functional abstractor for CMOS VLSI circuits. EDAC'92, pp. 22-27, 1992.

[15] Malik, S. Analysis of cyclic combinational circuits. ICCAD'93, pp. 618-625, 1993.

[16] McDougall, T., Parashkevov, A., Jolly, S., Zhu, J., Zeng, J., Pyron, C., and Abadir, M. SLV/ATPG: An Automated Flow for Test Model Generation from Switch Level Custom Circuits. Submitted to MTV '02

[17] Synopsys. Synopsys Online Documentation. Version 2001.08, August 2001.

[18] Verplex Systems. Logical Equivalence Checker Reference Manual version 3.0. August 2001.

A Practical and Efficient Method for Compare-Point Matching

Demos Anastasakis, Robert Damiano, Hi-Keung Tony Ma, Ted Stanion
Synopsys Inc.
{demosa, robertd, tonyma, teds}@synopsys.com

ABSTRACT
An important step in using combinational equivalence checkers to verify sequential designs is identifying and matching corresponding compare-points in the two sequential designs to be verified. Both non-function and function-based matching methods are usually employed in commercial verification tools. In this paper, we describe a heuristic algorithm using ATPG for matching compare-points based on the functionality of the combinational blocks in the sequential designs. Results on industrial-sized circuits show our methods are both practical and efficient.

Categories and Subject Descriptors
J.6 [**Computer-aided engineering**]: Verification, compare-point matching.

General Terms
Algorithms, Experimentation, Verification.

Keywords
Combinational verification, equivalence checking, latch mapping.

1. INTRODUCTION
Using combinational equivalence checkers to prove or disprove the functional equivalence of two sequential designs is a common and practical verification approach in today's design methodology. This verification approach is especially applicable when only combinational synthesis techniques are used for optimization. Corresponding combinational blocks from two sequential designs (one called the reference and the other the implementation) can be compared and verified using combinational formal techniques [1][2][3]. The sequential design can be a gate-level netlist or a register transfer level (RTL) source. No simulation vectors are required and the verification is complete over the entire vector space of the combinational blocks being verified. As designs become larger and more complex, verification methods utilizing combinational equivalence checkers and hence offering fast turnaround verification times and complete verifications, are rapidly gaining ground on the traditional simulation-based verification approaches.

An important step in applying combinational equivalence checkers is identifying and matching corresponding compare-points in the two sequential designs to be verified. A compare-point in a sequential design is a combinational logic endpoint during verification. A compare-point can be an output port, register, latch or black-box input pin. Compare-point matching (CPM) methods in commercial verification tools can be broadly divided into two classes: non-function-based and function-based.

Non-function-based methods use name or structural comparisons to match compare-points in the sequential designs. In most production verification flows, a large percentage of compare-points are usually matched using non-function-based methods. A significant number of compare points usually are left unmatched due to design transformations that do not preserve signal names or dramatically modify the circuit structure of some portion of the designs. Function-based methods are the only viable automatic methods for matching the remaining compare-points without requiring non-trivial effort from the designers.

Various function-based techniques [4][5][6][7] have been proposed to identify latch correspondences between two sequential designs. They are all exact methods based on fixed-point computations. Given N latches in each sequential design, there are $N!$ possible combinations to match them. All these exact methods in the worst case may have to enumerate all combinations, although in an implicit way. This is because they actually try to prove functional equivalence between latches in the two sequential designs.

Our function-based heuristic method does not attempt to prove functional equivalence between compare-points. It rather tries to establish inequivalence relationships between compare-points. The inequivalence information is then used to group compare-points that are most likely to be equivalent into pairs. Not only is this method practical and efficient for industrial-sized circuits, it also separates cleanly the compare-point matching process from the equivalence verification process that follows. A similar approach is described in [8], but it only uses random pattern simulation to discover inequivalence relationships rather than using an automatic test pattern generation (ATPG) program. It also does not handle cases where don't care conditions must be considered to show equivalence.

This paper is organized as follows. In Section 2, we first define the compare-point-matching (CPM) problem and describe how it can be solved as a partitioning problem. We then describe how we can use an ATPG program in conjunction with a search model to find solutions for the partitioning problem. Taken into consideration of don't cares, that are typical in the specification description (e.g. RTL source), to show equivalence between two designs are described in Section 3. Experimental results are presented in Section 4 and the paper is concluded in Section 5.

2. COMPARE-POINT MATCHING

2.1 Compare-Point-Matching Problem

We are given two sequential circuits; one called the reference (C_{ref}) and the other the implementation (C_{imp}). Without loss of generality, we assume the two circuits have the same number of latches N and the only compare-points considered are the latches. The CPM problem is to identify N latch-pairs such that each pair consists of one latch from C_{ref} and one from C_{imp} and the two latches are combinationally equivalent. The next-state functions of the latches in a latch-pair represent two compare-points to be verified. The outputs of the latches in a latch-pair are used as pseudo primary inputs that are always set to identical values during combinational verification of downstream compare-points. There are $N!$ possible way of creating the N latch pairs.

Most exact function-based methods [4][5][6][7] solve the compare-point matching and equivalence checking processes together and implicitly enumerate all $N!$ combinations. Our heuristic method on the other hand only tries to group latches that are most likely to be equivalent and let the equivalence checker that follows perform the verification task. Our heuristic method can also be used as an efficient front-end to any exact method.

2.2 Compare-Point Partitioning

We can identify and use inequivalence relationships between reference and implementation latches to solve the CPM problem. If we can show a reference latch in C_{ref} is not equivalent to $(N - 1)$ implementation latches in C_{imp}, then we can conclude it is most likely to be equivalent to the remaining implementation latch and hence group the two into a compare-point pair.

Theorem 1: Let C_{ref} and C_{imp} be two sequential circuits each containing N latches, and there exists a unique pairing of latches, one from each circuit, such that latches in each pair are combinationally equivalent. If it can be shown for each latch in C_{ref} there are $(N - 1)$ latches in C_{imp} that are not equivalent to it, then there is only one way left to group the latches into combinationally equivalent pairs.

Based on Theorem 1, we can devise an algorithm, similar to [6], to detect inequivalence between two latches, one at a time and eventually arrive at the final solution representing the unique grouping if it exists. The number of inequivalence checks can be exceedingly large, even though using functional or sequential random pattern simulation can help to reduce it. The inequivalence checking process can also inadvertently invoke a number of equivalence proofs.

The main idea behind our heuristic method is try to generate as much inequivalence information about latch pairs as possible in a single checking step. This is done by grouping potentially equivalent latches (compare-points) into a single partition and successively dividing each partition into two sub-partitions until all partitions contain only two latches, one from C_{ref} and one from C_{imp}. The heuristic method also makes use of the newly generated inequivalence information to facilitate the division process.

Let us assume we are given two sequential circuits, C_{ref} and C_{imp}, and M partitions of potentially equivalent latches. Each partition contains equal number of latches from C_{ref} and C_{imp}. A reference latch in a partition is potentially equivalent to any implementation latch in the same partition and is already proved to be inequivalent to any implementation latch in any other partition. If M is equal to 1, then any latch in C_{ref} is potentially equivalent to any latch in C_{imp}. M may be greater than 1 at the beginning if non-function-based CPM process has been run and hence some initial equivalent and inequivalent relationship information is already obtained.

Let us choose a partition P_i and assume it contains $2N$ latches. Let the next-state functions of the reference latches in the partition be denoted by RNS_1 to RNS_n and the next-state functions of the implementation latches be denoted by INS_1 to INS_n. The next-state functions of the latches correspond to compare-points to be matched. Let the outputs of the reference latches in the partition be denoted by RPS_1 to RPS_n and the outputs of the implementation latches be denoted by IPS_1 to IPS_n. The outputs of the latches correspond to their present-states and represent pseudo primary inputs used to verify the downstream compare-points combinationally.

Let us create a search model circuit with outputs RNS_1 to RNS_n and INS_1 to INS_n corresponding to the next-state functions of the reference and implementation latches in the partition respectively. Without loss of generality, let us assume the next-state functions only depend on the primary inputs I_1 to I_p and pseudo primary inputs RPS_1 to RPS_n and IPS_1 to IPS_n. Figure 1 shows the search model circuit.

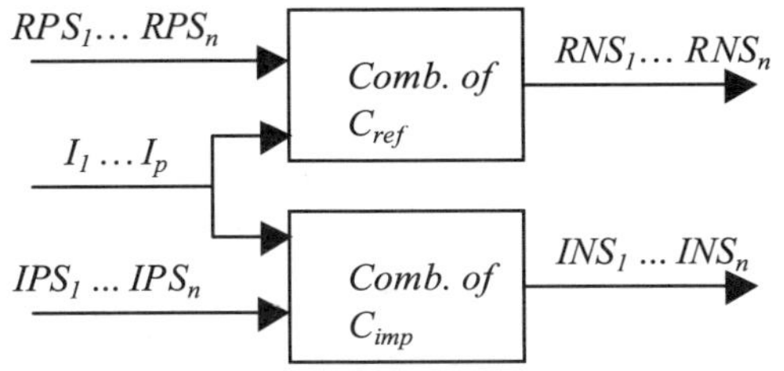

Fig. 1 Search Model Circuit.

Theorem 2: Assume there exists some pairings of latches in P_i such that latches in each pair are combinationally equivalent. Given two latches L_{ref} and L_{imp} in P_i, if we can find a test vector ($I_1 \ldots I_p, RPS_1 \ldots RPS_n, IPS_1 \ldots IPS_n$) with the constraints $RPS_1 = \ldots = RPS_i = \ldots = RPS_n = IPS_1 = \ldots = IPS_i = \ldots = IPS_n$ that shows the two latches are inequivalent, then L_{ref} and L_{imp} can never be shown equivalent with any latch-pairing combinations.

Based on Theorem 2, we can devise a heuristic CPM method using successive partitioning to create potentially equivalent latch-pairs. Figure 2 shows the partitioning procedure. Whenever a test vector is found that sets opposite values between some RNS_i's and some INS_i's, we can divide a partition into two, one containing RNS_i's and INS_i's with values 0 and the other containing RNS_i's and INS_i's with values 1. Reference latches in one sub-partition cannot be equivalent to any implementation latches in the other sub-partition. In general, the search model is driven by pseudo primary inputs coming from multiple partitions. The test vectors are constrained to have RPS_i's and IPS_i's from the same partition having identical logic values.

```
While (unprocessed partition with size > 2 still exits) {
   pick a partition;
   search for a test vector that sets opposite values at
     some reference and implementation latches;
   if (found such test vector) {
      simulate the test vector;
      create two new partitions, one containing latches
      with 0 values and the other with 1 values;
   } else {
      mark partition as processed and cannot be divided;
   }
}
```

Fig. 2 The Partitioning procedure.

An example to illustrate the working of the partitioning procedure is shown in Figure 3. The example has 2 inputs, I_1 and I_2, and six compare-points RNS_1, RNS_2, RNS_3, INS_1, INS_2 and INS_3. The pseudo primary inputs corresponding to the present states of the latches are RPS_1, RPS_2, RPS_3, IPS_1, IPS_2 and IPS_3. Initially, all compare-points are in a single partition. A partitioning vector consists of inputs I_1, I_2, RPS_1, RPS_2, RPS_3, IPS_1, IPS_2 and IPS_3. First, a vector (0,1,1,1,1,1,1,1), constrained with identical logic values at RPS_1, RPS_2, RPS_3, IPS_1, IPS_2 and IPS_3, is generated using random pattern generation or an ATPG program. This vector divides the single partition into 2. Partition [RNS_3 INS_3] consists of only two compare-points and is designated as a compare-point pair. The other partition is further divided into two sub-partitions with 2 elements by another vector (1,0,0,0,1,0,0,1) that is constrained with identical logic values at RPS_1, RPS_2, IPS_1 and IPS_2 and identical logic values at RPS_3 and IPS_3. Three pairs of potentially equivalent compare-points are thus generated. In the next section, we will describe how to augment the search model circuit so that an ATPG program can be used to generate the partitioning vectors.

Partitioning Vectors / Partitions

I_1 I_2 RPS_1 RPS_2 RPS_3 IPS_1 IPS_2 IPS_3 [RNS_1 RNS_2 RNS_3 INS_1 INS_2 INS_3]

0 1 1 1 1 1 1 1 — 0 0 1 0 0 1

↓

[RNS_1 RNS_2 INS_1 INS_2] [RNS_3 INS_3]

1 0 0 0 1 0 0 1 — 0 1 0 1 0 0

↓

[RNS_1 INS_1] [RNS_2 INS_2] [RNS_3 INS_3]

Fig. 3 A partitioning example.

2.3 Use of ATPG for generating vectors

During the partitioning process, we need to generate partitioning vectors that sets opposite values between at least one reference compare-point and one implementation compare-point. Random pattern generation (RPG) can be used initially and followed by ATPG. In order to use an ATPG program to generate partitioning vectors, we need to augment the search model circuit to facilitate the generation process. Figure 4 shows the augmented search model circuit.

Gates A and D in figure 4 are n-input NAND gates. Gates B and C are n-input OR gates. Gates E and F are 2-input AND gates and Gate G is a 2-input OR gate. A test vector that sets a value 1 at Gate G will guarantee to set opposite values between at least one reference compare-point (RNS_i) and one implementation compare-point (INS_i). We can use an ATPG program on the augmented search model circuit to generate such vectors with the constraints $RPS_1 = ... = RPS_i = ... = RPS_n = IPS_1 = ... = IPS_i = ... = IPS_n$.

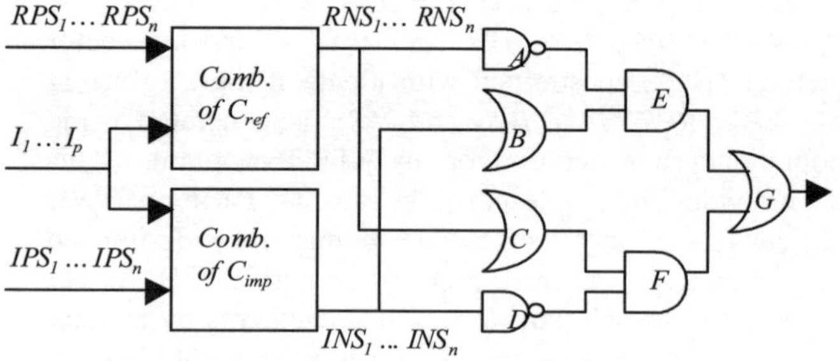

Fig. 4 Augmented Search Model Circuit for ATPG.

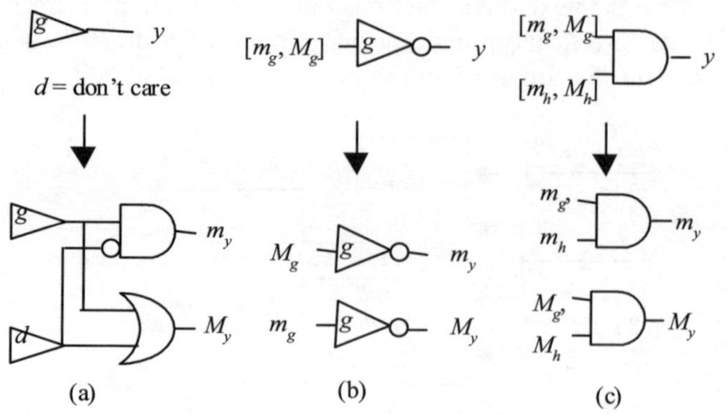

Fig. 5 Don't Care Circuit Modification Rules.

3. HANDLING OF DON'T CARES

Considering don't care conditions when comparing two designs is essential to avoid false negative verification results. Don't care conditions can come from the specification when performing RTL-To-Gate verification. They can also arise in hierarchical verification when external don't cares are derived from upstream modules for the verification of the downstream modules.

For our partitioning-based CPM method, there are two requirements for efficient handling of don't care conditions. First, we need to represent and model don't care conditions explicitly as part of the search model circuit so we can still readily apply an ATPG program to search for partitioning vectors. Second, the don't care representation circuit has to be size-efficient with respect to the original circuit. Our method of modeling don't care conditions is largely based on the work in [9].

3.1 Modeling don't cares

One way to represent an incompletely specified function that contains a don't care set is to use an interval. Given a function F with an on-set f and a don't care set d, we can represent the function with the interval $[f \,\&\, d', f + d]$. We interpret an interval $[m, M]$ as the set of functions $\mathcal{F} = \{f: m \subseteq f \subseteq M\}$. Any function in an interval is a valid implementation of the incompletely specified function f. A single don't care value x is represented by the interval [0,1]. Based on the interval concept, we can come up with a set of modification rules to change a circuit with don't cares to encode and represent explicitly the don't cares within the enhanced circuit.

Interval-based circuit modification rules to model don't cares within a circuit structure are shown in Figure 5. For each gate or input port with an external don't care function d, we create two gates which implement the two endpoints of the equivalent interval as shown in Figure 5a. If the inputs of a gate have an interval associated with them, two new gates are created to represent the endpoints of the interval for the gate. Examples for an inverter and an AND gate are shown in Figure 5b and 5c respectively. Given a circuit with don't cares associated with some of its gates/ports, the maximum number of gates added to encode the don't care conditions in the original circuit is $O(n)$, where n is number of gates in the original circuit.

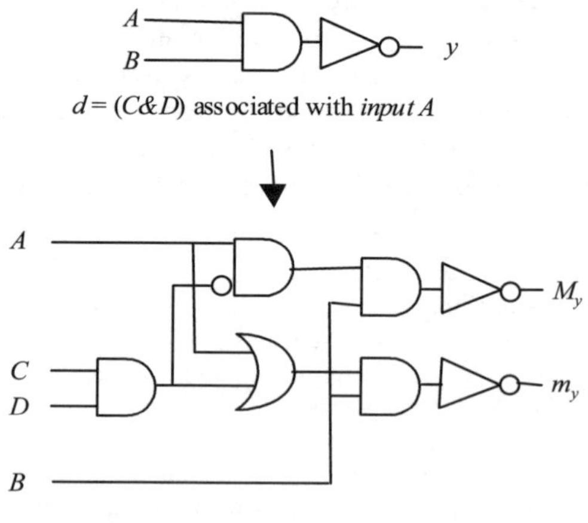

Fig. 6 An example for modeling don't cares.

An example to illustrate how to modify a circuit that has external don't cares associated with its primary input is shown in Figure 6. At the top of Figure 6 is the original circuit with 4 inputs *A, B, C* and *D*. There is a don't care condition *(C&D)* associated with the input *A*. The corresponding modified circuit encoded with the don't care information is shown at the bottom of the Figure 6.

The output y is split into two signals, m_y and M_y, that represent the two endpoints of the interval associated with it. If an input vector in the care set, e.g. C != 1 or D != 1, is applied to the primary inputs, the logical values at m_y and M_y are either all 0 or all 1. If any don't care vector, e.g. C = 1 and D = 1, is applied at the primary inputs, m_y and M_y are set to 0 and 1 respectively. By examining the logic values at the two corresponding interval endpoints for output y, we can determine whether an input vector is in the care/don't care set with respect to that particular output.

3.2 Impact on partitioning vector generation

Without loss of generality, we assume the reference circuit contains don't care conditions. In order to use an ATPG program to automatically generate partitioning vectors for circuits with don't cares, we first need to modify the original search model circuit (shown in Figure 1) based on the modification rules described in section 3.1. Two signals, RNS_{i1} and RNS_{i2}, are created for each compare-point, RNS_i, whose transitive fanin cone contains gates/ports with external don't cares. A vector that sets 0 and 1 at RNS_{i1} and RNS_{i2} respectively is a don't care vector with respect to the compare-point RNS_i. A care vector with respect to RNS_i will set either all 0 or all 1 at RNS_{i1} and RNS_{i2}.

Next, we need to modify the augmentation of the search model network. For every pair of RNS_{i1} and RNS_{i2}, we create one 2-input AND gate RNS_i_A and one 2-input OR gate RNS_i_O. The fanins to the two new gates are RNS_{i1} and RNS_{i2}. We replace the fanins of the n-input NAND gate A in Figure 4 with RNS_i_Os and replace the fanins of the n-input OR gate C with RNS_i_As. The augmentation circuit for a design with two reference compare points whose transitive fanin cones contain external don't cares is shown in Figure 7.

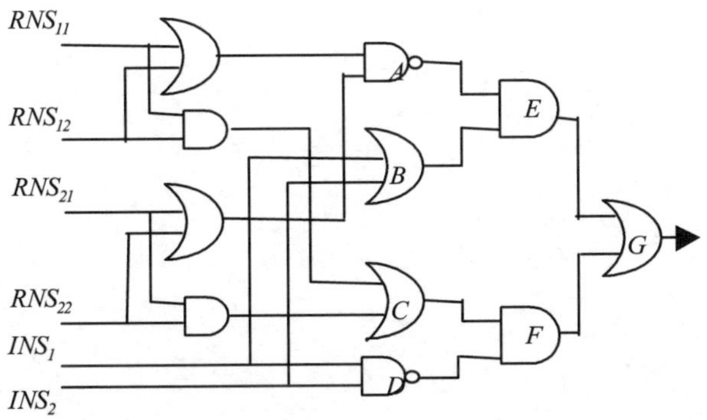

Fig. 7 Augmentation circuit for design with don't cares.

With the modifications described above, a generated vector that sets a value 1 at the output of the search model circuit will again guarantee to establish an inequivalent relationship between at least one reference compare-point and one implementation compare-point. However, multiple vectors may be needed to separate a partition into two as shown by the example in Figure 8. The first two vectors show INS_3 is not equivalent to RNS_1 and RNS_2. The last vector shows RNS_3 is not equivalent to INS_1 and INS_2. Based on the combined inequivalent information from the three vectors, we can then separate the partition into two.

Partitioning Vectors								Partitions					
I_1	I_2	RPS_1	RPS_2	RPS_3	IPS_1	IPS_2	IPS_3	[RNS_1	RNS_2	RNS_3	INS_1	INS_2	INS_3]
0	1	1	1	1	1	1	1	0	x	x	0	0	1
1	0	0	0	0	0	0	0	x	1	x	1	1	0
1	1	0	0	0	0	0	0	x	x	1	0	0	1

↓

[RNS_1 RNS_2 INS_1 INS_2] [RNS_3 INS_3]

Fig. 8 A partitioning example with don't cares.

4. EXPERIMENTAL RESULTS

We performed experiments on 7 industrial circuits with sizes ranging from tens of thousands to hundreds of thousands gates. Table 1 shows the results obtained on a 450 MHz Sun UltraSPARC-III machine. Compare-point pairs in all test cases were successfully proved equivalent by the verification process that follows. The second column indicates the total number of gates in the reference designs. The third column in the table indicates the total number of potential compare-point pairs. The fourth column indicates the number of compare points matched by non-function-based methods. The fifth the number of compare points matched by our heuristic partition-based method. The last column indicates the number of CPU seconds it took to match by our heuristic method. There are two main factors contributing to the efficiency of our heuristic method. First, equivalence checking is completely avoided during the compare-point matching process. Second, existing matching information generated by non-function-based methods, that are usually run a priori as part of the matching process, are effectively used in our heuristic method.

Table 1 Experimental Results

DESIGNS	Num. of gates	Num.of compare-point pairs	Matched by non-function-based methods	Matched by our heuristic method	CPU time (sec)
1	140k	3583	2762	821	50
2	112k	1843	667	1176	45
3	71k	859	411	448	32
4	86k	646	369	277	29
5	40k	718	417	301	6
6	156k	1975	1815	160	28
7	364k	4378	1795	2583	176

5. CONCLUSION

We have described a heuristic compare-point-matching method based on a partitioning algorithm. We have shown how to use an ATPG program and a search model to implement the partitioning algorithm. Extension of the CPM method to handle designs with don't cares was also presented. Experimental results on large industrial circuits show the method is very efficient in practice.

6. ACKNOWLEDGMENTS

The authors would like to thank Jim Kukula, Stan Liao, Brian Lockyear, Lisa McIlwain, and Carl Pixley for their participation in many insightful discussions.

7. REFERENCES

[1] D. Brand, "Verification of large synthesized designs," in ICCAD, 1993, pp. 534 – 537.

[2] S.M. Reddy, W. Kunz and D.K. Pradhan, "Novel verification framework combining structural and OBDD methods in a synthesis environment," in DAC 1995, pp. 414 – 419.

[3] A. Kuehlmann and F. Krohm, "Equivalence Checking Using Cuts and Heaps," in DAC 1997, pp. 263 – 268.

[4] T. Filkorn, "Symbolische methoden für die Verifikation endlicher Zustandssysteme," Ph.D. dissertation, Institut für Informatik der Technischen Universität München, Munich, Germany, 1992.

[5] C.A.J. van Eijk and J.A.G. Jess, "Detection of Equivalent State Variables in Finite State Machine Verification," in IWLS, 1995.

[6] S-Y. Huang, K-T. Cheng, K-C. Chen and U. Glaeser, "An ATPG-Based Framework for Verifying Sequential Equivalence," in ITC, 1996, pp. 865 – 874.

[7] J.R. Burch and V. Singhal, "Robust Latch Mapping for Combinational Equivalence Checking," in ICCAD, 1998, pp. 563 – 569.

[8] H. Cho and C. Pixley. Apparatus and method for deriving correspondence between storage elements of a first circuit model and storage elements of a second circuit model. U. S. Patent 5,638,381, June 1997.

[9] T. Stanion. Circuit synthesis verification method and apparatus. U. S. Patent 6,056,784, May 2000.

Self-referential verification of gate-level implementations of arithmetic circuits

Ying-Tsai Chang* and Kwang-Ting (Tim) Cheng
Department of Electrical and Computer Engineering
University of California
Santa Barbara, CA 93106
ytchang,timcheng@windcave.ece.ucsb.edu

ABSTRACT

Verification of gate-level implementations of arithmetic circuits is challenging due to a number of reasons: the existence of some hard-to-verify arithmetic operators (e.g. multiplication), the use of different operand ordering, the incorporation of merged arithmetic with cross-operator implementations, and the employment of circuit transformations based on arithmetic relations. It is hence a peculiar problem that does not fit quite well into the existing RTL-to-gate equivalence checking methodology. In this paper, we propose a self-referential functional verification approach which uses the gate-level implementation of the arithmetic circuit under verification to verify itself. Specifically, the verification task is decomposed into a sequence of equivalence checking subproblems, each of which compare circuit pairs derived from the implementation under verification based on the proposed self-referential functional equations. A decomposition-based heuristic using structural information is employed to guide the verification process for better efficiency. Experimental results on a number of implementations of the multiply-add units and the inner product units with different architectures demonstrate the versatility of this approach.

Categories and Subject Descriptors

B.5.2 [**Register-Transfer-Level Implementation**]: Design Aids—*verification*

General Terms

Algorithm, Verification

Keywords

Arithmetic circuit verification

1. INTRODUCTION

Combinational equivalence checking is a relatively mature field after over a decade's intensive research. Commercial combinational equivalence checking tools have advanced greatly in both speed and capacity, and have been successfully employed in the full-chip context for RTL functional validation sign-off. The key idea behind the solution to this intractable problem for complex industrial circuits is the effective use of structural similarities between circuits under comparison. Structurally similar circuits have an appreciable portion of functionally equivalent internal signal pairs between them, and these signal pairs could serve as cutpoints to facilitate the progression of equivalence checking in the miter approach [2]. With the aid of a robust RTL-to-gate equivalence checking tool, design functional validation and verification could be conducted more efficiently at the RTL level. A successful abstraction at a higher level of description could further facilitate design architecture exploration.

One class of circuits that proved to be difficult for RTL-to-gate equivalence checking is arithmetic circuits. Among them, the most familiar one is the multiplication operator. Arithmetic circuits play an important role in modern high-performance VLSI datapath design. The quest for performance in arithmetic circuits has stimulated the development of various architectures. However, these proliferated architectures have also posed new challenges to their implementation verification. Existing equivalence checking techniques are often incapable of handling some of these circuits due to either the lack of equivalent internal signal pairs when compared to a reference implementation with a different architecture, or the incapability of direct verification, e.g. BDD memory explosion or exponential number of backtracks for ATPG or SAT approaches. Since arithmetic operations are also the natural word level primitives in HDL, this issue is even more prevalent in modern RTL design methodology. Hence, the verification of gate-level implementations of arithmetic circuits is of practical importance. In this paper, we analyze this problem and propose a new and efficient solution. We believe that part of this problem has not been properly addressed before, and the proposed solution is the first to be able to efficiently perform gate-level verification for a large class of arithmetic circuits.

*Current affiliation: Novas Software, San Jose, CA 95110. E-mail: ytchang@novas.com

2. GATE-LEVEL ARITHMETIC CIRCUIT VERIFICATION - THE ISSUES

The difficulty of verifying gate-level implementations of arithmetic circuits is mainly due to the existence of some hard-to-verify constructs. These constructs include individual arithmetic operators (e.g. multiplication), the use of different operand ordering, merged arithmetic with cross-operator implementations, and circuit transformations based on arithmetic relations. Optimizations of arithmetic circuits for performance, which are often based on arithmetic relations, e.g. Booth's recoding or carry-save adder (CSA)

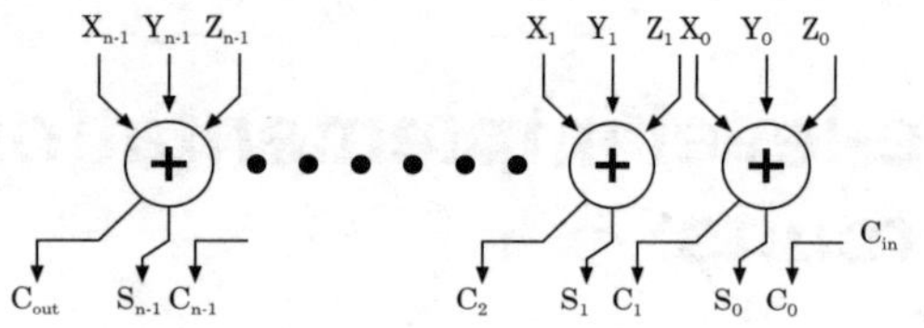

Figure 1: An n-bit carry-save adder.

making use of redundant number systems, inadvertently introduced some of these hard-to-verify constructs. In the modern HDL-based RTL design methodology, the capability of HDL design synthesis tools to perform some of these optimizations automatically and transparently further complicated their implementation verification problem.

The verification of arithmetic circuits has been addressed in a number of former researches, including backward construction of BMD [9], modular verification using BMD [6], implicit verification [16], residue BDD [11], the functional equation approach [8, 7], and the theorem-proving approach [1]. However, a more thorough analysis that across the entire spectrum of issues is still lacking. In the following, we list the key issues that are the major obstacles to an effective verification of arithmetic circuits. The listing is ordered from the more local and gate-level issues to the ones with an increasingly global and behavioral nature.

- *The difficulty of verifying individual arithmetic operators*

  It is well-known that the verification of multiplication operator is a hard one in the BDD-based verification paradigm due to its exponential spatial complexity in BDD [4, 5]. The difficulty of multiplication operator verification is closely related to the integer factorization problem [7]. Therefore it seems that this difficulty is quite independent of the verification paradigm. This difficulty is further complicated by the existence of multiple globally different architectures and, therefore, could not be easily verified at an equivalence checking setting. In the architectural space, the implementation of a multiplication operator could employ different addition reduction trees or different recoding schemes. Commonly encountered architectures of addition reduction trees include addition arrays (AA), carry-save-adder arrays (CS), Wallace tree (WA), and with Booth recoding (BR) [13, 6]. The division operator is another example that is known difficult to verify.

- *Operand ordering*

  The problem of operand ordering originates from the fact that implementations of arithmetic operators in general do not reflect the *commutative* and *associative* laws structurally. For example, the multiplication operator is in general implemented asymmetrically, i.e. in the addition reduction tree implementation, one operand is decomposed and the other one is kept intact in forming the partial products. Based on the commutative law $x \times y = y \times x$, if we use the second operand for decomposition, the implementation based on the left hand side (decomposing y) and the one based on the right hand side (decomposing x) will be significantly different in terms of circuit structure. Further considering the associative law $(x \times y) \times z = x \times (y \times z)$, the space for equivalent implementations of the same function is even larger. For example, $x \times y \times z$ could have 12 different implementations based solely on exploring the different ordering/grouping of the 3 operands.

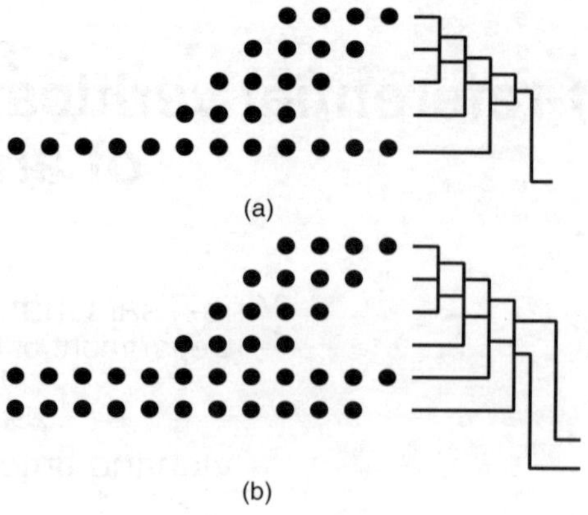

Figure 2: Two different multiply-add units implemented in carry-save-adder tree architecture.

- *Merged arithmetic with cross-operator implementation*

  Merged arithmetic refers to the direct implementation of multi-operator arithmetic expressions without breaking down into individual operators [17, 10, 18]. This approach often employs arithmetic relations to achieve a cross-operator implementation with a single addition reduction tree [14]. Merged arithmetic can achieve higher speed by reducing the number of carry propagation chain, and often has a smaller circuit size and lower power consumption. However, due to the lack of intermediate signals representing the individual operator decomposition, merged arithmetic circuits are an especially hard class of circuits for functional verification.

  One example is the multiply-add unit commonly encountered in the DSP design. To reduce the delay of the critical path, the multiply-add operation is implemented as one single addition reduction tree, i.e. the addition operator is merged into the addition reduction tree of the multiplication operator. This is often achieved by employing the CSA array (Figure 1). Implementations of two types of 4-bit multiply-add units based on the CSA array architecture are shown in Figure 2. In this figure, we use the dot notation and their skeleton addition reduction tree structure to represent the circuit implementation. The dots denote the bits of addenda in the addition reduction tree. The 3-to-2 and 2-to-1 reduction symbols represent the CSA and the binary full adder respectively. In multiply-add units, the addend typically has a larger size to prevent overflow or to reduce accumulation error. Figure 2 (a) implements the multiply-add function $S = a \times b + c$ directly, and Figure 2(b) implements another type of multiply-add unit with the addend and result in the CSA form for faster accumulation at successive operations, i,e, $S_0 + S_1 = a \times b + c_0 + c_1$. This figure shows two different multiply-add units implemented using the same CS architecture. In general, any addition reduction tree architecture utilized in the implementation of multiplier operators could also be used in merged arithmetic. Merged arithmetic hence introduced another rather large degree of freedom in the design implementation space.

  Another example, where this problem is more serious, is the inner product unit, e.g. $S = a_0 \times b_0 + a_1 \times b_1$ and higher di-

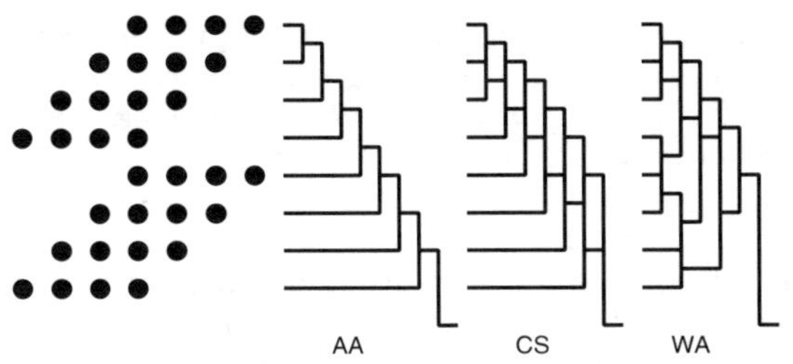

Figure 3: Designs of the inner product unit with different addition reduction tree structures

mensional generalizations. As shown in Figure 3, which follows the convention in Figure 2, the inner product unit could also have a number of different implementations in which the two multiplication operators are implemented in one single addition reduction tree. In this case, even the intermediate products are missing from the implementation, and its verification demands a new approach. The main issue for merge arithmetic is that operations of redundant number representation, such as CSA, do not have a suitable RTL description in modern HDL. Functional equivalence checking in this context is hence very difficult.

- *Circuit transformation by arithmetic relations*

  At a more global level, arithmetic circuits could be simplified by explicitly using more arithmetic transformations, such as the *distributive* law $x \times (y+z) = x \times y + x \times z$. Architectural changes by exploring common expression factoring and expansion often significantly change the datapath, e.g. $f = a-b,\ g = a+b,\ h = a^2 - b^2 \rightarrow h = fg$. Verification of circuits employing such transformations seems to be more tractable only at the RTL level [19]. Compared to the associative and commutative laws, the circuit transformations based on the distributive law is more difficult to prove, presumably because the distributive law bridges between two different arithmetic operations.

- *Floating point units.*

  The verification of floating point units as a class of arithmetic circuits is further complicated by the issue of rounding schemes and operand alignment. This class of circuits seems to be more suitably addressed with the assistance of the theorem-proving techniques [1], which requires more architectural information and human intervening. We do not consider this problem in this paper due to our main interest in the more common arithmetic operators encountered in modern HDL.

The above issues seriously impair our capability of gate-level arithmetic circuit verification at the common RTL-to-gate equivalence checking setting. In this paper, we present a self-referential functional verification approach which properly addresses the first three issues.

3. SELF-REFERENTIAL FUNCTIONAL VERIFICATION

The basic idea of self-referential functional verification is to use the gate-level implementation of the arithmetic circuit under verification to verify itself. More specifically, we decompose the arithmetic circuit verification problem into a sequence of mathematical equations, which entail the correctness of the arithmetic circuit when established. These mathematical equations are verified by checking the equivalence of similar circuit pairs derived from the circuit under verification based on these equations. Notice that the difficulty of gate-level arithmetic circuit verification is mainly due to the proliferation of globally different architectures of arithmetic circuits originated from the existence of arithmetic relations, while self-referential functional verification approach explicitly utilizes these arithmetic relations to facilitate their verification instead.

3.1 Verification by exploiting functional equations

Consider the verification of a circuit F implementing a binary arithmetic function f composed of $+, -$ and $\times$ operators in an expanded form, i.e. sum $(+,-)$ of product form as an arithmetic expression, with distinct l_i-bit binary operands x_i. The operands x_i of f are composed of Boolean bit variables x_{ij} such that the binary representation of x_i is $(x_{i(l_i-1)}x_{i(l_i-2)}..x_{i0})_2$. The total number of Boolean bit variables in f is L, i.e. $L = \sum_i l_i$. An ordering $(y_1, y_2..., y_L)$ of all bit variables $\{x_{ij}\}$ defines a sequence of sets $(A_0, A_1.., A_L)$ with A_k composed of the first k elements in this ordering, i.e. $A_0 = \phi$, and $A_k = A_{k-1} \cup \{y_k\}$. For $y_k = x_{ij}$, we use the notation $\tilde{y_k}$ to denote the set of bit variables composed of all x_{im} with $m \neq j$ and all bit variables of the operands that are not within the same product term with x_i in f, e.g. for $f = x_1 \times x_2 + x_3 \times x_4$, the set $\tilde{x_{12}}$ is composed of all $x_{1m} (m \neq 2)$, x_{3n}, and x_{4n}. We also use the notation $f|_{A=0}$, to denote the function f with the values of the bit variables in A restricted to zero. We have the following:

THEOREM 1. *(Soundness) For any ordering $(y_1, y_2.., y_L)$ of all bits $\{x_{ij}\}$, consider the finite sequence of sets $A_0..., A_L$ associated with this ordering as defined above. Then the set of $2L$ equations $(k = 1, 2.., L)$:*

$$F|_{A_{k-1}=0} = F|_{A_k=0} \pm F|_{(\tilde{y_k} \cup A_{k-1})=0} \quad (1)$$

$$F|_{(\tilde{y_k} \cup A_k)=0} = f|_{(\tilde{y_k} \cup A_k)=0} \quad (2)$$

where $\pm$ depends on whether the term containing $\tilde{y_k}$ in f is $+$ or $-$, and $F|_{A_L=0} = f|_{A_L=0}$ implies $F = f$.

PROOF. It follows from the definition of binary number representation, the distributive law, and induction on the total number of nonzero bits in all operands. □

Our self-referential functional verification of arithmetic circuits is based on this theorem. Notice this theorem is on a specific form of f rather than the function itself. However, even with this restriction, as will be shown in examples below, Theorem 1 still includes a lot of interesting cases. In essence, this theorem uses the distributive law to decompose the binary digits inside the arithmetic product terms of the mathematical expression f, and hence also the circuit F. For a gate-level implementation F of an arithmetic circuit f, the equations in (1) are verified by equivalence checking the circuits implemented at both sides of the equations, and the equations in (2) are verified by invoking Theorem 1 again on the new function $f|_{(y_k \cup A_k)=0}$. Note that the implementations of both sides of the equation (1) are derived from the same original circuit under verification F.

Intuitively, this sequence of equations successively reduces the original verification problem into the ones with smaller nontrivial circuit sizes and fewer inputs by bit-decomposition. Notice that the specification function f only enters in the form of specifying $\tilde{x_{ij}}$, $\pm$, and terminal cases, hence it is a framework of verification without an implementation standard. Although seemingly two copies of different F's appear in the right hand side of equation (1), as will be discussed below, a judicious choice of the ordering $y_0...,y_l$ will result in significant isomorphic portion in these two circuits that could be shared between them. Thus, the equivalence checking problems represented by the above equations are in fact comparing two F's with small difference in peripheral circuits and constraints; hence the name *self-referential* verification.

For example, to verify whether $F(x_{n-1}..,x_0,y_{n-1}..,y_0)$ implements $f = x \times y = (x_{n-1}x_{n-2}...x_0)_2 \times (y_{n-1}y_{n-2}...y_0)_2$, it suffices to prove the following equations ($k = 0,1,..n-1$):

$$\begin{aligned}
& F(0..0,x_{n-1-k},..,x_0,y_{n-1}..,y_0) \\
= \ & F(0..0,x_{n-2-k}..,x_0,y_{n-1}..,y_0) \\
& +F(0..0,x_{n-1-k},0..,0,y_{n-1}..,y_0) \\
& F(0..0,y_{n-1-k}..,y_0) \\
= \ & F(0..0,y_{n-2-k}..,y_0) \\
& +F(0..0,y_{n-1-k},0..0) \\
& F(0..0,x_{n-1-k},0..,0,y_{n-1}..,y_0) \\
= \ & f(0..0,x_{n-1-k},0..,0,y_{n-1}..,y_0) \\
& F(0..0,y_{n-1-k},0..0) \\
= \ & f(0..0,y_{n-1-k},0..0)
\end{aligned}$$

and $F(0..,0) = f(0..,0)$ on F. Notice that in this special case the ones with bit variables in one operand all equal to zero are often trivial to prove due to the fact that $0 \times y = 0$. To put it in a more familiar form, this sequence of equations essentially verifies the following equation for different k on F:

$$\begin{aligned}
& (0..0x_{n-1-k}x_{n-2-k}..x_0)_2 \times (y_{n-1}..y_0)_2 \\
= \ & (0..0x_{n-2-k}x_{n-3-k}..x_0)_2 \times (y_{n-1}..y_0)_2 \\
& +(0..0x_{n-1-k}0..0)_2 \times (y_{n-1}..y_0)_2
\end{aligned}$$

The verification of arithmetic circuits posed as equivalence checking problems also has the benefit of gracefully handling design error. A number of previous approaches were incapable of handling design error. In particular, word-level decision diagram based techniques [6, 9] share this problem due to the explicit word-level feature, which makes them incapable of representing bit-level errors appearing in a gate-level netlist. However, when posed in the equivalence checking setting, modern equivalence checking methodology is often capable of finding a counterexample for inequivalence quickly, and thus providing a convenient solution to this design error problem.

This proposed approach could not be easily applied to the division operator due to the fact that the decomposition functional equation for the division operator requires the information of the remainders, i.e. $(x+y)/z = x/z + y/z + (x\%z + y\%z)/z$, which might not be explicitly available in the circuit implementation under verification. Although it would still be desirable to be able to handle these cases, the division and remainder operations are not as often used in modern HDL design as the other operations handled by our approach.

3.2 Expediting equivalence checking problems

Efficient execution of these equivalence checking problems could in general be achieved by judiciously choosing the ordering $(y_1...,y_L)$ for bit-decomposition using structural information:

OBSERVATION 1. *(Effectiveness I) In verifying the sequence of equations in Theorem 1, choosing the bit with the smallest fanout cone size for decomposition first will result in easier equivalence checking problems.*

The fanout cone size of a signal denotes the number of signals within the fanout cone of that signal in the circuit. With this choice of bit-decomposition, the resulting equivalence checking problem will have a lot more equivalent internal signal pairs between the two circuits under equivalence checking. In fact, the resulting equivalence checking problem could be performed by first sharing common circuit portions between $F|_{A_{i+1}}$ and $F|_{\bar{y}_i \cup A_i}$, and then merging the isomorphic portions between both sides of the equation in the miter approach. The equivalence checking problem is then effectively reduced to verifying the equivalence of small portions of the two circuits under the constraint that these portions are driven by the merged isomorphic portion.

Choosing the bit for decomposition from the ones with small fanout cone sizes could also help the design error diagnosis. For example, if the largest i for which the equation in Theorem 1 is not satisfied is α, then there is an error in the fanout cone of y_α within circuit $F|_{A_{\alpha-1}=0}$. Notice that the operand ordering problem and the issue of merged arithmetic are automatically addressed by this decomposition strategy.

Although the application of theorem 1 could involve recursive invocation due to the requirement of verifying equation (2), it is often possible to reuse former equivalence checking results. For example, to verify whether $F(x_{n-1}..,x_0,y_{n-1}..,y_0,z_{n-1}..,z_0)$ implements $f = x \times y \times z = (x_{n-1}x_{n-2}...x_0)_2 \times (y_{n-1}y_{n-2}...y_0)_2 \times (z_{n-1}z_{n-2}...z_0)_2$ with ordering $(x_{n-1}..,x_0,y_{n-1}..,y_0,z_{n-1}..,z_0)$, it requires to prove the following equations ($k = 0,1,..n-1$):

$$\begin{aligned}
& F(0..0,x_{n-1-k},..,x_0,y_{n-1}..,y_0,z_{n-1}..,z_0) \\
= \ & F(0..0,x_{n-2-k}..,x_0,y_{n-1}..,y_0,z_{n-1}..,z_0) \\
& +F(0..0,x_{n-1-k},0..,0,y_{n-1}..,y_0,z_{n-1}..,z_0) \\
& F(0..0,y_{n-1-k}..,y_0,z_{n-1}..,z_0) \\
= \ & F(0..0,y_{n-2-k}..,y_0,z_{n-1}..,z_0) \\
& +F(0..0,y_{n-1-k},0..0,z_{n-1}..,z_0) \\
& F(0..0,z_{n-1-k}..,z_0) \\
= \ & F(0..0,z_{n-2-k}..,z_0) \\
& +F(0..0,z_{n-1-k},0..0)
\end{aligned}$$

and the following recursive invocation of Theorem 1:

$$\begin{aligned}
& F(0..0,x_{n-1-k},0..,0,y_{n-1}..,y_0,z_{n-1}..,z_0) \\
= \ & f(0..0,x_{n-1-k},0..,0,y_{n-1}..,y_0,z_{n-1}..,z_0) \\
& F(0..0,y_{n-1-k}..,y_0,z_{n-1}..,z_0) \\
= \ & f(0..0,y_{n-1-k}..,y_0,z_{n-1}..,z_0) \\
& F(0..0,z_{n-1-k},0..0) \\
= \ & f(0..0,z_{n-1-k},0..0)
\end{aligned}$$

These equations could be simplified by the equations:

$$\begin{aligned}
& F(0..0,x_{n-1-k},0..,0,y_{n-1}..,y_0,z_{n-1}..,z_0)|_{x_{n-1-k}=x_{n-1-k-l}} \\
= \ & F(0..0,x_{n-1-k-l},0..,0,y_{n-1}..,y_0,z_{n-1}..,z_0) \times 2^l
\end{aligned}$$

which will often result in more efficient equivalence checking in most implementations, and reduce to only one recursive invocation of theorem 1 in this case.

The appearance of possible false negatives in the direct application of the previous prescription should be properly addressed. It was pointed out in [7] that cutpoints inside a full adder is an intrinsic source of false negative. It could be avoided using the fanout cone of the adjacent bit in the ordering to move the support backward. Care should be taken to handle these false negative systematically to avoid excessive backward substitution in BDD-based equivalence checking.

3.3 Implicit verification of arithmetic circuit

One particularly serious problem for Wallace tree architecture using BDD-based equivalence checking in this approach is the large support variable size, which will result in excessive variable reordering. We found that the implicit verification approach [16] is capable of reducing the support variable size.

Implicit verification of arithmetic circuits is based on the fanin-based partitioning over all primary outputs of the arithmetic circuit [16]. It was observed that, under the setting of the equivalence checking of arithmetic circuits, the equivalence of lower significant bits could be used as a precondition to facilitate the equivalence checking of the next higher bit. More specifically, given two arithmetic circuits with binary value outputs $c = (c_{l-1}c_{l-2}..c_0)_2$ and $c' = (c'_{l-1}c'_{l-2}..c'_0)_2$, the equivalence checking of these two circuits starts from the least significant bits (LSB) c_0 and c'_0. The equivalence checking of the more significant bits c_k and c'_k are conducted by checking whether the expression

$$(c_k \oplus c'_k)\overline{(c_{k-1} \oplus c'_{k-1})}...\overline{(c_{k-b_k} \oplus c'_{k-b_k})}$$

equals to zero for each k, where the parameter b_h increases successively for an individual k if the equivalence is not proved until the LSB is reached. In proving this proposition using local BDD-based approach, c_k and c'_k are built using signals at the fanin cone boundary of c_{k-1} and c'_{k-1} as support variables, and the expressions above involving different contiguous bits are evaluated using composition. This prescription of local BDD construction could potentially avoid the intermediate explosion of BDDs corresponding to the above expression, and it turns out to be crucial in handling Wallace tree architecture.

4. EXPERIMENTAL RESULTS

Experimental results for different implementations of the following three arithmetic functions are shown in Table 1:

$$I: \quad S = a \times b + c$$
$$II: \quad S = a_0 \times b_0 + a_1 \times b_1$$
$$III: \quad S = a \times b \times c$$

where a's and b's are of size n bits, and c's are of size $3n$ bits in I and n bits in III. Some implementations of these arithmetic functions with different architectures are verified using the proposed approach with no *a priori* architectural information given to the verification program. The architectures are denoted in the subscripts using the abbreviations introduced in section 2. Function III essentially has two addition reduction tree and we denote their architectures in the subscript sequentially. The experiments are performed on a Pentium III 733MHz machine with 256MB memory running Linux OS. The results are shown with the upper field representing the verification time in seconds and the lower field representing memory usage in MB. The number appearing in the first row denotes the size parameter n within the corresponding column. The results for circuit size up to $n = 80$ is shown. A BDD-based equivalence checking procedure using CUDD package [15] with sifting reordering is used. We used implicit verification [16] to effectively reduce the support variable size. The results show that our approach is capable of verifying these examples with different architectures. The only one that we could not finish is I_{WA} for $n = 80$, which still suffers from excessive variable reordering even with the implicit verification technique. Notice that no prior work could be compared to our result due to the minimal assumptions made on the circuits in our approach. However, our results do show better or comparable performance to other former approaches even with weaker assumptions on the circuits. The low memory usage also suggests that structural similarities in each equivalence checking problem are effectively utilized. Our results support the following observation:

n	16	32	48	64	80
I_{AA}	0.96	16.7	86	258	617
	0.61	0.92	1.54	1.60	2.21
I_{CS}	1.59	17.7	87	259	619
	0.65	0.95	1.60	1.65	2.01
I_{WA}	1.72	42	251	1720	
	0.61	0.65	0.87	0.91	
I_{BR}	3.61	47	261	780	2476
	0.58	0.75	3.98	4.06	4.58
II_{AA}	1.87	33	161	503	1192
	0.41	0.71	1.02	1.23	1.61
II_{CS}	2.21	35	171	502	1196
	0.43	0.70	1.05	1.40	1.68
$III_{AA,AA}$	1.89	33	183	502	1215
	0.42	0.72	0.97	1.34	1.63
$III_{CS,CS}$	2.30	35	174	513	1237
	0.45	0.75	1.12	1.42	1.69

Table 1: Experimental results. Within each entry, the upper number is the verification time in seconds and the lower one is the memory usage in MB. (abbreviations: Addition array(AA), Wallace tree (WA), Carry-save-adder array (CS), Booth's recoded (BR))

OBSERVATION 2. *(Effectiveness II) Common architectures of arithmetic circuits have a bit-decomposition ordering that could result in effective execution of the equivalence checking problems in Theorem 1.*

This observation is a reflection of the underlying addition reduction tree structure. Theorem 1 by itself is merely a mathematical statement. Only the evidence from Observations 1 and 2 could justify the effectiveness of the proposed approach.

5. RELATED WORK

Verification of arithmetic circuits using functional equations was first introduced in [8], where the equation $x \times (y+1) = xy + x$ is used for the verification of multipliers. The verification problem of one multiplier is then posed as one single functional equation, but it is executed as a sequence of equivalence checking problems to break the carry-propagation chain in the $(y+1)$ term by cases. However, the circuits under equivalence checking could have quite few equivalent internal signal pairs due to the large fanout cone of the difference circuit, and the resulting equivalence checking problems could be quite hard.

Functional equation has also been used to verify multipliers in [7] with a similar procedure described in this paper. The solution proposed in the current paper offers a more general framework that could handle a larger class of arithmetic circuits through a structural-based bit-decomposition strategy. The inclusion of the implicit verification technique further extends the capability of this framework to handle Wallace tree architecture, which could not be handled effectively in [7].

Implicit verification [16] is the first technique that could handle equivalence checking over structurally dissimilar arithmetic circuits. In our context, the circuits under comparison are structurally similar, but implicit verification technique still introduced the benefit of reducing the support variable size, which is essential in some cases, e.g. Wallace tree architecture.

Self-referential verification is a novel idea that utilizes functional equations to use a circuit implementation to verify itself without referencing to a standard. However, similar idea has been proposed in quite different contexts that represent some even more surprising and elegant solution [12].

6. CONCLUSIONS AND FUTURE DIRECTIONS

We present a systematic analysis of the gate-level arithmetic circuit verification problem and propose a self-referential approach to solve this problem. Experimental results show that this approach is capable of handling a number of arithmetic circuits with different architectures that cannot be verified effectively by existing approaches, including the multiply-add units and the inner product units. This framework is capable of handling most arithmetic circuits composed of operators $+, -$, and $\times$ with different operand ordering and even with merged arithmetic implementations. It would be useful to integrate this approach into a generic equivalence checking engine.

This framework seems to be incapable of handling circuits derived by employing transformations using arithmetic relations other than the commutative and associative laws, especially factorization and expansion associated with the distributive law. We believe this problem should be best solved at the RTL level with the integration of a computer algebraic system [3]. Only a tight integration of an equivalence checker, a gate-level arithmetic circuit verifier and a computer algebra system could completely solve all the challenging issues of arithmetic circuit verification problem in modern HDL-based VLSI design.

7. ACKNOWLEDGEMENTS

This research was supported in part by SRC Task 835.001 and NSF International Research Center for SoC. The final version of this paper was prepared when the first author is at Novas Software.

8. REFERENCES

[1] Special issue on verification of arithmetic hardware. *Formal Methods in System Design*, 14(1), 1999.

[2] D. Brand. Verification of large synthesized designs. In *Proceedings of the International Conference on Computer-Aided Design (ICCAD)*, pages 534–537, November 1993.

[3] R. K. Brayton. The future of logic synthesis and verification. In *Logic Synthesis and Verification*, pages 403–434. Kluwer Academic Publishers, 2002.

[4] R. E. Bryant. Graph-based algorithm for boolean function manipulation. *IEEE Trans. on Computers*, C-35(8):677–691, 1986.

[5] R. E. Bryant. On the complexity of VLSI implementations and graph representations of boolean functions with application to integer multiplication. *IEEE Trans. on Computers*, 40(2):205–213, 1991.

[6] R. E. Bryant and Y. A. Chen. Verification of arithmetic circuits with binary moment diagrams. In *Proceedings of the 32nd ACM/IEEE Design Automation Conference (DAC)*, pages 535–541, 1995.

[7] Y.-T. Chang and K.-T. Cheng. Induction-based gate-level verification of multipliers. In *Proceedings of International Conference on Computer-Aided Design (ICCAD)*, pages 190–193, November 2001.

[8] M. Fujita. Verification of arithmetic circuits by comparing two similar circuits. In *Proceedings of the 8th International Conference on Computer Aided Verification (CAV)*, pages 159–168, 1996.

[9] K. Hamaguchi, A. Morita, and S. Yajma. Efficient construction of binary moment diagrams for verifying arithmetic circuits. In *Proceedings of International Conference on Computer-Aided Design (ICCAD)*, pages 78–82, November 1995.

[10] T. Kim, W. Jao, and S. Tjiang. Arithmetic optimization using carry-save-adders. In *Proceedings of the 35th ACM/IEEE Design Automation Conference (DAC)*, pages 433–438, 1998.

[11] S. Kimura. Residue BDD and its application to the verification of arithmetic circuits. In *Proceedings of the 32nd ACM/IEEE Design Automation Conference (DAC)*, pages 542–545, 1995.

[12] A. Migdall. Correlated-photon metrology without absolute standards. *Physics Today*, 52(1):41–46, 1999.

[13] B. Parhami. *Computer Arithmetic: Algorithms and Hardware Designs*. Oxford University Press, 1999.

[14] M. Potkonjak and J. Rabaey. Optimizing resource utilization using transformations. *IEEE Trans. on CAD*, 13(3):277–292, 1994.

[15] F. Somenzi. *CUDD: CU decision diagram package Release 2.3.1*. University of Colorado at Boulder, 2001.

[16] T. Stanion. Implicit verification of structurally dissimilar arithmetic circuits. In *Proceedings of IEEE International Conference on Computer Design (ICCD)*, pages 46–50, 1999.

[17] E. E. J. Swartzlander. Merged arithmetic. *IEEE Transactions on Computers*, 29(10):946–950, 1980.

[18] J. Um, T. Kim, and C.-L. Liu. Optimal allocation of carry-save-adders in arithmetic optimization. In *Proceedings of International Conference on Computer-Aided Design (ICCAD)*, pages 410–413, November 1999.

[19] Z. Zhou and W. Burleson. Equivalence checking of datapaths based on canonical arithmetic expressions. In *Proceedings of the 32nd ACM/IEEE Design Automation Conference (DAC)*, pages 546–551, 1998.

PANEL: Whither (or Wither?) ASIC Handoff?

Chair: Michael Santarini, EE Times, San Mateo, CA

Organizers: Sudhakar Jilla, Synopsys, Mountain View, CA and Mark Miller, Tera Systems, Campbell, CA

Abstract

The traditional ASIC netlist handoff is changing - but to what? Is RTL handoff finally a reality? Or, will a placement-based handoff model emerge? Are differences among underlying tool technologies and methodologies only cosmetic? Or, are there fundamental business and IP distinctions? These and other questions will be discussed as the panel examines the future of the designer - ASIC vendor - EDA vendor relationship.

Panelist Statements

Tommy Eng
Tera Systems, Campbell, CA

Complex ASIC designs are encountering production delays and multifaceted design closure issues, caused largely by the need to consider physical-design-induced performance factors brought on by DSM effects of the latest manufacturing processes. These factors must now be modeled in the front-end of the design flow to ensure design closure. The EDA industry has reacted by developing a new generation of Physical Synthesis tools taking physical modeling effects into account during synthesis and optimization. Unfortunately, its application is resulting in challenges for system designers and their ASIC vendors by driving mandatory methodology changes, i.e., moving the sign-off point up to the RT level, or down into the physical domain with the inclusion of placement data. This is further complicated by a shortage of physical design expertise within system design teams and the need for better RTL handover methodologies and tools for ASIC vendor support teams. One thing is abundantly clear: higher quality RTL will result in faster chips, sooner. Visibility into downstream implications of RTL, early in the design process, has enormous impact on design convergence and on-time schedule performance.

Sandeep Khanna
Synopsys, Mountain View, CA

Timing closure is becoming a mainstream problem for most designs for technologies at 0.18μm and below. Timing closure, the main value proposition of the new generation of Physical Synthesis tools such as Physical Compiler from Synopsys, eliminates the surprises between actual and expected interconnect loading and coupling at different stages in the design process that manifest themselves as chips failing to meet the target speeds. This is achieved by considering physical effects much earlier in the design process. The traditional netlist based signoff model has also similarly evolved to a placement based signoff model where customers hand off netlist and placement to the vendors, who complete detailed routing and are seeing a strong correlation with the earlier predictions made during Physical Synthesis. Customers are already beginning to experience benefits in the form of performance and productivity boosts. Vendors also see the benefit in being able to deliver silicon on time and eliminating the painful iterations between synthesis (customers) and routing (vendors). Some models are extending placement-based signoff to RTL or floorplan stages. By definition, a pure RTL signoff model involves handing off RTL instead of netlist and/or placement. The challenge is to develop tools that can give an indication of the ability to close timing without actually synthesizing or doing any physical design. If and when such tools become a reality, certain classes of designs can benefit from an RTL signoff. Placement signoff is here today while RTL signoff still remains a pipe dream.

Kamalesh Ruparel
Cisco, San Jose, CA

It is no longer an illusion. The verdict is in: timing closure is and continues to be a formidable challenge in today's ASIC designs. This is because as we increasingly soak ourselves in VDSM technologies, physics presents all manner of difficulties in physical design, signal integrity, wild defect mechanisms, etc. The sudden downturn in the economy has only exacerbated the situation by making these problems even costlier to surmount. High end (8 million gate and above) ASIC design is no longer an exclusive club – even medium sized ASICs (3 to 7 million gates) are getting to experience the spicy sauces of VDSM.

At a broad level, ASIC designs can be broadly bucketed into three major groups: the very high-end (die sizes of 18+ mm on a side, 500+ MHz clock frequency), mid-range (die sizes of 10 - 15mm on a side, 125 - 500 MHz), and the low end (die sizes of <10mm on a side, < 100 MHz).

In the low end ASIC design, RTL sign-off is possible, but ultimately ends up becoming an ASSP product. For mid-range to high-end ASIC designs, we see the RTL sign-off model as a complete misfit. Intricate dependencies between external versus internal IP, tools and methodologies, logical and physical design processes, functional verification and AC-signoff, ASIC/EDA/customer partnerships, die-size sweet spots and needed packaging technologies, TTM and yield/volume ramp-ups, and so on make it practically impossible for a true RTL-signoff to become a reality.

For these upper-tier ASIC designs, the only prayer to a reasonable turnaround time (TAT) and a predictable timing closure is some form of a placement based ASIC handoff. In order to achieve such a handoff with any significant value, all the necessary dependencies mentioned above need to be carefully managed while at the same time leveraging the best available technologies such as virtual prototyping and physical synthesis. Forcing designers and architects to view physical design as an integral and necessary part of the design process has now become inevitable. Wirability, congestion, power and clock planning have to be dealt with weeks before RTL gets frozen. What was a paradigm shift a couple of years ago is now becoming mainstream: physical synthesis has effectively conquered the land of the inadequate statistical WLM. As if this were not enough, a thorough SI analysis of the routed netlist is mandatory before final tapeout. All of this has not only increased the number of iterations between the logical and physical worlds, but has also squeezed the requirements of overall TAT windows available for design development.

RTL signoff, then, is another way of throwing the problem over the wall... just a little sooner. Unfortunately, there is a tendency on the part of the vendors to promise a single panacea to a customer's timing closure and other ASIC engineering challenges. The sheer nature and complexity of the problem prevents a single EDA or ASIC-process solution from achieving success across all designs. So, while a particular technical solution/tool may seem to provide a great solution for one class of designs, it will most likely fall apart in a different class of designs. The required results can only be achieved with a multitude of solutions, a disciplined methodology, steadfast execution and carefully managed three-way partnerships between EDA, ASIC and design houses.

Tom Russell
IBM ASIC, Burlington, VT

ASIC methodology must evolve to adapt to advances in process technology and EDA tools. The widespread acceptance of physical, timing/placement-driven synthesis, coupled with the advent of a new breed of powerful architectural RTL-stage design planning tools, is surely driving a shift away from today's traditional netlist signoff for ASICs. For many ASIC customers, this will drive them to take on more of the back-end, lower-level activities prior to sign-off, but for another class of customers - perhaps less cutting edge in their needs and unwilling to invest in mastering deep sub-micron effects – an RTL signoff model will be both efficient and economically attractive. For these reasons, signoff engagement models will be migrating in both directions and ASIC vendors are responding today to meet these changing customer needs by offering flexible engagement models.

Kazu Yamada
NEC Electronics Inc., Santa Clara, CA

Rapid and consistent changes in the technical landscape have forced the diminishing of the traditional definition of ASIC hand-off. It is critical and important to have a clear and comprehensive hand-off definition that defines boundary conditions among design stages in today's complex ASIC design projects. However, hand-off no longer means "event horizon" for both the ASIC user and the ASIC vendor. In a way, there is no hand-off any more, in that the absolute boundary has become a matter of job sharing. An ASIC user can "interface" his designs to the ASIC vendor at C language level or all the way down at GDSII level, depending on the user's preference. However, level of involvement or interaction between system/chip designer and ASIC provider always starts from design definition, and lasts to deep inside of the physical implementation stage. Only the magnitude of interaction varies from project to project, primarily due to the choice of balance between time-to-market and level of optimization to the silicon technology. ASIC vendors should be capable of handling this multi-level or continuous hand-off process in order to provide the most value to their customers. Besides, the higher the interface level, the more value-add from the ASIC vendor point of view, resulting in a better chance to lock in customers.

Software Synthesis from Synchronous Specifications Using Logic Simulation Techniques

Yunjian Jiang
wjiang@eecs.berkeley.edu

Robert K. Brayton
brayton@eecs.berkeley.edu

Department of Electrical Engineering and Computer Science
University of California, Berkeley CA 94720

ABSTRACT

This paper addresses the problem of automatic generation of implementation software from high-level functional specifications in the context of embedded system on chip designs. Software design complexity for embedded systems has increased so much that a high-level functional programming paradigm need to be adopted for formal verifiability, maintainability and short time-to-market. We propose a framework for efficiently generating implementation software from a synchronous state machine specification for embedded control systems. The framework is generic enough to allow hardware/software partition for a given architecture platform. It is demonstrated that the logic optimization and simulation techniques can be combined to produce fast execution code for such embedded systems. Specifically, we propose a framework for software synthesis from multi-valued logic, including fast evaluation of logic functions, and scheduling techniques for node execution. Experiments are performed to show the initial results of our algorithms in this framework.

Categories and Subject Descriptors

B.6.3 [**Logic Design**]: Design Aids—*Simulation*; D.2 [**Software**]: Software Engineering

General Terms

Algorithms Design

1. INTRODUCTION

We adopt a platform-based design methodology, with two major features: (a) using standardized or application specific processing platforms to execute intelligent software, and (b) raising the abstraction level of software designs with implementation code automatically generated from functional models.

By using standardized processors and moving intellectual properties into software design, the methodology reduces hardware manufacturing cost, increases flexibility of the design, hence extends product life cycle and shortens time-to-market. However, software design in such systems becomes extremely complex. High-level functional programming paradigms tackle this problem by abstracting away implementation details with code generated automatically from functional computation models.

We focus on control intensive embedded systems, for instance, automobile engine, airplane, and network protocol controls. The computation model used is a network of extended finite state machines (EFSMs). An EFSM is a system with a finite state controller interacting with an unbounded integer data-path [9]. Each transition of the controller is guarded by a predicate over the integer variables, and is associated with an action function which updates the values of the integer variables. This model has been studied for system design and verification. High-level synchronous languages like Esterel [7] can be used to program in such models.

A static synchronous model like EFSMs cannot model or verify dynamic aspects of the embedded software, like dynamic task and memory allocation, which may be important in some applications. We argue that in safety critical control applications, like avionics and automobile electronics, such synchronous models have to be used for formal verifiability and executable specifications. There are research efforts to combine the synchronous model with dynamic aspects to provide powerful system level modeling, which is not the focus of this paper.

Our design flow starts from high-level synchronous specification, textual or graphical, like Esterel, Argos, Lustre, or Matlab StateFlow. EFSMs are derived from these and optimized using multi-valued (MV) logic synthesis techniques. The internal representation, a network of functional nodes, supports abstract data variables (which may be implemented by integer or floating point data path) and black-box function calls. The computation involved with the abstract data variables are not interpreted by the tools, but the information flow is utilized for synthesis in the logic domain.

In the mapping phase, the functional model can be mapped to different architecture platforms. For a configurable system-on-chip platform, each node can be estimated for performance and implemented either in hardware or software, using techniques described in [4]. For pure software implementation, this paper presents a framework for generating efficient code using logic simulation techniques. This includes fast logic function evaluation and scheduling for node execution.

In the remainder of this paper, Section 2 reviews related work in this area; Section 3 describes the computation model we use for optimization and synthesis; Section 4 details our techniques on code generation and a scheduling scheme; Section 5 shows experimental results followed by conclusions.

2. RELATED WORK

The Polis project [2] uses EFSMs for intermediate representation and synthesis. A high-level design language, like Esterel [7],

Table 1: Node types in a control data network

node types	operation	input	output	example
control	logical	MV	MV	$a\{0\}b\{1,2\}+a\{1\}$
data	arithmetic	data	data	`x + y`
multiplexer	assignment	MV/data	data	$c\{0\}x+c\{1\}y$
predicate	predicate	data	MV	`x > y`

is compiled into a set of EFSMs, which are subsequently optimized and mapped into hardware and/or software [3], depending on system constraints. The communication model among these EFSMs is a Globally Asynchronous Locally Synchronous (GALS) mechanism. Each EFSM is represented by a single state transition table for the control path and a lookup table for the data-path. Binary Decision Diagrams (BDDs) are used to optimize the state transitions and synthesize the code. This representation cannot scale to very large designs. In addition, data-path information is not utilized for optimization.

The Esterel compiler from the Esterel team [11] has two ways of generating code from different representations of a FSM. The automata method [6] enumerates all possible states and generates the action code for each state. This can produce very fast code, but for large examples, the code size blows up with the state space. The circuit method [5] combines a multi-level Boolean circuit with a data-path table, and generates code based on the logic equations. The code size is scalable, but with compromises in execution speed. The communication model between processes is synchronous composition through a set of completely unordered synchronous events.

Edwards proposed compiling Esterel directly into sequential C code [10], by static scheduling and context switching. This approach applies only to statically schedulable applications. Other Esterel compilation techniques that have been proposed include modular compilation [14], and reaction acceleration [24]. By compiling Esterel directly into C code, these approaches do not easily support partitioning between hardware and software.

The area of code generation for control applications overlaps largely with discrete function simulation. Most closely related publications are in cycle simulation using decision diagrams [19, 1]. The scheduling techniques to be presented here also have some resemblance with event-driven and back-tracing simulation [23, 18]. However, these simulation techniques are tuned for implementation in high-performance computer servers, rather than cost sensitive and resource constrained embedded systems.

The literature is also rich in software synthesis for other application domains and computation models, for instance, code generation from synchronous data flow, process networks, and Petri Nets [22]. These are not studied or compared in our paper.

3. COMPUTATION MODEL

We use a network of EFSMs as the computation model. The techniques presented are for synthesis and code generation for a single EFSM, but they are easily extensible to interacting EFSMs with synchronous or asynchronous communication.

We adopt a functional representation that uses multi-valued variables and Boolean algebra. We believe this applies naturally to embedded software synthesis with advantages over binary logic.

- It explores bit level parallelism and thus saves execution instructions for updating variables and RAM usages.
- A larger optimization space is explored for finding an optimal implementation without being limited to a particular binary encoding.

We use MVSIS [13, 8] for optimization and synthesis. For EFSM minimization, it operates on a control-data network representation.

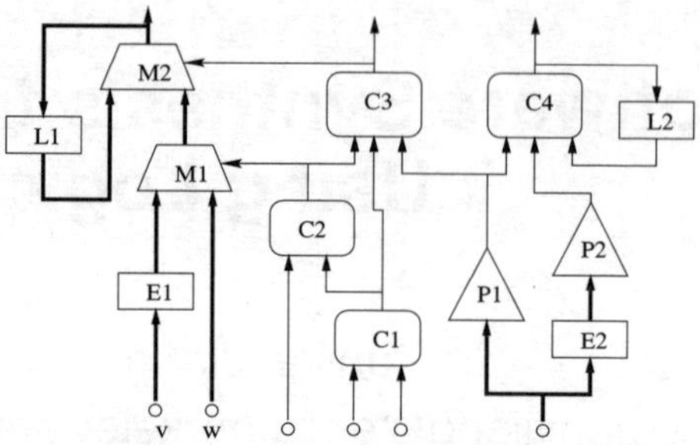

Figure 1: Control-data network

These have control nodes and data nodes interconnected with two types of variables: multi-valued variables with finite ranges and abstract data variables with unbounded ranges. There are four types of nodes: *control*, *multiplexer*, *data* and *predicate*. These are classified according to their input and output variables types, as shown in Table 1. In the examples in the table, a, b, c are multi-valued control variables, and x, y are data variables.

A multiplexer is defined as $f = f(y_c, y_0, \cdots, y_{n-1})$, where y_c is a MV-variable with n values, y_i, $(i \in [0, n-1])$ are data inputs. The output f is assigned to y_i if $y_c = i$. The data computation contained in predicate nodes and expression nodes are currently modeled as uninterpreted strings. The expressions must be arithmetics definable by the semantics of the C language.

There is a directed edge from node i to node j, if the function at node j syntactically depends on the output variable at node i. The network has a set of primary inputs and a set of nodes designated as the outputs of the network. There are also latches for both control and data variables. Figure 1 shows an example of a control-data network with two latches, where bold wires indicate data variables.

Each control node is a multi-valued function represented in Multi-valued Decision Diagrams (MDDs) and sum-of-products. In general, a variable x_i is multi-valued and takes on values from the set $P_i = \{0, 1, ..., |P_i| - 1\}$. A **literal** of a MV-variable x is associated with a subset of values for that variable and evaluates to 1 if the variable takes on any of the values in the set. A **product term** or **cube** is a conjunction of literals and evaluates to 1 if each of the literals evaluates to1. A **sum-of-products** (SOP) is the disjunction of a set of product terms and evaluates to 1 if any of the products evaluates to 1. For detailed definitions refer to [13].

A set of technology independent optimization methods have been implemented in MVSIS [8]. These include algorithms extended from binary logic: algebraic decomposition [12], don't care-based simplification [15], elimination, resubstitution; and also algorithms specifically tuned for multi-valued logic: node pairing and encoding [13].

4. SOFTWARE SYNTHESIS

The software synthesis problem is, given such a multi-level control-data network, generate an efficient software implementation in C (or target dependent machine language), such that appropriate constraints are satisfied. These may include resource constraints, like RAM and ROM usages, and timing constraints. This is a mapping problem from a parallel execution model, originally targeted for hardware, to a sequential execution model, where all tasks share the same processing unit. The goal is to generate fast sequential simulation code with constraints on the code size. This may be achieved by reducing the time spent on each node, and reducing the total number of nodes that has to be evaluated. We therefore solve two sub-problems: code generation for individual logic func-

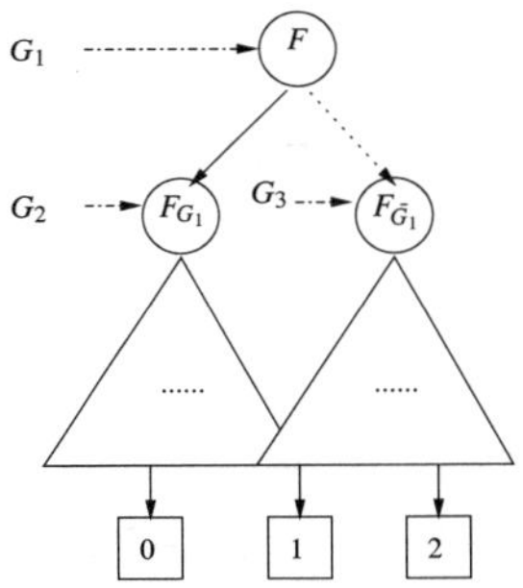

(a)

```
_N1:   if (! M ∈ G1) goto _N2;
       else goto _N3;
_N2:   if (! M ∈ G2) goto _N4;
       else goto _N5;
_N3:   if (! M ∈ G3) goto _N6;
       else goto _N7;
_N4:   ... ...
```

(b)

Figure 2: Generalized Cofactoring Diagram (a) and its generic software implementation (b)

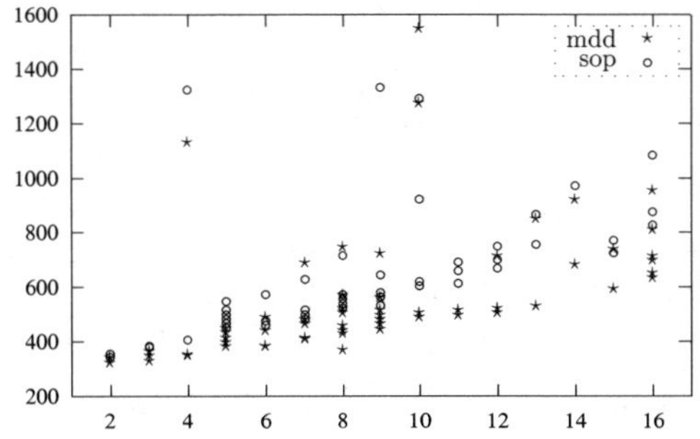

Figure 3: Comparison of code size w.r.t. the number of inputs

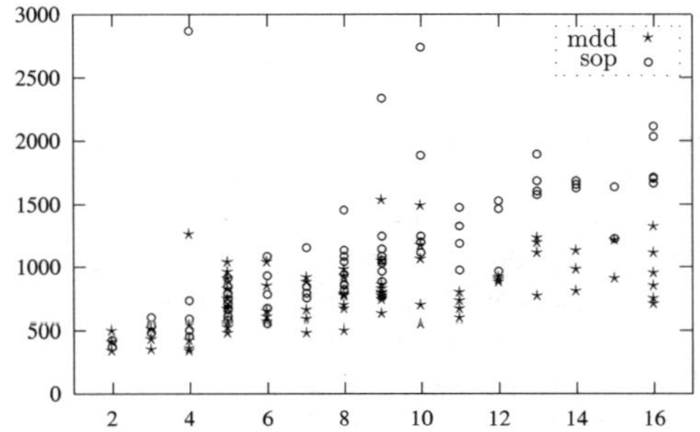

Figure 4: Comparison of run-time w.r.t. the number of inputs

tions, and scheduling of node execution, detailed in the remainder of this section.

4.1 Code Generation

Given multi-valued logic function $f(x_1,\ldots,x_n) : P_1 \times P_2 \times \ldots \times P_n \rightarrow Q$, where P_i is the domain of input variable x_i and Q is the domain of the output variable, find software implementation such that the code size and execution speed is optimized.

Different code generation schemes exist depending on various logic function representations: sum-of-products (SOP), multi-valued decision diagrams (MDDs), and lookup tables (LUT). We use a generic model called Generalized Cofactoring Diagram (GCD) to model and unify these different methods.

Similar to MDDs, GCD is a diagram composed of functional nodes and terminal nodes; different from MDDs, there is a set of general cofactoring functions $\{G_1, G_2, \ldots, G_n\}$ (rather than variables), one for each node in the structure, to be cofactored by the intermediate function nodes. As illustrated in Figure 2(a), the top node function F has two out-going edges, representing the cofactors F_{G_1} and $F_{\bar{G}_1}$. Generalized cofactoring for functions is defined as $F_G = F \cap G$. Note that in general, G_i can be multi-valued functions. In that case each value k of G_i is a binary function G_i^k. F would have as many output edges as the number of values for G, each representing the cofactoring of F on one of G_i's value functions: $F_{G_i^k}$.

The software derived from such a GCD takes an input minterm M and determines the output value based on the branching structure, as shown in Figure 2(b). Each branching point represents a sub-space of the input logic domain, and the cofactoring function G_i partitions this further into smaller domains, until the output value can be decided in the sub-space. The input minterm follows the GCD structure and makes branching decisions based on the result of $(M \in G_i)$. In the figure, label `_N1` refers to the code derived from the root node F; label `_N2` refers to the code derived from node F_{G_1}. Label `_N4` and beyond are not shown in the figure.

The code generation problem is to find the set of cofactoring functions $\{G_1, G_2, \ldots, G_n\}$ and build the cofactoring structure, such that the evaluation of the logic function in software is optimized. Long computation time can be tolerated for deriving the structure that gives the best performance, since the generated code is to be loaded in the final products and executed numerous times.

Note that unlike MDDs, GCDs need not be ordered so they are more like free MDDs. A similar optimization scenario based on free MDDs was proposed in [17]. The optimality is measured in code size, which is proportional to the total number of branching nodes, and speed, which is related to (a) the average depth of the diagram, and (b) the cost of the branching test $(M \in G_i)$.

Different code generation methods based on specific functional representations are variations of the GCD:

- **MDD**: Cofactoring functions G's are single variable functions. At each GCD node, the input minterm is tested for the value of a particular input variable. The evaluation speed is bounded by the number of inputs, but the size may be exponential depending on the logic function.

- **SOP**: Cofactoring functions are single cubes. At each GCD node, the input minterm is checked to see if it is contained in the cube associated with that GCD node. The containment test is performed by a bit-wise *AND* operation if the cubes and minterms are represented in bit vectors [16], or by evaluating the value of the cube through Boolean operations. Both the size and the evaluation speed may be exponential depending on the logic function.

- **LUT**: There is only one GCD node that performs a memory lookup. There are as many entries in the memory as the number of input minterms. The output value is retrieved by one memory operations. The code size is exponential, but can be managed by decomposing into cascading LUTs; the evaluation speed is potentially limited by the memory latency of the system architecture. Experiments show that the memory bottleneck of common processor architectures prevents this method from being competitive compared with the others.

Figure 3 and 4 shows code size and run-time comparison of MDD and SOP code for some 150 logic functions extracted from a common benchmark suite, with input variables ranging from 2 to 16. The codes are generated and compiled with `gcc -O2` on an Intel PIII platform running Linux. The code size is the number of bytes in the object code; the run time results are the average of 10 million simulations on pseudo random input vectors. We make the following observations: (a) Both methods demonstrate a linear run-time degradation as the number of inputs grows. The SOP code degrades faster than the MDD code. (b) The SOP code is in general slower then the MDD code, except for only very small logic functions (less than 3 inputs) and occasionally large logic functions with small cube counts. (c) There are cases when the SOP code is faster or smaller then the MDD code. This indicates that a hybrid approach (such as GCDs) which derives a specialized cofactoring structure for each logic function would produce consistently superior results.

The real runtime performance depends on the target processor architectures. Our result gives an approximate comparison of the code generation methods. New GCD structures based on optimal cube cofactoring is to be explored in future work. In the final experiments, we use MDD code for large logic functions and SOP code for small functions.

4.2 Scheduling

Given a network of nodes, as described in Section 3 (see Figure 1), find a scheduling scheme for the firing of the nodes, such that the overall execution time is minimized. A straight-forward approach is to statically schedule the nodes with a topological sort, similar to oblivious logic simulation. Every node in the network is always evaluated, even though sometimes they do not need to be.

Unlike hardware execution, where timing is characterized by the critical path delay, software execution time is associated with the total number of nodes that have to be evaluated for a particular input vector. A data expression node may not be evaluated if it is de-selected by its fanout multiplexer. (For example node `M1` and `E1` in Figure 1.) A control node or a predicate node may not be evaluated if its value is not observable at the primary outputs. In the remainder of this section, we develop theories for identifying these cases and develop techniques to make use of them in software scheduling.

4.2.1 *Triggerability*

To simplify the description, in the sequel if not specified otherwise, we use `inputs` to denote primary inputs and the output of latch variables; we use `outputs` to denote primary outputs and the inputs of latch variables.

DEFINITION 1. *A node is called a trigger node (or triggerable) if there exists at least one input vector, which can determine the output values without this node.*

The triggerability of a node can be tested by the existence its Maximal Observability Don't Cares (MODC). The $MODC(x_i, y_k)$ for an edge from node x_i to node y_k is the set of minterms in the domain of all intermediate variables, such that the values of x_i cannot be observed at the output of node y_k. The MODC of a node x_i is the intersection of the MODCs for all its fanout edges [15]. The MODC of node y_k is also inherited by all of its fanin nodes. Expressed in equation:

$$MODC_{x_i} = \cap_{y_k \in fanout(x_i)} (MODC(x_i, y_k) \cup MODC_{y_k})$$

THEOREM 1. *A node is a trigger node if and only if its MODC set is not empty.*

If the MODC set of a node is empty, it is always observable at the outputs for all possible input combinations. This means they have to be evaluated for all cases in order to obtain the output value. We call these nodes *static* nodes.

For control nodes with multi-valued logic, the MODC theory and algorithms have been developed in [15]. For a multiplexer $f = f(y_c, y_0, \ldots, y_{n-1})$, where y_c is the control variable and y_i are data input variables, let $MODC_f$ be the MODC set computed for the output f. Then the MODC set for its fanin edges are:

$$MODC(y_i, f) = (y_c \neq i) \cup MODC_f$$
$$MODC(y_c, f) = MODC_f$$

Intuitively, the control variable has to be evaluated in order to do the multiplexer assignment, therefore its MODC directly inherits from node f; a data variable is blocked from the output of f, if it is de-selected by the control variable.

For a predicate or expression node $f = f(y_1, \ldots, y_n)$, where all inputs are data variables, we have:

$$MODC(y_i, f) = MODC_f$$

In other words, an expression node is a trigger node if and only if all of its fanouts are trigger nodes. We use the following theorem to simplify the analysis of the data-path triggerability:

DEFINITION 2. *A trigger tree is a largest set of predicate and expression nodes, among which only one node fans out to the external world. This node is called the root node of this trigger tree.*

COROLLARY 1. *All nodes in a trigger tree share the same triggerability.*

A trigger tree is treated as a single node in our analysis and code generation. In the example in Figure 1, node `P2` and `E2` are combined as a trigger tree.

4.2.2 *Triggerized Scheduling*

After the triggerability of all nodes in the network are computed, we develop a quasi-static execution schedule based on this information. We first outline some features of the scheduling scheme and then give a general algorithm.

The generated code has two portions: (a) a static portion consisting of static nodes generated in a topological order from inputs

```
Algorithm [Network Code Generation]:
input: control-date-network
Partition network into trees and compute triggerability;
generate_header();
List1 = topological list of static trees;
Foreach tree i in List1
    code_gen_tree(i);
generate_tail();
List2 = topological list of trigger trees;
Foreach tree i in List2
    code_gen_tree(i);
End
code_gen_tree [Tree Code Generation]:
input: trigger tree
Foreach node k in tree in topological order
   If k is data node
      Generate expression; Continue;
   Build GCD for logic function k;
   Traverse each node g in GCD
      If g requires value from trigger node
        If cost(g) < threshold
           code_gen_tree(g);
        else
           Generate function call for g;
      Generate branching code for g;
End
```

Figure 5: Code generation with triggerized scheduling

to outputs; (b) a dynamic portion consisting of trigger nodes. A trigger node is generated as a "cheap" subroutine call (meaning the only overhead is storing the return address). They are executed *on-demand* as requested by their fanout nodes.

Because calling a trigger node has overhead, we use a simple heuristic to detect very small trigger nodes (less than three inputs), and generate the code on the spot, rather than using the subroutine call.

Trigger nodes may be invoked more than once depending on its number of fanouts. A trigger node is called *level_1* triggerable if it has only one fanout node; otherwise it is called *level_2* triggerable. We create a one bit flag for each level_2 trigger node to indicate the status of its execution. Before firing this trigger node, the associated flag is tested and the subroutine is called only if it has never been evaluated before. This again creates additional overhead for level_2 trigger nodes, which is also considered in the screening heuristic. The fact that data nodes are combined into trigger trees increases the size of triggered subroutines, which amortizes the overhead cost.

Figure 5 gives the overall procedure of the code generation scheme. Function call `generate_header()` refers to variable declaration and state initialization; `generate_tail()` refers to the update of latch variables and returning evaluation results. Generating expressions for data nodes is straight-forward for predicates and data expressions, but also involves trigger node testing for multiplexers. It is relatively simple, and hence not detailed in the figure.

A control node is treated as a tree by itself. The GCD structure for a control node is built from either an MDD or a SOP representation, depending on their estimated cost according to the size of the logic function. We use an estimation heuristic based on the number of inputs and number of product terms. At each GCD node, the cofactoring function (single variable for MDD and single product for SOP), is tested for triggerability. If the variables required are coming from trigger nodes, then they would be fired, either through cheap function call or on-the-fly computation. And then the bit-flag for level_2 trigger nodes would be updated.

4.3 Event Driven Simulation

The notion of triggerability explores statically the observability of a node with respect to its transitive fanouts, which extends the idea of back-tracing simulation [23]. Event driven simulation technique [18] tries to achieve the same goal (reducing the number of nodes that need to be evaluated), by dynamic scheduling according to input activities.

As a comparison we apply event-driven techniques to the static nodes in the network. Here a more conservative definition of static node is used instead of the one defined in the previous section. At a given node, if there is no activity among the static fanin nodes, the trigger fanin nodes will not be evaluated even if there might be activity in them. Therefore, for static node k, if k is a control node, all its fanins are labeled as static; if k is a multiplexer, only its control input is labeled as static.

In the generated code, we create a bit vector, one bit for each static node, indicating the activity at its local inputs. The evaluation of a static node then includes additional overhead of checking activity at local inputs and propagating activity to all fanout nodes, which proves to be useful in speeding up the code. Below shows the pseudo code of evaluating node η with input minter M.

```
eval_node (η, M):
if (!activity(η)) return;
if (result_changed( eval_logic (M) );
    foreach fanout k of η
        set_activity(k);
```

Table 2: Results on execution speed on 10 million pseudo random inputs

Examples	strl	E-auto	E-sort	E-opt	MVSIS
fuel-ctr	103	420	10100	6360	5880
pulse-ctr	132	430	12400	9890	7770
instr-ctr1	215	270	28770	8200	5800
instr-ctr2	240	275	30860	8800	6290
dma-ctr	455	400	114000	11900	8510

`eval_logic()` contains the actual code for evaluating the logic function (or data black-box). `activity()` checks the event bit of the corresponding node. These are implemented as fast inline function calls. As an illustration, the example in Figure 1 (without observing the logic in each node) produces the scheduling below, where `E1` and `M1` are trigger nodes:

```
update_latch(L1); update_latch(L2);
eval_node(C1); eval_node(C2); eval_node(P1); eval_node(C3);
eval_node(E2); eval_node(P2); eval_node(C4);
if (C3) { M2 = L1; }
else {
    if (C2) { eval_logic(E1); M1 = E1; M2 = M1; }
    M1 = w; M2 = M1;
}
```

5. EXPERIMENTS

In the experiment, we use Esterel as a high-level specification language, and the Esterel compiler to parse the input Esterel program and produce an intermediate circuit representation called DC. We translate DC into our intermediate control-data network representation in an extended BLIF-MV format. Implementation C code is generated after multi-valued logic optimization.

The examples consist of an electronic engine fuel controller, an instruction decoder (both acyclic and cyclic version), and a direct memory access controller. Table 2 shows the average execution speed of the compiled code on 10 million pseudo random input vectors. Column `strl` shows the lines of Esterel source code, indicating the size of these applications. `E-auto` shows the code generated by the Esterel compiler from an automaton representation, which for large examples tends to blow up; `E-sort` is the code from a binary circuit representation; `E-opt` is also circuit code but optimized by an extension of SIS [20] called `Basicopt`, which consists of binary combinational area optimizations, state encoding and latch removal [21]. If we treat the optimized circuits after `Basicopt` as input, and then optimize further in MVSIS to generate code, the results are shown in column `MVSIS`. The results are obtained on a Intel PIII platform with a Linux 6.1 operating system and 128MB RAM. We use GNU tool "`gprof`" to obtain the average run time consumed by the core evaluation code. As shown, the code generated from MVSIS has faster evaluation speed over `E-opt`. Both the triggerability and event driven approach are experimented and the best results are reported here. Table 3 shows the object code size in bytes after compiled by "`gcc -c -O2`". In general, the code size generated by our approach is larger due to the overhead of back-tracing and event propagation.

Discussion about the experiments:

- These examples are originated from hardware systems (microprocessors and DMAs). However the functionality is representative for controller designs (80-90% control).

- The DC circuit derived from the Esterel compiler is pure binary, and includes both combinational and sequential redundancies. MVSIS is able to rebuild MV variables through

Table 3: Results on code size compiled by `gcc -O2`

Examples	E-auto	E-sort	E-opt	MVSIS
fuel-ctr	2239	2565	2141	2514
pulse-ctr	1691	3525	2845	3400
instr-ctr1	65111	9997	2981	4380
instr-ctr2	65111	10905	3349	4460
dma-ctr	220783	23237	5665	5776

node pairing and merging, but it is hard to reconstruct the original program structure. Also MVSIS has limited sequential optimization capability.

- Automata-based code is very fast, because there is a piece of specialized code tailored for each reachable state. However, the code size explodes with the state space. The structural (circuit) representation is very compact with the state space encoded into a limited number of latches, but this has much slower evaluation speed. We experimented with merging binary latches to create multi-valued ones. This does give some run-time improvements, but exploring the state encoding space for fast simulation is still an open issue.

6. CONCLUSIONS AND FUTURE WORK

The methodology of high-level programming based on a functional model is inevitable in future embedded software applications, especially for mission critical discrete control as in avionics. For these applications, we adopt a design flow centered around a particular model of computation, EFSMs. This model supports high-level specification with synchronous languages (with data types and function calls), and supports hardware/software partitioning and mapping onto a set of architecture platforms.

We developed a framework for generating efficient software implementation from synchronous specifications. A generic setting called Generalized Cofactor Diagram (GCD) was proposed for evaluating individual logic functions. We developed the theory of triggerability for a quasi-static scheduling scheme, and gave an algorithm to generate code. Event-driven simulation techniques are also applied in the code generation phases, and prove to be useful in deriving faster code. Preliminary experiments are performed to analyze our first attempt on some of the algorithms in the framework.

A few algorithms need to be fine-tuned for the software synthesis framework. Logic evaluation of individual functions can be improved by searching for an optimal generalized cube cofactoring structure. Applications that are sequentially very deep (large number of latches) can benefit from exploring the state re-encoding space, or generalized cofactoring on the latch variables.

In the top-level design flow, we have experimented on mapping EFSMs to a reconfigurable architecture platform that has a specialized data-path augmented with reconfigurable FPGA instructions. The mapping onto other standardized system-on-chip architectures, especially the automatic synthesis of the communication scheme between hardware and software, seems to be a promising research direction.

Acknowledgement

The authors would like to acknowledge Max Chiodo for providing an intermediate format and its parser from Esterel DC. We are grateful for the support of the SRC under contract 683.004 and the California Micro program and our industrial sponsors, Fujitsu, Cadence, and Synplicity.

7. REFERENCES

[1] P. Ashar and S. Malik. Fast functional simulation using branching program. In *Proc. of the Intl. Conf. on Computer-Aided Design*, pages 408–412, Nov. 1995.

[2] F. Balarin, M. Chiodo, P. Giusto, H. Hsieh, A. Jurecska, L. Lavagno, C. Passerone, A. Sangiovanni-Vincentelli, E. Sentovich, K. Suzuki, and B.Tabbara. *Hardware-Software Co-Design of Embedded Systems: The Polis Approach*. Kluwer Academic Press, 1997.

[3] F. Balarin, M. Chiodo, P. Giusto, H. Hsieh, A. Jurecska, L. Lavagno, A. L. Sangiovanni-Vincentelli, E. M. Sentovich, and K. Suzuki. Synthesis of software programs for embedded control applications. *IEEE Trans. Comput.-Aided Design Integrated Circuits*, 18(6):834–49, June 1999.

[4] M. Baleani, F. Gennari, Y. Jiang, Y. Patel, R. K. Brayton, and A. Sangiovanni-Vincentelli. Hw/sw partitioning and code generation of embedded control applications on a reconfigurable architecture platform. In *Proc. of the Intl. Symposium on Hardware/Software Co-Design*, May. 2002.

[5] G. Berry. Esterel on hardware. *Philosophical Transactions of the Royal Society of London. Series A*, 1992.

[6] G. Berry. *The constructive semantics of pure Esterel*. Book in preparation, 1996.

[7] G. Berry and G. Gonthier. The Esterel synchronous programming language: Design, semantics, implementation. *Science of Computer Programming*, 1992.

[8] R. K. Brayton and et al. MVSIS. `http://www-cad.eecs.berkeley.edu/mvsis`.

[9] K. T. Cheng and A. Krishnakumar. Automatic functional test generation using the extended finite state machine model. In *Proc. of the Design Automation Conf.*, June 1993.

[10] S. Edwards. Compiling esterel into sequential code. In *Proc. of the Design Automation Conf.*, June 2000.

[11] The ESTEREL language. [On-line] http://www.esterel.org.

[12] M. Gao and R. K. Brayton. Semi-algebraic methods for multi-valued logic. In *Proc. of the Intl. Workshop on Logic Synthesis*, May. 2000.

[13] M. Gao, J. Jiang, Y. Jiang, Y. Li, S. Singha, and R. K. Brayton. MVSIS. In *Proc. of the Intl. Workshop on Logic Synthesis*, May. 2001.

[14] O. Hainque, L. Pautet, Y. L. Biannic, and E. Nassor. Cronos: a separate compilation toolset for modular esterel applications. *Formal Methods*, 1999.

[15] Y. Jiang and R. K. Brayton. Don't cares and multi-valued logic network minimization. In *Proc. of the Intl. Conf. on Computer-Aided Design*, Nov. 2000.

[16] Y. Jiang and R. K. Brayton. Logic optimization and code generation for embedded control applications. In *Proc. of the Intl. Symposium on Hardware/Software Co-Design*, Apr. 2001.

[17] C. Kim, L. Lavagno, and A. Sangiovanni-Vincentelli. Free MDD-based software optimization techniques for embedded systems. In *Proc. of the Conf. on Design Automation & Test in Europe*, Mar. 2000.

[18] P. M. Maurer. Event driven simulation without loops or conditionals. In *Proc. of the Intl. Conf. on Computer-Aided Design*, Nov. 2001.

[19] P. McGeer, K. McMillan, A. Saldanha, A. Sangiovanni-Vincentelli, and P. Scaglia. Fast discrete function evaluation using decision diagrams. In *Proc. of the Intl. Conf. on Computer-Aided Design*, pages 402–407, Nov. 1995.

[20] E. M. Sentovich, K. J. Singh, L. Lavagno, C. Moon, R. Murgai, A. Saldanha, H. Savoj, P. R. Stephan, R. K. Brayton, and A. L. Sangiovanni-Vincentelli. SIS: A System for Sequential Circuit Synthesis. Technical Report UCB/ERL M92/41, Electronics Research Laboratory, Univ. of California, Berkeley, CA 94720, May 1992.

[21] E. M. Sentovich, H. Toma, and G. Berry. Latch optimization in circuits generated from high-level descriptions. In *Proc. of the Intl. Conf. on Computer-Aided Design*, pages 428–35, Nov. 1996.

[22] M. Sgrio, L. Lavagno, Y. Watanabe, and A. L. Sangiovanni-Vincentelli. Synthesis of embedded software using free-choice petri nets. In *Proc. of the Design Automation Conf.*, June 1999.

[23] R. Ubar, J. Raik, and A. Morawiec. Back-Tracing and Event-Driven Techniques in High-Level Simulation with Decision Diagrams. In *Proc. of the Intl. Symposium on Circuits and Systems*, pages 208–211, May 2000.

[24] D. Weil, V. Bertin, E. Closse, M. Poize, P. Venier, and J. Pulou. Efficient compilation of Esterel for real-time embedded systems. In *Proc. of the Intl. Conf. on Compilers, Architecture and Synthesis for Embedded Systems*, Nov. 2000.

Complex Library Mapping for Embedded Software Using Symbolic Algebra

Armita Peymandoust
Computer Systems Laboratory
Stanford University
Stanford, CA 94305

Tajana Simunic
HP Labs & Stanford University
1501 Page Mill Rd., MS 3U-4
Palo Alto, CA 94304

Giovanni De Micheli
Computer Systems Laboratory
Stanford University
Stanford, CA 94305

{armita, tajana, nanni}@stanford.edu

ABSTRACT

Embedded software designers often use libraries that have been pre-optimized for a given processor to achieve higher code quality. However, using such libraries in legacy code optimization is nontrivial and typically requires manual intervention. This paper presents a methodology that maps algorithmic constructs of the software specification to a library of complex software elements. This library-mapping step is automated by using symbolic algebra techniques. We illustrate the advantages of our methodology by optimizing an algorithmic level description of MPEG Layer III (MP3) audio decoder for the Badge4 [2] portable embedded system. During the optimization process we use commercially available libraries with complex elements ranging from simple mathematical functions such as `exp` to the `IDCT` routine. We implemented and measured the performance and energy consumption of the MP3 decoder software on Badge4 running embedded Linux operating system. The optimized MP3 audio decoder runs 300 times faster than the original code obtained from the standards body while consuming 400 times less energy. Since our optimized MP3 decoder runs 3.5 times faster than real-time, additional energy can be saved by using processor frequency and voltage scaling.

Categories and Subject Descriptors

C.3 [**Special-Purpose and Application-Based Systems**]: *Microprocessor/microcomputer applications, Real-time and embedded systems, Signal processing systems.*

General Terms

Algorithms, Performance, Design, Experimentation, Theory.

Keywords

Embedded software optimization, Automated library mapping, Symbolic algebra, Polynomial representation, Computation intensive software.

1. INTRODUCTION

The market demand for portable multimedia applications has exploded in the recent years. Software optimization is critical for such systems due to time-to-market pressures. The software design process consists of translating a high-level specification into the optimized machine code for the target processor, often using compilers. There has been several research projects on optimizing compilers in last few years [5]. Prototype research compilers have shown impressive results [6]. Most optimizing compilers target high-performance and/or general-purpose computers. Relatively little effort has been dedicated to create powerful optimizing compilers for embedded processors. Even though several researchers are studying automatic code optimization techniques for embedded processors [7,8], currently, most embedded processors (or DSPs) are programmed directly by expert programmers and code optimization is mostly based on human intuition and skill.

Software engineers typically start with algorithmic level C code, sometimes developed by the standards groups such as MPEG, and manually optimize it to execute on the given hardware platform. Pre-optimized libraries for embedded system design are often available. For example, Intel recently released a library targeted at multimedia developers for SA-1110 embedded processor [10], and TI has a similar library for TI'54x DSP [11]. Embedded operating systems typically provide a choice from a number of mathematical and other libraries [12,13]. In addition, a library of more complex instructions, such as those developed by Tensilica tools [4], could be used. When a set of pre-optimized libraries is available, the designer has to choose the elements that perform best for a given section of code. Such manual optimization is error-prone and introduces undesired delay in the overall development process.

For example, consider a section of code that calls the `log` function. The library used in mapping consists of four different `log` implementations: double, float, fixed point using simple bit manipulation algorithm [14], and fixed point using polynomial expansion. Each implementation has a different accuracy, performance and energy trade-off. A designer would need to estimate which of the four functions would work best, test the hypothesis, and iterate until the best result is found. Designers are faced with an even more complex problem when attempting to map a software implementation of `IDCT` already present in MP3 standards code into an embedded software library. There are many ways to implement `IDCT` on a given processor, and it may be difficult for a designer to determine which library element is most appropriate. Clearly, the high-level arithmetic optimizations targeted at the use of complex library elements are currently left to the designers' ingenuity.

Our methodology automates the process of identifying the code sections that benefit from complex library mapping, and then performs the mapping using symbolic techniques. Similar symbolic techniques have been used for algorithmic level synthesis of data intensive circuits [20], and for software optimization of arithmetic functions by efficient mapping into processor's instructions set [15]. To illustrate the difference, let us go back to the examples discussed above. The work presented in [20] is concerned with mapping `log` to its hardware implementation, while in [15] the focus is on representing `log` and portions of `IDCT` with polynomials and then decomposing those into complex processor instructions, such as `MAC`. In contrast, the methodology presented in this work attempts to map `log` and `IDCT` into as complex software library element as possible, without resorting to decomposition into processor instructions when not necessary. In addition, while previous work reported only simulation results, in this work we present the measurement results by running the code on a hardware platform (Badge4 [2]).

The paper is organized as follows: Section 2 describes the target software and hardware platforms used in reported measurements. Section 3 describes our methodology, and gives an overview of each of its steps. The results of MP3 decoder optimization for Badge4 are presented in Section 4. We measured a significant performance increase and energy consumption decrease over the original executable specification from the standard body. Finally, Section 5 summarizes our contributions.

2. BACKGROUND

The main motivation for this work is the author's experience in porting MP3 audio decoder software available from the standards body [3] first to SmartBadge [2] with Angel operating system (OS), then to SmartBadge with eCos OS, and finally to the new version of SmartBadge, Badge4, running Linux OS. The manual optimization for MP3 decode on the SmartBadge [16] required the designer first to implement a fixed-point library and replace all floating-point operations with fixed point. Then, the designer needs to fully understand the details of the SmartBadge's design, so that the critical arithmetic operations could be manually optimized with inline assembly code. Each manual optimization process lasted several days.

While hand-optimizing code may be interesting the first time, it becomes increasingly tedious in the subsequent tries. This experience is common in typical industrial settings, where the software needs to be ported and optimized to the newer versions of hardware. Thus, the goal of this work is to automate mapping of complex arithmetic functions commonly occurring in portable system designs, into a set of pre-optimized library routines.

Badge4, as shown in Figure 1, is an embedded system powered by batteries through a DC-DC converter. It consists of StrongARM-1110 processor with SA-1111 companion chip that controls peripherals, audio CODEC with microphone and speakers, Lucent's WLAN card, sensors and three types of memory: SRAM, SDRAM and FLASH. Badge4 is the SmartBadge with a new processor and the addition of SDRAM and a companion chip.

Badge4 currently runs eCos [12] and an embedded version of Linux operating system [13]. In this work we use Linux OS since the libraries available to us are implemented only for Linux. Badge4's Linux has the main parts of the OS, including a small file system, residing in SRAM. The larger file system is remotely mounted from the server via the WLAN card. In our experiments, the MP3 files reside in the directory structure on the server and are streamed via wireless link to the Badge4 for decoding. The output can either be played back on the speakers, or streamed back and saved in a file for accuracy comparison purposes.

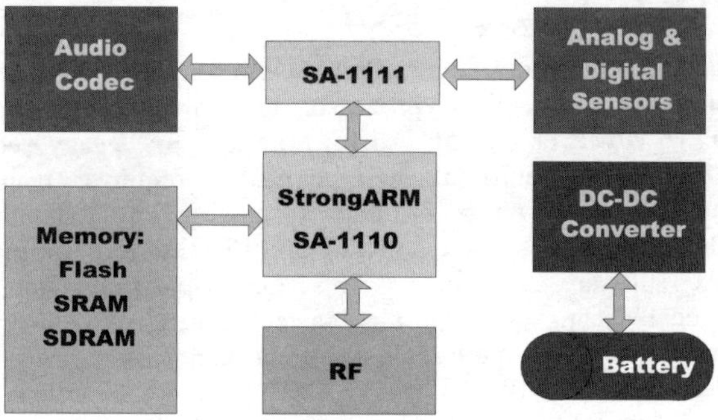

Figure 1. SmartBadge Architecture

We obtained the original MP3 audio decoder software from the International Organization for Standardization [3]. The first step in decoding MP3 stream is synchronizing the incoming bitstream and the decoder. Huffman decoding of the subband coefficients is performed before requantization. Stereo processing, if applicable, occurs before the inverse mapping which consists of an inverse modified cosine transform (`IMDCT`) followed by a polyphase synthesis filterbank. Compliance test provided by MPEG standard [17] is used to evaluate the accuracy of the optimizations. The range of RMS error between the original code's output and the samples produced by the code under test defines the level of compliance.

The outcome of our mapping algorithm is faster than real-time MP3 decoder software for Badge4. For an MP3 player, faster than real-time execution implies that lower voltage and frequency can still meet the real-time constraint. This in turn translates into longer battery life or lighter battery requirement for the embedded system. The next section gives an overview of our methodology for mapping source code into complex software library elements.

3. COMPLEX SOFTWARE LIBRARY MAPPING METHODOLOGY

Ideally, the software designer would write an algorithmic-level description of the software and have a compiler-like tool optimize it using software libraries available for the given platform. However, optimum implementation of calculation heavy routines for the particular hardware is not possible with traditional compiler optimizations. Commonly, the designer does most of such optimizations by hand. Automating even a portion of the optimization process can save much design time.

Our methodology automates most of the library mapping process. The mapping methodology consists of three main steps. The first step is to characterize the library elements. The characterization needs to include not only performance and energy consumption for a given architecture, but also the expected input and output format, accuracy and a polynomial representation. The next step identifies the target code for optimization. In this step, the critical functions are chosen via profiling. Traditional compiler techniques are used in representing the arithmetic section of the critical functions as polynomials. Finally, the target code represented by polynomials is mapped into the library elements. The mapping process uses symbolic techniques to decompose the target code into a set of library elements. The mapping process selects the solution that offers best performance with sufficient accuracy.

Our key contribution is a new method for mapping code into a library of complex software elements using symbolic polynomial manipulation. Since our methodology is compliant with other software optimization techniques, additional benefits are gained by combining it with traditional complier optimization algorithms. The next sections describe each part of our methodology in detail.

3.1 Library Characterization

The target library may consist of traditional embedded system library, such as IEEE floating-point mathematical library for Linux operating system [13], a commercial library available for the particular processor, such as Intel's integrated performance primitives library [10], and a set of in-house pre-optimized routines. Library characterization is done on element-by-element basis. Each element is labeled with the type of inputs and outputs, performance, accuracy, energy consumption, and finally the polynomial representation.

The format of library element inputs and outputs is determined from the library include files. Techniques discussed in the next section can be used to extract the polynomial from the source code if the code is available. Otherwise either the distributor needs to provide the equivalent polynomial representation or it might be obtained from the documentation.

Important part of library characterization is the determination of accuracy, performance and energy consumption. This information is used to guide the selection process when more than one library element has same functionality. Most embedded systems have OS timers that can be used for fine-granularity performance measurements on hardware. But often there is not an easy way to measure processor and memory power consumption. Alternatively, a cycle-accurate energy consumption simulator [16] easily provides energy and performance estimates of library elements.

Table 1. Sample Complex Library Elements

Library Element	Execution time	Execution time ratio
float SubBandSyn	0.95	1
fixed SubBandSyn	0.01	92
IPP SubBandSyn	0.002	479
float IMDCT	0.39	1
fixed IMDCT	0.014	27
IPP IMDCT	0.0002	1898

Examples of two complex library elements, SubBand Synthesis and `IMDCT`, are shown in Table 1. The library has three different versions of each library element: the open-source floating point version from the MP3 standards library [3], fixed-point in-house pre-optimized routine, and a version from Intel's integrated performance primitive (IPP) library for SA-1110 processor [10]. For each library element we have measured its performance on the Badge4 hardware. All entries in Table 1 are represented using polynomials. Since polynomials for complex library elements can be quite large, we show only a critical portion of `IMDCT` polynomial in Equation 1. Note that this is just a first order polynomial, since `cos(i,k,n)` can be calculated in advance for all `i,k` and `n`. Total of `n/2` windowed samples, y_k, are transformed into `n` x_i samples.

$$x_i = \sum_{k=0}^{\frac{n}{2}-1} y_k \cos\left(\frac{\pi}{2n}(2i+1+\frac{n}{2})(2k+1)\right) \tag{1}$$

3.2 Target Code Identification

The first step in target code identification is to identify the energy and performance critical procedures. This step can be done with either the energy profiler simulator [16], or by profiling directly on the hardware. Once the power and performance critical procedures are identified, they are formulated as polynomials suitable for mapping into library elements.

Critical procedures calculating an arithmetic or Boolean function can be easily represented as polynomials. The polynomial representation of a procedure can be directly extracted from the C code if it calculates a linear arithmetic function. If the procedure performs a series of bit manipulations or Boolean functions, previously developed algorithms based on interpolation [22] can be used to formulate its equivalent polynomial representation. When a section of the procedure implements a nonlinear function, we use an approximation, such as the Taylor or Chebyshev series expansion, as its polynomial representation.

The goal of this step is to formulate as large polynomials as possible, so the likelihood of finding a more complex library element that matches at least a portion of the formulated polynomial is increased. This can be accomplished by using code transformation techniques such as loop unrolling, constant and variable propagation, code motion, conditional expansion and model expansion.

3.3 Library Mapping Algorithm

This step decomposes the polynomial representations of the code blocks into available complex library elements while minimizing cost. Inputs to the library-mapping algorithm are a set of polynomial representations for critical code blocks, a characterized library of complex elements, and a routine that provides accuracy and cost (e.g. performance, energy) feedback. Note that a similar algorithm was used for algorithmic level synthesis of data intensive circuits [20], and for mapping basic blocks of arithmetic functions into complex processor instructions [15] (e.g. mapping of `log` into a series of `MAC`s). The core of the algorithm is a symbolic manipulation technique, known as *simplification modulo set of polynomials (simplify),* based on Gröbner basis fundamentals [20]. The polynomial representations of critical code blocks are simplified modulo a subset of polynomial representations of library elements.

Symbolic computer algebra is a set of algorithms capable of algebraic manipulation of expression containing undetermined values (symbols), such as variable `x` in `(x+1)*(x-1)`. Several commercial symbolic computer algebra systems are available on the market; Maple [18] and Mathematica [19] are most widely used. Most interesting symbolic polynomial manipulations are based on Gröbner bases [21]. Gröbner bases also solve variable elimination in a set of polynomials and ideal membership problems, which is the core of simplification modulo set of polynomials. We use the following set of symbolic techniques: factorization, expansion, Horner transform, multivariate polynomial substitution, and variable elimination.

Factor and expand are inverse operations. Consider using Maple to factor and expand the following polynomial:

```
>   S := x^2*(x^14+x^15+1);
>   P := expand(S);
                P := x^16+x^17+x^2
>   factor(P);
                x^2*(x^14+x^15+1)
```

Horner form of a polynomial is a nested normal form with minimal number of multiplications and additions. Any polynomial can be rewritten in Horner, or nested, form. An example of Horner form polynomial for multiple variables is shown below:

```
>   S:= y^2*x+y*x^2+4*x*y+x^2+2*x;
>   convert(S, 'horner', [x,y]);
                (2+(4+y)*y+(y+1)*x)*x
```

Elimination theory based on the Gröbner basis formalizes substitution and variable elimination for multivariate polynomials.

```
>   S:= x + x^3*y^2 -2*x*y^3
>   simplify(S, {p:= x^2-2*y}, [x,y,p]);
                x+y^2*x*p
```

Since evaluating all subsets of the library is exponentially expensive, the library-mapping algorithm uses the branch-and-bound method with performance and energy consumption as bounding functions to prune the search space. All previously described symbolic manipulation except *simplify* are used as guidelines in formulating different side relation sets to speed up the mapping algorithm. The symbolic manipulations result in various equivalent polynomials and thus provide more options to the mapping algorithm. Therefore, the speed and likelihood of finding the right match will increase. As with all branch-and-bound algorithms, in spite of using heuristics to speed up the mapping process, this algorithms worst-case complexity still remains exponential.

Table 2 shows the pseudo-code of the library-mapping algorithm. The target code to be mapped into a library *L* is represented with *S*. Mapping *S* into *L* is equivalent to simplifying *S* modulo elements of the library *L* as side relations (*sr*). Decision tree (*solution_tree*) implements the branch-and-bound algorithm. The results of each *simplify* step are also saved in the tree data structure. When a simplification result is within an acceptable tolerance of the polynomial representation of a library element, a possible solution is found and the corresponding tree node is marked accordingly. The algorithm also applies tree-height reduction, factorization, substitution, expansion, and Horner-based transform on *S*. As a result, there are several polynomials representing the target code (*exp_tree*), which can used to guide the initial side relation selection process. When all initial side relations are used and the result of simplify is not a library element, we decompose the result without further guidance from the expression tree. The algorithm is implemented in C with calls to Maple V for the symbolic manipulations. Typically, the algorithm takes only a few minutes to execute since we use heuristics to speed up the mapping process. However, its worst-case complexity is still exponential.

Table 2. Library Mapping Algorithm

```
function Decompose (exp_tree, boundVal) {
  #initialize a solution tree
  solution_tree ← tree (exp_tree)
  Depth ← 0
  Bound ← boundVal
  for all n ∈ tree with Depth {
     if Depth == 0
        choose sr ∈ L to preserve the exp_tree structure
     else for all sr ∈ L {
          result = simplify (n,sr);
          Add_child (n,result) #make result a child of node n
          Depth ← Depth + 1
          if result ∈ L
          { # solution is found
             Bound = cost of node result; }}}
  return the best solution with sufficient accuracy
end Decompose

procedure main (S,L)
  exp_tree [1 .. NoManipulations] = AllManipulations (S);
  for I = 1 to NoManipulations {
     boundVal[i]=Performance(exp_tree[i]);
     solution[i] = Decompose(exp_tree[i],boundVal[i]) }
  return the best solution in solutions[i]
end main
```

4. RESULTS

We illustrate the advantages of our methodology by performing library mapping on an algorithmic level description of the MPEG Layer III (MP3) audio decoder we obtained from the standards body [3]. The optimization target is the Badge4 portable embedded system shown in Figure 1. Badge4 currently runs an embedded version of Linux operating system [13]. During the optimization process we used a mathematical library available with Linux OS [13], Intel's integrated performance primitives (IPP) library for SA-1110 processor [10], and a library populated with in-house pre-optimized routines. The library elements ranged from simple mathematical functions such as `exp` to as complex elements as `IMDCT` routine.

Our library mapping methodology, as described in Section 3, consists of library characterization, target code identification and the final library mapping step. The library characterization step uses hardware measurements for performance and simulations for energy consumption [16]. The polynomial representation is obtained either from the source code (Linux mathematical and in-house libraries), or from documentation (IPP library).

The target code identification consists of two important steps: profiling the code and formulating polynomials to be mapped. All profiling is done using hardware measurements. Table 3 shows the results of profiling original MP3 decoder software we obtained from the standards body. The results are shown for one frame and represent only the most significant functions in terms of their performance impact.

Table 3. Original MP3 Profile

Function name	Execution time (s)	%
III_dequantize_sample	1.1754	45.33
SubBandSynthesis	0.9481	36.56
inv_mdctL	0.3872	14.93
III_hybrid	0.0670	2.58
III_antialias	0.0131	0.51
III_stereo	0.0010	0.04
III_hufman_decode	0.0007	0.03
III_reorder	0.0005	0.02
Total for one frame	2.5931	100.00

The next step is to represent portions of these functions as polynomials and then map them into available libraries. For the first mapping step we selected Linux mathematical (LM) and in-house libraries (IH). The resulting performance profile is shown in Table 4. Although the performance per frame is drastically reduced (by two orders of magnitude), we can see that still almost 85% of the execution time is spent in the `IMDCT` and subband synthesis functions.

Table 4. MP3 Profile after LM & IH mapping

Function name	Execution time (s)	%
inv_mdctL	0.0144	49.54
SubBandSynthesis	0.0103	35.30
III_dequantize_sample	0.0013	4.33
III_stereo	0.0008	2.83
III_reorder	0.0007	2.28
III_antialias	0.0006	2.15
III_hufman_decode	0.0007	2.48
III_hybrid	0.0003	1.10
Total for one frame	0.0291	100.00

The next step is to map to Intel's IPP library to further optimize the code. Here we find two primitives that match the two critical procedures shown in Table 4. The new performance profile is shown in Table 5. As shown, our method automatically uses two of the IPP routines. These two routines correspond to the two most time consuming sections of the code as shown in Table 4. While the new profile shows that subband synthesis still takes roughly 35% of the execution time for each frame, we see that `MDCT` is no longer a critical portion of the code. Notice that the execution of the IPP subband synthesis routine is one order of magnitude faster than the previous version and the total time for decoding one frame is reduced by a factor of 50.

Table 5. MP3 Profile after LM & IH & IPP mapping

Function name	Execution time (s)	%
ippsSynthPQMF_MP3_32s16s	0.00176	35.242
III_dequantize_sample	0.00124	24.79
III_stereo	0.00082	16.46
III_hufman_decode	0.00067	13.416
IppsMDCTInv_MP3_32s	0.00047	9.4113
III_get_scale_factors	3.4E-05	0.6808
Total time for one frame	0.00499	100.00

Table 6 summarizes the performance and the energy results of the overall mapping process. Both measurements were performed on the Badge4 while running at maximum processing speed and voltage. We started from the original source code that runs roughly two orders of magnitude slower than real time. The next two entries show the results of mapping only into Intel's IPP library; more specifically, we were able to automatically use IPP's SubBand Synthesis and `IMDCT` in the original code. However, the rest of the code remains intact and still operates on floating-point data. StrongARM-1110 cannot perform floating-point operations natively; therefore the code is still far from real-time execution. IH library entry represents the mapping from the original code, into Linux mathematical library and our in-house pre-optimized library. We save two orders of magnitude in both performance and energy for this mapping. This is due to changing most floating-point operations to fixed point. Fixed-point accuracy is verified through simulation. Additional saving of a factor of four is obtained by further optimizing the code and mapping to all three libraries: Linux mathematical, in-house and Intel's IPP library. The factor of four improvement is achieved

Table 6. Performance and Energy for MP3 library mapping

Code version	Perf (s)	Factor	Energy (J)	Factor
Original	503.92	1.0	509.6	1.0
IPP SubBand	301.43	1.7	292.5	1.7
IPP SubBand & IMDCT	211.27	2.4	199.1	2.6
IH Library	5.47	92.1	4.47	114.2
IH + IPP SubBand	3.33	151.4	2.78	182.3
IH + IPP SubBand & IMDCT	1.43	352.4	1.17	435.2
IPP MP3	0.41	1240.8	0.31	1626.

solely based on automatic use of complex library elements that have been pre-optimized for the given processor. Such optimization was not possible in previous work [15], since mapping was limited to the complex instructions available on the target processor (MAC). Full compliance to the standard of each version of MP3 code is ensured by checking the accuracy at each mapping step with MP3 compliance test [17].

The last table entry, IPP MP3, represents fully hand-optimized code available for MP3 from Intel. Our most optimized version (IH+IPP SubBand & IMDCT) is a factor of five worse than hand-optimized IPP MP3. Our approach only automates the process of mapping code that can be represented with polynomials into complex library elements. Since large portions of MP3 decoder software can be represented with polynomials, we were able to measure savings of a factor of 350 in performance and 435 in energy over the original source code obtained from the standards organization. Even larger energy savings are possible by using processor frequency and voltage scaling, because our most optimized MP3 code runs almost four times faster than real time. One should note that the improvement factors reported are algorithmic floating-point measurements divided by optimized code measurements. Improvements of this order are realistic when skilled designers hand optimize code for a given embedded system, as shown in the last column of Table 6. Our contribution is to automate most of this optimization process.

5. CONCLUSION

The contribution of this paper is a methodology for automatic mapping of critical code sections into complex library elements. Since our mapping methodology uses symbolic algebra methods, we focus only on code sections that can be represented with polynomials. There are three main steps in our methodology: library characterization, target code identification, and library mapping.

We have tested our methodology by mapping critical code sections of MP3 audio decoder into libraries we had available for the Badge4 portable embedded system. We have measured savings of a factor of 350 in performance and 435 in energy over the original specification while still keeping in full compliance with the MPEG standard. Currently, skilled software designers hand optimize code to achieve such improvement gains. Our proposed methodology automates this tedious process. Additional energy savings are possible by using frequency and voltage scaling since the final MP3 code runs a factor of four faster than real-time.

6. ACKNOWLEDGMENTS

This research is supported by ARPA/MARCO Gigascale Research Center, HP Labs, and Synopsys Inc. We would like to thank all preceding organizations for their support. We would also like to thank Dr. Yung-Hsiang Lu for his help with the data acquisition.

7. REFERENCES

[1] P. G. Paulin, C. Liem, M. Cornero, F. Nacabal, and G. Goossens, "*Embedded software in real-time signal processing systems: application and architecture trends*," Proc. IEEE, vol. 85, no. 3, pp. 419-435, Mar. 1997.

[2] G. Q. Maguire, M. Smith, H. W. Peter Beadle, "SmartBadges: a wearable computer and communication system", *6th International Workshop on Hardware/Software Codesign*, Invited talk, 1998.

[3] "Coded representation of audio, picture, multimedia and hypermedia information", *ISO/IEC JTC/SC 29/WG 11*, Part 3., May 1993.

[4] Albert Wang, Earl Killian, Dror Maydan, Chris Rowen, "Hardware/Software Instruction Set Configurability for System-on-Chip Processors", *Proceedings of the Design Automation Conference*, pp. 184-190, June 2001.

[5] S. Muchnick, *Advanced Compiler Design and Implementation*, Morgan Kaufmann, 1997.

[6] M. Hall, J. Anderson, S. Amarasinghe, B. Murphy, S. Liao, E. Bugnion, M. Lam, "Maximizing multiprocessor performance with the SUIF compiler", *IEEE Computer,* vol. 29, no. 12, pp. 84-89, Dec. 1996.

[7] P. Marwedel and G. Goossens. *Code Generation for Embedded Processors*. Kluwer Academic Publishers, 1995.

[8] R. Leupers, *Retargetable Code Generation for Digital Signal Processors*, Kluwer Academic Publishers, 1997

[9] F. Catthoor, S. Wuytack, E. De Greef, F. Balasa, L. Nachtergaele, A. Vanduoppelle, *Custom Memory Management Methodology: Exploration of Memory Organisation for Embedded Multimedia System Design*, 1998, Kluwer Academic Pub.

[10] Intel, "Integrated Performance Primitives for the Intel StrongARM SA-1110 Microprocessor", 2000.

[11] Texas Instruments, "TI'54x DSP Library", 2000.

[12] Cygnus Solutions, "eCos^{TM} Reference Manual", 1999.

[13] RedHat, "Linux-arm math library reference manual".

[14] J. Crenshaw, *Math Toolkit for Real-Time Programming*, CMP Books, Kansas, 2000.

[15] A. Peymandoust, T. Simunic, and G. De Micheli, "Low Power Embedded Software Optimization using Symbolic Algebra", *Proceedings of the Design Automation and Test in Europe Conference*, pp. 1052-1058, March 2002.

[16] T. Simunic, L. Benini, G. De Micheli, "Energy-Efficient Design of Battery-Powered Embedded Systems", *Special Issue of IEEE Transactions on VLSI*, pp. 18-28, May 2001.

[17] ISO/IEC JTC 1/SC 29/WG 11 13818-4, "Information Technology, Generic Coding of Moving Pictures and Associated Audio: Conformance", International Organization for Standardization, 1996.

[18] Maple, Waterloo Maple Inc., www.maplesoft.com, 1988.

[19] Mathematica, Wolfram Research Inc., www.wri.com, 1987.

[20] A. Peymandoust and G. De Micheli, "Symbolic Algebra and Timing Driven Data-flow Synthesis", *Proceedings of the International Conference on Computer Aided Design*, pp. 300-305, November 2001.

[21] T. Becker and V. Weispfenning, *Gröbner Bases*, Springer-Verlag, New York, NY, 1993.

[22] J. Smith and G. De Micheli, "High-Level Synthesis and Design Reuse using Polynomial Circuit Models", *IEEE Transactions on VLSI*, Vol. 9, No. 6, pp. 783-800, December 2001.

Retargetable Binary Utilities

Maghsoud Abbaspour, Jianwen Zhu
Electrical and Computer Engineering
University of Toronto, Ontario M5S 3G4, Canada
jzhu@eecg.toronto.edu

ABSTRACT

Since software is playing an increasingly important role in system-on-chip, retargetable compilation has been an active research area in the last few years. However, the retargetting of equally important downstream system tools, such as assemblers, linkers and debuggers, has either been ignored, or falls short of meeting the requirements of modern programming languages and operating systems. In this paper, we present techniques that can automatically retarget the GNU binutils tool kit, which contains a large array of production-quality downstream tools. Other than having all the advantages enjoyed by open-source software by aligning to a de facto standard, our techniques are systematic, as a result of using a formal model of instruction set architecture (ISA) and application binary interface (ABI); and simple, as a result of leveraging free software to the largest extent.

Categories and Subject Descriptors

D.3.4 [**Processors**]: Retargetable compilers

General Terms

Design, Languages

1. INTRODUCTION

New products in consumer electronics and telecommunications are characterized by increasing functional complexity and shorter design cycle. It is generally conceived that the complexity problem can be best solved by the use of system-on-chip (SOC) technology. And the design cycle problem can be best solved by pushing functionality as much as possible to software. However, the conventional wisdom here that "software is cheaper than hardware", is not necessarily true unless the software development platform, typically includes operating system, compiler, assembler, linker, and debugger are readily available. Unfortunately, all these tools depend intrinsically on the processor architecture, which in an SOC context is usually designed to adapt to the application. The development platform has to be *retargetted* to the new processor architecture and this task is by no means trivial.

DAC 2002, June 10-14, 2002, New Orleans, Louisiana, USA.

The field of retargetable compilation has evolved to the point where an architecture description language (ADL) can be used to model a processor micro-architecture, and a compiler can be generated automatically from such an architecture specification. While research in compiler generation is becoming mature, few efforts shed light on the automatic generation of other tools. This is partly due to the fact that these "downstream" tasks are perceived to be trivial engineering issues compared to optimizing compiler. While this perception has been persistent enough to be reflected in all computer science curriculum, it is no longer valid. Take the linker as an example, while the traditional linker does nothing but threading the object files together, the modern linker has to handle features such as shared libraries and dynamic linking as a result of modern operating system, static constructors and templates as a result of modern programming languages such as C++ and Java, and even inter-procedural optimization as a result of modern compiler theory. The most recent version of Free Software Foundation's `binutils` package, which delivers exactly the downstream development tool suite, has a daunting size of 250k lines of C code.

Neither the manual development nor the automatic generation of software with such complexity is reasonable to fit into an SOC development cycle. Most companies chose to *port*, or reuse the majority of an existing package, while manually rewriting the architecture dependent part of it. The de facto standard of such a package is the GNU `binutils` package, partly due to the fact that it is designed to be "portable", partly due to the fact that it is free software and accessible to everyone in the world. While this package has been ported to virtually every known processor in the world, it still has to be manually ported to every new processor ever created. Unfortunately, the black magic involved and the skill required to hack into this package is mastered only by a limited group of people who, by postulation [1], can be gathered all over the world and still packed in one room.

A tool that can automatically port this mature, robust and standard package then seems both ideal and feasible: the architecture-dependent part of the package is relatively small after all. It is not trivial however, since the interface between the architecture dependent and independent part is neither cleanly defined nor well documented.

In this paper, we focus on techniques that leads to the automatic porting of the GNU `binutils` package, which includes a suite of downstream software development tools such as assembler, linker, library manager, profiler, object file examiner and manipulator and C++ dismangler.

The contribution of this work is three fold: First, we present a formal, abstract model of instruction set architecture (ISA) to capture the architectural information required for retargetting. While this overlaps the goal of ADLs, it is a complementary effort since we can now formulate retargetting algorithms without concerning about the ADL syntax. Our tool can hence be trivially ported to compiler suites with different ADLs. Second, we introduce a formal model of application binary interface (ABI) as a new element of architectural model. While ABI is one of the essential information for retargetting, it hasn't been a subject of architecture specification in previous work. Third, to strike a balance between simplicity and standard compatibility, we have developed an automatic technique that can generate an implementation of obscure, poorly documented interface from a cleanly defined ISA and ABI model.

In the sequel, we will first review the related work in the area of retargetable compilation in Section 2. We will then describe a typical binary package, highlighting its architecture-dependent components. We will then present our model of ISA and ABI that is relevant to binary utility retargetting in Section 4 and Section 5 respectively. Finally, we describe our retargetting tool in Section 6 with experimental result.

2. RELATED WORK

The first step towards retargetable compilation is the establishment of architectural model and the definition of corresponding architectural description languages (ADLs). The earliest forms of ADLs are various code generator generators (CGGs) [2]. The CGGs typically use tree pattern specifications to drive the generation of the instruction selector. However, such specification is often tied to a particular compiler implementation, for example, a particular intermediate representation.

Architecture descriptions for embedded processors, for example, the DSP processors and application-specific instruction set processors (ASIPs), have also received intense interest in the recent years. MIMOLA [3] used a hardware description language to describe the structural model of the processor, code selection and register allocation can then performed by pattern matching. Since ILP can be concisely represented using a structural model, some recent work adopted a similar approach. For example, CHESS [4], [5] and MESCAL [6] use a graph-based model, while EXPRESSION [7] uses a network of abstract components including those that capture memory hierarchy. Other efforts, including Flexware [8], ISDL [9] and LISA [10], take a more traditional approach.

Many retargetable compiler suite generates assembler by merely "threading" C code segment attached to ADL instruction specification. Among the few exceptions that achieved a reasonable amount of automation for downstream tools, the New Jersey machine-code toolkit [11] can generate instruction encoding and decoding routines from an abstract ADL specification. However, its ADL does not allow a complete specification of ABI, and leave important issues such as the relocation closure organization to the application, which is not automated at all. While CHESS [5] and LISA [12] can generate assembler and linker, the detail of the retargetting process is not yet described and it is not clear if the generated tool can achieve the same level of versatileness as GNU `binutils` in terms of multiple object format and modern programming and OS feature support. Moreover, other useful binary tools, such as library and object file manipulators, are not supported.

3. GNU BINARY UTILITIES

Binary Utilities are tools that generate, examine and manipulate object files. The GNU `binutils` package contains two supporting libraries, `bfd` and `opcodes`, as well as the following tools: the assembler `gas` is used to convert assembly code into object file; the link editor `ld` is mainly used to group object files into executable; the library managers `ar` and `ranlib` are used to create and modify object archive files; `objdump, nm, strings` are used to examine the content of object files; `objcopy, strip` are used to manipulate the content of object files; `c++filt` is used to perform C++ name dismangling; `gprof` is used for program profiling.

While all tools are apparently architecture dependent, `binutils` is designed to be portable by limiting the architecture dependent function within three components, namely, the BFD library, which all tools depend on; the `opcodes` library, which both `gas` and `objdump` rely on, and `gas` itself. In addition, the BFD and `opcodes` library is used by another important GNU package, the GDB debugger. In the sequel, these three components are described in detail.

3.1 Binary File Descriptor Library (BFD)

GNU's BFD library is a package which contains a set of common routines to manipulate object files [?]. Due to historical reasons, object files present with different formats, called the binary file format (BFF). The most commonly used BFFs are `a.out`, `COFF` and `ELF`. The general structure of an BFF contains four major parts: a *file header* containing general information as well as pointers to other parts of the file, a number of *sections* holding code and raw data, *relocation tables* and *symbol table* information.

Since the processing of object files depends on different operating system, CPU target, and BFF configuration, the BFD package is designed with two layers: the frontend and backend. the unique frontend provides the interface to the application so that the differences between different CPU/OS/BFF configuration are abstracted away. The backend layer provides the concrete implementations for each of the CPU/OS/BFF configuration.

One difficulty in using the BFD library is its complexity. The library itself is very large, the number of functions offered in the front end are exceptionally many. The BFD front end was designed in mind to allow the programmer to be able to retrieve all type of information about any BFF, at least the existing ones. The BFD library can be integrated with disassemblers, decompilers, debuggers, etc. Due to this generality and hence its bulkiness, it is difficult to use it without spending a great deal of time learning how to use it. Perhaps because it is too general, it often contains information that is not needed for a particular application.

3.2 `opcodes` Library

For each target CPU `mycpu`, the `opcodes` library contains two C files: `mycpu-dis.c` and `mycpu-opc.c`. While the former provides the implementation for instruction disassembling, the latter constructs a data structure that describes the assembler syntax as well as other information for instructions, each of which is identified by a unique mnemonic. `binutils` defines a unique API for what is supposed to be implemented in `mycpu-dis.c`. On the other hand, the data structure defined `mycpu-opc.c` is rather arbitrary. The current GNU `binutils` distribution contains different representation for different targets.

3.3 GNU Assembler

The GNU tool `gas` is in fact a family of assemblers. `gas` is primarily intended to assemble the output of a C compiler for use by the linker. It can be configured to produce several alternative object formats and for several targets.

The assembler performs three tasks. In the first step, it reads the input file and performs preprocessing. In the second step, it parses and assembles each input line. While the processing of most directives and all expressions are common to all targets, the processing of some directives and all instructions are target-specific. In this step it also inserts relocations for unresolved symbols during the second pass. In the third step, it writes the object files including symbolic and debugging information. For each target `mycpu`, `gas` provides a C file `tc-mycpu.c` to implement the second step. In addition to other target-dependent code, `tc-mycpu.c` will also access BFD and `opcodes` libraries, which themselves are target-dependent. While the third step is both target-dependent and BFF-dependent, all the dependency is in fact encapsulated in the BFD library and therefore the implementation can be shared by all targets.

4. ISA MODELING

Before we can automatically generate the target-dependent parts of the binary utilities, a model that can be used to capture the instruction set of the target architecture is needed. To abstract away the irrelevant architecture information and the syntactical variances of different ADLs, we use the *formal algorithm notation* (FAN) to specify our model. The type system of FAN is based on sets, which allows us to formulate retargetting algorithms naturally with the architectural model.

As shown in Definition 1, our model of an ISA is characterized by stores, fields and instructions.

DEFINITION 1. *An **ISA architectural model** [1] is a member of*

ISA = **tuple** {
 S : $\langle\rangle^{ISAStore}$;
 F : $\langle\rangle^{ISAField}$;
 I : $\langle\rangle^{ISAInstrn}$;
}

*where S is a set of **stores**, F is a set of **fields** and I is a set of **instructions**.*

4.1 Stores

The stores of an ISA model provide an abstraction for register files contained in a target. As shown in Definition 2, a store is characterized by its granularity, its size, and the set of cells it contains. The first two define `size` number of physical storage cells, each of which has a bitwidth of `gran`. The latter defines a set of logical storage cells, each of which occupies a continuous range of physical cells.

DEFINITION 2. *A store is a member of set*

ISAStore = **tuple** {
 gran : $\mathcal{Z}$;
 size : $\mathcal{Z}$;
 cells : $\langle\rangle^{ISACell}$;
}

ISACell = **tuple** {
 index : $\mathcal{Z}$;
 size : $\mathcal{Z}$;
}

4.2 Fields

Fields in our ISA model provide an abstraction to concepts such as opcodes and addressing mode. Fields are essential for defining instruction formats, which can be shared by many instructions.

[1] In FAN, we use the notation $\langle\rangle^A$ to represent a power set of A, and the notation $[\,]^A$ to represent the set of all sequences over elements of A.

As shown in Definition 3, A field is first characterized by its `kind`, which can either be an opcode (`OPC`), or a register (`REG`), or a unsigned immediate value (`IMM`), or a signed immediate value (`SIMM`), or a symbol (`SYM`). `size` defines the number of bits the field occupies within an instruction. `argno` associates the field to a particular operand of the instruction behavior. `dest` signals if the field is the destination of the instruction. `reloc` indicates the relocation type of the field, which we will define in detail in Section 5.

DEFINITION 3. *An field is a member of set*

ISAField = **tuple** {
 kind : $\{OPC, REG, IMM, SIMM, SYM\}$;
 size : $\mathcal{Z}$;
 argno : $\mathcal{Z}$;
 dest : $\mathcal{B}$;
 reloc : *Reloc*;
}

4.3 Instructions

As shown in Definition 4, an instruction in our ISA model is characterized by its behavior, its assembly format and binary encoding format.

DEFINITION 4. *An instruction is a member of set*

Instrn = **tuple** {
 behavior : $\langle\rangle^{Signature}$;
 asmFormat : *string*;
 binFormat : *string*;
}

Every instruction is associated with a *behavior*, or computational task that transform values. Because behaviors are architecture-independent and hence can be shared by different architectural descriptions, we decide not to model the detail of the behavior in the ISA model, instead, behaviors appear as signatures. Because it is very hard to find a canonical behavioral representation, each instruction is associated with a set of signatures, each of which is one representation of the abstract behavior. Similar to a method signature of an object oriented language, a signature in our model consists of a class type, an identifier and a sequence of argument types. Since the behavior of an instruction is needed only for code selection or simulator generation, we omit the detail of instruction behavior modeling in this paper.

asmFormat specifies the assembly syntax of an instruction. While a general way of syntax specification needs a LEX/YACC type of BNF formalism, it is an overkill for this case. We choose instead a more intuitive approach based on usage: assembly syntax of an instruction is specified by a series of patterns starting with a common mnemonic. Another benefit of this approach is that the same specification can be used by the retargetable compiler for assembly emission. In each assembly pattern, instruction destinations or operands, called the *assembly fields*, are identified by the special character % for register fields, and # for immediate or symbol fields. These special characters are followed by a number: when the number value is zero, it indicates that the corresponding field is a destination; when it is non-zero, it indicates that it is an operand and the number specifies the argument number. The argument number can be used by the compiler to emit appropriate register, immediate or symbol values. It can also be used by the assembler to associate the operands to appropriate binary fields.

binFormat specifies the binary format of an instruction. The binary encoding of an instruction consists of a sequence of *binary fields*. Each field may be annotated inside the parenthesis pair with some additional information. For example, an opcode field is annotated with the opcode value of the instruction.

5. ABI MODELING

An ABI defines a binary interface for application programs that are compiled and packaged for a specific OS running on a specific hardware architecture. An ABI is a protocol between different software development tools, so that software created in different languages and compiled by different compilers can still be linked and interoperate with each other. An ABI is also a protocol between the application and the OS, so that the OS loader can create the correct process image from an executable file and possibly many shared object files. Rather than presenting a complete ABI model, in this paper we focus only on part of the ABI that is relevant to binary utilities and describe the modeling of *relocation* and *procedure linkage table* (PLT). An architecture description can be specified using this model and serve as the input to our automatic porting tool.

Since software programs are compiled separately into object files, each object file may contain data or instructions that reference symbols defined elsewhere. Even for the reference of local symbols, the actual address of these local symbols cannot be resolved at compile time since the enclosing sections can be moved to arbitrary locations at link or load time. The process of calculating the correct values of these external or local symbol references, called the *relocated values*, and adjusting the bits within the corresponding instructions or data, called the *relocation field*, is called relocation. Typically, an object file contains an array of *relocation entries* in a special relocation section, each of which points to the instruction or datum to be relocated. Depending on different BFF used, the relocation entry may contain a *relocation type*, which designates the calculation method of relocated values, and a *relocation addend*, which is an integer-sized storage that can help store useful information for the calculation in case the relocation field of the instruction or datum to be relocated is not large enough to hold the information. While the exact source of this information may vary, it is always an integer value that will be added to relocated value, and hence the name.

To support shared object and dynamic linking, position-independent code (PIC) whose instructions need no relocation should be supported. The linker usually creates a *global offset table* (GOT) that contains pointer to all the global data that the executable file addresses. GOT entries can be considered as global data themselves and therefore any reference to them need relocation. Similarly, the linker may also create a procedure linkage table (*PLT*) for procedure symbols. Due to the need for lazy evaluation, that is, not providing procedure addresses until they are called for the first time, each PLT entry contains a series of architecture-dependent instructions that call routines defined in the dynamic linker.

The calculation of the relocated value involves the following parameters, which are either information kept in the relocation field, or relocation entry, or values maintained by applications such as linker or loader:

- *A*: the addend, which can either be stored in the relocation field of the instruction or datum to be relocated, or the relocation entry;
- *B*: the base address at which a shared object is loaded into the memory during execution;
- *GOT*: the address of the global offset table;
- *G*: the offset into the global offset table at which the address of the referenced symbol resides during execution;
- *L*: the place (section offset or address) of the procedure linkage table entry for the reference procedure symbol;
- *P*: the place (section offset or address) of the instruction or datum to be relocated;
- *S*: the value of the referenced symbol.

The relocation model for ABI specification can thus be characterized by the calculation method as well as the identification of relocation field. Definition 5 gives our simple model of relocation type, which is on the other hand a complete one due to the specialty of the relocation calculation expression.

DEFINITION 5. *A* **relocation type** *is a member of*

```
Reloc = tuple {
    id          : Z;
    expCode     : Z;
    rightshift  : Z;
    bitsize     : Z;
    bitpos      : Z;
    complain    : {IGNORE, BIT, SIGN, UNSIGNED};
    }
```

where id is an unique integer identifier, expCode $= \langle C_7, C_6, ..., C_0 \rangle$ *encodes the expression* $\Sigma_i C_i P_i$*, with* $P_7, ...P_0$ *being* $-GOT, A, B, G, GOT, L, -P, S$ *respectively; rightshift represents the number of bits at the right side of the calculated* $\Sigma_i C_i P_i$ *that should be dropped; bitpos and bitsize represents the bit position as well as the size of the relocation field within the instruction or datum to be relocated; complain encodes the action to take when specific type of overflow occurs.*

Figure 1 shows an example of relocation expression.
Example 1 shows the specification of SPARC relocation types.

EXAMPLE 1. *SPARC relocation description.*

```
none          = ⟨0,0x00,0,0,0,IGNORE⟩;          // none
sparc8        = ⟨1,0x41,0,8,0,BIT⟩;             // S+A
sparc16       = ⟨2,0x41,0,16,0,BIT⟩;            // S+A
sparc32       = ⟨3,0x41,0,32,0,BIT⟩;            // S+A
sparcDISP8    = ⟨4,0x43,0,8,0,UNSIGNED⟩;        // S+A-P
sparcDISP16   = ⟨5,0x43,0,16,0,UNSIGNED⟩;       // S+A-P
sparcDISP32   = ⟨6,0x43,0,32,0,UNSIGNED⟩;       // S+A-P
sparcWDISP30  = ⟨7,0x43,2,30,0,UNSIGNED⟩;       // S+A-P>>2
sparcWDISP22  = ⟨8,0x43,2,30,0,UNSIGNED⟩;       // S+A-P>>2
sparcHI22     = ⟨9,0x41,10,22,0,IGNORE⟩;        // S+A>>10
sparc22       = ⟨10,0x41,0,22,0,BIT⟩;           // S+A
sparc13       = ⟨11,0x41,0,13,0,BIT⟩;           // S+A
sparcLO10     = ⟨12,0x41,0,10,0,IGNORE⟩;        // (S+A)&0x3ff
sparcGOT10    = ⟨13,0x10,0,10,0,IGNORE⟩;        // G&0x3ff
sparcGOT13    = ⟨14,0x10,0,13,0,BIT⟩;           // G
sparcGOT22    = ⟨15,0x10,10,22,0,IGNORE⟩;       // G>>10
sparcPC10     = ⟨16,0x43,0,10,0,IGNORE⟩;        // (S+A-P)&0x3ff
sparcPC22     = ⟨17,0x43,10,22,0,BIT⟩;          // S+A-P>>10
sparcWPLT30   = ⟨18,0x46,2,30,0,UNSIGNED⟩;      // L+A-P>>2
sparcCOPY     = ⟨19,0x00,0,0,0,IGNORE⟩;         // none
sparcGLOBDAT  = ⟨20,0x41,0,0,0,IGNORE⟩;         // S+A
sparcJMPSLOT  = ⟨21,0x00,0,0,0,IGNORE⟩;         // none
sparcRELATIVE= ⟨22,0x60,0,0,0,IGNORE⟩;          // B+A
sparcUA32     = ⟨23,0x41,0,0,0,IGNORE⟩;         // S+A
```

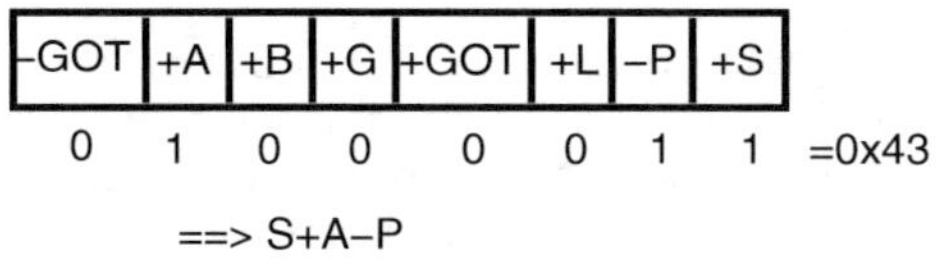

Figure 1: Relocation calculation expression.

6. RETARGETTING BINARY UTILITIES

As shown in Figure 2, our retargetting tool, called `rbinutils`, starts by compiling a target specification in an ADL into an architecture database. We have developed a *general purpose* Intellectual-Property (IP) specification language, called Babel, to capture the non-functional aspects (including architectural aspect) of an IP. Babel can be considered the textual form of FAN. Provided explicitly to the user, our ISA and ABI model is captured in Babel as types. The target information can then be captured in the form of typed values. The type inference engine of Babel can help check the validity of the target specification. While we have used Babel for our own experiments, target specifications in other ADLs can potentially be processed by our tool since our architectural database is independent of Babel. In fact, we have created a simple API to create and access the target information based on the presented ISA and ABI model.

Our tool also provides a set of template files, which help create the C headers, C files as well as configuration scripts to be retargetted. These templates include not only partial implementation, but also the placeholders which can direct our tool to generate codes at appropriate locations. The template files themselves, however, are architecture independent.

The code generator in `rbinutils` generates files from the template files that can be merged into the GNU `binutils` source tree. The user can subsequently configure the modified source tree and follow the normal building process to build *all* the tools.

To exercise the tool, we have specified the processor model for both SPARC and Intel 386, which exemplify the RISC and CISC architectures respectively. In the text that follows, we describe in detail the important components of our tool, while using data for the SPARC target (called `mysparc`) to demonstrate the result.

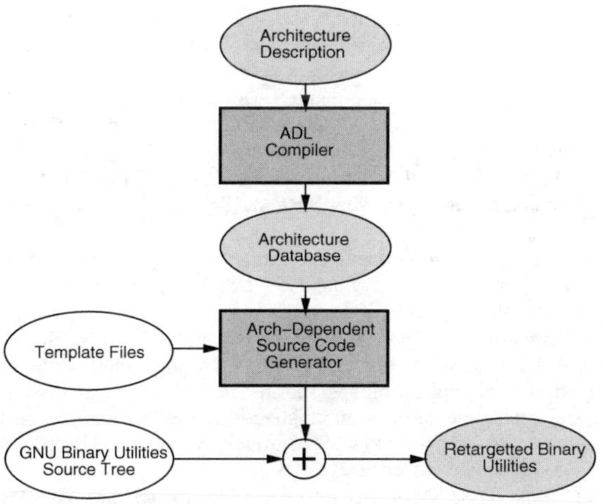

Figure 2: `rbinutils` block diagram.

Generated Files	#line (generated)
/bfd/config/mysparc-elf.mt	3
/bfd/archures.c	1483
/bfd/configure.host	112
/bfd/configure.in	286
/bfd/config.bfd	166
/bfd/elf32-mysparc.c	1482
/include/elf/common.h	229
config.sub	1014
/bfd/target.c	785
/bfd/cpu-mysparc.c	44
/bfd/elfcode.h	6582

Table 1: Generated and changed files for `bfd`.

6.1 Retargetting BFD Library

Since BFD is used by all binary tools, the first task of porting binary utilities is to port the BFD library. This unfortunately is a painstaking procedure since no clean interface has been defined by BFD implementers due to historical reasons. To make things worse, it is winded together with BFF-specific code and scattered in many different C files. `rbinutils` fully automates this task by generating an implementation of an obscure interface consisting of type declarations, macros, data and functions, from a clean target model. The generated implementation contains the following:

- type: It includes the definition of an enumeration type which defines the relocation type identifiers.
- data: It contains general information about the target processor, such as word size, address size and name. It includes the definition of a relocation *howto table*, an internal representation that characterizes the relocation calculation methods of each relocation type. It also contains an internal representation of the PLT entries to be generated.
- functions: It contains the functions for checking relocations as well as generating dynamic sections. The dynamic sections, such as *.dynamic*, *.hash*, *.got* and *.plt*, are used by dynamic linker for creating process image. They are created by the linker to hold various data, symbol table, global offset table and procedure linkage table respectively. It also contains the function to relocate all relocation entries of a section.

Table 1 shows the generated and changed BFD files for `mysparc` target.

6.2 Retargetting `opcodes` Library

As described in Section 3.2, the `opcodes` library creates an internal representation for each instruction, which can be accessed by a hash table keyed by instruction mnemonic. In the GNU distribution, such representation is specific to each target. In `rbinutils`, we have created a common representation for all targets. Therefore, the code generator can generate, in the same fashion for all targets, the initialized C struct data which can be derived from the ISA field and instruction definitions.

Table 2 shows the generated or changed opcodes files for `mysparc` target.

6.3 Retargetting Assembler

As described in Section 3.3, the target-dependent part of `gas` involves instruction parsing, assembling and relocation generation.

Generated Files	#line (generated)
/include/opcode/mysparc.h	60
/opcode/mysparc-opc.c	1552
/opcode/configure	262
/opcode/Makefile.in	1045

Table 2: Generated and changed files for `opcodes`.

Generated Files	#line (generated)
/gas/config/tc-mysparc.c	2480
/gas/config/tc-mysparc.h	368
/gas/configure.in	940
/gas/Makefile.in	2763

Table 3: Generated and changed files for `gas`.

Current distribution of GNU `gas` has ad hoc parser implementations for each target, partly due to the fact the `opcodes` library is add hoc. `rbinutils` uses a single algorithm to generate an assembly line parser for different targets. During parsing, the mnemonic of the instruction is first extracted from input assembly line. The internal instruction description is then searched and retrieved from the hash table maintained by the `opcodes` library. The remaining texts are then matched against the assembly patterns stored, during which symbol, immediate or register information is extracted and stored.

After collecting all assembly fields, the generated parser encodes the instruction according the the binary format specified in the target specification. As stated earlier, the matching between the assembly fields and the binary fields are achieved by matching the argument number in target specification. Opcodes and immediate values can thus be directly encoded by bit masking and shifting. Register values are encoded after retrieving its index from a register table, which is generated according to the store information specified in the target specification. For fields that need relocation, as indicated in the target specification, relocation entries are added in the object file.

Table 3 shows the generated `gas` files for target `mysparc`.

6.4 Retargetting Other Tools

Other tools depend only on the BFD and `opcodes` libraries. For example, the linker uses BFD to handle both dynamic or static linking. Retargetting the rest of the binary utilities is just a matter of changes in configuration files. Table 4 shows the files generated for `ld` retargetting.

6.5 Verification

To verify our tool, we used `gcc` to compile all SPEC2000 integer benchmarks including `gzip`, `mcf`, `eon`, `vortex`, `vpr`, `crafty`, `perlbmk`, `bzip2`, `gcc`, `parser`, `gap` and `twolf`, into assembly codes of the SPARC target. We then assembled, and linked them using the assembler and linker generated by `rbinutils`. We compared the generated executables with those generated by the original GNU binutils, and the results were exactly the same.

Generated Files	#line (generated)
/ld/configure.in	183
/ld/emulparam/elf32_mysparc.sh	9
/ld/Makefile.in	868
/ld/config/mysparc-elf.mt	1

Table 4: Generated and changed files for `ld`.

7. CONCLUSION

In conclusion, we have argued that the complexity involved in modern downstream tools such as assemblers and linkers have made development or automatic generation of these tools from scratch an impractical task. This has led to our effort which seeks to automatically port an existing tools that is powerful, robust and freely available. By augmenting the specification of instruction set information of a processor with ABI information, we are able to automatically port the GNU's `binutils` package, which itself consists of a quarter million lines of code. Our experiment shows that this approach is both feasible and practical. In future work, we will extend the same methodology to another extremely useful and freely available GNU development tool, the `gdb` source level debugger.

8. REFERENCES

[1] John R. Levine, *Linkers and Loaders*, Morgan Kufmann Publishers, 2000.

[2] C. W. Fraser, R. R. Henry, and T. A. Proebsting, "BURG—fast optimal instruction selection and tree parsing," *SIGPLAN Notices*, vol. 27, no. 4, pp. 68-76, April 1992.

[3] P. Marwedel, "The MIMOLA design system: Tools for the design of digital processors," in *Proceeding of the 21st Design Automation Conference*, June 1984, pp. 587-593.

[4] A. Fauth, J.V. Praet, and M. Freericks, "Describing instruction sets using nML," Tech. Rep., Technische Universiteat Berlin and IMEC, Berlin(Germany)/Leuven(Belgium), 1995.

[5] J. Van Prate, D. Lanneerand W.Geurts, and G. Goossens, "Processor modeling and code selection for retargetable compilation," *ACM Transaction on Design Automation of Electronic Systems*, vol. 6, no. 3, July 2001.

[6] *Mescal Architecture Description*, `http://www.ee.princeton.edu/MESCAL/mad.html`.

[7] A. Halambi, P. Grun, V. Ganesh, A. Khare, N. Dutt, and A. Nicolau, "Expression: A language for architecture exploration through compiler/simulator retargetability," in *Proceedings of the Design Automation and Test Conference in Europe*, March 1999.

[8] P. Paulin, C. Liem, T. May, and S. Sutarwala, "Flexware: A flexible firmware development environment for embedded systems," in *Code Generation for Embedded Processors*, P. Marwedel and G. Goossens, Eds. Kluwer Academic Publishers, 1995.

[9] G. Hadjiyiannis, S. Hanono, and S. Devadas, "ISDL: An instruction set description language for retargetability," in *Proceeding of the 34th Design Automation Conference*, June 1997.

[10] *LISA Language for Instruction Set Architectures*, Institute for Integrated Signal Processing System, ISS - RWTH Aachen, October 2000.

[11] N. Ramsey and M. Fernandez, "The New Jersey machine-code toolkit," in *Proceedings of the 1995 USENIX Technical Conference*, January 1995, pp. 289-302.

[12] A. Hoffmann, T. Kogel, A. Nohl, G. Braun, O. Schliebusch, O. Wahlen, A. Wieferink, and H. Meyr, "A novel methodology for the design of application-specific instruction-set processors (asips) using a machine description language," *IEEE Transactions on Very Large Scale Integration (VLSI) Systems*, vol. 20, no. 11, pp. 1338-1354, November 2001.

[13] S.Chamberlain, *libbfd: the Binary File Descriptor library.*, Cygnus Support, Free Software Foundation, Inc., first edition edition, April 1991.

Exploiting Operation Level Parallelism through Dynamically Reconfigurable Datapaths

Zhining Huang, Sharad Malik
Department of Electrical Engineering, Princeton University
Princeton, NJ 08544, USA
{znhuang, sharad}@ee.princeton.edu

ABSTRACT

Increasing non-recurring engineering (NRE) and mask costs are making it harder to turn to hardwired Application Specific Integrated Circuit (ASIC) solutions for high performance applications [12]. The volume required to amortize these high costs has been increasing, making it increasingly expensive to afford ASIC solutions for medium volume products. This has led to designers seeking programmable solutions of varying sorts using these so-called programmable platforms. These programmable platforms span a large range from bit-level programmable Field Programmable Gate Arrays (FPGAs), to word-level programmable application-specific, and in some cases even general-purpose processors. The programmability comes with a power and performance overhead. Attempts to reduce this overhead typically involve making some core hardwired ASIC like logic blocks accessible to the programmable elements. This paper presents one such hybrid solution in this space – a relatively simple processor with a dynamically reconfigurable datapath acting as an accelerating co-processor. This datapath consists of hardwired function units and reconfigurable interconnect. We present a methodology for the design of these solutions and illustrate it with two complete case studies: an MPEG 2 coder, and a GSM coder, to show how significant speedups can be obtained using relatively little hardware. The co-processor can be viewed as a VLIW processor with a single instruction per kernel loop. We compare the efficiency of exploiting the operation level parallelism using classic VLIW processors and this proposed class of dynamically configurable co-processors. This work is part of the MESCAL project, which is geared towards developing design environments for the development of application specific platforms.

Categories and Subject Descriptors

J.6 [**COMPUTER-AIDED ENGINEERING**]: *Computer-aided design (CAD)*

General Terms

Design, Performance

1. INTRODUCTION

Deep sub-micron issues of interconnect delay and signal integrity have significantly increased the design costs of ASICs – both due to the higher engineering costs resulting from longer design cycles, as well as due to the increasing cost of design tools. This increase in NRE costs has pushed up the breakeven volume point at which the per-device cost for the manufactured product is viable. As a consequence, for many products below this break-even volume mark there is a desperate need to satisfy the application performance requirements using specialized programmable solutions. These programmable solutions are increasingly referred to as programmable platforms, and their design (and use) is an exercise in trying to match application concurrency with architectural concurrency.

There is a vast range of these platforms, from bit-level programmable fine and coarse-grained FPGAs, to word-level programmable application specific processors. The application specific processors themselves span a fairly large range – from specialized VLIW processors with complete instruction sets tailored to the application domain [14], to simple RISC processors with a few specialized instructions implemented using hardwired logic blocks [15]. A relatively new class of platforms consists of simple processors with programmable co-processors, which are essentially accelerating datapaths with very little, if any, control [8, 16].

In this paper we present a solution in the last category. We introduce a dynamically reconfigurable co-processor with fixed hardware blocks and programmable interconnect and show how this can be used to accelerate the main computation loops (kernels) of compute intensive applications. The rest of the application runs on a relatively simple master processor. Specifically we provide a complete methodology and algorithm for the design of such a co-processor for a given application as well as the mapping of the kernel loops onto this co-processor.

This is illustrated through two complete non-trivial case studies, an MPEG2 encoder and a GSM encoder where we show significant speedups using relatively little hardware.

Interestingly, this co-processor can be viewed as a VLIW processor with a single instruction per kernel loop. In our effort to understand the matching of application concurrency to architectural concurrency, we compare the efficiency of exploiting the operation level parallelism using classic VLIW processors and this proposed class of dynamically reconfigurable co-processors with some interesting observations.

2. PREVIOUS RESEARCH

There have been several research efforts as well as commercial products that have tried to explore the use of configurable logic as acceleration co-processors.

Early research on reconfigurable computing focused on configurable logic in the form of fine-grained FPGAs. However, FPGAs failed to become a mainstream computing technology for complete applications because of reasons such as bit level logic granularity, hardware constraints and high configuration time [5]. Subsequently we saw the emergence of tightly coupled reconfigurable co-processor designs, for example, GARP [1] and RaPiD [6]. In these designs, kernels from applications are mapped to the co-processor for acceleration. However, as these co-processors still consisted of fine-grained architectures, the problems of reconfiguration bit size, hardware resource constraints, and slow clock rate were not fully solved. Coarse-

DAC 2002, June 10-14, 2002, New Orleans, Louisiana, USA.

grained reconfigurable computing, such as used in PipeRench [5] and Pleiades [7] differed in that they used coarser blocks of logic in the reconfigurable fabric, resulting in faster clock speeds and the ability to implement larger applications in the multimedia and communication domains. However, PipeRench is not targeted to specific applications and cannot be customized for specific kernel loops. Pleiades is a design for a specific application domain targeting low power. It has a hierarchical mesh interconnect between logic macros (ALUs, MEMs, etc). The interconnect is reconfigured for different kernels to enable the reuse of logic macros.

Recently, commercial products have emerged that include the ability to dynamically reconfigure the programmable logic during a single application. The Chameleon chip has a 32bit embedded processor with a dynamically reconfigurable fabric. Up to 84 datapath units and 24 multipliers reside in the fabric. The interconnection between them is dynamically reconfigurable. The Chameleon chip has a fixed architecture targeting communication applications. Kernels in applications are mapped to the reconfigurable fabric through hand written Verilog descriptions. The reconfigurable fabric is significantly general to permit a wide application for communication applications. This device is an example of the dynamically reconfigurable co-processors that our methodology targets. We describe how such a device can be designed and programmed for a given application (domain) by providing a complete methodology. This builds on our previous work on this, which focused on a specific algorithm that is part of this methodology. None of the previous efforts described here have presented a methodology for the design of application specific co-processors that use dynamically reconfigurable coarse-grained logic.

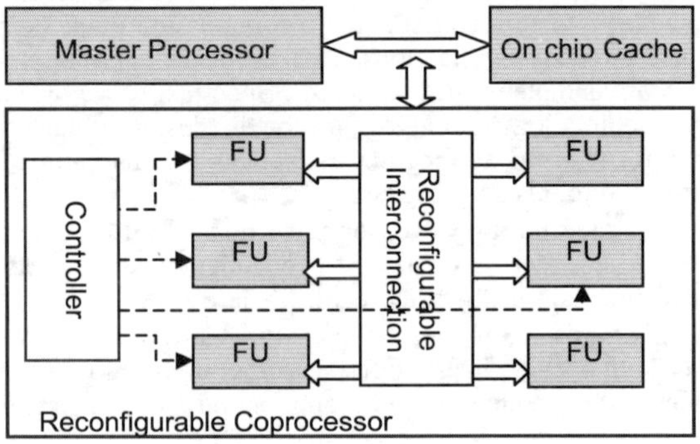

Figure 1. Architectural Model

3. METHODOLOGY OVERVIEW

The architectural model studied in this paper consists of a master processor and a reconfigurable co-processor as illustrated in Figure 1. The co-processor has the same access to the memory sub-system as the master processor. The co-processor mainly consists of a reconfigurable datapath and some control logic. The controller works as a state machine and controls the start and end of the datapath execution. The reconfigurable datapath has fixed ASIC-like function units and reconfigurable interconnections. This paper mainly describes the methodology of designing the reconfigurable datapath for an application, given its high-level language description.

The first step in the methodology is the identification of the computationally intensive loops in the application. For each such kernel loop a custom datapath is designed to maximize the parallelism that can be exploited. Datapaths designed for different loops in an application are then combined into a single reconfigurable datapath co-processor. The co-processor switches between the individual datapaths through changing the connections between the function units using programmable interconnect. The interconnect "program" bits for a loop datapath are referred to as the context of the loop. If the size of this program is small (as we will demonstrate) and the number of contexts is small (as we will also show), then the contexts can all be stored locally in the co-processor and need not be brought in from external memory. It also enables rapid switching of contexts – a switch can be accomplished in a single cycle.

4. DATAPATHS FOR KERNEL LOOPS

This section describes the first step in this methodology - how to generate the datapaths for kernel loops given high-level language descriptions of an application.

4.1 Extract Kernel Loops

We use the IMPACT compiler as the front end to do pre-processing, including performance profiling and loop detection. C codes are used as the description of applications and as the inputs to the front-end compiler. As we only map innermost loops to custom datapaths in the co-processor at this stage, the result of extracting kernel loops is a list of the most-executed innermost loops in Lcode (a meta assembly language), which is the intermediate representation of the IMPACT compiler.

In order to map kernel loops to customized pipelined datapaths, we need to do data dependence analysis to derive the following: register live-in set of the loop body; register live-out set of the loop body; data dependence between instructions within loop iterations; data dependence between instructions in different loop iterations.

The live-in and live-out sets are used during the switch between the master processor and the co-processor. When the execution of the application hits the loop, the general processor switches the execution to the co-processor. During the switch, the values of registers in the live-in set are copied into the co-processor. The live-out set contains registers whose value has changed during the loop execution and will be used later. After the loop execution, the co-processor switches execution back to the master processor and writes back the live-out register values.

The data dependence information helps in constructing the datapath for the loop body.

4.2 Direct Mapping of Kernel Loop Datapath

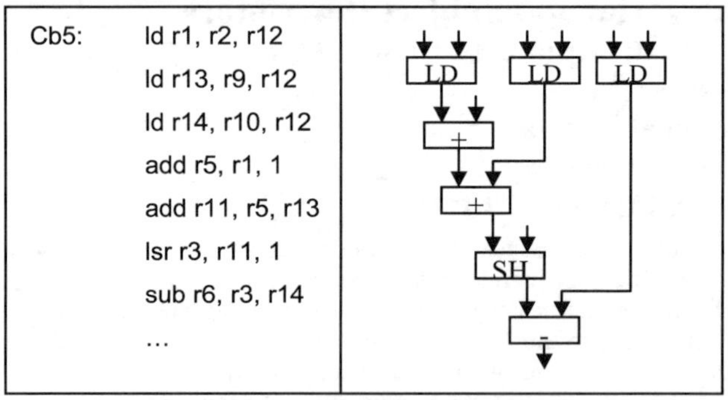

Figure 2. Direct mapping from IR to hardware

Given the Lcode intermediate representation for a kernel loop, our custom datapath design for that loop proceeds by starting with a direct hardware mapping from the Lcode to hardware blocks. Direct mapping here means that one instruction in software corresponds to one function unit in the hardware. By connecting all the function units in the hardware, we form the datapath for a loop body as illustrated in Figure 2.

Each function unit in the hardware executes an instruction in the software. Since the data goes through the datapath, no write back to the register file is needed. This further implies no register bandwidth limitation within loops. The advantage of direct mapping is the simplicity of the methodology, which makes it easy to implement automatically.

4.3 Branch Condition Transforms

If there is more than one basic block in the loop body, the two sides of branch instructions result in different datapaths, which are then merged using a multiplexer.

4.4 Pipelining the Execution

As the first step in pipelining the execution, we need to assign the function units into different pipe stages in the datapath. In order to do so, we insert registers in the datapath.

The scheduling of the pipeline stages for the datapath maximizes the parallelism of the loop body within the iteration. The scheduling of execution of consecutive loop iterations maximizes the parallelism between loop iterations. If there is no data dependence between loop iterations, at each clock cycle, we can fetch and begin to execute the next loop iteration until we finish all the iterations. If there is data dependence between loop iterations, we look at the pipeline stages of the instructions that cause the dependence to decide whether a delay or a bypass is needed. There are two kinds of data dependences, one is dependence from registers, and the other one is dependence from memory operations. The two kinds of dependences have different properties.

For register dependence, the register value is written in the first loop iteration and is used in the next one. The dependence always exists between adjacent iterations. Suppose the register value is written at pipeline stage i and is used at pipeline stage j. If $i \leq j$, the register value is always generated before it is used because of the pipelined execution. The second iteration is always one step behind the first one during the pipeline execution. So all we need is a bypass from the function unit that generates the register value to the function unit consuming the value. To synchronize the execution, (j-i) registers are inserted along the bypass. If $i > j$, the register value is generated after it is used if we pipeline the execution without adjustment. Obviously this is not correct. In this case, we need to delay the pipeline execution and add a feedback connection from the function unit which generates the register value to the function unit which consumes it. A total of (i-j) registers are inserted on the feedback connection for synchronizing the execution. An extra (i-j) clock cycle delay is needed besides the normal one clock cycle delay between the pipeline execution of the adjacent iterations. The total delay for this case is (i-j+1), which means that every (i-j+1) clock cycles a new loop iteration begins to execute.

For data dependence from memory operations (load and stores), if the dependence is store after load, we just need to reschedule the store into a pipeline stage that is late enough. If the dependence is store after store, the former store can be eliminated. There is never a dependence between two loads. If the dependence is load after store, the pair of conflicting instructions can be transformed into the instruction that generates the value to be stored, and the instruction which consumes the value load from memory. If this is the case, an extra execution delay or bypass is needed.

For memory data dependences, the dependence could happen beyond adjacent iterations. Since we fetch the iterations consecutively, iteration l and (l+k) have a normal delay of k clock cycles. If the dependent instructions are at the i and j pipeline stages, and the dependence happens between l and l+k iterations, the threshold value for delay or bypass is (j+k-i). If $(j+k-i)>0$, no extra delay is needed. Otherwise, [(i-j-k+1)/k] cycles of extra delay is needed. The delay between adjacent iteration execution is [(i-j+1)/k].

The assumption for pipelining the execution is that function units which take multiple cycles to complete can be internally pipelined. For example, an integer multiplier typically takes 3 cycles. So each multiplier occupies 3 pipeline stages in the datapath. Similarly, memory ports take 3 pipeline stages too.

4.5 Estimation of the Pipeline Execution Time

The pipeline execution of the datapath exploits the maximum parallelism that exists within the loop bodies and between loop iterations. If there is not much parallelism in a loop, it may not be worth switching the execution to the reconfigurable co-processor. If we store the reconfiguration context in a distributed cache on the co-processor to realize fast context switch as proposed, the number of loops to be mapped to the datapath is limited due to context memory size limits. It is important to measure the execution time and speed up we can have by mapping a loop to the datapath. Those loops that can bring maximum speed up should be selected.

Suppose the total number of pipeline stages of a datapath is S, the delay of the consecutive loop iterations is D (≥ 1) cycles, the loop iteration number is N, the execution switch (from general processor to co-processor) overhead is O cycles, and the write back time is W cycles, then the total execution time is given by:

$$T = [S + D*(N-1)] + O + W \quad \text{cycles} \qquad (1)$$

We use this estimation of the execution time for each kernel loop, to select the top kernel loops which can get the most speed up for the application after switching the execution to the reconfigurable co-processor.

5. RECONFIGURABLE DATAPATH

Using the direct mapping methodology and the scheduling algorithm for pipeline execution, we generate the datapath for each selected loop as described in the previous section. The datapath consists of function units and the interconnection between them. Given the multiple datapaths for top kernel loops, we need to generate the common datapath and then reconfigure it into the datapaths for different loops when necessary. The process for merging the multiple datapaths together and generating the common datapath is called datapath mapping. Our previous research [9] has addressed this specific problem in the methodology with a bipartite matching algorithm.

5.1 Datapath Mapping

This part of this methodology has been reported in our prior work and is presented here briefly for review. The datapath mapping algorithm essentially does a graph overlay of the topology graphs for each datapath. The vertices in the topology graph represent function units in the datapath. Vertices have the same color if they represent same type of function units. Directed edges represent the interconnections of function units. The goal is to overlay the graphs, by sharing same colored vertices and interconnections, into a single graph such that it has vertices and interconnections for all the individual graphs and the total number of interconnections is minimized. This combined graph represents the final datapath to be used, with the individual kernel loop datapaths being instantiated using reconfiguration of the interconnect.

5.2 Routing Box

The dotted block in Figure 3 is a function unit in the datapath as viewed from the outside. It consists of three parts. A routing box is placed before the actual function unit and is used to select from the multiple inputs. A register is inserted after the function unit for pipelining.

In Figure 3, the routing box selects two inputs to the function units from the five possible interconnections coming into this function unit. The more the interconnections, the more the configuration bits needed for this selection. The last element in Figure 3 is the control bits. These are needed during the configuration to specify the function for multi-function units, for example, add vs. subtract. In addition, some constant registers may need to be initialized during the configuration.

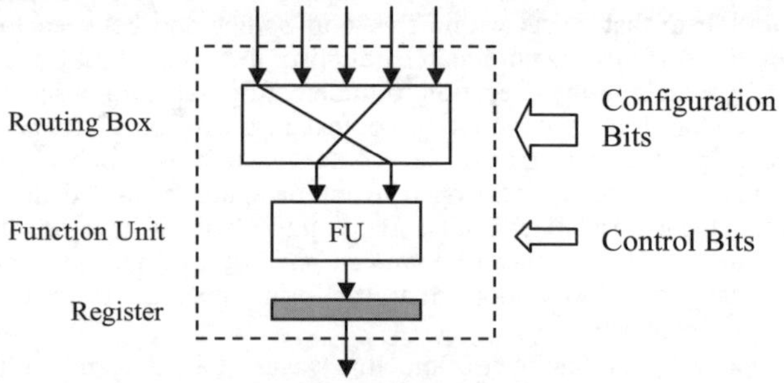

Figure 3. Routing Box for Function Units

5.3 Critical Path and Clock Speed

The critical path in the reconfigurable datapath determines the clock speed in the co-processor. In the reconfigurable datapath, the critical path of a pipeline stage is the critical path of the fixed function unit plus the delay of the routing box plus wire delays. Sometimes a complicated function unit is pipelined inside and executed in multiple clock cycles. Since the functional units are hardwired, their delay is the same as it would be if the functional unit was in the processor. The same is true for the wire delays. As the routing box just needs to select between its inputs, its delay is just one gate delay [9, 10], which is negligible. In general processors, the critical path of pipeline stages is no longer in the functional unit stage, but rather in the more sophisticated branch control and decoding stage. This may be significantly larger than functional unit delay. In fact, two Arithmetic Logic Units (ALUs) on the Pentium 4 processor are clocked at twice the core processor frequency [11]. Thus, it is reasonably safe to assume that the co-processor clock period is less than the master processor clock speed, with the potential for significant speed up.

5.4 Reconfiguration Overhead

Since the reconfigurable datapath is dynamically reconfigured at each context switch with transfer of execution control, the reconfiguration overhead is counted as part of the execution time and is crucial to speed up. The overhead consists of three parts: the reconfiguration context switch, execution switching to the datapath and execution switching back.

Typically a number (usually 8 to 16) of the loops are selected, so we can store the multiple contexts in a distributed cache in the co-processor. During execution, no reconfiguration bits need to be loaded from main memory. We only switch between the multiple contexts, which can be done in 1 clock cycle [2, 9].

Since the trivial part of the codes are executed in the master processor, when we switch the execution to the reconfigurable datapath, live-in registers are read from register files in the general processor to the datapath. After execution, live out registers are written back to register files in the general processor. Usually the 2 sets contain 10~20 registers for each kernel loop. The register file bandwidth in the master processor is the bottleneck for copying the registers. Assuming we have 4 register read/write ports, this results in a 5 cycle overhead.

6. BENCHMARK STUDIES

6.1 MPEG 2 Video Coder

The source code and test data files for this are from Mediabench [4]. The data file for test coding is a bitstream of YUV components (4 frames). The software is compiled and profiled using the IMPACT compiler. The IR we used to extract kernel loop datapaths is generated using the processor platform "EPIC_1G_1BL" without any modern compiler techniques such as superblock, hyperblock or execution predication. "EPIC_1G_1BL" is the EPIC 1-issue 1-branch processor, which we use to represent today's single-issue speculative processor.

The top 10 kernel loops are selected and listed in Table 1 after we do the compilation, profiling, datapath extraction, pipeline scheduling and execution time estimation. These loops constitute 86% of the total execution time.

Table 1. Top 10 kernel loops from MPEG

	Software Time	*Hardware Time*	*# of Memory Ports*	*Speed Up*
Loop 1	820	132.3	2	6.2
Loop 2	85.0	9.57	5	8.88
Loop 3	33.5	4.79	3	6.99
Loop 4	33.1	4.73	3	7.0
Loop 5	35.7	10.8	2	3.31
Loop 6	20.2	0.64	3	31.6
Loop 7	31.4	9.7	2	3.24
Loop 8	11.5	1.82	2	6.32
Loop 9	10.6	1.68	2	6.31
Loop 10	7.3	0.50	2	14.6
Rest	178.2	-	-	1
Overall	1266.3	354.73	-	3.57

(Number in Million cycles)

The software time is the execution time on the general processor in clock cycles. The hardware time accounts for the estimation of pipelined datapath execution time and reconfiguration overhead in clock cycles. Rest stands for the remainder of the code besides the 10 loops, which is always executed in the master processor. Compared with the single-issue processor, we can get a speed up of 3.57. The speed up is based on a conservative assumption of the same clock speed of the co-processor and general processor.

The 10 loops have 7 different datapaths - some of the loops share the same datapath. From the 7 different datapaths, we generate the reconfigurable datapath using the mapping algorithm described earlier. This results in 51 function units and 114 interconnections. Registers are inserted for pipelining and are viewed as part of the function units here. The 51 function units includes 24 registers, 8 adders, 5 memory ports, 3 shifters, 2 multipliers, 2 dividers, 2 logic ALUs, and 5 multiplexers.

An estimate of the total reconfiguration bit size for programming the interconnections is 221 bits. Including further the additional configuration bits for some function units such as adders, shifters and multiplexers, and the configuration bits for initializing some of the constant registers (no more than 8 x 16 bits), each reconfiguration context will be under 500 bits. We can view this as a single instruction for the co-processor needed for the entire loop. To store 5k bits for all 10 loops in the distributed cache in co-processor is very reasonable.

From the datapath mapping result, we see that the hardware resources needed to construct such a reconfigurable datapath are not significant. This makes it feasible to consider multiple copies of this if it can lead to further exploitation of operation level parallelism in the loop. If there is no data dependence between loop iterations, then having multiple copies of the datapath and executing loop iterations in parallel can bring significant speedup. However, this increases the required operand bandwidth, which translates to memory bandwidth. Having multiple copies of the datapath requires duplicating the memory ports too. However, memory bandwidth is likely to be constrained.

It is interesting to examine the relationship between datapath speed up with the memory bandwidth under the assumption of enough other hardware resources in the co-processor. Figure 4 shows the speedup obtained vs. the number of memory ports. The reason for the speed up saturation is the limit on the number of parallel iterations due to inter-iteration data dependency. In this study we have not considered the use of loop restructuring techniques to reduce inter-iteration data dependency. That is likely to further increase the operation parallelism and increase speedup [17]. This is one of the issues we intend to examine in further work.

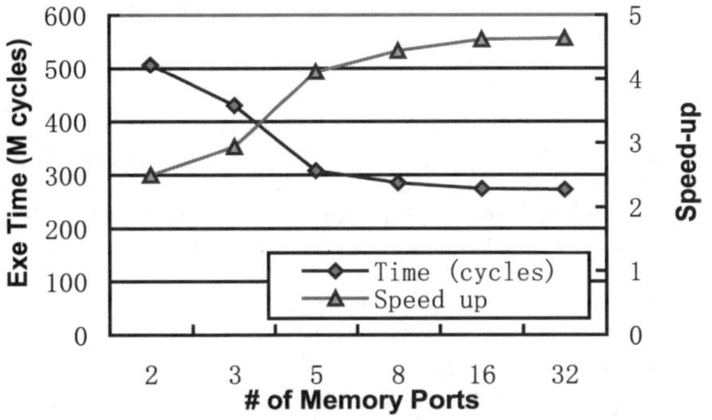

Figure 4. Speed-up vs. Memory Bandwidth for MPEG

Note however that before we see speedup saturation, the number of memory ports is a critical factor that affects the speed up. Usually this number cannot be very large in the current computer systems. Some applications may require a large number of memory ports in order to get a large speed up. We are currently examining techniques for reducing this.

Given that we can view the co-processor as a VLIW with a single instruction per loop, it is instructive to see how this speedup compares with using a conventional VLIW processor. This will provide some insight into the two different ways of harnessing operation level parallelism. The co-processor proposed in this work exploits it through customized datapath synthesis per loop, for the selected loops, and just a regular processor for the rest of the program. In comparison, a VLIW processor exploits it through instruction level parallelism through the entire program. Using the IMPACT compiler, we profiled and estimated the software execution time with different processor platforms. "EPIC_xG_1BL" means x-issue 1-branch EPIC processor. X (1, 4, 8 or 16) instructions can be fetched and issued at the same time. Hyperblocks and execution predication are also used to get the best execution time.

Figure 5 shows the execution times on different EPIC platforms, as well as the execution time of a single-issue processor with a co-processor with 8 memory ports. With 8-issue or 16-issue, a VLIW has almost the same speed for the whole application on the simple-processor, co-processor combination. However the cost of an 8-issue or 16-issue VLIW is large compared to the proposed co-processor. The VLIW processor size grows significantly due to the large number of register file ports, and large instruction decode logic. The program size increases significantly too. Hyperblock formation and predication execution will expand the code size, in this case from 17k instructions to 29k instructions. Large instruction width such as 8 or 16 would bring many empty slots into the VLIW instructions which further expand the code size. The cost of the reconfigurable datapath is much smaller. There is no complicated instruction decode logic, and no large register file ports. The program size can be even smaller since the kernel loop codes become reconfiguration contexts with small bit size. For each kernel loop, the reconfigurable datapath is faster than any EPIC processor. For example, the most executed kernel loop in the MPEG coder takes 128M cycles in an EPIC 8-issue processor while the reconfigurable datapath takes 71.7M cycles with 8 memory ports. By increasing the number of kernel loops mapped onto the co-processor, we can further increase the speedup at the cost of co-processor complexity and program size.

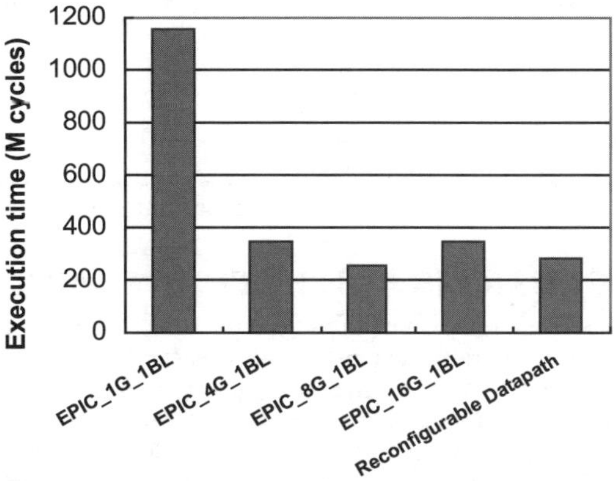

Figure 5. VLIW vs. Reconfigurable Datapath for MPEG

6.2 GSM

GSM (Global System for Mobile communication) is one of the most popular wireless communication standards. We now examine the implementation of a GSM coder, which compresses speech signals. More specifically, GSM 06.10 compresses frames of 160 16-bit linear samples into 33-byte frames. Both the source codes and test data for this are from Mediabench [4].

Table 2 gives the result of speedup of the co-processor against a single-issue speculative processor. An overall speed up of 2.78 for the application is achieved. In the GSM case, the top ten kernel loops account for 81% of the execution time.

The datapath mapping result of GSM shows that the reconfigurable datapath includes 67 function units and 197 interconnections. The 67 function units include 25 registers, 10 memory ports, 17 adders, 8 multipliers, 4 shifters, and 3 multiplexers. An estimate of interconnection reconfiguration size is 415 bits. With additional configuration bits for function units such as adders, shifters, and multiplexers and initialization of constant registers, the total reconfiguration size is expected to be under 1k bits. It is reasonable to store the 10k bits in the distributed cache in the co-processor for the 10 loops.

The speedup obtained vs. the number of memory ports under the assumption of enough other hardware resources so that we can make multiple copies of datapaths is similar to the MPEG case showed in Figure 4. In GSM, with 4 memory ports we could approach the limit of speed up very closely.

The result of comparing the reconfigurable datapath with different EPIC platforms is similar to the MPEG case showed in Figure 5. Again a significantly complex VLIW processor is needed to match the performance of this simple processor, co-processor combination.

Table 2. Top 10 kernel loops from GSM

	Software Time	*Hardware Time*	*# of Memory Ports*	*Speed up*
Loop 1	61.0	9.47	3	6.44
Loop 2	33.3	13.8	2	2.41
Loop 3	5.76	0.203	10	28.4
Loop 4	4.21	0.151	4	27.9
Loop 5	2.31	0.157	1	14.7
Loop 6	2.07	0.192	1	10.8
Loop 7	2.07	0.184	3	11.2
Loop 8	2.06	0.149	2	13.8
Loop 9	1.40	0.118	2	11.9
Loop	1.26	0.149	2	8.46
Rest	26.45	-	-	1
Overall	141.9	51.0	-	2.78

(Number in million cycles)

7. CONCLUSIONS AND FUTURE WORK

This paper describes a methodology for the design of a dynamically reconfigurable datapath co-processor for a specific application (domain). We describe how such a co-processor can be designed and an application mapped to it. Through the two case studies of an MPEG2 encoder and a GSM encoder, we demonstrate how significant speedups compared to a single processor execution can be obtained using relatively little hardware.

We also show how the co-processor can be viewed as a VLIW processor with a single (different) instruction per loop mapped to the co-processor. This then leads to comparing the exploiting of operation level parallelism using such a co-processor with regular VLIW processors. Here we see that significantly larger VLIW processors will be needed to match the performance of such a co-processor, with poorer area and power properties. In general, we can draw some conclusions about the relative applicability of such a co-processor vs. a classic VLIW processor. If the application has a few areas of localized operation level parallelism, such a co-processor provides for a cost (area and power) effective way of harnessing this parallelism without paying the price for it in the rest of the program. However, if the entire application code has operation level parallelism, a VLIW processor is likely to be able to exploit it just as well.

In the above discussion, we have been quite conservative about the parallelism and speed of the co-processor. We have assumed that the co-processor clock speed is the same as the main processor, a fact that we have argued is likely to be conservative. A detailed analysis is needed to determine how much faster the clock on the co-processor can be compared to the main processor. This can lead to further speedup. The VLIW processors we are comparing against have the benefit of the most sophisticated of compilation techniques (hyberblock formation and predicated execution), while our compilation is fairly straightforward. We intend to examine the use of loop restructuring techniques geared towards reducing intra and inter-iteration data dependencies. Further, the function blocks that we consider are in direct correspondence with operations in instructions. We know that many of these can be combined into a single complex function block as is done in application specific processors from Tensilica [15]. Again, this bears further investigation.

Based on the speedups we have seen, and the potential for significant additional speedup based on the directions we have outlined, we believe that these platforms are very interesting for a class of applications with localized sections of operation level concurrency.

8. ACKNOWLEDGEMENTS

This research is supported by the Gigascale Silicon Research Center (GSRC), sponsored by MARCO and DARPA. We would like thanks members of the MESCAL team for interesting discussions as part of this work.

9. REFERENCES

[1] J. R. Hauser and J. Wawrzynek. "Garp: A MIPS Processor with a Reconfigurable Coprocessor," In *Proceedings of the IEEE Symposium on Field-Programmable Custom Computing Machines* (FCCM '97, April 16-18, 1997), pp. 24-33.

[2] K. Furuta, T. Fujii, M. Motomura, K. Wakabayashi, M. Yamashina. "Spatial-Temporal Mapping of Real Applications on a Dynamically Reconfigurable Logic Engine (DRLE) LSI," CICC 2000.

[3] IMPACT research group, University of Illinois, at Urbana-Champaign, http://www.crhc.uiuc.edu/IMPACT.

[4] C. Lee, M. Potkonjak, W. Mangione-Smith, "MediaBench: A Tool for Evaluating and Synthesizing Multimedia and Communications Systems," Micro 30, 1997.

[5] S. C. Goldstein, H. Schmit, M. Moe, M. Budiu, S. Cadambi, R. Taylor, R. Laufer. "PipeRench: A Coprocessor for Streaming Multimedia Acceleration," *ISCA* 1999.

[6] D. Cronquist, P. Franklin, S. Berg, and C. Ebeling, "Specifying and Compiling Applications for RaPiD," IEEE workshop on FPGAs for Custom Computing Machines, 1998.

[7] M. Wan, H. Zhang, V. George, M. Benes, A. Abnous, V. Prabhu, J. Rabaey "Design Methodology of a Low-Energy Reconfigurable Single-Chip DSP System," Journal of VLSI Signal Processing, 2000.

[8] Chameleon Systems Inc., San Jose, CA, "CS2000 Reconfigurable Communications Processor Product BRIEF" from http://www.chameleonsystems.com.

[9] Z. Huang and S. Malik, "Managing dynamic reconfiguration overhead in system-on-a-chip design using reconfigurable datapaths and optimized interconnection networks", in DATE 2001.

[10] Product Data Sheets, "Virtex: DC and Switching Characteristics", from http://www.xilinx.com.

[11] Intel Data Sheet, "Product overview: Intel® Pentium® 4 processor-based workstations," from http://www.intel.com.

[12] "System Level Design: Orthogonolization of Concerns and Platform-Based Design", K. Keutzer, S. Malik, J. M. Rabaey, A. R. Newton and A. Sangiovanni-Vincentelli), IEEE Transactions on Computer-Aided Design, Vol. 19, No. 12, December 2000.

[13] T. J. Callahan and J. Wawrzynek, "Instruction-Level Parallelism for Reconfigurable Computing", *Field-Programmable Logic and Applications FPL'98*, Tallinn, Estonia, Aug-Sep 1998.

[14] TriMedia Technologies Inc., http://www.trimedia.com/.

[15] Tensilica Inc., http://www.tensilica.com/.

[16] Morphics Technology Inc., http://www.morphics.com/.

[17] M. Wolfe, "High Performance Compilers For Parallel Computing", Addison-Wesley Publishing Company, 1996.

Dynamic Hardware Plugins in an FPGA with Partial Run-time Reconfiguration

Edson L. Horta,*
Universidade de San Pãulo
Escola Politécnica - LSI
San Pãulo, SP, Brazil
edson-horta@ieee.org

John W. Lockwood,†
David E. Taylor
Applied Research Lab
Washington University
Saint Louis, MO 63130
{lockwood,det3}@arl.wustl.edu

David Parlour
Xilinx, Inc
2100 Logic Drive
San Jose, CA 95124
dave.parlour@xilinx.com

http://www.arl.wustl.edu/arl/projects/fpx/

ABSTRACT

Tools and a design methodology have been developed to support partial run-time reconfiguration of FPGA logic on the Field Programmable Port Extender. High-speed Internet packet processing circuits on this platform are implemented as Dynamic Hardware Plugin (DHP) modules that fit within a specific region of an FPGA device. The PARBIT tool has been developed to transform and restructure bitfiles created by standard computer aided design tools into partial bitsteams that program DHPs. The methodology allows the platform to hot-swap application-specific DHP modules without disturbing the operation of the rest of the system.

Keywords

FPGA, partial RTR, reconfiguration, hardware, modularity, network, routing, packet, Internet, IP, platform computing

Categories and Subject Descriptors

B.7.2 [**Hardware**]: Circuits—*Design Aids*; B.7.1 [**Hardware**]: Circuits—*VLSI*; B.4.3 [**Hardware**]: Input/Output and Data Communications—*Interconnections (Subsystems)*; C.2.1 [**Computer Systems Organization**]: Computer-Communication Networks—*Network Architecture and Design*

General Terms

Design, Experimentation

*Research is supported by CNPq (Brazil)

†Research is supported by NSF: ANI-0096052 and Xilinx

1. INTRODUCTION

FPGAs are frequently used in networking applications, where they offer both the performance of custom hardware and the flexibility of reprogrammability [1] [2]. Systems implemented with FPGAs can make use of their reprogrammability in one of two ways: Compile-Time Reconfiguration (CTR) or Run-Time Reconfiguration (RTR) [3]. CTR systems do not change the FPGA's configuration for the lifetime of the application, e.g. SPLASH [4] and PAM [5]. RTR systems change the FPGA configuration during the course of operation, either by full reconfiguration [6] [7] or partial reconfiguration [8] [9] [10].

The present research uses partial RTR in an FPGA to provide developers of hardware packet processing applications a capability similar to the Dynamically Linked Libraries (DLLs) which are used in software applications. Just as a DLL is a software module that can be attached to or removed from a running program as an application demands, we define a Dynamic Hardware Plugin (DHP) [11] as a module which can be loaded into or removed from a running FPGA without disturbing other circuits operating in it. The ability to change the hardware feature set in a running system is particularly useful in packet processing applications such as firewalls and routers, where it is not desirable to suspend the operation of a network during reprogramming. A practical system for implementing DHPs includes:

- A suitable FPGA test platform
- A well defined DHP interface specification
- A complete DHP design methodology
- Physical implementation tools (place & route)
- Configuration bitstream management tools

These five elements are analogous to an operating system platform, Application Programming Interface (API), modular programming methodology, compiler, and linker needed to implement DLLs in the software domain.

The Field Programmable Port Extender (FPX) [12] is the FPGA-based prototyping platform used in the Washington University Gigabit Switch (WUGS) [13]. Section 2 describes the well-defined DHP interfaces, the design methodology, the CAD tools which place and route DHP modules, and the PARBIT tool that manages bitstreams. The FPX is described in Section 3, along with an application targeted to this platform.

2. PARTIAL RECONFIGURATION

Partial reconfiguration allows an FPGA to implement multiple functions and to change those functions while the system is running. The target FPGA is logically partitioned into a static infrastructure region and a number of DHP sites. The infrastructure connects each DHP site to shared resources and/or other DHP sites. In the present system the FPGA Input/Output resources are owned by the infrastructure and all DHP sites are identical, although different design choices might be appropriate in other applications.

The layout and interface specification for the DHPs are influenced by both the architecture of the FPGA and its behavior during reconfiguration. The characteristics of the Xilinx VIRTEX-E family are described below. Following this description, the tool used to generate the partial configuration bitfile is presented, along with the requirements for the interface between the static infrastructure and the DHP sites, and the methodology to generate the configuration bitfiles used by PARBIT.

2.1 VIRTEX-E Architecture

The application presented in this paper targets a device from the Xilinx Virtex-E family [14]. Programmable Input/Output Blocks (IOBs) around the edge of the array are used to interface to off-chip resources. The interior consists of a matrix of Configurable Logic Blocks (CLBs) containing: lookup tables, flip-flops and programmable interconnect. The lookup tables in the CLB can be used as function generators, small distributed RAMs or programmable-length shift registers. A number of columns in the CLB matrix are replaced with Block SelectRAMs, which are dedicated dual-ported memories. A column of clock drivers used for global clock distribution run vertically though the center of the chip.

Virtex-E configuration bits are organized in columns corresponding to a column of the FPGA's logic resources [15]. The Center Column controls the global clock pins. The IOB Columns control the configuration for the left and right side IOBs. Each CLB Column controls one column of CLBs and two IOBs above and bellow these CLBs. Each column has n rows, with one CLB per row. The Block SelectRAM Interconnect Columns define the interconnection of each RAM column. The Block SelectRAM Content Columns define the contents of each RAM column.

To configure the VIRTEX-E FPGA, a series of bits, divided into fields of commands and data, are loaded into the device. Each one of the configuration columns are divided in smaller slices, called frames. A frame is the smallest part of the configuration memory that can be written or read.

2.2 PARBIT

In order to partially reconfigure a FPGA, it is necessary to isolate a specific area inside the device and download the configuration bits related to that area. A tool called PARBIT [16] has been developed to easily transform and restructure bitfiles to implement dynamically loadable hardware modules.

To restructure the configuration bitfile, the tool utilizes the original bitfile, a target bitfile, and parameters given by the user. These parameters include the block coordinates of the logic implemented on a source FPGA, the coordinates of the area for a partially programmed target FPGA, and the programming options. The tool reads the original configuration bitfile and copies to the partial bitfile only the configuration bits related to the area defined by the user.

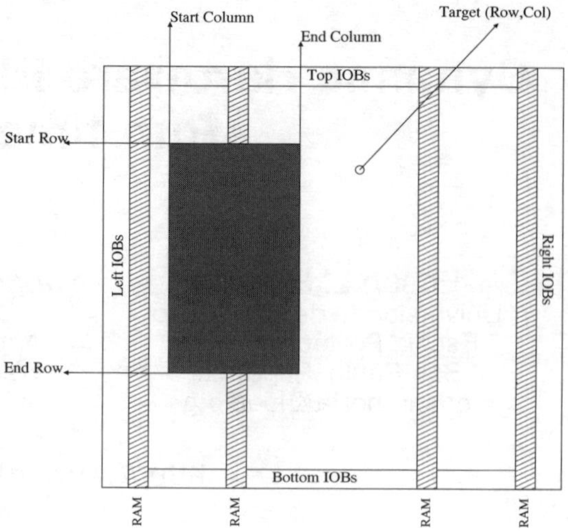

Figure 1: Block of logic selected from original bitfile

The target bitfile is used by PARBIT to copy the configuration bits that are inside a column specified by the user, but outside the partial reconfigurable area. This happens due to the fact that one frame takes all the rows of a column and the partial reconfigurable area is smaller than a whole column.

PARBIT allows arbitrary block regions of a compiled design to be re-targeted into any similar size region of an FPGA. It is possible to define an area inside the CLB columns of the chip, without the top and bottom IOB configuration bits. The tool generates the partial bitfile containing the area selected by the user (from the original bitfile) and this file will be used to reconfigure the FPGA.

To relocate blocks from the original bitfile, a user defines the start and end columns and rows for the block in the original design. Then, the user defines where to put this block into a target bitfile of the same device type. The tool then generates the partial bitfile containing the area selected by the user (from the original bitfile). This data is used to reconfigure the target device. A sample of a block within a VirtexE FPGA is shown in Figure 1.

It is important to note that the configuration bits for the top and bottom IOBs from the target device do not change after the partial bitfile is loaded. The configuration bits for the columns from the original and target bitfile are merged according to the rows defined by the user.

2.3 Gasket Interface

DHP modules that are downloaded into the FPX need fixed interconnection points to communicate with the infrastructure logic on the FPGA. These points are connected by special wires, called antennas.

Gaskets allow DHP modules to be built with only a minor modification to the standard FPGA design flow. When building a DHP, PAR (Place And Route) is made to limit routing to only the *included routes*. It may not use the *excluded routes*. By placing antennas between the DHP and the infrastructure region, the routing of signals through the gasket is fixed. Figure 2 shows how antennas cross out of the *DHP free routing zone*.

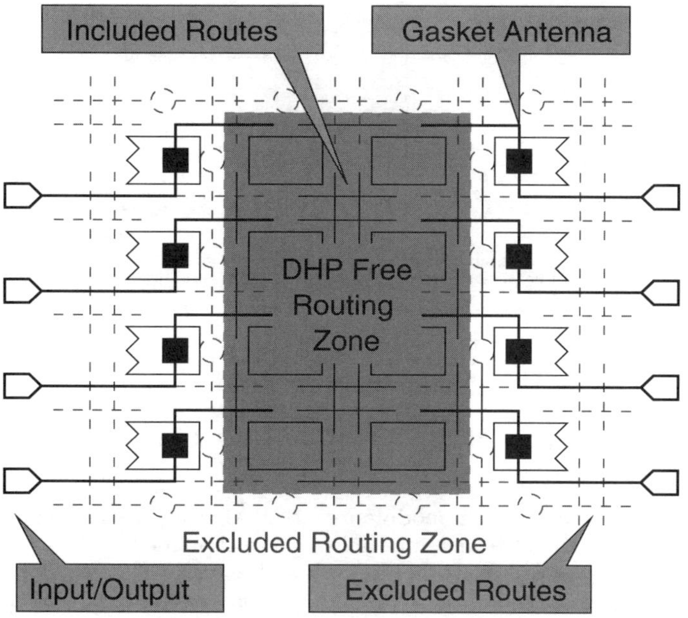

Figure 2: Gasket Antennas

2.4 Generation of Bitfiles

In order to generate bitfiles that can be read by PARBIT to create the partial bitfile, it is necessary to follow a few high-level rules regarding the logical names of the entities that make up the logic within the FPGA. These rules are applied during the synthesis, routing, and placement of the FPGAs that hold the infrastructure circuit and the DHP modules.

2.4.1 Synthesis

The bitfiles that are manipulated by PARBIT are generated from VHDL files containing the infrastructure and the user module descriptions (plugins), along with constraints commands used to define the regions of the FPGA that will be reconfigured.

The infrastructure bitfile contains the fixed logic area in the FPGA, encompassing all of the I/O pads, signals and flops that interface to the module, and logic that make up the on-chip system. In the VHDL file, there is one entity, called "INFRA", that contains the infrastructure, and one or more entities, called "GASKET", that contains the flops used to interface with each user module. Additional VHDL files are used to generate the physical constraints necessary to lock the gasket flops in fixed positions inside the FPGA. These constraints are also used to lock the infrastructure logic into a specific area, and to reserve space for the user modules inside the FPGA.

The bitfile for the DHP contains the description of a module and is generated in a similar way. The difference is that there is one user module entity, called "DHP", connected to one GASKET entity, as shown in Figure 3. For the GASKET entity in the design, the constraints are set so that this GASKET is placed in the same position as the first GASKET of the infrastructure design.

2.4.2 Placement

The placement of the infrastructure and DHPs can be accomplished with conventional FPGA placer tools, that have the ability to constrain logic to specific regions of the array. Infrastructure logic is kept out of the DHP sites, and DHP logic is confined to the appropriate rectangular area. These constraints can be expressed in the source VHDL as attributes or kept in a separate physical constraint file read by the placer.

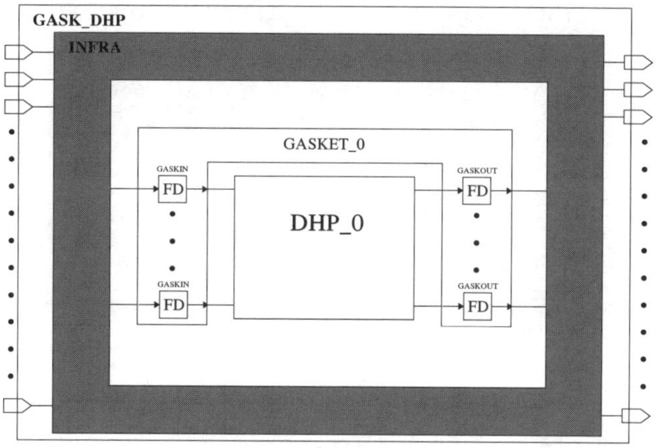

Figure 3: Logical Design Entities

2.4.3 Routing

The routing problem is more complex than placement, and requires a greater degree of control over the use of routing resources than is currently available in off-the-shelf FPGA CAD tools. To guarantee that there is no interference between DHPs and infrastructure as new DHPs are being configured, the nets in a design are sorted into one of three categories and then routed with special constraints as follows:

- All nets which are internal to the DHPs are routed on interconnect resources that do not cross the boundary of the DHP site. The Virtex-E architecture includes long lines which traverse the entire array, and hex lines which can reach six CLBs in any direction. These wires may not be used in a DHP since they could possibly cause interference when the DHP is loaded
- All nets which are completely contained in the infrastructure are routed using any wiring resource that is not used for use in routing DHPs.
- Nets which cross the the boundaries between the infrastructure and the DHP site are forced to follow the same path for each DHP. These are the gasket antennas which are identical for every DHP, since the DHP interface conforms to a fixed specification. These nets have identical configurations in every possible DHP implementation so the act of reconfiguration is guaranteed to cause no interference in the configuration or operation of the infrastructure or adjacent DHP sites

These routing constraints are satisfied using a modified version of the router in which the use of any individual Programmable Interconnect Point (PIP) in the array can be enabled or disabled. Individual nets can also have wire segments assigned to force the route solution to follow a predictable path as needed.

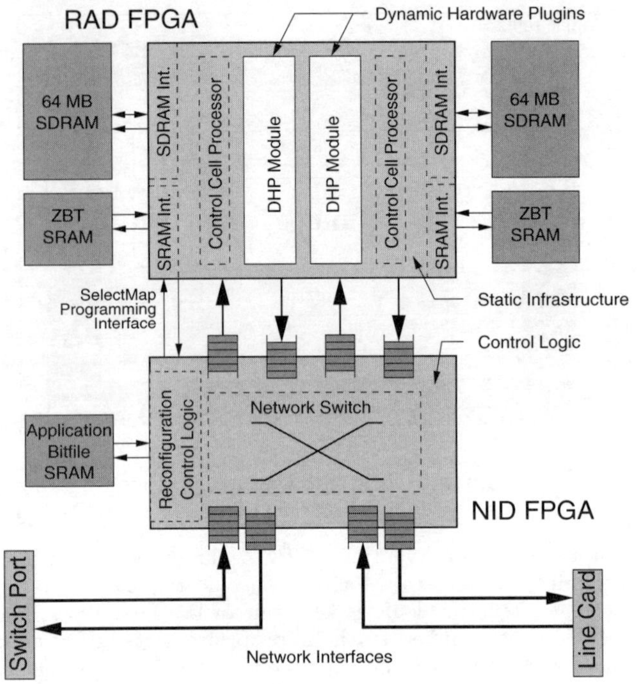

Figure 4: FPX System

3. THE FPX PLATFORM

The Field-programmable Port Extender enables the rapid prototype and deployment of hardware components for modern routers and firewalls [12]. The system allows new packet processing functions to be quickly prototyped as DHP modules in hardware, then downloaded into reconfigurable logic over the network [17]. New features are added by loading DHPs into well-defined DHP module interfaces.

All functions on the FPX are implemented with FPGAs. The core functionality of the FPX is implemented on the Networking Interface Device (NID) and on part of the Reprogrammable Application Device (RAD). The NID is a Xilinx XCV600E FPGA that contains the control logic to reconfigure regions of the RAD. The RAD is a Xilinx XCV-2000E FPGA that holds the DHP modules.

In order to reprogram a RAD module, the NID implements a reliable protocol to fill the contents of the an SRAM with configuration data that is sent over the network. A final control cell is sent to NID to initiate the reprogramming of RAD using the contents of the reprogram memory [18].

The NID also contains a network switch that forwards individual traffic flows between network interfaces and DHP modules on the RAD. This combination of partial reconfiguration control logic and per-flow routing circuits allow the FPX install new DHP modules without affecting the operation of the rest of the system.

3.1 Modular Logic

Dynamic Hardware Plugins (DHPs) are used to implement application-specific functionality on the FPX. Multiple DHPs can be loaded into the RAD and run in parallel on the single FPGA device. Data flows may pass through multiple hardware plugins. In order to support a broad spectrum of applications, DHPs can access off-chip memory resources.

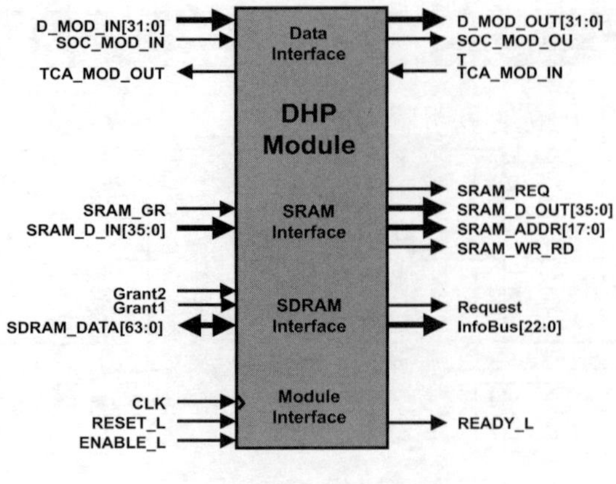

Figure 5: Modular Component of FPX

Hardware plugin modules on the RAD consist of a region of FPGA gates and internal memory, bounded by a well-defined interface to the network and external memory [19]. The modular interface of an FPX component is shown in Figure 5. Data arrives at and departs from a module over a 32-bit wide, Utopia interface. Data passes through modules in a cell. Large IP datagrams pass through the interface in multiple cells.

The module provides two interfaces to off-chip memory. The SRAM interface supports transfer of 36-bit wide data to and from off-chip SRAM. The Synchronous Dynamic RAM (SDRAM) interface provides a 64-bit wide interface to off-chip memory. In the implementation of the IP lookup module, the off-chip SRAM is used to store the data structures of an Internet route table [12].

3.2 DHP Implementation on the FPX

Figure 6 shows the infrastructure of the RAD when viewed by the Xilinx FPGA Editor. The logic for the CCP (Control Cell Processor), the SDRAM controller, and the SRAM interface is placed and routed in the left and the right sides of the chip. The center portion of the chip is reserved area for two hardware plugins modules.

The floorplan for the FPGA circuit that implements a DHP is shown on Figure 7. Here, the logic for the module is confined to a region the size of a DHP. Further, the I/O signals around the DHP include Flip/Flops that lock the location of the signals and provide timing isolation between DHP circuits. Inputs and outputs from the DHP are routed to the external I/O pins of the FPGA so that standard synthesis tools can be used for their implementation.

After generating these two bitfiles, PARBIT is run to generate a partial bitfile suitable for run-time reconfiguration. The parameters passed to PARBIT include the coordinates of the DHP module and the coordinate of the new position of this module inside the FPGA. The partial bitfile generated for a module has a size of 187 KBytes as compared to the complete XCV2000E configuration size of 1,270 KBytes.

Current efforts focus on the implementation of an enhanced on-chip gasket interfaces for additional types of signals. Some types of on-chip shared resources require tri-state busses that span the width of the device. By enhancing the gasket interface to support the use of horizontal long lines, it is possible to efficiently implement this type of interconnect.

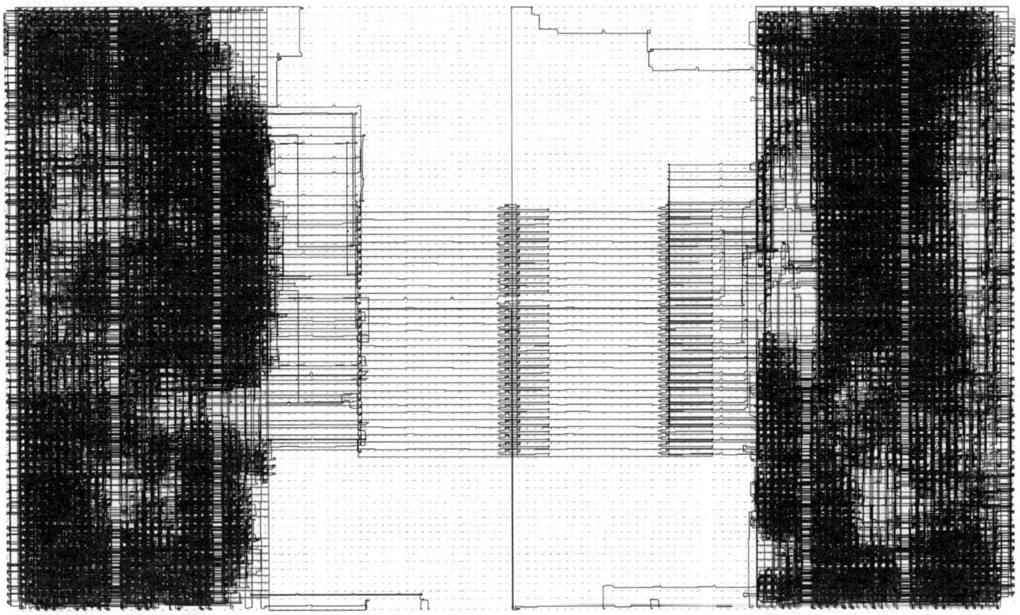

Figure 6: Infrastructure Floorplan of the RAD (A Xilinx Virtex 2000E FPGA). Note that Infrastructure logic, including the pair of SRAM controllers, the pair of SDRAM controllers, and the Control Cell Processor (CCP) are routed and placed on the left and right sides of the FPGA. Two regions have been reserved for DHP modules and are left empty when synthesizing the infrastructure.

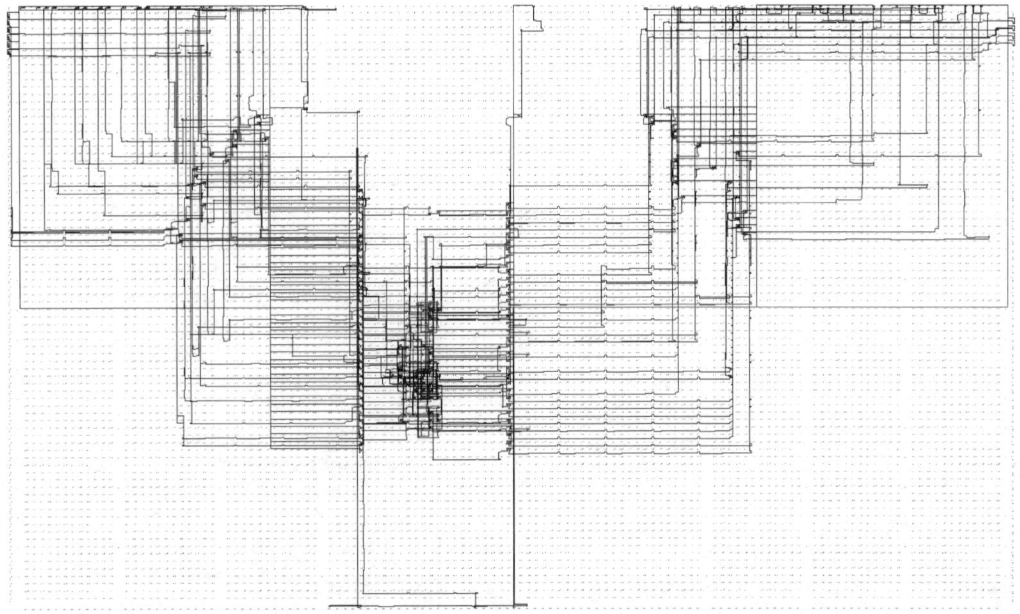

Figure 7: DHP Floorplan of the RAD FPGA. A single module has been routed and placed to fit within the area of one DHP module. Using PARBIT, this DHP can be placed into either location of the Infrastructure shown in the floorplan above. Note that input/output signals are routed to I/O pins so that standard design flows for synthesis and simulation of the FPGA circuit can be followed.

4. CONCLUSIONS

A technique has been demonstrated for designing partial RTR systems in a VIRTEX-E FPGA. The methodology uses PARBIT to generate the partial bitfiles. This tool can extract and relocate regions of logic in a compiled FPGA.

The interface between the DHP plugins and the infrastructure, called a gasket, is able to lock fixed interconnection points between infrastructure logic and dynamically reconfigured regions of an FPGA.

The approach shown in this paper reduces the time needed to download the new hardware module, due to the fact that these modules are pre-compiled and that the reconfiguration of DHP modules is performed in hardware (not software). This feature enables the implementation of network modules that perform a variety of networking functions and can be deployed at run time in a networking router, switch, or firewall.

5. FUTURE PLANS

Going forward, we are collaborating with application developers and network equipment vendors to design reconfigurable hardware into several routing, switching, and firewall platforms. As the installed base of reconfigurable platforms grows, it becomes easier to rapidly deploy new features in DHP modules. By following a standard and open modular interface, DHPs modules are fully interoperable.

6. ACKNOWLEDGMENTS

The authors would like to thank Richard Y. Sun and Robert Wells, from Xilinx, Inc., for their assistance in solving the routing problems associated with DHP implementation.

7. REFERENCES

[1] S. Hauck, "The roles of FPGAs in reprogrammable systems," *Proceedings of the IEEE*, vol. 86, pp. 615–638, Apr. 1998.

[2] W. Marcus, I. Hadzic, A. McAuley, and J. Smith, "Protocol boosters: Applying programmability to network infrastructures," *IEEE Communications Magazine*, vol. 36, no. 10, pp. 79–83, 1998.

[3] B. L. Hutchings and M. J. Wirthlin, "Implementation approaches for reconfigurable logic applications," in *Field-Programmable Logic and Applications (FPL'1995)* (W. Moore and W. Luk, eds.), (Oxford, England), pp. 419–428, Springer-Verlag, Berlin, Aug. 1995.

[4] D. T. Hoang, "Searching genetic databases on splash 2," in *IEEE Workshop on FPGAs for Custom Computing Machines* (D. A. Buell and K. L. Pocek, eds.), (Los Alamitos, CA), pp. 185–191, IEEE Computer Society Press, 1993.

[5] P. Bertin, H. Touati, and E. Lagnese, "PAM programming environments: Practice and experience," in *IEEE Workshop on FPGAs for Custom Computing Machines* (D. A. Buell and K. L. Pocek, eds.), (Los Alamitos, CA), pp. 133–138, IEEE Computer Society Press, 1994.

[6] J. M. Ditmar, "A Dynamically Reconfigurable FPGA-based Content Addressable Memory for IP Characterization," Master's thesis, KTH- Royal Institute of Technology, Stockholm, Sweden, 2000.

[7] D. Ross, O. Vellacott, and M. Turner, "An FPGA-based Hardware Accelerator for Image Processing," in *More FPGAs: Proceedings of the 1993 International workshop on field-programmable logic and applications* (W. Moore and W. Luk, eds.), (Oxford, England), pp. 299–306, 1993.

[8] J. D. Hadley and B. L. Hutchings, "Designing a partially reconfigured system," in *Field Programmable Gate Arrays (FPGAs) for Fast Board Development and Reconfigurable Computing, Proc. SPIE 2607* (J. Schewel, ed.), (Bellingham, WA), pp. 210–220, SPIE – The International Society for Optical Engineering, 1995.

[9] S. McMillan and S. Guccione, "Partial run-time reconfiguration using JRTR," in *Field-Programmable Logic and Applications / The Roadmap to Reconfigurable Computing (FPL'2000)*, (Villach, Austria), pp. 352–360, Aug. 2000.

[10] E. L. Horta and S. T. Kofuji, "The architecture of a reconfigurable ATM switch (RECATS)," in *Workshop de Computação Reconfigurável*, (Marília, SP, Brazil), Aug. 2000.

[11] D. E. Taylor, J. S. Turner, and J. W. Lockwood, "Dynamic Hardware Plugins (DHP): Exploiting reconfigurable hardware for high-performance programmable routers," in *IEEE OPENARCH 2001: 4th IEEE Conference on Open Architectures and Network Programming*, (Anchorage, AK), Apr. 2001.

[12] J. W. Lockwood, J. S. Turner, and D. E. Taylor, "Field programmable port extender (FPX) for distributed routing and queuing," in *ACM International Symposium on Field Programmable Gate Arrays (FPGA'2000)*, (Monterey, CA, USA), pp. 137–144, Feb. 2000.

[13] S. Choi, J. Dehart, R. Keller, J. W. Lockwood, J. Turner, and T. Wolf, "Design of a flexible open platform for high performance active networks," in *Allerton Conference*, (Champaign, IL), 1999.

[14] Xilinx Inc., "Virtex-E 1.8V Field Programmable Gate Arrays." Xilinx DS022, 2001.

[15] S. Kelem, "Virtex configuration architecture advanced user's guide." Xilinx XAPP151, Sept. 1999.

[16] E. Horta and J. W. Lockwood, "PARBIT: a tool to transform bitfiles to implement partial reconfiguration of field programmable gate arrays (FPGAs)," Tech. Rep. WUCS-01-13, Washington University in Saint Louis, Department of Computer Science, July 6, 2001.

[17] J. W. Lockwood, N. Naufel, J. S. Turner, and D. E. Taylor, "Reprogrammable Network Packet Processing on the Field Programmable Port Extender (FPX)," in *ACM International Symposium on Field Programmable Gate Arrays (FPGA'2001)*, (Monterey, CA, USA), pp. 87–93, Feb. 2001.

[18] J. W. Lockwood, "Evolvable internet hardware platforms," in *NASA/DoD Workshop on Evolvable Hardware (EH'2001)*, pp. 271–279, July 2001.

[19] D. E. Taylor, J. W. Lockwood, and N. Naufel, "Generalized RAD Module Interface Specification of the Field-programmable Port eXtender (FPX)," tech. rep., WUCS-01-15, Washington University, Department of Computer Science, July 2001.

A Reconfigurable FPGA-Based Readback Signal Generator For Hard-Drive Read Channel Simulator

Jinghuan Chen, Jaekyun Moon, and Kia Bazargan
Department of Electrical and Computer Engineering
University of Minnesota
Minneapolis, MN 55455, USA.
{jchen, moon, kia}@ece.umn.edu

ABSTRACT

A hard disk readback signal generator designed to provide noise-corrupted signals to a channel simulator has been implemented on a Xilinx Virtex^{TM}E FPGA device. The generator simulates pulses sensed by read heads in hard drives. All major distortion and noise processes, such as intersymbol interference, transition noise, electronics noise, head and media nonlinearity, intertrack interference, and write timing error, can be generated according to the statistics and parameters defined by the user. Reconfigurable implementation enables an update of the signal characteristics in run-time. The user also has the flexibility to choose from a set of bitstreams to simulate particular combinations of noise and distortion. Such customized restructuring helps reduce the area consumption and hence virtually increase the capacity of the FPGA device. The time to generate the readback signals has been reduced by four orders compared to its software counterpart.

Categories and Subject Descriptors

J.6 [**Computer Applications**]: Computer-aided Engineering—*Computer-aided design (CAD)*; B.5.1 [**Register-Transfer-Level Implementation**]: Design—*Data-path design, Control design*

General Terms

Measurement Performance Design Reliability

1. INTRODUCTION

In the hostile hard drive environment, advanced signal processing techniques are necessary in order to recover the user data from readback signals that are picked up by read sensors (read heads). Exploration of novel techniques are normally carried out through very intensive simulation. Conventionally, the simulation task is performed in software.

This work was supported by NSF under Grant CCR-9805195.

DAC 2002, June 10-14, 2002, New Orleans, Louisiana, USA.

The major drawback of the software approach is that it usually takes a very long time when low bit error rate (BER) is investigated. For example, a simulation of BER=10^{-8} with 10 errors that runs on a personal computer equipped with a 1GHz Pentium 4 processor can take days. In contrast, acceptable BERs in commercial hard drives normally go below 10 out of 10^{13} for desktop products and 10 out of 10^{15} for server-class products. Since it is very important to understand the behavior of many signal processing and coding techniques such as turbo codes and low density parity check (LDPC) codes at very low BERs, high-speed hardware-based or mixed hardware-software simulators are of great practical significance.

In view of the fact that noise sources and characteristics in hard drives are quite unique and different from those in other communication systems, a dedicated readback signal generator must be implemented in front of other signal processing modules in such a simulator. This signal generator should be accurate and, at the same time, easy to reconfigure. It shall incorporate all major noise processes and distortions and should be capable of generating very low probability events in accordance with the user-defined statistics. It should also provide the user with a flexibility of choosing from various combinations of noise and distortion characteristics. In this paper, we propose to implement the readback signal generator on a field programmable gate array (FPGA) utilizing its reconfigurability. This FPGA-based signal generator can also be used as part of a testbench for the data-recovery circuit (usually called the "read channel") since it is totally tunable compared to typical test spindle setups. The proposed FPGA-based signal emulator can provide a complete set of test conditions.

1.1 Recording Process and Noise/Distortion Categorization

In hard drives, the storage medium consists of one or more metal or glass disks coated with magnetic thin films. Binary information is stored in concentric circular tracks on the surface of disks. Each binary bit takes a very small portion of a track called a *bit cell.* During the write process, a write head magnetizes each bit cell to one of two tangential directions. Whether a 0 or a 1 is recorded is determined by the direction of magnetization. A read head picks up the magnetization flux emitted from the boundaries of bit cells and generates readback pulses (Fig. 1). These pulses will be passed through a pre-amplifier, analog/digital filters, and data-recovery modules and converted back to binary data.

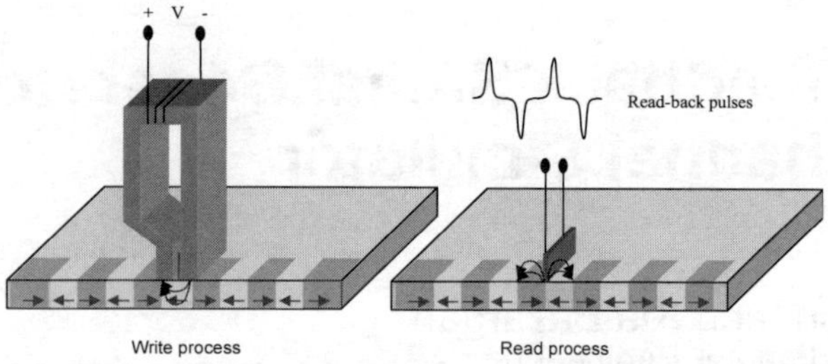

Figure 1: Write and read process in hard drives

The read channel includes the filters and the data-recovery modules.

The imperfect nature of the magnetic thin film results in a zigzag boundary between bit cells instead of an ideal straight edge. This will cause the pulse position to shift in time domain (*transition jitter*) and pulse-shape to deform (*width variation*). The noise introduced by this sort of phenomenon is called *transition noise (TN)* and it is data dependent in the sense that noise arises only in the presence of transitions that make up a data pattern. Almost all of today's read heads are *magneto-resistance (MR)* or *giant magneto-resistance (GMR)* heads which are not linear sensors. They introduce the *head nonlinearity (HNL)* that causes the positive and negative pulses to have different amplitudes. Two more types of media nonlinearity also occur during the write process: *nonlinear transition shift (NLTS)* and *partial erasure (PE)*. They generally create additional transition position shift and amplitude decline. Slow variant *write timing error (WERR)* stems from the irregularity of the disk spinning speed. A read head does not always position with 100% accuracy on the top of a track. It also senses the signal from the neighboring track and is influenced by *cross-talk* which is sometimes called *intertrack interference (ITI)*. Finally, *electronics noise (EN)* further corrupts the signal. Transition noise in hard-drive recording systems has been modeled in several ways. Moon[1] uses Tayler series to approximate transition jitter and width variation. Caroselli and Wolf [2] simplify the structure of recorded tracks by using a micro-track model. In our implementation, we mimic the physical reality by actually shifting the pulse position and changing the pulse-shape in runtime. The behavior of the transition noise and other noise processes as well as distortions follows the statistics laid down by the user. It should be noted that the statistics do not have to be Gaussian, as has been assumed in many previous modeling works.

1.2 Reconfigurable Signal Generation

Instead of running under static conditions, the signal generator may be required to change the noise properties during runtime. Furthermore, different kinds of noise and distortion characteristics mentioned above are not necessarily present in a system simultaneously. These features demand some degree of reconfiguration capability from the generator, which makes time-multiplexing FPGA the implementation method of choice.

In the next section, we will briefly describe how to model different types of noise processes and distortions observed in hard drives. Details of the modeling work is out of the scope of this paper. Further discussions will be given in another paper. We will deal with the design and implementation issues in Section 3. Then we will focus on pseudo-random number generators and structure and reconfigurability of the readback signal generator. Conclusions will be given in the last section.

2. MODELING OF READBACK SIGNALS

2.1 Distortion and Noise

A sequence of readback pulses are said to experience *intersymbol interference (ISI)* when the length of each pulse extends over a number of symbols. The pulses overlap and are distorted. Expression of noiseless readback signal is given in Eq. (1). The response $h(t)$ is the isolated pulse and b_k is the differentiation of data bit $a_k \in \{-1, 1\}$ (Eq. 2). The sequence b_k takes value of $\{-2, 0, 2\}$ and a non-zero b_k indicates the existence of a transition. ISI happens when the length of $h(t)$, L, is longer than the symbol period T.

$$y(t) = \sum_k b_k h(t - kT) \tag{1}$$

$$b_k = a_k - a_{k-1} \tag{2}$$

The *position jitter* and *width variation* are categorized as noise sources because they introduce noise when the readback signal is sampled. Both types of noise can be modeled as random variables. We rewrite Eq. (1) to reflect the transition noise. Now the isolated pulse $H(t)$ is not deterministic but a random process. The position jitter parameter δ_j is a random variable falling in the range $-T < \delta_j < T$.

$$y(t) = \sum_k b_k H(t - kT + \delta_j) \tag{3}$$

Eq. (1) to (3) are well known and straightforward. Head nonlinearity can be characterized by a nonlinear function $F(x)$, which can be measured from a read head. NLTS and PE can be modeled as a data dependent nonlinear distortion of the pulses. Simplistic assumptions have been made on the data dependence [4] and [5]. In this paper, only the 1*st* order effect is considered. We express HNL, NLTS, and PE as follows:

$$y(t) = \sum_k F\{r_k b_k h_k(t - kT + \delta_j - \frac{\varepsilon T}{4} b_k b_{k-1})\} \tag{4}$$

where r_k represents the amplitude loss due to PE and ε is the amount of NLTS in units of T. Note that NLTS component reduces to zero effectively unless two consecutive transition occurs, i.e., $b_k b_{k-1} \neq 0$.

Slowly varying WERR can be simplified and modeled as a slow-paced *"random walk"*. Intertrack interference is implemented by summing up weighted output signals from two uncorrelated generators that simulate two interfering tracks. *Electronics noise* normally is modeled as an additive white bandlimited gaussian noise (AWGN).

2.2 Signal-to-Noise Ratio (SNR)

The behavior of statistically defined noise should be determined by its probability distribution. However, when the noise pdf is assumed to be Gaussian, the first and second order statistics can provide all necessary information about the noise process. In this case, the user of the signal generator only needs to specify the SNR, which effectively fixes the variance of the noise. In addition, the user shall specify the norminal isolated pulse $h(t)$, the percentage of the transition noise power, and the composition of the position jitter and the width variation. In the remainder of this paper, we will focus on Gaussian noise. We note, however, that our simulator can handle any user-specified pdf for the noise process.

3. IMPLEMENTATION

3.1 Gaussian Pseudo-Random Number Generator (PRNG)

Transition noise, electronics noise, and write timing error are all randomly generated in a non-uniform fashion. These in-system pseudo-random number generators must be uncorrelated and bear good randomness. In our design, uniform pseudo-random numbers are first produced and then transformed to non-uniform random numbers. One effective and popular uniform PRNG is the linear feedback shift register (LFSR) [6]. Its output is a maximum-length sequence (*m-sequence*) with a period of $2^n - 1$, where n is the number of storage units (registers). However, if we need many sequences of random noise, the area consumed by LFSR will be large, something that we cannot afford in an FPGA implementation. For example, with five 6-bit wide PRNGs that can generate random numbers with a period longer than 1000 ($n = 10$), the number of registers will be 300. Facing area constraints in this design, we adopted an alternative method, linear hybrid cellular automaton (LHCA) (for example, see [7] and [8]). The principle of a cellular automaton is that the next value of each register is calculated by a Boolean function from the current values of immediate neighbors and itself. A cellular automaton is said to have a *null boundary* condition when the left- and right-most registers in the array connect to zeros. A null boundary is preferred to a feedback boundary because it avoids long feedback paths. It has been shown that when the Boolean functions are carefully selected and after the LHCA evolves for a number of initialization clock cycles, such a register array will output an m-sequence that has the same period as that of an LFSR [7]. The Boolean functions are called *computation rules* and categorized by Wolfram [11]. One of the setups that can generate m-sequences is a careful mix (hybrid) of two Boolean functions, *Rule 90* and *Rule 150*, in the calculation of register contents [7]. Rule 90 and Rule 150 are defined as follows:

$$\text{Rule 90: } a_i(t+1) = a_{i-1}(t) \oplus a_{i+1}(t) \tag{5}$$

and

$$\text{Rule 150: } a_i(t+1) = a_{i-1}(t) \oplus a_i(t) \oplus a_{i+1}(t) \tag{6}$$

where $a_i(t)$ is the content of register i at time t. How to determine the positions of Rule 90 and Rule 150 in a register array can be found in [9] and [10].

After a quick observation, we can find that not every output from registers in the array is a "good" bit because those from adjacent registers are correlated. In order to eliminate correlation, only one bit of every $\eta + 1$ bits will be used to form a random number where η is called the *site spacing parameter*. To compare with LFSR, if $\eta = 1$, the number of registers in a LHCA PRNG that can generate the same group of random numbers as mentioned above will be 60, a factor of 5 reduction in the number of registers.

Transformation from a uniform to a non-uniform pseudo-random number is illustrated in Fig. 2. An n-bit uni-

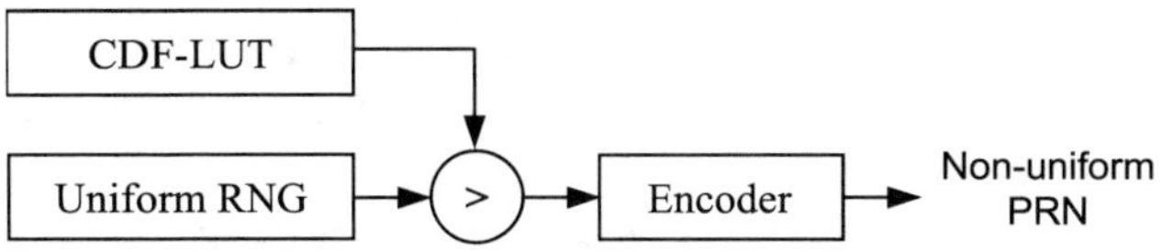

Figure 2: Generation of non-uniform pseudo-random number

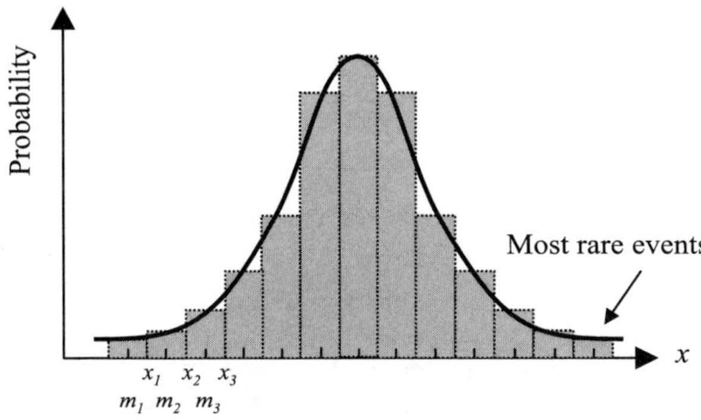

Figure 3: Splitting PDF into 2^m columns

form random number is compared with numbers in a user-defined *cumulative distribution function (CDF)* conversion table and then encoded to an l-bit non-uniform random number. This exercise is equivalent to randomly picking a point in the area under the corresponding pdf, grouping all points in a column, and substituting them with the one number m_i (Fig. 3). Attention shall be paid to the choice of n and l since they determine the accuracy and precision of the PRNG. The width n of a uniform random number limits the smallest probability we can simulate, which is $\frac{1}{2^n-1}$. Encoding from an n-bit to an l-bit random number can be considered as quantizing a continuous random variable X to an l-bit discrete random variable X_q, if $n \gg l$. The width l controls the mean square loss associated with the quantization process.

Max [13] shows that the optimal quantizer is nonuniform when the pdf of the random variable is not uniform. However, the digital nature of the FPGA device decides the quantization levels $m_j (j = 1, 2, \ldots, q)$ to be uniform where q is the number of quantization levels, which is 2^l. We need to determine the boundaries $x_j (j = 1, 2, \ldots, q-1)$ for the suboptimal quantizer in term of minimizing the quantization loss. The quantization loss N_q is given in Eq. (7) and $f_x(x)$ is the pdf of X.

$$N_q = E[(X - X_q)^2] = \sum_{i=1}^{q} \int_{x_{i-1}}^{x_i} (x - m_i)^2 f_x(x) dx \tag{7}$$

In order to minimize the quantization loss, let

$$\frac{\partial N_q}{\partial x_i} = 0, \qquad i = 1, 2, \ldots, q-1. \tag{8}$$

One can obtain:

$$x_j = \frac{1}{2}(m_j + m_{j+1}). \tag{9}$$

Eq. (9) indicates that the suboptimal quantizer is a uniform quantizer. If noise is assumed to have Gaussian distribution, N_q can be computed analytically. Since quantization noise can be seen as an additive noise to the random

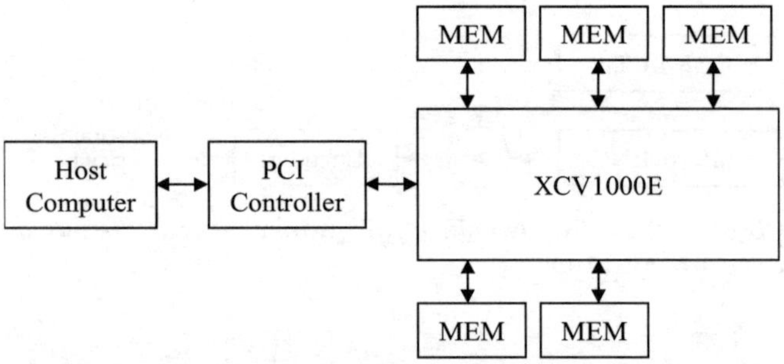

Figure 4: Simplified block diagram of an Annapolis Micro Systems FIREBIRD™ PCI board

variable, we can define a signal to quantization noise ratio (SQNR) to evaluate the loss due to quantization as follows:

$$SQNR = \frac{\sigma_{RV}^2}{N_q} \tag{10}$$

where σ_{RV}^2 is the variance of the continuous random variable. For example, when $q = 64$ and σ_{RV}^2=9, $SQRT$ is 20.4 dB. Hence, as long as we use sufficiently many quantization levels, the quantization loss can be suppressed down to a negligible level.

3.2 Implementation Platform

The design is synthesized by Synplicity's Synplify™ 6.2, simulated by ModelSim™ 5.5, and implemented by Xilinx Foundation™ 3.1 . The design has been implemented on a Xilinx Virtex™E XCV-1000E device which is mounted on an Annapolis Micro Systems FIREBIRD™ PCI board. A simplified block diagram of a FIREBIRD™ PCI board is shown in Fig. 4 [12]. The XCV1000E communicates with a host computer through PCI bus. The user's design, as well as some register space and a few interface blocks are implemented on the FPGA device (Fig. 5). By calling a group of application programming interface (API) functions, the user is able to command the board and establish communications between the host computer and the resister space as well as interface blocks. The user has access to 18 Mega bytes of RAM that is distributed in five modules on the board. Before the signal generator can produce readback signals, it has to be initialized with data pre-calculated by software. The initialization data is loaded into the on-board RAM before it is used to initialize the XCV1000E. The output signals are also stored into the RAM first and then transferred to the host computer. When the signal generator is integrated into the read channel simulator, however, the generated signal will be forwarded to the data-recovery module directly. Only recoverd binary data and BER will be sent back to the user. Clocks used by the XCV1000E and the RAM are synchronized and can run up to 150MHz.

The input interface of the generator in Fig. 5 bridges the host and the on-board RAM to the generator circuits and controls the timing of the initialization process. It reads the configuration values from the memory and from the host and initializes generator look-up tables. There are three types of tables need to be configured or reconfigured: pulse look-up tables, Gaussian conversion tables in PRNGs, and multiplier look-up tables.

3.3 Readback Signal Generator

The architecture of the readback signal generator is modularized and pipelined. Its structure is illustrated in Fig. 6. The data-track and interfering-track modules are almost identical except the random numbers from LHCA-PRNG are independent and the off-track (OT) ratio r is different. The output of data-track is multiplied by the ratio r while the interfering track is multiplied by a different ratio r', ($r' \leq 1 - r$). ITI to the data track is embodied through adding two weighted signals. The peak of the summed signal will be degraded by a ratio of $r + r' \leq 1$. The value of r and r' are determined by the off-track position of the read head [14]. In the diagram, the solid arrows mark the signal paths, the dash ones indicate the random number paths, and the dotted lines are the configuration data paths.

The remaining seven modules in each track process the noise corruption and distortions on the pulses that are described in Section 2. An L-bit long isolated pulse is chosen from the *pulse look-up table* that is implemented in RAM. The look-up table contains N pulses that have different pulse shapes and widths. The amplitude and the shape of the selected pulse are altered by HNL and PE multipliers. In NLTS module, nonlinear transition shift is integrated into the overall moves to be executed in a 2-level Barrel shifter. After passing through the shifter, the pulse is superimposed on a sequence of pulses that moves forward by T bits in every clock cycle. The write error is added to the pulse sequence in the WERR module. Electronics noise is then incorporated onto each point before the output signal is multiplied by the OT ratio. Since one operand to every embedded multiplier is a known parameter, it is practical to implement the multipliers in look-up tables to increase the run speed.

3.4 Reconfiguration

Reconfigurable computing in FPGAs provides a way to increase the device capacity and improve flexibility when solving a range of problems without duplicating hardware [15]. In the readback signal generator, two dimensions of

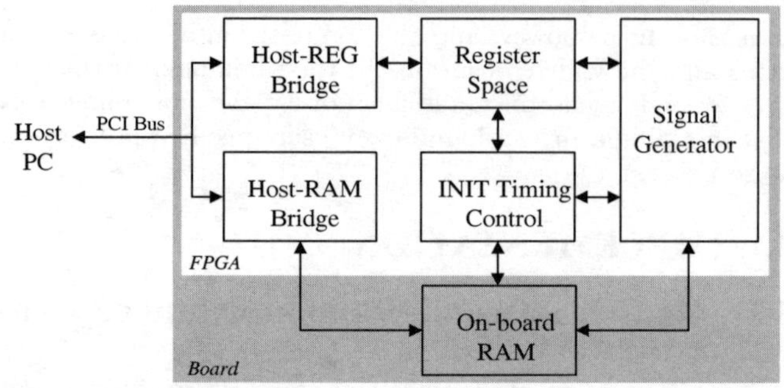

Figure 5: Interfaces between host, RAM, and the signal generator

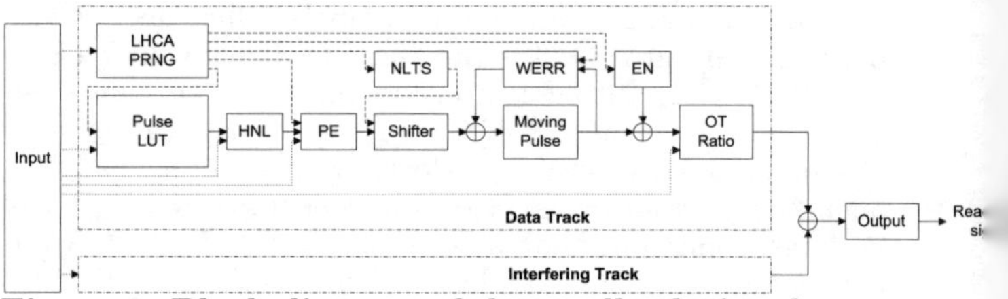

Figure 6: Block diagram of the readback signal generator

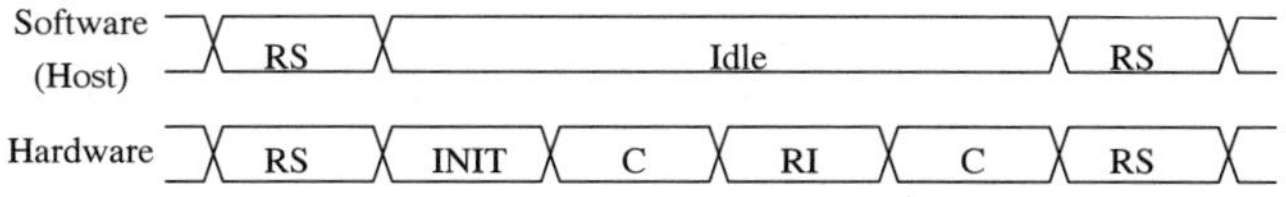

(RS: Restructuring; INIT: Initialization; C: Computation; RI: Reinitializing)

Figure 7: Time-duplexing

reconfigurations are considered: *reinitializing* and *restructuring*. They are accomplished jointly through hardware and software.

3.4.1 Reinitializing

When a read head moves from one track to another, the pulse-shape, noise statistics, noise amplitude, and ITI are likely to change. This reality motivates the reconfiguration of the signal generator at runtime. The execution time is duplexed to two phases: *initialization phase* and *computation phase*. Prior to the execution of these two phases, configuration data is calculated by host software and loaded to the on-board RAM. In the initialization phase, the initialization control unit redefines look-up table contents in the signal generator with configuration data stored in the RAM. The execution time of reinitialization ranges from about one thousand to less than a hundred thousand clock cycles depending on the scale of the generator while the number of clock cycles in the computation phase is significantly larger. Meantime, the computation circuit idles for the initialization to finish and contents in all registers stay unchanged. When the generator completes reinitialization, it will generate readback signals characterized by the new configuration (see Fig. 7).

3.4.2 Restructuring

Although the noise statistics or pulse-shape may change in time, the kinds of noise processes and distortions remains the same in many cases when a hard drive system is put together. For instance, PE may happen for one type of magnetic media but be absent on another. In other words, not every module in the design has to be involved in the generation of a particular sequence of readback signals. Therefore, a customized structure is preferred to virtually increase the device capacity. This becomes more critical when the signal generator is integrated into the whole simulator. It is normally expected to economize on the generator to preserve more resource for data-recovery modules. Additionally, since the complicated generator takes a large share of the area budget, optimizing the generator structure is an efficient way to reduce the overall area consumption. To serve the interests of different simulation emphasis, we have chosen six typical combinations of noise and distortion (See Table 1). These combinations are by no means conclusive. In theory, there are $2^7 = 128$ possible bitstreams that can be prepared in advance. The task of choosing which bitstream to download is done by the software according to the user's demand. In other words, the functionality of the generator now is split between the hardware and software. The consumed resources will be reported in Section 3.6.

In summary, reconfiguration through time-duplexing and hardware-software co-design helps achieve resource reusing in the FPGA device. Hence, we can simulate readback signals with different noise processes and distortions by using significantly less resources.

Table 1: Implemented noise/distortion combinations

#	TN	HNL	PE	NLTS	WERR	EN	ITI	slices
0	$\surd$	$\surd$	$\surd$	$\surd$	$\surd$	$\surd$	$\surd$	12286
1	$\surd$							5010
2	$\surd$					$\surd$		7649
3						$\surd$		5716
4	$\surd$					$\surd$	$\surd$	12235
5	$\surd$	$\surd$	$\surd$	$\surd$				6410

3.5 Discussion

In the cases when the restructuring is required to be done during runtime, we can store the register contents to on-board RAM and load a different bitstream from host in a very short time. A control unit then configures the new generator with data readback from the RAM. However, a better approach will be laying out each module into physically bounded and floorplanned blocks and dynamically reconnecting different blocks into a new circuit. An extra credit by doing so is the simplification of the initialization control unit by putting various initial conditions into blocks as well. In other words, the reinitialization and restructuring can be merged into one reconfiguration scheme and it can be performed on the fly. A capable tool to achieve this approach is Xilinx's JBits ([16], [17]). Unfortunately, by the time of this publication JBits and the board in use are not compatible; hence, we are not able to implement this scheme. However, we believe it can help further improve the flexibility in utilization of this generator and reduce its resource consumption.

3.6 Implementation Report

A set of sample implementations (Table 1) have been constructed with the parameters given in Table 2.

Table 2: Parameters of sample implementations

Parameter	Symbol	Value
Pulse look-up table size	N	16
ISI length	I	6
Symbol period	T	8
NLTS	εT	2
Signal-to-noise ratio	SNR	10 dB
Percentage of TN in total noise	α	90%
Percentage of jitter in TN	PJP	75%
LHCA length	l_{LHCA}	244
Site spacing parameter	η	1
Quantization levels	q	64
Uniform RN width	n	11
OT ratios	r and r'	0.9 and 0.1
Head nonlinearity	HNL	1 for pos. pulses 0.9 for neg. pulses

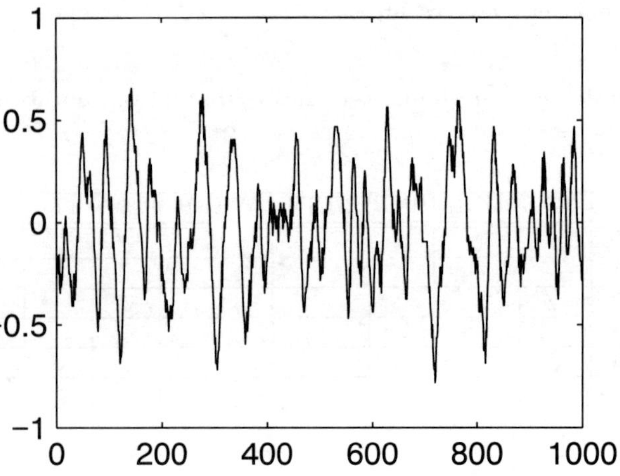

Figure 8: Generated readback signal

The total area consumption of the implementaion with all modules (bitstream 0) is 99% of 12288 slices. This extreme number clearly indicates that it is vital to utilize reconfiguration in this design. The resource consumption of other bitstreams is listed in the last column in Table 1. When the size of design approaches the limit of the device, the number of used slices may not reflect the actual area consumption by the generator very well. From the table, it shows that by doing customized restructuring, we can significantly avoid wasting resources on unused modules. The readback signal generator can run with clock speeds up to 70MHz. The runtime of generating 10^{12} pulses on a FPGA-based generator is 4 hours compared to about 50000 hours in software. The runtime on a software-based generator is estimated on a PC eqipted with one 1GHz Pentium^{TM} 4 processor and 256MB RAM. A sample output signal is shown in Fig. 8.

All the design parameters in Table 2 can be easily scaled up when a more precise generator is in demand and a larger FPGA device is available. Every module can be easily removed from the design and a new module can be inserted without modification of others.

4. CONCLUSIONS

A readback signal generator for hard-drive systems has been implemented on a Xilinx Virtex^{TM}E FPGA device. It is capable of simulating all major noise processes and distortions that have been observed in hard-drive read and write processes. The design is modularized to meet the requirements of reconfiguration. The noise and distortion sources can be reconfigured in runtime and a flexibility of choosing from various noise and/or distortion sources is provided. The resource consumption can be reduced by a significant amount through reconfiguration.

5. REFERENCES

[1] J. Moon, "Discrete-time modeling of transition-noise-dominant channels and study of detection performance", *IEEE Transactions on Magnetics*, vol. 27, no. 6, pp. 4573-4578, Nov. 1991.

[2] J. Caroselli and J. K. Wolf, "Applications of a new simulation model for media noise limited magnetic recording channels", *IEEE Transactions on Magnetics*, vol.32, no.5, pt.1, pp.3917-19, Sept. 1996.

[3] H. N. Bertram, *Theory of magnetic recording*, Cambridge University Press, 1994.

[4] T. Yamauchi and J. M. Cioffi, "A nonlinear model for thin film disk recording systems", *IEEE Transactions on Magnetics*, vol.29, no.6, pp.3993-95, Nov. 1993.

[5] B. Lin, E. Hild, and J. R. Cruz, "Magnetoresistive read/write channel models", *IEEE Transactions on Magnetics*, vol.35, no.6, pp.4528-31, Nov. 1999.

[6] W. G. Solomon, *Shift register sequences*, Aegean Park Press, 1982.

[7] P. D. Hortensius, R. D. McLeod, W. Pries, D. M. Miller, H. C. Card, "Cellular automata-based pseudorandom number generators for built-in self-test", *IEEE Transactions on Computer-Aided Design of Integrated Circuits & Systems*, vol.8, no.8, pp.842-59, Aug. 1989.

[8] P. D. Hortensius, R. D. McLeod, H. C. Card, "Parallel random number generation for VLSI systems using cellular automata", *IEEE Transactions on Computers*, vol.38, no.10, pp.1466-73, Oct. 1989.

[9] S. Zhang, D. M. Miller, J. C. Muzio, "Determination of minimal cost one-dimensional linear hybrid cellular automata", *Electronics Letters*, vol.27, no.18, pp.1625-7, Aug. 1991.

[10] K. Cattell, S. Zhang, "Minimal cost one-dimensional linear hybrid cellular automata of degree through 500", *Journal of Electronic Testing: Theory & Applications (Jetta)*, vol.6, no.2, pp.255-8, April 1995.

[11] S. Wolfram, "Statistical mechanics of cellular automata", *Reviews of Modern Physics*, vol.55, no.3, pp.601-44, July 1983.

[12] Annapolis Micro Systems, Inc., *FIREBIRD^{TM} Hardware Reference Manual*, 2001.

[13] J. Max, "Quantizing for minimum distortion", *IEEE Transactions on Information Theory*, IT-6(1), pp.7-12, March 1960.

[14] J. Chen and J. Moon, "Detection signal-to-noise ratio versus bit cell aspect ratio at high areal densities", *IEEE Transactions on Magnetics*, vol.37, no.3, pp.1157-67, May 2001.

[15] A. DeHon and J. Wawrzynek, "Reconfigurable computing: what, why, and implications for design automation", *Proceedings of 1999 Design Automation Conference*, pp.610-15, 1999.

[16] S. A. Guccione, D. Levi, and P. Sundararajan, "JBits: A Java-based interface for reconfigurable computing", *Proceedings of 2nd Annual Military and Aerospace Applications of Programmable Devices and Technologies Conference*

[17] C. Patterson, "High performance DES encryption in Virtex FPGAs using JBits", *Proceedings of 2000 IEEE Symposium on Field-Programmable Custom Computing Machines*, pp.113-21, 2000.

Embedded Software-Based Self-Testing for SoC Design

A. Krstic[1] W.-C. Lai[1] L. Chen[2] K.-T. Cheng[1] S. Dey[2]
[1] University of California, Santa Barbara, CA 93106, {angela, wlai, timcheng}@windcave.ece.ucsb.edu
[2] University of California, San Diego, CA 92037, {lichen, dey}@ece.ucsd.edu

ABSTRACT

At-speed testing of high-speed circuits is becoming increasingly difficult with external testers due to the growing gap between design and tester performance, growing cost of high-performance testers and increasing yield loss caused by inherent tester inaccuracy. Therefore, empowering the chip to test itself seems like a natural solution. Hardware-based self-testing techniques have limitations due to performance and area overhead and problems caused by the application of non-functional patterns.

Embedded software-based self-testing has recently become focus of intense research. In this methodology, the programmable cores are used for on-chip test generation, measurement, response analysis and even diagnosis. After the programmable core on a System-on–Chip (SoC) has been self-tested, it can be reused for testing on-chip buses, interfaces and other non-programmable cores. The advantages of this methodology include at-speed testing, low design-for-testability overhead and application of functional patterns in the functional environment. In this paper, we give a survey and outline the roadmap and challenges of this emerging embedded software-based self-testing paradigm.

Categories and Subject Descriptors

B.8.1 [**Integrated Circuits**]: Performance and Reliability – *reliability, testing and fault-tolerance.*

General Terms: Algorithms, Performance, Reliability.

Keywords: VLSI test, SoC test, functional test, microprocessor test.

1. INTRODUCTION

System-on-chip (SoC) has become a widely accepted architecture for highly complex systems on a single chip. An SoC contains a large number of complex, heterogeneous components that can include digital, analog, mixed-signal, radio frequency (RF), micromehanical and other systems on a single piece of silicon. The increasing heterogeneity and programmability associated with the system-on-chip architecture together with the rapidly increasing operating frequencies and technology changes are demanding fundamental changes in VLSI testing.

The test application using testers poses challenges because the testers performance is increasing at a slower rate than the device speed. This translates into an increasing yield loss due to external testing since guardbanding to cover tester errors results in a loss of more and more good chips. In addition, high-speed testers are very costly. Also, for mixed-signal testing, analog instrumentation needed for precise analog testing is not present on the digital logic tester. Having to test mixed-signal chips would require a mixed-signal tester that is even more expensive and implies a two-pass testing strategy. However, there are also cases (such as those where vast amount of digital data has to be pumped through an analog interface e.g., 3GIO, Infiniband, SATA to test the digital core) where neither of the two platforms would work.

Built-in self-test (BIST) solutions eliminate the need for high speed testers and offer the ability to apply and analyze at-speed test signals on chip with greater accuracy than that of the tester. Existing BIST techniques belong to the class of structural BIST. Structural BIST, such as scan-based BIST techniques [1][2][3], offer good test quality but require addition of dedicated test circuitry. Therefore, they incur non-trivial area, performance and design time overhead. Moreover, structural BIST applies non-functional, high-switching random patterns and thus, causes much higher power consumption than normal system operations. Also, to apply at-speed tests to detect timing related faults, existing structural BIST needs to resolve various complex timing issues related to multiple clock domains, multiple frequencies and test clock skews that are unique in the test mode.

A new *embedded software-based self-testing* paradigm [4][5][6] has a potential to alleviate the problems due to the use of external testers as well as embedded hardware tester problems described above. In this testing strategy, it is assumed that programmable cores on the SoC (such as processor and DSP cores) are first self-tested by running an automatically synthesized test program which can achieve high fault coverage. This allows application of functional tests in the normal functional operating environment of the design. Thus, it eliminates problems caused by application of non-functional patterns as well as problems caused by non-functional environment during test application. Next, the programmable core can be used as a pattern generator and response analyzer to test on-chip buses, interfaces between cores or even other cores including digital, mixed-signal and analog components on an SoC. This solution is sometimes also referred to as *functional self-testing*.

The concept of embedded software-based self-testing is illustrated in Figure 1 using a bus-based SoC. The IP cores in the SoC are connected to a Peripheral Interconnect (PCI) [7] bus via the virtual component interface (VCI) [8]. The VCI acts as a standard communication interface between the IP core and the on-chip bus. First, the microprocessor tests itself by executing a set of instructions. Next, the processor can be used for testing the bus and other non-programmable IP cores in the SoC. In order to support the self-testing methodology, the IP core has a test wrapper around it. The test wrapper contains test support logic needed to control shifting of the scan chain, buffers to store scan

This work was supported in part by the MARCO/DARPA Gigascale Silicon Research Center (http://www.gigascale.org). Their support is gratefully acknowledged.

data and support at-speed test, etc. In this example, the on-chip bus is a shared bus and the arbiter controls access to the bus.

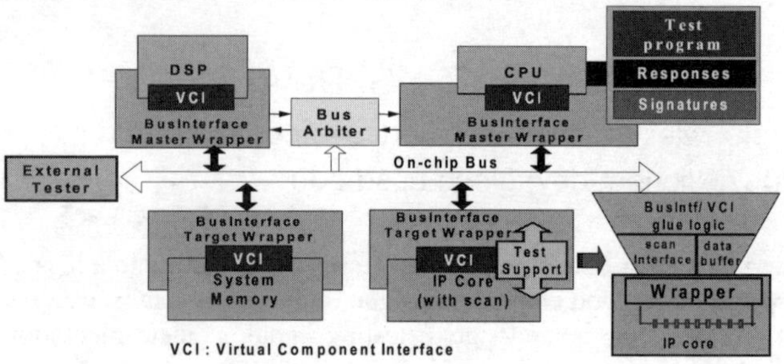

Figure 1: Embedded software-based self-testing for SoC.

There are several advantages of the embedded software-based self-test approach. First, it allows reuse of programmable resources on SoCs for test purposes. In other words, this strategy views testing as an application of the programmable components in the SoC and thus, minimizes the addition of dedicated test circuitry for design-for-testability or self-test.

Second, in addition to eliminating the need for costly high-speed testers, it can also reduce the yield loss due to tester accuracy problems. Self-testing offers the ability to apply and analyze at-speed test signals on chip with accuracy greater than that available with the tester.

Third, while the hardware-based self-test must be applied in the non-functional BIST mode, software-based self-test can be applied in the normal operational mode of the design, i.e., the tests are applied by executing instruction sequences as in regular system operations. This eliminates the problems created by application of non-functional patterns that can result in excessive power consumption when hardware BIST is used.

Also, functional self-test can alleviate some delay testing problems. Delay testing using scan techniques requires turning all state elements into scannable counterparts and they can be loaded serially through the scan port. The scan shifting is usually done at a lower frequency to make for easier Design-for-Testability (DfT) implementation. After the proper states are all scanned-in, the system clock is initiated for 2 or more cycles. The first clock is the *launch clock* that launches the output of the state elements into the combinational blocks. The subsequent clock (*capture clock*) captures the responses back into the scan chain for verification. During the brief launch/capture system clock, large current surges will happen affecting the circuit delays. Increasing the clock period would allow for defect based testing, but this would not guarantee that the circuit would meet the rated speed. Another possible solution would be to launch multiple system clocks after the scan shift so that the power grid has a chance to stabilize. This requires hundreds of cycles and presents a big problem for ATPG.

Functional testing can also relieve problems due to application of non-functional patterns during structural delay testing through AC-scan or BIST that can result in over-testing and yield loss. Experiments have shown that many structurally testable delay faults in the microprocessors can never be sensitized in the functional mode of the circuit [5]. This is because no functionally applicable vector sequence can excite these delay faults and propagate the fault effects into destination outputs/flip-flops at-speed. Defects on these faults will not affect the circuit performance and their testing is not necessary.

Testing of analog circuits has been a costly process because of the limited access to the analog parts and testers needed for functional testing. The situation has become worse due to integrating various digital and analog cores onto the SoC, in which testing the analog parts becomes the bottleneck of production testing. Self-testing of on-chip ADC/DAC and analog components using DSP-based approaches can alleviate these problems.

In this paper, we give a survey of the embedded software-based self-testing methods. We start by discussing processor self-test methods targeting stuck-at faults and delay faults. Next, we continue with self-testing of buses and global interconnects and well as other non-programmable IP cores on SoC. We also describe instruction-level DfT methods based on insertion of test instructions to increase the fault coverage and reduce the test application time and test program size. Finally, we summarize DSP-based self-test for analog/mixed-signal components.

2. EMBEDDED PROCESSOR SELF-TEST

While logic BIST may perform well on industrial application specific integrated circuits (ASICs), it is less feasible on microprocessors. First, the design changes needed for making a microprocessor BIST-ready (e.g., immune to problems such as bus contentions even when pseudorandom test patterns are applied) may come with unacceptable cost, such as substantial manual effort and significant performance degradation. In addition, microprocessors are especially random pattern-resistant. Due to timing-critical nature of microprocessors, test points may not be acceptable as a solution to this problem, as they could introduce performance degradation on critical paths. Deterministic BIST, on the other hand, may lead to unacceptable area overhead, as the size of the on-chip hardware for encoding deterministic test patterns depends on the circuit testability [9].

A number of approaches have been proposed to generate functional tests for microprocessors [10][11][12][13][14]. Some propose to apply the tests with external testers [10], while others allow the processors to tests themselves with self-test programs [11][12][13][14]. A common characteristic of approaches in [11][12] is application of randomized instructions to the processor under test. However, although processors are more amenable to random-instruction tests than to random-pattern tests, it is difficult to target structural faults by applying random instructions at the processor level. Approaches in [13][14] use structural ATPG to generate tests for stuck-at faults in the processor. They use the RTL information of the processor to form a set of RTL-module equations that can realize the generated gate-level test. Solving the set of equations specifies the required instruction sequences and operands. All of the above approaches target only stuck-at faults and the methods cannot be easily generalized for delay faults.

Unlike hardware-based self-testing, software-based testing is non-intrusive since it applies tests in the normal operational mode of the circuit. Moreover, software instructions have the ability of guiding the test patterns through a complex processor, avoiding the blockage of the test data due to non-functional control signals as in the case of hardware-based logic BIST.

Embedded software-based self-test methods for processors have been proposed in [4][5][6]. These methods consist of two steps: the test preparation step and the self-testing step. The test preparation step involves generation of *realizable* tests for components of the processor. Realizable tests are those that can be delivered using instructions. Therefore, to avoid producing undeliverable test patterns, the tests are generated under the constraints imposed by the processor instruction set. The tests can then be either stored or generated on-chip, depending on which method is more efficient for a particular case. A low-speed tester can be used to load the self-test signatures or the predetermined tests to the processor memory prior to the application of tests.

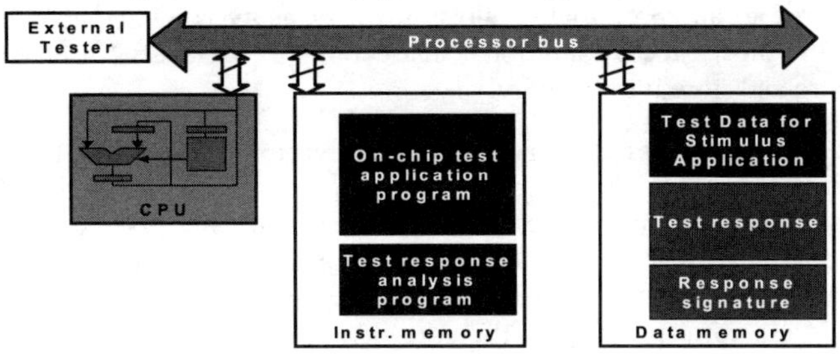

Figure 2: Embedded processor self-testing.

The self-testing step, illustrated in Figure 2, involves the application of these tests using a software tester. The software tester can also compress the responses into self-test signatures that can then be stored into the memory. The signatures can later be unloaded and analyzed by an external tester. The assumption here is that the processor memory has been tested with standard techniques such as memory BIST before the application of the test and the memory is assumed to be fault-free.

In the following, we describe the embedded software-based self-test method for testing stuck-at [4] and path delay faults [5][6] in microprocessors using their instruction set.

Stuck-at Fault Testing. As the first step, the realizable structural tests for components of the processor are developed. Component tests can either be stored or generated on-chip. If tests are generated on-chip, the test needs of each component are characterized by a self-test signature, which includes the seed S and the configuration C of a pseudorandom number generator as well as the number of test patterns to be generated N. The self-test signatures can be expanded on-chip into test sets using a pseudorandom number generation program. Multiple self-test signatures may be used for one component if necessary. Thus, this self-test methodology allows incorporation of any deterministic BIST techniques that encode a deterministic test set as several pseudorandom test sets [15][16].

By targeting the structural test needs of less complex components, the proposed method has the fault coverage advantage of deterministic structural testing. Since component test application and response collection are done with instructions instead of with scan chains, it requires no area or performance overhead and the test application is performed at-speed. Most importantly, by shifting the role of external testers to applying test programs and unloading responses, it enables at-speed testing of gigahertz processors with low-speed testers.

Using manually extracted constraints, the above scheme has been applied to a simple Parwan processor [17]. The results have demonstrated the feasibility and effectiveness of the software-based self-test method by generating a high-coverage test program for the simple processor.

Delay Testing. To synthesize a test program for self-test of path delay faults in a microprocessor using its instructions, first the spatial and temporal constraints between and at the registers and control signals are extracted (given the instruction set architecture and the micro-architecture of the processor core). Next, a path classification algorithm, extended from [18][19], implicitly enumerates and examines all paths and path segments. If a path cannot be sensitized with the imposed extracted constraints, the path is *functionally untestable* and thus, eliminated from the fault universe. This helps reduce the computational effort of the subsequent test generation process. As the experimental results in [5] show, a high percentage of the paths in Parwan processor [17] and DLX processor [20] are functionally untestable.

Next, a subset of long paths among the functionally testable paths are selected as targets for test generation. A gate-level ATPG for path delay faults is extended to incorporate the extracted constraints into the test generation process and it is used to generate test vectors for each target path delay fault. If the test is successfully generated, it not only sensitizes the path but it also meets the extracted constraints. Therefore, it is most likely to be deliverable by instructions (if the complete set of constraints has been extracted, the delivery by instructions could be guaranteed). In the test program synthesis process that follows, the test vectors specifying the bit values at internal flip-flops are first mapped back to word-level values in registers and values at control signals. These mapped value requirements are then justified at the instruction level. Finally, a pre-defined propagating routine is used to propagate the fault effects captured in the registers/flip-flops of the path delay fault to the memory. This routine compresses the contents of some or all registers in the processor, generates a signature and stores it in memory. The procedure is repeated until all target faults have been processed. The test program is generated off-line and later used to test the microprocessor at-speed.

To apply the synthesized test program, it is loaded into the on-chip memory by an external low speed tester. When the test program is being executed at-speed, a set of signatures is recorded in memory. At the end of the test program, a response analysis subroutine is called to further compress the recorded signatures in memory and finally, compare the compressed signature with the correct signature.

This test synthesis program has been applied to Parwan [17] and DLX [20] processors. On the average, 5.3 and 5.9 instructions were needed to deliver a test vector and the achieved fault coverage for *testable* path delay faults was 99.8% and 96.3% for Parwan and DLX, respectively.

3. SELF-TESTING OF BUSES AND GLOBAL INTERCONNECTS

In SoC designs a large amount of core-to-core communications must be realized with long interconnects. As gate delay continues to decrease, the performance of interconnect is becoming increasingly important in achieving a high overall performance. However, due to the increase of cross-coupling capacitance and mutual inductance, signals on neighboring wires may interfere with each other, causing excessive delay or loss of signal integrity. While many techniques have been proposed to reduce crosstalk, due to the limited design margin and unpredictable process variations, the crosstalk must also be addressed in manufacturing testing.

Due to its timing nature, testing for crosstalk effects needs to be conducted at the operational speed of the circuit-under-test. However, at-speed testing of GHz systems requires prohibitively costly high-speed testers. Moreover, with external testing, hardware access mechanisms are required for applying tests to interconnects deeply embedded in the system. This may lead to unacceptable area or performance overhead.

A BIST technique in which an SoC tests its own interconnects for crosstalk defects using on-chip hardware pattern generators and error detectors has been proposed in [21]. Although the amount of area overhead may be amortized for large systems, for small systems, the amount of relative area overhead may beb unacceptable. Moreover, hardware-based self-test approaches, as the one in [21], may cause over-testing and yield loss, as not all test patterns generated in the test mode are valid in the normal operational mode of the system.

The problem of testing system-level interconnects in embedded processor-based SoCs, which are the most dominant type of SoCs, has been addressed in [22][23]. In such SoCs, most of the system-level interconnects, such as the on-chip buses, are accessible to the embedded processor core(s). The proposed methodology is software-based and enables an embedded processor core in SoC to test for crosstalk effects in these interconnects by executing a software program. The strategy is to let the processor execute a self-test program with which the test vector pairs can be applied to the appropriate bus in the normal functional mode of the system. In the presence of crosstalk-induced glitch or delay effects, the second vector in the vector pair becomes distorted at the receiver end of the bus. The processor, can then store this error effect to the memory as a test response, which can be later unloaded by an external tester for off-chip analysis.

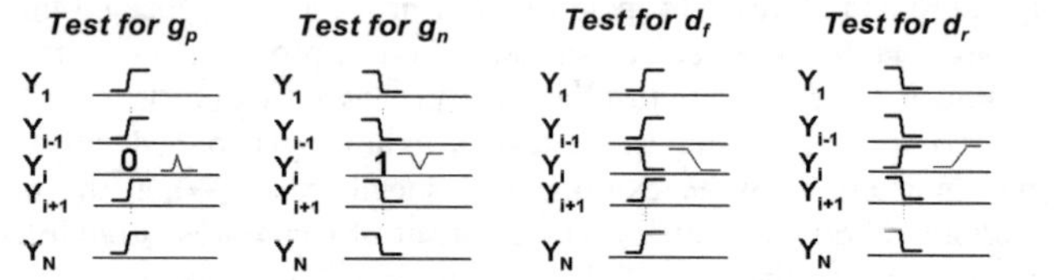

Figure 3: Maximal aggressor tests for victim Y_i.

Maximum Aggressor (MA) fault model proposed in [24] is suitable for modeling crosstalk defects on interconnects. It abstracts the crosstalk defects on global interconnects by a linear number of faults. It defines faults based on the resulting crosstalk error effects, including positive glitch (g_p), negative glitch (g_n), rising delay (g_r) and falling delay (g_f). For a set of N interconnects, the MA fault model considers the collective aggressor effects on a given victim line Y_i, while all other $N-1$ wires act as aggressors. The required transitions on the aggressor/victim lines to excite the four error types are shown in Figure 3. For a set of N interconnects, there are $4N$ MA faults, requiring $4N$ MA tests. It has been shown in [24] that these $4N$ faults cover all crosstalk defects on any of the N interconnects.

Chen *et al.* [22] concentrate on testing data and address bus in a processor-based SoC. The crosstalk effects on the interconnects are modeled using the MA fault model.

Testing Data Bus. For a bi-directional bus such as data bus, crosstalk effects vary as the bus is driven from different directions. Thus crosstalk tests need to be conducted in both directions [21].

To apply a test vector pair (*v1*, *v2*) for the data bus from an SoC core to the CPU, the CPU first exchanges data *v1* with the core. The direction of data exchange is irrelevant. For example, if the core is the memory, the CPU may either read *v1* from the memory or write *v1* to the memory. The CPU then requests data *v2* from the core (a memory-read if the core is memory). Upon the arrival of *v2*, the CPU writes *v2* to memory for later analysis.

To apply a test vector pair (*v1*, *v2*) to the data bus from the CPU to an SoC core, the CPU first exchanges data *v1* with the core. Then, the CPU sends data *v2* to the core (a memory-write if the core is memory). If the core is memory, *v2* can be directly stored to an appropriate address for later analysis. Otherwise, the CPU must execute additional instructions to retrieve *v2* from the core and store it to memory.

Testing Address Bus. To apply a test vector pair (*v1*, *v2*) to the address bus, which is a unidirectional bus from the CPU to an SoC core, the CPU first requests data from two addresses (*v1* and *v2*) in consecutive cycles. In the case of a non-memory core, since the CPU addresses the core via memory-mapped I/O, *v2* must be the address corresponding to the core. If *v2* is distorted by crosstalk, the CPU would be receiving data from a wrong address, *v2'*, which may be a physical memory address or an address corresponding to a different core. By keeping different data at *v2* and *v2'* (i.e., mem[*v2*] $\neq$ mem[*v2'*]), the CPU is able to observe the error and store it to memory for analysis.

The feasibility of this method has been demonstrated by applying it to test the interconnects of a processor-memory system. The defect coverage was evaluated using a system-level crosstalk defect simulation method.

Functionally Maximal Aggressor Tests. Even though the MA tests have been proven to cover all physical defects related to crosstalk between interconnects, Lai *et al.* [23] observe that many of them can never occur during normal system operation due to constraints imposed by the system. Therefore, testing buses using MA tests might screen out chips that are functionally correct under any pattern produced under normal system operation. Instead, *Functionally Maximal Aggressor (FMA) tests* meeting the system constraints and being possible to be delivered under the functional mode are proposed [23]. The tests can be synthesized by a software routine. The synthesized test program is highly modularized and very small. Experimental results have shown that a test program as small as 3K bytes can detect all crosstalk defects on the bus from the processor core to the target core.

The synthesized test program is applied to the bus from the processor core and the input buffers of the destination core capture the responses at the other end of the bus. Such responses need to be read back by the processor core to determine whether or not any faults on the bus occurred. However, because the input buffers of a non-memory core cannot be read by the processor core, a DfT scheme is suggested to allow direct observability of the input buffers by the processor core. The DfT circuitry consists of bypass logic added to each I/O core to improve its testability.

With the DfT support on the target I/O core, the test generation procedure first synthesizes instructions to set the target core to the bypass mode and then it continues with synthesizing instructions for the FMA tests. The test generation procedure does not depend on the functionality of the target core.

4. SELF-TESTING OF OTHER NON-PROGRAMMABLE IP-CORES

Testing non-programmable cores on an SoC is a complex problem with many unresolved issues [25]. Industry initiatives such as the IEEE P1500 Working Group [26] provide some solutions for IP core testing. However, they do not address the requirements of at-speed testing.

A self-testing approach for non-programmable cores on an SoC has been proposed in [25]. In this approach, a test program running on the embedded processor delivers test patterns to other IP cores in the SoC at-speed. The test patterns can be generated on the processor itself or fetched from an external ATE and stored in on-chip memory. This alleviates the need for dedicated test circuitry for pattern generation and response analysis. The approach is scalable to large size IP cores whose structural netlists are available. Since the pattern delivery is done at the SoC operational speed, it supports delay test. A test wrapper (shown in Figure 1) is inserted around each core to support pattern delivery. It contains test support logic needed to control shifting of the scan chain, buffers to store scan data and support at-speed test, etc.

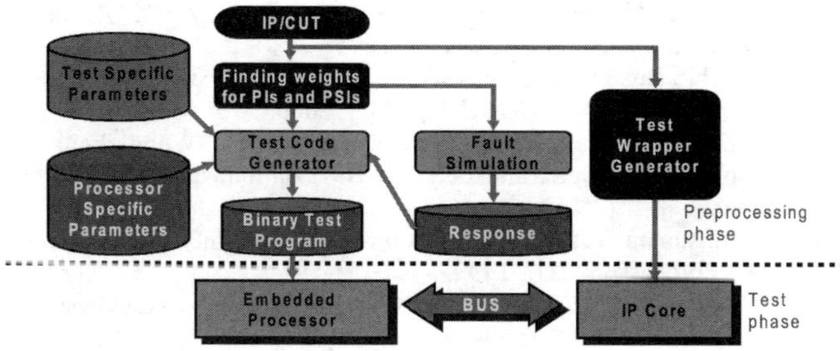

Figure 4: The test flow.

The test flow based on the embedded software self-testing methodology is illustrated in Figure 4. It offers tremendous flexibility in the type of tests that can be applied to the IP cores well as in the quality of the test pattern set without entailing significant hardware overhead. Again, the flow is divided into a pre-processing phase and a testing phase.

In the pre-processing phase, a test wrapper is automatically inserted around the IP core under test. The test wrapper is configured to meet the specific testing needs for the IP core. The IP core is then fault simulated with different sets of patterns. Weighted random patterns generated with multiple weight sets or using multiple capture cycles [3] after each scan sequence are used in [25]. Next, a high-level test program is generated. This program synchronizes the software pattern generation, start of the application of the test and analysis of the test response. The program can also synchronize testing multiple cores in parallel. The test program is then compiled to generate a processor specific binary code.

In the test phase, the test program is run on the processor core to test various IP cores. A test packet is sent to the IP core test wrapper informing it about the test application scheme (e.g., single or multiple capture cycle). Data packets are then sent to load the scan buffers and the PI/PO buffers. The test wrapper applies the required number of scan shifts and captures the test response for the programmed number of functional cycles. The results of the test are stored in the PI/PO buffers and the scan buffers and from there they are read out by the processor core.

5. INSTRUCTION LEVEL DFT

While self-testing manufacturing defects in an SoC by running test programs using a programmable core has many potential benefits, such a self-test strategy might require a lengthy test program and might not achieve a high enough fault coverage. These problems can be alleviated by applying a DfT methodology based on adding test instructions to an on-chip programmable core such as a microprocessor core. This methodology is called *instruction-level DfT*.

Instruction-level DfT inserts test circuitry in the form of test instructions and should be a less intrusive approach as compared to the gate-level DfT techniques which attempt to create a separate test mode somewhat orthogonal to the functional mode. If the test instructions are carefully designed such that their micro-instructions reuse the datapath for the functional instructions and do not require any new datapath, the overhead, which only occurs in the controller, should be relatively low.

Instruction-level DfT methods have been proposed in [11] [27]. The approach in [11] adds instructions to control the exceptions, e.g., microprocessor interrupts and reset. With the new instructions, the test program can achieve fault coverage between 87% and 90% for stuck-at faults. However, this approach cannot achieve a higher coverage because the test program is synthesized based on a random approach and it is not able to effectively control or observe some internal registers that have low testability.

The DfT methodology proposed in [27] systematically adds test instructions to an on-chip processor core to improve the self-testability of a processor core, reduce the size of the self-test program as well as reduce its run time (i.e., reduce the test application time). To decide which instructions to add, the testability of the processor is analyzed first. If a register in the processor is identified as hard-to-access, a test instruction allowing direct accessing of the register is added. In addition to these test instructions, test instruction can be also added to optimize the test program size and run time.

Adding test instructions to the programmable core does not improve the testability of other non-programmable cores on the SoC. Therefore, instruction-level DfT cannot increase the fault coverage of the non-programmable cores. However, the test programs for testing the non-programmable cores can be optimized by adding new instructions. In other words, the same set of test instructions added for self-testing the programmable cores can be used to reduce the size and run time of the test programs for testing other non-programmable cores.

The experimental results on two processors (Parwan [17] and DLX [20]) show that test instructions can reduce the program size and program run time by 20% at the cost of 1.6% area overhead.

6. ON-CHIP ADC/DAC AND ANALOG COMPONENTS

For mixed-signal systems integrating both analog and digital functional blocks onto the same chip, testing of analog/mixed-signal parts has become the bottleneck during production testing. Because most analog/mixed-signal circuits are functionally tested, analog/mixed-signal testing needs expensive automatic test equipment (ATE) for analog stimulus generation and response acquisition. One promising solution to this problem is BIST that utilizes on-chip resources (either shared with functional blocks or dedicated BIST circuitry) to perform on-chip stimulus generation and response acquisition. Under the BIST approach, the requirement on the external test equipment is less stringent. Furthermore, stimulus generation and response acquisition is more immune to environmental noise during the test process.

With the advent of the CMOS technology, DSP-based BIST becomes a viable solution for analog/mixed-signal systems as the required signal processing to make the pass/fail decision can be realized in the digital domain with digital resources.

An efficient BIST architecture for testing on-chip analog and mixed-signal components has been proposed in [28]. It employs the delta-sigma modulation technique for both stimulus generation

[29] and response analysis. Figure 5 illustrates this delta-sigma modulation-based BIST architecture.

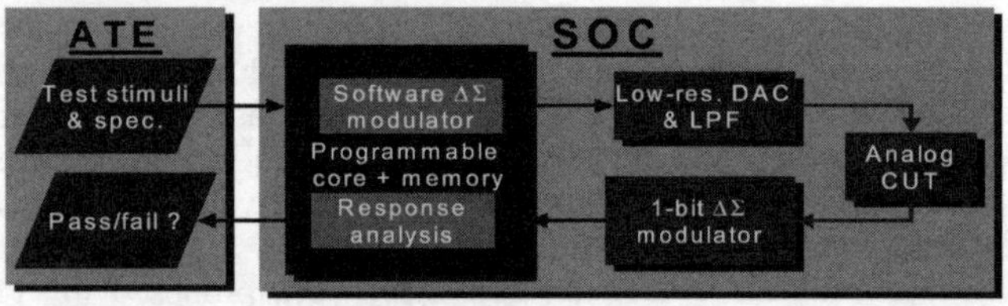

Figure 5: DSP-based self-test for analog/mixed-signal parts.

A *software* delta-sigma modulator converts the desired signal to one-bit digital steam. The digital 1's and 0's are then transferred to two discrete analog levels by one-bit DAC followed by a low-pass filter that removes the out-of-band high-frequency modulation noise, and thus restores the original waveform. In practice, one extracts a segment from the delta-sigma output bit stream that contains an integer number of signal periods. The extracted pattern is stored in on-chip memory, and periodically applied to the low-resolution DAC and low-pass filter to generate the desired stimulus. Similarly, for response analysis, a 1-bit $\Sigma-\Delta$ modulator can be inserted to convert the analog DUT output response into a 1-bit stream which is then analyzed by digital signal processing (DSP) operations performed by on-chip DSP/microprocessor cores. Among the 1-bit $\Sigma-\Delta$ modulation architectures, the 1st-order configuration is the most stable and has the maximal input dynamic range. However, it is not practical for high-resolution applications (as rather high over sampling rate will be needed), and suffers inter-modulation distortion. Compared to the 1st-order configuration, the 2nd-order configuration has a smaller dynamic range but is more applicable for high-resolution applications.

Note that the software part of this technique, i.e., the software $\Sigma-\Delta$ modulator and the response analyzer, can be performed by on-chip DSP/microprocessor cores, if abundant on-chip digital programmable resources are available (as indicated in Figure 5), or by external *digital* test equipment.

7. CONCLUSIONS

Embedded software-based self-testing has a potential to alleviate many of the current external tester-based and hardware BIST testing techniques for SoCs. In this paper, we give a summary of the recently proposed techniques for self-testing for system-on-chips. One of the main tasks in applying these techniques is extracting the functional constraints in the process of test program synthesis, i.e., deriving tests that can be delivered by processor instructions. Future research in this area must address the problem of automating the constraint extraction process in order to make the proposed solutions feasible for general processors. The software-based self-testing paradigm can be further generalized for analog/mixed-signal components through the integration of DSP-based testing techniques, $\Sigma-\Delta$ modulation principles and some low-cost analog/mixed-signal DfT.

ACKNOWLEDGMENTS

We would like to thank Dr. T. M. Mak, Intel on many stimulating discussions and useful insights on the topics in this paper.

References

[1] C.-J. Lin, Y. Zorian, and S. Bhawmik. Integration of Partial Scan and Built-in Self-Test. *JETTA*, Aug. 1995.
[2] K.-T. Cheng and C.-J. Lin. Timing-Driven Test Point Insertion for Full-Scan and Partial-Scan BIST. *ITC*, 1995.
[3] H.-C. Tsai, S. Bhawmik, and K.-T. Cheng. An Almost Full-Scan BIST Solution – Higher Fault Coverage and Shorter Test Application Time. *ITC*, 1998.
[4] L. Chen and S. Dey. Software-Based Self-Testing Methodology for Processor Cores. *TCAD*, Mar. 2001.
[5] W.-C. Lai, A. Krstic, and K.-T. Cheng. On Testing The Path Delay Faults of a Microprocessor Using its Instruction Set. *VTS*, 2000.
[6] W.-C. Lai, A. Krstic, and K.-T. Cheng. Test Program Synthesis for Path Delay Faults in Microprocessor Cores. *ITC*, 2000.
[7] PCI Special Interest Group, Hillsboro, Oregon, USA. *PCI Local Bus Specification, Revision 2.0*, Apr. 1994.
[8] On-Chip Bus Development Working Group. *Virtual Component Interface Standard (OCB 2 1.0)*, Mar. 2000.
[9] G. Kiefer, H. Vranken, E. J. Marinissen, and H.-J. Wunderlich. Application of Deterministic Logic BIST on Industrial Circuits. *ITC*, 2000.
[10] R. Rajsuman. Testing A System-on-Chip with Embedded Microprocessors. *ITC*, 1999.
[11] J. Shen and J. A. Abraham. Native Mode Functional Test Generation for Processors with Applications to Self Test and Design Validation. *ITC*, 1998.
[12] K. Batcher and C. Papachristou. Instruction Randomization Self Test for Processor Cores. *VTS*, 1999.
[13] J. Lee and J. H. Patel. Architectural Level Test Generation for Microporcessors. *TCAD*, 1994.
[14] J. Lee and J. H. Patel. Hierarchical Test Generation Under Architectural Level Functional Constraints. *TCAD*, Sep. 1996.
[15] S. Hellebrand and H.-J. Wunderlich. Mixed-Mode BIST Using Embedded Processors. *ITC*, 1996.
[16] R. Dorsch and H.-J. Wunderlich. Accumulator Based Deterministic BIST. *ITC*, 1998.
[17] Z. Navabi. *VHDL: Analysis and Modeling of Digital Systems.* McGraw-Hill, New York, 1997.
[18] K.-T. Cheng and H.-C. Chen. Classification and Identification of Nonrobustly Untestable Path Delay Faults. *TCAD*, Aug. 1996.
[19] A. Krstic, S. T. Chakradhar, and K.-T. Cheng. Testable Path Delay Fault Cover for Sequential Circuits. *EDAC*, 1996.
[20] M. Gumm. VHDL – Modeling and Synthesis of the DLXS RISC Processor. *VLSI Design Course Notes*, Univ. of Stuttgart, Germany, Dec. 1995.
[21] X. Bai, S. Dey, and J. Rajski. Self-Test Methodology for At-Speed Test of Crosstalk in Chip Interconnects. *DAC*, 2000.
[22] L. Chen, X. Bai, and S. Dey. Testing for Interconnect Crosstalk Defects Using On-Chip Embedded Processor Cores. *DAC*, 2001.
[23] W.-C. Lai, J.-R. Huang, and K.-T. Cheng. Embedded-Software-Based Approach to Testing Crosstalk-Induced Faults at On-Chip Buses. *VTS*, 2001.
[24] M. Cuviello, S. Dey, X. Bai, and Y. Zhao. Fault Modeling and Simulation for Crosstalk in System-on-Chip Interconnects. *ICCAD*, 1999.
[25] J.-R. Huang, M. K. Iyer and K.-T. Cheng. A Self-Test Methodology for IP Cores in Bus-Based Programmable SoCs. *VTS*, 2001.
[26] IEEE P1500 Web Site, http://grouper.ieee.org/groups/1500/.
[27] W.-C. Lai and K.-T. Cheng. Instruction-Level DFT for Testing Processor and IP Cores in System-on-a-Chip. *DAC*, 2001.
[28] J.L. Huang and K.T. Cheng. A Sigma-Delta Modulation Based BIST Scheme for Mixed-Signal Circuits. *ASPDAC*, 2000.
[29] B. Dufort and G. W. Roberts. Signal Generation Using Periodic Single and Multi-Bit Sigma-Delta Modulated Streams. *ITC*, 1997.

A Novel Wavelet Transform Based Transient Current Analysis for Fault Detection and Localization *

Swarup Bhunia
Purdue University
Electrical and Computer Engineering
Indiana, USA
bhunias@purdue.edu

Kaushik Roy
Purdue University
Electrical and Computer Engineering
Indiana, USA
kaushik@purdue.edu

Jaume Segura
Balearic Islands University
Physics Department
07071 Palma de Mallorca, Spain
jaume.segura@uib.es

ABSTRACT

Transient current (IDD) based testing has been often cited and investigated as an alternative and/or supplement to quiescent current (IDDQ) testing. While the potential of IDD testing for fault detection has been established, there is no known efficient method for fault diagnosis using IDD analysis. In this paper, we present a novel integrated method for fault detection and localization using wavelet transform based IDD waveform analysis. The time-frequency resolution property of wavelet transform helps us detect as well as localize faults in digital CMOS circuits. Experiments performed on measured data from a fabricated 8-bit shift register and simulation data from more complex circuits show promising results for both detection and localization. Wavelet based detection method shows superior sensitivity than spectral and time-domain methods. The effectiveness of the localization method in presence of process variation, measurement noise and complex power supply network is addressed.

Categories and Subject Descriptors

B.8.2 [**Hardware**]: Performance and Reliability—*Reliability, Testing, and Fault-Tolerance*

General Terms

Algorithms, Reliability, Experimentation

Keywords

Transient current (IDD), wavelet transform, fault localization

*This research was funded in part by MARCO GSRC under contract number SA3273JB.

1. INTRODUCTION

While many solutions have been proposed to deal with background leakage elevation in IDDQ testing, IDD or transient current based testing has emerged as an alternative and/or supplementary testing method. A number of research work [3, 4, 1, 13, 8, 7, 10, 11, 6, 5] have been targeted to establish that IDD waveform analysis is an effective technique to detect many of the defects that can occur in ICs, including defects such as resistive opens and weak transistor defects, which may not be detected by conventional IDDQ testing methods.

In 1987, Frenzel and Marinos [3] investigated a small TTL and described the complete power supply current as a signature of the DUT. Hasizume et al. [4] addressed the issue of analyzing the spectral content of the IDD current under normal and faulty conditions. Beasley et al. [1] applied simultaneous pulsing on the power supply rails and analyzed the temporal and/or spectral characteristics of the transient currents. Vinnakota [13] considered dynamic power dissipation of a circuit to detect faults. Plusquellic et al. [8] proposed the concept of Transient Signal Analysis (TSA) with distributed measurement points. Su et al. [12] applied dynamic current monitoring techniques on SRAMS using extensive DFT strategy. De Paul et al. [7] used the accumulated charge (computed by numerical integration of a current waveform) for signature comparison. In the method of Sachdev et al. [10] one sample per IDD test pattern at pre-determined instance is used as signature. Muhammad et al. [6] developed DFT based signature comparison to detect faults and an integrator based approach to extract delay information, which gives an idea about the depth where the fault resides. Their method works well only for levelized circuits and the localization is susceptible to problems like *aliasing*.

In this paper, we present a novel integrated approach for fault detection and diagnosis using wavelet transform of IDD signal. Wavelet transform has the potential to resolve a signal in both time and frequency domain simultaneously. For detecting fault, the complete set of wavelet coefficients for the IDD waveform can be used as signature to compare a faulty circuit with a fault-free one and to make pass/fail decisions. Once a fault is detected, we use time-domain information in the wavelet coefficients to localize it using delay measurement technique. Multiple number of input vectors are applied to the Device Under Test (DUT) to narrow down the faulty region.

Our experimental results are based on measured current data from a 8-bit shift register and transient response obtained from *HSpice* simulation of an 8 bit Arithmetic Logic Unit (ALU) and an 8x8 Wallace tree multiplier. Wavelet based method is shown to be much superior to other techniques in terms of detection sensitivity. Time domain information extracted from wavelet coefficients is used to efficiently locate fault in the test circuits. The effectiveness of the method on fabricated chip proves its potential for practical application.

The rest of the paper is organized as follows. Section 2 gives an overview of the wavelet transform and its basic properties; Section 3 deals with the fault detection and localization using wavelet analysis; Section 4 contains experimental results; Section 5 addresses some important issues concerning practical application of the method and Section 6 concludes the paper.

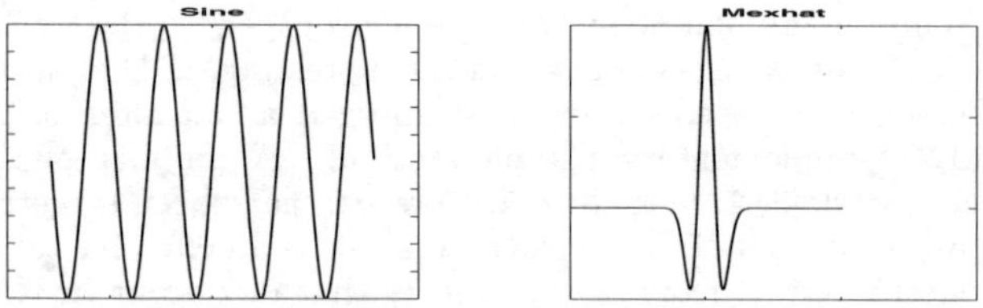

Figure 1: Basis functions for Fourier and wavelet transforms

2. AN OVERVIEW OF WAVELET TRANSFORM

Fourier analysis has a serious drawback since it transforms signal in frequency domain losing all information on how the signal is spatially distributed. Wavelet transform of a signal, on the other hand, decomposes signal in both time and frequency domain [2][9], which turns out to be very useful in fault detection and localization. In wavelet transform we take a real/complex valued continuous time function with two main properties - a) it will integrate to zero, b) it is square integrable. This function is called the mother wavelet or wavelet. Property (a) is suggestive of a function which is oscillatory or has wavy appearance and thus in contrast to a sinusoidal function, it is a small wave or wavelet (figure 1). Property (b) implies that most of the energy of the wave is confined to a finite interval. The CWT or the Continuous Wavelet Transform of a function $f(t)$ with respect to a wavelet $\Psi(t)$ is defined as:

$$W(a,b) = \int_{-\infty}^{\infty} f(t)\Psi^*_{a,b}(t)dt \quad (1)$$

$$\text{where } \Psi_{a,b}(t) = \frac{1}{\sqrt{|a|}}\Psi\Big(\frac{t-b}{a}\Big) \quad (2)$$

Here a,b are real and $*$ indicates complex conjugate. $W(a,b)$ is the transform coefficient of $f(t)$ for given a, b. Thus the wavelet transform is a function of two variables. For a given a, $\Psi_{a,b}(t)$ is a shift of $\Psi_{a,0}(t)$ by an amount b along time axis. The variable b represents time shift or translation. Since a determines the amount of time-scaling or dilation, it is referred to as *scale* or dilation variable. If $a > 1$, there is stretching of $\Psi(t)$ along the time axis whereas if $0 < a < 1$ there is a contraction of $\Psi(t)$ (figure 2). Each wavelet coefficient $W(a,b)$ is the measure of approximation of the input waveform in terms of the translated and dilated versions of the mother wavelet. Figure 1 compares the basis signals of DFT and wavelet transform. The mother wavelet shown in figure 1 is called *Mexican hat* wavelet. Figure 2 shows the translated and dilated mother wavelet used to approximate an IDD waveform of a test circuit.

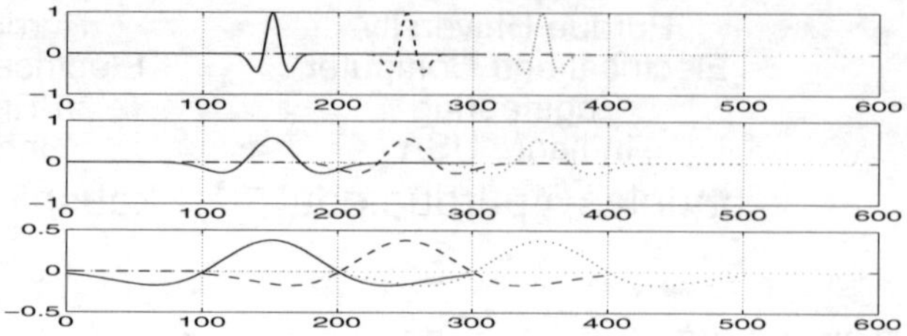

Figure 2: Translated and dilated mother wavelet (*mexican hat*) used in the wavelet decomposition

3. IDD ANALYSIS USING WAVELET TRANSFORM

3.1 Fault Detection

The Fault Detection strategy is based on comparing current signature of the DUT with the signature of the golden (fault-free) device. Input test stimuli are chosen randomly in our detection process. For each stimulus applied to the DUT, we compute the wavelet coefficients of the transient current and compare them with those for the golden device with the same input. The comparison in our case, is made by calculating the Mean Square Error (MSE) between the two sets of wavelet coefficients. Mean Square Error is chosen for comparison because it is a simple metric that can effectively detect faults. The pass/fail criterion can be decided by comparing the value of the MSE with a pre-selected test margin. Since the signature is based on wavelet coefficients, we take into account both the time and frequency components simultaneously in the transient current signature. This gives us better sensitivity in fault detection than methods based on only spectral [4] [10] [6] or only time-domain components [13] [6].

For comparing the effectiveness of our detection scheme with existing methods based on pure spectral and pure time domain contents of the IDD signal, we used a common metric referred as $normRMS$ which is defined in equation 3. It computes the root mean square of the difference between the coefficients for golden circuit response (G_i) and those for the DUT (F_i) as a fraction of the golden circuit coefficient(G_i) and can be considered as the direct measure of sensitivity. For the time domain method, we used the charge integration (as discussed in [7]) for comparison.

$$normRMS = \sqrt{\frac{1}{N}\sum_{i=1}^{N}\Big(\frac{F_i - G_i}{G_i}\Big)^2} \quad (3)$$

3.2 Fault Localization

An important advantage of wavelet based analysis of the transient current is its ability to efficiently localize fault.

For a faulty device, we can observe the wavelet coefficients and determine the delay at which the response of the faulty circuit deviates significantly from that of the fault-free one. We then use this information to identify region where the fault resides. There are two advantages of identifying the faulty region. First, it may be possible to improve the process and yield. Second, we can use this information either for fault tolerance, or to isolate a small portion of the circuit containing the fault.

Figure 3 depicts a simple test circuit consisting of a set of inverters in cascade. We use this circuit to explain how we measure delay and how it can be used to localize faults. First we apply input stimuli to the fault-free inverter chain as shown in the figure 3 and monitor the IDD waveform. We then introduce a metal bridge of resistance 1kΩ between T3 and supply line and obtain the IDD response. We repeat the same procedure for a similar fault at position T5. Figure 4 shows the plot of IDD for these three cases. Figure 5 shows the wavelet coefficients for the IDD waveform at four different scales. The faulty responses deviate from the fault-free case at some time instant after the transition in the primary input. We call this delay between the input transition and the point of deviation in the time axis as T_d (Figure 6). This is due to the propagation delay through the cells, since the fault is not activated until the effect of an input transition propagates to it. Since the propagation delay to T5 is more than that to T3, the point of deviation for fault at T5 is shifted right in time axis from the point for fault at T3.

One interesting observation we can make here is that T_d can be used as a measure of depth in the circuit at which the fault resides. Since T_d is the propagation delay to the point where the fault is located, we can calculate minimum and maximum propagation delay along each path in the DUT in which the effect of an input transition propagates and compare that to T_d to select a set of potential faulty cells.

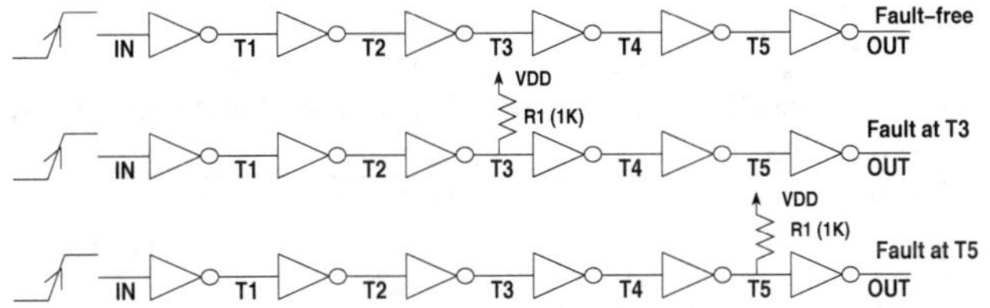

Figure 3: Inverter chain with bridging fault at T3 and T5

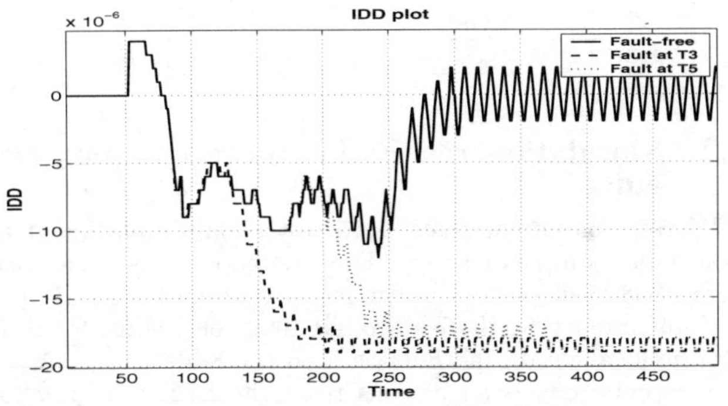

Figure 4: IDD Waveforms for the inverter chain with and without fault

Wavelet coefficients give us a more efficient way to calculate T_d than using simple integration [6] or a point-by-point comparison of the IDD signal. Wavelet transform can resolve a signal in time axis at different scales or frequencies. Hence we can compute T_d for multiple frequency components simultaneously using wavelet decomposition. This helps us to automatically get rid of the DC component in IDD. We can also avoid the *aliasing* effect, which may be present in the integration based method. By *aliasing* we mean that two sufficiently varying IDD waveform may have same area under the IDD curve which can lead to wrong delay computation. Since we can compute T_d at different scales in wavelet analysis, there is barely any chance of *aliasing*.

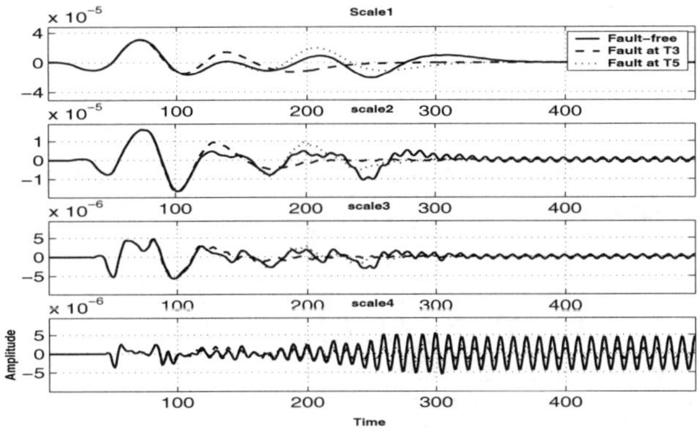

Figure 5: Plot of Wavelet coefficients at different scales for the IDD waveform in figure 4

Table 1: Fault localization steps

Locate-fault (DUT, Golden Device, Cell Delays)
1. Initialize set S with all cells in DUT 2. Apply a random transition $I_0 \to I_1$ to both the DUT and golden device. Check if a fault is detected by wavelet based method 3. If detected, compute the delay (T_d) by comparing wavelet coeffs of DUT and golden device, else go to step 2 4. If $T_d > 0$, *Partition* the circuit into set of faulty (Sf) and fault-free cells (Sff) for the input and delay, else go to step 2 5. Set $S = S \cap Sf$ 6. If terminating condition satisfied, exit, else go to step 2

The fault localization algorithm is depicted in table 1. We start by initializing S, the set of potential faulty cells at a particular iteration, with all cells in the DUT. If a fault is detected for an input transition $I_0 \to I_1$, we compute the delay. The delay computation unit compares wavelet coefficients of the fault-free and faulty response at different scales and determines the point in time axis where the wavelet components of faulty response vary by a pre-determined threshold from those of the fault-free response. We discard the high frequency components for delay computation.

Partition is a method for dividing the cells in the DUT into two non-overlapping sets to distinguish faulty cells from the non-faulty ones based on input transition and T_d. The partitioning algorithm traverses the cells in topological order and checks if a particular cell can be potentially faulty based on the min-max propagation delay to the cell. We can narrow down the faulty region by applying more input transitions and by taking intersection between previous faulty set and the current one.

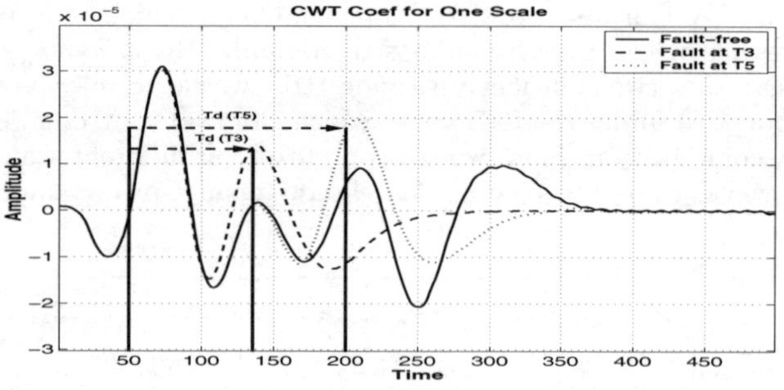

Figure 6: Plot of Wavelet coefficients for one scale showing measurement of delay

4. EXPERIMENTAL RESULTS

The detection and localization algorithm was implemented in *C*. We used *Matlab* toolbox for wavelet to perform wavelet decomposition of the IDD signl. Mother wavelet used in wavelet transform was *db2* [2]. The algorithms were tested on the measured data from the fabricated test chip for a 8-bit shift register. Detection and localization methods were also tested on simulation data from more complex circuits.

4.1 Results for a fabricated 8-bit shift register

The experiments used a 8-bit shift register in which several mask defects were designed. Transmission gates were connected to selected nodes of the circuit to emulate the presence of open defects. When a transmission gate is off, the corresponding open defect is activated. The circuit was designed with ES2 n-well dual metal 1.0μ technology. Current was measured by sensing the voltage drop at a very low inductive MP930 Caddok 300Ω resistor with a Tektronix P6247 1 GHz bandwidth differential probe. For testing the wavelet based detection and localization method, we worked with open defects preventing clock propagation from the defect site to the register.

Table 2 presents comparative results for the detection sensitivity of the wavelet, FFT and charge integration based method. The fault K_i implies that transmission gate between clock line and i-th register is open. It can be observed that the sensitivity for the wavelet based method is order of magnitude higher than either FFT or charge integration method. This difference is plotted in figure 7 in logarithmic scale. The superior sensitivity of wavelet can be attributed to its decomposition of the IDD signal in both time and frequency axis. Higher sensitivity makes wavelet a better candidate for fault detection in digital CMOS circuits, especially for parametric faults which are more difficult to detect.

Table 3 presents the localization results for the open defects in table 2. The delay T_d to each node K_i is computed from *Hspice* simulation of the extracted layout, which is then compared with the delay obtained from comparing

Table 2: Comparison of sensitivity of wavelet based detection with FFT and charge integration techniques

Fault	$normRMS$ (Wavelet)	$normRMS$ (FFT)	$normRMS$ (Charge)
K2	166.910	1.599	0.973
K3	109.921	1.508	0.948
K4	108.130	1.342	0.922
K5	215.780	1.293	0.872
K6	225.603	1.757	0.813
K7	313.904	5.097	0.698
K8	418.470	5.065	0.544

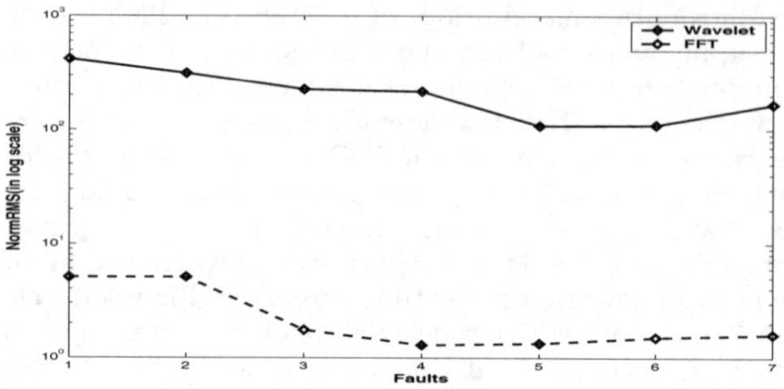

Figure 7: Comparison of fault detection sensitivity for the fabricated chip

the wavelet coefficients of the measured current and that obtained from simulation. The localization of fault, depends on how accurately we can measure the delay. The delay obtained from the measured data varies below 8% on average which is good enough to identify the fault location, i.e. which transmission gate is open.

Table 3: Localization results (in terms of delay) from the test chip

Fault	Delay from measured data (ns)	Delay from simulation (ns)	% Variation
K2	2.50	2.20	13.6
K3	7.00	7.60	7.8
K4	11.50	12.00	4.1
K5	14.20	15.53	8.6
K6	16.25	17.83	8.9
K7	19.50	19.88	1.9
K8	21.75	20.00	8.8

4.2 Simulation results for more complex circuits

Wavelet based methods were tested on simulation data from more complex circuits. Fault detection was performed for an 8x8 Wallace tree multiplier with around 800 cells and an 8-bit integer ALU with around 1000 cells, while for testing the localization method we used the 8-bit integer ALU. The circuits consisted of cells from the LEDA library for which the delay of each cell was known for different input conditions. We introduced different types of faults in the circuit and simulated the circuit with HSpice for a 0.25μ

TSMC technology library. Random input stimuli were applied to detect and localize faults.

Table 4 demonstrates the superiority of wavelet based detection compared to FFT and charge based technique. Resistive shorts and opens as well as parametric faults are used for comparison. For all the cases wavelet has a much higher sensitivity than current based techniques, proving its effectiveness for larger circuits.

Table 5 shows how our fault localization method converges as more and more test vectors are applied. The partitioning algorithm identifies a potential faulty set of cells for each test vector activating the fault. It uses the delay information obtained from wavelet coefficients. Concentrating on the common set of identified cells in two iterations of *Partition*, we norrow down our selection of cells. For the results in table 5, we inserted a metal bridge in the test circuit and applied a set of random vectors. Column 2 is the number of cells in the region identified as faulty at a specific run of partitioning. As we apply more number of vectors this region is narrowed down to almost 1% of total cells.

Table 4: Comparison of sensitivity for ALU and Wallace tree multiplier circuit

Design	Fault	$normRMS$ (Wavelet)	$normRMS$ (FFT)	$normRMS$ (Charge)
8-bit ALU	Bridge (Vss)	147.189	0.275	0.244
	Open	21.482	0.068	0.158
	Vth (10%)	255.378	0.093	0.189
	Bridge (Vss)	58.688	0.174	0.247
WTM (8x8)	Bridge (Vss)	46.566	0.215	0.283
	Bridge (Vdd)	304.322	0.276	0.209
	Open	477.436	0.034	0.083
	Vth (10%)	701.924	0.119	0.184

Table 5: Convergence of localization with increasing number of test vectors

# vectors	Cells in faulty region	% of total cells
1	398	39.25
3	112	11.05
7	59	5.82
12	33	3.25
17	10	0.99

Table 6 shows our experimental results for different types of faults injected in the test circuit at random locations. We successfully detected and localized faults in all the cases. How narrowly we can identify the faulty region largely depends on the number of input stimulus applied. We terminated our localization process if either we were able to localize the fault in 10% of the total cells or we have run 20 iterations of *Partition*. This was to limit the enormously slow *Hspice* run. Column 3 lists the number of random test vectors used (tests that detect the fault). Column 4 shows the number of cells in the identified region, which measures how accurately we can localize the faults. Column 5 lists the number of cells in the faulty region as percentage of total cells in the circuit.

Table 6: Fault Detection and Localization result for an 8-bit ALU

Fault	Detected/ Located	# vectors	Cells in faulty region	% of total cells
Metal bridge (with VDD)	Yes/Yes	13	71	7.00
Metal bridge (with GND)	Yes/Yes	7	91	8.88
Gate oxide short	Yes/Yes	16	52	5.13
Open	Yes/Yes	3	72	7.10
Metal bridge (with GND)	Yes/Yes	4	62	6.11

5. FACTORS AFFECTING LOCALIZATION

5.1 Effect of Power Supply Network

Our experiments assume that the circuit consists of only one module that is directly connected to the power supply pin. But in real chip, power supply network is usually designed as a mesh-like grid with different modules connected at different points on the grid. Wavelet based localization method can be extended to general mesh-like power-supply network. It can be observed that our localization method can be effective if we can ensure that the occurrence of a fault in a module does not significantly affect the current waveform in others. We simulated a power grid described as a RLC mesh with three different modules: an adder, a multiplexer and a comparator, connected to different points of the grid. It is observed that IDD waveform for fault-free modules does not suffer significant changes for a fault in another module, which verifies the applicability of localization for mesh-like power network.

5.2 Mother Wavelet Selection

Choice of mother wavelet is another issue that may affect computation of the T_d and thus localization. One of the advantages of wavelet transform is that it is adaptive i.e. we can selct a mother wavelet which can best approximate the input waveform. We experimented with a number of mother wavelets e.g. *db2*, *morlet*, *Mexican hat*, *haar* [2] [9] etc. and observed how T_d differs for different wavelets for a particular bridging fault in the ALU. Localization, as depicted in table 7, is best for *db2* and hence, we can use it as mother wavelet for testing the ALU.

Table 7: Localization results using different mother wavelets

Mother wavelet	# vectors	Cells in Faulty Region
db2	7	90
morl	7	101
mexh	12	98
haar	17	113

5.3 Effect of Sampling Frequency

The sampling rate at which the IDD waveform should be monitored is important because it affects the measurement noise and applicability of the method in real time. Ideally we need to sample the IDD waveform at above the Nyquist rate (i.e. twice the maximum frequency) to keep all the frequency components in the sampled data. However for detecting a fault, it has been observed that we do not need high sampling frequency [11]. This holds true for localization too because we can localize effectively without considering the high frequency spectral components of the IDD waveform. It is observed that we can sample the current waveform at as low as 50ps interval and still get a localization area less than 10% of the total cells for the test circuit.

5.4 Process Variation

The impact of process variation has to be taken into account in determining test margin for detection. Process variation also affects the delay along a path primarily because individual cell delays vary with process parameter changes. We modeled the process variation as transistor threshold (Vth) variation [14]. We have observed that if we compute test margins (the pass/fail limit and the delay threshold) based on 10% Vth variation, we can still detect and localize faults effectively. The impact of process variation on the localization is in terms of the resolution of the faulty region. Table 8 lists our experimental results with variations in Vth for a particular bridging fault in the ALU.

Table 8: Effect of process variation on localization

Vth change		# vectors	Cells in faulty region
NMOS	PMOS		
0%	0%	7	90
5%	5%	7	92
5%	-5%	7	92
-5%	5%	7	95

5.5 Effect of Measurement Noise

The hardware used to measure IDD waveform has some fixed resolution and introduces error in the measured waveforms. Typically the current waveform measured off-chip loses some high frequency components due to the presence of decoupling capacitance. As observed from the experimental results on the test chip, our method does not suffer significantly from measurement noise. This is also established from the results from simulation runs with the larger circuits, since, we consider only low frequency components of the waveform and still get efficient detection and localization.

6. CONCLUSIONS AND FUTURE WORK

A novel wavelet based IDD waveform analysis for fault detection and location is presented. Wavelet based delay computation method has been shown to withstand IDD waveform changes due to process and measurement noise due to its ability to do multi-resolution analysis of the IDD signal. Experimental results for measured data from a chip as well as for simulation data from larger circuits validates the technique's practical application.

Generation of optimal set of input stimuli for fault detection and localization is necessary to make the testing process more efficient. Test vectors for fault detection will be similar to that for IDDQ test. However, for localizing faults, automatic generation for optimal set of input vectors is non-trivial. Our approach to fault detection and localization can also be effective for pure analog or mixed-signal circuits. Currently, we are working on these aspects.

7. REFERENCES

[1] J. Beasley, H. Ramamurthy, J. Ramirez-Angulo, and M. DeYong. Idd pulse response testing of analog and digital cmos circuits. In *Proceedings of International Test Conference*, pages 626–634, 1993.

[2] I. Daubechies. *Ten Lectures on Wavelets.* Society for Industrial and Applied Mathematics, Philadelphia, Rutgers University, 1992.

[3] J. Frenzel and P. Marinos. Power supply current signature (pscs) analysis: A new approach to system testing. In *Proceedings of International Test Conference*, pages 125–135, 1987.

[4] M. Hasizume, K. Yamada, T. Tamesada, and M. Kawakami. Fault detection of combinational circuit based on supply current. In *Proceedings of International Test Conference*, pages 374–379, 1988.

[5] B. Kruseman, P. Janssen, and V. Zieren. Transient current testing of 0.25 μ cmos devices. In *Proceedings of International Test Conference*, pages 47–58, 1999.

[6] K. Muhammad and K. Roy. Fault detection and location using idd waveform analysis. *IEEE Design and Test of Computers*, 18(1):42–49, January-February 2001.

[7] I. D. Paul, J. L. Rossello, M. Roca, E. Isern, J. Segura, and C. F. Hawkins. Transient current testing based on current (charge) integration. In *Proceedings of the Workshop on IDDQ Testing*, pages 26–30, 1998.

[8] J. F. Plusquellic, D. M. Chiarulli, and S. P. Levitan. Digital integrated circuit testing using transient signal analysis. In *Proceedings of International Test Conference*, pages 481–490, 1996.

[9] R. Rao and A. Bopardikar. *Wavelet Transforms: Introduction to Theory and Applications.* Addison-Wesley, 1998.

[10] M. Sachdev, P. Janssen, and V. Zieren. Defect detection with transient current testing and its potential for deep sub-micron cmos ics. In *Proceedings of International Test Conference*, pages 204–213, 1998.

[11] H. Soeleman, D. Somasekhar, and K. Roy. Idd waveform analysis for testing of domino and low voltage static cmos circuits. In *Proceedings of 8th Great Lakes Symp. on VLSI*, February 1998.

[12] S. Su and R. Makki. Testing random access memory by monitoring dynamic power supply current. *Journal of Electronic Testing: Theory and Applications*, 3(4):265–278, 1992.

[13] B. Vinnakota. Monitoring power dissipation for fault detection. *Journal of Electronic Testing: Theory and Applications*, 11(2):173–181, October 1997.

[14] P. Yang, D. Hocevar, P. Cox, C. Machala, , and P. Chatterjee. An integrated and efficient approach for mos vlsi statistical circuit design. *IEEE Transaction on CAD*, 5(1):5–14, 1986.

Signal Integrity Fault Analysis Using Reduced-Order Modeling

Amir Attarha and Mehrdad Nourani
Center for Integrated Circuits & Systems
The University of Texas at Dallas
Richardson, TX 75083-0688
{attarha,nourani}@utdallas.edu

Abstract

This paper aims at analysis of signal integrity for the purpose of testing high speed interconnects. This requires taking into account the effect of inputs as well as parasitic RLC elements of the interconnect. To improve the analysis/simulation time in integrity fault testing, we use reduced-order modeling that essentially performs the analysis in the frequency domain. To demonstrate the generality and usefulness of our method, we also discuss its application for test pattern generation targeting signal integrity loss.

Categories & Subject Descriptors: *B.8.1 [Performance and Reliability]: Reliability, Testing, and Fault-Tolerance — signal integrity*
General Terms: *Design, Measurement, Reliability, Theory*
Keywords: *Integrity Fault, Locality Factor, Reduced-Order Model, Signal Integrity, Test Pattern Generation.*

1. INTRODUCTION

As we approach 100*nm* technology, noise and skew imposed by interconnects have emerged as main concerns in the interconnect design of gigahertz SoCs. Voltage distortion (noise) and delay violations (skew) contribute to signal integrity loss causing time-dependent dielectric breakdown (TDDB) [1], functional error [2], performance degradation [3] and reliability problems [4]. Therefore, new test development and coverage techniques for interconnects susceptible to noise and skew are substantial. Unfortunately, the existing methods analyzing the interconnect for integrity are either too slow or too inaccurate [5]. This motivated us to develop an analysis method that is not only accurate for purpose of integrity fault testing but also much faster than the existing methods.

Due to the complexity of the interconnect model, an efficient simulation method is essential for test pattern generation. We have adopted reduced-order model to alleviate computation complexity with slight loss of accuracy. Reduced-order model methods were developed as an alternative for circuit-level simulators to approximate the behavior of interconnect, power and clock networks [6].

On the application side of our method, we use the model order reduction methods such as the PVL [7] and SyMPVL [8] to evolve a test pattern generation approach for detecting intermittent failures due to integrity fault on long interconnects. These methods are chosen because of their high accuracy, numerical stability and ability to identify the true poles and zeros of the original system efficiently. Poles and zeros are the roots of denominator and numerator of the system transfer function, respectively [9].

This paper is organized as follows. In Section 2, we show how the reduced-order model can be used to partially analyze the interconnect network in frequency domain and how integrity loss problem (e.g. signal overshoot and skew) is formulated. In Section 3 we comment on few techniques that can further improve the integrity fault analysis in terms of running time. Section 4 discusses application of our analysis tool in a test pattern generation method that targets integrity faults. The experimental results are presented in Section 5. Finally, concluding remarks are in Section 6.

2. SYSTEM REPRESENTATION

2.1 Effect of Interconnect on Integrity

The origin of signal integrity problems lies in the circuit interconnects. In deep-submicron SoC's, an interconnect behaves like a lossy transmission line. In low and mid-range frequencies, common in the past, the *RC* delays have been the dominating factors in the global interconnect delay and distortion. As technology shrinks, the effects of coupling capacitances and mutual/self inductances will increase significantly, jeopardizing the efficiency of the SoCs.

We present a new formulation for the interconnect network which facilitates various test analysis targeting integrity loss including pattern generation [10]. The most important novelty of this formulation is the fact that the effects of inputs are parameterized into the behavior of the interconnect model. This leads to a unique formula which annotates the effects of input patterns within the behavior of the interconnect network.

2.2 Input Representation

In a System-on-Chip (SoC), the interconnect drivers are designed very strong to diminish any possible signal disturbance [11]. Therefore, the transitions on the input signals can be reckoned as step functions. On the other hand, some of the signals are quiescent at "1" or "0" depending upon the driver circuitry. We define an input $x_i(t)$ of the interconnect network as:

$$x_i(t) = a_i u(t) + b_i$$

where $u(t)$ is a step function; a_i and b_i are constants such that $a_i \in \{-1, 0, 1\}$ and $b_i \in \{0, 1\}$.

2.3 System Representation

An interconnect network can be represented by a set of transfer functions in the S-domain [9]. As shown in Figure 1, an interconnect network with n inputs and m outputs is represented by transfer

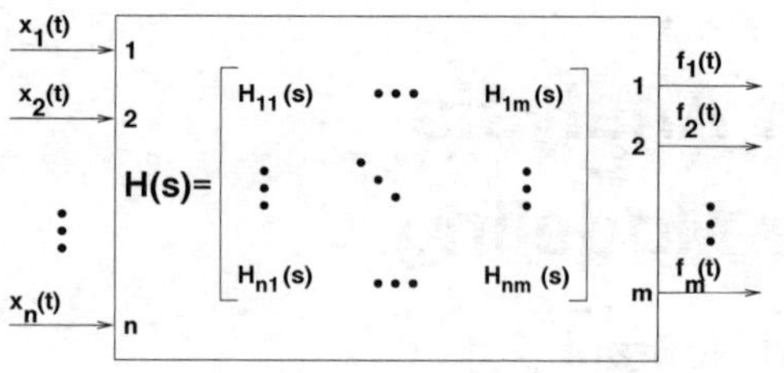

Figure 1: S-domain representation of interconnect network

function $H(s)$. $H(s)$ itself contains $n \times m$ partial transfer functions relating inputs to outputs. Note that, in general for an interconnect network, we have $m \geq n$ due to the possibility of fanout on some wires. The transfer function of a specific output f_r $(1 \leq r \leq m)$ can be represented as:

$$H_{f_r}(s) = \sum_{i=1}^{n} X_i(s) H_{ir}(s) \tag{1}$$

where $X_i(s)$ is the Laplace transform of $x_i(t)$ (the ith input).

Using an order reduction method (e.g. PVL [7] or ENOR [12]), the transfer function of output f_r becomes of order q:

$$H_{f_r}(s) = \sum_{i=1}^{n} \sum_{j=1}^{q} \frac{k_{ij}}{(s - p_{ij})} \tag{2}$$

where k_{ij} and p_{ij} are zeros and poles of the output f_r, respectively. Note that $H_{f_r}(s)$ is obtained based on the superposition of the effects of all n inputs on f_r. The output of a system can be computed in the frequency domain by multiplying the transfer function and the input function. After transferring it to time domain to observe the timing behavior of the output for $t > 0$, $f_r(t)$ is calculated by:

$$f_r(t) = \sum_{i=1}^{n} \left(a_i \sum_{j=1}^{q} \frac{k_{ij}}{p_{ij}} \left(e^{p_{ij}t} - 1\right) + b_i \sum_{j=1}^{q} k_{ij} e^{p_{ij}t} \right) \tag{3}$$

The practical use of Equation 3 is that if poles and zeros are available for a given set of inputs, $f_r(t)$ can be obtained efficiently using numerical methods. Therefore, it can be used to find patterns that cause maximal delay or overshoots. In providing this equation, we need to estimate q (the order of model reduction). Full discussion on empirical and heuristic methods to determine q is beyond the scope of this paper. We just point out that it plays an important role in the accuracy of the equation and computational complexity of the reduced model. We will briefly comment on such methods in the next section.

2.3.1 *Finding Input Patterns for Maximal Delay*

In Equation 3, all the timing information of the output signal is available and the impact of inputs $x_i(t)$ are considered within a_i and b_i coefficients. As the result, by defining delay as the time taken to reach certain voltage (e.g. 50% of V_{dd} or $V_d = V_{dd}/2$ as shown in Figure 2), the delay can be computed from the following equation:

$$f_r(t) - 0.5V_{dd} = 0 \tag{4}$$

Using numerical methods, we can solve Equation 4 to obtain the maximal delay t_d associated with an input set.

2.3.2 *Finding Input Patterns for Maximal Overshoot*

Overshoot occurs when a signal momentarily exceeds V_{dd}. In other words, overshoot is the global maximum of the output waveform of the system. Finding the global maximum of a nonlinear equation (e.g. Equation 3) is a complicated problem in general. However, overshoot is the maximum voltage value taking place in the neighborhood of the root of Equation 4. More specifically, the

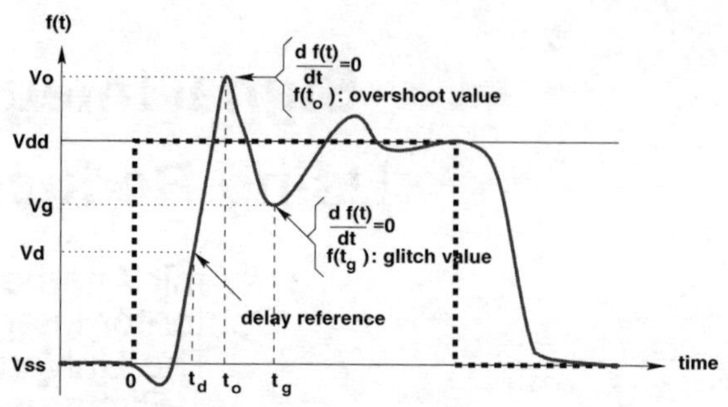

Figure 2: Calculating delay and overshoot values

neighborhood information (i.e. a point around t_d) has already been obtained by solving Equation 4. The following equation finds when the overshoot happens:

$$\frac{df_r(t)}{dt} = 0 \ni t \in (t_d, t_d + \Delta] \tag{5}$$

By solving Equation 5, t_o is found and the overshoot value ($f_r(t_o) = V_o$) is computed. Figure 2 graphically demonstrates the calculation of overshoot value.

2.3.3 *Finding Input Patterns for Glitches*

Glitches (ringings) occur when a signal momentarily falls below V_{Hmin}. Similar to what we explained for overshoots, glitches taking place in the neighborhood of t_o obtained already by solving Equation 5. The following equation finds when the glitch happens:

$$\frac{df_r(t)}{dt} = 0 \ni t \in (t_o, t_o + \Delta] \tag{6}$$

By solving Equation 6, t_g is found and the glitch voltage value ($f_r(t_g) = V_g$) is computed for further processing. Figure 2 demonstrates this graphically.

3. IMPROVING PERFORMANCE

Success of the integrity fault analysis model that we presented depends on numerical techniques that can solve the equations efficiently. In this section, we present few techniques that significantly enhance this analysis.

3.1 Numerical Methods to Solve Equations

Theoretically, we are able to analyze the output for any possible indication of signal integrity loss such as delay, overshoot, or undershoot. However, Equations such as 3, 4 and 5 must be solved efficiently. There are several numerical methods for solving nonlinear equations such as Bisection, Secant, and Newton-Raphson [13]. Although such numerical methods can solve these Equations, they may require long running time. Moreover, care must be given to the proper choice of a starting point that may not be easy to determine. If we are successful in choosing a good starting point, methods such as Newton-Raphson method converges quadratically to the root, as proven in [13].

3.2 Using SyMPVL Algorithm

The interconnect network is modeled as a large RLC circuit. Such a circuit can be described and characterized in terms of square and symmetric matrix transfer functions. Moreover, being composed of passive components, RLC networks are always stable and passive. In our work we used PVL/SyMPVL [7]/[8], an algorithm for the approximation of the symmetric multi-port transfer function of an RLC network. This algorithm takes advantage of these special structures of RLC networks. SyMPVL computes symmetric matrix-Pade approximations of the circuit matrix transfer function, using a symmetric block-Lanczos technique. It produces more accurate results compared to other models and even in special cases of RC and LC networks, guarantees stability and passivity at any order of approximation [8].

More specifically, in our application this method computes the reduced-order model for large multi-port RLC networks. This leads to considerably less execution time to provide Equation set 3. Instead of running PVL algorithm $n \times m$ (see Figure 1) and performing superposition, the desired matrices (e.g. poles and zeros) can be obtained by SyMPVL in one run. This significantly improves the preprocessing time to construct Equation set 3.

3.3 Defining the Order of Reduction

As discussed earlier, q (the order of model reduction) should be much smaller than the the order of original model. Choosing an appropriate q enhances accuracy of the model while keeping the computational complexity under control. Unfortunately, to the best of our knowledge, there is no analytical method to find the optimum value of q. In almost all works reported in the literature, the value of q is selected heuristically or based on trial and error [6] [8] [12]. The selection is done based on the application and desired level of accuracy and computation complexity. A key feature of the PVL [7] algorithm is that the accuracy of approximation can be guaranteed for a special frequency range. The authors in [7] have shown that a quality measure (Q) for the poles of $H(s)$ can be defined to determine q. Calculation of Q is based on defining bounds for eigenvalues of the reduced model. Finally, Q is used for determining the optimum order of reduction (q) so that to satisfy the desired accuracy. Details can be found in [7].

In our work, we have simply started from a very small q (e.g. 1/50 of the original order) and checked its accuracy by comparing to the simulation results of SPICE (i.e. the golden model). If the accuracy was above 8%, we have increased q and repeated the process. We acknowledge that such selection process is ad-hoc and may require few iterations. However, since the selection of q is done only once the effort, compared to the running time of the signal integrity analysis, is negligible. We are currently working on a mixed analytical-empirical heuristic method to choose the optimal (or near-optimal) value of q and we hope to report the outcome in near future.

3.4 Defining the Locality Metric

Considering the property of the interconnect network and the effect of parasitic elements, the search space of possible patterns can be restricted without decreasing the chance of finding effective patterns. To be more specific, we need to elaborate on the roles of different parasitic elements in the behavior of an interconnect network. In spite of coupling capacitances between all wires, the effect of capacitive coupling is considered local, in the sense that the coupling effects of adjacent wires are quite dominant compared to the capacitive coupling effects of far off wires [14] [15]. However, the inductance has larger range effect and thus the effect of mutual inductance could be significant. Furthermore, the effect of coupling inductances and capacitances on a wire oppose each other [14]. When the signal on a wire switches in one direction, the noise due to capacitive coupling affects other nearby signals in the same direction as that of switching. However, the noise due to inductive coupling is in the opposite direction.

To take advantage of above observation, we define the *locality* metric. Locality k is a parameter showing the proximity of neighboring wires in each side that actively affect one wire due to coupling capacitances. In other words, the victim is affected by the signals carried on k wires on each side of it mainly due to the repercussion effects of coupling capacitances. The signals traveling on wires beyond the locality border exert adverse effects on the victim mainly due to their mutual inductances.

Using the locality metric, we can effectively consider RC model for the victim wire and $2k$ wires next to it and RLC model for the rest. Practically, this reduces the complexity of the network and shortens the analysis time for constructing and solving Equation sets 3 and 5. Briefly, if the interconnect network contains n lines, the total number of patterns for integrity analysis shrinks from $n*4^n$ to $n*4^{n-2k-1}$, where 4 is the possible states for an input (falling, rising, quiescent at 0, and quiescent at 1). Also, the term n appears because the analysis of all patterns has to be done for all wires. Note that depending upon the targeted fault (overshoot/delay), the input of the specific line subject to fault is determined. We can further restrict the search space of patterns by exploiting the fact that the effects of mutual inductances oppose coupling capacitance effects. Beyond the locality region, the effects of mutual inductances are considered dominant. Therefore, the number of patterns can be significantly reduced, that is: $(n-2k)*4^{(n-2k-1)}+k\sum_{i=0}^{k-1}4^{(n-2k+i)}$, where, the summation term covers the asymmetry of k lines at the sides of interconnect network. Note that depending on the available information, test time budget and desired accuracy the candidate set of patterns will be chosen. For example, for application of test pattern generation, the method eventually selects a very small subset of the candidate set.

4. TEST GENERATION

The signal integrity model and analysis procedure presented in the previous section is a general technique that can be used in many signal integrity simulation and analysis applications. In this section we show how our technique can be used for signal integrity fault (unacceptable overshoot and skew) analysis. More specifically, we use this technique to find deterministic patterns that stimulate maximal integrity loss on the interconnects.

Conventional test pattern generation algorithms cannot be used for signal integrity faults due to fundamental differences between the nature of classic and integrity faults [10]. Classic faults such as stuck-at, bridging, and open faults, are often assumed to cause permanent adverse effects on system in terms of logic level of signals. On the other hand, integrity faults are intermittent, appear mostly as disturbances, heavily depend on the working frequency and are hard to categorize.

For generating the deterministic test patterns, accurate interconnect model is required which can be obtained using extraction tools such as NETAN [16]. Then, the locality factor k and the order of model order reduction q need to be decided. In this work, we use the PVL/SyMPVL algorithm [7]/[8] to get the reduced order model and find the poles and zeros. For a selected line, Equation 3 is formed. Then, the integrity loss regarding is calculated for an input pattern chosen from the candidate set. If the pattern causes integrity loss (i.e. overshoot and skew exceed the acceptable range) the pattern will be stored. Otherwise, another pattern will be tried until the whole candidate set is exhausted.

5. SIMULATION RESULTS

MATLAB [17] was used for generating the reduced order model and performing analytical computations. To test the accuracy, interconnects with various lengths and bitwidths are analyzed using our model and also simulated in SPICE [18]. The results of our reduced model were consistently within 8% of accuracy compared to SPICE simulation. Specifically, our method identifies the same

Line #	Time-Dependent Loss		Voltage-Dependent Loss	
	Avg. Iterations	*Delay [ps]*	*Avg. Iterations*	Overshoot [Volt]
1	424.6	25.4	208.7	2.76
2	320.0	32.9	132.3	2.69
3	129.0	25.8	159.3	2.91
4	106.3	31.3	84.4	3.04
5	13.7	52.7	61.0	3.19
6	276.3	33.5	131.7	2.73
7	180.3	41.5	70.7	2.85
8	168.3	28.3	365.4	2.63
9	1000	Not Found	330.7	2.71

Table 1: The results of integrity fault simulation

test patterns to stimulate the maximal integrity loss as SPICE finds but runs much faster by 2 to 3 order of magnitude.

We have employed an interconnect network containing nine parallel lines with RLC segments, coupling capacitances, and mutual inductances to perform the experimentations. Before starting the test pattern generation, we need to restrict the search space of possible test patterns for signal integrity faults. For instance, if maximum delay of $n = 9$ interconnect lines are targeted, the total number of possible patterns is equal to $9 * (4^9) = 2,359,296$ as we explained in previous section. While if we consider locality factor of $k = 2$, as described in Section 4, the number of patterns in candidate set will reduce to $(5 * 4^4) + (2 * 4^5) + (2 * 4^6) = 11,520$.

After defining k, the test pattern generation procedure is used. q is calculated along with the model order reduction. In our experimentation the selected order $q = 14$, while the original model is order of 100 since it contains more than 300 memory elements such as capacitances and inductances. The frequency and the source voltage are $f = 1GHZ$ and $V_{dd} = 2.5$ *Volt*. The thresholds for unacceptable skew, overshoot and ringing are assumed to be $T_d = 25$ *ps*, $V_o = 2.62$ *Volt* and $V_g = 1.75$ *Volt*, respectively.

Table 1 demonstrates the simulation results for capturing time-dependent and voltage-dependent integrity faults, respectively. As seen, the average number of MATLAB iterations to solve the equations numerically vary depending upon the layout information extracted and position of the line. The lines in the middle are more subject to coupling capacitance and mutual inductance interferences compared to the lines at (or close to) the side of interconnect network. Therefore, as a general trend, more efforts are required to find the desired test pattern at (or close to) sides because the candidate set is smaller. For instance, as Table 1 shows, MATLAB finds the test patterns for line 5 (in the middle) after only 14 iterations on the average while 424 iterations on average were needed to find the test patterns for line 1.

Note that in test generation, the patterns are chosen from the candidate set randomly. Additionally, the process is terminated after: 1) a test pattern causing unacceptable integrity loss (e.g. delay larger than 25 *ps* or overshoot larger than 2.62 *Volt*) is found; or 2) after 1000 MATLAB iterations. Obviously, such termination conditions can be revised to check more patterns from the candidate set or find more patterns violating the integrity thresholds. In general, there might be also some wires that no pattern can exceed T_d, V_o or V_g which means no unacceptable integrity loss will occur in that particular wire.

6. CONCLUSION

High-speed interconnects in deep-submicron need to be treated as transmission lines. This is a challenge for SoC testing due to the computational demand of distributed RLC model. We proposed an analysis technique based on the reduced-order model to improve the performance of signal integrity analysis. Our analysis method is quite general and can be employed by various integrity test methods. We applied this technique to find deterministic test patterns that can stimulate maximal integrity loss (overshoot and skew) in each line of the interconnect network.

Acknowledgements

This work was supported in part by the National Science Foundation CAREER Award #CCR-0130513.

REFERENCES

[1] W. Hunter, "The Statistical Dependence of Oxide Failure Rates on Vdd and Tox Variations with Applications to Process Design, Circuit Design and End Use," in Proc. *Reliability Physics Symposium*, pp. 72-81, 1999.

[2] H. Chen and L. Wang, "Design for Signal Integrity: The New Paradigm for Deep-Submicron VLSI Design," in Proc. *Intern. Symp. on VLSI Tech.*, pp. 329-333, 1997.

[3] P. Nordholz, D. Treytnar, J. Otterstedt, H. Grabinski, D. Niggemeyer and T. Williams, "Signal Integrity Problems in Deep Submicron Arising from Interconnects Between Cores," in *Proc. VLSI Test Symposium (VTS'98)*, pp. 28-33, 1998.

[4] M. Nourani and A. Attarha, "Detecting Signal Overshoots for Reliability Analysis in High-Speed SoCs," to appear in *IEEE Trans. on Reliability*, 2002.

[5] A. Kahng, S. Muddu and E. Sarto, "Interconnect Optimization Strategies for High-Performance VLSI Designs," in *Proc. Intern. Conference on VLSI Design*, pp. 464-469, Jan. 1999.

[6] A. Odabasioglu, M. Celik, and L. Pileggi, "PRIMA: Passive Reduced-Order Interconnect Macromodeling Algorithm," *IEEE Conf. on CAD* , 1997.

[7] Peter Feldmann, and Ronal W. Freund, "Efficient Linear Circuit Analysis by Pade Approximation via the Lanczos Process," *IEEE Trans. on CAD*, vol. 14, no. 5, pp. 639-649, 1995.

[8] R. Freund and P. Feldmann, "Reduced-Order Modeling of Large Linear Passive Multi-Terminal Circuits Using Matrix-Pade Approximation," in Proc. , 1998.

[9] Alan V. Oppenheim, Signals and systems, second Edition, Prentice Hall, 1997.

[10] A. Attarha, and M. Nourani, "Test Pattern Generation for Signal Integrity Faults on Long Interconnects," in *Proc. VLSI Test Symposium (VTS'02)*, April 2002.

[11] J. Rabaey, *Digital Integrated Circuits*, Prentice Hall, 1996.

[12] Bernard N. Sheehan, "ENOR: Model Order Reduction of RLC Circuits Using Nodal Equations for Efficient Factorization," *Design Automation Conference*, pp. 17-21, 1999.

[13] E. W. Cheney, *Numerical Mathematics and Computing*, Brooks Cole publishing Company, 1985

[14] C. Cheng, J. Lillis, S. Lin and N. Chang, *Interconnect Analysis and Synthesis*, John Wiley & Sons, 2000.

[15] L. He, N. Chang, S. Lin and O. Nakagawa, "An Efficient Inductance Modeling for On-chip Interconnects," *IEEE Custom Integrated Circuits Conference*, pp. 457-460, 1999

[16] NET-AN: Multi-net three-dimensional field solver extraction tool, Users reference manual, OEA International, Inc., 2001.

[17] MATLAB User's and reference manual, *2000 The MathWorks, Inc., Version 6.0.0.88*,2000.

[18] TI-SPICE3 User's and reference manual, *1994 Texas Instrument Incorporation*, 1994.

Enhancing Test Efficiency for Delay Fault Testing Using Multiple-Clocked Schemes

Jing-Jia Liou, Li-C. Wang and Kwang-Ting Cheng
Department of ECE, UC-Santa Barbara
jjliou@windcave.ece.ucsb.edu, {licwang, timcheng}@ece.ucsb.edu

Jennifer Dworak and M. Ray Mercer
Department of EE, Texas A&M University
jdworak@dropzone.tamu.edu, mercer@ee.tamu.edu

Rohit Kapur and Thomas W. Williams
Synopsys Inc., {rkapur, tww}@synopsys.com

ABSTRACT

In conventional delay testing, the test clock is a single pre-defined parameter that is often set to be the same as the system clock. This paper discusses the potential of enhancing test efficiency by using multiple clock frequencies. The intuition behind our work is that for a given set of AC delay patterns, a carefully-selected, tighter clock would result in higher effectiveness to screen out the potential defective chips. Then, by using a smarter test clock scheme and combining with a second set of AC delay patterns, the overall quality of AC delay test can be enhanced while the cost of including the second pattern set can be minimized. We demonstrate these concepts through analysis and experiments using a statistical timing analysis framework with defect-injected simulation.

Categories and Subject Descriptors

B.8.1 [**Hardware**]: Reliability, Testing, and Fault-Tolerance

General Terms

Experimentation, Measurement, Reliability

Keywords

Delay Testing, Statistical Timing Analysis, Transition Fault Model

1. INTRODUCTION

Suppose a circuit consists of n sites, numbered as $1 \dots n$. Without loss of generality, assume that by applying a set A of 100% transition fault patterns, $l_1 \dots l_n$ are the longest sensitized paths passing through site $1 \dots n$, respectively. Let C be the test clock period and let M be a timing analysis model where $M(l_i)$ represents the delay time returned by the model for path l_i. Then, after the application of A, for each site i, any delay defect with a size greater than $C - M(l_i)$ can be captured. For the whole circuit, any single-site delay defect with a size greater than $C - \min\{M(l_1) \dots M(l_n)\}$ can be captured. For enhancing the quality of transition fault testing, it is often suggested that we should maximize each $M(l_i)$.

In traditional timing analysis, model M is often based upon nominal or worst-case timing assumptions. In the deep sub-micron domain, factors that affect timing characteristics such as process variations, manufacturing defects and noise are often statistical in nature [1, 2]. These factors should be better captured and simulated using statistical models and methods [3, 4]. Hence, the outputs of $M(l_i)$ and $M(p_i)$ in the above analysis can be probability distributions rather than discrete delay values.

With M being a statistical timing analysis model, our focus does not lie in maximizing each $M(l_i)$ for transition fault patterns by selecting the longest l_i for explicit testing. Instead, we focus on achieving the following objective: We assume that two sets of transition fault patterns A and B are given (similarly, assume that A covers paths $T_1 = \{l_1^1 \dots l_n^1\}$ and B covers paths $T_2 = \{l_1^2 \dots l_n^2\}$). Test quality is enhanced by combining the two pattern sets. This is because for each site i, any delay defect with a size greater than $C - \max\{M(l_i^1), M(l_i^2)\}$ can be captured. Then, our goal is to find a clock scheme such that the test application cost is minimized relatively to the cost of applying only one set, and the quality is maximized relatively to the quality of applying both sets.

In our method, the application of tests is divided into two phases: *screening phase* and *confirmation phase*. In the screening phase, the goal is to use the first set of patterns with a tighter clock scheme to screen out *potential* defective chips. Then in the confirmation phase, both sets of patterns are applied with the normal clock to confirm which ones are actually bad. The key of this strategy is the selection of clock scheme in the screening phase in order to minimize the cost during the confirmation phase. The cost saving relies on the fact that majority of the chips are good and hence, only a few would be processed during the confirmation phase.

Several publications [5, 6, 7, 8] have tried to use variable or multiple clock schemes to improve the detection of delay defects. Unfortunately, there is one serious drawback of these schemes: the false negative outcomes. It is possible that most "bad chips" classfied by these schemes are functionally and timingly correct, because most delay defects that increase the delays of some paths do not actually locate at timing critical paths. This problem will result in unwanted increase in the defective part level and decrease of the total yield (yield loss).

2. CLOCK SCHEMES

It is important to note that given the two sets A and B, alternating the clock itself will not improve the quality beyond that obtained using both A and B with the normal clock. Hence, $A + B$ combined with the normal clock will represent an upper bound (in quality) of using any multiple-clocked scheme. This is because to avoid yield loss, it is necessary to apply again both A and B at the end with the normal clock to filter out good chips that were considered to be defective under a tighter clock. Hence, the total number of defective chips captured cannot be greater than that captured by all tests in $A+B$ combined with the normal clock.

2.1 Initial Clock Scheme

The above discussion implies a very simple 2-clock scheme. During the screening phase, a tighter clock $C-\delta$ is used with the test set A. During the confirmation phase, the normal clock is used with the combined test set $A+B$. It is important to note that path delay tests, if available, can be used for set B. However, set A has to target on transition faults. This is because a tighter clock in the screening phase does not help to capture a defect falling beyond the topological coverage of the paths sensitized by A. Hence, to ensure high quality, it is necessary to use a screening pattern set that can provide a complete topological (site) coverage.

Defect Miss A chip passing the screening phase may be defective and still can be captured if B is applied with the normal clock. However, it is guaranteed that the pass chip will also pass if A is applied with the normal clock. Based upon these points, it is possible that some defects are missed under the new test application scheme, but they can be captured by the original test method where $A+B$ is applied with the normal clock.

How likely will this happen? Consider a given site i. A true defect (which can affect circuit performance) on i will be missed if its size d falls into the range $[C-M(l_i^2),\ C-\delta-M(l_i^1)]$. This implies $M(l_i^1)+\delta < C-d < M(l_i^2)$. With a reasonably large δ, it is unlikely that $M(l_i^2) > M(l_i^1)+\delta$. Even with this happening, the chance of a defect occurrence with a size that falls exactly into the range is small. Hence, the problem of defect miss is not a serious concern.

Potential Yield Loss (Test Cost) Yield loss is defined as the percentage of chips which are considered to be defective by the tests but are actually good chips. To avoid yield loss, in the confirmation phase $A+B$ combined with the normal clock has to be applied again to test all potentially defective chips identified in the screening phase.

Again, how likely will potential yield loss happen? Since we have already use $A+B$ and the normal clock to prevent this from happening, the key here is that if a good chip is passed to the confirmation phase, it will increase our testing cost and hence, we want to minimize that as much as possible.

Similarly to the defect miss situation, a defect on site i will result in potential yield loss if its size d falls into the range $[C-\delta-M(l_i^1),\ C-\max(M(l_i^1),\ M(l_i^2))]$. This implies that $\max(M(l_i^1),\ M(l_i^2)) < C-d < \delta+M(l_i^1)$. Hence, to minimize potential yield loss after the screening phase (the test cost), δ should be as small as possible.

The Selection of δ

The above analysis suggests two contradictory objectives to select δ. On one hand, δ should be large to avoid defect miss. On the other hand, δ should be small to minimize the cost resulting from the effort to avoid yield loss during the confirmation phase. By combining the two objectives, it seems that the most logical solution is to set $\delta_i = D_i = |\ M(l_i^1) - M(l_i^2)\ |$ for each site i.

In practice, we can simply assume δ to be a small constant to begin with. Then, through statistical analysis and simulation, adjust δ to an acceptable level such that the cost to avoid yield loss during the confirmation phase is within the tolerance. From this perspective, δ is set based upon the cost we can afford. The larger the δ is, the better the quality will be and the higher the cost we will pay.

2.2 Revised Clock Scheme

If only one clock is used during the screening phase, the above iterative method provides a feasible way to approach the desired δ. Another alternative is to employ multiple clocks during the screening phase as well. Suppose we are allowed to use k clocks in the screening phase, what will be a good strategy to apply these k clocks? To answer this question, we first need to know how to partition the first test set. Then, we need to decide what clock to be used for each subset of patterns. Our overall strategy is illustrated in the following.

Test Set Partitioning Given a set of patterns, we first call the statistical timing analysis tool M [3] to characterize the delay distribution of the longest sensitized path passing through each site (again denote these results as $M(l_1^1)\dots M(l_n^1)$). Next we compute the mean value in each distribution (assume that the order is $m(l_1^1) < \dots < m(l_n^1)$). Then, the partition is based upon the difference $D = m(l_n^1) - m(l_1^1)$. The first partition is to divide $[0,D]$ into two ranges $D_1 = [0, \varepsilon D]$ and $D_2 = [\varepsilon D, D]$ where ε denotes the *ratio of partition* and $0 < \varepsilon < 1$ (normally, $0.5 < \varepsilon$ because short paths are less important than the long paths). For all paths whose m values falling into the first range will be grouped into the first set S_1. The rest will be grouped into the second set S_2. Then, the same idea can be applied recursively to partition S_2 further, and so on. In section 3.2, we will discuss the selection of ε so that the best performance can be achieved.

Clock Selection Given a subset S_i of paths, we extract the patterns to cover S_i. Then, for those patterns, we determine the path set S_i' being covered. Based upon the path set S_i', statistical timing analysis is performed to characterize the delay distribution of the longest path. This distribution is often normal and represented by its mean and standard deviation as (μ, σ). Then, the test clock for this subset of patterns will be set as $\mu + c\sigma$ where c is a small constant. Usually, it is reasonable to have $c = 3$.

2.3 Cost Evaluation

To enable us to compare different clock schemes, a cost evaluation method is required. Given the two sets of patterns A and B, intuitively we may define the cost as $A+\rho(A+B)$ where ρ represents the probability of a good chip being classified as being potentially defective during the screening phase.

As discussed before, this probability ρ depends on the clock(s) selected during the screening phase. However, the true cost is not entirely reflected in the simple equation $A+\rho(A+B)$. This is because we should also take defect size distribution into account. Without loss of generality, we assume defect distribution is a discrete function Γ. Then, for each defect size s, $\Gamma(s)$ gives the conditional probability that if a defect does occur, the chance of being with the size s. Then the true cost should be re-formulated as:

$$Cost = A + [\sum_{\forall s} \Gamma(s) \times \rho_s \times (A+B)] \qquad (1)$$

where s is the defect size and ρ_s is the probability of a good chip being classified as a defective chip during the screening phase due to a defect with size s. In essence, equation (1) calculates the cost by averaging across the defect distribution curve. If the defect distribution is uniform, then the equation will be reduced to $A + \rho(A+B)$ again where $\rho = \sum_{\forall s} \rho_s$.

2.4 Cost Reduction

The cost can be further reduced if we allow an extremely small yield loss tolerance level. For example, suppose for large s where $s > t$, $\Gamma(s) < \varepsilon$ where ε is a very small number such as 10^{-4}. Then, all short paths of which timing lengths less than $C - t$ (with high probabilities) in A can be removed during the confirmation phase. This is because these paths will provide minimal help in deciding that a potential defective chip is actually a good chip. Accordingly, we can remove all patterns that cover only those removed short paths. In the next section, we will demonstrate that if Γ is an exponential distribution, a large numbers of paths in A can be removed in the confirmation phase with very little increase (almost zero) of the yield loss probability. As a result, the cost of applying our proposed test scheme can be dramatically reduced.

3. EXPERIMENTAL RESULTS

3.1 Evaluation Framework

In order to evaluate the cost and effectiveness of each clock scheme with different pattern sets, we implemented an evaluation framework to estimate the test coverage based on defect injection in the statistical timing analysis model.

Figure 1 illustrates the complete procedure of our evaluation scheme for a particular clock and pattern set. Note that the pattern set is characterized using the set of paths S sensitized by the patterns. In each Monte Carlo sampling run, first a circuit instance with cell/interconnect delays is generated according to the delay distributions characterized through Monte Carlo SPICE. This instance will then be evaluated by "statistical analysis of S". The "statistical analysis of S" is to check if there is any path in S (on the given instance) longer than the testing clock C. If there is, then this instance is said to be faulty and covered by S (*Covered*). At the end, our scheme will calculate the probability of a faulty path captured by S based upon all the instances statistically produced. This *capture probability* is defined as $Covered/(\#\ of\ instances\ generated)$.

Note that at the beginning of each Monte Carlo run, we will inject a delay defect randomly at a chosen signal node. This is to simulate the randomness of spot defects that cause chips to fail. Also note that we use a statistical cell/interconnect delay model to simulate the variation of each path delay. For multiple-clocked scheme, the evaluation process is similar to the above, but for those circuit instances classified as *Covered* during the screening phase, they will further be tested in the confirmation phase. Then, at the output of the process we will obtain the number of instances which have to be further tested, and the final number of *Covered* after the confirmation phase.

In our experiments, we utilize a Monte Carlo SPICE (ELDO) [9] to extract the statistical delays of cells for 0.25μm, 2.5V CMOS technology. The input transition time and output loading of the cells are used as indices for building/accessing these libraries.

3.2 Results

In the following, we will first explain in detail our results obtained for circuit s5378. In Figure 2, we show the effectiveness of combining two transition pattern sets (Note that since our objective here is to compare the relative strength among different methods, the absolute values of these probabilities are not important and hence, to avoid confusion they are not marked). The quality level of the combined set is improved (relatively to the individual set alone) as expected. However, test cost will be dramatically increased by applying the two test sets together.

Figure 3 shows how the test quality is improved by using multiple-clocked schemes. When using the initial clock scheme, δ is set to

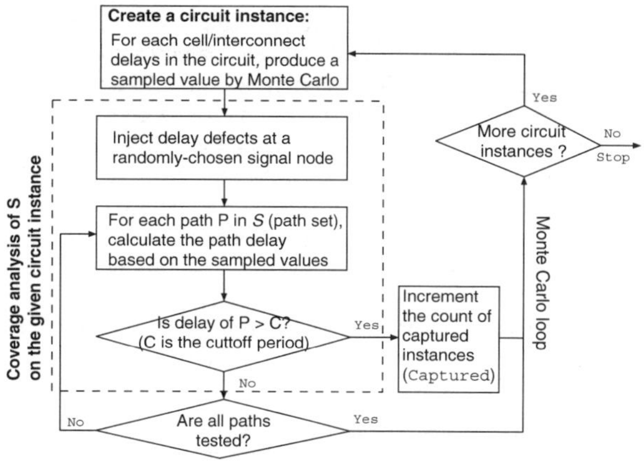

Figure 1: Flow Chart for Statistical Evaluation of *S*

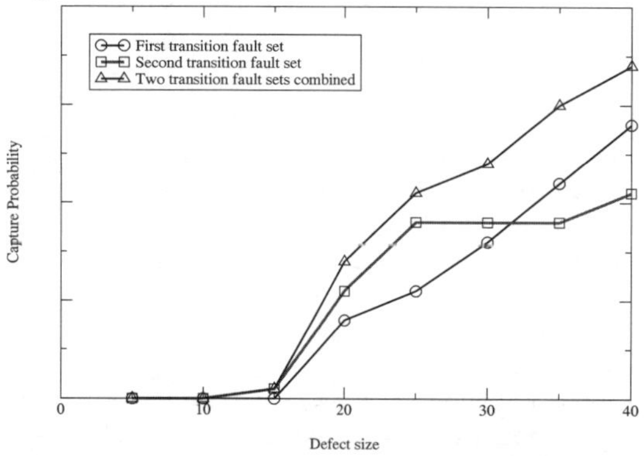

Figure 2: Defect Capturing Probabilities for Double Transition Pattern Sets

10 units. When using the revised clock scheme, the partition number is set to 3 (three different clocks are applied to three subsets during the screening phase). And the ratio of partition ε (as defined in section 2.2 above) is set to $1 - \frac{1}{e}$. Note that this particular ratio is obtained through experiments. We discovered that a smaller number such as 0.5 or a larger number such as 0.75 both would deliver inferior results.

In both figures, the first transition pattern set is used in the screening phase. As demonstrated in the figures, both the initial clock scheme and the revised clock scheme discussed in section 2 result in test quality levels that are bounded by the first transition pattern set and the *combined* transition set. However, the revised clock scheme provides a much better quality that is close to the combined set.

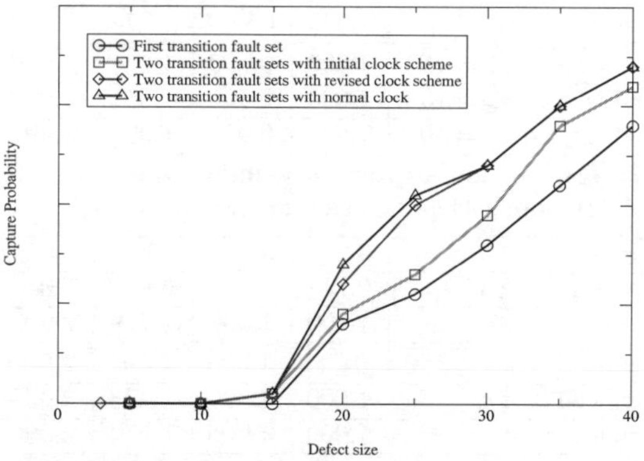

Figure 3: Defect Capturing Probabilities for Transition Pattern Sets Using Different Clock Schemes

Table 1 demonstrates the capture probabilities and costs for different testing schemes. The costs are calculated by the formula in section 2.3 and based on the assumption of a defect size distribution: $\lambda e^{-\lambda s}$ where s is the defect size and λ is a constant (we use 0.1 in the experiments). This exponential distribution for defect size (given that defects occur) has been studied in many publications [10, 11] and is a practical assumption to be used. Note that it is also possible to adopt other distributions. However, using other distributions in general does not invalidate the trends observed in our work.

To simplify the presentation, we will use the following notations in the table to represent different test schemes discussed above.

S1	first transition pattern set
S2	second transition pattern set
S3	first and second transition pattern sets with revised clock scheme
S4	first and second transition pattern sets combined with the normal clock

	S1	S2	S3	S4
capture prob.	0.60	0.77	**0.95**	1.0
cost (%)	1.000	1.006	1.274	2.006
reduced cost $t = 150$	1.000	1.006	1.113	2.006
reduced cost $t = 100$	1.000	1.006	1.027	2.006
reduced cost $t = 50$	1.000	1.006	**1.002**	2.006

Table 1: Test cost and capture probability comparison for different delay fault sets and clock schemes for s5378.

In the table, we also consider the cost reduction approach discussed in section 2.4. The t value in each case is shown in the table. Note that by assuming defect distribution as $\lambda e^{-\lambda s}$ ($\lambda = 0.1$), the probability of yield loss is close to 0.000676 for the worst case shown in the table (where $t = 50$ units). We emphasize that the absolute numbers in the table are less important than their trends.

Data in this table confirm various observations discussed above. For example, the quality of S3 (0.95) is close to that of S4 (1.0) and is greatly improved from S1 (0.60). Now given costs listed in the next three rows in the table, we can compare quantitatively the costs of different testing schemes. Here the capture probabilities are normalized with respect to that of S4, and the costs of schemes are normalized with respect to the cost of S1. Notice that in the case of cost reduction in S3, by linearly decreasing the t value, the cost can drop in terms of an order of magnitude.

The revised clock scheme S3 requires less testing effort than S4. Also S3 has the flexibility to reduce the cost by allowing a yield loss tolerance level. For the tolerance level of $t = 50$, the cost of S3 is only 1.002 which is almost the same as the cost from applying the first transition pattern set alone! We also have similar observations from Tables 2 and 3.

	S1	S2	S3	S4
capture prob.	0.75	0.8	**1.0**	1.0
cost (%)	1.000	0.990	1.274	1.990
reduced cost $t = 150$	1.000	0.990	1.221	1.990
reduced cost $t = 100$	1.000	0.990	1.061	1.990
reduced cost $t = 50$	1.000	0.990	**1.002**	1.990

Table 2: Test cost and capture probability comparison for different delay fault sets and clock schemes for s9234.

	S1	S2	S3	S4
capture prob. (%)	0.71	0.76	**0.95**	1.0
cost	1.000	1.000	1.779	2.000
reduced cost $t = 150$	1.000	1.000	1.406	2.000
reduced cost $t = 100$	1.000	1.000	1.233	2.000
reduced cost $t = 50$	1.000	1.000	**1.034**	2.000

Table 3: Test cost and capture probability comparison for different delay fault sets and clock schemes for s35932.

4. CONCLUSIONS

In this paper we discuss using multiple-clocked schemes to improve delay fault testing. We observed that applying a tighter clock alone does not help to improve delay test quality. Hence, we proposed a testing approach that consists of two test application phases: the screening phase and the confirmation phase. During the screening phase, the goal is to quickly identify those chips that are potentially defective. This goal is accomplished by using tighter clock(s) with an initial set of patterns that ensure a complete topological coverage. In the confirmation phase, a different pattern set is added to the original set for quality enhancement, and by testing with the normal clock, potential defective chips will further be confirmed as true defective chips.

Our two-phased multiple-clocked test application approach can deliver test quality close to that provided by combining the second pattern set and the original pattern set, with the cost close to that from the application of the original set alone. For the screening phase, we discuss two different clock schemes and conclude that the revised scheme is better than the initial scheme in quality. We studied our proposed methods based upon defect injection and simulation using a statistical timing analysis framework.

5. REFERENCES

[1] K. Baker, G. Gronthoud, M. Lousberg, I. Schanstra, , and C. Hawkins. Defect-Based Delay Testing of Resistive Vias-Contacts, A Critical Evaluation. *Proceedings of IEEE International Test Conference*, pages 467–476, September 1999.

[2] M. A. Breuer, C. Gleason, and S. Gupta. New Validation and Test Problems for High Performance Deep Sub-Micron VLSI Circuits. *Tutorial Notes, IEEE VLSI Test Symposium*, April 1997.

[3] J.-J. Liou, K.-T. Cheng, and D. Mukherjee. Path Selection for Delay Testing of Deep Sub-Micron Devices Using Statistical Performance Sensitivity Analysis. *Proceedings of IEEE VLSI Test Symposium*, pages 97–104, April 2000.

[4] J.-J. Liou, A. Krstić, K.-T. Cheng, D. Mukherjee, and S. Kundu. Performance Sensitivity Analysis Using Statistical Methods and Its Applications to Delay Testing. *Proceedings of Asian South Pacific Design Automation Coneference*, pages 587–592, January 2000.

[5] W. W. Mao and M. D. Ciletti. A Variable Observation Time Method for Testing Dlay Faults. *Proceedings of Design Automation Conference*, pages 728–731, June 1990.

[6] V. S. Iyengar and G. Vijayan. Optimized Test Application Timing for AC Test. *IEEE Transactions on Computer-Aided Design of Integrated Circuits and Systems*, 11(11):1439–1449, November 1992.

[7] D. Dumas, P. Girard, C. Landrault, and S. Pravossoudovitch. Effectiveness of a Variable Sampling Time Strategy for Delay Fault Diagnosis. *Proceedings of European Design and Test Conference*, pages 518–523, March 1994.

[8] W. B. Jone and Y. P. Ho. Delay Fault Coverage Enhancement Using Variable Observation Times. *Journal of Electronic Testing: Theory and Applications*, 11(2):131–146, October 1997.

[9] Anacad. Eldo v4.4.x User's Manual. 1996.

[10] N. N. Tendolkar. Analysis of Timing Failures Due to Random AC defects in VLSI moduels. *Proceedings of Design Automation Conference*, pages 709–714, June 1985.

[11] J. P. de Gyvez. *Integrated Circuits Defect-Sensitivity: Theory and Computational Models*. Kluwer Academic Publishers, Boston, MA, 1993.

Going Mobile: The Next Horizon for Multi-million Gate Designs in the Semi-Conductor Industry

Christian BERTHET
STMicroelectronics
BP217, 38019 Grenoble
Cedex,France

ABSTRACT

The complexity of a System-on-Chip design is not only in the million transistors packed in a square millimeter. The major challenge for technical success of a SoC is to make sure that millions lines of software fit in with millions gates.

In this paper, the problematic of multi-million gate design is illustrated from the viewpoint of a practical development of a complex digital system done at STMicroelectronics for a GSM/GPRS cellular application.

Categories and Subject Descriptors

C.3 [**Computer Systems Organisation**]: Special-purpose and application-based systems – *real-time and embedded systems.*

General Terms

Design.

Keywords

SoC Design, HW/SW co-design.

1. INTRODUCTION

For many people, the main complexity of a SoC is in the millions of transistors packed in a few square millimeters. This is partially true. The large majority of SoCs are made of both HW and SW: behind a multi-million gate chip, there are multi-million lines of C or assembly code...The former without the latter has less and less meaning! This general trend is confirmed by recent studies showing that out of a million engineers world wide working on embedded systems, there are eight times more engineers in SW design compared to HW design [1].

The content of the paper is organized as follows : In Section 2, we detail the general context of a multi-million gate SoC design. In Section 3, we present the problematic and raise some issues of SoC design based on our experience. In Section 4, we analyze some peculiarities of SoC design for mobile applications. Finally, in conclusion, we address the role of vertically integrated semi-conductor companies in this challenging context.

2. MULTI-MILLION GATE SOC

2.1 HW Components of a SoC

SoCs are generally made of several embedded processors, each one running control code or DSP code or both. They run either frozen code in ROM (sometimes called "firmware") or execute code located for example in external flash memories. It may happen that they form a true multi-processor system (i.e. with interactions, shared memories and/or shared peripherals). Each processor is usually associated with a set of peripherals such as DMA, interrupt controller, power and clock controller, etc... On-chip memories are ROM and RAM arrays such as caches and tightly coupled memories or slower memories. Systems-On-Chip also implement blocks of dedicated logic such as hard-wired accelerators and co-processors (e.g. Java), decoders (e.g. MPEG), dedicated functions (e.g. modem blocks), standard/proprietary IO functions (e.g. USB, UART, etc...). In general, these blocks are memory-mapped in a processor space.

2.2 SW Components of a SoC

The main categories of embedded software are OS (or kernels), drivers of HW blocks and IO peripherals, and applications such as communications (e.g. protocol stack and physical layers) or multimedia (e.g. MP3, MPEG4) software. The key factors of the software implementation are essentially the real-time behavior (in particular for OS, drivers, interrupt handlers, etc ..) and the performance of the processor and memory systems.

2.3 Complexity of SoC

For a SoC design using state-of-art .13 micron CMOS technology, gate density is in the range of 200Kg/mm^2 and SRAM density in the range of 500Kbit/mm^2. A complete SOC with say, 5Mgates and 8Mbits of SRAM, or about 100 millions transistors, will be in the range of 50 mm^2 (with pads, PLL, routing margin, etc ..).

In order to manage this complexity, a large part of the HW design work is based on reuse, up to 90% according to recent surveys [2] which confirm the effort of IP packaging and delivery methods in the semiconductor industry. In the SW area, reuse is still in its infancy, specially for low-level SW (physical layers).

2.4 Example of SOC Design: STEP-GSM

The STEP-GSM chip is a 20 Million transistor SoC design implemented in .18 micron technology. It is a digital base band device for 2.5G GSM/GPRS cellular applications, integrating two processor cores, a high-performance ST120 DSP/MCU core and an ARM720 MCU with cache. It also implements dedicated blocks of logic and high-speed SRAM banks for the cores (caches and fast memories). The complete HW and physical layer SW were developed by ST and the GSM and GPRS class 12 functionalities were validated using a commercial protocol stack.

As the saying goes, "today's SOC is tomorrow's IP", this base band will serve as a foundation when integrated in a more complex 3G system, together with a 3G modem, a multimedia engine, and other digital functions currently developed in isolation such as Bluetooth or GPS processors. Adding all these functionalities on a single die will give the kind of complexity mentioned above.

3. PRACTICAL ISSUES OF SOC DESIGN

Different characteristics of SoC design are detailed in the following paragraphs. Although they pertain to a given project, it is the feeling of the author that they can be of interest to others.

3.1 HW/SW Co-Design and Platforms

As for many SoCs, HW and SW are designed in parallel, no one can afford to wait for the HW before starting the SW development. However, they do not use the same vehicle all along the design process due to the different levels of description/validation of the SW and HW architectures, and their availability in time.

During the STEP-GSM design, HW and SW used two distinct flows, sometimes converging and sometimes diverging. Looking back at the chronology, three main phases took place: Phase 1 is the architecture definition with very strong HW/SW interactions, Phase 2 is the development phase using separate platforms with weak interaction, and Phase 3 is the convergence on a fast-prototyping platform with extremely strong interactions.

Combined with a very secure and safe silicon design methodology, this process has shown a very fast final integration time: the porting of the entire SW on the first STEP-GSM silicon took only 4 days to perform an over-the-air call, after assembly and testing of the silicon and application board.

3.1.1 HW/SW Architecture Definition

The initial specification phase requires many interactions between the SW and the HW teams. However the focus is not the same.

SW architects are mainly interested in the algorithmic aspects of the applications (communications, multimedia) using algorithmic tools such as Matlab, and consequently, in defining the range of performances required from the HW components.

In parallel, for HW architects, the emphasis is put on refining the required performances of the processor systems (core mips, bus bandwidth, latency of on-chip and off-chip memories, etc...) and also on estimating the resulting silicon in terms of speed, area, and power consumption. Typical tools are abstract models of the SoC architecture (e.g., C threads, which can be or not mapped on the targeted cores), and spreadsheets for the silicon estimates.

The decision process to allocate specific algorithms to HW blocks rather than SW is a kind of "meet-in-the-middle" compromise, based on application and system global performances, availability of existing IP's, availability of competent resources, silicon area and power consumption. Convergence occurs when an agreement is found on the HW/SW decomposition of the SoC translated into an architecture manual which is a kind of "wedding" contract. Then, after signature, "divorce" can really start... before reconciliation.

3.1.2 SW Development Platform

The SW development of STEP-GSM was basically done using a PC-based platform, thanks to the performances of PC processors: A 1GHz Pentium under Real Time Linux is easily able to emulate the execution of C code running on an embedded 52MHz CPU and a 104 MHz DSP (the upper bound of the application requirement, in our case GPRS Class 12).

The main advantages of this platform are that it is available sooner than a prototype, it is hardware independent (i.e., the hardware constraints are postponed to later steps: no need to care about memory footprint, power consumption, memory access times, external protocols, etc ..). Also it can be used to easily migrate pieces of code from one core to the other one, since SW allocation on cores is not yet done.

The Pentium processor also communicates with a DSP/FPGA off-the-shelf board on a PCI interface to handle real-time connections to the external world, e.g. radio interface, etc.... It also emulates the actual system tracing and application SW debuggers thru UART ports, with the exact same tools as on the final board.

At this stage the Pentium execution is faking the SW partitioning on the different cores, however the SW architecture is still an abstract view of the SW, it does not even use the target OS and the drivers are very different from the final ones. The platform is powerful enough to run the whole SW architecture including the protocol stack, and it was validated up to a GPRS call.

3.1.3 HW Development Platform

In parallel, the HW development of the STEP-GSM is an RTL model with a complete testbench, developed using VHDL and Verilog simulators on Unix workstations. It covers all the specification levels, from SoC architecture definition down to the RTL sign-off. This model targets first FPGAs for fast-prototyping, and second a CMOS library for the final silicon

Interaction with the SW was limited to the definition of tests scenarii and occasional (sometimes tough) discussions in front of the "solicitor" about possible changes in the "wedding" contract.

3.1.4 Fast-Prototyping Platform

The convergence of HW and SW occurred on the fast-prototype FPGA platform as soon as both HW and SW have reached an acceptable level of maturity. Two generations of platform were used: first, an Aptix machine and then, a dedicated board (more easily duplicated and portable). Each one basically uses bonded-out chips for the processor cores (when they are not synthesizable) and two Virtex-1000E FPGA from Xilinx for the rest of the logic, possibly with dedicated memory extensions.

This platform is the basic tool to find deep bugs in the HW. The turn-around time is very fast at this stage of development: once the bug is identified, it's a matter of few minutes for a SW fix and few hours for a HW fix, nothing compared to the cost of a mask set and the duration of a re-spin in foundry. More complex micro-architecture changes can be handled without major impact.

Main purpose of this platform is to perform the SW integration. At this stage, the SW is allocated on the different cores, using the target OS for each core, and this covers the partitioning assembly/C as well. Embedded processors' debuggers and trace tools for processors and system debug are extensively used (tracing dedicated events, bus access, interrupts) using JTAG and dedicated trace ports, in the same way as on the final silicon.

But it is also used to validate the HW architecture. The STEP-GSM implements an innovative SW partitioning that benefits from the MCU capabilities of the ST120 DSP. This partitioning lightens the performance burden on the other MCU and a direct consequence is that the overall system needs a limited on-chip SRAM memory to perform a GPRS class 12 connection. This major hypothesis of the HW architecture was validated on the fast-prototype platform, much before the availability of the first silicon and no other platform could have done it.

The fast-prototype guaranties the real-time execution (the dedicated board can even be used in a mobile environment !). Hardware constraints on the SW can be handled such as memory footprint, power optimization, memory access times, verification of external IO protocols.

As a matter of fact, the final silicon does not bring much more in terms of functionality for the HW/SW integration than the fast prototype, except that it is the final product in the adequate package which can be plugged on a target platform, and later on a real form-factor phone.

3.2 Trends in SoC Architecture

During the HW architecture phase, most of the main decisions are performed: processor cores, buses, peripherals, dedicated blocks of logic, etc .. based on different criteria such as their availability, their performances using actual memory system.

As systems get bigger and bigger, this phase becomes more and more critical. There is a trend to use standard cores, standard IP's and more recently, standard platforms. IP reuse already plays a fundamental role in the industry, obviously for time-to-market reasons, although the maturity of IP's and their verification level is not yet at the state-of-the-art. STEP-GSM makes an extensive use of pre-existing IP's and cores.

Platform-based approach is a consequence of the increasing cost of design NRE and mask set, and a way to make sure that the investments done in the industry for IP packaging will pay off. It is advocated as the only way to keep up with Moore's law.

In order to avoid the risk of slowing innovation down, as can do any standard, the right trade-off has to be found and new tools for SoC design must be invented. Bus architecture is a typical example: custom bus design and verification are error-prone tasks, and automatic bus synthesis is not yet a fully mature technology. A standard of bus interfaces will surely improve SoC design cycle, but standardization of a given bus will not suit all SoCs.

Another current trend is the possibility to use a common HW/SW platform all along the design cycle. During the STEP-GSM project, it has not been possible to do so, due to many reasons: First, SW designers are used to develop on PC and use C tools on PC whereas HW designers are used to workstations. Second, the cost of FPGA boxes and HW emulators remains prohibitive for extended usage for SW development. Dedicated boards are more affordable but, as explained above, they come later in the design cycle. Finally, the real-time aspect is a must quite soon in the SW development process justifying the need for platforms such as the PC-based one described above.

HW/SW co-design models (e.g. SystemC-based) with models of processor cores connected using BFM (Bus Functional Models) find their justification more as HW architecture exploration tools rather than true HW/SW co-design tools. Chip-level specification based on transaction-level is still to be widely accepted, as can be the RTL languages. A non-negligible issue with high-level models is to keep them updated with the architectural/micro-architectural changes which are obviously reflected in the RTL code. Even if it is the case, the use of Fast Prototyping will remain as the only reliable proof of the chip architecture concept before the real silicon is available.

3.3 ASIC Design Flow

ASIC design from RTL to layout relies on quite mature methods and flows for most of the design steps. We illustrate some characteristics of the flow and possible improvements.

3.3.1 Main Characteristics of the Flow

The STEP-GSM was designed using a flow based on RTL sign-off. This means that a lot of effort was done in RTL verification, including the early use of HW emulators based on a fast-track synthesis (at a million dollars the mask-set in the forth-coming technologies, nobody can afford for many silicon runs). RTL code is synthesizable, including the testbenches, therefore largely facilitating the design mapping onto fast-prototyping platforms. The tasks (and responsibilities) for RTL verification are separated from the RTL design. Similarly, tasks (and responsibilities) for functional RTL verification on one hand and timing/physical verification on the other hand are also kept apart.

3.3.2 Mature Technologies

Among the mature (or very mature) technologies let's mention: RTL code coverage techniques, RTL verification using HW emulators, physical design tools (budgeting, floor-planning, clock tree synthesis, gate-level synthesis and static timing analysis), combinational RTL-to-gate formal proof, DFT and ATPG for scanned designs, multi-site data management techniques and finally bug tracking analysis. The use of formal property checking is spreading and it will surely benefit a lot from a unified expression technique of temporal properties.

3.3.3 Wanted Technologies

Looking backwards at the STEP-GSM flow, it is possible to identify at least two areas where new tools could reduce the design time.

At the RTL phase, a large part of the verification effort is devoted to chip-level test creation: for STEP-GSM, 100K lines of C (MCU and DSP) were needed to simulate 20 millions clock cycles to cover 100% of the 100K lines of VHDL. Therefore, there is a lot of room for innovation in automatic test creation. It will not substitute for the intelligent task of hand-writing a test plan, but it will automate the tedious work of manually writing C programs to exercise RTL simulation. This will require that the very nature of the components behavior be expressed in a formal way, as was done for test generators for processors, such as Genesis from IBM. Such a tool could be combined with a coverage analysis tool and may give rise to new metrics for RTL coverage analysis.

Similarly, to improve the physical design time, effects such as interconnect cross-talk and voltage drop have to be addressed from the beginning [3]. Even if post-analysis tools exist today, these effects should be taken into account by synthesis and timing analysis tools, in order to avoid iteration loops.

4. SOC DESIGN FOR MOBILE SYSTEMS

Everything's getting worse when it comes to mobile systems. These products are defined to be small, portable, low-power, high volume, low cost and more and more complex! Compared to a simple voice-centric 2G phone, a 3G smart phone ("all-singing and all-dancing") means first, new 3G communication standards such as WCDMA (with increased band-width up to 2Mbits/s) and second, a merge with a lot of "computing-multimedia" technologies, with devices such as color LCD display and camera.

Furthermore, time-to-market pressure is constant and results from the consumer nature of the market, more and more shrinking the period of time between the system definition and the availability of a product (HW with SW validated on the field). This has a huge impact on the design of SoCs for theses applications. Methodologies allowing a rapid change in the design from a non-predictable request from the market, will definitely be favored. This pleads for the use of fast-prototyping.

4.1 Power-saving Design

With 1 Ah average, the integration degree of batteries for mobile devices is close to explosive materials and a technology breakthrough will be needed to go further. This dictates the generalization of power-saving design methods.

Dynamic power consumption of the SoC is minimized at all levels of design. At architecture level, power dissipation is taken as a major factor for the trade-off at HW/SW partitioning. A dedicated block for clock and power control is implemented at RTL level, so that voltage and frequency change mechanisms are dynamically controlled by SW. Power consumption is analyzed in detail at RTL. At design-level, low-power and low-leakage libraries are used and during synthesis, massive clock-gating is inserted

Power-down techniques are used to minimize static consumption. Leakage is a huge problem as device geometry scales down. Stand-by or sleep mechanisms are carefully tuned and power network are split in order to switch-off the current on parts of the SoC.

4.2 Limits of SoC Integration

Mobile appliances are a digest of today's most advanced semi-conductors technologies such as mixed (energy management), analog (RF), digital (base band and multimedia), and memories such as Flash and SRAM.

Integration on a single die is simply not feasible today. SoC design is already facing limits when integrating large amounts of SRAM on a CMOS device for different reasons among which yield and leakage. Doubling the size of the die to embed large SRAM banks has a major impact on the yield. Leakage is hundred times higher in a CMOS process compared to an SRAM process for the same memory size.

An alternative is to extend the SoC concept to a System-In-Package using dies from different technologies stacked in a single package. During the STEP-GSM project, a prototype has been developed with a triple stack of a CMOS SoC with a 32Mbits Flash die and a 4Mbits SRAM die [4]. If it is economically viable, this new concept can have a lot of impacts on the design, in particular at the system-level definition.

5. CONCLUSION

Vertically-integrated semi-conductors companies cover the whole range of technologies and processes from high-level system specification to foundry. We have given several examples of up-to-date technologies: fast-prototyping, power-saving and stacked packages which are complementary to address the challenges of complex mobile systems.

6. ACKNOWLEDGMENTS

I thank the HW, SW and integration teams of the STEP-GSM at STMicroelectronics, and The Who for their 1971 title.

7. REFERENCES

[1] Smith G. "The Embedded Software Design Crisis", Dataquest Presentation, VSI Alliance, Oct. 2001.

[2] Shandle, J., and Martin, G. "Making embedded software reusable for SoCs", EEdesign, March 1, 2002.

[3] Magarshack P. "Improving SoC Design Quality through a Reproducible Design Flow" IEEE Design and Test of Computers, Jan 2002.

[4] Matheron C., Damalix P. and Berthet C. "An ASIC/Flash/ SRAM triple stack package", submitted to Hot chips 2002.

HiPRIME: Hierarchical and Passivity Reserved Interconnect Macromodeling Engine for RLKC Power Delivery

Yahong Cao, Yu-Min Lee, Tsung-Hao Chen and Charlie Chung-Ping Chen
(ycao, yu-min, tchen)@cae.wisc.edu, chen@engr.wisc.edu
Department of Electrical and Computer Engineering
University of Wisconsin at Madison
Madison, WI 53706

ABSTRACT

This paper proposes a general hierarchical analysis methodology, HiPRIME, to efficiently analyze RLKC power delivery systems. After partitioning the circuits into blocks, we develop and apply the IEKS (Improved Extended Krylov Subspace) method to build the Multi-port Norton Equivalent circuits which transform all the internal sources to Norton current sources at ports. Since there is no active elements inside the Norton circuits, passive or realizable model order reduction techniques such as PRIMA can be applied. To further reduce the top-level hierarchy runtime, we develop a second-level model reduction algorithm and prove its passivity. Experimental results show 400-700X runtime improvement with less than 0.2% error.

1. CATEGORIES & SUBJECT DESCRIPTORS

B. Hardware
B.8 Performance and Reliability
B.8.2 Performance Analysis and Design Aids

2. GENERAL TERMS

Algorithms, Design, Measurement, Performance, Theory, Reliability, Verification

3. INTRODUCTION

With the UDSM (Ultra Deep Sub-Micron) technology, several features about today's chips (faster operating frequencies, larger amount of transistors, smaller feature size and lower power voltage) have pushed the power delivery noise analysis onto the list of the designer's high-priority concerns. Basically, power delivery noise consists of IR-drop, Ldi/dt drop and resonance fluctuations. IR-drop has been widely discussed and extensively studied in the literature [7], [8] and [9]. Due to the roaring clock frequency, increasing current consumption, and even the clock gating feature, Ldi/dt noise is quickly emerging into another power fluctuation concern [8]. Power delivery noise, which causes the power voltage greater or lower than the ideal value, can degrade the performance severely and even make the gate function wrong. Therefore, extensive analysis of RLKC power delivery systems are required to ensure they meet the targeted performance and reliability goals.

Generally speaking, one of the major difficulties for power delivery analysis is size explosion. Tens of millions of devices and parasitics are required to be modeled and simulated for over a long time period. The traditional circuit simulation engine such as SPICE [11] can not fulfill the demanding task in a timely manner. For this reason, hierarchical simulation algorithms have been proposed by David, Sachin, et al [8].

Model order reduction techniques have also been extensively studied in the literature [6]. Starting from AWE (Asymptotic Waveform Evaluation) [5] to PRIMA (Passive Reduction Interconnect Macromodeling Algorithm), model order reduction techniques have been successfully extended to consider inductance effects in a reasonable accuracy. Later, [7] develops an EKS (Extended Krylov Subspace) method to simulate large scale power delivery circuits with many current sources. To resolve the source waveform modeling issues, EKS has to perform moment shifting procedures to recover the proper moments.

In this paper, we propose a novel hierarchical power delivery analysis methodology. The contributions of our method are as follows: First, we establish a novel hierarchical power delivery macromodeling methodology, which integrates multiple-port Norton equivalent theorem with model order reduction algorithm to generate compact and accurate model and achieve significant runtime improvement. Secondly, we enhance the EKS method such that it no longer needs to perform moment shifting for source waveform modeling. Therefore, highly accurate simulation results are observed. Finally, to further reduce runtime, we develop a multiple level passive model reduction algorithm and prove its passivity.

The outline of this paper is as follows. In section 2, we introduce the basic power delivery network modeling, circuit formulations and model order reduction. In section 3, we present our hierarchical and passive order-reduced macromodeling methodology. In the last section, we present the experimental results.

4. PRELIMINARY

RLKC elements are applied to model power delivery system as shown in Figure 1. To reduce the simulation runtime, we decouple the linear simulation from the nonlinear

DAC 2002, June 10-14, 2002, New Orleans, Louisiana, USA.

simulation. Once the nonlinear simulation has been done, we use current sources and capacitors to model gate current consumption and the diffusion and gate capacitance, respectively. Therefore, the power grid analysis problem is reduced to simulate a linear RLKC network with linear time-varying current sources and measure the voltage drop at each grid. A linear RLKC circuit can be represented as a set of Modified Nodal Analysis (MNA) circuit equations as following:

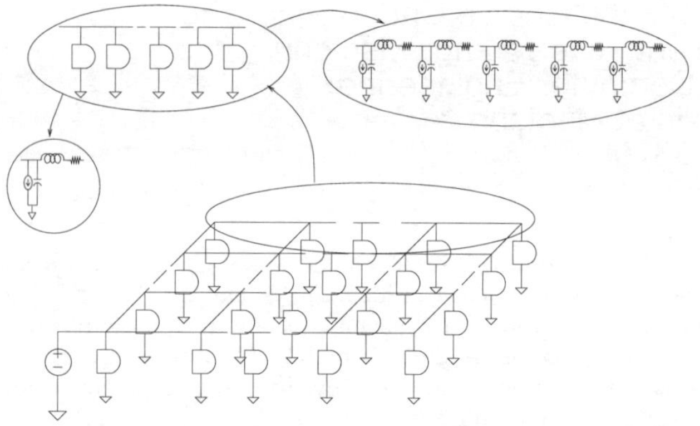

Figure 1: Modeling of the power grid network

$$Gx + C\frac{d}{dt}x = Bu, \quad (1)$$

where x is MNA variable vector, u denotes a vector of the port voltage sources and internal current sources.

$G = \begin{bmatrix} N & E \\ -E^T & 0 \end{bmatrix}$, $C = \begin{bmatrix} Q & 0 \\ 0 & H \end{bmatrix}$, $X = \begin{bmatrix} v \\ i \end{bmatrix}$ where v and i corresponds to the voltages at nodes and branch currents flowing through inductors and voltage sources. G and C matrices represent the conductance and susceptance matrices respectively. N, Q and H contain the stamping of the resistors, conductors and inductors. Note that H contains both self and mutual inductors (K elements). E corresponds to MNA current variables' contribution to the KCL equations.

Model order-reduction generates an analytic model which is a compact description of original circuits by matching their moments or poles. Circuit equations as shown in Equation 1 can be transformed to Laplace domain.

$$GX + CXs = BU \quad (2)$$

To illustrate the idea of moment matching, we expand both sides of the Equation (2) in a Taylor series around zero frequency

$$\begin{aligned}(G + sC)(m_0 + m_1 s + m_2 s^2 + \cdots) \\ = B(u_0 + u_1 s + u_2 s^2 + \cdots)\end{aligned}$$

where m_i and u_i, the coefficients of the i_{th} term in the Taylor series, are known as the i_{th} moment of x and u respectively. The basic idea of moment matching is to represent the finite unknown moments of the left hand side of the above equation in terms of the known moments of the right hand side. In PRIMA, the sources are assumed to be impulse sources in the moment matching process in order to preserve the input-output transfer characteristics. The impulse sources are constant in the frequency domain. So the Taylor expansion can be modified as

$$(G + sC)(m_0 + m_1 s + m_2 s^2 + \cdots) = Bu_0$$

This produces an iterative relationship between the moments of the x(s) and u(s): $Gm_0 = Bu_0$, $Gm_i + Cm_{i-1} = 0$. This explicit moment matching method is seldom used for the reason that it has numerical stability problem, especially in the higher order iterations. To avoid the numerical errors, a set of orthogonal bases is built to span the same subspace as spanned by the finite moments of x(s). The orthogonal bases V, or Krylov subspace, of matrix $A = -G^{-1}C$ and matrix $R = G^{-1}B$ is defined as $K_q(A, R) = colsp(R, AR, A^2R...A^{q-1}R)$. The dimension of the full description circuit(G,C,B) is reduced because the rank of V is much smaller than the original matrix A. This order-reduced model can be obtained by projecting the original model (G,C,B) onto this Krylov subspace using congruent transformation. The reduced MNA matrices are denoted as $\widetilde{G} = V^TGV$, $\widetilde{C} = V^TCV$ and $\widetilde{B} = V^TB$. Thus the compact model in the time domain are represented in MNA equations as:

$$\widetilde{G}\widetilde{x} + \widetilde{C}\frac{d}{dt}\widetilde{x} = \widetilde{B}u$$

5. HIERARCHICAL AND PASSIVE ORDER-REDUCED MACROMODELING

Our hierarchical and passive model order reduction analysis consists of three steps. First we partition power grid networks into multiple blocks. Each block may contain RLC interconnect network and multiple internal switching currents. The second step is to generate Norton equivalent order-reduced model for each block. This step is divided into three phases: The first phrase is to find the passive order-reduced model for the RLC interconnect network of this block. The second phase is to find the Norton equivalent current of the internal current sources. In the last phase of the second step, we attach the Norton equivalent currents at the ports of the order-reduced RLC model. At the third step, we apply integration algorithm to the macromodels generated by the first two steps and apply top level simulation.

The outline and flowchart of our algorithm are shown in Figure 2 and Figure 3, respectively. In the following subsections, we will discuss step $A2.1$, step $A2.2$ and step $A3$ in Figure 2.

Algorithm: HiPRIME (Hierarchical and Passivity Reserved Interconnect Macromodeling Engine)

A1. Partition the given circuit into multiple blocks

A2. For each block, we generate multiport Norton equivalent order reduced circuits by the following procedure:

- **A2.1** Zero all the active sources and perform passive model order reduction for the linear circuit using any passivity guaranteed model reduction algorithm such as PRIMA.
- **A2.2** Activate all sources and short all the ports nodes to ground and find out the Norton equivalent sources at each port by IEKS or SPICE simulation.
- **A2.3** Form the Norton Equivalent circuit by attaching the the Norton equivalent sources at each port to the reduced circuit generated by the model reduction algorithms in Step A2.1.

A3. Form the integrated circuit by combining all the reduced modules. Perform higher level of model order reduction such as IEKS or PRIMA when necessary.

Figure 2: HiPRIME Algorithm

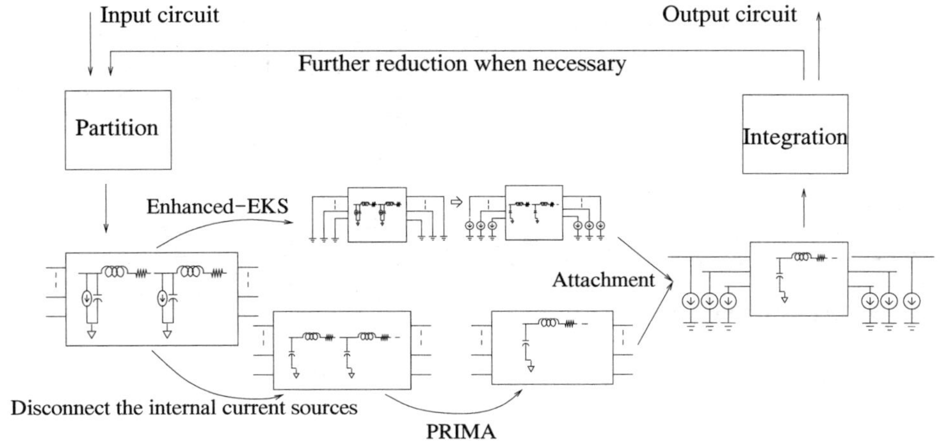

Figure 3: Flowchart for the hierarchical and passivity reserved interconnect macromodeling engine

5.1 Passive Reduced-Order Macromodeling of RLC Networks

After a power grid network is partitioned into multiple blocks, each block may contain passive RLC interconnect and internal switching current sources. In order to obtain a passive order-reduced model, we simply disconnect all the internal current sources to make this block a passive RLC network. The effect of these current sources on grid voltages will be considered later. We may apply a conventional passive model order reduction algorithm, PRIMA, to each simplified block. The MNA equations for the RLC interconnect network of the ith block is as:

$$G_iX_i + C_i\frac{d}{dt}X_i = B_iu_i \qquad (3)$$

where u_i is the port voltage vector of the i_{th} block. The result of PRIMA is that, by the transfer matrix V_i, G_i, C_i and B_i are transformed into $\tilde{G}_i, \tilde{C}_i$ and $\tilde{B}_i$, whose dimension are reduced. The compact MNA equation to describe this block is as:

$$\tilde{G}_i\tilde{X}_i + \tilde{C}_i\frac{d}{dt}\tilde{X}_i = \tilde{B}_iu_i$$

The benefits of model order reduction after partition include: 1) Model order reduction algorithms handle much smaller input circuit size, which means the limit of memory might be eased; 2) It also makes parallel order reduction for different blocks possible, by which the speed is improved.

5.2 Efficient Way of Finding the Norton Equivalent Current

In this section, we consider the effects of the internal current sources on grid voltages which are ignored in the previous procedure. Norton equivalent theory is used to find out the equivalent current sources at the ports and replace all the internal current sources so that the port responses of each block are preserved. To distinguish the port voltage sources from the internal current sources, we modify Equation (3) as

$$G_iX_i + C_i\frac{d}{dt}X_i = [B_iB_i'] \begin{bmatrix} v_i \\ i_{gi} \end{bmatrix} \qquad (4)$$

where v_i and i_{gi} denote the independent voltage sources and the internal switching current sources in the ith block respectively. B_i and B_i' denote the position of the voltage sources and the current sources relative to the whole network. The procedure of calculating the equivalent at the ports is illustrated in Figure 4. As shown, the port currents

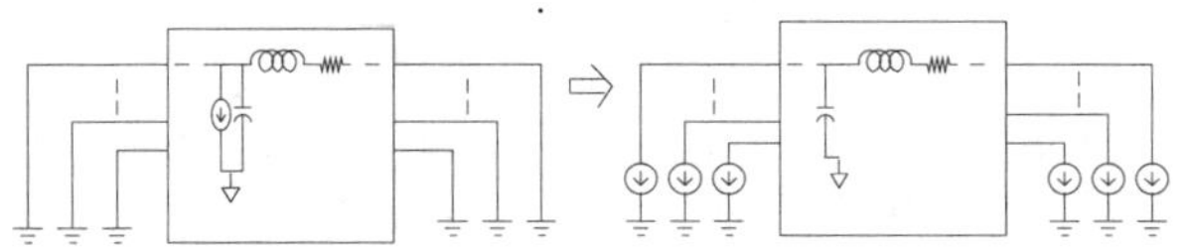

Figure 4: Finding the equivalent current of internal sources

with the port voltages set to zeros are the Norton equivalent current sources. Also as we know, port currents can be obtained by $i_{Ni} = B_i{}^TX_i$. To solve Equation (4) with the voltage sources u_i set to zeros, different algorithms can be applied. In our algorithm, we applied IEKS, an improved version of EKS such that no moment shifting is necessary which will be described in next session.

5.2.1 Improved Extended Krylov Subspace

Developed by Janet, et al. at [7], EKS directly calculates the orthogonalized moments of the response when multiple sources are turned on at the same time. Therefore, unlike PRIMA whose runtime is heavily dependent on the port number, the runtime of EKS is independent of that. EKS models a Piece-Wise-Linear(PWL) independent source as a sum of delayed ramps in the Laplace domain:

$$u(s) = \frac{1}{s^2}\sum_{i=1}^{K} r_i exp(-\beta_i s)$$

This expression contains $1/s$ and $1/s^2$ terms. Unfortunately the traditional Krylov subspace methods start the moment matching from the 0_{th} moment. EKS extends the Krylov subspace by shifting the moments toward right in the frequency spectrum.

This moment shifting in EKS is tedious and error-prone. Fortunately, we develop an improved moment calculation method which ensure the -1_{st} and -2_{nd} order moments are all zero for arbitrary finite time PWL waveform and hence the moment shifting process can be removed. Since for simulation purpose we are only interested for a specific time period, the finite-time assumption is quite general. We believe this procedure is numerically more sound than the original method.

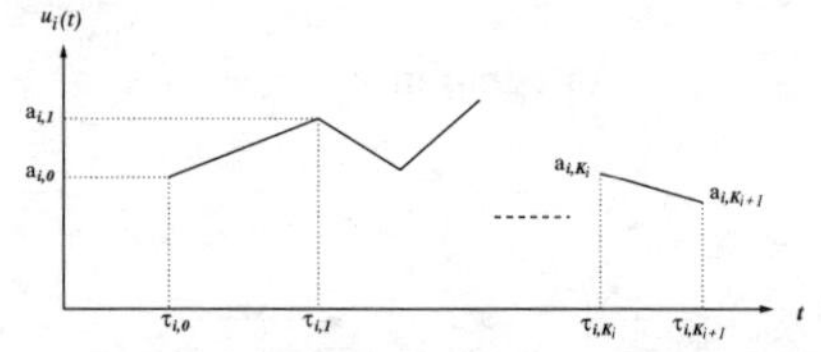

Figure 5: Waveform of the source

LEMMA 1. *Given a finite-time PWL source, IEKS calculates a moment representation with -1_{st} and -2_{nd} order moments to be zero.*

PROOF. Given an finite-time PWL source $u_i(t)$ as described in Figure 5, we have

$$u_i(t) = \sum_{j=0}^{K_i} \{[a_{i,j} + \gamma_{i,j}(t - \tau_{i,j})]\, \mathrm{E}_{(t-\tau_{i,j})} - [a_{i,j+1} + \gamma_{i,j}(t - \tau_{i,j+1})]\, \mathrm{E}_{(t-\tau_{i,j+1})}\} \qquad (5)$$

where $\gamma_{i,j} = (a_{i,j+1} - a_{i,j})/(\tau_{i,j+1} - \tau_{i,j})$, and $\mathrm{E}_{(t-\tau_{i,j})}$ is the unit-step function with $\tau_{i,j}$ delay. By taking the Laplace transform of Equation (5) and Taylor expansion, we now have

$$\begin{aligned}
&\mathcal{L}(u_i(t)) \\
= \;& \frac{1}{s^2}\sum_{j=0}^{K_i}\Bigg\{ a_{i,j}s\sum_{l=0}^{\infty}(-1)^l\frac{\tau_{i,j}^l}{l!}s^l + \gamma_{i,j}\sum_{l=0}^{\infty}(-1)^l\frac{\tau_{i,j}^l}{l!}s^l \\
& -a_{i,j+1}s\sum_{l=0}^{\infty}(-1)^l\frac{\tau_{i,j+1}^l}{l!}s^l - \gamma_i\sum_{l=0}^{\infty}(-1)^l\frac{\tau_{i,j+1}^l}{l!}s^l\Bigg\}
\end{aligned}$$

Let $\tilde{u}_{i,j}$ denote the coefficient of the s^j term. The moment representation of $\mathcal{L}(u_i(t))$ can be simplified to

$$\mathcal{L}(u_i(t)) = \{\tilde{u}_{i,-2}s^{-2} + \tilde{u}_{i,-1}s^{-1} + \tilde{u}_{i,0} + \tilde{u}_{i,1}s + \tilde{u}_{i,2}s^2 + \cdots + \tilde{u}_{i,m}s^m + \cdots\} \qquad (6)$$

When calculating the first two coefficients, we conclude

$$\begin{aligned}
\tilde{u}_{i,-2} &= \sum_{j=0}^{K_i}(\gamma_{i,j} - \gamma_{i,j}) = 0 \\
\tilde{u}_{i,-1} &= \sum_{j=0}^{K_i}(a_{i,j} - \gamma_{i,j}\tau_{i,j} - a_{i,j+1} + \gamma_{i,j}\tau_{i,j+1}) \\
&= \sum_{j=0}^{K_i}(a_{i,j} - a_{i,j+1} - \gamma_{i,j}(\tau_{i,j} - \tau_{i,j+1})) = 0
\end{aligned}$$

□

The first two terms are eliminated and Equation (6) can be rewritten as a moment representation starting from the 0_{th} moment:

$$\mathcal{L}(u_i(t)) = \{\tilde{u}_{i,0} + \tilde{u}_{i,1}s + \tilde{u}_{i,2}s^2 + \cdots + \tilde{u}_{i,m}s^m + \cdots\}$$

Table 1 summarizes the process of IEKS to calculate the first M moments.

Algorithm: IEKS Moments Calculating Algorithm
Input: N PWL sources, $\{a_{i,0\ldots(K_i+1)}, \tau_{i,0\ldots(K_i+1)}\}_{i=1}^{N}$
M, the number of moments to calculate
Output: $\mathbf{u}_{i,m} = \{[\; u_{i,1} \;\; u_{i,2} \;\; \cdots \;\; u_{i,M}\;]\}_{i=1}^{N}$,
the first M moments of the N PWL sources
Begin
for $i = 1 : N$
for $j = 0 : K_i$
$$\gamma_{i,j} = \frac{(a_{i,j+1} - a_{i,j})}{(\tau_{i,j+1} - \tau_{i,j})}$$
end
for $m = 1 : M$
for $j = 0 : K_i + 1$
$$\beta_{i,j}^{(m)} = \frac{(-\tau_{i,j})^m}{m!}$$
end
$$u_{i,m-1} = (a_{i,0} - \gamma_{i,0}\frac{\tau_{i,0}}{m})\beta_{i,0}^{(m-1)} - \sum_{j=0}^{K_i-1}(\gamma_{i,j} - \gamma_{i,j+1})\beta_{i,j+1}^{(m)} - (a_{i,K_i+1} - \gamma_{i,K_i}\frac{\tau_{i,K_i+1}}{m})\beta_{i,K_i+1}^{(m-1)}$$
end
end
End

Table 1: IEKS Moments Calculating Algorithm

5.2.2 *System solution by IEKS*

IEKS generates a system transform matrix V_i, by which the original system is transformed into a compact description.

$$\tilde{G}_i\tilde{X}_i + \tilde{C}_i\frac{d}{dt}\tilde{X}_i = [\tilde{B}_i\tilde{B}_i']\begin{bmatrix} 0 \\ i_{gi} \end{bmatrix} \qquad (7)$$

The compact form can be solved quickly in the time domain by standard integration algorithms. The solution of the original system can be recovered by $X_i = V\tilde{X}_i$. But we are only interested about the port currents in the solution. The port currents are directly obtained by $i_{Ni} = B^T X_i = \tilde{B_i^T}\tilde{X}_i$.

5.3 Macromodel Integration and Top Level Reduced-Model Simulation

After the $A2$ step of HiPRIME (Figure 2), we transform a block consisting of RLC segments and internal PWL currents into a passive order reduced block with current sources attached only at the ports. The new macromodel of each block is illustrated in Figure 6. The port response of the original block is preserved. We generate such macromodel

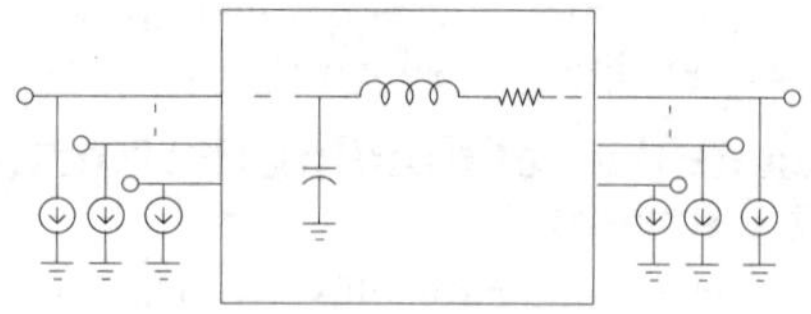

Figure 6: Equivalent circuit for each block

for each block and the entire network is integrated by combining these macromodels together. Considering each block as a node of the integrated network, we stamp them into the MNA equation of the entire network. The combination of ith block and jth are given in Equation (8).

$$\begin{bmatrix} \tilde{G}_i & 0 & -\tilde{B}_{ij} \\ 0 & \tilde{G}_j & -\tilde{B}_{ji} \\ \tilde{B}_{ij}^T & \tilde{B}_{ji}^T & 0 \end{bmatrix}\begin{bmatrix} \tilde{X}_i \\ \tilde{X}_j \\ u_{ij} \end{bmatrix} + \begin{bmatrix} \tilde{C}_i & 0 & 0 \\ 0 & \tilde{C}_j & 0 \\ 0 & 0 & 0 \end{bmatrix}\frac{d}{dt}$$

$$\begin{bmatrix} \tilde{X}_i \\ \tilde{X}_j \\ u_{ij} \end{bmatrix} = \begin{bmatrix} \tilde{B}_{ii} & 0 & 0 & 0 \\ 0 & \tilde{B}_{jj} & 0 & 0 \\ 0 & 0 & E_i{}^T & E_j{}^T \end{bmatrix}\begin{bmatrix} u_i \\ u_j \\ i_{Ni} \\ i_{Nj} \end{bmatrix} \qquad (8)$$

In this formulation, u_{ij} denotes the voltages at the common ports belonging to both the i_{th} and j_{th} blocks. u_i and u_j stand for the voltages at the ports of the i_{th} and j_{th} blocks, which are not connected with each other. B_{ii} and B_{jj} denote the connection of the internal nodes to the ports which are exclusive of each other. B_{ij} denotes the connection of internal nodes of ith block to jth block and B_{ji} denotes the connection of internal nodes of jth block to ith block. i_{Ni} and i_{Nj} stand for the equivalent port currents, which are extracted from the inside of the i_{th} and j_{th} block respectively. E_i and E_j record the connection of the internal nodes to the equivalent port currents for ith block and jth block.

Given this glued macromodels in Equation (8), we can further apply model order reduction and simulation techniques such as PRIMA or IEKS to the top level to save runtime. There may be more than 2 hierarchical levels and the higher level model order reduction is introduced as the following. Let us define some block system matrices for ith and jth blocks. Define $G' = \begin{bmatrix} \widetilde{G}_i & 0 & -\widetilde{B}_{ij} \\ 0 & \widetilde{G}_j & -\widetilde{B}_{ji} \\ \widetilde{B}_{ij}^T & \widetilde{B}_{ji}^T & 0 \end{bmatrix}$, $C' = \begin{bmatrix} C_i & 0 & 0 \\ 0 & C_j & 0 \\ 0 & 0 & 0 \end{bmatrix}$ and $B' = \begin{bmatrix} \widetilde{B}_{ii} & 0 \\ 0 & \widetilde{B}_{jj} \\ 0 & 0 \end{bmatrix}$, with the internal current sources disconnected, the system is described as

$$G' \begin{bmatrix} \widetilde{X}_i \\ \widetilde{X}_j \\ u_{ij} \end{bmatrix} + C' \frac{d}{dt} \begin{bmatrix} \widetilde{X}_i \\ \widetilde{X}_i \\ u_{ij} \end{bmatrix} = B' \begin{bmatrix} u_i \\ u_j \end{bmatrix} \quad (9)$$

Use $\begin{bmatrix} V_1 V_2 V_3 \end{bmatrix}^T$ represent the orthogonal basis of the subspace spanned by the moments of $\begin{bmatrix} \widetilde{X}_i \widetilde{X}_j u_{ij} \end{bmatrix}^T$ and denote $\widetilde{G'} = \begin{bmatrix} V_1 \\ V_2 \\ V_3 \end{bmatrix}^T G' \begin{bmatrix} V_1 \\ V_2 \\ V_3 \end{bmatrix}$, $\widetilde{C'} = \begin{bmatrix} V_1 \\ V_2 \\ V_3 \end{bmatrix}^T C' \begin{bmatrix} V_1 \\ V_2 \\ V_3 \end{bmatrix}$ and $\widetilde{B'} = \begin{bmatrix} V_1 \\ V_2 \\ V_3 \end{bmatrix}^T B'$, the MNA equation generated by the higher level order-reduction is noted as

$$\widetilde{G'}\widetilde{Z} + \widetilde{C'}\frac{d}{dt}\widetilde{Z} = \widetilde{B'} \begin{bmatrix} u_i \\ u_j \end{bmatrix}$$

5.4 Preservation of Passivity

In order to apply PRIMA or EKS to the already reduced model as described in previous session, we need to be sure that the higher level order-reduced models are also passive. The following theorem gives us the warrant.

THEOREM 1. *During the hierarchical order reduction, the passivity of the higher level order-reduced macromodel is also preserved. That is to say, the transfer function of the higher level order-reduced system $Y(s)$ satisfies*

1. $Y(s^\star) = Y^\star(s)$ for all complex s.

2. $Y(s)$ is a positive matrix, that means, $Z^{\star T}(Y(s) + Y^T(s^\star))Z \succeq 0$ for any complex s satisfying $Re(s) \succ 0$ and for any complex vector z.

PROOF. With the impulse voltages active at ports and applying Laplace transform to Equation (10),the transfer function can be obtained as

$$Y(s) = \widetilde{B'}^T(\widetilde{G'} + s \cdot \widetilde{C'})^{-1}\widetilde{B'}$$

For the system matrices are all real, the first condition is met naturally. To prove the second condition is also met, we start from

$Z^{\star T}(Y(s) + Y^T(s^\star))Z$

$= Z^{\star T}(\widetilde{B'}^T(\widetilde{G'} + s \cdot \widetilde{C'})^{-1}\widetilde{B'} + \widetilde{B'}^T(\widetilde{G'} + s^\star \cdot \widetilde{C'})^{-T}\widetilde{B'})Z$

Setting $w = (\widetilde{G'} + s^\star \cdot \widetilde{C'})^{-T}\widetilde{B'}z$ and $s = j\varpi + \sigma$ yields

$Z^{\star T}(Y(s) + Y^T(s^\star))Z$

$= w^{\star T}\left[(\widetilde{G'} + (j\varpi + \sigma) \cdot \widetilde{C'}) + (\widetilde{G'} + (j\varpi - \sigma) \cdot \widetilde{C'})^T\right] w$

$= w^{\star T}\left(\widetilde{G'} + \widetilde{G'}^T\right) w + w^{\star T} \cdot \sigma \cdot \left(\widetilde{C'} + \widetilde{C'}^T\right) w$

Since C_i and C_j are symmetric, $C_i^T + C_i = 2C_i$ and $C_j^T + C_j = 2C_j$. Along with the fact that C_i and C_j are nonnegative definite, it yields

$w^{\star T} \cdot \sigma \cdot \left(\widetilde{C'} + \widetilde{C'}^T\right) w$

$= \sigma \cdot w^{\star T} V_1^T V_i^T 2C_i V_i V_1 w + \sigma \cdot w^{\star T} V_2^T V_j^T 2C_j V_j V_2 w \succeq 0$

for any complex vector z and positive σ. Since N_i and N_j are symmetric nonnegative definite matrices, we have

$w^{\star T}\left(\widetilde{G'} + \widetilde{G'}^T\right) w$

$= w^{\star T} V_1^T V_i^T (G_i^T + G_i) V_i V_1 w + w^{\star T} V_2^T V_j^T \left(G_j^T + G_j\right) V_j V_2 w$

$= w^{\star T} V_1^T V_i^T \begin{bmatrix} 2N_i & 0 \\ 0 & 0 \end{bmatrix} V_i V_1 w$

$+ w^{\star T} V_2^T V_j^T \begin{bmatrix} 2N_j & 0 \\ 0 & 0 \end{bmatrix} V_j V_2 w \succeq 0$

for any complex vector z. From the above, we can conclude that the second passivity condition is satisfied. □

6. EXPERIMENTAL RESULTS AND FUTURE WORK

In this section, the speed and accuracy of HiPRIME and IEKS are demonstrated and compared with other methods. We use mesh networks to model the power delivery network, which consist of lumped RC/RLC segments with many current sources attached inside. The typical values for each lumped RC/RLC segment are $R = 0.2\Omega$, $L = 1.0pH$ and $C = 0.024fF$.

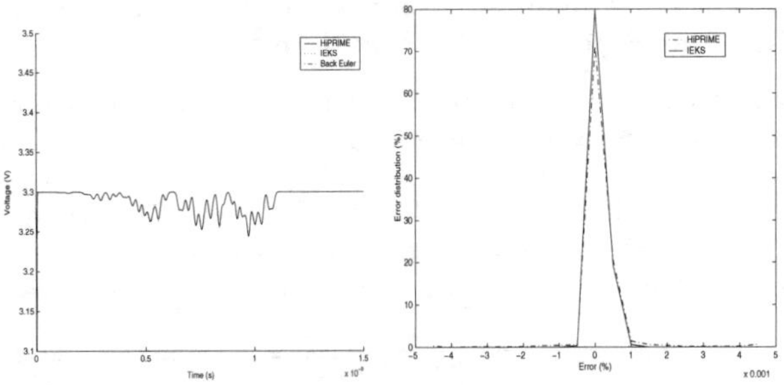

Figure 7: Accuracy analysis of RC circuit case: (a) waveform result of HiPRIME, IEKS and Back Euler (b) error spectrum of HiPRIME and IEKS

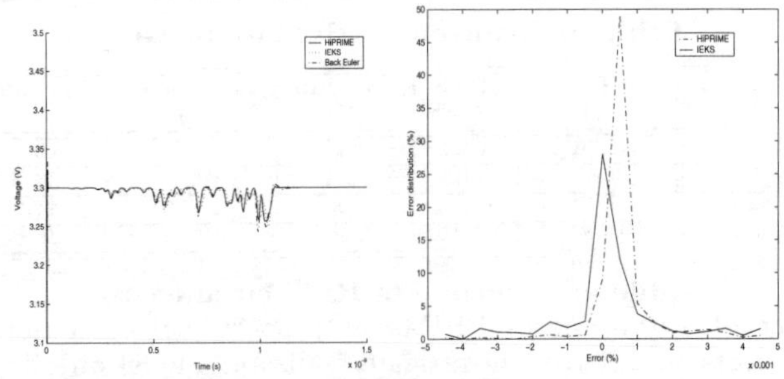

Figure 8: Accuracy analysis of RLC circuit case: (a) waveform result of HiPRIME, IEKS and Back Euler (b) error spectrum of HiPRIME and IEKS

The voltage waveforms of HiPRIME are compared with those of IEKS and Back Euler. The cases of RC circuit and RLC circuit are tested and the results are shown in Figure 7 and Figure 8 respectively. In the RC case, the waveforms are indistinguishable and the errors for 80% time intervals are within 0.001%. In the RLC case, the waveforms by HiPRIME and IEKS approximate the one by Back Euler method very well and the errors for 50% time intervals are within 0.001%.

We implement IEKS in C code and test it on a PIII 933MHZ machine. The result is compared with our time domain solver InductWise [10] and Spice. Table 2 summarizes the runtime results and the runtime comparisons are shown in Figure 9. Significant speed improvement, 700X faster than Spice, is observed and the same tendency that the speed up increases with larger circuit size is shown.

Circuit Size	IEKS (s)	InductWise (s)	Speedup (X)	Spice (s)	Speedup (X)
7861	1.46	14.76	10.1	697.13	477.48
14081	3.88	29.77	7.67	2728.18	703.14
43541	13.49	107.05	7.93	–	–
89201	35.33	244.95	6.93	–	–

Table 2: Runtime of IEKS, InductWise and Spice

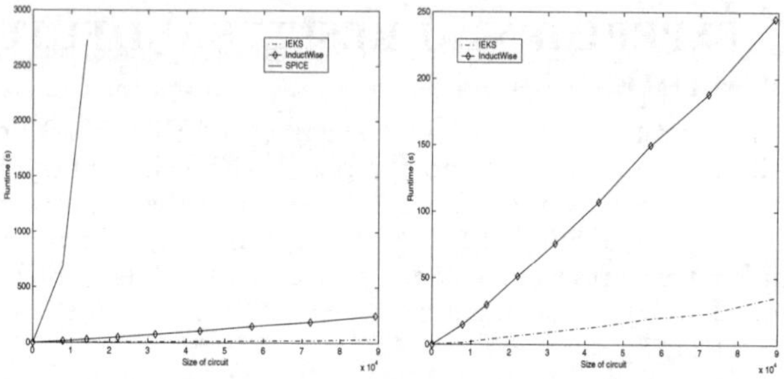

Figure 9: Runtime Analysis of (a) IEKS, InductWise and Spice (b) IEKS and InductWise

We also implement HiPRIME, flat IEKS and Back Euler in Matlab and test them on Sun Ultrasparc V. Table 3 and 4 summarize the runtime results and the runtime comparisons are shown in Figures 10 and 11 for RC and RLC circuits respectively. From the figures, we can see the tendency that the speed up is more impressive when the circuit size increases.

Circuit Size	HiPRIME (s)	IEKS (s)	Speedup (X)	Back Euler (s)	Speedup (X)
203	2.52	0.89	0.36	17.1	6.79
803	2.98	1.97	0.66	78.3	26.3
2403	4.39	6.17	1.40	288.9	65.8
5003	7.93	13.18	1.66	760.1	95.8

Table 3: Runtime of RC circuit case

Circuit Size	HiPRIME (s)	IEKS (s)	Speedup (X)	Back Euler (s)	Speedup (X)
443	2.76	1.29	0.47	29.5	10.7
1883	3.87	5.08	1.31	129.8	33.5
3843	6.72	12.94	1.92	276.9	41.2
5803	11.81	27.15	2.29	427.2	36.6

Table 4: Runtime of RLC circuit case

Since the runtime of PRIMA is proportional to number of ports, we plan to investigate realizable model order reduction algorithm whose runtime is ports size independent. Also we would like to point out that the focus of this paper is not on the partition algorithms. However, a good

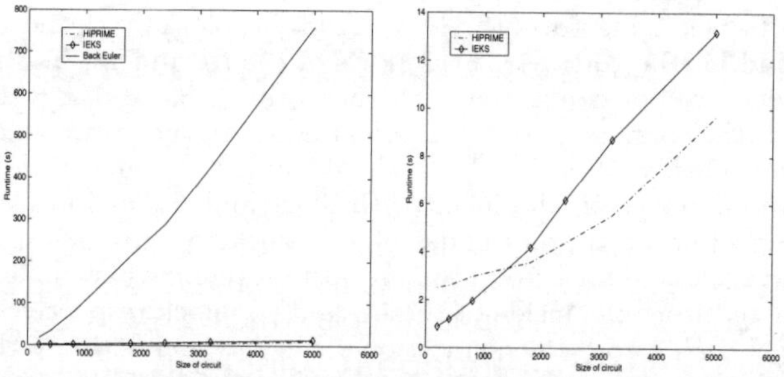

Figure 10: Runtime analysis of RC circuit case: (a) runtime of HiPRIME, IEKS and Back Euler (b) runtime of HiPRIME and IEKS

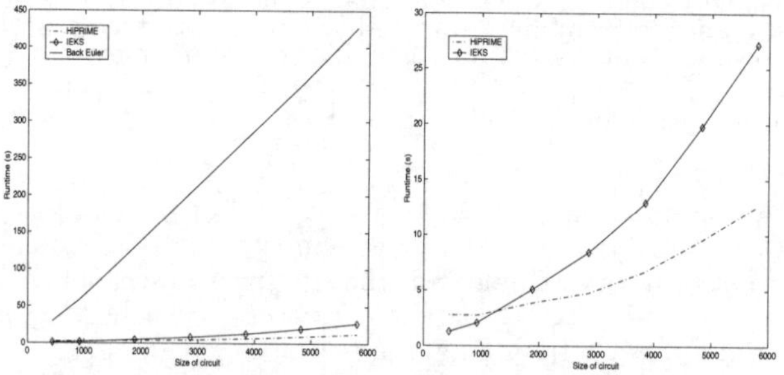

Figure 11: Runtime analysis of RLC circuit case: (a) runtime of HiPRIME, IEKS and Back Euler (b) runtime of HiPRIME and IEKS

partition algorithm is important for the performance of the hierarchical and passive model order reduction algorithm.

7. ACKNOWLEDGEMENTS

This work is partially supported by NSF Grant CCR-0093309, Intel Corp., and Faraday Inc.

8. REFERENCES

[1] L. O. Chua, C. Desoer and E. Kuh, *Linear and Nonlinear Circuits*, McGRAW-HILL Book Co., 1987.

[2] J. Cong and C. K. Koh, *Interconnect Layout Optimization Under Higher-Order RLC Model*, IEEE/ACM International Conference on Computer-Aided Design, pp. 713-720, CA, Nov. 1997.

[3] P. Feldman and R. W. Freund, *Efficient linear circuit analysis by Pade approximation via the Lanczos process*, IEEE Trans. Computer-Aided Design, vol. 14, no. 5, pp 639-649, May 1995.

[4] Q. J. Yu and E. S. Kuh, *Exact moment model of transmission lines and application to interconnect delay estimation*, IEEE Trans. on VLSI Systems. vol.3, pp. 311-322, June 1995.

[5] L.T. Pillage, R.A. Rohrer and C. Visweswariah, *Electronic Circuit and System Simulation Methods*, McGRAW-HILL Book Co.

[6] A. Odabasioglu, Mustafa Celik and L.T. Pillage, *PRIMA: Passive Reduced-order Interconnect Macromodeling Algorithm*, ICCAD, pp. 58-65, 1997.

[7] Janet M. Wang and Tuyen V. Nguyen, *Extended Krylov Method for Reduced Order Analysis of linear Circuits with Multiple Sources*, *DAC*, 2000.

[8] Min Zhao, Rajendran V. Panda, Sachin S. Sapatnekar, Tim Edwardsand, Rajat Chaudhry and David Blaauw, *Hierarchical Analysis of Power Distribution Networks*, *DAC*, 2000.

[9] Sani R. Nassif and Joseph N. Kozhaya, *Fast Power Grid Analysis*, *DAC*, 2000.

[10] Tsung-hao Chen and Charlei Chung-ping Chen, *Efficient Large-Scale Power Grid Analysis Based on Preconditioned Krylov-Subspace Iterative Methods*, *DAC*, 2001. http:// vlsi.ece.wisc.edu/Tools.htm

[11] Nagel, L.S., *SPICE2: A Computer Program to Simulate Semiconductor Circuits*, *ERL Memo ERL-M520*, *University of CAliforniaa*, *Berkeley*, 1975.

High-Level Current Macro-Model For Power-Grid Analysis *

Srinivas Bodapati
ECE Dept. and Coordinated Science Lab.
University of Illinois at Urbana-Champaign
Urbana, Illinois 61801, USA
bodapati@uiuc.edu

Farid N. Najm
ECE Department
University of Toronto
Toronto, Ontario, Canada M5S 3G4
f.najm@utoronto.ca

ABSTRACT

We present a frequency domain current macro-modeling technique for capturing the dependence of the block current waveform on its input vectors. The macro-model is based on estimating the *Discrete Cosine Transform* (DCT) of the current waveform as a function of input vector pair and then taking the inverse transform to estimate the time domain current waveform. The input vector pairs are partitioned according to Hamming distance and a current macro-model is built for each Hamming distance using regression. Regression is done on a set of current waveforms generated for each circuit, using HSPICE. The average relative error in peak current estimation using the current macro-model is less than 20%.

Categories and Subject Descriptors

B.7 [**Hardware**]: Integrated Circuits—*CAD*; B.7.2 [**Integrated Circuits**]: Design Aids—*Modeling*

General Terms

Algorithms

Keywords

Power grid, Current macro-model, DCT

1. INTRODUCTION

Integrated circuits are drawing increasingly large currents from the power grid. Such large currents in the power grid aggravate physical reliability of IC's due to electromigration. They also cause significant voltage drop leading to circuit slow-down or soft errors. As a result, power grid analysis and design is now an important concern during chip design, not only for power dissipation and reliability reasons, but for performance reasons as well. Therefore, it is important to do early design planning of the power grid, in order to reduce the chance of having to redesign large parts of it [4]. To enable this, in an environment where previously designed blocks are being reused (hard IP blocks) we propose a bottom-up current waveform macro-model for logic blocks. We work at a level of abstraction that may be called structural RTL: the circuit is described as an interconnection of flip-flops and Boolean (combinational) logic blocks. In this context, we develop a cycle-based model for the current waveform of each combinational block that captures the dependence of the current waveform per-cycle as a function of the *vector pair* applied at the block inputs. Previous work has targeted either the average power [11, 9, 13, 6, 2] or energy-per-cycle [12, 7].

Current waveform macro-modeling is difficult because of the large variations that are possible in current waveform shapes, and due to the very large number of possible vector pairs. To overcome this problem, we have developed a macro-modeling technique that is based on Discrete Cosine Transform (DCT). In this approach, we estimate the DCT of the current waveform, instead of the the time domain current waveform and then use the inverse DCT to get the time-domain current waveform. Unlike time-domain current waveforms, the DCT of current waveform is more regular. Specifically, large variations in waveform shapes in the time domain translate to variations only in the *parameters* of the DCT but not its overall shape. Therefore, in our modeling approach, we use a *template* function for the DCT of the per-cycle current waveform. The *template* parameters (discussed in section 4) depend on the input vector pair. We have found that one can use low-order polynomial models to capture this dependence. We use regression to generate these polynomials, based on a number of randomly generated vector pairs for which the circuit is simulated in HSPICE.

The form of this template function is not exactly *derived*, but *inferred* by examining the forms of the frequency transforms of two types of current waveform shapes that one commonly sees in practice. In the next section, we will describe the frequency transforms that will be used as the basis for inferring a DCT model template.

2. DISCRETE COSINE TRANSFORM

The current waveform obtained from circuit simulation (HSPICE) is a discrete-time signal, which can be considered

*This research was supported by the Semiconductor Research Corporation (SRC), under SRC 97-DJ-484 (with funds from Texas Instruments Inc.) and SRC 99-TJ-682 (with funds from IBM Corp.).

DAC 2002, June 10-14, 2002, New Orleans, Louisiana, USA.

as being obtained from the periodic sampling of a continuous-time current waveform, i.e, $i[n] = i_c(nT)$, $0 \leq n \leq N-1$, where N is the length of the current sequence, $i_c(\cdot)$ is the continuous-time current waveform and T is the *sampling period*, whose reciprocal is the *sampling frequency* [10]. In case of circuit simulation, T is equal to the (fixed) time step specified in the transient analysis (we have used a time step of 0.01 ns). The 1-dimensional Discrete Cosine Transform [8] (DCT) of a sequence $\{i[n], 0 \leq n \leq N-1\}$ is defined as:

$$I[k] = \alpha(k) \sum_{n=0}^{N-1} i[n] \cos\left[\frac{\pi(2n+1)k}{2N}\right], \; 0 \leq k \leq N-1 \quad (1)$$

where $\alpha(0) = \sqrt{1/N}$, $\alpha(k) = \sqrt{2/N}$, for $1 \leq k \leq N-1$, and the inverse transform is given by:

$$i[n] = \sum_{k=0}^{N-1} \alpha(k) I[k] \cos\left[\frac{\pi(2n+1)k}{2N}\right], \; 0 \leq n \leq N-1 \quad (2)$$

In order to gain some insight into the form that the DCT current macro-model should take, we will look at the analytical forms of the Fourier transform (FT) corresponding to simplified representations of typical current waveforms. For example, we will consider a piecewise-linear triangular current waveform and construct its FT as an analytical closed-form expression. The *form* of this FT will suggest what forms we should use in our DCT model template. The reason this makes sense is that there is a relationship between the two transforms, as follows: the DCT is closely related to the DFT (Discrete Fourier Transform), which is a sampled form of the DTFT (Discrete Time Fourier Transform), which is itself related to the FT. We will now explain these relationships.

The N-sample DCT is related to a $2N$-sample DFT [8], as follows. The DCT of an input sequence of N-samples can be obtained by extending the input to a $2N$-sequence sample with even symmetry, taking a $2N$-point DFT, and saving N terms of it. The even extension of $i[n]$ is defined as:

$$i'[n] = \begin{cases} i[n] & n = 0, 1, \ldots, N-1 \\ i[2N-1-n] & n = N, N+1, \ldots, 2N-1 \end{cases} \quad (3)$$

and the $2N$-point DFT of $i'[n]$, is given by:

$$\begin{aligned} I'[k] &= \frac{1}{\sqrt{2N}} \sum_{n=0}^{2N-1} i'[n] e^{-j(2k\pi/2N)} \\ &= \sqrt{\frac{2}{N}} e^{(jk\pi/2N)} \sum_{n=0}^{N-1} i[n] \cos\left[\frac{(2n+1)k\pi}{2N}\right] \quad (4) \end{aligned}$$

which shows that the $2N$-point DFT and the DCT are closely related. The relationship between the DFT and the DTFT [10], is as follows. The DFT of a sequence, $\{i'_n\}_{n=0}^{2N-1}$, is a set of evenly spaced samples of the DTFT over the frequency range 0 to 2π, multiplied by a constant factor to make the DFT an orthonormal transform. Thus, $I'[k] = \frac{1}{\sqrt{2N}} I_d(\omega)|_{\omega=\frac{2\pi k}{2N}}$ where $k = 0, 1, \ldots, 2N-1$ and $I_d(\omega)$ is the DTFT of the sequence $\{i'_n\}$, defined by: $I_d(\omega) = \sum_{n=0}^{2N-1} i'[n] e^{-j\omega n}$, where ω is the frequency in radians. Finally, if we consider that the sequence $\{i'_n\}$ is obtained by sampling from an even-extended continuous time current waveform, then the relationship between $I_d(\omega)$ and the Fourier transform (FT) of the continuous waveform, denoted by $I_c(\Omega)$, is given by [10]: $I_d(\omega) = \frac{1}{T} \sum_{n=-\infty}^{\infty} I_c\left(\frac{\omega+2\pi n}{T}\right)$ where Ω has been replaced by $\left(\frac{\omega+2\pi n}{T}\right)$. When $n = 0$, which corresponds to a DTFT between $-\pi \leq \omega \leq \pi$, we get $\Omega = \frac{\omega}{T}$ [10] . Thus, in summary, the DCT of a discrete time sequence is related to the Fourier transform of an even-extended version of the continuous-time function from which the given discrete-time samples were taken. In the next two sections, we discuss different possible approximations to the actual current waveform, present their corresponding continuous transforms, analyze them and then develop the DCT current macro-model templates based on them.

3. ANALYSIS OF CURRENT WAVEFORMS

A CMOS combinational logic circuit draws current over one cycle in response to a vector pair at its inputs, and these waveform shapes typically fall into three categories:

Approximately Triangular: In this case, the time domain current waveform has a triangular geometry, and the DCT of the current waveform is similar to the DCT of a regular triangular wave.

Approximately Trapezoidal: In this case, the time domain current waveform has a trapezoidal geometry, and the DCT of the current waveform is similar to the DCT of a regular trapezoid.

Multiple peaks: In this case the time domain current waveform has multiple peaks which are separated in time. Currently we use partitioning in time to convert such a waveform to a sequence of waveforms that are of one of the above forms.

In the following, we will explore the FT of the triangular and the trapezoidal current waveforms, which will be used to infer reasonable forms for our DCT model template.

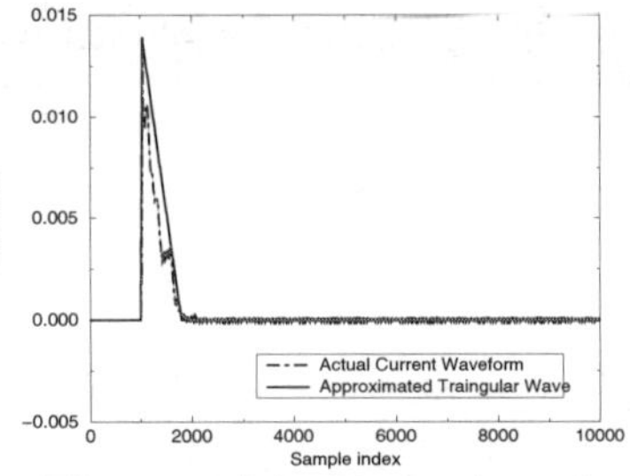

Figure 1(a). Typ. triangular current waveform and its approx.

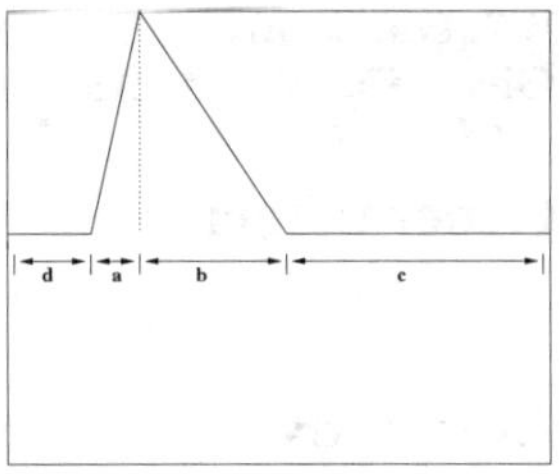

Figure 1(b). Explanation of $f_{\text{tri}}(t)$.

3.1 Triangular Current Waveform

A typical triangular current waveform and its triangular waveform approximation are shown in Fig. 1(a). Before taking the DCT, we increase the waveform time duration and assume zero values for the waveform over the time extension. This is referred to as "zero padding," and its main advantage is that it increases the resolution of the DCT (the corresponding DTFT is sampled more closely), which gives a smoother DCT curve. Zero padding in the time domain is equivalent to using a higher sampling frequency in the frequency domain of the DTFT to get the corresponding DFT/DCT [10]. Thus, even though the original current waveform ends at 2000 samples (which corresponds to 20 ns for our sampling period of 0.01 ns), we have shown 10,000 sample points.

We construct an even extension of this triangular waveform and derive the corresponding continuous Fourier transform. The triangular waveform shown in Fig. 1(a) can be analytically expressed as:

$$f_{\rm tri}(t) = \begin{cases} 0, & \text{for } 0 \le t \le d \\ A(t-d)/a, & \text{for } d \le t \le a+d \\ A(a+b+d-t)/b, & \text{for } a+d \le t \le a+b+d \\ 0 & \text{for } a+b+d \le t \le Q \end{cases} \quad (5)$$

where a, b, c, d are the dimensions shown in Fig. 1(b), and $Q = a+b+c+d$. The Fourier transform of the even extension of $f_{\rm tri}(t)$, denoted by $F_{\rm tri}(\Omega)$ is given by:

$$\begin{aligned} F_{\rm tri}(\Omega) &= Ae^{-j\Omega(Q)}\left[\frac{2\cos\Omega(b+c) - 2\cos\Omega(a+b+c)}{a\Omega^2}\right] \\ &+ Ae^{-j\Omega(Q)}\left[\frac{2\cos\Omega(b+c) - 2\cos\Omega(c)}{b\Omega^2}\right] \end{aligned} \quad (6)$$

where Ω is the frequency in the analog (continuous) domain and is related to ω, the digital frequency by $\omega = \Omega T$. In our case, we have $T = .01$ ns and $N = 10,000$ before the even extension. After extension, we have $2N$ samples therefore $\omega = 2\pi k/2N = \pi k/10,000$ and $\Omega = \pi k/100$ rad/ns. It is important to note that for $\Omega = \pi k/100$, $F_{\rm tri}(\Omega)$ is real for all integral values of k because $Q = a+b+c+d = 100$ ns.

In Fig. 2(a), we show the first 100 samples (for clarity) of the 10,00 point DCT of a current waveform, DCT of its triangular waveform approximation and the plot of the continuous transform. Fig. 2(a) is typical of all triangular current waveform shapes that we have seen. The DCT plot of the current waveform appears to be a decaying sinusoid. Therefore, instead of constructing a model for each point on the DCT (every frequency component), we use a generic function (a template) to model the entire DCT. In order to infer the form of this template, we compare the plot of continuous transform with the DCT plot. Except for a scale factor, it is obvious that $F_{\rm tri}(\Omega)$ has the right shape and thus may help us define the form of the current macro-model.

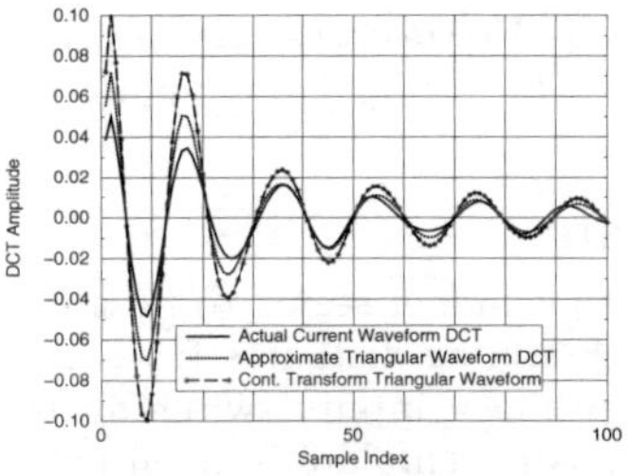

Figure 2(a). DCT of a typ. triangular current waveform.

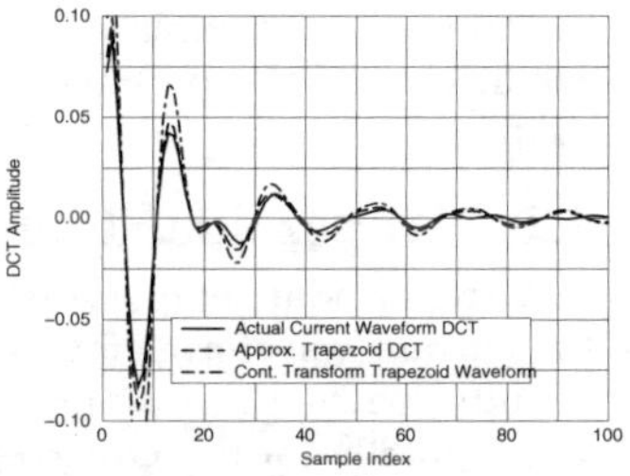

Figure 2(b). DCT of a typ. trapezoidal current waveform.

The most important aspect to observe is that the amplitude of $F_{\rm tri}(\Omega)$ decays as the square of the frequency, a fact which we will make use of in developing our DCT model template. Since the DCT looks like a decaying sinusoid, we can use a simple sinusoid which decays as a square of the frequency. But unlike a decaying sinusoid which has a constant time period, the plots in Fig. 2(a) show that in case of a current waveform and its approximations (both discrete and continuous transform plots) the time period is varying (we can use the difference between consecutive/alternate zero crossings to measure the time period). We use this fact in our model as well. It actually helps simplify our current macro-model, in the sense that we do not use multiple cosine terms as in $F_{\rm tri}(\Omega)$, instead we use a simplified expression with a variable time period, as discussed in section 4.

3.2 Trapezoidal Current Waveform

In some cases, the current waveform has *trapezoidal* shape, as shown in Fig. 3(a), which also shows a piecewise linear trapezoidal approximation to the current waveform. The equation of the trapezoidal wave is given by:

$$f_{\rm tra}(t) = \begin{cases} 0, & \text{for } 0 \le t \le a \\ A(t-a)/b, & \text{for } a \le t \le a+b \\ A, & \text{for } a+b \le t \le a+b+c \\ A(U-t)/d, & \text{for } a+b+c \le t \le U \\ 0 & \text{for } U \le t \le W \end{cases} \quad (7)$$

where the dimensions are illustrated in Fig. 3(b), $U = a+b+c+d$ and $W = a+b+c+d+e$. The Fourier transform of the even extension of $f_{\rm tra}(t)$, denoted by $F_{\rm tra}(\Omega)$ is given by:

$$\begin{aligned} F_{\rm tra}(\Omega) &= Ae^{-j\Omega(W)}\left[\frac{2\cos\Omega(c+d+e) - 2\cos\Omega(W-a)}{b\Omega^2}\right] \\ &+ Ae^{-j\Omega(W)}\left[\frac{2\cos\Omega(e+d) - 2\cos\Omega(e)}{d\Omega^2}\right] \end{aligned} \quad (8)$$

As before, it turns out that $F_{\rm tra}$ is real for the values of Ω under consideration. In Fig 2(b), we show the DCT of the current waveform, the DCT of the trapezoidal waveform approximation, and the plot of the continuous transform. Again, we can observe that, except for a scale factor, the continuous transform plot compares well with the actual current waveform DCT. Thus, we can use the continuous transform to infer the shape of the DCT model template. The most obvious inference we can draw from the continuous transform, is, as before, the quadratic decay in the amplitude of the DCT, in terms of frequency.

If we compare the plots in Fig. 2(a) and Fig. 2(b), we can see that the DCT plot of a trapezoid shows some deviation from the perfectly decaying sinusoid of triangular waves. The same observation holds true for the respective continuous transform as well, therefore we cannot use a decaying sinusoid with a varying time period to simplify our model as in the case of triangular waveforms (see section 4). Instead, we will use a slightly different analytical expression for the trapezoidal current waveform, which is based on $F_{\rm tra}(\Omega)$ to some extent, but also based on observations related to the plots. Specifically, for trapezoidal waveforms, we will use two cosine terms (the continuous transform has 4 cosine terms), and again use a varying time period. We use the difference between successive maxima (or minima) as as a measure of the time period. The DCT plots show that this difference varies for successive maxima (or minima). Details are given in section 4.

3.3 Waveform Partitioning

For multiple peak current waveforms we use a simple partitioning algorithm in the time domain, in which we detect the index of the first peak, and then traverse to a point where the current value is less 10% of the maximum value. We repeat this process and partition the time domain waveform into sub-intervals so that each sub-interval contains a single peak, and then use the same macro-modeling technique discussed in section 4. This technique is used only in cases where we have peaks which are very far apart. Partitioning also keeps the number of current macro-model templates to only two. Since in the current macro-model characterization phase we need to automatically detect which template to use depending on the error criterion (the error criterion

Table 1: Dummy Coding for transitions at a single node.

y	x_1	x_2	x_3	y
$0 \to 1$	0	0	1	1
$1 \to 0$	0	1	0	2
$1 \to 1$	1	0	0	3
$0 \to 0$	0	0	0	4

would detect which of the two templates is a better fit), it is good to minimize the number of possible templates.

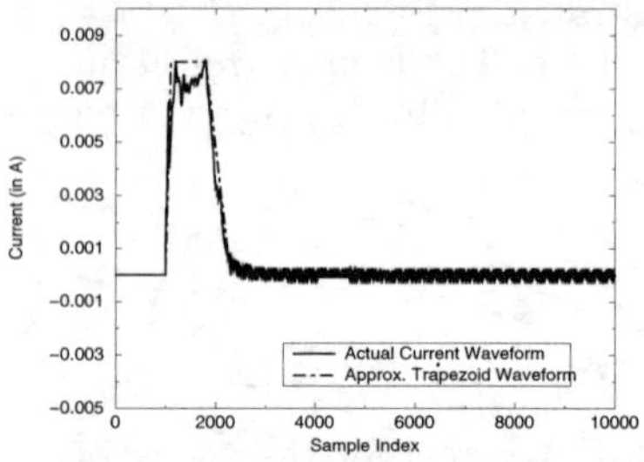

Figure 3(a). Typ. trapezoidal current waveform and its approx.

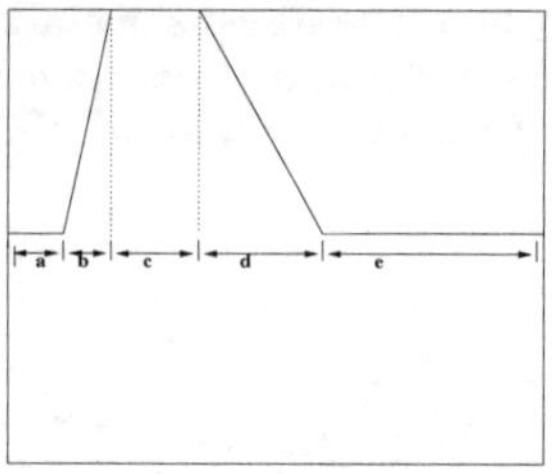

Figure 3(b). Explanation of $f_{\mathrm{tra}}(t)$.

4. MACRO-MODEL CONSTRUCTION

In order to formulate a macro-model, one needs to develop a mapping between the input Boolean vector pair and certain variables that can be the input variables to the macro-model. These variables are used as *predictor variables* in regression. For a given input vector, a single node at the primary input can undergo one of the following four transitions $\{0 \to 1, 1 \to 0, 1 \to 1, 0 \to 0\}$. We treat the set of four possible transitions at each primary input as a categorical variable [5]. In general, a categorical variable with k levels (in our case 4) is transformed into $k-1$ variables (in our case 3) each with 2 levels. This process of creating variables from categorical variables is called **dummy coding**. In our case, we use three variables x_1, x_2 and x_3 to code the categorical variables. The dummy coding for the 4 levels of our categorical variable is shown in the second, third and fourth columns of Table 1 (and also shown in [12]). These three columns are labeled as x_1, x_2, x_3 in Table 1. Basically, each input transition is mapped to a vector of $[x_1\ x_2\ x_3]$, but this has a clear disadvantage in that it increases the number of variables as well as the number of regression coefficients. But we have found that, depending on the Hamming distance of the vector pair, we can introduce a less expensive solution, as follows. For large Hamming distances (more than 60% of the inputs are switching), very few inputs undergo $0 \to 0$ and $1 \to 1$ transition, therefore we can assume that the categorical variable has just two levels, so that a single two-valued variable is required to represent a primary input. The mapping for large Hamming distance is shown in the fifth column of Table 1, labeled y.

4.1 Triangular Template

The triangular template is a simplified as well as a modified version of $f_{\mathrm{tri}}(\Omega)$. The simplification is achieved because we explicitly incorporate the effect of the changing time period (which happens because of multiple cosine terms in $f_{\mathrm{tri}}(\Omega)$). We estimate the first few time periods (in our case, five) directly and use them in a single cosine term, whose other parameters do not change. We can do this only because of the energy compaction achieved with orthogonal transforms like DCT. The template equation for the DCT of the triangular current waveform is given by: $I_{\mathrm{tri}}(k) = D(k)A\cos\left(\frac{2\pi(k-1)}{T_i}\right)$, $k = 1, 2, \ldots$, where k is the sample index, $D(k)$ is a decay factor, A is a scale factor, which we refer to as amplitude, and T_i, $i \in \{1, 2, 3, 4, 5\}$ is the variable time period. The DCT value corresponding to $I_{\mathrm{tri}}(k)|_{k=0}$ is called the DC Value. These terms are the *parameters* of the template, which we relate to the variables associated with the input vector pair. The triangular template is similar to what was presented in [1] (the trapezoidal template is new), where it was shown that the time period T_i does not vary significantly for a given Hamming distance of the input vector pair, prompting a partitioning of the model according to Hamming distance. Thus, T_i is estimated as in [1]. As for $D(k)$, we use the following template for it, which is motivated by the inverse square dependence on frequency seen above: $D(k) = \frac{f(x)k}{g(x)k^2+h(x)}$ where x is either y (a vector of length one) or the vector $[x_1\ x_2\ x_3]$ (refer to Table 1). This expression is similar to that defined in [1], except that we have modified $f(x), g(x)$ and $h(x)$, to account for possible interaction between the input variables, through cross-product terms, so that $f(x), g(x)$ and $h(x)$ are now of the form (where $\mathcal{H}$ stands for Hamming distance):

$$\Pi_{i=1}^{p}(\alpha_i y_i + \beta_i), \text{ for large } \mathcal{H}. \quad (9)$$

$$\Pi_{i=1}^{p}(\alpha_{i1}x_{1,i} + \alpha_{i2}x_{2,i} + \alpha_{i3}x_{3,i} + \beta_i), \text{ for small } \mathcal{H}. \quad (10)$$

where p is the number of primary inputs. The coefficients α and β are estimated using regression as in [1]. The Amplitude A, and the DC Value $I_{\mathrm{tri}}(0)$ are also modeled along the lines of $f(x), g(x)$ and $h(x)$. Their coefficients are also estimated using regression. The triangular template can provide a good model in cases of both large and small Hamming distances, but in some cases for small Hamming distance, the trapezoidal template is better. However, estimating the coefficients of the parameters of the triangular template is simpler.

4.2 Trapezoidal Template

A trapezoidal template equation model seems to best fit the current waveforms for low Hamming distances for some circuits, when the fraction of primary inputs switching is less than 60% or so. Our trapezoidal template is given by:

$$\begin{aligned} I_{\mathrm{tra}}(k) &= \frac{A}{D'(k)}\left[B(k)\cos\left(\frac{2\pi(k-1)}{T_i}\right)\right] \\ &+ \frac{A}{D'(k)}\left[C(k)\cos(2\pi(k-1)\omega)\right] \end{aligned} \quad (11)$$

where $k = 1, 2, \ldots$ is the sample index, T_i, $i \in \{1, 2, 3, 4, 5\}$, is the variable time period, A is the amplitude and $B(k) = p(x)k$, $C(k) = q(x)k$, $D'(k) = r(x)k^2 + s(x)$, $\omega = t(x)$, $A = w(x)$ where x is the vector $[x_1\ x_2\ x_3]$. The functions $p(x)$, $q(x)$, $r(x)$, $s(x)$, $t(x)$ and $w(x)$, all have the same functional form (with different coefficients), given by: $\Pi_{i=1}^{p}(\alpha_{i1}x_1 + \beta_{i2}x_2 + \gamma_{i3}x_3 + \delta_i)$, where the coefficients are estimated using non-linear regression. Since the number of parameters in the template is large compared to I_{tri}, we have a large number of coefficients and estimating the coefficients of the parameters of the trapezoid is computationally more expensive than estimating the coefficients of a triangular wave. The polynomial function used to model the DC value $I_{\text{tra}}(k)|_{k=0}$ is similar to that used for A. The coefficients of the function corresponding to $I_{\text{tra}}(k)|_{k=0}$ are estimated separately, using simple linear regression. The form of the above template is motivated by the functional form of the the continuous Fourier transform expression $F_{\text{tra}}(\Omega)$ for a trapezoid waveform, which was given above (8) and which involves four cosine terms, and decays as the square of the frequency. Since four cosine terms may be too many (computationally expensive model), and a single cosine term cannot capture the deviation from a simple decaying sinusoid, we have used only two cosine terms in our template, decaying with the square of frequency. In order to maintain a good fit with just two cosine terms, we have found that using separate frequency terms in the numerator, as we have done with $B(k)$ and $C(k)$, is advantageous.

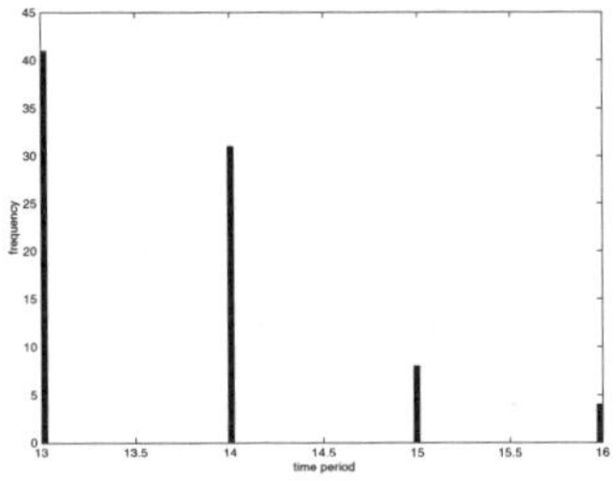

Figure 4. Dist. of the first time period for alu2.

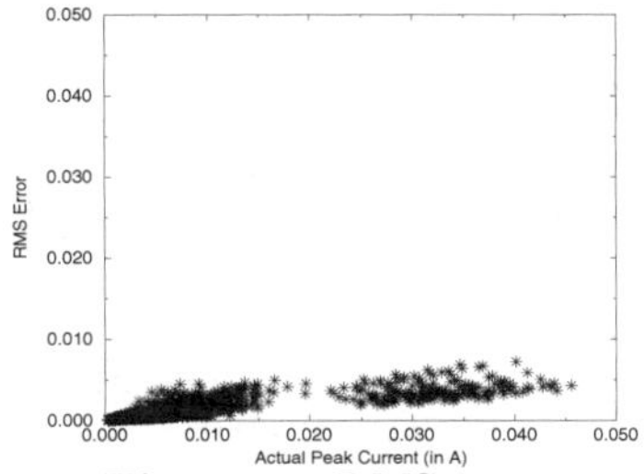

Figure 5. RMS error vs. peak current.

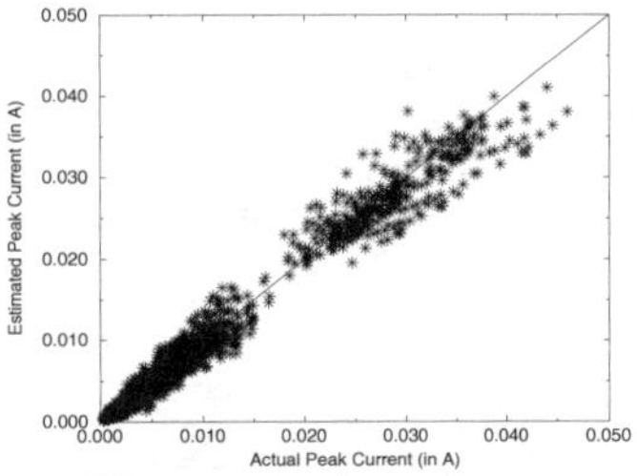

Figure 6. Actual vs est peak current.

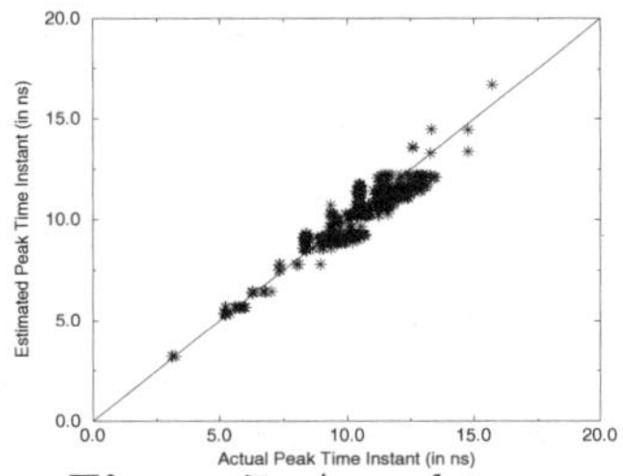

Figure 7. Actual vs. est time inst. when peak occurs.

The philosophy behind using two cosine terms is that one cosine term would produce a basic decaying sinusoid (as in the triangular case) and the other would account for the deviations from the decaying sinusoid that are seen at low Hamming distance. Therefore, the variable time period T_i is used with only one of the two cosine terms, the one which is aimed at generating the decaying sinusoid. The frequency ω of the other cosine term is estimated as a function of input vector pair using regression. The variable time period terms T_i are approximated as the difference between consecutive maxima (or minima) of the DCT. As in the triangular case and [1], the time period does not vary much for a given Hamming distance, as shown in Fig. 4. Thus, we use the same method as in [1] to get an estimate of the first five time periods. Basically, from the distribution of time periods for a given Hamming distance, we choose the time period value which occurs most often (in statistical terms, we choose the *mode* of the time period distribution as the time period for a given Hamming distance). The first five time periods are enough to get an estimate of the dominant terms of the DCT.

5. EXPERIMENTAL RESULTS

A set of randomly generated vectors were used to simulate various benchmark circuits in HSPICE, and the corresponding current macro-models were built for these circuits. The resulting macro-models were tested for accuracy on a different set of randomly generated vector pairs for each of the benchmark circuits listed in Table 2, which also shows the number of inputs and gates for each circuit. In Table 2, e_{pavg} (5th column) denotes the **average relative error** in peak current estimation, and e_{tavg} (6th column) denotes the **average relative error** in the estimation of the time instant at which the peak current occurs. The 4th column in Table 2, denoted by RMS_{avg} lists the **average Root Mean Square Error** [8] between the estimated and the actual current waveforms. The average error in all the three cases has been computed over the number of input vector pairs over which the macro-models have been tested. The last column in Table 2 shows the total number of simulations (number of vector pairs) needed to build the macro-model for all possible hamming distances, for each circuit. The number of simulations can be reduced by combining models for various hamming distances.

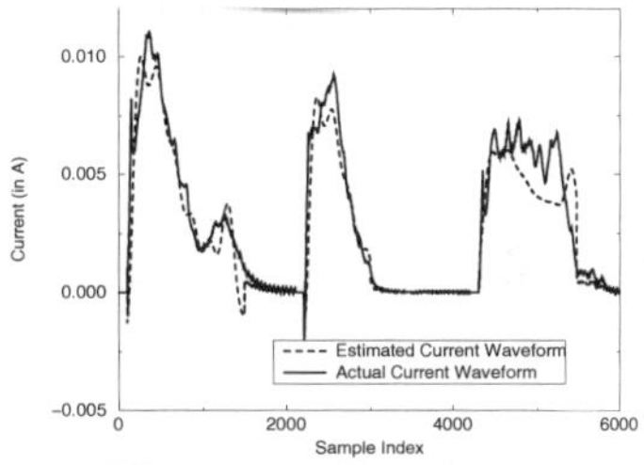

Figure 8. For alu2, ham dist =4.

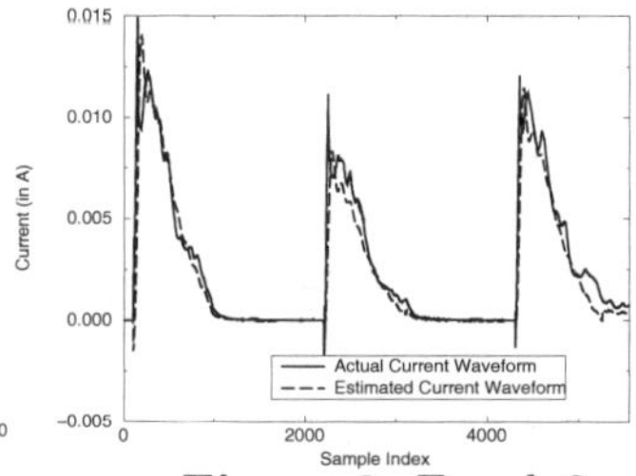

Figure 9. For alu2, ham dist =8.

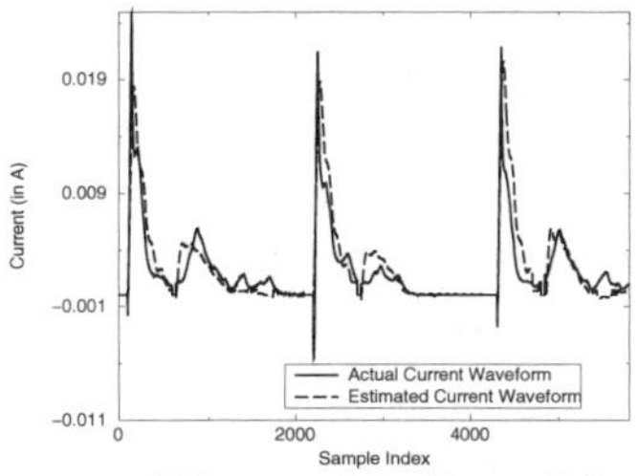

Figure 10. For c432, ham dist =32.

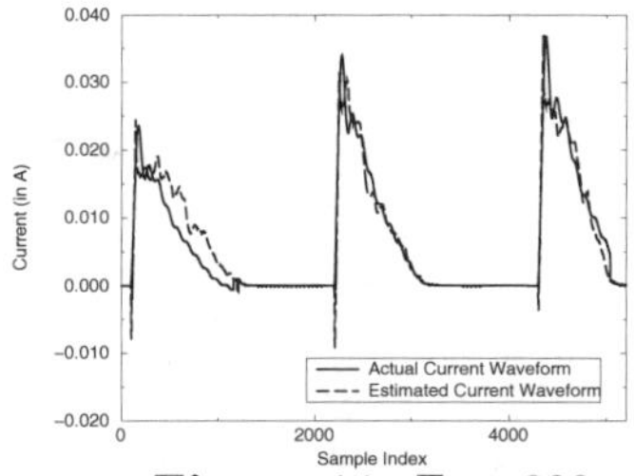

Figure 11. For c880, ham dist = 55.

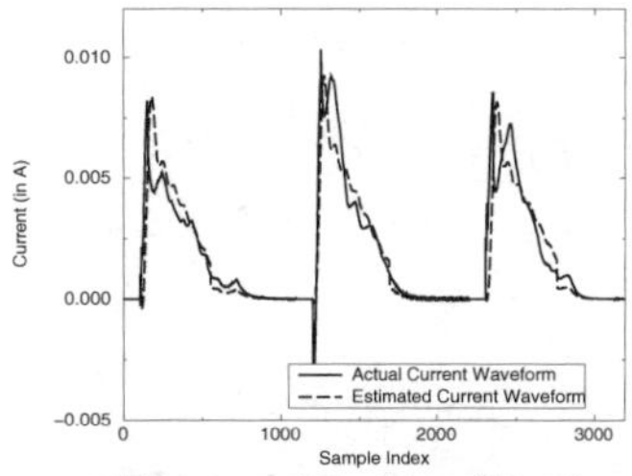

Figure 12. For f51ml, ham dist = 7.

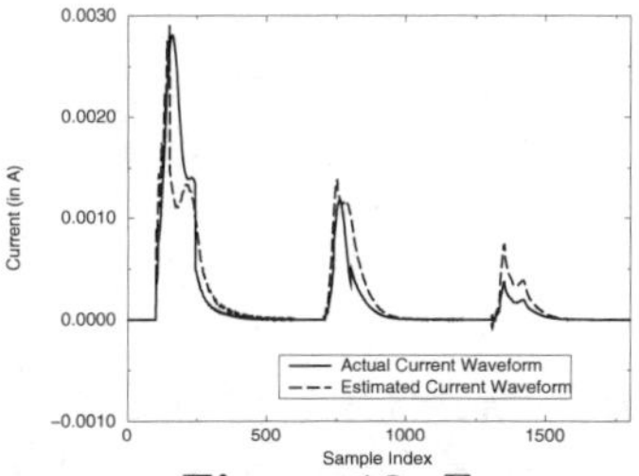

Figure 13. For cu, ham dist = 2.

In order to give some intuition as to the goodness of these average error numbers, and some measure of spread, we compared the RMS error for each vector pair with the actual peak current as shown in Fig 5. The plot shows that the

RMS error is more or less bounded across the various current peak values for the benchmark circuits, and increases very slowly with current peak. Figure 6 shows the correlation plot between actual and estimated peak current and similarly in Fig. 7 we show the correlation plot between actual and estimated *time instant* at which peak occurs. Both these correlation plots are mostly linear which implies that both the peak current and the time instant at which the peak occurs can be estimated fairly accurately using our macro-model.

Finally, in Figs. 8–17, we present a comparison of the actual and estimated current waveform for the benchmark circuits listed in Table 2, for some example Hamming distances, where we superimposed the the estimated current waveform on top of the actual current waveform. These plots show good accuracy in most of the cases, especially for large Hamming distances. This occurs because when a small fraction of inputs are switching, the current waveform significantly depends on what inputs are switching, which leads to large variations in the current amplitude. However, the current magnitude itself is much lower in case of low Hamming distance, so that the absolute error is actually small. Having said this, we must point out that the model in fact works very well, capturing the required current waveforms in a high-level black box macro-model. Finally, we should mention that even though our macro-model has been built using HSPICE, one can build current macro-models for much larger circuits using more efficient simulators, like PowerMill, or gate-level simulation techniques such as [3].

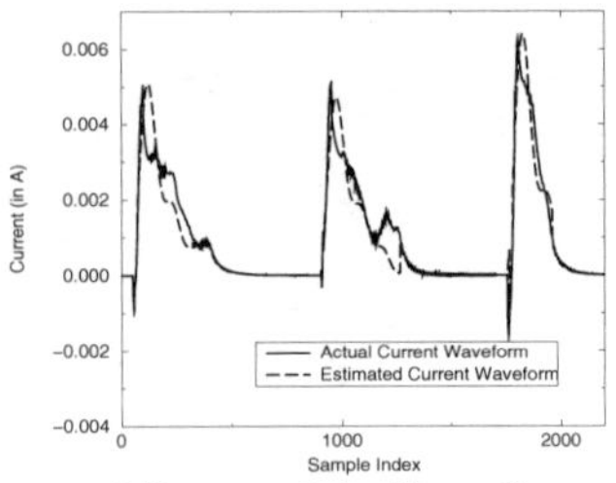

Figure 14. For x2, ham dist = 8.

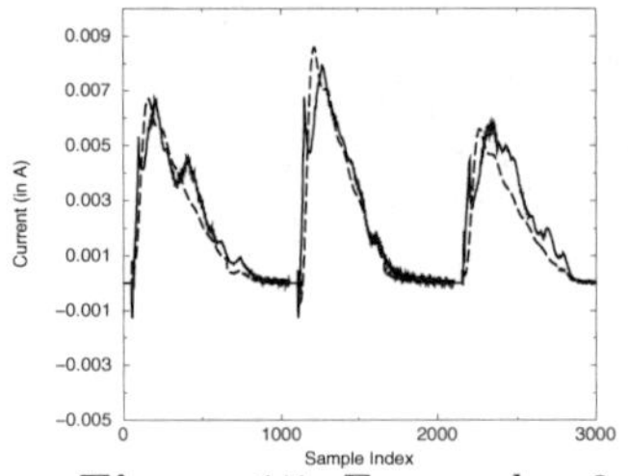

Figure 15. For random8, ham dist = 4.

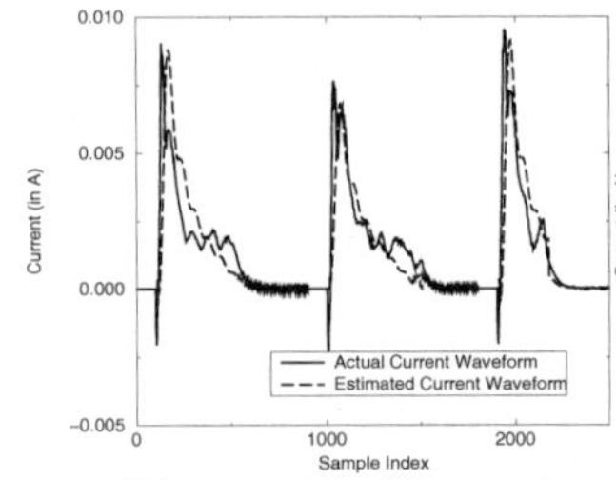

Figure 16. For parity, ham dist = 14.

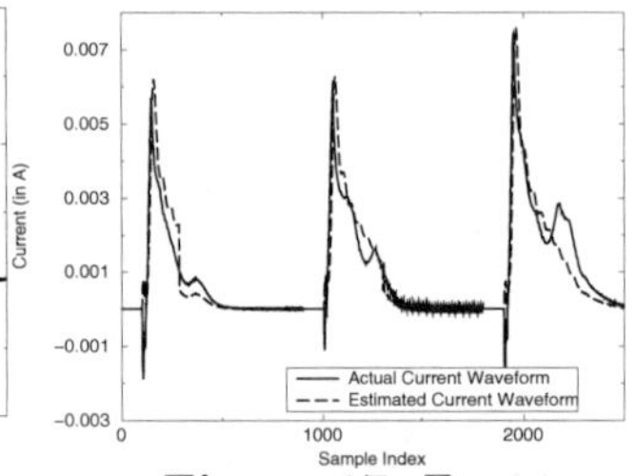

Figure 17. For mux, ham dist = 19.

6. CONCLUSION

In order to enable early block-level analysis of the power grid, when using hard IP blocks, we have proposed a cycle-based current waveform modeling technique that involves predicting the DCT of the current waveform from the input vector pairs, and using the inverse DCT to get back the time-domain waveform. The parameters of the DCT template function are expressed in terms of input vector pairs, using regression. While we continue to improve the estimation and validate this technique on more circuits, the data so-far show that this type of estimation is indeed possible, enabling early block level current estimation for power grid design.

Table 2: Benchmark circuits used in Figs. 5–17.

Circuit	#I	#Cells	RMS_{avg}	$e_{pavg}\%$	$e_{tavg}\%$	#sim.
c880	60	383	.0032	11.98	5.92	7000
alu2	10	368	.0023	17.70	7.51	700
c432	36	217	.0030	12.43	6.45	4500
cu	14	48	.00049	18.91	1.99	770
f51ml	8	105	0.0011	19.21	2.12	320
mux	21	91	.0004	17.86	3.00	2000
random8	8	158	0.0010	14.14	6.02	320
parity	16	68	.00077	15.37	3.98	640
x2	10	50	.00054	14.35	2.43	500

7. REFERENCES

[1] S. Bodapati and F. N. Najm. Frequency-domain supply current macro-model. In *ISLPED*, pages 295–298. IEEE, 2001.

[2] A. Bogliolo and L. Benini. Node sampling: A robust RTL power modeling approach. In *IEEE ICCAD*, pages 461–467. ACM/IEEE, 1998.

[3] A. Bogliolo, L. Benini, G. D. Micheli, and B. Ricco. Gate-level current waveform simulation of CMOS integrated circuits. In *ISLPED*, pages 109–112. IEEE, 1996.

[4] A. Dharchoudhury, R. Panda, D. Blaauw, R. Vaidyanathan, B. Tutuianu, and D. Bearden. Design and analysis of power distribution networks in PowerPC microprocessor. In *ACM/IEEE DAC*, pages 738 – 743. IEEE, 1998.

[5] R. F. Gunst and R. L. Mason. *Regression Analysis And Its Application, A Data-oriented approach.* Marcel Dekker, Inc, 1980.

[6] S. Gupta and F. N. Najm. Power macromodeling for high level power estimation. In *34th ACM/IEEE DAC*, pages 365–370. IEEE, 1997.

[7] S. Gupta and F. N. Najm. Energy-per-cycle estimation at RTL. In *ISLPED*, pages 121–126. IEEE, 1999.

[8] A. K. Jain. *The Fundamentals Of Digital Image Processing.* Prentice-Hall, Inc, 1989.

[9] P. E. Landman and J. M. Rabaey. Architectural power analysis: The dual bit type method. *IEEE Trans. on VLSI Systems*, 3:173–187, June 1995.

[10] A. V. Oppenheim and R. W. Schafer. *Discrete-Time Signal Processing.* Prentice-Hall, Inc, 1990.

[11] S. R. Powell and P. M. Chau. Estimating power dissipation of VLSI signal processing chips: The PFA technique. *VLSI Signal Processing IV*, pages 250–259, March 1990.

[12] Q. Qiu, Q. Wu, C.-S. Ding, and M. Pedram. Cycle-accurate macro-models for RT-level power analysis. *IEEE Tans. on VLSI Systems*, 6:420–528, December 1998.

[13] A. Raghunathan, S. Dey, and N. K. Jha. Register-transfer level estimation techniques for switching activity and power consumption. In *IEEE ICCAD*, pages 158 – 165. IEEE, 1996.

Macro-Modeling Concepts For The Chip Electrical Interface

Brian W. Amick
Sun Microsystems Inc.
Austin, TX
Brian.Amick@Sun.com

Claude R. Gauthier
Sun Microsystems Inc.
Sunnyvale, CA
Claude.Gauthier@Sun.com

Dean Liu
Sun Microsystems Inc.
Sunnyvale, CA
Dean.Liu@Sun.com

ABSTRACT

The power delivery network is made up of passive elements in the distribution network, as well as the active transistor loads. A chip typically has three types of power supplies that require attention: core, I/O, and analog. Core circuits consist of digital circuits and have the largest current demand. In addition to all of the system issues/models for the core, modeling the I/O subsystem has the additional requirement of modeling return paths and discontinuities. The analog circuits present yet a different challenge to the macro-modeling of the supply network because they place a tight demand on supply variations. This paper presents a design methodology on how to generate macro-models of the entire chip electrical interface. This methodology can be used by the chip, package, and system designers and is being used to design high-reliability servers.

Categories and Subject Descriptors

C.5.3 **[Computer System Implementation]**: VLSI Systems.

General Terms

Performance, Design, Reliability.

Keywords

VLSI Power Distribution, Inductance, High Speed Microprocessor Design, Analog and I/O Power Delivery.

1. INTRODUCTION

Delivering power to VLSI chips is an increasingly difficult challenge as voltage scaling decelerates, the supply current and power dissipation accelerate, and the device count increases. As voltage scales, the tolerances on the power supply get tighter, while the noise sources actually increase in magnitude. Processor power transients can be large in magnitude and occur at arbitrary frequencies though the largest power transients occur at relatively low frequencies, necessitating a model of the entire system. I/Os do not escape this challenge as I/O counts and frequencies increase. The high-frequency signal return current associated with each of these I/O circuits must travel through the entire network from chip to package to board.

A potential road-block to meeting this objective is the fact that most of the data required for design verification is available only when the design is completed, and the system is ready to ship. Also, when the data becomes available, the simulation problem can be intractable, owing to the amount of elements and information present.

Power-delivery models for core, IO, and analog supplies requires modeling of the entire system (typically circuit-board/package/die) to capture the basic first order effects. This paper will address the modeling constructs required to understand these effects early in the design cycle in such a way that the design can be influenced. Examples will be presented that illustrate the impact of the various components of the core, I/O, and analog supply models.

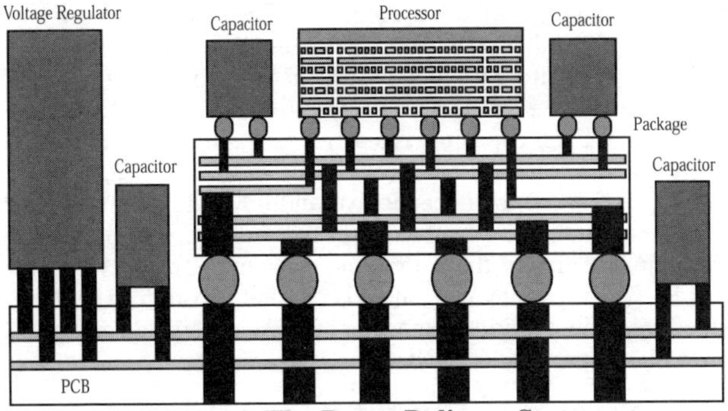

Figure 1. The Power Delivery System

2. CORE POWER MODELING

A microprocessor power delivery system is illustrated in Fig. 1. The major components include the voltage regulator, PCB stack-up, capacitor (type, quantity, and placement), the package type/stack-up, package capacitors, and on-die capacitance. These major issues are resolved by creating a system-level model of the core power distribution. For low/mid-frequencies, one can model this complex network as shown in Fig. 2. In Fig. 2 the parasitic inductance comes from: bumps, package and board vias, planes and perforated planes, decoupling capacitor parasitics, and the voltage regulator. The capacitance comes from on-chip capacitance (non-switching gates and decoupling capacitance) and discrete off-chip capacitors. The model is accurate in the 1 to 200MHz range and will serve to develop the system, package, and on-chip capacitor requirements. This type of model is not meant to see the effects of localized, very high-frequency on-chip noise, however it can be expanded into 2D/3D models to extend the effective frequency range upwards.

The off-chip inductance (L) and on-chip capacitance (C) form a LC tank circuit whose impedance peaks at a the chip/package resonant frequency (between 10MHz-100MHz in most systems). Depending on system configuration the peak can be quite large. The resonance can be excited by instruction patterns that cause large transients in power. However, the excitation is not limited to a sine or square wave, as any large power transient will contain energy at the frequencies of interest. If this resonance is not removed, then large power transients will degrade the supply significantly over multiple cycles, thereby reducing performance and reliability. The most substantial voltage droop in a microprocessor power delivery system is associated with this frequency band. The following sections addresses how to estimate the various parasitics in Fig. 2, and formulate a methodology for designing core power distribution networks using a macro-model of the chip electrical interface.

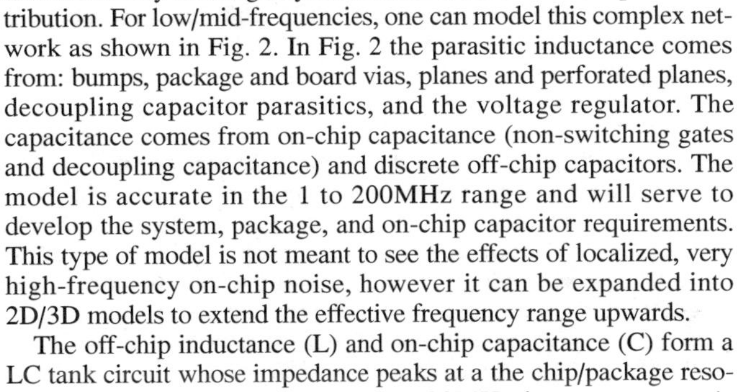

Figure 2. Core supply equivalent circuit (low/mid frequencies)

2.1 Decoupling Capacitors

Voltage regulators contain a feedback mechanism that allows them to respond to changes at the load. However the bandwidth of these regulators are typically limited to the 10-kHz to 1-MHz range [1], above which they behave like inductors. The inductor value can be obtained from the voltage regulator vendor and is highly dependent on the design of the connector. Therefore, a hierarchy of decoupling is used to provide power at higher frequencies.

Discrete capacitors are far from ideal as each capacitor contains an equivalent series resistance (ESR) and an equivalent series inductance (ESL). Each capacitor is a series resonant circuit that is most effective at its resonant frequency (f_O) where it appears resistive. Below f_O the response is capacitive and beyond f_O the inductance dominates. Capacitor effectiveness is limited by the capacitor's ESL and the inductance of the path to the capacitor [2]. The ESL can be reduced by using many capacitors is parallel. In order to explore the reduction of the path inductance, a model of the package and PCB power planes must be derived.

2.2 Package and Circuit Board Power Planes

A typical system contains both non-perforated and highly perforated power planes. The non-perforated sections of power planes are the sections between capacitors and the package. Perforated power planes are those that lie under the package where hundreds or thousands of vias cut through the planes. The loss of metal increases the inductance and resistance, which can be represented using a derating factor based on the amount of metal lost. Power planes have R, L, and C components, however for low/mid-frequencies the C can typically be ignored as this distributed capacitance will be relatively small compared to explicate capacitance. Different approaches to finding the spreading inductance can be found in [3] and [4].

Multiple planes in parallel can be treated as resistors in parallel, given an alternating power-ground stack-up. For example, two pairs of power planes have one third the inductance of a single pair because of the addition of a dielectric layer between the pairs of planes. Therefore, the key to managing inductance lies in using multiple pairs of planes, with thin spacing. A very important package consideration is the use of capacitors which will require a low impedance path through the horizontal package power planes. These planes are modeled just like perforated and unperforated planes on the PCB. With the use of package capacitors, it is possible to eliminate a high-Q chip/package resonance.

2.3 Vias, Pins, & Bumps

Hundreds or thousands of vias are used to bring the core power from the PCB power planes to the pads connecting to the package and between different layers in the package. The inductance is proportional to the via height and inversely proportional to the total number of vias. Typically core power is brought in through the middle section of the package pin array (to alleviate I/O routing issues). The layer location of PCB power planes controls the via lengths.

In general packages are a series of vertical via layers and pins that connect the PCB to the chip along with some horizontal planes of metal connecting the vias inside of the package. For the model shown in Fig. 2, the package planes can be ignored in the vertical path from chip to PCB. Along the way the via dimensions change to reflect the connection dimensions at the package ends. The total minimum thickness of the package is dominated by the number of routing layers needed to route the IO signals or by mechanical reasons. It often takes several layers for the IO signals to escape towards the edge of the package.

The vertical bump/via/pin inductances can be derived from approximations which relate the self and mutual inductance of a pair of parallel filaments [4]. The mutual inductance (M) of two equal parallel straight filaments is given by:

$$M = \frac{u_o}{2\pi} l \left[\ln\left(\frac{l}{sep} + \sqrt{\left(1 + \frac{l^2}{sep^2}\right)} \right) - \sqrt{\left(1 + \frac{sep^2}{l^2}\right)} + \frac{sep}{l} \right] \quad (1)$$

where l is the via length and sep is the separation distance between the center of the vias. To obtain the self inductance, we replace sep by the radius of the wire (or half the diameter, d) and add the internal inductance of the wire ($u_r/4$) we obtain the self inductance (L):

$$L = \frac{u_o}{2\pi} l \left[\ln\left(2\frac{l}{d} + \sqrt{\left(1 + \frac{4l^2}{d^2}\right)} \right) - \sqrt{\left(1 + \frac{0.25d^2}{l^2}\right)} + 0.5\frac{d}{l} + \frac{u_r}{4} \right] \quad (2)$$

It is of note that at higher frequencies, where the skin effect is predominant, the internal inductance term drops out.

Using EQ 2, the self inductance of each via/pin in the package can be calculated along with the mutual inductance, using EQ 1, between all pairs of vias/pins in the package. Once the inductive terms of the package vias/pins are computed the network is collapsed into a two-port network as illustrated in Fig. 3(a) by placing all power inductors in parallel and then in series with the parallel combination of ground inductors. The network can be excited with a 1-A_{AC} current source, and by measuring voltage at a given frequency, ω, we can solve for the equivalence inductance by applying Ohms law. Using this result, an equivalent single inductor and resistor can be found for arbitrary patterns such as the ones shown in Fig. 3(b) [5]. This methodology can be applied to the several alternatives for socketing the microprocessor including LGA, BGA, or μPGA. From a modeling perspective, the major difference is brought about by the pin/ball lengths and the pin/ball pattern used.

2.4 Chip Model

The processor determines the power transient behavior and contains a substantial amount of on-die capacitance. The on-die capacitance comes from non-switching logic, switching logic, bypass capacitance that is local to the logic, and global (in the channels) capacitance (as illustrated in Fig. 2). The local capacitance can be represented as a capacitor with a low ESR. Inductance is usually ignored on-chip in this analysis as off-chip inductances and on-chip resistance dominate at low/mid frequencies. The ESR for local capacitance is usually dominated by the channel resistance and increases as the channel length increases. Global capacitance resides in the channels, often far away from the logic. At this distance, the ESR is dominated by the power grid, and the capacitor is modeled with a much higher ESR than its local counterpart.

Digital logic is represented by a series pair of parallel capacitor and resistor. For non-switching logic, one of the two resistors is set to a very high value and the other is set to give a RC time constant equal to a fanout-of-four (FO4) inverter. Switching logic (including leakage and crow-bar current) can be represented by either switch-

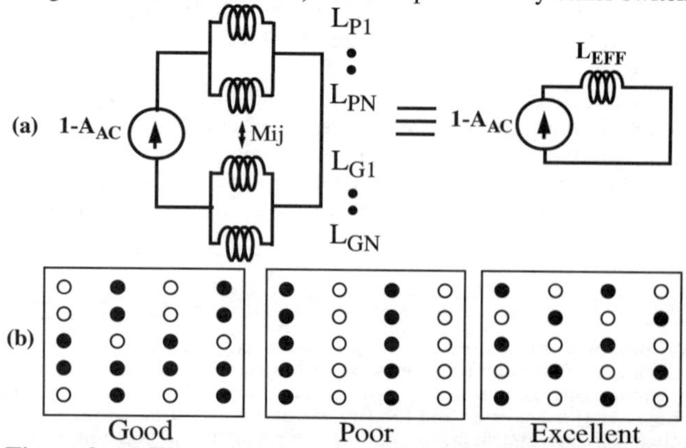

Figure 3. (a) The pattern can be simulated in the AC domain by mapping the L's, M's to a passive circuit. (b) Effective inductance is strong function of placement.

ing the resistors in another copy or copies of the circuit used for non-switching logic. Leakage current is modeled by making the higher "off" value resistor just large enough to still draw the correct amount of leakage. Crow-bar current is modeled by having both resistors in a low resistance state for a short amount of time. Finally by using multiple copies of the circuit, an expected current profile, from an architectural power model, can be modeled.

2.5 Capacitor Selection & Sizing Methodology

We have assembled all the pieces required for the sizing of capacitors at the various hierarchies. Several different types of capacitors are available, each with a different cost, area, and frequency response (due to ESR and ESL). The goal is to find a combination of the board, package and chip capacitors that reduces the overall impedance of the distribution network to within a tolerable limit, while fitting within the allotted area and for a reasonable total cost.

A model of the system similar to Fig. 2 is first constructed, with the parasitic inductance and resistance calculated based on physical geometry using the methodology described in the previous sections. The chip load is excited using an AC waveform with its frequency as the sweeping parameter. Using this model, the optimal capacitor size for each hierarchy can be found with a greedy approach. For board capacitors, the frequencies of interest range from 1-MHz to 10-MHz, while the package capacitors are responsible for frequencies from 10-MHz to 200-MHz.

Using the greedy approach, a unit quantity of the cheapest capacitor is added into the model. The impedance of the network across the frequency spectrum of interest is obtained by sweeping the frequency of the AC load using a circuit simulator. If the network impedance is above the target, another capacitor is added. Each additional capacitor lowers the impedance peak while possibly shifting its frequency. If the impedance is still above the target when the allotted decap area is full with the cheapest capacitor, then replace one capacitor with the next-most expensive type, assuming that the higher cost capacitor is more effective in lowering the overall impedance at higher frequencies. The search stops when the desired impedance is met. Algorithms for proper placement of the on-chip decoupling capacitance can be found in [6].

Fig. 4 shows the transient response to a step excitation. The absence of ringing in the design that meets the target impedance prevents even larger excitations from a repetitive source. The use of the methodology leads to a substantial increase in Vdd which would allow for higher frequency operation at a relatively low cost to the overall system.

3. I/O POWER MACRO-MODEL

I/O power modeling is an extension of the core power problem, but is slightly more complex because of the signals and return currents. Fortunately the PCB, vias, decoupling capacitors, and voltage regulator modeling constructs are the same as shown in the core power section. The current loops for I/O circuits are more complex

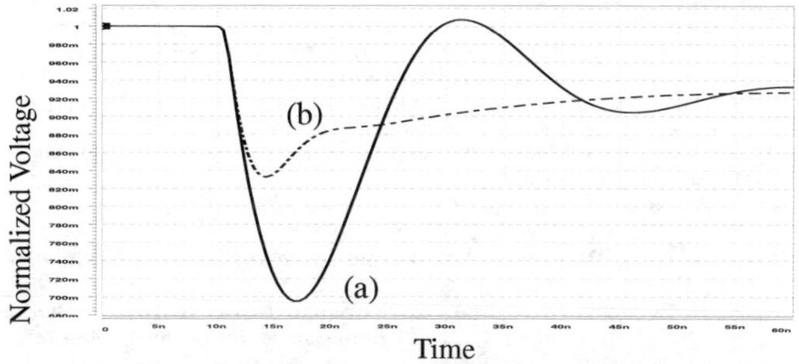

Figure 4. The effects of a 40A transient as seen from the chip (a) without a methodology for sizing capacitors and (b) using the proposed methodology to reduce the supply impedance

than that of the core because they include off-chip parasitics resulting in Vdd/Gnd bounces independent of each other across different chips. Therefore, careful attention must be paid to make sure return currents are modeled correctly. It is the flow of these return currents that causes large Vdd/Gnd bounces from either an overly inductive supply network or discontinuities in the signal return path.

In the presence of Vdd and Gnd bounces, the resulting timing and voltage margins will be reduced because the waveform will have increased jitter and reduced signal swing. In some cases the bounces, themselves, can be transmitted across a channel as noise. In addition, if the links are single-ended, the reference voltage may be inaccurate as well since it is often referenced to a different Gnd than the receiver.

To model these return currents correctly, signal traces must be referenced to the correct Vdd and/or Gnd nodes. This is done by using a 2D field solver, such as Ansoft, to extract the traces and reference planes. In addition the use of ideal ground (SPICE node 0) should be avoided in all locations except for one spot as this node is an infinite source/sink for current. This special node can allow current to avoid discontinuities in the system and jump from one point to another, masking important effects.

An example model that illustrates the major power supply effects that must be captured is shown in Fig. 5. The model represents two voltage mode drivers whose output signals are referenced to both Vdd and Gnd planes in the package, but in the PCB they are between Gnd planes only. PCB traces are typically only referenced to Gnd planes since Vdd is not routed through connectors. Note the signal being between Vdd and Gnd planes in the package is reflected in the model by the capacitors to Vdd and Gnd. Also, the extraction contains important capacitors between Vdd and Gnd, representing where signal traces are not present between reference planes.

While Fig. 5 is all that is required to accurately model the IO supply, it raises some possible issues that must be resolved. The first is the Vdd and Gnd bounce from the large supply inductance. The bounces from the supply network can be reduced by increasing the on-chip capacitance or by lowering the inductance to other decoupling capacitors (e.g., through thinner spaced planes and more pins).

The second issue is the discontinuity in the signal return when the signal travels from the package to the PCB. In the package, approximately half the return current flows in Vdd, and half through Gnd. However, in the PCB, all the return current flows through Gnd. Therefore half the return current must cross the discontinuity through the capacitance between Vdd and Gnd in the package,

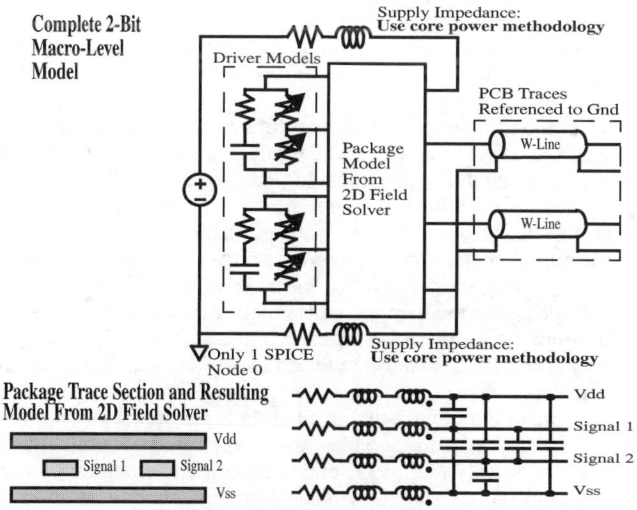

Figure 5. I/O Macro-Model

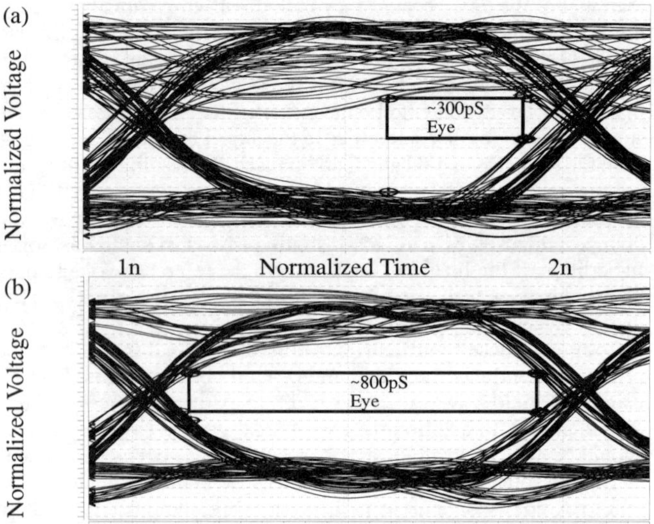

Figure 6. Effect of discontinuities to the received signal eye (a) and effect of using capacitor to satisfy discontinuities (b).

which will collapse the supply voltage in the package and on-chip. The effect, illustrated in Fig. 6a, can be avoided by placing low ESR and ESL capacitance between Vdd and Gnd near the return current discontinuity either on the PCB or on the package (illustrated in Fig. 6b). In contrast to Fig. 6a, which shows the data eye without the package capacitors, the eye opening in Fig. 6b is noticeably larger. This example model can be extended to reflect different numbers of signals by extracting more traces, scaling the non-package elements, and adding additional on-chip elements. Different configurations of capacitances, vias, and power planes, can be evaluated. The I/O macro-model allows us to quickly investigate all of the trade offs associated with the design.

4. ANALOG POWER MACRO-MODEL

Typically microprocessors have some analog circuits such as a PLL or DLL whose effectiveness is related to the stability of the power supply. PLL and DLL jitter is dominated by power supply noise [7]. The issues addressed by the analog supply macro-model include supply isolation, signal return current strategy, and decoupling capacitor methodology. The digital circuits on a microprocessor often cause significant supply fluctuation due to large switching current. Often the fluctuation is too large for analog circuits and it is advantageous to isolate the supply of the analog circuits. The system level approach to providing supply isolation is to route a separate quiet power supply from the PCB as a trace in the package to the processor. The macro-model for this very common approach is illustrated in Fig. 7.

The model includes a PCB filter, the pin and via impedance, a transmission line to represent the package trace, the bump and via impedance, and the analog transistor load. The PCB filter is used to isolate the board and chip. The impedance for the pin, bump, and vias are calculated using the parasitic estimation methods described in the earlier section on core power macro-modeling. In the package, the analog supply is routed as a signal sandwiched between ground planes and widely spaced from other traces in order to avoid coupling from other noisy signals. If the analog supply is to be routed near other signals, then the coupling between them must be included as well.

One important issue is that in almost all microprocessor applications, circuits on analog supplies have to communicate with circuits on another supply (e.g., PLL output driving the clock tree on the core supply). If the on-chip Gnd grids are not shorted, the current loop will include the package and substrate impedances. This could

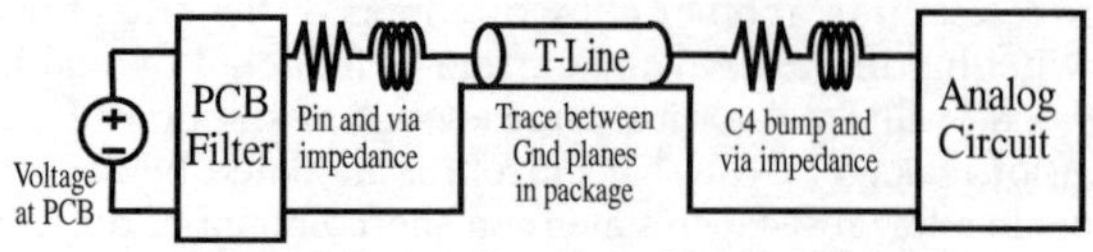

Figure 7. Analog Supply Macro-Model

result in a large portion of the signal being lost across these impedances as power supply bounce, a slowing of the signal rise/fall times, duty-cycle distortion, and jitter in the signal. Therefore the Gnd domains should be shorted in metal as metal results in a very small and controlled impedance.

The other critical issue with system level supply isolation is determining the amount of on-chip decoupling capacitance needed for the analog supply. This can be done by exciting the input to the macro-model in Fig. 7 with a noise source and finding the minimum amount of on-chip decap required for the circuits to meet required specifications. Decoupling capacitance and jitter performance can typically be traded for each other. As a final check a model should be simulated to make sure the decoupling capacitance is sufficient to limit supply noise from the return currents flowing through the supply decoupling capacitance when crossing power domains.

5. SUMMARY

Scaling trends in VLSI have created significant challenges for designers including chip power delivery and distribution as supply voltages have decreased, while transistor counts, frequency, and current have increased. A good model includes information about the entire distribution network, including the passive elements in the distribution network (decoupling capacitor hierarchies, board and package power planes, vias, and bumps), as well as the active transistor loads (core, I/O, and analog). This paper presented a methodology on how to construct the macro-models of the chip electrical interface, discussed the trade-offs involved in the design of various components, and provided some design solutions.

For I/O power, the same macro-models were used, however the model was extended to address signal return currents and potential discontinuities. Solutions to typical problems were presented within the context of the I/O macro-model. Finally, a macro-model for analog power was presented to address the major issues when analog and digital circuits reside on the same chip. Issues brought up by the model included the need for a supply isolation approach, decoupling capacitor methodology, and the interface issues associated with inter-power-domain communication. These models are currently being used to design high-reliability servers in industry.

6. ACKNOWLEDGMENTS

The authors would like to acknowledge the contributions of R. Wheeler, M. Blatt, S. Bobba, C. Durham, V. Ganesan, D. Greenhill, R. Heald, S. Pannala, P. Patel, L. Noujeim, S. Schmidt, B. Sen, G. Singh, L. Smith, J. Tam, P. Trivedi, R. Voelker, B. Wong, and G. Yee

7. REFERENCES

[1] L.D.Smith *et al.*, "Power Distribution System Design Methodology and Capacitor Selection For Modern CMOS Technology," *IEEE Transactions on Advanced Packaging*, August 1999, vol 22, no 3.

[2] Korony *et al*, "Controlling Capacitor Parasitics For High Frequency Decoupling Capacitors," *Proc. IMAPS Conference*, 2001.

[3] W. Dally and J. Poulton, "Digital Systems Engineering," Cambridge University Press, 1997.

[4] F. W. Grover, "Inductance Calculations Working Formulas and Tables," D. Van Nostrand Company, 1946.

[5] R. Wheeler, "Modeling Simultaneous Switching Noise (SSO) in the Z-axis Direction of VLSI Packages and PCB", www.wheeler.com,1999.

[6] H. H. Chen and D. D. Ling, "Power Supply Noise Analysis Methodology for Deep-Submicron VLSI Chip Design," *Proc. Design Automation Conference*, June 1997.

[7] V. Kaenel, *et al.*, "A 320 MHz, 1.5mW @ 1.35-V CMOS PLL for Microprocessor Clock Generation," *IEEE J. of Solid-State Circuits*, Vol. 31, No. 11, pp. 1715-1722, November 1996.

Modeling and Analysis of Regular Symmetrically Structured Power/Ground Distribution Networks†

Hui Zheng and Lawrence T. Pileggi
Carnegie Mellon University, Department of Electrical and Computer Engineering
5000 Forbes Avenue
Pittsburgh, PA 15213
{hzheng, pileggi}@ece.cmu.edu

ABSTRACT

In this paper we propose a novel and efficient methodology for modeling and analysis of regular symmetrically-structured power/ground distribution networks. The modeling of inductive effects is simplified by a folding technique which exploits the symmetry in the power/ground distribution. Furthermore, employment of susceptance [10,11] (inverse of inductance) models enables further simplification of the analysis, and is also shown to preserve the symmetric positive definiteness of the circuit equations. Experimental results demonstrate that our approach can provide up to 8x memory savings and up to10x speedup over the already efficient simulation based on the original sparse susceptance matrix without loss of accuracy. Importantly, this work demonstrates that by employing limited regularity, one can create excellent power/ground distribution designs that are dramatically simpler to analyze, and therefore amenable to more powerful global design optimization.

Categories and Subject Descriptors

B.7.2 [**Integrated Circuits**] Design Aids - *verification*

General Terms

Design, Verification

Keywords

Power/Ground Distribution, Susceptance, Folding Technique, Design Regularity

1 INTRODUCTION

The role of a power/ground distribution network is to supply stable voltage references to the on-chip circuitry. Due to the large amount of time-varying current surges, however, designing a reliable power/ground distribution is one of the most challenging design problems due to IR drops, Ldi/dt noise and electromigration [2-9]. Consequently, as the current requirement goes up due to the increasing density of VLSI chips, power/ground distribution networks require more precise design considerations and more stringent noise budgets -- particularly as voltage supply levels are scaled to control power dissipation.

†This work was supported by Semiconductor Research Corporation (SRC) under contract 2000-TJ-778.

DAC 2002, June 10-14, 2002, New Orleans, Louisiana, USA.

Traditionally, the on-chip power/ground distribution network is modeled as a resistive network for IR drop analysis. For static/DC analysis, only current sources modeling the average current consumption are attached to the network. For dynamic/transient analysis, there are time-varying current sources modeling the switching activity of circuit blocks, and capacitors that model decoupling and intrinsic capacitance to simulate the dynamic behaviors. To accommodate the millions of resistors in such power/ground network models, hierarchical approaches [5] and multigrid-like techniques [6] have been developed. An obvious solution to the IR drop problem is to size up the wires in a power/ground distribution network. However, as the width and thickness of the wires increase, the inductance effect becomes evident for the on-chip metal layers. Furthermore, for high-performance systems such as microprocessors, the C4 packaging technology dictates the combined analysis of on-chip power grid and packaging. There has been some work which attempts to address inductance for power grid analysis, but the power grid models are generally oversimplified by either assuming tree structures or only considering self inductance [7,8,9].

In this paper we propose a novel and efficient methodology for modeling and analysis of regular symmetrically structured power/ground distribution networks. Based on a folding technique and the susceptance concept [10,11], very simple circuit models can be used to accurately capture magnetic couplings for very large power/ground distribution networks. As we will explain in detail, such modeling and analysis simplification requires certain *regularity* of a power/ground distribution network. However, imposing *some* forms of design regularity may be just what is needed to address the increasingly difficult verification problem for large power/ground distribution networks. Such forms of regularity seem to be imperative in general for closing the ever increasing "design productivity gap" [1] that is making application-specific IC design less affordable. In addition, the simplifications brought on by regularity may facilitate powerful global optimizations -- which otherwise would be impossible for more complicated models -- that ultimately produce a superior design.

This paper is organized as follows. Section 2 presents a folding technique which exploits the symmetry in power/ground distribution and is applicable both for inductance and susceptance-based models. Section 3 shows that for certain power/ground configurations, based on the properties of the susceptance matrix, the S-based model can be further simplified to facilitate much more efficient simulation without compromising the accuracy. In Section 4, the validity and efficacy of our approach is verified by the experimental results and some discussions of regularity are also provided. Section 5 draws our final conclusions.

2 FOLDING TECHNIQUE

A typical circuit model of a power/ground distribution consists of two linear networks representing the Vdd and Gnd supply respectively. The circuit macroblocks between the two networks are modeled as time-varying current sources and capacitance (including decoupling and intrinsic capacitance.) Due to the inherent symmetry, when the model does not include mutual inductive couplings, it is easy to decouple the Vdd and Gnd networks through current source transportation and capacitance splitting. We will further show here, however, that even in the presence of mutual inductance, the symmetry in power/ground distribution still enables us to reduce the circuit size by half via a circuit folding technique.

Referring to Fig. 1, suppose L_{D1} and L_{G1} correspond to two symmetrical segments in the Vdd and Gnd networks, and L_{D2} and L_{G2} correspond to a second pair. Since the self inductances and mutual inductances of the segments can be evaluated with close-form formulas which are functions of geometrical parameters, the geometrical symmetry in power/ground distribution guarantees that $L_{D1} = L_{G1}$, $L_{D2} = L_{G2}$, $M_{DD} = M_{GG}$ and $M_{DG11} = M_{DG22}$. $M_{DG12} = M_{DG21}$ approximately holds under certain conditions. Our results will show that the slight difference between M_{DG12} and M_{DG21} does not affect the accuracy of our approach to any discernible level. Therefore, for notational simplicity, we assume M_{DG12} and M_{DG21} are exactly the same for now. In summary, the geometrical symmetry results in the circuit symmetry, which further leads to the symmetry in current distribution, i.e., $I_{D1} = -I_{G1}$ and $I_{D2} = -I_{G2}$. As a result, the voltage across L_{D1} can be represented as:

$$\begin{aligned} V_{D1} &= \frac{dI_{D1}}{dt}L_{D1} + \frac{dI_{D2}}{dt}M_{DD} + \frac{dI_{G1}}{dt}M_{DG11} + \frac{dI_{G2}}{dt}M_{DG21} \\ &= \frac{dI_{D1}}{dt}(L_{D1} - M_{DG11}) + \frac{dI_{D2}}{dt}(M_{DD} - M_{DG21}) \end{aligned}$$

which exemplifies that the magnetic couplings between Vdd and Gnd networks can be folded into the self-inductances and mutual inductances inside one of the networks.

More generally, the entire inductance matrix can be formed as:

$L = \begin{bmatrix} L_{DD} & L_{DG} \\ L_{DG}^T & L_{GG} \end{bmatrix}$, where L_{DD} and L_{GG} represent the inductance matrices for the Vdd and Gnd networks respectively and are identical; L_{DG} models the mutual inductances between the Vdd and Gnd networks and where ideally $L_{DG} = L_{DG}^T$. Thus, the folding technique for L can be understood as follows:

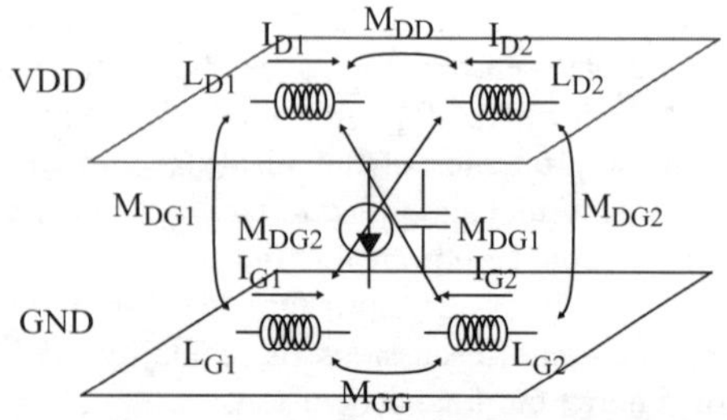

Fig. 1. Symmetry in Power/Ground Distribution

$$\begin{bmatrix} V_{DD} \\ V_{GG} \end{bmatrix} = \begin{bmatrix} L_{DD} & L_{DG} \\ L_{DG}^T & L_{GG} \end{bmatrix} \begin{bmatrix} i_{DD} \\ -i_{DD} \end{bmatrix} = \begin{bmatrix} L_{DD} - L_{DG} & 0 \\ 0 & L_{GG} - L_{DG}^T \end{bmatrix} \begin{bmatrix} i_{DD} \\ -i_{DD} \end{bmatrix} \quad (1)$$

Obviously, it holds that $V_{DD} = -V_{GG}$.

Importantly, this folding technique also works for the susceptance matrix [10, 11], since S is also a *block symmetric* matrix:

$$S = L^{-1} = \begin{bmatrix} L_{DD} & L_{DG} \\ L_{DG}^T & L_{GG} \end{bmatrix}^{-1} = \begin{bmatrix} S_{DD} & S_{DG} \\ S_{DG}^T & S_{GG} \end{bmatrix} \quad (2)$$

where $S_{DD} = S_{GG} = (L_{DD} - L_{DG}L_{DD}^{-1}L_{DG})^{-1}$ and $S_{DG} = S_{DG}^T = -L_{DD}^{-1}L_{DG}(L_{DD} - L_{DG}L_{DD}^{-1}L_{DG})^{-1}$ which can be derived from the formulas in [13]. Similarly, the decoupling of the Vdd and Gnd networks in terms of S can be shown as:

$$\begin{bmatrix} i_{DD} \\ i_{GG} \end{bmatrix} = \begin{bmatrix} S_{DD} & S_{DG} \\ S_{DG}^T & S_{GG} \end{bmatrix} \begin{bmatrix} V_{DD} \\ -V_{DD} \end{bmatrix} = \begin{bmatrix} S_{DD} - S_{DG} & 0 \\ 0 & S_{GG} - S_{DG}^T \end{bmatrix} \begin{bmatrix} V_{DD} \\ -V_{DD} \end{bmatrix} \quad (3)$$

Thus, a susceptance-based modeling flow can be devised, and the substantial benefits that it provides will be described in the following sections. First, the window technique [10,11] is applied to extract half of the whole S matrix, namely S_{DD} and S_{DG}^T. Then the folded S matrix is calculated as $S_{fold} = S_{DD} - S_{DG}$. We can further construct the circuit for the Vdd portion by adding the extracted resistors, transported current sources and split capacitors. The voltage waveforms on the Gnd nodes can be deduced from the simulation results using the corresponding nodes in the Vdd portion.

3 FOLDED S IS NEARLY DIAGONAL FOR REGULAR, SYMMETRICAL STRUCTURE

As presented in section 3, the folding technique can reduce the simulation effort by cutting the circuit size by half. However, this speed-up is not sufficient for the analysis improvement that is required for large power/ground distribution networks. Fortunately, we can further simplify the simulation formulation by exploiting the fact that the folded susceptance matrix is nearly diagonal when the symmetrical wires in the Vdd and Gnd distribution are very close. It should be further stressed that this is also a guideline for creating good power/ground distribution designs in general.

To illustrate our point we use the power/ground configuration shown in Fig. 4. In Fig. 2 and Fig. 3, the magnitude maps of the original and folded susceptance matrices extracted from a 5x4 power/ground distribution network are shown. In order to contrast the magnitudes more clearly, an S matrix is normalized with regard to its diagonal elements, and every entry in the matrix is then replaced by the absolute value of its logarithm. In this way, we can easily find out how many orders of magnitude the off-diagonal elements are smaller than the diagonals. In the magnitude map of the original S matrix, we can clearly see the four block matrices as in Eq. (2). It is worth noting that all the entries in S_{DG} and S_{DG}^T are negative, and because of the shielding effect of susceptance elements, the diagonal entries in S_{DG} and S_{DG}^T which model the susceptive couplings between physically close symmetrical Vdd and Gnd segments have much bigger magnitudes than the other off-diagonal elements in S.

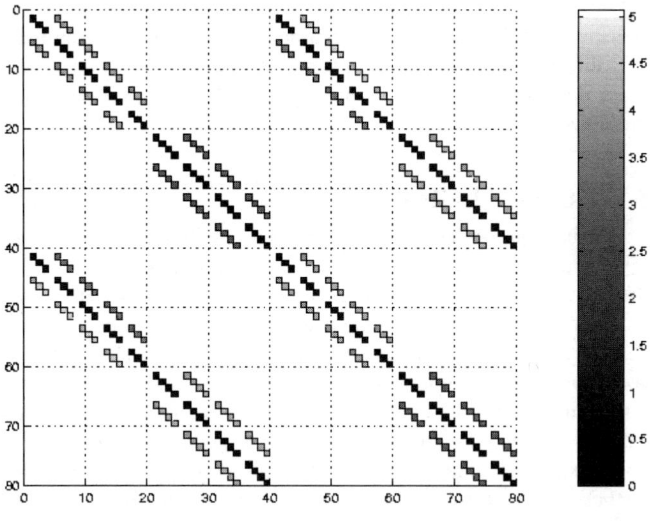

Fig. 2. Original S Matrix (normalized)

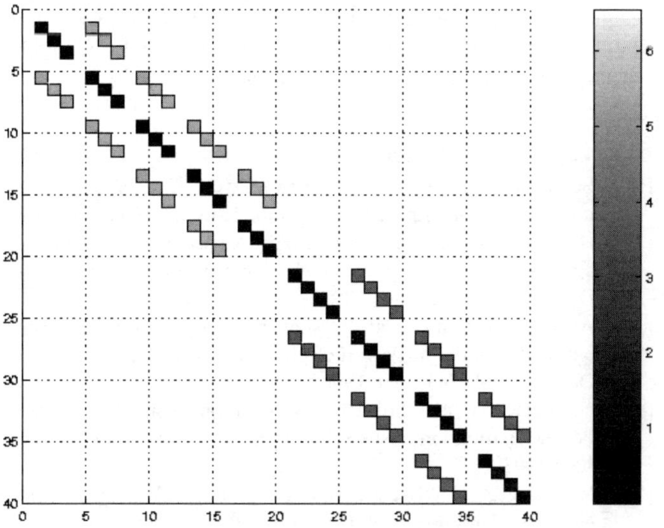

Fig. 3. Folded S Matrix (normalized)

The magnitude map in Fig. 3 shows that the folded S becomes extremely diagonally dominant. The reason is that the positive diagonal entries in s_{DD} become bigger after subtracting the negative diagonal entries in s_{DG}, while other off-diagonal elements become smaller. Therefore, it becomes possible to ignore the off-diagonal entries in the folded S without sacrificing too much accuracy for a symmetrical design such as the one in Fig. 4. However, the folding technique does not have the same effect on the inductance matrix as it does to the susceptance matrix. First of all, the original inductance matrix is not as sparse as an equivalent susceptance matrix, and it has been demonstrated that inductance is notoriously difficult to localize. Secondly, all of the terms in an inductance matrix are positive, therefore, the folding technique actually reduces the diagonal entries in the folded L matrix, which does not necessarily lead to a very diagonally dominant L matrix.

There are several benefits to using a diagonal S matrix in a transient simulation. In the extracted circuits modeling power/ground distributions, each segment is modeled as a resistor in series with a inductor or susceptor. For transient simulation, the Backward-Euler (BE) companion model for a segment, when there is no mutual susceptance, can be transformed into a combined companion model through a Norton-to-Thevinin equivalent circuit transformation. This enables elimination of one node for each segment. Also, it opens the door for further speedup through either a hierarchical [5] or a multigrid-like approach [6], since the symmetrical positive definiteness of the simulation formulation which previously existed only for RC circuits, is preserved for RCS circuits here [12].

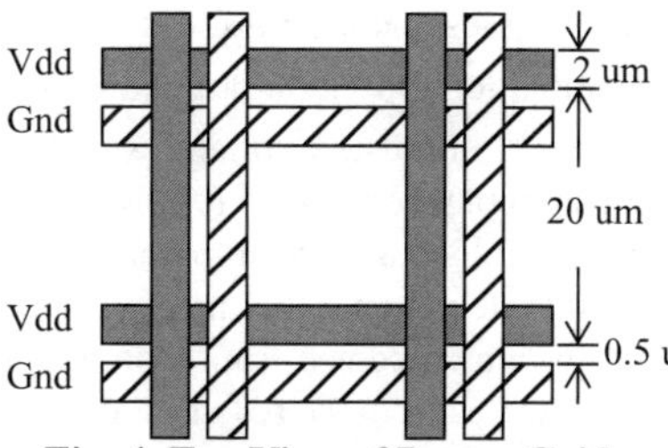

Fig. 4. Top View of Power Grid

4 EXPERIMENTAL RESULTS AND DISCUSSIONS

In order to test the methodology presented in Section 2 and 3, we created power grids following the configuration shown in Fig. 4. All of the wires have a width of 2 microns and a thickness of 1 micron, and the pitch for the Vdd and Gnd wire pairs in each layer is 20 microns. To generate power grids of different sizes, varying numbers of rails are used in each layer. Current sources modeling the circuit blocks and decoupling capacitors are inserted between the symmetric Vdd and Gnd nodes on the lower layer.

Note that the distance between two immediately adjacent Vdd and Gnd lines is very small (0.5 microns) compared to the pitch. As pointed out earlier, this choice is essential to the success of our methodology, since it not only guarantees the validity of the folding technique for s_{DG} is very close to s_{DG}^T as in Eq. (2), but also maximizes the shielding effects of susceptance elements which justifies the exclusion of the mutual coupling terms. After all, this configuration turns out to be a good design since the increased capacitance between Vdd and Gnd grids due to the small spacing becomes a valuable resource of decoupling capacitance.

Table 1. Comparison of Memory Usage and Runtime

	Original S			Folded S with Mutuals					Folded S without Mutuals				
	#nodes #nonzero	Memory (Mb)	Runtime (sec)	#nodes #nonzero	Memory (Mb)	Runtime (sec)	Memory Saving	Speed up	#nodes #nonzero	Memory (Mb)	Runtime (sec)	Memory Saving	Speed up
20x20	3003 102137	13.3	48.5	1722 26348	5.9	16.6	2.2x	2.9x	882 3444	3.3	4.8	4.0x	10x
40x40	11603 409937	49.0	249.1	6642 105508	18.0	91.8	2.7x	2.7x	3362 13284	6.7	22.2	7.3x	11.2x
60x60	25803 923337	110.8	650.1	14762 237468	40.3	249.2	2.7x	2.6x	7442 29524	12.5	56.9	8.9x	11.4x

With an extraction tool that implements the windowing technique [10], we generate the original and folded circuit netlists for power grids with size 20x20, 40x40 and 60x60. Then, a special simulator which supports susceptance elements [12] is used for simulating the power grid circuits. The runtimes and memory usages are tabulated in Table 1. For all simulations, a fixed time step is used and the number of time points is 500. All the statistics are from runs on an IBM Model 7042/7043 AIX machine. As we can see, the simulations based on folded and diagonal S models speed up by up to10 times and save up to 8 times memory usage compared to the simulations based on original S matrices.

Fig. 5 shows several simulated voltage waveforms on a node based on different susceptance models: (1) original S; (2) folded S with mutual terms; (3) folded S without mutual terms; (4) no S. We can see a very good match among the three waveforms. However, the fourth waveform, which comes from a simulation based on a RC model, clearly deviates from the results that include inductance effects. This demonstrates the necessity of modeling inductance for top-level on-chip power grid.

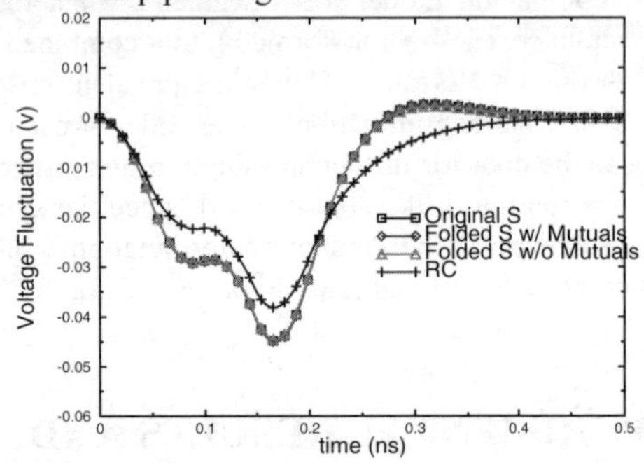

Fig. 5. Comparison of Voltage Waveforms (d = 0.5um)

To explore the efficacy of regular symmetrically structured designs, we further analyze this structure by attempting to violate the required regularity and symmetry by distances. We change the spacing between the Vdd and Gnd lines from 0.5 microns to 8.0 microns, which is the worst violation of the symmetry. As shown in Fig. 6, there is some visible discrepancy now among the new simulated waveforms from different susceptance models. Coincidentally, there seems to be a close match between the waveforms from the original S and the diagonal folded S, which is not true in general. Also note that the waveforms from the case of 0.5-micron distance show a smaller voltage drop, which can be attributed to the bigger coupling capacitance between the Vdd and Gnd lines.

Now the important question is: given a power/ground distribution network, how do we find out whether it is regular enough for our approach to work? We can find the answer by checking the diagonal dominance of the folded susceptance matrix. More importantly, having seen the great benefit from the simplified model, we may be able to turn around to ask a question from the design point of view: given a set of technology parameters, what regularity should we impose on the power/ground distribution design in order for it to be accurately analyzed by our methodology? Clearly such a methodology can be used to select proper design rules and spacings. By experimenting with susceptance matrices associated with small-scale power/ground distribution networks with different topologies, we can generate some design rules which can guarantee the accuracy of our modeling technique. In addition, the power/ground distribution designs, which follow those design rules, can be readily optimized by some of the existing powerful algorithms.

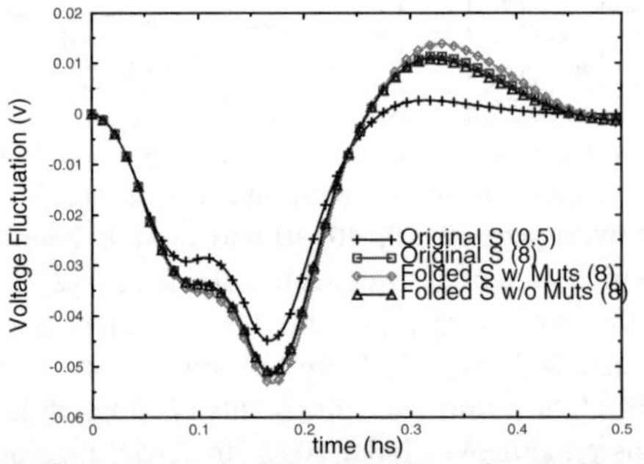

Fig. 6. Comparison of Voltage Waveforms (d = 8um)

5 CONCLUSIONS

We have presented a new and efficient methodology for modeling and analysis of regular symmetrically-structured power/ground distributions. Unlike other approaches which either ignore inductance or use oversimplified inductance models, our approach models the magnetic couplings in a realistic manner. The powerful combination of a folding technique and the susceptance concept results in extremely simple circuit models for regular symmetrically structured power/ground distributions, which not only speeds up transient simulation significantly, but also opens the door for further speedup through either a hierarchical [5] or a multigrid-like approach [6]. Furthermore, in order to take advantage of the greatly simplified verification offered by our approach, certain design rules can be generated for power/ground distribution design, and such designs can be further optimized which would be otherwise impossible with more complicated models.

6 REFERENCES

[1] Semiconductor Industry Association, "The International Technology Roadmap for Semiconductors: Design," 1999 Edition

[2] H. Chen and D. Ling, "Power Supply Noise Analysis Methodology for Deep-Submicron VLSI Chip Design," in Proc. 34-th DAC, pp. 638-643, June 1997.

[3] A. Dharchoudhury, et. al. "Design and Analysis of Power Distribution Networks in PowerPC Microprocessors," in Proc. 35-th DAC, pp. 738-743, June 1998.

[4] G. Steele, et. al. "Full-Chip Verification Methods for DSM Power Distribution Systems," in Proc. 35-th DAC, pp. 744-749, June 1998.

[5] M. Zhao, et. al. "Hierarchical Analysis of Power Distribution Networks," in Proc. 37-th DAC, pp. 150-155, June 2000.

[6] S. R. Nassif and J. N. Kozhaya, "Fast Power Grid Simulation," in Proc. 37-th DAC, pp. 156-161, June 2000.

[7] T. Chen and C. Chen, "Efficient Large-Scale Power Grid Analysis Based on Preconditioned Krylov-Subspace Iterative Methods," in Proc. 38-th DAC, pp. 559-562, June 2001

[8] H. Su, K. Gala and S. S. Sapatnekar, "Fast Analysis and Optimization of Power/Ground Networks," in Proc. IEEE ICCAD, pp. 477-480, Nov. 2000.

[9] S. Zhao, K. Roy and C-K Koh, "Frequency Domain Analysis of Switching Noise on Power Supply Network," in Proc. IEEE ICCAD, pp. 487-492, Nov. 2000.

[10] M. Beattie and L. Pileggi, "Efficient Inductance Extraction via Windowing," in Proc. DATE, pp. 430-436, March, 2001

[11] A. Devgan, H. Ji, and W. Dai., "How to Efficiently Capture On-Chip Inductance Effects: Introducing a New Circuit Element K," in Proc IEEE ICCAD, pp. 150-155, Nov. 2000

[12] H. Zheng, et. al. "Window-based Susceptance Models for Large-Scale RLC Circuit Analyses," in Proc. 2002 DATE, pp. 628 - 633, March, 2002

[13] W. H. Press, et. al. *Numerical Recipes in C*, 2nd Edition, pp. 77, Cambrige University Press, 1992

Clock Tree Optimization in Synchronous CMOS Digital Circuits for Substrate Noise Reduction Using Folding of Supply Current Transients

Mustafa Badaroglu[1,4], Kris Tiri[1,5], Stéphane Donnay[1], Piet Wambacq[1], Ingrid Verbauwhede[3], Georges Gielen[2], Hugo De Man[1,2]

[1]IMEC, Kapeldreef 75, B-3001 Leuven, Belgium
[2]ESAT - KU Leuven, Kasteelpark Arenberg 10, B-3001 Heverlee, Belgium
[3]EE Dept. - UCLA, 7440B Boelter Hall, P.O. Box 951594, Los Angeles, CA 90095-1594, USA
[4]Also Ph.D. Student at ESAT - KU Leuven, Leuven, Belgium
[5]Also Ph.D. Student at EE Dept. - UCLA, Los Angeles, USA
badar@imec.be

ABSTRACT
In a synchronous clock distribution network with zero latencies, digital circuits switch simultaneously on the clock edge, therefore they generate substrate noise due to the sharp peaks on the supply current. We present a novel methodology optimizing the clock tree for less substrate generation by using statistical single cycle supply current profiles computed for every clock region taking the timing constraints into account. Our methodology is novel as it uses an error-driven compressed data set during the optimization over a number of clock regions specified for a significant reduction in substrate noise. It also produces a quality analysis of the computed latencies as a function of the clock skew. The experimental results show >x2 reduction of substrate noise generation from the circuits having four clock regions of which the latencies are optimized.

Categories and Subject Descriptors
B.5.1 [**Register-Transfer-Level Implementation**]: Design – *data-path design.* B.6.1 [**Logic Design**]: Design Styles – *sequential circuits.* B.6.3 [**Logic Design**]: Design Aids – *optimization, simulation.* B.7.1 [**Integrated Circuits**]: Types and Design Styles-*VLSI.* B.8.2 [**Performance and Reliability**]: Performance Analysis and Design Aids.

General Terms
Algorithms, Design, Performance, Reliability.

Keywords
Substrate noise, di/dt noise, low-noise digital design, clock distribution networks, supply current shaping and optimization.

1. INTRODUCTION
There is a trend towards single-chip integration of more complex mixed-signal systems, higher speeds and lower supply voltages where the signal-integrity analysis becomes a challenging task. In mixed-signal ICs, substrate noise degrades the performance of the analog circuits due to noisy digital circuits in the same substrate.

A few publications concentrate on reducing the noise generation at the source. There are publications on low-noise logic cell design, such as low-voltage logic [1], current-mode logic [2], and CMOS gates with guard wiring and decoupling [3]. Speed degradation and lower noise margins are the drawbacks for low-voltage logic. Static power consumption increase is a major drawback of current-mode logic, not tolerable in large digital systems. The gates with decoupling and guard wiring have a drawback of increase in area and additional supply rails. Up till now, no good methodologies exist to reduce the substrate noise at its source without drawbacks such as increase in area and power increase, speed degradation.

Decreasing the peak and the slope of the supply current will reduce the substrate noise since a large part of the noise is generated due to the ringing of the damped LC-tank, which is formed by the on-chip capacitance and the package inductance with series resistance on the supply. Flattening the supply current profile requires an estimation of current waveforms, which is a complex task since:

- At a given time, one or more gates can switch simultaneously, depending on the inputs and the state of the circuit.
- For a gate, the supply current differs due to the gate, the load, the supply, the input transition time and the input state.
- The waveform width depends on the penetration depth of the switching into the combinatorial logic.

Designing regular blocks, where each block has the same delay and the same supply waveform pattern, can solve the problem. One must introduce a lot of logic redundancies to achieve this goal, which is not acceptable. Another possibility is to introduce different latencies in the different regions in a clock tree to make the supply current flatter. To optimize the latencies, the use of the total transient data of the supply current is not acceptable due to the complexity of an exhaustive search for optimum clock latencies. It is therefore necessary to find a representative current waveform of all clock cycles to reduce the number of points. There exist a number of techniques to find a representative supply current waveform. A pattern-independent algorithm, iMAX [4], generates a maximum instantaneous peak current for one clock cycle. The drawback of lacking logical constraints in iMAX has been solved by adding signal correlations [5]. Simulation based approaches have also been proposed to estimate either the maximum peak or RMS value of the supply current [6]. However, those techniques require not only an extensive simulation time due to the large number of input patterns but also the produced vectors are still large to be used for an exhaustive optimization. Besides this, none of the techniques described above gives an error interval on the estimation. The algorithm described in [7] compresses the supply currents of M current sources into C compression sets, with a higher compression than a single-cycle for a single compression set, assuring also a user-specified error bound. However, this algorithm can result in a high number of compression sets, which is not acceptable for our optimization, as the number of compression sets is highly dependent on the temporal locality and periodicity of the currents across different clock cycles.

In this paper we present a novel methodology for substrate noise reduction, which is based on an error-driven optimization of the clock tree latencies using supply current profiles taking timing constraints and the clock skew into account. There are techniques using clock latencies in order to reduce the peak current [8] and the ground bounce [9]. However, they suffer from a large number of the constraints, given as the total number of the flip-flops. Importantly they do not give a value for the number of the clock regions, which is set by the relation between the major resonance frequency of the circuit and the rise/fall time of the supply current.

First we describe the substrate noise simulation methodology that will be used to evaluate the results of the latency optimization. Next, the steps of the latency optimization are described together with their computational complexity. Then we define figure-of-merits for the clock skew sensitivity of the computed latencies. Finally we present our experimental results and draw conclusions.

2. OVERVIEW OF SUBSTRATE NOISE SIMULATION METHODOLOGY

For large designs, it is not feasible to use a transistor-level simulation of the substrate noise generation with detailed substrate models. We have proposed a methodology called SWAN [10] to simulate the substrate noise generation from large digital circuits. The accuracy of SWAN has been verified with measurements [11].

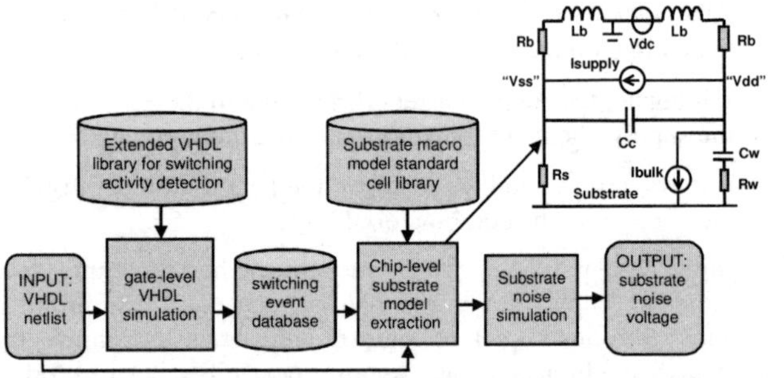

Figure 1. SWAN flow and chip-level substrate model [10]

In this paper, we will use SWAN to simulate the supply current and substrate noise. An overview of the methodology is shown in Figure 1. For every gate a substrate macro model is characterized once using transistor level model of each gate. It macro model contains two substrate noise injection mechanisms, the bulk and the supply current, together with the coupling impedances between VDD, VSS and the substrate [10]. A chip-level substrate model is extracted (see Figure 1) by using the macro models and the switching data generated by an event-driven simulator. Finally, we simulate the substrate noise using the chip-level substrate model.

3. CLOCK TREE LATENCY OPTIMIZATION METHODOLOGY

The coupling to the substrate from the supply current is due to the ringing of the supply voltage caused by the Ldi/dt noise and the resistive coupling via the ground contact resistance. Therefore, decreasing both the amplitude and the time derivative of the supply current will reduce the substrate noise generation. The RMS value of substrate noise is proportional to the integral of its spectrum, resulting from the multiplication of the supply current spectrum and the supply current transfer function to the substrate. Since most of the noise power is due to the major resonance frequency in the transfer function, reducing the spectrum power under this resonance will also reduce the substrate noise (see Figure 2).

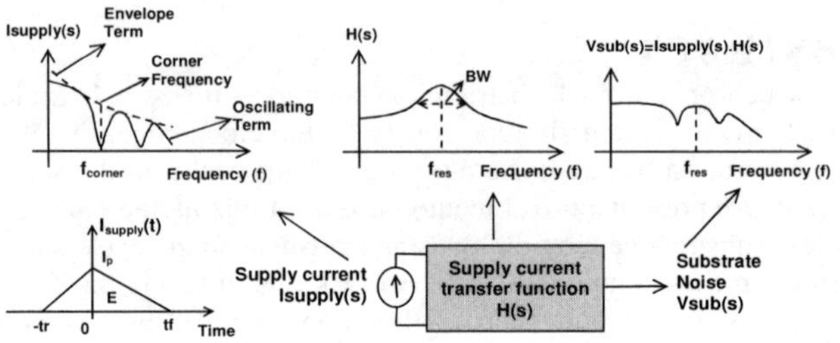

Figure 2. Effect of supply current on the substrate noise

For synchronous CMOS circuits, the total supply current in the time domain can be approximated by a triangular waveform as shown in Figure 2 where I_p, t_r, t_f and E are the peak current, rise time, fall time and total charge respectively. Fourier transformation of the supply current in Figure 2 gives: (1)

$$I_{supply}(f) = \frac{E}{2\pi^2.(t_r + t_f).f^2}\left[\frac{(1-\exp(j.2\pi f.t_r))}{t_r} + \frac{(1-\exp(-j.2\pi f.t_f))}{t_f}\right]$$

As the fastest oscillating harmonic in equation (1) determines the first local minimum on the oscillating term, we can define the corner frequency (f_{corner}) in the supply current spectrum as the minimum of $1/t_r$ and $1/t_f$. One can adjust the corner frequency, by modifying I_p, t_r, and t_f, in order to eliminate the major resonance frequency (f_{res}), which is set by the chip-level substrate model (see Figure 1). An optimum value for the rise/fall time is computed by:

$$\left(f_{corner} = \min\left(\frac{1}{t_r}, \frac{1}{t_f}\right)\right) < \left(f_{res} = \frac{1}{2\pi\sqrt{2L_b(C_w + C_c)}}\right) \quad (2)$$

The reduction becomes more by choosing the corner frequency as the notch points in (1) as:

$$\left(f_{corner} = \frac{K_r}{t_r} = \frac{K_f}{t_f}\right) < f_{res}, \quad K_r, K_f \in N \quad (3)$$

Without looking at the timing implications, the required minimum number of the clock regions, M, is found by the ratio of the actual rise/fall time over the optimum rise/fall time. The actual rise/fall time is computed after the triangular approximation of the total supply current of the circuit. Normally, the timing implications and the multiple peaks on the supply current will avoid having an optimum rise/fall time by using M or more clock regions. This requires an optimization of the latencies on these clock regions.

The clock latency optimization (see Figure 3) consists of 3 main parts: (1) assignment of every instance into M clock regions, (2) folding of supply current transients, (3) optimization of latencies. The next three sections will describe these steps in more detail.

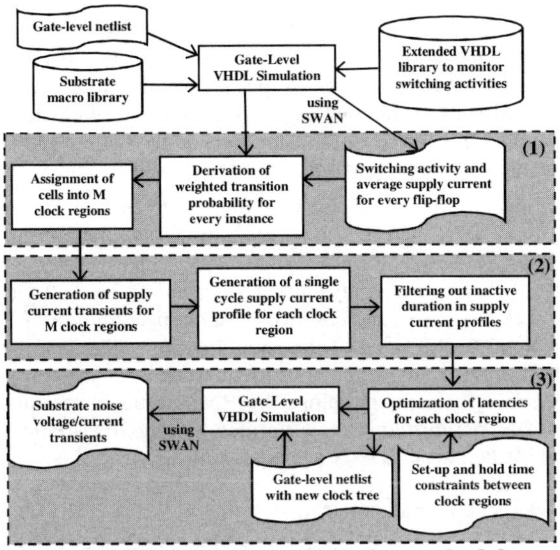

Figure 3. Clock latency optimization methodology

3.1 Clock region assignment

It is important to balance the instances over the clock regions for a significant reduction in the substrate noise generation. First, we divide the instances in sets FF_i, the set of all instances (u_{ij}) that have a data dependency on the driving flip-flop (ff_i) in FF_i. The set u_{ij} also contains the driving flip-flop (ff_i). Note that in general the intersection of two sets may not be empty. Instance assignment is based on transitive fan-out analysis from each flip-flop (ff_i). We define the weighted transition probability, a figure indicating switching statistics for all instances assigned to clock region-m (CR_m) weighted with individual contribution of the instances during a single switch, as follows:

$$P_{m,transition} = \frac{\sum_{i=1}^{N}\sum_{j=1}^{F(i)} a_{ij}.n_{ij,switch}.A_{ij,RMS}}{\sum_{i=1}^{N}\sum_{j=1}^{F(i)} n_{ij,switch}.A_{ij,RMS}} \qquad a_{ij} = 1 \text{ if } u_{ij} \in CR_m \qquad (4)$$

where $n_{i,switch}$ is the number of switching activities of the instance (u_{ij}), $A_{ij,RMS}$ is the average RMS value of the supply current for u_{ij} over all switching occurrences of u_{ij}. N is the total number of the flip-flops. F(i) is the total number of the instances in set FF_i. $A_{ij,RMS}$ factor reflects the average contribution of u_{ij} on the RMS value of the total supply current waveform during a single switching of u_{ij}. An instance u_{ij} is strictly an element of CR_m, if all sets (FF_i) where u_{ij} belongs are in CR_m. To derive $A_{ij,RMS}$ and $n_{ij,switch}$ values for each instance in the network, a gate-level simulation of the initial netlist using SWAN is performed. The sets of FF_i are joined into appropriate clock regions such that the weighted transition probabilities are balanced by 1/M. In a condition where $u_{ij} \in \{FF_{i1}, FF_{i2}\}$, FF_{i1} and FF_{i2} are assigned to different clock regions, e.g. $FF_{i1} \rightarrow CR_{m1}$ and $FF_{i2} \rightarrow CR_{m2}$, then u_{ij} is assigned to CR_{m1} if weighted transitional probability of u_{ij} is caused mostly by CR_{m1} or vice versa. It is vital to reduce the shared set of the cells as much as possible to reduce the possible glitches, which cause an increase in power, integrity problems and the error term in the supply current compression described in Section 3.2.

3.2 Folding of supply current transients

After the assignment of the instances to each clock region, the individual supply transient for each clock region is generated from the previously explained transient simulation of the supply current used during clock region assignment. This results in $M.(T_{clock}/\Delta t).n_{cycle}$ data points, where M is the number of clock regions, T_{clock} is the clock period (integer multiple of Δt), Δt is the unit time step of the simulation and n_{cycle} is the total number of clock cycles. We discretize a single clock cycle into K time intervals, where each time interval can be chosen as Δt resulting in $K=T_{clock}/\Delta t$. We define I(k,m,n) as the actual value of the supply current at time interval-k, clock cycle-n and clock region-m. For each clock region, the union of I(k,m,n) points is compressed into a set of supply current profiles of each having a single clock cycle representation. Every clock region contains at least one current profile defined over every time interval. The compression can create more than one profile in a clock region depending on the user error bound. The set of supply current profiles is defined as:

$$IP_m = \{ I_p(k,m,p) \} \quad \text{for } k = 1..K \quad \text{for } p = 1..P(m) \qquad (5)$$
$$IP = IP_1 \cup IP_2 \cup ... \cup IP_M$$

where P(m) is the number of elements in the set of supply current profiles, IP_m, defined for clock region-m. With tight assumption on temporal locality and periodicity of currents across different clock cycles, for the supply current profile one can choose a single supply current cycle with maximum peak-to-peak value, that is $I_p(k,m,p=1)= I(k,m,n_o)$ where n_o is the clock cycle number with maximum peak-to-peak value on the supply current. This choice of n_o leads to wrong results when the temporal locality assumption is violated. Better representation is to compute statistical properties such as mean, standard deviation and probability density function at each profile point $I_p(k,m,p)$ using all the points in the actual waveform. $I_p(k,m,p)$ contains a set of statistical functions given as:

$$I_p(k,m,p) = \{\mu(k,m,p), \sigma(k,m,p), h(k,m,p)\} \qquad (6)$$

where $\mu(k,m,p)$: mean of I(k,m,n),
$\sigma(k,m,p)$: standard deviation of I(k,m,n),
h(k,m,p): histogram of I(k,m,n) distribution,
over all clock cycles ($n=1..n_{cycle}$).

Each bin in the histogram, h(k,m,p), is created by the quantization of I(k,m,n) values. To compute the probability of achieving a supply current value in a certain interval, one can normalize the histogram, h(k,m,p), by dividing by the total number of actual transient points used for the construction of $I_p(k,m,p)$. The supply current profile is represented by $\mu(k,m,p)$ values where $\sigma(k,m,p)$ is a measure, how the actual values deviate from the supply current profile-p at time interval-k and clock region-m. For example, having $\sigma(k,m,p)$ zero means: "The supply current profile-p perfectly represents the actual supply current at time interval-k, clock region-m". Figure 4 shows the folding procedure.

The error interval on the supply current folding is derived for every time interval by using the derived probabilities from the normalized histogram h(k,m,p). For a confidence percentage, P_c (such as 98%), one has to find a minimum value of ε from:

$$P((\mu(k,m,p) - \varepsilon) < I_p < (\mu(k,m,p) + \varepsilon)) > P_c \qquad (7)$$

As $n_{cycle} \rightarrow \infty$, a normal distribution is likely to occur when the circuit contains a large number of independent processes and has an input vector having normal distribution. In this case, the error interval on $I_p(k,m,p)$ is given as $\mu(k,m,p) \pm 2.32\sigma(k,m,p)$ with 98% confidence. Using error interval analysis on $I_p(k,m,p)$, the set of supply current profiles, IP, are extracted within a given error interval (ε_{user}). For each I(k,m,n), the confidence percentage, P_c, on ε_{user} is tested using (7) by substituting $\varepsilon = \varepsilon_{user}$. If the inequality in (7) is not satisfied for all elements of IP_m set, an additional current profile, $I_p(k,m,p)$, is generated in IP_m using (6). P(m) is incremented by 1 else $I_p(k,m,p)$ in (7) is computed for the current profile-p in IP_m where P_c is maximum. The trade-off in this error driven profile extraction is the decrease of the compression ratio.

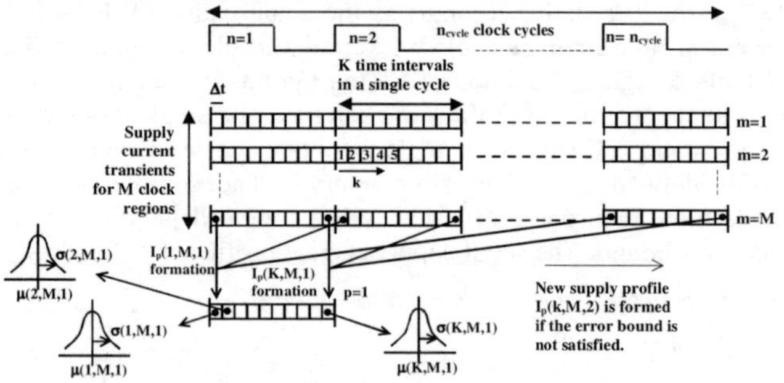

Figure 4. Folding algorithm for forming the current profiles

T_{clock} has to be chosen always larger than the maximum width of the supply current to satisfy timing constraints. For this reason, a filtering is performed on the supply current transients to increase the computational efficiency of the folding algorithm.

3.3 Clock latency optimization

Clock latency optimization is based on an exhaustive search of all latencies for a minimum on the cost function, which will be described later. An exhaustive search is necessary as the problem is NP-complete.

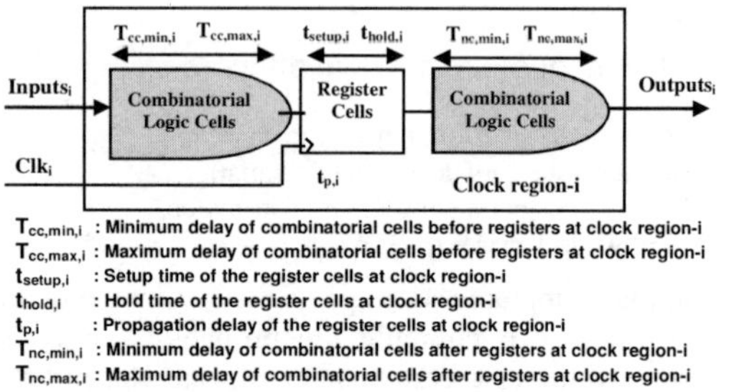

Figure 5. Timing values for clock region-i

The latencies have to be constrained with timing constraints defined in Figure 5. The clock region-i communicating with clock region-j has to satisfy the following constraints:

Setup time constraint: (8)

$$\Delta t_{clk,max}(i,j) < (T_{cycle} - T_{cc,max}(i) - t_{setup}(i) - t_p(j) - T_{nc,max}(j) - \delta)$$

Hold time constraint:

$$\Delta t_{clk,min}(i,j) > (t_{hold}(j) - T_{cc,min}(i) - t_p(j) - T_{nc,min}(j) + \delta)$$

where $\Delta t_{clk,max/min}(i,j)$ is the maximum/minimum allowed latency between Clk_i and Clk_j. δ is the clock uncertainty due to the unexpected skew coming from the clock interconnect respectively. The constraint for each clock latency becomes:

$$\Delta t_{clk}(i,j) > (\max(-\Delta t_{clk,max}(j,i), \Delta t_{clk,min}(i,j)) = t_{max}(i,j)) \qquad (9)$$
$$\Delta t_{clk}(i,j) < (\min(-\Delta t_{clk,min}(j,i), \Delta t_{clk,max}(i,j)) = t_{min}(i,j))$$

The optimization procedure is to find the best M latency bundle $(l_1, l_2, .., l_m, \ldots, l_M)$ that gives minimum value on the cost function computation using the total supply current shifted with the latencies where l_m is defined as the latency value of clock region-m. One can freely set one of the latencies to zero such as $l_1=0$ such that one of the clock regions is aligned to the edge of the clock.

At each latency value, the cost function is evaluated as the product of the peak value and the slope of the total supply current. This comes as a direct result from equation (1), which states the direct proportionality of the supply current spectrum to the multiplication of the peak, $2E/(t_r+t_f)$, and the inverse of the rise/fall time, $1/t_r+1/t_f$, proportional to the slope of the supply current. The optimization tries to minimize this factor in order to reduce the spectral energy of the supply current, therefore the RMS value of the substrate noise. The optimization is performed on the constraint space formed by the latencies $(l_1, l_2, \ldots, l_M)$ as follows:

$$\min_{l} f_{cost}(l_1, l_2, \ldots, l_M) = \max\left(\sum_{m=1}^{M} I_p([1,K] - l_m, m, 1)\right).$$
$$\mathrm{RMS}\left(\sum_{m=1}^{M} \left(I_p([1,K] - l_i, m, 1) - I_p([1,K] - l_i - 1, m, 1)\right)\right) \qquad (10)$$
$$\text{s.t. } l_1 = 0, \; t_{max}(i,j) < l_i - l_j < t_{min}(i,j) \quad \forall i,j \in [1,M]x[1,M]$$

4. COMPUTATIONAL COMPLEXITY

During the folding of the supply current transients of M clock regions, each having $K.n_{cycle}$, data points; the mean, the standard deviation and the probability histogram are computed recursively. Due to this recursion, the computational complexity is bound to $O(K.M.n_{cycle})$ whenever there is no error control. With error control, the complexity becomes $O(K.M.P_{MAX}.n_{cycle})$ where P_{MAX} is the maximum number of the supply profiles stored in IP_m. P_{MAX} increases by a decrease of the user error interval. The compression factor with and without error control becomes P_{MAX}/n_{cycle} and $1/n_{cycle}$ respectively. The complexity of the clock latency optimization is $O(K.(M+1).P_{MAX}.a^{M-1})$ where $a^{M-1} << K^{M-1}$ is the dimension of the search space of all M-1 latencies under set-up and hold time constraints while we set one of the latencies to zero. Overall complexity from the folding and latency optimization is $O(K.M.P_{MAX}.n_{cycle}) + O(K.(M+1).P_{MAX}.a^{M-1})$. If the number of compression sets in each clock region is bound to 1, computational complexity becomes $O(n)+O(a^{M-1}n/n_{cycle})$ where n is the number of the data points ($n=K.M.n_{cycle}$). The computational complexity approaches O(n), first order dependency on the number of the data, with increasing number of clock cycles (n_{cycle}).

5. CLOCK SKEW SENSITIVITY

Due to clock routing, the load balancing or other random effects within different clock regions there will be an uncertainty or skew at each clock region. As the high slew rate of the supply current used during the optimization, the optimum point can have a high sensitivity to clock skew. To analyze the quality of the results to skew, we construct a skew radius around the optimum point. In addition, we exhaustively search the space around the optimum for a given radius (δ). Within this radius, we introduce the following skew figure on the quality of the results:

$$SB_{MAX,RMS}(\delta) = MAX,RMS\left(\sqrt{f_{cost}(l_{opt} \pm r)/f_{cost}(0)}\right), \quad \forall r \in [0,\delta] \qquad (11)$$

where l_{opt} and r are the optimum latency bundle and the skew effect on the latencies respectively. $f_{cost}(0)$ is the value of the cost function (see equation (10)) before the optimization. The square root of the ratio is necessary to scale the cost function, which is given as $I_{peak}.(di/dt)_{RMS}$, appropriately in order to resemble the reduction factor of the substrate noise generation. $SB_{MAX,RMS}(\delta)$ is an indicator showing the maximum and RMS value of the reduction factor due to the clock skews, which is bound by δ.

6. EXPERIMENTAL RESULTS

The methodology is illustrated in a 4-bit Pseudo-Random-Noise-Generator (PRBS) implemented in a 0.35-μm CMOS process on an EPI-type substrate at 3.3V supply. Later, the results from ITC'99 benchmark circuits [12] and a test chip [13] are presented.

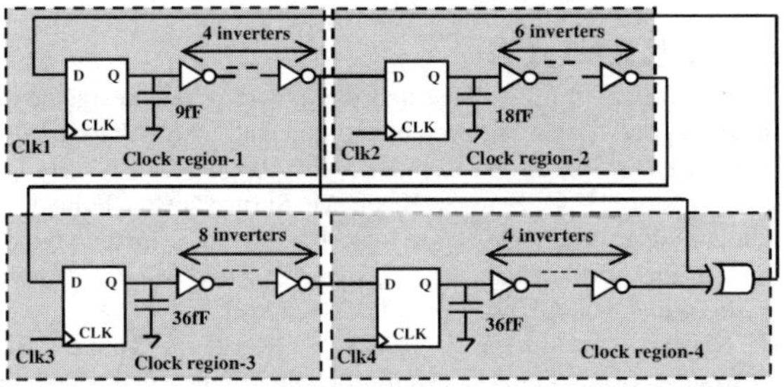

Figure 6. Pseudo-Random-Noise-Generator (PRBS) circuit

Figure 6 depicts the test-circuit and its division into four different clock regions. The supply current transfer function to the substrate has a resonance frequency of 2.3GHz. The 3dB bandwidth of the resonance stretches from 1.3GHz to 3.2GHz. The supply current has a corner frequency at 270MHz. So choosing 4 as the number of clock regions is appropriate as the initial corner frequency is already well below the resonance frequency. The design has a clock period of 4ns and a supply line parasitics of 5nH+0.5Ω. A single supply current profile has been constructed for each clock region using the actual supply current data from a total transient simulation of 105 clock cycles using SPICE. Choosing 105 clock cycles, considering the intrinsic periodicity of the 4-bit PRBS, results in an unbiased estimate.

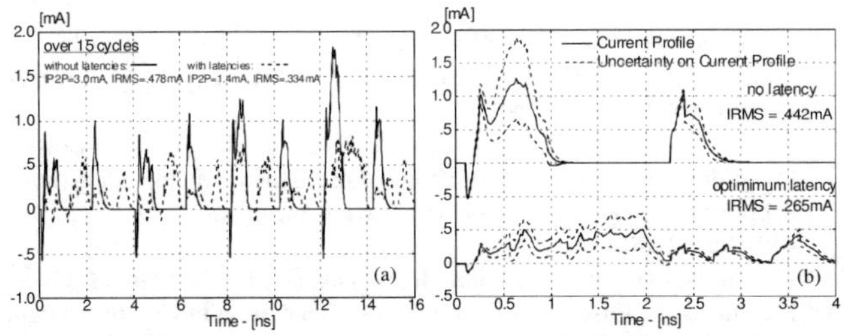

Figure 7. Simulated (SPICE) supply current (a) transients, (b) their profiles with/without latencies in the PRBS circuit

Figure 7 shows the transients (a) and the profiles (b) of the supply current with/without latencies. Figure 8 shows the simulated substrate noise transients (a) and the corresponding spectra (b) with/without latencies. The design with optimized latencies achieves factors of 2.10 and 1.75 reduction in the peak-to-peak and the RMS value of substrate noise respectively. The design with optimized latencies has a reduction of 8.9dB, 20dB, 18dB, 14dB, and 13dB at fundamental, 2nd, 3rd, 4th and 5th harmonics of the clock respectively. This is due to the reduction of the spectral power of the supply current as depicted in Figure 8. In this example, the spectral power of the supply current has been reduced by 6dB. Table 1 shows the clock skew sensitivity figures, explained in section 5, of the latency optimization results.

Table 1. Clock skew sensitivity of the latency results

Skew radius	0ps	50ps	100ps	150ps	200ps
SB_{MAX}	0.43	0.54	0.63	0.66	0.67
SB_{RMS}	0.43	0.47	0.49	0.50	0.51

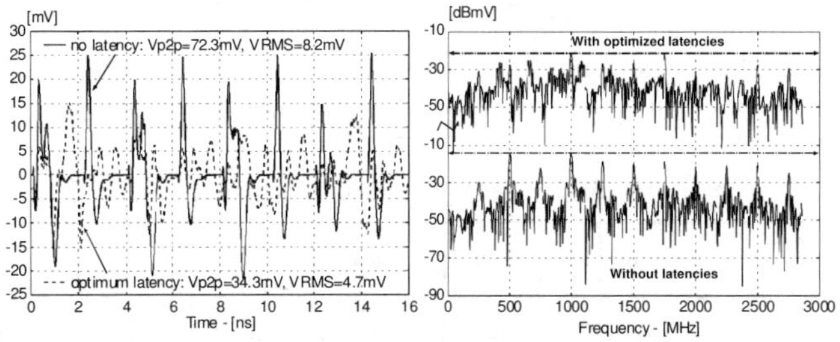

Figure 8. Simulated (SPICE) substrate noise (a) transients and (b) their spectra with/without latencies in the PRBS circuit

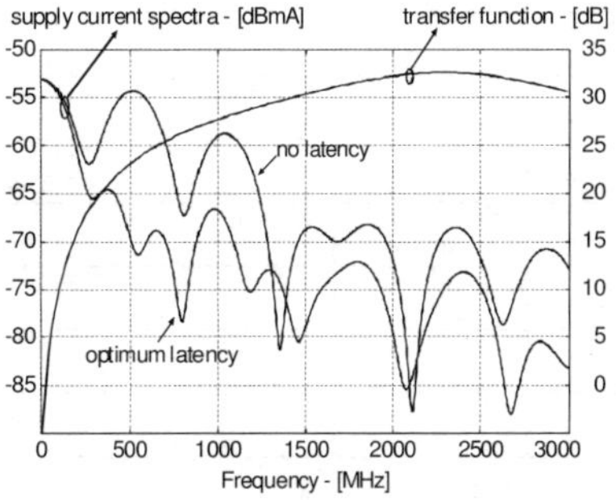

Figure 9. Supply current spectra with/without latencies and their transfer function to the substrate in the PRBS circuit

The reduction becomes more significant when the initial corner frequency is above the major resonance frequency and is shifted below the major resonance frequency. Figure 10 shows the transients and the corresponding spectra of a circuit, having a resonance at 950MHz, as a result of changing the amplitude and slope of its supply current. The supply current has been shaped from (50mA, 200ps) to (10mA, 1000ps), where we assume no timing constraints and equal rise and fall times. A significant reduction by a factor of 5.4 and 7.9 is achieved for the peak-to-peak and RMS value of the substrate noise respectively.

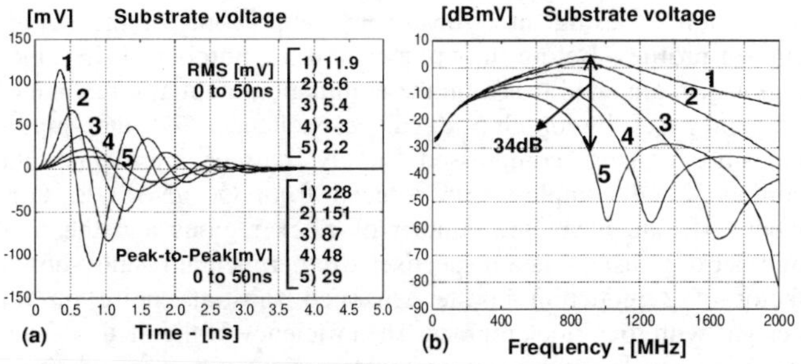

Figure 10. Simulated (SWAN) substrate noise (a) transients and (b) the corresponding spectra after supply current shaping in a circuit having a resonance frequency of 950MHz

We have also tested the introduction of the latencies on the substrate noise reduction figures of ITC99 benchmark circuits [12]. It is not anymore feasible to simulate those circuits using SPICE due to the increased complexity of the circuits relative to PRBS circuit illustrated above. Therefore, we use SWAN [10] to simulate supply current and substrate noise voltage transients using random test vectors as input. The circuits have been implemented in a 0.35-μm CMOS process on an EPI-type substrate using 1nH+0.1Ω package parasitics and 3.3V supply. The latencies of each design have been computed for the 4 clock regions. Table 2 shows the initial values of the substrate noise generation and the supply current together with the reduction percentages as a result of the introduced latencies. It is concluded that the reduction is around 37% in average for the RMS values of the substrate noise.

Table 2. ITC99 benchmark results

Circuit Name	#Gates/ FFs	Clk cycles/ Freq. [MHz]	#Sw. Act. Initial Final	Vs,rms [mV], % Vs,p2p [mV], %	Ipeak [mA], %
B04	606/66	199/50	131373 133242	22.30 [14%] 314.0 [15%]	49.70 [5%]
B05	977/34	199/50	56142 61704	6.875 [38%] 76.88 [30%]	17.57 [23%]
B06	61/9	199/50	9385 9663	4.204 [35%] 87.80 [43%]	4.96 [30%]
B07	422/49	199/50	32432 34206	7.186 [48%] 104.4 [2%]	27.80 [1%]
B08	168/21	199/50	10318 9813	4.942 [43%] 78.39 [16%]	8.89 [34%]
B09	160/28	199/50	22501 23746	7.525 [58%] 163.6 [43%]	12.44 [22%]
B10	190/17	199/50	7386 7381	0.420 [5%] 22.62 [32%]	1.98 [48%]
B11	484/31	199/50	77166 79405	7.095 [38%] 104.7 [5%]	19.97 [%13]
B12	1037/121	199/25	105424 108988	10.02 [58%] 123.2 [35%]	33.97 [18%]
B13	343/50	199/25	23962 22313	2.847 [30%] 112.8 [48%]	14.22 [39%]

We have also designed and measured a mixed-signal chip (see Figure 11), fabricated in a 0.35μm CMOS process on an EPI-type substrate, in order to compare several low-noise digital designs [13]. A comparison of two realizations, a reference circuit (REF) and a low-noise design with optimized clock latencies (LN1) in four clock regions, of a 5Kgate synchronous CMOS circuit, shows more than a factor of 2 reduction in the substrate noise generation.

7. Conclusions

Shaping the supply current is shown to be very effective as the coupling from the ringing of the supply into the substrate is dominant. In this paper, we have presented a methodology to optimize the clock tree latencies to reduce the substrate noise generation by using an error-driven compression of the supply current profiles. Before the optimization the number of the clock regions is computed based on the elimination the major resonance frequency set by on-chip circuit capacitance and the supply parasitics. Using compressed supply current transients, the computational complexity is reduced from $O((n/M)^M)$ to $O(n)$ where M, and n are the number of clock regions and the total number of transient data respectively. Experimental results show a factor of 2 reduction in the generated substrate noise on the designs with four clock regions. The efficiency of the methodology has been verified with measurements on a fabricated mixed-signal chip. The supply current shaping by the use of clock latencies is shown to be very effective if timing constraints allow shaping.

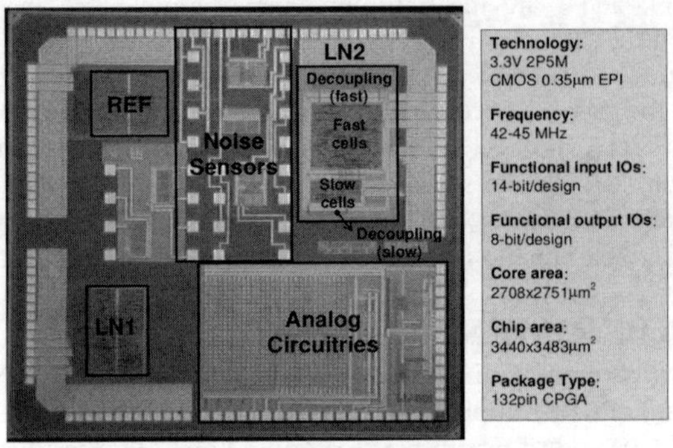

Figure 11. Microphotograph of the test chip

8. REFERENCES

[1] A. Matsuzawa, "Low-voltage and low-power circuit design for mixed analog/digital systems in portable equipment," *IEEE Journal of Solid State Circuits,* vol. 29, no. 4, pp.470-480, April 1994.

[2] D. J. Allstot, H. C. San, S. Kiaei, M. Shriyastawa, "Folded source-coupled logic vs. CMOS static logic for low-noise mixed-signal ICs," *IEEE Trans. on Circuits and Systems I: Fundamental Theory and Applications,* vol. 40, no.9, pp.553-563, Sep. 1993.

[3] M. Nagata, K. Hijikata, J. Nagai, T. Morie, A. Iwata, "Reduced substrate noise digital design for improving embedded analog performance," *ISSCC Dig. of Tech. Papers,* pp.224-225, Feb. 2000.

[4] H. Kriplani, F. Najm, I. Hajj, "Maximum current estimation in CMOS circuits," *Proc. of Design Automation Conf.,* pp.2-7, 1992.

[5] H. Kriplani, F. Najm, I. Hajj, "Pattern independent maximum current estimation in power and ground buses of CMOS VLSI circuits: Algorithms, signal correlations, and their resolution," *IEEE Tr. On Computer Aided Design of Integrated Circuits and Systems,* vol.14, no.8; Aug. 1995; pp. 998-1012.

[6] A. Krstic, T. C. Kwang, "Vector generation for maximum instantaneous current through supply lines for CMOS circuits," *Proc. of Design Automation Conference,* pp. 383-388, June 1997.

[7] R. Chaudhry, D. Blaauw, R. Panda, T. Edwards, "Current signature compression for IR-Drop Analysis," *Proc. of Design Automation Conference,* pp. 162-167, June 2000.

[8] P. Vuillod, L. Benini, A. Bogliolo and G. De Micheli, "Clock-skew optimization for peak current reduction," *IEEE Int. Symp. On Low Power Electronics and Design,* pp. 265-270, 1996.

[9] A. Vittal, H. Lia, F. Brewer, M. Marek-Sadowska, "Clock skew optimization for ground bounce control," *ICCAD Digest of Technical Papers,* pp. 395-399, 1996.

[10] M. van Heijningen, M. Badaroglu, S. Donnay, M. Engels, I. Bolsens, "High-level simulation of substrate noise generation including power supply noise coupling", *Proc. of Design Automation Conf.,* pp. 446-451, June 2000.

[11] M. van Heijningen, M. Badaroglu, S. Donnay, H. De Man, G. Gielen, M. Engels, I. Bolsens, "Substrate Noise Generation in Complex Digital Systems: Efficient Modeling and Simulation Methodology and Experimental Verification," *ISSCC Digest of Technical Papers,* pp.342-343, Feb. 2001.

[12] ITC99 benchmark circuits: http://www.cad.polito.it/tools/itc99.html

[13] M. Badaroglu, M. van Heijningen, V. Gravot, J. Compiet, S. Donnay, M. Engels, G. Gielen, H. De Man, "Methodology and Experimental Verification for Substrate Noise Reduction in CMOS Mixed-Signal ICs with Synchronous Digital Circuits," *ISSCC Digest of Technical Papers,* pp. 274-275, Feb. 2002.

Resynthesis and Peephole Transformations for the Optimization of Large-Scale Asynchronous Systems

Tiberiu Chelcea Steven M. Nowick*
Department of Computer Science
Columbia University
New York, NY, USA e-mail: {tibi,nowick}@cs.columbia.edu

ABSTRACT

Several approaches have been proposed for the syntax-directed compilation of asynchronous circuits from high-level specification languages, such as Balsa and Tangram. Both compilers have been successfully used in large real-world applications; however, in practice, these methods suffer from significant performance overheads due to their reliance on straightforward syntax-directed translation.

This paper introduces a powerful new set of transformations, and an extended channel-based language to support them, which can be used an optimizing back-end for Balsa. The transforms described in this paper fall into two categories: resynthesis and peephole. The proposed optimization techniques have been fully integrated into a comprehensive asynchronous CAD package, Balsa. Experimental results on several substantial design examples indicate significant performance improvements.

1. Introduction

Several approaches have been proposed for compilation of asynchronous circuits from high-level specification languages [1, 13, 4, 8]. Compilers such as Balsa [1] (from University of Manchester) or Tangram [13, 10] (from Philips Research Labs, and now used for commercial products) perform a syntax-directed compilation of high-level specifications into an intermediate representation using *handshake components*. These components are then synthesized into circuits using a template-based approach. An *occam*-based compiler [4] and an alternative translation approach from Caltech [8] have also been proposed.

Balsa and Tangram have been widely-used, but their syntax-directed translation methods introduce significant performance overheads [3, 15]. While these synthesis styles have the advantage of "transparency" (the designer is controlling the final results from the high-level program), they also have the disadvantage of avoiding the use of powerful back-end transformations, except for simple peephole optimizations. The Caltech approach uses only localized resynthesis techniques (such as handshake reshuffling and "guard symmetrization"), which are not systematically applied or captured at a higher language level. In contrast, this paper presents a much wider-ranging and more powerful set of transformations for the optimization of asynchronous systems.

This paper introduces three major contributions. The first contribution is a set of new transformations which can be used in a back-end optimizer for large asynchronous circuits. The transforms fall into two categories: peephole and resynthesis. A peephole optimization optimizes components in a sliding window; if the netlist of components in the window conform to a pattern, they are replaced in a template-based fashion by other existing components. In contrast, resynthesis optimizations attempt to replace components in a non-template-based fashion, by re-synthesizing them.

The second contribution of this paper is the extension to a component specification language, called CH. CH was introduced in [5] to model and manipulate only control components, and is extended here to model *datapath* components. CH is very important in the proposed approach: each transform is formalized as a simple language manipulation procedure in CH. The resulting specifications are thus independent of the synthesizable low-level specifications into which CH is translated. Burst-mode controller specifications are currently employed, but they are just one of several possible low-level specification styles. The final contribution of the paper is a new automated back-end for the Balsa synthesis system which incorporates the proposed transforms.

The proposed new transforms facilitate design-space exploration. By systematically applying them, the designer, starting with an unoptimized circuit, can obtain different improved implementations of the initial design. An additional advantage is that these transforms are formalized using an intermediate specification language, which allows the designer to observe tradeoffs between different related implementations at a higher level.

The integrated design flow was applied to three substantial asynchronous design examples. Pre-layout back-annotated Verilog simulations on technology-mapped implementations indicate up to 54% speed improvement over the unoptimized implementations. Improvements are significantly better than those obtained using previous optimizations proposed in [5].

The paper is organized as follows. Section 2 presents a brief overview of the proposed synthesis approach. Section 3 gives some asynchronous background, and Section 4 gives background information on the CH language. Section 5 introduces the new CH extensions, while Sections 6 and 7 present the proposed resynthesis and peephole optimizations, respectively. Finally, Section 8 presents experimental results, and Section 9 presents conclusions and directions for future work.

2. Overview of the Approach

Balsa [1, 2] is collection of programs that facilitate the description and synthesis of asynchronous systems. The original Balsa design flow takes a system description in the Balsa high-level description language and translates it into a netlist of handshake components. Each component is then mapped to actual circuits in a template-based fashion.

The improved back-end includes an *optimization step*, which consists of the proposed set of resynthesis and peephole transforms, as well as of the previous optimizations from [5]. The revised Balsa flow is shown in Fig. 1 (the shaded boxes indicate research contributions).

The new back-end takes the list of handshake components, and performs the proposed peephole and resynthesis transforms on them, to obtain a list of optimized components. It then partitions the list

*This work was supported by NSF ITR Award No. NSF-CCR-0086036 and NSF Award No. CCR-99-88241, and by a grant from the New York State Microelectronics Design Center.

DAC 2002, June 10-14, 2002, New Orleans, Louisiana, USA.

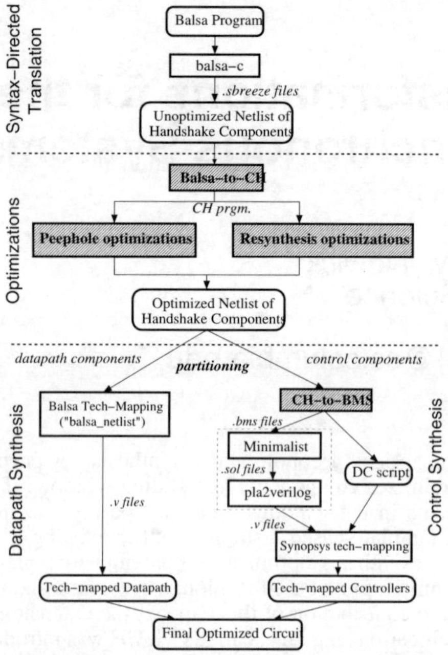

Figure 1: New Balsa System Design Flow

into datapath and control components. The datapath components are then synthesized using the existing Balsa system technology mapper (balsa-netlist). The control components are synthesized into hazard-free logic implementations using the Minimalist Burst-Mode CAD package [7], with speed-oriented scripts. The logic implementations are then technology mapped using a commercial tool (Synopsys Design Compiler). Hazards must not be introduced at this step. The proposed solution is to split the controllers into smaller modules (e.g. using Verilog, one module per level of a NAND-NAND implementation), technology map them separately, and then formally check the solutions for hazard freedom. In all cases, the technology-mapped circuits were correct (i.e. hazard-free). For more details on the technology-mapping step, see [5].

The current design flow incorporates some of the lower-level synthesis steps from the previously-proposed flow in [5]. The interactions between the optimizations and Minimalist, pla2verilog, Synopsys Design Compiler technology-mapping, and balsa-netlist are similar in both design flows. However, the major difference is in the optimization step, which is now enlarged and improved with the newly-proposed transforms.

2.1 Related Work

Several approaches for component modeling have been proposed ([2, 8, 13, 10]). However, some can only model a single behavior of a particular type (e.g. sequencing) rather than a desired set of closely-related variants (e.g. various "interleavings" of a basic sequencing protocol) [2, 10]. Others use low-level specifications where each signal transition is enumerated, which are cumbersome to manipulate [8, 13, 10].

Several peephole optimization and control resynthesis techniques for asynchronous systems have also been previously proposed. Previous peephole optimizations [13, 4] have typically been *behavior-preserving*, and most have dealt with simple components and improvements (such as redundancy removal). In contrast, unlike several of the previous techniques, all of the proposed transforms in this paper are *behavior-modifying*, and subsume many of the earlier ones.

Recent approaches to control resynthesis [11, 6] are mainly variants of component composition, using Petri-net or trace-theory formalisms, and are behaviorpreserving. In contrast, the proposed resynthesis techniques include much more powerful transformations for design-space exploration, including concurrency enhancement and protocol manipulation. In addition, the proposed transforms are integrated into a comprehensive CAD package for asynchronous synthesis. Finally, these previous resynthesis approaches did not report results on performance improvements, only on area.

3. Asynchronous Background

3.1 Asynchronous Communication Protocols

There are two key concepts in understanding the behavior of asynchronous components: their *communication protocol* and techniques to *interleave* two or more communication protocols.

Two asynchronous components communicating among themselves over one channel use a "communication protocol" (Fig. 2a). The

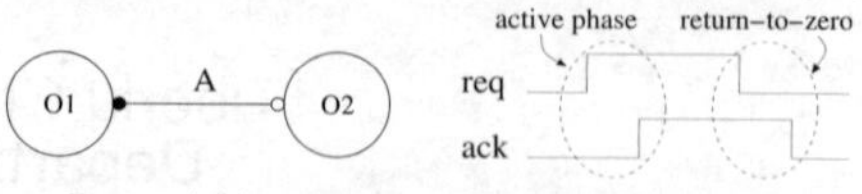

a) Communication channel b) Handshake expansion

Figure 2: Communication Protocol

most common protocol is called "4-phase handshaking": component O1 initiates the communication, and component O2 completes the communication. The component's interface to a channel is called a port. Each port has a type: component O1 initiates the communication, and thus has an *active* port A (indicated by the black dot); component O2 completes the communication, and thus has a *passive* port A (indicated by the hollow dot). The communication channel is implemented by two wires: a request and an acknowledge. A communication (Fig.2b) has the following handshake expansion: A_r↑ A_a↑ A_r↓ A_a↓ (where A is the channel name, and r/a are the request/acknowledge wires), and consists of two phases: the *active phase* (rising transitions on the request and acknowledge wires, in this order), and a *return-to-zero phase* (falling transitions in the same order).

When a component is connected to two or more channels, an important issue is how the communication protocols on the channels are *interleaved*. For example, a component communicating on channels A and B with two other components might *enclose* the handshake on B within the handshake on A (A_r↑ B_r↑ B_a↑ B_r↓ B_a↓ A_a↑ A_r↓ A_a↓), or, it might *sequence* the communication on B after the communication on A (A_r↑ A_a↑ A_r↓ A_a↓ B_r↑ B_a↑ B_r↓ B_a↓). Different interleavings give different tradeoffs (such as speed, area, or power) in the component's implementation. Therefore, a component description language such as CH must define operators that model different interleavings.

3.2 Burst-Mode Specifications

Burst-mode specifications are the target of the synthesis path. They are a commonly-used Mealy-type specification for asynchronous controllers [9]. The Minimalist CAD package [7] is used in the design flow to synthesize the optimized Burst-mode controllers.

A burst-mode specification (Fig. 8b) is a finite-state machine specification, and consists of a set of states and a set of arcs. An arc is labeled with an input burst (a set of input transitions, for example "B_a- C_a+"), followed by an output burst (a set of output transitions, for example "C_r-"), and connects two states. An input (output) burst describes a set of input (output) transitions: rising transitions are marked with a '+', and falling transitions are marked with a '-'. A BM machine waits for an input burst to arrive; transitions may come in any order and at any time. Once the complete input burst has arrived, the output burst is generated and the machine moves to the next specification state.

4. CH Language: Basics

The previously-proposed CH language [5] is an intermediate specification language. In the old version [5], CH could model only control handshake components; in this paper, CH is extended to model datapath components also. CH is important in the proposed approach: each optimization is formalized as a simple CH manipulation procedure on component specifications. This section introduces the original version of CH [5] through a series of examples, starting with channel modeling, and progressing through a series of component modeling examples. In the next section, the new extensions will be presented.

Channel Modeling. Channel communication protocols can be modeled easily and concisely using CH. Figure 3 shows two examples of components with only one port, each connected to a P-TO-P (point-to-point) channel. These channels are the most common ones, and they consist of one request and one acknowledge wire. Fig. 3a shows a point-to-point channel A connected to a *passive*

a) (p-to-p passive A) b) (p-to-p active A)

Figure 3: Channel Modeling in CH

port. The CH expression indicates the type of the channel (P-TO-P), the type of the port (active/passive), and the channel name, and it is a shorthand for the handshake expansion in Fig. 2b, where the handshake is initiated by an input transition on the request wire. In contrast, Fig. 3b shows a P-TO-P channel A connected to an *active* port. The new CH expression now indicates that the port has an active type, and has the same handshake expansion (Fig. 2b), but the handshake is initiated by an output transition on the request wire.

There are six more channel types in CH. Both MULT-REQ and MUX-REQ have multiple request wires, but during a MULT-REQ handshake there are transitions on *all* request wires, while during a MUX-REQ handshake there are transitions on *exactly one* request wire. In mirror, there are channels with multiple acknowledges (MULT-ACK, MUX-ACK), with a similar behavior. Finally, there are two special types of channels: VERB (which allows the designer to specify all transitions in a handshake) and VOID (used internally, during optimizations).

D-element. The first component example, a *D-element* (Fig. 4), was introduced by Martin [8]. The D-element has one passive port

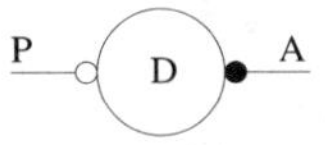


```
(rep
    (enc-early  (p-to-p passive P)
                (p-to-p active A)))
```

(a) Block Diagram (b) CH specification

Figure 4: D-Element

P, and one active port A (Fig. 4a). When activated on P, the D-element performs a full handshake on the active port A, before completing the handshake on P. The CH program (Fig. 4b) concisely captures this behavior. The top-level operator REP indicates that the behavior is repeated forever. Then, handshakes on the passive and active ports are interleaved. The particular interleaving is an *early-enclosure* (ENC-EARLY), which encloses the handshake on A in the middle of the active phase of P. Unlike the CH expression, the Martin-style specification (*[[P_r↑]; A_r↑; [A_a↑]; A_r↓; [A_a↓]; P_a↑; [P_r↓]; P_a↓]) for the same element indicates every signal transition.

Sequencer Component. The second example is a useful control handshake component: a *sequencer* (Fig. 5) [2, 14, 10], used to activate two processes in turn. The component has one passive port

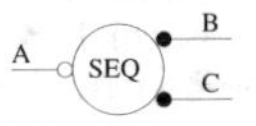


```
(rep (enc-early  (p-to-p passive A)
                 (seq   (p-to-p active B)
                        (p-to-p active C))))
```

(a) Block Diagram (b) CH program

Figure 5: Sequencer Component

(A) and two active ports (B and C); the active ports are connected to processes B and C on the respective channels. When activated on channel A, the component activates in turn each of the B and C channels (first B, then C) before completing the handshake on A. The CH expression (Fig. 5b) reads as follows. First, the top-level operator (rep) indicates that the behavior is repeated forever. Then, there is an early-enclosure (ENC-EARLY) interleaving between the handshake on A and the non-overlapped sequencing (SEQ) of handshakes on channels B and C. Unlike the CH expression, the Martin-style specification: *[[A_r↑]; B_r↑; [B_a↑]; B_r↓; [B_a↓]; C_r↑; [C_a↑]; C_r↓; [C_a↓]; A_a↑; [A_r↓]; A_a↓] is not hierarchical, and it is cumbersome for manipulation.

Call Component. The final example is a *call* component (Fig. 6) [2, 14, 10]. A CALL provides access to a resource whenever requested by one of several components, called callers, where the requests must be mutually-exclusive. The particular interleaving between a caller request and the resource access is a *middle-enclosure* (ENC-MIDDLE). The enc-middle operator takes two arguments (e.g. channels C and E) and returns an interleaving in which the active/return-to-zero phases of the E handshake are enclosed within the active/return-to-zero phases, respectively, of C (C_r↑ E_r↑ E_a↑ C_a↑ C_r↓ E_r↓ E_a↓ C_a↓). In addition, CH also defines a *late-enclosure* operator (enc-late), with the following handshake expansion: C_r↑ C_a↑

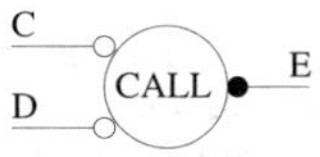


```
(rep (mutex
        (enc-middle  (p-to-p passive C)
                     (p-to-p active E))
        (enc-middle  (p-to-p passive D)
                     (p-to-p active E))))
```

(a) Block Diagram (b) CH program

Figure 6: Call Component

C_r↓ E_r↑ E_a↑ E_r↓ E_a↓ C_a↓.

Burst-Mode Aware Restrictions. Once a component is modeled in CH, it must be translated into a synthesizable Burst-Mode [7, 9] specification (the chosen low-level component specification language). Therefore, several restrictions (called "burst-mode aware") must be imposed on CH: only certain combinations of operators and argument types are allowed. For "burst-mode aware" interleavings, both the interleaving operator and the passivity/activity of relevant ports must be considered; all legal combinations are given in [5]. These restrictions are enforced in two steps. First, each initial handshake component is modeled only with "burst-mode aware" CH expressions. Then, each proposed transform is also restricted: starting with BM-aware CH specifications, their result (a CH expression) is checked to be BM-aware. If it is, the transform has succeeded, otherwise it has failed. For the peephole transforms, this check is performed only once: when the transform is first formalized. In contrast, for the resynthesis transforms, this check is performed each time the transform is applied, since the transforms do not assume fixed configurations of components.

5. New CH Extensions

CH, as defined in [5], can model a variety of control handshake components. In this section, CH is extended to handle datapath component modeling. In particular, CH is extended with channel types and datapath operators. There is one final useful extension to CH, a concurrent sequencer operator; however, this operator is used to model control (not datapath) components, and therefore is more suitable for presentation in the next section.

Data Channel Modeling: Formalism. A four-phase bundled-data channel consists of a pair of control wires (a request and an acknowledge), and a bundle of data wires on which a data item is passed [10]: the request is delayed until after the data item is valid. The communication protocol on a data channel is uniquely defined by the following parameters, which are also discussed in [10]:

- **Data Flow Direction**: as indicated, a data channel consists of two control wires and data wires. There are two options as to the direction in which data flows: in the same direction as the request ("push"), or in the same direction as the acknowledge ("pull").
- **Data Validity**: in asynchronous communication, the sender and the receiver must know when the data item being passed is valid. There are three common data-validity schemes: *broad*, *early*, and *late* (see [10] for formal definitions of these data-validity types).
- **Data Item Type**: each data item has an associated data type (such as byte, word, bool). To be compatible for communication, the sender's and the receiver's ports must both be able to exchange the same data type.
- **Data Width**: a data type is encoded on a number of data wires. To be compatible for communication, the sender's and receiver's ports must both have the same number of data wires encoding a data item.

The formal syntax for a data channel in CH is: ([push|pull] [b|e|l] name [data type] [data width]). The first parameter indicates the data-flow direction. Data validity is selected by the second parameter: broad, early, or late. The name of the channel is given by name, and the last two parameters describe the data item type and the number of data wires used to encode it. When a data channel is connected to a particular port, an additional port attribute (passive/active) is also used: ([passive|active] [?|!] [b|e|l] name [data type] [data width]). In this case, the data-flow attribute (push/pull) can be transformed into the simpler notation ?|!, which indicates whether data is input (?) or output (!).

Modeling of Datapath Operators: Formalism. A handshake datapath component consists of two parts: its interface communication protocol and its functional operation. Therefore, at the CH level, the component is modeled by two expressions: one that models the control, and which contains only interleaving operators and the newly-introduced data channels; and one CH expression for the datapath computation, which contains only *datapath operators*.

The two CH expressions for modeling a datapath component are used as follows. The CH control part, which models the behavior on the interfaces, is translated into a low-level specification (Burst-mode, in this approach) and then directly synthesized. In contrast, the CH expression for the datapath computation is used both to select the appropriate datapath circuit from a library of components, and to model a structural view of the connections to this datapath circuit.

Three new *datapath* operators are introduced to model the datapath computation:

- **func**: this operator models datapath circuits which compute a function (e.g. adder). The syntax is: (func circuit_type out1 ... outn inp1 ... inpn) where circuit_type indicates the type of computation performed (for example, addition, subtraction), inp1 ... inpn are the interfaces on which the inputs are received, and out1 ... outn are the interfaces on which the results are sent.
- **transf**: this operator describes datapath circuits that simply transfer their inputs to their outputs (wire connections). The syntax is: (transf out inp) where inp is the port on which data is received, and out is the port on which data is transferred.
- **var**: this operator describes a register datapath circuit (used to store data items). The syntax is: (var inp out1 ... outn) where inp is the channel name whose data wires are connected to the write port of the register, and out1 ... outn are the channel names whose data wires are connected to the read ports of the register.

Modeling of Datapath Components: Adder Example. To better understand the modeling of data channels and datapath components, an eight-bit adder handshake component (Fig. 7) is now presented. Notice that two CH expressions, E1 and E2, are needed to model this component. The adder's handshaking behavior is de-

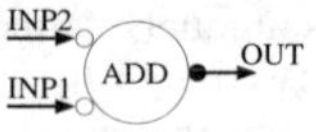

E1: (rep (enc-middle (passive ! b out byte 8)
(enc-middle (active ? b inp1 byte 8)
(active ? b inp2 byte 8))))

E2: (func add out inp1 inp2)

(a) Block Diagram (b) CH program

Figure 7: Eight-Bit Adder Component

scribed by expression E1. The adder's interfaces are described by CH data channel declarations: out is a passive output port (connected to a pull channel), and inp1, inp2 are active input ports (also connected to pull channels). The arrow indicates a data channel, and its direction indicates the direction of data flow (in on the two inputs, out on the output). The data validity on each channel is broad (b): during a communication, the data item passed on the channel is valid from before *req*↑ until after *ack*↓. The last two parameters in the channel declaration indicate the data item type (byte) and the number of wires which encode it (8). The component is activated on the out port, and then concurrently fetches the operands on inp1 and inp2; when the addition is complete, the result is output on out, and the interfaces are returned-to-zero in the same order.

Expression E2 describes the datapath portion of the asynchronous adder: the func CH datapath operator indicates that the inp1, inp2 data wires connect to the the inputs of the addition circuit, and that the out data wires are connected to its outputs. This expression can then be used to select an eight-bit adder from a library of datapath circuits.

6. Resynthesis Transforms

This section introduces two related transforms: Enc Replacement and Seq Replacement, which operate on individual components, and replace some of their interleaving operators by others. These transforms are simple, yet powerful: each one is formalized as a CH language manipulation procedure, and they allow for higher-level exploration of tradeoffs in asynchronous synthesis.

6.1 Seq Replacement

The Seq Replacement transform replaces a non-concurrent sequencing operator (seq) by a more concurrent sequencing operator (seq-concur). The potential benefit of the transform is increased throughput for the entire subsystem controlled by this operator. This subsection first introduces and formalizes the new seq-concur operator. It then illustrates its use in the new transform where sequencer components are replaced by concurrent sequencer components.

The new seq-concur operator defines a concurrent interleaving of handshakes on two channels. The operator takes two arguments (channels) and interleaves them such that the first channel completes its active phase; next, the return-to-zero phase of the first channel is concurrent with the active phase of the second channel; and, finally, the return-to-zero phase of the second channel is executed. More formally, if the seq-concur operator is applied to channels B and C, it returns the following handshake expansion: B_r↑; [B_a↑]; B_r↓, C_r↑; [B_a↓ ∧ C_a↑]; C_r↓; [C_a↓].

A sequencer component can be easily transformed into a concurrent sequencer using the Seq Replacement, as illustrated in Fig. 8[1]. Each component has one passive port A and two active ports (B

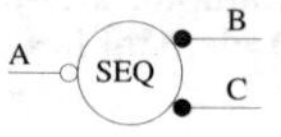

(rep
(enc-early
(p-to-p passive A)
(seq (p-to-p active B)
(p-to-p active C))))

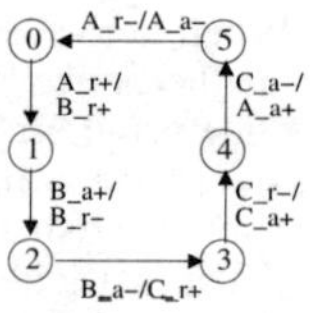

(rep
(enc-early
(p-to-p passive A)
(**seq-concur** (p-to-p active B)
(p-to-p active C))))

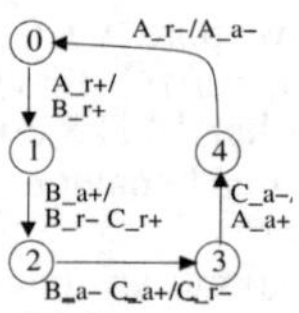

(a) Sequencer (b) Concurrent Sequencer

Figure 8: The Seq Replacement Transform

and C), whereas each of the active ports is connected to other components on the respective channels. When activated on channel A, the non-concurrent sequencer performs a non-overlapped activation of channels B and C in turn. In contrast, with the concurrent sequencer, the return-to-zero phase on channel B is overlapped with the active phase on channel C. At the CH level, a sequencer component is easily changed to the a concurrent sequencer by changing the seq operator into the newly-introduced seq-concur operator. The Burst-mode specifications for the two components are shown in Fig. 8.

There is an important correctness issue in the use of concurrent sequencers. The newly-introduced concurrency between processes may potentially result in data [12] and structural hazards. For example, data hazards may occur when two concurrent processes share a variable. Suppose a 4-way sequencer (activating four processes P1 ... P4) is to be made more concurrent, but there is a data hazard between processes P2 and P3. For example, P2 writes a variable, and P3 reads that variable; if a concurrent sequencer is used, so that P2 and P3 are concurrent, the variable might be read by P3 even before P2 finishes latching a new value.

The solution is to define *hybrid* sequencers, which result from the selective application of Seq Replacement. The CH expression for the hybrid sequencer to be used in the above example is:

(rep (enc-middle (p-to-p passive a)
(seq-concur (p-to-p active P1)
(**seq** (p-to-p active P2)
(seq-concur (p-to-p active P3)
(p-to-p active P4))))))

Effectively, a concurrent sequencer operator is used whenever there are no data hazards, and a non-concurrent sequencer operator whenever there are hazards (between P2 and P3). Now, using this hy-

[1]It should be noted, however, that the optimization is not limited to sequencers: any component that has a sequencing operator (for example, WHILE [1]) can be optimized with this transform.

brid sequencer, P2 finishes writing the variable before P3 is started; therefore, P3 correctly reads the new variable value.

6.2 Enc Replacement

The Enc-Replacement transform replaces one "enclosure" interleaving operator in a CH expression by another one (for example, replaces enc-early with enc-middle). Thus, the transform modifies the interleaving of communications on two or more channels connected to a given component's ports. This transform is now illustrated on two examples.

(L/A;R) element. The (L/A;R) element was introduced in [8] to implement the sequential execution of asynchronous processes. In this example, it is shown how this sequencing component is transformed into a Tangram *parallel component* [10].

The implementation of the (L/A;R) element corresponds to the following CH expression:

```
(rep (enc-middle (enc-late    (p-to-p passive L)
                              (p-to-p active R))
                 (p-to-p active A)))
```

Like a sequencer component, the (L/A;R) element is activated on a passive port, and then performs sequential handshakes on the active ports A and R, before completing the handshake on L. The Martin-style specfication for this element is: *[[L_r↑]; A_r↑; [A_a↑]; L_a↑; [L_r↓]; A_r↓; [A_a↓]; R_r↑; [R_a↑]; R_r↓; [R_a↓]; L_a↓].

Enc Replacement can be applied to the enc-late operator, which is simply changed to the middle-enclosure operator:

```
(rep (enc-middle (enc-middle (p-to-p passive L)
                             (p-to-p active R))
                 (p-to-p active A)))
```

Now, the behavior of the (L/A;R) component is radically changed: the active/return-to-zero phases of the A and R handshakes are *concurrently* executed in the middle of the active/return-to-zero phases, respectively, of the L handshake (*[[L_r↑]; A_r↑, R_r↑; [A_a↑ ∧R_a↑]; L_a↑; [L_r↓]; A_r↓, R_r↓; [A_a↓∧R_a↓]; L_a↓]). In effect, the (L/A;R) component is now a parallel component: the subsystems attached to each of the A and R channels are executed concurrently.

New Adder Component. The second example involves the adder datapath components. In Section 4 an adder component was modeled. Replacing the top-level middle-enclosure operator (see Fig. 7b) by an early-enclosure operator changes drastically the behavior of the adder component. The new CH expression is:

```
E1: (rep    (enc-early
                (passive ! b out byte 8)
                (enc-middle    (active ? b inp1 byte 8)
                               (active ? b inp2 byte 8))))
E2: (func add out inp1 inp2)
```

The new adder now performs all of its computation (requesting the operands, producing the result, and releasing the operands) in the middle of the active phase of the channel out handshake. The return-to-zero phase is extremely fast. In contrast, the previous adder releases its operands late, at the end of the return-to-zero phase on channel out. In effect, the new adder facilitates *resource sharing*: since the operands are now released early, they can be immediately used by other processes.

7. Peephole Transforms

This section introduces two classes of peephole transforms. The first class contains a single powerful transform (Protocol Reversal on Functional Units), which is applied to functional units, and replaces one port type by another. The second class is a collection of various template-based clustering transforms. Each transform is *behavior non-preserving*: the resulting component is equivalent up to a different ordering of the signal transitions of the initial components.

7.1 Protocol Reversal on Functional Units

This peephole transform changes the interface types (active/passive) of functional units and also of the components connected to these units. The goal of the transform is to decrease the latency of a subsystem. Very much like Enc/Seq Replacement, this transform manipulates CH constructs (in this case, the interface type parameter in channel declarations).

Before illustrating the transform in detail, it is worth giving some motivation for its application. In Balsa, functional units typically have a passive output port and several active input ports. A functional unit is activated on the output port when a result is needed; the unit then actively requests the operands on the input channels, receives the data, and performs the operation. The idea of the transform is to modify the unit's interfaces so that the *data operands* themselves activate the functional unit, thus computing a result even before requested on the output port. As a result, the latency to compute the result may be significantly decreased.

As an example, consider the initial peephole window shown in Fig. 9a. It consists of a unary functional unit, a variable, and a transferer [2, 14]. In general, the functional unit may have more than one input port. The variable component is used to store data

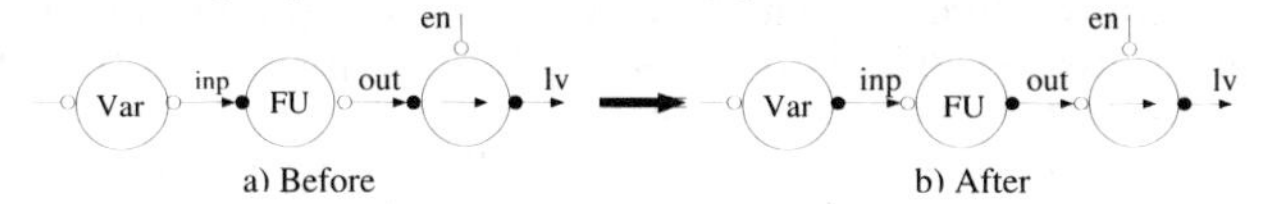

Figure 9: Protocol Reversal on Functional Units

items; it has a passive port (on which a data item is written), and an active port (on which a data item is read). First, the variable (VAR) latches a data input. Later on, the control enables the transferer (on channel en), which, in turn activates the functional unit (FU). The unit obtains the data input from the variable, computes the result, which is then transfered on lv to its destination. The whole cycle repeats with the variable latching the next data item.

In the optimized configuration (Fig. 9b), the port types in each component are now changed. This change corresponds to modifying the initial CH expression for the functional unit:

```
(rep (enc-middle (passive ! b out T1 L1) (active ? a inp T2 L2))) →
(rep (enc-middle (passive ? a inp T2 L2) (active ! b out T1 L1)))
```

The activity types on out and in are switched from passive/active to active/passive. Correspondingly, the inp port in the variable is now active, and the out port in the transferer is now passive.

The operation of the new configuration is now different. The variable, once written, immediately starts the computation in the functional unit. The result is sent to the new transferer. At some time later, the control will finally enable the transferer, which *immediately passes* the result on channel lv, thus decreasing the latency of the whole operation.

Unfortunately, it is not easy to generalize this peephole optimization for n-input functional units. For unary units, the pattern of new data items is fixed: the variable is written each cycle. However, for n-ary units, the pattern of new data items cannot be pre-determined. For example, within one cycle, a single variable can be written, while the others not; or, all but one variable can be written, while the last variable not. We are currently looking for ways to generalize this approach, perhaps using more information than just the connections of functional units.

7.2 Clustering Transforms

The final class of peephole optimizations uses template-based matching to identify and optimize targeted clusters of components. The three new transforms are graphically summarized in Fig. 10 by showing the initial and final configurations for each one. It is expected that a number of additional configurations can be added to this list in the future.

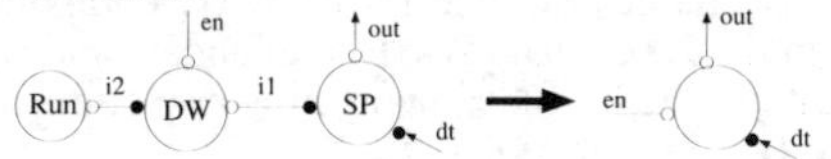

(a) Passive Output Port Transform

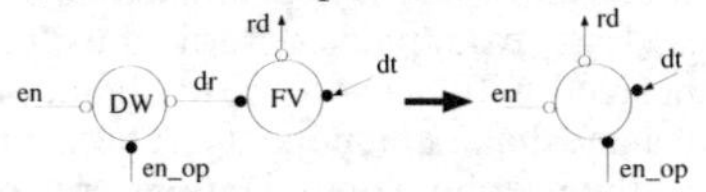

(b) Data Sampling Transform

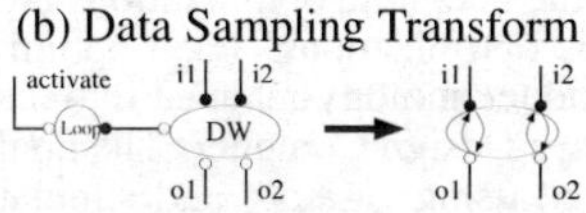

(c) Loop Enabling Transform

Figure 10: Clustering Peephole Transforms

Passive Output Port Transform. This transform optimizes a Balsa subsytem which outputs a data item to the environment, whenever the environment requests data. The transform takes three components, and clusters them into one.

The initial Balsa implementation (Fig. 10a) consists of three components: a SEQUENCE PULL [2], a DECISION WAIT [2], and a RUN [2, 13] component. The SEQUENCE PULL, when requested for data on channel OUT, first synchronizes with the DECISION WAIT on channel i1 before requesting a data item on channel dt; when the data item is received, it is passed on out. In turn, DECISION WAIT first synchronizes with the Balsa process on channel en, and then synchronizes with the RUN component on channel i2. The RUN component always completes any handshake started on its passive port. Thus, after it is requested for data, the subsystem performs four sequential operations: it synchronizes with the Balsa process, synchronizes with the RUN component, receives a data item on dt, and outputs it on out.

The result of this transform is a single clustered component, which has the same interfaces minus the i1 and i2 channels. The output cycle is now shortened: the component first concurrently synchronizes the data item request with the Balsa process, then receives a data item on channel dt, which is then output on channel out. The new component has a very simple implementation, just one C-element and wires between data inputs and data outputs.

Data Sampling Transform. This transform optimizes a Balsa subsystem which reads a data item from the environment. The transform takes two components, and clusters them into one.

The initial Balsa implementation (Fig. 10b) consists of only a FALSE VARIABLE component [2] and a DECISION WAIT component. An n-way FALSE VARIABLE component has one passive input port (dt – used to receive a data item), an active control port (dr – used to indicate data received), and n passive output ports used to sample the data item (Fig. 10b shows a one-way FALSE VARIABLE). The subsystem performs three sequential operations: it receives a data item, synchronizes with the Balsa process on channel en, and, finally, enables other processes (on channel en_op) to read the data item on channel rd.

The result of this transform is a single clustered component, which has the same interfaces, minus the eliminated dr channel. The data reading cycle is now shortened: the component first concurrently synchronizes the receiving of a data item with the Balsa process (on dt and en), and then enables other processes to read the data item. The implementation of the new component is very simple, just a C-element and wire connections on datapath signals.

Loop Enabling Transform. This transform operates on handshake components that synchronize two independent Balsa subsystems. The transform takes two components, and replaces them by just pairs of wires.

The transform takes as an input a LOOP component [2, 14] connected to both a special activate channel (automatically synthesized, used to start the whole system) and to a DECISION WAIT component (Fig. 10). The loop component, once started on the activate channel, enables forever the DECISION WAIT component to pass handshakes from a passive port to a corresponding active port.

The resulting component consists of just pairs of wires. The intuition behind the transform is simple: since the loop component *unconditionally* enables the DECISION WAIT component, the enabling logic in the DECISION WAIT is redundant and can be eliminated; in addition, the LOOP component is also eliminated.

8. Results

The entire synthesis flow has been evaluated on a number of substantial examples. Each example is described using the Balsa language, and synthesized with `balsa-c`, to obtain the initial unoptimized netlist of handshake components. The list of components is then optimized: the peephole optimizations are applied first, and then the resynthesis optimizations. Each optimized component is then synthesized and technology-mapped into the AMS 0.35μm library using Synopsys Design Compiler. The final implementations were back-annotated using `pearl`, and simulated with Cadence Verilog-XL. All steps, except the implementation of the enc/seq transforms, have been automated.

Three complete systems have been optimized and simulated: a four-handshake systolic counter [14], an eight-bit word three-place low-latency FIFO, and a programmable eight-bit counter.

Examples		Balsa	Previous Transforms [5]		All Transforms	
		Speed (ns)	Speed (ns)	Improvement	Speed (ns)	Improvement
Systolic Counter		24.81	16.06	35.26%	11.28	54.50%
FIFO	**latency**	17.33	15.19	12.36%	10.32	40.45%
	put cycle time	8.41	8.06	4.16%	6.22	26.04%
	get cycle time	11.78	9.91	15.87%	8.28	29.71%
Binary Counter		236.30	217.33	8.03%	126.96	46.26%

Table 1: Experimental Results

Table 1 shows, for each example, the throughput and latency improvements over the unoptimized Balsa circuits. The cycle time is reported for the systolic counter and programmable counter, while for the FIFO, the put and get cycle times, as well as the latency of a data item from input to output through an empty FIFO are reported. The "Previous Transforms" column indicates the results when only the previously-introduced transforms (Activation Channel Removal and Call Distribution [5]) are applied. The "All Transforms" column indicates the results when applying both the previous transforms and the newly-introduced transforms. For each example, the following transforms were successfully applied: *Systolic Counter* – Loop Enabling Transform; *Low-Latency FIFO* – Seq Replacement, Loop Enabling, Data Sampling, and Passive Output Port; and *Binary Counter* – Protocol Reversal on Functional Units, Seq Replacement. The performance improvements for on the simulated examples range up to 54%. Furthermore, the improvements using the new transforms are significantly better than those using only previous techniques [5].

9. Conclusions and Future Work

This paper introduces a powerful new set of transformations, and an extended channel-based language to support them, which can be used in an optimizing back-end for Balsa. The transforms described in this paper fall into two categories: resynthesis and peephole. Results on several substantial examples indicate significant performance improvements over unoptimized Balsa implementations, as well as over the implementations using only previous control-oriented transforms.

The current approach is still limited to supplying primitive transforms (somewhat analogous to the operations in `SIS` for logic optimization), and does not yet present systematic scripts or algorithms for their optimal application. This is a topic for future work.

10. REFERENCES

[1] A. Bardsley and D. Edwards, "Compiling the Language Balsa to Delay-Insensitive Hardware", *Hardware Description Languages and their Applications (CHDL)*, April 1997, pp. 89-91.
[2] A. Bardsley, "Implementing Balsa Handshake Circuits", PhD. Thesis, Department of Computer Science, University of Manchester, 2000.
[3] A. Bardsley, Personal Communication, 2000.
[4] E. Brunvand, "Translating Concurrent Communicating Programs into Asynchronous Circuits", Technical Report CMU-CS-91-198, Carnegie Mellon University, 1991.
[5] T. Chelcea, A. Bardsley, D. Edwars, and S.M. Nowick, "A Burst-Mode Oriented Back-End for the Balsa Synthesis System", Proceedings of *Design Automation and Test in Europe*, March 2002.
[6] T. Kolks, S. Vercauteren, and B. Lin, "Control Resynthesis for Control-Dominated Asynchronous Design", *IEEE ASYNC'96 Symposium*, pp. 233-243.
[7] R.M. Fuhrer and S.M. Nowick, "Sequential Optimization of Asynchronous and Synchronous Finite-State Machines: Algorithms and Tools", Kluwer Academic Press, 2001.
[8] A.J. Martin, "Programming in VLSI: From Communicating Processes to Delay-Insensitive Circuits", in C.A.R. Hoare, editor, *Developments in Concurrency and Communication*, UT Year of Programming Institute on Concurrent Programming, Addison-Wesley, 1990, pp. 1-64.
[9] S.M. Nowick, "Automatic Synthesis of Burst-Mode Asynchronous Controllers", Technical Report CSL-TR-95-686, Stanford University, March 1993.
[10] A.M.G. Peeters, "Single–Rail Handshake Circuits", PhD. Thesis, Department of Computer Science, Technical University of Eindhoven, 1996.
[11] M.A. Pena and J. Cortadella, "Combining Process Algebras and Petri Nets for the Specification and Synthesis of Asynchronous Circuits", *IEEE ASYNC'96 Symposium*, pp. 222-232.
[12] L.A. Plana and S.M. Nowick, "Architectural Optimization for Low-Power Nonpipelined Asynchronous Systems", *IEEE Transactions on Very Large Scale Integration Systems*, Vol. 6, No. 1, March 1998.
[13] K. van Berkel, "Handshake Circuits: an Intermediary between Communicating Processes and VLSI", PhD. Thesis, Department of Computer Science, Technical University of Eindhoven, 1992.
[14] K. van Berkel, "Handshake Circuits: an Asynchronous Architecture for VLSI Programming", *International Series on Parallel Computation*, Vol. 5, Cambridge University Press, 1993.
[15] K. van Berkel, Personal Communication, 1999.

Design of Asynchronous Circuits by Synchronous CAD Tools

Alex Kondratyev
Cadence Berkeley Laboratory
2001 Addison Street, 3rd floor
Berkeley, CA 94704, USA
kalex@cadence.com

Kelvin Lwin
Reshape Inc.
1255 Terra Bella Ave.
Mountain View, CA 94043, USA
klwin@reshape.com

ABSTRACT
The roadblock to wide acceptance of asynchronous methodology is poor CAD support. Current asynchronous design tools require a significant re-education of designers, and their features are far behind synchronous commercial tools. This paper considers a particular subclass of asynchronous circuits (Null Convention Logic or NCL) and suggests a design flow that is based entirely on commercial CAD tools. This new design flow shows a significant area improvement over known flows based on NCL.

Categories & Subject Descriptors
B.5.1 Design – Styles

General Terms: Design

1. INTRODUCTION
EDA flows, being industry-driven, use synchronous methodology as a de-facto standard. However, the implementation problems presented by imposing a synchronous model of operation in deep-submicron circuits motivates the investigation of other modes of operation, asynchronous in particular [1].

Asynchronous design has been proven capable of delivering:

- Higher speed due to of the average case performance versus worst case in synchronous circuits [2]
- Less power consumption due to the absence of clock and natural support of idle mode [3]
- Low EMI and noise due to even distribution of switching activity in time [4]

However the success stories in high-speed and low power asynchronous designs are somewhat controversial. To deliver the promised advantages they often rely on non-trivial timing assumptions that make verification difficult. Moreover, a lack of commercial CAD support for asynchronous synthesis is a major roadblock to wider acceptance of the methodology.

DAC 2002, June 10-14, 2002, New Orleans, Louisiana, USA.

Low EMI and noise coefficients are the only "free" advantages of asynchronous circuits. Getting rid of the clock results in a significantly flatter noise/EMI spectrum across the frequency domain (10DB drop according to [4]). Until recently, EMI and noise metrics were "second class citizens." The focus was on power and performance, but that is about to change. EMI and noise metrics are gaining significance because of two emerging applications: mixed signal and smart cards. For mixed signal designs, analog blocks are sensitive to clock correlated, digital switching noise. Reducing noise and EMI has an immediate impact, boosting both precision and performance significantly.

In a smart card domain, functionality is not sensitive to EMI itself, but the security is. Non-invasive security attacks are based on monitoring the power rail, or EMI signature, of a smartcard. Even distribution of circuit-switching activities vastly improves the security.

This paper suggests an automatic flow for the design of asynchronous circuits featuring: low EMI, high security and small flow turnover cost (HDL based methodology using commercial CAD tools) at a significantly reduced area penalty than the previous flow [5].

Section 2 introduces main theoretical notions. Section 3 describes a previously known HDL flow for NCL. Section 4 suggests a new way of NCL implementation. Section 5 presents experimental results for the suggested flow.

2. THEORTICAL BACKGROUND
2.1 Delay-Insensitive Combinational Circuits
The acknowledgement notion plays a key role in ensuring delay-insensitivity (DI). Informally, we say that the firing of gate g_i **acknowledges** the firing of gate g_j, if by looking at g_i switching, one can judge that g_j has already fired as well. In a delay-insensitive combinational circuit, any transition at the wire OR gate must be acknowledged by the primary outputs.

In practice, implementing the acknowledgment of every single wire in a circuit is costly. One can consider some wire forks in a circuit to be safe by making a timing assumption about their skew. These forks are called **isochronic** [6]. For an isochronic fork, it is sufficient to acknowledge a single wire from the fork while the acknowledgements for the rest of wires rely on the timing assumption. Circuits in which the only non-acknowledged wires come from isochronic forks are called quasi-delay-insensitive [6]. NCL circuits belong to the QDI class.

2.2 Null Convention Logic (NCL)

NCL is a specific way of implementing data communication based on **DI encoding.** Data changes from the spacer (**NULL**) to a proper codeword (**DATA**) in the set phase, and then back to NULL in the reset phase. NCL targets the simple DI encoding in which DATA codewords are **one-hot codes,** and the spacer NULL is represented by a vector with all entries equal to "0". For example in **dual-rail encoding** each signal *a* is represented by two wires *a.0* and *a.1* (i.e. *a*=1 encoded as *a.0*=0, *a.1*=1, and *a=0* encoded as *a.0*=1, *a.1*=0).

At an architectural level, NCL systems show a clear separation of sequential and combinational parts, much *in the same way as with synchronous systems* (see Figure 1).

NCL systems borrow the idea of organizing register interaction and data communication in DI fashion from micropipeline architectures [7].

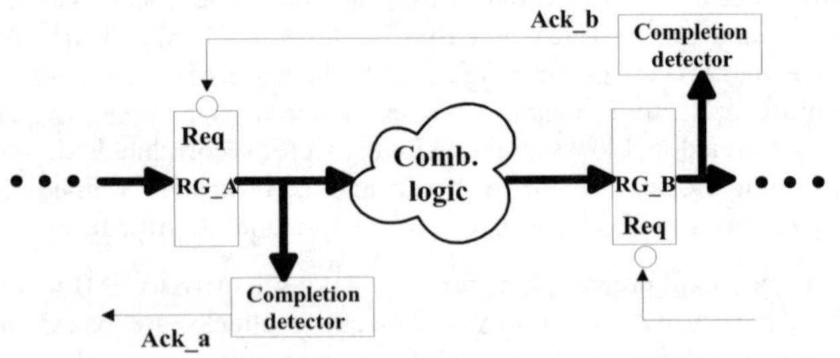

Figure 1. NCL system implementation

To explain how the NCL system functions, let us assume that all registers are initially in the NULL state and signals Ack are asserted to "0" (signals *Req* are asserted to "1"). When DATA arrives, the outputs of a register (e.g. RG_A) will change from NULL to DATA (the register stores the DATA value), and the DATA **wavefront** propagates through a combinational circuit to the inputs of the next register (RG_B). Simultaneously, a completion detector checks for a DATA codeword at its inputs, and replies by rising the *Ack* signal. This signal disables the request line of the previous register and prepares the register for storing the next NULL wavefront. The **request-acknowledgement mechanism** of register interaction [7] ensures a **two-phase discipline** in NCL system functioning and prevents collisions between different DATA wavefronts.

This behavior scales down to the level of NCL gates. Every gate implements the so-called threshold function and is represented as $g(x_1,\ldots,x_n) = S + g(x_1+x_2+ \ldots +x_n)$, where *S* is an unate set function. A gate *g* switches from NULL to DATA when its set function turns to 1, and it resets to NULL back when all inputs are reset to 0. A semi-static CMOS implementation of an NCL gate is shown in Figure 2(a), while Figure 2 (b)(c) show an implementation and notation for a particular NCL gate with the function $g=x1x2 + g(x_1+x_2)$, known from literature as a Muller's C-element.

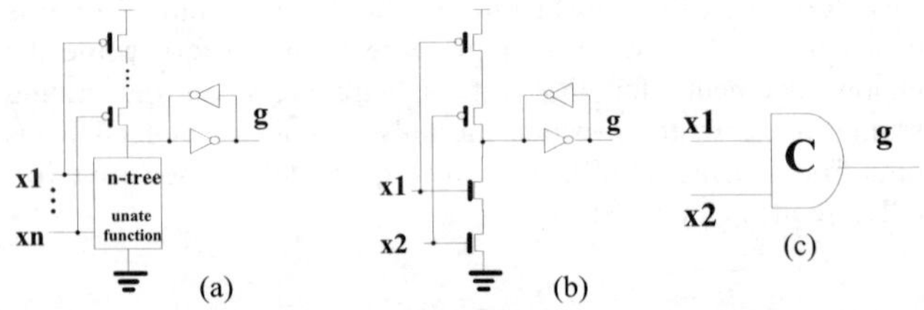

Figure 2. NCL gate implementations

3. Overview of HDL Design Flow

The NCL design flow uses off-the-shelf simulation and synthesis components (see Figure 3). The flow executes two synthesis steps. The first step treats NCL variables as single wires. The synthesis tool performs HDL optimizations and outputs a network in GTECH library as if it is a conventional Boolean RTL. The second step expands the intermediate GTECH netlist into a dual-rail NCL by making dual-rail expansions and mapping into the threshold library. The particulars of implementation on step1 and step 2 impact the quality of final results.

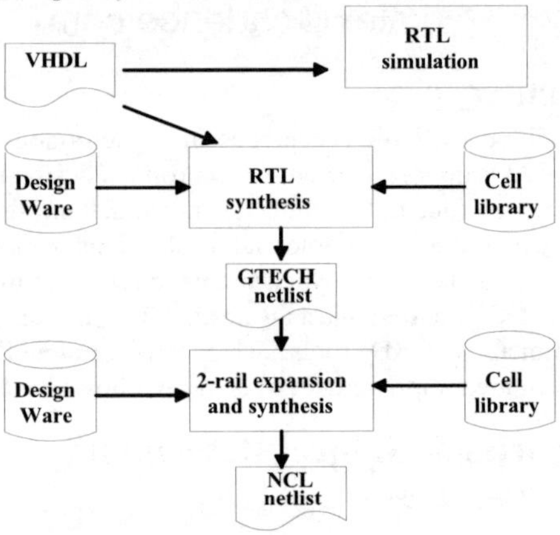

Figure 3 . RTL flow for NCL

In [5] a regular method for NCL implementation based on Delay Insensitive Minterm Synthesis (DIMS) [8] was suggested. In this method, steps 1 and 2 of the design flow are implemented as follows:

Step 1 performs a mapping of the optimized network into two-input NAND, NOR and XOR gates.

Step 2 first represents each wire *a* as a dual-rail pair *a.0* and *a.1* and then makes a direct translation of two-input Boolean gates into pairs of threshold gates with limited optimization of a threshold network.

Unfortunately DIMS-based implementations have significant overhead that comes from two main sources:

1. Overdesigning due to locality of ensuring DI (no sharing in the acknowledgement is allowed)
2. Little room for optimisation (optimisation can easily destroy DI properties)

4. NCL Flow with Explicit Completeness

The newly proposed flow exploits the idea of separate implementation of functionality and delay-insensitivity. A NCL circuit is partitioned on functional and completion parts with the possibility to optimize them independently from each other. This is achieved by a following modification of flow steps 1 and 2.

Step 1 performs a conventional logic synthesis (with optimization) from RTL specification of NCL. It maps an obtained network into GTECH library that consists of gates implementing set functions of threshold gates.

Step 2 consists of the following substeps:

2.1. Reduction of the logic network to unate gates (by using two different variables *a.0* and *a.1* for direct and inverse values of signal *a*). The obtained unate network implements *rail.1* of a dual-rail combinational circuit.

2.2. Dual-rail expansion of the combinational logic by creating for each gates in the *rail.1* network its corresponding dual gate in *rail.0* network.

2.3. Ensuring delay-insensitivity by providing local completion detectors (OR gates) for each pair of dual gates and connecting them into a completion network (multi-input C-element) with a single output *done.*

Implementation of the flow with explicit completion requires a minor modification of interfacing conventions within the NCL system. From now on we will assume that for each 2-rail primary input *a.0, a.1* there exists an explicit signal *a.go* such that $a.0 \neq a.1 \Rightarrow a.go=1$ (set phase), while $a.0=a.1=0 \Rightarrow a.go=0$ (reset phase). The modified organization of NCL system is shown in Figure 4. It is easy to see that it differs from the one in Figure 1 in having separate completion detectors for combinational logic and registers.

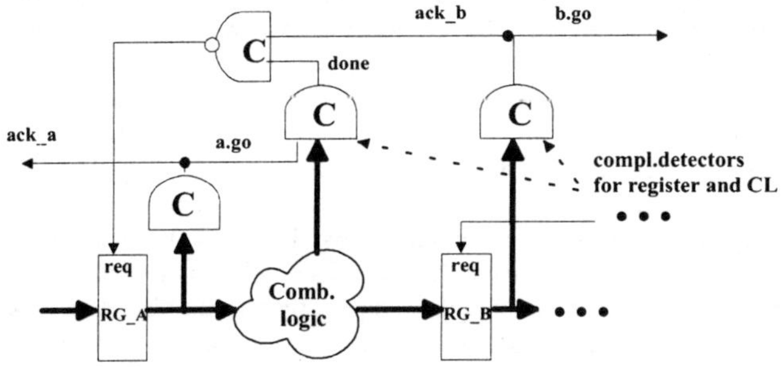

Figure 4. NCL system with explicit completion detection

Example. **Encoder 4 to 2.** The encoder is described by the following RTL specification:

```
encode : process(din)
  begin
    if din = "1000" then
      d <= "11";
    elsif din = "0100" then
      d <= "10";
    elsif din = "0010" then
      d <= "01";
    elsif din= "0001" then
      d <= "00";
    else
      d <= (others => '0');
    end if;
```

A mapping of encoder in GTECH library (step 1) is shown in Figure 5(a). After the reduction to unate gates (Figure 5(b)) the network is expanded into dual-rail implementation (Figure 5(c)). This is done locally by providing a dual function for each of the gates in unate representation from Figure 5(b). Each dual-gates pair contributes an output by a local completion detector (OR gate) that goes to multi-input C-element, which is providing signal *done* (Figure 5(d)). Information about the validity of input codewords is provided by the C-element, *I.go.* It combines all completion signals for primary inputs. The output *I.go* is connected to the C-element *done.*

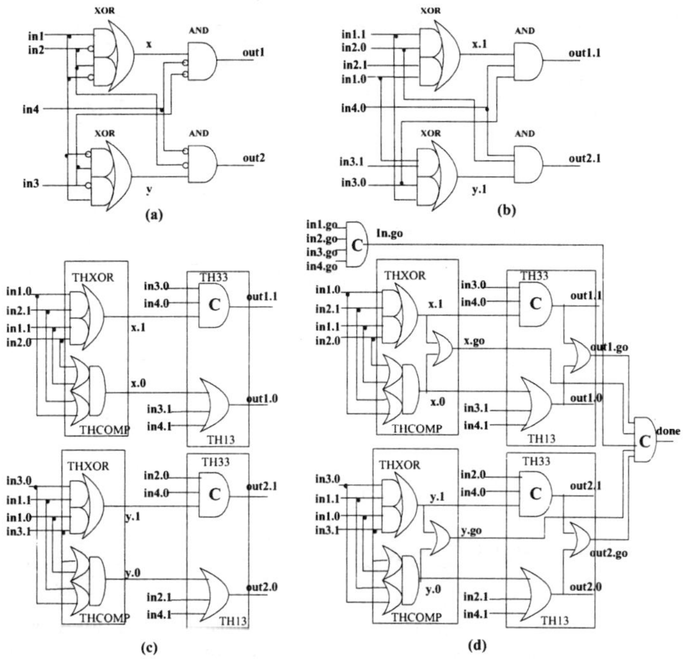

Figure 5. 4 to 2 encoder with explicit completion

Proposition. A NCL circuit implemented with explicit completion belongs to QDI class.

Proof for the proposition at hand is rather straightforward. In each phase (set and reset) exactly one gate out of the dual pair must switch. These transitions are acknowledged by local completion detectors. In their own turn all local detectors are acknowledged by the primary output *done.* Hence, at least one wire in every fork is acknowledged and a circuit is from QDI class.

5. EXPERIMENTAL RESULTS

This Section provides experimental data for comparison of the suggested flow with the previously known approach [5] according to area and power consumption. Speed estimation was less of an issue because the flow does not target high-speed applications. However preliminary data show that NCL_X circuits do not show degradation of performance with respect to [5].

Area. Implementation of NCL flow with explicit completion (NCL_X flow) kept the same front end RTL coding style so it is compatible with the known NCL flow based on DIMS [8] (NCL_D flow). This enables the same designs accomplished in NCL_D to be synthesized using NCL_X. Between these two, fair comparisons can be conducted. Table 1 illustrates the designs chosen as benchmark suite to compare the relative performance of NCL_X versus NCL_D. It shows the designs with large combinational clouds to have achieved the highest reduction in area (e.g. AND4) and registration elements dominant designs showing very little reduction (e.g. SET_CNT). The AND4 design is merely a 4-input AND function implementation. The circuit SET_CNT is a 3-stage ring register that counts up to a certain number. The area number used for the figures is based on the transistor count of physical cells in the threshold library. According to the design suite, the new flow, NCL_X, can achieve an average area reduction of 28%.

Table 1. Area Comparison

Circuit	NCL_D	NCL_X	%
AND16	480	186	-61%
AND4	96	30	-69%
IF_THEN_ELSE	72	38	-47%
HA	72	56	-22%
FA	176	118	-33%
CLIPPER	448	237	-47%
MUX_DECODER	456	352	-23%
SET_CNT	490	456	-7%
USHIFT	1264	762	-40%
NCL_ADDRCONV	1356	1045	-23%
SERIAL_CRC	2010	1936	-4%
SYNC_STATE	2402	1934	-19%
BIT_CNT	2779	2432	-12%
FSM_DATAPATH	10260	9290	-9%
NCL_X2	15338	13852	-10%
NCL_X1	18606	12724	-32%
VITERBI DECODER	45198	38130	-16%
Average	**-**	**-**	**-28%**

Power consumption. The wavelet design consists of two datapath blocks address and data) and one control block (FSM). Results from PowerMill simulations of NCL_D and NCL_X versions are shown in Table 2. The nominal supply voltage for the design is 2.5V. However the design is fully operational under the reduced supply voltages 1.8V and 1.1V. This is an additional advantage that comes from the asynchronous nature of the implementation in which reducing power results in performance degradation but not in design malfunction. Energy numbers are presented relative to the nominal supply voltage and NCL_D style of implementation. From Table 2 follows that NCL_D and NCL_X designs show approximately the same power numbers. This conclusion was also confirmed by analysis of switching activities for NCL_D and NCL_X design styles. For NCL_X circuit, the additional switching comes from completion network but in general offset by having smaller number of switching in the functional part.

Table 2. Energy Comparison

Design	Energy (V_{dd}=2.5V)		Energy (V_{dd}=1.8V)		Energy (V_{dd}=1.1V)	
	NCL-D	NCL-X	NCL-D	NCL-X	NCL-D	NCL-X
addr	100%	109%	46%	52%	14%	19%
data	100%	90%	55%	46%	12%	5%
wavelet	100%	102%	47%	50%	38%	29%

6. CONCLUSION

The paper has presented an efficient design flow for asynchronous circuits that is fully supported by commercial CAD tools. Implementations obtained through the flow are compared favorably to the ones reported in [5]: they show significantly lower area overhead, are faster and consume approximately the same power. To be fair we should admit that in comparison to synchronous circuits NCL_X implementations suffers from 2-2.5 times the area penalty and could consume more power. Power consumption might be less of the issue because it could be reduced due to natural support of idle mode and through the use of four-rail communication scheme instead of two-rail. Area penalty is unlikely to be further reduced because it is close to a theoretical lower bound of 2X coming from dual-rail nature of designs. However the advantages of NCL_X circuits with respect to synchronous are extremely low noise and EMI [4] and higher level of security and reliability during the circuit operation. Due to these advantages NCL_X circuits might be a good for emerging mixed signal and smart card applications.

7. ACKNOWLEDGMENTS

We wish to thank Alexander Taubin, Ross Smith and Michiel Lighthart for their invaluable help with the NCL_X flow implementation.

8. REFERENCES

1. International Technology Roadmap for Semiconductors, http://www.itrs.net/

2. Ivan Sutherland and Jon Lexau. Designing fast asynchronous circuits. In *Proc. International Symposium on Advanced Research in Asynchronous Circuits and Systems*, pages 184-193. IEEE Computer Society Press, March 2001.

3. S. B. Furber, D. A. Edwards, and J. D. Garside. AMULET3: a 100 MIPS asynchronous embedded processor. In *Proc. International Conf. Computer Design (ICCD)*, September 2000.

4. John McCardle, David Chester. Measuring an Asynchronous Processor's Power and Noise, SNUG Conference, April 2001.

5. Michiel Ligthart, Karl Fant, Ross Smith, Alexander Taubin, and Alex Kondratyev. Asynchronous design using commercial HDL synthesis tools. In *Proc. International Symposium on Advanced Research in Asynchronous Circuits and Systems*, pages 114-125. IEEE Computer Society Press, April 2000.

6. A.J. Martin. Programming in VLSI: From communicating processes to delay-insensitive circuits. In C. A. R. Hoare, ed., Developments in Concurrency and Communication, UT Year of Programming Series, pages 1-64. Addison-Wesley, 1990.

7. Ivan E. Sutherland. Micropipelines. Communications of the ACM, 32(6):720-738, June 1989.

8. Jens Sparso and Jorgen Staunstrup. Delay-insensitive multi-ring structures. Integration, the VLSI journal, 15(3):313-340, October 1993.

Implementing Asynchronous Circuits using a Conventional EDA Tool-Flow

Christos P. Sotiriou
Institute of Computer Science (ICS), Foundation for Research and Technology - Hellas (FORTH),
P.O Box 1385, Heraklion, Crete, GR 711 10 Greece.
sotiriou@ics.forth.gr

Abstract

This paper presents an approach by which asynchronous circuits can be realised with a conventional EDA tool flow and conventional standard cell libraries. Based on a gate-level asynchronous circuit implementation technique, direct-mapping, and by identifying the delay constraints and exploiting certain EDA tool features, this paper demonstrates that a conventional EDA tool flow can be used to describe, place, route and timing-verify asynchronous circuits.

Categories and Subject Descriptors

B7.1. [**Integrated Circuits**]: Types and Design Styles

General Terms

Design, Experimentation, Standardization

Keywords

Asynchronous, EDA, Tool-Flow

1. INTRODUCTION

Asynchronous design has often been quoted as an alternative design approach which can help overcome the implementation problems of Deep-Sub-Micron (DSM), high-performance synchronous circuits, *i.e.* maintaining low clock skew, efficient clock buffering, high-power consumption, high Electromagnetic Interference (EMI) [9][3]. However, the adoption of asynchronous design has been hindered by the lack of commercial support and inability to design asynchronous systems using conventional tools. In this paper we show how a class of asynchronous circuits can be manually described, timing-constrained and timing-verified and then automatically placed and routed using conventional EDA tools. Our emphasis is on asynchronous control circuits but the same principles can be applied to asynchronous datapaths.

2. ASYNCHRONOUS CIRCUITS IN EDA

A variety of approaches exist for the design and implementation of asynchronous circuits [5]. Transition-based approaches specify circuits and systems in terms of signal transitions using Petri-nets [11], then using Signal Transition Graphs (STGs) [2] map the specification into an implementable Asynchronous Finite State Machine (AFSM) form. Micropipeline circuits [16] are asynchronous pipelines that may incorporate logic between their stages. Programming/compilation approaches [8][19] start from a circuit specification in a formal language, then through a sequence of syntax-transformations derive circuit descriptions. Finally, FSM-based asynchronous design approaches include asynchronous Huffman machines [18], Burst-Mode AFSMs [15][21] and Direct-mapped AFSMs [6][14]. However, only a few of these approaches are suitable for EDA. The fundamental requirement is that they use standard logic gates. An important additional requirement is that the derived circuits possess easy to determine timing constraints, which can be supported throughout the stages of an EDA flow.

Two industrial-strength asynchronous EDA tool flows actually exist, the Tangram tool set which has been used to design the 80C51 microcontroller [20] and the Theseus Logic tool flow [7]. However, the former is based on proprietory tools, whereas the latter uses proprietory standard cell libraries.

In this work we fully-automate the timing verification, placement and routing of asynchronous circuits using standard EDA tools and standard cell libraries based the direct-mapping approach introduced by L. A. Hollaar [6], which uses set-reset flip-flops for generating and storing the states of a "one-hot" encoded AFSM [18]. The one-hot encoding eliminates general races between state variables and simplifies the circuit implementation as the logic that generates the state signals assumes a regular form. One-hot AFSMs have the attractive property that their timing constraints for correct operation are independent of the actual circuit specification, in contrast to other asynchronous design approaches. In Hollaar's implementation each state corresponds to a set-reset flip-flop. The condition for entering a state is fed to the set input of the state's flip-flop, whereas the condition for leaving the state is fed to its reset input. The set and reset conditions can be implemented by combinational logic. The reset condition is in most cases the setting of a successor state [6][13].

Figure 1 shows an set-reset flip-flop AFSM implementation of a sequential portion of a state diagram. Portions of states s1, s2 and s3 are shown. State s2 is entered on the transition of signal x. The circuit operates as follows: if input x is asserted, then the output of the NAND gate to which x is input will drop as both x and s1 are high. The consequence of this is the setting of state s2 and the resetting of state s1.

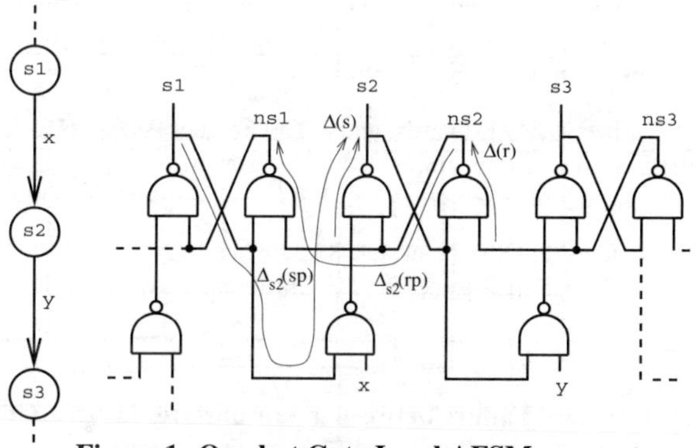

Figure 1: One-hot Gate-Level AFSM segment

2.1 One-Hot Gate-level AFSM Delay Analysis

In this section the delay assumptions for correct operation of one-hot gate-level AFSMs are identified. For a gate-level AFSM, the following delay paths may be defined:

- (a) Set input delay, *i.e.* delay between the set input of a state (the set input of the state flip-flop) and the state output, $\Delta(s)$.
- (b) Reset input delay, *i.e.*, delay between the reset input of a state (the reset input of the state flip-flop) and the inverted state output, $\Delta(r)$.
- (c) State setting delays (set paths), *i.e.* delays between the predecessor state signal outputs and the state output. This comprises the delays through combinational logic and the state flip-flop. For state n with m predecessors these delays can be represented as $\Delta_{n(1)}(sp)$, $\Delta_{n(2)}(sp)$, ... $\Delta_{n(m)}(sp)$ for m predecessors.
- (d) State resetting delays (reset paths), *i.e.* delays between the complementary output of a state (inverted version) and the state signal outputs of its predecessors. This comprises the delays through potentially combinational logic and the flip-flops of the predecessor states. For state n with m predecessors these delays can be represented as $\Delta_{n(1)}(rp)$, $\Delta_{n(2)}(rp)$, ... $\Delta_{n(m)}(rp)$ for m predecessors.

These critical delays are identified for state s2 in the simple gate-level AFSM segment of Figure 1. In order to ensure correct circuit operation one-hot critical races between two or more states must be avoided for all states of the AFSM.

2.1.1 *One-hot race between a pair of states.*

The correct order of state transitions is 10→11→01. The condition for correct operation is that the next state must be properly entered before the previous state is left. In CMOS one-hot AFSMs [14] this translates to selecting appropriate transistor sizes for the n-type transistors, used to enter a state, and p-type transistors, used to leave a state.

In gate-level AFSMs, the condition for correct operation depends on the threshold points of the set-reset flip-flops, *i.e.* the point where flip-flops change state. If all the NAND gates forming the set-reset flip-flops are identical, then it is nearly impossible to make the circuit fail for a 2 state transition, as the circuit is delay insensitive, *i.e.* will work correctly for arbitrary delays of wires and gates. However, if NAND gates of different thresholds are used (different $\frac{N}{P}$ transistor sizing for example) it is possible that the previous state's flip-flop changes state, before the next state's has flipped. This type of failure is illustrated in Figure 2, where the output waveforms of an HSPICE [10] simulation of two AFSM states s1 and s2 are shown. The top and bottom panels of Figure 2 show the two outputs of state flip-flops s1 and s2, *i.e.* s1 and ns1 and s2 and ns2 respectively. Failure has been achieved by moving the threshold point of the s1 flip-flop towards VDD and loading ns2.

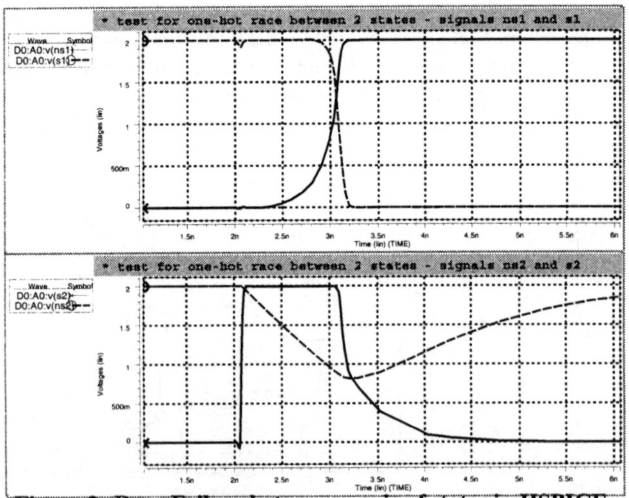

Figure 2: Race Failure between a pair of states in HSPICE

It is possible to express delay constraints which will ensure one-hot race free transitioning even if the NAND gates are non uniform. A state n is entered by a set path, $\Delta_n(sp)$, whereas it is left by a corresponding reset path, $\Delta_n(rp)$. Thus, for one-hot race free operation, the delay through the reset path and through the combinational logic leading back to the set input of the state must be longer than the delay of setting the state once the set input is being driven (for all possible set and reset paths), *i.e.*:

$$\Delta_n(rp) + (\Delta_n(sp) - \Delta(s)) > \Delta(s) \Rightarrow$$
$$\Delta_n(rp) + \Delta_n(sp) > 2\Delta(s) \quad (EQ1)$$

To ensure that this delay assumption holds it is possible to bias the $\Delta(s)$ and $\Delta(r)$ delays so that $\Delta(s) < \Delta(r)$ holds.

2.1.2 *One-hot race between three or more states.*

In the case of three states different transition orders are possible: 100→110→010→011→001 or 100→110→111→011→001. In any case, state 101 should not be reached. The prohibited 101 state can only be reached if the middle of the three states has been set and reset before the left state has been reset. This can only occur if the right state is set (thus resetting the middle state) and the middle state reset faster than the middle state can reset the state on the left, *i.e.* the situation to avoid is: Δ(sp of right state) + Δ(rp of right state) < Δ(rp of middle state). For two states, n-1 and n, the opposite condition must hold for correct operation (for all possible set and reset paths), *i.e.*:

$$\Delta_n(sp) + \Delta_n(rp) > \Delta_{n-1}(rp) \quad (EQ2)$$

Due to the nature of one-hot coding, ensuring that one-hot critical races do not occur for more than three states can be achieved by ensuring that two and three state races do not occur for every possible sequence of two and three states of the AFSM.

To summarise, correct circuit operation for a one-hot gate-level AFSM can be guaranteed if equations (EQ1) and (EQ2) above hold for all states of the AFSM. In the next sections we identify the features of the EDA tools that can be used to implement asynchronous circuits.

2.2 Asynchronous Circuit Specifications

In order to describe a gate-level asynchronous circuit into a synthesis tool, a circuit netlist form must be used, since synthesis is not (as yet) possible. One such form is the Synopsys GTECH [17] (Generic TECHnology) format, which allows for a circuit to be described in terms of generic technology-independent unsized gates. It is then possible to both map the GTECH specification into any technology and to specify path constraints which will appropriately size the technology mapped gates.

2.3 Delay Constraints and Hierarchy

In asynchronous circuit design, only true path delays are relevant to the circuit operation. For example, the delay constraints described in the previous section may be verified by measuring the true path delays for the corresponding paths. However, it is often the case that static timing analysis reports a path delay inappropriately, *i.e.* by combining arrival times of signals which will not change. In addition, timing analysis is particularly confused by the feedback of state registers. Careful hierarchical definition of the circuit can overcome this problem. Paths which are to be measured and constrained by the synthesis tool should not contain feedback. For example, the set and reset NAND gates may need to be constrained so that $\Delta(s) < \Delta(r)$ holds. If a cell is described so as to make feedback external to it, it is then possible for delay constraints to be fulfilled appropriately by the synthesis tool.

2.4 Constraint Specification

The delay constraints of direct-mapped AFSMs presented in Section 2.1 are relative, *i.e.* the delays of critical circuit paths are interrelated. There is currently little support for relative constraint specification in contemporary EDA tools, even though standards such as SDF (Standard Delay Format) v3.0 support them [12]. However, it is possible to specify absolute timing constraints on circuit paths through the commonly used GCF (General Constraint Format) timing constraints [1]. Experiments performed with Cadence's Pearl timing analyser demonstrated that timing analysis for GCF `path_delay` constraints considers the true paths between the specified points. GCF constraints are also compatible with Cadence's P&R (Placement and Routing) tools. However, as GCF constraints are designed to be imposed only on external, top-level pins, it was found necessary to introduce dummy pins to a physical design file (DEF format) to be able to constrain the appropriate paths.

2.5 Physical Design

Hierarchical P&R allows for timing constraints to be localised to the individual circuits and makes the process of constraining circuits manageable. The breakdown of a system into constituent parts is a very natural process for asynchronous systems, as these are generally composed of asynchronous blocks which communicate with each other via delay-insensitive asynchronous handshaking protocols. Therefore, it is straightforward to draw block boundaries at the handshake interfaces. In this way, constraints are localised, and at the top (block) level, where blocks communicate, it must be ensured that the handshake data are "bundled", *i.e.* synchronised, to their corresponding handshake control signals.

3. AN EXAMPLE DESIGN

We now demonstrate the process of realising a fully-asynchronous circuit using a conventional EDA flow by manually constraining an example circuit: an asynchronous, 5-stage, 32-bit pipeline, constructed out of fully-decoupled four-phase latch controllers [4][13]. The circuit was mapped to the VST-UMC 0.18μm technology.

3.1 Constraint Specification

Table 1 shows the delay analysis for the critical paths of the unconstrained version of the circuit. The timings of Table 1 are obtained by mapping the GTECH circuit specification to the technology library and then measuring the critical path delays. The numbers in bold represent the path delays reported by the Pearl `FindPathsBetween` command (with false paths blocked) on the critical paths, whereas the numbers in italics represent the path delays reported by the DC `report_timing` command. The DC timings are obtained by measuring the path delays at appropriate levels of hierarchy so that false paths are not considered. As Pearl allows for false paths to be blocked at the top level, the Pearl timings are more accurate as they include input and output loading.

State	Δ(s)	Δ(r)	Δ(sp)	Δ(rp)
ackin	**0.11**/*0.06*	**0.04**/*0.03*	**0.25**/*0.17*	**0.18**/*0.15*
waita	**0.07**/*0.06*	**0.04**/*0.03*	**0.19**/*0.16*	**0.14**/*0.15*
reqout	**0.11**/*0.06*	**0.04**/*0.03*	**0.24**/*0.16*	**0.18**/*0.15*
waitr	**0.04**/*0.03*	**0.06**/*0.05*	**0.18**/*0.15*	**0.15**/*0.15*
idle	**0.10**/*0.06*	**0.04**/*0.03*	**0.26**/*0.20*	**0.15**/*0.15*

Table 1: Unconstrained Latch Controller Delay Analysis (in ns)

Table 2 shows the delay analysis for the constrained circuit. This version of the circuit was technology mapped and then constrained so that the the delay of the reset gate of the state flip-flops was at least twice that of the set gate, *i.e.* $\Delta(\text{r}) \geq 2\Delta(\text{s})$, and the bundled-data assumption on the output request holds, *i.e* all data outputs are synchronised to signal `reqout`. Imposing the $\Delta(\text{r}) \geq 2\Delta(\text{s})$ synthesis constraint to all states ensures that after circuit synthesis the delay constraints necessary for correct circuit operation of the one-hot AFSM, *i.e.* $\Delta_n(\text{sp}) + \Delta_n(\text{rp}) > 2\Delta(\text{s})$ and $\Delta_n(\text{sp}) + \Delta_n(\text{rp}) > \Delta_{n-1}(\text{rp})$, will be fulfilled for all states.

State	Δ(s)	Δ(r)	Δ(sp)	Δ(rp)
ackin	**0.06**/*0.03*	**0.09**/*0.09*	**0.20**/*0.15*	**0.23**/*0.18*
waita	**0.04**/*0.03*	**0.11**/*0.09*	**0.17**/*0.14*	**0.20**/*0.18*
reqout	**0.06**/*0.03*	**0.09**/*0.09*	**0.20**/*0.14*	**0.24**/*0.19*
waitr	**0.04**/*0.03*	**0.10**/*0.09*	**0.17**/*0.14*	**0.20**/*0.23*
idle	**0.06**/*0.03*	**0.13**/*0.09*	**0.22**/*0.20*	**0.23**/*0.18*

Table 2: Constrained Latch Controller Delay Analysis (in ns)

By using the data of Table 2 the conditions for correct circuit operation can be calculated. In this way, constraints can be specified which can propagate throughout the physical part of the EDA flow. The values of the critical delays are shown in Table 3. As can be seen by this table, $\Delta_n(\text{sp}) + \Delta_n(\text{rp}) > \Delta_{n-1}(\text{rp})$ is a tighter constraint to $\Delta_n(\text{sp}) + \Delta_n(\text{rp}) > 2\Delta(\text{s})$.

State	$\Delta_{(n)}(\text{sp}) + \Delta_{(n)}(\text{rp})$	2Δ(s)	$\Delta_{(n-1)}(\text{rp})$
ackin	**0.43**	**0.12**	**0.23**
waita	**0.37**	**0.08**	**0.23**
reqout	**0.44**	**0.12**	**0.23**
waitr	**0.37**	**0.08**	**0.24**
idle	**0.45**	**0.12**	**0.20**

Table 3: Constrained Latch Controller Critical Delays (in ns)

From the figures of Table 3 an absolute upper delay bound can be obtained for the reset path of each state, *i.e.* Δ(rp). This value is equal to the sum of the set and reset paths of the next state. There is no upper bound on the set path, *i.e.* Δ(sp), but there is a lower bound for both the set and reset paths due to the same delay constraint equation. Table 4 shows appropriate minimum and maximum absolute delay constraints on the set and reset paths. These delay constraints can be specified in GCF file format, for them to be used in the physical parts of the EDA flow, by using the GCF `path_delay` construct.

State	Δ(sp) range	Δ(rp) range
ackin	**> 0.20**	**> 0.23, < 0.37**
waita	**> 0.17**	**> 0.20, < 0.45**
reqout	**> 0.20**	**> 0.24, < 0.37**
waitr	**> 0.17**	**> 0.20, < 0.45**
idle	**> 0.22**	**> 0.23, < 0.44**

Table 4: AFSM Delay Constraints (in ns)

When the circuit enters the physical design phase of the EDA flow and is P&R, the path delays will change. As the delay constraints are expressed as absolute numbers and not relatively, it is likely that although the absolute timings may have changed, the relative timings are still within the appropriate ranges and absolute timing violations reported by the tools are not affecting the circuit's operation. If contemporary EDA tool flows did support relative timing constraints that would be ideal for describing the timing constraints of asynchronous circuits. Then, it would be possible to use relative timing constraint specifications to ensure that asynchronous circuits operate correctly and absolute timing constraints for paths that impact on the circuit's performance.

3.2 Physical Design

Cadence's DesignPlanner was used for hierarchical design, Silicon Ensemble for block-level P&R and IC Craftsman for the top-level routing. The top-level floorplan and final layout are shown in Figure 3. The five stages correspond to the same synthesis-constrained latch controller circuit (as described in the previous section), only with different aspect ratios. After floorplanning, each stage was P&R separately in both the time-driven (with GCF constraints) and in the non time-driven (connection-driven) mode of the P&R tools.

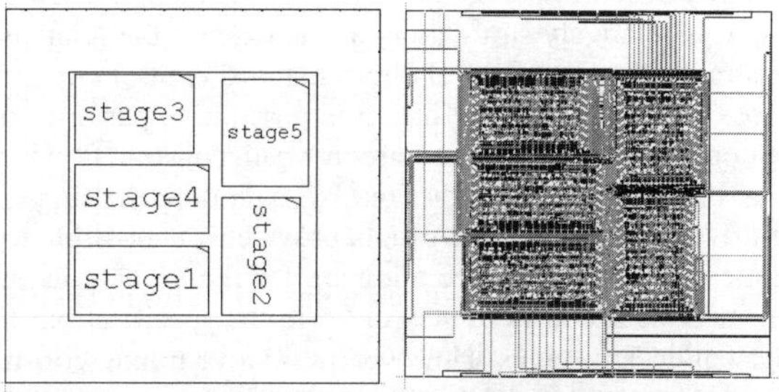

Figure 3: Asynchronous Pipeline Floorplan and Layout

Even in time-driven mode, small violations, less than 50ps, of the path constraints manifested. Figure 4 graphically represents the violations reported by Pearl after P&R for the non time-driven and time-driven versions of the 32-bit fully-decoupled latch controller circuits respectively. The violations shown are minimum path violations, *i.e.* the paths were faster than specified.

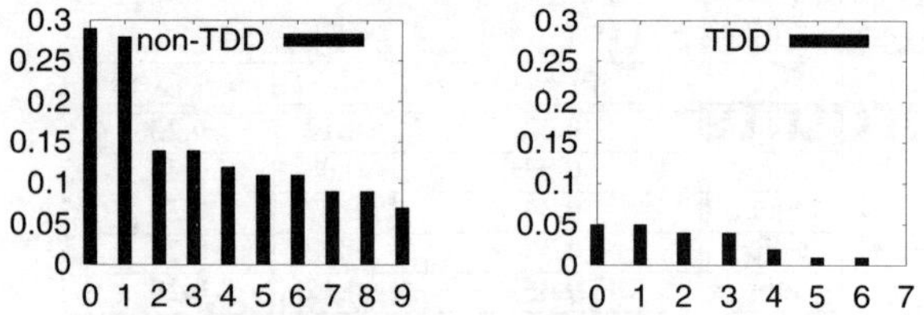

Figure 4: Range of Timing Violations (in ns)

Table 5 shows the timings of the post-layout critical paths of the time-driven design. As can be seen by these timings the circuit will operate correctly even though the timing analyser reports absolute timing violations. Contrasting this table with Table 3, which contains the critical post-synthesis delays, we can see that the defined absolute delays have controlled the placement and routing process effectively and have produced a circuit with delays very close to the specification.

State	$\Delta_{(n)}$(sp) + $\Delta_{(n)}$(rp)	2Δ(s)	$\Delta_{(n-1)}$(rp)
ackin	**0.42**	**0.12**	**0.18**
waita	**0.31**	**0.08**	**0.23**
reqout	**0.56**	**0.12**	**0.18**
waitr	**0.40**	**0.08**	**0.30**
idle	**0.39**	**0.12**	**0.17**

Table 5: Time-driven Post-Layout Path Delays (in ns)

State	$\Delta_{(n)}$(sp) + $\Delta_{(n)}$(rp)	2Δ(s)	$\Delta_{(n-1)}$(rp)
ackin	**0.22**	**0.04**	**0.09**
waita	**0.19**	**0.04**	**0.11**
reqout	**0.25**	**0.10**	**0.09**
waitr	**0.23**	**0.04**	**0.14**
idle	**0.22**	**0.04**	**0.09**

Table 6: Non Time-driven Post-Layout Path Delays (in ns)

Table 6 shows the timings of the post-layout critical paths of the non time-driven design. These timings demonstrate two important points. Firstly, they vary greatly from the post-synthesis timings (which estimate placement), indicating that the synthesis timings were conservative and affected the time-driven placement of the circuit negatively. The connection-driven algorithms of non time-driven mode have produced a faster circuit by minimising path lengths and packing the circuit. Secondly, despite the fact that the delays and ratios of the delays have changed, the relative path constraints still hold. The sizing of the set and reset gates of the state flip-flops, *i.e.* synthesis constraint Δ(r) $\geq$ 2Δ(s), and the bundled-data synthesis constraints make it hard for the placement and routing tool to violate the relative constraints for such a small circuit.

4. CONCLUSIONS

We have shown that it is indeed possible to exploit features of conventional synchronous EDA flows to implement asynchronous circuits. It is relatively straightforward to express the relative timing constraints that a gate-level direct-mapped control circuit must fulfill for correct operation. Thus, it is possible to perform timing verification of the asynchronous circuit's path constraints at various stages of the EDA flow. Further on, by exploiting the support for timing-driven design that EDA tools provide, it is possible to perform time-driven P&R, hence allowing for the asynchronous circuit's critical path delays to stay close to the specification during the physical design stages. However, as relative timing constraints are not as yet supported by commercial EDA flows, the timing constraints of an asynchronous circuit must be described as absolute constraints particular to a specific implementation technology. Such constraints are effectively stricter than the equivalent relative constraints which they represent and limit circuit performance.

5. REFERENCES

[1] Cadence Design Systems. *General Constraint Format Reference Manual*, June 1998.

[2] T. A. Chu. Synthesis of Self-Timed VLSI Circuits from Graph-Theoretic Specifications. Technical Report MIT/LCS/TR-393, Massachusetts Institute of Technology, June 1987.

[3] M. J. Flynn, P. Hung, and K. W. Rudd. Deep-Submicron Microprocessor Design Issues. *IEEE Micro*, 19, No. 4:11–22, 1999.

[4] S. B. Furber and P. Day. Four-Phase Micropipeline Latch Control Circuits. *IEEE Transactions on VLSI Systems*, 4(2):247–253, June 1996.

[5] S. Hauck. Asynchronous Design Methodologies: An Overview. Technical Report TR 93-05-07, Department of Computer Science and Engineering, University of Washington, Seattle, 1993.

[6] L. A. Hollaar. Direct Implementation of Asynchronous Control Units. *IEEE Transactions on Computers, Vol. C-31, No. 12*, December 1982.

[7] M. Ligthart, K. Fant, R. Smith, A. Taubin, and A. Kondratyev. Asynchronous design using commercial HDL synthesis tools. In *Proceedings of the International Symposium on Advanced Research in Asynchronous Circuits and Systems*, pages 114–125. IEEE Computer Society Press, Apr. 2000.

[8] A. J. Martin. Programming in VLSI: From Communicating Processes to Delay-Insensitive Circuits. In C. Hoare, editor, *Developments in Concurrency and Communication*, California Institute of Technology, 1990. Addison Wesley.

[9] D. Matzke. Will Physical Scalability Sabotage Performance Gains ? *IEEE Computer*, Sept. 1997.

[10] Meta-Software. *HSPICE User's Manual H9001*, 1990.

[11] T. Murata. Petri-Nets: Properties, Analysis and Applications. In *Proceedings of the IEEE*, volume 77-4, pages 541–580, 1989.

[12] Open Verilog International. *Standard Delay Format Specification Version 3.0*, 1995.

[13] C. P. Sotiriou. *Design of an Asynchronous Processor.* PhD thesis, Institute for Computing Systems Architecture, Division of Informatics, University of Edinburgh, 2001.

[14] C. P. Sotiriou. Direct-Mapped Asynchronous Finite-State Machines in CMOS Technology. In *Proceedings of the 14th International ASIC/SOC Conference*, September 2001.

[15] K. S. Stevens. *Practical Verification and Synthesis of Low Latency Asynchronous Systems.* PhD thesis, Department of Computer Science, The University of Calgary, September 1994.

[16] I. E. Sutherland. Micropipelines. *Communications of the ACM, Volume 32, Number 6*, 1989.

[17] Synopsys. *Designware - GTECH Library*, 2000.

[18] S. H. Unger. *Asynchronous Sequential Switching Circuits.* Wiley Interscience, Department of Electrical Engineering, Columbia University, 1969.

[19] K. van Berkel. *Handshake Circuits: An asynchronous architecture for VLSI programming.* Cambridge University Press, 1993.

[20] H. van Gageldonk, K. van Berkel, A. Peeters, D. Baumann, D. Gloor, and G. Stegmann. An Asynchronous low-power 80C51 Microcontroller. In *Proceedings of the International Symposium on Advanced Research in Asynchronous Circuits and Systems*, Apr. 1998.

[21] K. Y. Yun, D. L. Dill, and S. M. Nowick. Synthesis of 3D Asynchronous State Machines. In *Proc. International Conf. Computer Design (ICCD)*, Cambridge, Massachusets, 1992.

Transformation Rules for Designing CNOT-based Quantum Circuits

Kazuo Iwama
Sch. of Inform., Kyoto Univ.
QCI, ERATO, JST
iwama@kuis.kyoto-u.ac.jp

Yahiko Kambayashi
Sch. of Inform., Kyoto Univ.
yahiko@i.kyoto-u.ac.jp

Shigeru Yamashita
NTT Communication Science Labs.
QCI, ERATO, JST
ger@cslab.kecl.ntt.co.jp

ABSTRACT

This paper gives a simple but nontrivial set of local transformation rules for *Control-NOT(CNOT)*-based combinatorial circuits. It is shown that this rule set is complete, namely, for any two equivalent circuits, S_1 and S_2, there is a sequence of transformations, each of them in the rule set, which changes S_1 to S_2. Our motivation is to use this rule set for developing a design theory for *quantum circuits* whose Boolean logic parts should be implemented by CNOT-based circuits. As a preliminary example, we give a design procedure based on our transformation rules which reduces the cost of CNOT-based circuits.

Categories and Subject Descriptors

B.6.m [**LOGIC DESIGN**]: Miscellaneous

General Terms

Design, Theory

Keywords

Quantum Circuit, CNOT Gate, Local Transformation Rules

1. INTRODUCTION

It is widely considered that logic synthesis is a mature field in our community. However, this is only true for conventional AND-OR-NOT-based circuits or LSI's; new ideas must be needed if we face technology innovations. The main purpose of this paper is to introduce logic synthesis for *quantum Boolean circuits* [12] (QBCs for short). The key difference between conventional circuits and quantum ones is their base-family of logic gates, for the latter of which many people agree that *Control-Not (CNOT)* type logic gates will be a single possibility. Another (even more important) feature of QBCs is its severe restriction against wire-linking between gates: As we can see in a moment, only straight-line, parallel wiring is allowed. Thus, it looks obviously hard to apply conventional logic synthesis techniques for our present purpose.

In spite of such a new target, our basic strategy approaching to it is quite conservative. Namely, our logic synthesis proposed in this paper is based on *local transformations*, which has been constantly popular and successful for conventional circuits [2, 3]. (One can think of, for example, the DeMorgan's law, which has been used most often to make a local simplification of Boolean circuits.) More concretely, we give a set of local transformation rules. The rule set is complete, which means we can transform any QBC into any of its equivalent ones by applying these rules. We also make some concrete suggestions on how to use these transformation rules to simplify QBCs.

It should be noted that quantum algorithms are often described by using QBCs [12]. Designing a "good" QBC is thus plays a key role to the successful implementation of a quantum algorithm. A little surprisingly, however, relatively small attention has been paid for the design methodology of QBCs [1, 9]. [1] shows that any unitary transformation can be broken down into a sequence of basic quantum gates. The method is general, but it does not necessarily generate efficient circuits. [9] gives how to construct QBCs for Boolean functions by using CNOT gates but their resulting circuits are essentially the same as the elementary two-level AND-OR-NOT circuits.

To discuss how to design small QBCs, we should note that QBCs for quantum algorithms can be divided into two parts: One is a quantum specific part, for example, the Walsh-Hadamard transformation for making a quantum superposition and the Quantum Fourier transformation [6]. The other part, which is sometimes called *quantum Boolean oracles* [5], is for calculating (conventional) Boolean functions. To establish an efficient design methodology for QBCs, it is much more important to target the latter part (quantum Boolean oracles) since the structures of the latter part vary depending on each problem, whereas the structure of former part is usually fixed. Our ultimate goal is to develop a design theory for quantum Boolean oracles, and it is a nice start-up to have a set of local transformation rules for CNOT-based QBCs. It should be noted that a similar set of transformation rules for conventional circuits is given in [8].

We also introduce the canonical form, which is another fundamental concept in logic synthesis, for CNOT-based QBCs. Then we can prove the *completeness* of the rule set only by showing that there is a sequence of transformations from any circuit to its canonical form, since each transformation is bidirectional. Accordingly, we can assure that by using our transformation rules, we can modify a given circuit into another with a desirable property (e.g., of small cost) by executing a NP-type search, although the length of the search path might not be always short. The completeness of the rule set is quite non-trivial and theoretically interesting on its own. As mentioned before, however, our final goal is to use our rules for more practical purposes. Actually, we give a preliminary example of a design procedure which simplifies CNOT-based circuits.

2. QUANTUM BOOLEAN CIRCUITS

Although we need virtually no knowledge of quantum computation to understand the main result of this paper, let us start with a minimum introduction of the new technology: In quantum computing, a *quantum bit*, or *qubit* for short, plays an important role. There have been proposed various physical implementation objects for qubits [10], and they can be described as: $|x\rangle = \alpha|0\rangle + \beta|1\rangle$, where $|x\rangle$ means the quantum state of the qubit x, and α and β are

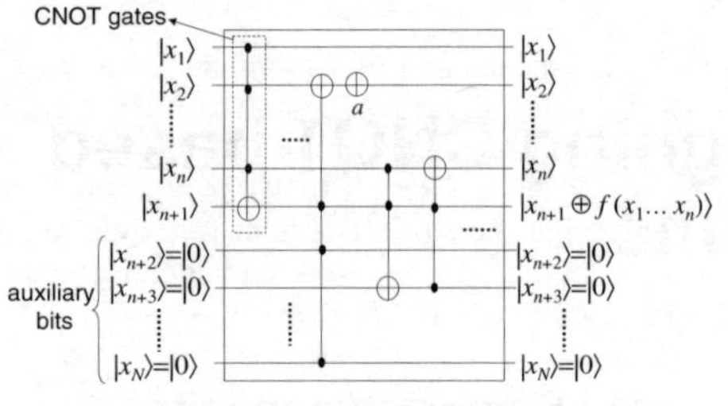

Figure 1: A Quantum Boolean Circuit

complex numbers which satisfy that (1) $|\alpha|^2 + |\beta|^2 = 1$ and (2) if we *measure* the quantum state, we get the states $|0\rangle$ and $|1\rangle$ with the probability $|\alpha|^2$ and $|\beta|^2$, respectively. This means that a qubit can store a *superposition* of the states $|0\rangle$ and $|1\rangle$ unlike a conventional bit. In the general framework of quantum computing, we apply a specific sequence of *unitary* operations to qubits, and get a desired solution by measuring the final quantum states of the qubits. Such a sequence of unitary operations is often described as a *quantum Boolean circuit* (QBC).

A QBC is a quantum system with N qubits, denoted by $|x_1\rangle|x_2\rangle\cdots|x_N\rangle$, as illustrated in Fig. 1 where we apply a specific unitary operation corresponding to each gate one by one to the qubits $|x_1\rangle$ to $|x_N\rangle$ from left to right. As an interaction of their qubits, we can only use CNOT gates whose functionality will be given later. The left-side $|x_1\rangle|x_2\rangle\cdots|x_n\rangle$ are used for the input, and their values should be restored finally. (This restoring is important when the circuit is used for quantum Boolean oracles as we mention later.) The $(n+1)$-st qubit $|x_{n+1}\rangle$, called a *work bit*, is changed into $|x_{n+1} \oplus f(x_1,\cdots,x_n)\rangle$, which is used to obtain the value of the Boolean function f. Furthermore, a circuit can use any finite number of *auxiliary qubits* which are reset to $|0\rangle$ initially and also restored to be $|0\rangle$ finally. More formally:

Definition 1. A *Control-NOT (CNOT) gate* is denoted by $[t, C]$, where t is an integer and C is a finite set of integers ($t \notin C$). $|x_t\rangle$ is called a *target bit* and $|x_k\rangle$ is called a *control bit* if $k \in C$.

This kind of gates is also called n-bit Toffoli gates [11] in *reversible computing*. Also the literature refers to $[t,C]$ with only one control bit as a Control-NOT gate, and to $[t,C]$ with two control bit as *Control-Control-NOT gate*, and so on, but we simply call both of them CNOT gates in this paper. In the figure, we use $\oplus$ for a target bit and $\cdot$ for a control bit. For example, the leftmost CNOT gate in Fig. 1 is given by $[n+1, \{1,2,n\}]$. As its symbol suggest, $[n+1, \{1,2,n\}]$ changes the state of $|x_{n+1}\rangle$ into $|x_{n+1} \oplus x_1 \cdot x_2 \cdot x_n\rangle$, where $\oplus$ is the conventional XOR. (See Definition 3 for details.)

We denote the set of N-bit basis vectors by $\{0,1\}^N$. The quantum state of N qubits $|x_1\rangle\cdots|x_N\rangle$ is a superposition (a linear combination) of those 2^N basis vectors. However, we often assume in this paper that the state is a single basis vector when describing the behavior of the system. Generalization to superposed states can be done simply by taking a linear combination of the results for each basis vector.

Definition 2. A *quantum Boolean circuit* of size M over qubits $|x_1\rangle,\cdots,|x_N\rangle$ is a sequence of CNOT gates $[t_1,C_1]\cdots[t_i,C_i]\cdots[t_M,C_M]$, where $1 \le t_i \le N, C_i \subseteq \{1,\cdots,N\}$. Note that unlike the conventional Boolean circuits we are only allowed to have a sequence of CNOT gates. (We cannot use so-called jump-wires which propagate the result of one gate to another gate placed far away. This is an important difference between our circuit model and the conventional circuit model.)

Definition 3. The state of the circuit after the i-th CNOT gate $[t_i, C_i]$ is denoted by $S_i = |x_1^i\rangle\cdots\cdots|x_N^i\rangle$ and defined as follows:

(1) $S_0 = |a_1\rangle|a_2\rangle\cdots|a_N\rangle$, where $|a_1\rangle|a_2\rangle\cdots|a_N\rangle \in \{0,1\}^N$ is an input state.

(2) For $1 \le k \le N$ and $1 \le i \le M$, $|x_k^i\rangle = |x_k^{i-1}\rangle$ if $k \ne t_i$, and $|x_{t_i}^i\rangle = |x_{t_i}^{i-1} \oplus X_{C_i}^k\rangle$ where $X_{C_i}^k = 1 \wedge x_{i_1} \wedge \cdots \wedge x_{i_j}$ if $C_i = \{i_1,\cdots,i_j\}$. Here $\wedge$ is the conventional AND and $\oplus$ is the conventional XOR operators.

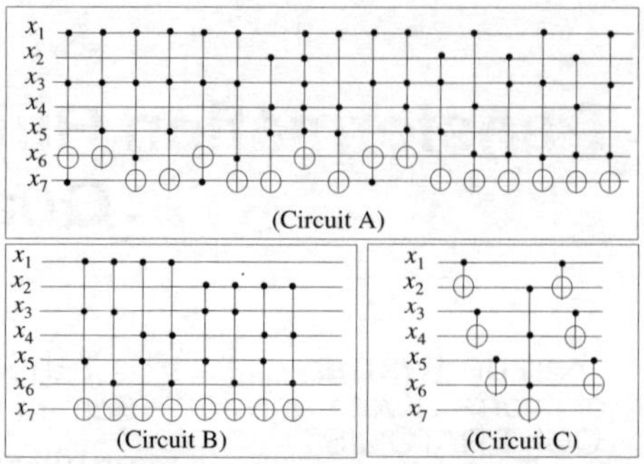

Figure 2: Equivalent Circuits

For example, we can calculate S_2 of Circuit A in Fig. 2 such that $|x_i^2\rangle = |x_i^1\rangle$ for $i \ne 6$, and $|x_6^2\rangle = |x_6^1 \oplus x_1^1 \cdot x_3^1 \cdot x_5^1\rangle$. Note that $X_{C_i}^k$ is 1 when C_i is an empty set, which means that the CNOT gate operates just as a negation (see gate a in Fig. 1). For better exposition, we often say "the state of $|x_1\rangle$ after the i-th gate" in stead of "the state of $|x_1^i\rangle$."

Definition 4. A quantum Boolean circuit is said to be *proper* and to *compute* a Boolean function $f(x_1,\cdots,x_n)$ iff (i) $S_0 = |a_1\rangle|a_2\rangle\cdots|a_{n+1}\rangle|0\rangle\cdots|0\rangle$, i.e., all the auxiliary bits are initially cleared, (ii) The final state $S_M = |a_1\rangle|a_2\rangle\cdots|a_{n+1} \oplus f(x_1,\cdots,x_n)\rangle|0\rangle\cdots|0\rangle$, i.e., the $(n+1)$-st state is XORed with the function value and all the others are the same as the input.

Remark. Note that our condition says that all auxiliary qubits must be reset finally. Otherwise, we cannot use the circuit as a Boolean oracle for the following reason: If the auxiliary qubits are not reset, the states of $|x_1\rangle\cdots|x_n\rangle$ may be *entangled* with some states of the auxiliary qubits. In such a case, we can no longer treat the states of $|x_1\rangle\cdots|x_n\rangle$ as one quantum system, and therefore quantum operations for the states of $|x_1\rangle\cdots|x_n\rangle$ might not work as desired. For example, the *f-controlled phase shift* used in the Grover's search algorithm [5] is defined as $|x_1\rangle\cdots|x_n\rangle|x_{n+1}\rangle = (-1)^{f(x_1,\cdots,x_n)}|x_1\rangle\cdots|x_n\rangle|x_{n+1}\rangle$ where $|x_{n+1}\rangle$ is initialized to $\frac{1}{\sqrt{2}}(|0\rangle - |1\rangle)$. If the circuit is not proper, we can no longer use the circuit as a primitive of the above unitary operation.

Fig. 2 shows three QBCs which are all equivalent, i.e., they compute the same Boolean function $f(x_1,\cdots,x_6) = (x_1 \oplus x_2)\cdot(x_3 \oplus x_4)\cdot(x_5 \oplus x_6)$. Note that x_7 is a work bit in these circuits. As will be shown later, we can always construct a circuit by only using CNOT gates whose target bit is x_{n+1}, i.e., those like Circuit B. However, different types of circuits are of course possible like Circuits A and C, where C is simpler than A and B.

3. TRANSFORMATION RULES

In this section, we introduce six *local transformation rules* which can be applied for a sequence of CNOT gates. Each transformation rule looks like $F \Leftrightarrow G$, where F and G are sequences of CNOT gates. If a (proper) circuit A is written as $A = A_1FA_2$, i.e., it includes a subsequence F, then A changes into $A' = A_1GA_2$ by applying the rule $F \Leftrightarrow G$.

Transformation Rule Set. In the followings, ε means the empty sequence, and we refer to a CNOT gate whose target bit is the i-th bit as $CNOT_i$.

(1) $[t_1,C_1]\cdot[t_1,C_1] \Leftrightarrow \varepsilon$. Namely, two adjacent identical gates cancel.

(2) $[t_1,C_1]\cdot[t_2,C_2] \Leftrightarrow [t_2,C_2]\cdot[t_1,C_1]$, if $t_1 \notin C_2$ and $t_2 \notin C_1$. The condition means that the two gates are "independent." If there is some influence between two gates, we cannot simply change the order of the two gates. Even in such cases, we can change the order by adding some gates, as we will see in the followings.

(3) $[t_1,C_1]\cdot[t_2,C_2] \Leftrightarrow [t_2,C_2]\cdot[t_1,C_1]\cdot[t_1,C_1 \cup C_2 - \{t_2\}]$, if $t_1 \notin C_2$ and $t_2 \in C_1$ (see Fig. 3).

(4) $[t_1,C_1]\cdot[t_2,C_2] \Leftrightarrow [t_2,C_1 \cup C_2 - \{t_1\}]\cdot[t_2,C_2]\cdot[t_1,C_1]$, if $t_1 \in$

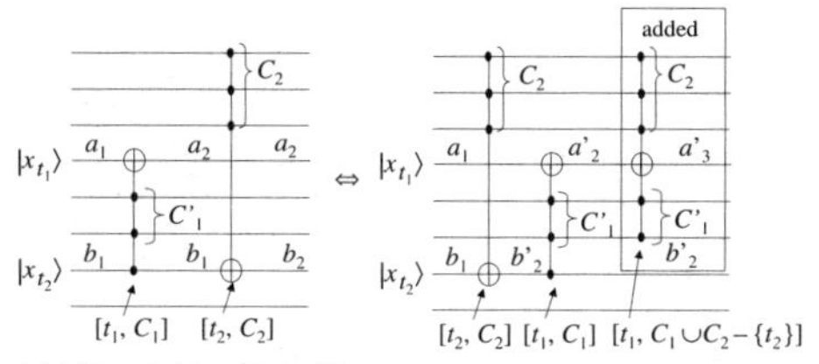

(a) lefthand-side of Rule (3) (b) righthand-side of Rule (3)

Figure 3: Transformation Rule 3

C_2 and $t_2 \notin C_1$. This rule is the dual of (3), i.e., the relationship between the two gates is opposite.

(5) $[t_1,\{c_1\}] \cdot [t_2, C_2 \cup \{c_1\}] \Leftrightarrow [t_1,\{c_1\}] \cdot [t_2, C_2 \cup \{t_1\}]$, if $(t_1 > n+1)$ and there is no $CNOT_{t_1}$ before $[t_1,\{c_1\}]$ (see Fig. 4).

(6) $[t,C] \Leftrightarrow \varepsilon$, if there is an integer i such that $i \in C, i > n+1$, and there is no $CNOT_i$ before $[t,C]$.

Lemma 1. Applying any one of Rules (1) to (6) does not change the Boolean function computed by the circuit.

Proof. (1) Rule (1) is obvious since $g \oplus f \oplus f = g$ for any Boolean functions f and g.

(2) Rule (2) is also easy because $[t_1, C_1]$ and $[t_2, C_2]$ do not affect each other if $t_1 \notin C_2$ and $t_2 \notin C_1$.

(3) See Fig. 3 again. We have two important qubits, i.e., $|x_{t_1}\rangle$ and $|x_{t_2}\rangle$ ($t_1 \neq t_2$ by the condition). In Fig. 3 (a), let a_1 and a_2 be the states of $|x_{t_1}\rangle$ before and after the gate $[t_1, C_1]$, respectively. Let a_1, a'_2 and a'_3 be the states of $|x_{t_1}\rangle$ before $[t_2, C_2]$, after $[t_1, C_1]$ and after $[t_1, C_1 \cup C_2 - \{t_2\}]$, respectively, in Fig. 3 (b). Also, b_1, b_2 and b'_2 are the states of $|x_{t_2}\rangle$ similarly defined. Furthermore let $C'_1 = C_1 - \{t_2\}$ (C'_1 and C_2 may not be disjoint). Note that the state whose index is in $C'_1 \cup C_2$ does not change throughout this portion of the circuit since there is no target bit index in $C'_1 \cup C_2$. Now let us calculate the states a_2, b_2, a'_3 and b'_2. First, both b_2 and b'_2 can be written as $b_1 \oplus X_{C_2}$, where X_{C_2} is a product term of all x_j such that $j \in C_2$. Similarly, let $X_{C'_1}$ be a product term of all x_j such that $j \in C'_1$, then we have $a_2 = a_1 \oplus X_{C'_1} \cdot b_1$, $a'_2 = a_1 \oplus X_{C'_1} \cdot b'_2 = a_1 \oplus X_{C'_1} \cdot (b_1 \oplus X_{C_2}) = a_1 \oplus X_{C'_1} \cdot b_1 \oplus X_{C_2} \cdot X_{C'_1}$ and $a'_3 = a'_2 \oplus X_{C_2} \cdot X_{C'_1} = a_1 \oplus X_{C'_1} \cdot b_1 \oplus X_{C_2} \cdot X_{C'_1} \oplus X_{C_2} \cdot X_{C'_1} = a_1 \oplus X_{C'_1} \cdot b_1$. Thus $b_2 = b'_2$ and $a_2 = a'_3$. The states of other qubits do not change as mentioned before and hence the transformation does not change the functionality of the circuit.

(4) Similar to the proof for Rule (3).

(5) If there is no $CNOT_{t_1}$ before $[t_1,\{c_1\}]$, $|x_{t_1}\rangle$ remains $|0\rangle$ just before the gate $[t_1,\{c_1\}]$. Therefore, after applying $[t_1,\{c_1\}]$, the state of $|x_{t_1}\rangle$ and that of $|x_{c_1}\rangle$ must be the same, which means that we can change t_1 to c_1 in the index set of the latter gate's control bits.

(6) If there is an integer i which satisfies the condition of the rule, $[t,C]$ has an auxiliary bit $|x_i\rangle$ as its control bit whose current state remains $|0\rangle$. Since one of its control bits is always 0, this CNOT gate is useless. □

4. COMPLETENESS OF THE RULE SET

In this section, we first introduce the canonical form for quantum Boolean circuits. Then it is shown that any circuit can be transformed into its canonical form using the transformation rules. The completeness of the rule set is its immediate consequence, since our transformation rules are bidirectional.

4.1 Canonical Form

Definition 4. A quantum Boolean circuit S is said to be *of the canonical form,* if (1) it has only $CNOT_{n+1}$ gates, namely CNOT gates whose target bit is the work bit, (2) it does not include two or more same $CNOT_{n+1}$ gates, (3) $CNOT_{n+1}$ gates are ordered lexicographically in terms of the indices of their control bits, and (4) no $CNOT_{n+1}$ gates have auxiliary qubits in their control bits, i.e., the circuit S uses no auxiliary qubits at all.

The condition (2) means that the canonical form must not be redundant in terms of Rule (1) in our transformation rule set. The condition (3) means that, for example, $[n+1,\{1,2,3\}]$ should be placed before $[n,\{2,3,4\}]$. Recall that Fig. 2 shows three different circuits computing the same Boolean function, and only Circuit B is of the canonical form. Now consider Boolean formulas of the following form: $a_0 \oplus a_1x_1 \oplus a_2x_2 \oplus \cdots \oplus a_nx_n \oplus a_{1,2}x_1x_2 \oplus a_{1,3}x_1x_3 \oplus \cdots \oplus a_{n-1,n}x_{n-1}x_n \oplus \cdots\cdots \oplus a_{1,2,\cdots n-1,n}x_1x_2\cdots x_{n-1}x_n$, where no negated literals are allowed and each a_i is 0 or 1. This form is called a *positive polarity Reed-Muller expression*, for which the following lemma is known:

Lemma 2. [4]. Any Boolean function can be expressed by a positive polarity Reed-Muller expression which is unique except for the order of terms.

Lemma 3. Any CNOT-based quantum circuit S has its unique canonical form.

Proof. By definition, S transforms $|x_i\rangle$ to $|x_i\rangle$ for $1 \leq i \leq n$ and $|x_{n+1}\rangle$ to $|x_{n+1} \oplus f(x_1, \cdots, x_n)\rangle$ for some formula f. Compute the positive polarity Reed-Muller expression of f, which naturally corresponds to a sequence of $CNOT_{n+1}$ gates. Since the order of such gates must be lexicographic with respect to their control bits in the canonical form, the resulting circuit is unique by Lemma 2. □

4.2 Transformation Procedure

Theorem 1. Any quantum circuit S can be transformed into its canonical form by using our transformation rules.

Proof. Our procedure, $Main(S)$, for such a transformation is illustrated in Fig. 8. $Main(S)$ first calls $Shift(S, n+1)$ which "shifts" all the $CNOT_{n+1}$ gates in S to the left portion of the circuit (see below for details). After all the $CNOT_{n+1}$ gates are placed in the left portion of the circuit, we let the remaining part be S' and then apply $Shift(S', n)$. Repeat this procedure until $Shift(S', 1)$. As one can see later, the *Shift* procedure also includes the lexicographically reordering operation and the operation of deleting redundant gates. Therefore, after $Shift(S', 1)$ is finished, all $CNOT_m$ for $1 \leq m \leq n$ have already vanished (see Lemma 4). Thus all we have to do more in line 9-15 of $Main(S)$ is to delete the gates whose target bits are auxiliary qubits.

We first concentrate ourselves on the *Shift* procedure. Let $k = [t_1, C_1]$ and $h = [t_2, C_2]$ be two consecutive gates, where $t_2 = n+1$ and $t_1 \neq n+1$. Then our basic idea is to switch the positions of k and h, i.e., kh into hk. If $t_1 \notin C_2$ or $t_2 \notin C_1$, then this switch can be done in a single step by using one of Rules (2)-(4). Unfortunately, that is not possible if $t_1 \in C_2$ and $t_2 \in C_1$; the following algorithm mainly targets this situation where we can no longer depend on a simple repetition of the above switching operations. Now here are details of the procedure $Shift(S, m)$ for $1 \leq m \leq n+1$:

Step 1. Suppose that the entire circuit S includes k CNOT gates whose control bits include the m-th bit. Let these gates be $g_1, g_2, \cdots g_k$, and $g_j = [t_j, C'_j \cup \{m\}]$. What we do first is to remove this m-th bit from the control bits of g_j by using an unused auxiliary bit, say the a_j-th bit. As an illustration, see Fig. 5 for the original circuit S and Fig. 7 for the circuit after Step 1 of $Shift(S, 3)$ is applied. In more detail:

Step 1.1. We add two identical gates $[a_j, \{m\}]$, right before g_j by Rule (1). Let these gates be g_{l_j} for the left one and g_{r_j} for the right one.

Step 1.2. We change the order of g_{r_j} and g_j by Rule (2), i.e., to the order of $g_{l_j} g_j g_{r_j}$.

Step 1.3. We apply Rule (5) to g_{l_j} and g_j so that $g_j = [t_j, C'_j \cup \{m\}]$ is changed to $[t_j, C'_j \cup \{a_j\}]$.

Step 2. We move each $g_{l_j} (1 \leq j \leq k)$ to the left as far as possible. Note that we can always do this transformation by applying Rules (2) to (4). We need to add a gate whose target bit is the a_j-th bit $(1 \leq j \leq k)$ when we apply Rule (4). We also move such added gates to the left part of the circuit by Rules (2) to (4). (One might have the concern that this chain-like addition does not stop.

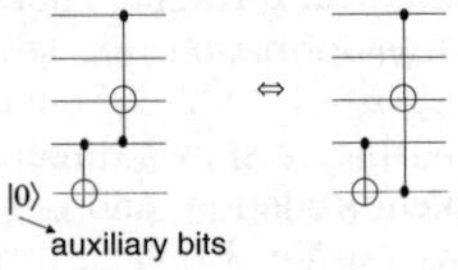

Figure 4: Transformation Rule 5

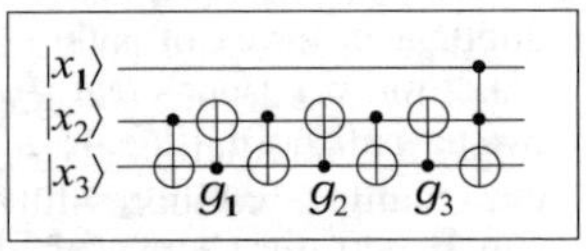

Figure 5: Initial Circuit

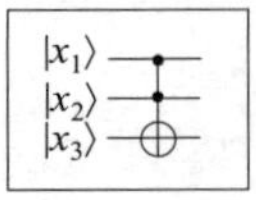

Figure 6: The Final Result

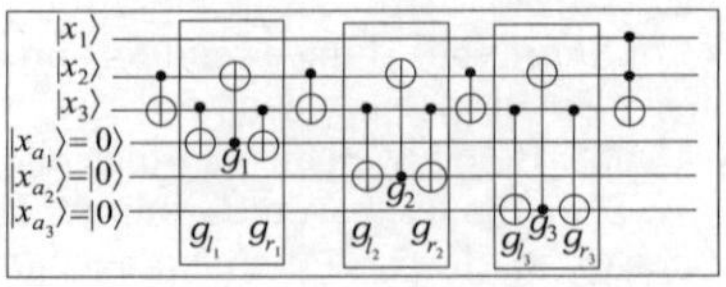

Figure 7: The Circuit after Step 1

```
Main(S)
{
    S' = S
    for (i = n+1 to 1) {
        Shift(S',i)
        C_i = S_L /* the left part of the result of Shift(S',i) which consist
                of only CNOT_i gates*/
        S' = S_R /* the remaining part of the result of Shift(S',i) */
    }
    while (gates g that have auxiliary qubits in its control bits exist) {
        Let g be the left most one of such gates.
        Move g to the next right position of CNOT_{n+1} by applying one
            of Rules (2) and (4) as many times as possible.
        Delete g by applying Rule (6).
    }
    Reorder the gates whose target bits are auxiliary qubits lexicographically
        by applying Rule (2) as many times as possible.
    Delete redundant pair of gates by applying Rule (1) as many times as possible.
}
```

Figure 8: $Main(S)$

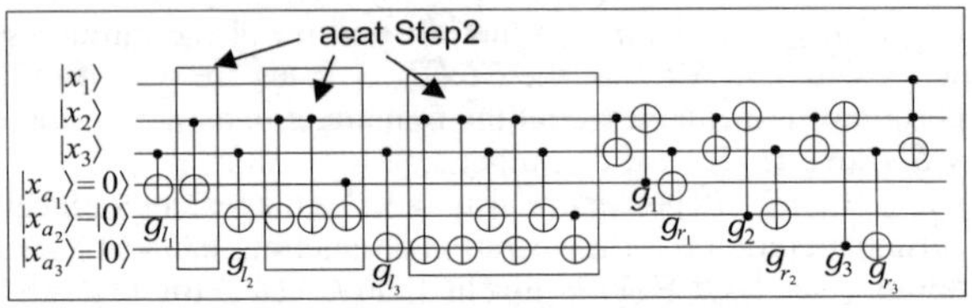

Figure 9: The Circuit after Step 2

This is actually not the case because a $CNOT_{a_j}$ is always added at the lefthand-side of the transformation of Rule (3).) Then we get a circuit where all $CNOT_{a_j}(1 \leq j \leq k)$ except for $g_{r_j}(1 \leq j \leq k)$ are placed in the left most part of the circuit. See Fig. 9 for a circuit after Step 2 in our example.

Step 3. We then move each $g_{r_j}(1 \leq j \leq k)$ to the opposite direction, i.e., to the right as far as possible. Again we can always do this by applying Rules (2) to (4) and exactly the same as before as for the added gates whose target bits are a_j-th $(1 \leq j \leq k)$. Thus we get the circuit where all $CNOT_{a_j}(1 \leq j \leq k)$ are placed in the left most part or in the right most part of the circuit.

Step 4. Without loss of generality we can assume that g_{l_1} is placed at the left-end position of the circuit after Step 2. As one can see later, we can overcome several difficulties we encounter when moving $CNOT_m$ gates to the left, by giving a special role to this gate g_{l_1}. Now, for each gate g which has the m-th bit in its control bits and placed in the lefthand-side of the $CNOT_m$ gates, we apply the following transformations: First we move g to the left so that it comes to the next right position of g_{l_1} by using Rule (2) or (3) as many times as possible. Then we apply Rule (5) to now consecutive g_{l_1} and g so that the control bit $|x_m\rangle$ of g_{l_j} is changed to the a_1-th bit. See Fig. 10 for a circuit after Step 4 in our example. At this moment no gates, except for the leftmost gate g_{l_1}, which have the m-th bit in their control bits exist if they are placed in the left side of the $CNOT_m$ gates. Note that the gates added in Step 2 must not have the m-th bit in their control bits, and therefore we can always perform the following Step 5.

Step 5. We move each $CNOT_m$ so that it comes to the next right position of g_{l_1} by applying Rules (2) to (4). If we need to add new $CNOT_m$ gates in the above transformation, we also move these added gates similarly. Now all $CNOT_m$ are placed consecutively, in the next right position of g_{l_1}. The left portion of our example looks like Fig. 11 (a) after this step (recall that $m = 3$).

Step 6. Thus we have almost moved the $CNOT_m$ gates to the left. The only remaining obstacle is g_{l_1}. In this step, we delete redundant $CNOT_m$ gates as follows.

Step 6.1. We reorder those $CNOT_m$ gates lexicographically by applying Rule (2) as many times as possible. Then we have three groups of $CNOT_m$; the first group does not include the a_j-th bit $(1 \leq j \leq k)$ in the control bits, the second group has only the a_1-th bit (no a_j-th bit $(2 \leq j \leq k)$) in the control bits, and the third group of the remaining gates has the a_j-th bit $(1 \leq j \leq k)$ in the control bits. Let these groups be Groups A, B and C, respectively. The three groups of $CNOT_3$ after this step are illustrated in Fig. 11 (b).

Step 6.2. We delete all gates in Group C by Rule (6).

Step 6.3. We apply Rule (1) to delete redundant pairs of gates in Groups A and B. We will claim that all the gates in Group B, and Group A as well if $m \neq n+1$, disappear at this moment in the next Lemma 4. Thus all $CNOT_m$ gates that have the a_1-th bit in their control bits, i.e., those in Groups B or C, have been deleted. Therefore, we can perform the following Step 7.

Step 7. We finally move g_{l_1} which remains in the left most position to the right side of all $CNOT_m$ by applying Rule (3) as many times as possible. We may again need to add $CNOT_{a_1}$, which are moved to the right side as well. Finally we get a circuit where all $CNOT_m$ are placed in the left most part of the circuit. That completes $Shift(S,m)$.

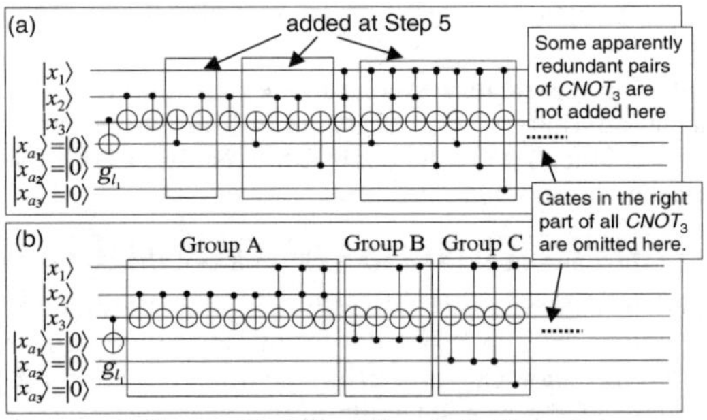

Figure 11: The Circuit in Step 6

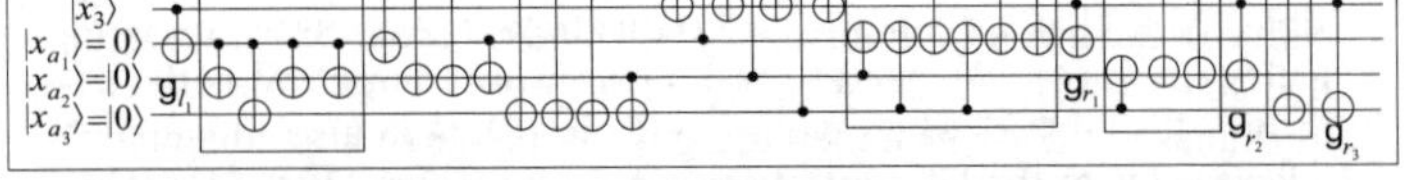

Figure 10: The Circuit after Step 4

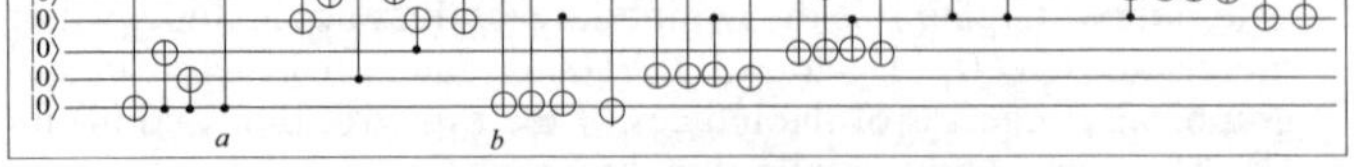

Figure 12: The Circuit after *Shift* Is Finished

Lemma 4. All the gates in Group B must disappear after Step 6.3 of *Shift*(S,m) if the circuit S is proper. All the gates in Group A must also disappear when $m \neq n+1$.

Proof. Below we consider the state of $|x_m\rangle$. Recall that we have only g_{l_1} in the left part of Group A. Therefore, before Group A the states of $|x_1\rangle$ to $|x_{n+1}\rangle$ remain unchanged, i.e., being equal to their initial states, and only $|x_{a_1}\rangle$ has changed to be equal to $|x_m\rangle$.

Suppose that some gates in Groups A or B have remained after Step 6.3, and let the i-th such gate in Group A be $[m, C_{A_i}]$, and the i-th gate in Group B be $[m, C_{B_i}]$. Furthermore, let $X_{C_{A_i}}$ be a product term of all x_j such that $j \in C_{A_i}$, and $X_{C_{B_i}}$ be a product term for C_{B_i}. Then the value of the state of $|x_m\rangle$ after Group A can be written as $x_m \oplus X_{C_{A_1}} \oplus X_{C_{A_2}} \oplus \cdots = x_m \oplus (X_{C_{A_1}} \oplus X_{C_{A_2}} \oplus \cdots) = x_m \oplus f_A$ because of associative laws for the XOR operator. An important point here is that $X_{C_{A_1}}, X_{C_{A_2}}, \cdots$ are all different since Step 6.3 includes the simplification procedure (by Rule (1)) after reordering the gates into the lexicographical order. This means that f_A is a positive polarity Reed-Muller form. Also note that f_A does not have the literal x_m since each C_{A_i} does not include m. We then can calculate the value of the state of $|x_m\rangle$ after Group B as $x_m \oplus f_A \oplus X_{C_{B_1}} \oplus X_{C_{B_2}} \oplus \cdots$. Recall that all the gates in Group B have the a_1-th bit as their control bits, therefore, all $X_{C_{B_i}}$ in the formula contain the literal x_{a_1} which is equal to x_m because of the leftmost gate g_{l_1}. Accordingly, we can rewrite the formula as $(x_m \oplus f_A \oplus x_m f_B')$, where f_B' does not have the literal x_m and is a positive polarity Reed-Muller form again. (The XOR has distributive laws, for example, $xyz \oplus x = x(yz \oplus 1)$.)

Since there are no $CNOT_m$ after Group B (recall that Group C's gates have already disappeared) and the circuit is proper, we have

$$|x_m \oplus f_A \oplus x_m f_B'\rangle = \begin{cases} |x_m\rangle & \text{if } 1 \leq m \leq n, \\ |x_m \oplus f\rangle & \text{if } m = n+1, \end{cases}$$

where f has only literals x_1 through x_n. Now we can conclude that $f_A = f_B' = 0$ if $1 \leq m \leq n$, and $f_B' = 0$ if $m = n+1$ for the following reason: The above formulas can be expanded with respect to x_m, i.e., $x_m \oplus f_A \oplus x_m f_B' = (f_A f_B' + \overline{f_A}\ \overline{f_B'}) x_m + f_A \overline{x_m}$, and $x_m \oplus f = \overline{f} x_m + f \overline{x_m}$. One can see that these two functions are equivalent iff $\overline{f} = f_A f_B' + \overline{f_A}\ \overline{f_B'}$ and $f = f_A$, which implies $f_B' = 0$. One can also see that $(f_A f_B' + \overline{f_A}\ \overline{f_B'}) x_m + f_A \overline{x_m} = x_m$ implies $f_A = f_B' = 0$. Also note that if $g = X_1 \oplus X_2 \oplus \cdots$ is a positive polarity Reed-Muller form, then $g = 0$ iff $X_1 = X_2 = \cdots 0$. Therefore, no product terms are included in f_A or f_B', i.e., there should be no gates in Group B, and neither in Group A when $m \neq n+1$. □

After we successfully get a circuit where all $CNOT_{n+1}$ are placed in the left most part of the circuit by *Shift*$(S, n+1)$, *Main* then calls *Shift*(S', n), by which all $CNOT_n$ gates are shifted to the left. They are placed next to $CNOT_{n+1}$ gates already shifted. Then we can delete all $CNOT_n$ by Lemma 4. This continues until *Shift*$(S', 1)$. In our example, after calling *Shift*$(S', 1)$, we get a circuit in Fig. 12.

After the 8th line in Fig. 8, we have $CNOT_{n+1}$ gates which are already of the canonical form at the left part of the circuit and then $CNOT_i$ gates $(i \geq n+2)$ after them. Now we execute the second half of *Main*, where we delete all the $CNOT_i (i \geq n+2)$ gates, i.e., those whose target bits are auxiliary ones, and finally there remain only $CNOT_{n+1}$ gates in the circuit.

The $CNOT_i$ $(i \geq n+2)$ gates are divided into two groups. The first group, denoted by A, has control bits in auxiliary bits and the second group, B, does not. We first move group A gates to the left until the next positions to $CNOT_{n+1}$ gates. This can be done by switching group A gates with group B gates, which is always possible by using Rules (2)-(4). (Switching two gates both in group A, like gates a and b in Fig. 12, is not so easy.) Then the group A gates can be deleted one by one from the leftmost one by Rule (6). Then we have only group B gates. Recall that when we introduced a $CNOT_i (i \geq n+2)$ gates, we always introduced another $CNOT_i$ which cancels the value of $|x_i\rangle$. Therefore the final values of the auxiliary qubits are all 0. Therefore we can delete the group B gates by reordering and applying Rule (1). (Otherwise, we can imply a contradiction just as we did in the proof of Lemma 4.) In our example, we successfully get to a circuit in Fig. 6 which is the canonical form for the initial circuit. □

4.3 Completeness of the Rule Set

Now our main result is almost immediate:

Theorem 2. Let S_1 and S_2 be any equivalent quantum Boolean circuits. Then there exists a sequence of transformation rules, each in the rule set given in Section 3, which transforms S_1 to S_2.

Proof. Since S_1 and S_2 are equivalent, their canonical forms are the same by Lemma 3. Let this canonical form be S. We can transform S_1 into S by Theorem 1. Let this sequence of transformation rules be r_1. We can also transform S_2 into S by sequence r_2. Since all of our transformation rules are bidirectional, we can get a sequence r_2' of rules that transforms S to S_2 simply by reversing r_2. Now the sequence r_1 followed by r_2' transforms S_1 into S_2. □

5. CONSTRUCTION OF EFFICIENT CIRCUITS

In this section, we propose an example of circuit design procedure based on our transformation rules. The design flow consists of the following three steps:

Step 1. We make an arbitrary initial CNOT-based circuit from a given formula (depending on the problem for the quantum algorithm).

Step 2. We transform the initial circuit to its canonical form by our procedure *Main*(S) in Section 4.

Step 3. A new transformation rule is given in Section 5.2, which can be realized by composing our transformation rules and guarantees a reduction if the cost of the circuit. We just apply this rule as many times as possible.

For Step 1, it should be noted that if we can use auxiliary qubits, we can simulate any type of formula by a CNOT-based circuit whose size is almost the same as the given formula. Here, we show how to simulate CNF formulas by CNOT-based circuits in the following section.

5.1 Circuits for CNF Formulas

For simplicity, we use an example. Suppose that $f = (x_1 + x_3)(\overline{x_1} + x_2 + x_3)$. Then we prepare two auxiliary qubits, $|x_5\rangle$ and $|x_6\rangle$ for the first and the second clause, respectively (see Fig. 13). Then we introduce a gate $g_1 = [5, \{1,3\}]$. (In general, for the i-th clause, we introduce $g_i = [n+1+i, C_i]$ where C_i includes j iff x_j or $\overline{x_j}$ appears in the clause. Thus we use one auxiliary qubit per clause.) Then, we place NOT_5, right after g_1 where NOT_i is a simpler notation for $[i, \{\}]$. NOT_i works just like a negation gate. Furthermore, we place two NOT_1 gates before and after g_1 since x_1 appears in the positive form. Similarly two NOT_3 gates are inserted. Let G_1 be the set of these gates. Then a similar set, G_2, of gates are introduced for the second clause, but no NOT_1 is inserted since x_1 appears in the negative form in the second clause. Then, we introduce one special gate $p = [4, \{5,6\}]$ to propagate the product value of the clauses into $|x_4\rangle$. After that we introduce G_2' and G_1', which are exactly the same as G_2 and G_1, respectively, in the mirror positions as shown in Fig 13.

Now it is easy to see that the value of $|x_5\rangle$ after G_1 is $\overline{0 \oplus \overline{x_1}\ \overline{x_3}} = (x_1 + x_3)$, which is equal to the first clause, and similarly for $|x_6\rangle$ and G_2. Thus the value of $|x_4\rangle$ after the special gate p simulates the value of the given CNF formula f. G_1' and G_2' are used for restoring the states of the auxiliary qubits. Thus our circuits simulate CNF formulas without much increasing the size, which immediately implies that the satisfiability problem for our circuits is NP-complete. Also, our transformation rules can simulate the Resolution system

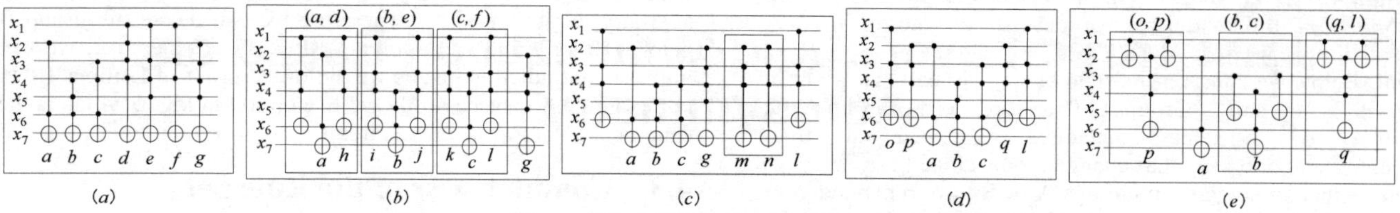

Figure 15: A Minimization Procedure

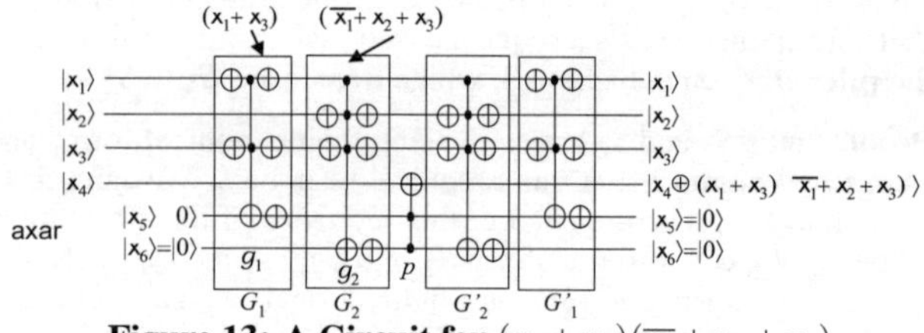

Figure 13: A Circuit for $(x_1+x_3)(\overline{x_1}+x_2+x_3)$

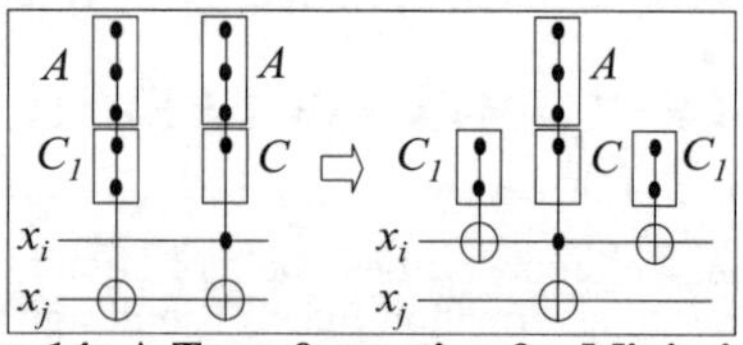

Figure 14: A Transformation for Minimization

to prove the unsatisfiability of CNF formulas with only a polynomial overhead (omitted in this paper).

5.2 Circuit Reduction from the Canonical Form

First we show a useful complex transformation rule which can be realized by applying a sequence of our transformation rules:

$$[j, A\cup C_1]\cdot[j, A\cup C\cup\{i\}] \Rightarrow [i, C_1]\cdot[j, A\cup C\cup\{i\}]\cdot[i, C_1],$$

if $C \subseteq C_1$.

Figure 14 illustrates this transformation. This transformation can be decomposed as follows: First we put two $[i, C_1]$ gates by Rule (1) right after $[j, A\cup C\cup\{i\}]$. Next we move $[j, A\cup C_1]$ and $[j, A\cup C\cup\{i\}]$ to the position between the two added $[i, C_1]$ gates by Rules (2) and (3), where a new $[j, A\cup C_1]$ gate is added. Finally, we remove two $[j, A\cup C_1]$ gates by Rule (1) and get the righthand-side of the transformation. We can generalize the transformation for more than two gates as follows:
$[j, A\cup C_1]\cdot[j, A\cup C_2]\cdots[j, A\cup C_k]\cdot[j, A\cup C\cup\{i\}] \Rightarrow [i, C_1]\cdot[i, C_2]\cdots[i, C_k]\cdot[j, A\cup C\cup\{i\}]\cdot[i, C_1]\cdot[i, C_2]\cdots[i, C_k]$,
if $C \subseteq C_l (1 \le l \le k)$.

It is of course hard to guess a reasonable definition of the cost of CNOT gates at this moment. However, as shown later, there is some evidence that the implementation of a CNOT gate with more inputs should be more expensive than with less inputs. If it is true, then the implementation cost of the circuit of the righthand-side of the above transformation is smaller than that of the lefthand-side.

It is easy to see that we can transform Circuit B to Circuit C in the previous Fig. 2 by applying this transformation many times. Below we present a more complicated and interesting example where we need some heuristics. Suppose we start from the circuit in Fig. 15 (a). First we apply the transformation to three pairs of gates, i.e., a and d, b and e, and c and f to get the circuit in Fig. 15 (b). Next we can simply delete two pairs of redundant gates, i.e., h and i, and j and k by Rule (1). Then we can optimize the circuit further by applying the above transformation to the gates b and c.

We have another way to optimize the circuit even more in this case: First we swap the order of l and g, and add two gates m and n as shown in Fig. 15 (c). Then we can apply the transformation to three pairs of gates, i.e., a and m, b and g, and c and n to get the circuit in Fig. 15 (d). Next we further apply the transformation to three pairs of gates, i.e., o and p, b and c, and q and l to get the final circuit in Fig. 15 (e).

Here we give a comparison between the implementation costs for the circuits in Fig. 15 (a) and Fig. 15 (e) based on the following cost assumption, where $Cost(n)$ means the cost of a CNOT gate with n control bits: (i) $Cost(1) = 1$, (ii) $Cost(2) = 14$, (iii) $Cost(3) = 56$, (iv) $Cost(4) = 140$, and (v) $Cost(m) = 112(m-3)$ when $m \ge 5$. Our assumption is based on [1], which shows that (i) a CNOT gate with two control bits can be decomposed to 14 basic quantum gates with fewer inputs, (ii) a CNOT gate with three and four control bits can be decomposed to four and ten CNOT gates with two control bits, respectively, and (iii) a CNOT gate with $m(\ge 5)$ control bits can be decomposed to $8(m-3)$ CNOT gates with two control bits. By a simple calculation we can see that the cost of the circuit in Fig. 15 (a) is 602, whereas the cost for Fig. 15 (e) is 188. Thus, we can usually make a circuit smaller if we apply our complex transformation to appropriate portions of the circuit. To find the best portion to which we should apply the transformation is a combinatorial problem similar to finding a best *divisor* in the conventional multilevel logic synthesis [7].

6. CONCLUDING REMARKS

As mentioned earlier, our ultimate goal is to develop a design theory for quantum Boolean circuits. For this purpose: (1) We will need more concrete "guide" or heuristics on how to change a given circuit into a *better* circuit. (2) During the course of this research, we may find "a counter example", i.e., a pair of circuits for which our rule set needs exponentially many steps. The rule set will then need to be modified to cope with such counter examples. (3) A design theory for *sequential* quantum circuits will also be interesting.

7. REFERENCES

[1] A. Barenco, C. H. Bennett, R. Cleve, D. P. DiVincenzo, N. Margolus, P. Shor, T. Sleator, J. A. Smolin, and H. Weinfurter. Elementary gates for quantum computation. *Physical Review A*, 52(5):3457–3467, Nov. 1995.

[2] R. K. Brayton, R. Rudell, A. Sangiovanni-Vincentelli, and A. R. Wang. MIS: A Multiple-Level Logic Optimization System. *IEEE Transactions on Computer-Aided Design of Integrated Circuits and Systems*, CAD-6(6):1062–1081, Nov. 1987.

[3] J. Darringer, W. Joyner, L. Berman, and L. Trevillyan. LSS: Logic synthesis through local transformations. *IBM J. Res. and Develop.,*, 25(4):272–280, July 1981.

[4] M. Davio, J.-P. Deshamps, and A. Thayse. *Discrete and Switching Functions*. McGraw Hill International, 1978.

[5] L. K. Grover. A fast quantum mechanical algorithm for database search. In *Proceedings of 28th ACM Symposium on Theory of Computing*, pages 212–219, 1996.

[6] J. Gruska. *Quantum Computing*. McGraw Hill, 1999.

[7] G. D. Hactel and F. Somenzi. *Logic Synthesis and Verification Algorithms*. Kluwer Academic Publishers, 1996.

[8] K. Iwama, K. Hino, H. Kurokawa, and S. Sawada. Random benchmark circuits with controlled attributes. In *Proc. European Design & Test Conference and Exhibition (ED&TC'97)*, pages 90–97, 1997.

[9] J.-S. Lee, Y. Chung, J. Kim, and S. Lee. A Practical Method of Constructing Quantum Combinational Logic Circuits. Technical Report http://arXiv.org/abs/quant-ph/9911053, LANL e-print, 1999.

[10] M. A. Nielsen and I. L. Chuang. *Quantum Computation and Quantum Information*. Cambridge University Press, 2000.

[11] T. Toffoli. Reversible computing. Technical Report MIT/LCS/TM-151, MIT LCS, Feb. 1980.

[12] A. Yao. Quantum circuit complexity. In *Proc. 34th Annual IEEE Symposium on Foudations of Computer Science*, pages 352–361, 1993.

Fast Three-Level Logic Minimization Based on Autosymmetry

Anna Bernasconi, Valentina Ciriani, Fabrizio Luccio, Linda Pagli
Department of Computer Science
University of Pisa, Italy
{annab, ciriani, luccio, pagli}@di.unipi.it

ABSTRACT

Sum of Pseudoproducts (SPP) is a three level logic synthesis technique developed in recent years. In this framework we exploit the "regularity" of Boolean functions to decrease minimization time. Our main results are: 1) the regularity of a Boolean function f of n variables is expressed by its *autosymmetry degree* k (with $0 \leq k \leq n$), where $k = 0$ means no regularity (that is, we are not able to provide any advantage over standard synthesis); 2) for $k > 1$ the function is *autosymmetric*, and a new function f_k is identified in polynomial time; f_k is "equivalent" to, but smaller than f, and depends on $n - k$ variables only; 3) given a minimal SPP form for f_k, a minimal SPP form for f is built in linear time; 4) experimental results show that 61% of the functions in the classical ESPRESSO benchmark suite are autosymmetric, and the SPP minimization time for them is critically reduced; we can also solve cases otherwise practically intractable. We finally discuss the role and meaning of autosymmetry.

Categories and Subject Descriptors

B.6.3 [**Logic Design**]: Design Aids—*Automatic synthesis, Optimization.*

General Terms

Algorithms, Design, Theory.

Keywords

Three-Level logic, Synthesis, Autosymmetry.

1. INTRODUCTION

The standard synthesis of Boolean functions is performed with Sum of Products (SP) minimization procedures, leading to two level circuits. More-than-two level minimization is much harder, but the size of the circuits can significantly decrease. In many cases three-level logic is a good trade-off among circuit speed, circuit size, and the time needed for the minimization procedure. Note that, in all cases, algorithms for exact minimization have exponential complexity, hence the time to attain minimal forms becomes huge for increasing size of the input.

Two level minimization is well developed [10, 4, 9]. Several techniques for three level minimization have been proposed for different algebraic expressions. Among them EX-SOP forms, where two SP forms are connected in EXOR [5, 6]; and Sum of Pseudoproducts (SPP) forms, consisting of an OR of pseudoproducts, where a pseudoproduct is the AND of EXOR factors [8]. For example $(x_0 \cdot x_1 + x_0 \cdot \overline{x}_2) \oplus (x_2 \cdot x_3 + x_1 \cdot x_4)$ is an EX-SOP form, and $(x_0 \oplus \overline{x}_1) \cdot x_2 + (x_0 \oplus x_2 \oplus x_3) \cdot (x_2 \oplus \overline{x}_4) + x_1 \cdot x_4$ is an SPP form. Experimental results show that the average size of SPP forms is approximately half the size of the corresponding SP, and SPP forms are also smaller than EX-SOP [2]. As a limit case each EXOR factor reduces to a single literal in SPP, and the SP and SPP forms coincide.

In this work we refer to SPP minimization. Initially this can be seen as a generalization of SP minimization, and in fact an extension of the Quine-McCluskey algorithm was given in [8] for SPP. In particular the pseudoproducts to be considered can be limited to the subclass of *prime* pseudoproducts, that play the same role of prime implicants in SP. The algorithm for SPP, however, was more cumbersome than the former, thus failing in practice in minimizing very large functions. A deeper understanding of the problem, together with the use of ad hoc data structures, has allowed to widely extend the set of functions practically tractable [2]. Still a number of standard benchmark functions can be hardly handled with this technique.

In fact the aim of this paper is to exploit the "regularity" of any given Boolean function, in order to decrease the time needed for its logical synthesis. Our main results are: 1) the regularity of a Boolean function f of n variables can be expressed by an *autosymmetry degree* k (with $0 \leq k \leq n$), which is computed in polynomial time. $k = 0$ means no regularity, that is we are not able to provide any advantage over standard synthesis; 2) for $k \geq 1$ the function f is said to be *autosymmetric*, and a new function f_k is identified in polynomial time. In a sense f_k is "equivalent" to, but smaller than f, and depends on $n - k$ variables only (a function f with $k = 2$, and its corresponding function f_2, are shown in Figure 1); 3) given a minimal SPP form for f_k, a minimal SPP form for f is built in linear time; 4) experimental results show that 61% of the functions in the classical ESPRESSO benchmark suite are autosymmetric: the SPP minimization

DAC 2002, June 10-14, 2002, New Orleans, Louisiana, USA.

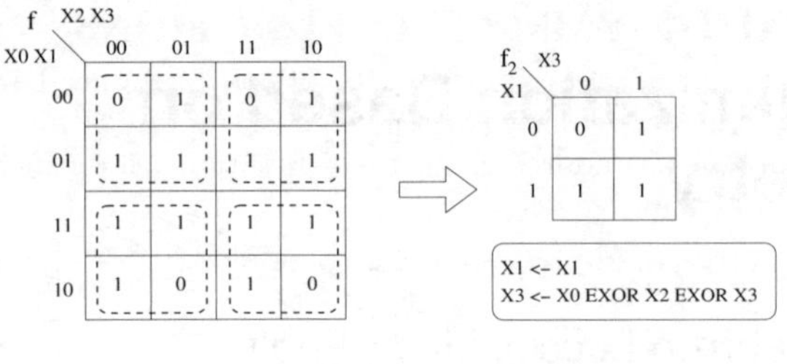

Figure 1: A 2-autosymmetric function f, in a Karnaugh map (on the left) of four variables. The corresponding reduced function f_2, which depends on the variables x_1 and x_3, and the linear substitutions are on the right.

	$\mathbf{c}_0$	$\mathbf{c}_1$	$\mathbf{c}_2$	$\mathbf{c}_3$	$\mathbf{c}_4$	$\mathbf{c}_5$
$\mathbf{r}_0$	0	1	0	0	0	1
$\mathbf{r}_1$	0	1	0	0	1	0
$\mathbf{r}_2$	0	1	1	1	0	1
$\mathbf{r}_3$	0	1	1	1	1	0
$\mathbf{r}_4$	1	1	0	1	0	0
$\mathbf{r}_5$	1	1	0	1	1	1
$\mathbf{r}_6$	1	1	1	0	0	0
$\mathbf{r}_7$	1	1	1	0	1	1

Figure 2: A canonical matrix representing a pseudocube P in $\{0,1\}^6$. The canonical columns are $\mathbf{c}_0$, $\mathbf{c}_2$, and $\mathbf{c}_4$.

time for them is critically reduced, and we can also solve cases otherwise intractable. Indeed, although autosymmetric functions form a subset of all possible Boolean functions, a great amount of standard functions of practical interest fall in this class. In the last section we speculate on the possible causes of this fact. Observe that even if we can study an autosymmetric function f in a $n-k$ dimensional space, f depends, in general, on all the n input variables, i.e., f is non degenerated. Whereas all degenerated function are autosymmetric.

In the next Section 2 we recall the basic definitions and results of SPP theory, and present a companion algebraic formulation later exploited for testing autosymmetry. In Section 3 we discuss the properties of autosymmetric functions, and how the problem of determining their minimal SPP forms can be studied on a reduced number of variables. In Section 4 we show how autosymmetry can be tested in polynomial time, and derive a new minimization algorithm that includes such a test in the initial phase. In Section 5 we present an extensive set of experimental results which validate the proposed approach, also proving that the number of benchmarks practically tractable is significantly increased. A final discussion on the role of autosymmetry is developed in Section 6.

2. THE UNDERLYING THEORY

The basic definitions and properties that we now recall were stated in [8] and extended in [2]. We work in a Boolean space $\{0,1\}^n$ described by n variables $x_0, x_1, \ldots, x_{n-1}$, where each point is represented as a binary vector of n elements. A set of k points can be arranged in a $k \times n$ matrix whose rows correspond to the points and whose columns correspond to the variables. Figure 2 represents a set of eight points in a space of six variables.

A Boolean function $f : \{0,1\}^n \to \{0,1\}$ can be specified with an algebraic expression where the variables are connected through Boolean operators, or as the set of points for which $f = 1$. $|f|$ denotes the number of such points.

From [8] we take the following. Let $\boldsymbol{u}$ be a (Boolean) vector; $\overline{\boldsymbol{u}}$ be the elementwise complementation of $\boldsymbol{u}$; and $\widehat{\boldsymbol{u}}$ denote $\boldsymbol{u}$ or $\overline{\boldsymbol{u}}$. The *constant* vectors $\mathbf{0}$ and $\mathbf{1}$ are made up of all 0's or all 1's, respectively. Vector $\boldsymbol{uv}$ is the concatenation of $\boldsymbol{u}$ and $\boldsymbol{v}$. A vector $\boldsymbol{u}$ of 2^m elements, $m \geq 0$, is *normal* if $m = 0$, or $m > 0$ and $\boldsymbol{u} = \boldsymbol{v}\widehat{\boldsymbol{v}}$ with $\boldsymbol{v}$ (hence $\widehat{\boldsymbol{v}}$) normal. For example all the columns in the matrix of Figure 2 are normal vectors.

A matrix M with 2^m rows is *normal* if all its rows are different and all its columns are normal. A normal matrix is *canonical* if its rows, interpreted as binary numbers, are arranged in increasing order (the matrix of Figure 2 is canonical). A normal vector is *k-canonical*, $0 \leq k < m$, if is composed of an alternating sequence of groups of 2^k 0's and 2^k 1's. In Figure 2, $\boldsymbol{c}_0$ is 2-canonical, $\boldsymbol{c}_2$ is 1-canonical, and $\boldsymbol{c}_4$ is 0-canonical. A canonical matrix M contains m columns $\boldsymbol{c}_{i_0}, \ldots, \boldsymbol{c}_{i_{m-1}}$ of increasing indices, called the *canonical columns* of M, such that $\boldsymbol{c}_{i_j}$ is $(m-j-1)$-canonical for $0 \leq j \leq m-1$. The other columns are the *noncanonical* ones.

If M represents a set of points in a Boolean space (where the rows are the points and the column c_i corresponds to the variable x_i), canonical and noncanonical columns correspond to *canonical* and *noncanonical variables*, respectively. We have:

DEFINITION 1. (From [8]) *A* pseudocube *of degree m is a set of 2^m points whose matrix is canonical up to a row permutation.*

The matrix of Figure 2 represents a pseudocube of 2^3 points in $\{0,1\}^6$. Note that a subcube in $\{0,1\}^n$ is a special case of pseudocube where the noncanonical columns are constant.

The function with value 1 in the points of a pseudocube P is a *pseudoproduct*, and can be expressed as a product of EXOR factors in several different forms, one of which is called the *canonical expression* (briefly *CEX*) of P. For the pseudocube P of Figure 2 we have:

$$CEX(P) = x_1 \cdot (x_0 \oplus x_2 \oplus \overline{x}_3) \cdot (x_0 \oplus x_4 \oplus x_5). \quad (1)$$

Refer to [8] for the nontrivial rule for generating *CEX(P)*. We simply recall that each EXOR factor of the expression contains exactly one noncanonical variable directed or complemented, namely the one with greatest index (x_1, x_3, and x_5 in the example), and each noncanonical variable is contained only in one EXOR factor; all the other variables in the expression are canonical (x_0, x_2, and x_4 in the example); and some canonical variables may not appear in the expression. Note that the minimal SP form for the above function is $\overline{x}_0x_1\overline{x}_2\overline{x}_3\overline{x}_4x_5+\overline{x}_0x_1\overline{x}_2\overline{x}_3x_4\overline{x}_5+\overline{x}_0x_1x_2x_3\overline{x}_4x_5+\overline{x}_0x_1x_2x_3x_4\overline{x}_5+x_0x_1\overline{x}_2x_3\overline{x}_4\overline{x}_5+x_0x_1\overline{x}_2x_3x_4x_5+x_0x_1x_2\overline{x}_3\overline{x}_4\overline{x}_5+x_0x_1x_2\overline{x}_3x_4x_5$, much larger than *CEX(P)*.

If P is in fact a subcube, each EXOR factor in *CEX(P)* reduces to a single noncanonical variable, the canonical variables do not appear, and the whole expression reduces to the well known *product* expression e.g. used for implicants in SP forms.

A general property of the algebraic representation of pseudocubes is given in the following:

THEOREM 1. *In a Boolean space* $\{0,1\}^n$*: (i) the EXOR factor of any subset of variables (directed or complemented) represents a pseudocube of degree* $n-1$*; (ii) the product of* $k \leq n$ *arbitrary EXOR factors represents either an empty set or a pseudocube of degree* $\geq n-k$.

Point (i) of the theorem can be easily proved by induction on the number of variables in the EXOR factor. Point (ii) then follows from a theorem of [8] which states that the intersection of two pseudocubes of degrees p, r is either empty, or is a pseudocube of degree $\geq p+r-n$.

For the example of Figure 2, the EXOR factors x_1, $x_0 \oplus x_2 \oplus \overline{x}_3$, and $x_0 \oplus x_4 \oplus x_5$ in the Expression (1) represent three pseudocubes of 2^5 points, and their product represents the pseudocube P of 2^3 points.

Note now that an equality of the form $\text{EXOR}_1 \cdot \text{EXOR}_2 \cdot \ldots \cdot \text{EXOR}_k = 1$, satisfied by all the points of a pseudocube, can be equivalently written as a system of k linear equations:

$$EXOR_i = 1, \quad 1 \leq i \leq k \tag{2}$$

that is an instance of a general linear system $A\boldsymbol{x} = \boldsymbol{b}$ with $\mathbf{b} \in \{0,1\}$, $\boldsymbol{x} = \{x_0, \ldots, x_{n-1}\}$, A is a $k \times n$ matrix of coefficients 0, 1, and the sum is substituted with EXOR. As known [3, 1] such a system specifies an *affine subspace* of the linear space $\{0,1\}^n$. Then, from the existence of *CEX(P)* for any pseudocube P, and from Theorem 1, we have:

COROLLARY 1. *In a Boolean space* $\{0,1\}^n$ *there is a one-to-one correspondence between affine subspaces and pseudocubes.*

This corollary allows to inherit all the properties of affine subspaces into pseudocube theory. In particular, a pseudocube containing the point (vector) **0** corresponds to a *linear* subspace.

The *structure* of a pseudocube P, denoted by STR(P), is CEX(P) without complementations. For the pseudocube P of Figure 2 we have: $STR(P) = x_1 \cdot (x_0 \oplus x_2 \oplus x_3) \cdot (x_0 \oplus x_4 \oplus x_5)$. Let us now extend the symbol $\oplus$ to denote the elementwise EXOR between two vectors. Then $\boldsymbol{\alpha} \oplus \boldsymbol{\beta}$ is the vector obtained from $\boldsymbol{\beta}$ complementing in it the elements corresponding to the 1's of $\boldsymbol{\alpha}$. For a vector $\boldsymbol{\alpha} \in \{0,1\}^n$ and a subset $S \subseteq \{0,1\}^n$, let $\boldsymbol{\alpha} \oplus S = \{\, \boldsymbol{\alpha} \oplus \boldsymbol{s} \mid \boldsymbol{s} \in S\}$. We have:

THEOREM 2. (From various results in [2]) *For any pseudocube* $P \subseteq \{0,1\}^n$ *and any vector* $\boldsymbol{\alpha} \in \{0,1\}^n$*, the subset* $Q = \boldsymbol{\alpha} \oplus P$ *is a pseudocube with* $STR(Q) = STR(P)$.

Finally recall that an arbitrary function can be expressed as an OR of pseudoproducts, giving rise to an SPP form.

For example if we add the two rows (points) $\mathbf{r}_8 = 1\,1\,0\,1\,0\,1$ and $\mathbf{r}_9 = 1\,1\,0\,1\,1\,0$ to the matrix of Figure 2, we have a new function f that can be seen as the union of two partially overlapping pseudocubes: namely P (already studied), and Q associated to the rows $\mathbf{r}_4, \mathbf{r}_5, \mathbf{r}_8, \mathbf{r}_9$. Note that Q is in fact a cube, and we have $CEX(Q) = x_0 x_1 \overline{x}_2 x_3$. In conclusion f can be expressed in SPP form as:

$$CEX(P) + CEX(Q) = x_1(x_0 \oplus x_2 \oplus \overline{x}_3)(x_0 \oplus x_4 \oplus x_5) + x_0 x_1 \overline{x}_2 x_3. \tag{3}$$

The minimal SP form for f contains 40 literals, while the SPP Expression (3) contains 11 literals. Passing from SP to SPP, however, amounts to passing from a two-level to a three-level circuit. This fact has always to be taken into account and will not be further repeated.

3. AUTOSYMMETRIC FUNCTIONS

The class of *autosymmetric functions* introduced in [8] seems to be particularly suitable for SPP minimization. The present work is addressed to these functions, for which we give an alternative definition.

DEFINITION 2. *A Boolean function* f *in* $\{0,1\}^n$ *is* closed under $\boldsymbol{\alpha}$*, with* $\boldsymbol{\alpha} \in \{0,1\}^n$*, if for each* $\boldsymbol{w} \in \{0,1\}^n$*,* $\boldsymbol{w} \oplus \boldsymbol{\alpha} \in f$ *if and only if* $\boldsymbol{w} \in f$.

Each function is obviously closed under the zero vector **0**. As proved in [8], if a function f is closed under two different vectors $\boldsymbol{\alpha}_1$, $\boldsymbol{\alpha}_2 \in \{0,1\}^n$, it is also closed under $\boldsymbol{\alpha}_1 \oplus \boldsymbol{\alpha}_2$. Therefore the set L_f of all the vectors $\boldsymbol{\beta}$ such that f is closed under $\boldsymbol{\beta}$ is a linear subspace of $\{0,1\}^n$. In fact, combining in EXOR, in all possible ways, k linearly independent vectors $\boldsymbol{\alpha}_1, \ldots, \boldsymbol{\alpha}_k$, we form a subspace of 2^k vectors that is closed under $\oplus$, and contains the vector **0** generated as $\boldsymbol{\alpha}_i \oplus \boldsymbol{\alpha}_i$, see for example [3]. L_f is called the *linear space* of f, and k is its *dimension*. By Corollary 1, L_f is a pseudocube, and we will refer to $CEX(L_f)$ and $STR(L_f)$.

DEFINITION 3. *A Boolean function* f *is* k-autosymmetric, *or equivalently* f *has* autosymmetry degree k, $0 \leq k \leq n$, *if its linear space* L_f *has dimension* k.

THEOREM 3. *Let* f *be a* k*-autosymmetric function. There exist* ℓ *vectors* $\boldsymbol{w}^1, \boldsymbol{w}^2, \ldots, \boldsymbol{w}^\ell \in f$*, with* $\ell = |f|/2^k$*, such that*

$$f = \bigcup_{i=1}^{\ell} (\boldsymbol{w}^i \oplus L_f) \tag{4}$$

and for each i, j, $i \neq j$, $(\boldsymbol{w}^i \oplus L_f) \cap (\boldsymbol{w}^j \oplus L_f) = \emptyset$.

PROOF. Let $\boldsymbol{w}^1$ be any vector in f. By Definition 2, $\boldsymbol{w}^1 \oplus L_f \subseteq f$. Consider the set $f^1 = f \setminus (\boldsymbol{w}^1 \oplus L_f)$. If $f^1 = \emptyset$, then $f = \boldsymbol{w}^1 \oplus L_f$ and f is a pseudocube of degree k with $|f| = 2^k$. Otherwise, let $\boldsymbol{w}^2$ be any vector of f^1. Again by Definition 2, $\boldsymbol{w}^2 \oplus L_f \subseteq f$. Observe that $(\boldsymbol{w}^1 \oplus L_f) \cap (\boldsymbol{w}^2 \oplus L_f) = \emptyset$ (by contradiction: let $\boldsymbol{v} \in (\boldsymbol{w}^1 \oplus L_f) \cap (\boldsymbol{w}^2 \oplus L_f)$, then $\boldsymbol{v} = \boldsymbol{w}^1 \oplus \boldsymbol{\alpha} = \boldsymbol{w}^2 \oplus \boldsymbol{\beta}$, with $\boldsymbol{\alpha}, \boldsymbol{\beta} \in L_f$; then $\boldsymbol{w}^2 = \boldsymbol{w}^1 \oplus \boldsymbol{\alpha} \oplus \boldsymbol{\beta}$, that is $\boldsymbol{w}^2 \in (\boldsymbol{w}^1 \oplus L_f)$ which is a contradiction). Therefore we have:

$$f = (\boldsymbol{w}^1 \oplus L_f) \cup (\boldsymbol{w}^2 \oplus L_f) \cup f \setminus ((\boldsymbol{w}^1 \oplus L_f) \cup (\boldsymbol{w}^2 \oplus L_f)),$$

and using the same argument on the set $f \setminus ((\boldsymbol{w}^1 \oplus L_f) \cup (\boldsymbol{w}^2 \oplus L_f))$ the theorem easily follows. □

From the proof above we see that the number of points of a k-autosymmetric function is a multiple of 2^k. Indeed, each affine subspace $\boldsymbol{w}^i \oplus L_f$ contains 2^k points. Recalling that L_f is a pseudocube, and from Theorems 2 and 3 we immediately have:

COROLLARY 2. *A* k*-autosymmetric function* f *is a disjoint union of* $|f|/2^k$ *pseudocubes* $\boldsymbol{w}^i \oplus L_f$ *of degree* k *all having the same structure* $STR(L_f)$*, and the same canonical variables of* L_f.

This corollary has two immediate consequences. First we can extend the definition of canonical variables to autosymmetric functions. Namely, the canonical (respectively: noncanonical) variables of L_f are designated as the canonical (respectively: noncanonical) variables of f. Second, note that the 2^k points of each pseudocube $\boldsymbol{w}^i \oplus L_f$ contain all the 2^k combinations of values of the canonical variables, hence in exactly one of these points all such values are 0. We have:

COROLLARY 3. *The vectors* $\boldsymbol{w}^1, \ldots, \boldsymbol{w}^\ell$ *of Theorem 3 can be chosen as the points of* f *where all the canonical variables have value 0. Moreover, for* $1 \le i \le \ell$, $CEX(\boldsymbol{w}^i \oplus L_f)$ *is obtained from* $STR(L_f)$ *complementing the noncanonical variables with value 0 in* $\boldsymbol{w}^i$.

EXAMPLE 1. *Consider the function* $f = \{00001, 00100, 00110, 01000, 01010, 01101, 10001, 10011, 10100, 11000, 11101, 11111\}$. *It can be easily verified that the linear space of* f *is* $L_f = \{00000, 01100, 10101, 11001\}$, *where each vector can be obtained as the EXOR of the other two.* f *is then 2-autosymmetric. We have* $STR(L_f) = (x_0 \oplus x_1 \oplus x_2)x_3 (x_0 \oplus x_4)$, *where* x_0, x_1, *and,* x_2, x_3, x_4, *are the canonical and noncanonical variables, respectively. From Corollary 3 we have* $\boldsymbol{w}^1 = 00001$, $\boldsymbol{w}^2 = 00100$, $\boldsymbol{w}^3 = 00110$, *hence:*

$$f = (00001 \oplus L_f) \cup (00100 \oplus L_f) \cup (00110 \oplus L_f)\,. \qquad (5)$$

From this we derive the following (nonminimal) SPP form for f, *where each term* $\boldsymbol{w}^i \oplus L_f$ *of Expression (5) is transformed into an EXOR factor derived from* $STR(L_f)$ *complementing the noncanonical variables with value 0 in* $\boldsymbol{w}^i$*:*

$$(x_0 \oplus x_1 \oplus \overline{x}_2)\overline{x}_3(x_0 \oplus x_4) + (x_0 \oplus x_1 \oplus x_2)\overline{x}_3(x_0 \oplus \overline{x}_4) + (x_0 \oplus x_1 \oplus x_2)x_3(x_0 \oplus \overline{x}_4)\,.$$

We will study any k-autosymmetric function f with $k \ge 1$ through a simpler function f_k. We state:

DEFINITION 4. *For a* $(k \ge 1)$*-autosymmetric function* f, *the* restriction f_k *consists of the* $|f|/2^k$ *points of* f *contained in the subspace* $\{0,1\}^{n-k}$ *where all the canonical variables of* f *have value 0.*

Note that f_k depends only on the $n-k$ noncanonical variables of f. Once L_f has been computed (see next section), the canonical variables of f are known, and f_k can be immediately determined applying Definition 4. For instance, for the function f with $k = 2$ of Example 1, f_2 depends on the noncanonical variables x_2, x_3, x_4. To build f_2 we take the subset $\{00001, 00100, 00110\}$ of the points of f for which the canonical variables x_0, x_1 have value 0, and then project these points into the $\{0,1\}^3$ subspace relative to x_2, x_3, x_4, where we have $f_2 = \{001, 100, 110\}$. f_2 is $\overline{x}_2\overline{x}_3x_4 + x_2\overline{x}_3\overline{x}_4 + x_2x_3\overline{x}_4$. Note that the same expression is obtained setting $x_0 = x_1 = 0$ in the last expression of Example 1. We now show that the SPP-minimal form of any autosymmetric function can be easily derived from the SPP-minimal form of its restriction. Note that finding the latter form is easier because the restriction has less variables and points. In the example above f depends on 5 variables and has 12 points, while f_2 depends only on 3 variables and has only 3 points (represented by the minterms $\overline{x}_2\overline{x}_3x_4$, $x_2\overline{x}_3\overline{x}_4$ and $x_2x_3\overline{x}_4$). An easy extension of Theorem 6 of [8] is the following

LEMMA 1. *A* $(k \ge 1)$*-autosymmetric function and its restriction have the same number of pseudoproducts in their minimal SPP forms.*

In fact we can prove a stronger property, namely not only f_k and f have the same number of pseudoproducts in their minimal forms, but a minimal form for f can be easily derived from a minimal form for f_k. Let $x_{z_0}, \ldots, x_{z_{n-k-1}}$ be the noncanonical variables of f, and let $STR(L_f) = p_0p_1\cdots p_{n-k-1}$, where p_i is the EXOR factor containing x_{z_i}, $1 \le i \le n-k-1$ (recall that each noncanonical variable is in an unique EXOR factor).

THEOREM 4. *If* $SPP(f_k)$ *is a minimal SPP form for* f_k, *then the form* $SPP(f)$ *obtained by substituting in* $SPP(f_k)$ *each variable* x_{z_i} *with the EXOR factor* p_i *is a minimal SPP form for* f.[1]

PROOF. By Lemma 1, the number of pseudoproducts in $SPP(f)$ is minimum, then we have only to prove that this form covers exactly all the points of f. When we transform f into f_k, we select the vector $\boldsymbol{u}^i$ with all canonical variables set to zero from each affine subspace $\boldsymbol{w}^i \oplus L_f$. Call $\boldsymbol{v}^i$ the vector $\boldsymbol{u}^i$ without the canonical variables, i.e., its projection onto a subspace $\{0,1\}^{n-k}$. When we apply the linear substitutions $x_{z_0} \leftarrow p_0, \ldots, x_{z_{n-k-1}} \leftarrow p_{n-k-1}$, we force any pseudoproduct that covers $\boldsymbol{v}^i$ in f_k to cover all the points in $\boldsymbol{w}^i \oplus L_f$ in f, and the thesis immediately follows. □

4. THE MINIMIZATION ALGORITHM

In the previous section we have shown that each Boolean function f is k-autosymmetric, for $k \ge 0$. If $k \ge 1$, f will be simply called *autosymmetric*. For minimization purposes we have an increasing advantage for increasing $k \ge 1$, as minimizing a k-autosymmetric function with n variables and ℓ points reduces to minimizing a different function with $n-k$ variables and $\ell/2^k$ points. Even for $k = 1$ we have to cover only one half of the original points.

Fortunately, for a given function f, finding the associated linear space L_f and computing the autosymmetry degree k is an easy task, because the required algorithm is polynomial in the number n of variables and in the number of points of f. By Definition 2, a function f is closed under $\boldsymbol{\alpha}$ if for any $\boldsymbol{u} \in f$ there exists $\boldsymbol{v} \in f$ such that $\boldsymbol{v} = \boldsymbol{u} \oplus \boldsymbol{\alpha}$. Thus, for all $\boldsymbol{u} \in f$, there exists a vector $\boldsymbol{v} \in f$ such that we can express $\boldsymbol{\alpha}$ as $\boldsymbol{\alpha} = \boldsymbol{u} \oplus \boldsymbol{v}$. In other words, $\boldsymbol{\alpha}$ must be searched within the vectors of the set $\{\boldsymbol{u} \oplus \boldsymbol{v} \mid \boldsymbol{u}, \boldsymbol{v} \in f\}$. More precisely:

THEOREM 5. *Let* f *be a Boolean function. Then* $L_f = \bigcap_{\boldsymbol{u} \in f} (\boldsymbol{u} \oplus f)$.

PROOF. Let $\boldsymbol{\alpha} \in \bigcap_{\boldsymbol{u} \in f} (\boldsymbol{u} \oplus f)$. Then

$$\boldsymbol{\alpha} \in \bigcap_{\boldsymbol{u} \in f} (\boldsymbol{u} \oplus f) \Leftrightarrow \forall \boldsymbol{u} \in f, \quad \boldsymbol{\alpha} \in \boldsymbol{u} \oplus f$$

$$\Leftrightarrow \forall \boldsymbol{u} \in f, \quad \exists \boldsymbol{v} \in f \mid \boldsymbol{\alpha} = \boldsymbol{u} \oplus \boldsymbol{v} \Leftrightarrow \boldsymbol{\alpha} \in L_f\,.$$

□

Based on Theorem 5 we state:

ALGORITHM 1. **Construction of L_f** (build L_f and find the autosymmetry degree k of a given function f)

1. for all $\boldsymbol{u} \in f$ build the set $\boldsymbol{u} \oplus f$;

2. build the set $L_f = \bigcap_{\boldsymbol{u} \in f} (\boldsymbol{u} \oplus f)$;

3. compute $k = \log |L_f|$.

The time complexity of Algorithm 1 is $\Theta(|f|^2 n)$, because we must build a set $\boldsymbol{u} \oplus f$ for all $\boldsymbol{u} \in f$, and the construction of each such a set requires $\Theta(|f|\,n)$ time.

Any SPP minimization algorithm can be easily extended for exploiting autosymmetry. For a given function f we first compute L_f and k with Algorithm 1. If $k = 0$ (i.e., f is not

[1]The resulting expressions may be reduced using some properties of EXOR, in particular $x \oplus x = 0$ and $0 \oplus x = x$.

autosymmetric) we proceed with regular minimization, otherwise we compute the restriction f_k of f, minimize it, and finally derive a minimal form $SPP(f)$ from $SPP(f_k)$. We propose the following A (for autosymmetry) minimization algorithm.

ALGORITHM 2. **A-Minimization** (build a minimal SPP form of a given function f)

1. build L_f and compute the value of k by Algorithm 1;
2. **if** $k = 0$ **then** minimize f with any SPP synthesis algorithm
3. **else**
   (a) determine the canonical variables of L_f and compute the restriction f_k as indicated in Section 3;
   (b) compute $STR(L_f) = p_0 p_1 \cdots p_{n-k-1}$;
   (c) compute the minimal form $SPP(f_k)$ for f_k with any SPP synthesis algorithm;
   (d) build $SPP(f)$ by substituting in $SPP(f_k)$ each noncanonical variable x_{z_i} with the EXOR factor p_i of $STR(L_f)$.

By the theory developed in the previous section, Algorithm 2 is correct. Note that the algorithm builds an SPP form minimal with respect to the number of pseudoproducts. To obtain the minimal SPP form with respect to the number of literals we must slightly rearrange steps *3(c)* and *3(d)*, executing the substitutions of all p_i for x_{z_i} in the prime pseudoproducts of f_k, before selecting such pseudoproducts in the set covering problem implicit in the minimization algorithm.

EXAMPLE 2. *Minimization of the function f of Example 1, using Algorithm 2.*

- *Derive L_f and k by Algorithm 1. For this purpose, for all $\boldsymbol{u} \in f$ compute the set $\boldsymbol{u} \oplus f$. (For example, for the point* 00100 *we obtain the set:* $00100 \oplus f = \{00101, 00000, 00010, 01100, 01110, 01001, 10101, 10111, 10000, 11100, 11001, 11011\}$*). The intersection of all the sets $\boldsymbol{u} \oplus f$ gives the linear space* $L_f = \{00000, 01100, 10101, 11001\}$. *We then have $k = 2$.*
- *Since $k = 2$ we proceed with the* **else** *branch of Algorithm 2. L_f has noncanonical variables x_2, x_3, x_4, hence f_2 is restricted to these variables and we have:* $f_2 = \{001, 100, 110\}$.
- *The minimization problem now consists of finding a minimal SPP cover of the points of f_2. Applying the algorithm of [2] we have the minimal form* $SPP(f_2) = x_2\overline{x}_4 + \overline{x}_3(x_2 \oplus x_4)$.
- *Compute:* $STR(L_f) = (x_0 \oplus x_1 \oplus x_2)x_3(x_0 \oplus x_4)$.
- *Derive the minimal SPP form for f by substituting x_2, x_3 and x_4 in $SPP(f_2)$ with the EXOR factors of $STR(L_f)$, respectively $(x_0 \oplus x_1 \oplus x_2)$, x_3 and $(x_0 \oplus x_4)$. With some immediate algebraic simplifications we obtain* [2]$SPP(f) = (x_0 \oplus x_1 \oplus x_2)(x_0 \oplus \overline{x}_4) + \overline{x}_3(x_1 \oplus x_2 \oplus x_4)$.

[2]E.g., in the first term of $SPP(f_k)$ we have: $\overline{x}_4[x_4 \leftarrow (x_0 \oplus x_4)] = \overline{(x_0 \oplus x_4)} = (x_0 \oplus \overline{x}_4)$. In the last term of $SPP(f_k)$ we have: $(x_2 \oplus x_4)[x_2 \leftarrow (x_0 \oplus x_1 \oplus x_2), x_4 \leftarrow (x_0 \oplus x_4)] = (x_0 \oplus x_1 \oplus x_2) \oplus (x_0 \oplus x_4) = (x_0 \oplus x_0 \oplus x_1 \oplus x_2 \oplus x_4) = (x_1 \oplus x_2 \oplus x_4)$.

function	k	#L	#PP	T_A	T_C
max512(0)	1	8	2	20.82	2618.02
Z9sym(0)	1	102	17	310.17	*
newtpla2(2)	2	17	4	1.71	236.89
radd(2)	3	16	3	0.22	5.43
addm4(6)	4	18	4	0.39	79.86
intb(0)	5	96	12	11420.80	*
addm4(7)	6	6	2	0.27	123.63
intb(5)	7	38	8	431.60	*
newtpla(4)	8	7	1	0.48	2893.04
opa(17)	10	37	7	503.01	*
alcom(5)	11	6	3	119.67	*

Table 1: Detailed results for a subset of autosymmetric functions. The last two columns report CPU times in seconds on a Pentium III 450 machine for Algorithms *A* and *C* (a star indicates non termination after 172800 seconds). The results are relative to single outputs. #L and #PP are the numbers of literals and of prime pseudoproducts in the minimal expression.

5. EXPERIMENTAL RESULTS

The new minimization Algorithm 2, also called Algorithm A, has been tested on a large set of functions without don't cares, taken from the ESPRESSO benchmark suite [11] (the different outputs of each function have been synthesized separately). The performance of Algorithm A has been compared with the performance of the best previous algorithm, that is the one proposed in [2], in the following indicated as Algorithm C (after Ciriani). In fact, also the minimization of function f_k in Algorithm A (step 3(c)) has been implemented with Algorithm C. For all the functions considered we have computed the values of the autosymmetry degree k with Algorithm 1, obtaining the results shown in the first two rows of Table 2. Surprisingly the overall percentage of autosymmetric functions ($k \geq 1$) is over the 61%. We have then attempted to run Algorithms A and C for all such functions, recording the CPU times whenever the computation terminated in less than 172800 seconds (2 days) on a Pentium III 450 machine. Results on program termination are given in the last three rows of Table 2.

Row **Avg(T_A/T_C)** of Table 2 shows the average reduction of computing time using Algorithm A instead of C, for all the benchmark functions for which both algorithms terminated (i.e., for the 546 functions of row **#AC**). Note the improvement introduced by the new algorithm for all the autosymmetric functions in the set, and how such an improvement drastically increases for increasing k. For $k = 0$, instead, we have $Avg(T_A/T_C) > 1$, and actually the ratio (T_A/T_C) is slightly greater than 1 for each such a function. This is because Algorithm A computes L_f in any case, then calls Algorithm C. The resulting slowdown is however always negligible because L_f is computed in polynomial time by Algorithm 1 (see previous section), while Algorithm C is exponential in nature. For all the functions in the table the forms obtained with the old and the new method coincide.

Finally Table 1 shows the CPU times for a small subset of the above functions with $k \geq 1$, and other relevant minimization parameters for them.

6. THE ROLE OF AUTOSYMMETRY

To understand the role of autosymmetric functions we must compare them with the set of all possible functions.

k=	0	1	2	3	4	5	6	7	8	9	10	11	12	Tot
#funct	365	116	72	95	41	43	39	27	16	24	58	28	18	942
%funct	38.9	12.3	7.6	10.1	4.3	4.6	4.1	2.9	1.7	2.4	6.2	3.0	1.9	100
#AC	264	82	55	54	23	35	21	10	2	0	0	0	0	546
#A	0	5	4	5	9	7	16	9	14	24	58	28	18	197
#*	101	29	13	36	9	1	2	8	0	0	0	0	0	199
Avg(T_A/T_C)	1.01	0.63	0.42	0.37	0.32	0.27	0.07	0.004	0.0001	-	-	-	-	-

Table 2: Distribution of k-autosymmetric functions in the Espresso benchmark suite. #funct are the total numbers of functions (single outputs) for any value of k, and %funct are the corresponding percentages. The other rows report the number of k-autosymmetric functions synthesized with the new Algorithm A, and with the previous best Algorithm C (#AC); synthesized with A only (#A), since C did not terminate; not synthesized at all since both algorithms did not terminate (#*). Avg(T_A/T_C) is the average of the ratio T_A/T_C, for the 546 functions of row #AC (k = 0 to 8). T_A and T_C are the times required by the two algorithms on the same function.

The total number of Boolean functions of n variables is $N_F = 2^{2^n}$, corresponding to all the ways a subset of points can be chosen in $\{0,1\}^n$. This is a huge number, however, due to the randomness of the above generating process very many of such functions do not correspond to any significant circuit. Autosymmetric functions are just a subset of the above. By a counting argument (omitted here for brevity) we can in fact prove that autosymmetric functions are $N_A = (2^n - 1)2^{2^{n-1}}$. Therefore, for increasing n, autosymmetric functions constitute a vanishing fraction of all the functions, as N_A/N_F goes to zero for n going to infinity. Still the question remains on how many significant functions are autosymmetric.

All what we can say at the moment is that most of the major benchmark functions are indeed autosymmetric, as shown in the previous section. The more so when n is small and the values of N_F and N_A are not too distant. The reason, we might argue, is that a function encoding a real life problem must exhibit a regular structure that can be reflected in some degree of autosymmetry. In fact, also degenerated functions (i.e., functions that do not depend on all the variables) are autosymmetric. Some of this functions are encountered among the benchmark ones; however, this property is not immediately evident and is not directly used in the standard minimization process.

This regularity may also allow to define an autosymmetric function f independently of the number of variables, and then to state a rule for deriving a minimal form for f valid for any n. Well known functions as, for example, the ones counting the parity of n bits, or giving the next-state values for an n bits Gray code, can be easily expressed in minimal form for an arbitrary number of variables just because they are autosymmetric (for the parity see [8]; for Gray codes elementary considerations suffice).

It might be relevant to examine the relation between autosymmetric functions, and functions which are simply *symmetric*. That is, functions which are invariant under any permutation of their variables (see for example [7]). The total number of symmetric functions is $N_S = 2^{n+1}$, hence $N_S << N_A$, but they are not a subset of the autosymmetric ones. In fact a symmetric function may be autosymmetric (e.g., the parity function), but there are symmetric functions that are not autosymmetric (e.g., any symmetric function with an odd number of points). At the moment we have no interesting results in this direction.

We observe that the *information content* of an autosymmetric function f is represented by its reduction f_k together with the linear transformation $x_{z_0} \leftarrow p_0, \ldots, x_{z_{n-k-1}} \leftarrow p_{n-k-1}$ (Section 3), so that the core of the synthesis problem is the minimization of f_k (Section 4). This suggests a formal generalization. For a given function $g \in \{0,1\}^m \rightarrow \{0,1\}$ we can define the *autosymmetry class* $\mathcal{C}_g$ as the class of all the autosymmetric functions $f \in \{0,1\}^t \rightarrow \{0,1\}, t \geq m$, such that $f_k \equiv g$. Since the information content of any given function f can be easily found, and a minimal SPP form for f can then be derived from $SPP(f_k)$, minimizing the function f_k corresponds to minimize the entire class $\mathcal{C}_g = \mathcal{C}_{f_k}$. Exploiting the full potential of such an approach is currently a matter of study.

Finally another interesting research direction is the generalization of the autosymmetry property to functions with don't care set. Since the autosymmetry is strictly related to the points of a function f, the major goal would be selecting a subset of don't cares as points of f to maximize the autosymmetry degree of the function.

7. REFERENCES

[1] A. Aleksanyan. Realization of Boolean Functions by Disjunctions of Products of Linear Forms. *Dokl. Akad. Nauk SSSR (Russian), (translation in Soviet Math. Dokl. 39(1), 1989)*, 304(4):781–784, 1989.

[2] V. Ciriani. Logic Minimization Using Exclusive OR Gates. In *ACM/IEEE 38th Design Automation Conference (DAC)*, pages 115–120, 2001.

[3] P. Cohn. *Algebra Vol. 1.* John Wiley & Sons, 1981.

[4] O. Coudert. Two-Level Logic Minimization: an overview. *INTEGRATION*, 17:97–140, 1994.

[5] D. Debnath and T. Sasao. An Optimization of AND-OR-EXOR Three-Level Networks. In *Asia and South Pacific DAC*, pages 545–550, 1997.

[6] E. Dubrova, D. Miller, and J. Muzio. AOXMIN-MV: A Heuristic Algorithm for AND-OR-XOR Minimization. In *4th Int. Workshop on the Applications of the Reed Muller Expansion in circuit Design*, pages 37–54, 1999.

[7] Z. Kohavi. *Switching and Finite Automata Theory.* McGraw Hill, 1970.

[8] F. Luccio and L. Pagli. On a New Boolean Function with Applications. *IEEE Trans. on Comp.*, 48(3):296–310, 1999.

[9] P. McGeer, J. Sanghavi, R. Brayton, and A. Sangiovanni-Vincentelli. ESPRESSO-SIGNATURE: A New Exact Minimizer for Logic Functions. *IEEE Transactions on VLSI*, 1(4):432–440, 1993.

[10] T. Villa, T. Kam, R. Brayton, and A. Sangiovanni-Vincentelli. *Synthesis of Finite State Machines: Logic Optimization.* Kluwer Academic Publishers, 1997.

[11] S. Yang. Synthesis on Optimization Benchmarks. User guide, Microelectronic Center of North Carolina, 1991.

An Efficient Optimization–based Technique to Generate Posynomial Performance Models for Analog Integrated Circuits

Walter Daems
K.U.Leuven, ESAT–MICAS
Kasteelpark Arenberg 10
B-3001 Leuven, Belgium
daems@esat.kuleuven.ac.be

Georges Gielen
K.U.Leuven, ESAT–MICAS
Kasteelpark Arenberg 10
B-3001 Leuven, Belgium
gielen@esat.kuleuven.ac.be

Willy Sansen
K.U.Leuven, ESAT–MICAS
Kasteelpark Arenberg 10
B-3001 Leuven, Belgium
sansen@esat.kuleuven.ac.be

ABSTRACT

This paper presents an new direct–fitting method to generate posynomial response surface models with arbitrary constant exponents for linear and nonlinear performance parameters of analog integrated circuits. Posynomial models enable the use of efficient geometric programming techniques for circuit sizing and optimization. The automatic generation avoids the time–consuming nature and inaccuracies of handcrafted analytic model generation. The technique is based on the fitting of posynomial model templates to numerical data from SPICE simulations. Attention is paid to estimating the relative 'goodness–of–fit' of the generated models. Experimental results illustrate the significantly better accuracy of the new approach.

Categories and Subject Descriptors

B.7.2 [**Integrated Circuits**]: Design Aids; B.8.2 [**Performance and Reliability**]: Performance Analysis and Design Aids; I.6.5 [**Simulation and Modeling**]: Model Development

General Terms

Performance, Design, Algorithms

Keywords

Performance Modeling for Analog Circuits, Posynomial Response Surface Modeling, Geometric Programming

1. INTRODUCTION

The sizing of transistor–level analog integrated circuits is a time–consuming and thus expensive step in the design of analog and mixed–signal circuits. Automation of this process is currently an important research target in the electronic design automation community.

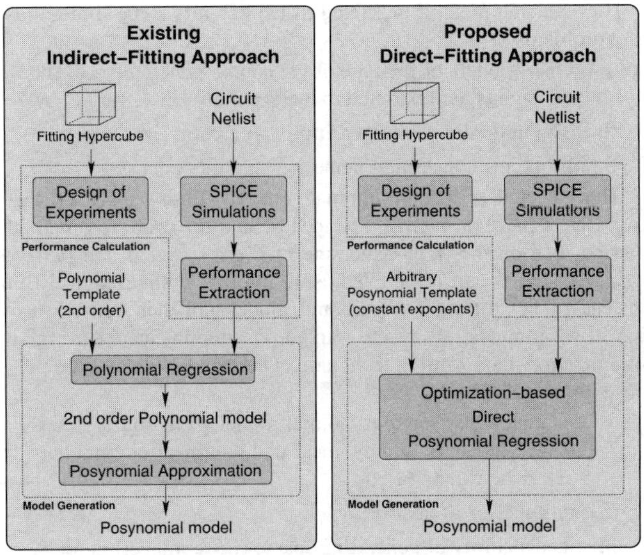

Figure 1: Comparison of the existing approach from [5] and the novel proposed direct–fitting approach

In the course of that research, it was demonstrated that the sizing of analog integrated circuits, like amplifiers, switched–capacitor filters, LC oscillators, etc., can be formulated as a geometric program [1, 2]. In that approach, the performance characteristics of the circuits are represented as symbolic equations written in posynomial format.

The advantages of a geometric program are multiple [3]:

1. The problem is convex and has only one global optimum.
2. The optimization is not dependent on the starting point.
3. Infeasible sets of constraints can be identified.
4. The geometric program's optimum can be found extremely efficiently, using interior point methods [4], even for relatively large problems.

The problem with the approaches in [1] and [2], however, is that the symbolic equations have to be derived manually, and that non-posynomial expressions have to be approximated manually into posynomial format. Both steps are time–consuming and can result in large inaccuracies, since e.g. also the device models have to be cast in posynomial format [1].

Recently, a method to automatically generate posynomial models for linear and nonlinear performance characteristics has been presented in [5]. The outline of the method has been sketched on the left–hand side of Fig. 1. The method first calculates the performance of a circuit for a set of samples that is composed using techniques from *design of experiments*. The data points can be generated using full–accuracy SPICE simulations. Then, a second–order polynomial template is fit to the sampled performance data. Finally, this model is approximated to render it posynomial. In this way it is possible to generate posynomial models of the form

$$p(X) = \sum_{k=-1,0,1} \; \sum_{\substack{l=-1,0,1 \\ \neg((k=-1)\wedge(l=-1))}} \left(\sum_{i=1}^{n} \sum_{j=1}^{n} \left(c_{i,k,j,l} x_i^k x_j^l \right) \right) \quad (1)$$

with n the number of design parameters and all $c_{i,k,j,l}$ positive constants.

However the method presented in [5] exhibits some limitations and problems:

1. Only models of the type of (1) can be generated, i.e. the exponents are restricted to integers between -2 and 2.
2. The fit quality for larger fitting hypercube sizes is poor.
3. Problems concerning numerical stability occur.

This paper describes an alternative direct–fitting approach that overcomes these problems. The idea behind the proposed method can be seen on the right–hand side of Fig. 1. Again, the performance of the circuit is sampled using numerical simulations. But instead of first fitting a polynomial model and then approximating it, we immediately fit the wanted posynomial model using an optimization–based fitting technique. The strong points of the proposed method are:

1. its generality: any posynomial model with arbitrary constant exponents for any linear and nonlinear circuit characteristic can be fitted;
2. its superior fit quality;
3. its capability of generating sparse (i.e. compact) models.

The paper is organized as follows. Section 2 will provide some theoretical background concerning *posynomial performance modeling*. In section 3, we will concisely indicate the principles of the indirect–fitting approach of [5], followed by a detailed description of our novel direct–fitting approach. To show the greatly improved performance, the experimental results obtained with our method will be compared in section 4 to the ones obtained in [5]. Finally, section 5 draws some conclusions and provides some ideas for future research.

2. POSYNOMIAL PERFORMANCE MODELING

2.1 Performance modeling

Consider a system S transforming an input signal E into an output signal Y (Fig. 2). The system S is governed by a set of design parameters X that influence its behavior.

$$Y = S(E, X) \quad (2)$$

The mathematical modeling of this input–output relationship is called *behavioral modeling*. In this paper, we are not directly interested in this relationship as such. However, the combination (Y, E) allows us to determine a particular performance p_i of the system, subject to excitation Y. Gathering a set of particular performances p_i into a vector gives us an idea of the total performance of the system.

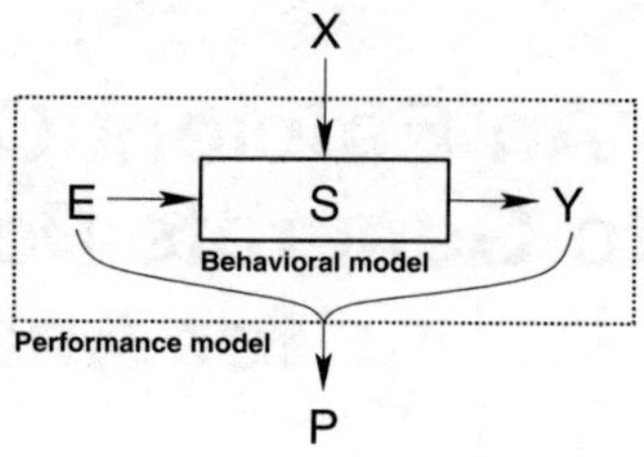

Figure 2: Electronic system S parametrized by X with excitation E and response Y leading to a performance P

Therefore, seen from a designer's point of view, the key relationship is:

$$P = F(X) \quad (3)$$

The mathematical modeling of this relationship is called *performance modeling*.

2.2 Posynomials & Geometric Programming

Let $X = (x_1, x_2, x_3, \ldots, x_n)^T$ be a vector of real, positive variables. An expression f is called *signomial* if it has the form

$$f(X) = \sum_{i=1}^{m} \left(c_i \prod_{j=1}^{n} \left(x_j^{\alpha_{ij}} \right) \right) \quad (4)$$

with $c_i \in \mathcal{R}$ and $\alpha_{ij} \in \mathcal{R}$. If we restrict all c_i to be positive ($c_i \in \mathcal{R}^+$), then the expression f is called *posynomial*. A posynomial consisting of a single term is called a *monomial*. The signomial fitting problem now can be defined as:

Signomial fitting problem:

Given a set of performance data samples

$$\{(X_1, p_{i,1}), (X_2, p_{i,2}), \ldots, (X_a, p_{i,a})\}$$

and a model template of type (4) with given exponents $\alpha_{ij} \in \mathcal{R}$,

determine the coefficients c_i, with $c_i \in \mathcal{R}$, such that

$$\left\| [f(X_1), f(X_2), \ldots, f(X_a)]^T - [p_{i,1}, p_{i,2}, \ldots, p_{i,a}]^T \right\|_2 \quad (5)$$

is minimal. ■

We use an Euclidean norm in (5) for three reasons:

1. it renders (5) smooth;
2. it causes a good centered spread of the sampling data around the model; and
3. it allows to solve the signomial modeling problem as the least–squares solution of an overdetermined system, which in turn reduces to solving a set of linear equations.

The posynomial fitting problem then reduces to:

Posynomial fitting problem:

Solve the signomial fitting problem with the extra constraint

$$c_i \geq 0, \quad \forall i \in [1:m] \quad (6)$$

■

Whereas the signomial form has better fitting properties, the posynomial form allows to formulate analog circuit sizing as a geometric program [1, 2].

A (primal) geometric program is the constrained optimization problem:

$$\begin{array}{lrcll} & \text{minimize } f_0(X) & & & \\ \text{with the constraints:} & f_i(X) & \leq & 1, & i = 1, \ldots, p \\ & g_j(X) & = & 1, & j = 1, \ldots, q \\ & x_k & \geq & 0, & k = 1, \ldots, n \end{array} \tag{7}$$

with all $f_i(X)$ posynomial and all $g_j(X)$ monomial. By substituting all variables x_i by $z_k = log\ (x_k)$ and taking the logarithm of the objective function f_0 and every constraint f_i, g_j, it can readily be seen that the transformed problem is a convex optimization problem. Because of this, it has only one global optimum. In addition, this optimum can be found very efficiently even for large problems, using interior point methods [4].

In view of the canonical geometric programming formulation, all performance constraints must be modeled in the normal form of (7). Therefore, we will scale the performance variables p_i in one of the two following ways:

- Linear scaling:

$$p_{i,scaled} = 1 \pm \frac{1}{W}\left(p_i - p_{i,spec}\right) \tag{8}$$

  with W an arbitrary weight factor.

- Logarithmic scaling:

$$p_{i,scaled} = 1 \pm \frac{1}{W}\log_{10}\left(\frac{p_i}{p_{i,spec}}\right) \tag{9}$$

  with W an arbitrary weight factor.

The scaling formulae (8) and (9) suggest a dependence of the models on the design specifications. This is only partly true. Indeed, the direction of the inequality may force us to fit $-p_i$ instead of the original performance parameter p_i.[1] The reason for this is obvious: changing the sign of a posynomial function does not preserve its posynomiality. However, the exact value of the specification (e.g. $GBW \geq$ 1GHz) is not an intrinsic part of the fitted model as the normalization does not harm the posynomiality of the models.

3. POSYNOMIAL MODEL GENERATION

In this section we will describe two techniques to fit posynomial models $p_i = f(X)$ (see (4)) with given constant exponents α_{ij} to a set of sample data $(X_k, p_{i,k})$:

1. the *indirect–fitting method*, as it is presented in [5], and
2. a *direct–fitting method*, the subject of this paper.

Both methods try to generate a posynomial response surface model [6] based on the generated performance samples. In section 4, the two methods will be compared using experimental results.

As the performance generation part of the two techniques is identical (see Fig. 1), we will briefly treat it in advance.

Fig. 3 illustrates how the set of performance samples is composed. The input samples X_k are generated using techniques from *Design of Experiments* in order to optimize the effectiveness of the sample set in the fitting process [7]. This input vector controls the values of (operating point) device currents and voltages. The *Operating Point Driven Solver* transforms these values into the corresponding transistor geometries and bias currents and voltages. Together with the analysis cards needed for the performance extraction, this results in a fully specified SPICE netlist. The consecutive

[1]This explains the $\pm$ sign in (8) and (9)

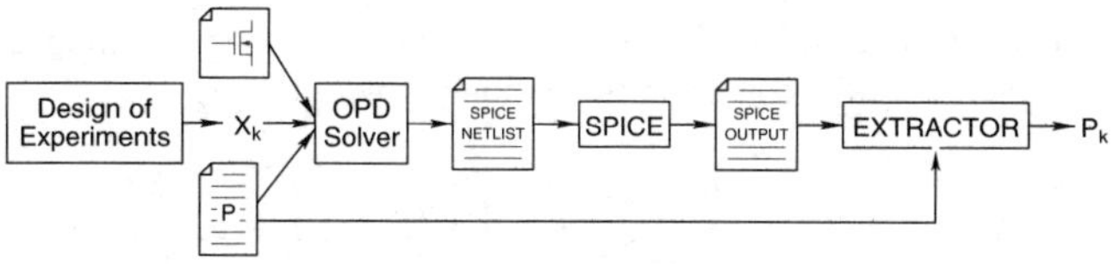

Figure 3: Performance generation outline

SPICE simulation results in an output file containing the simulation data. Afterwards, an automatically generated extractor script can distill the performance data. This concludes the part common to both methods. Note also that all these required simulations can be executed in parallel on a network of computers.

3.1 Indirect–fitting method

The indirect–fitting method is based on the fact that the signomial fitting problem reduces to solving an overdetermined set of linear equations in the least–squares sense when using an Euclidean norm in (5). The outline of the indirect fitting method is depicted in Fig. 4.

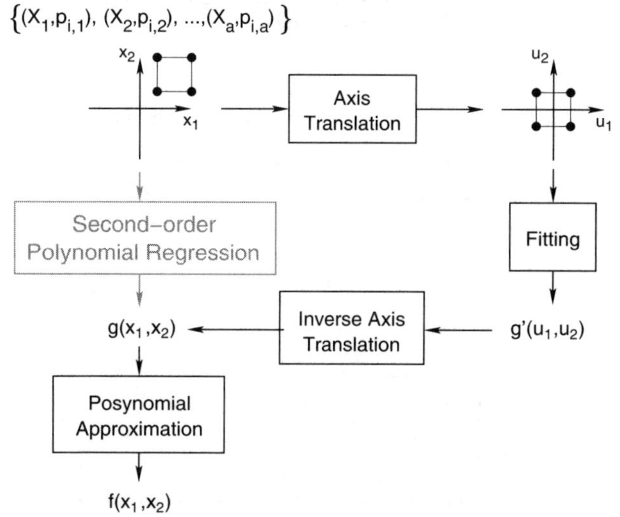

Figure 4: Outline of the indirect fitting method

The data set $\{(X_1, p_{i,1}), (X_2, p_{i,2}), \ldots, (X_a, p_{i,a})\}$ is first transformed into a data set that is located symmetrically around the origin of the U plane. Afterwards a second–order polynomial is fit to these data, such as to minimize the least squares error in the sampling points. Finally, the second–order polynomial is approximated by a posynomial form, i.e. the resulting model. The nature of the posynomial approximation step is to minimize the (nominal and first derivative) error in the center of the fitting hypercube. The details of this approximation can be found in [5].

Notice that the axis transformation was introduced to avoid the unstable solution of the least–squares fitting method that occurs in the case of asymmetrical data. Because of the axis translation, the indirect–fitting method can only generate models with integer exponents.

3.2 Direct–fitting method

The direct–fitting method solves the *posynomial fitting problem* directly. For this purpose, we developed a constrained minimization algorithm. The poysnomial fitting problem can be rewritten as a single–objective convex optimization problem:

$$\text{minimize } \psi(C) = \sum_{t=1}^{a}\left(\sum_{i=1}^{m}\left(c_i \prod_{j=1}^{n}\left(x_{j,t}^{\alpha_{ij}}\right)\right) - p_{i,t}\right)^2 \quad (10)$$

$$\text{subject to } c_i \geq 0, \quad \forall i \in [1:m] \quad (11)$$

with $x_{j,t}$ and $p_{i,t}$ respectively the value of x_j and the value of p_i at the tth experiment. The goal function (10) is a positive semi–definite second–order polynomial restricted to a convex constraint set. Therefore, the optimization problem is convex.

The entire fitting algorithm can be found in Fig. 5. This algorithm makes use of a boolean variable, *isposy*, a counter, *relcntr*, and a real variable, ψ_{prev}. It also assumes that c_1 is the coefficient of the model's constant term.

---

1. compose $\psi(C)$
2. set $\psi_{prev} = +\infty$
3. do
    - 3.1. do
        - 3.1.1. minimize $\psi(C)$ using a conjugate–gradient descent until C contains a negative component or until a local stop criterion has been fulfilled
        - 3.1.2. set *isposy* = *true*
        - 3.1.3. for $j = [2:m]$, do
            - 3.1.3.1. if ($c_j < 0$), then
                - 3.1.3.1.1. $c_j = 0$
                - 3.1.3.1.2. set *isposy* = *false*
        - 3.1.4. set *relcntr* = 0
        - 3.1.5. for $j = [2:m]$, do
            - 3.1.5.1. if $((c_j == 0) \wedge (\nabla_{C,j}(\Psi) < 0))$, then
                - 3.1.5.1.1. release c_j
                - 3.1.5.1.2. *relcntr* = *relcntr* + 1
            - 3.1.5.2. else
                - 3.1.5.2.1. fix c_j

      until *isposy*
    - 3.2. if (*relcntr* == 0), then
        - 3.2.1. stop
    - 3.3. if $((\psi(C) - \psi_{prev}) < \varepsilon)$, then
        - 3.3.1. stop
    - 3.4. else
        - 3.4.1. set $\psi_{prev} = \psi(C)$
        - 3.4.2. release c_j that offers the largest axis–wise descent

   forever

Figure 5: The direct–fitting algorithm

This algorithm has proven to be better than any other algorithm (interior–point algorithms with both external and internal barrier functions) we tested.

Besides the improved fitting capabilities (that will be demonstrated in section 4), this algorithm exhibits two clear advantages over the indirect approach:

- it is capable of fitting model coefficients of posynomial models with *arbitrary* real exponents, not only integers.
- the models it generates are sparse, i.e. a lot of the coefficients of the composing monomials are zero. This is advantageous for the needed CPU time when using these models in sizing applications afterwards.

3.3 Model quality assessment

In order to assess the fit–quality of the generated models, we use the quality–of–fit parameter q, defined in [5]. This allows a fair comparison between the indirect–fitting method and the direct–fitting method.

The starting point for this parameter is the root mean square of the deviation in the a sampling points. This parameter is then normalized by division with the performance range of the sampling points:

$$q = \frac{\sqrt{\sum_{k=1}^{a}\left(f(X_k) - p_k\right)}}{a\left[c + \left(\max_{k=1}^{a} p_k - \min_{k=1}^{a} p_k\right)\right]}. \quad (12)$$

In (12), c is a constant parameter to avoid error overestimation when the performance range approaches zero. If we reuse the sampling points used during the fitting process, then this figure is:

- computationally cheap (no extra simulations are needed), and
- easy to assess: a quality larger than 1 suggests a bad fit.

However, using verification points located within the fitting hypercube (not coinciding with the sampling points) yields a more reliable verification yardstick. In addition, the use of orthogonal arrays [8] (the sampling scheme that is used in [5]) locates the samples at the fitting hypercube's boundaries. This inherently disfavors the indirect–fitting approach.

We will therefore use the same three relative quality figures as proposed in [5]:

- q_{oc}: the relative model deviation in the center point,
- q_{tc}: the quality figure of (12) evaluated in sampling points located in the interior of the fitting hypercube, and
- q_{wc}: the quality figure of (12) evaluated in the original sampling points used for the model generation

The obvious drawback of using q_{tc} is the need for extra analyses (i.e. circuit simulations). Notice that for the generated posynomial models q_{wc} can be considered as the worst–case value (therefore the subscript *wc*), q_{tc} as the typical–case value and (especially for the indirect–fitting approach) q_{oc} as the optimal–case value.

4. EXPERIMENTAL RESULTS

As test case we use the same test circuit as used in [5], a high–speed CMOS OTA in a $0.7\mu m$ CMOS technology (see Fig. 6). The supply voltage is 5V. The nominal threshold voltages of this technology are $0.76V$ for NMOS–devices and $-0.75V$ for PMOS–devices. The circuit has to drive a load capacitance of 10pF.

Thirteen independent design variables can be chosen for the high–speed OTA of Fig. 6. The chosen design variables are gathered in Table 1. Note that currents and transistor drive voltages are chosen as variables, rather than transistor widths, since we use an operating–point driven formulation for analog circuit sizing [9]. The bounded range of variables $v_i \in [lb_i, ub]$ is logarithmically scaled onto $x_i \in [0, 1]$ using

$$x_i = \log\left(\frac{v_i}{lb}\right) \Big/ \log\left(\frac{ub}{lb}\right) \quad (13)$$

As a consequence all scaled variables are positive (as required for the posynomial formulation). The bounds have also been indicated in Table 1.

Our goal is to derive expressions for the low–frequency gain ($A_{v,LF}$), the unity–gain frequency (f_u), the phase margin (PM), the input–referred offset voltage (v_{offset}) and the positive and negative slew rate (SR_p, SR_n). In order to comply with the geometric programming formulation (which in its direct form only supports minimization and $\leq$ constraints), we will fit the inverse of the characteristics that need to be maximized or have a $\geq$ constraint (i.e. $-A_{v,LF}$, $-f_u$, $-PM$ and $-SR_p$). All characteristics are scaled lin-

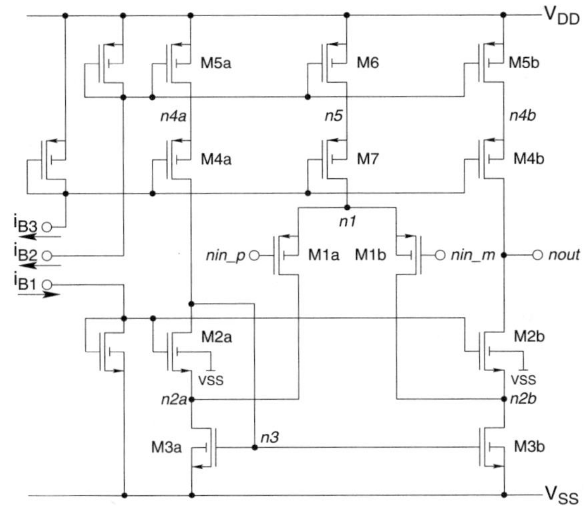

Figure 6: Schematic of a high–speed CMOS OTA

i	v_i	lb	ub	i	v_i	lb	ub
1	$v_{GS,M1}$	-0.75V	-4V	2	$v_{GS,M2}$	0.75V	4V
3	$v_{DS,M2}$	0.1V	4V	4	$v_{GS,M3}$	-0.75V	-4V
5	$v_{GS,M4}$	-0.75V	-4V	6	$v_{GS,M5}$	-0.75V	-4V
7	$v_{DS,M5}$	-0.1V	-4V	8	$v_{DS,M6}$	-0.1V	-4V
9	$i_{D,M1}$	-10uA	-10mA	10	$i_{D,M2}$	10uA	10mA
11	i_{B1}	1uA	100uA	12	i_{B2}	1uA	100uA
13	i_{B3}	1uA	100uA				

Table 1: Design variables chosen as model inputs

early according to (8), except for f_u which is scaled logarithmically according to (9).

For each of the characteristics to model, we will derive posynomial expressions using two different sampling hypercube widths ($dx = 0.1$, $dx = 0.01$).

The direct–fitting algorithm has been implemented in our posynomial model generation prototype called PRÎSM. Its TCP–based client-server simulation system schedules simulation and extraction jobs on remote workstations over the intra–net or the Internet (*resource parallelization*) and then fits the posynomial models on a single workstation. The core modules have been coded using C++, while the TCP–based client–server system has been coded in Perl. The total amounts to about 40 000 lines of code.

PRÎSM was run on an Intel Celeron 466MHz running GNU/Linux. The analysis servers ran on 16 UNIX workstations (attached to a standard 100Mbit/s TCP–IP network and under normal load circumstances), ranging from a SUN Ultra Sparc I (with a SPECfp95 of 9) to an HP B–1000 (with a SPECfp95 of 42) using their native OS. The simulations needed to obtain a full orthogonal hypercube of sampling points took approximately 3 minutes. Using these data the whole set of performance characteristics ($-A_{v,LF}$, $-f_u$, $-PM$, v_{offset}, $-SR_p$, SR_n) can be fitted. The time needed to fit each of the models using the algorithm of Fig. 5 is indicated in Table 2.

	generation time [s]		no. of coefficients		sparseness	
model	dx		dx		dx	
	0.1	0.01	0.1	0.01	0.1	0.01
$A_{v,LF}$	61	115	14	12	95.0%	95.8%
f_u	63	85	14	9	95.0%	96.9%
PM	125	130	22	20	92.3%	93.0%
v_{offset}	230	149	31	20	89.2%	93.0%
SR_p	116	116	32	12	88.9%	95.8%
SR_n	54	131	12	13	95.8%	95.5%

Table 2: Properties of the models generated using the direct–fitting technique

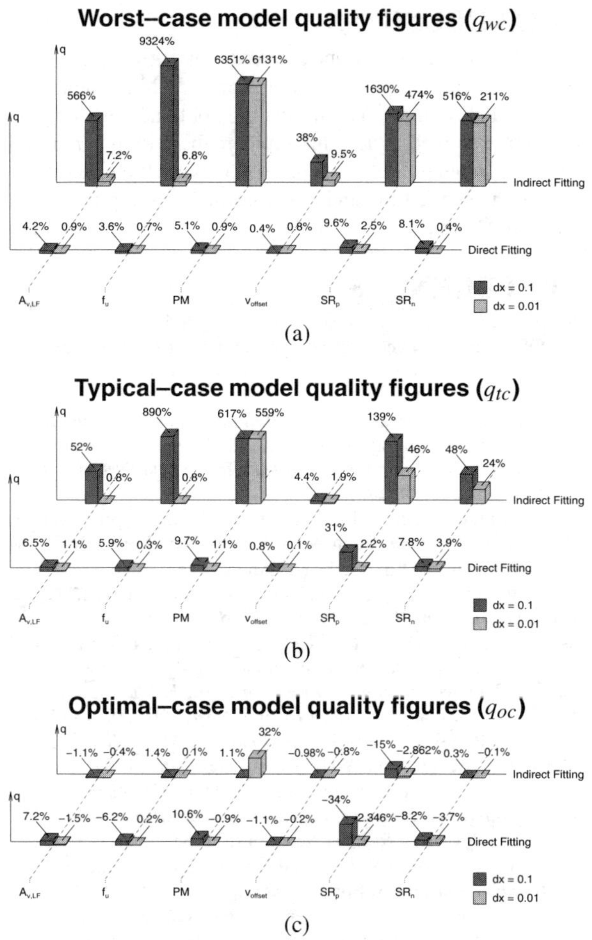

Figure 7: Model quality figures of the posynomial models: (a) worst case, (b) typical case, (c) optimal case

Table 2 demonstrates also clearly the sparseness (number of coefficients that are zero divided by the total number of template coefficients) of the models that are generated. Compare this to an average number of nonzero coefficients of 105 (corresponding to a model sparseness of 63.7%) for the models generated with the indirect–fitting technique. It can be concluded that concerning sparseness the indirect–fitting technique is outperformed by the new direct approach.

The quality of the generated models can be inspected in Fig. 7. The three subfigures, corresponding to the different quality measures q_{wc}, q_{tc} and q_{oc}, each contain two rows of data: the front row for the results of the direct–fitting technique and the back row for the results of the indirect–fitting technique. The improvement of the model qualities introduced by the direct–fitting technique is most considerable. On average an improvement with a factor of over 100 is observed, both for large as for small hypercubes.

5. CONCLUSIONS

In this paper, we presented a new direct–fitting method to fit posynomial model templates with arbitrary real exponents to numerical simulation data. The fit qualities obtained are very good, both for large as well as for small hypercube widths. In addition, the resulting models are very sparse. This method therefore effectively reduces the time and effort needed to setup and solve an analog circuit sizing problem in the form of a geometric program.

6. REFERENCES

[1] M. Hershenson, S. P. Boyd, and T. H. Lee, "Optimal design of a CMOS op–amp via geometric programming," *IEEE Transactions on Computer–Aided Design of Integrated Circuits and Systems*, vol. 20, no. 1, pp. 1–21, Jan. 2001.

[2] P. Mandal and V. Visvanathan, "CMOS op–amp sizing using a geometric programming formulation," *IEEE Transactions on Computer–Aided Design of Integrated Circuits and Systems*, vol. 20, no. 1, pp. 22–38, Jan. 2001.

[3] R. J. Duffin, C. Zener, and E. L. Peterson, *Geometric programming: theory and application*, John Wiley & Sons, New York, 1967.

[4] K. O. Kortanek, X. Xu, and Y. Ye, "An infeasible interior–point algorithm for solving primal and dual geometric programs," *Math. Programming*, vol. 76, pp. 155–181, 1996.

[5] W. Daems, G. Gielen, and W. Sansen, "Simulation–based automatic generation of signomial and posynomial performance models for analog integrated circuits," in *Proceedings IEEE/ACM International Conference on Computer Aided Design*, Nov. 2001, pp. 70–74.

[6] George E. P. Box and N. R. Draper, *Empirical Model Building and Response Surfaces*, John Wiley & Sons, New York, 1987.

[7] George E. P. Box, W. G. Hunter, and J. S. Hunter, *Statistics for experimenters: an introduction to design, analysis and model building*, John Wiley & Sons, New York, 1978.

[8] A. S. Hedayat, Sloane N. J. A., and J. Stufken, *Orthogonal arrays: theory and applications*, Springer–Verlag, New York, 1999.

[9] F. Leyn, G. Gielen, and W. Sansen, "An efficient dc root solving algoritm with guaranteed convergence for analog integrated cmos circuits," in *Proceedings IEEE/ACM International Conference on Computer Aided Design*, Nov. 1998, pp. 304–307.

Remembrance of Circuits Past: Macromodeling by Data Mining in Large Analog Design Spaces

Hongzhou Liu, Amit Singhee, Rob A. Rutenbar, L. Richard Carley
Dept. of ECE, Carnegie Mellon University
Pittsburgh, Pennsylvania, 15213 USA
{hongzhou, asinghee, rutenbar, carley}@ece.cmu.edu

ABSTRACT

The introduction of simulation-based analog synthesis tools creates a new challenge for analog modeling. These tools routinely visit 10^3 to 10^5 fully simulated circuit solution candidates. What might we do with all this circuit data? We show how to adapt recent ideas from large-scale data mining to build models that capture significant regions of this visited performance space, parameterized by variables manipulated by synthesis, trained by the data points visited during synthesis. Experimental results show that we can automatically build useful nonlinear regression models for large analog design spaces.

CATEGORIES AND SUBJECT DESCRIPTORS

B.7.2 [Integrated Circuits]: Design aids—verification

GENERAL TERMS: Algorithms

I. INTRODUCTION

The use of simplified macromodels for analog circuits, both to accelerate simulation and to enhance early design exploration, has a long history in mixed-signal design. The earliest techniques for macromodel construction relied on design expertise to create a suitably simplified circuit model, and the analytical equations needed to map the performance of the full circuit into parameters for the macromodel. More recent techniques mix design expertise about model structure with nonlinear regression (i.e., curve fitting) to fit macromodel parameters from samples of the full circuit's performance obtained from simulation. Two recent developments suggest a need to revisit this area: the recent standardization of analog/mixed-signal hardware description languages, and the recent introduction of commercial analog synthesis tools.

The introduction of standardized behavioral simulation languages for mixed-signal systems (*e.g.*, Verilog-AMS and VHDL-AMS [1],[2]) offers designers the ability to mix device-level models, analog behavioral models, and digital blocks, all in the same simulation environment. These languages are widely expected to be significant enablers for a more top-down mixed-signal design style. But to really exploit these AMS languages, we need a more rigorous approach to building macromodels for arbitrary analog blocks. For linear systems, the last decade has seen enormous progress in the construction of reduced-order models with mathematically guaranteed accuracy, [3]. For general nonlinear systems, we have no such mature techniques. Theoretical work on nonlinear reduced-order modeling is still quite new (e.g., [4]), and as yet limited in its applicability.

A different challenge is posed by the recent development of practical analog synthesis tools [5]. These tools take a fixed circuit topology and solve for the sizing/biasing parameters needed to meet a set of performance specifications. Today's synthesis techniques come in two broad styles. *Equation-based* synthesis techniques require that each circuit topology we seek to size be represented as a set of analytically tractable performance equations. The most successful of these use ideas from geometric programming [6]. The disadvantages of this approach are the need to build custom equations for each new topology (and the underlying device models as well), and the inability to capture all performance requirements in an accurate form. In contrast, *simulation-based synthesis* ([7]-[11]) uses simulator-in-the-loop numerical search techniques that fully simulate each visited solution candidate. They are capable of sizing any circuit that can be simulated, need no partial sizing starting solution, and reuse the same qualified verification environment used for manual design sign-off. Distributed workstation parallelism renders the overall approach tractable.

Simulation-based synthesis creates a wholly new opportunity for modeling: each synthesis run visits between 10^3 and 10^5 fully simulated samples of the design space for a given circuit topology. If we synthesize just one uniquely sized circuit per day for a year, we might easily visit a few million samples of a design space. *What can we do with all these circuit data points?*

Today's *ad hoc* model simplifications and low-dimensional curve fitting strategies seem ill-equipped to exploit this opportunity. Of course, building regression models from simulations is not new: response surface methods for design centering have been in widespread use for years [12], [13]. What is new here is the *scale* of the problem. In a typical centering scenario, we try to avoid simulating more than a few hundred different circuits to build a low-dimensional local response surface. In contrast, a single synthesis run may generate 10-100X more circuit samples, and range widely over the entire design space.

The question we address in this paper is the how to build models that can capture significant regions of this performance space, parameterized by variables manipulated by synthesis, trained by the data points visited during synthesis. We build such models by adapting recent ideas from large-scale *data mining* [14], which focuses on techniques for extracting patterns, predictive formulas, or classifiers from large amounts of high-dimensional data.

The remainder of the paper is organized as follows. Sec. II reviews relevant background work. Sec. III introduces our data mining formulation. Sec. IV presents a small illustrative example. Sec. V presents experimental results from real synthesis data. Finally, Sec. VI offers concluding remarks.

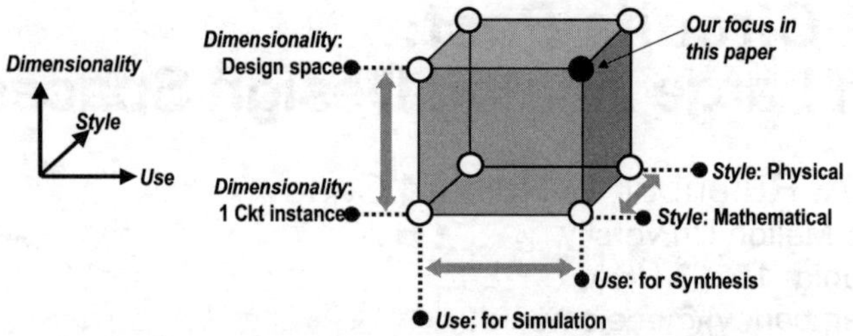

Fig. 1. A Simple Taxonomy of Analog Macromodels

II. Background: Macromodeling

The term "macromodel" has historically been used to refer to a variety of different analog modeling problems, with different associated model-building techniques, and intended end applications. Fig. 1 offers a simple taxonomy to categorize these different approaches, and clarify our own goals. We identify three separate axes in this taxonomy:

- *Use:* distinguishes between models built primarily to accelerate simulations, and those built to enable wider exploration of the design space in synthesis. Of course, these are not always mutually exclusive; the distinction usually appears on the other two axes.
- *Style*: distinguishes between models that are built explicitly as physical circuits, from those that are only intended to "fit" some of the original circuit's "true" responses, without any underlying physical model. Physical macromodels usually *are* circuits--albeit simpler circuits. Mathematical macromodels are usually obtained from some form of nonlinear regression, and offer no guarantees that all of their internal behaviors can be physically realized.
- *Dimensionality*: distinguishes between models that represent a single circuit (called an *instance*) from models intended to represent a parameterized range of circuits (called a *design space*). Instance models are built for simulation; design space models for optimization.

We can categorize relevant prior work using this simple taxonomy. Analog macromodeling initially focused on accelerating simulation. The earliest work built simplified physical macromodels for single circuit instances, e.g., the classical Boyle opamp model [15]. To handle more complex circuits and devices, later efforts turned to curve-fitting to obtain the parameters for these physical models from samples of the original instance's simulated performance [16]. Unfortunately, we often lack a good physical model to parameterize. Automated extraction of a reduced-order model is now easily done for linear systems [3]. This not the case for nonlinear systems, though recent efforts have appeared [4],[17]. The common strategy is to apply nonlinear regression techniques [18] to fit either a conjectured functional form, or use so-called *black box* modeling in which the regression itself chooses the functional form.

Synthesis-directed macromodeling usually focuses on design space models, since the goal is to explore or optimize over some space of circuit parameters. In the context of circuit tuning, response surface methods fall into this category [12], [13]. Similar ideas have been used for microwave synthesis [19]. Harjani [20] appears to be the first to explicitly build carefully constructed nonlinear regression models of analog design spaces, for use in hierarchical analog synthesis [21]. This work minimized the number of circuit simulations via careful design of experiments (DOE [22]), and proposed a regression formulation based on radial basis functions. More recently, Daems [23] developed a similar strategy for design space modeling, using a different DOE model, and regression via posynomials, motivated by [6].

In this paper, we focus on the problem of building models of analog design spaces, emphasizing synthesis rather than simulation applications. As a result, we are willing to trade some modeling accuracy for the ability to model a larger region of design space. Prior efforts here [19],[20],[23] relied on two critical assumptions: (1) simulations to obtain samples of the design space are expensive and must be minimized; and (2) model construction can control directly where to sample in this design space. In contrast, we start with many more fully simulated samples of the design space (e.g., 10,000 versus 100). However, we do not control where these samples are located, since our data is the by-product of circuit synthesis. This wealth of synthesis-produced data creates both opportunities and problems.

III. Macromodeling By Data Mining

In classical nonlinear regression, we have data points of the form (x, y), where $x \subseteq \Re^n$ and $y \in \Re$, which represent samples of some unknown, high-dimensional function $y = F(x)$. We seek to construct an approximation to this function $R(x)$, called a *regressor*, which minimizes some *error function* that penalizes the difference between the true function and our approximation (*e.g.*, sum of squared errors), measured over a population of sample data points $P = \{(x, y)_i\}$:

$$\sum_{(x,y) \in P} \text{error}[y, R(x)] \qquad (1)$$

To construct a suitable regressor we must solve three problems:

- *Model selection*: select an appropriate functional form for $R(x)$
- *Model fitting*: determine parameters for this model, so as to minimize overall fitting error (1), using the training data from *P*.
- *Model validation*: using a different set of testing data, verify how well the model fits data points outside of the training data points.

The essential problem is that for a large population of high-dimensional data points, it becomes extraordinarily difficult to find a suitable functional form for the regressor that can adequately fit the data, while remaining simple enough to allow us to solve for the fitting parameters. This is the principal motivation behind the radial basis functions in [20] and posynomial-approximated signomials in [23]; these are both forms with relatively tractable fitting procedures.

However, recent work in the data mining community suggests a different direction for coping with the problem of large, high-dimensional data sets. Rather than struggling to build a single regressor capable of fitting well across a very large sample space, we build instead a *committee of regressors*, each of which fits very well in some portions of the design space, and less well in others [14]. We predict the output value for any new data point by *combining* the individual regressors, in a procedure called *voting*. The mechanics of committee construction and voting are attractively simple. To build the committee of regressors, we use a powerful, relatively new statistical technique called *boosting* [14],[24]; to vote the regressors, we adapt ideas from *instance-based methods* [25]. We describe these ideas next.

A. Building a Committee of Regressors via Boosting.

The original idea was first developed by Freund and Schapire [24] and applied to so-called *categorical* classification problems, in which the model to be fit has a discrete two-element output range, e.g., $\{0, 1\}$. The idea is illustrated in Fig. 2. We begin with some data set S with N weighted training samples. Each sample $(x, y)_i$ has weight w_i, all samples equally weighted to start. We *sequentially* fit a set of regressors $R_t(x)$, each more heavily weighting those data points that were poorly fit in the previous iteration. Thus, after each iteration, we up-weight those points that were poorly fit, and down-weight points that were well fit. This has the interesting side effect of allowing well-fit samples to *vanish* from the population as boosting progresses: these points are so well fit by prior regressors that we no longer want to focus fitting efforts on them. The number of iterations t varies by application, but it is not uncommon to run hundreds of boosting cycles [14]. At the end, we have a committee of regressors, which we combine via weighted voting to construct the final output. The boosting method is easy to implement, and independent of the form of the individual regressors, $R_t(x)$. With a carefully constructed weighting update, boosting also comes with some surprisingly good theoretical bounds on its performance [24].

For building macromodels for large analog design spaces, a boosted committee of regressors is immediately attractive. However, boosting has to date been used primarily for *classification*--fitting regressors that map real-valued inputs to a small set of discrete category labels. The assumption of a discrete output set is critical to the construction of theoretically strong voting methods for boosting [24]. On the other hand, our interest is to apply boosting to quantitative nonlinear regression. Prior work here is more sparse, with Drucker [26] perhaps the most notable.

We use the boosting iteration from [26], illustrated in Fig. 3. However, we suggest a different mechanism for voting the individual regressors, discussed in the following subsection.

As a practical matter, note that the choice of form for individual regressors $R_t(x)$ will determine whether one can weight the training data for suitable effect. If not, the alternative is to *resample* the data. With suitable normalization, we can interpret weight w_i on each point as a probability. Given the data set S_t used to build regressor $R_t(x)$, construct S_{t+1} via sampling from S_t with w_i as the probability of selection for any point. Well-fit points, having small probabilities, may not be sampled. Poorly fit points, with large probabilities, may be *duplicated* in the population. When we cannot upweight an individual data point, we can replicate it in the population to increase its importance during fitting. To enhance this effect, we can choose a population size for each S_t larger than $N=|S|$, the size of the original data set.

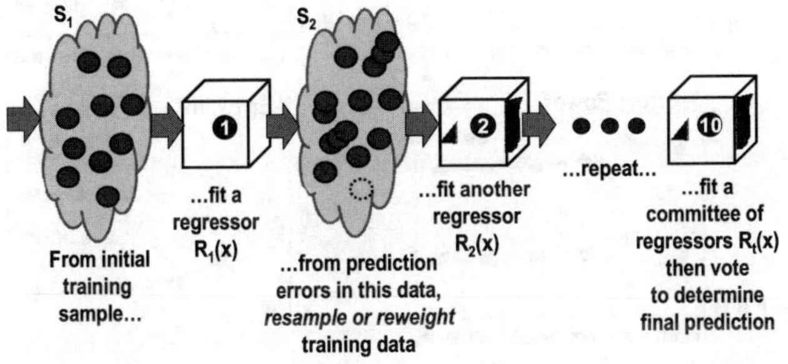

Fig. 2. Boosting for Regression

1. Given data set S with N training points of the form $(x, y)_i$, with x=<$x_1, x_2, ..., x_n$>. Set weight $w_i=1/N$ for each point.

2. Set t=0. Set $S_0 = S$.

3. Set $t=t+1$. Construct set S_t by sampling AN points (A>1)from S_{t-1} with the probability that we select point $(\mathbf{x}, y)_i = w_i$.

4. Fit a new nonlinear regressor $R_t(x)$ to data in S_t

5. Calculate (normalized) training error for each point $(x, y)_i$ as

$$L_i = \text{error}(y_i, R_t(x_i)) = \frac{|y_i - R_t(x_i)|}{\max_{(x, y)_j \in S_t} |y_j - R_t(x_j)|}$$

6. Calculate average weighted training error for entire data set S_t

$$\text{as } L = \sum_{(x, y)_j \in S_t} w_j L_j$$

7. Calculate *confidence parameter* for $R_t(x)$ as $\beta = L/(1-L)$

   Note: *smaller* β implies *higher* confidence in the fit for $R_t(x)$

8. Update weight for each data point $(x, y)_i$ as $w_i \Leftarrow w_i \beta^{(1-L_i)}$

9. If we want to boost another regressor, go to step **3**.

Fig. 3. Algorithm for Sequential Boosting Iteration [26]

B. Voting a Committee of Regressors via an Instance-Based Method

Prior efforts in both classification and regression have assumed that *all* the regressors in a boosted committee should be combined to create the final voted prediction. This fits well with categorical data, where we literally "outvote" the poor predictors. We argue that this is inappropriate when we know that our input data is likely to be sparse and of high dimension, and for circuits, is sampling very nonlinear behaviors. We want an alternative that selects the "best" set of regressors to combine.

Another technique from data mining can be applied here. In classical regression, we use our training data to build a suitable nonlinear regressor, but then we only use the analytical form of the regressor to make future predictions--the training data itself is abandoned. Like the scaffolding used to construct a building, we expect the scaffolding to be removed when the building is completed. This need not be the case: we might instead use the training data itself as an integral part of the final regressor. These are called *instance-based* methods; our strategy was influenced by the instance-based locally weighted polynomial regression ideas from Moore [25].

Fig. 4 shows our instance-based heuristic for combining regressors. For any new data point x', we look up the K nearest neighbors of x' in the training data, using ordinary Euclidean point-to-point distance. With each of these training samples $(x, y)_j$ we associate the index $t(j)$ of the best boosted regressor from among our committee of boosted regressors $R_t(x)$. Thus, the K nearest neighbor data points

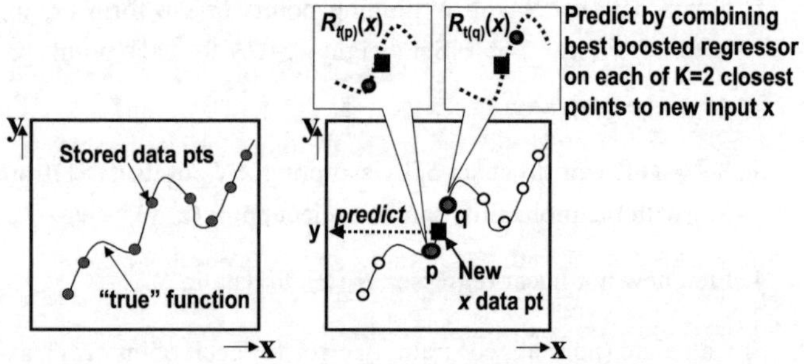

Fig. 4. Instance-Based Voting with K=2 Nearest Neighbors for 2-D Nonlinear Function

also specify (at most) K different regressors that we know fit *well* in this local region. This has two very useful advantages:

- *Efficient scaling to large data sets:* Although the size of the training data set itself may be quite large, we only need to use a few data points for any given prediction. Nearest-neighbor lookup in high dimensions is a well studied area, and there are several data structures that make look-up efficient, e.g., [27]. In our work, we use simple linear scans through all the data points.
- *Many regressors are fit, few are evaluated*: we are free to fit a large number of boosted regressors if this is necessary to "cover" our design space. But for any given prediction, we only need the regressors on the nearest K neighbors to compute our new data point. Of course, we also hope--but cannot guarantee--that we need fewer than K regressors in many well-fit regions of our design space.

These K nearest-neighbor points $(\boldsymbol{x}, y)_{n1}$, $(\boldsymbol{x}, y)_{n2}$, ..., $(\boldsymbol{x}, y)_{nK}$, specify K regressors $R_{t(n1)}(\boldsymbol{x})$ through $R_{t(nK)}(\boldsymbol{x})$. How might we combine these into a single numerical prediction? Simple averaging proves to be a workable scheme. However, when the data is of high dimension and sparse, even a few "nearest" neighbors may not be all that near. As well as an appropriate, empirically chosen value for K, it is also useful to weight each local regressor inversely with its distance from the point to be predicted. If the set of distances to the K nearest neighbors is $\{d_{n1}, d_{n2}, ..., d_{nK}\}$, then we vote our K local regressors as

$$y_{predicted} = \sum_{i=1,K}\left[\frac{(d_{ni})^{-1}}{D}\right]R_{t(ni)}(\boldsymbol{x}') \qquad (2)$$

where $D = \sum_{i=1,K}(d_{ni})^{-1}$. Note that if each neighbor has a unique best local regressor, then this simply favors the regressors on the closer points. But in the opposite case, where each neighbor chooses the same regressor, this weighting does not distort the prediction.

C. Fitting Each Local Regressor

The final component of our regression strategy is the choice of the individual regressors $R_t(\boldsymbol{x})$ for each boosting iteration. It is again worth emphasizing that both boosting and our instance-based voting heuristic can be applied to any choice for the regressor.

We use a standard feed-forward neural network for each regressor [28]. Neural nets have a reputation for successful fitting for highly nonlinear functions, and are reasonably close to "black box" models in that the structural choices needed to select a neural network model

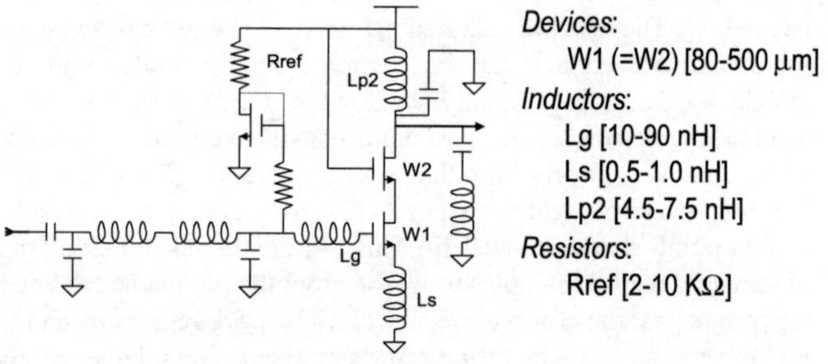

Fig. 5. Simple Example Circuit: RF LNA

are relative few. They have also been successfully used in other circuits-oriented modeling and optimization applications [19]. We use two hidden layers (10 sigmoid neurons per layer) and one output layer (1 linear unit) as the network architecture. Our implementation is based on the Matlab Neural Network toolbox [30].

IV. A Simple Macromodeling Example

We first illustrate our macromodeling formulation with a simple RF LNA circuit shown in Fig. 5. The circuit has 5 independent sizing parameters, variable over the ranges given in the figure The circuit is realized in 0.35μm TSMC CMOS.

Since the problem is rather small, we simply generate uniform random samples over the 5-dimensional hypercube defined by the input ranges. We generated 2000 total samples of the performance of this circuit (each of which required several SPICE runs), and used a randomly selected half to train our regressors, and reserved the other half to test (validate) our regressors. As does [23], we omit circuits whose random sizing renders them improperly biased (*e.g.*, transistors turned off). Fig. 6 shows the results using 10 boosting cycles to vote up to 10 boosted neural net regressors. We show both mean and maximum errors. (Note that in contrast to [23], this is a rather harsher set of metrics: there are *always* some very poorly predicted points for any regressor, and we evaluate on testing data *not* seen during model construction.) The data are interpreted as follows:

- **train1:** the leftmost column on each plot shows the error from fitting one single neural net regressor to the training data itself. We ex-

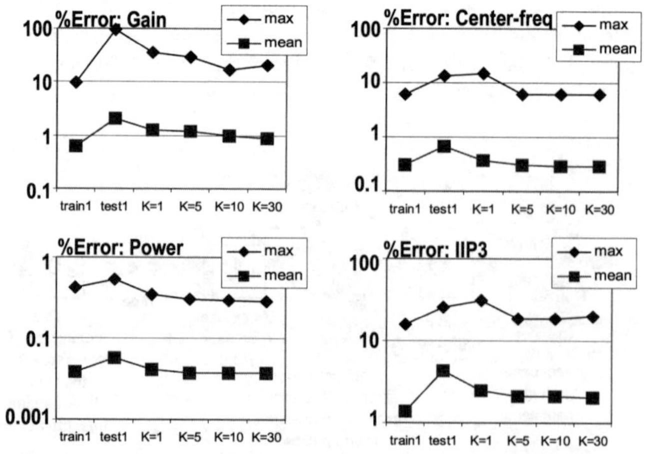

Fig. 6. Regression Results for Simple RF LNA Circuit

pect this to provide the best fit, since we evaluate only training data.

- **test1**: the second plot column to the right show the error from using only this single neural net regressor to predict the points in the testing data set. We expect this to be inferior, since this is data *unseen* during fitting. These are the points against which we need to compare and validate our new fitting techniques.
- **K=1,...K=30:** the rightmost 4 plot columns show the errors from predicting outputs over the testing data, using instance-based voting of the *K* nearest-neighbors from the training data, each labeled with the best from among 10 boosted neural net regressors. We evaluate K as 1, 5, 10, and 30.

The results are encouraging. For gain, center-frequency, power, and IIP3, boosting improves both the maximum and mean errors over the use of a single conventional neural net predictor. Usually, *K*=1 or *K*=5 nearest neighbors are sufficient: we select the local 1 or 5 best regressors and combine them, via inverse-distance weighting. Setting *K* too large usually has the effect of increasing the error: we are overfitting, trying to use "too much" model for the data. These extra neighbors are too far away to be of use in improving the local prediction accucuracy.

The log error scale of Fig. 6 obscures the impact that boosting has on the actual distribution of the errors; we illustrate this more directly in Fig. 7 by showing histograms of the prediction error at each testing point for the IIP3 fitting experiment of Fig. 6. Note how even *K*=1 (select the best boosted regressor on the *single* training point nearest the testing point) improves the average error--though not the worst-case error in this case. *K*=5 trims the worst errors from the right-side tail of the distribution. Subsequent boosting with *K*=10 has minimal effect, and boosting with *K*=30 neighbors starts to degrade the fitting, as shown in the increased spiking in the distribution.

V. Macromodeling a Synthesis Design Space

We turn now to a much larger experiment. Fig. 8 shows a circuit automatically sized in a commercial simulation-based analog synthesis tool [29]. The schematic has roughly 50 devices, and has been sized to achieve roughly 450MHz with power under 15mW. We have approximately 37,000 samples of the design space for this circuit, as a result of synthesis. Each data point has 27 independent design variables, and 12 circuit performance outputs. In contrast to prior fitting experiments of which we are aware, this is a much larger, much higher-dimensional case study.

It is also different for a reason fundamental to synthesis: synthesis tools are designed to *converge* to a good solution [7]-[11]. No particular invocation of synthesis need necessarily explore any region of the space beyond that needed to find a good solution. To emphasize this, we fit two portions of the this data set: the first 20% of the data points (time-ordered by synthesis) and the last 20%. This creates two populations of roughly 7500 samples of the design space.

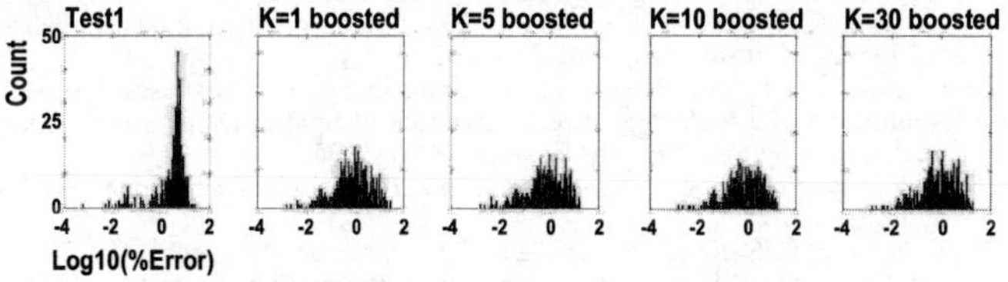

Fig. 7. Error Histograms for IIP3 Data from Fig. 6

One salient difference from the experiments of the last section is that these two data sets are very different in character. Samples from the first 20% of the run range more widely over the design space (as expected, since search ranges widely at the start). Samples from the last 20% of the run cluster more closely about the final solution. As a consequence, we see much larger worst-case prediction errors for the first 20% because the of existence of more "outlier" points.

We fit using the same techniques as in the previous section. We randomly select 3/4 of each data set to use for training, and reserve the remaining 1/4 for validation of our regression fit. We boost 10 cycles of regressors. Fitting the next neural network in each boosting cycle requires roughly 1 hour of CPU time, in our current Matlab implementation. We evaluate using instance-based nearest-neighbor voting for *K*=1,5,10,30. Results appear in Fig. 9. Let us briefly consider each experiment:

- **Gain**: boosting at *K*=1 or *K*=5 clearly helps for the first data set, but interestingly, does little to help the last data set. Mean error declines very slightly (but the fit is quite good in the first place), while maximum error increases.
- **Power**: boosting is more clearly helpful on the first data set. On the the later data, boosting helps the mean slightly, but at the cost of an increased maximum error. Bias/variance trade-offs are a common by-product of regression. A single regressor probably suffices.
- **Slew rate:** this is clearly a very challenging fit. Boosting improves the mean, but at the cost of some deterioration in maximum error.
- **Unity Gain Frequency (UGF):** as with slew, the trade-off is an improved mean, but a worsened maximum error.
- **Phase Margin:** also a very difficult fit like the slew, but in this case, boosting shows a clear advantage in both data sets.

This analysis should simply remind us of the need to regard these data mining techniques with an eye toward proper model selection and validation. A single trained regressor sometimes fits well, a committee of boosted regressors is sometimes "too much" fitting. On the other hand, a boosted committee is sometimes able to achieve significant fitting improvements. We regard this as a very satisfactory set of initial results for this very difficult new problem.

VI. Conclusions

Simulation-based synthesis tools routinely visit many fully simulated circuit solution candidates. We showed how to adapt ideas from large-scale data mining to build models that capture significant regions of these large, nonlinear high-dimensional spaces. Results are encouraging: there are spaces where our boosted regressors are clearly superior. However, as with all regression problems, no one technique fits best in all circumstances; in some cases a single regressor suffices, in others, our boosting techniques offer useful trade-offs between mean and worst-case error. Our regression architecture is a first attempt to apply these ideas to analog design space modeling; other constructions using these techniques also appear promising.

Acknowledgment: This work was funded by the Semiconductor Research Corp., the National Science Foundation under contract 9901164, the DARPA NeoCAD program managed by the Sensors Directorate of the Air Force Laboratory, USAF, Wright-Patterson AFB, and Texas Instruments. We thank Rodney Phelps and Akshat Shah for valuable discussions about this work.

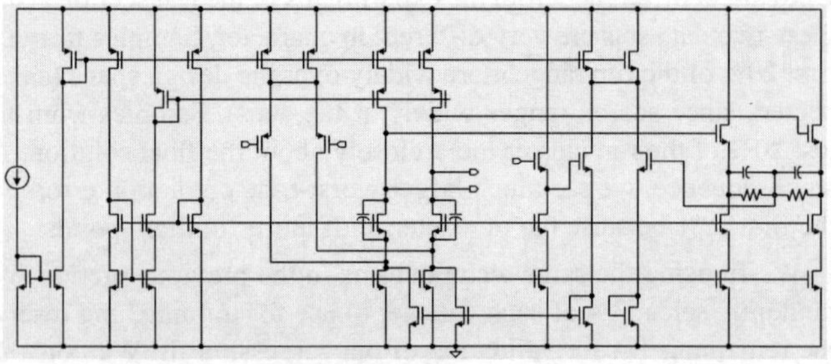

Fig. 8. Amplifier Circuit for Synthesis Design Space Macromodel Experiment.

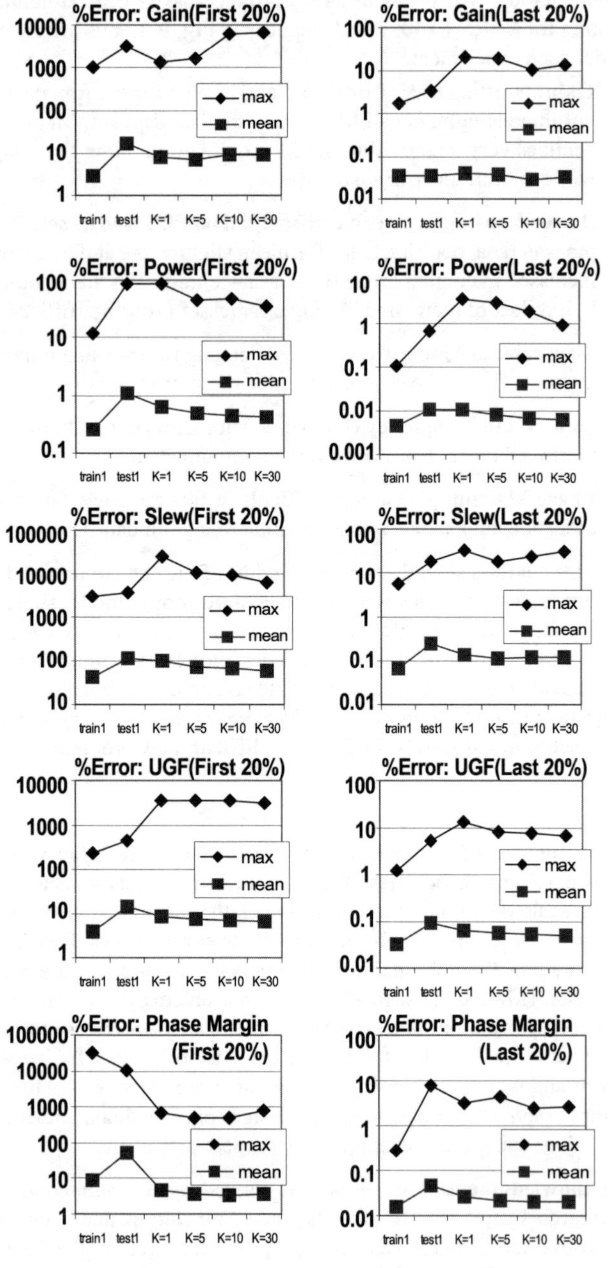

Fig. 9. Regression Results for Design Space of Circuit from Fig. 8.

REFERENCES

[1] P. Frey, D. O'Riordan, "Verilog-AMS: Mixed-Signal Simulation and Cross Domain Connection Modules," *Proc IEEE Behavioral Modeling and Simulation Conference (BMAS)*, 2000.

[2] E. Christen, K. Bakalar, "VHDL-AMS--A Hardware Description Language for Analog and Mixed-Signal Applications", *IEEE Trans. Circuits and Sys II: Analog and Digital Sig. Proc.*", vol. 46, no. 10, Oct. 1999

[3] A. Odabasioglu, M. Celik, and L. T. Pileggi, ``PRIMA: Passive Reduced-Order Interconnect Macromodeling Algorithm," *IEEE Trans. CAD*, vol. 17, no. 8, pp. 645-654, 1998.

[4] Joel Phillips, "Projection Frameworks for Model Reduction of Weakly Nonlinear Systems," *Proc. ACM/IEEE DAC*, June 2000.

[5] G.G.E. Gielen and R.A. Rutenbar, "Computer-Aided Design of Analog and Mixed-Signal Integrated Circuits, *Proc IEEE*, vol. 88, no. 12, Dec. 2000.

[6] M. Hershenson, S. Boyd, T. Lee, "GPCAD: a Tool for CMOS Op-Amp Synthesis", *Proc. ACM/IEEE ICCAD*, pp. 296-303, 1998

[7] M. Krasnicki, R. Phelps, R.A. Rutenbar, L.R. Carley, "MAELSTROM: Efficient Simulation-Based Synthesis for Analog Cells," *Proc. ACM/IEEE DAC*, June 1999.

[8] R. Phelps, M. Krasnicki, R. A. Rutenbar, L. R. Carley, J. R. Hellums, "Anaconda: Simulation-Based Synthesis of Analog Circuits via Stochastic Pattern Search," *IEEE Trans. CAD*, vol. 19, no. 6, June 2000.

[9] R Phelps, M. Krasnicki, R.A. Rutenbar, L.R. Carley, J. Hellums, "A case Study of Synthesis for Industrial-Scale Analog IP: Redesign of the Equalizer/ Filter Frontend for an ADSL CODEC," *Proc. ACM/IEEE DAC*, June 2000.

[10] M.J., Krasnicki, R. Phelps, J.R. Hellums, R.A. Rutenbar, L.R. Carley, "ASF: A Practical Simulation-Based Methodology for the Synthesis of Custom Analog Circuits," *Proc. ACM.IEEE ICCAD*, Nov. 2001.

[11] R. Schwenker, J. Eckmueller, H. Graeb, K. Antriech, "Automating the Sizing of Analog CMOS Circuits by Consideration of Structural Constraints,", *Proc DATE99*, March 1999.

[12] A. R. Alvarez, B. L. Abdi, D. L. Young, H. D. Weed, J. Teplik, and E. R. Herald, "Application of Statistical Design and Response Surface Methods to Computer-Aided VLSI Device Design", *IEEE Trans. CAD*, vol. 7, no. 2, pp. 272-288, Feb. 1988.

[13] K.K. Low and S.W. Director, "A New Methodology for the Design Centering of the IC Fabrication Process," *IEEE Trans. CAD*, vol. 10, no. 7, pp. 895-903, July 1991.

[14] E. Bauer, R. Kohavi, "An Empirical Comparison of Voting Classification Algorithms: Bagging, Boosting, and Variants". *Machine Learning* Vol 36, Nos. 1/2, pp. 105-139, July/August 1999.

[15] G.R.Boyle, B.M.Cohn, D.O.Pederson, and J.E.Solomon, "Macromodeling of Integrated Circuit Operational Amplifier," *IEEE J.Solid-State Circuits*, vol. SC-9, pp. 353-363, Dec.1974.

[16] Y-C Ju, V.B. Rao and R. Saleh, "Consistency Checking and Optimization of Macromodels", *IEEE Transactions on CAD*, vol. 10, no. 8, Aug. 1991.

[17] J. Phillips, "Automated Extraction of Nonlinear Circuit Macromodels," *Proc. IEEE CICC*, May 2000.

[18] T. Hastie and R. Tibshirani, *Modern Regression and Classification Short Course*, Washington DC, June 2000.

[19] M. H. Bakr, J. W. Bandler, K. Madsen, and J. Sondergaard, "Review of the Space Mapping Approach to Engineering Optimization and Modeling," *Optimization and Engineering*, vol. 2, 2001.

[20] R.Harjani and J. Shao, "Feasibility and Performance Region Modeling of Analog and Digital Circuits", *Analog Integrated Circuits and Signal Processing*, Kluwer Academic Publishers, 1996.

[21] R. Harjani, R.A. Rutenbar and L.R. Carley, "OASYS: a framework for analog circuit synthesis," *IEEE Trans. CAD*, vol. 8, no. 12, Dec. 1989.

[22] G. Box, W. Hunter and J. Hunter, *Statistics for Experimenters: and Introduction to Design Data Analysis and Model Building*, John Wiley, 1978.

[23] W. Daems, G.G.E. Gielen, W. Sansen, Gielen, "Simulation-Based Automatic Generation of Signomial and Posynomial Performance Models for Analog Integrated Circuits," *Proc. ACM/IEEE ICCAD*, Nov. 2001.

[24] Y. Freund and R.E. Schapire, "A Decision-Theoretic Generalization of On-Line Learning and an Application to Boosting," *Journal of Computer and System Sciences*, vol. 55, no. 1, 1997.

[25] A. W. Moore, J. Schneider, and K. Deng, "Efficient Locally Weighted Polynomial Regression Predictions", *Proc. 14th Int'l Conf on Machine Learning*, Morgan Kaufmann, pp 236-244, 1997.

[26] H. Drucker, "Improving Regressors Using Boosting Techniques," *Proc. 14th International Conference on Machine Learning*, ed. Douglas H. Fisher, Jr., Morgan -Kaufmann, pp. 107-215, 1997.

[27] Hjaltason, G.R. and Samet H., "Ranking in Spatial Databases," in M.J. Egenhofer and J.R.Herring, Eds., Advances in Spatial Databases - Fourth International Symposium, SSD'95, pp. 83-95, 1995.

[28] B. Ripley, *Pattern Recognition and Neural Networks*, Cambridge University Press, 1996.

[29] E. Hennig, R. Sommer, L. Charlack, "An Automated Approach for Sizing Complex Analog Circuits in a Simulation-Based Flow," *Proc. DATE02*, March 2002.

[30] Mathworks, Inc. http://www.mathworks.com.

Optimal Design of Delta-Sigma ADCs by Design Space Exploration

Ovidiu Bajdechi
Delft University of Technology
Mekelweg 4
2600GA Delft, The Netherlands
O.Bajdechi@ITS.TUDelft.NL

Georges Gielen
Katholieke Universiteit Leuven
Kasteelpark Arenberg 10
B-3001 Leuven, Belgium

Johan H. Huijsing
Delft University of Technology
Mekelweg 4
2600GA Delft, The Netherlands

ABSTRACT

An algorithm for architecture-level exploration of $\Delta\Sigma$ ADC design space is presented. The algorithm finds an optimal solution by exhaustively exploring both single-loop and cascaded architectures, with single-bit or multi-bit quantizer, for a range of oversampling ratios. A fast filter-level step evaluates the performance of all loop-filter topologies and passes the accepted solutions to the architecture-level optimization step which maps the filters on feasible architectures and evaluates their performance. The power consumption of each accepted architecture is estimated and the best top-ten solutions in terms of the ratio of peak SNDR versus power consumption are further optimized for yield. Experimental results for two different design targets are presented. They show that previously published solutions are among the best architectures for a given target but that better solutions can be designed.

Categories and Subject Descriptors

J.6 [**Computer Applications**]: Computer-Aided Engineering

General Terms

Design

Keywords

ADC, CAD, delta-sigma

1. INTRODUCTION

Among the many architectures of analog-to-digital converters (ADC), $\Delta\Sigma$ designs are used in a large class of applications ranging from low-frequency [1] and audio [2] to down-converted intermediate frequency [3] and digital video [4]. Their property to trade speed for accuracy makes them more attractive in the context of present CMOS technology evolution [5]. The spread of $\Delta\Sigma$ designs and the absence of an accurate analytical model for their non-linear behavior caused a rapid evolution of dedicated simulation software. There are a number of simulators readily available, some of them as free software toolboxes [6] and some developed in universities [7], [8]. However, even using fast, dedicated simulators, it is all but impossible for a designer to explore the entire range topology and design parameters that can yield the optimal solution for a target dynamic range (DR) or peak signal-to-noise+distortion ratio (SNDR).

This paper presents a global optimization approach which outputs a list of the best $\Delta\Sigma$ ADC architectures in terms of peak SNDR versus power consumption ratio. The optimality is guaranteed by thorough search of the entire design space, as opposed to other design automation software which only allow a user-specified architecture to be designed [8] or a set of user-designed architectures to be browsed in search for a set optimum [7], which compared to the entire design space is a local optimum.

The search is conducted in two steps. In a first step, the filter-level design, the loop filter(s) are analyzed using the linear model approximation [9] for all possible combinations of loop order, number of cascaded loops, number of bits in the quantizer, oversampling ratio and peak gain of the noise transfer function (NTF). The DR of each solution is evaluated [6] and qualifying solutions are delivered to the next step.

The second step, the architecture-level design, maps the filter-level qualified solutions on all possible system architectures. Each architecture is designed and its performance in terms of integral non-linearity (INL) [9], DR intercept, and conversion gain is analyzed from time-domain simulations. Accepted solutions are then subjected to power consumption estimation, based on the assumption of switched-capacitor (SC) differential circuit design using single-stage operational amplifiers [10], [2]. A reduced set of solutions is selected based on the peak SNDR versus power consumption ratio. Only this reduced set of solutions is subjected to Monte-Carlo analysis to optimize their yield performance with respect to process-induced mismatch of architecture coefficients.

The paper is organized as follows. Section 2 contains a description of the filter-level design step. Section 3 presents the architecture-design step with details on the performance tests and architecture-level power estimation. Experimental

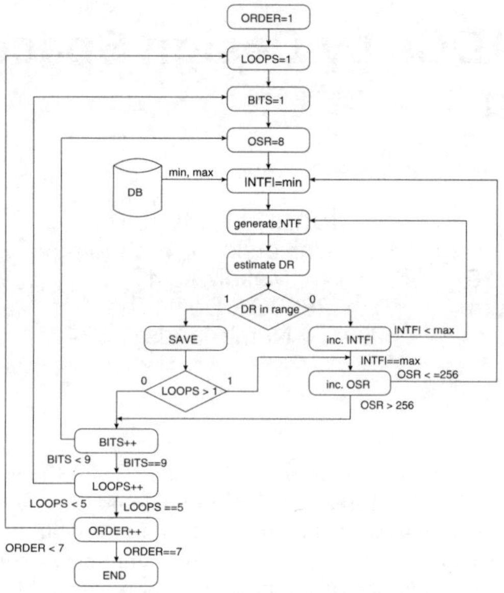

Figure 1: Design space exploration algorithm for filter level (linear model)

results are shown in Section 4 for two different targets, an audio and an xDSL ΔΣ ADC. The conclusions are presented in Section 5.

2. FILTER-LEVEL DESIGN

2.1 Exploration Algorithm

A fast design space exploration is performed first, using a linearized model of the ΔΣ ADC. The search algorithm is shown in Fig. 1. A separation is made between the *topology* design space defined by the *ORDER*, number of *LOOPS* and number of *BITS*, and the *parameter* design space defined by the oversampling ratio *OSR* and peak *NTF* magnitude. All these dimensions of the search space are browsed using constant stepping, linearly in *topology* space and exponentially in *parameter* space.

At each step a set of two filter transfer functions are generated, the NTF and the signal transfer function (STF). As a provision for direct sensor interfacing, it is possible to set the STF gain to values different from unity. Filter generation can also optimize in-band zeros for the NTF (to reduce total in-band noise power [6]) and off-band zeros for the STF (to reduce in-band gain ripple and increase off-band rejection). After the two filter transfer functions are generated the in-band noise power is calculated to estimate the DR. A test is applied with two thresholds derived from the target DR (in dB)

$$DR_{target} + DR_{lower} \leq DR \leq DR_{target} + DR_{upper} \quad (1)$$

The lower limit tests if, after the circuit (white) noise is added, the converter still reaches the target DR. The upper limit is set to reject power-hungry solutions which offer more quantization-noise DR than actually needed. If the test is passed, the solution is saved in a database.

For the single-loop solutions the algorithm stops when

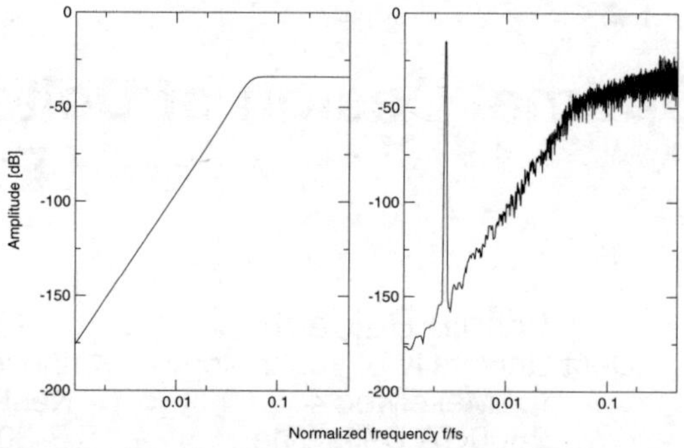

Figure 2: Estimated (*left*) and time-domain simulated (*right*) quantization noise spectrum for ORDER=4, BITS=1, LOOPS=1

a combination of minimal *OSR* and *NTF* is found. For cascaded designs, however, because significant architectural details are not available at the filter level, the algorithm continues searching for solutions even after finding the first valid one. This insures that no valid solution is prematurely rejected.

2.2 Dynamic Range Evaluation

A fast yet accurate method to estimate the DR is used. It overcomes the drawbacks of both time-domain simulations, which are slow, and classical DR estimation using a formula which is inaccurate especially for high-order ΔΣ loops:

$$DR = \frac{3}{2}\frac{2L+1}{\pi^{2L}} OSR^{2L+1}(2^B - 1)^2 \quad (2)$$

First the magnitude of NTF is calculated in n points equally distributed from DC to $f_s/2$

$$|NTF(Z)|_i = \left| \frac{\Sigma_{j=1}^{L} d_j Z^{-j}}{\Sigma_{k=1}^{L} n_k Z^{-k}} \right|, i = 1 \cdots n \quad (3)$$

using

$$Z^{-k}|_i = cos(k\pi i/n) - Jsin(k\pi i/n) \quad (4)$$

The total power of quantization noise is assumed to be [9]

$$P_{qn} = \frac{\Delta^2}{12} \quad (5)$$

which yields a (white) quantization noise amplitude in each bin i from DC to $f_s/2$, referred to $V_{ref} = \Delta(2^{B-1})/2$, as

$$S_{qn,Bbit} = 20log_{10}\left(\sqrt{\frac{2}{3n}}\right) - 20(B-1)log_{10}(2) \quad (6)$$

if the quantizer has B bits.

The value of the NTF magnitude, given by Eq. (3) expressed in dB, is added to the value of $S_{qn,Bbit}$ and a curve showing the quantization noise amplitude in each bin is drawn. Its integral in the band of interest yields the estimated DR value. Fig. 2 shows the estimated (*left*) and time-domain simulated (*right*) spectrum of the output of a fourth-order, one-bit, single-loop ΔΣ ADC. The two graphs show the good accuracy of this estimation method for high-order loops.

The DR of the fourth-order, single-loop, single-bit ΔΣ ADC operated at 64 times OSR is estimated to be 102dB,

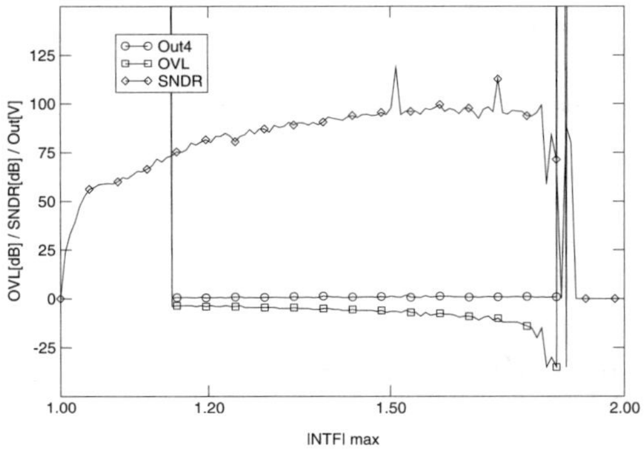

Figure 3: Peak NTF magnitude database entry example (ORDER=4, BITS=1, $V_{clip}/V_{Ref} = 1$)

which is only 3dB different from the time-domain simulated result of 105dB. As a comparison, the formula in Eq. (2) yields 134dB, due to the idealized NTF expression used to derive it [9]. The evaluation is fast, finding all 200-plus possible solutions which can yield 100dB DR in approximately 10 minutes on a 1GHz computer.

2.3 Peak NTF Magnitude Database

The one parameter of $\Delta\Sigma$ ADCs which cannot be predicted accurately by linear modeling is the overloading level (OVL) [9]. In the algorithm presented here, the overloading levels for each single-loop $\Delta\Sigma$ ADC in the entire range of loop orders and quantizer bits are computed once from time-domain simulations and stored in a database. The peak magnitude of NTF is varied in a range from 1.0 (0dB) to 16.0 (24dB). Each single-loop $\Delta\Sigma$ ADC is designed and optimized in a range of input signal values for each NTF peak magnitude. The overloading level is detected as the input signal which causes integrator clipping.

Fig. 3 shows the entry in the peak NTF database for the single-bit, fourth-order $\Delta\Sigma$ ADC. It shows the variation of overloading level *OVL*, peak output of the last integrator *Out4* and the peak *SNDR* in the NTF range where the loop can be stabilized.

Because there is a tight relationship between the *OVL* value and the ratio of the clipping voltage over reference voltage, the database has to be generated for each different ratio of V_{clip}/V_{Ref}. This takes about 24 hours on a 1GHz computer but the same database can be used for any target DR (SNDR) design as long as the V_{clip}/V_{Ref} ratio does not change.

3. ARCHITECTURE-LEVEL DESIGN

3.1 Exploration Algorithm

The architecture-level exploration algorithm evaluates the performance of the filter-level solutions mapped on a specific architecture, as shown in Fig. 4. If a single-loop solution is processed, an architecture is generated (the feedforward and feedback connectivity can be specified by the user) and its coefficients are calculated. A wide range of input signal amplitudes is then used to detect the overloading level. The input signal applied at this stage is a pulse (a busy signal [9]) with a fundamental frequency three times lower than the signal bandwidth.

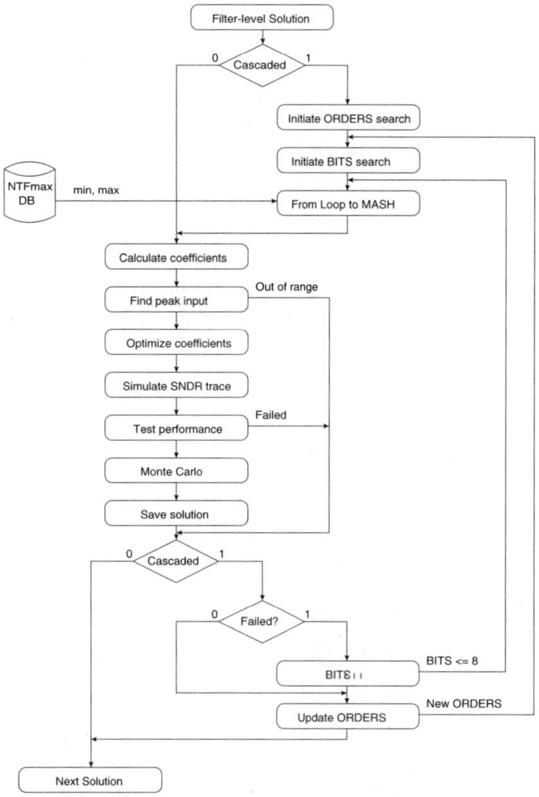

Figure 4: Design space exploration algorithm for architecture level

The next step is the coefficient optimization, which is performed at the OVL previously detected. The SNDR and DR variations as a function of the input signal are then simulated. The two curves are tested for performance and passing solutions are subject to a Monte-Carlo (MC) analysis to test the architecture's tolerance to coefficient mismatch. The solution is then saved for further processing.

For the cascaded $\Delta\Sigma$ ADCs the only architectural details predefined at filter level are the number of loops and the number of bits in the last loop. Therefore, the filter orders of individual cascaded loops are generated as one additional design space dimension named *ORDERS* in Fig. 4. Another new design space dimension is the number of *BITS* in the first loop of the cascade. For simplicity, the last $n-1$ loops in a cascade of n loops have the same number of bits as the last one. Each derivative of the input solution in the extended design space is analyzed as an independent solution.

3.2 Performance Test

Performance testing is based on statistics of SNDR and DR curves as functions of input signal level. Linear regressions are performed on each curve, from SNDR zero-crossing to the overloading level. The slope of SNDR is tested to be within 10% of the desired conversion gain (typically unity). A slope outside this range shows a strong dependency of quantization noise power of the input signal level, which is not desired. The mean of the DR curve is then tested against the target DR value to insure the target DR is attained. Finally, the peak regression residual of the DR curve is tested to be lower than 6dB (1 bit) to insure the required integral non-linearity (INL). Finally, the overloading level is tested to be larger than -15dB.

$SNDR(V_{in})$ and $DR(V_{in})$ curves for an architecture rejected by the performance test algorithm are shown in Fig. 5. The dotted lines are the linear regression fitted values for both simulated curves. The drop in DR at high input levels (larger than -20dB) shows that the peak NTF value needed to reach the target DR is too high so premature clipping occurs [6]. The peak SNDR which still keeps a good overall INL is about 85dB instead of almost 95dB, as shown by its absolute peak value. But the detection of this effect requires a set of linear regressions with each point in the curve used as higher limit, which would increase the computation time tenfold. Instead, because the decrease in peak SNDR already disqualifies this solution, the simple yet effective criterion of peak regression residual limiting is used as rejection reason. The slope of DR fitted line shows that the test for SNDR slope also works towards rejecting the solution, even if the curves would pass the DR mean value test.

3.3 Power Estimation

The power of each solution accepted by the performance tests explained in the previous sub-section is evaluated considering a switched-capacitor design approach. Fully differential circuits are considered, with independent paths for input and DAC signal integration, as shown in Fig. 6. The charge in the sampling capacitor C_s and in the feedback capacitor C_{fb} are both integrated on the integration capacitor C_i during half or the clock cycle (integration phase).

The power consumption of such an integrator can be expressed as a function of the amplifier's input stage transconductance g_m, which is designed considering the settling requirements for specific linearity performance [2], [10]. The capacitive load driven by g_m is derived from noise performance of each integrator. Considering a one-stage amplifier, the total noise power at the input of the integrator is

$$P_n = 4KT\left(\frac{1}{C_s}\left(1+\frac{C_{fb}}{C_s}\right)+\frac{1}{3\pi C_{load}}\right) \quad (7)$$

with

$$C_{load} = C_s + C_{fb} + C_l + \frac{C_l(C_s+C_{fb})}{C_i} \quad (8)$$

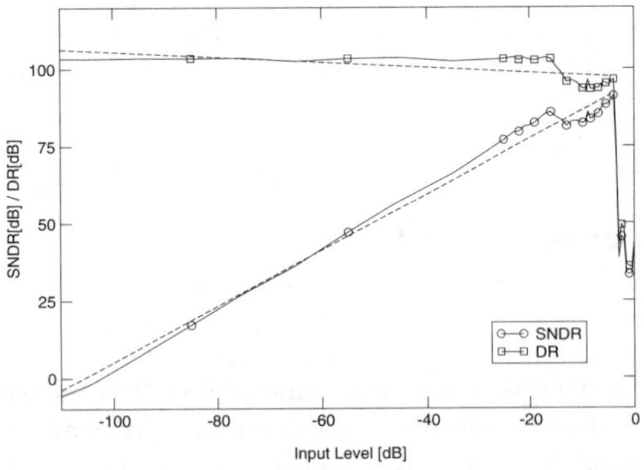

Figure 5: Example of SNDR and DR curves for a rejected architecture (ORDER = 3, OSR = 32, BITS = 2, LOOPS = 3 [1,1,1], 2 bits in the first loop)

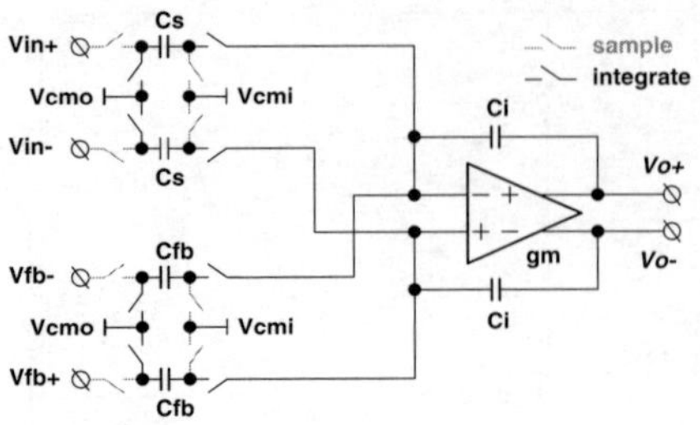

Figure 6: Schematic of a typical switched-capacitor integrator

where C_l is a fraction of the integration capacitor (parasitic capacitance of C_i) connected at the opamp's output. The first part of the noise power in Eq. (7) is the noise of the switch on-resistance and the second part is the noise of the g_m in the opamp. This power is referred at the input of the $\Delta\Sigma$ converter as

$$P_i = \left(P_n \frac{1}{\Pi_{j=1}^{i-1}\left(\frac{C_{sj}}{C_{ij}}\right)} \frac{\pi^{2(i-1)}}{(2i-1)OSR^{2i-1}}\right)_i \quad (9)$$

with i the order of the integrator in the $\Delta\Sigma$ converter.

All the capacitors are calculated from the required noise power of each integrator. The g_m is then calculated from the required settling performance, considering a slewing followed by settling model [2]

$$g_m = \frac{C_{load}\,(N_\tau - 1)}{\frac{T_{CK}}{2} - V_{fb}\frac{C_{load}}{I_0}} \quad (10)$$

The feedback voltage at the input of the opamp after the charge re-distribution phase [2], V_{fb}, can be calculated in the worst-case as

$$V_{fb} = V_{out,max}\left(1+\frac{C_l}{C_i}\right)\frac{C_s+C_{fb}}{C_{load}} \quad (11)$$

The number of needed time constants, N_τ in Eq. (10), is given by the settling required to reach B bits of linearity [10] as $N_\tau = Bln(2)$. The required linearity of an integrator is determined by allocating equal distortion power to each integrator and considering the loop gains to input-refer individual distortion powers, with the sum of the distortion powers set to 3dB (0.5 bit) below the target DR.

Considering MOS transistors operated in weak inversion are used in the input stage of the opamp, $g_m = 10I_0$ and a compact expression for g_m is obtained

$$g_m = \frac{2}{T_{CK}}C_{load}\left((N_\tau - 1) + 10V_{out,max}\left(1+\frac{C_l}{C_i}\right)\right) \quad (12)$$

The noise power of each integrator is allocated based on exploration of noise power distribution across the converter. A part of the noise power of the previous integrator is allocated to the next in the loop

$$(P_n)_i = R\,(P_n)_{i-1} \quad (13)$$

and the value of R is chosen to minimize total power consumption. Fig. 7 illustrates the dependence of power consumption of noise power allocation through the ratio R. The top curve is the power (expressed as the supply current) consumed for the total g_m for a fifth-order, 3-loop, 2-2-1 architecture, with 5 bits in the first loop and 3 bits in the

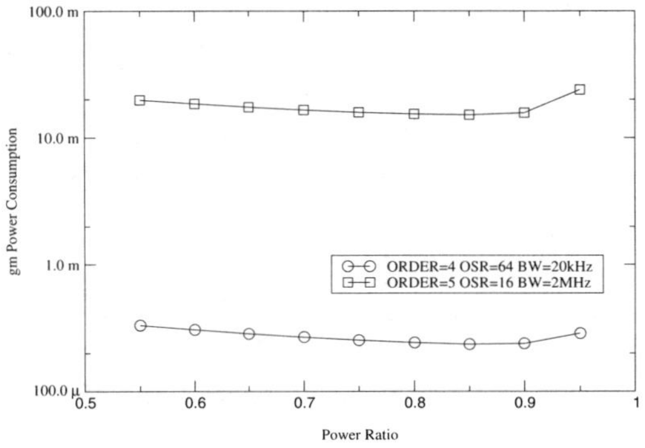

Figure 7: Power consumption for g_m as function of noise power ratio R

other two loops. It is operated at 16 times OSR for a signal bandwidth BW of 2MHz, as in [4]. It is worth noting the 25% reduction in current consumption by optimization of R and the fact that the minimum current, of 14mA, is well related to the reported consumption of 36mA if folded cascode amplifiers are used [2]. The second curve in Fig. 7 is the current consumption for a fourth-order, one-bit single-loop $\Delta\Sigma$ ADC for audio applications (BW=20kHz). The supply current estimated in the best case also matches the design reported in [2].

3.4 Yield-Based Optimization

The accepted solutions are passed to the MC analysis which varies the $\Delta\Sigma$ ADC coefficients using a user-supplied distribution. The user can specify, for each integrator, a spread 3σ value along with the coefficient-to-coefficient mismatch 3σ value. A required minimal capacitor is specified for the latter 3σ value. A few hundreds of Monte-Carlo simulation steps are run for the top-ten solutions and the performance tests explained in Subsection 3.3 are applied. The single-loop solutions can be designed from the early stages to give 100% yield with relaxed matching requirements. However, for cascaded solutions which do not attain an yield larger than 90%, the number of bits in the first loop in the cascade is increased and another Monte-Carlo yield analysis iteration is started.

4. EXPERIMENTAL RESULTS

Two examples are presented to show the effectiveness of global optimization through exhaustive design-space exploration. The first one is an audio $\Delta\Sigma$ ADC biased at 1.5V with rail-to-rail input and 1.5V reference voltage. A designed circuit has been reported [2] which consumes 0.95mW (only the analog part, without voltage reference buffers) for a DR of 98dB and peak SNR of 89dB at a signal bandwidth of 20kHz. The second one is an ADC for xDSL applications, biased at 2.5V, also with rail-to-rail input and reference voltage equal to the supply voltage. The signal bandwidth is 2MHz. A design has been reported which consumes 90mW in the analog circuits to attain 95dB DR and 90dB peak SNR.

The designs mentioned above are state-of-the art examples. The results presented here show that other architectures can offer better peak SNDR versus supply power consumption ratios (FOM) but these designs are still among the best options. The yield optimizations have been conducted assuming a 1% (fair capacitor/capacitor matching) coefficient-to-coefficient mismatch for single-loop architectures and 0.5% (good capacitor/capacitor matching) for cascaded architectures.

Table 1: Global solutions for Audio $\Delta\Sigma$ ADC

ORDER	LOOPS	BITS	OSR	SNDR	FOM
4	2 (2-2)	2 (6)	32	97.5	119.8
4	3 (2-1-1)	2 (5)	32	97.3	119.5
5	3 (2-1-2)	1 (7)	32	98.5	120.2
5	2 (2-3)	4 (7)	32	98.5	119.8
5	2 (2-3)	5 (7)	32	98.1	119.5

Table 2: Audio ADCs with LOOPS=1

ORDER	BITS	OSR	SNDR	FOM
3	6	16	95.3	117.6
3	8	16	96.3	117.8
3	4	32	95.5	118.4
4	8	8	95.6	116.4
4	5	16	96.0	117.8
5	4	16	95.4	116.5
5	6	16	96.2	115.9

4.1 Audio Delta-Sigma ADC

The search for an optimal audio $\Delta\Sigma$ has been first performed in the entire design space, to find the global optimum. In order to compare with previously reported state-of-the-art solution, restricted sets in terms on number of loops and number of bits have been analyzed. The global optimization results are shown in Table 1. It is worth noting the massive presence of cascaded solutions. The column *LOOPS* shows each cascaded loop's order in parentheses, while the column *BITS* contains the number of bits in the first loop in parentheses. The solutions have virtually the same figure-of-merit *FOM*. All solutions have OSR=32 and a large number of bits in the first loop, which increases the FOM value by increasing the overloading level with no power costs.

Table 2 contains the optimization results for a set of solutions restricted by the number of loops (LOOPS=1). The best solution is the third order, 4-bit loop, again working at OSR=32. This solution is also remarkable by its low number of bits compared to the other top performers. This shows that, for audio frequencies, high-OSR is still a good option.

Further design space restriction to 1-bit single-loop architectures yields only the state-of-the-art fourth-order, single-loop solution with OSR=64 reported in [2]. The 1-bit DAC is often needed for its inherent linearity but it requires a larger OSR. The solution is chosen from a set of four possible, three of which do not pass the yield test.

4.2 Delta-Sigma ADC for xDSL Applications

The results of global optimization for a 4MS/s $\Delta\Sigma$ ADC are shown in Table 4.

Again a third-order solution with OSR=32 and 4-bits quantizer has good performance, but most of the solutions operate at 16 times oversampling. They also have large number of bits in the (first loop) quantizer to attain high

Table 3: **Global solutions for xDSL $\Delta\Sigma$ ADC**

ORDER	LOOPS	BITS	OSR	SNDR	FOM
3	1	4	32	96.5	102.7
4	2 (2-2)	4 (8)	16	97.4	103.1
4	3 (2-1-1)	4 (7)	16	98.1	102.7
5	2 (2-3)	3 (8)	16	96.6	102.1
5	2 (2-3)	4 (6)	32	97.1	101.7
6	3 (2-2-2)	2 (8)	16	97.4	102.6

overloading levels. The FOM values are 20 dB smaller than the ones obtained with the audio ADCs because of increased sampling frequency.

To avoid solutions like the ones requiring 32 times OSR, the designer can conduct the search in a limited space, for example for (ORDER=5, OSR=16). Furthermore, the number of bits can be limited, for example to 6, to keep a low DAC complexity. The solutions for this search are shown in Table 4. They are all cascaded $\Delta\Sigma$ ADCs except two which, even with their low-FOM, can be good choices for low-voltage, mismatch-tolerant designs.

The best are the three-loops with a 2-2-1 configuration. The state-of-the-art solution reported so far [4] is among them. During initial optimization stages, only 3 bits were needed in the first loop, but yield optimization reached the 5-bit solution reported in [4], the increase being needed to accommodate capacitor/capacitor mismatch effects.

An overview of all fifth-order solutions operated at 16 times OSR is given in Fig. 8. The total biasing current g_m required by each of the 21 solutions is depicted with a different symbol for different number of loops. For the same number of cascaded loops there are multiple solutions with different combinations of orders for the loops in the cascade. A 3-bit solution needs the lowest power and generally the 2-loops solutions are the best performers. There are eleven 2-loops solutions, nine 3-loops and one 4-loops.

5. CONCLUSIONS

An exhaustive design-space exploration algorithm for global optimization of $\Delta\Sigma$ ADC design has been presented. The algorithm examines all possible solutions in two steps to increase overall speed. The large total number of possible solutions is only explored by the fast filter-level design step

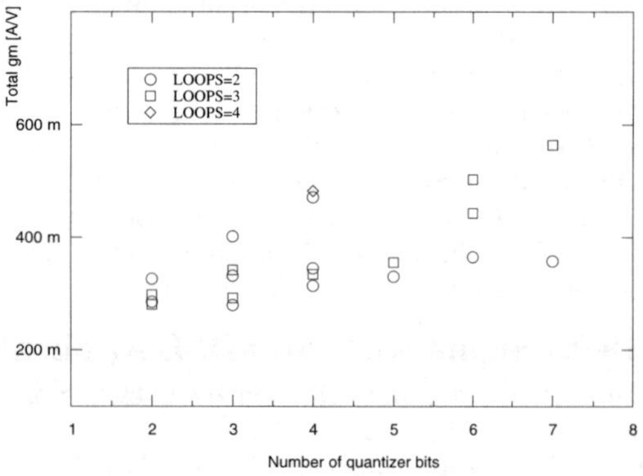

Figure 8: Total g_m for ORDER=5, OSR=16 solutions

Table 4: xDSL ADCs with ORDER=5, OSR=16

LOOPS	BITS	SNDR	FOM
2 (4-1)	2 (5)	93.1	98.0
3 (2-2-1)	2 (6)	97.1	102.6
3 (2-2-1)	3 (5)	95.7	101.0
3 (2-2-1)	4 (6)	97.8	102.5
1	5	96.6	100.4
3 (2-2-1)	5 (5)	98.1	102.6
1	6	96.5	99.6
3 (2-2-1)	6 (6)	98.2	101.7

which rejects the ones which can not attain the required dynamic range. A reduced set of solutions is forwarded to the architecture-level design step which evaluates their behavior by time-domain simulations. The power of each solution is estimated and only ten solutions are subjected to the time-consuming yield analysis. The generation of the entire set of possible solutions takes approximately 24 hours on a 1GHz computer. Yield analysis can be run on sub-sets of solutions to find only architectures with required properties. Experimental results show that previously published state-of-the-art solutions are among the best designs but not necessarily the best.

6. REFERENCES

[1] Y. Perelman, R. Ginosar, *A low-light-level sensor for medical diagnostic applications*, IEEE JSSC, Vol. 36, No. 10, Oct. 2001

[2] A.L. Coban, P.E. Allen, *A 1.5V 1.0 mW Audio $\Delta\Sigma$ Modulator with 98dB Dynamic Range*, ISSCC 1999 Digest of Technical Papers.

[3] L. Breems, E.J. van der Swan, J.H. Huijsing, *A 1.8-mW CMOS $\Sigma\Delta$ Modulator with Integrated Mixer for A/D Conversion of IF Signals*, IEEE Journal of Solid-State Circuits, Vol. 35, Apr. 2000.

[4] K. Vleugels, S. Rabii, B.A. Wooley, *A 2.5V Broadband Multi-Bit $\Delta\Sigma$ Modulator with 95dB Dynamic Range*, ISSCC 2001 Digest of Technical Papers, Feb. 2001.

[5] K. Bult, *Analog Design in Deep Submicron CMOS*, Proceedings of European Solid-State Circuits Conference, Sept. 2000.

[6] R. Schreier, *Delta-Sigma Toolbox for Matlab*, www.mathworks.com/support/ftp/controlssv5.html

[7] K. Francken, M. Vogels, G. Gielen, *Dedicated System-Level Simulation of $\Delta\Sigma$ Modulators*, Proceedings of the IEEE CICC, May 2001.

[8] V. Liberali, V.F. Dias, M. Ciapponi, F. Maloberti, *TOSCA: a simulator for switched-capacitor noise-shaping A/D converters* IEEE Transactions on Computer-Aided Design of Integrated Circuits and Systems, Vol. 12, Sept. 1993.

[9] S.R. Norsworthy, R. Schreier, G.C. Temes, *Delta-Sigma Data Converters*, IEEE Press, 1997.

[10] A. Marques, V. Piuri, M.S. Steyaert, W.M. Sansen, *Optimal Parameters for $\Delta\Sigma$ Modulator Topologies*, IEEE Transactions on Circuits and Systems II, Vol. 45, Sept. 1998.

Systematic Design of a 200 MS/s 8-bit Interpolating/Averaging A/D Converter

J. Vandenbussche, K. Uyttenhove, E. Lauwers, M. Steyaert and G. Gielen
Katholieke Universiteit Leuven, Dept. of Electrical Engineering, ESAT-MICAS
Kasteelpark Arenberg 10, B-3001 Heverlee, Belgium
Phone: +32 - 16 - 32 17 19, Fax: +32 -16 - 32 19 75
Email: georges.gielen@esat.kuleuven.ac.be

ABSTRACT

The systematic design of a high-speed, high-accuracy Nyquist-rate A/D converter is proposed. The presented design methodology covers the complete flow and is supported by software tools. A generic behavioral model is used to explore the A/D converter's specifications during high-level design and exploration. The inputs to the flow are the specifications of the A/D converter and the technology process. The result is a generated layout and the corresponding extracted behavioral model. The approach has been applied to a real-life test case, where a Nyquist-rate 8-bit 200 MS/s 4-2 interpolating/averaging A/D converter was developed for a WLAN application.

Categories and Subject Descriptors

B.7.m Integrated Circuits: miscellaneous

General Terms

Design

Keywords

A/D converters, Interpolating, Flash, Simulated Annealing.

1. INTRODUCTION

In the design of analog functional blocks as part of a large system on silicon, a number of phases are identified. These are depicted in Fig. 1. The first phase in the design is the specification phase. During this phase, the analog functional block is analyzed in relation to the surrounding system to determine the system-level architecture and the required block specifications. With the advent of analog hardware description languages (VHDL-AMS, VERILOG-A/MS), the obvious implementation for this phase is a generic analog behavioral model [1]. This model is parameterized with respect to the specifications of the functional blocks. A list of specifications is given in Table 3 later on. The next phase is the design (synthesis) of the functional block. It consists of sizing & layout and is shown in the center of Fig. 1. The design methodology used is top-down performance-driven [2]. This design methodology has been accepted as the de facto standard for systematically designing analog building blocks [2]. Finally, a behavioral model for the block is extracted from the sized circuit including (layout) parasitics. This allows verifying and efficiently simulating the block as part of a larger system. This methodology is now applied to an interpolating/averaging A/D converter.

The paper is organized as follows. Section 2 explains the chosen A/D converter architecture. Section 3 describes the systematic design methodology in detail and section 4 presents the measurement results. Finally, conclusions are drawn in section 5.

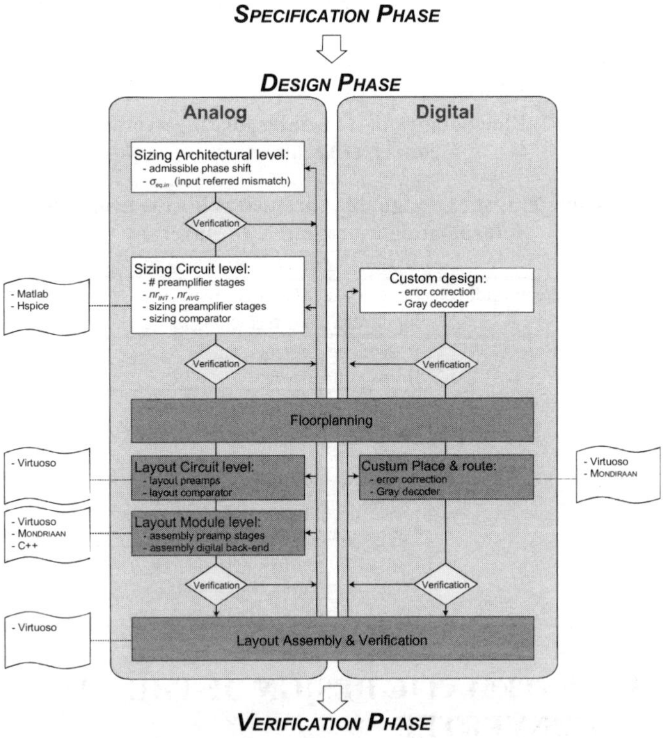

Figure 1: Presented systematic design flow for an interpolating/averaging A/D converter.

2. THE INTERPOLATING/AVERAGING ARCHITECTURE

The interpolating/averaging architecture is shown in Fig. 2. Just as in the flash architecture, processing is fully parallel resulting in high sampling rates [3]. The front-end is fully differential for improved dynamic performance. A Sample & Hold circuit (S/H) samples the differential input signal. The resulting signal is compared with the fully differential reference ladder network and amplified in the first amplification stage. The output of this

DAC 2002, June 10-14, 2002, New Orleans, Louisiana, USA.

preamplifier stage is interpolated $nr_{INT,st1}$ times. A second preamplifier stage is added, which is interpolated $nr_{INT,st2}$ times. Both preamplifier stages use averaging to improve static performance [4]. The outputs of the preamplifier stage(s) steer the regenerative comparators. A digital back-end performs additional error correction (e.g. against bubble errors) and encodes the thermometer coder output from the comparators in Gray code, which is synchronized at the output by a latch.

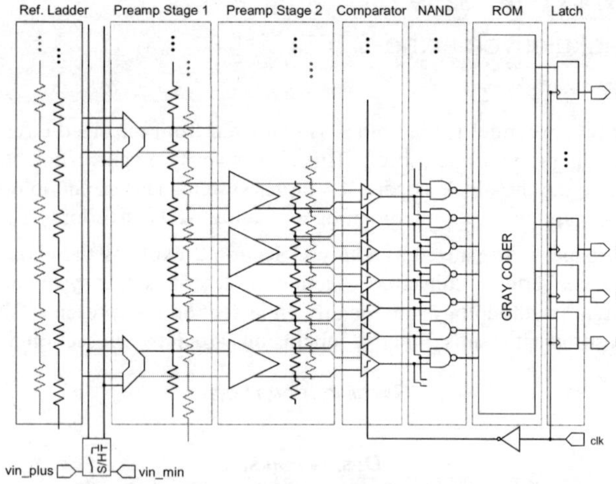

Figure 2: Block diagram of the interpolating/averaging A/D converter architecture.

Table 1: List of designable parameters for the proposed interpolating/averaging A/D converter.

	Designable parameters
Architectural level	Phase shift at Nyquist frequency
	Input referred offset $\sigma_{total,offset}$
Circuit level	Resistance reference ladder R_{ladder}
	Boost voltage S/H
	$(W,L)_i$ tors S/H
	Level of interpolation: $nr_{INT,st1}$ / $nr_{INT,st2}$
	Amount of averaging: $nr_{AVG,st1}$ / $nr_{AVG,st2}$
	Resistance for averaging: $R_{AVG,st1}$ / $R_{AVG,st2}$
	Gain preamps: $A_{preamp,st1}$ and $A_{preamp,st2}$
	$(W,L)_i$ transistors preamps
	Regeneration time constant τ_{reg}
	$(W,L)_i$ transistors comparator
	$(W,L)_i$ transistors digital back-end

3. SYSTEMATIC DESIGN OF THE A/D CONVERTER

3.1 Specification phase.

Two approaches are available for the statistical behavioral modeling of A/D converters: equation-based modeling [5] or macro modeling [1]. This equation-based approach has the advantage that Monte-Carlo simulations are no longer needed and thus simulations can be speeded up considerably. For timing verification/simulation though, macro models are better suited. To study the effects of clock jitter, signal dependent delay, etc. the designer needs to resort to macro-models as presented in [1]. In the presented approach these macro models were used.

3.2 Design phase

The specifications derived during the system-level specification phase are now the input to the converter design phase. The design of the converter is performed hierarchically. The design parameters are listed in Table 1. First, some architectural decisions have to be made. Both static and dynamic performance are taken into account, resulting in specifications for mismatch and admissible phase shift for the different building blocks.

3.2.1 Architectural-level sizing.

A Monte-Carlo simulation can be used to estimate the design yield as a function of the total equivalent input-referred offset [10]. For these simulations a targeted INL of 1.0 LSB and a targeted DNL of 0.5 LSB were used. Using averaging techniques, the DNL can be improved by a factor of nr_{AVG}, while the INL can be improved by $\sqrt{nr_{AVG}}$ [4], resulting in the plot shown in Fig. 3. The yield is plotted as a function of the offset, with the amount of averaging nr_{AVG} as a parameter varying from 1 (i.e. no averaging) to 9. With e.g. an averaging of 9 ($nr_{AVG}=9$) the simulations yield a constraint for the admissible total equivalent input-referred offset:

$$\sigma_{total,offset} \leq 0.7\ LSB \tag{1}$$

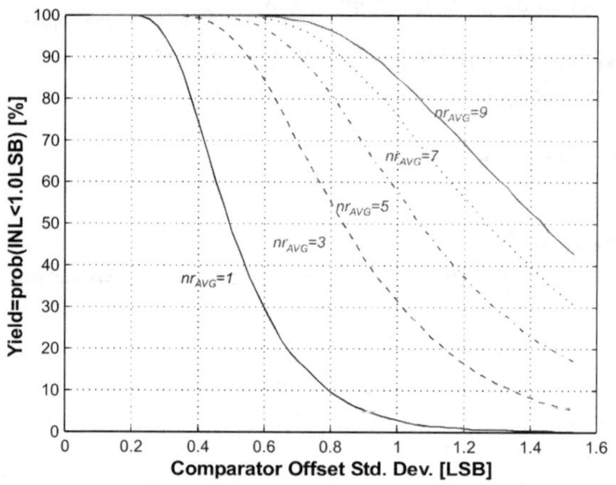

Figure 3: Estimated yield as a function of the total equivalent input referred offset for a targeted INL of 1.0 LSB.

From Fig. 2, the total equivalent input-referred offset $\sigma_{total,offset}$ can be calculated as:

$$\sigma^2_{total,offset} = \sigma^2_{preamp\_st1,offset} + \left(\frac{\sigma_{preamp\_st2,offset}}{A_{preamp\_st1}}\right)^2 + \left(\frac{\sigma_{comp,offset}}{A_{preamp\_st1} \cdot A_{preamp\_st2}}\right)^2 \tag{2}$$

where $\sigma_{preamp\_st1,offset}$ is the input-referred offset of the preamplifier stage 1, $\sigma_{preamp\_st2,offset}$ is the input-referred offset of the preamplifier stage 2, and $\sigma_{comp,offset}$ is the input-referred offset of the comparator stage. The latter term is negligible if the gain in the preamplifiers is high enough. From statistical behavioral modeling [1] and technological constraints of the process used (0.35 μm CMOS), it can be calculated that a total gain of 15 is sufficient for the comparator to have negligible contribution in the total equivalent input-referred offset.

$$A_{preamp} = A_{preamp\_st1} \cdot A_{preamp\_st2} = f(INL, technology) \geq 15 \quad (3)$$

In this design A_{preamp} was chosen 20. Thus mismatch and speed no longer have to be traded off for the comparator, allowing to optimize the comparator for speed.

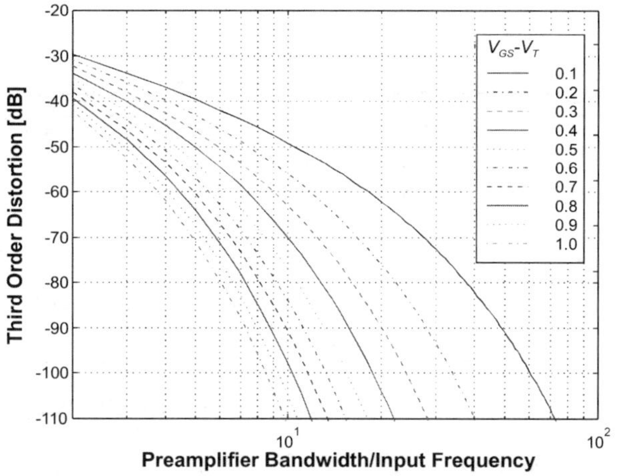

Figure 4: HD$_3$ as a function of the preamplifier bandwidth/input frequency ratio for a V_{fs} of 1.25 V.

Apart from the mismatch constraint, the admissible phase shift for the preamplifier is also determined at this stage of the design. In [3] a formula was derived for the resulting third-order distortion HD_3 as a function of the bandwidth of the preamplifiers:

$$HD_3 \approx \frac{2g\ f_{in}}{3\pi\ f_b}, \text{ where } \quad g \approx e^{-2b_n \frac{(V_{GS}-V_T)f_{in}}{V_{fs}\ f_b} - 1} \quad (4)$$

V_{fs} is the full-scale input range, f_{in} is the input frequency and f_b is the bandwidth of the preamplifier, g represents the normalized delay $\delta t_d/BW$ of the preamp, b_n is the relative output level. The normalized delay is worst-case around the mid-codes i.e. when $b_n = 0.5$. The results of equation (4) are depicted in Fig. 4: for this example the targeted 50 dB distortion would result in a constraint of 10° phase shift at Nyquist frequency for a V_{GS}-V_T of 0.3 V:

$$\varphi_{Nyquist} \leq \text{atan}(1/6) \approx 10° \quad (5)$$

3.2.2 Circuit-level sizing

The architectural-level design resulted in constraints in terms of gain ($A_{preamp} > 15$), bandwidth of the preamps (e.g. $\varphi_{Nyquist} \leq \text{atan}(1/6) \approx 10°$), and admissible input-referred offset (e.g. $\sigma_{total,offset} \leq 0.7\ LSB$) for the different building blocks. Using these constraints, each of the building blocks can be sized as will be discussed in detail in the following paragraphs for each block: S/H, fully differential ladder, 1st stage preamplifier, 2nd stage preamplifier, comparator and digital back-end.

Sample & Hold

The S/H was based on the architecture presented in [6] using the gain-boosting technique. Three modifications were done: (1) the gate was boosted with a fixed voltage, (2) special attention was paid to the clock recovery and timing and (3) a PMOS transistor was added in parallel with the NMOS switch transistor.

The S/H was designed to steer a load of 5 pF (worst-case estimate of the total input capacitance based on mismatch constraints) with an input swing of 0.8V (the input swing is chosen by the designer and fixed during the optimization of the preprocessing chain later on). The simulated 3rd harmonic is -68dB and the 5th harmonic is -83dB at a sampling rate of 200 MS/s.

Reference ladder network

The reference ladder has to be properly sized to avoid feed-through. A first-order estimation of the feedthrough to the midpoint (worst case) of the reference ladder is given by [7]:

$$V_{mid}/V_{in} = \tfrac{\pi}{4} f_{in} R_{ladder} C \quad (6)$$

In this formula f_{in} is the input frequency. R_{ladder} is the total resistance in the case of one ladder. C stands for the total coupling capacitance from the input to the reference ladder (the gate-source capacitance of the input transistors of the preamplifiers). The maximum resistance R_{ladder} is calculated to be 14 Ω.

Preamplifier stage 1

The schematic of the first stage preamplifier is shown in Fig. 5. The input referred offset was calculated using ISAAC [8]:

$$\sigma^2_{preamp\_st1} = 4\left[\sigma^2_{M_1} + \left(\sigma^2_{M_{33}}/\sqrt{2}\right) + \sigma^2_{M_3} + \sigma^2_{M_4}\right] \quad (7)$$

The gain A_{preamp_st1} of this preamplifier is a function of the number of averaging nr_{AVG} and the number of interpolations nr_{INT}. In [4] a new preamplifier topology was proposed which has high impedance load, which is beneficial for averaging. The presented preamplifier exhibits the same advantage of high intrinsic impedance load (formed by transistors M_{33} and M_4). Thus the gain in the preamplifier and the averaging effect no longer have to be traded off [4].

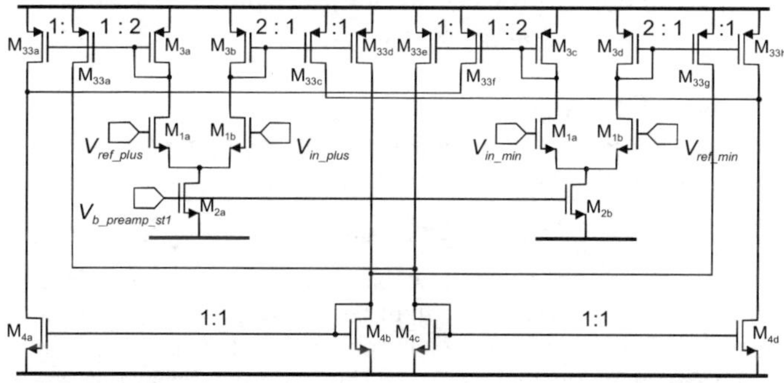

Figure 5: Schematic of the 1st stage preamplifier.

Using macro models for the amplifiers, a closed expression for the overall gain of the preamplifier as a function of the number of averaging nr_{AVG} and the number of interpolations nr_{INT}, was calculated using the ISAAC tool [8]. Fig. 6 shows the case where $nr_{AVG}=5$ and $nr_{INT}=2$. Similar macro models were used to derive equations for other values of nr_{INT} and nr_{AVG}. Comparing the different equations resulted in a closed expression for the gain:

$$A_{preamp\_st1} = -2 \cdot \frac{gm_{m1}}{g_{AVG}} \cdot \frac{nr_{INT}(nr_{AVG}+1)^2}{2^3} \quad (8)$$

This expression is a function of the amount of averaging nr_{AVG} and the number of interpolations nr_{INT}.

Not only the gain, but also the frequency behavior is affected by the averaging. Expressions were derived for the dominant pole as a function of both the number of averaging nr_{AVG} and the number of interpolations nr_{INT} :

$$f_{dominant\_st1} = \frac{1}{2\pi \cdot f(nr_{AVG}, nr_{INT}) \cdot R_{AVG} \cdot C_{load}} \quad (9)$$

where $f(nr_{AVG}, nr_{INT})$ is a fit factor extracted from simulations. This fit factor $f(\cdot)$ is a function of both the number of averaging nr_{AVG} and the number of interpolations nr_{INT}.

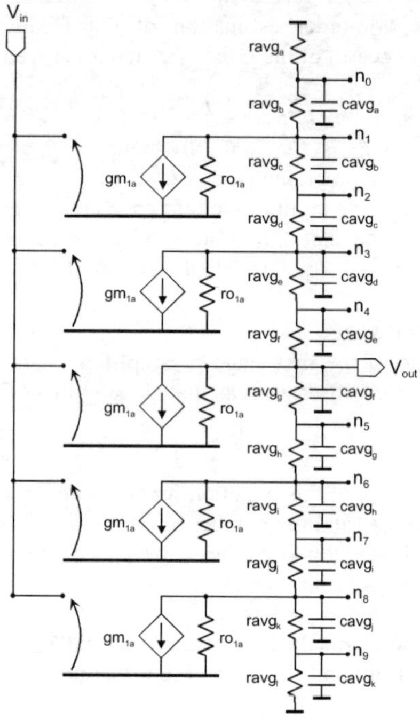

Figure 6: Simplified schematic for preamplifiers in case of nr_{AVG}=5 and nr_{INT}=2.

Preamplifier stage 2

For the second-stage preamplifier, the topology presented in [4] was used. The mismatch contribution is given by:

$$\sigma^2_{in\_st2} = \sigma^2_{m1} + 2\sigma^2_{m3}(gm_{m3}/gm_{m1})^2 \quad (10)$$

Also the 2nd stage preamplifier has a high output impedance. Equation (8) also gives the gain of the 2nd stage preamplifier but then as a function of the amount of averaging $nr_{AVG,st2}$ and the number of interpolations $nr_{INT,st2}$. The frequency behaviour was equally addressed. An equation for the dominant pole, similar to equation (9), has been derived.

Regenerative comparator

The comparator used in this A/D converter is a very fast regenerative structure, depicted in Fig. 7.

During the reset phase the regenerative nodes n_{r1} and n_{r2} are 'shorted' by the switch M_4. The reset time constant can be approximated by [9,10]:

$$\tau_{res} \cong \frac{C_{eq}}{g_{m5} + g_{m6} - g_{o5} - g_{o6} - 2g_{ds4}} \quad (11)$$

where C_{eq} is the total capacitance on the nodes n_{r1} and n_{r2}.

During the regeneration phase (clock is low) the injection of the current imbalance stops, and the conductance of the switch M_4 drops to zero. The regeneration speed is governed by a positive pole approximately given by [9,10]:

$$p_{reg} \cong \frac{g_{m5} + g_{m6} - g_{o5} - g_{o6}}{C_{eq}} \quad (12)$$

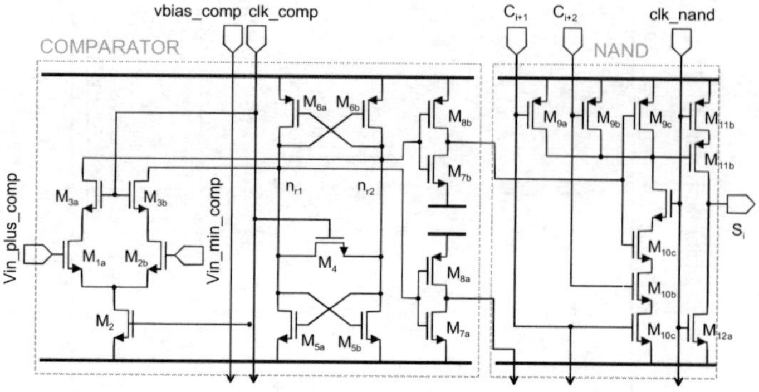

Figure 7: Schematic of the comparator & the digital back-end

Digital back-end logic

The outputs of the comparators form a thermometer code: all comparator outputs below the input level are '1' and vice versa. The thermometer code is then converted into Gray code in a ROM decoder. A flash converter with this type of structure typically suffers from two problems: bubbles in the thermometer code and metastability [10]. The simplest circuit that can detect a bubble is a 3 input and-gate to ensure only a single one drives the ROM, as shown in Fig. 7. To avoid metastability two asymmetric inverters (M_7/M_8 in Fig. 7) were inserted [10].

Sizing plan

Combining these equations with the set of constraints resulting from architectural-level synthesis, a full design plan for the converter was derived. The architectural design resulted in three constraints for the design of the preamplifier stages:

$$A_{preamp\_st1} = 10, \; A_{preamp\_st2} = 2 \quad (13a)$$

$$\varphi_{Nyquist} \leq \mathrm{atan}(1/6) \approx 10° \quad (13b)$$

$$\sigma^2_{preamp\_st1} \leq \tfrac{3}{4}(0.7LSB)^2, \sigma^2_{preamp\_st2} \leq \tfrac{1}{4}(0.7\,LSB)^2 \quad (13c)$$

With these constraints, and the complete set of design equations derived, all transistors can be sized. As the inter dependency of the different design variables is high, the sizing plan has been formulated in a (global) constraint optimization that has been resolved using advanced simulated annealing (ASA) where the power consumption and chip area of the circuit is minimized [11]. The phase shift constraint is evaluated using equation (4) during optimization. The offset constraint is implemented as a lookup table and checked as the amount of averaging nr_{AVG} evolves during optimization. The overdrive voltages V_{GS}-V_T of the preamplifiers, the lengths L of the transistors, the biasing currents and the averaging resistor values r_{AVG} are the input variables of the optimization. The input range was fixed during optimization as was the number of interpolations which was chosen $nr_{INT,st1}$=4 and $nr_{INT,st2}$=2.

The resulting global optimization problem is formulated as:

$$\underset{\underline{x}}{\text{minimize}}\, C(\underline{x}) = \sum_{i=1}^{k} w_i \cdot f_i(\underline{x}) + \sum_{j=1}^{l} w_j \cdot g_j(\underline{x}) \quad (14)$$

where $\underline{x}$ is the set of independent variables as listed in Table 2, $\underline{f}(\underline{x})$ is a set of k objective functions, and $\underline{g}(\underline{x})$ denotes a set of l constraints. Constraint functions are formulated such that a constraint is satisfied when $\underline{g}(\underline{x}) \leq 0$.

Table 2: Input variables x_i for the 1st & 2nd stage preamplifier and comparator.

	1st stage preamp	2nd stage preamp	comparator
x_i	L_1, L_2, L_3=L_{33}, L_4	L_1, L_2, L_3	L_1, L_2, L_5, L_6, L_4, W_4
	$(V_{GS}\text{-}V_T)_{M1}$, $(V_{GS}\text{-}V_T)_{M2}$, $(V_{GS}\text{-}V_T)_{M3}$, $(V_{GS}\text{-}V_T)_{M4}$	$(V_{GS}\text{-}V_T)_{M1}$, $(V_{GS}\text{-}V_T)_{M2}$, $(V_{GS}\text{-}V_T)_{M3}$	$(V_{GS}\text{-}V_T)_{M1}$, $(V_{GS}\text{-}V_T)_{M2}$, V_{reset}
	$I_{DS,2}$	$I_{DS,2}$	$I_{DS,2}$, $I_{DS,5}$
	$R_{AVG,st1}$, $L_{AVG,st1}$,, $nr_{AVG,st1}$	$R_{AVG,st2}$, $L_{AVG,st2}$,, $nr_{AVG,st2}$	

The cost function is built by a weighted sum of functions that force the optimization to evolve to *operational* (saturation/linear region), *functional* (design requirements fulfilled) and *applicable solutions* (specifications met). Within this last design subspace, trade-offs are optimized to result in a solution with minimal area and power. These four categories of cost terms have weighting terms which typically differ an order of magnitude in order to guide the optimization.

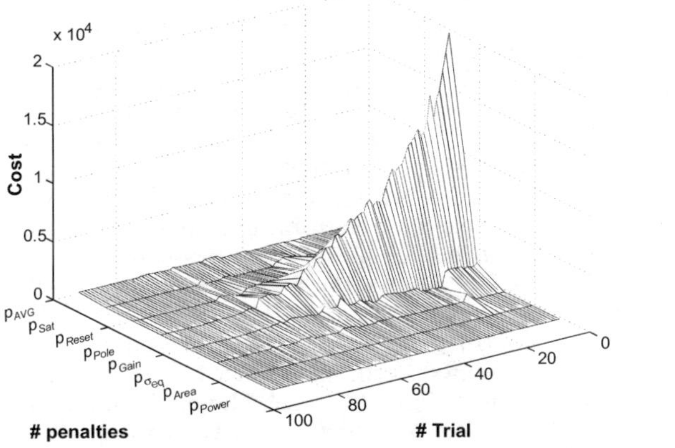

Figure 8: Global cost during sizing using ASA.

For the MOS transistors the input set is chosen to be:

$$\underline{x}_{Mi} = \{L, V_{GS} - V_T, I_{DS}\} \tag{15}$$

A level-1 Spice model was encapsulated in a separate Matlab routine. Given L, $V_{GS} - V_T$ and I_{DS}, the routine returns W, g_m, g_o, σ and the parasitic C_{GS}, C_{GD}. Other device-level evaluations (e.g. BSIM) could be used as well, by either a Matlab routine or an external C-code routine which can be invoked easily form within the Matlab environment.

For designing the averaging resistors, an additional routine was added in Matlab. The routine takes the resistive value R_{AVG} and the length L_{AVG} as inputs and decides in which layer to implement the resistor (e.g high-resistive poly, low-resistive poly) such as to minimize parasitic capacitance. The routine returns the width W_{AVG} and the parasitic capacitance C_{AVG}.

The amount of averaging nr_{AVG} is needed to calculate the averaging effect and thus the constraint on the admissible input-referred offset $\sigma_{total,offset}$. The amount of averaging nr_{AVG} is however not an independent variable, and is calculated from the V_{GS}-V_T, the gain and the linear output range of the preamplifier stages. This loop is resolved by choosing $nr_{AVG,st1}$ and $nr_{AVG,st2}$ as input variables and forcing them to be equal to the actual nr_{AVG} calculated.

The simulated annealing loop was implemented in the Matlab environment and uses the Advanced Simulated Annealing (ASA) C-routine as actual algorithm [11]. The evolution of the total cost term during sizing is depicted in Fig. 8. About 30 trials were needed to obtain the final result. One trial takes 14 min on a Sunblade 1000 workstation.

The sizing of the S/H, the reference ladder network and the digital back-end was not included in the global optimization. The S/H was designed using ELDO in the loop, while the digital back-end was designed manually.

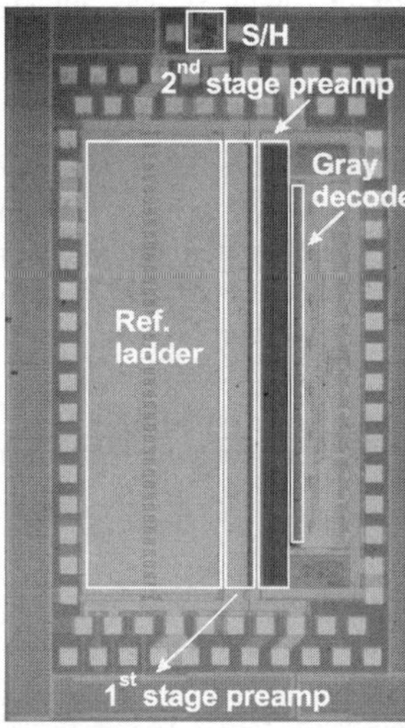

Figure 9: Micro photograph of the A/D converter.

3.2.3 *Layout*

The floorplan follows directly from the block diagram in Fig. 2; the result is depicted in Fig. 9. Around the perimeter of the chip 1 nF of decoupling capacitance has been integrated to provide stable power supplies.

The reference ladder was implemented in metal-1 layer. Dummies were added to provide identical surroundings. An additional decoupling capacitance of 10x30pF was added to each ladder to provide stable reference levels.

The layout of the preamplifiers and the routing was done manually: devices were generated using LAYLA [12], placement of the different modules (1st & 2nd stage preamplifier) was done using MONDRIAAN [13]. Internally an additional 500 pF of decoupling capacitance was added. Guard rings were used to reduce substrate (digital) noise coupling.

The layout of the digital back-end was done combining Virtuoso from Cadence and the MONDRIAAN tool [13]. The layout of the comparator was done manually as was the internal routing, transistors were generated using the LAYLA tool. The ROM cell is handcrafted, 8 ROM cell constitute a ROM line. MONDRIAAN is used to place the ROM cell and connect the cells using a listing of the Gray code as input. The ROM is generated within 1 minute.

The clock distribution is critical for analog design, and available digital tools cannot deal with the specific analog requirements. A buffered binary clock tree takes care of equal delay, which would otherwise deteriorate the dynamic performance. The design and layout of this clock buffer was done manually.

3.3 Verification Phase

After sizing the design was verified with device level simulations using ELDO. Layout parasitics are extracted and the static performance was verified with Monte-Carlo simulations considering process variations using the in-house developed tool MIMI [14]: a mismatch offset voltage and offset current are automatically added to the netlist.

The offset of the comparator was automatically extracted with an in-house developed tool presented in [1]. The offset is determined by narrowing down the input voltage interval for which the comparator toggles. A regeneration time constant τ_{reg} of 50 ps was simulated.

4. MEASUREMENTS

The A/D converter was processed in a 0.35µm CMOS process. The A/D converter was mounted on a ceramic substrate; all biasing was generated on the substrate to minimize incoupling noise. The analog power preprocessing chain runs from a 3.3 V power supply, the digital back-end runs at 2.5 V. All measurements were done at full speed of 200 MS/s [3]. The analog preprocessing chain consumes 285 mW, the reference ladder consumes 250 mW and the digital part consumes 120 mW worst case.

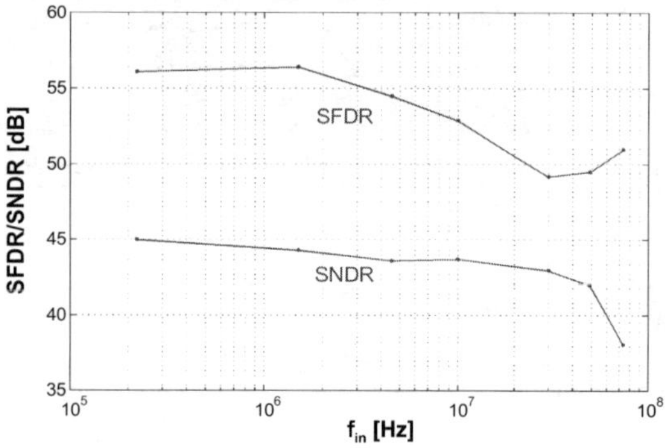

Figure 10: Measured dynamic performance: SFDR>50 dB and SNDR > 43 dB.

Static performance measurements show an INL < 0.95 LSB and a DNL < 0.8 LSB. The dynamic performance is shown in Fig. 10. A Signal-to-Noise-and-Distortion-Ratio (SNDR) of 44.3 dB is achieved at low frequencies; at 30 MHz a SNDR figure of 43 dB was measured.

5. CONCLUSIONS

The systematic design of an 8-bit interpolating/averaging 200 MS/s Nyquist-rate A/D converter has been presented. Using behavioral models the system specifications are translated in offset and phase shift constraints that steer the global optimization at the circuit level. The chip was processed in a standard 0.35 µm CMOS process. Measurements on the processed chip (see Table 3) yielded good results proving the viability off the presented approach.

Table 3: Specification list for a (interpolating/ averaging) A/D converter with target values and measured values.

	Specification	Unit	Target	Measured
Static	Resolution *N*	#bits	8	8
	INL / DNL	LSB	<1 / <½	0.8 / 0.9
	Yield	%	99.9	-
Dynamic	SFDR	dB	> 45	59.2
	SNDR	dB	> 40	44.3dB@1.5MHz 42.7dB@40MHz
	Sample freq.	MS/s	200	200
Environmental	Conversion rate	-	1 code/ clock cycle	1 code/ clock cycle
	Input capacitance	pF	< 5	4.8
	Input range	V_{ptp}	> 0.5	1.3 V
	Power supply	V	3.3	3.3/2.5
	Coding	-	Gray code	Gray code
Optimization	Power	mW	min	655
	Area	μm^2	min	1400x2400

6. REFERENCES

[1] G. Van der Plas, J. Vandenbussche, *et.al.*, "Statistical Behavioral Modeling for A/D Converters", *Proc. on the IEEE 1999 ICECS*, pp. 1713-1716, Cyprus, Sept. 1999.

[2] G.Gielen and R. Rutenbar, "Computer-Aided Design of Analog and Mixed-signal Integrated Circuits", *Proc. IEEE*, vol. 88, no. 12, pp. 1825-1852, Dec. 2000.

[3] R. Van de Plassche, "Integrated Analog-To-Digital and Digital-To-Analog Converters", *Kluwer Academic Publishers*, ISBN 0-7923-9436-4, p. 193-204, 1994.

[4] K. Bult, A. Buchwald, "Embedded 240-mW 10-b 50-MS/s CMOS ADC in 1 mm^2", *IEEE JSSC*, Vol. 32, No. 12, December, 1997, p. 1887-1895.

[5] E. Liu, *et.al.* " A behavioral representation for Nyquist Rate A/D Converters, in *Proc.* IEEE *International Conference on Computer-Aided Design*, 1991, pp. 386-389.

[6] T.L.Brooks, *et. al.*, "A 16b Sigma-Delta Pipeline ADC with 2.5MHz Output Data-rate", in *Proc. ISSCC*, Feb 1997, p 404.

[7] A. Venes *et. al.*, "An 80 MHz, 80 mW, 8 bit CMOS Folding A/D Converter with Distributed Track and Hold Preprocessing", *IEEE JSSC*, vol. 31, no. 12, Dec. 1996.

[8] G. Gielen, H. Walscharts and W. Sansen, "ISAAC: a symbolic simulator for analog integrated circuits", *IEEE JSSC*, vol.24, no. 6, December 1989, pp. 1587-1597.

[9] Geert Van der Plas, *et.al.*, "Symbolic Analysis of CMOS Regenerative Comparators", *European Conference on Circuit Theory and Design*, pp. 86-89, Italy, August 1999.

[10] K. Uyttenhove, A. Marques and M. Steyaert , "A6-bit, 1GHz Acquisition Speed ADC in 0.35 CMOS", in *Proc. IEEE CICC*, May 2000, Orlando, Florida, pp 249-252.

[11] This product includes software developed by Lester Ingber and other contributors: *http://www.ingber.com/*

[12] K. Lampaert, G. Gielen and W. Sansen, "A Performance-Driven Placement Tool for Analog Integrated Circuits", *IEEE JSSC*, pp. 773-781, July 1995.

[13] G. Van der Plas, J. Vandenbussche, *et. al.*, "Mondriaan: a Tool for Automated Layout Synthesis of Array-type Analog Blocks", in *Proc. IEEE CICC*, pp. 485-488, May 1998.

[14] W. Verhaegen, G. Van der Plas and G. Gielen, "Automated Test Pattern Generation for Analog Integrated Circuits", *VLSI Test Symposium,* April, 1997.

Petri Net Modeling of Gate and Interconnect Delays for Power Estimation

Ashok K. Murugavel and N. Ranganathan
Dept. of CSEE and Center for Microelectronics Research
University of South Florida
Tampa, Florida 33620

ABSTRACT

In this paper, a new type of Petri net called Hierarchical Colored Hardware Petri net, to model real-delay switching activity for power estimation is proposed. The logic circuit is converted into a HCHPN and simulated as a Petri net to get the switching activity estimate and thus the power values. The method is accurate and is significantly faster than other simulative methods. The HCHPN yields an average error of 4.9% with respect to Hspice for the ISCAS '85 benchmark circuits. The per-pattern simulation time is about 46 times lesser than PowerMill.

Categories and Subject Descriptors

B.7 [**Hardware**]: Integrated Circuits; B.7.2 [**Integrated Circuits**]: Design Aids—*simulation*

General Terms

Power Measurement

1. INTRODUCTION

As the integrated circuit design moves into deep sub-micron technology, it is possible to pack several million transistors in a small die area. Besides the traditional goals of area and time, power has become an important and critical parameter in the design of circuits and systems. Power estimation techniques, in general, represent a trade-off between accuracy and speed. With the advent of wireless and mobile high performance computing platforms and the limited operational lifetime of a battery, low power designs are required. With higher power consumption, packaging and cooling costs increase and these issues need to be addressed. Thus, it is obvious that with increased circuit integration, power dissipation assumes greater importance. An important open problem in the area of power estimation at the gate level, is the modeling of real delays (both intrinsic gate and interconnect delays). Until, the advent of deep sub-micron (DSM) technologies, the overall path delay between gates was dominated by gate delays, and the bus/interconnect delays could be ignored. In deep sub-micron IC's the interconnects introduce critical problems on the signal integrity and the interconnect delays dominate over the gate delays and so cannot be ignored. It has been observed that the existing power estimation methods are fast and accurate, but, do not consider real delay models that include interconnect effects [3].

Numerous works have appeared in the literature on average power estimation at the gate level under a zero delay model [9, 1]. The zero delay models ignore glitches completely, and can thus lead to underestimation of power. In [4, 12], techniques for real delay power estimation considering only the gate delays has been proposed. In deep sub-micron IC's the interconnects introduce critical problems on the signal integrity and the interconnect delays dominate over the gate delays and so cannot be ignored. It has been observed that the existing power estimation methods are fast and accurate, but, do not consider real delay models that include interconnect effects [3].

In [11], a fast simulation algorithm based on a standard logic simulator to estimate the gate-level power has been presented. In [2], a new accurate and efficient method to estimate the switching activity of a circuit has been proposed. The method involves a novel event driven delay simulation using a time parallel simulator. A fast and accurate glitch modeling and estimation technique has been proposed in [6], the average power estimation of the circuit is based on a toggle count mechanism has been presented. The toggle information obtained is not exact for circuits with high depth. This leads to discrepancies in the power estimate. The methodology has been shown to be faster than Hspice but with the loss of accuracy. The methodology has also been compared with PowerMill. The first RT-Level power estimation technique that includes the interconnect effects was presented in [3]. The model for switching activity estimation is based on the relationship between switching activity and entropy. The RT-level technique has an estimation error of about 25.8%.

In this paper, a new technique based on simulation is presented that models switching activity based on a complete delay model (considering both gate and interconnect delays). In order to model switching activity, we propose a new type of Petri net called the Hierarchical Colored Hardware Petri net (HCHPN) which incorporates accurately the spatial and temporal correlations with real delays. The logic circuit is first modeled as a Gate Signal Directed Acyclic Graph (GSDAG), which is then converted into the corresponding Hierarchical Colored Hardware Petri net (HCHPN) and simulated as a Petri net to get power values. While providing the same accuracy levels compared to existing simulators, the proposed strategy requires significantly less per-pattern simulation time.

2. PRELIMINARIES

In this section, we provide some definitions and terminology related to Petri nets, required for describing the proposed approach.

A Petri net model consists of a set of *places* (P) and a set of *transitions* (T) [5], in which the places correspond to the states of the model and the transitions represent the actions related to the states of the model. The places and transitions are connected by a set of directed *arcs* (A). An arc can only connect a transition with a place or vice-versa. An arc directed from a place to a transition is called an input-arc and an arc from a transition to a place is called an output-arc of the transition. A transition is said to be *enabled* if there are enough tokens in each of the input places as specified by the arcs connecting the input places to the transition. An enabled transition can *fire* if the other conditions associated with the transition are satisfied. The initial state of the Petri net is called its *initial-marking*. The current state of a Petri net is the distribution of tokens to the places in the Petri net.

The three basic elements of CPNs are places, transitions and arcs. The color set (Σ) determines the types, operations and functions that can be used in the net. It is assumed that the color sets have at least one element each. The arcs have expressions attached to them called *arc-expressions* (E). A transition in a CPN is enabled if it is possible to bind the variables in such a way that the arc-expressions of all the input arcs evaluate when tokens are present at the corresponding input places. The transition can, in addition, have Boolean expressions called *Guard-expressions* (G), that are needed to evaluate to "true" in-order that the transition fires. The *node function N* maps each arc into a tuple where the first element is the source node and the second element is the destination node, the *color function C* maps the place p to a color set, the *guard function* (G) maps the transition t to the boolean function and the *arc expression function* (E) maps the arc, a to an expression. The *initialization function* (I) specifies the initial state of the Petri net. A CPN, can be defined mathematically as in [5],

$$CPN = (\Sigma, P, T, A, N, C, G, E, I) \quad (1)$$

where, (i) Σ is the color set, (ii) P is a finite set of places, (iii) T is a finite set of transitions, (iv) A is a finite set of arcs such that, $P \cap T = P \cap A = T \cap A = \oslash$, (v) N is a node function, defined from A into PxT $\cup$ TxP, (vi) C is the color function defined from P into Σ, (vii) G the guard function, (viii) E is an arc expression function, (ix) I is the initialization function.

It is difficult to model large systems as a single large CPN. To alleviate this problem the concept of Hierarchical Colored Petri nets [5] is used. The following definitions and terminology from [5], are required for defining the HCHPN.

Definition 1 [5]: A *multi-set s*, over a set *S*, is a function from *S* to *N*, where *N* is the set of Natural numbers. If $a \in S$ then $s(a)$ is the number of appearances of element *a* in *S*. A_{MS} is the set of all multi-sets over S, defined as,

$$\sum_{a \in S} s(a)a \quad (2)$$

Definition 2 [5]: A Time set (TS) is defined as the set of all non-negative real numbers, $TS = x \in R$, where x represents the time instant and *R* is the set of all real numbers.

Based on the above, we propose a new type of Petri net called the HCHPN and develop some definitions and terminology required for modeling switching activity.

3. HCHPN BASICS

The concepts of the Hardware Petri net (HPN) [10] and Hierarchical Colored Petri net (HCPN) [5] can be modified to define a new kind of Petri net which we call as the Hierarchical Colored Hardware Petri net (HCHPN). In order to obtain the switching activity of a circuit under a real delay model, we need to add the factor of time into the Petri net. The features of HCHPNs are: (i) certain restricted places can store only one token at any time and a newly arriving token always overwrites the existing token, (ii) when a transition fires, the associated gate function is evaluated, (iii) besides the weights on the arcs, functions can be defined on the arcs as additional firing conditions; delays for the gates and the interconnects can be specified on the transitions as firing conditions, (iv) each token has a time stamp and is called a timed token, which is available only after the clock time of the simulator is greater than or equal to the time stamp on the token.

The HCHPN has a set of *substitution transitions*. The substitution transition allows the embedding of a small or sub-Petri net within a larger Petri net for higher level modeling. When the substitution transition needs to be fired, the underlying Petri net needs to be evaluated in order to determine the outcome of the substitution transition. We relate the individual CPN to a substitution transition, the individual non-hierarchical CPN in a HCHPN is called a *page*. The input and output places of the page are called *port nodes*. The port nodes specify the type and number of tokens that can be introduced into the page. Each of the substitution transitions is called a *super node* and the page that contains these super nodes are called a *super page*. In a super page, the places that are connected to the super nodes (substitution transitions) are called *socket nodes*. The socket nodes of a super node (substitution transition) are related to the port nodes of the page through the port assignment function. The HCHPN has in addition to the substitution transitions, a set of transitions called *gate transitions*. The gate transitions are present only in the sub-nets/pages of the HCHPN. An important function added to the HCHPNs is time. With the introduction of time to each of the tokens we can model highly complex real time systems. The token in a HCHPN is called a colored token. The colored token is a 3-tuple (p,d,t) where p is the current position (place) of the token $p \in P$, d is the color set of the token and t is the time of the firing of the token. The HCHPN consists of a set of places called *restricted places*, than can hold only a finite number of tokens. To impart additional conditions during the firing of a transition, functions can be added to the transitions, which are called *code segment functions*. The HCHPN can be defined mathematically as follows,

Definition 3: A HCHPN is defined as a 10-tuple,

$$HCHPN = (PG, S, GT, PO, PA, PR, TI, TO, TC, RP) \quad (3)$$

where,

(i) PG is a finite set of pages such that (a) each page $pg \in PG$ is a non-hierarchical CPN and, (b) none of the pages have any net element in common,

(ii) $S \subseteq T$ where S is a set of substitution transitions (super nodes),

(iii) ($GT \subseteq T$ and $GT \cap S = \emptyset$), where GT is the set of gate transitions,

(iv) $PO \subseteq P$ is the set of port nodes, the port nodes are the input and output places of a page,

(v) PA is a port assignment function, which defines the relation between the socket nodes of the substitution transition and the port nodes of the page,

(vi) $PR \in PG_{MS}$ is a multi-set of prime pages; it determines the number of instances each individual pages has,

(vii) TI is the time set defined in **Definition 2**,

(viii) TO is the set of all possible colored tokens, i.e., all 3-tuples (p,d,t), where $p \in P$, d is the value of the token and $t \in TI$,

(ix) TC is the code segment function. It is defined from T into the set of functions,

(x) $RP \subset P$ is a set of restricted places, which can hold only a finite number of tokens in them.

The 10-tuple (PG, S, GT, PO, PA, PR, TI, TO, TC, RP) specifies the structure of the HCHPN. The following definitions represent the various characteristics of the HCHPN.

Definition 4: A *binding B(t)* of a transition t is a function defined on Var(t) as, $(i)\forall v \in Var(t) : s(v) \in Type(v)$, $(ii) G(t) < s >$, where, B(t) denotes the set of all bindings for t as defined by the function s. $G(t) < s >$ denotes the evaluation of the guard expression G(t) in the binding b.

Definition 5: A *token element* is a tuple (p,c,t) where $p \in P$, $c \in C(p)$ and $t \in TI$, while a binding element is a pair (t,b), where $t \in T$ and $b \in B(t)$. The set of all token elements is denoted by TE while the set of all binding elements is denoted by BE.

Definition 6: A *state* is defined as a multi-set of colored tokens with a time-stamp. S is the set of all possible states. A *marking* is a multi-set over TE without the time-stamp. The initial marking M_0 is the marking which is obtained by evaluating the initialization expressions:

$$\forall (p,c,t) \in TE : M_0(p,c,t) = (I(p))(c)(t) \tag{4}$$

The set of all markings is denoted by $\mathbb{M}$.

Definition 7: *Event* (t,a,b) specifies the possibility of the firing of a transition t by removing tokens from the input places a and adding tokens to the output places b. The set of all events is the Event set ES.

Definition 8: A step S is *enabled* in a marking M iff

$$\forall p \in P: \sum_{(t,b) \in Y} E(p,t) < b > \leq M(p) \tag{5}$$

Let the step S be enabled in the marking M. When (t,b) ∈ S, we say that t is enabled in M for binding b. When $(t_1,b_1,),(t_2,b_2) \in Y$ and $(t_1,b_1) \neq (t_2,b_2)$ we then say that (t_1,b_1) and (t_2,b_2) are concurrently enabled.

After the transition is enabled, it is not necessary that the transition must fire, this is because of the time factor involved, which leads to the definition of enabling time.

Definition 9: The *enabling time E* of an event (t,a,b) is the maximum of the time-stamps of the tokens consumed by transition t from the input place set a.

$$E(t,a,b) = max_{(p,d,t) \in TO} x \tag{6}$$

Definition 10: If a transition t is enabled and the global clock is past the enabling time, it can *fire*. If more than one transition can fire at the same time, the priority of the transition firing is given such that the transition that was enabled first gets to fire first. The firing of transition t changes the marking **M** to marking **M'** such that,

$$M'(p) = \begin{cases} M(p) \pm aw & if (p,t) \in A, \\ M(p) & otherwise \end{cases}$$

where $M(p)$ and $M'(p)$ are the marking of place p before and after the firing of transition t. The value aw is the weight of the arc from place p to transition t.

Definition 11: The *firing delay* is defined as the difference between the firing time and the time-stamp of the token being produced as an output from a transition.

4. REAL DELAY MODELING

To model switching activity estimation using Petri nets, we first transform the gate-level combinational circuit into a GSDAG. We illustrate this using the example circuit c17, an ISCAS circuit with 6 NAND gates, shown in Figure 1.

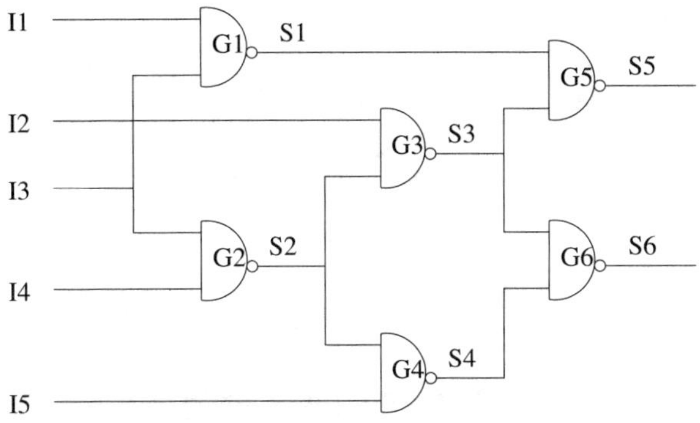

Figure 1: c17 ISCAS Combinational Circuit

The circuit has 6 gates marked {G1, G2, ..., G6} and we are interested in capturing the switching activity at the outputs of the gates marked as { S1, S2, ..., S6}. We need to determine the number of times the output of a gate switches from either 0 to 1 or 1 to 0. First, we represent the circuit as a DAG where each gate in the circuit is represented as a node and each signal in the circuit as an arc. The DAG thus obtained is converted into a GSDAG structure, by introducing additional nodes corresponding to the signals, {S1, S2, ..., S6} while maintaining connectivity by introducing arcs. The GSDAG for the circuit shown in Figure 1 is given in Figure 2. To represent multiple fan-outs, we introduce a node for each output fanning out from the gate. For instance, the output {S2} from gate {G2} is represented by two fan-out nodes {S21} and {S22} in the GSDAG. The GSDAG thus obtained can be represented as a Petri net by replacing the gates in the GSDAG by transitions and the signals by places. This Petri net does not contain the delay information of the circuit. The Petri net equivalent for the c17 circuit in Figure 1 is given in Figure 3.

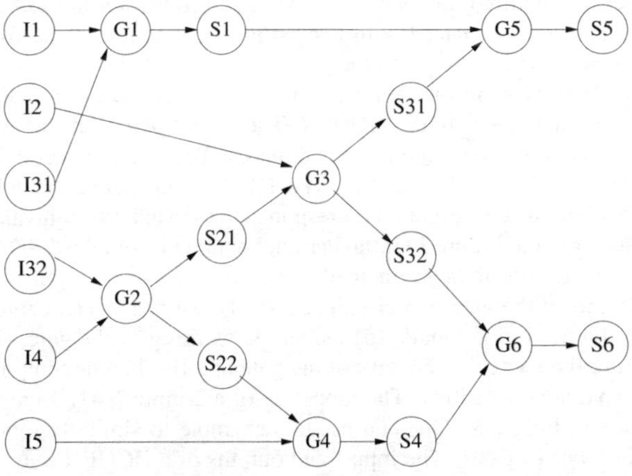

Figure 2: GSDAG structure of the c17 circuit

The important challenge in representing a synchronous system as a Petri net is in obtaining an exact correspondence between the propagation of signals in the circuit and the shifting of tokens in the Petri net. In a Petri net, when a transition fires, a token from

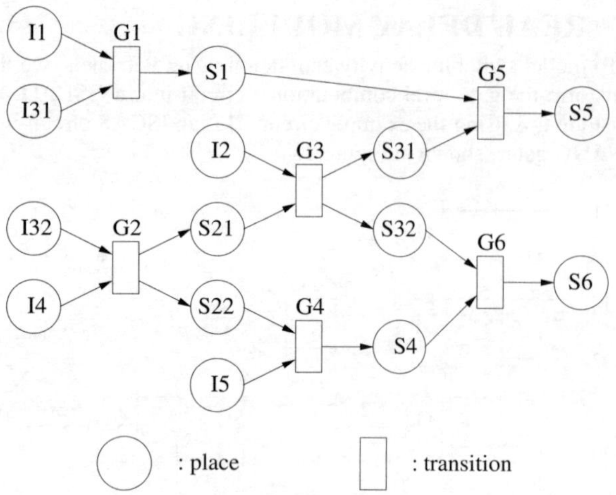

Figure 3: Petri net structure of the c17 circuit

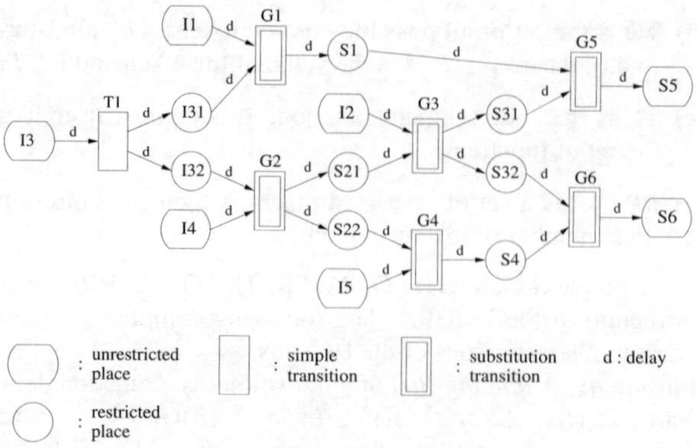

Figure 4: HCHPN structure of the c17 circuit

each input place to the transition is consumed and a token is introduced into each place originating from that transition. However, in a logic circuit, the logic signal changes value and does not disappear. The propagation of signals in the circuit can be modeled by using a Colored Petri net (CPN) and defining two different tokens labeled {one} and {zero}, to specify the two different states of a logic signal. However, the CPNs do not allow the modeling of inertial gate delays and interconnect delays in a circuit. It will be seen that the HCHPN introduced in the last section enables the accurate modeling of the various aspects of a digital circuit.

The Petri net obtained from the GSDAG is transformed into a HCHPN by adding type, color, hierarchy and delay information. A place in a HCHPN has a type function specifying the type of tokens the place is associated with. While in a Petri net, all places are of the same type which can store any number of tokens, in a HCHPN there are two kinds of places: (i) unrestricted places and (ii) restricted places. The restricted places can store only one token at any point in time, which is useful in modeling interconnects. Fanning out inputs are replaced by an input place and a transition, to get the required number of tokens. Each transition in the Petri net structure is replaced with a substitution transition to perform the action corresponding to the gate that is modeled by the transition. The places in the Petri net representing signals are replaced by restricted places in the HCHPN. The output places of the Petri net are not converted into restricted places. The interconnect delay values are added to the arcs of the HCHPN to incorporate the delay values into the net. Figure 4 corresponds to the HCHPN equivalent of the circuit in Figure 1. If the Petri net obtained from the GSDAG is large, it needs to be segmented.

To model the gate of a circuit accurately, we need to (i) capture any glitches in the inputs, (ii) model the working of the gate, (iii) capture the switching activity of the gate and (iv) handle temporal and spatial correlations. The modeling of a 2-input NAND gate is shown in Figure 5. One can use this example to similarly model other types of gates. The inputs and outputs of a HCHPN must be places. In Figure 5, places {PI1, PI2} are the input places and the place {PO} is the output place of the NAND gate. Glitches are of two types: (a) generated glitches, (b) propagated glitches. The generated glitches are due to the different arrival times of the input signals. The glitches already present in the input signals are further propagated by the gate and these are called propagated glitches. The input glitches are accurately captured in our model, by using the structure consisting of the set of places {P0, P1, P2} and the transitions {T0, T1}. The places, {P0, P1, P2} are restricted places that can store only one token at any time. A newly arriving token always overwrites the existing token. The *gate transition T2* models the gate function. The delay associated with the gate is incorporated in the form of a special data structure associated with the transition {T2}. An assumption made in this model is that the delay of the gate is independent of the input pattern and is always a constant. The gate function is given as a logical expression attached to the transition {T2}, as a function of the input tokens to the transition. The logical function is a code segment function (TC) defined in Definition 3 in Section 2.

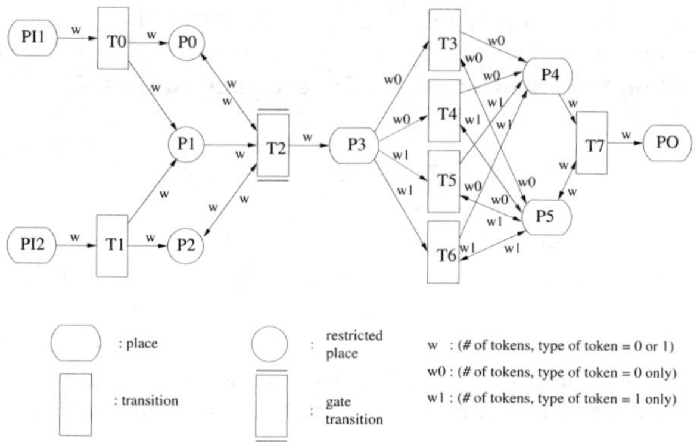

Figure 5: HCHPN structure of the NAND gate

The switching activity can be captured based on the previous and the current states of the gate. The current state of the gate is specified by the output token from the transition {T2}, stored in place {P3}. The previous state of the gate is the token stored in place {P5}, this token is initialized to 'zero' at the start of the simulation. The transitions {T3, T4, T5, T6} are used to distinguish the power consuming/dissipating transitions from the other transitions: (i) transition {T3} fires if the output of the gate remains at 0; (ii) transition {T4} fires if there is a switch from a 1 to a 0 (power dissipating transition); (iii) transition {T5} fires if there is a switch from a 0 to a 1 (power consuming transition); (iv) transition {T6} fires if the output of the gate remains at 1. The output tokens from the transitions {T3, T4, T5, T6}, are given to places {P4} and {P5}. It should be noted that in the HCHPN, *the transitions fire in the order in which they were enabled.* The temporal correlations are modeled using places {P4, P5} and transition {T7}. The arcs that connect the transitions {T3, T4, T5, T6} with places {P3, P4, P5} can only carry specific tokens. The other arcs in a HCHPN have arc weights specifying the type and the number of tokens that can be carried by the arcs. Thus, we have modeled the complete working of the NAND gate. The other gates such as *NOR, AND, OR, NOT, BUFFER and XOR* were also modeled similarly.

Now that we have seen the modeling of c17 as a HCHP-net, the working of the HCHP-net is shown here as a flow of tokens. The super page of the c17 HCHP-net is specified by Figure 4. The super page passes the tokens between the substitution transitions and handles the fanout from a gate. The output token from the substitution node is given to the places that are connected to the substitution transition. The tokens from the places are given to the sockets of other substitution transitions or the places themselves are port nodes of the super page. The sockets in the super page are of the same type as the port nodes of the page. The tokens are thus passed between the super page and the page. Since c17 consists of only NAND gates, the NAND gate in Figure 5 is used to explain the working of the HCHP-net. The port nodes of the NAND gate (page) are {PI1, PI2, PO}, {T2} is the gate transition, {P0, P1, P2} are restricted places and {P3, P4, P5} are unrestricted places of the HCHP-net. The tokens in the port nodes are fired by transition {T0, T1} into the restricted places {P0, P1, P2}. The restricted place {P1} ensures that the glitches that are smaller than the propagation time of the gate are not propagated through to the output. The gate transition {T2} has a code segment attached to it that specifies the working of the gate based on the token types that are input into the transition from the restricted places {P0, P2}. The transitions {T3, T4, T5, T6} are used to determine the type of transition on the output based on the current output from the gate transition and the previous output from the gate transition. The firing of these transitions are noted and the firing of transition {T4, T5} signify a switch in the output. The switching activity of all the gates are added to determine the average switching activity of the circuit. The previous state of the gate is stored in the unrestricted place {P5}. The transition {T7} puts the current token in the output place {PO} that is passed on to the super page through the socket node of the super page.

The HCHPN thus modeled is simulated in order to obtain the switching activity of the circuit. To calculate the toggle information, we need to calculate the various instances at which the gate switches from either 0 to 1 or from 1 to 0. A transition fires in a HCHPN only when an event occurs, i.e., the gate switches. The information obtained from the transition firing specifies the number of times the transition occurs at the output of a gate. Thus, we have a direct relationship between the transition firing in a Petri net and the switching activity of the circuit.

5. DELAY MODELING

To obtain a generalized delay model for deep sub-micron (DSM) circuits, the delays of both the gates and the interconnects have to be taken into consideration as the interconnect delays dominate over the gate delays in DSM circuits. Thus, in the real delay modeling of a circuit, the total delay of the gate is calculated as the sum of (i) intrinsic gate delay (T_{GD}) and (ii) gate load delay (T_{RD})as $t_d = T_{GD} + T_{RD}$ [7]. The *intrinsic gate delay* is the delay due to the physical devices like transistors that constitute the gate. We calculate the intrinsic gate delay assuming an infinite load at the output of the gate using Hspice. The *gate load delay* is the delay due to the RC network connected to the gates output.

There are different methods to calculate the delay of the gate load due to the RC network at the output of a gate. In this work, we use a Thevenin equivalent circuit as shown in Figure 6, to model the interconnect and thus, the delay of the signal in the circuit. The capacitance at node A and node B are known from the layout and so is the resistance. In [7], a one segment Π model has been proposed to model the gate output while still considering the resistance shielding effects. The model approximates the load by considering only the first three moments of the driving point admittance. This model approximates the entire interconnect load with a Π circuit. The delay formula [7] is given as,

$$T_{RD} \approx \frac{1}{|s_1|} \left| ln \left(\frac{(1+s_1 R_1 C_2)(e^{T_R |s_1|} - 1)}{b_2 s_1^2 (s_1 - s_2)(1 - v_{th}) T_R} \right) \right| \tag{7}$$

where $b_1 = R_S(C_1 + C_2) + R_1 C_2$, $b_2 = R_S R_1 C_1 C_2$, and $s_{1,2} = \frac{-b_1 \pm \sqrt{b_1^2 - 4b_2}}{2b_2}$, R_S is the series source resistance as seen in Figure 6, T_R is the source ramp input rise time, v_{th} is the voltage, R_1 is the resistance, C_1, C_2 the capacitances in the Π model used for modeling the interconnects as shown in Figure 6. This model is incorporated into our work.

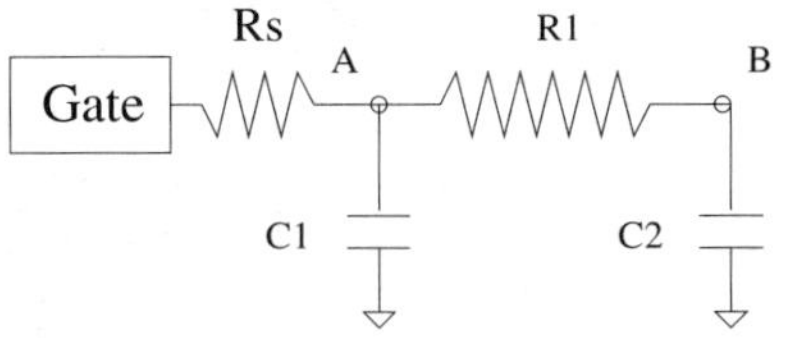

Figure 6: Thevenin equivalent circuit

Figure 7 gives the block diagram for the power estimation tool using HCHPNs. The input to the tool is the gate-level net-list, and the RC estimates (layout-level) at each node, the *intrinsic gate delay* estimates for each of the gates and the input vectors to be simulated. For, a gate delay model, we need only the gate-level net-list and the gate delay as inputs to the tool. The first step in the tool is the conversion of the input net-list into a GSDAG. The GSDAG is then converted into a HCHPN with the necessary code segments, substitution transitions, color functions, hierarchy and delay elements. The delay elements can be calculated based on the capacitance and resistance estimates obtained from the layout-level for each of the nodes in the circuit and equation 7. We have modified the Design/CPN tool [8], to perform the simulation. The toggle information thus generated is the switched capacitance of the circuit, it is used to calculate the power estimate as shown in Figure 7.

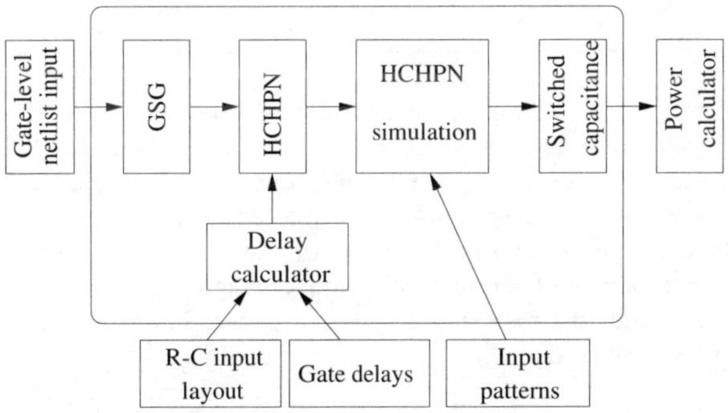

Figure 7: Block diagram of power estimation tool using HCHPNs

6. EXPERIMENTAL RESULTS

The experimental results for the ISCAS '85 combinational benchmark circuits. The ISCAS circuits consist of the basic gates like

AND, OR, NAND, NOR, NOT, EXOR, BUFF. The gates with larger number of inputs have been modified to include gates with only two, three and four inputs. The circuits were synthesized using MOSIS 0.4μ standard cell CMOS technology.

Since, there are no standard input vector sets available for these benchmark circuits, we generated a random pattern of 10,000 vectors for use with PowerMill and HCHPNs. Due to the time constraint placed by Hspice, it was not possible to simulate a large sequence of 10,000 vectors. Hence, a smaller sequence of 100 vectors was used to calculate the time taken per vector for all the three models in addition to the 10,000 vector sequence for HCHPNs and PowerMill. The rise and the fall times of the ramp input given to Hspice and PowerMill was $0.125ns$. The intrinsic gate delays for the HCHPNs were calculated using Hspice. The HCHPNs for large circuits need to be segmented, such that the number of arcs between the segments is kept to a minimum. The simulations were performed on a Sun Ultra 10 with 128 M RAM. The switching activity at each gate is used to calculate the power estimate at that gate, thus, an accurate power estimate of the circuit can be obtained. We used Design/CPN [8], a Colored Petri net simulation and analysis tool for performing the simulation of the HCHPNs. It supports the use of complex color sets and arc expressions for CPNs as explained in Section 3 through the use of a functional programming language *Standard ML (SML)*. The SML language is used to specify the delays of the gates in the gate transition and the interconnect delays in the arc functions. For a more detailed information about Design/CPN the reader is directed to [8]. The delay values can also be estimated based on a wire length estimation technique as proposed in [3]. Experiments were performed for a gate interconnect delay model, where in both the intrinsic gate delay and the interconnect effects are taken into account to calculate the power estimate.

Table 1: Accuracy of the HCHPN modeling

Circuit	Hspice (in uW)	PowerMill (in uW)	HCHPN (in uW)	% error over Hspice	
				PowerMill	HCHPN
C432	918.0	986.4	992.7	7.4	8.1
C449	1385.9	1233.2	1298.5	11.0	6.3
C880	608.3	569.1	634.8	6.4	4.4
C1355	630.0	651.4	652.3	3.3	3.5
C1908	422.0	449.5	441.5	6.5	4.6
C2670	448.0	492.1	458.4	9.8	2.3
C3540	719.7	735.6	759.0	2.2	5.5
C5315	1098.0	1126.3	1040.6	2.5	5.2
C6288	4509.0	4898.9	4329.3	8.6	4.0
C7552	7902.7	7491.0	7528.6	5.2	4.7
Average				6.3	4.9

Tables 1 and 2 compare the accuracy and speedup for modeling both gate and interconnect delays. It can be seen that the HCHPNs have an average speed-up of 46 times over PowerMill. The power estimates are within 4.9% of that of Hspice, while PowerMill results are within 6.3% of that of Hspice. A comparison of the performance of the real-delay models reported in [6] was not possible since the results were presented there in terms of charge with Joule as unit.

7. CONCLUSIONS

In this paper, we have introduced a new fast and accurate real delay simulator for power estimation at the gate-level using the proposed Hierarchical Colored Hardware Petri nets. The modeling is exact and captures all the correlations in the circuit. We have presented a model for real delay based power estimation considering both the intrinsic gate delays and the interconnect delays. The model considers only complete transitions, all partial or incomplete transitions are ignored. The state explosion problem of Petri nets also exists with this modeling and will need to be addressed when modeling huge circuits. Currently, we are investigating the modeling of sequential circuits using HCHPNs.

Table 2: Simulation times

Circuit	Hspice (in ms)	PowerMill (in ms)	HCHPN (in ms)	Speed-up of HCHPN	
				Hspice	PowerMill
C432	19180	138.5	2.8	6850	49
C449	31806	197.0	4.6	6914	42
C880	49258	316.2	7.3	6747	43
C1355	85286	423.6	11.7	7289	36
C1908	105286	558.0	13.6	7741	41
C2670	139720	991.7	19.8	7056	50
C3540	205751	1409.7	25.9	7944	54
C5315	486910	3019.4	68.0	7160	44
C6288	794164	5380.0	106.1	7485	50
C7552	929651	7753.7	143.5	6478	54
Average				7166.4	46.3

8. REFERENCES

[1] S. Bhanja and N. Ranganathan. Dependency preserving probabilistic modeling of switching activity using bayesian networks. In *Proc. of Design Automation Conf.*, pages 209–214, 2001.

[2] M. Buhler, M. Papesch, K. Kapp, and U. G. Baitinger. Efficient switching activity simulation under a real delay model using a bit-parallel approach. In *Proc. of the EURO-DAC*, pages 459–463, 1999.

[3] K. M. Buyuksahin and F. N. Najm. High-level power estimation with interconnect effects. In *Proc. of the Intl. Symp. on Low Power Electronic Devices*, pages 197–202, 2000.

[4] T.-L. Chou, K. Roy, and S. Prasad. Estimation of circuit activity for static and domino cmos circuits considering signal correlations and simultaneous switching. *IEEE Trans. on Computer Aided-Design*, 15(10):1257–1265, Oct. 1996.

[5] K. Jensen. *Colored Petri Nets: Basic Concepts*, volume 1. Springer, second edition, 1996.

[6] G. Jochens, L. Kruse, and W. Nebel. Application of toggle-based power estimation to module characterization. In *Proc. of Power and Timing Modeling of Integrated Circuits*, pages 161–170, 1997.

[7] A. B. Kahng and S. Muddu. Gate load delay computation using analytical models. In *Proc. of Asia-Pacific Conference on Circuits and Systems*, pages 433–436, 1996.

[8] Meta Software Corporation. *Design/CPN Reference Manual*, 2 edition, 1993.

[9] F. N. Najm. A survey of power estimation techniques in vlsi circuits. *IEEE Trans. on VLSI Systems*, 2(4):644–649, Dec. 1994.

[10] P. Rokyta, W. Fengler, and T. Hummel. Electronic system design automation using high level petri nets. In *Hardware Design and Petri Nets, ed. A. Yakovlev, L. Gomes and L. Lavagno*, pages 193–204. Kluwer Academic Publishers, 2000.

[11] T H. Krodel. Power play: Fast dynamic power estimation based on logic simulation. In *Proc. of the Intl. Conf. on Computer Design*, pages 96–100, 1991.

[12] C.-Y. Tsui, M. Pedram, and A. M. Despain. Efficient estimation of dynamic power consumption under a real delay model. In *Proc. of the Intl. Conf. on Computer Aided Design*, pages 224–228, 1993.

Power Estimation in Global Interconnects and its Reduction Using a Novel Repeater Optimization Methodology

Pawan Kapur
Dept. of Electrical Engineering
CIS 112, Stanford, CA-94305
kapurp@stanford.edu

Gaurav Chandra
Dept. of Electrical Engineering
CIS, Stanford, CA-94305
gchandra@stanford.edu

Krishna C. Saraswat
Dept. of Electrical Engineering
CISX 326, Stanford, CA-94305
saraswat@cis.stanford.edu

ABSTRACT

The purpose of this work is two fold. First, to quantify and establish future trends for the dynamic power dissipation in global wires of high performance integrated circuits. Second, to develop a novel and efficient delay-power tradeoff formulation for minimizing power due to repeaters, which can otherwise constitute 50% of total global wire power dissipation. Using the closed form solutions from this formulation, power savings of 50% on repeaters are shown with minimal delay penalties of about 5% at the 50 nm technology node. These closed-form, analytical solutions provide a fast and powerful tool for designers to minimize power.

Categories and Subject Descriptors

B6.3 [**Logic Design**] Design Aids – *Optimization*, B7.m [**Integrated Circuits**] Miscellaneous

General Terms

Design, Performance

Keywords

Global interconnects, Repeaters, Power dissipation

1. INTRODUCTION

The power dissipation in high performance integrated circuits (ICs) is quickly becoming a performance bottleneck. The scaling paradigm will exacerbate the power problem severely from many different angles. On one hand the device leakage power due to sub-threshold and gate leakage is likely to rise in the future. On the other hand, the dynamic power will also increase not only due to a larger number of devices and interconnects on the chip, but also owing to the burden of keeping the speeds of electrical interconnects compatible with increasing clock frequencies. Certain estimates, which merely include device power (both dynamic and static), have shown chip power densities to increase to about 200 W/cm^2 at 35 nm node [14]. Including the interconnect power would substantially increase these power density estimates. This may lead to a significant increase in the chip temperature, which would degrade both reliability, through greater susceptibility to electromigration failures, and performance, through increased interconnect resistance and poorer device characteristics. The situation presents an impending power crisis, which threatens to slow down the progress of the chip industry.

It is important to identify the major power consumption sources on a chip, quantify them, and focus on efficient technological, circuit and/or architectural solutions to minimize them. Toward this goal, we address a potentially large source of chip power: the power required for global signaling. We first deal with a realistic estimation of the power consumption in global wires including that due to repeaters, used to speed these wires. Having quantified this as a function of future technology nodes, we then introduce a novel, efficient, methodology which minimizes the power consumed by repeaters in global interconnects.

2. POWER CONSUMPTION IN GLOBAL SIGNALING

2.1 Sources of Power Increase in the Future

The power consumed in signaling over long on-chip, global interconnects will rise in the future owing to many reasons. Firstly, the number and the length of global signaling wires is likely to grow because of an increase in the complexity and the area of the chip, and because of a shrinkage in module sizes. This, along with wire scaling would lead to an increase in the total capacitance of global wires, and hence dynamic power dissipation, despite the introduction of lower dielectric constant materials. Secondly, deterioration of interconnect performance (delay and bandwidth) with scaling will lead to solutions, such as repeater insertion, which attempt to fix this problem, but burn power. Finally, communication across chip will require increasing number of clock cycles in the future. This would require deeper wire pipelining, hence larger power consumption. The first two sources of power increase will be modeled in detail in subsequent sections. The extent to which the third source of power is addressed in this work is limited to modeling the deterioration in the global interconnect latency with respect to the clock period. Since this will dictate the pipelining depth, the extra power consumption may be deduced.

2.2 Global Wire and Repeater Power Calculations

In this section we present the global wire and repeater power dissipation calculations. First, we address the repeater power. The power dissipated due to global wires can, subsequently, be deduced from repeater power, as will be elucidated later. Before performing any calculations related to interconnects we need realistic, technologically constrained, modeling of the interconnect parameters, such as resistance. This accurate resistance modeling will be even more important in the estimation of interconnect latency, as presented in a later section.

2.2.1 Realistic Copper Resistivity

Ideal, constant, copper (Cu) resistivity (ρ) may lead to overly optimistic results for both power and latency. Scaling of interconnects increases the effective copper ρ due to the electron surface scattering effect and the copper barrier effect [10]. The surface scattering effect depends on barrier/copper interface quality. This is characterized by a parameter P, which could range from 0 to 1 with P=0.5 being a reasonable value for Cu [10]. The barrier effect depends on the barrier thickness and its deposition technology. The possible barrier deposition technologies include Ionized Physical Vapor Deposition (IPVD), collimated PVD, PVD and Atomic Layer Deposition (ALD). ALD, because of its conformality, is most effective in lowering resistivity.

The rise in global wire effective Cu resistivity is shown in Fig. 1 [10]. The wire dimensions and other required parameters were taken from International technology roadmap for semiconductor (ITRS '99) [1]. With reasonable parameters defined as a chip temperature of 100 ^{0}C, a barrier thickness (BT) of 10 nm, P=0.5, and the ALD deposition technology, the copper effective resistivity rises to about 3.2μΩ-cm at the 35 nm node, much larger than the ideal value of 1.7μΩ-cm. The wire resistance was next calculated using above resistivity modeling for various technological conditions (Fig. 2). The effect of technology on resistance values at the 35 nm node is summarized in Table 1. It also indicates the error in resistance in parenthesis, if ideal ρ is used.

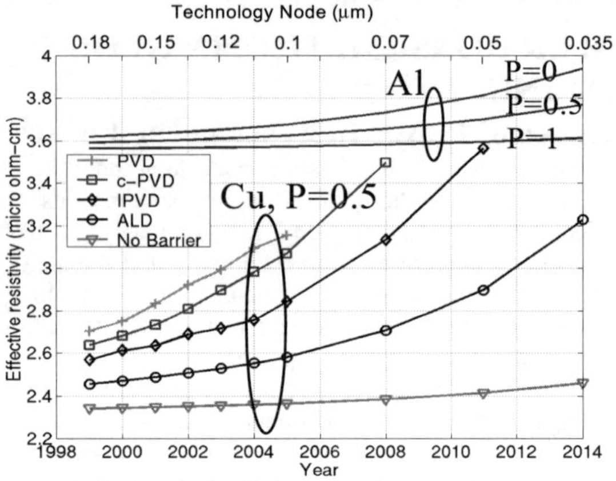

Figure 1: Cu resistivity rise (global wires) with scaling. Barrier thick.=10nm Temp.=100 ^{0}C. Al resistivity showed for comparison

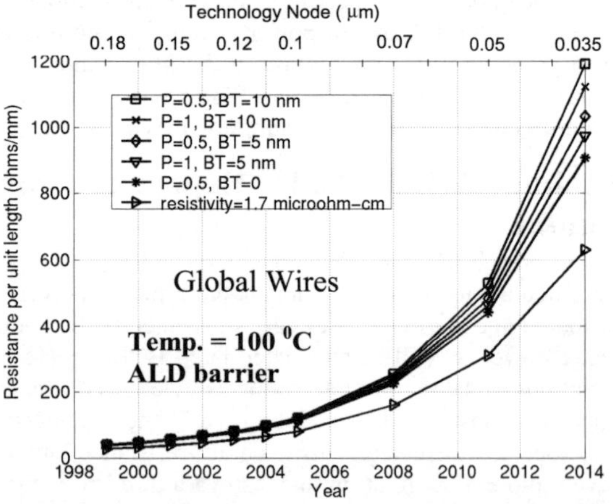

Figure 2: Global Cu wire resistance per unit length with ALD barrier with diff. P and barrier thickness

Table 1: Summary of resistance per mm at 35 nm techn. node. T=100^{0}C, ALD Barrier dep. Techn.

Practical Constraint	Global Resist. (Ω/mm)	Semi-global Resist. (Ω/mm)	Local Resist. (Ω/mm)
	Year 2014	Year 2014	Year 2014
None: ideal ρ=1.7μΩ-cm	628	1773	3275
P=0.5, BT=10 nm	1192 **(90%)**	4351 **(145%)**	9564 **(192%)**
P=1, BT=10 nm	1123 **(79%)**	3942 **(122%)**	8490 **(159%)**
P=0.5, BT=0	908 **(45%)**	2668 **(51%)**	5030 **(54%)**

2.2.2 Estimating the Number of Repeaters

The estimation of number of global wire repeaters requires 1) an accurate wire length distribution for various parts of the chip, 2) a demarcation method that decides the length boundaries between various wiring tiers (local, semi-global and global) as well as the length beyond which repeated wires are faster, and 3) the dimensions of the wires. In most previous approaches, which address the impact of scaling on wire length distribution, e.g. [7], the distribution is homogeneously calculated for the entire chip. However, a realistic wire length distribution necessitates a distinction between cache and logic areas and transistors. In our work, we partition the problem into logic and memory blocks and treat them separately. We get the area occupied by random logic by subtracting cache area from total chip area. A 6T SRAM cell occupies about 4.65 um^2 at 180 nm technology node and 3.45 um^2 at 150 nm technology node [2], giving a size of approximately 600λ^2. Here λ is the design parameter, such that, 2λ refers to the technology node in μm. Above memory size per bit along with the number of memory bits (taken from ITRS), is used to compute future cache area. Using the area occupied by random logic, and the number of logic transistors (from ITRS), we then obtain the wire length distribution based on the stochastic methodology proposed by Davis et. al. [5].

To obtain the length boundaries between local, semi-global, and global wires, we assume a modular design based approach proposed in [15]. In this approach, it is assumed that most modules will be on the order of 50,000 gates, within which the routing is done using local interconnects. The modules will be further grouped into isochronous regions connected by semi-global wires. The extent of these regions will be dictated by the distance traversed in 90% of the fast, local, clock cycle, over semi-global wires. Finally, fat global wires will connect various isochronous regions and communication between these regions will use a slower global clock. The idea of having isochronous regions operating on a local clock connected together by a network of global wires operating on a global clock parallels the present board level design scenario. Hence we think it is reasonable to believe that future on-chip designs would adapt to such design methodologies.

The third step for repeater number estimation is the knowledge of wire dimensions. We use the ITRS dictated future pitch and assume the width of wires to be half of that. There have been suggestions that the ITRS dictated pitch is only the minimum pitch, and in some cases, the global wires do not have to scale at all, staying at a constant pitch for future technology nodes [15]. However, this will lead to large number of metal levels. Concerns have already been made about the increasing signal wires leading to a rapid increase in the required number of metal layers [7]. We justify the use of ITRS dictated wire pitch by computing the number of metal layers required for just the signal wires with these pitches (Fig. 3). Two typical Rent's exponents of 0.55 and 0.6 are used for this calculation [6], [8], [13]. Additional layers would be needed to accommodate power/ground and clock wires. It is evident from the figure that even with the ITRS dictated aggressive wire scaling, the allocated metal levels by ITRS are not sufficient to accommodate the wires at far future nodes. For the present and near future technology nodes, the allocated metal layers appear to be in excess of the number required. However, owing to a large number of wires on the chip, slight increase in the pitch will lead to a rapid increase in metal levels. Thus, we don't expect a significant deviation in the average wire pitch from the ITRS dictated pitch even for near term technology nodes.

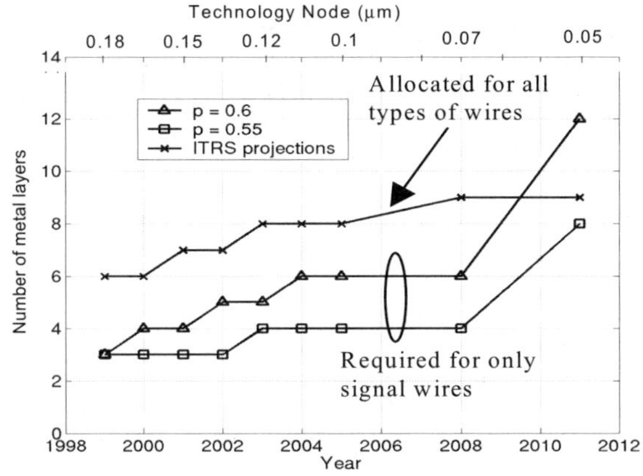

Figure 3: **No. of metal levels needed to accommodate only signal wires at future nodes. ITRS projection of total no. of metal levels also shown.**

The only remaining thing to be established is the fraction of the global wires, which will have repeaters on them. We assume that repeaters on a wire will be inserted when the repeated wire delay is smaller than the unrepeated delay. This occurs at the optimal distance between repeaters, (l_{opt}), originally formulated in [3] and expressed in the following form in [11]

$$l_{opt} = 3.24\sqrt{\frac{r_o C_{nmos}}{R_w C_w}} \quad (1)$$

Here, C_{nmos} and r_o are the capacitance and resistance of the minimum sized NMOS transistor, respectively. R_w and C_w are the resistance and capacitance per unit length of wires respectively. We find that l_{opt} for global wires is always less than the minimum global wire length. Hence, all global wires will have repeaters on them. We call the length, beyond which repeaters are inserted, as the crossover length. In our case, this length is the same as the minimum global wire length. Thus, for a wire of length l, the number of repeaters on that wire is:

$$n_{repeater}(l) = \begin{cases} 0, & \text{if } l < l_{crossover} \\ (\text{round}\left(\frac{l}{l_{opt}}\right)) - 1, & \text{otherwise} \end{cases} \quad (2)$$

Using the statistical wire length distribution, the minimum global wire length, and the number of repeaters at a given length from (2), we compute the total number of repeaters, $N_{repeater}$. The resulting number of repeaters, for two Rent's exponents of 0.55 and 0.6, are shown in Fig. 4, for realistic as well as ideal copper resistivity. The global signal wire repeaters are found to be as high as 5.5 million at the 50 nm technology node with reasonable copper resistivity and a Rent's exponent of 0.55. We compare our repeater number estimates with those obtained by other authors [4], [15] at the 70 nm technology node (Table 2). Our prediction of about 0.85 million repeaters, for a Rent's exponent of 0.55, lies between the two numbers predicted by references [15] and [4], where as, a Rent's exponent of 0.6 yields results which match well with [4]. The repeater estimate obtained in [15] is quite less because in this work the global wires are kept at a constant pitch at future nodes.

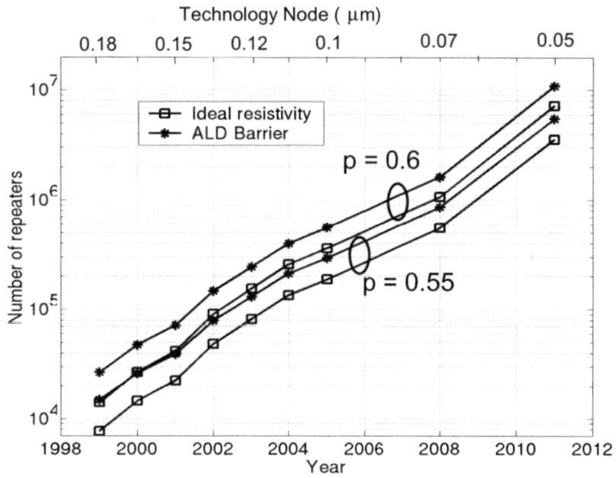

Figure 4: **Total no. of repeaters on global wires as a function of tech. node for different p (Rent's exponent)**

Table 2: Comparison of no. of repeaters of our approach with previous work. The numbers shown are for 70 nm technology

Number of repeaters estimated by [4]	Number of repeaters Estimated by [15]	Our approach, p=0.55	Our approach, p=0.6
1.6 million	0.2 million	0.85 million	1.61 million

2.2.3 Power Due to Delay optimized Repeaters

The short circuit power of repeaters is neglected in our analysis. For estimating dynamic power, the capacitance due to all the repeaters on global wires, $C_{repeater}$, is given by

$$C_{repeater} = N_{repeater}(6 S_{opt} C_{nmos}) \quad (3)$$

Where,

$$S_{opt} = 0.58\sqrt{\frac{r_o C_w}{R_w C_{nmos}}} \quad (4)$$

and

$$C_{nmos} = C_g(2\lambda) \quad (5)$$

Here, S_{opt} is the optimal sizing of the NMOS in the repeater [3], [11]. C_g is the NMOS gate capacitance per micron, and is expected to stay constant at about 1.75 fF/μm for future technology nodes [9]. For a repeater, PMOS is assumed to be twice as large as NMOS. Also the diffusion capacitance is assumed to be the same as the gate capacitance. This leads to 6 times the NMOS gate capacitance in (3). The total dynamic power dissipation due to repeaters is

$$P_{repeater} = s_w C_{repeater} V^2 f_{clock} \quad (6)$$

Where, s_w is the switching activity factor, and V and f_{clock} are supply voltage and clock frequency, respectively. For a reasonable switching activity of 0.15 [16], the power dissipation due to global wire repeaters for future technology nodes is shown in Fig. 5. It is evident that the added power dissipation due to repeaters is a serious problem in the future. At 50 nm technology node, with a reasonable Rent's exponent of 0.55 [13] and using ideal copper resistivity, the repeater power dissipation is about 50 Watts, and with realistic copper resistivity it is about 60 Watts. The resistance plays a role in repeater power as it dictates the crossover length beyond which repeaters are inserted. The power numbers are much worse for a Rent's exponent of 0.6.

2.2.4 Power Due to Global Wires

The power dissipation due to global wires themselves can be simply obtained from the repeater power by realizing an interesting fact regarding the total capacitance of all the repeaters placed

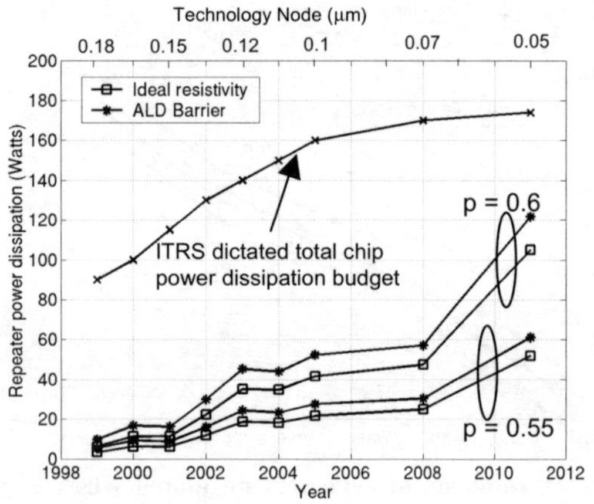

Figure 5: Repeater power dissipation as function of tech node. ITRS dictated total chip power budget also shown.

optimally on a single wire. This capacitance can be obtained by multiplying a single repeater capacitance ($6S_{opt}C_{nmos}$) by the number of repeaters on a wire (L/l_{opt}), where, l_{opt} and S_{opt} can be obtained using (1) and (4), respectively. The expression, thus obtained, is independent of the wire resistance, and is given by:

$$C_{repperline} = 1.07C_{line} \tag{7}$$

Where, $C_{repperline}$ is the total capacitance of all delay optimized repeaters on a single wire and C_{line} is the capacitance of a single wire. Since we established before that all global wires use repeaters, the total power dissipated due to global wires is approximately same as that due to all global repeaters. Hence, the total power dissipation approximately doubles, yielding about 120 Watts of power at 50 nm node with a Rent's exponent of 0.55. This can be a substantial fraction of the total chip power.

3. REPEATER POWER MINIMIZATION METHODOLOGY

The exorbitant power consumption due to delay-optimized repeaters at future technology nodes can be of serious concern. A simple method to reduce repeater power is to decrease the repeater size and/or space them further apart. Both these solutions lead to a delay penalty. In this section, we develop a novel formulation which optimizes the separation and sizing of the repeaters such that the power savings is maximized for a given delay penalty.

The expression for delay due to repeaters which are spaced distance l apart and whose NMOS transistor is sized, S, (channel width to length ratio) can be simply obtained by applying Elmore delay model to a simplified RC network for a stage (one repeater to the next) and is given by

$$\tau_{rp} = \frac{L}{l} t_{stage} \tag{8}$$

$$= L\left(\frac{b(1+e)(1+f)r_0C_{nmos}}{l} + aR_wC_wl + \frac{br_oC_w}{S} + b(1+e)R_wC_{nmos}S \right)$$

Here, L is the length of the wire, a and b are switching model dependent parameters. If we assume that the output of the repeaters switches when the input reaches half of the voltage swing, a and b are found to be about 0.4 and 0.7, respectively [12]. Parameter e is the ratio of the PMOS to the NMOS size and f is the ratio of the diffusion capacitance to the gate capacitance of the transistors. Equation (8) can be optimized independently with respect to S and l to give minimum delay. This yields

$$\tau_{rpopt} = 2L\left(\sqrt{ab(1+e)(1+f)} + b\sqrt{1+e}\right)\sqrt{R_wC_wr_oC_{nmos}} \tag{9}$$

$$l_{opt} = \sqrt{\frac{b(1+e)(1+f)}{a}}\sqrt{\frac{r_oC_{nmos}}{R_wC_w}} \tag{10}$$

$$S_{opt} = \sqrt{\frac{1}{1+e}}\sqrt{\frac{r_0C_w}{R_wC_{nmos}}} \tag{11}$$

For the typical value of e=2 (PMOS sized twice of NMOS), f=1 (diffusion capacitance is same as gate capacitance), and above stated a and b values, (10) and (11) reduce to (1) and (4), respectively. Now, in an attempt to reduce power, we decrease S and increase l, such that $S = x_sS_{opt}$ and $l = l_{opt}/x_l$. Here, x_s and x_l are less than one and denote the fractional change in sizing and spacing from delay optimal values. The total wire delay can be written as

$$\tau_{rp} = L\left(\sqrt{ab(1+e)(1+f)}\left(x_l + \frac{1}{x_l}\right) + b\sqrt{1+e}\left(x_s + \frac{1}{x_s}\right)\right)\sqrt{R_wC_wr_oC_{nmos}} \tag{12}$$

For x_s and x_l equal to 1, (12) reduces to (9). The delay penalty, β, expressed as a ratio of delay with sub-optimal (x_s and x_l not equal to 1) repeaters to that with delay optimized repeaters (x_s and x_l equal to 1) can be written as

$$\beta = \frac{\tau_{rp}}{\tau_{rpopt}} = \frac{\left(x_s + \frac{1}{x_s}\right) + A\left(x_l + \frac{1}{x_l}\right)}{2(1+A)} \tag{13}$$

$$\text{where,} \quad A = \sqrt{\frac{a(1+f)}{b}} \tag{14}$$

Next, we examine the power consumption of a single repeated wire due to its capacitance and the capacitance of repeaters on it. This power for the delay sub-optimal case (general form) is:

$$P = s_w\left[C_wL + (1+f)(1+e)C_{nmos}S_{opt}\frac{L}{l_{opt}}x_sx_l\right]V^2f_{clock}$$

$$= s_wV^2f_{clock}C_wL(1+x_sx_lA) \tag{15}$$

The first and the second terms in the parenthesis correspond to the wire and the repeater contributions, respectively. For delay optimal case, where $x_s=x_l=1$, the ratio of the capacitance of all the repeaters on a single wire to the wire capacitance becomes equal to A. For a reasonable value of f=1, A is 1.07 from (14), agreeing with (7).

The amount of power saving obtained per wire can be expressed as the ratio of the total power per wire in the power saving repeaters to that in the delay optimized repeaters (δ). This is easily obtained using (15) and is given by

$$\delta = \frac{P}{P_{opt}} = \frac{(1+x_sx_lA)}{(1+A)} \tag{16}$$

We propose that using the expressions for delay penalty, (13) and power savings, (16), one can find x_s and x_l, such that, for a required power saving, minimum delay penalty is incurred, or vice versa. This condition can be achieved by substituting x_l expressed in terms of δ and x_s from (16), into the expression for β, and minimizing β with respect to x_s. The minimum β and the corresponding x_{sopt} and x_{lopt} are obtained to be the following

$$\beta_{min} = \sqrt{\frac{A-1+\delta}{A+1-1/\delta}} \tag{17}$$

$$x_{lopt} = \frac{\sqrt{(A+1-1/\delta)(A-1+\delta)}}{A} \quad (18)$$

$$x_{sopt} = \delta\sqrt{\frac{A+1-1/\delta}{A-1+\delta}} \quad (19)$$

The minimum delay penalty (β_{min}) can also be expressed in terms of the fractional decrease in power due only to the repeaters on a single wire (η) instead of δ. η is given by the product of x_s and x_l. Using (16) to relate δ and η, and substituting in the expression for β_{min} leads to the following compact expression after simplification:

$$\beta_{min} = \frac{\sqrt{(1+\eta A)(1+A/\eta)}}{(1+A)} \quad (20)$$

Equations (17)-(20) can, thus, be used to minimize the delay penalty for a given power saving on repeaters, or vice versa.

To see the possibility of the most efficient delay-power trade off graphically, we first plot the delay penalty vs. desired power saving in Fig. 6 using (13). The power saving indicated by each curve represents saving only through size reduction at a fixed spacing. Different curves are for different spacings ranging from l_{opt} to $3l_{opt}$. The x-axis is the ratio of the repeater powers only. Different delay penalties for same desired power saving, with different combination of size and spacing, point to the possibility of optimization. In the past, a few attempts have been made to reduce repeater power dissipation [15]. In [15] the product of cube root of the width (proportional to S) and delay was taken to be the objective function for minimization. This minimization leads to an optimal spacing of l_{opt}, a size of $0.5S_{opt}$, a delay penalty, β of 12.5% and a 50% power saving on repeaters [15]. This particular data point is shown in Fig. 6. It is clear that for the same power saving there exists a different combination of S and l which gives a lower delay penalty of about 6%. Thus, the optimization we propose leads to a more efficient tradeoff, and provides a range in this tradeoff.

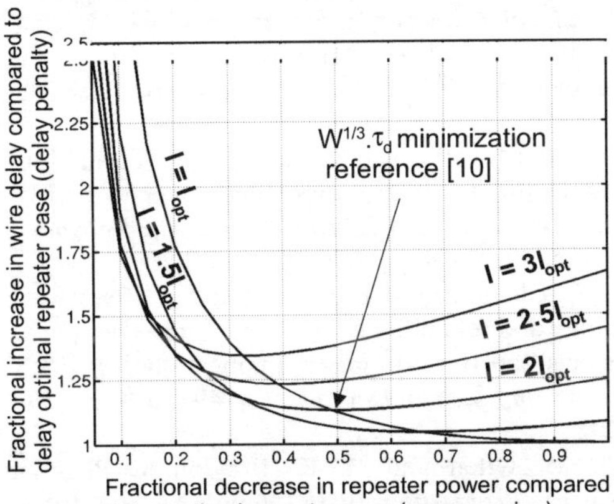

Figure 6: Plot showing the delay-power tradeoff. Each curve represents power saving only due to repeater size reduction

Fig. 7 shows the optimized curve for delay penalty-power saving tradeoff using (18)-(20). The figure also shows the corresponding fractional change in spacing and sizing required (compared to delay optimal repeaters) to achieve this optimization. We observe that with our proposed optimization it is possible to save large repeater power for moderate delay penalties.

Next, we apply the optimal trade off formulation to estimate the total power savings due to all the repeaters on a chip. Fig. 8 shows the total power dissipation due to repeaters as a function of technology node, for different tolerable delay penalties. Even for a small delay penalty of 5%, repeater power dissipation is reduced by approximately 50% (from 61 watts to about 30 watts at 50 nm node). The savings saturate rapidly for larger delay penalties and are 62% and 80% respectively for a 10% and 25% delay penalties. An important point to mention here is that the particular optimization for delay-power tradeoff is done using only a single wire. However, there can be a third source of power saving which comes about through a decrease in the number of repeaters in a different way. This reduction occurs if the crossover length, beyond which repeaters are inserted, is increased. For this to happen the crossover length has to be dictated by the repeater spacing and not by the minimum wire length at the global tier. While the methodology we choose does not lead to this possibility, we consider this case for the sake of completeness. In such a case, the constraining equation for power has to be modified to the following instead of (16).

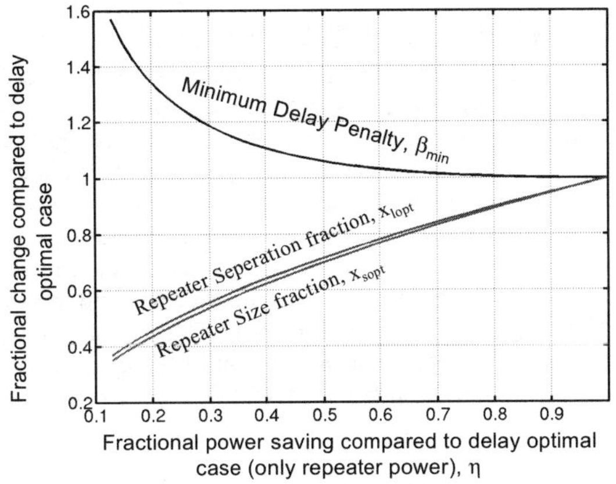

Figure 7: Optimized delay-power tradeoff curve with our methodology. Also shown repeater size and spacing to achieve the respective power saving.

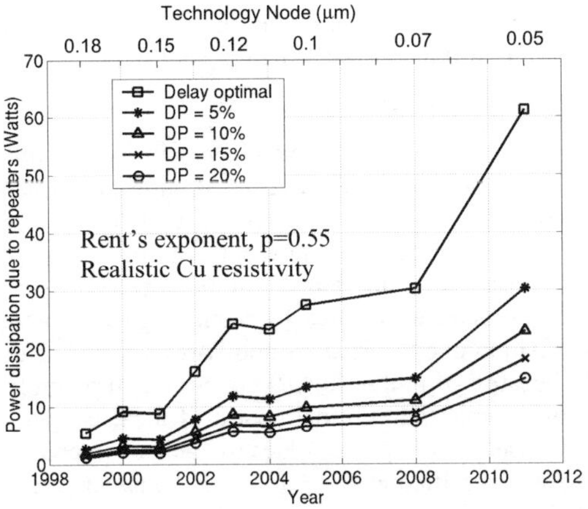

Figure 8: Plot showing the effectiveness of delay-power tradeoff. Power saving for various delay penalties (DP)

$$\delta_{global} = \frac{P_{global}}{P_{optglobal}} = \frac{(1+x_s x_l A)\int_{l_{opt}/x_l}^{L_{max}} xn(x)dx}{(1+A)\int_{l_{opt}}^{L_{max}} xn(x)dx} \quad (21)$$

Here, n(x) is the number of wires at length x and is given by wire distribution function obtained using the Rent's rule. If repeater spacing is greater than minimum global wire length, (21) along with (13) should be used for optimization of power-delay tradeoff. The solution would involve using numerical methods.

4. INTERCONNECT LATENCY

It is important to analyze interconnect latency with respect to clock period because a large interconnect latency for communication across the chip would require deeper pipelining, hence larger power dissipation. The second reason an analysis of wire latency is warranted, is to complete the discussion on repeater power-delay tradeoff by quantifying the latency penalty in absolute terms for power saving repeaters. Using the realistic resistance values, the latency calculations for global wires were performed and are shown in Fig. 9 [11]. Fig. 9 shows the chip-edge long global wire delay with and without the repeaters as a function of the clock period. It is seen from Fig. 9, that despite repeaters, the latency rises to about 7.5 times the clock period at 35 nm node, an increase of about 8X compared to the 180 nm node. This latency rise will cause a further increase in power consumption due to global wires.

Fig. 9 also depicts the latency curve relevant to power saving repeaters with a delay penalty of 25% compared to delay optimal case. It should be recalled that this small delay penalty results in a large power saving of approximately 80% over delay optimal repeaters. Further, it is obvious from the figure that the 25% delay penalty curve is still much faster than the non-repeated wires.

5. CONCLUSION

In this work, we have identified and quantified the future power expenditure on long distance signaling in high performance chips, and suggested an efficient power-delay trade-off formulation to lower this power. In an attempt to accurately model various sources of power dissipation in global signaling, we use practical interconnect resistance, dictated by technological constraints. The power dissipation due to repeaters and wires, as well as that arising indirectly because of multi-clock cycle across chip communication latency, can be misleadingly optimistic, if ideal copper resistivity values are used. The power due to global interconnects and repeaters was quantified and was found to be as large as about 120 Watts at 50 nm technology node. This could be a substantial fraction of the total chip power. Approximately, half of this power was due to repeaters. With the motivation to ameliorate the large repeater power, closed form solutions for an efficient repeater delay-power tradeoff were developed. As much as 50% repeater power can be saved with merely a 5% delay penalty using this formulation. The closed form solutions give designers a powerful tool for a quick back of the envelope calculation for saving power. Finally, interconnect latency was discussed to both partially address the third power dissipation source in global signaling (arising from multi-clock cycle interconnect latency), and to quantify the delay penalty in power-delay tradeoff, for power saving repeaters. On one hand, it showed that the even 25% latency penalty over delay optimized repeaters, which saved 80% of repeater power, was very less compared to non-repeated wires. On the other hand, it pointed to about 8 times clock period latency even with delay optimized repeaters at 35 nm, leading to a possibility of further power dissipation due to deeper pipelining.

6. ACKNOWLEDGEMENTS

The authors would like to thank the MARCO Interconnect Focus Center for providing support for this research.

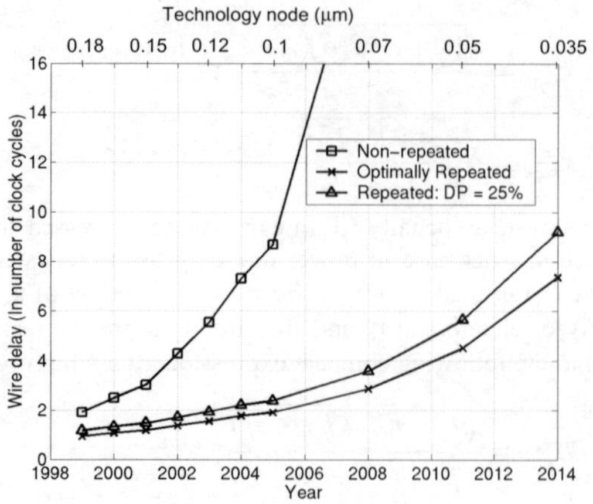

Figure 9: Chip -edge long global wire latency (in clock cycles) with and without repeaters. 25% delay penalty compared to delay-optimal repeaters also shown. ALD barrier assumed

7. REFERENCES

[1] *The International Technology Roadmap for Semiconductors* (ITRS), 1999.

[2] [Online]. TSMC manufacturing specifications. Available: http://www.tsmc.com/technology/index.html.

[3] H. B. Bakoglu, *Circuits, Interconnections and Packaging for VLSI*, Addison-Wesley, 1990

[4] J. Cong, "An interconnect-centric design flow for nanometer technologies," *Proceedings of the IEEE*, Vol. 89, No. 4, pp. 505-528, April 2001.

[5] J. A. Davis, V. K. De, and J. D. Meindl, "A stochastic wire-length distribution for gigascale integration (GSI)- Part I: derivation and validation," *IEEE Transactions on Electron Devices,* Vol. 45, no.3, pp. 580-589, March 1998.

[6] J. A. Davis, V. K. De, and J. D. Meindl, "A stochastic wire-length distribution for gigascale integration (GSI)- Part II: Applications to clock frequency, power dissipation, and chip size estimation," *IEEE Transactions on Electron Devices,* Vol. 45, no.3 March 1998.

[7] J. A. Davis, et. al., "Interconnect limits on gigascale integration (GSI) in the 21st century," *Proceedings of the IEEE,* Vol. 89, no.3, pp. 305-324. March 2001.

[8] P. Fisher and R. Nesbitt, "The test of time. Clock-cycle estimation and test challenges for future microprocessors," *IEEE Circuits and Devices Magazine*, pp.37-44, March 1998.

[9] R. Ho, K. W. Mai, and M. A. Horowitz, "The future of wires," *Proceedings of the IEEE*, Vol. 89, No. 4, pp. 490-504, April 2001.

[10] P. Kapur, J. P. McVittie, and K.C. Saraswat, "Technology and reliability constrained future copper interconnects-part I: Resistance modeling," *IEEE Transactions on Electron Devices*, Vol. 49, No. 4, pp. 590-597, April 2002

[11] P. Kapur, G. Chandra, J. P. McVittie and K. C. Saraswat, "Technology and reliability constrained future copper interconnects-part II: Performance implications," *IEEE Transactions on Electron Devices*, Vol. 49, no. 4, pp. 598-604, April 2002

[12] R. H. J. M. Otten and R. K. Brayton, "Planning for performance," *Proceedings of 35th Annual Design Automation Conference (DAC),* 1998, pp. 122-127.

[13] G. A. Sai-Halasz, "Performance trends in high-end processors," *Proceedings of the IEEE*, Vol.83, No. 1, pp. 20-36, January 1995.

[14] V. Swerdlov, Y. Naveh, and K. Likharev, "Nanoscale SOI Ballistic MOSFETs: An impending Power Crisis," *IEEE International SOI Conference,* pp. 151, 2001.

[15] D. Sylvester, and K. Keutzer, "Impact of small process geometries on microarchitectures in systems on a chip," *Proceedings of the IEEE*, Vol. 89, No. 4, pp. 467-489, April 2001.

[16] D. Sylvester and K. Keutzer, "Getting to the bottom of deep submicron," *Proceedings of International Conference on Computer-Aided Design,* pp. 203-211, November 1998.

Low-Swing Clock Domino Logic Incorporating Dual Supply and Dual Threshold Voltages

Seong-Ook Jung
University of Illinois
Urbana, IL 61801
sojung@uivlsi.csl.uiuc.edu

Ki-Wook Kim
Pluris Incorporation
Cupertino, CA 95014
kkim@pluris.com

Sung-Mo Kang
University of California
Santa Cruz, CA 95064
kang@soe.ucsc.edu

ABSTRACT

High-speed domino logic is now prevailing in performance critical block of a chip. Low Voltage Swing Clock (LVSC) domino logic family is developed for substantial dynamic power saving. To boost up the transition speed in proposed circuitry, a well-established dual threshold voltage technique is exploited. Dual supply voltage technique in the LVSC domino logic is geared to reduce power consumption in clock tree and logic gates effectively. Delay Constrained Power Optimization (DCPO) algorithm allocates low supply voltage to logic gates such that dynamic power consumed by logic gates is minimized. Delay time variations due to gate-to-source voltage change and and input signal arrival time difference are considered for accurate timing analysis in DCPO.

Categories and Subject Descriptors

B.6 [**Hardware**]: Logic Design; B.7 [**Hardware**]: Integrated Circuits

General Terms

Design

Keywords

domino logic, low swing clock, dual supply voltage, dual threshold voltage, low power

1. INTRODUCTION

Domino logic is faster than counterpart static logic since the output of domino logic gate drives only NMOS transistors while the output of static logic gate drives both PMOS and NMOS transistors [4]. Thus, domino logic has become a popular circuit design style to meet the high speed requirement. However, domino logic may be less power effective compared to static logic due to significant clock loading and high switching activity. The total dynamic power of domino logic is consumed mainly by both the clock tree and the domino logic gate. Since switching power consumption dominates dynamic power consumption, the total dynamic power consumption P_{total} of a circuit which is implemented by domino logic can be represented by

$$P_{total} = P_{clk} + P_{logic} \quad (1)$$

$$P_{clk} = \alpha_{clk} \cdot (C_{clk-line} + C_{clk-tr}) \cdot V_{clk-swing}^2 \cdot f_{clk} \quad (2)$$

$$P_{logic} = \sum_i \alpha_{logic\_i} \cdot C_{logic\_i} \cdot V_{logic-swing}^2 \cdot f_{clk} \quad (3)$$

where, P_{clk} and P_{logic} are the dynamic power consumption in the clock tree and the logic gates, respectively. α_{clk} and α_{logic_i} are the switching probability of the clock and each domino logic gate, respectively. $C_{clk-line}$, C_{clk-tr}, and C_{logic_i} represent the clock interconnection line capacitance, the clock load transistor capacitance, and the internal and output node capacitance of each domino logic gate. $V_{clk-swing}$ and $V_{logic-swing}$ are the voltage swing level of the clock and the logic gate. f_{clk} is the operating frequency.

Since switching power consumption is quadratically proportional to the supply voltage, supply voltage reduction is one of the most effective way to reduce power consumption. Dual supply voltage techniques have been proposed, which use high supply voltage (V_{ddH}) in the critical path to obtain high performance while using low supply voltage (V_{ddL}) in the non-critical path to reduce power consumption for static logic in [1, 3]. Rjoub in [6] proposed design technique to save power consumption by reducing the voltage swing only on the domino gate output nodes. Gate-level voltage scaling approach for domino logic was presented in [8]. However, it focuses only on the logic power minimization by replacing domino logic gates using V_{ddH} in the non-critical path with domino logic gates using V_{ddL}.

In this paper, we propose a new domino logic family, namely LVSC domino logic, in which low swing clock is used for domino logic gates to reduce clock tree power consumption. In order to improve performance, low threshold voltage (V_t) is applied to NMOS transistor in the critical path. The proposed LVSC domino logic is effectively incorporated with dual supply voltage technique. Delay Constrained Power Optimization (DCPO) algorithm is also proposed to replace logic gates using V_{ddH} with logic gates using V_{ddL}. In this algorithm, delay time variations according to gate-source voltage change and input arrival time difference are considered. The experiment results for ISCAS'85 benchmark show that proposed technique saves the total dynamic power consumption significantly while maintaining performance.

2. LVSC DOMINO LOGIC FOR DUAL SUPPLY VOLTAGE TECHNIQUES

2.1 Level Shifting Issue in Dual Supply Voltage Technique

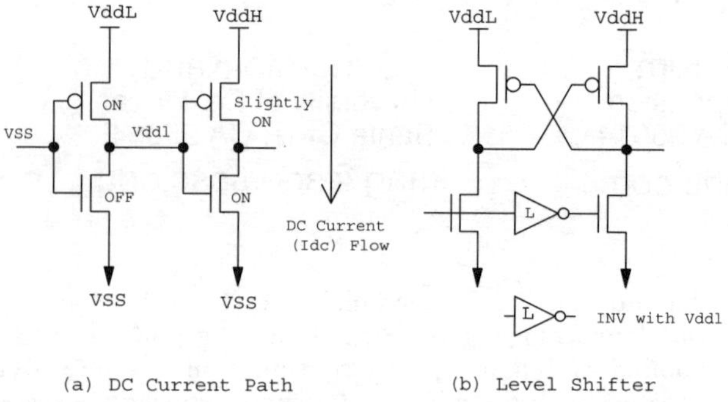

Figure 1: DC current and level shifter in dual supply voltage technique for static logic

The reduction of voltage swing level is the most effective way to reduce power consumption since dynamic power consumption is quadratically proportional to the voltage swing level. However, if low supply voltage is applied to all logic gates, performance is degraded since driving capability is lowered due to the reduced gate to source voltage. Dual supply voltage techniques for static logic use V_{ddH} for static logic gates in the critical path to achieve high performance while using V_{ddL} for static logic in the non-critical path to reduce power consumption. However, when the output of logic gate using V_{ddL} drives logic gate using V_{ddH}, PMOS transistor of logic gate using V_{ddH} does not completely turn off since gate to source voltage of PMOS transistor is less than 0 ($V_{GS_p} < 0$). Thus, DC current flows through PMOS and NMOS transistors as shown in Figure 1 (a). To prevent DC current, a level shifter should be inserted between the output of logic gate using V_{ddL} and the input of logic gate using V_{ddH}. Figure 1 (b) shows an example of the level shifter. Since the level shifter consumes much more power than an inverter, the effect of power saving varies with the number of added level shifters. Furthermore, the level shifter is much slower than normal gates. Thus, the level shifter can limit dual supply voltage technique for static logic [9].

On the other hand, DC current does not flow and the level shifter is not required in dual supply voltage technique for domino logic since the output of domino logic gate using V_{ddL} drives only NMOS transistor in the following domino logic gate. Thus, dual supply voltage technique is very efficient for domino logic to reduce power consumption while keeping performance.

2.2 Proposed LVSC Domino Logic

Shieh in [8] proposed the dual supply voltage technique for domino logic. The power consumption in the clock tree is large portion of the total power consumption in domino logic due to the large clock interconnection capacitance, the large clock load transistor capacitance, and the high switching probability. However, the previous dual supply voltage technique for domino logic does not consider on reducing power consumption in the clock tree. It uses V_{ddH} for the clock tree. AND all inputs of domino logic gates in the critical path should be driven by V_{ddH}, which limits high supply voltage replacement with V_{ddL}. The delay variations to gate-to-source voltage change and input arrival time difference are not considered either, which result in significant delay offset.

To reduce the power consumption in the clock tree, the clock can share the supply voltage with logic gates such that high voltage swing clock is used for high V_{dd} domino logic gates, while low voltage swing clock is used for low V_{dd} domino logic gates. We refer to this clock scheme as Dual Voltage Swing Clock (DVSC). In this scheme, two clock trees are required for both high and low supply voltage devices. It can complicate routing, and thus increase the effective load capacitance of clock tree. Eventually, power saving effect can be mitigated by increased interconnection line capacitance.

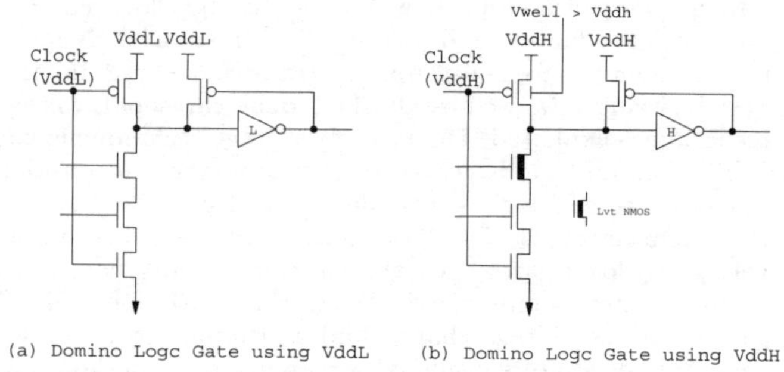

Figure 2: Low Voltage Swing Clock (LVSC) in dual supply voltage technique for domino logic

To further reduce the power consumption in the clock tree, we propose Low Voltage Swing Clock (LVSC) scheme, in which low voltage swing clock controls all domino logic gates regardless of supply voltage level of domino logic gate as shown in Figure 2. Since low voltage voltage clock controls PMOS precharge transistor in a high V_{dd} logic gate as shown in 2 (b), PMOS cannot completely turn off. As a result, DC current flows through PMOS transistor, which is similar to static logic without the level shifter. To resolve this problem by increasing V_t of PMOS precharge transistor, the well of PMOS precharge transistor is biased with higher voltage than V_{ddH}. Domino logic gate usually has smaller number of PMOS transistors than NMOS transistors. Thus, the well of PMOS precharge transistor can be readily separated from the other PMOS transistors with different well bias. Even though the precharge time may increase due to high V_t of PMOS precharge transistor, the precharge time of domino logic gate does not determine the circuit performance. Note that the evaluation time of domino logic gate using V_{ddH}, which is in the critical path, can also be degraded due to the low voltage swing clock. To compensate for such performance degradation, low V_t is applied to NMOS transistor in the critical path.

2.3 Simulation Results for LVSC Domino Logic

All simulation results in this section are based on a 0.18 μm CMOS process. The high and low supply voltage are 1.5V and 1.0V. Figure 3 shows the delay time variation according to the input arrival time difference for 2-input AND gate. VDDH_conv denotes the conventional domino logic

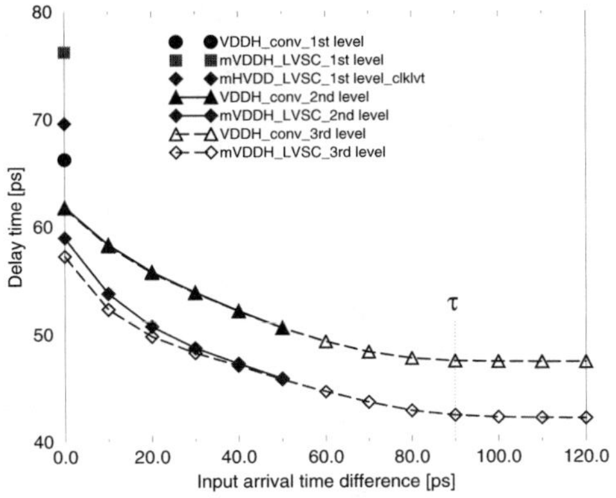

Figure 3: Delay time variations according to the input arrival time difference for 2-input AND gate

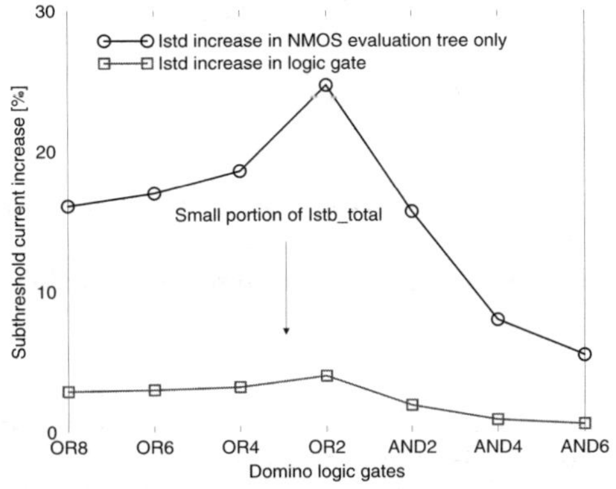

Figure 4: Subthreshold current increase in LVSC scheme for several domino logic gates

gate using V_{ddH} and mVDDH_LVSC denotes the domino logic gate using V_{ddH} in LVSC scheme. At the primary input stage, in which two input signals and the clock arrive at the same time, the delay time of domino logic gate in LVSC scheme increases by 15% compared to the conventional domino logic gate since clock gated NMOS transistor (MNCLK) of domino logic gate in LVSC scheme is driven by V_{ddL}. The delay overhead can be reduced to 5% by applying low V_t to MNCLK with small subthreshold current (I_{sub}) overhead.

From the next logic level, the LVSC scheme is faster than the conventional domino logic gate regardless of input arrival time difference. The reason is that the clock is already high and the driving capability improvement of low V_t NMOS transistor in the critical path overcomes low voltage swing clock effect. The delay time tends to be reduced as the input arrival time difference increases and it becomes invariant to the input arrival time difference beyond some specific value. We refer to the time difference as *the arrival time difference threshold*, or τ, of the gate.

Since low V_t is used for performance, the change in I_{sub} needs to be considered. In domino logic, all domino logic gates are in the precharge phase during the standby mode. Thus, I_{sub} in the standby mode is determined by both stacked NMOS evaluation transistors and the PMOS transistor of output inverter. Figure 4 shows I_{sub} increase in LVSC scheme for several domino logic gates. To drive large fanout, stage ratio between dynamic part and static part of domino logic gate is usually greater than 2 and the static output inverter of domino logic is usually high-skewed for high speed operation. Furthermore, I_{sub} of stacked transistor (NMOS evaluation transistor) is much less than that of non-stacked transistor [2]. Thus, the increase in I_{sub} by low V_t NMOS transistor in the critical path serves relatively small portion of the total static power consumption since large PMOS transistor of output inverter mainly contributes total I_{sub}. As a result, the performance of domino logic gate can be improved by applying low V_t to NMOS transistor in the critical path with small amount of increase in I_{sub}.

3. CIRCUIT DELAY CHARACTERIZATION

3.1 Gate delay

The gate delay $d(u)$ for a gate u can be represented by using alpha power law model [7].

$$d(u) \propto \frac{C_L \cdot V_{dd}}{K \cdot (V_{GS} - V_{th})^{\alpha}} \tag{4}$$

where C_L denotes load capacitance and V_{GS} is the gate-to-source voltage. K is a drivability factor and equals $\mu_0 C_{ox}(W/L)$. μ_0 is the zero bias mobility. C_{ox} is the gate oxide capacitance per unit area. α is the velocity saturation index.

The following expression for gate delay $d_0(u)$ of a domino logic gate u can be obtained from Equation (4) [5].

$$d_0(u) \propto \frac{C_{dyn} \cdot V_{ddH}}{K_n \cdot (V_{ddH} - V_{thn})^{\alpha}} + \frac{C_{out} \cdot V_{ddH}}{K_p \cdot (V_{ddH} - V_{thp})^{\alpha}} \tag{5}$$

where $d_0(u)$ is the intrinsic delay for high supply voltage (V_{ddH}), all fanin gates with V_{ddH} of V_{GS}, and no input arrival time difference effect on gate delay. C_{dyn} and C_{out} represent the capacitance of the dynamic node and the output node in domino logic gate. V_{thn} (K_n) and V_{thp} (K_p) denote the threshold voltage (the drivability factor) of the NMOS evaluation transistor and the PMOS transistor of the output inverter. The first term in Equation (5) represents the delay time from input to dynamic node and the second term represents the delay time from dynamic node to output node.

In our novel circuit scheme, timing analysis needs to be carried out in transistor level because supply voltage replacement, and V_{GS} and input signal arrival difference changes due to the replacement affect the delay time of each logic gate. The gate delay $d(u)$ and delay increase $\Delta d(u)$ of domino logic gate u are given by the following equations.

$$d(u) = d_0(u) + \Delta d_{V_{dd}}(u) + \Delta d_{V_{GS}}(u) + \Delta d_{z(u)}(u) \tag{6}$$

$$\Delta d(u) \triangleq d(u) - d_0(u) = \Delta d_{Vdd}(u) + \Delta d_{V_{GS}}(u) + \Delta d_{z(u)}(u) \tag{7}$$

$$\Delta d_{Vdd}(u) \propto \frac{C_{dyn} \cdot V_{ddL}}{K_n (V_{ddH} - V_{thn})^{\alpha}} + \frac{C_{out} \cdot V_{ddL}}{K_p (V_{ddL} - V_{thp})^{\alpha}} - d_0(u) \tag{8}$$

$$\Delta d_{V_{GS}}(u) \propto \frac{C_{dyn} \cdot V_{dd}}{K_n (V_{ddL} - V_{thn})^{\alpha}} - \frac{C_{dyn} \cdot V_{dd}}{K_p (V_{ddH} - V_{thp})^{\alpha}} \tag{9}$$

$$\Delta d_{z(u)}(u) = \begin{cases} 0 & \text{if } z(u) \geq \tau(u) \\ d_{ATO} & \text{otherwise} \end{cases} \tag{10}$$

The gate delay varies depending on (1) supply voltage of the gate u ($\Delta d_{V_{dd}}(u)$), (2) gate-to-source voltage of the NMOS transistor, namely the supply voltage of fanin gate ($\Delta d_{V_{GS}}(u)$), and (3) arrival time difference of input signal from fanin gates $(\Delta d_{z(u)}(u))$. $z(u)$ denotes arrival time difference between the latest signal and other signal and $\tau(n)$ denotes arrival time difference threshold. d_{ATO} represents increase in delay time due to the reduced input arrival time difference.

Suppose that the supply voltage of a gate u is scaled down. Then, the gate delay $d(u)$ increases by $\Delta d_{V_{dd}}(u)$. Since the supply voltage of a gate u is scaled down, the gate-to-source voltage of the NMOS transistor in fanout gates of the gate u reduces from V_{ddH} to V_{ddL}. As a result, the gate delay $d(v)$ for fanout gate v of u aggravates by $\Delta d_{V_{GS}}(v)$. Another determinant of the fanout gate delay is the variation of input signal arrival time difference that results from gate delay change in u, which is denoted by $\Delta d_{z(v)}(v)$. Such delay variations on the fanout are the integral parts of the timing analysis, which can be extracted from SPICE simulation.

3.2 Definitions and preliminaries

A combinational circuit is represented by a directed acyclic graph $G = (V, E)$. A node $u \in V$ represents a gate in the circuit. There is an edge $(u, v) \in E$ if the output signal of node u is fed directly as an input of node v.

Definition: The arrival time $AT(u)$ for gate u defines the latest path delay from the primary inputs to the gate output of u such that

$$\begin{aligned} AT(u) &= \max_{i \in FI(u)} AT(i) + d_0(u) + \Delta d_{V_{dd}}(u) \\ &\quad + \Delta d_{V_{GS}}(u) + \Delta d_{z(u)}(u) \end{aligned} \tag{11}$$

where $FI(u)$ is a set of fanin gates of gate u.

The required time $RT(u)$ is the latest time the signal must be available at the output of u such that

$$\begin{aligned} RT(u) &= \min_{j \in FO(u)} RT(j) - (d_0(j) + \Delta d_{V_{dd}}(j) \\ &\quad + \Delta d_{V_{GS}}(u, j) + \Delta d_{z(j)}(j)) \end{aligned} \tag{12}$$

where $FO(u)$ is a set of fanout gates of gate u. The slack time $ST(u)$ is defined as the time difference between required time and arrival time such that

$$ST(u) = RT(u) - AT(u). \tag{13}$$

If all the gates in a circuit have non-negative slack time, then the circuit is said to be safe.

Theorem 3.1. *For an edge $(u, v) \in E$ in G, with delay increase by $\Delta d_{V_{dd}}(u)$ in u, if $\Delta AT(v) \equiv \{(d_0(v) + \Delta d_{V_{dd}}(u) + \Delta d_{V_{GS}}(v) + \Delta d_{z(v)}(v)) - (AT(v) - AT(u))\} \geq 0$ then $AT(v)$ increases by $\Delta AT(v)$.*

Definition: Gate v is *arrival-time sensitive* to gate u, or $v \in ATS(u, \Delta d(u))$, if $\Delta AT(v) \geq 0$ with delay increase in u by $\Delta d(u)$.

Theorem 3.2. *A circuit is safe for delay variation $\Delta d(u)$ in gate u, if the slack time for all gates in the arrival time sensitive gate set is non-negative such that*

$$\forall v \in ATS(u, \Delta d(u)), \quad ST(v) \geq 0. \tag{14}$$

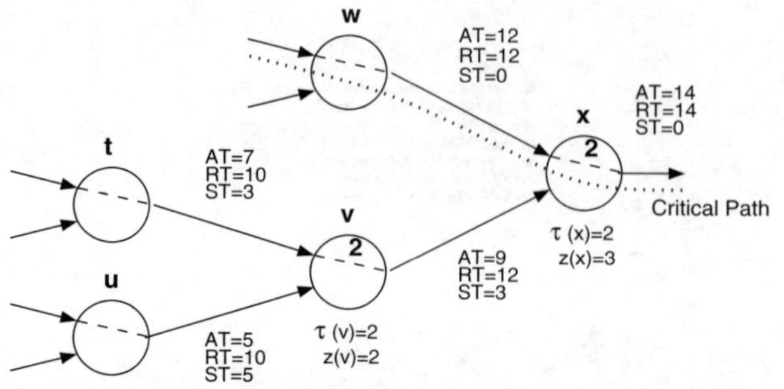

(a) Timing information for high Vdd configuration (circuit delay=15)

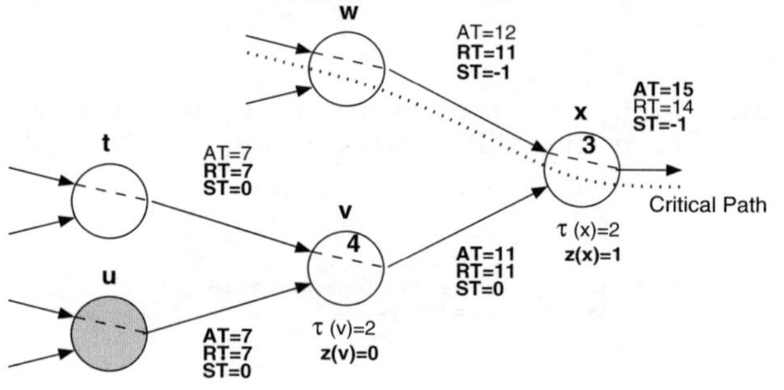

(a) Timing information when gate u has low Vdd (circuit delay=16)

Figure 5: Timing analysis example

Example 3.1.*:* This example shows the proposed timing analysis ensures accuracy by accounting timing variations due to gate-to-source voltage and input signal arrival time. In this example, let us suppose the following. First, for any gates (say u), V_{dd} scaling leads to gate delay increase by 2 (or, $\Delta d_{V_{dd}}(y) = 2$). Second, if a gate (say u) lowers its supply voltage, then the delay time of corresponding fanout gate (say v is a fanout gate of u) increases by 1 (or, $\Delta d_{V_{GS}}(v) = 1$). Third, if the arrival time difference of input signals $z(u)$ is less than arrival time difference threshold $\tau(u)$, then the delay time increases by 1 (or $\Delta d_{z(v) < \tau(v)}(v) = 1$).

Now let us investigate the gate u in Figure 5(b). Voltage scaling on u adds 2 on gate delay. The arrival time for gate u increases by $\Delta d_{V_{dd}}(u) = 2$ resulting in $AT(u) = 7$. Since the arrival time difference $z(v)$ becomes 0, which is less than $\tau(v)$ of 2, the gate delay increase by 1 (namely, $\Delta d_{z(v)}(v) = 1$). Furthermore, V_{GS} reduction on v increases $d(v)$ by $d_{V_{GS}}(v) = 1$. Thus, the arrival time for gate v increases from 9 to 11. Since the updated arrival time difference $z(x)$ of 1 is less than $\tau(x)$ of 2, the delay time of gate x increases by $\Delta d_{z(x)}(x) = 1$. In this particular example, circuit delay aggravates by 1 due to delay variations by gate-to-source voltage and input arrival time difference. Thus, the power supply voltage of gate u can not be scaled down. However, the previous dual supply voltage technique for domino logic does not consider those delay time variations. Thus, it replaces the supply voltage of gate u with V_{ddL}, which results in increasing the critical path delay. ■

```
procedure DCPO (circuit C, delay_constraint d(C))
  Build a circuit graph G for the circuit C;
  Create gate delay table for each gate in library with
    different V_GS and input arrival time difference z;
  Set circuit delay as d(C);
  Initialize all the gates using V_ddH in LVSC scheme;
  repeat
    foreach gate u using V_ddH
      if circuit C is safe after voltage scaling on u
      then
        add u to candidate gate set;
        set the weight of gate u to the gain σ(u);
      end
    end
    Build a timing sensitivity graph
      for the candidate gate set;
    Prune delay-orthogonal gates by applying
      maximal weight independent set algorithm
      to the timing sensitivity graph;
    foreach gate u in resulting delay-orthogonal set
      Update gate delay d(u) := d(u) + Δd_Vdd(u);
      Update gate delay of fanout gates
        according to V_GS change and
        input arrival time difference variation;
      Update arrival time for all gates
        in ATS(u, Δd_Vdd(u));
      Update slack time for all gates
        in ATS(u, Δd_Vdd(u));
    end
  until (candidate gate set is empty)
  return power optimal circuit C_power_opt
end
```

Figure 6: Delay optimization algorithm

4. DELAY CONSTRAINED POWER OPTIMIZATION (DCPO) ALGORITHM

The main idea for power optimization algorithm is to scale down the supply voltage of gates in the non-critical path with delay constraint $d(C)$ that is obtained from initial circuit with V_{ddH}. For each candidate gate u, the gain represents the amount of power-delay product reduction as a result of low V_{dd} application which is formally defined as

$$\sigma(u) = \frac{1}{(1 - \Delta P(u)/P_o)(1 + \Delta D(u)/D_o)} \quad (15)$$

where $\Delta P(u)$ represents saved power, P_o is original power before update, $\Delta D(u)$ is circuit delay overhead, and D_o is the original circuit delay. The more the gain is, the more favorable candidate gate is for V_{dd} reduction.

The pseudo-code of the algorithm is shown in Figure 6. Given a combinational circuit, a circuit graph G is built. For all the gates defined in cell library, generate a timing information base for keeping precharacterized delay variations according to V_{dd}, V_{GS}, and z. Then all the gates are configured to domino cells with high supply voltage in LVSC scheme.

For each iteration, we identify a candidate gate set that saves power while circuit delay overhead is within given delay constraint $d(C)$. If a gate satisfies delay constraint, assign the power-delay product gain in Equation (15) to the gate as a weight, which will be utilized as a priority in candidate gate selection step. Then, in order to prune out a set of gates that are independent of each other with respect to timing variation, timing sensitivity graph is set up. If a gate v is in $ATS(u, \Delta d_{V_{dd}}(u))$, then there is a directed edge from u to v in the timing sensitivity graph. Applying maximal weighted independent set algorithm to the timing sensitivity graph yields a set of temporally orthogonal gates. When two gates u and v are chosen as a result of this step, then gate delay variation of one gate does not affect timing information such as arrival time, required time, or slack time of the other gate. Thus, timing information of all the gates in this set can be updated at the same time. This iteration continues not until candidate gate set is empty. Finally, the procedure yields power-optimal circuit configuration with given delay overhead.

5. EXPERIMENTAL RESULTS

To simplify the analysis, technology mapping is carried out to map the logic to a library which contains AND and OR gates. For domino logic realization, all the internal inverters are eliminated by using duplication technique.

Initially, all logic gates use V_{ddH}. First, all logic gates are replaced with logic gates using V_{ddH} in LVSC scheme. The replacement results for ISCAS85 benchmark circuits are shown in column "mVDDH_LVSC" in Table 1. All the values shown in Table 1 are delay improvement and power reduction percentages with respect to the initial circuits with V_{ddH}. "Δt_D" denotes critical path speed-up. "ΔP_{logic}" and "ΔP_{std}" denote dynamic and standby power reductions, respectively. We assume that all domino logic gates are precharged in standby mode. Critical path delay time is improved up to 11.68% with small amount of increase in dynamic and static power consumptions. However, total dynamic power consumption can be saved significantly since low voltage clock swing is used. Then, DCPO algorithm is applied to further reduce the total dynamic power consumption within delay constraint. The results for ISCAS85 benchmark circuits are shown in column "LVSC_DualVDD" in Table 1. "Rep" denotes the percentages of high supply voltage replacement using low supply voltage. The dynamic power saving in logic gates ranges from 13.54% to 36.94% and the standby power saving in logic gates ranges from 5.63% to 27.61%. The dynamic power saving in logic tree ("ΔP_{clk}") is assumed to be 44.44% since clock swing voltage is reduced to 1.0V from 1.5V.

Table 2 shows the total dynamic power saving for ISCAS85 benchmark according to the ratio of P_{clk} over P_{total}. "Conv." and "LVSC" denote the total dynamic power saving without and with low swing clock, respectively. If P_{clk} is 20% of P_{total}, the total dynamic power saving of LVSC scheme increases up to 29.77% on average while the total dynamic power saving of conventional scheme is 20.81% on average. For "C432" and "C6288", power saving of LVSC scheme is around 2 times bigger than that of conventional scheme. If P_{clk} is 50% of P_{total}, the dynamic power saving of LVSC scheme is more than 2.5 times that of conventional scheme on average. Thus, the power saving of proposed LVSC scheme is significantly greater than that of conventional scheme.

Table 1: Delay speed-up and power saving for ISCAS85 benchmark circuits

Circuit	Gates	mVDDH_LVSC			LVSC_DualVDD				
		Δt_D	ΔP_{logic}	ΔP_{std}	Rep	Δt_D	ΔP_{logic}	ΔP_{clk}	ΔP_{std}
C432	171	9.34%	-2.05%	-2.73%	40.35%	4.18%	14.86%	44.44%	11.60%
C880	376	11.06%	-1.78%	-2.75%	80.21%	4.93%	31.76%	44.44%	23.30%
C1355	669	11.26%	-1.57%	-2.98%	36.31%	6.04%	13.54%	44.44%	5.63%
C1908	550	11.34%	-1.67%	-2.87%	63.46%	1.86%	24.77%	44.44%	17.84%
C2670	953	9.73%	-1.70%	-2.89%	92.67%	6.36%	36.94%	44.44%	27.61%
C3540	1492	11.68%	-1.78%	-2.84%	81.62%	0.57%	32.34%	44.44%	24.18%
C5315	2133	11.28%	-1.60%	-2.88%	86.00%	0.00%	34.19%	44.44%	25.50%
C6288	4588	10.33%	-1.78%	-2.84%	45.70%	0.00%	14.68%	44.44%	17.30%
C7552	3048	9.92%	-1.55%	-2.95%	91.57%	0.00%	31.85%	44.44%	27.51%
Avg.		10.66%	-1.72%	-2.86%	68.65%	2.60%	26.10%	44.44%	20.05%

Table 2: Total dynamic power saving for ISCAS85 benchmark circuits according to the ratio of P_{clk} over P_{total}

Circuit	20%		30%		40%		50%	
	Conv.	LVSC	Conv.	LVSC	Conv.	LVSC	Conv.	LVSC
C432	11.89%	20.78%	10.40%	23.73%	8.92%	26.69%	7.43%	29.65%
C880	25.41%	34.30%	22.23%	35.56%	19.06%	36.83%	15.88%	38.10%
C1355	10.83%	19.72%	9.48%	22.81%	8.12%	25.91%	6.77%	28.99%
C1908	19.82%	28.70%	17.34%	30.67%	14.86%	32.64%	12.39%	34.61%
C2670	29.55%	38.44%	25.86%	39.19%	22.16%	39.94%	18.47%	40.69%
C3540	25.87%	34.76%	22.64%	35.97%	19.40%	37.18%	16.17%	38.39%
C5315	27.35%	36.24%	23.93%	37.27%	20.51%	38.29%	17.10%	39.32%
C6288	11.74%	20.63%	10.28%	23.61%	8.81%	26.58%	7.34%	29.56%
C7552	25.48%	34.37%	22.30%	35.63%	19.11%	36.89%	15.93%	38.15%
Avg.	20.88%	29.77%	18.27%	31.60%	15.66%	33.44%	13.05%	35.72%

6. CONCLUSION

Domino logic has been widely used in high performance circuit design. Even though high performance and small area can be achieved using domino logic gate, the power consumption of domino logic is greater than that of static logic due to clock network and high switching probability. In this chapter, dual supply voltage technique for domino logic is proposed to save total power consumption. To reduce the power consumption in clock tree, low voltage swing clock domino logic family is proposed. Low voltage swing clock scheme utilizes low V_t transistor for NMOS evaluation transistor in the critical path to overcome performance degradation due to low voltage swing clock with the cost of tiny amount of I_{sub}. The delay variations according to the gate-to-source voltage and the input arrival time difference when supply voltage is replaced are elaborated in timing analysis. The delay constrained power optimization algorithm with proposed low swing clock domino logic yields significant power saving while maintaining the critical path delay.

7. ACKNOWLEDGMENTS

This work was supported in part by Semiconductor Research Corporation under Grant SRC891.

8. REFERENCES

[1] C. Chen, A. Srivastava, and M. Sarrafzadeh. On gate level power optimization using dual-supply voltages. *IEEE Trans. VLSI Systems*, 9(5):616–629, Oct. 2001.

[2] Z. Chen, M. Johnson, L. Wei, and K. Roy. Estimation of standby leakage power in CMOS circuits considering accurate modeling of transistor stacks . In *Proc. Int. Symp. Low Power Electronics and Design*, pages 239–244, 1998.

[3] M. Igarashi, K. Usami, K. Nogami, F. Minami, Y. Kawasaki, T. Aoki, M. Takano, C. Mizuno, T. Ishikawa, M. Kanazawa, and S. S. A Low-power design method using multiple supply voltages. In *Proc. Int. Symp. Low Power Electronics and Design*, pages 36–41, 1997.

[4] S. M. Kang and Y. Leblebici. *CMOS Digital Integrated Circuits: Analysis and Design*. McGraw-Hill, 2nd ed., 1999.

[5] B. Ramasubramanian, H. Schmit, and L. R. Carley. Mixed-swing QuardRail for low power dual-rail logic logic. In *Proc. Int. Symp. Low Power Electronics and Design*, pages 82–84, 1999.

[6] A. Rjoub, O. Koufopavlou, and S. Nikolaidis. Low power / low swing domino CMOS logic . In *Proc. IEEE Int. Symp. Circuits and Systems*, pages II.13–II.16, 1998.

[7] T. Sakurai and A. R. Newton. Alpha-power law mosfet model and its application to CMOS inverter delay and other formulas. *IEEE J. Solid State Circuits*, 29(6):655–662, June 1994.

[8] S. J. Shieh and J. S. Wang. Design of low-power domino circuits using multiple supply voltages. In *Proc. IEEE Int. Conf. Electronics, Circuits and Systems*, pages 711–714, 2001.

[9] Y. J. Yeh, S. Y. Kuo, and J. Y. Jou. Converter free multiple scaling techniques for low-power CMOS digital design. *IEEE Trans. Computer-Aided Design*, 20(1):172–176, Jan. 2001.

DRG-Cache: A Data Retention Gated-Ground Cache for Low Power[1]

Amit Agarwal, Hai Li, and Kaushik Roy
School of Electrical and Computer Engineering
Purdue University
West Lafayette, IN 47906, USA

<amita, hl, kaushik> @ ecn.purdue.edu

ABSTRACT

In this paper we propose a novel integrated circuit and architectural level technique to reduce leakage power consumption in high performance cache memories using single Vt (transistor threshold voltage) process. We utilize the concept of Gated-Ground [5] (NMOS transistor inserted between Ground line and SRAM cell) to achieve reduction in leakage energy without significantly affecting performance. Experimental results on gated-Ground caches show that data is retained (DRG-Cache) even if the memory are put in the stand-by mode of operation. Data is restored when the gated-Ground transistor is turned on. Turning off the gated-Ground transistor in turn gives large reduction in leakage power. This technique requires no extra circuitry; row decoder itself can be used to control the gated-Ground transistor. The technique is applicable to data and instruction caches as well as different levels of cache hierarchy such as the L1, L2, or L3 caches. We fabricated a test chip in TSMC 0.25μ technology to show the data retention capability and the cell stability of DRG-cache. Our simulation results on 100nm and 70nm processes (Berkeley Predictive Technology Model) show 16.5% and 27% reduction in consumed energy in L1 cache and 50% and 47% reduction in L2 cache with less than 5% impact on execution time and within 4% increase in area overhead.

Categories and Subject Descripters

B.3.2 [**Memory Structure**]: Design Styles --- *Cache memories;* B.3.1 [**Memory Structure**]: Semiconductor Memories --- *Static memory (SRAM)*; B.7.1 [**Integrated Circuits**]: Types and Design Styles --- *Memory technology.*

General Terms: Design, Performance and Experimentation.

Keywords: Gated-ground, SRAM, low leakage cache.

1 INTRODUCTION

Semiconductor devices are aggressively scaled each technology generation to achieve high integration density while the supply voltage is scaled to achieve lower switching energy per device. However, to achieve high performance there is need for commensurate scaling of the transistor threshold voltage [6]. Scaling of transistor threshold voltage is associated with exponential increase in sub-threshold leakage current [3]. To overcome the excessive leakage problem, advanced leakage control methods will become indispensable for future technologies.

State-of-the-art microprocessor designs devote a large fraction of the chip area to memory structures — e.g., multiple levels of instruction and data caches, translation look-aside buffers, and prediction tables. For instance, 30% of Alpha 21264 and 60% of StrongARM are devoted to cache and memory structures [4]. Caches account for a large (if not dominant) component of leakage energy dissipation in recent designs, and will continue to do so in the future. Recent energy estimates for 0.13μ process technology indicate that leakage energy accounts for 30% of L1 cache energy and as much as 80% of L2 cache energy [2]. To address the power inefficiency, we propose a single Vt (transistor threshold voltage) Data Retention Gated-Ground Cache (DRG-Cache) design and architecture, in which the unused portions of the memory core are set to low leakage mode to reduce power.

Many embedded designs [1], instead of gating the cell, use circuit only techniques [7] and primarily rely on a dual-threshold voltage (dual-Vt) process technology [8,9] to reduce leakage. Dual Vt makes sense till Vdd is 1V and Vt is around 0.25v due to Vt spread [8] and also it requires extra process cost. By providing an alternative solution, our single Vt integrated circuit/architecture approach to reduce leakage for high-performance designs offers a key advantage over the dual-Vt approach.

The Dynamically Resizable (DRI) I-cache presented in [5] varies the size of the L1 cache by turning off (using gated-Ground NMOS transistor between the Ground and SRAM cell) the unused section to reduce leakage. In this work we investigate the data retention capability of the DRG-cache having gated-Ground, its circuit level implications and architectural mechanisms. Our theoretical analysis and fabricated chip results suggest that data can be retained in DRG-Caches even when the gated-Ground transistor is turned-off. We propose that the data in the cache be in the data-retenting mode (gated-Ground transistor 'off') all the time unless we are reading/writing from that cell. Our approach is easily applicable to both data and instruction L1 as well as L2 cache with very little circuit overhead.

2. DRG-CACHE: CIRCUIT, ARCHITECTURE AND LAYOUT

To prevent the leakage energy dissipation in a DRG-Cache from limiting aggressive threshold-voltage scaling, we use a circuit-level mechanism called gated-Ground [5]. Gated-Ground enables a DRG-Cache to effectively turn 'off' the supply voltage and

---

[1]This research was funded in part by DARPA, MARCO GSRC, SRC, and by Intel and IBM.

DAC 2002, June 10-14, 2001, New Orleans, Louisiana, USA.

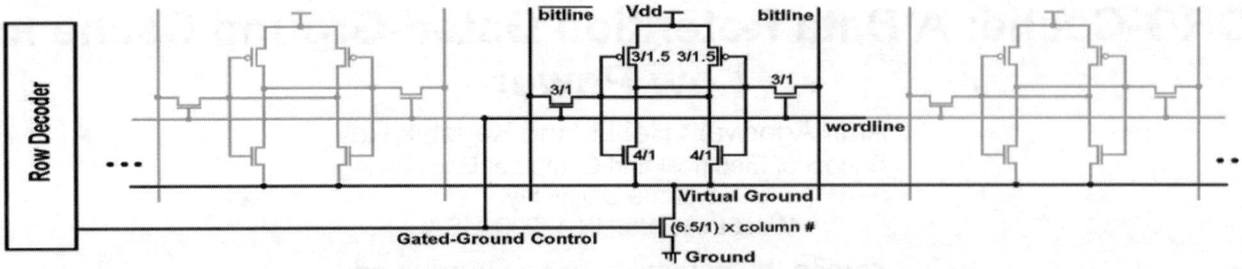

Figure 1. Anatomy of DRG-Cache : Data Retention Gated-Ground Cache.

virtually eliminate the leakage energy dissipation in the cache's unused (used section of the cache core is defined as the SRAM cells from which data is read/written) sections. The key idea is to introduce an extra NMOS transistor (Figure 1) in the leakage path from the supply voltage to the ground of the cache's SRAM cells; the extra transistor is turned 'on' in the used and turned 'off' in the unused sections, essentially "gating" the cell's supply voltage. Figure 1 shows the anatomy of DRG-Cache. Gated-Ground achieves significantly lower leakage because of the two off transistors connected in series reducing the leakage current by orders of magnitude; this effect is due to the self reverse-biasing of stacked transistors, and is called the stacking effect [10].

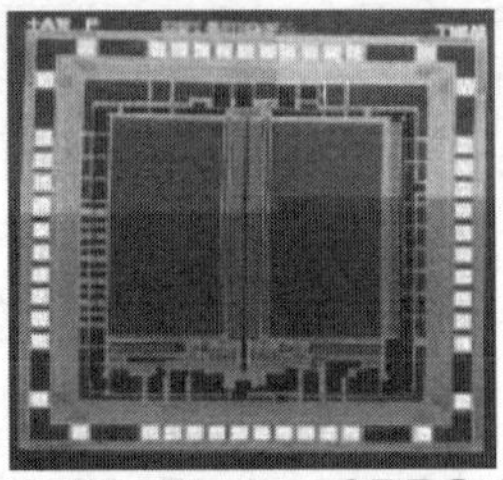

Figure 2. Die Photo of DRG-Cache.

Same as conventional gating techniques, the gated-Ground transistor can be shared among multiple SRAM cells from one or more cache blocks. It amortizes the overhead of the extra transistor. Because the size of gated-Ground transistor plays a major role in the data retention capability and stability (section 4) of the DRG-Cache, and also affects the power and performance savings (section 6), the gated-Ground transistor must be carefully sized (Figure 1) with respect to the SRAM cell transistors it is gating. While the gated-Ground transistor must be made large enough to sink the current flowing through the SRAM cells during a read/write operation in the active mode and to enhance the data retention capability of cache, too large a gated-Ground transistor may reduce the stacking effect, thereby diminishing the energy savings. Moreover, large transistors also increase the area overhead due to gating. In DRG-Cache the gated-Ground transistor is shared by a row of SRAM cells. The gated-Ground transistor is controlled by the row decoder logic of conventional SRAM. The cells are turned 'on' only when the row is being read from or when data is written into the row. However, this requires that the row decoder drives a larger gate capacitance associated with gated-Ground transistor unlike conventional caches. To maintain performance proper sizing of decoder is required.

Figure 2 shows the die photo of the fabricated DRG-cache along with a conventional cache (to the left) for comparison. The right most thin column in layout is the gated Ground transistor. To minimize the area overhead and optimize layout, we implemented gated-Ground transistor as rows of parallel transistors placed along length of the SRAM cells. This row of parallel transistor is placed at one end of row of SRAM cells. The metal line, which is used as a ground line in conventional cache, is used to connect the drain of gated-Ground transistor and the SRAM cell and acts as virtual ground. Wordline connecting each cell of a row to a row decoder is connected to the gate of the gated-Ground transistor. As it is clear from the layout, the area overhead due to gated-Ground transistor is about 4%. It is important to note that the DRG-Cache core is fully compatible with current cache design.

3. FACTORS AFFECTING THE STORED DATA IN DRG-CACHE

Conventional SRAM store the data as long as power supply is on. This is because the cell storage nodes at '0' and '1' are firmly strapped to power rails through conducting devices (by a pull-down NFET in one inverter and a pull-up PFET in the other inverter). Figure 3 shows a single cell schematic of our DRG-Cache. When the gated-Ground transistor is on, it behaves exactly like conventional SRAM in terms of storing data. Turning 'off' the gated-Ground nicely cuts-off the leakage path from the cell node that is at '1' to ground. However, it also cuts-off the opportunity to strap the cell node at '0' firmly to ground. This makes it easier for a noise source to write a '1' to that node. Node storing '1' remains firmly strapped to Vdd as long as input ($\overline{Q}$) to the pull-up PFET (M4) remains below the trip point of the inverter.

Secondly, there is an issue with data retention. If the sub-threshold current through pull-up PFET (M2) and through pass transistor (M6) is greater than sub-threshold current through gated-Ground via pull down NFET (M1) or in other words if resistance through the pull-up PFET (M2) and pass transistor (M6) be much less than the sub-threshold resistance of gated-Ground transistor, then a voltage divider may cause the node at '0' to rise high. This has a possibility of degrading the current drive of the PFET (M4) and increasing the leakage of the NFET (M3) in the inverter driven by '0' node. This case has potential, in worst case, to destroy data written into the cell and in the best case to degrade the stacking effect.

4. DATA RETENTION CAPABILITY OF DRG-CACHE

The leakage current in MOSFET depends on various process parameters, transistor size and the quiescent state of the circuit. We use leakage model in [10] for modeling our leakage current. This model uses the following simplified BSIM sub-threshold current equation.

$$I_{subth} = A e^{\frac{q}{nkT}(V_G - V_S - V_{TH0} - \gamma' V_{SB} + \eta V_{DS})} (1 - e^{-\frac{qV_{DS}}{kT}}) \quad (1)$$

where $A = \mu_0 C'_{ox} \frac{W}{L_{eff}} (\frac{kT}{q})^2 e^{1.8}$

V_G, V_D and V_S are gate voltage, drain voltage, and source voltage of the transistor, respectively. Body effect is represented by the term $\gamma' V_S$, where γ' is the linearized body effect coefficient. η is the DIBL coefficient, representing the effect of V_{DS} on threshold voltage. C_{ox} is the gate oxide capacitance. μ_0 is the zero bias mobility and n is the sub-threshold swing coefficient of the transistor.

We do our leakage modeling at the time when gated-Ground transistor is turned 'off' and DRG-Cache is in low leakage mode (standby leakage mode). Initially '1' is written at Q and '0' at $\overline{Q}$. In the Figure 3, the light color transistors are 'on', and dark color transistors are 'off'. We assume that the voltage at $\overline{Q}$ goes to saturation after sometime and this saturation value is Vg. This Vg is small enough to keep the pull-up PFET (M4) 'on'. Because M4 remains 'on', it firmly straps the node Q to Vdd supply rail. This in turn keeps the NFET (M1) 'on'. Because M1 is 'on', voltage at virtual ground follows the voltage at $\overline{Q}$ and goes to Vg. This makes both gate voltage and source voltage of pull-down NFET (M3) at Vg, and turns 'off' M3 because V_{GS} is '0'.

From Figure 3 we observe that the transistors responsible for charging $\overline{Q}$ node are M2 and M6. For considering the worst-case condition we assume that the $\overline{\text{bitline}}$ remains at Vdd even if pre-charging is not applied for a long period of time, and the leakage associated with pass transistor M6 is maximum (Eq 1).The voltage at $\overline{Q}$ is discharged through capacitance leakage and leakage through gated-Ground transistor. Equating all the current at time of voltage saturation we get

$$I_{subM7} = I_{subM3} + I_{subM6} + I_{subM2} - I_{cap}\ (\text{neglected})$$

where, the subthreshold leakage currents are given as

$$I_{subM7} = A_7 e^{\frac{q}{nkT}(-V_{TH0} + \eta V_g)}$$
$$I_{subM3} = A_3 e^{\frac{q}{nkT}(-V_{TH0} - \gamma' V_g + \eta(V_{DD} - V_g))} \quad (2)$$
$$I_{subM6} = A_6 e^{\frac{q}{nkT}(-V_{TH0} - (1 + \gamma' + \eta)V_g + \eta V_{DD})}$$
$$I_{subM2} = A_2 e^{-\frac{q}{nkT}(-V_{THOP} + \eta(V_g - V_{DD}))}$$

We use the parameters from TSMC 0.25μ technology library and sizes of the transistors from our cache layout in our leakage model. In our case, V_g turns out to be around 0.4V. Here we have neglected the capacitance leakage. However, if we consider the capacitance leakage, it will reduce the voltage at $\overline{Q}$. This in turn, because of the feedback, affects positively in turning on transistor M4. Because Vg is small enough to turn on pull-up PFET (M4), our initial assumptions are correct. Q is strapped to V_{DD} through low resistance, and hence it always remains at '1'. Turning on M7 will restore the data by pulling $\overline{Q}$ to '0' through M3 and M7.

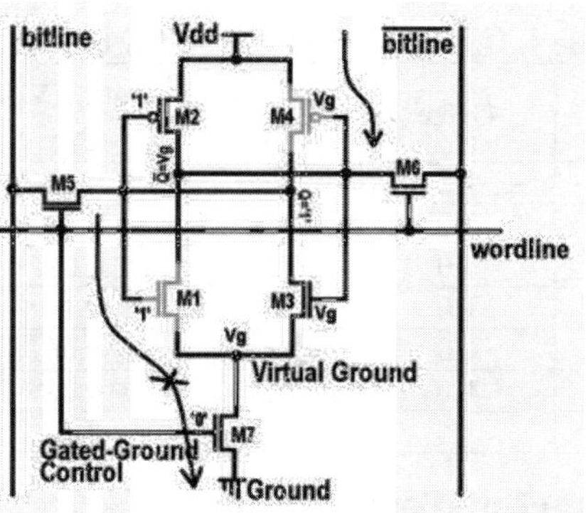

Figure 3. DRG-Cache: Data Retention Capability.

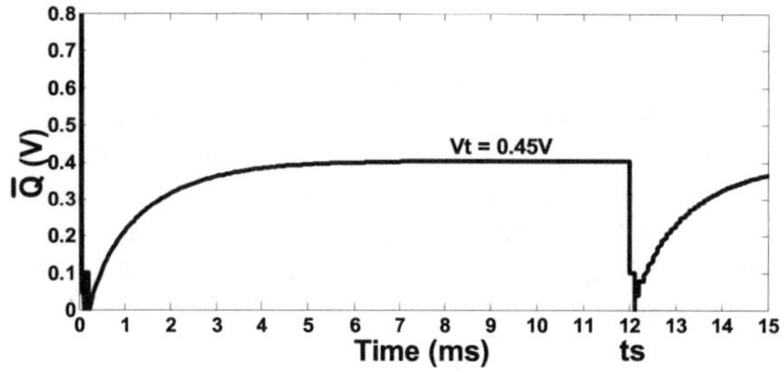

Figure 4. Voltage variation at node storing '0'.

We simulated the extracted net-list from layout of the DRG cache. After time t_s gated-Ground is again turned on. The voltage at node storing '1' always remains at Vdd. Figure 4 shows the voltage variation at $\overline{Q}$ (using TSMC 0.25μ process technology and supply voltage of 2.5V). As the gated-Ground is turned off the voltage at $\overline{Q}$ starts rising and gets saturated to 400mv. Turning on the gated-Ground transistor restores the data at $\overline{Q}$ to zero. This result shows that data is not lost even if we turn off the gated-Ground transistor for indefinite time.

Table 1 shows the simulation results for saturation voltage at node storing '0' using TSMC 0.25μ technology. Unless otherwise, mentioned the temperature for all simulations are considered as 25°C, which is worst case for data retention. Here Gsize is the relative size of gated-Ground transistor normalized with respect to the size of NMOS M3 shown in figure 3. Vtn_0 is the zero bias threshold voltage of the NMOS transistors of SRAM cell. Vtp_0 is kept at −0.50V. Sat $\overline{Q}$ is the saturation voltage at node storing 0($\overline{Q}$).

4.1 Impact of Widening the Gated-Ground Transistor

First three rows of Table 1 indicate that the saturation voltage of node storing '0' can be made as small as possible by increasing the size of gated-Ground transistor (Gsize). Increasing the size of gated-Ground increases the discharging current for node storing '0', which reduces the saturation voltage. This also degrades the power savings and improves the performance. Hence, there is a trade off between performance, stability and power dissipation. However, increasing the size of gated-Ground only affects leakage linearly. Recall, the effect of gated-Ground on leakage is exponential. Hence, we can get higher stability without losing much in terms of leakage reduction.

Table 1. Data Retention Capability of DRG-Cache

Temp (°C)	Gsize	Vtn_0 (V)	Sat $\overline{Q}$ (Vg) (V)	ΔVt_{max} (mV)
25	1.0	0.45	0.40	150
25	2.0	0.45	0.22	157
25	4.0	0.45	0.08	165
50	1.0	0.45	0.11	111
100	1.0	0.45	0.06	97
25	1.0	0.35	0.08	85
25	1.0	0.25	0.07	78

4.2 Impact of Temperature

The temperature of a chip can vary depending on the workload and power consumption. The leakage current increases exponentially with temperature (Eq. 1). However, this current increment is more in NFETs as compared to PFETs, because NFET parameters (for example μ, η, and γ) are stronger than PFET for contributing to leakage current. Hence, the discharging current is stronger than charging current in high temperature case. This makes Vg to saturate at a lower voltage. First, fourth and fifth rows of Table 1 indicate the impact of temperature on saturation voltage.

4.3 Impact of Lowering Vt

It is evident from the first, sixth and seventh rows of Table 1 that voltage at node storing '0' gets saturated at lower voltage with lower Vt process technology. This is due to fact that lowering the Vt increases the leakage current of all the transistors and since the discharging current is stronger than the charging current in the low Vt case, the saturation voltage Vg is lower.

4.4 Stability Analysis Under Vt Variation

We observed in the previous section that the voltage at the node storing '0' changes as we change the Vt of the cell. If this value crosses the trip point of the cell (which is close to Vdd), the cell may flip at the time of reading. In current process technology and on upcoming technology generations it is very difficult to fabricate transistors all having exact same Vt. The Vt mismatch may cause destructive read in the worst case. We simulated the DRG-cache to determine the effect of Vt variation on stored data. For worst-case analysis we assigned transistors M2, M3, M5 and M6 (Figure 3) low Vt (Vt - ΔVt) and transistors M1, M4 and M7 high Vt (Vt + ΔVt) transistors. We let the data to saturate for a long time (ts) and after that we try to read it (Figure 5). The saturation voltage rises as ΔV_t approaches close to 0.152V. However, we are able to read the data. Changing the ΔV_t to 0.153V flips the cell, i.e. the data cannot be read correctly from the cell ($\overline{Q}$ remains at Vdd after time ts).

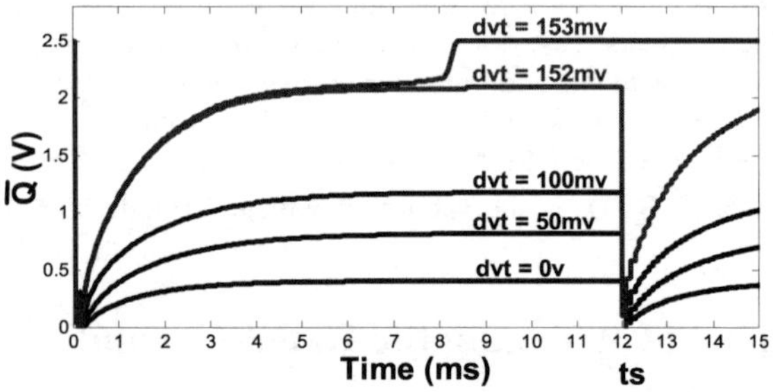

Figure 5. Stability analysis under Vt variation

The last column of Table 1 shows the limitations on V_t variation on various cases to avoid destructive read. Comparing the first row with the third and fourth row we observe that increasing the size of gated-Ground transistor has positive effect on stability. The stability of the cell is sensitive to lowering the Vt_0 of all the transistors including the gated-Ground transistor. We observe that ΔVt_{max} is more than 75mV (baseVtn_0 = 0.25V) in worst case and up to 165mv (base Vtn_0 = 0.45V) in best case.

4.5 Impact of Technology Scaling

We simulated DRG-cache using Berkeley Predictive Technology Model (BPTM) [11] for 70nm and 100nm technology after scaling our netlist. The Vdd used for these technologies are 1.0 and 1.2 V, respectively. Table 2 shows that even for these technologies, voltage at node storing '0' gets saturated to a value which is small compared to Vdd. This saturation voltage depends on supply voltage and leakage currents for respective technology.

Table 2. Impact of technology scaling on data retention capability of DRG-cache

Supply Voltage (V)	Technology	Sat $\overline{Q}$ (Vg) (V)	ΔVt_{max} (mV)
1.2	100nm	0.21	103
1.0	70nm	0.16	79

5. ENERGY AND DELAY ANALYSIS

A DRG-cache reduces leakage energy by gating ground to cache unused sections (gated-Ground turned off). However, it increases dynamic energy at the time of reading/writing due to overhead associated with extra gated-Ground transistor. The row decoder has to drive the gate capacitance associated with gated-Ground transistor, which in turn increases the delay and power.

In conventional cache, at the time of reading/writing, only one row (selected by row decoder) contributes to dynamic energy and the rest of the rows remain inactive (high temperature) leakage mode. If the cache is not accessed, all the rows leak actively. In our DRG-Cache architecture, only gated-Ground transistor associated with accessed row remains on, and all others remain off. Hence, all the rows except the accessed one remains in the standby leakage mode. When the cache is not accessed all the rows remains in standby leakage mode.

Figure 6. Breakdown of delay in cache.

Figure 6 shows the percentage contribution of all the four parts: decoder, wordline, bitline/sense amplifier, databus, in overall delay. This estimation is based on our simulation. The delay of overall cache is dominated by decoder and bitline/sense amplifier delay. The wordline contributes only 3% to overall delay.When we read from conventional cache, bitline gets discharged through pull-down NFET and pass-transistor NFET connected to node storing '0'. Putting an extra transistor (gated-Ground transistor) increases the resistance of discharging path, which in turns degrades the bitline/sense-amplifier delay in DRG-Cache compared to conventional cache. Secondly, the gated-Ground transistor is controlled by row decoder, which poses extra overhead on wordline driver. This increases the wordline delay of DRG-Cache compared to conventional cache. However, the contribution of wordline delay to overall cache delay is very small [12,14]. Hence, overall performance of DRG-Cache does not get much affected by increase in wordline delay.

6. RESULTS

In this section, we present experimental results on the energy and performance trade-off of a DRG-Cache as compared to a conventional-cache. Due to ease of fabrication we utilize the 0.25μ technology results to validate the data retention capability, stability and performance-leakage saving trade-off in DRG-Cache. We scale down our net-list for 100nm and 70nm technology for showing energy savings achieved by DRG-Cache.

6.1 Energy Performance Trade-off

Table 3 shows the leakage savings and relative read time for different gated-Ground implementations. Here relative read time is the read time of DRG-Cache with respect to corresponding conventional cache. Conv (conventional) cache leakage and DRG-Cache leakage shown are leakage energy normalized with respect to leakage energy of conventional cache at $Vtn_0 = 0.45V$.

Table 3. Energy Performance Trade-off

Gsize	Vtn_0 (V)	Normalized Conv. Cache leakage	Normailized DRG Cache Leakage	Relative Read Time
2.0	0.45	1.0	0.68	1.028
1.0	0.45	1.0	0.54	1.046
1.0	0.35	9.0	5.2	1.044
1.0	0.25	80.8	47.8	1.044

First two rows of Table 3 and Table 1 indicate that increasing the width of the gated-Ground transistor improves the read time, data retention capability and stability of cell but decreases energy savings and increases the area. The read time improves due to increase in drive current because of large sized gated-ground transistor. However, at the same time decoder driver has to drive large capacitance due to increase in capacitance associated with gated-Ground transistor, which increases the wordline delay and dynamic energy of cache. Because wordline delay is a small fraction of overall delay the overall read time is improved. Last two rows of Table 3 indicate that lowering the threshold voltage increases the leakage energy by several orders of magnitude. The DRG-Cache leakage energy is orders of magnitude smaller than the conventional cache leakage energy. The motivation behind lowering the threshold voltage is to get high performance cache, which at the same time improves the data retention capability but degrades the stability of the cell (Table 1).

6.2 Energy Savings and Power-Delay Products for 70nm and 100nm Technology

6.2.1 Utilization Factor

We simulated simplescalar [13] for getting the information about the cache utilization factor. In the context of aggressive modern out-of-order microprocessors, which exploit instruction level parallelism [15] we assume that L1 cache is accessed in each and every cycle (conservative assumption). The idle time of L2 cache depends on both number of load instruction in a cache and miss rates of both L1 and L2 cache. L2 accesses are generated by L1 misses, so lower the miss rate of L1, higher the idle time of L2 cache. L2 cache utilization is also affected by L2 miss rate and the number of load instruction in an application. Our simulation for spec95 benchmarks show that on an average L2 cache is not accessed 80% of time. This indicates that most of the time L2 remains in the idle mode, which in turn makes leakage energy of L2 to dominate over its dynamic energy. Hence, the DRG-Cache is ideal for reducing leakage in L2 cache.

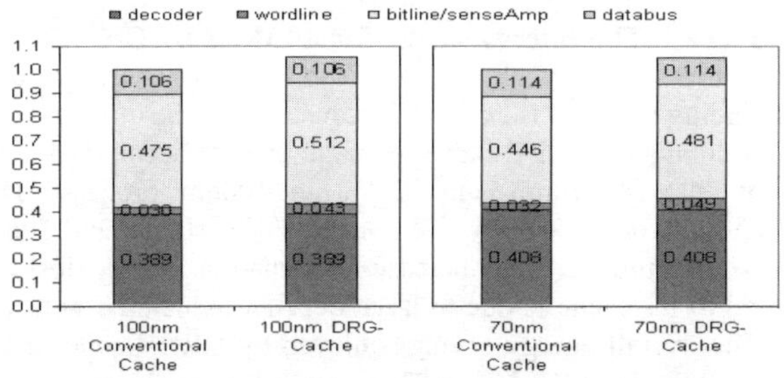

Figure 7. Performance loss in DRG-Cache.

6.2.2 Performance Loss

The DRG-Cache loses performance against conventional cache due to increase in wordline and bitline/sense amplifier delays. Figure 7 depicts the contribution of four stages of cache in overall delay. It also depicts the increase in delay in different stages with respect to conventional cache and its effect on overall delay. The graph shows that wordline delay increases by more than 43% in 100nm and 53% in 70nm technology, while the increase in bitline/sense amplifier delay is only 7.8% in both technologies. However, the contribution of wordline delay is only 3% to the overall delay causing overall performance degradation to be 5.2 – 5.3% in respective technology.

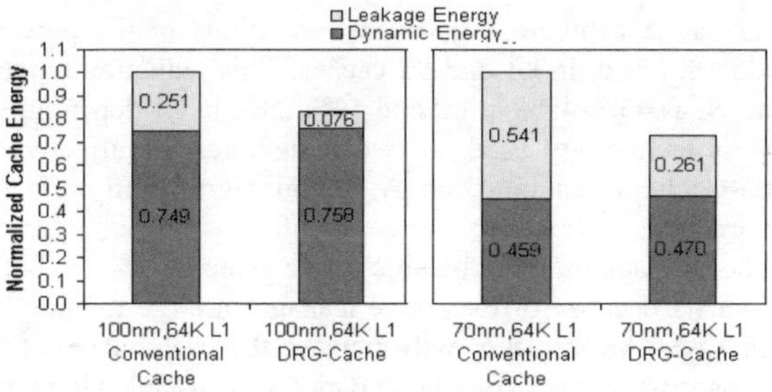

Figure 8. The energy savings for a 64K L1 DRG-Cache.

6.2.3 Energy Savings

Figure 8 shows the contribution of leakage and dynamic energy in L1 cache based on energy analysis given in section 5 and measured utilization factor. Assuming total energy of conventional cache is 1, we normalized leakage and dynamic energy consumption of DRG-Cache to the conventional cache. The graph indicates that leakage energy accounts for 25% in 100nm and 54% in 70nm technology for L1 caches. DRG-Cache saves around 70% leakage in 100nm and 51% in 70nm process. The energy overhead associated with decoder results in 1.2% and 2.3% increase in total dynamic energy of L1 cache for respective process technologies. The overall energy reduction achieved by DRG-Cache is as much as 27% in 70nm and 16.5% in 100nm process technology.

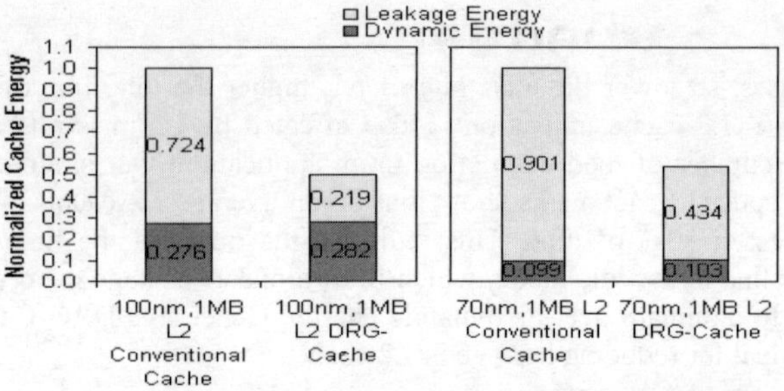

Figure 9. The Energy saving for a 1M L2 DRG-Cache.

Figure 9 shows the normalized (similar to Figure 8) energy results corresponding to 1 MB L2 Cache. Because L2 Cache utilization is small as compared to L1 cache, leakage energy accounts for as much as 90% in 70nm and 72.4% in 100nm process. The increment in dynamic energy is 2.1% and 4% in 100nm and 70nm process respectively. This increase in dynamic energy loss as compared to L1 cache is due to large decoder associated with L2 cache. The overall energy saving achieved by DRG-Cache in L2 is 50% in 100nm and 46.4 % in 70nm process technology.

In both L1 and L2 caches, the leakage power saving for 100nm technology is larger than 70nm technology, though it is shown in [16] that the leakage savings due to stacking is expected to increase with scaled technologies. However, for the Berkeley Predictive Technology Model [11] the subthreshold swing S is smaller in 100nm technology compared to 70nm, and hence we observe more leakage savings in 100nm. For L1 cache, the leakage energy contribution to the overall energy for 70nm is larger than the leakage contribution for 100nm process. For L2 it is approximately same for both the process technologies, which makes the overall energy savings to be larger for 70nm than for 100nm process in L1 cache. Hence, for L2 cache, the overall energy savings using DRG with 100nm process is larger than the energy savings using 70nm process, unlike L1 cache.

DRG-Cache achieves significant reductions in the power-delay product for both in L1 and L2 caches. This reduction ranges from as much as 43%-47% in L2 and 12%-23% in L1 depending on the process technology used. This demonstrates the effectiveness of DRG-Cache in reducing energy without significantly affecting the performance. However, one should note that for scaled technology one may not be able to use same low Vt for both logic and cache because of excessive leakage in large memory arrays. High Vt SRAMs in turn will increase the cache access time and process cost. Hence, our technology is a usable alternative for deep submicron design.

7 CONCLUSIONS

In this paper we explored an integrated circuit and architectural level approach to reduce leakage energy dissipation in deep-submicron cache memories while maintaining high performance. DRG-Cache utilizes inherent idleness of cache to save leakage by turning off the idle sections of the SRAM core. The key observation in this paper is that data is not lost when the gated-Ground transistor is turned off in the unused sections of the cache. A fabricated test chip in 250nm technology verified the feasibility of our approach.

Our simulations for 100nm and 70nm technologies using single V_t approach show that DRG-Cache saves around 16.5% and 27% energy in L1 and 50% and 47% energy in L2 cache, with approximately 5.2% and 5.3% impact on performance for 100nm and 70nm process, respectively, compared to conventional caches with same low Vt. The power-delay products show 12% and 23% improvement for L1 and 47% and 43% improvement for L2 compared to conventional design, for respective process technologies. This shows the effectiveness of DRG-Cache in saving leakage for contemporary as well as future cache designs.

8 REFERENCES

[1] J. Montanaro et. al. A 160-MHz, 32-b, 0.5-W CMOS RISC microprocessor. IEEE Journal of Solid-State Circuits, 31(11), 1703–1714, 1996.

[2] V. De. Private communication.

[3] S. Borkar. Design challenges of technology scaling. IEEE Micro, 19(4), 23–29, July 1999.

[4] S. Manne, A. Klauser, and D. Grunwald. Pipline gating: Speculation control for energy reduction. In Proceeding of the 25th Annual Int. Symp. on Comp. Archi., 32–141, 1998

[5] M. D. Powell, S.-H. Yang, B. Falsafi, K. Roy, and T. N. Vijaykumar. Gated-Vdd: A circuit technique to reduce leakage in cache memories. In Proceedings of ISLPED, (July 2000), ACM Press, 90–95.

[6] J. M. Rabaey. Digital Integrated Circuit. Prentice Hall, 1996.

[7] I. Fukushi, R. Sasagawa, M. Hamaminato, T. Izawa, and S. Kawashima. A low-power SRAM using improved charge transfer sense. In Proceedings of the 1998 Int. Symp. on VLSI Circuits, pages 142–145, 1998.

[8] L. Wei, Z. Chen, M. Johnson, K. Roy, and V. De. Design and optimization of low voltage high performance dual threshold CMOS circuits. In Proceedings of the 35th Design Auto. Conf., pages 489–494, 1998.

[9] F. Hamzaoglu, Y. Ye, A. Keshavarzi, K. Zhang, S. Narendra, S. Borkar, M. Stan, and V. De. Dual-Vt SRAM cells with full-swing single-ended bit line sensing for highperformance on-chip cache in 0.13um technology generation. In Proceedings of the 2000 Int. Symp. on Low Power Elect. and Design (ISLPED), July 2000.

[10] Z. Chen, L. Wei, M. Johnson, K. Roy. Estimation of Standby Leakage Power in CMOS Circuits Considering Accurate Modeling of Transistor Stacks. Int. Symp. on Low Power Electronics and Design, 1998, pp.239-244.

[11] http://www-device.eecs.berkeley.edu/~ptm/

[12] N. Shibata, M. Watanabe and Y. Sato. A 2-V 300-MHz 1-Nb Current-Sensed Double-Density SRAM for Low-Power 0.3-μm CMOS/SIMOX ASICs. IEEE Journal of Solid State Circuits, Vol. 36, No. 10, pages 1524-1537, Oct. 2001.

[13] D. Burger and T. M. Austin. The SimpleScalar tool set, version 2.0. Technical Report 1342, Computer Sciences Department, University of Wisconsin–Madison, June 1997

[14] T. Wada and S. Rajan. An Analytical Access Time Model for On-Chip cache Memories. IEEE Journal of Solid State Circuits, Vol. 27, No 8, pages 1147-1156,August 1992.

[15] J. L Hennessy and D. A. Patterson. Computer Architecture A Quantitative Approach. Morgan Kaufmann, 2nd edition

[16] Narendra S., et al. Scaling of Stack Effect and its Application for Leakage Reduction. in Proceeding of ISLPED'01 (Hiltington CA, Aug 2001), ACM Press, 194-200.

PANEL: Unified Tools for SoC Embedded Systems: Mission Critical, Mission Impossible or Mission Irrelevant?

Chair: Gary Smith, Dataquest

Organizers: Daya Nadamuni, Dataquest and Sharad Malik, Princeton University

Abstract

As designers struggle with developing application solutions consisting of complex systems-on-a-chip with a significant software component, they must deal with a diversity of tools with very different philosophies and assumptions, to help manage this task. On one hand are tools which assume a clean separation between the hardware and software parts of the design with an abstraction of the hardware available for software development. On the other hand are tools that try to handle the hardware and software parts of the design concurrently. What drives these different philosophies? Which of these is critical for emerging system designs? Which of these is viable going forward? Our panel of experts consisting of designers, embedded software tool providers, system design tool providers and an academic will answer these challenging questions.

Panelists

Rick Chapman
ST Microelectronics

John Fogelin
Wind River Systems

Kurt Keutzer
UC, Berkeley

Grant Martin
Cadence Design Systems

Brian Bailey
Mentor Graphics

Dynamic and Leakage Power Reduction in MTCMOS Circuits Using an Automated Efficient Gate Clustering Technique

Mohab Anis, Shawki Areibi*, Mohamed Mahmoud and Mohamed Elmasry

ECE Department, University of Waterloo, Waterloo, Ontario, Canada, N2L 3G1
email: manis,mohamedm,elmasry@vlsi.uwaterloo.ca
* School of Engineering, University of Guelph, Guelph, Ontario, Canada, N1G 2W1
email: sareibi@uoguelph.ca

ABSTRACT

Reducing power dissipation is one of the most principle subjects in VLSI design today. Scaling causes subthreshold leakage currents to become a large component of total power dissipation. This paper presents two techniques for efficient gate clustering in MTCMOS circuits by modeling the problem via Bin-Packing (BP) and Set-Partitioning (SP) techniques. An automated solution is presented, and both techniques are applied to six benchmarks to verify functionality. Both methodologies offer significant reduction in both dynamic and leakage power over previous techniques during the active and standby modes respectively. Furthermore, the SP technique takes the circuit's routing complexity into consideration which is critical for Deep Sub-Micron (DSM) implementations. Sufficient performance is achieved, while significantly reducing the overall sleep transistors' area. Results obtained indicate that our proposed techniques can achieve on average 90% savings for leakage power and 15% savings for dynamic power.

Categories & Subject Descriptors: B.7.1 [Integrated Circuits]: Types and Design Styles
General Terms: Design

1. INTRODUCTION

With the advent of technology, the reduction of the supply voltage V_{dd} has become vital to reduce dynamic power and to avoid reliability problems in Deep Sub-Micron (DSM) regimes. However, reducing V_{dd} alone causes serious degradation in the circuit's performance. One way to maintain performance is to scale down both V_{dd} and the threshold voltage V_{th}. However, reducing V_{th} increases the subthreshold leakage current exponentially. This problem escalates in DSM technologies. Multi-threshold CMOS (MTCMOS) technology has emerged as an increasingly popular technique to reduce leakage power during the standby mode, while attaining high speed in the active mode. Devices switching in the critical path are assigned low threshold voltage (LVT) while the others are high threshold voltage (HVT) to reduce leakage power [1],[2]. This technique requires accurate assignment of LVT and HVT, but has the advantage of preserved speed. Another way to implement MTCMOS technology is inserting a HVT device; called sleep transistor, in series to the normal LVT circuitry as shown in Figure 1(a) [3]. The sleep transistor is controlled by a controllable $SLEEP$ signal used for active/standby mode control ($SLEEP$=1,0 during standby and active modes respectively). Proper sleep transistor sizing is a key issue that affects the performance as well as the dynamic and leakage powers of the entire circuit. The design cycle is usually short, but at the expense of a slight speed loss.

In this paper we introduce two techniques that cluster logic gates at a fixed sleep transistor size, which will prove to be power efficient compared to the literature ([3] and [4]) while maintaining adequate performance.

2. BACKGROUND

During the active mode, the sleep transistor could be realized as a resistor **R** as shown in Figure 1(a) [5]. This generates a small voltage drop V_X equal to $I \times R$, where I is the current flowing through the sleep transistor. The voltage drop across R, reduces the gate's driving capability from V_{dd} to V_{dd}-V_X which in turn degrades the gate's performance. Therefore, the resistor should be made small and consequently the size of the sleep transistor large which comes at the expence of area and power overhead. On the other hand, if the resistor is made large meaning that the sleep transistor is sized small, the circuit speed will degrade. This trade-off between achieving sufficient performance and low power values will become even more severe in the DSM regime. In DSM technologies, the supply voltage is scaled down aggressively, causing the resistance of the sleep transistor to increase dramatically, requiring even larger size sleep devices. This will cause leakage and dynamic power to significantly mountain in the standby and active modes respectively. Therefore, an important design criterion is sizing the sleep transistor to attain sufficient performance. In other words, the current "I" flowing through the sleep transistor must be satisfactory to achieve the required speed.

The worst case design scenario takes place if all the gates supported by the sleep transistor are simultaneously switching in time (Figure 1(b)). The sleep transistor exhibits maximum current then (I=I_1+I_2+I_3)(Case I). The sleep transistor is thus sized up to contain the high current. If the gates are discharging mutually exclusive, the sleep transistor is sized according to the maximum current of the mutually exclusive discharging gates ($I = \max\{I_1, I_2, I_3\}$) (Case II). The sleep transistor is a lot smaller in this case. If a current-time graph is constructed of the discharged currents, I_1, I_2 and I_3 would overlap in time in Case I. On the other hand, no overlap in time occurs for Case II. An intermediate case occurs when the discharged currents "partially" overlap, if the LVT Logic Blocks have slightly different discharge times.

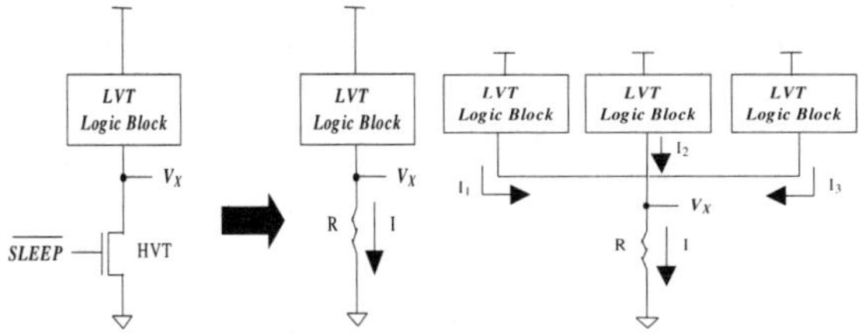

(a) Sleep Transistor modeled as resistor

(b) Worst case discharging scenario

Figure 1: Sleep Transistor in MTCMOS Circuits

A single sleep transistor to support the whole circuit was proposed in [3]. In another work [4], the sleep transistor was sized according to an algorithm based on mutual exclusive discharge pattern. In [4], cascaded gates are clustered together because simultaneous current discharge can never take place. This methodology may be efficient for balanced circuits with tree configurations, where mutually exclusive discharging gates are easily detected. However, this methodology would not be efficient for circuits with complicated interconnections and unbalanced structures. Sleep transistor assignments can therefore be wasteful, and would cause dynamic and leakage power to rise. Finally, the sets of sleep transistors in [4] are merged into a single large sleep transistor to accommodate the whole circuit as in [3]. In addition to the drawbacks listed above, sharing a single sleep transistor for the whole circuit would increase the interconnect resistance for distant blocks. As a result, the sleep transistor would be sized even larger than expected to compensate for the added interconnect resistance. Excessively large sleep transistors again augment dynamic and leakage power as well as area. This drawback would be even more severe in DSM regimes, where interconnects would have a large impact on the circuit's performance [6]. Our proposed methodology in Section 4 solves this problem, and not only clusters gates with exclusive discharge patterns, but with "partially" overlapping discharged currents as well. The first step in our technique is to calculate the size of the sleep transistor.

3. SIZING THE SLEEP TRANSISTOR

To estimate the size of the sleep transistor, the delay of a single gate (τ_d) at the absence of a sleep transistor can be expressed as

$$\tau_d = \frac{C_L V_{dd}}{(V_{dd} - V_{tL})^{\alpha}} \tag{1}$$

where C_L is the load capacitance at the gate's output, V_{tL} is the LVT=350mV, V_{dd}=1.8V and α is the velocity saturation index which is equal to ≈ 1.3 in 0.18μm CMOS technology. In the presence of a sleep transistor, the delay of a single gate τ_d^{sleep} can be expressed as

$$\tau_d^{sleep} = \frac{C_L V_{dd}}{(V_{dd} - V_X - V_{tL})^{\alpha}} \tag{2}$$

where V_X is the potential of the virtual ground. Assuming the circuit could tolerate a 5% degradation in performance due to the presence of the sleep transistor, therefore

$$\frac{\tau_d}{\tau_d^{sleep}} = 95\% \tag{3}$$

Substituting for τ_d and τ_d^{sleep}, and assuming $\alpha = 1$ for simplicity, we get

$$1 - \frac{V_X}{(V_{dd} - V_{tL})} = 95\% \tag{4}$$

Therefore V_X can be formulated as

$$V_X = 0.05(V_{dd} - V_{tL}) \tag{5}$$

The current flowing through the "linearly-operating" sleep transistor is expressed as:

$$\begin{aligned} I_{sleep} &= \mu_n C_{ox}(W/L)_{sleep}[(V_{dd} - V_{tH})V_X - {V_X}^2/2] \\ &\approx 0.05\mu_n C_{ox}(W/L)_{sleep}(V_{dd} - V_{tL})(V_{dd} - V_{tH}) \end{aligned} \tag{6}$$

where μ_n is the N-mobility, C_{ox} is the oxide capacitance and V_{tH} is the HVT=500mV. The size of the sleep transistor can be therefore expressed as

$$(W/L)_{sleep} = \frac{I_{sleep}}{0.05\mu_n C_{ox}(V_{dd} - V_{tL})(V_{dd} - V_{tH})} \tag{7}$$

I_{sleep} and consequently $(W/L)_{sleep}$ are chosen to exhibit low power dissipation. I_{sleep} is chosen to be 250μA, leading to a $(W/L)_{sleep}$ ≈ 6 for 0.18μm CMOS technology. This constant size $(W/L)_{sleep}$ = 6 will be used for both proposed methodologies i.e Bin-Packing (BP) and Set-Partitioning (SP) techniques. Agreeable delay, power and leakage values to analytical calculations were verified for the LVT HSPICE models, to ensure correct functionality. Leakage current increases by an order of magnitude for every 85mV reduction in V_{th}.

4. PROPOSED CLUSTERING TECHNIQUE

To illustrate our techniques, six benchmarks are used as test vehicles; a 4-bit Carry Look Ahead (CLA) adder, a 32-bit priority checker, a 6-bit array multiplier design, a 4-bit ALU/Function Generator (74181 ISCAS-85 benchmark), a 32-Single Error Correcting circuit (C499 ISCAS-85 benchmark) and finally a 27-bit Channel Interrupt Controller (CIC) (C432 ISCAS-85 benchmark). These benchmarks have been chosen to offer a variety of circuits with different structures employing various gates, with different fanouts. The 4-bit CLA adder will be first used to demonstrate our techniques, then the results to the other benchmarks will be provided later on.

Figure 2 shows a schematic diagram of the CLA adder, which consists of 28 gates (G_1-G_{28}). All gates are implemented in 0.18μm CMOS technology. In the next section a preprocessing stage of gate currents is described. This stage will be utilized in solving the BP problem.

4.1 Preprocessing of Gate Currents

The main objective of the preprocessing stage is to group gates into subclusters such that the combination would not exceed the max current of any gate within the cluster. Randomly chosen input vectors are applied and the highest discharging current at the output of every gate is monitored (worst case). The discharge current is only monitored because this is the current that flows through the sleep transistor and eventually *ground*. A load of 6fF is applied to the outputs of each circuit. The probability that discharging takes place (switching activity) at the output of each gate is

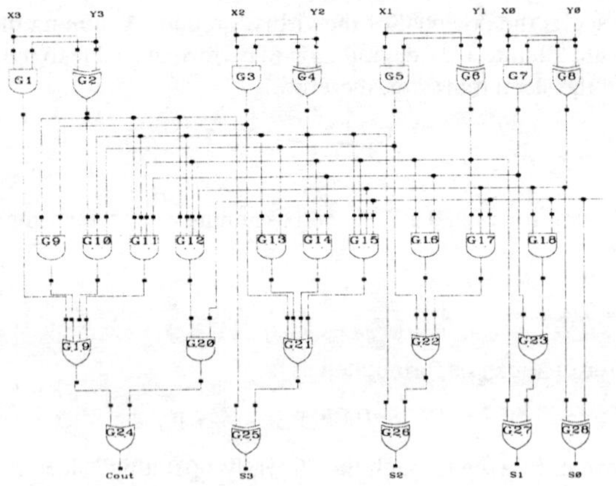

Figure 2: 4-bit Carry Look-Ahead Adder

calculated and multiplied by the corresponding discharge peak current. This current is composed of the discharge and the short-circuit currents that take place during switching. Sleep transistors should be sized to also accommodate the short-circuit currents, otherwise speed will degrade. The peak current value and time at which the switching occurs as well as its duration are monitored. The time the switching takes place depends on the gate's propagation delay and input pattern, while the current duration depends on the slope of the input signal as well as the fanout of the gate. The larger the input slope and/or gate fanout, the longer the switching duration. The discharge current of each gate takes a triangular shape, whose peak occurs at a time equal to the gate delay, and spans a time, mainly function in the fanout of the gate.

To facilitate vector comparisons and to offer an automated design environment, every discharge current at the output of a gate is represented by a vector. The time axis is divided into time slots each equal to 10psec as shown in Figure 3. A time slot of 10psec is sufficient in 0.18μm CMOS technology to offer relatively good accuracies. Each time slot holds a value that represents the magnitude of the discharge current at that specific time which constitutes an element in the vector. In order to illustrate this idea, Figure 3 shows a 2-input AND gate (G1) with a fanout of 2 driving a 2-input OR gate (G2) with a fanout of 4. The disharge currents of G1 and G2 (I_1 and I_2) are presented as a vector. Each element

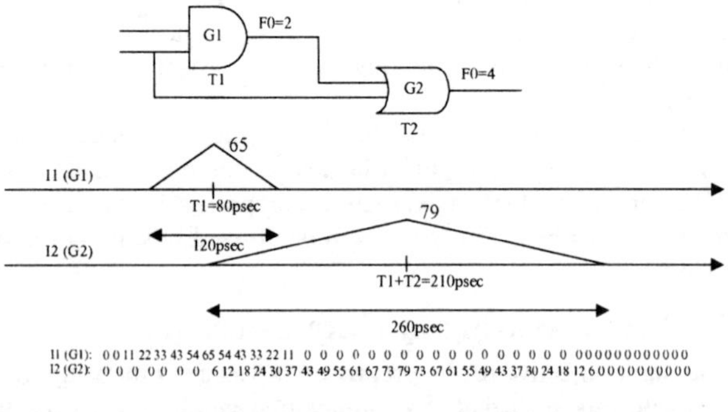

Figure 3: Timing Diagram

in the vector presents the magnitude of current at this 10psec time slot. The peak of the discharge current for G1 occurs at the gate's delay time ($T_1 = 80psec$), while the discharge current I_2 occurs at time ($T_1 + T_2 = 210psec$), because G2 will not discharge till G1 discharges. The peak currents of gates G1 and G2 are $65\mu A$ and $79\mu A$ respectively. The triangular shaped currents are converted into vectors as seen in Figure 3. Since G2 has a large fanout of 4, the duration of the discharge current is long (260 psec), while the duration of the discharge current in G1 is short due to the small fanout of 2 (120 psec). Therefore, for every gate in the circuit, a vector is constructed that carries information about the delay of the gate (when the peak occurs), the fanout of the gate (the duration at which the current lasts) and the magnitude of the current in each time slot. By constructing a vector for each gate, a series of vectors (28 in this case) are produced, that carry information about the whole circuit.

Figure 4 illustrates the used preprocessing heuristic that forms a set of subclusters of gates that when combined would not exceed the maximum current of any gate within the cluster. Table 1 shows

```
PREPROCESSING HEURISTIC
1. Initialize current vectors
2. Set all Gates free; to move to subcluster;
3. For all gates in circuit
        If gate i is not clustered yet
          assign gate i to new cluster k
          update cluster current vector
          calculate max current,start,end time
        For all other gates in circuit
          If (gate j is not clustered yet)
            add current of gate j to cluster k
            If (combination ≤ max current)
              append gate to cluster
              update cluster info
              set gate j locked in cluster k
        End For
   End For
4. Return all clusters formed.
```

Figure 4: Heuristic for Preprocessing

the results of applying the preprocessing heuristic to the 4-bit carry look-ahead adder that was presented in Figure 2. In this example seven subclusters were formed. For example the third column in Table 1 (I_{EQ_3}) represents a subcluster formed by combining Gates G5,G6 and G14 which has a maximum current of 110 of the partially overlapped discharging gates ($I_{G5}^{max} = 83$, $I_{G6}^{max} = 110$ and $I_{G14}^{max} = 30$). $I_{overlap}^{3} = max\{I_{G5}^{max}, I_{G6}^{max}, I_{G14}^{max}\}$. The objective is then to group as much current (gates) as possible without exceeding the current limit of the sleep transistor (250μA), while minimizing the number of sleep transistors used as will be shown in Table 2 of Section 4.2. This is analogous to the Bin-Packing problem in operations research.

Table 1: Results: Current Equivalence

I_{EQ_1} $I_{overlap} = 80$	I_{EQ_2} 80	I_{EQ_3} 110	I_{EQ_4} 90	I_{EQ_5} 50	I_{EQ_6} 30	I_{EQ_7} 50
$I_1, I_2, I_{10}, I_{11}, I_{12}, I_{19}, I_{20}, I_{21}, I_{22}, I_{24}, I_{25}, I_{26}, I_{27}$	$I_3, I_4, I_{13}, I_{15}, I_{28}$	I_5, I_6, I_{14}	I_7, I_8, I_{16}, I_{23}	I_9	I_{17}	I_{18}

4.2 The Bin-Packing Technique

The Bin-Packing (BP) problem [7] can be described as follows. Given n $items$ ($currents$ in this case) and m $bins$ ($sleep\ transistors$ in this case), with

$$I_{EQ_j} = \text{equivalent } current \text{ of gate } j,$$
$$I_{max} = capacity \text{ of each } sleep\ transistor = 250\mu\text{A}$$

The objective is to assign each I_{EQ} to one bin so that the total current in each bin does not exceed I_{max} and the number of bins used is minimized.

The mathematical formulation of the problem is as follows

$$Minimize\ z = \sum_{i=1}^{m} y_i \tag{8}$$

subject to

$$\sum_{j=1}^{n} I_{EQ_j} x_{ij} \leq I_{max} y_i, \quad i \in \{1, ..., m\},$$
$$\sum_{i=1}^{m} x_{ij} = 1, \tag{9}$$

where

$$y_i = \begin{cases} 1, & \text{if bin } i \text{ is used} \\ 0, & \text{otherwise} \end{cases} \quad x_{ij} = \begin{cases} 1, & \text{if items } j \in \text{bin } i; \\ 0, & \text{otherwise} \end{cases}$$

This model is a pure Integer Linear Programming problem (ILP). The objective function to be minimized; z, is analogous to the minimum number of sleep transistors used. y_i is analogous to the sleep transistors available. x_{ij} takes a value of "1" if current I_{EQ_j} is assigned to bin i. CPLEX 6.5; a commercial ILP solver, was used to solve this BP problem, to determine which currents should be grouped together, and to which sleep transistor they are assigned. A summary of the current assignments is shown in Table 2.

Table 2: Results: Current Assignments

Sleep Transistor (Cluster)	1	2
Equivalent Currents	$I_{EQ_3}, I_{EQ_4}, I_{EQ_7}$	$I_{EQ_1}, I_{EQ_2}, I_{EQ_5}, I_{EQ_6}$
Assigned Gates	G_5, G_6, G_7, G_8, $G_{14}, G_{16}, G_{18}, G_{23}$	$G_1, G_2, G_3, G_4, G_9, G_{10}, G_{11}, G_{12}, G_{13}, G_{15}, G_{17}, G_{19}, G_{20}, G_{21}, G_{22}, G_{24}, G_{25}, G_{26}, G_{27}, G_{28}$
$\sum$ Currents(μA)	250	240

It is clear from Table 2 that two sleep transistors will be needed to contain all the gates in the circuit ($z = 2$). It should be noted that the total current of any cluster must never exceed the maximum current limit of the sleep transistor, which is 250μA.

The BP technique was further applied to the other five benchmarks. Keeping the 5% speed degradation as a comparison basis (operational frequency 500MHz) , the BP technique is compared to [3] and [4]. The results are mentioned in Section 5 and summarized in Table 3 (Normalized to [3]).

The BP technique is particularly efficient when it is applied to small circuits that have unbalanced structures. One limitation is that the BP technique does not take the physical locations of the gates on the chip into consideration. For larger circuits this might cause two gates located far apart to be clustered together which will augment the routing complexity of the circuit, as discussed earlier. The Set-Partitioning technique solves this problem, and consequently reduces the routing complexity of the circuit unlike [3] and [4].

4.3 The Set-Partitioning Technique

The Set-Partitioning (SP) problem [7] can be described as follows: Similar to the BP problem, m currents (gates) are arranged into groups such that each element is included only once in a cluster. A cost function; c_j is associated with each group j (S_j). The cost function c_j is evaluated from the physical locations of the gates with respect to each other, which is related to the routing complexity of the circuit.

In order to evaluate the physical locations of the gates, the Cadence Virtuoso Placement and Route tool has been used to produce a compact layout from the schematic entry. Once the compact layout is constructed, the X,Y coordinates for every gate are extracted and the cost functions are evaluated. Figure 5 shows the floor-plan layout for the 4-bit CLA adder. The V_{dd} and gnd rails are shown and a cavity exists where the sleep transistors are located.The cavity of the sleep transistors has been taken into consideration when extracting the X,Y coordinates of every gate.

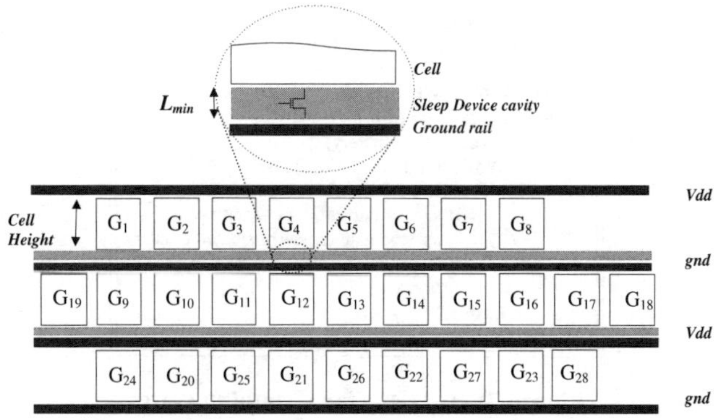

Figure 5: 4-bit CLA Adder Floorplan

In Figure 5 gates G1 to G28 are identified, and the relative distances are computed from the compact layout.

The cost function is formulated as follows:

$$c_j = (w_1 \times c_{j1}) + (w_2 \times c_{j2}) \tag{10}$$

where c_{j1} represents the difference between the maximum cluster capacity and the sum of all currents of gates within a cluster and c_{j2} is a distance function (i.e rectilinear distance between gates within a cluster).

$$c_{j1} = Sleep\_Transistor_{max\_current} - \sum current_i \ \ \forall i \tag{11}$$

The weights w_1 and w_2 are the weights associated with the cost of the two constraints i.e distance and capacity of the formed clusters. In this paper we have assigned equal values to the weights w_1 and w_2 respectively. This will allow the set partitioning modeling of the problem to favor clusters with full capacity (more gates within a cluster) and minimum distance as will be explained later on.

$$c_{j2} = \sum d_{uv} \text{ in a group } S_j \tag{12}$$

where d_{uv} is the distance between the centers of gates G_u and G_v. For example, referring to Figure 6, group S_j is composed of gates G_u,G_v and G_w. The value of the partial cost function of group S_j is: $c_{j2} = d_{uv} + d_{vw} + d_{wu}$

Gates are grouped, while meeting the constraint that the sum of currents does not exceed I_{max}=250μA. Figure 7 presents a very fast and efficient heuristic to form groups of clusters that will be used by the SP technique. The heuristic forms different types of clusters (i.e clusters consisting of single gates, two gates e.t.c). This will guarantee that the set partitioning technique will find a solution for the problem. The target is to select certain groups (clusters) to

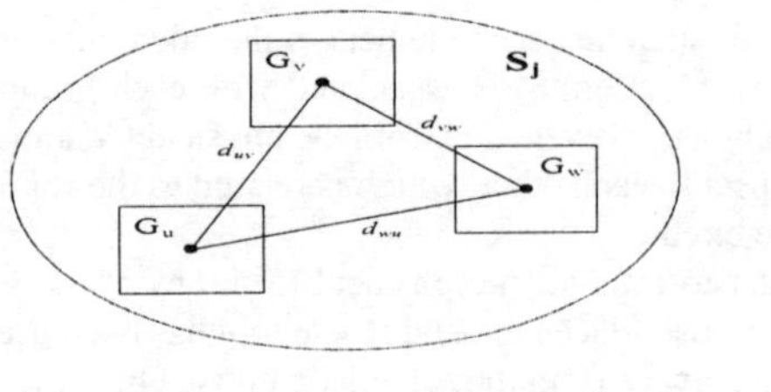

Figure 6: Cost Function Calculation Example

```
CLUSTERING HEURISTIC
Create_Clusters()
1. Calculate distances between all gates;
2. Initialize maxgates_per_cluster=n;
3. Create clusters with Single gates;
4. For cl=2; cl ≤ maxgates_per_cluster
       Create_n_Gate_Cluster(cl)
5. For all clusters created calculate_cost()

Create_n_Gate_Clusters(cl)
1. For cluster of type cl
       create_new_cluster()
       While not done
           Choose Gate with minimum distance
           If sum of currents ≤ capacity
               append gate to newly created cluster
           If total gates within cluster ≥ limit
               break;
       End
2. return newly created cluster
```

Figure 7: Heuristic for Grouping Gates

achieve lowest cost value, while maintaining the I_{max} constraint. The groups must also cover all gates with no repetition. The mathematical formulation of the set partitioning problem is as follows

$$Minimize \quad Z = \sum_{j=1}^{n} c_j S_j \tag{13}$$

subject to

$$\sum_{j} S_j = 1 \tag{14}$$

$$S_j = \begin{cases} 1, & \text{if the } j\text{th subset is formed into a group} \\ 0, & \text{otherwise} \end{cases}$$

n is the number of groups generated. The above model is also a 0-1 pure integer LP problem, which is again solved using CPLEX version 6.5. Figure 8 shows the solution of the SP technique by

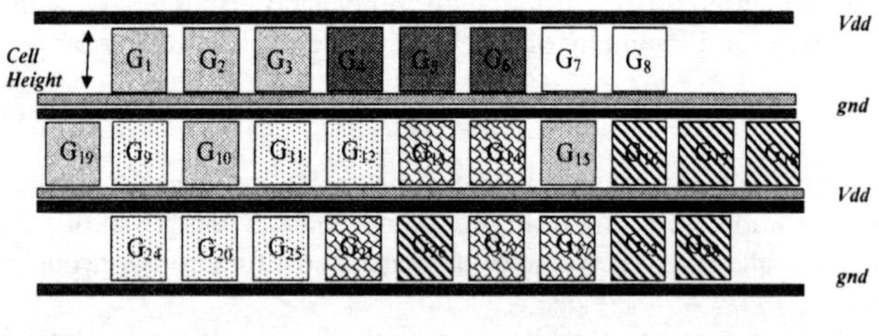

Figure 8: Results:4-bit CLA Adder Floorplan

highlighting the gates that are clustered together. It is evident from Figure 8 that gates that are placed closely were clustered together (i.e gates in two consecutive rows) with a specific sleep transistor therefore minimizing the wire-length. Figure 9 shows the CPU time involved in solving the benchmarks for both the BP and SP problems. It is evident from the figure that solving the SP problem involves more CPU cycles than solving the BP problem. This is due to the fact that the number of variables and constraints in the SP problem are much larger than that of the BP problem. It is important to point-out that as we increase the number of clusters generated for the SP technique the smaller the computation time involved. The BP preprocessing algorithm has a worst case complexity of $O(n^2)$, where n is the number of gates in the circuit. On the other hand, the SP algorithm complexity is $O(nk)$ where n is the number of gates in the circuit and k is the maximum gates to be appended in a cluster. For large circuits it is recommended that heuristic search techniques such as Genetic Algorithms would be used instead of the CPLEX solver.

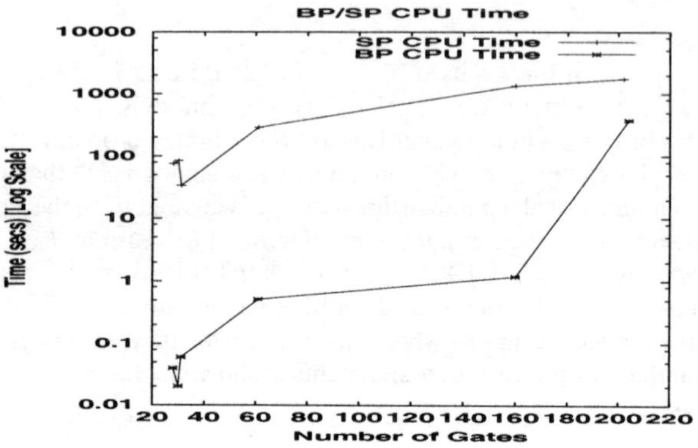

Figure 9: Computation Time for BP and SP

5. RESULTS AND DISCUSSION

Table 3 compares the SP and BP techniques to the literature, while keeping the 5% degradation in speed as a comparison basis. All results are normalized to [3]. The BP & SP techniques employ sleep transistors of equal size calculated as $((\frac{W}{L})_{sleep} = 6)$ to achieve only 5% degradation in circuit speed (frequency of operation is set at 500MHz). L_{min} for 0.18μm CMOS technology = 180nm leading to W_{sleep} = 1.1 μm. LVT=350mV and HVT=500mV.

From Table 3,[4] employs a smaller sized single sleep transistor containing the whole circuit compared to [3]. Consequently, a slight reduction in dynamic power is observed due to the reduction of the drain capacitance linked to the sleep transistor. [4] achieves an average of 50% reduction in leakage power compared to [3].

The highest leakage reduction occurs in the 27-bit CIC benchmark. This is due to the large reduction in sleep transistor area (3247 to 153). On the other hand, the BP technique produces large reductions in the sleep transistors total area. Although the number of sleep transistors is higher than [3] and [4], the size of every sleep transistor is much smaller achieving an overall reduction in sleep transistor area. Therefore, the BP technique offers significant dynamic power savings compared to [3] and [4] as shown in Table 3. On average the BP technique achieves 17% reduction over [3] and 14% dynamic power reduction over [4]. The main saving however is associated with the leakage power, due to the reduction of the sleep transistor size, which is directly proportional to the leakage power dissipation. On average the BP technique achieves 95% and 86% leakage power reduction compared to [3] and [4].

The SP technique is then compared to the BP technique, [3] and [4], while still keeping the 5% speed degradation as a compari-

Table 3: Algorithm Comparison

REF	Benchmark	4-bit CLA Adder	32-bit Parity Checker	6-bit Multiplier	4-bit 74181 ALU	32-bit Single Error Correcting C499	27-channel interrupt controller circuit C432
	No. of gates	28	31	30	61	202	160
[3]	Delay	1	1	1	1	1	1
	$P_{dynamic}$	1	1	1	1	1	1
	$P_{leakage}$	1	1	1	1	1	1
	# Sleep Trans	1	1	1	1	1	1
	ST_Area [$W_{sleep}(\mu$m)]	50	42	65	97	176	3247
[4]	Delay	1	1	1	1	1	1
	$P_{dynamic}$	0.98	0.97	0.89	0.97	0.99	0.98
	$P_{leakage}$	0.58	0.51	0.23	0.41	0.46	0.05
	# Sleep Trans	11→1	16→1	5→1	37→1	32→1	52→1
	ST_Area [$W_{sleep}(\mu$m)]	29.3	21.6	15	39.5	81	153
BP	Delay	1	1	1	1	1	1
	$P_{dynamic}$	0.86	0.82	0.69	0.83	0.80	0.98
	$P_{leakage}$	0.044	0.077	0.051	0.068	0.05	0.0054
	$P_{dynamic}$ savings to [3]	14%	18.4%	31.4%	17%	20%	2%
	$P_{dynamic}$ savings to [4]	12.2%	15.9%	23%	14.4%	19.2%	0%
	$P_{leakage}$ savings to [3]	95.6%	92.3%	94.9%	93.2%	95%	99.5%
	$P_{leakage}$ savings to [4]	92.5%	84.8%	77.8%	83.3%	89.1%	88.5%
	# Sleep Trans	2	3	3	6	8	16
	ST_Area [$W_{sleep}(\mu$m)]	2.2	3.3	3.3	6.6	8.8	17.6
SP	Delay	1	1	1	1	1	1
	$P_{dynamic}$	0.93	0.91	0.81	0.89	0.91	0.98
	$P_{leakage}$	0.13	0.15	0.15	0.14	0.13	0.011
	$P_{dynamic}$ savings to [3]	7%	9%	19%	11%	9%	2%
	$P_{dynamic}$ savings to [4]	5.1%	6.2%	9%	8.2%	8.1%	0%
	$P_{leakage}$ savings to [3]	87%	85%	85%	86%	87%	98.9%
	$P_{leakage}$ savings to [4]	77.7%	70.4%	34.8%	65.6%	71.1%	76.6%
	# Sleep Trans	6	6	9	12	22	33
	ST_Area [$W_{sleep}(\mu$m)]	6.6	6.6	9.9	13.2	24.2	36.3

son basis. The SP technique produces large reductions in the sleep transistors total area compared to [3] and [4], but higher than BP because an additional constraint to the objective function is added (i.e routing cost) and no preprocessing is incorporated as explained in Section 4.1. The SP technique reduces the dynamic power on average by 16% and 6% compared to [3] and [4] respectively. This is attributed to the reduction of capacitance due to the down-sizing of the sleep transistors.

Furthermore, the SP technique achieves 88% and 66% leakage reduction compared to [3] and [4]. The main advantage of the SP technique is taking into consideration the location of the blocks in order to reduce the overall interconnects, providing more optimization to the area. The advantages of the SP technique will be even more evident in the DSM regime when interconnects dominate circuit performance and dynamic power. More-over, equally sized sleep devices as for BP and SP facilitate design for other circuits and provides more regular layouts. The area of the sleep transistor (ST) is equal to $W_{sleep} \times L_{sleep}$. Keeping the length of the sleep transistor (L_{sleep}) constant in the 4 techniques mentioned in Table 3, the sleep transistor width (W_{sleep}) can now be used as the sleep transistor area representative. The reduction in dynamic power is dependant on the number and size of sleep transistors and how big the circuit is (ratio of ST capacitance to overall circuit capacitance), while leakage power is only dependant on the number and size of the sleep transistors. Therefore, it can be noticed from Table 3, that the savings in leakage power is directly proportional to the reduction in total sleep transistor area. Finally, the proposed technique offers minimal area overhead, with no perturbation to the layout. This is attributed to the very narrow cavity (Figure 5) that holds the sleep transistors, which is located at a fixed location parallel to either the supply or ground rails. This further guarantees that the sleep transistor will not change the overall floorplan of the circuit.

6. CONCLUSIONS AND FUTURE WORK

Two techniques are applied to efficiently cluster gates in MTCMOS circuits. The first gives the minimum number of sleep transistors to be employed, while the second takes the circuit's routing complexity into consideration. On average the BP technique reduces dynamic and leakage power by 15% and 90% respectively. The SP technique also reduces dynamic and leakage power on average by 11% and 77% respectively. Future work involves improving the computation time involved to solve the SP and BP problems by using heuristic search techniques in the form of Genetic Algorithms that are suitable for multi-objective optimization problems.

7. REFERENCES

[1] L.Wei et al., "Design and optimization of dual-threshold circuits for low-voltage low-power applications," *IEEE Trans. on VLSI Systems*, pp. 16–24, 1999.

[2] S.Sirichotiyakul et al., "Stand-by Power Minimization through Simulataneous Threshold Voltage Selection and Circuit Sizing," Proc. of the 36th DAC, pp. 436–441, 1999.

[3] S.Mutah et al., "1-V Power Supply High-Speed Digital Circuit Technology with Multi-Threshold Voltage CMOS," *IEEE JSSC*, pp. 847–853, 1995.

[4] J.Kao et al., "MTCMOS Hierarchical Sizing Based on Mutual Exclusive Discharge Patterns," Proc. of the 35th DAC, pp. 495–500, 1998.

[5] J.Kao et al., "Transistor Sizing Issues And Tool For Multi-threshold CMOS Technology," Proc. of the 34th DAC, pp. 409–414, 1997.

[6] M.Bohr and Y.Elmansy, "Technology for Advanced High-Performance Microprocessors," *IEEE Trans. on Electron Devices*, vol. 45, pp. 620–625, 1998.

[7] R.Rardin, Optimization in Operations Research, Prentice Hall, 1998.

Total Power Optimization By Simultaneous Dual-Vt Allocation and Device Sizing in High Performance Microprocessors

Tanay Karnik, Yibin Ye, James Tschanz, Liqiong Wei, Steven Burns¶,
Venkatesh Govindarajulu, Vivek De, Shekhar Borkar
Circuit Research, Intel Labs, 5200 NE Elam Young Parkway, Hillsboro, OR 97124.
¶Strategic CAD, Intel Labs, 5200 NE Elam Young Parkway, Hillsboro, OR 97124.
(503) 696 1975, 001
tanay.karnik@intel.com

ABSTRACT
We describe various design automation solutions for design migration to a dual-Vt process technology. We include the results of a Lagrangian Relaxation based tool, iSTATS, and a heuristic iterative optimization flow. Joint dual-Vt allocation and sizing reduces total power by 10+% compared with Vt allocation alone, and by 25+% compared with pure sizing methods. The heuristic flow requires 5x larger computation runtime than iSTATS due to its iterative nature.

Categories and Subject Descriptors
B.7 INTEGRATED CIRCUITS
B.7.1 Types and Design Styles – *Microprocessors and microcomputers, VLSI.*

General Terms: Algorithms, Performance, Design, Experimentation, Verification.

Keywords: Dual-Vt design, multiple threshold, sizing, optimization.

1. INTRODUCTION
Total power consumption of a high performance microprocessor (μP) is exceeding 100 Watts. Power density has exceeded that of a hot plate. μP's used for mobile applications have an additional battery life constraint. Total power consists of switching and leakage components. These two components can be traded off to minimize total active power consumption.

Leakage power minimization has prompted various IC manufacturers to employ multiple threshold voltage (Vt) processes. Most of the process technologies provide dual threshold voltage transistors (dual-Vt). Low-Vt transistors are used in performance-critical blocks to meet target clock frequency requirements, and high-Vt transistors are used in blocks with delay slacks to minimize overall leakage power. Various techniques [2,3,6-9,11,12] have been published for dual-Vt design, synthesis or process migration of libraries. To improve performance, upsizing of a high-Vt transistor, which increases switching power and die area, can be traded off against using a low-Vt transistor, which increases leakage power. Few techniques consider transistor sizing and Vt allocation as an integrated problem [4,5,10]. However, the reported techniques are either not evaluated for complex industrial designs [4,5], or focus on reduction of standby leakage power only [10], not total active power of μP's.

In this paper, we describe two methods for optimized migration of an existing single-Vt design database to a dual-Vt process technology: (1) a deterministic global optimizer for simultaneous Vt allocation and transistor sizing (iSTATS), and (2) a heuristic iterative optimization flow using an industry-standard sizing tool (AMPS) and a Transistor-level Dual-Vt Allocator (TA-DVT) tool. An existing μP database in single-Vt process is used for testing the two methods, and for understanding the trade-offs offered by the different Vt allocation and sizing options.

The paper is organized as follows. Section 2 provides a detailed description of our Lagrangian Relaxation (LR) based iSTATS tool with parameter estimation, constraints, assumptions, algorithms, implementations and results. Section 3 contains similar details of the iterative heuristic method. Results are discussed and analyzed in section 4, and conclusions are presented in Section 5.

2. SIMULTANEOUS VT ALLOCATION AND TRANSISTOR SIZING BY iSTATS

2.1 Estimation
The iSTATS optimizer performs gate-level timing analysis internally. The circuit blocks contain continuously sizeable CMOS logic gates of various types. The granularity of allocation is limited to half gate-level, that is, mixed Vt's are not allowed in the pull-up PMOS tree or pull-down NMOS tree inside a gate. This results in four different versions of a static CMOS logic gate - *hphn, hpln, lphn* and *lpln*. The first/third letter indicates whether the P/N tree is high-Vt (*h*) or low-Vt (*l*).

2.1.1 Gate Delays
The gate library is simulated for varying conditions of input slope, normalized load (output load capacitance divided by driving conductance) and P/N transistor skew. All library gates are designed to have two independent size variables per gate. Rise/fall delay and signal slope are modeled by separate equations for different Vt versions of a gate.

2.1.2 *Path Delays*

iSTATS includes a static timing analyzer that supports time borrowing across transparent latches. The results are correlated with Pathmill, an industry-standard static timing analyzer.

2.1.3 *Interconnect Parasitics*

The optimization is applied to a pre-existing microprocessor database. The interconnect capacitances are assumed to scale according to the new process technology. Interconnect resistances are assumed to be insignificant for this optimization.

2.1.4 *Power*

Block-level switching (P_A) and leakage (P_L) powers are:

$$P_A = a[c_T(W_{hV_t} + W_{lV_t}) + C_I]V^2 f; \quad P_L = (\lambda W_{hV_t} + W_{lV_t}) I_{off} \beta V$$

$$P_{total} = P_A + P_L$$

where P_{total} is total power, a is activity factor, c_T is capacitance per unit width, $W_{hVt}[W_{lVt}]$ is high[low]-Vt device width, V is supply voltage, f is target frequency, C_I is total interconnect capacitance, λ is h_{Vt} to l_{Vt} leakage ratio, I_{off} is l_{Vt} leakage current per unit width, β is an empirical factor which accounts for impacts of within-die L variations, stack effect, noise and fraction of "*off*" transistors on leakage power. Short-circuit power is modeled simply as an increase in a, since it is typically less than 10% of the switching power. Its dependencies on Vt and signal slopes are ignored.

2.2 Assumptions

The clock skew is assumed fixed and ideal clock waveforms are provided at the clock inputs. Furthermore, since migration of the design database to next generation process technology is in an early phase, we assume that very stringent delay requirements for few critical paths in some circuit blocks can be relaxed.

2.3 Optimization Algorithm

The optimization algorithm is based on the well-known Lagrangian Relaxation technique [1]. The problem formulation: Minimize $\alpha(P_{total}) + WNS + WS$

Subject to $\forall_{u \to v} A_u + d_{u \to v} \le A; \ \forall_{pi} A_{pi} = A^T_{pi} + adj_{pi};$

$\max_{u \to v}(s_{u \to v} - s_{\lim it}) = WS; \ \max_{po}(A_{po} - r_{po}) = WNS$

where α is power-delay tradeoff factor, WNS is worst negative slack, WS is worst slope, pi is primary input, po is primary output, A is arrival time, r is required time, d is delay, A^T is arrival time constraint, adj is adjustment based on changing load capacitance. Applying LR, the objective becomes

$$iSTATS(\mu, \sigma, w) = \alpha[W_{avg} + \lambda W_{avglV_t}] + \Sigma_{u \to v}\, \mu_{u \to v} d_{u \to v} + \Sigma_{po}\, \mu_{po}(-r_{po}) +$$
$$\Sigma_{pi}\, \mu_{pi}(a^T_{pi} + adj_{pi}) + \Sigma_{u \to v}\, \sigma_{u \to v}(s_{u \to v} - s_{\lim it})$$

where the multipliers μ and σ signify delay and slope criticality.

Figure 1 shows the algorithmic flowchart. The inner loop traverses the netlist in reverse topological order to numerically solve for width and Vt variables. As Vt is not a continuous variable, there is a slight quality degradation. Inner loop complexity is O(n). Timing analysis and multiplier update are performed during the outer loop. Lower bound is tracked during the outer loop to monitor convergence.

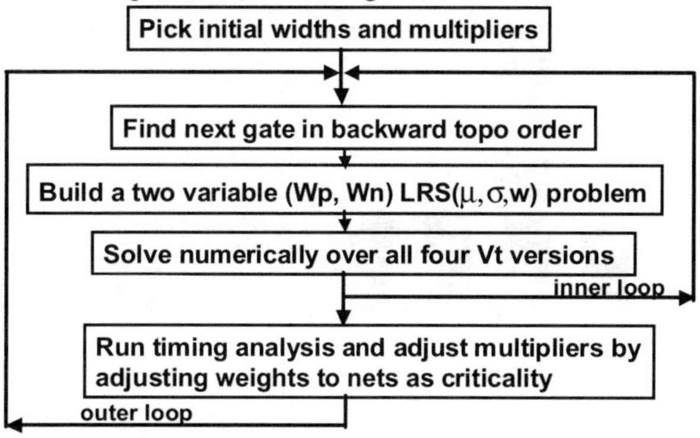

Figure 1: iSTATS algorithm flowchart

2.4 Implementation and Results

iSTATS is implemented as a modular and extensible C-code. The experiments are executed on HP-UX B.10.20. Various static CMOS circuit blocks in an existing microprocessor are selected as test cases. The design optimizations are performed in a 0.13μm dual Vt process for 2.2 Ghz target clock frequency.

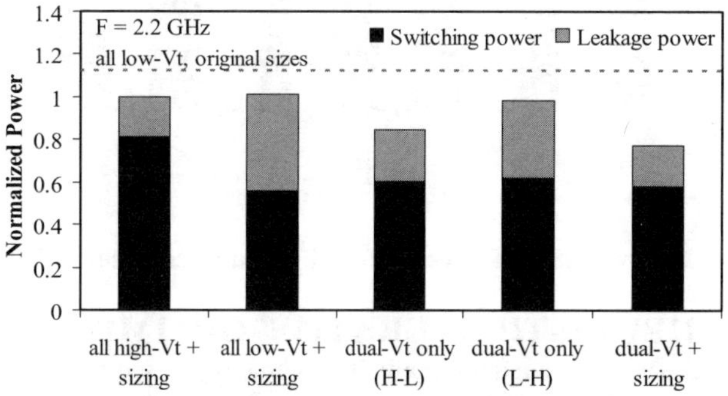

Figure 2: Switching and leakage power tradeoffs

Figure 2 shows P_{total}, with P_A and P_L components, for different design optimizations. The power is averaged over a number of circuit blocks for the following design options:

1. start with all high Vt, perform sizing optimization
2. start with all low Vt, perform sizing optimization
3. start with all high Vt, perform optimal low Vt allocation
4. start with all low Vt, perform optimal high Vt allocation
5. simultaneously optimize Vt allocation and sizing.

The total power achieved by simultaneous Vt allocation and sizing is at least 10% smaller than other design options that which use dual-Vt allocation without sizing, and is 20% smaller than either of the single-Vt design options where only sizing is used. Among "dual-Vt only" options, option 3 is better, indicating that high-to-low Vt conversion is preferred.

Figure 3 shows histograms of average device width and low-Vt usage in the different circuit blocks for options 3 and 5. Clearly, low-Vt usage and average width vary by large amounts from block to block since they are dependent on the inherent logic topology of the block and its timing requirement. This holds true for both design options. Therefore, design optimizations and guidelines for Vt allocation and sizing must be derived by comprehending the wide assortment of circuit block types in a complex μP design.

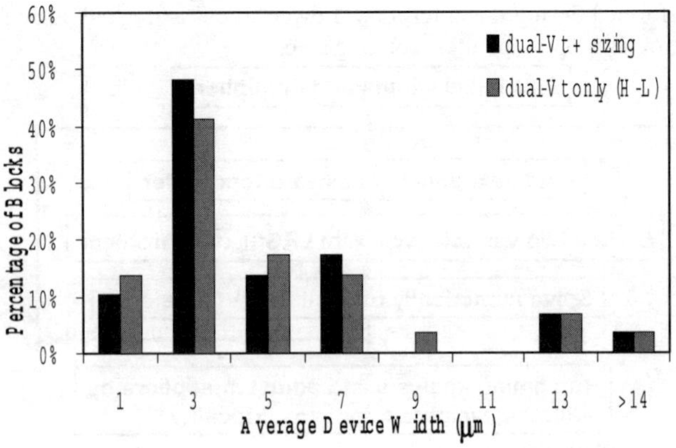

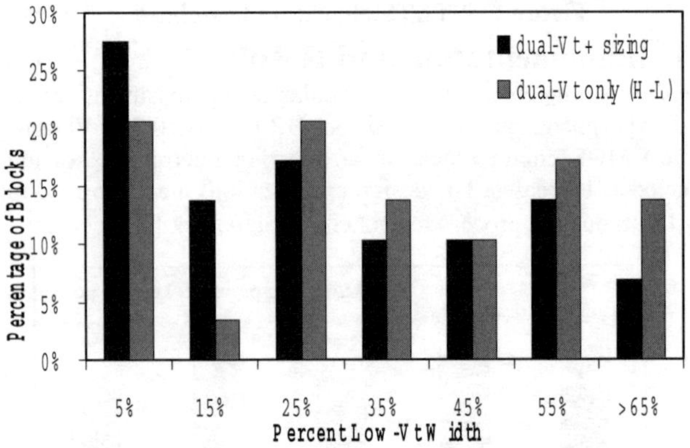

Figure 3: iSTATS results for different circuit blocks

3. HEURISTIC ITERATIVE OPTIMIZER

In IC design community, Pathmill is commonly used for static timing analysis. AMPS, based on pathmill, is widely used for sizing a circuit. There are several heuristic dual-Vt allocation algorithms reported in [4,5,10]. We developed an algorithm for priority-based transistor-level allocation of dual-Vt (TA-DVT). Both AMPS and TA-DVT are used in an iterative flow to achieve optimization by conjoint sizing and dual-Vt allocation. Parasitics for delay and power estimation are extracted from the original 0.18µm microprocessor design and scaled to 0.13µm process technology. After sizing and Vt allocation, we use the original interconnect parasitics, but re-estimate the transistor-related parasitics. Power estimations are the same as that described in the section 2 and include gate oxide leakage power. Some transistors are marked as "don't touch" during Vt allocation, and forced to be high-Vt. For example, many transistors in clock generation circuits and dynamic circuits are excluded for Vt change. We identify these "don't touch" transistors based on our design methodology for noise. Additionally, some transistors need to have the same Vt based on the design requirements. For instance, multiple array instances of the same circuit template may need to have the same Vt due to design and layout considerations.

3.1 Flows and Algorithms

We consider 4 flows that use AMPS for sizing and TA-DVT for Vt allocation. They are shown in Fig. 4(a). In flow 1, only sizing is done using AMPS. In flow 2, only dual-Vt allocation is performed. In flow 3, we do a heuristic iterative optimization with both sizing and dual-Vt allocation to minimize total power. The heuristics proposed in flow 3 introduce an intermediate performance target for each iteration, as illustrated in Fig. 4(b). AMPS is used to meet only the intermediate performance target and TA-DVT is then used to meet the final target. We use five different intermediate performance targets, relaxed by 0%, 7%, 15%, 22% and 30% from the final target. The design with the lowest total active power is then selected. We find that the best intermediate target depends strongly on the circuit block type. A second AMPS run is used after TA-DVT, as some paths may be faster than the target due to low-Vt allocations, creating opportunities for further downsizing. We also implement a flow 3x in which all transistors are converted to low-Vt first. Then we use AMPS which mostly downsizes the devices. TA-DVT is then used for selective high-Vt allocation.

3.2 TA-DVT Algorithm

The TA-DVT flow is illustrated in Figure 5. We first estimate parasitics of the initial circuit. The circuit can be all high-Vt, all low-Vt or dual-Vt to begin with. Pathmill is then used to identify critical paths. We extract slack S_p of all critical paths reported by Pathmill, and delay, d_{ip}, of all transistors in these paths. Note that the delay of a transistor is path dependent, i.e., the delay of the same transistor i can be different in path p and path q, or $d_{ip} \neq d_{iq}$. We also need to know the delay improvement Δd from high-Vt to low-Vt for all candidate transistors. If the initial circuit includes all high-Vt transistors, then we perform another Pathmill run with all low-Vt, and vice versa. For common paths found in both Pathmill runs, we extract the path-dependent delay improvement, Δd_{ip}, for transistor i in path p. When the corresponding path in the second pathmill run is not found for path q, the average transistor delay improvement Δd_i,, which is the difference between average high-Vt delay and low-Vt delay in the two Pathmill runs for transistor i, is used for path q. We find that the percentage delay change $\Delta d_i/d_i$,, varies significantly among transistors because the transistor delay is strongly dependent on the input signal slope, stack effect, the fraction of interconnect capacitance in the load, etc. However, the variation in delay change for a transistor among different paths running through it is small.

For low-Vt assignment, priority of each candidate transistor i is computed based on the following weight function:

$$\mathbf{W_i} = \Sigma_{\forall p \text{ covered by } i \text{ with } Sip < 0}\, \mathbf{MIN}(\Delta d_{ip}, |S_{ip}|)\, /w_i$$

where w_i is the width of transistor i and S_{ip} is the slack of transistor i in path p. Only paths with negative slack, covered by the transistor, are considered in the summation. Transistors with smaller width and higher impact on negative slack due to Vt change have higher weights. The candidate transistors are stored in a queue implemented as a binary heap: the transistor with the highest weight is placed at the top of the heap. In each round of low-Vt allocation, the top transistor in the heap is selected, converted to low-Vt, and removed from the heap. The slacks of all paths affected by the new low-Vt transistor are updated, along with their weights. Transistors that do not belong to critical paths are removed from the heap and the heap is updated with a new transistor on the top. The low-Vt allocation process continues until the heap is empty. The low-to-high Vt allocation works in a similar way as shown in Fig. 5. However, a different weight function is used.

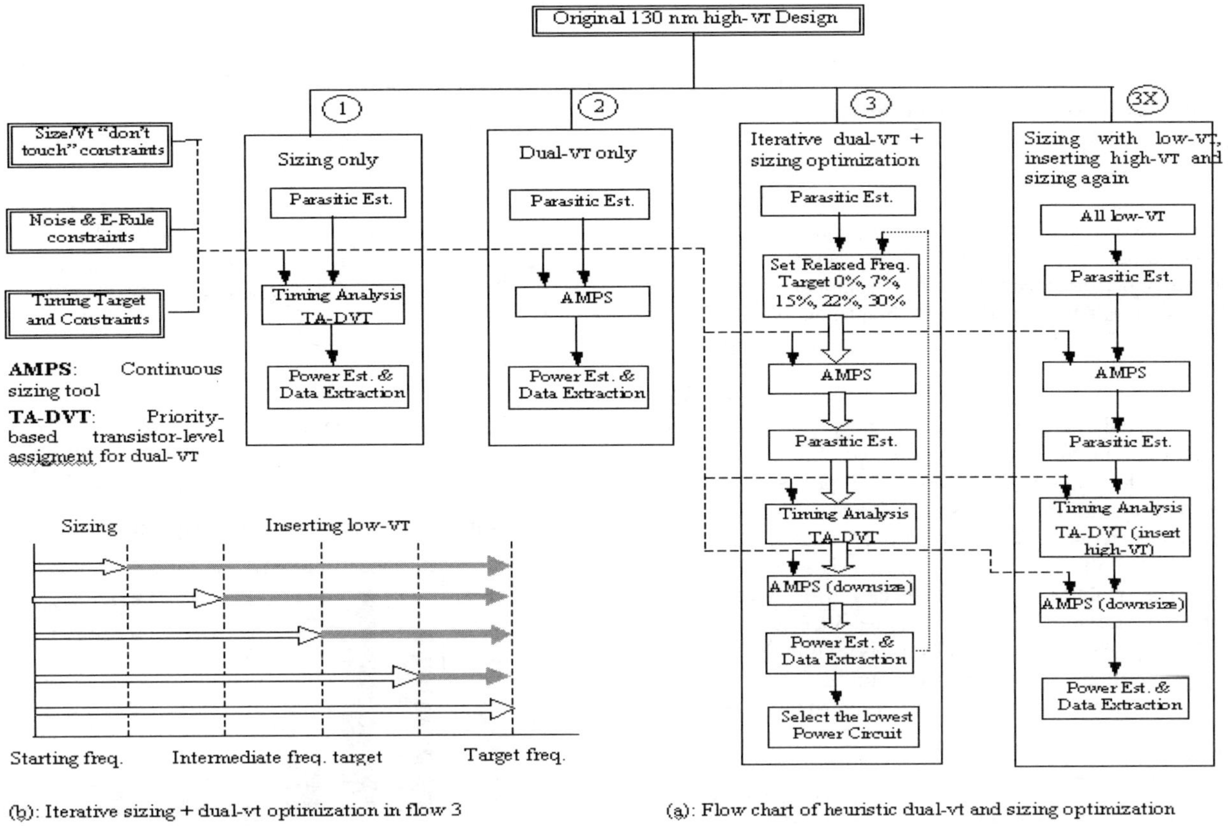

Figure 4: Flowcharts of heuristic iterative optimization (using AMPS and TA-DVT) and different design options

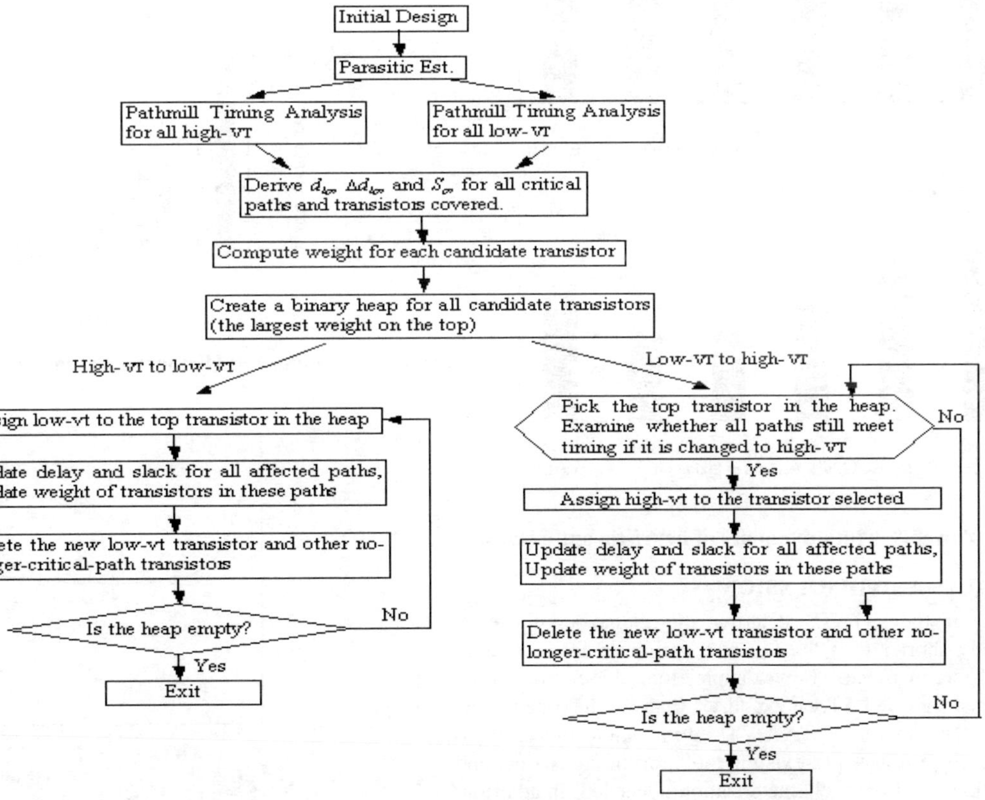

Figure 5: Flowchart of priority-based transistor-level allocation of dual-Vt (TA-DVT)

3.3 Implementation and Results

In Fig. 6, we use a circuit block as an example to illustrate the dynamics of flow 3. The lowest total power is obtained when the intermediate target is relaxed by 15%. As expected, both low-Vt percentage and leakage power increase as the intermediate target is relaxed further, since more low-Vt transistors are needed to meet the final performance target. Fig. 7 summarizes total power of 100 circuit blocks for different frequency targets for all techniques. The "iterative sizing + dual-Vt allocation" produces the best solution. However, "sizing with all lowVt + low-to-high dual-Vt" provides approximately the same power. "Sizing only" and "dual-Vt only" are comparable in power, but they cannot meet the high performance targets achievable by the other techniques. Fig. 8 shows histograms of percentage low-Vt device width and average transistor width for all circuit blocks at 2GHz clock frequency for "dual-Vt only" and "sizing + dual-Vt". The dual-Vt optimization varies significantly across circuits. The number of circuit blocks with large low-Vt percentage is substantially smaller in "dual-Vt + sizing" than in "dual-Vt only". An activity factor of 0.1 is used in the dual-Vt + sizing optimizations. However, the optimization is not very sensitive to activity factor assumption, as shown in Fig. 9.

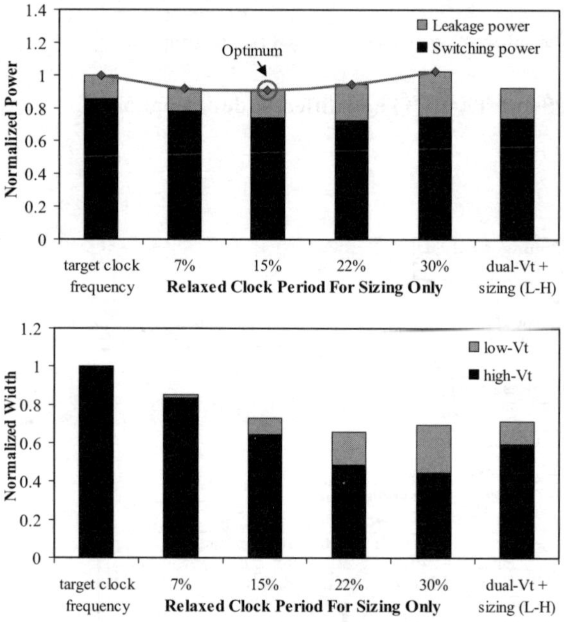

Figure 6: Power & width dynamics of heuristic optimization

4. RESULTS AND DISCUSSION

Run time of the iSTATS deterministic global optimizer is significantly shorter than the iterative optimizer across a large sample of circuit blocks. For example, for a block containing 10000 transistors, iSTATS is 5x faster than the heuristic flow. However, the AMPS + TA-DVT optimization is easier to implement since it uses pre-existing tools for timing, sizing, and Vt allocation. No library characterization is needed. In addition, transistor-level Vt allocation and continuous sizing are possible to provide the best spatial granularity for the optimization.

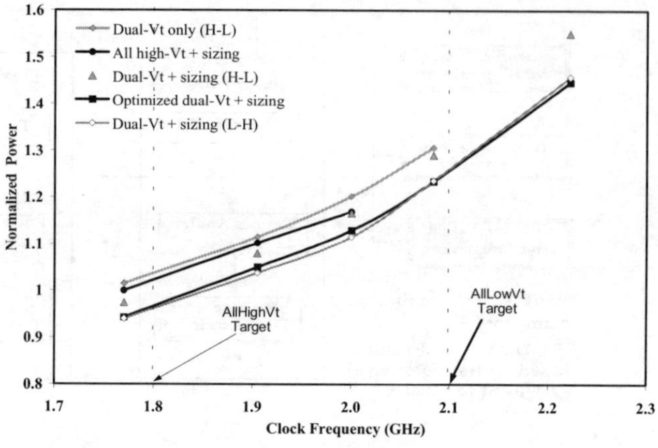

Figure 7: Total power versus frequency for 5 techniques

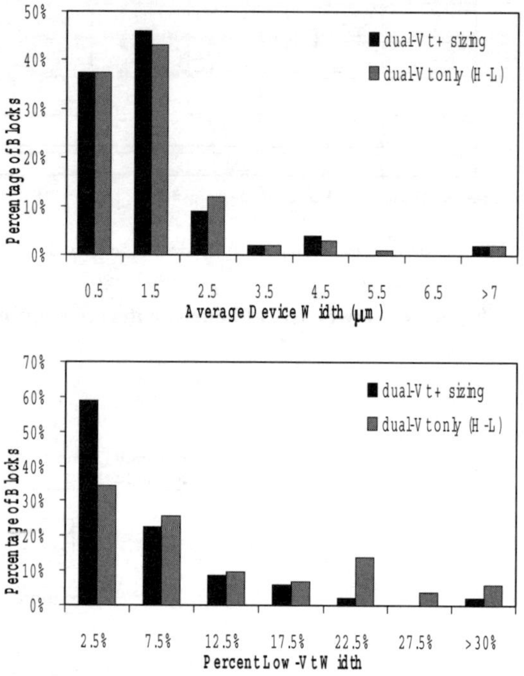

Figure 8: Histograms of low-Vt usage and device width

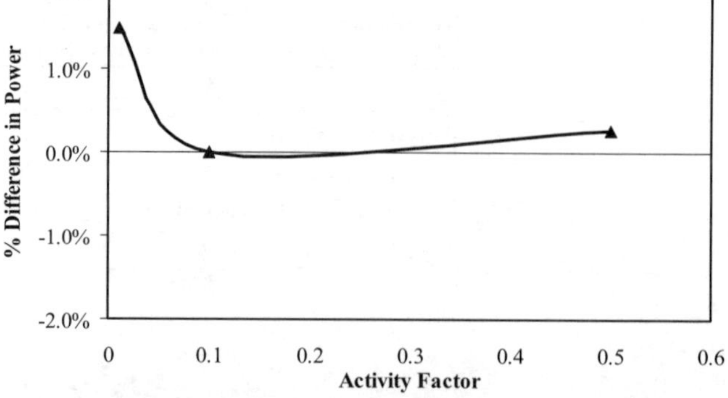

Figure 9: Sensitivity of power sensitivity to activity factor for dual-Vt+sizing optimization

Figs. 10 and 11 show results of the optimization across a range of target clock frequencies and activity factors for 100 functional unit blocks containing 500-50000 transistors per block. With all high-Vt devices and original sizes, typical achievable clock frequency would be 1.8GHz. If all devices are converted to low-Vt, 2.1GHz frequency is achievable without any resizing. Target 2GHz clock frequency can be achieved by sizing only with all high-Vt. The total power for 0.1 activity factor is 5% smaller than a design with dual-Vt only.

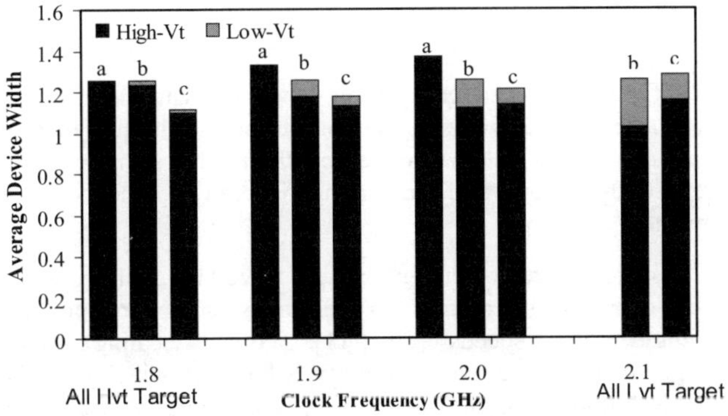

(a) All high-Vt + sizing (b) dual-Vt only (H-L) (c) dual-Vt + sizing

Figure 10: Average device width vs. frequency for various design options

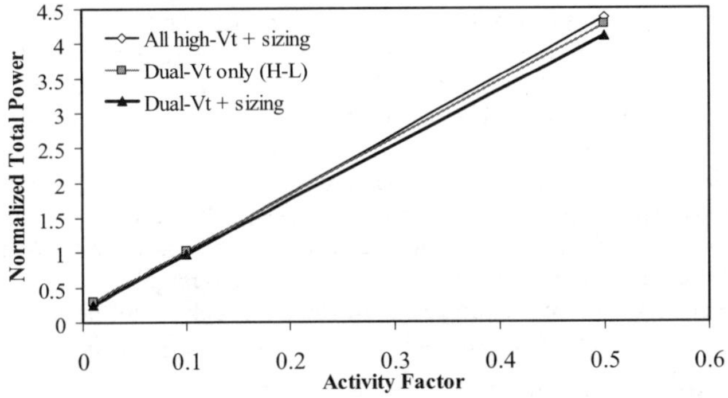

Figure 11: Power vs. activity for various design options at 2.0Ghz clock frequency

Even though using low-Vt devices prevents upsizing of transistors resulting in smaller switching power and reduced average device width than the "sizing only" option, leakage power is significantly larger. As a result, total power is increased. Conjoint dual-Vt and sizing achieves a more optimal balance between low-Vt usage and transistor upsizing needed to achieve the target 2GHz clock frequency. As a result, total power and average device width are 5-10% smaller than the previous two design options. A design with only high-Vt devices fails to achieve 2.1GHz clock frequency target by sizing only, whereas designs containing dual-Vt devices are able to meet this aggressive performance target. The total power is 5-10% smaller when both sizing and dual-Vt are optimized jointly, when compared to optimal dual-Vt allocation only with no transistor resizing. Leakage power is significantly smaller for the design with joint dual-Vt and sizing, since it reduces low-Vt usage from 40% to 15% to achieve proper balance between switching and leakage power components that minimizes total power. The average device widths are approximately the same for both. At higher values of activity factors, leakage power has less impact on total power. As a result, techniques that favor more low-Vt usage become more beneficial. Downsizing of transistors allowed by more low-Vt usage reduces switching power, and thus total power, significantly without incurring major impacts of additional leakage from the low-Vt devices. Thus, benefits of joint dual-Vt and sizing over "dual-Vt only" or "sizing only" are more pronounced when activity is higher.

5. CONCLUSIONS

Various design automation solutions for single Vt design migration to a dual-Vt process technology have been presented. The results are generated using a Lagrangian Relaxation based tool, iSTATS and a heuristic iterative optimization flow. Dual-Vt allocation and sizing reduces total power by at least 10% compared with pure Vt allocation methods and at least 25% compared with pure sizing methods.

6. REFERENCES

[1] Chen, C.P., Chu, C., and Wong, D.F., Fast and exact simultaneous gate and wire sizing by LR. IEEE TCAD, Vol. 18, No. 7, July 1999, 1014-1025.

[2] Ishihara, T., and Asada, K., **A** system level memory power optimization technique using multiple supply and threshold voltages**.** IEEE/ACM DAC 2001, 456-461.

[3] Ko, U., Pua, A., Hill, A., and Srivastava, P., Hybrid dual-threshold design techniques for high-performance processors with low-power features. ISLPED 1997, 307-311.

[4] Pant, P., Roy, R., and Chatterjee, A., Dual-Threshold Voltage Assignment with Transistor Sizing for Low Power CMOS Circuits. TVLSI, Vol. 9, No. 2, 4, 2001, 390-394.

[5] Sirichotiyakul, S., et al, Stand-by power minimization through simultaneous threshold voltage selection and circuit sizing. DAC 1999, 436-441.

[6] Sundararajan, V., and Parhi, K. K., Low power synthesis of dual threshold voltage CMOS VLSI circuits. IEEE ISLPED 1999, 139-144.

[7] Tripathi, N., Bhosle, A., Samanta, D., and Pal, A., Optimal assignment of high threshold voltage for synthesizing dual threshold CMOS circuits. VLSI Design, India, 2001, 227-232.

[8] Wei, L., et al, Design and Optimization of Dual-Threshold Circuits for Low-Voltage Low-Power Applications. IEEE TVLSI, Vol. 7, No. 1, March 1999, 16-23.

[9] Wei, L., Chen, Z., and Roy, K., Ye, Y., De, V., Mixed-Vth (MVT) CMOS Circuit Design Methodology for Low Power Applications. ACM/IEEE DAC 1999, 430-435.

[10] Wei, L., Roy, K., and Koh, C. K., Power Minimization by Simultaneous Dual-Vth Assignment and Gate-sizing. IEEE CICC 2000, 413-416.

[11] Wong, Q., and Vrudhula, S.B.K., Static power optimization of deep submicron CMOS circuits for dual V/sub T/ technology. IEEE/ACM ICCAD 1998, 490-496.

[12] Wong, Q., and Vrudhula, S.B.K., An investigation of power delay trade-offs for dual V/sub t/ CMOS circuits. ICCD 1999, 556-562.

An Optimal Voltage Synthesis Technique for a Power-Efficient Satellite Application

Dong-In Kang, Jinwoo Suh, and Stephen P. Crago
University of Southern California/Information Sciences Institute
3811 N. Fairfax Dr. Suite 200, Arlington, VA 22203
703-248-6164, 703-248-6160, 703-812-3729
{dkang, jsuh, crago}@isi.edu

ABSTRACT

This paper presents an optimal voltage synthesis technique for a satellite application to maximize system performance subject to energy budget. A period of a satellite's orbit is partitioned into several independent regions with different characteristics such as type of computation, importance, performance requirements, and energy consumption. Given a periodic energy recharge model, optimal voltages for the regions are synthesized such that the overall performance is maximized within the energy budget in the period.

Categories and Subject Descriptors

C.4 [**PERFORMANCE OF SYSTEMS**] Design studies, Modeling techniques, Performance attributes.

General Terms

Algorithms, Management, Performance, Design.

Keywords

Power-aware design, power-efficient design, satellite application, queueing.

1. INTRODUCTION

A satellite such as an LEO (Low-Earth Orbit) satellite circles Earth and captures information of the events on Earth and is used for remote sensing and weather monitoring. Its behavior with respect to energy charge/consumption is highly periodic. Energy is recharged every period using solar panels. The computation is also periodic. A satellite flies over precalculated regions every period and does pre-determined computations depending on the region that it is flying over.

Due to the increasing demand of computations and limited energy source of the satellite, energy-efficient usage of the resources is becoming more important. However, low-power design is not always the best solution because any unused energy within a period is simply wasted due to the periodic recharge. Power-aware design that maximizes system performance within the energy budget is superior to low-power design. The sampling rate of a remote sensing satellite can be increased when there is extra energy left unless the sensor limits the sampling rate. Likewise, the accuracy of the computation can be enhanced with extra energy.

A major source of energy savings is voltage scaling, which scales operational voltages of resources and corresponding maximum clock speeds [1][3][11]. Voltage scaling affects throughput and latency in a nonlinear fashion [2], which makes it difficult to find an optimal trade-off between energy consumption and throughput. Maximum clock frequency depends on the supply voltage. As supply voltage increases, both the maximum clock frequency and energy consumption increase.

In this paper, we solve the optimal voltage synthesis problem for power-aware satellite system design. The voltage synthesis problem is to find the (optimal) voltages of resources in the system that satisfy system-level energy constraints and maximize overall performance. The period of a satellite's orbit is partitioned into several independent regions. Each region has its own event rate, computation demand per input, importance, and energy consumption per computation. We assume the event rate and computation demand are not deterministic but stochastic. With those inputs, our technique synthesizes an optimal voltage for each region that maximizes aggregated throughput in the period while total energy consumption is within the energy budget.

Motivational example:
As an example of an LEO satellite, the weather satellite NOAA10 tracked the following regions on January 26th, 2001 at around 11:00am. Figure 1 is downloaded from the J-Track by NASA [7]. The satellite covered seven different regions during the period. Region R_1 corresponds to the Pacific Ocean, R_2 to North America, R_3 to the Arctic and Greenland, R_4 to the Atlantic ocean, R_5 to Europe and Africa, R_6 to the Atlantic ocean, and R_7 to Antarctica. Based on the characteristics of the similar satellite NOAA12, its period is 101.3 minutes and it's in polar orbit.

The synthetic characteristics in each region are shown in Table 1. The duration d_i at region R_i is the length of the satellite's stay over the region. The arrival rate λ_i at region R_i is the average arrival rate of inputs per millisecond. The service rate $\mu_{i,ref}$ at region R_i is the average number of inputs that can be processed per millisecond at the reference voltage V_{ref} shown in Table 2. The weight w_i of an input for region R_i denotes how valuable an output is when compared to the outputs in other regions. Energy consumption $e_{i,ref}$ for region R_i is the average energy at the reference voltage V_{ref} consumed to process an input in milli-Joules. Energy consumption is assumed to depend on the service time.

We assume that computation in a region may differ from the computation in the other regions. Likewise, the size of input data can be different for each region. The buffer size B denotes how many inputs can be stored in the buffer. The maximum input loss probability Λ_i for region R_i is the maximum allowed input loss probability for the region.

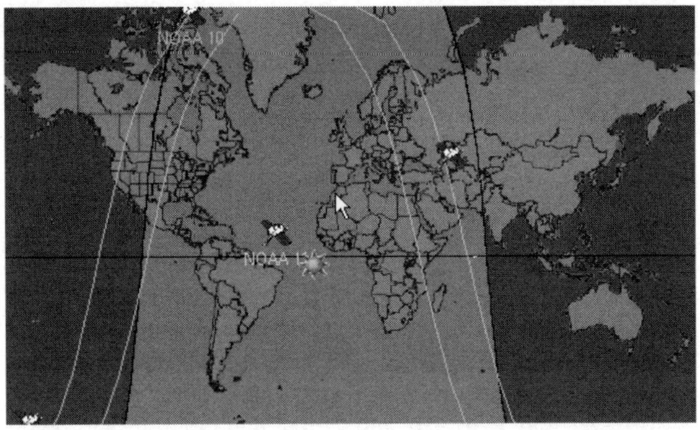

Figure 1. A screenshot of Jtrack

Table 1. Characteristics of the system per region

Region	R_1	R_2	R_3	R_4	R_5	R_6	R_7
d_i(min.)	26.1	11.6	11.6	8.7	14.5	11.6	20.3
λ_i	0.4	1	0.2	0.4	1.2	0.4	0.2
$\mu_{i,ref}$	1	1.5	1	1	1.8	1	1
w_i	1	1.5	0.9	0.7	1.2	0.7	0.7
$e_{i,ref}$(mJ)	2.1	2.5	1.8	2.1	2.5	2.1	1.8
B	100	50	80	100	70	100	80
Λ_i	0.1	0.02	0.1	0.05	0.01	0.05	0.1

Table 2. System constraints

V_{ref}	V_t	E_{Total}
3.3V	0.6V	3300 J

As an example, the second column labeled R_1 in Table 1 is interpreted as follows. The satellite flies over the Pacific Ocean for 26.1 minutes. During that time 0.4 inputs arrive per millisecond on the average. The satellite can process one input per millisecond on the average in the region at the reference voltage. The average energy consumption for processing one input is 2.1 milli-Joules in the region. The system can hold up to 100 inputs in its buffer. Any more inputs are discarded. Due to the limited buffer space, input loss is expected. However, there is a maximum allowable input loss probability for each region. Up to 10 percent of inputs gathered during the flight over Pacific Ocean can be discarded.

System-wide constraints are described in Table 2. According to Table 2, the satellite can use up to its maximum energy budget E_{Total}, which is 3300 Joules, for computation within a period. The system is assumed to be voltage schedulable between its reference voltage V_{ref}, 3.3 Volts, and its threshold voltage V_t, 0.6 Volt, shown in Table 2.

Our interest lies in how to derive the optimal voltage for each region that produces maximum performance with a given energy constraint E_{Total}. Performance is defined as the sum of inputs processed multiplied by their weights. The solution to the example is derived using our optimal algorithm in Section 3.3.

The rest of this paper is organized as follows. Section 2 presents our model and an overview of the problem. In Section 3, the power optimization techniques and voltage synthesis algorithms are discussed. Simulation results are presented in Section 4. Related work is presented in Section 5. We conclude in Section 6.

2. MODEL AND PROBLEM OVERVIEW

A set of regions $\{R_i \mid i = 1 .. N\}$ exists within a period T. Each region R_i has its own characteristics: data arrival rate (λ_i), service rate at the reference voltage ($\mu_{i,ref}$), energy consumption per input processing at the reference voltage ($e_{i,ref}$), duration (T_i), buffer size (B_i), and weight of an output (w_i) in that region. The weight value associated with a region reflects the importance of an output produced in the region. The performance (Ω_i) of a region R_i is the product of the weight and the number of outputs in the region, which is $\Omega_i = T_i w_i \lambda_i (1 - p_{i,loss})$, where $p_{i,loss}$ denotes the probability of input loss due to the limited buffer space. Performance in a period is described as the sum of the performance of the regions in the period, which is $\sum_i \Omega_i = \sum_i T_i w_i \lambda_i (1 - p_{i,loss})$. Our goal is to find an optimal voltage assignment of each region so that overall performance is maximized within the energy budget.

Periodic Energy Recharge: In the case of satellites with solar panels, the battery is recharged periodically. Both battery charge and discharge occur throughout the entire orbit, although the charging amount at any time may vary. We assume that the amount of the energy usable during a period is predictable and available at the start of a period. This simple model may not capture dynamic behavior of energy supply and consumption.

Heterogeneous computation patterns within a period: As a satellite orbits around the earth, its covering areas change: from land to sea, from desert to mountains, etc. Different areas may have different computation needs and input characteristics. We assume that within a period, there may be multiple regions that have their own input arrival rates, computation requirements, data sizes and energy consumptions per data processing. Those patterns of regions in a period may change as the earth rotates on its axis. We assume that those patterns and their characteristics are known at the start of a period, which is reasonable for most satellites.

Input Arrival and Service Patterns: In this paper we assume inputs arrive in accordance with a Poisson process of rate λ_i within region R_i. The probability distribution of the service time is assumed to be exponential with parameter μ_i. Maximum system capacity is limited to K_i. We assume no correlation among the inputs, and one input produces one output. The system in region R_i is modeled with an M/M/1/K_i queueing system. To simplify the static analysis, we ignore the dependencies between different regions, and treat each region independently. The interference between regions is studied using simulation in Section 4.

Dynamic power consumption of a CMOS digital circuit: In CMOS digital circuits, latency and energy consumption by a task are given by the following equations (Eq. 1) and (Eq. 2) [2]. In the equations, v denotes supply voltage, V_t and V_{ref} denote threshold voltage and reference voltage respectively, which are inherent to the implementation technology, C is a technology dependent constant, and C_{eff} is the effective capacitance.

$$Latency = C\frac{v}{(v-V_t)^2},\quad V_t < v \le V_{ref} \qquad \text{(Eq. 1)}$$

$$Energy \approx (\#\text{ of switches}) \times C_{eff} \times v^2 \qquad \text{(Eq. 2)}$$

When latency $l_{i,ref}$ and average energy consumption $e_{i,ref}$ per input in region R_i at reference voltage $V_{i,ref}$ of the resource are given, latency $l_i(v_i)$ and average energy consumption $e_i(v_i)$ at the supply voltage v_i can be driven by the following equations.

$$e_i(v_i) = e_{i,ref} \times (\frac{v_i}{V_{i,ref}})^2 \qquad \text{(Eq. 3)}$$

$$l_i(v) = l_{i,ref} \times \frac{v \times (V_{i,ref} - V_{i,t})^2}{V_{i,ref} \times (v - V_{i,t})^2} \qquad \text{(Eq. 4)}$$

As the supply voltage decreases, latency increases and the corresponding operating frequency decreases. Likewise, the service rate μ_i at v_i decreases as the voltage v_i decreases shown in (Eq. 5), where $\mu_{i,ref}$ denotes the service rate at the reference voltage $V_{i,ref}$.

$$\mu_i = \mu_{i,ref}\frac{V_{i,ref}(v_i - V_{i,t})^2}{v_i(V_{i,ref} - V_{i,t})^2} \qquad \text{(Eq. 5)}$$

Design Constraints and Problem Overview: Design constraints consist of performance constraints and energy constraints. Performance constraints are given for each region R_i and they are described as (1) maximum input loss probability Λ_i, shown in (Eq. 6) and (2) the sum of the products of weight and the number of outputs per region shown in (Eq. 7). $p_{i,loss}$ denotes the probability of input loss in the region R_i due to the limited buffer space. The weight of an output w_i in region R_i denotes the importance of an output compared with the outputs of other regions. T_i denotes the duration of region R_i. The energy constraint E_{Total} is the total amount of energy that can be used for the processing in a period shown in (Eq. 8), where e_i denotes energy consumption in region R_i. The design goal is to find feasible voltages for the regions that maximize performance while the energy consumption is within the energy budget E_{Total} by tuning voltages and corresponding clock frequencies in the regions.

$$\forall R_i,\ p_{i,loss} \le \Lambda_i \qquad \text{(Eq. 6)}$$

$$Performance = \sum_i \Omega_i = \sum_i T_i \times w_i \times \lambda_i \times (1 - p_{i,loss}) \qquad \text{(Eq. 7)}$$

$$\sum_i E_i = \quad_i T_i \times \lambda_i \times (1 - p_{i,loss}) \times e_i \le E_{Total} \qquad \text{(Eq. 8)}$$

3. STATIC VOLTAGE SYNTHESIS FOR MAXIMUM PERFORMANCE

In this section, we present an analytical framework for our optimal voltage synthesis technique. We first present the technique for a single region model, and then we extend it for a multiple region model.

3.1 A Period with a Single Region

Throughout a period, we assume the system is homogeneous and has fixed parameters: arrival rate λ, service rate μ_{ref} at the reference voltage V_{ref}, system capacity K, maximum input loss probability Λ and average energy consumption per input e_{ref} at the reference voltage V_{ref}.

The system can be modeled with a single M/M/1/K queue. The steady state probability p_n that there are n data in the M/M/1/K queue is described in (Eq. 9) where ρ is defined in (Eq. 10) [10]. The input loss probability p_{loss} is equal to the steady state probability.

$$p_n = \begin{cases} \frac{(1-\rho)\rho^n}{1-\rho^{K+1}}, & \text{if } \rho \ne 1, n = 0,..,K \\ \frac{1}{K+1}, & \text{if } \rho = 1, n = 0,..,K \end{cases} \qquad \text{(Eq. 9)}$$

$$\rho = \frac{\lambda}{\mu} = \frac{\lambda}{\mu_{ref}\frac{V_{ref}(v-V_t)^2}{v(V_{ref}-V_t)^2}} = \frac{\lambda v(V_{ref}-V_t)^2}{\mu_{ref}V_{ref}(v-V_t)^2} \qquad \text{(Eq. 10)}$$

(Eq. 9) and (Eq. 10) show that the input loss probability depends on the supply voltage. With a higher supply voltage, the input loss probability becomes smaller. Maximum performance is achieved at the smallest p_{loss} value within the energy budget. If the smallest p_{loss} value within the energy budget is larger than the maximum input loss probability Λ, there is no feasible solution. The system produces the largest throughput at the largest v value which satisfies both (Eq. 11) and (Eq. 12). It is trivial to find an optimal solution. We first find v using (Eq. 10), (Eq. 11), and $p_{loss} = \Lambda$. If the resulting v value is not within the feasible range, no feasible solution exists. If there is a feasible v value, the optimal supply voltage is the largest v value satisfying (Eq. 11) and (Eq. 12)

$$E = T \times \lambda \times (1 - p_{loss}) \times e_{ref} \times (\frac{v}{V_{ref}})^2 \le E_{Total} \qquad \text{(Eq. 11)}$$

$$p_{loss} \le \Lambda \qquad \text{(Eq. 12)}$$

3.2 A Period with Multiple Regions

Period T is partitioned into several sub-intervals, and we call them regions denoted by R_1, ..., R_N. Each region R_i has its own parameters: λ_i, μ_i, K_i, T_i, w_i, e_i and Λ_i. We assume no correlation among the inputs, and one input produces one output. The system in region R_i is modeled with a M/M/1/K_i queueing system. We ignore the dependencies between different regions. The assumption of the independence between regions is investigated using simulation in Section 4.

Each region R_i is modeled with an M/M/1/K_i queue. The steady state probability $p_{i,n}$ that there are n data in the M/M/1/K_i queue is described in (Eq. 13) and (Eq. 14). The input loss probability $p_{i,loss}$ is equal to $p_{i,Ki}$ and it must be equal to or smaller than maximum loss probability Λ_i for all regions while the sum of the energy consumption of all regions must be within energy budget, which is shown in (Eq. 7) and (Eq. 8).

$$p_{i,n} = \begin{cases} \frac{(1-\rho_i)\rho_i^n}{1-\rho_i^{K_i+1}}, & \text{if } \rho_i \ne 1, n = 0,..,K_i \\ \frac{1}{K_i+1}, & \text{if } \rho_i = 1, n = 0,..,K_i \end{cases} \qquad \text{(Eq. 13)}$$

$$\rho_i = \frac{\lambda_i}{\mu_i} = \frac{\lambda_i}{\mu_{i,ref}\dfrac{V_{i,ref}(v_i - V_{i,t})^2}{v_i(V_{i,ref} - V_{i,t})^2}} = \frac{\lambda_i v_i (V_{i,ref} - V_{i,t})^2}{\mu_{i,ref} V_{i,ref}(v_i - V_{i,t})^2} \qquad \text{(Eq. 14)}$$

It is a trivial process to find the voltages that satisfy the maximum allowable loss probabilities for all regions. However, it is not trivial to find voltages that maximize system performance. We propose an optimal algorithm that finds voltages of the regions producing maximum system throughput within energy constraints. We first define a function $\phi_i(v_i)$ of each region R_i, which is shown in (Eq. 15).

$$\phi_i(v_i) = \frac{dE_i}{d\Omega_i} = \frac{dE_i}{dv_i}\frac{dv_i}{d\Omega_i} = \frac{dE_i}{dv_i}\frac{1}{\dfrac{d\Omega_i}{dv_i}} \qquad \text{(Eq. 15)}$$

Function $\phi_i(v_i)$ of each region R_i shows how quickly energy consumption changes as performance changes. Function $\phi_i(v_i)$ increases monotonically as v_i increases in the interval $V_{i,t} < v_i \le V_{i,ref}$, which is shown by Lemma 2. The proofs of the following lemmas can be found in [9][8].

Lemma 1. When the value of v_i is not bounded, $\phi_i(v_i)$ values should be equal for all regions in the optimal solution.

Lemma 2. Function $\phi_i(v_i)$ increases monotonically as v_i increases where $\lambda_i > 0$, $\mu_i > 0$, $V_{i,t} < v_i \le V_{i,ref}$.

Lemma 3. Let n regions $R_1, R_2, R_3, .., R_n$ exist in period T, and their $\phi_i(v_i)$ values be equal. Let energy consumption in period T be E, and performance be Ω. When more energy ΔE is given for more performance and reference voltages are assumed to be infinite, the performance increase $\Delta\Omega$ is maximized where the $\phi_i(v_i)$ values of all regions are the same.

From Lemma 2 and Lemma 3, the maximum throughput gain compared to the setting where their original $\phi(v)$ values are the same is obtained at the same $\phi(v)$ value for all resources. The assumption of infinite reference voltage guarantees the existence of a voltage at the resulting $\phi(v)$ value.

Based on this observation, we propose an optimal voltage assignment algorithm that produces maximum performance subject to energy constraints.

Algorithm 1. Optimal Voltage Assignment for regions in a period

(1) **S** ← $\{i \mid R_i\}$; $E \leftarrow E_{Total}$
(2) For $i \in$ **S** [
(3) Calculate v_i, $\phi_i(v_i)$ and $E_i(v_i)$ where $p_{i,loss} = \Lambda_i$
(4) $V_{i,bot} \leftarrow v_i$
(5) If ($v_t > V_{i,ref}$) [No feasible solution exists!! Exit.]]
(6) If ($E < \sum_{i \in S} E_i(v_i)$)[No feasible solution exists!! Exit.]
(7) If ($E \ge \sum_{i \in S} E_i(V_{i,ref})$)[Solution found!! Exit.]
(8) **Q** ← $\{i \mid v_i = V_{i,ref}\}$; **S** ← **S** – **Q**; $E \leftarrow E - \sum_{i \in Q} E_i$
(9) **T** ← $\{(i, \phi_i(v)) \mid i \in$ **S** and ($v = V_{i,ref}$ or $v = V_{i,bot}$)$\}$
(10) Sort **T** in descending order of $\phi_i(v)$
(11) While (**S** is not empty) [
(12) m ← number of unique $\phi_i(v)$ values in **T**
(13) if ($m = 0$) [
(14) Maximize $\sum_{i \in S} \Omega_i$ subject to $E = \sum_{i \in S} E_i(v_i)$ using the method of Lagrange multipliers.]
(15) Φ ← ($\lfloor m/2 \rfloor$+1)-th largest $\phi_i(v)$ value in **T**
(16) For $i \in$ **S** [
(17) $v'_i \leftarrow v$, where $\phi_i(v'_i) = \Phi$
(18) if ($v'_i > v_{i,ref}$) then [$v'_i \leftarrow v_{i,ref}$]
(19) If ($\psi \ge \sum_{i \in S} E_i(v'_i)$)[/* Upper Half */
(20) For ($i_i \in$ **S**) [$v_i \leftarrow v'_I$]
(21) If ($E = \sum_{i \in S} E_i(v'_i)$) [Solution found!! Exit]
(22) $Low \leftarrow \Phi$
(23) **Q** ← $\{i \mid v_i = V_{i,ref}\}$; **P** ← $\{(i, \phi_i(v)) \mid \phi_i(v) \le \Phi\}$]
(24) Else ($E < \sum_{i \in S} E_i(v'_i)$)[/* Lower Half */
(25) $High \leftarrow \Phi$
(26) **Q** ← $\{i \mid v'_i = V_{i,bot}\}$; **P** ← $\{(i, \phi_i(v)) \mid \phi_i(v) \ge \Phi\}$]
(27) **S** ← **S** – **Q**
(28) $E \leftarrow E - \sum_{i \in Q} E_i(v_i)$
(29) **T** ← **T** – $\{(i, \phi_i(v)) \mid i \in$ **Q** and ($v = V_{i,bot}$ or $V_{i,ref}$)$\}$
(30) **T** ← **T** – **P**]

The algorithm first finds an initial solution that satisfies input loss constraints of all regions. Each region is evaluated independently and has a single solution. If the resulting solution does not satisfy energy constraints, there is no feasible solution, which is shown in lines (1) through (6). The while loop in lines (11) through (32) is a binary search algorithm that reduces the search space by half, as shown in line (15). The size of the search space is at most $2N$ where N is the number of regions in the period. At each iteration of the while loop, the median $\phi_i(v)$ value Φ in set **T** is chosen and the corresponding voltage is estimated in lines (15) through (18). If the $\phi_i(v)$ value of a region at the reference voltage is smaller than Φ, its voltage is set at the reference voltage. When the energy consumption at Φ is smaller than the energy constraint E_{Total}, Φ is set to the lower bound of the search space and the regions with current voltages equal to the reference voltage are removed from further consideration, which is shown in lines (22) – (23) and (27) – (30). When the energy consumption is equal to the energy constraint, the optimal solution is found in line (21).

Otherwise, Φ is set to the upper bound, and the regions with current voltages equal to the initial voltages are removed from further consideration as shown in lines (25) – (26) and (27) – (30). When set **T** becomes empty as shown in line (13), the upper bound 'High' and the lower bound 'Low' of $\phi(v)$ value are given. Lagrange's method is used to find an optimal solution as shown in line (14).

We present a simple example to show the behavior of the algorithm. Assume that there are two regions in a period R_1, R_2, $\phi_1(V_{1,bot}) = a_1$, $\phi_1(V_{1,ref}) = b_1$, $\phi_2(V_{1,bot}) = a_2$, $\phi_2(V_{2,ref}) = b_2$, $a_1 < a_2 < b_1 < b_2$. In its first iteration of the while loop, there are four unique ϕ values in **T** and the third largest ϕ value in **T** is chosen as Φ that is a_2. Energy consumption at Φ is smaller than the energy constraint. According to line (23), **Q** = {} and **P** = {(1, a_1),(2, a_2)}, and **T** becomes {(1, b_1), (2, b_2)}. In the second iteration, the algorithm probes b_1. Now, the energy consumption at b_1 is larger than the energy constraint. According to line (26), **Q** = {} and **P** = {(1, b_1), (2, b_2)}. Since set **T** is empty, the algorithm finds optimal voltages $v_{1,opt}$ and $v_{2,opt}$ using Lagrange's method, which is shown in line (14). More explanation of the example and the proof of Theorem 1 are found in [9].

Theorem 1. When there is a feasible design that meets energy constraints and input loss constraints for all regions, Algorithm 1 finds an optimal solution with maximum performance.

When we assume the existence of the inverse function of $\phi_i(v_i)$, the complexity of Algorithm 1 is O(N log N), where N denotes the number of regions in a period and solving for the *Lagrange* multiplier is assumed to be done in O(N) time [4].

3.3 Solution of the motivational example

We used Algorithm 1 to solve the motivational problem shown in Table 1 and Table 2. We used binary search algorithm to find Lagrange's multiplier in line (14) of Algorithm 1 with 0.0001 percent error bound. It takes less than 8 milliseconds to find the solution on a Pentium III-733MHz PC running Linux.

	R_1	R_2	R_3	R_4	R_5	R_6	R_7
V_d	1.795	2.528	1.374	1.844	2.544	1.844	1.340
Λ_i	0.1	0.02	0.1	0.05	0.01	0.05	0.1
$p_{i,loss}$	0.1	0.02	0.02	0.05	0.01	0.05	0.075

4. SIMULATION

The analytical framework assumes there is no interference between two neighboring regions. However, this may not be true. Region R_i can interfere with R_{i+1} when the queue is not empty or the processor is not idle at the time of the region change. In this situation, the processor is working on data collected in region R_i even though the satellite is flying over region R_{i+1}. This situation gets worse when region R_i is highly loaded and the duration of region R_{i+1} is short.

To observe interference between regions, we simulated the behavior of the motivational example and its variants. Three queue management policies at the boundaries of the regions are simulated which include (1) no management, (2) drop all inputs in the queue except the one that is being processed, and (3) drop all inputs including the one that is currently processed.

The simulated result of the example is shown in Table 3. In Table 3(a), $p_{i,loss}$ denotes the loss probability given by the analysis and $p_{i,loss(i)}$ denotes the average packet loss probability observed by the simulation using the *i-th* queue management policy at the boundaries of the regions. Likewise, $\sigma_{i,loss(1)}$ denotes standard deviation of the packet loss probability using the *i-th* queue management policy.

Our analytic result does not depend on how large the arrival rate and service rate are but does depend on their ratio. To observe the influence of a large and small arrival rate and service rate, we scaled the service rate and arrival rate by 10, 0.1 and 0.01 while their ratio remained the same. The results are summarized in Table 4, Table 5, and Table 6. From those tables, we observe that our analysis technique predicts well for high arrival and service rates. At low arrival and service rates, our analytical result deviates from simulation results. As the arrival and service rates decrease as shown in Table 6, interdependency among the regions at the boundaries between regions gets bigger and the technique predicts better for queue management policies (2) and (3) than no management policy (1).

Table 3. Simulation result of the example.

	R_1	R_2	R_3	R_4	R_5	R_6	R_7
$p_{i,loss}$	0.1	0.02	0.02	0.05	0.01	0.05	0.075
$p_{i,loss\ (1)}$	0.10	0.02	0.02	0.05	0.01	0.05	0.075
$p_{i,loss\ (2)}$	0.10	0.02	0.02	0.05	0.01	0.05	0.076
$p_{i,loss\ (3)}$	0.10	0.02	0.02	0.05	0.01	0.05	0.076
$\sigma_{i,loss\ (1)}$	2e-3	1e-3	3e-3	3e-3	7e-3	3e-3	3e-3
$\sigma_{i,loss\ (2)}$	2e-3	1e-3	3e-3	3e-3	7e-4	3e-3	3e-4
$\sigma_{i,loss\ (3)}$	2e-3	1e-3	3e-3	3e-3	7e-4	3e-3	3e-4

(a)

	Energy	σ_{energy}	Performance	$\sigma_{performance}$
(1)	3.3e6	1748	3.4e6	1750
(2)	3.3e6	1754	3.4e6	1764
(3)	3.3e6	1752	3.4e6	1763

(b)

Table 4. Simulation result (10 times service rate and arrival rate)

	R_1	R_2	R_3	R_4	R_5	R_6	R_7
$p_{i,loss}$	0.1	0.02	0.02	0.05	0.01	0.05	0.075
$p_{i,loss\ (1)}$	0.1	0.02	0.02	0.05	0.01	0.05	0.075
$p_{i,loss\ (2)}$	0.1	0.02	0.02	0.05	0.01	0.05	0.075
$p_{i,loss\ (3)}$	0.1	0.02	0.02	0.05	0.01	0.05	0.075
$\sigma_{i,loss\ (1)}$	6e-4	3e-4	8e-4	9e-4	2e-4	9e-4	7e-4
$\sigma_{i,loss\ (2)}$	6e-4	3e-4	8e-4	9e-4	2e-4	9e-4	7e-4
$\sigma_{i,loss\ (3)}$	6e-4	3e-4	8e-4	9e-4	2e-4	9e-4	7e-4

(a)

	Energy	σ_{energy}	Performance	$\sigma_{performance}$
(1)	33.0e6	5627	34.3e6	5404
(2)	33.0e6	5785	34.3e6	5666
(3)	33.0e6	5785	34.3e6	5666

(b)

Table 5. Simulation result (0.1 times service rate and arrival rate)

	R_1	R_2	R_3	R_4	R_5	R_6	R_7
$p_{i,loss}$	0.1	0.02	0.02	0.05	0.01	0.05	0.075
$p_{i,loss\ (1)}$	0.1	0.02	0.02	0.05	0.01	0.05	0.074
$p_{i,loss\ (2)}$	0.1	0.02	0.02	0.05	0.01	0.05	0.076
$p_{i,loss\ (3)}$	0.1	0.02	0.02	0.05	0.01	0.05	0.076
$\sigma_{i,loss\ (1)}$	5e-3	3e-3	8e-3	9e-3	2e-3	8e-3	8e-3
$\sigma_{i,loss\ (2)}$	5e-3	3e-3	9e-3	9e-3	2e-3	8e-3	8e-3
$\sigma_{i,loss\ (3)}$	5e-3	3e-3	9e-3	9e-3	2e-3	8e-3	8e-3

(a)

	Energy	σ_{energy}	Performance	$\sigma_{performance}$
(1)	0.33e6	552	0.34e6	562
(2)	0.33e6	564	0.34e6	564
(3)	0.33e6	565	0.34e6	563

(b)

Table 6. Simulation result (0.01 times service rate and arrival rate)

	R_1	R_2	R_3	R_4	R_5	R_6	R_7
$p_{i,loss}$	0.1	0.02	0.02	0.05	0.01	0.05	0.075
$p_{i,loss\ (1)}$	0.11	0.05	8e-3	0.05	0.03	0.03	0.064
$p_{i,loss\ (2)}$	0.10	0.02	0.04	0.06	0.01	0.06	0.080
$p_{i,loss\ (3)}$	0.10	0.02	0.04	0.06	0.01	0.06	0.081
$\sigma_{I,loss\ (1)}$	2e-3	1e-2	2e-2	3e-2	7e-3	2e-2	0.026
$\sigma_{I,loss\ (2)}$	2e-3	1e-2	1e-2	3e-2	3e-2	2e-2	0.026
$\sigma_{I,loss\ (3)}$	2e-3	1e-2	2e-2	3e-2	7e-3	2e-2	0.026

(a)

	Energy	σ_{energy}	Performance	$\sigma_{performance}$
(1)	32.4e3	198	33.7e3	201
(2)	32.9e3	175	34.2e3	177
(3)	32.9e3	175	32.9e3	177

(b)

5. RELATED WORK

There have been many run-time techniques for system-level power reduction. Dynamic voltage scheduling techniques change the voltages of resources and corresponding clock frequencies [5][6][14]. Dynamic power management techniques shut down resources at idle time for additional power savings [11]. Dynamic power management techniques have been extended with a dynamic voltage scheduling technique [13]. In real-time systems, exploiting resource slack time has been studied for reducing power consumption in a single CPU system [1][11] and a distributed system [8]. Low-power techniques have been an important topic for CAD research [1][12]. However, most of them are chip-level VLSI design techniques, such as multiple voltage scheduling among functional units in a chip [3][11]. These techniques address the problem of assigning a supply voltage from a finite number of pre-known supply voltages to each operation in a data flow graph so that the resulting schedule minimizes power consumption and satisfies timing constraints.

Our technique is similar to a dynamic voltage scheduling technique in that it synthesizes the supply voltage of a resource in a region based on stochastic event prediction. However, our technique is unique in solving the power-aware optimization problem subject to a performance constraint rather than the low-power scheduling problem. While low-power scheduling techniques consider only power consumption, our technique considers both system performance and energy consumption and synthesizes optimal voltages that maximize system performance.

6. CONCLUSION

We presented an analytical optimal voltage synthesis technique for a satellite application. The technique optimizes voltages for each region of a satellite's period so that the system performance is maximized under an energy budget. We presented simulation results showing the quality of the packet loss probabilities at the regions estimated by the analysis.

7. ACKNOWLEDGMENTS

This effort is sponsored by Defense Advanced Research Projects Agency (DARPA) through the Air Force Research Laboratory, USAF, under agreement number F30602-00-2-0548.

8. REFERENCES

[1] L. Benini and G. D. Micheli, "System-Level Power Optimization: Techniques and Tools," International Symposium on Low Power Electronics and Design, pp. 288-293, 1999.

[2] A. P. Chandrakasan, S. Sheng, and R. W. Brodersen, "Low-Power CMOS Digital Design," IEEE Journals of Solid-State Circuits, pp. 473-484, vol. 27, no. 4, April 1992.

[3] J. Chang and M. Pedram, "Energy Minimization Using Multiple Supply Voltages," IEEE Transactions on Very Large Scale Integrated (VLSI) Systems, vol. 5, no.4, Dec. 1997.

[4] K. Govil, E. Chan, and H. Wasserman, "Comparing Algorithms for Dynamic Speed-Setting of a Low-Power CPU," Mobile Computing and Networking, November 1995.

[5] J. Dey, J. Kurose, and D. Towsley, "On-Line Scheduling Policies for a Class of IRIS (Increasing Reward with Increasing Service) Real-Time Tasks," IEEE Transactions on Computers, Vol. 45, No. 7, Jul. 1996.

[6] I. Hong, G. Qu, M. Potkonjak, and M. B. Srivastava, "Synthesis Techniques for Low-Power Hard Real-Time Systems on Variable Voltage Processors," IEEE Real-Time Systems Symposium, pp. 178-187, Dec. 1998.

[7] JTRACK, NASA http://liftoff.msfc.nasa.gov/realtime/JTrack/, 2001

[8] D. Kang, S. Crago, and J. Suh, "Power-Aware Design Synthesis Techniques for Distributed Real-Time Systems," ACM Workshop on Languages, Compilers, and Tools for Embedded Systems (LCTES) '01, Jun. 2001.

[9] D. Kang, J. Suh, and S. Crago, "An Optimal Voltage Synthesis Technique for a Power-Efficient Satellite Application," unpublished USC/ISI technical report, also found at http://www.east.isi.edu/publications/satellite.pdf.

[10] Leonard Kleinrock, "Queueing Systems," volume I, John Wiley & Sons, 1975.

[11] Q. Quiu and M. Pedram, "Dynamic Power Management Based on Continuous-Time Markov Decision Processes," Design Automation Conference, June 1999.

[12] D. Singh, J. Rabaey, S. Rajgopal, N. Sehgal, and T. Mozdzen, "Power Conscious CAD Tools and Methodologies: A Perspective," Proceedings of IEEE, pp. 570-594, vol. 83, no. 4, April 1995.

[13] T. Sinumic, L. Benini, A. Acquaviva, P. Glynn, and G. D. Micheli, "Dynamic Voltage Scaling and Power Management for Portable Systems," Design Automation Conference, June 2001.

[14] M. Weiser, B. Welch, A. Demers, and S. Shenker, "Scheduling for Reduced CPU Energy," Operating Systems Design and Implementation, Nov. 1994.

Fast and Accurate Behavioral Simulation of Fractional-N Frequency Synthesizers and other PLL/DLL Circuits

Michael H. Perrott
Microsystems Technology Laboratory, MIT
perrott@mit.edu, http://www-mtl.mit.edu/~perrott

ABSTRACT

Techniques for fast and accurate simulation of fractional-N synthesizers at a detailed behavioral level are presented. The techniques allow a uniform time step to be used for the simulator, and can be applied to a variety of phase locked loop (PLL) and delay locked loop (DLL) circuits beyond fractional-N synthesizers, as well as to a variety of simulation frameworks such as Verilog and Matlab. Simulated results from a custom C++ simulator are shown to compare well to measured results from a prototype fractional-N synthesizer using a Σ-Δ modulator to dither its divide value.

Categories and Subject Descriptors

I.6.5 [**Simulation and Modeling**]: Model Development

General Terms

Algorithms

Keywords

fractional-N,frequency,synthesizer,sigma,delta,PLL,DLL

1. INTRODUCTION

Fractional-N frequency synthesizers provide high speed frequency sources that can be accurately set with very high resolution, which is of high value to many communication systems. Figure 1 illustrates a fractional-N synthesizer, which consists of a phase-frequency detector (PFD), charge pump, loop filter, voltage controlled oscillator (VCO), and a frequency divider that is dithered between integer values to achieve fractional divide ratios. This paper will focus on a class of fractional-N synthesizers known as Σ-Δ frequency synthesizers [12], for which the divide value is dithered according to the output of a Σ-Δ modulator [8].

Dithering of the divide value by the Σ-Δ modulator allows high frequency resolution to be achieved [12], but also has the negative side effect of introducing quantization noise that degrades the overall PLL noise performance. It is highly desirable to be able to simulate the effects of this quantization noise, along with other noise sources in the PLL shown in Figure 2, on the overall PLL performance. It is also desirable to simulate the dynamic response of the synthesizer in response to variations of the Σ-Δ input in order to evaluate stability and characterize the performance of the system when it is used as a transmitter [10].

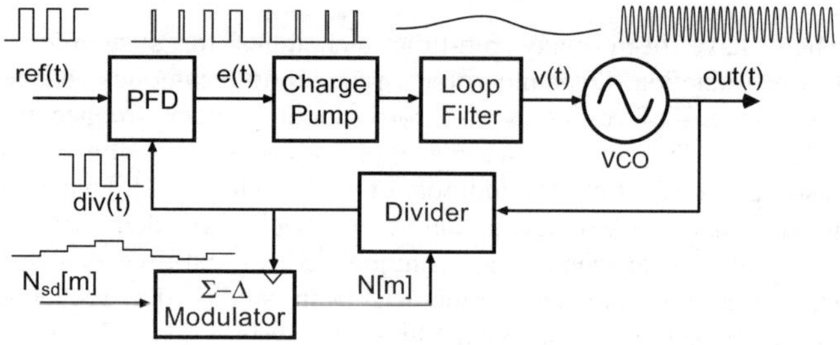

Figure 1: Σ-Δ synthesizer and associated signals.

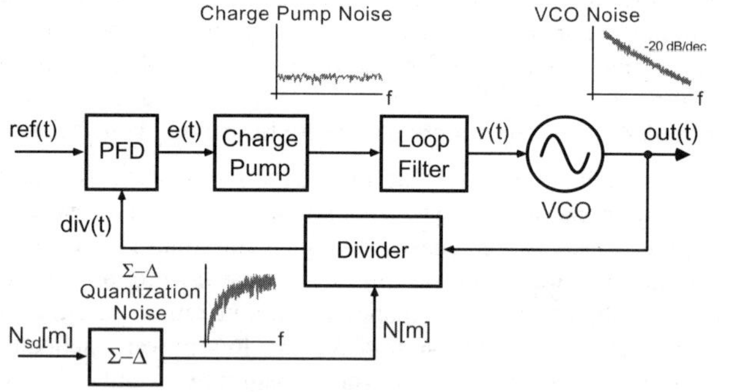

Figure 2: Spectral densities of PLL noise sources.

Simulation of fractional-N synthesizers is particularly challenging for a variety of reasons. First, the high output frequency of the synthesizer (often in the GHz range) imposes a high simulation sample frequency for traditional simulators. Unfortunately, the overall PLL dynamics have a bandwidth that is typically three to four orders of magnitude lower in frequency than the output frequency (often 100 kHz to 1 MHz bandwidth compared to a GHz output frequency). Thus, traditional simulators take a long time to compute the dynamic response of the system since many simulation samples are required. This is the classical problem that is encountered with the simulation of PLL circuits. For noise simulation, the fractional-N synthesizer adds the additional constraint that its behavior is non-periodic in steady-state due to the dithering action of the divide value, which prevents the use of methods developed for periodic steady-state conditions [7] as used with simulators such as SpectreRF.

In contrast to the above approaches, two techniques are presented in this paper that allow fast and accurate simulation of both dynamic and noise performance of fractional-N synthesizers at a detailed behavioral level. The first provides accurate representation of the continuous-time (CT) PFD output with a discrete-time (DT) sequence using an area conservation principle. The second allows a dramatic reduction of the simulation sample frequency, and therefore a longer sample period, by including the divider implementation in the VCO simulation module. Both of these methods allow a uniform time sample period to be used, and also allow non-iterative computation of the sample values of the various signals within the system. The uniform time sample period allows the results of the simulator to be readily examined in the frequency domain without resampling, and the non-iterative computation allows the technique to be easily used in mainstream simulators such as Verilog, VHDL, Matlab, and custom C/C++ programs.

An outline of the paper is as follows. Section 2 provides an overview of the discretization technique for representation of the CT PFD output with a DT sequence, and presents the corresponding mathematical analysis. Section 3 describes, at a high level, how the discretization technique can be implemented with the PFD described in terms of basic building blocks such as registers and logic gates. Section 4 focuses on the method of dramatically reducing the required simulation sample rate by combining the VCO and divider functions into one simulation block. Section 5 compares simulated results from a custom C++ simulator to measured results of an actual circuit implementation described in [10]. Finally, Section 6 concludes.

2. PFD DISCRETIZATION TECHNIQUE

For simulation based on uniform time sampling, a straightforward approach of converting the CT PFD output to a DT sequence is to apply the sampling operation shown in Figure 3 [6]. Unfortunately, this approach effectively quantizes the location of the PFD edges according to the simulation sample period, T_s. A reasonable assessment of the dynamic performance of the PLL can be achieved if T_s is made sufficiently small. However, the resulting quantization noise overpowers the true noise characteristics of the signals, and prevents proper noise analysis of the overall PLL.

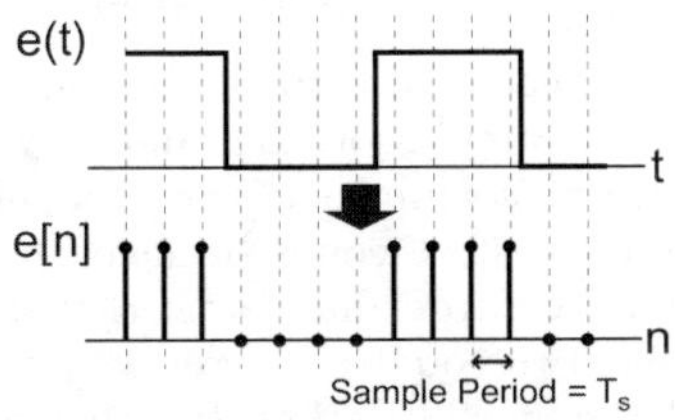

Figure 3: Classical uniform time sample method.

To solve the quantization noise issue, event-driven simulation methods have been developed for classical frequency synthesizers that align simulation samples precisely to the edges of the PFD output [2, 5, 1]. Although higher accuracy can be achieved with such methods, they are generally more complicated than uniform time sampling methods. Either closed-form calculation of the loop filter step response must be developed and then inserted into the simulation, or iterative methods, as used in SPICE or Verilog-A, must be incorporated into the simulator to calculate the loop filter response with varying time steps. The former approach is tedious and typically restricted to a low loop filter order, so that most of the recent methods focus on the latter approach [2, 5, 1]. In this case, the up-front work of the designer is minimized, but the simulation time is often longer due to the iterative calculations that are performed at each time step. Unfortunately, for either case, event-driven simulators have not yet been applied successfully to the noise analysis of fractional-N frequency synthesizers in which the divide value is dynamically varied.

In contrast to the above approaches, a constant time step method is proposed in this paper that applies an area conservation principle when converting the CT PFD output to the DT domain. This approach allows non-iterative computation of the loop filter dynamics by allowing them to be converted from CT to DT using either impulse invariance or bilinear transform methods [9]. Figure 4 illustrates an example of the resulting DT PFD signal, along with the corresponding DT loop filter impulse response. The charge pump is ignored in this analysis for simplicity; its effect can be included by simply scaling the PFD output by the value of the charge pump current. In the example, we see that the DT PFD output takes on values at its transitions that vary between 0 and 1 depending on the location of the transition edge. The DT version of the loop filter simply consists of a DT filter whose impulse response corresponds to samples of the CT impulse response of the loop filter, $h(t)$.

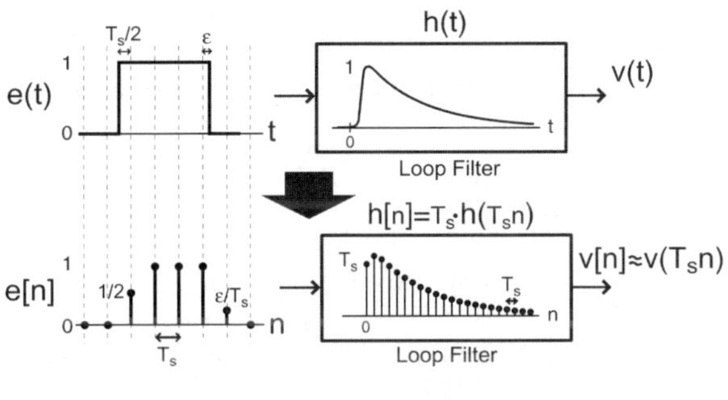

Figure 4: Proposed discretization method.

The discretization procedure is now discussed in more detail. As illustrated by Figure 5, we can view the CT PFD output, $e(t)$, as a series of rectangular pulses with height of one or zero and a width and time offset that varies according to the location of PFD edges. For rectangular pulses not associated with edges, the width corresponds to the sample period of the simulation, T_s. For rectangular pulses at edge boundaries, the width of the pulse varies between 0 and T_s as shown in the figure. In either case, these pulses look like impulses to the loop filter so that, from an intuitive standpoint, their influence can be characterized by two parameters — their area and time offset. Therefore, in line with this intuition, the corresponding DT PFD signal, $e[n]$, is chosen as samples that have amplitude proportional to the *area* of the respective rectangular pulse in that time sample interval. The area of each pulse corresponds to its associated timing parameter ϵ shown in the figure — the method of calculating ϵ for each pulse will be discussed in Section 4. It will be shown that the proposed discretization procedure yields highly accurate results, fast computation, and a simple implementation framework.

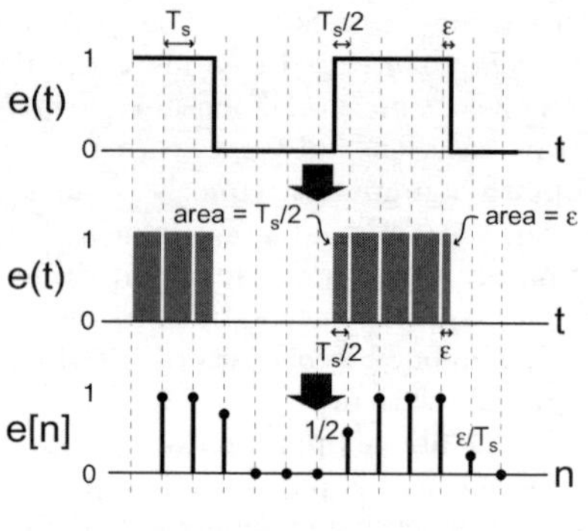

Figure 5: Details of PFD discretization technique.

To mathematically justify the technique, let us begin by specifying a notation for the rectangular pulses composing $e(t)$ that are shown in Figure 5, namely

$$\text{rect}(t, \epsilon_k) = 1 \text{ for } -\epsilon_k/2 \le t \le \epsilon_k/2, \quad 0 \text{ elsewhere.}$$

We can describe the loop filter output as

$$v(t) = \sum_{k=-\infty}^{\infty} \text{rect}(t - kT_s - \Delta t_k, \epsilon_k) * h(t), \tag{1}$$

where $*$ denotes convolution, and Δt_k shifts its associated pulse in time according to the value of ϵ_k and whether the transition edge is rising or falling. Note that ϵ_k and Δt_k are constrained to $0 \le \epsilon_k \le T_s$ and $-T_s < \Delta t_k < 0$.

Taking the Fourier Transform of both sides of Equation 1, we obtain

$$V(jw) = \sum_{k=-\infty}^{\infty} e^{-jw(kT_s+\Delta t_k)} \frac{2\sin((\epsilon_k/2)w)}{w} H(jw), \tag{2}$$

where $H(jw)$ is the loop filter frequency response.

One might be bothered that the Fourier Transform is being taken with sequences that are stochastic in nature, namely ϵ_k and Δt_k. This step is justified by noting that these sequences will be defined and finite in duration for a specific simulation run. However, we cannot infer statistical properties from Equation 2 or the analysis that follows.

Equation 2 can be simplified in light of the following assumptions:

- $H(jw)$ is a lowpass filter such that $|H(jw)| \approx 0$ for $|w| > w_o$,
- The sample frequency, $1/T_s$, is much higher than the bandwidth of $|H(jw)|$, so that $w_oT_s \ll 1$. Since $|\epsilon_k| < T_s$, we have $\sin((\epsilon_k/2)w_o) \approx (\epsilon_k/2)w_o$.

The second assumption is well justified in practice since it is typical for $w_oT_s \ll 1/100$. For instance, the author recommends sampling at a rate that is greater than a factor of 10 above the reference frequency, which is, in turn, at least a factor of 10 higher in frequency than the loop filter bandwidth in Hz, f_o, to achieve stable PLL dynamics [11]. In this case, $f_oT_s < 1/100$, so that $w_oT_s < 1/(2\pi 100)$.

Based on the above assumptions, Equation 2 is approximated as

$$V(jw) \approx \sum_{k=-\infty}^{\infty} \epsilon_k e^{-jw(kT_s+\Delta t_k)} H(jw).$$

The inverse Fourier Transform of the above expression is

$$v(t) = \sum_{k=-\infty}^{\infty} \epsilon_k h(t - kT_s - \Delta t_k). \tag{3}$$

We are now ready to develop the DT model of the PFD/loop filter that we are seeking. We begin by sampling Equation 3:

$$v(nT_s) = \sum_{k=-\infty}^{\infty} \epsilon_k h((n-k)T_s - \Delta t_k). \tag{4}$$

The above formulation requires nonuniform sampling of $h(t)$ due to the inclusion of Δt_k — it is preferable to remove this parameter if it can be shown that its influence is negligible. We will examine this issue using a specific form for $h(t)$, and then comment on the extension of the analysis for more general forms of $h(t)$.

Let us assume that the loop filter corresponds to a lead/lag network with transfer function of the form:

$$H(jw) = K\frac{jw + w_z}{jw(jw + w_o)}.$$

Using the method of partial fractions [9], it is straightforward to show that the corresponding loop filter impulse response is of the form

$$h(t) = K_1 e^{-w_o t}u(t) + K_2 u(t) = (K_1 e^{-w_o t} + K_2)u(t),$$

where $u(t)$ is the unit step (0 for $t < 0$, 1 for $t \ge 0$), and K_1 and K_2 are constant scale factors. Plugging the above expression into Equation 4, we obtain

$$v(nT_s) = \sum_{k=-\infty}^{\infty} \epsilon_k (K_1 e^{-w_o((n-k)T_s - \Delta t_k)} + K_2) u((n-k)T_s - \Delta t_k). \tag{5}$$

We note that:

$$u((n-k)T_s - \Delta t_k) = u((n-k)T_s) \quad \text{since } -T_s < \Delta t_k < 0,$$

and

$$\begin{aligned} e^{-w_o((n-k)T_s - \Delta t_k)} &= e^{w_o \Delta t_k} e^{-w_o(n-k)T_s} \\ &\approx (1 + w_o\Delta t_k)e^{-w_o(n-k)T_s}. \end{aligned}$$

Therefore, the effect of the time shift operation by Δt_k has no influence on samples of $u(t)$, and only slightly modulates the amplitude of samples of the exponential response $e^{-w_o t}$. Although its effect could be incorporated into the numerical model, the author has found that it can be safely ignored given that two conditions are met. The first condition is that w_oT_s be much less than 1 so that $1 + w_o\Delta t_k \approx 1$. This condition is satisfied in practice; it was argued earlier in this section that we can typically expect that $w_oT_s \ll 1/100$. The second condition is that the sample rate of the simulator, $1/T_s$, be chosen as an integer multiple of the nominal frequency of the pulses associated with the CT PFD output. The effect of violating either of these conditions is the introduction of false spurs in the output phase noise of the synthesizer, as will be demonstrated in Section 5.

Given that the above conditions are satisfied, we can simplify Equation 5 as

$$v(nT_s) = \sum_{k=-\infty}^{\infty} \epsilon_k (K_1 e^{-w_o(n-k)T_s} + K_2) u((n-k)T_s),$$

so that we have

$$v(nT_s) = \sum_{k=-\infty}^{\infty} \frac{\epsilon_k}{T_s} T_s h((n-k)T_s). \qquad (6)$$

Equation 6 is the conclusion of our effort, and matches the picture representation of the method illustrated in Figure 4 when $e[n] = \epsilon_n / T_s$.

Although the above analysis was performed for a simple lead/lag loop filter, higher order filters can be analyzed in similar fashion using the partial fraction expansion method. Specifically, high order filters have impulse responses that consist of a sum of exponentials, with each exponential corresponding to a distinct pole in the loop filter. The impact of e_k and Δt_k on each of these exponentials can be assessed in the same manner as derived above.

3. IMPLEMENTATION OF PFD

Now that it has been established that the PFD output signal can be accurately represented as a DT sequence using a principle of area conservation, let us examine the practical issue of representing the PFD topology in simulation code. As revealed by Figure 6, computation of $e[n]$ for a given PFD topology requires that transition information be passed along and processed by primitive elements such as registers and logic gates. Basic operations such as complementing signals must also be supported.

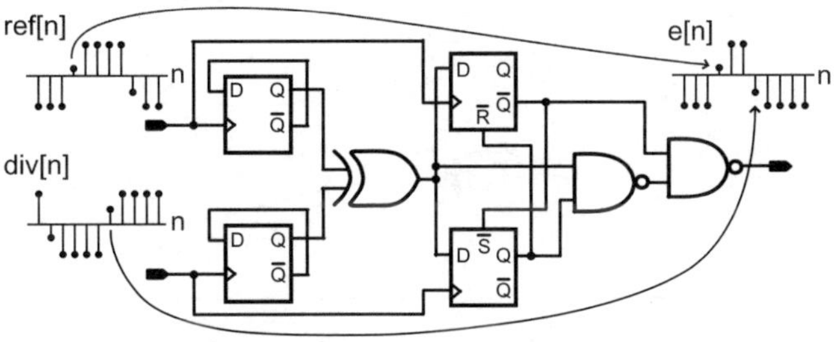

Figure 6: XOR-based PFD.

As illustrated in Figure 7, the complement operation is easily achieved as a sign change by slightly modifying the representation of $e[k]$ such that it alternates between -1 and 1 as opposed to 0 and 1. The new representation is achieved through the transformation $e[n] \Longrightarrow 2e[n] - 1$.

The transfer of transition information through primitives is illustrated in Figure 8 for a register and a representative logic gate, namely the 'and' gate. In the case of the register, the relevant timing information is contained in the clock signal. Specifically, whenever there is a transition at the output of the register, the location of that transition in time is set by the location of the rising (or falling) edge of the clock. As shown in the figure, this information is transfered to the register output by simply passing on the clk transition value when the output transitions in the same direction, and passing on the complement of the clk transition value when the output transitions in the opposite direction. In the case of the 'and' gate, either input can cause the output to transition. As gleaned from the figure, it is straightforward to determine which input is causing the transition, and appropriately pass its edge location value to the output of the 'and' gate. Similar arguments can be made for more complicated registers that include set and reset functions, and other primitives such as 'or' and 'xor' gates.

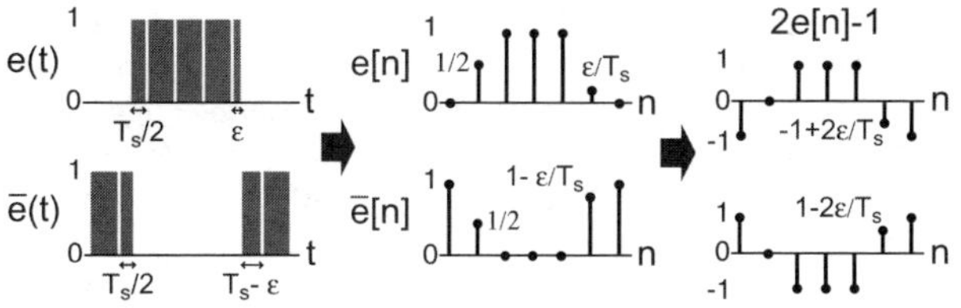

Figure 7: A better signal representation.

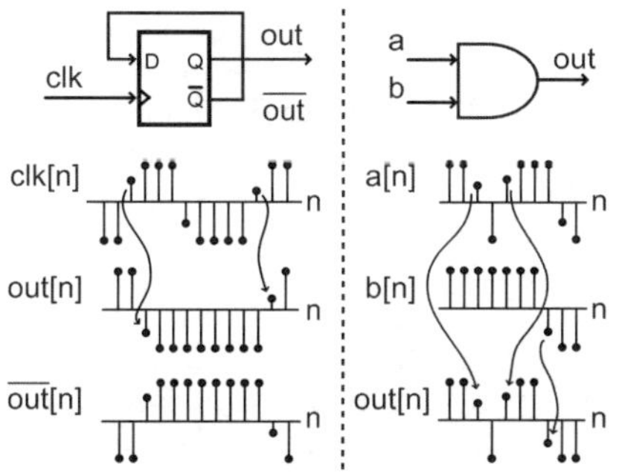

Figure 8: Example of register and logic gate signals.

4. VCO AND DIVIDER

Simulation of the VCO and divider portions of the PLL is now discussed, and a technique illustrated whereby the simulation sample period, T_s, can be set according to the reference frequency rather than the much higher VCO frequency. This technique typically allows more than two orders of magnitude speedup in simulation time of the PLL since the VCO frequency is typically more than two orders of magnitude higher than the reference frequency. Unlike a previous method that provided speedup for a PLL with no divider and a memoryless phase detector by modeling the VCO entirely in the phase domain [13], the presented technique accommodates fractional-N synthesizers that have dividers with dynamically varying value and digital PFD topologies as described in the previous sections. The key idea behind the technique is to combine the VCO and divider into one computation block.

To begin, let us define the phase of the VCO, $\Phi_{vco}(t)$, as the integral of its output frequency. Since the output frequency of the VCO is varied about its nominal frequency by its input voltage, we have:

$$\Phi_{vco}(t) = \int_{-\infty}^{t} 2\pi(K_v v(\tau) + f_c) d\tau + \Phi_{vn}(t), \qquad (7)$$

where $v(t)$ is the VCO input voltage, K_v is the VCO gain (Hz/V), f_c corresponds to the nominal VCO frequency when $v(t) = 0$, and $\Phi_{vn}(t)$ is VCO noise as illustrated in Figure 2. To model a nonlinear relationship from input voltage to VCO frequency, the VCO input would be multiplied by a polynomial gain expression rather than just K_v. In general, $\Phi_{vco}(t)$ looks like a ramp in time, and rising edges of the VCO output occur each time it increments by 2π radians.

Simulation of the VCO is performed by simply discretizing Equation 7 as

$$\Phi_{vco}(nT_s) = \sum_{k=-\infty}^{n} 2\pi T_s(K_v v(kT_s) + f_c) + \Phi_{vn}(nT_s). \quad (8)$$

To prevent loss of information in the CT to DT conversion, $1/T_s$ must be higher than twice the highest frequency content of $v(t)$ and $\Phi_{vn}(t)$, as stated by the Nyquist theorem [9]. From a practical perspective, this condition will often be satisfied by meeting the sampling requirements for the PFD output. Choosing a sample rate such that $w_o T_s \ll 1/100$ is obviously sufficient for $v(t)$ since it is the output of the loop filter with bandwidth w_o rad/s. $\Phi_{vn}(t)$ is also bandlimited since it rolls off at -20 dB/decade, or more, before eventually hitting a low valued noise floor [3].

Rising edges of the divider output occur every $N[m]$ rising edges of the VCO output, where $N[m]$ corresponds to the instantaneous divide value. Therefore, as illustrated on the left side of Figure 9, the VCO phase, $\Phi_{vco}(t)$, completely specifies the location of the divider edges. As such, we can determine the value of ϵ_k at the transition points of the divider output based entirely on computed VCO phase, as shown on the right side of Figure 9 for first-order interpolation [2]. Note that the expressions assume that phase is wrapped every $2\pi N[m]$ radians. It suffices to choose a sample rate for the VCO phase computation according to the divider frequency, which equals the reference frequency, rather than the much higher VCO frequency.

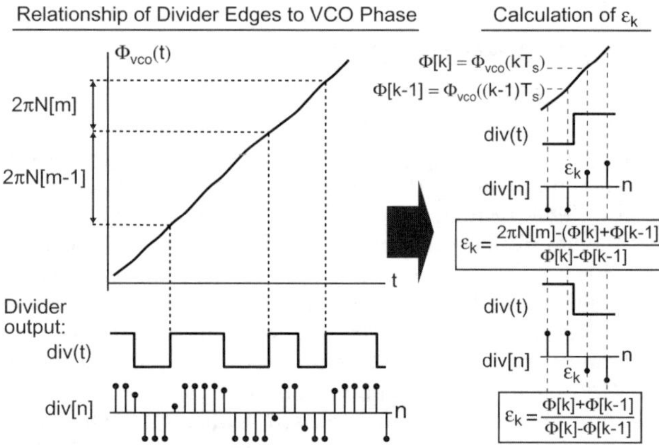

Figure 9: VCO and divider signals.

5. RESULTS

The results of simulating the dynamic behavior and noise performance of a prototype synthesizer described in [10] using a custom C++ simulator employing the presented techniques are now presented, and simulated noise compared to measured results. Figure 1 provides a block diagram of the prototype system; the reader is referred to [10] for more details. Both dynamic and noise simulations will include the noise sources depicted in Figure 2, with VCO noise being input referred as a white noise source as described in [10]. Parameters associated with the noise sources are shown in Figure 10, which were computed from Hspice simulations and VCO measurements.

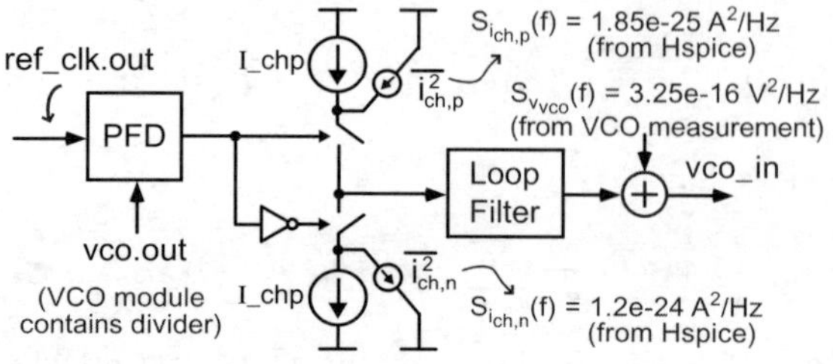

Figure 10: Model of charge pump and VCO noise.

Relevant characteristics of the prototype include a reference frequency of 20 MHz, a VCO with $f_c = 1.84$ GHz and $K_v = 30$ MHz/V, a second order Σ-Δ modulator, a charge pump that outputs ±1.5 μamps, a nominal divide value of 92.3, a PFD topology as shown in Figure 6 [4], and a lead/lag filter with transfer function

$$H(jw) = \frac{1 + jw/(2\pi f_z)}{C_3 jw(1 + jw/(2\pi f_p))},$$

where

$$f_z = 11.6 \text{ kHz}, \quad f_p = 127.2 \text{ kHz}, \quad C_3 = 30\text{e-}12.$$

Figure 11 shows the measured synthesizer phase noise of the prototype taken from [10], along with the measured open loop VCO noise from which the input referred VCO noise variance in Figure 10 was computed.

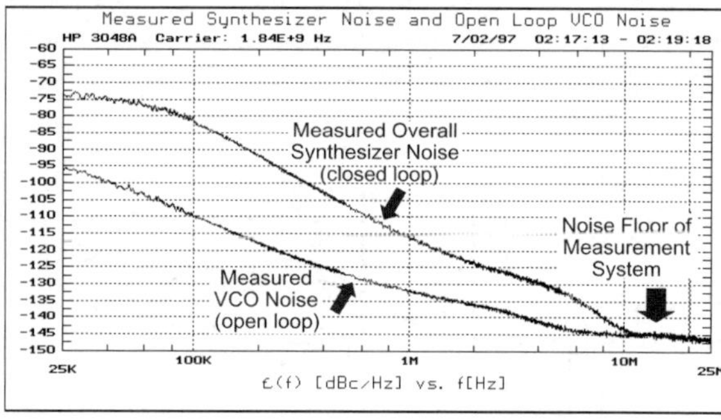

Figure 11: Measured synthesizer noise.

The simulation sample frequency was chosen as $1/T_s =$ 400 MHz, which is a factor of 20 higher than the reference frequency. The CT loop filter was converted to DT using the bilinear transform [9]. All simulations were run on a 650 MHz Pentium III laptop computer.

Beginning with dynamic behavior, Figure 12 shows the simulated VCO output frequency (constructed from the simulated VCO input) in response to variations at the input of the Σ-Δ modulator that include step and ramp functions. The step size is chosen to be large enough to knock the synthesizer out of frequency lock — the corresponding oscillations in the VCO output frequency are a result of cycle slipping before the VCO becomes frequency locked again. The subsequent ramp in divide value illustrates the high resolution of the synthesizer as its output frequency is varied over a 40 MHz range. For this simulation, 260 thousand time steps were computed in less than 5 seconds.

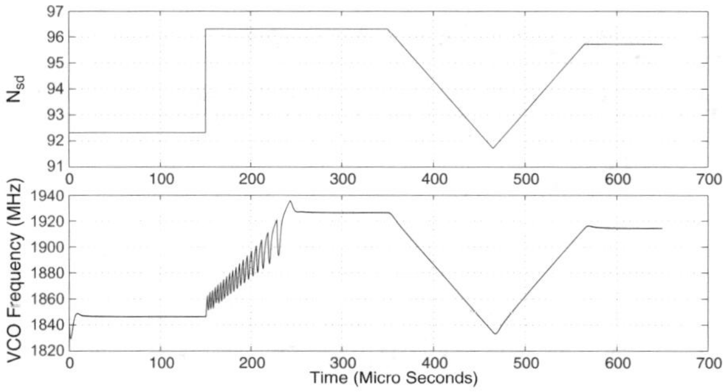

Figure 12: Simulated synthesizer dynamics.

Noise simulations of the prototype (constructed from the simulated VCO input) are shown in Figure 13 with the input to the Σ-Δ modulator being held constant. The top plot shows the simulated output noise spectral density with the simulation sample frequency, $1/T_s$, set to an integer multiple of the reference frequency, as recommended to reduce the effects of Δt_k in Equation 4. The bottom plot shows the impact of choosing $1/T_s$ to be a non-integer multiple of the reference frequency. We see that, in both cases, the simulated phase noise agrees quite well with measured results. The larger discrepancy at frequencies close to 100 kHz is probably due to non-ideal characteristics of the charge pump, such as duty cycle offset and transient dynamics, not being modeled. Such effects could be included within the given framework, but it is useful to observe that, despite ignoring such effects, the simulation results are still quite accurate for this prototype. As observed in the bottom plot, the impact of choosing $1/T_s$ to be a non-integer multiple of the reference frequency is that the reference spur at 20 MHz offset is aliased to other frequency values. This aliasing occurs due to the presence of harmonics above 402 MHz of the 20 MHz reference spur in the CT PFD output. For each simulation, 5 million time steps were computed in 80 seconds.

6. CONCLUSION

Two techniques were presented in this paper that allow fast and accurate simulation of fractional-N synthesizers at a detailed behavioral level using a uniform time sample period. The first provides accurate representation of the CT PFD output with a DT sequence using an area conservation principle. The second allows a dramatic reduction of the simulation sample frequency by including the divider implementation in the VCO simulation module. The techniques were incorporated into a custom C++ simulator, which was used to simulate the dynamic and noise performance of a prototype Σ-Δ frequency synthesizer. The simulated noise performance was shown to agree quite well with measured results. The techniques can also be applied to other phase locked loop circuits, and be implemented in other simulation frameworks such as Verilog and Matlab.

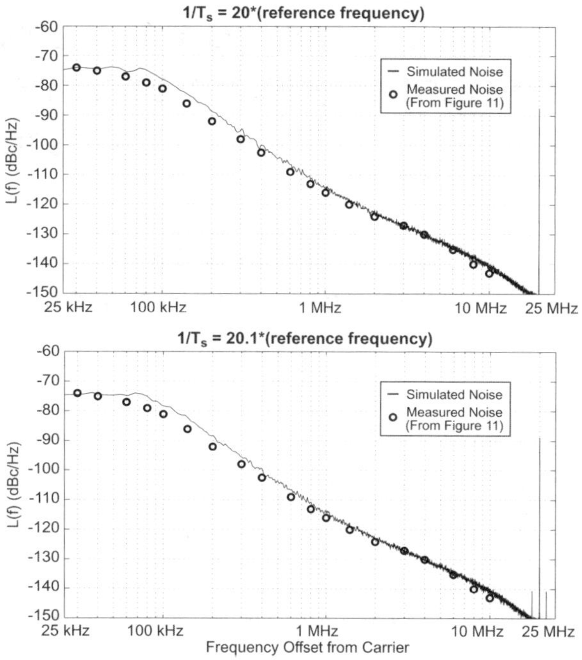

Figure 13: Simulated synthesizer phase noise.

7. REFERENCES

[1] B. De Smedt and G. Gielen. Nonlinear Behavioral Modeling and Phase Noise Evaluation in Phase Locked Loops. In *CICC*, pages 53–56, 1998.

[2] A. Demir, E. Liu, A. L. Sangiovanni-Vincentelli, and I. Vassiliou. Behavioral Simulation Techniques for Phase/Delay-Locked Systems. In *Custom Integrated Circuits Conference (CICC)*, pages 453–456, 1994.

[3] A. Hajimiri and T. Lee. A General Theory of Phase Noise in Electrical Oscillators. *IEEE Journal of Solid State Circuits (JSSC)*, 33(2):179–194, Feb. 1998.

[4] A. Hill and A. Surber. The PLL Dead Zone and How to Avoid It. In *RF Design*, pages 131–134, Mar. 1992.

[5] M. Hinz, I. Konenkamp, and E.-H. Horneber. Behavioral Modeling and Simulation of Phase-locked Loops for RF Front Ends. In *43rd Midwest Symp. on Circuits and Systems*, pages 194–197, 2000.

[6] D. Johns and K. Martin. *Analog Integrated Circuit Design.* Wiley, 1997.

[7] K. Kundert, J. White, and A. Sangiovanni-Vincentelli. *Steady-State Methods for Simulating Analog and Microwave Circuits.* Kluwer, Boston, 1990.

[8] S. Norsworthy, R. Schreier, and G. Temes. *Delta-Sigma Data Converters: Theory, Design, and Simulation.* IEEE Press, New York, 1997.

[9] A. V. Oppenheim and R. W. Schafer. *Discrete Time Signal Processing.* Prentice Hall, N.J., 1999.

[10] M. Perrott, T. Tewksbury, and C. Sodini. A 27 mW CMOS Fractional-N Synthesizer using Digital Compensation for 2.5 Mb/s GFSK Modulation. *JSSC*, 32(12):2048–2060, Dec. 1997.

[11] B. Razavi. *Monolithic Phase-Locked Loops and Clock Recovery Circuits: Theory and Design.* IEEE Press, New York, 1996.

[12] T. A. Riley, M. A. Copeland, and T. A. Kwasniewski. Delta-Sigma Modulation in Fractional-N Frequency Synthesis. *JSSC*, 28(5):553–559, May 1993.

[13] P. Van Halen and G. Boyle. SPICE-Compatible Behavioral Phase-Space Simulation Techniques for Phase-Locked Systems. In *38th Symposium on Circuits and Systems Conference*, volume 1, pages 53–56, 1996.

Time-domain steady-state simulation of frequency-dependent components using multi-interval Chebyshev method

Baolin Yang
Cadence Design Systems*
byang@celestry.com

Joel Phillips
Cadence Berkeley Laboratories
jrp@cadence.com

ABSTRACT

Simulation of RF circuits often demands analysis of distributed component models that are described via frequency-dependent multi-port Y, Z, or S parameters. Frequency-domain methods such as harmonic balance are able to handle these components without difficulty, while they are more difficult for time-domain simulation methods to treat. In this paper, we propose a hybrid frequency-time approach to treat these components in steady-state time-domain simulations. Efficiency is achieved through the use of the multi-interval Chebyshev (MIC) simulation method and a low-order rational-fitting model for preconditioning matrix-implicit Krylov-subspace solvers.

Categories and Subject Descriptors

I.6 [**Computing Methodologies**]: Simulation and Modeling

General Terms

Algorithms

Keywords

RF circuit simulation, frequency dependent, S parameter

1. INTRODUCTION

Special-purpose RF circuit simulators exploit the sparsity of the spectrum in order to make the computations tractable[1]. RF communication circuits often contain components (particularly passive components) such as transmission lines, integrated inductors, and SAW filters, where distributed effects are important. Distributed components are those that are not conveniently described by a finite-dimensional (i.e. lumped) state-space model, for example because they have an infinite-dimensional state space. We will use the term to mean any component more conveniently described by a frequency-domain or convolutional representation.

*The author is now with Celestry Design Technologies, Inc.

DAC 2002, June 10-14, 2002, New Orleans, Louisiana, USA.

Circuit simulation methods solve circuit equations in either the time domain or frequency domain. Time-domain simulation methods discretize circuit equations using finite difference methods such as the second-order Gear method. The advantage of the time-domain methods is that they can select time-points based on localized error estimates and as a result can easily handle strongly nonlinear phenomena and sharp transitions in circuit waveforms. Frequency-domain simulation methods, such as the harmonic balance method, are popular for RF simulation and have the advantage of attaining spectral accuracy for smooth waveforms. Development of matrix-free Krylov-subspace algorithms [2, 3, 4] has made dedicated RF simulation tools even more popular as they can now be used to analyze circuits with thousands of devices.

In contrast to frequency-domain techniques, which have no more difficulty simulating distributed devices than lumped devices, a drawback of time-domain methods is that the simulation of distributed or passive components is more difficult. There are generally two approaches. The first approach is to apply a model generation tool[5, 6] to generate finite-dimensional state-space models for those components. These state-space models are easily simulated in the time domain. The second approach is to compute the impulse response of the distributed element, and apply a convolution approach at each timestep to obtain the time-domain response of the element. In transient analysis, one may use direct convolution approaches to simulate these components[7] without significant difficulty.

In RF simulation, however, such as periodic steady-state analysis via shooting-Newton methods[3], problems arise due to the fact that distributed components have infinite-dimensional state. First, it is not sufficient to calculate the state of the circuit at a single point of the periodic time interval in order to describe the periodic steady-state, the distributed state of the devices must be computed[8]. This in turn implies that the sensitivity calculation in shooting methods involves more than the two ending points of the periodic interval. Second, the distributed components destroy the block-banded structure of the Jacobian that is exploited in preconditioner computation[3].

In this paper, we propose a novel hybrid frequency-time method to include these distributed passive components in an RF simulator. Because we will take a matrix-implicit approach to equation solution, the distributed devices can be treated as black-boxes with a port-to-port operator given most naturally by frequency-dependent S/Y/Z-parameters. Since it is only necessary to evaluate this operator, not to write an explicit matrix, we propose an approach that is a dual to traditional harmonic balance. We formulate the equations in the time-domain, then use Fourier analysis to switch between candidate time-domain waveform solutions and the frequency-domain spectrum where we evaluate the frequency-domain devices. With this combined frequency/time-domain approach, we can inherit the advantages of both time-domain methods in describing hard non-linearities, and harmonic balance methods in conveniently treating frequency-domain devices.

In principle, any time domain method can be used, but for traditional low-order polynomial methods such as second-order Gear, the Fourier transformation step is expensive. Therefore we advocate the multi-interval Chebyshev (MIC) method [9] as our time-domain simulation method, as we can preserve the accuracy of the frequency-domain model description, but with less computational time. Within the multi-interval Chebyshev context, the remainder of the paper gives details of the formulation, describes optimal partitioning of distributed device evaluation between time and frequency domains, and considers practical aspects such as effective preconditioner construction.

2. CIRCUIT EQUATIONS

2.1 Formulating Distributed Component Equations

Lumped circuit behavior is usually described by a set of N non-linear differential algebraic equations (DAEs) that can be written, without loss of generality, as

$$\frac{dQ(v(t))}{dt} + I(v(t)) + u(t) = 0\,, \tag{1}$$

where $Q(v(t)) \in \mathbb{R}^n$ represents contribution of the dynamic elements such as inductors and capacitors, $I(v(t) \in \mathbb{R}^n$ is the vector representing static elements such as resistors, $u(t) \in \mathbb{R}^n$ is the vector of inputs, and the vector $v(t) \in \mathbb{R}^n$ contains state variables such as node voltages. Distributed components must be described using a more general form. Suppose that there are K linear, time-invariant, but distributed devices, each of which possess a convolutional (impulse response) representation. Then we may formulate the time-domain circuit equations as

$$\frac{dQ(v(t))}{dt} + I(v(t)) + \left[\sum_{k=1}^{K} P_k \int_{-\infty}^{t} H_k(t-\tau) P_k^T v(\tau)\,d\tau\right] + u(t) = 0\,, \tag{2}$$

Here $H : (-\infty, t] \to \mathbb{R}^{p_k \times p_k}$ is the $p_k \times p_k$ matrix of impulse responses for the kth distributed component. It is related to the rest of the circuit equations via a matrix $P_k \in \mathbb{R}^{n \times p}$. P_k^T is an operator that extracts the relevant portion of the state vector. That is, P_k is one in the (i,j) entry if the jth connection to the kth device is through the state variable with index i.

In RF simulation, we are usually interested in solving steady-state problems. The periodic steady-state problem is the simplest and serves as a model for all the others. It is also important to study because it forms the basis of more advanced analyses, such as cyclostationary noise analysis, multi-frequency distortion analysis, or envelope simulation. The periodic steady-state solution is the solution of Eq. (2) that also satisfies the two-point boundary condition $v(t+T) - v(t) = 0$ for some period T and all t. It is therefore sufficient to calculate the periodic steady-state solution in the interval $(0, T]$, i.e. we seek $v \in (0, T] \times \mathbb{R}^n$. This is useful because we may utilize the (discrete) Fourier transform $\mathcal{F} v : \mathbb{Z} \to \mathbb{C}$, and its inverse $\mathcal{F}^{-1} = \mathcal{F}^T$, in representing and manipulating signals. $\mathcal{F}$ maps v on $[0, T)$ to its frequency-domain representation, the Fourier series coefficients.

Without loss of generality, we can describe any physically-realizable distributed linear element as an N-port component described by an $N \times N$ frequency-dependent scattering (S) parameter matrix. As the convolution operation is diagonalized by the Fourier transform, we may write the circuit equations in terms of the frequency-dependent matrix $H(\omega)$ as

$$\frac{dQ(v(t))}{dt} + I(v(t)) + \left[\sum_{k=1}^{N} P_k \mathcal{F}^T H_k(\omega) \mathcal{F} P_k^T v\right] + u(t) = 0\,. \tag{3}$$

Here $H_k(\omega)$ is the (continuous) Fourier transform of the kth impulse response matrix. The relation between the scattering parameter matrix $S(\omega)$ of the underlying element and the matrix $H(\omega)$ depends on the way the circuit equations are stamped. For simplicity of presentation, we consider a 2-port device with S-parameter description to be stamped in the MNA formulation. We denote the potential at the first port as e_1 and e_1' and the second port as e_2 and e_2'. The currents at the two ports are denoted as I_1 and I_2. The port resistance is R_1 and R_2. If we ignore contributions from other circuit elements (*), formally the circuit equation that describes the behavior of the 2-port and defines $H(\omega)$ is:

$$\begin{bmatrix} 0 & 0 & 0 & 0 & 1 & 0 \\ 0 & 0 & 0 & 0 & -1 & 0 \\ 0 & 0 & 0 & 0 & 0 & 1 \\ 0 & 0 & 0 & 0 & 0 & -1 \\ 1-S_{11} & S_{11}-1 & -S_{12}\sqrt{\frac{R_1}{R_2}} & S_{12}\sqrt{\frac{R_1}{R_2}} & -(S_{11}+1)R_1 & -S_{12}\sqrt{R_1 R_2} \\ -S_{21}\sqrt{\frac{R_2}{R_1}} & S_{21}\sqrt{\frac{R_2}{R_1}} & 1-S_{22} & S_{22}-1 & -S_{21}\sqrt{R_1 R_2} & -(S_{22}+1)R_2 \end{bmatrix}$$
$$\cdot \begin{bmatrix} e_1 & e_1' & e_2 & e_2' & i_1 & i_2 \end{bmatrix}' = \begin{bmatrix} * & * & * & * & 0 & 0 \end{bmatrix}' \tag{4}$$

2.2 Discretization

In frequency-domain simulation methods one solves for the Fourier spectrum of the waveform instead of function values at a set of time points. It is straightforward to include distributed or passive components that may be described using frequency-dependent tabulated S-parameter. With the S-parameter information, one knows how to load the Jacobian of the circuit equation in frequency domain and set up the linear system. A linear solver, such as a Krylov-subspace iterative solver, can be used to solve the linear system.

In time-domain methods, on the other hand we solve for the waveform values, $u(t_j)$, at a set of time steps: $t_j,\ j = 0, 1, \cdots, M$. To perform the discretization at $t = t_k$, we seek an interpolating polynomial of degree p based on $p+1$ time points. The time derivative is approximated by evaluating the derivative of the interpolating polynomial at $t = t_k$. Thus the time-derivative of the approximate solution is required to match the differential equation at the point t_k. The time steps are not uniform and their variation can be big. It is difficult to simulate these components on this set of time points without loss of efficiency. We propose an approach that makes this happen. In our novel approach, advanced Krylov-subspace iterative solver and multi-domain Chebyshev method are used to achieve this purpose.

Here we give the basics of MIC method. Given the values, $u(t_j)$, at the Chebyshev Gauss-Lobatto collocation points, $t_j = \cos\frac{\pi j}{M},\ j = 0, \ldots, M$ in one subinterval, we seek an interpolating polynomial of degree M:

$$I_M u(t) = \sum_{k=0}^{M} \tilde{u}_k T_k(t)\,, \tag{5}$$

and the derivative of the polynomial can be computed by using backward recurrence relation for Chebyshev polynomials $T_k(t)$. We can also obtain a differentiation matrix D by noting that the interpolating polynomial can be expressed in terms of a Lagrange interpolation polynomial $g_j(t)$. The entries of the matrix D are computed by taking the analytical derivative of $g_j(t)$, $D_{kj} = g_j'(t_k)$. Refer to

[9] for the exact formulation of the matrix. As in harmonic balance, high accuracy is achievable because in each approximation interval all the desirable approximation properties of Chebyshev polynomials can be exploited.

3. HYBRID FREQUENCY-TIME DOMAIN SIMULATION

3.1 Matrix-Implicit Krylov Subspace Solution Technique

The major computational task for large-scale RF simulation is to solve the large linear systems generated from the circuit Jacobian matrix. In the nonlinear steady-state computation, a system must be solved at each step of Newton's method. The Jacobian J is composed of three pieces, the dynamic piece J_C, the static piece J_G, and the distributed piece J_D. With the Chebyshev discretization and

$$G(t_k) = \left.\frac{\partial I}{\partial v}\right|_{v(t_k)}, \quad C(t_k) = \left.\frac{\partial Q}{\partial v}\right|_{v(t_k)}, \tag{6}$$

we have

$$J_C = \begin{bmatrix} D^1 \odot I^1 & & & D_0^1 \odot I^1 \\ & & \ddots & \\ & D_0^2 \odot I^2 & D^2 \odot I^2 & \\ & & & \ddots \end{bmatrix} \begin{bmatrix} C(t_1) & & & \\ & C(t_2) & & \\ & & C(t_3) & \\ & & & \ddots \end{bmatrix} \tag{7}$$

where D^k is part of the Chebyshev differentiation matrix D in the kth interval ($[D_{ij}]|_{i,j=1,\cdots,M}$), D_0^k is $[D_{ij}]|_{i=1,\cdots,M;j=0}$, and I^k the identity matrix of dimension equal to the number of timepoints in the kth interval,

$$J_G = \begin{bmatrix} G(t_1) & & & \\ & G(t_2) & & \\ & & G(t_3) & \\ & & & \ddots \end{bmatrix} \quad \textit{and } J_D = \left[\sum_{k=1}^{N} P_k \mathcal{F}^T H_k(\omega) \mathcal{F} P_k^T\right]. \tag{8}$$

In small-signal and noise analysis, a related system must be solved at each frequency point of analysis. One notes that because of the frequency-dependent property of the distributed components, the transformation $\mathcal{F}$ is dependent on all time points in the simulation time interval. From Equation 3, it is clear that every p-port distributed element introduces p rows and columns into the Jacobian that are full. This can be seen by noting that the matrices $P\mathcal{F}^T$ and $\mathcal{F} P^T$ are dense. It is not possible to form and solve large systems of such a form efficiently using direct linear algebra. In our approach, as is now standard, a Krylov-subspace iterative solver such as GMRES is used. The use of these methods in circuit simulation are treated in detail elsewhere [2, 3, 4, 10].

We focus on the matrix-implicit treatment of the distributed components. To evaluate the term J_D, we need a procedure that takes time-domain waveforms as input, applies the Fourier transformation to transform them into the frequency domain, applies the transfer function of the distributed components, in terms of frequency-dependent S-parameters, to the frequency-domain result, and inverse-transforms to the time-domain. Such a procedure has as much flexibility in dealing with distributed-components as harmonic balance methods. In fact, there is more flexibility. We can extract any part of the transfer function that is best suited to evaluation in the time-domain and perform that part of the port-to-port transformation in time domain. For example, we can perform the delay transformation of transmission lines in the time domain and obtain much more accurate answers than possible by representing the delay using a sum of Fourier harmonics. All these operations fit in the matrix-vector product calculation of iterative solvers.

3.2 Time-Frequency Decomposition

Taking a two-port system as an example, we now explain the partitioning of the distributed component properties between the time- and frequency- domain representations. For each port in the distributed component, there is a port equation in the circuit equations. For a two-port system, we have

$$(1-S_{11})e_1 + (S_{11}-1)e_1' - S_{12}\sqrt{\frac{R_1}{R_2}}e_2 + S_{12}\sqrt{\frac{R_1}{R_2}}e_2' - (S_{11}+1)R_1 i_1 - S_{12}\sqrt{R_1 R_2}\, i_2 = 0\,, \tag{9}$$

for one port and a similar equation for the other port. Note that the S-parameter in this formula is frequency dependent. Hence the corresponding equations in time domain are more complicated and can be obtained using frequency-time-domain transformations. For periodic steady state analysis, the spectrum of solution, which we show how to compute in the next section, is at integer multiples of the fundamental frequency. We need the S-parameters at those frequencies and then we can obtain the equations in time domain.

Now we discuss decomposing the port equations into a frequency-independent part and a frequency-dependent S-parameter part. We decompose the first port equation into

$$A_1 = (1-\bar{S}_{11})e_1 - (1-\bar{S}_{11})e_1' - \bar{S}_{12}\sqrt{\frac{R_1}{R_2}}e_2 + \bar{S}_{12}\sqrt{\frac{R_1}{R_2}}e_2' - (1+\bar{S}_{11})R_1 i_1 - \bar{S}_{12}\sqrt{R_1 R_2}\, i_2 \tag{10}$$

and

$$B_1 = (\bar{S}_{11}-S_{11})e_1 + (S_{11}-\bar{S}_{11})e_1' - (S_{12}-\bar{S}_{12})\sqrt{\frac{R_1}{R_2}}e_2 + (S_{12}-\bar{S}_{12})\sqrt{\frac{R_1}{R_2}}e_2' - (S_{11}-\bar{S}_{11})R_1 i_1 - (S_{12}-\bar{S}_{12})\sqrt{R_1 R_2}\, i_2 \tag{11}$$

and enforce $A_1 + B_1 = 0$. Likewise we can decompose the second port equation.

In general, s-parameters with bar's in the split equations are the S-parameters of some time-domain effect, such as delay, DC offset, or linear scale. These time-domain effect can be calculated in time domain separately. One simple example is to choose constants for the approximation when S-parameters at high frequencies approach a constant. Thus, the frequency-dependent parts, B_1 and B_2, do not have any high-frequency effects, which can be calculated in frequency domain through Fourier transforms without exhibiting oscillations in sharp-transition regions of time-domain waveform, known as Gibbs phenomena.

3.3 Chebyshev Fourier Quadratures

It should be clear that, while the above procedure can be performed with time-domain periodic-steady-state solution method enforce periodicity explicitly, one key to efficient implementation is efficient evaluation of the Fourier transforms. The Chebyshev discretization achieves high accuracy with relatively few timepoints. With this property, one can afford to apply Fourier transformation and inverse Fourier transformation to transform the waveforms between frequency domain and time domain. The computational cost is $O(M^2)$, where M is the number of time steps. Because of the non-uniform timepoint distribution, we cannot base an implementation on the standard FFT, though we could use more sophisticated $O(M \log M)$ non-uniform Fourier analysis techniques[11].

In order to use the frequency-dependent S-parameters directly, the time-domain waveform is transformed into frequency domain. We use Chebyshev quadrature formulas to obtain the Fourier spectrum with high-order accuracy. The order of accuracy is basically equal to the number of Chebyshev collocation points, which in turn given by the number time steps in one Chebyshev time interval. Unlike low-order polynomial approximations such as Gear methods, the Chebyshev quadratures maintain spectral accuracy in this process. Assume the time steps we have are $t_i, i = 0, \cdots, i = M$. The integration formula is

$$\int_{t_0}^{t_M} v(t)dt = \sum_{i=0}^{M} w_i v(t_i) \tag{12}$$

where

$$w_i = \begin{cases} \frac{1}{M^2-1} & i = 0 \\ \frac{2}{M}(1 + \frac{(-1)^(i-1)}{M^2-1} + \sum_{k=1}^{\frac{M}{2}-1} \frac{2}{1-4k^2}\cos(2ki\frac{\pi}{M})) & i \le \frac{M}{2} \\ w_{M-i} & \frac{M}{2} < i \le M \end{cases} \tag{13}$$

Note that we are considering a periodic function. The Fourier spectrum can be calculated using the following formula:

$$\int_{t_0}^{t_M} v(t)e^{jk\frac{2\pi}{T}t}dt = \sum_{i=0}^{M} w_i v(t_i)e^{jk\frac{2\pi}{T}t_i}. \tag{14}$$

With the Fourier spectrum known, it is possible to perform the necessary part of the Jacobian and equation evaluation in the frequency domain, just as in harmonic balance. After we obtain the result of the operation in frequency domain, we then transform it back to the time domain using inverse Fourier transformation. Our time-domain simulation needs to have the result at M time points $t_i, i = 0, \cdots, M$. For all these transformations, the computational complexity of the multi-interval Chebyshev method and traditional low-order methods are the same, generally proportional to M or M^2. Because M for the Chebyshev methods is much less than that of traditional time-domain methods, the Chebyshev methods have a large advantage over traditional low-order methods.

This Fourier weight calculation method is subject to aliasing error. When a large k is required, the integration formula (14) may not have enough resolution, because M needs to be at least $2k$ for the formula to have any accuracy. Spectral accuracy is achieved when $\frac{M}{k}$ is large. However, one can interpolate the solution at $t_i, i = 0, \cdots, M$ to a finer Chebyshev grid. This interpolation keeps the high-order accuracy of the time-domain solution. Then one can perform integration of (14) on the fine grid to guarantee the accuracy. In practice, it is not necessary to really interpolate the solution. One only needs to pre-correct the integration weights, $w_i, i = 0, \cdots, M$, to realize this refinement in integration. This situation can occur when the time-domain waveform in an interval is much smoother than the Fourier harmonic to be evaluated.

4. PRECONDITIONING

4.1 Block-Lower Triangular Preconditioners

Rapid convergence of Krylov-subspace iterative methods requires construction of effective preconditioners. Shooting-based time-domain RF simulation methods[2, 3, 10] typically construct preconditioners by observing that that block-lower-triangular portion of the Jacobian is easily inverted, given that the diagonal blocks can be efficiently factored via sparse direct techniques. That is, to solve the preconditioned system $P^{-1}Jx = P^{-1}b$, the Jacobian is first decomposed into upper- and lower- block triangular pieces, $J = L + U$, with

$$U = \begin{bmatrix} 0 & & D_0^1 \otimes I^1 \\ & \ddots & \\ & & 0 \end{bmatrix} \begin{bmatrix} C(t_1) & & & \\ & C(t_2) & & \\ & & C(t_3) & \\ & & & \ddots \end{bmatrix}, \tag{15}$$

$L = \tilde{J}_C + J_G$, where

$$\tilde{J}_C = \begin{bmatrix} D^1 \otimes I^1 & & & \\ & & \ddots & \\ & D_0^2 \otimes I^2 & D^2 \otimes I^2 & \\ & & & \ddots \end{bmatrix} \begin{bmatrix} C(t_1) & & & \\ & C(t_2) & & \\ & & C(t_3) & \\ & & & \ddots \end{bmatrix}. \tag{16}$$

The preconditioner is then $P = \tilde{J}_C + J_G = L$. As P is block-lower-triangular, inversion is easy as long as the diagonal blocks can be factored or inverted efficiently, which is the case here. The problem with applying a similar technique, when distributed elements are present, is the term J_D. In the matrix-implicit context, it is not even clear how to construct the lower-diagonal piece, nor how to invert it, much less that this would be an effective strategy.

In general, solutions at nodes of the distributed components are coupled in time because of the Fourier transformation done in the periodic time interval. Existing preconditioning techniques for periodic steady-state simulation can not solve this difficulty. In this paper, we propose to use low-order lumped models to construct the distributed part of the preconditioner.

4.2 Rational Fitting

The first idea in our preconditioning strategy is to first construct an approximate $\tilde{H}(\omega)$ that is close to $H(\omega)$. One possibility to construct an approximate $\tilde{H}(\omega)$ is to use rational approximation[5] of the underlying device parameters $S(\omega)$. We propose constructing a rational approximation $\tilde{S}(\omega)$ of $S(\omega)$. Since, because the rational approximation will be used for a preconditioner, it does not have to be extremely accurate, we may approximate $S(\omega)$ with a low-order model, which is cheap to construct. This approach has the advantage that we may use the block-lower-triangular preconditioners discussed in the previous section. Constructing a rational function model is equivalent to finding a state-space model of the form

$$\frac{dx}{dt} = Ax + Bu(t), \quad y = Cx + Du \tag{17}$$

such that the transfer function $\tilde{S}(\omega) = D + C(i\omega I - A)^{-1}B$ approximates $S(\omega)$. We do this by requiring that the mean-square error

$$E = \sum_k W(\omega_k)^2 ||\tilde{S}(\omega_k) - S(\omega_k)||_2 \tag{18}$$

be minimized[5]. In the above, $W(\omega_k)$ represent the weighting, at frequency point k, for the least-squares minimization.

4.3 Lumped Model Preconditioner

Ideally, the preconditioner would then be $P = \tilde{J}_C + J_G + \tilde{J}_D$, where

$$\tilde{J}_D = \left[\sum_{k=1}^{N} P_k \mathcal{F}^T \tilde{H}_k(\omega) \mathcal{F} P_k^T v. \right] \tag{19}$$

Unfortunately, even with the rational model technique, we cannot invert matrices of this form efficiently. However, as state-space models have finite-dimensional state space, the rational function models may be included in an extended version of the circuit equations. The extended circuit equations will include the states of the rational fitting model, and so the Jacobian of the extended system will be the Jacobian of the original system treating the distributed component as the rational fitting model. Adopting the J, L, U notation, let

$$J_R = L_R + U_R \tag{20}$$

be the Jacobian for the circuit equations including the rational fitting model, where R denotes rational fitting, and L_R and U_R denote the upper- and lower-, respectively, block-triangular pieces as defined above. If there are M timesteps in the Chebyshev discretization of the periodic interval, and n circuit equations, we have $J \in \mathbb{R}^{Mn \times Mn}$ but $J_R \in \mathbb{R}^{M(n+q) \times M(n+q)}$ where q is the total number of additional state space variables introduced by the rational approximation of the frequency response of all the distributed components. Thus we cannot simply take $P = L_R$, as L_R and J have different dimensions.

Define an operator $Q : \mathbb{R}^{Mn} \to \mathbb{R}^{M(n+q)}$ that embeds the dimension $M(n+q)$ space into the Mn one – it is a section of the identity matrix. Q^T likewise extracts the original state space from the extended one. The relation that the Jacobian of the extended system will be the Jacobian of the original system treating the distributed component as the rational fitting model can be expressed as

$$J = Q^T J_R Q. \tag{21}$$

To use L_R as the preconditioner for Jacobian J, we may use Q^T to extract the relevant components, so that

$$P^{-1} = Q^T L_R^{-1} Q. \tag{22}$$

Note that the inverse of L_R can be easily computed because it is a block lower-triangular system. Note that the preconditioner will also include an approximation to $\tilde{H}$. If the rational fitting model is a fairly good approximation of the original system, the GMRES solver should be efficient whenever the block-lower-triangular strategy is also effective.

Note that, because we use the rational model as a preconditioner, its accuracy does not affect the accuracy of the final solution. If we were to use rational approximation to construct an equivalent lumped model, instead of evaluating the distributed components, the rational model would need to very closely track the frequency response, meaning it would often be of very high order. We would also have to introduce physical constraints on the rational model – such as passivity – that are computationally expensive to enforce for high-order multi-port models. Both the model generation process, and the later time-domain simulation, would be more computationally expensive than the strategy proposed here, where rational approximations of orders two or three are often enough to give effective preconditioners.

5. NUMERICAL RESULTS

In this section we show the effectiveness of the hybrid frequency-time multi-interval Chebyshev (MIC) method in periodic steady state simulation of distributed elements including a spiral inductor, a transmission line, and a SAW filter. We compare the new approach with the pure rational-fitting model generation approach. Efficiency of the low-order rational fitting model in building a preconditioner for the hybrid frequency-time approach is shown through comparing the convergence history of GMRES solvers with and without the preconditioner.

In the first example, we simulate the periodic steady state of a spiral inductor driven by a pulse source. The spiral inductor is described with an S-parameter table. The frequency range of the S-parameters is from $50MHz$ to $10GHz$. The fundamental frequency of the periodic steady state is $1GHz$. We show the numerical result, waveform at the output, of three simulations. First, we use

Figure 1: Comparison of low-order rational fitting, high-order rational fitting, and hybrid frequency-time MIC with low-order rational fitting preconditioner.

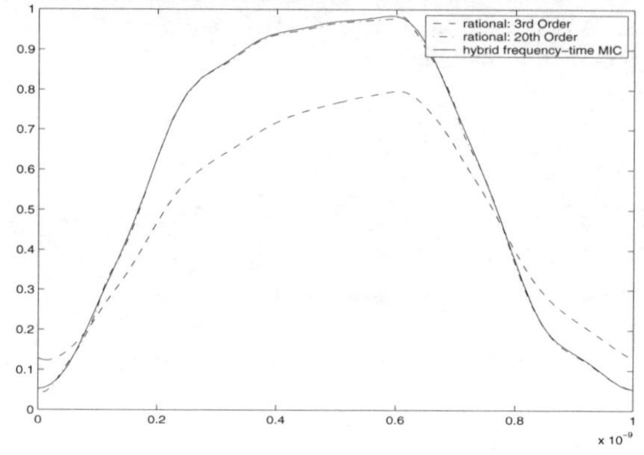

Figure 2: Comparison of GMRES convergence with and without preconditioner.

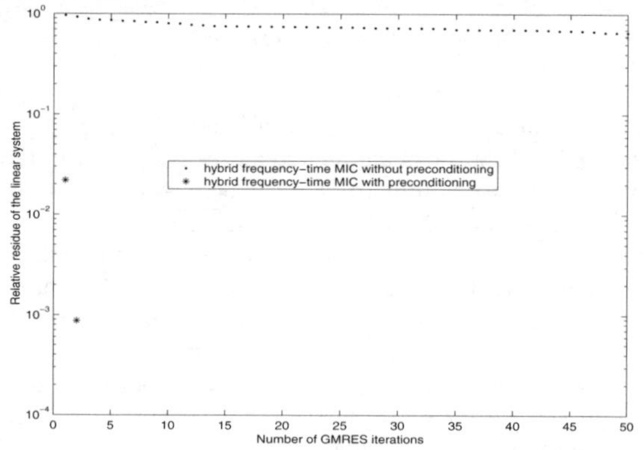

a third-order rational fitting model and shooting-Newton method to obtain the periodic steady state. The solution is very inaccurate compared to other results. Clearly, the third-order rational fitting model is inaccurate by itself. Second, we use a $20th$ order rational fitting model instead. The result is much more accurate than that of the first simulation. Last, we use the hybrid frequency-time MIC method with the third-order rational fitting model as the preconditioner. With no regard to the inaccuracy of the $3rd$-order model, the result of the new method is satisfactory. There is only a small discrepancy between the second and third simulation result, which may be due to the fact that the treatment of the S-parameters is done in fundamentally different ways and there is no guarantee that the two methods will agree in the frequency range outside of $[50MHz, 10GHz]$.

Although the accuracy of rational fitting model is satisfactory in this experiment, it is inefficient because of the high order used in doing rational fitting. The additional state-space equations added when the high-order rational fit model is used means that the circuit equations we need to solve are of much larger size. Indeed, the overall simulation time of the hybrid frequency-time domain method is 5 seconds while that of the 20th-order rational fitting method is 30 seconds. In Fig. 2 we plot the convergence history of the GMRES solver. We compare GMRES convergence with the $3rd$-order fitting model preconditioner and without the preconditioner. One can clearly see that GMRES convergence without the preconditioner is very slow while GMRES converges in a few itera-

Figure 3: Comparison of rational fitting and hybrid frequency-time in transmission line simulation.

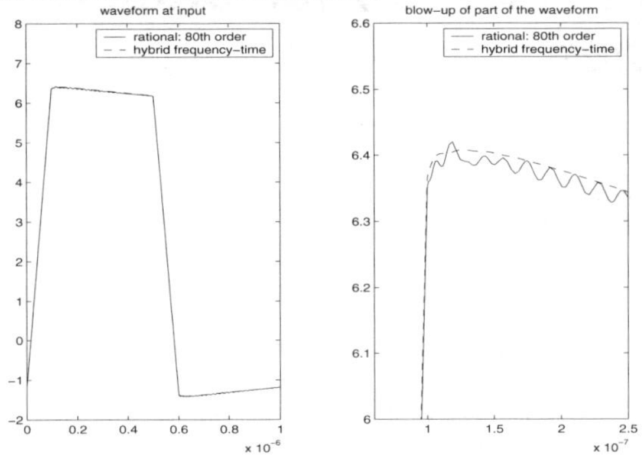

tions with the preconditioner. The convergence of our linear solver is satisfactory in this and other experiments.

Figure 4: IF output waveforms of a mixer circuit.

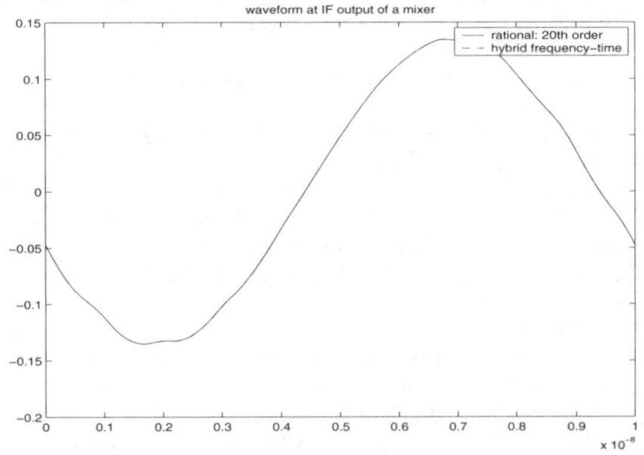

In the second example, we apply the new method to simulate a transmission line with loss and dispersion due to skin effect and finite conductor thickness. We show the waveform of both the rational fitting based simulation and the hybrid frequency-time simulation. In Fig. 3, one can clearly see from the expanded plot on the right that the rational fitting model exhibits high-order oscillation in the waveform, while the waveform of hybrid frequency-time method does not have this problem. The spurious oscillation of rational fitting models is often present in this and other experiments as well as those given in the literature. The hybrid frequency-time approach is promising in that it gives more accurate results.

In the next example, we calculate the periodic steady-state of a mixer circuit. This is a nonlinear circuit that contains 4 bipolar transistors and a SAW filter described by measured S-parameters. We again compare the high-order rational fitting simulation and the hybrid frequency-time simulations. From Fig. 4 we observe that the waveforms of the two simulations are basically on top of each other, verifying the correctness of the result.

Finally, we have performed simulations of a mixer designed with dozens of microstrip transmission lines. In our simulations, we observe that the hybrid frequency-time domain method, using low-order rational fitting model only as preconditioner, has better convergence properties than the pure rational fitting method, which is what we would expect. We verified the simulation result with that of a long transient simulation.

In summary, the new method is promising in terms of accuracy, convergence, and efficiency in all of our numerical experiments.

6. CONCLUSIONS

The hybrid frequency-time domain approach proposed in this paper for the treatment of distributed components, such as described by frequency-dependent tabulated S-parameters, is an effective technique for steady-state time-domain simulation with such components. Because of the spectral accuracy of the Chebyshev discretization in time domain, efficiency and accuracy is achieved in the transformation between frequency and time domains for device evaluation involving frequency-dependent S-parameters. A matrix-implicit Krylov-subspace method is naturally and conveniently used as the linear solver. Numerical experiments show that the new method gives more accurate results and has better convergence properties, compared to the pure rational-fitting lumped model approach. Efficiency of the low-order rational fitting model as a preconditioner for the hybrid frequency-time method is also shown. The method proposed in this paper can be conveniently extended to other time-domain RF analyses, such as quasi-periodic steady-state analysis[10] and small-signal or noise analysis[2].

7. REFERENCES

[1] Ken Kundert, "Introduction to RF simulation and its applications," *IEEE J. Sol. State Circuits, to appear September 1999*, September 1999.

[2] R. Telichevesky, J. White, and K. Kundert, "Efficient AC and noise analysis of two-tone RF circuits," in *Proc. 33rd Design Automation Conference*, June 1996.

[3] R. Telichevesky, J. White, and K. Kundert, "Efficient steady-state analysis based on matrix-free Krylov-subspace methods," in *Proc. 32rd Design Automation Conference*, June 1995.

[4] D. Long, R. Melville, K. Ashby, and B. Horton, "Full chip harmonic balance," in *Proc. Custom Integrated Circuits Conference*, May 1997.

[5] Carlos P. Coelho, Joel R. Phillips, and L. Miguel Silveira, "Robust rational function approximation algorithm for model generation," in *Proceedings of the 36th Design Automation Conference*, New Orleans, June 1999, pp. 207–212.

[6] K. Nabors, Tze-Ting Fang, Hung-Wen Chang, K. S. Kundert, and J. K. White, "Lumped interconnect models via gaussian quadrature," in 34th *ACM/IEEE Design Automation Conference*, 1997, pp. 40–45.

[7] J. E. Schutt-Aine and R. Mittra, "Scattering Parameter Transient Analysis of Transmissions Lines loaded with Nonlinear Terminations," *IEEE Transactions on Microwave Theory and Techniques*, vol. MTT-36, pp. 529–536, 1988.

[8] K. Engelborghs, T. Luzyanina, K. J. in 't Hout, , and D. Roose, "Collocation methods for the computation of periodic solutions of delay differential equations," *SIAM J. Sci. Comput.*, vol. 22, pp. 1593–1609, 2000.

[9] B. Yang and J. Phillips, "A multi-interval chebyshev collocation method for efficient high-accuracy rf circuit simulation," in *Proc. 37th Design Automation Conference*, June 2000, pp. 178–183.

[10] D. Feng, J. Phillips, K. Nabors, K. Kundert, and J. White, "Efficient computation of quasi-periodic circuit operating conditions using a mixed frequency/time approach," in *Proc. 36th Design Automation Conference*, June 1999.

[11] A. Dutt and V. Rokhlin, "Fast fourier transforms for nonequispaced data II," *Appl. Comput. Harmon. Anal.*, vol. 2, pp. 85–100, 1995.

A Time-domain RF Steady-State Method for Closely Spaced Tones

Jaijeet Roychowdhury
University of Minnesota
jr@ece.umn.edu

Abstract

Verifying circuits with two or more closely-spaced driving frequencies is important in RF and wireless communications, *e.g.*, in the design of down-conversion mixers. Existing steady-state calculation methods, like harmonic balance, rely on Fourier series expansions to find the difference-frequency components typically of interest. Time-domain methods are, however, better suited for circuits with strong nonlinearities such as switching. Towards this end, we present a purely time-domain method for direct computation of difference tones in closely-spaced multi-tone problems. Our approach is based on multiple artificial time scales for decoupling the tones driving the circuit. Our method relies on a novel multi-time reformulation that expresses circuit equations directly in terms of time-scales corresponding to difference tones. We apply the new technique to an RF-CMOS mixer to predict baseband bit-streams and down-conversion gain and distortion, in two orders of magnitude less CPU time than traditional time-stepping simulation.

Categories and Subject Descriptors

B.7.2 [**Integrated Circuits**]: Design Aids—*Analog Verification*

General Terms

Algorithms, Design, Verification

Keywords

MPDE, difference-frequency time scales, multi-time PDEs, harmonic balance, shooting, envelope, analog, mixed-signal, analog/RF simulation, RF switching mixers, artificial time scales, homotopy, continuation methods

1. INTRODUCTION

RF sections of wireless communication systems typically involve one or more stages of frequency conversion, with frequency tones that are either widely separated, or very closely spaced. While frequency up-conversion in transmitters involves widely separated frequencies, down-conversion circuits in receivers are typically driven by two or more very closely spaced frequencies. For example, a direct-conversion cellphone could feature tones at 1.8GHz, spaced only a few Mhz or 100s of kHz apart. It is the difference frequencies that are of primary interest in receiver design and verification, since they carry the information being received; hence design and verification tools must necessarily deal with the small frequencies together with the large ones.

As is well known, simulating circuits with disparate frequencies using time-stepping methods, such as those used for transient simulation in SPICE [5, 8] and similar tools, can be very inefficient, particularly when the time-varying steady-state solution is desired. Better methods exist for both widely-separated and closely-spaced driving tones. An important technique that is useful for both situations is harmonic balance (HB) (*e.g.*, [2–4]). HB expands all time-varying waveforms in the circuit in Fourier series components featuring the driving tones, their harmonics and mixes . Since sum and difference frequencies appear explicitly in the mix components, HB is able to accommodate both widely-separated and closely-spaced driving tones naturally. Unfortunately, Fourier series expansions are also the Achilles' heel of HB, for they are ill-suited to waveforms with sharp corners or peaks. Such waveforms arise in modern integrated-RF designs, such as switching RF circuits, desirable for their low power and noise characteristics. In these situations, *i.e.*, when strong nonlinearities are encountered, time-domain methods are usually preferred over frequency-domain ones like HB.

Purely time-domain methods for steady-state and RF calculations have, until recently, been limited to single-tone problems, with shooting and its variants commonly used (*e.g.*, [1, 6, 10]). Recently, artificial time scale approaches [2, 9] have addressed the multi-tone case, resulting in purely time-domain methods for quasi-periodic steady state and envelope simulation. So far, however, these techniques have concentrated only on the case of widely-separated driving tones.

In this work, we extend multi-time approaches to the case of closely-spaced driving tones. The key enabler towards this end is the use of *difference-frequency time scales* (or simply *difference time scales*). In contrast to the widely-used concept of difference frequencies, which is linked to sinusoidal basis functions and quasi-periodic waveforms, difference time scales capture the essential intuition of slow variations at difference frequencies without being limited to sinusoidal bases or quasi-periodicity. This makes it possible to obtain, for example, the shape of baseband or down-converted information signals (such as bit streams) directly in the time domain, without at any point involving frequency-domain representations or calculations.

We present applications of the new technique to balanced and unbalanced switching mixer circuits for direct-conversion receivers, similar to those reported recently in [7, 11]. We are able to obtain, directly, time-domain waveforms for sections of bit streams downconverted to baseband. Using pure-tone driving excitations, we are also able to obtain down-conversion gain and distortion figures. The new method is faster by about two orders of magnitude than periodic steady-state computation using single-time shooting.

Although we demonstrate the new technique on RF communication circuits in this paper, we note that the proposed method can be applied generally to other systems featuring closely-spaced tones, such as power conversion circuits and electro-optical communication systems.

In the remainder of the paper, we describe schemes for reformulating and solving the circuit equations in terms of difference-frequency time scales in Section 2. In Section 3, we describe the application of the new technique to a balanced switching mixer, and provide interpretations of the results obtained.

2. DIFFERENCE TIME SCALES

Multi-time circuit equations

We first recall the basic concepts of multi-time circuit formulations, following the approach of [9]. The equations of the circuit are represented as

$$\dot{q}(x(t)) + f(x(t)) + b(t) = 0, \quad (1)$$

where $q(\cdot)$ represents the linear and nonlinear charge/flux terms, $f(\cdot)$ the conductive terms, $b(t)$ the excitation, and $x(t)$ the unknowns (voltages and currents) in the circuit. When the excitation is multi-tone (whether closely or widely spaced), $b(t)$ can be expressed in terms of multi-variate functions involving several artificial time scales. For example, a two-tone signal may be expressed as

$$b(t) = \hat{b}(t,t), \quad (2)$$

where

$$\hat{b}(t_1,t_2) = \hat{b}(t_1+T_1, t_2+T_2), \quad (3)$$

is periodic in each of its two arguments, for some periods T_1 and T_2. The two tones of excitation are $f_1 = \frac{1}{T_1}$ and $f_2 = \frac{1}{T_2}$; they are widely separated if $f_1 \gg f_2$ (or vice-versa). If the tones are closely spaced, then typically, $f_1 \sim f_2 \gg |f_1 - f_2|$.

In [9], it has been shown that when the tones are widely separated, the multi-time representation (3) can be represented far more compactly, in terms of numerical samples, than the normal single-time form (2). This fact is applied towards solving the circuit equations (1), by assuming that the circuit unknowns $x(t)$ can also be expressed in a numerically compact multi-time form similar to (3). Finding this multi-time solution directly involves solution of the following *multi-time partial differential equation* (MPDE) corresponding to (1):

$$\frac{\partial q(\hat{x}(t_1,t_2))}{\partial t_1} + \frac{\partial q(\hat{x}(t_1,t_2))}{\partial t_2} + f(\hat{x}(t_1,t_2)) + \hat{b}(t_1,t_2) = 0. \quad (4)$$

In (4), $\hat{x}(t_1,t_2)$ is the multi-time representation of $x(t)$ that is sought. It has been shown that if $\hat{x}(t_1,t_2)$ satisfies (4), then $x(t) = \hat{x}(t,t)$ solves the original circuit equations (1). Regardless of whether the tones in the circuit are separated or not, this basic fact always holds.

When the driving tones are in fact widely separated, *e.g.*, $T_1 \ll T_2$, solving the MPDE (4) is preferred over solving (1) for two reasons. First, $\hat{b}$ and $\hat{x}$ can be represented much more efficiently than their single-time counterparts; and second, the t_1 and t_2 variations of the solution $\hat{x}(t_1,t_2)$ directly provide information of interest, namely the slowly-varying envelope, and the detailed shape of the fast variations at any point under the envelope.

Closely-spaced tones

When the tones are closely spaced, *i.e.*, $T_1 \sim T_2$, the MPDE (4) as it stands is still perfectly valid, however it is not immediately useful as in the widely separated case, even though $\hat{b}$ and $\hat{x}$ are still numerically far more compact than their single-time counterparts. The reason for this is that the frequencies of interest are typically *difference frequencies*, such as $f_1 - f_2$, $2f_1 - f_2$, *etc.*. The difference frequencies are much smaller than f_1 or f_2, and manifest themselves as baseband signals or as slow envelopes riding on much faster waveforms. The multi-time solution $\hat{x}$ does not directly provide the slow waveforms at these frequencies. As a concrete example, consider the ideal mixing operation

$$z(t) = x(t)y(t), \quad (5)$$

where $x(t) = \cos(2\pi f_1 t)$ and $y(t) = \cos(2\pi f_2 t)$, with $f_1 = 1\text{GHz}$ and $f_2 = f_1 - 10\text{kHz}$. In other words, the two tones f_1 and f_2 are very closely spaced, with the difference tone $f_1 - f_2 = 10\text{kHz}$ being much smaller. The mixing operation (5) generates this difference tone explicitly, since

$$z(t) = \frac{1}{2}\left[\cos(2\pi \times 10\text{kHz} \times t) + \cos(2\pi(f_1+f_2)t)\right], \quad (6)$$

and the high frequency component at $f_1 + f_2$ is usually eliminated by separate filtering. The multi-time representation of $z(t)$, given by

$$\hat{z}_1(t_1,t_2) = \cos(2\pi f_1 t_1)\cos(2\pi f_2 t_2), \quad (7)$$

does not directly provide this difference-frequency component.

For a multi-time method to be practically useful for the case of closely-spaced driving tones, it must be capable of directly solving for these slow waveforms in the time domain. We now outline how this can be achieved. The key insight is to realize that a given $\hat{b}$ satisfying (2) is not unique, *i.e.*, there are many functions $\hat{b}$, with *different* T_1 and T_2 values in (3), providing the same $b(t)$ by (2). We illustrate this with our ideal mixing example. For convenience, we first define

$$\hat{z}_s(t_{1s},t_{2s}) = \cos(2\pi t_{1s})\cos(2\pi t_{2s}) \quad (8)$$

to be a scaled multi-time representation of $z(t)$, periodic with period 1 in both its arguments. $\hat{z}_1$ in (7) can be expressed in terms of this as

$$\hat{z}_1(t_1,t_2) = \hat{z}_s(f_1 t_1, f_2 t_2). \quad (9)$$

Note that $\hat{z}_1$ above, with periods $T_1 = \frac{1}{f_1}$ and $T_2 = \frac{1}{f_2}$ very similar to each other, does not provide difference-frequency information directly; but as always required, it satisfies $z(t) = \hat{z}(t,t)$.

If we now define a difference frequency of interest to be

$$f_d = f_1 - f_2 = 10\text{kHz} \quad (10)$$

and the corresponding period $T_d = \frac{1}{f_d} = 0.1\text{ms}$, we can devise a new multi-time representation

$$\hat{z}_2(t_1,t_2) = \hat{z}_s(f_1 t_1, f_1 t_1 - f_d t_2). \quad (11)$$

Note that $\hat{z}_2$ is periodic with respect to t_1 and t_2, with periods T_1 and T_d, respectively, hence its changes along t_2 capture variations corresponding to difference-frequency time scales. Furthermore, it continues to satisfy the requirement $z(t) = \hat{z}_2(t,t)$, hence is as relevant to the underlying one-time problem as $\hat{z}_1(t_1,t_2)$. $\hat{z}_1$ and $\hat{z}_2$ are plotted in Figure 1 and Figure 2, respectively; note the explicit variation in the difference time scale, spanning 0.1ms, in the latter.

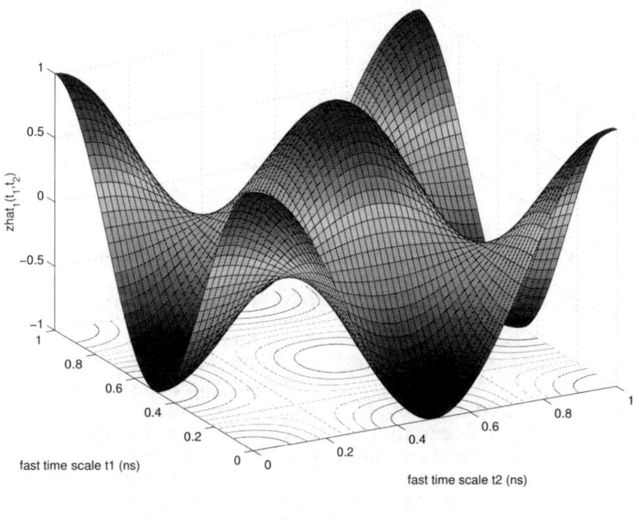

Figure 1: $\hat{z}_1(t_1,t_2)$

Note that the arguments $(f_1 t_1, f_1 t_1 - f_d t_2)$ in (11) can be interpreted geometrically as scaling and shearing of the original time scales (t_{1s}, t_{2s}) in (8). Furthermore, numerical compactness of representation is not affected by the shearing, as can be seen in Figure 2. Hence time-sheared multi-time representations for $\hat{b}$ and $\hat{x}$ can be applied to the MPDE (4), and a solution obtained directly in terms of the slow difference-frequency time scale of interest.

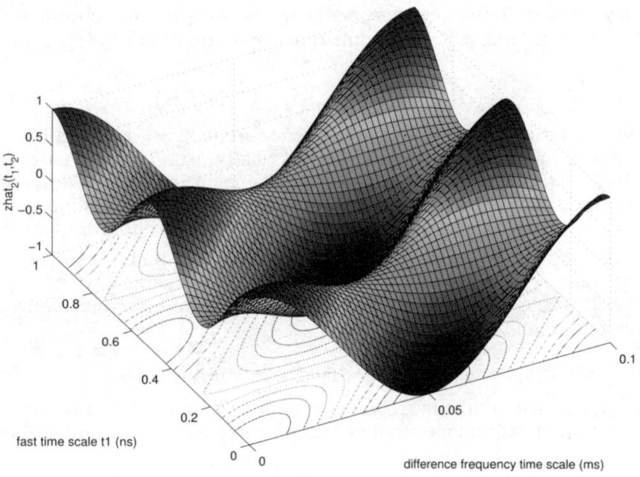

Figure 2: $\hat{z}_2(t_1,t_2)$

We also note that (10) is only one way of defining a difference-frequency time scale; depending on the application, other difference frequencies should be used. For example, in the balanced mixer circuit in Section 3, the local oscillator frequency $f_1 = 450$MHz is doubled internally within the circuit and then mixed with the information-carrying tone close to 900MHz. Hence the difference frequency of interest (at baseband) is

$$f_d = 2f_1 - f_2, \tag{12}$$

with a corresponding change to (11),

$$\hat{z}_2(t_1,t_2) = \hat{z}_s\left(f_1 t_1, 2f_1 t_1 - f_d t_2\right), \tag{13}$$

restoring the property $z(t) = \hat{z}_2(t,t)$.

An important point to note is that the signals need not be sinusoidal or near-sinusoidal. Indeed, high-frequency 'tones' involving modulation by bit streams can be used to drive a circuit. For example, the normalized-to-period-1 information-carrying tone driving the balanced mixer in Section 3 is given in multi-time form by

$$\hat{b}_i(t_{1s},t_{2s}) = 2\cos(2\pi \times 2t_{1s})\,\mathrm{pulse}\left(2(2t_{1s} - t_{2s})\right). \tag{14}$$

Once reformulated in terms of difference time scales, the MPDE (4) may be solved by any of the time-domain numerical methods in [9].

3. APPLICATION AND EXAMPLES

Balanced LO-doubling down-conversion mixer

A CMOS balanced down-conversion mixer, adapted from a circuit proposed in [11], was analyzed using the above ideas. An important feature of this circuit is that the lower pair of MOSFETs constitutes a frequency doubler, generating a current at twice the LO frequency. This current feeds the differential pair formed by the upper two MOSFETs, resulting in mixing and down-conversion. The supplied LO signal in this case is a sinusoid at 450MHz; the RF signal 'tone' is as for the previous circuit, a modulated bit stream close to 900MHz. For this circuit, the natural choice for the difference frequency is given by (12), again resulting in a baseband frequency of 15kHz. The time-shearing function is given by (13). The multi-time form of the differential output voltage between the drains of the upper MOSFETs is shown in Figure 3. Once again, the time-domain shape of the bit-stream is evident from the variation along the difference-frequency time scale. Figure 4 shows the envelope along the difference-frequency time scale, corresponding to the actual baseband voltage of the output.

The sharp waveforms created by the action of the frequency doubler in the circuit are clearly visible in Figure 5, which depicts the

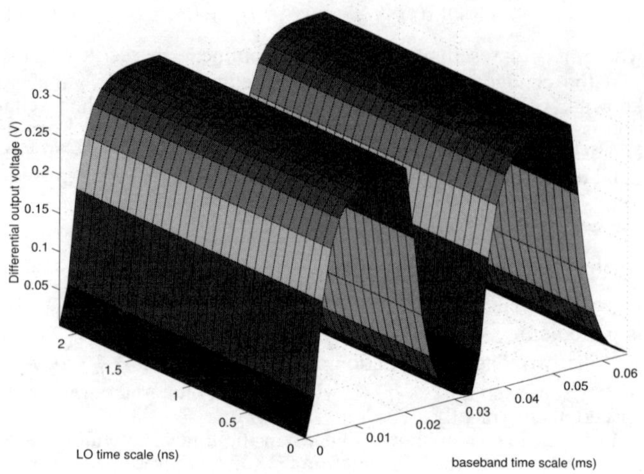

Figure 3: Balanced mixer: differential output voltage

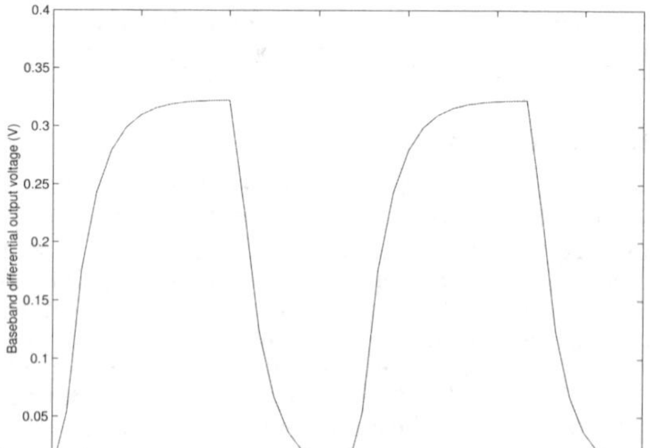

Figure 4: Balanced mixer: baseband differential output

multi-time voltage at the sources of the differential pair. The advantages of fully time-domain solution methods are best utilized in such situations. A small section of the actual voltage waveform, over a period of 5 LO cycles, is depicted in Figure 6.

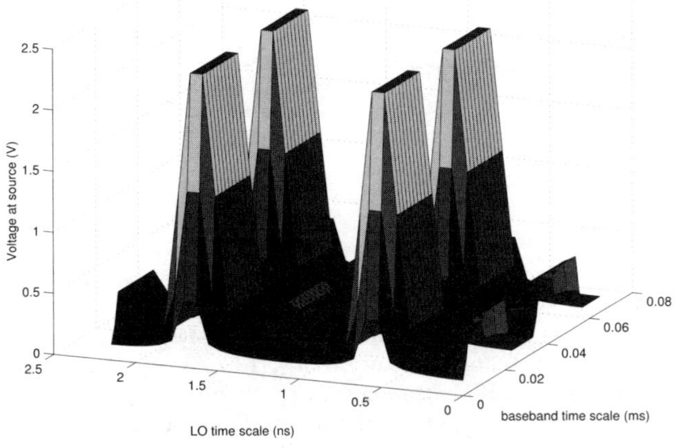

Figure 5: Balanced mixer: voltage at MOSFET sources

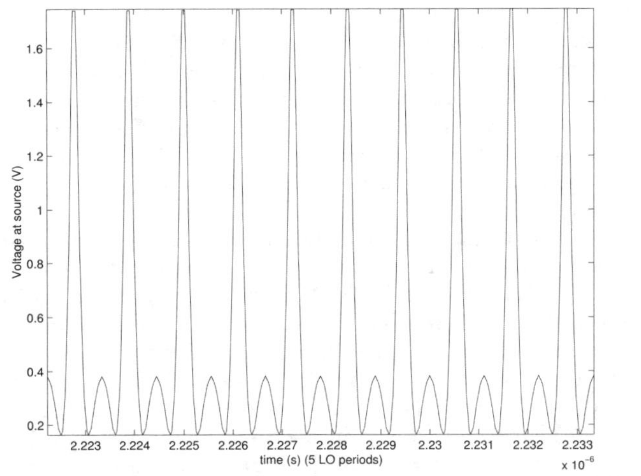

Figure 6: Balanced mixer: voltage at MOSFET sources

Computational speedup

The computational speedup of the new method stems from that relatively few grid points in the multi-time plane are sufficient to capture solution waveforms, compared to the number of time-points needed for the normal time representation. For the balanced mixer circuit above, we employed a 40×30 grid, resulting in 1200 grid points. Wall-clock time for the longest run (26 iterations) of Newton-Raphson on the balanced mixer circuit (given a good starting guess) was 1m3s[1]. In cases where Newton-Raphson did not converge, using continuation reliably obtained solutions in 10-20m.

The closest comparable traditional time-domain approach is shooting (or time-discretization) across one period of the difference frequency, but with time-steps small enough to capture the LO variations with sufficient accuracy, *i.e.*, 10 or more time-steps per LO period. This amounts to some 300000 or more time-steps over the difference-frequency period, resulting in an equation system more than $250\times$ larger. Using iterative linear solution methods and in the presence of similar nonlinear convergence behaviour, this results in a computational advantage of more than two orders of magnitude in favour of the new method described here. However, because it can be significantly more difficult to obtain nonlinear solution convergence for traditional shooting than for the multi-time approach in the presence of widely separated time scales, the real user-experienced speedup can be considerably larger.

We note that the speedup depends roughly linearly on extent of disparity between the LO time scale and the difference-frequency time scale. The break-even point (in terms of frequency separation) for computational speedup over single-time shooting is strongly dependent on implementation. We have noted that frequency disparities of 200 and above confer a speed advantage to multi-time methods.

Conclusion

Through the use of sheared, difference-frequency time scales, we have extended multi-time circuit solution formulations, previously useful only for widely-separated driving tones, to situations where driving signals are close in frequency. A purely time-domain method for solving the resulting equations has been presented and applied to compute the time-domain shapes of baseband bit-streams in down-conversion mixers. The new method is well-suited for estimating effects such as ISI and ACI in communication symbol streams. When baseband variations are much slower than those of the driving signals, it is also more than two orders of magnitude faster than comparable single-time methods.

4. REFERENCES

[1] T.J. Aprille and T.N. Trick. Steady-state analysis of nonlinear circuits with periodic inputs. *Proc. IEEE*, 60(1):108–114, January 1972.

[2] H.G. Brachtendorf, G. Welsch, R. Laur, and A. Bunse-Gerstner. Numerical steady state analysis of electronic circuits driven by multi-tone signals. *Electrical Engineering (Springer-Verlag)*, 79:103–112, 1996.

[3] R.J. Gilmore and M.B. Steer. Nonlinear circuit analysis using the method of harmonic balance – a review of the art. Part I. Introductory concepts. *Int. J. on Microwave and Millimeter Wave CAE*, 1(1), 1991.

[4] K.S. Kundert, J.K. White, and A. Sangiovanni-Vincentelli. *Steady-state methods for simulating analog and microwave circuits*. Kluwer Academic Publishers, 1990.

[5] L.W. Nagel. *SPICE2: a computer program to simulate semiconductor circuits*. PhD thesis, EECS Dept., Univ. Calif. Berkeley, Elec. Res. Lab., 1975. Memorandum no. ERL-M520.

[6] J. Parkhurst and L. Ogborn. Determining the steady-state output of nonlinear oscillatory circuits using multiple shooting. *IEEE Trans. CAD*, 14(7):882–889, 1995.

[7] J. Pihl, K.T. Christensen, and E. Braun. Direct Downconversion with Switching CMOS Mixer. In *Proc. IEEE ISCAS*, pages I–117–I–120, 2001.

[8] Thomas L. Quarles. *SPICE 3C.1 User's Guide*. University of California, Berkeley, EECS Industrial Liaison Program, University of California, Berkeley California, 94720, April 1989.

[9] J. Roychowdhury. Analysing circuits with widely-separated time scales using numerical PDE methods. *IEEE Trans. Ckts. Syst. – I: Fund. Th. Appl.*, May 2001.

[10] R. Telichevesky, K. Kundert, and J. White. Efficient steady-state analysis based on matrix-free krylov subspace methods. In *Proc. IEEE DAC*, pages 480–484, 1995.

[11] Z. Zhang, Z. Chen, and J. Lau. A 900MHz CMOS Balanced Harmonic Mixer for Direct Conversion Receivers. In *Proc. IEEE Radio and Wireless Conference (RAWCON)*, pages 219–222, 2000.

[1] All times are on a single-CPU, 1.4GHz AMD Athlon system with 512MB of 133MHz DDR memory, running Linux 2.4.15.

An Algorithm for Frequency-Domain Noise Analysis in Nonlinear Systems

Giorgio Casinovi
School of Electrical and Computer Engineering
Georgia Institute of Technology
Atlanta, GA 30332-0250
Giorgio.Casinovi@ece.gatech.edu

ABSTRACT

Many mixed-technology and mixed-signal designs require the assessment of the system's performance in the presence of signals that are best characterized as having continuous frequency spectra: for instance, the simulation of a circuit's behavior in the presence of electrical noise. This paper describes an algorithm for frequency-domain simulation of nonlinear systems capable of handling signals with continuous spectra. The algorithm is based on an orthogonal series expansion of the signals in the frequency domain. Thus it is, in a sense, the dual approach to frequency-domain simulation with respect to harmonic balance, which relies on a time-domain series expansion of the signals. Signal spectra are obtained from the solution of a system of nonlinear algebraic equations whose dimension increases with the desired spectral accuracy. Numerical results obtained on an optical amplifier are presented.

Categories and Subject Descriptors

J.6 [**Computer-Aided Engineering**]: Computer-Aided Design; I.6.8 [**Simulation and Modeling**]: Types of Simulation—*continuous*

General Terms

Algorithms, Verification

Keywords

Circuit simulation, frequency-domain analysis

1. INTRODUCTION

The steady increase in the number of systems based on mixed-technology designs, combined with the continuing trend in the decrease of their physical dimensions, is bringing renewed attention to algorithms and tools capable of assessing circuit or system performance in the presence of noise.

Most methods developed so far to perform noise analysis in nonlinear systems assume that the perturbation created by the noise is sufficiently small for the system to respond in a linear way to noise interference [1, 2, 3, 4]. Hence they cannot be used if the changes in the system response caused by the interfering signal are not small enough to justify the linearization of the system equations.

The only existing frequency-domain simulation methods that can fully handle nonlinearities rely on the principle of harmonic balance [5], and a number of noise-analysis algorithms based on this approach have been published [6, 7]. In harmonic balance, the solution to the system equations is expressed as a periodic or quasi-periodic Fourier series in the time domain. The drawback intrinsic in this approach is that, by construction, it can only handle signals with discrete frequency spectra. There are many situations, however, in which it is necessary to simulate the behavior of a system in the presence of signals that do not satisfy this requirement.

This paper presents an algorithm suitable for the simulation of nonlinear systems in the frequency domain when the signals involved have continuous spectra. The algorithm is based on a Fourier series expansion of the signals in the frequency domain. Thus it can be considered, in a sense, the dual approach to frequency-domain simulation with respect to harmonic balance. The mathematical formulation of the algorithm is given in Section 2, while Section 3 contains an analysis of the error introduced in the signal spectra computed by the algorithm. Finally, results obtained from numerical simulations are presented in Section 4.

2. MATHEMATICAL FORMULATION

Consider a nonlinear system (e.g. an integrated circuit) that is described by the following system of equations:

$$\frac{d}{dt}\mathbf{f}[\mathbf{x}(t), t] + \mathbf{g}[\mathbf{x}(t), t] = \mathbf{u}(t). \tag{1}$$

In addition to electrical circuits, this equation describes many other types of dynamical systems, with the variables taking on various physical meanings.

In the frequency domain, this equation corresponds to the following equivalent equation [5]:

$$j\omega\mathbf{F}[\mathbf{X}(\omega)] + \mathbf{G}[\mathbf{X}(\omega)] = \mathbf{U}(\omega), \tag{2}$$

where $\mathbf{X}(\omega)$ and $\mathbf{U}(\omega)$ represent the Fourier transforms of $\mathbf{x}(t)$ and $\mathbf{u}(t)$. The traditional approaches to frequency-domain analysis, which are based on harmonic balance, perform sinusoidal expansions of the signals in the time domain. This type of expansion, however, is applicable only to signals that have discrete spectra. This section describes an algorithm developed specifically for frequency-domain simulation of nonlinear systems when the signals have continuous spectra. It is based on a sinusoidal expansion of the signals *in the frequency domain.*

It will be assumed that both the input $\mathbf{u}(t)$ and the solution $\mathbf{x}(t)$ have a frequency spectrum that has a finite upper bound ω_M, as shown in Fig. 1. In practice, this condition can always be satisfied by choosing ω_M sufficiently large. Moreover, it will also be assumed that the spectra of $\mathbf{f}[\mathbf{x}(t),t]$ and $\mathbf{g}[\mathbf{x}(t),t]$ satisfy the same condition.

In many instances, $\mathbf{u}(t)$ and $\mathbf{x}(t)$ contain a DC component, which creates a spike[1] in their spectra at $\omega = 0$. The presence of spikes makes it problematic from a numerical standpoint to expand the spectrum in a sinusoidal series. For this reason, it is more convenient to rewrite (1) so that the DC component is separated from the signals. For this purpose, let $\mathbf{u}_0, \mathbf{x}_0, \mathbf{f}_0, \mathbf{g}_0$ be the DC components of $\mathbf{u}(t), \mathbf{x}(t), \mathbf{f}[\mathbf{x}(t),t]$ and $\mathbf{g}[\mathbf{x}(t),t]$, respectively, and define the following quantities:

$$\Delta\mathbf{x}(t) = \mathbf{x}(t) - \mathbf{x}_0, \quad \Delta\mathbf{f}[\Delta\mathbf{x}(t),t] = \mathbf{f}[\mathbf{x}_0 + \Delta\mathbf{x}(t),t] - \mathbf{f}_0,$$
$$\Delta\mathbf{u}(t) = \mathbf{u}(t) - \mathbf{u}_0, \quad \Delta\mathbf{g}[\Delta\mathbf{x}(t),t] = \mathbf{g}[\mathbf{x}_0 + \Delta\mathbf{x}(t),t] - \mathbf{g}_0.$$

Since $\mathbf{u}_0 = \mathbf{U}(0)$ and $\mathbf{g}_0 = \mathbf{G}[\mathbf{X}(0)]$, setting $\omega = 0$ in (2) shows that $\mathbf{g}_0 = \mathbf{u}_0$. Therefore the following equation is equivalent to (1):

$$\frac{d}{dt}\Delta\mathbf{f}[\Delta\mathbf{x}(t),t] + \Delta\mathbf{g}[\Delta\mathbf{x}(t),t] = \Delta\mathbf{u}(t). \tag{3}$$

Note that this is *not* a linearization of (1) around the DC component of the signals: it is easy to verify that $\mathbf{x}(t) = \mathbf{x}_0 + \Delta\mathbf{x}(t)$ solves (1) exactly, if $\Delta\mathbf{x}(t)$ is the solution of (3).

By assumption, the spectra of all the terms in (3) are contained in the interval $[-\omega_M,\ \omega_M]$. Therefore they can be expanded in an orthogonal series over that interval. For example:

$$\Delta\mathbf{U}(\omega) = \sum_{k=-\infty}^{\infty} \mathbf{U}_{-k} e^{-jk\pi\omega/\omega_M} r(\omega/\omega_M), \tag{4}$$

where the function $r(x)$ is defined as:

$$r(x) = \begin{cases} 1, & |x| \le 1 \\ 0, & |x| > 1. \end{cases}$$

Taking the inverse Fourier transform of (4) yields the following series expansion for $\Delta\mathbf{u}(t)$:

$$\Delta\mathbf{u}(t) = \frac{\omega_M}{\pi} \sum_{k=-\infty}^{\infty} \mathbf{U}_{-k}\, \mathrm{sync}[\omega_M(t - t_k)], \tag{5}$$

where: $\mathrm{sync}\, x = \sin x / x$, and $t_k = k\pi/\omega_M$. Evaluating this expression at $t = t_n$ shows that coefficients $\mathbf{U}_k$ in (4) are related to the values of $\Delta\mathbf{u}(t)$ by the following relationship:

$$\Delta\mathbf{u}(t_n) = \frac{\omega_M}{\pi}\mathbf{U}_{-n}. \tag{6}$$

Equations (5) and (6), taken together, express the well-known sampling theorem [8].

[1] Theoretically, a Dirac impulse.

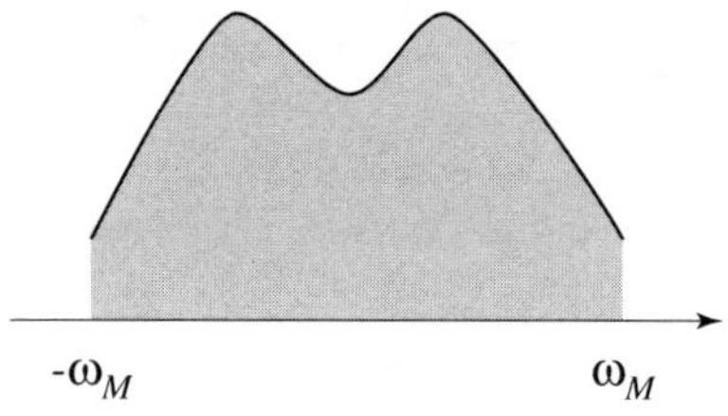

Figure 1: Continuous frequency spectrum

A similar series expansion can be obtained for the left-hand side of (3):

$$\frac{d}{dt}\Delta\mathbf{f}[\Delta\mathbf{x}(t),t] + \Delta\mathbf{g}[\Delta\mathbf{x}(t),t]$$
$$= \sum_{k=-\infty}^{\infty} (\Delta\dot{\mathbf{f}}[\Delta\mathbf{x}(t_k),t_k] + \Delta\mathbf{g}[\Delta\mathbf{x}(t_k),t_k])\, \mathrm{sync}[\omega_M(t - t_k)],$$

where: $\Delta\dot{\mathbf{f}}[\Delta\mathbf{x}(t_k),t_k] = \left.\frac{d}{dt}\Delta\mathbf{f}[\Delta\mathbf{x}(t),t]\right|_{t=t_k}$. Consequently, in order for (3) to be satisfied the following equality must be satisfied at all points t_k:

$$\Delta\dot{\mathbf{f}}[\Delta\mathbf{x}(t_k),t_k] + \Delta\mathbf{g}[\Delta\mathbf{x}(t_k),t_k] = \Delta\mathbf{u}(t_k). \tag{7}$$

An expression for $\Delta\dot{\mathbf{f}}[\mathbf{x}(t_k),t_k]$ can be obtained starting from the following series expansion:

$$\Delta\mathbf{f}[\Delta\mathbf{x}(t),t] = \sum_{i=-\infty}^{\infty} \Delta\mathbf{f}[\Delta\mathbf{x}(t_i),t_i]\, \mathrm{sync}[\omega_M(t - t_i)].$$

Differentiating both sides of this equation yields:

$$\Delta\dot{\mathbf{f}}[\Delta\mathbf{x}(t_k),t_k] = \omega_M \sum_{i=-\infty}^{\infty} \Delta\mathbf{f}[\Delta\mathbf{x}(t_i),t_i]\, \mathrm{sync}'[\omega_M(t_k - t_i)].$$

But:

$$\begin{aligned} \mathrm{sync}'[\omega_M(t_k - t_i)] &= \mathrm{sync}'[\pi(k-i)] \\ &= \begin{cases} 0, & i = k \\ \dfrac{(-1)^{k-i}}{\pi(k-i)}, & i \ne k. \end{cases} \end{aligned}$$

Therefore:

$$\Delta\dot{\mathbf{f}}[\Delta\mathbf{x}(t_k),t_k] = \frac{\omega_M}{\pi} \sum_{\substack{i=-\infty \\ i\ne k}}^{\infty} \Delta\mathbf{f}[\Delta\mathbf{x}(t_i),t_i]\frac{(-1)^{k-i}}{k-i}, \tag{8}$$

where the value $i = k$ is excluded from the summation. Substituting this expression for $\Delta\dot{\mathbf{f}}[\Delta\mathbf{x}(t_k),t_k]$ in (7), and truncating the infinite series to a finite number of terms (which is dependent upon the accuracy that one wishes to achieve), yields the following equation:

$$\frac{\omega_M}{\pi} \sum_{\substack{i=-N \\ i\ne k}}^{N} \Delta\mathbf{f}[\Delta\mathbf{x}(t_i),t_i]\frac{(-1)^{k-i}}{k-i} + \Delta\mathbf{g}[\Delta\mathbf{x}(t_k),t_k] = \Delta\mathbf{u}(t_k). \tag{9}$$

This is a set of $2N+1$ algebraic equations in the $2N+1$ unknowns $\Delta\mathbf{x}(t_{-N}), \Delta\mathbf{x}(t_{-N+1}), \ldots, \Delta\mathbf{x}(t_N)$, which can be solved using Newton's method. The spectrum of $\Delta\mathbf{x}(t)$ can then be computed from relationships analogous to those in (4) and (6):

$$\Delta \mathbf{X}_N(\omega) = \frac{\pi}{\omega_M} \sum_{k=-N}^{N} \Delta \mathbf{x}(t_k) e^{-jk\pi\omega/\omega_M} r(\omega/\omega_M). \quad (10)$$

3. ERROR ANALYSIS

The algorithm described in the previous section replaces $\Delta\dot{\mathbf{f}}[\Delta\mathbf{x}(t_k), t_k]$ with a linear combination of the values of $\Delta\mathbf{f}[\Delta\mathbf{x}(t), t]$ at other timepoints, as shown in (8). This is the same type of approximation that is made in linear multistep numerical integration methods. This suggests that an in-depth analysis of the error introduced by the algorithm when the system is linear can already provide useful insight into the algorithm's numerical performance.

If the system is linear, equation (1) becomes:

$$\mathbf{F}\frac{d\mathbf{x}}{dt} + \mathbf{G}\mathbf{x}(t) = \mathbf{u}(t), \quad (11)$$

or, equivalently, in the frequency domain:

$$(j\omega\mathbf{F} + \mathbf{G})\mathbf{X}(\omega) = \mathbf{U}(\omega). \quad (12)$$

The exact solution of this equation is:

$$\mathbf{X}(\omega) = (j\omega\mathbf{F} + \mathbf{G})^{-1}\mathbf{U}(\omega). \quad (13)$$

Let $F(\omega)$ be a generic function vanishing outside the interval $[-\omega_M, \omega_M]$ of Fig. 1. Let $P_N F$ denote the function obtained by truncating the Fourier expansion of F to the N-th term:

$$P_N F(\omega) = \sum_{k=-N}^{N} F_k e^{jk\pi\omega/\omega_M} r(\omega/\omega_M),$$

and let: $R_N F(\omega) = F(\omega) - P_N F(\omega)$. The following result can be established.

THEOREM 1. *The spectrum $\mathbf{X}_N(\omega)$ obtained by applying (9) and (10) to (11) satisfies the following equation:*

$$P_N[(j\omega\mathbf{F} + \mathbf{G})\mathbf{X}_N(\omega)] = P_N\mathbf{U}(\omega). \quad (14)$$

Proof: Because of space limitations, the proof of this theorem is omitted. ■

An equation for the error: $\mathbf{E}_N(\omega) = \mathbf{X}(\omega) - \mathbf{X}_N(\omega)$ can now be obtained by subtracting (14) from (12):

$$\begin{aligned} R_N\mathbf{U} &= \mathbf{U} - P_N\mathbf{U} \\ &= (j\omega\mathbf{F} + \mathbf{G})\mathbf{X} - P_N[(j\omega\mathbf{F} + \mathbf{G})\mathbf{X}_N] \\ &= (j\omega\mathbf{F} + \mathbf{G})(\mathbf{X} - \mathbf{X}_N) + (j\omega\mathbf{F} + \mathbf{G})\mathbf{X}_N \\ &\quad - P_N[(j\omega\mathbf{F} + \mathbf{G})\mathbf{X}_N] \\ &= (j\omega\mathbf{F} + \mathbf{G})\mathbf{E}_N + R_N[(j\omega\mathbf{F} + \mathbf{G})\mathbf{X}_N]. \end{aligned}$$

Consequently:

$$\begin{aligned} &(j\omega\mathbf{F} + \mathbf{G})\mathbf{E}_N(\omega) \\ &\quad = R_N\mathbf{U}(\omega) - R_N[(j\omega\mathbf{F} + \mathbf{G})\mathbf{X}_N(\omega)]. \quad (15) \end{aligned}$$

Equation (15) is similar to (12), but its right-hand side consists of the sum of two terms, both of which represent errors introduced by truncating an infinite Fourier series to a finite number of terms. The first term derives from the truncation of the Fourier series expansion of $\mathbf{U}(\omega)$, and the second from that of $(j\omega\mathbf{F} + \mathbf{G})\mathbf{X}_N(\omega)$. In particular, $\mathbf{E}_N$ may be non-zero even if $R_N\mathbf{U} = 0$, i.e. even if the right-hand side of (12) is expressed exactly by a finite Fourier sum.

Further insight into the error introduced by truncating an infinite Fourier series can be obtained by deriving an explicit expression for $P_N F(\omega)$ in the following way. From the well-known formula for the Fourier coefficients:

$$F_k = \frac{1}{2\omega_M}\int_{-\omega_M}^{\omega_M} F(w) e^{-jk\pi w/\omega_M} dw$$

it follows that:

$$P_N F(\omega) = \sum_{k=-N}^{N} F_k e^{jk\pi\omega/\omega_M} =$$

$$\frac{1}{2\omega_M}\int_{-\omega_M}^{\omega_M} F(w)\left(\sum_{k=-N}^{N} e^{jk\pi(\omega - w)/\omega_M}\right) dw.$$

But:

$$\sum_{k=-N}^{N} e^{jkx} = \frac{\sin\left(N + \frac{1}{2}\right)x}{\sin(x/2)}.$$

Therefore:

$$P_N F(\omega) =$$

$$\frac{1}{2\omega_M}\int_{-\omega_M}^{\omega_M} F(w)\frac{\sin\left(N + \frac{1}{2}\right)\pi(\omega - w)/\omega_M}{\sin[\pi(\omega - w)/2\omega_M]} dw.$$

This expression for $P_N F(\omega)$ shows that truncating an infinite Fourier series to a finite number of terms is equivalent to performing a convolution of the original function $F(\omega)$ with the following function:

$$W_N(\omega) = \frac{1}{2\omega_M}\frac{\sin\left(N + \frac{1}{2}\right)\pi\omega/\omega_M}{\sin(\pi\omega/2\omega_M)}.$$

The oscillatory nature of $W_N(\omega)$ can potentially create spurious oscillations in the computed spectrum (Gibbs phenomenon). This effect, known as *windowing* in the spectral analysis of signals, occurs whenever the Fourier transform of a signal that is theoretically infinitely long is computed numerically using a finite length FFT. In particular, the function $W_N(\omega)$ obtained above corresponds to the application of a rectangular window to the time-domain signal. This observation suggests that the numerical performance of the algorithm described in Section 2 can be improved using windowing techniques developed for the spectral analysis of signals [9].

4. NUMERICAL RESULTS

The numerical performance of the algorithm described in the previous sections was verified by performing simulations of a number of test circuits. The results obtained on a CMOS optical amplifier, shown in Fig. 2, are presented next.

This circuit amplifies the output of a photodetector, modeled by a current source applied to its input. The amplifier load is modeled by a 30 kΩ resistor at the amplifier's output. A signal having a uniform power spectral density within a 1 GHz bandwidth (white noise) was applied to the amplifier's input. The spectrum of the resulting output signal, computed by the algorithm with $N = 128$, is shown in Fig. 3. The computation required approximately 32 minutes of CPU time on a Sun Ultra 60 workstation running Solaris 8. The ripple that is clearly visible in the graph is a manifestation of the Gibbs phenomenon mentioned earlier,

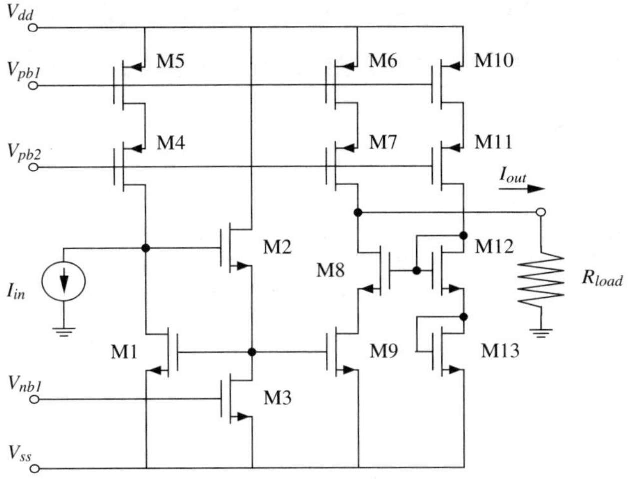

Figure 2: A CMOS optical amplifier

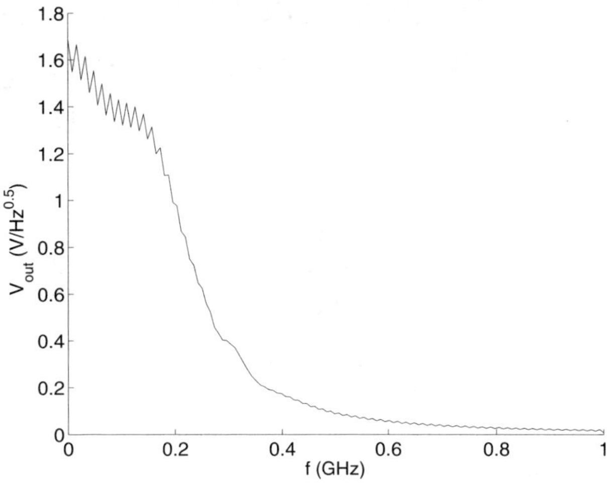

Figure 3: Output spectrum of optical amplifier

and it is caused by the abrupt truncation of an infinitely long signal to one of finite length. It can be reduced or eliminated by using appropriate windowing or curve fitting techniques.

5. CONCLUSION

The approach to frequency-domain simulation described in this paper is, in a way, the dual of the well-known analysis method based on harmonic balance, in that the unknown signals are expanded in a sinusoidal series in the frequency domain, instead of the time domain. Taking the inverse Fourier transform of the series expansion transforms the differential equation describing the system into a finite set of algebraic equations in the time domain (instead of the frequency domain, as with harmonic balance). The number of equations is chosen based on the desired accuracy, just as in harmonic balance the number of harmonics to be solved for is determined by the desired accuracy. Thus the two methods are one the mirror image of the other, but, unlike harmonic balance, the algorithm described in this paper is suitable for handling signals with continuous spectra. As with harmonic balance, the computational effort required for the analysis increases with the desired accuracy: in this case, the resolution in the computed spectrum. Specifically, the size of the system generated by (9) is $(2N+1)D$, where D is the dimension of $\mathbf{x}(t)$. Thus the size of this system increases linearly with the number of terms in (10), and can become very large even for circuits of relatively modest size. This problem can be alleviated by the use of iterative solution methods that do not require matrix inversion [10].

6. ACKNOWLEDGEMENTS

Partial support for this work by DARPA under contract F33615-01-C-1971, and by the Yamacraw Initiative of the State of Georgia is gratefully acknowledged.

7. REFERENCES

[1] A. Demir, E. W. Y. Liu, and A. L. Sangiovanni-Vincentelli, "Time-domain non-Monte Carlo noise simulation for nonlinear dynamic circuits with arbitrary excitations," *IEEE Trans. Computer-Aided Design*, vol. 15, pp. 493–505, May 1996.

[2] M. Okumura, H. Tanimoto, T. Itakura, and T. Sugawara, "Numerical noise analysis for nonlinear circuits with a periodic large signal excitation including cyclostationary noise sources," *IEEE Trans. Circuits Syst. I*, vol. 40, pp. 581–590, Sept. 1993.

[3] J. Roychowdhury, D. Long, and P. Feldmann, "Cyclostationary noise analysis of large RF circuits with multitone excitations," *IEEE J. Solid-State Circuits*, vol. 33, pp. 324–336, Mar. 1998.

[4] R. Telichevesky, K. S. Kundert, and J. White, "Efficient AC and noise analysis of two-tone RF circuits," in *Proc. 33rd Design Automation Conf.*, pp. 292–297, June 1996.

[5] K. S. Kundert and A. Sangiovanni-Vincentelli, "Simulation of nonlinear circuits in the frequency domain," *IEEE Trans. Computer-Aided Design*, vol. CAD-5, pp. 521–535, Oct. 1986.

[6] K. S. Kundert, J. White, and A. Sangiovanni-Vincentelli, *Steady-State Methods for Simulating Analog and Microwave Circuits*. Norwell, MA: Kluwer Academic Publishers, 1990.

[7] V. Rizzoli and A. Neri, "State of the art and present trends in nonlinear microwave CAD techniques," *IEEE Trans. Microwave Theory Tech.*, vol. 36, pp. 343–365, Feb. 1988.

[8] A. V. Oppenheim and R. W. Schafer, *Discrete-Time Signal Processing*. Englewood Cliffs, NJ: Prentice Hall, 1989.

[9] F. J. Harris, "On the use of windows for harmonic analysis with discrete Fourier transform," *Proc. IEEE*, vol. 66, pp. 51–83, Jan. 1978.

[10] Y. Saad and M. H. Schultz, "GMRES: A generalized minimal residual method for solving nonsymmetric linear systems," *SIAM J. Sci. Stat. Comput.*, vol. 7, pp. 856–869, July 1986.

System-level performance optimization of the data queueing memory management in high-speed network processors

Ch. Ykman-Couvreur
J. Lambrecht, D. Verkest
F. Catthoor*
IMEC, Kapeldreef 75
B3001 Leuven, Belgium
ykman@imec.be

A. Nikologiannis
ELLEMEDIA technologies
Syggrou Avenue 223
EL17121 Athens, Greece
anikol@ellemedia.com

G. Konstantoulakis
Inaccess Networks SA
Sygrou Avenue 230
EL17672 Athens, Greece
gkonst@inaccessnetworks.com

ABSTRACT

In high-speed network processors, data queueing has to allow real-time memory (de)allocation, buffering, retrieving, and forwarding of incoming data packets. Its implementation must be highly optimized to combine high speed, low power, large data storage, and high memory bandwidth. In this paper, such data queueing is used as case study to demonstrate the effectiveness of a new system-level exploration method for optimizing the memory performance in dynamic memory management. Assuming that a multi-bank memory architecture is used for data storage, the method trades off bank conflicts against memory accesses during real-time memory (de)allocation. It has been applied to the data queueing module of the PRO^3 system [8]. Compared with the conventional memory management technique for embedded systems, our exploration method can save up to 90% of the bank conflicts, which allows to improve worst-case memory performance of data queueing operations by 50% too.

Categories and Subject Descriptors

B.3.3 [**Memory Structures**]: Performance Analysis and Design Aids

General Terms

Memory performance

Keywords

System-level exploration, memory management

1. INTRODUCTION

The rapid growth in the dimension and diversity of networks and telecom systems, along with the ever increasing user demand in networking services, has imposed the need for efficient protocol processing and the development of high capacity systems running OC3 (155 Mbps), OC12 (622 Mbps), OC48 (2.488 Gbps) and sometimes OC192 (10 Gbps) speeds. One of the fundamental functionalities, and also one of the most difficult parts in high-speed network processors offering quality of service, is *data queueing*. This has to allow real-time memory (de)allocation, buffering, retrieving, and forwarding of incoming data packets. Its implementation must be highly optimized to combine high speed, low power, large data storage, and high memory bandwidth. It must also be optimized to support worst-case situations as efficiently as possible. First, to minimize large data copying, pointers to these data are better passed as arguments of the data queueing operations. Secondly, to increase the memory bandwidth, a combination of SRAMs (for pointers, flags, ..., which are small and frequently accessed) and DRAMs (for data packets, which are large and not frequently accessed) in the memory architecture becomes a must. Hence the memory performance is affected by two factors: the huge amount of memory accesses and the effectiveness of access to memory. E.g. accesses to non-busy banks in a multi-bank memory architecture can be performed much faster than successive accesses to different pages insisde the same bank.

In this paper, such data queueing is used as case study to demonstrate the effectiveness a new system-level exploration method for optimizing the memory performance in dynamic memory management. Assuming that a multi-bank memory architecture is used for data storage, this method trades off bank conflicts (due to successive accesses to the same bank) against memory accesses during real-time memory (de)allocation. It has been applied to the data queueing module of the PRO^3 system [8]. Compared with the conventional memory management technique for embedded systems, our exploration method can save up to 90% of the bank conflicts, which improves worst-case memory performance of data queueing operations by 50% too. It also fits in Matisse [19], a system-level synthesis approach supporting the memory management of dynamic (de)allocation of data structures in embedded systems.

The remainder of the paper is organized as follows. Section 2 summarizes the design characteristics of existing data queueing implementations. Section 3 presents the abstract

*Also prof. at Katholieke Univ. Leuven

specification of the data queueing module in the PRO^3 system. This module is used to illustrate our method. Section 4 analyzes the system-level parameters that affect the memory performance the data queueing module. These parameters are taken into account in Section 5, that describes our exploration method. Experimental results for the considered data queueing module are discussed in Section 6.

2. RELATED WORK

Data queueing from the *algorithmic* point of view is already well established. To guarantee quality of service, important policies use per-flow queuing, where data packets are organized into queues, one queue per active connection/flow. These policies are described in [6, 12, 18]. In addition to large data storage, to implement the queues, these policies are characterized by real-time (de)allocation and frequent updates of large data structures.

Hence the implementation of the data queueing module (DQM) must also be combined with an appropriate *memory management technique to organize the virtual memory, to allow real-time memory (de)allocation, and to control memory overflow.* Conventionally, as in memory management of embedded systems, the virtual memory is partitioned into blocks to store the data packets and all dynamic data structures implementing the queues [4, 5, 15, 16, 17]. All these blocks that are free in the virtual memory are maintained in a single linked list, called *free list* [14]. An alternative approach is to simulate the DQM with different memory management techniques and then to choose the most convenient one, relatively to power, area, and performance. Another more promising approach is to develop a *cost-driven exploration method* that takes into account the network characteristics and its real-time requirements, and that systematically provides the best memory management technique for that DQM. This approach is followed in [19], where the complete design space is characterized by identifying all its orthogonal decision trees, and where an exploration methodology, driven by the number of memory accesses, is outlined. However nowadays, none of the existing approaches *exploit the features of the used memory architecture.*

Relatively to the selected *memory architecture*, on-chip memories (limited by the storage capacity) are used in [5, 16]. Content addressable memories (limited by both performance and power) are used in [4]. In [7, 15], the memory architecture only consists of off-chip DRAMs. By storing data structures with data packets in the same DRAMs, instead of storing data structures in separate off-chip SRAMs, the number of off-chip memories are economized and the number of chip pins is minimized. However the latency of accessing the data structures is increased, since DRAMs are slower than SRAMs. Additionally storing data structures with data packets in the same DRAMs increases the number of DRAM accesses per segment enqueueing or dequeueing. Hence nowadays to combine high speed, low power, large data storage, and high memory bandwidth, the *use of off-chip DRAMs (with multi banks) combined with off-chip SRAMs* becomes a must.

To increase the performance of data queueing, not only the memory bandwidth, but also the access to memory must be highly optimized. Research has already been carried out to reduce the number of bank conflicts in a multi-bank memory architecture. [2, 9] proposes data placement schemes suitable for static data dominated applications. [10, 11] propose local optimizations for memory access scheduling. However nowadays *minimizing at the system level the number of bank conflicts during real-time memory (de)allocation* has not been given much attention.

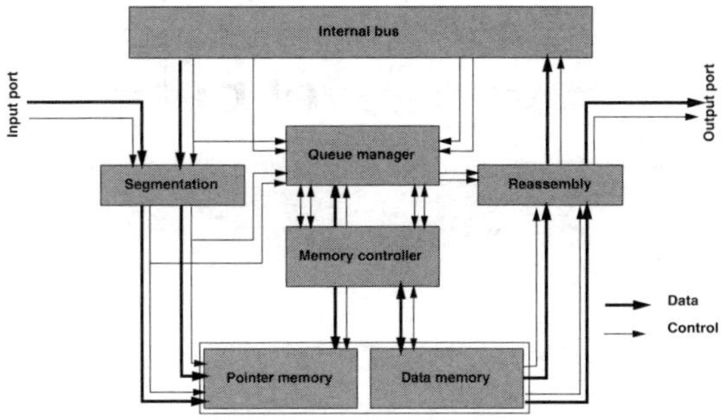

Figure 1: DQM block diagram

What we propose in this paper is to optimize at the system level, where the impact on the memory performance is the most important, the dynamic memory management in data queueing. This is realized as follows:

- Select a memory architecture combining off-chip DRAMs with off-chip SRAMs. This can access data structures in parallel with data packets, and thus to aim at keeping the rate of DRAM accesses to one per segment enqueueing or dequeueing.
- Systematically explore possible memory management techniques, based on network characteristics, and on its real-time requirements, and trading off bank conflicts in DRAMs against memory accesses in SRAMs during real-time memory (de)allocation.

3. DEMONSTRATOR

As previously mentioned, DQMs in network processors are responsible for real-time memory (de)allocation, buffering, retrieving, and forwarding of incoming data packets. These data packets can be of variable length. They are organized into queues, one queue per active connection. The DQM has to perform operations like: create or destroy a queue, enqueue a new packet in a queue, dequeue a packet from a queue and forward it, dequeue a packet from a queue and append it to another queue, delete a packet from a queue. The major bottleneck in the DQM implementation is not in the data path and the controller, but in the communication and the memory organization [3]. In this paper we focus on the worst-case memory performance of the DQM.

The DQM block diagram is shown in Figure 1. The virtual memory is split into the pointer memory and the data memory. All pointers, flags, ..., in the data structures implementing the queues, which are small and frequently accessed, are stored in the *pointer memory.* All data packets, which are large and not frequently accessed, are stored in the *data memory.*

The DQM has to segment the incoming data packets being of variable length into fixed-size segments. Similarly, the outgoing segments from the data memory are reassembled to form the outgoing data packets.

To simplify the functionality and increase the modularity of the DQM, the DQM is divided into two main blocks: the queue manager and the memory controller. The queue manager handles the control information from the network pro-

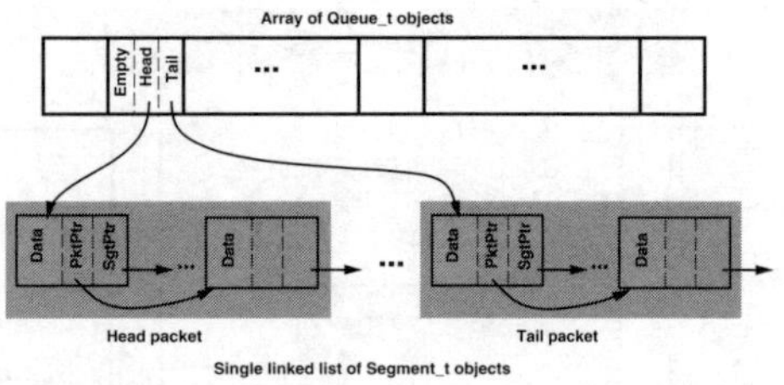

Figure 2: Queue manager abstract specification

cessor for transferring data packets to/from the data memory. It handles the queues and organises the data packets in the queues. It also manages the memories for dynamically (de)allocating the data structures implementing the queues and for detecting memory overflow.

In the PRO^3 system, the DQM has to agree with the following network characteristics. It must support up to 0.5M connections simultaneously. In congestion situations, it must support a data memory of 256M bytes. It must preserve the OC48 (2.488 Gbps) line rate of the incoming packets. However, taking into account that bandwidth is wasted due to packet segmentation/reassembly, and that the DQM must support 2 input ports, 2 output ports, and 1 internal bi-directional port, the DQM needs to meet a bandwidth of at least 15 Gbps (around $6 * 2.488$ Gbps).

The abstract specification of the queue manager is now detailed. This queue manager is the most critical block of the DQM and an intelligent implementation is needed. It also has to exploit the parallelism not only of the DQM operations, but also of the huge amount of memory accesses in order to meet the strict real-time requirements. In this paper, we assume that an abstract specification has already been selected for the queue manager. This is depicted in Figure 2 and described below. Each queue consists of a single linked list of segments, and it is characterized by a `Queue_t` object consisting of the following main fields: an `Empty` bit indicating whether the queue is empty or not, a `Head` pointer to the first segment of the head packet in the queue, and a `Tail` a pointer to the first segment of the tail packet in the queue. This `Queue_t` object is stored in an array whose index respresents the queue identifier generated by the network processor. *This array is statically allocated.*

Each segment is characterized by a `Segment_t` object consisting of the following main fields: `Data` storing the segment data bytes, a `SgtPtr` pointer to the next segment in the queue, and a `PktPtr` pointer to the last segment of the current packet. *These* `Segment_t` *objects are dynamically (de)allocated.* A memory management technique is then required to organize the virtual memory, to allow real-time memory (de)allocation of these objects, and to control memory overflow. This is selected by our exploration method described in Section 5.

4. MEMORY PERFORMANCE

As mentioned in Section 3, at the system level, all `Data` fields from the `Segment_t` objects are stored in the data memory, which is mapped later during system synthesis to DRAMs. All other fields are stored in the pointer memory, which is mapped later to SRAMs. In the following, we analyze the system-level parameters that affect the performance of both data and pointer memories, in order to take them into account in our exploration method.

4.1 Data memory

To accommodate the requirements of the data memory (both memory storage and bandwidth), a module of double data rate synchronous DRAMs [13] can be used. It provides large memory space and high memory bandwidth, but suffers from bank conflicts. This module has a total capacity of 256 Mbytes and is organized into 8 banks. The memory data bus of this module is 64-bit wide and runs at 133 MHz. Taking into account that both edges of the clock are used to access a double data rate memory, the module offers a bandwidth of 17 Gbps in best case. To effectively utilize the module, this should work as much as possible in the burst mode. The burst has a length of 64 bytes. Hence to optimize memory bandwidth utilization, a segment size of 64 bytes too is selected. In a non-busy bank, a segment can be read/written in 4 clock cycles. In a busy bank, there is a *bank conflict* and the access is delayed (8 clock cycles, i.e. around 60 ns) until the bank becomes available.

Hence at the system level, the data memory performance heavily depends on the number of bank conflicts. A bank conflict happening during a read access cannot be controlled. However a bank conflict happening during a write access can be controlled by storing new data in a non-busy bank.

4.2 Pointer memory

For the pointer memory, a module of zero bus turnaround SRAMs [13] can be used. These memories employ a high-speed and low-power design using an advanced CMOS process, and they eliminate any turnaround clock cycle between read and write memory accesses. Each SRAM has a capacity of 0.5M 32-bit words, and it can run at 166 MHz. To agree with the assumed network characteristics, the DQM requires to support a data memory of 256M bytes, which corresponds to 4M segments of 64 bytes each. Hence 22 bits are needed to specify a `segment_t` pointer. The DQM also requires to support 0.5M connections. Hence within the selected queue manager specification (see Section 3), a pointer memory of 9M `Segment_t` pointers[1] is needed, and it must be mapped to a module of 18 SRAMs. To minimize the number of pins, the number of parallel memories is restricted to 2, and the memory data bus of the module is 64-bit wide. Since data pipelining is possible, a 64-bit word (that is, any `Queue_t` object or the part of any `Segment_t` object stored in the pointer memory) can be accessed every clock cycle. Taking into account that both data and pointer memories can be accessed in parallel, 5 (resp. 15) 64-bit word accesses to the SRAM module can be performed in parallel with a segment reading/writing in the DRAM module without (resp. with) bank conflict.

Hence at the system level, the pointer memory performance can be further optimized by minimizing the number of `Queue_t` and `Segment_t` accesses during `Segment_t` dynamic (de)allocation. These operations are required in the following DQM operations: enqueue new segments in a queue during packet storage, and delete a packet from a queue.

4.3 Conclusion

[1] 1M pointers for the `Queue_t` objects, 8M pointers for the `Segment_t` objects.

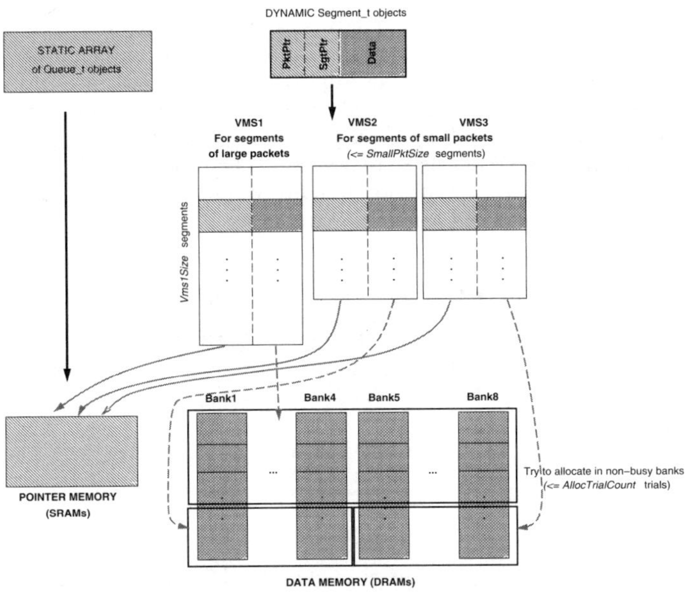

Figure 3: Our DQM memory management, parameterized by $AllocTrialCount$**,** $SmallPktSize$**, and** $Vms1Size$

Our method described in the next section explores the design space of memory management techniques, taking this previous analysis into account, and efficiently trading off: (1) write accesses to non-busy banks in the data memory during packet storage, against (2) pointer memory accesses during packet storage and deletion.

5. MEMORY MANAGEMENT

As mentioned in Section 3, the selected queue manager specification uses: (1) the **Queue_t** data structure whose objects are stored in a static array; (2) the **Segment_t** data structure whose objects are dynamically (de)allocated in the virtual memory. Hence a memory management technique is required to organize the virtual memory, to allow real-time memory (de)allocation of the **Segment_t** objects, and to control memory overflow.

Our exploration method to derive an efficient memory management technique for the DQM is depicted in Figure 3 and is described below. It fits in the context of embedded system synthesis, as our Matisse approach [19] based on fast and successive explorations at all levels of the complete design flow, in order to achieve efficicent system designs. Since optimizing the dynamic memory management is one of the first system-level synthesis steps, exploration in this step must be simple and fast.

5.1 Virtual memory organization

The virtual memory is first partitioned into virtual memory segments (VMSs) of fixed-size blocks. In this paper, 4 VMSs are considered. A first VMS stores the static array of **Queue_t** objects, and it is assigned to the pointer memory. Three other VMSs consist of a reserved amount of memory to store **Segment_t** objects. **VMS1** stores segments from large packets (containing $> SmallPktSize$ segments), it has a capacity of $Vms1Size$ segments, and it is mapped later to the 8 banks of the DRAM module (described in Section4.1). **VMS2** and **VMS3** store segments from small packets (containing $\leq SmallPktSize$ segments), and they have a capacity of $(4M - Vms1Size)/2$ segments each[2]. **VMS2** (resp. **VMS3**) is mapped later to the first (resp. last) 4 banks of the DRAM module. To control the bank conflicts at the system level, each block (whose size is **Segment_t** size) in these three VMSs is also initially associated with one of the 8 banks. Also the free blocks of each VMS are maintained in a single linked list by using the **SgtPtr** pointer to specify the next free block in the list. Such lists of free blocks are called *free lists* [14]. As mentioned in Section 4, the pointers in the **Segment_t** objects are assigned to the pointer memory, whereas the **Data** fields are assigned to the data memory.

5.2 (De)allocation implementation

For these three VMSs (**VMS1, VMS2** and **VMS3**), a memory management technique is required, where dynamic (de)allocation is implemented as follows.

5.2.1 Allocation in segment enqueueing

Whenever a **Segment_t** object is created, a free block to store it is allocated as follows, so as to avoid bank conflicts as much as possible. In the DQM, **Segment_t** creation is required in segment enqueueing.

1. If the segment belongs to a large packet (containing $> SmallPktSize$ segments), **VMS1** is selected. If the head block in its free list is associated with a busy bank, it is pushed at the free list tail. This is repeated until a block associated with a non-busy bank is found. We limit the number of trials to $AllocTrialCount$. The current head block is then allocated.
Hence for each trial one pointer memory read access is required to access the next free block in the free list. One pointer memory write access is required to link the rejected head blocks to the free list tail. Another pointer memory read access is required to update the free list head. Hence the total number of pointer memory accesses is less than $AllocTrialCount + 1$.
2. If the segment belongs to a small packet, and if the head block in the free list of **VMS2** is associated with a non-busy bank, this head block is allocated.
Hence only one pointer memory read access is required to update the free list head.
3. Otherwise the head block in the free list of **VMS3** is allocated and only one pointer memory read access is required too.

Hence *worst-case segment enqueueing* either implies a bank conflict in the data memory or requires more than one access in the pointer memory during segment allocation.

5.2.2 Deallocation in packet deletion

Whenever a **Segment_t** object is deleted, the block storing it is pushed to the free list it came from. In the DQM, **Segment_t** deletion is only required in packet deletion.

1. For large packets, all segments are guaranteed to be stored in blocks allocated from the same free list, the one of **VMS1**.
In this case, only one pointer memory write access is required to update the tail of this free list, independently from the number of deleted segments in the packet.
2. For small packets, for each segment to be deleted, the block storing it is pushed to the free list it came from, that

[2] To agree with the network characteristics, the DQM requires to support 4M segments (see Section 4.2).

is, the one of either VMS2 or VMS3.

One pointer memory read access is required to access each next segment of the packet in the queue. For each segment, one pointer memory write access is also required to link the freed block to the free list. Hence the total number of pointer memory accesses per packet deletion is always less than $2 * SmallPktSize - 1$.

Hence *worst-case packet deletion* requires more than one access in the pointer memory.

5.3 Exploration

In our exploration method, the values of the three most important parameters are explored: $AllocTrialCount$, $SmallPktSize$, and $Vms1Size$. Larger $AllocTrialCount$ is, less bank conflicts in the data memory are generated, but more pointer memory accesses are required during allocation of segments belonging to large packets. Larger $SmallPktSize$ is, more segment allocations are performed without bank conflict in the data memory and without extra pointer memory access, but more pointer memory accesses are required during packet deletion. $Vms1Size$ has to trade-off memory overflows during segment allocation between large packets stored in VMS1 and small packets stored in both VMS2 and VMS3.

Within this design space, the conventional memory management technique, consisting in using one free list, is characterized by: $AllocTrialCount = 1$, $SmallPktSize = 0$, and $Vms1Size = 4M$ (i.e. $2^{22} = 4194304$) segments. In contrast, the best memory management technique is the one that minimizes the memory overflows in the three VMSs of the virtual memory organization, together with the memory performance in worst-case segment enqueueing and packet deletion. However, as illustrated in Section 6, these cost parameters must be traded off against each other. To this end a 5-dimension Pareto curve is generated as follows, from which the designer can select an optimized memory management technique meeting the required trade-off:

• Each possible memory management technique, parameterized by the explored $AllocTrialCount$, $SmallPktSize$, and $Vms1Size$ values, is simulated using some real packet traces. Each simulation profiles worst-case situations and derives the following 3 parameters:

– The percentage of segment allocation failures in order to estimate the number of memory overflows.

– $OverheadMemCycle/WorstCases$ in order to estimate the memory performance overhead in worst-case segment enqueueing, where $OverheadMemCycle$ denotes the total number of overhead memory cycles (in terms of SRAM clock cycles) in all worst-case segment enqueueing executions, and $WorstCases$ denotes the number of these executions.

– The memory performance overhead in worst-case packet deletion is similarly estimated.

• A technique for which all profiled parameters are larger than in another explored alternative is eliminated.

• The set of remaining techniques is called the *Pareto curve* of the considered DQM relatively to the exploration.

Practically, $AllocTrialCount$ which has the larger impact is greedily explored, then $SmallPktSize$, and finally $Vms1Size$. For each parameter, the exploration stops when no better technique can be reached.

6. RESULTS

Results of our exploration method are reported for IP networks receiving large IP packets. These results illustrate that the conventional memory management technique used in embedded systems is far from being well-suited in DQM modules. To this end, simulations are performed on real packet traces found in [1]. First we analyse the impact of each parameter in the exploration. Then we describe the resulting Pareto curve and compare the selected optimized techniques with the conventional one.

Table 1: $AllocTrialCount$ impact

Alloc Trial Count	Mem. acc. Per seg. alloc. Avg/Max	Bank conflict percentage In write ops Wo/with page ctrl	In read ops Wo/with page ctrl
1	1/1	15.8/11.9	12.4/12.4
2	1/3	6.2/1.9	6.7/6.7
3	1/4	5.5/1.9	6.2/6.2
4	1/5	5.3/1.7	6.1/6.1
5	1/6	5.4/1.7	6.1/6.1

First, $AllocTrialCount$ is explored, while $SmallPktSize$ and $Vms1Size$ are kept constant. Results are reported in Table 1 for $SmallPktSize = 0$, i.e. for using only one VMS. From this exploration, we make the following observations. (1) The conventional memory management technique corresponds with $AllocTrialCount = 1$. Its number of pointer memory accesses is constant per segment allocation. However its average number of bank conflicts is the highest. (2) Athough the performance of segment enqueueing depends on $AllocTrialCount$ (since the number of pointer memory accesses per segment allocation is bounded by $AllocTrialCount + 1$), it is not affected by it on average (since this number of accesses remains constant on average). (3) Supporting page control in the DRAM memory controller is a must, since it allows an additional reduction of up to 70% in the number of bank conflicts during DRAM write accesses. (4) Already with $AllocTrialCount = 2$, 84% of the number of bank conflicts are avoided during DRAM write accesses, and this also yields a reduction of 46% in the number of bank conflicts during DRAM read accesses, which are not controlled though. (5) From $AllocTrialCount = 3$, no further reduction in the number of bank conflicts is observed, and the greedy exploration can stop.

Table 2: $SmallPktSize$ impact

Small Pkt Size	Mem. acc. Per pkt deletion Avg/Max	Bank conflict percentage In write ops Wo/with page ctrl	In read ops Wo/with page ctrl
0	1/1	5.5/1.9	6.2/6.2
1	1/1	3.5/1.2	7.3/7.3
2	1/3	3.6/1.5	6.8/6.8
3	1/5	3.4/1.4	6.6/6.6
4	1/7	3.2/1.4	6.6/6.6

Second, $SmallPktSize$ is explored, while $AllocTrialCount$ and $Vms1Size$ are kept constant. Results are reported in Table 2 for $AllocTrialCount = 3$. From this exploration, we make the following observations. (1) Although the performance of packet deletion depends on $SmallPktSize$ (since its number of pointer memory accesses is bounded by $2 * SmallPktSize - 1$), it is also not affected by it on average (since this number of accesses also remains constant on average). (2) Considering three VMSs instead of only one does not allow to further reduce the number of bank conflicts for these considered network characteristics. However as illustrated in Table 4, this allows to trade off the worst-case performance in segment enqueueing against the ones in packet deletion. (3) The greedy exploration can stop with $SmallPktSize = 4$.

Table 3: $Vms1Size$ impact

Vms1 Size	Memory Overflow Probability	Bank conflict percentage In write ops Wo/with page ctrl	In read ops Wo/with page ctrl
2097152	1.85	3.6/1.5	6.8/6.8
3145728	2.28	4.2/1.8	5.6/5.6
3495232	2.39	4.0/1.8	5.1/5.1
3670016	2.41	3.8/1.7	4.7/4.7
3774848	2.43	3.8/1.7	4.5/4.5
3984576	2.46	3.5/1.6	3.9/3.9
4054528	2.49	3.4/1.5	3.7/3.7
4089472	2.5	13.5/11.5	12.5/12. 5

Finally,, $Vms1Size$ is explored, while $AllocTrialCount$ and $SmallPktSize$ are kept constant. Results are reported in Table 3 for $AllocTrialCount = 3$ and $SmallPktSize = 2$, and for a particular trace of DQM operations. We observe that this exploration still allows to reduce the number of bank conflicts during DRAM read accesses by 43%, but at the expense of the number of memory overflows. The greedy exploration stops as soon as a degradation in the number of bank conflicts is observed.

Table 4: Optimized techniques compared with the conventional one

AllocTrCt/ SmPkSize/ Vms1Size	Bank conflict percentage With page ctrl In write/read ops	Memory Overflow Percentage	Seg. enq. Overhead Perf/Energy
3/1/2097152	**1.2/7.4**	**1.7**	**4/164**
4/1/3145728	**1.5/5.9**	**2.2**	**4/160**
5/1/3774848	**1.6/4.9**	**2.4**	**4/158**
5/1/4054528	**1.3/4.0**	**2.5**	**4/152**
4/0/4194304	**1.7/6.1**	**2.5**	**3/122**
1/0/4194304	11.9/12.4	2.45	8/337

As mentioned in Section 5, to help the designer to select an optimized memory management technique meeting the required trade-off, a Pareto curve is generated. In this experiment, this consists of 16 techniques out of the 180 explored ones. In these optimized techniques, we observe that: (1) In all packet deletion executions, the memory performance is minimal; (2) Per worst-case segment enqueueing execution, the memory performance overhead is between 3 and 4 SRAM clock cycles; (3) The memory overflow percentage is between 1.7 and 2.5. Some of these optimized techniques are reported in Table 4 and compared with the conventional technique. They can save up to 90% (resp. 68%) of the bank conflicts during DRAM write (resp. read) accesses. This bank conflict reduction allows to improve both worst-case memory performance in segment enqueueing by 50%.

7. CONCLUSION

One of the most difficult part in high-speed network processors is data queueing. In this paper, such data queueing was used as case study to demonstrate the effectiveness of a new system-level exploration method for optimizing the memory performance in dynamic memory management. This method also fits in Matisse [19], a system-level synthesis approach supporting the memory management of dynamic (de)allocation of data structures in embedded systems. In the future we intend to adapt this new method to the memory management of look-up tables, another bottleneck in the design of high-speed network processors.

7.1 Acknowledgements

This paper describes work undertaken in the context of the IST-1999-11419 Protocol Processor Project (PRO^3), partially funded by the Commission of the European Union. The authors would like to acknowledge the contributions of all their collegues involved in the project from Lucent Technologies, Hyperstone electronics, IMEC, National Technical University of Athens and Ellemedia Technologies.

8. REFERENCES

[1] Traces available in the internet traffic archive. http://ita.ee.lbl.gov/html/traces.html.

[2] A.A. Bertossi and M.C. Pinotti. Mappings for conflict-free access of paths in elementary data structures. *COCOON*, p.351-361, July 2000.

[3] F. Catthoor et al.. *VLSI Signal Processing*, volume VII, chapter Global Communication and Memory Optimizing Transformations for Low Power Signal Processing Systems, p.178-187. IEEE Press, New York, 1994.

[4] H.J. Chao et al. Design of a generalized priority queue manager for ATM switches. *IEEE Journal on Selected Areas in Communications*, 15(5):867-880, June 1997.

[5] A.K. Choudhury and E.L. Hahne. New implementation of multi-priority pushout for shared memory ATM switches. *Computer Communications*, 19(3):245-256, March 1996.

[6] V. Firoiu and M. Borden. A study of active queue management for congestion control. *Infocom*, Tel Aviv, Israel, March 2000.

[7] G.Doumenis et al. A parallel VLSI video/communication controller. *The Journal of VLSI Signal Processing, Special Issue on Parallel VLSI Architectures for Image and Video Processing*, July 2000. Kluwer Academic.

[8] C. Georgopoulos et al. A protocol processing architecture backing TCP/IP-based security applications in high speed networks. *Interworking Conference*, Bergen, Norway, 2000.

[9] L.K. John. Data placement schemes to reduce conflicts in interleaved memories. *Computer Journal*, 43(2):138-151, 2000.

[10] M. Kandemir, U. Sezer, and V. Delaluz. Improving memory energy using access pattern classification. *ICCAD*, p.201-206, November 2001.

[11] K.A. Khare, P.R. Panda, N.D. Dutt, and A. Nicolau. High-level synthesis with SDRAMs and RAMBUS DRAMs. *IEICE Trans. on Fundamentals of Electronics, Communications, and Computer Sciences*, 11:2347-2355, 1999.

[12] V. Kumar, T. Lakshman, and D. Stiliadis. Beyond best effort: Router architectures for the differentiated services of tomorrow's internet. *IEEE Communications Magazine*, p.152-164, May 1998.

[13] Micron. http://www.micron.com.

[14] N. Murphy. Safe memory usage with dynamic allocation. *Embedded Systems*, p.49-57, May 2000.

[15] A. Nikologiannis and M. Katevenis. Efficient per-flow queueing in DRAM at OC-192 line rate using out-of-order execution techniques. *Int. Conf. on Communications*, p.2048-2052, 2001.

[16] J.L. Rexford, A.G. Greenberg, and F.G. Bonomi. Hardware-efficient fair queueing architectures for high-speed networks. *Infocom*, p.638-646, March 1996.

[17] D.N. Serpanos and P. Karakonstantis. Efficient memory management for high-speed ATM systems. *Design Automation for Embedded Systems*, 6(2):207-235, April 2001.

[18] B. Suter, T.V. Lakshman, D. Stiliadis, and A.K. Choudhury. Buffer management schemes for supporting TCP in gigabit routers with per-flow queueing. *IEEE Journal on Selected Areas in Communications*, August 1999.

[19] S. Wuytack et al. Memory management for embedded network applications. *IEEE Trans. on CAD*, 18(5):533-544, 1999.

Analysis of Power Consumption on Switch Fabrics in Network Routers

Terry Tao Ye
Computer Systems Lab
Stanford University
taoye@stanford.edu

Luca Benini
DEIS
University of Bologna
lbenini@deis.unibo.it

Giovanni De Micheli
Computer Systems Lab
Stanford University
nanni@stanford.edu

ABSTRACT

In this paper, we introduce a framework to estimate the power consumption on switch fabrics in network routers. We propose different modeling methodologies for node switches, internal buffers and interconnect wires inside switch fabric architectures. A simulation platform is also implemented to trace the dynamic power consumption with bit-level accuracy. Using this framework, four switch fabric architectures are analyzed under different traffic throughput and different numbers of ingress/egress ports. This framework and analysis can be applied to the architectural exploration for low power high performance network router designs.

Categories and Subject Descriptors

C.4 [**Performance of Systems**]: Design studies, Modeling techniques

General Terms

Design, Experimentation, Performance

Keywords

Networks on Chip, Interconnect Networks, Systems on Chip, Power Consumption

1. INTRODUCTION

With the fast development of Internet applications, the network bandwidth and capacity are moving quickly to the Giga-bit and Tera-bit domains. The switch fabric circuit is the fundamental building block inside a network router, it distributes all network traffic from ingress ports to egress ports. The performance of switch fabrics is very critical in network applications.

While most attention is focused on speed and capacity issues of switch fabrics, power consumption is becoming a more and more serious problem, especially for single chip network routers, where switch fabric circuit contributes a significant part of the total power consumption.

There are many different switch fabric architectures used in network routers. They have different characteristics in terms of bandwidth, throughput and delay [1]. In this paper, we will focus on the power consumption analysis of these different architectures, and estimate how the power consumption scales with the network capacity. Particularly, the following questions are to be addressed and answered:

1. What is the power consumption of different switch fabric architectures under different network traffic loads?
2. How does the power consumption scale with different numbers of ingress and egress ports?

In a switch fabric circuit, the power is dissipated on three components: 1) the internal node switches, located on the intermediate nodes between ingress and egress ports; 2) the internal buffers, used to temporarily store the packets when contention between packets occurs; and 3) the interconnect wires that connect node switches. The power consumption on these three components changes differently under different traffic loads and configurations. Therefore, they need to be analyzed with different modeling methodologies.

In this paper, we first propose different power consumption models for each of the above three components. The total power consumption is then analyzed and simulated on four widely-used switch fabric architectures; 1)Crossbar, 2)Fully Connected 3)Banyan and 4)Batcher-Banyan. A Simulink based multi-threading simulation platform is created for this analysis. It performs a time domain simulation with bit level accuracy. The power consumption of every bit in the traffic flow is traced as the bit moves among the components inside switch fabrics.

Previous approaches for switch network power estimation are either based on statistical traffic models [2], or analytical models [3] [4]. The simulation is performed on gate or circuit levels, it is time consuming and not practical for large switch fabric designs. Furthermore, these approaches are not suitable for tracing the power consumption dynamically in real-time network traffic conditions. For example, the power consumption on internal switch buffers depends on the dynamic contention between packets. Compared with previous approaches, our modeling is an architectural-level estimation with bit level accuracy. It traces the power consumption based on dynamic packet dataflows. This approach is ideal for architectural design exploration as well as application specific power analysis.

The paper is organized as follows. Section 2 describes the network router architectures and functions of switch fabrics. Section 3 proposes the power modeling techniques for switch fabrics. Using these techniques, four switch fabric architectures are analyzed in Section 4. Experiment benchmarks are described in Section 5 with results in Section 6.

2. ROUTERS AND SWITCH FABRICS

A network router consists of four parts: 1) the ingress packet process unit, 2) the egress packet process unit, 3) the arbitration unit (arbiter), and 4) the switch fabrics. Fig. 1 shows the block diagram of a network router architecture.

The ingress packet process unit parallelizes the serial dataflow on the transmission line into bus dataflow, and inspects the header and the content (e.g. in content-aware routing algorithms) of the incoming packets. The egress process unit re-assembles the processed

DAC 2002, June 10-14, 2002, New Orleans, Louisiana, USA.

Figure 1: Block Diagram of a Network Router

packets and delivers the packets to their destination ports. The arbitration unit (the arbiter) determines when and where a packet should be routed from the ingress ports to the egress ports. It uses the packet information from the ingress inspection and makes the decision based on arbitration algorithms.

A switch fabric circuit is an interconnect network that connects the ingress ports to the egress ports. Different switch fabric architectures have different impacts on the network router performance, e.g. throughput, delay, power, etc. In the following sections, we will only analyze the power consumption issues on switch fabrics and estimate the power consumption of different switch fabric architectures with different numbers of ingress and egress ports.

3. POWER MODELING WITH BIT ENERGY

A switch fabric circuit is an on-chip interconnect network. The power consumption on switch fabrics comes from three different sources: 1)the internal node switches; 2) the internal buffer queues; and 3) the interconnect wires. Inside the switch fabrics, different packets travel on different data paths concurrently, and the traffic load on each data path may change dramatically from time to time. To estimate the dynamic power consumption in this multi-process interconnect network, we propose a new modeling approach: the *Bit Energy* E_{bit}.

The bit energy E_{bit} is defined as the energy consumed for each bit when the bit is transported inside the switch fabrics from ingress ports to egress ports. The bit energy E_{bit} is the summation of the bit energy consumed on node switches, $E_{S_{bit}}$, on internal buffers, $E_{B_{bit}}$ and on interconnect wires, $E_{W_{bit}}$. We will analyze these three bit energies in details in the following sections.

3.1 Node Switch Power Consumption

Node switches are located on the intermediate nodes inside switch fabrics. They direct the packets from one intermediate stage to the next stage until reaching the destination ports. In different switch fabric topologies, node switches may have different functions and node degrees. For example, in Banyan switch fabrics, the node switch is a 2 × 2 switch and has degree of 4.

When a data bit travels through a node switch, The logic gates on the data path inside the node switch consume power as they toggle between power rails. We will analyze the 2 × 2 switch used in Banyan switch fabrics as an example (Fig. 2).

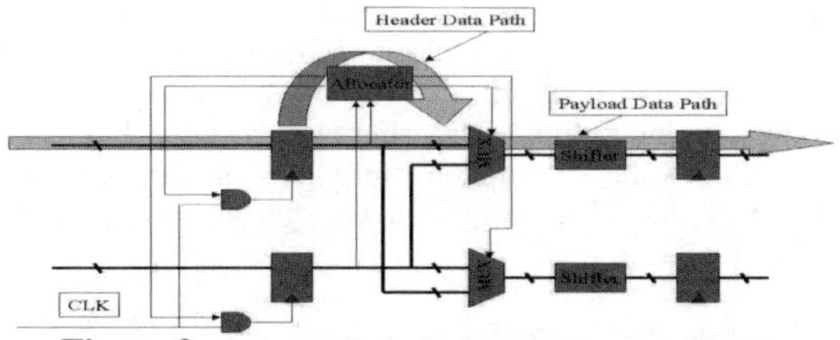

Figure 2: A 2 × 2 Node Switch in Banyan Switch

The 2 × 2 switch directs the packets from its two inputs to the outputs, according to the destination addresses of the packets. The ingress process unit had already parallelized the serial dataflow on the transmission line into a parallel bus dataflow (16-bit or 32-bit wide), so the destination address can be read out at one clock cycle. The destination bits are first detected by the allocator. If the destination port is available, the allocator allocates the output port to the packet and preserves the allocation throughout the packet transmission. The allocation process and packet transmission process are denoted in Fig. 2 as "header data path" and "payload data path" respectively. Both processes consume energy when packet bits travel through the data paths. The energy consumed by the header bits is actually different from the payload bits. However, the header normally occupies a small portion of the entire packet. Without loss of generality, we will use the payload bit energy as the node switch bit energy $E_{S_{bit}}$ in our analysis.

In reality, the bit energy $E_{S_{bit}}$ also depends on the presence or absence of packets on other ports of the node switch. For example, the switch will consume more power to process two packets at the same time, but the power consumption is not necessarily twice as much as that of processing a single packet. Therefore, the bit energy $E_{S_{bit}}$ is an input state-dependent value and should be expressed in an input vector indexed look-up table. For a switch with n input ports, there will be 2^n different input vectors with different $E_{S_{bit}}$ values.

In our analysis, the look-up table is pre-calculated from Synopsys Power Compiler simulation. The node switch circuit is first simulated based on input vectors. Then, the switching activities on every gate are also traced. Last, the power consumption of the entire circuit is estimated. We simulate different combinations of the input vectors and the results are shown in details in Section 5. Using the look-up table, the power consumption on the switch node can be estimated under different traffic conditions.

3.2 Internal Buffer Power Consumption

When contention occurs between packets, internal buffers are needed to temporarily store the packets with lower priorities (Fig. 3). There are two different types of contention between ingress packets, namely, the *destination contention* and *interconnect contention*.

- *Destination contention* – When there are two or more packets in the ingress ports requesting the same destination port at the same time, *destination contention* will occur. This type of contention is application dependent. In our analysis, we assume the arbiter had already solved this type of contention before the ingress process unit delivers the packets to the switch fabrics.

- *Interconnect contention (Internal Blocking)* – Inside switch fabric circuits, the same interconnect link may be shared by packets with different destinations. The contention on the shared interconnects is called *interconnect contention* or *internal blocking*. It happens inside switch fabric interconnect networks and it is architectural dependent.

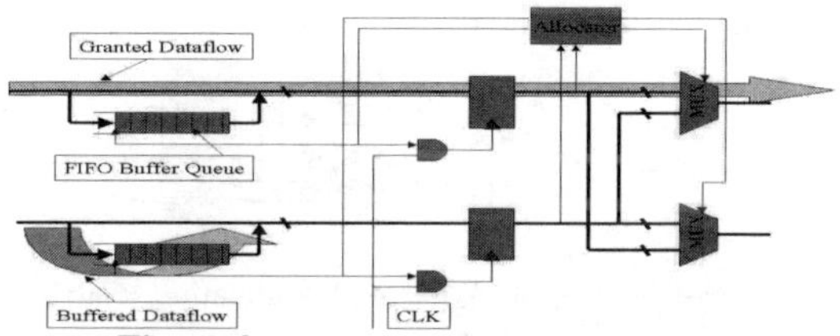

Figure 3: Buffers in a 2 × 2 Node Switch

In switch fabric circuits, the buffers are normally implemented with shared SRAM or DRAM memories. The energy consumption in buffers comes from two sources: 1) the data access energy, consumed by each READ or WRITE memory access operation, and 2) the refreshing energy, consumed by the memory refreshing operation (in the case of DRAM). The bit energy on the internal buffers $E_{B_{bit}}$ can be expressed by the following equation(Eq. 1):

$$E_{B_{bit}} = E_{access} + E_{ref} \qquad (1)$$

where E_{access} is the energy consumed by each access operation and E_{ref} is the energy consumed by each memory refreshing operation. In reality, memory is accessed on word or byte basis instead of a single bit, the E_{access} is actually the average energy consumed for one bit.

The energy consumed by memory access is determined by the contentions between the ingress packets. As discussed earlier, *destination contention* is application dependent, regardless of what switch fabric circuits are used. In comparison, *interconnect contention* is switch fabric architecture dependent. Different architecture topologies result in different contention occurrence. In this paper, we are interested in comparing the power consumption on different switch fabric architectures under the same network traffic, therefore, we assume the *destination contention* has already been resolved by the arbiter before the ingress packets are delivered to the switch fabrics. We only compare the internal buffer energy consumption occurred from *interconnect contention*.

3.3 Interconnect Wires Power Consumption

When the node switch delivers one bit with flipped polarity to the interconnect wires, the signal on the wire will toggle between logic "0" and logic "1". Energy is dissipated in this charging or discharging process. Only bits with flipped polarity consume energy, namely, $E_{W_{bit0\to1}}$ or $E_{W_{bit1\to0}}$ have bit energy values, $E_{W_{bit0\to0}} = 0$, $E_{W_{bit1\to1}} = 0$.

Assuming a rail-to-rail toggling, the bit energy on the interconnect wires $E_{W_{bit}}$ for bit 1→0 and 0→1 can be described by the following equation (Eq. 2).

$$E_{W_{bit}} = \frac{1}{2}C_{wire}V^2 + \frac{1}{2}C_{input}V^2 = \frac{1}{2}C_W V^2 \qquad (2)$$

Here C_{wire} is the wire capacitance on the interconnect, C_{input} is the total capacitance of the input gates connected to the interconnect. $C_W = C_{wire} + C_{input}$ is the total load capacitance propagated by that bit. The rail-to-rail voltage is denoted by V, assuming the CMOS gates are switching from Vdd to GND.

The bit transmitted on interconnect wires consumes energy only when its polarity is flipped from previous transmitted bit. The switching activities on interconnects can be traced by our simulation approach proposed in Section 5.

3.4 Interconnect Wire Length Estimation

The wire capacitance C_{wire} is a function of wire-length and coupling capacitance between adjacent wires [5]. Estimation of interconnect wire-length of the switch fabric network is essential for the bit energy calculation on wires.

Here we adopt the Thompson model [6] for wire-length estimation. The estimation is based on a graph embedding process as described below and also illustrated in Fig. 4.

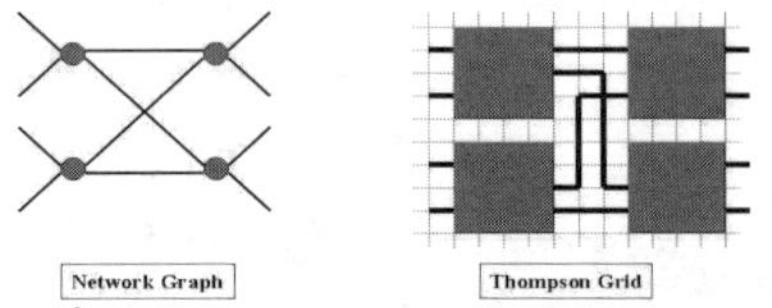

Figure 4: Thompson Wirelength Estimation Model

We are given a source graph $G(V_G, E_G)$, where $V_{G_i}(0 < i \leq n)$ are the vertices of graph G and $E_{G_i}(0 < i \leq m)$ are the edges. The source graph represents the network topology. The target graph is denoted by $H(V_H, E_H)$, where V_H and E_H are the vertices and edges of H. Graph H is a 2-dimensional grid mesh consisting of p columns and q rows. An embedding of graph G into graph H is performed as follows. Each vertex in G is mapped into a $d \times d$ square of vertices in H, where d is the degree of vertex $v_i \in V_G$ and no more than one vertex in V_G occupies the same vertex in V_H. Each edge in G is mapped into one or more edges of graph H, and no more than one edge in E_G occupies the same edge in graph H. The optimal Thompson embedding of graph G into graph H is to find the minimum number of columns p_{min} and rows q_{min} in H that are needed for this embedding. The interconnect wire-length is defined as the number of grids that an edge E_G covers.

In our approach, we manually map the switch fabric topologies into Thompson grids and estimate the interconnect wire-length by counting the number of grids the interconnect covers. The detailed mapping of each particular switch fabric topology will be shown in Section 4.

The Thompson model is only a global wire-length estimation. It is not as accurate as detailed wire-routing, but it is an effective way of architectural planning for interconnect networks, especially when the network topology is regular. In the case of switch fabrics, it is straightforward to map the regular switch nodes and interconnects into a 2-dimensional mesh of regular rows and columns. Therefore, the Thompson model is an ideal model for switch fabric wire-length estimation.

For an interconnect wire of length equal to one Thompson grid, we define the wire bit energy consumption as $E_{T_{bit}}$. If an interconnect wire has its length equal to m Thompson grids, its wire bit energy is $E_{W_{bit}} = m \times E_{T_{bit}}$.

4. SWITCH FABRIC ARCHITECTURES

With the above proposed power consumption model, we will analyze four widely-used switch fabric architectures in this section.

4.1 Crossbar Switch Fabrics

An $N \times N$ crossbar network connects N input ports with N output ports. Any of the N input ports can be connected to any of the N output ports by a node switch on the corresponding crosspoint (Fig. 5).

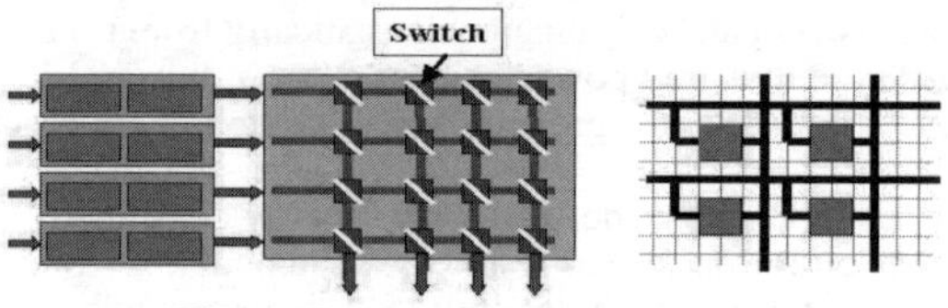

Figure 5: Crossbar Switch Fabrics

Crossbar topology uses space division multiplexing for input-output connection. Every input-output connection has its own dedicated data path, therefore, crossbar is *interconnect contention* free.

The node switch on the crosspoint of crossbar network can be a simple CMOS pass gate, or a tri-state CMOS buffer. Both are relatively simple compared to the node switches used in other network topologies.

Every bit will propagate throughout the long interconnect wires that connect to the input port (the row interconnect, in Fig. 5) and output port (the column interconnect). It also toggles the input gates of all the node switches connected to the same row. The load on the input port is the total of the wire capacitance and the sum of all input capacitances of N switches.

Some crossbar switch networks use buffers at every crosspoint to solve the *destination contention* problems. As discussed earlier in this paper, we assume the *destination contention* is already resolved by the arbiter, i.e., there are no buffers needed in the power modeling of crossbar network.

A Thompson embedding of crossbar switch network is also shown in Fig. 5. Under the Thompson model, the mapping is straightforward and the total bit energy for the crossbar switch fabrics is described in Eq. 3.

$$E_{bit_{crossbar}} = N \times E_{S_{bit}} + 8N \times E_{T_{bit}} \qquad (3)$$

where $E_{S_{bit}}$ is the bit energy for the switch and $E_{T_{bit}}$ is the bit energy of a Thompson grid wire. Each crossbar node switch has degree of 4, however, two of the ports are used as feed-through ports, so we assume it occupies 2×2 Thompson grids. Two extra grids are also needed for horizontal and vertical interconnects for each node switch. Each bit traveling from input i to output j will propagate both the interconnect wires connected to the input port i and output port j, each of the interconnect has length of $4N$ of a Thompson grid.

Crossbar has the benefit of being free of *interconnect contention*. However, the bit energy will increase linearly with the number of input and output ports N. The power consumption will be very high for switch fabrics with large port numbers.

4.2 Fully-Connected Network

An $N \times N$ fully-connected network uses MUXes to aggregate every input to the output (Fig. 6). Each MUX is controlled by the arbiter that determines which input should be directed to the output.

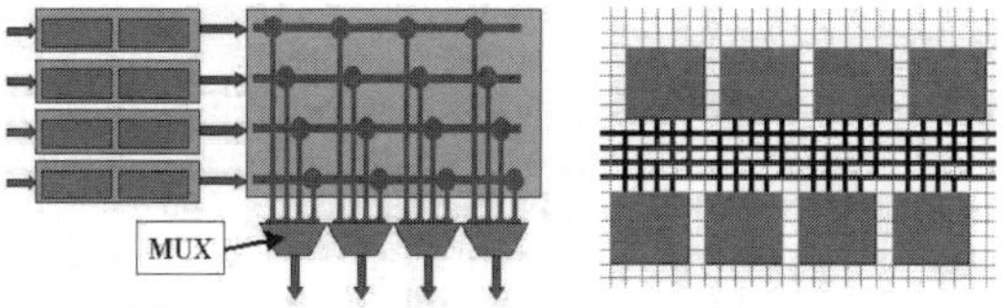

Figure 6: Fully Connected Switch Fabrics

Similar to the crossbar network (fully connected switch network is also often referred as crossbar), in fully connected switch network, each source-destination connection has its dedicated data path. The network is also free of *interconnect contention*. There are no internal buffers needed in its power modeling.

The bit energy for a fully connected switch network is consumed on the interconnect wires and the MUXes. A Thompson embedding is shown in Fig. 6, where the MUXes are placed in a double-row fashion. The bit energy can be estimated with the following equation (Eq. 4).

$$E_{bit_{fullyconn}} = E_{S_{bit}} + \frac{1}{2} N \times N \times E_{T_{bit}} \qquad (4)$$

where $E_{S_{bit}}$ is the bit energy on the MUX. Compared with crossbar switch, each bit only consumes energy on one of the MUXes, instead of N switches as in the case of crossbar. However, the N-input MUX has more complicated logic gates, and its power consumption and complexity scale up with the number of inputs N.

4.3 Banyan Network

Banyan network (Fig. 7) is an isomorphic variation of Butterfly topology. It has $N = 2^n$ inputs and $N = 2^n$ outputs, where n is called the dimension of Banyan. It has total of $\frac{1}{2} N \log_2 N$ switches in n stages, each stage is referred as stage i where $0 \leq i < n$ [1].

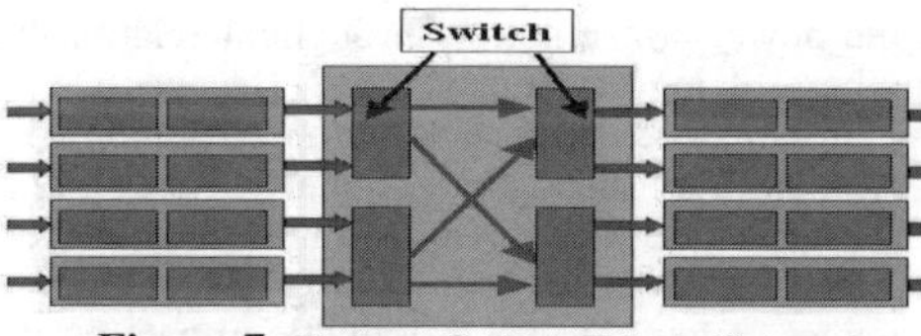

Figure 7: Banyan Switch Fabric Network

The switch used in a Banyan network is a binary switch, as described in Section 3. It has two inputs and two outputs. The input packet with destination bit *"0"* or *"1"* goes to output *"0"* and *"1"* respectively. Stage i in the Banyan network checks the ith bit of the destination address of the packet, therefore, the packet will be routed automatically from stage 0 to stage $n-1$. This routing scheme is also called self-routing switch fabrics. If two input packets have the same destination bit coming at the same time, one of the input packets will be buffered.

Banyan network has *interconnect contention* problems [1]. The same interconnect might be shared by different data paths. A buffer is needed at each internal node switch.

The binary switches in Banyan network have more complex logic as compared to the crosspoint node switches in crossbar network. A binary switch also consumes more power when bit is switched from the input port to the output port.

A simple Thompson embedding of Banyan network is shown in Fig. 4. Detailed analysis of Thompson embedding of Banyan isomorphic networks can be found in [7]. Using the Thompson model, the longest interconnect wire-length for stage i of Banyan network can be estimated as 4×2^i Thompson grid. Different input/output connections have different interconnect data paths. The worst case bit energy of Banyan network (the longest interconnects a bit may travel) can be estimated using Eq. 5 below.

$$E_{bit_{Banyan}} = \sum_{i=0}^{n-1} q_i E_{B_{bit}} + 4 \sum_{i=0}^{n-1} 2^i E_{T_{bit}} + n E_{S_{bit}} \qquad (5)$$

where $N \geq 2$ and $n \geq 1$. q_i has the value of 0 or 1. When there is a contention at stage i, q_i is 1, otherwise it is 0. The value of q_i is determined by the contentions between packets on the interconnect.

4.4 Batcher-Banyan Network

To solve the *interconnect contention* problem, the Batcher-Banyan network architecture is introduced, as shown in Fig. 8. It consists of a Batcher sorting network, followed by the Banyan network. After sorting network, each input-output connection will have its own dedicated path, therefore there is no *interconnect contention* [1].

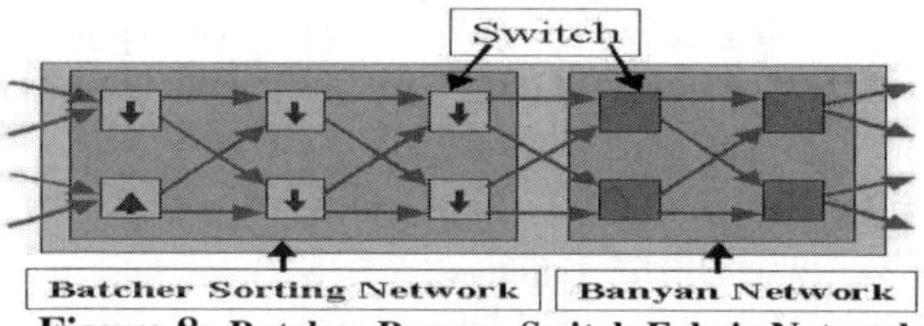

Figure 8: Batcher-Banyan Switch Fabric Network

Although Batcher-Banyan network solves the *interconnect contention* problem, it pays the price by increasing the number of stages between the inputs and outputs. It has total of $\frac{1}{2}(log_2 N)(log_2 N + 1)$stages, which will in turn increase the bit energy consumed on switches and interconnect wires.

The Thompson embedding of Batcher-Banyan is very similar to Banyan network, except it has more stages. The worst case bit energy for Batcher-Banyan network can be expressed in the following equation (Eq. 6).

$$E_{bit_{Batcher}} = 4 \sum_{j=0}^{n-1} \sum_{i=0}^{j} 2^i E_{T_{bit}} + 4 \sum_{i=0}^{n-1} 2^i E_{T_{bit}} \qquad (6)$$

$$+ \frac{1}{2} n(n+1) E_{SS_{bit}} + n E_{SB_{bit}} \qquad (7)$$

where $N \geq 4$ and $n \geq 2$. Because the sorting switch used in Batcher sorting network is different from the binary switch used in Banyan network. we denote $E_{SS_{bit}}$ for the bit energy of sorting switches and $E_{SB_{bit}}$ for the bit energy of binary switches.

5. BIT-ENERGY CALCULATION AND EXPERIMENTS

In this paper, we present a framework to estimate switch fabric power consumption. The calculation and results are based on case studies. The accurate values of each parameter for a specific switch fabric implementation depend on particular circuit design techniques and technologies. However, the methodology introduced here can be applied in other cases.

5.1 Bit Energy Calculation

1. Bit energy of Node Switches

As discussed in Section 3, the bit energy of node switches is state-dependent, it depends on the presence or absence of the packets on other input ports. For a node switch with n-input ports, there are 2^n different input vectors with different bit energy values. Normally switch fabrics with a large input/output ports are constructed from node switches with smaller number of degrees (2×2 or 4×4), therefore, the vector number 2^n is not prohibitively large. In this paper, the bit energy is pre-calculated from Synopsys Power Compiler simulation. We build each of the node switches with 0.18μm libraries, apply different input vectors and calculate the average energy consumption on each bit. The circuits for each node switch range from a few hundred gates to 10K gates, therefore, the simulating time is very quick (tens of seconds, in most cases). The bit energy results for crossbar switches, the N-input MUXes, the Banyan binary switches, the Batcher sorting switches are listed in Table 1.

Table 1: Bit Energy Under Different Input Vectors

Switch Fabric Architectures	Input Vector	Bit Energy	Input Vector	Bit Energy
		$10^{-15} joule$		$10^{-15} joule$
Crossbar 1×1	[0]	0	[1]	220
Banyan 2×2	[0,0]	0	[0,1]	1080
	[1,0]	1080	[1,1]	1821
Batcher 2×2	[0,0]	0	[0,1]	1253
	[1,0]	1253	[1,1]	2025
N-input MUX	N = 4	431		
	N = 8	782		
	N = 16	1350		
	N = 32	2515		

The input vectors for the 2×2 switches in the above table indicate presence or absence of the packets on the corresponding input ports, e.g. [1,0] means only input port 0 has packet coming in. For the N-input MUXes, bit energy values are very close among different input vectors, but they increase with the number of inputs N, as shown in the table.

2. Bit Energy of Buffer Queues

We use SRAM as the shared buffers inside Banyan switch fabrics. We adopt techniques similar to those proposed by [8] and [9] to estimate the memory access power consumption. SRAMs with different sizes have different access time and current, therefore they have different memory access power dissipation. The buffer size at each node switch greatly affects the performance of Banyan switch. The trade-off analysis between buffer size and switch throughput is beyond the scope of this paper. Researches in [10] and [11] show that buffer size of a few packets will actually achieve ideal throughput under most network traffic conditions. In our experiments, we use 4K bit buffer queue for each Banyan node switch. Based on the buffer size of each switch, we calculate the size of the shared memory. The memory access energy consumption can then be estimated based on selected memory size.

An off-the-shelf $0.18\mu m$ 3.3V SRAM is used as a reference for bit energy estimation. The calculation of access energy is based on 133MHz operation with access time specified in the data sheet. The results are shown in Table 2.

Table 2: Buffer Bit Energy of $N \times N$ Banyan Network

In/Out Size	Number of Switches	Shared SRAM Size	Bit Energy ($10^{-12} joule$)
4×4	4	16K	140
8×8	12	48K	140
16×16	32	128K	154
32×32	80	320K	222

3. Bit Energy of Interconnect Wires

The length of the Thompson grid can be estimated from the bus pitch distance of the interconnect. In Thompson model, each interconnect is a signal bus and occupies one grid square. Assuming the bus width is 32 bit, in 0.18μm technology, the wire pitch of global buses is around 1μm, therefore, the Thompson grid is around 32μm. The interconnect wire capacitance can be calculated from the method presented in [5]. For a global wire in 0.18μm technology, the wire capacitance is around 0.50fF/μm. Using these estimations, under 3.3V, the bit energy on interconnect wire of a Thompson grid length $E_{T_{bit}} = 87 \times 10^{-15}$ joule.

Comparing the buffer bit energy $E_{B_{bit}}$ values in Table 2 with the interconnect wire bit energy $E_{T_{bit}}$ value calculated above, we can see that storing a packet in buffer consumes far more energy than transmitting the packet on interconnect wires. This "buffer penalty" indicates that energy consumed in buffers is a significant part of total energy consumption of switch fabrics, and the buffer energy will increase very fast as the packet flow throughput increases. Our experiments in Section 6 will show this result.

5.2 Simulation Platform

The bit energy introduced in Sections 3 and 4 is the energy consumption for one bit. In switch fabric interconnect networks, different packets will travel on different data paths. To calculate the total power consumption of the entire switch fabrics, we need to trace the dataflow of every packet and summarize the bit energy of every bit on nodes switches, internal buffers and interconnect wires. In this paper, we introduce a Simulink based bit-level multi-threading simulation platform.

The complete network router system is implemented in Simulink. The ingress process units, the egress process units, the global arbiter and different switch fabric architectures are all written in C++ and then compiled into Simulink S-functions. The switch fabric architecture is constructed hierarchically. The activities of every bit in the packet are traced at every node switches, every buffers as well as interconnect wires (all implemented in S-functions).

A TCP/IP packet traffic flow is generated as inputs for the network router. Because we only need to simulate the switching activities inside the switch fabrics, the packet payloads are random binary bits. The IP address of each packet has been translated into destination port address by the ingress process unit. The arbiter uses the first-come-first-serve arbitration with round robin policy. The destinations of the TCP/IP packets are random. We use input buffer scheme to store the packets when there is *destination contention*. The input buffers are located at each ingress process unit. Because the input buffers are outside the switch fabric network, they are not counted for switch fabric power consumption. The line data rate is assumed to be 100BaseT.

Power consumption is measured under three metrics: 1) different traffic throughput, 2) different switch fabric architectures, and 3) different numbers of ingress and egress ports.

We implement four switch fabric architectures with different numbers of input/output ports, namely, 4×4, 8×8, 16×16 and 32×32. Packet dataflow is generated at each input port. The throughput of the packet dataflow can be adjusted by controlling the packet generation intervals. The throughput is measured at the egress process units. The throughput indicates the traffic loads that flow through the switch fabric networks.

6. RESULTS AND ANALYSIS

Fig. 9 shows the power consumption of different switch fabric architectures under traffic throughput from 10% to 50%. The figure also shows the power consumption/throughput relationship under different numbers of ingress/egress ports. Because we use input buffering scheme to store the packets with *destination contention*, the theoretical maximum throughput is 58.6% (measured at egress ports). In reality, the 58.6% throughput is not achievable [1].

From these results, we have the following observations:

1) *Interconnect contention* has a dramatic impact on the power consumption of Banyan switch. Banyan switch has the lowest power consumption under low traffic throughput, as the throughput increases, the power consumption increases exponentially. This is caused by the "buffer penalty" as discussed in Section 5. However, as the number of ingress and egress ports increases, the interconnect wire-length and bit energy also increase, the "buffer penalty" domination will have less impact. This can be seen by comparing the "Banyan curve" in the figures with the curves of other architectures. Actually, in the 32×32 configuration, Banyan had the lowest power consumption when the traffic throughput is less than 35%.

2) *Fully connected switch* has the lowest power consumption among all four architectures with different numbers of ports. But the difference with Batcher-Banyan narrows down as the number of ports increases. This is because in switch fabrics with a small number of ports, the power consumption on the internal node switches dominates, as the switch fabrics is getting bigger (more ports), interconnect power consumption will dominate.

The impact of number of ports on power can be better seen from Fig. 10. In this figure, power consumption of each architecture is compared with different number of ports. The traffic throughput is 50%. The power consumption difference between fully con-

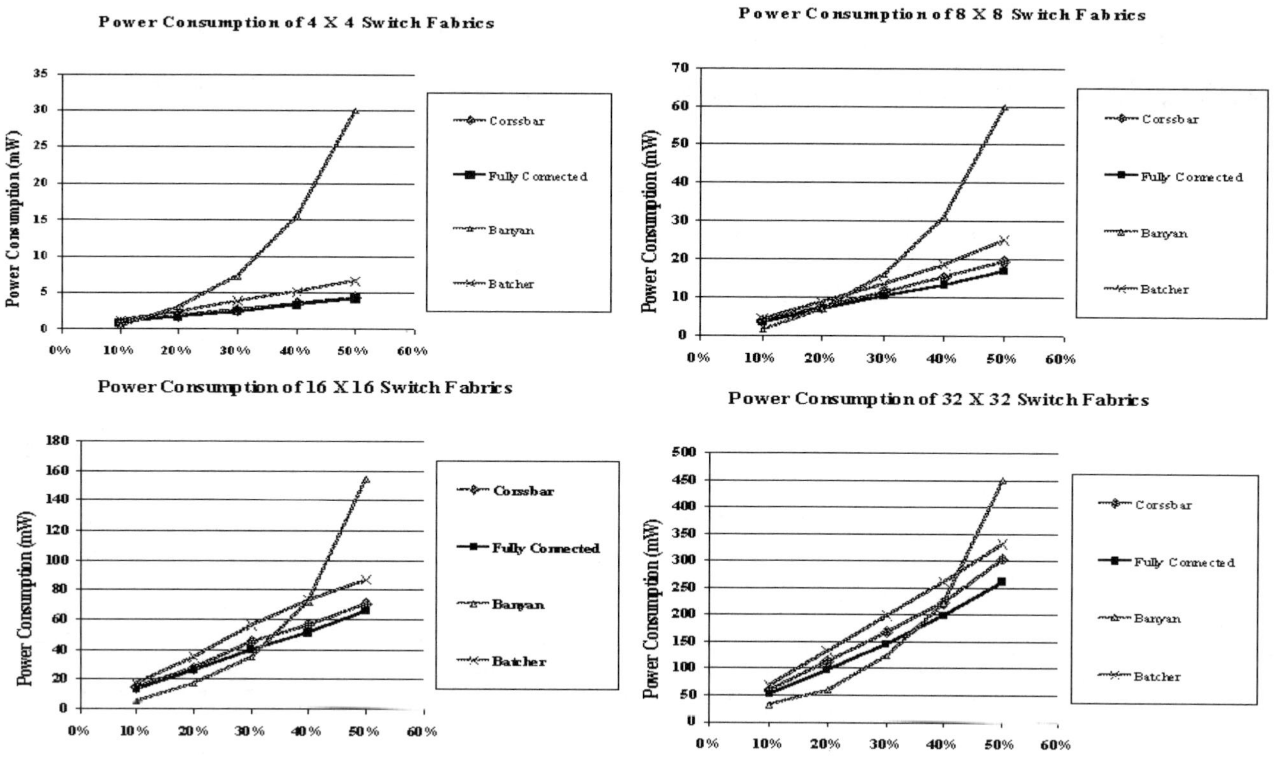

Figure 9: Power Consumption Under Different Traffic Throughput

nected switch and Batcher-Banyan switch decreases from 37% in 4×4 switches to 20% in 32×32 switches.

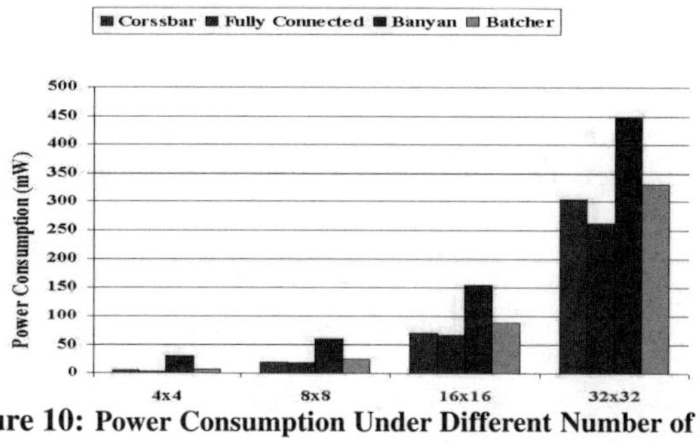

Figure 10: Power Consumption Under Different Number of Ports

3) The power consumption of crossbar, fully connected and Batcher-Banyan networks increases almost linearly with the increase of the traffic throughput, except the Banyan network, which is dominated by the power consumption on internal buffers. This observation is not surprising because data flow with higher throughput needs more power to process the bits along the data path.

7. CONCLUSION

A power consumption estimation framework is proposed in this paper. Using the framework, we analyze the power consumption of different switch fabric architectures used in network routers. From our analysis, we have the following conclusions:

1) *Interconnect contention* (internal blocking) induces significant power consumption on internal buffers, and the power consumption on buffers will increase sharply as throughput increases.

2) For switch fabrics with a small number of ports, internal node switches dominate the power consumption, for switch fabrics with a larger number of ports (e.g. beyond 32×32), interconnect wires will gradually dominate the power consumption.

The power consumption estimation in the results only represents the power dissipated on switch fabrics, the total power consumption for the entire network router needs to include other components as well, e.g. the ingress units, egress units, arbitration logics, peripheral logics, and IOs. The analysis and comparison above are only based on case studies with specific technology parameters. Different implementations of switch fabrics will have different comparison results. However, the methodology presented in this paper is general, it can be applied to different switch fabric designs.

8. ACKNOWLEDGMENTS

This research was supported by GSRC/MARCO.

9. REFERENCES

[1] H. Jonathan Chao, Cheuk H. Lam, Eiji Oki "Broadband Packet Switching Technologies: A Practical Guide to ATM Switches and IP Routers" *Wiley-Interscience 2001*

[2] Wassal, A.G.; Hasan, M.A. "Low-power system-level design of VLSI packet switching fabrics" *CAD of Integrated Circuits and Systems, IEEE Transactions on* , June 2001

[3] C. Patel, S. Chai, S. Yalamanchili, D. Shimmel, "Power constrained design of multiprocessor interconnection networks," *Int'l Conference on Computer Design*, 1997.

[4] Langen, D.; Brinkmann, A.; Ruckert, U. "High level estimation of the area and power consumption of on-chip interconnects," *IEEE Int'l ASIC/SOC Conference, 2000.*

[5] R. Ho, K. Mai, M. Horowitz, "The Future of wires," *Proceedings of the IEEE*, April 2001.

[6] C. D. Thompson, "A Complexity Theory for VLSI, PhD thesis", *Carnegie-Mellon University*, August 1980

[7] Andre DeHon, "Compact, Multilayer Layout for Butterfly Fat-free" 12th ACM Sym. on Parallel Algorithms and Architectures, 2000

[8] E. Geethanjali, N. Vijaykrishnan, M. Kandemir, M. J. Irwin, "Memory System Energy: Influence of Hardware-Software Optimizations", Int'l Sym on Low Power Design and Electronics, July 2000.

[9] Wen-Tsong Shiue; Chakrabarti, C. "Memory exploration for low power, embedded systems", 36th Design Automation Conference, 1999. Proceedings.

[10] Moustafa A. Youssef, M. N. El-Derini and H. H. Aly "Structure and Performance Evaluation of a Replicated Banyan Network Based ATM Switch" IEEE Sym. on Computers and Communications 1999

[11] Oktug, S.F.; Caglayan, M.U. "Design and performance evaluation of a banyan network based interconnection structure for ATM switches", Selected Areas in Communications, IEEE Journal on , June 1997

Memory Optimization in Single Chip Network Switch Fabrics *

David Whelihan, Herman Schmit
Department of Electrical and Computer Engineering
Carnegie Mellon University, Pittsburgh,PA 15213
{whelihan,herman}@ece.cmu.edu

ABSTRACT

Moving high bandwidth (10Gb/s+) network switches from the large scale, rack mount design space to the single chip design space requires a re-evaluation of the overall design requirements. In this paper, we explore the design space for these single chip devices by evaluating the ITRS. We find that unlike ten years ago when interconnect was scarce, the limiting factor in today's designs is on-chip memory. We then discuss an architectural technique for maximizing the effectiveness of queue memory in a single chip switch. Next, we show simulation results that indicate that a more than two order of magnitude improvement in dropped packet probability can be achieved by re-distributing memory and allowing sharing between the switch's ports. Finally, we evaluate the cost of the optimized architecture in terms of other on-chip resources.

Categories and Subject Descriptors

B.3.2 [**Harware**]: Design Styles

General Terms

Design, Performance

1. INTRODUCTION

This paper explores the architecture of single chip network switch fabrics. Single-chip network switches are less costly and consume less power than today's rack-level and board-level switches, which will mean higher bandwidth and profitability for network users and suppliers. The levels of integration achievable by modern process technologies significantly alter the design space for network switches, however. This paper first evaluates the new realities for switch design. We conclude that unlike a rack-level system, wires are abundant. The limited resource in a single-chip switch is memory, which is used for queuing data throughout the chip.

*This work was funded by the Pittsburgh Digital Greenhouse under contract EDT-99-1.

This work was inspired by Dally's exploration of topology in message-passing networks[1]. Since our research goal was the exploration of the design space for single chip network switches, we re-evaluated the resources available to an SOC switch designer, establishing which are critical for single a chip network switch now and in the next 7 years. We accomplish this by analyzing data provided by the International Technology Roadmap for Semiconductors(ITRS)[10] with the needs of a scalable, high bandwidth switch architecture in mind. We then describe an architectural technique by which a designer could optimize the efficiency of critical on-chip resources. Based on this, we present a parameterizable class of network switch fabrics and a simulator capable of evaluating instances of this class of fabrics. We then use this simulator to determine the optimal memory organization, using dropped packets caused by overflowing queues as an optimization metric. Finally, we show simulation data supporting our design choices.

2. CRITICAL RESOURCES

A network switch, in its most basic form, can be viewed as a large conglomeration of inputs and outputs. The inputs absorb data packets of either fixed length(as in ATM) or variable length(IP)[7], and send them to the proper output. Further, such a system would require that a memory be placed at each output port to buffer in the case where two or more input packets are switched to the same output. Figure 1 illustrates this

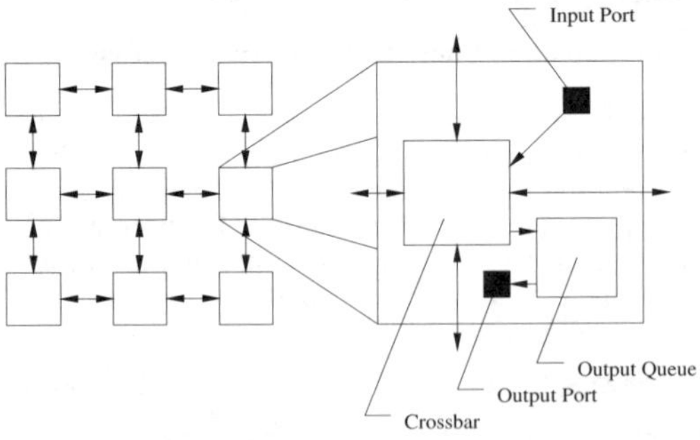

Figure 1: Array model of a single-chip network switch comprised of individual *nodes*

model, in which input and output blocks are grouped together with output queue memory and a switch-box unit. These *nodes* are then grouped in a grid with metal interconnect between them. This is consistent with earlier work[1] and due to its regularity supports scalability. For our chip resource analysis, we concentrate on three main areas: interconnect, memory, and I/O bandwidth.

2.1 I/O Bandwidth

The amount of on-chip bandwidth is irrelevant if the switch cannot be fed data from off chip. The number of I/O pads allocated to a particular port will depend on the incoming data rate, the maximum speed of an I/O Buffer, and the requirements of the signaling protocol used. As for the latter, we assumed Low Voltage Differential Signaling(LVDS)[6], in which each individual input or output requires two I/O pads. This was chosen because it is currently capable of GHz speeds, and it allowed us to be conservative in our IOB utilization estimates. Figure 2a shows the maximum number of nodes for ports running at both 2.4Gb/s(OC-48) and 10Gb/s(OC-192).[1] These values indicate the absolute bandwidth ceiling. Though this does not take into account control and strobe inputs, it is still true that by 2008 the IOB bandwidth will allow chips with over 100 bi-directional ports running at 10Gb/s[10].

2.2 Interconnect

The object of this particular analysis was to determine to a reasonable degree of accuracy the maximum *bi-section bandwidth* of a routing channel (in this case, 1mm was chosen). In this analysis, the first two metal layers are assumed to be used for local routing, and power and ground rails are routed on high level auxiliary layers. The intermediate layers are allocated evenly to north-south and east-west routing with the following rule:

$$N_{layers} = \left\lfloor \frac{N_{int} - 2}{2} \right\rfloor \qquad (1)$$

where N_{int} is the total number of intermediate wiring layers predicted by the ITRS. Figure 2b shows the bandwidth ceiling for a 1mm bisection of a chip through 2008. We have also graphed two demand lines, calculated by multiplying the number of ports per year from Figure 2a by their respective line speeds. The demand lines show that a 1mm wide routing channel could transmit all of the required data bandwidth for such a chip. This means that we are IO limited and not bound by on-chip wiring bandwidth. This has been made possible by the substantial decrease in wire pitch, the increase of on-chip speeds (easily 2500 MHz by 2008), and the increase in the number of metal layers (9 intermediate layers by 2008).

2.3 Memory

Currently, due to advances in embedded DRAM technology, it is possible to efficiently integrate both DRAM and processing logic efficiently on the same die. In [11], a 1MB embedded DRAM with a footprint of $19mm^2$ is described. This device has a large line width(1024 bits) as well as a 3.7ns address access time, putting the per bit access time well within the limits for 10Gb/s port speeds. The DRAM cells, while about 2.5 times larger than a comparable cell made in a DRAM only process[14][15], still achieve roughly seven times the density of an SRAM cell.

Using this work as a baseline, to first order we can graph memory sizes for the maximum chip size predicted by the ITRS by extrapolating the cell scaling for both the DRAM and the hypothetical SRAM that is seven times larger. Figure 2c shows curves for both SRAM and DRAM on an $800mm^2$ die for four process nodes.[2] The cell size in future processes was extrapolated by simply scaling based on the shrinking feature size.

[1]The jump in 2008 is due to the IOB speed increasing such that it is almost an exact multiple of the data frequency.

[2]The ITRS specifies 800 mm^2 as the maximum chip size through 2008.

2.4 Analysis

While the absolute IOB bandwidth limit does provide a hard cap on the throughput of our device, it still allows for a high aggregate bandwidth (1Tb+). Since this is a hard technology barrier, it does not lend itself well to architectural improvement.

As for on-chip interconnect, it is apparent that the situation of ten years ago is no longer the case. Cross-chip metal interconnect running at well over 2 GHz will be in abundance. While it is true that this is only a first order approximation, it seems obvious that even with wires at greater than minimum pitch, and with wire speeds halved the bi-section bandwidth still outpaces the bandwidth capability of the IOBs.

We have determined that *memory* is the most critical chip resource today. We justify this by pointing out that commercial high bandwidth switches routinely have more than 1 MB per port of output queue memory[18]. Also, the worst case traffic at any one output port will increase with the number of input ports in the system, which in today's embedded DRAM technology would occupy the majority of an $800mm^2$ chip. The memory requirement for queueing under this assumption is plotted as a dashed line in figure 2c for all process nodes through 2008 [3]. This line was calculated by multiplying the number of possible ports from Figure 2a by 1MB. We see that while there is always enough DRAM availability, 1MB per port generally requires at least half of an $800mm^2$ chip. This problem is compounded by the fact that memory storage is required for other tasks on a switch:

- Keeping track of packet priority when service guarantees have been made.
- Keeping track of packet time-stamps to ensure that the oldest packet goes out first.
- Routing tables. Storage for upwards of 500k routes is not unreasonable in a high bandwidth switch.

3. ARCHITECTURES

Within the last few years, several research teams have devised and fabricated single chip network switches and message routers [5][8][9][17]. Most of the previous work has used a shared buffer approach, in which a centralized block of memory is partitioned into multiple *output queues*. This is the most efficient use of memory, as a particular output does not truly have a fixed queue allocation (up to the limits of the physical memory). One could envision this as one output queue *borrowing* unused memory from other, less heavily utilized queues. An unfortunate side affect of this is that an output that fills the entire memory with its data at time t could starve the other outputs for queue memory at time $t+1$ unless per port maximum allocations are enforced. Scalability is yet another problem that the shared buffer approach must contend with. As the number of Input/Output ports increases on a switch, so too does the chance of contention for the shared memory. One solution to this problem is to add more ports[4] to the shared memory, allowing multiple simultaneous accesses. Unfortunately, the area and power dissipation increase prohibitively when many simultaneous memory accesses are necessary.[12].

The other extreme involves the allocation of separate output queue memory to every output port. This has the benefit of

[3]2000:180nm,2002:130nm,2005:100nm,2008:70nm

[4]this is different from the definition of *port* previously discussed.

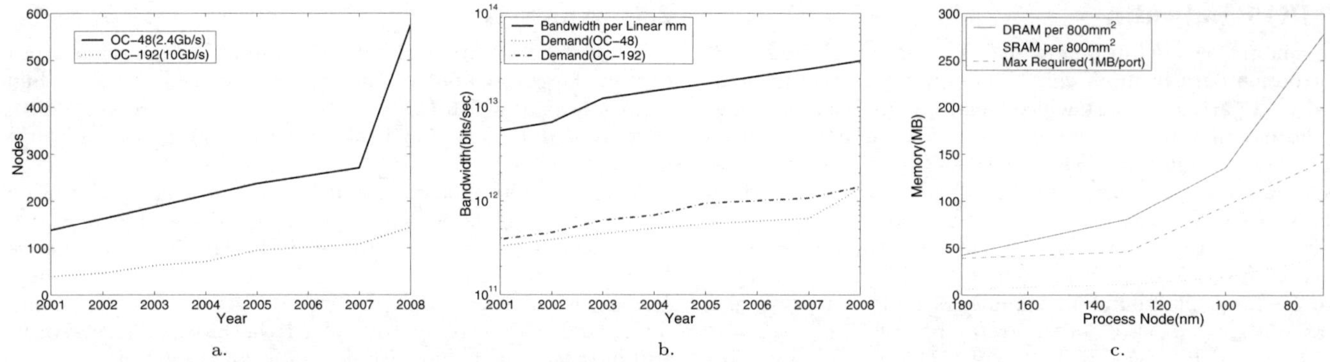

Figure 2: a) Maximum number of bi-directional ports. b) Maximum bandwidth over a 1mm bi-section. c) Memory limits for an 800mm^2 chip.

more predictable output performance, since each output port is guaranteed a certain amount of memory. The major problem with this scheme is that one must add enough memory to *all* of the queues to ensure that any *one* can handle the target peak load, resulting in tremendous inefficiency.

We decided that a regular structure of identical nodes (k-ary n-cubes) best fit our scalability requirement, but these structures necessitate spreading out memory resources[5], leading to inefficiency in silicon real estate allocation. We would rather devote on-chip space to more advanced Quality Of Service(QOS) hardware[6].

The memory organization we used in our test architecture was based on the concept of wormhole routing[2][3][4]. In wormhole routing, flows of data are broken up into blocks called *flits*. A flit is defined as the smallest bufferable chunk of data. Wormhole routing is a circuit switched protocol in which a path from the input port to the output port is established, and persists until the packet has been received by the destination port in the network. The path across chip is comprised of a number of *hops* in intermediate nodes. The nodes only buffer data if a blockage occurs along the path, either at the terminal end (due to a lack of buffer space), or along the route (due to a lack of routing channels). In either case, the flits are buffered along the route in the intermediate nodes. The tendency to buffer en route makes wormhole routing optimal for our situation. If the output queue memory at the terminal node overflows, a packet need not be dropped, since it can temporarily share memory with the ports along its path. Our model therefore incorporates *transit buffers*, which allow wormhole buffering along a packet's path through the fabric. The buffers are only accessible to packets whose destination is *not* the current node.

It is impractical to allocate memory on a flit-by-flit basis, as the line widths for an embedded memory would most likely be larger than a flit. This is especially true for embedded DRAM, which would require a large line size to mitigate its slower speed. We would therefore allocate memory in blocks sized as an integer number of flits. Unfortunately, if the minimum block size is 20 flits, and a packet is only 10 flits long, another 10 flits worth of memory are wasted. Another problem is the choice of the amount of memory that any node can grant a transitory(not destined for that node) packet. Intuitively, since the goal of this architecture is to maximize memory sharing across a chip, the optimum situation would be one in which every packet was strung out evenly across its path length:

$$N_{bph} = \frac{L_{packet} - B_{oq}}{N_{hops}} \qquad (2)$$

where N_{bph} is the number of flits allocated per hop, L_{packet} is the packet length in flits, B_{oq} is the number of flits allocated in the output queue, and N_{hops} is the number of intermediate nodes a packet's circuit traverses.

4. SIMULATION

We used a software simulator written in C++ to explore the memory architecture/allocation space. Our simulator is capable of running systems with an arbitrary number of ports(*switch nodes*) and connectivity between the ports. The simulator currently only supports toroidal(k-ary n-cube[1]) topologies of arbitrary degree, though the overall simulator architecture does not preclude other topologies[16]. The simulator can achieve speeds of 50-100k cycles per minute on a 64 node switch.

The amounts and ratios of the two types of memory in our system(output queue and transit buffer) are parameterizable, as is the minimum block allocation for each. Figure 3 illustrates the functionality of our prototype network. The bottom of the figure indicates each node's position in the network. The spaces between the lightly shaded node boxes on the upper part of the figure represent interconnect with data in flight between nodes. In this example time moves forward as one moves down the figure. Packets are routed adaptively via minimal paths.

At time t_0, a header is received at node 1. After processing(assumed to take zero time for simplicity), the incident node chooses one of a number of *inter-node channels* to send the incoming packet out on. It is important to note that two adjacent nodes may have many such channels between them, but only one is used per packet. The incident node sends a *connect* message (along with the header) to the next node in this packets on-chip route. This message is propagated along the shortest path to the target node(8), until at t_3, a connection message reaches the destination node. At this point a circuit is established that will remain while the packet is being transmitted to the destination node. At t_1, the first data block or *flit* arrives at the input port. This flit is blocked pending the reception of an *unblock* message from the next node in the path, which is received at time t_2.

On reception of the unblock message, the first node transmits a data flit, while buffering the arriving data flit which is waiting for the connection to unblock. The transmitted flit arrives

[5]Unless memory is shared between clusters of nodes, which is not explored in this paper

[6]By this we mean structures such as per-session queues and class based scheduling

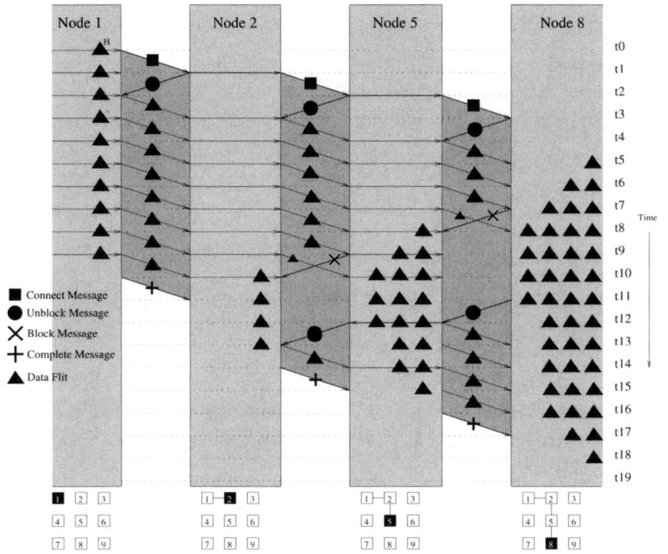

Figure 3: Network functionality

at the first intermediate node(2) before the entire connection is established through to the destination node. In this example, the packet is being buffered at the output node(8) because another packet is ahead of it in line to exit. Therefore, the first data flit is received and buffered at time t_5. This buffering continues until time t_7, when the packet is one flit away from filling its output queue allocation(in this case 4 flits) in node 8. At this time, the destination port(8) sends a *block* message to the second to last node in the path(5). Though it subsequently receives another flit at t_8, fully utilizing its output queue allocation, further transmissions are stopped and the next to last node begins to buffer in its transit memory. The second to last node(5) in turn buffers until its allocation is filled at time t_{10} when the second node in the path(2), after receiving a block message, begins to buffer.

One cycle later, at time t_{11}, node 2 receives a *completion* message from the first node(1). The first node has at that point finished transmitting data, and will subsequently deallocate the *channel* the packet was using, freeing it for use with other packets. The dark gray areas of the diagram represent the lifetime of the inter-node channels.

In this example, the output port on the last node(8) is freed simultaneously with the arrival of the completion message at the second node(2). An unblock message is propagated backwards to the first active node in the chain, while flits are being sent out the output port in order. If this had not been simultaneous, the packet and all of its allocated resources(channels and buffer memory) would have remained static until the output port was freed. At time t_{14}, the second node(2) transmits its final flit, which is followed one cycle later by a completion message, and the freeing of all resources allocated to the packet. The packet is transmitted in this manner until the last flit has passed through the output queue.

This example does not cover packet dropping. In our simulator, the packet in the previous example would have been dropped if the packet had been blocked until the first node had run out of transit space. At that time, there is no previous node to block, and the packet is dropped by sending a *drop* message forward from the incident port all the way to the end of the circuit. The simulator currently produces data such as the number of flits in, out and dropped, system utilization, average memory used per cycle, average latency, as well as detailed statistics on the characteristics of dropped packets (average length, number of hops, and memory allocation). The later statistics will be integral to the following analysis.

5. ANALYSIS

The purpose of this section is to answer the following three questions:

- What is the optimal ratio of output queue memory to transit buffer memory?
- What is the optimal policy for distributing transit memory to packets blocked at intermediate nodes?
- What do these policies cost us in terms of interconnect?

The first question aims to discover the amount of each of the two types of memory that must be at each node. To answer this, we ran simulations in which the total node memory stayed constant, but the ratio of output memory to transit memory changed. For the second question, we will show results for two different policies: *Distributed Packets*, in which the switch makes its best effort to spread a buffered packet across the chip, and *Greedy Packets*, in which a hard allocation is granted to each packet in every node it passes through, regardless of the packet's length. We will analyze the results of the two aforementioned studies to determine optimal points from the memory utilization, and dropped packet probability perspective. The third question will be answered by looking at wire utilization metrics produced by our simulator.

As we have been assuming that the the usage model for a chip of this type will be that of a high bandwidth IP switch, we used a packet length distribution from the Cooperative Association for Internet Data Analysis(CAIDA)[13], which is based on real network traces. Packet arrival rates were drawn from a Poisson process, and the destination distribution in the network of an incoming packet was uniform. The network for the following analysis is a 4-ary 3-cube(64 node 3-d torus), and the system utilization is 93%[7].

In the following analysis, we concentrate on the *hop length* of dropped packets. We define the hop length as the number of nodes that a packet must traverse before it reaches its output port. This value is important because we expect that a packet with a large number of intermediate nodes, and therefore a high probability of finding buffer space, will be less likely to have been dropped than a packet with a short path length when transit buffers are implemented.

Figure 4 shows baseline data for a system in which nodes only have dedicated output queue memory(there are no transit buffers, and all buffering for an output port takes place within that port). In this simulation, as with all the rest, interconnect bandwidth is considered infinite, therefore the amount of dropped packets is related only to the output queue depth. Figure 4b shows the aggregate packet drop rate for the 64 node system for output queue depths ranging from 7000 flits(28kB) to 17000 flits(68kB). As we would expect, the packet drop rate decreases as the amount of available output queue memory increases. Figure 4a breaks down the dropped packets by their on-chip path length. Since our destination distribution is uniform, the longest path possible in a 4-ary 3-cube is six, and

[7] System utilization is calculated by dividing the number of flits into the system by the number of nodes times the number of cycles run

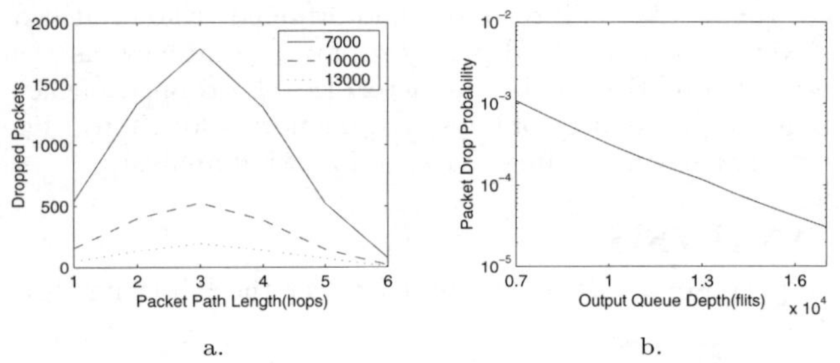

Figure 4: Baseline simulation; a) Dropped packets by path length for three different output queue depths(measured in flits). b) Dropped packet probability vs. output queue depth.

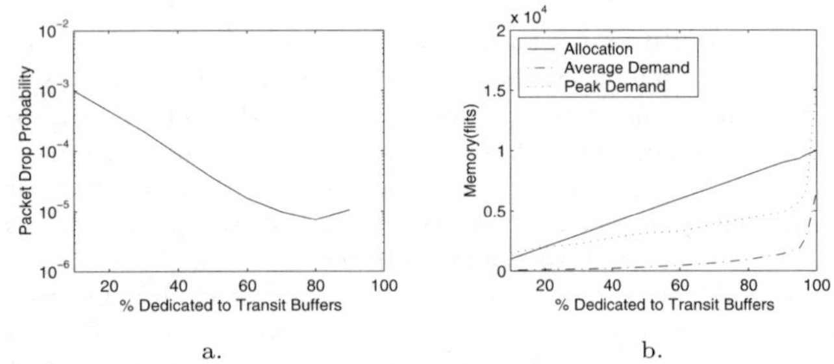

Figure 5: Distributed Packet Simulation; a) Dropped packet probability by memory ratio when total node memory is limited to 5k flits. b) Transit buffer utilization.

zero hop packets(packets whose ingress node is the same as their egress node) are disallowed, we see a Gaussian distribution centered at 3 hops for all three output queue depths.

5.1 Distributed Packets

To verify the assertion of equation (2), we allow intermediate nodes along a packet's path some extra information: the number of hops preceding them in the packet's path. When the terminal node in a 4 hop path blocks, the second to last node will only give enough buffer space to buffer 1/4 of the flits that have yet to be transmitted[8]. It is expected in this case that when the second to last node blocks, the preceding one will buffer 1/3 of the remaining traffic and so on, until the ingress node is required to buffer all of the rest of the packet in transit buffers (ideally 1/4 of the packet overflow). Figure 5a shows the dropped packet probability for different ratios of output queues to transit buffers when the *total* node memory is kept steady at 5k flits. We see a steady decline in the dropped packet probability until the 90% transit buffers point, where the probability has dropped down to the order of 10^{-6}. We see however, that a ratio of 1:4 output queues to transit buffers seems to be optimal. Interestingly, going beyond the 1:9 point, toward the 100% transit buffers point (no output buffering), the dropped packet probability sky-rockets to over 20%. Figure 5b shows both the peak and average transit buffer demands for this simulation, as well as the transit memory ceiling for each point defined by system parameters. For the majority of this plot, the peak remains below the maximum allocation. However, near the right side of the graph, we see a sharp, exponential increase in both the average and peak demands, ending with the peak at 15000 flits. It was at this point in the simulation that the packet dropping increased due to switch saturation.

[8]Since a four hop path will cross five nodes including the terminal one

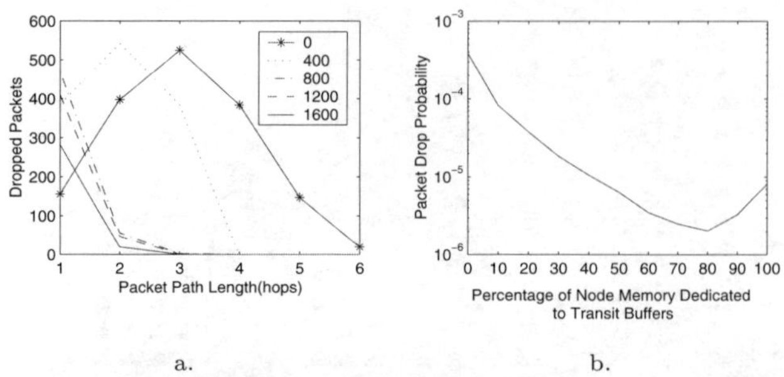

Figure 6: Greedy Packet Simulation; a) Packet drop rate vs. path length for 5 different per node transit buffer allowances at the 9:1 memory ratio. b) Packet drop probability for 11 memory ratios with total node memory 10k flits.

5.2 Greedy Packets

Though the memory allocation technique just discussed reduced the dropped packet probability significantly, the requirement that intermediate nodes store the number of nodes preceding them in each packet chain might be a prohibitive factor in the design. Therefore, we sought to investigate the effectiveness of fixed limits at each node.

Figure 6a shows the results of this experiment where each node has 10k flits of memory instead of the 5k of the previous study. For comparison, we have included the data for the case in which there are only output queues (no transit buffers). The other four curves represent progressively larger transit allocations to each packet that passes through a node. We find that as we increase the allocation, effectively allowing packets to become greedier, clustering closer to their destination ports, the dropped packet rate decreases significantly. Also, we notice that the mean path length for dropped packets shifts to the left until nearly all dropped packets have traversed only one hop. This is a good indicator that this scheme is working properly, as the probability of finding transit buffer space increases with path length. The greatest percent decrease in dropped packets occurs when 200 flits are allocated per node per packet.[9]

Figure 6b shows the overall dropped packet probability for 11 ratios of output queue memory to transit memory. In this experiment, the maximum allocation per node is set to be an entire packet(400 flits or 1600 bytes). We see a 2+ orders of magnitude improvement over the case with only output queues when the ratio is 1:4 output buffers to transit buffers. The graph also trends upward at ratios beyond 1:4. All dropped packets in this region only traveled one hop. These packets were dropped due to an inopportunity to get enough queue memory along their path caused by their short path length and not due to saturation, as in the distributed study.

5.3 Interconnect Cost

One downside to circuit-switched transmission protocols is their tendency to hold wire resources longer. In a Store-and-

[9]The packet length distribution specifies that this is larger then 80% of injected packets

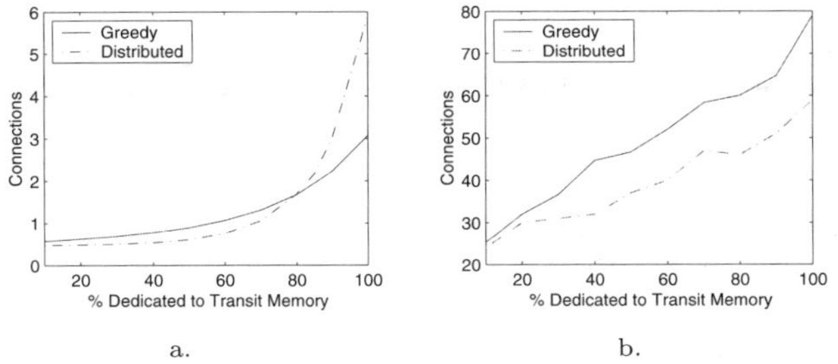

Figure 7: a)Average inter-node connections. b)Peak inter-node connections for greedy packet study.

Forward network, a path that is 3 hops long will hold 1/3 of the path at any one time, since packets are fully buffered at intermediate nodes before being granted a channel to the next node in the path.

Figure 7 shows the peak and average number of connections for the inter-node interfaces on-chip for both the distributed and greedy packet studies. When no transit buffers are allocated, and only output queues are used, we see a peak utilization of less than ten channels. As the memory ratio increases, so to does the peak channels used metric. The average number of connections(Figure 7a) increases exponentially as one moves in the direction of more transit buffers. It is important to note however, that the average utilization is quite low.

6. CONCLUSION

In this paper, we showed that wormhole routed architectures using distributed *transit buffers* for packet buffering allow us to attain a better packet drop probability by more than as 2 orders of magnitude over the case in which ports(nodes) only use output queues. This is due to the tendency to share transit memory between ports. We discussed two ways of allocating memory resources in this environment: *Distributed* and *Greedy*. We found that while the Distributed method achieved a better packet drop probability with less memory than the Greedy method, the distributed method did not react well to the memory saturation seen when the switch is using only transit buffers, and not output queues. In both cases, a switch with only transit memory did not perform as well as mixed designs. In both the Distributed and Greedy packet models, we saw that a 1:4 ratio of output queues to transit buffers is optimal. We also showed that the cost of such optimizations is more interconnect. However, the the number of physical/virtual channels actually *passing* data at any one time is quite small.

7. FUTURE WORK

The intent of this study was to begin with first principles(basic silicon constraints), rising up to a level in which on-chip structures were created to realize network switch functionality. For this reason, the study concentrated only on memory utilization. Obviously, many other variables exist, such as the traffic model. A uniform distribution with Poisson arrival rates is good for a first order survey, but what about bursty models(representing FTP), or hot-spot models? We believe based on extrapolation of existing data that such traffic will require more memory, but that the transit buffers should still be better utilized than output queues due to the variability of these models.

We would like to experiment with topologies other than k-ary n-cubes. How does a mesh topology(non-toroidal) affect the overall performance? Higher and lower dimensional systems(2-d or 4-d) in addition to other topologies are also interesting avenues for exploration. We have shown that longer paths and high connectivity can utilize transit buffers effectively, but how does this affect message latency? If the head of all messages reaches the output queue without blocking due to sufficient inter-node bandwidth, the latency is just a function of the load and queue depth on that particular output port, resulting in a seemingly ideal fabric.

8. REFERENCES

[1] W.J. Dally *Performance Analysis of k-ary n-cube Interconnection Networks*, IEEE Trans. Comput., vol. 39 no. 6, pp. 775-785, Jun 1990.

[2] W.J. Dally and C.L. Seitz *The Torus Routing Chip,* J. Distributed Computing, vol. 1, no. 3, pp. 187-196, 1986.

[3] L.M. Ni and P.K. McKinley *A Survey of Wormhole Routing Techniques in Direct Networks*, IEEE Computer, vol. 26, issue 2, pp. 62-76, Feb 1993.

[4] J. Duato, et. al. *Interconnection Networks.*, Los Alamitos: IEEE Computer Society Press, 1997.

[5] J. Rexford, et. al. *A Router Architecture for Real-Time Communication in Multicomputer Networks*, IEEE Trans. Comput., vol. 47, no. 10, pp. 1088-1100, Oct 1998.

[6] A. Boni, et. al. *LVDS I/O Interface for Gb/s-per-pin Operation in 0.35-μm CMOS,* IEEE Journ. Solid-State Circ., vol. 36, no. 4, pp. 706-711 Apr 2001.

[7] A. Tanenbaum *Computer Networks*, Englewood Cliffs:Prentice Hall, 1989.

[8] G. Kornaros, et. al. *Atlas I: Implementing a Single-chip ATM switch with Backpressure,* IEEE Micro, vol. 19, issue 1, pp. 30-34, Jan-Feb 1999.

[9] D. Pnevmatikatos and G. Kornaros *Atlas II: Optimizing a 10Gbps Single-Chip ATM switch,* Proc. IEEE Intl. ASIC/SOC Conf., pp. 142-146, 1999.

[10] Semiconductor Industry Association, *International Technology Roadmap for Semiconductors,* 1999.

[11] O. Takahashi, et. al. *1-GHz Fully Pipelined 3.7-ns Address Access Time 8kx1024 Embedded Synchronous DRAM Macro,* IEEE Journ. Solid-State Circ., vol. 35, issue. 11, pp. 1673-1679, Nov. 2000.

[12] S. Rixner, et. al. *Register Organization for Media Processing,* Proc. Intl. Symp. on High-Perf. Comp. Arch., pp. 375-386, 2000.

[13] Cooperative Association for Internet Data Analysis, http://www.caida.org

[14] P. W. Diodato *Embedded DRAM: More than Just a Memory,* IEEE Communications, vol. 38, issue 7, pp. 118-126, Jul. 2000.

[15] D. Keitel-Schultz and N. Wehn *Embedded DRAM development: Technolgy, physical design, and application issues,* IEEE Design & Test of Comp., vol. 18, issue. 3, pp. 7-15, May-Jun. 2001.

[16] F. Karim, et. al. *On-chip communication architecture for OC-768 network processors,* Proc. Design Autom. Conf. Las Vegas, NV, pp. 678-683, Jun. 2001.

[17] G.J. Jeong et. al. *Reconfigurable Shared Buffer ATM Switch,* Proc. IEEE APC Conf. on ASICs, pp. 327-330, 1999.

[18] Cisco Systems http://www.cisco.com/warp/public /cc/pd/ifaa/192/prodlit/cc1_ds.htm

Behavioral Modeling of (Coupled) Harmonic Oscillators

Piet Vanassche, Georges Gielen and Willy Sansen
Katholieke Universiteit Leuven - ESAT/MICAS
Kasteelpark Arenberg 10, B-3001 Leuven, Belgium
piet.vanassche@esat.kuleuven.ac.be

ABSTRACT

A new approach is presented for the construction of behavioral models for harmonic oscillators and sets of coupled harmonic oscillators. The models can be used for system-level simulations and trade-off analysis. Besides the steady-state behavior, the model also takes the transient behavior of the oscillation amplitudes and phase differences into account. This behavior is modeled using a set of linear differential equations. Their extraction from a netlist description is based upon the ideas of perturbation analysis and stochastic averaging. The modeling equations are valid in the neighbourhood of the oscillator's steady-state operating point and can be evaluated at very low computational cost. Another major advantage of the approach is that it explicitely separates the fast-varying steady-state behavior and the slow-varying transient behavior. This allows for straightforward application of multi-rate simulation techniques, greatly boosting simulation speeds. The technique is illustrated for a quadrature oscillator.

Categories and Subject Descriptors

G.1.7 [**Mathematics of Computing**]: Ordinary Differential Equations; I.6.5 [**Computing Methodologies**]: Model Development—*modeling methodologies*

General Terms

Algorithms, Theory

Keywords

Averaging, Behavioral modeling, Coupled harmonic oscillators, Perturbation theory

1. INTRODUCTION

Compact behavioral models of system-level building blocks, like filters, mixers and oscillators, are needed for several reasons. From a bottom-up point of view, they allow for efficient system-level verification. From a top-down point of view, they can be used for trade-off analysis and specification translation. They also assist in understanding the interactions that take place between the building block and its surroundings. To this account, behavioral models should both be accurate and allow efficient evaluation.

This paper presents an approach for behavioral modeling of harmonic oscillators and sets of coupled harmonic oscillators. The term harmonic oscillator means that the oscillation principle is based upon energy resonating between a magnetic field (an inductor) and an electric field (a capacitor). Harmonic oscillators are used in many RF-applications [12] because of their good linearity and phase noise performance. Coupled harmonic oscillators occur for example in zero- and low-IF receiver systems where one needs both an in-phase and a quadrature oscillation [2]. The modeling approach presented in this paper allows to capture the steady-state, transient and noise behavior of this type of oscillators.

When looking at the behavioral oscillator models published in litterature, oscillators are often treated as simple phase-integrating elements [3, 10]. In these models, amplitudes and phase differences are treated as constants. They neglect all of the oscillator's transients. Also, there is no systematic way of incorporating the influence of noise into the behavioral modeling equations. In recent years, more systematic modeling approaches have emerged as people investigated the oscillator's phase noise behavior [4, 6, 7]. That work, however, has only been concerned with the behavior of the (common-mode) phase θ. Although [7] also discusses amplitude variations, the behavior of the phase differences of coupled oscillators is never adressed. Another interesting approach towards oscillator analysis was presented in [11], where one uses a multivariate method to separate the fast-varying, steady-state, time constants and the slow-varying, transient, time constants. That work, however, only addresses the simulation issue and not the extraction of a compact behavioral model.

The work presented in this paper builds upon the multi-variate idea to generate separate models describing the fast-varying, steady-state, and the slow-varying, transient, behavior. More precisely, it is shown that the behavior of a set of coupled harmonic oscillators can be modeled as

$$v_{osc}(t) = A_k(t)\cos(\omega_{osc}t + \theta(t) + \Delta\phi_k(t)) \qquad (1)$$

where the amplitudes $A_k(t)$, the common-mode phase $\theta(t)$ and the phase differences $\Delta\phi_k(t)$ –measured with respect to the phase of some reference oscillator– are slow-varying compared to the oscillation time constant. In a first-order approximation, these slow-varying processes are shown to be governed by a system of linear ordinary differential equations (ODEs). This provides a very compact model which is evaluated at low computational cost. The model extraction procedure builds upon the ideas of perturbation theory [8] and averaging [1, 8, 9, 13]. It is found to be especially well suited for oscillators with a moderate to high Q-factor of the

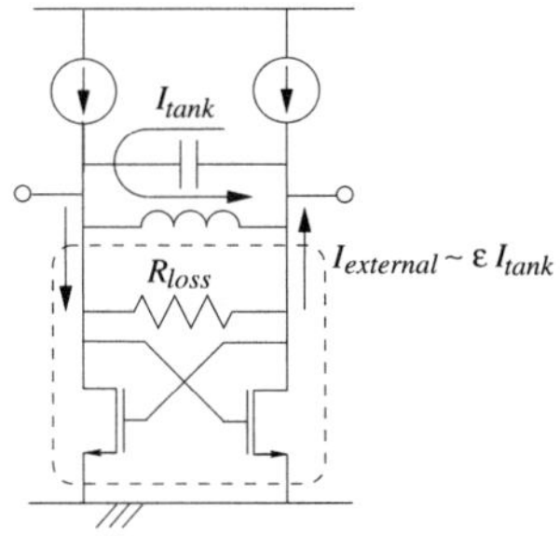

Figure 1: Decomposition of a harmonic oscillator into a lossless resonance tank and external sources changing the tank energy. Inductor and capacitor losses are considered to be part of these external sources and are modeled using the resistor R_{loss}.

resonance tank. Current testing has shown the approach to work for $Q \geq 5$. By separating fast-varying and slow-varying behavior, it allows for straightforward application of multi-rate simulation techniques, greatly boosting simulation speeds.

The remainder of this paper is organized as follows. Section 2 outlines the general ideas upon which the modeling approach is founded. In section 3, perturbation analysis of (coupled) harmonic oscillators allows us to construct a set of ODEs governing the behavior of the oscillation amplitudes and phase differences. In section 4, we apply the averaging technique to separate the fast-varying, steady-state, and slow-varying, transient, time constants. In section 5, we linearize the equations in the neighbourhood of the oscillator's operating point. Section 6, discusses how to extract the behavioral model from a given netlist, while section 7 experimentally demonstrates our approach on a harmonic quadrature oscillator. Finally, conclusions are drawn in section 8.

2. FUNDAMENTAL CONCEPTS

Separating the steady-state and transient behavior of a harmonic oscillator basically comes down to decomposing the oscillator circuit into a set of lossless resonance tanks and a set of "external" sources modifying the energy contained in the tanks. This decomposition is illustrated in Fig. 1 for a single harmonic oscillator. The part encircled by the dashed line is the collection of sources modifying the resonance tank energy. These sources also include inductor and capacitor losses which can be modeled using a parallel resistor R_{loss}.

In practical realizations, the energy injections by the external sources are kept as limited as possible. Typically, larger sources generate more noise, contributing for example to bad phase noise behavior. In most cases, the active elements are designed to be just large enough to compensate for the tank losses due to R_{loss}. This design implies that the currents $I_{external}$ injected by the external sources into the resonance tank are an order of magnitude smaller than the current already circulating within the tank, or

$$I_{external} \sim \epsilon I_{tank} \qquad (2)$$

with $\epsilon \ll 1$.

When left unperturbed, the resonant tank voltages and currents will settle onto a circular orbit in phase space. Such a solution of the unperturbed system is completely characterized by its amplitude and initial phase, or equivalently by the initial voltage v_0 over the capacitor and the initial current i_0 through the inductor. It can hence be represented as a single point in a two-dimensional space, for example S_1 in Fig. 2. Injection of the currents $I_{external}$ will however disturb the energy in the resonance tank, causing changes

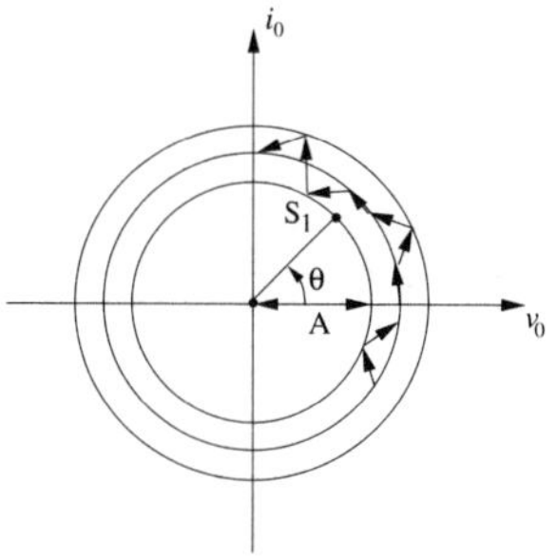

Figure 2: For a single harmonic oscillator, each solution of the *unperturbed* system is characterized by an amplitude A and an initial phase θ. Such a solution can hence be represented as a single point in a 2-dimensional space. The transient behavior of the *perturbed* system can then be considered as the oscillator "slipping" from one solution of the unperturbed system to another one as indicated by the arrows.

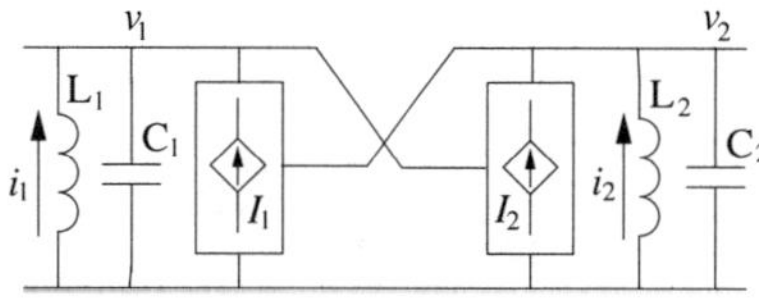

Figure 3: A system of two coupled harmonic oscillators can be decomposed into two independent resonance tanks which are coupled through voltage-controlled current sources.

in its oscillation amplitude and phase. It is hence the interaction between the resonance tank and the external sources which causes the oscillator's transient behavior. This transient behavior can be interpreted as the oscillator slipping from one solution of the unperturbed system to another. This behavior is illustrated in Fig. 2. Since the energy injected per oscillation period T is small, these transient changes will be slow compared to the period of oscillation. More precisely, these changes occur on a time scale T/ϵ, with ϵ determined by equation (2).

From the discussion above, it follows that the behavior of the oscillator is completely determined by the characteristics of the unperturbed system, the fast-oscillating resonance tank, and the equations characterizing the slow-varying behavior of its amplitude and phase. As will be discussed in section 4, the ODEs governing this slow-varying behavior can be separated from the fast steady-state oscillation using the averaging technique. This technique relies on the assumption that, over a single oscillation period, amplitudes and phases can be considered constant. This yields a model for the transient behavior that can be solved on a time scale T/ϵ.

The discussion above can easily be extended to a system of K coupled harmonic oscillators. In this case, the system is separated into K independent resonance tanks which are mutually coupled through controlled current sources. Fig. 3 illustrates this for $K = 2$. This template will serve as a starting point for our analysis in the next sections. In section 6, we will discuss how to map a given circuit topology on this template. Note that it is also possible to deal with other types of resonance tanks, like capacitor-inductor series connections. Adjustments are straightforward but are not discussed within this text.

3. PERTURBATION ANALYSIS

A system of K coupled harmonic oscillators, as depicted in Fig. 3

for $K = 2$, is described by the normalized system of equations

$$\begin{aligned} \frac{d\bar{i}_k}{d\bar{t}} &= \bar{v}_k + \epsilon\delta_k \frac{d\bar{i}_k}{d\bar{t}} \\ \frac{d\bar{v}_k}{d\bar{t}} &= -\bar{i}_k - \epsilon \bar{I}_k\left(\bar{v}_1, \frac{d\bar{v}_1}{dt}, \bar{v}_2, \frac{d\bar{v}_2}{dt}, \ldots, \bar{t}\right) + \epsilon\delta_k \frac{d\bar{v}_k}{d\bar{t}} \end{aligned} \quad (3)$$

for $k = 1, \ldots, K$. The normalized quantities are related to the actual voltages and currents through $\bar{v}_k = \frac{v_k}{V_k}$, $\bar{i}_k = \frac{i_k}{\omega_{0,k} C_k V_k}$, $\bar{I}_k\left(\bar{v}_1, \frac{d\bar{v}_1}{dt}, \ldots, t\right) = \frac{I_k(v_1, \frac{dv_1}{dt}, \ldots, t)}{I_{k,max}}$ and $\bar{t} = \omega_0 t$. Furthermore,

$$\epsilon = \frac{I_{k,max}}{\omega_{0,k} C_k V_k} \quad (4)$$

Here, $\omega_{0,k} = 1/\sqrt{L_k C_k} = \omega_0 (1 + \epsilon\delta_k)$ is the resonance frequency of the k-th tank. We assume these resonance frequencies to be clustered around their geometric mean $\omega_0 = \left(\prod_k \omega_{0,k}\right)^{1/k}$, or $\delta_k \leq 1$. Furthermore, V_k is an estimate for the magnitude of the voltage swing over the capacitor C_k. It should be chosen such that (4) yields the same ϵ for all k. $I_{k,max}$ is a measure for the maximum current delivered by the active elements. Using vector notation, the system (3) looks like

$$\frac{d\mathbf{x}}{dt} = \mathbf{L}\mathbf{x} + \epsilon \mathbf{g}\left(\mathbf{x}, \frac{d\mathbf{x}}{dt}, t\right) \quad (5)$$

where $\mathbf{x}(t) = \begin{bmatrix} \bar{i}_1(t) & \bar{v}_1(t) & \bar{i}_2(t) & \bar{v}_2(t) & \cdots \end{bmatrix}^T \in \mathbb{R} \to \mathbb{R}^{2K}$, $\mathbf{L} \in \mathbb{R}^{2K \times 2K}$ and $\mathbf{g}(\cdot, \cdot, \cdot) \in \mathbb{R}^{2K} \times \mathbb{R}^{2K} \times \mathbb{R} \to \mathbb{R}^{2K}$. The nature of the oscillation is characterized by the parameter ϵ, defined by equation (4). If $\epsilon \ll 1$, the oscillation is essentially one where energy resonates between an electric and a magnetic field, with the active elements only weakly involved to compensate for dissipative losses. If $\epsilon \gg 1$, the active elements dominate the oscillating behavior, and we get a relaxation type oscillator. For harmonic oscillators with a sufficiently high Q-factor, i.e. the type of oscillators being considered in this text, it holds that $\epsilon \sim \frac{1}{Q} \ll 1$.

In constructing a compact behavioral model equivalent to the original system (5), we proceed along the same lines of reasoning as outlined in [1, 9]. For $\epsilon \to 0$, i.e. for the unperturbed oscillator, (3) is solved by

$$\bar{i}_k(t) = A_k \sin(t + \theta_k) \quad (6)$$

$$\bar{v}_k(t) = A_k \cos(t + \theta_k) \quad (7)$$

with $\bar{i}_k(t)$ and $\bar{v}_k(t)$ 2π-periodic processes. In vector notation this becomes

$$\mathbf{x}(t) = \mathbf{x}_s(t; \mathbf{p}) \quad (8)$$

where $\mathbf{x}_s \in \mathbb{R} \times \mathbb{R}^{2K} \to \mathbb{R}^{2K}$ and

$$\mathbf{p} = \begin{bmatrix} A_1 & \cdots & A_K & \theta_1 & \cdots & \theta_K \end{bmatrix}^T \in \mathbb{R}^{2K} \quad (9)$$

For unperturbed oscillators, amplitudes and phases are determined by the initial conditions and remain constant with respect to time, i.e. they can be considered as constants of motion characterizing a particular periodic solution.

Under the influence of weak perturbations, i.e. for $\epsilon \ll 1$, the oscillation amplitudes and phases will no longer remain constant, but will slowly change values with time. This is the "slipping" behavior as outlined in section 2 and illustrated in Fig. 2. This implies that the solution of the perturbed system can be modeled as

$$\mathbf{x}(t) = \mathbf{x}_s(t; \mathbf{p}(t)) \quad (10)$$

where the vector $\mathbf{p}(t)$ is no longer a constant, but is evolving with time. The equations governing this "slipping" behavior are found by substituting the signal model (10) into the ODEs (5). Up to first order in ϵ, this yields

$$\frac{d\mathbf{p}}{dt} = \epsilon \mathbf{h}_p(t; \mathbf{p}) \quad (11)$$

with $\mathbf{h}_p(t; \mathbf{p}) \in \mathbb{R} \times \mathbb{R}^{2K} \to \mathbb{R}^{2K}$ determined by

$$\mathbf{h}_p(t; \mathbf{p}) = \frac{\partial \mathbf{x}_s}{\partial \mathbf{p}}(t; \mathbf{p})^{-1} \mathbf{g}\left(\mathbf{x}_s(t; \mathbf{p}), \frac{\partial \mathbf{x}_s}{\partial t}(t; \mathbf{p}), t\right) \quad (12)$$

As the Jacobian matrix $\frac{\partial \mathbf{x}_s}{\partial \mathbf{p}}(t; \mathbf{p})$ in equation (12) is concerned, it is easy to show that it has a block-diagonal structure with

$$\mathbf{J}_k(t; \mathbf{p}) = \begin{bmatrix} \sin(t + \theta_k) & A_k \cos(t + \theta_k) \\ \cos(t + \theta_k) & -A_k \sin(t + \theta_k) \end{bmatrix} \quad (13)$$

being the (block)-diagonal elements. Its inverse is also block-diagonal with the diagonal elements equal to

$$\mathbf{J}_k(t; \mathbf{p})^{-1} = \frac{1}{A_k} \begin{bmatrix} A_k \sin(t + \theta_k) & A_k \cos(t + \theta_k) \\ \cos(t + \theta_k) & -\sin(t + \theta_k) \end{bmatrix} \quad (14)$$

In a final step, we change variables, reformulating (12) in terms of one common-mode phase θ and the phase differences $\Delta\phi$ between the oscillators. The reason for this reformulation lies in the fact that, for most practical applications, the phase differences between the oscillators are of greater interest than their absolute values. These phase differences are often required to converge to some fixed value, a property we will use later on in our model extraction procedure. Introducing the vector

$$\mathbf{q} = \begin{bmatrix} A_1 & \cdots & A_K & \Delta\phi_2 & \cdots & \Delta\phi_K & \theta \end{bmatrix}^T \quad (15)$$

$$= \mathbf{T}_q^{-1} \mathbf{p} \quad (16)$$

with $\theta = \theta_1$, $\Delta\phi_k = \theta_k - \theta_1$ and $\mathbf{T}_q \in \mathbb{R}^{2K \times 2K}$, the ODE (12) can be recast into

$$\frac{d\mathbf{q}}{dt} = \epsilon \mathbf{h}_q(t; \mathbf{q}) \quad (17)$$

where

$$\mathbf{h}_q(t; \mathbf{q}) = \mathbf{T}_q \left[\frac{\partial \mathbf{x}_s}{\partial \mathbf{p}}(t; \mathbf{T}_q^{-1}\mathbf{q})\right]^{-1} \mathbf{g}\left(\mathbf{x}_s, \frac{\partial \mathbf{x}_s}{\partial t}, t\right) \quad (18)$$

Here, $\mathbf{x}_s$ stands short for $\mathbf{x}_s\left(t; \mathbf{T}_q^{-1}\mathbf{q}\right)$. In what follows, (17) serves as a starting point for extraction of a compact behavioral model for the dynamic behavior of a set of coupled harmonic oscillators.

4. AVERAGING

Equation (17) learns that the transient changes in $\mathbf{q}(t)$ occur on a time scale t/ϵ. This means that, over one oscillation period, $\mathbf{q}(t)$ can be considered a constant. This observation is the foundation of the averaging technique in [1, 9], where it was first discussed for deterministic systems, and [5], where it is extended to include noisy inputs. In what follows, we use the results obtained in [5] to extract a set of equations describing the oscillator's transient behavior.

In order to apply the averaging technique properly, we need to separate the deterministic and stochastic processes at the oscillator input. To this account, we assume that the perturbation part in the right-hand side of equation (5) can be written as

$$\mathbf{g}\left(\mathbf{x}, \frac{d\mathbf{x}}{dt}, t\right) = \mathbf{g}_d\left(\mathbf{x}, \frac{d\mathbf{x}}{dt}, \mathbf{u}\right) + \mathbf{G}_n\left(\mathbf{x}, \frac{d\mathbf{x}}{dt}, \mathbf{u}\right) \mathbf{n}(t) \quad (19)$$

Here, $\mathbf{u}(t) \in \mathbb{R} \to \mathbb{R}^U$ represents a vector of slow-varying, deterministic inputs and $\mathbf{n}(t) \in \mathbb{R} \to \mathbb{R}^N$ is a vector of uncorrelated, standard (normalized) white noise sources. The assumption of the deterministic inputs $\mathbf{u}(t)$ being slow-varying, i.e. varying on a time scale t/ϵ, seems justified by the fact that these inputs are typically control signals which should never vary faster than the transients of the oscillator they try to control. Combining the equations (18) and (19), the right-hand side $\mathbf{h}_q(t;\mathbf{q})$ in equation (17) can be split into a determistic and a stochastic component, or

$$\mathbf{h}_q(t;\mathbf{q}) = \mathbf{h}_d(t;\mathbf{q},\mathbf{u}) + \mathbf{H}_n(t;\mathbf{q},\mathbf{u})\,\mathbf{n}(t) \tag{20}$$

where both $\mathbf{h}_d(t,\cdot,\cdot) \in \cdot \rightarrow \mathbb{R}^{2K}$ and $\mathbf{H}_n(t;\cdot) \in \cdot \rightarrow \mathbb{R}^{2K\times N}$ are 2π-periodic in t. Using furthermore the fact that the last component of the vector $\mathbf{q}$, i.e. the reference phase θ, will always appear in combination with the time variable t as $t+\theta$, and with $\mathbf{q}_1$ being defined by $\mathbf{q} = \begin{bmatrix} \mathbf{q}_1^T & \theta \end{bmatrix}^T$, The ODE (17) is transformed into

$$\frac{d}{dt}\begin{bmatrix} \mathbf{q}_1 \\ \theta \end{bmatrix} = \epsilon\left(\mathbf{h}_d(t+\theta;\mathbf{q}_1,\mathbf{u}) + \mathbf{H}_n(t+\theta;\mathbf{q}_1,\mathbf{u})\,\mathbf{n}(t)\right) \tag{21}$$

which, because of the noise terms, is a system of stochastic differential equations.

Using the averaging principle [1, 5], it can be shown that for ϵ being small, the solution of (21) converges to that of the "averaged" equation

$$\frac{d}{dt}\begin{bmatrix} \mathbf{q}_1 \\ \theta \end{bmatrix} = \epsilon\left(\overline{\mathbf{h}}_d(\mathbf{q}_1,\mathbf{u}) + \overline{\mathbf{H}}_n(\mathbf{q}_1,\mathbf{u})\,\overline{\mathbf{n}}(t)\right) \tag{22}$$

with

$$\overline{\mathbf{h}}_d(\mathbf{q}_1,\mathbf{u}) = \frac{1}{2\pi}\int_0^{2\pi} \mathbf{h}_d(t;\mathbf{q}_1,\mathbf{u})\,dt \tag{23}$$

and

$$\overline{\mathbf{H}}_n(\mathbf{q}_1,\mathbf{u}) = \sqrt{\frac{1}{2\pi}\int_0^{2\pi} \mathbf{H}_n(t;\mathbf{q}_1,\mathbf{u})\,\mathbf{H}_n(t;\mathbf{q}_1,\mathbf{u})^T\,dt} \tag{24}$$

In equation (22), $\overline{\mathbf{n}}(t) \in \mathbb{R}\times\mathbb{R}^N$ again is a vector of uncorrelated, standard white noise processes.

5. LINEARIZATION

With its right-hand side independent of θ, the system (22) can be separated into two subsystems, one modeling the transient behavior of the vector $\mathbf{q}_1$, i.e. the amplitudes A_k and the phase differences $\Delta\phi_k$, and another one modeling the behavior of the common phase θ. These systems are given by

$$\frac{d\mathbf{q}_1}{dt} = \epsilon\left(\overline{\mathbf{h}}_{d,1}(\mathbf{q}_1,\mathbf{u}) + \overline{\mathbf{H}}_{n,1}(\mathbf{q}_1,\mathbf{u})\,\overline{\mathbf{n}}(t)\right) \tag{25}$$

$$\frac{d\theta}{dt} = \epsilon\left(\overline{\mathbf{h}}_{d,2}(\mathbf{q}_1,\mathbf{u}) + \overline{\mathbf{H}}_{n,2}(\mathbf{q}_1,\mathbf{u})\,\overline{\mathbf{n}}(t)\right) \tag{26}$$

where $\overline{\mathbf{h}}_{d,1}$, $\overline{\mathbf{h}}_{d,2}$ are the first $2k-1$ rows, respectively the last row of the vector $\overline{\mathbf{h}}_d$ and $\overline{\mathbf{H}}_{n,1}$, $\overline{\mathbf{H}}_{n,2}$ are likewise extracted from $\overline{\mathbf{H}}_n$.

For a properly designed, stable oscillator and for a constant input $\mathbf{u}(t) = \mathbf{u}_{op}$ (the nominal control voltage of a VCO), the solution of the system (25) converges to the neighbourhood of a fixed point $\mathbf{q}_{1,op}$ determined by

$$\overline{\mathbf{h}}_{d,1}(\mathbf{q}_{1,op},\mathbf{u}_{op}) = 0 \tag{27}$$

A good approximation of the oscillator's transient behavior in the neighbourhood of such a fixed point is found by linearizing (25) and (26) around the desired operating point $\mathbf{q}_{1,op}$. With $\mathbf{q}_1(t) = \mathbf{q}_{1,op} + \Delta\mathbf{q}_1(t)$ and $\mathbf{u}(t) = \mathbf{u}_{op} + \Delta\mathbf{u}(t)$, this yields

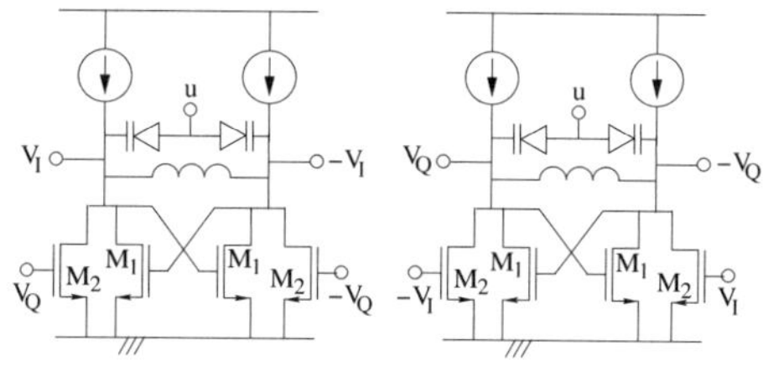

Figure 4: Basic quadrature oscillator topology used to illustrate the numerical methods involved in extracting the behavioral model.

$$\frac{d\Delta\mathbf{q}_1}{dt} = \mathbf{A}_1\Delta\mathbf{q}_1 + \mathbf{B}_{u,1}\Delta\mathbf{u}(t) + \mathbf{B}_{n,1}\overline{\mathbf{n}}(t) \tag{28}$$

with $\mathbf{A}_1 = \epsilon\frac{\partial\overline{\mathbf{h}}_{d,1}}{\partial\mathbf{q}_1}(\mathbf{q}_{1,op},\mathbf{u}_{op}) \in \mathbb{R}^{(2K-1)\times(2K-1)}$, etc. Note that for the operating point $(\mathbf{q}_{1,op},\mathbf{u}_{op})$ to be stable, all eigenvalues of the matrix $\mathbf{A}_1$ should be located in the left half plane.

In the neighbourhood of the operating point $(\mathbf{q}_{1,op},\mathbf{u}_{op})$, the behavior of the common phase θ is likewise determined by

$$\frac{d\theta}{dt}(t) = \Delta\omega_0 + \mathbf{A}_2\Delta\mathbf{q}_1(t) + \mathbf{B}_{u,2}\Delta\mathbf{u}(t) + \mathbf{B}_{n,2}\overline{\mathbf{n}}(t) \tag{29}$$

where $\Delta\omega_0 = \epsilon\overline{\mathbf{h}}_{d,2}(\mathbf{q}_{1,op},\mathbf{u}_{op}) \in \mathbb{R}$, $\mathbf{A}_2 = \epsilon\frac{\partial\overline{\mathbf{h}}_{d,2}}{\partial\mathbf{q}_1}(\mathbf{q}_{1,op},\mathbf{u}_{op}) \in \mathbb{R}^{1\times(2K-1)}$, etc. For a free-running oscillator in steady-state regime, i.e. with $\Delta\mathbf{q}_1(t) = 0$ and $\Delta\mathbf{u}(t) = 0$, equation (29) implies that

$$\theta(t) = \Delta\omega_0 t + \sqrt{\mathbf{B}_{n,2}\mathbf{B}_{n,2}^T}\,B(t) \tag{30}$$

with $B(t)$ being a standard Brownian motion. In (30), $\Delta\omega_0$ represents the deviation of the actual oscillation frequency from the natural frequency of the resonance tank. The second term is the phase-noise component as discussed in [4].

Equations (28)-(29) specify a behavioral model that completely characterizes the transient behavior of a set of coupled harmonic oscillators in the neighbourhood of the operating point $(\mathbf{q}_{1,op},\mathbf{u}_{op})$. Once $\mathbf{A}_1$, $\mathbf{A}_2$, $\mathbf{B}_{u,1}$, $\mathbf{B}_{u,2}$, $\mathbf{B}_{n,1}$, $\mathbf{B}_{n,2}$ and $\Delta\omega_0$ have been determined, something which needs to be done only once, evaluation of the model comes at a minimal computational cost. This is particularly interesting when long (or repetitive) simulations need to be performed.

6. NUMERICAL COMPUTATIONS

In extracting the model equations (28)-(29) from a given circuit-level description, we need to map the circuit topology on the template of Fig. 3 (where it is illustrated for $K = 2$). Here, the current sources I_k should include all energy dissipating elements, also the ones arising from capacitor and inductor losses. Furthermore, we must find a way to evaluate the functions $\overline{\mathbf{h}}_d(\mathbf{q}_1,\mathbf{u})$ and $\overline{\mathbf{H}}_n(\mathbf{q}_1,\mathbf{u})$. In what follows, we outline the computational procedures for the example of the harmonic quadrature oscillator shown in Fig. 4. Adjusting this approach for other types of (coupled) harmonic oscillators is straightforward.

The characteristics of the oscillator resonance tanks, more specifically the effective capacitance and inductance values C_k and L_k, are determined by fitting the frequency characteristic of the admittance $G(j\omega)$ of an RLC tank to the one measured using the simulation setup shown in Fig. 5. This setup is simply a single oscilla-

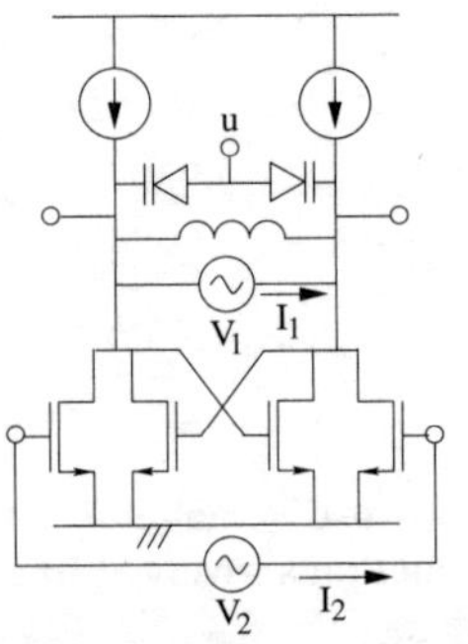

Figure 5: Measurement setup used to extract the resonance tank characteristic and the currents $I_k\left(v_1, \frac{dv_1}{dt}, v_2, \frac{dv_2}{dt}, \ldots, t\right)$ exchanged between the resonance tank and the remainder of the circuit.

tor section, where a given in-phase and quadrature voltage are enforced using the voltage sources V_1 and V_2. To obtain the circuit's $G(j\omega)$, we use the results of an AC analysis in the neighbourhood of $V_1 = V_2 = 0$. This involves applying

$$V_1(t) = \Delta v \cos(\omega t) \quad (31)$$
$$V_2(t) = 0 \quad (32)$$

with Δv a sufficiently small voltage. Measuring the (steady-state) current $I_1(t)$ through the voltage source V_1 allows us to compute $G(j\omega) = I_1/V_1$. The fitting is performed for frequencies ω in the neighbourhood of the resonance frequency ω_0, detemined by

$$\omega_0 = \arg\min_{\omega} |G(j\omega)| \quad (33)$$

All of this needs to be done for the external (control) inputs set to their nominal DC value, or $\mathbf{u} = \mathbf{u}_{op}$. Note that use of the setup in Fig. 5 implies that the transistor capacitances are taken into account as part of the effective tank capacitance C_k.

Computation of $\overline{\mathbf{h}}_d(\mathbf{q}_1, \mathbf{u})$ and $\overline{\mathbf{H}}_n(\mathbf{q}_1, \mathbf{u})$, for a given set of values for $\mathbf{q}_1$ and $\mathbf{u}$, basically requires evaluation of the currents I_k. Evaluating, for example, $I_1\left(v_1, \frac{dv_1}{dt}, v_2, \frac{dv_2}{dt}, \ldots, t\right)$ is done by applying

$$V_1(t) = A_1 \cos(\omega_0 t + \theta) \quad (34)$$
$$V_2(t) = A_2 \cos(\omega_0 t + \theta + \Delta\phi_2) \quad (35)$$

in the measurement setup of Fig. 5. $I_1\left(v_1, \frac{dv_1}{dt}, v_2, \frac{dv_2}{dt}, \ldots, t\right)$ then equals the (steady-state) current $I_1(t)$ flowing through the voltage source V_1. In computing this current I_1, the (slow-varying) input signals $\mathbf{u}(t)$ should be kept at a constant value. The deterministic component $\overline{\mathbf{h}}_d(\mathbf{q}_1, \mathbf{u})$ is separated from the stochastic component $\overline{\mathbf{H}}_n(\mathbf{q}_1)$ by first setting all noise sources to zero, which leaves only the deterministic component to be measured. The stochastic component then follows from (time-varying) linearization in the neighbourhood of the "large-signal" deterministic component.

7. EXPERIMENTAL RESULTS

The model extraction procedure was implemented in Matlab for the harmonic quadrature oscillator shown in Fig. 4. We compared both the behavioral model's steady-state and transient behavior with the results obtained from SPICE-like simulation algorithms. In Matlab, transistor modeling was done based upon the MOST level-1 current equations, as no link between Matlab and more sophisticated SPICE transistor models was available. All computations were done starting from the normalized system of equations (3). Extraction of the behavioral model takes a few minutes on a Sun Ultra30.

	$\gamma = 0.5, m = 0.3$ Original	Model	$\gamma = 0.5, m = 0.5$ Original	Model
A_1	2.51	2.51	2.62	2.62
A_2	2.51	2.51	2.62	2.62
ω_{osc} (rad/sec)	1.010	1.008	1.010	1.011
$\Delta\phi$	90.1^o	90°	90.0^o	90.0^o

Table 1: Comparing operating point values as computed by solving (27) with the values extracted from Spice-like simulations. Computations were done for $\epsilon = 0.1$.

Besides the characteristics of the resonance tank, there are three other important parameters specifying the design of the oscillator in Fig. 4. So different values of these parameters correspond to different designs. Firstly, we have

$$\gamma = \frac{1/R_{loss}}{g_{m,1}/2} \quad (36)$$

being the ratio between the parallel conductance $G_{loss} = 1/R_{loss}$, modeling capacitor and inductor losses, and the transconductance of the restoring active element (as seen by the resonant tank). Here, g_{m1} is the transconductance of the transistor M_1. A second design parameter is given by

$$m = \frac{g_{m2}}{g_{m1}} \quad (37)$$

which is the ratio of the transconductances of the transistors M_1 and M_2. It is a measure for the strength with which the two oscillators are coupled. Finally, there is ϵ, the ratio between the maximum current which can be delivered by the active elements and the current circulating in the tank. It roughly equals

$$\epsilon = \frac{K}{\gamma Q} \quad (38)$$

where Q equals the quality factor of the inductor coil and where K can be shown to be a constant in the range $0.25 - 1$. This makes ϵ of the same order of magnitude as $1/Q$. For typical Q-factors and choices of γ, ϵ is in the range $0.05 - 0.2$. In what follows, we perform most computations for $\epsilon = 0.1$.

In a first stage, we solve equation (27) for the operating point $\mathbf{q}_{1,op} = \begin{bmatrix} A_1 & A_2 & \Delta\phi \end{bmatrix}$ and $\omega_{osc} = \omega_0 + \Delta\omega_0$. These results are compared with those extracted from SPICE-like simulations. Table 1 summarizes the results for two sets of values for γ and m. All operating point values were found to be at least 1% accurate. This even holds for larger values of ϵ (up to 0.2).

In a next step, we compare the model's transient behavior, in the neighbourhood of the selected operating point, with results obtained from SPICE-like simulations. This is done by starting simulations with initial conditions for amplitudes and phase-differences deviating about 10% from their steady-state values. Fig. 6 shows the resulting amplitudes of the in-phase and quadrature oscillations as computed using the behavioral model (solid line) and as extracted from the envelope of the SPICE-like simulations (x-marks). In order to evaluate the accuracy of the transient model, we computed the ratio of the energy contained in the error component, i.e. the difference between the result obtained from the behavioral model and from the SPICE-like simulations, and the energy contained in the transient itself. The accuracy was computed for two

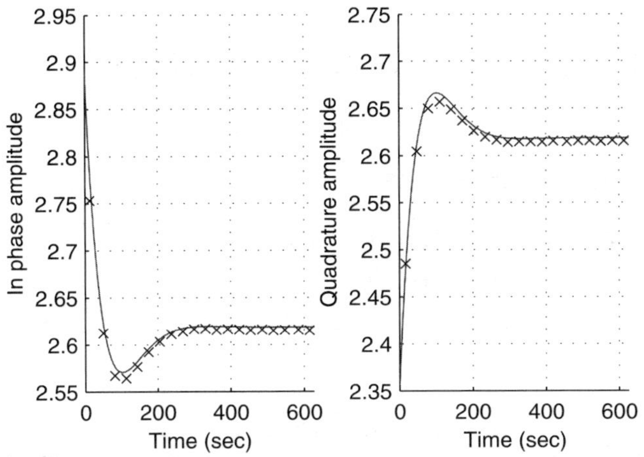

Figure 6: Amplitudes of the in-phase and quadrature oscillations as computed using the behavioral model (solid line) and as extracted from the envelope of the SPICE-like simulations (x-marks). These result were obtained for $\epsilon = 0.1$, $\gamma = 0.5$ and $m = 0.5$.

	Relative error (dB) $\gamma = 0.5$,$m = 0.3$	Relative error (dB) $\gamma = 0.5$,$m = 0.5$
$\epsilon = 0.05$	-22.5	-22.1
$\epsilon = 0.1$	-21.9	-20.4
$\epsilon = 0.15$	-18.4	-19.3
$\epsilon = 0.2$	-18.7	-16.4

Table 2: Ratio of the energy contained in the error component, i.e. the difference between the result obtained from the behavioral model and from the SPICE-like simulations, and the energy contained in the transient itself.

sets of values for γ and m and for several values of ϵ, i.e. for several values of the Q-factor of the tank. These results are summarized in Table 2, which shows that, even for larger values of ϵ, the modeling accuracy is still within 1%.

Table 3 compares the computational complexity involved in the evaluation of the original oscillator equations and the behavioral model equations. Both sets of equations were solved with a SPICE-like solver with variable time step. Results are shown for the two extreme values of the parameter ϵ and indicate a speed gain per oscillation period T ranging from 20-40. As expected, speed gain increases with decreasing ϵ due to the larger timestep that can be used in integrating the behavioral model equations. Note that this gain in simulation speed will even further increase when compared to circuit simulations involving more complex transistor models than the level-1 models used in our Matlab experiments. This is due to an increase in computational cost in evaluating the original circuit equations, while the cost in evaluating the behavioral model remains unchanged once it has been computed.

	$\epsilon = 0.05$		$\epsilon = 0.2$	
	Original	Model	Original	Model
CPU-time/T (sec)	0.18	4e-3	0.17	8e-3
KFlops/T	42.5	1.60	42.8	1.93

Table 3: Computational complexity per oscillation period for 2 extreme values of ϵ.

8. CONCLUSIONS

This paper has presented a novel approach for the construction of behavioral models for (coupled) harmonic oscillators. The models can be used for system-level simulations and trade-off analysis. Besides the steady-state behavior, the model also takes transient behavior into account. The modeling equations are valid in the neighbourhood of the oscillator's steady-state operating point. Once being constructed, they can be evaluated at very low computational cost. A major advantage of the approach is that the fast-varying, steady-state, behavior and the slow-varying, transient, behavior are explicitely separated. This allows for straightforward application of multi-rate simulation techniques, boosting simulation speeds with a factor 20 and more. The modeling approach has been applied to a harmonic quadrature oscillator. Results, both as steady state and transient behavior are concerned, were shown to be accurate within 1% as compared to SPICE-like simulations.

9. ACKNOWLEDGEMENTS

This work has been supported in part by the Flemish IWT.

10. REFERENCES

[1] N.N. Bogoliubov and Y.A. Mitropolsky, *Asymptotic methods in the theory of non-linear oscillations*, Hindustan Pulishing Corp., Delhi, 1961

[2] J. Crols and M. Steyaert, *CMOS Wireless Transceiver*, Kluwer Academic Publishers, June 1997

[3] A. Demir et al., "Behavioral Simulation Techniques for Phase/Delay-Locked Systems", In *Proc. IEEE CICC*, pp. 21.3.1-4, 1994

[4] A. Demir, A. Mehrotra and J. Roychowdhury, "Phase Noise in Oscillators: A Unifying Theory and Numerical Methods for Characterization", In *IEEE Trans. Circ. and Syst.-I*, vol. 47, no. 5, pp. 655-674, May 2000

[5] M.I. Freidlin and A.D. Wentzell, *Random perturbations of dynamical systems*, 2nd ed., Springer, 1998

[6] A. Hadjimiri and T.H. Lee, "A General Theory of Phase Noise in Electrical Oscillators", In *IEEE J. Solid-State Circuits*, vol. 33, no. 2, pp. 179-194, February 1998

[7] F.X. Kaertner, "Analysis of White and $f^{-\alpha}$ Noise in Oscillators", In *Int. J. Circuit Theory Appl.*, vol. 18, pp 485-519, 1990

[8] J. Kevorkian and J.D. Cole, *Perturbation Methods in Applied Mathematics*, Springer-Verlag, New York, 1981

[9] N. Kryloff and N. Bogoliubov, *Introduction to non-linear mechanics, Princeton University Press*, New York, 1947

[10] M.F. Mar, "An Event-Driven PLL Behavioral Model with Applications to Design Driven Noise Modeling", In *Proc. BMAS*, Session 1.1, 1999

[11] O. Narayan and J. Roychowdhury, "Multi-Time Simulation of Voltage-Controlled Oscillators", In *Proc. IEEE DAC*, Session 36, 1999

[12] M. Steyaert et al., "A 2-V CMOS Cellular Transceiver Front-End", In *IEEE Journ. Solid-State Circ.*, vol. 35, no. 12, pp. 1895-1907, December 2000

[13] F. Verhulst, *Nonlinear Differential Equations and Dynamical Systems*, 2nd ed., Springer 1996

Model Checking Algorithms for Analog Verification

Walter Hartong, Lars Hedrich, Erich Barke
Institute of Microelectronic Circuits and Systems, University of Hannover
Appelstrasse 4, 30167 Hannover, Germany, www.ims.uni-hannover.de
hartong, hedrich, barke@ims.uni-hannover.de

ABSTRACT

In this contribution we present the first method for model checking on nonlinear analog systems. Based on digital CTL model checking algorithms and results in hybrid model checking, we have developed a concept to adapt these ideas to analog systems. Using an automatic state space subdivision method the continuous state space is transfered into a discrete model. In doing this, the most challenging task is to retain the essential nonlinear behavior of the analog system. To describe analog specification properties, an extension to the CTL language is needed. Two small examples show the properties and advantages of this new method and the capability of the implemented prototype tool.

Categories and Subject Descriptors

I.6.m [**Computer Methodologies**]: Simulation and Modeling

General Terms

Algorithms, Language, Theory, Verification

Keywords

Model Checking, Formal Methods, Nonlinear Analog Systems

1. INTRODUCTION

In recent years formal methods, like equivalence checking and model checking, have been successfully introduced into the digital design flow, indicated by the growing number of commercial vendors. However, these methods are still almost unknown in analog verification. Symbolic analysis is a promising technique but it cannot cope with the power of formal methods in the digital world. As far as we know, there has been only one approach published on equivalence checking for analog circuits [5].

On the other hand, digital tools have been extended to special hybrid systems, being digital systems connected to some analog blocks or to an analog environment. These techniques are focused on the digital part of the system. The analog behavior is mostly restricted and the verification results are not appropriate to assess the functionality of the analog part, because the analog behavior is assumed to be correct during the verification process. The result does therefore not include useful information for analog designers. Furthermore, model checking languages used are not able to describe analog system properties. Thus, formal methods for analog systems are still lagging far behind the digital tools.

In this contribution a discrete model for the nonlinear system will be presented. Further, model checking algorithms known from digital model checking tools will be adapted to analog systems.

This paper is organized as follows. In Section 2 a general comparison between digital, hybrid and analog systems is given. The next two sections describe the main algorithms developed, namely the generation of the discrete model and the model checking methods. Finally, some experimental results are given, showing the use of the developed prototype.

DAC 2002, June 10-14, 2002, New Orleans, Louisiana, USA.

2. SYSTEM DESCRIPTION

The most general system class to be considered in this context is the transition system. Most digital and hybrid model checking tools are based on such systems.

Definition 1. A state transition system $T = (Q, Q_0, \Sigma, R)$ consists of

- a set of states Q and a set of initial states Q_0,
- a set of generators or events Σ and
- a state transition relation $R \subseteq Q \times \Sigma \times Q$.

The state transition system is a suitable model for digital as well as for analog systems [11]. In the analog case, the set of states Q can be represented by an Euclidean space (state space) spanned by the system's state variables. The number of states is infinite, due to the continuous definition of the state variables. The initial state Q_0 is a single point in the state space. Often, but not necessarily, it is the DC operating point. There are only two generators Σ causing state transitions, namely, the time t and the input values $u(t)$. The state transition relation R is a continuous function with respect to time. It can be visualized by a vector field in the state space.

Unfortunately, the representation of states and state transitions in the discrete digital case is totally different. Published model checking tools for hybrid systems, require discrete or discrete and partly linear system descriptions [1]. Our focus is on nonlinear analog behavior, which cannot be described by these simple approximations. Linear phase-portrait approximation [6] is an encouraging method, but the calculation of proper state space intersection planes seems to be complicated, especially for high dimensional and strongly nonlinear problems. There are two aspects which make a different approach necessary. Firstly, model checking for analog systems requires extensions of the language to define analog system properties in a reasonable way. Secondly, the analog system model used in the model checker needs to retain the essential analog dynamics to get useful results.

2.1 Analog System

An analog system description consists of a nonlinear first order differential algebraic equation system

$$f(\dot{x}(t), x(t), u(t)) = 0 \quad (1)$$

with the input variable vector $u(t)$ and the vector $x(t)$ of n system variables. In general, the function f is arbitrarily nonlinear. However, in practice the nonlinearities are restricted by the device models used. There are several ways to build such equations from a transistor netlist or a behavioral model, for example the modified nodal analysis (MNA).

The number of state variables is less than or equal to the number of energy storage elements (e.g. capacitors and inductors). The definition of state variables spanning the state space depends on the way to build up Equation (1) and is not unique for one system [4]. Moreover, the link between state variables and system variables is not necessarily obvious. Mathematically, this is known as initial value problem for differential algebraic equations [8]. However, since we focus on basic model checking methods, these issues will not be discussed here. To take the input values into account, the state space can be extended by the number of inputs. We call this extended state space. Since, the input values are not determined by the system itself but rather by environmental conditions, it is not possible to model their behavior within the state space.

3. DISCRETE MODEL GENERATION

Some of the following algorithms have been inspired by research in the area of approximating dynamical behavior [3]. Despite the similarities, there are a lot of differences, mainly caused by the overall target of the algorithms. Therefore, the algorithms have not been used, but some of the ideas have been adopted.

As we have seen, the continuous variables in an analog system - state values and time - have to be transfered into a discrete state space description and a state transition relation. The next three sections illustrate this process.

3.1 Discrete Time Steps

The transition relation R for an analog system is given by the state variables' time derivation $\dot{x}(t)$, which is a continuous function. The actual state transition can be calculated by integrating this function.

$$x(t) = x_0(t_0) + \int_{t_0}^{t} \dot{x}\left(x(t), u(t)\right) dt \quad (2)$$

In practice $\dot{x}(t)$ may possibly not be derived as explicit function, however, the relation is given implicitly by Equation (1). In general, $x(t)$ can only be calculated using numerical integration. This problem is well known in analog simulators, like Spice, Spectre, Saber, etc. During transient simulation the differential equation system is solved using discrete time steps. Given a small time step $\Delta t = t_1 - t_0$, the transition between the actual state $x(t_0)$ and the next state $x(t_1)$ is determined by a numerical integrator, using e.g. the backward Euler formula:

$$x(t_1) = \left\{ x \,\middle|\, f\left(\frac{x - x(t_0)}{t_1 - t_0}, x, u(t_1)\right) = 0 \right\} \quad (3)$$

A step size control uses the second derivation with respect to time for a local measurement of the integration error. If the given error threshold is exceeded, step size is reduced otherwise the transient step is accepted.

This method can be directly used in an analog model checking tool. An arbitrary test point in the state space is mapped to its successor state, depending on the actual Δt used. In contrast to transient simulation, there is no temporal predecessor state for a test point. A second time step has to be calculated for each point to determine the second derivation, enabling a local error control.

In general, the time step Δt will vary throughout the state space, due to the step size control. As we will see later, this makes the checking of explicit time dependencies difficult. To make this, easier the time step is chosen to be equal for each test point within one state space region (see Section 3.2). By this algorithm, every state space point can be mapped to its successor point including a local step length control. The resulting tuple of test and target point is represented by a successor-vector in the state space. Due to the error control, the vector length is restricted to reasonable values in terms of the second derivation. Thus, a time discrete transition relation for single points is given: $s(x(t)) = x(t + \Delta t)$.

3.2 State Space Subdivision

To get a discrete and finite state description, the continuous and infinite state space has to be bounded and subdivided. The restriction to a finite region is simply done by a user defined start area, comprising the considered system behavior. This causes special border problems, which will be discussed later. However, it does not impact the correctness of the model checking result because in real world systems there is alway a natural bound for the state variable values.

Since we do not have a digital environment, a natural subdivision for the start area, given for example by threshold values of digital state transitions, is missing [1]. Furthermore, to retain the analog system's behavior correctly, a sufficient number of subdivisions is needed, especially in state space regions with highly nonlinear behavior. On the other hand, the number of discrete regions should be as small as possible to reduce the total runtime.

The subdivision is done by rectangular boxes or hyper boxes, which are not necessarily the best choice [6]. However, for implementation reasons boxes are the far most convenient data structure. Other subdivision geometries might be considered during future improvements.

At first, we start with a user controlled uniform subdivision in all state space dimensions. Secondly, an automatic subdivision strategy is used to react on different system dynamics, depending on the actual state space region. The main target is to get a uniform behavior in each state space box. The uniformity is measured by the variation of the successor-vectors ($sv(x) = s(x) - x$), calculated in the state space (Section 3.1). Namely, vector length l_m and angle a_m between different vectors are considered. Equations (4) and (5) give the definition of these values. The function $L(\cdot)$ gives the length of the delivered vector or vector component.

$$l_m = 1 - \frac{\min_{x \in box} L(sv(x))}{\max_{x \in box} L(sv(x))} \quad (4)$$

$$a_m = \max_{\substack{x \in box \\ i \in dim}} \frac{L(sv(x_i))}{L(sv(x))} - \min_{\substack{x \in box \\ i \in dim}} \frac{L(sv(x_i))}{L(sv(x))} \quad (5)$$

Box subdivision is continued recursively until l_m and a_m drop under a given threshold or a given subdivision depth is exceeded. Within the expected accuracy, all boxes fulfilling the l_m and a_m thresholds do not contain fix points. This is, because fix points are always surrounded by regions with ununiform behavior in terms of Equation (4) and (5) (see midpoint in Figure 1). This information is stored and used in the connection relation algorithm (Section 3.3).

Additional subdivisions are applied, if the successor-vectors in a region are too short in relation to the box size. This occurs mainly in regions where the system is strongly nonlinear, which implies Δt to be very small. Each box in the state space will represent a single state in the discrete model. Thus, the set of i states is given by $Q = \{box_1, box_2, ..., box_i\}$.

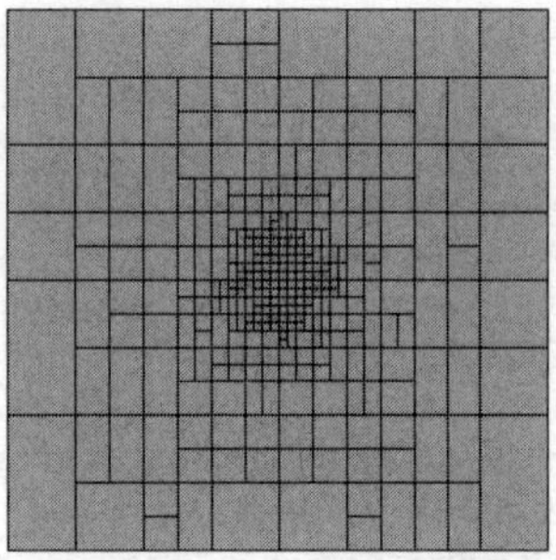

Figure 1: Automatic state space subdivision.

The result of automatic subdivision for a simple example, given by $\{\dot{x}_1 = x_1, \dot{x}_2 = x_2\}$, is shown in Figure 1. According to the uniformity values, box size is getting smaller in the middle of the picture, until the maximum subdivision depth is reached.

3.3 Connection Relation

The last step in getting a discrete system model is the connection relation between state space regions. In Section 3.1 successor states for single state space points have been defined ($s(\cdot)$). Using this point to point relation, the target region r_{target} is given by the set of all target points associated with a test point within the state space region r_{test} (see Equation (6)), as illustrated by the gray regions in Figure 2. We call this exact transformation $T_1(\cdot)$.

$$r_{target1} = \{s(y) \mid y \in r_{test}\} = T_1(r_{text}) \quad (6)$$

Since it is not practical to calculate a huge or - mathematically - infinite number of successor-vectors for each box, a good estimation or inclusion of the target region is needed. Three different approaches will be discussed.

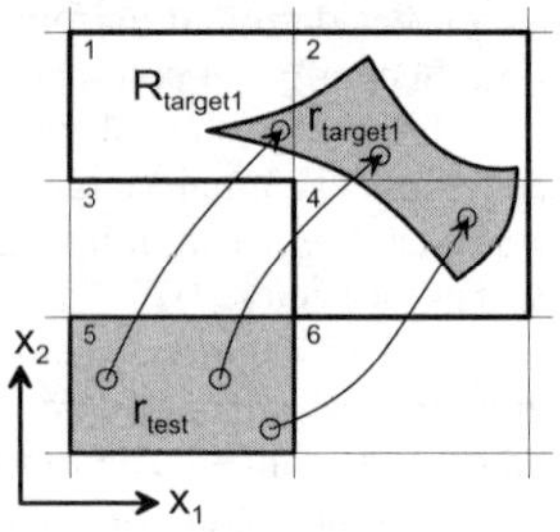

Figure 2: State space connection relation using T_1.

An inclusion $r_{target2}$ can be calculated using interval analysis [10]. This approach provides an exact and overestimated solution. That means, the correct solution $r_{target1}$ is fully included in $r_{target2}$. However, $r_{target2}$ might be much larger than $r_{target1}$. This effect is called overestimation and might be a serious problem for large systems [5]. Moreover, this approach is very time consuming.

$$r_{target2} = T_{interval}(r_{test}) = T_2(r_{test}) \supseteq r_{target1} \quad (7)$$

A more practical but also less accurate way to approximate the target region is to choose a number of test points p_{test} within the test region and to calculate the dedicated target points p_{target}. The target region $r_{target3}$ can be approximated using an appropriate inclusion of these points. As we will see below, an inclusion operation is also needed while expanding the target regions to the actual state space regions. I.e. these two steps can be combined. Even only a few test points may give a reasonable target approximation, but the region $r_{target3}$ might be under- or overestimated.

$$\begin{aligned} p_{test3} &= \{s_1, s_2, ..., s_n\},\ s_i \in r_{test} & (8) \\ r_{target3} &= \{\text{inclusion}(s(y)) \mid y \in p_{test3}\} & (9) \\ &= \mathrm{T}_3(r_{test}) \simeq r_{target1} & (10) \end{aligned}$$

It is shown in [7] that T_3 is surely overestimated if all corner values are used as test points and if $s(\cdot)$ can be assumed to be monotonic. Following the argumentation in [3], T_3 can also be made rigorous using Lipschitz constants L in each state space dimension. Using a grid of test points, spaced by h, one can calculate a extension diameter $d_{ex} = Lh$ for the target points. Expanding each target point by this diameter d_{ex} in each dimension gives a set of boxes. The union of these boxes is an overestimated target approximation $r_{target4}$.

$$\begin{aligned} r_{target4} &= \left\{\text{expand}_{(Lh)}(s(y)) \mid y \in \text{grid}(h, r_{test})\right\} & (11) \\ &= \mathrm{T}_4(r_{test}) \supseteq r_{target1} & (12) \end{aligned}$$

All discussed target regions do not fit into the state space subdivisions used. Therefore, a second step is needed to extend these regions to legal sets of boxes. For example, region $R_{target1}$ (hatched areas in Figure 2) is given by the set of all boxes having contact with the target region $r_{target1}$. Fortunately, this operations is always an overestimation and does therefore not impact the correctness of the above results. Until now, only the third operation $T_3(\cdot)$ has been implemented. It has been shown that a number of 5-8 random test points are sufficient for the target box approximation.

Some additional steps are needed to optimize the connection relation for some corner cases. Namely, these are prevention of long successor-vectors, resulting in a box over-jump, boxes with self-connection and boxes with no connections to other boxes due to short successor-vectors. The last two conditions are unphysical if the box does not contain fix points (Section 3.2). As we have already mentioned in Section 3.1, no explicit time relations are considered. It might be useful or necessary in future implementations to store not only the connection relation $R \subseteq Q \times Q$ but rather this relation combined with the related transition time delays.

Until now, we have not discussed how to model the input values in the discrete extended state space. There are mainly two extreme assumptions for the transition behavior: Firstly, the input values do not change at all. That means, the model is build up as described before for several constant input values. There will be no connection between states with different input values. The second assumption is that the input value can change instantaneously over the whole input value range. A state space region has therefore not only connections to regions at the same input level but additional connections to the neighbor boxes in terms of input values. As we will see later, both of these input models are useful for certain conditions to be checked.

4. ANALOG CTL MODEL CHECKING

Given the subdivided state space and the connection relation, the continuous problem has been transfered into a discrete model. In this way, it is possibly treatable by digital model checking tools. However, these tools and also the model checking languages are not well suited for the generated models and the description of analog properties. In particular, the intensively used BDD structures are not helpful for this kind of models, because set of states can not efficiently be described by binary state variable combinations. Therefore, a modified model checker has been developed, based on the basic CTL algorithms described in [2]. There are some algorithmic modifications due to the special need of the analog model, explained below. The language has been extended by a minimal set of operations enabling the work on analog models. Additionally, the results are visualized graphically. The meaning of the CTL operators is the same as in digital model checkers. The following table gives a syntax overview on the classical CTL language (13):

$$\phi := a \mid \phi \circ \phi \mid \neg\phi \mid \triangleright \diamond \phi \mid \triangleright \phi \, \mathsf{U} \, \phi \tag{13}$$

$$\phi := b * v \mid \phi \circ \phi \mid \neg\phi \mid \triangleright \diamond \phi \mid \triangleright \phi \, \mathsf{U} \, \phi \tag{14}$$

a	boolean variable			
b	continuous variable			
v	real value			
$*$	analog operators	$>$	$\rightarrow$	greater
		$<$	$\rightarrow$	smaller
$\circ$	boolean operators	$\vee$	$\rightarrow$	or
		$\wedge$	$\rightarrow$	and
$\neg$		$\neg$	$\rightarrow$	not
$\triangleright$	path quantifiers	E	$\rightarrow$	on some path
		A	$\rightarrow$	on all paths
$\diamond$	temporal operators	X	$\rightarrow$	next-time
		F	$\rightarrow$	eventually
		G	$\rightarrow$	always
U		U	$\rightarrow$	until

For example the formula $\Theta = \mathsf{AF}(state_1)$ can be read as follows: All paths starting in a state within Θ will eventually reach a state in which $state_1$ is true. It is obvious, that this language is not sufficient for analog model checking, since it is not possible to describe sets of states in a continuous state space. Therefore, the language has been expanded (14) by operators to describe half planes in the state space, e.g. $(x_1 > -13.2546)$. In combination with the boolean operators this enables the definition of arbitrary Manhattan polytopes. If the threshold values used in the CTL formula are not already subdivision values in the state space, they have to be added before executing the CTL formula. This makes the discrete model not only depend on the analog system but also on the CTL formula used, which is another difference to digital model checkers.

Some CTL operations are not clearly defined for analog systems. For example, the time quantifier "X - at the next time step" depends on the time step Δt used. As we have seen in Section 3.1 the time step is not always the same for the whole state space. Therefore, this operation is not reasonable in analog model checking. The other operations can be used in the same manner as in digital.

Obviously, this language is not very powerful describing analog design specifications. This has two reasons. Firstly, we want to follow the digital model checking ideas as far as possible without major changes, secondly, useful or necessary expansions have not been applied yet in the prototype tool. However, it will be shown below that it is possible to check some analog properties even with this minimal set of operations.

4.1 Border Problems

Due to the restricted state space, the border boxes get a special status, because at least one box face is connected to the outside area. Since this area is not considered in the system model, there are no connections from or to the outside area in the connection relation, even if there should be some in the none restricted case. During the CTL evaluation, the border boxes will behave incorrectly.

Consider for example differential equation (15) and CTL formula (16) in the restricted state space ($x = ([-5 \,..\, 5], [-5 \,..\, 5])$).

$$\{\dot{x}_1 = 1, \dot{x}_2 = 0\} \tag{15}$$

$$EG((x_2 > 1.0) \,\&\, (x_2 < 2.0)) \tag{16}$$

Theoretically, the region defined in (16) is not restricted in dimension x_1, but in our case it is, due to the given state space area. The theoretical result of (16) is $((x_2 > 1.0) \,\&\, (x_2 < 2.0))$ whereas we get $\emptyset$ in the restricted case, because some border boxes have no successor box. To avoid these problems, the border boxes are treated specially, depending on the CTL formula. As a result, the border boxes are not part of the model checking result and have to be omitted during interpretation.

4.2 Experimental Results

Three small nonlinear examples are used to show the capability of the proposed tool. The first one in a simple Schmitt trigger circuit. Secondly, a Biquad lowpass filter including two state variables is shown and the last example is a tunnel diode oscillator.

4.2.1 Schmitt Trigger Example

The inverting Schmitt trigger consists of an opamp behavioral model, two resistors, and an output capacitance shown in Figure 3. The opamp has an open loop gain of 10000. The output voltage restriction is ± 5 V and the maximum output current is 80 mA.

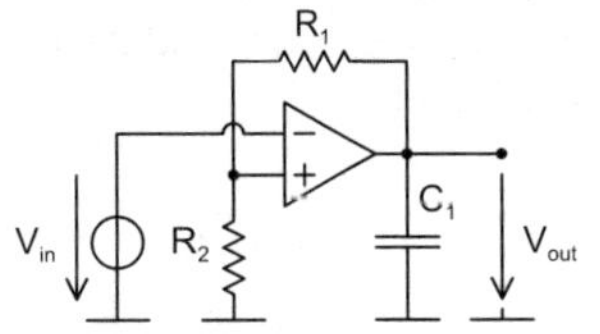

Figure 3: Schmitt trigger circuit.

Resistors R_1 and R_2 are both set to 10 kΩ. Thus, the switching threshold is about ± 2.5 V and the output voltage varies between +5 V and -5 V. This circuit has only one state, namely, the output voltage. Additionally, it has one input signal V_{in}. To consider all states that might occur in the circuit, the extended state space is chosen to be $V_{out} = [-7.7 \,..\, 7.7]$ and $V_{in} = [-7.7 \,..\, 7.7]$. The most interesting feature of the Schmitt trigger function are the switching properties for one output state to the other. Formulating this by CTL results in: $\Phi_1 = \mathsf{EF}(V_{out} < -4.5)$. Φ_1 is the set of states in which a path exists that will eventually reach the region $V_{out} < -4.5$. We choose the constant input value model for this calculation. The collection of boxes fulfilling this condition is shown in Figure 4 in light gray.

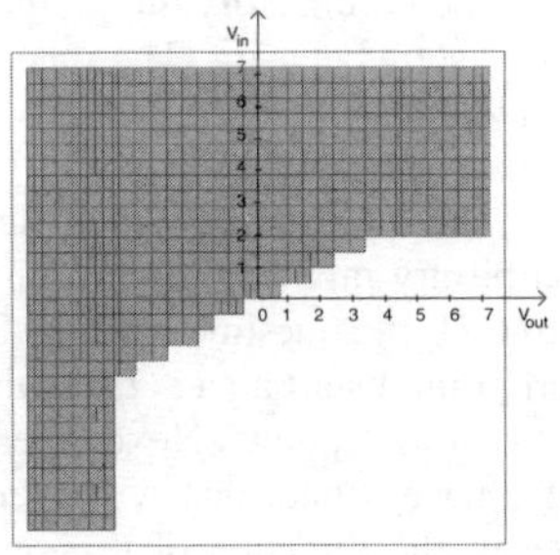

Figure 4: Schmitt trigger model checking result.

It is clearly seen, that above $V_{in} \simeq 2$V the circuit always switches to the negative output state. Below this point the switching depends on the output state V_{out}, but since this will be either 5 V or -5 V there will occur no switching in this region. An equivalent formula can be applied to find the positive switching conditions.

For analog designers such graphical output might be useful. In general however, one would expect only *true* or *false* as output for a CTL formula. To derive this, the checking condition can be expanded by an additional statement. For example

$$\Phi_2 = \mathsf{EF}(V_{out} < -4.5) \;\&\; ((V_{out} > -1) \;\&\; (V_{in} < -1)). \qquad (17)$$

In this case the output is an empty set, which means there is no path form region $((V_{out} > -1) \;\&\; (V_{in} < -1))$ to the negative output state. CTL formula (17) is *false* for the whole state space. Every CTL formula can be changed in the same manner to get a binary result. However, for this contribution we prefer graphical results since they gives more insight into the algorithms.

4.2.2 Biquad Lowpass Filter

The second example is a 2nd order Biquad lowpass filter, shown in Figure 5. The opamp model is the same as in the Schmitt trigger example but the maximum output voltage range is restricted to ± 1.5 V. This circuit has two state variables, namely the capacitor voltages V_{c1} and V_{c2}. Using a charge oriented capacitor model will normally lead to the two charges as state variables. We changed this, because it seems us more convenient to think in voltages than in charges. The corner frequency ω_c and the damping factor d are given by Equations (18) and (19).

$$\omega_c = \frac{1}{\sqrt{R_1 R_2 C_1 C_2}} \qquad (18)$$

$$d = 0.5 C_1 \omega_c (R_1 + R_2) \qquad (19)$$

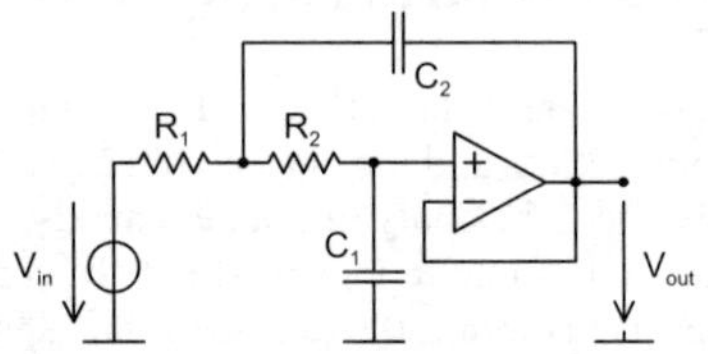

Figure 5: 2nd order Biquad lowpass filter.

We use two different value sets for the resistors and capacitors, one with $\omega_c = 100\ s^{-1}$ and $d = 0.5$ and the second at the same frequency but with $d = 2$. The property to be checked in this example is the occurrence of overshooting in the two filters. Since these properties should be proved for arbitrary input signals, the appropriate input value model is chosen. The input signal range is $V_{in} = [-2 \,..\, 2]$, so that the nonlinearity due to the restricted output voltage will effect the system behavior. The state space is restricted to $V_{c1} = [-4 \,..\, 4]$ and $V_{c2} = [-2.5 \,..\, 2.5]$.
The initial state in this example is assumed to be the DC operating point at $V_{in} = 0$. The question is, which states are reachable from this point for arbitrary input signals within the given range? Directly transfered to CTL the question reads as Equation (20). Instead of a single start point a start area is used, that is for example a box surrounding the initial point. Next, operation EF is used to check which states have a path that will eventually reach this starting area. However, the direction of this operation is wrong. Inverting time (iv) gives the correct formula (Equation (20)).

$$\Phi_3 = \mathsf{iv}(\mathsf{EF}((V_{c2} < 0.5) \;\&\; (V_{c2} > -0.5) \;\&\; (V_{c1} < 0.5) \;\&\; (V_{c1} > -0.5) \;\&\; (V_{in} < 0.5) \;\&\; (V_{in} > -0.5))) \qquad (20)$$

This equation is applied to both circuits. The results are shown in Figure 6 and Figure 7. The black box indicates the state space borders. The input voltage axis is perpendicular to the paper. As expected, in the highly damped circuit state V_{c1} remains within a range of ± 2 V whereas it reaches higher levels in the less damped case. It turns out that the opamp output restriction of ± 1.5 V does not have an impact on this result. This is because states V_{c1} and V_{c2} are not restricted by the opamp output.

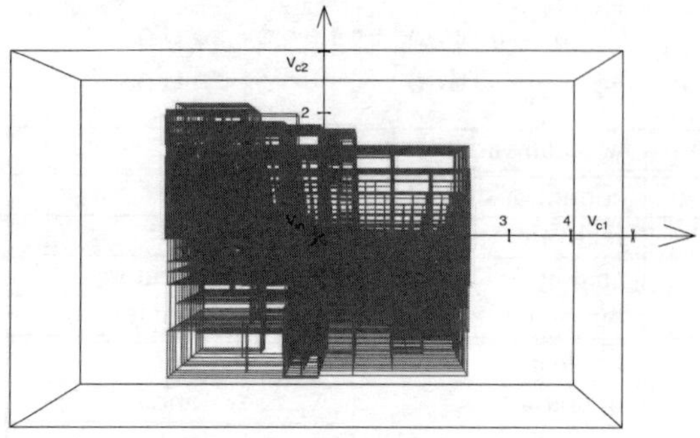

Figure 6: Result of Equation (20) for the highly damped circuit.

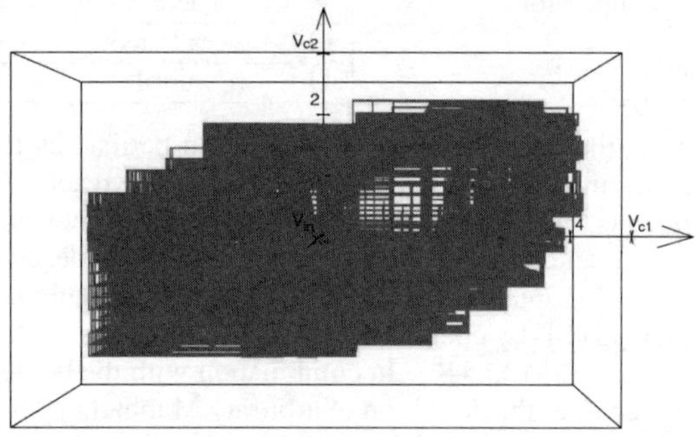

Figure 7: Result of Equation (20) for the less damped circuit.

To check the situation in more detail, one might be interested in the circuit's fix points. This is achieved by formula $\Phi_4 = \mathsf{iv}(\mathsf{EG}(1))$, where 1 denotes the whole state space.
To achieve the requested result, the input model has to be changed to constant input values, because it does not make sense to look for fix points while the input value may change arbitrarily. The graphical result is shown in Figure 8. The turning point due to the output voltage saturation is clearly visible.

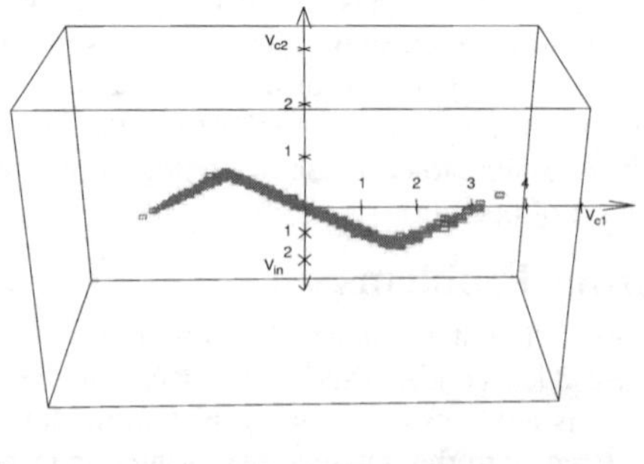

Figure 8: Fix points in the highly damped lowpass circuit.

The calculation of the discrete model in this example takes about three hours on a Sun ULTRA-II 300 MHz. Once the model is generated the calculation of CTL formulas is quite fast. It takes only seconds or a few minutes. Most of the runtime is spent in the numerical equation solver. Until now, a self written Newton solver

is used, so that a large improvement can be expected by using a commercial analog simulator for this task.
However, due to the n-dimensional subdivision algorithm, the total runtime complexity is exponential in the number of state variables, which is a serious problem for all state space exploration algorithms. Exponential runtime complexities are also known from digital model checking tools, but the use of efficient algorithms and data structures (e.g. BDDs) has extended the applicability for a wide range of circuits. Since similar techniques have not been utilized in the presented tool, a successful use of analog model checking in the future is likely, especially because the size of analog parts in todays designs is small compared to digital modules.

4.3 Tunnel Diode Oscillator

The analog system used in our third example is a simple tunnel diode oscillator circuit shown in Figure 9. The input voltage V_{in} is set to 2.6 V. In this operating point the circuit starts an oscillation automatically. The bounded state space is given by $V_C = [-0.2 .. 4.4]$ and $I_L = [-0.2 .. 4.0]$.

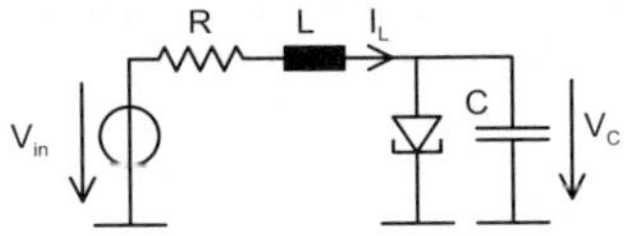

Figure 9: Tunnel diode oscillator circuit.

In a digital system a stable oscillation might be proved by $\mathsf{AG}(\mathsf{AF}(state_1))$ & $\mathsf{AG}(\mathsf{AF}(\overline{state_1}))$ [9]. This means: "On every computational path $state_1$ is *true* at some future time ($\mathsf{AF}(state_1)$). This condition is true for all future time ($\mathsf{AG}(\cdot)$). Simultaneously, this expression holds for $\overline{state_1}$". By this, the expression states that $state_1$ switches between *true* and *false* for all future time. The model checking algorithm will find all points in the state space which fulfill this condition. We choose the same expression for the analog verification tool, except of the inner terms. They are replaced by a translation to the analog world, that is, the current I_L is larger or smaller than some value.

$$\Phi_5 = \{\mathsf{AG}(\mathsf{AF}(I_L > 2.2))\ \&\ \mathsf{AG}(\mathsf{AF}(I_L < 1.6))\} \qquad (21)$$

The collection of boxes fulfilling this condition is shown in Figure 10 in light gray. Except of some border boxes and the middle region, the whole state space fulfills formula (21).

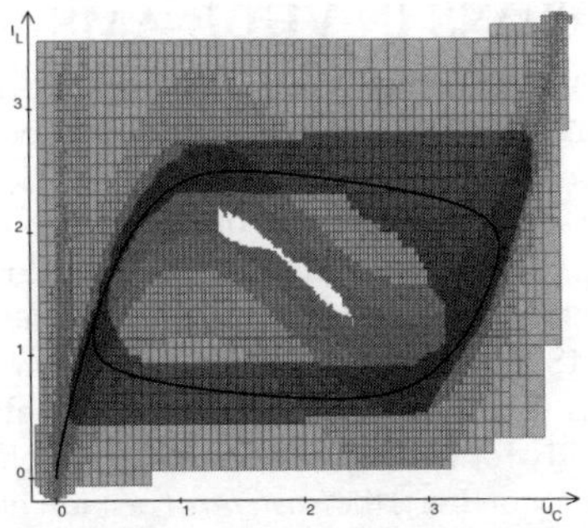

Figure 10: Model checking results Φ_5 and Φ_6.

We can conclude that nearly the whole plane will float into an stable orbit. The next question might concern the possible orbit geometry. We generate this by applying $\Phi_6 = \mathsf{iv}(\mathsf{EG}(\Phi_5))$;. By inverting the time, we exclude all points floating out of the orbit region. Since the orbit is stable, there is always a way out of the orbit at negative time. Therefore, operator EG is used instead of AG (resulting in an empty plane). The result Φ_6 contains the whole orbit calculated by an ordinary simulation (black line in Figure 10).

5. CONCLUSION

To apply digital model checking ideas to analog systems a discrete system model is needed. The main algorithmic task is to develop a state transition model in such a manner that the main nonlinear and dynamic properties of the analog system are retained. This is done using an automatic state space subdivision method and an algorithm developing the connection relation.
The implemented CTL model checker is based on digital algorithms, but it is extended by some operations, enabling the definition of analog properties in CTL. The capability of this language is shown by two experimental results. However, it is clear that the language is not sufficient for all analog properties. Especially, explicit time dependent properties, like slew rates or delays, are not covered until now.
The nonlinear examples show the use of the tool in analog design. However, applying CTL to analog systems, seems even more uncommon than doing so in digital. The result of CTL formulas can either be shown graphically or - even more formally - as a binary $true/false$ decision. The runtime is quite high for the discrete model generation, but there is a big optimization potential due to the prototype implementation.
As far as we know, the presented tool is the first approach to model checking for nonlinear analog systems. That opens a wide range of possibilities in applying formal methods not only to digital and hybrid systems but also to analog systems. Therefore, it is a step towards a more formalized analog design flow.

6. REFERENCES

[1] R. Alur, T. Henzinger, G. Lafferriere, and G. Pappas. Discrete abstractions of hybrid systems. *Proceedings of IEEE*, (88):971–984, 2000.

[2] J. Burch, E. Clarke, D. Long, K. McMillian, and D. Dill. Symbolic model checking for sequential circuit verification. *IEEE Transactions on Computer-Aided Design of Integrated Circuits and Systems*, 13(4):401–424, 1994.

[3] M. Dellnitz, G. Froyland, and O. Junge. The algorithms behind gaio - set oriented numerical methods for dynamical systems. In B. Fiedler, editor, *Ergodic Theory, Analysis, and Efficient Simulation of Dynamical Systems*, pages 145–174. Springer-Verlag, Berlin, 2001.

[4] M. Günther and U. Feldmann. Cad-based electric circuit modeling in industry, part i: Mathemetical structure and index of network equations. *Suveys on Mathematics for Industry*, 8(2):97–129, 1999.

[5] L. Hedrich and W. Hartong. Approaches to formal verification of analog circuits. In P. Wambacq, editor, *Low-Power Design Techniques and CAD Tools for Analog and RF Intergrated Circuits*, pages 155–191. Kluwer Academic Publishers, Boston, 2001.

[6] T. Henzinger and P.-H. Ho. Algorithmic analysis of nonlinear hybrid systems. *CAV '95: International Conference on Computer-Aided Verification, LNCS*, 939(7):225–238, 1995.

[7] R. Kurshan and K. McMillan. Analysis of digital circuits trough symbolic reduction. *IEEE Transactions on Computer-Aided Design of Integrated Circuits and Systems*, 10(11):1356–71, 1991.

[8] R. März. Numerical methods for differential algebraic equations. *Acta Numerica*, pages 141–198, 1991.

[9] K. McMillian. Symbolic model checking. *Kluwer Academic Publishers, Boston*, 1993.

[10] A. Neumaier. Interval methods for systems of equations. *Cambridge University Press, Cambridge*, 1990.

[11] A. Puri. Theory of hybrid systems and discrete event systems. *Dissertation, University of California ar Berkeley*, 1995.

Regularization of Hierarchical VHDL-AMS Models using Bipartite Graphs

Jochen Mades
Infineon Technologies AG
Munich, Germany
Jochen.Mades@vhdl-ams.de

Manfred Glesner
Institute of Microelectronic Systems
Darmstadt University of Technology, Germany
glesner@mes.tu-darmstadt.de

ABSTRACT

The powerful capability of VHDL-AMS to describe complex continuous systems in form of differential algebraic equations (DAEs) often leads to problems during numerical simulation. This paper presents a discrete algorithm to analyze unsolvable DAE systems and to correct the underlying hierarchical VHDL-AMS description automatically in interaction with the designer, avoiding time-consuming manual error correction.

Categories and Subject Descriptors

G.2.2 [**Discrete Mathematics**]: Graph algorithms

General Terms

Algorithms, Design

Keywords

DAEs, VHDL-AMS, Regularization, Structural Solvability, Bipartite Graphs

1. INTRODUCTION

VHDL-AMS is established as an IEEE standard (1076.1) designed to describe discrete as well as continuous behavior in a unique language. VHDL-AMS is very valuable for the simulation of complex mixed Systems-on-Chip (SoC), due to the capability to describe mixed-signal systems in one common language and the possibility to simulate the SoC with a single simulator. Therefore, the time consuming coupling of discrete simulators with continuous simulators gets unnecessary.

Apart from the simulation of SoCs, VHDL-AMS is also of great interest in the world of analog design, since it allows the description of analog systems or subsystems on a very high level of abstraction, which can lead to enormous simulation speed-ups (factor 10 to 100).

In contrast to SPICE-based simulators the intention of VHDL-AMS and especially of VHDL-AMS simulator packages is not to have only a variety of basic library elements, which are interconnected by topology descriptions like structural VHDL or SPICE-like netlists to get the final circuit behavior. The power of VHDL-AMS is to have the possibility to describe the behavior of a complex system by formulating directly the corresponding differential algebraic equations (DAEs) of the system.

However, this universality implies that standard topological tests (based on simulator libraries) like the ones utilized to check DC-solvability are not sufficient anymore. Thus, for VHDL-AMS new diagnostic tools have to be developed to identify problems caused by unsolvable systems.

For the purpose of supporting designers using VHDL-AMS to describe complex systems on a high level of abstraction, we developed an algorithm that localizes critical equations in a DAE system. Furthermore, if it is not solvable, the algorithm checks whether the system can be regularized by a special VHDL-AMS language construct - the break statement. In that case the description of the DAE system is automatically regularized in interaction with the designer, insuring that the corrected VHDL-AMS source code then leads to a solvable system.

Section 2 initially gives an overview over different types of equations appearing in VHDL-AMS. Then Section 3 introduces standard analysis methods for structural solvability before in Section 4.1 a short overview of hierarchical modeling in VHDL-AMS is given. Section 4.2 introduces the *break statement* which is used in Section 5 where the algorithm for the regularization of structural unsolvable DAE systems, described in hierarchical VHDL-AMS, is presented. Finally, the algorithm will be illustrated in Section 6 on an abstract description of a $\Sigma\Delta$-analog to digital converter and first results of a prototype implementation will be shown in Section 7.

2. EQUATIONS IN VHDL-AMS

During the pre-simulation phase of a VHDL-AMS model, the so-called elaboration [1], the DAE system needed for numerical simulation is created. Therefore, different VHDL-AMS language constructs like *simple simultaneous statements, component instantiations* and *quantity declarations* have to be evaluated and corresponding characteristic expressions have to be generated.

The VHDL-AMS standard differentiates between three different sets of equations. The first set, the *explicit set*, contains all the expressions explicitly formulated in the model. The characteristic expressions of the second set, the *structural set*, are generated implicitly. These so-called *structural expressions* represent the conservative laws KCL and KVL in case of *terminal ports*, while they represent the equality of internal and external quantities in case of *quantity ports*. The third set, the *augmentation set*, provides characteristic expressions defining the explicitly used derivatives of the VHDL-AMS model. For each derivative of the VHDL-AMS model there is exactly one characteristic expression in the *augmentation*

DAC 2002, June 10-14, 2002, New Orleans, Louisiana, USA.

set and the derivative is called the *tag* of the characteristic expression. During DC-Analysis these expressions set the derivatives to zero (*quiescent state augmentation set*).

3. STRUCTURAL SOLVABILITY

A necessary condition for the solvability of a system of equations is the so-called *structural solvability*. The *structural solvability* focuses on the structure of the matrix (Jacobian), considering only matrix entries unequal to zero. This structure is normally represented by a bipartite graph B where the nodes of the first partition of B represent the rows (expressions) and the nodes of the second partition of B represent the columns (quantities) of the matrix [2, 3, 4, 5]. An edge between a node i of the first partition and a node j of the second partition exists if the entry (i, j) of the matrix is unequal to zero. If there exists a *perfect matching* in B, which can be determined by efficient matching algorithms [6], then the matrix is structural solvable.
In the following we denote by a *structural DC-solvable DAE system*, a DAE system for which the corresponding DC system is structural solvable.

4. PROPERTIES OF VHDL-AMS

4.1 Hierarchical Modeling

VHDL-AMS allows to model in a hierarchical manner, meaning that subsystems, the so-called *design units*, can be developed independently. A *top-level description* connects all the subsystems together to a complete *design*.
Each declaration in such a *design unit* has its own *scope*. A scope is the area inside the VHDL-AMS source code, where a declaration is visible and can be used.
The quantities of the characteristic expressions of the *explicit set* belong to the same scope, in contrast to the quantities of the *structural set* belonging to different scopes. Therefore, the presented algorithm distinguishes between nodes in the bipartite graph representing *structural expressions* and others.

4.2 The Break Statement

The *break statement* is an advanced VHDL-AMS language construct which allows to initialize the DAE system at the beginning of a transient simulation and to mark discontinuities during such a transient simulation. In either case a DC-Analysis of the system is performed.
The application of a *break statement* substitutes a characteristic expression of the *current augmentation set* (see Section 2) selected by the *selector quantity* of the *break statement*[1] by an expression

$$Q_{break} = |E|, \tag{1}$$

where Q_{break} denotes the *break quantity* and $|E|$ the evaluated *expression* of the *break statement*. The *selector quantity* and the *break quantity* are restricted to belong to the same scope.
In the bipartite graph of the DAE system, the characteristic expression of the *augmentation set* is represented by a node of the first partition. The substitution of this characteristic expression by the application of the *break statement* is reflected in the replacement of the edge in the bipartite graph, which is incident to the node representing the characteristic expression of the *augmentation set* with an edge between that node and the node of the second partition of the bipartite graph representing the *break quantity* Q_{break} of the *break statement*.

[1] There is a one to one correspondence between the *selector quantity* and the *tag* of the characteristic expressions of the augmentation set [1].

5. REGULARIZATION-ALGORITHM

The presented algorithm analyzes a structural DC-unsolvable DAE system described in VHDL-AMS and checks if the DAE system is regularizable, meaning if it is possible to transform it into a structural DC-solvable DAE system. To this end, a first bipartite graph $B_1 = \{V_1^1 \cup V_1^2; E\}$ of the DAE system is set up as described in Section 3. The nodes of the first partition of B_1 are divided into three different categories belonging to the type of the characteristic expression they represent:

$$V_1^1 = V_e \cup V_s \cup V_a$$

The nodes $v \in V_e$ represent the characteristic expressions of the *explicit set*, nodes $v \in V_s$ characteristic expressions of the *structural set* and nodes $v \in V_a$ represent characteristic expressions of the current *augmentation set* (see Section 2).
If the analyzed DAE system is structural DC-unsolvable, then the *maximal matching* in B_1 is not a *perfect* one (see Section 3). The aim of the presented algorithm is to replace specific edges in B_1 in a way that a *perfect matching* in B_1 can finally be determined. This edge-replacement can be performed with *break statements* (see Section 4.2). To formulate adequate *break statements* the algorithm analyzes B_1 in order to detect valid *break-* and *selector quantities* in the following manner.
First, the algorithm removes all nodes $v \in V_s$ from B_1 representing *structural expressions*, since the *break-* and *selector quantities* are restricted to belong to a common scope and because of *structural expressions* may contain unknowns from different scopes (see Section 2). Edges incident to these nodes are also removed from B_1.
Then the *connected components* [7] of the modified bipartite graph are considered and standard matching algorithms [6] determine a *maximal matching* in every connected component in a way that no *alternating path* from an *unmatched* node $v \in V_e$ representing a characteristic expression of the *explicit set* to a *matched* node $v \in V_a$ representing a characteristic expressions of the *augmentation set* in B_1 exists. An *alternating path* is a path in a bipartite graph with edges belonging alternately to a given matching M starting with an edge $e \notin M$ [8]. Furthermore the matching algorithm avoids *alternating paths* from *unmatched* nodes $v_u \in V_1^2$ representing quantities which are not contained in *structural expressions* to *matched* nodes $v_m \in V_1^2$ representing quantities which are contained in *structural expressions*.
In the next step the algorithm considers *matching edges* incident to nodes $v \in V_a$. If there exists an *alternating path* from the node $v \in V_1^2$, incident to such a *matching edge*, to a node representing a quantity which is contained in a *structural expression* then the *matching edge* is removed from the *connected component*. This removement leads to an *unmatched* node in both partitions of the *connected component*. The reason for that is to give the algorithm a higher degree of freedom in detecting valid *selector-* and *break quantities* as described below.
In the following step the algorithm starts building a second bipartite graph $B_2 = \{V_2^1 \cup V_2^2; E\}$ by analyzing the nodes of the *connected components* of B_1. Each *unmatched* node $v \in V_a$ of the first partition of the analyzed *connected component* serves for the determination of a valid *selector quantity*. Therefore, for each determined *unmatched* node $v \in V_a$ a corresponding node $sel \in V_2^1$ of the first partition of B_2 is created. If the algorithm detects an *unmatched* node $v \in V_e$ in the first partition of the *connected component* then the algorithm suspends without having a chance to regularize the system by specifying *break statements*. Additionally, the algorithm creates a node $brk \in V_2^2$ in the second partition of B_2 for every *unmatched* node $v \in V_1^2$ of the second partition of the analyzed *connected component*. The nodes $brk \in V_2^2$ serve for the determination of valid *break*

quantities needed for the specification of regularizing *break statements* as described later. Furthermore the nodes $v \in V_s$ representing the *structural expressions* removed at first from the first bipartite graph B_1 are represented by nodes $sce \in V_2^1$ of the first partition of the second bipartite graph B_2.
The edges of the second bipartite graph B_2 between nodes of the first partition and nodes of the second partition are created in the following manner. A Node $sel \in V_2^1$ is connected via an edge to a node $brk \in V_2^2$ if the *unmatched quantity* represented by the node *brk* belongs to the same scope as the *selector quantity*[2] represented by the node *sel*. A node $sce \in V_2^1$ of the first partition of B_2 representing a *structural expression* is connected to a node $brk \in V_2^2$ of the second partition of B_2 if either the *unmatched quantity* represented by *brk* is contained in the *structural expression* or if there exists an *alternating path* in the first bipartite graph B_1 from the node representing the *unmatched quantity* to a node representing a *quantity* which is contained in the *structural expression*.
In the next step standard matching algorithms determine a *maximal matching* in B_2. If this matching is a *perfect matching*, then the structural DC-unsolvable DAE system is regularizable by the specification of different *break statements*. These *break statements* are added to the VHDL-AMS source code in order to regularize the structural DC-unsolvable DAE system.
The *break statements* are determined by the evaluation of the *matching edges* of the second bipartite graph B_2 incident to nodes $sel \in V_2^1$. The evaluation of such a *matching edge* is performed in the following way. First the algorithm determines all nodes $v \in V_a$ in the first bipartite graph B_1 which can be reached by *alternating paths* in B_1 starting from the *unmatched* node $v \in V_a$ represented by the node *sel*. The *tags* of these determined nodes together with the *tag* of the *unmatched* node represented by the node *sel* are the valid *selector quantities* for the regularizing *break statement*. Then the algorithm evaluates the node $brk \in V_2^2$ incident to the *matching edge* in order to detect the valid *break quantities* of the *break statement*. The valid *break quantities* are the *quantities* represented by nodes $v \in V_1^2$ of the second partition of B_1 which can be reached on *alternating paths* starting from the *unmatched* node represented by *brk* and the *quantity* represented by that *unmatched* node itself.
In the next step the algorithm interacts with the designer asking to choose one of the determined valid *selector quantities* as well as one of the determined valid *break quantities* and to specify an expression for the corresponding *break statement* (see Section 4.2).
Then the first bipartite graph B_1 has to be updated in the following manner. The edge incident to the node $v \in V_a$ in B_1 representing the *characteristic expression tagged* by the chosen *selector quantity* is replaced by an edge between that node $v \in V_a$ and the node $v \in V_1^2$ representing the chosen *break quantity*. If either the node $v \in V_a$ selected by the *selector quantity* or the node $v \in V_1^2$ representing the *break quantity* is a *matched* node in B_1, then the *alternating path* from the *unmatched* node represented by the node *sel* or *brk* of B_2 to the selected node has to be *augmented* [8], meaning that the edges $e \in M$ of the *alternating path* are removed from the given maximal matching M and the edges $e \notin M$ are added to the maximal matching M. This enlarges the maximal matching of B_1, which was originally not a perfect one. Then the algorithm continues evaluating the next *matching edge* in B_2.
Finally, after having evaluated all the *matching edges* the matching in the first bipartite graph B_1 is a *perfect matching* which represents the structural DC-solvability of the DAE system. The proof and a more detailed description of the algorithm are given in [9].

[2]The derivative of the *selector quantity* is the *tag* of the *characteristic expression* of the current *augmentation set* represented by the *unmatched* node $v \in V_a$ in the first bipartite graph B_1.

6. EXAMPLE

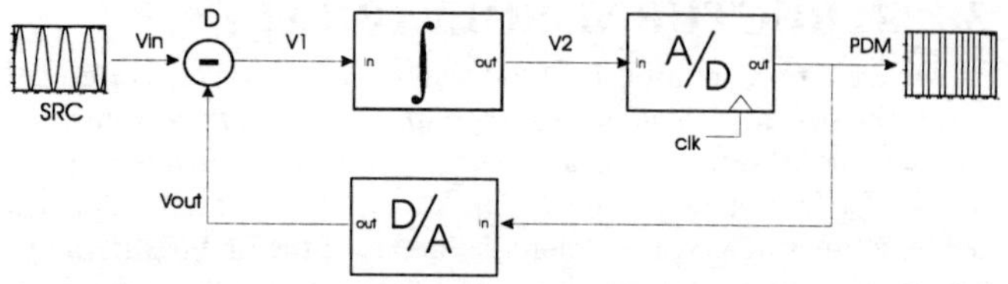

Figure 1: Sigma-Delta Analog to Digital converter

Figure 1 shows the block diagram of a sigma-delta analog to digital converter (ΣΔ-ADC). Such a mixed signal system can conveniently be described in VHDL-AMS. To demonstrate how the presented algorithm works we described the ΣΔ-ADC hierarchically. The continuous behavior of the sinus source *SRC*, the difference building part *D*, the integrator *I* and the 1-Bit digital to analog converter *DA* can be described with equations (e1) to (e4), respectively.

$$
\begin{aligned}
SRC &= 2.0 * sin(2.0\pi * 5.0 \cdot 10^7 * t) && \text{(e1)}\\
D_{out} &= D_{in1} - D_{in2} && \text{(e2)}\\
\tfrac{d}{dt} I_{out} &= I_{in} && \text{(e3)}\\
DA_{out} &= \pm V_{dd} && \text{(e4)}
\end{aligned}
$$

The *structural expressions* (s1) to (s8) result from the elaboration of the complete hierarchical design of the ΣΔ-ADC.

$$
\begin{aligned}
V_{in} &= SRC \quad \text{(s1)} & V_1 &= I_{in} \quad \text{(s5)}\\
V_{in} &= D_{in1} \quad \text{(s2)} & V_2 &= I_{out} \quad \text{(s6)}\\
V_{out} &= D_{in2} \quad \text{(s3)} & V_2 &= AD_{in} \quad \text{(s7)}\\
V_1 &= D_{out} \quad \text{(s4)} & V_{out} &= DA_{out} \quad \text{(s8)}
\end{aligned}
$$

For the calculation of the DC operating point the *quiescent state augmentation set* [1] is determined setting the derivatives of the DAE system to zero. In our example there is just one derivative appearing in the description of the Integrator (see equation (e3)):

$$\tfrac{d}{dt} I_{out} = 0.0. \quad \text{(a1)}$$

The subsystem formed by equation (s6) and (s7) leads to an unsolvable DAE system, because of the three unknowns V_2, AD_{in} and I_{out} are not contained in further equations. Figure 2(a) shows the corresponding bipartite graph with a maximal matching. Due to the maximal matching being not a perfect one - s_2 as well as AD_{in} are unmatched - the structural unsolvability of the DAE system is confirmed.

The presented algorithm starts analyzing the VHDL-AMS description of the structural DC-unsolvable DAE system by removing the nodes representing *structural expressions* from B_1 (see Figure 2(b)) and by determining a *maximal matching* for each *connected component*. Then the algorithm the *matching edges* incident to nodes of the current *augmentation set* are removed under the conditions explained in Section 5. The next step of the algorithm creates the second bipartite graph B_2 (see Figure 3[3]) by evaluating the *unmatched* nodes of the *connected components*. There is one *unmatched* node a_1 representing the characteristic expression (a1) of the *augmentation set*. This node is represented as node sel_1 in the first partition of B_2 (shadowed node). Together with the eight *structural*

[3]First partition - upper lane, second partition - lower lane.

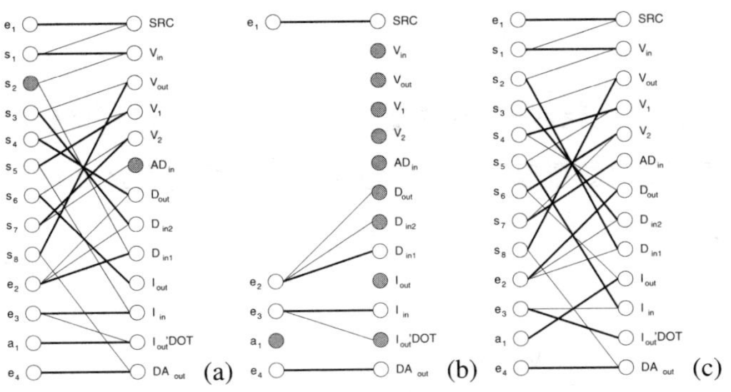

Figure 2: Bipartite graphs of DAE Systems

expressions sce_1 to sce_8 this node completes the first partition of the second bipartite graph B_2, while the second partition is formed by nodes representing the *unmatched* nodes of the second partition of B_1.

To determine whether the structural DC-unsolvable DAE system is

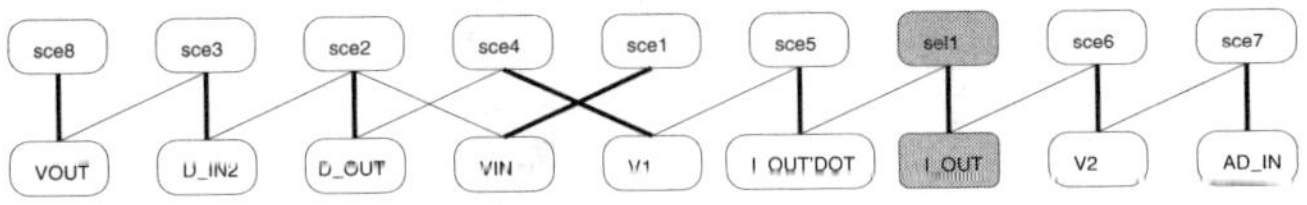

Figure 3: Second bipartite graph used for regularization

regularizable with *break statements* the algorithm determines a maximal matching. Because of the existence of a *perfect matching*[4] in B_2, the DAE system of the ΣΔ-ADC is regularizable.

The evaluation of the *matching edge* $e = \{a_1, I_{out}\}$ leads to the following *break statement*. Because there is no *alternating path* in B_1 from the node representing (a1) to another node representing a characteristic expression of the *augmentation set*, no choice for the *selector quantity* can be given. Therefore the *selector quantity* I_{out} is determined from the *tag* of (a1). Furthermore in our example there is no *alternating path* in B_1 from the node representing the quantity I_{out}. Due to that, the *break quantity* in our example is I_{out}. Finally the algorithm interacts with the designer to specify an expression $|E|$ for the following *break statement*:

```
break for Iout use Iout => |E|;
```

The application of this *break statement* finally will substitute the equation (a1) with equation

$$I_{out} = |E|. \tag{2}$$

This substitution can be seen by the replacement of the edge being incident to the node representing (a1). Figure 2 (c) shows the bipartite graph corresponding to the regularized DAE system described in VHDL-AMS. The perfect matching shows the structural DC-solvability.

7. EXPERIMENTAL RESULTS

The presented algorithm was implemented to extend our VHDL-AMS Simulation Framework [10] to support the diagnostic of hierarchical VHDL-AMS models. Analyzing the ΣΔ-ADC example from Section 6, the algorithm interacts with the designer who specifies e.g. $|E| = 2.3 \cdot 10^{-3}$ for the *expression* of the proposed *break statement*. Then the *break statement* is added to the VHDL-AMS source code of the integrator (see Figure 4).

[4]Matching edges are illustrated bold.

```
entity integrator is
   port (quantity I_in : in  real;
 quantity I_out : out real);
end integrator;

architecture behave of integrator is
begin -- behave

  I_out'DOT == I_in;

end behave;
```

```
entity integrator is
   port (quantity I_in : in  real;
 quantity I_out : out real);
end integrator;

architecture behave of integrator is
begin  -- behave

  I_out'DOT == I_in;
  -- inserted by BA (05.12.2001)
  break for I_OUT use I_OUT => 2.3e-3;
end behave;
```

Figure 4: Original and corrected VHDL-AMS description

Finally the algorithm asks to recompile the design unit and to restart the simulation with the corrected design unit of the integrator.

8. CONCLUSION

The presented algorithm supports the development of high level models of mixed-signal systems using VHDL-AMS. Unsolvable DAE systems resulting from VHDL-AMS descriptions are analyzed and the designer is supported in correcting the underlying VHDL-AMS source code.

A first prototype of the algorithm extends the environment of the Infineon Technologies inhouse circuit simulator TITAN [11] and was efficiently used at the Technical University of Darmstadt in the summer course "High-Level Modeling of Mixed-Signal VLSI Circuits".

9. REFERENCES

[1] The Institute of Electrical and Electronics Engineers, *IEEE Standard 1076.1-1999. VHDL Language Reference Manual*, IEEE, 1999.

[2] Kazuo Murota, *Matrices and Matroids for Systems Analysis*, Algorithms and Combinatorics 20. Springer Verlag, 2000.

[3] Andras Recski, *Matroids Theory and its Applications in Electric Network Theory and in Statics*, Springer Verlag, 1989.

[4] J. Unger, A. Kröner, and W. Marquardt, "Structural analysis of differential-algebraic equation systems - theory and application," in *Comput. Chem. Engng.*, 1995, vol. 19.

[5] B. Petersen, "Investigating solvability and complexity of linear active networks by means of matroids," in *IEEE Transactions on Circuits and Systems*, 1979, vol. 26.

[6] B. V. Cherkassky, A. V. Goldberg, P. Martin, J. C. Setubal, and J. Stolfi, "Augment or push? a computational study of bipartite matching and unit capacity flow algorithms," in *Technical Report 98-036R*. NEC Research Institute, Inc., 1998.

[7] O. Melnikov, R. Tyshkevich, V. Yemelichev, and V. Sarvanov, *Lectures on Graph Theory*, B.I. Wissenschaftsverlag, 1994, ISBN 3-411-17121-9.

[8] A. V. Aho, J. E. Hopcroft, and J. D. Ullmann, *Data structures and algorithms*, Addison-Wesley, 1983.

[9] Jochen Mades, *Structural Solvability of hierarchical VHDL-AMS Models*, PhD. Thesis, Darmstadt University of Technology, 2002, (in preparation).

[10] T. Schneider, J. Mades, M. Glesner, A. Windisch, and W. Ecker, "An Open VHDL-AMS Simulation Framework," in *2000 IEEE/ACM International Workshop on Behavioral Modeling and Simulation (BMAS 2000)*, Orlando, Florida, USA, October 16-18 2000, (pages missing).

[11] G. Denk, U. Feldmann, C. Hammer, M. Kahlert, R. Neubert, G. Reissig, and A. Windisch, "Erweiterung eines Standard-Schaltkreissimulators in Richtung VHDL-AMS," in *Proceedings of the 5. ITG/GMM-Diskussionssitzung, Analog 99*. Informationstechnische Gesellschaft im VDE, 1999.

Improving the Generality of the Fictitious Magnetic Charge Approach to Computing Inductances in the Presence of Permeable Materials

Yehia Massoud
Synopsys Inc.
Mountain View, CA

Jacob White
MIT
Cambridge, MA

ABSTRACT

In this paper we present an improvement to the fictitious magnetic charge approach to computing inductances in the presence of permeable materials. The improvement replaces integration over a "non-piercing" surface with a line integral and an efficient quadrature scheme. Eliminating the need to generate non-piercing surfaces substantially simplifies handling problems with general geometries of permeable materials. Computational results are presented to demonstrate the accuracy and versatility of the new method.

Categories and Subject Descriptors

G.4 [**Mathematical Software**]: Algorithm design and analysis; G.1.2 [**Approximation**]: Approximation of surfaces and contours

General Terms

Algorithms

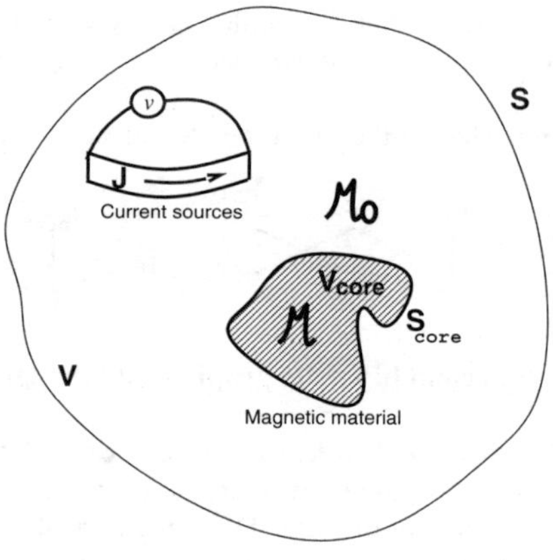

Figure 1: Current sources are outside the magnetic material

1. INTRODUCTION

Improvements in fabrication technology is making it possible to use permeable materials in a variety of micromachined devices (or MEMS), and designers are exploiting these materials to increase magnetic force or inductance while preserving small device size [1, 2, 3, 4, 5, 6, 7]. However, the planar nature of most micromachining technology implies that designers can not simply wrap coils of wire around highly permeable cores. Instead, the geometric freedom available results in designs that are difficult to analyze analytically, making numerical extraction tools essential.

One approach to computing inductance in the presence of permeable materials is the fictitious magnetic charge method [8, 9], in which permeable material interfaces are replaced with fictitious sheets of magnetic charge. The fictitious magnetic charges are then determined by enforcing continuity of normal magnetic fields at the permeable material interfaces. The method is accurate, even for highly permeable materials, because fields are never computed inside those permeable materials. However, there is also a difficulty. In order to use the fictitious magnetic charge method it is necessary to construct a surface which bounds the current carrying conductors and does not penetrate the permeable material. Constructing "non-piercing" surfaces for arbitrary geometries of conductors and permeable material structures is very difficult.

In this paper we show that the fictitious magnetic charge method can be modified, eliminating the need for the non-piercing surfaces. In Section 2, we briefly describe the integral formulation of the fictitious magnetic charge method and its use of non-piercing surfaces. In Section 3, we present a new method which uses line integrals and quadrature schemes. In Section 4, we give computational results to demonstrate the accuracy of our approach. Finally, in Section 5, we summarize the work presented in the paper.

2. FORMULATION BACKGROUND

For many problems in MEMS, one can assume that the permeable materials do not conduct current and are separated from conductors, as shown in Figure 1. For such problems it is possible to derive an integral formulation for the conductor currents based on introducing fictitious magnetic charge at permeable material interfaces [8]. Using this

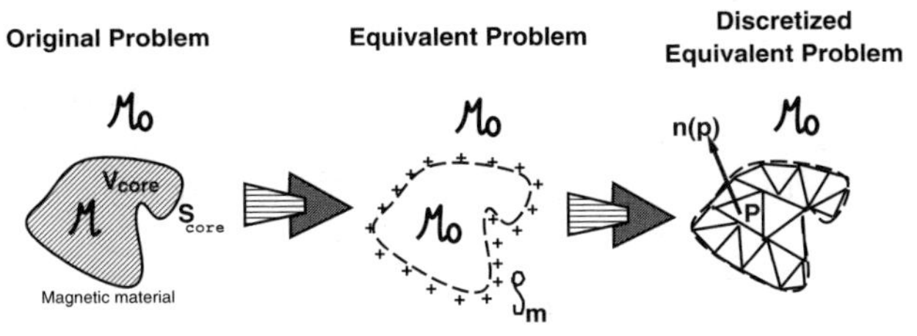

Figure 2: Linear magnetic material can be represented with an equivalent free space problem with magnetic surface charges distributed on the magnetic material interfaces. The interface is discretized into panels on which charge is assumed constant

approach the magnetic problem becomes equivalent to the free space problem, as shown in Figure 2. The fictitious magnetic charge can be determined from the equation for the jump in the normal magnetic field at the permeable material interface as in,

$$\rho_m(r)\frac{2\pi(\mu_r+1)}{(\mu_r-1)} = \nabla \times \int_v \frac{J(r')dv' \cdot n(r)}{| r - r' |} - \int_s \rho_m(r')n(r) \cdot \nabla \frac{1}{| r - r' |} dS'_{core} \quad (1)$$

where ρ_m is the fictitious surface charge density, S_{core} is the surface of the magnetic material, μ_r is the magnetic material's relative permeability, J is the current density, and n(r) is the unit vector normal to the magnetic material surface calculated at point r. Note that the second term of (1) represents the normal magnetic field due to currents, and the third term represents the normal magnetic field due to fictitious magnetic charges.

The conductor currents satisfy an integral equation derived using a vector potential generated by the conductor currents and the fictitious magnetic charges,

$$\frac{J(r)}{\sigma} + \frac{j\omega\mu}{4\pi}\int_V' \frac{J(r')}{| r - r' |} dv' + \frac{j\omega\mu}{4\pi} \nabla_r \int_{S'_m} \frac{-\rho_m(r')}{| r - r' |} dS'_m = - \nabla \phi(r) \quad (2)$$

where S_m is the surface of the permeable material, and ϕ is the scalar potential. Note that the second term and the third term of (2), divided by $j\omega\mu$, represent the vector potential due to the currents and the the vector potential due to fictitious magnetic charges, respectively.

In order to solve these two coupled integral equations (1) and (2), the surface of the magnetic material is discretized into panels and the current carrying conductors are sliced into filaments that are then combined into loops. Using these so-generated loop and panel basis functions converts the integral equations (1) and (2) to a linear system of equations [8]:

$$\begin{bmatrix} R(\omega) + j\omega L_J(\omega) & j\omega L_\rho(\omega) \\ Hn_J & (Hn_\rho - I) \end{bmatrix} \begin{bmatrix} \overline{I}_M \\ \overline{q}_m \end{bmatrix} = \begin{bmatrix} V \\ 0 \end{bmatrix} \quad (3)$$

where I_M, V, and q_m are vectors of loop currents, loop voltages, and fictitious magnetic charge densities, respectively. The sub-matrices R, L_J, Hn_ρ and Hn_J are computed using techniques described in [8, 9].

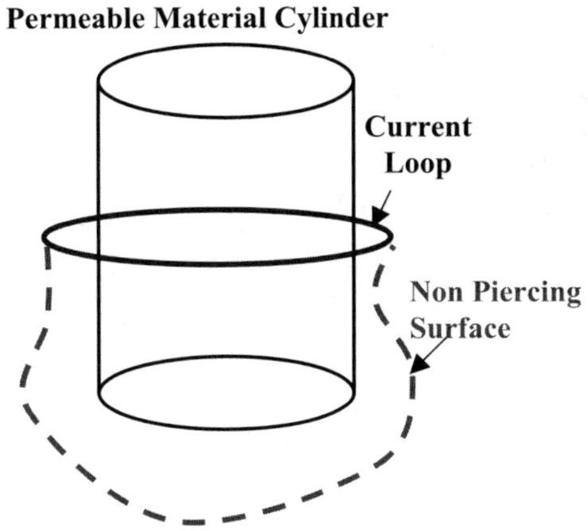

Figure 3: A non Piercing current loop surface. The surface should avoid cutting the permeable material in order to avoid numerical cancellation errors.

2.1 Evaluating $[L_\rho]_{ki}$

The matrix element $[L_\rho]_{ki}$ corresponds to the impact of magnetic charge on panel k on the current in loop i and is given by:

$$[L_\rho]_{ki} = -\frac{j\omega\mu_0}{4\pi}\int_{S_{loop\ i}} \nabla \frac{1}{|r - r'|} \cdot n(r)dS, \quad (4)$$

where n(r) is the normal to the surface of current loop i at point r and $S_{loop\ i}$ is the non-piercing surface bounded by the filaments in loop i. In order to avoid numerical cancellation errors, the surface should be non-piercing. That is, the surface should not penetrate the magnetic material, as shown in Figure 3.

The integral in (4) comes from integrating 1/r over the surface bounded by the filaments in current loop i and evaluating the result at a point at the permeable material surface. For simple configurations of current carrying conductors and permeable materials, one can find a non-piercing surface like the tented one shown in Figure 4. This tented surface can then be divided into triangles, and therefore, $[L_\rho]_{ki}$ can be expressed as:

$$[L_\rho]_{ki} = -\frac{j\omega\mu_0}{4\pi}\sum_{\triangle}\int_{S_\triangle} \nabla_r \frac{1}{| r - r' |} \cdot n(\triangle)dS_\triangle \quad (5)$$

where $S_\triangle$ is a triangle on the current loop surface, and $n(\triangle)$ is the normal to triangle on the current loop (loop) surface. The integral inside the summation in (5) is the potential due to a dipole charge distribution on a triangles which can be evaluated analytically [10].

3. IMPROVING THE EVALUATION OF $[L_\rho]_{KI}$

In this section, we show the improvement we have made to calculation of the $[L_\rho]_{ki}$ integral using line integration. This avoids the problem of finding and constructing non-piercing surfaces which can be very difficult and sometimes impossible for a range of MEMS devices such multiple turn spiral inductors. Also, this method facilitates the automation of inductance extraction using the fictitious magnetic charge no matter how complicated is the configuration of the current carrying conductors or the permeable structures.

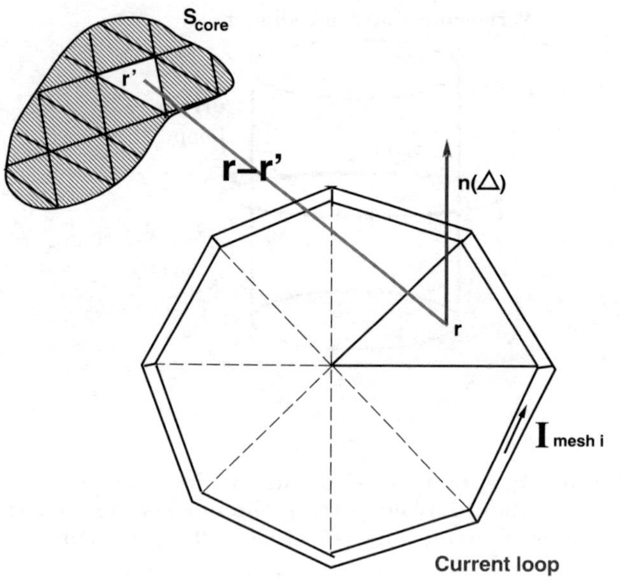

Figure 4: Evaluating $[L_\rho]_{ki}$, the effect of the magnetic charge on panel k on the current in loop i. Note that the surface of the current loop i is divided into triangles.

Consider translating the surface of the loop to a new coordinate system such that the center of panel k is the origin of the new coordinate system, P_0. Thus, the integral in (4) becomes

$$[L_\rho]_{ki} = \frac{j\omega\mu_0}{4\pi} \int_{S_{loop\ i}} \frac{1}{r^2} [u_r \cdot n(r)]\, dS \qquad (6)$$

where $n(r)$ is the normal to the surface of current loop i at point r and u_r is the unit vector along the r direction. The integral in (6) is equal to $\frac{j\omega\mu_0}{4\pi} I_\Omega(P_0)$, where $I_\Omega(P_0)$ is the solid angle integral. The solid angle integral is dependent only on the contour of the surface [11]. The solid angle integral $I_\Omega(P_0)$ is given by

$$I_\Omega(P_0) = \frac{1}{4\pi} \oint_C \left[\frac{Z(l)}{R(l)} - 1\right] \frac{(\hat{\rho}\cdot\hat{v})\,(\hat{t}\cdot\hat{\eta})}{\rho} dL \qquad (7)$$

where R and Z are the two spherical coordinates of the point l on contour C, ρ is the projection of vector R on the x-y plane, $\hat{\rho}$ is the unit vector along ρ, $\hat{t}$ and $\hat{\eta}$ are the unit tangential vectors of the contour C and the projection curve of C on x-y plane, respectively, and $\hat{v} = \hat{\eta} \times \hat{z}$ with $\hat{z}$ being a unit vector along the z-axis. The geometry is shown in Figure 5.

Now we use the idea of the solid angle to get an efficient way to calculate $[L_\rho]_{ki}$. After translating the surface of the loop to a new coordinate system such that the center of panel k is the origin of the new coordinate system, P_0, $[L_\rho]_{ki}$ can be calculated as in

$$[L_\rho]_{ki} = \sum_{f=1}^{\#filaments\ in\ loop\ i} E_{kf} =$$

$$\frac{j\omega\mu_0}{4\pi} \sum_{f=1}^{\#fil.in\ loop\ i} \int_{L_{fil_f}} \left[\frac{Z(l)}{R(l)} - 1\right] \frac{(\hat{\rho}\cdot\hat{v})\,(\hat{t}\cdot\hat{\eta})}{\rho} dL \qquad (8)$$

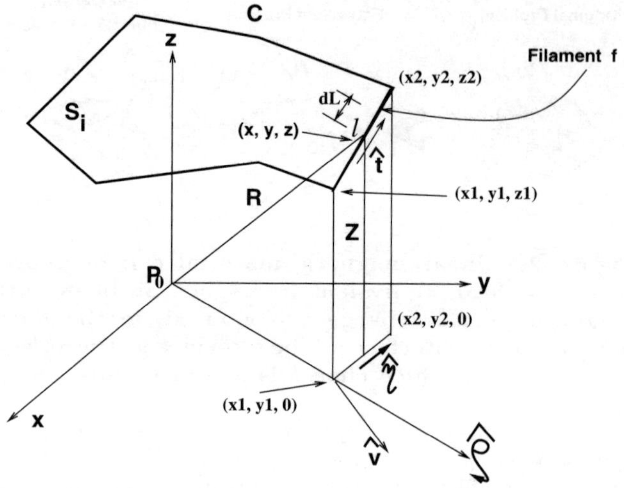

Figure 5: Evaluating of the solid angle integral at the origin P_0

After some mathematical manipulation, we get an expression for E_{kf} as in

$$E_{kf} = \frac{j\omega\mu_0}{4\pi} \int_l \left[\frac{z}{\sqrt{x^2+y^2+z^2}} - 1\right] * \left[\frac{x(y_2-y_1) - y(x_2-x_1)}{(x^2+y^2)}\right] * \left[\frac{1}{(x_2-x_1)^2+(y_2-y_1)^2+(z_2-z_1)^2}\right] dL \qquad (9)$$

where dL is a line integral along filament f,

$$x = x_1 + \left[\frac{(x_2-x_1)*L}{(x_2-x_1)^2+(y_2-y_1)^2+(z_2-z_1)^2}\right],$$

$$y = y_1 + \left[\frac{(y_2-y_1))*L}{(x_2-x_1)^2+(y_2-y_1)^2+(z_2-z_1)^2}\right],$$

and

$$z = z_1 + \left[\frac{(z_2-z_1))*L}{(x_2-x_1)^2+(y_2-y_1)^2+(z_2-z_1)^2}\right]$$

Note that (x_1, y_1, z_1) is the translated starting point of filament f and (x_2, y_2, z_2) is the translated end point of filament f, as shown in Figure 5.

The integral in (9) has a smooth integrand and can be computed easily using numerical quadrature [12]. We, thus, efficiently evaluate the elements of L_ρ in the linear system (3).

4. COMPUTATIONAL RESULTS

In order to demonstrate the accuracy and the versatility of the improved fictitious magnetic surface charge method, we used the improved method to solve for the frequency dependent inductance of a realistic spiral inductor over a permeable material substrate, as shown in Figure 6.

The new technique computes integrals using the line integral over the length of the spiral and avoids generating non-piercing surfaces. The tested realistic spiral inductor in Figure 6 has a diameter of 280μ, with a 10μ by 10μ cross section. The permeable material substrate is 1500μ by 1500μ,

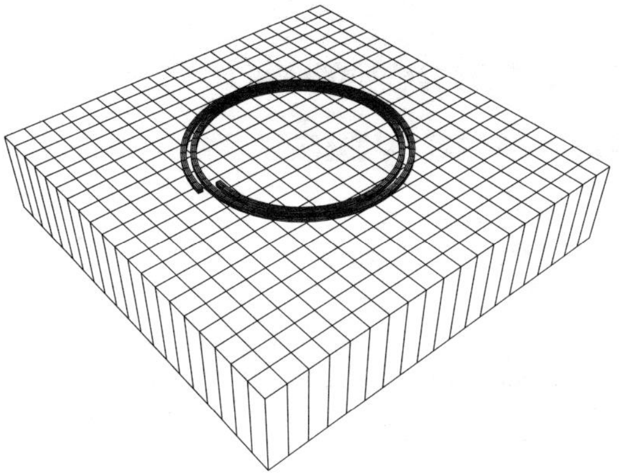

Figure 6: A spiral inductor over a magnetic substrate.

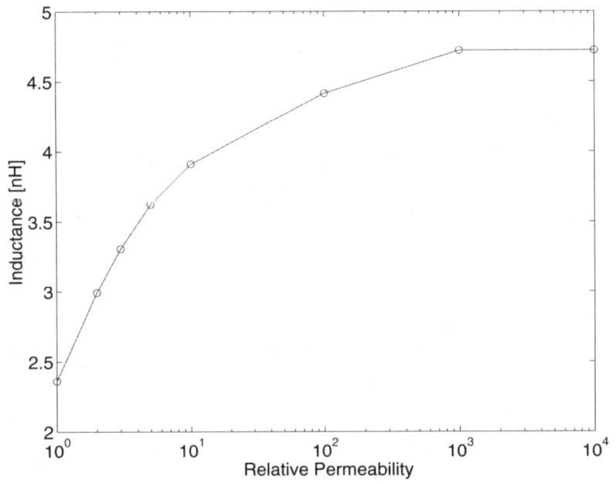

Figure 7: Variation of the inductance of spiral inductor with relative of the magnetic substrate.

with 200μ as its thickness. Figure 7 shows the variation of extracted inductance with the permeability of the permeable substrate. The inductance of the spiral inductor over a magnetic substrate increases as the permeability increases, till it reaches an upper limit which is almost double the value of the spiral inductance without a substrate. This agrees very well with the theoretical analysis in [13]. Figure 8 shows the frequency response of the inductance of the spiral inductor in Figure 6. Note the high frequency inductance is lower than the low frequency one, due to the skin effect.

5. CONCLUSIONS AND ACKNOWLEDGMENTS

An improved method for extracting frequency dependent inductances in the presence of permeable materials has been presented. The method eliminates the need to generate non-piercing surfaces, and is easily applied to general three-dimensional problems. Results were presented to demonstrate the accuracy and versatility of the new method.

The authors would like to acknowledge the support of the Advanced Technology Group at Synopsys, as well as support from the DARPA MEMS program, the National Science Foundation, and the Semiconductor Research Corporation.

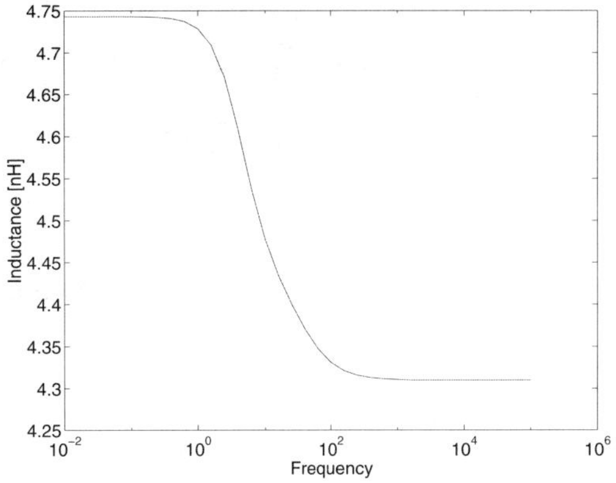

Figure 8: Inductance frequency response of the spiral inductor over a magnetic substrate example. The substrate has a relative permeability of 2000.

6. REFERENCES

[1] H. Guckel, T. Christenson, K. Skrobis, T. Jung, J. Klein, K. Hartojo, and I. Widjaja, "A first functional current excited planar rotational magnetic micromotor," *IEEE Transactions on Micro-Electromechanical Systems*, pp. 7–11, January 1993.

[2] C. Ahn and M. Allen, "Micromachined planar inductors on silicon wafers for mems applications," *IEEE Transactions on Industrial Electronics*, pp. 866–876, December 1998.

[3] W. Taylor, O. Brand, and M. Allen, "Fully integrated magnetically actuated micromachined relays," *Journal of Microelectromechanical Systems*, pp. 181–191, June 1998.

[4] C. Ahn and M. Allen, "A planar micromachined spiral inductor for integrated magnetic microactuator applications," *Journal of Micromechanics and Microengineering*, pp. 37–44, June 1993.

[5] C. Sullivan and S. Sanders, "Design of microfabricated transformers and inductors for high-frequency power conversion," *IEEE Transactions on Power Electronics*, pp. 228–238, February 1996.

[6] T. Sato, H. Tomita, A. Sawabe, T. Inoue, T. Mizoguhci, and M. Sahashi, "A magnetic thin film inductor and its application to a MHZ switching dc-dc converter," *IEEE Transactions on Micro-Electromechanical Systems*, pp. 217–223, March 1994.

[7] K. Yamaguchi, E. Sugawara, O. Nakajima, and H. Matsuki, "Load characteristics of a spiral inductor type thin film microtransformers," *IEEE Transactions on Magnetics*, pp. 3207–3209, June 1993.

[8] Y. Massoud and J. White, "Fast inductance extraction of 3-d structures with non-const permeabilities," in *Proceedings of the International Conference on Modeling and Simulation of Microsystems*, April 1998.

[9] Y. Massoud, J. Wang, and J. White, "Accurate inductance extraction with permeable materials using qualocation," in *Proceedings of the International Conference on Modeling and Simulation of Microsystems*, April 1999.

[10] J. N. Newman, "Distributions of sources and normal dipoles over a quadrilateral panel," *Journal of Engineering Mathematics*, vol. 20, no. 2, pp. 113–126, 1986.

[11] Y. Liu, "Analysis of shell-like structures by the boundary element method based on 3-d elasticity: formulation and verification," *International Journal for Numerical Methods in Engineering*, vol. 41, pp. 541–558, 1998.

[12] D. Engels, *Numerical Quadrature and Cubature*. New York: Academic Press, 1980.

[13] W. Roshen and D. Turcotte, "Effect of finite thickness of magnetic substrate on planar inductors," *IEEE Transactions on Magnetics*, pp. 270–275, January 1990.

A General Probabilistic Framework for Worst Case Timing Analysis

Michael Orshansky and Kurt Keutzer
Department of Electrical Engineering and Computer Sciences
University of California, Berkeley

1 ABSTRACT

The traditional approach to worst-case static-timing analysis is becoming unacceptably conservative due to an ever-increasing number of circuit and process effects. We propose a fundamentally different framework that aims to significantly improve the accuracy of timing predictions through fully probabilistic analysis of gate and path delays. We describe a bottom-up approach for the construction of joint probability density function of path delays, and present novel analytical and algorithmic methods for finding the full distribution of the maximum of a random path delay space with arbitrary path correlations.

Categories and Subject Descriptors

J.6.1 [**Computer-Aided Engineering**]: Computer-aided design.

General Terms: Algorithms

2 INTRODUCTION

Over the years it has been widely acknowledged that the uncertainty about the true design and manufacturing conditions is a major cause of unnecessary over-design and resulting underperformance of circuits [1][2]. The sources of this uncertainty are manifold, and are due to the limitations of the actual design practices, uncertainty about the environmental design characteristics (cross-talk noise, temperature and supply voltage variation), and the inherent variation of the underlying process parameters. With the advance of deep sub-micron technologies, process variability and, in particular, intra-chip variation, has been increasing. This is due to various processing and device physics factors such as random dopant placement in the channel, spatially correlated and proximity-caused Lgate variation, and interconnect metal thickness variation [2].

The emergence of intra-chip parameter variability as a dominant source of uncertainty and circuit degradation requires a new set of approaches to circuit timing analysis, whose role is to guarantee that the predicted maximum clock speed is as close as possible to the actual (silicon) timing behavior. Industrial experience shows that the gap between the worst-case timing constraints predicted by the tools, and the final silicon performance is routinely greater than what can be tolerated and is sometimes as high as 30% [3].

What is wrong with the existing tools and approaches? Circuit-dependent parametric yield loss is predicted to become a key issue in nanometer silicon technologies [4]. The fundamental problem is that the standard timing techniques are incapable of accurately predicting parametric yield of a circuit due to their non-probabilistic formulation. One particular result of this failing is the well-known conservatism of traditional worst-case modeling techniques. We may distinguish at least two levels of conservatism. The first is the practice of defining the worst-case timing behavior of a cell by performing circuit analysis in SPICE that simultaneously sets all the device model parameters to their worst-case values. Several approaches to reduce this type of conservatism have been proposed [5]. When we move to the level of cell-based static timing analysis, an additional level of conservatism is created by the non-probabilistic delay computation of the traditional analysis. This conservatism is relatively new but rapidly growing in importance, and it arises due to the breakdown of key assumptions regarding the correlation between the timing responses of the various delay elements, implicit in the traditional worst-case timing tools.

In the past, several attempts have been made to introduce statistical computations into the domain of gate-level timing analysis. Hitchcock [6] describes a Monte-Carlo based technique for computing the distributions of gate delays. Jyu [7] proposes a faster approach that is capable of dealing with false paths. The deficiency of these techniques is that they are still computationally very expensive. A table look-up algorithm, considered by Berkelaar [8], is faster but fails to account for correlation between gate delays as well provide a way to compute the maximum of more than two variables.

In contrast to the earlier work, we propose an *analytical* theoretical framework for statistical timing analysis. Only the analytical (as opposed to computationally expensive sampling-based techniques, such as Monte-Carlo) methodology has a chance to be used in the timing analysis of real-sized circuits. The proposed approach is entirely probabilistic, seeking to construct the probability distribution of an achievable clock period for a given circuit. We show how to construct the joint probability density function (*jpdf*) of path delays, and, specifically, how to form the covariance matrix of *jpdf* starting from the statistical gate models, and statistical process variation models. The most difficult theoretical problem that has to be solved next is finding the distribution of the maximum of a multivariate vector of path delays with arbitrary covariance structure, described by the just-derived *jpdf*. We introduce a set of previously unavailable analytical estimates and algorithmic solutions that allow constructing tight upper and lower bounds on the distribution of the maximum.

We begin in the next section by motivating the need for a new probabilistic approach. Section 4 describes a modeling strategy for derivation of the joint probability density function of individual delay elements (both gates and wires) and path delays. Then, in section 5.1, the bounds for the distribution of the maximum of path delays, and, in section 5.2, an algorithm to compute them, are developed. Finally, in section 6, the circuit results and a set of quantitative comparisons are provided.

DAC 2002, June 10-14, 2002, New Orleans, Louisiana, USA.

3 WHY TRADITIONAL WORST-CASE TIMING IS OVERLY-PESSIMISTIC

Before we dive into the technical details of the new approach, let us look at why the traditional worst-case timing analysis is bound to result in an unreasonable level of conservatism. Let us for simplicity consider the gate-level static timing analysis. For the sake of simplicity we ignore the orthogonal issue of false paths. Worst-case static timing analysis proceeds by setting each gate to its worst-case timing value and performing a longest path computation to arrive at the worst-case critical path delay. The assumption that is implicit in this approach is that delay elements (gates and wires) are perfectly correlated with each other. The assumption of perfect correlation is reasonable only if the following two conditions hold: (a) intra-chip variation is negligible, and (b) the sensitivity of delay response of *each gate* and *each wire segment* to variation in *every parameter* is perfectly correlated with all the other delay elements (e.g. *both* gate and wire delays). We already know that with the rise of intra-chip variation the first assumption is getting less and less reliable. It is also known that the delay response of the different cells is not perfectly correlated [9]. Lastly, with the rise of the contribution of interconnect delay the fact of its poor correlation with gate delays cannot be ignored.

The failure to consider the validity of the assumptions makes the probability of finding a manufactured chip with the characteristics, assigned to it by the worst-case timing analysis, extremely small. The majority of manufactured chips exhibit a significantly higher performance, as confirmed by practice [3].

4 A PROBABILISTIC FOUNDATION FOR TIMING ANALYSIS

The fundamental problem of the traditional approach to worst-case timing analysis is that it is essentially formulated in a non-probabilistic manner. *The delays of gates, and later, of the paths are treated as fixed numbers, not random variables.* The inherently probabilistic problem is reduced to a purely arithmetical one, and once this transition is made, the ability to probabilistically quantify the likelihood of timing estimates is irreversibly lost. A different formulation of the timing problem is required that would do justice to the probabilistic nature of the problem. Such a completely general formulation is now advanced.

4.1 Problem Formulation

The clock cycle of a chip is constrained by the maximum path delay, $\max\{D_1...D_N\} \le T_{clock}$, where D_i is the delay of the i_{th} path in the circuit. The delay of each path is a random variable, described by a probability distribution. Because $\{D_1...D_N\}$ is a random vector, the value of $\max\{D_1...D_N\}$ is also a random variable. Then, in order to estimate the statistical properties of the chip's timing, we must find the distribution of $\max\{D_1...D_N\}$. The cumulative probability function of $\max\{D_1...D_N\}$ is given by $F(t) = P\{\max\{D_1...D_N\} \le t\}$, or equivalently:

$$F(t) = P\{D_1 \le t, D_2 \le t, ..., D_N \le t\} \qquad (1)$$

where *F(t)* is the cumulative probability function defined over the path delay probability space. In general, we can find the cumulative probability function by direct integration:

$$F(t) = \int_{-\infty}^{t} (N-1) \int_{-\infty}^{t} f(D_1, D_2, ...D_N) dD_1 dD_2 .. dD_N \qquad (2)$$

where $f(D_1, D_2, ...D_N)$ is the joint probability density function (*jpdf*) of $\{D_1...D_N\}$. Unfortunately, the direct evaluation of an *N*-dimensional integral for an arbitrary $f(D_1, D_2, ...D_N)$ is extremely difficult for large *N*.

Given that it is impractical to solve (2) directly for large *N*, we are faced with the task of finding the distribution of $\max\{D_1...D_N\}$ by some other means. While the exact analytical expressions for the distribution of $\max\{D_1...D_N\}$ are still not available, in section 5, we introduce a set of tight analytical bounds for the distribution of $\max\{D_1...D_N\}$. But, first, we describe the derivation of the *jpdf* of path delays, and specifically, show how to construct the path delay covariance matrix on the basis of individual delay elements (gates and wires).

4.2 Path Delay Distribution

We limit our analysis to a combinational circuit containing *N* paths. The analysis can be easily extended to cover sequential circuits. A path is defined by a set of gates and wires that this path traverses, but let us for simplicity talk about gates only, with an understanding that the wire delays can also be covered by the notation introduced below. Let there be a gate set *GS* of gates belonging to a path p_i: $GS\{p_i\} = \{g_i^1 ... g_i^{m_i}\}$, where m_i is the number of gates along the path *i*. In the presence of process variability and other sources of circuit variation, delay of each gate, and thus, each path, can be described by a random variable. Together, all path delays form a random vector $D = \{D_1...D_N\}$, whose joint probability density function, and especially, covariance structure, we want to establish.

We assume that, as is common practice, a Gaussian distribution best describes the process variation. Then, under the mild assumptions that we consider below, the individual gate delay and path delay distributions are also Gaussian. The *jpdf* of the *multivariate normal* vector of path delays is fully characterized by the vector of means, $E\{D\}$, and the covariance matrix, Σ_D: $D \sim N(E\{D\}, \Sigma_D)$. The mean vector of path delay *jpdf* is simply the *N*-dimensional vector of nominal delays of each path, e.g. $\mu = \{E\{D_1\}, ..., E\{D_N\}\}$. The mean delay of a path is given by $E\{D_j\} = \sum E\{d_g(i)\}$, for $i \in GS(p_j)$, where $d_g(i)$ is the delay of the i_{th} gate of path *j*. The *NxN* covariance matrix, Σ_D, is fully characterized by the pair-wise covariance terms between path delays, $\Sigma_{(i,j)} = \mathrm{cov}\{D_i, D_j\}$, where D_i is the random delay of path *i*.

Path covariances $\Sigma_{(i,j)}$ can be constructively computed on the basis of pair-wise gate delay covariances. In the analysis that follows, we assume that statistical interactions between gate delays can be neglected in computing the total path delay response to process variations. This assumption is usually justified. Then,

$$\mathrm{cov}\{D_i, D_j\} = \sum_{k_i=1}^{m_i} \sum_{k_j=1}^{m_j} \mathrm{cov}\{d_g(i,k_i), d_g(j,k_j)\} \qquad (3)$$

where m_i is the number of gates along path *i*, and $d_g(i,k_i)$ is the delay of gate k_i of path *i*. This equation can also be modified to account for the effect of gate delay dependence on the output slew of the previous gate within the path, by propagating variance through a path using a chain rule.

We now derive gate delay covariances. First, we introduce a general notation that will give the ability to include a wide range of gate delay dependences on the design-level, and ultimately, layout-level parameters. Let the attributes specifying a gate be contained in a vector *L*. Among other characteristics, *L* can incorporate the geometric properties of each gate (vertical/horizontal, proximity, left/right), gate sizes, threshold voltages, and so on. Also, let *M* be subset of the elements of the vector *L* that are affected by the process and manufacturing variations, for example, $M = \{L_{gate}, V_{th}\}$. Then, *M* is a random vector, and for simplicity we assume that it has a zero mean vector, $M \sim N(0, \Sigma_M)$. In other words, *L* would describe the nominal values of gate parameters, while *M* would describe the deviations from the nominal.

Let the delay of a gate be given by an arbitrary function $d_g = f(L)$. In order to establish an expression for the pair-wise covariances of gate delays, we assume the linearity of delay response to the localized variation of process parameters. In other words, we assume that a first order Taylor expansion of the gate delay function is adequate. Then,

$$d_g = d_g(L_o) + \varphi(L_o)^T M \quad (4)$$

Here $\varphi(\cdot)$ is the Jacobian of the first-order derivatives of the delay function to *M*. For example, $\varphi(\cdot) = \left(\partial d_g / \partial L_g, \partial d_g / \partial V_{th}\right)$. Under the expansion of Eq. 4, the gate delay distribution is Gaussian. Letting $d = \dim\{M\}$, the covariance of two gate delays *i* and *j* is given by:

$$\operatorname{cov}\{d_g(i), d_g(j)\} = \sum_{t_i=1}^{d} \sum_{t_j=1}^{d} \varphi_{t_i} \varphi_{t_j} \operatorname{cov}\{M_{t_i}, M_{t_j}\} \quad (5)$$

Let us emphasize that our derivation has been completely general with respect to the particular nature of process variability and the types of delay elements we want to include in the analysis. Therefore, using this analysis, the *jpdf* (both the vector of means and the covariance matrix) of the random delay vector can reflect any design or layout dependence. And, again, while we talked only about 'gates', the wire delays as well as gate delays can be naturally accommodated into this analytical framework.

5 ANALYSIS OF GENERAL MULTINORMAL DISTRIBUTIONS

The probabilistic formulation of timing analysis advanced in section 4.1 requires us to estimate the distribution of $\max\{D_1 \ldots D_N\}$. The biggest obstacle to the analytical approach to statistical timing analysis, however, is the lack of a closed-form solution for $\max\{D_1 \ldots D_N\}$ of a system of random variables. While the various non-analytic ways of addressing this problem in the context of worst-case timing analysis have been explored, the analytic estimates have not been available for $N > 2$. In this section we derive tight analytical approximations for the distribution of $\max\{D_1 \ldots D_N\}$.

5.1 Finding Bounds of Gaussian Processes

It must be stated clearly that finding the exact distribution of the maximum of a multivariate distribution with an arbitrary covariance structure is an open mathematical problem. Only for certain types of covariance matrices such a solution can be found. Since the covariance of our multivariate path delay vector is unlikely to have any special symmetry, we can only derive bounds of the form $LB \le f(\max\{D_1 \ldots D_N\}) \le UB$, where *LB* and *UB* are correspondingly the lower and upper bounds. We establish the bounds by employing the powerful results from the study of the general Gaussian processes.

A brief explanation of notation used in this section is helpful. For a random variable *X*, *EX* is the expected (mean) value of *X*, and $EX^2 = E(X^2)$ is the variance of *X*. Note also that we use $\max\{X_1 \ldots X_N\}$ and $\max\{X\}$ interchangeably.

The first step in approximating the distribution $f(\max\{X\})$ is to find the location of the expected value of $\max\{X\}$ - $E\max\{X\}$. The following theorem is the fundamental result of the theory of Gaussian processes, and is key to our analysis [10].

Theorem 1. *Let X and Y have centered multivariate normal distributions, such that* $EX_i^2 = EY_i^2$ *for all i, j, and* $E(Y_i - Y_j)^2 \le E(X_i - X_j)^2$. *Then for all real* λ

$$P\{\max\{Y\} > \lambda\} \le P\{\max\{X\} > \lambda\}$$

In other words, for two multivariate distributions with constant variance, *the maximum of a more correlated distribution is stochastically smaller than the maximum of a less correlated distribution.* (Note that if $E(X_i - X_j)^2 \ge E(Y_i - Y_j)^2$, $\operatorname{cov}(X_i, X_j) \le \operatorname{cov}(Y_i, Y_j)$, i.e. *Y* has a more correlated distribution.) The importance of this inequality lies in the fact that it allows to deduce inequalities for multivariate normal distributions with complex covariance structures by comparing them with simpler distributions. First, a corollary of Theorem 1 for the relation between the expected maximums [10]:

Corollary 1. *Under conditions of Theorem 1*

$$E\max\{Y\} \le E\max\{X\}$$

$$\textit{Proof}: E\max\{Y\} = \int_0^\infty P\{\max\{Y\} > \lambda\} d\lambda - \int_{-\infty}^0 P\{\max\{Y\} < \lambda\} d\lambda$$
$$\le \int_0^\infty P\{\max\{X\} > \lambda\} d\lambda - \int_{-\infty}^0 P\{\max\{X\} < \lambda\} d\lambda = E\max\{X\}$$

Corollary 2. *If X and Y have centralized multivariate normal distributions, such that* $EX_i^2 = EY_i^2$ *for all i, j, and* $\operatorname{cov}(X_i, X_j) = 0$ while $\operatorname{cov}(Y_i, Y_j) \ne 0$ *for all i, j. Then*

$$E\max\{Y\} \le E\max\{X\}$$

The last corollary simply states that if two distributions have the same and constant variance, but one is a correlated distribution while the other is not, *the expected maximum of the un-correlated distribution establishes an upper bound for the expected maximum of the correlated distribution.* As we show in the next section, we can compute the exact value of the expected maximum of the un-correlated distribution. Combining it with Corollary 2, we get an upper bound on the expected maximum of the actual correlated distribution.

Theorem 1 can be extended to cover non-centralized multivariate normal distributions, e.g. distributions for which $EX_i \ne const$. Due to the lack of space, we present the result without a proof but it can be found in [11].

Theorem 2. *Let X and Y have multivariate normal distributions such that* $EX_i \neq const$, $EY_i \neq const$, *and* $EX_i = EY_i$. *Now, if* $EX_i^2 = EY_i^2$ $\text{cov}(X_i, X_j) = 0$, and $\text{cov}(Y_i, Y_j) \neq 0$ *for all i, j, then*

$$E\max\{Y\} \leq E\max\{X\}$$

Among other results, Theorem 1 leads to the useful lower bound on the expected maximum of a multivariate distribution. While parametric yield of a circuit, the percentage of chips that function properly, is determined by the upper bound, the lower bound helps to assess the amount of conservatism potentially contained in the upper bound [12]:

Theorem 3. *Let* $X = (X_1, ..., X_N)$ *have a centered multivariate normal distribution, and let* $E(X_i - X_j)^2 \geq \delta^2$, *then*

$$E\max\{X\} \geq \frac{E|N(0,1)|}{\sqrt{8}} \delta \sqrt{\log_2 N} ,$$

where $E|N(0,1)| = 0.798$. For the sake of interpretation, note that $E(X_i - X_j)^2 = 2\sigma_X^2(1-\rho_{ij}) \geq \delta^2$, if the variance is constant. Thus, the lower bound is determined by the largest correlation between the elements of the multivariate normal vector.

So far our analysis has been limited to finding the expected value of the maximum. Ultimately, however, we would like to get an estimate of the probability distribution of $\max\{X_1...X_N\}$, or, at least, an estimate of some higher moments of its probability density function. This is due to the fact that in order to accurately predict parametric yield we need to be able to find the value of the maximum at different quantiles of the clock cycle distribution. Another fundamental inequality that comes from the study of general Gaussian processes helps us establish the bound for the distribution of the maximum. It states that the spread of $\max\{X_1...X_N\}$ about its mean is no worse than the spread in the distribution of X_i with largest variance. Moreover, the distribution of the maximum above its mean value is such that the likelihood of a deviation is bounded from above by the probability of the same deviation for a centralized normal variable with the variance equal to the largest variance [14].

Theorem 4. *Let* $X = (X_1, ..., X_N)$ *have a centralized multivariate normal distribution. Let* $S = \max(X_i)$ *and* $\sigma_{\max}^2 = \max(\text{var}(X_i))$, *then for any* $r \geq 0$,

$$P\{(S - E\max\{X\}) > r\} \leq P\{N(0, \sigma_{\max}^2) \geq r\}$$

While Theorem 4 is formulated for a centralized distribution, the claim can be extended to an arbitrary mean vector. Due to the lack of space, we present the result without a proof but it can be found in [11].

Theorem 5. *Let X have a multivariate normal distribution such that* $EX_i \neq const$. *Then, under conditions of Theorem 4,*

$$P\{(S - E\max\{X\}) > r\} \leq P\{N(0, \sigma_{\max}^2) \geq r\}$$

Using this theorem we can directly establish the conservative bounds for the quantiles of the distribution of $\max\{X\}$.

Corollary 3. *Under conditions of Theorem 5, the value of* $\max\{X\}$ *at the* k_{th} *percentile of the distribution is:*

$$\max\{X\} \leq E\max\{X\} + z_k \sigma_{\max}$$

where z_k *is the value of the standard normal at the* k_{th} *percentile of the normal distribution.*

This result finally gives us all the necessary tools to bound $\max\{X\}$ of a general multivariate probability distribution. Later we provide some examples for the use of this result in estimating the parametric yield of a circuit.

5.2 An Algorithm for Exact Evaluation of Expected Maximum

In order to use the result of Corollary 3 to conservatively estimate the clock cycle, we first need to find $E\max\{D\}$ for a general path delay vector $D = \{D_1...D_N\}$. We know from Corollary 2 that the expected (mean) maximum of an un-correlated distribution establishes an upper bound for the expected (mean) maximum of the correlated distribution. In this section we describe an algorithm for exact evaluation of $E\max\{X\}$ for an un-correlated multivariate normal distribution with arbitrary mean and variance vectors.

The algorithm to find $E\max\{X\}$ is based on the following probabilistic idea: It is possible to find a point T_1, such that in the repeated drawings of N standard normal random variables, on average, at least one random variable out of N would exceed it. Because T_1 is the point, which is *guaranteed to be exceeded* by at least one variable, $T_1 \leq E\max\{X\}$. Let us now consider the single variable that is guaranteed to exceed T_1. We know that for this variable it is possible to find $T_1 \leq T_2$, such that $P\{\max\{X\} \geq T_2\} = P\{\max\{X\} \leq T_2\} = 1/2$. Then, by definition $T_2 = median(\max\{X\})$. Since by Theorem 4 the distribution of $\max\{X\}$ approaches normal distribution, the median approaches the mean, and we finally have $E(\max\{X\}) \to T_2$.

Let us consider a point T_2, and the probability p_i that a random variable X_i, distributed as $N(\mu_i, \sigma_i^2)$, would exceed it. Then,

$$p_i = p_i(T_2, \mu_i, \sigma_i) = P\{X_i > T_2\} = 0.5 - \Phi((T_2 - \mu_i)/\sigma_i) \quad (6)$$

where $\Phi(t) = 1/\sqrt{2\pi} \int_0^t \exp(-z^2/2)dt$ is the Laplace function.

For each component of the random vector $X = (X_1, ..., X_N)$, we create a mapping to a Bernoulli random variable, whose probability of success, - the probability of X_i exceeding T_2, - is p_i. Let $n(N) = Z_1 + ... + Z_N$ denote the number of successes among N Bernoulli variables. The expected value of the number of successes is

$$E\{n(N)\} = EZ_1 + ... + EZ_N = \sum_{i=1}^{N} p_i \quad (7)$$

since for a Bernoulli variable $EZ_i = p_i$. As was observed above, in order to find $E\max\{X\}$, we need to find the point, at which the expected number of random variables exceeding it is ½. To find this point, T_2, we must solve the equation:

$$E\{n(N)\} = \sum_{i=1}^{N} p_i(T_2, \mu_i, \sigma_i) = 1/2 \quad (8)$$

with $p_i(T_2, \mu_i, \sigma_i)$ given by Eq. 6. If X is a centralized multivariate distribution with constant variance, we have $p_i = const$, and the equation simplifies to $N \cdot p(T_2, \mu, \sigma) = 1/2$. In this case, a closed form solution can be found with

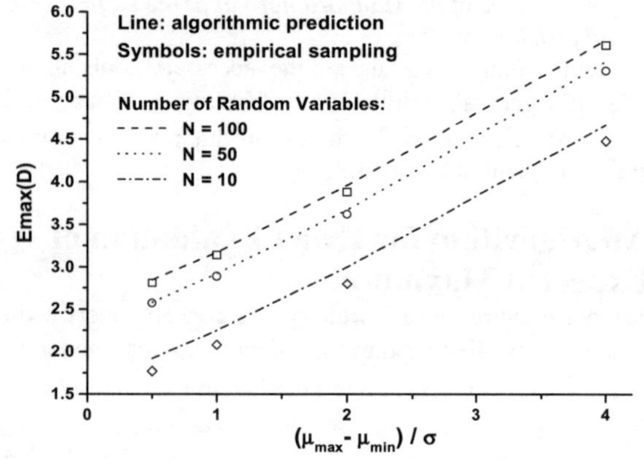

Figure 1. The algorithm is reasonably accurate for different ranges of the mean. Variance is fixed in this example. The results are shown for several values of $(\mu_{\max} - \mu_{\min})/\sigma$.

$E\max\{X\} = \eta$, where $\Phi(\eta) = (2N-1)/2N$. For X with a non-constant mean vector and/or non-constant variance, the equation (8) has to be solved iteratively.

The algorithm has been implemented in Matlab. We verified its accuracy by predicting $E\max\{X\}$ for different configurations of mean and variance vectors. The true value of $E\max\{X\}$ was found by generating sufficiently large random data sets. First we verified accuracy across the range of mean values by varying the ratio of $R = (\mu_{\max} - \mu_{\min})/\sigma$, under fixed variance, as shown in Figure 1. The algorithm always generates slightly conservative estimates but the accuracy improves, as N gets larger. For N=50 and N=100, the conservatism is very small 1-2%. The conservatism gets larger for smaller N and smaller R: for N=10 and R=0.5, it is about 8%. Because our approach is targeted for use with large N, this is an acceptable behavior of the algorithm.

We also considered accuracy of the algorithm for three different variance relations among the elements of the random vector, as shown in Figure 2: (a) homogenous (constant) variance,

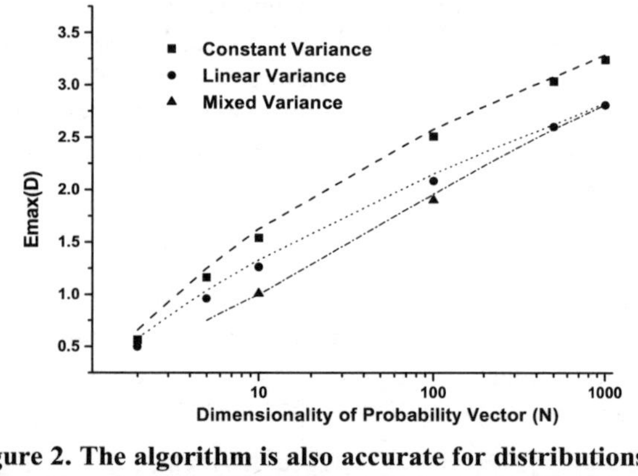

Figure 2. The algorithm is also accurate for distributions with different variance profiles. The mean is assumed constant.

(b) linear reduction of variance, and (c) a mixture of high and low variance. This time we assumed identical mean values. Accuracy is again very reasonable, as shown in Table 1. The values of $E\max\{X\}$ grow roughly logarithmically with N. $E\max\{X\}$ gets larger for the distributions with the larger number of elements with high variance (e.g. the constant variance distribution).

Table 1. The true values of $E\max\{X\}$, and the error of the algorithm's prediction. The algorithm is always conservative.

N	Constant Variance		Linear Variance	
	$E\max\{X\}$	Error (%)	$E\max\{X\}$	Error (%)
100	2.51	2.8	2.08	3.3
500	3.04	1.4	2.60	0.9
1000	3.24	1.5	2.81	0.7

6 RESULTS AND COMPARISONS

We now consider a simple circuit example that will illustrate several features of the probabilistic framework for worst-case timing analysis proposed in the previous sections. To highlight the important features of the approach, we constructed a topologically simple circuit that consists of a repeated pattern of 4 uniquely constructed paths. The four unique paths have similar but non-identical mean delays, $E\{D\}$. Because paths contain different number and types of cells, the variance of path delays, σ_D^2, is also different. (This way, path 4 whose $E\{D_4\} < E\{D_1\}$ may be stochastically slower than path 1, because $\sigma_4 > \sigma_1$.)

We use SPICE to analyze the circuit for a generic 0.18um CMOS technology. To evaluate the statistical properties of the circuit, we superimpose a variation of the gate length (L_{gate}) and the threshold voltage (V_{th}) on the nominal technology values. Both L_{gate} and V_{th} are assumed to be normally distributed; the standard deviations are given by $3\sigma_L = 0.15L_o$ and $3\sigma_{Vth} = 0.1V_{th\,o}$. We assume that both inter-chip and intra-chip components of variation are non-negligible, but that intra-chip component dominates. This may be expressed by setting $cor(L_i, L_j) = cor(V_{th\,i}, V_{th\,j}) = 0.3$. Equations 3-5 are used to estimate gate delay covariance, using SPICE-evaluated delay sensitivities. Path variance and covariance is then computed using Eq. 5.

Table 2. We consider a set of paths with different means and variances. The differences are due to path compositions.

Path	Path composition	# gates	$E\{D\}$ (ps)	σ_D (ps)	D^{wc} (ps)
1	AND2+NAND2+XOR2 +NOR2+NOR4	5	435	24.2	597
2	AND2+4NAND2 +NOR2+NOR4	7	430	18.5	580
3	INV+AND2+NAND2 +XOR2+ NOR4	5	424	23.6	579
4	2XOR+NOR4	3	425	27.9	590

Path delay distribution is clearly correlated. We estimate the upper bound for $E\max D$ using the exact algorithm of section 5.2. We also estimate the lower bound on $E\max D$ using the result of Theorem 3, and compare the newly derived bounds with the predictions of traditional timing analysis. The standard timing would find the *average* clock period given by 435ps, the largest $E\{D\}$ in Table 2. The results of probabilistic estimation of T_{clock} are in Table 3. For N=16, the average T_{clock} is predicted to be bounded by $443ps \le ET_{clock} \le 473ps$. In this case, standard timing would underestimate the mean of the correlated path delay distribution by 1.5-8%. For N=400, the error would be 2-15%.

On the other hand, standard timing overestimates the worst-case behavior of a circuit. The standard timing analysis would conclude that the clock period is given by the largest worst-case delay (D^{wc}) among the paths of Table 2, giving $T_{clock}^{wc} = 597ps$. The results of probabilistic timing analysis are derived for the 98th percentile of T_{clock} distribution. Using the results of Corollary 3:

$$T_{clock}(k) \le E\max\{D\} + z_k \sigma_{\max D}$$

where z_k is the value of the standard normal at the k_{th} percentile of the distribution. From Table 2 we have $\sigma_{\max D} = 27.9ps$. Then, probabilistic timing analysis would conclude, for N=16, that $T_{clk}(0.98) \le 529ps$. Thus, it overestimates the worst-case timing behavior by 6-12%, Table 3.

Table 3. The upper and lower bounds on $E\max\{D\}$ of correlated path delay distribution, the probabilistic T_{clock}, and the error comparison.

N	Bounds on $E\max\{D\}$, ps		$T_{clk}(0.98)$ (ps)	T_{clock}^{wc} (ps)	$\frac{T_{clock}^{wc} - T_{clk}(0.98)}{T_{clock}^{wc}}$ (%)
	Lower	Upper			
16	443	473	529	597	11.5
40	444	483	539	597	9.8
100	445	492	548	597	8.4
400	446	504	560	597	6.3

The difference between the standard and probabilistic timing is significantly larger for circuits with more balanced mean and variance vectors, and with more levels of logic in the paths. In the analysis of another circuit example, we found the difference in the worst-case timing behavior to be 7-21% [11].

The probabilistic formulation is vital, however, for reasons other than the difference in the timing estimates considered above. Among other things, it allows us to reveal the parametric yield curve. The true distribution of T_{clock} is tighter than given by the standard timing analysis, and is shifted towards the slower clock periods, (Figure 3). When the difference in T_{clock} distributions is mapped onto the parametric yield curve (which is just the cumulative distribution function of T_{clock}), one sees that a dramatic drop in parametric yield occurs at the high end of the clock frequency range. For example, the standard analysis predicts 50% yield for chips running at, or faster, than the average T_{clock}, while the probabilistic analysis shows that the yield is, at most, 8%.

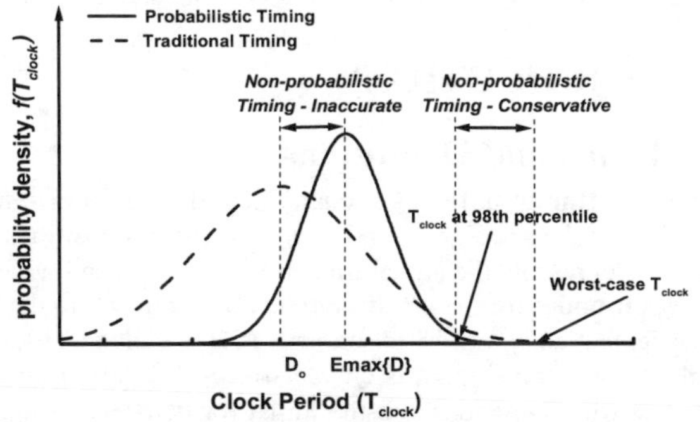

Figure 3. Traditional timing underestimates typical (mean) T_{clock} and overestimates worst-case T_{clock}.

Because the manufacturers of high-speed circuits get most of their revenue from the top percentile of the clock speeds, even the small decrease in parametric yield is very significant for them.

6 CONCLUSION AND FUTURE WORK

The innovation of this work is to define the problem of statistical timing analysis as the problem of finding the distribution of the maximum of circuit path delays with arbitrary covariance structure. The distribution of the maximum was estimated via the theoretical estimates and algorithms, which show a very good level of accuracy. Preliminary results indicate that traditional timing fails in several respects. It underestimates the value of the typical clock period, and overestimates the worst-case timing behavior, requiring expensive redesigns. It is also incapable of providing accurate parametric yield information. In the future, as the magnitude and complexity of process variation grows, we expect these deficiencies of non-probabilistic timing analysis to become even more prominent.

Overall, the results are encouraging, and indicate large potential for use of probabilistic timing analysis. More understanding of the computational behavior of the proposed algorithms for the analysis of large circuits is required. The other direction for substantial improvement is to further tighten the bounds on the probability distribution of the maximum of an arbitrary multivariate space. We believe that the development of such tighter bounds will show that the difference between the standard and probabilistic timing estimates is even greater.

7 REFERENCES

[1] Boning, D., and Nassif, S., "Models of Process Variations in Device and Interconnect", in *Design of High-Performance Microprocessor Circuits*, A. Chandrakasan (ed.), 2000.

[2] M. Orshansky, "Increasing Circuit Performance through Statistical Design Techniques," in *Closing the Gap Between ASIC and Custom*, Kluwer, D. Chinnery, K. Keutzer (ed.), 2002.

[3] Chinnery, D., and Keutzer, K., "Closing the Gap Between ASICs and Custom", *Proc. of DAC*, 2000.

[4] Semiconductor Industry Association, *International Technology Roadmap for Semiconductors*, 2001.

[5] S. Nassif, "Statistical worst-case analysis for integrated circuits," *Statistical Approaches to VLSI*, Elsevier Science, 1994.

[6] R. Hitchcock, "Timing Verification and the Timing Analysis Program", *Proc. of DAC*, 1982.

[7] H.-F. Jyu, S. Malik, S. Devadas, K. Keutzer, "Statistical timing analysis of combinational logic circuits," *IEEE Trans. on VLSI Systems*, vol.1, (no.2), June 1993. p.126-37.

[8] M. Berkelaar, "Statistical Delay Calculation", *International Workshop on Logic Synthesis,* 1997.

[9] S. Zanella et al, "Statistical Timing Macromodeling of Digital IP Libraries", *Workshop on Statistical Metrology*, 2000.

[10] R. Adler, *An Introduction to Continuity, Extrema, and Related Topics for General Gaussian Processes*, 1990, p. 49.

[11] M. Orshansky, *DAC 2002 Technical Materials*, www-device.eecs.berkeley.edu/~omisha/dac2002.htm

[12] D. Pollard, *A User's Guide to Measure Theoretic Probability*, Cambridge University Press, 2001.

False Timing Path Identification Using ATPG Techniques and Delay-Based Information

Jing Zeng, Magdy Abadir
High Performance Tools and Methodology
Motorola Inc.
Austin, TX78729
Jing.Zeng,M.Abadir@motorola.com

Jacob Abraham
Computer Engineering Research Center
University of Texas at Austin
Austin, TX78712
jaa@cerc.utexas.edu

ABSTRACT

A well-known problem in timing verification of VLSI circuits using static timing analysis tools is the generation of false timing paths. This leads to a pessimistic estimation of the processor speed and wasted engineering effort spent optimizing unsensitizable paths. Earlier results have shown how ATPG techniques can be used to identify false paths efficiently [6],[9], as well as how to bridge the gap between the physical design on which the static timing analysis is based and the test view on which ATPG technique is applied to identify false paths [9]. In this paper, we will demonstrate efficient techniques to identify more false timing paths by utilizing information from an ordered list of timing paths according to the delay information. More than 10% of additional false timing paths out of the total timing paths analyzed are identified compared to earlier results on the MPC7455, a Motorola processor executing to the PowerPC^{TM} [1] instruction set architecture.

Categories and Subject Descriptors

J.6 [**Computer-Aided Engineering**]: --Computer-aided design; B.8.2 [**Performance and Reliability**]: Performance Analysis and Design Aids; F.2.3 [**Analysis of Algorithms and Problem Complexity**]: Tradeoffs between Complexity Measures; B.6.3 [**Logic Design**]: Design Aids-Optimization

General Terms

Algorithms, Performance

Keywords

static timing analysis, false timing paths, ATPG, timing slack

1. MOTIVATION

Static timing verification is a crucial step in the design methodology of high performance VLSI circuits. It handles the exponential number of paths in the circuit using efficient linear structural algorithm which ignores the circuit functionality. It is generally used iteratively to allow the timing issues with a certain number of paths to be fixed before it is run again. The cost of optimization and iteration is expensive and optimizing unsensitizable paths does not help to improve the performance. Thus effective techniques for identifying false paths and shortening the cycle of the iterative process are needed.

There have been different approaches at solving the false path identification(**FPI**) problem. Chen and Du used satisfiability(SAT)-based algorithms to check if the sensitizability functions can be satisfied[5]. Chang and Abraham proposed a path sensitization method[4]. Once false path sections have been identified, there are algorithms identifying them from the timing graph [2] [3]. Binary Decision Diagram(BDD)-based model checking and bounded SAT techniques were used at resolving false paths in synthesized logic blocks in [7]. In [1], ATPG techniques are used to remove false paths during timing analysis while expensive circuit modification technique is performed to resolve re-convergent fanout. In [8], a path delay fault identification prototype tool is used to determine circuit stabilization time. We make use of commercial ATPG tools at our design center. The tools can handle very large design models on the order of hundred million transistors.

Earlier results in [6] [9] demonstrate the effectiveness of ATPG techniques. Since they are not BDD-based, the problems with BDD blow-ups while analyzing timing paths for a full chip are not encountered.

Section 2 describes our terms, assumptions and our previously developed FPI technique. We then present our new FPI technique which utilizes information from timing report. We present our experimental results in Section 4, and our conclusions in Section 5.

2. INTRODUCTION

2.1 Terms and Definitions

A **critical timing path** (P) is characterized by a set of n nodes $x_1, x_2, ..., x_n$ and a set, $T = \{t_1, t_2, ..., t_n\}$, of signal transitions such that $t_i \in T$ represents the signal transition on x_i. We call nodes x_1, x_2, ..., x_n **on-nodes** for path P. If a circuit node connects to the same circuit element as x_i, but is not on path P, it is called a **side-node** of path P. Each transition t_i is characterized by a pair of booleans $< b_i, a_i >$ where b_i and a_i are the initial (or ***b***efore) and final (or ***a***fter) boolean values at node x_i, respectively. Note that b_i and a_i are always complementary to each other. We call the set $\{b_1, b_2, ..., b_n\}$ the **before** set and the set $\{a_1, a_2, ..., a_n\}$ the **after** set and test for their satisfiability.

Timing paths start from primary inputs or outputs of sequential

[1]PowerPC is a trademark of the International Business Machines Corporation.

elements, often called the **launch points**, and end with primary outputs or inputs to sequential elements, which are called the **capture points**. Timing paths with outputs of sequential elements as their launch points and inputs of sequential elements as their capture points are called **latch to latch timing paths**.

A path is **sensitizable** if a two-pattern test (v_1, v_2) activates a transition at its launch point and propagates to its capture point. We call the time frame during which v_2 is applied the **current time frame**, while that during which v_1 is applied the **previous time frame**.

Timing paths are generally listed in the timing reports based on their timing slacks. **Timing slack** is defined as the required arrival time minus the actual arrival time at a capture point of a timing path. Static time analysis can be configured to run so that only the single worst case timing path is generated for each capture point in the circuit. We call these paths **main paths**. Main paths can be ordered based on their timing slacks at the capture points with the path with the worst slack showing up first in the report. Besides the main path for a particular capture point, paths converging to the same capture point, with differences in timing slacks from that of the main path within a given threshold, can be generated. These paths are called **subpaths**. Subpaths and their corresponding main path form a group of **converging paths**.

For the scope of this paper, we restrict our analysis to identifying false timing paths which violate logic constraints within the combinational blocks in a design. We will be using the terms *timing paths* and *paths* interchangeably.

2.2 Previous Work: FPI using Logic Info

Previously developed FPI techniques took into consideration of the following logic conditions.

1. e_a = True iff $x_i = a_i$, for all i can be justified simultaneously when evaluated in the current time frame.

2. e_n = True iff respective non-controlling values can be assigned simultaneously at all side-nodes in the current time frame.

3. e_b = True iff $x_i = b_i$ for all i can be justified simultaneously in the previous time frame.

The subscripts of e_a, e_n and e_b indicate the evaluation of different criteria. e_a describes the *e*valuation of *a*fter value criteria. e_n describes the *e*valuation of *n*on-controlling value criteria while e_b describes the *e*valuation of *b*efore value criteria.

The following algorithm, which we will refer to as algorithm *A*, was presented in [6], [9].

Given a path P with e_a, e_b and e_n
if $e_a = false$, then
    P is a false path(f_a)
else
    if $e_n = true$ then
        P is an active critical path
    else if
        $e_b = false$ where $e_n = false$
        then P is a false path(f_b)
    else P is an active critical path

The labels in parenthesis represent the group of paths which are identified false at a particular step in the algorithm. This notation will be used to describe other algorithms later on. We also call the *false paths* identified in algorithm A *logically false paths* since only logic information is utilized for false path identification. If a path is not identified as logically false, we call it a *logically active path.*

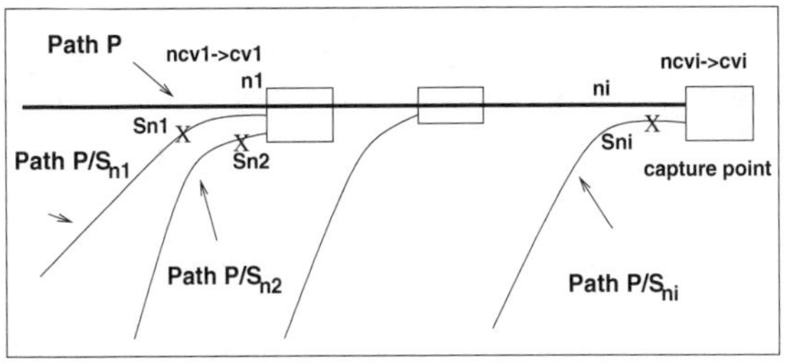

Figure 1: General Multiple Race Condition

3. SLOW PATH CONDITION: FPI USING LOGIC AND DELAY INFORMATION

Due to reconvergent fanouts in the circuits, race conditions between timing paths exist. In a **race condition**, more than one inputs of a gate on the path under consideration(PUC) transition from NCV(non-controlling value) to CV(controlling value). Here e_a and e_b are satisfied for the inputs/output of the gate, but e_n is violated. The transition at the output of the gate is associated with that of the input with earlier timing than others. Unless the arrival times of the transitions associated with the side-nodes of a gate are slower than that associated with the on-node, path PUC is not sensitized.

Any input of the gate in a race condition which is a side-node of the PUC is called S_{n_i}. All the S_{n_i} for the PUC form a set **SN**. The path which forms race condition with PUC at S_{n_i} is called P/S_{n_i}. All the S_{n_i} corresponding to P/S_{n_i} with longer delay than the PUC form a set **SSN**. *SSN* is a subset of *SN*.

One way to decide whether a PUC with race conditions is sensitized, is to keep track of the transition arrival time at each of its on-node and S_{n_i} node. This can be an expensive approach. It is more efficient to utilize the delay information of converging timing paths from the timing report. So for a timing path *P*, which may or may not be a main path, we have the following.

Theorem I. A timing path *P* with race condition, is false if there is not a logically active P/S_{ni} of longer delay than P at any S_{ni}.

Proof: In Figure 1, an "x" indicates a S_{n_i} node.

Take for example, two side-nodes S_{n1}, S_{n2} and the corresponding on-node n_1 in a race condition. If either of the two side paths P/S_{n1}, P/S_{n2} corresponding to S_{n1}, S_{n2} has been identified as logically false, then it would not help to sensitize the transition of path P. If either of the two side paths P/S_{n1}, P/S_{n2} corresponding to S_{n1}, S_{n2}, is of shorter delay than path *P*, it would kill the propagation of the transition on node n_1.

The same condition needs to be true for all S_{ni}s *simultaneously*, otherwise P is a false path □

We call the FPI condition in Theorem I the *Slow Path Condition*.

The comparison of timing slacks of paths depends on how accurately delays are estimated. A threshold is needed to estimate how close the slacks of two timing paths can be and still be differentiated. In a race condition, if one timing path has a slightly better slack than another one, it can activate the other one, if it ends up with worse timing in real silicon. We take this into consideration in our experiments. The false paths identified based on the slow path condition are not logically false paths. Whether they can be sensitized depending on the delays of other logically active paths.

To check Theorem I, we can identify the exact locations of race conditions where e_a, e_b are satisfied, but e_n is violated. We can then check if there are logically active paths of longer delays associated with these nodes. This can be expensive. Instead, we identify the upper bound of the locations of race conditions while allowing the PUC sensitization using efficient structural analysis as follows.

Corollary I. A timing path P with race conditions is false if race conditions happen at nodes which are not intersections between path P and every one of its converging paths with longer delays.

The converging paths of P with longer delays can be identified from the delay information in the timing report. Their intersections with path P can be identified by comparing path nodes. Corollary I avoids explicitly stating exactly where the race conditions are, but provides an upper limit for them for path P to be sensitized.

We can make the upper bound of race condition locations tighter and further identify false paths. If a converging path P' with longer delay is logically false, then its associated intersection with path P cannot be a location for race condition. A path P with race conditions is false if none of its converging paths with longer delays is logically active. Thus we have Corollary II.

Corollary II. A timing path P with race condition(s) is false if none of its converging paths with longer delays is logically active.

We call the FPI condition defined by Corollary I and II a *revised slow path condition* since it does not attempt to identify the exact locations of race conditions, only the upper bound. The identified upper bound is called **PRN**.

Note if a path P violates e_n, it could, 1). violate e_a; 2). violate e_b where e_n is violated; 3). have a race condition. For the first two items, P is false by definition. In general, if e_s and e_n can be checked to eliminate the third possibilty, then P is false whatever the exact cause of its being false. This allows earlier and more efficient identification of false paths.

A main path with a race condition violates Corollary I.

Corollary III. A main path *P* with a race condition is false.

3.1 Improved FPI Algorithm

Let e_s = True iff the set *PRN* is non-empty. e_s describes the evaluation of the revised slow path condition. Given a path *P* with e_a, e_s, e_n and e_b, Figure 2 shows our improved algorithm *B*. The main advantages over algorithm A are the following.

- Additional false paths(f_s and f_n) are identified.
- The checking of the satisfiability of e_n and e_a simultaneously makes sure that a specific transition can be sensitized.
- The locations where e_b is checked are identified more efficiently using the corollaries which do not take additional ATPG run time, compared to being checked exactly where e_n is violated in algorithm A.
- The checking of e_n after e_s is done at a node $\notin$ PRN which is identified efficiently using the corollaries, while e_n is checked everywhere in algorithm A.
- Paths which fail e_b can be identified false as early as during the checking of e_s.

3.2 A More Efficient FPI Algorithm

The checking of e_a, e_n and e_s in algorithm B can be done either in parallel or in serial where only the paths not identified false by e_a need to be checked further. Either way it requires duplication of

if e_a = false, then P is false(fa)
else
  if e_a and e_n are true, then P stays active
else
    if e_s = false, then P is false(fs)
    elsif e_n = false at node $x_i \notin$ PRN
      then P is false(fn)
    elsif e_b = false at node $x_i \in$ PRN
      then P is false(fb)
    else P stays active

Figure 2: Algorithm B

if e_a and e_n are true, then P stays active
else
  if e_s = false, then P is false(fs)
  elsif e_n = false at node $x_i \notin$ PRN
    then P is false(fn)
  elsif e_a = false then P is false(fa)
  elsif e_b = false at node $x_i \in$ PRN
    then P is false(fb)
  else P stays active

Figure 3: Algorithm C

resources or a lot of run time of an ATPG tool. To minimize the amount of logical checking, we observe the following.

Observation: the set of timing paths which fail the simultaneous satisfaction of e_a and e_n is the upper bound of all the false timing paths we identified in algorithm B.

With e_a, e_s, e_n and e_b, we propose algorithm C(Figure 3) for path *P* with the following main efficiency over algorithm B.

- The upper bound of all the false paths is identified by e_a and e_n. This reduces the amount of paths that need be checked by further logic conditions, especially e_a. As in algorithm B, the upper bound is further reduced by checking e_s.

Note in algorithm B, we check e_s on the timing paths which fail e_a and e_n together, but not e_a by itself. Now we perform the check on *ALL* the timing paths which fail e_a and e_n. This tradeoff is minimal for structural analysis. Also, paths which fail e_a by itself and e_b can be identified as early as during the checking of e_s.

4. Experimentation Results

A circuit made of boxes and RC networks has been analyzed. The timing behaviors of the boxes are precharacterized under numerous environmental parameters using transistor-level simulation. The RC nets are precharacterized for the estimation of interconnet timing behavior using extraction tool. Timing analysis is then performed on the whole chip. The output of the timing analysis consists of a set of critical paths.

Our FPI engine translates timing paths into delay paths for ATPG tools[9]. It checks the satisfiability of different criteria for the path by setting the corresponding values at the nodes along the path simultaneously using the ATPG tool commands. Based on the status returned after running ATPG tool with command files, we use our

latch to latch timing paths	Algorithms									FPI improv of B,C over A
	A	B				C				
	fa	fa	fs	fn	fb	fs	fa	fn	fb	
61	6	6	9	5	0	12	3	5	0	3.3
332	26	26	36	6	0	54	8	6	0	2.6
566	54	54	56	6	0	95	15	6	0	2.1

Figure 4: False Path Identified using Different Algorithms

FPI engine again to identify a list of false paths. Additional command files are fed to the ATPG tool and the log files from the ATPG tool are analyzed for further logic checking.

# of Transistors	# of IO pins	# of latches
33 million	281	90k

Table 1: Statistics for MPC7455

We ran our experiments on MPC7455 microprocessor. All runs were performed on a 400MHz Ultra60 running Solaris 5.6 with 1GB memory. Three sets of most critical timing paths were generated using the timing analyzer. We simplified the issue by analyzing only most critical latch to latch timing paths, but it is straightforward to extend the analysis to other kinds of paths. The cycle time target is 950ps. The threshold for generating subpaths was 3ps. The threshold for differentiating between the timing slacks of two timing paths was set to be 0.01ps. Any two timing paths with timing slacks of less than 0.01ps difference are considered to be of comparable delay. False paths identified in different groups following the convention described earlier are shown in Figure 4.

Our additional checking of f_s and f_n was effective in identifying false paths. This highlights the need of taking into consideration delay-based information of the paths besides logic value justification when performing FPI. The last column in Figure 4 shows the number of false paths identified using algorithms B and C, both utilizing f_s and f_n, over algorithm A. At least twice as many false paths are identified using f_s and f_n, or at least 10% additional false timing paths out of the total timing paths analyzed in all three data groups. Note only f_a is identified in algorithm A since identifying f_b without structural analysis information is expensive.

We can see that the number of false paths identified using algorithm C changed from those identified using algorithm B. e_s identifies false paths under category e_a in algorithm B efficiently during the structural analysis. This results in the number under the f_a using algorithm C being much less than that using algorithm B. The number of timing paths needing to be checked using f_a is also reduced. In general, algorithm C performs much less logic checking than algorithm B, thus algorithm C takes much less time than algorithm B since the run time of the algorithms mainly come from running ATPG tool for logical checking. The run time for the 3 sets of timing paths is listed in Table 2. The ATPG abort limit is set to 100. We can see algorithm C takes comparable amount of time to algorithm A, which is about 50% of what it takes for algorithm B. With a set abort limit, in the worst case, the overall run time scales linearly with the total number of paths. If the logic checking for most timing paths takes much less time than the abort limit, the ratio of overall run time for two sets of timing paths can be smaller than the ratio of timing path number. From Table 2, we see between the first two data groups, the ratio of timing path number is 5.4(332/61), while the ratio of run time is 2.3. The same is true for comparison between all the data groups. The loading of the gate-level model takes around 1 hour is not included in the run time since it is an one-time cost and is the same for all algorithms.

latch to latch timing paths	A(mins)	B(mins)	C(mins)
61	30	63	33
332	68	141	73
566	94	195	101

Table 2: Comparison of Algorithms

5. Conclusions

New techniques are demonstrated to perform FPI utilizing additional delay-based information of the timing paths. We are able to identify at least 20% of the total timing paths within the different data groups as false. The new techniques can efficiently identify at least 10% more false paths out of all the timing paths compared to the earlier results. We experimented with a single threshold value for differentiating between slacks of timing paths for this paper. More threshold values will be experimented with for future experiments. Their impacts on the number of false timing paths identified and run time will be explored. So far we have worked on identifying false paths using scan test vectors. Future work will include identifying functionally false paths taking into consideration of the sequential/functional behavior of the circuits.

4. REFERENCES

[1] P. Ashar, S. Malik. "Functional timing analysis using ATPG", in *IEEE Transactions on Computer-Aided Design of Integrated Circuits and Systems*, vol. 14, no. 8, pp. 1025-1030, Aug. 1995.

[2] K. P. Belkhale, A. J. Suess. "Timing Analysis with known False Sub-graphs" in *Proceedings of ACM/IEEE International Conference on Computer-Aided Design*, pp. 736-739, 1995.

[3] D. Blaauw, R. Panda and A. Das. "Removing user-specified false paths from timing graphs", in *Proceedings of 36th ACM/IEEE Design Automation Conference*, pp. 270-273, 2000.

[4] H. Chang and J. Abraham "VIPER: An Efficient Vigorously Sensitizable Path Extractor"in *Proceedings of 30th ACM/IEEE Design Automation Conference*, pp. 112-117, 1993.

[5] H. C. Chen, D. and H. C. Du. "Path Sensitization in Critical Path Problem," in *Proceedings of IEEE International Conference on Computer-Aided Design*, pp. 208-211, 1991.

[6] J. Bhadra, M. S. Abadir, J. A. Abraham. "A Quick and Inexpensive Method to Identify False Critical Paths Using ATPG Techniques: an Experiment with a PowerPC Microprocessor" in *Proceedings of IEEE 2000 Custom Integrated Circuits Conference*, pp. 71-74, May. 2000.

[7] R. Raimi and J. A. Abraham. "Detecting False Timing Paths: Experiments on PowerPC(TM) Microprocessors", in *Proceedings of 36th ACM/IEEE Design Automation Conference*, pp. 737-741, 1999.

[8] A. Sivaraman, A. J. Strojwas. "Timing Analysis Based on Primitive Path Delay Fault Identification", in *Proceedings of IEEE/ACM International Conference on Computer-Aided Design*, pp. 182-189, 1997.

[9] J. Zeng, M. S. Abadir, J. Bhadra, J. A. Abraham. "Full chip false timing path identification: applications to the powerPC microprocessors " in *Proceedings of Design, Automation and Test in Europe*, pp. 514-518, 2001.

False-Path-Aware Statistical Timing Analysis and Efficient Path Selection for Delay Testing and Timing Validation*

Jing-Jia Liou, Angela Krstic, Li-C. Wang and Kwang-Ting Cheng
Electrical and Computer Engineering Department,
University of California, Santa Barbara,
{jjliou,angela,licwang,timcheng}@windcave.ece.ucsb.edu

ABSTRACT

We propose a false-path-aware statistical timing analysis framework. In our framework, cell as well as interconnect delays are assumed to be correlated random variables. Our tool can characterize statistical circuit delay distribution for the entire circuit and produce a set of true critical paths.

Categories and Subject Descriptors

B.8.2 [**Hardware**]: Performance Analysis and Design Aids

General Terms

Algorithm, Performance, Reliability

Keywords

Critical path selection, false path, statistical timing analysis

1. INTRODUCTION

With the continuous shrinking of VLSI devices, accurate prediction of circuit performance is becoming an increasingly difficult problem. Factors such as process variations, noise sources, modeling errors can significantly affect the performance of deep-submicron designs, and these factors are all statistical in nature. Therefore, the use of statistical methods for timing analysis [10, 11] seems to be inevitable in the near future. The use of statistical models is equally important for path selection tools for maximizing the probability of covering critical paths and for detecting defects caused by delay disturbances [1, 2, 3]. These models allow considering various sources of delay variations and can result in a more realistic set of critical paths for delay testing and timing validation. The importance of including power supply noise induced delay variations into the statistical timing models and into the critical path selection process has been demonstrated in [4].

Even though the potential benefits of using statistical models are clear, a significant effort is required to develop models, tools and algorithms for statistical analysis. One of the main difficulties is to include false path analysis. The traditional timing analysis methodologies with false path analysis [5, 6, 7, 8, 9] are not easily extendable into the statistical domain. This is because, with statistical timing models, a path can be true in some circuit instances and false in other instances. By definition, *false paths* are paths that cannot be sensitized under any input vector pair and hence including false paths in circuit timing calculation may lead to unrealistic results. The false paths can also result in waste of engineering resources during the path selection process. In this paper, we present an efficient false-path-aware statistical framework that can minimize the negative effects of false paths on circuit performance estimation and critical path selection for delay testing and timing validation.

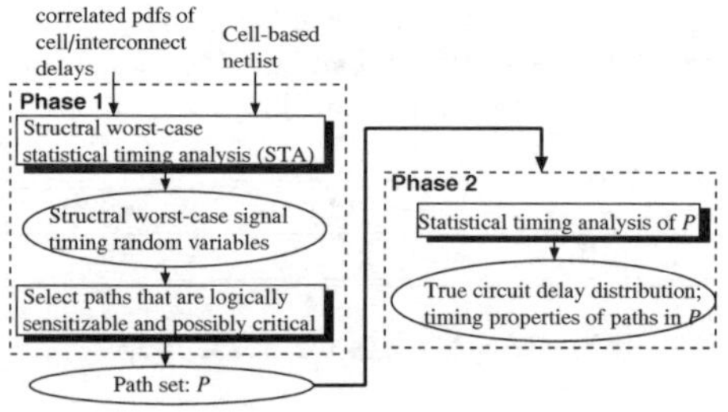

Figure 1: False-path-aware statistical timing analysis.

Given the correlated delay pdfs (pin-pin cell or interconnect delays), our goal is to estimate the true circuit delay distribution and deliver a set of *logically sensitizable* (or simply "sensitizable"), *timingly critical* (or simply "critical"), and *timingly true* (or simply "true") paths. We differentiate between a path being sensitizable and being true because the former considers only the functional (logic) criteria for sensitizing a path while the later considers both logic and timing (statistical timing) criteria for a path to be a long path. For each path, our tool can report its probability of being critical, being true as well as being true *and* critical. This information can be used to improve path selection for timing optimization and delay testing. It also allows ranking the selected paths according to their critical and true probabilities.

2. THE FRAMEWORK

The complete flow of our false-path-aware statistical timimg analysis and path selection framework is shown in Figure 1. It consists of two phases. In the first phase, the objective is to select all logically sensitizable long paths. Worst-case statistical timing information is used to select the long paths [1]. In the second phase, we process the paths in P to obtain their true timing information. With this information, we can determine the circuit delay distribution and further eliminate timingly false paths The outputs of our framework are the true signal timings of the refined set P and the circuit delay distribution.

2.1 First Phase: Identifying Logically Sensitizable Critical Paths

Critical paths are defined as paths whose delays exceed a certain percentage of the longest path delay (or exceed a predefined

*This work was supported in part by the MARCO/DARPA Gigascale Silicon Research Center (http://www.gigascale.org). Their support is gratefully acknowledged.

cutoff period). Although there are several path selection tools [13, 14] available, these methods are based upon nominal or worst-case cell/interconnect delays and cannot capture the statistical factors. As a result, they may fail to select the right paths for devices whose cells/interconnects are subject to delay variations.

To find the critical paths for statistical timing models, we first find the critical nodes. In the beginning, our statistical timing analysis produces the worst-case signal timing random variables including arrival times, required times and slacks (Figure 1). The slack is used to determine if the node is critical or not. Since in the statistical framework, the arrival times, required times and slacks are all random variables, a node could be critical in some circuit instances but non-critical in others. Actually, the area of the slack pdf in the negative region is the probability of the node being a critical node. If the probability of being critical is 1 (with a certain confidence level), this node has a negative slack. It means for all sampled circuit instances, this node is critical and hence it has to be considered when selecting structurally long paths. On the other hand, if the probability is 0, the node has a positive slack. There is also a mixed-type node whose probability can be a non-zero value (other than 1). Mixed-type nodes with slacks of higher critical probabilities than the threshold set by the user will be included in the path selection. Once slack distributions are produced by the structurally worst-case statistical timing analysis tool, we can construct paths from critical nodes such that all nodes on the paths have non-zero probabilities of being critical. The above analysis finds all structurally long paths in the statistical domain. Next, we want to find the logically true paths by checking the functional sensitizability condition [12] for these paths. The outcomes of the first phase are the set of logically sensitizable and structurally long (critical) paths. However, a logically sensitizable path might not be sensitizable in the timing sense, i.e., no vector can sensitize this path such that it always dominates all on-path cell's timing behavior. Furthermore, the true delay of a structurally long path might not be critical because the path is constructed from the worst-case slack distributions. It is the responsibility for the second phase to handle these two problems. In the following, we will discuss the implementation details of the first phase process.

2.1.1 Statistical Timing Analysis

The statistical timing analysis is the core engine of our framework.We adopt a cell-based approach in building our framework. It requires pre-characterization of cells, i.e., extracting random variables for cell delays and output transition times. The input transition time and output loading of the cells are used as indices for building these libraries. Since the goal of statistical timing analysis is to describe the timing behavior of the circuit regardless of the applied input patterns, we use the worst-case analysis in building the cell libraries. The interconnect delay is also modeled as a random variable and is pre-characterized once the RCs are extracted. The random variables of the signal arrival times at cell/interconnect outputs are computed using the information on the arrival and transition times of the cell fanins as well as the information on the cell/interconnect delays [15]. Our framework uses Monte-Carlo-based techniques to approximate the pdfs of the signal arrival times for all internal signals and primary outputs. The convergence criteria are decided based on desired accuracy of results.

2.1.2 Handling Correlations

In Monte Carlo sampling, we need to deal with the fact that the cell/interconnect delay random variables are strongly correlated. Our technique [1] to support correlations is to partition circuit into several groups. We assume a common correlation factor among

```
SearchTrueCriticalPaths()
  for each PO with a falling/rising transition
    if (the prob. of the negative area of the PO slack > threshold
      push the gate connected to PO into the stack
    while (stack is not empty)
      pop a gate from the stack
      if (all fanins of the gate have been searched)
        Backtrack to the previous gate
      else // Extend a partial path
        bool Successful = ExtendBySensitization(gate);
        if (Successful) Successful=Implications();
        if (Successful)
          if (the current gate is a PI)
            Print out the current found path
            and its probability of being true and critical
            Backtrack to the previous gate
          else
            for each fanin of the current gate:
              if (the negative area of the slack > threshold)
                push this fanin into the stack
        else // Extend a partial path unsuccessfully
          Backtrack to the previous gate
```

Figure 2: Identifying logically true and critical paths.

random variables of each group. Our method is to add an independent random variable into the group of variables which share the same correlation factor r. In the following, we use the case of two random variables as an example. Let $X = A + t_1 C$ and $Y = B + t_2 C$, where A, B, C (C is standard normal pdf) are independent random variables. It can be proved that $t_1 = \sqrt{r}\sqrt{\bar{X^2}}$ and $t_2 = \sqrt{r}\sqrt{\bar{Y^2}}$. In the sampling process, we can simply sample values from A, B and C (using sampling techniques for independent random variables) and hence a sample of X and Y can be obtained.

2.1.3 Selection Algorithm for Logically True Paths

In this algorithm, The circuit is processed in a depth-first manner from POs towards PIs level-by-level. At each level, the current partial path is extended (if contains mixed-type slacks) and the sensitization criteria (functional sensitization [12]) are checked. If the criteria are satisfied, the process continues by further extending the current partial path. If they are not satisfied, it means that none of the paths that include the current partial path are true and the algorithm backtracks to the previous level. The process, then, continues by trying a different partial path. Once a primary input has been reached, a sensitizable and structurally true path is found. The pseudo code for our path search algorithm is shown in Figure 2.

2.2 Second Phase: Identifying Timingly True Critical Paths

A path is defined as a timingly true path for a circuit instance if the transitions on the path cannot be invalidated by any other off-input paths. In the statistical environment, the status of timing conditions could be different from instance to instance. Therefore, a path could be a true path in some instances and a false path in others. Such a path would have a probability of being true between 0 and 1. Then the goal of this phase is to derive various problility values by performing statistical analysis on P set (Figure 1): (1) the probability of a path being critical, i.e., the critical probability, p_c, (2) the probability of a path being true (functionally and timingly sensitizable), i.e., true probability, p_t and (3) the probability of a path being true *and* critical, p_{tc}. One possible method for timing analysis in Phase 2 is to modify the statistical analysis method in [16] to obtain the signal timing distributions (in the paper the analysis is used to obtain diagnostic infomation). However, probability values, p_t and p_{tc}, are not available in this scheme.

It is often useful to know what are the most critical true paths among the selected paths, i.e., orders of long paths. Traditionally, long paths are defined according to a fixed path delay (one circuit instance). In our framework, the probability values p_c, p_t and p_{tc}

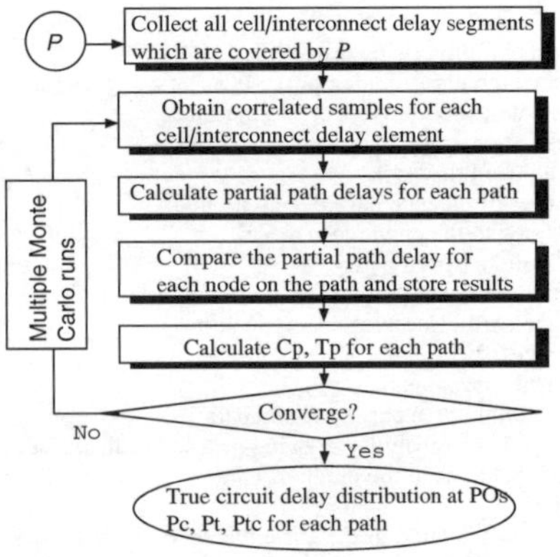

Figure 3: Statistical analysis of *P*.

can be used to rank the selected paths among themselves in a statistical sense. The longest path in the statistical domain can be defined as a path with the largest p_c and nonzero p_t. That means that it is statistically a true longest path. In the case that there are paths with the same p_c, we propose to use p_{tc} for further ordering of the paths. Choosing a path with higher p_{tc} will yield a higher probability to sensitize a critical path.

2.2.1 Statistical Analysis of P

Details of the second phase algorithm is illustrated with the flow chart in Figure 3. It is basically a Monte Carlo sampling technique. In the beginning, the procedure collects all covered segments (cell and interconnect delay elements) in *P*. This is to avoid getting two samples for the same segment in the Monte Carlo run and also help to maintain the correlations between segments using techniques described in section 2.1.2. Maintaining the correlations is particularly important because the relative comparison of arrival times will be incorrect if the correlations are not kept during calculation.

For a circuit instance, the delay of partial paths in *P* that converges at a node will be compared and both the minimum and maximum of the partial delays will be stored for each node. Also, we can obtain a true circuit delay sample using the same technique for POs. It is to compare and store all complete path delays in *P*. By the method described in the next section, all true timing properties (p_c,p_t and p_{tc}) can also be calculated.

2.2.2 Calculation of Probability Values

First, we define two events C_P and T_P for a circuit instance. The C_P event of which a path is critical is defined as $C_P = \{\sum D_i - cutoff\ time\}$, where *i* is the index running over all delay elements on *P*, and D_i is a sample of the distribution of the delay element *i*. The event of T_P of which a path being true for an instance is also defined as follows. Assume that *a* and *b* are inputs to gate *g* in path *P* and both *a* and *b* have transitions. And *a* is the on-path input (*b* is a side-input to *g*). T_P is the event that all gates on *P* satisfy each of the following condition:

$T_g = \{\sum D_i < max_arrival_time(b)\}$if *a* is a controlling input,

$T_g = \{\sum D_i > min_arrival_time(b)\}$if *a* is a non-controlling input.

where *i* is the index running over all delay elements of the partial path from PI to *g* on *P*. For gates that have more than two inputs, this method is applied for each side-input. Counting these event occurrence for paths, we can derive the probability values for a Monte Carlo sampling scheme. Note that for the event of a path being true and critical is the conjuncture of the two events: $\{C_P, T_P\}$.

3. EXPERIMENTAL RESULTS

As described in the section 2, the framework (Figure 1) is a cell-based statistical timing analysis. It requires pre-characterization of cells. We use a Monte-Carlo-based SPICE (ELDO) [17] to extract cell delays for a 0.25μm, 2.5V CMOS technology. The input transition time and output loading of the cells are used as indices for building these libraries. The interconnect delay is also modeled as a random variable. We perform 1000 runs of Monte Carlo analysis for each phase.

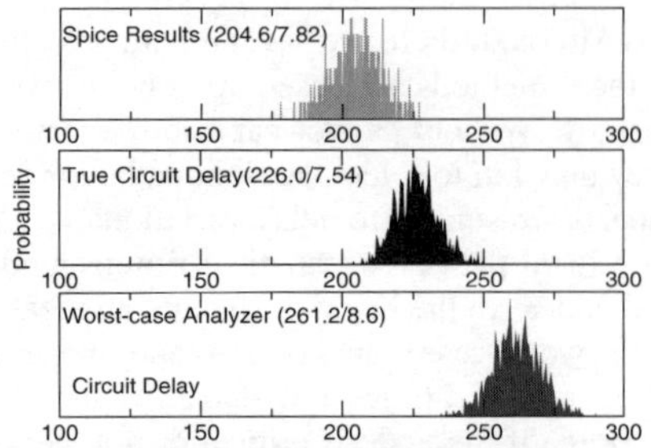

Figure 4: Delay distributions.

Figure 4 demonstrates the results of our algorithm comparing to a Monte-Carlo-based SPICE on a benchmark circuit, s1196. In the Figure, the random variables are expressed as mean/σ, e.g., 204.6 is the mean and 7.82 is the σ for the spice results. The true circuit delay distribution obtained by our algorithm (the second frame) represents a better statistical upper bounds of the results of the SPICE. On the other hand, the results of the statistical worst case analysis (third frame) give a pessimistic estimation of the circuit performance. It shows that our algorithm has a better estimation for this circuit than the traditional statistical timing analysis tools which use only structural information. Note also that typical static timing analysis tool will predict the circuit delays beyond the 3σ bound (261+3*8.6=286.8) of the structural statistical worst case in the figure because these tools only use fixed worst-case cell/interconnect delays. Even enhancing traditional tools with false path pruning, the best figure can be obtained is still beyond the 3σ bound (226+7.54*3=248.6) of our "True Circuit Delay". Moreover, traditional timing analysis tools suffer from inability to target the correct paths [2, 4]. Still the results in Figure 4 is about 12% behide the SPICE results. It is because "static" analysis is applied and thus only "worst-case" random variables for cells can be used. If test vectors are generated for select paths and dynamic simulation is performed, it is possible to match SPICE results more closely.

Circuit delays estimated by our algorithm and traditional worst-case analysis under different correlation assumptions for a set of ISCAS89 circuits are shown in Table 1. There is notable reduction of the estimated circuit delays for most circuits, especially for s1196 and s38417. The mean of the true circuit delay for s5378 is slightly larger than the one obtained by structural worst-case analysis, but on the other hand the σ is smaller. This means that we still have better bound estimation for the circuit delay. It is interesting to note that when correlation factors increase, the standard deviations (σ) increase accordingly while the means decreases. This kind of analysis and comparison is only available by our new algorithm.

From Table 1, one interesting observation can be made: some σ values of true circuit delay are significantly smaller than those of structural worst-case, especially when assuming independent delay random variables. For s38417, σ=0.9 in the case of true circuit delay is smaller than σ=2.1 for the structural worst case. Further analysis reveals that the true circuit delay in this example is actually the collection distribution of 899 paths. All 899 paths have non-zero probabilities of becoming the longest path and they are almost equally long. On the other hand, for structural worst case the circuit delay is only the composite of 64 paths. A larger number of equally long paths means a higher probability for circuit delay to concentrate in a smaller range and hence, a smaller σ.

Table 1: Comparison of random variables for circuit delays with three different correlation factors.

	independent cell/interconnect delays				cell/interconnect delays with a correlation factor of 0.5				cell/interconnect delays with a correlation factor of 0.9			
	true circuit delay		structural worst-case circuit delay		true circuit delay		structural worst-case circuit delay		true circuit delay		structural worst-case circuit delay	
circuits	mean	σ	mean	σ	mean	σ	mean	σ	mean	σ	mean	σ
s1196	226.1	2.7	261.6	2.8	226.0	5.7	261.3	6.7	226.0	7.5	261.2	8.6
s5378	242.8	1.8	242.3	2.2	241.0	7.3	241.3	7.1	239.2	9.0	239.7	9.3
s9234	180.4	1.1	179.4	1.8	178.2	4.6	179.1	4.3	178.3	5.9	179.0	5.6
s13207	482.0	2.6	483.4	2.5	481.3	11.2	482.4	11.7	480.2	14.9	481.3	15.6
s15850	372.3	1.8	374.2	2.4	370.7	7.7	373.6	7.8	370.0	10.4	373.0	10.3
s35932	193.5	0.8	209.3	0.6	192.7	4.2	208.1	4.6	190.7	5.6	206.3	6.2
s38417	299.0	0.9	363.7	2.1	297.3	7.1	363.4	8.5	296.0	10.0	362.7	11.3
s38584	559.2	4.2	566.7	4.4	559.2	12.6	566.6	12.6	558.7	16.1	566.6	16.4

Our framework (Figure 1) can also be used to select paths for performance validation and delay testing. It will output comprehensive timing information associated with each selected path. We compiled the number of selected paths in Table 2 for the outputs of both phases in our framework. In the table, the third column ("logically sensitizable path") is the number of paths (P in Figure 1) after the first path selection phase. This is the number of paths which are functionally sensitizable and possibly critical. And the fourth column ("true and critical path") is the number of paths after the complete process is finished. This is the number of paths that satisfy the following timing conditions: $p_{tc}(P) > 0$ for each path in P. It means that these paths are both timingly true and critical with respect to the cutoff period. We also listed in the second column ("structurally long paths") the number of paths which will be outputed at the end of the first phase in Figure 1 without any functional analysis to eliminate functionally unsensitizable paths, i.e., they are only possibly critical paths. As it can be seen, while the number of the logically sensitizable paths (column 3) is only a fraction of the total possibly critical paths (column 2), the number of the true critical path (column 4) is even smaller. This emphasizes the importance of eliminating the false paths. This table also shows that our algorithm runs efficiently for the benchmark circuits.

4. CONCLUSIONS

We have proposed an efficient statistical timing analysis algorithm to analyze the true timing behavior of a circuit and select true critical paths based on statistical delay modeling. Our framework will provide a better estimation of the circuit delay by considering the statistical cell/interconnect delay models and also the correlations between the pdfs of the delay elements. Furthermore, for applications of path selection the framework can eliminate false paths which are functionally unsensitizable and/or timingly invalidated by other paths. The experimental results show that there is significant reduction of paths if these false paths are eliminated. This tool also provides the capability to order paths according to their probabilities of being true and critical. For deep submicron designs with large variations of different circuit parameters, this information is crucial for circuit optimization, test generation for delay faults and dynamic timing simulation.

Table 2: The number of selected paths by different methods and CPU/MEM consumption of our algorithm.

circuits	# structurally long paths	# logically sensitizable path	# true and critical path	cutoff time	CPU time (s)	MEM (Mbytes)
s1196	534	255	98	200	12.7	2.8
s5378	144	144	114	240	11.8	6.8
s9234	14	7	5	179	8.8	5.4
s13207	3571	926	221	470	55.9	19.8
s15850	8730	1728	281	370	96.9	11
s35932	56339	1664	576	190	207	39.5
s38417	29622	6506	1235	298	181	50
s38584	110	97	66	500	102	41

Note that the CPU time includes the time for both the path selection and and analysis phase. The benchmarks are done on a PentiumIII 733MHz with 256M RAM.

5. REFERENCES

[1] J.-J. Liou, A. Krstić, K.-T. Cheng, D. Mukherjee, and S. Kundu. Performance Sensitivity Analysis Using Statistical Methods and Its Applications to Delay Testing. *Proc. ASP-DAC*, pp 587-592, Jan 2000.

[2] J.-J. Liou, K.-T. Cheng, and D. Mukherjee. Path Selection for Delay Testing of Deep Sub-Micron Devices Using Statistical Performance Sensitivity Analysis. *Proc. of VTS*, pp 97-104, Apr 2000.

[3] J.-J. Liou, K.-T. Cheng, S. Kundu, and A. Krstic. Fast Statistical Timing Analysis by Probabilistic Event Propagation. *Proc. of DAC*, Jun 2001.

[4] J.-J. Liou, A. Krstić, Y.-M. Jiang, and K.-T. Cheng. Path selection and pattern generation for dynamic timing analysis considering power supply noise effects. *Proc. of ICCAD*, pp 493-496, Nov 2000.

[5] D. Brand and V. Iyengar. Timing Analysis Using Functional Analysis. *IBM Thomas J. Watson Center, Technical Report*, 1986.

[6] P. McGeer and R. Brayton. Efficient Algorithms for Computing The Longest Viable Path in A Combinational Network. *Proc. of DAC*, pp 561-567, Jun 1989.

[7] S. Devadas, K. Keutzer, and S. Malik. Computation of Floating Mode Delay in Combinational Circuits: Theory and Algorithms. *IEEE Trans. on CAD*, 12(12):1913-1923, Dec 1993.

[8] H.-C. Chen and D. H. C. Du. Path Sensitization in Critical Path Problem. *IEEE Trans. on CAD*, 12(2):196-207, Feb 1993.

[9] H. Chang and J. A. Abraham. An Efficient Critical Path Tracing Algorithm for Designing High Performance VLSI Systems. *JETTA*, 11:119-129, 1997.

[10] D. R. Tryon, F. M. Armstrong, and M. R. Reiter. Statistical Failure Analysis of System Timimg. *IBM Journal of Research and Development*, 28(4):340-355, Jul 1984.

[11] H.-F. Jyu, S. Malik, S. Devadas, and K. Keutzer. Statistical Timing Analysis of Combinational Logic Circuits. *IEEE Trans. on VLSI*, 1(2):126-137, Jun 1993.

[12] A. Krstić and K.-T. Cheng. *Delay Fault Testing for VLSI Circuits*. Kluwer Academic Publishers, Boston, MA, 1998.

[13] W.-N. Li, S. M. Reddy, and S. K. Sahni. Long and Short Covering Edges in Combinational Logic Circuits. *IEEE Trans. on CAD*, 9(12):1245-1253, Dec 1990.

[14] S. Tani, M. Teramoto, T. Fukazawa, and K. Matsuhiro. Efficient Path Selection for Delay Testing Based on Partial Path Evaluation. *Proc. of VTS*, pp 188-193, May 1998.

[15] H. Edamatsu, K. Homma, M. Kakimoto, Y. Koike, and K. Tabuchi. Pre-Layout Delay Calculation Specification for CMOS ASIC Libraries. *Proc. of ASP-DAC*, pp 241-248, Feb 1998.

[16] M. Sivaraman and A. J. Strojwas. Path delay fault diagnosis and coverage-a metric and an estimation technique. *IEEE Trans. on CAD*, 20(3):440-457, Mar 2001.

[17] Anacad. Eldo v4.4.x User's Manual. 1996.

A Fast, Inexpensive and Scalable Hardware Acceleration Technique for Functional Simulation

Srihari Cadambi
cadambi@nec-lab.com

Chandra S Mulpuri
chandi@nec-lab.com

Pranav N Ashar
ashar@nec-lab.com

C&C Research Laboratories
NEC, Princeton NJ 08540

ABSTRACT

We introduce a novel approach to accelerating functional simulation. The key attributes of our approach are high-performance, low-cost, scalability and low turn-around-time (TAT). We achieve speedups between 25 and 2000x over zero delay event-driven simulation and between 75 and 1000x over cycle-based simulation on benchmark and industrial circuits while maintaining the cost, scalability and TAT advantages of simulation. Owing to these attributes, we believe that such an approach has potential for very wide deployment as replacement or enhancement for existing simulators. Our technology relies on a VLIW-like virtual simulation processor (SimPLE) mapped to a single FPGA on an off-the-shelf PCI-board. Primarily responsible for the speed are (i) parallelism in the processor architecture (ii) high pin count on the FPGA enabling large instruction bandwidth and (iii) high speed (124 MHz on Xilinx Virtex-II) single-FPGA implementation of the processor with regularity driven efficient place and route. Companion to the processor is the very fast SimPLE compiler which achieves compilation rates of 4 million gates/hour. In order to simulate the netlist, the compiled instructions are streamed through the FPGA, along with the simulation vectors. This architecture plugs in naturally into any existing HDL simulation environment. We have a working prototype based on a commercially available PCI-based FPGA board.

Categories and Subject Descriptors

B.6.3 [**Logic Design**]: Design Aids - *Simulation, Verification*

General Terms

Algorithms, Performance, Experimentation, Verification

Keywords

Functional simulation, hardware acceleration, VLIW, FPGA

1. INTRODUCTION

Functional simulation plays a critical role in determining the overall time-to-market of a design. The amount of functional simulation that designers have to perform before they incur the time and expense of manufacture is large. More than 60% of human and computer resources are used for verification in a typical design process, of which more than 85% are for functional simulation[2].

The need for faster functional simulation has been addressed to some extent by better simulators as well as specialized hardware emulators. Levelized compiled code simulators and cycle-based simulators specifically address the needs of functional verification by ignoring timing[9, 12]. They offer an inexpensive method of accelerating functional simulation. Specialized emulators, on the other hand, offer a considerable performance gain over software simulators, but at a much higher cost.

In this work, we present a novel and inexpensive hardware acceleration scheme for functional logic simulation. Functional simulation is a good candidate for hardware acceleration since it has considerable concurrency (or instruction-level parallelism) that cannot be exploited by traditional processors. We use a standard "off-the-shelf" PCI-board with a single FPGA as our accelerator. Instead of emulating the circuit directly on the FPGA, we introduce the notion of an intermediate *simulation processor* called SimPLE (Simulation Processor for Logic Evaluation). SimPLE is mapped onto the FPGA, while the netlist to be simulated is compiled to SimPLE. This makes the technique scalable, i.e., independent of the netlist size. Using the single FPGA board reduces the cost, while the FPGA itself offers flexibility in the SimPLE architecture.

Figure 1 compares our solution against software simulators and emulators. SimPLE, with its inexpensive hardware and software, costs the same as software simulators, but is orders of magnitude faster. In addition, the turn-around-time for simulating a design with SimPLE is similar to software simulators and much faster than emulators. Thus, it is a viable replacement or enhancement to software simulators and a cheaper alternative to emulators.

2. RELATED TECHNOLOGY

FPGAs are frequently used in emulators. A typical emulation board consists of several FPGAs interconnected to-

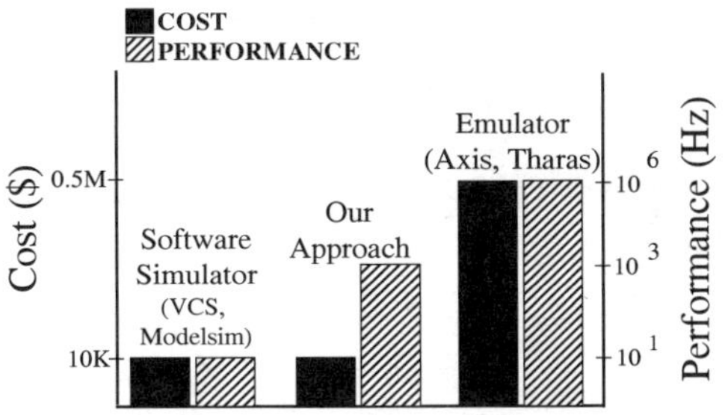

Figure 1: SimPLE vs. simulators and emulators.

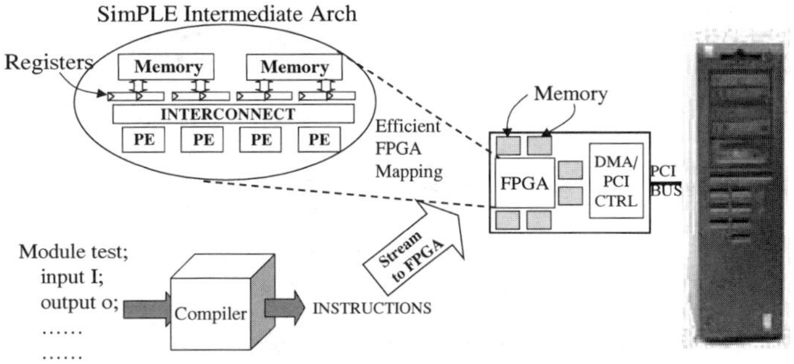

Figure 2: Simulating using SimPLE.

gether. [3, 4] demonstrated the Virtual Wires scheme to time-multiplex wires on physical pins connecting the FPGAs, thereby improving their utilization. However, such large emulation systems are neither cost-effective nor scalable. Several emulation vendors use multiple FPGAs and specially designed hardware within emulation systems costing hundreds of thousands of dollars.

Another approach to emulation is to time-multiplex large designs onto smaller FPGAs. The circuit is not emulated as a whole, but in portions: each portion fits inside the single FPGA, which is repeatedly reconfigured. While this does not have the high cost of the multi-FPGA solution, it's performance is adversely affected by the FPGA's reconfiguration overhead. Generic FPGAs dedicate only a small number of I/O pins for configuration purposes (for example, the SelectMap interface in Xilinx FPGAs [13]). Thus they have a very small configuration bandwidth resulting in significant delays during reconfiguration. Specialized FPGA architectures with extra on-chip storage for multiple configuration contexts have been devised to address this issue[10, 11]. However, such architectures are not commercially available.

3. OUR ACCELERATION SYSTEM

We solve the FPGA configuration bandwidth problem by introducing the notion of a virtual simulation processor called SimPLE. Netlists to be simulated are transformed into instructions for SimPLE by a compiler. These instructions use the data I/O pins of the FPGA and are therefore not affected by the small configuration bandwidth. Once it is configured on the FPGA, different netlists may be simulated on SimPLE using the instructions.

3.1 The Overall System

Our hardware acceleration system consists of a generic PCI-board with a commercial FPGA, memory and PCI and DMA controllers, so that it naturally plugs into any computing system. The board is assumed to have direct access to the host's memory, with its operation being controlled by the host. Thus, the host can direct DMA transfers between the main memory and the memory on the board, which the FPGA can access. Further, with our scheme, the board memory need only be single-ported with either the FPGA or the host (via the PCI interface) accessing it at any time.

Figure 2 shows our simulation methodology. The VLIW-stype instructions compiled for SimPLE are transferred to the on-board memory along with a set of simulation vectors using DMA. Each instruction represents a slice of the netlist. Executing all instructions simulates the entire netlist for one simulation vector. For each vector therefore, all instructions are streamed from the board memory to the FPGA after which the result vector is stored back in the board memory. If the SimPLE instruction is wider than the FPGA-memory bus on the board, it is time-multiplexed into smaller pieces that are reorganized using extra hardware on the FPGA. When all the simulation vectors are done, the result vectors are DMA'ed back from the board to the host. More simulation vectors may now be simulated if required. The host controls the entire simulation using an API.

In order to quantify the simulation speed, we define *user cycles*, *processor cycles*[10] and *FPGA cycles*. The FPGA cycle is the clock period of the FPGA with SimPLE configured on it. A processor cycle is the rate at which SimPLE operates. It is defined as the time taken to complete a single SimPLE instruction. Note that if the instruction is time-multiplexed (i.e., when the SimPLE instruction is wider than the FPGA-memory bus on the board), the processor cycle is larger than the FPGA cycle. Finally, a user cycle is the time taken to fully simulate the netlist for a single simulation vector, i.e., process all the instructions.

We can now quantify the simulation rate. Assume the SimPLE compiler produces N instructions for a netlist when targeting a SimPLE architecture whose instruction width is I_W. If the FPGA-memory bus width on the board is B_W and the FPGA clock cycle is F_C, then the user cycle U_C and simulation rate R are given by

$$U_C = N * \lceil \frac{I_W}{B_W} \rceil * F_C \quad (1)$$

$$R = \frac{1}{U_C} \quad (2)$$

Thus the simulation rate can be increased by reducing (i) the number of instructions produced by the compiler, (ii) the instruction width and (iii) the FPGA clock cycle.

If a very large circuit compiles to too many instructions that do not fit in the on-board memory, the instructions are broken up into smaller portions and DMA'ed separately. This affects the overall performance but maintains the scalability of SimPLE. By upgrading the on-board memory however, we can achieve scalability with no loss of performance. Reasonable amounts of memory allow very large netlists to be simulated: a board with 256MB of SDRAM, for instance, can hold all instructions for a 40-million gate netlist.

3.2 The SimPLE Architecture

SimPLE is based on the VLIW architectural model. Such an architecture can take advantage of the abundant inherent parallelism present in functional logic simulations. A template of SimPLE is shown in Figure 3. It consists of a

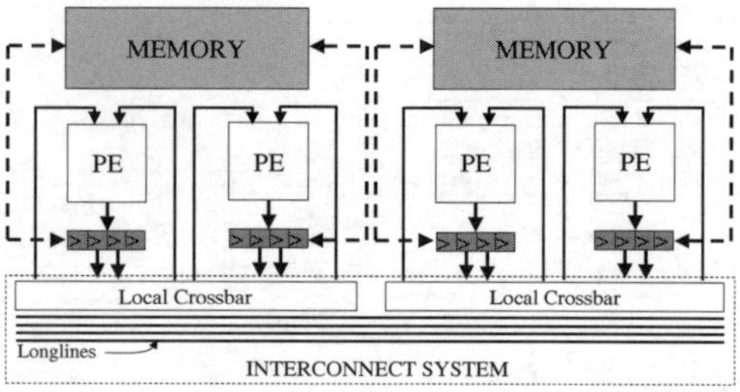

Figure 3: Architectural model of SimPLE.

large array of very simple interconnected functional units or processing elements (PEs)[1]. Each processing element is a 2-input look-up table (LUT) that can evaluate any 2-input Boolean function. The array of PEs can evaluate a large number of gates every cycle. In order to store intermediate signal values, SimPLE has a distributed register file system that provides considerable accessibility at high clock speeds. Each PE has its own register file, with a small number of read ports. Since the number of registers is limited by hardware considerations (FPGAs are not register-rich), there is a second-level of memory hierarchy in the form of a distributed memory system. Registers may be spilled into this memory, and loaded again later. The presence of multiple memory banks permits fast simultaneous accesses. Each memory bank services a set of processing elements and their register files. The PEs perform (i) a Boolean operation during which a result is written to their register files (ii) a memory store during which a register from the register file is spilled into memory and (iii) a memory load during which a value from memory is loaded into a register. Each register file has a single write port that may be used by the PE or the memory, a read port for memory stores and a small number of general read ports connected to other PEs by means of an interconnect system.

The interconnect system in SimPLE consists of local crossbars and long lines. A local crossbar connects a subset of PEs, i.e., it connects read ports of a PE's register file to the inputs of any PE within the subset. The long lines span the entire range of PEs, that is, they enable a register to be read by any PE. If there is a single local crossbar with no long lines, the interconnect system becomes a full crossbar switch. For this paper, we restrict ourselves to a single crossbar switch with a fixed latency.

The SimPLE architecture is parameterizable using the number of PEs, the register files (size and number of ports), the memory size and configuration (i.e., how many PEs each memory bank spans, etc.) and memory and interconnect latencies. All parts of SimPLE, including the memory, are mapped to an FPGA. The VLIW instruction format of SimPLE, along with further architectural details, is shown in Figure 4.

3.2.1 Advantages of SimPLE

SimPLE has the following advantages over software simulators and hardware emulators, whether FPGA-based or otherwise. (i) **Parallelism:** Parallelism present in functional simulations is effectively exploited by the wide processing datapath of SimPLE. (ii) **Register and Memory Access:** The architecture offers easy access to a large number of registers, much larger than what is possible in traditional CPUs. The memory banks are also within close proximity, permitting fast memory access times (1-2 FPGA cycles). (iii) **Configurability:** Since SimPLE is parameterizable, several different configurations may be premapped to FPGAs and stored in a library. Depending on the target FPGA or on the netlist at hand, an appropriate architecture may be chosen. (iv) **Scalability:** SimPLE is transparent to the size of the netlist, much like a software solution. A netlist is compiled into a set of instructions, any number of which may then be executed. Larger versions of SimPLE provide better performance, but smaller ones will still simulate the netlist. (v) **Configuration Bandwidth:** We avoid the small configuration bandwidths of FPGAs by using the data I/O pins for instructions.

[1]The PEs currently target gate-level netlists. We are in the process of extending them to RT-level netlists.

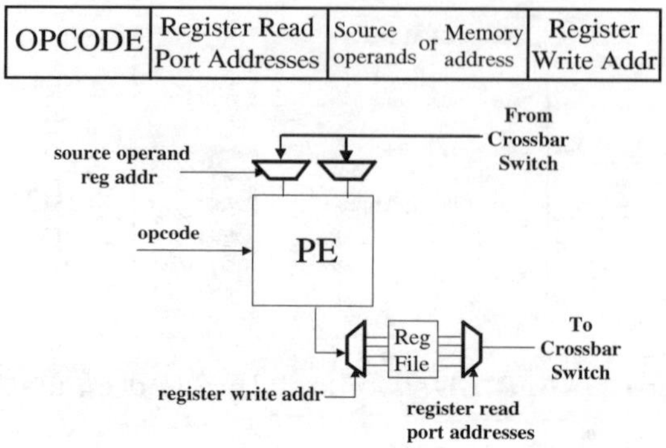

Figure 4: SimPLE instruction format.

4. THE COMPILER

Prior to simulation, a compiler translates netlists into SimPLE instructions. The compiler for SimPLE accepts as input a gate-level netlist described in structural verilog or the Berkeley Logic Interchange Format (BLIF) and the parameters of the target SimPLE architecture. The initial netlist is optimized using standard Boolean optimization techniques and then mapped to lookup-tables using FlowMap[5]. Subsequently, SimPLE instructions are generated. Instruction generation involves PE, register and routing resource allocation as well as memory management. We now focus on the instruction generation procedure.

4.1 The Scheduling Algorithm

The compiler schedules each operation to PEs and manages registers and memory according to the architectural constraints of SimPLE. It maps gates to PEs and wires interconnecting the gates to registers. The registers are allocated such that overall register usage is minimized. The inputs of all PEs are connected to source registers using the register ports and the interconnect system. When a register file is full, a register is selected to be spilled and stored into memory, to be loaded upon demand later. The scheduling algorithm is deterministic and fast, similar to [7].

At every stage of the compilation, the set of graph nodes that have their sources scheduled comprise the *ready-front*. Instruction generation proceeds by selecting nodes from the ready-front and packing them into instructions. Figure 5 shows the instruction generation process. If resources (i.e., PEs and registers) are available, a node is selected to be

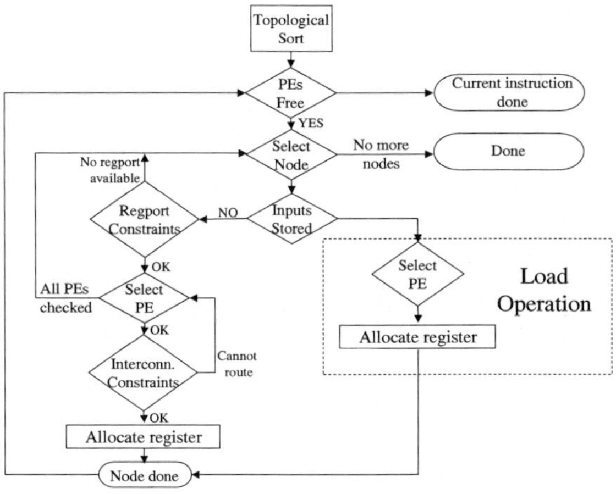

Figure 5: Scheduling and instruction generation.

scheduled from the ready-front using a *selection heuristic.* Our heuristic selects a node N based on (i) its fanout and (ii) number of registers freed by scheduling N. A node with a large fanout opens up more possibilities for scheduling nodes in future time-steps, and is therefore prioritized, while prioritizing nodes that free a large number of registers is a simple, greedy strategy to minimize register usage[7]. On the other hand, if N has one or more inputs stored in memory, it's priority function is based on how many of the stored inputs' fanout are in the ready-front.

If no PEs are available to schedule a node, the current instruction is completed, and a new one started. If no registers are available, *store* operations are scheduled to spill registers. The node whose output is stored is chosen based on the number of its fanout in the ready front.

If the operands of a selected node correspond to registers that have been spilt, *load* operations are scheduled. Once the operands are in registers, the PE with the most free registers is selected to perform the operation. Subsequently, the operands are routed to the PE's inputs using the readports of register files and the interconnect system. Then a register is allocated for the output of the node. This continues until all PEs are occupied, after which new instruction is started.

4.1.1 Handling Registers Specified by the User

A register in the netlist needs to be handled in a special manner with our scheme. We simulate the behaviour of a register using memory loads and stores. Each register R is broken up into its input D and output Q. When an input to a gate G is Q, then the value that must be used to evaluate the gate is the value of the register from the *previous* user cycle (defined in Section 3.1). When the output of a gate G is connected to D, the value that must be written to the register is the value computed by G in the *current* user cycle. Thus, the SimPLE compiler transforms D into a memory load and Q into a memory store to the same memory location. Additionally, Q is constrained to be scheduled after D.

4.2 Compilation Results and Analysis

We analyze results using a combination of industrial, ISCAS and other representative benchmarks. For every result in this paper, we use 4 industrial benchmarks (Industry1-4), the integer and the microcode units of the PicoJava processor (IU and UCODE), and 6 large gate-level combinational and sequential netlists selected from ISCAS89, ITC99[6], and from common bus and USB controllers. The benchmarks range in size from 31,000 to 430,000 2-input gates.

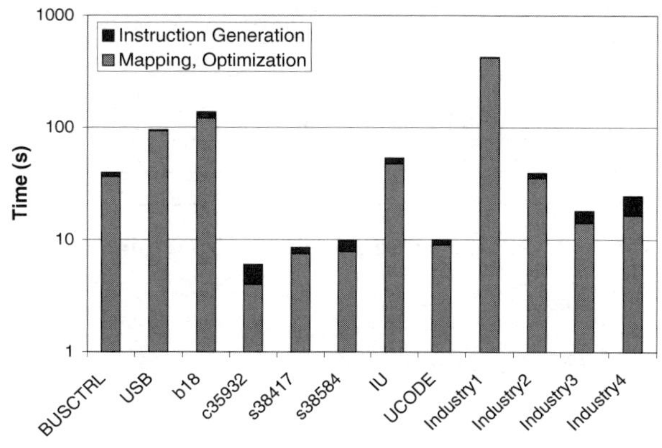

Figure 6: Compilation speed.

Figure 6 illustrates the speed of our compiler on a 440MHz UltraSparc10. A fast compiler facilitates debugging, and is therefore important for simulation. We incorporate efficient data structures and make the instruction generation complexity approximately linear with the netlist size.

We also found that the storage requirement for intermediate values was very modest when using the SimPLE compiler. Further, we found that more than 2 register ports make little difference to the compilation efficiency when there are 32 or more processors. This is explained by the fact that all netlists are mapped to 2-LUTs during compilation, and sufficient parallelism exists with 32 processors to minimize overlap of values on the same processor (overlapping values on a single processor require the use of multiple readports). Henceforth we confine ourselves to SimPLE architectures with 2 readports.

5. FPGA SYNTHESIS

Prior to simulation, SimPLE must be configured onto the FPGA. This is done *only once,* after which an arbitrary number of simulations may be performed. The configuration bits for several SimPLE architectures may be produced beforehand and stored in a library. Thus, the time taken to place and route SimPLE on the FPGA does not affect the simulation speed. However, the FPGA clock speed does (Equation 2). Therefore, it is important to place and route SimPLE on an FPGA and achieve a high clock speed. This section describes our FPGA place and route procedure.

An HDL generator generates a behavioral description of SimPLE with a specific set of parameters, namely the number of processors, memory size, etc. It can also generate extra hardware to time-multiplex the SimPLE instruction if required. This description is synthesized using Synopsys' FPGA Express and mapped, placed and routed on a Virtex-2 FPGA using the Xilinx Foundation 4.1i.

5.1 FPGA Place and Route Methodology

Placement on an FPGA is extremely important in order to achieve good routability. It has been shown that correct placement of modules prior to routing can reduce congestion and enhance the clock speed considerably[8, 1]. We

SimPLE Arch		Clock Speed (MHz)	
Processors	Registers	No Macros	With Macros
32	32	61	124
32	64	56	120
48	32	48	112
48	64	43	112

Table 1: Effect of regularity-driven placement.

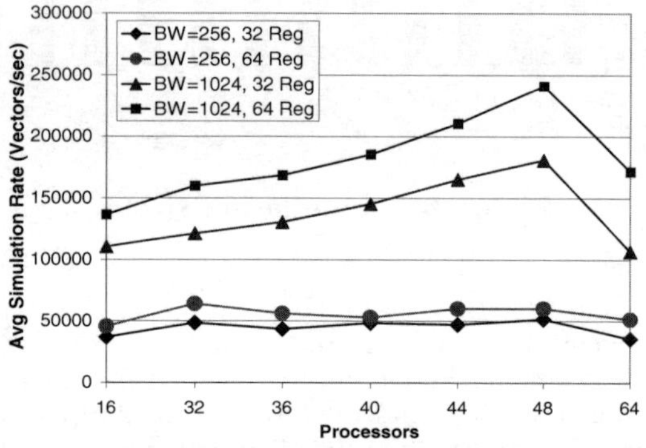

Figure 7: Simulation rate.

use a regularity-driven scheme to obtain a good placement. Every instance of SimPLE inherently has a high degree of regularity since the processing elements, memory blocks and register files are all identical to each other. The hierarchy of SimPLE, including all the regular units, is clear from Figure 3.

Our FPGA place and route methodology involves the following four steps: (i) identification of the best repeating unit in the design, (ii) compact pre-placement of the repeating unit as a single (relatively placed) hard macro, (iii) placement of the entire design using the macros and (iv) overall final routing.

From among the several macros possible in Figure 3, we experimentally found that the largest one (i.e., the top-level macro) was the best, since it had the best compaction ratio and relatively less IO. Once identified, a macro is synthesized, mapped to the FPGA CLBs and then placed. The overall description of SimPLE is instantiated in terms of the macro and then mapped, placed and routed. A more detailed description is beyond the scope of this paper.

Table 1 compares FPGA clock speeds with and without our macro strategy. All experiments were performed using the latest Xilinx Foundation 4.1i. We see clock speed improvements of up to 3x with our approach.

Using the FPGA clock cycle, along with the number of compiled instructions and the instruction width, we can compute the simulation rate using Equation 2. Figure 7 shows the simulation rate in vectors per second for various SimPLE architectures for two values of the width of the FPGA-memory bus on the board: 256 and 1024. The architecture with 48 processors is clearly the best when the FPGA-memory bus is 1024 bits wide. Wider architectures have wider instructions that need to be time-multiplexed more, and are therefore not necessarily better. With a smaller FPGA-memory bus width, several architectures were close. This indicates that the instruction width offsets gains provided by the wider architectures when the FPGA-memory bus width is small.

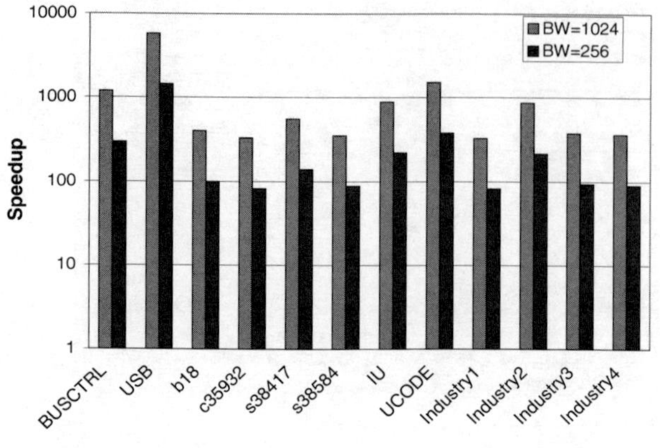

Figure 8: SimPLE vs. cycle-based simulation.

6. SIMULATION SPEEDUPS

In this section, we present actual speedups resulting from an implementation of SimPLE on a large Virtex-II FPGA as well as our first prototype on a generic board.

6.1 Speedup on Virtex-II

Based on the results in Section 5, we synthesized a version of the SimPLE processor with 48 PEs, 64 registers per PE, 2 register read ports per register file, a distributed memory system consisting of banks of 16Kbits each spanning two PEs, a memory word size of 4 bits and an interconnect latency of 2 on an 8-million gate Virtex-II FPGA (XV2V8000). We used Xilinx's Foundation tools as described in Section 5.

6.1.1 Comparison to Cycle-based Simulation

We used the publicly available cycle-based simulator Cyco, which produces straight line C code from structural verilog. We compiled and ran the C code on an UltraSparc 10 system with 1GB RAM containing a SparcV9 processor running at 440MHz. It may be noted that the time for compiling the generated C code is large (around a few hours). This is another advantage of SimPLE which has small compile times.

Figure 8 shows the speedup obtained by SimPLE with 48 processors and 64 registers running at 100MHz (restricted since most boards run at 100 MHz) over a cycle based simulator running on an UltraSparc 440MHz workstation. The right column for each benchmark indicates the speedup if the width of the FPGA-memory bus on the board is 1024 bits, while the smaller left column indicates the speedup for a FPGA-memory bus width of 256 bits. The speedups are 200-3000x for a FPGA-memory bus width of 1024 bits and decrease to 75-1000x for a FPGA-memory bus width of 256 bits.

6.1.2 Comparison to Event-driven Simulation

For this comparison, we used a commercial event-driven simulator with zero-gate delays on a 440MHz UltraSparc-10. Figure 9 shows the speedups obtained for the same benchmarks. The speedups are 100-10000x for a FPGA-memory bus width of 1024 bits and decrease to 25-2000x when the FPGA-memory bus width reduces to 256 bits[2].

[2]For examples with very low activity rates, event-driven simulation is faster than cycle-based simulation.

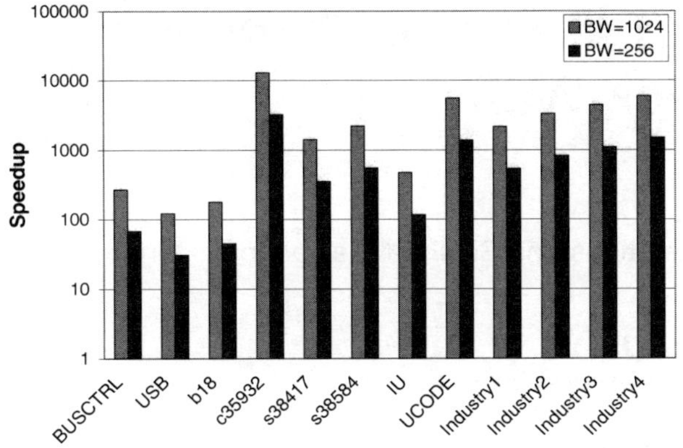

Figure 9: SimPLE vs. event-driven simulation.

6.2 Speedup using the prototype

We implemented a prototype using a generic FPGA board (ADC-RC1000) from AlphaData (www.alphadata.co.uk). The board had a Xilinx Virtex-E 2-million gate FPGA with an FPGA-memory bandwidth of 128 bits at 66 MHz. We have a fully working simulation environment that allows the user to compile and simulate a netlist, and view selected signals. We measured speedups obtained on the small prototype board for two designs. One was a 400,000-gate sequential benchmark, and the other a portion of the pipeline datapath of the PicoJava processor. For both of these, the prototype board was about 30x faster than the event-driven simulator, and 12x faster than the cycle-based simulator. These speedups are expected to increase by a factor of 6 with the Virtex-II board currently being targeted.

6.3 Where does the speedup come from?

The primary reasons for the speedups are (i) the parallelism (ii) large number of registers and memory in SimPLE (iii) high bandwidth between the FPGA and board memory and (iv) high FPGA clock speed. Superscalar processors typically execute 2-3 instructions per cycle. However, this does not usually translate to 2-3 gate evaluations per cycle; memory loads and stores when register files overflow, cache misses and long access times to main memory lower the simulation rate. In SimPLE however, the large number of registers reduces memory operations, while the array of processors permits more parallel gate evaluations than possible in a superscalar processor. The high bandwidth between the FPGA and the board memory allows quick transfer of the wide SimPLE instructions.

6.4 SimPLE in a larger system

SimPLE may be used as a transparent back-end to commercial software simulators. Portions of testbenches that consist of a large number of uninterrupted vectors may be simulated on SimPLE. The designer annotates such parts of a testbench, and compiles the design to the SimPLE architecture. During simulation, control switches from the software simulator to SimPLE using an interface such as Verilog-PLI. The SimPLE tools convert the testbench into vectors and download them (along with the compiled instructions) onto the accelerator board. Thus, SimPLE is best used as an enhancement for commercial software simulators.

7. CONCLUSIONS

In this paper, we introduced a novel approach to accelerating functional simulation. The key attributes of our approach are high-performance, low-cost, scalability and low turn-around-time (TAT). We reported speedups of up to 2000x over zero delay event-driven simulation and up to 1000x over cycle-based simulation on benchmark and industrial circuits while maintaining the cost, scalability and TAT advantages of software simulation.

8. REFERENCES

[1] J. Abke and E. Barke. A New Placement Method for Direct Mapping into LUT-based FPGAs. In *International Conference on Field Programmable Logic and Applications*, pages 27–36, August 2001.

[2] S. I. Assn. International Technology Roadmap for Semiconductors. *ITRS*, 1999. http://public.itrs.net.

[3] J. Babb, R. Tessier, and A. Agarwal. Virtual Wires: Overcoming Pin Limitations in FPGA-based Logic Emulators. In *Proceedings of the IEEE Workshop on FPGAs for Custom Computing Machines*, April 1993.

[4] J. Babb, R. Tessier, M. Dahl, S. Hanano, D. Hoki, and A. Agarwal. Logic Emulation with Virtual Wires. In *IEEE Transactions on CAD of Integrated Circuits and Systems*, June 1997.

[5] J. Cong and Y. Ding. An Optimal Technology Mapping Algorithm for Delay Optimization in Lookup-Table based FPGA Designs. In *IEEE Transactions on CAD*, pages 1–12, January 1994.

[6] F. Corno, M. S. Reorda, and G. Squillero. RT-level ITC99 Benchmarks and First ATPG Results. In *IEEE Design and Test of Computers*, pages 44–53, July 2000.

[7] S. C. Goldstein, H. Schmit, M. Moe, M. Budiu, S. Cadambi, R. R. Taylor, and R. Laufer. PipeRench: A Coprocessor for Streaming Multimedia Acceleration. In *The 26th Annual Internation Symposium on Computer Architecture*, pages 28–39, May 1999.

[8] C. Mulpuri and S. Hauck. Runtime and Quality Tradeoffs in FPGA Placement and Routing. In *International Symposium on Field Programmable Gate Arrays*, pages 29–36, February 2001.

[9] E. Shriver and K. Sakallah. Ravel: Assigned-delay Compiled-code Logic Simulation. In *International Conference on Computer-Aided Design*, pages 364–368, November 1992.

[10] S. Trimberger. Scheduling Designs into a Time-multiplexed FPGA. In *Proceedings of the 1998 ACM/SIGDA Sixth International Symposium on Field Programmable Gate Arrays*, February 1998.

[11] S. Trimberger, D. Carberry, A. Johnson, and J. Wong. A Time-multiplexed FPGA. In *IEEE Symposium on FPGAs for Custom Computing Machines*, February 1997.

[12] K. Westgate and D. McInnis. Reducing Simulation Time with Cycle Simulation. *Quickturn White Paper*, 2000. http://www.quickturn.com/tech/cbs.htm.

[13] Xilinx. Virtex-II 1.5v Field Programmable Gate Array: Advance Product Specification. *Xilinx Application Databook*, October 2001. http://www.xilinx.com/partinfo/databook.htm.

PANEL: Formal Verification Methods: Getting around the Brick Wall

Chair: David Dill, Stanford University

Organizers: Nate James, Prover Technology/KVO Public Relations and Shishpal Rawat, Intel Corporation

Abstract

Do formal verification tools and methodologies require a drastic overhaul to move beyond equivalence checking? Equivalence checking catches errors in synthesis and local hand-modifications to designs. However, powerful formal verification technologies are emerging to combat "behavioral" errors, which represent today's biggest verification problems. Nonetheless, formal verification experts are split on how formal tools should adapt to this challenge. Some of our panelists feel that designers can sufficiently benefit from new formal verification technologies by making incremental changes to current methodologies. Others, however, argue that major changes are required to reap meaningful benefits from these new technologies. Just how much change is enough, what is the capacity of our current tools and what is limiting the full deployment of FV technology?

Our panel of experts, consisting of users, tool providers, and core engine builders, will answer these challenging questions. The panel will debate these issues while discussing real life examples from the user base. They will provide a perspective of how the progression of technology will bring the real promise of formal verification to the user base.

Position Statements

Gérard Berry
Esterel Technologies

Thanks to continuous progress in Boolean and arithmetic computation engines, designers can now use formal verification techniques to validate their designs. These techniques systematically explore design behavior and especially corner cases in a way inaccessible to human reasoning. They are applicable to module or module interconnection verification at various abstraction levels. They can automatically check dynamic properties of designs. They can also analyze the true behavior coverage of test suites, fully dealing with complex concurrent component interaction coverage in a way that traditional code coverage cannot reveal.

The current design flow mostly operates at RTL-level and can be called Design for Synthesis. Since many existing verification tools rely on internal RTL descriptions, they are immediately usable within this flow. However, we think that matching the correctness constraints for larger future designs will require one more step up, Design for Verification, where automatic techniques will play a major role. The current HDLs may rapidly run out of steam since they work at too low an abstraction level to fully optimize the use of verification engines. New or enhanced HDLs will be needed to help the designer by providing higher behavioral description levels, simpler paths to synthesis, and built-in support for verification.

Limor Fix
Intel Corporation

Formal Property Verification, targeted at either catching "behavioral" errors or for certifying the correct behavior of a design, has been used in our industry, and in particular at Intel, for almost ten years. Our findings indicate that formal property verification is a validation vehicle that brings value not achievable in simulation. Complex control and data-intensive circuits that cannot be exhaustively checked by simulation can achieve high level of coverage using formal verification methods. However, the deployment of formal property verification is still limited due to (1) lack of specification and (2) limited capacity of FV technology.

Both roadblocks can be overcome by changing the design methodology. Unfortunately, small (large) changes will yield small (large) benefits. For example, if designers will specify embedded assertions in the RTL and designs employ parameterized RTL blocks and arrays, then these incremental changes only partially solve the lack of specification and the limited capacity of the tools that design teams continuously face.

While we fully support these changes, we hope to utilize the full power of the FV technology by allowing more drastic changes to the front-end design methodology. For example, by unifying the efforts of developing Cluster-Test-Environments and Formal Specification, or by developing high level models which are written in a formal language which can serve as formal specification for the RTL.

Harry Foster
Verplex Systems, Inc.

Proving functional correctness using formal techniques (e.g., model-checking) has established its worth with leading-edge companies. Yet, transitioning the technology into mainstream design environments (beyond the bounds of research and early adopters) has been limited by the following factors: (a) lack of a universally accepted formal property language, (b) lack of commercial tools—until recently, (c) lack of methodology guidelines for effective use (e.g., addressing scalability), and (d) unpersuasive ROI—*what's the gain for the pain?* The lack of a formal property language has recently been addressed through the efforts of the Accellera standardization organization. Likewise, commercial tool offerings are proliferating as the technology matures. Yet, to clearly demonstrate the value of formal or semi-formal techniques requires a change in design methodology. That is, a formal interface specification is required for any individual block we wish to verify. Many argue that in real designs, the block-level interfaces are changing too rapidly to maintain up-to-date interface specifications. However, a counter-argument is that complexity of block-level interfaces could be reduced if serious effort were placed on developing and maintaining interface specifications. The additional *gain for the pain* is an improved simulation process (e.g., the block-level interface specifications can serve as monitors during system integration).

Rajeev Ranjan
Real Intent, Inc.

The increasing high premium of functional correctness in designs with growing complexity demands that verification needs to be concurrent with the design process. Formal verification can facilitate this objective while making only incremental changes to the mainstream HDL based design methodology. In this incremental methodology, automatically identified assertions as well as user specified design assertions within RTL are exhaustively analyzed before testbenches are developed. The formal analysis of these assertions can either confirm desired behavior or eliminate undesired behavior as RTL is being developed.

While the capacity of formal analysis is substantially lower than completely constrained simulation, use of innovative methodologies for adequate and easy constraint specification and for seamless assume-guarantee reasoning enables meaningful formal analysis to obtain 100% confidence for assertions verified.

The biggest barrier to adoption is the absence of clear metrics for performance and quality of the design process. It ranges from cobbling up RTL and throw it over the wall for the verification engineers to developing proper specification and developing RTL from it. A clean guideline on enforcing a quality sign-off criteria for every designer would proliferate usage of formal techniques. The tool vendors can help by reducing the designer's effort needed to get tangible benefits from formal techniques by providing easy specification and debugging methods.

Gunnar Stålmarck
Prover Technology

Although formal verification has made progress in recent years, several factors continue to limit its use in mainstream design flows. Proof of concept remains a problem for the industry, as does a lack of standardized FV languages. Also, one cannot overlook insufficient proof power when considering the reasons why formal verification has not progressed beyond standard applications such as equivalence checking and model checking.

If proof power were greater today, say 100 times greater, formal verification would already be a critical part of every design flow. Instead, designers have been forced to bridge the gap between proof power available and proof power needed by adapting design flows to apply formal verification only where absolutely critical. Increasing the proof power of formal verification tools is the key to unlocking formal verification's potential as a powerful high-level tool for designers. Once the issue of proof power is addressed, revolutionary new standards and methodologies will emerge and be adopted for use on a large scale. Although both incremental and revolutionary paths will be explored, the benefits of revolutionary new methodologies with increased proof power will prevail, enabling adventurous new designs, faster time to market, and complete verification coverage.

Curt Widdoes
0-In Design Automation

Formal verification technologies today are vastly more thorough than simulation in finding corner-case problems in RTL designs. However, hardware design teams have historically been reluctant to abandon established methodologies until they have no choice. In the case of verification, mainstream design teams will continue to rely on simulation and will adopt formal technology incrementally only where it can reduce verification time and effort within the existing simulation flow.

Assertion-based verification (ABV) is an effective way to bring formal verification into the traditional simulation flow. In ABV, the design team instruments the RTL design with statements of design intent (assertions), and simulates with the assertions in place. Then, they use formal technology to directly target the assertions. Constraints, which are necessary for formal verification, are simply assertions that check for illegal interface behavior; in ABV, constraints are simulated along with all the other assertions.

Assertion-based verification fits into the simulation flow with minimal disruption, and the resulting assertions improve the effectiveness of simulation. By enabling formal technology to be used in the simulation flow, ABV increases the thoroughness of verification and decreases the verification time and effort. As a result, ABV is being used today by many companies to verify large designs with complex interfaces. Gradually, as users achieve success with incremental use of formal verification, they will adopt new languages and flows that are designed to maximize its benefits. Eventually, formal verification will play as large a role in mainstream verification as simulation does today.

S-Tree: A Technique for Buffered Routing Tree Synthesis

Milos Hrkic
mhrkic@cs.uic.edu

John Lillis
jlillis@cs.uic.edu

University of Illinois at Chicago, CS Dept., Chicago, IL 60607

ABSTRACT

We present the S-Tree algorithm for synthesis of buffered interconnects. The approach incorporates a unique combination of real-world issues (handling of routing and buffer blockages, cost minimization, critical sink isolation, sink polarities), robustness and scalability. The algorithm is able to achieve the slack comparable to that of buffered P-Tree [7] using less resources (wire and buffers) in an order of magnitude less cpu time.

Categories and Subject Descriptors

B.7.2 [**Hardware**]: Integrated Circuits—*Design Aids*

General Terms

Algorithms

Keywords

timing optimization, Steiner trees, routing, buffer insertion.

1. INTRODUCTION

In the deep submicron era effective performance driven interconnect synthesis has become crucial for achieving chip-level timing closure. When synthesizing an interconnect structure for a timing critical net, there are a number of degrees of freedom which may be exploited including *buffer insertion*, *wire tapering* and *topology*. In the past ten years we have seen the growth of a fairly substantial body of work in the area. Many of the practical buffer insertion techniques in use today can be traced to the seminal work of van Ginneken [10] which proposed a dynamic programming algorithm for inserting buffers into a *given* topology. In addition, the buffer insertion techniques for two-pin nets has received some attention (e.g., [11], [5]). In the area of routing topology construction, there are several works of particular interest to this paper. These include the *P-Tree* based methods [8], methods for timing driven routing of two-pin nets [4], [11] and methods which combine buffer insertion and topology construction (e.g., [7], [9], [3]).

This work was supported in part by NSF CAREER Award CCR-9875945 and in part by SRC under contract 2001-TJ-914.

DAC 2002, June 10-14, 2002, New Orleans, Louisiana, USA.

In this paper we address the interconnect synthesis problem with a number of criteria and objectives in mind which we believe to be of practical importance but not adequately addressed in the literature. An overview of the issues emphasized is as follows.

Simultaneous Buffer Insertion and Tree Construction
We believe that substantial overall improvements in solution quality can be achieved by not viewing the interconnect synthesis problem as a two phase process (routing topology construction followed by buffer insertion) but as a simultaneous approach. While some past works have attempted such a unification (e.g., [7], [3]), those methods do not meet some of our other objectives (e.g., scalability and critical sink isolation).

Handling Routing and Buffer Blockages
In a real design there are typically limitations on where wires can be routed and where buffers can be inserted. We address these issues explicitly in the proposed algorithms by adopting a general graph model for our routing target.

Effective Isolation of Critical Subtrees
One potential weakness of some topology constructions is that they are oblivious to sink criticality. For example in the P-Tree method, a sink permutation is formed where consecutive subsequences of sinks are candidates for subtrees. Since the sink permutation is determined by the relative *physical locality* of the sinks and is oblivious of such criteria as sink criticality (i.e. *temporal locality*), some very high performance and/or cost effective solutions may fall outside the solution space. It is instructive to consider the case in which buffers are used to decouple a non-critical group of sinks. In such a case (in order to conserve scarce buffering resources) it may be preferable to disregard (to a limited degree) the physical locality of such sinks at the expense of additional wire-length so that they may be "tapped-off" with a single buffer. A similar phenomenon occurs when sinks have specified signal polarity requirements as pointed out recently in [1]. The algorithm presented herein is designed to deal with both issues.

Ability to Handle Relatively High Fanout Nets
It is often the case that in a typical modern design flow we see some signal nets with relatively high fanout (e.g., 15 pins or more). Such nets may result from the removal of buffers introduced by logic synthesis producing nets that will be re-buffered with more physical information. Such nets are often

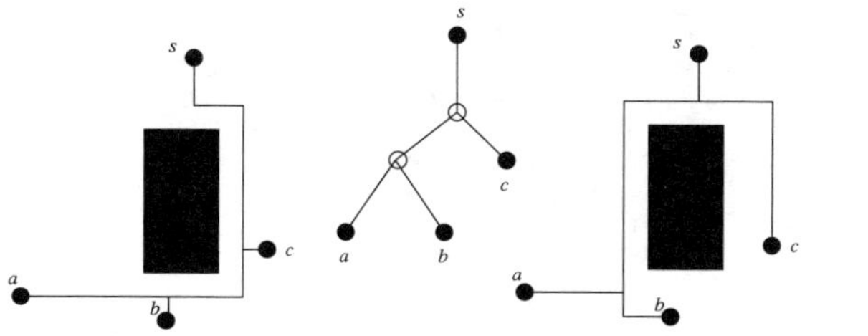

Figure 1: A topology for a 3-sink net and two physical embeddings of that topology. Pin s is the driver.

crucial to overall system performance. Prior methods which exhibited significant robustness and generality of problem formulation tended to not scale well to such size of nets (e.g., [7], [3] which are high-degree polynomial and exponential respectively).

Cost/Performance Tradeoffs

We recognize that while our algorithms focus on the one-net problem, a real CAD system's objective is to drive an entire design to timing closure. Thus, multiple nets must compete for wiring and buffering resources and it is clearly not sufficient to, for example, maximize the performance of a particular net without paying attention to some measure of cost incurred. Further, when one considers that recent estimates indicate that, in the near future, perhaps more than 700K buffers will be required simply to buffer global or near-global interconnects [2], it becomes clear that the over-use of buffering resources will have dramatic consequences.

With these issues in mind we have developed an algorithm called *S-Tree*. There are three degrees of freedom exploited by S-Tree: *Topology*, *Embedding* (placement of Steiner nodes in a target graph) and *Buffer placement*. An overview of how these degrees of freedom are exploited is given in the next section. We first focus primarily on the solution space covered by the algorithm and then sketch some of the implementation details.

2. SOLUTION SPACE DESCRIPTION

We begin with the problem of mapping the Steiner nodes of a given topology to the vertices of a target graph (typically a partial grid graph). Fig. 1 illustrates the corresponding solution space; for the given topology in the middle, we show two possible embeddings. Even though the topology is fixed, there is clearly some flexibility in the embedding space which may be useful particularly in timing driven applications. This flexibility however is certainly limited and precludes certain (potentially useful) solutions (e.g. it may be useful depending on timing requirements to isolate sink a completely with its own path from the driver s; this is not possible for this topology).

Finding an optimal embedding (and buffer assignment) for a given topology is in itself an interesting problem. However, we have generalized the concept in S-Tree to provide more flexibility in the topology space (in fact, exponential flexibility).

Given this notion of topology embedding, the S-Tree algorithm provides additional flexibility in the topology space. Suppose that in addition to a topology T, we are also given a partitioning of the sinks into two disjoint sets S_1 and S_2 (this partitioning may be arbitrary, in our implementation we partition sinks based on timing criticality). Now consider a subtree in T rooted at vertex u and with left and right children $l(u)$ and $r(u)$ (if they exist). Some sinks in the subtree belong to S_1 and others to S_2. If we have two sets of topologies L and R (covering disjoint sets of sinks), then $L \times R$ is the set of all topologies where a root has a member of L as its left child and a member of R as its right child (basically a cross-product; if one of the sets is empty, the output is the other non-empty set). Given these notions, let $P_{12}(u)$ be the set of topologies in the S-Tree space for the subtree rooted at u and covering all sinks in the subtree. Let $P_1(u)$ be the set of topologies for the subtree rooted at u and covering only those sinks in the subtree which are in partition S_1. Similarly $P_2(u)$ is the set of topologies rooted at u and covering sinks from S_2. Since u must be either in S_1 or S_2, the base case is trivial: If $u \in S_1$ and u is a sink then $P_{12}(u) = P_1(u) = \{u\}$ and $P_2(u) = \emptyset$. Case where u is a sink in S_2 is handled similarly. The following recurrence relations establish these sets:

$$P_1(u) = \{P_1(l(u)) \times P_1(r(u))\}$$
$$P_2(u) = \{P_2(l(u)) \times P_2(r(u))\}$$
$$P_{12}(u) = \{P_{12}(l(u)) \times P_{12}(r(u))\} \cup \{P_1(u) \times P_2(u)\}.$$

The expansion in solution space comes from $P_{12}(u)$. It allows us to "promote" one of the subsets and *stitch* it to the root (giving rise to the S-Tree name).

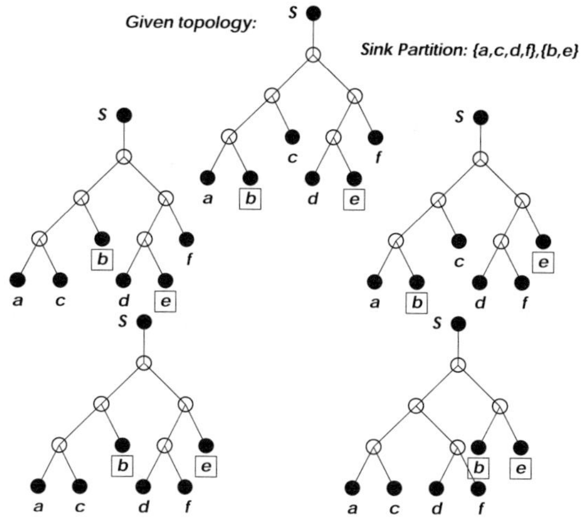

Figure 2: S-Tree topology solution space example.

Fig. 2 illustrates the solution space for a 6-sink topology with a given sink partitioning. Naturally, the given topology (on top) is included in the space in addition to the other four shown. Note that while three topologies are isomorphic to the given topology, some of the sinks have been re-labeled (b and e now being closer to the root and c and f farther).

Two notions motivate this idea. First, in a dynamic programming framework, the optimal solution in the expanded solution space can be found with only a small amount of additional work vs. the totally fixed topology case. Second, it is well-suited to timing related issues where it is often desirable for groups of critical or non-critical sinks to be in the same subtree. If S_1 and S_2 are critical and non-critical sinks, then the stitching operation enables this quite naturally.

To illustrate the second point, consider Fig. 3. In the figure we see a given topology and a sink partitioning. Now suppose that sink b is highly critical. It is then likely that

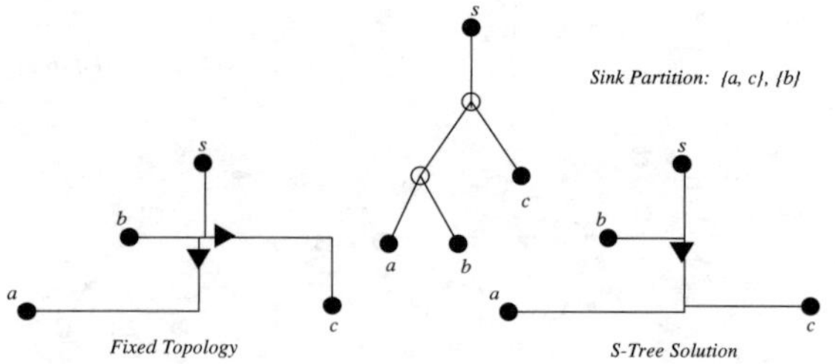

Figure 3: Illustration of critical sink isolation and buffer savings in S-Tree.

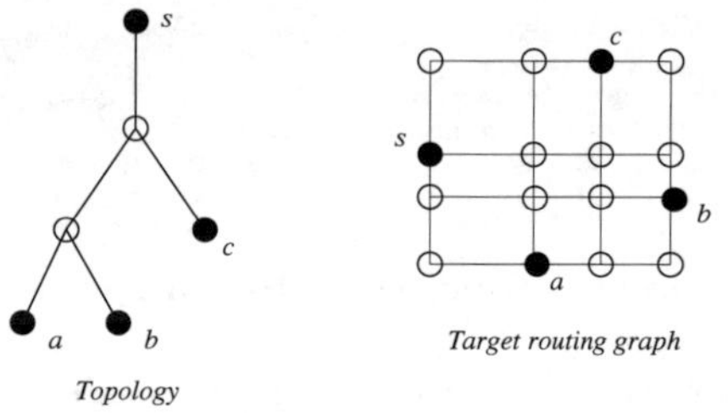

Figure 4: Abstract topology and target graph.

we desire an unbuffered path from s to b which decouples all off-path capacitance with one or more buffers. For the given topology, the best we can do is illustrated on the lower-left. In S-Tree however, the solution on the lower-right becomes possible. Not only will the S-Tree solution have lower delay to sink b, it also uses just a single buffer. We expect that such savings will be increasingly important as more and more buffers are needed in typical designs and conservation of buffering resources becomes crucial.

Given this overview of S-Tree, we can now informally define the optimization problem: Given technology parameters, sink timing requirements, a buffer library, a target routing graph, a topology and a sink partition (S_1, S_2), find a topology in the corresponding space, its embedding and buffer assignments which minimizes cost (e.g., number of buffers, total area, total wire length) subject to timing constraints being met.

2.1 Dynamic Programming Overview

The Dynamic Programming algorithm used for S-Tree is similar in flavor to several past works (e.g., [6], [8]). A candidate solution for a buffered subtree rooted at some vertex in the target routing graph will be represented by its *signature* (p, c, q) indicating that this candidate subsolution incurs cost p, has upward capacitance c and has required arrival time q at its root. Given this notion of a signature, a subsolution is *non-dominated* if no other solution is superior in *all three* dimensions. Any dominated solution may be discarded. Note that while, as many papers have noticed, only c and q are necessary to assure a maximum q solution, all three parameters appear necessary if we want to avoid excessive cost overhead. This increases run-time complexity, but we believe it is truly necessary for practical solutions.

Now let u be a vertex in the given topology T and v be a vertex in the target routing graph G (Fig. 4). We then define the following sets.

$A_1(u, v)$: the set of non-dominated solutions for subtree rooted at u in T, with root embedded at v in G *and* connecting only sinks in S_1 contained in that subtree.

```
Subroutine: GenDijkstra(A^b, G)
    A^b: Joined solutions
    G: Target routing graph
    A ← ∅
    for each solution in A^b(v)
      Insert A^b(v) into Queue
    endif
    while Queue ≠ ∅
      v1 ← Top Queue
      if Solution(v1) is not suboptimal
        A(v1) ← A(v1) ∪ Solution(v1)
        for each egde (v1, v2) adjecent to v1
          v2 ← Augment(v1, v2)
          Insert v2 into Queue
        endfor
      endif
    endwhile
    return A
Subroutine: ComputeA(T, G)
    T: Topology subtree
    G: Target routing graph
    for each vertex v ∈ G
      A^b_1 ← Join(A1.left, A1.right)
      A^b_2 ← Join(A2.left, A2.right)
    endfor
    A1 ← GenDijkstra(A^b_1, G)
    A2 ← GenDijkstra(A^b_2, G)
    for each vertex v ∈ G
      A^b_12 ← Join(A12.left, A12.right) ∪ Join(A1, A2)
    endfor
    A12 ← GenDijkstra(A^b_12, G)
    return A
Subroutine: ComputeAll(T,G)
    T: Topology subtree
    G: Target routing graph
    if(leaf(T))
      A(T) ← ComputeInitial(G)
    else
      A(T.left) ← ComputeAll(T.left, G)
      A(T.right) ← ComputeAll(T.right, G)
      A(T) ← ComputeA(T, G)
    endif
    return A(T)
Algorithm: S-Tree(T, G, s, S1, S2)
    T: Topology
    G: Target routing graph
    s: source node
    S1, S2: sink prtitions
    A(T) ← ComputeAll(T, G)
    Final ← AugmentForDriver(A(T, s))
    return Final
```

Figure 5: Basic S-Tree Algorithm.

$A_2(u, v)$: Similarly defined except limited to sinks in S_2.

$A_{12}(u, v)$: Similarly defined, but topologies must connect *all* sinks in the subtree.

We apply dynamic programming techniques to compute these sets in bottom-up fashion (an algorthm overview appears in Fig. 5). When all sets are established, the candidate solutions in $A_{12}(u_s, v_s)$ (where u_s is the root of the topology and v_s is the location of the driver in the target graph) are augmented to consider the effect of the driver (additional load-dependent delay) which then results in the overall set of non-dominated solutions with respect to cost p and required-time q. These form a tradeoff curve from which a solution can be selected.

In order to compute these sets it is useful to define a similar group of sets in which the vertex v in the target graph is constrained to be a branching point (basically, a Steiner). Let these sets be $A_1^b(u, v)$, $A_2^b(u, v)$ and $A_{12}^b(u, v)$.

Proceeding bottom-up in the topology, suppose we are at some vertex u; we visit each graph vertex v and first compute $A_1^b(u, v)$ and $A_2^b(u, v)$ by joining solutions from the

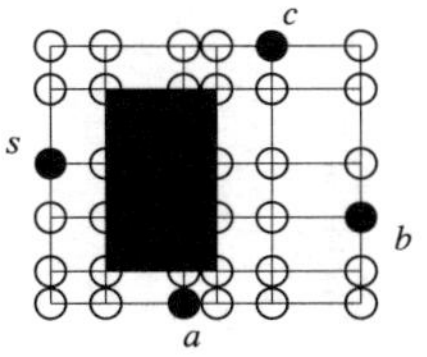

Figure 6: Target routing graph construction.

appropriate subtrees (a solution in $A_1^b(u,v)$ is formed by joining a solution in $A_1^b(l(u),v)$ with one from $A_1^b(r(u),v)$ where $l(u)$ and $r(u)$ are u's left and right children in T).

We then compute $A_{12}^b(u,v)$ by first joining $A_{12}(l(u),v)$ with $A_{12}(r(u),v)$ and then $A_1(u,v)$ with $A_2(u,v)$; the non-dominated solutions in the resulting union is retained. It is worth mentioning that in *Join* step (line c2 in Fig. 5) to guarantee optimality we have to construct a cross product of all costs p and join sets in respect of that cross product. Also pruning dominated solutions is not trivial any more. Methods and data-structures from [6] have been used.

As a preprocessing step we construct a graph on which topology will be embedded. We extend grid lines from the driver and every sink in all directions (4 in single layer, 3-6 in multilayer environment). Grid line intersections become candidates for branch points (Fig. 4). In the similar way we introduce grid lines that follow the contours of routing and buffer blockages (Fig. 6). Then we remove vertices and edges covered by routing blockages and mark vertices covered by buffer blockages as infeasible for buffer insertion.

Once we have a target graph and A^b set of branching solutions we must "augment" them to form "single-stem" solutions $A_1(u,v)$, $A_2(u,v)$ and $A_{12}(u,v)$. Such solutions eventually must reach a branch point computed in the previous step. This is accomplished efficiently through what can be viewed as a generalized version of Dijkstra's algorithm. We let the branching point solutions form the initial wavefront and expand in a manner similar to [11] and [4]. In order to do that we first introduce a virtual vertex x in target graph and add directed edges from that vertex to every other vertex labeling them with appropriate A^b solution (Fig. 7). So if we are generating solutions for vertex u in topology then the edge from x to v in target graph will have label $A^b(u,v)$. Starting from vertex x we update labels on all other vertices, insert them into priority queue, expand them in best cost first order and continue to update neighboring vertices until the queue becomes empty.

When we run our algorithm in trade-off mode, instead of a single 'best' solution we generate a set of non-dominated solutions, and instead of a single label we have a list of non-dominated labels which are expanded by best cost first.

3. DISCUSSION

Implementation Issues

Some important implementation details have not been mentioned in the interest of space. For instance, efficient determination of the dominance property can be non-trivial and the method and data-structures of [6] have been employed. In addition, sometimes a subtree's sinks are from only one set. Careful bookkeeping can exploit this situation to avoid redundant computations.

Initial Topology

Although initial topology is a parameter to S-Tree algorithm, it has to be generated somehow. In our implementation we used MST based approach and a P-Tree$_A$ topology, both discussed in [8].

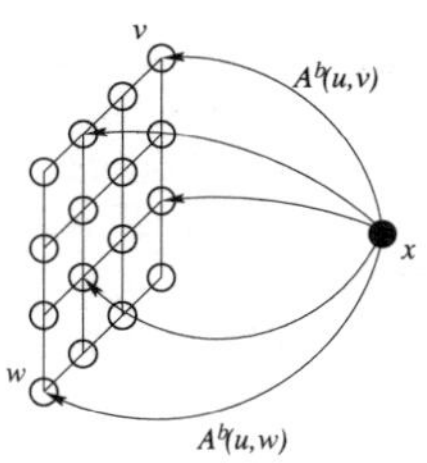

Figure 7: Illustration of generalized Dijkstra's algorithm. From vertex x directed edges with weight of $A^b(u,v)$ go to every vertex v in the target graph.

Speedup Techniques

It is often the case that a candidate solution, while non-dominated can be discarded by computing various lower-bounds or putting a threshold on the global cost allowable. We have incorporated such strategies and achieved substantial speedups. Also a buffer library pruning (discarding buffers which are dominated by other buffers) as a preprocesing step improves runtime.

Solution Quality

Careful choice of buffer candidate locations may improve solution quality. Segmenting long wires and/or inserting additional grid lines from buffer blockages may yield better solutions. These simple modifications do have impact on runtime but do not affect overall algorithm complexity.

Sink Partitioning

Currently we use simple heuristics to partition critical from non-critical sinks. We compute estimated delays from the source to each sink, adjust the given required time by these estimates and then rank the estimated achievable slacks. Partitioning with respect to sink polarity is trivial.

Generalizations

We have presented the algorithm where there are two sets of sinks. However there is no reason (except computational complexity as there is a run-time term which is exponential in the number of sets) that we cannot use three or more sets. In fact, in the limit where we have n singleton sets, the topology becomes irrelevant and we obtain optimality.

Modifications

There are a number of basically trivial modifications to the algorithm which are important in practice. For instance it is desirable to be able to handle inverters as buffers. This is easily done through previously studied techniques in buffer insertion. Also buffer insertion step is done in generalized Dijkstra's algorithm in a way similar to [5] which implicitly allows buffer cascading.

Algorithm Complexity

Due to limited space we constrain our algorithm complexity analysis to minimum wire length mode only. Assume that we have n sinks. Then topology still has $O(n)$ vertices including sinks. For every vertex u in topology and every vertex v in target graph we construct $A^b(u,v)$ and $A(u,v)$ set of solutions. Size of the target graph is $O(n^2)$ (if we have b blockages then it is $O((n+4b)^2)$). Computing $A^b(u,v)$ $\forall v$

takes $O(n^2)$ time. Computing $A(u,v)$ $\forall v$ involves running generalized Dijkstra's algorithm which runs in $O(E \log V)$ time where E is the number of edges and V number of vertices in the graph. In our case $E \approx O(n^2)$ (sparse graph) and $V \approx O(n^2)$ we have $(O(n^2 \log n^2))$ what is $O(n^2 \log n)$. Computing these sets for every u we have $O(n(n^2+n^2 \log n))$ what is $O(n^3 \log n)$. In trade-off mode we have an additional multiplicative factor because we do not keep a single solution for every (u,v) pair but a set of non-dominated solutions, yielding a pseudo-polynomial complexity.

4. EXPERIMENTS

We have implemented the S-Tree algorithm and performed some initial experiments in Linux environment on a PC with a 766MHz CPU. The three main criteria of interest are (1) run-time scalability, (2) solution quality in terms of required-time and (3) solution quality in terms of cost (wire length and buffer usage). Our experiments are compared with a P-Tree based approach [7] which is known to produce high quality solutions particularly for uniform required times.

In our experiments we used two different ways to derive the initial topology fed to S-Tree. First is induced by the minimum spanning tree (same tree used to construct initial permutation in P-Tree [8]). For the other we used the P-Tree$_A$ (wire-length only) algorithm to derive the initial topologies (because of the pseudo-polynomial nature of timing-driven versions of the various algorithms, this is reasonable as S-Tree still dominates the overall run-time). In test results the first method is referred to as S-Tree, and the second one as S-Tree+ (the runtime of P-Tree$_A$ is included). Wire length is reported in microns, slack in pico seconds, and execution time in seconds of CPU time.

We compared performance of all 3 timing driven algorithms with simultaneous tree construction and buffer insertion on 2 sets of nets. Nets in the first set (6-24 'a' and 'b') have uniform random distribution of required arrival times for sinks (Tab. 1). This is the unfavorable setup for S-Tree since P-Tree's larger solution space (which in general does not cover S-Tree's solution space) allows more exploration when it is not obvious which paths will become critical at the end. S-Tree was able to achieve maximum slack very close to the one of P-Tree on all nets using same number of buffers, less wire, and an order of magnitude less cpu time. On net 6b it gave exactly the same solution and on net 12a slightly faster one. Only on net 12b, 3% more wire was used, but solution was generated 4 times faster. For net sizes 21 and 24, P-Tree was not able to give buffered solutions in reasonable amount of time while S-Tree was able to complete and meet sink timing requirements.

Second set of nets (6-16 crit) contains a few highly critical sinks among non-critical ones (Tab. 2). Here we compared S-Tree against P-Tree (both with buffer insertion) on how much resources they require to achieve given slack. For a better understanding of wire usage, minimum wire length achievable by both algorithms is also included. S-Tree is a clear winner in all 3 dimensions. It was able to achieve same slack using less wire, less buffers and using an order of magnitude less cpu time.

5. EXAMPLE

To demonstrate how S-Tree works in practice we have constructed a simple example instance. Some of the solution topologies are shown in Figure 8. Driver is represented by a triangle, sinks by circles and inserted buffers by squares. Sink that is critical for that particular solution is circled. Also each solution's slack is reported in pico seconds, total wire length in microns, and total number of used buffers. In this example, due to timing requirements sinks are partitioned in the following way: $S_1 = \{1,3,4\}$ and $S_2 = \{2,5\}$. Solution on the top left obeys initial topology, which is in this case constructed using geometric MST approach. As the solutions move on trade-off curve towards more expensive (and faster) ones, it can be clearly seen how that initial topology "breaks" and at the end we have sinks 2 and 5 grouped in the same subtree, isolated from the others. Also note how current critical sink changes from one solution to the other.

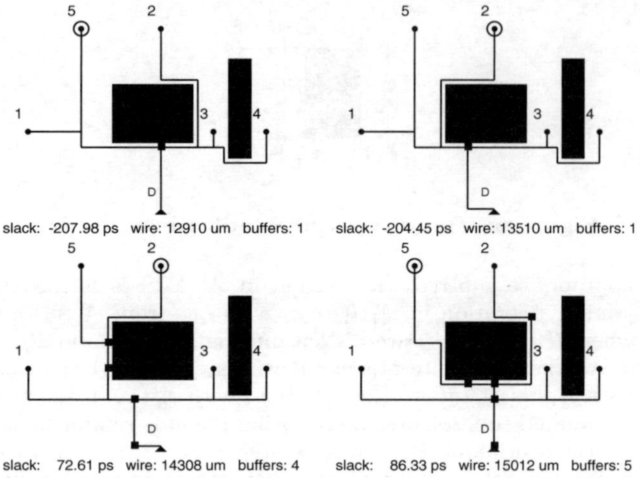

Figure 8: Example topology.

6. CONCLUSIONS

We have presented the S-Tree algorithm for synthesis of buffered interconnects. The approach incorporates a unique combination real-world issues (handling of routing and buffer blockages, cost minimization, critical sink isolation, sink polarities), robustness and scalability. The algorithm is able to achieve the slack comparable to that of buffered P-Tree using less resources (wire and buffers) in an order of magnitude less cpu time. The saving on resources becomes even larger for nets with non-uniform timing requirements.

7. REFERENCES

[1] C. Alpert, et. al. "Buffered Steiner Trees for Difficult Instances," ISPD-01, pp. 4-9.

[2] J. Cong, "Challenges and Opportunities for Design Innovations in Nanometer Technologies," *Frontiers in Semiconductor Research: A Collection of SRC Working Papers.* SRC, 1997.

[3] J. Cong, X. Yuan, "Routing Tree Construction Under Fixed Buffer Locations," DAC-2000, pp. 368-373.

[4] S.-W. Hur, A. Jagannathan, J. Lillis, "Timing-Driven Maze Routing," *IEEE Transactions on Computer Aided Design* Feb. 2000, vol. 19, no. 2, pp. 234-241.

[5] A. Jagannathan, S.-W. Hur, J. Lillis, "A Fast Algorithm for Context-Aware Buffer Insertion," DAC-2000, pp. 368-373.

Table 1: Performance of different algorithms on 6-24 pin nets

Alg.	net 6a							net 6b						
	0 buf		3 buf		5 buf		cpu	0 buf		1 buf		3 buf		cpu
	wl	slk	wl	slk	wl	slk		wl	slk	wl	slk	wl	slk	
PTree	18034	-682	20310	517	20014	608	0.126	20440	-945	25396	183	17902	488	0.071
STree	17799	-646	19397	541	19718	606	0.045	20440	-945	17902	174	17902	488	0.033
STree+	17799	-646	19397	541	19718	606	0.045	20440	-945	17902	174	17902	488	0.033

Alg.	net 9a							net 9b						
	0 buf		4 buf		9 buf		cpu	0 buf		3 buf		6 buf		cpu
	wl	slk	wl	slk	wl	slk		wl	slk	wl	slk	wl	slk	
PTree	22157	-941	28917	605	37468	704	1.943	29819	-2891	40156	-14	49972	83	1.336
STree	22157	-941	28508	575	32859	696	0.411	34401	-3495	34282	-117	30128	48	0.528
STree+	23764	-946	26830	598	32859	696	0.357	34401	-3495	30128	-161	29450	23	0.554

Alg.	net 12a							net 12b						
	0 buf		3 buf		7 buf		cpu	0 buf		5 buf		8 buf		cpu
	wl	slk	wl	slk	wl	slk		wl	slk	wl	slk	wl	slk	
PTree	36220	-2947	32173	-110	50694	299	11.98	31714	-2245	28664	415	30370	461	9.000
STree	35703	-2929	36949	38	36949	303	1.529	31382	-2849	35959	294	36125	435	2.251
STree+	35703	-2929	36949	38	36949	303	1.543	31145	-2684	29549	322	31654	440	2.052

Alg.	net 15a							net 15b						
	0 buf		5 buf		8 buf		cpu	0 buf		3 buf		7 buf		cpu
	wl	slk	wl	slk	wl	slk		wl	slk	wl	slk	wl	slk	
PTree	41410	-3650	46366	312	49111	399	58.69	32076	-1839	36306	525	45358	652	45.83
STree	49424	-3451	39501	209	46802	385	6.283	35303	-2009	43003	487	46691	641	7.041
STree+	52536	-3638	38701	134	45731	353	7.481	33980	-1930	42843	487	42700	641	6.813

Alg.	net 18a							net 18b						
	0 buf		5 buf		11 buf		cpu	0 buf		5 buf		8 buf		cpu
	wl	slk	wl	slk	wl	slk		wl	slk	wl	slk	wl	slk	
PTree	42817	-3650	48972	265	53062	464	209.6	41265	-3544	49166	120	54590	181	187.4
STree	35325	-3784	43696	62	49776	450	14.28	43004	-3757	46037	36	49840	180	17.01
STree+	35287	-3760	43696	62	49776	450	14.72	42384	-3720	48283	32	47738	177	17.98

Alg.	net 21a							net 21b						
	0 buf		3 buf		7 buf		cpu	0 buf		5 buf		9 buf		cpu
	wl	slk	wl	slk	wl	slk		wl	slk	wl	slk	wl	slk	
PTree	-	-	-	-	-	-	-	-	-	-	-	-	-	-
STree	61769	-3376	49528	159	50387	783	46.14	55245	-4478	47434	29	45281	183	50.37
STree+	45708	-2836	45480	409	47421	1137	38.75	49930	-4094	48422	61	49898	184	38.89

Alg.	net 24a							net 24b						
	0 buf		5 buf		11 buf		cpu	0 buf		6 buf		12 buf		cpu
	wl	slk	wl	slk	wl	slk		wl	slk	wl	slk	wl	slk	
PTree	-	-	-	-	-	-	-	-	-	-	-	-	-	-
STree	69602	-5960	62186	-146	62176	143	84.66	56775	-4385	64101	99	60947	305	76.32
STree+	69602	-5960	60760	-482	60094	-292	101.24	52377	-4254	62822	-98	62891	239	77.93

Table 2: Usage of resources for critical nets PTree vs STree

nets	slack	minWL	PTree			STree		
			wl	buf	cpu	wl	buf	cpu
6crit	521	10500	15500	6	0.026	10500	5	0.011
8crit	680	7000	10000	6	0.027	10000	4	0.004
10crit	673	10000	18000	6	0.125	14000	5	0.026
12crit	784	11000	18000	9	0.643	15000	5	0.033
14crit	614	12000	20000	7	0.702	18000	5	0.059
16crit	1567	14400	23200	7	4.982	21200	6	0.382

[6] J. Lillis, C.-K. Cheng, T.-T. Y Lin, "Optimal Wire Sizing and Buffer insertion for Low Power and a Generalized Delay Model," *IEEE Journal of Solid State Circuits*, 31 (3): pp. 437-447, March 1996.

[7] J. Lillis, C.-K. Cheng, T.-T. Y. Lin, "Simultaneous Routing and Buffer Insertion for High Performance Interconnect," *Proc. 6th IEEE Great Lakes Symposium on VLSI*, Ames, Iowa, Mar. 1996, pp. 148-153.

[8] J. Lillis, C.-K. Cheng, T.-T. Y Lin, "New Performance Driven Routing Techniques With Explicit Area/Delay Tradeoff and Simultaneous Wire Sizing," DAC-96.

[9] T. Okamoto, J. Cong, "Buffered Steiner Tree Construction with Wire Sizing for Interconnect Layout Optimization," ICCAD-96, pp. 44-49, 1996.

[10] L.P.P.P. van Ginneken, "Buffer Placement in Distributed RC-tree Networks for Minimal Elmore Delay," ISCAS-90, pp. 865-868, 1990.

[11] H. Zhou, D.F. Wong, I.M. Liu, A. Aziz, "Simultaneous Routing and Buffer Insertion with Restrictions on Buffer Locations," DAC-99, pp. 96-99.

An Algorithm for Integrated Pin Assignment and Buffer Planning *

Hua Xiang†, Xiaoping Tang†‡, and D.F. Wong†
†University of Texas at Austin, Austin, TX 78712
‡Silicon Perspective, A Cadence Company, Santa Clara, CA 95054
{xiangh, tang, wong}@cs.utexas.edu

ABSTRACT

The *buffer block* methodology has become increasingly popular as more and more buffers are needed in deep-submicron design, and it leads to many challenging problems in physical design. In this paper, we present a polynomial-time exact algorithm for integrated pin assignment and buffer planning for all two-pin nets from one macro block (source block) to all other blocks of a given buffer block plan as well as minimizing the total cost $\alpha \cdot W + \beta \cdot R$ for any positive α and β where W is the total wire length and R is the number of buffers. By applying this algorithm iteratively (each time pick one block as the source block), it provides a polynomial-time algorithm for pin assignment and buffer planning for nets among multiple macro blocks. Experimental results demonstrate its efficiency and effectiveness.

Categories and Subject Descriptors

B.7.2 [**Integrated Circuits**]: Design Aids—*Placement and routing*; J.6 [**Computer Applications**]: Computer-Aided Engineering—*Computer-aided design*

General Terms

Algorithms, Design, Performance, Theory

Keywords

buffer insertion, pin assignment, min-cost maximum flow

1. INTRODUCTION

As chip size grows larger and minimum feature size is reduced, the capacitance and resistance of wires increase dramatically. This makes interconnects play a critical role in achieving high performance and reducing circuit complexity in deep-submicron design. Many techniques have been proposed to reduce interconnect delay. One effective way is buffer insertion[3][9] and it is heavily used in chip design. It is estimated that the amount of inserted buffers for on-chip interconnect may be up to 800,000 in 50*nm* technology[4]. At the same time, the introduction of so many buffers raises many challenging problems in physical design. In a hierarchical approach, buffers are clustered together as a buffer block to facilitate floorplan and routing. Cong et al.[5], Tang and Wong[12] and Sarkar et al.[11] proposed algorithms for buffer block planning problem to minimize the chip area and the number of buffer blocks. In two recent works, Dragan et al.[7][8] gave algorithms for global buffered routing problem with fixed pins. Their work is based on multicommodity flow which is an NP-hard problem. In this paper, we address the problem of simultaneous Pin assignment and Buffer planning (PB) for a given buffer block plan. Our algorithm uses min-cost flow computation which is solvable in polynomial time.

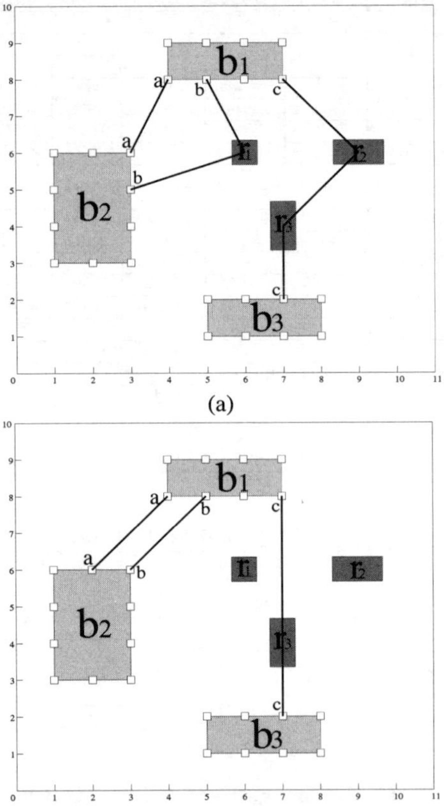

Figure 1: (a) Three nets use 3 buffers and the total wire length is 20. (b) An optimal solution with 1 buffer and wire length 14.

Informally, the problem can be described as follows: given a placement of macro blocks and buffer blocks, assign pins and plan buffer insertions for the given set of nets subject to the required lower and upper bounds of connection intervals (i.e., the range of allowable distance between two buffers or two pins or a pin and a buffer) as well as minimizing the total cost $\alpha \cdot W + \beta \cdot R$ where W

*This work was partially supported by the National Science Foundation under grant CCR-9912390.

is the total wire length and R is the number of buffers. Figure 1 shows an example. The placement includes 3 macro blocks, and 3 buffer blocks. Buffer blocks may have different capacities, i.e., the number of buffers in a buffer block can be different. r_1 has a capacity 1 while the capacities of r_2 and r_3 are 2. A net set includes 2 nets between b_1 and b_2 and 1 net between b_1 and b_3. The range of allowable distance between two buffers or a buffer and a pin is bounded by Manhattan distance 2 and 4. Also if the distance of two pins is longer than 5, buffers have to be inserted, i.e., the lower and upper bounds for the allowable distance between two pins are 0 and 5. The purpose is to assign pins and plan buffers for the 3 nets while minimizing the number of buffers and the total wire length.

The goal of pin assignment is to find the exact locations of pins on macro blocks. Buffer planning is to decide buffer usages along net connections to maintain required delay constraints. The two tasks are closely related. Pin assignment alone may neglect many important factors since interconnect is hard to predict, and a global view of net connections is always helpful. Figure 1(a) illustrates a solution by a two-step approach (i.e., pin assignment following buffer planning). The pin assignments are decided according to the shortest Manhattan distance. Totally 3 buffers are used and the wire length is 20. Figure 1(b) shows an optimal solution with only one buffer and the wire length is 14.

In this paper, we present a polynomial-time exact algorithm for simultaneous pin assignment and buffer planning for all two-pin nets from one macro block (source block) to all other blocks for a given buffer block plan such that each net satisfies the lower and upper bounds on connection intervals as well as minimizing the total cost $\alpha \cdot W + \beta \cdot R$ for any positive constants α and β where W is the total wire length and R is the number of buffers. By applying this algorithm iteratively (each time pick one block as the source block), it provides a polynomial-time algorithm for pin assignment and buffer planning for nets among multiple macro blocks. Experimental results demonstrate its efficiency and effectiveness.

The rest of the paper is organized as follows. Section 2 defines the PBO (Pin assignment and Buffer planning for One source block) problem which simultaneously assigns pins and plans buffers for nets between one macro block and all other blocks. In section 3, we present a network flow formulation to solve the PBO problem. Then we extend PBO problem to PB(Pin assignment and Buffer Planning) problem which considers all nets among multiple macro blocks in section 4. In section 5, we also provide a node clustering method to speed up the computation. Finally, we show the experimental results in section 6 and conclude the paper in section 7.

2. PIN-ASSIGNMENT AND BUFFER PLANNING FOR ONE SOURCE BLOCK (PBO)

Given a placement of $m+1$ macro blocks $B = \{b_s, b_1, ..., b_m\}$ with n buffer blocks $R = \{r_1, ..., r_n\}$. (For convenience, we call b_s source block and other blocks sink blocks.) Each buffer block r_i $(i = 1, ..., n)$ is associated with a positive integer c_i denoting the capacity of r_i, i.e., buffer block r_i can hold c_i buffers.

Let $N = N_1 \cup N_2 \cup ... \cup N_m$ where N_i $(i = 1, ..., m)$ is the set of nets between block b_s and b_i; $P = P_s \cup P_1 \cup ... \cup P_m$ where P_i $(i = s, 1, ..., m)$ is the set of available pin locations of block b_i.

The distance of two points u and v on a planar region is denoted as d_{uv}. Let *buffer interval* be the allowable distance range between two buffers or a pin and a buffer; and *pin interval* be the allowable distance range between two pins. For convenience, let *connection interval* refer to buffer interval or pin interval.

Suppose the lower and upper bounds for buffer intervals are $\tilde{L}$ and $\tilde{U}$ respectively; and those for pin intervals are $\bar{L}$ and $\bar{U}$ respectively. A valid path means:

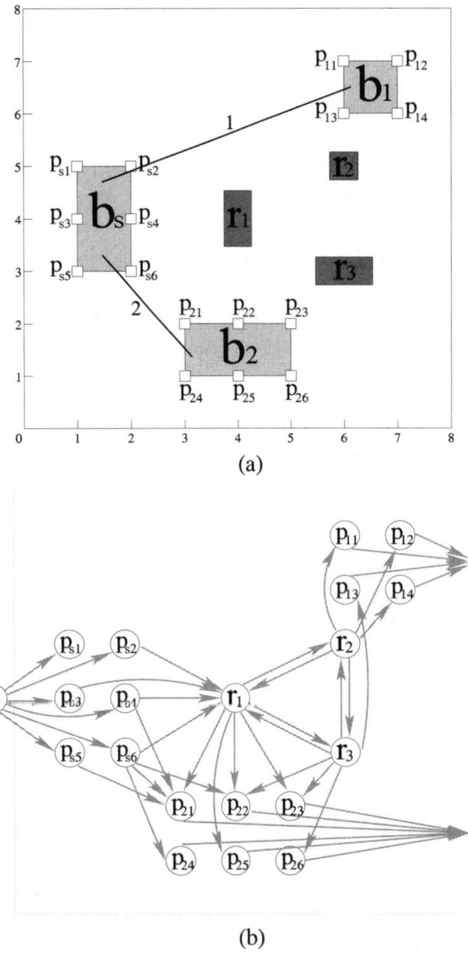

Figure 2: (a) A PBO problem with 3 macro blocks and 3 buffer blocks. The lower and upper bounds of buffer intervals are 2 and 3 respectively. And the pin intervals are bounded 0 and 3. The tiny squares p_{xy} on the boundaries of macro blocks are available pin locations. The net set includes one net between b_s and b_1 and two between b_s and b_2. (b) The corresponding flow network graph.

1. If the path is $p = (p_s, r'_1, ..., r'_k, p_t)$ where $p_s, p_t \in P$ and $r'_i \in R$ $(i = 1, ..., k)$, then the distance of each path segment is bounded by $\tilde{L}$ and $\tilde{U}$, i.e., $\tilde{L} \le d_{p_s r'_1} \le \tilde{U}$, $\tilde{L} \le d_{r'_i r'_{i+1}} \le \tilde{U}$ $(i = 1, ..., k-1)$ and $\tilde{L} \le d_{r'_k p_t} \le \tilde{U}$;
2. If the path is $p = (p_s, p_t)$ where $p_s, p_t \in P$, then $\bar{L} \le d_{p_s p_t} \le \bar{U}$.

The length of a path is the sum of the distances of all path segments.

For any given positive constants α and β, the PBO problem is to find a set of valid paths connecting b_s and all other macro blocks as well as minimizing the total cost $\alpha \cdot W + \beta \cdot R$ where W is the total length and R is the number of buffers. Each path corresponds to a net in N and the two end points of the path are assigned pin locations for this net.

Figure 2(a) gives a simple example. There are 3 macro blocks and 3 buffer blocks. The capacities of the three buffer blocks r_1, r_2

and r_3 are 2, 1 and 2 respectively. For any two points u and v on the planar region, their coordinates are (u_x, u_y) and (v_x, v_y) respectively. Define distance $d_{uv} = |u_x - v_x| + |u_y - v_y|$. The required lower and upper bounds for buffer intervals are 2 and 3 respectively; and the bounds for pin intervals are 0 and 3 respectively. The purpose is to decide pins and buffer locations for 1 net between b_s and b_1 and 2 nets between b_s and b_2 with a minimum cost $\alpha \cdot W + \beta \cdot R$.

3. THE ALGORITHM

To solve the PBO problem, we first construct a network graph, then apply a min-cost flow algorithm to get the solution.

Given a PBO problem, we construct the network graph $G = (V, E)$ with capacity U and cost C as follows:

1. $V = \{s, t, t_1, t_2, ..., t_m\} \cup R \cup P$, where s is the source node, t is the sink node, and t_i $(i = 1, ..., m)$ is a subsink node.

2. $E = E_s \cup E_t \cup E_{\tilde{t}} \cup E_b \cup E_r \cup E_{br} \cup E_{rb}$, where
   $E_s = \{(s, p_s) | p_s \in P_s\}$,
   $E_t = \{(t_i, t) | i = 1, 2, ..., m\}$,
   $E_{\tilde{t}} = \bigcup_{i=1}^{m} \{(p, t_i) | p \in P_i\}$,
   $E_b = \bigcup_{i=1}^{m} \{(p_s, p) | p_s \in P_s, p \in P_i, \bar{L} \le d_{p_s p} \le \bar{U}\}$,
   $E_r = \{(r_i, r_j) | i \ne j, i, j = 1, ..., n, \tilde{L} \le d_{r_i r_j} \le \tilde{U}\}$,
   $E_{br} = \bigcup_{i=1}^{n} \{(p_s, r_i) | p_s \in P_s, \tilde{L} \le d_{p_s r_i} \le \tilde{U}\}$,
   $E_{rb} = \bigcup_{i=1}^{n} \{(r_i, p) | p \in P_j, j = 1, ..., m, \tilde{L} \le d_{r_i p} \le \tilde{U}\}$

3. Edge Capacity:
   for edges $e(t_i, t)$, $U(e) = |N_i|, (i = 1, ..., m)$;
   other edges e, $U(e) = 1$.

4. Node Capacity:
   for $r_i \in R$, $U(r_i) = c_i$;
   for $p \in P$, $U(p) = 1$;
   other nodes are incapacitated.

5. Edge Cost:
   for $e \in E_s \cup E_t \cup E_{\tilde{t}}$, $C(e) = 0$;
   other edges $e(u, v)$, $C(e) = \alpha \cdot d_{uv}$.

6. Node Cost:
   for $r \in R$, $C(r) = \beta$;
   other nodes v, $C(v) = 0$.

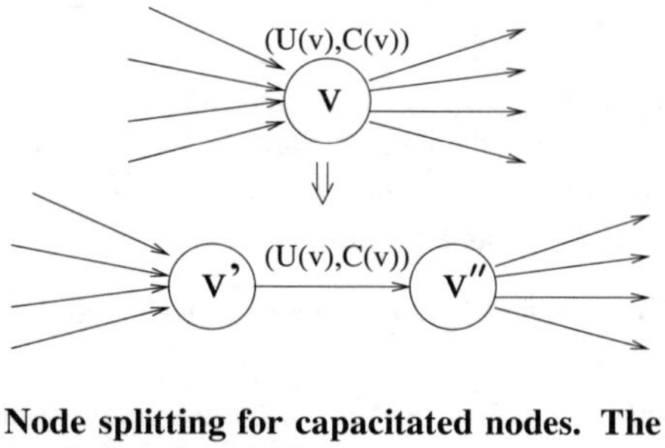

Figure 3: Node splitting for capacitated nodes. The new edge has capacity $U(v)$ and cost $C(v)$.

Figure 2(b) illustrates the constructed network graph for the PBO problem in Figure 2(a). Note that whether an edge (u, v) $(u, v \in P \cup R)$ should be added to G depends on whether d_{uv} falls in the range $[\bar{L}, \bar{U}]$ (or $[\tilde{L}, \tilde{U}]$) or not. For example, the distance between p_{13} and r_2 is 1, which is less than the lower buffer interval bound $\tilde{L} = 2$. Thus there is no edge between p_{13} and r_2 in the constructed flow network. Similarly, the distance between p_{s2} and p_{13}

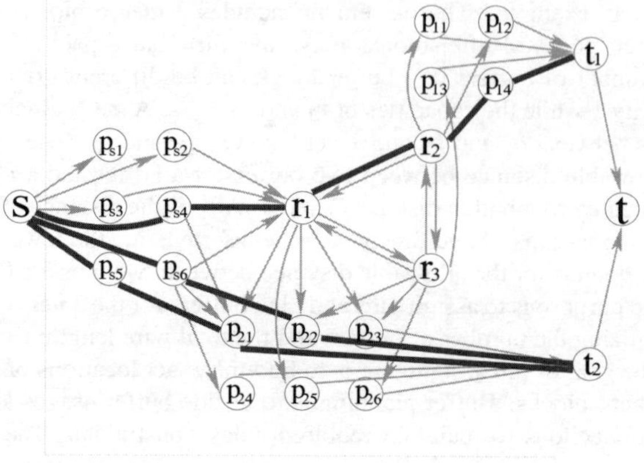

(a)

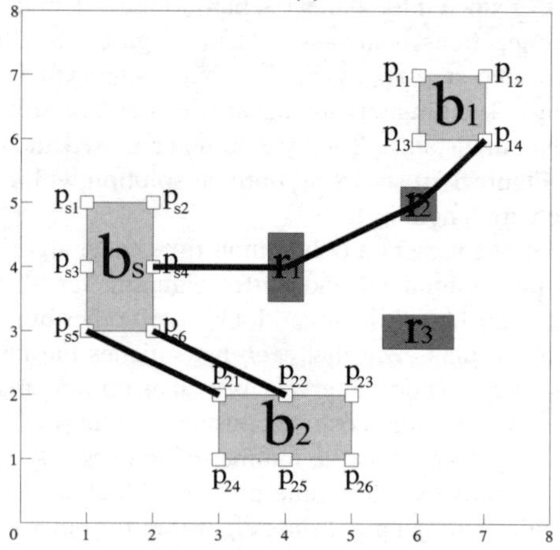

(b)

Figure 4: (a) a flow f in the network in Figure 2(b), $|f| = 3$. (b) The corresponding solution of pin assignment and buffer planning to the PBO problem of Figure 2(a).

is 5 which is larger than the upper pin interval bound $\bar{U} = 3$, thus the edge (p_{s2}, p_{13}) is not included.

In the constructed flow network, every node in $P \cup R$ has a cost and a capacity. However, classical network flow problem only assigns cost and capacity to edges. This can be solved by splitting the capacitated node v into two nodes v' and v''. A new edge (v', v'') is added with a capacity $U(v)$ and a cost $C(v)$. Then change the original edges (u, v) and (v, w) into edges (u, v') and (v'', w) respectively (refer to Figure 3).

Any flow in G can be mapped to a pin assignment and buffer planning solution for a subset of the given nets. The used capacities of R nodes in the flow are the number of buffers needed in the solution. Figure 4(a) illustrates a flow f, $|f| = 3$. And Figure 4(b) is a solution of pin assignment and buffer planning derived from the above flow. If a flow f exists and $|f| = |N|$, then we can find a feasible solution of pin assignment and buffer planning for all of the nets in N. On the other hand, given a pin assignment and buffer planning solution for n nets, a flow f $(|f| = n)$ can always be found on the constructed flow network. Since the total capacities of edges going into sink node t are $\sum_{i=1}^{m} U(t_i, t) = \sum_{i=1}^{m} |N_i| = |N|$, the maximum flow f_{max} in G, $|f_{max}| \le |N|$. Thus if a flow $|f| < |N|$, then there is no feasible solution to the original PBO problem, which requires considering all of the nets between b_s and all other macro blocks.

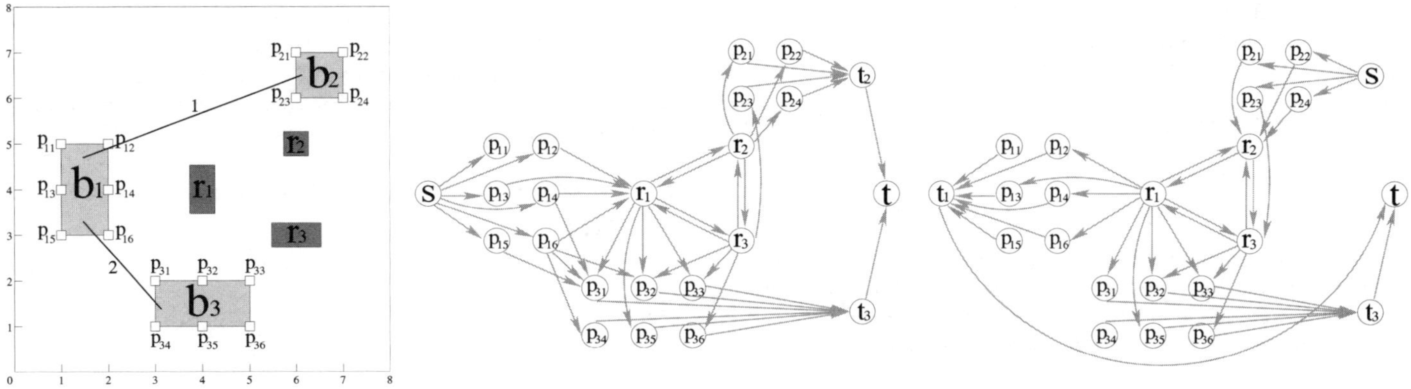

(a) (b) (c)

Figure 5: (a) A PB problem with 3 macro blocks and 3 buffer blocks. (b) The corresponding flow network when b_1 is the source block. (c) The corresponding flow network when b_2 is the source block.

Furthermore, the cost of the flow is also the cost of pin assignment and buffer planning solution. Therefore min-cost maximum flow assigns pins and plans buffers for as many nets as possible with minimum total cost.

The following theorem shows that the PBO problem can be exactly solved by min-cost flow computation on G.

THEOREM 1. *A min-cost flow f, $|f| = |N|$, in G corresponds to a pin assignment and buffer planning solution to PBO problem for all nets in N with minimum total cost $\alpha \cdot W + \beta \cdot R$ for any given α and β where W is the total wire length and R is the number of buffers. If the size of the max-flow, $|f_{max}| < |N|$, then there is no feasible solution to the PBO problem. A min-cost maximum flow assigns pins and plans buffers for the maximum number of nets with minimum total cost.*

The algorithm PBO-Flow can be summarized as the following:

Algorithm PBO-Flow(B, R, N, P, C, α, β)
1. Construct the network graph $G(V,E)$
2. Assign capacities U and costs C
3. Apply min-cost maximum flow algorithm on G
4. Derive the pin assignment and buffer planning solution

Finding a min-cost maximum flow in a network is a classical problem for which several polynomial-time optimal algorithms are available[2][6]. Deriving a solution of PBO from a flow in G can be done in $O(E)$ time. Thus, if we adopt the double scaling algorithm in[1], we get the following time complexity for the PBO problem.

THEOREM 2. *PBO-Flow algorithm optimally solves the PBO problem in $O(VE \log\log U_{max} \log(VC_{max}))$ time for $G = (V,E)$, U_{max} is the maximum value of U, and C_{max} is the maximum value of C.*

Note that the complexity of our PBO-Flow algorithm is mainly dependent on the size of the constructed network graph $G(V,E)$. According to the graph construction rules, $|V| = 2 + m + 2(|P| + |R|)$ (consider node splitting), and the upper bound of edges is $m + |P| + |P_s| \cdot |P| + |P| \cdot |R| + |R|^2 + |P| + |R|$ (in fact, the number of edges is much smaller due to distance constraint) which is bounded by $(|P| + |R|)^2$.

In applications, we may put more effort on reducing the number of buffers. In this case, we can set a large weight to buffer nodes, i.e., a large β. For example, let β larger than the upper distance bound $\tilde{U}$. If we set β large enough, we tend to get a solution with the minimum number of buffers.

Further in some circuits, some locations on a block may not be allowed for pin assignment. In this case, these kinds of locations will not appear in the pin set P. Obviously, our network-flow based algorithm will not assign a pin to these kinds of locations.

4. PIN ASSIGNMENT AND BUFFER PLANNING (PB)

In the above section, we discuss how to solve PBO problem using min-cost maximum flow computation. PBO problem only consider net connections between one source block and other blocks. In reality, we need to deal with net connections among all of the macro blocks, called PB(Pin assignment and Buffer Planning) problem. The definition of PBO problem can be easily extended to PB problem as the following:
Given:

1. a placement of m macro blocks $B = \{b_1, b_2, ..., b_m\}$ and n buffer blocks $R = \{r_1, r_2, ..., r_n\}$; buffer block r_i has a capacity c_i.
2. a set of available pin locations $P = P_1 \cup P_2 \cup ... \cup P_m$ where P_i $(i = 1, ..., m)$ is a set of available pin locations of macro block b_i.
3. a set of nets $\bar{N} = \bar{N}_1 \cup \bar{N}_2 \cup ... \cup \bar{N}_m$ where $\bar{N}_i$ $(i = 1, ..., m)$ is the set of nets between block b_i and all other blocks.
4. two non-negative numbers $\tilde{L}$ and $\tilde{U}$ denoting the lower and upper bound of buffer intervals.
5. two non-negative numbers $\bar{L}$ and $\bar{U}$ denoting the lower and upper bound of pin intervals.

Goal:
For any given positive α and β, find a set of valid paths corresponding to the net set $\bar{N}$ as well as minimizing the total cost $\alpha \cdot W + \beta \cdot R$ where W is the total wire length and R is the number of buffers.

PBO-Flow algorithm solves pin assignment and buffer planning for nets between one block and all other blocks. Naturally, if we treat each macro block as the source block and apply PBO-Flow algorithm repeatedly, we can get a solution for all nets among multiple macro blocks. This is the basic idea of PB-Flow algorithm.

Algorithm PB-Flow(B, R, $\bar{N}$, P, C, α, β)

1. Let b_1 be the source block;
2. Apply PBO-Flow(B, R, $\bar{N}_1$, P, C, α, β)
3. For i = 2 to m
4.   Let b_i be the source block;
5.   Adjust the network graph $G(V,E)$
6.   Apply PBO-Flow(B, R, $\bar{N}_i$, P, C, α, β)
7. Derive the pin assignment and buffer planning solution

Note that when different blocks are selected as the source block, their constructed flow networks are slightly different, i.e., the edges incident from/to the pin nodes of the two blocks are different. So in PB-Flow algorithm, at each iteration, we need to do some adjustment to transform the existing network graph to the one corresponding to the next source block(line 6).

Now suppose two different macro blocks b_i and b_{i+1} are selected as the source block sequentially. Let G_i be the constructed flow network when b_i is chosen as the source block. The following steps change G_i to b_{i+1}'s corresponding flow network G_{i+1}:

1. delete edges connecting the pin nodes of P_i and the pin nodes of other blocks;
2. add edges connecting the pin nodes of P_{i+1} and the pin nodes of other blocks as well as maintaining distance constraints.
3. reverse the direction of edges (p,r) where $p \in P_i$ and $r \in R$;
4. reverse the direction of edges (r,p) where $r \in R$ and $p \in P_{i+1}$;
5. reverse the direction of edges (s,p) where $p \in P_i$;
6. reverse the direction of edges (p,t_{i+1}) where $p \in P_{i+1}$;
7. let t_{i+1} be the source node s and the original source node become a subsink node t_i. Thus remove the edge (t_{i+1},t) and add one new edge (t_i,t).

As illustrated in Figure 5, (a) is a PB problem. (b) shows the constructed flow network when b_1 is the source block. When the source block is switched to b_2, t_2 becomes the source node and the original one becomes a subsink node t_1. But the edges connecting two buffer nodes remain unchanged.

For a newly added edge (u,v), the cost is $\alpha \cdot d_{uv}$ and the capacity is 1. For other edges (u',v'), the cost is unchanged and the capacity is 1 if no flow flows through the corresponding edge in the previous iteration. Otherwise, the capacity is 0. As to the capacities of nodes, similarly, if a node v has c capacities left after pushing flow in G_i, then the capacity of v is c in G_{i+1}.

G_i and G_{i+1} have the same node set. Also it is easy to show that the number of changed edges is bounded by $(|P_i|+|P_{i+1}|)\cdot(|P|+|R|)$. When taking the distance constraint into consideration, this bound can be further greatly reduced. Thus the adjustment to the graph can be efficiently accomplished.

A two pin net connecting b_i and b_{i+1} belongs to both net sets $\bar{N}_i$ and $\bar{N}_{i+1}$. Suppose we can get a feasible solution to the PB problem with PB-Flow algorithm. After applying PBO-Flow on G_i, the nets belong to $\bar{N}_i \cap \bar{N}_{i+1}$ should have already been found. Thus we only need to consider $\bar{N}_{i+1} - \bar{N}_i - \ldots - \bar{N}_1$ while pushing flows on G_{i+1}.

On the other hand, as we notice that a net belonging to $\bar{N}_i \cap \bar{N}_{i+1}$ should correspond to a path $(p_u, r'_1, \ldots, r'_k, p_v)$ ($p_u \in P_i$, $p_v \in P_{i+1}$, $r'_i \in R, i = 1, \ldots, k$) in G_i. Then in the modified flow network G_{i+1}, the path $(p_v, r'_k, \ldots, r'_1, p_u)$ must exist since the edge connections among buffer blocks are not changed. Thus we can remove all paths connected to nodes in P_{i+1} by increasing the capacities along all these paths, and applying PBO-Flow with nets $\bar{N}_{i+1}$. The optimality of PBO-Flow guarantees that the cost will not increase.

Further even if the algorithm doesn't return a feasible solution, (e.g, no feasible solution exists), the second way still outperforms the first one since the optimality of PBO-Flow assures that the new solution won't become worse: either connecting more nets or reducing the cost.

5. IMPROVEMENT WITH NODE CLUSTERING

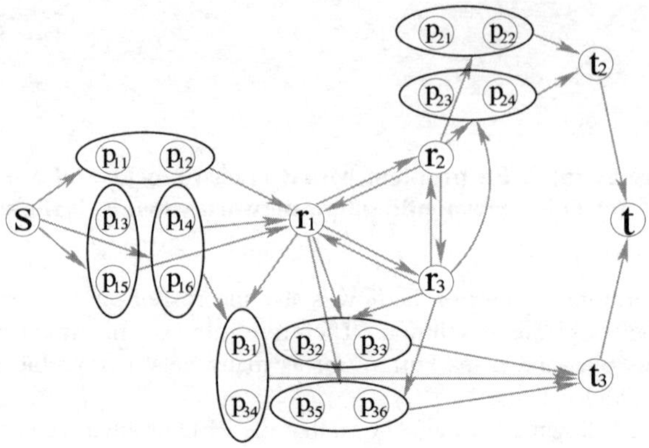

Figure 6: The corresponding flow network of the PB problem in Figure 5(b) using node clustering method.

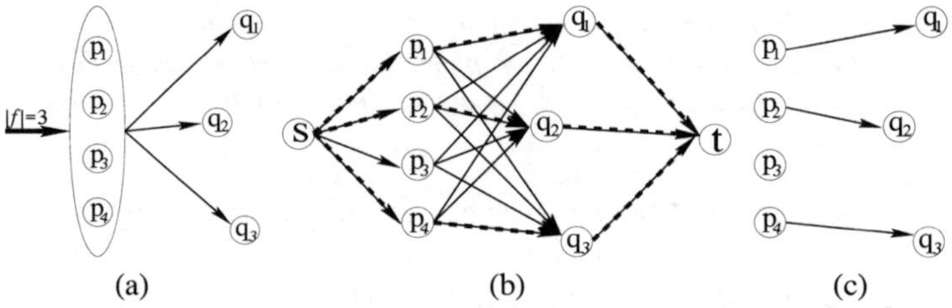

Figure 7: (a) A flow f, $|f| = 3$ flows through a super-node. (b) The corresponding network and a flow solution. (c) Deriving connections for original pin nodes.

When we handle a big circuit which may include a huge amount of nets, the corresponding flow network might be quite large. In order to facilitate the process of huge PB problems, we propose the following node clustering method to speed up the computation.

For any $p \in P$, if p is inside a macro block, it must be connected to some pin outside the block. Without loss of generality, we may assume that a pin is only on the boundary of a macro block. Then by grouping neighbor pin nodes together, we can greatly reduce the number of nodes, consequently reducing the number of edges. Once several nodes are grouped together, we can use the average coordinate as the location of the new "super-node". And the capacity of the super-node is the number of nodes it includes. Figure 6 illustrates a constructed flow network for Figure 5(b) when 2 neighbor pins are clustered to one super-node. Actually, we can set different scale rates to blocks according to their sizes or other requirements.

Once we get a solution from the super-node flow network, we need to map the flow to a solution of the original PB problem, i.e, distributing the flows through one super-node to its pin nodes. Of

Table 1: Average results of PB-Flow for 5 times. All nets are found using PB-Flow algorithm

File		A33n	X40	H80	S100	F110	M200
Grid		107x106	146x160	235x230	232x227	367x401	587x560
Blocks		33	40	60	91	110	100
Buffer Blocks		30	40	120	90	90	120
Nets		640	1021	1317	2021	4180	5659
Time (second)	Net-Net	27.80	34.49	75.46	119.67	91.91	135.77
	PB-Flow	41.91	27.32	67.94	83.12	220.45	326.9
Found nets	Net-Net	528.4	940.2	1316	1994	4167	5578
	PB-Flow	640	1021	1317	2021	4180	5659
Buffers	Net-Net	391.8	1731.8	2259	6400.2	9964.4	15463
	(per net)	(0.741)	(1.842)	(1.717)	(3.210)	(2.391)	(2.772)
	PB-Flow	327	1439	1949.8	5627	8631.4	13873.4
	(per net)	(0.511)	(1.409)	(1.480)	(2.784)	(2.065)	(2.452)
	reduced	31.04%	23.51%	13.80%	13.27%	13.63%	11.54%
Wire Length	Net-Net	920.2	2672	3575	8394.2	14131.4	21041
	(per net)	(1.741)	(2.842)	(2.717)	(4.210)	(3.391)	(3.772)
	PB-Flow	967	2460	3266.8	7648	12811.4	19532.4
	(per net)	(1.511)	(2.409)	(2.480)	(3.784)	(3.065)	(3.452)
	reduced	13.21%	15.24%	8.72%	10.12%	9.61%	8.48%

course, we hope the mapping has the minimum connection length. In Figure 7(a), a flow f, $|f| = 3$ flows through a super-node which includes 4 pin nodes. 7(c) shows a feasible mapping. Still it can be solved with min-cost maximum flow as shown in Figure 7(b). Each node p_i $(i = 1,2,3,4)$ in the super-node is connected to the destination nodes q_j $(j = 1,2,3)$ if $d_{p_iq_j}$ satisfies the distance constraints. The capacity of each edge is 1. The cost of edge (p_i,q_j) is the distance of two nodes $d_{p_iq_j}$. All other edges have a cost 0. Min-cost maximum flow guarantees to find an optimal mapping.

6. EXPERIMENTAL RESULTS

Our algorithms were implemented in C++ on Sun Sparc Ultra 5 (360MHz) with 128M memory. We tested PB-Flow on 6 circuits which were generated randomly. For all files, we adopted the node clustering method in Section 5 to reduce runtime.

We compared PB-Flow algorithm with a two-step approach (first assign pins, then plan buffers). We used a classical way[10] to assign pins as follows: by connecting the centers of two macro blocks, we got two crossing points on the boundaries of each block, then assign pins around the crossing points for the nets between these two macro blocks. After pin assignment was done for all nets, we used a net-by-net approach for buffer planning. The net-by-net approach considered only one net each time and found the min-cost path between the two pins of the net for buffer insertion.

For each test circuit, we repeated both approaches 5 times. Table 1 listed the average results of these 5 times. Using PB-Flow, we could find a feasible solution of pin assignment and buffer planning for all of the nets with a significant improvement on both the total wire length and the number of buffers. The last two rows show the comparison of the number of buffers and the total wire length. For both methods, we listed the test results and the values per net. The percentage was calculated according the values per net since the numbers of found nets are different. The last two testing files $F110$ and $M200$ include more than 4000 nets. The net-by-net approach can not handle such large files. For comparison purpose, we first apply the node clustering strategy to group pins, then use net-by-net for buffer insertion.

7. CONCLUSION

In this paper, we present a polynomial-time algorithm for simultaneous pin assignment and buffer planning for all 2-pin nets between a source macro block and all other blocks such that each net satisfies the lower and upper bound of connection intervals as well as minimizing the total cost $\alpha \cdot W + \beta \cdot R$. By applying this algorithm iteratively (each time pick one block as the source block), it provides a polynomial-time algorithm for pin assignment and buffer planning for nets among multiple macro blocks. Experimental results demonstrate that the algorithm is very efficient and effective.

8. REFERENCES

[1] R.K. Ahuja, A.V. Goldberg, J.B. Orlin, and R.E. Tarjan, "Finding minimum-cost flows by double scaling", Mathematical Programming 53, pp. 243-266, 1992.

[2] R.K. Ahuja, T.L. Magnanti, and J.B. Orlin, Network Flows, Prentice Hall, 1993.

[3] H.B.Bakoglu, Circuits, Interconnections, and Packaging for VLSI, Addison-Wesley, 1990.

[4] J.Cong, "Challenges and opportunities for design innovations in nanometer technologies", SRC Working Papers, http://www.src.org/prg_mgmt/frontier.dgw, Dec 1997.

[5] J.Cong, T.Kong, and D.Z. Pan, "Buffer block planning for interconnect driven floorplanning", Proc. ICCAD, pp. 358-363, Nov.1999.

[6] T.H. Cormen, C.E. Leiserson, and R.L. Rivest, Introduction to Algorithms, The MIT Press, 1992.

[7] F.F. Dragan, A.B.Kahng, I.I. Mandoiu, S. Muddu, and A. Zelikovsky, "Provably good global buffering using an available buffer block plan", Proc. ICCAD, pp. 104-109, 2000.

[8] F.F. Dragan, A.B.Kahng, I.I. Mandoiu, S. Muddu, and A. Zelikovsky, "Provably good global buffering by multiterminal multicommodity flow approximation", Proc. ASPDAC, pp. 120-125, 2001.

[9] R.Otten, "Global wires harmful?", Proc. ISPD, pp. 104-109, 1998.

[10] B.Preas and M.Lorenzetti, "Physical Design Automation of VLSI Systems", Benjamin/Cummings, Menlo Park, CA 1988.

[11] P.Sarkar, V. Sundararaman and C.K. Koh, "Routability-Driven Repeater Block Planning for Interconnect-Centric Floorplanning", Proc. ISPD, April 2000.

[12] X.Tang and D.F. Wong, "Planning buffer locations by network flows", Proc. ISPD, pp. 180-185, April 2000.

An Efficient Routing Database

Narendra V. Shenoy
Synopsys Inc.
700 E. Middlefield Rd.
Mountain View CA 94043
nshenoy@synopsys.com

William Nicholls
Synopsys Inc.
1907 NW Cornelius Pass Rd.
Hillsboro OR 97124
nicholls@synopsys.com

ABSTRACT
Routing is an important problem in the process of design creation. In this paper, we focus on the problem of designing a database for the non-partitioned routing problem. New technology libraries describe constraints that are hard to manage in grid-based approaches to the routing database. While general region query based data-structures have been proposed, they typically suffer from speed problems when applied to large blocks. We introduce an interval-based approach. It provides more flexibility than grid-based techniques. It exploits the notion of preferred direction for metal layers to manage the memory efficiently. It supports efficient region queries. We finally present a comparison study for real industrial designs on this database.

Categories and Subject Descriptors
J.6 [**Computer-Aided Engineering**]: Computer-aided design

General Terms: Algorithms, Design

Keywords: Physical design, routing, database.

1 INTRODUCTION
Managing data in any software tool is a challenge when the data sizes become large. This is particularly true in the routing problem. Routing has been a subject of fertile research, particularly from the algorithmic point of view. Less emphasis on the database has been found in the research community. In this paper, we investigate the key issues that arise in the design and management of a routing database. We propose an efficient data-structure that meets the demands of new technologies. In the rest of Section 1 we discuss the motivation, requirements and previous work in this field. In Section 2, we present our approach and discuss relevant issues that a routing database must address. Section 3 deals with experimental data that shows the merits of our approach. Section 4 provides some concluding remarks.

1.1 Motivation and previous work
We believe that the design and implementation of an efficient routing database is an interesting and challenging problem for the following reasons. With the recent advances in combining placement and synthesis[21][22], the next focus on design closure will be in the area of routing. In recent technologies, capacitance prediction using routing topology (which is done during placement today) is insufficient. Post-route timing results will be less predictable unless a feedback loop to a timer and extractor is present during routing. This means (assuming a single process) that the memory available for routing will be reduced. A flat database will make this simple to manage. Another compelling reason to investigate new routing database representations is the lack of flexibility in grid-based schemes. Newer technology libraries describe complex pin and obstruction geometry. Abstracting such structures onto a fixed grid representation means that a certain part of the solution space is lost. Finally, recent technology constraints such as wire widths and wire spacing vary across layers. Grid based schemes often require these numbers to be multiples of a fixed number so that a common grid can be created for all layers. This can lead to inefficient use of routing resources.

A routing database requires the following essential ingredients to be successful. It should support fast region query, as the search engine will make millions of queries. A classic search such as Dijkstra's [1] approach or variants such as Lee's [2][3] approach operates in the following fashion. A heap is used to maintain the current frontier of the search. The least cost point is popped off the heap and its neighbors are examined for possible expansion. It is this step that needs fast access to the database. An expansion can be prohibited due to metal geometry or obstructions at or near the expansion point. In addition it should be easy to modify and manipulate the database during routing. It should be memory efficient. The simplest routing database is the grid-point database. The area under concern is populated by a set of grid-points. Each point stores the information required to respond to the queries made during expansion. Each point also stores memory for the cost of the search and for backtracking.

An alternative routing database relies on data-structures tailored for efficient region queries. There is a large body of literature in this area[4][5][6][7][8][9][10][11][12]. Finkel and Bentley[4] develop quad-trees as a data structure for higher dimensional query. Later Bentley[5] proposes the kd-tree as an improvement over the quad-trees. The book by deBerg et al.[13] provides an excellent reference to this literature. Kedem[6] describes the use of quad-trees with bisector lists for storing/retrieving objects using minimal bounding boxes. Ousterhout[7] designs the corner-stitching technique for fast incremental local operations. Rosenberg[8] conducts a detailed study of data structures that support area query in two dimensions. Later work by Brown[9] eliminates bisector lists in the quad-trees. Fontayne and Bowman[10] describe a radix hash tree for region query. Marple et al.[11] extend corner-stitching to manage trapezoidal objects in the database. Lai et al.[12] introduce VH/HV-trees for region queries. If we choose to design a database relying on any of these structures, the following issues need to be kept in mind:

DAC 2002, June 10-14, 2002, New Orleans, Louisiana, USA.

- Data-structures that rely on trees can become quickly unbalanced as objects are inserted. Explicit balancing schemes need to be designed and implemented. This can add a significant level of complexity to the implementation.
- A routing database will contain millions of rectangles. Any structure whose memory growth is not linear in the input size will be hard pressed to handle large designs. Data structures such as corner-stitching can cause severe fragmentation are also ill suited.
- A common approach to speed up queries is to maintain shadows (regions of effect in the neighborhood) of metal objects along with the original geometry so that all region query can be reduced to a simple point query.

The ability to implement efficient area based search algorithms on the database is also important. In this context, we can find various approaches in [2][3][15][16][14]. Lee's approach and Rubin's variant [2][3] implement a search on a grid graph. These often require the routing problem to be partitioned. We do not favor the partitioning approach because it leads to a local view of routing; making global properties of a design such as timing, and antenna rules harder to determine. Line search techniques[15][16] provide a mechanism to speed up the search at the expense of sub-optimal solutions. Both these techniques rely on a grid graph that restricts the search space for a router. The grid-graph coincides with the grid-point database. This can be expensive in terms of memory as shown by Cong et al.[17] and Hetzel [14]. One approach to solve this is to ignore the grid and build the search graph on the fly[17][18][19]. Memory is saved but time is consumed to construct the graph for every search. Hetzel[14] proposes a data-structure relying on intervals for routing. He describes a fast search engine that can deal with large graph sizes. Since we are seeking a balance between performance, capacity and more flexibility than a grid based solution, our approach is strongly motivated by his work. In [14] the author focuses on a new search algorithm for large grid graphs. We design a database that coincides with the search graph required by him. Our main contribution is to extend the graph on which Hetzel describes his search to support all the requirements of a routing database.

1.2 Terminology

The routing problem is defined on a set of planes, numbered from zero (layer zero referring to the metal layer closest to the substrate). Each plane has a preferred direction of routing which alternates between horizontal and vertical directions. A layer k has a default metal width denoted by mw_k. The spacing between any two metal geometries is required to be at least sp_k units apart. Often metal objects with large geometries may require other objects to be more than sp_k units apart (width based spacing). So the spacing rule can actually be a function of the size of the objects. Each adjacent layer pair will have a set of vias used for connection. The geometries of the vias can be often complex with metal overhangs and can also be asymmetric. Each adjacent layer pair has a set of default vias. These are used to connect default metal on adjacent layers. Vias can be placed on top of one another to jump across multiple layers. The actual via invoked to implement this is called a stack via whose geometry need not be the same as a combination of default vias. For the sake of the discussion we will assume that a default via has the same shape as a square piece of default metal on the appropriate layer. Extending the database to other vias is conceptually straightforward, but requires a fair amount of book-keeping.

The routing database is populated by three kinds of objects.

1. Objects such as pins and pre-routes that need to be connected by the router. For sake of discussion, we only consider **pins**.
2. Objects that appear as **obstructions** to the router. The router needs to satisfy spacing constraints with respect to these objects.
3. In addition we have to be able to annotate **route geometry**. This includes routes with different widths and different vias.

Pins and obstructions are input constraints and hence fixed for a routing problem instance. Route geometry is constructed by the router and hence is changeable. We call a route in the non-preferred direction of a plane as a jog.

2 OUR APPROACH

We have mentioned the fact that grid-point based schemes are expensive in terms of memory requiring a partitioning of the design. General point query structures are too expensive in terms of speed. We shall demonstrate both these facts in the experimental section. We naturally look for an approach that provides the middle ground. We define a data representation that depends on intervals. From an intuitive sense, an interval captures more than a grid point but less than a rectangle that describes the true geometry.

Consider the data on a given plane for routing. Most technologies define a preferred direction for routing. This means that a large set of metal geometry will run parallel to the preferred direction (except in the case of lower layers such as layer zero and layer one possibly). The database is designed to make use of this natural bias. We construct a set of gridlines that run the span of the chip in the preferred direction. The spacing between the gridlines can be non-uniform. Since an interval is constrained to lie on a gridline, its co-ordinates in the direction of the gridline can take on any value (i.e. are not restricted to be grid points as in the case of a grid based system).

2.1 Interval database

Consider a plane in the routing database with a preferred routing direction. We use the term "xy" to denote the axis of the preferred direction and "yx" to denote the axis of the non-preferred direction for a plane. We create a set of gridlines parallel to the "xy" axis, whose "yx" co-ordinates are described by the user. The gridlines as managed as an array. Each gridline holds a set of intervals. There are two ways to store the intervals.
- If we choose to distinguish between the impact of different metal geometries at a point, we can use an interval-tree to represent the set of intervals. An interval tree requires linear memory and provides logarithmic response to query. However the tree requires explicit balancing.
- If we choose not to distinguish between the impact of different metal geometries at a point, we can use a simpler representation. In the case of the routing problem this distinction is not necessary and we prefer this approach. See Section 2.2 for further details.

The intervals are **non-overlapping** and maintained in a tree such as a skip-list[20] or a red black tree[1]. Skip-lists are an interesting alternative to balanced trees, as they do not require explicit balancing. Preliminary experiments show that red-black trees are better than skip-lists by a constant. The main advantage of a skip-list is that you get a pointer to the "next" interval for free. Hence skip-lists are more efficient for neighborhood query at a point. Skip-lists use randomization to provide amortized O(log(n)) insertion, deletion and query, where n is the number of intervals in the skip-list. The intervals have a next and previous pointer, which make traversal along a gridline very efficient.

Let us now focus on the query response time. Managing the gridlines as a static array of size N provides O(log(N)) access for a gridline with a relationship (equal, less than, greater than) to a given "yx" value. Querying for the object at a point in a plane p takes O(log(N) + log(n)). Obtaining all the objects in a region requires a series of logarithmic queries. If n is the maximum number of intervals in any gridline, and the query range intersects G gridlines, and the number of intervals in the region is k, it is easy to show that the complexity of a query is O(log N + G log n + k). Neighborhood queries of a point (p, p_{xy}, p_{yx}), where p is the plane, p_{xy} and p_{yx} are the co-ordinates in the "xy" and "yx" axis respectively, are accomplished as follows. Let g be the index for the gridline at p_{yx} on plane p. The next and previous intervals can be efficiently obtained. The interval in the gridline in the positive yx direction is obtained by querying the gridline at $g+1$ with location p_{xy}. Similarly the neighbor in the lower gridline is obtained by querying the gridline at $g-1$ with location p_{xy}. The neighbor in the above(below) plane is obtained as follows. We query for the gridline on $p+1$ ($p-1$) with a yx value of p_{xy}. We then query that gridline for the interval at p_{yx}. Thus neighbor queries require O(log N + log n). Note that it is possible that there is no gridline on $p+1(p-1)$ with a yx value of p_{xy}.

Let us now compare the memory consumption of the proposed routing database with that of a uniform grid. Assume that we reserve one word for database annotation. Consider a row of n grid points. This requires n words. A skip-list requires three words on an average per object for its manipulation (akin to each node in a binary tree having a pointer to its parent and the two children). Each interval also has a pointer to the previous interval, the xy value of the high end point of the interval and a pointer to the gridline that contains it. Hence the total memory required per interval is seven words: for k intervals the memory used will be $7k$. However we know that the sum of the lengths of the intervals in grid units must be n (as the intervals are contiguous). Hence as long as the average length of the interval (n/k) exceeds 7 grid units, the memory in the skip-list will be less than the uniform grid. This requirement is easily met in the higher layers. On the lower layers, the metal pins and obstructions in the standard cells cause a fair amount of fragmentation of the gridline and this requirement may not be met.

2.2 Manipulating the database

In our implementation, we reserve two words on each interval for annotating the database. Each object that is to be inserted in the database has a descriptor that describes it. One word on the interval is reserved to store a pointer to this descriptor, called its "owner". There are two notions of owning an interval. The first one is the due to actual metal belonging to an object intersecting the interval. The second notion is a weaker one, which expresses a design rule to the interval in question without actual metal overlaying it. The second word is used to encode the nature of "ownership". Thus the effect of each object in the routing database is mapped onto the set of gridlines. We term the intent of an object to express interest on an interval as the "color" of an object. We encode the following "ownership" rules using different bits:

1. Interval owner is permanent metal (cannot be ripped up).
2. Interval owner is temporary metal (can be ripped up).
3. Interval owner permanently blocks interval for other route metal.
4. Interval owner temporarily blocks interval for other route metal.

These consume four bits on the second word. The differentiation between temporary and permanent geometry is required to facilitate search during the rip-up phase. The metal block shadow will include the original geometry or polygon expanded by the spacing rule (sp_k for layer k) plus half of the default metal width ($mw_k/2$). This assumes a default metal routing width per layer and a default via for each pair of routing layers. This scheme of marking the shadow of the blockage, will allow us to mark the impact of an object in the database once. During the expansion step of a search, the region query is replaced by a point query.

Figure 1 : Pin marking in interval database

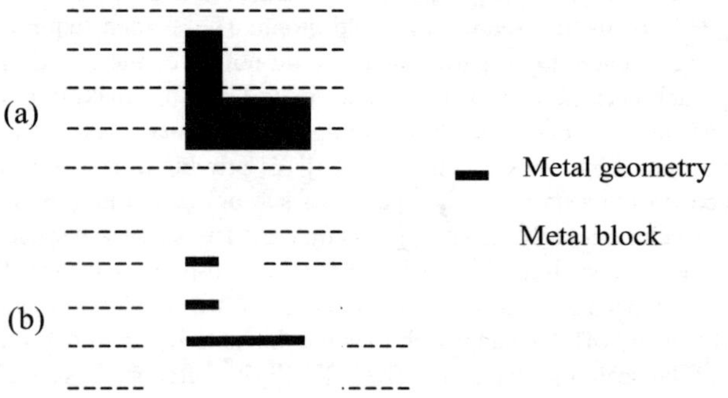

We now describe the mechanism to annotate the database for the three kinds of objects that populate the routing database. Pin geometry marking marks all intervals in the region that it occupies. In Figure 1, we show an L shaped pin over the gridlines in the database (a). In (b), the impact of the pin is marked in the database. Each interval that overlaps the rectangle is marked with the pin as the owner. In addition there is a metal block due to the pin in the same region. The metal block region extends outside the pin. Often via geometry is larger than metal half-width and hence via block regions can extend outside the metal block region.

Obstruction geometry marking is the same as pin geometry except for two points of note. We set the ownership pointer to a value, which will not be a legal pointer and use it for all obstruction geometry. Secondly there is no need to mark the permanent metal color, instead we simply mark the metal block space. This reduces fragmentation without affecting the information required for routing. Figure 2 shows the marking of an L shaped polygonal obstruction. So far we have presented a very simplistic view of pin and obstruction marking. Extensions to pin marking that enable centerline marking of routes and to block marking for Euclidean spacing rules and to permit obstructions to lie between gridlines have been omitted due to space constraints.

Routes are maintained as centerline objects, where each route segment is an interval on the gridline. These intervals have the ownership pointer set to the descriptor of the route. The temporary metal bit is set. Appropriate encoding on the second word of the interval represents vias and jogs. To better explain this, let us examine the possible neighbors of an interval. A route segment can have up to ten neighbors as shown in Figure 3. We enumerate the neighbors at the positive end as (index refers to neighbor in Figure 3):

1. Via up at positive end of interval (pos-via-pos)
2. Via down at positive end of interval (pos-via-neg)

Figure 2: Obstruction marking in interval database

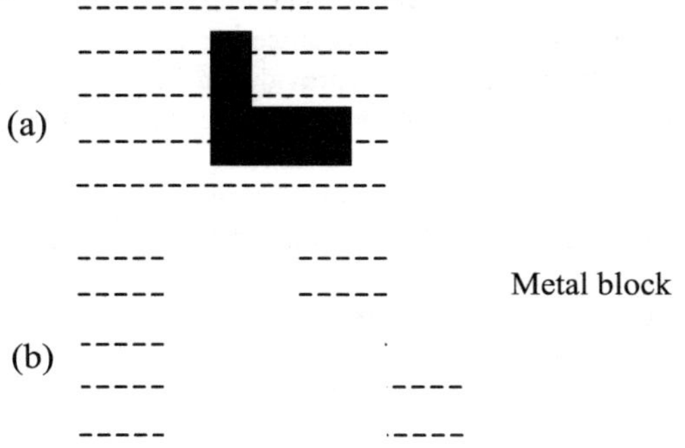

3. Jog (wrong way segment) in the negative direction at positive end of interval (pos-jog-neg)
4. Jog in positive direction at positive end of interval (pos-jog-pos)
5. Next interval at positive end of interval

Indices 6-10 are similarly neighbors at the negative end of the segment. These consume ten bits on the second word. This allows us to do two things efficiently. Given an interval we can enumerate all neighbors to the interval on the same net or (route) efficiently. In addition it is possible to generate a routing path given a point on it. The marking of route geometry differs from pins and obstructions. The centerline of the path is annotated on the intervals. To obtain the metal geometry for an interval segment in plane *k*, we simply bloat the segment by $mw_k/2$. To compute the geometry due to a via, we look at the neighbor bits to decide the direction of a via. Since we only support a default via per adjacent layer pair, the via geometry can be easily recovered. To mark the metal blockage due route objects, we follow the procedure of bloating the geometry (of a route segment or via) by the spacing rule (sp_k for layer *k*) plus half of the default metal width ($mw_k/2$).

We present a simple example in Figure 4. We have a vertical route on layer two with a via up at the high end. The route on layer three is a horizontal segment with a jog in the positive direction at the high end. The centerline markings and neighbor markings for these connections are shown in the Figure. Note that the jog on the center gridline on layer three causes a singleton interval. Also observe that the jog on layer three can appear at any X co-ordinate and is not constrained to happen at a grid-point. The hatched segments indicate the metal blockage (some blockage intervals and gridlines have been omitted for clarity).

Often multiple objects in a neighborhood will express interest on a common interval. The intervals in the common neighborhood will inherit the "tinted" color of these pins. Tinting is the process of resolving the resultant color when multiple objects express interest in an interval with their own colors. When multiple objects do interact, the owner word of the interval is set to a value that is not a legal pointer; unless the multiple objects belong to the same net. In this case any one of the owners is maintained. In the first case the interval

Figure 3: Interval and its neighbors

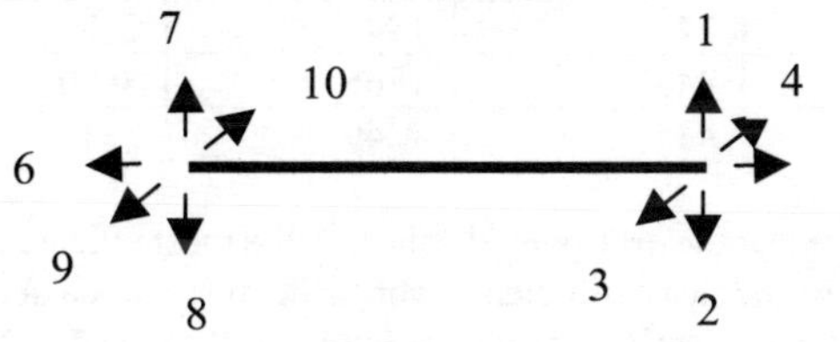

Figure 4: Route geometry marking in the interval database

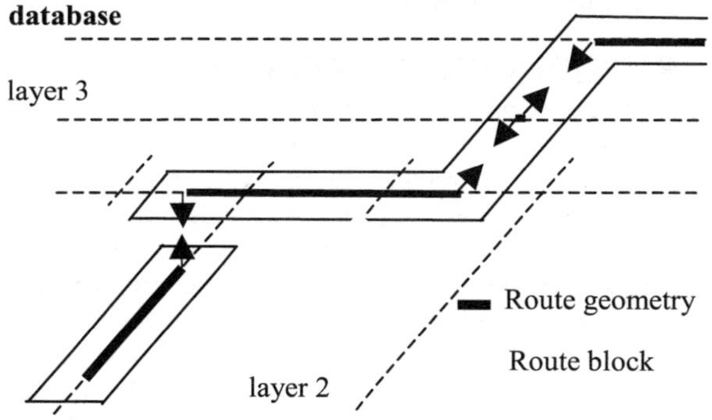

is orphaned. The nature of ownership word on the interval is set to be the union of the bits. Once an interval is orphaned, it will continue to remain so for all future tints. Consider the example in Figure 5. We have two pins whose blockage markings interact in the region between them. Clearly no net (neither the net that owns pin 1 nor the net that owns pin 2, nor any other net) can occupy the common interval (marked as tinted block) without causing a design rule violation. Should we seek to preserve the original interactions, then instead of maintaining the intervals as non-overlapping (or contiguous) objects, we use a segment tree.

An interesting problem arises when route geometry interacts with other geometry (pin, obstruction or route geometry) leading to orphaned intervals. During the rip-up phase, if such route geometry is deleted, then we must clear the effect of blockage markings due to the deleted object. This is easily accomplished in our system by a simple two step approach. First, we erase all blockage markings in a region where the deleted objected could have had an impact. Next, we rebuild all blockages in this region. In order to do this we need the geometries of pins, obstructions and routes that impact this region. The route geometries are calculated by region queries in the interval database. The pin and obstruction geometries are obtained by querying an on-disk version of kd-trees.

We are now in a position to discuss some of the natural advantages of the interval database over the traditional grid-based schemes. The interval database provides the search engine with the flexibility of making jogs that need not be on a grid point. This is extremely useful in finding connections to pins that are hard to reach on congested layers such as layer zero and layer one.

Since we only need to determine the "yx" co-ordinates for the grid-lines on a layer, we can support routing of default widths and spacing that vary across layers easily. Traditional grid-based schemes require the grid-points on adjacent layers to align up. This means that

Figure 5: Interaction of colors

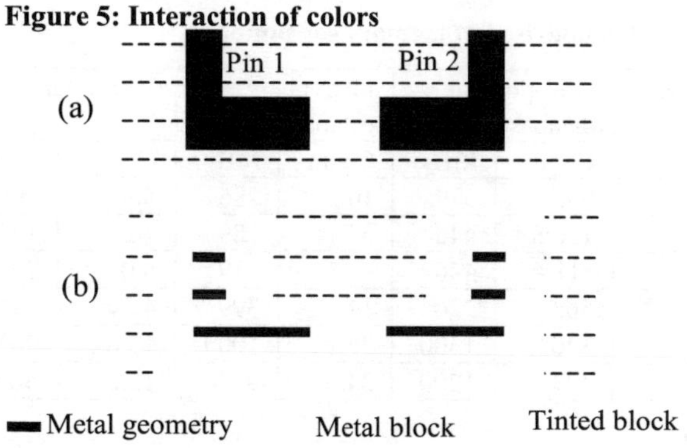

if the grid separations are relatively prime numbers, only a small subset of feasible via locations is captured.

3 EXPERIMENTAL RESULTS

We now present experimental results that justify the efficiency of the interval database. In Table 1 we describe characteristics of some industrial problems for routing. These problems use a fixed area standard cell methodology. All designs include macros along with standard cells. Power routing is also done, so these appear as constraints to the routing problem. Columns 1-3 give the design name, number of nets to be routed (K being used to represent 1000) and number of layers. Columns 4 and 5 describe layers with the same metal width and spacing as the layers within braces. For example, in the case of D5, the metal widths are 190 for layer zero, 220 for layer one and 240 for layers two to five. Similarly the spacing rules are 200 for layer one, 220 for layer two, and 240 for the rest. Observe that design rules vary across layers for all but one design. The technology libraries for all designs are different and are at least 0.25m or below. The technologies describe vias, which have metal overhangs and are not inline with the routes.

Table 1: Experimental designs

Design	# nets	# layers	Metal width rules	Metal spacing rules
D1	107K	5	{0}{1-3}{4}	{0}{1-3}{4}
D2	115K	4	{0-3}	{0-3}
D3	148K	6	{0-3}{4-5}	{0-3}{4-5}
D4	150K	6	{0}{1-4}{5}	{0}{1-4}{5}
D5	319K	6	{0}{1}{2-5}	{0}{1}{2-5}
D6	373K	5	{0}{1-3}{4}	{0}{1}{2-5}

In Table 2 we consider the memory impact for three different database schemes.

1. Grid-point database: We calculate the number of grid-points implied by the intersection of the user-defined tracks on all the layers. We assume that 1 word (4 bytes) per grid point is a reasonable memory cost.
2. Interval database: We take the fully routed designs (almost – see Table 4) and calculate the total number of intervals in the database. Note that this includes fragmentation due to via blockages as well. We calculate the memory as 7 words (28 bytes) per interval.
3. Tree database: We count the total number of rectangles due to pins, pre-routes and metal routes. We use 10 words (40 bytes) per rectangle as the cost of an object. We ignore the impact of complex via-geometry to be conservative.

Table 2 : Comparison of memory consumption

Design	Grid-point database		Interval database		Tree database	
	Count	MB	Count	MB	Count	MB
D1	94.7	379	10.2	286	4.9	200
D2	110.8	443	6.7	189	4.2	172
D3	111.4	446	11.3	319	5.0	200
D4	56.0	224	14.2	399	4.7	190
D5	340.2	1300	39.2	1000	17.6	705
D6	323.6	1200	34.5	967	15.2	608

In Table2 the columns labeled "Count" (columns 2, 4 and 6) count the number of objects (grid-points in column 2, intervals in column 4 and rectangles in column 6). The object counts are in millions. The memory required to manage these is provided in the columns denoted by "MB" (in megabytes). We observe that the grid-point system is most expensive in memory, the tree-based system consumes least memory and the interval database is in the middle. A point to note, the search algorithm can be directly implemented on the grid-point database or on the interval database. However any search on a tree-based system typically requires the construction of a search graph on the fly, or repeated queries into the tree database. The latter is computationally expensive (as shown in the next paragraph). So the comparison with a tree based system must be done, keeping in mind that an adjoining structure for the search needs to be built. Also note that the interval database is sensitive to the actual geometries and the spacing rules. So for a testcase like D4 although the grid-point database is smaller than that for D1, the interval database is larger.

We now study the CPU time for point queries, as these are indicative of the speed of a search. Once again we compare three different database schemes populated with the objects after full detailed routing.

1. Grid-point database: We implement this as a simple two-dimensional array per layer. Given an (x, y), we need to find the indices which define that grid point. The query simply returns the appropriate entry in the array.
2. Interval database: The implementation is as described in Section 2. It includes all the extensions to handle vias correctly. A query to at a point (x, y), returns the interval in the database that contains the point.
3. Tree database: We build a kd-tree on a per layer basis. All pin geometry, obstruction geometry and route geometry arising from segments are added to the appropriate trees. We balance the kd-trees before any query. Note that the geometry for vias is not added to the kd-trees. We believe these should be handled separately for the following reasons. First a design such as D5 has 3.6 million vias at the end. This corresponds to roughly a fifth of the rest of the geometry. Second, via geometry can be inferred from a point query in a two-dimensional kd-tree (where points are inserted instead of rectangles). This will consume less memory.

We query each grid-point (defined by the intersection of horizontal and vertical tracks on a plane) on all the layers. We report the total CPU time required to query all the grid-points on a SUN 400MZ, 4G machine.

Table 3: Comparison of query times

Design	Grid-point database (seconds)	Interval database (seconds)	Tree database (seconds)
D1	119	287	460
D2	134	354	689
D3	132	420	654
D4	74	194	278
D5	443	1866	3600
D6	435	1249	2517

We observe that the grid-point database is the most efficient. The tree database is the least efficient, while the interval database is in between. Quite clearly it would be impractical to have a router that operates on the trees alone and capable of handling large search regions. Also recall that via geometry will need to be queried in addition. In addition, the trees have been balanced statically with all the route geometries added. In reality the route geometries will be added as the routing proceeds and the trees will not be as well balanced. So the times in column 4 are more generous than reality permits.

Finally we present preliminary results for a routing system based on the interval database. This demonstrates the viability of the interval database for routing. All routes have been verified to be design rule clean for spacing and width rules using an independent design rule checker. We have a few opens as we are still refining our rip-up strategy.

Table 4: Routing system with intervals

Design	Open nets/Total nets	Time (seconds)
D1	1/107K	2458
D2	0/115K	3009
D3	4/148K	4769
D4	1/150K	3137
D5	1/319K	11686
D6	18/373K	12071

The run time is for the total run and includes time to build the database, do global routing and detail routing. We use a single CPU for all examples. Bulk of the time is spent in detail routing and rip-up routing.

4 CONCLUSIONS

We have proposed a routing database using intervals as the basic structure instead of grid-points or rectangles. We have shown that it is flexible to support recent trends in design rules for spacing and widths. Our approach is more efficient in memory management than the grid-point based approach. Its performance is superior to a kd-tree based approach. We are quite confident that the interval-based scheme will outperform region queries on any tree-based approach. Our database has handled designs as large as 370K nets in a flat manner.

5 REFERENCES

[1] T. Cormen, C. E. Leiserson and R. L. Rivest, Introduction to Algorithms, MIT Press and McGraw Hill, New York, 1990.

[2] C. Y. Lee, "An algorithm for path connection and its applications", IRE Transactions on Electronic Computers, EC-10:346-365, 1961.

[3] F. Rubin, "The Lee path connection algorithm", Transactions on Computers, C-23:907-914, 1974.

[4] R. A. Finkel and J. L. Bentley, "Quad trees: a data structure for retrieval on composite keys", Acta Inform. 4:1-9, 1974.

[5] J. L. Bentley, "Multidimensional binary search trees used for associative searching", Commun. ACM, 18:509-517, 1975

[6] G. Kedem, "The Quad-CIF tree: a data structure for hierarchical on-line algorithms", Proceedings of the 19th Design Automation Conference, pages 352-357, 1982.

[7] J. K. Ousterhout, "Corner Stitching: a data structuring technique for VLSI layout tools", IEEE Trans. On Computer-Aided Design, Vol 3, No 1, January 1984.

[8] J. B. Rosenberg, "Geographical data structures compared: a study of data structures supporting region queries", IEEE Trans. On Computer-Aided Design, Vol. 4, No. 1, pages 53-67, January 1985.

[9] R. L. Brown, "Multiple storage quad trees: a simpler faster alternative to bisector list quad trees", IEEE Transactions on Computer-Aided Design, Vol 5, No. 3, July 1986.

[10] Y. D. Fontayne and R. J. Bowman, "The Multiple Storage Radix Hash Tree: An Improved Region Query Data Structure", International Conference on Computer-Aided Design, pages 302-305, 1987.

[11] D. Marple, M. Smulders and H. Hegen, "Tailor: A layout system based on trapezoidal corner stitching", IEEE Trans. On Computer-Aided Design, Vol. 9, No. 1, pages 66-90, January 1990.

[12] G. G. Lai, D. Fussel, D. F. Wong, "HV/VH Trees: A New Spatial Data Structure for Fast Region Queries", Proceedings of the 30th ACM/IEEE Design Automation Conference, pages 43-47, June 1993.

[13] M. de Berg, M. van Kreveld, M. Overmars, O. Schwarzkopf, "Computational Geometry Algorithms and Applications", Springer Verlag 1997.

[14] A. Hetzel, "A Sequential Detailed Router for Huge Grid Graphs", European Design Automation Conference, pages 332-338, 1998.

[15] D. W. Hightower, "A solution to line-routing problem on the continuous plane", Proceedings of the 6th Design Automation Workshop, pages 1-24, 1969.

[16] W. Heyns, W. Sansen, and H. Beke, "A line expansion algorithm for the general routing problem with a guaranteed solution", Proceedings of the 17th Design Automation Conference, pages 243-249, 1980.

[17] J. Cong, J. Fang, and K.-Y. Khoo, "An Implicit Connection Graph Maze Routing Algorithm for ECO Routing", Proceedings of the International Conference on Computer-Aided Design, pages 163-167, 1999.

[18] K. L. Clarkson, S. Kapoor, and P. M. Vaidya, "Rectilinear shortest paths through polygonal obstacles in $O(n (\log n)^2)$ time", Proceedings of the 3rd Annual Symposium on Computational Geometry, pages 251-257, 1987.

[19] S. Q. Zheng, J. S. Lim and S. S. Iyengar, "Finding Obstacle-Avoiding Shortest Paths Using Implicit Connection Graphs", IEEE Transactions on Computer-Aided Design of Integrated Circuits and Systems, Vol. 15, No. 1, Jan. 1996.

[20] W. Pugh, "Skip Lists: A Probablistic Alternative to Balanced Trees", Communications of the ACM, pages 668-676, June 1990, Volume 33, Number 6.

[21] S. Hojat and P. Villarubia, "An Integrated Placement Synthesis Approach for Timing Closure of PowerPC Microprocessor", Proceedings of the International Conference on Computer Design, pages 206-210, 1997.

[22] N. Shenoy et. al, "A Robust Solution to the Timing Convergence Problem in High-Performance Design", Proceedings of the International Conference on Computer Design, pages 250-257, 1999.

Automatic Generation of Embedded Memory Wrapper for Multiprocessor SoC

Ferid Gharsalli Samy Meftali Frédéric Rousseau Ahmed A. Jerraya
TIMA laboratory
46 av. Felix Viallet
38031Grenoble cedex (France)
Tel: (33) 4 76 57 46 41
{ferid.gharsalli, samy.meftali, frederic.rousseau, ahmed.jerraya}@imag.fr

ABSTRACT

Embedded memory plays a critical role to improve performances of systems-on-chip (SoC). In this paper, we present a new methodology for embedded memory design in the case of application specific multiprocessor system-on-chip. This approach facilitates the integration of standard memory components. The concept of memory wrapper allows automatic adaptation of physical memory interfaces to a communication network that may have a different number of access ports. We give also a generic architecture to produce this memory wrapper. This approach has successfully been applied on a low-level image processing application.

General Terms

Design, Experimentation.

Categories and Subject Descriptors

M1.9 and T4.5: High level and architectural synthesis

Keywords

Embedded Memory, System-on-Chip, Memory access, Memory Wrapper Generation.

1. INTRODUCTION

Embedded memory is becoming the new paradigm allowing entire systems to be built on a single chip. In the near future, new developments in process technology will allow the integration of more than 100 Mbits of DRAM and 200 millions gates of logic onto the same chip. Hence, several applications types (networking, computer graphics and multimedia applications) that require a large number of embedded memories (DRAM, flash, etc) can exploit this evolution. Unfortunately, nowadays, there is not yet a complete and automatic method allowing the designers to integrate all these memory types (particularly the global memory) into a multiprocessor heterogeneous systems-on-chip (SoC) from a high abstraction level. The memory integration process in such systems is still hand-made. This is time-consuming and error prone.

This evolution is creating several breaking points in the memory design process.

- The SoC specification consists of heterogeneous components in terms of communication protocols and abstraction levels. To integrate existing memory components in such systems, the physical memory ports should be adapted to the number of accesses required by the application.
- During the design space exploration, designers need to try different kind of memories (in terms of type, size, and consumption). This exploration should not modify the rest of the architecture. In this case, memory interface flexibility is needed.
- SoC architectures allow the integration of several master CPUs sharing a global memory. One problem is the parallel access to the memory. Multiple memory access may require sophisticated synchronization scheme.

This paper presents a systematic approach for memory integration into multiprocessor SoC. Section 2 explains the main problems of integrating memory into multiprocessor SoC. In section 3, we describe the basics of our methodology and our architectural models at different abstraction levels. In section 4, we present a generic wrapper architecture as well as its automatic generation. Section 5 highlights this methodology applied to an image processing application for a digital camera. Finally, section 6 concludes this paper.

2. MEMORY FOR MULTIPROCESSOR SYSTEM-ON-CHIP

To reduce the complexity of memory design, most of current system design methods adopt design reuse [1, 2, 3, 4, 5, 6]. In such design methods, system architecture specification includes heterogeneous components in terms of communication protocols and abstraction levels.

Figure 1 shows an example of a conventional multiprocessor SoC composed of CPU, DSP and/or memory components.

DAC 2002, June 10-14, 2002, New Orleans, Louisiana, USA.

Compared with the design of conventional embedded systems, the implementation of system communication becomes much more complicated in multiprocessor SoC design, since (1) heterogeneous processors are involved in communication, (2) complex communication protocols and networks are used, and (3) standard memory components need to be connected to processors and/or network [7,8].

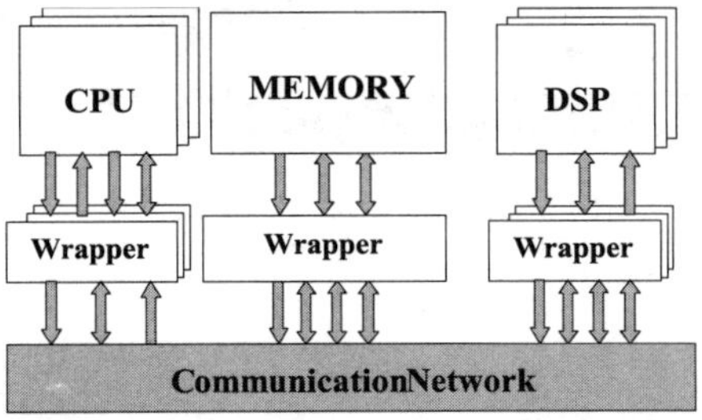

Figure 1: A typical Multiprocessor SoC

Existing memory components may have a fixed access structure (e.g. number of ports). To integrate this memory component into a heterogeneous system, we often need a different number of access ports. The key issue is then to adapt the physical memory interface (with fixed number of access ports and protocols) to the logical interface. Wrappers have been widely used to resolve this problem for simulation and synthesis [4, 9, 10, 11, 12]. For simulation, BFM (bus functional model) encapsulates a memory functional model with a cycle accurate interface [4, 9]. In [10], a wrapper for mixed-level cosimulation between FIFO channel and cycle-accurate models is presented. In [1, 3, 4], a protocol wraaper is used to adapt a communication protocol of IP to the communication protocol of on-chip bus. Most previous approaches using wrappers have the following limitations: (1) there is no generic architecture model for memory wrappers and (2) there is no automatic method for memory wrapper generation. Our contribution is (1) the introduction of a generic memory wrapper architecture that can be applied to several types of memory and (2) the definition of a method to generate automatically these memory wrappers.

3. MEMORY DESIGN FLOW

This work is an extension of the work presented in [1] and [5] to cover memories. The overall flow allows component based design for multicore SoC. The system is initially described as a set of virtual components interconnected via channels. Virtual component consists of a wrapper and internal module. In this environment an internal model may correspond to a software task or hardware module. The wrapper adapts the interface of the module to the communication network. This section introduces the concept of virtual memory module in the component based design flow.

3.1 Memory design based on separation between behavior and communication

Memory flexibility is ensured by the separation between the high-level memory access code and the physical memory access protocols. From the processor view, memory access must be completely independent of the memory type (synchronous, asynchronous, static, dynamic, etc). Accordingly, physical memory accesses are abstracted to logical accesses at the system level. A logical access may hide the memory protocol, several operations (address decoding, row and column selection, refresh management, etc) that depend on the type of memory and control signals. The operations on these ports are fixed data type assignments (e.g. real, integer). On the other side, physical memory access needs explicit signals and explicit protocols.

To adapt the logical interface to the physical interface, we use a virtual memory module (Figure 2). The latter consists of behavior and interface. The interface, which encapsulates the memory, is composed of two kinds of ports: internal ports and external ports. External ports correspond to logical ports that allow a logical access to the memory, whereas the former correspond to physical ports and allow a physical access to the memory. In the remainder of the paper, we will use the expressions "physical ports" and "logical ports", and the correspondence between them is made by a wrapper.

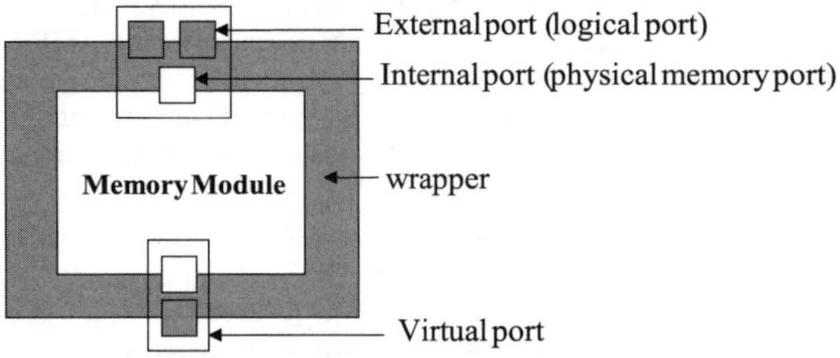

Figure 2: Logical and physical port concept

The wrapper introduces the required flexibility in the design. It allows the designer to decide quite late during the design process which component will be used for the implementation of the memory. Flexibility is achieved thanks to the separation between behavior and interface.

The memory module and its wrapper compose a virtual memory module. We can access to the virtual memory module through its virtual ports. The virtual port is hierarchical and has internal as well as external ports.

3.2 Architectural model at different abstraction levels for SoC design

In order to master the complexity, the architecture model may be described at different abstraction levels. We distinguish two main levels: virtual architecture level and micro-architecture level.

3.2.1 Virtual architectural model

The virtual architectural model (also called macro-architecture model) represents an abstract architecture. This abstract architecture is composed of a set of virtual modules interconnected through logical wires (Figure 3.a). Each module may represent a software processor (e.g. DSP or a micro-controller executing software), a hardware processor (specific hardware) or an existing memory module in the final architecture. The logical wires are abstract channels that transfer fixed data types (e.g. integer, real) and may hide low-level protocols (e.g. handshake or memory mapped I/O).

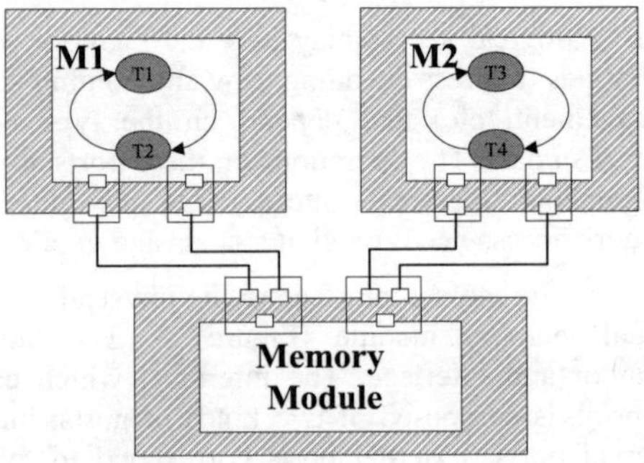

(a) Virtual architecture model

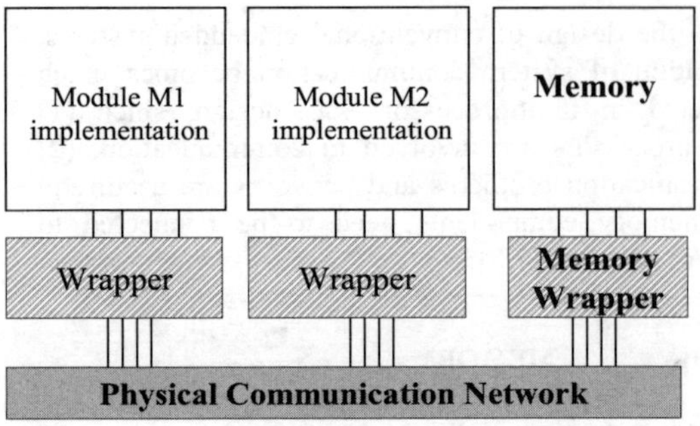

(b) Micro-architecture model

Figure 3: Architectural models

Each virtual module consists of a module and its wrapper, and they communicate with each other through abstract channels connected to their virtual ports. For instance, FIFO communication is realised using high-level communication primitives. In our design flow, the virtual architecture is described using an extension of SystemC. More details about this model are given in [1].

The memory wrapper may include controllers and buffers when needed. At this level, the time unit becomes the clock cycle and the memory wrapper behavior corresponds to finite state machines (FSM), where each transition is realized in a clock cycle. To resolve the heterogeneity problem (in terms of communication protocols and abstraction levels), the memory wrapper has channel adapters that perform conversion protocol and abstraction level adaptation between logical ports and physical ports. The architecture of the memory wrapper is more detailed in the next section.

3.2.2 Micro-architecture model

The micro-architecture abstraction level gives the detailed RTL architecture (Figure 3.b). In this model, existing global memory modules are encapsulated within an interface (memory wrapper) in order to accommodate the final protocol and to isolate memory from the communication network. This wrapper adapts memory protocols to the communication network. The communication between the memory module and other modules of the architecture is made through the memory wrapper by using physical wires that implement the final protocols.

4. AUTOMATIC MEMORY WRAPPER GENERATION BY ASSEMBLING LIBRARY COMPONENTS

The key idea behind this work is to allow the automatic memory wrapper generation. In order to achieve this, we use a generic wrapper architecture that can be customized according to the architecture under design.

4.1 Generic memory wrapper architecture

Figure 4 shows a generic model of a memory wrapper that connects a global memory shared by several processors through the communication network. Its generic architecture is based on the wrapper architecture described in [13]. It is composed of two main parts. The first part, called Memory Port Adapter (MPA) is specific to the memory. The second part depends on the communication protocol used by the communication network and it contains several modules called channel adapters (CA). These two components communicate through an internal communication bus.

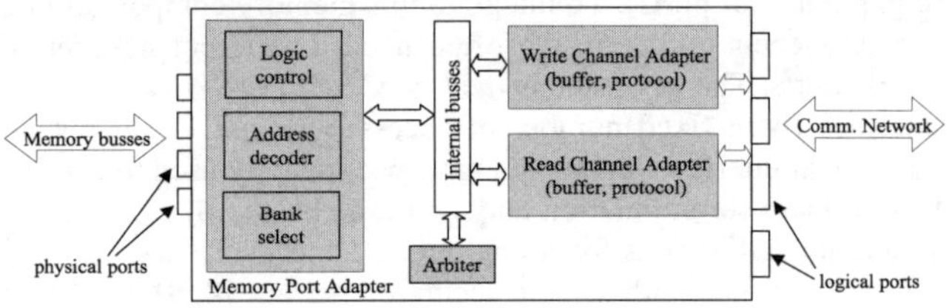

Figure 4: Generic memory wrapper architecture

- Memory port adapter: the memory dependent part includes a controller and several memory-specific functions such as control logic, address decoder, bank controller and other functions which depend on the type of memory (refresh management in DRAM). In addition, the MPA performs type conversion and data transfer between internal communication bus and memory bus. The complexity of the MPA behaviour depends on the memory used.

The communication network dependent part is made of three types of basic elements:

- Channel adapter (CA): CA implements the communication protocol. Its implementation depends on many parameters such as communication protocol (FIFO, burst, etc), channel size (int, short, etc), and port type (in, out, master, slave, etc). The channel adapter manages read and write operations requested by the other modules accessing the memory. The number of channels gives the number of CA.
- Internal communication bus interconnects the two parts of the memory wrapper (MPA part and the CA part). This internal bus is usually composed of address, data and control signals. The size of this internal bus depends on the memory bus size and the channel size. This size is determined for each instance of memory wrapper at the wrapper generation step.
- Arbiter: the arbiter manages the conflicts generated by multiple connection of CAs to the internal bus of memory wrapper. In the case of multi-access to a shared memory port, the arbiter must give the access permission to only one channel. In our model, this is managed using a priority access list.

4.2 Basic elements needed for wrapper generation

In order to generate memory wrapper implementation at the RT Level from the virtual architecture level and to simulate it, three basic libraries are used.

1. The library of physical memories contains several memory models (SRAM and SDRAM abstract memory models).
2. The memory port adapter library contains several specific MPA in relation with memories from the library of physical memories.
3. The communication protocols library contains several types of channel adapters (read and write channel adapter with handshake, FIFO or burst protocol).

4.3 Memory wrapper generation flow

We assume that the virtual architecture is fixed by the designer, or following a high level design flow [14, 15, 16]. The virtual architecture is annotated with parameters that will guide the wrapper generation step.

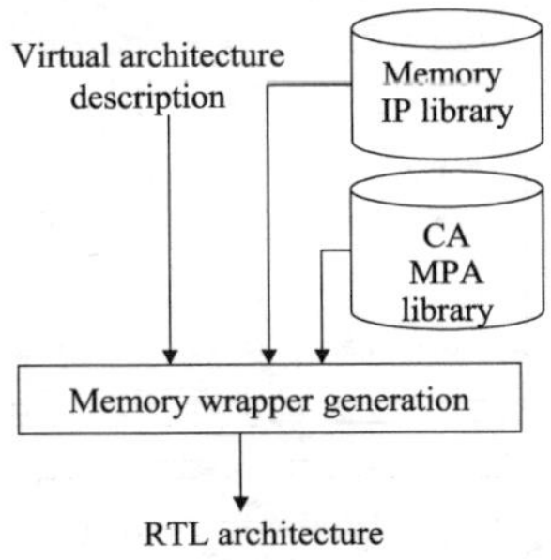

Figure 5: Memory wrapper generation flow

The memory wrapper generation step consists of choosing (1) the channel adapter and (2) the memory port adapter from the libraries (Figure 5). We use parameters to configure selected components. For instance, CA configuration uses the following allocation parameters: read/write type, master/slave operation, type of transmitted data, access mode. Parameters used for the MPA configuration correspond to the address/data bus size, the memory interface signals and static or dynamic. After configuration, customized components are instantiated. The CA component(s) is (are) connected to the MPA through internal busses. Finally, the entire wrapper is connected to the rest of the system. Every time we add a new component in the library of memory IP, we have to write all the specific parts related to this new memory, such as MPA and some other functionalities (refresh, bank selection, …).

5. LOW LEVEL IMAGE PROCESSING FOR DIGITAL CAMERA

5.1 Principle and System Specification

Our application is about a JamCam digital camera. We use the version detailed in [17]. This system architecture is memory rich and processor-poor. It contains 2 Mbytes Flash image storage, 2 Mbytes ROM and 256 Kbytes static RAM. It uses an 8-bit RISC processor with embedded USB and other interface functionalities.

The input image is stored in the memory system before the image-processing step done by the processor. We are interested in the filtering image part that consists of blurring the image by applying a Gaussian convolution mask. It has optimal properties in a particular sense: it removes some aberrations introduced by the sensor, including noise.

To accelerate the image processing, we use two ARM7 processors sharing a memory module. Each processor communicates with three communication channels: address and data channels, and a buffer channel for read operations. In the rest of this section, we study the architecture generation of the image processing. To show the flexibility of the approach, we tried two memory types (a single port SDRAM and a double port SRAM) in two different experiments. The goal is to show that we can change a memory module without changing the rest of the architecture.

5.2 Virtual architecture model

The virtual architecture of the application includes three virtual modules named VM1, VM2 and Virtual memory module. The basic module behaviors are described as SystemC processes (thread). Figure 6 shows the specification for the two experiments. The only difference between these models is the content of the memory module. The specifications of the two other modules are the same.

5.2.1 Experiment 1: using a double port memory

In Figure 6.a, the memory module corresponds to a SRAM memory that has **two physical ports**. Both module VM1 and VM2 access the virtual memory module simultaneously. Each module (VM1 and VM2) is connected to the virtual memory module via one abstract channel.

The virtual memory module is composed of the SRAM module and its wrapper. It contains **two virtual ports**. One is connected to VM1 port, the other is connected to the VM2 port. Each virtual port is composed of the SRAM physical port (internal ports) and one logical port (external port). Operations on these ports are fixed data type assignments (integer 32 bits).

5.2.2 Experiment 2: using a single port memory

The memory module (Figure 6.b) corresponds to a single port SDRAM memory that has **only one physical port**. The virtual port is composed of one physical port connected to the SDRAM and two logical ports (external ports) connected to the virtual ports of VM1 and VM2 via two abstract channels. VM1 and VM2 may access the logical ports in the same time, but only one access to the physical SDRAM port is accepted. So, the SDRAM wrapper performs the arbitration of the multi-access to its single port.

5.3 Micro-architecture model

The micro-architectures corresponding to Figure 6.a and Figure 6.b are shown in Figure 7.a and Figure 7.b respectively. To obtain the RTL architecture, the wrapper is generated. The wrapper establish the RTL communication between the two ARM7 processors and the memory. The physical memory access needs several explicit signals (RAS, CAS, BS, etc), numerous operations and protocols (data conversion, address decoding, refresh, burst protocol, etc).

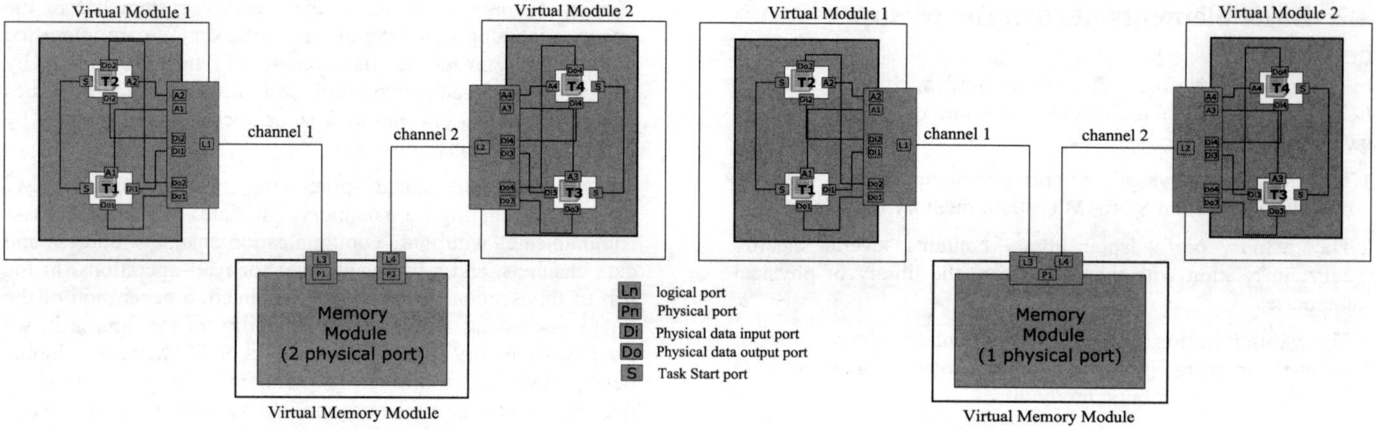

Figure 6: The virtual architecture of the image filtering application using (a) double port memory (b) single port memory

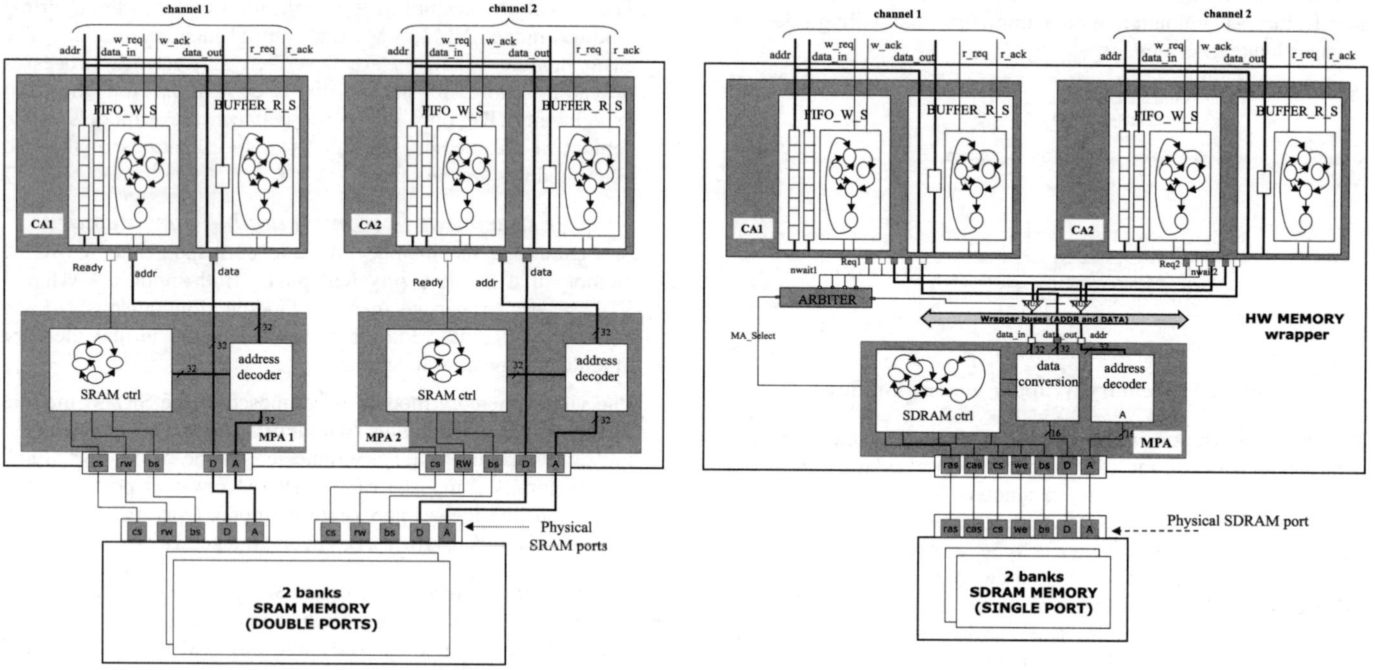

(a) double port memory

(b) single port memory

Figure 7: RTL generated architecture of the filtering application

Communication protocols are refined:

- For write operations, we use two FIFOs (for the address and data) that allows processors to send several write requests to the memory. The size of each FIFO is 32 words of 32 bits wide. Both FIFOs are controlled by only one FSM.
- For read operations, the processor gives the data address, and we use a buffer (one word of 32 bits wide) to send back the read data.

5.3.1 Experiment 1: using a double memory port

In this experiment, we use a dual port SRAM with two banks (Figure 7.a). The size of the memory buses is 32 bits.

The generated memory wrapper ensures the interconnection between the SRAM and the ARM7 wrappers. To adapt the memory protocol to the point-to-point network protocol, we instantiated two channel adapters. To adapt the two SRAM physical ports to the logical ports of the system, we instantiated two MPA specific to the SRAM. MPAs are composed of memory controller which generates control signals and address decoder.

5.3.2 Experiment 2: using a single memory port

In the Figure 7.b, we use a single port SDRAM. The size of the SDRAM busses are 16 bits (data and address). To interconnect this memory to the network, we are not obliged to modify all the wrapper. Indeed, we instantiate only one MPA specific to the SDRAM. MPAs are composed of 3 parts: the memory controller which generates control signals, address decoder and data converter. The data converter performs address and data size conversion between the wrapper internal bus (32 bits) and the memory bus (16 bits). The CAs used for the first architecture (Figure 7.a) and all the other part of the wrapper remain intact. As multi-access is not accepted, we implement an arbiter to control the memory multi-access. We use a HSK protocol to synchronize the memory accesses.

5.4 Results

The automatic generation of these wrappers allows a fast design space exploration of various types of memories. We generate SystemC model (for cosimulation) and we are currently working on the VHDL generation for synthesis. Memory models are also written in SystemC and VHDL.

The memory architecture and the generated wrapper have been validated with a cycle accurate co-simulation approach based on SystemC. Two ISSs of ARM7 core (40 MHz) are used.

We note that there is a small difference in the code size of the memory wrapper in the two RTL architecture models. In fact, CAs are not changed. Only the MPA is changed (10% of the wrapper code), all the rest remains intact. As the SDRAM requires complex control signals, the two controllers implemented in MPAs are more complex than the one implemented in the SRAM wrapper and it explains the small difference in the code size. This easy integration of memory proves the flexibility of the architecture.

6. CONCLUSION

This paper described the advantages of using memory wrapper to connect a shared memory to processors or IP in a system-on-chip. This memory wrapper separates the logical access (by processor or IP) to the physical access to the memory, which depends on the memory type. It allows designers to try easily several kinds of memory without modifying the rest of the system. We have given also a generic architecture of the memory wrapper in order to generate it automatically. We proved the applicability of this method by applying it on an image processing system.

REFERENCES

[1] W. Cesário, A. Baghdadi, L. Gauthier, D. Lyonnard, G. Nicolescu, Y. Paviot, S. Yoo, A.A. Jerraya, M. Diaz-Nava "Component-Based Design Approach for Multicore SoCs", DAC 2002.

[2] J.A Rowson and A. Sangiovanni-Vincentelli, "Interface-Based Design", Proc. of DAC, pp. 178-183, 1997.

[3] C.K Lennard, "Standards for System-Level design: Practical Reality or Solution in Search of a question?", Proc. DATE, Mar. 2000.

[4] D.D. Gajski, J. Zhu, R. Dömer, A. Gerstlauer, and S.zhao, "SpecC: Specification Langage and Methodology, Kluwer Academic Publisher., 2000.

[5] Synopsys, Inc., http://www.systemc.org/.

[6] Coware, Inc., http://www.coware.com/cowareN2C.html.

[7] D.A Culler, J. P Singh, "Parallel Computer Architecture", Morgan Kaufmann Publishers, 1999.

[8] D.A. Patterson, J.L Hennessey, "Computer Organization and Design-The Hardware/Software Interface", Morgan Kaufmann Publishers.

[9] L. Sémira and A. Ghosh, "Methodologie for Hardware/Software Co-verification in C/C++", Proc. ASP DAC, Jan. 2000.

[10] K. Takemura, M. Mizuno, and A. Motohara, "An approach to System-Level Bus Architecture validation and its Application to digital Still Camera Design", Workshop SASIMI, pp. 195-201, Apr. 2000.

[11] J-Y. Brunel, W.M. Kruijtzer, H.J.H.N. Kenter, F.Petrot, and L. Pasquier, "COSY Communication IP's", Proc. DAC, pp. 406-409, June 2000.

[12] S. Yoo, G. Nicolescu, D. Lyonnard, A. Baghdadi, and A. A. Jerraya, " A Generic Wrapper Architecture for Multi-Processor SoC Cosimulation and Design ", CODES/CASHE 2001.

[13] D. Lyonnard, S. Yoo, A. Baghdadi, A.A. Jerraya, "Automatic Generation of Application-Specific Architectures for Heterogeneous Multiprocessor System-on-Chip", Proc. of DAC 2001, June 2001.

[14] F. Catthoor & al, Custom Memory Management Methodology, Kluwer Academic Publishers, 1998.

[15] P. R. Panda, N. Dutt, A. Nicolau, "Memory Issues in Embedded Systems-on-chip, Optimization and exploration", Kluwer Academic Publishers, 1999.

[16] S. Meftali, F. Gharsalli, F. Rousseau, A.A. Jerraya, "An Optimal Memory Allocation for Application-Specific Multiprocessor SoC", proc. of ISSS 2001.

[17] http://www.pixelphysics.com

A Novel Synthesis Technique for Communication Controller Hardware from declarative Data Communication Protocol Specifications

Robert Siegmund, Dietmar Müller
Chemnitz University of Technology
Professorship Circuit and Systems Design
09107 Chemnitz, Germany
{rsie,mueller}@infotech.tu-chemnitz.de

ABSTRACT

An innovative methodology for the efficient design of communication controller hardware for popular protocols such as ATM, USB or CAN is proposed. In our approach, controller hardware in form of RTL models is synthesized from a formal specification of a communication protocol. The difference to previously published work related to hardware synthesis techniques from protocol specifications is that in our approach a complete communication architecture consisting of *both* the interacting transaction producer and the consumer controllers, as well as the interconnect between them, are synthesized from *one single* protocol specification in the *same* synthesis tool run, thus ensuring conformity of all producer and consumer controllers to the protocol specification while tremendously reducing the modeling effort for the controller specifications. The formalism used for protocol specification and a corresponding hardware synthesis algorithm from such specifications are presented. The methodology has been applied to the design of various communication controllers including IEC14443 Wireless SmartCard, ATM and CAN. The novelty and efficiency of our methodology is demonstrated through comparison to State-of-The-Art protocol synthesis tools such as [10].

Categories and Subject Descriptors

B.4.4 [**Hardware**]: Input/Output and Data Communications Devices—*Performance Analysis and Design Aids*; B.5.2 [**Hardware**]: Register-Transfer-Level-Implementation—*Design Aids*

General Terms

Algorithms, Design, Languages

Keywords

Protocol Specification, Controller Hardware Synthesis, Interface-based Design

1. INTRODUCTION

Today's digital systems are communication-dominated designs. They comprise a large number of modules of high functional complexity which communicate with each other by means of high-performance data protocols. Especially in order to minimize the physical interconnect between system modules, bit-serial protocols using a single wire for data transmission such as USB, CAN or FireWire are extensively applied. However, corresponding protocol controllers are control-oriented designs which include complex state machines. The manual design of such protocol controllers is thus a tedious, error-prone and time-consuming task.

In this paper, a novel methodology for the efficient design of communication controller hardware which is especially suited for (but not limited to) complex, bit-serial protocols is presented. The methodology is based on synthesis of controller hardware from a formal high-level specification of the protocol. Compared to previous work [6],[8],[4],[5],[10], our approach is unique in two aspects: First, the formalism that is used for protocol specification has been realized as an extension to the system description language SystemC, called SystemC^{SV} . This ensures that protocol specifications become an integral part of the system specification and can be simulated and verified directly in the system context. State-of-the-Art approaches published to date require the generation of corresponding behavioural models for transaction producer and consumer in a HDL or C from the protocol specification which have then to be inserted manually into the system model prior to simulation. The major drawback is that errors in the protocol description are only reflected in the simulated behaviour of these modules so that from the erroneous behaviour the necessary corrections of the protocol specification must be concluded.

The second aspect is related to the fact that in any kind of communication between two or more system modules usually two distinct types of interacting controllers are involved: Transaction producing controllers, which structure and encode the information to be transmitted into a signal sequence on a physical communication medium, and transaction consuming controllers which assemble this signal sequence and extract the relevant information (see Fig 1). The approach presented in this paper is to the author's knowledge the first one that generates *both* transaction producer and consumer controllers in form of RTL models from *one single* protocol specification in the same run of the synthesis algorithm. Beside a tremendous raise in design efficiency and reduced modeling effort for the protocol specifications, this approach ensures that the behaviours of both transaction producer and consumer conform to the protocol specification, a property which has to be otherwise verified

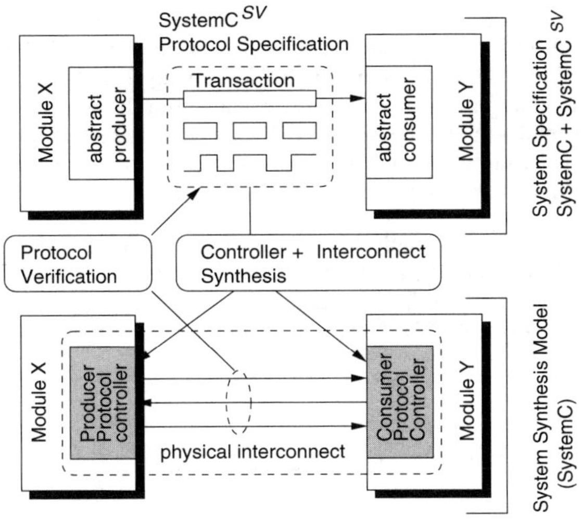

Figure 1: Overview over the proposed Methodology

in extensive simulations using protocol reference models. However, in case verification of the controller implementation against the protocol specification is required, in the presented approach the protocol specification also serves as a protocol checker during system simulations (Fig. 1). Details on this can be found in [9].

1.1 Related Work and Motivation

High-level communication modeling concepts are an integral part of a large number of State-of-the-Art system modeling languages: So do SystemC [11] and SpecC [2] provide the concept of interfaces and channels which are an abstraction of the signal level communication and enable the specification of hierarically structured communication protocols. These modeling concepts support an interface-based design style [7] through separation of communication and computation. However, the communication modeling concepts in these languages have been primarily designed for simulation purposes and to date no feasible hardware synthesis methodologies have been presented for related interface protocol descriptions.

A number of publications have addressed synthesis of communication controller hardware from protocol descriptions. In [8] the system Clairvoyant was presented which generates controller hardware from production-based protocol specifications. An extension to this system is the Protocol Compiler [10], which uses a graphical hierarchical regular expression based language for protocol specification. In [4],[5] regular grammars are used for protocol descriptions which can be synthesized into a VHDL description of a corresponding protocol controller by means of the PRO-GRAM system.

The drawback of these synthesis approaches is that the formalisms used to capture the protocol specification are not part of a system modeling language. Therefore, no verification support for the protocol is offered by the standard simulation tools, and the methods are complicated to integrate into existing design flows. Another disadvantage that is common to all formalisms for communication protocol specification presented so far is that in order to synthesize a transaction producer/consumer pair, the protocol must be described from two different views: the transaction producer view and the transaction consumer view. Therefore, for the same protocol two different specifications have to be developed which results in a decrease in design efficiency and increases verification effort.

The rest of the paper is organized as follows: In Sect. 2 the formalism used for protocol specification is presented. Section 3 describes our approach to synthesize controller hardware from such specifications. The synthesis results obtained for various protocols such as IEC14443 SmartCard, ATM and CAN are given in Sect. 4. Section 5 concludes the paper.

2. SPECIFICATION OF DATA COMMUNICATION PROTOCOLS

The formalism which is used to capture specifications of data communication protocols is an extension to SystemC and named SystemC ^{SV} [9]. In SystemC ^{SV} , system modules (which themselves can be either hardware or software components) communicate through transmission of **communication items** over abstract channels. The information payload carried by an item is specified by a set of item parameters. Further item attributes describe the direction and duration of item transmission [9]. Items are used to describe communication at different levels of abstraction. So, taking the USB bus as an example, a communication item may describe a complete USB transaction, the data field within this transaction or just a single bit of information in the serial USB bit stream. SystemC ^{SV} provides three types of interface items which are distinguished by the level of communication abstraction: **TRANSACTION** items are used to describe multi-directional communication such as transfers with acknowledge phase. **MESSAGE** items describe unidirectional information transfers from a source to a target. Finally, the **PHYMAP** item provides a means for clock-synchronous mapping of transactions and messages and their parameters to states of physical signals which eventually represent the interconnect between hardware modules. Sample specifications for these items are given in Lst. 1, which is a description of the ISO/IEC14443 serial protocol used for wireless communication between a SmartCard transponder and a terminal[1].

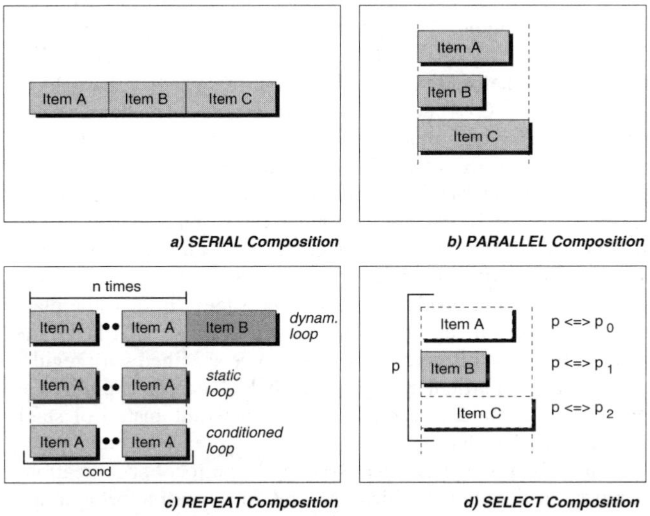

Figure 2: Item Composition Schemes used for Protocol Specification in SystemC ^{SV}

For transaction and message items a *transmission protocol* can be defined in form of a **COMPOSITION** of constituting lower level items. For that purpose SystemC ^{SV} provides four different composition schemes (Fig. 2), which are used to describe a transmission protocol of an item in terms of an execution schedule of a set of constituting lower level items. The **SERIAL** and **PARALLEL** schemes execute a number of items sequentially or in parallel, respectively. In addition to that there are two parameter-controlled schemes which allow for alternative (**SELECT** scheme) or repeated item execution (**REPEAT**). The latter two schemes enable the specification of protocols which involve data-dependent branching and looping. Extensive investigations into State-of-the-Art serial communication protocols showed that virtually any non-pipelined, point-to-point or single-master broadcast protocol can be modeled through combination of these four schemes.

```
SV_TRANSACTION(IEC14443Frame) {

  /* parameters */
  SV_Param<sc_uint<8> >  framelen;
  SV_ParamArray<sc_uint<8>, 8> data;

  SV_TRANSACTION_CTOR(IEC14443Frame) {
     SV_FROM << "TX"; SV_TO << "RX";
     ...
     /* protocol declaration */
     SV_COMPOSITION(
        SV_SERIAL(
           SoF_M(),
           SV_REPEAT(z,framelen,
                     Data_M(framelen,data)),
           EOF_M()));
  } };

/* specifications of SoF,Data,EoF left out*/

SV_PHYMAP(PHYBit) {
  SV_Param<bool> b;
  SV_SignalRef<bool> XD;
  SV_PHYMAP_CTOR(PHYBit) {
  ...
  SV_ASSOCIATE(XD-b);
};
```

Listing 1: Specification of the ISO/IEC14443 Protocol

A further SystemC^{SV} language construct used for protocol specification is the `SV_SEQBEHAVIOUR` statement which is used in composition schemes to embed the execution of sequential behaviours given as a C++ method in between item executions. Such sequential behaviours must execute in zero time and are generally used to modify local state of communication items, so that with the aid of parameter-controlled composition schemes which evaluate the item state the execution of items can be made dependent on previous executions of this or other items. Example applications for embedded sequential behaviours are CRC computations or automatic bit stuffing inside bit transfer messages. In fact, the class of protocols which can be described with this approach is a a superset of the class of protocols which can be specified with regular expressions. In particular all protocols which can be expressed as context-free grammars recognized by a stack automaton of stack size 1 can be modeled.

It has to be pointed out that SystemC^{SV} protocol specifications are purely declarative specifications and, in contrast to behavioural specifications, are not directly executable. Instead, it is the task of a corresponding simulation or synthesis algorithm to derive in the elaboration phase two *distinct, executable behaviours* from such a protocol specification:

Definition 1. The **Decompositional Behaviour** defines how a communication item is decomposed over time into a set of constituting lower level items according to its protocol. It furthermore defines how the set of item parameters are mapped to the formal parameters of these lower level items. For synthesis, it specifies the behaviour of the *transaction producer* controller.

Definition 2. The **Compositional Behaviour** corresponds to the *inverse* decompositional behaviour and defines how a set of constituting lower level items are assembled over time according to the protocol in order to generate a higher level item. It furthermore defines how the information payload contained in the parameters of the assembled item is reconstructed from the parameter values of the constituting items. For synthesis, it specifies the behaviour of the *transaction consumer* controller.

Modules which communicate using items such as IEC14443 Frames would then be specified in SystemC^{SV} as follows:

```
SC_MODULE(Transponder) {
   SV_InterfacePort<IEC14443> RX;
   SV_InterfacePort<IEC14443> TX;

   void send_thread() {
     TX.sv_send(TX->IEC14443Frame(32,data));
   }
   void receive_thread() {
     RX.sv_receive(RX->IEC14443Frame(len,data));
   }
}
```

Listing 2: Module Specification with Item Communication

3. CONTROLLER SYNTHESIS

3.1 Overview

Fig. 3 visualizes the steps required to automatically transform a SystemC specification containing SystemC^{SV} protocol specifications into a model which can be further synthesized with e.g. the Synopsys CoCentric Compiler. The input to our synthesis algorithm is a complete SystemC system model, consisting of two or more SystemC modules which communicate using one or more SystemC^{SV} interface items. The item transmission protocols are specified in terms of a SystemC^{SV} protocol description. Module and interface descriptions are parsed and transformed into an intermediate representation. For the protocol description a protocol flow analysis is then performed which results in a protocol flow graph (PFG). In the next step both transaction producer and consumer are generated from this PFG in form of extended FSM models. These models replace the abstract channel interfaces in the communicating modules. The last step is the generation of synthesizable SystemC code for the communicating modules which have the descriptions of the protocol controllers included and which are connected by physical wires and signal level interfaces.

3.2 Generation of Protocol Flow Graphs

In the protocol analysis step, a protocol flow graph (PFG) is constructed from a SystemC^{SV} item protocol description.

Definition 3. A transaction protocol flow graph is a directed cyclic graph defined by the tuple

$$PFG := < S, T, P, C, I, F >$$

where S denotes the set of protocol states with $I \in S$ being the initial state and $F \in S$ the final state, $T \subseteq S \times S$ the set of state transitions, P the set of transaction parameters and C the set of signals which constitute the physical communication channel.

PFG transitions $t \in T$ are labeled with composition and decomposition guards [C/D Cond], composition and decomposition actions [C/D Action] and signal associations χ, denoted by the symbol $<=>$. The latter describe an unambiguous mapping of a parameter $p \in P$ or a constant value to a channel signal $c \in C$. Transition guards

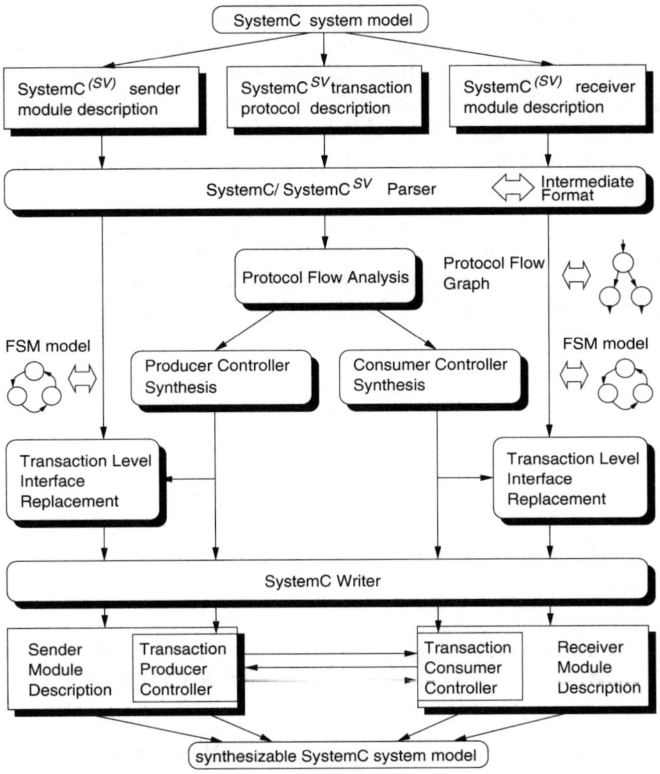

Figure 3: Transformation of a SystemC^{SV} model into a synthesizable SystemC model

condition the transitions between graph states. Conditions can be described in terms of values of parameters or state variables. Transition actions describe actions to be performed when a transition is executed. Such actions include the modification of local state variables or the execution of embedded sequential behaviours. If a transition has multiple actions attached, the order of execution is not determined. Guards and actions are classified into composition specific, denoted by *C*, and decomposition specific ones, denoted by *D*, which accounts for the fact that a PFG represents both compositional and decompositional behaviour of a transaction protocol. For PFG generation from a SystemC^{SV} transaction protocol specification we have chosen a template-based construction approach. For each SystemC^{SV} composition scheme as well as for each item type, a corresponding PFG template is stored in a template library. The contents of this library is shown in Fig. 4. For the **SELECT** scheme two different templates exist. Which one is chosen depends on the selection parameter being a reversible parameter (e.g. during composition the parameter value is established depending on the item matched by the SELECT scheme) or being a constant (during composition a certain item is expected to be matched). PFG generation is done simply through appropriate recursive nesting of these templates, starting with the transaction/message template. This recursive nesting process completes with the insertion of PHYMAP item templates containing the channel signal associations. Care must be taken when transaction item protocols involve a turnaround in the direction of information flow (e.g. handshaking protocol). This is indicated by the fact that lower level items differ in their `FROM,TO` direction attributes (see Lst. 1). In this case, for all items which differ in their logical transmission direction from the direction of the first item in the transaction composition, in the further PFG construction the *C* and *D* labels in the guards and actions

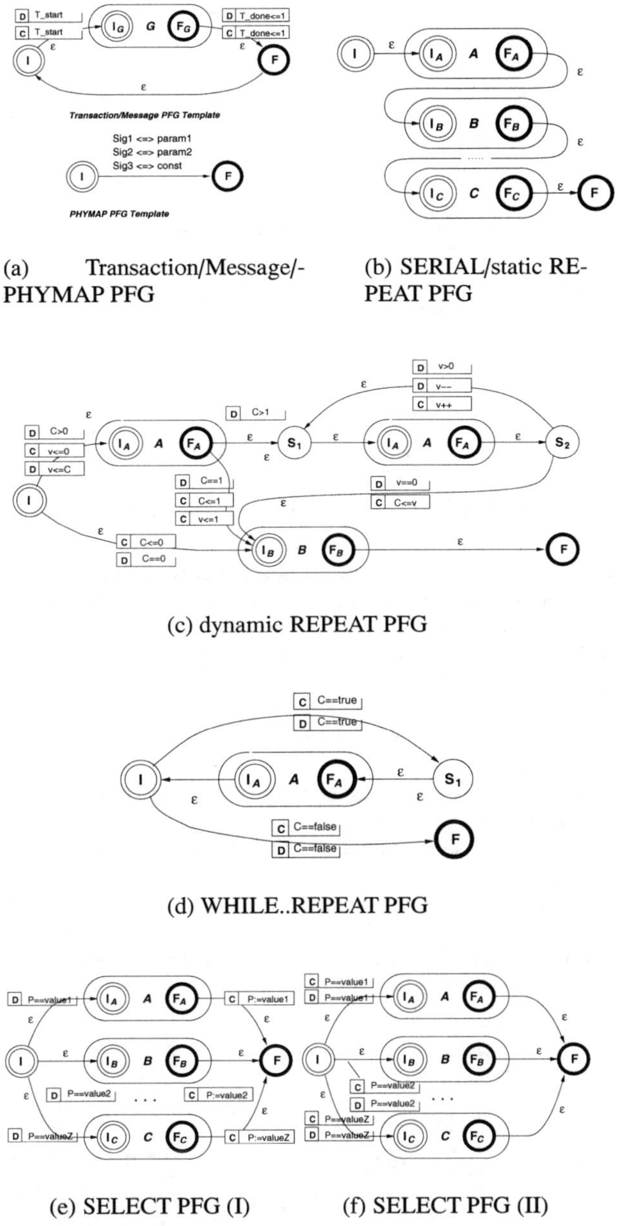

(a) Transaction/Message/-PHYMAP PFG

(b) SERIAL/static REPEAT PFG

(c) dynamic REPEAT PFG

(d) WHILE..REPEAT PFG

(e) SELECT PFG (I)

(f) SELECT PFG (II)

Figure 4: Protocol Flow Graph Templates used for PFG construction

must be swapped for this item instance. Finally, when PHYMAP items are inserted, all signal associations are labeled with *C* if such swapping has been performed for the instancing higher level item. Otherwise they are labeled with *D*. In order to be able to generate PFG's also for non-deterministic protocols (specified through dynamic REPEAT or SELECT compositions), PFG templates are modeled using ε-transitions which corresponds to the Thompson construction of NDFA-based pattern matchers [3]. In our case, ε-transitions describe PFG state transitions which are executed in zero time while non-ε transitions in the PFG take exactly one clock cycle to execute. In Fig. 5 an example PFG is shown for IEC14443 frame protocol which was given in Lst. 1.

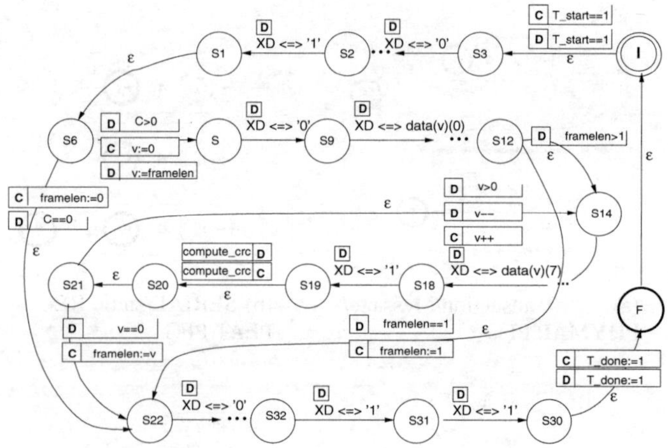

Figure 5: PFG for the IEC14443 Frame Protocol

3.3 Controller Generation from PFG

After successful generation of the PFG from a SystemC^{SV} protocol description, the two distinct controllers representing transaction producer and transaction consumer have to be synthesized from the PFG. The first step is to generate two distinct finite automata for producer FA_D and consumer FA_C. FA_D is generated in the following way: (FA_C is obtained in an analog fashion with C and D labels swapped)

- erase all PFG composition guards and actions (labeled C) from the PFG
- transform signal associations labeled with D into a channel signal assignment
- transform signal associations labeled with C into a transition condition.

The resulting automaton graphs still contain ε-transitions which must not exist in the final controller RTL description in order to be synthesizable into hardware. Thus, in the next step a transformation is applied to the two automaton graphs which generates equivalent graphs without ε-transitions. A corresponding algorithm used in DFA construction for lexicographical scanners would compute the ε-closure of a state, e.g. the set of all states reachable from this state by ε transitions and then simply re-target non-ε transitions ending in this state to all states in the ε-closure. However, since in our case ε-transitions may have guards and actions attached which need to be handled appropriately, a different approach is chosen. The idea is to traverse the FA graph from each state having outgoing ε-transitions in order to find all paths of the form $Q_S \xrightarrow{\varepsilon} Q_A \xrightarrow{\chi} Q_B \xrightarrow{\varepsilon} Q_T$. For each such path, a new transition $Q_S \xrightarrow{\chi} Q_T$ is created. Furthermore all guards and actions along the path are collected and annotated on this transition. After all such paths have been identified, ε-transitions and consequently isolated states are removed from the automaton graph. The result is an equivalent condensed graph as depicted in Fig. 6. The condensed graphs for FA_C,FA_D are in most cases non-deterministic e.g. they contain states which have outgoing transitions to different target states with the same transition conditions. Non-determinism results from the use of the dynamic REPEAT or SELECT composition in protocol descriptions. The construction of a deterministic FSM from the possibly non-deterministic automaton graphs poses no problem since a deterministic FSM can be obtained using the subset construction algorithm [3]. This algorithm computes all clusters of target states reachable from a source state for a certain transition condition. We extended this algorithm for our purposes in order to correctly handle transition guards and reassign actions to the resulting FSM graph. An example for this transformation is shown in Fig. 7. The non-deterministic FA has two transitions leaving State $Q0$, where one transition is conditioned with guard G1 and the other is unconditioned. Furthermore are actions A1, A2 attached to these transitions. In case the condition of guard G1 evaluates to true, both states $Q1$ and $Q2$ are reachable from $Q0$, resulting in the state subset cluster $Z1 = \{Q0, Q1\}$. When this transition is taken, both A1 and A2 must be executed which are consequently attached to the transition $Z0 \longrightarrow Z1$. If G1 evaluates to false, the only reachable state is $Q2$, thus $Z2 = Q2$ and A2 must be executed which is annotated to transition $Z0 \longrightarrow Z2$.

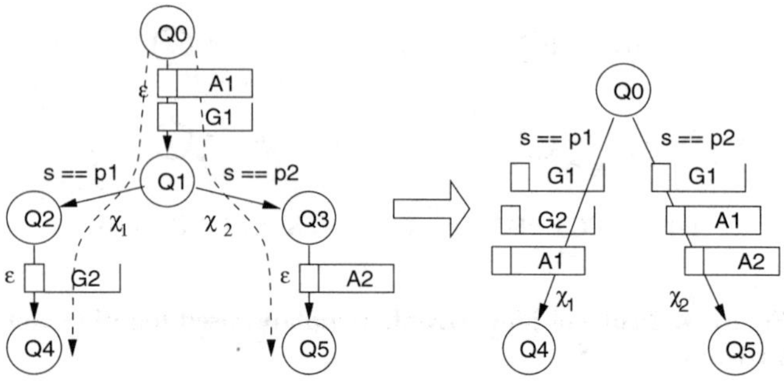

(a) FA graph with labeled ε edges

(b) condensed FA graph without ε edges

Figure 6: Transformation of FA graph with ε transitions into a condensed FA without ε transitions

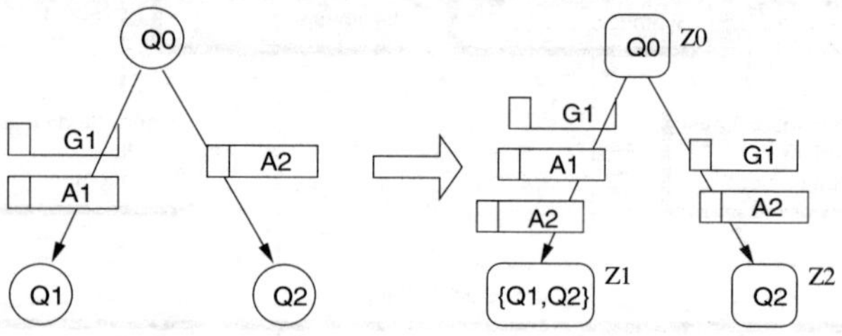

(a) non-deterministic FA graph

(b) equivalent deterministic FSM graph

Figure 7: Transformation of non-deterministic FA into deterministic FSM

3.4 Controller Integration

The last step in the synthesis process is the integration of the generated controller models into the models of the communicating modules. In this step, also the transaction level interfaces of these modules, represented by the `sv_InterfacePort` objects (see Lst. 2), are replaced by a signal level interface with a set of physical pins which are then interconnected between the modules. (see also Fig 1). The `sv_send()`, `sv_receive()` methods used for communication in the SystemC system specification are replaced by a conditional loop containing a single SystemC `wait()` state-

Table 1: Synthesis Results for various Protocol Specifications (Area in Virtex Slices, Clock Frequency in MHz)

Protocol	COSYNE								Protocol Compiler							
	Producer Controller				Consumer Controller				Producer Controller				Consumer Controller			
	Stats	Tran	f_{max}	A	Stats	Tran	f_{max}	A	Stats	Tran	f_{max}	A	Stats	Tran	f_{max}	A
IEC 14443	43	46	85.3	94	77	131	78.5	126	49	116	83.6	91	256	410	77.2	134
ATM	56	62	85.3	52	9	25	63.8	56	56	62	85.9	51	9	22	65.7	54
CAN	452	1022	58.2	271	421	1217	51.7	334	512	1132	50.3	258	476	1342	48.2	321

ment. This loop tests the `T_done` signal of the generated controllers and blocks the execution of the current process until the corresponding protocol controller which now performs the functionality of the former send() or receive() statement has either sent or received a transaction.

4. EXPERIMENTAL RESULTS

The proposed protocol synthesis algorithm has been implemented with about 25000 lines of C++ code in a tool called **COSYNE** (**Co**ntroller **Syn**thesis **E**nvironment) and has been applied to synthesis of hardware controllers for transaction producer/consumer pairs for three different protocols of sufficient complexity. The synthesized protocol examples include the IEC14443 SmartCard communication protocol as well as popular peripheral device interconnect protocols such as CAN. Furthermore was an ATM protocol specification synthesized as an example for a byte-serial protocol. For the ATM example, the protocol compiler models available as demos in the Synopsys distribution were used and the corresponding SystemC^{SV} protocol description has been developed on their basis. Finally, for the CAN example controllers were generated from a protocol description which defines the four basic CAN messages (Data/Remote/Overload/Errorframe).

Table 1 shows the results of the controller synthesis with COSYNE in terms of the number of controller states and transitions in order to give an impression of the complexity of the generated RTL controller models. Furthermore are area and clock frequency measures shown which were obtained through further logic synthesis of the controller models for a XILINX Virtex V100CS144-4 using Synopsys FPGA II compiler. For that purpose, the SystemC controller models generated by COSYNE had to be manually converted to equivalent VHDL RTL models since SystemC synthesis tools were not available yet to the authors. Functional equivalence of the VHDL models resulting from COSYNE and from Protocol Compiler, respectively, was verified through pairwise simulation of a COSYNE generated producer controller in combination with a Protocol Compiler generated consumer controller and vice versa. For comparison purposes, the synthesis results obtained for the same protocols using Synopsys Protocol Compiler are listed. In order to get comparable results, Protocol Compiler options were set to generation of single process controller architectures with binary min-encoded state machines. The results suggest that the controllers synthesized with COSYNE are in terms of performance and area consumption comparable to those generated with protocol compiler. In some cases (IEC14443 and CAN producers) even a smaller number of states and transitions were obtained with COSYNE, resulting in performance and area improvements of the controller implementations.

5. CONCLUSIONS AND FUTURE WORK

We presented a novel design methodology for communication controller hardware based on synthesis from declarative protocol descriptions. A formalism for protocol specification was presented as integral part of the system modeling language SystemC which enabled the specification, verification and synthesis of protocols in the context of the system description. Furthermore we presented a synthesis algorithm which generates from a single SystemC^{SV} protocol specification all interacting transaction producer/consumer controllers. The approach could be especially interesting for the design of future *Networks on Chips* which are protocol-dominated designs and integrate transaction producer and consumer of a protocol onto a single chip. The feasibility of our methodology was shown by means of various protocol examples which were captured in SystemC^{SV} and synthesized into RTL controller models using the COSYNE tool. Further work will address the problem of modeling and synthesis of pipelined protocols such as AMBA as well as protocols which involve split transactions. Other work will address the extension of the proposed methodology to the synthesis of interfaces between software and HW/SW components.

6. REFERENCES

[1] Fujitsu FME GmbH, Dreicheich/Buchschlag. *BabyFace2 Chip Specification*, 1999.

[2] D. Gajski, J. Zhu, R. Doemer, A. Gerstlauer, and S. Zhao. *SpecC: Specification Language and Methodology*. Kluwer Academic Publishers, Boston/Dordrecht/London, 2000.

[3] J. Hopcroft and J. Ullman. *Introduction to Automata Theory, Languages, and Computation*. Addison-Wesley, Reading, MA, 1986.

[4] J.Oberg, A. Kumar, and A. Hemani. Grammar-based hardware synthesis of data communication protocols. In *Proceedings of the 9th International Symposium on System Synthesis*, La Jolla, CA, November 1996.

[5] J. Oberg, A. Kumar, and A. Hemani. Grammar-based Hardware Synthesis from Port Size Independent Specifications. In *IEEE Transactions on VLSI*, volume Vol. 8, pages pp. 184–194, April 2000.

[6] P. Probert and K. Salesh. Synthesis of communication protocols: survey and assessment. In *IEEE Transactions on Computers, Special Issue on Protocol Engineering*, volume Vol. 40, pages pp. 468–476, 1991.

[7] J. Rowson and A. Sangiovanni-Vincentelli. Interface-based design. In *Proceedings of the 34th Design Automation Conference*, Anaheim, CA, 1997.

[8] A. Seawright. *Grammar-based Specifications and Synthesis for Synchronous Digital Hardware Design*. PhD thesis, University of California, Santa Barbara, June 1994.

[9] R. Siegmund and D. Mueller. SystemC-SV: Extension of SystemC for Mixed Multi Level Communication Modeling and Interface-based System Design. In *Proceedings of the Conference on Design, Automation and Test in Europe (DATE 2001)*, Munich, Germany, 2001.

[10] SYNOPSYS Inc. *Protocol Compiler User Manual*, 1998.

[11] Synopsys Inc., CoWare Inc., Frontier Design, Inc., http://www.systemc.org. *SystemC User's Guide*, 2000.

An Integrated Algorithm for Memory Allocation and Assignment in High-level Synthesis

Jaewon Seo
Dept. of EECS, KAIST, Korea

Taewhan Kim
Dept. of EECS, KAIST, Korea

Preeti R. Panda
Synopsys Inc., USA

Abstract - With the increasing design complexity and performance requirement, data arrays in behavioral specification are usually mapped to fast on-chip memories in behavioral synthesis. This paper describes a new algorithm that overcomes two limitations of the previous works on the problem of memory-allocation and array-mapping to memories. Specifically, its key features are *(1) a tight link to the scheduling effect*, which was totally or partially ignored by the existing memory synthesis systems, and supporting *(2) non-uniform access speeds* among the ports of memories, which greatly diversify the possible (practical) memory configurations. Experimental data on a set of benchmark filter designs are provided to show the effectiveness of the proposed exploration strategy in finding globally best memory configurations.

Categories and Subject Descriptors : B.5.2 [Register-Transfer-Level Implementation]: Design Aids

General Terms : Algorithms, Performance, Design

Keywords : memory allocation, memory assignemt, memory design, scheduling effect

1. INTRODUCTION

Behavioral synthesis, which is the process of automatically generating an RTL design from an algorithmic description, is an important research area in design automation. Many behavioral descriptions for manipulating large amounts of data computations use *array variables* to represent data storages. Consequently, behavioral synthesis is required to allocate memory modules for implementing the array variables. However, many synthesis systems either do not support this process or simply rely on designer's specification of particular memory modules. Thus, for memory intensive applications, it is necessary to automate the process of finding the best memory configuration in terms of both total memory cost and design performance.

There are number of research works which have addressed the problem of memory exploration in behavioral synthesis. Shiue *et. al* proposed an ILP model and heuristic-based algorithm for solving the problem of determining memory configuration with minimum area under power constraint or minimum power consumption under area constraint [1]. However, the approach does not take into account the scheduling effect, which could lead to a major limitation in the design space exploration. An exploration strategy under the assumption of a fixed hierarchical model (with different access speeds) of memories was proposed by Holmes and Gajski [2]. However, since the memory granularity is determined by the memory hierarchy used, its application to general (diverse) types of memories in library is not supported. Ramachandran *et. al* proposed a bottom-up cluster-based memory synthesis algorithm that produces an array variable configuration [3]. They assumed that for each array group, any memory module can be allocated only if it satisfies the array size requirement. This clearly limits the exploration of memories with different (read/write) access speeds and different number of ports. Panda proposed a memory bank exploration algorithm which exploits the memory access sequences to minimize page misses in a bank [4]. One limitation is that the exploration has excluded the memories with multiple ports. Wuytack *et. al* addressed the problem of storage bandwidth optimization, by which subsequent memory allocation and assignment tasks produce an architecture with less memories and memory ports [6]. Panda *et. al* provides a full survey on memory and array optimization in high-level synthesis [7].

In this paper, we propose a novel memory exploration strategy. Specifically, our key contributions are *a tight link to the scheduling effect*, which was hardly taken into account by the existing approaches, and supporting *non-uniform access speed* among the ports of memories, which greatly diversifies the possible memory configurations. We integrate the two features in one framework of our algorithm to explore memory configurations more fully and effectively.

2. MOTIVATING EXAMPLES

In this section, we motivate the key features of our algorithm with examples. Our algorithm is based on the following observations.

- *Scheduling effects* during memory exploration can significantly enhance the quality of the memory configuration. (Example 1)
- *Different speed of memory (read/write) accesses* during memory exploration significantly extends the scope of the exploration of the memory configurations. (Example 2)

Example 1: (scheduling effects)
Figure 1(a) shows a segment of behavioral description that manipulates four arrays where each size is 16 (bits) × 1024 (words). Suppose we have a library of memory modules available to use, as summarized in Figure 1(b). It specifies the number of (read-only, write-only, read/write) ports and the area[1] (i.e., cost) of each memory. In the library, we temporarily assumed that every memory access operation takes one clock cycle and all the operations in Figure 1(a) should be completed within three clock cycles. Figure 2(a) shows a schedule for the DFG, called *schedule-1*, and its possible allocations and assignments of memory modules using the library in Figure 1(b). We can easily find that arrays A, B and D cannot be grouped to allocate one memory module because three (read) access operations must be executed in parallel at clock step 1 and none of the memories support the parallelism. The highlighted configuration in the table of Figure 2(a) indicates the one with minimum total memory cost. However, when we reschedule the read operation $read\_D[i]$ from clock step 1 to 2, as shown in Figure 2(b), we can further reduce the total memory module cost, that is, from 26.55 to 22.46. This simple example indicates that scheduling is one of the major factor which drastically affects the memory exploration and the quality of the resulting memory configuration.

[1]In our work, the area (also energy consumption) of each memory was calculated according to the model proposed in [1].

```
reg [15:0] A[1023:0];
reg [15:0] B[1023:0];
reg [15:0] C[1023:0];
reg [15:0] D[1023:0];
...
C[i] = A[j] + B[k];
t = D[i];
...
```

(a)

mem.	size (bits×words)	num. of ports r	w	rw	area (mm^2)
M1	16×1024	0	0	1	7.94
M2	16×1024	1	1	0	8.66
M3	16×2048	0	0	1	11.23
M4	16×2048	1	1	0	12.25
M5	16×2048	1	0	1	15.32

(b)

Figure 1: (a) A segment of input behavioral description, (b) A memory module library.

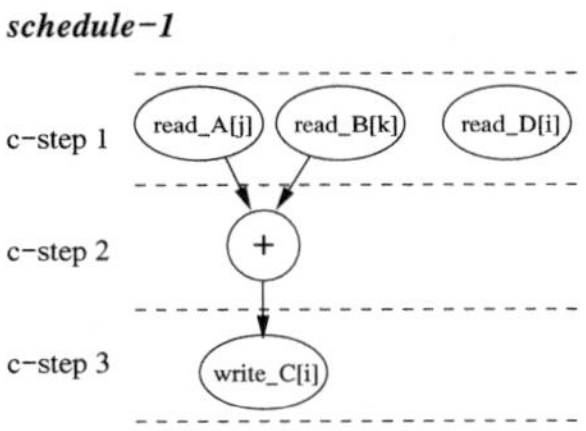

mem. configuration (array group : memory)	area (mm^2)
A:M1, B:M1, C:M1, D:M1	31.76
A:M1, B:M1, CD:M3	27.11
A:M1, C:M1, BD:M5	31.20
...	...
AB:M5, CD:M3	**26.55**
...	...

(a) *schedule-1* for Figure 1(a) and its possible mem. configurations

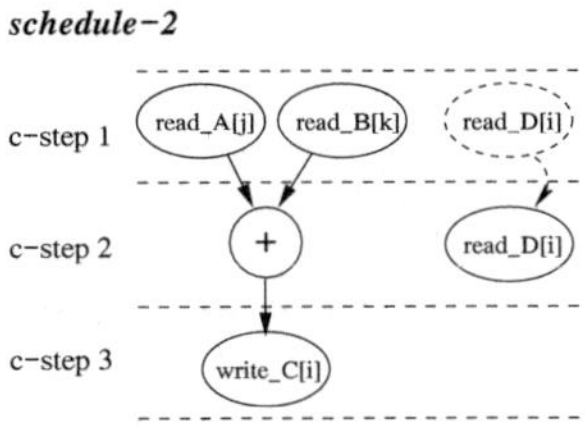

mem. configuration (array group : memory)	area (mm^2)
A:M1, B:M1, C:M1, D:M1	31.76
A:M1, B:M1, CD:M3	27.11
A:M1, C:M1, BD:M5	31.20
C:M1, D:M1, AB:M3	27.11
...	...
AC:M3, BD:M3	**22.46**
AB:M5, CD:M3	26.55
...	...

(b) *schedule-2* for Figure 1(a) and its possible mem. configurations

Figure 2: An example of showing the effect of scheduling on the exploration of memory configuration.

Example 2: (non-uniform speeds of memory accesses)
Table 1 shows a list of memory modules with detailed description of the number of clock cycles taken by each type of memory access. For example, memory module $M2$ has one read-only and one write-only ports, each consuming 2-clock cycle time, whereas $M2'$ has one read-only port having 2-clock cycle time and one write-only port having 3-clock cycle time. Clearly, integrating the non-uniform access times of memory modules and scheduling in one framework of memory exploration is essential to find globally best memory configurations.

3. THE MEMORY EXPLORATION ALGORITHM

The proposed algorithm accepts an unscheduled *data-flow graph* (DFG) as input. Let L be the memory library, and T be the number of clock cycles (i.e., latency constraint) allowed to execute all the operations in DFG. Then, the optimization problem is to select memories from L and assign them to arrays in DFG so that the total memory cost[2] is minimized while satisfying the latency constraint.

Our algorithm consists of three major steps : In Step 1, an initial solution is generated. In Step 2, the previous solution is refined iteratively. In Step 3, array clustering is performed.

Array grouping is a partition on the set of arrays, where arrays that reside in a same memory module are grouped together. Initially in Step1 each array is mapped to a separate group, which means that each array resides in a distinct memory. During Step 2 the grouping of arrays is not altered. After completion of Step 2, in Step 3 the grouping is modified by merging any two groups, and then with the

[2]The cost usually refers to the total area of the allocated memories. However, our approach can easily change the cost to the 'total energy' consumed by the memories. (We provide a set of data on the energy consumptions in Table 3 of section 4.)

mem.	size (bits×words)	number of ports r	w	r/w	access cycles r	w	r/w	area (mm^2)
M1	16×1024	0	0	1	-	-	2	7.94
M1′	16×1024	0	0	1	-	-	3	7.94
M2	16×1024	1	1	0	2	2	-	8.66
M2′	16×1024	1	1	0	2	3	-	8.66
M3	16×2048	0	0	1	-	-	2	11.23
M3′	16×2048	0	0	1	-	-	3	11.23
M4	16×2048	1	1	0	1	2	-	12.25
M4′	16×2048	1	1	0	2	2	-	12.25
M5	16×2048	1	0	1	2	-	2	15.32
M5′	16×2048	1	0	1	3	-	3	15.32

Table 1: An example of memory library with access timings.

resultant grouping, Step 2 is performed again. The whole process is repeated until a single large cluster is formed and the memory configuration and corresponding schedule with the minimum cost is chosen as a result. This decoupling of memory allocation and array grouping is based on the fact that the considering them together is computationally very expensive and it also enables the manual specification of the grouping by the designer, if needed.

• Step 1 (*Generation of an initial solution*): We generate an initial schedule and a corresponding memory configuration that will be refined iteratively during Step 2. We use Force-Directed Scheduling (FDS) algorithm [8] to balance memory operations across clock steps for minimizing the requirements for memory read/write ports. FDS is one of the most representative scheduling algorithms whose objective is to minimize the hardware resources under a certain latency constraint. In our case, we use FDS to schedule the memory operations so that the maximum number of concurrent reads/writes is minimized. From the schedule obtained by FDS, we determine, for each array, the maximum concurrent accesses. Then, among the memories in L that supports this concurrent accesses, we select the one with the minimum cost, and assign it to the array. During FDS, access time of each memory operation is specified by the designer or simply assumed to be the minimum in the library to avoid the requirement for excessive number of ports

• Step 2 (*A simultaneous rescheduling and reassignment*): We refine the schedule and memory configuration obtained in Step 1. Our objective is to reduce the total cost of memory configuration by attempting to iteratively perform rescheduling (operations) and reassignment (arrays to memories) together. To support a large number of iterations, the task of rescheduling and reassignment performed at each iteration should be fast.

In scheduling, we place our primary importance on the optimization of memories to be used. Consequently, we extract all the memory access operations from DFG, and construct a graph, called the Memory Dependency Graph (MDG), which maintains a partial order between the memory access nodes in the DFG.

MDG has two dummy nodes called *source* and *sink*. An arc is created from *source* to every memory access node with no predecessor, and from every memory access node with no successor to *sink*. We assume *source* and *sink* are scheduled at clock step 0, $T+1$, respectively. The MDG concept is similar to that proposed in [4], but ours is more general in usage, including a minimum delay information between nodes.

Definition (*MI: minimum interval*): An arc $v \rightarrow w$ with *minimum interval* $MI(v \rightarrow w)$ in the MDG indicates the minimum number of clock cycles required for the execution of arithmetic/logical operations along the arc between the two memory access nodes v and w.

For example, $v \xrightarrow{MI=3} w$ indicates that at least three clock cycles are required for the period between the completion of the execution of v and the start of w. We calculate the MI values for each arc during the process of generating an MDG by using a depth-first graph traversal procedure. For a schedule S, a memory configuration M for

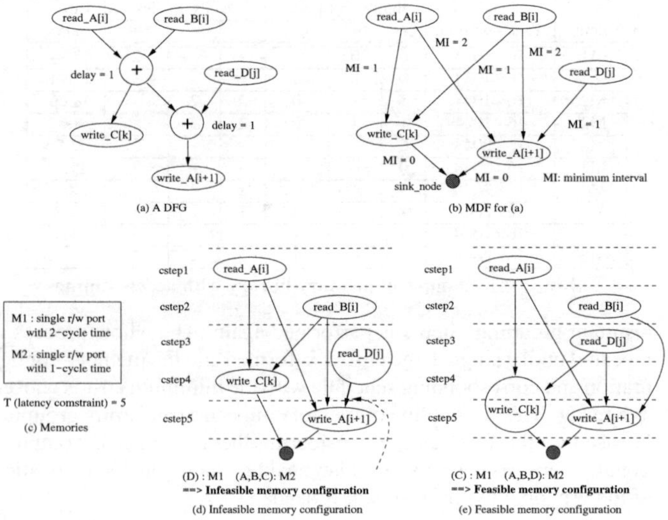

Figure 3: An example showing MDG construction and infeasible/feasible memory configurations.

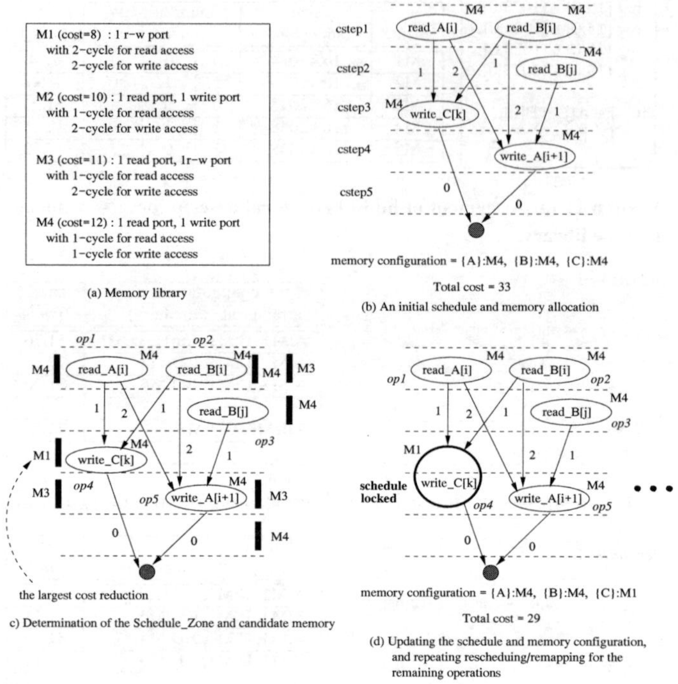

Figure 4: An example illustrating our incremental step of simultaneous rescheduling and remapping.

S is called a *feasible* memory configuration if for every arc $v \longrightarrow w$ in MDG, it satisfies

$$cstep(v) + delay(v) + MI(v \longrightarrow w) \leq cstep(w) \qquad (1)$$

where $cstep(v)$ is the clock step at which memory access operation v starts to execute and $delay(v)$ is the execution delay (in the number of clock cycles) of v. Consequently, *our goal is to find a schedule with latency constraint of T that leads to a feasible memory configuration with minimum total memory cost.*

An example of DFG and its MDG are shown in Figures 3(a) and (b), respectively. The number assigned to each arc in Figure 3(b) indicates the MI value. Figure 3(d) shows the memory operations for a schedule of the MDG when arrays A, B and C each uses a memory of type *M2* in Figure 3(c) and array D uses *M1*. Consequently, as indicated with dotted arrow in Figure 3(d), the arc from $read\_D[j]$ to $write\_A[i+1]$ violates the inequality relation in Eq.(1), resulting in an infeasible memory configuration for the schedule. On the other hand, Figure 3(e) shows a feasible memory configuration for the same schedule (i.e., all arcs satisfying the inequality in Eq.(1)), where A, B and D each uses a memory of type *M2* while C uses a type *M1* memory.

The example in Figure 4 shows how operation rescheduling and memory remapping are performed at an iteration of our algorithm. Suppose we have an initial schedule and its memory configuration shown in Figure 4(b) using the memories in Figure 4(a). For each operation, we determine *schedule_zone*, which is the set of clock steps at which the operation can be scheduled for execution without causing any timing violation of Eq.(1). The heavy lines next to the operations represent *schedule_zone*. For example, *schedule_zone* for $op1$ in Figure 4(c) is $\{cstep1\}$ since scheduling $op1$ to *cstep2* causes MI violations between $op1$ and $op4$ and between $op1$ and $op5$, while *schedule_zone* for $op4$ is $\{cstep3, cstep4\}$.

Then, we check for each memory in L, if the selected array can be scheduled within its *schedule_zone* without causing any MI violation. As a result, we obtain a new schedule for that array. We use a simple list scheduling heuristic for the checking using the selection guideline that the most timing-critical operation is scheduled first. Then, among the memories that lead to feasible memory configurations, we choose the one with the smallest memory cost. For example, the symbol attached to each interval of *schedule_zone* in Figure 4(c) indicates the selected memory.

Let $M(a_i)$ be the memory assigned to array a_i for a certain memory configuration R, and M_j be the memory selected according to the procedure described above when all operations of a_i are rescheduled. Then, among all array-memory pairs (a_i, M_j), we choose to reschedule the one that maximizes the quantity

$$\Delta C = C(M(a_i)) - C(M_j) \qquad (2)$$

where $C(M)$ represents the memory cost of type M. From the example in Figure 4(c), we selected op_4 and rescheduled (actually the same clock step as before) it at clock step 3 because its cost reduction in Eq.(2) (i.e.,$C(M4)$-$C(M1)$=12-8=4) is the largest. op_4 is then locked for rescheduling (but may change the assigned memory by subsequent iterations), and the process of rescheduling and reassignment for the remaining operations is repeated until all operations are rescheduled and reassigned.

The overall procedure of Step 1 and Step 2, called MEM-exp, works as follows. First, we generate an initial schedule for an input DFG and a memory configuration for the schedule using a memory library L. We refine the initial solution iteratively in the loop of our algorithm. At each iteration, we select a pair of array and memory that lead to a maximum decrease of memory cost (i.e., the value in Eq.(2)) without violating the latency constraint of T and causing an infeasible memory configuration. We then reschedule all the operations of that array and at the same time, reassign the memory determined using the procedure described in the previous paragraph to the selected array. Once the rescheduling and remapping are done, any attempt to reschedule the operation is disallowed for the rest of the execution of the inner loop of our algorithm. The loop continues if we find a new configuration whose total cost is less than the minimal cost found so far or there is an overall reduction of the memory cost in the current inner loop (i.e., $\sum \Delta C < 0$).

• Step 3 (Array Clustering): The exploration of memory configurations in Step 2 has been performed with the restriction that the grouping of arrays is fixed. In Step 3, the grouping is changed to further reduce the total memory cost, by bottom-up array clustering. Arrays are clustered by merging two different groups into a single group and reallocating a new memory module to that group. Storing two or more arrays into the same memory module may require additional operations for address translation. After adding proper operations

design	$\|L\|$	MEM-exp		*Exhaustive*		diff.
		cost (mm^2)	time (sec)	cost (mm^2)	time (sec)	(%)
kalman	180	10.772	0.3	10.718	77	0.5
sor	200	371.134	0.1	371.134	20	0.0
phoneme	180	132.050	0.7	-	-	-
heat	180	17.103	0.2	17.103	27	0.0
ddpoly	200	8.349	0.1	8.349	32	0.0
tridag	200	24.816	0.1	23.309	24	6.1
avg.						1.32

Table 2: Comparisons of memory area for the memory configurations produced by MEM-exp and an exhaustive searching method.

into DFG, rescheduling is performed in the same way as in Step 2. Among all possible merging of two groups the one that reduces the memory cost maximally is chosen. After completion of Step 3, current memory configuration is refined iteratively without changing the grouping of arrays (Step 2). The whole process terminates when only a single large group remains.

Extension to DRAM Memories : When L includes DRAM memories, MEM-exp is extended to exploit one important feature of DRAM architecture. The architecture enables a significant speed-up in executing subsequent memory accesses that refer to the words in the same page because additional steps of row decoding and precharging are not needed. Such memory access mode is called *page mode* [4].

We extend MEM-exp to exploit the page mode accesses: If two memory access operations in MDG are executed successively and they use the same DRAM memory, we check the MI violation between the two operations and *schedule_zone* violation of the operations by determining the number of clock steps required for executing them based on the page mode accesses, and comparing the number with the MI value and *schedule_zone*.

It should be noted that our memory exploration method can easily incorporate other types of DRAM access modes (e.g., those described in [5]) with similar ways.

4. EXPERIMENTAL RESULTS

The proposed memory exploration algorithm MEM-exp was implemented in C++ and executed on a Sun Sparc20 workstation. We tested a set of memory intensive applications to demonstrate the effectiveness of the proposed memory exploration algorithm. kalman, sor and heat are the kalman filter design, the successive over relaxation design, and the differential heat release computation design respectively [9]. phoneme is the phoneme recognition [10]. trida is the tridiagonal systems of equations and ddpoly is the polynomial and its derivative evaluation [11].

The experiments were performed in two respects: (1) to check the effectiveness of memory exploration for minimizing the total memory area and (2) to check the effectiveness for minimizing the total energy consumption by memories.

The latency constraint in each experiment is set to the median value in the range between the lower and upper bound of latency constraints. With latency constraint which is lower than lower bound, no feasible solution can be obtained, while, with latency constraint higher than upper bound, no more improvements can be achieved.

• **Effectiveness of minimizing total memory area**: Table 2 shows a comparison of results produced by our algorithm and an exhaustive searching method (*Exhaustive*), in which we used a numberof pruning techniques to speed up the exhaustive exploration without losing optimality. $|L|$ is the number of memories in memory library L. In terms of total memory area, our solutions are on average 1.3% more than the optimal solutions. In terms of run time, our method is significantly fast. *Exhaustive* slows down considerably as the number of arrays become large, and for design phoneme which contains 28 arrays, it was not able to produce a solution within 6 hours.

design	$\|L\|$	MEM-exp cost (nJ)	*Greedy* cost (nJ)	red. (%)
kalman	180	6.95	8.09	14.0
sor	200	1158.89	3827.38	69.7
heat	180	14.29	20.91	31.7
phoneme	180	362.66	1293.56	72.0
ddpoly	200	16.75	18.64	10.1
tridag	200	25.38	28.38	10.6
avg.				34.6

Table 3: Comparisons of energy consumption for the memory configurations produced by MEM-exp and a greedy heuristic.

• **Effectiveness of minimizing total energy consumption**: We compare our results, in terms of energy consumption, with those produced by *Exhaustive* and a greedy heuristic (called *Greedy*). The results are summarized in Table 3. The idea in *Greedy* was taken from [1], in which it tries to assign arrays that are more frequently accessed to memories with less energy consumption. Since the heuristic in [1] does not take into account the latency constraint, we modified the heuristic so that it allocates memories in a greedy manner while satisfying the latency constraint. The comparisons in Table 3 indicate that MEM-exp allocates memories more selectively according to the access patterns and frequencies in the behavioral specification, with 34.6% less energy consumption than that produced by *Greedy*.

5. CONCLUSIONS

In this paper, we proposed a new (SRAM/DRAM) memory exploration algorithm to solve two important problems: *(1) a tight link to the scheduling effect*, which was totally or partially ignored by the existing memory synthesis systems, and supporting *(2) non-uniform access speeds* among the ports of memories, which greatly diversify the possible practical memory configurations. We solved the problems in *an integrated fashion* to explore memory configurations more fully and effectively by devising efficient cost evaluation and timing-violation checking techniques. Experimental results showed that the run time of our algorithm is significantly faster than that of the exhaustive search based method while the quality, in terms of memory area, of the memory configurations was within 1.3% difference (on average) with the optimal solutions. In term of energy consumption, our algorithm was able to find schedules and memory configurations with 34.6% less energy consumption (on average) compared to the existing method [1].

Acknowledgment : This work was supported by the Korea Science and Engineering Foundation (KOSEF) through the Advanced Information Technology Research Center(AITrc).

6. REFERENCES

[1] W.-T. Shiue, *et. al*, "Low Power Multi-Module, Multi-Port Memory Design for Embedded Systems", *Proc. of Signal Processing Systems*, 2000.

[2] N. Holmes and D. Gajski, "Architectural Exploration for Datapaths with Memory Hierarchy", *Proc. of EDAC*, 1994.

[3] L. Ramachandran, *et. al*, "An Algorithm for Array Variable Clustering", *Proc. of the EDAC*, 1994.

[4] P. R. Panda, "Memory Bank Customization and Assignment in Behavioral Synthesis", *Proc. of ICCAD*, 1999.

[5] P. R. Panda, *et. al*, "Incorporating DRAM Access Modes into High-level Synthesis", *IEEE Trans. on CAD*, 1998.

[6] S. Wuytack, *et. al*, "Minimizing the required memory bandwidth in VLSI system realizations.", *IEEE Trans. on VLSI Systems*, 1999.

[7] P. R. Panda, *et. al*, "Data and Memory Optimization Techniques for Embedded Systems", *ACM Trans. on Design Automation of Electronic Systems*, 2001.

[8] P. G. Paulin and J. P. Knight, "Force-Directed Scheduling for the Behavioral Synthesis of ASIC's", *IEEE Trans. on CAD*, 1989.

[9] N. D. Dutt, "High-Level Synthesis Design Repositories", *http://www.ics.uci.edu/ dutt*.

[10] Schmit, H., Thomas, D.E. "Synthesis of application-specific memory designs", *IEEE Trans. on VLSI Systems*,, 1997.

[11] W. H. Press, *et. al*, *Numerical Recipes in C*, Cambridge University Press, 1988.

High-Level Synthesis of Multiple-Precision Circuits Independent of Data-Objects Length*

M. C. Molina, J. M. Mendías, R. Hermida
Dpto. Arquitectura de Computadores y Automática
Universidad Complutense de Madrid
Avda. Complutense s/n, 28040 Madrid (SPAIN)
{cmolinap, mendias, rhermida}@dacya.ucm.es

ABSTRACT
This paper presents a heuristic method to perform the high-level synthesis of multiple-precision specifications. The scheduling is based on the balance of the number of bits calculated per cycle, and the allocation on the bit-level reuse of the hardware resources. The implementations obtained are multiple-precision datapaths independent of the number and widths of the specification operations. As a result impressive area savings are achieved in comparison with conventional algorithms implementations.

Categories and Subject Descriptors
B.5.2 [**Register-transfer-level Implementation**]: Design Aids – *automatic synthesis, optimization.*

General Terms: Algorithms, Design.

Keywords
High-level synthesis, multiple-precision, scheduling, allocation.

1. INTRODUCTION
Conventional high-level synthesis (HLS) scheduling algorithms set the execution cycle in which every specification operation starts its execution. They try to balance the number of operations of each different type scheduled per cycle, independently of their widths. Figure 1a) shows a possible scheduling provided by a conventional algorithm for a multiple-precision specification formed by three additions with no data dependencies. However, it is possible to obtain circuits with smaller area by uniformly distributing the number of bits calculated among the cycles. In this case, an operation may be executed during a set of non-necessarily consecutive cycles starting from the least significant bits. The carry-out produced in one cycle must be stored to be supplied as the carry-in during the next execution cycle of the operation. Figure 1b) shows another possible scheduling for the specification in which the least significant bits of the operation $(A=B+C)$ are calculated in cycle i, and the remaining ones in cycle $i+1$.

Conventional HLS allocation algorithms provide datapaths where every operation is executed over one functional unit (FU) of the same or greater width. The number of FUs is determined by the maximum number of operations scheduled in the same cycle, with their respective widths depending on the number of bits of the wider operations. When these wider operations are not scheduled in such *busy* cycle, a considerable waste of area may result. Figure 1c) shows the conventional solution for the scheduling in Figure 1a). However, it is possible to obtain circuits with smaller area by following other design strategies, like the simultaneous execution of several operations over one FU, or the execution of one operation over several linked FUs. Figures 1d) and 1e) show these novel solutions for the scheduling in Figure 1b).

The argumentation made for operations and FUs can be extrapolated to variables and storage resources, and to data transfers and routing resources.

In this paper we propose an algorithm able to offer solutions like the ones presented above. These scheduling and allocation methodologies lead to smaller datapaths where the number and widths of the HW resources are independent of the number and widths of the specification operations and variables.

2. RELATED WORK
Multiple-precision behaviours appear both in SW and HW, especially in the development of DSP applications.

The problem to be solved in DSP SW development consists in the transformation of the original multiple–precision specification into another one with a unique width. This obliges to use the truncation and extension operators to make uniform the operations widths, with the consequent loss of precision in the first case, and waste of HW in the second one. Recent research in multiple-precision SW derives fixed-point implementations from floating-point or infinite-precision descriptions [1] [8].

Within the HW field there has been little research in HLS for multiple-precision specifications. Up to now most systems have adopted trivial solutions, balancing the number of operations executed per cycle, and binding operations to FUs with identical widths [4] [5]. More efficient systems execute operations over wider FUs, extending the operands when necessary, and discarding some bits of the solutions produced [2] [3]. Nevertheless, some authors admit that these trivial solutions are not good enough and suggest (but not implement) other alternatives, like the execution of an operation during several cycles [3].

DAC 2002, June 10-14, 2002, New Orleans, Lousiana, USA.

* Supported by Spanish Government Grant CICYT TIC-99 0474

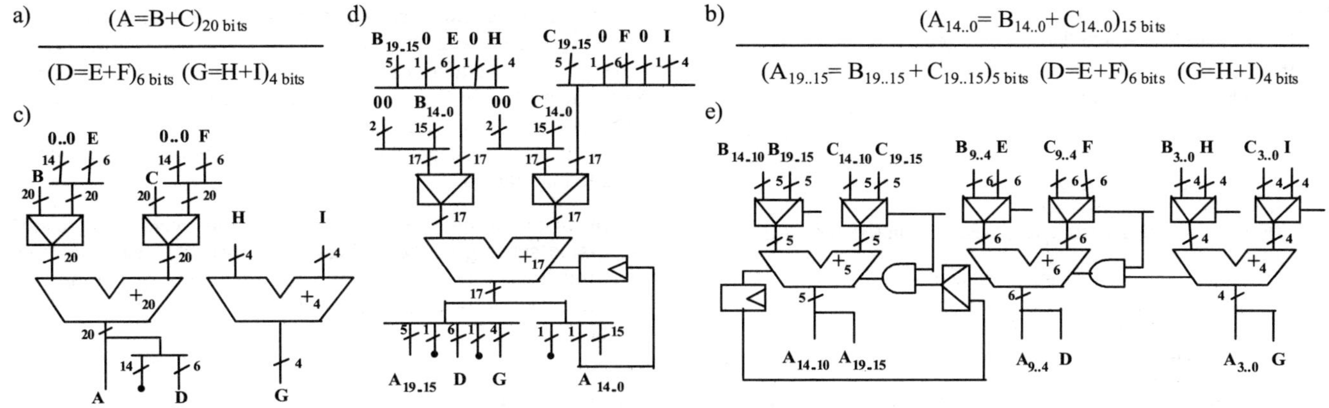

Figure 1. a) Scheduling proposed by a conventional algorithm, b) more efficient scheduling, c) implementation proposed by a conventional algorithm, d) and e) more efficient implementations for the 2nd scheduling.

3. PROPOSED ALGORITHM

Our algorithm performs the scheduling and allocation of multiple-precision specifications by using the design strategies shown in the introduction to deal with additions and those operations with an *additive* kernel. Remaining operations are treated in a conventional way.

The input is a multiple-precision specification whose *additive* operations (comparisons, maximum/minimum, subtractions, etc) are previously transformed into additions. The outputs are a datapath and a controller. In the datapaths provided the sum of all adder widths and the sum of all register widths is equal to, respectively, the maximum number of bits added and stored simultaneously in a cycle. And the sum of all multiplexer widths is less or equal to the maximum number of bits transmitted simultaneously in a cycle.

Figure 2 shows a schema of the proposed algorithm, and in the next subsections its central parts are explained.

3.1. Scheduling

We propose a variant of the popular and classical force-directed scheduling algorithm [7]. The intent of the original method is to minimize HW cost subject to a given time constraint (λ) by balancing the number of operations executed per cycle. The algorithm successively selects, among all operations and all execution cycles, an (operation, cycle) pair according to an estimation of the circuit cost called *force*.

Since the area of multiple-precision datapaths depends on the maximum number of bits computed simultaneously in a cycle, the intent of our variant is to minimize HW cost by balancing this number. Our method successively selects, among all operations and all execution cycles, a pair formed by an operation or an operation fragment, and an execution cycle according to a new *force* definition that takes into account the operations widths. As a result of the possible fragmentation an operation may be scheduled in a set of non-necessarily consecutive cycles.

In order to calculate the new *force* measure, the *time frame* of every operation (set of cycles an operation may start its execution) is calculated first:

$$\mathrm{TF}_{op} = \left\{ c \in \mathrm{N} \;\middle|\; \sigma_{\mathrm{ASAP}}(op) \le c \le \sigma_{\mathrm{ALAP}}(op) \right\}$$

Assuming that all bindings to feasible execution cycles have equal probability, operation *op* covers execution cycle *c* with probability:

$$\mathrm{P}(op,c) = \begin{cases} \dfrac{1}{\left\| \mathrm{TF}_{op} \right\|} & \text{if } c \in \mathrm{TF}_{op} \\ 0 & \text{otherwise} \end{cases}$$

The *estimated operation cost* in cycle *c*, is the average value of the number of operation widths covering execution cycle *c*:

$$\mathrm{EOC}(c) = \sum_{op \in \mathrm{EOP}_c} \left(\mathrm{width}(op) \cdot \mathrm{P}(op,c) \right)$$

$$\text{where} \quad \mathrm{EOP}_c = \left\{ op \in \mathrm{OP} \;\middle|\; c \in \mathrm{TF}_{op} \right\}$$

Force is a rather intuitive measure for how desirable a certain operation to cycle binding is, and its basic formula is:

$$\mathrm{F}(op,c) = \mathrm{EOC}(c) - \sum_{i \in \mathrm{TF}_{op}} \frac{\mathrm{EOC}(i)}{\left\| \mathrm{TF}_{op} \right\|}$$

The smaller or more negative this expression becomes, the more preferable is an operation to cycle binding. Of course, an operation to cycle binding results in consequences that need to be considered. Fixing the time frame of an operation possibly affects the time frames of data dependent operations, resulting in additional changes of the distribution graph. In our complete new *force* definition this has been taken into account as in the complete original *force* definition [7].

Our algorithm binds operations (or fragments) to cycles subject to the following *upper bound*:

$$\frac{\sum_{op \in \mathrm{OP}} \mathrm{widths}(op)}{\lambda} + \mathrm{error\_margin}$$

which corresponds to the most uniform operation bits distribution among cycles plus a certain error margin. The error margin is necessary because due to the data dependencies this perfect

```
INPUT: DFG
OUTPUT: Datapath & Controller
BEGIN
  AdditiveOperations2Additions(OP);
  ---------------------------------------------------------------
  Unscheduled = OP;                                  Scheduling
  REPEAT                                                  Phase
    F = CalculateForces(Unscheduled);
    BestOC = SelectMinForce(F);
    IF Bound < (BitsSched(BestOC.cycle)+Width(BestOC.op))
    THEN
      width1 = Bound - BitsSched(BestOC.cycle);
      width2 = Width(BestOC.op) - width1;
      Fragment(BestOC.op, width1, width2, op1, op2);
      Schedule(op1, BestOC.cycle);
      Add(Unscheduled, op2);
      FOR ope IN (Suc(BestOC.op) ∪ Pred(BestOC.op)) DO
        Fragment(ope, width1, width2, op1, op2);
        Remove(Unscheduled, ope);
        Add(Unscheduled, op1);
        Add(Unscheduled, op2);
    ELSE Schedule(BestOC.op, BestOC.cycle);
    END IF;
    Remove(Unscheduled, BestOC.op);
    UpdateDependencies(Unscheduled);
  UNTIL Unscheduled = Ø;
  ---------------------------------------------------------------
  REPEAT                                             Allocation
    C = CalculateCandidates(REG);                         Phase
    VL = SelectValidLocations(C, FU, REG);
    IS = CalculateIS(C, VL, FU, REG);
    BestCL = SelectBestCandidate(C, VL, IS);
    Allocate(BestCL, FU, REG);
    Remove(C, BestCL);
  UNTIL C = Ø;
  CompactDatapath(FU, REG);
  CreateInterconnections(FU, REG, Datapath);
  ---------------------------------------------------------------
END;
```

Figure 2. Algorithm to perform the HLS of multiple-precision specifications.

distribution is not always reachable. Experimental results have measured this margin in the 15% to 20% increase range.

The scheduling algorithm consists in a loop which finishes when all the operations and all the operation fragments have been scheduled. In every step an operation and one of its time frame cycles are selected using the *force* measure. If the number of bits already scheduled in the selected cycle plus the width of the selected operation does not reach the *upper bound*, then the operation is scheduled in the cycle. Otherwise the operation is fragmented into two new operations. The width of one of these fragments is *upper bound* minus number of bits already scheduled in the selected cycle. This fragment is scheduled in the cycle, and the other one remains unscheduled with its original time frame. In order to avoid reductions in the time frames of the predecessors and successors of the fragmented operation, they are also fragmented. After scheduling an operation, or an operation fragment, the data dependencies of its successors and predecessors need to be updated.

When an operation is scheduled in several cycles, the result is calculated starting in the earliest cycle from the least significant bits, and storing the partial results until the latest execution cycle. An additional 1-bit register is also needed to store the partial carry signal.

3.2. Allocation

This phase is performed in three steps, first a virtual resources selection and binding of operations and variables take place; next, the partial datapath is compacted; and finally, it is completed with the routing resources. It should be highlighted that the set of operations and variables the algorithm allocates is not the original one, but the obtained after the scheduling phase.

The aim of the first step is to construct a datapath of as many 1-bit adders and 1-bit registers as, respectively, the maximum number of bits added and stored simultaneously in a cycle.

During this phase the algorithm works with a virtual array of linked 1-bit adders. The carry-out of each adder is connected, through an *and* gate, to the carry-in of the one on its left. These *and* gates are used to enable the carry propagation when necessary.

Before explaining this phase of the algorithm some concepts need to be defined:

Candidate: set of operations scheduled in different cycles and aligned in a particular way (different alignments result in different candidates). Figures 3a) and 3b) show respectively, a scheduling of a multiple-precision specification, and all the possible *candidates* formed by two of its operations.

Location of a candidate: set of contiguous 1-bit FUs which are not busy during the cycles in which the *candidate* operations are scheduled. Figure 3c) shows the two possible *locations* of one of the *candidates* drawn in Figure 3b).

Valid location: *location* that leaves enough contiguous regions of 1-bit FUs to allocate the still not allocated operations. Only the first *location* shown in Figure 3c) is a *valid location.*

Interconnection saving: cost function used to select the best pair (*candidate, location*). The *interconnection saving* of allocating a *candidate* in a *location* takes into account the number of bits of the *candidate* operands and results which can be stored in the

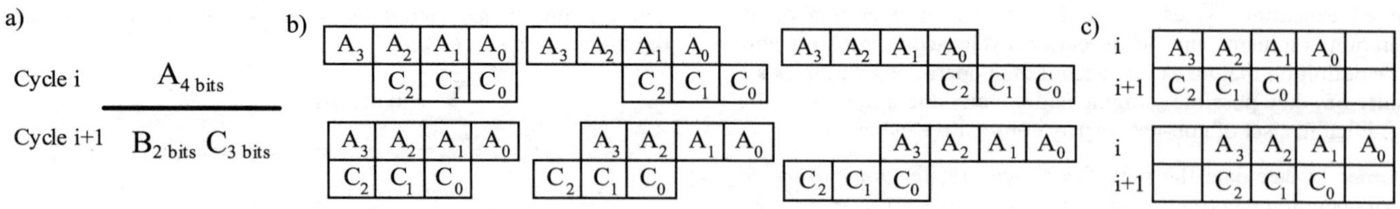

Figure 3. a) Possible scheduling for a multiple-precision specification, b) all possible *candidates* formed by operations *A* and *C*, c) all possible *locations* of one of these *candidates*.

a)

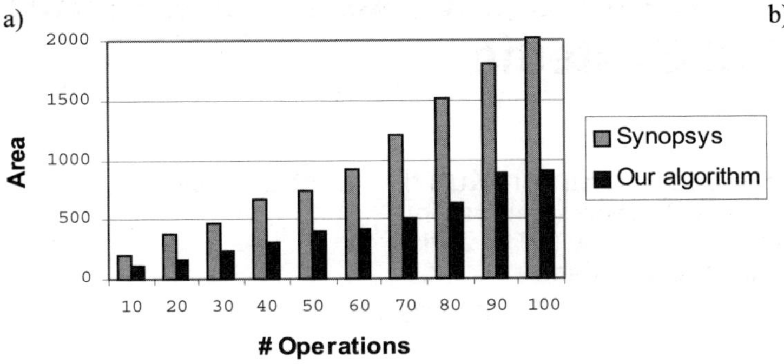

b)

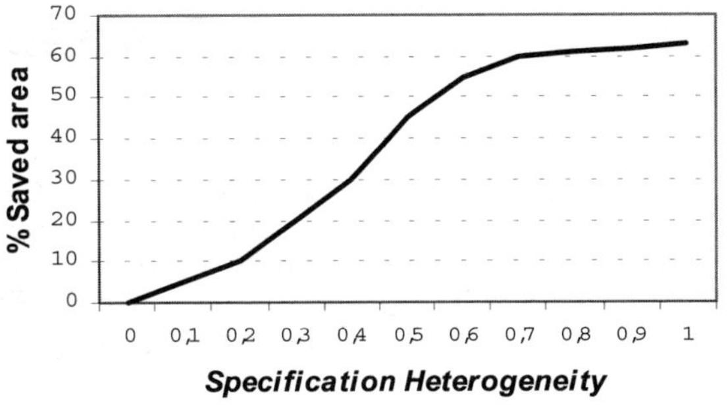

Figure 4. a) Average area of some circuits synthesized by Synopsys and our algorithm, b) percentage of area saved by the proposed algorithm in comparison with Synopsys implementations.

same registers, and the number of bits of the *candidate* operands and results which can be stored in the registers used by the operations already allocated in the selected *location*.

This phase of the algorithm consists in a loop which finishes when all the operations and all the variables have been allocated. In every step a subset of the *candidates* and a subset of their *valid locations* are created. The criteria to select these subsets are explained in [6]. Next the algorithm calculates the *interconnection saving* of each *candidate* in every *valid location*, and finally, the *candidate* with the maximum *interconnection saving* is allocated.

The allocation of a *candidate* consists in the individual allocation of each of its constituent operations to a set of contiguous 1-bit FUs, and the allocation of their respective operands and results to a set of 1-bit registers (not necessarily contiguous). In order to reduce the interconnection area, the algorithm tries to allocate all the *candidate* right operands to the same set of registers, and the same applies to the left operands and the results. It also tries first the registers used to store the variables of the operations already allocated in the selected *location*.

Once all the specification operations have been allocated, the algorithm joins those 1-bit-width FUs which must remain linked to propagate the carry signal during all the cycles to form the final FUs. And those 1-bit-width registers with compatible load signals are also joined to form wider ones.

4. EXPERIMENTAL RESULTS

In order to measure the quality of the implementations obtained by our algorithm we have compared them with the ones proposed by Synopsys Behavioral Compiler. We have synthesized a wide collection of randomly generated multiple-precision specifications, which ranges from 10 to 100 *additive* operations with latencies from 4 to 30 cycles. Results show that the area of the implementations obtained by our approach is always smaller than the area of the circuits proposed by the commercial tool. For the circuits synthesized the average saved area ranges between 45% and 55%. Figure 4a) shows graphically some of these results.

In order to explain these results it is necessary to identify the factors which influence the quality of the implementations obtained by both methods. The area of our implementations only depends on the specification size, the data dependencies, and the given time constraint. However, the area of the implementations obtained by conventional algorithms also depends on the number and widths of the specification operations. Figure 4b) shows the amount of area saved by our algorithm in function of the *specification heterogeneity* (ratio of the number of different widths in the specification to the number of operations).

5. CONCLUSION

This paper presents novel scheduling and allocation algorithms for multiple-precision specifications, which balance the number of bits calculated per cycle and guarantee a maximum bit-level reuse of adders and registers. The implementations obtained are multiple-precision datapaths, where the number and widths of the resources are independent of the specification operations.

Experimental results show that the implementations given by our algorithm have smaller area than the ones offered by commercial tools, and also that the amount of saved area grows with the *specification heterogeneity*.

Future work will include the generalization of the algorithm to increase the reuse of all kinds of FUs, and the integration of the scheduling and allocation phases, what we expect could even report better results.

6. REFERENCES

[1] Cmar, R., Rijnders, L., Schaumont, P., Vernalde, S., and Bolsens, I. A methodology and design environment for DSP ASIC fixed point refinement. In Proceedings of DATE, 1999.

[2] Constantinides, G.A., Cheung, P.Y.K., and Luk, W. Heuristic datapath allocation for multiple wordlength systems. In Proceedings of DATE, 2001.

[3] Ercegovac, M., Kirovski, D., and Potkonjak, M. Low-power behavioural synthesis optimization using multiple precision arithmetic. In Proceedings of DAC, 1999.

[4] Huang, C., Chen, Y., Lin, Y., and Hsu, Y. Data path allocation based on bipartite weighted matching. In Proceedings of DAC, 1990.

[5] Küçükçakar, K., and Parker, A. Data Path tradeoffs using MABAL. In Proceedings of DAC, 1990.

[6] Molina, M.C., Mendías, J.M., and Hermida, R. Multiple-Precision Circuits Allocation Independent of Data-Objects Length. In Proceedings of DATE, 2002.

[7] Paulin, P.G., and Knight,J.P. Force-directed scheduling for the behavioral synthesis of ASICS. IEEE Trans. on CAD, 1989.

[8] Willens, M., Bürsgens, V., Keding, H., Grötker, T., and Meyer, M. System-level fixed-point design based on an interpolative approach. In Proceedings of DAC, 1997.

Schedulability of Event-Driven Code Blocks in Real-Time Embedded Systems

Samarjit Chakraborty Thomas Erlebach Simon Künzli Lothar Thiele
Computer Engineering and Networks Laboratory
Swiss Federal Institute of Technology (ETH) Zürich
CH-8092 Zürich, Switzerland
{samarjit,erlebach,kuenzli,thiele}@tik.ee.ethz.ch

ABSTRACT

Many real-time embedded systems involve a collection of independently executing event-driven code blocks, having hard real-time constraints. Tasks in many such systems, like network processors, are either not preemptable or have restrictions on the number of preemptions allowed. All the previous work on the schedulability analysis of such systems either have exponential complexity, or allow unbounded number of preemptions and are usually based on heuristics. In this paper we present the exact necessary and sufficient conditions under EDF, for the schedulability of such a collection of code blocks in a non-preemptive environment, and give efficient algorithms for testing them. We validate our analytical results with experiments and show that the schedulability analysis problem in such systems can be exactly and efficiently solved in practice.

Categories and Subject Descriptors

C.3 [**Computer Systems Organization**]: Special-purpose and application-based systems—*Real-time and embedded systems*

General Terms

Algorithms, Performance, Verification

1. INTRODUCTION

Real-Time systems are generally modeled as a collection of independent *tasks*, where each task generates a sequence of *jobs*, each of which is characterized by a *ready-time*, an *execution requirement*, and a *deadline*. The schedulability analysis of such a system is concerned with determining whether it is possible to assign to each job a processor time equal to its execution requirement, between its ready-time and its deadline. In the context of most real-time embedded systems, each such task is required to model an event-driven block of code, parts of which are triggered by external events and require to be executed within a given deadline from the triggering time. Such blocks of code are usually represented by their control flow graphs on some appropriate level of abstraction, where the vertices have associated deadlines and represent portions of code implementing some functionality, and the edges represent the flow of control. The vertices when triggered by external (or internal) events have to be executed within their associated deadlines. The schedulability analysis of such a collection of control flow graphs answers whether it is possible to execute all the vertices of all graphs within their deadlines, under all possible event triggering sequences.

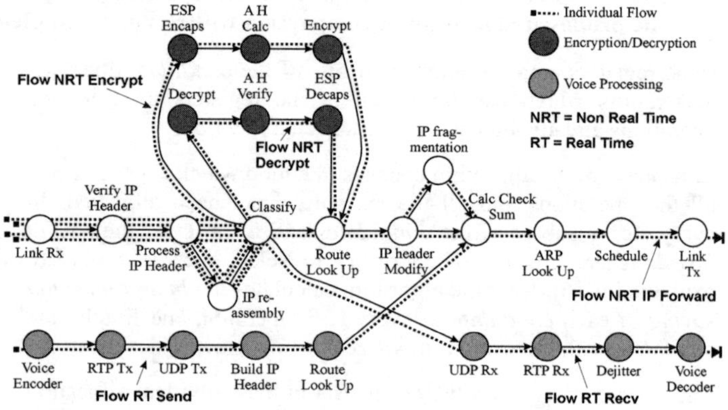

Figure 1: Control flow graph of a code implementing parts of a network packet processor

As an example, Figure 1 shows on a very high level of abstraction, the control flow graph of a code, implementing parts of an embedded network packet processor. The vertices in this graph represent different packet processing functions and get triggered by incoming packets (external events) or by preceding vertices when they complete execution and are ready to forward the packet for further processing. Such a packet processor may have several input ports through which packets flow in, and after being processed they are put out on the appropriate output ports. The code (whose control flow graph is shown in the figure) corresponding to each input port is responsible for handling packets flowing through that port, and all such blocks of code execute concurrently. The schedulability analysis of a collection of such graphs is necessary to determine if all packets belonging to real-time flows (such as voice) can be processed within their respective deadlines.

In this example and also in many other embedded systems scenarios, due to constraints on memory and also due to efficiency reasons the number of preemptions allowed is restricted. This is because of the usually large overhead

DAC 2002, June 10-14, 2002, New Orleans, Louisiana, USA.

involved in preemptions. Hence, once a vertex of the control flow graph in the above example starts executing on a processor, it is required to continue till completion before another vertex (probably from a different graph) can be scheduled for execution. The advantages of preemption in such cases are usually offset by the large overhead involved in preempting a process which has only partially processed a packet.

Conditional real-time code. The main difficulty in the schedulability analysis (both for preemptive and non-preemptive cases) of a collection of such control flow graphs lies in the fact that for general directed acyclic graphs, what constitutes a worst case event triggering sequence for an individual graph can not be determined in isolation, due to the presence of conditional branches. To illustrate this, consider our next example.

```
while (external event) do
  execute code block B0          /* (e0, d0) */
  if (C) then
    execute code block B1        /* (e1, d1) */
  else
    execute code block B2        /* (e2, d2) */
  end if
end while
```

In the above code, for each code block B, the tuples (e, d) enclosed within the comments indicate the execution requirement and the deadline of B. Now, if the condition C depends on some external event, or on the value of a variable which can not be determined at compile time, then the worst case branch here would depend on the other blocks of code executing concurrently with this one. Let $e_1 = 2$, $d_1 = 2$, $e_2 = 4$ and $d_2 = 5$. If another code block is simultaneously executing with $e = 1$ and $d = 1$ then the (e_1, d_1) branch corresponds to the worst case, whereas if $e = 2$ and $d = 5$ then the (e_2, d_2) branch corresponds to the worst case. Hence the usual method followed for the feasibility analysis of hard-real-time systems, of approximating a piece of code by its worst case behavior does not work in the presence of conditional branches. The alternative, which involves enumerating all possible execution paths in the control flow graph, leads to exponential complexity.

Previous results. There is a large body of work on modeling real-time embedded systems and on answering scheduling theoretic questions arising in these models (see [1] for an overview). Whereas most of the previous work considered independently executing tasks, of late there has been a considerable amount of work on trying to model data and control dependencies between tasks and addressing scheduling issues in such models (see [11] and the references therein). Very recently, a new model called the *recurring real-time task model* was proposed in [2, 4] for modeling codes with conditional branches as shown in the examples above. It generalizes many of the previous models like the sporadic [9], multiframe [10], generalized multiframe [5], and recurring branching [3]. However, the algorithms presented in [2] for the feasibility analysis problem in this model for the preemptive uniprocessor case, have a running time which is exponential in the number of vertices of the task graphs, and the complexity of this problem was undecided. Recently it was proved that this problem is NP-hard [7].

Our results. Relatively little is known about the non-preemptive version of the problem, where tasks are triggered by external events and there exist control dependencies between them as shown with our network packet processor example. All the work in this area (see [8] and the references therein) is based on heuristics and no exact test for schedulability is known till now. Moreover, in view of the recent NP-hardness result for the preemptive recurring real-time task model, the non-preemptive version is very likely to be NP-hard as well.

In this paper we study the non-preemptive version of the recurring real-time task model, for modeling a set of concurrently executing event-driven code blocks with real-time constraints. Our main contributions are that we give an exact necessary and sufficient condition for schedulability under Earliest-Deadline-First (EDF), and show that this condition can be efficiently tested. Towards this, we give a pseudo-polynomial time algorithm and efficient approximation schemes which involve a trade-off between the running time of the algorithms and the quality of the results produced. We validate our analytical results using experiments and show that for all practical purposes the schedulability analysis of a collection of such code blocks can be accurately and efficiently done. For the ease of presentation, the model we consider here is slightly simpler than that of [2] in the sense that we do not consider the recurring behavior of the code blocks. Using techniques described in [2] it is possible to extend our results to incorporate this recurring behavior and we postpone the details of this to a full version of this paper.

Apart from the results that we present here for EDF, it is also possible to derive a sufficient condition for fixed priority scheduling and efficiently test this condition. Due to space restrictions, this result and some other proofs are omitted here and can be found in [6]. In the next section we formally describe the model. Section 3 presents the schedulability test for EDF, following which we describe our approximation schemes in Section 4. Finally, Section 5 describes our experimental results.

2. THE TASK MODEL

A task modeling a block of code is represented by a directed acyclic graph with a unique source and a sink vertex. Associated with each vertex v is its execution requirement $e(v)$ (which can be previously determined, for example at compile time), and deadline $d(v)$. Whenever the vertex v is *triggered*, the code corresponding to it has to be executed (which takes $e(v)$ amount of time) within the next $d(v)$ time units. Since we consider a non-preemptive environment, once a vertex has started execution it can not be preempted, and continues executing till completion. After it completes, another vertex which has already been triggered, possibly belonging to a different task graph, can be scheduled for execution.

Each directed edge (u, v) in the graph is associated with a minimum intertriggering separation $p(u, v)$, denoting the minimum amount of time that must elapse before the vertex v can be triggered after the triggering of the vertex u, and $p(u, v) \geq d(u)$.

The semantics of the execution of such a task graph state that the source vertex can be triggered at any time, and once a vertex u is triggered then the next vertex v can be triggered only if there exists a directed edge (u, v) and at least $p(u, v)$ amount of time has elapsed since the triggering of u. If there are directed edges (u, v_1) and (u, v_2) from the vertex u (representing a conditional branch) then only one among v_1 and v_2 can be triggered, after the triggering of u. Therefore, a sequence of vertices $v_1, v_2, \ldots, v_k$ getting triggered at time instants $t_1, t_2, \ldots, t_k$ is legal if and only if there are directed edges (v_i, v_{i+1}) and $t_{i+1} - t_i \geq p(v_i, v_{i+1})$ for $i = 1, \ldots, k-1$. The real-time constraints require that the code corresponding to vertex v_i be executed within the time interval $[t_i, t_i + d(v_i)]$.

Task sets and schedulability analysis. A task set $\mathcal{T} = \{T_1, T_2, \ldots, T_k\}$ consists of a collection of task graphs, the vertices of which can get triggered independently of each other. A triggering sequence for such a task set $\mathcal{T}$ is legal if and only if for every task graph T_i, the subset of vertices of the sequence belonging to T_i constitutes a legal triggering sequence for T_i. In other words, a legal triggering sequence for $\mathcal{T}$ is obtained by merging together (ordered by triggering times, with ties broken arbitrarily) legal triggering sequences of the constituting tasks.

The schedulability analysis of a task set $\mathcal{T}$ is concerned with determining whether for all possible legal triggering sequences of $\mathcal{T}$, the codes corresponding to the vertices of the task graphs can be scheduled such that all their associated deadlines are met. As already mentioned before, here we are interested in the non-preemptive uniprocessor version of this problem.

3. SCHEDULABILITY ANALYSIS FOR EDF

An EDF scheduler always selects a ready job with the shortest deadline for execution and is known to be optimal under preemption. In the non-preemptive case EDF is known to be optimal for independently executing jobs if the scheduler is work conserving or non-idle (i.e. if a job is ready then it has to be scheduled if the processor is empty). In this section we derive an exact necessary and sufficient condition for the schedulability of a set of task graphs under EDF. Not surprisingly, we show that for our task model EDF is also an optimal non-preemptive work conserving scheduler.

3.1 Demand-Bound Function ($T.dbf(t)$)

Our schedulability analysis is based on an abstraction of a task, represented by a function called the *demand-bound function*. The demand-bound function of a task T, denoted by $T.dbf(t)$, takes as an argument a real number t and returns the maximum possible cumulative execution requirement by vertices of T that have been triggered by a legal triggering sequence and have both their ready times and deadlines within a time interval of length t. Intuitively, $T.dbf(t)$ denotes the maximum possible execution requirement that can possibly be demanded by T within any time interval of length t, if all its vertices are to meet their deadlines. As

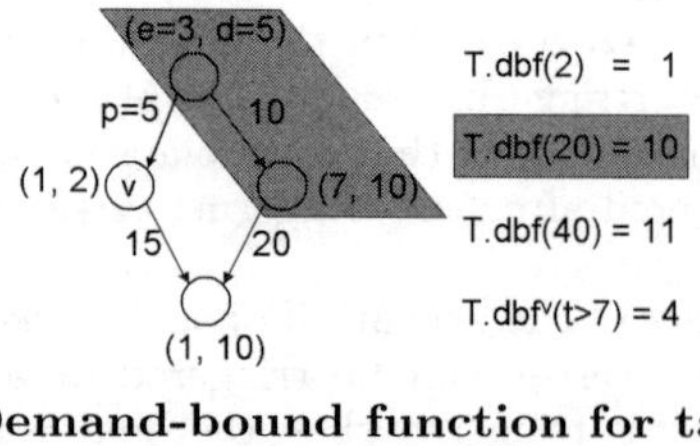

Figure 2: Demand-bound function for task graph T

an example, consider the task graph T shown in Figure 2. For this graph, $T.dbf(2) = 1$ because the vertex having an execution requirement of 1 and deadline 2, can trigger at the beginning of any time interval of length 2 and has an execution requirement of 1 if it has to meet its deadline. Similarly, $T.dbf(20) = 10$ because of a possible triggering of the two shaded vertices in the graph within any interval of length 20.

In addition to $T.dbf(t)$, we denote by $T.dbf^v(t)$, the maximum execution requirement demanded by T within any time interval of length t, due to any triggering sequence ending at the vertex v.

3.2 Conditions for Schedulability

In this section we give a necessary and sufficient condition for the schedulability of a set of task graphs under EDF scheduling. This condition is specified by Algorithm 1, which uses the two functions $T.dbf(t)$ and $T.dbf^v(t)$ introduced in the last subsection.

Algorithm 1 Algorithm for schedulability analysis under EDF

Input: Task set $\mathcal{T}$
$decision \leftarrow YES$
for all tasks $T_i \in \mathcal{T}$ and **for all** vertices $v \in T_i$ and **for all** $\hat{\tau} \geq 0$ **do**
  Let $\tilde{\mathcal{T}} \leftarrow \mathcal{T} \backslash \{T_i\}$
  $\mathcal{T}_{dbf=0} \leftarrow \{T \in \tilde{\mathcal{T}} \mid T.dbf(\hat{\tau} + d(v)) = 0\}$
  $e_{max} \leftarrow \max_{v'}\{e(v') \mid v'$ is a vertex of a task $T \in \mathcal{T}_{dbf=0}\}$
  Let $\mathcal{T}_{dbf>0} \leftarrow \{T \in \tilde{\mathcal{T}} \mid T.dbf(\hat{\tau} + d(v)) > 0\}$ and $q \leftarrow |\mathcal{T}_{dbf>0}|$
  $index \leftarrow 0$
  for $p = 1$ to q **do**
    Let $e'_{max} \leftarrow \max\{e(v') \mid v' \in T_p,\ d(v') > \hat{\tau} + d(v)\}$
    if $index = 0$ **then**
      if $e'_{max} > (T_p.dbf(\hat{\tau} + d(v)) + e_{max})$ **then**
        $e_{max} \leftarrow e'_{max}$
        $index \leftarrow p$
      end if
    else
      if $e'_{max} + T_{index}.dbf(\hat{\tau} + d(v)) > (T_p.dbf(\hat{\tau} + d(v)) + e_{max})$ **then**
        $e_{max} \leftarrow e'_{max}$
        $index \leftarrow p$
      end if
    end if
  end for
  if $index \neq 0$ **then**
    $\hat{\mathcal{T}} \leftarrow \mathcal{T}_{dbf>0} \backslash \{T_{index}\}$
  end if
  if $\hat{\tau} + d(v) < (T_i.dbf^v(\hat{\tau} + d(v)) + \sum_{T \in \hat{\mathcal{T}}} T.dbf(\hat{\tau} + d(v)) + e_{max})$ **then** /* Condition (†) */
    $decision \leftarrow NO$
  end if
end for
return $decision$

THEOREM 1. *A task set $\mathcal{T}$ is schedulable under EDF if and only if Algorithm 1 returns YES.*

PROOF. Let v be any vertex of a task graph $T_i \in \mathcal{T}$. The vertex v has an execution requirement of $e(v)$ and a deadline equal to $d(v)$. Let v be triggered at time t and it completes execution at time $t + \delta$.

Let $R_{\mathcal{T}}^{\leq d}[t, t+\tau]$ denote the sum of the execution requirements of the vertices of any task graph $T \in \mathcal{T}$ which have been triggered in the time interval $[t, t+\tau]$ and which have their deadlines less than or equal to d. Let $W^{v,t}(t+\tau)$ $(0 \leq \tau \leq \delta)$ denote the total execution requirement at time

$t + \tau$ that was generated by all the tasks in $\mathcal{T}$, and which must be met by the processor (under EDF scheduling) before the vertex v that was triggered at time t can complete its execution. $W^{v,t}(t+\tau)$ includes the execution requirement $e(v)$ of the vertex v as well. We assume that the processor was idle before time 0.

If we look back in time, let $t-\hat{\tau}$ be the first time before the time instant t when the processor does not have any vertex to execute with deadline less than or equal to $t+d(v)$ (i.e. the deadline of the vertex v). Hence, during the entire interval $[t-\hat{\tau}, t+\delta)$, the processor always has some vertex to execute with deadline less than or equal to $t + d(v)$. $W^{v,t}(t + \tau)$ for any $0 \leq \tau \leq \delta$ is therefore composed of the following: (1) The remaining execution requirement of the vertex that is in execution at time $t - \hat{\tau}$, denoted by $P(t - \hat{\tau})$. By our assumption of $\hat{\tau}$, the deadline of this vertex is greater than $t + d(v)$. (2) The execution requirement generated by the vertices of the task T_i during the time interval $[t - \hat{\tau}, t]$. This includes the vertex v. Clearly, all these vertices have a deadline less than or equal to $t + d(v)$. Therefore, this equals to $R_{T_i}^{\leq t+d(v)}[t - \hat{\tau}, t]$. (3) The execution requirement generated by vertices with deadlines less than or equal to $t + d(v)$, from all tasks belonging to a set, say $\hat{\mathcal{T}}$, where $\hat{\mathcal{T}} \subseteq \mathcal{T}\backslash\{T_i\}$, during the time interval $[t-\hat{\tau}, t+\tau]$. Therefore, this is equal to $\sum_{T\in\hat{\mathcal{T}}} R_T^{\leq t+d(v)}[t-\hat{\tau}, t+\tau]$. (4) The execution requirement served by the processor during the time interval $[t - \hat{\tau}, t + \tau]$.

Since we are considering a non-preemptive environment, the vertex which is in execution at the time $t - \hat{\tau}$ has to finish executing before any vertex having a deadline less or equal to $t + d(v)$ can be executed. Therefore, the processor always executes some vertex having a deadline less than or equal to $t + d(v)$ during the interval $[t - \hat{\tau} + P(t - \hat{\tau}), t + \tau]$. Hence,

$$W^{v,t}(t+\tau) = P(t-\hat{\tau}) + R_{T_i}^{\leq t+d(v)}[t-\hat{\tau}, t] + \sum_{T\in\hat{\mathcal{T}}} R_T^{\leq t+d(v)}[t-\hat{\tau}, t+\tau] - (\hat{\tau}+\tau) \quad (1)$$

Now, note that if there exists a τ $(0 \leq \tau \leq d(v))$ such that $W^{v,t}(t+\tau) = 0$, then the vertex v completes execution on or before its deadline. Substituting $\tau = d(v)$ in Equation (1), we obtain:

$$W^{v,t}(t+d(v)) = P(t-\hat{\tau}) + R_{T_i}^{\leq t+d(v)}[t-\hat{\tau}, t] + \sum_{T\in\hat{\mathcal{T}}} R_T^{\leq t+d(v)}[t-\hat{\tau}, t+d(v)] - \hat{\tau} - d(v)$$

Following our definition of the *demand-bound functions* (*dbf* and *dbf*^{v}), clearly,

$$W^{v,t}(t+d(v)) \leq P(t-\hat{\tau}) + T_i.dbf^v(\hat{\tau}+d(v)) + \sum_{T\in\hat{\mathcal{T}}} T.dbf(\hat{\tau}+d(v)) - \hat{\tau} - d(v) \quad (2)$$

To compute an upper bound on $W^{v,t}(t + d(v))$ we would like to maximize the right hand side of the above inequality (2). For this, note that if a vertex v' of a task T contributes to the term $P(t - \hat{\tau})$, then T can not belong to the set $\hat{\mathcal{T}}$. Following this constraint, for any task T_i and any vertex $v \in T_i$, Algorithm 1 computes $P(t - \hat{\tau}) = e_{max}$ and the task set $\hat{\mathcal{T}}$ which maximizes the right hand side of Inequality (2). Therefore, if the algorithm returns YES, then we have (from Condition (†) of the algorithm),

$$W^{v,t}(t + d(v)) \leq \hat{\tau} + d(v) - (\hat{\tau} + d(v)) = 0$$

Hence, there exists a $\tau \leq t+d(v)$ such that $W^{v,t}(t+d(v)) \leq 0$ and therefore the vertex v completes execution before its deadline.

The proof of necessity is based on similar arguments. The interested reader is referred to [6]. □

Note that Step 2 in Algorithm 1 involves a loop over all possible values of $\hat{\tau} \geq 0$. However, it suffices to consider only a finite set of $\hat{\tau}$s and this is explained at the end of Section 4.2. The optimality of EDF in this model follows from the fact that the proof of necessity makes no assumptions about the scheduling discipline.

4. APPROXIMATE SCHEDULABILITY ANALYSIS

The demand bound function of a task graph can clearly be computed by enumerating all possible paths in the graph and computing the execution requirement and deadline corresponding to each path. In the worst case, since the number of such paths can be exponential in the number of vertices in the graph, this procedure will incur an exponential running time. It can be shown by a reduction from the knapsack problem that computing $T.dbf(t)$ for a task graph T is NP-hard, implying that our algorithm for schedulability analysis can have a worst case running time which is exponential in the number of vertices in any task graph.

In this section we first show that $T.dbf(t)$ for any task graph can be efficiently approximated. Towards this we give a fully-polynomial time approximation scheme (FPTAS) for computing $T.dbf(t)$. Using this result we then give approximate decision algorithms for schedulability analysis.

4.1 Approximating the Demand-Bound Function

Given a task graph T we first give a pseudo-polynomial time algorithm for computing $T.dbf(t)$ for any $t \geq 0$, based on dynamic programming. Let there be n vertices in T denoted by $v_1, \ldots, v_n$, and without any loss of generality we assume that there can be a directed edge from v_i to v_j only if $i < j$. Following our notation described in Section 2, associated with each vertex v_i is its execution requirement $e(v_i)$ which here is assumed to be integral (a pseudo-polynomial algorithm is meaningful only under this assumption), and its deadline $d(v_i)$. Associated with each edge (v_i, v_j) is the minimum intertriggering separation $p(v_i, v_j)$.

Let $t_{i,e}$ be the minimum time interval within which the task T can have an execution requirement of exactly e time units due to some legal triggering sequence, considering only a subset of vertices from the set $\{v_1, \ldots, v_i\}$, if all the triggered vertices are to meet their respective deadlines. Let $t_{i,e}^i$ be the minimum time interval within which a sequence of vertices from the set $\{v_1, \ldots, v_i\}$, and ending with the vertex v_i, can have an execution requirement of exactly e time units, if all the vertices have to meet their respective deadlines. Lastly, let $E = \max_{i=1,\ldots,n} e(v_i)$. Clearly, nE is an upper bound on $T.dbf(t)$ for any $t \geq 0$. It can be trivially shown by induction that Algorithm 2 correctly computes $T.dbf(t)$, and has a running time of $O(n^3E)$.

Given this algorithm, any $t \geq 0$, and an $0 < \varepsilon \leq 1$, let T_t be the subgraph of T consisting only of those vertices v_i for

Algorithm 2 Computing $T.dbf(t)$

Input: Task graph T, and a real number $t \geq 0$

for $e \leftarrow 1$ to nE **do**

$t_{1,e} \leftarrow \begin{cases} d(v_1) & \text{if } e(v_1) = e \\ \infty & \text{otherwise} \end{cases}$

$t^1_{1,e} \leftarrow t_{1,e}$

end for

for $i \leftarrow 1$ to $n-1$ **do**

for $e \leftarrow 1$ to nE **do**

Let there be directed edges from the vertices $v_{i_1}, v_{i_2}, \ldots, v_{i_k}$ to v_{i+1}

$t^{i+1}_{i+1,e} \leftarrow \begin{cases} \min\{t^{i_j}_{i_j, e-e(v_{i+1})} - d(v_{i_j}) + p(v_{i_j}, v_{i+1}) + \\ d(v_{i+1}) \mid j = 1, \ldots, k\} \text{ if } e(v_{i+1}) < e, \\ d(v_{i+1}) \text{ if } e(v_{i+1}) = e, \text{ and } \infty \text{ otherwise} \end{cases}$

$t_{i+1,e} \leftarrow \min\{t_{i,e}, t^{i+1}_{i+1,e}\}$

end for

end for

$T.dbf(t) \leftarrow \max\{e \mid t_{n,e} \leq t\}$

which $d(v_i) \leq t$, and let E_t denote the maximum execution requirement of a vertex from among all vertices of T_t. Now we scale all the execution requirements associated with the vertices of T_t by $K = \varepsilon E_t/n$ i.e. $e'(v_i) = \lfloor e(v_i)/K \rfloor$ and run Algorithm 2 with the new $e'(v_i)$s and the graph T_t. Let V be the set of vertices (with the scaled execution requirements) that result in the computation of $T.dbf(t)$ in this algorithm. We claim that the summation of the original (unscaled) execution requirements of these vertices is greater than or equal to $(1-\varepsilon)$ times the actual demand-bound function for the task graph for this value of t. Further, since this algorithm now runs in time $O(n^4/\varepsilon)$, (with the scaled execution requirements), it is an FPTAS for computing $T.dbf(t)$. We denote this approximate value of $T.dbf(t)$ computed by this algorithm by $T.dbf'(t)$.

4.2 Approximate Decision Algorithms

Our approximate decision algorithms use the approximate demand-bound function ($T.dbf'(t)$) introduced in the last subsection, instead of the exact values $T.dbf(t)$, in Algorithm 1. The decision algorithms are parameterized by $0 < \varepsilon \leq 1$, where the number of wrong answers and the running time depend on the value of ε chosen. For smaller values of ε the percentage of possible wrong answers decrease, but at the expense of the running time of the algorithm.

Using the same notation as in the last subsection, note that for all possible values of $t \geq 0$, there can be at most n distinct values of E_t for any task graph. For each such E_t, we consider the corresponding subgraph that gives rise to this E_t as described above, and scale the execution requirements of the vertices of this subgraph by $K = \varepsilon E_t/n$. In each such subgraph T_t, the number of values of time intervals t' at which the value of $T_t.dbf'(t')$ changes is bounded by $O(n^2/\varepsilon)$, and hence the number of values of time intervals t at which the value of $\sum_{T \in \mathcal{T}} T.dbf'(t)$ changes is bounded by $O(|\mathcal{T}|n^3/\varepsilon)$.

It follows that in Step 2 of Algorithm 1 it is sufficient to run the loop only for $O(|\mathcal{T}|n^3/\varepsilon)$ values of $\hat{\tau}$, since the value of $\sum_{T \in \mathcal{T}} T.dbf'(\hat{\tau})$ can change at most this number of times. Therefore, the loop in Step 2 executes for a total of $O(|\mathcal{T}|^2 n^4/\varepsilon)$ times.

For each task $T \in \mathcal{T}$, computing the $t_{n,e}$ values for each of its subgraphs T_t, using Algorithm 2 and the scaled execution requirements requires $O(n^4/\varepsilon)$ time, and these values are stored in a table. Hence computing all such values for all the task graphs in $\mathcal{T}$ takes $O(n^5|\mathcal{T}|/\varepsilon)$ time. The Step 25 in the algorithm dominates the running time among all the steps inside the loop (Steps 3-27), and requires a computation of $T_i.dbf'^v(t) + \sum_{T \in \hat{\mathcal{T}}} T.dbf'(t) + e_{max}$ where $t = \hat{\tau} + d(v)$. To compute this, note that computing $T.dbf'(t)$ for any $T \in \mathcal{T}$ requires a binary search to identify the appropriate table corresponding to a subgraph T_t, and then a linear search through this table. Therefore, this requires $O(n^2\varepsilon^{-1}\log n)$ time. The exactly same time is for computing $T.dbf'^v(t)$. Hence, computing the value of $T_i.dbf'^v(t) + \sum_{T \in \hat{\mathcal{T}}} T.dbf'(t) + e_{max}$ for $t = \hat{\tau} + d(v)$ in Step 25 requires a total of $O(|\mathcal{T}|n^2\varepsilon^{-1}\log n)$ time. Therefore, the total run time of Algorithm 1 using the approximate demand-bound functions is $O(|\mathcal{T}|^3 n^6 \varepsilon^{-2} \log n)$.

Since $T.dbf'(t) \leq T.dbf(t)$ for any $t \geq 0$, this algorithm is overly *optimistic*, in the sense that for certain task sets which are not schedulable, the algorithm might still return a YES. However, for task sets where some vertices might miss their deadlines by a large time lengths, the algorithm always returns a NO. So the algorithm errs only for task sets where some vertices might miss their deadlines by "small" amounts of time and this can be parametrized by ε. Therefore, any $0 < \varepsilon \leq 1$ characterizes a class of task sets for which the algorithm errs. Decreasing ε reduces this class of such task sets for which the algorithm errs, at the cost of increasing the running time quadratically in $1/\varepsilon$, and therefore this gives a fully polynomial-time approximate decision scheme for approximate feasibility testing.

It may be shown that for any $t \geq 0$, $T.dbf'(t) + \varepsilon E_t \geq T.dbf(t)$. Hence, using $(T.dbf'(t) + \varepsilon E_t)$ instead of $T.dbf(t)$ in Algorithm 1 results in a *pessimistic* algorithm with analogous properties as in our optimistic algorithm.

A Pseudo-Polynomial Time Algorithm. Lastly, it may be noted that Algorithm 1 along with the pseudo-polynomial time algorithm for computing the demand-bound function of a task graph also implies a pseudo-polynomial time algorithm for schedulability analysis. To see this, let for any task $T \in \mathcal{T}$, t^T_{max} denote the maximum amount of time elapsed among all execution sequences starting from the source vertex of T and ending at the sink vertex, if every vertex is triggered at the earliest possible time (respecting the minimum intertriggering separations). Let $t_{max} = \max_{T \in \mathcal{T}} t^T_{max}$. Clearly, it is sufficient to test the Condition (†) in Algorithm 1 only for $\hat{\tau} = 1, \ldots, t_{max}$. Both $T.dbf^v(\hat{\tau} + d(v))$ and $T.dbf^v(\hat{\tau} + d(v))$ in the Step 25 of the algorithm for any $\hat{\tau}$ can be determined in pseudo-polynomial time by Algorithm 2 and clearly, t_{max} is pseudo-polynomially bounded, implying a pseudo-polynomial algorithm for schedulability analysis.

5. EXPERIMENTAL RESULTS

In spite of the theoretical guarantees in our algorithms the experiments reported here are interesting because of two reasons. Firstly, many approximations schemes are exceedingly difficult to implement and in practice might have running times which are comparable or even worse than the equivalent simpler exponential time algorithms, for practical input instances. Secondly, the parameter ε in our algorithms represents a trade-off between the quality of the results obtained and the running time. Hence, it is interesting to identify a suitable value of ε for any realistic input instance.

We have implemented the pseudo-polynomial exact algo-

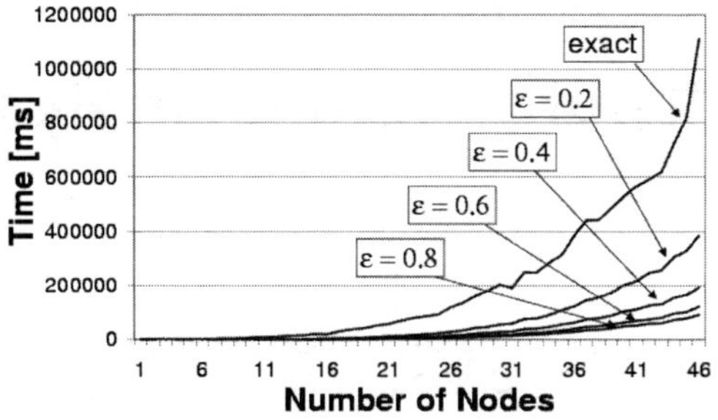

Figure 3: Running time versus the number of vertices in the task graphs when $E = 500$

rithm for feasibility analysis, and also the approximation scheme. For our experiments, we have randomly generated synthetic task graphs, using two parameters. The first is the maximum execution requirement associated with any vertex of the graph, E, which effects the running time of the pseudo-polynomial time algorithm and the quality of the results generated by the approximation scheme. We call the second parameter the *connectivity factor*. If $v_1, \ldots, v_n$ are the vertices of a task graph such that there is an edge from v_i to v_j only if $j > i$, then for each vertex v_j we construct an edge from v_i to v_j with a probability equal to the connectivity factor of the graph, for $i = 1, \ldots, j-1$.

Figure 3 shows the running time of the exact pseudo-polynomial algorithm and the approximation scheme for four different values of ε on a task set consisting of three graphs, when the number of vertices in each of these graphs is gradually increased. The maximum execution requirement (E) associated with any vertex was set to 500. The connectivity factor in all the graphs was set to 0.4; this generates realistic control flow graphs [6]. The CPU time was measured on a moderately loaded Sunblade 1000 running SunOS 5.8 with 750 MHz CPU and 2 GB RAM. All the algorithms were implemented in Java.

Note that for any set of task graphs, the optimal choice of ε depends on the maximum execution requirement E, associated with any vertex. For instance, in the above example if $E = 100$ then the performance of the exact algorithm is better than the approximation scheme with $\varepsilon = 0.2$. To give an example of the values of E that might occur in practice, typical execution requirements associated with the vertices in Figure 1 (for the voice flow) are 300, 190, 640 and 500 μsecs for UDP Tx, Build IP Header, Route Look Up, and ARP Look Up respectively when these are implemented on a DSP like TMS320C620x. These values are 270, 130, 420 and 330 μsecs respectively when implemented on a RISC processor like ARM9TDMI.

Figure 4 shows the exact value of the demand-bound function $T.dbf(t)$ computed by the pseudo-polynomial algorithm, and its upper and lower bounds ($T.dbf'(t)+\varepsilon E$ and $T.dbf'(e)$ respectively) computed by the approximation scheme. It should be noted that the value of $T.dbf'(t)$ for all values of t is almost equal to $T.dbf(t)$, and this is better than the worst case theoretical bound. The values shown in this graph are for a task graph with $E = 1000$, $\varepsilon = 0.6$ and the number of vertices in the task graph and the connectivity being equal to 30 and 0.4.

Lastly, the percentage of wrong answers returned by the approximation scheme for different values of ε are given by $(0.2, 5\%)$, $(0.4, 6\%)$, $(0.6, 11\%)$ and $(0.8, 15\%)$. Each of the tuples indicate the value of ε and the percentage of wrong answers. For this we have used 100 task sets, each consisting of three task graphs with 30 vertices in each graph and having a connectivity factor of 0.4. The maximum execution requirement of any vertex was set to 100. All the task sets considered here either almost fully load the processor, or when not schedulable, the vertices miss their deadlines only by small amounts of time, and therefore these represent the difficult cases for our approximation algorithms.

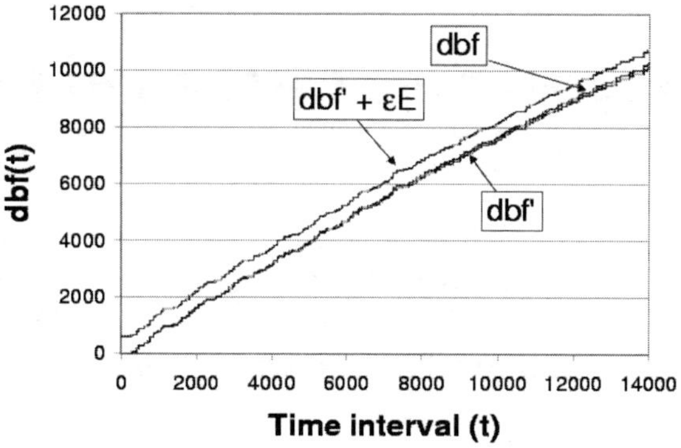

Figure 4: The demand-bound function ($T.dbf(t)$) and the upper and lower bounds on its approximation

6. CONCLUSION

In this paper we have considered the schedulability analysis of a collection of event-driven real-time code blocks. Although this problem is very likely to be computationally difficult (NP-hard) we have shown that for all practical purposes it can be efficiently solved. For fixed priority schedulers only a sufficient condition for schedulability is known [6]. Here it would be interesting to come up with a test which is both necessary and sufficient.

7. REFERENCES

[1] F. Balarin, L. Lavagno, P. Murthy, and A. Sangiovanni-Vincentelli. Scheduling of embedded real-time systems. *IEEE Design and Test of Computers*, pages 71–82, January-March 1998.

[2] S. Baruah. Dynamic- and static-priority scheduling of recurring real-time tasks. To appear in Real-Time Systems.

[3] S. Baruah. Feasibility analysis of recurring branching tasks. In *Proc. 10th Euromicro Workshop on Real-Time Systems*, pages 138–145, 1998.

[4] S. Baruah. A general model for recurring real-time tasks. In *Proc. IEEE RTSS*, pages 114–122, 1998.

[5] S. Baruah, D. Chen, S. Gorinsky, and A.K. Mok. Generalized multiframe tasks. *Real-Time Systems*, 17(1):5–22, 1999.

[6] S. Chakraborty, T. Erlebach, S. Künzli, and L. Thiele. Schedulability of event-driven code blocks in real-time embedded systems. Technical Report TIK 130, ETH Zürich, 2002.

[7] S. Chakraborty, T. Erlebach, and L. Thiele. On the complexity of scheduling conditional real-time code. Technical Report TIK 107, ETH Zürich, 2001.

[8] P. Eles *et al.* Scheduling of conditional process graphs for the synthesis of embedded systems. In *Proc. Design, Automation and Test in Europe*, 1998.

[9] A.K. Mok. *Fundamental Design Problems of Distributed Systems for the Hard-Real-Time Environment.* PhD thesis, LCS, MIT, 1983.

[10] A.K. Mok and D. Chen. A multiframe model for real-time tasks. *IEEE Transactions on Software Engineering*, 23(10):635–645, 1997.

[11] P. Pop, P. Eles, and Z. Peng. Schedulability analysis for systems with data and control dependencies. In *Proc. 12th Euromicro Conference on Real-Time Systems*, 2000.

Associative Caches in Formal Software Timing Analysis

Fabian Wolf*
Volkswagen AG, Wolfsburg, Germany
fabian.wolf@volkswagen.de

Jan Staschulat
Technical University of Braunschweig, Germany
jans@ida.ing.tu-bs.de

Rolf Ernst
Technical University of Braunschweig, Germany
ernst@ida.ing.tu-bs.de

ABSTRACT

Precise cache analysis is crucial to formally determine program running time. As cache simulation is unsafe with respect to the conservative running time bounds for real-time systems, current cache analysis techniques combine basic block level cache modeling with explicit or implicit program path analysis. We present an approach that extends instruction and data cache modeling from the granularity of basic blocks to program segments thereby increasing the overall running time analysis precision. Data flow analysis and local simulation of program segments are combined to safely predict cache line contents for associative caches in software running time analysis. The experiments show significant improvements in analysis precision over previous approaches on a typical embedded processor.

Categories and Subject Descriptors

C.3 [**Special-Purpose and Application-Based Systems**]: [Real-time and embedded systems]; D.2.4 [**Software Engineering**]: Software/Program Verification—*formal methods*; C.4 [**Performance of Systems**]: Performance Attributes

General Terms

Algorithms, Performance, Verification

Keywords

Cache Analysis, Embedded Software, Real-Time, Timing Analysis

1. INTRODUCTION

Embedded software running time is typically not unique but is bound by intervals which result from data dependent behavior, environment timing and target system properties. In system design automation, these intervals are used in many ways. In some cases, only the worst case is of interest, e.g. for single processor schedulability analysis, in another context both best and worst cases are relevant, such as for multiprocessor scheduling analysis or for variable life time analysis as used for memory assignment. In all these cases, running time intervals of the individual software processes are fundamental data needed to analyze system behavior. Profiling and simulation tools such as [9] are the state-of-the-art in industry but since exhaustive simulation is impractical regarding test patterns selection for the conservative running time bounds, simulation results can only cover part of the system behavior.

With growing importance of embedded software, formal analysis of running time intervals has met increasing interest in the EDA and real-time systems communities. Major improvements were the introduction of implicit path enumeration and the inclusion of cache analysis [8]. In [2, 10], data cache analysis has been included using abstract interpretation. While all approaches are conservative, i.e. all possible timing behavior is included in the resulting intervals, the main differences are in the architecture features that are covered by the hardware model and the width of the conservative interval. The closer this interval to the real running time bounds, the higher is the practical use of formal analysis.

Cache hits or misses depend on the mapping of programs and data to memory, and dependencies can span a whole process or even several process executions. This makes simulation particularly risky since the user cannot anticipate the effects of optimization and address mapping in compilers and linkers and, therefore, cannot provide the necessary simulation patterns to test critical cases. On the other hand, cache performance is too significant to be neglected in running time analysis of software on advanced target architectures. Previous research approaches for running time and cache analysis work on basic blocks as elements. These approaches identify the cache lines used by a basic block and compare the different basic block cache tables to identify upper and lower bounds for the number of cache line replacements, hits and misses per basic block execution. Then, based on implicit or explicit program path analysis, bounds on the total cache hits and misses are determined. These approaches may get too expensive because the number of basic blocks can be very high.

In earlier papers [13], we have demonstrated how to extend analysis elements from basic blocks to larger program segments thereby significantly improving both path analysis and architecture modeling precision. The approach is based on local program segment execution, i.e. cycle true simulation. These program segments may contain loops where cache lines are repeatedly replaced by others. Therefore, table comparison alone is not sufficient. In [12] we have presented a preliminary technique which combines local cache tracing with global data flow analysis. In this previous approach we have only determined bounds for the number of cache

*This work was performed for the Technical University of Braunschweig, Germany. Since September 2001, Fabian Wolf has been working for Volkswagen AG, Wolfsburg, Germany.

misses for simple direct mapped caches. In this paper, we present significant improvements to this approach. We extend it to set associative caches and consider the impact of the determined cache hits and misses on the running time interval of a process. We consider the majority of effects of compiler optimizations in static software analysis and present detailed experimental results.

The rest of this paper is structured as follows: Previous work is reflected in section 2. Local cache simulation for program segments is explained in section 3 before the methodology for data flow analysis based cache line content prediction is introduced in section 4. The experiments are presented in section 5 before we conclude in section 6.

2. PREVIOUS WORK

Trace based cache simulators have been used in modeling computer architecture for many years. Due to the well-known drawbacks of simulation for real-time guarantees, research approaches focus on static cache analysis. Some of the most important approaches are explained in the following.

In the running time analysis framework in [8], the actual state of the instruction cache depending on the flow of the program is modeled. A simple conflict graph for the cache lines referenced by basic blocks is sufficient for direct mapped caches while associative cache analysis requires a more complex cache state transition (CSTG) graph. The CSTG is used in a so called implicit program path enumeration, an efficient technique that maps timing analysis to a min/max optimization problem which is then solved by integer linear programming (ILP). For higher associativities, the problem becomes prohibitively complex because the problem size grows more that exponentially with the number of basic blocks mapped to a set. The integer linear programming (ILP) problem that computes the running time interval on the process level is already NP-complete. Moreover, implicit enumeration does not exclude many of the false paths of a program and their resulting cache reference sequences, such that this method tends to result in wider bounds than necessary. This approach presented in [8] does not consider data caches.

In [5], a timing analysis tool focusing on the instruction cache analysis part is presented. The problem size is reduced by a clustering of basic blocks in the cache conflict graph mapped to the same cache line. This is an improvement, but cache lines have a very limited size, e.g. 16 instruction words, so the clusters cannot be very large. Data caches and access addresses are not analyzed.

The approach presented in [7] uses data flow analysis for the determination of data dependent access addresses. This can be very useful for the prediction of data cache behavior. The main assumption is that not all data accesses with complex address expressions have to be treated as cache misses. "Use-/define chains" [1] for access addresses are derived to determine whether address expressions are just depending on constants. In this case, the data access address can be predicted reducing the number of potential cache misses. The framework this approach is embedded in uses an architecture where the instruction and data word size are equal to the size of a cache line, so the improvements through spatial locality have a limited impact.

For many applications, the instruction cache achieves high hit rates for repetitive loops because the loop body is small enough to fit entirely into the cache. Only the first iteration will lead to cache misses while subsequent iterations will lead to hits. In the static cache analysis methodology in [3], a method to analyze the control flow of a program by statically categorizing the cache behavior of each instruction to *always hit*, *always miss*, *first hit* or *first miss* is introduced. The approach looks at the loop bodies starting at the innermost loop nest level. At each level, if the loops fits entirely into the cache, then only the first iteration leads to misses while the subsequent iterations lead to hits. If the loop does not entirely fit into the cache, portions of the loop will always miss the cache due to conflicts. The running time bounds for the loop can be determined. After analyzing the innermost loop, the next loop nest level can be considered. For input data dependent control structures, first miss scenarios have to be assumed for the following instructions. This can have a significant impact on loop nests. Path information relating different loop nests of the program is not considered, so the predicted bounds may be too conservative.

The approach in [2] describes static instruction and data cache analysis using abstract interpretation. It has been refined in [10]. Starting from the memory accesses of the program and their influence on the caches a collecting semantic is built. This collecting semantic assigns all possible cache states to every given memory access. The *program analyzer generator (PAG)* builds the control flow graph from the executable program under investigation. For every node in the control flow graph two analyzes are implemented. The *must*-analysis for the node in the control flow graph determines all memory blocks that must be in the cache for every possible execution of the program. The *may*-analysis for the node delivers all memory blocks that may be in the cache at this point. The memory accesses can be categorized to *always hit (ah)*, *always miss (am)* or *not classified (nc)*. For conservative running time analysis, (nc) are treated as (am) for the upper bound and as (ah) for the lower bound. This methodology is still based on the analysis granularity of single instructions or basic blocks.

3. LOCAL CACHE SIMULATION

In [12], an analysis technique for direct mapped caches has been proposed. It delivers tight bounds on the number of cache hits and misses for a single process on a given target architecture. We review and refine this basic approach, extend it to set associativity and describe the impact of the determined cache hits and misses on the process-level running time intervals. These are crucial extensions to the previously presented technique that enable an application in embedded system design automation.

3.1 Local Running Time Model

Previous approaches like [8] are based on the execution count and the running time of the basic blocks of a program which are then combined in some global path analysis, such as implicit path enumeration. The basic block running time can be determined by adding up the running times for each instruction in a basic block, possibly with upper and lower bounds in case of data dependent instruction running time. This leads to wide intervals in case of architectures with pipelines and caches because it is still assumed that all executions of one instruction take the same time. Precise modeling of individual basic blocks by local simulation only solves part of the problem since cache effects, pipelining and register allocation extend over basic block borders. This results in a dependency of the running time on the program path through a sequence of basic blocks. A startup overhead covering overlapping execution for all possible entry paths into the basic block has to be considered.

To obtain higher precision, the analysis should be extended from basic blocks to longer program segments consisting of sequences of basic blocks whenever possible. On the left side of figure 1, the analysis methodology based on the granularity of basic block running times is shown. Overheads representing conservative assumptions for cache misses, pipeline stalls and register spills have to be assumed for every basic block leading to high overestimations but the control flow between these basic blocks is predictable.

A closer look at the program segment shows that the example has a unique, data independent execution path only. A program segment with such a property can be treated like a generalized basic block. On the right side of figure 1 this program segment is shown. Functional constraint annotation becomes obsolete for this program segment. As we apply local *program segment simulation*, this approach only needs a startup overhead for the segment running time while intermediate basic block transitions can exactly be modeled. In summary, the basic block overhead as a main source of overestimation can be significantly reduced using an identification of suitable longer program segments and the extension of the basic block running time model to program segments.

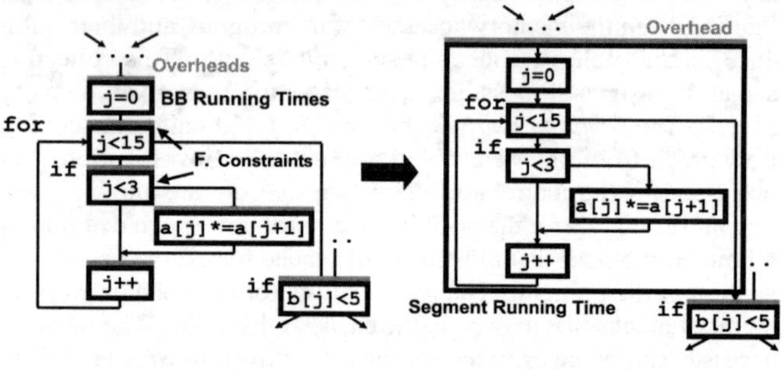

Figure 1: From basic blocks to program segments

3.2 From Basic Blocks to Program Segments

In previous research, we have developed efficient techniques to identify *program segments (PrS)* with unique paths and exploit such segments to reduce analysis problem size and safely reduce timing intervals [11, 13]. Large parts of typical embedded system programs have a single path that is independent of input data [14]. An example is given in the graph in figure 2, other examples may be an FFT or an FIR filter. The control flow of the loop is independent of input data. This path may wrap around many loops, conditional statements and even function calls that are used for source code structuring and compacting. A PrS has a *Single Feasible Path (SFP)*, when paths through the segment are not depending on input data [11, 13].

The key to finding SFP program segments (SFP-PrS) is to distinguish between input data dependent control flow and source code structuring aids. When data dependencies are more difficult to determine than in figure 2, a depth first search algorithm on the syntax graph combined with a symbolic simulation of basic blocks can be applied. In [8], path analysis may be accurate, but it requires much designer interaction even for SFP-PrS while it still does not deliver the precision for program segment running time and cache behavior implied in local simulation. For SFP-PrS, local cache simulation automatically chooses the one correct path and exploits the basic block sequence and memory footprint as well as the resulting cache access sequence without any designer interaction. Local simulation can cover several adjacent or nested SFP-PrS.

Most practical systems also contain non-SFP parts. A program segment has *Multiple Feasible Paths (MFP)*, when paths through its control structure depend on input data. This is the state-of-the-art in basic block based analysis and the according methodologies [8] can be used. An ILP solver computes the process-level running time intervals on the analysis granularity of SFP-PrS instead of basic blocks. The running times for the SFP-PrS nested in the potential paths through the MFP-PrS can be determined using local simulation in a hierarchical path analysis approach [11, 13].

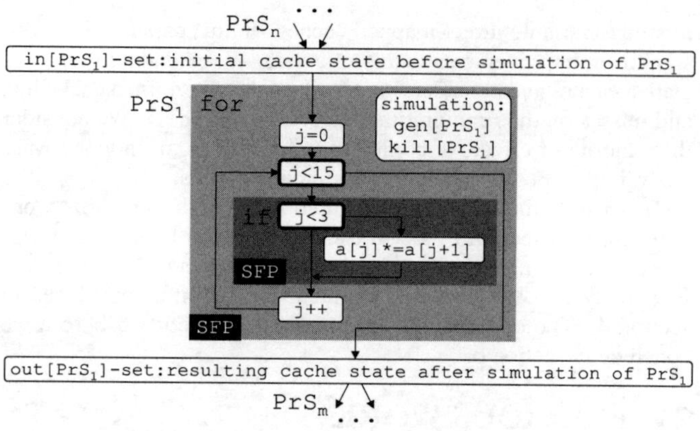

Figure 2: Cache set replacement in SFP simulation

3.3 Set Associativity in Cache Modeling

Instruction and data caches consist of lines that can contain more than one word. The cache line is treated as a whole on every reference. For an associativity of n, n cache lines are summarized to a cache set. The set is referenced by the cache index. Due to the calculation method of the cache index by using the least significant bits of the access address, cache sets are independent. Memory can be divided into line blocks [8], according to the cache lines the parts of the memory segments are mapped to. When more than one line block is mapped to a cache set, n line blocks can be stored before the replacement strategy decides which cache line is discarded.

3.4 Data Caches

Data access addresses are often given as an offset to a fixed base address [7]. When the offset is input data dependent, two misses are assumed, one for the loading of the data of the unknown address and one for the potential replacement of cache line contents that could have been leading to a hit. If the data access address is composed by the base address, constants and variables resulting from a local access sequence in an SFP-PrS. e.g. as for the array *a*[ ] in figure 2, this is referenced to as a *Single Data Sequence (SDS)* [14] which is quite common in embedded systems. Data cache behavior of SDS can be determined by local simulation of the program segment.

3.5 Program Segment Cache Evaluation

Simulation for SFP-PrS starts with the best case and worst case assumptions being first hit/miss regarding the contents at the beginning of the PrS. Best case and worst case bounds for an n-way set associative cache are delivered by the established cache simulator DINERO III [6]. It models the associative instruction and data cache behavior of the local address sequence.

In data flow analysis [1], the $gen[bb]$-set contains the generated definitions for a variable in a basic block (bb) while the $kill[bb]$-set contains the destroyed definitions. The $in[bb]$-set is the set of definitions reaching and the $out[bb]$-set the set of definitions leaving the basic block. We apply this methodology to complete cache sets consisting of n lines in the following.

Definition 1 *A cache definition is a content modification of the cache set.* □

A definition for a cache set can result from a miss of the instruction cache, a miss when reading the data cache and every data cache write. On a cache hit on either cache, a definition occurs when the priorities of the cache lines in a set are changed.

Definition 2 *The $gen_{set}[PrS]$-sets are the sets containing the last definitions for each cache set when executing the PrS.* □

This means on local conflict misses, the definition generated by a replaced line block is removed. The $gen_{set}[PrS]$-sets can be computed by local simulation.

Definition 3 *The $kill_{set}[PrS]$-sets are the sets containing the destroyed definitions for the cache sets of the disjoint PrS.* □

A definition from a disjoint PrS is destroyed when the same cache set is defined by the current PrS and the line block from the disjoint PrS is removed from the set as it was chosen by the replacement strategy. The $kill_{set}[PrS]$-sets can be computed by local simulation.

Definition 4 *The $in_{set}[PrS]$-sets are the sets containing the cache set definitions reaching the PrS.* □

The $in_{set}[PrS]$-sets potentially reduce worst case overhead assumptions for the beginning of the local simulation in further analysis.

Definition 5 *The $out_{set}[PrS]$-sets are the sets containing the cache set definitions leaving the PrS.* □

In figure 2, we can see the execution of an SFP program segment containing several basic blocks. The cache contents the simulation starts from corresponds to the $in[PrS]$-set while the cache set replacement during execution defines $gen[PrS]$-set and $kill[PrS]$-set. The $out[PrS]$-set corresponds to the cache state after execution. Running times for the PrS can be computed by local simulation before a process-level solution is computed with an ILP solver. This is described in section 4.4.

4. GLOBAL FLOW ANALYSIS

Even for the first reference of a cache set in the simulation of a program segment (PrS), the assumption of a cache miss is too pessimistic. The line block might already be in the cache set from a previous execution of the current PrS or from a loading of the line block by a PrS mapped to the same line block. This can lead to cache hits even for the first reference in a PrS. This propagation of definitions is referred to as the *Set Definition Propagation* of a program segment $SDP[PrS]$. We utilize well-established work in global flow analysis from compiler design [1] for computation.

4.1 Cache Set Content Prediction

In figure 3, the mapping of the address space of a program to line blocks referring to the cache sets (CS) according to [8] can be seen. Path analysis has found four SFP-PrS where cache behavior can be simulated while the control flow from PrS_1 to PrS_2 and PrS_3 cannot be predicted (MFP-PrS). However, line block 1 will be present in the set at the beginning of PrS_2 because of the previous execution of PrS_1 that loads the complete cache line including parts of the code for PrS_2. Line block 2 may be present because of a previous execution of PrS_2 itself or an execution of PrS_3, when it has not been replaced. This effect is valid for direct mapped caches as well as for higher associativities. In general, overlapping cache sets lead to predictable cache line contents.

The effect reduces the worst case assumption of empty $in_{set}[PrS]$-sets. An analysis technique combining local simulation and static propagation analysis in three steps is proposed. The major differences of this methodology to *must/may* analysis [2, 10] are a conservative path analysis for SFP-PrS and the data flow analysis granularity of program segments instead of single instructions or basic blocks.

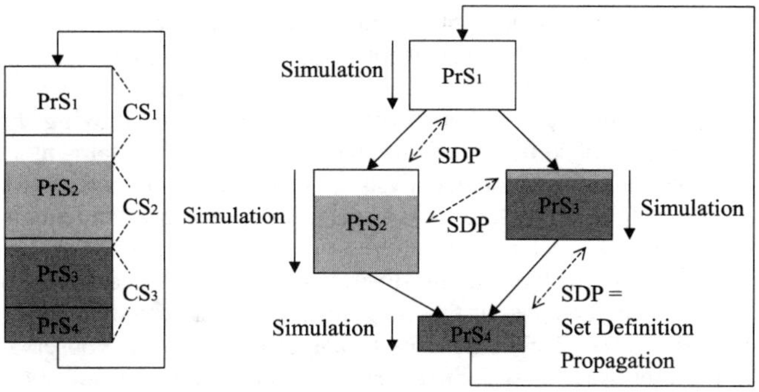

Figure 3: Set Definition Propagation

4.2 Hybrid Prediction Approach

In a **first step**, cache behavior is simulated for every PrS starting with a first miss scenario meaning empty $in_{set}[PrS]$-sets. A first hit scenario is needed for the best case analysis. Local cache hits and misses for the address sequence of the PrS are found by local simulation. The $gen_{set}[PrS]$-sets and $kill_{set}[PrS]$-sets can be computed.

In the **second step**, the data flow analysis defines $out_{set}[PrS]$-sets from $gen_{set}[PrS]$-sets and $kill_{set}[PrS]$-sets given by simulation and $in_{set}[PrS]$-sets.

$$out_{set}[PrS] = gen_{set}[PrS] \cup (in_{set}[PrS] - kill_{set}[PrS])$$

This means the set references leaving the PrS $out_{set}[PrS]$ are composed by the set references within the PrS plus the references entering the PrS which are not replaced. A second set of equations is given by the fact that the $in_{set}[PrS]$-set is defined from the predecessor $out_{set}[PrS]$-sets.

$$in_{set}[PrS] = \bigcap_{P \in pred(PrS)} out_{set}[P]$$

For worst case analysis, only definitions occurring on all previous PrS are propagated, so the intersection operator $\cap$ is used. For best case analysis, the union operator $\cup$ needs to be applied. The equations for every cache set can be solved by iterative forward data flow analysis algorithms [1]. The $in_{set}[PrS]$-sets are assumed to be empty in the first iteration. When data flow analysis is finished, the $in_{set}[PrS]$-sets either contain a subset of the propagated definitions for worst case analysis or a superset of the propagated definitions for best case analysis.

In the **third step**, worst case analysis finds unique definitions in the $in_{set}[PrS]$-sets that PrS simulation has classified as misses for the first reference due to the conservative cache state, namely empty $in[PrS]$-sets at the beginning of the PrS. Similarly, best case analysis finds the maximum number of definitions reaching the PrS. The initial first hit assumptions not being in the $in[PrS]$-sets can be replaced by misses improving the best case analysis bound.

4.3 Evaluation of Flow Analysis Results

A definition in the $in_{set}[PrS]$-set represents an instruction or data word that is present in the set. When the cache line referenced by the definition in the $in_{set}[PrS]$-set is classified as a first miss by simulation, the assumption in the first step is too conservative.

When the definition in the $in_{set}[PrS]$-set is from another program segment (PrS) whose execution would load the current line block — an execution of PrS_3 loads CS_3 for PrS_1 in figure 3 — the first miss (compulsory) for this line and PrS can be removed because the memory block is loaded in the other PrS and not killed.

When the unique definition in the $in_{set}[PrS]$-set is from the memory block of the program segment (PrS) itself, all misses (compulsory) for this line and PrS can be removed, except for the miss of the first execution. For associative caches, just removing the first miss is not sufficient because the referenced set contains more than one line. The global analysis only propagates definitions for a set, not for a cache line, so the replacement strategy can be considered in the local simulation of the program segments. Per predicted cache hit, the running time of one cache miss can be subtracted from the conservative program segment startup with empty $in_{set}[PrS]$-sets. This is done by means of replacing the conservative compulsory misses with the newly predicted hits in a new local simulation where pipeline and register set are still modeled conservatively assuming a cold start for the beginning of the program segment. When this approach is used, global cache effects can be integrated into the overheads of the local program segment running time intervals determined by simulation. For this purpose, a slight modification of the cache simulation tool is necessary. DINERO III has been extended with the possibilities to preset, dump or directly modify the cache behavior of a given address sequence. It can model least recently used and round robin replacement.

4.4 Running Time Calculation

For the MFP-PrS and the process-level result, the approach from [8] can be applied on the simulated lower level SFP-PrS running times including several basic blocks. Implicit path enumeration can bound the longest path through the MFP-PrS. The startup overhead has been reduced using the flow analysis results. In further work, more analysis techniques from the real-time research community can be added to further improve the precision of the SFP program segment startup overhead with respect to pipeline and register set behavior. The complex cache analysis using a transition graph is not needed because the cache behavior is modeled using the startup overheads.

4.5 Compiler Optimization and Code Motion

As the presented analysis approach is based on the source code level, local and global compiler optimization can have significant effects on the running time intervals. It has been shown that most compiler optimizations are confined to the level of basic blocks [4], so they do not influence the results when the same optimizations are used for analysis. Code motion, e.g. removing loop invariant code from a loop body and inserting it before or after the loop introduces more problems. For the analysis granularity of basic blocks, these optimizations have to be avoided to deliver correct results. This leads to wider running time intervals to stay conservative. For the analysis granularity of program segments, code motion from a loop body that stays in the same program segment is allowed. Instruction implementation via assembly library functions can be treated as data dependent instruction running time. Data dependent instruction running time must be treated by formal subtraction or addition of the best case or worst case timing as described in [14]. Running time intervals for the library functions need to be provided by their vendors. Instructions implemented as exceptions and interrupt effects are not included in the program segment running time but must be treated on the process level.

5. EXPERIMENTS

Running time analysis based on the granularity of basic blocks [8] is compared to running time analysis based on SFP-PrS identification and local simulation of PrS. In this experiment, the original intention is to free the designer from functional constraint annotation. For the basic block based approach, this would lead to infinite running time intervals. For this reason, the loop bounds have been annotated for the basic block based approach which already implies designer knowledge about the program under investigation. This is not necessary in the approach based on program segments when loop bounds are input data independent (SFP). For the local simulation of the SFP-PrS as well as the basic blocks BB, a StrongARM (SA-110) simulator with 160 MHz core frequency, 40 MHz bus frequency and 85 ns memory cycle time has been used. The parameters in local cache simulation are selected according to the real parameters of the StrongARM processor which is an associativity of 32, a block size of 32 bytes, 16 kbytes instruction cache and 16 kbytes data cache. This leads to a cache miss penalty of 32 bus cycles. In [8], this penalty for the i960 has been much lower, so overestimations would seem lower for the i960 instead of the StrongARM. The round robin replacement strategy is modeled. No cache state transition graph (CSTG) for the basic block based approach has been used. A CSTG for an associativity of 32 would have been much too complex. Cache analysis has only been done by means of local simulation which leads to high overestimations in a basic block based approach. Cache analysis supported by set definition propagation has only been applied on the granularity of program segments. Both instruction and data cache behavior have been considered.

Four configurations exploring the effects of program segment source code translation from isolated files or connected program segments in one single file as well as the effects of compiler optimizations (GNU -O1) are presented. In the first column of each table, the benchmark program is provided. The exact running time intervals given by an application of exact worst case and best case input data annotated by the designer are shown in the second column. This would not be possible for more complex programs. In the presented cases, these have been determined manually only to serve as a reference for the evaluation of the static analysis approaches. These running time intervals are different for the same input data when compiler optimizations are applied. The third column shows the running time intervals determined on the analysis granularity of basic blocks (ILP on BB) while the next column shows the number of loop bounds (lb) being the functional constraints that have been annotated to avoid infinite running time intervals. In the next column, the running time intervals for the approach using SFP identification (ILP on SFP) are given. In the last column, the running time intervals for the extension of the SFP identification approach with cache analysis using set definition propagation (ILP on SDP) are shown.

In the first configuration, the running time intervals have been determined using isolated local simulation of program segments, i.e. in different source code files after path analysis. No compiler optimizations have been applied. The results are given in table 1.

Table 1: Intervals for isolated PrS, no optimization

Benchmark [μs]	Exact	ILP on BB	lb	ILP on SFP	ILP on SDP
arrcalc	[19.45,20.37]	[4.805,206.9]	4	[3.339,29.92]	[9.20,28.93]
chkdata	[15.62,20.72]	[1.082,226.0]	2	[1.233,152.2]	[9.039,39.82]
bsort	[58.69,104.6]	[13.48,3046]	2	[8.696,1316]	[15.09,846.2]
circle	[47.96,151.1]	[4.269,622.1]	1	[4.287,154.4]	[5.962,153.5]
FIRFilter	[72.15,100.0]	[38.53,2566]	4	[42.99,158.9]	[60.17,136.5]
countsort	[38.10,41.47]	[15.77,1079]	2	[16.28,475.9]	[29.50,290.5]
exchsort	[43.18,43.96]	[17.46,1164]	2	[19.40,237.9]	[30.51,49.34]

In the second configuration, the running time intervals have been determined using local simulation of all PrS and cache analysis in one single file. Worst case startup overheads for the program segments or basic blocks according to the applied analysis granularity have been inserted afterwards. No compiler optimizations have been applied. The results are given in table 2.

Table 2: Intervals for connected PrS, no optimization

Benchmark [μs]	Exact	ILP on BB	lb	ILP on SFP	ILP on SDP
arrcalc	[19.45,20.37]	[2.54,195.7]	4	[3.189,27.30]	[9.186,23.43]
chkdata	[15.62,20.72]	[0.69,172.1]	2	[0.864,78.74]	[8.089,31.77]
bsort	[58.69,104.6]	[12.61,2802]	2	[8.831,1076]	[14.80,459.8]
circle	[47.96,151.1]	[4.26,622.1]	1	[4.287,154.4]	[5.962,153.5]
FIRFilter	[72.15,100.0]	[38.44,2552]	4	[42.99,150.5]	[60.21,122.8]
countsort	[38.10,41.47]	[10.49,788.9]	2	[11.16,295.2]	[20.04,126.2]
exchsort	[43.18,43.96]	[14.62,1116]	2	[19.75,153.2]	[31.67,46.49]

In the third configuration in table 3, the running time intervals have been determined using isolated local simulation of program segments in different input files again. Compiler optimizations with -O1 have been applied.

Table 3: Intervals for isolated PrS, with optimization -O1

Benchmark [μs]	Exact	ILP on BB	lb	ILP on SFP	ILP on SDP
arrcalc	[11.95,12.97]	[1.188,124.7]	4	[1.262,29.71]	[3.881,14.03]
chkdata	[8.792,10.58]	[1.199,167.0]	2	[1.225,116.1]	[2.826,98.72]
bsort	[23.58,27.52]	[4.714,1017]	2	[1.572,693.3]	[13.22,682.7]
circle	[39.27,100.9]	[3.292,480.6]	1	[3.295,123.6]	[5.901,101.4]
FIRFilter	[30.97,37.59]	[13.52,1874]	4	[20.87,353.1]	[29.68,210.0]
countsort	[15.31,15.50]	[6.554,746.5]	2	[7.003,369.8]	[14.48,188.3]
exchsort	[17.20,17.20]	[5.932,622.2]	2	[6.322,250.9]	[9.863,250.6]

In the final configuration presented in table 4, the running time intervals have been determined using local simulation of all PrS and cache analysis in one single source code file. Worst case startup overheads have been inserted afterwards. Compiler optimizations with -O1 have been applied.

Table 4: Intervals for connected PrS, with optimization -O1

Benchmark [μs]	Exact	ILP on BB	lb	ILP on SFP	ILP on SDP
arrcalc	[11.95,12.97]	[1.324,122.7]	4	[1.412,29.40]	[3.881,14.03]
chkdata	[8.792,10.58]	[0.306,94.14]	2	[0.312,48.14]	[3.834,23.61]
bsort	[23.58,27.52]	[5.070,703.9]	2	[2.894,129.2]	[14.69,35.61]
circle	[39.27,100.9]	[3.292,480.6]	1	[3.295,123.6]	[5.901,101.4]
FIRFilter	[30.97,37.59]	[11.28,988.8]	4	[14.35,296.0]	[26.78,198.1]
countsort	[15.31,15.50]	[2.045,208.5]	2	[3.409,75.22]	[8.949,20.75]
exchsort	[17.20,17.20]	[6.067,322.1]	2	[5.644,26.66]	[7.903,24.23]

The experiment reveals that the approach using SFP identification delivers significantly tighter running time intervals than the approach on basic block level. This is an expected result because local simulation for SFP-PrS can consider overlapping basic block effects instead of assuming worst cases for caches at basic block beginnings. These have a significant impact when the StrongARM processor with its high miss penalty is assumed as the target architecture. When cache analysis is applied in combination with SFP identification, the intervals get tighter because many startup overheads for SFP-PrS, especially in nested loops, can be reduced.

Due to the high cache miss penalty compared to the running time for a hit in these configurations, nested loops with embedded MFP can still cause high overestimations because empty caches for the isolated segments in the MFP have to be assumed for every iteration. When SDP cannot predict all of these cache contents, the interval stays relatively wide, e.g. for the *bsort* algorithm. Nested loops with an embedded MFP require the assumption of several cache misses for every iteration. ILP analysis of connected PrS with a later addition of overheads according to the selected granularity delivers tighter intervals than for isolated PrS. More information on cache states is available. The consideration of compiler optimizations leads to more realistic running time intervals with respect to the final solution on the embedded target architecture.

6. CONCLUSION

We have shown how to combine local cache tracing with global data flow analysis in formal cache analysis. The technique matches a processor modeling and path analysis technique which extends architecture modeling from basic blocks to larger program segments thereby increasing the overall analysis precision. The approach has been extended from direct mapped caches to set associative caches and the impact of compiler optimizations has been considered. We have demonstrated that our approach considering both instruction and data cache behavior can significantly increase the analysis precision of state-of-the-art approaches.

7. REFERENCES

[1] A. V. Aho, R. Sethi, and J.D. Ullman. *Compilers: Principles, Techniques and Tools*. Addison-Wesley, Reading, GB, 1988.

[2] C. Ferdinand and R. Wilhelm. Efficient and precise cache behavior prediction for real-time systems. *Journal of Real-Time Systems, Special Issue on Timing Analysis and Validation for Real-Time Systems*, November 1999.

[3] C. Healy, R. Arnold, F. Müller, D. Whalley, and M. Harmon. Bounding pipeline and instruction cache performance. *IEEE Transactions on Computers*, pages 53–70, January 1999.

[4] J. L. Hennessy and D. A. Patterson. *Computer Architecture: A Quantitative Approach*. Morgan Kaufmann Publ., 1996.

[5] A. Hergenhan and W. Rosenstiel. Static timing analysis of embedded software on advanced processor architectures. In *Proceedings of Design, Automation and Test in Europe*, pages 552–559, Paris, March 2000.

[6] M. Hill. DINERO III Cache Simulator. www.ece.cmu.edu/ ece548/tools/dinero/src/, 1998.

[7] S.-K. Kim, S. Min, and R. Ha. Efficient worst case timing analysis of data caching. In *Proceedings of IEEE Real-Time Technology and Applications Symposium*, 1996.

[8] Y. Li and S. Malik. *Performance Analysis of Real-Time Embedded Software*. Kluwer Academic Publishers, 1999.

[9] Mentor Graphics. *Seamless Co-Verification Environment*. http://www.mentorg.com/seamless/.

[10] H. Theiling, C. Ferdinand, and R. Wilhelm. Fast and Precise WCET Prediction by Seperate Cache and Path Analyses. *Journal of Real-Time Systems*, 18(2/3):157–179, May 2000.

[11] F. Wolf and R. Ernst. *Behavioral Intervals in Embedded Software*. Technical University of Braunschweig, Germany, To be published by Kluwer Academic Publishers in 2002.

[12] F. Wolf and R. Ernst. Data flow based cache prediction using local simulation. In *Proceedings of the IEEE High Level Design Validation and Test Workshop*, pages 155–160, Berkeley, USA, November 2000.

[13] F. Wolf and R. Ernst. Intervals in software execution cost analysis. In *Proceedings of the IEEE/ACM International Symposium on System Synthesis*, pages 130–135, Madrid, Spain, September 2000.

[14] W. Ye and R. Ernst. Embedded program timing analysis based on path clustering and architecture classification. In *Proceedings of the IEEE International Conference on Computer-Aided Design (ICCAD '97)*, pages 598–604, San Jose, USA, 1997.

Compiler-Directed Scratch Pad Memory Hierarchy Design and Management*

M. Kandemir
Microsystems Design Lab
Pennsylvania State University
University Park, PA 16802, USA
kandemir@cse.psu.edu

A. Choudhary
ECE Department
Northwestern University
Evanston, IL 60208, USA
choudhar@ece.nwu.edu

ABSTRACT

One of the primary challenges in embedded system design is designing the memory hierarchy and restructuring the application to take advantage of it. This task is particularly important for embedded image and video processing applications that make heavy use of large multi-dimensional arrays of signals and nested loops. In this paper, we show that a simple reuse vector/matrix abstraction can provide compiler with useful information in a concise form. Using this information, compiler can either adapt application to an existing memory hierarchy or can come up with a memory hierarchy. Our initial results indicate that the compiler is very successful in both optimizing code for a given memory hierarchy and designing a hierarchy with reasonable performance/size ratio.

Categories and Subject Descriptors

B.3 [**Hardware**]: Memory Structures; D.3.4 [**Programming Languages**]: Processors—*Compilers;Optimization*

General Terms

Design, Experimentation, Performance

Keywords

Data Reuse, Scratch Pad Memory, Memory Hierarchy

1. INTRODUCTION

Embedded system design has undergone a major renaissance in the last five years. An important characteristic of this change is that software is playing an ever increasing role in system design. Consequently, automatic compiler support for optimizing embedded software is of critical importance. Classical compiler methods alone, however, may not be sufficient for attaining the highest levels of performance from an embedded platform. Instead, embedded system-specific issues should also be accounted for. Software-managed memories (called scratch-pad memories [9]) are an important part of design process, particularly in image and video processing applications that make heavy use of large multi-dimensional arrays of signals and nested loops. Compiler can restructure a given code to adapt it to a given memory hierarchy; but, it can also help designer to come up with a memory hierarchy for a given application. This hierarchy can then be used as a starting point for a more sophisticated memory hierarchy optimization.

While it might be possible (for some applications) to adopt a plain (single level) scratch-pad memory architecture and obtain large savings in performance and energy [9, 7], many access patterns encountered in embedded image and video applications exhibit data reuse in multiple loop levels, thereby necessitating a multi-level memory hierarchy if we are to obtain the best results. Therefore, we believe that as embedded codes get more and more complex, memory architectures that contain multi-level scratch-pad memories will be more widely employed. This will also make it extremely important to provide compiler support for optimizing applications in such hierarchies.

In this paper, we present a compiler-directed memory hierarchy design and code optimization strategy. More specifically, we make the following major contributions:

- We show how an optimizing compiler can manage the flow of data to/from a scratch-pad memory (SPM) hierarchy using a mathematical abstraction based on reuse vectors and reuse matrices.
- We present an algorithm that, given an input code, tries to come up with an SPM hierarchy. This hierarchy might be used as an input for a more detailed systematic analysis and optimization framework.
- We present an algorithm that enforces a memory hierarchy for a given code. For example, using this algorithm, it might be possible to transform a given code in such a way that a majority of array references in the code would work best with a specific depth of memory hierarchy.
- We report experimental data showing the effectiveness of our approach. Our results indicate that the compiler is very successful in both optimizing code for a given hierarchy and designing a hierarchy with reasonable performance/size ratio.

Our approach is different from many previous studies on software-directed memory management. First, in contrast to studies in [9, 10, 7, 2] where the main focus is a single-level memory management, we focus on a memory hierarchy. We focus on both designing a memory hierarchy and exploiting a given hierarchy for current data access pattern. As compared to the design methodology in [3], our approach is simpler as it uses a reuse vector/matrix abstraction (which can be extracted by an optimizing compiler during data access pattern/data dependence analysis). By using this abstraction, we can easily decide the flow of data through memory hierarchy. In addition, this abstraction also allows us to come up with an initial hierarchy design very quickly.

The rest of this paper is organized as follows. Section 2 presents background material and gives details of our optimization strategy. Section 3 discusses our implementation, introduces our experimental framework, and presents performance and energy consumption data. Section 4 summarizes our major results.

*This work is supported in part by NSF Career Award #0093082 and by a grant from PDG.

2. OUR APPROACH

2.1 Background

The iteration space $\mathcal{I}$ of a loop nest contains one point for each iteration. An iteration vector can be used to label each such point in $\mathcal{I}$. We use $\vec{i} = (i_1, ..., i_n)$ to represent the iteration vector for a loop nest of depth n, where each i_k corresponds to a loop index (or its value at a specific point in time), starting with i_1 for the outermost loop index. In fact, an iteration space $\mathcal{I}$ can be viewed as a polyhedron bounded by the loop limits.

The subscript function for a reference to an array U is a mapping from the iteration space $\mathcal{I}$ to the data space $\mathcal{D}$. The data space can also be viewed as a polyhedron bounded by array bounds. A subscript function defined this way maps an iteration vector to an array element. In this paper, we assume that subscript mappings are affine functions of the enclosing loop indices and symbolic constants. Many array references found in embedded image and video processing codes fall into this category. Under this assumption, a reference to an array U can be represented as $L^U\vec{i} + \vec{b}^U$ where L^U is a linear transformation matrix called the access (reference) matrix, $\vec{b}^U$ is the offset vector, and $\vec{i}$ is the iteration vector [12]. If the loop is n-deep and the array U is m-dimensional, L^U is $m \times n$ and $\vec{b}^U$ has m elements. For example, the reference $V[k][j]$ in Figure 1(a) can be written as:

$$L^V\vec{i} + \vec{b}^V = \begin{pmatrix} 0 & 0 & 1 \\ 0 & 1 & 0 \end{pmatrix} \begin{pmatrix} i \\ j \\ k \end{pmatrix} + \begin{pmatrix} 0 \\ 0 \end{pmatrix}.$$

According to Li [8], if $\vec{i}$ and $\vec{j}$ are two iteration vectors that access the same memory location, $\vec{r} = \vec{j} - \vec{i}$ is called a reuse vector (assuming that $\vec{j}$ is lexicographically greater than $\vec{i}$). For a loop of depth n, the reuse vector is n dimensional. The loop corresponding to the first non-zero element from top is the one that carries the corresponding reuse. If there are more than one reuse vector in a given direction, they are represented by the lexicographically smallest one. The reuse vector gives us useful information about how array elements are used by different loops in a given nest. In this paper, we mainly focus on self reuses (that is, the reuses originating from individual references) as in very rare circumstances group reuse (that is, the reuse between different references to the same array) brings additional benefits that cannot be exploited by self reuse [8]. There is a temporal reuse due to reference $L^U\vec{i} + \vec{b}^U$ if and only if there exist two different iterations $\vec{i}$ and $\vec{j}$ such that $L^U\vec{i} + \vec{b}^U = L^U\vec{j} + \vec{b}^U$; that is, $\vec{j} - \vec{i} \in Ker\{L^U\}$. This last expression indicates that the temporal reuse vector $\vec{r} = \vec{j} - \vec{i}$ belongs to the kernel set (null set) of L^U [12].

As an example, let us focus on the matrix multiply nest shown in Figure 1(a). The reuse vector due to $U[i][k]$ is $(0, 1, 0)$. Informally, what this means is that, for fixed values of loops i and k, the successive iterations of the j loop access the same element from this array. That is, the j loop carries (exhibits) temporal reuse for this reference.

The reuse vectors coming from individual references make up a reuse matrix, R. In considering the matrix multiply code again, we find that:

$$R = \begin{pmatrix} 0 & 0 & 1 \\ 0 & 1 & 0 \\ 1 & 0 & 0 \end{pmatrix}.$$

The first, second, and third columns in this matrix correspond to the reuse vectors for $W[i][j]$, $U[i][k]$ and $V[k][j]$, respectively. As will be discussed shortly, each column indicates which array sections should be brought into (and discarded from) the SPM at what time.

A linear one-to-one mapping between two iteration spaces can be represented by a square, non-singular matrix T. Using such a transformation, each element $\vec{i}$ of the original iteration space $\mathcal{I}$ is mapped to $\vec{i'} = T\vec{i}$ on the transformed iteration space $\mathcal{I}'$. Converting the original loop nest to the transformed nest is a two-step process [12]. First, each array reference $L^U\vec{i} + \vec{b}^U$ is transformed to $L^U T^{-1}\vec{i'} + \vec{b}^U$. Second, the loop bounds are transformed accordingly. Finding the new loop bounds can necessitate the use of Fourier-Motzkin elimination, details of which can be found in [12]. It can be shown that if T is a linear transformation matrix, $\vec{r}$ is a reuse vector, and R is a reuse matrix, $T\vec{r}$ and TR are the transformed reuse vector and the transformed reuse matrix, respectively [8].

In Sections 2.2 through 2.5, we assume the existence of an SPM hierarchy and focus on how to manage data flow through this hierarchy. In Section 2.6, we show how to design a memory hierarchy based on data reuse. Till Section 2.4, we exclusively focus on a two-level memory hierarchy which contains a large (slow and energy-consuming) memory and a small (fast and less energy-consuming) memory. The problem is then to decide what data to bring to the fast memory at what time and how to decide when data in the fast memory are not useful anymore. We use the terms (software-controlled) memory hierarchy and SPM hierarchy interchangeably.

2.2 Reuse Based Data Transfers

Recall that $U[i][k]$ in Figure 1(a) leads to a reuse vector of $(0, 1, 0)$. Suppose that we have a section of array U in Figure 1(a), $U[i][1 : N]$ (the ith row), residing in the fast memory. Since all loop iterations

$$(i, 1, 1), (i, 1, 2), ..., (i, 1, N), (i, 2, 1), (i, 2, 2), ..., (i, 2, N),$$

$$......(i, N, 1), (i, N, 2), ..., (i, N, N)$$

access this same array section, this section can be kept in the fast memory (in the fast SPM) during the entire execution of these iterations; and, following the execution of the iteration (i, N, N), it can be discarded from the fast memory. It should also be noted that this section can be brought into the fast memory just before the iteration $(i, 1, 1)$ starts execution (i.e., just before the j loop is entered). For clarity, we represent these iterations collectively using $(i, *, *)$, where $*$ denotes all iterations in the corresponding loop level; we also represent this array section using $U[i][*]$, where $*$ indicates all elements in the corresponding dimension. To sum up, by just considering the value of 1 in this reuse vector (the second entry), we can decide the point in the code at which an array section (in our case, this is the ith row of U) should be brought into the fast memory and the point at which it should be removed from the fast memory. This brings us to the following conclusion:

> *An array section should remain in the fast memory from the start of the loop that carries reuse for the corresponding array reference to the termination of the same loop.*

Let us consider the following generic n-deep loop nest that accesses an m-dimensional array:

$$for(i_1 = i_1{}^l; i_1 \le i_1{}^u; i_1 ++)$$
$$for(i_2 = i_2{}^l; i_2 \le i_2{}^u; i_2 ++)$$
$$for(i_3 = i_3{}^l; i_3 \le i_3{}^u; i_3 ++)$$
$$...$$
$$for(i_n = i_n{}^l; i_n \le i_n{}^u; i_n ++)$$
$$U[f_1(i_1, i_2, .., i_n)][f_2(i_1, i_2, .., i_n)]...[f_m(i_1, i_2, ..., i_n)]$$

In this nest, each $f_j(.)$ $(1 \le j \le m)$ is an affine function corresponding to a subscript expression. $i_k{}^l$ and $i_k{}^u$ are the lower and upper bounds for loop i_k. Assume that $\vec{r} = (r_1, r_2, ..., r_n)$ is the corresponding reuse vector. Assume further (for now) that there is only a single non-zero element (say r_k) in $\vec{r}$; all other elements are assumed to be zero. In other words, the loop i_k is the one that carries the reuse.

In this case, the compiler can make use of the fast memory as follows. Let us use $f_l(i_1, i_2, .., i_{k-1}, *, *, .., *)$ to denote all indices enumerated in a dimension l by fixing the first $k - 1$ loops at specific values $(i_1, i_2, .., i_{k-1})$ and considering all values (within the loop bound ranges) for all other loop index positions. Just before executing the loop i_k, the compiler brings the array section containing all elements represented by $U[f_1(i_1, i_2, .., i_{k-1}, *, *, .., *)]$ $[f_2(i_1, i_2, .., i_{k-1}, *, *,$

$.., *)] ... [f_m(i_1, i_2, .., i_{k-1}, *, *, .., *)]$ into the fast memory, and keeps them there while executing all iterations in $(i_1, i_2, .., i_{k-1}, *, *, .., *)$. When the last iteration in this set has been executed, the section can be removed from the fast memory. If this is done, the said set of elements would reside in the fast memory as long as they are reused, and as soon as the last iteration that reuses the elements is completed, the set can be removed from the fast memory. It should be emphasized that if any element in this set is updated while it is in the fast memory, it should be written back to slow memory.

If there are multiple references to the same array, the reuse vectors are considered together and the corresponding data transfers are combined as much as possible. Consider the nested loop in Figure 1(b). We have two reuse vectors: $(0, 1, 0)$ (due to reference $U[i][k]$) and $(0, 0, 1)$ (due to reference $U[i][j]$). We note that the data transfer indicated by $(0, 0, 1)$ is subsumed by the transfer indicated by $(0, 1, 0)$; that is, once the transfer of the ith row to the fast memory is complete (just before entering the j loop), there is no need for performing an extra transfer for the element $U[i][j]$. Let us now focus on the code given in Figure 1(c). In this case, the reuse vectors are $(0, 1, 0)$ and $(0, 1, 0)$. Consequently, both data transfers should be performed before the j loop is entered. However, one data transfer contains the ith row of the array whereas the other contains the ith column. Except for one element, these two transfers do not overlap; so, they should be performed separately. Our current implementation uses the following rule to decide whether the transformations required by different references to the same array can be combined. Let $L_1{}^U\vec{i} + \vec{b_1}^U$ and $L_2{}^U\vec{i} + \vec{b_2}^U$ be two references to array U. If $L_1{}^U = L_2{}^U$, there is a good chance that there exists significant amount of data reuse between these two references (even if $\vec{b_1}^U \neq \vec{b_2}^U$). Consequently, in this case, the data transfers due to these references can be combined and performed together. On the other hand, if $L_1{}^U \neq L_2{}^U$, these two references are treated as if they belong to different arrays. If a nest contains references to multiple arrays, each array is treated independently and a separate data transfer is performed for each array.

2.3 Transformations

In some cases, it may not be clear which loop carries the reuse. To make the loop carrying the reuse explicit, it may be necessary to apply loop transformations. As an example, consider the loop nest shown in Figure 1(d). When considered alone, none of the references in this nest has temporal reuse. However, there is a group reuse in this nest with a reuse vector of $\vec{r} = (1, -1)$. Consequently, it is not trivial to find the point in the code to perform memory transfers so that we can exploit the fast memory. However, if we transform this nest using the loop transformation matrix

$$T = \begin{pmatrix} 1 & 1 \\ 1 & 0 \end{pmatrix},$$

we obtain the nest in Figure 1(e). We now see that the reuse vector for this transformed nest is $\vec{r} = (0, 1)$. This means that only the innermost loop carries reuse, and we can make use of the fast memory. To see how the reuse is exploited in this case, let us list the array elements accessed by a few initial iterations:

$$\begin{array}{lcl}
(i' = 3, j' = 2) & \Rightarrow & U[2][1] \text{ and } U[1][2] \\
(i' = 4, j =' 2) & \Rightarrow & U[2][2] \text{ and } U[1][3] \\
(i' = 4, j' = 3) & \Rightarrow & U[3][1] \text{ and } U[2][2] \\
(i' = 5, j' = 2) & \Rightarrow & U[2][3] \text{ and } U[1][4] \\
(i' = 5, j' = 3) & \Rightarrow & U[3][2] \text{ and } U[2][3] \\
(i' = 5, j' = 4) & \Rightarrow & U[4][1] \text{ and } U[3][2]
\end{array}$$

We observe that (for a specific i') the iterations in $(i', *)$ cause the reuse of elements $U[j'][i' - j']$, where $max(2, i' - N + 1) \leq j' \leq min(N, i' - 1) - 1$. Therefore, just before entering the j' loop (in the transformed nest), these elements can be brought into the fast memory. As compared to the matrix multiply case, this example illustrates two interesting points. First, in some cases, loop transformations might be necessary for identifying where (in the code) to perform data transfers. Second, the data transfers for different values of the outer loop are of different sizes. So, if we are to design a memory hierarchy, in determining the size of the fast memory, we should consider the maximum number of reused elements over all i' values.

Another question in this example is how to determine the transformation matrix to use. We can determine the transformation matrix using the default and desired reuse vectors. For our current example, the default reuse vector is $\vec{r_d} = (1, -1)$. Since we have only two loops in the nest, if want to make sure that the reuse will be carried by one of them, the desired reuse vector should $\vec{r_s} = (0, 1)$. Then, from $\vec{r_s} = T\vec{r_d}$, we can determine the entries of T. Note that we used loop transformations in this section to modify the group temporal reuse vector so as to exploit a two-level memory hierarchy. As will be discussed shortly, loop transformations can also be used for modifying the temporal reuse vectors.

2.4 Multi-Level Hierarchy

When there are multiple non-zero elements in a given reuse vector, we can use this fact for exploiting a multi-level memory hierarchy. In the most general case, we can exploit a memory level for each non-zero entry in the reuse vector. To illustrate this, let us consider the four-deep nest in Figure 1(f). For $U[i][k]$, the reuse vector is $(0, 1, 0, 1)$ which indicates that there are reuses in j and l loops (although the reuse itself is carried by the j loop). We can exploit these reuses as follows. Before entering the j loop, the i row of this array can be transferred from the slow memory (called Level 1) to the fast memory (called Level 2). In addition to that, if we have an even faster memory (called Level 3), the element $U[i][k]$ can be transferred to this memory (just before the l loop) and can be used from there throughout the execution of the l loop. In other words, at a given time, one row of array U can be in Level 2 SPM and one element from this row can be in Level 1 SPM (the fastest level). This is depicted in Figure 2. A similar data transfer scheme is valid for reference $V[j][l]$ as well. Before entering the i loop, the entire array can be transferred to Level 2 (if we have that much space in Level 2), and a row the array can be transferred to Level 1 before the k loop.

It is important to note that just the fact that there are two non-zero entries in the temporal reuse vector does not mean that we can use the nest in question only with a three-level memory hierarchy. In fact, we can still use it with a two-level memory hierarchy and the fact that there are two non-zero entries in the reuse vector gives the compiler flexibility in scheduling data transfers. For example, focusing on reference $U[i][k]$ in Figure 1(f), we have two options assuming a two-level memory: (i) we can perform data transfer before the j loop, or (ii) we can perform data transfer before the l loop. Note that transfer before the j loop requires a larger fast memory space, but it is also expected to give better results. It is also important to emphasize that loop transformations can make a nested loop written for a two-level of memory hierarchy suitable for a three-level of memory hierarchy, and vice versa. For example, the loop transformation matrix

$$T = \begin{pmatrix} 0 & 1 & 0 & 0 \\ 1 & 0 & 0 & 0 \\ 0 & 1 & 1 & -1 \\ 0 & 0 & 0 & 1 \end{pmatrix}$$

can transform a reuse vector $(0, 1, 0, 0)$ to $(1, 0, 1, 0)$. Note that while $(0, 1, 0, 0)$ is suitable for a two-level of memory hierarchy, $(1, 0, 1, 0)$ is suitable for a three-level of memory hierarchy.

2.5 Imperfectly Nested Loops

While all examples given so far have focused on perfect nests only, our reuse vector/matrix based strategy can work with imperfectly-nested loops as well. An imperfect nest is the one that contains assignment statements between loops, whereas in a perfect nest all assignment statements are nested within the innermost loop. In our framework, imperfectly-nested loop nests in general present opportunities for min-

```
for(i = 1; i <= N; i++)
 for(j = 1; j <= N; j++)
  for(k = 1; k <= N; k++)
   W[i][j] += U[i][k] * V[k][j];
```
(a)

```
for(i = 1; i <= N; i++)
 for(j = 1; j <= N; j++)
  for(k = 1; k <= N; k++)
   c += U[i][k] + U[i][j];
```
(b)

```
for(i = 1; i <= N; i++)
 for(j = 1; j <= N; j++)
  for(k = 1; k <= N; k++)
   c += U[i][k] + U[k][i];
```
(c)

```
for(i = 2; i <= N; i++)
 for(j = 1; j <= N - 1; j++)
  k += U[i][j] + U[i - 1][j + 1];
```
(d)

```
for(i' = 3; i' <= 2N - 1; i'++)
 for(j' = max(2, i' - N + 1); j' <= min(N, i' - 1); j'++)
  k += U[j'][i' - j'] + U[j' - 1][i' - j' + 1];
```
(e)

```
for(i = 1; i < N; i++)
 for(j = 1; j <= N; j++)
  for(k = 1; k <= N; k++)
   for(l = 1; l <= N; l++)
    d += (U[i][k] * U[i][k]) - V[j][l];
```
(f)

```
for(i = 1; i <= N; i++)
 for(j = 1; j <= N; j++)
  for(k = 1; k <= N; k++)
   c += U[i][k] * V[j][k];
 for(j = 1; j <= N; j++)
  for(k = 1; k <= N; k++)
   d = c + U[k][i] + U[i][k];
```
(g)

```
for(jt = 1; jt <= N; jt = jt + S)
 for(i = 1; i <= N; i++)
  for(j = jt; j <= jt + S - 1; j++)
   for(k = 1; k <= N; k++)
    W[i][j] += U[i][k] * V[k][j];
```
(h)

Figure 1: Different code fragments.

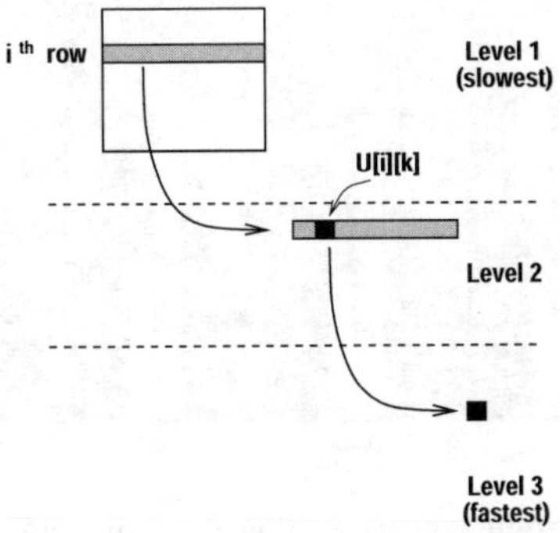

Figure 2: Exploiting three-level memory.

```
compute reuse matrix R = [r1, r2, ..., rs]
level=1;
for i = 1, s
  k = number of non-zeros in ri;
  if (k > level) then level = k; endif
endfor;
for l = 2, level
  capacity[level] = 0;
endfor;
for i = 1, s
  current_level = 2;
  let ri be [r1i, r2i, ..., rni];
  for j = 1, n
    if (rji != 0) then
      amount = F(reference(ri));
      capacity[current_level] = capacity[current_level] + amount;
      current_level = current_level + 1;
    endif;
  endfor;
endfor;
return(level,capacity[]);
```

Figure 3: Memory hierarchy design algorithm.

imizing data transfers by overlapping or combining them with each other. Consider, for example, the code fragment shown in Figure 1(g). Let us focus on a two-level memory hierarchy and on data transfers due to array U. Since both the sub-nests in this code has the same reference $U[i][k]$, the data transfer due to this reference can be performed just before the first j loop; there is no need to perform data transfer (due to this reference) before the second j loop. The second sub-nest has another reference to array U: $U[k][i]$. The data transfer due to this reference can be performed along with that due to $U[i][k]$ before the first j loop. After the execution of the second j loop, the elements of array U which reside in the fast memory can be discarded (as their reuse ends). Our current implementation handles perfectly-nested and imperfectly-nested loops of any depth. But, it does not attempt to optimize the hierarchical SPM usage across multiple, independent imperfectly-nested loops. That is, it does not try to take advantage of inter-nest data reuse. However, it should also be noted that handling imperfectly-nested loops means that our approach can be extended operate on an entire program as the entire code can be thought of as an imperfectly nested loop enclosed by an imaginary outermost loop that iterates only once.

2.6 Memory Design Algorithm

So far, we have assumed that a memory hierarchy already exists and tried to determine how this hierarchy can be exploited by capturing the reuse of data using reuse vectors/matrices. Our reuse vector/matrix based analysis however can also be used for designing a suitable SPM hierarchy for a given code. In this subsection, we address this problem. Note that after obtaining an initial memory hierarchy using the technique presented in this subsection, more sophisticated techniques [3] can be used, if desired, to refine the hierarchy.

2.6.1 Hierarchy Design for a Given Code

We first present a strategy to determine the cache configuration given a loop nest. A sketch of our approach is given in Figure 3. Our approach tries to build a separate hierarchy for each nest and then combines these hierarchies into a single hierarchy by choosing the largest memory required for each level. The algorithm first builds the reuse matrix and then it constructs the memory hierarchy incrementally. It starts with the first reuse vector and determines how much SPM space we need in each level. It then moves to the second reuse vector. Considering this vector, the algorithm can increase either the number of levels or the space needed by existing level(s) or both. These incremental updates stop once the last reuse vector has been processed.

As an example, consider once more the matrix multiply code in Figure 1(a) and its reuse matrix in Section 2.1. Considering the first reuse vector, $(0, 0, 1)$, we decide that the hierarchy should be at least two levels and that the smaller (and faster) memory should be able to hold $W[i][j]$. Considering the second reuse vector, $(0, 1, 0)$, we can see that the number of levels does not need to be increased (as this reuse vector has the same number of non-zero elements as the previous one). However, it should be large enough to hold an entire row of U. Finally, the last reuse vector $(1, 0, 0)$ implies that the fast memory should be able to hold the entire array V. Based on these, our algorithm decides that we need a two-level SPM hierarchy and that the faster level should have a capacity of at least N^2+N+1 elements. By default, we assume that the slower level has a capacity to hold the entire working set (N^3 elements). In the algorithm in Figure 3, we first compute the number of levels in the hierarchy and then find the capacity of each level. The function 'reference(.)' gives the array reference corresponding to the reuse vector under consideration, and the '$F(.)$' function computes the SPM space demanded by that reference (see Section 2.2).

We next discuss how to modify this strategy if the underlying system has some memory space constraints. For example, in case we may not be able to store the entire V array (in our current example) in the fast memory, we need to modify the code slightly. We note that strip-mining, a loop transformation that divides the iteration space of a given loop into stripes [12], can be useful in this context. In our example, we can strip-mine the j loop as shown in Figure 1(h). In this code, S is the stripe size. Now, the total space requirement for the fast memory (considering only i, j, and k loops) is $NS + N + 1$. By choosing a suitable value for S, the space demand (put on the fast memory) can be kept under control.

2.6.2 *Enforcing a Hierarchy*

The algorithm presented in the previous subsection does not try to modify the code to enforce it to adopt a pre-specified memory hierarchy; that is, it does not modify the reuse matrix; it just builds a hierarchy considering the current reuse matrix. In many cases, however, we might want to modify the reuse matrix to control the number of levels in the hierarchy. A reuse matrix can be changed using two types of transformations:

- The transformations that change the number of zeroes and non-zeroes in the reuse vectors.
- The transformations that re-order zeroes and non-zeroes in the reuse vectors (without changing their numbers).

Since in general the second type of transformation can be reduced to the first type, we focus only on the first types of transformations here. Our approach can take a reuse matrix and transform it to another reuse matrix such that in the resulting matrix most reuse vectors demand a t-level memory hierarchy, where t being a value between 2 and n (the number of loops in the nest). For the clarity of presentation, however, we present the algorithm for only the case when $t = 3$ (see Figure 4); that is, for each reuse vector in R, we would like to make all the entries in the vector zero, except two entries (these two entries will represent the loops that exhibit reuse). Note that in general the order in which the reuse vectors in a given reuse matrix are processed might also be important. Our current approach orders reuse vectors from left to right considering the number of times each reuse vector occurs in the code. The most frequently used reuse vector occupies the leftmost column whereas the least frequently used one sits in the rightmost column. Then, for the most frequent reuse vector, the algorithm makes sure that only two non-zero elements are obtained (after the transformation). It achieves this by selecting suitable vectors for the first $n - 2$ rows of T. In doing so, data dependences are also checked using the approach in [8] (this is not shown in the code for clarity). It then processes the remaining reuse vectors and selects suitable vectors for the remaining two rows. As soon as all rows of T have been determined, the algorithm returns the transformed reuse matrix (TR) and terminates.

3. EXPERIMENTS

We used array-based versions of four embedded image processing applications to evaluate the effectiveness of our strategy. `wood` is a

```
compute reuse matrix R = [r1, r2, ..., rs]
let T be the transformation matrix and t1, t2, ..., tn be the rows of T;
finished=false;
select t1, t2, ..., tn-3, tn-2 such that
  t1 ∈ Ker{r1}, t2 ∈ Ker{r1} ... tn-3 ∈ Ker{r1}, tn-2 ∈ Ker{r1};
j = 2;
while (not finished)
  bool1 = (t1 ∈ Ker{rj});
  bool2 = (t2 ∈ Ker{rj});
  ....
  booln-2 = (t3 ∈ Ker{rj});
  k = number of trues in {bool1,bool2,...,booln-2};
  if (k ≤ (n-4)) then
    set tn-1 to vector in Ker{rj};
    set tn to vector in Ker{rj};
    finished = true;
  else
    if (k = (n-3)) then
      if (tn-1 is not set) then
        set tn-1 to vector in Ker{rj};
      else
      if (tn is not set) then
        set tn to vector in Ker{rj};
        finished = true;
      endif;
    endif;
  endif;
  j = j + 1;
endwhile;
return(TR);
```

Figure 4: Memory hierarchy transformation algorithm.

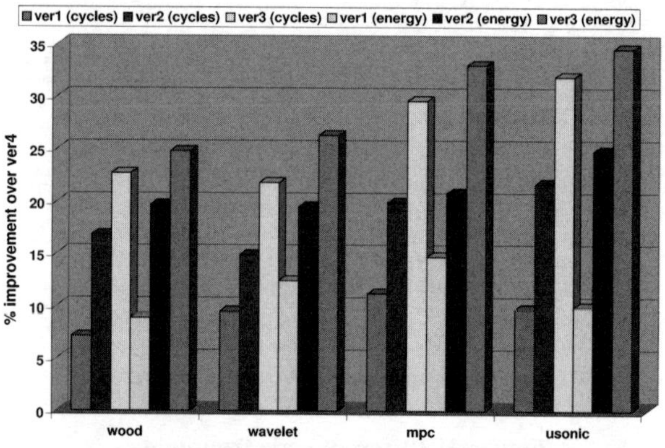

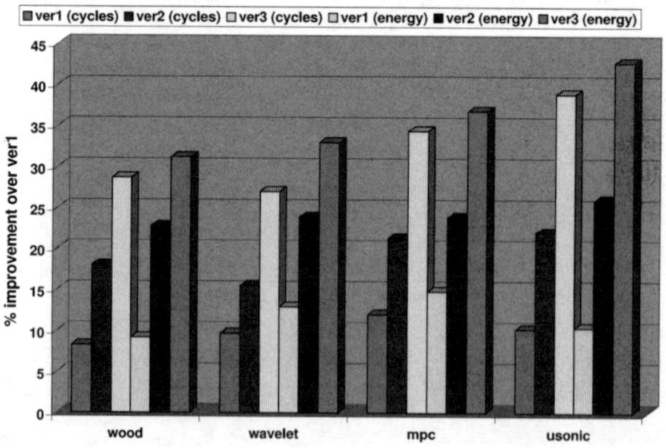

Figure 5: Percentage energy and performance improvements over `ver4`. Top: two-level memory hierarchy; Bottom: three-level memory hierarchy.

color-based visual surface inspection method for wood properties. It builds the color histogram percentile features of the image to recognize wood surface defects with relatively low complexity. wavelet is a wavelet-based noise reduction and restoration application. mpc is a medical pattern classification application. It is used mainly for choosing therapy for peptic ulcers. Finally, usonic is a feature-based estimation algorithm for ultrasonic image sequences. The total input sizes used for wood, wavelet, mpc, and usonic are 271KB, 306KB, 586KB, and 410KB, respectively. For each benchmark code, we experimented with four different versions. ver1 is a static SPM optimization strategy. It profiles the code and selects a set of array elements that has the highest reuse. It then puts these elements in the fast memory (at the beginning of the program) and keeps them there throughout the execution. ver2 is a nest-based dynamic SPM management strategy. It changes the contents of SPM depending on the current access pattern; its details can be found elsewhere [7]. ver3 is the reuse vector/matrix based SPM management strategy discussed in this paper. ver4 is a version that uses a cache memory (instead of SPM) in conjunction with the locality-optimized version of each application. All versions tested here have been implemented in a source-to-source translator using the SUIF infrastructure [1]. Our compiler takes a C code as input and optimizes it taking into account the memory hierarchy. The compiler output (which is also a C code) is then fed into a custom simulator. The simulator models a memory hierarchy (which is constructed using SPMs) and assumes a simple, single-issue embedded core (running on 200MHz). For cache experiments, the simulator directly calls Dinero [4].

We performed two sets of evaluations. In the first set, we fixed the SPM hierarchy and evaluated the performance and energy consumption of each version. In the second set, for each version, we determined the total fast level memory size and energy consumption under a given performance bound. We calculated the access latencies for different sizes of SPMs and data caches using the CACTI tool [6]. To compute the per access energy costs for different sizes of memories, we employed a slightly modified form of the model proposed by Shiue and Chakrabarti [11]. While our simulator also gives the energy spent in interconnect, we found this value very small for all benchmarks used.

Figure 5 gives the performance (execution cycles) and energy results for two-level (top) and three-level (bottom) memory hierarchies. All values presented are normalized with respect to the corresponding value of ver4 (the cache version). In the two-level memory, the simulated faster memory (SPM) size is 4KB; the cache used in the ver4 version is also 4KB (two-way associative with a block size of 32 bytes). In the three-level memory hierarchy, the fastest SPM size is 2KB and the middle level memory is 6KB. In all cases, the first level memory (the slowest memory) is 1MB. These results show that our strategy outperforms both ver1 and ver2 in both energy and performance. We also note that the gains are larger with the three-level memory (as our reuse vector based strategy utilizes all the fast levels fully). The average performance improvements due to ver1, ver2, and ver3 are 9.4%, 18.3%, and 26.5%, respectively, for the two-level memory. The corresponding values for the three-level memory are 10.0%, 19.2%, and 32.3%. Figure 6 shows the percentage increase in memory capacity and energy consumption (with respect to our approach) when all version generate the same performance results. Our experimentation methodology is as follows. First, we ran the algorithm in Figure 3 and determined a memory hierarchy for our version (ver3). We then recorded the total memory size (excluding the slowest memory, the size of which is always fixed at 1MB) used by our version and its execution time. Then, we ran other versions using several different levels and capacities, and found (for each version) the configuration that gave the same execution time as ver3 (within an error margin of 2%). We observe from these results that in order to catch the performance of ver3, the other versions demand much larger memory space and (because of this) significantly larger energy consumption. Even ver2 (which is the best among the remaining ones) increases the memory capacity (omitting the slowest one) by 35.4% and energy consumption by 16.5% on the average.

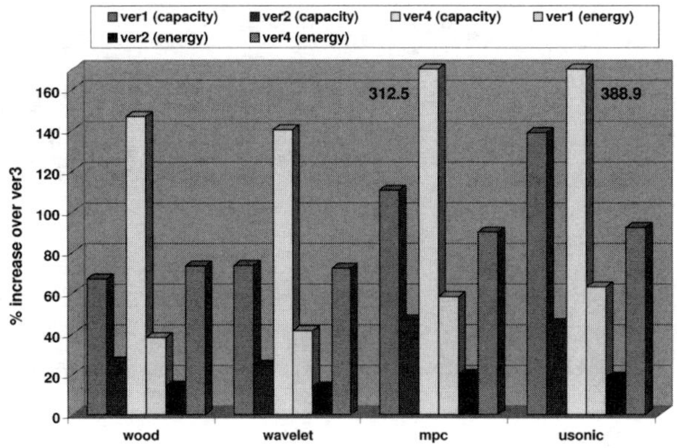

Figure 6: Percentage increase in memory size and energy over ver3 under constant performance.

4. CONCLUSIONS

We presented and discussed a scratch-pad memory (SPM) hierarchy design and optimization framework. In this framework, the compiler has a central role in the sense that it manages the flow of data across a given hierarchy (by staging computation and data). Given an access pattern, the compiler can also come up with a memory hierarchy which might be used as a starting point for further analysis and synthesis. The effectiveness of this strategy was measured using four complete applications from the embedded image processing domain. Our results reveal that this optimization framework is very successful in reducing both energy and execution cycles and that it outperforms two previous approaches to SPM management.

5. REFERENCES

[1] S. P. Amarasinghe, J. M. Anderson, M. S. Lam, and C. W. Tseng. The SUIF compiler for scalable parallel machines. In Proc. *the Seventh SIAM Conference on Parallel Processing for Scientific Computing,* February, 1995.

[2] L. Benini, A. Macii, E. Macii, and M. Poncino. Increasing energy efficiency of embedded systems by application-specific memory hierarchy generation. *IEEE Design & Test of Computers,* pages 74–85, April-June, 2000.

[3] F. Catthoor, S. Wuytack, E. D. Greef, F. Balasa, L. Nachtergaele, and A. Vandecappelle. Custom memory management methodology – exploration of memory organization for embedded multimedia system design. *Kluwer Academic Publishers,* June, 1998.

[4] Dinero IV Trace-Driven Uniprocessor Cache Simulator. URL: http://www.cs.wisc.edu/~markhill/DineroIV/

[5] J. Eyre and J. Bier. DSP processors hit the mainstream. *IEEE Computer Magazine,* pp. 51–59, August 1998.

[6] N. P. Jouppi and S. J. E. Wilton. An enhanced access and cycle time model for on-chip caches. *Research Report 93/5,* Compaq WRL, Palo Alto, CA, July 1994.

[7] M. Kandemir, J. Ramanujam, M. Irwin, N. Vijaykrishnan, I. Kadayif, and A. Parikh. Dynamic management of scratch-pad memory space. In Proc. *the 38th Design Automation Conference,* Las Vegas, NV, June 2001.

[8] W. Li. *Compiling for NUMA Parallel Machines.* Ph.D. Thesis, Computer Science Department, Cornell University, Ithaca, NY, 1993.

[9] P. R. Panda, N. D. Dutt, and A. Nicolau. Efficient utilization of scratch-pad-memory in embedded processor applications. In Proc. *European Design and Test Conference,* Paris, March 1997.

[10] P. R. Panda, N. D. Dutt, and A. Nicolau. Architectural exploration and optimization of local memory in embedded systems. In Proc. *ISSS'97,* Antwerp, September 1997.

[11] W-T. Shiue and C. Chakrabarti. Memory exploration for low power, embedded systems. In Proc. *Design Automation Conference,* New Orleans, Louisiana, 1999.

[12] M. Wolfe. *High Performance Compilers for Parallel Computing,* Addison-Wesley Publishing Company, 1996.

Unlocking the Design Secrets of a 2.29 Gb/s Rijndael Processor

Patrick R. Schaumont
UCLA Dept of EE
Los Angeles, CA
schaum@ee.ucla.edu

Henry Kuo
UCLA Dept of EE
Los Angeles, CA
henrykuo@ee.ucla.edu

Ingrid M. Verbauwhede
UCLA Dept of EE
Los Angeles, CA
ingrid@ee.ucla.edu

ABSTRACT
This contribution describes the design and performance testing of an Advanced Encryption Standard (AES) compliant encryption chip that delivers 2.29 GB/s of encryption throughput at 56 mw of power consumption. We discuss how the high level reference specification in C is translated into a parallel architecture. Design decisions are motivated from a system level viewpoint. The prototyping setup is discussed.

Categories and Subject Descriptors
C.3 [**Special-Purpose and Application-based Systems**]

General Terms: Measurement, Performance, Design, Security.

Keywords: Rijndael, Encryption, Domain-Specific, Low-Power.

1. INTRODUCTION
Recently we developed and tested an AES compliant encryption processor that implements Rijndael [1] at 2.29 Gb/s and 56 mW of power consumption. This device can be used to instrument a digital broadband platform like a router or a wireless base station with security services. The processor is programmable and supports Rijndael in any combination of key length (128,192,256 bits) and data size (128,192,256 bits). It is integrated into the host platform through a 16 bit data bus and operated through a small instruction set. The implementation uses 173 KGates of 1.8V 0μ18 CMOS standard cell technology, and has been verified operational at up to 154 MHz clock frequency in a prototype setup.

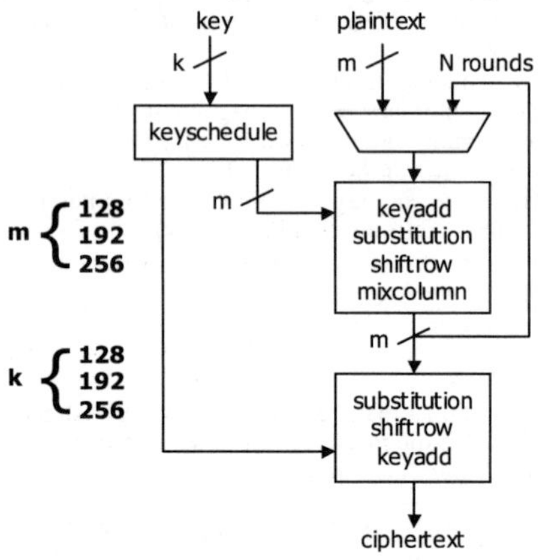

Figure 1: Rijndael Encryption

The paper is organized as follows. In section 2, we will review the system level architecture. This architecture was designed starting from the NIST reference implementation in C [2]. We will discuss the transformations that are needed on this description to create the system architecture. In addition we touch upon integration issues for our architecture.

In section 3 we provide motivation for some design decisions with a system level look at the chip. In addition, we also touch upon related work and other Rijndael implementations that have been done by academia and industry.

We used an HDL based design flow for this chip, which will be discussed in section 4. We will also discuss design statistics with respect to design efficiency. Finally the prototyping approach and measurements are shown in section 5, and conclusions and outlook are given in section 6.

2. SYSTEM SPECIFICATION AND ARCHITECTURE
The Rijndael encryption algorithm is a block cipher that converts cleartext data blocks of 128, 192 or 256 bit into ciphertext blocks of the same length. It uses a key of selectable length (128, 192, 256 bit). The algorithm is organized as a set of iterations called *rounds* as illustrated in Figure 1. The number of rounds is dependent on the data block length. For each round, a subkey is created out of the original key by means of a key schedule. The operations performed on the data blocks include byte substitution by means of a lookup table, transpositions and rotations, modulo-2 addition of a subkey as well as Galois field operations. They always affect a complete 128, 192 or 256-bit data block at a time.

2.1 System Architecture
The approach we took was to design a high-speed hardware accelerator of the Rijndael algorithm. As will be demonstrated, the energy efficiency of this approach is orders of magnitude better than using a general-purpose programmable platform.

The system architecture of our implementation is shown in Figure 2. The central block of the architecture, `encrypt`, implements one round of a Rijndael encryption in a fully parallel, non-pipelined fashion. A Rijndael encryption can be completed at one

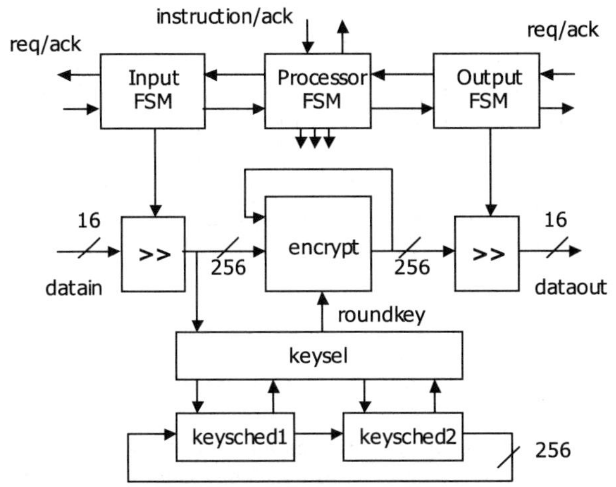

Figure 2: System Architecture

clock cycle per round. Thus, following the standard we need from 10 clock cycles for 128 bit blocks up to 14 clock cyles for 256 bit blocks.

The architecture can support any combination of key-length and data-block length. As a result the key-scheduling process speed must be adjusted relative to the encryption process speed when data-blocks and sub-keys have different lengths. For example, when 256 bit data blocks and 128 bit subkeys are needed, then 2 key schedule iterations are needed for each data block. With a dual implementation of key schedulers, this double rate can be supported. Double-rate key scheduling is only a worst case situation, but also non-integral rates can occur. The combination of 192-bit data blocks with 128-bit keys for instance requires 1.5 key schedule iterations per data block.

The processor has three controllers, two for I/O interfacing and one for instruction sequencing. They communicate through request/acknowledge protocols with the host system. This asynchronous interfacing method allows the chip to be clocked much faster than the bus it is connected to. It also brings considerable simplification of the performance testing process. Separation of I/O controllers from the instruction sequencing controller makes this block easily portable to a different context, were a different data-bus or I/O interface protocol would be used.

Figure 3 shows the internals of the central encrypt block and illustrates the nature of the data processing that is done by this processor. Data words are 256 bit wide. The data is organized conceptually in a 4 by 8 matrix of bytes, totaling 32 bytes. If the processor is working in 128 or 192 bit mode, then the leftmost 2 resp. 4 columns of the matrix are unused. The processing starts at the bottom `keyadd` block which performs modulo-2 addition of a subkey to each data byte. Next, each data byte is fed into an S-box, implemented as a lookup array in the `substitution` block. This happens in parallel for all 32 bytes, so 32 lookup tables (of 256 8-bit entries) are used. In the `shiftrow` block, the matrix rows are circularly shifted with a parameter- and row-dependent amount of positions. Finally in `mixcolumn` each column is transformed linearly using galois-field constant multiplications. All of these operations are performed within one clock cycle. Yet by careful design the critical path of this block is

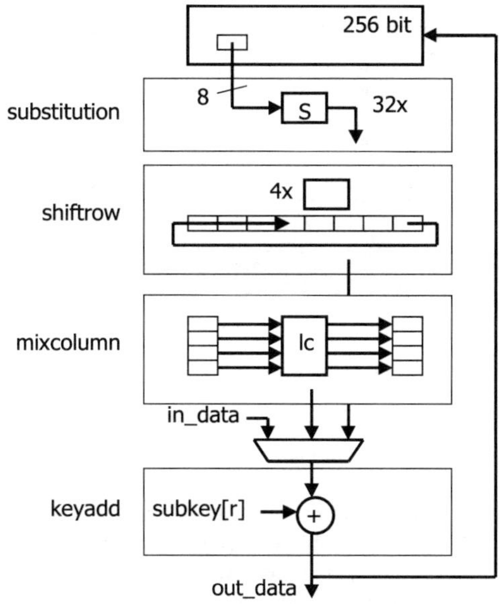

Figure 3: Encrypt Block

6ns in 0µ18 technology. The overall critical path of the design is 10 ns and resides in the key schedule block. A detailed discussion of the architecture optimization has been published [3].

2.2 System Specification

The Rijndael encryption algorithm is available as a C program on the NIST website [2]. For hardware implementation, the reference implementation in C presents a number of issues that need to be resolved before architecture design can start. For a hardware designer, these are best described as *ambiguities*. We enumerate those below to demonstrate what problems would be encountered by a C-to-hardware compiler, assuming it would start from the reference implementation.

First, a parallel version of the sequential specification is needed. This requires dataflow analysis that crosses the boundaries of function calls. The hardware in Figure 3 for example is the result of merging several C function calls, and unrolling and merging the for-loops in each function. It is nontrivial because the loop bounds in C are variable and parameter-dependent. In addition, not all C functions are treated the same way. For example, the evaluation of sub-keys can be done online (in parallel with the encryption process) or offline. In C, the notion of parallelism is absent, and the distinction between on-line and off-line is merely an assumption reflected in the ordering of function calls. In hardware however, it has major impact on the memory architecture as all subkeys need to be stored first in offline mode.

A second aspect involves parameterization. The introduction of runtime parameters like selectable key-length has a direct impact on the implementation overhead of the hardware. The C reference code implements parameterization with lookup arrays and control constructs such as `if-then`. In hardware, an efficient implementation requires us to get rid of runtime variability as much as possible and for example apply constant propagation whenever a parameter is known constant. Also if the parameter range is restricted, we can simplify the implementation. One particular example is shown in Figure 4 and demonstrates the implementation of the Galois Field multiplications. In the

Original

```
word8 Alog[255] = { ..};
word8 Log[255]  = { ..};
word8 mul(word8 a, word8 b) {
if (a&&b)
  return Alog[(Log[a]+Log[b])% 255];
}
```

Optimized

```
/* we know that only mul(a,2) and mul(a,3)
   are used. Do both of them in this
   function without lookup tables */
void mul23(word8 a, word *m2,
                    word *m3) {
  *m2 = (a & 0x80) ? (a<<1) ^ 0x1C
                   : (a<<1);
  *m3 = *m2 ^ a;
}
```

Figure 4: An example of code specialization

reference code, the Galois Field multiplications are implemented using a Log-Alog lookup table (with two such multiplications per matrix row en per round). The Log-Alog lookup table is elegant to describe and easy to understand but expensive as it uses large lookup tables. On closer inspection it turns out that for encryption only two different constant multiplications are used, with 2 and with 3. We thus can replace the lookup-table method with a dedicated structure.

A third aspect is memory architecture, which is richer and more elaborate in a hardware implementation then in software. The C reference implementation uses arrays for both the storing of intermediate results, the storing of keys and data blocks, as well as lookup tables. In the hardware implementation, these translate to wiring, registers, random logic tables and dedicated computation blocks. The problem that the designer is facing is that the initial description in C gives no clue about how to treat each array. This problem is also worsened by the lack of parallelism in the initial description.

A fourth aspect involves using transformations that cross memory-computation boundaries. An example is the implementation of the key scheduling. Subkeys are created in our architecture online. Because of variable key and datablock length, a setting can occur where two subkeys per round are needed, each 128 bit, for one datablock of 256 bit. As a result the key scheduling hardware is implemented twice, as illustrated in Figure 2. This way the hardware can evaluate up to two subkeys per clock cycle. In the C reference implementation, the key scheduling function initializes one array that contains all subkey material for one data block. It uses a parameter-dependent `while` loop to ensure enough subkeys have been evaluated. Clearly such a while loop cannot be transformed into a fixed rate hardware implementation. The kind of transformation that is needed here crosses the boundaries of memory and computation space.

Summarizing, we observed the need for the following analysis techniques of the C code.

- Dataflow analysis that can cross the borders of function calls, and that gives feedback to the designer as to what strategies can be used for parallelization of the code, both at function-level as well as block-level. Such feedback includes memory-access and operation profiling.
- Source-level loop transformations and analysis of the storage architecture alternatives [5].

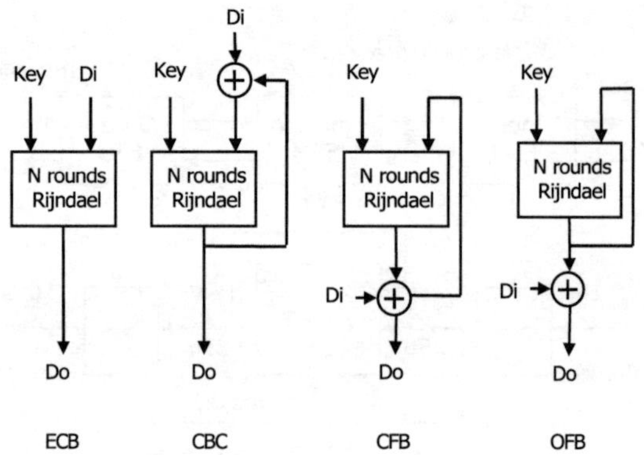

Figure 5: Feedback Modes prevent Pipelining

- Source-level constant propagation and folding, also for variables for which a constant input can be guaranteed. Reduction of source-level control-flow complexity as a result of this.

These techniques are known in the high-performance software compiler community. The existing C-to-hardware technology however focuses on datapath- and sequential controller synthesis. High level synthesis support is available for fixed architecture templates such as VLIW or ASIP with customizable instruction sets. Since our architecture could not be classified as one of those, we choose to base our implementation flow on HDL and do manual architecture design.

2.3 Integration Issues

The interface that attaches the accelerator processor to the host system is used to transfer blocks of plaintext and keys onto the processor, and extract ciphertext out of it. It also presents an instruction set to the host system to control the execution of the Rijndael algorithm. Two factors were taken into account here.

First, the interface operation was designed to be independent of the processor clock. Typically the interface clock speed is decided by the system communication bus present in the host. The Rijndael processor clock speed on the other hand is as fast as critical path and power consumption constraints allow. Therefore, two-way handshakes were used on both the data bus as well as the instruction bus (Figure 2).

The second issue is that the interface should allow easy porting over a range of host processors, with potentially different buswidths. With separate I/O controller FSMs, the amount of code that needs to be revised in a new context is kept minimal.

3. SYSTEM ISSUES AND RELATED WORK

We now motivate several design decisions in our architecture and indicate some related work along the way.

3.1 Use of Pipelining

In hardware implementation, pipelining is used as a throughput enhancing technique. Considering the system architecture in Figure 2, one possible transformation is to unroll the iterative encryption algorithm (`encrypt` block) and next insert pipeline stages. This leads to very high throughputs (25 Gbps is claimed by some vendors of 0μ18 CMOS softcores). Such pipelined implementations have only limited application for a block cipher. These ciphers are always used in well defined *mode of operation*

(Figure 5). The modes include Electronic Code Book, Cipher-Block Chaining, Cipher Feedback and Output Feedback. Three of these four modes rely on feedback of the ciphertext to the input. For these modes, pipelining of the Rijndael block is not useful to increase throughput. In addition, the mode that can be pipelined (ECB), is also cryptographically weakest since it translates identical plaintext blocks to identical ciphertext blocks. An ECB cipher will for example not hide recurring data patterns [7]. For this reason, our implementation does not use pipelining. The 2 Gb/s performance figure of our chip holds for any encryption mode.

3.2 Online Subkey Computation

The subkeys that are created by the key scheduling process need to be evaluated only once per key. After that the subkeys can be reused for all data blocks as long as the key is not changed. Therefore some implementations choose to evaluate all subkeys offline before the encryption process starts, and to store the subkeys in a lookup table. This approach is good for batchmode encryption applications, like for instance secure document storage, but not for applications where keys are changed frequently.

In our implementation, subkeys are calculated online as needed and can be changed at each data block. This situation occurs for example in Internet routers with IPSEC support [4]. In an IPSEC router, the key and subkey material (1280 bits for 128-bit data blocks) is potentially different for each secured packet route. Due to the large number of active routes, subkeys cannot be stored in on-chip memory. They need to be either calculated on-line or else moved together with a packet payload onto chip. However, half of the Internet packets are only 64 bytes in length (512 bits) [6]. With offline calculation and off-chip storage of subkeys the router will use more bus bandwidth for the communication of subkeys than for the communication of useful data payloads. The effective solution for this is online computation of subkeys such as is done in our architecture.

3.3 Decryption and Authentication

While our chip does not contain a Rijndael decryption algorithm, it can still be used to build a decryption unit. The OFB and CFB modes shown in Figure 5 create ciphertext from plaintext by means of a simple XOR operation. Since this operation is its' own inverse we can use the same Rijndael encryption block for both encryption and decryption. In communication systems, the use of the OFB mode is popular for securing data streams because it prevents the propagation of communication errors into multiple blocks of data.

Another application where only an encryption algorithm is used is message authentication. Here a sender creates a message signature by encrypting the message with a CBC-mode cipher, and sending the last encryption output as the signature. The signature is hard to compute without knowing the encryption key. A receiver who shares this secret key, can then verify the authenticity of the message by recomputing the signature using the same CBC encryption on the received data.

3.4 Software Implementations

Several software implementations have been created with handcrafted assembly for highest performance. The best implementation we found requires 227 cycles for 10 rounds of

Table 1: Comparison based on Energy Efficiency

AES 128bit key 128bit data	**Throughput**	**Power**	**Figure of Merit (Gb/s/W)**
This work (0u18 CMOS)	1.28 Gbits/sec	56 mW	22.8 (100%)
This work (on FPGA [i])	640 Mbit/s	1.63 W	0.39 (1.7%)
Pentium III [ii],[iii]	645 Mbits/sec	41.4 W	0.015 (0.06%)

[i] Xilinx Webpack + Xilinx Virtex Power Estimator on Domain Spec Processor Design
[ii] Helger Lipmaa, http://www.tcs.hut.fi/~helger/aes/rijndael.html: PIII assembly handcoded
[iii] Intel Pentium III Datasheet 1.13 GHz (Vcc=1.8V, Icc=23A)

128 bit data blocks and 128 bit keys on a Pentium III, and 124 cycles on IA64 [8]. Since these run on processors clocked in the GHz range, this performance eventually approaches the efficiency of a dedicated hardware implementation. With respect to power consumption however, the software solution on general purpose processors is three orders of magnitude higher then a dedicated hardware implementation (Table 1).

3.5 Hardware Implementations

Initial designs from academia [9,10,11] and industry seem to have focused on the design of Rijndael cores in reconfigurable hardware. These implementations show that contemporary reconfigurable platforms with their rich distributed memory architecture are well suited for Rijndael prototype implementations. Most of them use pre-computed subkeys. For pipelined ciphers, very high performances have been obtained. For non-pipelined operation in FPGA, the 1.5 Gbps limit is to our knowledge still unbroken.

While almost no published power consumption figures exist, we estimate that our processor is at least two orders of magnitude better then an FPGA solution (Table 1).

Several softcore implementations have been designed as well, which advertise clock speeds ranging from 200 to 350 MHz in

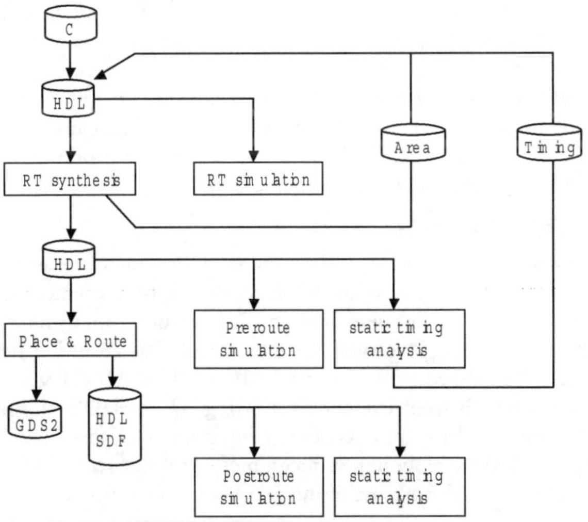

Figure 7: HDL Design Flow

0μ18 CMOS technology. One published ASIC design [12] implements a fully enrolled encryption unit in in 613 Kgates of 0u35 CMOS technology. An overview based on pre-tapeout performance figures can be found in [13].

3.6 AES standard implementation

Since we started design of the chip, NIST has finalized the standard for AES encryption. While the architecture of Figure 2 supports a superset of the AES parameters, it can also be simplified when only AES compliant settings are needed. AES restricts the data block size to 128 bits only. This means that the encrypt block of Figure 2 can be halved. In addition, one key scheduling block can be eliminated, since the keys are always larger then the data blocks. This would roughly reduce the chip gate count by half (85 Kgate). The critical path that currently resides in the key scheduler would be reduced as well. We estimate that the resulting design would reach at least 200 MHz with the same HDL synthesis methodology.

4. DESIGN AND IMPLEMENTATION

4.1 HDL Design

From the design point of view, translating the architecture of Figure 2 into HDL is a straightforward process, once the RT level architecture is designed.

One aspect that was found to be difficult was the design of the control architectures. This is because there is no elegant HDL mechanism by which control structures (finite state machines) can be captured. A coding style based on `case` statements quickly becomes confusing and complex, and tends to increase the amount of coding as shown in Figure 6.

This figure shows the relative weights of design elements used to implement hierarchy, controller, datapath and lookup tables. The four categories are non-commented lines of HDL code, number of HDL files, number of instances in the design hierarchy and finally the silicon area. It is easily seen that less then 50% of the lines of HDL occupy more then 90% of the chip area (lookup tables and datapath). This observation is important when considering design productivity issues. Closer inspection reveals that the use of HDL hierarchy is very effective in datapath design, but not in the design of controllers. Each of the three controllers in our design has their own HDL design, and each translate to a single instance in the design hierarchy. Yet, they require one third of the coding. They also represent a semantically hard part because of the way FSM are encoded in HDL. While there are tools that capture FSM design in a graphic formalism, these require partioning of the design in a graphic and a non-graphic part.

4.2 Design Flow

The chip was implemented with an HDL based design flow as shown in Figure 7. The inital RT level description, created out of the C reference implementation by careful analysis, was synthesized to gate level with Synopsys DC. The resulting netlist was then processed with Avanti's SIlicon Ensemble P&R. The verification path uses Cadence's Verilog-XL as HDL simulator and Synopsys' Primetime as static timing analysis tool. The chip was processed by National Semiconductor. Of the 16 test samples that were received from the foundry, 14 were operational.

5. TEST AND MEASUREMENTS

The test setup, shown in Figure 8, allows creating complete schmoo plots and at-speed testing of the chip. The test setup was made using readily off the shelf equipment, connected as shown in Figure 9. A vector generator provides testbench stimuli at a rate of 20 Mword/s. Only 9 bits of testvector input data are needed; by using simple keys and data patterns some of the input bits could be tied to ground. The testvectors are processed by our AES chip which is clocked by a clock generator at a high rate (100 MHz nominally). The outputs of the chip are observed by a logic analyser. In order to save logic analyser memory the analyser is operating synchronously and clocked out of the vector generator. This setup works because the interfaces at the chip boundaries are implemented using two way handshakes. Once the test vectors are entered in the chip, the encryption process runs at full speed. The relevance of this testing approach is that it operates our core at full speed in order to demonstrate 2.29 Gb/s encryption throughput, without requiring expensive equipment that can feed test vectors at such rates into the chip. The AES chip also uses a variable voltage supply. In combination with the variable clock generator this allows to determine the power consumption at each critical clock frequency. This is illustrated in Figure 10. The chip characteristics are summarized in Table 2.

Figure 8: Test Setup

Table 2: Chip Characteristics

Feature		Value		Unit
CMOS Stcell Technology		0.18		um
Core Voltage (VDD)		1.8		V
IO Ring (VDDIO)		3.3		V
Package		68		PLCC
Min. Clock Period	(1.95 V)	6.5		ns
	(1.9 V)	7.5		ns
	(1.8V)	8		ns
Power Consumption	(1.95 V)	82		mW
	(1.9 V)	67		mW
	(1.8V)	54		mW
Gate Count		173		Kgate
Throughput	**Data/Key**	**128**	**192**	**256**
	128	1.6	1.33	1.14
	192	2.0	2.0	1.71
	256	2.29	2.29	2.29
Interfaces	16 bit data-in with req-ack handshake			
	16 bit data-out with req-ack handshake			
	4 bit instruction with req-ack handshake			
	reset, clock and processor status signals			

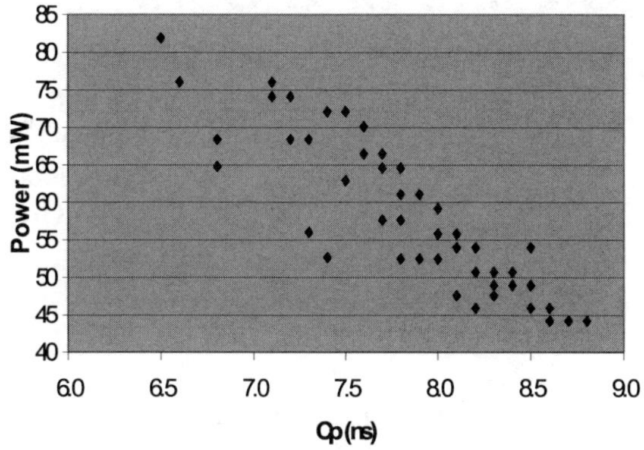

Figure 10: Power Consumption and Critical path

6. CONCLUSIONS AND OUTLOOK

In this contribution, we presented a 173 Kgate Rijndael encryption core that was verified at 2.29 GB/s encryption speed. The architecture was designed for best performance over several different cryptographic modes, and several different host systems. Our current research efforts will further improve this design in two areas. First, power consumption will remain a critical factor, especially when cryptographic applications will move into embedded context. This leads to the concept of power-optimized domain specific processors, which we expect to see much more in the future. Second, while state of the art compilers cannot yet take arbitrary high level code into implementation, there is a clear need for design at higher abstraction level. We are also developing a design environment that can tackle the design of domain specific processors at higher abstraction level.

7. ACKNOWLEDGEMENTS

The authors would like to acknowledge the support of National Semiconductor for chip processing, as well as the funding support of NSF and UC Micro.

8. REFERENCES

[1] J. Daemen, V. Rijmen, The block cipher Rijndael, Smart Card Research and Applications, LNCS 1820, J.-J. Quisquater, B. Schneier, Eds, Springer Verlag, 2000, pg. 288-296.

[2] NIST Advanced Encryption Standard, http://www.nist.gov/aes/

[3] H. Kuo, I. Verbauwhede, Architectural Optimization for a 1.82 Gb/s VLSI Implementation of the AES Rijndael algorithm, Cryptographic Hardware and Embedded Systems (CHES 2001), LNCS 2162, pp. 51-64, Springer-Verlag.

[4] R Oppliger, Security at the Internet Layer, IEEE Computer, September 1998.

[5] F. Catthoor, S. Wuytack, E. De Greef, F. Balasa, L. Nachtergaele, A. Vandecappelle, "Custom Memory Management Methodology", ISBN 0-7923-8288-9, Kluwer Academic Publishers, 1998.

[6] D. Whiting, B. Schneier, S. Bellovin, AES Key agility issues in high speed IPsec implementations, Public Comments on AES Candidate Algorithms Round 2, http://www.nist.gov/aes

[7] A. Menezes, P. van Oorschot, S. Vanstone, Handbook of Applied Cryptography, CRC Press, 1997.

[8] H. Lipmaa, AES Candidates: A survey of implementations, http://www.tcs.hut.fi/~helger/aes/rijndael.html

[9] M. McLoone, J. McCanny, High Performance Single Chip FPGA Rijndael Algorithm Implementations, Workshop on Cryptographic Hardware and Embedded Systems, Paris, 2001.

[10] V. Fischer, M. Drutarovsky, Two Methods of Rijndael Implementations in Reconfigurable Hardware, Workshop on Cryptographic Hardware and Embedded Systems, Paris, 2001.

[11] P. Chodowiec, K. Gaj, P. Bellows, B. Schott, Experimental Testing of the Gigabit IPSec-Compliant Implementations of Rijndael and Triple DES Using SLAAC-1V FPGA Accelerator Board, Proc. Information Security Conference, Malaga, Spain, Oct 1-3, 2001.

[12] T. Ichikawa et al, Hardware Evaluation of the AES Finalists, in Proc. 3th AES Candidate Conference, New York, April 13-14, 2000.

[13] A. Satoh, S. Morioka, K. Takano, S. Munetoh, A Compact Rijndael Hardware Architecture with S-Box Optimization, Proc. ASIACRYPT 2001, LNCS 2248, pp. 239-254, 2000.

The iCORE^{tm} 520 MHz Synthesizable CPU Core

Nick Richardson, Lun Bin Huang, Razak Hossain, Tommy Zounes, Naresh Soni
STMicroelectronics, Inc
Advanced Designs, Central R&D
4690 Executive Dr. Ste. 200, San Diego, CA 92121
(858) 452-7715 x 270
nick.richardson@st.com; lun-bin.huang@st.com; razak.hossain@st.com; tommy.zounes@st.com; naresh.soni@st.com

Julian Lewis
STMicroelectronics, Ltd.
Advanced Designs, Central R&D
1000 Aztec West - Almondsbury
BRISTOL, BS32 4SQ - UK
+44 145 446 2647
julian.lewis@st.com

ABSTRACT

This paper describes a new implementation of the ST20-C2 CPU architecture. The design involves an eight-stage pipeline with hardware support to execute up to three instructions in a cycle. Branch prediction is based on a 2-bit predictor scheme with a 1024-entry Branch History Table and a 64 entry Branch Target Buffer and a 4-entry Return Stack. The implementation of all blocks in the processor was based on synthesized logic generation and automatic place and route. The full design of the CPU from microarchitectural investigations to layout required approximately 8-man years. The CPU core, without the caches, has an area of approximately 1.5 mm^2 in a 6-metal 0.18m CMOS process. The design operates up to 520 MHz at 1.8V, among the highest reported speeds for a synthesized CPU core [1].

Categories and Subject Descriptors

M.2.4 [**Design Methodologies**]: High performance ASIC design: *Optimizing microarchitecture to layout.*

General Terms: Performance, Design.

Keywords: Synthesis, ASIC, CPU, embedded, st20, pipeline, cache, branch-prediction, high-frequency, microarchitecture

1. INTRODUCTION

The goal of the iCORE project was to show that a synthesizable core could be capable of execution speeds typically only achievable by complex and time-consuming custom designs. This was to be done by using a customized ASIC design flow to create a very high performance version of STMicroelectronics' ST20-C2 embedded CPU core.

The ST20-C2 architecture is a superset of the acclaimed inmos transputer, which was first introduced in 1984 (inmos was acquired by STMicroelectronics in 1989). The ST20-C2 added an exception mechanism (interrupts and traps), extensive software debug capability and improved support for embedded systems. It is now used in high-volume consumer applications such as chips for set-top boxes.

For the project to be considered successful, however, we had to demonstrate optimization in all the following areas, in approximate order of importance:

- Frequency
- Execution efficiency (IPC)
- Portability
- Core size
- Power consumption

2. OPTIMIZING THE MICROARCHITECTURE

To achieve good instruction-execution efficiency without excessive design complexity, previous implementations of the ST20-C2 architecture employed relatively short pipelines to reduce both branch-delays and operand/result feedback penalties, thus producing good IPC counts over a wide range of applications. In the case of iCORE, however, the aggressive frequency target dictated the use of a longer pipeline in order to minimize the stage-to-stage combinatorial delays. The problem was how to add the extra pipeline stages without decreasing instruction execution efficiency, since there would be little point in increasing raw clock frequency at the expense of IPC. Following analysis using a C-based performance model, a relatively conventional pipeline structure was chosen, but one that had some important variations targeted at optimizing instruction flow for the unique ST20-C2 architecture.

The pipeline microarchitecture is shown in Figure 1. It comprises four separate units of two pipeline stages each. The **IFU** (Instruction Fetch Unit) comprises the IF1 and IF2 pipeline stages and is responsible for fetching instructions from the instruction cache and performing branch predictions. The **IDU** (Instruction Decode Unit) comprises the ID1 and ID2 pipeline stages and is responsible for decoding instructions, generating operand addresses, renaming operand-stack registers and checking operand/result dependencies, as well as maintaining the program counter, known in this architecture as the Instruction Pointer (IPTR), and a Local Workspace Pointer (WPTR). The **OFU** (Operand Fetch Unit) comprises the OF1 and OF2 pipeline stages and is responsible for fetching operands from the data cache, detecting load/ store dependencies, and aligning and merging data supplied by the cache and/or buffered stores and data-forwarding buses. Finally, the **EXU** (Execute Unit) comprises the EXE and WBK pipeline stages and performs all arithmetic operations other than address generation, and stages results for writing back into the Register File Unit (RFU) or memory.

The RFU contains the 3-deep register-stack, conceptually organized as a FIFO, comprising the *A*, *B*, and *C* registers (in top-to-bottom order). In practice, it is implemented using a high-speed multi-port register file. The memory-write interface is coupled with a store buffer (SBF) that temporarily holds data while waiting for access to the data cache.

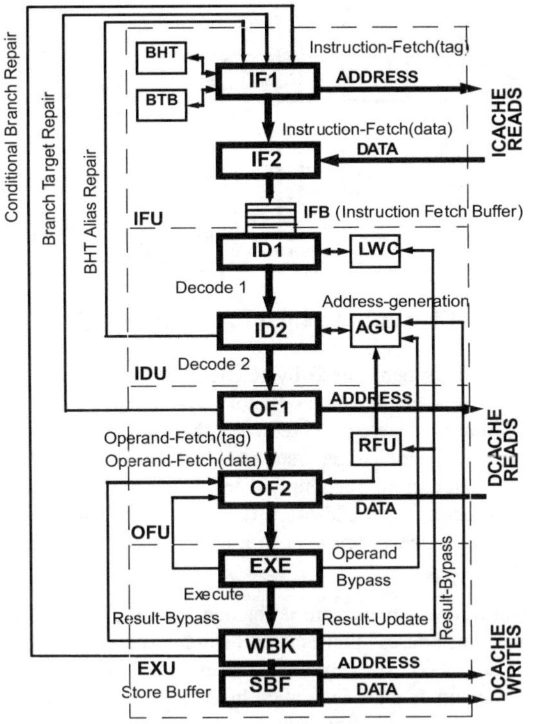

Figure 1. iCORE Pipeline Structure

The instruction-fetch portion of the pipeline (the IFU) is coupled to the execution portion of the pipeline (the IDU, OFU, and EXU) via a 12-byte Instruction Fetch Buffer (IFB).

To ensure that the pipeline is not continually stalled waiting for memory operands, two in-line operand caches are tightly integrated into the pipeline, both capable of being accessed in a single pipeline throw.

The first operand cache is for local variables, which are accessed relative to a Local Workspace Pointer (WPTR). It is called the Local Workspace Cache (LWC) and accelerates references to the first 16 words in the current local workspace, as pointed to by WPTR. The size of the LWC was determined by the results of running a number of benchmark programs on the C-based statistical performance model. It is placed early in the pipeline (in the ID1 stage) so that local variables used for non-local address calculations can be supplied to the Address Generation Unit (AGU) in the next pipeline stage (ID2) for early calculation of the full non-local address.

The second operand cache is an 8 KB in-line data cache that accelerates references to non-local variables, which can be anywhere in the 4 GB address space, as well as local variables that miss the LWC. It uses non-local addresses calculated by the AGU in the ID2 stage, and supplies operands to the Execute Unit, thus occupying the two pipeline stages before the EXU. For critical path optimization, its tag RAM and data RAM are placed in two different pipeline stages. Accessing the tag RAM before the data RAM eliminates the cache's set-associative output multiplexing that would have been needed if the tag and data RAMs were placed in the same stage, and also allows cache reads and writes to share the same control flow, thus simplifying the design.

Together, these two highly integrated operand caches are complementary to the ST20-C2 architecture's *A*, *B*, and *C* registers, effectively acting like a very large register file.

2.1 Instruction Folding

An important effect of the in-line operand caches is to increase the opportunities for instruction folding. This is an instruction decoding technique that combines two or more machine instructions into a single-throw pipeline operation. Since the ST20-C2 has a stack-based architecture, most of its instructions specify very simple operations such as loading and storing operands to and from the top of the 3-deep register-stack, or performing arithmetic operations on operands already loaded onto the register-stack, but for simplicity, memory and ALU operations are never combined in the same instruction. Without such compound instructions that allow memory and arithmetic operations to occur as single-slot pipeline operations, it would not be possible to take advantage of iCORE's in-line memory structure. With folding, however, up to three successive instructions can be merged into one operation that occupies a single execution slot and which fully uses iCORE's pipeline resources.

As an example, take the following very common three-instruction sequence:

ldl <n>
ldnl <m>
add

The first instruction, **load local** (*ldl*) loads from memory a variable that is addressed by an offset <n> from the current value of the Local Workspace Pointer (WPTR) and pushes it onto the top of the register-stack. The second instruction, **load non-local** (*ldnl*) loads from memory a variable that is addressed by an offset <m> from the value on top of the register-stack (in this case the value that was just placed there by the *ldl* instruction). Finally the *add* instruction adds the top two values on the register-stack (*A*, *B*) and pushes the result back onto it.

Without folding, this instruction sequence would occupy three distinct pipeline slots, and could be subject to stalling due to data-dependency between the *ldl* and the *ldnl* instructions. However, with folding and iCORE's in-line memory structure they can be executed in a single pipeline slot, even though they require two memory reads. To understand how this is done, consider the stage- by-stage execution:

IF1: Instruction fetch (instruction cache tag access)

IF2: Instruction fetch (instruction cache data RAM access, load IFB)

ID1: execute the *ldl* by reading local variable from LWC at offset <n>; issue *ldl, ldnl, add* from IFB as one operation if *ldl* hits the LWC

ID2: Using the AGU, generate address to be used by the *ldnl* instruction by adding the local operand obtained by the *ldl* in ID1 to the offset <m>.

OF1: Fetch data requested by *ldnl* from data cache (tag RAM)

OF2: Fetch data requested by *ldnl* from the data cache (data RAM); fetch second data operand from register file unit (pop top of stack)

EXE: Add the data obtained by the *ldnl* in OF2 to the second operand from register-stack.

WBK: Push result onto the top of the register-stack.

Note how the three folded instructions are executed separately by different pipeline stages (the *ldl* in ID1, the *ldnl* in ID2, OF1, OF2, and the *add* in EXE, WBK), but because only one pipeline stage at a time is occupied, the three instructions are apparently executed in a single clock cycle.

There are many such folding sequences supported by iCORE, but they all follow the same principle as described above.

2.2 Data Forwarding

In a deeply pipelined design, it is important to mitigate the effects of delays caused by pipeline latencies. One significant source of latency-based performance degradation is caused by data dependency between successive instructions. If the dependency is a true one (such as when an arithmetic operation uses the result of another instruction immediately preceding it) then there is no way of eliminating it, although microarchitecture techniques can be used to reduce the pipeline stalls it causes. This is generally done by providing data forwarding paths between potentially inter-dependent pipeline stages. In iCORE this includes data bypass paths from WBK to OF2, EXE to OF2, WBK to ID2, and EXE to ID2, as shown in Figure 1.

2.3 Branch Prediction

iCORE's relatively long pipeline gives rise to the danger of branches causing significant performance degradation compared to previous ST20-C2 implementations. After performance simulations showed that the effect of the longer pipeline on some benchmarks was significant, a low-cost branch prediction scheme was incorporated into iCORE's microarchitecture.

In a machine without branch prediction, the source of most branch penalties is two-fold:

Firstly, the target address of a taken branch or other control-transfer instruction must be determined. This cannot be done until after the branch has been decoded, since it must first be identified and then (usually) an arithmetic operation performed, such as adding an offset to the current Instruction Pointer value, to obtain the address of the next valid instruction in the program flow. By the time its target address can be applied to the instruction cache to make it start fetching instructions from the new destination, several instructions have been fetched from the wrong address. These must be cancelled, creating a penalty of several clock cycles.

Secondly, conditional branches, i.e. those that are dependent on the results of previous instructions to determine their direction, usually must progress all the way to the end of the pipeline before the condition on which their action depends is resolved. Only then can instruction fetches from a new target address proceed, causing an even greater branch penalty than that incurred by the target address calculation alone.

iCORE implements a branch prediction mechanism that reduces the penalties in both these cases. A two-bit predictor scheme is used to predict branch and subroutine-call instruction behaviour (taken or not taken), while a Branch Target Buffer is used to predict target addresses for taken branch and call instructions. Also, a 4-deep return stack is used to predict the special case of *ret* instruction return addresses.

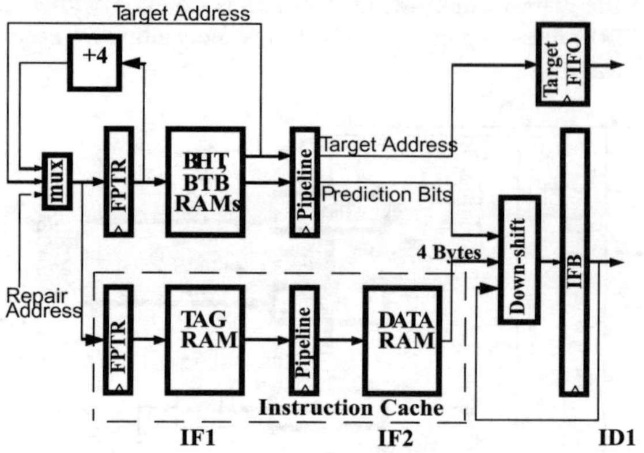

Figure 2. Branch Prediction Logic

The prediction scheme used by iCORE is shown in Figure 2. It employs a 1024-entry Branch History Table (BHT), each entry of which contains a two-bit predictor that implements a branch-hysteresis algorithm (weakly/strongly taken/not-taken). The BHT keeps the most current branching history of a set of instruction cache words. Predicted fetch-addresses (i.e. the target addresses of predicted-taken branches) are kept in a 64-entry Branch Target Buffer (BTB).

Whenever a new 4-byte word is fetched from the instruction cache, the instruction address that is used to index the cache also indexes the BHT and BTB. The retrieved information from the BHT is used to predict whether there is a taken branch at the current instruction address. If a taken branch is predicted, the address retrieved from the BTB is loaded into the Instruction- Fetch Pointer and used to fetch the next instruction word. Otherwise, the sequential value of the Instruction-Fetch Pointer is used. All this takes place in the very first pipeline stage, ID1, so that predicted target addresses can be applied immediately to the instruction cache, and the performance impact of successfully predicted branches is negligible.

As branches are predicted, mispredictions may occur. One type of misprediction is caused by the cost-saving measures of using only the low-order bits of the instruction-fetch address to index BHT and BTB entries while ignoring the upper address bits, and by providing only one BHT/BTB entry per 4-byte cache word, even though there could be more than one branch in a word. These factors can cause a single BHT or BTB entry to be shared by many fetched instructions, even though the entry is likely to be correct only for the one instruction that was originally used to update it; the instructions at other fetch addresses that share the entry may get wrong predictions or wrong target addresses, or both.

Other sources of misprediction include mispredicted direction of conditional branches, and incorrect return addresses. Mispredictions are detected and resolved in later pipeline stages. The instruction decoder in the ID1 stage can detect many mispredictions caused by BHT aliasing. For instance, if a *taken* condition is predicted for an instruction that is not a branch, or an *untaken* condition is predicted for an unconditional branch, the prediction is clearly in error and a repair operation can start immediately with only a small penalty. To check predicted targets, an adder in the OF1 stage calculates the correct target address of a taken branch and compares it with the predicted one (which is fed forward from the BTB). If there is a mismatch due to a missing aliased BTB entry, then recovery can start from OF1. Finally, a conditional branch must wait until it reaches the EXE stage before the condition on which its action was determined can be checked, and thus has the longest repair time when incorrectly predicted.

The steps needed to repair a branch depend on the type of misprediction. In the case where BHT aliasing causes a non-branch instruction to be taken, then all instructions fetched after that instruction are flushed from the pipeline, and instruction fetching is restarted from the *source* address of that instruction to ensure that it is correctly fetched, while the BHT is updated to indicate *not taken.* Conversely, in the case where a target address of a taken branch is incorrectly predicted, then instructions from the bad target are flushed from the pipeline and instruction fetching is restarted from the correct *target* address computed in the OF1 stage, while the BTB is updated with the new target address.

Finally, repairs needed to correct a mispredicted *conditional* branch flush all subsequent instructions from the pipeline and begin fetching again either from the target address of the branch or the next-sequential address of the branch, depending on whether the branch is taken or not.

2.4 Cache Subsystem

Features of the cache controller's design were focused on the need for simplicity (thereby enabling high frequency) and the requirement for very tight coupling to the CPU pipeline. These requirements resulted in a two-stage pipelined approach, the first stage of which handled the tag access and tag comparison, and the second stage that handled the data access and data alignment. This gave a very good balance of delays, and had the advantage of saving both power and delay by eliminating the need for physically separate data RAMs for each associative cache-bank (Way), since the data RAM's Way-number could be pre-determined in the tag cycle and then simply encoded into extra data RAM address bits. To further reduce delays, integrated pipeline input registers were built into both the tag and data RAMs to eliminate input wire delays from their access times.

A simplified block diagram of the IFU and OFU cache controllers is shown in Figure 3. One controller occupies the IF1 and IF2 pipeline stages in the IFU and the other occupies the OF1 and OF2 pipeline stages in the OFU. The same cache controller design is used in each case, so its main pipeline stages corresponding to IF1/IF2 in one case and OF1/OF2 in the other are generically known as CH1 and CH2.

Before the CH1 phase, the cache controller receives the request (read or write), address, and data from the CPU and loads them into its input pipeline registers.

In the CH1 phase, the tag RAMs are accessed and the tag comparison is performed. The first version of iCORE has 8 KB caches and a 16-byte line size, so 512 tag entries are required. At the end of CH1, the result of the tag comparison (hit/miss), and the CH1 address and data fields from the original CPU request are forwarded to the CH2-phase pipeline input registers.

In CH2, the data RAMs are accessed and the data is returned to the CPU's Instruction Fetch Buffer (IFB) in the case of the instruction

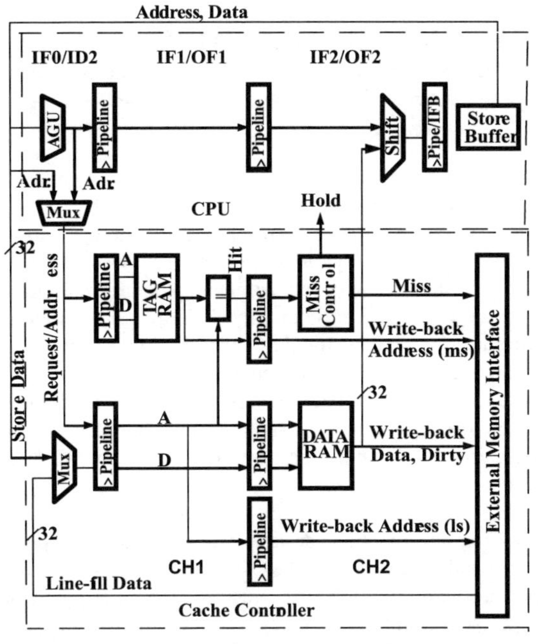

Figure 3. Cache Controller

cache, or the data aligner on the inputs to the EXE stage in the case of the data cache.

As well as returning data on a cache hit, the CH2 stage also initiates a cache-miss sequence if the requested data is not in the cache, asserting the *hold* signal to freeze the CPU pipeline until data fetched from external memory is returned.

2.5 Effects of Microarchitecture on IPC

Most of the microarchitectural enhancements' effect on IPC were first studied through the use of a C-based statistical performance model, which "executes" various benchmark program traces produced by an ST20-C2 Instruction Set Simulator (ISS). They include basic instruction folding, the Local Workspace Cache and its alternative designs, enhanced instruction folding enabled by the use of LWC, and branch prediction schemes and designs. The architecture of some of the features was fine-tuned during RTL implementation, as more accurate performance and area trade-off analyses were made possible. A summary of the effects of these microarchitecture features on IPC is listed in Table1.

Table 1: IPC Enhancements on Dhrystone 2.1

Features	IPC Effect	Overall IPC Improvements
Add basic fold	10%	10%
Add LWC	10%	20%
Enhance instruction folding	7%	27%
Add LWC rotating index	8%	35%
Add branch prediction	5%	40%
Enhance branch prediction transition algorithm	1%	41%
Add Return Stack	3%	44%
Add LWC dynamic disable	4%	48%

3. OPTIMIZING THE IMPLEMENTATION

Implementation of the design was based on a synthesis strategy for the control and datapath logic. Only the RAMs used custom design methods.

3.1 Synthesis Strategy

The synthesis process was divided into two steps. The first step employed a bottom-up approach using Design Compiler™ in which all top-level blocks were synthesized using estimated constraints. Blocks were defined as logic components generally comprising a single pipeline stage. The execute unit of the processor included a large multiplier, for which Module Compiler™ was used as it was found to provide better results than Design Compiler™. As the blocks were merged into larger modules, which were finally merged into the full processor, several incremental compilations of the design were run to fine tune the performance across the different design boundaries.

The only region of the design where synthesis provided unacceptable path delays was in the OF2 and the IF2 stages. In both cases the problems involved the use of complex multiplexer functions. Custom multiplexer gates were designed for these cases and logic was manually inserted with a *dont_touch* attribute to achieve the desired results. Other complex multiplexer problems were resolved by re-writing the HDL to guide the synthesis tool towards the preferred implementation.

The second step of the synthesis strategy employed a top-down approach, with cell placement using Synopsys' Physical Compiler™, which optimized gates and drivers based on actual cell placement. Physical Compiler™ proved to be effective in eliminating the design iterations often encountered in an ASIC flow due to the discrepancy between the gate load assumed by the wire-load model and that produced by the final routed netlist. The delays based on the placement came within 10% of those obtained after final Place and Route. Physical Compiler™ can be run with either an "RTL to placed gates", or a "gates to placed gates" flow. The "gates to placed gates" flow was found to provide better results for this design.

3.2 Standard Cell Library

A higher performance standard cell library (H12) was developed for iCORE to improve circuit speed compared to a supplied generic standard library. Speed improvement of more than 20% was observed in simulation and on silicon when the H12 library was used.

The conventional CMOS static designs used in the generic library were used in the H12 library, but with the following differences:

- Reduced P/N ratio
- Larger logic gates (vs. buffered)
- More efficient layout
- Increased cell height

The P/N ratio for each cell was determined by the best speed obtained when the gate was driving a light load. It was observed that when the gate-load increased, the P/N ratio needed to be increased to get optimal speed. However, the physical-synthesis tool minimizes the fan-out and line loading on the most critical paths, so the H12 library was optimized for the typical output loading of those paths.

Delays were also reduced by creating larger, single level cells for high-drive gates. The original library added an extra level of buffering to its high-drive logic gates to keep their input capacitance low and their layout smaller, but at the expense of increasing their delay. Eliminating that buffer by increasing the cell size also increased the input capacitance, but still the net delay was significantly reduced. These cells were only used in the most critical paths, so their size did not have much impact on the overall layout size.

The H12 cells generally had larger transistors that required the layouts to be taller. However, by using more efficient layout techniques, many of the cells widths were reduced compared to the generic library. Also, the larger height of the cells allowed extra metal tracks to be used by the router that in turn increased utilization of the cell placement.

4. PHYSICAL DESIGN STRATEGY

Physical Compiler was used to generate the placement for the cells. The data cache and instruction cache locations were frozen during the placement process and the floorplanning was data-flow driven. Clock lines were shielded to reduce delay uncertainty by using Wroute in Silicon Ensemble™ by Cadence. To minimize the power-drops inside the core, a dense metal5/Metal6 power grid mesh was designed.

The balanced clock tree was generated with CTGen™. The maximum simulated skew across the core was 120 ps under the worst-case process and environmental conditions. Typically this would be significantly less. The final routed design was formally compared to the provided netlist using Formality™. The Arcadia™ extracted netlist delay was calculated using Primetime™.

5. RESULTS

The iCORE processor was fabricated in STMicroelectronics' 0.18m HCMOS8D process. A GDSII plot of the test-chip is shown in Figure 5. Excluding the memories, the entire chip is seen to have the distinctive random layout of synthesized logic. The large memories on the right and on the top and bottom of the plot are the data and instruction RAMs. The other small memories seen in the plot are the Local Workspace Cache, the Branch History Table and Branch Target Buffer.

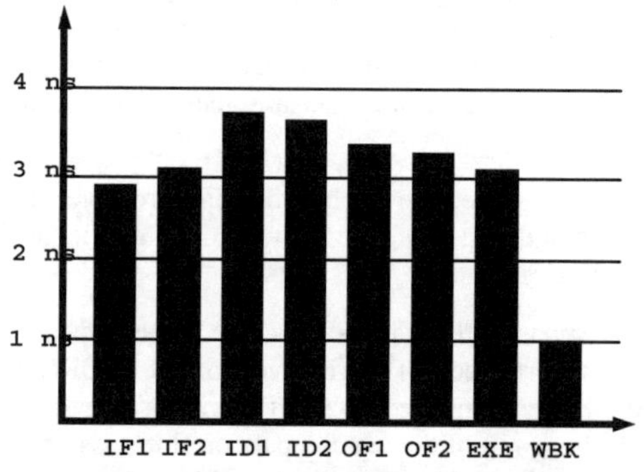

Figure 4. Static Analysis: Worst-case Pipeline Paths

Timing simulation indicated that a good balance of delays between the pipeline stages was achieved, as shown in Figure 4.

Silicon testing showed functional performance of the design from 475 MHz at 1.7 V to 612 MHz at 2.2 V and 25 degrees Celsius, ambient.

Analysis of the Dhrystone 2.1 benchmark showed that iCORE achieved an IPC (Instructions Per Cycle) count of about 0.7, which met the goal of being the same or greater than that of previous ST20-C2 implementations, indicating that the various pipeline optimizations were functioning correctly.

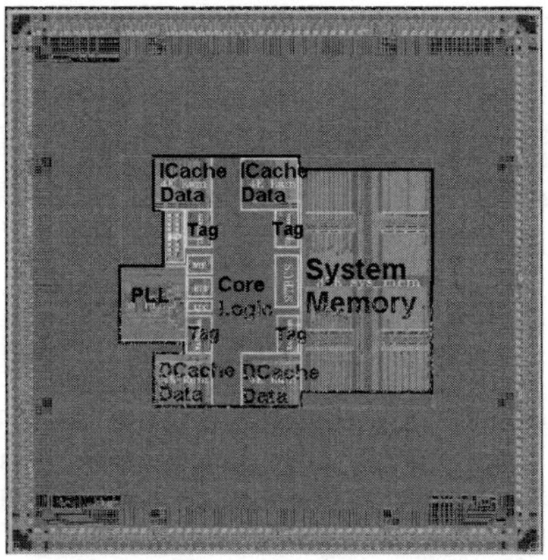

Figure 5. iCORE Test-chip GDSII Plot

6. CONCLUSIONS

The performance gap between custom and synthesized embedded cores can be closed by using deep, well-balanced pipelines, coupled with careful partitioning, and mechanisms that compensate for increased pipeline latencies. This enables the use of simple and well-structured logic functions in each pipeline stage that are amenable to highly optimal synthesis[3]. The performance gains are consolidated by use of placement-driven synthesis, and careful clock-tree design.

This paper explored the use of advanced logic design to implement a high performance CPU core using an ASIC methodology. By avoiding custom circuit design, we have been able to reduce the design time and high costs traditionally associated with it, while still achieving excellent performance.

The final chip has been demonstrated to be functional in silicon at up to 520 MHz under typical conditions.

7. REFERENCES

[1] C. D. Snyder, "Synthesizable core makeover: Is Lexra's seven-stage pipeline core the speed king?," *Cahner's Microprocessor Report*, July 2, 2001.

[2] "ST20-C2 Instruction Set Reference Manual*", STMicroelectronics, 72-TRN-273-02,* November 1997

[3] D. G. Chinnery and K. Keutzer, "Closing the gap between ASIC and custom: An ASIC perspective," *Proceedings of the 38th Design Automation Conference*, Las Vegas, NV, pp. 420-425, June 2001.

A Flexible Accelerator for Layer 7 Networking Applications

Gokhan Memik and William H. Mangione-Smith
Department of Electrical Engineering
University of California, Los Angeles
{memik, billms}@ee.ucla.edu

ABSTRACT

In this paper, we present a flexible accelerator designed for networking applications. The accelerator can be utilized efficiently by a variety of Network Processor designs. Most Network Processors employ hardware accelerators for implementing key tasks. New applications require new tasks, such as pattern matching, to be performed on the packets in real-time. Using our proposed accelerator, we have implemented several such tasks and measured their performance. Specifically, the accelerator achieves 25-fold improvement on the performance of pattern matching, and 10-fold improvement for tree lookup, over optimized software solutions. Since the accelerator is used for different tasks, the hardware requirements are small compared to an accelerator group that implements the same set of tasks. We also present accurate analytic models to estimate the execution time of these networking tasks.

Categories and Subject Descriptors

B.2.4 [**Arithmetic and logic structures**]: High-Speed Arithmetic, Cost/Performance.

General Terms

Algorithms, Measurement, Performance, Design.

Keywords

Application-Specific Processor, Networking Applications, Network Processor, Accelerator, Table Lookup, Pattern Matching.

1. INTRODUCTION

Traditionally, the processing elements in networks are either ASICs or variations of general-purpose processors. Both schemes have their advantages and shortcomings. ASICs generally provide better performance, but they have higher manufacturing costs and lack the flexibility of programmable processors. If there is a change in the protocol or application, it is hard to implement the change in the design. General-purpose processors, on the other hand, are not optimized for networking applications and hence do not provide satisfactory performance for many networking applications.

Network Processors (NPUs) are a new type of processor optimized for networking applications that combine the advantages of ASIC and general-purpose processors. By utilizing specially designed hardware, these designs achieve performance comparable to ASIC. Since they are software programmable, they have the flexibility comparable to general-purpose processors.

Soon after their introduction [2, 10], the NPU market became one of the fastest growing segments of the microprocessor industry. Currently there are more than 30 companies with a variety of NPU designs [8].

New networking applications coupled with the higher link speeds demand new and more complex tasks to be performed efficiently in the processing elements. A significant portion of these new applications search or modify Layer 7 application payload information in the packet that was not accessed by the traditional network processing elements. In this paper, we present design details of an accelerator used to implement key networking operations such as tree lookup and pattern matching. These operations are expected to be in demand in the near future by applications that must access and modify Layer 7 payloads. For example, pattern matching is performed by several classification engines to categorize packets for security, QoS, or similar purposes.

Accelerators have been widely used by traditional routers and NPUs. However, due to their proprietary nature, design details of these accelerators have not been publicly discussed. We provide design details of such an accelerator and provide a thorough study of the performance of the accelerator. Specifically, we

- present a novel accelerator design and discuss the effect of the applications on our design decisions,
- discuss specialized algorithms used to implement different tasks on the accelerator,
- present analytical models to estimate the execution times of these algorithms,
- compare the performance of the accelerator against software solutions.

Currently implemented accelerators in the NPUs are designed for a single task. For example, each protocol processor in IBM PowerNP [6] has eight separate coprocessors for tasks such as accessing tree data structures and calculating CRCs. Our proposed accelerator can efficiently implement several different tasks. Hence, it reduces the chip area required to implement a set of accelerators.

The next section summarizes the related work. In Section 3, we explain the networking applications and present how frequently the accelerator can be used by these applications. In Section 4, we explain the accelerator design and the algorithms implemented in detail. Section 5 presents experimental results. We conclude the paper with Section 6, which summarizes the paper.

2. RELATED WORK

There is a wide variety of Network Processor design methodologies, which can be grouped into three major categories: VLIW-based processors, Simultaneous Multithreaded (SMT) processors, and single-chip multiprocessor systems. Crowley et al. [5] evaluate different design methodologies (a VLIW, an SMT, a single-chip multiprocessor, a fine-grain multiprocessor) for NPUs. Nie et al., on the other hand, discuss a RISC-based Network Processor that has hardware support for thread switching and bit-wise data manipulation [11].

Vendors such as Clearwater [4] implement smart memory management units for bulk packet data transfers. Wuytack et al. discuss a memory-oriented synthesis methodology and uses embedded network applications as an example [15].

Several NPU vendors implement accelerators to enhance the performance of the processor. Although the design details of these accelerators have not been disclosed, it emphasizes the importance of accelerators in Network Processors. Design details of ClearSpeed Table Lookup Engine are also not discussed, but the performance of the engine is presented for different key sizes used for tree lookup [3]. In IBM PowerNP [6] two processors share eight different accelerators such as checksum calculator that calculates and verifies frame; control access bus controller that controls access to data structures; string copier that accelerates data movement between coprocessors and the shared memory pool; tree search engine that performs analysis through tree searches; and semaphore manager that assists in controlling access to shared resources. C-port's C-5 employs six accelerators for up to 15 processing cores [2]. The accelerators perform tasks such as buffer and queue management and tree lookup. The table lookup engines in both PowerNP and C-5 are based on hashing techniques.

3. NETWORKING APPLICATIONS

Network Processors are special-purpose architectures designed for networking applications. Hence, we need to understand the networking applications in detail to be able to judge the implications of design decisions for these processors. For this study, we have examined several applications from the NetBench [9] suite. NetBench is designed for NPUs and contains ten applications listed in Table 1. CRC performs the 32-bit cyclic redundancy checking. DH, MD5, and SSL are security applications. DH implements the Diffie-Hellmann public key encryption/decryption algorithm. MD5 is the message digest algorithm version 5, which generates secure signatures for packets. SSL implements the secure sockets layer, which first performs RSA and DSA authentication and then uses the blowfish and 3DES encryption/decryption algorithms. DRR is the deficit-round-robin scheduling algorithm used to achieve fair share of the available bandwidth between the connections going through the switch or router. Ipchains is a common firewall application used to filter malicious packets from a network. Nat implements the network address translation application used by several service providers to increase the utilization of the available IP addresses. Route implements the IPv4 routing. Tl is the table lookup routine implemented as a radix-tree search and URL implements the URL-based switching, which performs intelligent switching according to the contents of the packet.

Table 1 also presents the fraction of the execution cycles spent in functions that can be accelerated by the proposed accelerator (*accel*), the remaining cycles executed excluding the memory stalls (*rest*), and the number of instructions executed per packet processed (*inst/packet*). To find the execution fraction that can be accelerated, we have performed a thorough analysis of the applications with the provided input sets. We have examined each function in the applications and decided whether the function (or a sub-task in it) can be implemented using the accelerator proposed in this work. Then we have measured the fraction of the execution that is spent in these functions. The fraction of the execution time that can utilize the proposed accelerator varied from 0% (for CRC) to 81.2% (for URL) with an arithmetic mean of 37.2%.

Table 1. Application information: Input parameters used for applications (*input*), the portions of the executed instructions that can be accelerated (*accel*), remaining execution in the applications excluding memory stalls (*rest*), and number of instructions executed per packet (*inst/packet*)[1].

App.	Input	Accel [%]	Rest [%]	Inst/packet [K]
CRC	50000	0	98.6	23.9
DH	20	27.3	71.4	121700[1]
DRR	128 50000	32.4	55.7	6.1
Ipchains	9 64 1000	37.1	53.9	7.4
MD5	50000	44.5	49.5	20.4
Nat	128 50000	33.4	57.7	2.1
Route	128 50000	35.1	49.5	1.8
SSL	1	34.7	62.6	408000[1]
Tl	128 50000	46.1	37.4	1.2
URL	128 50000	81.2	12.6	17.1

4. ACCELERATOR DESIGN

In this section, we explain the design details of the accelerator and present novel algorithms that utilize it.

Figure 1 shows the overall accelerator design. The design of the unit resembles a barrel shifter. In a traditional barrel shifter, each module selects either the shifted input bit or the unshifted input bit to perform the conditional shifting operation. In our design, however, the input to each module is a byte and all control signals affect complete bytes. In addition, modules are capable of performing some logical operations on their input values. Specifically, the module is selected to perform one of the following functions on their input signals:

- output = left input (shift right)
- output = right input (pass without shift)
- output = (left input AND right input)
- output = (left input OR right input)
- output = (left input XNOR right input)
- output = (left input BYTE-WISE-OR right input)
- output = (left input BYTE-WISE-AND right input)

The utility of each these operations will be examined during the discussion of the applications in the following subsections. AND, OR, and XNOR are bit-wise operations, performing the logical operation on the corresponding bits of the two input signals.

---

[1] For DH this column corresponds to instructions per key generation and for SSL it corresponds to instructions for performing all supported key generations and performing encryptions on the packet.

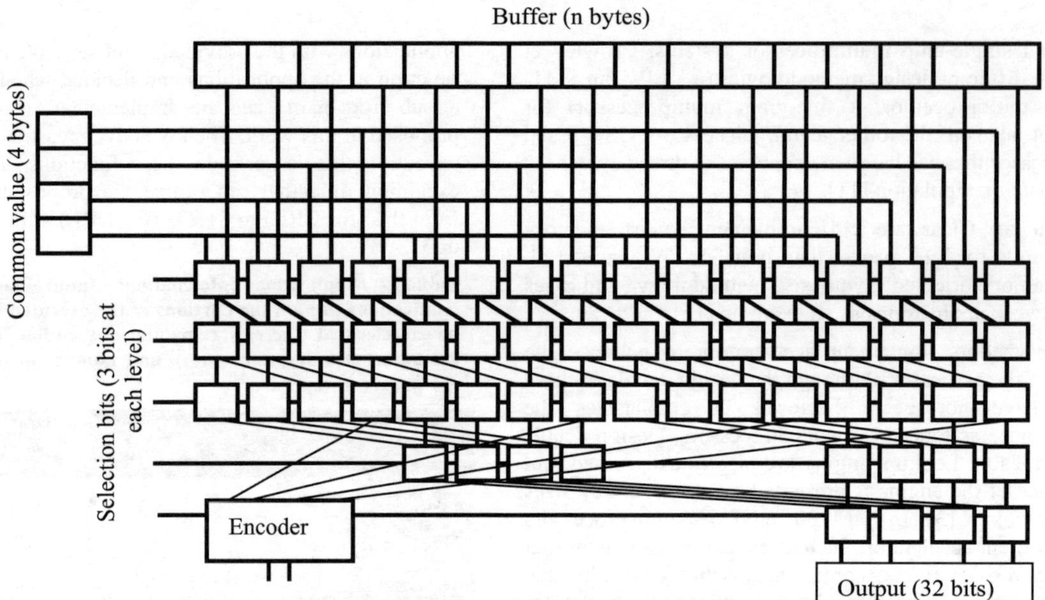

Figure 1. Accelerator design (the figure is plotted 16-bytes wide for illustrative purposes)

Byte-wise-or results 0xFF if one of its inputs is equal to 0xFF, 0x0 otherwise. Similarly, byte-wise-and results 0xFF if both of its inputs are equal to 0xFF, 0x0 otherwise. Figure 2 illustrates the design of the modules in the accelerator. All modules are equivalent. The leftmost AND gates in the figure are 8-bit AND gates taking all their input either from left input or from right input byte and performing the AND operation on these bits. Hence, they output 1 if the corresponding input equals to 0xFF. The outputs of these gates are connected to 2-bit and and or gates. The results of these 2-bit logic gates are rippled to form 8-bit inputs to the multiplexer. Hence, the signals on their output are 0xFF or 0x00.

The accelerator uses a common value register and a wide buffer (n bytes) to store the packet information. The common value is 4 bytes wide. We have selected this size, because it is an efficient width for many tasks. For example, this size is equal to the size of an IP address in IPv4. Hence, many address related tasks (lookup, match, etc.) can be efficiently performed. We have performed several simulations with different widths (n) of the accelerator. The results for the simulations with the accelerator widths ranging from 64 to 512 bytes are given in Section 5. Using the selection bits, the operations on each level is selected. Using these selection bits and appropriate setting of the common value and the buffer, a wide variety of tasks can be accomplished by the accelerator.

The accelerator also uses an encoder, which is embedded in the third level of the modules. It has (n / 4) inputs chosen one bit from every four modules starting with the fourth leftmost module as shown in Figure 1. This encoder gives the index of the leftmost input that is high.

For illustrative purposes, the accelerator in Figure 1 is drawn 16-byte wide. In our design, however, we have selected the unit to be 128-byte wide, which gives a good performance / cost ratio. The results are explained in Section 5. This architecture directly supports the extraction of any 4-bytes from the input buffer without requiring that they be contiguous. This function is very useful for traditional packet analysis operations.

We have implemented an RTL level description of the accelerator in VHDL and synthesized it using Synopsis design compiler [11]. We have used a library representing a .25µ technology. The accelerator with 128-byte buffer required 115K transistors and has a total delay of 6.65 ns.

In the following subsections, we describe algorithms to implement key emerging tasks for networking applications and present analytical models to estimate the execution time of these tasks.

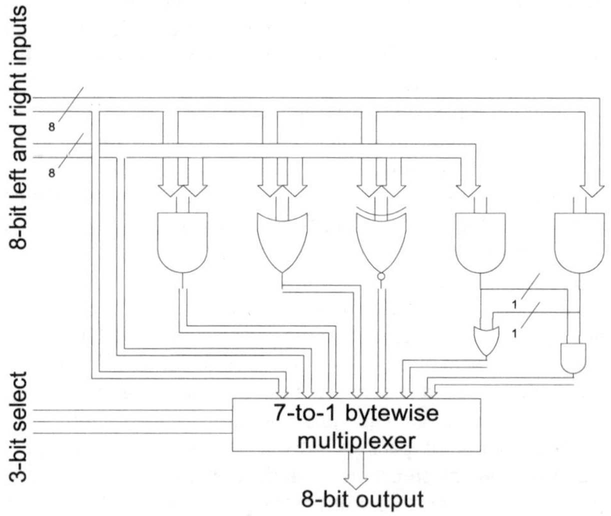

Figure 2. Block diagram of the modules

4.1 Pattern Matching

One of the common tasks in the networking applications is pattern matching. For example, in URL-based switching, a known pattern is searched in the payload to decide the type of the HTML request encoded in the packet. Similarly, several security applications search for known patterns in the payload to detect malicious packets [12].

Consider an example, in which we are searching for the pattern '*.jpg*' in the payload. The pseudo-code in Figure 3 illustrates the pattern-matching algorithm. We first load the common value with the '*.jpg*' value and load the payload into the buffer. Then, we configure the accelerator to perform 'xnor' in the highest level, 'and' in the next two levels and 'byte-wise-or' in the following levels. If there is no match in the payload, the 'xnor' results will not be equal to 0xFF and this will in turn make the output equal to zero. If the payload contains '*.jpg*' and the first character ('.') is aligned with the first character of the common value ('.'), then there will be four consecutive 0xFF values at the end of the first level ('xnor') and at least one 0xFF after the third level ('and'). Hence, the result will not be equal to zero. Note that, to guarantee whether there is a match or not, we have to perform the operation with different alignments. This is achieved by rotating the common value. After four iterations, if there is not a match, we can conclude that the payload section examined does not contain the pattern. Note that the pattern matching code contains two unique instructions: load-buffer and op. 'load-buffer r, add' results in loading r bytes from address add into the accelerator input buffer. 'Op r1, r2, r3' activates the accelerator with the selection bits stored in r2 and the common value stored in r3. The result of the operation is written to r1.

```
load                $2,ACCELERATOR_WIDTH
load                $3,'.jpg'
load-buffer         $2,payload
load                $4,0x2db6da93 #conf.
op                  $2,$3,$4
branch-not-equal    $2,$0,$match
rotote-right        $3,$3,8
op                  $2,$3,$4
branch-not-equal    $2,$0,$match
rotote-right        $3,$3,8
op                  $2,$3,$4
branch-not-equal    $2,$0,$match
rotote-right        $3,$3,8
op                  $2,$3,$4
branch-not-equal    $2,$0,$match
#no match
```

Figure 3. Code for pattern matching

If the pattern of interest is larger than four bytes, we first divide it into four byte sub-patterns. Then each sub-pattern is searched in the buffer. If all sub-patterns are found, we return the buffer to the processor to be examined for the whole pattern. If one of the sub-patterns is not matched, we indicate a mismatch to the processor. Note that, if the pattern of interest is smaller than four bytes, we can use the accelerator directly by modifying the selection bits.

We have developed an analytical model to estimate the execution time of the pattern matching algorithm. The number of cycles required can be estimated using:

$$\#cycles = \lceil haystack / n \rceil * (Mem(max(n, haystack)) + exec)$$

Equation 1. Cycles required for pattern matching

In Equation 1 haystack corresponds to the size of the buffer that is searched, Mem (x) corresponds to the number of cycles spent in reading x amount of data from memory and exec is number of cycles required by the accelerator when the buffer data is available. For example, assuming a delay of 4 cycles for the accelerator, exec equals to 26 (1 cycle for setting common value, 1 cycle for setting the select bits and 24 cycles to produce the results, because we have to perform 4 iterations each lasting for 6 cycles). In Section 5.2, we compare the estimations for the model against simulation numbers and show that the model estimates the number of execution cycles within 22% in worst case.

We have performed several simulations to see the performance effect of the accelerator. The accelerator is able to improve the performance by as much as 25 times over a software solution. The experiments and the results are explained in Section 5.3.

4.2 Tree Lookup

Most networking applications contain significant amount of table lookup operations. For example, in traditional IP routing, routing tables must be traversed to decide what rules to apply on the packet. An efficient method for implementing these tables is in the form of trees. Particularly, radix-tree lookup is commonly used in several UNIX systems and routing architectures [14].

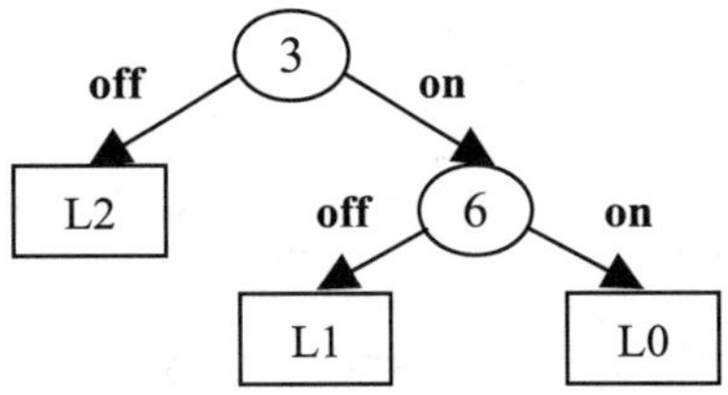

Figure 4. A radix-tree example

In radix-tree lookup, the tree is traversed from the root to the leaves. Each node has a specified offset or position. If the position of the address (the bit at the position) is set, then the right sub-tree is traversed, otherwise, the lookup continues on the left sub-tree (the selection of right or left for the active bit is implementation dependant). An example tree is given in Figure 4. For the example tree, if the third bit of the searched address is zero, the lookup will return leaf node 2. Otherwise, the sixth bit will be examined.

The radix-tree algorithm for the accelerator uses the fact that the "on" locations used to reach a leaf uniquely determines the resultant leaf node of the lookup. For example, leaf 0 in Figure 4 will be reached if and only if both the third and sixth bits of the searched address are equal to one. Our algorithm first examines the tree to find the correct buffer setting for the accelerator. This process is given in Figure 5. The leaf nodes are sorted from right to left and a mask value for each leaf node is found. These masks have a 0 at the on positions to reach the leaf and 1 at all other positions. For our example tree, this will result in the following buffer values: 0xFFFFFFB7, 0xFFFFFFF7, 0xFFFFFFFF. This processing must be performed whenever there is a change in the network route (hence the tree). Since in several networking applications the tree is mostly stable (for example, in the traditional routing it changes if there is a topology change in the network, which occurs infrequently) the overhead is small.

```
void
setup_bufs (struct tree *tree, unsigned int masks)
{
   if (tree->type == leaf)
     {
       bufs[order] = masks;
       order++;
       return;
     }
    setup_bufs (tree->right, masks ^ (1 << tree-
    >position));
    setup_bufs (tree->left, masks);
}
```

Figure 5. C code for preparation of the buffer value for the for radix-tree lookup

After the buffer values are set, the common value in the accelerator is set as the address from the packet. The highest level is set to perform 'or' and the following two levels perform 'byte-wise-and'. If there is a 0xFF value at the result of the third level, the index of the module will determine the result of the lookup operation. This index will be equal to the output of the encoder. This value is later used for retrieving the leaf information from the array of leaf nodes.

If the required buffer size exceeds the width of the accelerator, an indirect addressing is used. The first accelerator operation identifies the buffer to be retrieved, for which the accelerator gives the index of the resultant leaf node.

We have developed an analytical model to estimate the number of execution cycles for a tree lookup:

$$\#cycles = \lceil \log_{(n/4)} leaves \rceil * (Mem(avg.) + exec)$$

Equation 2. Cycles required for table lookup

In Equation 2 leaves correspond to the number of leaf nodes, Mem (avg.) corresponds to the number of cycles spent in reading data from memory and exec is number of cycles required by the accelerator when the buffer data is available. For example, assuming the delay of the accelerator is 4 cycles, exec equals to 7 (1 cycle for setting common value, 1 cycle for setting the select bits and 4 cycles to produce the results and 1 cycle for reading the encoder value). In Section 5.2, we compare the estimations for the model against simulation numbers and show that the model estimates the number of execution cycles within 3% in worst case.

We have performed several simulations to measure the effectiveness of the accelerator for tree-lookup. The accelerator is able to improve the performance of software solution by as much as 12-times. The detailed results are explained in Section 5.3.

4.3 Security

Many encryption/decryption algorithms use permutation to find unique signatures for a packet or to change the original packet payload. For example, the MD5 algorithm use the permutations listed in Figure 6. Such permutations can be directly mapped into the accelerator. For example, to implement 'F' from Figure 6, the buffer is loaded with x, y, ~x, and z series. Then, in the first two levels the buffer values are forwarded. In the third level, 'and' and in the fourth level 'or' operation is performed. The results at this stage are propagated one by one to the output register to get the desired section from the series.

```
#define F(x, y, z) (((x) & (y)) | ((~x) & (z)))
#define G(x, y, z) (((x) & (z)) | ((y) & (~z)))
#define H(x, y, z) ((x) ^ (y) ^ (z))
#define I(x, y, z) ((y) ^ ((x) | (~z)))
```

Figure 6. MD5 code segment for permutations

5. EXPERIMENTS

5.1 Experimental Setup

In the simulations performed, the programmable core is an out-of-order execution processor much like Alpha 21264 [7]. We have modified the SimpleScalar [1] simulator by implementing the accelerator as a function unit in the processor. Three different accelerator widths are studied: 64 bytes, 128 bytes and 512 bytes. The delay for the 64 and 128-byte accelerators is set to 4 cycles and the delay for the 512-byte accelerator is set to 5 cycles. The simulated core has 32 KB of separate L1 data and instruction caches and a 512 KB unified L2 cache. The latencies for L1 and L2 caches are set to 1 and 10 cycles, respectively.

5.2 Validating Analytical Models

In Sections 4.1 and 4.2, we have presented analytical models to estimate the execution latency for pattern matching and tree lookup algorithms. In this section, we compare the estimations made using these models and simulation values from the SimpleScalar simulator.

Figure 7 summarizes the estimations for the pattern matching task. The worst estimation is made for 512-byte wide accelerator for the haystack size of 640 bytes, for which the error is 21.85%.

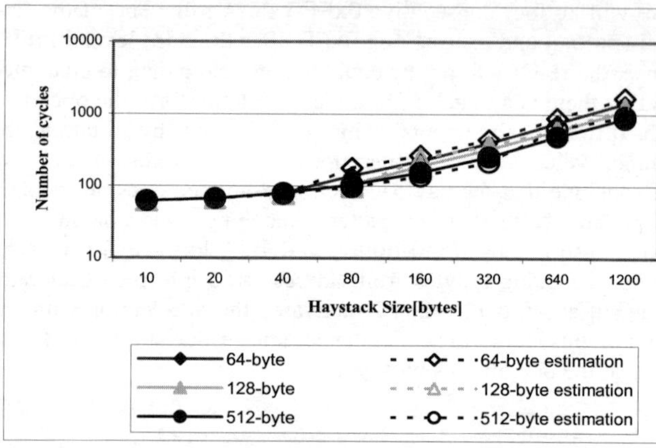

Figure 7. Estimations for pattern matching

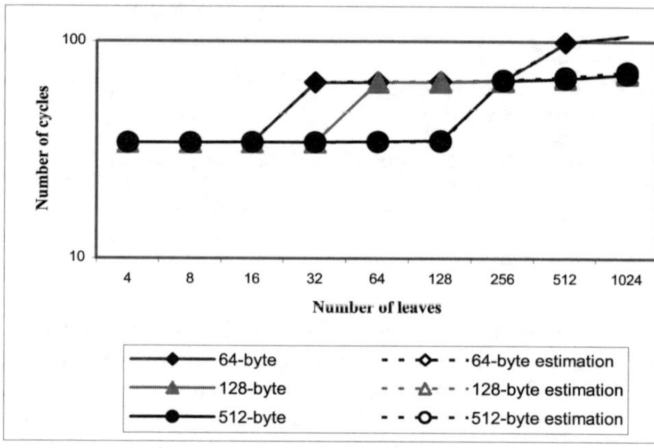

Figure 8. Estimations for tree lookup

The results for tree lookup are summarized in Figure 8. In all the configurations, the analytical model for tree lookup almost exactly matches the simulation numbers. The largest error is 2.6% for lookup in a tree with 1024 rules with 512-byte wide accelerator.

5.3 Performance Simulations

The simulated applications in this section are kernels that perform the radix-tree lookup and pattern matching tasks. Our goal in these simulations is to measure the performance of the accelerator for different input parameters.

Figure 9 summarizes the results for pattern matching. The improvement achieved by the accelerator increases with the increased string size. For large strings, the accelerator achieves up to 25-fold improvements (e.g., the accelerator with 512-byte width achieves 24.2-fold improvement for 1200-byte haystack).

The results for tree lookup are summarized in Figure 10. As the table size is increased the improvement of the accelerator increases. However, the speed-up is non-monotonic. It can be seen that these jumps exactly occur when the buffer is not large enough to hold all the leaves. At these sizes, the accelerator has to perform multiple accesses, creating the saw tooth pattern observed in the results. In addition, the overall trend in the improvement for 64-byte wide accelerator is downward, whereas it is upwards for the 512-byte wide accelerator and is almost constant for the 128-byte design.

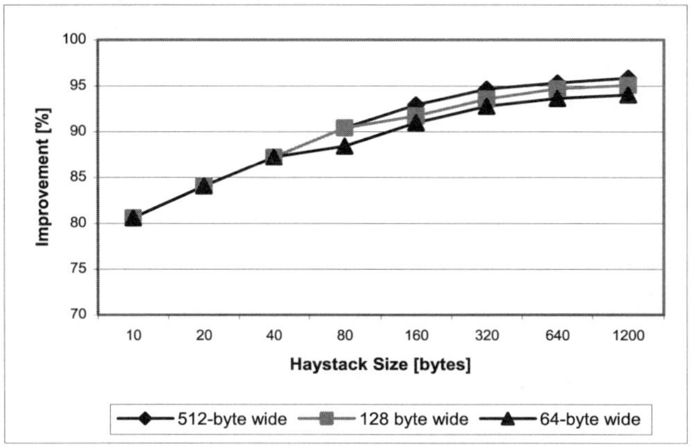

Figure 9. Improvement by accelerator for pattern matching

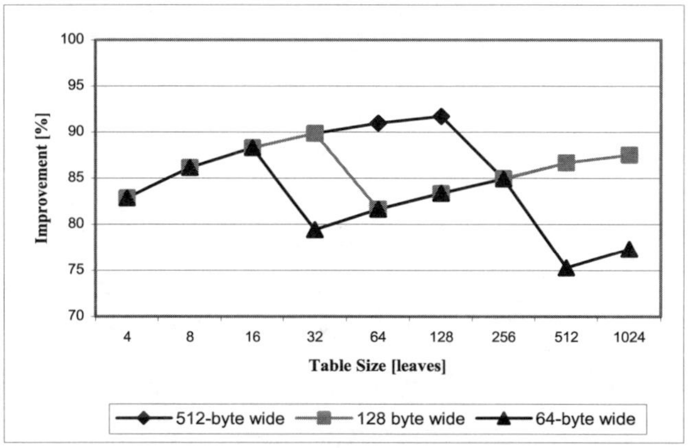

Figure 10. Improvement by accelerator for tree lookup

We have also performed simulations to measure the effectiveness of the accelerator for security applications. We have modified the MD5Transform routine (including the Decode sub-routine) to utilize the function unit. The accelerator was able to increase the performance of MD5Transform by 58.54%.

6. CONCLUSIONS

New networking applications coupled with the higher link speeds demand new and more complex tasks to be performed efficiently in the processing elements. A significant portion of these applications searches or modifies Layer 7 information in the packet that was not accessed by the traditional network processing elements. Hence, implementing accelerators for such tasks is a necessity to achieve high performance. In this paper, we have presented design details of such an accelerator. We have also presented novel algorithms to implement such key tasks. We have shown that our proposed accelerator can perform tasks such as tree lookup and pattern matching efficiently. Specifically, it achieves up to 25-fold improvements for pattern matching and 10-fold improvement for tree lookup over optimized software solutions. Finally, we have performed a study for the utilization of the accelerator in a multi-core environment.

REFERENCES

[1] Burger, D. and Austin, T. The SimpleScalar Tool Set, Version 2.0. Technical Report CS-TR-97-1342, University of Wisconsin, June 1997.

[2] C-port Corporation. C-5 Network Processor Architecture Guide. C-port technical library C5NPD0-AG/D, May 2001.

[3] ClearSpeed Technology Ltd. The ClearSpeed Table Lookup Engine. White paper 02-WP-1021 v3.0. http://www.clearspeed.com/papers/tle_book.pdf

[4] Clearwater Networks. Introducing the CNP810 Family. http://www.clearwaternetworks.com/clearwater-overview.pdf

[5] Crowley, P., Fiuczynski, M. E., Baer, J. L, Bershad, B. N. Characterizing Processor Architectures for Programmable Network Interfaces. In *Proc. of International Symposium on Supercomputing*, Santa Fe / NM, 2000.

[6] International Business Machine Corporation. IBM PowerNP NP4GS3 Network Processor Datasheet. IBM technical library, np3_DLTOC.fm.08, May / 2001.

[7] Kessler, R. The Alpha 21264 Microprocessor. In IEEE Micro, 19(2), Mar/Apr 1999.

[8] Mangione-Smith, and W. H. and Memik, G. Network Processing: Applications, Architectures and Examples. Tutorial at *International Symposium on Microarchitecture*, Austin / TX, Dec. 2001.

[9] Memik, G. and Mangione-Smith, W. H. NetBench: A Benchmarking Suite for Network Processors. In Proc. of *International Conference on Computer-Aided Design*, San Jose / CA, Nov. 2001.

[10] MMC Networks, Inc. Leading the Network Processor Revolution. http://www.mmcnet.com/Solutions

[11] Nie, X., Gazsi, L., Engel, F., Fettweis, G. A New Network Processor Architecture for High-Speed Communications. In Proc. of *IEEE Workshop on Signal Processing Systems*, Taipei / Taiwan, Oct 1999.

[12] Roesch, M. Snort: The Open Source Network Intrusion Detection System Web Site. http://www.snort.org

[13] Synopsys, Inc. Synopsys design compiler – Overview. http://www.synopsis.com/products/logic/design_comp_cs.html

[14] Villamizar, C. OSPF Optimized Multipath (OSPF-OMP). Internet Draft ietf-ospf-mpp-02, Feb, 1999.

[15] S. Wuytack, J.L. da Silva, Fr. Catthoor, G. de Jong and Ch. Ykman. Memory Management for Embedded Network Applications. *IEEE Transactions on Computer-Aided Design,* Volume 18, Number 5, pp. 533-544, May 1999.

PANEL: What's the Next EDA Driver?

Chair: Jan Rabaey, UC Berkeley

Organizers: Joachim Kunkel, Synopsys and Dennis Brophy, Model Technology

Abstract

The PC industry was the major consumer of silicon in the 80's and 90's. It defined the requirements for EDA. In a world dominated by PC's, clock frequency was the ultimate measure of performance. Times have changed. Today the internet, wireless communications and consumer applications such as high end gaming stations have replaced the PC as the primary driver. How we measure 'cutting edge' has also changed. Raw computing speed measured in terms of clock speed has given way to alternate forms of measuring performance – MBits/s, MMACS, and Polygons/sec now capture the differing computation requirements for these domains. Depending on the application domain, weight and power are just as important if not more important than raw computational power. How do these changes drive technology development in EDA? Which of these is the dominant EDA driver? How much of this drive is a function of EDA following the money? How much of this drive is a function of unique technical challenges posed by these domains? This diverse set of panelists with representation from various application segments and the EDA industry will attempt to answer these questions.

Panelists

Raul Camposano
Synopsys

Davoud Samani
Wireless ICs, Infineon

Larry Lerner
Agilent Technologies

Rick Hetherington
Nexsi

Estimation of the Likelihood of Capacitive Coupling Noise

Sarma B. K. Vrudhula*
ECE Dept., Univ. of Arizona
Tucson, AZ
sarma@ece.arizona.edu

David Blaauw† ‡
EECS Dept., Univ. of Michigan
Ann Arbor, MI
blaauw@ece.umich.edu

Supamas Sirichotiyakul†
Sun Microsystems
Boston, MA
supamas@sun.com

Categories and Subject Descriptors

B.7.2 [**Hardware**]: Integrated Circuits—*Design Aids,Reliability and Testing*

General Terms

Preformance, Reliability, Verification

Keywords

Noise, Signal Integrity, Deep Submicron

ABSTRACT

The corruption of signals due to capacitive and inductive coupling of interconnects has become a significant problem in the design of deep submicron circuits (DSM). Noise simulators, based on worst-case assumptions, are overly pessimistic. As a result, when they are used on industrial ICs with hundreds of thousands of nets, thousands of nets are reported as having potential noise violations. There is a need to prioritize the problem nets based on the likelihood of the noise and possibly even eliminate them from further consideration if the likelihood is *negligable*. In this paper, a probabilistic approach is described which allows for a quantitative means to prioritize nets based on the likelihood of the reported noise violation. We derive upper bounds on the probability that the total noise injected on a given victim net by a specific set of aggressors exceeds a threshold. This bound is then used to determine a lower bound on the *expected number of clock cycles* (ENC) before the first *violation* occurs on a given net. Nets can be prioritized based on the ENC. We demonstrate the utility of this approach through experiments carried out on a large industrial processor design using a state-of-the-art industrial noise analysis tool. A significant and interesting result of this work is that a substantial portion (25%) of the nets were found to have an ENC of more than five years. If five years is deemed to be sufficiently long time, then these could be eliminated from further consideration.

*Work done while on sabbatical with Advanced Design Tools group at Motorola, Austin TX. Also suported by NSF Center for Low Power Electronics (CLPE) (NSF Grant No. EEC-9523338, State of Arizona and Industrial consortium).

†Formerly with Advanced Design Tools group at Motorola, Austin TX.

‡Supported by SRC grant 2001-HJ-959

DAC 2002, June 10-14, 2002, New Orleans, Louisiana, USA.

1. INTRODUCTION

A frequent cause of signal corruption in DSM circuits is capacitive and inductive coupling of interconnects [2, 5, 6, 8, 14, 16]. Their effects can lead to a malfunction of the circuit or degrade its performance. *Functional noise* refers to the situation where a glitch occurs on an otherwise stable net (victim) due to switching of its neighboring nets (aggressors), and propagates to a storage element or a dynamic node, possibly altering its state. *Delay noise* occurs when the victim net is also switching, and can lead to a significant increase in the delay *uncertainity* [2]. This would translate into a reduced clock frequency. The impact of signal interference due to interconnect coupling has become quite significant, and therefore techniques for estimating its magnitude and methods for reducing its effects are critical for the design of DSM circuits [1, 4, 5, 7, 9, 10, 12, 13, 15, 17]. An excellent summary of DSM design issues appears in [16].

Noise analysis tools are designed to be pessimistic due to the consequences of missing a potential malfunction. In large industrial designs, this results in hundreds or even thousands of reported violations, which would be too expensive to fix in terms of chip area, performance and design time. Often the number of aggressors with significant coupled noise can be large, exceeding 5 to 10 aggressors. In such a case, the likelihood of a worst-case composite noise, where all aggressors are required to switch at exactly the right time, is small. Logic correlations and timing windows have been used to reduce the pessimism in noise analysis [3, 7, 10, 13]. In general, only a limited number of logic implications can be derived due to the high computational cost. Moreover, for noise analysis, logic correlations have not been very effective since the relation between *physical proximity* and *logical dependence* gets weaker as circuit sizes increase due to a larger number of globally routed nets. Timing intervals, obtained from static timing analysis, can also be used to identify aggressors that cannot inject noise on the victim net simultaneously. In practice, the intervals are too wide and eliminate only a small portion of the reported noise violations.

In this paper, we present a method that will allow designers to prioritize the potential noise violations based on their likelihood of occurance. If this likelihood is sufficiently small, it is possible that even during operation of the part for a number of years, the probability of a noise violation on the net is negligible, and the net can be assigned a lower priority for the application of noise avoidance strategies or even eliminated from further consideration. The proposed method is based on estimating the probability that the noise on a victim will exceed a given threshold. This is used to prioritize nets based on the expected number of clock cycles (ENC) before the first violation occurs on a specific victim. Based on this, a net with a reported noise violation can be assigned a lower priority if its ENC exceeds say, five years or ten years.

In Section 2 a description of the noise waveforms is presented. Sections 3 and 4 contain the main technical contribution, namely the upper bounds on the tail probabilities and estimates of ENC. Results of experiments carried out on a large, high performance PowerPC microprocessor are presented in Section 5.

2. A PROBABILISTIC MODEL OF NOISE

The focus here is on *functional noise* and hence, without loss of generality, we assume that the victim is stable at logic 0, and an aggressor switches from logic 0 to logic 1. A victim with a set of n aggressors will be referred to as a *cluster of size n*. As in the case with most other work on noise analysis, we assume *linearity*, i.e., the composite noise waveform is obtained by taking the sum of the noise waveforms resulting from each aggressor. The general form of the noise waveform seen on a victim due to an aggressor switching is a sum of weighted, decaying exponentials, the number of terms being equal to the order of the circuit. To simplify the algebraic work, we take a linear approximation for the *rising* and *falling* portions of the waveform and consequently, the noise waveform resulting from each aggressor is approximated by a triangular pulse. In the interest of being conservative one can construct a triangular pulse that is an enevelope of a given exponental pulse, i.e. contains most of the exponential pulse.

Associated with each aggressor i, is a timing interval $[a_i, b_i]$ (obtained from static timing analysis), where a_i and b_i denote the earliest and latest possible arrival times of the transition. Thus, the noise waveform resulting from aggressor i, which is denoted by $Z_i(t)$, is represented by $(h_i, r_i, d_i, a_i, b_i)$, where h_i is the peak noise voltage and r_i and d_i are the slopes of the rising and falling edges of the noise waveform, respectively.

The *randomness* or variability of the noise on a victim net is over different clock cycles and arises due to the fact that an aggressor may or may not switch during a clock cycle, as well as the time when an aggressor switches over its interval. Let τ_i be the random variable that denotes the time instance in $[a_i, b_i]$ at which aggressor i switches. We assume that τ_i is uniformly distributed over $[a_i, b_i]$, i.e.,

$$F_{\tau_i}(x) = \text{Prob}(\tau_i \leq x) = \frac{(x-a_i)_+ - (x-b_i)_+}{b_i - a_i}, \quad (1)$$

where $(x)_+ = x$ if $x \geq 0$, otherwise $(x)_+ = 0$. In the absence of any other information, the uniform distribution is the meaningful choice. However, this assumption is not very restrictive and the approach can be extended to other distributions of switching events, e.g., the triangular distribution. For a given value of τ_i, $Z_i(t)$ is expressed as

$$Z_i(t) = \begin{cases} r_i(t-\tau_i) & \tau_i \leq t \leq \tau_i + \frac{h_i}{r_i} \\ d_i(\frac{h_i}{r_i} - (t-\tau_i)) + h_i & \tau_i + \frac{h_i}{r_i} \leq t \leq \frac{h_i}{r_i} + \frac{h_i}{d_i} + \tau_i \\ 0 & \text{elsewhere.} \end{cases} \quad (2)$$

With τ_i being a random variable, $Z_i(t)$ is a stochastic process, and for a fixed value of t, $Z_i(t)$ is a random variable. Let $F_{Z_i(t)}(z) = \text{Prob}(Z_i(t) \leq z)$ denote the distribution function of $Z_i(t)$. For a fixed value of t, if N waveforms given by (2) are generated corresponding to N sample observations of τ_i, then $F_{Z_i(t)}(z)$ represents the fraction of those N waveforms that at time t have a value $\leq z$. For a fixed t,

$$\{Z_i(t) \leq z\} = \{\tau_i \geq t - \frac{z}{r_i}\} \cup \{\tau_i \leq t - \frac{h_i}{r_i} - \frac{h_i}{d_i} + \frac{z}{r_i}\}. \quad (3)$$

This leads directly to the distribution function of $Z_i(t)$, expressed in terms of the distribution function of τ_i.

$$F_{Z_i}(z) = \begin{cases} 1 - F_{\tau_i}(t - \frac{z}{r_i}) + F_{\tau_i}(t - \frac{h_i}{r_i} - \frac{h_i}{d_i} + \frac{z}{d_i}) & \\ & 0 \leq z \leq h_i \\ 1 & z \geq h_i \\ 0 & \text{otherwise.} \end{cases} \quad (4)$$

Let $S_n(t)$ represent the waveform of the total noise on a given victim net in a cluster of size n. As stated earlier, the use of linear models for the victim and aggressor drivers, allows us to represent the total noise on a victim as a sum of the noise due to each aggressor. We assume that the random variables $Z_i(t)$, $i = 1, 2, \ldots, n$, are independent. Under the zero delay model, the determination of logic correlations (which is an intractable problem in itself) is done separately and only those aggressors that would be switching in the same direction would be included in a cluster. These have been accounted for in the experiments. However, temporal correlations are far more difficult to model and no effective solution to include them and still maintain analytical tractability is known. To account for the fact that an aggressor may or may not switch within a clock period, we introduce a binary random variable variable X_i associated with aggressor i, where $X_i = 1$ with probability p_i and $X_i = 0$ with probability $(1 - p_i)$. p_i is called the switching probability of aggressor i. The random variable that represents the total noise on the victim is:

$$S_n(t) = Z_1(t) * X_1 + Z_2(t) * X_2 + \cdots + Z_n(t) * X_n. \quad (5)$$

3. CRITERION FOR PRIORITIZING NETS

A simple criterion that designers can use to prioritize nets that are identified as having potential noise violations is the *expected* number of clock cycles (ENC) before the first violation occurs. A noise violation is said to occur on a victim net if the total noise $S_n(t)$ on that victim exceeds a given threshold α. The threshold is calculated based on the characteristics of the victim's receiver gate. On each clock cycle, we observe a realization of $S_n(t)$. Let the random variable $N(t)$ denote the number of clock cycles required to observe the first noise violation at time t. Assuming independence of noise waveforms over different clock cycles, the probability that the noise on this victim will exceed α for the first time on kth cycle is given by

$$\text{Prob(N(t) = k)} = [P(S_n(t) \leq \alpha)]^{k-1} * (1 - P(S_n(t) \leq \alpha)). \quad (6)$$

Let $\mathcal{E}$ denote the expectation operator. The expected number of clock cycles before the first violation occurs at time t within a clock period, is given by

$$\text{ENC}(t) = \mathcal{E}(N(t)) = \frac{1}{P(S_n(t) > \alpha)}. \quad (7)$$

When assigning a victim net a very low priority or even discarding it from further consideration because its ENC(t) is very large (e.g. 5 years), we should ensure that t is selected so that at no other value of t, the ENC(t) is smaller. Thus we have

$$t^* = \{t \mid \text{ENC}(t) \text{ is minimum}\} \quad (8)$$

$$\text{ENC}^* = \text{ENC}(t^*). \quad (9)$$

To compute ENC(t), we need the distribution function of $S_n(t)$. An analytical form does not appear to be possible. The alternative, which is to carry out numerical convolution for each t, would be computationally prohibitive. Therefore, we proceed with the next best alternative, which is to derive bounds on $P(S_n(t) > \alpha)$.

4. BOUNDS ON NOISE PROBABILITY

A common strategy to construct a upper bound on the probability that a non-negative random variable exceeds a given value is to construct a parametric family of upper bounds and then find the value of the parameter that minimizes the upper bound. This approach is based on the Chernoff bound [11], which states that

$$P(S_n(t) > \alpha) \le e^{-\theta\alpha}\Phi_{S_n(t)}(\theta)\ \forall\theta \ge 0, \tag{10}$$

where $\Phi_{S_n(t)}(\theta)$ is the moment generating function (mgf) of $S_n(t)$, and θ is an unknown parameter to be determined. From Equation 5 and from the properties of the mgf, we have

$$\Phi_{S_n(t)}(\theta) = \prod_{i=1}^{n}(p_i\Phi_{Z_i(t)}(\theta) + 1 - p_i). \tag{11}$$

By definition of the mgf, we have

$$\Phi_{Z_i(t)}(\theta) = \sum_{k=0}^{\infty}\frac{\mathcal{E}(Z_i^k(t))\theta^k}{k!}. \tag{12}$$

Let $B(\theta,t,\alpha) = e^{-\theta\alpha}\Phi_{S_n(t)}(\theta)$. The value of θ that minimizes $B(\theta,t,\alpha)$ is the solution to the equation [11]

$$\frac{\mathcal{E}(S_n(t)e^{\theta S_n(t)})}{\mathcal{E}(e^{\theta S_n(t)})} = \frac{\Phi'_{S_n(t)}(\theta)}{\Phi_{S_n(t)}(\theta)} = \alpha. \tag{13}$$

Equation (13) can be simply expressed as

$$\frac{d\log(\Phi_{S_n(t)}(\theta))}{d\theta} = \alpha. \tag{14}$$

Using (11) and (14), the value of θ that minimizes the bound given by the RHS of (10) is the solution to the equation

$$\sum_{i=1}^{n}\frac{p_i\Phi'_{Z_i(t)}(\theta)}{p_i\Phi_{Z_i(t)}(\theta) + 1 - p_i} = \alpha. \tag{15}$$

We have now established a method to compute a lower bound on ENC. First, we have to determine t^* (see (8)). Then we have to minimize $B(\theta,t^*,\alpha)$ with respect to θ. Before we proceed with this task, we need expressions for $\mathcal{E}(Z_i^k(t))$ (see Equations 11 and 12).

4.1 Moments of Total Noise

By definition, $\mathcal{E}(Z_i^k(t)) = \int_{-\infty}^{\infty} z^k dF_{Z_i(t)}(z)$ and $F_{Z_i(t)}(z)$ is given by (1). Integrating by parts and noting that $F_{Z_i(t)}(z)$ has a finite jump at $z = 0$, we obtain the following recurrence relation.

$$\mathcal{E}(Z_i^k(t)) = h_i^k - k\int_0^h z^{k-1}F_{Z_i(t)}(z)dz. \tag{16}$$

Equation (16) can be solved exactly. The result is

$$\begin{aligned}
&\mathcal{E}(Z_i^k(t)) = 0^k F_{Z_i(t)}(0) + \\
&\frac{1}{(b_i-a_i)(k+1)!r_i}\Big[(h_i + r_i(t-b_i-h_i/r_i)_+)^{k+1} - \\
&(h_i + r_i(t-a_i-h_i/r_i)_+)^{k+1} + \\
&(r_i(t-a_i)_+)^{k+1} - (r_i(t-b_i)_+)^{k+1}\Big] + \\
&\frac{1}{(b_i-a_i)(k+1)!d_i}\Big[(h_i - d_i(t-b_i-h_i/r_i)_+)^{k+1} - \\
&(h_i - d_i(t-a_i-h_i/r_i)_+)^{k+1} + \\
&(-d_i(t-a_i-h_i/r_i-h_i/d_i)_+)^{k+1} - \\
&(-d_i(t-b_i-h_i/r_i-h_i/d_i)_+)^{k+1}\Big],
\end{aligned} \tag{17}$$

where $F_{Z_i(t)}(0)$ (using (4) and (1)) is given by

$$\begin{aligned}
F_{Z_i(t)}(0) = 1 - \frac{1}{b_i-a_i}[(t-a_i)_+ - (t-b_i)_+] + \\
\frac{1}{b_i-a_i}\left[(t-(\frac{h_i}{r_i}+\frac{h_i}{d_i}+a_i))_+ - (t-(\frac{h_i}{r_i}+\frac{h_i}{d_i}+b_i))_+\right].
\end{aligned}$$

Equation (17) expresses the kth order moment of the noise waveform due to aggressor i, in terms of its descriptors, $(h_i, r_i, d_i, a_i, b_i)$. The moments of $S_n(t)$ can be obtained from Equations 17 and 5. To see how the theoretical and sample moments compare, the first four moments were computed for many clusters taken from a PowerPC microprocessor. Using the timing intervals for each aggressor that were obtained from static timing analysis, and the noise estimates produced by the simulator [10], a MonteCarlo simulation was carried out by varying the switching point of each aggressor. For each selection of switching points, the composite waveform was computed, and this was repeated 5000 times. This corresponds to 5000 clock cycles. Figures 1 and 2 show plots of the theoretical (using Equation (17)) and sample mean, and standard deviation. The timing intervals associated with each aggressor are shown at the bottom of each plot. There is also equally good agreement between the higher order theoretical and sample moments.

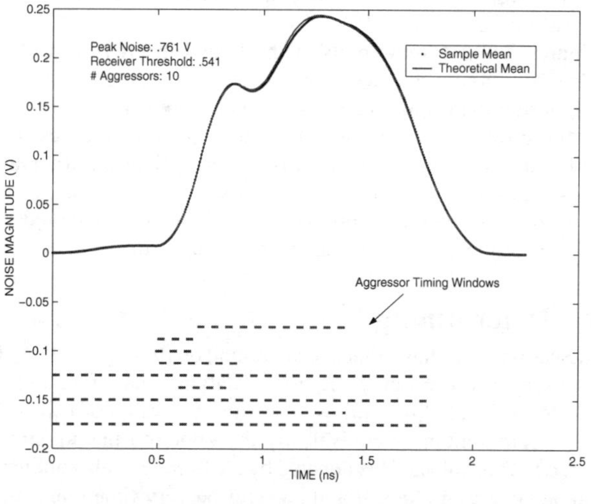

Figure 1: Theoretical and Sample Mean of Total Noise of one victim with 10 aggressors

Since the ordinary moments involve only simple powers of t, the mgf $\Phi_{Z_i(t)}(\theta)$ can be obtained almost by inspection. Using the

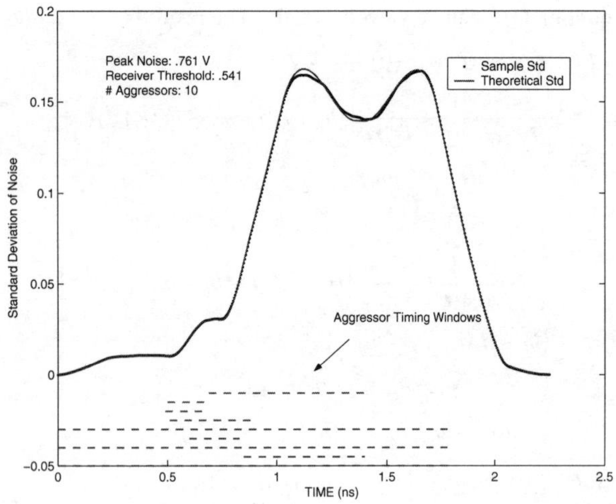

Figure 2: Theoretical and Sample Std. of Total Noise of one victim with 10 aggressors

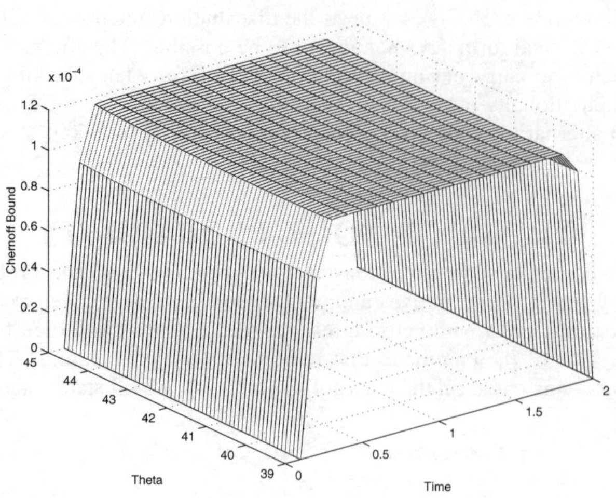

Figure 3: Chernoff Bound of a Net with 10 aggressors without timing intervals.

definition of $\Phi_{Z_i(t)}(\theta)$ given in (12) and substituting the expression for $\mathcal{E}(Z_i^k(t))$ given in (17) into (12), we obtain

$$\begin{aligned}\Phi_{Z_i(t)}(\theta) &= F_{Z_i(t)}(0) + \\ &\frac{1}{(b_i-a_i)r_i\theta}[\exp(\theta(h_i+r_i(t-b_i-h_i/r_i)_+)) \\ &-\exp(\theta(h_i+r_i(t-a_i-h_i/r_i)_+)) + \\ &\exp(\theta r_i(t-a_i)_+)-\exp(\theta r_i(t-b_i)_+] + \\ &\frac{1}{(b_i-a_i)d_i\theta}[\exp(\theta(h_i+d_i(t-b_i-h_i/r_i)_+)) \\ &-\exp(\theta(h_i+d_i(t-a_i-h_i/r_i)_+)) + \\ &\exp(-\theta d_i(t-(h_i/r_i+h_i/d_i+a_i)_+))- \\ &\exp(-\theta d_i(t-(h_i/r_i+h_i/d_i+b_i))_+)]. \qquad (18)\end{aligned}$$

$\Phi_{S_n(t)}(\theta)$ (the mgf of $S_n(t)$) is computed by substituting Equation (18) into Equation (11).

Figures 3 and 4 show plots of the bound in (10) for a cluster of size 10, without and with timing intervals, respectively. Note that ignoring the timing intervals of aggressors simply means that the timing intervals span the entire clock period. Figures 5 and 6 show that the bound (10) as a function of θ in the interval of interest. From these plots, it is clear that once the desired value of t is obtained, then the optimal value of θ can be obtained very quickly by either gradient or direct search techniques.

4.2 Determining t^*

In Section 3 we stated that a lower bound on the expected number of clock cycles to observe the first violation (total noise exceeding a given threshold) requires determining a value of t at which ENC(t) is minimum, or equivalently the value of t that maximizes $B(\theta,t,\alpha)$. That value of t is denoted by t^*. Since θ is also unknown, an iterative search along θ and t would be very time consuming, and prohibitively so, given the large number of clusters that have to processed. Moreover, convergence is not guaranteed. We now describe a procedure that identifies t^* first without having to know θ and then the bound $B(\theta,t*,\alpha)$ can be easily minimized with respect to θ using gradient search.

The solution to finding t^* is based on the monotonicity properties of the mgf $\Phi_{Z_i(t)}(\theta)$. These properties allow us to identify a finite set ($4n$ for a cluster of size n) of distinguished time points $\mathcal{T}$, which we refer to as *breakpoints*. Now for a fixed θ, the maximum value of $\Phi_{S_n(t)}(\theta)$ will occur at the breakpoints or in between two breakpoints. The latter may occur because at some breakpoint, the mgf of some of the aggressors are increasing, while the mgf of others are decreasing. In such a situation, we would have to iteratively search, between every pair of such breakpoints, for the value of t where $\Phi_{S_n(t)}(\theta)$ attains a maximum. Moreover, the point within such an interval where the maximum of $\Phi_{S_n(t)}(\theta)$ occurs will generally depend on θ. To avoid this, we construct a modified mgf $\tilde{\Phi}_{S_n(t)}(\theta)$ such that $\tilde{\Phi}_{S_n(t)}(\theta) \geq \Phi_{S_n(t)}(\theta)$, for all θ, and so that only the breakpoints in $\mathcal{T}$ need to be examined in order to determine the value of t where $\tilde{\Phi}_{S_n(t)}(\theta)$ is maximum. Furthermore, many of the points in $\mathcal{T}$ can be discarded. Finally, at each time point in the reduced set of breakpoints, we solve (15) to find θ using gradient search and choose that value of θ that minimizes the bound. Note

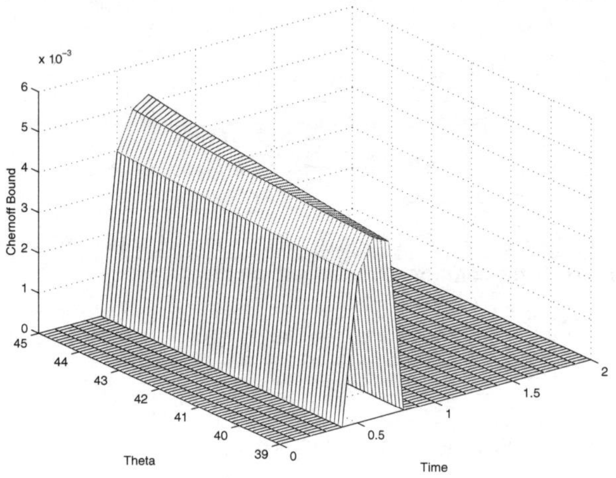

Figure 4: Chernoff Bound of a Net with 10 aggressors with timing intervals.

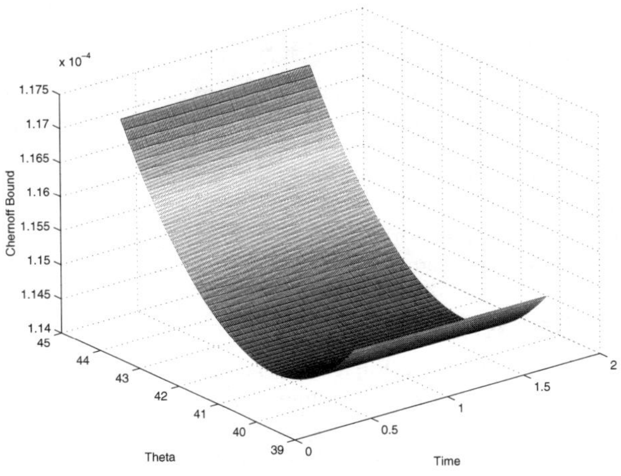

Figure 5: Plot in Figure 3 restricted to smaller range of t.

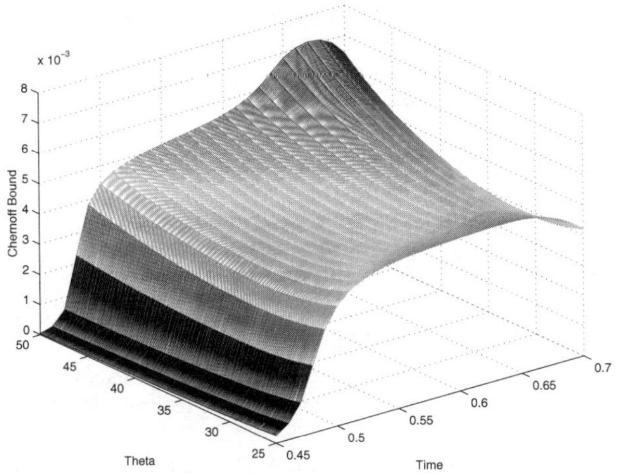

Figure 6: Plot in Figure 4 restricted to smaller range of t.

that this is a conservative solution in that it will lead to a smaller lower bound on ENC*.

Consider the mgf of $Z_i(t)$ given in (18). For a fixed θ, its behavior as a function of t can be classified into two cases.

Case 1. $\mathbf{b_i - a_i \geq h_i/r_i + h_i/d_i}$. This situation occurs when the width of the timing interval is greater than the width of the noise pulse. In this case, $\Phi_{Z_i(t)}(\theta)$ is

- monotonically increasing for $t \in [a_i, h_i/r_i + h_i/d_i + a_i)$,
- constant for $t \in [h_i/r_i + h_i/d_i + a_i, b_i)$, and
- monotonically decreasing for $t \in [b_i, h_i/r_i + h_i/d_i + b_i]$.

Case 2. $\mathbf{b_i - a_i < h_i/r_i + h_i/d_i}$. In this case, $\Phi_{Z_i(t)}(\theta)$ is

- monotonically increasing for $t \in [a_i, \tilde{t}_i]$, and
- monotonically decreasing for $t \in (\tilde{t}_i, h_i/r_i + h_i/d_i + b_i]$,

where $\tilde{t}_i$ is given by

$$\tilde{t}_i = \frac{d_i}{r_i + d_i}\left(\frac{h_i}{r_i} + \frac{h_i}{d_i} + a_i\right) + \frac{r_i}{r_i + d_i} b_i. \quad (19)$$

The points where $\Phi_{Z_i(t)}(\theta)$ changes direction are the *breakpoints* of $Z_i(t)$. The breakpoints of all the aggressors are collected into a list $\mathcal{T}$ and sorted in increasing order. Formally, $\mathcal{T} = \cup_i \tau_i$, where τ_i is defined as

$$\tau_i = \begin{cases} \left\{a_i, \frac{h_i}{r_i} + \frac{h_i}{d_i} + a_i, b_i, \frac{h_i}{r_i} + \frac{h_i}{d_i} + b_i\right\} & \text{if case 1} \\ \left\{a_i, \tilde{t}_i, \frac{h_i}{r_i} + \frac{h_i}{d_i} + b_i\right\} & \text{otherwise.} \end{cases}$$

With n aggressors, the maximum possible number of points in $\mathcal{T}$ is $4n$. However, most of the points in $\mathcal{T}$ can be eliminated from further consideration. This is done by associating a direction $\delta_{i,j}$ with $\Phi_{Z_i(t_j)}(\theta)$ for each $t_j \in \mathcal{T}$.

$$\delta_{i,j} = \begin{cases} -1 & \text{if } \Phi_{Z_i(t_j)}(\theta) \text{ is decreasing} \\ 0 & \text{if } \Phi_{Z_i(t_j)}(\theta) \text{ is constant} \\ 1 & \text{if } \Phi_{Z_i(t_j)}(\theta) \text{ is increasing.} \end{cases}$$

Now $\tilde{\Phi}_{Z_i(t)}(\theta)$ can be constructed from $\Phi_{Z_i(t)}(\theta)$ as follows. Let $\mathcal{T} = \{t_1, t_2, \ldots, t_m\}$.

$$\tilde{\Phi}_{Z_i(t_1)}(\theta) = \Phi_{Z_i(t_1)}(\theta)$$

$$\tilde{\Phi}_{Z_i(t_j)}(\theta) = \begin{cases} \Phi_{Z_i(t_{j-1})}(\theta) & \text{if } \delta_{i,j-1} = -1, \\ & \text{and } \exists k, \delta_{k,j-1} = 1 \\ \Phi_{Z_i(t_j)}(\theta) & \text{otherwise} \end{cases}$$

and $\tilde{\Phi}_{S_n(t)}(\theta) = \prod_{i=1}^{n} \tilde{\Phi}_{Z_i(t)}(\theta)$.

As stated above, many of the points in $\mathcal{T}$ can be discarded. Let t_j be the first point in $\mathcal{T}$ such that $\delta_{i,j} = -1$, for some i. Let t_k be the last point in $\mathcal{T}$ such that $\delta_{i,k} = 1$, for some i. Then points in $(t_1, \ldots, t_{j-1})$ and $(t_{k+2}, \ldots, t_m)$ can be discarded. This is because, t_j and t_k are the the first and last points where one of the modified mgfs $\tilde{\Phi}_{Z_i(t)}(\theta)$ has reached its peak. Note that this reduction is possible only with $\tilde{\Phi}_{Z_i(t)}(\theta)$ and not with $\Phi_{Z_i(t)}(\theta)$. Once a time point $t^* \in \mathcal{T}$ where $\tilde{\Phi}_{S_n(t)}(\theta)$ is maximum is identified, then the value of θ that minimizes $\tilde{B}(\theta, t^*, \alpha)$ is computed by numerically solving (15) with $\tilde{\Phi}$ replacing Φ.

5. EXPERIMENTAL RESULTS

The experimental results reported in this paper were generated by performing noise analysis for a high performance microprocessor. The noise simulator, called ClariNet [10], was developed to analyze large, high performance processor designs. It embodies several features that help speedup noise analysis, allowing it to process hundreds of thousands of nets in a few hours. The total number of nets analyzed were nearly 200,000. Each net was analyzed twice - once for *low overshoot* (victim net is to be stable at logic 0 and its aggressors switch from logic 0 to logic 1) and once for *high undershoot* (victim net is stable at logic 1 and aggressors switch from logic 1 to logic 0). As a result, the total number of violations was 2501. Although the cluster size (number of aggressors per victim) can be quite large, the number that contributed a significant amount of noise was found to be less than 10. For this reason, we set the the maximum cluster size to 10. For each net, the noise report indicates the peak height and width of the noise injected on the net by each aggressor, and the threshold of the receiver gate. This is data that was used for the probabilistic analysis as described in this paper.

For each cluster, the optimal value of the bound was computed, with α set to the receiver's threshold. Of the 2501 reported violations, 634 or 25.35% of them had an expected number of clock cycles for the first violation exceeding 5 years for a continuously running 555 MHz processor. If 5 years is an acceptable level then these can be eliminated from further consideration. Figure 7 shows a plot of the percentage of nets that have an ENC greater than or equal to the value on the abscissa. It is important to note that the set of all nets when using timing intervals (lower plot in Figure 7) is much smaller than the set of nets when timing intervals are ignored.

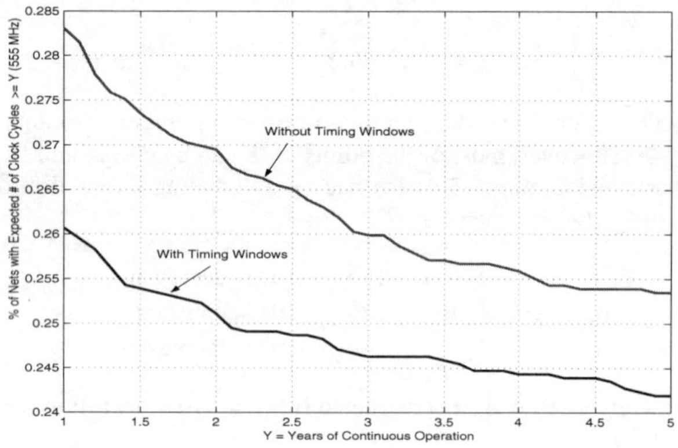

Figure 7: Expected # of Clock Cycles for 1st Violation

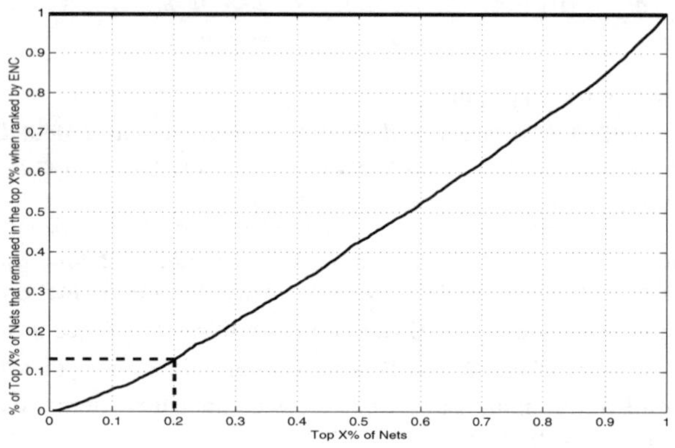

Figure 8: Ranking Nets by Peak Noise and ENC

Next we deleted the nets whose ENC was greater than or equal to 5 years. The number of nets remaining was 1867. For these nets, we computed the ENC* values. This provides an alternative ranking based on the expected number of clock cycles required to observe a noise violation for the first time, rather than simply the sum of the all the noise peaks injected by each aggressor. Two sorted lists L_{PEAK} and L_{ENC} of the nets were generated. L_{PEAK} is the set of nets in decreasing order of the magnitude of the peak noise of the composite waveform. L_{ENC} is the set of nets in increasing order of their ENC* values. Figure 8 shows a plot of the percentage of nets in the original list L_{PEAK} that retain their ranking in the second list L_{ENC}, as each list is traversed. The abscissa represents the top $X\%$ of the nets. The ordinate represents the percentage of the nets in the top $X\%$ of list L_{PEAK} that remained in the top $X\%$ of the nets in list L_{ENC}. For example, from the top 20% of the list L_{PEAK} which represents 373 nets, only 12.5% of them ($\approx$ 46 nets) remained in the top 20% of L_{ENC}. If the ranking of the nets did not change then the plot would be a horizontal line at 1. The important conclusion here that the probabilistic approach identifies nets based not only on the noise magnitude but also on the likelihood of occurrence. Nets for which the simulator reports a relatively small noise violation may become more important, resulting in corrective action being taken on them before nets with larger noise magnitudes.

6. CLOSING REMARKS

In this paper we presented an approach to analyzing functional noise. The objective was to use estimates of the probability that the noise will exceed a specified threshold as a means to prioritize nets or even eliminate them from further consideration if the likelihood is small. Results of experiments carried out on a large industrial high performance microprocessor, using a recently developed state-of-the-art noise simulator were also presented.

7. REFERENCES

[1] R. Arunachalam, K. Rajagopal, and L. T. Pileggi. TACO: Timing Analysis with Coupling. In *Proc. ICCAD*, pages 1–8, Nov. 1998.

[2] Y. Cao, T. Sato, X. Huang, C. Hu, and D. Sylvester. New Approaches to Noise-Aware Static Timing Analysis. In *Proc. TAU*, pages 8–13, Dec. 2000.

[3] P. Chen and K. Keutzer. Towards True Crosstalk Noise Analysis. In *Proc. ICCAD*, pages 132–137, Nov. 1999.

[4] J. Cong, D. Z. Pan, and P. V. Srinivas. Improved Crosstalk Modeling for Noise Constrained Interconnect Optimization. In *Proc. TAU*, pages 14–20, Dec. 2000.

[5] A. Devgan. Efficient Coupled Noise Estimation for On-Chip Interconnects. In *Proc. ICCAD*, pages 147–151, Nov. 1997.

[6] K. Gala, V. Zolotov, R. Panda, B. Young, J. Wang, and D. Blaauw. On-Chip Inductance Modeling and Analysis. In *Proc. DAC*, pages 63–68, June 2000.

[7] A. Glebov, S. Garrilov, D. Blaauw, S. Sirichotiyakul, C. Oh, and V. Zolotov. False-Noise Analysis using Logic Implications. In *Proc. ICCAD*, pages 515–521, Nov. 2001.

[8] H. Hu and S. Sapatnekar. Circuit-Aware On-Chip Inductance Extraction. In *Proc. CICC*, 2001.

[9] A. B. Kahng, S. Muddu, and E. Sarto. Switch Factor Based Analysis of Coupled RC Interconnects. In *Proc. ASIC/SoC Conf.*, pages 3–8, 1999.

[10] R. Levy, D. Blaauw, G. Braca, A. Dasgupta, A. Grinshpon, C. Oh, B. orshav, S. Sirichotiyakul, and V. Zoltov. ClariNet: A noise analysis tool for deep submicron design. In *Proc. DAC*, pages 223–238, June 2000.

[11] J. Proakis. *Introduction to Digital Communications.* McGraw-Hill, New York, 1983.

[12] S. S. Sapatnekar. On the chicken-and-Egg Problem of Determining the Effect of Crosstalk on Delay in Integrated Circuits. In *Proc. IEEE 8th Topical Meeting on Electrical Performance of Electronic Packaging*, pages 245–248, 1999.

[13] Y. Sasaki and G. De Michelli. Crosstalk delay analysis using relative window method. In *Proc. ASIC/SoC Conf.*, pages 9–13, 1999.

[14] K. L. Sheppard and V. Narayanan. Noise in Deep Submicron Digital Design. In *Proc. ICCAD*, pages 524–531, Nov. 1996.

[15] S. Sirichotiyakul, D. Blaauw, R. Levy, C. Oh, S. Shams, V. Zolotov, and J. Zuo. Driver Modeling and Alignment of Worst-Case Delay Noise. In *Proc. TAU*, pages 1–7, Dec. 2000.

[16] D. Sylvester and K. Keutzer. Getting to the Bottom of Deep Submicron. In *Proc. ICCAD*, pages 203–211, Nov. 1998.

[17] A. Vittal, L. H. Chen, M. Marek-Sadowska, K.-P. Wang, and S. Yang. Crosstalk in VLSI Interconnections. *IEEE Trans. on Computer Aided Design*, pages 1817–1824, Dec. 1999.

Crosstalk Noise Estimation for Noise Management

Paul B. Morton
University of California at Santa Cruz
Dept. of Computer Engineering
Santa Cruz, CA 95064
morton@cse.ucsc.edu

Wayne Dai
University of California at Santa Cruz
Dept. of Computer Engineering
Santa Cruz, CA 95064
dai@cse.ucsc.edu

ABSTRACT

One of the main challenges in developing an effective crosstalk noise management strategy is to develop a crosstalk noise estimate which is both accurate and leads to a tractable optimization problem that can be used to optimally redistribute uncommitted routing resources to resolve crosstalk noise violations. Devgan's [4] estimate comes very close to meeting these objectives, however, it is extremely pessimistic for nets with long couplings or aggressor nets with short rise times. The increased complexity of more sophisticated estimates, such as that proposed by Vittal et. al. [13] lead to a crosstalk management problem that is very hard to solve. In this paper we develop two estimates, similar in form to Devgan's estimate, that are based on local approximations of Vittal's estimate. Our estimates are substantially more accurate than Devgan's estimate while still allowing us to formulate the crosstalk management problem in a form that can be solved efficiently.

Categories and Subject Descriptors

B.7.2 [**Integrated Circuits**]: Design Aids—*Layout, Placement and routing*

General Terms

design, performance

Keywords

crosstalk, noise, estimation, local approximation, optimal spacing, noise management

1. INTRODUCTION

One of the main challenges in developing and effective crosstalk noise management strategy is to develop a crosstalk noise estimate which is both accurate and leads to a tractable optimization problem that can be used to optimally redistribute uncommitted routing resources [10] to resolve crosstalk noise violations. To date, the vast majority of work that has been done on crosstalk noise management, see for example [11, 5, 12, 6], has been based on a simple capacitive coupling estimate of crosstalk noise. However, coupling capacitance alone leads to an extremely unreliable estimate of crosstalk noise. In particular, a capacitive coupling estimate ignores the multiplicative effects of aggressor net driver rise time, and ignores the moderating effects of interconnect resistance and self capacitance.

Devgan's estimate [4] is a more sophisticated crosstalk noise estimate that comes close to meeting our objectives. It is easily computable, and it incorporates capacitive coupling, victim net interconnect resistance and aggressor net driver rise time. The increased sophistication of Devgan's estimate necessitates a more complex formulation of the crosstalk noise management problem. In particular, it can be formulated as a separable, convex, nonlinear programming problem [7]. However, even with the increased sophistication of Devgan's estimate, it still does not account for self capacitance and aggressor net resistance, making it overly pessimistic for nets with long couplings or aggressor nets with short rise times.

Recently, Vittal et. al. [13] have developed a crosstalk noise estimate that takes all of these variables into consideration while maintaining the same computational complexity as Devgan's estimate. While advances in crosstalk noise estimation can be directly and easily applied to the task of crosstalk noise analysis, they are much more difficult to effectively apply to the task of crosstalk noise management, since changes that do not increase the computational complexity of crosstalk noise estimation and analysis can dramatically increase the complexity of the crosstalk noise management problem. In particular, when Vittal's estimate is applied to the crosstalk management problem, the resulting nonlinear programming problem is no longer separable, and is, in general, nonconvex. It is unlikely that a programming problem of this form can be directly solved in a reasonable amount of time.

In this paper we develop two estimates, each of which combines the best properties of Devgan's and Vittal's estimates. To do this we take advantage of the fact that, for the purpose of the crosstalk noise management problem, the crosstalk noise estimate need only be accurate over a relatively small range of net spacings. This small range allows us to develop two estimates based on local approximation of Vittal's estimate, which are similar in form to Devgan's estimate. Both of our estimates are significantly more accurate than Devgan's estimate, while they still allow us to formulate the crosstalk management problem as separable, convex, nonlinear programs. In addition, we will demonstrate that one of these programs can be solved in $n\,log(n)$ time.

2. VITTAL'S NOISE ESTIMATE

In this paper we represent the routing for a net, v, as a tree where the nodes of the tree for net v form a set, N_v. Each wire segment of a net, v, connects two nodes in $v's$ routing tree. All information describing the wire segment connecting node i to its parent, $Par(i)$, is associated with the node i, that is, all information is stored in the down stream node for a wire segment.

We base our estimate of the crosstalk noise seen at any node of a net on Vittal's crosstalk noise estimate. This estimate takes into account capacitive coupling, driver rise times, net resistance, and self capacitance. Vittal's estimate, for unit amplitude excitation, can be written as

$$V_n = \sum_{x \in N_{net(n)}} \sum_{m \in CP(x,n)} R_m \sum_{y \in Adj(x)} \frac{C_{xy}}{T_{net(y),net(n)}} \quad (1)$$

where: V_n is the estimated crosstalk noise voltage induced on node n by all nets adjacent to $net(n)$; $net(n)$ is the net containing node n; $N_{net(n)}$ is the set of all nodes of $net(n)$; $CP(x,n)$ is the set of nodes, of $net(n)$, that are common to the path from the root to node x and the path from the root node to node n; R_m is the resistance of the wire segment connecting node m to node $Par(m)$; $Adj(x)$ is the set of adjacent nodes capacitively coupled into node x; C_{xy} is the capacitive coupling between nodes x and y. $T_{net(y),net(n)}$ is the "effective rise time" of $net(y)$ as seen by $net(n)$. The effective rise time is defined as

$$T_{net(y),net(n)} = D_{net(n)} + D_{net(y)} + \frac{T_{net(y)}}{2} \quad (2)$$

where $T_{net(y)}$ is the rise time of the driver for $net(y)$, and $D_{net(n)}$ is the "total delay" associated with $net(n)$. The total delay associated with $net(n)$ is defined as

$$D_{net(n)} = \sum_{i \in N_{net(n)}} \sum_{m \in CP(i,i)} R_m \left[C_i^a + C_i^s + \sum_{j \in Adj(i)} C_{ij} \right] \quad (3)$$

where C_i^a is the capacitance associated with the area covered by the wire segment connecting node i to $Par(i)$, and C_i^s is the capacitance associated with any sink pin connected to node i.

3. PROBLEM FORMULATION

Crosstalk noise violations can be resolved by using uncommitted routing resources to modify wire spacings and/or wire widths. In this paper we will concentrate on the wire spacing approach, and note that a similar development can also be carried out based on modifications to wire widths. To formulate the underlying optimal spacing problem (OSP) for a net, v, with crosstalk noise violations, we write the coupling capacitance between two adjacent nodes, x and y, in terms of the adjacency length and adjacency spacing

$$C_{xy} = C_T \frac{L_{xy}}{S_{xy}} \quad (4)$$

where C_T is a proportionality constant determined by the technology, L_{xy} is the length of the adjacency between nodes x and y, and S_{xy} is the spacing between nodes x and y. Substituting (4) into (1) and (3) we can express the crosstalk noise at any node of a net, v, in terms of a set of spacing variables

$$S_v = \{S_{xy} \mid x \in N_v,\ y \in Adj(x)\} \quad (5)$$

The routing resources consumed by this spacing for net v can be expressed as

$$A_v = \sum_{x \in N_v} \sum_{y \in Adj(x)} L_{xy} S_{xy} \quad (6)$$

Using (1), (3), (4), (5), and (6) we can formulate the OSP as the following constrained minimization problem:

$$\min_{S_v} (A_v) \quad (7)$$

subject to

$$S_{xy} \geq S_{xy\_min} \quad \forall\, x \in N_v,\ y \in Adj(x) \quad (8)$$

$$S_{xy} \leq S_{xy\_max} \quad \forall\, x \in N_v,\ y \in Adj(x) \quad (9)$$

$$V_x \leq M_x \quad \forall\, x \in Pins(N_v) \quad (10)$$

where (8) and (9) are routing resource constraints, and (10) are the crosstalk noise constraints. Specifically, S_{xy_min} is the spacing between nodes x and y prior to optimization, S_{xy_max} is the maximum spacing allowed between nodes x and y, and M_x is the noise margin for node x.

4. LOCAL APPROXIMATION

While (7), (8), and (9) are linear in S_v, (10) is nonlinear, nonseparable, and, in general, nonconvex (see [9] appendix A). Since most of the routing resources on a routed design have already been committed and the remaining uncommitted routing resources are dispersed throughout the design, we expect S_{xy_max} to be a small multiple of S_{xy_min}. Further more, because nonconvex and/or inseparable programming problems are, in general, difficult to solve in a reasonable amount of time, we would like to make a convex, separable, local approximation for (1).

4.1 Taylor Series Estimate

One common approach to making a local approximation is to use several of the low order terms of a Taylor series expansion. Since we are interested in a separable approximation, we are limited to the first two terms of the series. This would give us a linear approximation, however, while the range of spacings is relatively small, it is still too large for this to be an accurate approximation. To improve the accuracy of the approximation we take advantage of the fact that crosstalk noise is primarily a function of capacitive coupling and is therefore a strong function of $1/\overline{S}_v$, where $\overline{S}_v$ is the vector of all spacing variables for net v. With this in mind, we can express (1) as

$$V_n(\overline{S}_v) = g\left(\frac{1}{\overline{S}_v}\right) = g(\overline{Q}) \quad (11)$$

where $\overline{Q} = 1/\overline{S}_v$. Using the first two terms of the Taylor series expansion of $g(\overline{Q})$ around $\overline{Q}_1$ we can approximate $g(\overline{Q}_2)$ as

$$g(\overline{Q}_2) \approx g(\overline{Q}_1) + (\overline{Q}_2 - \overline{Q}_1)^T \nabla_{\overline{Q}}\, g(\overline{Q})|_{\overline{Q}=\overline{Q}_1} \quad (12)$$

Substituting for $\overline{Q}$, $\overline{Q}_1$, and $\overline{Q}_2$, we have

$$g\left(\frac{1}{\overline{S}_{v2}}\right) \approx g\left(\frac{1}{\overline{S}_{v1}}\right) + \left(\frac{1}{\overline{S}_{v2}} - \frac{1}{\overline{S}_{v1}}\right)^T \nabla_{\frac{1}{\overline{S}_v}}\, g\left(\frac{1}{\overline{S}_v}\right)\bigg|_{\frac{1}{\overline{S}_v}=\frac{1}{\overline{S}_{v1}}}$$

and from (11) we see that we can approximate $V_n(\overline{S}_{v2})$ as

$$V_n(\overline{S}_{v2}) \approx V_n(\overline{S}_{v1}) + \left(\frac{1}{\overline{S}_{v2}} - \frac{1}{\overline{S}_{v1}}\right)^T \nabla_{\frac{1}{\overline{S}_v}}\, V_n(\overline{S}_v)|_{\overline{S}_v=\overline{S}_{v1}} \quad (13)$$

We can now define the local optimal spacing problem (LOSP1) using (7) through (10), where (10) is defined using (13), expanded around $\overline{S}_{v1}$.

While (13) is separable, and similar in form to Devgan's estimate, it is not, in general, convex, since one or more of the gradients could be negative. However, using the simple substitution of variables $\overline{S}_v = 1/\overline{q}$, this problem can be easily converted to a convex programming problem that can be solved using sequential quadratic programming, as described in [7].

From (1) we see that Vittal's estimate can be written such that it has at most $|S_v|$ terms, one for each C_{xy}. Additionally, since each term contains a reference to every C_{xy} through $D_{net(n)}$, the gradient in (13), and thus the Taylor series estimate, have a computational complexity of $O(|S_v|^2)$. From [7] it can be seen that the computational complexity of LOSP1 is roughly cubic in the number of spacing variables, $|S_v|$. Given these computational complexities, we see that this estimation technique would be useful for problems with a moderate number of spacing variables, but would be impractical for use on problems with large numbers of spacing variables.

4.2 MERT Estimate

To develop an estimate for (1) that can be used for problems containing large numbers of spacing variables we again take advantage of the fact that we expect S_{xy_max} to be a small multiple of S_{xy_min}. Because of this we can assume that $D_{net(n)}$ and $D_{net(y)}$ can be reasonably approximated by their minimum values, $D^{min}_{net(n)}$ and $D^{min}_{net(y)}$, over the range $S_{xy\_min} \leq S_{xy} \leq S_{xy\_max}$. Defining the "minimum effective rise time" (MERT) as

$$T^{min}_{net(y),net(n)} = D^{min}_{net(n)} + D^{min}_{net(y)} + \frac{T_{net(y)}}{2} \tag{14}$$

we can define the local approximation as

$$V_n \approx \sum_{x \in N_{net(n)}} \sum_{m \in CP(x,n)} R_m \sum_{y \in Adj(x)} \frac{C_{xy}}{T^{min}_{net(y),net(n)}} \tag{15}$$

This gives us an approximation, identical in form to Devgan's estimate, which is an upper bound on Vittal's estimate over $S_{xy\_min} \leq S_{xy} \leq S_{xy\_max}$. Substituting (4) into (15) we have

$$V_n \approx \sum_{x \in N_{net(n)}} \sum_{m \in CP(x,n)} R_m \sum_{y \in Adj(x)} \frac{C_T L_{xy}}{S_{xy} T^{min}_{net(y),net(n)}} \tag{16}$$

which is both convex and separable in S_v. We can now define the second local optimal spacing problem (LOSP2) using (7) through (10), where (10) is defined using (16).

5. OPTIMAL SPACING SOLUTION

For the nontrivial cases, the LOSP2 forms a convex programming problem with at least one strictly feasible solution. These nontrivial cases can be solved using Lagrange relaxation [1]. Additionally, this approach can easily incorporate timing considerations, as illustrated in [8]. To begin, we associate with each sink pin, p, a Lagrange multiplier, λ_p, which is used to relax that pin's noise constraint into the objective function. This gives us the following Lagrange subproblem (LSP):

$$\min_{S_v} \left(\sum_{x \in N_v} \sum_{y \in Adj(x)} L_{xy} S_{xy} + \sum_{p \in Pins(N_v)} \lambda_p (V_p - M_p) \right) \tag{17}$$

subject to

$$S_{xy} \geq S_{xy\_min} \quad \forall\, x \in N_v,\ y \in Adj(x) \tag{18}$$

$$S_{xy} \leq S_{xy\_max} \quad \forall\, n \in N_v,\ y \in Adj(x) \tag{19}$$

From this we define the Lagrange dual problem (LDP) as

$$\max_{\lambda} (LSP) \tag{20}$$

subject to

$$\lambda_p \geq 0 \ \forall\, p \in Pins(N_v) \tag{21}$$

where $\lambda = \{\lambda_p \mid p \in Pins(N_v)\}$.

Since the LOSP2 is convex, we know that the optimal objective values of the LOSP2 and the LDP are equal, and therefore we can solve the LOSP2 indirectly by solving the LDP.

1. $\lambda_p \leftarrow 0\ \forall\, p \in Pins(N_v)$
2. $i \leftarrow 1$
3. Repeat
4. Solve LSP
5. $d_p \leftarrow V_p - M_p\ \forall\, p \in Pins(N_v)$
6. Set θ_i s.t. $\lim_{i \to \infty} \theta_i = 0$ and $\sum_{j \leftarrow 1}^{i} \theta_j \to \infty$
7. $\lambda_p \leftarrow \lambda_p + \theta_i d_p\ \forall\, p \in Pins(N_v)$
8. $\lambda_p \leftarrow max(0, \lambda_p)\ \forall\, p \in Pins(N_v)$
9. $i \leftarrow i + 1$
10. Until convergence

Figure 1: Subgradient optimization algorithm.

We solve the LDP using the subgradient optimization technique outlined in Fig 1. This is a generalized ascent algorithm which takes into account the possibility that the search direction computed in line 5 may not be an ascent direction, due to the fact that the LSP may not, in general, be differentiable. As a consequence, the search direction computed in line 5 is based on the subgradient of the LDP. Since the search direction is not necessarily an ascent direction, we cannot rely on a line search to determine the optimal step length, thus, in line 6, we use a predefined sequence of step lengths which will guarantee the convergence of the subgradient optimization algorithm. To enforce the constrains imposed by (21), line 8 projects any infeasible solution back onto the LDP's feasible region.

To determine when the subgradient optimization algorithm has converged, we take advantage of the fact that the optimal values of the LOSP2 and LDP are equal. This implies that the algorithm converges when

$$\sum_{p \in Pins(N_v)} \lambda_p (V_p - M_p) = 0 \tag{22}$$

In practice, however, this criteria is too strict and needs to be loosened to

$$\left| \frac{\sum_{p \in Pins(N_v)} \lambda_p (V_p - M_p)}{\sum_{x \in N_v} \sum_{y \in Adj(x)} L_{xy} S_{xy}} \right| \leq error\ bound \tag{23}$$

However, in loosening the convergence criteria we have to be mindful of the fact that feasible, nonoptimal solutions to the LDP do not necessarily generate feasible solutions to the LOSP2. We overcome this difficulty by adding the following feasibility constraints to the convergence criteria:

$$\frac{V_p - M_p}{M_p} \leq error\ bound \ \ \forall\, p \in Pins(N_v) \tag{24}$$

In order for the subgradient optimization to be useful in practice, we need to be able to efficiently solve the LSP and efficiently compute a new set of spacings and Lagrange multipliers for each iteration of the algorithm. To efficiently solve the LSP, we take advantage of the fact that (17) is separable in S_v and can be written as

$$\min_{S_v} \left(\sum_{x \in N_v} \sum_{y \in Adj(x)} g_{xy}(S_{xy}) \right) \tag{25}$$

where

$$g_{xy}(S_{xy}) = K_{1xy} S_{xy} + \frac{K_{2xy}}{S_{xy}} + K_{3xy} \tag{26}$$

From the Karush-Kuhn-Tucker optimality conditions [1] we find that the LSP can be solved as

$$S_{xy} = \min\left(S_{xy\_max},\ \max\left(S_{xy\_min},\ \sqrt{K_{2xy}/K_{1xy}}\right)\right) \quad (27)$$

where

$$K_{1xy} = L_{xy} \quad (28)$$

To determine K_{2xy} we collect all terms of (17) which contain $1/S_{xy}$. This gives us

$$K_{2xy} = \sum_{p \in Pins(N_v)} \lambda_p \sum_{m \in CP(x,p)} \frac{R_m C_T L_{xy}}{T^{min}_{net(y),v}} \quad (29)$$

Rearranging the order of summation we have

$$K_{2xy} = \sum_{m \in Ans(x)} R_m \sum_{p \in DesPins(m)} \frac{\lambda_p C_T L_{xy}}{T^{min}_{net(y),v}} \quad (30)$$

where $Ans(x)$ is the set of nodes which contain x and all of its ancestors. $DesPins(m)$ is the set of m's descendant nodes which contain sink pins and the node m if it contains a sink pin. This can be written as

$$K_{2xy} = \frac{R^V_x C_T L_{xy}}{T^{min}_{net(y),v}} \quad (31)$$

where R^V_x is the "weighted up stream resistance" [3] seen from node x and is computed as

$$R^V_x = \sum_{y \in Ans(x)} \Lambda_y R_y \quad (32)$$

where the "weighting factor", Λ_y, associated with node y is computed as

$$\Lambda_y = \sum_{p \in DesPins(y)} \lambda_p \quad (33)$$

From (33) we see that the set of weighting factors can be computed in one bottom up traversal of the routing tree [2]. From (32) we see that the set of weighted up stream resistances can be computed in a single top down traversal of the routing tree. Further, once the weighted up stream resistance for a node has been computed we can immediately compute the node's spacings using (31), (28), and (27). From these three observations we have the following:

THEOREM 1. *The LSP can be solved in $O(|S_v| + |N_v|)$ time.*

From Fig. 1, we see that each iteration of the solution of the LDP must compute a new set of Lagrange multipliers. This can be accomplished by two traversals of the routing tree. One bottom up traversal to compute the down stream capacitance seen by each node, and one top down traversal to compute the noise seen at each sink pin. From these observations, Theorem 1, and the fact that the number of sink pins is bounded above by the number of nodes in the routing tree, we have the following:

THEOREM 2. *Each iteration in the solution of the LDP requires $O(|S_v| + |N_v|)$ time.*

6. DELAY ESTIMATION

The total delay for a net consists of three components, area capacitance delay, sink pin capacitance delay, and coupling capacitance delay. Of these three, only the coupling capacitance delay is affected by changes in S_v. To compute the coupling component of $D^{min}_{net(n)}$ we use the maximum allowed spacing, S_{xy_max}, for each spacing variable in S_v. To compute the coupling component of each $D^{min}_{net(y)}$ we need to divide the set of spacing variables for each $net(y)$

$$S_{net(y)} = \{S_{ab} \mid \forall\, a \in N_y,\ b \in Adj(a)\} \quad (34)$$

into two sets, the set of spacing variables common to both net v and $net(y)$

$$S^v_{net(y)} = S_v \cap S_{net(y)} \quad (35)$$

and the set of spacing variables of $net(y)$ which are not common to net v

$$S^{\overline{v}}_{net(y)} = S_{net(y)} - S^v_{net(y)} \quad (36)$$

To compute the coupling component of $D^{min}_{net(y)}$ we use the maximum allowed spacings, S_{xy_max}, for each spacing variable in $S^v_{net(y)}$ and for each spacing variable in $S^{\overline{v}}_{net(y)}$ we use the spacing determined by the layout prior to optimization.

7. RESULTS

To illustrate the improved accuracy of our estimates compared to Devgan's estimate, we computed the worst case errors between our two estimates, as determined by (13) and (16), and Vittal's estimate. We compared these errors to the error between Devgan's estimate and Vittal's estimate. In order to be consistent with the MERT estimate we chose to expand our Taylor series estimate around the maximum allowable spacing for each spacing variable. The results of our comparison are summarized in the plots shown in Fig. 2 through Fig. 4. In each of these figures we have plotted three surfaces, the upper surface represents the error associated with Devgan's estimate, the middle surface represents the error associated with our MERT estimate, and the lower surface represents the error associated with our Taylor series estimate.

The plots in Fig. 2 through Fig. 4 have been generated based on a pair of adjacent single segment nets, one representing the victim net, and one representing the aggressor net. The length of each of these nets ranges from $1\mu m$ to $1mm$. Each net has a source resistance of 500Ω and a load capacitance of $50fF$. Each has a self capacitance of $0.051fF/\mu m$ and a resistivity of $0.25\Omega/\mu m$. The minimum spacing between the nets is $0.33\mu m$ and the maximum spacing is three times the minimum spacing. The coupling capacitance at minimum spacing is $0.116fF/\mu m$. The wire resistance and capacitance have been lumped at every $0.1\mu m$.

In order to capture the worst case errors for our approximations we calculated all results using the minimum spacing between the victim and aggressor nets. In order to illustrate the behaviors of our estimates with respect to aggressor net driver rise time we have plotted the results for three different rise times ranging from $1ps$ to $100ps$. From Fig. 2 through Fig. 4 we can clearly see the dramatic improvements in accuracy obtained by our estimates. The improvements are particularly dramatic for aggressor nets with short rise times, and when there is a long coupling length between the aggressor and victim nets. Additionally, we can see that our Taylor series estimate is also an upper bound on the Vittal estimate, and, as can be seen in Fig. 3 and Fig. 4, produces an estimate that is roughly twice as accurate as the MERT estimate.

In Fig. 5 we have plotted the peak noise voltages predicted by our estimates (using the same values stated above), for an aggressor net driver rise time of $0s$, over a wide range of aggressor and victim net lengths. From this plot we see that, unlike Devgan's estimate, our estimates remain bounded as the rise time goes to zero, and also remain bounded as the coupling length increases.

To evaluate the performance of our optimal spacing algorithm for the MERT estimate we have tested it on a large set of randomly

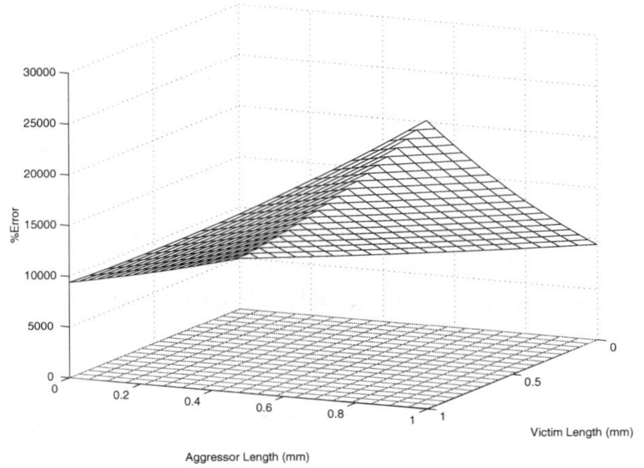

Figure 2: Error comparison for Devgan (upper surface), MERT (middle surface), and Taylor series (obscured lower surface) estimates with an aggressor net driver rise time of $1ps$.

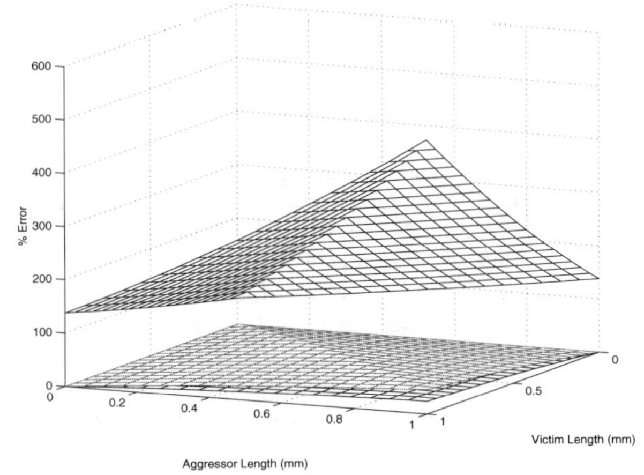

Figure 3: Error comparison for Devgan (upper), MERT (middle), and Taylor series (lower) estimates with an aggressor net driver rise time of $50ps$.

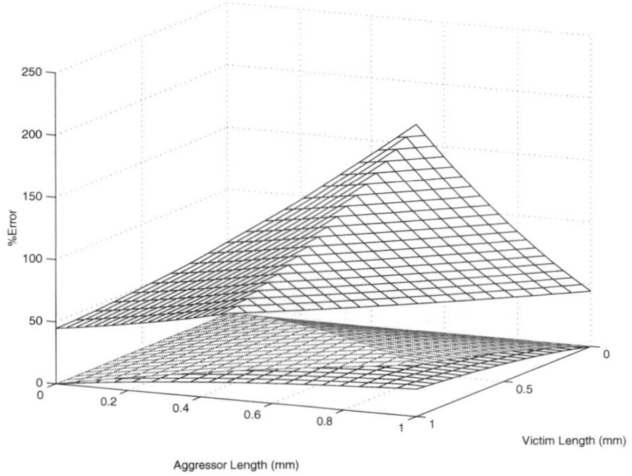

Figure 4: Error comparison for Devgan (upper), MERT (middle), and Taylor series (lower) estimates with an aggressor net driver rise time of $100ps$.

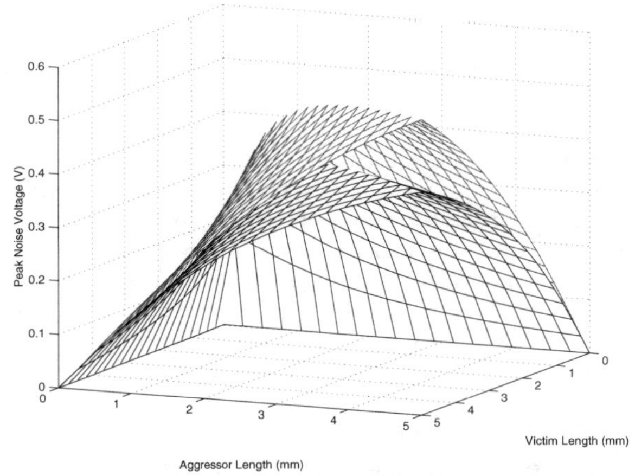

Figure 5: Peak noise predicted by MERT (upper), and Taylor series (lower) estimates with an aggressor net driver rise time of $0s$.

generated multisegment nets. The capacitance and resistance characteristics of these nets are the same as that outlined earlier. Our results have been generated on a Sun Enterprise 450 with a $300\,MHz$ Ultra SPARC-II CPU and $1\,GB$ of memory. The minimum effective rise time of each adjacent net was randomly selected between $20\,ps$ and $500\,ps$. The upper bound for each spacing variable was randomly selected between $0.35\,\mu m$ and $1.0\,\mu m$. The sequence of step lengths used in line 6 of the subgradient optimization algorithm, outlined in Fig. 1, was determined as

$$\theta_i = 1000(0.999)^{(i-1)} \tag{37}$$

For each of the randomly generated nets, we measured the CPU time for our algorithm, coded in C++, to find an optimal set of spacings for the net. We considered a solution to be optimal when the error bounds in (23) and (24) were set to 0.5%.

The results of our experiments are summarized in Fig. 6. In particular, we show two sets of results, one set forming a "cloud" of points close to the x-axis, and one set of points forming a bounding curve above the cloud. The cloud of points were obtained when the noise margin, M_x, for each pin, x, on a randomly generated net, was randomly selected from

$$M_x^{min} + 0.001(M_x^{max} - M_x^{min}) \leq M_x \leq M_x^{max} \tag{38}$$

where M_x^{min} and M_x^{max} are the minimum and maximum feasible noise margins for the pin. From our experiments we found that when the noise margin, M_x, for a pin, x, was less than approximately $M_x^{min} + 0.0005(M_x^{max} - M_x^{min})$ our algorithm occasionally had trouble converging. Note that these cases start to approach the trivial case where $M_x = M_x^{min}$. Because of this, the Theorem of Strong Duality [1] may not apply, and thus we cannot guarantee that our algorithm will converge. However, this difficulty does not pose any problems in practice since these cases can be easily identified and solved by inspection. In particular, if the noise margin for any pin on a net is approximately equal to its minimum feasible

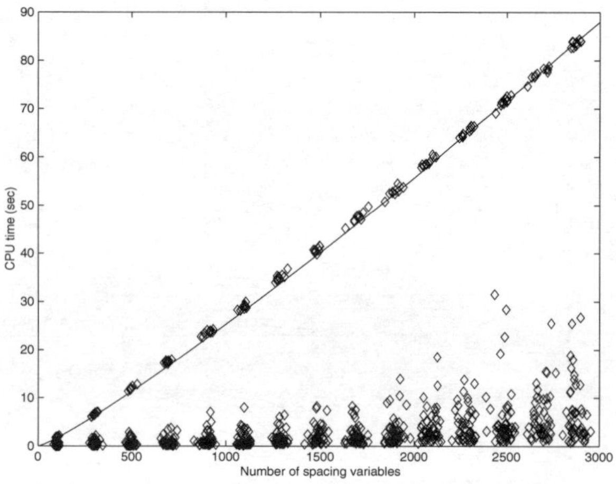

Figure 6: CPU time to compute optimal spacing based on MERT estimate for randomly selected noise margins (lower "cloud"), and for worst case performance (upper curve).

noise margin

$$M_x^{min} \leq M_x < M_x^{min} + 0.001(M_x^{max} - M_x^{min}) \qquad (39)$$

then each of the nets spacing variables should be set to its maximum value.

From our experiments we also found that as we narrowed the range from which the noise margins were selected, the performance of the algorithm converges to the worst case performance indicated by the points on the bounding curve in Fig. 6. In particular, we found that the worst case performance occurred when we set the noise margin, for each pin on a randomly generated net, to approximately $M_x^{min} + 0.75(M_x^{max} - M_x^{min})$. From these results we found that the performance of our algorithm can best be described as $n\,log(n)$, as indicated by the line in Fig. 6.

8. CONCLUSION

In this paper we have presented two crosstalk noise estimates for use in efficiently solving the crosstalk noise management problem. We have presented results that demonstrate that these estimates are significantly more accurate than Devgan's estimate and, in particular, they remain bounded with increasing coupling length and/or decreasing aggressor net rise time. The more accurate of these estimates allows us to formulate the crosstalk noise management problem as a mathematical programming problem that can be solved using sequential quadratic programming, which operates in roughly cubic time. We have shown that the less accurate estimate allows us to formulate the crosstalk noise management problem as a separable, convex, nonlinear programming problem. We have presented a Lagrange relaxation algorithm which can solve this mathematical program and present results that demonstrate that this algorithm operates in $n\,log(n)$ time.

9. ACKNOWLEDGMENTS

The authors would like to thank Eli Chiprout and Bing Zhang for their helpful comments and suggestions during the preparation of this paper.

10. REFERENCES

[1] M. Bazaraa, H. Sherali, and C. Shetty. *Nonlinear Programming*. John Wiley and Sons, New York, 1993.

[2] C. Chen and D. Wong. A Fast Algorithm for Optimal Wire-Sizing Under Elmore Delay Model. In *Proc. IEEE ISCAS*, pages 412–415, 1996.

[3] C. Chu and D. Wong. Greedy Wire-Sizing Is Linear Time. *IEEE Trans. Computer-Aided Design*, 18(4):398–405, 1999.

[4] A. Devgan. Efficient coupled noise estimation for on-chip interconnects. In *Proc. IEEE/ACM Int. Conf. on Computer Aided Design*, pages 147–151, 1997.

[5] T. Gao and C. Liu. Minimum Crosstalk Channel Routing. *IEEE Trans. Computer-Aided Design*, 15(5):465–474, 1996.

[6] H.-R. Jiang, Y.-W. Chang, and J.-Y. Jou. Crosstalk-Driven Interconnect Optimization by Simultaneous Gate and Wire Sizing. *IEEE Trans. Computer-Aided Design*, 19(9):999–1010, 2000.

[7] P. Morton and W. Dai. An Efficient Sequential Quadratic Programming Formulation of Optimal Wire Spacing for Cross-Talk Noise Avoidance Routing. In *Proc. Int. Symposium on Physical Design*, pages 22–28, 1999.

[8] P. Morton and W. Dai. Optimal Spacing for Managing Cross-talk Induced Noise and Delay in Resistive VLSI Interconnections. Technical Report UCSC-CRL-00-06, UCSC, 2000.

[9] P. Morton and W. Dai. Crosstalk Crosstalk Noise Estimation for use in an Efficient Noise Management Strategy. Technical Report UCSC-CRL-01-09, UCSC, 2001.

[10] P. Morton and W. Dai. Routing Resource Management for Post-route Optimization. Technical Report UCSC-CRL-02-12, UCSC, 2002.

[11] A. Onozawa, K. Chaudhary, and E. Kuh. Performance Driven Spacing Algorithms using Attractive and Repulsive Constraints for Submicron LSI's. *IEEE Trans. Computer-Aided Design*, 14(6):707–719, 1995.

[12] P. Saxena and C. Liu. Crosstalk Minimization using Wire Perturbations. In *Proc. Design Automation Conf.*, pages 100–103, 1999.

[13] A. Vittal, L. H. Chen, M. Marek-Sedowska, K.-P. Wang, and S. Yang. Crosstalk in VLSI Interconnections. *IEEE Trans. Computer-Aided Design*, 18(12):1817–1824, 1999.

Variable Frequency Crosstalk Noise Analysis: A Methodology to Guarantee Functionality from dc to f_{max}

Byron Krauter
IBM Corporation
Austin, TX 78758-3493
krauter@us.ibm.com

David Widiger
IBM Corporation
Austin, TX 78758-3493
widiger@us.ibm.com

Abstract - One means of reducing pessimism in crosstalk analysis is to consider timing orthogonality. While earlier works have addressed the temporal alignment of timing windows [1, 2, 3, 4], these treatments have overlooked one key point. Crosstalk noise failures are frequency dependent. A chip that functions at one frequency can fail due to crosstalk noise at faster and slower frequencies. Moreover, because system developers and manufacturers need chips that operate over a wide range of frequencies, noise analysis tools should guarantee a wide range of operation. In this paper, we explain this phenomenon and propose a simple procedure to include timing information and guarantee functionality from dc to f_{max}.

Categories and Subject Descriptors

B.7.2 [**Integrated Circuits**]: Design Aids - *simulation, verification*

General Terms

Algorithms, Design, Verification

Keywords

crosstalk, noise analysis, timing windows, timing orthogonality, GCD frequency, LCM window, frequency-dependent noise

1. Introduction

While it is paramount that a crosstalk noise analysis tool should never under-predict noise, it is also important that these tools not be too pessimistic since undue pessimism can produce an uncompetitive product (either a larger die size or a schedule slip fixing false failures). One way to reduce pessimism in crosstalk noise analysis is to consider timing windows and timing orthogonality.

Timing windows come in two flavors, aggressor switching and victim sensitivity windows. Aggressor switching windows define when aggressor nets can switch, and victim sensitivity windows define when victim pins are sensitive to noise. Aggressor switching windows are specified at source pins, and range from the earliest to latest source pin arrival times. (The aggressor source to victim sink pin noise delays are calculated during noise analysis.)

Victim sensitivity windows are specified at the victim sink pin, and extend from the pin set-up to the pin hold time. While the concepts of set-up and hold time are commonly understood for clocked circuits - clock gates, latches, and dynamic circuits - these concepts also apply to non-clocked circuits. The set-up and hold times of a non-clocked circuit are found by backwards propagating clocked circuit set-up and hold times back to the victim sink pin. To be more specific, by subtracting the long and short path delays from the clocked circuit set-up and hold times respectively.

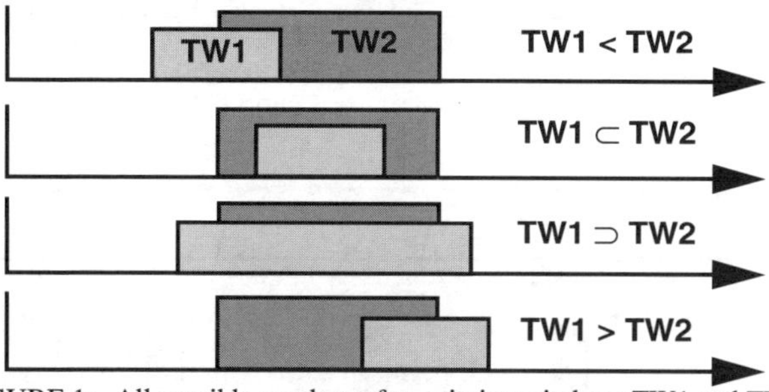

FIGURE 1: All possible overlaps of two timing windows TW1 and TW2

For single frequency operation, the recipe for examining timing orthogonality is straightforward. Timing windows are gleaned from timing analysis results, and two windows overlap when their minimum and maximum edges satisfy the following inequalities:

$$TW1_{min} < TW2_{max} \quad (1)$$

and

$$TW2_{min} < TW1_{max}. \quad (2)$$

In Fig. 1, we've depicted every possible overlap of timing windows TW1 and TW2, and indeed, every case satisfies (1) and (2).

An accurate noise analysis at a single frequency begins by considering only those aggressor nets whose noise pulse windows overlap the victim sensitivity window. Accuracy can be further enhanced by not superimposing the individual peak noises but instead by reanalyzing the coupled network with a nonlinear victim driver model when all the aggressors whose noise pulse windows overlap the victim sensitivity window, switch simultaneously. [1] describes such a procedure as well as how to include the effects of propagated noise from the previous stage.

However, chips are not run at a single frequency. Systems are brought up and debugged at many frequencies. Manufacturing test may require functionality over a range of frequencies. And finally, the final product frequency may be different than the design target frequency. While the requirement to operate at slower frequencies doesn't affect timing analysis, it does affect noise analysis. That is, a chip that meets timing requirements at f_{max} will also meet timing requirements at slower frequencies, but a chip that operates at f_{max} can fail at a slower frequency due to noise.

2. Why Frequency Affects Noise Analysis

Timing window edges cannot be assigned to absolute times. An individual timing window edge is temporally fixed to one (or more) clock edges. Crosstalk noise is frequency dependent because the temporal position of two timing window edges fixed to different clock edges, changes with frequency. The following paragraphs demonstrate this phenomenon with a few examples.

2.1 Single Clock Example

Even the simplest designs - static CMOS chips using Master-Slave latches and a single clock - exhibit this frequency-dependent behavior. Windows that don't overlap at one frequency can overlap when the chip is run either faster or slower. Consider the example depicted in Fig. 2. Both the aggressor launch and the victim cap-

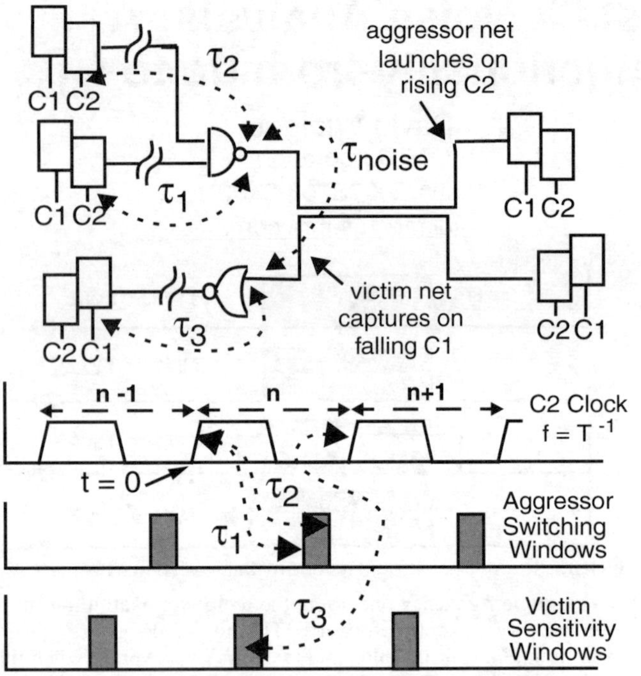

FIGURE 2: Changing Temporal Alignment of Aggressor Switching and Victim Sensitivity Window in a Single Clock Edge Design

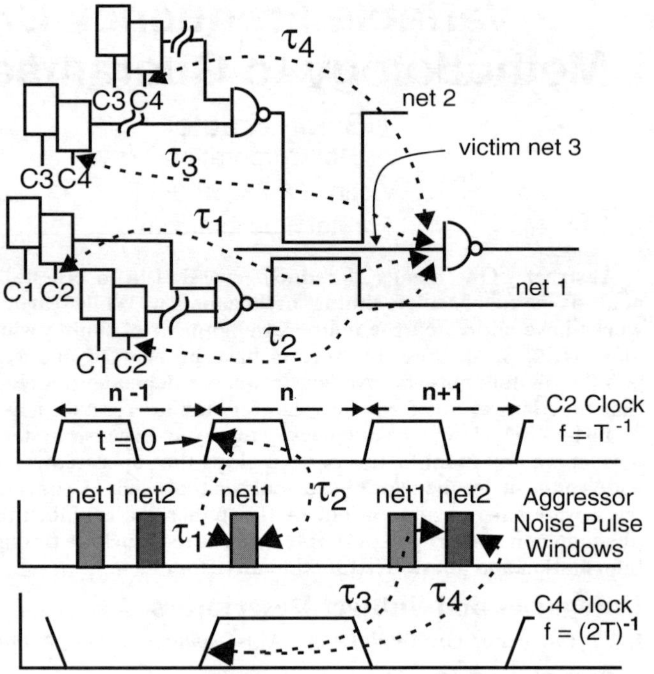

FIGURE 3: Changing Temporal Alignment of Aggressor Switching Windows with Clock Frequencies f and f/2.

ture are triggered on a rising C2 (or falling C1) clock edge. For convenience t=0 has been assigned to the rising C2 clock edge in the *nth* cycle. At the clock frequency depicted, the aggressor switching window and victim sensitivity window do not overlap.

In the *nth* cycle, the aggressor switching window at the nand gate output ranges from τ_1 to τ_2. When noise can appear at the nor gate input then ranges from ($\tau_1 + \tau_{noise}$) to ($\tau_2 + \tau_{noise}$) due to the noise pulse delay τ_{noise}. And finally, the victim sensitivity window at the nor gate input ranges from ($T - \tau_{set\text{-}up} - \tau_3$) to ($T + \tau_{hold} - \tau_3$) where $\tau_{set\text{-}up}$ and τ_{hold} are the setup and hold times at the downstream latch. To summarize

$$\tau_1 < \begin{array}{c}\text{Switching Window}\\ \text{at Nand Gate Output}\end{array} < \tau_2, \tag{3}$$

$$\tau_1 + \tau_{noise} < \begin{array}{c}\text{Noise Pulse}\\ \text{Window at Nor}\\ \text{Gate Input}\end{array} < \tau_2 + \tau_{noise}, \tag{4}$$

and

$$T - \tau_{setup} - \tau_3 < \begin{array}{c}\text{Victim}\\ \text{Sensitivity}\\ \text{Window}\end{array} < T + \tau_{hold} - \tau_3. \tag{5}$$

Applying (1) and (2) to the windows defined by (4) and (5), yields a range of cycle times where the two windows will overlap.

$$\tau_1 + \tau_{noise} - \tau_{hold} + \tau_3 < T < \tau_2 + \tau_{noise} + \tau_{setup} + \tau_3 \tag{6}$$

So the non-overlap in Fig. 2 is a consequence of the frequency depicted. If the period T is increased to satisfy (6), the windows will overlap, potentially causing a noise failure at a slower frequency.

2.2 Multiple Synchronous Clocks Example

In the previously-discussed example where all signals are launched and captured on the same clock edge, frequency dependent crosstalk failures result from the relative movement between aggressor switching and victim sensitivity windows. With multiple synchronous clocks, frequency dependent failures can also result from the relative movement of switching windows when aggressor nets are launched on different, yet synchronous, clock frequencies.

Consider the simple example in Fig. 3. Nets 1 and 2 are two synchronous aggressors nets that couple to a third victim net 3. (For the purpose of this discussion, victim net 3 need not be synchronous to nets 1 and 2.) Net 1 is launched on a rising C2 clock running at a frequency f, and net 2 is launched on a rising C4 clock running at a frequency f/2.

At the frequencies depicted in Fig. 3, the (aggressor clock pin to victim sink pin) noise pulse delays τ_1 and τ_2 on net 1 and τ_3 and τ_4 on net 2 disallow overlap of the noise pulse windows. But when the common source clock frequency f is decreased, the overlap can and will occur every other cycle of the C2 clock.

Net 1 can produce a noise pulse at the victim sink every C2 clock cycle. In cycle n, the net 1 noise pulse window is given by

$$\tau_1 < \begin{array}{c}\text{Net 1 Noise Pulse}\\ \text{Window in Cycle n}\end{array} < \tau_2 \tag{7}$$

and in cycle n + 1, the window is

$$\tau_1 + T < \begin{array}{c}\text{Net 1 Noise Pulse}\\ \text{Window in Cycle n+1}\end{array} < \tau_2 + T. \tag{8}$$

Net 2 produces a noise pulse at the victim sink every other cycle of the C2 clock. In cycle n of the C2 clock, the noise pulse window is

$$\tau_3 < \begin{array}{c}\text{Net 2 Noise Pulse}\\ \text{Window in Cycle n}\end{array} < \tau_4. \tag{9}$$

While the windows defined by (7) and (9) will never overlap, applying (1) and (2) to the timing windows defined by (8) and (9) yields a cycle time range where the noise pulse windows overlap:

$$\tau_3 - \tau_2 < T < \tau_4 - \tau_1 \tag{10}$$

So while aggressors nets 1 and 2 do not simultaneously couple onto victim net 3 at the frequency depicted in Fig. 3, simultaneous coupling will occur at a slower frequencies. And again, this overlap could potentially cause a noise failure at a slower frequency.

3. Crosstalk Noise Analysis from dc to f_{max}

The following paragraphs will explain the general analysis procedure by first discussing timing windows and their relationship with clock edges, introducing new concepts that are critical to the

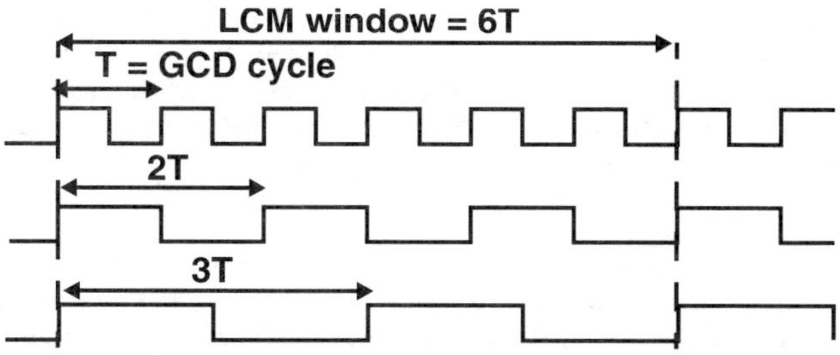

FIGURE 4: Least Common Multiple and Greatest Common Divisor Cycle Times for Synchronous Clocks with T, 2T, and 3T Cycle Times

analysis, and then discussing more specific analysis procedures that naturally build up to the general case. This section will conclude with a discussion of the more general analysis procedure that handles both asynchronous and synchronous nets whose frequencies range from dc to f_{max}.

3.1 Timing Windows and Clock Edges

Timing window edges are controlled by clock edges. The timing windows presented in Fig. 2 and 3 were single-clock-edge windows. That is, the leading and trailing edges were controlled by the same clock edge. In general, static logic circuits and master-slave latches produce single-clock-edge windows, and dynamic circuits, clock gates, and more complicated latches create windows that are controlled by multiple clock edges.

Transparent latches, for example, can create timing windows where the controlling clock edges change with frequency. At high frequencies when data flushes through the latch, leading and trailing edges are controlled by different clock edges. But at slower frequencies when the latch relaunches every signal, both edges are controlled by the same clock edge. Other timing window subtleties that aren't apparent in Fig. 2 and 3 include: When a upstream gate logically combines signals launched from different clocks, the downstream aggressor nets have multiple timing windows. When a victim sink feeds multiple downstream latches that are clocked with different, but synchronous clocks, the victim sink has multiple sensitivity windows.

3.2 Least Common Multiple (LCM) and Greatest Common Divisor (GCD) Cycle Times

For a group of synchronous clocks, the least common multiple and the greatest common divisor cycle times, T_{lcm} and T_{gcd}, are the cycle times of the beat and "source" clocks. Consider the three clocks depicted in Fig. 4. With periods of T, 2T, and 3T, the relative position of these clocks repeats every $T_{lcm} = 6T$, the least common multiple of 1, 2, and 3. And every clock frequency can be expressed as an integer multiple of $T_{gcd} = T$, the greatest common divisor of 1, 2, and 3. Moreover, every clock edge can also be expressed as a half integer multiples of T_{gcd}.

The importance of T_{lcm} and T_{gcd} is that all possible alignments between any two timing windows can be found inside an arbitrary LCM window, and every timing window inside the LCM window can expressed as a linear function of the GCD cycle time, T_{gcd}. Because every edge is temporally pinned to one (or more) GCD clock edges, these expressions are of the form

$$\tau_{edge} = \frac{n}{2}T_{gcd} + \tau_{offset} \qquad (11)$$

where $0 \le n < 2T_{lcm}/T_{gcd}$.

3.3 Fixed Frequency Analysis with Multiple, Yet Synchronous, Clocks

We now describe here a new procedure for examining all possible window alignments that can occur among signals derived from multiple, yet synchronous, fixed-frequency clocks. This procedure will make use of the previously-discussed LCM window concept.

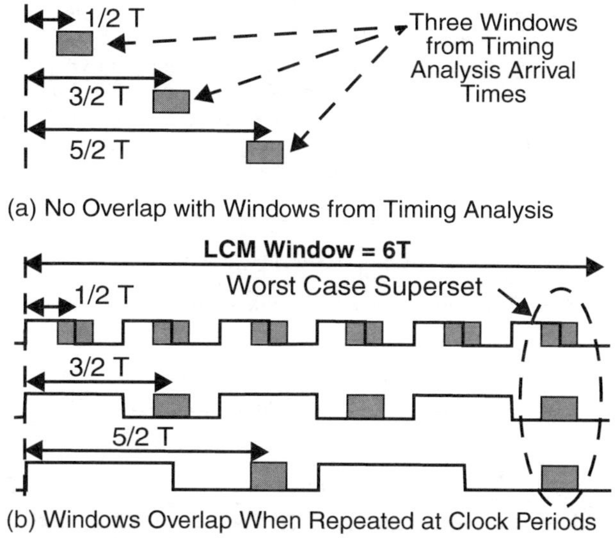

FIGURE 5: Three Aggressor Switching Windows Clocked at T, 2T, and 3T and How They Repeat Inside Their 6T LCM Window

Consider a group of aggressor nets and a victim sink that are synchronous but do not all share the same clock period. We would begin a fixed frequency noise analysis by obtaining, from timing analysis, switching windows for every aggressor net, sensitivity windows for the victim sink, and the clock frequencies at which these windows repeat. Because we are considering fixed clock frequencies, timing window edges could be treated as absolute times, and not functions of T_{gcd} like that shown in (11).

Next we would determine via circuit simulation, the noise waveforms caused by each aggressor net switching by itself. The so-obtained noise pulse delays would be added to the aggressor switching windows to obtain aggressor noise pulse windows.

We would then determine the group LCM cycle time and build a list of all timing windows inside an arbitrary LCM window. This means repeating every timing window at its own fixed clock frequency across the LCM window. Timing windows that overlapped the LCM window endpoints would be split, and those portions outside the LCM window would be discarded.

Every entry in the list would have a window type (aggressor noise pulse or victim sensitivity), a net or pin identity, and the timing window endpoints. After sorting the list by timing window start times, we would traverse the list to determine which windows were simultaneously active. And in particular, we would determine via superposition, the set of noise pulse windows overlapping the sensitivity window that would result in the worst noise. Accuracy could be further enhanced by re-analyzing the coupled network with a nonlinear victim driver model when all the aggressors identified in the previous step, switch simultaneously [1].

The need for a LCM window analysis can be demonstrated with an example. Imagine timing analysis reported the three windows centered about 1/2, 3/2, and 5/2 T in Fig 5-a. When viewed as single windows, these windows obviously do not overlap. But if these windows were repeated at the clocks periods T, 2T, and 3T shown in Fig 5-b, every possible combination of timing window overlap would occur over the LCM window of 6T.

3.4 Variable Frequency Analysis with Multiple, Yet Synchronous Clocks

For the same set of synchronous aggressor nets and victim sink, a variable frequency noise analysis would also determine which clock edges controlled the leading and trailing timing window edges, and express each edge as a function of the group GCD cycle time, T_{gcd}. The analysis would then determine the GCD clock ranges where pairs of timing windows overlapped. This determination can be best described as a double nested `for` loop that examines (1) and (2) for every pair of timing windows and finds a range of T_{gcd}, if one exists, that satisfies both inequalities.

For example, given a timing window TW1 with a leading edge

$$TW1_{min} = \frac{n_1}{2}T_{gcd} + \tau_1 \qquad (12)$$

and another timing window TW2 with a trailing edge

$$TW2_{max} = \frac{n_2}{2}T_{gcd} + \tau_2, \qquad (13)$$

a test of (1) ($TW1_{min} < TW2_{max}$) above yields

$$T_{gcd} < \frac{\tau_2 - \tau_1}{n_1 - n_2}. \qquad (14)$$

If this and the subsequent test of (2) ($TW2_{min} < TW1_{max}$) both yield a GCD cycle time greater than the minimum GCD cycle time, timing windows TW1 and TW2 would, at some viable clock frequency overlap. Of course, (14) is only valid when $n_2 \neq n_1$, but determining whether two timing windows overlap when $n_1 = n_2$ is trivial, and their relative position does not change with T_{gcd}.

The algorithm to find the cycle times when timing windows overlap can be improved by recognizing that a timing window can never overlap a periodic copy of itself, and that sub-LCM windows can be examined for efficiency. For example, if the synchronous group had T, 2T, and 3T clocks, a comparison between a window clocked at T and a window clocked at 2T need only look over a 2T region. Likewise a comparison between a window clocked at T and and window clocked at 3T need only look over a 3T region.

Next we would build a list of GCD cycle times representing when pairs of timing windows overlap. Each member of the list would contain a beginning and ending GCD cycle time and the two windows that overlap in this cycle time range. Two windows that always overlap would be given beginning and ending GCD cycle times of 0 and ∞. The GCD cycle time list would be sorted by beginning GCD cycle times. Finally a series of fixed frequency analyses would be performed.

The first analysis would be performed at the maximum GCD clock frequency using (11) to build the timing windows and the previously-discussed fixed frequency analysis procedure in Section 3.3. To determine worst case noise at the slower clock frequencies, the sorted GCD cycle time list would be traversed (starting at the minimum GCD cycle time) looking for changes in timing window overlap. That is, every new beginning GCD cycle time in the list would indicate that a new pair of timing windows overlapped relative to the preceding noise analysis. And with this change would come the possibility of a yet noisier set of aggressor nets. So at each new beginning GCD cycle time another fixed frequency analysis would be performed, again using (11) to build the timing windows at the new frequency.

3.5 General Analysis

The general analysis would begin by dividing the controlling clocks into synchronous groups. Clocks belong to the same group when they have a fixed and known phase relationship. The C1, C2, C3, and C4 clocks in Fig. 3 belong to the same synchronous clock group because the phase relationship between these clocks is both fixed and known. A source synchronous I/O clock, however, can't be grouped with its source clock because the phase relationship between the clocks, although fixed, is nevertheless unknown.

Next, the timing windows themselves would be identified and grouped into these synchronous clock groups. All timing window edges would be expressed as functions of their synchronous clock group GCD cycle time, T_{gcd}. Note that while the leading and trailing edges of a single timing window can be controlled by different clocks, both controlling clocks will belong to the same synchronous clock group.

Now if any of the nets in the total set of aggressor nets and victim sink is asynchronous to another net in the set, the processing proceeds as follows: If the victim sink is also synchronous to one of these groups, it is included in that group, and the peak noise for the group is analyzed using the previously-discussed process in Section 3.4. Next, the peak noise of every synchronous group that doesn't include the victim net is computed. This computation also uses the variable frequency analysis process, but now we find the peak noise as if the victim sensitivity window extended over the entire LCM cycle. Because the noise from these groups is asynchronous to the victim and asynchronous to the other groups, the noise from each such synchronous group is summed to the synchronously-determined noise (the noise from the aggressor nets in the victim sink's synchronous group) without further pruning to find worst case noise that the victim net can sample. Again accuracy can be enhanced by re-analyzing the coupled network with nonlinear models and simultaneous switching aggressor nets.

4. Other Considerations

Noise affects timing, and vice versa, timing affects noise. An accurate determination of either, even when clock frequencies are fixed, is an iterative process [3]. The procedure presented here addresses one critical aspect of the noise/timing relationship, but it is not an iterative process and it will miss second order effects. While the timing windows can always be padded to prevent these escapes, this added margin will not appeal to every design team. But some design teams, prodded by more demanding manufacturing, test, and system debug requirements, will opt for a methodology that guarantees functionality over a wider frequency range.

5. Conclusions

In this paper we have shown that crosstalk noise is dependent on clock frequency, and have proposed a methodology using timing windows pinned to GCD (Greatest Common Divisor) clock edges that guarantees functionality from dc to f_{max}. Furthermore, we have proposed a more accurate treatment of noise derived from multiple synchronous clocks using LCM (Least Common Multiple) windows for even fixed clock frequency analyses.

References

[1] L.H.Chen and M.Marek-Sadowska, "Aggressor Alignment for Worst-Case Crosstalk Noise", *IEEE Trans. on Computer-Aided Design of Integrated Circuits and Systems*, vol. 20, no. 5, pp. 612-621, May 2001

[2] R.Levy, D.Blaauw, G.Braca, A.Dasgupta, A.Grinshpon, C.Oh, B.Orshav, S.Sirichotiyakul, and V.Zolotov, "ClariNet: A Noise Analysis Tool for Deep Submicron Design", *Proc. Design Automation Conf.*, June 2000, pp. 233-238.

[3] K.L.Shepard, V.Narayanan, P.C.Elmendorf, and G.Zheng, "Global Harmony: Coupled Noise Analysis for Full-Chip RC Interconnect Analysis", *Proc. Intl. Conf. Computer-Aided Design*, Nov. 1997, pp. 139- 146.

[4] D.L.Rude, "Statisical Method of Noise Estimation in a Synchronous System", *IEEE Trans. on Components, Packaging, and Manufacturing Technology - Part B*, vol. 17 no. 4, pp. 514-519, Nov. 1994.

Towards Global Routing With RLC Crosstalk Constraints *

James D. Z. Ma and Lei He
Department of Electrical and Computer Engineering
University of Wisconsin, Madison, WI 53706

Abstract

Conventional global routing minimizes total wire length and congestion. Experiments using large industrial benchmark circuits show that up to 24% of nets in such routing solutions may have RLC crosstalk violations at 3GHz clock. We develop an extremely efficient length-scaled K_{eff} (LSK) model that has a high fidelity for long-range RLC crosstalk. We formulate an extended global routing problem (denoted as GSINO) to consider simultaneous shield insertion and net ordering with RLC crosstalk constraints, then propose an effective three-phase GSINO algorithm. The GSINO algorithm *completely* eliminates the RLC crosstalk violations, and has small area and wire length overhead compared to conventional routing.

Categories and Subject Descriptors

B.7.2 [Design Aids] – Placement and Routing

General Terms

Algorithms, Design, Theory

1. INTRODUCTION

As VLSI technology advances, crosstalk becomes increasingly critical. Global routing has been studied to consider track assignment [1] or separated layer/track assignment and shield insertion [2] for capacitive crosstalk constraints. A few recent works have addressed crosstalk avoidance techniques for both capacitive and inductive crosstalk. Examples include shielding [3], simultaneous shield insertion and net ordering (SINO) [4], twisted bundle layout structure [5], and differential signaling [6]. However, there has been no in-depth study on automatic global routing that is able to consider both capacitive and inductive crosstalk constraints.

The rest of the paper is organized as follows: Section 2 introduces the preliminaries and proposes a length-scaled K_{eff} (LSK) model. Section 3 formulates an extended global routing ($GSINO$) problem and develops a three-phase algorithm. Section 4 presents the experimental results. Section 5 concludes the paper with discussions on future work. A full version of this paper is available as a technical report [7], including details of LSK model development and verification, in-depth GSINO algorithm description, and more experimental results.

2. RLC NOISE MODELING

2.1 Preliminaries

We consider over-the-cell routing for global interconnects in this paper. We denote a set of signal nets as $\mathcal{N} = \{N_1, N_2, \cdots\}$. Each net N_i has a number of pins ($p_{i0}, p_{i1}, \cdots$), where p_{i0} is the source and p_{ij} ($j > 0$) is the sink. We assume a pair of routing layers, one for horizontal wires and the other for vertical wires. The routing layers are divided by pre-routed power/ground (P/G) networks into routing *regions* $\mathcal{R} = \{R_1, R_2, \cdots\}$. The number of horizontal (vertical) tracks available for region R_k is the horizontal (vertical) capacity $HC(R_k)$ ($VC(R_k)$). A track can be occupied by a segment of either a net or a shield. A shield is a wire directly connected (without through devices) to P/G networks. One convenient way to connect shields is to add vias between shields and P/G networks. We also assume that all global interconnects have the same driver resistance and loading capacitance, and all wires (except P/G wires) have the same width, spacing, and thickness. Moreover, P/G wires are wide enough so that there is *no* crosstalk (coupling) between regions separated by P/G wires.

According to [4], two signal nets N_1 and N_2 are *sensitive* to each other if a switching event on N_1 causes N_2 to malfunction (due to extraordinary crosstalk or delay variation). In this case we call N_1 an aggressor for N_2 and N_2 a victim of N_1. The *sensitivity rate* of N_i is defined as the ratio of the number of aggressors for N_i to the total number of signal nets. Further, we assume that there is no coupling between different routing regions. The simultaneous shield insertion and net ordering (SINO) problem [4] has been studied to find the minimum area solution with the following RLC crosstalk constraints: all signal nets are capacitive crosstalk free (i.e., no sensitive nets are adjacent to each other) and have inductive crosstalk (or equivalently, the total crosstalk) less

*This research is partially supported by NSF CAREER Award CCR-0093273 and Intel. We used computers donated by HP and SUN Microsystems. Lei He is now Assistant Professor of Electrical Engineering Department at UCLA. Address comments to lhe@ee.ucla.edu.

than a given threshold. An accurate and extremely efficient model for long-range RLC crosstalk, such as the LSK model to be developed below, is of paramount importance for a routing algorithm considering SINO at the full chip level.

2.2 LSK Model

A formula-based K_{eff} model has been proposed in [4] to characterize the inductive coupling between two signal nets. It uses a formula to calculate the coupling coefficient K_{ij} between two signal nets N_i and N_j. The total amount of inductive coupling K_i induced on N_i is $\sum_{j \neq i} K_{ij}$ for all signal nets that are sensitive to N_i. The K_{eff} model is easy to compute and convenient to use at a higher design level or an earlier design stage. Furthermore, it has a high fidelity in the following sense [8]: for a SINO solution of multiple nets with a fixed wire length, a signal net with a higher K_i value given by the K_{eff} model also has a higher SPICE-computed noise voltage. Moreover, empirical evidence [7] has shown that the noise voltage is roughly a linearly increasing function of the wire length and this observation holds for a large number of SINO solutions in our experiments considering different design and fabrication technologies.

Therefore we propose the following length-scaled K_{eff} model (in short, LSK model). We first compute the LSK value as

$$LSK = \sum_j l_j \cdot K_i^j \quad (1)$$

where K_i^j is the total inductive coupling for the net N_i under study in region R_j and l_j is the length of N_i in R_j. We then compute the RLC crosstalk voltage from the LSK value by looking up a table with two columns, one for LSK and the other for the corresponding crosstalk voltage. To build such a table, we generate a number of SINO solutions for a single routing region, and compute the LSK values and corresponding crosstalk voltages via SPICE simulations for different wire lengths. Our table used in the paper contains 100 entries, with crosstalk voltage values from 0.10V to 0.20V, which is about 10% $\sim$ 20% of the supply voltage Vdd (1.05V for the ITRS 0.10μm technology [9]).

Note that the LSK model can only be applied to a net with the following property: none of the neighboring wires of N_i switches simultaneously with N_i. All nets in a SINO solution satisfy this property. We also assume uniform driver and receiver for all interconnects, and the aforementioned table should be re-computed for different combinations of driver and receiver. Our future work includes generalizing the LSK model for non-uniform drivers and receivers

3. GSINO PROBLEM FORMULATION AND ALGORITHM

We formulate the GSINO problem as follows:

FORMULATION 1. *The optimal GSINO problem decides a Rectilinear Steiner Tree RST_i for each net N_i in the context of global routing, and finds a SINO solution within each region such that the given RLC crosstalk constraint is satisfied for each sink, and the total wire length and routing area are minimized.*

The GSINO problem has a high complexity, as even its sub-problem SINO is NP-hard [4]. Therefore, we propose the following three-phase heuristic algorithm. In Phase I, we *uniformly* partition the crosstalk bounds among routing regions for each signal net, and perform global routing with consideration of allocating and minimizing shielding area. In Phase II, we find a SINO solution within each routing region under partitioned crosstalk bounds. In Phase III, we carry out local refinement to eliminate the remaining (but very limited) crosstalk violations and further reduce the routing congestion. We only present details of Phases I and III below, and refer readers to [4] for the SINO algorithm used in Phase II.

3.1 Phase I Algorithm

For simplicity of presentation, we assume a uniform constraint of crosstalk voltage for all sinks. Both our algorithm and program implementation, however, can handle non-uniform crosstalk constraints. For crosstalk bound partitioning we first map the crosstalk voltage into an LSK value by the table look-up approach described in Section 2. We then use $L_{e,ij}$, the Manhattan distance between the source p_{i0} and a specific sink p_{ij} of net N_i, to approximate the wire length in the final routing solution, and compute the inductive coupling bound for each net segment on the path from the source to the sink as $K_{th} = \frac{LSK}{L_{e,ij}}$. For the net segment on the common paths from the source to multiple sinks, K_{th} is the minimum of those bounds determined for individual paths.

After the above uniform partitioning of the crosstalk bounds, K_{th} is available and fixed for each net segment in global routing phase. We then apply the iterative deletion (ID) algorithm [10] to synthesize a global routing solution with shielding area (i.e., a number of tracks) properly reserved and more importantly minimized. We choose the ID algorithm because it is able to consider all the signal nets simultaneously and thus is independent of the net routing order. It is less efficient but may lead to better solutions compared to other order-dependent routing approaches. Nevertheless, our algorithm framework can be applied to other routing algorithms.

We first define the *net connection graph* for net N_i as an undirected graph $G_i = (V_i, E_i)$, where V_i is the set of regions within the bounding box for the pins of net N_i and there is an edge $e_{i,jk} \in E_i$ if $V_{i,j}$ and $V_{i,k}$ are adjacent regions. Let Ω be the set of the connection graphs we construct for all the nets. We iteratively delete the edge with the largest weight from Ω until Ω is a net connection forest, i.e., the connection graphs for all nets are reduced to connection trees. The ID algorithm is summarized in Figure 1, with details in [10] and [7].

ID Algorithm
For each net N_i, construct G_i
Let Ω be the set of G_i's.
Repeat
  e = the edge with the maximum weight in Ω.
  remove e from the corresponding G_i.
  update the weight of affected edges.
Until Ω is a net connection forest.

Figure 1: ID algorithm.

The weight of a horizontal edge is computed as

$$w(e) = \alpha \cdot f(WL) + \beta \cdot HD(R_i) + \gamma \cdot HOFR(R_i) \quad (2)$$

where $f(WL)$ is the normalized wire length with respect to

the estimated wire length of the Rectilinear Steiner Minimum Tree ($RSMT$) for the current net. $HD(R_i)$ is the routing density defined as $\frac{HU(R_i)}{HC(R_i)}$, where $HC(R_i)$ is the horizontal capacity of region R_i and $HU(R_i) = N_{ns} + N_{ss}$ is the utilization of horizontal tracks, with N_{ns} being the number of net segments in the region and N_{ss} being the number of shields needed by the min-area SINO solution to satisfy the K_{th} constraint for each net segment in the region. Moreover, $HOFR(R_i)$ is the relative horizontal overflow, i.e., the number of overflow net segments over the routing capacity. Finally, α, β, and γ are three constants and can be tuned for different objectives. Generally, γ is much larger than α and β so that virtually no overflow is allowed in the final global routing solution. We set α=2, β=1, and γ=50 in this paper. The weight for a vertical edge can be defined by a formula similar to Formula 2.

Note that given the fixed K_{th} for each net segment, the number of shields in each region is a function of the number of net segments N_{ns} in the region and their sensitivities (S_i's) [8, 7]. I.e.,

$$\begin{aligned} N_{ss} &= a_1 \cdot \sum_{i=1}^{N_{ns}} S_i^2 + a_2 \cdot \frac{1}{N_{ns}} \cdot \sum_{i=1}^{N_{ns}} S_i^2 + a_3 \cdot \sum_{i=1}^{N_{ns}} S_i \\ &\quad + a_4 \cdot \frac{1}{N_{ns}} \cdot \sum_{i=1}^{N_{ns}} S_i + a_5 \cdot N_{ns} + a_6 \qquad (3) \end{aligned}$$

In [7] we present the values of the coefficients and verify the formula using a large range of N_{ns} and S_i. The estimates differ by at most 10% from the min-area SINO solutions.

It is worthwhile to emphasize that according to Formula (3), the number of shields in a region is proportional to the density of sensitive nets in that region. Because the net segments in different regions are not sensitive to each other by definition and the weight of an edge considers the density of sensitive nets, our new ID algorithm is able to evenly distribute the sensitive nets across the chip, avoid routing too many sensitive nets in the same region, and reduce the total number of shields.

3.2 Phase III Algorithm

During the crosstalk budgeting in Phase I, we use the Manhattan distance from source to sink to approximate the wire length of the final route. A detour may actually occur, cause under-estimation of the crosstalk, and lead to crosstalk violations. Even though such violations are very limited in our experiments, we use the algorithm shown in Figure 2 to *completely* eliminate the crosstalk violations and further reduce the routing congestion. Phase III contains two passes of iterative local refinement (denoted as LR) in a greedy fashion, considering crosstalk bound re-distribution which is different from the uniform distribution used in Phase I.

There are two loops in the first pass to eliminate the crosstalk violations. The outer loop picks the net N_i with the most severe crosstalk violation, and the inner loop locates the least congested routing region R_j through which N_i is routed. Further, the inner loop allows exactly *one* more shield to be added into R_j by using Formula (3) to decide how much the K_{th} can be reduced for the net segment of N_i. Finally, SINO is invoked again in R_j with respect to the reduced K_{th}. This inner loop is iterated until we eliminate the crosstalk violation for N_i. The outer loop is stopped when there are no more crosstalk-violating nets.

There are also two loops in the second pass to reduce the congestion. In the outer loop, we start with the most congested region R_j and compute the *slack* as K_i - K_{th} for all the nets that are routed through region R_j. To avoid being too greedy, in the inner looper we allow only *one* shield to be removed at a time. Again, Formula (3) is used to decide how much increase on K_{th}'s is needed. The inner loop iteration is stopped when one shield can be removed. We then re-run SINO under the increased K_{th}'s. We only accept the new SINO solution if it has no crosstalk violations, and in this case update the congestion of region R_j and the slacks of affected nets. The outer loop is stopped when no reduction on slacks is possible without causing crosstalk violations.

```
Pass 1: Eliminate crosstalk violations
Repeat
    N_i = net with most severe crosstalk violation
    Repeat
        R_j = the least congested region containing N_i
        decrease K_th for N_i's segment by
        allowing one more shield in R_j
        re-do SINO in R_j
    Until N_i has no crosstalk violation
Until no nets have crosstalk violations

Pass 2: Reduce routing congestion
Repeat
    R_j = the most congested region
    compute the slacks for nets routed through region R_j
    Repeat
        N_i = the net with the largest slack
        increase K_th for N_i's segment in R_j by N_i's slack
    Until one shield can be removed from R_j
    new_SINO = re-do SINO in R_j
    if (new_SINO causes no crosstalk violations) then
        SINO=new_SINO
        update the congestion of region R_j
        update the slacks of affected nets.
Until no reduction on the slacks is possible without
        causing crosstalk violations
```

Figure 2: LR algorithm.

4. EXPERIMENTAL RESULTS AND DISCUSSIONS

We have implemented the GSINO algorithm using C/C++ on a UNIX workstation, and applied it to the ISPD'98/IBM benchmark circuits at 3GHZ clock (see [7] for detailed description of the circuits and the derived technology parameters). The placement is generated by DRAGON [11] and the crosstalk constraint is set to 0.15V, around 15% of the supply voltage Vdd, for all sinks.

We compare GSINO with the following two approaches: (1) ID+NO: ID-based global router to minimize wire length and congestion only, followed by net ordering (NO) within each region to eliminate as much capacitive coupling as possible; and (2) *i*SINO: ID-based global router to minimize wire length and congestion only, followed by SINO within each routing region. In order to make fair comparisons, we still consider the weight function (Formula (2)) but without N_{ss} in $HU(R_i)$ for both ID+NO and *i*SINO (i.e., no shielding area reservation or minimization is considered).

Both the *i*SINO and GSINO approaches apply SINO to meet the crosstalk bounds so that no crosstalk violations are observed. In Table 1 we only present the numbers of crosstalk-violating nets in ID+NO solutions. Since ID+NO is not aware of an RLC crosstalk constraint during global routing, up to 24% of nets may have crosstalk violations.

	sensitivity rate = 30%	sensitivity rate = 50%
ibm01	1907 (14.60%)	2583 (19.78%)
ibm02	3254 (16.87%)	4275 (22.16%)
ibm03	4920 (18.85%)	6056 (23.20%)
ibm04	5143 (16.42%)	5928 (18.92%)
ibm05	4361 (14.71%)	7135 (24.07%)
ibm06	4802 (13.96%)	6573 (19.11%)

Table 1: Numbers of crosstalk-violating nets for ID+NO solutions. The data in the parentheses are the percentages with respect to the total numbers of signal nets.

Because applying SINO within each region after global routing does not change the wire length, *i*SINO has the same wire length as ID+NO. On the other hand, GSINO has an average of 7% wire length overhead compared with ID+NO for 30% sensitivity rate and 13% overhead for 50% sensitivity rate (see Table 2). It is worthwhile to point out that the SINO solution has a relatively smaller delay per unit length as no neighboring wires switch simultaneously [12]. Therefore, the performance penalty due to the increase on wire length should be less than the wire length penalty.

	ID+NO	GSINO	ID+NO	GSINO
	sensitivity rate = 30%		sensitivity rate = 50%	
ibm01	639	683 (6.89%)	639	706 (10.49%)
ibm02	724	796 (9.94%)	724	829 (14.50%)
ibm03	647	717 (10.82%)	647	753 (16.38%)
ibm04	748	815 (8.96%)	748	868 (16.04%)
ibm05	695	741 (6.62%)	695	784 (12.81%)
ibm06	769	827 (7.54%)	769	860 (11.83%)

Table 2: Average wire lengths (μm) of ID+NO and GSINO solutions. The data in the parentheses are the average increase on wire lengths compared to ID+NO solutions.

In Table 3, we calculate the routing area by the product of the maximum row and column lengths, and compare the three routing approaches. *i*SINO has large routing area overhead compared to ID+NO, 18% on average for 30% sensitivity and 23% on average for 50% sensitivity. GSINO reduces the area overhead to 7% and 9%, respectively.

	ID+NO	*i*SINO	GSINO
	sensitivity rate = 30%		
ibm01	1533 × 1824	1658 × 1974 (17.04%)	1589 × 1866 (6.04%)
ibm02	3004 × 3995	3306 × 4283 (17.99%)	3105 × 4087 (5.74%)
ibm03	3178 × 3852	3492 × 4108 (17.18%)	3285 × 3950 (6.00%)
ibm04	3861 × 3910	4157 × 4241 (16.78%)	4120 × 3932 (7.31%)
ibm05	9837 × 7286	11529 × 7443 (19.73%)	10526 × 7404 (8.74%)
ibm06	5002 × 3795	5412 × 4107 (17.09%)	5283 × 3890 (8.26%)
	sensitivity rate = 50%		
ibm01	1533 × 1824	1714 × 2048 (25.53%)	1591 × 1872 (6.51%)
ibm02	3004 × 3995	3476 × 4329 (25.39%)	3187 × 4125 (9.54%)
ibm03	3178 × 3852	3662 × 4139 (23.82%)	3318 × 4050 (9.77%)
ibm04	3861 × 3910	4278 × 4322 (22.47%)	4115 × 3951 (7.67%)
ibm05	9837 × 7286	11772 × 7489 (23.00%)	10428 × 7406 (7.75%)
ibm06	5002 × 3795	5527 × 4206 (22.46%)	5418 × 3889 (11.00%)

Table 3: Routing areas ($\mu m \times \mu m$) of ID+NO, *i*SINO, and GSINO solutions. The data in the parentheses are the increase on routing areas compared to ID+NO solutions.

In Tables 1–3, we consider two sensitivity rates 30% and 50%. In the case of 30%, a signal net is sensitive to random 30% of other signal nets in the netlist. We observe that when the sensitivity rate decreases from 50% to 30%, GSINO results in reduced wire length (26% on average) and reduced routing area overhead (20% on average). The actual sensitivity rate in real designs depends on logic and physical implementation, and hence our assumption on random sensitivity rate is only an approximation. However, we expect the overall sensitivity rate in real designs to be less than 50%, implying that the wire length and routing area overhead will be less than what we report in this work.

5. CONCLUSIONS AND FUTURE WORK

In this paper we have formulated a GSINO problem and developed an effective three-phase algorithm to satisfy the RLC crosstalk constraints, with consideration of SINO. Experimental results have demonstrated that the GSINO algorithm is able to completely eliminate the RLC crosstalk violations, with small routing area and wire length overhead.

The majority of running time in the current three-phase GSINO algorithm is consumed by the ID-based global routing phase. A more efficient global router will be developed or be integrated into the GSINO framework. Furthermore, we plan to explore alternative crosstalk budgeting approaches.

6. ACKNOWLEDGMENT

The authors would like to thank Dr. Prashant Saxena at Intel, and Mr. Kevin Lepak and Mr. Jinjun Xiong at University of Wisconsin-Madison for helpful discussions.

7. REFERENCES

[1] T. Xue and E. S. Kuh, "Post global routing crosstalk synthesis," *TCAD*, pp. 1418–1430, Dec. 1997.

[2] H. Zhou and D. F. Wong, "Global routing with crosstalk constraints," *TCAD*, November 1999.

[3] L. He, N. Chang, S. Lin, and O. S. Nakagawa, "An efficient inductance modeling for on-chip interconnects," in *CICC*, 1999.

[4] L. He and K. M. Lepak, "Simultaneous shielding insertion and net ordering for capacitive and inductive coupling minimization," in *ISPD*, 2000.

[5] G. Zhong, H. Wang, C.-K. Koh, and K. Roy, "A twisted bundle layout structure for minimizing inductive coupling noise," in *ICCAD*, 2000.

[6] Y. Massoud, J. Kawa, D. MacMillen, and J. White, "Modeling and analysis of differential signaling for minimizing inductive cross-talk," in *DAC*, 2001.

[7] J. D. Ma and L. He, "Towards global routing with RLC crosstalk constraints," in *University of Wisconsin, Technical Report, ECE-02-1*, 2002.

[8] J. D. Ma, A. Parihar, and L. He, "Pre-routing estimation of shielding for RLC signal integrity," in *ICCD*, 2001.

[9] Semiconductor Industry Association, *International Technology Roadmap for Semiconductors*, 1999.

[10] J. Cong and B. Preas, "A new algorithm for standard cell global routing," *Integration, the VLSI Journal*, November 1992.

[11] M. Wang, X. Yang, and M. Sarrafzadeh, "DRAGON2000: Standard-cell placement tool for large industry circuits," in *ICCAD*, 2000.

[12] J. D. Ma and L. He, "Formulae and application of interconnect estimation considering shield insertion and net ordering," in *ICCAD*, 2001.

Reduction of SOC Test Data Volume, Scan Power and Testing Time Using Alternating Run-length Codes*

Anshuman Chandra and Krishnendu Chakrabarty
Dept. Electrical and Computer Engineering
Duke University
Durham, NC 27705, USA.
{achandra, krish}@ee.duke.edu

ABSTRACT

We present a test resource partitioning (TRP) technique that simultaneously reduces test data volume, test application time and scan power. The proposed approach is based on the use of alternating run-length codes for test data compression. Experimental results for the larger ISCAS-89 benchmarks and an IBM production circuit show that reduced test data volume, test application time and low power scan testing can indeed be achieved in all cases.

Categories and Subject Descriptors

B.7.3 [**Integrated circuits**]: Reliability and testing

General Terms

Algorithms, design, experimentation, reliability, verification

1. INTRODUCTION

Intellectual property (IP) cores are now commonly used in large system-on-a-chip (SOC) designs. Although IP cores help reduce design cycle time, they pose several test challenges. The system integrator is confronted with the problems of test data volume, test application time and power consumption during test. New techniques based on test resource partitioning (TRP) that reduce test data volume, testing time and power during testing are therefore necessary to facilitate plug-and-play SOC test automation.

Power consumption in test mode is considerably higher than in normal mode [1]. Test data volume and test application time are two additional problems faced in SOC test integration. Current techniques do not provide a unified solution to the three problems of test data volume, test application time and test power. These techniques typically provide point solutions that target at most two out of the above three objectives.

Structural methods for reducing test data volume and testing time typically require design modifications. For example, the Illinois scan architecture (ILS) [2, 3] uses a single scan-in pin to simultaneously feed multiple scan chains during broadcast mode. However, a drawback of the ILS architecture is that it does not address the problem of reducing power consumption during scan testing.

In a different approach, which can be described as an algorithmic strategy, the precomputed test set T_D provided by the core vendor is compressed (encoded) to a much smaller test set T_E and stored in ATE memory. An on-chip decoder is used for pattern decompression to generate T_D from T_E during pattern application [16, 17]. These compression techniques are typically based on run-length codes and their variants, e.g. Golomb and FDR codes. A particularly attractive feature of TRP based on compression methods is that it does not require any redesign of IP cores.

Another way to reduce test data volume and testing time is to use BIST [4]. However, BIST can only be applied to SOCs if the IP cores in them are BIST-ready. Since most currently-available IP cores are not BIST-ready, the incorporation of BIST in them requires considerable redesign.

A number of techniques to control power consumption in test mode have also been presented in the literature. These can be broadly classified as (a) structural (b) algorithmic, and (c) tester-based.

1. *Structural methods*: These methods, which do not address test data volume or testing time, involve gated scan chains [5], tailored test generation circuits [6, 9], modified scan latch and vector inhibition [7, 8], and scan chain organization [10].

2. *Algorithmic methods*: These include automatic test pattern generation (ATPG) under power constraints, test data compression, and test scheduling algorithms. ATPG techniques for generating vectors for low power testing usually leads to an increase in the number of test vectors [11]. On the other hand, static compaction of scan vectors causes significant increase in power consumption during testing [13]. While compacted vectors are useless if they exceed power constraints, uncompacted vectors cannot be used as they require excessive tester memory. In order to address this problem, power minimization based on test data compression was presented in [19]. Test scheduling techniques for system integration attempt to reduce testing time by applying scan/BIST vectors to several cores simultaneously [14, 15]. Test scheduling is typically carried out under power constraints since multiple cores are tested in parallel.

3. *Tester frequency*: Reduction in power dissipation can be achieved by running the tester at a slower frequency. Although this method offers the simplest way to reduce power consumption, it leads to unacceptable testing times and is therefore impractical.

We note that structural methods for reducing test power in SOCs require modification to the embedded cores, e.g. via scan latch reordering [12], scan chain and scan cell redesign. This is usually not feasible for IP cores. ATPG techniques are also infeasible for IP

*This research was supported in part by the National Science Foundation under grant number CCR-9875324.

DAC 2002, June 10-14, 2002, New Orleans, Louisiana, USA.

Group	a=0 Run-length of 0s	a=1 Run-length of 1s	Group prefix	Tail	Codeword
A_1	0	0	0	0	00
	1	1		1	01
A_2	2	2	10	00	1000
	3	3		01	1001
	4	4		10	1010
	5	5		11	1011
A_3	6	6	110	000	110000
	7	7		001	110001
	8	8		010	110010
	9	9		011	110011
	10	10		100	110100
	11	11		101	110101
	12	12		110	110110
	13	13		111	110111
...	...	...	...	...	...

Figure 1: The alternating run-length code.

cores since they require gate-level structural models. We therefore focus on test data compression for reducing test power, test data volume, and testing time simultaneously.

It was shown in [19] that test data volume and test power can be reduced simultaneously using Golomb coding. The key idea is to map the don't-cares in the test vectors to zero. This results in long runs of zeros that can be efficiently compressed using Golomb code [17]. The resulting fully-specified test set also reduces switching activity during scan shifting. Hence, significant reduction in scan power is accompanied with test data compression. However, the test power can be reduced further if we map the don't-cares to derive test sets that minimize switching activity. Unfortunately these test sets are not amenable for compression by run-length codes.

In this paper, we present a new class of codes, called alternating run-length codes, for test data compression. These codes are particularly effective for compressing test sets that lead to minimum switching activity . They also reduce testing time due to the reduction in the amount of test data that needs to be transported from the tester to the SOC.

2. ALTERNATING RUN-LENGTH CODE

We first review FDR coding and its application to test data compression. The FDR code is a data compression code that maps variable-length runs of 0s to variable-length codewords. The reader is referred to [18] for a detailed discussion and motivation for the FDR code.

It was shown in [18] that the FDR code is very efficient for compressing data that has few 1s and long runs of 0s. However, for data streams that are composed of both runs of 0s and runs of 1s, the FDR code is rather inefficient. In fact, in our initial experiments, the sizes of encoded test sets obtained for such test sets were larger than the sizes of uncompressed test sets. This provides the motivation to develop a code that can efficiently compress both runs of 0s and 1s.

Figure 1 illustrates the encoding procedure for the new alternating run-length code. The alternating run-length code is also a variable-to-variable-length code and consists of two parts—group prefix and tail. The prefix identifies the group in which the run-length lies and the tail identifies the member within the group. An additional parameter associated with this code is the alternating binary variable a. The encoding produced by the alternating run-length code for a given run-length depends on the value of a. If

Input data stream:	000000111111000001 (18-bits)
FDR encoded data:	1100000000000000011010 (22-bits)
Alternating FDR encoded data:	11000010111010 (14-bits)
	| a = 0 | a = 1 | a = 0 |

Figure 2: An example to compare FDR coding with alternating run-length coding.

$a = 0$, the run-length is treated as a run of 0s. On the other hand, if $a = 1$, the run-length is treated as a run of 1s. Note that the values of a for the different runs are not added to the encoded data stream.

Figure 2 shows the encoded data obtained using the two codes for a data stream composed of interleaved runs of 0s and 1s. We observe that the size of the FDR-encoded data set (22 bits) is larger than the size of the input data set (18 bits); hence the FDR code provides no compression for this case. On the other hand, the size of the alternating run-length-encoded data set (14 bits) is smaller than the size of the input data set. Therefore, we are able to achieve compression with the new code. We also note that $a = 0$ is used for compressing a runs of 0s, $a = 1$ is used for compressing a runs of 1s, and $a = 0$ is then used for compressing the next run of 0s. Hence, a is inverted after each run is encoded and it keeps alternating between 0 and 1 thereafter. In this paper, we assume a default initial value of $a = 0$, i.e., we assume that the input data stream starts with a run of 0s.

2.1 Power estimation

We use the weighted transitions metric (WTM) introduced in [13] to estimate the power consumption due to scan vectors. The WTM metric models the fact that the scan power for a given vector depends not only on the number of transitions in it but also on their relative positions. For example, consider a scan vector $v_1v_2v_3v_4v_5 = 01000$, where v_1 is first loaded into the scan chain. The 0-to-1 transition between v_1 and v_2 causes more switching activity in the scan chain than the 1-to-0 transition between v_2 and v_3.

WTM is also strongly correlated to the switching activity in the internal nodes of the core under test during scan operation. It was shown experimentally in [13] that scan vectors with higher weighted transition metric dissipate more power in the core under test. In addition, experimental results for an industrial ASIC [5] confirm that the transition power in the scan chains is the dominant contributor to test power.

Consider a scan chain of length l and a scan vector $t_j = t^\star_{j,1}t^\star_{j,2} \ldots t^\star_{j,l}$, with $t^\star_{j,1}$ scanned in before $t^\star_{j,2}$, and so on. As shown in [19], the *weighted transitions metric* for t_j, denoted WTM_j, is given by $WTM_j = \sum_{i=1}^{l-1}(l-i)\cdot(t^\star_{j,i} \oplus t^\star_{j,i+1})$. If the test set T_D contains n vectors $t_1, t_2, \ldots, t_n$ then the average scan in power P_{avg} and peak scan in power P_{peak} are estimated as $P_{avg} = \sum_{j=1}^{n} WTM_j/n$ and $P_{peak} = \max_{j\in\{1,2,\ldots,n\}}\{WTM_j\}$.

It was shown in [19] that Golomb coding can be used to simultaneously reduce the volume of test data, P_{avg} and P_{peak}. The don't-care bits were mapped to 0s and the resulting test data was compressed using the Golomb code. While this approach provides significant reductions in power consumption, and at the same time, decreases the test data volume considerably, it does not minimize either P_{avg} or P_{peak}.

Here we improve upon [19] by using the alternating run-length code. Table 1 shows a partially-specified scan vector $t_i = 01XXX$

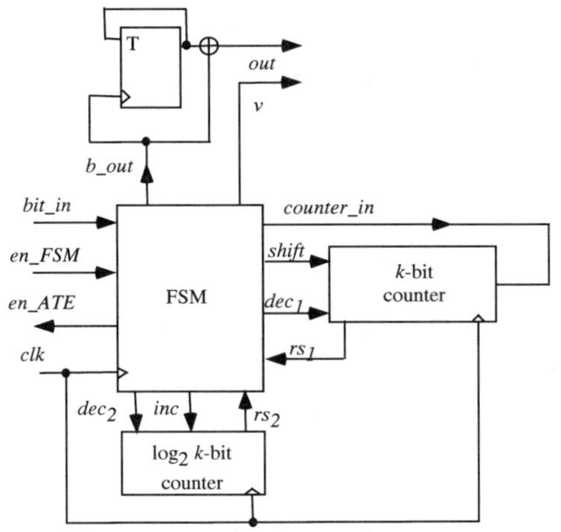

Figure 3: The decoder block diagram for the alternating run-length code.

$10XXX01$ with scan chain length $l = 12$, where X denotes a don't-care bit. If the don't-cares are mapped to appropriate binary values to minimize the weighted transition metric, then a sequence $dXXXXd'$, $d \in \{0,1\}$, must be mapped to $ddddd d'$. Similarly, a sequence $dXXXX$ must be mapped to $ddddd$. This ensures that the few unavoidable transitions occur "late" during scan in.

Table 1 shows the impact of don't-care mapping on the weighted transitions metric WTM_i and compression for a given test vector t_i. Since the FDR code has been shown to be more efficient than the Golomb code [18], we only consider the FDR code here. The WTM value is clearly higher if the don't-cares are always mapped to 0. However, FDR coding is much more effective in reducing test data volume if this strategy is used. On the other hand, while FDR coding is ineffective for the fully-specified test vector that minimizes WTM, the alternating run-length code provides the same compression as achieved with the FDR code with all don't-cares mapped to 0s.

2.2 Decompression architecture

An on-chip decoder decompresses the encoded test set T_E and produces T_D. Let r_{max} be the longest run of 0s in T_D and let $k = \lceil \log_2 r_{max} \rceil$. As discussed in [18], the FDR decoder can be efficiently implemented by a k-bit counter, a $\log_2 k$-bit counter and a finite-state machine (FSM), and it is independent of the precomputed test set and the circuit under test. The synthesized decode FSM circuit contains only 4 flip-flops and 34 combinational gates. The decoder for alternating run-length code can be implemented by making a small modification to the FDR decoder. The block diagram of the alternating run-length decoder is shown in Figure 3. An additional toggle flip-flop and an exclusive-OR gate are required to switch between $a = 0$ and $a = 1$.

3. TESTING TIME ANALYSIS

We now analyze the testing time when a single scan chain is fed by the alternating run-length decoder. Test data compression decreases testing time, and allows the use of a low-cost ATE running at a lower frequency to test the core without imposing any penalties on the total testing time. Let the ATE frequency and the on-chip scan frequency be f_{ATE} and f_{scan}, respectively, where $f_{ATE} < f_{scan}$. Since the ATE and the scan chain operate at two different frequencies, the decoder also consists of two parts—one operating at f_{ATE} and the other operating at f_{scan} such that $f_{ATE} = f_{scan}/\alpha$, $\alpha > 1$. The parameter α should ideally be a power of 2 since it is easier to synchronize the ATE clock with the scan clock for such values of α [21]. The proposed TRP scheme therefore decouples the internal scan chain(s) from the ATE via the use of a decoder interface. This decoupling implies that the scan clock frequency is no longer constrained by the ATE clock frequency limitation. Thus f_{scan} can now be made much larger than f_{ATE}.

For the alternating run-length code, let $t(k,i)$ be the total time required to decompress a codeword that is the ith member of the kth group, and $t_{shift}(k,i)$ and $t_{decode}(k,i)$ be the time required to transfer the data from the ATE to the chip and to decode the codeword, respectively. An upper bound on $t(k,i)$ can be obtained by assuming that decoding begins after the complete codeword is transfered from the ATE. This implies that $t(k,i) \leq t_{shift}(k,i) + t_{decode}(k,i)$.

The prefix length and the tail length of the codeword belonging to the kth group is each equal to k bits; see Figure 1. Since data is transfered from the ATE to the chip at the tester frequency, the time required to transfer any codeword of the kth group is given by $t_{shift}(k,i) = 2k/f_{ATE}$.

For any codeword, the prefix is identical to the binary representation of the run-length corresponding to the group's first element. As shown in Figure 1, the number of 0s in the prefix of a codeword belonging to the kth group is equal to $2^k - 2$. The decoder has to output $(2^k - 2)$ 0s before the tail decoding starts. The time $t_{prefix}(k)$ required to decompress the prefix of any codeword from the kth group is therefore given by $t_{prefix}(k) = (2^k - 2)/f_{scan}$.

Similarly, the time $t_{tail}(k,i)$ required to decompress the tail of the ith member of the kth group is equal to the sum of time required to output $(i-1)$ 0s and a single 1. Hence, $t_{tail}(k,i) = ((i-1)+1)/f_{scan} = i/f_{scan}$.

Therefore, the total decoding time $t_{decode}(k,i)$ is given by

$$\begin{aligned} t_{decode}(k,i) &= t_{prefix}(k) + t_{tail}(k,i) \\ &= (2^k - 2)/f_{scan} + i/f_{scan}. \end{aligned}$$

The total time needed to decompress the codeword is given by

$$\begin{aligned} t(k,i) &\leq t_{shift}(k,i) + t_{decode}(k,i) \\ &= 2k/f_{ATE} + (2^k - 2)/f_{scan} + i/f_{scan} \\ &= \frac{1}{f_{ATE}}(2k + \frac{2^k - 2 + i}{\alpha}) \end{aligned} \quad (1)$$

where $f_{scan} = \alpha\, f_{ATE}$.

Let $q(k,1), q(k,2), q(k,3), \ldots, q(k,2^k)$ be the absolute frequencies of the members of the kth group. Therefore, the decompression time $\tau(k)$ for the runs belonging to k^{th} group is given by

$$\begin{aligned} \tau(k) &= \frac{1}{f_{ATE}} \sum_{i=1}^{2^k} (2k + \frac{2^k - 2 + i}{\alpha}) q(k,i) \\ &= \frac{1}{f_{ATE}} (2k \sum_{i=1}^{2^k} q(k,i) + \frac{1}{\alpha} \sum_{i=1}^{2^k} (2^k - 2 + i) q(k,i)). \end{aligned}$$

Let us assume that k_{max} is the largest group. The test application time TAT_{SSC} for the entire test set with a single scan chain (SSC) is given by

$$TAT_{SSC} \leq \sum_{k=1}^{k_{max}} \tau(k) = \frac{|T_E| + \frac{1}{\alpha} \sum_{k=1}^{k_{max}} \sum_{i=1}^{2^k} (2^k - 2 + i) q(k,i)}{f_{ATE}}$$

where, $|T_E|$ is the size of the encoded test set. Next, to derive a lower bound on the testing time, suppose the tail bits are shifted in

Partially-specified scan vector (test cube)	Fully-specified vector (Minimum WTM)	Fully-specified vector (Don't-cares mapped to 0s)
$t_i = 01XXX10XXX01$ (12-bits)	011111000001 FDR code length: 14 bits Alternating FDR code length: 10 bits $WTM_i = 18$	010001000001 FDR code length: 10 bits $WTM_i = 25$

Table 1: Mapping of don't-cares in a test cube to binary values.

Circuit	No. of bits in T_D	Size of Mintest test set (bits)	FDR coding using T_D: Comp-pression (percent)	FDR coding using T_D: No. of bits in T_E	Alternating FDR coding using T_D: Comp-pression (percent)	Alternating FDR coding using T_D: No. of bits in T_E	Alternating FDR coding using T_D: Improvement over Mintest (percent)
s9234	39273	25935	43.59	22152	44.96	21612	16.66
s13207	165200	163100	81.30	30880	80.23	32648	79.98
s15850	76986	57434	66.22	26000	65.83	26306	54.19
s38417	164736	113152	43.26	93466	60.55	64976	42.57
s38584	199104	161040	60.91	77812	61.13	77372	51.95
Average	—	—	57.21	—	60.57	—	48.16

Table 2: Experimental results on test data compression using the FDR and the alternating run-length code.

while the prefix is being decompressed. Since, the tail bits are now shifted in parallel while the prefix bits are decoded, a lower bound on decoding time using (1) is given by:

$$t(k,i) \geq k/f_{ATE} + (2^k - 2)/f_{scan} + i/f_{scan} = \frac{k + \frac{2^k-2+i}{\alpha}}{f_{ATE}}.$$

Therefore,

$$TAT_{SSC} \geq \frac{1}{f_{ATE}}\left(\frac{|T_E|}{2} + \frac{1}{\alpha}\sum_{k=1}^{k_{max}}\sum_{i=1}^{2^k}(2^k - 2 + i)q(k,i)\right).$$

We next compare the testing time using the proposed TRP scheme with that for an ATPG-compacted test set with p patterns and an external tester operating at frequency $f^{\star}_{ATE}$. Let the length of the scan chain be n bits. The size of the ATPG-compacted test set is pn bits and the test application time TAT^{ATPG}_{SSC} equals $pn/f^{\star}_{ATE}$. Experimental results presented in Section 4 show that the testing time is reduced considerably using the proposed method if $f^{\star}_{ATE} = f_{ATE}$.

4. EXPERIMENTAL RESULTS

In this section, we evaluate the effectiveness of alternating run-length coding for reducing test data volume, testing time and power consumption during scan testing. We carried out experiments for the larger ISCAS-89 benchmark circuits and a production circuit from IBM. The experiments were conducted on a Sun Ultra 10 workstation with a 333 MHz processor and 256 MB of memory.

Table 2 presents the experimental results for test sets obtained from the Mintest ATPG program [20]. We compare the compression obtained using the FDR code and the alternating run-length code. In order to compare with [19], we first assigned all don't-cares to 0s and compressed T_D using the FDR code. We then carefully mapped the don't-cares to minimize WTM and compressed the resulting T_D using the alternating run-length code. Table 2 shows the sizes of T_D, the size of the smallest encoded test set obtained after static compaction using Mintest, the size of compressed test set obtained with all don't-cares mapped to 0 ($|T_{E1}|$) and size of compressed test set obtained with an optimal mapping of don't-cares to minimize WTM ($|T_{E2}|$).

As is evident from Table 2, the alternating run-length code yields better compression than the FDR code for four out of the six benchmark circuits. This is particularly remarkable since, as we show later in this section, high compression with the alternating run-length code is accompanied with significant reduction in testing time and test power. In all cases, the size of the encoded test set is less than the smallest ATPG-compacted test sets known for these circuits. On average, the size of T_E is 48.16% less than that of the compacted test sets obtained using Mintest.

Circuit	f_{ATE} (MHz)	α	Lower bound on TAT_{SSC} (ms)	Upper bound on TAT_{SSC} (ms)	TAT^{ATPG}_{SSC} (ms)
s9234	20	4	0.589	0.881	1.037
		8	0.785	1.326	1.296
		16	0.663	1.203	1.296
s13207	20	4	2.881	3.697	8.155
		8	1.848	2.664	8.155
		16	1.332	2.148	8.155
s15850	20	4	1.619	2.277	2.871
		8	1.138	1.796	2.871
		16	0.898	1.555	2.871
s38417	20	4	3.683	5.308	5.657
		8	2.654	4.278	5.657
		16	2.139	3.763	5.657
s38584	20	4	4.423	6.357	8.052
		8	3.178	5.113	8.052
		16	2.556	4.490	8.052

Table 3: Comparison of testing time using the proposed TRP method with traditional scan-based testing.

Table 3 presents test application time for the proposed method using the alternating run-length codes and for traditional scan-based testing with $f^{\star}_{ATE} = f_{ATE}$. We note that in all the cases the upper bound on test application time using the proposed scheme is lower than that for scan-based external testing. For example, the test application time for s13207 with $\alpha = 8$, and $f^{\star}_{ATE} = f_{ATE} = 20$ MHz lies between 1.848 ms and 2.664 ms, which is lower than the time of 8.155 ms required for conventional scan testing using ATPG-compacted patterns derived using Mintest. Furthermore, let us assume that the desired testing time for s38417 is the average of lower and upper bounds i.e., 2.256 ms. In this case, an external tester operating at $f^{\star}_{ATE} = 72.29$ MHz will be required, as opposed to a 20 MHz tester with the proposed TRP scheme.

Circuit	Uncompacted test sets with don't-cares mapped to 0s				Uncompacted test sets with don't-cares mapped to minimize *WTM*			
	Peak power P^F_{peak}	Peak power reduction (percent)	Average power P^F_{avg}	Average power reduction (percent)	Peak power P^A_{peak}	Peak power reduction (percent)	Average power P^A_{avg}	Average power reduction (percent)
s9234	12994	25.72	5692	61.09	12060	31.06	3466	76.30
s13207	101127	25.42	12416	89.82	97606	28.02	7703	93.68
s15850	81832	18.35	20742	77.18	63478	36.66	13381	85.27
s35932	172834	75.56	73080	87.47	125490	82.25	46032	92.11
s38417	505295	26.10	172665	71.31	404617	40.82	112198	81.35
s38584	531321	7.21	136634	74.50	479530	16.25	88298	83.52
Average	—	28.98	—	75.89	—	37.72	—	84.32

Table 4: Impact of the mapping of don't-cares to binary values on scan-in power consumption.

Circuit	Mintest test sets after static compaction		Uncompacted test sets with don't-cares mapped to minimize *WTM*			
	Peak power P^C_{peak}	Average power P^C_{avg}	Peak power P^G_{peak}	Peak power reduction (percent)	Average power P^G_{avg}	Average power reduction (percent)
s9234	16777	14555	16927	-0.89	12957	10.97
s13207	132129	116695	120992	8.42	97004	16.87
s15850	99647	88385	88445	11.24	67064	24.12
s35932	745282	390622	125298	83.18	50351	87.11
s38417	619929	553491	549950	11.28	311072	43.79
s38584	577365	521882	529315	8.32	437339	16.19
Average	—	—	—	20.25	—	33.17

Table 5: Experimental results on peak and average scan-out power consumption.

We next present results on the peak and average power consumption during the scan-in operation. These results show that test data compression can also lead to significant savings in power consumption. As discussed in Section 2, we estimate power using the weighted transitions metric. Let P^C_{peak} (P^C_{avg}) be the peak (average) power with compacted test sets obtained using Mintest, and let P^F_{peak} (P^F_{avg}) be the peak (average) power when the FDR code is applied to T_D after mapping the don't-cares to 0s. Similarly, let P^A_{peak} (P^A_{avg}) be the peak (average) power when the alternating run-length code is applied to T_D after mapping the don't-cares to minimize the weighted transitions metric. Table 4 compares the average and peak power consumption when FDR coding and alternating run-length coding are used.

Peak power and average power are significantly less if the alternating run-length code is used for test data compression and decompressed patterns are applied during testing. On average, the peak (average) power is 37.72% (84.32%) less in this case than for the Mintest test sets. (Note that the power values for the Mintest test sets are shown in Table 5.) Thus our results demonstrate that substantial reduction in test data volume and testing time are also accompanied by significant reduction in power consumption during scan testing.

We next present results on the peak and average power consumption during the scan-out operation. Table 5 shows that the peak power and average power are significantly less in five out of six cases if alternating run-length coding is used for test data compression. The peak power during scan out operation was only slightly higher for the s9234 benchmark circuit. On average, the peak (average) power is 20.25% (33.17%) less than for the Mintest test sets. Thus our results demonstrate that the substantial reduction in test data volume is also accompanied by significant reduction in power consumption during scan testing. The reduction in scan-out power is an important added advantage since we do not directly target scan-out power in our compression scheme.

Table 6 presents the experimental results when the FDR code and the alternating run-length code are applied to scan vectors for a production circuit from IBM. This circuit contains 1.2 million gates, 32200 latches and a total of 30 scan chains. We were provided with four sets of scan vectors for this circuit. Each vector set consists of 32 patterns (a total of 1031072 bits of test data per vector set). We find that the compression obtained using the alternating run-length code is comparable to that obtained using the FDR code. The slight (1-2%) decrease in compression is offset by the significant savings in test power as shown next.

Table 7 presents the average and peak power values for the IBM circuit when the FDR code and the new alternating run-length code are applied to the scan vectors. We find that the alternating run-length code is extremely efficient for reducing test power. Compared to the FDR code, as much as 43.36% greater reduction is obtained in peak power and as much as 42.38% greater reduction is obtained in average power.

Finally we present results on test application time for the IBM circuit when the TRP scheme based on the new alternating run-length codes is applied to the scan vectors. We assume that the circuit consists of a single scan chain. The minimum (maximum) testing time for the four vectors using TRP and a tester with $f_{ATE} =$ 50 MHz is 12.757 ms (15.205 ms). On the other hand, the test application time based on traditional scan-based testing using the same external tester is 61.863 ms.

5. CONCLUSIONS

We have shown that test resource partitioning (TRP) based on test data compression can be used to reduce SOC test data volume, testing time and test power simultaneously. The proposed TRP method is based on the use of a new code, which we call the

Scan vector set	No. of bits in T_D	FDR coding using T_D Comp-pression (percent)	No. of bits in T_E	Alternating run-length coding using T_D Comp-pression (percent)	No. of bits in T_E
1	1031073	95.08	50648	93.90	62862
2	1031073	94.71	54476	93.98	62030
3	1031073	95.02	51286	93.75	64354
4	1031073	95.63	45026	94.60	55596
Average	—	95.11	—	94.05	—

Table 6: Results on test data compression for an IBM circuit.

Scan vector set	Uncompacted test sets with don't-cares mapped to 0s: Peak power P^F_{peak}	Average power P^F_{avg}	Uncompacted test sets with don't-cares mapped to minimize *WTM*: Peak power P^A_{peak}	Peak power reduction (percent)	Average power P^A_{avg}	Average power reduction (percent)
1	9825669	3587758	6113059	37.78	2101363	41.42
2	10964254	3892459	6255619	42.94	2151415	44.72
3	8996102	3611662	5171926	42.50	2187382	39.43
4	7712574	3161122	3836810	50.25	1771358	43.96
Average	—	—	—	43.36	—	42.38

Table 7: Power reduction associated with the compression of IBM test data.

alternating run-length code. Experimental results for the ISCAS-89 benchmark circuits and for an IBM production circuit show that a slower ATE can often be used with no adverse impact on testing time. Therefore, the proposed approach not only decreases test data volume and the amount of data that must be transfered from the ATE, but it also reduces test power and testing time, and it allows the use of slower ATEs.

Acknowledgment

We thank Brion Keller of IBM Corporation for providing scan vectors for the production circuit.

6. REFERENCES

[1] Y. Zorian, "A distributed BIST control scheme for complex VLSI devices", *Proc. VLSI Test Symp.*, pp. 4-9, 1993.

[2] I Hamzaoglu and J. H. Patel, "Reducing test application time for full scan embedded cores", *Proc. Int. Symp. Fault-Tolerant Comp.*, pp. 260-267, 1999.

[3] F. F. Hsu, K. M. Butler and J. H. Patel, "A case study on the implementation of Illinois scan architecture", *Proc. Int. Test Conf.*, pp. 538-547, 2001.

[4] S. Hellebrand, J. Rajski, S. Tarnick, S. Venkataraman and B. Courtois, "Built-in test for circuits with scan based on reseeding of multiple-polynomial linear feedback shift registers", *IEEE Trans. Computers*, vol. 44, pp. 223-233, February 1995.

[5] J. Saxena, K. Butler and L. Whetsel, "An analysis of power reduction techniques in scan testing", *Proc. Int. Test Conf.*, pp. 670-677, 2001.

[6] P. Girard, L. Guiller, C. Landrault, S. Pravossoudovitch and H. -J. Wunderlich, "A modified clock scheme for a low power BIST test pattern generator", *Proc. VLSI Test Symp.*, pp. 306-311, 2001.

[7] S. Gerstendörfer and H.-J. Wunderlich, "Minimized power consumption for scan-based BIST", *Proc. Int. Test Conf.*, pp.77-84, 1999.

[8] P. Girard, L. Guiller, C. Landrault and S. Pravossoudovitch, "A test vector inhibiting technique for low energy BIST design", *Proc. VLSI Test Symp.*, pp. 407-412, 1999.

[9] F. Corno, M. Rebaudengo and M. S. Reorda, "Low power BIST via non-linear hybrid cellular automata", *Proc. VLSI Test Symp.*, pp. 29-34, 2000.

[10] L. Xu, Y. Sun and H. Chen, "Scan solution for testing power and testing time", *Proc. Int. Test Conf.*, pp. 652-659, 2001.

[11] S. Wang and S. K. Gupta, "ATPG for heat dissipation minimization during scan testing", *Proc. DAC*, pp. 614-619, 1997.

[12] V. Dabholkar, S. Chakravarty, I. Pomeranz and S. M. Reddy, "Techniques for minimizing power dissipation in scan and combinational circuits during test application", *IEEE Trans. CAD*, vol. 17, pp. 1325-1333, Dec. 1998.

[13] R. Sankaralingam, R. R. Oruganti and N. A. Touba, "Static compaction techniques to control scan vector power dissipation", *Proc. VLSI Test Symp.*, pp. 35-40, 2000.

[14] V. Iyengar and K. Chakrabarty, "Precedence-based, preemptive, and power-constrained test scheduling for system-on-a-chip", *Proc. VLSI Test Symp.*, pp. 368-374, 2001.

[15] K. Chakrabarty, "Test scheduling for core-based systems using mixed-integer linear programming", *IEEE Trans. CAD*, vol. 19, pp. 1163-1174, October 2000.

[16] A. Jas and N. A. Touba, "Test vector decompression via cyclical scan chains and its application to testing core-based design", *Proc. Int. Test Conf.*, pp. 458-464, 1998.

[17] A. Chandra and K. Chakrabarty, "System-on-a-chip test data compression and decompression architectures based on Golomb codes", *IEEE Trans. CAD*, vol. 20, pp. 355-368, March 2001.

[18] A. Chandra and K. Chakrabarty, "Frequency-directed run-length (FDR) codes with application to system-on-a-chip test data compression", *Proc. VLSI Test Symp.*, pp. 42-47, 2001.

[19] A. Chandra and K. Chakrabarty, "Combining low-power scan testing and test data compression for system-on-a-chip", *Proc. DAC*, pp. 166 -169, 2001.

[20] I. Hamzaoglu and J. H. Patel, "Test set compaction algorithms for combinational circuits", *Proc. Int. Conf. CAD*, pp. 283-289, 1998.

[21] D. Heidel, S. Dhong, P. Hofstee, M. Immediato, K. Nowka, J. Silberman and K. Stawiasz, "High-speed serializing/de-serializing design-for-test methods for evaluating a 1 GHz microprocessor", *Proc. VLSI Test Symp.*, pp. 234-238, 1998.

Embedded Test Control Schemes for Compression in SOCs

Douglas Kay and Sung Chung
Cisco Systems
170 West Tasman Dr.
San Jose, CA 95134
408 525 8409
dkay@cisco.com, schung@cisco.com

Samiha Mourad
Santa Clara University
500 El Camino Real
Santa Clara, CA 95053
408 554 4163
smourad@scu.edu

ABSTRACT
This paper presents novel control schemes for testing embedded cores in a system on a chip (SOC). It converts a traditional BIST scheme into an externally controllable scheme to achieve a high test quality within optimal test execution time without inserting test points. The scheme promotes design and test reuse without revealing IP information.

Categories
Testing; Test generation and debugging

General Terms
Algorithms, Performance, Design, Experimentation

Key words:
BIST, Data Compression, Test Resource Allocation

1. INTRODUCTION
Test costs represent a major and rapidly increasing percentage of IC's manufacturing costs. Also, as the technology features became smaller, the number of gates per chip has increased. Although the I/O pins count increased as well, the gate to pin ratio has decreased. As a consequence, the test data bandwidth is becoming extremely large. Testing is further complicated by the fact that most ICs are in effect whole systems on a chip (SOC). Each SOC consists of several cores, thus accessing the cores within the chip is another challenging problem to tackle. To alleviate some of these problems, embedded test is presently favored.

Built-In Self-Test (BIST) is a promising approach for testing embedded core-based systems because of its flexibility in adapting a single LFSR to generate patterns for different cores [3], [4]. Furthermore, BIST makes chip testing less dependent on the external tester and reduces the test data volume and bandwidth between the chip and the tester. The effectiveness of BIST is dependent on the fault coverage achieved, the hardware overhead, and the test execution time. While pseudorandom test pattern generation (PRTPG) costs lower than deterministic test pattern generation, its limitations need to be addressed to improve their effectiveness in testing.

LFSR generated pseudo-random (PR) test patterns are usually longer than deterministic tests to achieve comparable fault coverage. In addition, because of the linear dependency and auto-correlation of these patterns, they are not as effective in detecting faults. Although, in general, PR test sets may detect up to 80% of the faults within reasonable test application time, some faults require test patterns that have a low probability of being generated. These faults are known as *random pattern resistant* faults (RPR). Complete fault coverage may be achieved by steering the LFSRs toward increasing the probability of generating test patterns for these RPR faults. They include: 1) re-seeding techniques [11], [5], [19], [16], 2) weighted random pattern [12], [13], and 3) methods for altering the output sequence of LFSRs [15],[17]. For the third approach, the sequence needed to alter the test patterns is generated on demand by some circuitry. Our method is to generate the sequence in a programmable way and apply it interleaved during BIST operation [8, 9].

In a deterministic BIST environment, the method of altering the test patterns to detect RPR faults requires that the control data be generated by an on-chip circuit. In this paper, we concentrate on a methodology that minimizes test data volume by compressing the data set that controls the PRPG. Compression technique in testing, in the past, has been applied to the raw test patterns to reduce test data volume in the tester. This in turn reduces the data download time from the workstation to the tester. [18], [6]. Recently, it has been used for the full scan test patterns, which will be decompressed on the chip. [7] [1, 2]. Different compression schemes have been used for this purpose. In our case, we applied our special coding [10] to the control data to achieve better compression effectiveness as compared with the one applied to the original raw test data. The decompression is performed on the chip during BIST. We call our test scheme as interactive BIST (iBIST). One of our possible general schemes is outlined in Fig. 1.

To minimize the size of U, we need to minimize the cycles of controlling the LFSR and to carefully apply the appropriate bit values. In this paper, for a given code, we examine different schemes for controlling the LFSR on the overall compaction effects of the data. We formulate the problem in Section 2. In Section 3, we define the control data attributes that are used in comparing the various LFSR control schemes described in Section 4. In Section 5, actual BIST controller implementation and its area overhead will be discussed. And in the last section, we conclude with a summary. First, however, we formulate the problem in the next section.

2. PROBLEM FORMULATION
Since a standard LFSR is used in BIST as a PRTPG, it may seem strange to ask the question whether producing desired test patterns at will would be possible. If it is possible, this will lead to generating tests for faults other than stuck-at-fault. Figure 2

shows the situation where the desired set of target test pattern is given for a scan testing. Consider the case where the LFSR should provide the desired patterns $t_t = (0\ x\ 0\ x\ 1)$ starting from its initial state value Si = (0 0 1).

Given a sequential circuit C:

1. Generate a deterministic test D for C and let its length be L
2. Generate a pseudo-random test R for C using a primitive polynomial LFSR
3. Compare test sets D and R and replace unspecified bits, usually indicated by x(don't care), to match the D patterns with R patterns. The subset of unmatched patterns in D represents the patterns needed for RPR faults.
4. Apply a signal set U that controls the LFSR to generate the patterns for RPR faults. Let Lu be the length of U.
5. Compress U to Uc with length Lc and store (transmit) Uc on (to) the Chip.
6. At testing time, decompress Uc to U on chip and apply U to the LFSR.

Figure 1 The General Testing Scheme

In this configuration, LFSR will produce the test vector, $\mathbf{t}_p$ = (1 1 0 0 1) after five shift clocks and arrive at the final state $\mathbf{S}_f$ = (1 0 1) via intermediate states highlighted in the set of contiguous LFSR states, **S** = {(0 0 1), **(1 0 0), (1 1 0), (1 1 1), (0 1 1)**, (1 0 1)}. We can tell the last bit of each state will be shifted into the scan chain to produce $\mathbf{t}_p$. When $\mathbf{t}_p$ is compared with $\mathbf{t}_t$, the difference vector can be derived from the equation $\mathbf{t}_d = \mathbf{t}_t \oplus \mathbf{t}_p = \{1\ 0\ 0\ 0\ 0\}$.

Adding control tap (a 2-input XOR gate) into the different locations of LFSR to flip the output value of the flip-flop located in front of the control tap can make $\mathbf{t}_d$ become all zero vector. Where should one insert the control tap? There are different positions within the LFSR, as we will mention later. For now, we place it right at the output of the LFSR as shown in Fig. 3: the control vector, **U** = (0 0 0 0 1), transforms $\mathbf{t}_d$ to a null vector. The vector be applied from left to right in the time domain. Next, we can think of other positions anywhere inside LFSR.

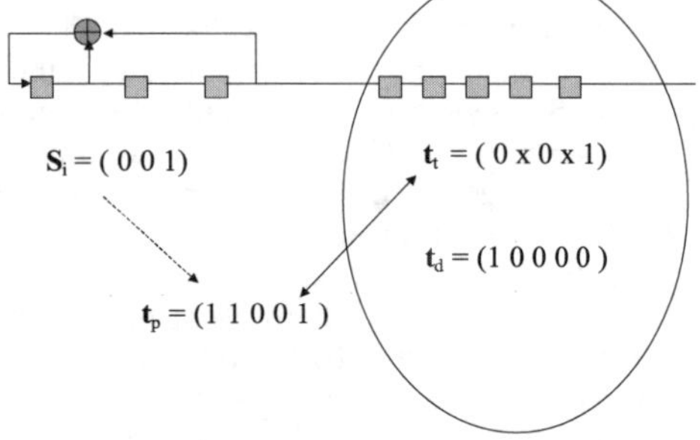

Figure 2 Generic Scan-BIST

As you move the tap position to the left, you will find out that you need to begin the control vector application as many clock cycles early as the number of your move from the beginning position, i.e. output of least significant flip-flop. As long as you apply the control vectors with the proper timing, you are able to produce desired patterns at will. To describe the problem formally, we define some terminology:

Definition: An LFSR consists of several flip-flops, say *N*. The LFSR states are the binary values extracted from its flip-flops. We partition these states into two types: S0 and S1. S0 (S1) states are such that their least significant bit (LSB) is 0 (1).

Theorem: Assume the flip-flops are labeled in the descending order from left to right, *N* to 1. Then, the control input u(n) applied at n^{th} clock cycle of any tap located in front of k-th flip-flop, where $0 < k < N + 1$, can steer the LFSR state, S, into S0 or S1 at $(n+k)^{th}$ clock cycles.

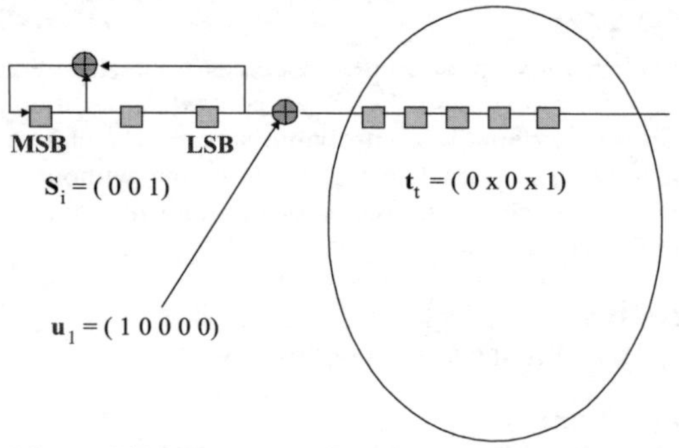

Figure 3 iBIST: Controlling the test pattern stream

Proof: If the tap is located at the front of Kth flip-flop, the input value, $\mathbf{I}_k$ of k-th flip-flop, at a certain discrete time point n , can be set to either 0 or 1 by supplying the input u(n) such that $\mathbf{I}_k(n) = R_{k+1}(n) \oplus u(n)$, where $R_{k+1}(n) \in (0, 1)$, $u(n) \in (0,1)$, and n = 0, 1, 2, …, L, where L is any positive integer. The input value, $\mathbf{I}_k(n)$, can be propagated to the flip-flop R_1 in k clock cycles because $\mathbf{I}_k(n) = \mathbf{I}_{k-1}(n+1) = \ldots = \mathbf{I}_1(n+k-1) = R_1(n+k)$.

Since $R_1(n+k)$ can be set to 0 or 1, the LFSR state belongs to S0 or S1 at the $(n+k)^{th}$ discrete time.

Corollary. The control input can be calculated as the following equation.

$U(n) = R_{k+1}(n) \oplus \mathbf{I}_k(n) = R_{k+1}(n) \oplus R_1(n+k)$

We targeted only pattern $\mathbf{t}_t$ so far for the sake of explanation. In the case of a whole test of m vectors, $\boldsymbol{T}_t = \{\mathbf{t}_1, \mathbf{t}_2, \ldots, \mathbf{t}_m\}$, we repeat the process for each pattern **ti** and develop *m* vectors each of dimension *n*. All such vectors form the control vector **U.** Therefore, the problem that we are solving can be stated concisely as follows.

The problem: To find the vector **U** that maps the PR test set, Tp into the deterministic test Ti, $\boldsymbol{f}(\mathbf{U})$: $\mathbf{T}_p \Rightarrow \mathbf{T}_t$ such that the size of **U** is minimal.

The above problem statement can be rephrased as: Find the vector **U** that transforms a projected test set $\mathbf{T}_p$ into target set $\mathbf{T}_t$ while optimizing the associated cost function. The cost function may be the overall size of *U*, its energy level, or the potential compression gain which would be obtained from using a particular encoding scheme. When we think about a single scan chain, the tap position in Fig. 3 is the easiest because we do not need to worry about the lead time for the start of control data application. However, in reality, the circuits usually contain multiple scan chains and the LFSR drives those multiple scan chains from each output of flip flops through spread network (in IBM terminology) or Phase Shifters (in Mentor Graphics terminology). Therefore, putting a control tap in front of LFSR, shown in Fig. 4, will provide the capability of controlling all of the scan chains without adding multiple control taps in front of each scan chain. However, we can put the control tap anywhere in the LFSR, for experimentation, without loss of generality.

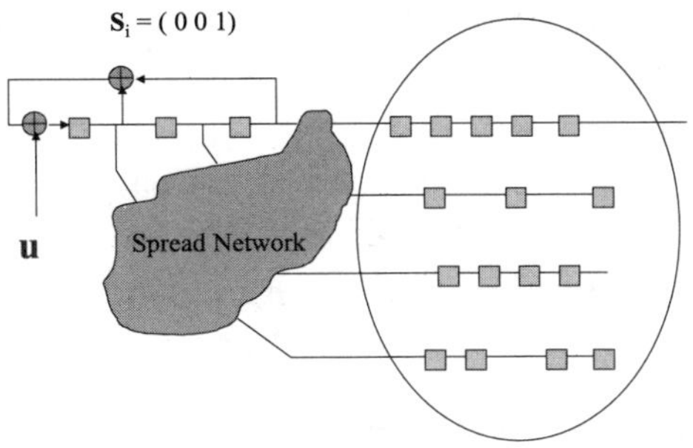

Figure 4. Ideal control tap positions

The control of LFSR may be applied for every pattern of the deterministic test, or if this test is too long, for a selection of the patterns. For example, it is possible to target patterns that are at a certain interval from each other. We refer to this distance as the control interval and denote it by *CI*.

Before describing the LFSR control algorithm, we will define some data attributes that will be used in the formulation and the evaluation of these algorithms.

3. ATTRIBUTES OF THE CONTROL DATA

Data compression consists of mapping a group of bits in the stream, i.e., a source word of length n_i into a code word of length m_i, where m_i is desired to be less than n_i and we define the difference $n_i - m_i$ as the gain g_i, for $i = 1, 2,..N$, where N is the total number of words in the source data stream. The size n_i and m_i do not have to be constant for the whole data stream, implying variable to variable coding. The *gain* of the compression scheme should be positive for the majority of words; otherwise, the scheme is not effective in compressing the data. The result of the compression is largely dependent on the characteristics of raw data and the code used to encode it.

Consider the binary stream of test data, length L, consisting of runs of 0s and 1s and let these different *runs* be of lengths L_i, with $1 \le i \le N$, then $L = \sum_{i=1}^{i=N} L_i$. To characterize the data and the code, we use the following attributes: *energy*, *sparsity,* and *statistical polarization.*

A test data stream may include an 'x' (don't care) in addition to '0' and '1'. We denote the energy by ε and it is defined as the fraction of the 1s in the data stream; the fraction of 0s is denoted by $\bar{\varepsilon}$. But, since this alphabet of the data may include xs (don't care), the fraction of xs represents the sparsity, s. This is the fraction of the data that can be adjusted to increase or decrease the energy for the data. These three attributes are thus related by the expression: $\varepsilon + \bar{\varepsilon} + \sigma = 1$.

The x bits can be replaced with 0 or 1 in a random or algorithmic manner. The data stream can be decomposed into the words, the succession of 0s and 1s. However, if the energy is low, then the majority of words will be succession of 0s in different length by assuming x is converted into 0. Then, instead of looking at the words of the data stream to be the both successions of 0s or 1s, we can define the words to be the successions of 0s, ending with single 1 without loss of generality. This is a conversion into uni-phase source words. These observations led us to develop a code that is suitable for the test data [10]. That is, a code yielding a good compression gain for the type of data. However, we are more interested in finding the overall compression effect of the coding scheme on the source data. That is, we would like to assess the percentage of data reduction after compression to the original deterministic test. We will define the compression effectiveness, *CE*, as : $CE = \frac{L - L_C}{L} 100\%$ where L is the length of the deterministic test and Lc that of the compressed data (compressed $\boldsymbol{U}$). From this definition, we realize that the higher *CE* is, the better the compression scheme. The compression process consists of two step process, preprocessing (steps 1-4 of Fig. 1), and compression (step 5 of the same figure). We can express *compression effectiveness* in terms of the compression through the two processes in the form $CE = 1 - \frac{L_C}{L_u}\frac{L_u}{L} 100\%$.

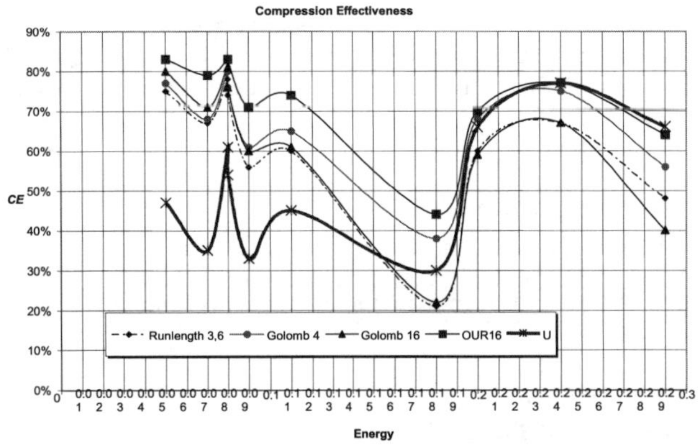

Figure 5. Compression Effectiveness versus Energy.

The compression gain is defined as $G = (L_u - L_c)$. This gain can also be evaluated for each pattern and in this case the Ls refer to the length of the pattern before and after compression. We will use the compression effectiveness to compare the different algorithms presented in the next section.

The code used in this study has been proven to be effective compared to Golomb and Runlength codes. The compression effectiveness for the different codes are plotted against the energy level of the control vectors in Fig. 5. The heavy line represents the *CE* for the vector $\boldsymbol{U}$. Useful results correspond to points above this line. Among the results, OUR coding was the best. However, we realize that the scheme used to control the LFSR may be further improved over the results obtained. In the next section we present then several algorithms for controlling the LFSR and analyze their efficiency.

4. ALGORITHMS

In the theorem of Section 2, we showed that the LFSR could be controlled to produce a target pattern from any initial state. Controlling the LFSR for a particular target pattern of the deterministic test, changes the state of LFSR and thus affects the control of the remaining patterns. Determining the best control vector requires an exhaustive search. It is an NP complete problem. To make the problem manageable, we developed several efficient algorithms and compared them. The algorithms are all based on a greedy strategy but use different cost functions. The cost function may be (1) the Hamming distance between the target pattern and the projected pattern, (2) the energy level of control vector, ε, (3) the compression gain of encoded control vector, ($L_u - L_c$).

4.1 Minimum Hamming distance

This algorithm applies to the configuration of Fig. 3. It generates the control vector **u** of the control tap located at the output of the LFSR, the output of its least significant flip-flop. Since the application of control vector does not modify the natural sequence of LFSR states, the control vector **u** is equivalent to the difference vector $\mathbf{t}_d = \mathbf{t}_t \oplus \mathbf{t}_p$, where $\mathbf{t}_t$ is the target pattern from the deterministic test and $\mathbf{t}_p$ is the PR pattern generated by the LFSR. The energy of the $\mathbf{t}_d$ is the Hamming distance between the two patterns.

Minimum Hamming distance algorithm

[A greedy algorithm that creates a control data set U with minimal energy level starting with LFSR state si with control interval CI]

1. Td <= D – R ∩D ; [set target pattern set as the remaining patterns after running pseudorandom testing. The size of matrix T_d is m x n. R is the pseudo random pattern. D is the original deterministic pattern]
2. [Initialization]
    - 2.1 U <= ∅;
    - 2.2 s <= si; [LFSR initial state]
    - 2.3 k <= m;
    - 2.4 i <= 0;
    - 2.5 D <= n; [the maximum distance possible for a given pattern]
    - 2.6 T <= Td;
3. While T ≠ ∅ do

```
  if ( i / CI) ≠ 0, then
        s[i]=next(LFSR(s)), tp[i]=out(LFSR(s)) for i = 1 to n; [shift LFSR n times, update s, and generate tp]
        T <= T – {tp} and k <= k -1 if tp ∈ T;
  otherwise
        si = s; [ Save the current LFSR state]
        for j = 1 to k [Try for remaining target patterns]
              s[i]=next(LFSR(s)), tp[i]=out(LFSR(s)) for i = 1 to n; [shift LFSR n times, update s, and generate tp]
              tx = tj ⊕ tp; [Hamming operation];
              if tx < D then
                    (a) D <= tx;
                    (b) t <= tj;
                    (c) Ssave=s;
              end-if
              s <= si [reset the state]
        end_for
        T <= T – {t}; U = U ∪ {u}; [u is derived for target pattern t using LFSR with its initial state s using the corollary in section 2]
        s=Ssavei
  end_if
  i <= i + 1;
end_while
```

4. Output U

4.2 Minimum energy of control vector

If the control tap is located inside the LFSR as indicated in Fig. 6, the applied control vector modifies the LFSR state sequence. This prevents us from using the Hamming distance between the target and projected patterns because the projected pattern will change as every control vector is applied.

We solve the problem by generating the control vector for each pattern of the remaining test set. We Calculate the energy level of all control vectors. Then, we choose the one with the minimum energy. Once the control vector is selected, we remove the target pattern from the remaining test set. The process is repeated until the test set is empty.

4.3 Direct compression gain of control vector

This algorithm is the same as the previous one except that the cost function use is the compression gain, CG, that was defined in Section 3. The calculation of compression gain depends on which type of encoding scheme it uses. In this paper, we use the OUR Coding [Author 2001].

4.4 Progressive Seeding

So far, for the above three algorithms, the control bit is assumed to be a zero whenever the corresponding target pattern bit is a "don't-care". From corollary in Section 2, $U(n) = R_{k+1}(n) \oplus \mathbf{I}_k(n) = R_{k+1}(n) \oplus R_1(n+k)$. $R_1(n+k)$ corresponds to the specific bit of the target pattern at timestamp n + k. If the bit in the target pattern is "don't care", U(n) was automatically set to 0 to lower the control vector's energy. If instead, we allow the choice of setting the control bit to be either 0 or 1, then we have to decide, each time we encounter a don't-care whether to set it to 0 or 1.

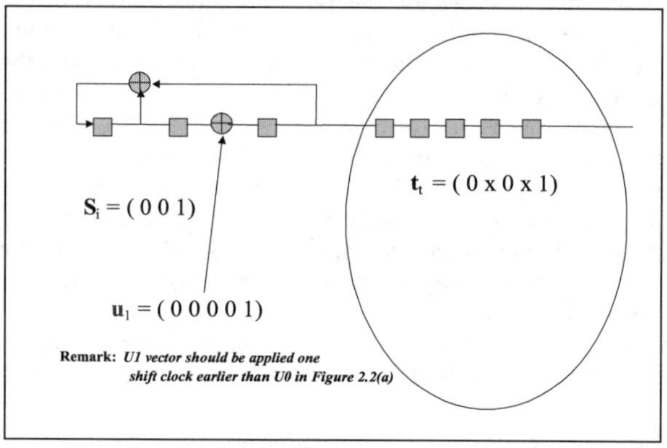

Figure 6 Minimum energy control

This choice allows the control of the LFSR, the activity of the *U* vector in a progressive manner, on the bit-level instead of the pattern-level. The scheme will provide more granularity of controlling LFSR although it may be more computationally complex. The cost function used for this algorithm continue to be the compression gain, CG.

4.5 Polarity inversion and static value control

We have shown previously, that the compression effectiveness is higher the lower is the energy of the data compressed [10]. Also, in section 3, we showed that the data attributes satisfy the equation $\varepsilon + \bar{\varepsilon} + \sigma = 1$. Thus when ε, is high, then the inverted data will have low energy. Since we cannot invert the logic for the original pattern, this would be possible if we invert the output of LFSR, instead.

Whether to invert the logic at the output of LFSR can be achieved by adding an extra bit at the first location of each control pattern. This bit should not be counted as part of control vector.

Rather, it is used as a flag bit to set or reset the logic inversion by BIST controller.

Also, some RPR patterns consist of mostly series of 1's or 0's. For these cases, we can supply static high or low input by bypassing LFSR with control gates show in Fig 7. This extra hardware is really very minute with respect to any modern circuit.

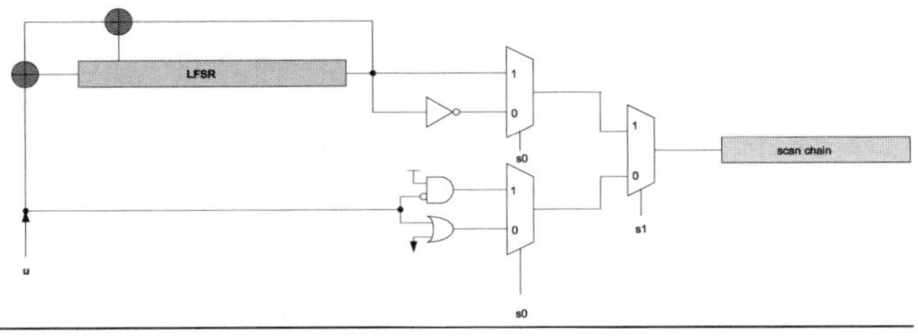

Figure 7 Polarity inversion and static control logic

Assigning {s0, s1} value to the first and the second bit of control vector {ui0, ui1} for ith control pattern, the BIST controller can multiplex the proper logic signal to the target scan chain at the beginning of applying ith control pattern signals.

4.6 Results comparison

iBIST provides the benefit of having compression effect on the original test pattern **D** in two ways. One way is to screen easy to hit test patterns with pseudorandom testing and the other is to apply the control vector to LFSR to hit the remaining or other types of deterministic test patterns. Screening the easy to detect patterns is performed in the same manner by running the same pseudorandom testing for all different control algorithms.

The difference from the traditional BIST fault grading is that we did not use the fault simulation, but used the method of matching between the random pattern, R, and the deterministic pattern, D. We applied coding scheme to the control pattern to compress the control pattern further.

Summing up, we see compression effectiveness in two phases: (1) During pseudorandom testing by filtering easy to detect test pattern, (2) During actual encoding on the control pattern. We applied the different control schemes to several benchmark circuits [14] whose profiles are listed in Table 1.

For each benchmark we calculated the corresponding overall compression effectiveness, CE, of the vector U and plotted the results versus the energy, ε, of the original deterministic pattern. The plots are shown in Fig. 8. As defined in Section 3, the higher the value of CE, the more efficient is the compression scheme.

The polarity inversion and static algorithm yields the best results for 8 out of the 10 circuits. For the other two circuits, it is the progressive seeding algorithm that gave the best results. Also, we would like to draw your attention that most of the algorithm did not perform well on s838 circuit. However, the polarity inversion and the static control scheme brought CE level up to above 70%. Overall, we conclude that the algorithm of Section 4.5, polarity inversion and the static control, yields the best result algorithm

Table 1 Profiles of the Benchmark Circuits

Design	In	Out	scan length	D-test length	PR-test length	PR-fault Cov.	U-test length
ALU	48	18	57	71	10K	66.2	24
Firewire	18	166	258	223	16K	34.5	14
IIR	35	10	243	814	10K	45.0	448
MMX	155	140	298	1883	50K	32.9	1264
Router	70	67	263	576	41.2K	54.3	263
S838	43	3	67	141	30.1K	29.8	99
S1238	22	15	32	205	50K	76.6	48
S13207	39	122	677	231	50K	46.8	123
Sdram	101	86	241	174	18.2K	60.9	68
USB	22	51	118	61	10K	65.6	21

5. PERFORMANCE OVERHEAD OF IBIST

Although BIST testing has been adopted in by the test community, its general acceptance by the design community was quite limited due to several concerns: (1) area overhead, (2) performance degradation during functional operation, and (3) negative impact on project schedule. As the technology feature size plunged to the submicron range, the concern about the area overhead has been eased. Allocating 10 to 20 K gates to the BIST logic became a norm these days. The impact on the performance along the critical path is still challenging because of the need for test point insertion. Our iBIST scheme avoids test point insertion

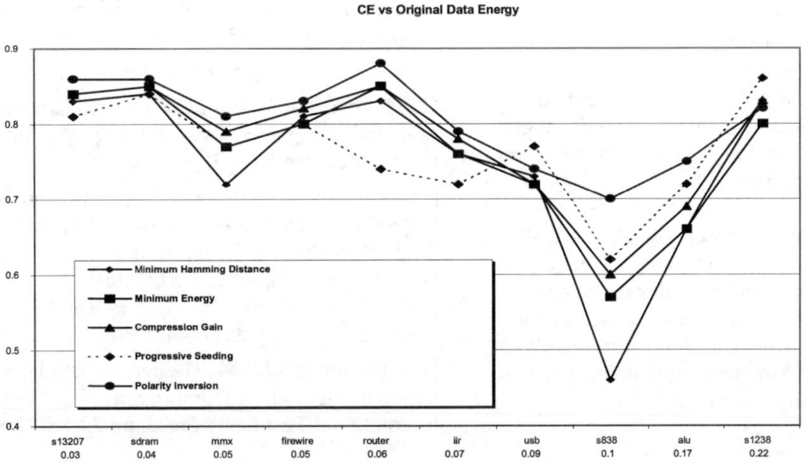

Figure 8 Compression effectiveness

This paper proposes the new BIST method of not using test point insertion, but of using control vectors to target the hard to detect fault in a programmable way. Therefore, the impact on the functional performance is expected no more than the ordinary internal scan implementation. The area overhead and the speed during test mode are listed in Table 2.

The iBIST configuration has been compared to the ordinary BIST implementation. This configuration consists of 5 major parts: (1) BIST controller, (2) LFSR, (3) MISR, (4) shift counter, and (5) pattern counter. The difference between BIST and iBIST design is in the LFSR and the pattern counter module due to their size differences. We used 48 bit for the LFSR and used 24 bit for pattern counter in ordinary BIST implementation. In iBIST implementation case, we could use 16-bit sizes for both LFSR and pattern counter, still satisfying full fault coverage. The synthesis of the designs has been done using Synopsys DC compiler and LSI 0.35 u process library (lcbg10) with worst case set up. We used the 50MHz clock speed, maximal signal transition on the IOs with medium map effort. Both BIST and iBIST could meet the 20 ns clock cycle constraint (for 50 MHz operation) effortlessly with no problem. The major slow down is in the shift and pattern counter. Speeding up these parts, we can run the test in much faster frequency. Using the assumption on the size of LFSR and the pattern counter, iBIST controller area actually came out smaller than the ordinary BIST even after including the decoder module.

Table 2 Comparing iBIST to traditional BIST and controllers

	IBST		iBIST	
Module	Area	Speed	Area	Speed
BIST_control	201.32	2.16ns	201.32	2.16ns
LFSR	1666.83	1.89ns	**553.38**	1.85ns
MISR	527.35	1.95ns	527.35	1.95ns
Shift Counter	900.60	6.32ns	900.60	6.32ns
Pattern Cnter	1572.50	9.15 ns	**1085.29**	8.01ns
Decoder	N/A	N/A	344.20	2.73ns
Total	4868.60	9.15ns (worst)	3612.14	**8.01ns** (worst)

6. SUMMARY AND CONCLUSIONS

We presented several control schemes for maximizing the compression effectiveness and reducing the test data in iBIST environment. Benchmark results show that the control schemes using selective polarity inversion and static control is superior to all the presented control schemes. In addition, we described the overall iBIST controller implementation including decoder logic. The results indicate that iBIST uses lower overhead area and produces higher performance than traditional BIST.

iBIST is a viable solution for overcoming the test data bandwidth gap between external I/Os and the internal circuit. It also promotes the design and test reuse and provides the protection of IP cores via control vectors. Its main benefit is achieving a high test quality within optimal test execution time without inserting test points. The decompression technique is independent of the design and promotes reusability at low test overhead.

8. REFERENCES

[1] A. Chandra and K. Chakrabarty, "Test data compression for system-on-a-chip using Golomb codes," *Proc. IEEE VLSI Test Symposium*, pp. 2000.

[2] A. Chandra and K. Chakrabarty, "On Using Golomb Codes and Internal Scan Chains for Test Data Compression / Decompression in a System-on-a-Chip," *Proc. of Testing Embedded Core-based System-Chips Workshop*, pp. 2.2-1 ~ 2.2-7, 2000.

[3] Joseph Desposito. "SOC and Deep-Submicron Technology Drive New DFT Strategies," *Electronic Design*, August 3, 1998. pp. 49-62.

[4] Rajesh K. Gupta and Yervant Zorian, "Introducing Core-Based System Design," *IEEE Design and Test of Computers*, October-December, 1997.

[5] Hellebrand S., et al., "A Mixed Mode BIST Scheme Based on Reseeding of Folding Counters," IEEE Proc. of Int'l Test Conference, pp 778-984, Oct. 2000

[6] M. Ishida, D.D. Ha, T. Yamaguchi, "COMPACT: A Hybrid Method for Compressing Test Data," Proc. Of VLSI Test Symposium, pp.62-69, 1998

[7] A. Jas, J. Ghosh_Dastidar, and N.A. Touba, "Scan vector compression/decompression using statistical coding," Proc. IEEE VLSI Test Symposium, pp. 114-120, 1999

[8] D. Kay and S. Mourad, "Controllable LFSR for Built-In Self-Test" Proc. of IEEE Instrument and Measurement Technology Conference, pp 223-229, May 1-4, 2000

[9] D. Kay and S. Mourad, "Controllable LFSR for Embedded Core BIST", Proc. Of Testing Embedded Core-based System-Chips Workshop, pp. 2.4-1 ~ 2.4-6, 2000

[10] D. Kay and S. Mourad, "Compression Technique for Interactive BIST Application", Proc. Of VLSI Test Symposium, pp. 9-14 2001

[11] Koenemann, B., "LFSR-Coded Test Patterns for Scan Designs," Proc. of European Test Conference, pp. 237-242, 1991

[12] Muradali, F., V.K. Agrawal, B. Nadeau-Dostie, "A New Procedure for Weighted Random Built-In Self-Test," proc. of International Test Conference, pp. 660-668, 1990.

[13] Bahram Pouya and Nur A. Touba. "Modifying user-defined logic for test access to embedded cores." International Test Conference, 1997 IEEE pp.60-68.

[14] SCU-RTL Benchmarks. http://www.engr.scu.edu/mourad

[15] Nur A. Touba and Edward J. McCluskey, "Altering a pseudo random bit sequence for scan based bist" International Test Conference, 1996 IEEE pp.167-175

[16] Venkataraman, S., et al. "An Efficient BIST Scheme based on Reseeding of Multiple Polynomial Linear Feedback Shift Registers," Proc. of International Conference on Computer-Aided Design (ICCAD), pp 572-577, 1993

[17] H.-J. Wunderlich, G. Kiefer: "Bit-Flipping BIST", Proc. Int. Conf. On CAD, pp 337-343, 1996

[18] T. Yamaguchi, M. Tilgner, M. Ishida, and D.S. Ha, "An Efficient Method for Compressing Test Data," Proceedings of International Test Conference, pp.79-88, November 1997

Wrapper/TAM Co-Optimization, Constraint-Driven Test Scheduling, and Tester Data Volume Reduction for SOCs*

Vikram Iyengar[1] Krishnendu Chakrabarty[1] Erik Jan Marinissen[2]

[1]Electrical & Computer Engineering, Duke University, Durham, NC 27708, USA
vik@ee.duke.edu krish@ee.duke.edu
[2]Philips Research Laboratories, 5656 AA Eindhoven, The Netherlands
erik.jan.marinissen@philips.com

ABSTRACT

This paper describes an integrated framework for plug-and-play SOC test automation. This framework is based on a new approach for wrapper/TAM co-optimization based on rectangle packing. We first tailor TAM widths to each core's test data needs. We then use rectangle packing to develop an integrated scheduling algorithm that incorporates precedence and power constraints in the test schedule, while allowing the SOC integrator to designate a group of tests as preemptable. Finally, we study the relationship between TAM width and tester data volume to identify an effective TAM width for the SOC. We present experimental results for non-preemptive, preemptive, and power-constrained test scheduling, as well as for effective TAM width identification for an academic benchmark SOC and three industrial SOCs.

Categories and Subject Descriptors

B.7.3 [**Integrated circuits**]: Reliability and testing

General Terms

Algorithms,design,experimentation,reliability,verification

1. INTRODUCTION

Test automation is widely recognized as an important part of the overall system-on-chip (SOC) design and integration cycle. In order to reduce design cycle time and provide greater functionality, a large number of intellectual property (IP) cores are stitched into an SOC. To facilitate plug-and-play test, an embedded core must be isolated from surrounding logic, and test access must be provided from the I/O pins of the SOC. Test wrappers are used to isolate the core for modular test application, while test access mechanisms (TAMs) transport test patterns and test responses between SOCs pins and core I/Os [18]. In addition, core tests must be scheduled such that precedence and power constraints are met and conflicts in the TAM are avoided.

To reduce cost and ensure quality, testing must make effective use of SOC test resources [6]. The rapidly increasing size of SOCs has spurred an enormous growth in test resource usage, leading to complex test hardware, long test application times, and large test data sets. An integrated framework for SOC test automation that increases the utilization of test resources is therefore necessary to increase production capacities and reduce test cost. This paper describes the design of such a framework that integrates the following three design processes into the test automation flow.

1. Wrapper/TAM co-optimization. Test wrapper design and TAM optimization are of critical importance during system integration since they directly impact hardware overhead, testing time and tester data volume on the tester. The new approach for TAM optimization is based on a generalized version of rectangle packing [8, 14].

2. Constraint-driven preemptive test scheduling. The objective of the proposed scheduling algorithm is to minimize testing time, while addressing the following issues: (a) resource conflicts between cores arising from the use of shared TAMs and on-chip BIST engines, (b) precedence constraints among tests, and (c) power dissipation constraints. Testing time is decreased further through the selective use of test preemption.

3. Tester data volume reduction. The third component is used to identify an SOC TAM width that reduces testing time as well as tester data volume. Test data for large SOCs now require several Gigabits of tester memory, which is a significant contributor to test cost [3, 6]. The impact of TAM design and test schedules on tester memory requirement has not been directly studied before. We show here that TAM widths that minimize testing time do not always lead to minimum tester data volume. We investigate the correlation of TAM width to the tester data volume, and determine TAM widths that minimize a cost function involving both the testing time and the amount of tester data. This allows the system integrator to trade-off testing time with data volume.

2. PRIOR WORK

Most prior research in test access for SOCs has either studied wrapper design and TAM optimization as independent problems, or not addressed the issue of sizing TAMs to minimize SOC testing time [1, 4, 18]. Alternative approaches that combine TAM design with test scheduling [9, 16, 19] do not address the problem of wrapper design and its relationship to TAM optimization.

The first integrated wrapper/TAM co-optimization methodologies were proposed in [12, 13]. A drawback of [12] is that the problem of wrapper/TAM co-optimization as formulated is intrinsically intractable; the compute time for the exact method of [12] increases exponentially with the number of TAMs. Efficient heuristics to reduce CPU time for wrapper/TAM design were presented in [13]. However, the approaches in [12, 13] are limited to test access architectures based on "fixed-width" TAMs. In fixed-width TAMs, the total TAM width is explicitly partitioned among a finite number of TAMs, and each core is assigned to exactly one TAM.

*This research was supported in part by the National Science Foundation under grant number CCR-9875324 and by an IBM Graduate Fellowship.

DAC 2002, June 10-14, 2002, New Orleans, Louisiana, USA.

Such architectures are inflexible and they generally lead to inefficient usage of TAM wires [14]

Several techniques for test scheduling of SOCs have been proposed [5, 7, 11, 15, 16, 18]. Methods to incorporate precedence and power constraints in a preemptive test schedule were presented in [11]. These methods assume that a pre-designed TAM for the SOC is provided. We address the design of an integrated framework for SOC test automation where TAM optimization and test scheduling are performed in conjunction. An integrated TAM design and test scheduling method based on flexible-width TAMs was presented in [14]. However, precedence and power constraints in the test schedule and test preemption were not considered.

Tester data volume reduction methods are either based on built-in self test (BIST) [2] or test data compression [6]. While both methods have been shown to be useful for test data volume reduction, there has not been a concerted effort to examine their use coupled with TAM design and test scheduling in the form of an integrated framework for test automation.

In this paper, we present a novel approach to integrated wrapper/TAM co-optimization and test scheduling based on a generalized version of rectangle packing [8]. A rectangle representation of core tests, similar to the one presented in [9], is used in our co-optimization and scheduling framework. We partition the TAM width among the set of of cores being tested during each interval of time. This partition is varied with time to match the TAM width to each core's test data needs. TAMs can fork and merge between cores to improve TAM wire utilization. Precedence constraints among tests can be embedded into the schedule by the system integrator, and the maximum power constraint that must not be exceeded during test can be easily incorporated. Moreover, the system integrator can identify certain tests as preemptable, as well as specify limits on the number of preemptions allowed for each test. Finally, the relationship between TAM width and tester data volume is investigated to identify an effective choice of total TAM width for the SOC.

3. TAM DESIGN AND TEST SCHEDULING

The general integrated wrapper/TAM co-optimization and test scheduling problem that we address in this paper is as follows. We are given the total SOC TAM width W, and the test set parameters for each core, i.e., the number of primary I/Os, test patterns, scan chains, and the scan chain lengths. Unlike in [1], we assume that the lengths of scan chains are fixed. The goal is to determine the TAM width and a wrapper design for each core, and a test schedule that minimizes the testing time for the SOC such that the following constraints are satisfied.

1. The total number of TAM wires utilized at any moment does not exceed W;
2. Precedence and concurrency constraints are met;
3. The maximum power dissipation value is not exceeded;
4. Selective preemption of tests is allowed.

Additionally, we would like to determine a value of W for the SOC to trade-off testing time with tester data volume.

We formulate this problem as a progression of three problems of increasing complexity that lead up to the general problem. These three problems are as follows:

Problem 1. Wrapper/TAM co-optimization and test scheduling.
Problem 2. Wrapper/TAM co-optimization and test scheduling with selective preemption, and precedence and power constraints.
Problem 3. Wrapper/TAM co-optimization, test scheduling with selective preemption, and precedence and power constraints, and identification of a TAM width to trade-off testing time with tester data volume.

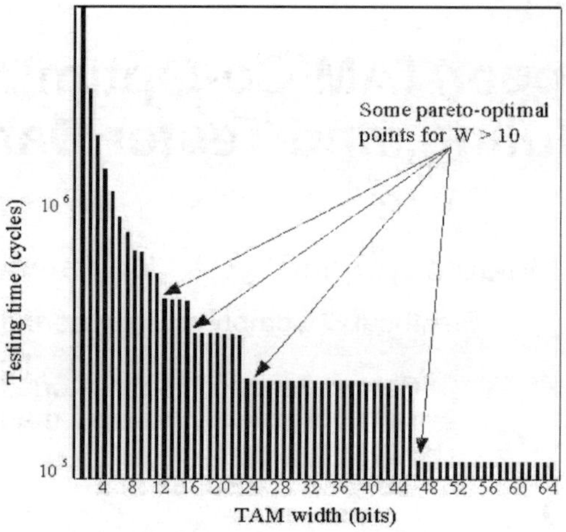

Figure 1: **Relationship between testing time and TAM width for Core 6 in Philips SOC p93791.**

In this section, we address *Problem 1*, and show how wrapper/TAM co-optimization can be integrated with test scheduling. In Section 4, we show how this problem is generalized to include precedence, preemption and power-constraints-*Problem 2*. Finally, in Section 5, we study *Problem 3*, i.e., we identify TAM widths that provide a trade-off between test time and tester data volume.

Problem 1: Given W and the test set parameters for each core, determine the TAM width and a wrapper design for each core, and a test schedule for the SOC that minimizes the total testing time, such that the total number of TAM wires utilized at any moment does not exceed W. □

We solved the problem of wrapper design for cores in [12] using the *Design_wrapper* algorithm based on the Best Fit Decreasing (BFD) heuristic. In order to solve the problem of assigning TAM width to cores and scheduling tests, we represent core tests by rectangles. The *Design_wrapper* algorithm is used to construct a set of rectangles representing the different testing times for each value of TAM width for a core, such that the height of the rectangle corresponds to the TAM width and the width of the rectangle represents the core test application time. The use of rectangles for core test representation during test scheduling has previously been studied in [7, 9, 10, 16]. A bin packing approach based on rectangle representation was used in [10] for test scheduling.

For a given core, the testing time decreases only at Pareto-optimal points when the TAM width exceeds core-specific thresholds [12]. Pareto-optimal points are formally defined in [14]. For example, in Figure 1, a TAM width of 46 results in a testing time of 115850 cycles, while all TAM widths from 47 up to 64 result in the same testing time of 114317 cycles. Hence 47 is a Pareto-optimal TAM width, and rectangles of height between 48 and 64 can be ignored. The TAM width assigned to cores is thus always the minimal value required to achieve a specific testing time. The extra TAM wires can be used for other cores in the SOC, thereby yielding a more efficient TAM design.

We now formulate *Problem 1* as a generalized version of the rectangle packing problem [8]. Consider an SOC having N cores, and let $\mathbf{R}_i$ be the set of rectangles for core i, $1 \leq i \leq N$. Problem $\mathcal{P}_{\mathrm{GRP}}1$ (generalized rectangle packing) is stated as follows.
$\mathcal{P}_{\mathrm{GRP}}1$: Select one rectangle $R_{ij} \in \mathbf{R}_i$ from each set $\mathbf{R}_i$, $1 \leq i \leq N$, and pack the selected rectangles into a bin of fixed height and unbounded width, such that no two rectangles overlap, and the width to which the bin is filled is minimized. Furthermore, during packing, each rectangle selected is allowed to be split vertically into several non-adjacent pieces, each having the same co-ordinates

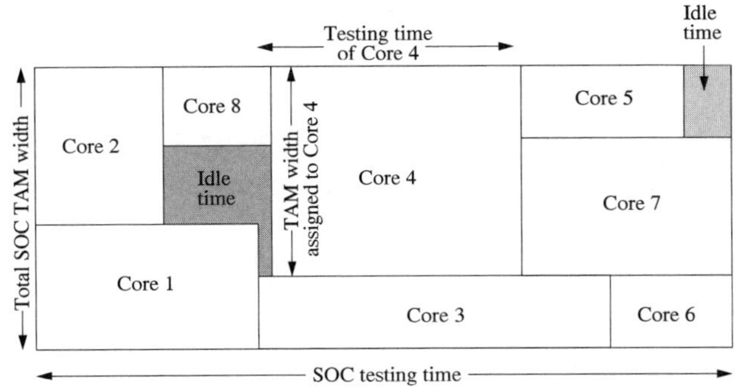

Figure 2: **Example test schedule using rectangle packing.**

on the horizontal axis. □

In Problem $\mathcal{P}_{\mathrm{GRP}}1$, during packing, the rectangle selected for a core can be vertically split into several non-adjacent rectangles having the same width. This is because it is possible to assign a group of non-contiguous TAM wires to a single core, using fork-and-merge of TAM wires.

Problem $\mathcal{P}_{\mathrm{GRP}}1$ relates to *Problem 1* as follows; see Figure 2. The height of the rectangle for each core corresponds to the TAM width assigned to the core, while the rectangle width corresponds to the testing time. The height of the bin corresponds to the total SOC TAM width, and the width to the which the bin is filled corresponds to the SOC testing time. The unfilled area of the bin corresponds to the idle time on TAM wires during test. Furthermore, the distance between the left edge of each rectangle and the left edge of the bin corresponds to the beginning time of each core test. Thus a one-to-one correspondence exists between the packed bin and the final test schedule.

Problem $\mathcal{P}_{\mathrm{GRP}}1$ can be shown to be $\mathcal{NP}$-hard by a restriction argument. A special case of $\mathcal{P}_{\mathrm{GRP}}1$ in which the cardinality of each set $\mathbf{R}_i$, $1 \leq i \leq N$ equals one directly corresponds to the rectangle packing problem in [8]. Since the rectangle packing problem was shown to be $\mathcal{NP}$-hard in [8] (by restriction to Bin Packing), $\mathcal{P}_{\mathrm{GRP}}1$ is also $\mathcal{NP}$-hard.

4. CONSTRAINED TEST SCHEDULING

In this section, we first detail *Problem 2* (integrated TAM design and constraint-driven test scheduling), and then formulate Problem $\mathcal{P}_{\mathrm{GRP}}2$, a generalized version of $\mathcal{P}_{\mathrm{GRP}}1$ that is equivalent to *Problem 2*.

Problem 2: Solve *Problem 1*, such that:

1. Precedence and concurrency constraints are met;

2. The maximum power dissipation value P_{max} is not exceeded;

3. Selective preemption of tests is allowed.

Precedence constraints reduce test cost and increase test efficiency. For example, "abort at first fail" strategies order tests such that either the cores most likely to fail are tested first, or tests that detect more faults are applied first [15]. Furthermore, memories are often tested and diagnosed earlier so that they can be used later for system test. Concurrency constraints seek to avoid conflicts in the test hardware. For example, a hierarchical parent core cannot be tested at the same time as the child cores lying within it. This is because the wrappers of the child cores must be in Extest mode while the parent core is being tested in Intest mode. Finally, power constraints must be incorporated in the schedule to ensure that the power rating of the SOC is not exceeded during test.

Problem 2 can be expressed in terms of rectangle packing as follows. Consider an SOC having N cores, and:

1. Let $\mathbf{R}_i$ be the set of rectangles for core i, $1 \leq i \leq N$;

2. Let precedence constraints between tests be defined, e.g., $i < j$ (test i must complete before test j is begun);

3. Let concurrency constraints between tests be defined, e.g., $i <> j$ (test i must not be applied at the same time as test j);

4. Let the test for Core i have a power dissipation of P_i;

5. Let each core be assigned a maximum number of allowed preemptions.

$\mathcal{P}_{\mathrm{GRP}}2$: Solve Problem $\mathcal{P}_{\mathrm{GRP}}1$, while allowing each rectangle to be horizontally split into several smaller, non-adjacent rectangles, each having the same height (the rectangles are split along the time axis). The number of such partitions for a rectangle must not exceed the maximum number of preemptions allowed for the test. For each precedence constraint $i < j$, the rectangle for Core i can be packed only after all partitions of the rectangle for Core j are packed. Furthermore, the rectangle representing test i must not overlap (in time) the rectangle for test j, if there is a specified concurrency constraint $i <> j$. Finally, at any moment of time the sum of the P_i values for the rectangles selected must not exceed the maximum specified value P_{max}. □

Problem $\mathcal{P}_{\mathrm{GRP}}2$ is a generalized version of $\mathcal{P}_{\mathrm{GRP}}1$, and can therefore be shown to also be $\mathcal{NP}$-hard by a restriction argument.

Data structure *Schedule*
1 $width_p(i)$ /* preferred TAM width for Core i */
2 $width(i)$ /* TAM width assigned to Core i */
3 $first\_begin(i)$ /* first begin time of Core i */
4 $end(i)$ /* end time of Core i */
5 $scheduled\_times(i)$ /* times during which Core i is scheduled */
6 $time\_left(i)$ /* testing time remaining for Core i */
7 $begun(i)$ /* boolean indicates Core i has begun at least once */
8 $scheduled(i)$ /* boolean indicates Core i is scheduled */
9 $complete(i)$ /* boolean indicates test for Core i has completed */
10 $preempts(i)$ /* number of times Core i has been preempted */
11 $max\_preempts(i)$ /* max. number of preemptions allowed */

Figure 3: **Data structure for the test schedule.**

Next, we describe our solution to Problem $\mathcal{P}_{\mathrm{GRP}}2$. Our algorithm first identifies a "preferred TAM width" ($width_p(i)$) for each Core i such that the core's testing time is within a small percentage of its testing time at a maximum allowable TAM width W_{max}. (In this paper, W_{max} is chosen to be 64.) A group of tests is then selected to be scheduled, such that precedence and power constraints are met, and BIST-scan test conflicts are not created. If the number of available TAM wires is insufficient to add a new test to the group, the resulting idle time is filled using several heuristics that seek to insert tests to minimize the idle time. If idle time is inevitable, the algorithm waits until the first currently-running test completes, and then repeats the scheduling process for the remaining tests. Tests may be preempted and resumed again such that the number of preemptions does not exceed a designer-specified value for each test. Next, we explain the rationale behind the heuristic decision-making in our algorithm, and show how these decisions minimize system testing time.

Data structure. The data structure in which we store the TAM width and testing time values for the cores of the SOC is presented in Figure 3. This data structure is updated with the assigned TAM width, begin and end times, and preemption count for each core as the test schedule is developed.

Preferred TAM widths. The pseudocode for our TAM optimization and test scheduling algorithm is presented in Figure 4. The inputs to our algorithm are the set **C** of cores, total TAM width W, set **PC** of precedence constraints, set **CC** of concurrency constraints, power constraint P_{max}, and user-input parameters p and d (explained later). In Procedure *Initialize* (Line 1), we calculate the collection **R** of Pareto-optimal rectangles as described in Section 3, and the preferred TAM width values for each core from the input percent value p. Recall that the core testing time varies with TAM width w as a staircase function that drops rapidly at first for small

Procedure *TAM_schedule_optimizer*(**C**, W, **PC**, **CC**, P_{max}, d, p)
1 *Initialize*(**C**, d, p);
2 Set $w\_avail = W$; $current\_time = 0$;
3 **While C** $\neq \emptyset$
4 **If** $w\_avail > 0$
5 **If** a Core $i \in$ **C** can be found, such that $begin(i) < current\_time$ AND $scheduled(i) = 0$ AND $preempts(i) = max\_preempts(i)$
6 *Assign*(i, $width(i)$, 0);
7 **If** a Core $i \in$ **C** can be found, such that $begun(i) = 1$ AND $width(i) \leq w\_avail$ AND $time\_left(i)$ is maximum AND (*Conflict*(i, **PC**, **CC**, P_{max})=0)
8 **If** $end(i) < current\_time$
9 *Assign*(i, $width(i)$, 1);
10 **Else** *Assign*(i, $width(i)$, 0);
11 **If** a Core $i \in$ **C** can be found, such that $begun(i) = 0$ AND $width_p(i) \leq w\_avail$ AND $time\_left(i)$ is maximum AND (*Conflict*(i, **PC**, **CC**, P_{max})=0)
12 *Assign*(i, $width_p(i)$, 0);
13 **If** a Core $i \in$ **C** can be found, such that $begun(i) = 0$ AND $width_p(i) \leq w\_avail + 3$ AND $width_p(i)$ is minimum AND (*Conflict*(i, **PC**, **CC**, P_{max})=0)
14 *Assign*(i, w_avail, 0);
15 **If** a Core $i \in$ **C** can be found, such that $first\_begin(i) = current\_time$ AND $T_i(width(i)) - T_i(width(i) + w\_avail)$ is maximum;
16 *Assign*(i, $width(i) + w\_avail$, 0);
17 **Else** *Update*(**C**);
18 **Return** *Schedule*;

Figure 4: Algorithm for solving $\mathcal{P}_{\text{GRP}}2$.

Procedure *Initialize*(**C**, d, p)
1 Compute collection **R** of rectangles using *Design_wrapper*;
2 **For** each Core $i \in$ **C**
3 Calculate $T_{ip} = T_i(W_{max}) + \frac{p}{100} \times (T_i(1) - T_i(W_{max}))$;
4 Set $width_p(i) = w$, such that $T_i(w) - T_{ip}$ is minimum;
5 Calculate highest Pareto-optimal width w_h;
6 **If** $w_h - width_p(i) \leq d$ **then** $width_p(i) = w_h$;

Figure 5: Preferred widths initialization subroutine.

values of w and less rapidly after that. For example, for Core 6 in p93791 (Figure 1), at $w = 10$ the testing time reaches within 10% of its value at $w = 64$, and at $w = 15$, the testing time is within 5% of its value at $w = 64$. Hence, instead of attempting to assign the highest Pareto-optimal width 47 to this core, a considerable savings in system TAM width can be realized by assigning a pre-calculated value, such that the testing time of the core reaches within a small percent value p of its testing time at $w = W_{max}$. The value of p is usually between 1 and 10.

Initializing subroutine. In subroutine *Initialize* (Figure 5), we initialize $width_p(i)$ to the preferred TAM width T_{ip}. ($T_i(w)$ denotes the testing time of Core i, with a TAM width of w.) In Lines 5 and 6 of *Initialize*, we allow $width_p(i)$ to be set to the highest Pareto-optimal width w_h if the difference between $width_p(i)$ and w_h is less than the input difference value d. This heuristic aids significantly in minimizing system testing time, when it is beneficial to assign a few ($\leq d$) extra TAM wires to a bottleneck core. For example, when using $p = 2$ for example SOC p34392 (presented in Section 6), we noticed that Core 18 was assigned 9 bits, leading to $T_{18}(9) = 622163$ cycles. The testing time for the SOC was also found to be 622163 cycles, from which we noted that Core 18 is a bottleneck core. Furthermore, Core 18's highest Pareto-optimal width is 10 bits, at which the testing time for Core 18 reaches its minimum value of 544579 cycles. Hence, providing an extra TAM wire to Core 18 reduced its testing time as well as the overall SOC testing time to $T_{18}(10) = 544579$ cycles. Thus the minimum testing time for SOC p34392 could be achieved using the heuristic in Lines 5 and 6 of *Initialize* with $d = 1$.

TAM width assignment and test scheduling. Line 2 of Figure 4 initializes the main rectangle packing loop. While executing the main **While** loop (Line 3), if there are w_avail TAM wires available for assignment, cores are assigned to the TAM using a three-

Procedure *Assign*(i, $width$, $preempt$)
1 Let i be the core to be scheduled;
2 Set $width(i) = width$; $w\_avail = w\_avail - width$;
3 Set $scheduled(i) = 1$; $preempts(i) = preempts(i) + preempt$;
/* s_i, s_o are the longest wrapper scan-in and scan-out lengths [12]*/
5 **If** $preempt = 1$, Set $time\_left(i) = time\_left(i) + \min\{si, so\}$;
6 **If** $begun(i) = 0$
7 Set $begun(i) = 1$; $first\_begin(i) = current\_time$;
9 Set $time\_left(i) = T_i(width(i))$;
10 Set $end(i) = current\_time + time\_left(i)$;
11 **Return** to Line 4 of *TAM_schedule_optimizer*;

Figure 6: The core assign algorithm.

Procedure *Conflict*(i, **PC**, **CC**, P_{max})
1 Let i be the core to be scheduled;
2 **For** all precedence constraints $j < i$
3 **If** $complete(j) = 0$, **Return** 1;
4 **For** all concurrency constraints $i <> j$
5 **If** $scheduled(j) = 1$, **Return** 1;
6 Let $P = 0$;
7 **For** all cores $j \neq i$
8 **If** $scheduled(j) = 1$, $P = P + P_j$;
9 **If** $P > P_{max}$, **Return** 1;
10 **For** all cores $j \neq i$
11 **If** there is a BIST-scan test conflict between Cores i and j, **Return** 1;
12 **Return** 0;

Figure 7: Precedence, concurrency and power constraints.

priority selection mechanism:
Priority 1: Line 5 searches for a core i whose test has already been preempted $max\_preempts(i)$ times, but has not completed. If such a core is found it is scheduled using the *Assign* subroutine (Figure 6). After *Assign* completes, control is returned to Line 4 of Figure 4, and the process of selecting a core begins again. Thus only if no core is found by *Priority 1*, does*Priority 2* come into play.
Priority 2: If a core is found, whose test has begun earlier, whose assigned TAM width is less than w_avail, and whose remaining testing time is largest among all such cores, then it is scheduled using *Assign* in Lines 7 to 10.
Priority 3: If a core is found, whose test has not begun earlier, whose preferred TAM width is less than w_avail, and whose remaining testing time is largest among all such cores, then it is scheduled using *Assign* in Lines 11 to 12.
Priority 1 is motivated by the need to complete the test for cores that cannot be preempted further, while *Priorities 2* and *3* seek to assign the preferred TAM width to each core.

Precedence, concurrency and power constraints. During selection of a core to be scheduled in Lines 7, 11, and 13 of Figure 4, the *Conflict* subroutine (Figure 7) is invoked to ensure that (i) precedence conflicts, (ii) concurrency conflicts, and (iii) power constraint conflicts are avoided.

Rectangle insertion in idle time. If there is no core found in Lines 5 to 12, rather than let the w_avail TAM wires remain idle, *TAM_schedule_optimizer* attempts to insert the rectangle for some unscheduled core into the available time. In Line 13, we find a core that has not been scheduled and whose preferred TAM width is within 3 bits of w_avail. This core is then scheduled using *Assign*. The 3-bit limit was found to be the most useful after extensive experimentation. This is because, in general, as long as the width of the rectangle chosen is within 3 bits of the preferred width, the gains to testing time reduction are greatest. However, if a value other than 3 is found to be more useful for another group of SOCs, this can be easily entered into our program by the system integrator during test automation.

Increasing TAM widths to fill idle time. If no rectangle is available to fill in the idle time, then the heuristic in Lines 15 to 16 is used to determine which of the cores currently scheduled to begin at $current\_time$ will benefit the most, in terms of testing time decrease, from an extra w_avail TAM wires. If such a core can

```
Procedure Update(C)
Find Core i such that time_left(i) is minimum;
Set next_time = current_time + time_left(i);
For all Cores i such that scheduled(i) = 1
  Set scheduled(i) = 0; end(i) = next_time;
  Set time_left(i) = time_left(i) − (next_time − current_time);
  If time_left(i) = 0
    Set complete(i) = 1; C = C − i;
  Update scheduled_times(i);
Set current_time = next_time; w_avail = W;
```

Figure 8: Test schedule update algorithm.

be found, then its currently-assigned $width(i)$ is increased to the highest Pareto-optimal width less than $width(i) + w\_avail$.

Finally, if the heuristics in Lines 4 to 16 fail to find a core to assign, the value of w_avail is set to 0 and the loop beginning at Line 4 is repeated. When w_avail is found to be 0 in Line 4, the execution proceeds to Line 17, where the process of updating $current\_time$ and w_avail is begun. This is presented in subroutine *Update* in Figure 8. Once $current\_time$ is incremented, the widths assigned to packed rectangles (or parts of rectangles) are fixed and cannot be changed later on in the schedule. The resulting test schedule is output in Line 18.

Test preemption. Tests can be selectively preempted each time *Update* is executed. At each execution of *Update*, the value of w_avail is reset to W, and all the incomplete tests contend for the available TAM width in Lines 4 to 16 of *TAM_schedule_optimizer*. If the maximum limit on preemptions for a certain Core i is reached in Line 5, then Core i is continuously scheduled until it completes. On the other hand if the limit on preemptions is not reached, and Core i is preempted (Line 9), then $time\_left(i)$ is incremented by $\min\{s_i, s_o\}$ in Line 5 of *Assign*. Here, s_i (s_o) is the length of the longest wrapper scan in (out) chain [18]. This increment in testing time is necessary because each time a preemption occurs, an extra scan in or scan out must be performed.

5. TESTER DATA VOLUME REDUCTION

In this section, we present the third part of our SOC test automation framework - identification of a value of W that minimizes a weighted cost function involving both the testing time and the tester data volume.

We first motivate the need for identifying such a TAM width. The cost of testing SOCs is closely related to the testing time and the volume of test data. While the time required to apply digital patterns to the SOC is a relatively small fraction of the total test time, the time required to transfer several Gigabytes of data from a workstation to the tester memory is significant if performed frequently [3]. Techniques to ensure that the test data required per pin is contained to a single tester buffer, are therefore vital. The motivation for trading-off TAM width with testing time and data volume lies in multisite testing, in which several ICs are tested in parallel by a single tester. Reduced TAM widths that contain test data per pin to buffer sizes will allow a larger number of ICs to be tested concurrently, thereby decreasing testing time for the entire production batch. Reducing TAM widths also leads to lower routing complexity. It is therefore important to develop techniques that can identify a low number of TAM wires, and to trade-off testing time and data volume.

We begin by plotting the variation of testing time $\mathcal{T}$ with W. This is shown for Philips SOC p22810 in Figure 9(a). (Results for the remaining example SOCs are shown in Table 2 in Section 6.) Note the progressive decrease in $\mathcal{T}$ at Pareto-optimal points. Next, we plot the variation of tester data volume $\mathcal{M}$ with W (Figure 9(b)). $\mathcal{M}$ varies as a non-monotonic function in W, achieving local minima at the Pareto-optimal W values of the $\mathcal{T}$ curve of Figure 9(a).

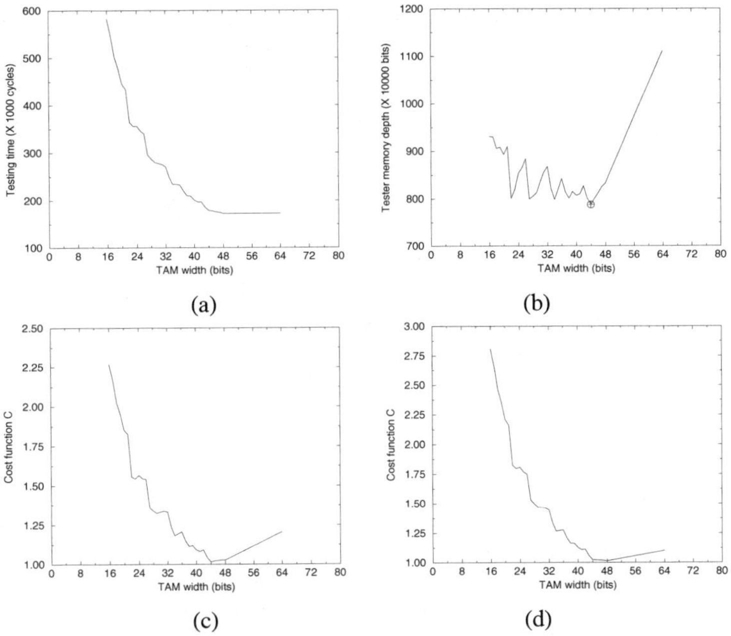

Figure 9: Relationship between (a) $\mathcal{T}$ and W, and (b) $\mathcal{M}$ and W for SOC p22810; The normalized cost function $\mathcal{C}$ for (c) $\alpha = 0.5$, and (d) $\alpha = 0.75$ for SOC p22810.

The global minima (marked in Figure 9(b)) is achieved at $W = 44$. However, $W = 44$ does not provide the lowest testing time for the SOC. $\mathcal{T}$ can be decreased by increasing W from 44 to 48 (at which point there is an increase in $\mathcal{M}$). Therefore, by varying W the system integrator can trade-off testing time with tester data volume.

We incorporate this feature in our framework by defining the normalized cost function $\mathcal{C} = \alpha\frac{\mathcal{T}}{\mathcal{T}_{min}} + (1-\alpha)\frac{\mathcal{M}}{\mathcal{M}_{min}}$, where $\mathcal{T}_{min}$ ($\mathcal{M}_{min}$) is the minimum value of $\mathcal{T}$ ($\mathcal{M}$), and $0 \leq \alpha \leq 0$ is a user-input parameter to control the trade-off. As α is varied from 0 to 1, the shape of the $\mathcal{C}$-curve changes from the $\mathcal{M}$-curve to the $\mathcal{T}$-curve. The $\mathcal{C}$-curve is "U" shaped in general having a single minima, illustrated for $\alpha = 0.5$ in Figure 9(c), and for $\alpha = 0.75$ in Figure 9(d). These values of W that minimize $\mathcal{C}$ for various values of α provide the system integrator with effective choices of TAM width for tester data volume reduction. Note that the choice of W will also be affected by the core type and testing method used, e.g., for a core tested only using BIST, the TAM may be very narrow to transport only the BIST control signals. In this paper, we consider only tests that use the TAM for transporting the entire test set.

6. EXPERIMENTAL RESULTS

In this section, we present experimental results for four example SOCs: d695 (an academic SOC from Duke University), p22810, p34392, and p93791 (industrial SOCs from Philips). These four SOCs are part of the ITC 2002 SOC test benchmarks initiative [17]. The experimental results were obtained using a 333 Mhz Sun Ultra 10 with 256 MB memory.

Table 1 presents results on integrated TAM design and test scheduling. We considered all possible integer values of the parameters p and d in the range $1 \leq p \leq 10$, $0 \leq d \leq 4$, and tabulated the best results. A lower bound on the testing time for an SOC can be expressed as $\max\left\{\max_i T_i(W_{\max}), \left\lceil \left(\sum_{i=1}^{N} T_i(1)\right)/W \right\rceil\right\}$. Lower bound values on the testing time for each SOC for several values of W are presented in Table 1. We compare the testing times obtained using non-preemptive and preemptive scheduling. For preemptive testing, the value of $max\_preempts(i)$ was

Table 1: **Wrapper/TAM co-optimization and test scheduling.**

SOC	W	Testing time (cycles): Lower bound	Non-preemptive	Pre-emptive	Preemptive + power-constrained
d695	16	41232	43410	43423	47574
	32	20616	22229	21757	29039
	48	13744	15698	15499	28441
	64	10308	11285	11354	20004
p22810	16	421473	466383	459951	527573
	32	210737	243779	243978	277151
	48	140491	164420	162554	213845
	64	105369	140222	134732	176076
p34392	16	936882	1071043	1082065	1180187
	24	624588	728986	702322	1075971
	28	544579	617018	615126	1075242
	32	544579	544579	544579	1075242
p93791	16	1749388	1860752	1860752	1966092
	32	874694	929311	929311	1247221
	48	583130	637717	643605	656214
	64	437347	503661	492095	631840

set to 2 for the larger cores. Preemptive scheduling obtains lower or equal testing times in most cases. However, in a few cases, non-preemptive schedules are shorter. This is because each preemption adds $\min\{s_i, s_o\}$ cycles to the length of a test, which can increase the overall SOC testing time for SOCs having a large number of short tests that are preempted several times. A careful investigation of the effects of preemption and the use of the $max\_preempts(i)$ parameter considering test lengths is therefore warranted. For p34392, we present testing time results only for $W \leq 32$, since at $W = 32$, we achieve the minimum testing time for p34392, 544579 cycles.

In Table 1, we also present the testing times obtained with power-constraints. We assigned a hypothetical power value P_i to the test for each Core i based on the number of test data bits per test pattern for Core i. The value of P_{max} was set to $\max_i\{P_i\}$ for test scheduling. The increases in testing times for power-constrained scheduling reflect this concurrency constraint. The CPU time of our new algorithm is less than 5 seconds in all cases; this is several orders of magnitude lower than the CPU times required by the method in [12].

Next, Table 2 presents results on effective TAM widths for tester data volume reduction. We present the minimum values of $\mathcal{T}$ and $\mathcal{M}$ obtained for the four SOCs, as well as the values of W at which they occur. We then present the values of $\mathcal{C}_{min}$ and the resulting effective TAM widths W_e obtained for several different values of α. Finally, the corresponding values of $\mathcal{T}$ and $\mathcal{M}$ obtained for these values of W_e are shown. It is clear from Table 2 that the system integrator can trade-off testing time with tester data volume by varying α between 0 and 1. For example, for SOC p22810, the minimum value of $\mathcal{T}$ (14022 cycles) is achieved at $W = 63$. However, the minimum values of $\mathcal{M}$ (7202214 bits) is actually achieved at $W = 18$. By setting $\alpha = 0.3$, the system integrator obtains a TAM width of 48 bits, at which $\mathcal{T} = 164420$ cycles and $\mathcal{M} = 7892160$ bits.

7. CONCLUSION

We have presented new techniques based on rectangle packing for wrapper/TAM co-optimization and test scheduling. TAMs have been tailored to the test data needs of cores through the use of Pareto-optimal widths. We have also presented several heuristics that minimize the idle time on TAM wires, thereby leading to a fast and efficient algorithm for TAM width allocation and test scheduling. The new rectangle packing algorithm is scalable for large industrial SOCs. Finally, the proposed approach allows the system integrator to determine an SOC-level TAM width to trade off testing time with tester data volume.

Table 2: **TAM widths for tester data volume reduction.**

$\mathcal{T}_{min}$ (cycles)	$^1W_1$ (bits)	$\mathcal{M}_{min}$ (bits)	$^2W_2$ (bits)	α	$\mathcal{C}_{min}$	^3W_e (bits)	$\mathcal{T}$ at W_e (cycles)	$\mathcal{M}$ at W_e (bits)
d695								
11285	63	675554	22	0.1	1.031	60	11612	696720
				0.3	1.031	60	11612	696720
				0.5	1.026	63	11285	710955
p22810								
$\mathcal{T}_{min}$	$^1W_1$	$\mathcal{M}_{min}$	$^2W_2$	α	$\mathcal{C}_{min}$	^3W_e	$\mathcal{T}$ at W_e	$\mathcal{M}$ at W_e
140222	63	7377480	44	0.01	1.018	44	167670	7377480
				0.3	1.103	48	164420	7892160
				0.5	1.093	55	148005	8140275
p34392								
$\mathcal{T}_{min}$	$^1W_1$	$\mathcal{M}_{min}$	$^2W_2$	α	$\mathcal{C}_{min}$	^3W_e	$\mathcal{T}$ at W_e	$\mathcal{M}$ at W_e
544579	32	16659486	27	0.2	1.000	27	617018	16659486
				0.25	1.033	27	617018	16659486
				0.3	1.032	32	544579	17426528
p93791								
$\mathcal{T}_{min}$	$^1W_1$	$\mathcal{M}_{min}$	$^2W_2$	α	$\mathcal{C}_{min}$	^3W_e	$\mathcal{T}$ at W_e	$\mathcal{M}$ at W_e
503661	62	29399656	22	0.5	1.012	22	1336348	29399656
				0.95	1.000	62	503661	31226982
				0.99	1.000	62	503661	31226982

$^1W_1$: W at $\mathcal{T}_{min}$; $^2W_2$: W at $\mathcal{M}_{min}$; ^3W_e: W at $\mathcal{C}_{min}$

8. REFERENCES

[1] J. Aerts and E.J. Marinissen. Scan chain design for test time reduction in core-based ICs. *Proc. Int. Test Conf.*, pp. 448-457, 1998.

[2] V.D. Agrawal, C.R. Kime and K.K. Saluja. A Tutorial on Built-In Self-Test, Part 1: Principles. *IEEE Design & Test of Computers*, pp. 73-82, March 1993.

[3] C. Barnhart et al. OPMISR: The foundation for compressed ATPG vectors. *Proc. Int. Test Conf.*, pp. 748-757, 2001.

[4] K. Chakrabarty. Optimal test access architectures for system-on-a-chip. *ACM Trans. Design Automation of Electronic Systems*, vol. 6, pp. 26-49, January 2001.

[5] K. Chakrabarty. Test scheduling for core-based systems using mixed-integer linear programming. *IEEE Trans. CAD*, vol. 19, pp. 1163-1174, Oct 2000.

[6] A. Chandra and K. Chakrabarty. Test resource partitioning for SOCs. *IEEE Design & Test of Computers*, vol. 18, pp. 80-91, Sept-Oct 2001.

[7] R.M. Chou, K.K. Saluja and V.D. Agrawal. Scheduling tests for VLSI systems under power constraints. *IEEE Trans. VLSI Systems,* vol. 5, no. 2, June 1997.

[8] E.G. Coffman, Jr., M.R. Garey, D.S. Johnson, and R.E. Tarjan. Performance bounds for level-oriented two-dimensional packing algorithms. *SIAM J. Computing*, vol. 9, pp. 809-826, 1980.

[9] Y. Huang et al. Resource allocation and test scheduling for concurrent test of core-based SOC design. *Proc. Asian Test Symp.*, pp. 265-270, 2001.

[10] Y. Huang et al. On concurrent test of core-based SOC design. *J. Electronic Testing: Theory and Applications*, vol. 18, Aug. 2002, to appear.

[11] V. Iyengar and K. Chakrabarty. Precedence-based, preemptive and power-constrained test scheduling for system-on-a-chip. *Proc. VLSI Test Symp.*, pp. 368-374, 2001.

[12] V. Iyengar, K. Chakrabarty, and E.J. Marinissen. Test wrapper and test access mechanism co-optimization for system-on-chip. *J. Electronic Testing: Theory and Applications*, vol. 18, pp. 213-230, April 2002.

[13] V. Iyengar, K. Chakrabarty, and E.J. Marinissen. Efficient wrapper/TAM co-optimization for large SOCs. *Proc. Design Automation and Test in Europe (DATE) Conf.*, pp. 491-498, 2002.

[14] V. Iyengar, K. Chakrabarty, and E.J. Marinissen. On using rectangle packing for SOC wrapper/TAM co-optimization. *Proc. VLSI Test Symp.*, 2002, to appear.

[15] W. Jiang and B. Vinnakota. Defect-oriented test scheduling. *Proc. VTS,* pp. 433-438, 1999.

[16] E. Larsson and Z. Peng. Test scheduling and scan-chain division under power constraint. *Proc. Asian Test Symp.*, pp. 259-264, 2001.

[17] E.J. Marinissen, V. Iyengar and K. Chakrabarty. ITC 2002 SOC test benchmark initiative. *http://www.extra.research.philips.com/itc02socbenchm*

[18] E.J. Marinissen. The role of test protocols in automated test generation for embedded-core-based system ICs. *J. Electronic Testing: Theory and Applications*, vol. 18, Aug. 2002, to appear.

[19] M. Nourani and C. Papachristou. An ILP formulation to optimize test access mechanism in system-on-chip testing. *Proc. Int. Test Conf.*, pp. 902-910, 2000.

Communication Architecture Based Power Management for Battery Efficient System Design *

Kanishka Lahiri
Dept. of ECE
UC San Diego
klahiri@ece.ucsd.edu

Anand Raghunathan
C&C Research Labs
NEC USA
anand@nec-lab.com

Sujit Dey
Dept. of ECE
UC San Diego
dey@ece.ucsd.edu

ABSTRACT

Communication-based power management (CBPM) is a new battery-driven system-level power management methodology in which the system-level communication architecture regulates the execution of various system components, with the aim of improving battery efficiency, and hence, battery life. Unlike conventional power management policies (that attempt to efficiently shut down idle components), CBPM may delay the execution of selected system components even when they are active, in order to adapt the system-level current discharge profile to suit the battery's characteristics.

In this paper, we present a methodology for the design of CBPM based systems, which consists of system-level performance and power profiling, battery discharge analysis, instrumentation of system components to facilitate CBPM, definition of CBPM policies, and generation of the CBPM-based system architecture. We present extensive evaluations of CBPM, and demonstrate its application to the design of an IEEE 801.11 Wireless LAN MAC processor system. Our results indicate that CBPM based systems are significantly more battery efficient than those based on conventional power management techniques. Further, we demonstrate that CBPM enables design-time as well as run-time tradeoffs between system performance and battery life.

Categories and Subject Descriptors

C.5.4 [**Computer System Implementation**]: VLSI Systems

General Terms

Algorithms, Design, Experimentation

Keywords

Embedded Systems, Communication Architectures, Battery Efficiency, Power Management, Low Power Design

1. INTRODUCTION

Improvements in wireless communication and semiconductor fabrication technologies are key factors driving a rapidly expanding market for battery powered electronics. Unfortunately, projections of the complexity, functionality and performance of these electronic systems far exceed projected improvements in battery technologies, leading to a widening "battery gap" [1, 2, 3]. Bridging this gap is a challenge that system designers must face for the foreseeable future.

The need to improve battery life has, in large part, driven research and development of low power design techniques for electronic circuits and systems [4, 5, 6]. Low power design successfully reduces the energy that is drawn from the battery, and hence improves battery life to some extent. However, truly maximizing battery life requires an understanding of both the source of energy and the system that consumes it. It has been shown that the amount of energy that can be supplied by a battery varies significantly, depending on the manner in which the energy is drawn [3],[7]-[12]. Consequently, there is a need for "battery efficient", (not merely energy efficient) design techniques that are specifically tailored towards the characteristics and limitations imposed by battery-powered systems.

In this paper, we present *Communication Based Power Management* (CBPM), a new *battery driven* design methodology that aims at improving battery lifetime beyond what is achievable using state-of-the-art techniques for low power design. CBPM is implemented in the system-level communication architecture, and utilizes the communication architecture's ability to regulate the time profile of the system's power consumption.

1.1 Paper Overview and Contributions

Communication based power management (CBPM) is a system-level power management methodology that tailors the run time system power profile to ensure efficient battery discharge, in order to maximize battery life. CBPM policies exercise dynamic, regulatory control over the system components, by proactively blocking execution of selected components to prevent inefficient use of the battery, while minimally impacting performance. In this paper, we present a methodology and algorithms for the design of CBPM based systems. We illustrate the benefits of CBPM by applying it to the design of an IEEE 802.11 MAC processor system. We present results that indicate that CBPM results in a degree of battery efficiency superior to those obtained using traditional power management approaches. We also demonstrate that CBPM policies can be configured to allow for flexible and fine-grained trade-offs between battery life and performance.

1.2 Related Work

Recent work in the area of battery efficient design includes modeling techniques to efficiently and accurately capture important electro-chemical effects during battery discharge (*e.g.,* [8]-[11]. These models estimate battery lifetime and capacity under a variety of discharge scenarios, and are a key part of a battery driven design environment. Techniques for designing battery-efficient systems include battery driven speed and voltage setting [12, 13]. These approaches are typically aimed at general purpose computing systems, and hence do not exploit characteristics of the application. Battery driven voltage setting and task scheduling for application specific systems are described in [7, 14]. While the latter can be applied only to systems for which a static schedule exists, neither approach lends itself to dynamic control over the system power profile. Extensions to conventional dynamic power management techniques have been suggested in [15], to take battery constraints into account. Their approach is a reactive one, based on monitoring the remaining charge in the battery. Battery scheduling and management for multi-battery systems ([15]-[18]) do not address system power consumption, but optimize the battery subsystem to best satisfy power requirements. The Smart Battery System [19], an emerging industry standard, aims at standardizing communication between batteries and the systems they power, to facilitate interoperability and improved battery efficiency.

To our knowledge, ours is the first contribution to battery efficient design that provides techniques to dynamically regulate the system power profile, by exploiting the system-level communication architecture.

1.3 Paper Organization

In the next section, we provide background on batteries and system-level communication architectures. In Section 3, we present basic concepts of CBPM, and illustrate its benefits through an example. In Section 4, we present a methodology for designing CBPM based systems. Finally, in Section 5, we present results of experiments conducted while applying CBPM to the design of an IEEE 802.11 MAC processor.

*Supported by the National Science Foundation, grant 99-12414

2. BACKGROUND

In this section, we describe battery characteristics that are taken into account by the CBPM methodology. We also describe relevant concepts associated with system-level communication architectures.

2.1 Battery Background

Terminology: The *standard capacity* of a battery is the energy that can be extracted from a battery when discharged under standard load conditions. For example, a typical lithium ion battery may have standard capacity 500 *mAh* when discharged at a constant current of 125 *mA*, at $25^\circ C$. The *actual capacity* of a battery is the amount of energy that the battery delivers under a given load, and is usually used (along with battery life) as a metric to judge the battery efficiency. While the actual capacity may differ from the standard capacity (depending on the load conditions), it cannot exceed a certain *theoretical capacity*. In this paper, *battery capacity* is used to refer to the actual capacity of the battery.

Rate Capacity Effects: Several electro-chemical effects make battery capacity sensitive to the discharge current profile. Significant among them are *rate capacity effects*, which result in a dependency between battery capacity and the magnitude of the discharge current [11].[1] Battery life depends largely on the availability of active reaction sites throughout the cathode. During discharge, chemical reactions cause the creation of inactive cathode sites, eventually leading to a state of discharge. During time intervals when the current drawn is small, inactive reaction sites are uniformly distributed throughout the material of the cathode. However, during intervals when the discharge current is larger, the outer surface of the cathode gets covered with inactive sites, making many internal active sites unreachable, effectively decreasing battery capacity. This rate capacity effect contributes to a significant dependence of battery capacity on the system's power profile [11].

2.2 System-level Communication Architectures

System-level communication architectures serve as a fabric to integrate various system components, and provides a medium for interacting system components to communicate. The *topology* of communication architectures can range from a single channel (shared bus), to a network of channels interconnected by *bridges*. System components include those that can initiate a communication transaction, called *bus masters* (*e.g.*, CPUs, DSPs, *etc.*), and those that merely respond to transactions initiated by a master, called *slaves* (*e.g.*, memories). Since channels are often shared by several masters, *communication protocols* include arbitration mechanisms to ensure that exactly one master has control of the channel at any given time. Typical arbitration mechanisms include *round-robin* access, *priority based* selection, *etc.*, which are implemented in centralized or distributed *arbiters*. The communication protocol also defines other communication functions, such as handshaking, burst transfers, and split transactions.

3. COMMUNICATION BASED POWER MANAGEMENT

In this section, we first explain the basic concept of communication based power management (CBPM), and then illustrate, through an example, the impact of CBPM on battery efficiency.

3.1 Basic Concept

Communication based power management makes use of the system-level communication architecture to regulate the execution of system components in order to obtain desirable system-level power profiles and improved battery efficiency. CBPM functionality differs significantly from conventional dynamic power management (DPM) techniques, which only aim to prevent wastage of energy during periods when the system or its components are idle. CBPM also differs from techniques that merely target an overall reduction in average power consumption. CBPM exercises dynamic and proactive control over system components, delaying their execution even if they are not in an idle state. CBPM based designs include mechanisms to block the execution of certain system components and force them to idle. Each component in a CBPM based system is internally power managed so as consume significantly less power when blocked than when active. By dynamically blocking the execution of certain components, CBPM can regulate the power profile of the system.

[1] Though CBPM does not explicitly target other effects, such as *charge recovery*, the battery life improvements that we report are generated using battery models that do account for these effects.

CBPM Rationale: While the aim of proactively controlling the execution of components in order to regulate the system power profile could be realized in many ways, we chose to use the system-level communication architecture for the following reasons:

Minimal hardware overhead: The communication architecture is a fabric that physically and logically integrates all the system components together. Embedding power management functionality in the communication architecture eliminates the need to include a separate power management unit and connect each component to it.

System-level visibility: To exercise control over the system power profile, power management policies need access to system wide information (*e.g.*, current execution states of different system components). By virtue of its connectivity, the communication architecture lends itself naturally to (i) effectively monitoring and detecting changes in the execution states of system components, and (ii) providing an efficient delivery mechanism for power management decisions to any system component.

Significant impact on power profile: The communication architecture directly controls the timing of system-level communications, and hence the execution of system components. In particular, the communication architecture can be used to control the timing of component idle and active states, impacting the component's power profile. Since the system power profile is the sum of the component power profiles, the communication architecture can be used to significantly impact the system power profile, as shown in the next subsection.

In summary, due to its system-level connectivity, and its potential to control the power profile of the system, the communication architecture naturally lends itself to use in a battery efficient system design methodology.

3.2 Example: IEEE 802.11 MAC Processor

We next illustrate the impact of CBPM by considering its application to an IEEE 802.11 MAC processor [20]. We compare the execution of the original performance optimized MAC processor with a CBPM based MAC processor. Note that, the original MAC processor incorporates DPM to minimize power consumed by idle components. For each design, we analyze representative power profiles, and battery efficiency, as the system processes a sequence of MAC frames.

The details of the MAC processor are presented in Section 5. However, for the sake of our illustration in this section, it is sufficient to consider only the system power profiles. Figure 1(a) shows a snapshot of the current discharge profile of the MAC processor as it processes 3 MAC frames.[2] The first is of length 54 bytes, and the next two (which arrive back to back) are of length 1414 bytes each. This snapshot is part of a longer profile obtained by simulating the MAC processor as it processes frame arrivals during 1.5 minutes of streaming video over a wireless LAN.

Using a stochastic model of a Lithium-ion battery (250 *mAh* capacity, rated discharge current of 125 *mA* [9]), we measured battery life and ca-

[2] Current discharge profiles are obtained by dividing the power profiles by the supply voltage (assumed to be constant).

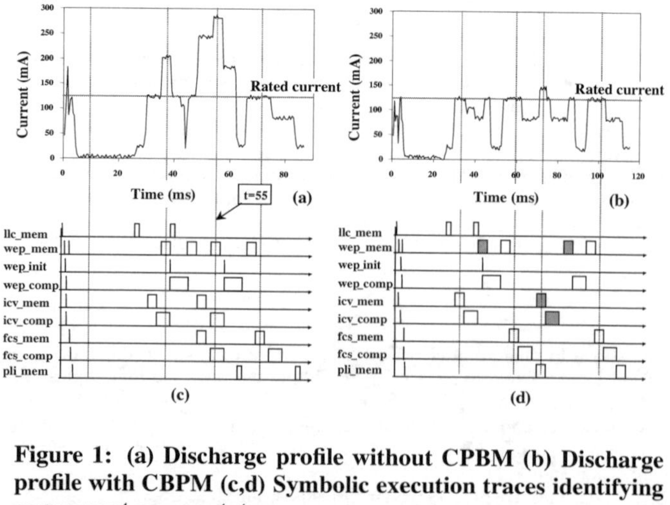

Figure 1: (a) Discharge profile without CPBM (b) Discharge profile with CBPM (c,d) Symbolic execution traces identifying component macro-states

pacity under the given profile. Under these conditions of discharge, the battery capacity and lifetime obtained are 154.3 *mAh* and 3209.2 *sec* respectively. We observe from Figure 1(a) that the current profile frequently exhibits regions where the current drawn is significantly higher than the rated discharge current of 125 *mA*.

To understand how CBPM improves battery efficiency, consider a symbolic execution trace of the system. An example of such a trace is shown in Figure 1(c), corresponding to the current profile of Figure 1(a). The trace depicts 9 "macro-states", which indicate the specific sub-tasks that each of 5 components (`LLC, WEP, ICV, FCS, PLI`) perform at any given time. For example, `WEP`, which is the component responsible for encrypting frame data, may be in one of several macro-states depending on whether it is reading a frame body from memory (`wep_mem`), initializing the key state array (`wep_init`), or computing an encrypted frame body (`wep_comp`).

The state of the system at any point of time is determined by examining the macro-states of individual components. From Figures 1(a) and (c), it is clear that different system states exhibit wide variation in the current drawn, depending on the individual macro-states of the system components. For example, Figure 1(c) indicates that, at $t = 55$ *ms*, the system state is one where two HW components (`ICV, FCS`) are engaged in lengthy iterative computations (`icv_comp, fcs_comp`), while a third (`WEP`) is accessing memory over a shared bus (`wep_mem`). Figure 1(a) indicates that this system state leads to a drawn current of 280 *mA*.

We next consider the processing of the same sequence of MAC frames, now with the CBPM policies enabled. A snapshot of the new current profile is shown in Figure 1(b), while the corresponding symbolic execution trace is shown in Figure 1(d). By comparing the symbolic execution traces in Figures 1(c) and (d), it can be seen that the CBPM policy effectively delays the execution of certain macro-states (indicated by shading), which would otherwise have led to a violation of the rated discharge current. For example, Figure 1(d) shows that CBPM delays the occurrence of macro-state `wep_mem` to $t = 42$ *ms* so as to avoid overlap with the macro-state `icv_comp`. From the current discharge profile in Figure 1(a), it is clear that the CBPM policy greatly reduces the occurrence of system states that violate the rated current. Battery life of the new system was measured to be 10199 *sec* (a 3.2X improvement) and more importantly, battery capacity was measured to be 225.4 *mAh* (a 1.5X improvement). This example illustrates the significant impact CBPM can have on the system power profiles, and hence overall battery efficiency.

4. CBPM DESIGN METHODOLOGY

In this section, we first present a structured methodology highlighting the various steps involved in designing CBPM based systems. Next, we present a detailed description of each step, including the various algorithms and tools used.

4.1 CBPM Methodology: Overview

The inputs to our methodology (shown in Figure 2) are (i) a simulatable system specification mapped to a system architecture consisting of HW/SW components (ii) a system-level communication architecture to integrate the various components, (iii) typical environment stimuli or traces, (iv) performance and battery life objectives. The output of the methodology is a CBPM policy that is capable of dynamically regulating the system power consumption profile so as to ensure efficient battery discharge. From a hardware perspective, the result is a system with a communication architecture enhanced with new battery aware communication protocols.

In **Step 1**, HW/SW co-simulation is performed to generate execution traces and system power profiles under provided input stimuli. In **Step 2**, the specification mapped to each system component is decomposed into a set of macro-states, whose execution is under the control of the communication architecture. In **Step 3**, power profiling tools are used to estimate the average power consumption of each macro-state. In **Step 4**, performance analysis is used to prioritize the macro-states in terms of their performance criticalities. In **Step 5**, the CBPM policy is defined. In **Step 6**, system-level performance analysis and battery life estimation are performed, accounting for the defined CBPM policy. Incorporation of CBPM policies effectively change the timing of certain macro-states. This may lead to a change in the relative criticalities of the macro-states, requiring recalculation of macro-state criticalities each time the CBPM policy is modified. Hence, if battery life and performance objectives are not met, **Steps 4, 5, 6** are repeated in an iterative manner. Finally, **Step 7** is executed, where the CBPM policy is implemented in the system-level communication architecture.

4.2 CBPM Methodology: Details

In this sub-section, we consider, in turn, the various steps outlined above.

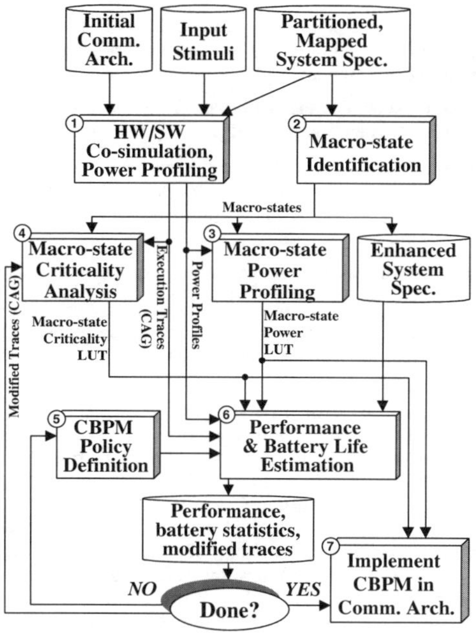

Figure 2: Communication Based Power Management (CBPM) design methodology

Step 1 — HW/SW co-simulation, trace generation: HW/SW co-simulation and power co-estimation of the entire system is performed, under provided input stimuli. The resulting execution traces and power profiles are used in **Steps 3, 4** and **6**. In our implementation, we used the framework described in [21].

Step 2 — Macro-state identification: A component's impact on system performance, and its contribution to the system power profile may exhibit significant variation, depending on which part of its specification it is currently executing. To take this dependence into account, we introduce the concept of a component macro-state. For each component C, we define a set of macro-states $s_0, \ldots, s_{n_C}$, and modify the specification of the component to make these macro-states identifiable at run-time. The procedure for identifying macro-states is as follows:

2A. For each component, label each basic block in the Control Flow Graph (CFG) of its specification as a macro-state s_i.

2B. Use performance analysis (**Step 4**) to measure $delay(s_i)$, the execution time of each macro-state s_i under provided inputs.

2C. The macro-states so obtained must satisfy:

- **Battery Constraint:** $k_1\tau < delay(s_i) < k_2\tau, i = 1, \ldots, n_C$, where τ is the battery time constant, n_C is the number of macro-states of component C, $k_1 < 1$ and $k_2 > 1$ are positive constants. This constraint provides a means to control the granularity at which CBPM regulates the power profile.[3]
- **Complexity Constraint:** $n_C < N$, where N is a constant. This constraint can be used to control the run-time overhead of keeping track of component macro-states.

If these conditions are satisfied, **Step 2E** is executed. Otherwise, **Step 2D**, (a clustering step), and **Steps 2B, 2C** are repeated.

2D. For each s_i for which $delay(s_i) < k_1\tau$, an attempt is made to combine s_i with neighboring macro-states. This is performed in a greedy manner, where s_i is combined with its neighbor t_i that has maximum delay. Due to control flow requirements, the clustering of s_i with t_i may force expansion of the cluster to include other macro-states.

2E. For each macro-state s_i, a blocking communication event e is identified such that s_i is dependent on e (unless e executes, s_i cannot execute). If no such events can be found, an explicit blocking communication is added at the beginning of s_i. Additionally, these communication events are modified to carry a macro-state identifier.

[3] The benefits of fine grained control over the power profile are limited by (typically large) battery time constants.

The output of **Step 2** is an enhanced specification for each component, containing specific communications that carry information about the current macro-state of the component.

Step 3 — Macro-state Power Profiling: The average power consumption of each macro-state (identified in **Step 2**) is estimated as follows:

3A. From the system execution traces, construct symbolic traces for each component, which indicate timing of each macro-state;

3B. From the power profiles and symbolic traces, calculate $pow(s_{ij})$, the average power consumed by the jth instance of s_i. [4]

3C. Calculate $pow(s_i)$, the average power consumed by macro-state s_i, by averaging over all instances of s_i.

The output of **Step 3** is a *macro-state power table*, consisting of tuples, $\langle s_i, pow(s_i)\rangle$ representing macro-states, and their respective average power consumptions.

Step 4 — Macro-state Criticality Analysis: The sensitivity of overall system performance to the execution times of each macro-state is estimated using a trace based system-level performance analysis framework [22]. The performance analysis proceeds in two phases. In a pre-processing phases, execution traces (obtained in **Step 1**) are analyzed to construct a Communication Analysis Graph (CAG), an abstract yet accurate representation of the system execution traces, consisting of time stamped vertices that represent computations and communications, and edges that represent control flow dependencies. In the second phase, the tool provides estimates of system performance for any specified communication architecture by rapidly transforming the CAG, to take into account effects of the communication architecture (*e.g.*, arbitration, burst transfers, CBPM policies, topology *etc*). The output of the tool is a transformed CAG (Figure 3) that provides useful performance statistics for any specified communication architecture. Next, the CAG is analyzed to determine the criticalities of each macro-state. Note that, when **Step 4** is executed for the first time, the input is a CAG which represents system execution under the initial communication architecture. For subsequent executions, the input CAG represents system execution under a CBPM enhanced communication architecture (Figure 2).

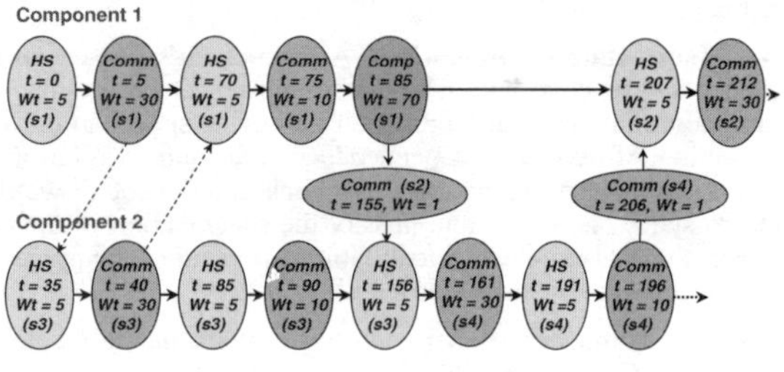

Figure 3: Modified communication analysis graph (CAG) that reflects effects of a priority based shared bus

4A. Each vertex in the CAG is labeled with the macro-state it is associated with (Figure 3).

4B. For each occurrence s_{ij} of a macro-state s_i, the start time of the first vertex in s_i is delayed by δ.

4C. The CAG is analyzed[22] to estimate the effect of this delay on system performance, $crit(s_{ij})$. The results of this analysis are the values of $crit(s_{ij})$ for all i, j.

4D. The average criticality of macro-state s_i, $crit(s_i)$ is obtained by averaging $crit(s_{ij})$ over all j.

The output of **Step 4** is a *macro-state criticality table*, which consists of tuples, $\langle s_i, crit(s_i)\rangle$, representing macro-states, and their respective criticalities.

Step 5 — CBPM Policy Definition: The two aims while defining the CBPM policy are (i) minimize violations of battery rate capacity constraints, and (ii) minimize the extent to which the execution of performance critical macro-states are blocked. The basic CBPM policy is as follows. Let C = set of currently executing macro-states. Let S = set of pending macro-states (those awaiting a grant from CBPM). At each CBPM arbitration, select a set of macro-states $S' \subseteq S$ such that the following conditions hold:

$$\sum_{x \in S' \cup C} pow(x) < P_{max} \quad (1)$$

$$crit(x) > crit(y), \forall (x,y) \mid x \in S', y \in S - S' \quad (2)$$

Clearly the power constraint P_{max} in Eqn 1 is a critical determinant of the extent to which performance is traded off for battery life improvements. A larger value of P_{max} leads to more macro-states being included in S', resulting in fewer blocked components, enhanced performance, and decreased battery efficiency. A smaller value of P_{max} constrains the size of S, inhibiting the number of simultaneously active components, potentially trading battery efficiency for performance. Eqn 2 ensures that if all the pending macro-states cannot be granted for a given value of P_{max}, the policy grants the most performance critical subset from the set of pending macro-states.

A natural question that arises when blocking the execution of system components, is that of deadlock [23]. To see that CBPM policies cannot result in deadlock, note that, in a CBPM based system, communications are blocking, and components are internally power managed. Hence, when components are blocked, they consume very little power, and hence, free up the power "resource" for other components to use. This prevents circular wait, a necessary condition for deadlock.

Step 6 — Performance and Battery Life Estimation: Detailed analysis is carried out in order to examine the impact of the CBPM policy on system performance and battery life. For performance analysis, we use the technique described in **Step 4**, which also generates a CAG with effects of the CBPM policies incorporated. The battery efficiency of the system is estimated as follows:

6A. The CAG derived via performance analysis is used to generate a symbolic execution trace of the system indicating the timing of various macro-states, considering the effects of the new communication protocols.

6B. A new system power profile is generated from the symbolic execution traces and the macro-state power table.

6C. Battery discharge is simulated using the power profiles to obtained battery capacity and lifetime estimates. In our work, we used a stochastic model of a lithium ion battery of 250 *mAh* capacity, with rated discharge current 125 *mA* [9].

If desired performance and battery goals are met, **Step 7** is executed. Otherwise, the CBPM policy is modified by adjusting the power constraint P_{max}, and **Steps 4, 5, 6** are executed iteratively. Note that, in these iterations, macro-state criticality analysis (**Step 4**) is driven by a CAG that includes effects of the CBPM policy.

Step 7 — CBPM Implementation: In this step, the CBPM policy is implemented in the communication architecture. Figure 4 shows a shared bus based communication architecture enhanced to implement CBPM. Requests from the components to the bus arbiter optionally carry a macro-state identifier. Requests to the arbiter are of two types (i) Type I requests: these

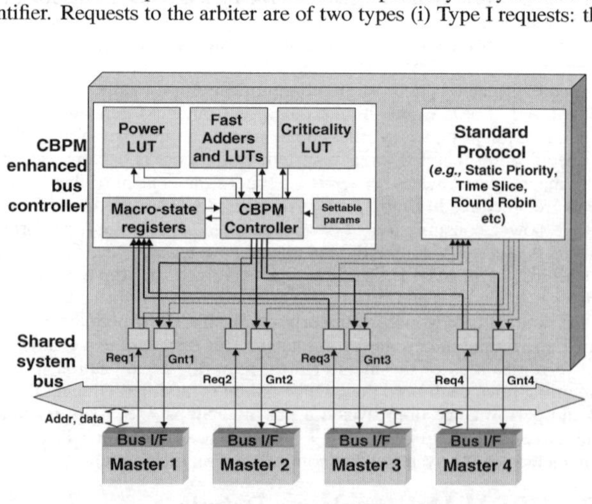

Figure 4: CBPM based communication architecture

[4]Note that, this estimate includes the power consumed by all the components associated with a macro-state s_i. For example, when s_i represents a large volume data transfer, the estimate includes the power consumed by the master and slave components, bus interface logic, and bus lines.

occur at macro-state boundaries, and are requests to transition to the next macro-state, and (ii) Type II requests: these represent conventional bus access requests. Type I requests (which occur at macro-state time-scales), are handled by the CBPM circuitry. Type II requests occur at small time-scales (potentially every bus cycle) and are handled by the underlying bus protocol (*e.g.*, priorities, TDMA, *etc*). Multiplexers are used to route requests to the appropriate logic in the arbiter. The CBPM circuitry includes the following:

- Macro-state registers that store a current and pending macro-state identifier for each component;
- Power LUT that stores macro-state power table;
- Criticality LUT that stores the macro-state criticality table;
- CBPM controller, which implements the arbitration algorithm for Type I requests. The algorithm is based on the policy defined in **Step 5**. At each arbitration, the total power consumed by the currently executing macro-states is calculated. This is obtained either by (i) using fast adders to add power numbers obtained by looking up the Power LUT, or (ii) using LUTs which store precomputed results for commonly occurring macro-state combinations. In parallel, the set of pending macro-states are sorted into descending order of criticality by looking up the Criticality LUT. Starting with the most critical, the controller grants pending states till either (i) there are no more pending macro-states, or (ii) granting another macro-state violates the power constraint.
- Settable parameters are stored in a register to provide for run-time configuration of the CBPM policy. For example, the power constraint P_{max} can easily be made run-time settable, permitting dynamic flexibility in trading off performance for battery life.

5. EXPERIMENTS

In this section, we describe (i) the IEEE 802.11 MAC processor system used for our experiments (full details of our implementation are available in [24]), (ii) the experimental methodology, and (iii) experimental results, including battery versus performance trade-offs, comparison with conventional power management techniques, and HW implementation.

5.1 IEEE 802.11 MAC Processor System

HW/SW Architecture: The MAC processor architecture (Figure 5(b)) consists of two embedded SPARCLite CPUs [25], 3 HW co-processors, 2 memories, and 3 queue buffers. The component `llc_c` is a HW implementation of the Logical Link Control Interface (LLC) task. It receives frames from the LLC layer, writes them in `mem_1`, and enqueues frame addresses in `Queue1`. Addresses are read from the queue by (i) `wep_c`, a HW co-processor implementing the Wired Equivalent Privacy (WEP) task, which encrypts frame data, and (ii) `sparc_1`, which computes an Integrity Checksum Vector (ICV) for each frame. The ICV and WEP tasks proceed in parallel. Encrypted frames are written to `mem_2`, and frame addresses to `Queue2`. The embedded CPU `sparc_2` implements tasks HDR, which generates the MAC header, FCS, which checksums the entire frame, and MAC_CTRL, which implements CSMA/CA, the channel access algorithm [20]. Once processing is complete and the frame can be transmitted, `pli_c`, a HW implementation of the Physical Layer Interface (PLI) dequeues frame addresses from `Queue3`, retrieves frames from `mem_2`, and passes them to the physical layer hardware.

System-level Communication Architecture: The communication architecture consists of dedicated channels for low volume communications (*e.g.*, synchronization between `sparc_1` and `wep_c`), and two 32-bit shared busses for data transfers between system components. The busses are interconnected by a `bridge`, and a static priority based protocol [26] is employed on each bus. Components `llc_c, wep_c, sparc_1`, and the `bridge` contend for access to `bus_1`, while components `bridge`, `sparc_2` and `pli_c` contend for access to `bus_2`. Note that, the CBPM methodology is general, and can be applied to systems with other communication architectures as well.

5.2 Experimental Methodology

The MAC processor was designed in the POLIS [27] environment, using a combination of Esterel and C to specify the system tasks. For performance analysis, we used a tool described in [22], while system power profiling was performed using a modified version of the power estimation framework described in [21]. On average, three macro-states were identified for each component. For example, the identified macro-states for `wep_c` were `wep_init` (state array initialization), `wep_mem` (block transfer of data from memory) and `wep_comp` (frame encryption). Other macro-states are listed in Figure 1(c,d). Battery life estimation was performed using a stochastic model of a Lithium-ion battery [9], with standard capacity 250 *mAh*, at a rated current of 125 *mA* (typical values for batteries commonly found in cell phones, PDAs *etc* [28]). Simulations were driven by actual 802.11 traffic, obtained by running ETHEREAL [29], a packet capture software on a laptop PC connected to a wireless LAN, under different applications. Battery efficiency was measured in terms of battery capacity and lifetime. Performance (maximum data rate) was estimated using the PTOLEMY simulation environment [30].

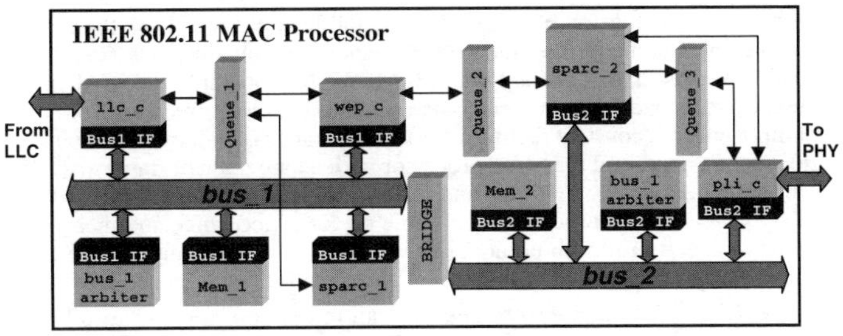

Figure 5: IEEE 802.11 MAC Processor Architecture

5.3 Battery Efficiency and Performance Trade-Off

To evaluate the effectiveness of the CBPM techniques, we experimented with different configurations of the MAC processor, each with a different setting of the power constraint. Varying the power constraint effectively changes the subset of macro-states that the CBPM policy chooses to grant. Figure 6 shows (i) battery capacity (*mAh*) as well as (ii) data rate (*Kbps*) of the MAC processor under different configurations of the CBPM policy. The trace consists of 4818 frames generated by 1.5 minutes of streaming video over an 802.11 based wireless network on a laptop running the Microsoft Windows Media Player [31]. From Figure 6 we can make the following observations:

- The CBPM based MAC processor shows a substantial increase in battery efficiency. Battery lifetime improvements from 3209.2 seconds to 10199.9 seconds are observed (3.2X), while more importantly, capacity improves from 154.3 *mAh* under configuration 0 (CBPM disabled) to 225.4 *mAh* under configuration 6 (1.5X).
- Improvements in battery capacity come at a price in terms of system performance. For example, configuration 6 provides for an improvement in battery lifetime of 3X, while degrading the data rate by 2.3X. However, as shown in the next experiment, the performance/battery trade off resulting from CBPM is significantly superior to what can be achieved using conventional power management techniques.

5.4 Comparison with other Power Management Techniques

For this experiment, we consider a design of the MAC processor that supports clock frequency scaling, a commonly used power management

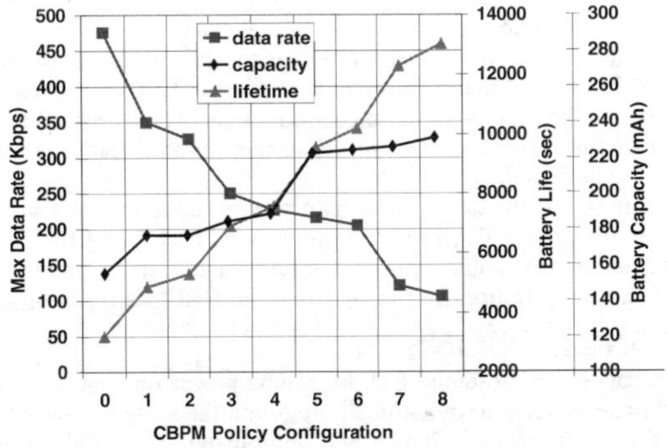

Figure 6: Battery efficiency *vs* performance for different configurations of CBPM policy

technique for extension of battery life in mobile appliances [12], the objective being to compare the trade-off characteristics obtained via frequency scaling versus those obtained via CBPM. We first operated the MAC processor over a wide performance range, scaling the clock frequency of each component by a constant factor over a discrete range of values. For each frequency, we exercised the MAC processor with a long trace of frame arrivals, and evaluated battery lifetimes and battery capacity. Next, we operated the CBPM based MAC processor over same performance range and measured battery capacity for each point. The two trade-off curves thus obtained are shown in Figure 7, based on which we make the following observations:

- CBPM is superior to frequency scaling as a tool for trading off battery efficiency for performance, because (i) it incurs a smaller performance penalty than frequency scaling for equivalent improvements in capacity, and (ii) it provides more battery-efficient solutions for equivalent performance. For example, to improve the battery capacity from 128 mAh to 182 mAh (an improvement of 42%), frequency scaling incurs a performance degradation of 22%, while that of the CBPM based system is only 7%.
- A major problem with the frequency scaling approach in improving battery efficiency is observed. In some cases, as the frequency is reduced, (decreasing performance), battery capacity actually decreases. This occurs because, decreasing the clock frequency results in the overlapping of various power hungry macro-states (which were disjoint at higher frequencies), leading to rated current violations. No such anomalies are observed with the CBPM trade-off curve.

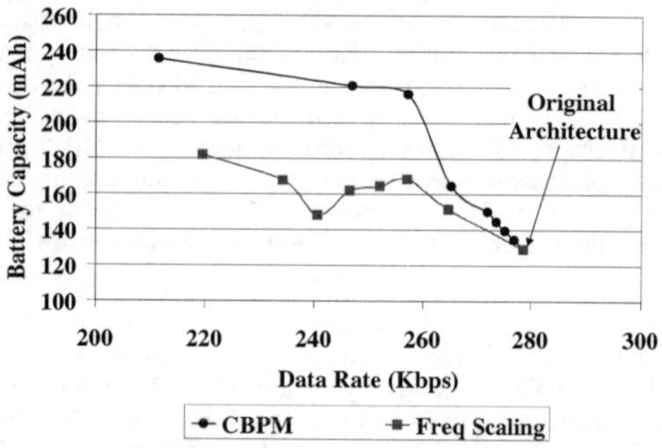

Figure 7: Comparison of the trade-off characteristics of the CBPM policy with frequency scaling

The experiment highlights the benefit of using a battery driven power management policy rather than one that merely aims at reducing average power.

A common dynamic power management (DPM) technique is idle shut down, where idle components are forced into a low power state. Note that, our original architecture (without CBPM) already incorporates DPM. Hence, the reported battery efficiency of the CBPM based MAC processor is over and above what is achievable using DPM.

5.5 HW Complexity

The CBPM enhanced bus arbiter was synthesized and mapped to a commercial 0.18μ standard cell library [32]. The area of the CBPM enhanced arbiter was found to be $15014\mu^2$, a 56% increase over the area of the original arbiter. To investigate the impact of this at the system-level, we estimated the total chip area by using fast synthesis for the HW co-processors, and data sheets for the memory and embedded CPUs. The overhead at the system level due to CBPM enhancements to the communication architecture was found to be less than 0.5%.

The delay of the enhanced bus arbiter was measured to be 4.6ns, implying that single cycle CBPM arbitrations can be performed for bus speeds upto 218Mhz. Significantly higher bus speeds can be supported by supporting multi-cycle arbitration for the less frequent CBPM arbitrations.

6. CONCLUSIONS

In this paper, we presented CBPM, a new power management methodology that can be used to dynamically regulate the system power profile to ensure efficient battery discharge, and demonstrated its application to the design of a battery-efficient IEEE 802.11 MAC processor. Results of our experiments indicate that CBPM can be used to design systems that are significantly more battery efficient than those based on conventional power management.

7. REFERENCES

[1] S. H. Gunther, F. Binns, D. M. Carmean, and J. C. Hall, "Managing the Impact of Increasing Microprocessor Power Consumption." Intel Technology Journal, First Quarter, 2001, available online at http://developer.intel.com/technology/itj/

[2] I. Buchmann, "Batteries in a Portable World." http://www.cadex.com.

[3] K. Lahiri, A. Raghunathan, S. Dey, and D. Panigrahi, "Battery driven system design: a new frontier in low power design," in *Proc. ASP-DAC/Int. Conf. VLSI Design*, pp. 261–267, Jan. 2002.

[4] A. R. Chandrakasan and R. W. Brodersen, *Low Power Digital CMOS Design*. Kluwer Academic Publishers, Norwell, MA, 1995.

[5] J. Rabaey and M. Pedram (Editors), *Low Power Design Methodologies*. Kluwer Academic Publishers, Norwell, MA, 1996.

[6] L. Benini and G. De Micheli, *Dynamic Power Management: Design Techniques and CAD Tools*. Kluwer, 1998.

[7] M. Pedram and Q. Wu, "Design considerations for battery-powered electronics," in *Proc. Design Automation Conf.*, pp. 861–866, June 1999.

[8] D. Rakhmatov and S. B. K. Vrudhula, "Time to failure estimation for batteries in portable systems," in *Proc. Int. Symp. Low Power Electronics & Design*, pp. 88–91, Aug. 2001.

[9] D. Panigrahi, C. F. Chiasserini, S. Dey, R. R. Rao, A. Raghunathan, and K. Lahiri, "Battery life estimation for mobile embedded systems," in *Proc. Int. Conf. VLSI Design*, pp. 55–63, Jan. 2001.

[10] L. Benini *et.al.*, "A discrete-time battery model for high-level power estimation," in *Proc. Design Automation & Test Europe Conf.*, pp. 35–39, Mar. 2000.

[11] M. Doyle, T. F. Fuller, and J. S. Newman, "Modeling of galvanostatic charge and discharge of lithium/polymer/insertion cell," *J. Electrochem. Soc*, vol. 140, pp. 1526–1533, June 1993.

[12] T. L. Martin, *Balancing Batteries, Power and Performance: System Issues in CPU Speed-Setting for Mobile Computing*. PhD thesis, Department of Electrical and Computer Engineering, Carnegie Mellon University, 1999.

[13] "Mobile Intel Pentium III Processor featuring Intel SpeedStep (TM) Technology Performance Brief." http://developer.intel.com/design/mobile/

[14] J. Luo and N. K. Jha, "Battery-aware static scheduling for distributed real-time embedded systems," in *Proc. Design Automation Conf.*, pp. 444–449, June 2001.

[15] L. Benini *et.al.*, "Battery-driven dynamic power management," *IEEE Design & Test of Computers*, vol. 18, pp. 53–60, Apr. 2001.

[16] L. Benini *et.al.*, "Extending lifetime of portable systems by battery scheduling," in *Proc. Design Automation & Test Europe Conf.*, pp. 197–201, Mar. 2001.

[17] C. F. Chiasserini and R. R. Rao, "Energy Efficient Battery Management," *IEEE J. on Selected Areas in Communications*, vol. 19, pp. 1235–1245, July 2001.

[18] Q. Wu, Q. Qiu, and M. Pedram, "An interleaved dual-battery power supply for battery powered electronics," in *Proc. Asia and South Pacific Design Automation Conference (ASP-DAC)*, pp. 387–390, Jan. 2000.

[19] "Smart Battery System Implementers Forum (http://www.sbs-forum.org).".

[20] "Wireless LAN Medium Access Control (MAC) and Physical Layer (PHY) Specifications." IEEE Computer Society LAN MAN Standards Committee, IEEE Std 802.11-1999 Edition.

[21] M. Lajolo, A. Raghunathan, S. Dey, and L. Lavagno, "Efficient Power Co-Estimation Techniques for System-on-Chip Design," in *Proc. Design Automation & Test Europe Conf.*, pp. 27–34, Mar. 2000.

[22] K.Lahiri, A.Raghunathan, and S.Dey, "System level performance analysis for designing on-chip communication architectures," *IEEE Trans. Computer-Aided Design*, vol. 20, pp. 768–783, June 2001.

[23] A. Silberschatz, P. Galvin, and G. Gagne, *Operating System Concepts*. Wiley and Sons, 2001.

[24] K. Lahiri, A. Raghunathan, and S. Dey, "Battery efficient architecture for an 802.11 MAC processor," in *Proc. IEEE Int. Conf. on Communications*, Apr. 2002.

[25] Fujitsu Microelectronics Inc., "Sparclite series mb86934 addendum." 1994.

[26] "Peripheral Interconnect Bus Architecture." http://www.omimo.be.

[27] F. Balarin *et.al.*, *Hardware-software Co-Design of Embedded Systems: The POLIS Approach*. Kluwer Academic Publishers, Norwell, MA, 1997.

[28] "Panasonic Lithium Ion Batteries: Individual Datasheet CGR 17500." http://www.panasonic.com/industrial/battery/oem/chem/lithion/

[29] "Ethereal 0.8.18." http://www.ethereal.com.

[30] J. Buck *et.al.*, "Ptolemy: A framework for simulating and prototyping heterogeneous systems," *International Journal on Computer Simulation, Special Issue on Simulation Software Management*, vol. 4, pp. 155–182, Apr. 1994.

[31] "Microsoft Windows Media Player." http://www.microsoft.com/windows/windowsmedia/

[32] "UMC 0.18 μ Copper Process Low-Voltage SAGE(TM) Standard Cell Library." http://www.umc.com/english/process/

Scheduler-Based DRAM Energy Management

V. Delaluz, A. Sivasubramaniam, M. Kandemir, N. Vijaykrishnan and M. J. Irwin
The Pennsylvania State University, University Park, PA 16802

ABSTRACT

Previous work on DRAM power-mode management focused on hardware-based techniques and compiler-directed schemes to explicitly transition unused memory modules to low-power operating modes. While hardware-based techniques require extra logic to keep track of memory references and make decisions about future mode transitions, compiler-directed schemes can only work on a single application at a time and demand sophisticated program analysis support. In this work, we present an operating system (OS) based solution where the OS scheduler directs the power mode transitions by keeping track of module accesses for each process in the system. This global view combined with the flexibility of a software approach brings large energy savings at no extra hardware cost. Our implementation using a full-fledged OS shows that the proposed technique is also very robust when different system and workload parameters are modified, and provides the first set of experimental results for memory energy optimization with a multiprogrammed workload on a real platform. The proposed technique is applicable to both embedded systems and high-end computing platforms.

Categories and Subject Descriptors

D.4.0 [**Operating Systems**]: General; D.4.1 [**Operating Systems**]: Process Management—*Scheduling*; B.3.1 [**Memory Structures**]: Dynamic Memory (DRAM)

General Terms

Management, Design, Experimentation

Keywords

Energy Management, DRAM, Operating Systems, Scheduler, Energy Estimation.

1. INTRODUCTION AND MOTIVATION

Optimizing power consumption of computing components has become as important as performance. Reducing the overall energy expended in sustaining computing demands on embedded devices is commonly accepted as being important for conserving battery energy. At the same time, power optimization is also taking center-stage on high end systems due to thermal and packaging considerations. To quote a recent Intel press release [7], 'Power and heat are the biggest issues for this decade'. Energy management is important at different levels of the system architecture. Further, hardware and software power optimization techniques are both equally important, and often go hand-in-hand to complement each other. Adhering to this philosophy, this paper presents a software-based strategy that can be easily incorporated in the operating system, to optimize the energy consumption of DRAMs by exploiting their hardware capabilities of operating in different power modes.

Energy optimization is important for all components including the processor datapath, caches, memory, peripherals, and buses [2]. Optimizing just one may shift the bottleneck to another component. Realizing this, Intel has recently announced a target power budget for each component as a percentage of overall system consumption [6]. Our focus in this paper is on the DRAM energy consumption, whose target consumption in [6] has put a limit of not more than 5% of overall system consumption. It has been observed [2, 8, 16, 18] that the memory system is a dominant consumer of the overall system energy and is an important candidate for software and hardware optimizations. This is especially true for many embedded applications that are typically memory-intensive (i.e., array-dominant such as signal and video processing). In addition, applications are gradually becoming more data-centric with stringent memory requirements (for both storage and speed), causing vendors to incorporate large storage capacities into their offerings. Typically, a computer system contains several DRAM chips (organized in rows/banks and columns), with each of them consuming power even if it is not being currently used. It would be extremely valuable to explore techniques for selectively transitioning the unused memory modules into lower energy consumption modes whenever possible.

As with other hardware components, DRAM modules have started providing power mode control capabilities [13, 14], where individual modules can be selectively transitioned into a low-power operating mode when inactive. There is a performance (and perhaps energy) cost to be paid in exiting this low-power operating mode and coming back to the active mode before a memory request can be serviced (called *resynchronization cost*). In the hope of exploiting these mode control capabilities, there has been recent interest in hardware-based techniques [3, 9] and compiler-directed techniques [3] to explicitly transition individual modules to a low-power operating mode during periods of idleness.

However, each of these approaches has its own limitations. With the compiler-directed approach, we are limited to optimizations for statically analyzable situations, and have to often resort to conservative options. Further, these approaches can reach decisions based on reference patterns for only one program at a time, and do not have information on a more global scale, e.g., across processes running on that system. While we can be much more aggressive/optimistic (perhaps overly optimistic in some circumstances) in hardware-based solutions, there is the cost of the additional hardware that is required to keep track of historical access patterns and to reach decisions based on these patterns.

In this paper, we present an alternate software-based solution, where the transitioning is effected by the operating system (OS).

The operating system has the physical frame usage information across the processes, and can often base its frame allocation policies to each process for energy optimization. Further, it has this information on a global scale, across the processes. This information can be used by the operating system to effect mode transitions at appropriate points when it has a good idea that a module will not be needed for a certain period. For example, if the OS can predict that a given module will not be used by the next process to be scheduled, the module can be transitioned into a low-power operating mode. This prediction can be made through information that can be gleaned by the OS by examining the reference bits in the TLB/page-table. If this turns out to be inaccurate, a resynchronization cost would be paid. Tracking resource access patterns across processes within the operating system has been used in prior research [12] in the context of powering down peripherals mainly, and this is the first investigation of such issues for DRAM power-mode control. Previous OS-based power management studies also considered voltage and frequency scaling [1, 4, 5, 11, 15, 17].

DRAM mode control by the operating system at re-scheduling points attempt optimization at a much coarser granularity than a compiler or a pure hardware based approach, where transitions can potentially be effected at memory access granularities. However, it is our belief (and our experimental results do confirm this) that a substantial portion of the energy savings that those techniques can provide can be attained with our OS scheduler-based mode transition strategy, without requiring extensive compilation support, or any additional hardware support (other than mode control capability). In fact, we go even further to show that if there is additional hardware support such as those that have been proposed in prior research [3, 9], our scheduler-based mode control strategy can be used in conjunction with these schemes to further the energy savings.

Previous studies [3, 9] that have attempted DRAM power optimization have used a simulation-based strategy, using a set of applications running in isolation (dedicated mode). In contrast, this work conducts an evaluation on an actual platform, running a full-fledged operating system with a multiprogrammed workload. This study modifies an off-the-shelf open-source operating system (Linux Redhat 6.2) running on Sun Sparc hardware, to incorporate our energy optimizations. Further, memory references from a multiprogrammed workload running on this system are traced. The Linux scheduler has also been augmented to track the idle times for different modules with our mode-control mechanism. Using several applications on this experimental platform, this paper makes the following contributions:

- We present the first results for the energy savings through power-mode control for a multiprogrammed workload and a real operating system during an actual execution.
- We demonstrate that our scheduler based (software-only) DRAM power mode control strategy can provide the bulk of the energy savings of a pure hardware based approach, without requiring the extra hardware.
- We also show that our scheduler based strategy, can further the energy savings even if a hardware based DRAM energy optimizer is available.

The rest of this paper is organized as follows. Section 2 reviews basic concepts related to DRAM power-mode management and summarizes important characteristics of hardware-based power-mode management strategies. Section 3 presents details of our scheduler-based power-mode management strategy. Section 4 introduces our experimental methodology and presents our results. Section 5 summarizes our major conclusions.

2. MEMORY ARCHITECTURE AND LOW-POWER OPERATING MODES

We focus on an architecture where the memory system is composed of multiple memory modules organized into banks (rows) and columns. Accessing a data in such an architecture would require activating the corresponding modules. There are several ways of saving power in such a memory organization. The approach adopted in this paper is to put the unused memory banks into one of several low-power operating modes. In all our experiments, we use one module per bank; consequently, the terms *bank* and *module* are used interchangeably.

	Energy Consumption (nJ)	Re-synchronization Time (cycles)
Active	3.570	0
Standby	0.830	2
Nap	0.320	30
Power-Down	0.005	9,000

Figure 1: Energy consumption and resynchronization times for different operating modes.

Each memory bank operates independently, and when not in active use, it can be placed into a *low-power operating mode* to conserve energy. Each low-power operating mode works by activating only specific parts of the memory circuitry such as column decoders, row decoders, clock synchronization circuitry, and refresh circuitry (instead of all parts of the circuitry) [13]. The model for our memory system associates an energy consumption value per cycle for four different operating modes together with a resynchronization time that is needed to bring the module back to the active state (which is the only state where a memory request can be serviced). The energy consumptions and resynchronization times for our operating modes are given in Figure 1. Earlier studies [3, 9] have used similar models for the DRAM modules. The energy values shown in this figure have been obtained from the measured current values associated with memory modules documented in memory data sheets (for a 3.3 V, 2.5 nanoseconds cycle time, 8 MB memory) [13]. The re-synchronization times are also obtained from data sheets. Based on trends gleaned from data sheets, the energy values increase by 30% when module is doubled in size.

A bank can be placed into a low-power mode using a hardware-based, software-based, or a hybrid approach [3, 9]. Previous research [3] has shown that hardware-based schemes, particularly a history-based mechanism (called HBP), achieves the best energy savings. To demonstrate that our approach can give energy savings close to those obtained through sophisticated hardware mechanisms, we compare our proposed scheduler-based strategy with two previously proposed hardware based mode control schemes — called Constant Threshold Predictor (CTP) and History Based Predictor (HBP).

Hardware-based mode control mechanisms, in general, monitor the memory references to each DRAM bank, and based on the history, effect appropriate power-mode transitions. The rationale behind CTP is that if a memory bank has not been accessed in a while, then it is not likely to be accessed in the near future (that is, inter-access times are predicted to be long). A (constant) threshold is used to determine the idleness of a bank, and if the next access to this bank does not come within this time, the bank is transitioned to a lower power mode. As in previous studies (e.g., [3]), after 10 cycles of idleness, the corresponding bank is put in standby mode in our experiments. Subsequently, if the bank is not referenced for another 100 cycles, it is transitioned into the napping mode. Finally, if the bank is not referenced for a further 1,000,000 cycles, it is put into power-down mode. Whenever a bank in the low-power mode is referenced, it is brought directly back into the active mode incurring the corresponding re-synchronization costs (based on what mode it was in). HBP, as discussed in [3], estimates the next inter-access time for a bank to be the same as the previous inter-access time. With this estimate, it calculates what would be the ideal power mode that it can transition the bank to after the current access, and at what time it should bring it back to the active mode (so that the processor does not see resynchronization overheads).

3. OUR APPROACH

Adhering to the philosophy that software-based solutions are more flexible, and less expensive, we propose a power-mode control strategy within the OS. The OS not only has physical frame allocation information for each executing process, but also has access to those that are actually being referenced (by sampling the reference bits in the page table/TLB). Further, it has this information not just for one process, but across all processes. Based on this information, it can determine points in the system execution when one or more banks may remain idle for extended periods, and consequently power them down.

We propose performing such power-mode transitions within the OS scheduler, when context switching from one process to the next. With time quantums (the period of time that a process normally runs before being preempted) typically being tens of milliseconds, one can hope to get substantial energy savings if a bank is not used during a time quantum. Also, if a bank is used only by a single process, additional energy savings can be obtained by placing the bank in a low-power mode until the next time quantum allocated to that process. Further, if the degree of multiprogramming increases (i.e., the number of process pseudo-concurrently executing), then the time gap between successive quanta for a process would further go up, increasing the window of opportunity for energy savings with this strategy. If we are able to track what banks are actually being used, then we can expect the same set of banks to be used the next time this application is scheduled. The other banks can be put into low-power mode.

Based on this principle, we describe a scheduler-based power-mode control scheme, that tracks memory banks that are being used by different applications and selectively turning on/off these banks at context switch points.

3.1 Bank Usage Table

The OS tracks memory bank usage of applications with a Bank Usage Table (BUT). Figure 2 shows an abstract view of a BUT which gives the bank usage information for `n` banks (`B1` through `Bn`) and `m` processes (daemons would also be included in this table). An entry in ith row and jth column marked using × in this table indicates that process i used bank j the previous time it was scheduled. This information can help us in selecting which banks to transition to a low-power mode when the scheduler picks this process the next time. We next discuss how to mark the × in the BUT.

When a process is selected to run, we put the banks that do not have an × into *power-down* mode (see Figure 1), and turn on the ones (to *active*) that do. The × markers for that row are then wiped out. We next need to find out whether for the current quantum, the process actually refers to a given bank. This can be achieved in different ways:

• Whenever the operating system schedules a process, it can wipe out the permissions (read/write/execute) of its virtual pages that have been mapped in (to frames in memory banks) to invalid in the page table, so that any reference to a virtual page by the process would result in a fault (segmentation fault in this case). Note that pages that do not have a physical frame mapping will page fault anyway. At the time of the fault (whether a segmentation or page fault), the OS can update the corresponding column of the BUT based on where the physical frame resides. It can also give back the permission for the remaining virtual pages mapping into the same physical bank so that references to those pages do not fault again in the same time quantum. With this approach, there is at most 1 extra fault per bank — which does not incur the disk overheads but only updates permissions in the page table entry — to track the BUT within each quantum. Further, setting page table entry permissions before scheduling a process is not a significant overhead (e.g., for an 8MB memory bank with 8KB pages, we need to update at most 8MB/8KB (= 1024) entries per memory bank).

• Instead of taking faults, the OS could use reference bits (that are maintained for each virtual page in the page table) to track what is being referenced. Before scheduling a process, the OS could wipe out the reference bits of all pages that have been physically mapped in. When context switching out this process, it could re-

Process Id	B1	B2	B3	...	Bn
1		×	×	×	
2	×	×			×
3				×	×
4					×
5	×	×			×
...	...	...	...	...	...
m	×		×		×

Figure 2: Bank usage table (abstract view).

examine all these reference bits and mark the corresponding columns of the BUT for whatever banks are referenced. Note that the page replacement algorithm (second chance LRU in Linux) also samples these reference bits, resetting them and moving on to the next candidate when searching for the physical frame. This code also needs to be extended to update the BUT when resetting the reference bits. While this approach avoids any extra page faults, it can affect the approximate LRU page replacement algorithm unless that is also made aware of the BUT updates that are taking place.

• The last approach that we identify is actually not an exact mechanism, but tries to approximate bank usage pattern. The advantage with this approach is that it can be much more efficient than the others. In this scheme, when a process is being context switched out, we simply mark the banks in the BUT for which there is an entry referencing a physical frame in that bank in the TLB. If the TLB does a good job of capturing the application's working set, then this scheme is ideally suited for capturing bank access behavior. The only situations when this approach fails to mark all banks being accessed is when the working set is larger, and there is no TLB entry associated with a bank that is currently being used. In this case, TLB misses need to be handled in software (as in our experimental platform)

All these schemes are quite involved, requiring extensive modifications of the operating system. While a comprehensive experimental evaluation of all these approaches is really needed to evaluate their pros and cons, in this study we use the first approach to mark the BUT table entries since our focus here is mainly to demonstrate the potential of our scheduler-based DRAM mode control strategy. All the energy and performance overheads for implementing this approach are reflected in our experimental results.

3.2 Evaluation Strategy

Our current implementation and evaluation platform uses an actual Sun UltraSparc 5, running Linux RedHat 6.2, kernel version 2.2.14. The physical memory available on this machine is 128 MB, and we can give parameters to the system specifying the bank organization. The default number of banks used in our experiments is sixteen (each of which is 8 MB), and the default OS time quantum is 200 milliseconds. However, we also studied the sensitivity of our energy savings to the bank organization.

We have implemented the BUT table maintenance within the Linux kernel by modifying the scheduler and page fault handler. Instead of having a separate global table, we kept each row of the table in the corresponding Linux *task* structure [19] for that process. We also enhanced the page allocation/faulting mechanism to keep track of banks that are being referenced by a process.

As was mentioned earlier, this is the first study to use an actual implementation-based approach for evaluation purposes, considering a full-fledged system operating in a multiprogrammed fashion. While this gives us a very realistic evaluation, one of its drawbacks is the inability to be able to track energy savings/consumption. For our energy calculations, we dump traces of memory references of the system (not just for one application, but across applications) and then use a post-processing strategy to compute memory-energy consumption (based on the memory model discussed earlier). We have ensured that the trace collection facility does not itself perturb the results significantly by using extensive buffers for accumulating logs and flushing them to disk rather infrequently. The memory reference trace contains whether the reference is a load or store, the

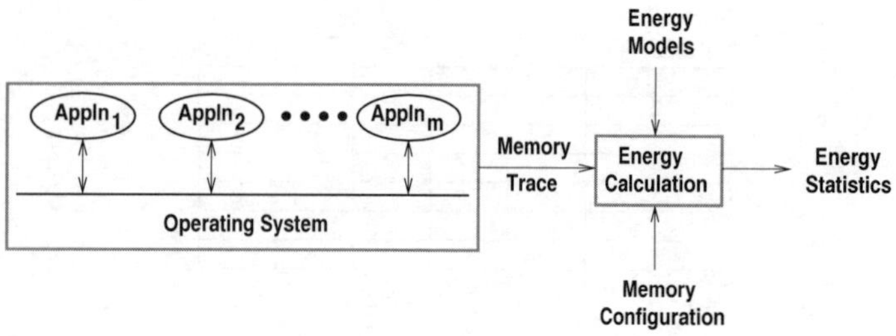

Figure 3: Trace collection and DRAM energy computation.

virtual page numbers, the corresponding physical frame number, and the timestamp of the reference. The timestamp was generated using the high resolution TICK register available on the UltraSparc processor. A schematic of our overall evaluation strategy is given in Figure 3.

3.3 Advantages and Disadvantages

Having presented our approach, we now qualitatively discuss its pros and cons, specifically in relation to hardware-based strategies that have been shown to give good energy savings in prior research [3].

Hardware-based approaches such as CTP or HBP require additional hardware that needs to monitor the ongoing accesses to each bank and then additional logic to make predictions based on this history. This not only incurs additional cost, but may itself consume some amount of energy. Another important concern with hardware-based schemes is that they are not very flexible, and it is not very easy to customize or adapt them based on the current environment/application.

Our scheduler based software strategy, on the other hand, does not require any additional hardware, except the mode transitioning mechanisms that DRAM chips provide for power management. It also offers us the flexibility of determining what banks to power manage, when to initiate transitions, and can adapt to application behavior dynamically. It can also track DRAM usage patterns across applications to base its decisions. For instance, let us say a process that has been accessing certain banks solely dies. The OS can, perhaps, initiate transitions for those banks right away, instead of incurring some delays which are necessary in certain hardware based schemes. In addition, the OS knows when the process is going to be scheduled again after the current epoch. Consequently, to avoid resynchronization costs, it can bring the corresponding banks to active state before the process may need them. Our current implementation does not, however, perform this optimization. An earlier study [12] has also pointed out the benefits of taking actions at the operating system level using task information rather than micromanaging the resource at the device level.

There is a downside to the scheduler-based strategy, which is a consequence of the granularity at which we operate. The mode control decisions are made at milliseconds granularity (at context switch points). On the other hand, hardware-based approaches operate at a much finer resolution, and can effect transitions at an access granularity. However, we think that, in the big picture, extended periods of idleness is what really matters, and our scheduler-based strategy is expected to exploit those periods.

We would like to point out that there is nothing to prevent our scheduler-based approach to be used in conjunction with the hardware-based mechanisms. With such an integrated approach, the scheduler manages the power consumption between one quanta of execution and the next, while the hardware-based mechanism manages the power consumption within each quanta.

We pictorially illustrate the pros and cons of the hardware-based (CTP in particular in this example), scheduler-based and integrated strategies in Figure 4. In this example, we assume two application processes, and only one memory bank being accessed (process 2 does not access this bank during its time quantum). The resynchronizations activities are not explicitly shown for the sake of clarity.

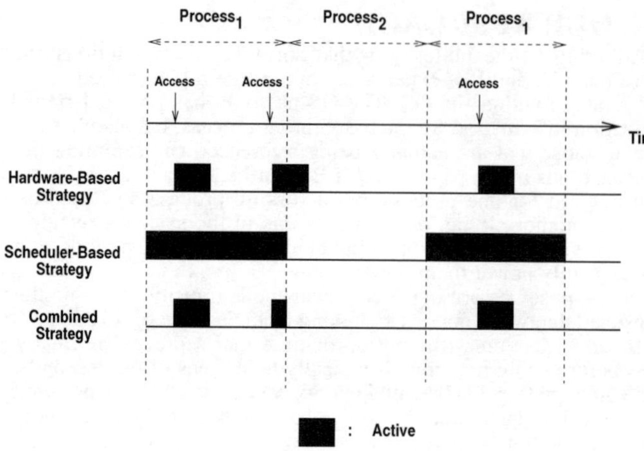

Figure 4: Comparison of hardware-based, scheduler-based, and combined strategies.

Benchmark	Source	Input	Energy (mJ)
`rawcaudio`	MediaBench	clinton.cpm	123.2
`rawdaudio`	MediaBench	clinton.adpcm	112.0
`polyphase`	MediaBench	polyphase.INT polyphase.INT.COE	85.7
`md5`	MediaBench	-	40.4
`cordic`	MediaBench	-	44.2
`paraffins`	Trimaran	15 entries	16.5
`g721encode`	MediaBench	clinton.pcm	6063.4
`mcf`	Spec2000	inp.in	3172.4

Figure 5: Benchmark characteristics.

The hardware-based strategy activates a bank upon an access, and following a threshold (after the access has been complete) the bank is transitioned into a low-power operating mode. Note that this threshold period could either be completely contained during the quantum for process 1, or some of it may spill over into the quantum for process 2. Note that this may not happen in the scheduler-based strategy since it can put this bank into a low-power mode when process 1 is preempted (if it knows that process 2 does not access this bank). However, it pays the penalty of keeping the bank active during the entire quantum of process 1. The advantage of hardware based vs. scheduler based schemes depends on whether the former or latter factor is more significant. On the other hand, the combined strategy can get the best of both worlds, and in this example gives the maximum energy savings as shown in the figure. This example has been given only to illustrate where each scheme gets its savings, and a realistic evaluation with actual applications is needed to really compare the schemes as is done next.

4. EXPERIMENTS AND RESULTS

4.1 Benchmarks

To evaluate the effectiveness of our scheduler-based strategy in saving energy, we used eight complete benchmarks. Six of these benchmarks are from MediaBench [10], a suite of applications that stream data. One of the remaining applications is `mcf`, a representative benchmark (which is very memory-intensive) from the SpecInt 2000 suite and the other is `paraffins` that comes with the Trimaran system. The salient characteristics of these benchmarks are summarized in Figure 5. The last column gives the *memory system energy consumption* for each benchmark when all 16 banks are kept in the fully active (normal operating mode) during the entire execution of the application (assuming no cache memory). All energy numbers reported in the rest of this paper are values normalized to those in the last column.

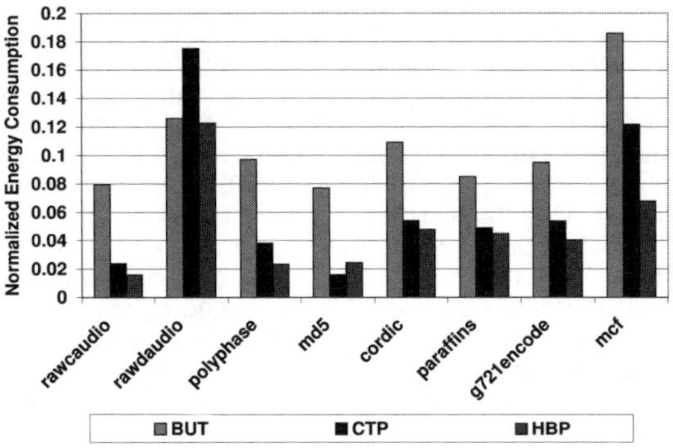

Figure 6: Normalized memory energy consumption.

4.2 Energy Savings and Performance with Single Application Execution

Figure 6 shows the normalized energy consumption for our benchmarks in an execution environment where only one application is running in the system with no cache memory present (results with a cache are presented later). Recall that there are always some daemons that execute. The energy savings obtained using CTP and HBP are also given. We can observe that CTP and HBP give very good energy savings, as was observed in previous studies. More noteworthy is the observation that our OS scheduler based strategy (denoted as BUT in the figure) is giving most of the energy savings provided by the hardware mechanisms. Please note that while the differences may look deceptively larger, the scale of the graph is quite amplified (for example, BUT gives 92% energy savings for `rawcaudio` while CTP/HBP give around 98-99% energy savings). More importantly, our strategy provides these benefits without incurring any additional hardware costs. This overhead of HBP scheme in particular can account to up to 10% additional energy consumption (consuming 241pJ every cycle) [3]. This overhead is not included in our comparison as it could vary based on the implementation. On the average, we have about 90% savings in DRAM energy with our scheduler-based strategy. We would like to point out that we did not put a bank into low-power mode during the execution of an OS daemon (while the hardware-based mechanisms could). Providing this enhancement can further lessen the differences between the hardware and our software strategies.

As mentioned earlier, we also considered the option of integrating our scheduler-based strategy with the hardware-based mechanism (both CTP and HBP), with the former providing savings across time quantum and the latter providing savings within a time quantum. We find that such a combined strategy can further the energy savings that are obtained with just the hardware-based mechanisms. More specifically, combining CTP with BUT improved the energy consumption of CTP by 22.7% and combining HBP with BUT improved the energy consumption of HBP by 73.0%.

We also found that the performance overhead of our scheduler-based approach is negligible (less than 1% for all benchmarks), assuming a resynchronization latency of 9000 cycles for the power-down mode. In another set of experiments, we increased the resynchronization time ten-fold, anticipating a large relative resynchronization latency for future memory chips (due to increasing microprocessor frequencies). We found that, even under this scenario, the increase in execution cycles is less than 7%.

It should be noted that this set of experiments gives the bias more towards the hardware-based schemes since there is not too much gap between successive time quanta for an application (the only activities between successive quanta for an application is the routine bookkeeping work for the OS). Even in this case, we find that our scheduler-based strategy comes close to the hardware mechanisms. In the rest of the discussion, we focus on our scheduler based strategy and examine its behavior for different workload and hardware parameters.

4.3 Energy Savings with Multiple Applications

We performed another set of experiments where we measured the effectiveness of our strategy under a multiprogrammed workload that imposes a higher demand on the memory system. No previous related study has studied this influence on the effectiveness of the DRAM power management schemes. To conduct this study, we simultaneously ran several instances of the same benchmark in the system. The results are depicted in Figure 7 for `mcf`, with the number of instances varying from 1 to 40. As the multiprogramming level increases, the time gap between successive executions of an application increases, thereby saving more energy (we observed that the OS page allocation provides some amount of insulation between processes by allocating them pages from different banks). However, beyond a certain point (4 in this case), the number of processes becomes so high that their memory demands cause a single bank to be shared by several processes. This is evidenced by the average number of banks used by an instance (process). We found that the average number of banks used at any instance increases from 3.4 when 10 instances are running to 7.6 when 40 instances are running. When an instance is scheduled, the scheduler ends up transitioning fewer banks into the low-power mode. This consequently reduces the energy savings at high multiprogramming levels, making the energy consumption increase almost linearly beyond 10 instances. Nevertheless, we observe that even with forty instances of the application concurrently running, we achieve nearly a 40% saving in memory energy consumption. Similar observations were made when we conducted tests with multiple instances of other benchmarks in our experimental suite.

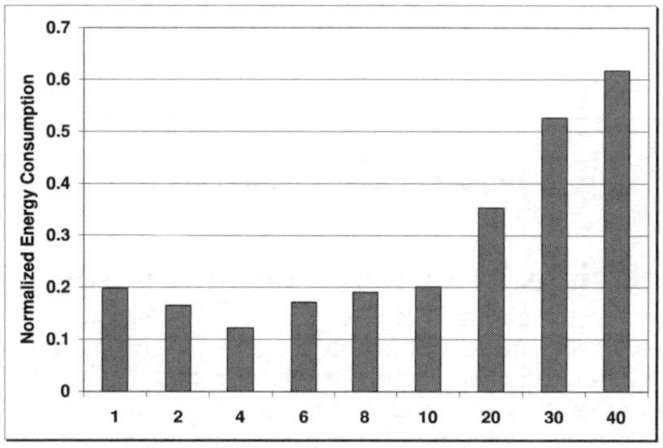

Figure 7: Normalized memory energy consumption (multiple instances of `mcf`).

4.4 Sensitivity to the Number of Banks

In the next set of experiments, we changed the number of banks in the system keeping total memory size fixed at 128MB. Recall that the original bank configuration has sixteen 8MB banks. We observe from Figure 8 that increasing the number of banks improves the energy savings. This is because a larger number of banks allows our approach to manage power-mode transitions for smaller portions of memory (at a finer granularity) and place more memory space into low-power operating mode when inactive. However, our other experiments with different number of banks also reveal that increasing the number of banks beyond a point does not increase the savings further. This is due to the fact that when a certain number of banks is reached, the access pattern of the application does not allow to transition more banks into the low-power mode, i.e., the working set spans that many banks. We also observed that this point is highly application-dependent. Another interesting observation from Figure 8 is that even with only four memory banks, our scheduler-based strategy achieves nearly a 50% saving in memory energy of `mcf`.

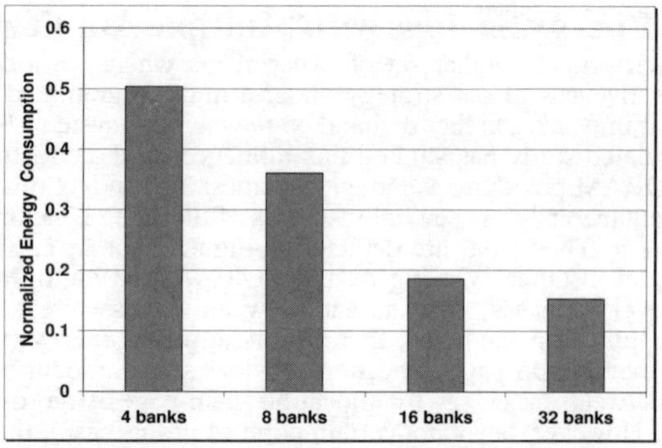

Figure 8: Sensitivity of energy savings to the number of banks (single instance of `mcf`).

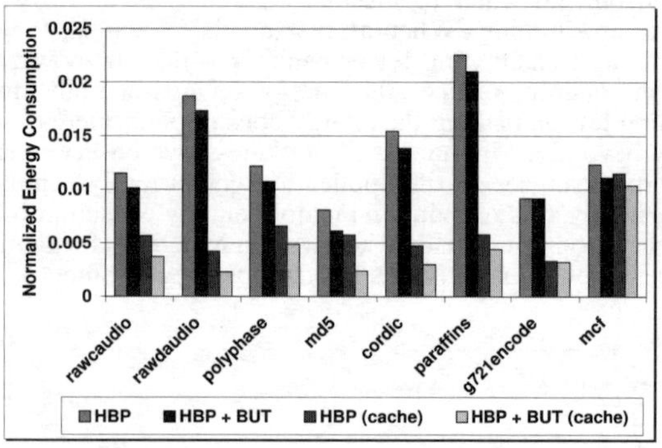

Figure 9: Impact of L1-L2 data cache hierarchy.

4.5 Savings in a Cache-Based Environment

We performed another set of experiments to measure the energy impact of power-mode control in a cache based environment. It should be noted that the behavior of our scheduler-based strategy is oblivious of the cache architecture in the system (if any). Therefore, we report results only for HBP and combined strategies. In this set of experiments, we use a two-way L1 data cache of 16 KB and a direct-mapped L2 cache of 32 KB. The line (block) sizes for L1 and L2 are 32 bytes and 128 bytes, respectively. We present the results in Figure 9. We see from these results that an L1-L2 cache hierarchy can help to improve the energy consumption further, as temporal locality helps the memory banks to transition to more aggressive energy-saving modes (the idleness duration for main memory banks increases). We see that for the `mcf` benchmark the additional savings obtained from the cache are not significant. This is because this benchmark exhibits poor locality, and consequently, has a high miss rate. For the other benchmarks, on the other hand, the improvement ranges from 47% (for `polyphase`) to 80% (for `rawdaudio`). We also observe from these results that even with a cache-based environment, our scheduler-based strategy helps increase the effectiveness of HBP in saving memory energy.

5. CONCLUSIONS

With power/energy optimization taking center-stage together with performance, it is becoming essential to optimize energy consumption at all levels of the system architecture. Such optimizations are not only important in embedded environments, but in high-end systems as well for reliability, packaging, and cooling constraints. Memory energy consumption has been pointed out as playing an important role in many applications, many of which are becoming data centric. At the same time, the small target budget for memory power [6] makes this an important hardware component to optimize using hardware and software techniques.

Earlier research on memory energy optimization has mainly used hardware mechanisms to detect bank idleness, and effect transitions to low-power modes accordingly. Apart from the inflexibility of such mechanisms, hardware cost is an important consideration.

Addressing these drawbacks, this paper has presented a new software mechanism for energy management within the operating system. We have proposed a scheduler-based mode transitioning mechanism for DRAMs that can keep track of banks referenced by different application processes, and use this information to power down banks that will not be referenced in the next time quantum during context switch points. Apart from this new proposal, we have also implemented and evaluated this mechanism on an actual Linux based system using realistic/multiprogrammed workloads. Performance results clearly indicate that our software-based strategy can give most of the savings of more expensive hardware-based schemes. Further, we have also shown that the scheduler-based mode control can even be integrated with an available hardware strategy to further the memory energy savings.

6. ACKNOWLEDGEMENTS

This work was supported in part by NSF grants 0082064, 0093082, 0093085, 0103583, 9701475, 9988164, 0130143 and by GSRC and PDG grants.

7. REFERENCES

[1] L. Benini, A. Bogliolo, and G. D. Micheli. A survey of design techniques for system-level dynamic power management. *IEEE Transactions on VLSI Systems,* 8(3), June 2000.

[2] F. Catthoor, S. Wuytack, E. D. Greef, F. Balasa, L. Nachtergaele, and A. Vandecappelle. *Custom Memory Management Methodology – Exploration of Memory Organization for Embedded Multimedia System Design.* Kluwer Academic Publishers, June 1998.

[3] V. Delaluz, M. Kandemir, N. Vijaykrishnan, A. Sivasubramaniam, and M. J. Irwin. Hardware and software techniques for controlling DRAM power modes. *IEEE Transactions on Computers,* November 2001 (Vol.50,No.11).

[4] K. Govil, E. Chan, and H. Wasserman. Comparing algorithms for dynamic speed-setting of a low-power CPU. In *Proc. the ACM International Conference on Mobile Computing and Networking,* pages 13–25, 1995.

[5] I. Hong, D. Kirovski, G. Qu, M. Potkonjak, and M. B. Srivastava. Power optimization of variable voltage core-based systems. In *Proc. Design Automation Conf.,* 1998.

[6] Intel announcement. http://developer.intel.com/design/mobile/intelpower/int_mpg.htm.

[7] Intel announcement. http://www.intel.com/pressroom/archive/releases/20011126tech.htm

[8] M. Kandemir, N. Vijaykrishnan, M. J. Irwin, and W. Ye. Influence of compiler optimizations on system power. In *Proc. the 37th Design Automation Conference,* Los Angeles, California USA, June 5-9, 2000.

[9] A. R. Lebeck, X. Fan, H. Zeng, and C. S. Ellis. Power aware page allocation. In *Proc. ASPLOS'00,* November 2000.

[10] C. Lee, M. Potkonjak, and W. H. Mangione-Smith. MediaBench: a tool for evaluating and synthesizing multimedia and communication systems. In *Proc. the International Symposium on Microarchitecture,* 1997.

[11] J. R. Lorch and A. J. Smith. Scheduling techniques for reducing processor energy use in MacOS. *Wireless Networks,* 3(5):311-324, 1997.

[12] Y.-H. Lu, L. Benini, and G. De Micheli. Operating system-directed power reduction. In *Proc. ISLPED'00,* Rapallo, Italy, 2000.

[13] Rambus Inc. http://www.rambus.com/.

[14] 128/144-MBit Direct RDRAM Data Sheet, Rambus Inc., 1999.

[15] Y. Shin and K. Choi. Power-conscious fixed priority scheduling for hard real-time systems. In *Proc. Design Automation Conference,* pages 134–139, 1999.

[16] N. Vijaykrishnan, M. Kandemir, M. J. Irwin, H. Y. Kim, and W. Ye. Energy-driven integrated hardware-software optimizations using SimplePower. In *Proc. the International Symposium on Computer Architecture,* June 2000.

[17] M. Weiser, B. Welch, A. Demers, and S. Shenker. Scheduling for reduced CPU energy. In *Proc. Symposium on Operating Systems Design and Implementation,* pages 13–23, 1994.

[18] W. Ye, N. Vijaykrishnan, M. Kandemir, and M. J. Irwin. The design and use of SimplePower: a cycle-accurate energy estimation tool. In *Proc. the 37th Design Automation Conference,* Los Angeles, California USA, June 5-9, 2000.

[19] M. Beck. *Linux Kernel Internals* (2nd Edition). Addison-Wesley, 1999.

An Integer Linear Programming Based Approach for Parallelizing Applications in On-Chip Multiprocessors*

I. Kadayif
Microsystems Design Lab
Pennsylvania State University
University Park, PA 16802
kadayif@cse.psu.edu

M. Kandemir
Microsystems Design Lab
Pennsylvania State University
University Park, PA 16802
kandemir@cse.psu.edu

U. Sezer
ECE Department
University of Wisconsin
Madison, WI 53706
sezer@ece.wisc.edu

ABSTRACT

With energy consumption becoming one of the first-class optimization parameters in computer system design, compilation techniques that consider performance and energy simultaneously are expected to play a central role. In particular, compiling a given application code under performance and energy constraints is becoming an important problem. In this paper, we focus on an on-chip multiprocessor architecture and present a parallelization strategy based on integer linear programming. Given an array-intensive application, our optimization strategy determines the number of processors to be used in executing each nest based on the objective function and additional compilation constraints provided by the user. Our initial experience with this strategy shows that it is very successful in optimizing array-intensive applications on on-chip multiprocessors under energy and performance constraints.

Categories and Subject Descriptors

D.3.4 [**Programming Languages**]: Processors—*Compilers, Optimization*

General Terms

Design, Experimentation, Performance

Keywords

Embedded Systems, Loop-Level Parallelism, Constraint-Based Compilation

1. INTRODUCTION

System-on-a-Chip architectures are increasingly employed to solve a diverse spectrum of problems in embedded and mobile systems domain. For example, GVPP [1], a system-on-a-chip architecture manufactured by a French company, executes 20 billion instructions per second and models the human perceptual process at the hardware level by mimicking the separate temporal and spatial functions of the eye-to-brain system. GVPP sees its environment as a stream of histograms regarding the location and velocity of objects. Those objects could be the white lines on a highway, the football in a televised game or the annotated movement of enemy ground forces from satellite telemetry. The system is also capable of learning-in-place to solve a variety of pattern recognition problems. It has automatic normalization for varying object size, orientation, and lighting conditions, and can function in daylight or darkness.

As the applications loaded into system-on-a-chip architectures become more and more complex, it is extremely important to have sufficient compute power on the chip. One way of achieving this to put multiple processor cores in a single chip. This strategy has advantages over an alternate strategy which puts a more powerful and complex processor in the chip. First, the design of an on-chip multiprocessor composed of multiple simple processor cores is simpler than that of a complex single processor system [8, 13]. Note that this simplicity also helps reduce the time spent in verification and validation [12]. Second, an on-chip multiprocessor is expected to result in better utilization of the silicon space. The extra logic that would be spent on register renaming, instruction wake-up, and register bypass on a complex single processor can be spent for providing higher bandwidth on an on-chip multiprocessor. Third, the on-chip multiprocessor architecture can exploit loop-level parallelism at the software level in array-intensive applications. In contrast, a complex single processor architecture needs to convert loop-level parallelism to instruction-level parallelism at runtime (that is, dynamically) using sophisticated (and power-hungry) strategies. During this process, some loss in parallelism is inevitable.

An on-chip multiprocessor improves execution time of applications using on-chip parallelism. An application program can be made to run faster by distributing the work it does over multiple processors on the on-chip multiprocessor. There is always some part of the program's logic that has to be executed serially, by a single processor; however, in many applications, it is possible to parallelize some parts of the program code. Suppose, for example, that there is one loop in the code where the program spends 50% of its execution time. If the iterations of this loop can be divided across two processors so that half of them are done in one processor while the other half are done at the same time in a different processor, the whole loop can be finished in half the time, resulting in a 25% reduction in execution time. While this argument favors increasing the number of processors as much as possible, there is a limit beyond which using more processors might actually degrade performance. This is because in many cases parallelizing a loop entails interprocessor communication. Increasing the number of processors in general increases the frequency and volume of this communication. After a specific number of processors is reached, increasing the number of processors further increases communication costs so significantly that additional benefits due to more parallelism may not be able to offset this.

The preceding discussion indicates that even one focuses only on performance, selecting the number of processors to use in parallelizing loop nests may not be trivial. When we consider multiple objective functions at the same time, the problem becomes even

*This work was supported in part by NSF Career Award #0093082.

DAC 2002, June 10-14, 2002, New Orleans, Louisiana, USA.

more difficult to address. Suppose, for example, that we would like to minimize the energy-delay product of a given loop using parallelization. To decide the number of processors to use, we need to evaluate the impact of increasing the number of processors on both energy consumption and execution cycles. If using more processors does not bring significant reductions in execution time, the energy-delay product may suffer as using more processors means that more processors should be powered on. Evaluating the impact of increasing the number of processors on both energy and performance is extremely difficult if one restricts itself only to static (compile-time available) information. Finally, the possibility that each loop in a given application may demand a different number of processors to generate the best results makes the overall parallelization problem highly complex.

In this paper, we focus on a constraint-based parallelization problem and propose an integer linear programming (ILP) based strategy to select the number of processors to use in parallelizing each loop nest in an application code to be run on an on-chip multiprocessor. Our strategy has two components. First, there is a profiling phase, where we run (simulate) each loop with different number of processors and collect performance and energy data. In the second phase, called the optimization phase, we formulate the problem of selecting a processor size for each loop as an integer linear programming (ILP) problem and solve it. Our ILP-based approach can take into account multiple constraints that involve execution time and energy consumption on different hardware components. To the best of our knowledge, this is the first constraint-based application parallelization study for on-chip multiprocessors.

The remainder of this paper is organized as follows. Section 2 gives an overview of the on-chip multiprocessor architecture considered in this paper. Section 3 demonstrates why parallelization for minimum execution time and parallelization for minimum energy consumption demand different strategies. Section 4 presents details of our constraint-based parallelization strategy and Section 5 presents experimental results. Finally, Section 6 concludes the paper by summarizing our major contributions.

2. ON-CHIP MULTIPROCESSOR

In this paper, we focus on an on-chip multiprocessor architecture. This architecture contains multiple processors (each with its own instruction and data caches), a shared on-chip SRAM memory, and an interprocessor synchronization logic. This is a shared memory architecture; that is, all interprocessor communication occurs through reading from and writing into the shared SRAM memory. A bus-based on-chip interconnect is used to perform interprocessor synchronization. This synchronization is necessary for the processors to get synchronized at the beginning and end of each nest they execute.

As mentioned in the previous section, an on-chip multiprocessor architecture has some advantages compared to superscalar and VLIW machines. Maybe the most important of these is the simplicity and independence of the processors. The processors we assume in this study are single-issue, five-stage pipelined architectures without any branch prediction or data speculation logic. This brings an important side-advantage in terms of execution time predictability as it is easier to predict execution time with simple processors without sophisticated prediction/speculation logic. Also, each processor can operate independently from each other and processors engage in synchronization only to maintain data integrity during parallel execution.

On-chip multiprocessors are very suitable for array-intensive embedded applications. Many array-intensive embedded applications are composed of multiple independent loop nests that operate on large, multidimensional arrays of signals. An important issue in such applications is to map the application code to the embedded architecture in question. Note that, in the on-chip multiprocessor case, this is relatively simple as, in the source-code level, we can parallelize each loop nest and distribute iterations across the processors in the on-chip multiprocessor. If the loop in question is parallelizable, high execution time benefits can be achieved using this strategy. In contrast, a superscalar or a VLIW architecture should map parallel loop iterations to parallel (and sometimes irregular) functional units, which may not form a good match for the application structure.

Simulation Parameter	Value
Processor Speed	400MHz
Number of Processors	8
Instruction Cache	4KB 2-way associative 32 byte blocks
Data Cache	4KB 2-way associative 32 byte blocks
Memory	1 MB
Cache Dynamic Energy Consumption	0.6 nJ
Memory Dynamic Energy Consumption	1.17 nJ
Leakage Energy Consumption for 32 bytes Active State Sleep State	 4.49 pJ 0.92 pJ
Resynchronization Time	20 msec

Figure 1: Simulation parameters.

Figure 1 gives the simulation parameters used in this paper. We have used Cacti [17] to obtain cache and memory dynamic energy values. We assume that the leakage energy per cycle for 4KB SRAM is equal to the dynamic energy consumed per access to a 32 byte data from that SRAM. This assumption tries to capture the anticipated importance of leakage energy in the future as leakage becomes the dominant part of energy consumption for 0.10 micron (and below) technologies for the typical internal junction temperatures in a chip [3]. Since not all processors are used in executing a given loop nest, the unused processors and their instruction and data caches can be placed into a power-down (sleep) mode (state). In the power-down state, the processor and caches consume only a small percentage of their original (per cycle) leakage energy. However, when a processor and its data and instruction caches in the sleep state are needed, they need to be reactivated (resynchronized). This resynchronization costs extra execution cycles as well as extra energy consumption. Note also that since the SRAM main memory is shared between processors, it is never placed into power-down state. In this paper, we use the terms 'number of processors' and 'processor size' interchangeably.

It should be mentioned that on-chip multiprocessors are gaining popularity in commercial world as well. A number of architectures such as IBM's Power4, Sun's MAJC-5200, and MP98 contain multiple processors (currently between two and four) on a single die [9, 11, 5, 4].

Tuning processor's resources according to the needs of application has been studied by previous research in the context of superscalar processors [6, 10, 2]. In contrast, in this work, we focus on optimizing a given on-chip multiprocessor application in a constrained-based environment using integer linear programming.

3. IMPACT OF ON-CHIP PARALLELIZATION ON ENERGY

To see whether parallelization for energy and parallelization for performance are the same thing or not, we conducted an initial set of experiments where we run each nest of each code in our experimental suite using different number of processors (ranging from 1 to 8), and measured the energy consumption and execution cycles. Since the trends for different benchmarks were similar, we focus here only on tomcatv, one of our array-intensive benchmarks.

Figure 2 gives the execution time and energy consumption for each of the nine nests of this benchmark. For each nest, the eight bars from left to right correspond to different number of processors from 1 to 8. Also each bar represents a result *normalized* with respect to the single processor case (that is, the energy consumption or execution cycles when the nest is executed on a single processor). The details of our benchmarks and simulation environment will be given in a later section.

From these graphs, we can make several observations. First, en-

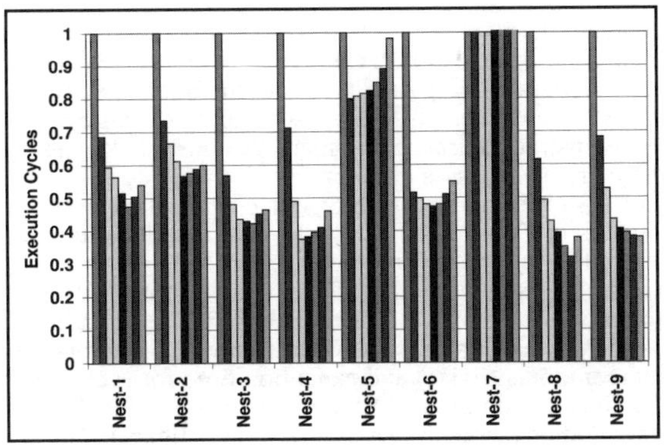

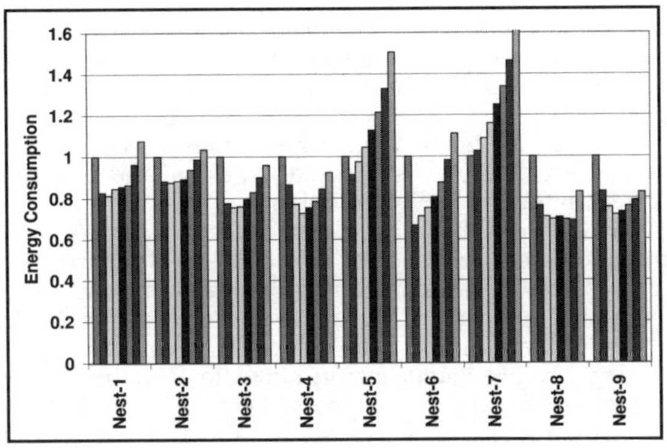

Figure 2: Normalized execution cycles and energy consumptions for tomcatv with different processor sizes.

ergy and performance trends are different from each other. This is because in many cases increasing the number of processors beyond a certain point may continue to improve performance, but the extra energy to power on the additional processor and its caches may offset any potential energy benefits coming from the reduced execution cycles. Second, for a given nest, increasing the number of processors does not always reduce execution cycles. This occurs because there is an extra performance cost of parallelizing a loop, which includes spawning parallel threads, interprocessor synchronization during loop execution if there are data dependences, and synchronizing threads after the loop execution. If the loop does not have sufficient number of iterations to justify the use of a certain number of processors, this extra cost can dominate the performance behavior. Third, the best number of processors from energy or performance perspectives depend on the loop nest in question.

4. OUR APPROACH

Our approach has two phases: profiling phase and optimization phase. In the profiling phase, we run each nest of the application being optimized for each possible processor size and record the energy consumption and execution time. These data are then fed to the second phase where we use integer linear programming (ILP) to determine the number of processors for each nest taking into account the objective function and compilation constraints. Therefore, the number of processors that will be used for executing each nest in the final optimized code is decided at compile time. However, the processor deactivations/reactivations (that is, placing a processor and its caches into a sleep state and later transitioning them to active state) take place at runtime.

To be more specific, if we have M different nests in a given array-intensive code and N different processors in our on-chip multiprocessor, we prepare two tables (one for energy values and the other one for execution cycles) in which each nest has N entries (that is, a total of NM energy values and NM execution time values). An entry i, j in these tables denote the energy consumption (or execution cycles) when nest i is executed using j processors. While in this paper we experimented with an architecture that contains eight processors, it is not difficult to modify this strategy to work with a machine with different number of processors. After building such tables (using the energy simulator that will be discussed shortly), our approach takes into account the energy and performance constraints (given as input) and selects the best number of processors for each nest. It should be noted that the number of processors selected for two consecutive nests might be different from each other, and in switching from one processor size to another, the associated time and energy overhead has to be accounted for. To select the processors sizes for the nests, our approach formulates the problem as a zero-one ILP (ZILP) and uses a publicly-available solver [14].

Let us first make the following definitions, assuming that $1 \leq i \leq M$ and $1 \leq j \leq N$, where N is the total number of processors and M is the number of nests in the code being optimized:

- $E_{i,j}$: the estimated energy consumption for nest i when j processors are used to execute it.
- $X_{i,j}$: the estimated execution time for nest i when j processors are used to execute it.

We use zero-one integer variables $s_{i,j}$ to indicate whether j processors are selected for executing nest i or not. Specifically, if $s_{i,j}$ is 1, this means that j processors are selected to execute nest i in the final solution. Obviously,

$$\sum_{j=1}^{N} s_{i,j} - 1 \quad (1)$$

should be satisfied for each nest i. Consequently, the total energy consumption (E) and total execution time (X) for the application can be expressed as:

$$E = \sum_{i=1}^{M}\sum_{j=1}^{N} s_{i,j}E_{i,j} \quad (2)$$

$$X = \sum_{i=1}^{M}\sum_{j=1}^{N} s_{i,j}X_{i,j} \quad (3)$$

It should be noted, however, that these energy and execution time calculations do not include any overhead for changing the number of processors between nest boundaries.

In our experiments, we assumed that a fixed (constant) amount time (denoted R) is required to make an inactive processor (i.e., a processor that has been placed into the sleep state along with its caches) active. We can compute the execution time overhead as follows:

$$n_{i,j} \geq s_{i,j} - \sum_{k=j}^{N} s_{i-1,k} \quad (4)$$

$$y_i = \sum_{j=1}^{N} n_{i,j} \quad (5)$$

$$Y = R\sum_{i=2}^{M} y_i \quad (6)$$

where $2 \leq i \leq M$ and $1 \leq j \leq N$. In this formulation, $n_{i,j}$ and y_i are zero-one variables. Expression (4) restricts $n_{i,j}$ to be 1 if nest i has j active processors and nest $i-1$ has less than j active processors; otherwise, $n_{i,j}$ is restricted to be 0. Expression (5), on the other hand, sets y_i to 1 if nest i has more active processors than nest $i-1$. Note that only when nest i uses more processors than nest $i-1$ we incur a resynchronization penalty in going from nest $i-1$ to nest i. Finally, expression (6) computes the total time overhead for activating processors between nest boundaries. Note that R is constant (20 msec in our experiments).

Suppose that in going from one nest to another, we need to activate new processors. These new processors plus the ones used to execute the former nest are considered to consume leakage energy for the time period of R. The energy consumption including overheads can be computed as follows:

$$t_{i,j} = \sum_{k=1}^{j} s_{i,k} \quad (7)$$

$$z_i = \sum_{j=1}^{N} t_{i,j} - 1 \quad (8)$$

$$u_i \geq z_{i-1} - z_i \quad (9)$$

$$U = \sum_{i=2}^{M} u_i \quad (10)$$

$$Z = r(U + \sum_{i=1}^{M-1} \sum_{j=1}^{N} j s_{i,j}) \quad (11)$$

where $1 \leq i \leq M$ and $1 \leq j \leq N$ except for expression (9) where $2 \leq i \leq M$. Also, r is the leakage energy consumption of a single processor during the time period of R and is constant. In this formulation, $t_{i,j}$ is a zero-one variable. Expression (7) here sets $t_{i,j}$ to 1, if nest i has j or less than j active processors. Expression (8) gives the number of inactive (powered-down) processors at nest i. Expression (9) forces u_i to take 0 or a positive value and, determines the number of processors activated in going from nest $i-1$ to nest i. Expression (10) gives the total number of processors that need to be activated between nest boundaries. Finally, expression (11) gives the total leakage energy overhead for the processors activated between nest boundaries and processors already active during activation periods.

Note that, taking into account these energy and performance overheads, $X+Y$ gives the execution time including overheads and $E+Z$ gives the energy consumption including overhead energy. It should also be noticed that expressions for calculating time overhead and energy overhead are different in a sense that while we overlap the time overhead for activated processors between loop boundaries, we consider each activated processor to consume leakage energy during the activation period.

In some cases, the optimization problem involves constraints or objective function built from energy consumptions of different hardware components. One example would be minimizing main memory energy only (instead of the overall system energy) under a specific execution time and a specific overall energy consumption constraints. To capture the energy breakdown across different components, we use $E^d_{i,j}$, $E^{dc}_{i,j}$, $E^{ic}_{i,j}$, and $E^m_{i,j}$ to denote (for nest i executing using j processors) the estimated energy consumptions in datapath, data cache, instruction cache, and main memory, respectively. If these are the only hardware components of interest, we have:

$$E_{i,j} = E^d_{i,j} + E^{dc}_{i,j} + E^{ic}_{i,j} + E^m_{i,j}.$$

We can find the overall main memory (E^m), instruction cache (E^{ic}), data cache (E^{dc}), and datapath energies (E^d) as:

$$E^m = \sum_i \sum_j E^m_{i,j} s_{i,j}$$
$$E^{ic} = \sum_i \sum_j E^{ic}_{i,j} s_{i,j}$$
$$E^{dc} = \sum_i \sum_j E^{dc}_{i,j} s_{i,j}$$
$$E^d = \sum_i \sum_j E^d_{i,j} s_{i,j}.$$

5. EXPERIMENTS AND RESULTS

We made experiments with several array-intensive benchmarks. The important features of our benchmarks are listed in Figure 3. Among these codes, `3step-log`, `full-search`, `hier`, and `par allel-hier` are different implementations for motion estimation. The remaining codes are representative array-intensive benchmarks from Spec and Perfect Club suites. The fourth column gives execution times in msecs and the last two columns give overall dynamic and leakage energy consumptions, in millijoules, for each benchmark when they are executed on a single processor of our on-chip multiprocessor. As shown in the table, leakage energy dominates dynamic energy since whole memory is active regardless of the number of active processors, and at each cycle, active data cache, instruction cache and memory consume leakage energy even there are no accesses to them.

Figure 4 shows the structure of our optimization system. The input code is first optimized using classical data locality optimizations as well as loop transformations that enhance loop-level parallelism (e.g., loop skewing) [16]. Then, this optimized code is fed to our simulator (see below). The simulator simulates each nest using all possible processor sizes and records energy consumption (both total and component-wise) and execution time for each nest. After that, this information along with the code itself is fed to the parallelization module (shown as Parallelizer in the figure). This module builds up ILP formulation (in a form suitable for the solver) and passes it to the ILP solver. The solver also takes the compilation constraints and the objective function as input. The parallelization module then obtains the solution from the solver and parallelizes the application (i.e., each nest) accordingly. If it is not possible to parallelize the application under given constraints and objective function, an error message is signaled. If desired, the user can relax some constraints and retry compilation.

Our simulation methodology is as follows. We simulate an on-chip multiprocessor architecture that contains 8 identical single-issue processors. Our simulation model has four parts: datapath, instruction and data caches, (shared) main memory, and interprocessor synchronization. For each processor, datapath energy is obtained using a cycle-accurate energy simulator developed for a five-stage, single-issue pipelined processor [15]. This simulator employs transition-sensitive energy models for datapath components such as functional units and register file. It has been validated using a commercial low-power microprocessor and its accuracy has been shown to be within 10% (of real current measurements). For cache memories and main memory (in our case, an SRAM), we used the analytical energy model proposed by Kamble and Ghose [7]. Since in our on-chip multiprocessor interprocessor synchronization is through a bus-based mechanism, we modeled its energy consumption as that of placing 8 bits into the interconnect (one for each processor) and the associated control register update (to set/reset the synchronization bit). Note that processors share data through main memory; that is, there is no extra interconnect for explicit interprocessor data communication. All the results presented in this section were obtained using the simulation parameters given in Figure 1. Also, all energy values reported in this paper are based on parameters for 0.10 micron, 1V technology.

To see the energy and performance benefits of our approach, for each benchmark, we performed experiments with two different versions. The first version is the original (unoptimized) code that uses the maximum number of processors (8) for each nest. The second version, is the optimized code where the number of processors for each nest is selected using the ILP-based strategy detailed in the previous section. Apart from the number of processors used for parallelizing loops, both the original and optimized codes have exactly the same structure (i.e., both of them are loop and data transformed for the best execution). After showing that the optimized codes exhibit much better energy behavior than their unoptimized counterparts, we demonstrate how our strategy can be used for compiling a given application under different energy/performance constraints.

Our presentation is in three parts. First, we give overall energy and performance results when our strategy is used for optimizing the total energy and performance. In our context, total energy is the sum of energies expended in datapath, instruction and data caches, main memory, and interprocessor synchronization bus. These results are given in Figure 7. The first graph in this figure gives the execution time of the optimized code as a fraction of the corresponding original code when the latter uses all 8 processors for all the nests. We see that the savings are modest (6.9% on the average). The second graph shows the overall energy consumption of the optimized code as a fraction of the original code when the latter uses all 8 processors for all the nests. We observe that the energy savings are much higher than performance savings. This is because, as

Benchmark	Input Size (KB)	Number of Nests	Exec. Time	Dynamic Energy	Leakage Energy
3step-log	814.16	3	1271.8	139.0	1709.3
adi	242.60	2	1187.6	129.5	1596.1
aps	548.92	3	764.4	98.8	1019.2
bmcm	525.44	4	1236.2	128.7	1661.4
btrix	212.04	7	3076.3	418.7	4134.7
eflux	780.30	2	268.8	42.5	361.3
full-search	814.16	3	4546.1	464.0	6109.9
hier	814.16	7	6456.7	529.8	8677.7
lms	320.00	4	584.2	106.0	785.2
n-real-upd.	80.00	3	2640.9	456.0	3549.4
par.-hier	814.16	5	5784.4	746.4	7774.4
tomcatv	633.60	9	2534.2	359.6	3406.0
tsf	468.18	4	2934.0	351.8	3943.3

Figure 3: Array-intensive benchmarks.

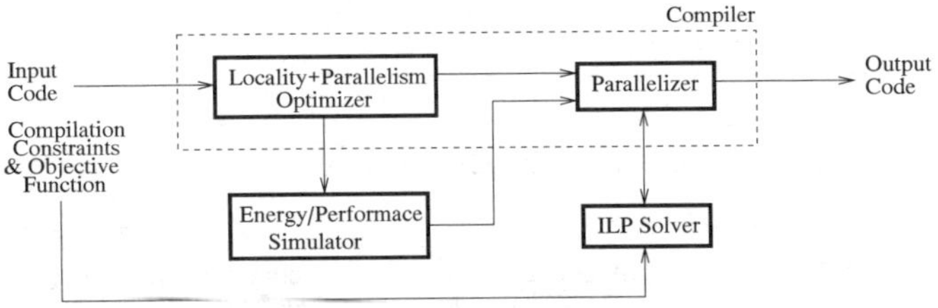

Figure 4: Our optimization system.

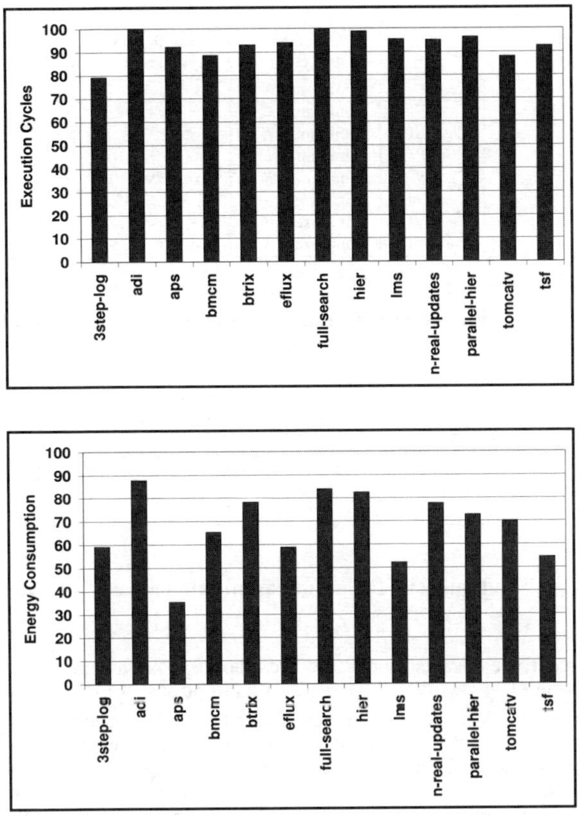

Figure 7: Normalized execution cycles and energy consumptions.

can be seen from the tomcatv example in Figure 2, optimizing (i.e., generating the best result) for performance generally requires large number of processors than required for optimizing for energy consumption. The largest energy savings in Figure 7 occur with `aps` (35.3%), `lms` (52.1%), and `tsf` (54.4%). This is because each of these codes contains a small number of costly nests that demand a small number of processors for the best energy behavior. The results presented in Figure 7 also reveal that it is not a good idea from the energy perspective to work always with the maximum number of processors available. In other words, for the best energy results, the number of processors should be changed adaptively.

In the second part of our presentation, we focus on tomcatv and give detailed results when this application is compiled under different energy and performance constraints. Let us assume for the moment that the overhead of dynamically changing the number of active processors is zero. We consider different compilation strategies whose objective functions and performance/energy constraints (if any) are given in the second and third columns, respectively, of Figure 5. First, let us focus on two simple cases: Case-I and Case-II. In Case-I, we try to minimize execution time; note that this is the classical objective of a performance-oriented compilation strategy. We see from columns four through twelve that in this case the average number of (selected) processors per nest is 4.67. Case-II represents the other end of the spectrum where we strive for minimizing the energy consumption (without any performance concerns). The average number of processors per nest is 3.22. The last two columns in Figure 5 give the energy consumption (in millijoules) and execution time (in msecs) for each scenario. We observe that optimizing only for performance results in an energy consumption which is 11.5% higher than that of the best energy-optimized version. Similarly, optimizing only for energy consumption leads to an execution time which is 14.5% higher than that of the best performance-optimized version. These results clearly show that parallelization for energy and performance are different objectives.

Let us now focus on other compilation strategies. Case-III and Case-IV correspond to minimizing energy consumption under performance constraints. We note that the processor sizes selected in these cases are different from those selected for Case-II. The average number of processors per nest are 3.88 and 4.11 for Case-III and Case-IV; that is, including a performance constraint in the compilation increases the average number of processors selected. In addition, we see that a more stringent performance constraint (e.g., $X \leq 925$) results in higher number of processors. Case-V corresponds to optimizing execution time under energy constraint. Note that, in this case, the number of processors selected for some nests is different from the corresponding numbers in Case-I. To sum up, when we consider Cases III, IV, and V, we see that including constraints during compilation can change the number of processor selected for executing nests. Note also that without an ILP-based framework such as the one presented in this paper, it would be extremely difficult to decide on the number of processors for each nest when a compilation scenario such as Case-V, Case-II, or Case-III is encountered.

Up to this point, we have assumed that the overhead of dynamically changing the number of processors between nests is zero. The remaining cases in Figure 5 correspond to scenarios with non-zero energy and performance overhead for reactivating an inactive processor. Specifically, we assumed a resynchronization penalty of 20 msec (which is a highly conservative [large] value even for powerful leakage optimization techniques such as power supply gating [3]) during which all processors (except the ones in low power) are assumed to consume leakage energy. We see from Cases VI, VII, and VIII that when the overhead is taken into account, the ILP solver tends to keep the number of processors the same between the neighboring nests as much as possible. We also see that even in these cases, optimizing for performance and optimizing for energy generate different number of processors for most of the nests.

So far, we focused on overall energy consumption without considering individual hardware components. In our last set of experiments, we investigate how targeting energy consumptions of different components might affect the selected number of processors. As before, we focus only on tomcatv. Our observations for other benchmarks are similar. Figure 6 gives the number of loops

Case	Objective Function	Compilation Constraints	Selected Number of Processors									Energy Consumption	Execution Time
			n1	n2	n3	n4	n5	n6	n7	n8	n9		
Case-I	minimize X	none	6	5	6	2	2	5	1	7	8	3065	908
Case-II	minimize E	none	3	3	3	4	2	2	1	7	4	2750	1040
Case-III	minimize E	$X \leq 975$	5	5	4	4	2	2	1	7	4	2814	974
Case-IV	minimize E	$X \leq 925$	6	3	4	4	2	4	1	7	7	2944	921
Case-V	minimize X	$E \leq 2975$	6	5	5	4	2	4	1	6	8	2974	915
Case-VI	minimize $E+Z$	$X+Y \leq 975$	6	4	4	4	2	2	1	4	4	2999	975
Case-VII	minimize $E+Z$	$X+Y \leq 925$	6	6	6	4	4	4	1	7	7	3253	924
Case-VIII	minimize $X+Y$	$E+Z \leq 2975$	6	3	3	3	2	2	1	3	3	2974	998

Figure 5: Different compilation strategies and the selected number of processors for each nest (tomcatv).

Case	Objective Function	Compilation Constraints	Selected Number of Processors									Energy Consumption			
			n1	n2	n3	n4	n5	n6	n7	n8	n9	E	E^m	E^{ic}	E^{dc}
Case-II	minimize E	none	3	2	2	2	2	2	1	3	3	2887	1025	628	446
Case-IX	minimize E^m	none	6	5	6	5	2	5	1	7	8	3354	856	899	690
Case-X	minimize E^{ic}	none	1	1	1	1	1	1	1	1	1	3765	1845	622	451
case-XI	minimize E^{dc}	none	1	1	1	1	1	1	1	1	1	3765	1845	622	451

Figure 6: Different compilation strategies and the selected number of processors for each nest (tomcatv).

selected for each nest and the corresponding overall energy consumption (E) as well as the energy consumptions in main memory (E^m), instruction cache (E^{ic}), and data cache (E^{dc}) when different objective functions are used (Case-II is repeated here for comparison purposes; it differs from the one in Figure 5 as E here includes overhead energy as well). We make several observations from these results. First, optimizing energy consumption of a specific component generates very different results than optimizing overall (total) energy consumption. For example, when the objective function is minimizing the main memory energy consumption (that is, Case-IX), the resulting overall energy consumption becomes 16.2% higher than the case when we try to minimize overall energy. This is despite the fact that Case-IX reduces the main memory energy by 16.5% with respect to Case-II. Second, similar observations can be made when we consider Case-X and Case-XI. Third, it is also interesting to see that if one targets only instruction cache energy or data cache energy, the ILP solver favors single processor execution for each nest. This is because in a single processor execution seven data caches and seven instruction caches can be placed in the sleep state saving significant leakage energy. As can be seen from Figure 6, however, such a single processor execution strategy increases the overall energy consumption by more than 30% compared to Case-II. It should also be mentioned, however, that if main memory resides in a different chip, then objective functions such as 'minimize E^m' or 'minimize $E^{ic}+E^{dc}+E^{d}$' might be more desirable. These example scenarios show how flexible our technique is in compiling a given code under constraints.

6. CONCLUSIONS

This paper presented an integer linear programming (ILP) based strategy for compiling a given array-intensive application in an on-chip multiprocessor under energy/performance constraints. Our strategy has two parts: profiling and ILP formulation. At the end of the optimization process, the compiler determines the most suitable number of processors that will be used in executing each nest in the code. Our preliminary results are encouraging, indicating up to 54% savings in energy compared to a static scheme that uses the maximum number of available processors for each nest. We also demonstrated that our approach can optimize a given application targeting energy consumptions of individual hardware components.

7. REFERENCES

[1] System-On-A-Chip Mimics Human Vision. http://content.techweb.com/wire/story/ TWB20000309S0002

[2] R. I. Bahar and S. Manne. Power and energy reduction via pipeline balancing. In Proc. *the 28th Annual International Symposium on Computer Architecture,* pages 218-229, 2001.

[3] A. Chandrakasan, W. J. Bowhill, and F. Fox. *Design of High-Performance Microprocessor Circuits.* IEEE Press, 2001.

[4] Chip Multiprocessing. http://industry.java.sun.com/javanews/stories/ print/0,1797,32080,00.html

[5] Chip Multiprocessing. *ITWorld.Com,* http://www.itworld.com/Comp/1092/CWSTO54343/

[6] T. M. Conte, K.N. Menezes, S. W. Sathaye, and M. C. Toburen. System-level power consumption modeling and tradeoff analysis techniques for superscalar processor design. *IEEE Transactions on VLSI Systems,* vol 8, no. 2, April 2000.

[7] M. Kamble and K. Ghose. Analytical energy dissipation models for low power caches. In Proc. *International Symposium on Low Power Electronics and Design,* August 1997.

[8] V. Krishnan and J. Torrellas. A chip multiprocessor architecture with speculative multi-threading. *IEEE Transactions on Computers, Special Issue on Multi-threaded Architecture,* September 1999.

[9] MAJC-5200. http://www.sun.com/microelectronics/MAJC/5200wp.html

[10] D. Marculescu. Profile-driven code execution for low power dissipation. In Proc. *the International Symposium on Low Power Electronics and Design,* pp. 253-255, 2000.

[11] MP98: a mobile processor. http://www.labs.nec.co.jp/MP98/top-e.htm.

[12] B. A. Nayfeh, L. Hammond, and K. Olukotun. Evaluating alternatives for a multiprocessor microprocessor. In Proc. *the 23rd Intl. Symp. on Computer Architecture,* pp. 66-77, Philadelphia, PA, 1996.

[13] K. Olukotun, B. A. Nayfeh, L. Hammond, K. Wilson, and K. Chang. The case for a single chip multiprocessor. In Proc. *the 7th Int'l Conference on Architectural Support for Programming Languages and Operating Systems,* ACM Press, New York, 1996, pp. 2-11.

[14] H. Schwab. *lp_solve Mixed Integer Linear Program Solver,* ftp://ftp.es.ele.tue.nl/pub/lp_solve/

[15] N. Vijaykrishnan, M. Kandemir, M. J. Irwin, H. Y. Kim, and W. Ye. Energy-driven integrated hardware-software optimizations using SimplePower. In Proc. *International Symposium on Computer Architecture*, June 2000.

[16] M. Wolfe. *High Performance Compilers for Parallel Computing*, Addison-Wesley Publishing Company, 1996.

[17] S. Wilton and N. Jouppi. Cacti: An enhanced cache access and cycle time model. *IEEE Journal of Solid-State Circuits,* May 1996.

Embedding Infrastructure IP for SOC Yield Improvement

Yervant Zorian
Virage Logic
46501 Landing Parkway
Fremont, CA 94538
+1-510-360-8035
zorian@viragelogic.com

ABSTRACT

In addition to the functional IP cores, today's SOC necessitates embedding a special family of IP blocks, called Infrastructure IP blocks. These are meant to ensure the manufacturability of the SOC and to achieve adequate levels of yield and reliability. The Infrastructure IP leverages the manufacturing knowledge and feeds back the information into the design phase. This paper analyzes the key trends and challenges resulting in manufacturing susceptibility and field reliability that necessitate the use of such Infrastructure IP. It also describes several examples of such embedded IPs for detection, analysis and correction.

Categories and Subject Descriptors

B.8.1 [**Performance and Reliability**]: Reliability, testing and fault tolerance.

General Terms: Measurement, Performance, Design, Reliability, and Verification.

Keywords: Semiconductor IP, Embedded Test & Repair, Yield Optimization, Test Resource Partitioning.

1. INTRODUCTION

Every new semiconductor technology node provides further miniaturization and higher performance, thus increasing the functions that electronic products could offer. Although adding such new functions do benefit the end-user, but they also necessitate finer and denser semiconductor fabrication processes, which make chips more susceptible to defects. Today's very deep-submicron semiconductor technologies of 130 nanometers and below are reaching defect susceptibility levels that result in lowering the manufacturing yield and reliability, and hence lengthening the production ramp-up period, and therefore the time to volume (TTV). The very deep submicron impact on yield, reliability and TTV is very critical for the semiconductor industry. It puts the conventional IC realization flow at an impasse.

In fact, every single phase in the IC realization flow Figure (3) impacts yield and reliability. This includes the design phase, prototyping or production ramp up, volume fabrication, test, assembly, packaging, and even the post-production life cycle of the chip. In order to optimize yield and reach acceptable TTV levels, the semiconductor industry needs to adopt advanced yield optimization solutions. These solutions need to be implemented at different phases of the chip realization flow.

The conventional semiconductor manufacturing infrastructure, i.e. the external equipment and processes, alone are insufficient to handle such advanced yield optimization solutions; supplemental on-chip infrastructure is needed. To optimize yield and reliability, the industry has recently introduced a range of embedded intellectual-property (IP) blocks, called infrastructure IP [6]. These are meant for inclusion into IC design and utilized during the different phases of product realization.

Semiconductor IP is well known for the last decade. Most of the known IP blocks, though, are functional ones, such as embedded processor, embedded memory, embedded analog, or embedded FPGA cores. Whereas, Infrastructure IP (I-IP) is not functional, i.e. does not contribute to the normal functionality of a given IC. Rather, I-IP is embedded in an IC solely to ensure its manufacturability and lifetime reliability [6]. This role is similar to the infrastructure elements of a building, such as wiring networks or plumbing, which are independent from the actual function of the building.

This paper introduces the basic types of infrastructure IP and presents their effectiveness in improving yield and reliability. The paper is organized in the following way. Section 2 discusses the trends and challenges of today's very deep submicron technologies concentrating on the ones that have major impact on yield and reliability. Section 3 introduces the architecture of I-IP. Sections 4-8 demonstrate examples of I-IP solutions used at different phases of the IC realization and use flow. These solutions are based on on-chip resources for embedded diagnosis, timing measurement, debugging, test, repair, and fault tolerance.

2. TRENDS AND CHALLENGES

Every phase in the product realization flow affects yield. The following subsections describe the key trends and challenges of today's very deep submicron technologies that have considerable impact on yield.

2.1 Yield Learning and Time to Volume

The semiconductor fabrication process is constantly evolving to implement the advances necessary to realize the improvements in fabrication technology. This introduces new materials and techniques into the fabrication process, which lead to a new generation of yield limiting faults [1]. Furthermore, the increase in design complexity and the shrinking geometries in very deep submicron technologies made devices more susceptible to systematic and random defects. Typically, yields are low for the initial lots of an IC design. This is called the ramp up period, as shown in Figure (1). During which the yield problems are learned (identified and diagnosed). As a result, the yield slowly ramps to a mature level. Then traditionally the volume production starts.

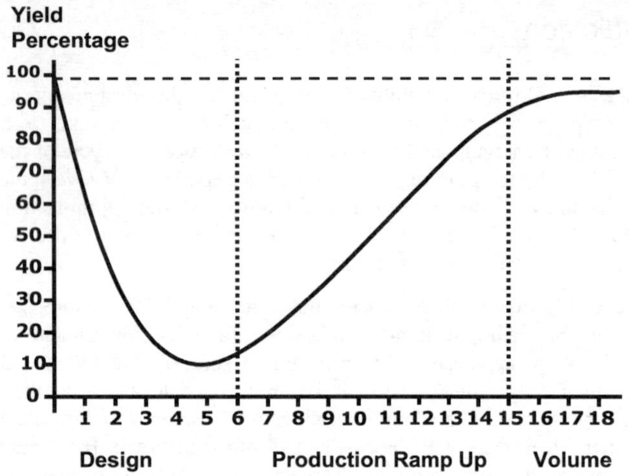

Figure (1) Yield Learning Curve (Source: HPL)

However due to the increase in time-to-market pressures, foundries are often forced to start volume fabrication on a given semiconductor technology before reaching the traditional defect densities, and hence the yield maturity levels necessary prior to volume production. Hence, improving yields as quickly as possible is an important component in lowering costs and improving profitability. The yield learning curve can be considerably improved, if the yield optimization can start at the design stage. This can happen by applying knowledge from the fabrication process into design. To optimize the design and obtain better yield, foundries should diagnose yield problems during the early development process. Collecting the manufacturing data, like defect distribution data, is done using special infrastructure IP blocks called embedded process monitoring IP, which will be discussed in Section 4.

Another trend is the supply chain disaggregation in the semiconductor industry due to the fabless model. The resulting supply chain includes fabless houses, functional-IP suppliers, pure-play foundries, and contract manufacturers for test, assembly and packaging services. In this disaggregated model, functional-IP providers must supply SoC designers with IP blocks optimized for functionally as well as manufacturability and yield. As a result, the advanced IP suppliers have begun assuming yield optimization responsibilities for their corresponding functional IP blocks, and thus embedding the necessary I-IP to ensure its manufacturability.

2.2 Memory Dominance

There is a clear trend to integrate large quantities of memory on a chip. Memories are designed with aggressive design rules and tend to be more prone to manufacturing defects. The overall yield of an IC or System-on-Chip (SoC) design relies heavily on the memory yield. Even though intrinsic or native memory yield may be unsatisfactory, memory and therefore, overall die yield, can be improved. The lower curve on Figure (3) shows the memory yield as a function of aggregate memory bit count. Traditionally embedded memories were testable but not repairable because embedded memories did need redundant elements for repair. The yield challenge in today's memories is addressed by offering memories with redundancy, i.e. spare elements. However, having redundancy only does not resolve the problem. The know-how of how to detect the defects in a memory and how to allocate the redundant elements require manufacturing know-how in terms defect distributions. In order to achieve the higher curve in Figure (2), i.e. optimized yield solution, one needs to utilize an infrastructure IP to contain the know-how, without which the yield can only be improvement to a limited level, as in Figure (2). An infrastructure IP to achieve optimized yield is discussed in Section 5, and is known as embedded test and repair IP.

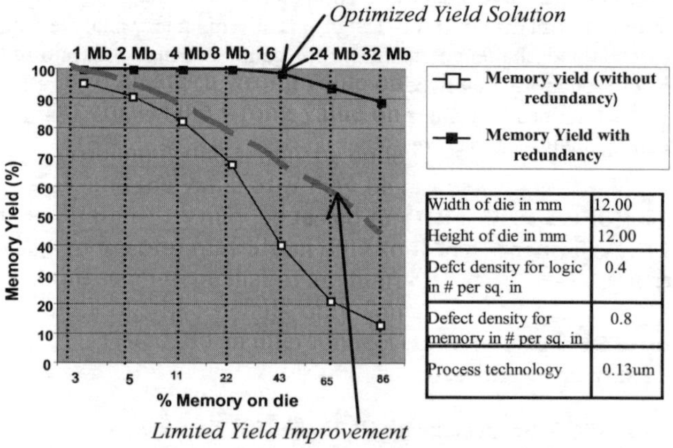

Figure (2) Memory Yield Optimization Curve (Source: Virage Logic Corp)

2.3 Diagnosis and Failure Analysis

The traditional failure analysis is comprised of fault localization, silicon deprocessing and physical characterization and inspection steps. The migration towards smaller geometries severely challenges this physical failure analysis process. Because of the increased sensitivity to failure, which requires finding smaller, more subtle defects; tighter pitches, which require greater spatial resolution; and increasing numbers of metal layers, which along with flip-chip packaging force the use of backside analysis [itrs]. These trends combined make the physical failure analysis process difficult to rely upon [6]. The key alternative is gathering failure data by using embedded diagnosis I-IP, such as signature analyzers, dedicated test vehicles or on-chip test processors, and then analyzing the obtained data by off-chip fault localization methodologies and tools. Examples of such embedded diagnosis I- IP are described in Section 6.

2.4 High Performance Circuits

Increasing performance in SOC designs require increased accuracy for proper resolution of timing signals. While semiconductor off-chips speeds have improved at 30% per year, tester accuracy has improved at rate of 12% per year. The tester timing errors are approaching the cycle time of the fastest device [5]. Yield losses due to tester inaccuracy are becoming a problem when using a traditional functional test methodology. The yield loss in this case is due to additional guard-bending performed during the test stage. The alternate solution to address this challenge is to use Infrastructure IP for measuring and analyzing timing specifications. This alleviates the yield loss on high-speed devices due to tester timing accuracy. Section 7 describes an effective embedded timing IP.

2.5 Transient Errors

Drastic decreases in device dimensions and power supply have significantly reduced noise margins and challenged the reliability of very deep-submicron chips. Soft errors, timing faults, and crosstalk are major signal integrity problems [2]. Hence, both the logic block and the embedded-memories require self-correcting intelligence, such as an infrastructure IP for robustness. Section 8 introduces examples of embedded Fault Tolerance IP.

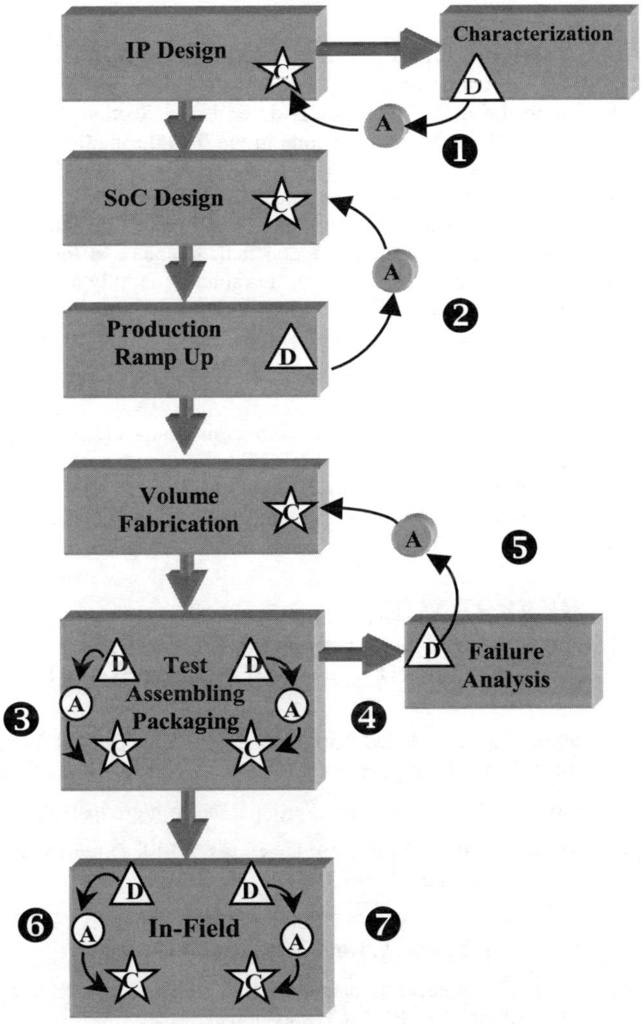

Figure (3) Yield Feedback Loops in IC Realization Flow

3. I- IP ARCHITECTURE

The trends and challenges described in Section 2 affect different phases of the IC realization flow. In order to improve yield and reliability a number of yield feedback loops need to be leveraged. Typically, a feedback loop, i.e. a yield enhancement solution, encompasses three consequent components: Detection, Analysis and Correction (DAC). Naturally, all three components are needed to achieve yield improvement. Due to the challenges of very deep submicron technology several of the feedback loop components (DACs) are embedded into the SOC design as Infrastructure IP. In fact, some of the feedback loops are completely embedded, such as loops 4, 5, 6, and 7; whereas others such as 1, 2, and 3 have only their detection monitors (D) only on-chip.

In the first wave, the infrastructure IP blocks were embedded mainly at the top level chip design and provided direct monitoring and control to the peripheries of different functional blocks. However, the trend has been towards integration of infrastructure IP with the functional IP. The following five sections will discuss examples of I-IP used in the yield feedback loops of Figure (3).

4. EMBEDDED PROCESS MONITOR IP

In loops 1 and 2, the D components are specialized I-IP solutions that monitor the process characteristics and collect device attributes. These process monitors are also called test vehicles or test chips. Such an I-IP can be used during the process development phase, or later during production, as shown in Figure (4). It may be incorporated in the IP design, as in loop 1 or in the SOC design, as in loop 2. It may be a full chip or an IP in a product chip [1]. The attributes collected by I-IP are fed to external analysis engines (A) and the results are utilized for modifications (C) to optimize yield. The A and C in loops 1 and 2 are performed not by an I-IP, but by external resources.

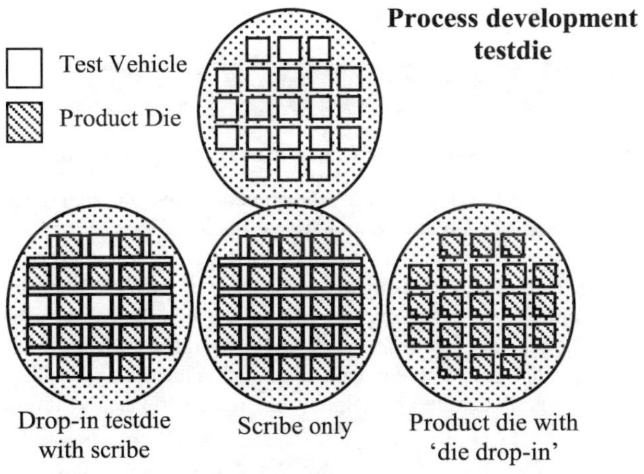

Figure (4) Embedded Process Monitor IP Application Modes (Source: HPL)

5. EMBEDDED TEST & REPAIR IP

The feedback loops 3 and 6 in Figure (3) correspond to this embedded test and repair IP. This I-IP is meant to address the embedded memory yield challenges, described in Section 2.2. The function of this I-IP is to perform embedded test, diagnosis, redundancy allocation and repair signature generation. This I-IP is a comprehensive one, which includes the whole feedback loop, i,e. D-A-C. During the testing phase loop 3 is operational, which means that the embedded memories achieve optimized yields during manufacturing. Loop 6 becomes operational upon power up, i.e. for in-field test and repair. The type of I-IP requires integration the functional IP, i.e. the embedded memory. This composite IP, in addition to the embedded memories with redundancy, consists of a processor for embedded test and repair, intelligent wrappers associated with each memory, and a fuse box to permanently store the repair information on-chip.

6. EMBEDDED DIAGNOSIS IP

The feedback loop 5 of Figure (3) is meant to detect the root cause of failures using embedded diagnosis IP, and then to transfer the gathered data to external resources in order to perform analysis and correction steps to the fabrication process. Several examples of embedded diagnosis IPs can be mentioned here. In case of embedded memories, the dedicated processor in Section 5 can gather the failure data at every error occurrence and transfer it to external analysis software, which builds the failed bit map of the memory and performs statistical and graphical analysis on it. Using this information, the fabrication process may be corrected.

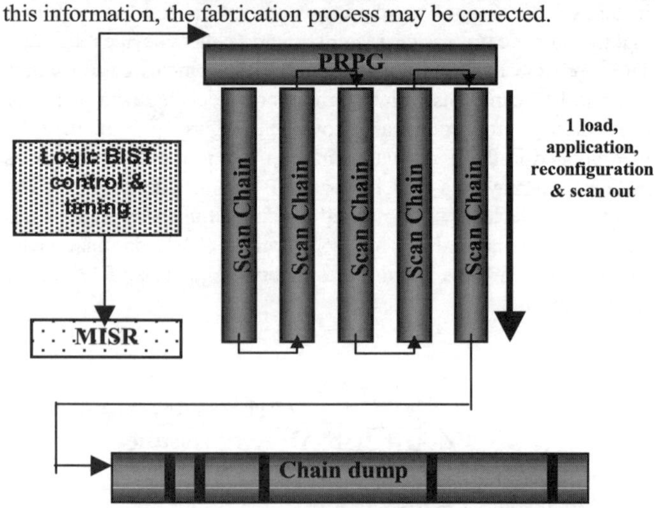

Figure (5) Diagnosis for Logic Blocks (Source: LogicVision)

In the case of random logic blocks, the embedded test and diagnosis IP is comprised of scan chains and test points incorporated into the random logic block itself, and an embedded signature analyzer on the peripheries of the random logic IP, as in Figure (5). This is another example demonstrating the integration of infrastructure IP with functional IP. The infrastructure IP operates with external software to allow interactive diagnostic modes. This allows flip-flop level diagnosis. A detailed description of this scheme can be found in [4].

The need to locate faults with further granularity is rising. Using dedicated test vehicles as, described in Section 4 and Figure (4), provides defect distribution data for each process layer and silicon structure [1]. Localization of performance fails is especially important. In addition, certain I-IP and external analysis tools do jointly locate defects to single transistor level. Some also handle realistic physical defects, including resistive bridges, resistive contacts/vias and opens.

This type of infrastructure IP may also use the IEEE proposed standard, P1500, which allows standard accessibility and isolation to individual functional IP blocks [3]. Designers implement this basic infrastructure IP around the peripheries of their individual functional IP blocks.

7. EMBEDDED TIMING IP

Because the timing specifications are often very stringent in today's SoCs, external instrumentation is not enough to ensure accurate measurement, as discussed in Section 2.4. An example of embedded timing IP is introduced in [5]. In this article, the I-IP achieves effective accuracy and is implemented via the feedback loop 4, as shown in Figure (3). This IP distributes multiple probes over different parts of an SOC to collect the necessary timing information. A central I-IP core controls the probes and transfers the information to a timing processor for analysis.

8. EMBEDDED FAULT TOLERANCE IP

The growing environmental susceptibility increases the reliability risks during the SOC life cycle, as described in Section 2.5. An Infrastructure IP is needed to operate in the field during the normal mode operation of the SOC. With this type of I-IP, the transient errors, including the soft errors, are detected, analyzed and corrected on-line, as shown in the feedback loop 7 of Figure (3). One well-known solution to perform embedded fault tolerance at the block level is described in [2]. The I-IP in this solution is fully integrated with the functional IP or SOC design. This allows for timing and area optimization and provides protection throughout the life cycle.

9. SUMMARY

This paper discusses the very deep submicron trends and challenges affecting manufacturing yield and reliability. It covers a wide range of infrastructure IP solutions to address these challenges — some used in the design phase, some in manufacturing; and others, in the field.

10. REFERENCES

[1] Bordelon, J. et al. "A Strategy for Mixed Signal Yield Improvement", IEEE Design & Test of Computers, Vol 19, No 3, May-June, 2002.

[2] Dupont, E., "Embedded Robustness IP", IEEE Design & Test, Vol 19, No 3, May-June, 2002.

[3] IEEE P1500 Web Site. http://grouper.ieee.org/groups/1500/.

[4] Pateras, S., "IP for Embedded Diagnosis", IEEE Design & Test of Computers, Vol 19, No 3, May-June, 2002.

[5] Tabatabaei, S., Ivanov, A., "Embedd Timing Analysis", IEEE Design & Test, Vol 19, No 3, May-June, 2002.

[6] Zorian, Y., "What is Infrastructure IP?", IEEE Design & Test of Computers, Vol 19, No 3, May-June, 2002.

Using Embedded FPGAs for SoC Yield Improvement

Miron Abramovici
Agere Systems
Murray Hill, NJ 07974
miron@agere.com

Charles Stroud
University of North Carolina
Charlotte, NC 28223
cestroud@uncc.edu

Marty Emmert
Wright State University
Dayton, OH 45435
marty.emmert@wright.edu

Abstract: In this paper we show that an embedded FPGA core is an ideal host to implement infrastructure IP for yield improvement in a bus-based SoC. We present methods for testing, diagnosing, and repairing embedded FPGAs, for which complete testability is achieved without any area overhead or performance degradation. We show how an FPGA core can provide embedded testers for other cores in the SoC, so that cores designed to be tested with external vectors can be tested with BIST, and the entire SoC can be tested with a low-cost tester.

Categories and Subject Descriptors: B.8.1 [Performance and Reliability]: Reliability, Testing, and Fault-Tolerance

General Terms: Algorithms, Design, Reliability

1. Introduction

Reusing embedded IP cores in System-on-Chip (SoC) design has become the primary means of improving the productivity of designers faced with very large and complex circuits. One problem in manufacturing a SoC with millions of transistors using deep-submicron technologies (0.13μm and below), is an increase in the probability of defects in silicon, which results in decreasing manufacturing yield. The economic consequences of low yield for expensive SoC devices can be disastrous. To effectively deal with this increased defect density, we need efficient methods for fault detection, location, and fault tolerance implemented on-chip. Typically, such methods are implemented by embedded IP blocks referred to as *infrastructure IP*. In this paper we will show that an embedded FPGA core is an ideal vehicle for infrastructure IP.

But an embedded FPGA is also a very useful functional SoC component. A SoC design where at least part of the user-defined logic is implemented in an embedded FPGA provides the following benefits:

1) Usually fixing logic design errors in an ASIC requires a "respin" to change a set of manufacturing masks. The cost of a respin is around $1M, and it also causes a 2-to-6 months hit in the time-to-market. In contrast, a fix that can be done in the FPGA can be completed in one day.

2) The embedded FPGA may be reused for different functions at different times. For example, a cell phone may work with different protocols (CDMA, GSM, etc.) in different geographical areas, and the logic implementing each protocol can be downloaded in the FPGA on a demand basis.

3) Although FPGA clock speeds are slower than ASIC clock speeds, the massive parallelism of an FPGA can be exploited to implement many algorithms (such as image processing, signal processing, cryptography, etc.) at higher performance levels than on μPs or DSPs. Usually such FPGA implementations consume less power than μP or DSP implementations, because they can be run at lower clock frequency.

4) The same SoC can be used as a platform for different products, or for different versions of the same product, which are created by different FPGA configurations.

5) A good strategy is to use the embedded FPGA to implement protocols and algorithms likely to change in the future (for example, hardware support for a standard not yet officially adopted). In this way, future changes will be easy to implement without changing the SoC.

6) The embedded FPGA can support remote and Internet-based field-upgrades, which may be very important features for system manufacturers.

This material is based upon work supported in part by the DARPA ACS program under contract F33615-98-C-1318, and by Lucent Technologies.

DAC 2002, June 10-14, 2002, New Orleans, Louisiana, USA.

2. Using an FPGA Core for Infrastructure IP

In addition to the functional features described above, an embedded FPGA can also provide great benefits in implementing infrastructure IP. SoC yield-improvement requires detecting faults, locating them, and repairing them so that the SoC will work correctly in their presence. Methods for detecting, locating, and repairing faults in RAMs are described in [9]. The problems we address in this paper are detecting, locating, and repairing faults in an embedded FPGA; some of our previous work dealt with the same problems in the context of on-line testing [4], and some of the solutions developed for on-line testing can be directly applied to, or adapted for, manufacturing testing of an embedded FPGA. Another contribution of this paper is novel ways to use an embedded FPGA to test other cores in the SoC.

We assume a bus-based SoC architecture, where the embedded cores are connected by a common system bus. The FPGA core can become the bus master. The SoC is tested by automatic test equipment (ATE) for manufacturing testing, or by a maintenance processor for in-system testing. In either case, the tester also controls the configuration process for the embedded FPGA core via the boundary-scan access mechanism of the SoC [22]; the same mechanism can be used both for manufacturing test and for in-system test. Various configurations for the FPGA are stored on disk. We will refer to the external tester as *test and reconfiguration controller* (TREC). If BIST is used within the SoC, TREC initiates the internal

BIST controller and processes the results. TREC also performs diagnosis and fault tolerance functions that will be described in the following sections. Later we will show that the embedded FPGA allows TREC to be a low-cost tester (possibly PC-based).

Our basic principle for using reconfigurable logic for infrastructure IP is "***create the infrastructure only when needed***." Hence instead of having the infrastructure for test always present in the circuit, we download it in the embedded FPGA only when we want to test the SoC. In most cases, this infrastructure will be used only once for manufacturing testing, so having it permanently in the system is clearly wasteful (we can recreate it for maintenance testing, if and when needed). Even if the FPGA implements user-defined system functions during normal operation, this logic is not needed when the SoC is under test and may be temporarily replaced by test logic. This is a generalization of our "free lunch" FPGA BIST [24]. Typically, logic BIST for ASICs introduces about 20% area overhead and some performance degradation. In contrast, BIST for FPGAs is a "free lunch" with *no area overhead or delay penalties*, since the entire FPGA is configured only for BIST, and all the BIST logic disappears when the FPGA is no longer under test.

Our basic principle can have surprising consequences. For example, the proposed IEEE P1500 Standard for Embedded Core Test [16] specifies that embedded cores should provide a wrapper for test purposes. An embedded FPGA core, however, can safely ignore this requirement, since the wrapper can be downloaded in the FPGA only during the SoC test.

Taking advantage of FPGA features, such as reconfigurability and regular structure, allows FPGA testing to achieve features that are impossible for ASICs: for example, FPGAs can be tested pseudo-exhaustively [20] and faults can be very precisely located. Pseudo-exhaustive testing results in *practically complete fault coverage* without the use of computationally expensive ASIC test tools, such as fault simulation and automatic test-pattern generation (ATPG).

Two types of fault tolerance exist for FPGAs. The first, *manufacturer-level fault-tolerance*, relies on providing spare resources that can be used to replace the faulty ones. For example, some FPGAs have a spare column and a hardware mechanism that allows any column to be replaced by an adjacent one [8]; the replacement for each column is controlled by a fuse. After the faulty column is diagnosed, all columns between the faulty one and the spare are "shifted" by one position, so that the faulty one is completely avoided, and the change is invisible to the outside world. The costs of this approach - the additional area needed for the spare resources and the performance penalty introduced by the column selection hardware - are paid by all chips, including the non-defective ones.

The second approach, *user-level fault-tolerance*, avoids these costs since it relies on the spares naturally available in an FPGA, where any application uses only a subset of the existing resources. Knowing the user circuit to be implemented in the FPGA, and the exact location of the faulty resources, one can modify the existing implementation (mapping, placement, routing) to replace the faulty resources with fault-free spares. For example, this approach allowed the Teramac custom computer [10] to be build from 864 FPGAs, out of which 75% were defective. Note that configuration data for each faulty FPGA should be separately maintained. Unlike manufacturing-level fault tolerance, which is independent of the target circuit, user-level fault tolerance can ignore any fault that does not affect the circuit. User-level fault tolerance requires testing and diagnosis of FPGAs to be done by users, and this may be unacceptable for off-the-shelf FPGAs typically tested by their manufacturers.

In contrast, an embedded FPGA core is tested only as part of an newly fabricated SoC, and the SoC designer is also responsible for its testing. In this environment, user-level fault-tolerance is clearly the right solution for yield improvement.

The remainder of the paper is organized as follows. Section 3 explains why a commonly used approach - testing the user's circuit - is not good for FPGA test. Section 4 presents techniques for testing and diagnosing logic cell faults in an embedded FPGA. Section 5 describes techniques for detecting and locating faults in the programmable interconnect network. Section 6 shows how to use an embedded FPGA to create the infrastructure for testing other cores. Section 7 presents fault tolerance methods that allow the embedded FPGA to work correctly in the presence of the located faults. Section 8 concludes the paper.

3. A Wrong Approach: Testing User's Circuit

Many FPGA design and test flows generate tests for the circuit implemented in the FPGA using ATPG tools for ASICs. However, such an approach will not completely test the FPGA hardware. The first problem is testing of a configuration multiplexer (MUX), which is a commonly used hardware mechanism to select subcircuits for various modes of operation. A configuration MUX is controlled by configuration memory bits to select one input to be connected to its output. In Figure 1a, assume that we set the configuration bit *S* to 0 to connect subcircuit *C0* to *X*. Then subcircuit *C1* disappears from the circuit model seen by the user. This is correct from a design viewpoint, because the value *V1* produced by *C1* can no longer affect the MUX output in the current configuration. But from a testing viewpoint, in any test for the MUX, we need to set *V0* and *V1* to complementary values. In general, for a MUX with *k* inputs, if *V* is the value of the selected input, all the other *k*-1 inputs should be set to value $\overline{V}$.

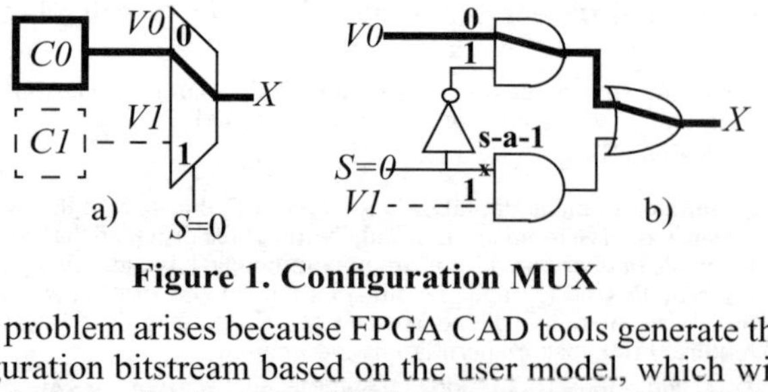

Figure 1. Configuration MUX

The problem arises because FPGA CAD tools generate the configuration bitstream based on the user model, which will never include the functionally inactive subcircuits (called "invisible logic" in [1]). Thus in Figure 1a, when S=0, $V0$ will be set to both 0 and 1, but $V1$ cannot change. Similarly, the user logic cannot control $V0$ in any configuration where S=1. Hence the testing of the MUX may not be complete. For example, the s-a-1 fault in the gate-level MUX model in Figure 1b is detected only when S=0, $V0$=0, and $V1$=1. But this pattern may never be applied if $V1$ cannot be controlled when S=0.

In most previous work dealing with testing FPGAs, the problem of testing a configuration MUX is either not addressed or it is "solved" functionally, by connecting every input in turn to the output, and providing both 0 and 1 values to the selected input. However, the invisible logic driving the inactive inputs is incorrectly ignored. Since an FPGA has tens of thousands of configuration MUX structures, such a test is likely to have a poor quality.

Our solution relies on separately configuring the invisible logic so that it will generate the proper values needed for the inactive MUX inputs. Then we "overlay" the resulting configuration files over the main configuration file with the active logic, and we "merge" them without changing any MUX setting done in the main configuration. This process is conceptually simple, but its implementation requires knowledge of the FPGA configuration stream structure.

A second problem with testing only the user's circuit is that spare resources are never tested, and when the need arises to replace a faulty resource with a spare one, we cannot be sure that the spare is fault-free. Our approach is to test the entire FPGA and to construct tests that are independent of the applications to be implemented in the device.

4. Testing the Logic Cells

4.1 BIST for Logic Cells

The embedded FPGA is composed of an array of programmable logic cells connected by programmable interconnect resources. Usually the logic and the interconnect of an FPGA are separately tested. The goal of logic tests is to detect any faults (single or multiple) affecting the operation of a cell, and also any combination of multiple faulty cells.

Figure 2 illustrates our logic BIST architecture [1]. Some of the cells are configured as *test-pattern generators* (TPGs) and *output-response analyzers* (ORAs), while another group of cells are configured as *blocks under test* (BUTs). The two TPGs provide identical test patterns to alternating rows of

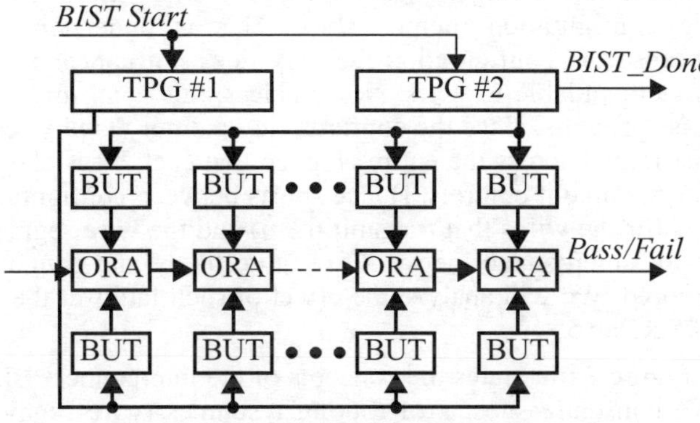

Figure 2. Logic BIST architecture

identically configured BUTs. The outputs of the BUTs are compared by the ORAs which latch any mismatches. The *Pass/Fail* result flip-flops of all ORAs are connected in a scan chain. The BUTs are then repeatedly reconfigured so that they are tested in all of their modes of operation. Each reconfiguration of the embedded FPGA to test a different cell mode of operation is referred to as a *test phase*. A *test session* is a collection of test phases that completely test the BUTs in all of their modes of operation. The number of test phases is a function of the cell architecture. Once the BUTs have been tested, the roles of the cells are reversed, so that in the next test session the previous BUTs become TPGs or ORAs, and vice versa. Figure 3 shows the cell functions in the two test sessions for an 8×8 FPGA. Since half of the cells are BUTs during each test session, only two test sessions are needed to test all cells in the embedded FPGA.

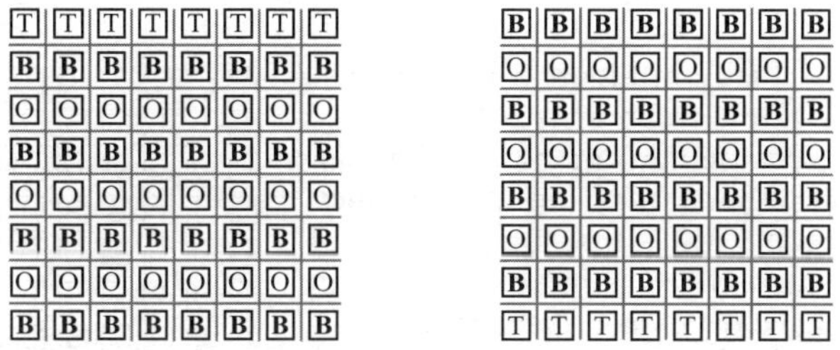

Figure 3. Cell functions for the two test sessions

In every test phase, TREC performs the following steps: 1) reconfigure the embedded FPGA with a BIST configuration retrieved from the disk, 2) initiate the BIST sequence, and 3) read the *Pass/Fail* results from the ORAs. The downloading of the BIST configuration and the control of the BIST process are done via the boundary-scan access mechanism of the SoC [22]. Since the test application time for any phase is dominated by the embedded FPGA reconfiguration time, an important goal of the BIST approach is to minimize the total number of test configurations.

Figure 4 illustrates the typical structure of a cell, consisting of a memory block that can function as a look-up table (LUT) or RAM, several flip-flops, and multiplexing output logic. The LUT/RAM block may also contain special-purpose logic for arithmetic functions (counters, adders, multipliers, etc.) The RAM may be configured in various modes of operation (synchronous, asynchronous, single-port, dual-port, etc.). The flip-flops can also be configured as latches, and may have programmable clock-enable, preset/clear, and data selector functions. Our TPG applies exhaustive patterns to every subcircuit of a cell for each one of its modes of operation, except for the RAM modes, where the memory block is checked with RAM March tests which cover most RAM-specific faults. Exhaustive testing of every subcircuit is feasible since their number of inputs is reasonably small. This results in ***practically complete cell fault coverage*** without explicit fault model assumptions and without fault simulation.

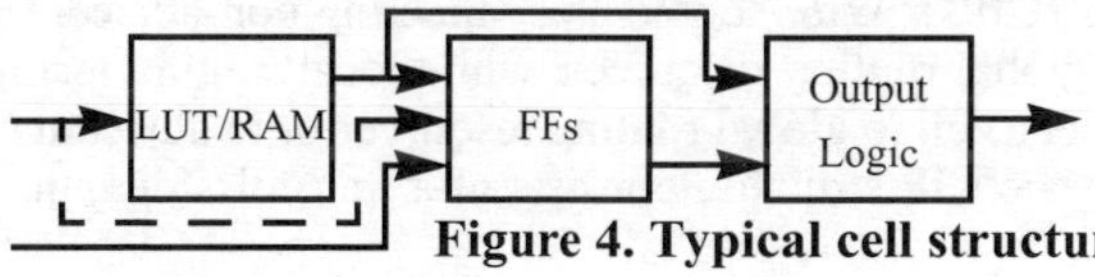

Figure 4. Typical cell structure

Although there are situations that make detection of multiple faulty cells more complicated (a pair of compared BUTs have equivalent faults and do not cause a mismatch; a faulty ORA does not detect a mismatch; a faulty TPG skips patterns that detect a faulty BUT), *almost any combination of multiple faulty cells is guaranteed to be detected* by our two test sessions. We have shown that the conditions that would allow a group of faulty cells to escape detection are extremely restrictive, so that they are very unlikely to occur in practice [1].

4.2 Logic Cell Diagnosis

Figure 5 illustrates a test session where two faulty BUTs (denoted by black squares) are detected. Errors are observed at the ORAs bordering the faulty cells (shaded squares). Clearly, the set of failing ORAs is unique for every faulty BUT, and therefore it can be used to locate single faulty cells. (As a consistency check, the two failing ORAs observing the same BUT must fail in the same phases [1].)

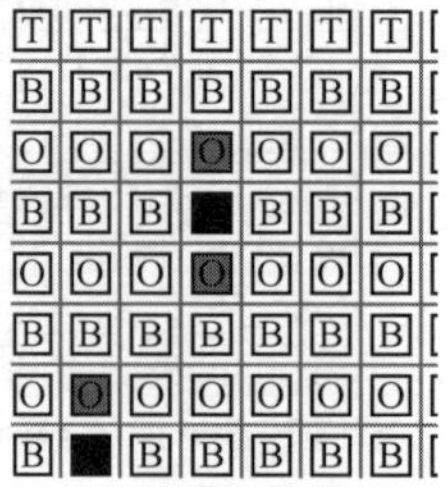

Figure 5.

Actually in Figure 5 we are dealing with a group of two faulty cells, but because they are observed at disjoint ORAs, they do not interact. Locating BUTs that are observed at common ORAs is more difficult, but we have developed diagnostic procedures [1] that can locate almost any combination of multiple faulty cells likely to occur in practice. These procedures rely on analyzing the sets of failing test phases obtained at different ORAs, and by applying additional BIST configurations when needed.

In addition to diagnosing the faulty cells, we can also identify the faulty subcircuits (LUTs, flip-flops) or the defective modes of operation (e.g., count-up, multiply) within a faulty cell [2].

Although our tests are application-independent, we can take advantage of knowing the circuits to be implemented in the embedded FPGA. Specifically, TREC may avoid applying additional diagnostic configurations when the suspected cells are not used in any of the target circuits. However, accurate diagnosis of unused cells may be required later to allow these cells to replace other defective cells for fault tolerance.

If a configuration memory bit that controls a resource in a logic cell is stuck, this fault is detected by the tests for the corresponding resource. However, the faulty resource and the stuck fault in its controlling configuration bit cannot be distinguished, unless the configuration memory has a readback feature that can detect the stuck bit.

5. Testing the Interconnect Network

5.1 BIST for Interconnect

The programmable interconnect network consists of wire segments of different lengths that can be connected via programmable switches referred to as *configurable interconnect points* (CIPs). Wire segments connecting non-adjacent cells form global routing resources, while local routing resources connect a cell to global routing resources or to adjacent cells. The basic CIP structure consists of a transmission gate controlled by a configuration memory bit (Figure 6a). There are three types of CIPs which we refer to as the *cross-point CIP* (Figure 6b), the *break-point CIP* (Figure 6c), and the *multiplexer (MUX) CIP* (Figure 6d) [18]. While a cross-point CIP connects wire segments located in disjoint planes (a horizontal segment with a vertical one), a break-point CIP connects two wire segments in the same plane. The MUX CIP comes in two varieties: decoded and non-decoded. A decoded MUX CIP is a group of 2^k cross-point CIPs sharing a common output wire and controlled by k configuration bits, such that the input wire being addressed by the configuration bits is connected to the output wire. A non-decoded MUX CIP contains a configuration bit controlling each transmission gate; here only one of the configuration bits is active for any configuration. There is also a *compound CIP* (Figure 6e), which is a combination of four cross-point and two break-point CIPs, each separately controlled by a configuration bit [31].

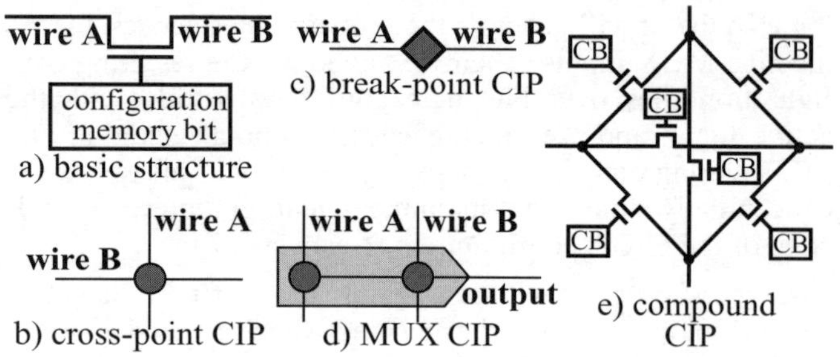

Figure 6. Configurable interconnect points

The fault model we consider is typical for interconnect: CIPs stuck-closed (stuck-on) and stuck-open (stuck-off), wires stuck at 0 or 1, open wires, and shorted wires. A stuck-closed CIP creates a short between its two wires. We assume that no detailed layout information is available regarding adjacency relations between segments. Instead, we use only "rough" physical data (available in data books) to determine adjacent "bunches" of wires, where a bunch is a group of wires that *may* have pair-wise shorts, but not every wire is necessarily adjacent with every other wire in the bunch. This treatment makes our method layout-independent.

Detecting a CIP stuck-open (closed) fault also detects the stuck-at-0 (1) fault in the configuration memory bit that controls that CIP. However, most previous work in FPGA interconnect testing has ignored the problem of shorts involving configuration memory bits. This is understandable, because the configuration memory does not appear in any model available to users. Nevertheless, this is an important class of faults, since the configuration memory is physically distributed across the entire chip, so that each bit is close to the resource it controls. Hence shorts between configuration bits (or the wires that transmit them) and the wire segments in the interconnect network are quite likely and cannot be ignored. We will analyze the effect of such faults at the end of Section 5.

Figure 7 illustrates the concepts of the interconnect BIST. We configure subsets of routing resources (wire segments and CIPs) to form two groups of *wires under test* (WUTs) that receive exhaustive test patterns from a TPG, while the values at the other end of the WUTs are compared by an ORA [26]. A WUT consists of several wire segments connected by closed CIPs. To check local routing resources, WUTs may also go through cells configured to pass their inputs directly to outputs. In Figure 7, the WUTs are shown by bold lines, and the activated (closed) CIPs are shown in gray, while the open CIPs are white. To test the open CIPs for stuck-on faults, when the TPG drives a value v on the WUTs (0 in Figure 7), $\bar{v}$ must be applied to the wire segments on the opposite side of the open CIPs.

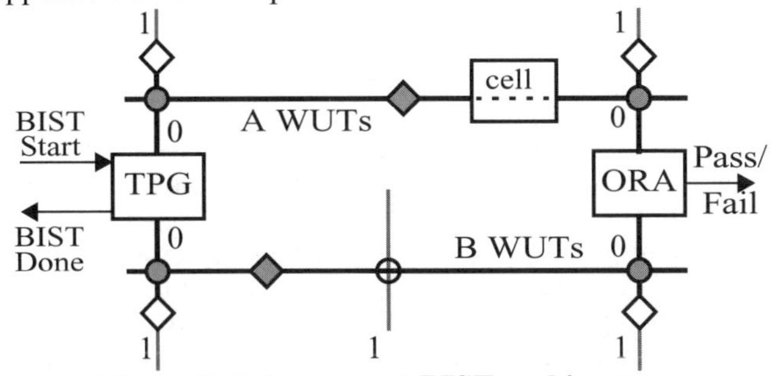

Figure 7. Interconnect BIST architecture

The TPG applies exhaustive patterns to the WUTs, all possible 2^n test vectors to every group of n WUTs. For reasonably small n, the application time for 2^n test vectors is still negligible compared to the reconfiguration time of the FPGA. The TPG also provides test patterns to the opposite sides of open CIPs that isolate the WUTs from the rest of the interconnect.

If the two compared sets of WUTs have identical (or equivalent) faults, then no mismatch will be detected at the ORA [15]. For example, each set of WUTs could have a different CIP stuck-open in its ith wire. To overcome this problem, we test every set of WUTs (at least) twice, each time being compared with a different set of WUTs [25].

The basic routing BIST architecture shown in Figure 7 is implemented within *self-testing areas* (STARs), where each STAR spans two columns or two rows of the embedded FPGA to test vertical or horizontal interconnect resources, respectively [4][5]. Figure 8 illustrates the structure for testing the horizontal interconnect. Testing within STARs is done concurrently.

While most of the interconnect resources are tested either in an horizontal or in a vertical STAR, testing the cross-point CIPs connecting global horizontal and vertical busses requires both a vertical and an horizontal STAR, where the cross-point CIPs at the intersection of the STARs are under test (Figure 9).

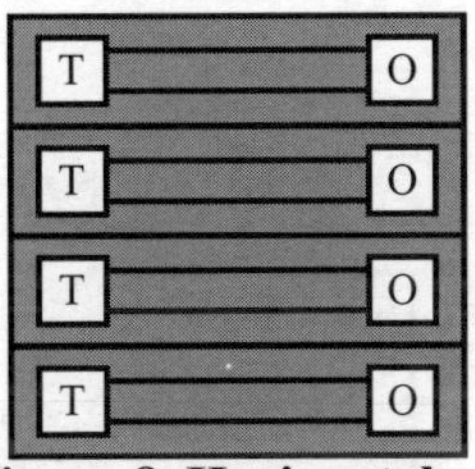

Figure 8. Horizontal interconnect testing

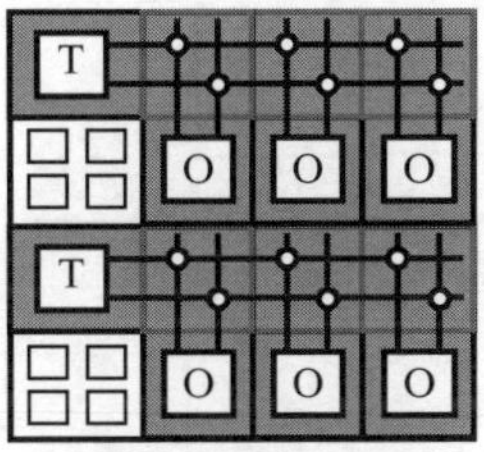

Figure 9. Global cross-point CIP test

5.2 Interconnect Fault Diagnosis

Our diagnosis procedure begins by analyzing the BIST results to narrow the search area, and proceeds only in the failing STARs. We use an adaptive strategy, where the next diagnostic phase is determined based on the results obtained so far. Figure 10 illustrates several diagnosis techniques (failing ORAs are darkened).

In Figure 10a, after the first phase we do not know which group of WUTs is faulty. During the second test phase, the WUTs are swapped between TPGs, and the faulty group can be identified by checking whether the failure moves to the other ORA or stays at the first ORA.

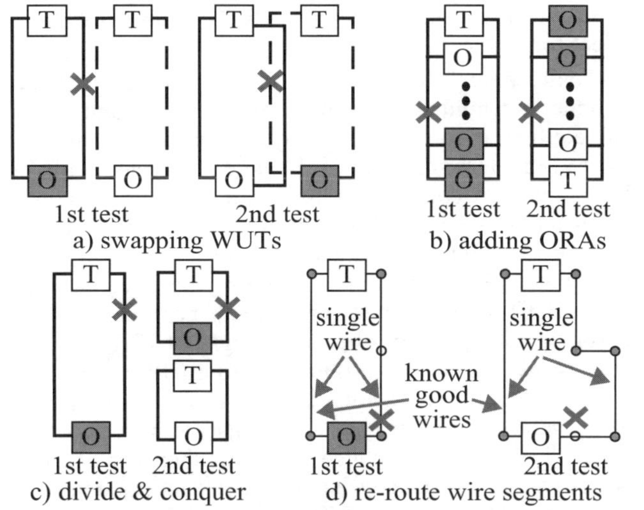

Figure 10. Diagnostic configurations

Figure 10b depicts one technique used to identify the faulty region in a set of WUTs. We add several ORAs to observe the WUTs at different points and use two phases with opposite directions of the test pattern flow through WUTs. Faults are located between the highest failing ORA in the first phase and the lowest failing ORA in the second phase. An alternative technique is to divide the WUTs into pairs of shorter WUTs and retest (Figure 10c). This "divide and conquer" method can be recursively repeated to obtain a very high diagnostic resolution.

Identifying the faulty WUT in a group can be accomplished by either testing one wire at a time (and comparing with a known-good wire), or by making different ORA connections with the faulty set of WUTs and observing the movement of ORA failures (a similar idea to that of Figure 10a). The faulty wire segment or CIP is then identified by re-routing to eliminate a segment (or CIP) at a time during the subsequent tests (Figure 10d).

Diagnostic phases can be either precomputed and stored on disk, or generated "on the fly" by TREC. In general, the diagnostic phases illustrated in Figure 10a through 10c can be precomputed, while the detailed diagnostic phases implied by Figure 10d would be generated "on the fly" for most FPGA interconnect architectures.

A tool like JBITS [29] that can efficiently generate configuration streams without relying on CAD tools, would be ideal to use during testing. Since no such tools are commercially available, we rely on commercially available place-and-route tools used incrementally, followed by configuration-generation software; these may run for several minutes to generate multiple diagnostic phases, but this time could be acceptable if TREC is a low-cost tester. A limit on the time allowed for computing diagnosis configurations can be established based on factors such as the cost of a SoC, the production volume, the current yield, the probability of repairing the defective SoC given the number of faults found so far, the target resolution, etc. The use of precomputed diagnostic phases in conjunction with an adaptive diagnostic procedure minimizes the total time spent on TREC, while maintaining defect-tolerant capabilities.

To illustrate the diagnostic resolution that can be obtained with these techniques, we will assume a single fault in a typical configuration of interconnect resources illustrated in Figure 11a. Here segment *X* is bounded by break-point CIPs that control its connections to *Y* and *Z*, while segment *W* shares no CIPs with *X*. For example, if a cross-point CIP is stuck-open (Figure 11b), this is the only condition that would allow the vertical and the horizontal segments connected by the that CIP to be driven to complementary values. A similar reasoning identifies an open wire segment (Figure 11c) In these cases, diagnostic resolution is to a CIP or a segment. However, we cannot distinguish between an open in a segment and a stuck-open fault in its adjacent break-point CIP (Figure 11d). Figure 11d also shows another pair of equivalent faults - a short between two segments separated by a break-point CIP and that CIP being stuck-on. On the other hand, shorts between two segments that do not share a CIP, such as *W* and *X* in Figure 11e, can be diagnosed by checking that each segment is fully functional while the other segment it is shorted to is unused, but when both segments are driven to opposite values, then we detect a failure on at least one of them.

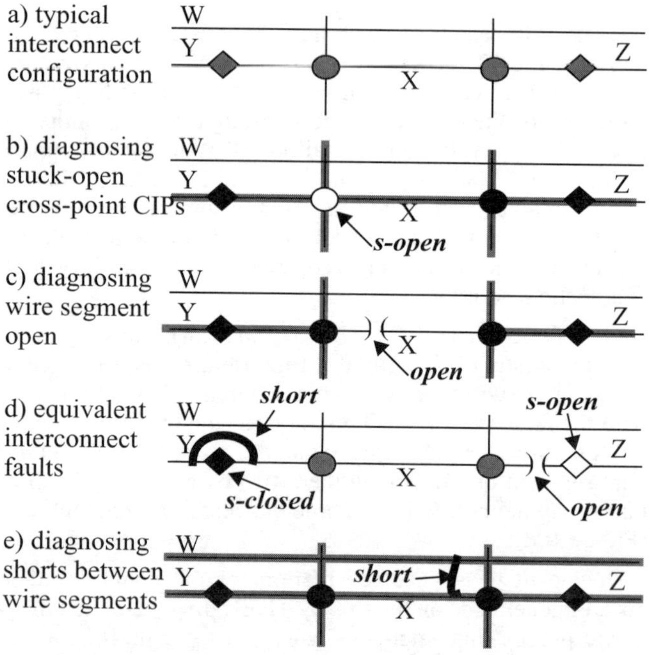

Figure 11. Diagnostic resolution examples

For multiple faults in the same area, there are cases where we will not always be able to identify the faulty resources among a set of suspects. These cases arise when some faults interfere with the routing configurations needed to diagnose other faults. In such cases, we will "play it safe" by labeling the entire set of suspects as faulty, so that all of them will be bypassed by fault tolerant techniques.

To illustrate the problem of detecting and diagnosing shorts with configuration memory bits, consider a short between a configuration bit *B* and an interconnect segment *S*, and let us assume that *B* controls a CIP *C*. Note that the value *b* of *B* is constant during a phase. This short may be detected in any configuration where *S* is used and at least once is set to value $\overline{b}$. The detection will occur in one of the two cases: 1) the short creates a wrong value on *B* and *C* is under test; 2) the short creates the wrong value on *S* and *S* is under test.

The problem is that *C* may be under test in different configurations, and the value of *B* may be affected in only some of them. Similarly, *S* may be tested in different configurations, and the value of *B* may affect only some of them. This behavior would seem to be that of an intermittent fault, as the same resource looks like it fails only some of the configurations where it is tested. So an interconnect diagnosis procedure that looks for consistency in all cases is doomed to fail. The practical solution in such situations is to label all the suspected resources as faulty. Avoiding all the potential defects is a reasonable strategy for yield enhancement.

6. Testing Other Cores with the FPGA Core

The embedded FPGA can also be used to test any other embedded core connected to the same system bus. We will separately analyze *BIST-cores*, whose testing is based on in-core BIST, and externally-tested-(*ET*)*-cores*, which rely on vectors external to the core.

6.1 BIST-Cores

For a BIST-core, its ***BIST logic may be moved from the core to the embedded FPGA***, as illustrated in Figure 12. (Similar techniques have been described for using an FPGA installed on a board to test other devices on the board

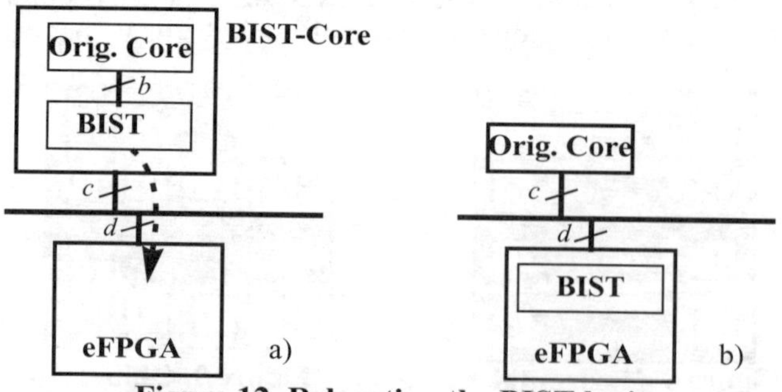

Figure 12. Relocating the BIST logic

[22][27].) In practice, this means that the core providers do not have to implement BIST in embedded cores; instead, they provide the BIST logic as an additional IP to be implemented in the FPGA core. Several conditions must be satisfied to make this transformation possible. First, the BIST logic should not be intrusive, that is, it should be easy to separate from the rest of the logic. Let *b*, *c*, and *d* denote, respectively, the number of connections between the BIST logic and the original core (without BIST), between the core (with BIST) and the system bus, and between the embedded FPGA and the system bus. The relocation of the BIST logic is possible if $b \leq c$ and $b \leq d$. Another condition is for the FPGA to supply test patterns at the speed required to test the core. Of course, the BIST logic implemented in the core would be faster than its FPGA implementation, but this speed may be higher than the one used in normal operation, when the core is accessed over the system bus. So if the FPGA can apply the test at the same speed the core is accessed over the system bus, and if the bus width conditions are also satisfied, the BIST logic may be relocated in the FPGA.

We will illustrate this scheme for testing an embedded RAM core. Figure 13 illustrates a typical RAM BIST architecture. *Data In*, *Address*, *Control*, and *Data Out* are included in the system bus. During BIST, the inputs *Data In*, *Address*, and *Control* are replaced with the signals produced by an integrated BIST controller and TPG, which generate a March test for the RAM block. The controller/TPG also produces the expected output responses to be compared with the RAM output data by the ORA, which latches any mismatches observed to produce a failure indication at the end of the BIST sequence. (Note that separate tests are needed to check the connections between the RAM block and the I/Os of the RAM core.)

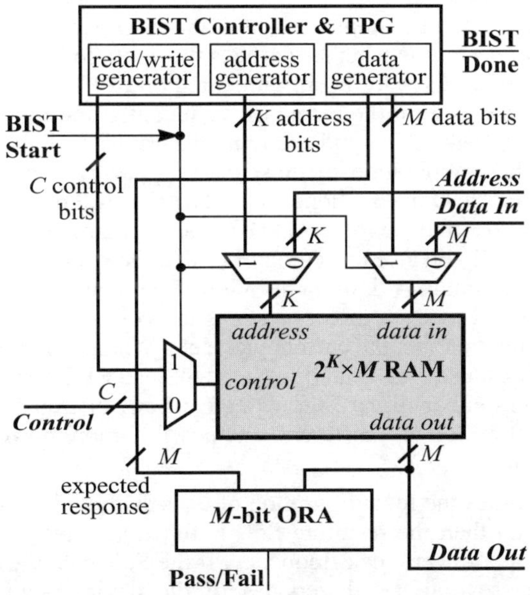

Figure 13. Typical RAM BIST architecture

Using the embedded FPGA, we can follow the scheme in Figure 12, remove all the BIST logic from the RAM core, and download the BIST controller/TPG and the ORA logic in the FPGA only when we need to test the RAM. This results in several advantages. First, the RAM core is smaller, being reduced to only the shaded RAM block, and faster, since we no longer need multiplexers to select between system and BIST data (this selection is now done by the system bus). Second, when the BIST logic is in the same core with the RAM (Figure 13), we also need to test the BIST logic to make sure that it will correctly test the RAM; this is no longer necessary when the RAM BIST logic is implemented in the FPGA core, because the self-test of the embedded FPGA (described in the previous sections) completely tests all its resources. Third, while defects in the BIST logic of a RAM core cannot be repaired and will likely cause the entire SoC to be discarded, defects in the embedded FPGA core can be easily tolerated (see Section 7).

If we have several RAM cores in the SoC, one drawback of this approach is that it can test only one RAM at a time, because the FPGA can observe the output of only one core at a time. In contrast, with the BIST logic in the same core with the RAM, all embedded RAMs may be tested in parallel, provided that the power dissipation constraints of the SoC are not violated.

A compromise solution, illustrated in Figure 14, moves only part of the BIST logic in the embedded FPGA core, while allowing several RAMs to be tested concurrently. We move only the BIST controller/TPG in the embedded FPGA, and we leave the ORA and a separate data generator block in the RAM core to produce only the expected data. The FPGA provides the same *Data In*, *Address*, and *Control* inputs to all RAMs under test. All embedded RAMs can be tested concurrently, since here results analysis is independently done within each RAM core. This scheme still avoids the BIST multiplexers which cause the performance degradation.

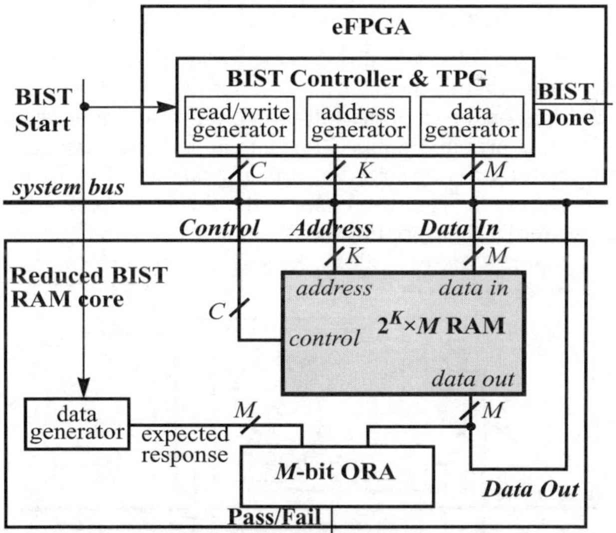

Figure 14. Partial move of RAM BIST logic

Similarly, the embedded FPGA core can be used in turn to implement (at least part of) the BIST logic for most embedded BIST-cores that are connected to the system bus. Moving the BIST logic in the embedded FPGA requires analyzing the trade-offs involving the area overhead and the performance penalties introduced by in-core BIST, the total test time, and the SoC power constraints. This analysis is the responsibility of the SoC designer; to make the choice possible, the core provider should supply both BIST and non-BIST (and, where applicable, partial-BIST) versions of the same core.

6.2 ET-Cores

In this section we consider embedded ET-cores that are tested with externally applied vectors. Typically, these cores use scan DFT and the vectors are generated by an ATPG program. Applying an external test to an embedded core is a difficult problem (for example, the core may have more pins than the SoC); the solution usually consists of inserting DFT structures to allow the input stimuli and the core responses to be transported between core and SoC pins [16]. If the ET-core is connected to the system bus, our approach is to ***change the core testing mode from external to BIST and implement the BIST circuit in the embedded FPGA.*** Note that ***the ET-core does not have to be changed***, since the source for its input vectors and the sink for its output vectors are still external to the core under test. But from the SoC's perspective, this is a significant change.

Given a full-scan circuit *C*, many procedures exist to create a TPG that produces a pseudo-random test of reasonable length that always achieves 100% fault-coverage for *C*; [7][28][30] are more recent techniques of this type. The key concept is generating vectors whose signal probabilities are dynamically changed so that the random-pattern-resistant faults [6] of *C* not yet detected will be more likely (or even certain) to be detected by the subsequent vectors. Although some of these techniques require a large area overhead for circuits with multiple scan-chains, the *area overhead is no longer a primary concern when BIST is implemented in reconfigurable logic* (the only constraint is to fit within the embedded FPGA).

Figure 15 illustrates this scheme for an ET-core with a single scan-chain. The circuit under test (CUT) is the combinational part of the full-scan circuit. An input vector is applied at primary inputs (PIs) and via the scan chain, and an output vector is observed at primary outputs (POs) and via the scan chain. The scan chain in the ET-core is extended by a register RI that supplies the PI values, and by a register RO that captures the PO values; RI and RO are implemented in the embedded FPGA, along with a TPG and an ORA. (A separate TPG per chain is needed for multiple scan chains.) The TPG provides the input sequence to the extended scan-chain, and the ORA processes the CUT outputs captured in the extended scan-chain. The ORA is a signature analyzer [6]. Note that the FPGA must have access to the serial data input, serial data output, and the scan-enable pins of the scan chain.

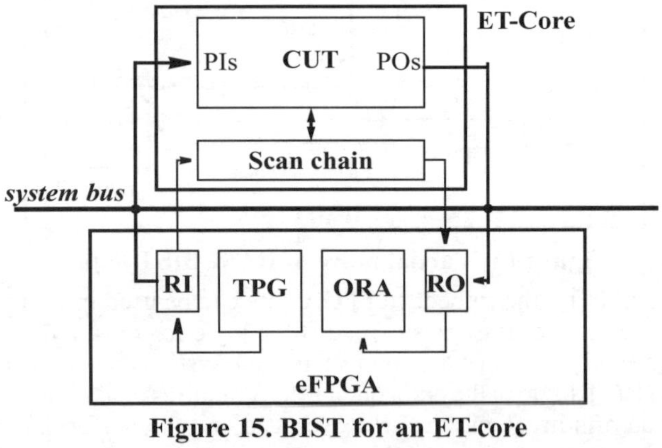

Figure 15. BIST for an ET-core

If the synthesis of the TPG is based on the structure of the CUT, which is not available to the SoC designer, then the tool that generates the TPG will be run by the core provider, and the TPG may be offered as an additional IP to be implemented in the FPGA core. Alternatively, if (non-compacted) ATPG vectors for the CUT are provided with the core, the TPG can be constructed to generate a pseudo-random test set that includes the given vectors. This solution allows the TPG to be created without access to the CUT implementation.

If this approach is feasible for every full-scan embedded core, then *the entire SoC can be tested without external vectors. The FPGA core takes turns to implement an embedded tester for every ET-core* connected to the system bus. If TREC does not need to store and apply vectors, then it can be a low-cost PC-based tester. The total test time will increase because most cores will not be concurrently tested, but this becomes less important when TREC is a low-cost tester.

7. Fault Tolerance Techniques

Our approach for user-level fault-tolerance for the embedded FPGA starts with the circuit implemented in the reconfigurable logic and with the location and type of every diagnosed fault. Note that we may have different circuits residing in the FPGA core at different times, either to perform different functions, or to implement various BIST circuits for other cores; the fault-tolerance analysis is separately done for each such circuit.

Given a circuit *C* implemented in the FPGA and a fault *f*, the first question is whether *f* affects the operation of *C*. If it does, then we have to reconfigure a region of *C* to bypass *f*. We will try to limit the extent of the reconfigured region as much as possible, to avoid changing the critical timing paths of *C*. The computation of most *fault-bypassing configurations* (FABCOs) is done by TREC. Although this computation may involve CAD tools, they are used only incrementally on a small region of *C*, and their run-times are reasonably small. A limit on the allowed computation time is established based on factors such as the cost of a SoC, the production volume, the current yield, etc. When computing a FABCO exceeds the time limit, we will let TREC test other chips, and either discard the defective SoC, or transfer the computation job to a different, and possibly more powerful, processor.

Sometimes the reconfiguration of *C* does affect its critical paths, and then the resulting SoC is functional, but it must operate at a lower clock frequency. If the SoC manufacturer provides several speed versions of the device, selling a repaired SoC as a slower device is much better than discarding it as defective.

In the process of bypassing faults in a defective FPGA core, we have created new FPGA configuration files for the defective SoC; these files should be used only with the defective chip they were generated for. A good method to maintain the association between a faulty SoC and its configuration files is to create a unique *faulty chip serial number*, and to label both the defective SoC and its configuration files using this identifier. The serial number for all the fault-free chips will be zero. The serial number will be recorded in a small non-volatile memory within the SoC, and it will be readable using a *USERCODE* boundary-scan instruction [22]. The serial number should also appear in the header of every configuration file for this SoC, and the configuration process will be allowed to proceed only if there is a match between the serial number in the SoC and the one in the configuration file.

If fault f does not affect C, then no reconfiguration for tolerating f is required. This is clearly the case when *f* affects FPGA cells or interconnect resources not used by *C*. In addition, we have developed techniques for ***using defective resources whenever possible***. This increases the effective yield without having to change *C*, and in most cases without impact on performance. Since using defective resources avoids using spare resources for fault tolerance, we will be able to tolerate more faults in the same area. Using defective resources is a major departure from most previous work in fault tolerance, whose goal is to bypass every located fault.

7.1 Compatibility

If the function of a circuit *C* implemented in the FPGA is not changed by fault *f*, *C* and *f* are said to be ***compatible***. A compatible fault does not require *C* to be reconfigured. Note that this definition allows *f* to change the timing of *C*. We will define compatibility separately for logic faults and interconnect faults.

7.1.1 Compatibility for Logic Faults

We say that a logic fault *f* in a logic cell is *compatible with the function F* to be implemented in that cell, iff *F* does not change in the presence of *f*. For example, if *F* is a combinational function using only the LUTs, any fault in flip-flops is compatible with *F*. Similarly, a fault affecting only the multiplication operation is compatible with any *F* not using multiplication. In general, any fault that affects only modules or modes of operation of the cell that are not used by *F* is compatible with *F* [3][12]. Then any such *F* can be implemented in the faulty cell, and the configuration file of the fault-free FPGA core is also applicable to this faulty SoC.

Higher diagnostic resolution can allow additional compatibility relations. For example, suppose that testing the RAM module of the cell has identified *f* as a single bit in the RAM being stuck-at-0(1). Consider any function *F* using the RAM module only as a LUT. If the value needed for the faulty bit is 0(1), *f* is compatible with *F*.

7.1.2 Compatibility for Interconnect Faults

We define a *layout* of a programmable interconnect network as the set of positions (open or closed) for all its CIPs, needed for a specific routing of the netlist. Similar to [21], we say that an interconnect fault *f* is *compatible with a layout L* iff *f* does not change the connections implemented by *L*. If *f* and *L* are compatible, then *L* can tolerate *f* and we don't have to reconfigure to bypass *f*.

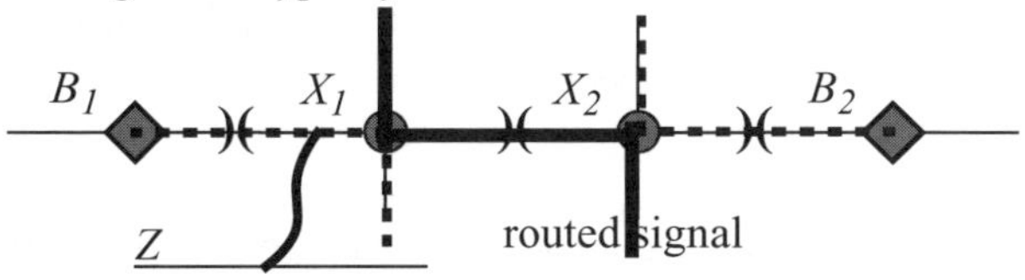

Figure 16. Illustration for fault compatibility

For example, in Figure 16, assume a layout *L* where breakpoint CIPs B_1 and B_2 are open, crosspoint CIPs X_1 and X_2 are closed, and segment *Z* is not used. Then an open on B_1X_1 is compatible with *L*, while an open in X_1X_2 is not. More interestingly, the short between B_1B_2 and *Z* is also compatible with *L*, since *Z* does not carry any signal. Thus the routed signal net can use the shorted segment B_1B_2. Table 1 gives detailed conditions that need to be met for single-fault compatibility.

Table 1. Single-fault compatibility conditions

f	Condition for *f*~*L*
segment open ***or*** CIP stuck-open ***or*** segment stuck-at	*L* does not transmit values through the faulty resource
CIP stuck-closed	*L* uses the CIP ***or*** *L* drives values only on at most one of the shorted segments
short between segments	*L* drives values only on at most one of the shorted segments ***or*** *L* uses the shorted segments for the same signal

Note that the compatibility relation between a fault *f* and a layout *L* may be affected by the presence of another fault *g*. This happens because *g* changes the structure of the programmable network on which *L* is implemented. This interaction between faults is not considered in [21]. For example, in Figure 17, assume a layout *L* where crosspoint CIPs X_1 and X_2 and are closed, and breakpoint CIPs B_1 and B_2 are open. Segment *A* is used by signal S_1 via crosspoint X_1, segment *C* is used by signal S_2 via crosspoint X_2, and segment *B* is not used. Let *f* be the short between segments *A* and *B*, and let *g* be the breakpoint B_2 stuck-closed. The single fault *f* is compatible with *L*, as is the single fault *g*. However, the multiple fault {*f*, *g*} shorts the signals S_1 and S_2, so it is not compatible with *L*.

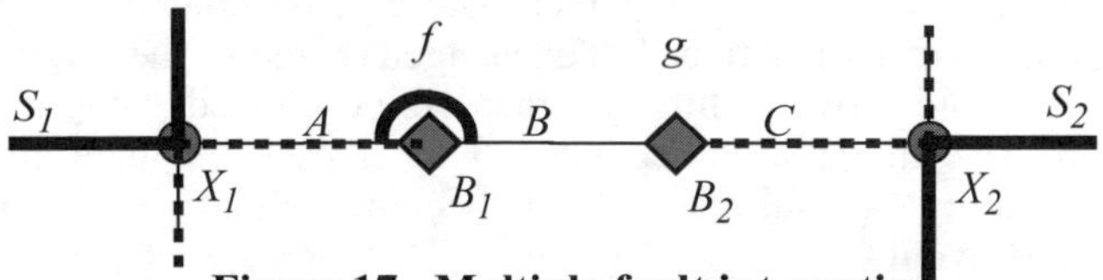

Figure 17. Multiple fault interaction

Our procedures rely on a network model coupled with a layout data base. The *network model* is a graph that represents the topological relations between the components of the programmable interconnect network - wire segments, cell pins, FPGA pins, and CIPs. The *layout data base* defines the CIP settings required to route all the signal nets in the circuit.

From the layout data base we can retrieve all the segments and CIPs used by a given signal, and determine which signal is routed through a given segment or CIP. In a fault-free network, a segment may be driven by only one signal.

To record faults in the network model, every CIP and segment has an additional field that indicates its fault status: a CIP can be fault-free, stuck-on, or stuck-off, and a segment can be fault-free, open, stuck, or shorted. For a shorted segment we also record the identity of its *partner* - the other segment involved in the short.

We determine the faulty resources that may be reused (and the ones that need to by bypassed) indirectly, by *finding the signal nets that have to be rerouted*. We say that a *signal net S is compatible with fault f* iff S can connect its source to all of its sinks in the presence of f, and no other signal drives any segment of S. (Note that f may be a multiple fault.)

The procedure to find the incompatible nets works as follows. We first insert all the located faults in the network model. Next, we process the faults to find all the signal nets affected by faults. Then we trace every affected signal S from its source to all its sinks, keeping track of the segments used. Tracing stops at open segments, at stuck-open CIPs, and at stuck segments, but it does go through stuck-on CIPs, and it jumps from a shorted segment to its partner. If faults prevent S from reaching at least one of its destinations, then S is incompatible with the faults. We record all the segments reached by every traced net. After all signals are traced, we can identify the segments driven by more than one signal; the signals that drive these segments are shorted, and only one of them may be allowed to continue to drive the shorted segments, while the other ones will be marked as incompatible. To determine which signal to leave in place, first we exclude the ones already found incompatible because the trace has not reached some of their destinations. To select among the remaining signals, we analyze the timing margins (slacks) of their paths, and chose the one with the least slack. In this way, we minimize the impact of rerouting on the system timing, since the signals to be rerouted have more slack. We break ties by selecting the longest net as the one left in place, since shorter nets are likely to be easier to reroute.

For example, in Figure 17, tracing S_1 will go through A, B, C, and continue along S_2, while tracing S_2 will go through C, B, A, and continue along S_1. Since the traced segments are recorded as being driven by both S_1 and S_2, these signals are shorted. Assuming that neither S_1 nor S_2 have other faults, we select one of them to be left unchanged (based on the analysis described above), while the other will be allowed to continue to drive the shorted segments. This is possible since after rerouting, the multiple fault will become compatible with the new layout.

In Figure 18, assume that the source of S_1 is on the left. Then tracing S_1 stops at the open segment X, and tracing S_2 reaches B via the short between B and C. Because of the open, B is driven by only one signal (S_2), so that there is no other signal shorted to S_2, which is hence compatible with the multiple fault, and does not have to be rerouted. However, S_1 needs to be rerouted.

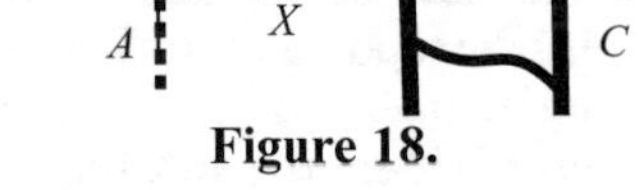

Figure 18.

7.2 Reconfiguration

A tool like JBITS [29] would be ideal to generate FABCOs by directly modifying configuration files without involving any CAD tools. Since no such tools are commercially available, we rely on commercially available place-and-route tools used incrementally. The concept of a fault, however, is alien to these tools, and we need modeling workarounds to make them avoid incompatible faulty resources. To avoid a defective logic cell, we program it to implement a dummy logic function whose inputs and outputs are unconnected. To avoid a defective CIP or wire segment, we program a dummy signal net with no source or sink to occupy the faulty resource and its immediate neighbors.

When we consider reusing a wire segment S which is shorted to an unused segment, the delay of S increases because of the additional capacitance provided by its partner. However, a conventional path tracing tool that reports the delay along the traced path, will not see the additional delay caused by the short. Our solution is to add the delay of the unused shorted segment to the delay of S, and if the new delay causes any path going through S to become too slow to operate at the specified clock speed, we give up on reusing S and mark the short as incompatible. Although this solution may be conservative in overestimating the delay of S, it guaranteed that the short will not affect the critical paths in the circuit.

7.2.1 Reconfiguration for Logic Cells

Many times the function of a logic cell does not utilize all its internal resources. In such cases, when a cell fault f is incompatible with the cell function F, it is often possible to avoid f by small changes in the way F is implemented. For example, if one of the LUTs in the cell is faulty, but F is not using all the LUTs, we remap F to use only fault-free LUTs, which makes f compatible and the cell reusable. Of course, we need to incrementally reroute the signals that connected to the inputs and output of the faulty LUT, to bring them to the replacement LUT.

Another example is a 3-input combinational logic function mapped to a 4-input LUT. This is implemented by setting identical values to every pair of RAM bits whose addresses differ only in the value of the "don't care" address bit, so that the RAM has two identical halves. In this case we can tolerate any combination of faults that affect only one of the two halves, by simply setting the previously unused bit to the value that selects the fault-free half. This also requires an incremental reconfiguration.

If a defective cell has a logic fault incompatible with the function assigned to this cell, we relocate the function to a compatible unused cell, preferably one adjacent to the defective cell (see Figure 19). If we cannot move the function to a compatible cell adjacent to the defective one, we set up a

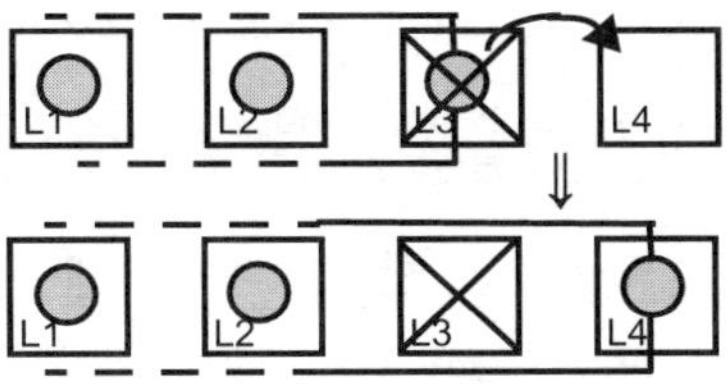

Figure 19. Cell replacement

chain of cell replacements [11], as illustrated in Figure 20. This approach averages the overall change in signal disturbance, but increases the number of signals that must be rerouted. Note that in our approach, no replacement cell (including the unused one) is required to be fault-free, but only compatible with the desired function. After cell functions are relocated, we use an incremental router to reroute the signal nets connected to the moved cells. Also we need to analyze the changes in the timing of the rerouted signals, to determine whether the chip should operate at the original clock frequency or a reduced one.

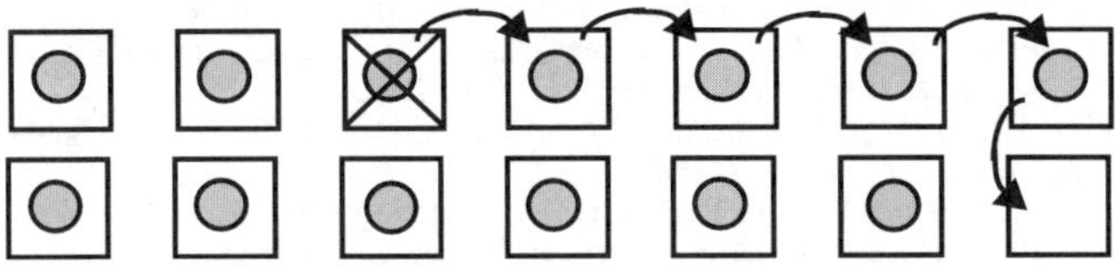

Figure 20. Chain of cell replacements

As a preprocessing step, we prepare a FABCO for each one of the used cells in the target circuit. This not a computationally expensive job, since all the circuit transformations are only incremental. These precomputed configurations are stored on the TREC disk, and are applied by TREC based on the diagnosis results. Note that the same unused cell can be designated as replacement for several working cells. After applying one FABCO, several other precomputed configurations may no longer be valid, since the circuit for which they were derived has been changed. It is another TREC task to keep track of the validity of precomputed FABCOs. If the FABCO for a defective cell has been invalidated, TREC computes a new one, by incrementally rerouting the signal nets connected to the moved function.

Another major advantage of our technique over conventional techniques is that we handle groups of faults in a tight area. Other techniques limit the number of faults they can handle by providing a limited number of spare logic resources in a region of the FPGA. When these local spares are exhausted, their fault tolerant techniques fail. We provide a global reconfiguration approach that uses *minimax matching* [19] to match defective cells to compatible cells, so that the maximum distance between any defective cell and its replacement cell is minimized.

7.2.2 Reconfiguration for Interconnect

Our experimental results show that in most practical circuits implemented in FPGAs, a large number of interconnect faults are compatible with the circuit [13]; this happens because many interconnect resources are unused, even in circuits where more than 70% of the cells are used. Every such fault can be tolerated without reconfiguration.

Next we describe our multi-step approach for avoiding incompatible interconnect faults. In the following procedure, S is the set of nets that have to be rerouted.

Stage 1: First we process incompatible faults that do not allow some signal nets to connect to inputs or outputs on a cell, because such faults may need reconfiguration for both logic and interconnect. If an input or an output of a cell is blocked, and the cell is not fully used, then we try to move the function of the blocked resource to an unused resource in the same cell. Figure 21 shows an example of a stuck-open CIP denying access to an input pin on LUT B. If LUT A is not used, we first try to move the function from LUT B to LUT A; keeping the LUT function within the same cell is less likely to introduce delays. For this, we reconfigure the cell logic and we rip-up the signal nets going to the inputs and outputs of LUT A [17]. If such a transformation is not possible within the blocked cell, we rip-up all the signal nets attached to the cell, and move the cell function a new compatible unused location. In either case we add the ripped-up signals to S

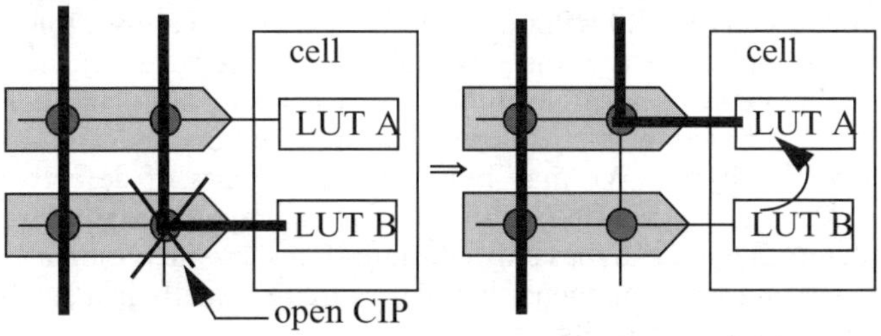

Figure 21. Reconfiguring a cell to bypass a CIP fault

Stage 2: We determine all the signal nets incompatible with the remaining faults using the procedure described in Section 7.1. For every such net we perform a partial rip-up, trying to maintain as much as possible from its original routing, and add the signal to S. For example, Figure 22a shows a signal net that was originally routed through segment X_1X_2 that is now open. Here we rip-up only the faulty segment.

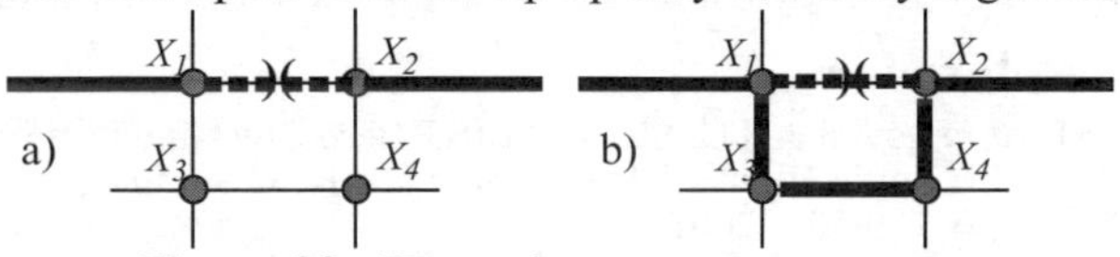

Figure 22. Bypassing an open segment

Stage 3: In this stage we attempt to incrementally reroute each signal net in S, without changing the routing of other non-affected signal nets. For example, in Figure 22a, if X_3 and X_4 are not used, the signal may be easily rerouted as shown in Figure 22b. Every successfully rerouted net is removed from S. If S becomes empty, we stop.

Stage 4: Here we use a more global transformation, completely ripping-up all the signal nets within a variable-sized window in the area around the signal nets in S, adding all the ripped-up signals to S, and rerouting all signals in S. If we are not successful, we increment the window size and repeat the process. If S becomes empty, we stop.

Stage 5: Here we do a new place-and-route for the entire circuit (with the faults marked by dummy signals).

Because many of interconnect faults are compatible with the layout, the last two stages are unlikely to be used in practice. Since each stage is computationally more expensive than its preceding one, this procedure will be subject to run-time bounds as described at the beginning of Section 7.

8. Conclusions

In this paper, we have shown that an embedded FPGA core is an ideal host to implement infrastructure IP for yield improvement in a bus-based SoC. The reconfigurable logic allows the infrastructure IP to be created only when needed, and enables BIST without any area overhead or performance degradation. Our BIST techniques achieve practically complete fault coverage for both logic cells and interconnect, and our adaptive diagnosis techniques can locate almost any combination of faulty cells or interconnect faults, and can also identify faults within defective cells. Such accurate diagnosis is a key factor in achieving effective repair for the embedded FPGA. Instead of the conventional approach of reconfiguring to bypass every located fault, our fault tolerance approach reuses defective resources whenever possible. This increases the effective yield without reconfiguration, and allows tolerating more faults in the same area. Reuse is based on new techniques to determine compatibility between faults in a cell and its function, or between a group of interconnect faults and the layout. We may reuse fault-free parts of defective cells and even shorted segments. For incompatible faults we determine fault-bypassing configurations by incrementally changing the implementation of the circuit, usually in a small area around the faults.

An embedded FPGA core makes significant changes to the SoCs testing paradigm. After the FPGA core is fully tested and repaired if needed, we can use it to provide embedded testers in turn for other cores in the SoC. This allows non-intrusive BIST logic to be removed from cores, reducing area overhead and delay penalties. It also enables cores designed to be tested with external vectors to be tested with BIST, and the entire SoC to be tested with a low-cost tester.

9. References

[1] M. Abramovici and C. Stroud, "BIST-Based Test and Diagnosis of FPGA Logic Blocks," *IEEE Trans. on VLSI Systems*, Vol. 9, No. 1, pp. 159-172, 2001

[2] M. Abramovici, C. Stroud, B. Skaggs, and J. Emmert, "Improving On-Line BIST-Based Diagnosis for Roving STARs", *Proc. IEEE Intn'l. On-Line Test Workshop*, pp. 31-39, 2000

[3] M. Abramovici, C. Stroud, S. Wijesuriya, C. Hamilton, and V. Verma, "Using Roving STARs for On-Line Testing and Diagnosis of FPGAs in Fault-Tolerant Applications," *Proc. Intn'l. Test Conf.*, pp. 973-982, 1999

[4] M. Abramovici, J. Emmert, and C. Stroud, "Roving STARs: An Integrated Approach to On-Line Testing, Diagnosis, and Fault Tolerance for FPGAs in Adaptive Computing Systems," *Proc. Third NASA/DoD Workshop on Evolvable Hardware*, pp. 73-92, 2001

[5] M. Abramovici, C. Stroud, and J. Nall, "BIST-Based Diagnosis of FPGA Interconnect," submitted to *Intn'l. Test Conf.*, 2002

[6] P. H. Bardell, W. H. McAnney, and J.Savir, *Built-In Testing for VLSI: Pseudorandom Techniques*, John Wiley and Sons, 1987

[7] N. Basturkmen, S. M. Reddy, and I. Pomeranz, "Pseudo Random Patterns Using Markov Sources for Scan BIST," submitted to *Intn'l. Test Conf.*, 2002

[8] R. Cliff *et al.*, "Programmable logic devices with spare circuits for replacement of defects," *U.S. Patent 5,485,102*, Jan. 1996

[9] K. Chakraborty and P. Mazumder, Fault-Tolerance and Reliability Techniques for High-Density Random-Access Memories, Prentice Hall, 2002

[10] B. Culbertson *et al.*, "Defect Tolerance on the Teramac Custom Computer," *Proc. 5th IEEE Symp. on Field-Programmable Custom Computing Machines*, pp. 140-147, 1997

[11] J. Emmert and D. Bhatia, "Partial Reconfiguration of FPGA Mapped Designs with Applications to Fault Tolerance and Yield Enhancement," *Proc. Intn'l Conf. on Field-Programmable Logic*, pp. 141-150, 1997

[12] J. Emmert, C. Stroud, B. Skaggs, and M. Abramovici, "Dynamic Fault Tolerance in FPGAs via Partial Reconfiguration," *Proc. IEEE Symp. on Field-Programmable Custom Computing Machines*, pp. 165-174, 2000

[13] J. Emmert, S. Baumgart, P. Kataria, A. Taylor, C. Stroud, M. Abramovici, "On-line Fault Tolerance for FPGA Interconnect with Roving STARs," *IEEE Intn'l. Symp. on Defect and Fault Tolerance in VLSI Systems*, pp. 445-454, 2001

[14] I. Harris and R. Tessier, "Interconnect Testing in Cluster-Based FPGA Architectures", *Proc. AMC/IEEE Design Automation Conf.*, pp. 49-54, 2000

[15] I. Harris and R. Tessier, "Diagnosis of Interconnect Faults in Cluster-Based FPGA Architectures", *Proc. IEEE Intn'l Conf. on Computer Aided Design*, pp. 472-476, 2000.

[16] IEEE P1500 Standard for Embedded Core Test, *http://grouper.ieee.org/groups/1500/*

[17] V. Lakamraju and R. Tessier, "Tolerating Operational Faults in Cluster Based FPGAs", *Proc. ACM Intn'l. Symp. on FPGAs*, pp. 197-194, Febr. 2000

[18] Lattice Semiconductor Co., *http://www.latticesemi.com/products/fpga*

[19] T. Leighton and P. Shor, "Tight Bounds for Minimax Grid Matching with Applications to Average Case Analysis of Algorithms," *Proc. Symp. on Theory of Computing*, pp. 91-103, 1986

[20] E. McCluskey, "Verification Testing - A Pseudoexhaustive Test Technique," *IEEE Trans. on Computers*, Vol. C-33, No. 6, pp. 541-546, June, 1984.

[21] K. Roy and S. Nag, "On Routability for FPGAs Under Faulty Conditions," *IEEE Trans. on Computers*, Vol. 44, pp. 1296-1305, 1995

[22] A. Russ and C. Stroud, "Non-Intrusive Built-In Self-Test for FPGA and MCM Applications," *Proc. IEEE Automatic Test Conf.*, pp. 480-485, 1995

[23] "Standard Test Access Port and Boundary-Scan Architecture," *IEEE Standard P1149.1*, 1990

[24] C. Stroud, S. Konala, P. Chen, and M. Abramovici, "Built-In Self-Test for Programmable Logic Blocks in FPGAs (Finally, A Free Lunch: BIST Without Overhead!)", *Proc. IEEE VLSI Test Symp.*, pp. 387-392, 1996

[25] C. Stroud, M. Lashinsky, J. Nall, J. Emmert, and M. Abramovici, "On-Line BIST and Diagnosis of FPGA Interconnect Using Roving STARs," *Proc. IEEE Intn'l. On-Line Test Workshop*, pp. 31-39, 2001

[26] C. Stroud, S. Wijesuriya, C. Hamilton, and M. Abramovici, "Built-In Self-Test of FPGA Interconnect," *Proc. Intn'l. Test Conf.*, pp. 404-411, 1998

[27] C. Stroud, *A Designer's Guide to Built-In Self-Test*, Kluwer Academic Publishers, 2002.

[28] N.A. Touba and E.J. McCluskey, "Bit-Fixing in Pseudo-Random Sequences for Scan BIST", *IEEE Trans. on CAD*, Vol. 20, No. 4, pp. 545-555, Apr. 2001.

[29] P. Sundararajanand and S. A. Guccione, "Run-Time Defect Tolerance using JBits," *Proc. ACM Intn'l. Symp. on FPGAs*, pp. 193-200, Febr. 2001.

[30] S. Wang, "Low Hardware Overhead Scan Based 3-Weight Weighted Random BIST," *Proc. Intn'l. Test Conf.*, pp. 868-877, 2001

[31] Xilinx, Inc., *http://www.xilinx.com/products*

A Proof Engine Approach to Solving Combinational Design Automation Problems

Gunnar Andersson, Per Bjesse, Byron Cook
Prover Technology
{guan,bjesse,byron}@prover.com

Ziyad Hanna
Intel Corporation
ziyad.hanna@intel.com

ABSTRACT

There are many approaches available for solving combinational design automation problems encoded as tautology or satisfiability checks. Unfortunately there exists no single analysis that gives adequate performance for all problems of interest, and it is therefore critical to be able to combine approaches.

In this paper, we present a proof engine framework where individual analyses are viewed as strategies—functions between different proof states. By defining our proof engine in such a way that we can compose strategies to form new, more powerful, strategies we achieve synergistic effects between the individual methods. The resulting framework has enabled us to develop a small set of powerful composite default strategies.

We describe several strategies and their interplay; one of the strategies, variable instantiation, is new. The strength of our approach is demonstrated with experimental results showing that our default strategies can achieve up to several magnitudes of speed-up compared to BDD-based techniques and search-based satisfiability solvers such as ZCHAFF.

Categories and Subject Descriptors

B.6.3 [**Hardware**]: Logic Design—*Design Aids*

General Terms

Algorithms, Design, Verification

1. INTRODUCTION

As many combinational design automation problems are NP- or coNP-hard, one fundamental approach to their solution is to model the problems as propositional logic formulas, and to apply proof methods to decide whether the formulas are tautologies or not.

However, there exists a plethora of techniques for discharging propositional proof conditions, including (but not limited to) the DPLL method [5], Stålmarck's method [11], Binary Decision Diagrams [2], and rewriting. Each proof technique has a particular characteristic in terms of space and time behavior, sensitivity to the size of the system, and the particular domain it will work the best for. As a result, it is not unusual to combine techniques in different ways when solving particular classes of combinational design automation problems.

At Prover Technology, we develop and maintain a plug-in proof engine that is used to solve combinational logic problems in design automation tools. This proof engine, PROVER CL, is delivered with default composite analyses for a number of domains.

As developers and maintainers of PROVER CL, we need to experiment with new variations of the default analyses so that we can keep them updated. We also need to develop new analyses that work well in new problem domains, and on problems with particular characteristics. It is thus important that our framework is constructed in such a way that (1) we can get the maximum amount of synergy between the individual methods, and (2) we can construct composite analyses with a minimum amount of effort.

In this paper, we present an approach to addressing these issues that we have been developing for several years—initial results were reported in early 2000 [4].

Our approach builds on two key ideas. First of all, we structure our proof engine in such a way that we view individual proof techniques as *strategies*; that is, functions between different proof states. Second, we define our strategies to be compositional in the sense that they can be combined in any order to form more powerful proof-search strategies.

As our experimental results indicate, the resulting compositional approach allows us to get synergy effects that provide order of magnitude speedups compared to individual analyses such as the ZCHAFF [8] search method, and state-of-the-art BDD-based analyses.

2. RELATED WORK

What we describe in this paper is an industrially proven framework for creating composite analyses for arbitrary combinational design automation problems (as opposed to just test pattern generation, for example). Our framework is carefully structured in such a way that any analysis from a large set can be composed in any order. We are not aware of any previous attempt to come up with such a general framework for solving problems in all of the many subdomains of combinational design automation.

However, a number of previous approaches to solving problems in particular subdomains of combinational design automation can be viewed as composite strategies in the sense that we will present. This is especially true in the domain of combinational equivalence checking, where combinations of BDDs, SAT and rewriting have been suggested by a number of different researchers [3, 6, 7, 9]. The most compelling example is probably the filter-based analysis for combinational equivalence checking that was proposed by Jain and coworkers in [7]. In this paper, the overall analysis is a sequential composition of a small set of *filters*. Each filter inspects a set of conjectured equivalences, and selects a few of them that it tries to prove using some technique. The filter-based analysis is hence a particular simple composition of individual analyses, each of which we can see as a strategy.

The view of strategies that we adopt in this paper is similar to the notion of tactics and tacticals [10] developed for LCF-style theorem provers in the early 1970s. The ideas behind tactics and tacticals have since proved to be fruitful enough that most theorem provers for first order and higher-order logics use variations on this approach.

In essence, our contributions in this paper are (1) to take the research from the higher-order domains and apply it in the domain of completely automated correctness proofs for combinational logic, (2) to describe how some classical proof methods and a novel method for simplifying SAT problems can be viewed as strategies, and (3) to present our experiences and some experimental results that illustrate the power of the approach.

3. PRIMITIVE STRATEGIES

The core problem that we are interested in solving is as follows. Given a propositional logic formula ϕ, we want to decide whether this formula is a *tautology*—that is, we want to decide whether the formula evaluates to true for all assignments to the variables in ϕ. We solve this problem by trying to prove ϕ using a *reductio ad absurdum* argument, where we use a combination of different analyses to check whether the negation of the formula is unsatisfiable.

We model individual analyses as strategies; that is, as functions

$$f : (\texttt{PState} \times \texttt{Parameters}) \rightarrow ([\texttt{PState}] \times \texttt{Res})$$

that take a proof state and some parameters, and return a list of new proof states together with information about the result of the analysis. The result is one of the values `Sat`, `Unsat`, and `Indet`, and we require that all strategies return `Indet` precisely when the returned list of proof states is nonempty.

A proof state contains a number of objectives $\texttt{Obj}_1 \ldots \texttt{Obj}_k$, and a set of current-time assumptions `Asmpt`. The objectives are formulas, and the assumptions are equivalence classes of negated and unnegated subformulas of the objectives.

Intuitively, a proof state represents the problem of deciding whether there is an assignment to the variables in the objectives that (1) makes all the formulas in each assumption equivalence class evaluate to the same value, and that (2) makes all of the objectives evaluate to true. We model the situation at the start of the overall proof search as a proof state that contains no assumptions together with the single objective $\neg\phi$; this represents the notion that we want to prove that ϕ is a tautology by deciding whether there is a model of the negation of ϕ.

Now, assume that *AllSat* is a mathematical function that decides problems represented as proof states. A strategy is *correct* if it fulfills two criterias:

1. If the list of returned proof states l is empty, then the returned outcome agrees with the result of $AllSat(p)$.

2. If the list of returned proof states l is nonempty, then $AllSat(p)$ holds precisely when $AllSat(p_2)$ holds for some proof state p_2 in l.

We require that all strategies in our framework are correct.

There are a number of additional properties of strategies that we can define as follows. Assume that the strategy *strat* is given the proof state p and returns the list of new proof states l together with the evaluation result *res*. The call to the strategy *strat* is

- *decisive* if l is an empty list.
- *refining* if l is a singleton list containing the original proof state with some extra assumptions.
- *splitting* if l is a list of length at least 2.

Most strategies are decisive in some cases, refining and splitting in others. However, if we want to guarantee that we can solve our top level goal, we need to construct an overall strategy that always is decisive.

Let us now consider how three previously known and one novel algorithm for satisfiability solving can be cast as strategies in our framework.

3.1 Stålmarck's saturation method

Stålmarck's *saturation method* is an algorithm for propositional proof [11] that has been applied to solve a variety of problems in software and hardware verification.

An application of the saturation algorithm takes three parameters as input: (1) a set of formulas $\phi_1, \ldots, \phi_n$, (2) an equivalence relation over negated and unnegated subformulas of the set of formulas, and (3) a number k, called the *degree of saturation*. Given these inputs, the saturation algorithm generates an extension of the equivalence relation that provably must hold if the original equivalence relation holds. The key idea in the saturation algorithm is to extend the relation with new equivalences that can be inferred from the original relation using *short* proofs. A proof is k-short if not more than k simultaneous assumptions are needed in the Dilemma [11] proof system.

Saturation is normally used for generating propositional proofs by contradiction: We first construct an equivalence relation where the negation of the formula is put in the same class as the value *True*, and then apply the saturation algorithm. If the formula is provable, there will exist a degree of saturation for which the resulting equivalence relation is *contradictory*, in the sense that the values *True* and *False* are in the same equivalence class.

In the present context, we are interested in constructing a strategy that we can use to process problems represented as proof states. The approach we take is to construct a wrapper around the saturation procedure that takes a proof state as argument together with a saturation degree. From the proof state, we construct an initial equivalence relation from the assumptions Asmpt, and add the formula $\texttt{Obj}_1 \wedge \ldots \wedge \texttt{Obj}_k$ to the class containing *True*. We then apply the saturation algorithm to the equivalence relation, with the given degree of saturation, and compute the strategy result in the following way: If the result of the saturation is contradictory, we return the result ([ ], Unsat). If the equivalence relation is noncontradictory, and all variables in the formula are in the equivalence classes containing *True* and *False*, then we return the result ([ ], Sat). Otherwise we return ([PState′], Indet), where PState′ is the original proof state augmented with assumptions corresponding to the resulting equivalence classes. Our saturation strategy is thus either decisive or refining.

3.2 DPLL

The Davis-Putnam-Loveland-Logemann algorithm [5] is a popular search-based satisfiability method. The DPLL procedure takes as input a formula in Conjunctive Normal Form (CNF). A CNF formula is a conjunction $\bigwedge_{k=1}^{n} C_k$, where each *clause* C_k is a disjunction $\bigvee_{i=1}^{m_k} l_i$ of literals—negated or unnegated variables.

The DPLL procedure attempts to find a satisfying valuation for the variables in the CNF formula by recursively selecting a variable x and case-splitting on its value. If either of the problem instances where x is set to true or false is satisfiable, then the whole problem must be satisfiable. Equivalently, if none of the instances are satisfiable, then the whole problem is unsatisfiable.

A naive implementation of the DPLL procedure that only uses case splitting to search for a satisfying assignment is theoretically both sound and complete. However, it is rarely computationally feasible to apply it to real life formulas as too many case splits will be required to decide satisfiability. As a consequence, most DPLL implementations use a number of optimizations.

One of the most important optimizations is the generation of *conflict clauses*. Whenever it is detected that the currently explored partial assignment makes some clause evaluate to false, it is possible to find a subset of the partial assignment which is the reason for this conflict. Assume for example that this subset is the valuation $(x := 0, y := 1, z := 0)$. If the set of clauses at hand is satisfiable, we know that none of its models can contain this particular subvaluation. We can express this piece of information as a constraint—a conflict clause—of the form $x \vee \neg y \vee z$. Now, as this clause is a consequence of the satisfiability of the CNF formula, it is sound to add it to our set of clauses, where it will restrict the remainder of the search.

Let us now consider how to make a strategy out of the DPLL procedure. Our strategy takes a proof state and a timeout value as arguments, encodes the proof state in clausal form, and runs our DPLL implementation on the resulting CNF formula. If the algorithm terminates without timing out, then we can decide the problem at hand. However, even if our search times out, valuable information will have been generated in the form of conflict clauses that we can use to augment the proof state we return. There are a number of choices for how we can do that. Here, we go with the simplest possible solution, and investigate the generated conflict clauses to see if we have any clauses containing a single literal. If so, then we add a new assumption for each one literal clause in such a way that the variable is forced to the value indicated by the literal. Just as in the case of the saturation strategy, the DPLL strategy is hence either decisive or refining.

3.3 BDD-based cut

A standard approach to deciding the satisfiability of a propositional formula is to construct its Binary Decision Diagram (BDD) and check whether there is at least one path from the BDD's top node to the constant node *True*. However, it is often not feasible to build BDDs for complex formulas. We therefore adopt the ideas in [1] and construct our BDD-based strategy in such a way that the user can set a limit on the size of the formula that we build a BDD for.

We define the *construction depth* of a node in a parse tree to be the number of nodes between it and the root of the tree. Hence, the root of a parse tree has construction depth 0. Our BDD strategy takes a proof state as argument together with a construction depth k, and makes a cut in the parse tree for $\texttt{Obj}_1 \wedge \ldots \wedge \texttt{Obj}_k$ at depth k. Assume that this cut intersects the nodes $\texttt{Node}_1 \ldots \texttt{Node}_k$. We substitute fresh variables $\texttt{Var}_1 \ldots \texttt{Var}_k$ for each of the intersecting nodes, and build a BDD for the top node of the resulting simplified parse tree.

If the cut depth of the parse tree is equal to or greater than the maximum construction depth of a variable in the original formula, then the resulting BDD allows us to decide the problem represented by our proof state. However, if the parse tree is cut at a lower level, what do we know?

Consider the paths in the generated BDD that end up at the *True* terminal. Each such path represents a valuation of a subset of the original variables and internal nodes that guarantees that the top node becomes true. Note that such a path may or may not correspond to a satisfying assignment: Given a set of values for internal nodes in the formula, we can not be sure that there exists an extension of the assignment to the original variables that generates this particular combination of internal values. However, we do know that if there exists a satisfying assignment, then it will be covered by one of the paths to *True*. We therefore define our BDD strategy to be splitting when it is not decisive by letting it generate one derived subproblem for each path to *True* in the following way.

Assume that we have a path to *True* that encodes the assignment $(\texttt{Var}_1 := 0, \texttt{Var}_2 := 1, \texttt{Var}_3 := 0)$, and that $Def(x)$ is an operator that returns the formula that is assigned the name x. Then the new proof state that we generate to represent the derived subproblem contains the objectives $\neg\, Def(\texttt{Var}_1)$, $Def(\texttt{Var}_2)$ and $\neg\, Def(\texttt{Var}_3)$, together with the previous assumptions.

3.4 Variable instantiation

The previous strategies are all based on previously known algorithms. We now present *variable instantiation*, one of the many in-house strategies that we have developed.

Let us define two formulas to be *equisatisfiable* if one is satisfiable precisely if the other is satisfiable. Now, assume that we want to check the satisfiability of a formula ϕ that contains the variable x. Also assume that we know that ϕ is equisatisfiable with the result of substituting the value *val* for every occurrence of x in ϕ. Then it clearly suffices to check the instantiated, simpler, formula for which there are only half as many assignments possible.

We can decide whether it is safe to instantiate a given variable as follows. Consider the case where we want to know whether it is safe to instantiate x in ϕ with the value *True*. If $\phi(x{:=}\,True)$ is satisfiable, it is clear that ϕ is satisfiable. However, in order to safely be able to set x to *True*, we also need to know that ϕ is unsatisfiable when $\phi(x:=True)$ is unsatisfiable. We can check whether this holds by trying to prove the formula $\phi(x{:=}\,False) \rightarrow \phi(x{:=}\,True)$: The provability of this formula implies that if $\phi(x{:=}\,True)$ is unsatisfiable, then there can be no models at all for ϕ, regardless of the value of x.

Let us now consider how to construct a strategy based on variable instantiation. Our variable instantiation strategy takes a proof state containing a set of objectives $\texttt{Obj}_1 \ldots \texttt{Obj}_k$ as input, and a set of assumptions `Asmpt`. For each variable that is not in the equivalence classes of *True* and *False* we apply variable instantiation in the following way.

Assume that the current variable under consideration is x. To check whether x can be set to *True*, we construct a new proof state containing the original assumptions, and the single objective

$$\neg(\bigwedge_{i=0}^{k} \texttt{Obj}_i(x{:=}\,False) \;\rightarrow\; \bigwedge_{i=0}^{k} \texttt{Obj}_i(x{:=}\,True))$$

We then invoke some fast but potentially incomplete strategy, say saturation of degree 0, and check the result. If the result is `Unsat`, then it is safe to set x to *True*. If not, we check whether it is safe to set x to *False* in the same way. Once we know that we can set a variable to some value, we substitute this value in all the original objectives and assumptions and move on to the next variable. If there are no variables left after an iteration over the variables, then we have solved the problem. Otherwise we return the resulting proof state together with the result `Indet`. The variable instantiation strategy is thus decisive or refining.

4. COMPOSING STRATEGIES

The primitive strategies in Section 3 are all carefully designed to behave in a similar fashion: They take a parameter as argument that directly or indirectly controls how much time will be spent on the problem, and then apply some analysis to the given proof state. This analysis either decides the problem given the resource bounds, or generates one or more simpler problems.

However, so far we have no way to compose the individual analyses to form new, more powerful, analyses. We now explore this.

Recall our definition of individual strategies as functions

$$f : (\texttt{PState} \times \texttt{Parameters}) \rightarrow ([\texttt{PState}] \times \texttt{Res})$$

that takes a proof state and some parameters `Parameters` as inputs. In the case of the saturation strategy, the single parameter is the saturation degree, and in the case of the DPLL strategy the parameter is the cutoff time. Informally, the way we will do composition is to allow `Parameters` to also contain an analysis to use if the problem is indeterminate given the resource constraints. After the post-processing analysis has been applied to every generated subproblem, the primitive strategies computes the overall result as follows. If one of the subproblems is diagnosed as satisfiable by the post-processing analysis, then the overall problem is satisfiable. If all are contradictory, then the overall problem is contradictory. Otherwise the result is indeterminate.

In order for every strategy to be able to use any other strategy to post-process indeterminate results, we need to settle on a standardized strategy interface. We define the type `StratI` of *strategy instances* to be the type of functions of the form

$$f_{inst} : \texttt{PState} \rightarrow ([\texttt{PState}] \times \texttt{Res})$$

and provide a trivial strategy instance `triv` that takes a proof state ps and returns $([ps], \texttt{Indet})$. We can now modify the signature of our example primitive strategies to take a post-processing strategy instance as an additional argument:

$$\texttt{sat} : (\texttt{PState} \times \texttt{Degree} \times \texttt{StratI}) \rightarrow ([\texttt{PState}] \times \texttt{Res})$$
$$\texttt{cut} : (\texttt{PState} \times \texttt{Level} \times \texttt{StratI}) \rightarrow ([\texttt{PState}] \times \texttt{Res})$$
$$\texttt{dpll} : (\texttt{PState} \times \texttt{Time} \times \texttt{StratI}) \rightarrow ([\texttt{PState}] \times \texttt{Res})$$
$$\texttt{inst} : (\texttt{PState} \times \texttt{StratI} \times \texttt{StratI}) \rightarrow ([\texttt{PState}] \times \texttt{Res})$$

In the case of the variable instantiation strategy, this results in two strategy instance arguments. The first is used to decide instantiation safety, and the second is used for post-processing.

As things stand, the only strategy instance we can provide for post-processing is the trivial strategy instance `triv`, and this is not very interesting. However, it is easy to turn a strategy into a strategy instance by wrapping it in a function that provides particular parameter values (including the strategy instance that is to be used for the proof states resulting from an indeterminate result). We can hence achieve sequential composition by using appropriate wrappers and `triv` to chain together individual strategies from the bottom up.

As an example, assume that we are interested in writing a composite strategy instance that applies DPLL for a hundred seconds only. We can do this by writing a wrapper that passes `dpll` the given proof state together with the value 100 and the strategy instance `triv`. After the DPLL invocation has finished, the wrapper takes the result from the DPLL strategy, and returns it. The resulting composite strategy instance can then be used to build a new strategy instance that first does a BDD cut on construction depth 20, and then applies DPLL to the resulting problems for a hundred seconds:

```
dpllI(ProofState) :=
  return dpll (ProofState, 100, triv)

cutDpll(ProofState) :=
  return cut (ProofState, 20, dpllI)
```

Note that we are not restricted to constructing strategy instances that are *static* in the sense that they always execute a set of strategies in a particular order, without any flexibility. We are writing the wrappers ourselves, so it is simple to construct strategy instances that examines the result returned from a strategy and acts based on the outcome. For example, assume that we want to use a BDD cut to generate a list of new problems, and apply `dpll` for some period of time on the results that fulfill some predicate, and saturation on the others. The flexibility of our framework makes this easy to do by providing an appropriate instance wrapper in the call to `cut`.

In our framework, we can thus construct complex composite analyses with a few lines of programming. However,

Instance	PROVER CL (s)	ZCHAFF (s)
c1355	0.3	2.4
c1908_bug	0.0	3.2
c1908	0.0	3.5
c2670_bug	0.2	0.0
c2670	0.1	1.8
c3540_bug	1.0	0.0
c3540	1.0	51.6
c432	0.0	0.1
c499	0.1	4.5
c5315_bug	1.7	1.1
c5315	0.8	46.4
c6288	4.2	[>60 min]
c7552_bug	2.2	0.4
c7552	2.2	84.5
c880	0.1	1.4

Table 1: ISCAS'85 benchmark results

when we are experimenting with different compositions (for example, when developing default analyses), it is desirable to avoid writing code and recompiling altogether. To address this issue, we have developed a domain specific language for describing composite strategy instances, which we interpret using a command line tool called CAPTAIN PROVE. This tool takes a formula as an argument, together with a text file that describes a strategy instance that it should apply to decide provability of the formula. Internally, CAPTAIN PROVE constructs an initial proof state, and then applies the described instance.

A simple description of a composite analysis in our strategy language looks as follows:

```
cut  10.
instantiate (dpll 10.).
sat  1.
dpll 100.
```

This strategy first does a cut at level 10. The resulting proof states are then refined by instantiation, where DPLL for ten seconds is used to decide whether a variable can be instantiated. Next, saturation of degree one is performed, followed by a hundred seconds of DPLL.

5. EXPERIMENTAL RESULTS

In order to demonstrate the power of the strategy approach, we have run three sets of experiments. Our objective is to demonstrate that the synergy effects between the many analyses that we combine can generate order of magnitude speedups, compared to individual techniques.

In the first two sets of experiments, we solve problems from the public benchmark collections ISCAS'85 and Satlib. The two suites represent different classes of design automation problems: The ISCAS'85 suite contains combinational equivalence checking problems, and the Satlib suite consists of bounded model checking problems. In the experiments, we compare the result from PROVER CL's default analyses against results from ZCHAFF [8], a state-of-the-art DPLL solver. The version of PROVER CL we are using here is version 4.1α10. All measurements for the public suites are done on a 1.5 GHz P4 with 2 GB of memory, running Red Hat Linux 7.1.

Instance	PROVER CL (s)	ZCHAFF (s)
ibm-1	0.5	2.5
ibm-2	0.0	0.0
ibm-3	0.8	0.9
ibm-4	1.2	4.3
ibm-5	0.2	0.4
ibm-6	5.2	14.6
ibm-7	0.2	0.2
galileo-8	7.0	92.3
galileo-9	8.6	140.2
ibm-10	7.1	404.2
ibm-11	3.5	34.8
ibm-12	16.4	165.4
ibm-13	7.4	10.1

Table 2: SATLIB benchmark results

In the third set of experiments we present results that have been generated in-house at Intel. Here we compare the performance of PROVER CL version 4.1α9, ZCHAFF, and a BDD-based tool developed internally at Intel. The measurements for this suite were done on a 1 GHz PIII with 1 GB of memory, running Red Hat Linux 6.2.

A note on strategies used in the experiments: For the ISCAS'85 and Intel experiments, we use the default strategy for combinational equivalence checking. For the Satlib experiments, we use the default strategy for bounded model checking. These default strategies were not devised to work well on these particular examples, and no changes to the default parameters were made in any of the experiments.

ISCAS'85

The ISCAS'85 suite consists of 15 combinational equivalence checking problems. Here, the objective is to show that all output signals from a pair of circuits are equivalent. One of the circuits in the pair is the original ISCAS'85 circuit and the other circuit is an optimized version. Due to erroneous optimizations, it may be the case that the optimized circuit is not equivalent to the original circuit.

We have run the experiments as follows. Using PROVER CL, we verify that individual pairs of outsignals from the circuits are equivalent. The time given for each circuit is the total time for deciding equivalence for every single outsignal. For ZCHAFF we use the CNF files produced by João Marques da Silva. Each CNF file corresponds to the problem of deciding whether the conjunction of individual equivalences for the outsignals is true or not. Note that this is an easier problem than that solved by PROVER CL.

We present the results in Table 1. As can be seen, it is hard to say anything about the relative performance of the engines for trivial examples that take less than a few seconds. However, for problems that take more than a couple of seconds, the trend is uniform: PROVER CL is one or more orders of magnitudes faster than ZCHAFF. Moreover, not only is PROVER CL always the faster engine, but the distribution of runtimes is also more uniform than for ZCHAFF.

Satlib

The Satlib benchmark suite contains 13 industrial model checking problems from IBM and Galileo (a telecommunication hardware manufacturer). Here, the objective is to decide whether there exists a run of a particular length from

Problem	PROVER CL (s)	ZCHAFF (s)	BDDs (s)
1	0.1	18.3	[> 10^7 nodes]
2	0.1	0.8	[> 10^7 nodes]
3	0.2	222.6	[> 10^7 nodes]
4	0.9	49.9	[> 10^7 nodes]
5	0.1	9.5	[> 10^7 nodes]
6	0.1	0.3	[> 10^7 nodes]
7	0.8	[>60 min]	[> 10^7 nodes]
8	9.9	101.5	[> 10^7 nodes]
9	0.3	15.5	[> 10^7 nodes]
10	450.2	[>60 min]	110.0
11	2.8	[>60 min]	[> 10^7 nodes]
12	139	232.5	[> 10^7 nodes]

Table 3: Intel benchmark results

the initial states of a circuit that ends up in a state violating a design invariant. For all of the problems, there exists such a run.

We have run the experiments as follows. The Satlib benchmarks are only available as pregenerated conjunctive normal form formulas optimized for CNF solvers like ZCHAFF. Unfortunately, PROVER CL operates best on formulas in full propositional logic; the conversion to CNF destroys valuable structural information. Since our objective is to compare performance on design automation problems, this situation is very troublesome—we can only do a partial translation back to full propositional logic for PROVER CL (this translation takes negligible time). Nevertheless, as can be seen in Table 2, PROVER CL outperforms ZCHAFF on all examples.

Intel benchmarks

The ISCAS and Satlib benchmark suites are relatively easy both for ZCHAFF and PROVER CL. We now consider another set of benchmarks where this is not the case.

The Intel problem set contains twelve combinational equivalence checking problems. Here, the objective is again to check that an optimized and an unoptimized version of a net behave the same. In ten of the cases, there is a bug, and in two cases the two nets are equivalent.

The experiments have been run as follows: We compare the performance of PROVER CL, ZCHAFF, and a BDD-based method developed internally at Intel. The problem instances are generated in full propositional logic for PROVER CL, and in clausal form for ZCHAFF. In the experiments, a timeout of one hour is used for all methods. For BDDs, we also abort when ten million BDD-nodes have been built.

We present the results in Table 3, where problems 8 and 9 are the two equivalent pairs of circuits. As can be seen, the BDD-based method runs out of memory in all but a single case. However, for the example where it finishes, it is the fastest method. ZCHAFF is always slower than PROVER CL, and times out in three cases. PROVER CL completes all benchmarks (in all cases except two in less than ten seconds), and performs between one and several orders of magnitude better than ZCHAFF on examples that are not trivial for both methods.

6. CONCLUSIONS

In this paper, we have taken an idea that has proven fruitful in other domains—semi-interactive theorem proving in rich logics—and transferred it to the domain of propositional proof search, with the aim to speed up combinational design automation problems. We have introduced the underlying ideas, presented some particular strategies, one of which is new, and discussed how they are integrated in Prover Technology's combinational proof engine plug-in PROVER CL.

The end result of the research we have presented here provides three important benefits. First of all, many of the users of our technology need state-of-the-art performance without modifying parameters, and we provide this in our default strategies. Second, the customizability of our engine makes it possible for power users to construct analyses that are tuned for the particular set of in-house problems that need to be solved. Third, for many customers, it is vital that our tool makes it easy to expand the existing set of strategies with new analyses.

Ongoing and future work includes to add more primitive analyses to the current set of basic building blocks, as they become known. Up to now, every time we have expanded our set of primitive building blocks, we have been able to extend the range of our default strategies substantially. Moreover, we are continuing development of default analyses for new problem domains.

7. REFERENCES

[1] D. Brand. Verification of large synthesized designs. In *Proc. Int. Conf. on Computer Aided Verification*, 1993.

[2] R. Bryant. Graph-based algorithms for boolean function manipulation. *IEEE Trans. on Computers*, C-35(8):677–691, Aug. 1986.

[3] J. R. Burch and V. Singhal. Tight integration of combinational verification methods. In *Proc. Int. Conf. on Computer Aided Verification*, 1998.

[4] K. Claessen and G. Stålmarck. A framework for propositional proof strategies. Technical report, Chalmers University of Technology, Computing Sciences Dept., January 2000.

[5] M. Davis, G. Logeman, and D. Loveland. A machine program for theorem-proving. *Communications of the ACM*, 5(394–397), 1962.

[6] S.-Y. Huang and K.-T. Cheng. *Formal Equivalence Checking and Design Debugging.* Kluwer Academic Publishers, 1998.

[7] J. Jain, R. Mukherjee, and K. Takayama. An efficient filter-based approach for combinational verification. *Fujitsu Scientific and Technical Journal*, 36(1):17–23, 2000.

[8] M. Moskewicz, C. Madigan, Y. Zhao, L. Zhang, and S. Malik. Chaff: Engineering an efficient SAT solver. In *Proc. 39th Design Automation Conference*, 2001.

[9] V. Paruthi and A. Kuehlmann. Equivalence checking using a structural SAT-solver, BDDs, and simulation. In *Proc. Int. Conf. on Computer Design*, 2000.

[10] L. Paulson. Tactics and tacticals in Cambridge LCF, 1979.

[11] G. Stålmarck. A system for determining propositional logic theorems by applying values and rules to triplets that are generated from a formula. Swedish Patent No. 467 076 (approved 1992), US patent No. 5 276 897 (1994), European Patent No. 0403 454 (1995).

Solving Difficult SAT Instances in the Presence of Symmetry

Fadi A. Aloul, Arathi Ramani, Igor L. Markov and Karem A. Sakallah
Department of EECS, University of Michigan, Ann Arbor 48109-2122
{faloul,ramania,imarkov,karem}@eecs.umich.edu

ABSTRACT

Research in algorithms for Boolean satisfiability and their implementations [23, 6] has recently outpaced benchmarking efforts. Most of the classic DIMACS benchmarks [10] can be solved in seconds on commodity PCs. More recent benchmarks take longer to solve because of their large size, but are still solved in minutes [25]. Yet, small and difficult SAT instances must exist because Boolean satisfiability is NP-complete.

We propose an improved construction of symmetry-breaking clauses [9] and apply it to achieve significant speed-ups over current state-of-the-art in Boolean satisfiability. Our techniques are formulated as pre-processing and can be applied to any SAT solver without changing its source code. We also show that considerations of symmetry may lead to more efficient reductions to SAT in the routing domain.

Our work articulates SAT instances that are unusually difficult for their size, including satisfiable instances derived from routing problems. Using an efficient implementation to solve the graph automorphism problem [18, 20, 22], we show that in structured SAT instances difficulty may be associated with large numbers of symmetries.

Categories and Subject Descriptors

I.1.2 [**Algorithms**]: Algebraic algorithms.

General Terms

Algorithms, experimentation, verification.

Keywords

SAT, CNF, faster, search, symmetry, difficult, instances, speed-up.

1. INTRODUCTION

Boolean satisfiability (SAT) is a pivotal problem in Computer Science and has numerous applications in Design Automation that range from microprocessor verification [25] to FPGA layout [19]. A one-million-dollar prize is offered by the Clay Institute for Mathematical Sciences for a complete polynomial-time SAT solver or a proof that such an algorithm does not exist (the P-vs-NP problem). Neither is likely to be found. Nevertheless, industrial applications motivate intensive research in SAT algorithms that quickly solve real-life instances. The fundamental framework for state-of-the-art SAT algorithms was laid out in the 1960s, but a number of recent improvements in algorithms and implementation techniques [23, 6] have led to performance breakthroughs. Most DIMACS challenge benchmarks [10] from the early 1990s are now solved in seconds on commodity PCs. Recently posted SAT benchmarks [25] take somewhat longer to solve (minutes), but that is primarily due to their enormous size (50MB+ files, etc). With the exception of artificially constructed families of benchmarks, it appears that SAT can be solved in polynomial time "for practical purposes".

It is well known that the dominant back-track solvers, such as GRASP [23] and CHAFF [6] do not perform well on randomly-created 3-SAT instances with ≈ 4.3 clauses per variable. However, such instances do not arise in Design Automation because application-derived SAT instances are typically structured. Attempts to explain the ease of structured instances were successful for certain applications [21], and generic ways to exploit certain types of structure were proposed [1].

Our work addresses both benchmarking and algorithmic aspects of SAT research. Given the excellent performance of existing SAT solvers, there is no room for improvement on easy benchmarks, and we focus on difficult instances.[1] Since the works of Haken and Urquhart [24] on lower bounds for resolution and back-tracking algorithms for SAT, several instance families have been known to require exponential time for DP/DLL (Davis-Putnam and Davis-Logemann-Loveland) solvers. For example, a recent lower bound for the pigeon-hole problem is $\Omega(2^{n/20})$ [2] where n is the number of pigeons. Another such family was constructed by Urquhart in terms of expander graphs and with considerable use of randomization [24]. Indeed, state-of-the-art SAT solvers, such as CHAFF, take a long time to solve those instances (see Table 1), but the relevance of such pathological cases to Design Automation is questionable. While lower bounds for SAT are often proven for unsatisfiable instances, it remains to be seen whether satisfiable instances can be difficult for the best solvers. We demonstrate CAD-related SAT instances, both satisfiable and unsatisfiable, that are very difficult for their size. Moreover, an easy instance of any size can be made difficult by adding a small difficult instance to it and connecting the two by inconsequential clauses to defeat partitioning.

Over many years, empirical research in algorithms for Design Automation identified a number of fundamental problem formulations, such as Boolean satisfiability, and mustered significant efforts to solve them efficiently. State of the art is gauged by optimized solver implementations ("engines"). Performance break-throughs are often due to novel algorithmic ideas, leaner implementations or the ability to apply a highly optimized engine in a novel way. In this work, we suggest that graph automorphism engines can be applied to the satisfiability problem in certain cases. Given that the graph automorphism problem is thought to not be NP-complete (thus potentially easier than SAT) and that very little CAD research was done on high-performance engines for graph automorphism (one such work is [16]), there may be significant room for future improvement. To be precise, we will be dealing with the colored variant of the graph automorphism problem that can be easily extended to hypergraphs (see definitions in Section 2).

Several works suggested that "breaking symmetries" in CNF formulae can speed up SAT solvers [3, 4, 5, 9, 16]. Symmetries of a CNF formula include clause-preserving permutations of variables. Such permutations may involve arbitrarily many variables at once, e.g., a complete cyclic shift. In this work, we do not address permutations that change the CNF formula but leave unchanged the Boolean function it represents.[2] However, if such symmetries are detected by other techniques [14], our proposed methods can process them in the same way as

[1] In practice, the difficulty of domain-specific classes of SAT instances is often known, and adequate SAT algorithms can be chosen. Otherwise, one can run several SAT solvers in parallel until one of them finishes. On a single processor, this may buy exponential speed-ups at the cost of a constant-factor slow-down.

[2] Such permutations can be called "semantic" symmetries versus "syntactic" symmetries that leave the CNF formula unchanged.

symmetries of the CNF formula. Similarly, many of the works we cite do not deal with symmetry detection, but rather assume that symmetries of the Boolean function are given. Using this assumption, two main directions were explored: (a) preprocessing the original CNF formula by adding symmetry-breaking clauses that do not affect satisfiability but speed up search [9], (b) extending SAT solvers, particularly those based on back-tracking, to dynamically use symmetries during the search process [5]. In this paper we pursue the pre-processing approach due to its simplicity, but will outline how our techniques can be applied within a back-tracking solver for increased efficiency.

Prior works on symmetries in SAT predate recent breakthroughs in SAT solvers and typically use several carefully constructed instances to illustrate their approach. E.g., Crawford et al. [9] suggest that symmetry-based techniques allow the pigeon-hole instances to be solved in polynomial time, but their empirical data [9, Figure 3] do not support this suggestion. Also it remains unclear whether the performance of leading-edge SAT solvers can be improved via the use of symmetries. In principle, the overhead due to symmetry detection and usage may outweigh the benefits, and it remains to be seen that useful CNF formulae have many symmetries. Pólya (1937), Erdös and Rényi (1963) proved that a random graph on n vertices has *no symmetries* with probability $1-\binom{n}{2}2^{-n-2}(1+o(1))$ [12, p. 1461]. This claim can be extended to CNF formulae using constructions in Section 2, but structured real-world instances may have richer symmetries. Indeed, Boolean functions from synthesis applications may have many symmetries [14]. If exponentially many symmetries exist, adding all possible symmetry-breaking clauses can be disastrous [9]. Despite these pitfalls, symmetry-based approaches have been useful in model checking [13, 7], verification [16], logic synthesis [15] and DSP algorithms [11].

In this work, we propose an automated flow that starts with a CNF formula in the DIMACS format and detects all of its symmetries (not just pairwise swaps). In this flow, all symmetries are captured implicitly, with exponential compression. The CNF formula is then preprocessed adding symmetry-breaking clauses that do not affect satisfiability. A black-box SAT solver is applied to the preprocessed CNF instance to produce the final answer; any satisfying assignment to this instance is (or corresponds to) a satisfying assignment of the original instance, and if the preprocessed instance is unsatisfiable then so is the original instance.

Our construction of symmetry-breaking clauses is novel. It is more economical and provides better coverage than that in [9]. Additionally, it directly applies to the compressed representation of all symmetries in the format produced by graph automorphism software [17, 18, 20, 22]. Our empirical results show significant improvements on CNF instances arising in Design Automation applications as well as highly randomized provably-difficult Urquhart benchmarks [24]. Two extensions are proposed to speed up symmetry detection. One is opportunistic symmetry detection, where only some symmetries are found. The other extension pursues domain-specific symmetries and leads to improvements of SAT formulations by adding domain-specific symmetry-breaking clauses. Thus generic symmetry detection is avoided by creating symmetry-less SAT instances that can be solved quickly.

The remaining material is organized as follows. Symmetry detection is described in Section 2 and symmetry-breaking in Section 3. Section 4 discusses constructions of SAT benchmarks. Our empirical results are presented in Section 5 and further extensions in Section 6.

2. FINDING SYMMETRIES

In general, a symmetry of a discrete object is a permutation of its components that leaves the object unchanged. Every discrete object has at least one symmetry — the "do-nothing" permutation. It is easy to see that a composition of two symmetries is a symmetry, and that composition with the do-nothing permutation does not change a symmetry. The composition of symmetries is associative, and every symmetry has an inverse. Composition is often *not* commutative. Abstract algebraic structures defined axiomatically in terms of such a composition operation (multiplication) are commonly called *groups*. In this work we will only deal with groups of symmetries whose elements can be thought of as permutations. A permutation can be represented by cycles, e.g., $(23)(567)$ represents a permutation on a set of at least 7 mark (elements). This permutation swaps marks 2 and 3, it cycles marks 5, 6 and 7 in that order. All other marks, e.g., 1 and 4, are left unchanged.

Computational group theory is approximately 25 years old, and great strides were made in the last decade with the development of the GAP package ("Groups, Algebra and Programming") [22]. A major efficiency in computational group theory comes from the notion of irredundant sets of generators of a group. A set of generators is made of group elements such that any other group element can be composed of generators and their inverses (no uniqueness required). Elementary group theory implies that any irredundant set of generators for any group with $N>1$ elements contains *at most* $\log_2 N$ elements, e.g., the $k!$ permutations on k marks can be generated by (12) and $(12..k)$. Thus, representing groups by sets of generators *always ensures exponential compression*. Computational group theory provides efficient algorithms for manipulating groups represented by sets of generators, without decompression. Therefore, an intelligent algorithm for symmetry detection may return a small set of generators rather than list all symmetries.

COLORED AUTOMORPHISM PROBLEMS. Given a graph, a *symmetry* is a permutation of its vertices that maps edges to edges. In case of directed graphs, edge orientations must be preserved. In the Graph Automorphism problem one seeks all symmetries of a given graph, e.g., in terms of group generators. It is known that all graphs except for an exponentially small family have *no symmetries* [12, p. 1461]. No worst-case polynomial-time algorithms are known for this problem, but it is commonly believed not to be NP-complete unless P=NP. Polynomial-time algorithms are available in many special cases [12, p. 1511]. Generic algorithms [17, 16] are based on linear-time partition refinement passes; a simple version finishes in three passes for all but an exponentially small family of graphs [12, p. 1513].

The Graph Automorphism problem may be constrained by vertex labels — symmetries must map each vertex into a vertex with the same label. Label constraints are computationally easy and can be formally reduced to plain graph automorphism. Labels are often expressed by integers and called colors (no relation to *graph coloring*). Another extension is to colored *hypergraphs* — symmetries must map hyperedges to hyperedges (of the same cardinality because no two vertices can map to one). The colored hypergraph automorphism problem reduces to the colored graph automorphism via the bipartite graph of the hypergraph. This graph contains a vertex for each hypergraph vertex and hyperedge, and connects them with edges according to the hypergraph's incidence relation. Graph vertices in the hyper-edge part are painted with a new color, and other vertices retain their original colors.

Brendan McKay implemented a practical algorithm for Graph Automorphism [17] in a software package called NAUTY [18], which has been continually improved for the last 20 years (version 2.0 released in 2001). NAUTY has been integrated into the computational group theory system GAP [22] by means of the GRAPE package [20]. This integration enables efficient group-theoretic operations on the results returned by NAUTY and facilitates some of our proposed algorithms. In 1998, Manku et al. [16] claimed speed-ups over a pre-2.0 version of NAUTY in the context of hardware verification. However, their code is not generic (built into a larger system) and is no longer supported.

CNF SYMMETRIES VIA GRAPH AUTOMORPHISM. The problem of finding symmetries of a CNF formula is reduced to colored graph automorphism, similarly to the reduction from the hypergraph automorphism outlined above. Every variable is represented by two vertices that correspond to the positive and negative literals. Every clause is represented by a vertex, and bipartite edges connect those vertices to vertices of relevant literals. Clause vertices are painted with color 1

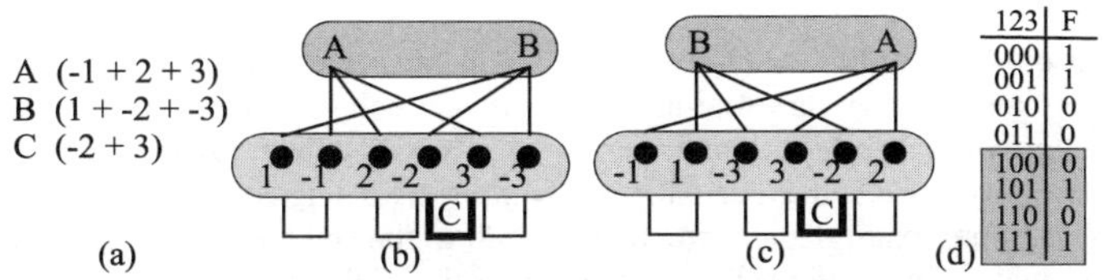

Figure 1: A CNF formula with three clauses A, B and C and three variables (a) is converted into a bipartite graph (b) for symmetry detection purposes. The two-literal clause C is represented by one edge (bold) while larger clauses A and B are represented by a vertex and three edges each. Any symmetry must map $C \to C$ therefore this instance has only one non-trivial symmetry (1 -1)(2 -3)(-2 3)(A B) shown in (c). The first cycle yields a symmetry-breaking clause $(\bar{1})$ which reduces the search space by half (d). Alternatively, the clause (2 + 3) corresponding to the third cycle can be added.

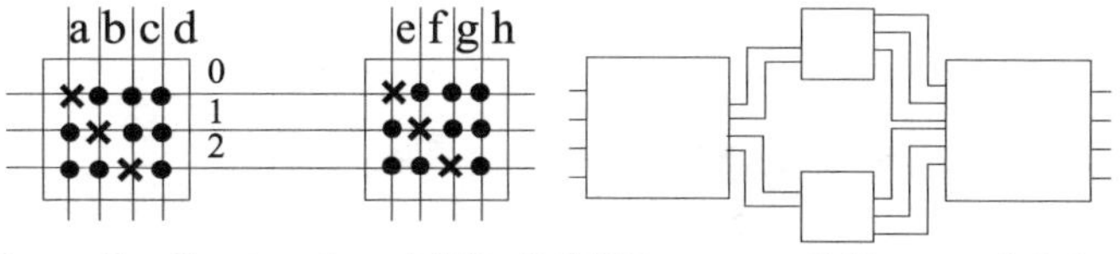

Figure 2: Construction of difficult SAT instances: (left) two switch-boxes in common FPGA architectures, (right) similar *N*-by-*M* switch-boxes are used to build hard satisfiable instances.

and literal vertices are painted with color 2. To ensure Boolean consistency, vertices of opposite literals are mated by direct edges. An additional simplification, motivated by [9, footnote 6], is to represent each two-literal clause by an edge directly connecting their two literals rather than by two edges and a vertex. In the final colored graph, $2 * Vars$ vertices represent the positive and negative literals and the remaining $Clauses - 2 * LitClauses$ vertices represent clauses. Figure 1 shows an example. This reduction has the advantage of detecting phase-shift symmetries $(a \to \bar{a})$ and their compositions with permutational symmetries.

3. SYMMETRY-BREAKING

Symmetries induce equivalence classes in the solution space (in group theory, they are called *orbits*). Given a satisfying truth assignment, all other truth assignments to which it can be mapped by symmetries, must also be satisfying. Similarly, symmetries always map unsatisfying assignments to unsatisfying assignments. Therefore, for a complete SAT solver it suffices to reason about one representative from each such class. This restriction can be implemented by selecting unique representatives from every equivalence class and adding clauses that are only satisfied on those representatives. An earlier construction of such symmetry-breaking clauses [9] is based on a given ordering of variables. Its main idea is (i) to order all elements from the solution space lexicographically, and (ii) to select the lexicographically smallest element from each equivalence class as its representative.

The construction described in [9] is applied to every given symmetry and generates many redundant clauses. To prune redundant clauses, the authors propose the concept of a symmetry tree, but it is not well supported by efficient algorithms for permutation groups, does not always prevent redundant clauses and is itself not always prunable to polynomial size [9]. Phase-shift symmetries were not addressed in that work.

We compute symmetry-breaking clauses on a per-symmetry basis [9], but consider only irredundant sets of symmetry generators, returned by graph automorphism programs. By breaking generator symmetries only, one does not necessarily break all symmetries, except for some cases [9]. So far, the power of such partial symmetry-breaking has not been evaluated, but we believe that a reasonable coverage is often achieved because an irredundant set of generators contains "maximally independent" symmetries — none of them can be expressed in terms of others.

Our construction is formulated in terms of cycles of a permutation (cf. [9]). For the variable swap (ab) the construction in [9] entails one additional variable and six symmetry-breaking clauses. Our construction below entails only one clause. First observe that if the cycle (ab) is a symmetry, whenever there is a satisfying assignment with $a = 0, b = 1$, there should be a symmetric (equivalent) satisfying assignment with $a = 1, b = 0$ and other variables unchanged. To allow only the first assignment, we add the symmetry-breaking clause $(\bar{a} + b)$, which can also be interpreted as $(a \leq b)$. Similarly, to "break" a cycle of length three (abc), we add $(\bar{a}+b)(\bar{b}+c)$, i.e., $(a \leq b)(b \leq c)$. To prevent transitivity violations, one has to choose an ordering of all variables at the beginning, and always use the $\leq$ sign consistently with that ordering. Longer cycles require more complex symmetry-breaking clauses, but one can always improve on the construction from [9]. We observed that symmetry generators produced for CNF formulae often consist of just 2-cycles and only rarely have 3-cycles. A cycle $(\bar{a}a)$ means that the value of a can be fixed arbitrarily, and this can be expressed by a one-literal symmetry-breaking clause. The construction in [9] does not address such phase-shift symmetries and never results in one-literal clauses.

If no cycles generate single-literal clauses (which achieve maximal pruning if all clauses have length ≤ 2), we can produce symmetry-breaking clauses from any one 2- or 3-cycle of a symmetry. That significantly speeds up SAT solvers in many cases. However, clauses of the form $(\bar{a} + b)$ achieve no pruning in areas of the solution space where the variables involved have identical values. A key idea in that case, similar to that in [9] is to process another cycle. Namely, for a symmetry $(ab)(cd)(ef)$..., we first add $(\bar{a} + b)$, then $(a = b) \Rightarrow (c \leq d)$, then $((a = b)(c = d)) \Rightarrow (e \leq f)$, etc. This construction can be efficiently implemented with one additional variable per cycle to indicate the equality of all variables in the cycle. A sample clause with new variables looks like $(\bar{x}_{a=b} + \bar{x}_{c=d} + \bar{e} + f)$. To ensure consistency, we sort marks within cycles and sort cycles by their first marks.

The construction of symmetry-breaking clauses is dwarfed by the symmetry-detection time. However, with every cycle processed, we add larger and larger symmetry-breaking clauses. Since large clauses typically do not have a great effect on the behavior of SAT solvers, we optionally limit symmetry-breaking clauses to the first 10 cycles of every symmetry. For the price of incomplete coverage, this technique considerably reduces the overhead of symmetry-breaking clauses. In our experiments it often performed better than the addition of symmetry-breaking clauses for all cycles. Moreover, extending back-track algorithms for SAT to dynamically check conditions of the form $((a = b)(c = d)...(u = v))$ may lead to improvements over pure pre-processing.

4. DIFFICULT SAT INSTANCES

FPGA ROUTING INSTANCES. A recent comparative study of two Boolean formulations of FPGA detailed routing constraints [19] showed that problem encoding affects the difficulty of SAT instances. Our work uses the better formulation, but still produces difficult instances. Two such constructions are shown in Figure 2 in terms of FPGA switch-boxes (see [19] for details on SAT formulations). The one on the left entails routing $N + k$ connections through N tracks and yields unsatisfiable instances that for $k = 1$ resemble the well-known pigeon-hole benchmarks. Empirical results in Table 1 are shown for six routing configurations (`chnl`) in which one tries to route (a) 11, 12 or 13 connections through 10 tracks, and (b) 12, 13 or 20 connections through 11 tracks. These instances are extremely difficult for the leading-edge SAT solver CHAFF [6] and also have many symmetries. They can appear as sub-instances in larger routing instances, and such sub-instances may be difficult to find.

From the benchmarking point of view, it is natural to expect *unsatisfiable* instances among the most difficult to solve. Indeed, randomized restarts used by CHAFF [6] typically allow it to avoid difficult regions of the search space and to quickly find satisfying solutions if they exist. However, our second construction is designed to create difficult *satisfiable* instances that trap even the best solvers in hopeless regions of their solution space for a long time before a satisfying solution can be found. The main idea is to create a satisfiable instance with a large number of hard-to-avoid unsatisfiable sub-instances. If the number of unsatisfiable

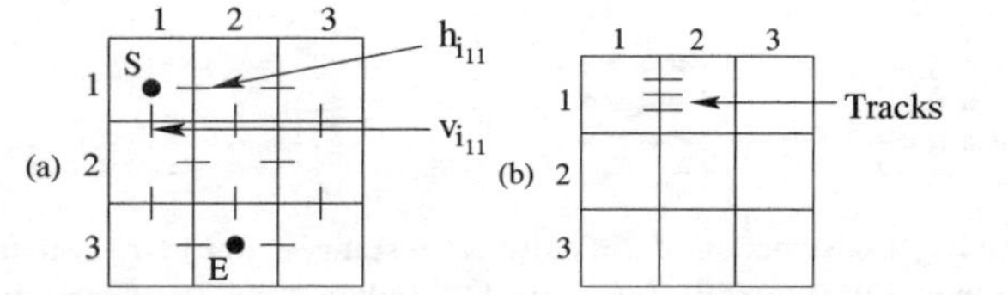

Figure 3: Construction of difficult SAT instances (global routing).

branches is much larger than the number of satisfiable branches, then random restart will keep on jumping from one unsatisfiable branch to another for a long time. Solvers without random restarts will, too, need to prove the unsatisfiability of many branches.

Our second construction entails routing a number of wires through four *N*-by-*K* FPGA switch-boxes of the type used in the first construction. The rightmost switch-box in the configuration in Figure 2 has several redundant outgoing tracks that are divided into two channels. Each channel is connected to a smaller switch-box with an insufficient number of outgoing tracks. The two groups of tracks that leave the smaller switch-boxes are connected to the left-most switch-box. When routing connections through tracks right-to-left, connections must be split between switch-boxes subject to the throughput constraints of switch-boxes. However, to a SAT solver, the throughput constraints are obscured by the pigeon-hole principle. SAT solvers first partition the connections between the two channels and back-track from every partition that does not lead to a satisfying assignment. If the capacities of the two channels leading to the smaller switchboxes are greater than the throughput of those switchboxes, an overwhelming majority of partitions will lead to unsatisfiable pigeon-hole instances. On average, at least several such instances must be solved before a good partition is found. SAT solvers without random restarts will also need to solve many pigeon-hole sub-instances before finding satisfying solutions. Empirical results for these satisfiable instances (`fpga`) in Table 1 show that they are very difficult for CHAFF. We observe that these instances become more difficult when the difference between the throughput of the small switchboxes and the capacities of the channels that lead to them is increased. This is consistent with our observations for the unsatisfiable `chnl` instances.

GLOBAL ROUTING INSTANCES. We propose a new construction of difficult randomized *satisfiable* instances unrelated to pigeonholes. They express routing two-pin connections in a grid with edge capacity constraints. To ensure that an instance is satisfiable but difficult, we use *randomized flooding*. Namely, we create a routing configuration by adding shortest possible routes while unused routing resources (edge capacities) remain. Shortest routes are created by breadth-first-search between pairs of randomly chosen grid cells or, if that fails, by finding a maximal shortest route starting at a given grid cell with unused routing resources. After a routing configuration is created, routes are erased and their end-points are used to formulate a SAT instance.

Our SAT encoding of routing instances has two components. One deals with *route definition* and captures possible ways to route each connection. The other addresses *capacity constraints* and restricts the number of connections that can be routed across a grid cell boundary.

Route definition. Routes are specified in terms of edges across cell boundaries in a grid. For each connection, there are routing tracks across each cell boundary on the grid. In the SAT formulation, each track is treated as a variable. Figure 3 (a) illustrates routing tracks in a 3-by-3 grid. Horizontal tracks for connection i are labeled $h_{i_{r,c}}$, where r and c are the row and column indices of the cell whose boundary the track crosses. Vertical tracks are labeled $v_{i_{r,c}}$. In Figure 3 (a), let the points marked S and E be the terminals of some two-terminal connection i. The SAT formulation proceeds as follows. Consider the terminal marked S. A route for this connection must pass through $h_{i_{1,1}}$ or $v_{i_{1,1}}$. Therefore, we add the clause $(h_{i_{1,1}} + v_{i_{1,1}})$. Since these two track selections are incompatible, we add the mutual exclusion clause $(\bar{h}_{i_{1,1}} + \bar{v}_{i_{1,1}})$.

We now push the cells reachable from the possible tracks into a queue. The queue contains cells reachable from those already visited. A list of visited cells is also maintained so that a cell is not pushed on the queue twice. While the queue is not empty, cells are popped off it and new clauses are introduced for the route tracks across the cell boundaries. In our example, assume that the cell to the right of S is popped off the queue. Since this cell is not an endpoint of the connection, exactly two of its boundaries must be selected. The cell boundaries in this case are $h_{i_{1,2}}, h_{i_{1,1}}$ and $v_{i_{1,2}}$. We therefore introduce the clauses $(h_{i_{1,1}} + v_{i_{1,2}})$, $(h_{i_{1,1}} + h_{i_{1,2}})$, and $(h_{i_{1,2}} + v_{i_{1,2}})$. However, again it is not possible for more than two tracks to be selected. Therefore, we add clauses of the form: $(h_{i_{1,1}} \wedge v_{i_{1,2}}) \Rightarrow \bar{h}_{i_{1,2}}$. This procedure is repeated for every cell popped off the queue until the queue is empty.

Capacity constraints. Each grid cell boundary has a capacity associated with it to restrict the number of connections that can be routed through it. The capacity limits are intended to prevent congestion. If C is the capacity limit for an edge of a grid cell, we include C variables per edge for each connection. In other words, each connection can be routed through one of C tracks across a cell boundary as shown in Figure 3 (b).

Consider two connections i and j. Consider horizontal route tracks for each connections, $h_{i_{r,c}}$, and $h_{j_{r,c}}$ for some row r and column c. Let $i_{r,c_1}, i_{r,c_2}, \ldots i_{r,c_C}$ and $j_{r,c_1}, j_{r,c_2}, \ldots j_{r,c_C}$ be the C extra variables introduced in the SAT formulation for the horizontal track in question. Then clearly, for any $i_{r,c_k}, 1 \leq k \leq C, i_{r,c_k} \Rightarrow h_{i_{r,c}}$, and also $h_{i_{r,c}} \Rightarrow (i_{r,c_1} + \ldots + i_{r,c_C})$. Clauses of this form are added to the SAT instance. Another restriction is that a route cannot pass through two tracks in the same channel (edge of a grid cell), i.e., if for some $k, 1 \leq k \leq C$, if i_{r,c_k} is true, then for all $l, 1 \leq l \leq C, l \neq k, (i_{r,c_k} \Rightarrow -i_{r,c_l})$. These clauses are also added. Finally, two connections cannot be routed through the same track, i.e. for all $k, 1 \leq k \leq C$, $(i_{r,c_k} \Rightarrow -j_{r,c_k})$ for all $j \neq i$, where j represents another connection. By combining the aforementioned techniques, we are able to express routing instances as SAT problems.

We created ten routing configurations by randomly flooding a 3-by-3 routing grid with connections subject to edge capacity constraints of 3. Then we applied the SAT encoding above. Table 1 shows empirical results for the five most difficult instances (`grout`).

5. THE EFFECT OF SYMMETRY-BREAKING

Our computational experiments were performed on PCs with AMD Athlon processors @1.2GHz and 1Gb of RAM. All codes were compiled with `g++ 2.95.4 -O3` and ran on Debian Linux. In addition to the instances described in Section 4 (`chnl` and `fpga`) and (`grout`), Table 1 lists six standard pigeon-hole instances (`hole`), five families of artificially constructed randomized Urquhart benchmarks (`Urq`) [24] and seven recent benchmarks from the micro-processor verification domain [25]. CHAFF runtimes in Table 1 are averages of (up to) 20 independent starts because CHAFF uses randomization internally and results of different runs may vary significantly. All runs that did not complete in 1000 seconds were aborted and did not contribute to averages. The percent of time-outs is shown for each instance.

To detect symmetries in CNF formulae, we converted them into colored graphs (see Section 2). We then used the NAUTY program [17, 18]. At each run, the result was a list of permutation generators of the group of symmetries. Permutation generators are specified by cycles. For each SAT instance, Table 1 lists NAUTY runtime in seconds excluding I/O. the total number of symmetries and the number of permutation generators. Those symmetry detection implementations are deterministic and not affected by re-ordering of vertices in the input graph. For some benchmarks we built symmetry-breaking clauses only for ten cycles per symmetry. The first ten cycles typically capture most of the speed-up provided by "breaking" a given symmetry. After new clauses were added, the preprocessed CNF instance was solved with CHAFF. Table 1 lists average runtimes of 20 independent runs of CHAFF for each instance. Pre-processed CNFs never timed out in our experiments.

The last column in Table 1 shows relative speed-up ratios due to the

Table 1: CHAFF runtime on original SAT instances is compared to the combined runtime of symmetry detection and CHAFF on instances with symmetry-breaking clauses (the right-most column). The full name of benchmark `2dlx_ca_mc` is `2dlx_ca_mc_ex_bp_f`. The numbers of symmetry generators and max cycles used per generator are shown (10 or all). Pure search speed-up (that does not take symmetry detection into account) is also given. Results for opportunistic window-based symmetry-finding are also given and in most cases discover all or a large fraction of all symmetries. All benchmarks that we generated for these experiments are available at `http://www.eecs.umich.edu/~faloul/benchmarks.html`

Instance	Satis-fiable?	#variables and #clauses	Plain Chaff sec	Time -out %	Symmetries Finding sec	Number of	#generators cycles		Chaff sec	Speed-up ratios: total ; search only
hole07	UNS	56;204	0.37	0%	0.1	2.03e8	all	13	0.01	3.32; 36.50
hole08	UNS	72;297	1.27	0%	0.07	1.46e10	all	15	0.01	15.22; 94.15
hole09	UNS	90;415	3.79	0%	0.1	1.32e12	all	17	0.02	32.00; 204.97
hole10	UNS	110;561	22.44	0%	0.15	1.45e14	all	19	0.02	132; 1122
hole11	UNS	132;738	212.73	0%	0.18	1.91e16	all	21	0.03	1229.6; 7090.9
hole12	UNS	156;949	>1000	100%	0.24	2.98e18	all	23	0.04	— ; —
Urq3_5	UNS	46;470	232.44	10%	0.48	2.32e6	all	29	0.0	484.16; —
Urq4_5	UNS	74;694	250.01	25%	1.35	2.50e6	all	43	0.0	185.18; —
Urq5_5	UNS	121;1210	>1000	100%	13.15	>1e7	all	72	0.0	—; —
Urq6_5	UNS	180;1756	>1000	100%	62.93	>1e7	all	109	0.0	—; —
Urq7_5	UNS	240;2194	>1000	100%	176.62	>1e7	all	143	0.0	—; —
grout3.3-01	SAT	864;7592	19.01	0%	4.79	8.71e9	10	26	0.67	3.48; 28.37
grout3.3-03	SAT	960;9156	44.35	0%	8.94	6.97e10	10	29	0.40	4.75; 110.89
grout3.3-04	SAT	912;8356	19.36	0%	6.81	2.61e10	10	27	0.36	2.70; 53.79
grout3.3-08	SAT	912;8356	21.30	0%	7.14	3.48e10	10	28	0.67	2.73; 31.80
grout3.3-10	SAT	1056;10862	28.18	0%	10.65	3.48e10	10	28	0.85	2.45; 33.15
chnl10x11	UNS	220;1122	22.17	0%	0.45	4.20e28	all	39	0.11	39.91; 210.13
chnl10x12	UNS	240;1344	81.88	0%	0.61	6.04e30	all	41	0.12	111.63; 663.00
chnl10x13	UNS	300;2130	657.61	25%	1.28	4.50e37	all	47	0.17	454.78; 3961.4
chnl11x12	UNS	264;1476	207.37	0%	0.75	7.31e32	all	43	0.15	231.31; 1415.5
chnl11x13	UNS	286;1742	788.32	20%	1.08	1.24e35	all	45	0.16	633.45; 4792.2
chnl11x20	UNS	440;4220	>1000	100%	4.4	1.89e52	all	59	0.31	—; —
fpga10.08	SAT	120;448	7.56	0%	0.63	6.00e71	all	62	0.05	11.15; 157.56
fpga10.09	SAT	135;549	3.80	0%	0.88	6.33e77	all	68	0.03	4.16; 113.39
fpga12.11	SAT	198;968	694.00	50%	3.76	7.18e77	all	95	0.06	181.63; 11377.0
fpga12.12	SAT	216;1128	80.20	0%	5.31	7.44e77	all	104	0.13	14.74; 616.92
fpga12.08	SAT	144;560	246.70	10%	1.23	8.41e77	all	72	0.08	188.39; 3103.14
fpga12.09	SAT	162;684	885.00	80%	1.7	2.25e77	all	79	0.05	504.56; 16388.8
fpga13.09	SAT	176;759	550.00	85%	2.57	2.56e77	all	84	0.06	208.81; 8593.75
fpga13.10	SAT	195;905	>1000	100%	4.04	5.76e77	all	93	0.08	— ; —
fpga13.12	SAT	234;1242	>1000	100%	6.9	8.85e77	all	110	0.08	— ; —
2dlx_ca_mc*	UNS	3250;24640	6.54	0%	38.36	9.36e77	10	66	6.30	0.15; 1.04
2pipe.cnf	UNS	892; 6695	2.08	0%	10.74	2.26e45	10	38	1.56	0.17; 1.33
2pipe_1_ooo	UNS	834; 7026	2.55	0%	9.37	8	10	3	1.80	0.23; 1.41
2pipe_2_ooo	UNS	925; 8213	3.43	0%	11.14	32	10	5	2.82	0.25; 1.22
3pipe	UNS	2468;27533	36.44	0%	463.57	7.29e77	10	85	19.65	0.08; 1.85
4pipe	UNS	5237;80213	337.61	0%	>1000	—	—	—	—	—; —
5pipe	UNS	9471;195K	325.92	0%	>1000	—	—	—	—	—; —
WINDOW-BASED SYMMETRY FINDING (1000 variables per window)										
2dlx_ca_mc*	UNS	3250;24640	6.54	0%	3.17	2.34e77	10	64	5.42	0.76; 1.21
2pipe	UNS	892; 6695	2.08	0%	10.47	2.26e45	10	38	1.30	0.18; 1.63
2pipe_1_ooo	UNS	834; 7026	2.55	0%	9.02	8	10	3	1.80	0.24; 1.41
2pipe_2_ooo	UNS	925; 8213	3.43	0%	11.09	32	10	5	2.80	0.25; 1.23
3pipe	UNS	2468;27533	36.44	0%	3.63	1.42e77	10	78	36.20	0.91; 1.01
4pipe	UNS	5237;80213	337.61	0%	9.32	1.03e78	10	142	334.00	0.98 ; 1.01
5pipe	UNS	9471;195K	325.92	0%	29.42	3.64e78	10	227	290.50	1.02; 1.12

use of symmetry-breaking clauses. For a given CNF instance, the first number is the ratio of (i) CHAFF runtime on original instance, and (ii) the total runtime of symmetry detection and CHAFF on preprocessed instances. The second number is produced similarly, except that symmetry detection runtime is ignored. This is the maximal possible speed-up if symmetries are detected instantaneously or provided as domain-specific knowledge. We make several observations: **(1)** the proposed SAT instances are only a fraction of the size of recent micro-processor verification benchmarks [25], but are more difficult to solve; **(2)** some difficult SAT instances have astronomical numbers of symmetries; this includes the randomized `Urq` and `grout` benchmarks; **(3)** symmetry-breaking clauses often speed up the best available SAT solver CHAFF [6]; **(4)** symmetry-breaking clauses typically do not slow down CHAFF and often speed it up, even when few symmetries are present; **(5)** CHAFF runtime and symmetry detection runtime are not correlated; either step may be a bottleneck. **(6)** among the `chnl` instances, the hardest to solve was routing of 20 connections through 11 tracks. Adding extra unrouted connections consistently increased difficulty.

6. OPPORTUNISTIC SYMMETRY-FINDING

The use of symmetry-breaking clauses does not require finding *all* symmetries; symmetry detection can be performed *opportunistically*. An algorithm that does not guarantee to find all symmetries may finish sooner. Some symmetries may be found using domain-specific knowledge, and new clauses can be added during the creation of SAT instances. This may speed up the detection of other symmetries.

Window-based Symmetry Finding. We observed that a variable would sometimes be symmetric to another variable connected by a clause (one hop) or through a chain of two clauses (two hops). When this is not true for all symmetries of a CNF formula, many symmetries may be composable from permutation generators of that kind. We therefore focus on "local" symmetries that permute small subsets of variables and fix all other variables. We define the subsets by sliding a window of fixed size along a given linear ordering of the variables — either original variable ordering of the CNF formula or the connectivity-based MINCE ordering [1]. For a window, we consider the left and right cuts, as in Figure 4 (b). To find symmetries local to a given window, the standard

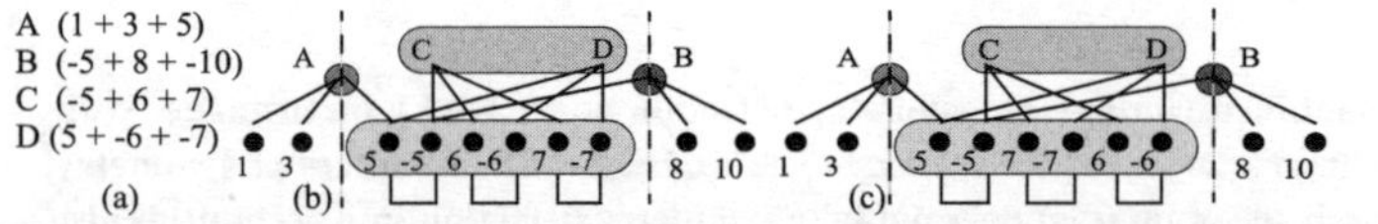

Figure 4: (a) window-based opportunistic symmetry detection for a CNF instance with ten variables and four clauses; (b) a colored graph for detecting only symmetries local to a window; (c) the "local" symmetry (6 7) (-6 -7).

construction of colored graph is applied to clauses and literals that are entirely inside the window. Each cut clause is represented by a vertex of unique color that is connected to literals inside the window. Since the size of this graph increases with cut size, a min-cut ordering improves runtime. We concatenate lists of permutation generators produced for different windows, consider the group generated by all those and use GAP [22] to produce an irredundant list of generators of this "global" group. Symmetry-breaking clauses are constructed from those generators. Producing symmetry-breaking clauses independently from each window and concatenating them may cause considerable redundancy. The trade-off between runtime, coverage and redundancy among windows depends on their overlap. Similarly, the window size affects the trade-off between runtime and coverage. We observe good empirical performance with windows of size 1000. Results in Table 1 show that our window-based technique found all or a significant portion of all symmetries for the micro-processor verification benchmarks [25] in a fraction of the runtime spent by complete symmetry-finding. If a randomized variable ordering is used, one could combine local permutation generators found for different orderings.

Improving SAT Formulations One way to reduce the runtime of symmetry finding is to learn how to detect (or predict) symmetries from domain-specific knowledge. Given the well-understood structure and symmetries of the `hole`, `chnl` and `fpga` benchmarks, we evaluated this approach on (randomized) `grout` benchmarks. We noticed that permuted variables in many cases correspond to neighboring tracks, e.g., if two connections are routed in parallel through several grid cells, there is considerable freedom (symmetry) in track assignment. To break this symmetry, we added clauses that preserve the relative order of tracks taken by every pair of connections routed through the same two edges of a grid cell. In other words, if one connection is routed through track 2 when entering the cell, and another connection is routed through track 3 when entering the cell, then the connections are allowed to leave the cell through tracks 2 and 3 resp., 1 and 2 resp. or 1 and 3 resp. Such constraints speed-up CHAFF: each `grout` instance is now solved in **0.50-0.80 seconds versus 19-45 seconds**. More dramatic speed-ups are achieved for `grout` instances built with larger routing grids.

7. CONCLUSIONS

We describe an automated flow that finds symmetries in CNF instances and uses them to speed up SAT search. This flow dramatically speeds up the solution of two well-known provably-difficult benchmark families — pigeon-hole problems and Urquhart benchmarks. We propose constructions of realistic satisfiable and unsatisfiable SAT instances, arising in routing applications, that are unusually difficult for their size. Unlike most existing SAT benchmarks, our benchmark families enable studies of the asymptotic performance of SAT solvers.

Since symmetry-finding is a bottleneck, we speed it up using opportunistic approaches. In one, we only look for symmetries that permute small groups of variables. Those groups are determined by sliding a fixed-sized window along a given variable ordering. The second approach attempts to improve the construction of SAT instances by detecting symmetries in domain-specific terms so that new clauses can be added during construction. We find astronomically many symmetries in randomized Urquhart and `grout` benchmarks, showing that randomization is compatible with symmetries. We explain symmetries in `grout` benchmarks and break them using domain-specific knowledge.

Our proposed flow does not require source code modifications and should work with all complete SAT solvers. In experiments described in Table 1 but performed with GRASP [23] instead of CHAFF [6], our flow demonstrated speed-ups 1.5-5 times even for the micro-processor verification benchmarks. The proposed flow may not give improvement on arbitrary SAT benchmarks — many difficult SAT instances do not have symmetries [8]. While many DIMACS benchmarks [10] have large numbers of symmetries, they are easy and can be solved faster than their symmetries can be detected.

Acknowledgements. This work is funded by an Agere Systems/SRC Research fellowship, a DAC fellowship and DARPA/MARCO GSRC.

8. REFERENCES

[1] F. Aloul, I. Markov and K. Sakallah, "Faster SAT and Smaller BDDs via Common Structure", *ICCAD 2001*, pp. 443-448.

[2] P. Beame, R. Karp, T. Pitassi and M. Saks, "The efficiency of Resolution and Davis-Putnam Procedure", to appear in *SIAM Journal on Computing*. `http://www.cs.washington.edu/homes/beame/papers/resj.ps`

[3] B. Benhamou and L. Sais, "Tractability through symmetries in propositional calculus". *Journal of Autom. Reasoning*, vol. 12, (no.1), Feb. 1994. pp. 89-102.

[4] L. Brisoux, E. Gregoire, L. Sais, "Improving backtrack search for SAT by means of redundancy", *Foundations of Intelligent Systems. 11th Intl. Symp., ISMIS'99*. Warsaw, Poland, June '99. Berlin, Germany: Springer 1999. p.301-9.

[5] C. A. Brown, L. Finkelstein, and P. W. Purdom. "Backtrack searching in the presence of symmetry". (T. Mora, editor), *Applied algebra, algebraic algorithms and error correcting codes, 6th intl. conf.*, pages 99- 110. Springer-Verlag, 1988.

[6] M. Moskewicz, C. Madigan, Y. Zhao, L. Zhang and S. Malik, "Chaff: Engineering an Efficient SAT Solver", *DAC 2001*.

[7] E.M. Clarke et al., (Edited by: Hu, A.J.; Vardi, M.Y.) "Symmetry Reductions in Model Checking", CAV'98, pp.159-71.

[8] S. A. Cook, D. G. Mitchell, "Finding Hard Instances of the Satisfiability Problem: A Survey", *DIMACS Ser. Discr. Math. & Theor. Comp. Sci.*, '97.

[9] J. Crawford, M. Ginsberg, E. Luks and A. Roy, "Symmetry-breaking predicates for search problems", *5th Intl Conf. Principles of Knowledge Representation and Reasoning (KR'96)*, Cambridge, MA, pp. 148-159.

[10] DIMACS Boolean Satisfiability Challenge Benchmarks: `ftp://dimacs.rutgers.edu/pub/challenge/sat/benchmarks/cnf`

[11] C.A.J. van Eijk, E.T.A.F. Jacobs, B. Mesman and A.H.Timmer, Identification and Exploration of Symmetries in DSP Algorithms, *DATE '99*, March 1999, pp. 602-608.

[12] L. Babai, "Automorphism Groups, Isomorphism, Reconstruction", Chapter 27, pp. 1447-1541, In (R. L. Graham, M Grötschel and L. Lovász, eds, Handbook of Combinatorics, vol. 2, MIT Press, 1995).

[13] C. Norris Ip and D. L. Dill, "Better verification through symmetry" , *Formal Methods in System Design*, 9(1/2):41-75, 1996.

[14] V. Kravets and K. Sakallah, "Generalized Symmetries of Boolean Functions", *ICCAD 2000*, pp. 526-532.

[15] V. Kravets and K. Sakallah, "Constructive Library-Aware Synthesis Using Symmetries", *DATE 2001*, pp. 208-213.

[16] G.S. Manku, R. Hojati and R. Brayton, "Structural symmetry and model checking", *Proc. Intl. Conf. Comp.-Aided Verific. (CAV '98)*, pp. 159-171.

[17] B. D. McKay, "Practical Graph Isomorphism", *Congressus Numerantium*, 30 (1981), pp. 45-87.

[18] B. D. McKay, "Nauty user's guide" (version 1.5), Technical report TR-CS-90-02, Australian National University, Computer Science Department, ANU, 1990. `http://cs.anu.edu.au/~bdm/nauty/`

[19] G. Nam, F. Aloul, K. Sakallah, and R. Rutenbar, "A Comparative Study of Two Boolean Formulations of FPGA Detailed Routing Constraints", in *Proc. Intl. Conf. on Physical Design (ISPD 2001)*, pp. 222-227.

[20] L. H. Soicher, "GRAPE: A System For Computing With Graphs and Groups", *in "Groups and Computation"* (L. Finkelstein and W.M. Kantor, eds), *DIMACS Ser. in Discr. Math. & Theor. Comp. Sci.* 11, pp. 287-291. `www-groups.dcs.st-andrews.ac.uk/~gap/Share/grape.html`

[21] M. Prasad, P. Chong, K. Keutzer, "Why is ATPG easy?", *DAC '99*.

[22] E. L. Spitznagel, "Review of Mathematical Software, GAP", *Notices Amer. Math. Soc.*, 41 (7), (1994), pp. 780-782. `http://www-groups.dcs.st-andrews.ac.uk/~gap/gap.html`

[23] J. P. M. Silva and K. A. Sakallah, "GRASP: A New Search Algorithm for Satisfiability", *IEEE Trans. On Computers*, vol. 48, no. 5, May 1999.

[24] A. Urquhart, "Hard Examples for Resolution", *JACM*, vol. 34, 1987.

[25] M.N. Velev, and R.E. Bryant, "Effective Use of Boolean Satisfiability Procedures in the Formal Verification of Superscalar and VLIW Microprocessors", *DAC 2001*, pp. 226-231. `http://www.ece.cmu.edu/~mvelev/#BENCHMARKS`

Satometer: How Much Have We Searched?

Fadi A. Aloul, Brian D. Sierawski, Karem A. Sakallah
Department of Electrical Engineering and Computer Science
University of Michigan
{faloul, bsieraws, karem}@eecs.umich.edu

ABSTRACT

We introduce Satometer, a tool that can be used to estimate the percentage of the search space actually explored by a backtrack SAT solver. Satometer calculates a normalized minterm count for those portions of the search space identified by conflicts. The computation is carried out using a zero-suppressed BDD data structure and can have adjustable accuracy. The data provided by Satometer can help diagnose the performance of SAT solvers and can shed light on the nature of a SAT instance.

Categories and Subject Descriptors

I.1 [**Symbolic and Algebraic Manipulation**]: Expressions and Their Representation, Algorithms.

General Terms

Algorithms, Measurement, Experimentation, Verification.

Keywords

BDDs, ZBDDs, SAT, CNF, backtrack search, conflict diagnosis, search space coverage, search progress.

1. INTRODUCTION

The last few years have seen significant algorithmic advances in, and carefully-crafted implementations of, Boolean Satisfiability (SAT) solvers [3, 13, 15, 19]. This has led to their successful application to a wide range of large-scale EDA problem instances consisting of thousands of variables and millions of clauses [4, 16, 18]. Despite these remarkable developments, SAT solvers cannot escape the underlying worst-case exponential complexity of their search space and must sometimes be aborted after a certain time-out limit has been reached. Typically, when a solver aborts it provides very little data about how much progress it had achieved up to that point. Such data can be quite useful. Knowing, for instance, that the solver had managed, after several hours, to explore only 1% of the search space might suggest a very hard problem instance and the need, perhaps, to try a different approach. If, on the other hand, the solver reports exploring more than 99% of the search space without finding a solution, it may be reasonable to assume that the instance has very few satisfying assignments or is possibly unsatisfiable.

Satometer (pronounced like barometer) is an accessory that can be used with any backtrack search SAT solver to report the percentage of search space actually explored by the solver. It requires the solver to emit the set of clauses corresponding to the conflicts encountered during the search. It can be used dynamically, while the SAT solver is running, to indicate progress in the search for a solution. It is more useful, however, as a postprocessor to analyze the result of an aborted or completed search.

The paper is organized as follows. In Section 2 we introduce our measure of search progress. We then describe, in Section 3, how this measure can be computed using BDDs and ZBDDs. In Section 4 we illustrate the utility of this measure in a variety of experimental scenarios and conclude, in Section 5, with a summary of the paper's main contributions.

2. MEASURING SEARCH PROGRESS

Despite the considerable activity in SAT research, the question of measuring the progress of a search process does not seem to have attracted much attention. The only relevant work that we were able to find is that of Kokotov and Shlyakhter [10]. They describe a progress bar that can be integrated into a backtrack SAT solver to measure its progress. The bar is updated based on either historical or predictive estimates of the size of the decision tree maintained by the SAT solver. They reported that the bar is able to predict progress with an accuracy of 80-90% without significantly impacting the solver's run time.

In our approach, we view the search process as a sequence of moves that continually (and systematically) modify a (partial) variable assignment until 1) a satisfying assignment (a solution) is found, 2) the formula is proven to be unsatisfiable (has no solution), or 3) a time-out limit is reached. Along the way, many assignments that are explored will correspond to zeros of the function represented by the formula and will cause the search process to backtrack. Every time such a "conflict" occurs, it identifies a portion of the search space that can be regarded as having been *explored* and found to contain no solutions.

Let $A_1, A_2, \ldots A_i$ denote the assignments that correspond to the first i conflicts. We can measure how much of the search space has been explored by counting the number of minterms[1] covered by the function $A_1 + A_2 + \ldots + A_i$. Normalizing this count by the total size of the space yields the percentage of the space that has been explored up to this point. We will use the notation $\|f\|$ to express the normalized number of minterms of the function f. Thus, $\|a+b\| = 75\%$, $\|a \cdot b\| = 25\%$, and $\|a \oplus b\| = 50\%$[2]. In the sequel, we will refer to $\|f\|$ as the *size* of f.

This measure can be equivalently computed by considering the *conflict clauses* identified at each conflict. Let $C_1, C_2, \ldots, C_j$ denote the conflict clauses identified after the first i conflicts. In general, $j \geq i$ as one or more conflict clauses may be identified at each conflict. The portion of the search space that would have been explored after processing the *ith* conflict can now be computed as $1 - \|C_1 \cdot C_2 \cdot \ldots \cdot C_j\|$.

1. Complete truth assignment that sets the function to 1.
2. Assuming that the number of variables is 2.

Decisions		Implications	Conflicts		Explored Space	
			Y/N	Clause	Minterms	%
1	a		N			
2	ab		N			
3	abc	d'	Y	$(a'+b'+c'$	2	12.5
4	abc'	d	Y	$(a'+b'+c)$	4	25
5	ab'	$c'd$	Y	$(a'+b)$	8	50
6	a'		N			
7	$a'b$		N			
8	$a'bc$	d'	Y	$(a+b'+c')$	10	62.5
9	$a'bc'$	d	N	Solution!		

(a) Execution trace of a basic backtrack SAT Solver

Decisions		Implications	Conflicts		Explored Space	
			Y/N	Clause	Minterms	%
1	a		N			
2	ab		N			
3	abc	d'	Y	$(b'+c')$	4	25
4	ab	$c'd$	Y	$(a'+b')$	6	37.5
5	a	$b'c'd$	Y	(a')	10	62.5
		a'	N			
6	b	$a'c'd$	N	Solution!		

(b) Execution trace of a conflict-based backtrack SAT Solver

Figure 1. Execution traces of two different SAT solvers on the formula in (1) illustrating how search progress is measured.

An illustration of these computations is shown in Figure 1 for the 4-variable formula

$$\varphi = (a+b+c)(a+b+c')(a'+b+c')(a+c+d)\cdot (a'+c+d)(a'+c+d')(b'+c'+d')(b'+c'+d) \quad (1)$$

3. COMPUTING SPACE COVERAGE

When the conflicting assignments are disjoint (i.e., when $A_k \cdot A_l = 0$ for $k \neq l$), space coverage can be simply calculated by the formula:

$$\|A_1 + A_2 + \ldots + A_i\| = \sum_{1 \le k \le i} \|A_k\| \quad (2)$$

Equivalently, if the conflict clauses are disjoint, i.e. if $C_k + C_l = 1$ for $k \neq l$, then space coverage is simply

$$1 - \|C_1 \cdot C_2 \cdot \ldots \cdot C_j\| = \sum_{1 \le k \le j} (1 - \|C_k\|) \quad (3)$$

In other words, if conflicts identify non-overlapping pieces of the search space, then the size of the explored space can be found by simply adding the sizes of the different pieces. In general, this will not be the case except for standard backtrack algorithms that do not employ conflict diagnosis to prune the search space. To compute the size of the explored space in such cases we have no choice but to build some type of symbolic representation for the disjunction of conflict assignments or the conjunction of conflict clauses. We describe below the two representations we examined and show how we used them to measure space coverage. Without loss of generality, we restrict the discussion to building representations for conjunctions of conflict clauses.

3.1 Using BDDs

The conflict clauses can be symbolically "anded" using a reduced ordered binary decision diagram (ROBDD or BDD for short) [5]. BDD semantics allow us to write the function f at a node labeled with variable x using Boole's expansion:

$$f = x' \cdot g + x \cdot h \quad (4)$$

where g and h are the functions associated with the 0- and 1-children of that node (see Table 1.) This immediately leads to the following formula for the size of f:

$$\|f\| = \frac{1}{2}(\|g\| + \|h\|) \quad (5)$$

The size of the function represented by a BDD can now be obtained by sweeping the BDD from the terminal nodes towards the top node and applying (5) at each visited node. The sweep is initialized by setting $\|0\| = 0$ and $\|1\| = 1$ for the constant functions of the terminal nodes.

3.2 Using ZBDDs

The problem with the BDD representation, of course, is that it quickly runs out of memory. An alternative that has lower memory requirements is the zero-suppressed BDD (ZBDD) originally proposed by Minato [14] for manipulating large combination sets, including sets of Boolean cubes. A combination set S can be regarded as a *set of sets*, e.g. $\{\{a, b\ , \{c, d, e\}, \{a, d\}, \{b\}\}$. Recently, Chatalic and Simon [6] demonstrated that ZBDDs can be an effective implicit representation of large CNF formulas and showed how they can be used to perform "multi-resolution" to solve some large structured SAT instances. In this scenario, the above example set corresponds to the CNF formula $(a+b)(c+d+e)(a+d)(b)$, i.e. each combination is viewed as an OR term (a clause) and the entire set (a union of combinations) as an AND term. Such an interpretation allows the semantics of Boolean algebra to be layered on top of the semantics of set algebra to obtain further compression of the ZBDD structure. In particular, Chatalic and Simon extended the standard ZBDD set-union operation to a subsumption-free union that automatically removes any clause that is completely subsumed by another clause. In the above example, combination $\{a, b\}$ is subsumed by combination $\{b\}$ yielding the *logically equivalent* set $\{\{c, d, e\}, \{a, d\}, \{b\}\}$. Additional reduction rules based on literal absorption, i.e. $(a)(a'+b+c) = (a)(b+c)$, were subsequently described in [1].

The semantics of ZBDD nodes were first articulated by Lobbing et al. in [11]. Given a set of atoms $\{a, b, c, \ldots\}$, a ZBDD node labeled with atom x represents a combination set f constructed according to the formula:

Table 1. Semantics of Decision Diagrams

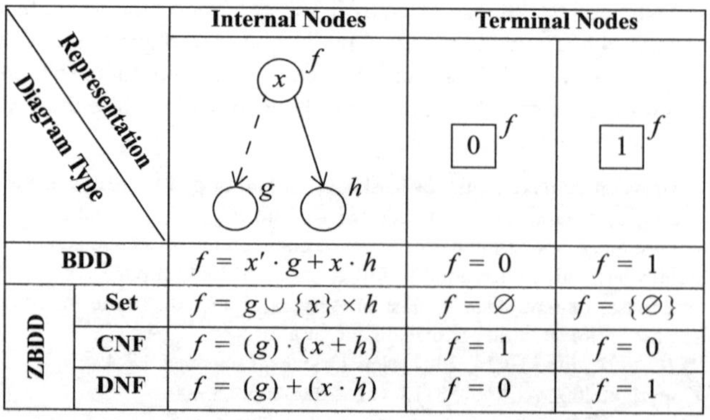

Diagram Type \ Representation		Internal Nodes	Terminal Nodes	
BDD		$f = x' \cdot g + x \cdot h$	$f = 0$	$f = 1$
ZBDD	Set	$f = g \cup \{x\} \times h$	$f = \varnothing$	$f = \{\varnothing\}$
	CNF	$f = (g) \cdot (x+h)$	$f = 1$	$f = 0$
	DNF	$f = (g) + (x \cdot h)$	$f = 0$	$f = 1$

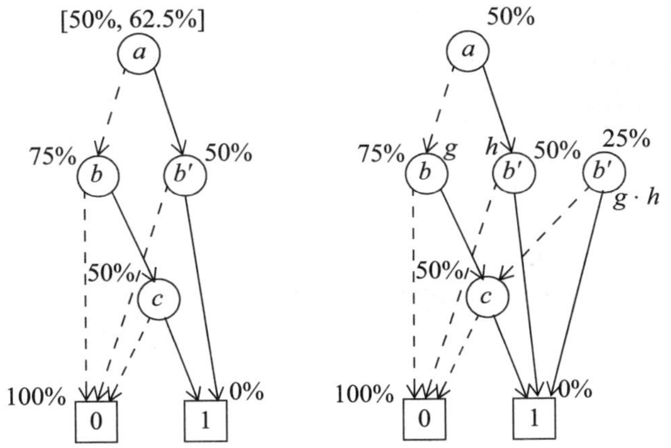

(a) Using bound in (12) (b) By computing $g \cdot h$

Figure 2. Computation of $\|(a+b')(b+c)\|$ using (11) and (12).

$$f = g \cup \{x\} \times h \tag{6}$$

where g and h are the combination sets associated with the 0- and 1-children of that node (see Table 1.) The terminal 0 and 1 nodes correspond, respectively, to the empty set (set of no combinations) and to the set of consisting of the empty combination. The "product" in (6) is similar to the Cartesian product of two sets and is defined by

$$S \times T = \bigcup_{s \in S,\, t \in T} \{s \cup t\} \tag{7}$$

For example, given the combination sets $S = \{\{a, b\}, \{b, c\}\}$ and $T = \{\{a, d\}, \{e\}\}$, their product is[3]

$$S \times T = \{\{a, b, d\}, \{a, b, e\}, \{a, b, c, d\}, \{b, c, e\}\} \tag{8}$$

When used to represent a CNF formula, the formula f associated with a ZBDD node labeled by variable x follows the same template of (6) except that the union of atoms in a combination is viewed as logical OR and the union of the combinations is viewed as logical AND yielding

$$f = (g) \cdot (x + h) \tag{9}$$

where g and h are the formulas associated with the 0- and 1-children of that node (see Table 1.) The terminal 0 and 1 nodes, correspond, respectively, to the constant 1 and constant 0 functions.

To represent CNF formulas with ZBDDs, the set of atoms is taken to be the set of literals over which the formula is defined. In addition, the positive and negative literals of each variable are grouped together so that they are adjacent in the total order used in constructing the ZBDD. This restriction facilitates, among other things, the identification and automatic removal of tautologies, i.e. combinations that have the form $(x + x' + \ldots)$, to further reduce the size of the ZBDD [6].

To determine the size of the function represented by the CNF formula associated with a ZBDD node, we must first re-write (9) as the disjoint sum of two terms

$$f = (g) \cdot (x + h) = x \cdot g + x' \cdot (g \cdot h) \tag{10}$$

This immediately leads to

$$\|f\| = \frac{1}{2}(\|g\| + \|g \cdot h\|) \tag{11}$$

which, unlike (5) for BDDs, requires that we compute the size of the product of the two child formulas. This is not a problem if one or both of the children is a terminal node, but does pose a serious complication if they are both internal nodes. One way to resolve this complication is to (recursively) create additional ZBDD nodes for such products until one of the children becomes terminal. This will provide us with the *exact* answer, but may exponentially increase the size of the ZBDD. Some of that increase can be ameliorated with caching and garbage collection. In particular, created nodes can be eliminated as soon as they have been used to tighten the bound of their parent.

An alternative to computing $\|g \cdot h\|$ exactly is to *bound* it. The upper bound is easily established as $\min(\|g\|, \|h\|)$ and occurs when either $g \le h$ or $h \le g$. The lower bound can be determined by noting that $\|g \cdot h\| = 1 - \|g' + h'\|$. Thus $\|g \cdot h\|$ is smallest when $\|g' + h'\|$ is largest which occurs when g' and h' are disjoint. This gives a lower bound of $\|g\| + \|h\| - 1$ and yields the interval

$$\|g \cdot h\| \in [\max(0, \|g\| + \|h\| - 1), \min(\|g\|, \|h\|)) \tag{12}$$

where the max in the lower bound insures that the estimate remains non-negative.

An illustration of these computations is given in Figure 2 for the example formula $(a + b') \cdot (b + c)$. The percentages annotating the ZBDD nodes denote the function sizes of their corresponding formulas as computed by (11) and (12). The uncertainty in the size at the top node is resolved, in part b of the figure, by creating a node for the product of its children.

Between the two extremes of an *exact* count and a *bound* computed according to (12) we can produce a range of approximations that trade accuracy with speed and memory consumption. Specifically, when a given level of accuracy, say 10%, is exceeded by the bound computed from (12), additional ZBDD nodes are created for the product formulas until the desired level of accuracy is achieved.

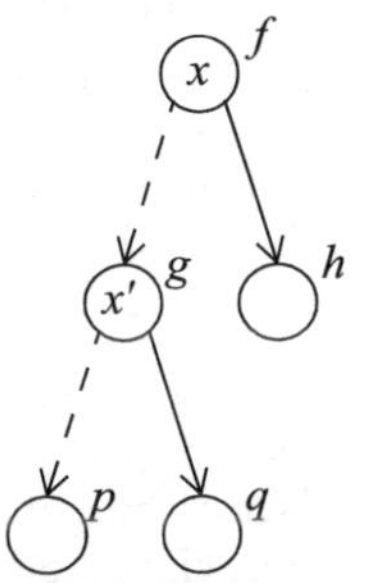

Figure 3. The special case when g is not vacuous in x.

We must finally note that (11) is correct only when g is vacuous in x. The only situation when this is not true is depicted in Figure 3 where g's node is labeled by the literal x'.[4] Substituting $g = (p) \cdot (x' + q)$ in (10) produces the disjoint sum

$$f = x \cdot (p \cdot q) + x' \cdot (p \cdot h) \tag{13}$$

which readily leads to

$$\|f\| = \frac{1}{2}(\|p \cdot q\| + \|p \cdot h\|) \tag{14}$$

Figure 4 illustrates the three possible modes of our approach on the bridging-fault *bf2670-001* instance. Despite setting an error limit of 20%, on average, the restricted bound and the unrestricted bound methods reported results within 7% and 24%, respectively, of the exact answer.

4. EXPERIMENTAL EVALUATION

Satometer is implemented in C++ using the CUDD package [17]. It incorporates the ZBDD enhancements described in [1] and [6] for symbolic manipulation of CNF formulas. In this section we demon-

3. Note that $S \times T \neq S \cup T$. For this example, $S \cup T = \{\{a, b\}, \{b, c\}, \{a, d\}, \{e\}\}$.

4. Note that h's node cannot be labeled by x' as this would create a tautology that is automatically eliminated.

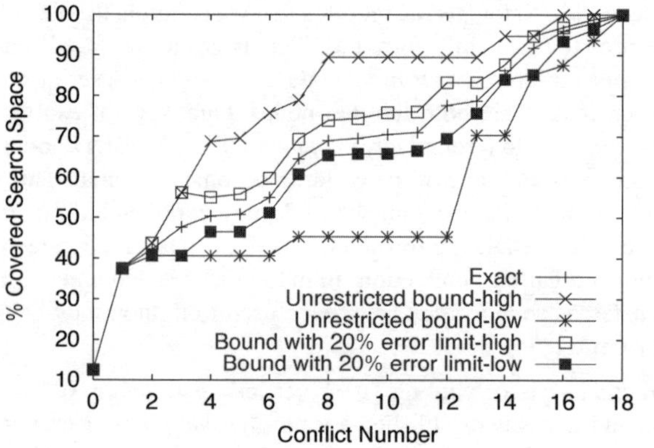

Figure 4. Applying three modes of the proposed approach to *bf2670-001.cnf* instance

strate its utility by applying it in a number of experimental scenarios. We configured it to report the size of the explored search space to within 20% of the exact answer; in many cases it was able to achieve a higher level of accuracy or to even report the exact answer. In the tables to follow, a single number in the explored space columns indicates that an exact answer was reported; ranges are indicated as intervals. All experiments were performed on an AMD Athlon 1.4 GHz machine with 1GB of RAM running the Linux operating system.

4.1 Effect of Preprocessing the CNF Formula

A variety of preprocessing techniques have been proposed to modify a CNF formula before submitting it to a SAT solver. These techniques generally add clauses to the formula in order to increase the number of potential implications or perform stylized algebraic simplifications to reduce the number of variables. We used Satometer to study the effectiveness of such techniques. In each case we compare the size of the space explored by a standard DLL algorithm[5] [8] (i.e. without conflict analysis) on the original as well as on the modified formula. The time-out limit in these experiments was set to 10 seconds; *Satometer's run time was negligible*. The results of these experiments are given in Table 2, Table 3, and Table 4.

Addition of consensus clauses. In [2] the authors report that augmenting a CNF formula with clauses identified using consensus can reduce search time. To avoid generating an exponential number of clauses, they proposed a truncated iterative consensus procedure that augments the original formula with clauses whose size (number of literals) is limited by a small user-specified constant. They report speedups on the *aim* benchmarks from the DIMACS set [9] when the size of added clauses are limited to 3 or fewer literals.

A sampling of results on some unsatisfiable instances from this suite is shown in Table 2. Column 1 lists the name of the benchmark; columns 2 and 3 give the number of variables (V) and clauses (C) in the original formula; column 4 gives the number of consensus clauses that are added to the formula; and columns 5 and 6 indicate the size of explored space reported by Satometer. The data in this table clearly show the effectiveness of these added clauses. For the two smaller instances, the search algorithm was actually able to explore the entire search space, and thus prove the unsatisfiability of the

Table 2. Addition of consensus clauses

Benchmark	Original		Modified	Explored Space, %	
	V	C	Extra C	Original	Modified
aim-50-1_6-no-4	50	80	54	57.06	100
aim-100-1_6-no-3	100	160	73	0.015	100
aim-200-1_6-no-3	200	320	233	0.049	[97.72, 100]
aim-200-2_0-no-1	200	400	191	0	[87.75, 100]

Table 3. Addition of symmetry-breaking predicates

Benchmark	Original		Modified	Explored Space, %	
	V	C	Extra C	Original	Modified
hole-7	56	204	14	100	100
hole-8	72	297	16	79.2	100
hole-9	90	415	18	37.5	100
hole-10	110	561	20	18.75	[99.98, 100]
hole-11	132	738	22	9.39	[99.96, 100]
hole-12	156	949	24	4.68	[99.96, 100]

Table 4. Algebraic Simplification

Benchmark	Original		Modified		Explored Space, %	
	V	C	V	C	Original	Modified
longmutl7	3319	10335	2184	7635	0.280	0.341
queinvar20	2435	20671	2343	28438	50	50.1
barrel7	3523	13765	800	3447	51.02	62.46
dlx2_cc_bug08	1515	12808	1486	13875	0	9.38

modified formula. In all cases, the addition of these clauses helped the SAT solver explore a significantly larger portion of the search space in the allotted amount of time.

Addition of symmetry-breaking predicates. In [7] the authors propose analyzing a CNF formula 1) to identify its symmetries, and 2) to augment it with clauses that break those symmetries. The intuition here is that the symmetry-breaking clauses act by allowing only one of many equivalent variable assignments to be a potential solution to the formula. If the original formula is satisfiable, the number of solutions may considerably decrease after preprocessing, clearly indicating that the search space was reduced. However, even if the original instance was not satisfiable, "the number of equivalent roads leading nowhere" would be reduced, and a generic SAT solver is likely to conclude much faster that no solution exists.

This intuition is confirmed by the data in Table 3 (whose layout is identical to that of Table 2.) The benchmarks in this experiment are members of the unsatisfiable *hole* suite [9] (which relates to the Pigeonhole principle.) The augmentation of each instance by a small number of symmetry-breaking clauses drastically enhances the ability of the SAT solver to prove unsatisfiability. This trend is clearly accentuated as instance sizes increase.

Algebraic simplification. Another formula preprocessing technique is based on formula simplification rules aimed at reducing the number of variables or clauses in the formula [12]. We studied this approach on some large hard bounded model checking [4] and microprocessor verification [18] instances. Results on a representative sample are given in Table 4.

Unlike the earlier experiments, the performance of the SAT solver on the modified formulas is not significantly better than its performance on the original formulas. The best improvement is in the *barrel7* benchmark and can be attributed to the simplifier's ability to drastically reduce the number of variables (from 3523 to 800.)

5. The solver uses a fixed decision heuristic, chronological backtracking, and implements BCP as implemented in Chaff.

Table 5. Percentage of Explored Search Space for Various SAT Solver and Decision Heuristics

Benchmark				Space Explored,%		
Family	**Name**	**V**	**C**	**DLL**	**Chaff–Fixed**	**Chaff–VSIDS**
uP Verification	**2dlx_cc**	4524	41704	0	[81, 100]	[99.06, 100]
	3pipe	2392	27533	0.098	[47.23, 62.63]	[80.41, 100]
	4pipe	5096	80213	0.025	[69.68, 88.15]	[77.77, 95.46]
	9vliw	19148	179492	0	[28.91, 35.16]	[99.97, 100]
DIMACS	**par32-1-c**	1315	5254	0	[78.64, 89.39]	[82.72, 100]
Bounded Model Checking	**barrel6**	2306	8931	52.77	[60.94, 63.83]	100
	barrel7	3523	13765	51.02	[60.95, 68.79]	[98.34, 100]
	barrel9	8903	36606	50.11	[58.59, 58.84]	[99.94, 100]
	longmult6	2848	8853	0.40	[72.39, 80.43]	[99.93, 100]
	longmult8	3810	11877	0.21	[80.27, 87.78]	[90.48, 100]
	queuin18	2081	17368	0	[96.57, 100]	100
	queuin20	2435	20671	50	[92.3, 100]	[97.59, 100]
FPGA Routing	**alu2_gr_rcs_w7**	3570	73478	2.36	[29.99, 36.55]	[50, 58.75]
	k2fix_gr_rcs_w8	10056	271393	1.18	[0.665, 7.65]	[0.798, 9.03]
	k2fix_gr_rcs_w9	11313	305160	0.59	[0.393, 5.147]	[0.400, 3.34]
	vda_gr_rcs_w8	5776	116522	0	[0.615, 6.65]	[0.819, 9.75]

The low coverage in this experiment is also an indication of the difficulty of these instances.

4.2 Analysis of Dynamic Techniques

In this set of experiments, we report on the application of Satometer to various SAT solvers with a variety of parameters. Our experiments involve two different SAT solvers: a simple DLL solver [8] and Chaff [15]. Chaff represents an efficient implementation of the basic DLL solver and is currently known as the leading DLL-based SAT solver. The goals of this experiment are to determine a) the best of two black-box SAT solvers, in which each solver's description is hidden, b) the best of a variety of decision heuristics, c) the difficulty of CNF instances, and d) an estimate of the number of satisfying assignments in a satisfiable instance.

Black box A vs. black box B experiment. In the following experiment, several SAT solvers are provided. However, the user has no knowledge of the internals of any of the SAT solvers. Given a set of hard instances, the user is required to identify the best solver in the shortest possible time. In general, the user will need to run each SAT solver for a specified time or randomly select a solver and hope that it is the best among all others. Using the proposed method, however, can give an insight to which solver performs best within the specified run time limit. Table 5 shows several results for various hard instances from bounded model checking [4], microprocessor verification [18], FPGA routing [16], and the DIMACs set [9]. We tested each instance for 10 seconds using the following three SAT solvers and options: standard DLL solver, Chaff with a fixed decision heuristic, and Chaff with the default cherry.smj heuristic. The results clearly indicate the superiority of the third solver among the other two solvers for almost all benchmarks, due to the significantly high search space coverage achieved in the given time limit. Figure 5 shows a detailed space coverage analysis of the *barrel5* instance for all three solvers.

Comparison of decision heuristics. As shown in Table 5, the proposed method can also be used to classify decision heuristics and rate their performance on various SAT instances. We show the results for two decision heuristics: a) static fixed [8]: unresolved variables with minimum index are selected first for decisions; b) dynamic VSIDS [15]: variables that appear in the highest number of clauses are selected first. (Some weight is given to variables appear-

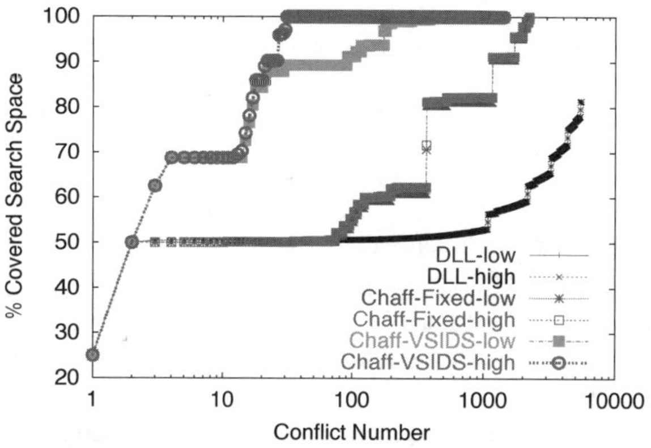

Figure 5. Search Space Coverage for *Barrel5.cnf*

ing in recent conflict-induced clauses). Again, the results show the effectiveness of VSIDS as opposed to the fixed decision heuristic. Nevertheless, the *k2fix_gr_rcs_w9* instance show a larger upper bound of the explored search space using the fixed decision order as opposed to VSIDS. However, since the ranges for both heuristics overlap, its hard to identify the optimal decision heuristic.

Hard problem prediction. Table 5 also shows the difficulty of solving the FPGA routing instances as opposed to other hard instances for the given decision heuristics and SAT solvers. Figure 6 shows a detailed space coverage analysis of the *k2fix_gr_rcs_w9* instance after unsuccessfully trying to solve it with Chaff for up to 150 seconds. Perhaps, this method can be used as a metric to rate the difficulty of SAT instances and assist SAT solver developers in improving their SAT tools.

Number of satisfiable assignments. As mentioned earlier, the search space will never be totally explored in "satisfiable" instances, as SAT solvers typically abort after identifying the first satisfying assignment. However, in some cases, several satisfying assignments, if not all, are needed. An example is to identify all possible primary input assignments for a circuit that would minimize the total gate delay. An insight into the number of possible satisfying assignments can be very helpful. A satisfiable instance in which a satisfying assignment is identified at an early stage of the search process is likely to have many satisfying assignments. In contrast,

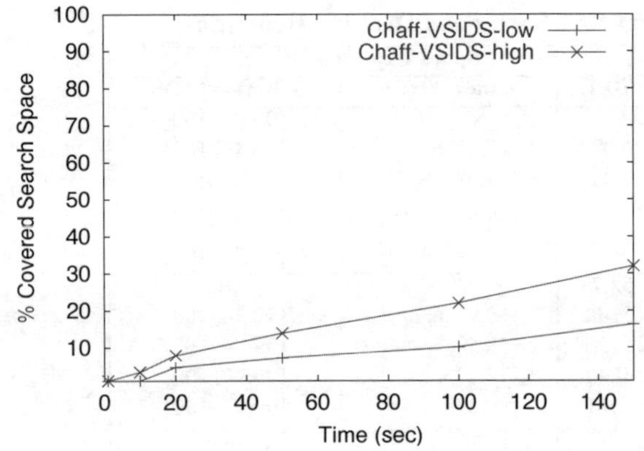

Figure 6. Search Space Coverage for *k2fix_gr_rcs_w9.cnf*

an instance that identifies a satisfying assignment after exploring almost the complete search space probably has a few satisfying assignments only. In order to test our assumption, we selected two satisfiable instances from the DIMACS set [9], namely the *aim-200-1_6-yes1-1.cnf* and *ssa7552-160.cnf*. The former is known to have a single satisfying assignment only, whereas the latter represents a stuck-at-fault problem with many satisfying assignments. Both instances were solved by Chaff in less than a second. We measured the explored search space after the search was completed for a single satisfying assignment. As expected, the percentage of the search space explored by the *aim** instance (99.99%) was tremendously larger than the *ssa** instance (28.13%).

Again, as in the experiments in Section 4.1, the accuracy of our results are significant. Although a user specified error limit of 20% is set, out of the 78 runs, 47, 6, 16, 8, reported results with 100%, >99%, 90%~99%, 80%~90% accuracy.

In terms of run time and memory consumption, constructing the ZBDDs is fast and is usually dependent on the size of the clauses. Furthermore, the high compression power of the ZBDD data structure utilizes less memory than a list data structure. As mentioned in Section 3.2, computing the search space coverage with an unrestricted bound is done by a single traversal of the ZBDD. On the other hand, the restricted bound and the exact count methods are slower, since additional ZBDD nodes are created during the ZBDD traversal. The size of the ZBDD, however, doesn't grow exponentially since the additional ZBDD nodes are removed as soon as the function sizes of their corresponding formulas are computed.

One way to reduce the run time and memory consumption is to only analyze conflict-induced clauses of size k or less. In general, smaller clauses are more useful in measuring the explored search space and require less ZBDD construction time and fewer ZBDD nodes. This approach, however, can only be used to measure the lower bound of the explored search space. For the instances reported in Table 5, Satometer was able to compute the search space coverage for almost all instances in less than a second each.

5. SUMMARY AND CONCLUSIONS

We described Satometer, a tool that measures the percentage of search space explored by a SAT solver. The tool can provide helpful diagnostic information, either during or at the conclusion of a SAT run. We believe that tools such as this are needed to complement the powerful SAT engines that have been developed in recent years. We plan to identify other metrics that can help characterize a search process (e.g., the maximum number of satisfied clauses encountered at any point during the search), to look for ways to further improve the efficiency of Satometer (e.g., by caching computation results), and to use it to analyze the performance of solvers on hard SAT instances. We are also planning to integrate Satometer into known SAT solvers and use the search space information to improve decision and restart heuristics.

6. ACKNOWLEDGMENTS

This work is funded by the DARPA/MARCO Gigascale Silicon Research Center and an Agere Systems/SRC Research fellowship.

7. REFERENCES

[1] F. Aloul, M. Mneimneh, and K. Sakallah, "Backtrack Search Using ZBDDs," *in Int'l Workshop on Logic Synthesis*, 293-297, 2001.

[2] F. Aloul, J. Silva, and K. Sakallah, "An Experimental Study of Satisfiability Search Heuristics," *in the Proc. of Design, Automation and Test in Europe*, 2000.

[3] R. Bayardo Jr. and R. Schrag, "Using CSP look-back techniques to solve real world SAT instances," *in Proc. of the 14th National Conference on Artificial Intelligence*, 203-208, 1997.

[4] A. Biere, A. Cimatti, E. Clarke, M. Fujita, and Y. Zhu, "Symbolic Model Checking using SAT procedures instead of BDDs," *in Proc. of the Design Automation Conf*erence, 317-320, 1999.

[5] R. Bryant, "Graph-based algorithms for boolean function manipulation," *in Proc. of IEEE Trans. on Computers*, 35(8), 677-691, 1986.

[6] P. Chatalic and L. Simon, "Multi-Resolution on Compressed Sets of Clauses," *in Proc. of the Int'l Conference on Tools with Artificial Intelligence*, 2-10, 2000.

[7] J. Crawford, M. Ginsberg, E. Luks, and A. Roy, "Symmetry-Breaking Predicates for Search Problems," *in Knowledge Representation: Principles of Knowledge Representation and Reasoning,* 148-159, 1996.

[8] M. Davis, G. Logemann, and D. Loveland, "A Machine Program for Theorem Proving," *in Comm. of the ACM*, 5(7), 394-397, 1962.

[9] DIMACS Challenge benchmarks in *ftp://Dimacs.rutgers.EDU/pub/challenge/sat/benchmarks/cnf.*

[10] D. Kokotov, I. Shlyakhter, "Progress Bar for SAT Solvers," unpublished manuscript, *http://sdg.lcs.mit.edu/satsolvers/progressbar.html,* 2000.

[11] M. Lobbing, O. Schroer, and I. Wegner, "The Theory of Zero-Suppressed BDDs and the Number of Knight's Tours," *in IFIP WG 10.5 Workshop on Applications of the Reed-Muller Expansion in Circuit Design, 1995.*

[12] J. P. Marques-Silva, "Algebraic Simplification Techniques for Propositional Satisfiability," in *Proc. of the Int'l Conference on Principles and Practice of Constraint Programming, 2000.*

[13] J. Marques-Silva and K. Sakallah, "GRASP: A Search Algorithm for Propositional Satisfiability," *in IEEE Trans. on Computers*, 48(5), 506-521, 1999.

[14] S. Minato, "Zero-Suppressed BDDs for Set Manipulation in Combinatorial Problems," *in Proc. of the Design Automation Conf.*, 1993.

[15] M. Moskewicz, C. Madigan, Y. Zhao, L. Zhang, and S. Malik, "Chaff: Engineering an Efficient SAT Solver," *in Proc. of the Design Automation Conference*, 530-535, 2001.

[16] G. Nam, F. Aloul, K. Sakallah, and R. Rutenbar, "A Comparative Study of Two Boolean Formulations of FPGA Detailed Routing Constraints," *in the Proc. of the Int'l Symp. on Physical Design*, 2001.

[17] F. Somenzi, CUDD: CU Decision Diagram Package, University of Colorado at Boulder, *ftp://vlsi.colorado.edu/pub/.*

[18] M. Velev and R. Bryant, "Boolean Satisfiability with Transitivity Constraints," *in Proc. of the Conference on Computer-Aided Verification*, 86-98, 2000.

[19] H. Zhang, "SATO: An Efficient Propositional Prover," *in Int'l Conference on Automated Deduction*, 272-225, 1997.

SAT with Partial Clauses and Back-Leaps

Slawomir Pilarski and Gracia Hu
Synopsys, Inc.
2025 NW Cornelius Pass Rd.
Hillsboro, OR 97124

Abstract

This paper presents four new powerful SAT optimization techniques: *partial clauses, back-leaps, immediate implications,* and *local decisions*. These optimization techniques can be combined with SAT mechanisms used in Chaff, SATO, and GRASP to develop a new solver that significantly outperforms its predecessors on a large set of benchmarks. Performance improvements for standard benchmark groups vary from 1.5x to 60x. Performance improvements measured using VLIW microprocessor benchmarks amount to 3.31x.

Categories and Subject Descriptors

J.6 [Computer-Aided Engineering]: Computer-Aided Design.

General Terms

Algorithms, performance, verification.

Keywords

Boolean satisfiability, SAT, design verification.

A Introduction

Boolean Satisfiability (SAT) solvers have been successfully applied to solve various problems in EDA [8]. They are useful in formal verification for combinational and sequential equivalence checking [5], property checking, and microprocessor verification [16]. They are known to be effective in automatic test pattern generation (ATPG) [6, 7, 14], which includes delay fault testing [1] and redundancy removal. SAT solvers were used in logic synthesis [2], timing analysis, and FPGA routing [12, 11].

The wide range of successful SAT applications in EDA is due to the tremendous progress in the performance of SAT solvers [9, 17]. A recently developed solver, Chaff, demonstrated further dramatic performance gains [10, 16, 18]. The speed of Chaff is due to very efficient implementation of all mechanisms used in a standard backtracking search algorithm.

This paper introduces four new powerful SAT optimization techniques: *partial clauses, back-leaps, immediate implica tions,* and *local decisions.* It demonstrates that the new optimization techniques can be successfully implemented together with SAT mechanisms used in most efficient SAT solvers. The resulting new solver significantly outperforms its predecessors on a large set of benchmarks.

The paper is organized as follows. Improvements to boolean constraint propagation (BCP) are introduced in Section B. New SAT decision optimization techniques are presented in Sections C.1 and C.2. Benchmark results are given in Section D. Concluding remarks are presented in Section E.

B Improvements to BCP

Efficient boolean constraint propagation based on watched literals [17, 10] can be further improved by using *immediate implications* and *partial clauses.* Both concepts are presented below.

B.1 Immediate Implications

EDA-related CNF formulae are usually derived from logic circuits. For example, a simple AND gate, $x = AND(y, z)$, can be expressed as $(\overline{x} + y)(\overline{x} + z)(x + \overline{y} + \overline{z})$. It is easy to note that CNF for AND, NAND, OR and NOR gates are very similar, and they are dominated by two-literal clauses. In a typical circuit-derived CNF formula, 55%-85% of all clauses have two literals. Now, note that a 0-assignment to a watched literal in a two-literal clause is an *immediate implication*, and there is no need to search the clause for non-watched literals, as it happens in Chaff or SATO. The notion of immediate implications can be used to speed up BCP. In the case of an already highly optimized BCP, immediate implications can noticeably

Table 1: Average BCP time per implication.

benchmark	no optim.	imm. impl.	speedup
5pipe	800 cycles	599 cycles	1.33x
bug001	686 cycles	606 cycles	1.13x

improve performance [13]. Table 1 presents examples of average time in machine cycles spent on BCP per implication with and without immediate implications. Note that various benchmarks may have very different speedups - not always as significant.

We also found out that SAT performance can be further improved if BCP processes immediate implications first. This, however, appears to be important for conflict clause selection, not for BCP speed.

B.2 Partial Clauses

Conflict clauses are usually much longer than the original benchmark clauses. The average conflict clause size often exceeds 200 or even 300 literals. Since most literals in an undeleted conflict clause are assigned to logic 0, a significant fraction of BCP time is spent on finding the next watched literal. This overhead could be reduced if, during BCP, unsatisfied clauses are temporarily replaced with clauses consisting of only unassigned literals. Such a transformation would not affect SAT correctness since 0-assigned literals do not affect BCP, except for slowing it down. Our observation leads to the concept of a *partial clause*. A partial clause is created by elimination of most 0-assigned literals in its parent clause. It can be used instead of its parent clause only when all eliminated literals are assigned to logic 0. For example clause
$(v_1{}^{=0} + v_2{}^{=0} + v_3{}^{=0} + v_4{}^{=0} + v_5{}^{=?} + v_6{}^{=?} + v_7{}^{=?})$,
where literals v_1, v_2, v_3, v_4 are assigned to logic 0 and literals v_5, v_6, v_7 are unassigned, can be replaced with partial clause
$(v_4{}^{=0} + v_5{}^{=?} + v_6{}^{=?} + v_7{}^{=?})$.

In our implementation, validity of a partial clause is linked to the highest decision level associated with eliminated literals. Partial clause inherits watched literals from its parent clause. Only conflict clauses are replaced with partial clauses.

Partial clauses together with immediate implications provide effective BCP optimization. Table 2 presents examples of average time in machine cycles spent on BCP per implication; column "no optim." shows data for BCP without optimizations; column "optim." shows data for BCP with immediate implications and partial clauses. Note that various benchmarks may have very different speedups - not always as significant. In general, the speedup due to partial clauses depends on the average size of a conflict clause; it also depends on the fraction of conflict clauses in the database, not to mention quality of partial clause implementation.

Table 2: Average BCP time per implication.

benchmark	no optim.	optim.	speedup
5pipe	800 cycles	330 cycles	2.42x
bug001	686 cycles	369 cycles	1.86x

C Improvements to Decisions

Dynamic decision making mechanism used in Chaff can be improved by using *local decisions* and *back-leaps*. Both concepts are presented below.

C.1 Local Decisions

In general, as in many other SAT solvers, the decision process used in Chaff selects a variable regardless of its relation to currently assigned variables. This means that non-chronological backtracking often invalidates many decisions and BCP computations unrelated to the conflict. Such an observation suggests that a significant performance gain is possible if the decision process could be focused on variables related to currently assigned variables.

In EDA formulae conflicts often express a correlation between variables that are close in terms of circuit connectivity. This suggests that the decision process should be biased towards unassigned variables that share clauses with already assigned variables. Relatively strong correlation between variables distant in terms of circuit connectivity can also be observed. Because of this, too much focus on circuit structure may lead to suboptimal decisions and conflict clauses that prune the search space relatively slowly.

Our decision strategy has two components:

- *Local:* select the "best" variable that shares a clause with already assigned variables; and
- *Global:* select globally "best" variable.

The mechanism of ranking variables in both groups is similar, though not identical, to that used in Chaff. We use two priority queues one for local and one for global decisions. We put less emphasis on literals in two-literal clauses - they carry smaller weight than other literals. About 80% of decisions select the locally "best" variable.

Our structure-oriented decisions appear to be especially effective for hard-to-satisfy and unsatisfiable formulae. Table 3 reports the impact of local decisions on SAT run time measured for CMU benchmarks. Note that our implementation also used immediate implications, but did not use partial clauses.

C.2 Back-Leaps

Standard non-chronological backtracking is a powerful mechanism used to prune the search space, but most backtracks

Table 3: CMU benchmark [15] results (in seconds).

benchmark	#tests	zChaff	local d.	speedup
sss 1.0	48	99	**33**	2.96x
sss 1.0a	8	37	**19**	1.86x
sss-sat 1.0	100	672	**358**	1.88x
fvp-unsat-1.0	4	1480	**839**	1.76x
fvp-unsat-2.0	22	218268	**55081**	3.96x
vliw-sat 1.0	100	13626	**9237**	1.48x

are are still chronological (see Table 3 in [9]). This observation suggests that a SAT solver would benefit from a mechanism that guides the decision sequence out of chronological backtrack regions. A crude version of such a mechanism is provided by restarts. A more sophisticated version of such a mechanism would monitor a number of most recent backtracks; if the majority of them are chronological, cancellation of some recent decisions, *a back-leap*, could be forced. Unlike backtracking, back-leaping does not flip the value of the variable truth assignment. This means that introduction of back-leaping does not compromise SAT solver completeness.

In our implementation we monitor ten most recent backtracks; if the total number of backtracked levels is smaller than a threshold value, we consider a back-leap to the second most recent decision level found in the the conflict clause; a back-leap takes place only if we can back-leap at least five decisions. For example, conflict clause
$(v_1{}^{=0,level=3} + v_2{}^{=0,level=7} + v_3{}^{=0,level=15} + v_4{}^{=0,level=22} + v_5{}^{=0,level=23} + v_6{}^{=0,level=33} + v_7{}^{=0,level=45})$
may trigger back-leaping to level 33 (since $45 - 33 > 5$).

Back-leaping seems to be especially effective for relatively hard-to-satisfy formulae, but not very effective for some unsatisfiable formulae.

A combination of back-leaping, local decisions, immediate implications and partial clauses translate into significant performance gains.

D Experimental Results

This section presents performance results for a SAT solver that combines effective SAT techniques used in GRASP, SATO and Chaff with the powerful SAT optimizations we introduced in Sections B and C. Table 4 summarizes results for the DIMACs benchmark suite [4]. Each row presents benchmark name, number of tests in the benchmark, run time (in seconds) for GRASP, SATO, and Chaff, run time for the new solver, and the speed up factor with respect to Chaff. Table 5 summarizes results for the CMU Benchmark Suite [15]. Both sets of benchmarks were used in [10] to illustrate performance gains offered by Chaff. We use the same benchmarks to demonstrate performance gains due to our SAT optimizations.

Note that all experiments reported in Tables 4 and 5 were run on the same 4CPU/400MHz/4GB Ultra-SPARC-4 machine. We used original unmodified GRASP, SATO, and zChaff packages with the following settings: GRASP(+T100 +B10000000 +C10000000 +S10000 +g20 +rt4 +dDLIS), SATO(default), and zChaff(default). Best results are highlighted. A dash means that at least one test was timed out.

E Conclusions

This paper demonstrates that the new four SAT optimization techniques introduced in Sections B can be successfully implemented together with efficient SAT mechanisms used in Chaff, SATO, and GRASP. The resulting new SAT solver is optimized for EDA applications. Benchmark results show performance improvements with respect to Chaff ranging from 1.5x to 60x. It appears that a revised variable scoring mechanism (e.g., the one used in [3]) could amplify already significant performance gains.

References

[1] C.-A. Chen and S. K. Gupta. A satisfiability-based test generator for path delay faults in combinational circuits. In *Proc.Design Automation Conference*, pages 209–214, June 1996.

[2] L. Entrena and K.-T. Cheng. Sequential logic optimization by redundancy addition and removal. In *Proc. International Conference on Computer-Aided Design*, pages 310–315, November 1993.

[3] Evgueni Goldberg and Yakov Novikov. Berkmin: a fast and robust sat-solver. In *Proc. Design Automation and Test in Europe*, pages 142–149, March 2002.

[4] D. S. Johnson and M. A. Trick eds. Second DIMACS implementation challenge. In *DIMACS benchmarks available at ftp://dimacs.rutgers.edu/pub/challenge/sat/benchmarks*, 1993.

[5] Andreas Kuehlmann, Malay K. Ganai, and Viresh Paruthi. Circuit-based boolean reasoning. In *Proc. ACM/IEEE Design Automation Conference*, pages 232–237, June 2001.

[6] T. Larrabee. Test pattern generation using boolean satisfiability. *IEEE Trans. CAD*, 11(1):4–15, January 1992.

[7] J. Marques-Silva and K. A. Sakallah. Robust search algorithms for test pattern generation. In *Proc. of the Fault-Tolerant Computing Symposium*, pages 152–161, June 1997.

[8] J. Marques-Silva and K. A. Sakallah. Boolean satisfiability in electronic design automation. In *Proc. ACM/IEEE*

Table 4: DIMACs benchmark results (in seconds)[4].

benchmark	# tests	GRASP	SATO	zChaff	new Solver	speed up w.r.t. zChaff
ii16	10	-	2.18	67.71	**1.62**	41.80x
ii32	17	4.47	2.47	2.75	**1.06**	2.59x
ii8	14	2.03	0.48	0.53	**0.04**	13.25x
aim200	24	5.81	0.77	1.11	**0.37**	3.00x
aim100	24	0.70	0.20	0.20	**0.04**	5.0x
pret	8	4.34	**0.07**	1.13	0.14	8.07x
par8	10	0.25	0.07	0.02	**0.01**	2.00x
ssa	8	3.30	3.78	0.28	**0.27**	1.04x
jnh	50	5.69	1.04	0.72	**0.47**	1.53x
dubois	13	0.35	0.17	0.14	**0.05**	2.08x
hole	5	-	142.73	**42.76**	48.61	0.88x
par16	10	-	-	31.77	**28.51**	1.11x
hanoi 5	1	-	-	76705	**1299**	59.05x

Table 5: CMU benchmark results (in seconds)[15].

benchmark	# tests	zChaff	new Solver	speedup
sss 1.0	48	99.14	**40.39**	2.45x
sss 1.0a	8	36.72	**8.56**	4.29x
sss-sat 1.0	100	671.75	**278.76**	2.41x
fvp-unsat-1.0	4	1480.24	**216.97**	6.82x
fvp-unsat-2.0	22	218268	**92987**	2.35x
vliw-sat 1.0	100	13626	**4115**	3.31x

Design Automation Conference, pages 675-680, June 2000.

[9] Joao P. Marques-Silva and Karem A. Sakallah. GRASP: A search algorithm for propositional satisfiability. *IEEE Trans. Comput*, 48(5):506–520, May 1999.

[10] Matthew W. Moskewicz, Conor F. Madigan, Ying Zhao, Lintao Zhang, and Sharad Malik. Chaff: Engineering an efficient SAT solver. In *Proc.Design Automation Conference*, pages 530–535, June 2001.

[11] G.-J. Nam, K.A. Sakallah, and R. Rutenbar. Satisfiability based FPGA routing. In *Proc. of the International Conference on VLSI Design*, pages 574–577, January 1999.

[12] G.-J. Nam, K.A. Sakallah, and R. Rutenbar. Satisfiability-based layout revisited: Detailed routing of complex FPGAs via search-based boolean SAT. In *Proc. of the International Symposium on Field-Programmable Gate Arrays*, pages 167–175, February 1999.

[13] Slawomir Pilarski and Gracia Hu. Speeding up SAT for EDA. In *Proc. Design Automation and Test in Europe*, page 1081, March 2002.

[14] P. R. Stephan, R. K. Brayton, and A. L. Sangiovanni-Vincentelli. Combinational test generation using satisfiability. *IEEE Trans. CAD*, 15(9):1167–1176, September 1996.

[15] M. N. Velev. CMU benchmark suite. In *benchmarks available at htpp://www.ece.cmu.edu/m̃velev.*

[16] Miroslav N. Velev and Randal E. Bryant. Effective use of boolean satisfiability procedure in the formal verification of superscalar and VLIW microprocessors. In *Proc. Design Automation Conference*, pages 226–231, June 2001.

[17] H. Zhang. SATO: An efficient propositional prover. In *Proc. of the International Conference on Automated Deduction*, pages 272–275, July 1997.

[18] L. Zhang, C. F. Madigan, M. H. Moskewicz, and S. Malik. Efficient conflict driven learning in a boolean satisfiability solver. In *Proc. International Conference on Computer-Aided Design*, pages 279–285, November 2001.

Combining Strengths of Circuit-based and CNF-based Algorithms for a High-Performance SAT Solver

Malay K Ganai, Lintao Zhang[1], Pranav Ashar, Aarti Gupta, and Sharad Malik[1]
NEC USA CCRL, Princeton NJ
{malay,ashar,agupta}@nec-lab.com
[1]EE Dept, Princeton University
{lintaoz,malik}@ee.princeton.edu

Abstract

We propose Satisfiability Checking (SAT) techniques that lead to a consistent performance improvement of up to 3x over state-of-the-art SAT solvers like Chaff on important problem domains in VLSI CAD. We observe that in circuit oriented applications like ATPG and verification, different software engineering techniques are required for the portions of the formula corresponding to learnt clauses compared to the original formula. We demonstrate that by employing the same innovations as in advanced CNF-based SAT solvers, but in a hybrid approach where these two portions of the formula are represented differently and processed separately, it is possible to obtain the consistently highest performing SAT solver for circuit oriented problem domains. We also present controlled experiments to highlight where these gains come from. Once it is established that the hybrid approach is faster, it becomes possible to apply low overhead circuit-based heuristics that would be unavailable in the CNF domain for greater speedup.

Categories and Subject Descriptors

J.6 [**Computer Applications**]: Computer-aided Engineering – *Computer-aided design.*

General Terms

Algorithms, Performance, Experimentation, Verification.

Keywords

Boolean Satisfiability (SAT), Boolean Constraint Propagation (BCP), Conjunctive Normal Form (CNF), Bounded Model Checking (BMC).

1. Introduction

The Boolean Satisfiability (SAT) problem has extensive applications in VLSI CAD. Recent advances in SAT solvers based on Conjunctive Normal Form (CNF) representation have resulted in significantly improved performance. In particular, innovative techniques for decision variable selection [1], Boolean constraint propagation (BCP) [1, 2], and backtracking with conflict analysis based learning [1, 3] have led to high-performance CNF-based SAT solvers like Chaff [1]. For circuit application domains such as Automatic Test Pattern Generation (ATPG) [4], equivalence checking [5], and Bounded Model Checking (BMC) [6], the Boolean reasoning problem is typically derived from the circuit structure. This has also led to interest in circuit-based SAT solvers [4, 5, 7, 8], which use circuit specific knowledge to guide the search. In general, attempts to include circuit structure information into CNF-based solvers have been unsuccessful due to significant overhead. Furthermore, it is also difficult to integrate CNF-based solvers with other useful techniques like BDD sweeping and dynamic circuit transformation [5]. On the other hand, CNF-based solvers are better than circuit-based solvers in handling conflict analysis and learned clauses [1].

In this paper, we propose techniques that combine the strengths of circuit-based and CNF-based SAT solvers. In particular, we describe a *hybrid SAT solver* where the original logic formula is processed in circuit form, and the learned clauses are processed separately in CNF. We discuss important differences between the two approaches, and highlight how we reap benefits from both by employing state-of-art innovations in BCP, decision variable selection, and backtracking. Our technique leads to a consistent performance improvement of up to 3x over state-of-the-art SAT solvers like Chaff on representative problem applications.

2. Motivation for Hybrid SAT

Boolean Constraint Propagation (BCP) forms the core of deduction in many successful and widely studied SAT algorithms based on the DPLL algorithm [9]. When a variable is assigned, BCP is used to find the consequences of the assignment, like unit clauses produced and whether a conflict exists. It constitutes about 80% of the total SAT time in our experience. Therefore, any improvement in BCP significantly benefits the overall performance of a SAT solver. We start by describing the details of the circuit-based BCP and CNF-based BCP that we use in our approach. Their relative strengths motivate the specific hybrid scheme we use.

2.1 Circuit-Based BCP

Existing circuit-based SAT solvers [4, 5, 7, 8] use a circuit representation based on AND and OR gate vertices, with INVERTERs either as separate vertices, or as attributes on the gate inputs. Constant propagation across AND/OR gates is, of course, well known, but the speed tends to be very implementation dependent. We use a lookup table for fast implication propagation [5]. Based on the current values of the inputs and output of the vertex, the lookup table determines the

DAC 2002, June 10-14, 2002, New Orleans, Louisana, USA.

next "state" of the gate where the state encapsulates any implied values and the next action to be taken for the vertex. The algorithm **imply** (from [5]), shown in Figure 1 for a generic vertex type, iterates over the circuit graph. For each vertex, it determines new implied values and the direction for further processing. As an example, Figure 2 (from [5]) shows some cases from the implication lookup table for a two-input AND gate. For Boolean logic, only one case, a logical *0* at the output of an AND vertex, requires a new case split to be scheduled for justification. All other cases either cause a conflict and backtracking, or further implications, or a return to process the next element to be justified. Due to its low overhead, this implication algorithm is highly efficient. As an indication, on a 750MHz Intel PIII with 256 MB, it can execute over one million implications per second.

```
Algorithm imply (vertex, value) {
  assign (vertex, value);
  lvalue = get_value (vertex->left);
  rvalue = get_value (vertex->right);
  next_state = lookup (value, lvalue, rvalue);
  switch (next_state) {
   case CONFLICT:
    return 0;
   case CASE_SPLIT:
    add_vertex(vertex, justification_queue);
    return 1;
   ...
   case PROP_LEFT_AND_RIGHT:
    if (imply (vertex->left, next_state->lvalue) &&
      imply (vertex->right, next_state->rvalue)) {
       return 1;
    }
    return 0;
    ...
  }
  return 1;
}
```

Figure 1: Circuit-based BCP Procedure

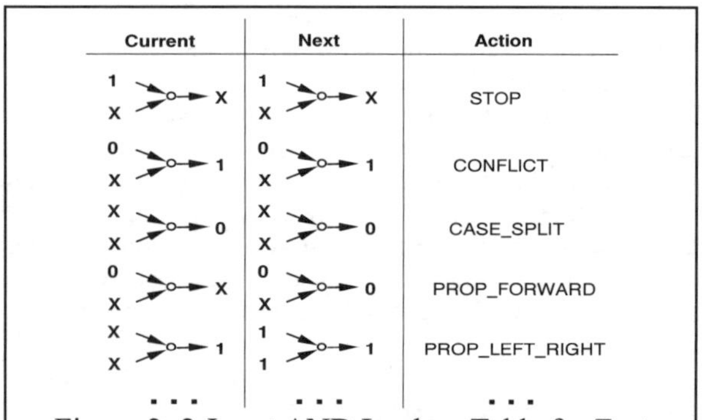

Figure 2: 2-Input AND Lookup Table for Fast Implication Propagation

2.2 CNF-Based BCP

For CNF, we use the BCP scheme called *lazy update* proposed by Chaff [1] :

1. For each clause, only two literals are monitored for state change.
2. The clause state is updated lazily when a variable is assigned, i.e., only when the two monitored literals coincide.
3. It does not require state change for clauses during the backtracking process, thus unassigning a variable takes constant time.

For clauses with many literals, lazy update works significantly better than other schemes like SATO [2], and GRASP[3]. It is especially useful for large clauses where it avoids unnecessary traversal.

2.3 Comparison of BCP Algorithms

There is an inherent overhead built into the translation of circuit gates into clauses. A two-input gate translates to three clauses in the CNF approach, while in the circuit-based approach a gate is regarded as a monolithic entity. Therefore, in the circuit approach an implication across a gate requires a single table lookup, while in the CNF approach it requires processing multiple clauses. In addition, the CNF-based BCP in Chaff does not keep track of the clauses that have been satisfied in order to reduce overheads. However, there is an inherent cost associated with visiting the satisfied clauses. Specifically, even if a clause gets satisfied due to an assignment to some un-watched literal, the watched literal pointers could still get updated. Overall, for the generally small clauses arising from circuit gates, these differences translate to significant differences in BCP time, usually in favor of the circuit-based approach.

On the other hand, learned clauses arising from conflict analysis are typically much larger than those arising from two-input gates. Adding a large learned clause as a gate tree can lead to a significant increase in the size of the circuit. This in turn, can increase the number of implications, thereby negating any potential gains obtained from circuit-based BCP. For such clauses, it is more appropriate to maintain them as monolithic clauses and take advantage of CNF-based 2-literal watching and lazy update to process them efficiently.

Based on these observations, we propose a hybrid approach in which we maintain the circuit-based logic expressions using a uniform-gate data structure, and the learned clauses as CNF. We process them separately using the appropriate BCP approach.

2.4 BCP Results

Our experimental results for BCP are shown in Table 1. The times shown are in seconds *per million implications* on a 750 MHz Intel PIII with 256 MB. The examples used are large logic formulas derived from the BMC application on a large industrial circuit. The columns `Hybrid` and `Chaff` show BCP times for the hybrid and CNF approaches, respectively, on learned clauses added during SAT as well as the original circuit formula. The size of the formula in terms of the number of primary inputs and gates is indicated in the columns `pi` and `gate`. The column `CH` is the ratio between the `Chaff` and `Hybrid` BCP times. It is clear that the hybrid approach is consistently faster than the CNF-based approach in Chaff on these large problems.

Logic Formula	pi	gate	BCP Time (sec/million implications)					
			Hybrid	Chaff	CH	Structure	CS	HS
230U	5016	228346	1.154	1.589	1.37	0.9	1.76	1.28
252U	5244	242170	1.168	1.586	1.35	0.914	1.73	1.28
272U	5244	238755	1.101	1.411	1.28	0.894	1.578	1.23
274S	3863	179217	1.11	1.52	1.369	0.908	1.67	1.22
275U	5472	255979	1.162	1.566	1.347	0.955	1.64	1.22
276U	5472	256000	1.172	1.571	1.34	0.91	1.72	1.29
405U	6612	325069	1.236	1.605	1.29	0.92	1.74	1.34
406U	6612	325090	1.207	1.639	1.35	0.94	1.74	1.28
407U	6612	325090	1.232	1.694	1.375	0.942	1.798	1.31
409U	6612	325090	1.245	1.639	1.316	0.923	1.775	1.35
435U	6840	338899	1.233	1.668	1.35	0.995	1.676	1.24
2dlx_100	557	33848	1.82	2.22	1.21	0.931	2.38	1.95

Table 1: BCP Time per million implications

To demonstrate the effect of learned clauses, we also present the BCP time per million implications for the circuit-based method on *only the circuit formula for the same problems*, shown in the Column `Structure`. Note that this is, in a sense, the best case possible, since BCP on the learned clauses is not strictly necessary for the search. Now, the column `CS (HS)` is the ratio between the `Chaff (Hybrid)` time and the `Structure` time, and reflects the overhead due to the extra work in performing BCP on the learned clauses as well. Clearly, large learned clauses introduce a significant overhead in comparison to the circuit formula alone. Our hybrid approach uses the better-suited CNF approach for large learned clauses.

3. Hybrid SAT Solver

A faster BCP with the hybrid approach is only the partial story. The additional benefit of a hybrid approach is that circuit-based decision heuristics can be easily applied, which are otherwise unavailable in a pure CNF approach. Thus, the overall speedup we observe with our hybrid solver can be attributed to the improved BCP in the deduction engine, and additional use of circuit-based heuristics in the decision engine. In this section we demonstrate the benefits derived from them. The diagnosis engine in our hybrid solver is similar to that in Chaff [1], originally proposed by GRASP [3]. In the hybrid approach, conflict driven learning records both clauses *and circuit nodes* as the reasons for the failure of search.

Our experimental results are presented in Table 2. The times are on a 750 MHz Intel PIII with 256 MB. The logic formulas are derived from the application of BMC on three large industrial circuits (bus, arbiter, and controller) and some public domain benchmarks[1] [10] for which circuit descriptions were available. (Note that use of our hybrid approach requires a circuit description, which is not available for most public benchmarks for SAT.) In order not to pollute the results, we ran the hybrid approach on more than 70 logic formulas, but report results only on those requiring more than 40s of CPU time. The formulas are distributed between unsatisfiable and satisfiable instances. The size of the formula can be determined from the `pi` and `gate` columns indicating the number of primary inputs and gates.

[1] In the table, 2dlx* is 2dlx_cc_mc_ex_bp_f_new

3.1 Comparison of Chaff and Hybrid SAT for Identical Heuristics

Our first comparison is between the hybrid solver and Chaff for exactly the same heuristics, i.e. without use of any circuit-based decision heuristics. Apart from the different BCP algorithms on the circuit formula, the two use identical algorithms for order of processing of implications, conflict-based learning, backtracking, and decision variable selection. In spite of the same heuristics, a minor difference can creep in due to the uncontrolled choice of the conflict node when several nodes are in conflict. This difference may have a pronounced effect in satisfiable instances, for which one of the two solvers may get lucky in hitting upon a solution early. However, it has relatively less effect in unsatisfiable instances since the entire search space must be explored. With this in mind, we consider only the unsatisfiable instances as reasonable data for this controlled experiment. The columns `Chaff` and `H` indicate the times for the Chaff and hybrid solvers, respectively. Due to unavailability of a circuit-based solver in public domain, we could not provide any comparison results with such solvers at this time.

It is clear that the overall performance of the hybrid solver is much better than Chaff. The typical ratio of `Chaff` time to `H`

Example	pi	gate	Chaff	H	H-jft	C/H	C/H-jft
Unsatisfiable Instances							
53U	6384	546779	1893.43	1629.05	639.20	1.16	2.96
55U	6612	571185	1585.08	1402.80	715.20	1.13	2.22
51U	6156	524116	1283.82	1010.90	508.80	1.27	2.52
47U	5700	477047	1190.05	863.60	391.87	1.38	3.04
49U	5928	499710	1072.64	738.30	606.70	1.45	1.77
435U	6840	338899	922.61	326.50	317.00	2.83	2.91
3pipe	198	10810	920.50	245.60	566.90	3.75	1.62
45U	5472	452641	888.86	415.27	443.40	2.14	2.00
434U	6840	338899	880.84	532.10	318.00	1.66	2.77
407U	6612	325090	736.49	367.20	588.00	2.01	1.25
438U	6840	338899	647.71	427.20	283.00	1.52	2.29
437U	6840	338899	621.17	281.10	419.00	2.21	1.48
439U	6840	338899	599.14	441.80	453.00	1.36	1.32
436U	6840	338899	559.08	369.60	384.00	1.51	1.46
105U	1017	94528	547.68	248.90	506.30	2.20	1.08
409U	6612	325090	496.31	437.20	329.00	1.14	1.51
406U	6612	325090	486.86	353.90	317.00	1.38	1.54
405U	6612	325069	471.50	585.00	445.00	0.81	1.06
43U	5244	429978	471.13	353.87	181.25	1.33	2.60
109U	1055	101180	441.33	173.70	229.00	2.54	1.93
408U	6612	325090	433.21	358.10	347.00	1.21	1.25
41U	5016	405572	364.03	230.27	233.80	1.58	1.56
39U	4788	382909	307.59	208.87	120.20	1.47	2.56
101U	979	88088	230.97	141.00	117.00	1.64	1.97
276U	5472	256000	193.24	132.30	104.00	1.46	1.86
37U	4560	358503	174.47	146.96	139.12	1.19	1.25
252U	5244	242170	165.46	118.30	93.50	1.40	1.77
35U	4332	335840	165.20	128.90	76.02	1.28	2.17
275U	5472	255979	162.97	132.10	129.00	1.23	1.26
230U	5016	228346	132.54	100.60	77.00	1.32	1.72
33U	4104	311434	98.64	80.10	42.10	1.23	2.34
31U	3876	288771	92.22	51.30	29.30	1.80	3.15
2dlx*	414	25075	89.07	49.30	72.40	1.81	1.23
272U	5244	238755	47.54	27.70	8.00	1.72	5.94
Satisfiable Instances							
109S	2374	144590	811.12	330.4	182.5	2.45	4.44
105S	2286	134654	432.22	196.4	72.9	2.20	5.93
108S	2367	142198	423.63	766.8	138.5	0.55	3.06
101S	2198	125070	388.35	399.38	111.9	0.97	3.47
102S	2235	127558	344.34	312.7	144.4	1.10	2.38
103S	2242	129818	343.95	481.8	86.7	0.71	3.97
107S	2330	139578	286.1	256.4	178.2	1.12	1.61
104S	2279	132350	270.01	384.4	49.1	0.70	5.50
106S	2323	137230	226.57	198.58	94.4	1.14	2.40
100S	2191	120305	153.41	374.8	97.8	0.41	1.57
2dlx_100	557	33848	120.12	76.8	17	1.56	7.07

Table 2: Comparison of Hybrid SAT Solver and Chaff

time, shown in the Column `C/H`, is greater than 1.3, with the maximum being 3.75. The ratio of the total time spent in Chaff to the total time spent in the hybrid solver for all the unsatisfiable instances is 1.48. Due to possibly a bad choice for a conflict node, we do see a larger time with the hybrid solver in a few cases, e.g., 405U.

For the satisfiable instances shown in Table 2, the Chaff to Hybrid ratio is distributed evenly on either side of 1.0 with a large standard deviation. This is most likely due to the randomness in choice of a conflict node, as explained earlier.

3.2 Incorporation of Circuit-Based Heuristics

Once it is established that the hybrid approach is, in fact, faster, it becomes possible to apply low overhead circuit-based heuristics that would be unavailable in the CNF domain to obtain an even higher speedup. Information regarding gate fanout/fanin, paths, and signal direction becomes readily available without overhead. We find that such an approach works much better than the circuit-oblivious decision heuristics used in the pure CNF-based methods. As an example, the column `H-jft` in Table 2 shows the time in our hybrid SAT solver with the use of the justification frontier heuristic [7]. This heuristic restricts the decision nodes to be those that justify the values on their fanout node. The use of this heuristic further improves performance in 24 of the 34 unsatisfiable instances. The ratio of the total time spent in Chaff to the total time spent in the hybrid solver for the unsatisfiable instances is now 1.89 versus the earlier 1.48, with the maximum ratio being 5.94 versus the earlier 3.75.

Now consider the satisfiable instances in Table 2. Where there was wide variation in the speedups (or lack thereof) earlier, we now have consistent and significant speedup over the CNF approach. The ratio of the total time spent in Chaff to the total time spent in the hybrid solver is now 3.24 for the satisfiable instances with the maximum ratio being 7.07. Clearly, the circuit heuristics lead to a satisfying solution much faster than the decision heuristics in Chaff.

Based on these data, there appears to be a clear benefit in using the circuit-based heuristics for both the satisfiable and unsatisfiable instances. In a sense, the hybrid approach allows us to exploit the benefits of the circuit-based as well as the CNF-based heuristics.

3.3 Impact on Verification Applications

SAT is a core engine in many verification applications like equivalence checking [5], and BMC [6]. In fact, it is typical for a long BMC run to last for multiple days, requiring thousands of applications of the SAT solver. A speed up by a factor of two in the core engine, therefore, can prove to be very significant since it would lead to a large absolute saving in the run time. Another practical aspect of our hybrid approach is that it is unnecessary to incur the overhead of copying the entire circuit into a CNF data structure for the purpose of SAT. This has the benefit of reducing the memory requirement of these applications, allowing them to scale to larger circuits, or in the case of BMC, also to a larger number of time frames.

4. Conclusions

We have presented a hybrid SAT solver that combines the strengths of circuit-based and CNF-based SAT solvers. We demonstrated it to be the highest performing SAT solver in circuit-based application domains. Our approach is based on an analysis of the source of efficiencies in state-of-art techniques for BCP, decision variable selection, and backtracking. We apply this understanding to develop SAT techniques leveraging off the circuit nature of these application domains while maintaining advantages arising from the CNF representation. Moreover, any future improvements in the performance of circuit-based and CNF-based SAT solvers can directly translate into improvements of the hybrid SAT solver as well.

References

[1] M. Moskewicz, C. Madigan, Y. Zhao, L. Zhang, and S. Malik, "Chaff: Engineering an Efficient SAT Solver," in *Proceedings of Design Automation Conference*, 2001.

[2] H. Zhang, "SATO: An efficient propositional prover," in *Proceedings of International Conference on Automated Deduction*, vol. 1249, *LNAI*, 1997, pp. 272-275.

[3] J. P. Marques-Silva and K. A. Sakallah, "GRASP: A Search Algorithm for Propositional Satisfiability," *IEEE Transactions on Computers*, vol. 48, pp. 506-521, 1999.

[4] M. Schulz, E. Trischler, and T. Sarfert, "SOCRATES: A highly efficient ATPG System," *IEEE Transactions on Computer-Aided Design of Integrated Circuits and Systems*, vol. 7, pp. 126-137, 1988.

[5] A. Kuehlmann, M. Ganai, and V. Paruthi, "Circuit-based Boolean Reasoning," in *Proceedings of Design Automation Conference*, 2001.

[6] A. Biere, A. Cimatti, E. M. Clarke, and Y. Zhu, "Symbolic Model Checking without BDDs," in *Proceedings of Workshop on Tools and Algorithms for Analysis and Construction of Systems (TACAS)*, vol. 1579, *LNCS*, 1999.

[7] H. Fujiwara and T. Shimono, "On the Acceleration of Test Generation Algorithms," *IEEE Transactions on Computers*, vol. C-32, pp. 265-272, 1983.

[8] P. Goel, "An implicit enumeration algorithm to generate tests for Combinational circuits," *IEEE Transactions on Computers*, vol. C-30, pp. 215-222, 1981.

[9] M. Davis, G. Longeman, and D. Loveland, "A Machine Program for Theorem Proving," *Communications of the ACM*, vol. 5, pp. 394-397, 1962.

[10] M. N. Velev, "Benchmark Suites. http://www.ece.cmu.edu/~mvelev," October 2000.

A Solenoidal Basis Method For Efficient Inductance Extraction *

Hemant Mahawar
Department of Computer Science
Texas A&M University
College Station, TX 77843
mahawarh@cs.tamu.edu

Vivek Sarin
Department of Computer Science
Texas A&M University
College Station, TX 77843
sarin@cs.tamu.edu

Weiping Shi
Department of Electrical Engineering
Texas A&M University
College Station, TX 77843
wshi@ee.tamu.edu

ABSTRACT

The ability to compute the parasitic inductance of the interconnect is critical to the timing verification of modern VLSI circuits. A challenging aspect of inductance extraction is the solution of large, dense, complex linear systems of equations via iterative methods. Accelerating the convergence of the iterative method through preconditioning is made difficult due to the non-availability of the system matrix. This paper presents a novel algorithm to solve these linear systems by restricting current to a discrete solenoidal subspace in which Kirchoff's law is obeyed, and solving a reduced system via an iterative method such as GMRES. A preconditioner based on the Green's function is used to achieve near-optimal convergence rates in several cases. Experiments on a number of benchmark problems illustrate the advantages of the proposed method over FastHenry.

Categories and Subject Descriptors

B.7.2 [**Integrating Circuits**]: Design Aids—*placement and routing, simulation, verification*

General Terms

ALGORITHMS

Keywords

Inductance extraction, interconnect, solenoidal basis, iterative methods, preconditioning

1. INTRODUCTION

The effect of inductance is increasingly felt on-chip, mainly due to long interconnect and high operation speed. Therefore fast and accurate inductance extraction is increasing important to the design and verification of VLSI circuits. There are three types of inductance extraction algorithms: *loop inductance*, *partial inductance* and *shape based*. The loop inductance algorithms are the most accurate but slowest, while the shape-based algorithms are the least accurate but the fastest. FastHenry [8] computes the loop inductance. Due to its high accuracy, FastHenry is often used as a reference for all other extraction algorithms. But FastHenry is very slow, and to improve its speed is one of the most challenging problems. Partial inductance was first proposed by Rosa and introduced to circuit design by Ruehli [13]. A number of algorithms have been proposed, such as Krauter [10] and He [5]. Partial inductance algorithms are faster than loop inductance algorithms. However it is shown that partial inductance without current return paths is inaccurate [3]. Shape-based algorithms, such as [9, 15], are fast but inaccurate for complex structures. For multilayer dielectric, a mixed potential integral equation method was proposed by Daniel [2] and a multilayer Green's function method was proposed by Liu [11].

In this paper, we study the extraction of loop inductance of 3D electrical conductors in uniform dielectric. We present a solenoidal basis method that gives a better formulation of the linear system compared to FastHenry, and we also present a novel preconditioning method that works with the solenoidal basis method. FastHenry and other 3D extraction algorithms for loop inductance are slow mainly due to the ill-conditioning of the linear systems. Therefore, our new formulation and preconditioning techniques address the key issue of this problem. Experimental results show that our techniques work well, and the new algorithm has several advantages over FastHenry.

*This work has been supported in part by National Science Foundation grants NSF-CCR 9984400 and NSF-CCR 0113668 and the Texas Advanced Technology Program grant 000512-0266-2001.

2. MATHEMATICAL PRELIMINARIES

The impedance of an s-conductor geometry can be summarized by an $s \times s$ impedance matrix $\tilde{\mathbf{Z}}$. The lth column of the impedance matrix is determined as follows: a *unit* current is sent through conductor l, and zero current is sent through other conductors. The numerical value of the potential difference between the two ends of conductor k gives values of the matrix element $\tilde{\mathbf{Z}}_{kl}$. The above procedure is repeated s times to compute all columns of $\tilde{\mathbf{Z}}$.

The current density $\mathbf{J}$ at a point r is related to the potential ϕ by the following integral equation [8]

$$\rho\mathbf{J}(\mathbf{r}) + j\omega \int_V \frac{\mu}{4\pi} \frac{\mathbf{J}(\mathbf{r}')}{||\mathbf{r}-\mathbf{r}'||} dV' = -\nabla\phi(\mathbf{r}), \tag{1}$$

where μ is the magnetic permeability, ρ is the resistivity, $\mathbf{r}$ is a three-dimensional position vector, ω is the frequency, $||\mathbf{r}-\mathbf{r}'||$ is the Euclidean distance between $\mathbf{r}$ and $\mathbf{r}'$, and $j = \sqrt{-1}$. The volume of conductors is denoted by V and incremental volume with respect to $\mathbf{r}'$ is denoted by dV'.

A numerical solution of (1) can be obtained by discretizing the conductors into n filaments $V_1, V_2, \ldots, V_n$. Assuming current flows along the length of the filament and current density is constant within each filament, a linear system of the following form is obtained:

$$[\mathbf{R} + j\omega\mathbf{L}]\,\mathbf{I} = \mathbf{V},$$

where $\mathbf{R}$ is an $n \times n$ diagonal matrix of filament resistances, $\mathbf{I}$ is the vector of filament currents, and $\mathbf{V}$ is the vector of the difference of potential between ends of each filament. The kth diagonal element of $\mathbf{R}$ is given by $\mathbf{R}_{kk} = \rho l_k / a_k$, where l_k is the length of the kth filament and a_k is the cross-sectional area of the kth filament. Let $\mathbf{u_k}$ denote the unit vector along the kth filament V_k. The elements of the inductance matrix $\mathbf{L}$ are given by

$$\mathbf{L}_{kl} = \frac{\mu}{4\pi}\frac{1}{a_k a_l}\int_{r_k \varepsilon V_k}\int_{r_l \varepsilon V_l}\frac{\mathbf{u}_k \cdot \mathbf{u}_l}{\|\mathbf{r}_k - \mathbf{r}_l\|}\,dV_k dV_l. \quad (2)$$

The current satisfies Kirchoff's current law at each node according to the constraints

$$\mathbf{B}^T\mathbf{I} = \mathbf{I}_d,$$

where $\mathbf{B}^T$ is the $m \times n$ branch index matrix and $\mathbf{I}_d$ is the vector of external source currents. Here, m and n denote the number of nodes and filaments, respectively, in the mesh. The vector $\mathbf{I}_d$ has non-zero values only for those nodes where external current source is applied. The potential difference across the filaments is expressed as

$$\mathbf{V} = \mathbf{B}\mathbf{V}_d,$$

where $\mathbf{V}_d$ is the node potential vector of length m. To compute the unknown filament currents and node potentials, the following linear system must be solved

$$\begin{bmatrix} \mathbf{R} + j\omega\mathbf{L} & -\mathbf{B} \\ \mathbf{B}^T & 0 \end{bmatrix}\begin{bmatrix} \mathbf{I} \\ \mathbf{V}_d \end{bmatrix} = \begin{bmatrix} \mathbf{0} \\ \mathbf{I}_d \end{bmatrix}. \quad (3)$$

A straightforward approach to solve this system involves removal of $\mathbf{I}$ by a block-step of Gaussian elimination. The resulting system is defined in terms of the unknowns $\mathbf{V}_d$ only:

$$\mathbf{B}^T[\mathbf{R} + j\omega\mathbf{L}]^{-1}\mathbf{B}\mathbf{V}_d = \mathbf{I}_d.$$

When solving this system by an iterative method, each iteration involves a matrix-vector product with the system matrix. In practice, this matrix is never computed explicitly. Instead, the matrix-vector product is computed as a sequence of three steps: (i) product with $\mathbf{B}$, (ii) solution of the system $[\mathbf{R} + j\omega\mathbf{L}]z = d$, and (iii) product with $\mathbf{B}^T$. The second step may require an inner iterative scheme, resulting in expensive outer iterations. Furthermore, with this kind of a complicated system matrix, it is difficult to find efficient preconditioners for the outer solver.

Alternately, one can compute a subspace for current that satisfies Kirchoff's law at each node. The linear system in (3) is transformed to a reduced system when current is restricted to this subspace, which is then solved by an iterative method. This approach is competitive only when the iterative scheme is preconditioned effectively. In the next section, we discuss how to compute the appropriate subspace and precondition the reduced system without explicitly constructing the matrix. Experiments in Section 4 demonstrate the superior properties of our approach over competing techniques.

3. THE SOLENOIDAL BASIS ALGORITHM

The difficulty in solving the system (3) arises partly due to the constraints imposed by Kirchoff's law on the flow of current at each node. These constraints are identical to divergence-free constraints in incompressible fluids that arise from conservation laws. A *solenoidal* basis is a basis for divergence-free functions that automatically satisfy conservation laws such as Kirchoff's law. The solenoidal basis algorithm constructs a basis for functions that satisfy Kirchoff's law. Expressing current in this basis leads to a reduced system which is solved by a preconditioned iterative method such as GMRES.

3.1 A Solenoidal Basis Approach

The solenoidal basis method is used to solve the problem

$$\begin{bmatrix} \mathbf{Z} & -\mathbf{B} \\ \mathbf{B}^T & 0 \end{bmatrix}\begin{bmatrix} \mathbf{I} \\ \mathbf{V}_d \end{bmatrix} = \begin{bmatrix} \mathbf{F} \\ \mathbf{0} \end{bmatrix}, \quad (4)$$

where $\mathbf{Z} = \mathbf{R} + j\omega\mathbf{L}$ is the filament impedance matrix, $\mathbf{L}$ is an $n \times n$ dense, symmetric and positive definite matrix, $\mathbf{R}$ is a diagonal matrix, and $\mathbf{B}$ is an $n \times m$ sparse matrix. The filament current vector $\mathbf{I}$ has size n, the node potential vector $\mathbf{V}_d$ has size m, and the right hand side vector $\mathbf{F}$ has size n.

The main difference between (3) and (4) is that the boundary conditions are specified for current in the first system whereas they are specified for potential in the second system. To convert (3) to (4), we let $\mathbf{I} = \mathbf{I}' + \mathbf{I}_p$, where $\mathbf{I}_p$ is a current vector that satisfies the constraints $\mathbf{B}^T\mathbf{I}_p = \mathbf{I}_d$. The following linear system

$$\begin{bmatrix} \mathbf{Z} & -\mathbf{B} \\ \mathbf{B}^T & 0 \end{bmatrix}\begin{bmatrix} \mathbf{I}' \\ \mathbf{V}_d \end{bmatrix} = \begin{bmatrix} -\mathbf{Z}\mathbf{I}_p \\ \mathbf{0} \end{bmatrix}$$

is solved for the unknown current $\mathbf{I}'$. The current vector $\mathbf{I}_p$ can easily be found by number of techniques. For instance, when the known branch current has unit magnitude, one can assign a unit current to filaments on an arbitrary path from node with input source current to the node with output source current (see Fig. 1). This approach can be extended to more general boundary conditions in a straightforward manner.

Since a unit current flowing in a closed loop in the mesh satisfies Kirchoff's law, it also satisfies the constraints imposed by $\mathbf{B}^T$. The solenoid basis method for (4) uses these mesh currents to represent the unknown current $\mathbf{I}$. This scheme is similar to the mesh current approach proposed in [6, 7] with the exception of the treatment of source current. In addition, the preconditioning proposed in section 3.2 is more powerful than those suggested in [7].

The null space of $\mathbf{B}^T$ represents a basis for current that obeys Kirchoff's law. Given a full-rank matrix $\mathbf{P} \in R^{n\times(n-m)}$ such that $\mathbf{B}^T\mathbf{P} = \mathbf{0}$, a current vector computed as follows

$$\mathbf{I} = \mathbf{P}x, \qquad x \in R^{(n-m)}$$

will satisfy the constraint $\mathbf{B}^T\mathbf{I} = 0$ for all x. A purely algebraic approach such as QR factorization of $\mathbf{B}$ cannot be used to compute $\mathbf{P}$ due to the prohibitive cost of computation and storage. We define a *unit* current flow in a closed loop as a *local* solenoidal function. Each such mesh current is represented as a vector and the set of these vectors forms the columns of $\mathbf{P}$. The local nature of these mesh currents leads to efficient computation and storage schemes for $\mathbf{P}$.

Since the current vector $\mathbf{I} = \mathbf{P}x$ automatically satisfies the constraint $\mathbf{B}^T\mathbf{I} = 0$, we only need to solve

$$\mathbf{Z}\mathbf{P}x - \mathbf{B}\mathbf{V}_d = \mathbf{F}.$$

After eliminating the branch potential unknowns $\mathbf{V}_d$ by multiplying this equation with $\mathbf{P}^T$, we get

$$\mathbf{P}^T\mathbf{Z}\mathbf{P}x = \mathbf{P}^T\mathbf{F}. \quad (5)$$

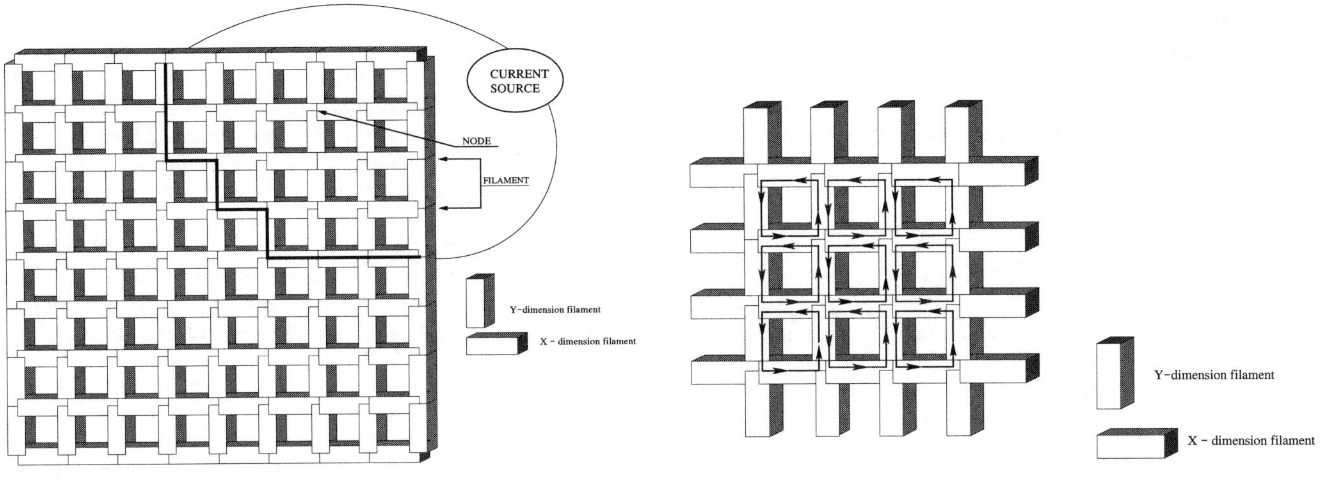

Figure 1: Discretization of a ground plane with a mesh of filaments. Current flowing through the filaments must satisfy Kirchoff's law at each node in the mesh. The bold line indicates a path for current that satisfies boundary conditions. The current is made up of two components: constant current along the bold line shown in the left figure and a linear combination of mesh currents as shown in the partial mesh on the right.

This reduced system can be solved via a suitable iterative scheme such as the GMRES method [14]. Once x is obtained, current is computed as $\mathbf{I} = \mathbf{P}x$. The potential difference across each filament is given by

$$\mathbf{V} = \mathbf{ZI} - \mathbf{F}.$$

The potential difference between two nodes is computed by adding the potential difference across filaments on any path connecting the nodes. This allows computation of impedance between the points where source current is applied.

3.2 Preconditioning the Reduced System

To accelerate the convergence of the iterative method, one must devise robust and effective preconditioning schemes for the system matrix in (5). The task of designing effective preconditioners is made especially challenging due to the unavailability of $\mathbf{L}$. Popular techniques based on incomplete factorizations of the system matrix compute parts of the system matrix selectively, followed by incomplete factorizations. These schemes tend to be expensive and limited in their effectiveness [7].

The success of the solenoidal basis method depends on effective preconditioning of the reduced system (5). A preconditioner is said to be optimal when the condition number of the preconditioned system, computed as the ratio of the largest and the smallest singular values, is bounded by a constant. In such a case, iterative methods may converge to the solution in fixed number of iterations regardless of the frequency ω and mesh width h of the mesh used for discretizing the conductor. The system matrix $\mathbf{P}^T\mathbf{ZP}$ can be analyzed using the observation that the matrices $\mathbf{P}$ and $\mathbf{P}^T$ are equivalent to discrete curl operators. Furthermore, the matrix $\mathbf{P}^T\mathbf{P}$ is equivalent to a discrete Laplace operator in the divergence-free space. We approximate $[\mathbf{P}^T\mathbf{P}]^{-1}$ by the matrix $\tilde{\mathbf{L}}^2$, where $\tilde{\mathbf{L}}$ is defined as follows:

$$\tilde{\mathbf{L}}_{kl} = \frac{\mu}{4\pi}\frac{1}{a_k a_l}\int_{r_k \varepsilon V_k}\int_{r_l \varepsilon V_l}\frac{1}{\|\mathbf{r}_k - \mathbf{r}_l\|}dV_k dV_l. \quad (6)$$

Here, $\tilde{\mathbf{L}}_{kl}$ gives the mutual inductance between parallel filaments placed at the centers of loop k and l.

The reduced system (5) can be approximated by

$$\mathbf{P}^T\mathbf{ZP} \approx \tilde{\mathbf{L}}^{-1}\left[\tilde{\mathbf{R}} + j\omega\tilde{\mathbf{L}}\right]\tilde{\mathbf{L}}^{-1}$$

where $\tilde{\mathbf{R}}$ is a diagonal matrix of resistance to mesh currents. The preconditioner for the reduced system (5) is defined as

$$M^{-1} = \tilde{\mathbf{L}}\left[\tilde{\mathbf{R}} + j\omega\tilde{\mathbf{L}}\right]^{-1}\tilde{\mathbf{L}}. \quad (7)$$

The matrix M^{-1} must be multiplied to the reduced system before the iterative method is employed. At each iteration, the preconditioning step consists of the matrix-vector product $z = M^{-1}r$ which can be computed in the following three steps

$$u = \tilde{\mathbf{L}}r, \qquad v = \left[\tilde{\mathbf{R}} + j\omega\tilde{\mathbf{L}}\right]^{-1}u, \qquad z = \tilde{\mathbf{L}}v.$$

The matrix-vector products in the first and third steps use approximate hierarchical techniques identical to those used for $\mathbf{L}$. The second step is implemented via an inner iterative solver which is used to solve the system

$$\left[\tilde{\mathbf{R}} + j\omega\tilde{\mathbf{L}}\right]v = u$$

to obtain v.

At very low frequency operation, the preconditioner takes the form

$$M_{\text{low}}^{-1} = \tilde{\mathbf{L}}\tilde{\mathbf{R}}^{-1}\tilde{\mathbf{L}},$$

in which inverting the diagonal matrix $\tilde{\mathbf{R}}$ is inexpensive. At very high frequency operation, the preconditioner takes the form

$$M_{\text{high}}^{-1} = -\frac{j}{\omega}\tilde{\mathbf{L}}$$

which is also inexpensive to apply to a vector. In each case, the preconditioning step is relatively cheap since it doesn't involve an inner solve. For intermediate frequencies, however, one must use the preconditioner (7).

There are several advantages of our preconditioning approach. The preconditioning step requires a matrix-vector product which is relatively inexpensive compared to incomplete factorization based preconditioners. The latter involve incomplete factorizations of partially computed $\mathbf{P}^\mathrm{T}\mathbf{ZP}$ and triangular solves which are expensive, especially on parallel platforms. In addition, experimental evidence suggests that the preconditioner is robust and very effective for the a wide range of frequencies.

3.3 The Algorithm

An outline of the solenoidal basis method to compute $\tilde{\mathbf{Z}}$ is given below.

ALGORITHM 1. *Solenoidal Basis Method.*

1. Generate a uniform mesh for each conductor in the circuit.
2. Compute the solenoidal basis matrix **P**.
3. For each conductor $l = 1, \ldots, s$,
   (a) Compute the particular solution $\mathbf{I}_p^{(l)}$ for a unit current flow through conductor l and the corresponding induced potential drop vector $\mathbf{F}^{(l)}$.
   (b) Solve the preconditioned system
   $$\mathbf{P}^T\mathbf{Z}\mathbf{P}Mx = \mathbf{P}^T\mathbf{F}^{(l)}, \qquad x^{(l)} = Mx$$
   to determine the mesh current vector $x^{(l)}$ and compute filament current vector
   $$\mathbf{I}^{(l)} = \mathbf{P}x^{(l)} + \mathbf{I}_p^{(l)}.$$
   Use GMRES algorithm with variants of the preconditioner (7). Use approximate hierarchical methods such as FMM or Barnes-Hut to compute matrix-vector products with **L** and $\tilde{\mathbf{L}}$ at each iteration.
   (c) For each conductor $k = 1, \ldots, s$, determine the conductor impedance matrix element $\tilde{\mathbf{Z}}_{k,l}$ by computing the potential difference between the two ends of the conductor. This is done by adding the potential drop across all the filaments along a path from one end to the other end of the conductor due to the current $\mathbf{I}^{(l)}$.

An efficient implementation of this algorithm can use a number of optimizations. The matrix **P** is never computed explicitly. A matrix-vector product with **P** is used to compute filament currents from mesh currents. Since this computation is defined locally, it can be performed by accumulating the contribution of each mesh current to the four filaments that comprise the mesh or loop. Knowledge of the structure of the discretization mesh is sufficient to develop an implementation in which **P** is not computed and stored explicitly. Similarly, matrix-vector products with $\mathbf{P}^T$ are used to compute mesh currents from filament currents. These products can also be computed without explicitly computing $\mathbf{P}^T$. This approach leads to significant saving in storage without increase in computation.

The most expensive operations in the algorithm are matrix-vector computations with the system matrix $\mathbf{P}^T\mathbf{Z}\mathbf{P}$ and preconditioning steps involving matrix-vector products with $\tilde{\mathbf{L}}$. A matrix-vector product with $\mathbf{P}^T\mathbf{Z}\mathbf{P}$ is computed as a sequence of three products:

$$u = \mathbf{P}x, \qquad v = [\mathbf{R} + j\omega\mathbf{L}]\,u, \qquad y = \mathbf{P}^T v.$$

Among these, the product with the dense matrix **L** is by far the most time-consuming computation. Since the matrix $\tilde{\mathbf{L}}$ used in the preconditioning step is similar to **L**, improving the computational complexity of this matrix-vector product can reduce the overall computation time significantly. There are several approximate hierarchical methods to compute products with such matrices in which accuracy is traded for a reduction in computational complexity.

Accurate matrix-vector products with the dense matrices **L** and $\tilde{\mathbf{L}}$ require $O(n^2)$ operations where n is the size of these matrices. A number of techniques have been developed to exploit the rapid decay of the kernel in (2) and (6) with distance to compute approximate matrix-vector products in $O(n \log n)$ or $O(n)$ operations. These include hierarchical techniques such as Barnes-Hut [1] and Fast Multipole Method (FMM) [4, 12] which use a truncated series approximation of filament currents within a localized region to estimate impact on well-separated sets of filaments. The method of Barnes and Hut relies only on filament-cluster interactions to achieve an $O(n \log n)$ computational bound for uniform filament distributions. The Fast Multipole Method uses both filament-cluster and filament-filament interactions to achieve an $O(n)$ bound for uniform distributions. In each case, reduction in computational complexity is associated with decrease in accuracy of the matrix-vector product.

We use a variant of the Barnes-Hut algorithm developed specifically for the inductance extraction problem. A hierarchical quadtree is used to partition the filaments in the mesh. Internal nodes of the tree represent hypothetical filaments carrying current equal to the sum of currents in that subtree. The coordinates of these hypothetical filaments are computed as a weighted sum of coordinates of the subtree filaments in which the weights correspond to the magnitude of current flowing in those filaments. This strategy exploits the fact that the combined inductance effect of a cluster of filaments can be approximated by the inductance effect of a single filament placed at the "center of current" of that group. This approximation is used to compute inductance on a filament which is well-separated from the cluster. This is analogous to the center of mass concept in Barnes-Hut algorithm. The leaf nodes of the quadtree are filaments of the mesh. Inductance on a leaf node is computed by traversing the tree in a top down manner. A subtree is traversed only if the corresponding box is not well-separated from the leaf node. Well-separatedness is established by using a distance metric to determines if the leaf node is sufficiently "far off" such that the error in the associated truncated expansion is below a threshold. The inductance due to filaments in a well-separated subtree is computed by direct interaction with the corresponding hypothetical filament. Remaining interactions consist of leaf-leaf interactions which consist of mutual inductance computation between pairs of filaments. Care should be taken to ensure that the algorithm doesn't check the well-separatedness criteria for the box containing the leaf node itself.

4. EXPERIMENTS

4.1 Ground Plane

We consider the problem of computing impedance of a $1\text{cm} \times 1\text{cm}$ ground plane with unit inflow current from the left edge and unit outflow current at the right edge (see Fig. 1). This benchmark problem allows discretization by uniform two-dimensional meshes with varying mesh width h. This problem exhibits the dependence of the condition number of the system matrices on h via the growth in iterations of the unpreconditioned GMRES algorithm. Such convergence behavior is ideal for testing the effectiveness of the preconditioning approach for different levels of refinement as well as operating frequencies.

Table 1 shows the number of iterations needed by the preconditioned GMRES to solve the linear system. The width of each filament is one-third of its length, and thickness is 2^{-13}cm. A tolerance of 10^{-3} was specified on the relative residual norm. The table shows that the rate of convergence is independent of the frequency ω and the mesh width h indicating that the preconditioner is optimal for the benchmark problem .

Table 1: **Ground plane: Iterations for convergence of preconditioned GMRES.**

Mesh Size	Filament Length (cm)	Frequency (in GHz)			
		1	10	100	1000
33×33	2^{-5}	6	5	5	5
65×65	2^{-6}	6	6	5	5
129×129	2^{-7}	8	7	7	6
257×257	2^{-8}	11	9	8	8

4.2 Spiral Inductor

This is a more complicated problem consisting of a conductor in the shape of a coil (see Fig. 2) which is discretized by a two-dimensional mesh. The coil is contained within a square region of size 1 cm×1 cm. Table 2 shows the number of iterations needed by the preconditioned GMRES to solve the linear system with a tolerance of 10^{-3} on the relative residual norm. Filament width and thickness are identical to the previous experiment. The number of iterations required by the solver are nearly independent of the mesh width for the range of frequencies considered here.

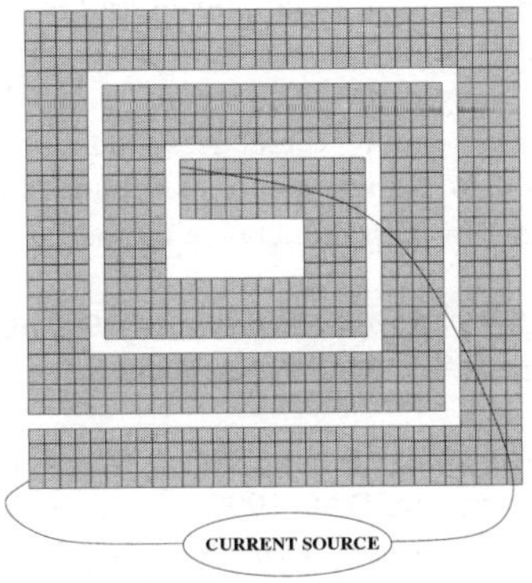

Figure 2: Spiral inductor.

Table 2: Spiral inductor: Iterations for convergence of preconditioned GMRES.

Mesh Size	Filament Length (cm)	Frequency (in GHz)			
		1	10	100	1000
33×33	2^{-5}	7	6	6	6
65×65	2^{-6}	8	7	7	7
129×129	2^{-7}	10	9	9	9
257×257	2^{-8}	16	12	11	11

4.3 Wire Over Ground Plane

We consider a 3D problem of determining impedance of a horizontal wire of length 0.25cm suspended at a distance of 0.1cm over a 1cm×1cm ground plane (see Fig. 3). The wire is discretized along its length. The ground plane is discretized by a two-dimensional mesh similar to the one in the ground plane problem discussed earlier. Filament width and thickness are identical to previous experiments. Table 3 shows the number of iterations needed by the preconditioned GMRES. Solver parameters were identical to earlier experiments. These experiments also indicate that the number of iterations required by the solver are nearly independent of the mesh width for the range of frequencies considered here.

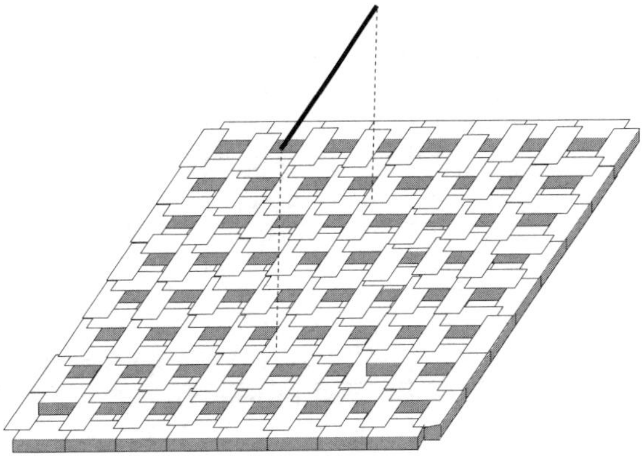

Figure 3: Wire over ground plane.

Table 3: Wire over ground plane: Iterations for convergence of preconditioned GMRES.

Mesh Size	Filament Length	Frequency (in GHz)			
		1	10	100	1000
33×33	2^{-5}	5	4	4	4
65×65	2^{-6}	6	5	5	5
129×129	2^{-7}	8	6	6	6
257×257	2^{-8}	12	8	8	7

4.4 Comparison with FastHenry

The performance of the solenoidal basis method was compared to FastHenry on two problems: the 2D ground plane problem and the 3D problem of a wire over the ground plane. FastHenry uses mesh currents to generate a reduced system which is then solved by the preconditioned GMRES method. Fast Multipole Method is used to compute matrix-vector products with dense matrices. Preconditioners are derived from incomplete factorizations of approximations to the reduced system. These approximations are computed by various techniques to sparsify the inductance matrix **L**. The software allows preconditioners such as CUBE and SHELL, in which the off-diagonal non-zeros in each column of **L** are restricted to those resulting from mutual inductance between filaments in the same box at a specific level in the quad tree and mutual inductance between filaments within a specific radius, respectively. The DIAG preconditioner corresponds to the case when all mutual inductances are ignored.

For the solenoidal basis method, the Barnes-Hut variant described in Section 3 was used to compute products with the system matrix **L** and preconditioner $\tilde{\mathbf{L}}$ directly without explicitly computing these matrices. The resulting implementation is a matrix-free code in which neither the system matrix nor the preconditioner matrix is ever computed. This reduces the storage requirement considerably, thereby allowing larger problems to be solved.

Ground Plane

Table 4 shows the number of iterations needed by preconditioned GMRES for 10GHz frequency. A tolerance of 10^{-3} was specified on the relative residual norm of both the methods. FastHenry was allowed to use default values for all other parameters. In these experiments, the inductance computed by the solenoidal basis method was within 3% of that obtained by FastHenry.

FastHenry requires significant amount of memory to construct the preconditioner matrix and compute its LU factorization. The entries marked "-" indicate the inability of FastHenry to solve the

Table 4: Ground plane problem.

Mesh	FastHenry-DIAG			FastHenry-CUBE			Solenoidal Method		
	Iter.	Time (s)	Mem. (MB)	Iter.	Time (s)	Mem. (MB)	Iter.	Time (s)	Mem. (MB)
33×33	13	2.04	10	13	2.36	10	5	2.03	1
65×65	16	12.70	42	17	16.59	42	6	11.91	5
129×129	21	95.30	177	19	142.42	177	7	67.61	17
257×257	26	835.66	734	28	1364.19	734	9	409.21	69
513×513	-	-	-	-	-	-	14	2924.58	298

Table 5: Wire over ground plane problem.

Mesh	FastHenry-DIAG			FastHenry-CUBE			Solenoidal Method		
	Iter.	Time (s)	Mem. (MB)	Iter.	Time (s)	Mem. (MB)	Iter.	Time (s)	Mem. (MB)
33×33	13	2.03	10	11	2.20	10	4	1.43	1
65×65	13	11.60	42	14	16.01	42	5	9.20	4
129×129	13	79.06	178	12	123.96	178	6	55.12	15
257×257	3	718.93	735	3	2732.73	735	8	351.19	61
513×513	-	-	-	-	-	-	12	2426.73	260

problem within the available system memory. These experiments were conducted on a 1.5GHz Pentium Dell Workstation with 1GB of memory. Furthermore, a growth in the number of iterations with mesh size indicates a sub-optimal preconditioning scheme which contributes an additional factor towards the increase in cost of solving these systems as the mesh size is increased. In contrast, the solenoidal basis method is able to solve the systems in almost fixed number of iterations.

Wire Over Ground Plane

Table 5 shows the number of iterations needed by preconditioned GMRES for 10GHz frequency. Again, a stopping tolerance of 10^{-3} was used for the relative residual norm of both the methods. FastHenry was allowed to use default values for all other parameters. These results also demonstrate the comparative advantage of the solenoidal basis method. The performance is similar to the ground plane problem.

Discussion

The modest performance of the preconditioners in FastHenry come with a significant cost of computing the preconditioners themselves as well as storing them. While these storage requirements can be reduced by computing incomplete factorizations, often this results in a weak preconditioner. The slower convergence rates associated with ineffective preconditioning may lead to overall higher computational cost. The comparative advantage of the solenoidal method is expected to grow with larger mesh sizes.

5. CONCLUSIONS

This paper presents a solenoidal basis method for inductance extraction of VLSI circuits. The proposed approach solves a linear system of equations to compute the mutual inductance effect on a filament mesh using eddy currents defined on the mesh loops. The resulting reduced system is solved iteratively by the preconditioned GMRES method. The preconditioner suggested for the system matrix exhibits convergence in a relatively few iterations irrespective of the mesh refinement and operational frequency. Experimental results indicate that the algorithm has several advantages over FastHenry. Since our algorithms is kernel independent, it can be applied to multi-layer dielectric by using multi-layer Green's function.

6. REFERENCES

[1] J. Barnes and P. Hut. A hierarchical O(n log n) force calculation algorithm. *Nature*, Vol. 324, 1986.

[2] L. Daniel, A. Sangiovanni-Vincentelli, and J. White. Techniques for including dielectrics when extracting passive low-order models of high speed interconnect. ICCAD 01.

[3] K. Gala, et. al. On-chip inductance modeling and analysis. DAC 2000.

[4] L. Greengard. *The Rapid Evaluation of Potential Fields in Particle Systems*. The MIT Press, Cambridge, Massachusetts, 1988.

[5] Z. He, M. Celik and L. Pileggi. SPIE: Sparse partial inductance extraction. DAC 1997.

[6] M. Kamon. *Fast Parasitic Extraction and Simulation of Three-dimensional Interconnect via Quasistatic Analysis*. PhD thesis, EE & CS Dept., MIT, 1998.

[7] M. Kamon and J. R. Philips. Preconditioning techniques for constrained vector potential integral equations, with application to 3-D magnetoquasistatic analysis of electronic packages. In *Proceedings of the Colorado Conference on Iterative Methods*, Breckenridge, Colorado, 1994.

[8] M. Kamon, M. J. Tsuk, and J. White. FASTHENRY: A multipole-accelerated 3D inductance extraction program. *IEEE Trans. MTT*, Vol. 42:1750–1758, 1994.

[9] B. Krauter and S. Mehrotra. Layout based frequency dependent inductance and resistance extraction for on-chip interconnect timing analysis. DAC 1998.

[10] B. Krauter and L. T. Pileggi. Generating sparse partial inductance matrices with guaranteed stability. ICCAD 95.

[11] M. Liu, T. Yu and W. W.-M. Dai. Fast 3D inductance extraction in lossy multi-layer substrate. ICCAD 01.

[12] V. Rokhlin. Rapid solution of integral equations of classical potential theory. *Journal of Computational Physics*, Vol. 60:187–207, 1985.

[13] A. E. Ruehli. Inductance calculation in a complex integrated circuit environment. *IBM J. Res. Dev.*, Sept. 1972.

[14] Y. Saad. Iterative methods for sparse linear systems. PWS publishing company, 1996.

[15] K. L. Shepard and Y. Zhang. Return-limited inductances: A practical approach to on-chip inductance extraction. *IEEE Trans. CAD*, 19(4):425–436, 2000.

On the Efficacy of Simplified 2D On–Chip Inductance Models†

Tao Lin Michael W. Beattie‡ Lawrence T. Pileggi

Department of Electrical and Computer Engineering, Carnegie Mellon University
5000 Forbes Avenue, Pittsburgh, PA 15213
{tl, beattie, pileggi}@ece.cmu.edu

ABSTRACT

Full three–dimensional (3D) inductance models of on-chip interconnect contain an extremely large number of forward coupling terms. It is therefore desirable to use a two–dimensional (2D) approximation in which forward couplings are not included. Unlike capacitive coupling, however, truncating mutual inductance terms can result in loss of accuracy and even instability. This paper investigates whether ignoring forward couplings is an acceptable choice for all good IC designs or if full 3D models are necessary in certain on-chip interconnect configurations. We show that the significance of the forward coupling inductance depends on various aspects of the design.

Categories & Subject Descriptors

B7.2 [**Integrated Circuits**] Design Aids – *simulation, verification*

General Terms

Theory, algorithms

Keywords

On-chip inductance, PEEC, sparsified model

1. Introduction

Although accurate analysis of the RLC interconnects can be performed based on the Partial Element Equivalent Circuit (PEEC) method [10], the full PEEC models obtained by three dimensional field solvers (refer to [6][7][8][9]) often produce an intractable complexity problem for subsequent analyses. Several approaches have been proposed to tackle the problem by extracting *sparsified* inductance matrices [11][12][13][14][15]. Recently, novel approaches for modeling the inductive effect based on a new type of circuit element *K (*susceptance*)* were proposed [16][17] that can be used to facilitate sparse, localized window–based inductance extraction models [17].

Inductance sparsification approaches based on truncation or approximation of forward coupling elements have been applied recently for both modeling and analysis [3][4][5]. In [3], a partial circuit model was proposed in which mutual inductance elements are included only if they are between parallel wires. In [4], the authors utilized a similar truncation approach for inductance modeling of shielded clock trees. An interesting 2D model was proposed for parallel wires in [5]. The authors first extract the total self and coupling inductance of the whole wires. A fraction of the total inductance value is then assigned to each bundle of segments as self and parallel inductance. Since in these circuit models, inductance coupling only exists in the two dimensions perpendicular to the current flow, we refer to them as two–dimensional (2D) models. Clearly, the complexity of the inductance modeling problem can be substantially reduced by using 2D models. However, the accuracy of the 2D models is not adequately justified, and only some experimental results have shown a good match for the specific types of circuits in [4][5].

In this paper we present a theoretical analysis of the 2D modeling error. We show that for long on-chip busses, where the inductance effects are typically most pronounced, the magnetic field generated resembles that of dipoles due to close-by current return paths. Analytical formulas are derived for the forward coupling inductance in terms of both partial and loop (effective) inductance. Furthermore, realistic RLC circuit models are used to analyze the forward coupling effect in real-world on-chip interconnects. Finally, we propose a combined two–dimensional inductance methodology that can provide a better accuracy and efficiency trade-off and, therefore, a practical solution to the modeling and analysis of large interconnect systems.

2. Motivation

Accurate three-dimensional PEEC models for general shaped on-chip interconnects can be obtained via field solver based extractors such as gismo [6], FastCap [7], FastHenry [8], and Raphael [9]. Usually the capacitance and inductance models are extracted separately as matrices, then they are combined to form the PEEC model. A full three-dimensional PEEC model includes a large number of resistors, capacitors, and inductors. Before sparsification, the coupling capacitors and the mutual inductors constitute a dominant portion of the model.

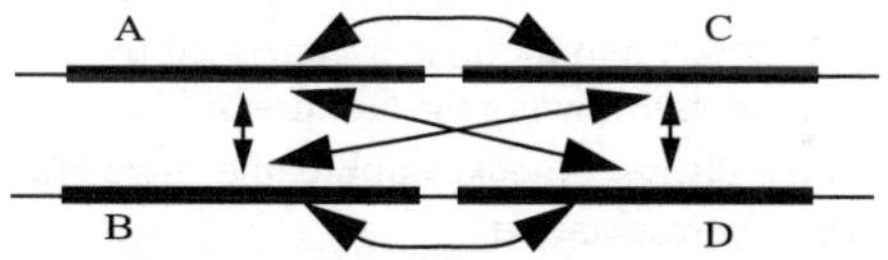

We distinguish three types of coupling (mutual) terms as illustrated using the two wire/two segment example above. The parallel segments in the two wires are perfectly aligned.

- *Parallel coupling:* the capacitive and inductive coupling between the parallel, aligned segments of the conductors. For example, the coupling between segment A and B.
- *Forward self coupling:* the coupling between segments of conductors belonging to the *same* line. For example, the coupling between segment A and C.

† This work was supported in part by Semiconductor Research Corporation contract 2000-TJ-778 and a grant from Intel Corporation.

‡ Michael W. Beattie was with Carnegie Mellon University, Dept. of Electrical and Computer Engineering. He is now with IBM Corporation, Austin, TX.

DAC 2002, June 10-14, 2002, New Orleans, Louisiana, USA.

- *Forward mutual coupling:* the coupling between non-aligned segments of conductors belonging to *different* lines. For example, the coupling between segment B and C.

As we know, capacitive coupling is a short range effect. Particularly, the forward self capacitive coupling and the forward mutual capacitive coupling terms are several orders of magnitude smaller than the self and parallel coupling terms for long wires. Moreover, the forward self coupling does not change the total capacitance of a conductor. Therefore, the forward capacitive coupling is known to be negligible. In many situations, even the parallel coupling capacitors beyond immediate neighbors are often discarded due to insignificant relative magnitude.

Unlike capacitance, coupling inductance between conductors reduces very slowly with distance. The long range magnetic coupling presents a vast number of non-negligible mutual terms in full 3D inductance extractions. For a system of N lines with M segments each, a full inductance model includes NM self terms, $(N-1)NM/2$ parallel coupling terms, and $N(M-1)NM/2$ forward coupling terms. The number of the forward coupling terms are no less than the parallel terms as long as each line contains at least two segments. When the number of segments for each wire is increased to improve the accuracy of the discretized model, apparently forward coupling inductance terms will become the major obstacle for runtime and storage efficiency.

3. Two–Dimensional Inductance Models

Two types of methods have been used in practice for the 2D model formulation.The first method is a partitioning and cascading based approach that has been reported in recent publications [3][4]. The circuit area is first partitioned into several sections. The self inductance and the mutual inductance are extracted for the parallel wire segments in every section and then cascaded to form the inductance model for the entire area.

The second method is applicable for aligned parallel interconnect structures [5]. Instead of dividing the wires into segments before extraction, this method first extracts the partial inductance between the entire wires. Then the wires are divided into equal length sections and the inductance values are uniformly distributed to the parallel segments in each section.

Note that the 2D models obtained from the above two methods are numerically different, even though they assume the same circuit topology. We refer to the model generated by the first method as the *cascaded* model, and the model generated by the second as the *normalized* model, following the authors of [5].

For the two wire/two segment example, the *cascaded* 2D model has the following representation,

$$\begin{bmatrix} L_A & M_{AB} & 0 & 0 \\ M_{AB} & L_B & 0 & 0 \\ 0 & 0 & L_C & M_{CD} \\ 0 & 0 & M_{CD} & L_D \end{bmatrix}, \tag{1}$$

where L_A stands for the partial self inductance of segments A, M_{AB} stands for the partial coupling inductance between segment A and B.

The second method, in contrast, generates the 2x2 inductance matrix for the entire wires first:

$$\begin{bmatrix} L_A + L_C + 2M_{AC} & M_{AB} + M_{BC} + M_{AD} + M_{CD} \\ M_{AB} + M_{BC} + M_{AD} + M_{CD} & L_B + L_D + 2M_{BD} \end{bmatrix} \tag{2}$$

Suppose the two stages of segments have equal length, splitting (2) renders the *normalized* 2D model.

$$\begin{bmatrix} L_A + M_{AC} & M_{AB} + M_{BC} & 0 & 0 \\ M_{AB} + M_{BC} & L_A + M_{AC} & 0 & 0 \\ 0 & 0 & L_A + M_{AC} & M_{AB} + M_{BC} \\ 0 & 0 & M_{AB} + M_{BC} & L_A + M_{AC} \end{bmatrix} \tag{3}$$

Note that the forward coupling terms, e.g., M_{AC}, M_{BC}, are not discarded. Instead, they are components of the 2D self and parallel terms ("folded" into these terms).

An important advantage of the 2D models as a sparsification to the original full 3D model is the *guaranteed stability* of these models. Note that the representations of both the cascaded models and the normalized models are block diagonal matrices. Furthermore, all the blocks are positive definite because they are the full matrix representations for real systems. For the cascaded model, each block is a full inductance matrix of the corresponding section. For the normalized case, every block is the scaled version of the full matrix for the unpartitioned system. Therefore, the block diagonal matrix for the 2D models is always positive definite.

Given the obvious benefits of using 2D inductance models, the primary consideration is the accuracy of these models. Critical to this problem is the magnitude of the forward coupling inductance.

4. Analysis of Forward Coupling Inductance

4.1 Basic Formulas for Partial Inductance

Although 3D field solver based extraction programs are generally required to compute the inductance model of arbitrary shaped objects, accurate analytical formulas are available for interconnects of regular shape [1][2]. In this section we explore the significance of forward coupling inductance using the following formulas.

- The mutual partial inductance between two parallel conductors with constant cross section (including the self partial inductance, which can be viewed as the mutual inductance between a conductor and its self),

$$M(l, d) = \frac{\mu}{2\pi} l\left(ln\left(\frac{2l}{R}\right) - 1 + \frac{d}{l}\right) \tag{4}$$

  where d is arithmetic mean distance between the two conductors and R is the geometric mean distance between them.

The above expression is the expansion form of the exact formula. It is accurate when d/l is small [1]. It is not limited to the conductors with rectangular cross section, nor does it require the two conductors to have identical cross section. Thus our discussion will be valid to a wide range of interconnects including wires of different aspect ratios. Two specialized forms of (4) are also used later in this paper.

- The self partial inductance of a rectangular bar

$$L(l) = \frac{\mu}{2\pi} l\left(ln\left(\frac{2l}{B + C}\right) + \frac{1}{2} - ln(\varepsilon)\right) \tag{5}$$

where B and C are the thickness and width of the wires respectively, and $\ln(\varepsilon)$ is a small positive constant depending solely on the ratio B/C, $\ln(\varepsilon) < 0.0025$.

- The mutual partial inductance between two parallel equal rectangular bar

$$M(l, d) = \frac{\mu}{2\pi} l\left(ln\left(\frac{2l}{d}\right) - ln(k) - 1 + \frac{d}{l}\right) \tag{6}$$

where d is the pitch between the traces and $ln(k)$ is another constant determined by the cross section of the conductors and d. The value of $ln(k)$ and $ln(\varepsilon)$ can be obtained from lookup tables in [1].

4.2 Partial Forward Coupling Inductance

The forward self coupling inductance between two adjacent segments of length l_1, l_2 on the same straight line is illustrated as:

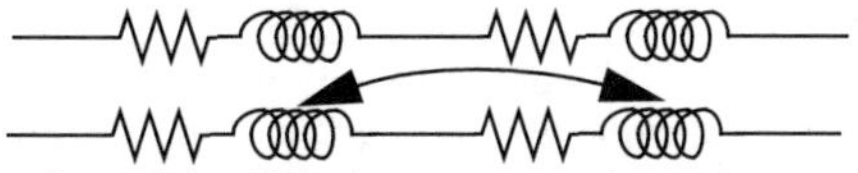

Since the total partial self inductance can be written as the sum of the self terms for each segments and the coupling terms between them,

$$L(l_1 + l_2) = L(l_1) + L(l_2) + 2M_f(l_1, l_2) \tag{7}$$

the coupling term can be computed by solving (7) for M_f and replacing the self terms by (5):

$$M_f(l_1, l_2) = \frac{\mu}{4\pi}\left(l_1 ln\left(\frac{l_1 + l_2}{l_1}\right) + l_2 ln\left(\frac{l_1 + l_2}{l_2}\right)\right) \tag{8}$$

The second type of the partial forward coupling inductance exists between two segments of length l_1, l_2 located on two parallel wires with distance d, but adjacent in the length direction:

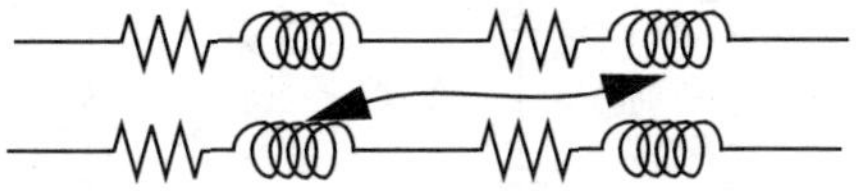

This term can be computed similarly by decomposing the mutual inductance between the two equal parallel segments of length l_1+l_2 into parallel coupling terms and forward coupling terms.

$$M(l_1 + l_2, d) = M(l_1, d) + M(l_2, d) + 2M_f(l_1, l_2, d). \tag{9}$$

Again, the forward coupling term can be solved from (9) using the formula for mutual inductance between parallel wires (6). We have,

$$M_f(l_1, l_2, d) = \frac{\mu}{4\pi}\left(l_1 ln\left(\frac{l_1 + l_2}{l_1}\right) + l_2 ln\left(\frac{l_1 + l_2}{l_2}\right) - d\right). \tag{10}$$

To assess the relative significance of forward coupling, we compare the forward coupling terms (8) and (10) to the self inductance (5) and the parallel coupling inductance (6) respectively. For simplicity, we assume the two segments of concern are of equal lengths. The ratio of the forward coupling self term (8) to the partial self term (5) is

$$\frac{M_f}{L} = \frac{ln(2)}{ln\left(\frac{2l}{B + C}\right) + \frac{1}{2} - ln(\varepsilon)}, \tag{11}$$

while the ratio of the forward coupling mutual term (10) to the parallel mutual term (6) has a similar form.

$$\frac{M_f}{M_p} = \frac{ln(2) - \frac{d}{2l}}{ln\left(\frac{2l}{d}\right) - ln(k) - 1 + \frac{d}{l}} \tag{12}$$

There are two obvious characteristics of (11) and (12). First, for the region when the long line assumptions of $l \gg d$ and $l \gg B + C$ are valid, both ratios are always smaller than 1. Secondly, the ratios asymptotically decrease to zero as the segment length increases. Most importantly, we have the following observation:

Observation 1: *Due to the logarithmic denominators of (11) and (12), the forward coupling partial inductance are **NOT** negligible.*

For example, assuming an 1μm by 1μm cross section, 2μm pitch and 1000μm segment length, the above ratios are 9.4% and 11.7% respectively.

Since usually long wires contain more than two segments, each self and parallel coupling inductance should be compared with the total effect of several forward coupling terms to the up-stream and down-stream segments. The total forward coupling inductance is even larger than the numbers shown above. Clearly, the forward coupling inductance terms are not negligible numerically, though it is favorable to do so in terms of efficiency.

4.3 Loop Inductance

Inductance is ultimately a property of the current loops. In this section, we study the mutual inductance between loops using a simple model. In this model, the current loop consists of only two parallel wire segments with one delivering current and the other returning current.

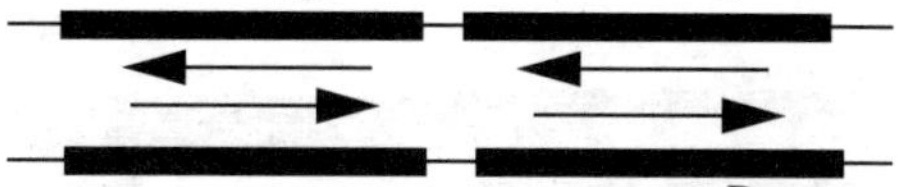

First we consider the case using the two wire/two segment example. Segment A and segment B form a current loop, while segment C and segment D form the other loop as shown above. Assume the wire distance, d, is much smaller than the length of the segments l_1, l_2.

Note that each of the small current loops consists of two identical currents flowing closely in the opposite directions. Such a current pair resembles the behavior of a ***dipole***. The induced electrical field generated by the current variation on one segment due to Faraday's Law counteracts the field generated by the other. The resulting induced field drops much faster with distance than that of a monopole (each segment current alone). Consequently, we should expect the couplings between the loops to be much weaker than the coupling between the partial elements.

The forward coupling between the two loops (dipoles), also shows the cancellation effect in its expression (13).

$$M_{loop}(l_1, l_2) = (M_{AC} - M_{BC}) + (M_{BD} - M_{AD}) \tag{13}$$

For the typical case where the segment lengths of A, B, C, D are much larger than the distance of the ideal return path d, we can apply the formulas (8)(10) derived in the previous section to the partial terms in (13). It results in the following astonishing expression:

$$M_{loop}(l_1, l_2) = \frac{\mu}{2\pi}d \tag{14}$$

The above formula shows that the (dipole) forward coupling between the loops is (approximately) a small constant when the segment lengths are long enough. In other words, the forward coupling inductance between the loops is well approximated by a constant which only depends on the distance of the current return; when the segment lengths increase, the forward coupling inductance approaches this constant as the limit value.

For comparison, we compute the self inductance for a loop, e.g., the loop formed by segment A and B,

$$L_{loop}(l_1) = L_A + L_B - 2M_{AB} \tag{15}$$

$$= \frac{\mu}{\pi} l_1 \left(ln\left(\frac{d}{B+C}\right) + \frac{3}{2} - ln(\varepsilon) + ln(k) \right) - \frac{\mu}{\pi} d$$

Note that the loop self inductance grows linearly with the wire length. When $l \gg d$, the loop forward coupling inductance becomes negligible compared with the loop self inductance. Assume that each trace has an 1μm by 1μm cross section and the pitch between the traces is 2μm. The ratio of (14) to (15) is 0.68% when the loop length is 100um and it decreases to 0.067% when the loop length is 1000um.

More generally, in the case when the two loops are not exactly aligned in the current direction as shown below, the forward coupling between the loops is given by $\frac{\mu}{2\pi}\tilde{d}$, where $\tilde{d}$ is the overlap of the two loops in Y direction. When $\tilde{d}$ is zero, the forward coupling is so small that a more accurate high-order formula has to be used in place of (4) to observe it.

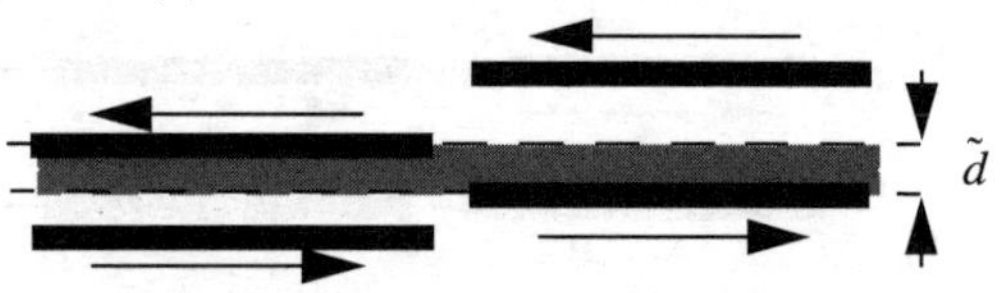

We can also derive the loop parallel coupling inductance in a similar fashion. For example, the coupling inductance between the loop formed by A and B and the loop formed by C and D in the figure below, is given by (16).

$$M_{loop}(l_1) = \frac{\mu}{2\pi} l_1 ln\left(\frac{R_{AD}R_{BC}}{R_{AC}R_{BD}}\right) \tag{16}$$

R_{AD}, R_{BC}, R_{AC}, R_{BD} are the geometric mean distance between the traces A, B, C, D respectively. The loop parallel coupling inductance between loops is a linear function of the wire (loop) length.

Therefore, we make the following observations.

Observation 2: *For the current loops defined above, the loop forward coupling inductance is a negligible constant* $\frac{\mu}{2\pi}\tilde{d}$.

Observation 3: *For the current loops defined above, the loop self inductance and the parallel coupling inductance between the loops are linear functions of the wire length, therefore the unit length inductance can be defined as the linear coefficients in equation (15)(16).*

Note that for partial inductance, the unit length inductance concept is invalid because the partial inductance grows superlinearly with the segment length as shown by the formulas in the previous section.

In summary, since the forward coupling terms between the loops are negligible, it doesn't matter whether they are discarded or approximated. Both 2D models will work fine in this case.

5. Impact of Current

In this section, we take a closer look at the expected on-chip current return paths and analyze the impact of the current distribution on the induced voltage due to forward coupling. Generally, for on-chip interconnects, there is no *dc* path for the current to return. The current delivered by the signal lines has to return via *ac* paths. The *ac* return paths are comprised of coupling capacitors among the parallel lines and even the cross–over metal strips in the neighboring layers.

Normally, a fully extracted RLC circuit contains two types of capacitors, coupling capacitors and grounding capacitors. The differences in the formulation of the capacitances can play an important role in the accuracy of the two–dimensional model.

Consider the current flowing to the common ground via the grounding capacitors. It virtually returns from the infinity and behaves like a monopole. On the other hand, the current can also return via the *ac* path formed by the lateral coupling capacitance and the neighboring wires or low resistive shields, and behaves like a dipole.

The reality is a mix of these two situations. Below we discuss the composite effect of the different current returns paths based on the coplanar bus structure in Fig. 1.

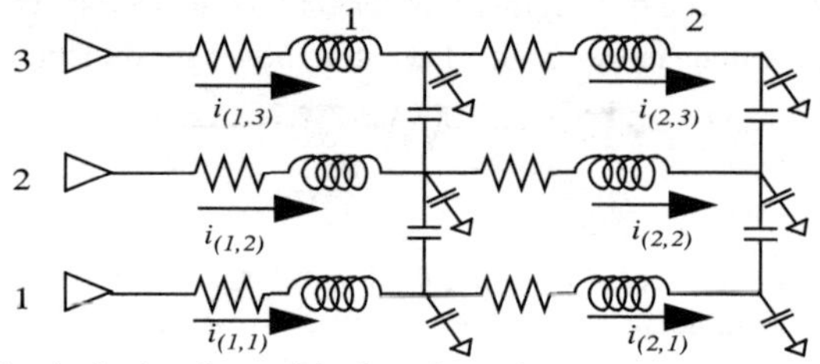

Fig. 1: A circuit model of coplanar busses

Assume that the circuit contains N wires and M segments. Denote the current flowing over the segment at the mth column and the nth row, *(m, n)* in Fig. 1, by $i_{(m,n)}$. This term can be decomposed into two parts,

$$i_{(m,n)} = i_{int(m,n)} + i_{ext(m,n)} \tag{17}$$

where $i_{int(m,n)}$ is the part of the current injected into *(m,n)* that returns from its parallel peers, while $i_{ext(m,n)}$ is the current returning via the ground capacitance paths.

It follows that the total induced voltage on a segment *(j,k)* by the current in stage *m,* $V_{(j,k),m}$ can be considered as the sum of the induced voltage by these two types of current.

$$V_{(j,k),m} = \tag{18}$$

$$\sum_{n=1}^{N} M_{(j,k)(m,n)} \frac{di_{int(m,n)}}{dt} + \sum_{n=1}^{N} M_{(j,k)(m,n)} \frac{di_{ext(m,n)}}{dt}$$

We compare the voltage induced by the forward coupling from its adjacent segments with the voltage induced by the self and parallel coupling inductance on this segment based on the above decomposition. Applying the general formula in terms of geometric mean distance(4), we obtain the induced voltage due to the *self and parallel coupling* inductance as follows

$$VP_{(j,k)} = VPI_{(j,k)} + VPE_{(j,k)} \tag{19}$$

$$VPI_{(j,k)} = \frac{\mu}{2\pi}\sum_{n=1}^{N}\left((-l_j ln(R_{k,n}) + d_{k,n})\frac{di_{int(j,n)}}{dt}\right) \tag{20}$$

$$VPE_{(j,k)} = \frac{\mu}{2\pi}\sum_{n=1}^{N}\left(l_j\left(ln\left(\frac{2l_j}{R_{j,n}}\right)-1+\frac{d_{k,n}}{l_j}\right)\frac{di_{ext(j,n)}}{dt}\right) \tag{21}$$

where $VPE_{(j,k)}$ and $VPI_{(j,k)}$ are the voltage responses due to parallel couplings generated by either ground capacitance (***E***xternal) current or coplanar capacitive coupling (***I***nternal) current. The above expressions are after simplification based on the Kirchoff's Current Law (KCL).

The voltage induced by the *forward coupling* inductance from the adjacent stage $m=j+1$, has the form

$$VF_{(j,k)} = VFI_{(j,k)} + VFE_{(j,k)} \tag{22}$$

$$VFI_{(j,k)} = \frac{\mu}{4\pi}\sum_{n=1}^{N}\left(-d_{k,n}\frac{di_{int(m,n)}}{dt}\right) \tag{23}$$

$$VFE_{(j,k)} = \tag{24}$$

$$\sum_{n=1}^{N}\left(\frac{\mu}{4\pi}\left(l_j ln\left(\frac{l_j+l_m}{l_j}\right)+l_m ln\left(\frac{l_j+l_m}{l_m}\right)\right)\frac{di_{ext(m,n)}}{dt}\right)$$

Comparing (20) and (21) with (23) and (24), we have the following observations.

Observation 4: *If the return current via the grounding capacitance is* negligible, *the induced voltage due to forward coupling (23) has a much smaller inductance coefficient than that due to the self and parallel inductance (20). Forward coupling inductance is comparatively negligible. The error of using 2D inductance models is small.*

Observation 5: *If the return current via the grounding capacitance is* significant, *the forward coupling impact (24) is* not *negligible compared with the self and parallel inductance impact (20) and (21). In this case, discarding the forward coupling inductance, as done in the cascaded model, will introduce error.*

Note that the return path is not the only influence on the inductance effect. The actual waveform of the current and its propagation in the wires also have an impact.

Consider the special case when the normalized model is applicable, the system consists of equal length parallel wires. For signals with large risetime and small propagation delay, due to the continuity of the current distribution along the line, at any given time the neighboring segments on the line have the same current flow approximately, i.e., $i_{(j,n)} \approx i_{(m,n)}$. The currents in the equations (23) and (24) can be approximated by the currents in the jth segments themselves. Thus the induced voltage due to the forward coupling inductance can be viewed as if the forward coupling inductance terms are "folded" into the self and parallel inductance. When the assumption on the current is true, the normalized model will be a good approximation to the original 3D model regardless of the current distribution in the return paths.

6. Experiments on RLC Interconnects

The overall impact of the forward coupling inductance should be considered in the combined RLC system. Since the lumped RLC circuit is an approximation of the original distributed multiconductor system, the total error of the 2D based RLC circuit can be viewed as the composite effect of two components, the error between the original distributed system and the finite 3D PEEC model, and the error between the 3D circuit model and the 2D based models. Both errors vary with the degree of discretization, e.g., the number of the segments used to represent each conductor or the size of the segments. In this section, We compare the response characteristics of the RLC circuits based on the 2D inductance models with the same system based on the 3D model, when the degree of discretization is varying.

A 5-bit bus with one shield at each side is used as an example in this experiment. The wiring material is copper. The wires are 1μm thick and 3000μm long. The width of the signal wires is 0.5μm while the width of the shields is 1μm. The space between the neighboring wires is 0.5μm. Each signal wire is driven by a 100Ω resistor and loaded by a capacitance of 40ff. The two shields on the sides are grounded at the near end. The bus is extracted using field solver based extraction tools. The signal delay obtained via HSpice simulation of the 2D and 3D models under a 50ps finite ramp input is shown against the number of segments used to represent each conductor.

First, we force the current to return via the coupling capacitance and the in-plane return paths by zeroing the grounding capacitors. In Fig. 2a, we find the delay curves given by the 2D models fit consistently well with the delay curve obtained from the full 3D model, as expected in the previous section.

In the second experiment, the grounding capacitors are included in the simulation. Since the current is no longer limited to return within the bus, the curve from the cascaded model starts to deviate from that of the 3D model (Fig. 2b). The discrepancy increases with the number of segments (when the segment length decreases). In this case, the normalized model is still able to provide a fair approximation to the 3D model.

To further investigate the efficacy of the normalized model, we repeat the second experiment with the input signal risetime set to 10ps. On the contrary to the assumption in the previous section, the response signal has a sharp rising edge but a larger LC propagation delay. The results are shown in Fig. 3. We find that the normalized model tends to produce overestimated delay values. However, even in this case, the relative error of the normalized model is still less than 5%.

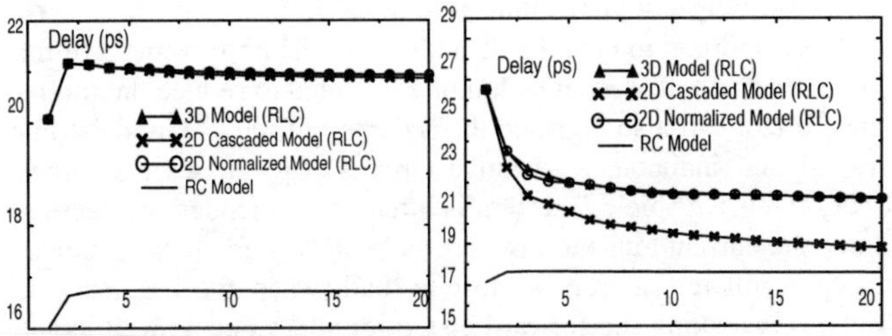

Fig. 2: Delay vs. # of segments (a) without grounding capacitors (b) with grounding capacitors

Fig. 4 shows the total extraction and simulation runtime for the full 3D model and one of the 2D model (the normalized model). Obviously, the latter is much more efficient.

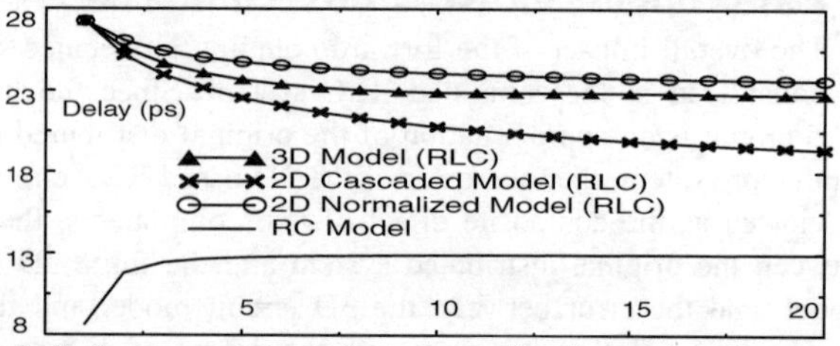

Fig. 3: Delay vs. number of segments (fast input signal)

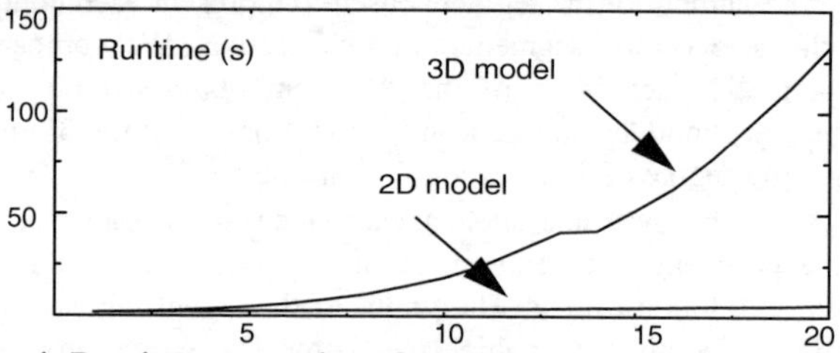

Fig. 4: Runtime vs. number of segments.

7. A Combined Methodology

Although the normalized 2D model has good consistency with the 3D model for a wide range of discretization, it is only applicable to aligned parallel wires with equal length. The cascaded model is more general and suitable for arbitrary interconnect topologies. However, from the experimental results, we observe that the cascaded 2D model could introduce significant error when the return current is not limited to close-by return paths. To solve this dilemma, we propose a general 2D inductance modeling methodology that combines the advantage of the cascaded model and the normalized model. Furthermore, it can be incorporated with other screening and sparsification techniques([16][17]) to reduce the number of parallel coupling terms.

We suggest the following methodology of RLC modeling for large on-chip interconnect systems:

1. Proceed along the X direction, divide the circuit area into sections so that all of the horizontal wires within each section span the width of the section.
2. The inductance of the equal length wires in each section is extracted using a field solver or a table-lookup based extractor.
3. Build a normalized model for each section based on the inductance matrix for the entire section.
4. Cascade the normalized models to form the 2D inductance model of the horizontal wires in the target area.
5. Repeat the above process in Y direction. Combine the results to form the complete inductance model.

It is important to note that in addition to the improvement of the modeling method, design techniques invented to reduce the inductance effect can also enhance the accuracy of the 2D models. For critical and inductance sensitive circuits, e.g. busses and clock trees, extensive shield insertion is often recommended in practice. Since the current tends to flow via the path with the lowest impedance, it will return from a close-by path when the frequency is high, thus making the forward coupling inductance as well as the effective loop inductance minimized. In other words, for well designed circuits, the on-chip interconnect structure is often of a form that can be analyzed via 2D models.

8. Conclusion

The significance of forward coupling inductance in on-chip interconnect models has been studied in detail. The impact of sparsifying the inductance matrix by truncating the forward coupling inductance was analyzed analytically and through numerical examples.

As expected, the impact of the forward coupling inductance heavily depends on the actual current distribution. The 2D inductance models are applicable whenever the return current can be restricted to local return paths. For aligned coplanar structure, the normalized model provides better accuracy then the cascaded model. The two types of 2D inductance models can be combined to provide more accurate and efficient approximations for full 3D inductance models.

9. References

[1] F. Grover, "Inductance Calculations: Working Formulas and Tables", Dover Publications, 1973.

[2] C. Hoer and C. Love, "Exact inductance equations for rectangular conductors with applications to more complicated geometries", *Journal of Research of the National Bureau of Standards,* 69C, No. 2, April-June 1965.

[3] K. Gala, *et al.*, "On-chip Inductance Modeling and Analysis", *Proc. 37th DAC,* pp. 63, June 2000.

[4] N. Chang, *et al.*, "Clocktree RLC Extraction with Efficient Inductance Modeling", *Proc. DATE,* pp. 522-526, 2000.

[5] M. Xu and L. He, "An Efficient Model for Frequency-based On-Chip Inductance", *GLSVLSI*, March 2001.

[6] M. Beattie and L. Pileggi, "Electromagnetic Parasitic Extraction via a Multipole Method with Hierarchical Refinement", *Proc. ICCAD,* Nov. 1999.

[7] K. Nabors and J. White, "FastCap: A Multipole Accelerated 3D Capacitance Extraction Program", *IEEE Trans. CAD*, vol. 10, pp. 1447-1459, Nov. 1991.

[8] M. Kamon, M. Tsuk, and J. White, "FastHenry: A Multipole Accelerated 3D Inductance Extraction Program", *IEEE Trans. Microwave Theory and Techniques,* 42, pp. 1750-1758, Sept. 1994.

[9] *Raphael Interconnect Analysis Program V98.4 Reference Manual*, Avant! Corporation, Nov. 1998.

[10] A. Ruehli, "Equivalent Circuit Models for Three-Dimensional Multiconductor Systems", *IEEE Trans. Microwave Theory and Techniques*, MTT-22, No. 3, pp. 216-221, March 1974.

[11] B. Krauter and L. Pileggi, "Generating Sparse Partial Inductance Matrices with Guaranteed Stability", *Proc. ICCAD,* Nov., 1995.

[12] Z. He, M. Celik, and L. Pileggi, "SPIE: Sparse Partial Inductance Extraction", *Proc. 34th DAC,* June 1997.

[13] A. Dammers and N. van der Meijs, "Virtual Screening: A step towards a sparse partial inductance matrix", *Proc. ICCAD,* pp. 445, November 1999.

[14] K. Shepard, D. Sitaram, and Y. Zheng, "Full-Chip, Three-Dimensional, Shape-based RLC Extraction", *Proc. ICCAD,* pp. 142, Nov. 2000.

[15] M. Beattie, *et al.*, "Equipotential Shells for Efficient Inductance Extraction", *IEEE Trans. CAD*, 20, No. 1, pp. 70-79, January 2001.

[16] A. Devgan, H. Ji, and W. Dai, "How to efficiently capture on-chip inductance effects: Introducing a new circuit element K", *Proc. ICCAD,* pp. 125, Nov. 2000.

[17] M. Beattie and L. Pileggi, "Efficient Inductance Extraction via Windowing", *Proc. DATE,* March 2001.

A Physical Model for the Transient Response of Capacitively Loaded Distributed *rlc* Interconnects

Raguraman Venkatesan, Jeffrey A. Davis and James D. Meindl

School of Electrical and Computer Engineering, Georgia Institute of Technology
791 Atlantic Drive NW, Atlanta, GA 30332-0269, USA
Phone: 1-404-894-9910 Fax: 1-404-894-0462 E-mail: vragu@ee.gatech.edu

ABSTRACT

Rapid approximation of the transient response of high-speed global interconnects is needed to estimate the time delay, crosstalk, and overshoot in a GSI multilevel wiring network. This paper outlines a rigorous and physical solution for transients in a capacitively-loaded distributed *rlc* interconnect using a convergent series of modified Bessel functions. Compact models for time delay and repeater insertion are derived. The single-line model is extended to evaluate crosstalk in two-coupled lines. These solutions are validated by HSPICE simulation, and have potential applications to rapid *rlc* timing analysis, global wire sizing and repeater insertion, signal integrity estimation, and reliability modeling (e.g. voltage overshoot and high current density concerns).

Categories and Subject Descriptors

B.4.3 **[Input/Output and Data Communications]**: Interconnections (Subsystems) – *physical structures (e.g. backplanes, cables, chip carriers).*

General Terms

Performance, Design, and Verification.

Keywords

Interconnects, distributed *rlc* lines, transient response, time delay, crosstalk, overshoot, repeaters.

1. INTRODUCTION

Being able to rapidly pinpoint interconnect noise and signal reliability problems caused by inductance in a GSI multilevel interconnect network is becoming extremely important for future multi-gigahertz systems. Direction of current research has focused on approximate models for the transient response of distributed *rlc* lines. For example, Ismail and Friedman used curve-fitted empirical models to extract expressions for time-delay of distributed *rlc* lines [7]. In addition Cao et al. [5] developed 2-pole and second order waveform models while Banerjee and Mehrotra [3], [4] used second and fourth order Pade approximations for the transfer function, the solutions of which require numerical iterations to calculate line characteristics such as time delay and crosstalk.

This paper focuses on models that describes transient waveforms on the state-of-the-art global interconnect structures that have one or two signal lines sandwiched between co-planar ground lines as in Fig. 1 [1], [8]. This paper significantly extends the work in [6] by rigorously solving the Laplace domain transfer function of a *capacitively loaded* distributed *rlc* line (Fig. 2) using a convergent series of modified Bessel functions to obtain a highly accurate and explicit solution for the transient response. These solutions are used to develop unified models for time delay and repeater insertion in *rc* and *rlc* interconnects. The complete model is extended to two-coupled lines, and the solutions are verified against HSPICE simulations.

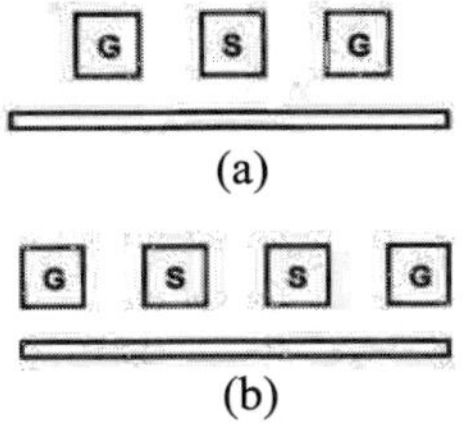

Fig 1. State-of-the-art global interconnect structures

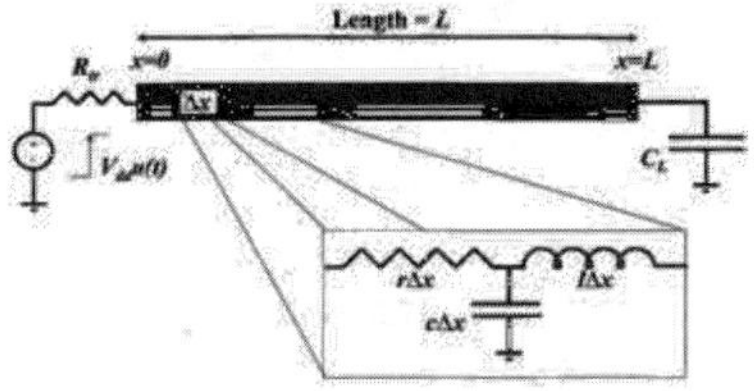

Fig 2. Distributed *rlc* line with load capacitance.

2. SINGLE LINE TRANSIENT MODEL

The voltage at the end of the line of length L, $V_{fin}(L,s)$, is described in the Laplace domain as [6]

$$V_{fin}(x=L,s) = (1+\Gamma_L)V_{inf}(L,s) + (1+\Gamma_L)\sum_{n=1}^{g}\left(\Gamma_L^n \Gamma_S^n V_{inf}\left((2n+1)L,s\right)\right) = V_1(s)+V_2(s) \quad (1)$$

where n and g are the number of reflections, $V_{inf}(x,s)$ is the Laplace domain voltage at x for a distributed *rlc* line of infinite length given by

$$V_{inf}(x,s) = \frac{V_{dd}}{s}\frac{Z_o\sqrt{(s+r/l)/s}}{Z_o\sqrt{(s+r/l)/s}+R_{tr}}e^{-x\sqrt{lc}\sqrt{s(s+r/l)}}. \quad (2)$$

Γ_L and Γ_S are the reflection coefficients at the load and source ends respectively, defined as:

$$\Gamma_L = \frac{Z_L-Z(s)}{Z_L+Z(s)} = \frac{1-sC_LZ_o\sqrt{1+\frac{r}{sl}}}{1+sC_LZ_o\sqrt{1+\frac{r}{sl}}}, \text{ and } \Gamma_S = \frac{Z_S-Z(s)}{Z_S+Z(s)} = \frac{R_{tr}-Z_o\sqrt{1+\frac{r}{sl}}}{R_{tr}+Z_o\sqrt{1+\frac{r}{sl}}} \quad (3)$$

where $Z_o = \sqrt{l/c}$ is the characteristic impedance of the line. The following two inverse laplace transform identities [9] will be used to obtain the time domain solution of (1) :

$$\left(\frac{s+\sqrt{s^2-\sigma^2}}{\sigma}\right)^{-v} \frac{e^{-k\sqrt{s^2-\sigma^2}}}{\sqrt{s^2-\sigma^2}} \rightarrow \left(\frac{t-k}{t+k}\right)^{\frac{v}{2}} I_v\left(\sigma\sqrt{t^2-k^2}\right)u_o(t-k), \quad v \geq 0 \tag{4}$$

where $I_v()$ is the modified Bessel function of order v.

$$\left(\frac{s+\sqrt{s^2-\sigma^2}}{\sigma}\right)^{v} \frac{e^{-k\sqrt{s^2-\sigma^2}}}{\sqrt{s^2-\sigma^2}} \rightarrow \left(\frac{t-k}{t+k}\right)^{\frac{v}{2}} I_v\left(\sigma\sqrt{t^2-k^2}\right)u_o(t-k) \tag{5}$$

$$2\sum_{r=0}^{\left[\frac{v}{2}\right]}\sum_{m=0}^{\left[\frac{v}{2}\right]}\left[\begin{array}{l}\binom{v}{2m}\binom{\frac{2m-1}{2}}{r}(-1)^r\sigma^{2r-v} \\ \times\sum_{y=1}^{\left[\frac{v-2r-1}{2}\right]+1} C_{v-2r-1,y}\dfrac{\sigma^{v-2r-y+1}kt^{v-2r-2y+1}}{\left(\sqrt{t^2-k^2}\right)^{v-2r-y+1}} I_{v-2r-y+1}\left(\sigma\sqrt{t^2-k^2}\right)\end{array}\right] u_o(t-k), \quad v>0$$

where the coefficient $C_{x,y}$ is defined recursively as :

$$\begin{array}{ll} C_{x,y} = 0 & y \leq 0, \text{ for all } x \\ C_{x,y} = 1 & x = 0,\ y = 1 \\ C_{x,y} = 0 & x = 0,\ y > 1 \\ C_{x,y} = (x-2y+3)C_{x-1,y-1} + C_{x-1,y} & \text{for all other x and y} \end{array} \tag{6}$$

i. First order reflection (n=0 or t≤3t_f)

Using algebraic expansions and inverse laplace transforms, the time domain solution for the first reflection (i.e. $t_f<t<3t_f$, t_f=time of flight) can be written as

$$V_{fin}(L,t_f<t\leq 3t_f) = V_{DD}\frac{Z_o}{Z_o+R_{tr}}e^{\frac{-rt}{2l}} \tag{7}$$

$$\times\left\{\begin{array}{l} P\sum_{i=0}^{\infty}(-M)^i\left(\dfrac{t-L\sqrt{lc}}{t+L\sqrt{lc}}\right)^{\frac{i+1}{2}} I_{i+1}\left(\sigma\sqrt{t^2-L^2lc}\right) \\ +Q\sum_{i=0}^{\infty}(-N)^i\left(\dfrac{t-L\sqrt{lc}}{t+L\sqrt{lc}}\right)^{\frac{i+1}{2}} I_{i+1}\left(\sigma\sqrt{t^2-L^2lc}\right) \\ +\dfrac{P}{1-G_s}\sum_{i=0}^{\infty}\sum_{k=1}^{\infty}(-M)^i\left(4-(1+G_s)^2G_s^{k-1}\right)\left(\dfrac{t-L\sqrt{lc}}{t+L\sqrt{lc}}\right)^{\frac{i+k+1}{2}} I_{i+k+1}\left(\sigma\sqrt{t^2-L^2lc}\right) \\ +\dfrac{Q}{1-G_s}\sum_{i=0}^{\infty}\sum_{k=1}^{\infty}(-N)^i\left(4-(1+G_s)^2G_s^{k-1}\right)\left(\dfrac{t-L\sqrt{lc}}{t+L\sqrt{lc}}\right)^{\frac{i+k+1}{2}} I_{i+k+1}\left(\sigma\sqrt{t^2-L^2lc}\right) \end{array}\right\}$$

$$\times\ u_o\left(t-L\sqrt{lc}\right)$$

where

$$P = \frac{B\left(B+\sqrt{B^2+4}\right)-2}{\sqrt{B^2+4}},\quad Q = \frac{-B\left(B+\sqrt{B^2+4}\right)-2}{\sqrt{B^2+4}},\quad M = \frac{B+\sqrt{B^2+4}}{2} \tag{8}$$

$$N = \frac{B-\sqrt{B^2+4}}{2},\quad B = \frac{4l}{rC_LZ_o},\quad G_S = \frac{R_{tr}-Z_o}{R_{tr}+Z_o},\quad \sigma = \frac{r}{2l}$$

When (7) no longer converges, a different set of algebraic expansions are used to write the first reflection laplace domain solution as

$$V_1^{'}(s) = V_{DD}\frac{Z_o}{Z_o+R_{tr}}\frac{e^{-L\sqrt{lc}\sqrt{s^2-\sigma^2}}}{\sqrt{s^2-\sigma^2}}\left\{\begin{array}{l}\left(P\left(\sum_{i=0}^{\infty}M^{-i-1}(-1)^i\left(\dfrac{s+\sqrt{s^2-\sigma^2}}{\sigma}\right)^i\right)+Q\sum_{i=0}^{\infty}(-N)^i\left(\dfrac{s-\sqrt{s^2-\sigma^2}}{\sigma}\right)^{-i-1}\right) \\ \times\left(1+\dfrac{1}{1-G_s}\sum_{k=1}^{\infty}\left(\dfrac{s+\sqrt{s^2-\sigma^2}}{\sigma}\right)^{-k}\left(4-(1+G_s)^2G_s^{k-1}\right)\right)\end{array}\right\} \tag{9}$$

Using (4) and (5), the equivalent time domain solution can be written as $V_{fin}(L,t_f<t\leq 3t_f) = e^{\frac{-rt}{2l}}V_1^{'}(t),\quad V_1^{'}(t) = L^{-1}\left[V_1^{'}(s)\right]$ (10)

ii. Second and higher order reflections (n>0 or t>3t_f)

For second and higher order reflections (i.e. $t>3t_f$), the time domain solution can be written as

$$V_{fin}(L,t>3t_f) = e^{\frac{-rt}{2l}}V_{DD}\frac{Z_o}{Z_o+R_{tr}} \tag{11}$$

$$\times\sum_{n=1}^{g}\left(\begin{array}{l}\left(S_C^{'}+S_C^{''}\right)\left(\dfrac{t-(2n+1)L\sqrt{lc}}{t+(2n+1)L\sqrt{lc}}\right)^{\frac{2i+2j+k+l+m+p+q+r}{2}} I_{2i+2j+k+l+m+p+q+r}\left(\sigma\sqrt{t^2-(2n+1)^2L^2lc}\right) \\ +\dfrac{1}{1-G_s}\left(S_C^{'}+S_C^{''}\right)\sum_{u=1}^{\infty}\left[\begin{array}{l}D\left(\dfrac{t-(2n+1)L\sqrt{lc}}{t+(2n+1)L\sqrt{lc}}\right)^{\frac{2i+2j+k+l+m+p+q+r+u}{2}} \\ \times I_{2i+2j+k+l+m+p+q+r+u}\left(\sigma\sqrt{t^2-(2n+1)^2L^2lc}\right)\end{array}\right]\end{array}\right)$$

$$\times\ u_o\left(t-(2n+1)L\sqrt{lc}\right)$$

where $S_C^{'}$ and $S_C^{''}$ are notations used for compactness. They represent the following series summation functions :

$$S_C^{'}\left(a^{-2i-2j-k-l-m-p-q-r}\right) \tag{12}$$

$$=\sum_{i=0}^{n}\sum_{j=0}^{\infty}\sum_{k=0}^{\infty}\sum_{l=0}\sum_{m=0}^{n}\sum_{p=0}\sum_{q=0}^{n}\sum_{r=0}\left[\begin{array}{l}\dfrac{n(n+j-1)!}{i!j!(n-i)!}\dfrac{n(n-p-1)!}{m!p!(n-m)!}\dfrac{n(n+1-r)!}{q!r!(n-q)!}(-1)^{i+k+n+m+p+q} \\ \times\binom{-i}{k}\binom{-j}{l}B^{k+l+m+p}G_s^{n+r-q}a^{-2i-2j-k-l-m-p-q-r}\end{array}\right]$$

$$S_C^{''}\left(a^{-2i-2j-k-l-m-p-q-r}\right) \tag{13}$$

$$=\sum_{i=0}^{n+1}\sum_{j=0}^{\infty}\sum_{k=0}\sum_{l=0}^{\infty}\sum_{m=0}^{n}\sum_{p=0}\sum_{q=0}^{\infty}\sum_{r=0}^{n}\left[\begin{array}{l}\dfrac{(n+1)(n+j)!}{i!j!(n-i+1)!}\dfrac{(n+1)(n-p)!}{m!p!(n-m+1)!}\dfrac{n(n-1-r)!}{q!r!(n-q)!}(-1)^{i+k+n+m+p+q+1} \\ \times\binom{-i}{k}\binom{-j}{l}B^{k+l+m+p}G_s^{n+r-q}a^{-2i-2j-k-l-m-p-q-r}\end{array}\right]$$

and, $D = 4-(1+G_s)^2G_s^{k-1},\ a = \left(s+\sqrt{s^2-\sigma^2}\right)/\sigma$. (14)

When (11) no longer converges, a different set of algebraic expansions are used to write the laplace domain solution for higher order reflections as

$$V_2^{'}(s) = \sum_{n=1}^{g}\left[\left(S_D^{'}\left(a^{i+j-2m-p-q}\right)+S_D^{''}\left(a^{i+j-2m-p-q}\right)\right)V_{inf}^{'}\left((2n+1)L,s\right)\right] \tag{15}$$

where $S_D^{'}$ and $S_D^{''}$ are the following functional notations used for compactness;

$$S_D^{'}\left(a^{i+j-2m-p-q}\right) = \sum_{i=0}^{n}\sum_{j=0}^{\infty}\sum_{m=0}^{i+j}\sum_{p=0}^{n}\sum_{q=0}\left[\begin{array}{l}G_s^{n+q-p}\dfrac{n(n-1+j)!}{i!j!(n-i)!}\dfrac{n(n+1-q)!}{p!q!(n-p)!} \\ \times(-1)^{i+j+p+m}B^{-i-j}\binom{i+j}{m}a^{i+j-2m-p-q}\end{array}\right] \tag{16}$$

$$S_D^{''}\left(a^{i+j-2m-p-q}\right) = \sum_{i=0}^{n+1}\sum_{j=0}^{\infty}\sum_{m=0}^{i+j}\sum_{p=0}^{n}\sum_{q=0}^{\infty}\left[\begin{array}{l}G_s^{n+q-p}\dfrac{(n+1)(n+j)!}{i!j!(n+1-i)!}\dfrac{n(n-1-q)!}{p!q!(n-p)!} \\ \times(-1)^{i+j+p+m}B^{-i-j}\binom{i+j}{m}a^{i+j-2m-p-q}\end{array}\right]. \tag{17}$$

Using (4) and (5), the corresponding time domain solution can be written as

$$V_{fin}(L,t>3t_f) = e^{\frac{-rt}{2l}}V_2^{'}(t),\quad V_2^{'}(t) = L^{-1}\left[V_2^{'}(s)\right]. \tag{18}$$

iii. Comprehensive model

The comprehensive model consists of three separate models which are tabulated in Table I. The choice of model depends on conditions which are developed based on simple convergence rules applicable to summations of infinite series.

Model 1 converges if $t \leq t_1$, where $t_1 = \dfrac{64}{M\sigma}+t_f$ (19)

while Model 2 converges if $t \le t_2$, where $t_2 \quad \frac{100}{11}\frac{2}{B\sigma}+3t_f$ (20)

Using (19) and (20), the comprehensive model can be described as in Table II. The comprehensive model is compared to HSPICE simulations for two values of the load capacitance in Fig. 3.

iv. Basis variables

Four dimensionless basis variables R_{ratio}, C_{ratio}, T_{ratio} and R_T, are defined as follows :

$$R_{ratio}=\frac{rL}{Z_o}\frac{\text{line resistance}}{\text{characteristic impedance}},\quad C_{ratio}=\frac{C_L}{cL}\frac{\text{load capacitance}}{\text{line capacitance}} \tag{21}$$

$$T_{ratio}=\frac{t}{t_f}=\frac{\text{time}}{\text{time of flight}},\quad R_T=\frac{R_{tr}}{Z_o}=\frac{\text{driver output resistance}}{\text{characteristic impedance}}$$

The complete transient model can be expressed in terms of just these 4 basis variables. These basis variables not only simplify the equations but also lend physical insight into the interplay of the various line parameters.

3. UNIFIED TIME DELAY MODEL

The 50% rise time t_d can be obtained by equating the transient voltage at the end of the line to $V_{DD}/2$. However, a simplified expression yields more physical insight and is therefore valuable. Using exponential series expansion, the 50% rise time in distributed *rlc* lines is simplified to

$$t_{d,ratio}=t_d/t_f \quad 1 \quad 0.693C_{ratio}\left(R_{ratio} \quad 0.65R_T \quad 0.36\right) \tag{22}$$

Distinction between *rlc* lines (with considerable inductive effects) and *rc* lines (with dominant resistive effects) is achieved by evaluation of the following condition :

Condition I : "*If* $0.377R_{ratio}+0.693R_T \quad 1$, *use rlc models; else use rc models.*"

Therefore the *unified* time delay model for *rc* and *rlc* lines is :

$$t_{d,ratio}=t_d/t_f \quad \max\left(1,\ 0.377R_{ratio} \quad 0.693R_T\right.$$
$$\left.+0.693C_{ratio}\left(R_{ratio}+0.65R_T+0.36\right)\right. \tag{23}$$

The rise time in (23) can be interpreted as consisting of two parts: (a) time for the signal to reach the load end of the line and (b) time to charge up the load capacitance. The time for signal propagation through the interconnect is dictated by time-of-flight for *rlc* lines and by time to charge up the line capacitance in *rc* lines. Fig. 4 compares the model in (23) with the time delay calculated by HSPICE; for $0.1 \le R_{ratio} \le 5$, $0 \le C_{ratio} \le 0.1$ and $0.25 \le R_T \le 1$, the error is less than 2%.

4. OPTIMUM REPEATER INSERTION

Based on an analysis similar to that in [2], expressions for repeater insertion in *rlc* lines are derived. For an interconnect with k repeaters, each h times the size of a minimum feature-size repeater, the condition for quantitative differentiation between *rc* and *rlc* lines is

Condition II : "*If* $0.377R_{ratio}/k+0.693R_T/h \le 1$, *use rlc models; else use rc models.*"

The unified expression for time delay is (Fig. 5)

$$t_{d,ratio}=t_d/t_f \quad \max\left(1,\ 0.377R_{ratio}/k \quad 0.693R_T/h\right)$$
$$+0.693C_{ratio}\left(hR_{ratio}+0.65kR_T+0.36\right). \tag{24}$$

Total time delay is minimized when each interconnect segment operates at time of flight (ToF) i.e. Condition II is satisfied in the equality for each segment – providing a relation between number and size of repeaters. Using this condition, the total time delay is minimized to calculate the optimum number (k_{opt}) and size of repeaters (h_{opt}) and optimum time delay ($t_{d,opt}$):

$$k_{opt}=0.9513R_{ratio},\ h_{opt}=1.208R_T$$
$$\text{and } t_{d\_opt,ratio}=t_{d\_opt}/t_f=1+1.5532C_{ratio}R_{ratio}R_T \tag{25}$$

The equivalent *rc* optimal repeater expressions from [2] are:

$$k_{opt,rc}=\sqrt{(0.377R_{ratio} \ / \ 0.693R_TC_{ratio}},\ h_{opt,rc} \quad \sqrt{R_T/\left(R_{ratio}C_{ratio}\right)} \tag{26}$$
$$\text{and } t_{d\_opt,ratio,rc}=2.5\sqrt{C_{ratio}R_{ratio}R_T}$$

Using *rc* models for repeater insertion [2] may sometimes violate ToF operation, whereas *rlc* repeater models ensure compliance with ToF operation and do away with the excessive number of repeaters, resulting in a decrease in repeater count. Substituting (26) in Condition II gives the expression for differentiation between *rc* and *rlc* lines for optimum repeater insertion :

Condition III : "*If* $1.332\sqrt{R_{ratio}C_{ratio}R_T} \le 1$, *use optimum rlc models (25); else use optimum rc models (26).*"

The new models result in a fewer number of repeaters as well as smaller size of repeaters compared to [6] and [3], as can be seen from Fig. 6 and Fig. 7, decreasing total area needed for repeaters.

Table I. Composition of the 3 models

	$t \le 3t_f$	$t > 3t_f$
Model 1	Eqn (7)	Eqn (11)
Model 2	Eqn (10)	Eqn (11)
Model 3	Eqn (10)	Eqn (18)

Table II. Comprehensive model

	Comprehensive Model		
	For $t \le t_1$	For $t_1 \le t \le t_2$	For $t_2 < t$
If $t_1 < t_2$	Model 1	Model 2	Model 3
If $t_1 > t_2$	Model 1		Model 3

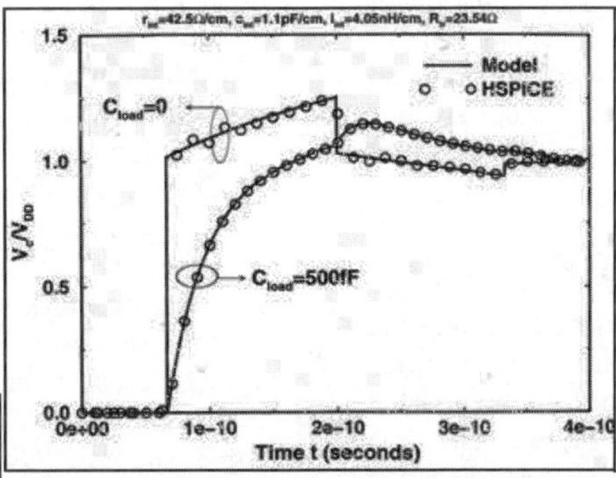

Fig 3. Transient response of a 1cm long interconnect - complete model vs HSPICE simulation.

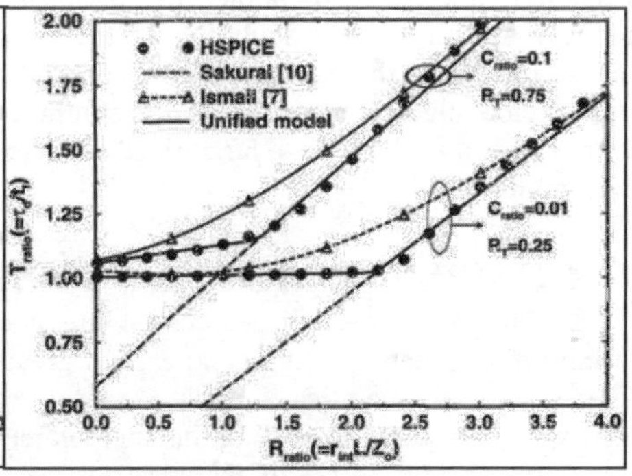

Fig. 4. Normalized time delay vs. R_{ratio} for different values of R_T and C_{ratio}.

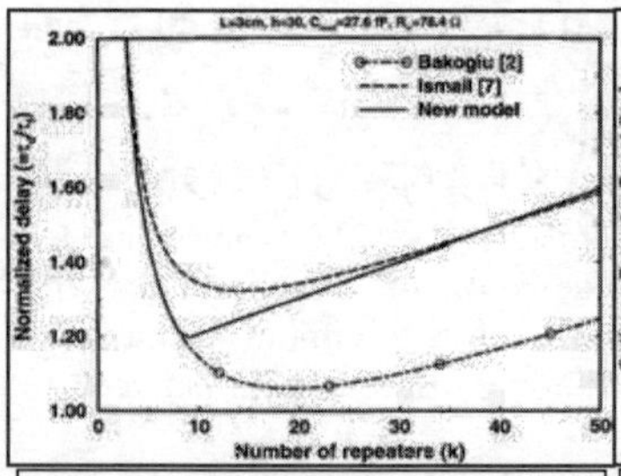

Fig. 5. Normalized delay vs. no. of repeaters, L=3 cm.

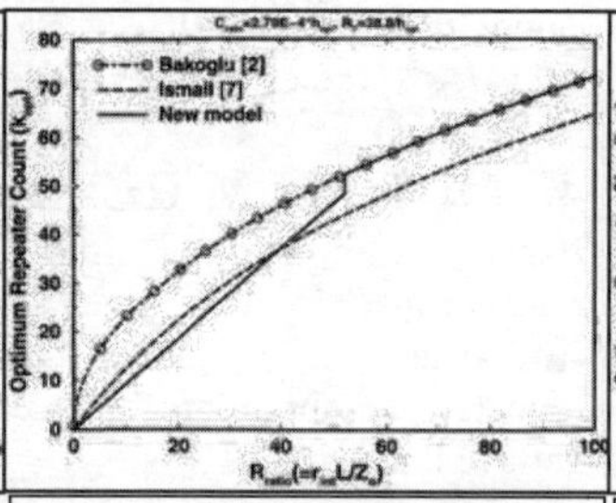

Fig. 6. Optimum number of repeaters vs line resistance.

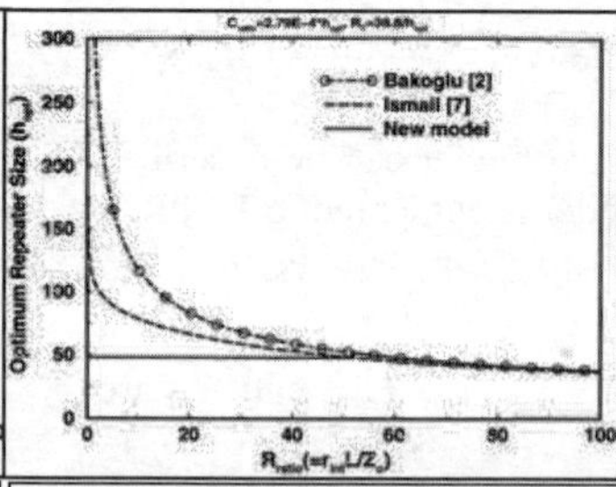

Fig. 7. Optimum size of repeaters vs line resistance.

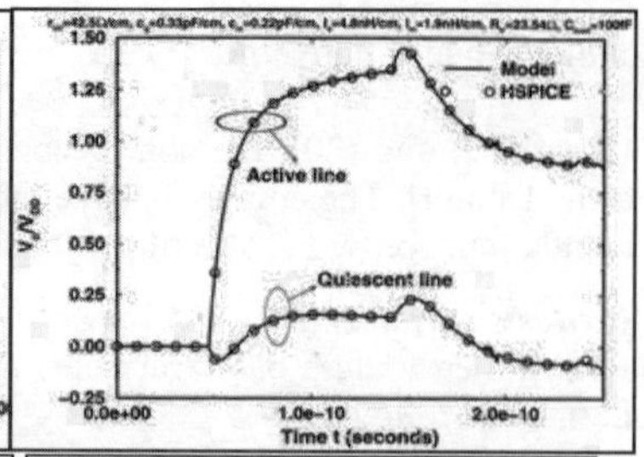

Fig. 8. Complete model vs HSPICE for 2-coupled lines.

5. CROSSTALK IN TWO COUPLED LINES

An important concern in high speed interconnect designs is the peak crosstalk induced in a quiescent line by an active neighbor (Fig 1 b). For two coupled lines, the transient voltage in the active and quiescent lines can be obtained based on an analysis similar to [6]. The differential equations describing signal propagation down two coupled lines is given by

$$\frac{\partial^2 \hat{V}}{\partial x^2} = r[C]\frac{\partial \hat{V}}{\partial t} + [L][C]\frac{\partial^2 \hat{V}}{\partial t^2}, \quad \hat{V} = \begin{bmatrix} V_A(x,t) \\ V_Q(x,t) \end{bmatrix} \tag{27}$$

The solutions for the active and quiescent line are given by

$$V_A(x,t)= 0.5[V_{fin}(r,\ l_s+l_m,\ c_g,x,t) + V_{fin}(r,\ l_s-l_m,\ c_g+2c_m,x,t)] \tag{28}$$

$$V_Q(x,t)= 0.5[V_{fin}(r,\ l_s+l_m,\ c_g,x,t) - V_{fin}(r,\ l_s-l_m,\ c_g+2c_m,x,t)] \tag{29}$$

where $V_A(x,t)$ and $V_Q(x,t)$ are the voltages at the end of the active and quiescent line. Fig. 8 compares the models in (28)–(29) with HSPICE simulations of the transient voltages in two-coupled lines. The presence of a load capacitor causes the peak crosstalk to occur after the second reflection (i.e. $t>3t_f$), whereas it occurs exactly at $t=3t_f$ in the no load case.The worst case time delay occurs when the two coupled lines are switching with opposite polarity; the transient response of the active line for the worst case time delay is

$$V_A(x,t)= V_{fin}(r,\ l_s-l_m,\ c_g+2c_m,x,t) \tag{30}$$

The worst case crosstalk occurs when both the lines are initially uncharged, and the active line charges to V_{DD}; the transient response of the quiescent line for the worst case crosstalk is given by (29). This two line solution is valid for ideal and non-ideal ground return paths because there are no constraints on the relation between the inductance and capacitance matrices.

6. PEAK OVERSHOOT

The single-line and two-coupled line transient expressions enable an easy and accurate calculation of the peak overshoot in the active line. Since the expressions are completely analytical, a fast and efficient algorithm can be used to compute the peak overshoot From Fig 3, it is clear that a larger load capacitance decreases peak overshoot.

7. COMPLETE MODEL VS HSPICE

HSPICE approximates a distributed rlc line by using large number of lumped elements, but at the expense of greater simulation time. The complete analytical model exactly solves the partial differential equation describing the interconnect system. The time stepping algorithm in HSPICE forces it to perform an exhaustive calculation of the transient response over a large range in order to pinpoint characteristics like time delay, crosstalk and overshoot. The ability of this new model to accurately calculate the transient response at an any time instant has the potential to utilize efficient algorithms to rapidly estimate desired interconnect characteristics.

8. CONCLUSIONS

A highly accurate and explicit solution has been rigorously derived for the transient response of single and two-coupled distributed *rlc* interconnects with a *capacitive load* termination. These solutions are validated by HSPICE simulation, and have been used to derive a unified model for rise time and repeater insertion. These new solutions have potential applications in CAD for rapid timing analysis, global wire sizing and repeater insertion, signal integrity estimation, and reliability modeling (e.g. voltage overshoot and high current density concerns).

9. REFERENCES

[1] F.E.Anderson, J.S.Wills and E.Z.Berta, "The core clock system on the next generation Itanium microprocessor," in Proc. of ISSCC, pp. 146-147, 2002.

[2] H.B.Bakoglu, "Circuits, interconnections and packaging for VLSI," Reading, MA: Addison Wesley, 1990.

[3] K.Banerjee and A. Mehrotra, "Analysis of on chip inductance effects using a novel performance optimization methodology for distributed RLC interconnects", in Proc. of DAC, pp.798-803, 2001.

[4] K.Banerjee and A. Mehrotra, "Accurate analysis of on chip inductance effects and implications for optimal repeater insertion and technology scaling," in Proc. of symposium on VLSI circuits, pp. 195-198, 2001.

[5] Y.Cao, X.Huang, D.Sylvester and C.Hu, "A new analytical delay and noise model for on-chip *RLC* interconnect," Proc. of IEDM, pp. 823-826, Dec. 2000.

[6] J.A.Davis and J.D.Meindl, "Compact distributed *rlc* interconnect models - Part I: Single line transient, time delay and overshoot expressions," IEEE Trans. Elect. Dev., Nov 2000, pp. 2068-2077.

[7] Y.I.Ismail and E.G.Friedman, "Effects of inductance on the propagation delay and repeater insertion in VLSI circuits," IEEE Trans. VLSI Sys., Apr 2000, pp.195-206.

[8] A.Kowalczyk et al., "First generation MAJC dual microprocessor," in Proc. of ISSCC, pp. 236-237, 2001.

[9] G.E.Roberts and H.Kaufman, *Table of Laplace Transforms*, W.B.Saunders Company, 1966.

[10] T.Sakurai, "Closed form expressions for interconnect delay, coupling and crosstalk," TED, Jan 1993, pp. 118-124.

HSpeedEx: A High-Speed Extractor for Substrate Noise Analysis in Complex Mixed-Signal SOC

Adil Koukab, Catherine Dehollain, and Michel Declercq
Swiss Federal Institute of Technology (EPFL), Electronics Laboratory CH-1015 Lausanne, Switzerland
adil.koukab@epfl.ch

ABSTRACT: The unprecedented impact of noise coupling on Mixed-Signal Systems-On-a-Chip (MS-SOC) functionality, brings a new set of challenges for Electronics Design Automation (EDA) tool developers. In this paper, we propose a new approach which combines a thorough physical comprehension of the noise coupling effects with an improved Boundary-Element-Method (BEM) to accelerate the substrate model extraction and to avoid the dense matrix storage. The low computational efforts required, as well as speed and accuracy reached, makes this method a highly promising alternative to verify complex MS-SOCs.

Categories and Subject Descriptors

J. [Computer Applications]: J.6 Computer-Aided Engineering – Computer-aided design (CAD).

General Terms: Algorithms, Design, Theory, Verification.

Keywords: Substrate coupling, noise, substrate noise, Mixed-Signal noise, Boundary-Element-Method, Numerical analysis, switching circuits, supply noise.

1. INTRODUCTION

The relentless move to a single-chip integration of such heterogeneous designs as Digital, Analog, RF front-ends, etc. has opened the door to a host of challenging substrate noise coupling effects. In fact, the large cross-talks and the power supply noise generated by digital blocks, can be transmitted to sensitive analog sections through the substrate, resulting in malfunction of the system and even its instability.

A most precise solution for substrate modeling certainly comes from the use of solver based on finite difference method (FDM) [1-4]. However, despite the improvements recently realized in this domain [1-4], the computational effort involved in computing by field solvers is still considerable. In addition, even if the size of the circuits enable the substrate modeling, the storage and the manipulation of the enormous N×N matrix of coupling resistances, is impractical [5]. Several approaches were proposed to deal with the complexity encountered when simulating large dense coupling networks [6-9]. Recently [6], an approach inspired by wavelets to sparsify the dense conductance matrix, was proposed. Other approaches, which are worthy for further investigation, are presented in [5]. Boundary-Element Methods (BEM) have been successfully adapted to the substrate-modeling problem [10-12].

This work has been supported by the Swiss Office Fédérale de l'Education et de la Science (N OFES 99-0336-2)

Storage and inversion of the resulting impedance matrix, however, make the required computational effort often prohibitive and full-chip MS-SOC analysis intractable.

In this paper, we propose a new approach which combine a thorough physical comprehension of the noise coupling effects with an improved Boundary-Element-Method (BEM) to accelerate the substrate model extraction and to avoid the dense matrix storage. Our substrate analysis package HspeedEx uses the algorithms of the method in an efficient and design-oriented methodology. The unprecedented low computational efforts required as well as speed and accuracy reached by this method, make that currently it can be one of the most promising alternative, able to verify complex MS-SOCs.

2. BACKGROUND

Consider a system of M planar contacts, decomposed into N panels on the top of a substrate. For such a system, the relation between the N total panels potentials (Φp ε □^{N}) and N total currents through each panel (Ip ε □^{N}) is given by $\Phi p = Zp.Ip$. Where Zp ε □$^{N\times N}$ is the impedance matrix. Obtaining each entry z_{ij} in this impedance matrix requires computing an integral involving the Green's function over the appropriate panel surfaces as shown by Eq. 1.

$$z_{ij} = \frac{\overline{\varphi_i}}{I_j} = \frac{1}{S_i S_j} \int_{s_i}\int_{s_j} G(s_i, s_j) ds_j ds_i \; ; G = \sum_{m,n=0}^{\infty} \left\{ \begin{matrix} f_{mn}\cos\left(\frac{\alpha x}{a}\right)\cos\left(\frac{\alpha x'}{a}\right) \\ \cos\left(\frac{\beta y}{b}\right)\cos\left(\frac{\beta y'}{b}\right) \end{matrix} \right\} \quad (1)$$

Where $\alpha = m\pi/a$, $\beta = n\pi/b$, a and b are the substrate lateral dimensions, s_i and s_j the surfaces of the panels i and j. f_{mn} is computed using a recursion formulas as shown in [10]. It is also demonstrated in [10] that z_{ij} can be expressed as a function of 64 2–D discrete cosine transform (DCT) coefficients (K(p,q)), with

$$K(p,q) = \sum_{m=0}^{P-1}\sum_{n=0}^{Q-1} k_{mn}\cos\left(m\pi\frac{p}{P}\right)\cos\left(n\pi\frac{q}{Q}\right) \quad (2)$$

k_{mn} is a function of f_{mn}, and the 64 (p , q) terms are determined from the ratio of contacts coordinates and substrate dimensions. For more details see [10] and [11]. A high-speed computation of those coefficients can be made using the fast Fourier transform, FFT [10]. Once the impedance matrix Zp is computed, one needs to invert it in order to generate the admittance matrix Yp.

The inversion of the impedance matrix and its storage is the time and memory hungriest stage of BEM . Iterative algorithms such as Generalized Minimum Residual algorithm, GMRES [13] can be used to speed up the computation. Another approach consists of the use of the sparsification via eigendecomposition technique [12], to enhance the scope of applicability of this technique to more large circuits. Despite these improvements, the extractions process remains too slow to perform a practical full-chip analysis accurately [12]. It's worth noting that the substrate model generated by BEM, as any electromagnetic model, often introduces physical and numerical modeling errors. Some precautions can be taken to limit these errors. For instance, we can, force GEMRES to converge to very tight tolerances. On the other hand, numerical instability problems can be avoided by using a particular green function [11]. Despite all these precautions, BEM continues to have difficulties to handle the large circuits.

3. SUBSTRATE MODELING FLOW

In this section, an efficient hierarchical modeling methodology is proposed. The substrate coupling verification flow is broken into a set of independent modeling stages as shown in Fig. 1.

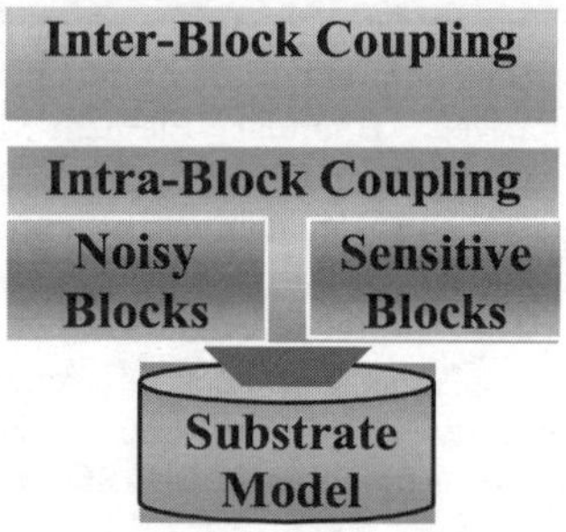

Fig. 1 Substrate Modeling Flow

The crucial result, we demonstrate here, is that the accuracy parameters governing the global substrate coupling (i.e., the inter-block coupling) are different from those governing the local transistor level coupling (i.e. intra-block coupling). As result, we can accurately analyze the inter-block substrate coupling with a coarse local intra-block coupling representation and conversely.

3.1 Inter-block Coupling fundamentals

Several investigations of the substrate noise coupling process are performed in order to capture their fundamental characteristics. The designed substrate noise evaluation chip (Fig. 2) include N inverters (N varies from 12 to 1200). Vin1, Vin2, and Vout are respectively the on-chip ground, the on-chip Vcc of the circuit and an on-chip ground (GND) node representing a sensitive node. A simulation Vout/Vin1 is performed for a chip with 12 to 1200 inverters and shown in Fig. 3. For the Simplified substrate Model (SM) only ground contacts of the layout are considered. For the Full substrate Model (FM), the whole layout (considering PMOS, NMOS, Wells, Vcc and Ground contacts) is used. In both cases the netlist of the circuits are added to the substrate models generated by FDM, to simulate the transfer functions.

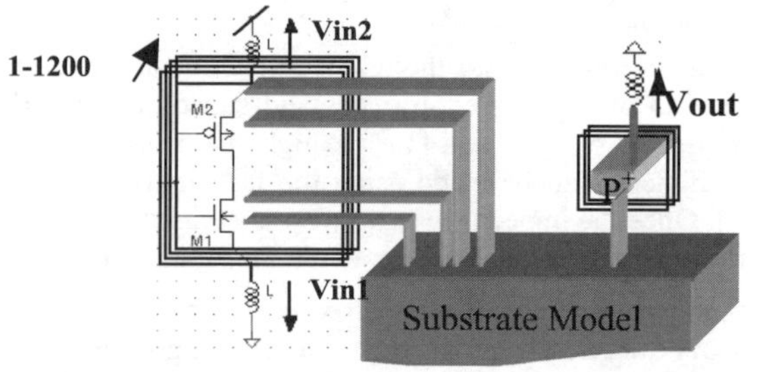

Fig. 2 Overview of substrate coupling evaluation chip with a typical package parasitic (wire inductance L =5nH).

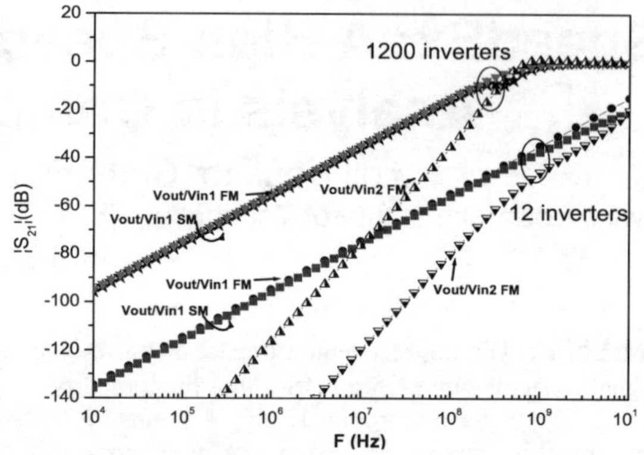

Fig. 3 The transfer function Vout/Vin1 and Vout/Vin2 for 12 and 1200 inverters: using simplified (SM) and full substrate model (FM)

As shown in Fig.3, the simplified and full substrate models show an excellent agreement for all frequencies and numbers of inverters considered. This is due to the fact that: the coupling path from Vin1 to Vout can be decomposed into N parallel paths, and each path can also be fragmented into an indirect path through the NMOS (R_t $+1/j\omega C_t$) in parallel with a direct path through the substrate (R_s). The term ($1/j\omega C_t+R_t$) is much larger than R_s for two reasons: the low value of C_t (~fF) and high value of R_t (indirect path). At high frequency, however, we should consider also the Vcc to Vcc and Vcc to ground coupling. In fact, in high frequency range the ground to ground isolation (Vout/Vin1) is in the same order of magnitude as the Vcc to ground isolation (Vout/Vin2) (see Fig. 3).

3.3 Inter-block Coupling: Numerical analysis

In this section, we describe the algorithm proposed to extract the substrate model. As demonstrated in the last section, a layout with wells, Vcc , and ground contacts, is sufficient to have an accurate representation of the inter-block substrate coupling. Despite these simplifications of the layout, the network to generate remains too dense to enable the targeted full-chip MS-SoC analysis. Therefore, further progress in numerical methods is needed in order to reduce computational efforts. The crucial observation we make here, is that the ground substrate contacts (respectively Vcc contacts) of each block of the chip are linked by metal lines and thus all substrate coupling paths between them are shorted. We can, thus, consider the ground contacts (respectively Vcc contacts) of each block (supposed to have its own on-chip ground bond-path) as the same contact when we perform inter-block substrate coupling. Consequently, the discretization explained in Fig. 4 is sufficient for an accurate inter-block coupling representation. In fact, since we focus on the coupling path between the various blocks of the chip, we can consider that the current at the contacts positioned in the edges of each block is very high, compared to the currents of the contacts situated in its center. Therefore, the edge contacts are the most dominant paths between blocks. This is the reason why a fine partition should be used in the edge region. More we go towards the center of the block, more the role of the contacts in the inter-block coupling is becoming weak, and thus, more we can use a coarse partitions. The currents at the ground contacts (respectively Vcc contacts) of each partition are considered to be constant.

The question that emerges now is: how to exploit this partition to speed up the numerical computation?. Let us consider two partitions of the chip i and j, having respectively as number of contacts L and M (Fig. 4).

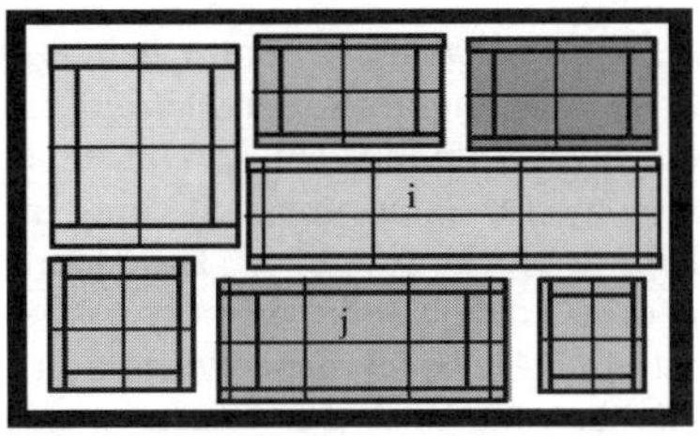

Fig. 4 MS-SOC partitioning example for inter-block coupling analysis

Since we assume that the currents of the L contacts (respectively the M contacts) is constant, the impedance representing the substrate coupling between the two partitions can be defined as

$$z_{ij} = \frac{\overline{\phi_i}}{I_j} = \frac{\overline{\phi_j}}{I_i} = \frac{1}{\sum_{l=1}^{L} S_l \sum_{m=1}^{M} S_m} \int_{\sum_{l=1}^{L} S_l} \int_{\sum_{m=1}^{M} S_m} G(s,s')ds\,ds' \qquad (3)$$

S_l and S_m are the surfaces of each contact in the partitions i and j ($l \in [1,L]$ and $m \in [1,M]$). From a decomposition of the integrals on the sum of integrals on each contact of partitions we obtain

$$z_{ij} = \left(\sum_{l=1}^{L} S_l \sum_{m=1}^{M} S_m \right)^{-1} \sum_{l=1}^{L} \sum_{m=1}^{M} \int_{S_l} \int_{S_m} G(s,s')ds\,ds' \qquad (4)$$

The double integrals in this equation represent the impedance between the contact l of partition i and contact m of partition j. Thus Z_{ij} can be represented by a sum of impedances between the contacts of each partition as

$$z_{ij} = \sum_{l=1}^{L} \sum_{m=1}^{M} z_{lm} s_l s_m \Big/ \sum_{l=1}^{L} \sum_{m=1}^{M} s_l s_m \qquad (5)$$

Using the same procedure with some algebra, we can relate z_{ii} to the impedance between the contacts of the partition i. Thus, we obtain

$$z_{ii} = \left(\sum_{l=1}^{L} s_l^2 \right)^{-1} \left(2 \times \sum_{l=1}^{L} \sum_{j=l+1}^{L} z_{lj} s_l s_j + \sum_{l=1}^{L} z_{ll} s_l^2 \right) \qquad (6)$$

The gain in computational cost is evident, since we transform an impedance matrix of M^2 elements, with M the number of contacts, to an impedance matrix of P^2 elements, with P the number of partitions. Suppose that the chip of Fig. 4 is composed of one million of contacts. The memory requirements is O (10^{12}) space to store the impedance matrix entries. If each block is decomposed into 1000 partitions, with a chip of 7 blocks, the memory requirement is O ($(7.10^3)^2$). The memory gain is, thus, considerable. The inversion requirement by LU factorization, for example, is O (10^{18}), and the gain in execution time by our procedure is more important.

3.4 Intra-block transistor-level coupling

For the intra-block coupling, we will distinguish between the noisy blocks modeling and the sensitive blocks modeling.

34.1 Noisy blocks

In addition to power supply noise, the switching activity of digital blocks injects spurious signals in the substrate through the reverse junction capacitances of transistors or by impact ionization (hot carriers). Transistor Neighboring substrate contacts pick up the most portion of these signals. On the other hand each transistor is linked to the substrate ground by (Rs+1/jCω). As C is of about few fF, the substrate resistances play a negligible role in this coupling. In conclusion, for all noisy circuits, we can connect the bulk of NMOS transistors directly to ground (respectively PMOS to Vcc) without any significant loss of accuracy.

3.4.2 Sensitive analogue blocks

The sensitive analogue blocks may require a more accurate substrate model, since the former can have a direct effect on their performances. For instance, the noise figure of an LNA can be affected by the substrate resistance thermal noise. The substrate resistances can also result in a change in the input/output matching of LNA, reducing its gain and its reverse isolation. The methodology for an accurate transistor level substrate modeling in analogue blocks is the following one: once the inter-block coupling is finished, we eliminate all the Vcc to GND resistances, of the targeted sensitive circuit, generated by this process. We keep only resistances representing the inter-block coupling. The partitioning process used for transistor-level coupling analysis is exactly the inverse of the one used for inter-block coupling, that is, we use a partition which becomes increasingly coarse as we go far from the handled analogue block. An example is shown in Fig.5.

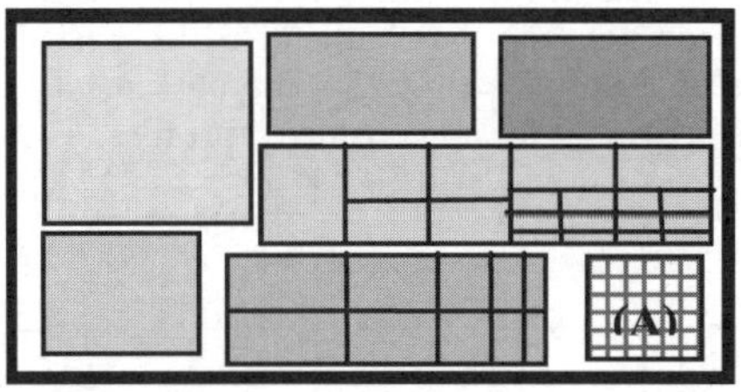

Fig. 5 MS-SOC Partitioning example for fast intra-block coupling analysis: (A) is the targeted sensitive block

The partition around the handled parts are only used for modeling the effects of distant contacts(i.e. the inter-block resistances coupling) on the intra-block substrate coupling. As for inter-block strategy, the currents in the partitions around the handled circuit are assumed to be constant, and the same analytical formula (Equ. 5 and 6) will be used to speed up the computation and avoid a large ill-conditioned matrix.

4. EXPERIMENTAL RESULTS

In this section, we present examples that show the accuracy and the efficiency of the methodology proposed in this paper. We focus on the inter-block substrate modeling, since it is the crucial process, enabling to evaluate the efficiency of the isolation strategy adopted, such as the flooplaning, the block placement, the guard ring distribution, the multiple bond-path assignment etc [14-15].

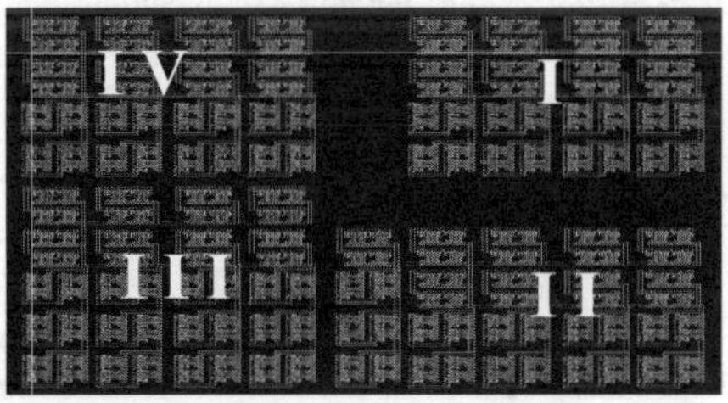

Fig. 6 Overview of substrate coupling evaluation chip: Test structure [4 blocks(I,II,III,VI) - 864 Inverters]

The test layout used for preliminary verifications is presented in Fig.6. Several experiments, using three substrate doping profiles (P1, P2, P3 in Fig. 7), has been conducted. The small thickness of P3 will enable a fine meshes discretization, for FDM analyses.

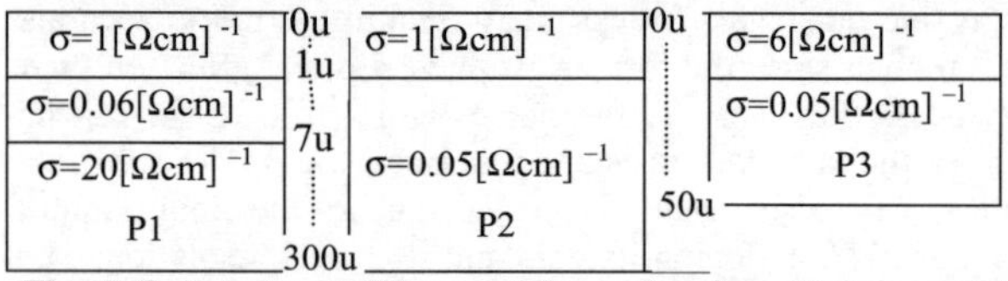

Fig. 7 Substrate profiles used for extraction: P_1 low resistivity; P_2 high-resistivity; P_3 high_resistivity and low-thickness substrate

For each of the substrate profiles indicated, extraction was performed between the GND and the Vcc contacts of the blocks. Table I (a) and I (b) show the inter-block computed resistances with BEM and the proposed method (FastBEM), of respectively substrate P_1 and P_2 of Fig. 7. R I-0, for instance, is the computed resistance between the ground contacts of block I and the backplane of the substrate and R I-II is the resistances between the ground contacts of block I and those of block II. Table II shows a comparaison between the results of BEM, FastBEM, and FDM in the case of P_3 profile (Fig. 7). The agreement between the three methods is evident for all considered profiles.

R	BEM	FastBEM
I-0	95	95
II-0	76	76
III-0	75	75
VI-0	94	94
I-II	9119	9180
I-III	10642	10694
I-VI	12325	12743
II-III	6829	6930
II-IV	10670	10740
III-IV	7932	7919

(a) low-res. profile P_1

R	BEM	FastBEM
I-0	791	801
II-0	721	717
III-0	746	746
VI-0	835	836
I-II	1587	1712
I-III	7412	8113
I-VI	1763	1822
II-III	825	856
II-IV	7330	7836
III-IV	956	1133

(b) high-res. profile P_2

TableI Inter-block substrate resistances of the test-chip

R	FDM	BEM	FastBEM
I-0	117	115	115
II-0	104	104	104
III-0	105	107	108
VI-0	121	121	121
I-II	1172	1036	1067
I-III	-	17069	20050
I-VI	1183	1123	1204
II-III	352	328	349
II-IV	-	14087	15615
III-IV	430	407	412

TableII Inter-block substrate resistances of the test-chip using P_3

Profil(P_3)	K(p,q)	BEM	FastBEM	FDM
Runtime user+sys	1mn 14s	4mn 30s 3mn 16s (*)	1mn 18s 4s (*)	1h 3mn
MaxMem	9328K	17M	9328K	1200M

Table III: Performances comparison between the extractions methods in the case of P_3 profile. (*) i.e. without K(p,q) computation time

Table III summarizes the computational cost of the three methods. "Runtime line" represents both User (the time to execute the command run_extraction_job) and System (the additional system time to complete the job) time of the extraction including post-processing. "MaxMem line" represent the peak memory usage over the total runtime, including input data processing, parasitic extraction, and writing outputs. For run time as for peak memory usage, the supremacy of the FastBEM is evident. It should be noted also that this supremacy is increasingly important as the test cases will include a more larger and more complex designs.

5. CONCLUSION

In this paper, we have presented a new approach, based on an improvement of Boundary-Element-Method (BEM), to accelerate the substrate model extraction and to avoid the dense matrix storage. We have shown that the accuracy parameters governing the global substrate coupling are different from those governing the local transistor level coupling. This has enabled us to build an efficient methodology for inter-block coupling as well as for transistor level intra-block coupling. The supremacy of the method in terms of extraction time, memory usage and accuracy has been demonstrated. The weak dependence of its efficiency on the complexity of the chip, makes it one of the most attractive method to verify MS-SOCs.

References

[1] S. Kapur and D. Long "Large scale capacitance extraction" IEEE/ACM Design Automation Conference 2000, pp. 744-749.

[2] K. Nabros and J. White "Fast capacitance extraction of general three-dimensional structures" IEEE Transactions on Microwave Theory and Techniques, June 1992 vol. 40, pp. 1496-1507.

[3] S. Kpur and D. E. Long "IES^3: Afast integral equation solver for efficient 3-dimensional extraction" IEEE/ACM Int. Conf. On Computer-Aided Design,1997, pp. 448-455

[4] F. Clément ''Computer Aided Analysis of Parasitic Substrate Coupling in Mixed Digital-Analog CMOS Integrated Circuits '' Ph.D. Dissertation No.1449. Swiss Federal Institute of Technology in Lausanne, 1996. ("SubtrateStorm User's Guid" Simplex).

[5] J. R. Phillips and L. Miguel Silveira "Simulation Approaches for Strongly Coupled Interconnect Systems" IEEE/ACM Computer Aided Design International Conf. 2001 pp. 430 -437

[6] J. Kanapka, J. Phillips, and J. White "Fast Methods for Extraction and Sparsification of Substrate Coupling" IEEE/ACM Design Automation Conference 2000, pp. 738-743

[7] M. Chou "Fast Algorithms for Ill-Conditioned Dense-Matrix Problems in VLSI Interconnect and Substrate Modelling" Ph.D. Dissertation, Massachusetts Institute of Technology, 1998.

[8] R. Gharpurey and S. Hosur "Transform domain techniques for efficient extraction of substrate parasitics " IEEE/ACM Int. Conf. On Computer-Aided Design,1997, pp. 461-467

[9] J. P. Costa, M. Chou, and L. M. Silveira "Precorrected-DCT techniques for modeling and simulation of substrate coupling in mixed-signal IC's" IEEE Int. Symposium on ciruits and Systems, 1998, pp. 358-362

[10] R. Gharpurey and R. G. Meyer "Modeling and Analysis of Substrate Coupling in Integrated Circuits" IEEE J. Solid-State Circuits, March 1996, vol .31, pp.344-353

[11] A. M. Niknejad, R. Gharpurey, and R. Meyer "Numerical Stable Green Function for Modeling and Analysis of Substrate Coupling in Integrated Circuits" IEEE Trans. On Computer-Aided Design of Integrated Circuits and Systems, 1998, pp. 305-315

[12] J. P. Costa, M. Chou, and L. M. Silveira "Efficients Techniques for Accurate Modeling and Simulation of Substrate Coupling in Mixed-Signal IC's" IEEE Trans. On Computer-Aided Design of Integrated Circuits and Systems, 1999, pp. 597-607

[13] Y. Saad, and M. H. Schultz "GMRES: A generalized minimal residual algorithm for solving nonsymmetric linear systems" SIAM I. Scientif., Stat. Computing, vol. 7 July 1986, pp. 856-869.

[14] A. Koukab, M. Declercq, and C. Deholain "Analysis and Improvement of the Noise Immunity in a Single-Chip Super-Regenerative Tranceiver" IEE Proc.-Circuits Devices Syst., October 2001, vol. **148**, pp 250-254

[15] S. Mitra, R. A. Rutenbar, L. R. Carley, and D. J. Allstot "A Methodology for Rapid Estimation of Substrate-Coupling Switching Noise" Custom Integrated Circuit Conference 1995 pp. 129-132

Combined BEM/FEM Substrate Resistance Modeling

E. Schrik
Delft University of Technology
DIMES / Circuits & Systems Group
Mekelweg 4, 2628 CD, Delft, The Netherlands
eelco@cas.et.tudelft.nl

N.P. van der Meijs
Delft University of Technology
DIMES / Circuits & Systems Group
Mekelweg 4, 2628 CD, Delft, The Netherlands
nick@cas.et.tudelft.nl

ABSTRACT

For present-day micro-electronic designs, it is becoming ever more important to accurately model substrate coupling effects. Basically, either a Finite Element Method (FEM) or a Boundary Element Method (BEM) can be used. The FEM is the most versatile and flexible whereas the BEM is faster, but requires a stratified, layout-independent doping profile for the substrate. Thus, the BEM is unable to properly model any specific, layout-dependent doping patterns that are usually present in the top layers of the substrate, such as channel stop layers. This paper describes a way to incorporate these doping patterns into our substrate model by combining a BEM for the stratified doping profiles with a 2D FEM for the top-level, layout-dependent doping patterns, thereby achieving improved flexibility compared to BEM and improved speed compared to FEM. The method has been implemented in the SPACE layout to circuit extractor and it has been successfully verified with two other tools.

Categories and Subject Descriptors

I.6.5 [**Simulation and Modeling**]: Model Development—*Modeling Methodologies*; B.7.2 [**Integrated Circuits**]: Design Aids—*Verification*; I.6.4 [**Simulation and Modeling**]: Model Validation and Analysis

General Terms

Verification, Design

Keywords

substrate noise, modeling, finite element method, boundary element method

1. INTRODUCTION

In present-day micro-electronic designs, substrate noise can have a significant impact on the functionality of the design. For example, in a mixed-signal design, the noise injected into the substrate by the digital part can seriously influence the functionality of the analog part. Thus, accurately modeling the behaviour of the substrate as a "noise-propagator" is becoming ever more important in the field of layout-to-circuit extraction [1, 2, 3, 4, 5, 6, 7, 8]. Basically, there are two principal ways of obtaining such a model: the Finite Element Method (FEM; applied in e.g. [6]) and the Boundary Element Method (BEM; described in e.g. [5]). The FEM, which makes a full, in-depth 3D discretization of the substrate, is usually slow, but very versatile and flexible, whereas the BEM, which only discretizes the contact areas on the boundary of the substrate, is much faster but requires a uniform, stratified substrate.

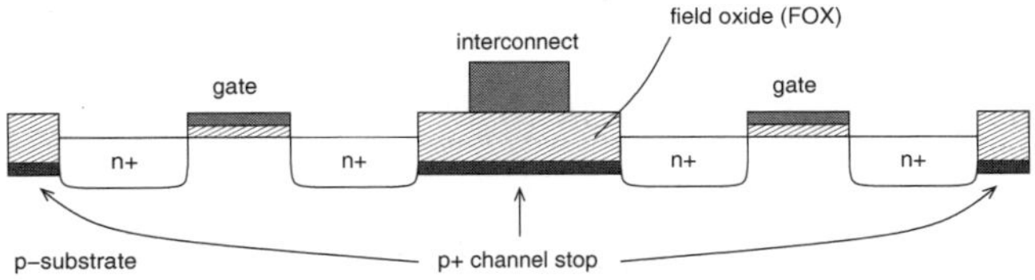

Figure 1: Highly doped (p+) channel-stop layer increases V_T of the parasitic FET being formed under the Oxide.

However, in many cases, a uniform stratified substrate is not a realistic assumption, because specific, layout-dependent doping profiles in the top-layers of the substrate are widely applied. An example of this is presented in Figure 1. It actually shows two layout-dependent doping patterns, i.e. the channel-stopper and the source/drain diffusions. Other examples include trenches, sinkers and buried layers.

While FEM based modeling could accurately model such structures, it can often be too slow. On the other hand, the BEM might not always be accurate enough. Therefore, we have developed a combination of BEM and FEM that is faster than FEM but can be more accurate than BEM based methodologies. We describe a new combination of BEM and FEM for substrate resistance extraction, but note that other BEM/FEM combinations can be found in the literature, see e.g. [9]. Reference [10] presents a combined BEM/FEM approach to capacitance extraction.

In fact, we have developed a 3D BEM/2D FEM combination. That is, we treat the top layer of the substrate as 2 dimensional. This will even be faster than a 3D BEM/3D FEM combination, and we show, by comparison of results obtained by our method to those obtained with Momentum [11] and Davinci [12], that in many practical situations the

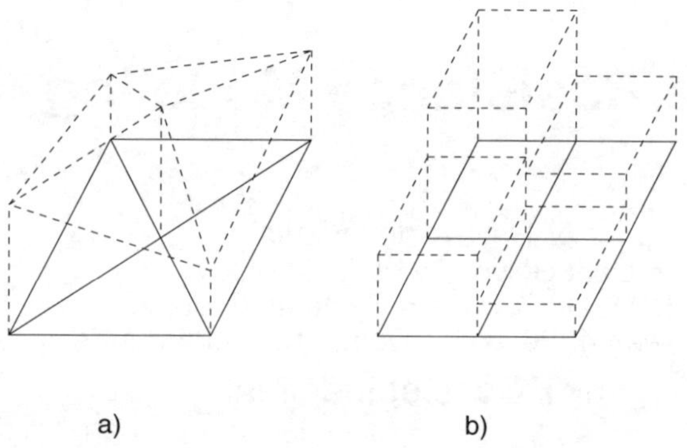

Figure 2: A triangular mesh with a piecewise linear potential distribution for the FEM (a) and a rectangular mesh with a piecewise constant current distribution for the BEM (b).

accuracy is still good enough. The method has been implemented in the SPACE layout-to-circuit extractor [13].

This paper is structured as follows. Section 2 will briefly present some relevant background information on the FEM and BEM methodologies. Section 3 describes our new method. Section 4 presents comparisons of our new method, as implemented in the SPACE layout-to-circuit extractor, to the two other tools mentioned above. Finally, Section 5 states our conclusions.

2. BEM AND FEM BACKGROUND

The finite element method in its simplest form subdivides the conductor domain into triangles (or tetrahedrons in the 3D case) and approximates the potential distribution in the conductor domain by a piece-wise planar (1st order) function that is defined over these triangles (see Figure 2a). The minimization of a functional that relates to the consumed energy in the conductor domain is then used as an alternative formulation for the Laplace equation. This is equivalent to modeling each triangle by a delta-type resistor network [14, 15] and solving the final resistances from the relation between the potentials and currents in the total resistor network.

The so-obtained resistor network is sparse. That is, the number of resistors is proportional to the number of nodes in the network. The network solution is typically accomplished by Gaussian elimination, or equivalently, star-delta transformation. All nodes but the terminal nodes (connected to circuit or device ports) are subsequently eliminated. By exploiting the properties of VLSI interconnect patterns, this can be done efficiently [16, 17] without destroying the sparsity of the network too much.

The BEM for resistance extraction [5] is derived from an integral equation form of the Laplace equation. Note that we will here consider the 3D case. The BEM only requires a discretization of that part of the boundary of the domain where (inhomogeneous) Dirichlet conditions hold. These Dirichlet conditions hold at those parts of the boundary where current can enter/leave the substrate (i.e. at the substrate contacts, which can be formed in many ways, see [4, 1]). At the remainder of the boundary, homogeneous Neumann conditions hold.

The BEM uses a so-called Green's function as the kernel in the integral equation. This Green's function can be interpreted as the potential in a certain point due to a unit current injected somewhere else. In order to approximate an arbitrary current distribution over the Dirichlet part of the boundary, the substrate contacts are discretized into panels and a shape or basis function for the current is assumed. In its simplest form, this models the current as a piecewise constant (0-th order) current distribution (see Figure 2b). That is, the current is constant on each panel but discontinuous between panels. Because of the Dirichlet condition, the potential over a panel must also be constant — the panels form an equipotential region.

Based on this discretization, the collocation method or the Galerkin method [18] can be used to obtain a system of equations with $\mathbf{U}$ as the vector of N panel potentials, $\mathbf{X}$ as the vector of (unknown) panel currents and $\mathbf{G}$ as the elastance matrix describing the potential at panel i due to a unit current in panel j:

$$\mathbf{U} = G\mathbf{X} \tag{1}$$

The BEM then continues by defining an incidence matrix $\mathbf{F}$ relating panels and contacts. By denoting $\mathbf{V}$ as the contact potential vector and $\mathbf{I}$ as the contact current vector, we can write

$$\mathbf{U} = F\mathbf{V} \tag{2}$$

$$\mathbf{I} = F^T\mathbf{X} \tag{3}$$

It follows that

$$\mathbf{I} = F^T G^{-1} F\mathbf{V} = Y\mathbf{V} \tag{4}$$

where $\mathbf{Y}$ is an admittance matrix for the resistive substrate with the substrate contacts as ports.

Thus, we see that also with the BEM the result is a resistance network. Compared to the FEM, however, this is a full resistance network, and each node is in principle connected to every other node. All the nodes are port nodes; there are no internal nodes. Model order reduction should be used to simplify this network, and/or a windowing technique such as [5] can be used to a-priori extract a reduced order model.

3. A COMBINED FEM/BEM METHOD

We will consider layout-dependent doping patterns with the following properties

1. they are situated in the top layer of the substrate
2. they are thin compared to e.g. the thickness of the epi-layer or their horizontal dimensions
3. their resistivity is low compared to the resistivity of the surrounding substrate

Under these conditions, the noise voltages in the neighborhood of such structures are mainly dominated by the shape and resistivity of these structures and the noise current is flowing mainly horizontally along these structures. Thus, we could approximate the substrate behavior locally with a 2D model. Globally, we would still need a 3D model.

The above observation motivates the combination of a local 2D method (i.e. 2D FEM) with a global 3D method (i.e. 3D BEM). In Section 4 we will experimentally investigate the validity of the 2D approach.

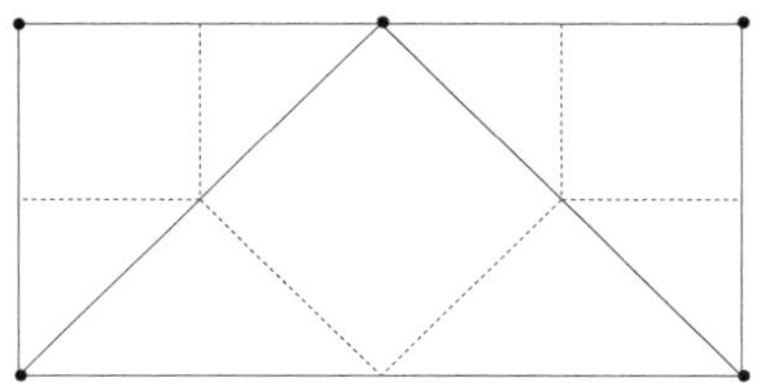

Figure 3: Illustration of a dual FEM/BEM mesh.

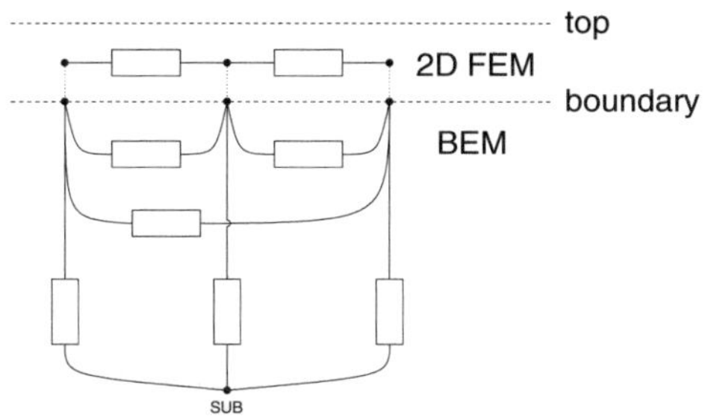

Figure 4: Schematic representation of the combined BEM/2D FEM modeling approach.

For the development of this method, we will consider 1st order elements for the FEM and 0th order elements for the BEM, as they are more common.

Equation 4 actually forces the potential of the panels on the same contact to be equal; contacts form an equipotential region. However, we are free to treat each panel as a separate 'subcontact', also when they touch each other. This corresponds to **F** being the unity matrix. The result is a full network of resistances with all the panels as ports.

Thus, we can identify nodes in the FEM mesh with panels in the BEM mesh, and it is natural to make the meshes each other's dual. Stated otherwise, with each node of the FEM discretization exactly one BEM panel is associated, which geometrically consists of all the points in the plane closer to this FEM node than to any other FEM node. Thus, a BEM panel is actually the Voronoi polygon [19] of a FEM node. In fact, the dual of the Voronoi diagram, with an edge between each node and all the nodes "closest" to it, is a so-called Delaunay triangulation which is in a certain sense optimal for FEM modeling as it avoids sharp angles.

Figure 3 illustrates two dual meshes. The solid lines form the FEM mesh, while the dashed lines form the dual BEM mesh. Note that, in general, the edges from the Voronoi polygons (i.e. BEM panels) are constructed from the perpendicular bisectors of the Delaunay triangulation on the set of nodes.

Given this correspondence of FEM and BEM discretizations, the resistance networks from both procedures may be connected together, as is illustrated in Figure 4. The result is a mixed sparse/dense resistance network, that will model the current flow in the top and deeper layers of the substrate respectively.

Figure 3 illustrates that the resulting BEM panels are, in general, not rectangular. This is allowed from a theoretical point of view, although it makes the numerical calculation of the entries of the matrix **G**, using either the collocation or the Galerkin method, less efficient. One practical way of performing these computations might be by decomposing

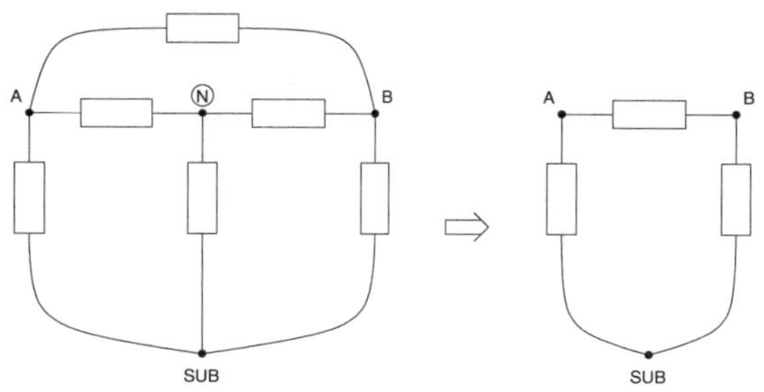

Figure 5: Simple example of the star-delta transformation; by eliminating node N, a simpler but equivalent network remains.

the irregularly shaped BEM panels into independent rectangular and triangular panels and connecting them with the incidence matrix. An alternative solution, but probably less accurate, would be to construct independent FEM and BEM meshes, and to join the associated resistance networks based on proximity of the FEM nodes and, say, the center of gravity of the BEM panels. That is, each BEM panel is associated with the closest FEM node.

The FEM/BEM combination produces a combined resistance network, of which the FEM part is sparse and the BEM part is full. Nearly the simplest of such a combined FEM/BEM network is illustrated in Figure 5a. Here, the resistances between A and N and B and N are parallel connections of a FEM and a BEM resistance, the others are from the BEM only. We assume that nodes A and B are the port nodes, connected to the rest of the circuit. Node N is an internal node.

This network can be simplified using Gaussian elimination or, equivalently, star-delta transformation. The result is shown in Figure 5b. Care should however be taken when interpreting the results quantitatively. This is explained as follows.

Consider the numerical example as in Figure 6. Figure 6a and 6b show the two originally unconnected FEM and BEM networks respectively, where the >> notation means that the value of the resistor is very large compared to the other values. Figure 6c shows the resulting combined network before star-delta transformation. Figure 6d shows how the delta network from eliminating the star node N is connected in parallel to the other resistors, and Figure 6e shows the final result.

Now, it is important to note that the resulting direct resistance between A and B (i.e. 41k) is actually larger than the 28k from the FEM method alone. Thus, by connecting a BEM network to the FEM network, the direct resistance is actually increased. This possibly counter-intuitive fact can be understood by noting that there is also a path from A to B via node SUB. After elimination of SUB, the resistance between A and B would actually be smaller than the 28k obtained from the FEM. In addition, a more detailed analysis shows that when adding more internal nodes the value between A and B actually converges to still greater values.

Furthermore, note that it may not be allowed to eliminate node SUB when it is connected to a bias or ground supply. In that case, the coupling between A and B in Figure 6e would actually be weaker than in the pure FEM case of Figure 6a. This is however precisely what is intended and accomplished by fixing the substrate potential.

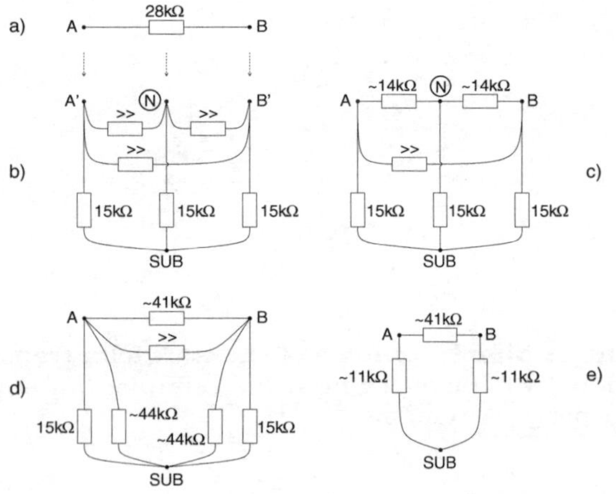

Figure 6: Illustration of star-delta transformation for combined FEM/BEM network

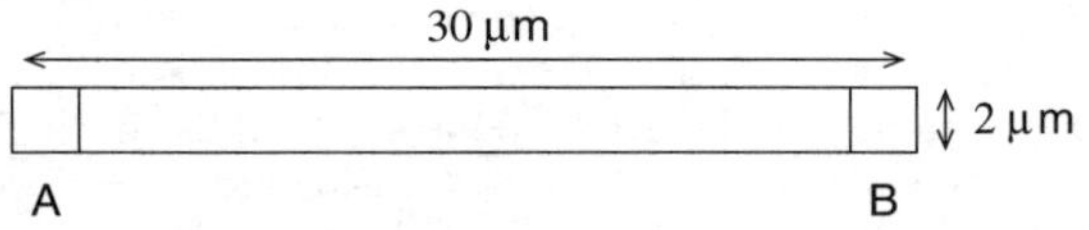

Figure 7: A simple strip of channel-stop with terminals A and B at the ends.

4. RESULTS

We have implemented the new method as described above in the SPACE layout-to-circuit extractor and performed a number of experiments to assess the validity and efficiency of the method.

The convergence behavior as discussed in Section 3 is actually confirmed in the following experiment. Consider the layout shown in Figure 7. It represents a line of channel-stop which has terminals A and B at the ends. The line is $30\,\mu m$ long, $2\,\mu m$ wide and $0.5\,\mu m$ thick, and it has a sheet resistivity of $2000\,\Omega/sq$. The substrate is uniform with conductivity of 10 S/m and a thickness of $250\,\mu m$, with a backplane metalization. The corresponding terminal in the netlists is called 'SUB'. The FEM resistance between A and B (i.e. for the channel stop alone) is 28K. Table 1 shows the result of the combined FEM/BEM method for an increasing number of discretization sections, as indicated in the first column labeled '#'. Columns 2-4 give the resistor values in the extracted netlist. The last column presents the effective resistance value if node SUB would be eliminated. Both columns 2 and 5 clearly show convergence behavior, as also illustrated in Figure 8.

We have verified the results above with Momentum [11], a 2.5D EM simulator for passive circuit analysis. The results

Table 1: Resistances in kΩ.

#	R(A,B)	R(A,SUB)	R(B,SUB)	R_{eff}(A,B)
1	16.93	8.145	8.146	8.302
2	26.90	9.632	9.785	11.28
3	34.09	10.17	10.28	12.78
4	37.63	10.42	10.50	13.45
5	41.07	10.58	10.64	13.99
9	44.91	10.81	10.85	14.61

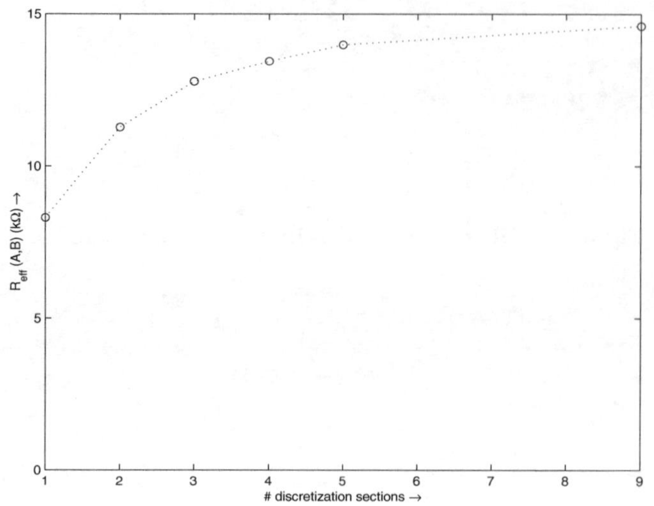

Figure 8: The effective resistance between nodes A and B under increasing refinement.

were as follows:

R(A,B)	R(A,SUB)	R(B,SUB)	R_{eff}(A,B)
40.3 kΩ	11.0 kΩ	11.2 kΩ	14.3 kΩ

Clearly, this data is consistent with the SPACE data for the finer discretization levels.

Figure 9 compares the results of our method, as implemented in SPACE, with those of Momentum as a function of channel stop resistivity. The results were obtained using a 9 section discretization for SPACE and we ensured that the Momentum mesh was comparable (by visual comparison) to the SPACE mesh. Upon varying the resistivity of the channel stop layer from $100\,\Omega$/sq until $10,000\,\Omega$/sq, we see a slight increase of the error, but the difference stays within 6%.

Figure 10 shows another layout where the channel stop region is only located in the middle between A and B but not directly connected. A and B are still of size $2\,\mu m \times 2\,\mu m$ and $30\,\mu m$ apart (center to center). The channel stop region is of size $w \times w$ and has resistivity $1000\,\Omega$/sq.

We have varied w between $0\,\mu m$ (no island present) and $24\,\mu m$, and we have calculated the resistance network between A, B and SUB again, with both SPACE and Momentum. The results can be found in Figure 11. We can see clearly that SPACE and Momentum both show similar behaviour; the difference is typically within 8%. As it appears, there is some offset between SPACE and Momentum. At this moment, we are not yet able to fully explain this offset, but it might be caused by a minor difference in the handling of the terminals (A and B).

Both our new method as well as Momentum both model the channel-stop layer *on top* of the substrate and cannot accurately model an actual doping pattern *in* the substrate. Therefore, we have to investigate how good this is as an approximation. This investigation was performed by comparing our method to Davinci [12], a 3D FEM device simulator.

Davinci, because of its adaptation to the task of device simulation, severely limits the size of the structure that can

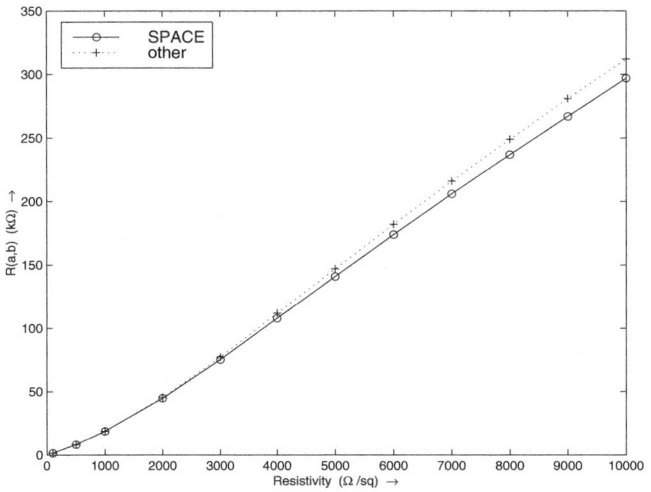

Figure 9: The direct resistance between terminals A and B for increasing values of the resistivity of the channel-stop strip (see Figure 7).

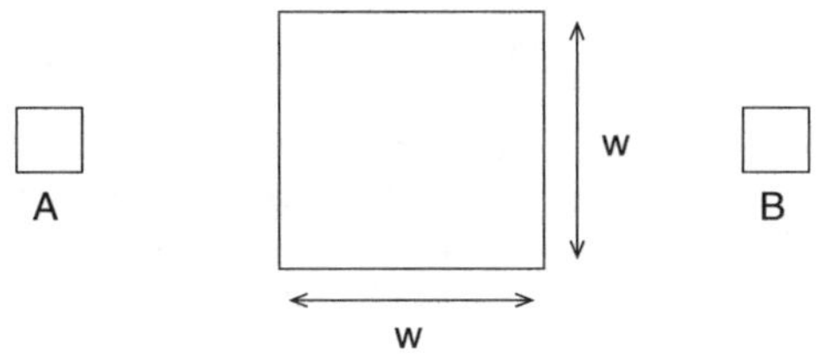

Figure 10: Basic setup of the "island" structure. The parameter "w" determines the size of the island, while keeping it square.

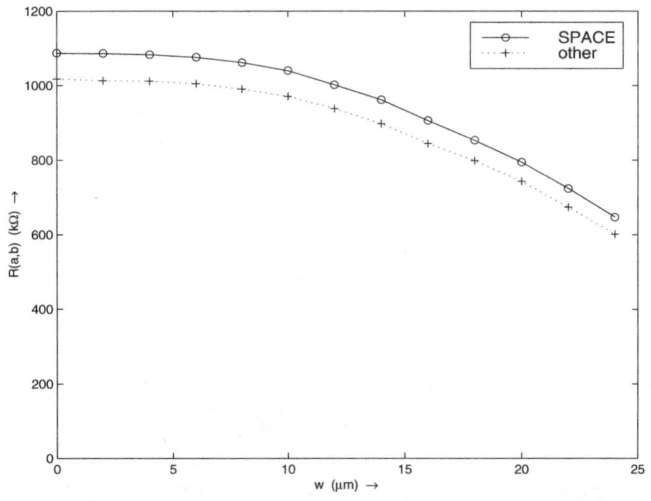

Figure 11: Direct resistance between terminals A and B for increasing values of w (see Figure 10).

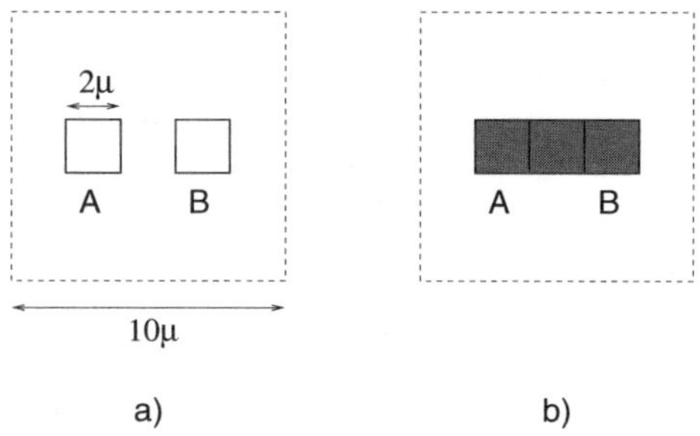

Figure 12: Top view of the experiment setup for the comparison between SPACE and Davinci. a) 2 separate terminals, b) 2 terminals "connected" by highly doped pattern

Table 2: Resistance values and computation times for different experiments. R_{AB} = resistance between terminals A and B, R_{AS} = resistance between terminal A and the backplane.

Experiment	Figure 12a		Figure 12b	
method	SPACE	Davinci	SPACE	Davinci
R_{AB} ($k\Omega$)	187	177	5.22	4.08
R_{AS} ($k\Omega$)	22.2	38.0	21.1	33.0
time (s)	31	351	58	382

be analyzed. We have used a version that is restricted to 30k grid points, and for simplicity we have applied a uniform grid.

The basic layout is shown in Figure 12a. It has two terminals A and B on top of a uniform substrate. All dimensions are (multiples of) $2\,\mu m$, the substrate has a conductivity of 10 S/m and is $5\,\mu m$ thick. Again, there is a backplane metallization. The dashed box in Figure 12 corresponds to the FEM domain boundary. For Davinci, we have assumed n-type silicon with $N_d = 4.6 * 10^{14} cm^{-3}$, which should correspond to the conductivity of 10 S/m.

Figure 12a shows the layout without, and Figure 12b with a channel-stop doping. This doping ($0.5\,\mu m$ thick and with $N_d = 4.6 * 10^{16} cm^{-3}$ which corresponds to a conductivity of 1000 S/m) will "connect" A and B like in the example of Figure 7.

The results can be found in Table 2, which was obtained using suitable mesh settings such that the results from both methods had converged (Davinci used 26011 grid points). The table clearly shows that the results of the FEM method and the FEM/BEM method are reasonably close. There can be many reasons for the remaining differences, including the fundamental fact that the FEM (Davinci) needs a bounded domain (a cuboid) and the BEM (SPACE) assumes a domain laterally extending to infinity.

The bottom row of the table shows that the new method can be considerably faster.

5. CONCLUSIONS

In this paper, we have addressed the problem of how to model layout-dependent doping patterns in the top layers of the substrate. A full 3D FEM method would easily be able to incorporate these doping patterns into a model, but this method is not fast enough for large-scale problems. To increase the speed, a BEM method would be more convenient to use, but this method is not capable of dealing with the layout-dependent doping patterns.

This paper describes a method that makes a combination between a 2D FEM method for the layout-dependent doping patterns and a 3D BEM method for the lower levels of the substrate. Our main assumptions upon which the validity of this approach is based, are that the layout-dependent doping patterns are thin, and that their resistivity is significantly lower compared to the underlying substrate.

A number of experiments have shown that the results of our method compare very well to those generated by Momentum and Davinci, and that the new method will be a valuable addition to the repertoire of methods/models that are useful for layout-to-circuit extraction.

Our future research will concentrate on determining the limitations of our method and how we can circumvent them. In particular, we will investigate the behaviour of our method when varying the thickness and the resistivity of the specific doping patterns and we will investigate the possibilities of extending our 2D FEM/BEM method to a 3D FEM/BEM method for more accuracy and flexibility. Also, the problem of model reduction for the resulting large, mixed sparse/dense networks will have to be addressed as well.

6. REFERENCES

[1] R. Gharpurey and R.G. Meyer, "Modeling and Analysis of Substrate Coupling in Integrated Circuits," *IEEE Journal of Solid-State Circuits*, vol. 31, pp. 344–353, Mar. 1996.

[2] R.B. Merrill, W.M. Young, and K. Brehmer, "Effect of Substrate Material on Crosstalk in Mixed Analog/Digital Integrated Circuits," in *IEEE International Electron Devices Meeting*, pp. 433–436, Dec. 1994.

[3] R. Singh, "A Review of Substrate Coupling Issues and Modeling Strategies," in *Proc. CICC*, pp. 491–498, May 1999.

[4] X. Aragones, J.L. Gonzalez, and A. Rubio, *Analysis and Solutions for Switching Noise Coupling in Mixed-Signal ICs*. Boston: Kluwer Academic Publishers, 1999.

[5] T. Smedes, N. P. van der Meijs, and A. J. van Genderen, "Extraction of Circuit Models for Substrate Cross-talk," in *Proc. Int. Conf. on Computer-Aided Design*, (San Jose, California), pp. 199–206, Nov. 1995.

[6] D.K. Su, M.J. Loinaz, S. Masui, and B.A. Wooley, "Experimental Results and Modeling Techniques for Substrate Noise in Mixed-Signal Integrated Circuits," *IEEE Journal of Solid-State Electronics*, vol. 28, pp. 420–430, Apr. 1993.

[7] F.J.R. Clement, E. Zysman, M. Kayal, and M. Declercq, "LAYIN: Toward a global solution for parasitic coupling modeling and visualization," in *Proc. IEEE Custom Integrated Circuits Conference*, pp. 537 – 540, May 1994.

[8] N.K. Verghese, T. Schmerbeck, and D.J. Allstot, *Simulation techniques and solutions for mixed-signal coupling in ICs*. Boston: Kluwer Academic Publishers, 1995.

[9] C.A. Brebbia, *The Boundary Element Method for Engineers*. Plymouth: Pentech Press, 1978.

[10] P. M. Dewilde and E. B. Nowacka, "Circuit Models for the Hybrid Element Method," in *Proc. Int. Symp. on Circuits and Systems*, (Atlanta, Georgia), pp. IV 616–619, May 1996.

[11] *Momentum, a 2.5 D EM Simulator by Agilent EEsof, see http://eesof.tm.agilent.com.*

[12] *Davinci, a 3D Device simulator by Avanti, see http://www.avanticorp.com.*

[13] F. Beeftink, A.J. van Genderen, N.P. van der Meijs, and J. Poltz, "Deep-Submicron ULSI Parasitics Extraction Using Space," in *Design, Automation and Test in Europe Conference 1998, Designer Track*, pp. 81–86, Feb. 1998.

[14] J.E. Hall, D.E. Hocevar, P. Yang, and M.J. McGraw, "SPIDER - A CAD System for Modeling VLSI Metallization Patterns," *IEEE Transactions on CAD*, vol. 6, pp. 1023 – 1031, Nov. 1987.

[15] A. J. van Genderen, *Reduced Models for the Behavior of VLSI Circuits*. PhD thesis, Delft University of Technology, Delft, The Netherlands, Oct. 1991.

[16] N. P. van der Meijs and A. J. van Genderen, "Delayed Frontal Solution for Finite-Element based Resistance Extraction," in *Proc. 32nd Design Automation Conf.*, (San Francisco, California), pp. 273–278, June 1995.

[17] A. J. van Genderen and N. P. van der Meijs, "Using Articulation Nodes to Improve the Efficiency of Finite-Element based Resistance Extraction," in *Proc. 33rd Design Automation Conf.*, (Las Vegas, Nevada), pp. 758–763, June 1996.

[18] R.F. Harrington, *Field Computation by Moment Methods*. New York: The Macmillan Company, 1968.

[19] F.P. Preparata and M.I. Shamos, *Computational Geometry: An Introduction*. Springer Verlag, 1985.

System Design Methodologies for a Wireless Security Processing Platform

Srivaths Ravi, Anand Raghunathan, Nachiketh Potlapally and Murugan Sankaradass
C & C Research Labs, NEC USA, Princeton, NJ 08540
{sravi,anand,nachiketh,murugs}@nec-lab.com

Abstract

Security protocols are critical to enabling the growth of a wide range of wireless data services and applications. However, they impose a high computational burden that is mismatched with the modest processing capabilities and battery resources available on wireless clients. Bridging the security processing gap, while retaining sufficient programmability in order to support a wide range of current and future security protocol standards, requires the use of novel system architectures and design methodologies.

We present the system-level design methodology used to design a programmable security processor platform for next-generation wireless handsets. The platform architecture is based on (i) a configurable and extensible processor that is customized for efficient domain-specific processing, and (ii) layered software libraries implementing cryptographic algorithms that are optimized to the hardware platform. Our system-level design methodology enables the efficient co-design of optimal cryptographic algorithms and an optimized system architecture. It includes novel techniques for algorithmic exploration and tuning, performance characterization and macro-modeling of software libraries, and architecture refinement based on selection of instruction extensions to accelerate performance-critical, computation-intensive operations. We have designed a programmable security processor platform to support both public-key and private-key operations using the proposed methodology, and have evaluated its performance through extensive system simulations as well as hardware prototyping. Our experiments demonstrate large performance improvements (e.g., 31.0*X for DES,* 33.9*X for 3DES,* 17.4*X for AES, and upto* 66.4*X for RSA) compared to well-optimized software implementations on a state-of-the-art embedded processor.*

Categories and Subject Descriptors

C.0 [**Computer Systems Organization**]: General- *System architectures*; C.1.0 [**Computer Systems Organization**]: Processor architectures- *General*; C.2.0 [**Computer Systems Organization**]: Computer-Communication Networks- *General, Security and protection*; C.5.3 [**Computer Systems Organization**]: Computer System Implementation- *Microcomputers, Portable devices*; E.3 [**Data**]: Data encryption- *DES, Public key cryptosystems*

General Terms

Security, Performance, Design, Algorithms

Keywords

Security, Security processing, Encryption, Decryption, Wireless, Handset, Embedded system, Performance, DES, 3DES, AES, RSA, SSL, IPSec, Design methodology, Platform, System architecture

1. INTRODUCTION

A large fraction of the applications and services that are of interest to Internet users involve access to, and transmission of, sensitive information (*e.g.*, e-commerce, access to corporate data, virtual private networks, online banking and trading, multimedia conferencing, *etc.*), making security a serious concern [1, 2]. The deployment of high-speed wireless data and multimedia communications ushers in even greater security challenges. Wireless communication relies on the use of a public transmission medium, making the physical signal easily accessible to malicious entities. Surveys of current and potential users of mobile commerce (m-commerce) services have indicated security concerns as the single largest bottleneck to their adoption [3].

Several security mechanisms have been developed for wired and wireless networks, based on providing security enhancements to various layers of the

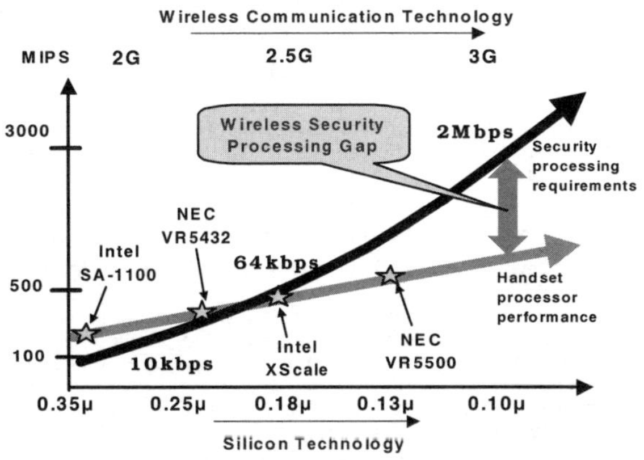

Figure 1: The security processing gap: Projected trends in security processing requirements and embedded processor performance

protocol stack (*e.g.*, IPSec at the network layer, SSL/TLS and WTLS at the transport layer, SET at the application layer, *etc.*) [4, 5]. While the above mechanisms provide satisfactory security if utilized appropriately, there is a critical bottleneck that impedes their use to address security concerns in wireless networks. Wireless clients (*e.g.*, smart phones, PDAs) are, and will always be, much more resource (processing capability, battery) constrained than their wired counterparts. On the other hand, security protocols significantly increase computational requirements at the network clients and servers [6, 7, 8] to levels that exceed the capabilities of wireless handsets. For example, a PalmIIIx handset requires 3.4 minutes to perform 512-bit RSA key generation, 7 seconds to perform digital signature generation, and can perform (single) DES encryption at only 13kpbs, assuming that the CPU is completely dedicated to security processing [8]. Further, security processing has been reported to rapidly drain the Palm's batteries [8].

The increase in data rates (due to advances in wireless communication technologies), and the use of stronger cryptographic algorithms (to stay beyond the extending reach of malicious entities) threaten to further widen the gap between security processing requirements and embedded processor performance (the "*security processing gap*"). Figure 1 compares the projected trends in computational requirements (MIPS) for security processing, and the increase in embedded processor performance (enabled by improvements in fabrication technology and innovations in embedded processor architecture). The inadequate performance of embedded processors in processing security protocols leads to high network transaction latencies, and low effective data rates. Another critical bottleneck to security processing on wireless handsets is battery capacity, whose growth (5-8% per year) is far slower than the growth in processing requirements or processor performance [9]. In practice, various metrics such as performance, power, and cost, need to be considered together and it is their interaction that poses the toughest challenges to the system designer. For example, power and cost are the main reasons why embedded processors for wireless handsets are slower than their desktop counterparts. The proposed system design methodology and security processing platform architecture result in large improvements in performance as well as energy efficiency. However, space restrictions dictate that the discussions in this paper be limited to performance issues.

Algorithm-specific custom hardware implementations can always provide the highest levels of efficiency [10, 11, 12, 13]. However, in practice, the need for efficiency in security processing has to often be considered together with, and traded off against, the need for flexibility. Each security protocol standard typically specifies a wide range of cryptographic algorithms that the network servers and clients need to execute in order to facilitate inter-operability [4, 5]. Further, a security processor is often required to execute multiple distinct security protocol standards in order to support (i) security processing in different layers of the network protocol stack (*e.g.*, WEP, IPSec, and SSL), or (ii) inter-working among different networks (*e.g.*, an appliance that needs to work in both 3G cellular and wireless LAN environments). Finally, programmability is desirable in order to allow easy adaptation to future security protocols and evolving standards. Hence, novel technologies to alleviate the computational burden of security processing while maintaining sufficient programmability are required.

1.1 Paper overview and contributions

We are developing a programmable security processor platform to enable secure data and multi-media communications in next-generation wireless handsets. The objective is to enable secure communications at data rates provided by 3G cellular (100 kbps - 2 Mbps) and wireless LAN (10 - 55 Mbps) technologies, while allowing for easy programmability in order to support a wide range of current and future security protocol standards. As explained above, the growth in computational requirements for security processing outstrips improvements in embedded processor performance, resulting in a significant performance gap. We believe that the use of novel system architectures and system-level design methodologies is critical to bridge this gap.

The system architecture of our security processing platform consists of

- A state-of-the-art commercial configurable and extensible processor (the Xtensa processor from Tensilica Inc. [14]) that is customized for efficient domain-specific processing, while retaining sufficient programmability, and
- Layered software libraries implementing cryptographic algorithms that are optimized and tuned to the underlying hardware platform.

Our system-level design methodology is based on the co-design of optimal cryptographic algorithms and an optimized system architecture. It allows the system designers to efficiently match the software to the characteristics of the hardware platform, and vice-versa. Our methodology includes novel techniques for algorithmic exploration and tuning as well as architecture refinement.

Concurrent development of the security algorithms and the underlying hardware architecture requires that the performance of algorithms be evaluated using either hardware models or instruction set simulation (ISS) models. In such a scenario, algorithmic exploration may be infeasible due to the size of the algorithm space, and the amount of time required to simulate realistic network transactions with hardware models. For example, simulating a single transaction of the SSL handshake protocol over a space of 495 RSA algorithm configurations would require over a month of simulation time with ISS models of the Xtensa processor, on a 440MhZ Sun Ultra 10 workstation with 1 GB memory. We propose a novel methodology to enable efficient and accurate exploration of the algorithm space, based on automatic performance characterization and macro-modeling of software functions that implement the various atomic steps in the cryptographic algorithm.

Architecture exploration is performed in our design flow through the generation and selection of custom instructions that accelerate performance-critical, computation-intensive operations. For programs where several distinct parts (*e.g.* functions) need to be accelerated through custom instructions, the large number of candidate sets of custom instructions make it difficult to evaluate all possibilities explicitly. The problem is further complicated by the fact that, it is often possible to have several different alternative custom instructions for accelerating a single sub-program, which present a tradeoff between the performance improvement and the overheads incurred by the hardware additions. We have developed techniques to automate the selection of custom instructions from a given candidate set, while considering the performance *vs.* hardware overhead tradeoffs.

We have designed a programmable security processor platform to support both private-key (*e.g.*, DES, 3DES, AES) and public-key (*e.g.*, RSA, El-Gamal) operations, using the proposed methodology. We have evaluated the performance of the security processor through extensive system simulations, and through hardware implementation using a prototyping platform. Our experiments demonstrate large performance improvements (*e.g.* 31.0*X* for DES, 33.9*X* for 3DES, 17.4*X* for AES, and upto 66.4*X* for RSA) compared to well-optimized software implementations on a state-of-the-art embedded processor. We believe that system-level design methodologies, such as the one proposed here, are critical to overcoming the challenges encountered in security processing on wireless handsets.

2. OVERVIEW OF THE SECURITY PROCESSING PLATFORM

Figure 2 presents an overview of the target system architecture for our security processor platform. Efficient security processing is attained in this architecture through (i) the use of a configurable and extensible processor that is customized through the selective addition of custom instructions, co-processors, and peripherals, which implement performance-critical, computation-intensive operations, and (ii) optimized software libraries that are derived through extensive algorithmic exploration and tuning of the cryptographic algorithms they implement.

2.1 HW Platform architecture

The hardware platform is based on the Xtensa T1040 processor from Tensilica, Inc. [14]. The Xtensa features a 32-bit RISC-like base processor architecture. It offers a wide range of options to configure the base processor, including selection of generic instructions (*e.g.*, hardware multiplier, MAC, floating point unit, *etc.*), exceptions and interrupt mechanisms, endianness, register window customization, cache and memory interface configuration, debug and test hardware, *etc.* In addition to its configurability, the Xtensa also provides the designer with the ability to extend the instruction set through the addition of custom instructions that execute on designer-specified custom hardware units, which are tightly integrated into the processor execution pipeline. In our work, we exploit the customizability offered by the Xtensa processor platform in order to meet our performance objectives for security processing. HW/SW partitioning at the granularity of custom instructions often results in satisfactory performance improvements. However, in some cases, the characteristics of the application require that more coarse-grained functions be mapped to custom hardware. In such cases, one option is to use a HW co-processor that interfaces to the Xtensa's single-cycle cache interface. Alternatively, HW units that do not require high-performance communication with the processor core can be implemented as peripherals connected to the processor bus. In our work, we attempt to use custom instruction extensions to the maximum extent possible since they allow for easier integration, and facilitate higher levels of programmability and HW re-use, compared to the co-processor and peripheral options.

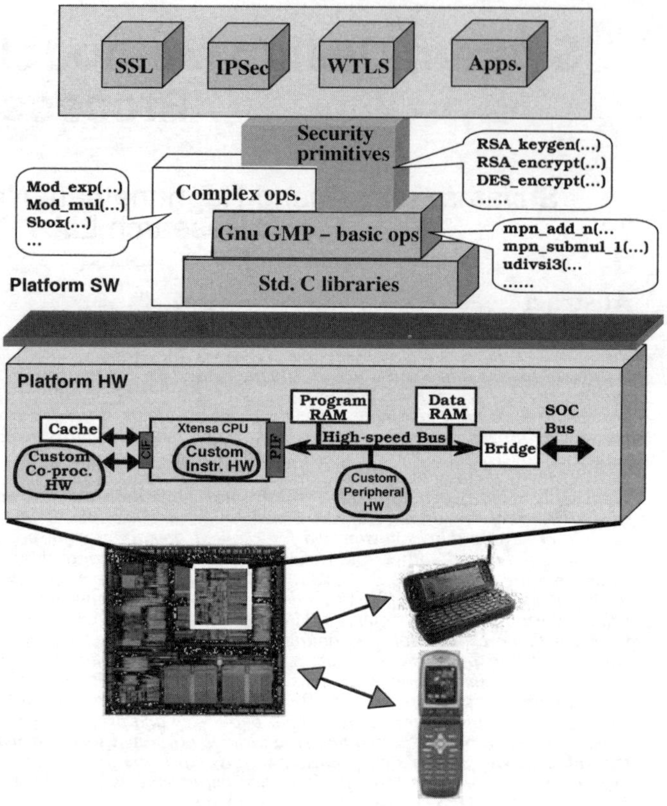

Figure 2: Overview of the target security processing system architecture

2.2 SW architecture

The choice of a suitable software architecture is critical to enable an efficient system design methodology. The software architecture for our security processor platform was designed using a layered philosophy, much like the layering used in the design of network protocols [15]. At the top level, the SW architecture provides a generic interface (API) using which security protocols and applications can be ported to our platform. This API consists of security primitives such as key generation, encryption, or decryption of a block of data using a specific public- or private-key cryptographic algorithm (*e.g.* RSA, ECC, DES, 3DES, AES, *etc.*). The security primitives are implemented on top of a layer of complex mathematical operations such as modular exponentiation, prime number generation, Miller-Rabin primality testing *etc.* [4]. These complex operations are in turn decomposed into basic mathematical operations, including bit-level operations (typically used in private-key algorithms) and multi-precision operations on large integers (typically used in public-key algorithms). The advantages of using the layered SW architecture approach include:

- The API interface at each software layer was fixed before implementation, allowing the design of each layer, and the porting of security protocols to our platform, to proceed concurrently. This reduced design time significantly, and enabled the use of more realistic application workloads to drive the design of each SW layer early in the

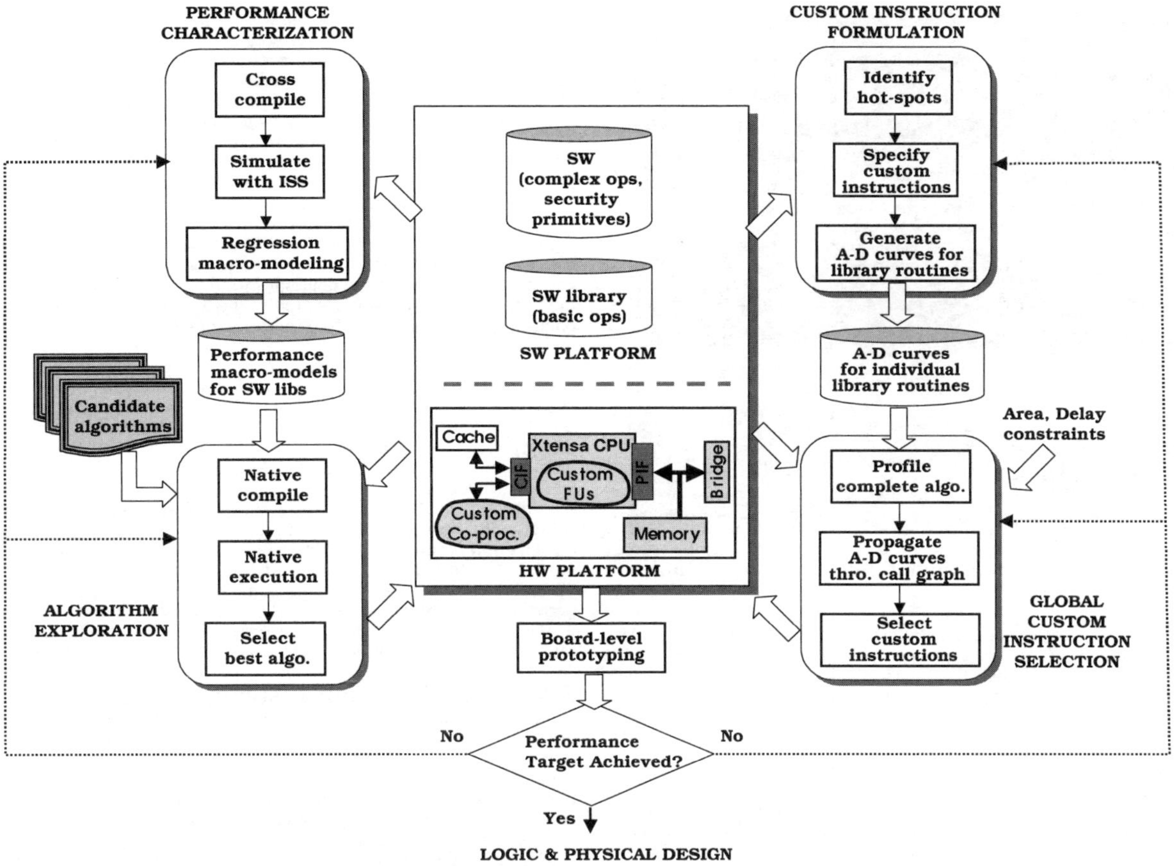

Figure 3: Overview of the security processing system design methodology

design process.

- The generation of candidate custom instructions could proceed once the software layer implementing basic operations was available (*i.e.*, without waiting for the entire SW implementation), since computations of the desired granularity are exposed in the basic operations.
- The separation of the top-level algorithms from the primitives or building blocks that are used to implement them enabled us to characterize the primitives and derive high-level performance macro-models, which were then used for efficient algorithmic exploration As illustrated in Section 3.2, this novel performance characterization methodology enabled the efficient exploration of a large number of candidate algorithms, which would have required several months of simulation time using ISS models.

3. DESIGN METHODOLOGIES

In this section, we present the methodology used to design the architecture of our wireless security processing platform. Section 3.1 presents an overview of the design methodology. Section 3.2 details the selection of the software constituents of the platform, while Sections 3.3 and 3.4 describe the steps involved in customizing the hardware platform.

3.1 Overview

Figure 3 outlines the system-level design flow used for our security processing platform. There are four major phases in the flow: (i) performance characterization of software libraries, (ii) algorithm exploration, (iii) formulation of candidate custom instructions to accelerate individual library routines, and (iv) global custom instruction selection to generate the required performance for each security algorithm. The methodology exploits the layered SW architecture in order to separate the above steps in a clean manner. Specifically, only implementations of the lower SW layers (standard libraries, basic operations) are required for performance characterization and formulation of custom instruction candidates, while algorithm exploration and global custom instruction selection are performed using the higher SW layers (complex operations, security primitives) while regarding the lower SW layers as a black box.

We now briefly describe the salient steps of our methodology, details of which are found in later subsections.

- The simulation time required for performance estimation is a significant bottleneck in algorithm design space exploration (in our context, several hours to few days per candidate algorithm). The performance macro-modeling phase effectively addresses this problem by enabling performance estimation through native compilation and execution, which can be orders of magnitude faster than Instruction Set Simulation. During the performance macro-modeling phase, we characterize the software library routines that constitute the basic steps of the algorithm, using a cycle-accurate ISS. We use statistical regression techniques to build macro-models that express the execution time of each routine as a function of parameters characterizing its input variables. The performance macro-modeling phase is explained in further detail in Section 3.2.
- The algorithm exploration phase attempts to identify optimal algorithmic implementations of security processing algorithms such as RSA, AES, 3DES *etc.* For each algorithm candidate, we instantiate the performance macro-models for library routines in the source code, and replace ISS runs with native compilation and direct execution on a host workstation, resulting in large speedups in simulation time. In our context, that allows exhaustive exploration of the algorithmic design space to be performed.
- In most scenarios, the optimized algorithm running on the base hardware platform does not achieve the target performance. Therefore, it becomes necessary to customize the underlying HW architecture, through custom instruction extensions in our case. During the custom instruction formulation phase, we focus on speeding up individual software library routines. That allows our designers to focus on small problem instances, where they best apply their creativity, leaving the global tradeoffs to the subsequent phase. The routine under consideration is profiled using traces derived from simulation of the entire algorithm. The computation-intensive parts of the routine are specified as a custom instruction. The hardware resources (functional units, register files, lookup tables, *etc.*) used in the custom instruction are varied to create a local area *vs.* delay tradeoff for the individual library routine. Having a rich set of alternatives is critical to achieving a high-quality solution in the global custom instruction selection phase. The custom instruction formulation phase is discussed further in Section 3.3.
- The global custom instruction selection phase determines a combination of (possibly several) custom instructions to result in maximum

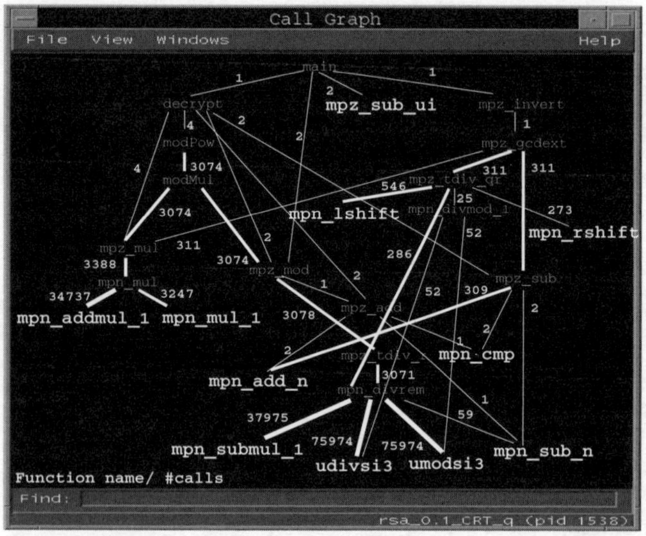

Figure 4: Call graph for an optimized modular exponentiation algorithm

speedup for the entire security algorithm subject to any applicable area constraints. This phase proceeds by propagating A-D curves for library routines through the function call graph of the entire algorithm. The potential explosion in the number of instruction combinations is contained using several techniques. The global custom instruction selection phase is described in detail in Section 3.4.

The optimized security processor platform is evaluated in the context of the target environment (*e.g.*, SSL and IPSec protocol processing, real-time video encryption/decryption, *etc.*) through board-level prototyping. Inadequacies in performance are addressed through further refinements to the HW or SW parts by iterating the steps described above with either relaxed area constraints, additional candidate algorithms, or additional custom instruction candidates.

3.2 Performance Macro-modeling for Algorithm-level Design Space Exploration

In this section, we introduce performance macro-models and describe their use in algorithmic design space exploration. A performance macro-model is a function that expresses the number of cycles incurred by the actual run of a library routine in terms of parameters that characterize the routine's input variables. For example, the performance of a routine *mpn_add_n* that adds two arbitrary length integers *in*1 and *in*2 can be expressed as a function of the bitwidths of the two inputs. The characterization process proceeds as follows. The routine under consideration is invoked in a test program that exercises it with a wide range of pseudo-randomly generated input stimuli. This test program is simulated using the cycle-accurate ISS for the target HW to generate performance data that consists of (i) the value of the parameters (*e.g.*, input variable bitwidths), and (ii) the number of execution cycles, for each invocation of the routine. A statistical regression is performed to fit the above data, resulting in the performance macro-model for the library routine. Note that, characterization is a one-time process, and results in acceleration of the overall performance estimation process.

Since the input space for a library routine can potentially be infinite, the input values used for characterization are generated to lie within a bounded super-space of the input space used by the application. For example, the GNU GMP library [16] provides a wide variety of functions that can perform arbitrary precision arithmetic on integers, rationals and floats. However, a 1024-bit RSA algorithm only requires operations restricted to (less than or equal to) 1024-bit arithmetic. Therefore, we characterize the library routines for this restricted domain only.

The performance profiles of arithmetic functions typically show a regular behavior (piecewise linear, quadratic, *etc.*) over input bit-width subspaces. Therefore, we can derive the performance model for a library routine fairly easily and accurately using regression-based approaches. All library routines instantiated in the source code of an algorithm can now be augmented with their respective performance models to allow performance estimation through native code execution on any host workstation. Further details of the performance macro-modeling technique can be found in [17].

3.3 Formulating custom instruction candidates and A-D curves

Figure 4 shows the profile statistics of an optimized modular exponentiation algorithm as a function call graph, with nodes representing function names, and edges weighted by the number of calls made to each function. For example, the function *decrypt* makes 4, 4, 2, 2 and 2 calls, to functions *mpz_mul*, *modPow*, *mpz_mod*, *mpz_add* and *mpz_sub*, respectively. Each node in the call graph may have more than one parent, since a function may be invoked by multiple higher-level functions. For example, *mpz_mul* is called by three functions - *decrypt*, *modMul*, and *mpz_gcdext*. For the sake of simplicity, the call graph in Figure 4 is truncated at functions that are highlighted with bold text, *i.e.*, calls to lower-level functions are not shown. The leaf nodes of the call graph in Figure 4 correspond to the library routines for which custom instructions are added in an interactive manner with the designer's involvement. It bears mentioning that, the granularity of the leaf nodes is a critical choice that determines the effectiveness of the custom instructions. Ideally, a function chosen to be a leaf node should contain sufficient amount of computation so as to provide scope for optimization, while being small enough that it is easy for a designer to understand and optimize. Our methodology contains heuristics for the choice of the leaf node based on the function's size and the fraction of the total program execution time it accounts for. However, we also provide the designer with an option to override automatic choices and manually specify the leaf nodes.

Since the added custom instructions can be provided with a variable number of hardware resources, we can associate an area-performance trade-off curve (also called A-D curve) with each custom instruction. The lower-most set of points in Figure 5(a) shows the A-D curve for a sample library routine *mpn_add_n* that performs the addition of two vectors. The original library routine is represented by the design point that has a zero area overhead and a performance of 202 cycles, as shown. All other design points are derived through custom instruction additions with varying number of adder resources, and hence, have non-zero area overheads. For example, the second design point is achieved by adding custom load/store instructions *load_UR*1, *load_UR*2 and *store_UR*3, and an addition instruction *add_*2 that uses two 32-bit adder resources. When the number of adders is changed to 4 (*add_*4), performance improves at increased area costs, creating the next design point in the A-D curve. At some point, additional resources bring diminishing returns (*e.g.*, due to limits on parallelism or memory bottlenecks).

3.4 Global Custom Instruction Selection

In this section, we describe our methodology for selecting custom instructions using A-D curves of software library routines and the annotated call graph of the entire algorithm. Our procedure for selecting custom instructions involves combining and justifying A-D curves in a bottom-up fashion to derive a composite A-D curve for the root node of the call graph. The area and performance constraints for the platform can then be applied at the root node to pick the final custom instruction(s).

For any subgraph rooted at a node f, with children given by the set $children(f)$, the performance of f is governed by the following equation

$$cycles(f) \quad = \quad local\_cycles(f) + \sum_{g \in children(f)} cycles(g) \qquad (1)$$

In the above equation, $local\_cycles(f)$ refers to the number of cycles spent in computations local to f, which do not involve calls to any of its children. The above equation can be directly applied when all members of the set $children(f)$ have a single performance number associated with them (*i.e.*, no A-D curves). However, when A-D curves of one or more functions in $children(f)$ need to be combined, there are a few issues involved, as illustrated below. When the root node of a subgraph in the call graph has multiple children, the A-D curve computation simply degenerates to repeated application of the following cases.

Two child nodes - one child with an A-D curve and another with no A-D curve: Figure 5(a) illustrates this case for the graph rooted at node *root*, with one child *mpn_add_n* (which has an A-D curve), and a second child *other* (which requires 10 cycles per call). In this case, for every design point in the A-D curve of *mpn_add_n*, we have a corresponding design point in the A-D curve of *root*, with the performance computed using Equation (1).

Two child nodes with A-D curves: Figure 5(c) illustrates this case using a graph rooted at node *root* with two children, *mpn_add_n* and *mpn_addmul_1*, whose A-D curves are shown in Figures 5(a) and 5(b), respectively. As in the previous case, the performance of *root* is the sum of the performances of its children, each weighted by the number of calls made to them. In general, every combination of design points (Cartesian product) from the A-D curves of *mpn_add_n* and *mpn_addmul_1* must be represented as a distinct point in the A-D curve of *root*. However, it turns out that whenever instructions are shared or dominated between design points, the number of design points in the composite A-D curve can be significantly reduced, as explained next.

Figure 6 shows the Cartesian product of the points on the A-D curves for *mpn_add_n* and *mpn_addmul_1*. Each entry corresponds to the union of the custom instructions that constitute the individual design points (we ignore load/store instructions, which are shared across both the children). For example, the shaded entry *add_2*, *mul_1* is the union of custom instructions *add_2*, *mul_1* for function *mpn_addmul_1*, and *add_2* for function *mpn_add_n*. The symbol ϕ is used to denote the null set, *i.e.*, no custom instructions. Observe that the shaded entry *add_2*, *add_4*, *mul_1* in Figure 6 is equivalent with many other design points. This is possible (i) when entries have the same custom instructions or (ii) when entries reduce to the same custom instructions. For example, the entry *add_2*, *add_4*, *mul_1* has two add instructions *add_2* and *add_4*, which differ only in the number of adder

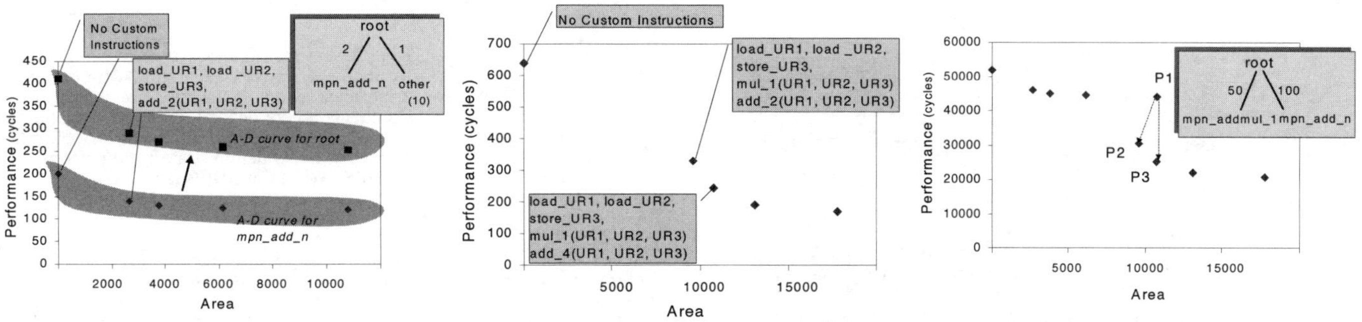

(a) (b) (c)

Figure 5: (a) A-D curve for library routine *mpn_add_n* and its propagation through an example call graph, (b) A-D curve for *mpn_addmul_1*, and (c) computing the A-D curve for a node with two children that have A-D curves

resources available while realizing the same functional capabilities. Given that *add_4* can be used to perform *add_2* with equal or better performance, we say that *add_4 dominates add_2*, and reduce *add_2*, *add_4*, *mul_1* to *add_4*, *mul_1*. Figure 6 contains 25 candidate design points, which can be reduced to only 9 points corresponding to the shaded entries in Figure 6. The reduced set of 9 points are represented in the A-D curve for *root*, as shown in Figure 5(c).

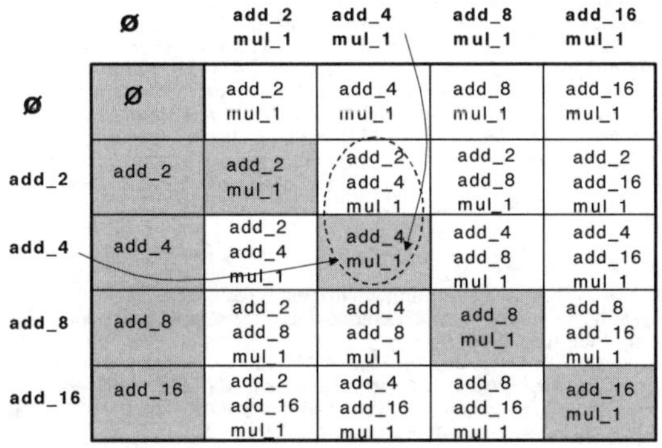

Figure 6: Combining the design spaces of two area-delay (A-D) curves

Note that, at the root node of the entire call graph, the standard notion of Pareto-optimality can be applied to eliminate inferior points. In Figure 5(c), we can prune away design point *P*1 which has inferior performance while incurring more area with respect to design points *P*2 and *P*3.

4. EXPERIMENTAL RESULTS

We used the design methodology presented in this paper to build a security processing platform for wireless handsets that supports popular network-layer and transport-layer security protocols (*e.g.*, IPSec, SSL, WTLS, *etc.*). Section 4.1 describes the different software and hardware tools used to carry out the various steps of the methodology. Section 4.2 presents an overall evaluation of the security processing platform, including its performance in speeding up the secure socket layer (SSL) protocol. Section 4.3 discusses the results of the algorithmic design space exploration methodology, focusing on the efficiency and accuracy of the macro-modeling based performance estimation technique.

4.1 Experimental Methodology

For algorithmic design space exploration, each algorithm candidate was implemented as a highly modular, optimized C implementation using library routines from two well-known software libraries: (i) The GNU MP library [16] provides a wide variety of functions that can perform arbitrary precision arithmetic on integers, rationals and floats, and (ii) a hash library that provides a reliable means for creating hash tables. The GNU based cross-compiler, and the instruction set simulator for the target processor (an Xtensa processor core running at 188 MHz in 0.18 micron technology) were used to profile the different library routines. Performance macro-models were constructed using the statistical modeling tool S-Plus [18]. Native simulation was then performed on a SUN Ultra 10 440 MHz workstation with 1 GB of memory to select the best algorithm configuration for the given target hardware.

Custom instructions for the different library routines were implemented as Tensilica Instruction Extension (TIE) descriptions and parameterized for generating A-D curves. The TIE descriptions were compiled using the TIE compiler [14], which generates both C-stubs and synthesizable RTL Verilog descriptions. The C-stubs were then instantiated as intrinsics in test programs to derive the performance numbers in the A-D curves. The RTL descriptions were subject to logic synthesis using Synopsys Design Compiler [19] and technology mapped to the NEC CB-11 0.18 micron technology library [20] to determine the area numbers. The global instruction selection procedure described in Section 3.4 was then used to evaluate the different TIE candidates. The TIE solutions determined were combined with the base Xtensa processor core using the Xtensa processor generator [14] to build the enhanced target hardware.

Figure 7: Functional prototype of the security processing platform

4.2 Evaluation of the security processing platform

We evaluated the performance of our security processor platform using standard implementations of private-key algorithms such as DES, 3DES, and AES, as well as the public-key algorithm RSA. The optimized HW platform and SW implementation resulting from our system design methodology were used to build a a board-level prototype implementation of the security processing platform, which is shown in Figure 7. The prototype was built using the Xtensa XT-2000 emulation board [21] with an EPSON graphics controller card [22] interfacing with an NEC LCD panel [23] (see Figure 7). The system prototype was used to demonstrate security processing performance improvements under various application scenarios, including real-time video decryption and SSL transaction acceleration.

Table 1: Performance speed-ups for popular security processing algorithms

Sec. Algo.	*Processing Rates*		
	Orig. (cycle/byte)	Final (cycle/byte)	Speedup
DES enc./dec.	476.8	15.4	31.0X
3DES enc./dec.	1426.4	42.1	33.9X
AES enc./dec.	1526.2	87.5	17.4X
RSA enc.	$34.29 * 10^3$	$3.16 * 10^3$	10.8X
RSA dec.	$12658 * 10^3$	$190.78 * 10^3$	66.4X

Table 1 illustrates the performance speed-ups for the individual security processing algorithms: 31.0X for DES, 33.9X for 3DES, 17.4X for AES,

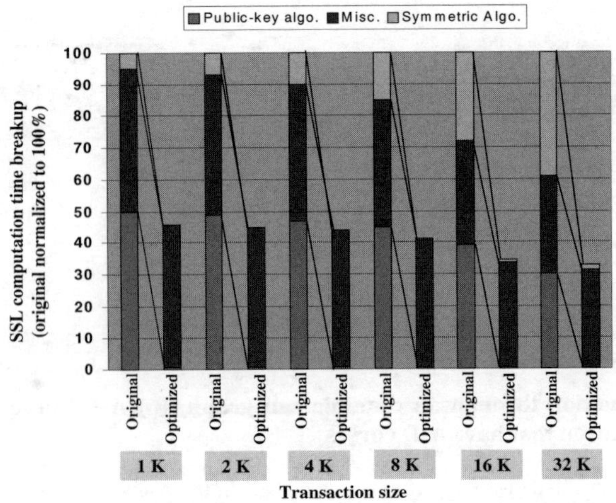

Note: Due to large speedups in the optimized case, the public-key and private-key components are not always visible in the above graph

Figure 8: Estimated speedups for SSL transactions

and upto 66.4X for RSA. Note that, these improvements are obtained compared to already optimized software implementations. We next see how the enhancements made to these security algorithms help in speeding up the popularly used transport layer security protocol, SSL [5]. SSL uses a combination of private-key and public-key algorithms to secure the data transferred between a client and a server. The SSL handshake first allows the server and client to authenticate each other, using public-key techniques such as RSA. Then, it allows the server to create symmetric keys, which are exchanged and used for rapid encryption and decryption of bulk data transferred during the session. Figure 8 shows the estimated speedup of SSL transactions through the use of our security processing platform. The breakup of the computation workload for SSL processing between the private-key algorithm, public-key algorithm, and other miscellaneous computations, is also indicated in Figure 8. Note that, the breakup depends on the session size, hence we considered various session sizes ranging from 1KB to 32KB. For small data transactions (where public-key algorithm computations in the SSL handshake dominate), our platform contributes to an overall transaction speedup of around 2.18X. In the case of large transactions, (where the private key algorithm starts to dominate the overall computation) our platform achieves an overall transaction speedup of 3.05X.

4.3 Algorithm design space exploration

We illustrate our algorithm design space exploration technique through the example of modular exponentiation, which is used for encryption and decryption in several public-key algorithms. Over 450 candidate algorithms were considered for evaluation due to the permutations arising from five modular multiplication algorithms, five input block sizes, three Chinese Remainder Theorem implementations, two radix sizes and three different software caching options [24]. Performance macro-model based evaluation of all the algorithm candidates completes in under 4 hours and 40 minutes. In comparison, only six algorithm candidates could be evaluated in nearly 66 hours of CPU time, using actual ISS runs. On an average, macro-model based performance estimation was found to be 1407 times faster than actual ISS runs. The performance estimated using the macro-models accurately tracked the performance profile determined by actual target simulation. The mean absolute error in the macro-model based estimates was only 11.8 %, and the relative accuracy was more found to be than adequate for the purpose of algorithm exploration.

5. RELATED WORK

Most of the efforts towards improving the efficiency of security processing have been targeted at addressing performance issues in e-commerce servers, network routers, firewalls, and VPN gateways [7, 25, 26, 27]. The fact that public key algorithms often dominate security processing requirements has driven the recent development of alternative public-key algorithms that offer reduced computational complexity [28, 29].

Various companies offer commercial security processor ICs to improve the performance of transaction servers and network routers [30, 31, 32, 33, 34, 35]. Architectural enhancements to high-end microprocessor systems to improve their performance in security processing have been investigated [25, 26]. Embedded processor designers have also developed security extensions to their products, typically based on the addition of application-specific co-processors and/or peripherals [36, 37]. Computer architects have researched domain specific instructions for security processing, with an aim to maximize efficiency without compromising programmability [38, 39]. Our target architecture and the system-level design methodologies presented here are complementary to most of the above efforts, and can enable high efficiency in security processing while maintaining programmability.

6. CONCLUSIONS

We presented the system-level design methodology used to design a programmable security processor platform for next-generation wireless handsets. The methodology was constructed using off-the-shelf commercial tools as well as novel in-house components where needed, in order to enable the efficient co-design of optimal cryptographic algorithms and an optimized HW platform architecture. Our experiments demonstrate large performance improvements (*e.g.* 31.0*X*, for DES, 33.9*X* for 3DES, 17.4*X* for AES, and upto 66.4*X* for RSA) compared to software implementations on a state-of-the-art embedded processor. We believe that system-level design methodologies, such as the one proposed here, are critical to meeting the challenging objectives and constraints encountered in security processing.

Acknowledgments: We acknowledge all brand or product names that are trademarks or registered trademarks of their respective owners. We would like to thank the members of the Tensilica support team for their invaluable assistance with the use of the Xtensa processor and tools.

7. REFERENCES

[1] U. S. Department of Commerce, *The Emerging Digital Economy II*. http://www.ecommerce.gov/ede/report.html, 1999.

[2] W. W. W. Consortium, *The World Wide Web Security FAQ*. http://www.w3.org/Security/faq/www-security-faq.html, 1998.

[3] *ePaynews*. http://www.epaynews.com/statistics/ecappstats.html.

[4] B. Schneier, *Applied Cryptography: Protocols, Algorithms and Source Code in C*. John Wiley and Sons, 1996.

[5] W. Stallings, *Cryptography and Network Security: Principles and Practice*. Prentice Hall, 1998.

[6] S. K. Miller, "Facing the Challenges of Wireless Security," in *IEEE Computer*, pp. 46–48, July 2001.

[7] G. Apostolopoulos, V. Peris, P. Pradhan, and D. Saha, "Securing Electronic Commerce: Reducing SSL Overhead," in *IEEE Network*, pp. 8–16, July 2000.

[8] D. Boneh and N. Daswani, "Experimenting with Electronic Commerce on the PalmPilot," in *Proc. Financial Cryptography*, pp. 1–16, 1999.

[9] K. Lahiri, A. Raghunathan, and S. Dey, "Battery-driven system design: A new frontier in low power design," in *Proc. Joint Asia and South Pacific Design Automation Conf. / Int. Conf. VLSI Design*, pp. 261–267, Jan. 2002.

[10] A. G. Broscius and J. M. Smith, "Exploiting parallelism in hardware implementation of DES," in *Proc. CRYPTO'91*, pp. 367–376, 1991.

[11] A. Curiger, H. Bonnenberg, R. Zimmermann, N. Felber, H. Kaeslin, and W. Fichtner, "VINCI: VLSI implementation of the new secret-key block cipher IDEA," in *Proc. IEEE Custom Integrated Circuits Conf.*, pp. 15.5.1–15.5.4, May 1993.

[12] C. K. Koc, "RSA hardware implementation," Tech. Rep. TR-801 (available online at http://security.ece.orst.edu/koc/ece575/rsalabs/tr-801.pdf), RSA Data Security Inc., Apr. 1996.

[13] T. Ichikawa, T. Kasuya, and M. Matsui, "Hardware evaluation of the AES finalists," in *Third Advanced Encryption Standard (AES) Conference*, Apr. 2000.

[14] *Xtensa application specific microprocessor solutions - Overview handbook*. Tensilica Inc. (http://www.tensilica.com), 2001.

[15] A. S. Tanenbaum, *Computer Networks*. Prentice-Hall, Englewood Cliffs, NJ, 1989.

[16] T. Granlund, *The GNU Multiple Precision Arithmetic Library*. http://www.gnu.org, 2000.

[17] N. Potlapally, S. Ravi, A. Raghunathan, and G. Lakshminarayana, "Algorithm exploration for efficient public-key security processing on wireless handsets," in *Proc. DATE Designers Forum*, pp. 42–46, Mar. 2002.

[18] W. N. Venables and B. D. Ripley, *Modern Applied Statistics with S-PLUS*. Springer-Verlag, 1998.

[19] "Design Compiler, Synopsys Inc. (http://www.synopsys.com).".

[20] *CB-11 Family 0.18um CMOS Cell-based IC Design Manual*. NEC Electronics, Inc., December. 1999.

[21] *Xtensa Microprocessor Emulation Kit XT 2000 - User's Guide*. Tensilica Inc. (http://www.tensilica.com), 2001.

[22] *S1D13806 Embedded Memory Display Controller*. Epson Research & Development Inc. (http://www.erd.epson.com).

[23] *NL6448BC33-31 10.4 inch digital VGA LCD display*. NEC Electronics Inc. (http://www.necel.com).

[24] N. Potlapally, S. Ravi, A. Raghunathan, and G. Lakshminarayana, "Optimizing Public-Key Encryption for Wireless Clients," in *Proc. IEEE Int. Conf. Communications*, May 2002.

[25] Intel Corp., *Enhancing Security Performance through IA-64 Architecture*. http://developer.intel.com/design/security/rsa2000/itanium.pdf, 2000.

[26] K. Kant, R. Iyer, and P. Mohapatra, "Architectural Impact of Secure Sockets Layer on Internet Servers," in *Proc. Int. Conf. Computer Design*, pp. 7–14, 2000.

[27] A. Goldberg, R. Buff, and A. Schmitt, "Secure Server Performance Dramatically Improved by Caching SSL Session Keys," in *ACM Wksp. Internet Server Performance*, June 1998.

[28] M. Rosing, *Implementing Elliptic Curve Cryptography*. Manning Publications Co., 1998.

[29] *NTRU Communications and Content Security*. http://www.ntru.com.

[30] *Broadcom Corporation, BCM5840 Gigabit Security Processor*. http://www.broadcom.com.

[31] Corrent Inc. http://www.corrent.com.

[32] *HIFN Inc.* http://www.hifn.com.

[33] *Motorola Inc., MC190:Security Processor*. http://www.motorola.com.

[34] *NetOctave Inc.* http://www.netoctave.com.

[35] *Securealink USA Inc.* http://www.securealink.com.

[36] *ARM SecurCore*. http://www.arm.com.

[37] *SmartMIPS*. http://www.mips.com.

[38] Z. Shi and R. Lee, "Bit Permutation Instructions for Accelerating Software Cryptography," in *Proc. IEEE Intl. Conf. Application-specific Systems, Architectures and Processors*, pp. 138–148, 2000.

[39] J. Burke, J. McDonald, and T. Austin, "Architectural Support for Fast Symmetric-Key Cryptography," in *Proc. Intl. Conf. ASPLOS*, pp. 178–189, Nov. 2000.

Constraint-Driven Communication Synthesis

Alessandro Pinto*　　Luca P. Carloni　　Alberto L. Sangiovanni-Vincentelli

ABSTRACT

Constraint-driven Communication Synthesis enables the automatic design of the communication architecture of a complex system from a library of pre-defined Intellectual Property (IP) components. The key communication parameters that govern all the point-to-point interactions among system modules are captured as a set of arc constraints *in the* communication constraint graph. *Similarly, the communication features offered by each of the components available in the IP communication library are captured as a set of* feature resources *together with its cost figures. Then, every communication architecture that can be built using the available components while satisfying all constraints is implicitly considered (as an* implementation graph *matching the constraint graph) to derive the optimum design solution with respect to the desired cost figure. The corresponding constrained optimization problem is efficiently solved by a novel algorithm that is presented here together with its rigorous theoretical foundations.*

Categories and Subject Descriptors

J.6.1 [**Computer Applications**]: Computer-Aided Engineering—*CAD*.

General Terms

Algorithm.

Keywords

Communication Synthesis, Systems-on-Chip, Network Design.

1. INTRODUCTION

In this work we propose a novel approach to design the communication architecture for a system of computational modules whose interaction is specified from an abstract point of view as a collection of communication requirements on a set of point-to-point unidirectional "virtual" channels. By abstracting away the specific functionality of each module, we can focus on exploring the various communication topologies that can be built composing a set of library elements that include "passive elements" (links) as well as active ones (repeaters, switches), each of them coming with a fixed cost function that captures an application-specific optimality criterion. The proposed approach lies on top of a mathematical model that allows us to fully separate computational issues from communication ones. While each computational module acts on the data streams that travel within the system (reading from input channels and writing new data onto the output channels according to its functionality) the communication elements limit themselves to transfer the data between two points (the links), two receive and re-transmit the same data (repeaters) or to route in the proper direction (the switches). This clear task division allows us to focus on the complete exploration of all the possible communication architecture topologies that can be built composing these primitive building blocks. In fact, we limit ourselves to three main composition types to physically implement the virtual channels: the segmentation of long channels by inserting repeaters between shorter links, the duplication of bandwidth-challenged channels by adding extra parallel links together with a pair of mux/demux switches, and the merging of distinct channels (which generally involves segmentation and duplication). Different from previous approaches to the problem of communication synthesis, we rely on the definition of a fine-grain library whose elements are combined to derive a communication topology that tightly match the system structure. The proposed algorithm discards all the sub-optimal local solutions, while generating a core set of candidate channel implementations from which it picks the optimum-cost subset based on the library cost functions. By composing this subset the algorithm returns the detailed topology of the final optimum-cost architecture that is guaranteed to satisfy all the original requirements.

Among the several related papers published in recent years, the authors of [3] split the development of the communication architecture in two steps: channel binding and channel mapping. The former binds virtual communication units to high-level communication channels, while the latter associates to each unit a tree of alternative physical implementations from a library. Then a depth-first search strategy is used to derive an optimal solution. In [6] and [7] the design of the communication architecture is done with an exploration of different solutions validated by a fast performance simulation (based on a detailed characterization of the library components). The authors of [9] assume that the network topology is given and find an implementation that allows to achieve very high performance by sending control signals and deadlines on the delivery of the message in advance to the corresponding data. The approach of [2] is similar to the present one, but specialized to ATM networks: the problem is to select the topology of a network composed of links specified in a library with their speed and cost. Differently from our approach this paper assumes that the location of the intermediate communication nodes is fixed and the optimization is limited to link selection.

2. THE MODEL

The abstract model to specify a communication system is represented in Figure 1 and consists of a set of computational modules communicating through point-to-point

*The authors are with UC Berkeley EECS Department, Berkeley, CA 94720-1772. E-mail: {apinto,lcarloni,alberto}@ic.eecs.berkeley.edu. This research was supported partially by the SRC and the GSRC/Marco Center at Berkeley.

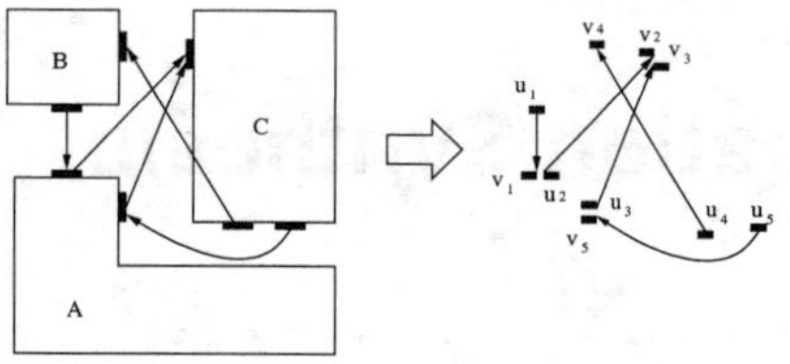

Figure 1: Model of Communication Requirement.

unidirectional communication "virtual" channels that are connected to the modules by means of input/output ports. A module may communicate with another module through multiple unidirectional channels (in both directions). For each entering (leaving) channel connected to the module there is a corresponding dedicated input (output) port. Communication requirements are specified for each channel as a set of two parameters: the distance to be covered and the required bandwidth. Our intent is to use this model as a basic common starting point to define the communication specifications of various kinds of systems, such as a "System-on-Chip", a multi-chip multi-processor system, or a local area network (LAN). Naturally, the basic model will be appropriately extended/refined for each particular application. For instance, in case of a "System-on-Chip", a channel could represent the set of wires implementing the address bus that a processor uses to access a cache memory and a certain required channel bandwidth could be specified in gigabyte per second. Furthermore, for each port of every computational module on the chip a certain location could be specified, thus making it possible to compute the Manhattan distance between any two communicating ports. On the other hand, if we are studying how to implement a LAN and we want to evaluate whether to realize it as a fiber-optic network or a wireless network, (or a combination of the two), the set of channels could just capture all the specified links among the clients and the servers. Here the Euclidean distance among all these components could be sufficient, while for each channel, the bandwidth is usually specified in gigabit per second.

Independently from the specific application, we follow the principle of orthogonalization of concerns [1, 5] and we derive from the network a *communication constraint graph* that allows us to focus on the design of the communication architecture while disregarding the functionality of each computational module in the system. Working with the constraint graph, we define the *constraint-driven communication synthesis problem* as the task of finding the communication architecture which satisfies all the constraints specified as communication requirements on the channels, while minimizing a predefined cost function that captures an optimality criterion which must be defined for the specific application.

DEFINITION 2.1. *A communication constraint graph, or simply* constraint graph, $\mathcal{G} = G(V, A)$ *is a directed graph, where each vertex v is associated to a port of computational module of the system and each directed arc a (also called, simply,* constraint arc*) represents a point-to-point communication channel between two modules. A position $p(v)$ is assigned to each vertex $v \in V$, while the following quantities are associated to each directed arc $a = (u, v)$ and referred to as* arc properties*: (1) the arc* length *(or,* distance*) $d(a)$ between vertex u and vertex v, and (2) the communication bandwidth $b(a)$ on the constrained arc a.*

The right-end side of Figure 1 illustrates the communication constraint graph that is derived from the network on the left-end side. The previous definition doesn't specify whether the position of the vertices should be considered on the plane or in space, nor which type of distance is used to compute the arc length. However, for all arcs $a = (u, v)$ in the graph, the values of $d(a)$ must be consistent with the positions $p(u)$ and $p(v)$. For instance, in the case of a "System-on-Chip", the position of vertex v (corresponding to a module port) will be given by its coordinates $p(v) = (x_v, y_v)$ and the length of the arc $a = (u, v)$ may be computed using the Manhattan distance between the coordinates of its two vertices $a = |x_u - x_v| + |y_u - y_v|$. In the sequel, we will rely on the notion of geometric norm $||p(u) - p(v)||$ to identify the generic distance between two vertices $u, v \in \mathcal{G}$.

The set of all arc properties (lengths and bandwidths) represent the set of design constraints that need to be satisfied while deriving a communication architecture that implements all point-to-point communication channels of the system. As discussed in the introduction, we assume that this communication architecture is realized by putting together elements taken from a communication library. In particular, the library may contain several kinds of communication links, repeaters, and switches. For instance, a communication link guarantees that a certain flow of information can be transferred with up to a specified bandwidth between two ports as long as they lie within a specified distance. Examples of communication links are optical fiber connections, wireless links, or metal lines on a chip that can sustain up to a certain bandwidth given a certain distance. A repeater is used to connect to links (that are able to sustain a certain bandwidth) to cover a distance that they would not to be able to cover stand-alone. A switch, while being able to act as a repeater, enables the connection of multiple links that share a specified bandwidth. A multiplexer is a switch that takes multiple incoming links and "merges" them into one outgoing link whose bandwidth is larger than the sum of the incoming one. A de-multiplexer does the inverse function. In the sequel we define more formally the notion of communication library and we show how putting together these basic elements and defining a few simple operation to combine them we are able to build a rich set of heterogeneous communication architectures having various topologies and bandwidth characteristics.

DEFINITION 2.2. *A communication library $\mathcal{L} = L \cup N$ is a collection of communication links and communication nodes. Each node $n \in N$ has a cost $c(n)$. Each link $l \in L$ is characterized by a set of* link properties*: (1) the link* length *(or,* distance*) $d(l)$ corresponds to the length of the longest communication channel that can be realized by this link, (2) the link* bandwidth $b(l)$ *corresponds to the bandwidth of the fastest communication channel that can be realized by this link, and (3) the link* cost $c(l)$*, that is defined with respect to the other links in the library based on an optimality criterion that varies with the type of application.*

The realization of a communication architecture that satisfies the requirements specified by a constraint graph can be modeled as a set of graph transformations (including the addition of new arcs and vertices). This leads us to define a new graph, called implementation graph whose set of vertices is an extension of the set of vertices of the constraint graph. In particular, each vertex in the implementation graph is either a "computational vertex" (corresponding to a vertex in the original constraint, i.e. a port of a computational module of the original system) or a newly added "communication vertex", corresponding to an instance of a communication node from the library. Also, every arc in the graph is mapped to a library link.

DEFINITION 2.3. *Given a graph $G = G(V, A)$, a path $q = (v_1, a_1, v_2, a_2, \ldots, v_{Q-1}, a_{Q-1}, v_Q)$ is an alternating sequence of distinct vertices and arcs in G, with $V(q, G)$ and $A(q, G)$ denoting respectively the set of vertices and arcs touched by q. Furthermore, we define the* sub-path *of p up to vertex $v_j \in V(q, G)$ as $sub(q, v_j) = (v_1, a_1, v_2, a_2, \ldots, a_{j-1}, v_j)$. As for an arc we can define the following* path properties*: the length $d(q) = \sum_{i=0}^{Q-1} d(a_i)$, the path bandwidth $b(q) = \min_{(i=0,\ldots,Q-1}\{b(a_i)\}$, and the cost $c(q) = \sum_{i=0}^{Q-1} c(a_i)$, with $c(a_i)$ denoting the arc cost as specified in the following definition.*

DEFINITION 2.4. *Given a constraint graph $\mathcal{G} = G(V, A)$ and a communication library $\mathcal{L} = L \cup N$, an* implementation graph *$\mathcal{G}'(\mathcal{G}, \mathcal{L}) = G(V' \cup N', A')$ is a directed graph s.t.:*

- *for each vertex in V there is a corresponding vertex in V' and vice versa (and they have the same positions), i.e. there is a bijective mapping function $\chi : V \to V'$ s.t. $\forall v \in V, \exists v' \in V'(v' = \chi(v) \wedge v = \chi^{-1}(v') \wedge p(v) = p(v'))$.*
- *for each vertex in N' there is a corresponding communication node in N, i.e. there is a surjective mapping function $\psi : N' \to N$ s.t. $\forall n' \in N', \exists n \in N(n = \psi(n'))$. The elements of N' are called communication vertices.*
- *for each arc in A' there is a corresponding communication link in L and they share the values of their properties, i.e. there is a surjective mapping function $\phi : A' \to L$ s.t. $\forall a' \in A', \exists l \in L(l = \phi(a') \wedge d(a') = d(l) \wedge b(a') = b(l) \wedge c(a') = c(l)$.*
- *for each arc $a = (u, v)$ in the constraint graph there is a set of paths $\mathcal{P}(a)$ in the implementation graph connecting $\chi(u)$ to $\chi(v)$ without passing through any other computational vertex (but only, possibly, through communication vertices) that together satisfy the bandwidth constraint $b(a)$ as the sum of the bandwidth $b(q)$ of each path $q \in \mathcal{P}$. Formally, $\forall a = (u, v) \in \mathcal{G}, \exists \mathcal{P}(a) \in \mathcal{G}'$ s.t. $\forall q = (n_1, \ldots, n_Q) \in \mathcal{P}$:*
  1. *$n_1 = \chi(u) \wedge n_Q = \chi(v) \wedge \forall m \in [2, Q-1](n_m \in N')$.*
  2. *$b(a) \leq \sum_{q \in \mathcal{P}} b(q)$.*

  The set of paths $\mathcal{P}(a)$ is called the constraint arc implementation *(or, simply,* arc implementation*) and its cost is $C(\mathcal{P}(a)) = \sum_{q \in \mathcal{P}} c(q)$.*

DEFINITION 2.5. *The* cost *of an implementation graph $\mathcal{G}'$ is defined as [1]:*

$$C(\mathcal{G}') = \sum_{n' \in N'} c(n') + \sum_{a' \in A'} c(a') \qquad (1)$$

where $c(n') = c(\psi(n'))$ and $c(n')$ are as of definition 2.4.

[1] Note that the "computational vertices" in V are not part of the cost equation, they may be though as having null cost.

Generally, for a given library there are many possible implementation graphs that satisfy the requirements expressed by the constraint graph while having different costs. In particular, one implementation graph, the optimum point-to-point implementation graph, is guaranteed to exist and it is derived by implementing a single arc constraint independently from all the others present in the constraint graph.

DEFINITION 2.6. *Given a constraint graph $\mathcal{G} = G(V, A)$ and a communication library $\mathcal{L} = L \cup N$, a* optimum point-to-point implementation graph *$\mathcal{G}'(\mathcal{G}, \mathcal{L}) = G(V' \cup N', A')$ is an implementation graph such that $\forall a_i \in G, \mathcal{P}(a_i)$ has the minimum cost $C(\mathcal{P}(a_i))$ while being subject to the constraint that $\bigcap_{a_i \in G} \mathcal{P}(a_i) = \emptyset$, i.e. its arc implementations are disjoint.*

The following definition gives a characterization of all possible structures for the arc implementations in an optimum point-to-point implementation graph.

DEFINITION 2.7. *Given a constraint graph $\mathcal{G} = G(V, A)$ a communication library $\mathcal{L} = L \cup N$, and an* implementation graph *$\mathcal{G}'(\mathcal{G}, \mathcal{L}) = G(V' \cup N', A')$ the arc implementation $\mathcal{P}(a)$ of $a(u, v) \in A$ is called:*

- *an* arc matching *iff $\mathcal{P}(a) = \{p = (\chi(u), \chi(v))\}$, i.e. the implementation is exactly one library link.*
- *a* K-way arc segmentation *iff $\mathcal{P}(a) = \{p = (\chi(u), n_1, \ldots, n_{K-1}, \chi(v))\}, \forall k \in [1, K-1](n_k \in N')$, i.e. the implementation is the concatenation of K library links interleaved by $k-1$ repeaters.*
- *a* K-way arc duplication *iff $\mathcal{P}(a) = \{p_1 = (\chi(u), \chi(v)), \ldots, p_K = (\chi(u), \chi(v))\}$, i.e. the implementation is made by K library links in parallel.*

Clearly, the optimum point-to-point implementation graph can be seen as the representation of a communication architecture that is built considering sequentially the implementation of each constraint arc a as a stand-alone task, performed according to the following steps: (1) if it exists, the minimum cost link l in the library that satisfies the constraints $d(l) \geq d(a) \wedge b(l) \geq b(a)$; if such a link exists an arc matching is the desired arc implementation [2]; (2) if $d(l) < d(a)$ for all library links l (while $b(l) \geq b(a)$ is satisfied by some l), then arc segmentation will lead to an implementation; (3) conversely, if $b(l) < b(a)$ for all library links l (while $d(l) \geq d(a)$ for some l), then arc duplication will lead to an implementation; (4) in case both constraints can not be satisfied by any link in the library, then a combination of arc segmentation and arc duplication will lead to an implementation.

LEMMA 2.1. *For all constraint graphs $\mathcal{G} = G(V, A)$ and all communication libraries $\mathcal{L} = L \cup N$, there exists an* optimum point-to-point implementation graph *$\mathcal{G}'(\mathcal{G}, \mathcal{L}) = G(V' \cup N', A')$ and $C(\mathcal{G}') = \sum_{n' \in N'} c(n') + \sum_{a' \in A'} c(a') = \sum_{a \in A} C(\mathcal{P}(a))$.*

On the other hand, by analyzing the definition of implementation graph it is clear that some of its arc implementations may share paths (i.e. links and/or communication vertices). In fact, in general the cost of an implementation

[2] See also the assumption 2.1 defined below.

graph is smaller than the sum of the costs of its arc implementations, i.e., re-considering equation 1, we have:

$$C(\mathcal{G}') = \sum_{n' \in N'} c(n') + \sum_{a' \in A'} c(a') \leq \sum_{a \in A} C(\mathcal{P}(a)) \quad (2)$$

As a consequence, we are forced to analyze the interactions between point-to-point constraint arc implementations and the task of finding the optimum implementation graph becomes more challenging.

DEFINITION 2.8. *Given a constraint graph $\mathcal{G} = G(V, A)$ a communication library $\mathcal{L} = L \cup N$, and an* implementation graph *$\mathcal{G}'(\mathcal{G}, \mathcal{L}) = G(V' \cup N', A')$, the union of $K \in [2, |A|]$ arc implementations $\mathcal{P}(a_1), \ldots, \mathcal{P}(a_K)$ is called a K-way* arc merging *when $\exists q^\star$ s.t. $\bigcap_{k=1}^{K} \mathcal{P}(a_k) = q^\star$. The path $q^\star$, whose bandwidth $b(q^\star) \geq \max_{(k=1,\ldots,K)}\{b(a_k)\}$ is called the* common path *of the merging transformation.*

After considering the possibility of K-way mergings, it becomes natural to define a constrained optimization problem aimed to find that implementation graph whose cost (expressed as the sum of the cost of all its components mapped to a library element) is minimum.

PROBLEM 2.1. *Given a constraint graph $\mathcal{G} = G(V, A)$ and a communication library $\mathcal{L} = L \cup N$, minimize the cost $C(\mathcal{G}')$ over all* implementation graphs *$\mathcal{G}'(\mathcal{G}, \mathcal{L}) = (V' \cup N', A')$.*

Clearly, this problem can be seen as a special case of 0-1 integer linear programming (ILP). In the sequel, we will present an exact algorithm to find the solution of this constrained optimization problem when the following assumption holds:

ASSUMPTION 2.1. *Given a constraint graph $\mathcal{G} = G(V, A)$ and a communication library $\mathcal{L} = L \cup N$, for each constraint arc $a = (u, v)$, $C(\mathcal{P}(a)) > 0$ and for all pairs of arcs $a = (u, v), a' = (u', v') \in A$ and for all corresponding minimum-cost constraint arc implementations $\mathcal{P}(a), \mathcal{P}(a')$:*

$$((d(a) \leq d(a') \wedge b(a) \leq b(a')) \Leftrightarrow (C(\mathcal{P}(a)) \leq C(\mathcal{P}(a'))) \quad (3)$$

This assumption is justified from a practical point in most application domains: for instance an optical fiber supporting a given bandwidth is priced per meter; similarly, for radio link covering a fixed distance the higher is the desired bandwidth the more expensive is the cost of the equipment.

3. THE ALGORITHM

To solve exactly the constrained optimization problem defined in the previous section we developed an algorithm that is based on a sequence of two steps, namely *local solution generation* and *global solution derivation*: (1) We efficiently generate the set $\mathcal{S}$ of all those alternative distinct implementations of each arc in the constraint graph that are not "dominated" by other less expensive implementations. The set $\mathcal{S}$ includes all $|A|$ arc implementations in the optimum point-to-point implementation graph (see definition 2.6) together with a minimal set of arcs implementations that are built applying k-way arc mergings ($k \in [2, |A|]$) (possibly combined with some arc segmentation/duplication). The elements of $\mathcal{S}$ are said "local" since they generally provide an implementation only for a subset of the constraint arcs, and, in the case of ($k = |A|$)-way mergings (where all arcs are implemented), the implementation may only represent a "local minimum" that is encountered while searching the whole solution space. (2) After computing the cost of each element of $\mathcal{S}$, we solve an instance of the weighted Unate Covering Problem(UCP) to find that subset of $\mathcal{S}$ providing the minimum cost global implementation for all arcs of the constraint graph.

```
GenerateCandidateArcImplementations (𝒢, ℒ)
{Compute set 𝒮 of candidate arc implementations}
𝒮 ⇐ ∅
{Get optimum point-to-point arc implementations}
for all arc a ∈ 𝒢 do
  𝒫(a) ⇐ findBestPointToPointImplementation (a, ℒ)
  𝒮 ⇐ 𝒮 ∪ σ(𝒫(a))
end for
{Find non-dominated candidate k-way mergings}
B ⇐ ComputeBandwidthVector (𝒢)
Γ ⇐ ComputeConstrainedDistanceSumMatrix (𝒢)
Δ ⇐ ComputeMergingDistanceSumMatrix (𝒢)
k ⇐ 1
foundCanditateMapping ⇐ TRUE
while (foundCanditateMapping) do
  for all column j ∈ col(Γ) do
    kWayMergingNotFound ⇐ TRUE
    for all row subsets I = {i_1, ..., i_k} ⊆ row(Γ) do
      if (Σ_{i=1}^{k} Δ[i, j] < Σ_{i=1}^{k} Γ[i, j]) then
        b_min ⇐ min{B[j], min_{i∈[1,k]}{B[i]}}
        if (∃l ∈ ℒ s.t.(b(l) + b_min) < Σ_{i=1}^{k} B[i] + B[j]) then
          𝒮 ⇐ 𝒮 ∪ findKMergingImplementation(j, I)
          kWayMergingNotFound ⇐ FALSE
        end if
      end if
    end for
    if (kWayMergingNotFound) then
      col(Γ) ⇐ col(Γ) \ {j}
    end if
  end for
  if col(Γ) = ∅ then
    foundCanditateMapping ⇐ FALSE
  else
    k ⇐ k + 1
  end if
end while
```

Figure 2: Deriving candidate arc implementations.

When combined, the two steps correspond to implicitly considering all possible communication sub-architectures that can be generated by putting together communication library components while satisfying the constraint graph requirements. To be effective, this approach must rely on the ability of generating the smallest possible set of local solutions $\mathcal{S}$ that must necessarily be considered to guarantee that the entire solution space is explored as part of the exact search of the global optimum. One could naively think to generate all possible solutions and leave the responsibility of finding the global optimum to state-of-the-art UCP solvers [4, 8], which provide sophisticated techniques to prune away suboptimal local solutions. In practice, this would turn out to be quite expensive from a computational point-of-view during both steps. Instead, we present a set of theoretical results that guide us during the first step, thus enabling the pruning of many useless branches during the search of the tightest $\mathcal{S}$. Since, the proliferation of alternative arc implementations (on top of those of the optimum point-to-point implementation graph) is due to the possibility of realizing K-way arc mergings, it is natural to focus on defining criteria that establish when a subset of K arcs should not be merged.

DEFINITION 3.1. *Let $\mathcal{G} = G(V, A)$ be a constraint graph and $\mathcal{L} = L \cup N$ a communication library. Also $\forall k \in [2, |A| - 1]$, let A^k denote a subset of A with cardinality $|A^k| = k$ and V^k the vertices connected by the arcs of A^k. Let $\mathcal{G}^k = G(V^k, A^k)$ be the projection of $\mathcal{G}$ onto A^k. A^k is said to be* k-way mergeable *iff the union of the arc implementations of the minimum-cost implementation graph $\mathcal{G}^\star(\mathcal{G}^k, \mathcal{L})$ is a k-way merging. The set of sets of k-way mergeable arcs is denoted as $\mathcal{M}^k$.*

The following lemma provides a sufficient condition to detect when a pair of arcs is not 2-way mergeable [3]. As for all the remaining results in this section, this condition is valid independently from the characteristics of the chosen communication library $\mathcal{L} = L \cup N$, as long as the library satisfies Assumption 2.1.

LEMMA 3.1. *Given a constraint graph $\mathcal{G} = G(V, A)$, let $A^2 = \{a, a'\} \subseteq A$, with $a = (u, v), a' = (u', v')$. If $d(a) + d(a') \le ||p(u) - p(u')|| + ||p(v) - p(v')||$ then $A^2 \notin \mathcal{M}^2$*

The following lemma can be seen as an extension of the result of the previous lemma to the case of k arcs because it provides a sufficient condition that guarantees that they are not k-way mergeable.

LEMMA 3.2. *Let $A^k = \{a_1, \dots, a_k\} \subseteq A$ be a subset of k arcs $a_1 = (u_1, v_1), \dots, a_k = (u_k, v_k)$ of $\mathcal{G} = G(V, A)$. If*

$$\left((k-1) \cdot d(a_k) + \sum_{i=1}^{k-1} d(a_i)\right) \le \sum_{i=1}^{k-1} ||p(u_i) - p(u_k)|| + ||p(v_i) - p(v_k)||$$

then $A^k \notin \mathcal{M}^k$

The following theorem guarantees that if an arc a can not be part of any k-way merging then it also can not be part of any $(k+h)$-way mergings with $h \in [k+1, |A|]$.

THEOREM 3.1. *Let $\mathcal{G} = G(V, A)$ be a constraint graph. For all arc $a \in A$, if $\forall k \in [2, |A| - 2], \forall A^{k-1} \subset A \setminus \{a\}$, $(A^{k-1} \cup \{a\} \notin \mathcal{M}^k)$, then $\forall h \in [k, |A| - 1], \forall A^h \subset A \setminus \{a\}$, $(A^h \cup \{a\} \notin \mathcal{M}^{h+1})$.*

Given a constraint graph $\mathcal{G} = G(V, A)$ and a communication library $\mathcal{L} = L \cup N$, the following result provides a sufficient condition to establish that a subset A^k of A is not k-way mergeable, i.e., formally, that $A^k \notin \mathcal{M}^k$.

THEOREM 3.2. *Let $A^k = \{a_1, \dots, a_k\} \subseteq A$ be a subset of k arcs $a_1 = (u_1, v_1), \dots, a_k = (u_k, v_k)$ of a constraint graph $\mathcal{G} = G(V, A)$ and $\mathcal{L} = L \cup N$ be a communication library. Then,*

$$\left\{\sum_{i=1}^{k} b(u_i, v_i) \ge \left(\max_{l \in L} \{b(l)\} + \min_{j \in [1,k]} \{b(u_j, v_j)\}\right)\right\} \Rightarrow A^k \notin \mathcal{M}^k$$

Figure 2 illustrates the algorithm to generate a minimal set of candidate arc implementations that is based on the above results. First, it is convenient to define two distinct symmetric matrices (the *Constrained Distance Sum Matrix* Γ and the *Merging Distance Sum Matrix* Δ) to capture key quantities related to each pair of arcs in the constraint graph $\mathcal{G} = G(V, A)$. In particular, for any two arcs $a_i = (u_i, v_i), a_j = (u_j, v_j)$, $\Gamma(a_i, a_j) = d(a_i) + d(a_j)$ and $\Delta(a_i, a_j) = ||p(u) - p(u')|| + ||p(v) - p(v')||$. Notice that since the two matrices are symmetric, we only need to scan the values of their upper diagonal part.

After having saved into $\mathcal{S}$ the optimum point-to-point arc implementation associated to each constraint arc, the algorithm proceeds by subsequently considering all possible k-way mergings for incrementing value of k. Using the result of Theorem 3.1, as soon as no k-way mergings are possible for an arc a_j, the corresponding column (and row) is removed from the matrix Γ. The algorithm leaves the main loop, when the set of columns of Γ becomes empty. The algorithm terminates returning the set $\mathcal{S}$ of candidate arc implementations, whose exact structures (i.e the exact topology, communication node position, number of links,...) are later obtained solving a simple nonlinear optimization problem, which computes also their costs. Finally, a unate covering matrix is built by associating to each row a constraint arc, to each column a candidate implementation and setting each entry i, j to 1 if the implementation j implements the arc i, to 0 otherwise. Each columns has also a weight corresponding to the implementation cost. The selection of the optimum global solution corresponds to the solution of this instance of weighted Unate Covering Problem (UCP) and can be found by using state-of-the-art UCP solvers [4, 8].

[3] The theorem proofs can be found in [10].

	a_1	a_2	a_3	a_4	a_5	a_6	a_7	a_8
a_1		10.38	14.05	102.02	105.18	103.61	8.60	8.60
a_2			14.44	102.40	105.56	104.00	8.99	8.99
a_3				106.07	109.23	107.67	12.66	12.66
a_4					197.20	195.63	100.62	100.62
a_5						198.79	103.78	103.78
a_6							102.22	102.22
a_7								7.21
a_8								

Table 1: $\Gamma(a_i, a_j) = d(a_i) + d(a_j)$.

	a_1	a_2	a_3	a_4	a_5	a_6	a_7	a_8
a_1		9.05	14.05	102.02	97.02	102.40	200.09	200.17
a_2			5	103.61	98.61	104.00	201.69	201.58
a_3				98.61	103.61	107.67	198.61	198.42
a_4					5	9.05	100.00	100.63
a_5						5.38	103.07	103.78
a_6							101.40	102.22
a_7								7.21
a_8								

Table 2: $\Delta(a_i, a_j) = ||p(u) - p(u')|| + ||p(v) - p(v')||$.

4. DOMAIN APPLICATION EXAMPLES

The algorithm presented in the previous section is illustrated here by means of two examples that are taken from different application fields. The first example represents a simple wide-area network (WAN), while the second example shows how the current approach can be adapted to attack the on-chip communication synthesis problem, Figure 3-(a) reports the diagram of a wide-area communication network, where the length of the arcs suggests that the"computational" nodes A,B,C are fairly close to each other as well as nodes D and E, while the two groups are separated by a distance which is relatively much larger. We assume that every channel presents the same bandwidth requirement, namely $10 Mbps$. Figure 3-(b) shows the corresponding communication constraint graph where only the ports and the channels are retained. In this case, it is reasonable to adopt the approximation that all the ports of a computation node have the same position. The library that is available for implementing the communication architecture consists of two types of links, whose cost is a function of the supported channel length: a radio link $l_r = (11 Mbps, l, \$2 \times meter)$, and an optical link $l_o = (1 Gbps, l, \$4 \times meter)$. Table 1 and Table 2 report respectively the values of the Constrained Distance Sum Matrix Γ and the Merging Distance Sum Matrix Δ expressed in kilometers. By running the algorithm reported in Figure 2, it is easy to determine that arc a_8 is not mergeable with any other arc and, therefore, will have to be implemented as a minimum-cost point-to-point link. Due to its distance, this links turns out to be the radio

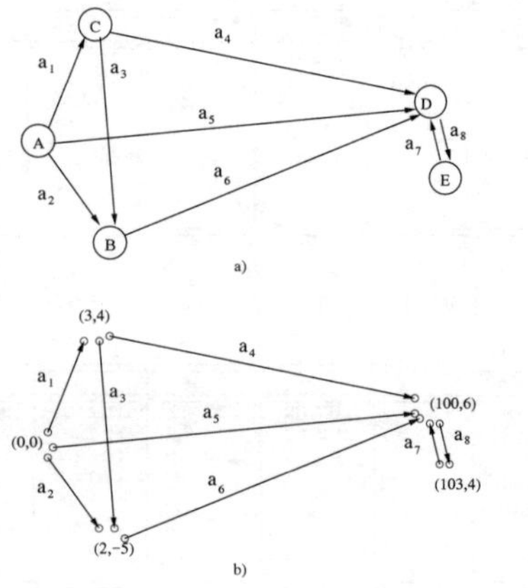

Figure 3: A Simple WAN and its Constraint Graph

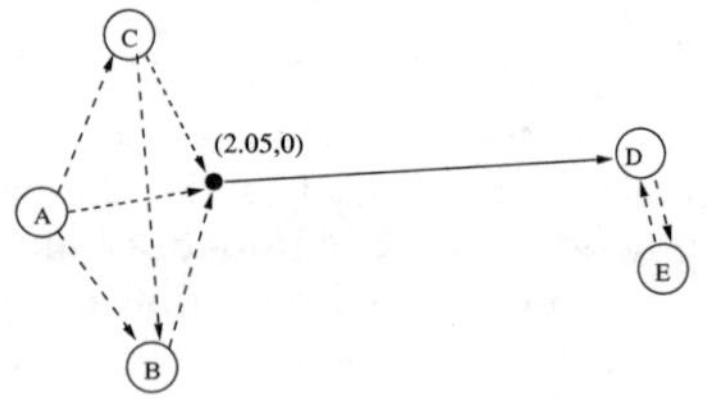

Figure 4: Example 1: Implementation Graph

link. The algorithm also determines that arc a_7 cannot be involved in any 4-way arc merging (nor, therefore, in any k-way mergings with $k > 4$). Besides the 8 optimum-cost point-to-point implementations, the set $\mathcal{S}$ contains thirteen 2-way, twenty-one 3-way, sixteen 4-way, and five 5-way candidate arc mergings. For each candidate implementation the following minimization problem is solved to derive its cost as well as the position of its communication nodes. implementation: *minimize the cost* $C(x)$ *subject to* $K \cdot x = d$. The K matrix is derived by writing the equation forcing that the sum of the lengths along x and y axes must be equal to the difference in position between source and destination points. Finally, after solving the weighted UCP, we learn that the minimum cost solution is obtained by merging the arcs a_4 with a_5 and a_6 in an optical link and implementing each of the other arcs with a dedicated radio link. The result is shown in Figure 4 where the dash-dot lines indicate a radio link and the solid line indicates an optical link.

If we change the application domain by moving to the problem of deriving an architecture for an on-chip communication network, the characteristics of the constraints and the cost function are quite different. Still the proposed approach can be used to find for instance the minimum number of repeaters (stateless buffers) that it is necessary to insert on a metal line while performing an optimum segmentation using the notion of critical length (l_{crit}) as defined in [11]. For this application, a first-cut library $\mathcal{L}$ can be considered as composed by only one link (a metal wire of length l_{crit} that is only dependent on the technology process) and three communication nodes (an inverter, a multiplexer and a demultiplexer, all optimally sized). By using the Manhattan distance as the appropriate measure for the length of the links, the cost of each arc in the implementation graph is given by $\lfloor (|x_v - x_u| + |y_v - y_u|)/l_{crit} \rfloor$. Figure 5-(a) illustrates an example of this application, where we have studied the most critical channels on a multi-processor MPEG 4 decoder implemented in a 0.18μ technology. The final communication architecture, reported in Figure 5-(b), has a total number of 55 required repeaters (with $l_{crit} = 0.6mm$). It is important to notice that this result is valid as long as the assumption that all links on the chip have a delay smaller than the clock period. Naturally, with the advent of deep sub-micron (DSM) process technology (0.13μ and below), this will be true for fewer wires. Still the approach presented in this work can be combined with the recently proposed latency-insensitive methodology [1], after making sure to define a cost function centered on the minimization of both stateless (buffers) and stateful (latches) repeaters.

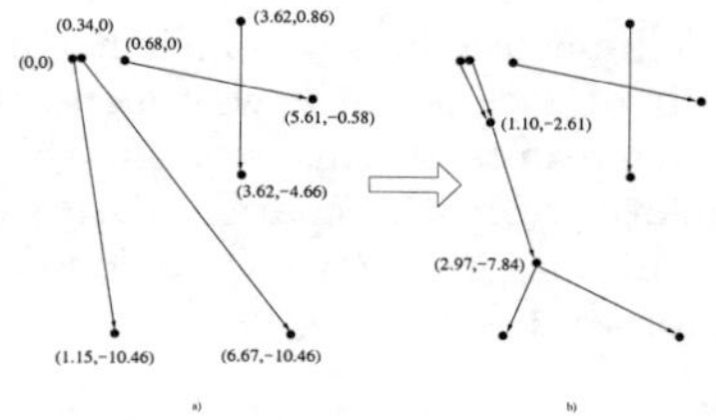

Figure 5: Final Implementation (Ex. 2).

5. CONCLUSIONS

We introduced a novel algorithm for the automatic synthesis of a communication architecture among a set of computational blocks once their relative positions and required pairwise communication bandwidth is provided. The algorithm is the result of a new way of modeling the problem (embodied by the notion of communication constraint graph) and it is based on a series of theoretical results that give simple conditions to detect when a possible merging between point-to-point communication channels should be avoided because it is guaranteed to be part of a suboptimal solution.

6. REFERENCES

[1] L. P. Carloni, K. L. McMillan, A. Saldanha, and A. L. Sangiovanni-Vincentelli. A Methodology for *"Correct-by-Construction"* Latency Insensitive Design. In *Proc. Intl. Conf. on Computer-Aided Design*, pages 309–315. IEEE, Nov. 1999.

[2] C. Chang, P. Kermani, and A. Kershenbaum. Multi-link-speed network topology design. In *International Phoenix Conference on Computers and Communications*, pages 299–306, 1992.

[3] J. Daveau, T. B. Ismail, and A. Jerraya. Synthesis of system level communication by an allocation based approach. In *International Symposium on System Synthesis*, pages 150–155, 1995.

[4] E. I. Goldberg, L. P. Carloni, T. Villa, R. K. Brayton, and A. L. Sangiovanni-Vincentelli. Negative Thinking in Branch-and-Bound: the Case of Unate Covering. *IEEE Transactions on Computer-Aided Design*, 19(3):281–294, Mar. 2000.

[5] K. Keutzer, S. Malik, A. R. Newton, J. Rabaey, and A. Sangiovanni-Vincentelli. System level design: Orthogonolization of concerns and platform-based design. *IEEE Transactions on Computer-Aided Design of Integrated Circuits and Systems*, 19(12), December 2000.

[6] P. V. Knudsen and J. Madsen. Integrating Communication Protocol Selection with Hardware/Software Codesign. *IEEE Transactions on Computer-Aided Design*, 18(8):1077–1095, Aug. 1999.

[7] K. Lahiri, A. Raghunathan, and S. Dey. Efficient exploration of the soc communication architecture design space. In *Proc. Intl. Conf. on Computer-Aided Design*, pages 424–430, 2000.

[8] S. Liao and S. Devadas. Solving covering problems using LPR-based lower bounds. In *Proc. of the Design Automation Conf.*, June 1997.

[9] L.-S. Peh and W. J. Dally. Flit reservation flow control. In *International Symposium on High-Performance Computer Architecture*, pages 74–84, 1999.

[10] A. Pinto, L. P. Carloni, and A. L. Sangiovanni-Vincentelli. Constraint-Driven Communication Synthesis. Technical Report available at www-cad.eecs.berkeley.edu/~lcarloni, Apr. 2002.

[11] R. H. J. M. Otten and R. K. Brayton. Planning for Performance. In *Proc. of the Design Automation Conf.*, pages 122–127, June 1998.

Component-Based Design Approach for Multicore SoCs[1]

W. Cesário, A. Baghdadi, L. Gauthier, D. Lyonnard, G. Nicolescu, Y. Paviot, S. Yoo, A.A. Jerraya
TIMA Laboratory, SLS Group
46, av. Félix Viallet
38031 Grenoble Cedex - France
{Wander.Cesario,Ahmed.Jerraya}@imag.fr

M. Diaz-Nava
STMicroelectronics
AST Grenoble Lab
12, rue Jules Horowitz, BP217
38019 Grenoble Cedex - France
Mario.Diaznava@st.com

ABSTRACT
This paper presents a high-level component-based methodology and design environment for application-specific multicore SoC architectures. Component-based design provides primitives to build complex architectures from basic components. This bottom-up approach allows design-architects to explore efficient custom solutions with best performances. This paper presents a high-level component-based methodology and design environment for application-specific multicore SoC architectures. The system specifications are represented as a virtual architecture described in a SystemC-like model and annotated with a set of configuration parameters. Our component-based design environment provides automatic wrapper-generation tools able to synthesize hardware interfaces, device drivers, and operating systems that implement a high-level interconnect API. This approach, experimented over a VDSL system, shows a drastic design time reduction without any significant efficiency loss in the final circuit.

Categories and Subject Descriptors
C.3 [**SPECIAL-PURPOSE AND APPLICATION-BASED SYSTEMS**]: *Microprocessor/microcomputer applications, Real-time and embedded systems.*

General Terms
Design, Experimentation, Standardization.

Keywords
Multicore System-on-Chip, Component-based design, HW/SW interfaces abstraction.

1. INTRODUCTION
The ITRS roadmap [1] predicts that by 2004 70% of ASICs will include at least one embedded instruction-set processor, which means that most ASICs will be Systems-on-Chip (SoC). Some existing designs confirm and strengthen this prediction, many applications include several processors with different instruction-sets: mobile terminals (e.g. GSM [2]), set-top boxes (e.g. PNX8500 from Philips [3]), game processors (e.g. PlayStation2 from Sony [4]) and network processors [5]. Most system and semiconductor houses are developing platforms involving several cores (MCU, DSP, IP, etc) and sophisticated communication interconnects (hierarchical bus, TDMA-based bus, point-to-point connections and packet-routing switches) on a single chip. The trend is moving now to interconnect standard components in the same fashion performed for boards few years ago. This evolution is creating several breaking points in the design process and new challenges for the EDA industry:

1. Systems are the more and more complex: communication cannot be specified manually at the register-transfer level (RTL) as before without introducing many errors. At this level, bus structures (data, address and control) and clock cycles need to be detailed to verify logical and electrical constraints. However, this level is not well adapted to check communication protocols because this would be too much time consuming. By these reasons, higher abstraction levels, to model and verify the interconnection between components, are required to master the complexity: reduce design time and errors.

2. SoCs will include many processors with different instruction-sets to execute dedicated functions in order to increase the flexibility of the whole system. The complexity of the code will become higher than the hardware part and will require, in the near future, several hundreds or thousands person-years [1]. This software will not be programmed at the assembler level as today.

3. Complex, on-chip, hardware/software interfaces are required to implement an application-specific communication interconnect. The hardware (microprocessor interfaces, bank of registers, memories) and software (drivers, operating systems) elements required to perform the communication protocols need to be adapted to the communication interconnect according to the type of core.

This paper presents a systematic approach for the assembling of application-specific SoC and the definition of a composition model to abstract the interconnection between hardware (cores) and software (tasks) components. Section 2 introduces the basic concepts for complex SoC design and the component-based approach. Section 3 details a new component-based specification model and the design flow that enables multicore SoC design at a higher level than RTL. Section 4 presents the application of this flow onto the design of a VDSL circuit and the analysis of the results. Finally, section 5 concludes this paper.

[1] This research was partially funded by MEDEA+ project A508.

2. BASICS: SYSTEM-ON-CHIP DESIGN

The expression SoC is used to designate an ASIC that combines heterogeneous components (CPUs, DSPs, IPs, memories, buses, etc.) on the same chip. This is similar to what we were used to see on system boards.

2.1 SoC Architecture

The classic literature about multiprocessor systems provides all the basic concepts used in this work [7][8], but it is too general to be efficient for SoCs. Figure 1 shows a typical multicore SoC architecture with heterogeneous processors and the on-chip communication interconnect. A key difference with classic computer architecture is that this model distinguishes (based on their utilization) two kinds of processors (CPUs): those used to run the end application and those dedicated to execute specific functions that could have been designed in hardware. The programming and interfacing of these two kinds of processors are quite different, as we will explain later.

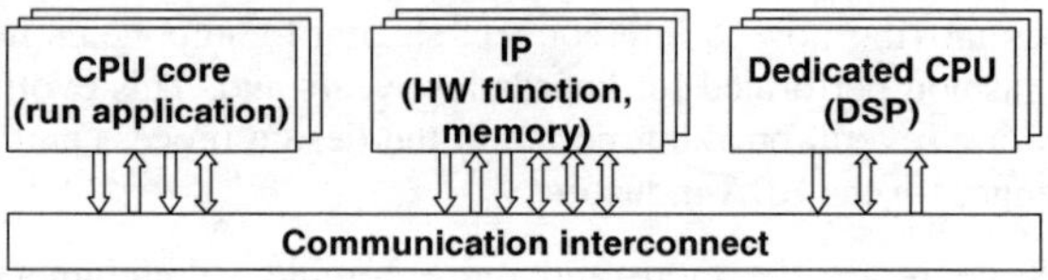

Figure 1. A typical multicore SoC architecture

Additionally, dedicated processors require application-specific communications and memory schemes due to performance optimization reasons. These specific architecture optimizations are generally related to the application domain. For instance, application-specific interfaces are required to respect tight resource/performance constraints (e.g. area, power, runtime, etc.).

2.2 Software Organization

Each kind of embedded processor employs a different software organization. The application software is generally organized as a stack of layers on top of the hardware as shown in Figure 2a.

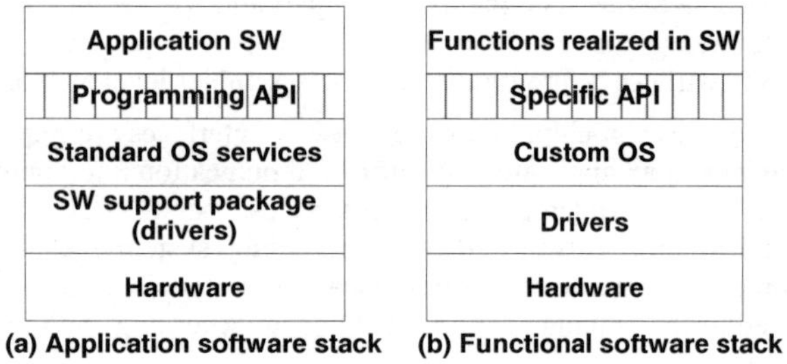

Figure 2. Software organization

The lowest software layer provides the drivers and the low-level architecture controllers. This is also called the software support package, and is generally provided by the SoC design team. The upper layers, OS services and SW application layer are provided by the application designer. The OS layer generally reuses an existing embedded OS, which may be customized to the application. The top layer is application-specific software that may be designed using high-level languages and code generation techniques. Application coding is not part of the SoC design and will be omitted in the rest of this paper.

Presently, the software running on the dedicated processors is written in Assembly/C. When multitasking and complex I/O are required, specific code is provided to perform scheduling and to manage I/O. This code is generally mixed with the functional code, and there is no separated OS layer. This low-level programming is one of the main bottlenecks in SoC design. Additionally, this scheme lacks flexibility/portability and makes it difficult to change the hardware part without a complete redesign.

One of the main contributions of this work is to consider higher-level programming approach for dedicated software (often-called firmware). Each dedicated CPU will use a software stack (see Figure 2b). Of course, for code size and performance reasons, it will not be realistic to use any existing generic OS as isolation layer. In this case, a custom OS supporting an application-specific program interface (API) is required. Custom OS design automation is a new research area brought by SoC design [9]. With this scheme, both dedicated and application software can be written independently from the hardware implementation. Another benefit of this approach is that dedicated software can be independent from hardware/software frontiers on architecture components and OS choices. Finally, the main advantage of the work presented in this paper is to allow this level of flexibility/portability without any loss of performance.

2.3 SoC Design Methods

Currently, most SoC designs start at the RT-level for the hardware interfaces and at the ISA level for the dedicated software components. Validation of the overall design uses co-simulation techniques that generally combine one hardware simulator and one or several instruction-set simulators (ISS). As explained in [10], two design automation approaches are competing to improve productivity: top-down and bottom-up approaches. The most popular top-down approaches are: synthesis from system level models [11][12][13] and platform-based design [14][15]. These approaches start with an architectural solution, target architecture or architecture platform. In contrast, the bottom-up approach, also called *component-based design*, starts with a set of components and provides a set of primitives to build application-specific architectures and communication APIs. The key idea of component-based design is to increase the abstraction level when designing component interconnections. Even if this approach does not provide much help on automating the architecture exploration phase, it may provide a considerable reduction of design time for hardware/software communication refinement and component integration, and facilitate IP reuse.

The key point in such a flow is the use of an abstract architecture where communication is separated from the component on the hardware side and from the functions on the software side. This abstract architecture may be used by the software programmer as an API. On the other side, the abstract architecture may use hardware components through an abstract API. This is generally called separation between communication and computation for component-based design [15].

2.4 Component-based Approaches

In component-based SoC design [16], the goal is to enable the integration of heterogeneous processors and communication protocols by using abstract interconnections. Behavior and communication must be separated in the system specification. Thus, system communication can be described at a higher-level and refined independently of the behavioral of system.

There are two component-based design approaches: usage of a standard bus protocol [17] and usage of a standard component protocol [18][19]. IBM defined a bus architecture called CoreConnect that adopts the first approach [17]. To connect heterogeneous components to the bus architecture, a wrapper is designed to adapt the protocol of each component to CoreConnect protocols. VSIA promotes the second approach with VCI and FI as standard component protocols [18]. In this case, the designer can choose a bus protocol and then design wrappers to interconnect using this protocol. [20] also present methods for on-chip network communication design in the case of multicore SoCs. This paper introduces a new concept, called virtual architecture, to cover both approaches listed above.

Recently, several commercial tools are trying to deal with the component-based design concept. CoWare presents the tool N2C [21], Cadence presents VCC (Virtual Component Codesign) [14], Sonics presents Silicon Backplane µNetwork [19]. Still, designers are not able to obtain significant reduction in system design cycle and optimized multicore architectures. This is mainly due to the non-separation of functional and application software and to the non-availability of custom OS generation tools.

3. COMPONENT-BASED DESIGN

This section introduces a high-level system design methodology for multicore SoC design using a communication refinement approach for virtual components. The system is described as a set of virtual components interconnected via channels. A virtual component consists of a wrapper and an internal component (or module). The internal component corresponds to a software task or a hardware function. The wrapper adapts accesses from the internal component to the external channels connected to the virtual component. The internal component and external channel(s) can be different in terms of: (1) communication protocol, (2) abstraction level, and (3) specification language. Depending on the difference, the functionality/structure of the wrapper is determined and automatically generated.

3.1 Virtual Architecture Model

The virtual architecture, also called macro architecture, represents a system as an abstract netlist of virtual components (see Figure 3). Each virtual component has a wrapper that is composed of a set of virtual ports [22]. A virtual port has internal and external ports. There could be an n to m (n and m are natural numbers) correspondence between the internal and the external ports. Internal and external ports are not directly connected because they can use different communication protocols and/or transmit signals at different abstraction levels. For this reason, in some cases the wrapper will do a conversion between different internal/external communication protocols and/or abstraction levels.

Virtual channels hide many details of communication protocols, for instance, FIFO communication is realized using high-level communication primitives (e.g. *put* and *get*). In our design flow, the virtual architecture is described using an extension of SystemC. Three new concepts are used:

1. virtual module: consists of a module and its wrapper;

2. virtual port: groups the corresponding internal and external ports having a conversion relationship. A wrapper may be composed of an hierarchy of several virtual ports;

3. virtual channel: grouping several channels having a logical relationship (e.g. multiples nets belonging to the same communication protocol).

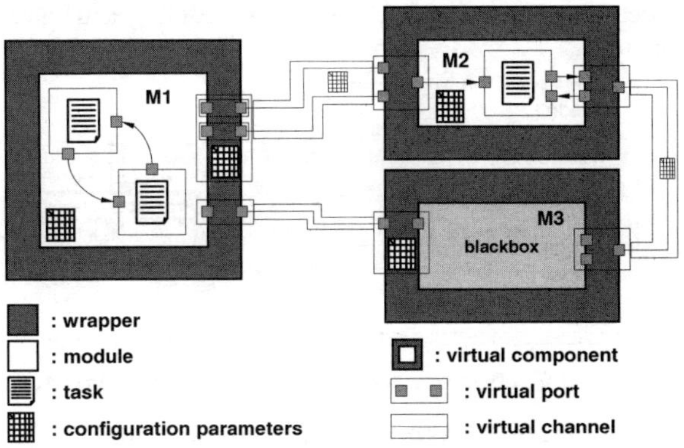

Figure 3. The virtual architecture model

For system refinement, this model needs to be annotated with architecture configuration parameters (e.g. protocol and physical addresses of ports). This model is not synthesizable/executable because the behavior of wrappers is not described. The main goal of this design methodology is to generate automatically these wrappers, in order to produce a detailed architecture that can be both synthesized and simulated.

3.2 The Generic Architecture Model

When defining this model our goal was to have a generic model where both computation and communication may be customized to fit the specific needs of the application. For computation, we may change the number and kind of components and, for communication, we can select a specific communication scheme. The architecture model is suitable to a wide domain of applications; more details can be found in [24].

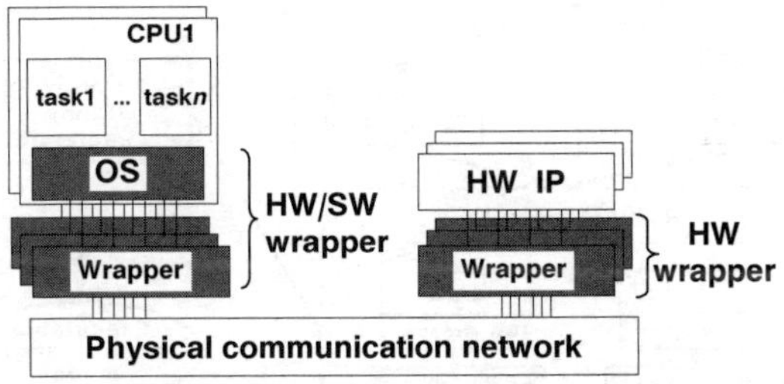

Figure 4. Generic architecture model for multicore SoC

We use a generic multicore SoC architecture where processors are connected to communication networks via wrappers (see Figure 4). In fact, HW processors are separated from the physical communication network by wrappers that act as communication coprocessors or bridges. Such a separation is necessary to free the processors from communication management and it enables parallel execution of computation tasks and communication protocols. Software tasks need also to be isolated from hardware through an OS that plays the role of SW wrapper.

As shown in Figure 5, the wrapper is made of a software part and a hardware part. On the hardware side, the internal architecture of the wrapper consists of a processor adapter, a channel adapter, and an internal bus. The number of channel adapters depends on the number of channels that are connected to the corresponding virtual module. On the software side, wrappers provide the implementation of high-level communication primitives (API) used in the software module and the drivers to control the hardware. If required, the software wrapper will also provide sophisticated OS services such as task scheduling and interrupt management.

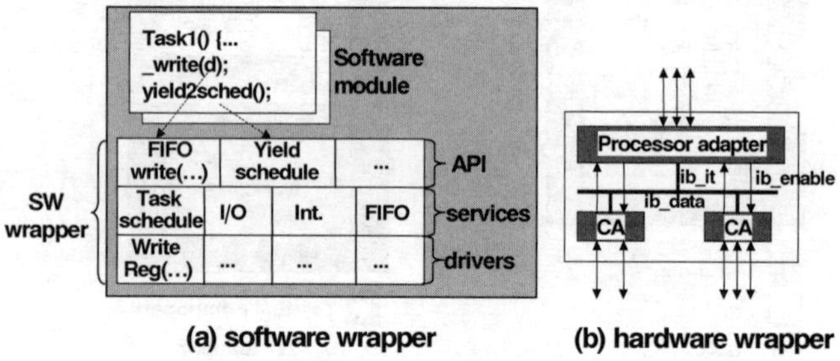

Figure 5. HW/SW wrapper architecture

3.3 Automatic Wrapper Generation

The overall view of our design environment is shown in Figure 6. An initial internal model is obtained from a translation of the extended SystemC specification, which is a virtual architecture annotated with configuration parameters. There are three tools for automatic wrapper generation: co-simulation wrapper generator, hardware wrapper generator, and the software wrapper generator. Details about these tools can be found in [9][23][24], only their principle will be discussed here.

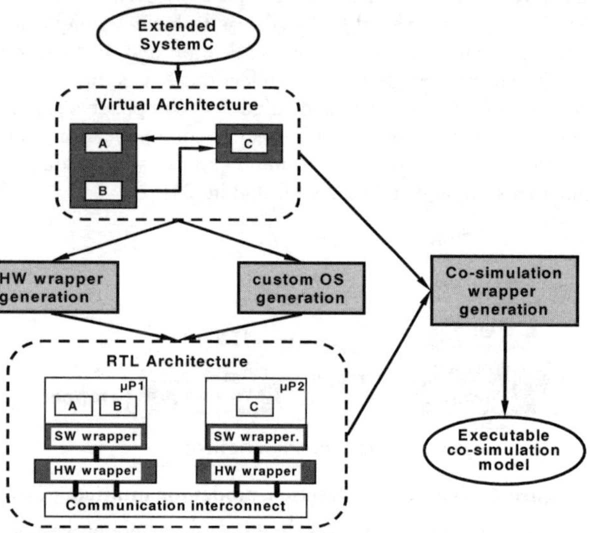

Figure 6. Design environment for multicore SoC

The co-simulation wrapper generator [23] produces an executable model that is used to validate the internal model. This executable model is composed of a SystemC simulator that acts as a master for other simulators. A variety of simulators can participate in this co-simulation: SystemC, VHDL, Verilog, and Instruction-set simulators. In the co-simulation wrapper library, there are simulation adapters for the different simulators supported. There are also channel adapters that implement all supported communication protocols in different languages.

The software wrapper generator [9] produces operating systems streamlined and pre-configured for the software module(s) that run(s) on each target processor. It uses an operating system library that is organized in three parts: APIs, communication/system services, and device drivers. Each part contains elements that will be used in a given software layer in the generated OS. This library contains a dependency graph between services/elements that is used to determine the minimal set of elements necessary to implement a given OS service. This mechanism is used to keep the size of the generated OS at a minimum, by avoiding inclusion of unnecessary elements from the library.

The library used by the hardware wrapper generator [24] has two parts: the processor library and the protocol library. All models stored in this library are synthesizable; they are instantiated and configured using the architecture configuration parameters. The processor library contains processor adapters, template architectures for processors, processor cores, local memories and peripherals. For instance, the following configuration parameters are used for processor adapter instantiation: port addresses, number of interrupts and their priorities. The protocol library contains channel adapters and communication network models. Channel adapter configuration uses the following configuration parameters: input/output type, master/slave operation, type of data transmitted, buffer size, and interrupt parameters.

4. COMPONENT-BASED DESIGN OF A VDSL APPLICATION

This section demonstrates the application of the high-level component-based methodology using a VDSL application as a design example.

4.1 The VDSL Modem Specification

The design we present in this section was taken from the implementation of a VDSL modem using discrete components [6]; the block diagram for this prototype implementation is shown in Figure 7. The subset we will use in the rest of this paper is shaded in Figure 7. It is composed of two ARM7s and part of the datapath: the TX_Framer, described at the RT-level and was used as a blackbox in this experiment.

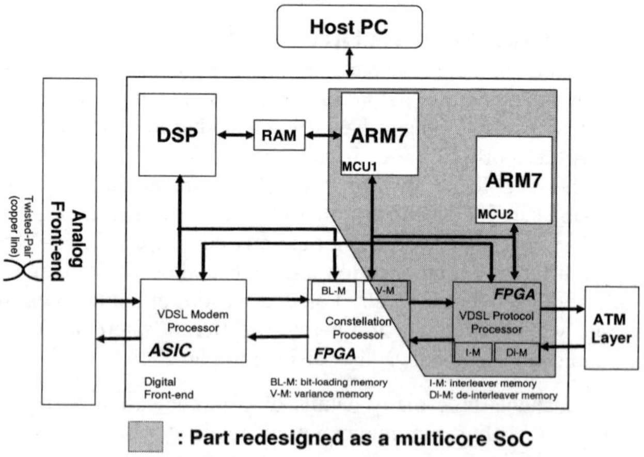

Figure 7. VDSL multicore SoC architecture

Despite all simplifications, the design of the selected subset remains quite challenging. In fact, this application uses two processors executing parallel tasks. The control over the three modules of the specification is fully distributed. All three modules act as masters when interacting with their environment. Additionally, the application includes some multipoint communication channels requiring sophisticated OS services.

4.2 VDSL Virtual Architecture

Figure 8 shows the virtual architecture model that captures the VDSL specification. Modules **VM1** and **VM2** correspond to the ARM7s on Figure 7 and module **VM3** represents the TX_Framer block (only the interface is known so it is represented as a blackbox). The virtual architecture can be mapped onto different architectures depending on the configuration parameters annotated in modules, ports, and nets. For instance, the three point-to-point connections (**VC1**, **VC2**, and **VC3**) between **VM1** and **VM2** can be mapped onto a bus or onto a shared memory if the designer changes the configuration parameters placed on these virtual channels and virtual ports.

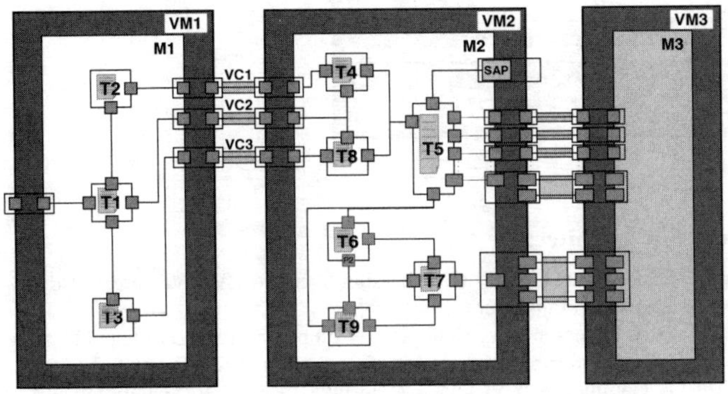

Figure 8. VDSL virtual architecture specification

4.3 Results

The manual design of a full VDSL modem requires several person-years; the presented subset was estimated as a more than five persons-year effort. When using this high-level component-based approach, the overall experiment took only one person during four months (not counting the effort to develop library elements and debug design tools). This corresponds to a 15-fold reduction in design effort. Running all wrapper generation tools takes only a few minutes on a Linux PC 500 MHz, most of the time was spent in writing the virtual architecture model with all the necessary configuration parameters. The behavior of each task was described using the SystemC C++ library [25].

Figure 9 shows the RTL architecture model obtained after HW/SW wrapper generation: two ARM7 cores with their local architectures, the TX_Framer block. Each hardware wrapper acts as a communication co-processor for the ARM7, it contains an ARM7 processor adapter that bridges the ARM7 local bus to the communication adapters (CAs). There is a CA for each virtual channel in the virtual architecture specification. For instance, module **VM1** reads test vectors from the environment through a simple register using the *Pooling* CA, and communicates with **VM2** through asynchronous FIFOs using 3 *HNDSHK* CAs (corresponding to virtual channels **VC1**, **VC2**, and **VC3**).

Each software wrapper (custom OS) is customized to the set of tasks executed by the processor core. For example, software tasks running on **VM1** access the custom OS using an API composed of two functions: *Pipe* for communication with **VM2**, and *Signal* to modify the task scheduling on runtime. The custom OS contains a round-robin scheduler (*Sched*) and resource management services (*Sync*, *IT*). The driver layer contains low-level code to access the CAs (e.g. *Pipe LReg* for the *HNDSHK* CA), and some low-level kernel routines.

The hardware wrapper for processor **VM2** includes a timer module because task **T5** (see Figure 8) must wait 10ms before starting its execution. A hardware interrupt is generated by the TIMER block, the task can configure this block using the *Timer* API provided by the service access port **SAP** (see Figure 8). The custom OS for **VM2** provides a more complex API: the *Direct* API is used to write/read to/from the configuration/status registers inside the TX_Framer block; *SHM* and *GSHM* are used to manage shared-memory communication between tasks.

Application code and generated OS are compiled and linked together to execute on each ARM7 processor. The HW wrapper can be synthesized using RTL synthesis. Table 1 presents the results regarding the generated OSs. Part of the OS is written in Assembly, it includes some low-level routines (e.g., context switch and processor boot) that are specific to each processor.

Table 1. Results for OS generation

OS results	# of lines C	# of lines Assembly	Code size (bytes)	Data size (bytes)
VM1	968	281	3829	500
VM2	1872	281	6684	1020
Context switch (cycles)				36
Latency for interrupt treatment (cycles)				59(OS) + 28(ARM7)
System call latency (cycles)				50
Resume of task execution (cycles)				26

If we compare the numbers presented in Table 1 with configurable commercial embedded OSs, the results are still very good. Generally, the minimum size for commercial OSs is around 4kbytes; but with this size, few of them could provide the required functionality. Performance was also very good: context-switch takes 36 cycles, latency for hardware interrupts is 59 cycles (plus 4 to 28 cycles needed by the ARM7 to react), latency for system calls is 50 cycles, and task reactivation takes 26 cycles.

Table 2 shows the numbers obtained after RTL synthesis of the HW wrappers using a CMOS 0.35μm technology. These results are good because wrappers account for less than 5% of the ARM7s core surface and have a critical path that corresponds to less than 15% of the clock cycle for the 25MHz ARM7 processors used in this case study.

Table 2. Results for the hardware wrapper generation

HW interfaces	# of Gates	Critical path delay (ns)	Max. freq. (MHz)
VM1	3284	5.95	168
VM2	3795	6.16	162
Latency for read operation (clock cycles)			6
Latency for write operation (clock cycles)			2
Number of code lines (RTL VHDL)			2168

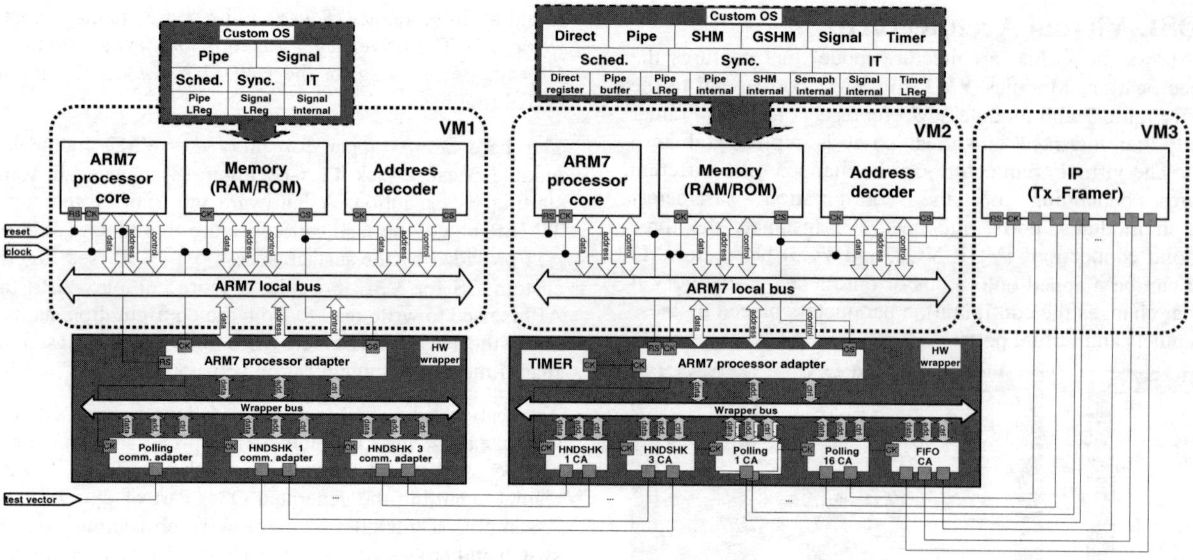

Figure 9. Generated multicore SoC architecture

4.4 Evaluation

The results extracted from the RTL model show that our approach can generate hardware/software interfaces and operating systems that are as efficient as manually coded/configured ones. Wrappers are build-up from library components so the HW/SW frontier in wrapper implementation can be displaced. This choice is transparent to the final user since everything that implements the interconnect API (hardware interfaces and OS) is generated automatically. Designers do not need to rewrite the application code because the API does not change (only its implementation does). Furthermore, correctness and coherence can be verified inside tools and libraries against the API semantics without having to impose fixed boundaries to the hardware/software frontier (in contrast to standardized interfaces and buses).

5. CONCLUSION

This paper presented a high-level component-based methodology and design environment for application-specific multicore SoC architectures. The system specification is a virtual architecture annotated with configuration parameters described in a SystemC-like model. Our component-based design environment has automatic wrapper-generation tools able to synthesize hardware interfaces, device drivers, and operating systems that implement a high-level interconnect API. Results show that wrappers generated automatically by these tools have performances close to the commercial/handcrafted equivalents.

6. REFERENCES

[1] ITRS, available at http://public.itrs.net/

[2] A. Nagari, et al., "A 2.7V 11.8 mW Baseband ADC with 72 dB Dynamic Range for GSM Applications," 21st Custom Integrated Circuits Conference, San Diego, 1999.

[3] http://www.semiconductors.philips.com/platforms/nexperia/

[4] Oka and Suzuoki, "Designing and Programming the Emotion Engine," IEEE Micro, vol. 19:6, pp. 20-28, Nov/Dec 1999.

[5] P. Paulin, F. Karim, and P. Bromley, "Network Processors: A Perspective on Market Requirements," Proc. of DATE, 2001.

[6] M. Diaz-Nava, G.S. Okvist, "The Zipper prototype: A Complete and Flexible VDSL Multi-carrier Solution", ST Journal special issue xDSL, September 2001.

[7] D.E. Culler, J. Pal Singh, "Parallel Computer Architecture," Morgan Kaufmann Publishers, 1999.

[8] D.A. Patterson , J.L. Hennessy, "Computer Organization and Design - The Hardware/Software Interface," Morgan Kaufmann Pub., 1998.

[9] L. Gauthier, S. Yoo, and A.A. Jerraya, "Automatic Generation and Targeting of Application Specific Operating Systems and Embedded Systems Software", IEEE TCAD, Vol. 20 Nr. 11, November 2001.

[10] K. Keutzer, "A Disciplined Approach to the Development of Platform Architectures," Synthesis and System Integration of Mixed Technologies, Nara, Japan, October 18 - 19, 2001.

[11] R. Ernst, et al., "The COSYMA environment for hardware/software cosynthesis of small embedded systems," Microprocessors and Microsystems, 1996.

[12] F. Balarin, et al., "Hardware-Software Co-design of Embedded Systems: The POLIS approach," Kluwer Academic Press, 1997.

[13] D. Gajski, et al., "SpecC Specification Language and Methodology," Kluwer Academic Publishers, 2000.

[14] Cadence Design Systems, Inc., Virtual Component Co-design: http://www.cadence.com/products/vcc.html

[15] K. Keutzer, et al., "System-level design: orthogonalization of concerns and platform-based design," IEEE TCAD, Dec. 2000.

[16] M. Sgroi, et al., "Addressing the System-on-Chip Interconnect Woes Through Communication-Based Design," Proc. of 38th Design Automation Conference, Las Vegas, June 2001.

[17] IBM Inc., Blue Logic Technology, http://www.chips.ibm.com/bluelogic/

[18] Virtual Socket Interface Alliance, http://www.vsi.org.

[19] D. Wingard, "MicroNetwork-Based Integration for SOCs," Proc. of DAC, Las Vegas, June 2001.

[20] J. A. J. Leijten et al., "PROPHID : A Heterogeneous Multi-Processor Architecture for Multimedia," Proc. of ICCD, 1997.

[21] Coware Inc., N2C: http://www.coware.com/

[22] W.O. Cesário, et al., "Colif: A design representation for application-specific multiprocessor SOCs," IEEE Design & Test of Computers, Vol.: 18, Issue: 5, pp. 8-20, Sept.-Oct. 2001.

[23] S. Yoo, G. Nicolescu, D. Lyonnard, A. Baghdadi, A. A. Jerraya, "A Generic Wrapper Architecture for Multi-Processor SoC Cosimulation and Design," Int. Symposium on HW/SW Codesign (CODES) 2001.

[24] D. Lyonnard, S. Yoo, A. Baghdadi, A. A. Jerraya, "Automatic Generation of Application-Specific Architectures for Heterogeneous Multiprocessor System-on-Chip," Proc. of DAC, Las Vegas, 2001.

[25] OSCI: http://www.systemc.org/

Traffic Analysis for On-chip Networks Design of Multimedia Applications[†]

Girish Varatkar Radu Marculescu

Department of Electrical and Computer Engineering
Carnegie Mellon University
Pittsburgh, PA 15213-3890

{gvv,radum}@ece.cmu.edu

Abstract: *The objective of this paper is to introduce self-similarity as a fundamental property exhibited by the bursty traffic between on-chip modules in typical MPEG-2 video applications. Statistical tests performed on relevant traces extracted from common video clips establish unequivocally the existence of self-similarity in video traffic. Using a generic communication architecture, we also discuss the implications of our findings on on-chip buffer space allocation and present quantitative evaluations for typical video streams. We believe that our findings open up new directions of research with deep implications on some fundamental issues in on-chip network design for multimedia applications.*

Categories and Subject Descriptors: C.4 [**Performance of systems**]: Modeling techniques; B.8.2 [**Performance and reliability**]: performance analysis and design aids.

General terms: performance, design

Keywords: system-level design, on-chip networks, communication analysis, self-similarity, long-range dependence.

1. Introduction and objectives

A fundamental issue in system-level design consists of selecting the optimal mechanism of communication between different on-chip modules [16][17]. For complex systems composed of many heterogeneous components, the on-chip traffic produced among different modules has very diverse characteristics. Since the traffic patterns depend so much on the target application, it is necessary to judiciously allocate the communication resources, especially since the on-chip buffer space is usually very limited.

Recently, Dally and Towles [1] proposed a novel on-chip interconnection network (Fig.1(a)) which can be used instead of the classical ad-hoc global wiring structure. What makes this generic architecture very attractive is that it offers well-controlled electrical parameters which enables high-performance circuits to reduce latency and increase bandwidth.

[†]Research supported by NSF CCR-00-93104, DARPA/Marco Gigascale Research Center (GSRC), and SRC 2001-HJ-898.

DAC 2002, June 10-14, 2002, New Orleans, Louisiana, USA.

As shown in Fig.1(a), a chip employing such a communication architecture consists of several network *clients* (e.g. processors, memories, and custom logic) which are connected to a network that routes *packets* between them. Each client is placed on a tile and communicates with other clients (not only its neighbors) via the on-chip network. A *router* is needed for each tile and it consists of several input-output controllers and their associated *buffers* (Fig.1(b)). From a practical point of view, the success of such an architecture depends on the ability to keep the overall area overhead to a *minimum*[1]. Since the area of the router is heavily dominated by the space occupied by the on-chip buffers, the problem of *optimal buffer sizing* becomes an issue of critical importance. Indeed, dropping or misrouting packets because of inappropriate buffer sizing reduces the overall performance and significantly increase the on-chip power dissipation. We also point out that this severe limitation of the on-chip buffer space comes in deep contrast with real data networks where there is ample room for very large buffers. This makes the on-chip network design problem unique and challenging.

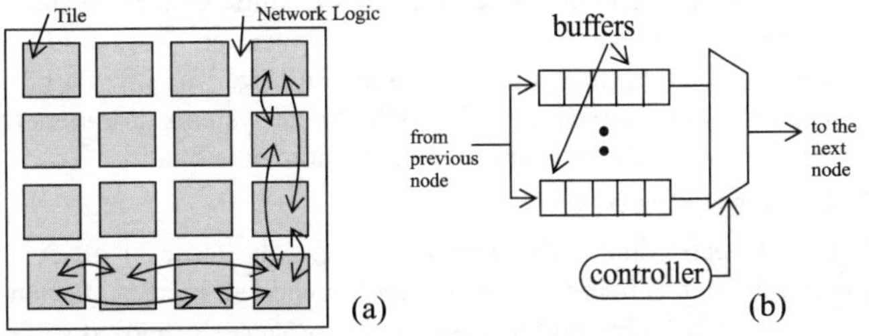

Fig.1(a) Die module tiles and network logic, Fig.1(b) A generic input controller and its buffers.

The objective of this paper is to propose a novel traffic analysis approach as a precise way to characterize the on-chip communication pattern of multimedia applications. More precisely, we propose a technique for traffic modeling based on *self-similar* or *Long-Range Dependent* (denoted as LRD[2]) stochastic processes. By analyzing the statistical properties of the arrival process at different points in a generic architecture like the one in Fig.1, for a standard MPEG-2 application, we first demonstrate that the self-similarity is a characteristic behavior of the on-chip traffic. Second, we characterize *quantitatively* the degree of self-similarity of the on-chip traffic using standard techniques based on Hurst parameter [3]. Knowing the Hurst parameter helps the designer to choose the *minimal* buffer size for the router at each tile in Fig.1 which will guarantee a certain *Quality of Service* (QoS).

1. The authors in [1] suggest about 6% area overhead of network logic for each tile.
2. We use interchangeably Self-Similarity and Long-Range Dependence (LRD).

The analysis we propose is especially relevant to the large class of *portable embedded multimedia systems* where the QoS requirements vary considerably from one media to another (e.g. video connections require consistently high throughput, but tolerate reasonable levels of jitter or packet errors) and buffer space is very limited. Consequently, the ability to explore several communication schemes while trying to satisfy QoS requirements is of crucial importance. As we show later in the paper, making use of the knowledge of traffic pattern for achieving a certain QoS with optimal resources turns out to be extremely helpful.

1.1. Contributions of the paper

The contributions of this paper are threefold:

• First and foremost, we propose a completely new way to address the problem of on-chip network design. To this end, we show how *self-similar* (or *LRD*) processes can be used effectively to model the bursty traffic behavior at chip-level.

• Second, we provide evidence about the presence of self-similar phenomena in on-chip traffic generated by multimedia applications. This has very important consequences since self-similar processes have properties which are completely different from traditional *short-range dependent* autoregressive (ARMA) or Markovian processes which have been mostly used in system-level analysis [15][18-20].

• Third, knowing the Hurst parameter which characterizes the traffic pattern for a particular application can be used to generate *synthetic traces* with statistical properties similar to the original ones [30]. These synthetic traces can be used to dramatically speed up the simulation process for multimedia applications where tens of hours of simulation are typically required to gather useful information for on-chip network design.

Taken together, our proposed technique allows media systems designers implementing on-chip communication networks to choose the appropriate buffer sizes and use large multimedia data benchmarks more effectively. This will enable systems designers to optimally trade-off performance metrics and media quality.

1.2 Related work

In recent years, due to the advent of SoCs, the issue of efficient communication schemes - at chip level - received increased attention [21-25][28]. One problem with the approaches proposed so far for on-chip network exploration is that they rely entirely on explicit simulation. Consequently, due to the huge amount of data contained in multimedia applications, the simulation-based techniques tend to become prohibitively expensive in practice [14][21][27]. Typically, tens of hours are needed to simulate a few minutes of video data. Moreover, simulating randomly video data, without a precise (quantitative) measure of traffic characteristics, is dangerous since the actual implications of traffic on the system performance may be obscured by using inappropriate data. These issues prompted our attention towards a more *formal* approach for on-chip communication analysis with emphasis on precise characterization of multimedia traffic.

As such, our paper is an attempt to bridge conceptually two very different worlds: data networks and on-chip networks. To this end, we first identify, at chip-level, a phenomenon discussed so far only in the context of traffic for real data networks [2][29]. Second, we analyze the traffic of a multimedia application which targets a novel packet-based SoC implementation and illustrate the impact of our analysis on on-chip network design.

We hope that beyond its practical implications, the connection we create between these apparently so different domains will stimulate further research on formal methods for on-chip network design.

1.3 Organization of the paper

Section 2 describes the motivation behind self-similarity and its definition. In Section 3, we present a detailed analysis of traffic for the MPEG-2 video decoder and show the results for four different video clips. In Section 4, we illustrate the implications of the LRD on the on-chip network design process. Finally, we conclude by summarizing our main contribution.

2. What is Self-similarity?

Self-similarity and fractals are concepts pioneered by Mandelbrot [11]. They describe the phenomenon where a certain property of an object - for instance, a natural image or a time series - is preserved with respect to *scaling* in space and/or time. If an object is self-similar (or fractal), then its parts, when magnified, resemble - in a suitable sense - the shape of the whole. For example, a two dimensional (2D) *deterministic* Cantor set is obtained by starting with a black unit square, scaling its size by 1/3, then placing four copies of the scaled square at the four corners, and repeating this process recursively ad infinitum (Fig.2).

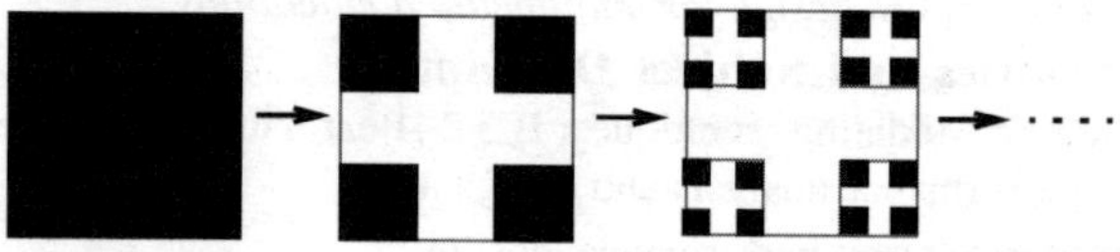

Fig.2. Deterministic fractal example: Two-dimensional Cantor set

The one-dimensional Cantor set can be obtained by projecting the 2D Cantor set onto a horizontal time axis. This can be further interpreted as an ON/OFF time series which model data traffic [26].

Stochastic self-similarity admits the infusion of probabilistic behavior. Unlike the deterministic fractals, the objects do not possess the *exact* resemblance of their parts at finer levels of detail. If we think, for instance, in terms of time series which may characterize some real data traces and relax a little bit the measure of resemblance, say, by focusing on certain statistics of rescaled time series, then it may be possible to expect an *approximate* similarity with respect to these relaxed measures. Second-order (or temporal) statistics are the statistical properties that capture burstiness (or variability) in time series which characterize, for instance, traffic patterns in real networks [2][5]. In particular, the *autocorrelation function*, as a function of the time lag, decreases *polynomially* rather than exponentially. The existence of such non-trivial correlation "at a distance" is referred to as LRD.

In the case of video traces, we concentrate on traffic characteristics at the *macroblock level*. If we look within each frame of a video, there are certain objects. All the macroblocks within a single object carry similar amounts of information and hence they get coded using almost the same number of bits. Since the macroblocks within an object generally occur next to each other in a frame, this leads to long range correlations. This may be the intuitive reason behind observing the LRD phenomenon in video traffic, at macroblock level.

We also note that, from a practical point of view, LRD has a considerable impact on queueing performance of the communication architecture. The traditional *short-range dependent* (or Markovian) processes have an autocorrelation function which decays *exponentially* fast. But the LRD processes exhibit a much slower decay of correlations; that is, their correlation functions typically obey some power-law decay. Intuitively, the presence of LRD indicates that while long-range correlations are individually small, their *cumulative effect* is non-negligible. This produces scenarios which are drastically different from those experienced with traditional short-range dependent models such as Markovian processes. This is the subtle point where the long-range dependence analysis we propose surpasses classical Markovian analysis and proves its practical value.

2.1 Definition of Long-Range-Dependence

The mathematical definition of long-range-dependence is given as follows. Let $X = (X_t : t = 0,1,...)$ be a wide-sense stationary stochastic process with mean *m*, variance σ^2 and autocorrelation function *r(k)*, $k \geq 0$. According to [2] *X* is said to exhibit *long-range dependence* if

$$r(k) \sim k^{-\beta} L_1(t) \quad \text{as } k \to \infty \qquad (1)$$

where $0 < \beta < 1$, $L_1(t)$ is a slowly varying function and ~ denotes the "asymptotically close" condition; that is, $\lim_{t \to \infty} L_1(tx)/L_1(t) = 1$, for all $x > 0$.

From equation (1) we see that LRD is characterized by an autocorrelation function that decays *hyperbolically* rather than exponentially fast. It also implies that the spectral density obeys a *power-law* function near the origin (also called *1/f* - noise). This captures the intuition behind LRD, namely that while high-lag correlations are individually small, their *cumulative effect* matters and gives rise to features which are very different from those of short-range dependent processes. In what follows, we describe two methods for testing LRD in any time series *X*.

2.2 Variance-Time Analysis

Let *X* be a wide-sense stationary time series. For each $m = 1,2,3,...$ let $X^{(m)} = X_k^m : k = 1, 2, 3,...$ denote the new wide sense stationary time series obtained by averaging the original time series *X* over non-overlapping blocks of size *m*. That is, for each $m = 1,2,3,...$; X^m is given by $X_k^m = \frac{1}{m}(X_{km-m+1} + \ldots + X_{km})$, $k > 0$.

The variances of X^m, $m = 1,2,3,...$ for short-range dependent processes will eventually decrease *linearly* in log-log plots against *m* with a slope equal to -1. On the other hand, for processes with LRD, the variances of the aggregated processes X^m, decrease linearly (for large *m*) in log-log plots against *m* with slopes arbitrarily flatter than -1.

Cox [4] shows how a specification of the sequence (var(X^m): $m > 0$) is equivalent to a specification of the autocorrelations given by (1). More importantly, for a constant *c*, we have

$$\text{var } X^{(m)} \sim cm^{-\beta} \text{ as } m \to \infty, \qquad (2)$$

with $0 < \beta < 1$. Actually, this value of β is related to the rate at which autocorrelations decay for large values of the lag. From equation (1), we can see that the autocorrelations decay *hyperbolically* with decay constant β.

2.3 R/S Analysis

Historically, stochastic processes with long-range dependence are important because they provide an elegant explanation of an empirical law that has been observed in many naturally occurring time series. What has since come to be known as the *Hurst effect* can be described as follows. Given the observations $(X_k : k = 1,2,...,n)$ with sample mean $\bar{X}(n)$ and variance $S^2(n)$, the *rescaled adjusted range statistic* (denoted as *R/S*) statistics is given by

$$\frac{R(n)}{S(n)} = \frac{1}{S(n)}[Max(0,W_1,W_2,...,W_n) - min(0,W_1,...,W_n)] \qquad (3)$$

where $W_k = (X_1 + X_2 + \ldots + X_k) - k\bar{X}(n)$, $1 \leq k \leq n$. In his study of the rescaled adjusted range [3], Hurst found that many historical records appeared to be well represented by

$$E[R(n)/S(n)] \sim cn^H, \text{ as } n \to \infty, \qquad (4)$$

with Hurst parameter *H* about 0.7. On the other hand, if X_k's are Gaussian pure noise or *short range dependent*, then $H = 0.5$ in equation (4) and the discrepancy is referred to as the Hurst effect. The Hurst effect is fully accounted for by stationary stochastic processes with long-range dependence. The relation between Hurst parameter and the rate at which the autocorrelation decays is given by $H = 1 - \beta/2$.

In practice, *R/S* analysis is based on a heuristic graphical approach, originally described in [11][12]. Formally, given a sample of *N* observations, $(X_k : k = 1, 2,..., N)$, one subdivides the whole sample into *K non-overlapping* blocks and computes the rescaled adjusted range $R(t_i, d)/S(t_i, d)$ for each of the new starting points $t_1 = 1, t_2 = N/K + 1, t_3 = 2(N/K) + 1, \ldots$ which satisfy $(t_i - 1) + d \leq N$. Here $R(t_i, d)$ is defined as in (3) with W_k replaced by $W_{t_i+k} - W_k$ and $S^2(t_i, d)$ is the sample variance of $X_{t_i+1}, X_{t_i+2}, \ldots, X_{t_i+d}$. The slope of the least square fit line fitting the set of values of *R/S* gives the asymptotic value of parameter *H*.

3. Traffic analysis for MPEG-2 video decoder

Our main observation is that, the traffic between different modules for a MPEG-2 decoder exhibits LRD. This is explained through the example of an MPEG-2 video decoder (Fig.3a) [9]. The decoder consists of the VLD (Variable Length Decoder), IQ (Inverse Quantization), IDCT (Inverse Discrete Cosine Transform), Motion Compensation (MC) units, and the associated buffers.

3.1 Modelling and measurement setup

We model the MPEG-2 Video decoder using the Stateflow component of Matlab which uses the semantics of Statecharts, formally proposed by Harel [10]. To create the Stateflow model of the MPEG-2 video decoder, the *sequential* C-code of the decoder was split into several processes and the communication among processes made explicit by using synchronization signals. We model the process graph obtained from the application in Fig.3a following the *Producer-Consumer* paradigm; that is, we describe the VLD process as the *Producer* and the IDCT/IQ unit as the *Consumer*. The VLD decodes the input stream, generates macroblocks, and

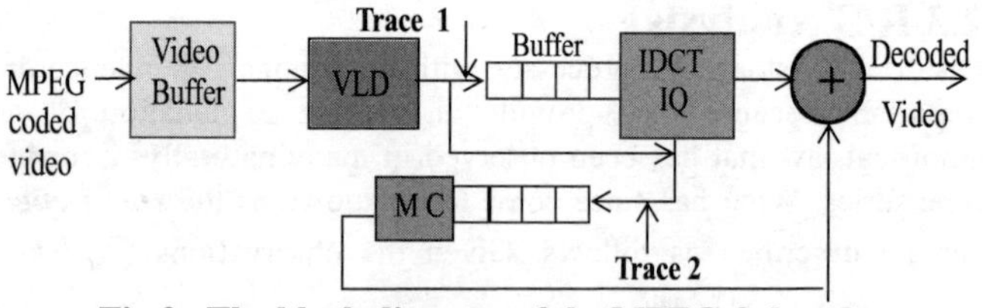

Fig.3a The block diagram of the MPEG-2 decoder

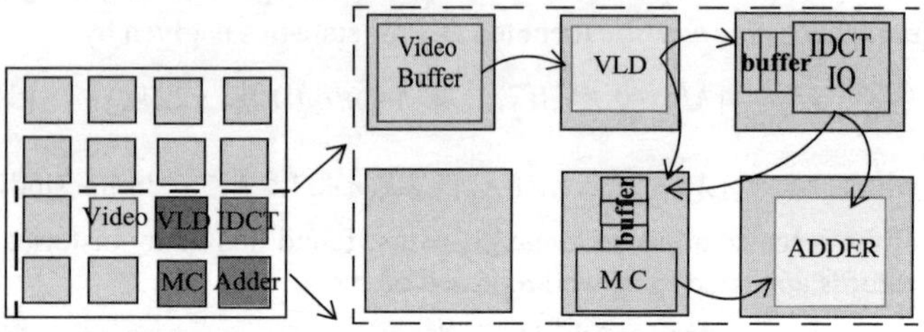

Fig.3b A possible mapping of the MPEG-2 decoder onto architecture in Fig. 1

puts them into the buffer. These are picked up by the *Consumer* to compute IDCTs and output data to reconstruct the frames. We assume that all computing processes are mapped onto the architecture discussed in Section 1 as shown in Fig.3b. The remaining unused tiles can be used to map other applications (e.g. audio, encryption, etc.)

Using the *Mpegstat* tool [13], we analyze a MPEG-2 video stream and find the detailed information about the macroblocks in the frames of a video. Depending upon whether a frame is I, P or B type, the macroblocks are processed differently and they follow different paths in the block diagram and then take different times to process. This results in various traffic patterns for different videos.

We monitored the arrival processes at the IDCT and MC modules recording their corresponding traces (that is, *Trace 1* and *Trace 2* in Fig.3a). The corresponding traces obtained were further evaluated using the variance-time method and the *R/S* method mentioned in Section 2. Using these methods, we were able to obtain the variance-time plots and R/S plots for the two traces. These results are discussed in the following section.

3.2 Results and discussion

Our approach to traffic modeling is "data driven". We rely upon four video sequences (*Clouds, Simpsons, Disc_ir, Hawaii*) of different video screen sizes ranging in length from 27 seconds (88000 macroblocks) to 1 second (43000 macroblocks). This represents all kinds of different scenes as shown in Table 1 by the statistics of I, P and B frames.

Video Clip	I frames	P frames	B frames	Macroblocks per frame
Clouds	24	12	0	1200
Simpsons	136	136	542	108
Disc_ir	18	9	0	1024
Hawaii	195	96	0	300

Table 1. Statistics for different clips

We focus on long sequences (X_i: $i = 1,2,..., N$) of data, where X_i represents the *number of bits* which contain the compressed and coded information for a macroblock in a frame of an MPEG video. Based on statistical analysis of the sequences, our main finding is that LRD *is* a characteristic of the MPEG-2 video traffic traces between different modules of a MPEG-2 decoder.

The monitored trace file consists of two columns. The first one records the *time* measured from the beginning of the trace at which a block of the video stream arrived at a module in the system. The second column gives the integer *size* in bits of the macroblock. The actual traffic therefore consists of alternating sequence of macroblock arrivals and silence periods. We consider the discrete version of the process where the process is averaged within a window of size δ [8]. To compute H, we perform two tests[1]:

- The ***first test*** corresponds to the variance-time analysis of the time series X [2]. As an illustration, Fig.4 shows the plots corresponding to *Simpsons* video clip. We plot the least square fit line in the graph. The slope of the line gives the value of the parameter β, from which we find out the Hurst parameter as $H = 1 - \beta/2$.
- The ***second test*** corresponds to the R/S analysis of the time series [3]. The plots corresponding to the rescaled adjusted range statistics for *Simpsons* video clip are shown in Fig.5.

Video Clip	Trace 1 H by Variance-time method	Trace 1 H by R/S plot method	Trace 2 H by Variance-time method	Trace 2 H by R/S plot method
Clouds	0.7240	0.7646	0.7603	0.7639
Simpsons	0.6874	0.7432	0.7407	0.7943
Disc_ir	0.8108	0.8180	0.8421	0.8131
Hawaii	0.7238	0.7453	0.5455	0.6839

Table 2. The Hurst parameter values for different clips by two methods

For convenience, a summary of the estimated Hurst parameters is also given in Table 2. As we can see, the values of H lie between 0.5 and 1 clearly indicating the presence of LRD. Also, the values of H obtained from *both* methods are sufficiently close to each other to further support the claim about the presence of LRD.

4. Implications of LRD traffic in designing on-chip networks

Beyond its statistical significance, long-range dependence has considerable impact on queueing performance of on-chip network. Only a small number of analytical queueing results are available for LRD traffic [6]. In traffic analysis, we typically deal with time series with hundreds of thousands of observations. If we try to fit the best ARMA model to such a process, then the number of parameters will tend to infinity[2]. Using an excessive number of parameters is undesirable as it increases the uncertainty of the statistical inference and parameters are difficult to interpret. So we need to model these processes with parametrically parsimonious models (that is, using the minimal number of parameters).

Norros in [7] used Fractional Brownian Motion (FBM) model which parsimoniously captures the LRD effects. This model finds out the *lower bound* for the probability that the queue length Q *exceeds* a certain buffer size x, under the assumption of having an infinite buffer. Mathematically:

$$P(Q > x) \sim \exp[-cx^{2-2H}] \tag{5}$$

1. We present the complete set of results for both tests in [31].
2. It will be as good as approximating a hyperbolically decaying function by a sum of exponentials.

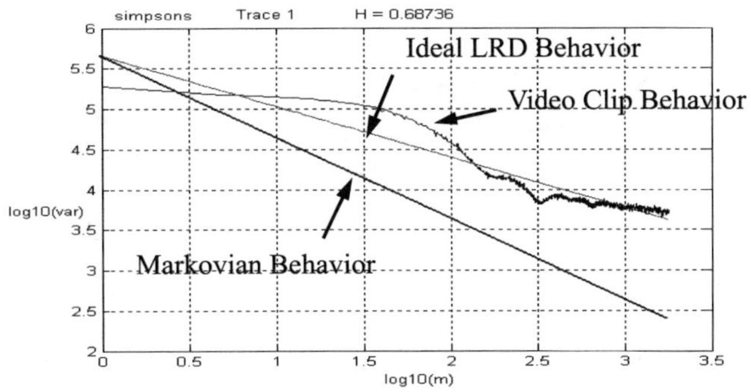

Fig.4 Variance-time plots for *Simpsons* at the IDCT module (Trace 1 in Fig.3a) and at the MC module (Trace 2 in Fig.3a).

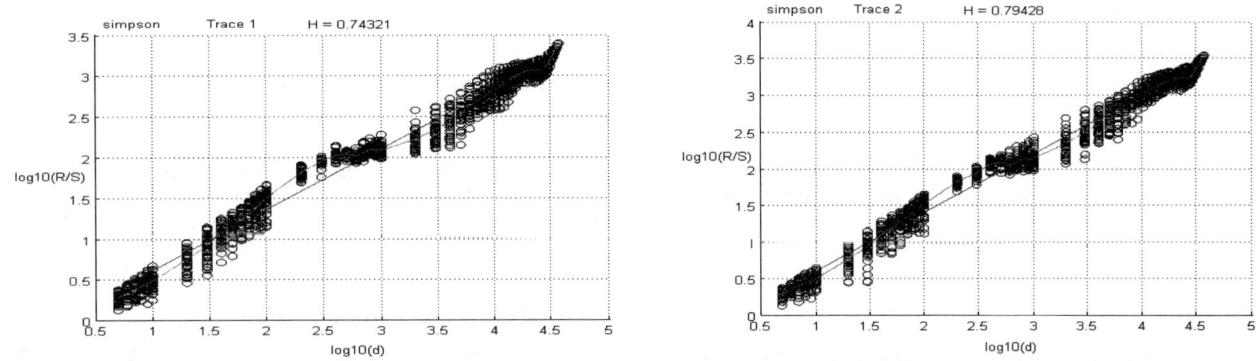

Fig.5 *R/S* plots for *simpsons* video clip at IDCT module (Trace 1 in Fig.3a) and at the MC module (Trace 2 in Fig.3a).

with

$$c = \frac{m^{2H-1}}{2a}\left(\frac{1-\rho}{\rho}\right)^{2H}\left[\left(\frac{1-H}{H}\right)^{H}+\left(\frac{H}{1-H}\right)^{1-H}\right]^{2} \quad (6)$$

where: m is the mean input rate, ρ is the utilization of the queue (that is, the ratio of average service time to average interarrival time); H and a are the Hurst parameter and the "peakedness" values which can be obtained from plots like those in Figs.4-5.

To asses the *accuracy* and *impact* of our predictions on the overall performance of the on-chip network, the complementary buffer length distributions for two different video traces are shown in Fig.6. (The values of H and a from the Fig.4 were used to plot these graphs). In these graphs, the dashed curves indicate the *predicted* probability values given by eqs. (5) and (6) while the continuous curves indicate the results obtained by *simulation*. There are a few conclusions which can be derived from these plots:

• First, the predicted and simulated curves show a very good agreement as a function of buffer length. That is, the small difference between them is because the simulation corresponds to just one instance of the arrival process while the analytical formula gives the result averaged over *many* traces. That is the reason why simulation curves which represent just one instance of the stochastic process with that particular value of Hurst parameter lie close to the curve predicted by the deterministic formula and sometimes overestimate or underestimate the buffer length.

• Second, the dash-dot lines (obtained for H = 0.5) correspond to short-range dependent (SRD) models (like the Markovian ones). From plots in Fig. 6, we can see that Markovian models significantly underestimate (typically 1-2 orders of magnitude) the buffer overflow probabilities since it assumes the distribution of the arrival process to be short-range dependent i.e. exponential. This may cause severe performance degradation at chip-level.

• Third, we 'enriched' the *Disc_ir* trace by concatenating *Clouds* followed by *Simpsons* followed by *Disc_ir* and the second graph in Fig.6 corresponds to the simulation of this edited trace. Again, we can see a very good agreement between the buffer overflow prediction by the LRD formula (5) and the simulation values.

There is also another way of using the plots in Fig. 6 (and therefore eqn. (5)) for on-chip network design. For instance, if the QoS needed by the target application asks for not more than 1% of lost macroblocks, then from the first plot in Fig.6, one can easily see that we need a buffer length of 9000 bits at the IDCT module. This way, we have a theoretical basis for choosing the buffer length. On the other hand, the Markovian analysis will predict a buffer length of 3000 bits and that will result in around 10% bits lost (instead of the target 1%). This represents a very serious performance degradation for an MPEG-2 video decoder. More generally, a multimedia system designer may have a set of typical video clips (e.g. news reader, weather channel etc.) that are expected to run using a MPEG-2 decoder. We can find the Hurst parameter for each of these clips and use it to predict different buffer lengths corresponding to the same degree of lost bits. We can then choose the maximum predicted buffer length and have a theoretical support for assuring the QoS guarantee to the consumer.

Last but not least, compared to simulation-based buffer sizing, the LRD analysis framework offers a *fast* approach for buffer sizing. More precisely, for the simulation-based approach, if we want to asses the impact of changing the speed of the processors in the on-chip network, then we need to rerun the simulations all over again for all the typical video clips. This is extremely time-consum-

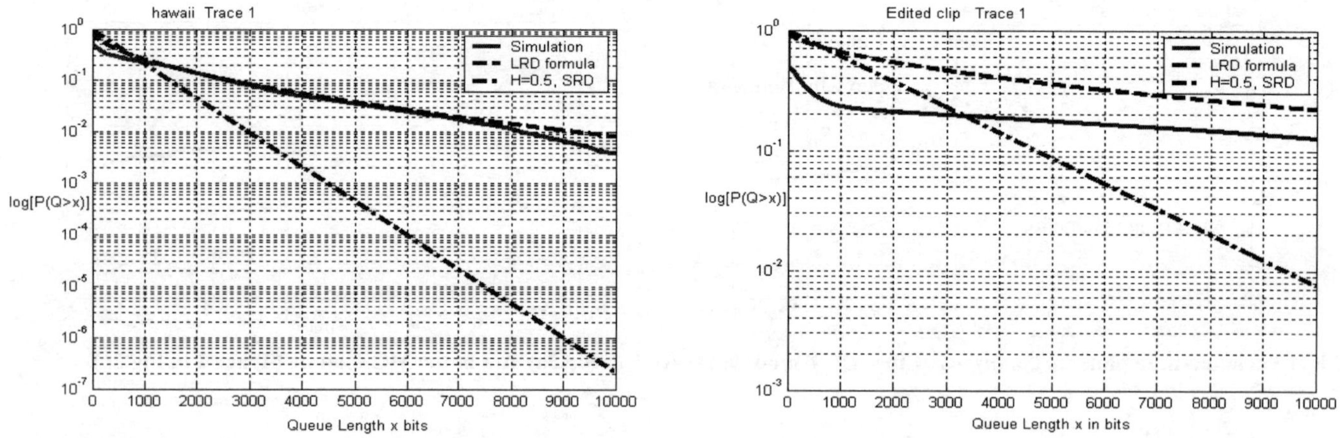

Fig.6 Complementary queue length distribution plots predicted by eqs. (5) and (6) for different video clips. The dashed curves indicate simulation results for an infinite queue with the arrival process following empirical trace (the server utilization is 0.5). The straight lines indicate the prediction by a SRD model (H = 0.5 in eqs.(5) and (6)).

ing since tens of hours of simulation may be needed. Conversely, in the LRD-based analysis, the value of H will *not* be affected as it depends only on the underlying video clip statistical properties. Consequently, we just need to change the value of utilization factor ρ in eq.(6) and get instantaneously the new buffer length.

5. Conclusion

We have presented a technique for on-chip traffic analysis using self-similar processes. For a recently proposed communication architecture based on packet switching, we have shown that, under various input traces, the arrival process at different nodes, for an MPEG-2 video application, exhibits self-similar phenomena. Characterizing the degree of self-similarity via the Hurst parameter helps in finding the optimal buffer length distribution which turns out to be the critical issue for the routers at each node in the on-chip communication network. We plan to explore the application of traffic characterization for other systems in the video domain. We believe that our findings open up new directions of research with deep implications on fundamental issues in on-chip network design.

6. Acknowledgements: We would like to thank Alberto Sangiovanni-Vincentelli for stimulating discussions on this topic and anonymous reviewers for providing excellent feedback.

7. References

[1] W. Dally, B. Towles, 'Route Packets, Not Wires: On-chip Interconnection Networks,' *Proc. DAC, Las Vegas, NV*, June 2001.

[2] W. E. Leland, et al. 'On the self-similar nature of ethernet traffic,' *IEEE/ACM Trans. on Networking*, Vol.2, No.1, Feb.1994.

[3] J. Beran, 'Statistics for Long-Memory Processes,' Chapman & Hall, 1994.

[4] D. R. Cox, 'Long-Range dependence: A Review,' in *Statistics: An Appraisal*, H.A.David and H.T.David, Eds., The Iowa State University Press, 1984.

[5] J. Beran, et al., 'Long-Range Dependence in Variable-Bit-Rate Video Traffic,' *IEEE Trans. Commun.*, Vol. 43, No. 2/3/4, 1995.

[6] A. Erramilli, O. Narayan and W. Willinger, 'Experimental Queueing Analysis with Long-Range Dependent Packet Traffic,' in *IEEE/ACM Trans. on Networking*, Vol.4, No.2, April 1996.

[7] I. Norros, 'A storage model with self-similar input,' in *Queueing Systems* Vol. 16, 1994.

[8] P. Abry, D. Veitch, 'Wavelet Analysis of Long-Range Dependent Traffic,' in *IEEE Trans. on Info. Theory*, Dec. 1997.

[9] Sikora T., 'MPEG Digital Video Coding Standards,' in *IEEE Signal Processing Magazine*, Sept. 1997.

[10] D. Harel, 'Statecharts: A visual formalism for complex systems,' in *Sci. Comp. Prog*, Vol. 8, 1987.

[11] B. B. Mandelbrot and J. R. Wallis, 'Computer Experiments with Fractional Gaussian Noises', *Water Resources Research*, vol.5, 1969.

[12] B. B. Mandelbrot and M. S. Taqqu, 'Robust R/S Analysis of Long Run Serial Correlation', *Proc. 42nd Session ISI*, Book 2, 1979.

[13] *http://bmrc.berkeley.edu/ftp/pub/mpeg/stat/*

[14] K. Lahiri, A. Raghunathan, S. Dey, 'Evaluation of the Traffic-Performance Characteristics of System-on-Chip Communication Architectures', *Proc. Intl. Conf. on VLSI Design*, Bangalore, India, Jan. 2001.

[15] A. Kalavade, P. Moghe, 'A tool for performance estimation of networked Embedded End-Systems,' *Proc. DAC*, San Francisco, CA, June 1998.

[16] K. Keutzer et al., 'System-Level Design: Orthogonalization of Concerns and Platform-Based Design,' *IEEE Trans. on CAD*, Vol.19, No.12, Dec. 2000.

[17] S. Edwards, L. Lavagno, E. A. Lee, A. Sangiovanni-Vincentelli, 'Design of embedded systems: formal models, validation, and synthesis,' *Proc. IEEE*, Vol.85, No.3, March 1997.

[18] A. Mathur, A. Dasdan, R. Gupta, 'Rate Analysis for Embedded Systems,' in *Trans. on Design Automation of Electronic Systems*, Vol. 3, No. 3, July 1998.

[19] A. Nandi, R Marculescu, 'System-level Power/Performance Analysis for Embedded Systems Design,' in *Proc. DAC*, Las Vegas, NV, June 2001.

[20] A. Nandi, R. Marculescu, 'Probabilistic Application Modeling for System-Level Performance Analysis,' *Proc. DATE*, Munich, March 2001.

[21] K. Lahiri, A. Raghunathan, S. Dey, 'Fast Performance Analysis of Bus-based System-on-chip Communication Architecture,' *Proc. ICCAD*, Nov. 1999.

[22] M. Gasteir, M. Glesner, 'Bus-based Communication Synthesis on System Level,' in *Trans. Design Automation Electronic Systems*, Jan. 1999.

[23] T. Yen, W. Wolf, 'Communication Synthesis for Distributed Embedded Systems,' in *Proc. ICCAD*, Nov. 1995.

[24] J. A. Rowson and A. Sangiovanni-Vincentelli, 'Interface Based Design,' in *Proc. DAC.*, June 1997.

[25] J. Daveau, T. B. Ismail, A. A. Jerraya, 'Synthesis of System-level Communication by an Allocation Based Approach,' in *Proc. ISSS*, 1995.

[26] K. Park, W. Willinger, (Eds.), 'Self-Similar Network Traffic and Performance Evaluation,' J. Wiley and Sons, 2000.

[27] M. Sgroi, et al., 'Addressing the System-on-a-chip Interconnect Woes Through Communication-Based Design,' in *Proc. of DAC.*, Las Vegas, 2001.

[28] F. Karim, A. Nguyen, S. Dey, R. Rao, 'On-chip Communication Architecture for OC-768 Network Processors,' in *Proc. of DAC.*, Las Vegas, June 2001.

[29] V. Paxson, S. Floyd, 'Wide-Area Traffic: The Failure of Poisson Modeling,' *Proc. ACM SIGCOMM'94*, London, UK, 1994.

[30] G. Varatkar, R. Marculescu, 'Modeling and Synthesis of On-chip Multimedia Traffic', *Packet Video Workshop*, Pittsburgh, PA, April 2002.

[31] G. Varatkar, R. Marculescu, 'On-chip Traffic Analysis for Multimedia Applications', *Technical Report*, CMU CAD-01-25, Sept. 2001.

Deriving a Simulation Input Generator and a Coverage Metric From a Formal Specification

Kanna Shimizu
Computer Systems Laboratory
Stanford University
Stanford, CA 94305, USA
kanna.shimizu@stanfordalumni.org

David L. Dill
Computer Systems Laboratory
Stanford University
Stanford, CA 94305, USA
dill@cs.stanford.edu

ABSTRACT

This paper presents novel uses of functional interface specifications for verifying RTL designs. We demonstrate how a simulation environment, a correctness checker, and a functional coverage metric are all created automatically from a single specification. Additionally, the process exploits the structure of a specification written with simple style rules. The methodology was used to verify a large-scale I/O design from the Stanford FLASH project.

Categories and Subject Descriptors

B.5.2 [**RTL Implementation**]: Design Aids

General Terms

Documentation, Performance, Design, Verification

Keywords

testbench, input generation, BDD minimization, coverage

1. INTRODUCTION

1.1 Motivation

Before a verification engineer can start simulating RTL designs, he must write three *verification aids*: input testbenches to stimulate the design, properties to verify the behavior, and a functional coverage metric to quantify simulation progress. It would be much easier if the three can be automatically derived from the interface protocol specification. Not only will this save a great deal of work, but it will also result in fewer bugs in the test inputs and the checking properties because they are mechanically derived from a verified specification. Motivated by these advantages, we developed a methodology where the three aids are automatically generated from a specification. Furthermore, we demonstrate how a specification structured by certain style rules allows for more memory efficient simulation runs. The primary contributions of this paper are:

- *a presentation of how a bus protocol specification can be used for simulation to automatically*
  - *generate the most general set of legal input sequences to a design*
  - *produce properties to check interface protocol conformance.*
  - *define a new functional coverage metric*
  - *bias inputs to increase the probability of hitting uncovered cases*
- *a new input generation scheme that greatly reduces the size of the BDD (binary decision diagram) [1] used*
  - *by dynamically selecting relevant constraints on a cycle-by-cycle basis and not translating the entire specification into a BDD*
  - *and by extracting out the pure "constraining" logic and discarding the conditional logic ("the guard") when building the BDD*
- *a report on the successful application of the methodology to verify a fabricated working design, and a description of the new bugs found*

1.2 Background

Many of today's digital circuit designs depend on the tight integration of multiple design components. These components are often designed by different engineers who may have divergent interpretations of the interface protocol that "glue" the components together. Consequently, designs may be incompatible and behave incorrectly when combined. Thus, functional interface specifications are pivotal for successful module integration and should be accordingly solid and precise. However, specifications widely in use today are still written informally, forfeiting an opportunity for automated analysis and logical clarity. In many cases, specifications such as those of standard bus protocols are buggy, ambiguous, and contradictory: all problems that can be resolved by formal specification development.

The advantages of formal specifications seem clear but they are nonetheless avoided due to a perceived cost-value problem; they are often considered too costly for the benefits they promise. Specifically, they are criticized for their lengthy development time and the need for formal verification training. For many, the value of a correct specification does not justify these costs, and precious resources are allocated for more pressing design needs. Thus, to counter these disincentives, we are developing a formal specification methodology that attacks this cost-value problem from two angles.

Cost Side The earlier papers [9, 8] for this project focus on the first half of the problem: minimizing the cost by making the specification process easier. We have directed our attention to signal-level

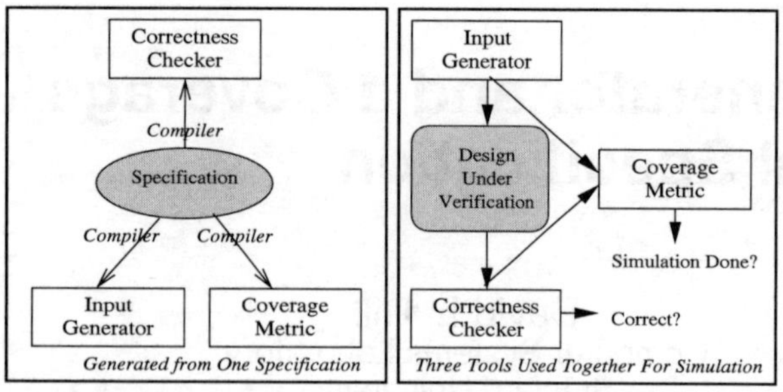

Figure 1: The Trio of Verification Aids

bus protocol descriptions for this aim since they are both important and challenging to specify. Using the PCI (Peripheral Component Interconnect) bus protocol and the Intel® Itanium™ Processor bus protocol as examples, we developed a specification style that produces correct, readable, and complete specifications with less effort than free-form, *ad hoc* methods. The syntactic structuring used in the methodology is language-independent and can be applied to many specification languages from SMV [6] to Verilog. This is a reflection of our belief that methodology, as opposed to tool or language development, is the key to achieving the stated goal.

Value Side As the latest work in this series, this paper focuses on the second angle of the cost-value problem: to increase the value of a formal specification beyond its role as a documentation. It is based on the idea that once a correct, well-structured specification is developed, it can be exploited in a way that a haphazard and incomplete specification cannot be. In particular, we investigate ways to use the specification *directly* to generate inputs, check behavior, and monitor coverage. Because our goal is to verify large designs common in industry, the methodology is specifically tailored for simulation-based verification.

1.3 The Problem and Our Approach

Given an HDL (hardware description language) component design to verify, an engineer needs various additional machinery (Figure 1, right).

1. Input Logic There must be logic to drive the inputs of the design. One method uses random sequences, which are not guaranteed to comply with the protocol, and consequently, it is difficult to gauge correctness of the design because its inputs may be incorrect. A more focused method is directed testing where input sequences are manually written, but they are time-consuming to write and difficult to get correct.

2. Output Check Logic to determine the correctness of the component's behavior is needed if manual scrutiny is too cumbersome. With the methodology presented here, the scope of correctness checking is limited to interface protocol conformance. It cannot check higher-level properties such as whether the output data from one port correctly corresponds to another port's input data.

3. Coverage Metric Because complete coverage with all possible input sequences tested is not possible, there must be some metric that quantifies the progress of verification coverage. The verification engineer would like to know whether the functionalities of the design have been thoroughly exercised and all interesting cases have been reached. This is an open area of research, and currently, most practitioners resort to methods with little theoretical backing.

Our Approach At the foundation of our methodology is a *unified framework approach* where the three tools are generated from a single source specification (Figure 1). This is possible because all three are fundamentally based on the interface protocol, and consequently, the interface specification can be used to automatically create the three. In current practice, the verification aids are each written from scratch; this requires a tremendous amount of time and effort to write and debug. By eliminating this step, our methodology enhances productivity and shortens development time.

Also, a thoroughly debugged, solid specification invariably leads to correct input sequences, checking properties, and coverage metrics. The correctness of the core document guarantees the correctness of the derivatives. In contrast, with current methods, each verification aid needs to be individually debugged. The advantages of this are most pronounced for standard interfaces where the correctness effort can be concentrated in the standards committee and not duplicated among the many implementors. Furthermore, when a change is made to the protocol (a frequent occurrence in industry), one change in the protocol specification is sufficient to reflect this because the verification aids can be regenerated from the revised document. Otherwise, the engineer would have to determine manually the effect of the change for each tool.

The derivation of a behavioral checker is the most straightforward of the three. The checker is on-the-fly; during simulations, it flags an error as soon as the component violates the protocol. As addressed in the first paper[9], the specification is written in a form very close to a checker. Furthermore, the specification is guaranteed to be executable by the style rules (described in section 2.1), and so the translation from it to a HDL checker requires minimal changes [9].

The bulk of the current work addresses the issue of automatically generating input sequences. Our method produces an input generator which is dynamic and reactive; the generated inputs depend on the previous cycle outputs of the design under verification. In addition, these inputs always obey the protocol, and the generation is a one-pass process. The mechanism relies on solving boolean constraints by building and traversing BDD structures on every clock cycle. Although input generation using constraint solvers is not by itself novel, our approach is the first to use and exploit a complete and structured specification.

Finally, a new simulation coverage metric is introduced, and the automatic input biasing based on this metric is also described. Although more experiments are needed to validate this metric's effectiveness, its main advantage (currently) is that it is specification-based and saves time: extra work is not needed to write out a metric or to pinpoint the interesting scenarios for they are gleaned directly from the specification document.

Previous Works Clarke et al. also researched the problem of specifications and generators in [3] but the methodology is most closely related to the *SimGen* project described in the 1999 paper [11] by Yuan et al. As with the SimGen work, we are focusing on practical methods that can be used for existing complex designs. Within that framework, there are mainly two features that differentiate our approach from SimGen.

First, the SimGen software uses a statically-built BDD which represents the entire input constraint logic; in contrast, our framework dynamically builds the appropriate BDD constraint on every clock cycle. This results in a dramatically smaller BDD for two reasons. One, only a small percentage of the protocol logic is relevant on each cycle, and so the corresponding BDD is always much smaller than the static BDD representing the entire protocol. Two, the BDD *contains only the design's input variables and do not contain state variables or the design's outputs*. Thus, for the PCI example, instead of a BDD on 161 variables, we have a BDD on 15 variables, *an order of magnitude difference*. Consequently, our input generation uses exponentially less memory. This reduction is based on the observation that the sole role of the state variables and the design's outputs is to determine which parts of the protocol are relevant (and thus required in the BDD) for a particular cycle. Otherwise, these variables are not needed to calculate the inputs. We believe that these two reasons for smaller BDDs would hold for many interfaces and therefore allow input generation for a large interface that may otherwise be hindered by BDD blowup.

A second difference with SimGen is that, unlike our framework, it requires the users to provide the input biases. A unique contribution of our work is the automated process of determining biases. It is noted that all these advantages are possible because our method exploits the structure of a stylized specification whereas SimGen is applicable for more general specifications.

2. METHODOLOGY

2.1 Specification Style

The specification style was introduced in [9] and is summarized here for the reader. It is based on using multiple *constraints* to collectively define the signalling behavior at the interface. The constraints are short boolean formulas which follow certain syntactic rules. They are also independent of each other, rely on state variables for historic information, and when AND-ed together, define exactly the correct behavior. This is similar to using (linear or branching time) temporal logic for describing behavior. However, our methodology allows and requires only the most basic operators for writing the constraints, and it aims for a complete specification as opposed to an *ad hoc* list of properties that should hold true.

This decomposition of the protocol into multiple constraints has many advantages. For one, the specification is easier to maintain. Constraints can be added or removed and independently modified. It is also believed that it is easier to write and debug. Since most existing natural-language specifications are already written as a list of rules, the translation to this type of specification requires less effort and results in fewer opportunities for errors. For debugging, a symbolic model checker can be easily used to explore the states allowed by the constraints.

Style Rule 1 The first style rule requires the constraints to be written in the following form.

$$prev(signal_0 \ldots \wedge \neg signal_j \ldots \vee variable_0 \ldots \wedge \neg variable_k) \\ \rightarrow signal_i \vee \ldots \wedge \neg signal_n$$

where "$\rightarrow$" is the logical symbol for "implies". The antecedent, the expression to the left of the "$\rightarrow$", is a boolean expression containing the interface signal variables and auxiliary state variables, and the consequent, the expression to the right of the "$\rightarrow$", contains *just the interface signal variables.* The allowed operators are AND, OR, and NEGATION. The *prev* construct allows the value of a signal (or the state of a state machine) a cycle before to be expressed. The constraints are written as an implication with the past expression as the antecedent and the current expression as the consequent. In essence, the past history, when it satisfies the antecedent expression, requires the current consequent expression to be true; otherwise, the constraint is not "activated" and the interface signals do not have to obey the consequent in the current cycle. In this way, the *activating* logic and the *constraining* logic are separated. For example, the PCI protocol constraint, $prev(trdy \wedge stop) \rightarrow stop$ means "if the signals $trdy$ and $stop$ were true in the previous cycle (the "activating" logic), then $stop$ must be true in the current cycle (the "constraining" logic)" where a "true" signal is asserted and a "false" signal is deasserted. This separation is what identifies the relevant (i.e."activated") constraints on a particular cycle. Also, it allows the BDD to be an expression purely of the "constraining" logic (as explained in the next section, *2.2.1*).

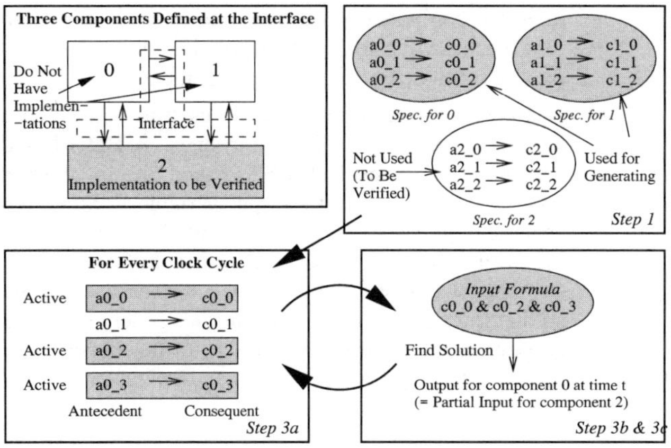

Figure 2: The Input Generation Algorithm

Style Rule 2 The second style rule, the *separability* rule, requires *each constraint to constrain only the behavior of one component.* Equivalently, because the constraining part is isolated from the activating part (due to the first style rule), the rule requires *the consequent to contain only outputs from one component.*

Style Rule 3 The third rule requires that the specification is dead state free. This rule effectively guarantees that an output satisfying all of the constraints always exists as long as the output sequence so far has not violated the constraints. There is a universal test that can verify this property for a specification. Using a model checker, the following CTL (computation tree logic)[2] property can be checked against the constraints, and any violations will pinpoint the dead state: $AG(all\ constraints\ have\ been\ true\ so\ far \rightarrow EX(all\ constraints\ are\ true))$

Although abiding by the style rules may seem restrictive, it promises many benefits. Furthermore, the style is still powerful enough to specify the signal-level PCI and Intel® Itanium™ Processor bus protocols.

2.2 Deriving an Input Generator

2.2.1 Basic Algorithm

Based on the following algorithm, input vectors are generated from the structured specification (Figure 2).

1. Group the constraints according to which interface component they specify. (This is possible because of style rule 2, the separability rule.) If there are n interface components, there will be n groups.
2. Remove the group whose constraints are for the component under verification. These will not be needed. Now, there are $n-1$ groups of constraints.
3. For each group of constraints, do the following on every clock cycle of the simulation run. The goal is to choose an input assignment for the next cycle.
   (a) For each constraint, evaluate *just the antecedent half.* The antecedent values are determined by internal state variables and observed interface signal values. For antecedents which evaluate to *true*, the corresponding constraints are marked as *activated.*
   (b) Within each group, AND together *just the consequent halves* of the activated constraints to form the *input formula.* As a result, there is one input formula for each

(c) A boolean satisfiability solver is used to determine a solution to each of the input formulas. A BDD-based solver is used instead of a SAT-based one in order to control the biasing of the input variables. Since the specification is nondeterministic and allows a range of behaviors, there will most likely be multiple solutions. (In section 2.3, we discuss how a solution is chosen so that interesting simulation runs are generated.) The chosen solutions form the input vector for this cycle.

(d) Go back to step 3(a) on the next clock cycle.

The significance of the style rules become clear from this algrotihm. The "activating" - "constraining" division is key to alowing for a dramatically smaller expression (just the consequent halves) to solve (rule 1). The separability rule also allows for smaller expressions by enforcing strict orthogonality of the specification along the interface components (rule 2). Finally, the lack of dead states guarantees the existence of a correct input vector assignment for every clock cycle (rule 3).

2.2.2 *Implementation*

A compiler tool, which reads in a specification and outputs the corresponding input generation module, has been designed and implemented. There are two parts to the input generator: the Verilog module which acts as the frontend and the C module as the backend (Figure 3).

The Verilog module contains all the antecedents of the constraints, and based on its inputs (the component's outputs) and its internal state variables, determines which constraints are activated for that clock cycle. Then, the indices of the activated constraints are passed to the backend C module. The C module will return an input assignment that satisfies all the activated constraints, and the Verilog module will output this to the component under verification. The choice of Verilog as a frontend allows many designs to be used with this framework.

The C module contains the consequent halves of the constraints. It forms conjunctions (ANDs) of the activated consequents, solves the resulting formula, and returns an assignment to the Verilog module. It is initialized with an array of BDDs where each BDD corresponds to a constraint consequent. On every clock cycle, after the activation information is passed to it, it forms one BDD per interface component by performing repeated BDD AND operations on activated consequents in the same group. The resulting BDD represents an (aggregated) constraint on the next state inputs from one component, and by traversing the BDD until the "1(TRUE)" terminal node is reached, an assignment can be found. Once an assignment is determined for each interface component, the complete input assignment to the component under verification has been established. The CUDD (Colorado University Decision Diagram) package [10] version 2.3.1 was used for BDD representation and manipulation, and Verilog-XL was used to simulate the setup.

2.3 Biasing the Inputs

2.3.1 *Coverage Metric*

We use the specification to define *corner cases*, scenarios where the required actions are complex. These states are more problematic for component implementations, and thus, simulations should drive the component through these scenarios. Consequently, whether

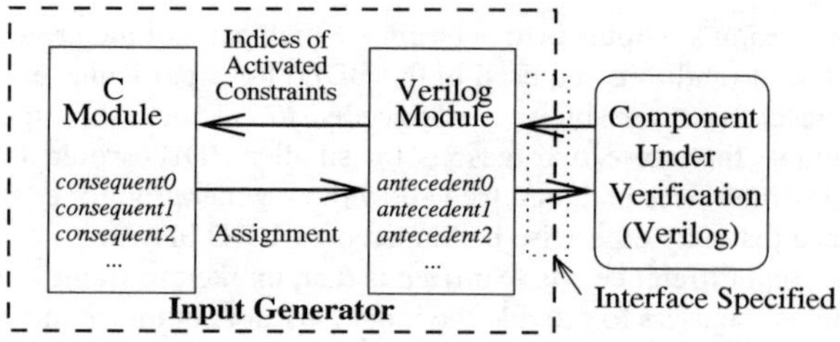

Figure 3: Implementation Details of the Input Generator

a corner case has been reached or not can be used to measure simulation progress, and missed corner cases can be used to determine the direction of further simulations.

As a first order approximation of corner cases, the antecedents of the constraints are used. This is because only when the antecedent clause is true does the implementation have to comply with the constraint clause. As an example, consider the PCI constraint, "master must raise irdy within 8 cycles of the assertion of frame." The antecedent is "the counter that starts counting from the assertion of frame has reached 7 and irdy still has not been asserted" and the consequent is "irdy is asserted." Unless this antecedent condition happens during the simulation, compliance with this constraint cannot be completely known. For a simulation run which has triggered only 10% of the antecedents, only 10% of the constraints have been checked for the implementation. In this sense, the number of antecedents fired during a simulation run is a rough coverage metric.

There is one major drawback to using this metric for coverage. The problem is intimately related to the general relationship between implementation and specification. By the process of design, for every state, a designer chooses an action from the choices offered by the nondeterministic specification to create a deterministic implementation. As a result, the implementation will not cover the full range of behavior allowed by the specification. Thus, some of the antecedents in the specification will never be true because the implementation precludes any paths to such a state. Unless the verification engineer is familiar with the implementation design, he cannot know whether an antecedent has been missed because of the lack of appropriate simulation vectors or because it is structurally impossible.

2.3.2 *Deriving Biases for Missed Corner Cases*

To reach interesting corner cases, verification engineers often apply biasing to input generation. If problematic states are caused by certain inputs being true often, the engineer programs the random input generator to set the variable true n% instead of the neutral 50% of the time. For example, to verify how a component reacts to an environment which delays its response, *env_response*, the engineer can set the biasing so that the input, *env_response*, is true only 5% of the time. 0% is not used because it may cause the interface to deadlock. With prevailing methods, the user needs to provide the biasing numbers to the random input generator. This requires expert knowledge of the design, and the biases must be determined by hand. In contrast, by targeting antecedents, interesting biasing can be derived automatically. The algorithm works as follows:

1. Gather the constraints that specify the outputs of the component to be verified. The goal: the antecedents of these constraints should all become true during the simulation runs.
2. Set biases for all input signals to neutral (50% true) in the input generator described in section 2.2. (Exactly how this is done will be explained in the following subsection.)

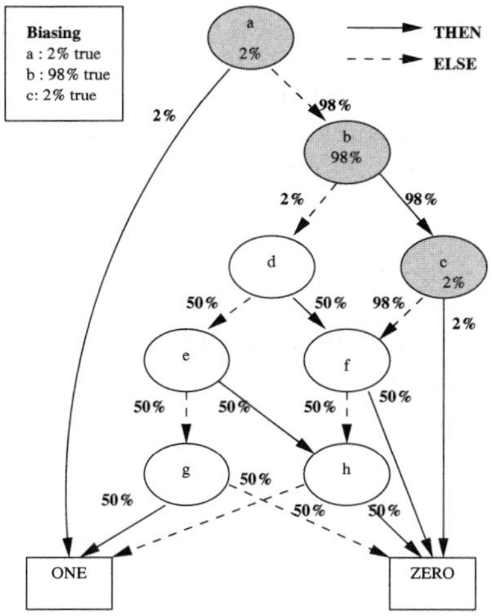

Figure 4: The Biased BDD Traversal

3. Run the simulation for some number of cycles.
4. Determine which antecedents have not fired so far.
5. Pick one missed antecedent, and use it to determine the variable biasing. If, for example, antecedent $\neg a \wedge b \wedge \neg c$ has not been true, set the following biases: a is true 2% of the time, b for 98%, c for 2%.
6. Re-run the simulation and repeat from step 4. Continue until all antecedents have been considered.

There are a number of interesting conclusions. First, although effort was invested in determining optimal bias numbers exactly, biases that simply allowed a signal to be true (or false) "often" was sufficient. Empirically, interpreting "often" as 49 out of 50 times (98%) seems to work well. Second, an antecedent expression contains not only interface signal variables but also counter values and other variables that cannot be skewed directly. Just skewing the input variables in the antecedent is primary biasing, and a more refined, secondary biasing can be done by dependency analysis. This was done manually. For example, many hard-to-reach cases are states where a counter has reached a high value, and by dependency analysis, biases that will allow a counter to increment frequently without resetting were determined.

2.3.3 *Implementing Biasing*

The actual skewing of the input variables is done during the BDD traversal stage of the input generation. After the input formula BDD for a component has been built, the structure is traversed according to the biases. If variable b is biased to be true 49 out of 50 times, the THEN branch is taken 49 out of 50 times (Figure 4). If this choice of branching forces the expression to evaluate to false (i.e. the traversal inevitably leads to the "ZERO" leaf), the algorithm will backtrack and the ELSE branch will be taken. As a result, even if b is biased to be true 49 out of 50 occurrences, the protocol logic can force b to be false most of the time. What is guaranteed by the biasing scheme is that whenever b is allowed to be true by the constraint, it will most likely be true.

An extra step is added to the input generation algorithm to accommodate the biasing. The variables need to be re-ordered so that the biased variables are at the top of the BDD, and their truth value are not determined by the other variables. In Figure 5, variable c is intended to be true most of the time. However, since c is buried towards the bottom of the BDD, if $\{a = 0, b = 1\}$ is chosen, c is forced to be false to satisfy the constraint. In contrast, if c is at the top of the BDD, the true branch can be taken as long as the other variables are set accordingly (for example, $a = 1$). Fortunately, since the number of BDD variables is kept small, reordering for this purpose does not lead to BDD blowup problems.

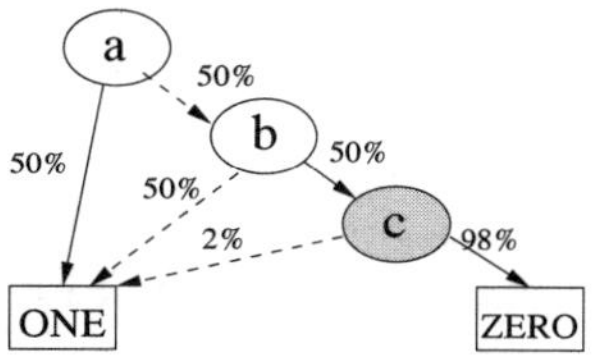

Figure 5: Incorrect Ordering

Compared to the biasing technique used in SimGen, the biasing used in this framework is coarse. With SimGen, branching probabilities, which take into account variable ordering, are calculated from the desired biases. In contrast, this method directly uses the biasing as the branching probabilities; it requires no calculations and compensates for possible distortions by reordering. Although implementing the SimGen calculations is not difficult, the advantages of achieving more precise biasing are not clear from the examples attempted.

3. EXPERIMENTAL RESULTS

To demonstrate the methodology on a meaningful design, we chose the I/O component from the Stanford FLASH [5] project for verification. The I/O unit, along with the rest of the project, had been extensively debugged, fabricated, and tested and is part of a working system in operation. The methods are evaluated on the PCI interface of the component.

The design is described by 8000 lines of Verilog and contains 283 variables which range from 1-bit to 32-bit variables: a complexity which renders straightforward model checking unsuitable. Approximate model checking was used by Govindaraju et al [4] to verify this design but no bugs were found because the design inputs were overly constrained and only a small state space was explored. Our simpler and more flexible simulation-based checking proved to be more effective by finding new bugs.

The Setup A formal PCI specification was used to constrain the inputs and check the outputs at the PCI interface of the design. A simulation checker that flags PCI protocol violations was generated from the specification using a compiler tool written in OCAML [7]. The same compiler tool was modified to output the constrained random simulation generator which controls the PCI interface inputs of the I/O unit. The I/O unit (the design under verification), checker, and input generator are connected and simulated together, and results are viewed using the VCD (Value Change Dump) file. The inputs were skewed in different configurations for each simulation run in order to produce various extreme environments and stress the I/O unit.

Verification Results Using the 70 assertions provided by the interface specification, nine previously unreported bugs have been found in the I/O unit. Most are due to incorrect state machine design. For example, one bug manifested itself by violating the protocol constraint, "once *trdy* has been asserted, it must stay asserted until the completion of a data phase." Because of an incorrect path in the state machine, in some cases, the design would assert *trdy* and then, before the completion of the data phase, deassert *trdy*. This can deadlock the bus if the counterparty infinitely waits for the assertion of *trdy*. The bug was easily corrected by removing the problematic and most likely unintended path. The setup makes

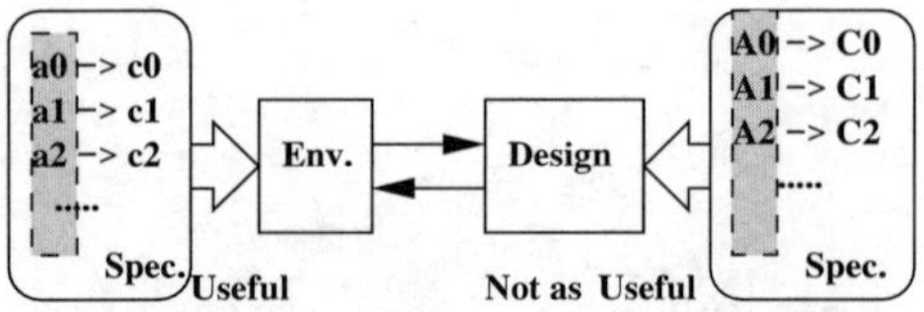

Figure 6: The Two Types of Simulation Coverage Metric and their Effectiveness

the verification process much easier; the process of *finding* signal-level bugs is now nearly automated, and so, most of the effort can focus on *reasoning* about the bug once it is found.

Coverage Results Unfortunately, the original intended use of the coverage metric proved to be fruitless for this experiment. Using antecedents of the constraints that *specify the component* was meaningless because the FLASH PCI design is conservative and implements a very small subset of the specification. For example, the design only initiates single data phase transactions, and never initiates multiple data phase transactions. Thus, most of the antecedents remained false because it was structurally impossible for them to become true.

However, using the metric to ensure that the environment is maximally flexible proved to be much more powerful. The motivation is to ensure that the design is compatible with any component that complies with the interface protocol. The design should be stimulated with the most general set of inputs, and so, using the missed antecedents from the constraints that *specify the environment* (in Figure 6, "a0, a1, ...") to determine biases was extremely fruitful; most of the design bugs were unearthed with these biasings.

Performance Results Performance issues, such as speed and memory usage, did not pose to be problems, and so, we were free to focus on generating interesting simulation inputs. However, to demonstrate the scalability of the method for larger designs, performance results were tabulated. The simulations were run on a 4-Processor Sun Ultra SPARC-II 296 MHz System with 1.28Gbytes of main memory. The specification provided 63 constraints to model the environment. These constraints required 161 boolean variables, but because of the "activating" - "constraining" logic separation technique, only 15 were needed in the BDDs. Consequently, the BDDs used were very small; the peak number of nodes during simulation was 193, and the peak amount of memory used was 4Mbytes.

Furthermore, speed was only slightly sacrificed in order to achieve this space efficiency. The execution times for different settings are listed in Table 2. With no constraint solving, where inputs are randomly set, the simulation takes 0.64s for 12,000 simulator time steps. If the input generator is used, the execution time increased by 57% to 1.00s; this is not a debilitating increase, and now the inputs are guaranteed to be correct. The table also indicates how progressively adding signal value dumps, a correctness checker module, and coverage monitor modules, adds to the execution time.

4. FUTURE WORK

Better coverage metrics can probably be deduced from the specification. A straightforward extension would be to see whether pairs of antecedents become true during simulations. Exploiting a structured formal specification for other uses is also of interest. Perhaps incomplete designs can be automatically augmented by specification constraints for simulation purposes. Or, useful synthesis information can be extracted from the specification. Also, experiments to determine whether designs that are too big for SimGen-type algorithms can be handled by ours would further validate the methodology. Furthermore, more extensive experiments to quantify the speed penalty for the dynamic BDD building should be done.

	Number
Boolean Vars in Spec	161
Boolean Vars in BDD	15
Constraints on Env	63
Assertions on Design	70
Peak Nodes in BDD	193
BDD Memory Use	4 Mbytes
Bugs Found in Design	9

Table 1: Interface Specification Based Generation Details for the FLASH Example

Settings	User Time	System Time	Total
Random	0.53s	0.11s	0.64s
Constrained	0.77s	0.23s	1.00s
with Dump	0.77s	0.26s	1.03s
with Monitor	1.33s	0.29s	1.62s
with Coverage	1.54s	0.25s	1.79s

Table 2: Time Performance of the Methodology on FLASH Example (for 12000 simulator time steps)

Acknowledgement This research was supported by GSRC contract SA2206-23106PG-2.

5. REFERENCES

[1] R. Bryant. Graph-based Algorithms for Boolean Function Manipulation. In *IEEE Transactions on Computers*, volume C-35, pages 677–691, August 1986.

[2] E. Clarke and E. Emerson. Synthesis of synchronization skeletons for branching time temporal logic. In *Logic of Programs: Workshop*, volume 131 of *Lecture Notes in Computer Science*, May 1981.

[3] E. Clarke, S. German, Y.Lu, H.Veith, and D.Wang. Executable Protocol Specification in ESL. In *Proceedings of FMCAD*, November 2000.

[4] S. G. Govindaraju and D. L. Dill. Counterexample-guided choice of projections in approximate symbolic model checking. In *Proceedings of ICCAD*, November 2000. San Jose, CA.

[5] J. Kuskin, D. Ofelt, M. Heinrich, J. Heinlein, and et al. The Stanford FLASH Multiprocessor. In *Proceedings of ISCA*, pages 302–313, April 1994.

[6] K. McMillan. http://www-cad.eecs.berkeley.edu/~kenmcmil/smv/.

[7] OCAML. http://caml.inria.fr.

[8] K. Shimizu, D. L. Dill, and C.-T. Chou. A Specification Methodology by a Collection of Compact Properties as Applied to the Intel Itanium Processor Bus Protocol. In *Proceedings of CHARME*, September 2001.

[9] K. Shimizu, D. L. Dill, and A. J. Hu. Monitor-Based Formal Specification of PCI. In *Proceedings of FMCAD*, November 2000.

[10] F. Somenzi. http://vlsi.colorado.edu/~fabio/CUDD/cuddIntro.html.

[11] J. Yuan, K. Shultz, C. Pixley, H. Miller, and A. Aziz. Modeling Design Constraints and Biasing in Simulation Using BDDs. In *Proceedings of ICCAD*, November 1999.

Hole Analysis for Functional Coverage Data

Oded Lachish
odedl@il.ibm.com

Eitan Marcus
marcus@il.ibm.com

Shmuel Ur
ur@il.ibm.com

Avi Ziv
aziv@il.ibm.com

IBM Research Laboratory in Haifa
Haifa University
Haifa, 31905
Israel

ABSTRACT

One of the main goals of coverage tools is to provide the user with informative presentation of coverage information. Specifically, information on large, cohesive sets of uncovered tasks with common properties is very useful. This paper describes methods for discovering and reporting large uncovered spaces (*holes*) for cross-product functional coverage models. Hole analysis is a presentation method for coverage data that is both succinct and informative. Using case studies, we show how hole analysis was used to detect large uncovered spaces and improve the quality of verification.

Categories and Subject Descriptors

B.6.3 [**Logic Design**]: Design Aids—*Verification*; D.2.4 [**Software Engineering**]: Software/Program Verification; D.2.5 [**Software Engineering**]: Testing and Debugging—*Testing tools*

General Terms

Verification, Measurement, Algorithms, Experimentation

Keywords

Functional verification, Coverage analysis

1. INTRODUCTION

Functional verification comprises a large portion of the effort in designing a processor [5]. The investment in expert time and computer resources is huge, as is the cost of delivering faulty products [3]. In current industrial practice, most of the verification is done by generating a massive amount of tests using random test generators [1, 2]. The use of advanced random test generators can increase the quality of generated tests, but it cannot detect cases in which some areas of the design are not tested, while others are tested repeatedly.

The main technique for checking and showing that the testing has been thorough is called test coverage analysis [9]. Simply stated, the idea is to create, in a systematic fashion, a large and comprehensive list of tasks and check that each task was covered in the testing phase. Coverage can help monitor the quality of testing and direct the test generators to create tests that cover areas that have not been adequately tested.

Coverage tools collect coverage data and then aid in the analysis of that data. Our experience shows that informative presentation of the coverage information is as important as efficient collection of that information.

Most coverage tools provide two types of reports — status reports and progress reports. Progress reports show how coverage progresses over time. These reports are useful in detecting when the testing is "running out of steam" and predicting its potential coverage value [7]. Status reports show which tasks are covered and which are not, and how many times each task was covered. This information helps the verification team direct the testing to untested, or weakly tested, areas in the design [6].

This paper describes *hole analysis*, a method for discovering and reporting large uncovered spaces for cross-product functional coverage models [6]. In a cross-product functional coverage model, the list of coverage tasks comprises all possible combinations of values for a given set of attributes. A simple example of a coverage model is all the possible combinations of requests and responses sent to a memory subsystem. In this case, a coverage task is the pair <`request, response`>, where `request` is any of the possible requests that can be sent to the memory (e.g., `memory read, memory write, I/O read, I/O write`) and `response` is one of the possible responses (e.g., `ack, nack, retry, reject`).

Hole analysis groups together sets of uncovered tasks that share some common properties, thus allowing the coverage tool to provide shorter and more meaningful coverage reports to the user. For example, it is much more informative to report that a `reject` response never occurred, than it is to include all possible cases of a request with a `reject` response in the list of uncovered tasks. This grouping of uncovered tasks into meaningful sets, in addition to shortening the list of uncovered tasks reported to the user, provides vital information on the cause of the hole. For example, investigation of the hole described above may lead to the simple conclusion that requests are never rejected because they do not arrive fast enough. In this case, increasing the rate of requests to the memory subsystem may be all that is needed to generate rejects.

Our hole analysis technique is based on finding uncovered tasks that have common values in some of the attributes, and finding some common denominator in the attribute values that distinguish them. The simplest case is when all the values of a specific attribute are not covered when each of the other attributes has a particular value. The hole above is an example of such a case. None of the values of the attribute `request` are covered when the `response` attribute equals `reject`. Another possibility is to find a meaningful subset of the values of a certain attribute, for example, I/O requests (including I/O read and I/O write). This type of hole detection leads to the discovery of rectangular holes that are parallel to the attribute axis. This analysis is similar to classification methods used in AI, such as ID3 [11].

Hole analysis has been implemented in a number of coverage tools — Comet [6], its successor Meteor, and FoCus [13] — developed at IBM, and has become an integral part of the verification process for users of these tools. In the paper, we provide two examples of real coverage models and show how hole analysis can provide meaningful reports to the user. In each case, there was a significant improvement in the testing quality. The first example is a coverage model for results of floating point instructions. The second example is a micro-architectural coverage model for register interdependency in the pipelines of a super-scalar PowerPC processor [10].

The rest of the paper is organized as follows. In Section 2, we explain cross-product coverage models and how they should be used during verification. In Section 3, we describe our hole analysis techniques using the floating point coverage model as an example. Section 4 illustrates the usefulness of hole analysis with the interdependency coverage model example. Finally, Section 5 concludes the paper.

2. FUNCTIONAL COVERAGE MODELS

Functional coverage focuses on the functionality of an application. It is used to check that all important aspects of the functionality have been tested. Unlike code coverage models that are program based, functional coverage is design and implementation specific [12].

When using functional coverage, coverage models containing a large set of verification (coverage) tasks are systematically defined by the user. These are used to measure, monitor, and redirect the testing process. Holes in the coverage space expose gaps in the testing, and new tests are created to "plug" these gaps. This is an iterative process in which the quality of the testing is continuously measured, and improvements to the testing are suggested when analyzing the results of the measurement.

Functional coverage models are composed of four components. The main component is a semantic description (story) of the model. The second component is a list of the attributes mentioned in the story. The third component is a set of all the possible values for each attribute. The last component is a list of restrictions on the legal combinations in the cross-product of attribute values.

Table 1 shows the attributes and their values for a functional coverage model taken from the floating point domain. The model consists of four attributes - Instr, Result, Round Mode, and Round Occur - each with the possible values shown. The semantic description of the functional coverage model is: test that all *instructions* produce all possible *target results* in the various *rounding modes* supported by the processor both when *rounding* did and did not occur.

Each attribute may be partitioned into one or more disjoint sets of semantically similar values. This provides a convenient way for users to conceptualize their model and for analysis tools to report on coverage data. In Table 2, which gives several partitioning examples, the instruction attribute is partitioned into arithmetic instructions and non-arithmetic instructions, or according to the number of input operands. The result attribute is partitioned according to the sign of the result, and so on.

Restrictions for this model are shown in Table 3. They describe combinations of attribute values that should never occur, for example, the result of an `fabs` instruction cannot be negative. The coverage model illustrated here has been used to check the coverage of tests generated for the floating point unit of several PowerPC processors.

Attr	Description	Values
Instr	Opcode of the instruction	fadd, fadds, fsub, fmul, fdiv, fmadd, fmsub, fres, frsqrte, fabs, fneg, fsel, ...
Result	Type of result	± 0, $\pm$MinDeNorm, $\pm$DeNorm, $\pm$MaxDeNorm, $\pm$MinNorm, $\pm$Norm, $\pm$MaxNorm, $\pm\infty$, SNaN, QNaN, none
Round Mode	Rounding mode	toward nearest, toward 0, toward $+\infty$, toward $-\infty$
Round Occur	Did rounding occur?	True, False

Table 1: Attributes of the floating-point model

Attribute	Partition	Values
Instr	Arith	fadd, fsub, fmul, fdiv, ...
	Non Arith	fabs, fneg, fmr, fsel, ...
Instr	1 Operand	fabs, fneg, fres, frsqrte, ...
	2 Operands	fadd, fsub, fmul, fdiv, ...
	3 Operands	fmadd, fmsub, fsel, ...
Result	Positive	$+0, \ldots, +$Norm$, \ldots, +\infty$
	Negative	$-0, \ldots, -$Norm$, \ldots, -\infty$
	Not A Number	SNaN, QNaN

Table 2: Attribute partitions (partial list)

Functional coverage models come in many flavors. Models may cover the inputs and outputs of an application (black box), or may look at internal state (white box). Functional coverage models may be either snapshot models, which check the state of the application at a particular point in time, or temporal models, whose tasks correspond to the application's state as it evolves.

2.1 Creating a Functional Coverage Model

The first step in the functional coverage process is the design of the coverage model. This is typically done by isolating "interesting" attributes of the application and building a story around them. For coverage to impact positively on the testing process, models should cover areas that are considered risky or error prone. The size of the model (number of legal tasks) should be chosen in accordance with testing resources. The model size should not be so large as to make it difficult to cover or impractical to analyze. Constructing a good coverage model requires functional coverage skills and domain specific knowledge.

The part of the model most difficult to define correctly is the model restrictions. Often a deep understanding of the specification, design, and implementation is needed to create correct restrictions.

Name	Description
Arith-SNaN	Instr $\in$ Arith $\Rightarrow$ Result $\neq$ SNaN
NonArith-Round	Instr $\notin$ Arith $\Rightarrow$ Round Occur $= 0$
Abs-Nabs	Instr $=$ fabs $\Rightarrow$ Result $\notin$ Negative AND Instr $=$ fnabs $\Rightarrow$ Result $\notin$ Positive

Table 3: Restrictions for the floating-point model

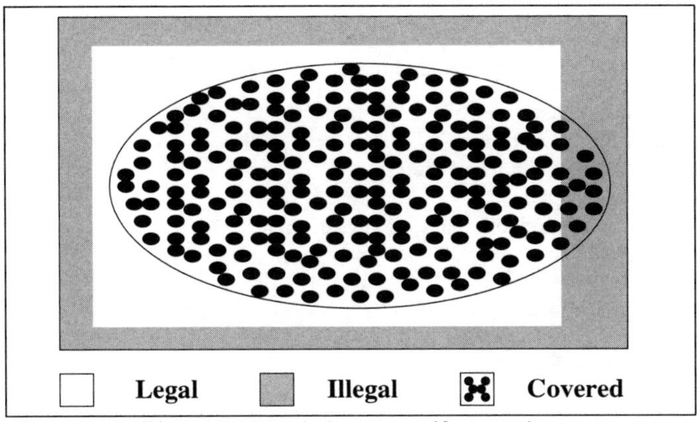

Figure 1: Restrictions are self-correcting

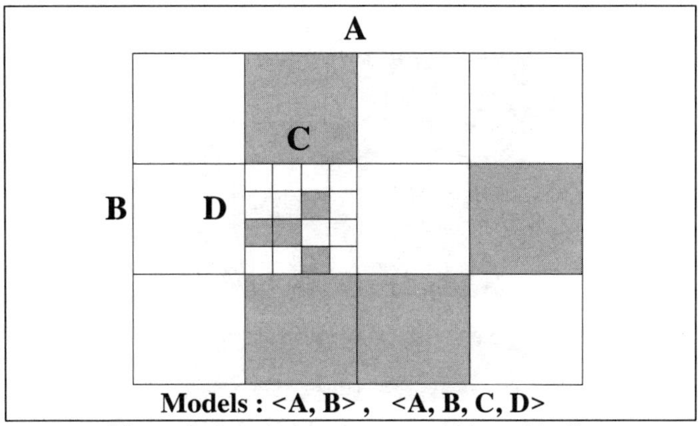

Figure 2: Restrictions are created hierarchically

	X	Y
→	0	2
	0	3
→	1	2
	1	4
	2	1
→	2	2
	2	6
→	3	2
	3	7
→	4	2

	X	Y	
	4	4	
→	5	2	
	5	8	
→	6	2	
	6	6	←
	6	7	←
	6	8	←
→	7	2	
	7	3	
	7	4	

	X	Y	
	7	6	←
	7	7	←
	7	8	←
→	8	2	
	8	6	←
	8	7	←
	8	8	←
	8	9	
→	9	2	
	9	5	

Figure 3: List of uncovered tasks

While creating restrictions, designers may need to think of how the system operates in ways not previously considered. This process turns out to be useful in itself, as it increases their understanding of the application.

2.2 Refining the Functional Coverage Model

Every task in the cross-product of a model's attribute values is either legal or illegal, and has either been covered or remains uncovered, as shown in Figure 1. Uncovered legal tasks are holes in the coverage model. Sometimes after careful analysis, it is determined that a hole is in fact an illegal task, and additional restrictions need to be introduced into the model to eliminate the hole. Conversely, illegal tasks that have been covered are either actual bugs in the application, or reflect a coverage model that is over-restrictive. Thus, the definition of a coverage model is dynamic, with tasks occasionally drifting between being legal and illegal as the application is better understood and the model is refined. In practice, actual bugs in the application are rare, and covered illegal tasks usually reflect problems in the model definition. Holes, in contrast, are more common, and are as likely to result from inadequate testing as from missing restrictions in the model.

The size of a model's cross-product is determined by the number of its attributes and the number of values for each attribute. Consequently, the number of tasks in a functional coverage model can be quite large making the model difficult to cover and analyze. One way to reduce the size of a model is to define restrictions that eliminate illegal tasks.

An alternative approach, suggested by Figure 2, is to start with a coverage model that has fewer attributes. In the floating point domain, for example, an initial model may consider just the two attributes, Instr and Result. Restrictions for such models are simpler both to write and refine. Once these smaller coverage models have been satisfactorily defined, we can apply their restrictions to more comprehensive models. This hierarchical approach to modeling makes it quicker to define restrictions and to adjust the coverage model to the simulation resources.

3. HOLE ANALYSIS

Once the functional coverage model has been defined and coverage of tests has been measured, the next step in the verification process is to analyze the coverage data. A critical aspect of the coverage analysis is to determine those areas of the application that have not been adequately tested. While most coverage tools are able to report on individual uncovered tasks, it remains the function of hole analysis to discover and concisely present information about meaningful holes. For us, holes are considered meaningful either because they represent a relatively large number of uncovered tasks, or because they correspond to a set of tasks that have some semantic coherency. In this section, we present a number of techniques for discovering such quantitatively and conceptually meaningful holes.

Consider a model with just two integer attributes, X and Y, each capable of taking on values between 0 and 9. Figure 3 shows the individual uncovered tasks. This is typical of the results that can be obtained from most tools. Not immediately obvious are the two meaningful holes that exist in the coverage data, shown with arrows in the figure. One hole occurs whenever Y equals 2, and a second hole exists when both attributes have values 6, 7, or 8. This is, however, readily seen in Figure 4. Such graphs, which can be obtained using our tools, provide a convenient way to present holes that are clustered along ordered values in models with a small number of attributes. The challenge for hole analysis is to discover more complex holes in arbitrarily large models, and to present these holes in such a way that their root cause may be more easily discerned.

3.1 Aggregated Holes

Meaningful holes can be automatically discovered between uncovered tasks that are quantitatively similar. A simple metric for similarity is the Hamming distance between two holes. This corresponds to the number of attributes on which the two differ. The distance will be one for holes that differ in only one attribute, two for holes that differ in two attributes, and so on, up to the number of attributes in the coverage space. We aggregate together any two holes whose Hamming distance is one. Thus, the two uncovered tasks <0,2> and <0,3> from Figure 3 can be aggregated into the single hole <0,{2,3}>.

Hamming distances can be also computed on aggregated holes.

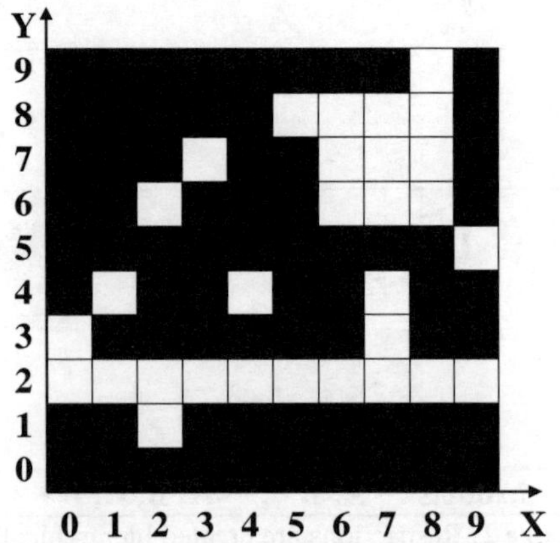

Figure 4: Holes visualization example

$$\begin{array}{c} <0,2> \\ <0,3> \\ <7,2> \\ <7,3> \\ <7,4> \end{array} \Longrightarrow \begin{array}{c} <\{0,7\},2> \\ <\{0,7\},3> \\ <0,\{2,3\}> \\ <7,\{2,3,4\}> \end{array} \Longrightarrow \begin{array}{c} <\{0,7\},\{2,3\}> \\ \\ <7,\{2,3,4\}> \end{array}$$

Figure 5: Aggregated holes calculation

Again, the distance between any two holes is equal to the number of differing attributes, but now the comparison is done for aggregated sets as well as atomic values. The process can be applied iteratively until no more new aggregated holes are discovered. Figure 5 shows how the five tasks <0,2>, <0,3>, <7,2>, <7,3>, and <7,4> can be aggregated together until only the holes <{0,7},{2,3}> and <7,{2,3,4}> remain. This technique is similar to Karnaugh binary mapping [8] and is useful for much the same reasons.

3.2 Partitioned Holes

A second technique for performing hole analysis is to group together holes that have been defined to be conceptually similar. Grouping is done according to semantic partitioning of the attribute values provided by the user when defining the coverage model. Attribute values that are partitioned together are frequently tested together, and holes in one value may occur for the whole group.

Grouping is used to describe larger holes and to provide more focus on the meaning of the holes. For example, the second hole in Table 4 is a grouping of many smaller holes. This grouping allows us to detect the common denominator (and perhaps the common cause) of all the sub-holes, that none of the instructions in the SquareRoots group produced denormalized results. After some analysis, it was determined that square root instructions in fact should not produce such results, and the hole was converted into a restriction in the coverage model.

Both aggregated and partitioned holes look to combine individual, uncovered tasks into larger holes that in turn can suggest reasons for the observed gaps in coverage. In both cases, similarity between tasks is used to determine which tasks should be combined. The difference between the two is that the similarity for partitioned holes is semantic and based upon groupings provided by the user, whereas for aggregated holes it is quantitative and discovered automatically by the tool.

3.3 Projected Holes

Consider the first hole described in Table 4, which has the specific value **fadd** for the instruction attribute and wild cards in all the other attributes. This hole denotes that all the tasks with **fadd** are not covered and it describes 152 tasks. We call this a projected hole. The dimension of a projected hole is the number of attributes in it that do not have specific values.

```
Data    : set T of covered tasks
Result  : list of projected holes

// mark phase
foreach t ∈ T do
    if t is not marked then
        mark t
        recursively mark all ancestors of t
    end
end

// report phase
for i = k downto 0 do
    foreach subspace v with dimension i do
        if v is not marked then
            report v as hole
            recursively mark all descendants of v
        end
    end
end
```

Figure 6: Projected holes algorithm

Any coverage model of n attributes can be viewed as a set of tasks or points in an n-dimensional space. Viewed as such, a projected hole of dimension 1 contains tasks that lie on a line in hyperspace, a projected hole of dimension 2 describes tasks that lie on a plane, and so on. A projected hole of a higher dimension subsumes all holes of a lower dimension that it contains. For example, the hole $p = < *, *, x_3 \ldots, x_n >$ of dimension 2 contains all the tasks described by the subspace $q = < x_1, *, x_3 \ldots, x_n >$. In such a case, we say that p is the *ancestor* of q, and that q is the *descendant* of p. In general, the higher the dimension, the larger and, therefore, more meaningful the hole it describes. Since a hole of higher dimension is both more informative and more concise than the holes it subsumes, it becomes unnecessary, and in fact redundant, to report on the latter.

One goal of hole analysis is to find projected holes of the highest dimensionality. The algorithm shown in Figure 6 first iterates over all covered events and marks them and all their ancestors. Next, it iterates over all potential holes, starting with those of the highest dimensionality ($< *, *, \ldots, * >$). Any subspace that is unmarked is reported as a significant hole and its descendants are recursively marked.

A projected hole is considered *lightly covered* if the percentage of tasks subsumed that have been covered is less than some given threshold. It is fairly easy to adapt the projected holes algorithm to report on lightly covered holes.

Another effective approach is to use a classification algorithm, such as ID3 [11], to find projected holes. Most classification algorithms use some kind of entropy measurement on the data to guide the construction of the decision tree used in the classification. In our case, the target function is the coverage data and ID3 is used to find a large cluster of uncovered tasks.

Table 4 shows some of the holes found in the floating point model when the model was implemented at one of the design laboratories as part of their floating point test plan. The first hole indicates that none of the tasks related to the **fadd** instruction were covered. The hole was caused by a bug in the specification to the test

Instr	Result	Round Mode	Round Occur	Hole Size
fadd	$*$	$*$	$*$	152
Square Roots	+DeNorm +MaxDeNorm +MinDeNorm	$*$	$*$	72
Estimates	$*$	$*$	True	144
$*$	±MaxDeNorm	$*$	True	263

Table 4: Coverage holes report

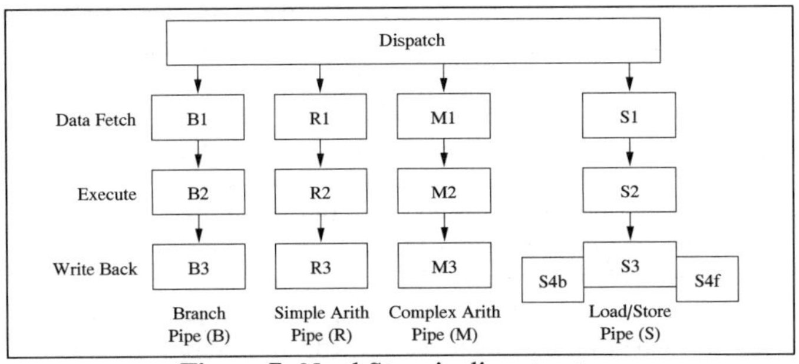

Figure 7: NorthStar pipeline structure

generator that omitted the **fadd** instruction from the list of instructions that needed to be generated. The next two holes show two groups of tasks that cannot be achieved. The first points to square root instructions (fsqrt, fsqrts, frsqrte) with small results, while the second points to estimation instructions (fres, frsqrte) with inexact results. Both holes are covered by restrictions that eluded the developer of the model. After their detection, the two holes were converted to restrictions. The last hole in the list corresponds to a group of tasks with specific results and rounding that was hard to generate, and was therefore only lightly covered (less than 10 percent). This hole was filled by enhancing the capabilities of the test generator.

A significant feature of our coverage tools is that the language they use to report holes is equivalent to the one used by the user for defining restrictions. The hole <Arith,SNaN,*,*>, for example, is expressed by the restriction $Instr \in Arith \Rightarrow Result \neq SNaN$. This means that holes can be easily translated into restrictions if necessary. In many cases, the translation can be automated.

4. INTERDEPENDENCY MODEL

In this section, we describe the coverage model and hole analysis that was done for checking register interdependency in the pipelines of the NorthStar processor. NorthStar is a member of the multi-threaded, super-scalar PowerPC family of processors developed by IBM. As depicted in Figure 7, the processor consists of four execution pipelines: branch (BPipe), simple arithmetic (RPipe), complex arithmetic (MPipe), and load/store (SPipe). Single cycle, fixed point arithmetic instructions can be handled either by RPipe or MPipe. All other arithmetic instructions, including floating point instructions, are handled by MPipe. Each pipeline consists of a data fetch, execution, and write back stage. The load/store pipe has two additional stages that are used for thread switching. The NorthStar processor does not allow out-of-order execution, so that all instructions progress through their respective pipelines in the same order in which they appear in the test.

A coverage model was defined to check for register interdependency between two instructions that appear together in the pipeline.

Attr	Descr	Values
I_1	Instr 1 type	FP-ld-simple, FP-st-simple, fix-ld-simple, fix-st-simple, fix-ld-complex, fix-st-complex, FP-M, fix-M, fix-R, branch, cache, condition, sync-call-return
P_1	Instr 1 pipe	BPipe, RPipe, MPipe, SPipe
S_1	Instr 1 stage	1-5
I_2	Instr 2 type	same as I_1
P_2	Instr 2 pipe	same as P_1
S_2	Instr 2 stage	same as S_1
Dep	Dependency	RR, RW, WR, WW, None

Table 5: Attributes of the interdependency model

$I_1(I_2)$ is a simple fixed point instruction $\Rightarrow P_1(P_2)$ is RPipe or MPipe
$I_1(I_2)$ is a complex fixed point or an FP instruction $\Rightarrow P_1(P_2) = $ MPipe
$I_1(I_2)$ is a load/store instruction $\Rightarrow P_1(P_2) = $ SPipe
$I_1(I_2)$ is a branch instruction $\Rightarrow P_1(P_2) = $ BPipe
$P_1 = P_2 \Rightarrow S_1 \neq S_2$
$P_1(P_2) \neq SPipe \Rightarrow S_1(S_2) <= 3$
$I_1(I_2)$ is a branch instruction $\Rightarrow Dep = $ None
$I_1(I_2)$ is a cache or complex store instruction $\Rightarrow Dep \neq$ WR (RW)
$S_1 = 2$ and I_1 is not a complex load instruction $\Rightarrow Dep \neq$ WR
$S_2 = 2$ and I_2 is not a store or mtspr instruction $\Rightarrow Dep \neq$ WR
$S_1 >= S_2$

Table 6: Restrictions for the interdependency model

The attributes of the model are shown in Table 5. The model consists of seven attributes: the type, pipeline, and stage of one instruction; the type, pipeline, and stage of a second instruction; and the register interdependency that may exist between the two. We say that two instructions in a pipeline have a register interdependency if they both access the same register during execution. The model distinguishes between read/read (RR), read/write (RW), write/read (WR), and write/write (WW) dependencies depending if the first/second instruction was reading or writing to the register. The model story looks for coverage of all possible interdependencies (including none) between all combinations of instructions in any possible pipeline stage.

An example of a coverage task in the model is the task <fix-R, RPipe, 2, fix-ld-simple, SPipe, 1, RR>, which corresponds to a simple fixed point instruction in stage 2 of the RPipe which has a read/read dependency with a fix-ld-simple instruction in stage 1 of the SPipe.

Restrictions in the model arise for various reasons: pipelines are dedicated to handle only certain types of instruction, limitations that exist on the pipeline architecture, certain instructions never read or write to registers, instructions are restricted in what dependencies they can have at certain stages of execution, and out-of-order execution is not supported by the processor. A partial list of restrictions for the interdependency model is shown in Table 6.

After performing hole analysis on the interdependency model,

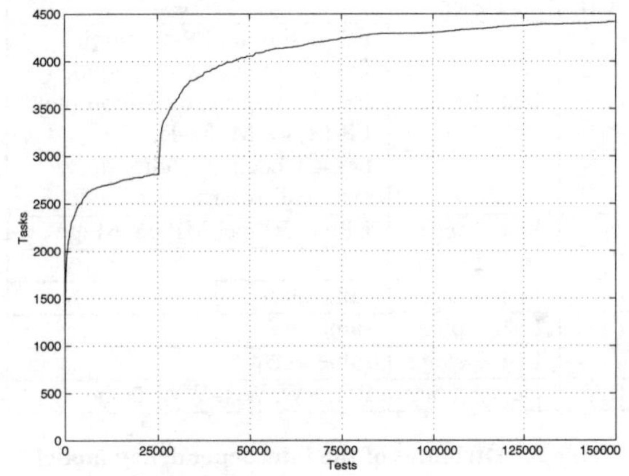

Figure 8: Interdependency coverage progress

two holes of significance were discovered. The first was the result of a bug in the test generator that was used to produce the test inputs. The test generator did not consider the fact that a store and update instruction also writes to a register, and therefore, such events were generated with low probability. The bug was fixed with the help of the hole that was discovered.

The second hole was due to the fact that some tasks need a rare combination of events to occur before they can be observed. When not enough thread switches are created, for example, the logic that checks dependency between the foreground and the background threads, about a third of the total pipeline logic, is not sufficiently tested. The solution was to reduce the timeout between threads and increase the mix of instructions which cause thread switches. While the test generator was capable of generating such events, the likelihood of doing so unintentionally was low. To improve coverage, the test generator was instructed by the verification engineer to concentrate on these corner cases. Figure 8 shows that a significant improvement in coverage was observed after about 25,000 tests were measured when these changes were made.

5. CONCLUSIONS

In this paper, we described hole analysis, a method that provides concise information about uncovered and lightly covered areas in cross-product functional coverage models. We presented a number of classification techniques used in holes analysis. Using two case studies, we showed how hole analysis can assist users of coverage tools to improve the quality of their verification.

We illustrated that effective coverage models are built hierarchically and through iterative refinement. We also showed that holes and restrictions can be similarly expressed, leading to two main advantages. First, the learning curve for writing functional coverage models and for using hole analysis is reduced. Also, the translation of holes into missing restrictions can be automated in many cases.

We are currently investigating several other techniques for increasing the analysis capabilities of our coverage tools. Items on which we are working include using machine learning and data-mining [4] techniques to detect coverage holes that are not rectangular or not parallel to the attribute axis (e.g., hole such as $x+y<7$). We are also looking at automatic and semi-automatic techniques for discovering relationships between the input to the simulation (e.g., test cases, test cases specification) and the coverage data. This type of analysis can identify which inputs contribute more towards coverage and should therefore be utilized more often.

6. REFERENCES

[1] A. Aharon, D. Goodman, M. Levinger, Y Lichtenstein, Y. Malka, C. Metzger, M. Molcho, and G. Shurek. Test program generation for functional verification of PowerPC processors in IBM. In *Proceedings of the 32nd Design Automation Conference*, pages 279–285, June 1995.

[2] A.M. Ahi, G.D. Burroughs, A.B. Gore, S.W. LaMar, C.R. Linand, and A.L. Wieman. Design verification of the HP9000 series 700 PA-RISC workstations. *Hewlett-Packard Journal*, 14(8), August 1992.

[3] B. Beizer. The Pentium bug, an industry watershed. *Testing Techniques Newsletter, On-Line Edition*, September 1995.

[4] P. Cabena, P. Hadjinian, R. Stadler, J. Verhees, and A. Zanasi. *Discovering Data Mining, from Concept to Implementation*. Prentice Hall, 1997.

[5] L. Fournier, Y. Arbetman, and M. Levinger. Functional verification methodology for microprocessors using the Genesys test-program generator. In *Proceedings of the 1999 Design, Automation and Test in Europe Conference (DATE)*, pages 434–441, March 1999.

[6] R. Grinwald, E. Harel, M. Orgad, S. Ur, and A. Ziv. User defined coverage - a tool supported methodology for design verification. In *Proceedings of the 35th Design Automation Conference*, pages 158–165, June 1998.

[7] A. Hajjar, T. Chen, I. Munn, A. Andrews, and M. Bjorkman. High quality behavioral verification using statistical stopping criteria. In *Proceedings of the 2001 Design, Automation and Test in Europe Conference (DATE)*, pages 411–418, March 2001.

[8] M. Karnaugh. The map method for synthesis of combinational logic circuits. *Transactions of the American Institute of Electrical Engineers*, 72(9):593–599, November 1953.

[9] B. Marick. *The Craft of Software Testing, Subsystem testing Including Object-Based and Object-Oriented Testing*. Prentice-Hall, 1985.

[10] C. May, E. Silha, R. Simpson, and H. Warren, editors. *The PowerPC Architecture*. Morgan Kaufmann, 1994.

[11] J. R. Quinlan. Inductions of decision trees. *Machine Learning*, 1:81–106, 1986.

[12] S. Ur and A. Ziv. Off-the-shelf vs. custom made coverage models, which is the one for you? In *proceedings of STAR98: the 7th international conference on software testing analysis and review*, May 1998.

[13] Focus. http://www.alphaworks.ibm.com.

Effective Safety Property Checking Using Simulation-Based Sequential ATPG *

Shuo Sheng
Department of ECE
Rutgers University
Piscataway, NJ, 08854
shuo@ece.rutgers.edu

Koichiro Takayama
Fujitsu Labs. of America Inc.
595 Lawrence Expressway
Sunnyvale, CA, 94086
ktakayam@fla.fujitsu.com

Michael S. Hsiao
Bradley Department of ECE
Virginia Tech.
Blacksburg, VA, 24061
mhsiao@vt.edu

ABSTRACT

In this paper, we present a successful application of a simulation-based sequential Automatic Test Pattern Generation (ATPG) for safety property verification, with the target on verifying safety property of large, industrial-strength, hardware designs for which current formal methods fail. Several techniques are developed to increase the effectiveness and efficiency during state exploration and justification of the test generator for verification, including (1) incorporation of a small combinational ATPG engine, (2) reset signal masking, (3) threshold-value simulation, and (4) weighted Hamming distance. Experimental results on both ISCAS89 benchmark circuits and real industry circuits have shown that this simulation-based verifier achieves better or comparable results to current state-of-the-art formal verification tools BINGO and CHAFF.

Categories and Subject Descriptors

B.7.2 [**INTEGRATED CIRCUITS**]: Design Aids, Verification, Simulation ; B.8.1 [**PERFORMANCE AND RELIABILITY**]: Testing

General Terms

Algorithms, Verification

Keywords

Verification, Simulation-Based, Sequential ATPG

1. INTRODUCTION

The ever-increasing complexity and circuit size have made verification and testing bottlenecks in the VLSI design cycle. In this paper, we address the problem of safety property verification in model checking and propose a simulation-based solution. The safety property verification problem involves checking if a property is held or violated in the design, and the current solutions essentially involve extensive state traversal in the finite-state-machine (FSM) of the design. The FSM traversal problem targeting property verification can be further decomposed into three subproblems: (1) how to effectively represent the states; (2) how to effectively represent the state transition function; (3) how to effectively traverse the state space. The main difficulty in all three problems lies in the exponential gap between design size and the number of states to explore.

Verification based on functional simulation is a very popular approach due to its simplicity. Its effectiveness, however, largely depends on the quality of vectors that are simulated. For large designs (which often implies large state space), exhaustive simulation would not be feasible, and some corner-case bugs can often be missed. On the other hand, significant advances in *formal methods* to design verification have been published. These formal methods can rigorously prove if the (small to medium-sized) design is correct.

Model checking and equivalence checking are two verification problems for which successful formal methods have shown promise. Model checking determines whether or not the design satisfies a given set of properties. If a property is proven to be violated, a counterexample (or error-trace) can be generated. Equivalence-checking, on the other hand, attempts to rigorously prove if the implementation circuit is functionally equivalent to the specification circuit (golden model). Formal methods for both model-checking and equivalence-checking frequently involve FSM traversal. In this paper, we target model-checking of safety properties. It is noted, however, that our method may also be applicable to equivalence checking as well.

The current model checking methods fall into three categories: (1) explicit methods, (2) symbolic methods, (3) ATPG or SAT based methods. Explicit methods take the finite state model of the design and enumerate the states explicitly. The example is $Mur\phi$ [2]. The limitation of this approach is its lack of scalability to larger design due to the explicit storage of the enumerated states. Symbolic model checking [1] utilizes Reduced Ordered Binary Decision Diagrams (ROBDDs) to compactly represent and manipulate the set of reachable states and the state transition functions. ROBDDs represent the sets of state and transition functions implicitly by its canonical structure. The attractive feature

*This work was done while the first author was a summer intern at Fujitsu Labs. of America, Inc. 2001, and is supported in part by NSF Grant CCR-0196470, a grant from NJ Commission on Science & Technology, and Fujitsu Labs of America.

of OBDDs is that the OBDD size may be much smaller than the states represented (with a given variable order). Thus, it is able to handle more states than explicit methods. Example verifiers in this class are SMV [3] and BINGO [18]. Though more efficient than explicit methods, the primary limitation of BDD-based approaches is the "memory explosion" problem, which occurrs when the BDDs for representing the state transition functions of the circuits may not be constructed entirely [7]. As a result, methods other than BDDs have been proposed, such as SAT [4] and ATPG based approach for model checking [8, 15, 14]. The ATPG based approach does not explicitly store the complete state space or build a canonical representation of the Boolean state transition function, instead, the state transition is done by forward simulation and the state information is learned on the fly. A solution is derived when sufficient knowledge has been acquired. However, ATPG could potentially take a long time when a hard property for a large circuit is encountered. In essence, BDD approaches are limited in space (memory) and could be limited in time if variable ordering is needed, while ATPG approaches are limited in time.

In terms of sequential ATPG, two approaches are possible: deterministic and simulation-based. Currently, almost all previous work on using ATPG for verification are deterministic [8, 15, 14] because they can prove *correctness* (i.e., when the modeled fault is untestable). In this case, a branch-and-bound algorithm is employed to exhaust the search space progressively so that no input vector could be found to detect the targeted fault; thus, we can safely claim that this property is held as the search space has been exhausted. If the procedure succeeds in finding a sequence to detect the target property, a counter example is generated to invalidate the property. As ATPGs are limited in time, the procedure aborts on difficult properties because of limited time, which is restricted in practice by a given limit in the number of backtracks, time-limit, etc. When the target objective is aborted, we can claim nothing about the property. Although various techniques have been developed to prune the search tree (e.g. given the recent significant progress in Boolean Satisfiability (SAT) solver [5, 6]) so that many unnecessary backtracks are reduced, aborts still occur because the deterministic ATPG intrinsically tries to exhaust the search space. The major difference between SAT and deterministic sequential ATPG are, though both are branch-and-bound, ATPG can target sequential time-frames implicitly, while SAT generally targets it by explicitly unrolling frames.

The other category, simulation-based ATPG for verification, is investigated in this work. Unlike the deterministic counterpart which tries to exhaust the complete search space, simulation-based ATPG does a "random walk" in the state space, hoping to hit the "target" in a limited number of steps. The walk is guided by some heuristics, thus it is not an entirely random traversal. At each step, a number of trial steps are made and evaluated by a *cost function* which measures how far the state of the circuit is from the target state. The cost function is computed through either logic simulation or fault simulation. The step with the least cost is picked as the next step and the process is repeated until the target is reached or certain limit is exceeded. Such ATPGs have the advantage of scaling to large designs because of its simulation-based nature, where only forward simulation is used and backtracks never occur. However, the major drawbacks are: (1) it can not prove correctness, due to simulation-based ATPG's inability to identify untestable faults; (2) a good cost function is needed to guide the search, otherwise a lot of time will be wasted on traversing the useless states. Although the simulation-based methods cannot prove correctness in verification, it fits quite well the objective of "debugging". In practice, for a large industry design, the verification objective is usually concerned less with providing formal proofs of correctness than finding bugs. In other words, designers are more concerned with finding a counter example to falsify the *large* design, which is generally easier than proving if it is 100% correct.

Our main contribution in this work is on elimination of the two "drawbacks" of simulation-based ATPG for verification. First, we incorporate a small deterministic engine, which is used to generate justification objective for the simulation engine as well as identify certain number of untestable faults. Second, we developed several techniques to refine the cost function in the simulation engine to efficiently guide the state traversal during the "bug hunt". Experimental results on both large ISCAS89 benchmark and real industrial circuit show our simulation-based approach achieved comparable or better performance than the existing formal approaches including CHAFF [6].

The organization of the paper is as following: Section 2 present some preliminaries such as problem formulation and introduction of sequential ATPG tool STRATEGATE; the four of our approaches for using sequential ATPG for verification are described in Section 3; Section 4 gives the overall procedures and discusses the "local minima" problem encountered; experimental results are presented in Section 5 and followed by conclusion in Section 6.

2. PRELIMINARIES

The prototype of the sequential ATPG we use for verification is STRATEGATE [9, 10]. It is a simulation based ATPG that uses genetic algorithm (GA) as the core algorithm for test generation. In GA, a *population* contains a set of *individuals* or *strings*. Every string is a candidate solution for the problem at hand. Each is associated with a *fitness*, which measures the quality of this individual for solving the problem. In test generation context, this fitness measures how good the candidate sequence (bit string) is for exciting or propagating the target faults. The fitness evaluation is done by fault simulation. Based on the evaluated fitness, evolutionary process of *selection, crossover and mutation* are used to generate a new population from the existing population. The process is repeated until the fitness of the best individual is good enough.

In the application context, we focus on verifying a class of safety properties which are expressed in CTL as $EF(p)$. The property to be verified is converted to a stuck-at-0 fault by constructing a monitor circuit and add it to the original circuit. The output of the monitor becomes one of the primary output (PO) of the transformed circuit, which is shown in Figure 1. Then the problem of verification becomes test generation for this stuck-at-0 fault. Because this is a PO fault, all we need is fault excitation. STRATEGATE performs test generation for this target fault in the following framework: first, a combinational ATPG is performed to derive a test (a vector of primary inputs I_p plus state S_{target}) which can excite the fault in one time frame, assuming full control of the state; second, justification process – a search is done for a sequence $\{ I_0, I_1, ..., I_n \}$ that can justify the

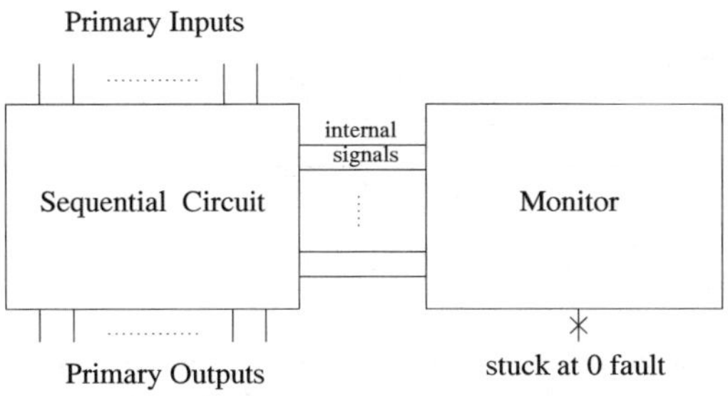

Figure 1: Transform Safety Property to S-@ Fault

state S_{target} from a given reachable state. If the justification process succeed, we then obtain a test sequence { I_0, I_1,..., I_n , I_p } that can detect the fault, which falsifies the property. If it failed, the only conclusion we can made is that we can not prove this property to be wrong. This step is hard because it involves essentially sequential ATPG.

3. OUR APPROACH TO VERIFICATION

Our work focuses on developing several techniques to build a strong state justification engine into the simulation-based test generator. The techniques we developed are described in the following subsections.

3.1 Incorporation of Deterministic ATPG

The first work is embedding a small combinational deterministic test generator into the simulation-based framework. The original STRATEGATE uses single-time-frame GA to do a simulation-based combinational test generation. However, it can not guarantee a solution even if one exists. Therefore, we put in a PODEM-based deterministic engine to help STRATEGATE. Since this deterministic engine is only called for combinational test generation, its cost is low even for large circuits. We modified the PODEM algorithm so that it can generate *multiple* solutions (if they exist) for a target fault. This is desirable because each solution is associated with a target state to be justified. When justification fails for one (unreachable) state, we can switch to the next solution. This deterministic engine also lends another advantage: it can identify some untestable faults. Because our deterministic engine is mainly combinational oriented, it is not powerful enough to identify sequentially untestable faults. However, it can identify all combinational untestable faults (given enough backtrack limits) and thus can prove the correctness of some properties.

3.2 Reset Signal Masking

A technique called *reset signal masking* is incorporated in the GA generation process to avoid revisiting previously visited states (thus forming cycles) during state traversal. Usually, designers will put in reset (or partial) reset signals into circuit for Design for Testability (DFT) purposes. Assertion of these signals will force the circuit state to return to certain reset or partial-reset states, thereby repeating cycles or loops in the FSM traversal. This will significantly reduce the number of new states visited and limit the depth the test generator can reach in the State Transition Graph (STG). The hard properties to be verified usually require a long sequence of states be traversed before the target states are reached. Therefore, we developed the reset signal masking technique in the GA to avoid resetting or partially resetting the state.

Definition 1. Reset Power 0(1) *for a given PI is the number of FFs that can be reset to a specified value by applying a 0(1) at this PI and X at all other PIs, with the initial state of circuit being all X.*

Reset power measures how powerful a single PI is in resetting the Finite State Machine (FSM) of the circuit. If a PI can reset all the FFs, then its reset power is n_{FF}, the total number of flip-flops in the circuit. Reset power for each PI can be identified by parallel logic simulation, starting from the all-unknown state, which takes negligible time. From the reset power, we define the following *Reset Mask Probability.*

Definition 2. Reset Mask to 1 Probability: if the reset power 0 of PI_j is greater than 0, then its reset mask to 1 probability is

$$prob(PI_j == 1) = \frac{ResetPower0_j \times k}{n_{FF}},$$

where k is a weight used to scale the probability value. Sometimes n_{FF} is far greater than the reset power value, which makes the corresponding reset mask probability near to zero. In this case, we use k to scale the value to a noticeable range. Similarly, we can also define *reset mask to 0 probability,* which is analogous to its reset mask to 1 probability counterpart. After we identify the reset power, reset mask probability is calculated and applied in GA after crossover and mutation operation.

Table 1 reports the reset signals we identified from the circuits we were using in the experiments. For each circuit, the number of FFs in this circuit is first given. Next, each subsequent column reports the number of PIs that can reset the corresponding number of flip-flops. For example, the value "1" under the column "1452" for circuit *s38584* indicates that there is one PI that can reset 1452 FFs for this circuit. This is actually a global reset signal since all FFs are reset. For circuit *s5378*, the value "4" under column "3" indicates that there exist 4 PIs that can reset groups of 3 FFs. We noticed from the table that some PIs have stronger resetting powers than others, e.g. in the same circuit of *s38584*, there is another PI that can reset 103 FFs.

3.3 Threshold-Value Simulation

Threshold-value model of logic gates was first proposed by Cheng and Agrawal [11]. They developed a test generator based on threshold value simulation [12]. Threshold value model is an analog model of logic gates which not only keeps the Boolean logic value information but also carries the dynamic controllability information with the corresponding threshold value. A logic '0' is represented by a value within the real value range [0,0.1] and a logic '1' is represented by a value within [0.9,1]. Therefore, it leaves space for the concept of "strong" and "weak" 0 and 1. The closer the thresh value to 0, the stronger the logic '0' is. Conversely, the closer the threshold value to 1, the stronger the logic '1' is. The evaluation of a gate is done by mapping the average input value of the gate to a output value within range [0,1] through the threshold function of the gate. The detail of the threshold function of each gate type is defined in [12]. In implementation, floating value is represented as integer to eliminate floating-point operation.

In our work, we use the threshold-value simulation to refine the GA fitness function for state justification. The

Table 1: Reset Signal Statistics

circuit	# of FF	1636	1452	327	179	103	15	11	5	3	2	1
s5378	179	-	-	-	1	-	-	-	2	4	5	2
s38417	1636	1	-	-	-	-	-	1	2	-	2	23
s38584	1452	-	1	-	-	1	1	-	-	-	-	1
mma	327	-	-	1	-	-	-	-	-	-	-	1

original STRATEGATE uses the number of FFs of the current state that matches the target states as the fitness function. However, it can not distinguish two different sequences where both justify the same subset of FFs but neither of them justifies the entire state. Threshold value simulation, in this case, will help us further distinguishing the two by measuring how strong or weak the 0s or 1s are for those unmatched (failed) FFs. Figure 2 shows a simple example of how threshold value model differentiate two cases which have the same logic value. Suppose we want to justify a FF, whose input is a 3-input AND gate. Assume the objective is to justify the FF to be 1. In figure2(a) and figure2(b) the objective both failed because a logic '0' is obtained at the output of the AND gate. However, it is observed that the causes of the logic '0' at the AND gate are different in the two cases. In figure2(a), we have two inputs to the AND gate being '1' and one input being '0'. In figure2(b), we have all three inputs to the AND gate being '0'. This implies that (a) has a weaker '0' at the output of AND gate than (b) and hence is more preferable because we just need to flip one of the three inputs to get a 1 to justify the objective FF while for case (b), we need three. This is reflected by a output threshold value the AND gate (obtained through threshold value simulation) being 0.074 for (a) and 0 in case (b). As a result, (a) is favored instead of (b) in GA evolution.

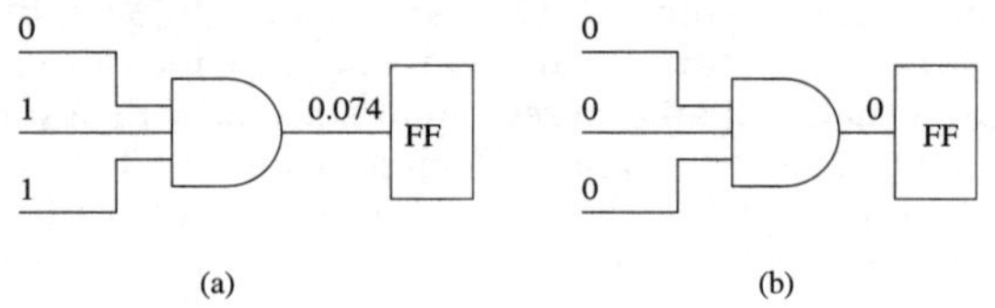

Figure 2: Two cases in threshold value simulation

With threshold value simulation, we use the following fitness function in GA for state justification

$$fitness = \sum_{i=1}^{N} \frac{1}{|FF_{thresh}(i) - FF_{obj}(i)|} \quad (1)$$

where $FF_{thresh}(i)$ is the threshold value of the ith unjustified FF in the target state, $FF_{obj}(i)$ is the objective value of that FF (0 or 1, generated by PODEM). N is the total number of unsatisfied objectives. For the justified FFs in the target state, $FF_{thresh}(i) = FF_{obj}(i)$, we do not include them into the fitness function because otherwise the denominator will become zero. The fitness is defined in the reciprocal form because GA will maximize it.

3.4 Weighted Hamming Distance

The fourth technique is the employment of weighted Hamming distance in calculating the distance between two states. In estimating how far the current circuit state is from the target, we constantly need to compute the distance of two states. Previously, Hamming distance has been used, which counts the number of bits differed in the two given state. However, reported results on this metric suggested that the Hamming distance does not necessarily correlate with the actual distance in state space [16]. Therefore, we enhance it by weighting the bits in the state by their controllability. Intuitively, the difference in a hard-to-control FF will contribute more than a easy-to-control FF to the overall distance of the two states. There are two cases to be differed here when calculating the difference:

- the bit in candidate state is 0 and in target state is 1;
- the bit in candidate state is 1 and in target state is 0.

In the first case, we need a 0 to 1 transition in the FF bit to justify the FF value and a 1 to 0 transition is needed in the second case. Therefore, the distance is target-state-related. We define the *state distance weight* according to the target state's value:

Definition 3. Distance weight for i^{th} FF in the state is $DSW_i = (1 - FFCC_i) * n_{FF}$.

where $FFCC_i$ is the controllability of FF_i. If the target state value in this bit is 0, then it is the 0 controllability of this FF. If target state value in this bit is 1, then it is the 1 controllability. n_{FF} is the total number of FFs. The 0 and 1 controllability used here are the probabilistic based measurement instead of SCOAP so that their value are always between 0.0 and 1.0. In practice, we use the frequency of each FF switching to 0 and 1 during simulation as its 0 and 1 controllability. Therefore, the easier the FF is to control 0 or 1, the close its corresponding swithing frequency value is to 1.0, and from definition 3, the small the weight for this FF. n_{FF} serves here as the upper bound for the weight. The worse case is that a FF is never switched to 0/1, then its corresponding $FFCC_i$ is 0, then its weight is the max weight – n_{FF}. Based on the weight defined for each FF, the overall weighted Hamming distance between state A and target state B is computed using the following formula:

$$WHD(A, B) = \sum_{i=1}^{n_{FF}} |FF_i^A - FF_i^B| * DSW_i \quad (2)$$

where FF_i^A and FF_i^B are the bit value for the two states respectively. WHD denotes "Weighted Hamming Distance". From equation (2) we can see that the original Hamming distance is the case in the weighted Hamming distance that all $DSW_i = 1$.

4. OVERALL PROCEDURE

We describe the overall procedure of the simulation-based ATPG for safety property verification in this section. We assume that each safety property has been transformed to a "target fault" using the monitor circuit introduced in Section 2. The procedure is given below.

1. Compute reset power, calculate reset mask probability
2. Build initial state table by simulating 1000 random vecs
3. State table enrichment phase: enrich the state table by generating & simulating another 1000 vecs, using GA & reset signal masking to maximize # of state visited.
4. For each target fault mapped from the target property
   4.1 Call PODEM to generate a combinational test;
   4.2 Justify the state part of the combinational test S_{tar}:
     nFail = 0; nLocalFail = 0;
     while ($nFail \leq nLimit$), start from the current state,
       4.2.1 Initialize GA population with random seqs;
             Seed the GA any past moves that took the ckt to a state S_p, such that $WHD(S_{tar}, S_p) \leq d$;
       4.2.2 Use GA w/ threshold value simulation to evolve this population and picks the best candidate;
       4.2.3 Make the next move: simulate the best candidate, update the current state;
       4.2.4 If (current state == target state),
               success: goto next property.
             else
               fail: nFail++;
               if fail at the same local minima
                 nLocalFail++;
               end if
               if (nLocalFail == nLocalLimit)
                 generate a random seq as next move,
                 nLocalFail = 0, goto 4.2.1
               end if
             end if
     end while

The main purpose of introducing steps 2 and 3 is to enrich the state table. The test generator maintains a state table which stores every visited state and the vector sequence ("move") that took the circuit to this state. All moves are concatenated to form the final test set. Every time after GA optimization, the test generator will append the best candidate in GA as the "next move" to the test set. The enrichment of state table will help because step 4.2.1 depends on the richness of this table. In step 4.2.1, some past moves are selectively seeded into GA. Their role is as an educated guess of initial values for the optimization algorithm. The more information we have in the state table, the more seeds we would have and hence the greater the chance that the GA will succeed. This state table and the associated moves are the important knowledge learned on-the-fly by the test generator.

4.1 Local Minima

As noticed from the procedure, our state space search treats the state justification problem as a large-scale optimization problem. Since the objective generally cannot be reached in a few steps, heuristics (threshold value fitness, weighted Hamming distance) are used to guide the search to approach the target step by step. GA is in charge of local optimization to decide the next move. As in all greedy algorithms for solving large-scale optimization problems, our procedure can be stuck into local minima as well, which is manifested by reaching a state closest to the target state and has a fitness higher than all its nearby neighbors.

To solve the local minima problem, we introduce in step 4.2.4 a counter *nLocalFail*. If the search has been trapped into the same local minima for a number of times *nLocalLimit*, then a random sequence is accepted as the next move, although its fitness might be much lower than the best one in GA. With this random walk, we hope the search can escape from the local minima. This technique is also known as *hill climbing* in optimization methods.

5. EXPERIMENTS AND RESULTS

Table 2: Circuit Characteristics

circuit name	total # of gates	total # of FFs	total # of properties
s5378	3043	179	9
s38417	26274	1636	50
s38584	23008	1452	50
mma	7166	327	1

We implemented the aforementioned techniques in a total of 21,000 lines of C++. We called this new tool "Threshold value sImulation guided VErifier" (THRIVE). THRIVE was tested on some large ISCAS89 benchmark circuits *s5378, s38417, s38584* and an industrial circuit *mma*. All these circuits have large numbers of FFs which can potentially make the formal verification methods that involve BDDs difficult to handle. The safety properties to be verified for ISCAS89 circuits are random conjunctures of internal signals of the circuits. The one property for *mma* is, however, a real property from the designer, which is known to be false. We compared our results with four tools: HITEC [17], BINGO [18], CHAFF [6] and the original STRATEGATE [9, 10]. HITEC is the most widely used deterministic ATPG tool in academia. BINGO is an efficient symbolic model checking tool for verifying realistic properties of sequential circuits and protocols. CHAFF is a SAT solver. Due to a careful engineering of all aspects of search, CHAFF has achieved significant performance gains over conventional SAT solver and is considered as the state-of-the-art. To apply CHAFF to verification, each property to be verified is converted to a CNF formula by unrolling the circuit k time frames. We did the unrolling with an incremental step of 20 time frames each time and up to 100 time frames in total: if the property can not be verified within k time frames, then $k + 20$ is tried. STRATEGATE is currently the state-of-the-art simulation-based ATPG tool, which has achieved significant performance gain over HITEC in test generation. The HITEC and BINGO experiments were run on a UltraSparcII, 200MHz with 256M RAM. All other experiments were run on a Pentium III 1GHz clock, 1024M bytes RAM machine installed with Redhat Linux 7.1.

Table 2 reports the characteristics of the circuits, including the total number of gates, total number of FFs and total number of properties to be verified for each large circuit experimented. Verification results are shown in table 3, in which we compared the results for HITEC, BINGO, CHAFF, STRATEGATE with THRIVE, proposed in this paper. For each circuit, the number of properties is first listed, followed by the number of verified properties and the execution time taken to verify them. As shown in the table, THRIVE achieved impressive performance over other techniques. For example, in *s5378*, HITEC aborted all nine properties. BINGO can verify all the properties, however, with large computational costs. CHAFF verified all nine of them in a very short time (10.3 sec). STRATEGATE can verify only 1 of the 9 properties to be false while THRIVE can verify all 9 properties to be false, though taking longer

Table 3: Verification Results

ckt	prop.	HITEC verified	HITEC time(s)	BINGO verified	BINGO time(s)	CHAFF verified	CHAFF time(s)	STRATEGATE verified	STRATEGATE time(s)	THRIVE verified	THRIVE time(s)
s5378	9	0	aborted	9	36835	9	10.3	1	58	9	2063
s38417	50	2	2913	fail to	build BDD	12	8805	9	4178	18	8957
s38584	50	2+10	4563	fail to	build BDD	31	11797	27	3852	27+2	5771
										28+2	12656
										29+2	19852
										30+2	59114
mma	1	0	aborted	1	7442	1	374	0	aborted	1	189

time than CHAFF. For *s38417* and *s38584*, BINGO could not build the BDD for the state transition function due to memory limitation. THRIVE, on the other hand, verified 18 properties, which is 16 more properties than HITEC and 9 more than STRATEGATE and 6 more than CHAFF. For *s38584*, HITEC proved 2 of the properties to be false and 10 to be correct (shown as 2+10). CHAFF proved 31 to be false. STRATEGATE proved 27 to be false and could not prove any to be correct. This is because it is purely simulation-based. THRIVE took a long time to successfully prove 30 to be false and 2 to be correct. The 2 came from the embedded deterministic engine – they are proved to be combinational untestable faults. We gave multiple check points in test generation time for this circuit in Table 3. We can see that most of the time was spent on verifying the last three properties. For the real industry circuit *mma*, both HITEC and STRATEGATE failed. BINGO succeeded after some extra work in restricting the search space was applied [19]. However, it took significantly more time (7442s). CHAFF verified this property as well but took longer time than THRIVE (374 seconds vs. 189 seconds).

6. CONCLUSION

We have presented an effective technique on safety property checking using simulation-based ATPG. We proposed several heuristics, namely, small deterministic ATPG, reset signal masking, threshold value simulation and weighted Hamming distance, to effectively guide the simulation engine in the large state space to reach the target state. These enhancements help the simulation engine to avoid revisiting previously visited states, deterministically obtain a combinational solution, and refine the cost function during search. Experiments on large benchmark and real industry circuits showed that our simulation-based verifier can verify more properties which previous techniques failed in comparable execution time.

7. REFERENCES

[1] Kenneth L. McMillan, *Symbolic Model Checking, Kluwer Academic Publishers*, Boston, 1993.

[2] "Murϕ Description Language and Verifier", *http://verify.stanford.edu/cgi-bin/wrap/dill/Murphi*

[3] "The SMV System", *http://www.cs.cmu.edu/ modelcheck/smv.html*

[4] A.Biere, A.Cimatti, E.M.Clarke, M.Fujita, and Y.Zhu, "Symbolic Model Checking Using SAT Procedures Instead of BDDs," *Proc. DAC*, 1999, pp317-20.

[5] J. P. Marques-Silva and K. A. Sakallah, "GRASP: A Search Algorithm for Propositional Satisfiability", *IEEE Trans. Computers*, vol. 48, no.5, May, 1999.

[6] M. W. Moskewicz, C. F. Madigan, Y. Zhao, L. Zhang, and S. Malik, "Chaff: Engineering an Efficient SAT Solver", *Proc. DAC*, pp. 530-535, June 2001.

[7] J. Jain, A. Narayan, M. Fujita and A. Sangiovanni-Vincentelli, "A Survey of Techniques for Formal Verification of Combinational Circuits", *Proc. Int. Conf. Comp. Design (ICCD)*, 1997, pp 445-454.

[8] V. Boppana, S. P.Rajan, K. Takayama, and M. Fujita, "Model Checking Based on Sequential ATPG," *Proc. CAV*, 1999.

[9] M. S. Hsiao, "Sequential circuit test generation using genetic techniques," *Ph. D. Dissertation, Department of Electrical and Computer Engineering, University of Illinois*, Oct., 1997.

[10] M. S. Hsiao, E. M. Rudnick, and J. H. Patel, "Dynamic state traversal for sequential circuit test generation," in the ACM Trans. Design Automation Electronic Systems, vol. 5, no. 3, pp. 548-565, July, 2000

[11] K.-T. Cheng, V. Agrawal, "A Simulation-Based Directed-Search Method for Test Generation," *Proc. Int. Conf. Comp. Design (ICCD)*, 1987, pp 48-51.

[12] K.-T. Cheng, V. Agrawal, E. S. Kuh, "A Sequential Circuit Test Generator Using Threshold-Value Simulation," *Proc. Fault Tolerance Computing*, 1988, pp 24-29.

[13] S.-Y. Huang, K.-T. Cheng and K.C. Chen, "Verifying Sequential Equivalence Using ATPG Techniques", *ACM Trans. Design Automation Electronic Systems*, pp244-275 , April, 2001.

[14] C.-Y. Huang and K.-T. Cheng, "Assertion Checking by Combined Word-level ATPG and Modular Arithmetic Constraint-Solving Techniques", *Proc. ACM/IEEE Design Automation Conference* , June 2000.

[15] M. K. Ganai, A. Aziz, and A. Kuehlmann, "Enhancing Simulation with BDDs and ATPG", *Proc. DAC*, 1999.

[16] A. Kuehlmann and K. L. McMillan, "Probabilistic State Space Search", *Proc. ICCAD*, 1999.

[17] T. M. Niermann and J. H. Patel, "HITEC: A Test Generation Package for Sequential Circuits," *Proc. European Conf. Design Automation*, pp. 214-218, 1991.

[18] H. Iwashita and T. Nakata, "Forward Model Checking Techniques Oriented to Buggy Designs", *Proc. ICCAD*, pp.400-404, Nov. 1997.

[19] K. Takayama, T. Satoh, T. Nakata, and F. Hirose, "An Approach to Verify a Large Scale System-on-a-chip Using Symbolic Model Checking", *Proc. ICCD*, pp.308-313, October 1998.

A Comparison of Three Verification Techniques: Directed Testing, Pseudo-Random Testing and Property Checking

Mike G. Bartley[1]
Elixent Ltd
Castlemead
Lower Castle Street
Bristol BS1 3AG
Mikebartleyuk@Yahoo.co.uk

Darren Galpin
Infineon Technologies UK Ltd
Infineon House, Great Western Court
Hunts Ground Road, Bristol BS34 8HP
+44(0)117 952 8741
Darren.Galpin@Infineon.com

Tim Blackmore
Infineon Technologies UK Ltd
Infineon House, Great Western Court
Hunts Ground RoadBristol BS34 8HP
+44(0)117 952 8745
Tim.Blackmore@Infineon.com

ABSTRACT
This paper describes the verification of two versions of a bridge between two on-chip buses. The verification was performed just as the Infineon Technologies Design Centre in Bristol was introducing pseudo-random testing (using Specman) and property checking (using GateProp) into their verification flows and thus provides a good opportunity to compare these two techniques with the existing strategy of directed testing using VHDL bus functional models.

Categories and Subject Descriptors
B.5.2 [**Register-Transfer-Level Implementation**]: Design Aids – *Verification.*

General Terms: Verification.

1. INTRODUCTION
The verification of two versions of a bridge between two on-chip busses is described. The verification methodology incorporated directed testing, pseudo-random testing and property checking at module level, as well as in-system testing. Only the module-level verification is considered in detail. The incorporation of the three techniques into a verification flow is described in Section 3. In Section 4, results are used to discuss the relative strengths of each verification technology, as well as the strength of the verification provided by all three combined. Since verification is not an exact science, rather than trying to draw hard and fast conclusions, recommended use of these technologies is presented based on the experience gained in Section 5. Some ideas as to how the three technologies may more easily be used in conjunction are included.

2. BACKGROUND
2.1 The Design-Under-Test (DUT)
The LFI is a bus bridge that connects a 32-bit address, 64-bit data local memory bus to a 32-bit address, 32-bit data flexible peripheral bus. The LFI is approximately 30K gates in size. Two versions were verified - the first version (LFI-IBC32) required that the DUT could work with LMB:FPI clock ratios of 2:1 and 1:1. The second version (LFI-S) extended the LFI-IBC32 functionality to support any clock frequency ratio so long as the LMB frequency is equal to or greater than the FPI frequency, and as long as the positive edges of the clock signals align. In addition, power saving features were incorporated into the LFI-S version.

2.2 Methodology
Four main techniques were applied to the verification of the LFI:

- Directed testing using VHDL BFM's for the test bench and transaction-based tests for directed tests.
- Pseudo-random constraint driven testing using Specman from Verisity.
- Formal property checking using GateProp from Siemens.
- The LFI was placed within a larger system and verified as part of that system. This system is a reference system used solely for the purpose of verifying system level aspects of a DUT that are hard to verify at module level. For example, the LFI can go into power down mode with other parts of the system when a certain set of conditions apply and the LFI then engages in a complex handshake with the CPU and other blocks. The complete operation is better verified at system level than at module level.

This paper is more concerned with module level verification and so will concentrate on the first three techniques.

Figure 1 indicates when each technique was applied during the project. There follows a brief explanation of why these techniques were applied at these times and how they were applied.

- When the LFI-IBC32 project started the only infrastructure available was the VHDL Bus Functional Models (BFM's). Hence a directed test bench and directed tests was developed first.

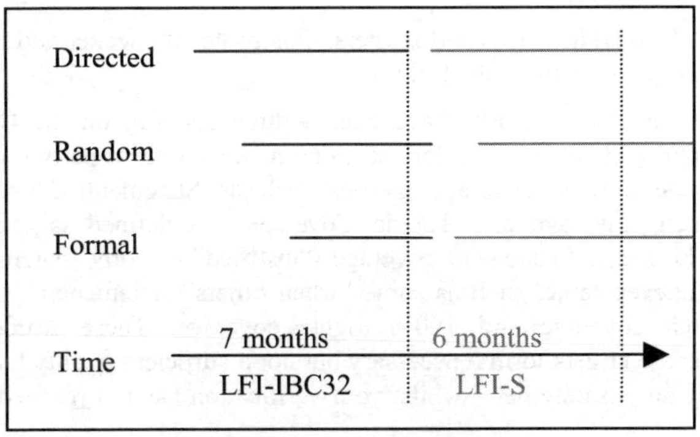

Figure 1: Verification time-line

[1] Mike Bartley was with Infineon Technologies when this work was carried out

DAC 2002, June 10-14, 2002, New Orleans, Louisiana, USA.

- The directed tests were completed about halfway through LFI-IBC32 but were continually run in regression and were enhanced to achieve the structural coverage targets.
- The LFI-IBC32 directed test bench and tests were ported to LFI-S and used for regression and structural coverage there too.
- e language Verification Components (eVCs) for the FPI and LMB became available during the LFI-IBC32 project and were then used to build a LFI Specman test bench (this accounts for starting the random testing later for LFI-IBC32). This test bench was then ported to the LFI-S project.
- GateProp was a new tool to the Infineon Technologies Design Centre in Bristol and some training and infrastructure development was required before it could be used (this accounts for starting the property checking later for LFI-IBC32). The GateProp environment and properties were however very quickly ported to LFI-S.

3. VERIFICATION TECHNOLOGIES

3.1 Directed testing

Directed testing started with the writing of a test specification document and a test bench description document by a separate verification team to give the independent interpretation discussed in Bergeron (2000). The test specification is written using a "black box" approach from the verification engineers understanding of the DUT spec. The document aims to test fully all of the stated functionality in the DUT spec, and it is from this that the directed tests are written. The writing of the test spec forms a very good review of the DUT spec and can uncover discrepancies or holes in the DUT spec. The test bench spec is aimed at describing the test bench features such that the tests to be run on it are able to exercise all of the design features and to allow for the implementation of the test specification. The design and verification teams reviewed both specifications.

The directed test bench is built from internally supplied BFM's, written in VHDL. These are driven by a simple test language that allows the test to perform read and write transactions plus idle cycles. The test is thus written at the transaction level and the BFM translates the transactions to the appropriate signal behaviour on the bus. The BFM's allow the DUT to be stimulated across the full range of functionality specified in the bus protocol. Adherence to the protocol by all parties is monitored by a protocol checker - another VHDL module which plugs directly onto the bus.

The directed test bench used consisted of:

- Four BFM Masters and Slaves, plus a protocol checker on the LMB bus.
- Two BFM slaves and masters, plus protocol checker and local memory unit on the FPI bus.

Once the directed tests have been written and run on the DUT, structural coverage is performed using a suitable third party tool to provide defined coverage metrics such as Statement Coverage, Branch Coverage and Toggle Coverage. Pre-defined targets of 100% *justified* statement coverage ("justified" as some statements are unexecutable, such as some "when others" statements), >95% branch coverage and 100% toggle coverage. These structural coverage targets form a necessary but not a sufficient quality target. They are a safety net – whilst your verification has not reached the targets you know you haven't done enough, but reaching the targets does not mean completion of verification. Experience shows that the tests derived from the black box test spec are insufficient to reach these structural coverage targets and so extra tests are specified and implemented using a "white box" approach. The white box approach works by looking at the coverage holes with a designer and specifying specific transaction sequences that need to be performed in order to hit the previously uncovered holes. This results in a set of very targeted tests.

As well as relying on structural coverage metrics, directed testing also aims to demonstrate bus compliance. The bus protocol specifies certain scenarios that have to be performed correctly in order for bus compliance to be achieved. A compliance document with tick boxes is used, and for each sequence the test which demonstrates this is supplied.

3.2 Random testing

Random testing was performed using Specman Elite and dedicated in-house *e*VC's (*e* language verification components – comparable to BFM's). The testbench was constructed using configurable numbers of *e*VC masters and slaves on both busses, together with an *e*-protocol checker on each bus. The system clocks were driven from the VHDL top level in order to improve simulation performance, but otherwise all stimuli were generated from the *e*-code itself.

A test specification is again written in advance based on the DUT spec - the document describes the scenarios that need to be covered. Similarly a test bench specification is written. Both are reviewed with the designers who may suggest additional scenarios.

Coverage scenarios fell into two main types for both of the DUT's considered in this paper.

1) Transaction coverage. On a per-master and per-slave basis, the types of transactions applied and received are recorded, along with associated information such as acknowledge codes, byte lanes used, wait states observed. Cross-coverage of these transaction types is then performed, in order to ensure that for each opcode type we have observed every error condition, every type of wait state and so on.
2) FSM transition coverage. We hook Specman to the internal state machines within the DUT itself, and define what the legal transitions are. During the course of the testing, we can measure coverage of every legal transition, and check that we did not cause any illegal transitions.

The combination of ensuring that we apply every possible cross-coverage point and FSM transition ensures that we achieve a high functional coverage of the DUT, and any areas that we initially miss during random testing are rapidly highlighted in the coverage results. The creation of new test scenarios with modified transaction constraints can then be added to ensure that we achieve 100% of the coverage that we have defined.

Cross-coverage of all opcode types with all byte lanes and all errors is impossible, as it can generate impossible to hit combinations. All impossible combinations and FSM transitions were removed from the coverage targets, so that a simple check for 100% coverage can be made at sign-off.

3.3 Formal

Formal verification was carried out using the Siemen's property checker, GateProp (see Bormann and Spalinger (2001) for details of this tool). The main tasks involved in property checking are writing and testing constraints and writing and running properties.

Earlier property checkers could not easily deal with large designs, suffering 'state space explosion'. One of the ways that GateProp overcomes this is not to do any reachable state analysis[1]. Hence it is the responsibility of the user to ensure that there are no false negatives due to internal signals in the design being in an unreachable state. This involves checking each failing property and developing a set of constraints on the internal signals so that they remain in a suitable superset of all possible reachable states. Both of these are very time consuming. Also, the constraints are not going to be re-usable on other designs, and in fact may be difficult to maintain across changes in design. It was therefore decided to start all properties from reset, this being the simplest valid state to define. It should be stated that more general approaches have successfully been applied in other industrial applications of GateProp, where a very high functional coverage has been achieved. In most of these projects the always-start-from-reset approach would not have touched the most interesting functionality, e.g. in telecommunication systems where you have frame lengths of thousands of cycles with corresponding serial synchronisation protocols. Admittedly, these approaches make limited use of some internal states, yet important ones only which make sense at an abstract level. They allow at least a certain degree of reusability of properties.

Given our always-start-from-reset approach, to ensure that the design did not reach an invalid state it was only necessary to constrain the inputs to model the DUT's environment. The goal here is to allow the inputs to take any legal combinations and sequences of values, in order to find corner cases. Of course, random testing also chooses inputs from inside a set of constraints. The difference here is that the property checker 'intelligently' tries to find a counter-example, or else prove, a property. However since the property checker is potentially exhaustive, it moves through clock cycles considerably slower than directed or random testing.

We split properties into two types - 'bus protocol properties' and 'transaction properties'. Transaction properties drove explicit transactions across the bridge, checking that addresses, control signals and data passed through correctly, and that the bridge responded appropriately to the various acknowledgements from bus slaves. The property checker has the advantages that all legal behaviour around the transaction was allowed, and the address and data were checked symbolically - in effect testing all possible values. The slowness of moving through clock cycles was a big limitation however. The transaction properties were sensitive to timing (e.g. the number of clock cycles it took each type of transaction to pass through the bridge). This made them more time consuming to write, with a degree of trial and error being necessary, and also to maintain since cycle timings were not specified and so could change with design updates.

Bus protocol properties checked that the LFI obeyed the protocols of the FPI and LMB. Since the bus protocol properties were checking that certain behaviour always, or never, happened they were not sensitive to timing. This also meant that they could exercise the design further just by running for longer. Another advantage they have is that, like the input constraints, they are re-usable on any module with a similar interface (discussed later).

The transaction properties and protocol properties each found bugs. Exact numbers are given later in Section 4.2.

3.4 Combining the techniques

Directed Testing, Random Testing and Formal Verification are not completely orthogonal techniques. During both the LFI projects the random test bench was used to seed both property checking and directed testing.

When a failure occurs in directed or random testing, a specific set of circumstances have caused the observed failure, and it may be possible that there are other similar signal combinations which expose the same or similar errors. Thus when the bug is "fixed" and the test re-run, the bug may no longer be observed, but a similar or new bug may still be present in the code but it is reachable only via another set of input stimuli. Consequently it is useful to test a wider set of signal inputs using property checking. If the bug failure is carefully observed, a set of properties can be deduced which must be satisfied for the bug to occur, and thus we can write a property to prove that the bug cannot possibly occur. In this way we can prove that the bug has been fixed over all circumstances, rather than the one circumstance observed in the directed or random test. However, not all directed test bench failures may be able to be analysed in the manner needed, so we may not be able to prove all bugs in all cases.

Thus, many failing tests were converted to properties. These properties were used to check the fix and often found new bugs or that the fix had introduced a new bug. This raises the question as to why such properties were not specified initially. The reason for this is that the set of all possible properties is too large – so using this approach helps to prioritise the properties to be written and run.

In addition, a random test may exercise a piece of code that has not been exercised by directed testing before, and it may prove to be a particularly problematic piece of code. The random test gives the verification engineer a set of input conditions which need to be applied to the DUT in order to reach that piece of code, so these can then be applied via the directed test bench to explore the behaviour of this previously untested code. This can free up the random test bench resource to explore and verify other areas of the DUT while the directed test bench is used to fix the bugs that cause the random test bench to fail every time it enters this area.

4. RESULTS

4.1 Coverage

The coverage achieved from the three approaches enabled us to release a DUT with a high degree of confidence in its design. Our directed testbench achieved the following structural coverage metrics:

- 100% Justified Statement Coverage
- 98.4% Branch Coverage
- 95.2% Condition Coverage
- 84.2% Trigger Coverage
- 100% Toggle Coverage

The random testing achieved 100% FSM transition coverage, but did not attain 100% cross-coverage of the FSM's. This proved to be impossible because of the design of the DUT. The bus transaction

[1] Note that versions of GateProp subsequent to that used on the LFI include approximation of the reachable state space and analysis of unreachable counter-traces.

coverage targets were met (the target was 100% after removal of impossible transactions as discussed earlier).

Coverage for Property Checking is far harder to measure as GateProp has no coverage metric. A review of the properties by the verification team established that we were convinced that all the practical transactions and all protocol issues had been covered. Indeed it is possible to establish full protocol compliance via review of the properties, or better still mathematically (see Barrett and McIsaac (1997)).

4.2 Bug Analysis

4.2.1 Bug Graph

Bugs were recorded using an issue tracking system that allowed statistics to be gathered such as the cumulative number of bugs found, the number fixed, and whether they occurred in the design or test bench. Bug find rates are often expected to follow an S-curve shape, as the number of bugs found rapidly increases as the verification effort is increased, and gradually reduces as the DUT reaches maturity. Thus if the levelling out in the curve is not seen, it can be an indication that more time is needed to debug the design.

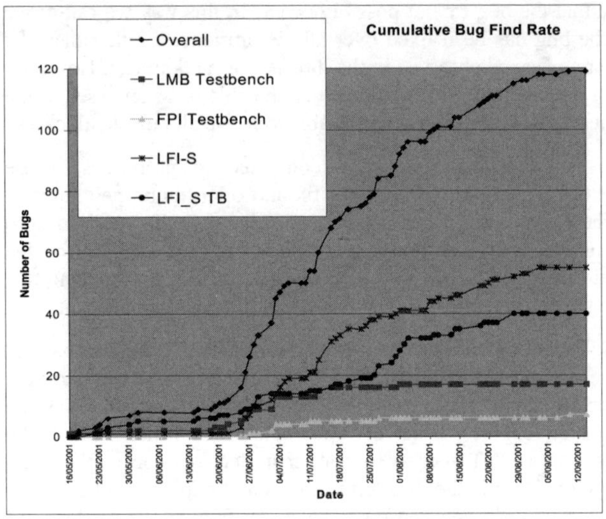

Figure 2 – LFI-S Bug Curves

Figure 2 shows the bug curve seen during the LFI-S project. It clearly shows the progression of the project through its life cycle. At the beginning of the project the test bench is constructed and debugged, then testing of the DUT begins and the bug curve rapidly moves upwards. Bugs are found in all parts of the design, and as new parts of the DUT functionality are tested, new areas of the verification IP are also tested, resulting in more test bench bugs. Eventually the test bench reaches maturity, and the DUT reaches maturity shortly afterwards. The cumulative total flattening out completely for a sufficient time can thus be used as a sign-off criteria, as it suggests that maturity has been reached.

4.2.2 What technique found what bug

Table 1 gives a breakdown of bugs found in the LFI-S by type, and by the method used to find them.

Bug Type	Prop.Chk	Direct.	Rnd.
Bus Protocol violations	11	4	8
Internal DUTqueue deallocation issues	0	2	4
Livelock/Deadlock	0	5	2
Other DUT issues	7	3	8
Total	18	14	22

Table 1 : Breakdown of LFI-S bugs

When reading these figures the reader is reminded that the various verification techniques were started at different times in the project. As stated at the start of the paper, for the LFI-S project the directed tests from LFI-IBC32 were used to prove initial functionality, and once basic functionality of the DUT had been demonstrated, random testing was started. However, property checking was started almost immediately, although time was required to modify the transaction properties due to changes in internal DUT timing.

In theory all of the bugs found by random testing could be found by directed testing. Random testing generates large sequences of transactions that can easily be replicated in the directed test bench. However, in practise there is not enough resource (to generate such large numbers of self-checking sequences) or ingenuity (to think of the corner cases that random generation creates).

Similarly, most of the bugs found by random testing could have been found using property checking given the resource to write an extensive set of properties (less ingenuity is required because property checking probes corner cases automatically). However, some bugs did require a number of cycles to cause the failure (e.g. those caused by filling up internal queues and then trying to add further transactions). The number of simulated cycles needed to get to this point can result in an excessively long run time for properties. For the same reason, property checking is unlikely to find the livelock or deadlock bugs, although if we create a livelock or deadlock by initial conditions we can prove that we cannot escape.

Property checking was most effective at demonstrating that the DUT does not violate bus protocols. Many of the violations would require a directed test to be written with a knowledge of what is happening cycle by cycle on the bus, and are generally found more by luck than judgement during random testing and than during directed testing.

Thus, during the two LFI projects, the verification strategies proved to be complementary – all stressed different areas of the DUT in different ways, and speeded the finding of bugs. They also allowed the verification manager to have several quantitative and qualitative measures on which to assess confidence in the quality of the DUT. Indeed, the following were all used in the sign-off criteria:

- directed and random test bench and test specifications reviewed and signed off by design and verification teams;
- structural and functional coverage targets met;
- all directed tests passing;
- a certain number of random transactions run successfully since last bug found;
- property specification reviewed and signed off by design and verification teams;

- all properties passing;
- bug curve flattened off and no bugs found for a given period of time.

4.3 Combined Effectiveness

Both the LFI-IBC32 and LFI-S have now been manufactured in silicon as part of larger SOC's and no bugs have reported against either.

5. RECOMMENDATIONS AND FUTURE USE

5.1 Directed testing vs. random testing

The combination of approaches improved the quality of the verification, but also increased the resources applied, especially as two test benches were built and maintained. One future course of action will be to become less reliant on directed testing (and thus more reliant on random testing and functional coverage) and to use the random test bench to perform the directed testing. However, this does result in compromise, as it requires the verification engineer to heavily constrain the random testbench and results in a loss of controllability compared to a pure directed testbench. This raises the question of whether to start with unconstrained random tests and constrain them to more directed tests over the project or vice-versa (Verisity recommend the former).

Also, without a large set of directed tests, a repeatable regression run of tests which can demonstrate DUT functionality becomes more problematic. Although the use of seeds in Specman allows repeatability, we require a list of seeds per scenario in order to guarantee full coverage. This also requires that the *e*VC's used are stable, and that the same version of Specman is also used each time the regression is run. If this requirement is not met, then the repeatability of the regression run cannot be guaranteed.

Finally, the structural coverage targets still need to be met and this requires Specman to be run with an appropriate structural coverage tool.

5.2 Productisation of verification IP

If users are to gain the benefits of verification IP reuse, then the productisation of the verification IP itself is very important.

The random test bench elements have been developed into re-usable IP (using the eVC approach recommended by Verisity). Indeed, these have been re-used successfully on a number of other projects with big reductions in test bench construction times

Some of the property checking deliverables have also been turned into re-usable verification IP. As already noted the property checking effort can be divided between the writing of constraints and the writing of properties. Each of these can be further divided. Constraints can be internal, i.e. constraining the internal signals of the design to avoid unreachable states, or external i.e. constraining the inputs to the design to reflect its environment. We've already divided the properties into bus protocol and transaction properties.

Clearly the internal constraints are not going to be re-usable, and may in fact be difficult to maintain across changes in a design during development. This was a major factor in our decision not to develop such constraints (and rely on reset to ensure properties used valid internal states). However the external constraints can be re-used, with appropriate modifications to signal names, on other modules that have a similar interface. The most commonly used interfaces are the busses, so constraints modeling busses give the largest opportunity for re-use.

Similarly the transaction properties will not be re-usable and may take some effort to maintain, due to small changes in timing. However the bus protocol properties are easily re-usable. A little care should be taken to monitor run times to ensure that the design is really being exercised, since some designs may have a ramp up time after reset where they do little and so are unlikely to break any protocols.

So it is the constraints on, and properties for, bus protocols that are re-usable. Indeed those developed for the LFI have been used on other modules that interface with the FPI and LMB busses, namely a peripheral control processor and an external bus unit. This experience of re-use will be used to develop bus protocol packages for these busses, using a simple wrapper to standardise signal names. These packages will be easy to use for someone with no experience of property checking. Of course such packages are applicable to any standardised protocol and there are also plans to develop them for other busses, such as the AHB.

5.3 Easier re-use between dynamic and static verification

During the LFI projects a set of constraints and properties were developed and maintained. There is much scope for re-use from random/directed testing to property checking and vice-versa. For example:

- The random test bench and property checking both use constraints to first define the legal sequences of inputs and then to further restrict them to a smaller set of sequences of interest to test or property in question.
- Properties should also hold during simulation.
- Dynamic verification can get the design into legal internal states from where it would be useful to start a property check (as opposed to always starting it from reset).

Fortunately, all three of these areas are being addressed in various ways. Accellera (www.accellera.org) are trying to define an industry standard formal language that can be used in both formal and dynamic verification. This should allow the first two to be realised. There are also tools appearing that enable the third (so-called semi-formal tools), such as 0-In Search.

6. REFERENCES

[1] Bergeron, J. Writing Testbenches; functional verification of HDL models (2000), Kluwer Academic Publishers

[2] Bormann, J. and Spalinger, C. (2001) Formale Verifikation fuer Nicht-Formalisten (Formal Verification for Non-Formalists*)*, Informationstechnik und Technische Informatik, volume 43, issue 1/2001, Oldenbourg Verlag

[3] Barrett, G. and McIsaac, A. (1997) Model Checking in a MicroProcessor Design Project*,* CAV'97 (Proceedings of the 9th International Conference on Computer-Aided Verification, Haifa, Israel 1997, Orna Grumberg (Ed.), Springer Lecture Notes in Computer Science 1254)

Energy-Efficient Communication Protocols

Carla F. Chiasserini
Dip. di Elettronica Politecnico di Torino
C.so Duca degli Abruzzi 24
10129 Torino, Italy
chiasserini@polito.it

Pavan Nuggehalli
Dept. of ECE
UCSD
9500 Gilman Dr.
La Jolla, CA 92093
pavan@cwc.ucsd.edu

Vikram Srinivasan
Dept. of ECE
UCSD
9500 Gilman Dr.
La Jolla, CA 92093
vikram@cwc.ucsd.edu

ABSTRACT

Wireless networking has experienced a great deal of popularity, and significant advances have been made in wireless technology. However, energy efficiency of radio communication systems is still a critical issue due to the limited battery capacity of portable devices. In this paper, we deal with the charge recovery effect that takes place in electrochemical cells and show how we can take advantage of this mechanism to increase the energy delivered by a battery. Then, we present energy-aware traffic shaping techniques, as well as scheduling and routing algorithms, which exploit the battery recovery effect.

Categories and Subject Descriptors

C.2.2 [**Computer-Systems Organization**]: Network Protocols

General Terms

Algorithms

Keywords

Wireless networks, energy efficiency, battery charge recovery

1. INTRODUCTION

Wireless devices have become an integral part of modern society. These devices are employed in cellular, ad hoc and sensor networks to name a few. A common thread restraining the true potential of these applications is the limitation in energy supply which are predominantly provided by batteries. For example, in cellular and ad hoc networks, the addition of wireless interfaces to portable devices allows for connectivity to the Internet. In order to provide ubiquitous and continuous access to Internet-based services, a significant energy supply is required from batteries. The same requirement exists in sensor networks, which are often deployed in inaccessible locations. In this case, the nodes batteries must have a long lifetime since it may be impossible to replace them.

The major contributions to energy consumption in wireless devices are: (i) the power consumed by the digital part of the device circuitry; (ii) the power consumed by the electronics of the transceiver in transmitting and receiving mode; (iii) the output transmission power.

For devices that operate in the ISM bands, typical values of the power amplifier output range over the interval [1 mW,1000 mW], while the transceiver power consumption is of the order of tens to hundreds of mW [1]. The total power consumption depends on the amount of transmitted, received, and processed traffic. It follows that the greater the energy capacity of the battery, the longer the run time of a wireless device. On the other hand, battery features such as long lifetime, light weight and small size are highly desirable in portable wireless devices. An obvious solution to the problem would be to increase the energy delivered by a battery by improving battery design. However, in the last few years, advances in battery technology have not kept pace with the wireless revolution. An alternative strategy consists in developing energy-efficient protocols and architectures which enhance the devices lifetime.

Various MAC protocols [2, 3] and schemes for power management control during transmissions [5, 6] have been proposed in the literature. Dynamic power management policies have been proposed in [7, 8]. The work in [9] focuses on data link layer mechanisms and show that, if energy is a constraint, then retransmissions should be minimized. In [10, 11], energy aware routing protocols have been proposed. Traditional routing protocols typically implement the shortest path algorithm; thus, it is likely that a few nodes will lie on many routes resulting in their quick death and loss of network connectivity. On the contrary, the solutions presented in [10, 11] distribute the load more evenly among the available routes leading to network longevity.

In this paper, we present a different approach to energy saving. Our goal is to understand the battery behavior and leverage this understanding to develop new energy-efficient protocols. In Section 2, we show that battery life is dependent on its usage pattern. Batteries recover charge during idle periods. This phenomenon can be exploited by the upper layers to design battery aware scheduling and traffic shaping algorithms. Moreover in general, all levels of the protocol stack may contribute to energy saving by intelli-

DAC 2002, June 10-14, 2002, New Orleans, Louisiana, USA.

gently using some knowledge about the behavior of the traffic, the radio channel and the network. A scheduling scheme and a routing protocol exploiting the battery recovery effect are described in Section 3.

2. THE RECOVERY MECHANISM IN BATTERIES

Our research focuses on the understanding of the intrinsic behavior of batteries and on the use of this understanding to develop a stochastic battery model and new energy-efficient protocols.

A battery consists of one or more electrochemical *cells*, organized in an array. Each cell consists of an anode, a cathode and the electrolyte that separates the two electrodes. The electrical current obtained from a cell results from electrochemical reactions occurring at the electrode-electrolyte interface [12, 13, 14]. At zero current, the concentration of the active species in the cell is uniform. As the discharge current increases, the active species are consumed at the electrode-electrolyte interface by electrochemical reactions, and replaced by new active materials that move from the electrolyte solution to the electrode through *diffusion*. As the intensity of the current is increased, the deviation of the concentration from the average becomes more significant and the state of charge as well as the cell voltage decrease. This phenomenon is called *rate capacity effect*. Beyond a threshold value, called the *limiting* current, the diffusion phenomenon is unable to compensate for the depletion of active materials and the cell voltage drops below the usable value even though the *theoretical* available energy stored in the cell may not have been exhausted. The amount of electrical energy that can be obtained from a battery cell is fundamentally limited by the type and mass of the electrodes and electrolytes. In practice, the delivered specific energy (ratio of the total delivered energy to the weight of the cell) depends on the intensity of the discharge current, the power level drained from the cell and whether the discharge is constant or pulsed.

The typical way to drain power from cells is at a constant level of current (constant current discharge). In this case, the relationship between specific energy and specific power (ratio of the delivered power to the weight of the cell) of different battery technologies, is displayed in the so-called Ragone plot, as shown in Fig. 1. The fact that in the plot the curves "lean" to the left shows that a high specific energy can be obtained only if the discharge is at low power levels. However, even if one settles for a low rate of discharge, the energy delivered by a battery under constant discharge is typically only 10-30% of the theoretical value. Although improvements in battery technology are being made, they tend to lag behind the demand.

Some of the adverse consequences of constant current discharge can be overcome when the discharge is pulsed. If a cell is allowed to relax long enough after delivering a pulse, the diffusion process, mentioned above, compensates for the depletion of the active materials that takes place during the current drain. The concentrations gradient of the electrolyte species decreases, and charge recovery takes place at the electrode. Thus, by using a pulsed discharge instead of a constant current discharge, it turns out that: (i) for a fixed power level, the delivered specific energy is greater; (ii) for a constant delivered specific energy, a higher specific power can be drained from the cell. Several findings [12, 15, 16, 17] quantify these advantages. For instance, in batteries characterized by a relatively low conductivity, e.g., lithium-polymer cells [18], pulsed discharge can increase the delivered specific power. Lithium-polymer cells have lightweight and flat formats and fit well in extra-thin cellular phones, but ion diffusion through the conducting polymers is slow and this limits the rate at which current can be withdrawn from a cell. However, in [19] a lithium-polymer cell was discharged at up to 50 mA/cm^2, about fourteen times the typical constant discharge rate, using pulses of 10 ms duration followed by a 50 ms rest period.

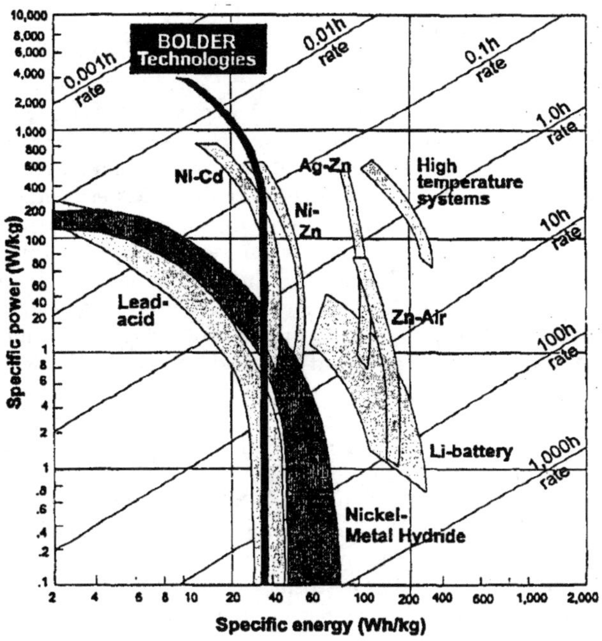

Figure 1: Ragone plots for various battery chemistries.

These facts suggest that the burstiness of the battery discharge process may be a more significant determinant of the delivered energy than the initial charge stored in the cell. That being so, we are led to explore the possibility of discharge shaping to maximize the battery energy efficiency. In order to investigate this possibility systematically, we need models for representing the battery behavior.

2.1 Modeling Battery Recovery

We start by considering an electrochemical battery model [20] that represents the underlying electrochemical phenomena in great detail. The electrochemical model has been modified to let the discharge of the cell be driven by a stochastic process, and can be numerically solved by using a program developed by Newman *et al.* [21]. Results are obtained assuming that the time scale is divided into time slots with duration equal to 1 ms and that the discharge process is composed of pulses occurring at stochastic time instants according to a Bernoulli distribution. Pulses have a constant duration equal to 1 ms (i.e., equal to one time slot). We take the current discharge at each pulse as a varying parameter of the system. The specific delivered energy is derived as a function of the number of discharge pulses per time slot. By fixing the current discharge, we obtain that as the number of current pulses per time slot decreases, the specific delivered energy dramatically increases since the chance to recover for the cell increases. This proves that a significant improvement in performance of real batteries is possible when a stochastic pulsed discharge is used instead of a constant current discharge. By varying the current discharge, results show that higher current pulses may degrade performance, even if the number of higher current pulses is really small. This phenomenon takes place whenever the current drained from a cell exceeds the specified limiting value and, therefore, the concentration gradient of active materials becomes significant. Thus, applications that involve the use of different levels of power (e.g., routing in ad hoc networks, power control in CDMA systems, etc.) should carefully administer the available battery energy. The set of results that we could derive through the electrochemical model however is limited because the computation time for the model solution is exceedingly large. Since our objective is to define a battery model for studying the energy efficiency of communication protocols, we need a more tractable model [22, 23].

We therefore develop a stochastic model, which is much simpler and represents the battery behavior mathematically in terms of parameters that can be related to physical characteristics of the electrochemical cell. The stochastic model captures the recovery effect that is observed when a pulsed discharge is applied, and neglects all the other phenomena represented in the electrochemical model. The model characterizes a fully charged cell by a certain theoretical energy and by a nominal energy, the latter representing the energy that could be extracted using a constant discharge profile. Notice that the nominal energy is much less than the theoretical energy. The recovery effect is modeled as a decreasing exponential function of the state of charge and delivered energy. To more accurately model real cell behavior, the exponential decay coefficient is assumed to take different values as a function of the delivered energy. In each time slot, some energy units are lost if any request of current supply arrives and has to be satisfied, otherwise the battery may recover or remain in the same state. The resulting cell behavior is a transient stochastic process that starts from the state of full charge and terminates when the cell is completely discharged, or the theoretical available energy is exhausted. The stochastic model has been matched to the electrochemical model for the particular case of a lithium-ion battery, and a comparison between the curves obtained from the stochastic and the electrochemical model has showed a very good agreement [22].

2.2 Application: An Example

By using the stochastic battery model, we are able to explore ways in which the energy efficiency of mobile wireless communications can be enhanced through the use of improved energy-efficient battery management techniques. We consider an array of L cells connected in parallel, that can be selectively scheduled. Fig. 2 illustrates the case for L=2. A delivered energy equal to N can be drained from each cell by means of a constant current discharge. However, by using a proper pulsed discharge technique, the region of delivered energy can be increased. The goal is to find methods to extend the region of delivered energy up to the maximum limit, i.e., the theoretical available energy T. Cells can be scheduled based on the discharge demand and their own status, so that the demand can be distributed in a manner that provides adequate rest periods for individual cells. We apply load balancing algorithms to battery management and propose both a *delay-free* and *delayed delivery* approach.

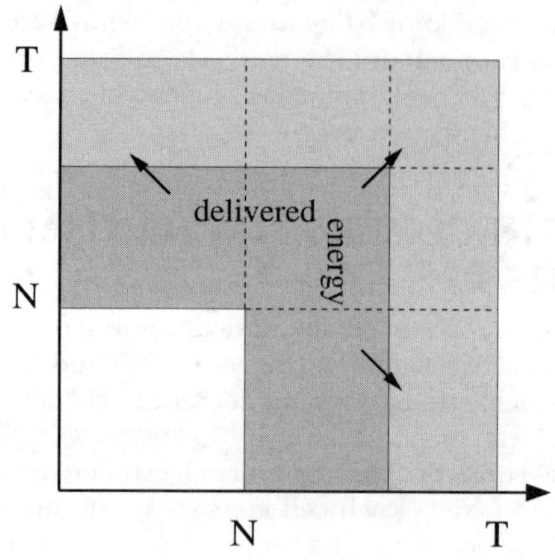

Figure 2: Margin of improvement for the energy delivered by an array of L=2 cells.

The former implements a scheduling scheme among the cells and provides the power supply as soon as required. A technique based on such an approach is the round robin scheduling, i.e., the discharge request arrivals are directed to the cells by switching from a cell to the next one. By using the stochastic battery model, we can evaluate the improvement in delivered battery energy that is achieved.

The latter approach uses scheduling algorithms in conjunction with discharge shaping, so that an additional delay is introduced in the power supply. In this case, the battery is always able to deliver its maximum available energy even for high discharge demand. This approach involves a coordination among the cells of the array and drains current from the cells according to their state of charge. The goal is to monitor the cells status and make them recover as much as they need to obtain the maximum available energy from the discharge process. We define a lower threshold for the cell state of charge such that, whenever the state of charge drops to this value, the cell is considered *not active* and current cannot be drawn off until recovery occurs. The event of the cell state of charge dropping to a certain threshold can be easily revealed by the control apparatus present in smart battery packages; the acquisition of the information is therefore considered instantaneous. A scheme, that can be implemented to schedule a cell among the set of the active ones, consists in selecting the cell that currently has the best state of charge. Whenever the state of charge of a cell drops to the value of threshold, the cell is removed from the set of active cells and allowed to recover. If none of the cells is active, the discharge requests that arrive are buffered and served as soon as a sufficient amount of charge is recovered. By applying this scheme, a battery is able to deliver the maximum available energy at the cost of a fairly small additional delay and complexity.

3. EXPLOITING THE RECOVERY EFFECT IN COMMUNICATION PROTOCOLS

Typically, energy efficiency can be obtained at the expense of other performance measures, such as delay and throughput. We provide two examples in which we show how communication protocols can exploit battery behavior to increase the delivered energy. The first example concerns scheduling and can be incorporated in the physical layer. The second example deals with routing, allowing for energy efficiency at the network layer.

3.1 Scheduling

We address the problem of designing an energy aware transmission schedule for a wireless node subject to some delay constraints [24]. We do so by combining two unconnected concepts, namely the battery recovery mechanism and channel coding [4]. In [4], it was shown that channel coding can be used to conserve energy by transmitting at reduced power levels over longer durations. The channel coding idea favors extending transmission durations for as long as possible. On the other hand, battery dynamics require the transmitter to be idle to allow for recovery. In other words, the device saves energy by transmitting in bursts and recovering in the idle periods. Delay constraints prohibit the use of arbitrarily large transmission or idle times. Clearly the two energy conserving ideas above are at variance with each other. Thus, a strategy which is based entirely on either one or the other idea is not optimal. We pose the following question: can these two concepts be amicably merged to devise a delay constrained energy-efficient transmission schedule?

In [4], they observe that, in a variety of channel coding situations, the energy required to transmit a bit is a strictly convex and decreasing function of its transmission time. Using this insight, they propose an optimal off-line lazy scheduling policy. The policy is off-line in the sense that the arrival times of all the packets in a bounded interval $(0, T)$ are known *a priori*. The aim of the policy is to schedule all departures before the deadline T. They contend that, for energy consumption to be minimized, a transmitter should attempt to transmit a packet for as long as possible subject to the deadline constraint, T. Motivated by the off-line analysis, they propose an online algorithm which they show to be near optimal. They then consider the problem of scheduling packets over an infinite time horizon with an average queuing delay constraint. They favorably compare a certain heuristically derived scheduling policy with a constant service policy that results in the same average delay. However they leave open the problem of devising an optimal scheduling policy for an arbitrary average delay constraint.

We first tackle the deadline constraint problem by adapting the optimal off line scheduling policy in [4] to a discrete-time scenario. We then assume a simple battery recovery model and derive the optimal transmission strategy. Our results show that energy savings of more than 50% are feasible by accounting for energy recovery.

Next, we consider the average delay case. In this model, packets arrive at a transmitter's queue according to a Poisson process. Packets are served on a first come, first serve basis. We use constrained dynamic programming (CDP) [25], to devise an optimal transmission strategy which uses only the channel coding concept and ignores battery dynamics. The resulting transmission policy is called the non-battery aware policy. We then consider a stochastic model of battery behavior incorporating battery recovery. This allows the transmitter to conserve energy by scheduling idle periods after a transmission. However, incorporating a realistic model of energy recovery along with the channel coding concept in the CDP framework is computationally prohibitive. We resolve this by making simplifying assumptions about battery recovery. This leads to policies which may be sub-optimal. We call these the battery aware policies. For a given average delay constraint, we evaluate the battery aware policy and the non-battery aware policy against a constant service policy using simulation in Fig. 3. Note that with tight delay constraints, the non-battery aware policy can save as much as 40% over a constant service policy. With the battery aware policy, this figure can improve up to 90% for low arrival rates. As the delay constraint is relaxed, the constant service policy is seen to do almost as well as the non-battery aware policy. However, the gains from the battery aware policy become more substantial for low arrival rates.

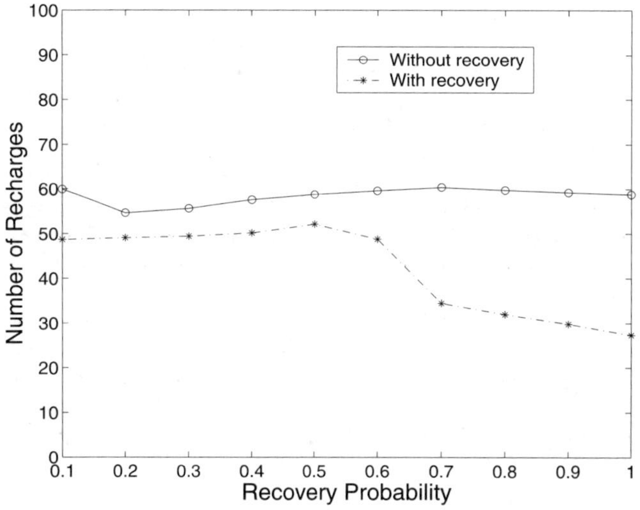

Figure 3: Arrival Rate = 0.1, Average Delay Constraint = 1.5. Each curve shows the number of recharges required per 100 constant service policy recharges. As arrival rate increases, the battery aware policy outperforms the non-battery aware policy for high values of β.

3.2 Routing

We present a routing protocol that aims at maximizing the network lifetime, i.e., the time period from the instant when the network starts functioning to the instant when the first network node runs out of energy [11]. We consider the battery capacity of the nodes as a common resource of the system and we improve the network lifetime by efficiently exploiting the available battery capacity.

We develop a routing protocol, named BEE (Battery Energy-Efficient) protocol [26], that aims to balance the battery consumption among all network nodes. The scheme selects routes whose links have a low energy cost; e.g., a route with 3 hops may be preferred to a route with 2 hops if the maximum level of power per link is lower. Also, the BEE protocol favors the nodes with low battery status by routing the multihop traffic through nodes with high battery capacity and allowing the others to benefit of the recovery effect.

We define K as the set of the nodes in the network, S as the set of traffic sources and G as the set of destination nodes. Nodes belonging to S can direct the traffic to any of the destination nodes using the other nodes as relays. We consider that time is discretized into unitary intervals corresponding to the packet transmission time. Nodes are assigned an initial amount of energy, denoted by B_i; the instantaneous battery status is denoted by b_i with $i \in K$. The transmission range of each node is limited to a certain value, indicated by ρ, so that a radio link can be established between any pair of nodes (i, j) only if the distance d_{ij} is less than or equal to ρ. We indicate as R_i the set of nodes whose distance from i is less than ρ. The energy spent for each data transmission from node i to node j is [11]: $e_{ij} = (d_{ij}/\rho)^4$ if $j \in R_i$, and infinite otherwise. The energy consumed by node j ($j \in R_i$) to receive a data unit is considered to be ten times lower than the transmission energy e_{ij}.

We discretize the transmission energy to any node in R_i ($i \in K$) into few levels ranging from e_{min} to e_{max}. To represent the benefit of the recovery phenomenon, the battery status of a node is increased whenever the node's transmission rate is sufficiently low. We model the energy accrue at node i as a function of the transmission rate λ_i and of the mean energy necessary to node i to transmit a packet ($\overline{e_i}$). We indicate the energy increase function as $\Gamma(\lambda_i, \overline{e_i})$ with $\overline{e_i} = \frac{1}{R_i}\sum_{j \in R_i} e_{ij}$. $\Gamma(\cdot,\cdot)$ is obtained through the stochastic battery model described in Section 2.1. The stochastic model was matched to the electrochemical model [21] for the case of impulses energy equal to $\overline{e_i}$ drained from the battery at rate λ_i.

Similarly, whenever a node needs to use a transmission energy level $e_{ij} > e_{min}$, the node's battery status is decreased by $e_{ij} + \Phi(e_{ij} - e_{min})$, where $\Phi(\cdot)$ takes into account the rate capacity effect and is derived from the electrochemical model results. Thus, at each time unit the evolution of the battery status is,

$$b_i = \begin{cases} b_i + \Gamma(\lambda_i, \overline{e_i}) & \text{if node } i \text{ is idle} \\ b_i - [e_{ij} + \Phi(e_{ij} - e_{min})] & \text{else} \end{cases} . \quad (1)$$

The routing algorithm assigns to each route a cost function taking into account both energy transmission and battery behavior. Given the generic source s and the generic destination g, the cost function associated to the k-th route, r^k_{sg}, is

$$F_k = \sum_{l_{ij} \in r^k_{sg}} [\Psi(\lambda_i)e_{ij} + p_{ij}] - \min_{i \in r^k_{sg}} b_i \quad (2)$$

where: (i) l_{ij} is the link between nodes i and j belonging to route r^k_{sg}. (ii) $\Psi(\lambda_i)$ is a weighting function that emphasizes the energy loss of the source node and depends on the source transmission rate; we have: $\Psi(\lambda_i) = A \cdot \lambda_i$ for $i = s$, with A being a proper constant value greater than 1, and $\Psi(\lambda_i) = 1$ otherwise. (iii) p_{ij} is the energy penalty due to the rate capacity effect experienced at node i when a power level higher than the mean power value is required to transmit a packet over link l_{ij}; we have $p_{ij} = max\{0, e_{ij} - \overline{e_i}\}$. (iv) $\min_{i \in r^k_{sg}} b_i$ is the minimum value of battery status among the nodes belonging to route r^k_{sg}.

Whenever source s has a block of data to transmit toward destination g, s evaluates the cost function for all the possible loop-free routes and selects the route r^m_{sg} such that: $F_m = \min_{r^k_{sg}} F_k$. We notice that every time the cost function has to be computed the information about the nodes battery status have to be updated.

If the set of routes over which we have to minimize the cost function is very large, the complexity of this scheme may become unacceptable for networks whose topology varies often. A simpler and faster algorithm is obtained as follows. We consider that the node in charge of selecting a route (i.e., either the source or the destination) chooses c routes independently and uniformly at random from all the available routes. The cost function is then evaluated over the c routes, and the selected path is the one among the c routes which minimizes F.

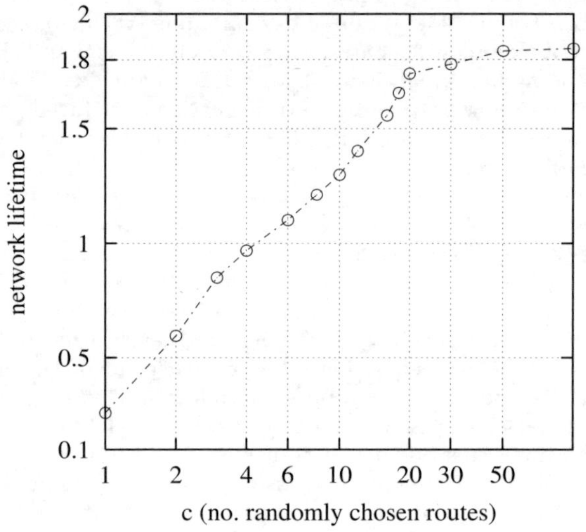

Figure 4: Network lifetime derived through the BEE scheme as a function of c, normalized to the value obtained through the MTE algorithm.

Fig. 4 shows the network lifetime obtained through the BEE scheme. We assume that the network consists of 15 nodes randomly scattered all over a circular area. The maximum transmission range is ρ=2.0 and the length of the routes connecting any pair source-destination can be equal to four hops at most. The initial battery status is set equal to 1 for all the nodes in the network. There are five source nodes with traffic rate $\lambda_0 = \lambda_1 = \lambda_2 = 0.4$ and $\lambda_3 = \lambda_4 = 0.3$, and two fixed destination nodes. A source node updates the cost function and selects a new route to a destination every time it finishes to transmit a block of five packets. Results show the network lifetime obtained through the BEE scheme as c varies, normalized to the value of lifetime derived through the MTE algorithm [27]. For $c > 4$, the BEE algorithm outperforms the MTE technique; the maximum value of network lifetime is obtained when the cost function is evaluated over all the possible routes. In this case, the factor of improvement with respect to the MTE scheme is equal to 1.8. These results suggest that, depending on the level of acceptable complexity, an optimal value of c can be selected.

4. CONCLUSIONS

We addressed the problem of energy efficiency in wireless networks. We proposed strategies that exploit battery dynamics to increase the energy delivered by a battery. We found that the longer the idle times of a battery-powered device, the greater the charge recovery effect and the delivered energy. In order to introduce idle times in the devices' activity, energy-aware communication algorithms can be developed at each level of the protocol stack. In this paper, we focussed on scheduling and routing algorithms, and showed the improvement in performance that can be achieved by implementing such schemes.

5. ADDITIONAL AUTHORS

Additional authors: Ramesh R. Rao (Dept. of ECE, UCSD, email: rao@cwc.ucsd.edu).

6. REFERENCES

[1] V. Rodoplu and T.H. Meng, "Minimum energy mobile wireless networks," *IEEE J. Selected Areas in Commun.*, vol. 17, no. 8, pp. 1333–1344, August 1999.

[2] J.-C. Chen, K.M. Sivalingam, P. Agrawal, and R. Acharya, "Scheduling multimedia services in a low-power MAC for wireless and mobile ATM networks," *IEEE Transactions on Multimedia*, vol. 1, no. 2, pp. 187–201, June 1999.

[3] A. K. Salkintzis and Christodoulos Chamzas, "An In-Band Power-Saving Protocol for Mobile Data Networks," *IEEE Trans. on Commun.*, vol. 46, pp. 1194–1205, Sept. 1998.

[4] B. Prabhakar, E. Uysal, and A. El Gamal, "Energy-efficient transmission over a wireless link via lazy packet scheduling," *Proc. of Infocom 2001*, April 2001.

[5] N. Bambos and J.M. Rulnick, "Mobile power management for maximum battery life in wireless communication networks," *Proc. of IEEE Infocom'96*, pp. 443–450, March 1996.

[6] B. Narendran, J. Sienicki, S. Yajnik, and P. Agrawal, "Evaluation of an adaptive power and error control algorithm for wireless systems," *Proc. of IEEE ICC'97*, pp. 349–355, 1997.

[7] T. Simunic, L. Benini, P. Glynn, and G. De Micheli, "Dynamic power management for portable systems," *Proc. of IEEE/ACM MobiCom'2000*, Boston, MA, USA, August 2000.

[8] L. Benini, A. Bogiolo, and G. De Micheli, "A survey of design techniques for system-level dynamic power management," *IEEE Transactions on VLSI Systems*, vol. 8, no. 3, pp. 299–316, June 2000.

[9] M. Zorzi and R.R. Rao, "Error control and energy consumption in communications for nomadic computing," *IEEE Transactions on Computers*, vol. 46, no. 3, pp. 279–289, March 1997.

[10] S. Singh, M. Woo, C. S. Raghavendra, "Power-aware Routing in Mobile and Ad Hoc Networks," *Proc. of 4th Annual ACM/IEEE International Conference on Mobile Computing and Networking*, Dallas, TX, pp. 181–190, October 1998.

[11] J.-H. Chang and L. Tassiulas, "Energy conserving routing in wireless ad-hoc networks," *Proc. of Infocom 2000*, Tel Aviv, Israel, March 2000.

[12] H.D. Linden, *Handbook of batteries*, 2nd ed., McGraw-Hill, New York, 1995.

[13] B. Nelson, R. Rinehart, and S. Varley, "Ultrafast pulse discharge and recharge capabilities of thin-metal film battery technology," *11th IEEE International Pulsed Power Conference*, Baltimore, pp. 636–641, June 1997.

[14] M. Doyle and J.S. Newman, "Analysis of capacity-rate data for lithium batteries using simplified models of the discharge process," *J. Applied Electrochem.*, vol. 27, no. 7, pp. 846–856, July 1997.

[15] T.F. Fuller, M. Doyle, and J.S. Newman, "Relaxation phenomena in lithium-ion-insertion cells," *J. Electrochem. Soc.*, vol. 141, no. 4, pp. 982–990, April 1994.

[16] R.M. LaFollette, "Design and performance of high specific power, pulsed discharge, bipolar lead acid batteries," *10th Annual Battery Conference on Applications and Advances*, Long Beach, pp. 43–47, January 1995.

[17] R.M. LaFollette and D. Bennion, "Design fundamentals of high power density, pulsed discharge, lead-acid batteries. II Modeling," *J. Electrochem. Soc.*, vol. 137, no. 12, pp. 3701–3707, December 1990.

[18] J.W. Halley and B. Nielsen, "Simulation studies of polymer electrolytes for battery applications," *MRS Symposium Proceedings*, vol. 496, pp. 101–107, December 1997.

[19] H.S. Choe and K.M. Abraham, "Synthesis and characterization of $LiNiO_2$ as a cathode material for pulse power batteries," *MRS Symposium Proceedings*, vol. 496, pp. 303–308, December 1997.

[20] M. Doyle, T.F. Fuller, and J.S. Newman, "Modeling of galvanostatic charge and discharge of the lithium/polymer/insertion cell," *J. Electrochem. Soc.*, vol. 140, pp. 1526–1533, 1993.

[21] J.S. Newman, *FORTRAN programs for simulation of electrochemical systems*, http://www.cchem.berkeley.edu/~jsngrp/ (Accessed July 1999).

[22] C.F. Chiasserini and R.R. Rao, "Energy efficient battery management," *IEEE J. Selected Areas in Commun.*, vol. 19, no. 7, pp. 1235–1245, July 2001.

[23] C.F. Chiasserini and R.R. Rao, "Improving battery performance by using traffic shaping techniques," *IEEE J. Selected Areas in Commun.*, vol. 19, no. 7, pp. 1385–1394, July 2001.

[24] P. Nuggehalli, V. Srinivasan, and R.R. Rao, "Delay Constrained Energy Efficient Transmission Strategies for Wireless Devices," *Proc. of Infocom 2002*, New York City, NY, June 2002.

[25] L. I. Sennott, "Computing Average Optimal Constrained Policies in Stochastic Dynamic Programming," *Probability in the Engineering and Informational Sciences*, vol. 15, pp. 103–133, 2001.

[26] C.F. Chiasserini and R.R. Rao, "Routing protocols to maximize battery efficiency," *Proc. of IEEE MILCOM 2000*, Los Angeles, USA, October 2000.

[27] T. Shepard, "Decentralized channel management in scalable multihop spread spectrum packet radio networks," *Tech. Rep. MIT/LCS/TR-670*, Massachusetts Institute of Technology Laboratory for Computer Science, July 1995.

Reliable and Energy-Efficient Digital Signal Processing

Naresh Shanbhag
Department of ECE & Coordinated Science Laboratory
University of Illinois at Urbana-Champaign
1308 West Main Street, Urbana, IL-61801
shanbhag@uiuc.edu

ABSTRACT

This paper provides an overview of *algorithmic noise-tolerance* (ANT) for designing reliable and energy-efficient digital signal processing systems. Techniques such as prediction-based, error cancellation-based, and reduced precision redundancy based ANT are discussed. Average energy-savings range from 67% to 71% over conventional systems. Fluid IP core generators are proposed as a means of encapsulating the benefits of an ANT-based low-power design methodology. CAD issues resident in such a methodology are also discussed.

Categories and Subject Descriptors

B.7 [**Hardware**]: Integrated Circuits

General Terms

DESIGN

Keywords

reliability, energy-efficiency, low-power, deep submicron, noise, noise-tolerance, broadband, communications

1. INTRODUCTION

Next generation wireless communication standards such as wireless LAN (IEEE 802.11b and a), access (IEEE 802.16) and 3G (UMTS) reflect the rapid growth in demand for portable and wireless services. Such systems demand higher functionality with extremely low levels of energy consumption. These conflicting requirements have been met to a great extent via scaling of feature sizes in modern CMOS technologies. However, technology scaling has also resulted in the emergence of deep submicron (DSM) noise [13] raising concerns regarding the ability of the semiconductor industry to extend Moore's law well into the deep submicron regime.

At the heart of the issue mentioned above is the problem of *achieving energy-efficiency in the presence of noise*, referred to as the *reliable low-power design* [12] problem. This problem has two components: 1.) determining bounds on energy-efficiency in the presence of noise, and 2.) developing design techniques for approaching these bounds. Since 1995, our research has directly addressed these two subproblems.

First, we developed an information-theoretic paradigm for DSM VLSI systems [11, 6] and employed it to determine bounds on energy-efficiency of these systems in the presence of noise. In [11], we proposed the idea of viewing DSM VLSI systems as communication networks. Recently, other researchers have employed information-theoretic considerations to determine bounds on the switching energy of binary transition [9], bounds on energy for deep submicron busses [15], and on biological systems [1].

Second, we developed key elements of a design philosophy based on noise-tolerance, including noise-tolerant circuit design [2], and algorithmic noise-tolerance (ANT) [7] for broadband communication systems. Indeed, the recent 2001 International Technology Roadmap for Semiconductors [8] (ITRS2001) has described *error-tolerance* as a cost-effective approach for achieving reliability. The ITRS2001 further asserts the need to view systems-on-a-chip as a communication network.

In this paper, we provide an overview of ANT as a effective technique that provides significant gains in energy-efficiency in the presence of noise over conventional systems. The key idea behind ANT is to permit errors to occur in a signal processing block and then correct it via a separate error control block. This approach of error/noise-tolerance is fundamentally superior to mitigating noise and achieves energy-efficiencies beyond what is achievable via present day techniques. Note, ANT techniques are somewhat orthogonal to known low-power techniques [3, 10] in that ANT can be applied after/along with known power reduction techniques. A key difference between ANT and related work in the information-theoretic and coding communities [4] is that the latter do not account for energy-efficiency.

In this paper, we focus on ANT techniques as an attractive approach for jointly addressing reliability and energy-efficiency issues for broadband communications and digital signal processing systems.

2. ALGORITHMIC NOISE-TOLERANCE

In this section, we first define an ANT-based system in section 2.1 followed by a motivational example in section 2.2. In section 2.3, we describe the key elements of an ANT-based system and in section 2.4 we provide a rationale for the effectiveness of ANT in enhancing energy-efficiency.

Table 1: Motivational Example

i	1	2	3	4	5	6	7	8
x	0.875	-0.875	0.875	-0.875	0.875	0.875	-0.875	0.875
η	0.000	0.000	0.875	0.000	0.000	0.875	0.000	0.000
y	0.547	0.219	0.328	-0.328	0.328	0.766	0.766	0.328
y'	0.547	0.219	1.203	-0.328	0.328	1.641	0.766	0.328
y_r	0.438	0.000	0.438	-0.438	0.438	0.438	0.438	0.438
y	0.547	0.219	0.438	-0.328	0.328	0.438	0.766	0.328

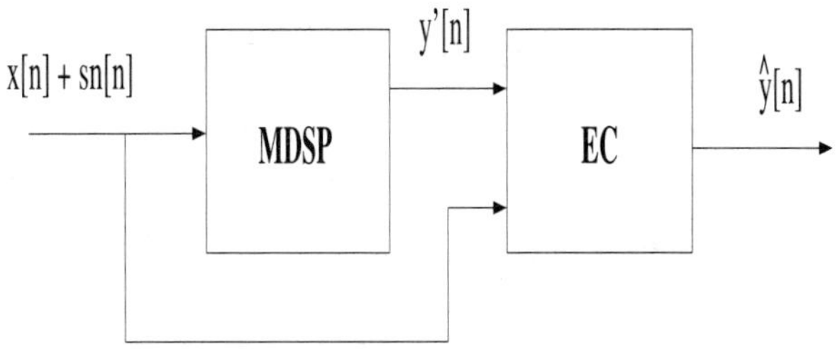

Figure 1: ANT-based DSP System.

2.1 Formal Definition of ANT

An ANT-based DSP system (see Fig. 1) has a main DSP (**MDSP**) block that computes in an energy-efficient manner but makes intermittent errors. It accepts as its input the signal $x[n] + sn[n]$, where $x[n]$ is the input signal and $sn[n]$ is the input signal noise. The noisy output of the **MDSP** block is denoted by $y'[n]$. In essence, the **MDSP** block sacrifices noise-immunity for energy-efficiency. The **EC** block observes the noisy output $y'[n]$, and the input $x[n] + sn[n]$ and perhaps certain internal **MDSP** signals to detect and correct errors. The final corrected output of the ANT-based system is denoted as $\hat{y}[n]$. The error-control (**EC**) block operates in an error-free manner but consumes significantly more energy per operation than the **MDSP** block.

The assumption of an error-free **EC** block can be easily justified as follows. Due to energy-efficiency considerations, the **EC** block will be designed to be much simpler/smaller than the **MDSP** block. Thus, the constraints on delay and power will be significantly relaxed for the **EC** block and therefore can be easily met with a much higher noise-immunity. For example, one can use a highly noise-tolerant circuit technique [2] to design the **EC** block and pay a price in terms of extra power. As the **EC** block is designed to be much smaller than the **MDSP** block, this extra power will have minimal impact on the overall power consumption.

If the **MDSP** block is an FIR filter, then its noisy output $y'(n)$ can be written as,

$$\begin{aligned} y'[n] &= \sum_{k=0}^{N-1} h_k x[n-k] + \eta[n] \\ &= y[n] + \eta[n] \end{aligned} \tag{1}$$

where $y[n]$ is the error-free output and $\eta[n]$ represents the manifestation of DSM noise at the algorithmic level on a per (output) sample basis. We now describe an example to illustrate the ANT concept.

2.2 Motivational Example

Consider the error-free output $y[n]$ of a finite-impulse response (FIR) filter (main filter) given by

$$y[n] = 0.625x[n] + 0.875x[n-1] + 0.625x[n-2], \tag{2}$$

where $x[n]$ is a 4-bit two's complement uncorrelated input that is equally likely to take on a value of either 0.875 or -0.875. Note that the coefficients also have a 4-bit two's complement representation.

We model the impact of a DSM noise source with the signal $\eta[n]$ that takes on a value of ± 0.875 with probability 0.1 and 0 with a probability of 0.9. We denote the noisy output of the main filter as $y'[n]$, where

$$y'[n] = y[n] + \eta[n] \tag{3}$$

Note that $\eta[n]$ is non-zero quite infrequently but that the non-zero value (0.875) is quite large, i.e., comparable to the maximum value of the output $y[n]$ (1.86). Table 1 shows the values of $x[n]$, $y[n]$, $\eta[n]$ and $y'[n]$ for 8 sample periods.

In order to detect and correct these errors, we implement a half-precision filter (i.e., a filter whose coefficient precision is half of that of the main filter in (2)) whose input is $x[n]$ (i.e., same as that of the main filter), and whose output $y_r[n]$ is given by

$$y_r[n] = 0.5x[n] + 0.5x[n-1] + 0.5x[n-2] \tag{4}$$

Note that the coefficients of the half-precision filter can be obtained by truncating the 4-bit two's complement representation of the main filter by two LSBs.

We flag an error when the absolute value of the difference between the two filter outputs $|y'[n] - y_r[n]|$ is greater than a pre-specified threshold E_{th}. In this example, we set $E_{th} = max(|y[n] - y_r[n]|) = 0.547$, where the maximization is over all input combinations. Let $\hat{y}[n]$ be the final (corrected) output of this system. If an error is flagged then we correct it via the assignment: $\hat{y}[n] = y_r[n]$. Otherwise, we set $\hat{y}[n] = y'[n]$.

Table 1 shows the values of $y_r[n]$ and $\hat{y}[n]$. A run of approximately 1000 samples shows that the output signal-to-noise ratios (SNRs) as

$$SNR_n = 10log_{10}(\frac{\sigma_y^2}{\sigma_n^2}) = 8.0474dB \tag{5}$$

$$SNR_{ANT} = 10log_{10}(\frac{\sigma_y^2}{\sigma_{\hat{y}-y}^2}) = 13.2667dB \tag{6}$$

where σ_y^2, σ_n^2, and $\sigma_{\hat{y}-y}^2$ are the variances of the error-free output $y[n]$, the noise signal $\eta[n]$ and the residual noise signal $\hat{y}[n] - y[n]$, respectively. Note that SNR_n is the output SNR of the noisy main filter *before* error control and SNR_{ANT} is the output SNR of the ANT-based system, i.e., after error correction. It can be seen that a very simple error correction scheme can improve the output SNR by more than $5dB$.

From this example, we conclude that the key elements of an ANT-based system are: 1.) the DSM noise model $\eta[n]$, 2.) the statistical nature of the metrics (e.g. SNR) employed to evaluate the algorithmic performance, and 3.) statistical properties of the signals being processed, and the algorithm itself. These are described next.

2.3 Elements of ANT

2.3.0.1 DSM Noise models and Architecture.

The signal $\eta[n]$ in (1) models the impact of DSM noise on an output sample. In general, noise may be random (e.g., alpha particle hits on a dynamic node) or may be systematic (e.g., critical path violations or delay faults due to numerous causes including variations in process, temperature and supply voltage). For practical reasons, in certain cases we may deliberately take $\eta[n]$ to be random even if the corresponding physical noise source is systematic. For example, supply bounce is a systematic source of noise because of its dependence on the input. This dependence is however extremely complex for large designs. Therefore, we may choose an $\eta[n]$ to be random when modelling the impact of supply bounce on the output of a DSP block.

DSM noise in general is hard to model especially at the algorithmic level as needed by ANT based systems. For this reason, in the past, we have proposed the concept of *voltage overscaling* (VOS) which extends the idea of voltage scaling and provides a 'well understood' source of noise. Assume that $V_{dd-crit}$ is the supply voltage at which the critical path delay just meets the delay constraints imposed by the application and architecture. VOS implies scaling of the supply voltage to a value below $V_{dd,crit}$, i.e., $V_{dd} = V_{dd,crit}/k_{vos}$, where $k_{vos} \geq 1$ is referred to as the voltage overscaling factor (VOSF). VOS results in one or more critical paths violating their delay constraints. This in turn will result in the output $y'(n)$ being in error when ever an appropriate input sequence is applied. We refer to such errors as soft errors because these disappear when the input sequence no longer excites the critical path.

Note that the statistics of $\eta[n]$ in a VOS scenario is a strong function not only of the input statistics but also the architecture. This is because the distribution of critical paths a function of the delay is dependent on the architecture. For example, an architecture utilizing ripple-carry structures and array multipliers will have a skewed distribution, i.e., a few paths will determine the overall delay of the system. This *delay imbalance* favors ANT because a skewed distribution implies a low frequency of errors when the architecture is voltage overscaled.

Therefore, in a VOS scenario and perhaps others, the noise model $\eta[n]$ has a dependence on the input signal $x[n]$. Later in the paper, we will present ANT techniques that exploit this dependence and those that do not.

2.3.0.2 Algorithmic Performance Metrics.

ANT techniques exploit the fact that the functional performance metrics for DSP and communication systems are statistical in nature. In reference to Fig. 1, the output SNR of a critically scaled (i.e., $V_{dd} = V_{dd,crit}$) (or noiseless) filter is given as

$$SNR_o = 10\log_{10}\left(\frac{{\sigma_y}^2}{\sigma_{sno}^2}\right) \tag{7}$$

where σ_y^2 and σ_{sno}^2 are the powers of the output signal $y[n]$ and the output *signal noise* $sno[n]$ (not shown in Fig. 1), respectively. The term 'signal noise' refers to the noise that is present in the input signal $x[n]$ and which is denoted in Fig. 1 as $sn[n]$. Therefore, $sno[n]$ is a filtered version of $sn[n]$. Note, a signal noise source will exist even though the critically scaled filter is error-free thereby providing a finite value for SNR_o.

Now, the output SNR for a voltage overscaled or noisy filter is given by

$$SNR_n = 10\log_{10}\left(\frac{{\sigma_y}^2}{{\sigma_{sno}}^2 + {\sigma_n}^2}\right) \tag{8}$$

where ${\sigma_y}^2$, ${\sigma_{sno}}^2$ and ${\sigma_n}^2$ are the powers of the output signal $y[n]$, output signal noise $sno[n]$, and soft errors due to DSM noise $\eta[n]$, respectively.

Similarly, the output SNR of an ANT-based filter is given by

$$SNR_{ANT} = 10\log_{10}\left(\frac{{\sigma_y}^2}{{\sigma_{sno}}^2 + {\sigma_r}^2}\right) \tag{9}$$

where ${\sigma_r}^2$ is the power in the residual soft error $r[n] = \hat{y}[n] - y[n]$.

Note that the presence of σ_{sno}^2 in the denominator in (9) provides room for imperfect error control and hence greater energy-efficiencies. This implies that in an ANT-based system, it is sufficient to reduce the DSM noise source $\eta[n]$ below the output signal noise floor represented by σ_{sno}^2 in order to minimize the impact on the output SNR.

2.3.0.3 Signal and Algorithmic Properties.

Signals being processed in DSP and communication systems can be modelled as random processes. The power spectrum of these signals are modulated by transmit filters, the physical channel and the receive filters (equalizers). Thus, one has a good idea of what the statistical properties of signals should be in a well-behaved or error-free system. DSM noise sources tend to disturb the statistical structure of signals. ANT techniques are extremely effective when the disturbance is large but correctable. Large disturbances make it easy to detect errors and, if correctable, the overall impact on SNR can be made minimal, i.e., $SNR_{ANT} \approx SNR_o$.

2.4 Energy Savings due to ANT

We now derive the conditions under which an ANT-based system leads to energy savings in a VOS scenario over a critically voltage-scaled system (defined as a system operating at $V_{dd-crit}$). The dynamic energy dissipation per clock cycle $\mathcal{E}_{orig}$ of such a system is given by

$$\mathcal{E}_{orig} = C_{orig}{V_{dd-crit}}^2 \tag{10}$$

where C_{orig} is the average switching capacitance. Note that $\mathcal{E}_{orig}$ is the minimum energy dissipation that conventional voltage scaling can achieve.

In comparison, the dynamic energy dissipation per clock cycle $\mathcal{E}_{ANT}$ of the corresponding ANT-based system is given by

$$\mathcal{E}_{ANT} = C_{orig}\left(\frac{V_{dd-crit}}{k_{vos}}\right)^2 + C_{ANT}{V_{dd-ant}}^2 \tag{11}$$

where C_{ANT} represents the overhead complexity due to ANT, V_{dd-ant} is the critical supply voltage for **EC** block, and

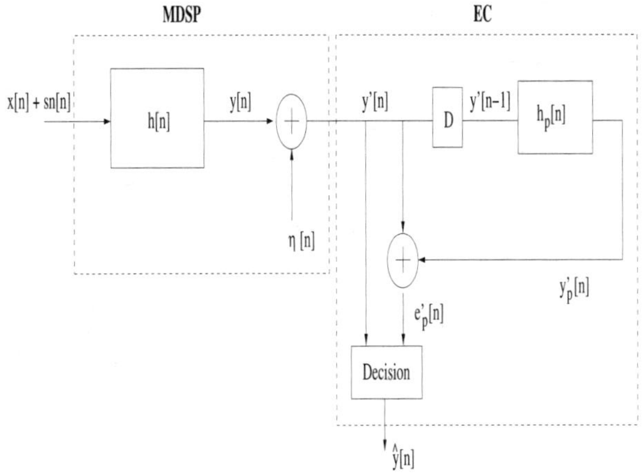

Figure 2: Prediction-based ANT technique.

$k_{vos} > 1$ is the VOS factor (VOSF). From (10)−(11), it can be easily shown that $\mathcal{E}_{ANT} < \mathcal{E}_{crit}$ provided that

$$C_{ANT} {V_{dd-ant}}^2 < C_{orig} {V_{dd-crit}}^2 (1 - \frac{1}{{k_{vos}}^2}) \quad (12)$$

In practice, the condition in (12) is easily satisfied by making C_{ANT} as small as possible and/or by making k_{vos} as large as possible. There is indeed an interesting trade-off between k_{vos} and C_{ANT}. When k_{vos} is increased, the performance degradation becomes larger as more critical paths and other longer paths start to fail. This requires increasingly sophisticated and hence complex ANT techniques which would increase C_{ANT}.

3. PREDICTION-BASED ANT

In this section, we describe a technique referred to as prediction-based ANT [7] (see Fig. 2) that employs the filter output $y'[n]$ to detect and correct errors.

Consider the output $y'[n]$ of a VOS filter that uses a least significant bit (LSB) first architecture. LSB first architectures are in fact the most commonly employed in practice. If the filter is sufficiently narrowband then the error-free output signal $y[n]$ is highly correlated, i.e., the MSBs do not change from one sample to the next. A linear predictor, which is another filter, therefore can be used to statistically predict the output $y[n]$. The output $y_p[n]$ of a linear predictor of $y[n]$ is given by

$$y_p[n] = \sum_{k=0}^{N_p-1} p_k y[n-k-1] \quad (13)$$

where N_p is the number of predictor taps and p_k are the predictor coefficients. The prediction error $e[n] = y[n] - y_p[n]$ can be minimized via a proper choice of the coefficients p_k.

Under VOS, we do not have access to $y[n]$ (the error-free output) but only to $y'[n]$ (the noisy output). We will assume however that the errors occur with a frequency which is sufficiently less than $1/(2N_p)$. In that case, the predictor output under VOS $y'_p[n]$ equals $y_p[n]$ (the error-free output).

Also, under VOS, the MSBs will fail first, i.e., $\eta[n]$ in (1) is large. The prediction error under VOS $e'_p[n]$ is therefore given by

$$\begin{aligned} e'_p[n] &= y'[n] - y'_p[n] \\ &= y[n] - y_p[n] + \eta[n]. \end{aligned} \quad (14)$$

In this case, error detection is accomplished via a simple threshold on $e'_p[n]$. Error correction on the other hand is also achieved by assigning the final corrected output $\hat{y}[n]$ as

$$\hat{y}[n] = y_p[n]. \quad (15)$$

In addition, we also assign $y_p[n]$ as the next N_p corrected outputs. This is because once $y'[n]$ is in error (i.e., $y'[n] \neq y[n]$) then $y_p[n] \neq y'_p[n]$ for the next N_p samples while the incorrect output $y'[n]$ is being flushed out from the data buffer in the predictor (see (13).

We summarize the prediction-based ANT scheme consists of the following steps:

- Error detection: if $|e'_p[n]| > E_{th}$ an error is declared.
- Error correction: If an error is declared then $\hat{y}[n] = y'_p[n]$ else $\hat{y}[n] = y'[n]$.

The assumptions made in deriving the prediction-based ANT technique are:

- magnitude of noise $\eta[n]$ is relatively large, i.e., comparable to the maximum value that $y[n]$ can achieve.
- the probability of $\eta[n] \neq 0$ is less than $1/(2N_p)$.

In addition, for obtaining large energy savings, we require that the bandwidth of the output $y[n]$ is sufficiently small so that the predictor complexity is much smaller than that of the **MDSP** block. We have shown through measurements conducted on an integrated circuit designed in 0.35μm, 3.3V CMOS process [5] that the prediction-based ANT technique provides up to 67% energy savings over a critically scaled filter.

The prediction-based ANT technique is surprisingly effective even when the output bits are flipped randomly with a probability p_{err}. We have shown [7] that ANT technique reduces the drop in SNR over a noiseless system from 11dB to 2dB for a value of p_{err} as high as 10^{-3}. In other words, the ANT based system improves the SNR by 9dB when each filter output bit is being flipped at an average rate of once every 1000 samples independent of each other.

4. ERROR CANCELLATION-BASED ANT

The error cancellation-based ANT technique shown in Fig. 3 exploits any correlation that may exist between $\eta[n]$ and the input $x[n]$. Note that such correlation certainly exists under a VOS scenario.

The error cancellation-based ANT requires a separate filter called the error canceller $h_e[n]$ that generates an estimate of $\eta[n]$ (denoted as $\hat{\eta}[n]$) from $x[n]$. This error canceller needs to be trained first in order to learn the correlation structure between $\eta[n]$ and the input $x[n]$. Figure 3, shows how this is done. During the training phase, a known input sequence $x[n] + sn[n]$ is provided at the input to the **MDSP** block. At the same time, the multiplexer provides the corresponding precomputed error-free output $y[n]$. Therefore, during the training phase, the output $y"[n]$ is given by,

$$y"[n] = y'[n] - y[n] = y[n] + \eta[n] - y[n] = \eta[n] \quad (16)$$

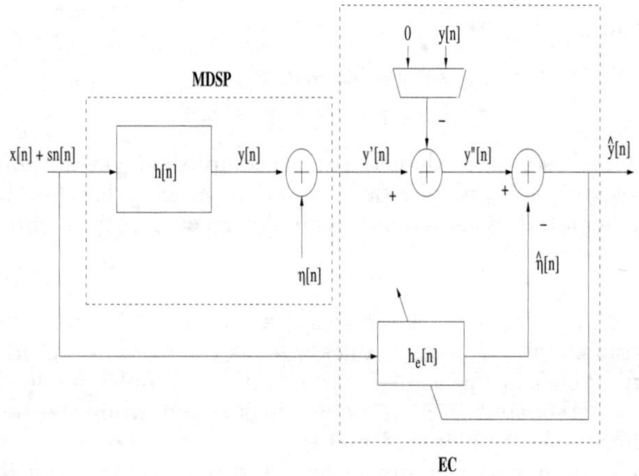

Figure 3: Error cancellation-based ANT technique.

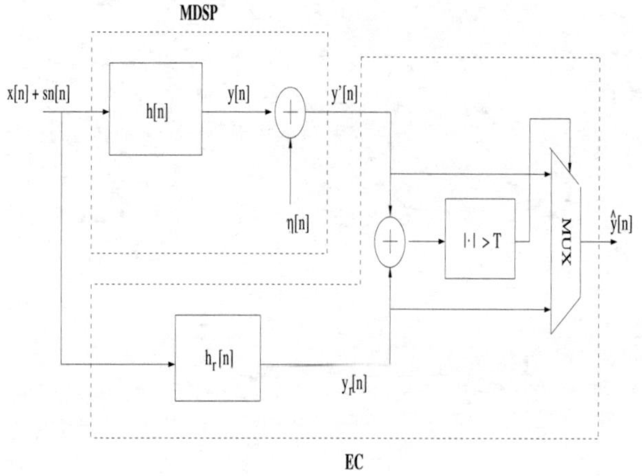

Figure 4: Reduced precision redundancy-based ANT technique.

and therefore the final output $\hat{y}[n]$ is given by

$$\hat{y}[n] = \eta[n] - \hat{\eta}[n] = e[n] \tag{17}$$

where the estimation error $e[n]$ is used to adapt the error canceller. Any of the well-known adaptive filtering algorithms such as the least-mean squared (LMS) algorithm can be employed here.

Normal operation commences once the error canceller has been trained, i.e., the variance of $e[n]$ has been minimized. In this phase, the input sequence is unknown and the multiplexer outputs a zero. In addition, the error canceller stops adapting. The output $y"[n]$ is therefore given by,

$$y"[n] = y'[n] - 0 = y[n] + \eta[n] \tag{18}$$

and hence the final output $\hat{y}[n]$ is given by

$$\begin{aligned} \hat{y}[n] &= y"[n] - \hat{\eta}[n] \\ &= y[n] + \eta[n] - \hat{\eta}[n] = y[n] + e[n] \end{aligned} \tag{19}$$

Equation (19) indicates that if the error canceller has been trained correctly then during normal mode of operation, the final output $y"[n]$ will be very close to the error free output $y[n]$.

When applied to digital FIR filters, we have shown that error-cancellation works best for broadband filters and energy savings of up to 71% can be achieved over critically scaled systems.

5. REDUCED PRECISION REDUNDANCY (RPR) BASED ANT

In the RPR technique [14], we exploit the fact that MSBs are the critical bits and hence need to be protected from noise.

The RPR technique shown in Fig. 4 has a low precision replica $h_r[n]$ of the filter $h[n]$ in the **MDSP** block. In this case, $h_r[n]$ takes the same input $x[n] + sn[n]$ as $h[n]$ but computes only the MSBs of the error-free output $y[n]$. Note that the output of the RPR filter $y_r[n]$ will not equal the error-free output $y[n]$ of the **MDSP** filter due to quantization noise. Quantization noise properties for DSP systems are well-understood. Thus, the error control scheme in case of RPR is as follows:

- compute the difference metric: $d(y', y_r) = |y'[n] - y_r[n]|$.
- Error detection: if $d(y', y_r) > E_{th}$, where E_{th} is a predefined threshold value, then flag an error.
- Error correction: if an error has been flagged then $\hat{y}[n] = y_r[n]$ otherwise $\hat{y}[n] = y'[n]$.

Note that the threshold E_{th} will need to be set at a value above the quantization noise floor. An ideal value for E_{th} would be $max(y'[n] - y_r[n]$, where the maximization is carried out over all noise scenarios. In case of VOS, this maximization needs to be done over all possible input combinations.

When applied to digital filtering [14], RPR provides up to 67% energy savings. We have also shown that RPR-based multipliers provide up to 44% energy savings in an FFT.

6. CAD ISSUES

The design of ANT-based DSP systems presents numerous opportunities for CAD researchers. This section describes some of the challenges.

6.1 Noise-Tolerant Design Methodology

New techniques for DSM noise analysis are required that lead to a better understanding of the energy penalty involved in controlling noise at the physical level. Also, algorithms and techniques need to be developed that propagate the impact of various noise sources to the algorithmic level, and development of algorithmic noise models. This will enable designers to evaluate the energy-efficiency benefits of ANT-based DSP systems under general noise scenarios and contribute greatly to the goals stated in ITRS2001[8].

Determining energy-optimal trade-offs between circuit level noise-tolerance and ANT is an open problem. The VOS scenario also provides a nice test case for exploring these trade-offs because the above mentioned noise modelling problem for VOS is much more tractable than for the general DSM noise case.

While any noise-tolerance based design paradigm would be effective in achieving energy-efficiency in the presence of noise, a question still remains as to what the bounds on energy-efficiency in the presence noise are? The information-theoretic view proposed in [11] has been applied to simple gates, where ANT techniques do not apply. There is

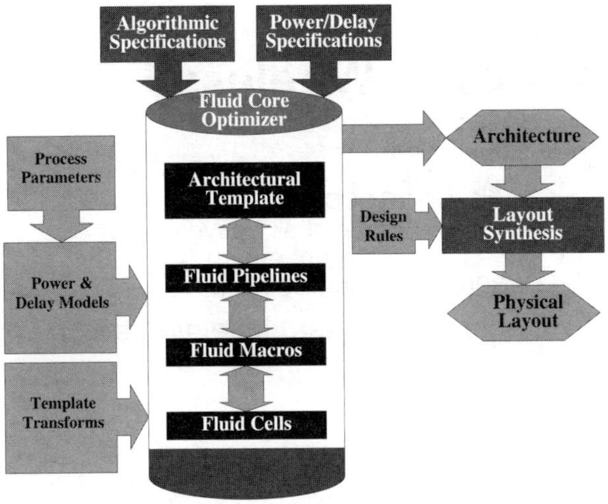

Figure 5: A fluid IP core generator.

currently a gap between our ability to compute achievable bounds on energy-efficiency and being able to compare those with the efficiencies achieved via techniques such as ANT, which by default are effective for complex systems. Computational techniques are required for determining the bounds on energy-efficiencies for complex VLSI systems.

6.2 Fluid IP Core Generators

In order to make the benefits of an ANT-based design methodology available to the designer, we propose the development of a fluid IP core generator as shown in Fig. 5.

A fluid IP core generator synthesizes a custom-quality layout of an algorithm-specific block (such as an equalizer) from high-level specifications without going through a synthesis and place & route step. The term 'fluid' refers to the fact that the core generator while optimizing the architecture can reach down into the circuit fabric and tune transistor sizes while generating layouts. Core generators are useful in broadband communication systems where energy and throughput efficiency limits need to be approached and where canonical blocks (such as equalizers, filters, FFTs, Reed-Solomon decoders etc.) are employed repeatedly though with varying parameters across multiple applications.

The key features/benefits of a fluid IP core generator are: 1.) process and device scalability, 2.) custom quality design within a synthesis quality design cycle, 3.) predictability of hard cores and flexibility of soft cores, and 4.) encapsulation of cross-domain optimization techniques such as ANT. An inherent drawback of such a core generator is that these are designed for algorithm-specific blocks.

An IP core generator has two major components: 1.) an architecture optimizer, and 2.) a layout synthesizer, and it accepts as inputs: 1.) power and delay models, 2.) library of template transforms, 3.) algorithmic specifications and 4.) power and delay specifications.

Design of a fluid IP core generator presents numerous problems. Development of process scalable power and delay models is one. Such development seems feasible given the fact that the functionality of algorithm-specific blocks do not change over process generations. Hence, models that require incremental changes from one process generation to the next would be useful.

7. ACKNOWLEDGMENTS

The author would like to acknowledge support from National Science Foundation grants CCR-000987, CCR 99-79381 and DARPA for the work reported in this paper.

8. REFERENCES

[1] P. Abshire and A. Andreou. Capacity and energy cost of information in biological and silicon photoreceptors. *Proceedings of IEEE*, 89(7):1052–1064, July 2001.

[2] G. Balamurugan and N. R. Shanbhag. The twin-transistor noise-tolerant dynamic circuit technique. *IEEE Journal of Solid-State Circuits*, 36(2):273–280, February 2001.

[3] A. Chandrakasan and R. Brodersen. Minimizing power consumption in digital cmos circuits. *Proceedings of the IEEE*, 83(4), April 1995.

[4] B. Hajek and T. Weller. On the maximum tolerable noise for reliable computation by formulas. *IEEE Transactions on Information Theory*, 37(2), March 1991.

[5] R. Hegde and N. R. Shanbhag. A low-power digital filter ic via soft dsp. In *Proceedings of Custom Integrated Circuits Conference*, May 1991.

[6] R. Hegde and N. R. Shanbhag. Towards achieving energy-efficiency in the presence of deep submicron noise. *IEEE Transactions on VLSI*, 8(4):379–391, April 2000.

[7] R. Hegde and N. R. Shanbhag. Soft digital signal processing. *IEEE Transactions on VLSI*, 9(6):813–823, December 2001.

[8] ITRS2001. International technology roadmap for semiconductors. *http://public.itrs.net/Files/2001ITRS/Home.htm*, 2001.

[9] J. D. Meindl and J. A. Davis. The fundamental limit on binary switching energy for terascale integration (tsi). *IEEE Journal of Solid-State Circuits*, 36:1515–1516, October 2000.

[10] J. Rabaey and M. Pedram. *Low Power Design Methodlogies*. Kluwer Academic Publishers, 1996.

[11] N. R. Shanbhag. A mathematical basis for power reduction in digital vlsi systems. *IEEE Transactions on Circuits and Systems: Part II*, 44(11):935–951, Nov. 1997.

[12] N. R. Shanbhag and et. al. Reliable low-power design in the presence of deep submicron noise. In *Proceedings of the Intl. Symposium on Low-Power Design*, pages 295–302, July 2000.

[13] K. L. Shepard. Conquering noise in deep submicron digital ics. *IEEE Design and Test of Computers*, pages 51–62, Jan.-Mar. 1998.

[14] B. Shim and N. R. Shanbhag. Reduced precision redundancy for low-power digital filtering. In *Proceedings of Asilomar Conference*, 2001.

[15] P. Sotiriadis, A. Chandrakasan, and V. Tarokh. Maximum achievable energy-reduction using coding with applications to deep sub-micron busses. In *Proc. of Intl. Symp. on Circuits and Systems*, 2002.

CMOS: A Paradigm for Low Power Wireless?

Michiel Steyaert
Katholieke Universiteit Leuven
Kasteelpark arenberg 10
3001 Heverlee, Belgium
michiel.steyaert@esat.kuleuven.ac.be

Peter Vancorenland
Katholieke Universiteit Leuven
Kasteelpark arenberg 10
3001 Heverlee, Belgium
peter.vancorenland@esat.kuleuven.ac.be

ABSTRACT

An overview and comparison of different topologies for wireless architectures are discussed, where the main focus lies on the power consumption and possibilities towards integration and reduction of external components. Architectures with reduced number of building blocks (both internal and external) are presented where the main benefits are the low costs, both in the CMOS technology as well as the power.

Categories and Subject Descriptors

B.7 [**Hardware**]: Integrated Circuits

General Terms

Design

Keywords

CMOS Wireless receivers Low-power

1. INTRODUCTION

The super heterodyne receiver has had a good run. It has been the basic workhorse for most RF designs since the 1920s. But many RF designers have been wooed by the idea of dispensing with the cumbersome bundle of filtering, oscillator and frequency-synthesis parts that make up a superhet. Those parts are needed to convert the signal received at the antenna to an intermediate frequency (IF), and ultimately to the baseband level. Direct conversion receivers are freed of the IF stage and, in principle, of the circuitry that comes with it. The transition to direct conversion has picked up pace across a number of wireless application areas. The technique has been easy enough to deploy in systems such as RF ID products, for example garage-door controls and pagers. But they use simple modulation schemes, with relatively easy-going requirements for interference rejection and error rates. There are far more performance related barriers to break through when it comes to modern wireless standards such as GSM. Direct-conversion technology has been slower to penetrate here, although it now accounts for roughly a quarter of GSM phones currently in use.

Zero and low-IF designs are a far simpler proposition in the wireless LAN arena because standards such as Bluetooth and 802.11b have looser requirements in terms of channel filtering compared with GSM. No-one seems to be developing superhets for wireless LANs. Developers of third-generation handsets are expected to pursue direct conversion aggressively, largely because the handsets will need to support both GSM and either flavour of CDMA.

A parallel move to direct conversion is the use of CMOS RF circuitry. It is a research avenue that entered the frame once CMOS geometries had shrunk enough to support gigahertz frequencies. Since "sub-quarter-micron" technologies now come readily available from the main foundries, the amount of full CMOS wireless products available on the market is also clearly rising.[7, 8, 9, 10]

Both price and power consumption are major driving forces for the commercial breakthrough of wireless products. This paper addresses the feasibility of the cheaper CMOS technology as the choice for low power wireless circuit implementation.

DAC 2002, June 10-14, 2002, New Orleans, Louisiana, USA.

2. WIRELESS ARCHITECTURES

The choice of a suitable wireless architecture can have a direct impact on the overall power consumption. On one hand there are the popular and reliable multi-stage down conversion architectures. On the other hand architectures with a reduced number of down conversion stages have gained interest due to their power saving potential.

Heterodyne receiver

The best known and most widely used architecture is the heterodyne receiver, depicted in Fig. 1. This receiver employs one to multiple stages of an image-reject filter, a mixer and a variable gain amplifier block. Due to the required high quality factors of the image reject mixers, these are generally implemented off-chip. The drawback of implementing off-chip components, apart from the cost increase, is the increased power consumption. These blocks have to be driven with a characteristic impedance (typically 50Ω). Driving signals at or near the RF frequency with such an impedance requires power hungry driver blocks. The implementation of a increasing number of cascaded down conversion stages will generally also be reflected in the overall power consumption.

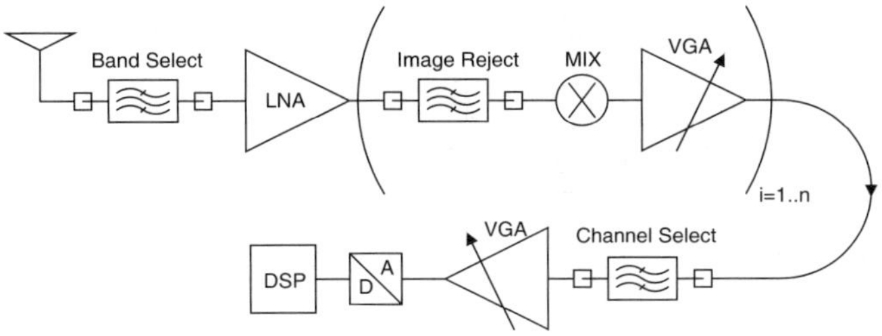

Figure 1: General schematic of the heterodyne receiver

Zero IF receiver

The zero-IF receiver has also been around for a long time [5]. This receiver, also known as a "homodyne" or "direct conversion" receiver, as depicted in Fig. 2 has proven to be a good alternative to the previously described heterodyne receiver. In this architecture, the frequency of the local os-

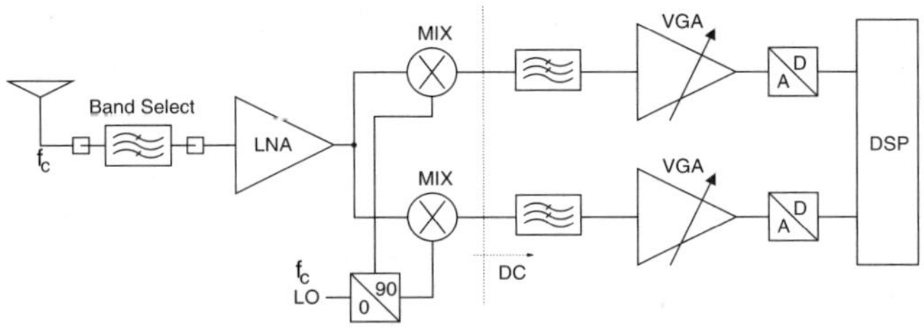

Figure 2: General schematic of the zero IF receiver

cillator signal is equal to the carrier frequency. The advantage being that the wanted signal and the image signal are the same, seriously relaxes the Image Rejection Ratio requirements. The use of an off chip image reject filter is no longer required and hence the LNA no longer needs to drive a 50 Ohm load. The channel select filters of the heterodyne receiver are replaced by low-pass filters that can be implemented on chip. All these properties lead to an architecture with a reduced power consumption.
The major drawback of this architecture, however, is the generation of DC offsets in the downconverted signal, due to self-mixing in the down-conversion mixers, even order distortions and 1/f noise sources. These offsets can easily saturate the amplifiers and filters at the end of the receive chain.
Different strategies, including tuning, trimming, adaptive offset corrections, frame-by-frame corrections,... have been employed to counter this problem [1].

Low IF receiver

A number of other architectures has been developed [13, 2], but only one has found its way from the academics into the commercial market in a number of designs.

The low IF receiver, while having different possible implementations, can generally be represented by Fig. 3. This architecture combines the advantages of the previously discussed receivers into a single analog down conversion stage with reduced power consumption without the DC offset problems. The IF frequency is slightly higher than DC, so the

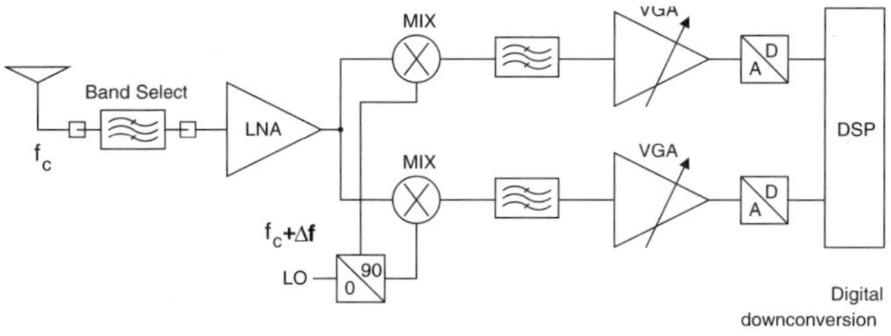

Figure 3: General schematic of the low IF receiver

wanted signal is not corrupted by DC offsets, nor by 1/f noise. The RF image rejection is performed by the I/Q down conversion mixers, which can be implemented as quadrature down converters or double quadrature down converters[4]. The channel selection is performed by the polyphase filter, which also aids the image rejection of the final down conversion to DC, as well as relaxing the dynamic range requirements on the AD converter stage.

3. LOW POWER CMOS WIRELESS RECEIVER IMPLEMENTATION

This section discusses a CMOS implementation example of a low power wireless receiver for GPS applications. The low IF architecture has been chosen for its good power performance.
The implemented receiver topology is shown in Fig. 4. The

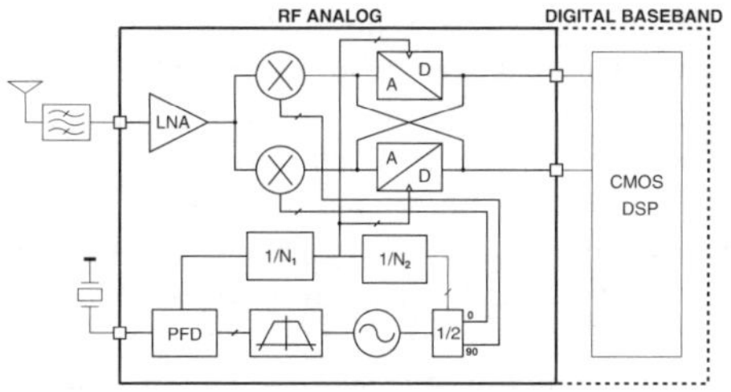

Figure 4: High Level receiver topology

receiver consists of an LNA connected to a switched mixer followed by the AD converter. By using a wide dynamic range AD converter, the VGA function used in many receivers can be shifted to the digital domain. A $\Delta\Sigma$ AD converter can achieve a sufficiently high dynamic range while not exaggerating the power consumption. The front-end converts the RF-signal to digital I and Q signals in a 2 MHz BW centered at 4MHz. The RF mixers are switched with a quadrature LO-signal derived from a VCO running at 3.14GHz followed by a divide-by-2 block. The RF-signal is quadrature down converted by mixers connected directly to the input of the A/D converter. The loop filter of the AD converter is a complex bandpass loop filter which has an asymmetric frequency response for its noise shaping function. The output of the receiver is a digital I and Q bitstream at a bit-rate of 128 MHz.
An anti alias filter in front of the ADC is not necessary anymore because the continuous time loop filter attenuates out of band signals before any signal sampling occurs, as opposed to the switched implementations [17], and is therefore left out.

The different building blocks are discussed next.

3.1 LNA

As seen in Fig. 4 the LNA is only preceded by the antenna and an external blocking-filter. This means no amplification of the signal has been performed before it reaches the input of the LNA. Therefore, the LNA is designed to have a very low noise figure since it sets a lower bound for the total receiver noise figure. The circuit schematic of the LNA is shown in Fig. 5.

A high voltage gain is necessary to sufficiently reduce the noise contribution of following mixer block. The low noise figure and high gain are provided by a single-ended common-source amplifier with inductive source degeneration. The cascode transistor reduces the Miller-effect, increasing the stability of the amplifier. It also increases the reverse isolation so that LO leakage to the antenna is minimized. The load of the LNA consists of an inductor which was designed for a high equivalent load resistance R_p, in order to have sufficient voltage gain without jeopardizing the linear operation of the mixer. A patterned ground shield was placed beneath the inductor in order to avoid noise injection and to increase the Q of the parasitic capacitance. To further boost the gain without a penalty in power consumption, the LNA input impedance was designed slightly lower than 50Ω [11]. A ground shield was used for the input bonding pad in order to minimize its capacitance and maximize the Q, which improves noise and gain performance. The input of the LNA is protected against ESD by two reverse biased diodes and a supply clamp.

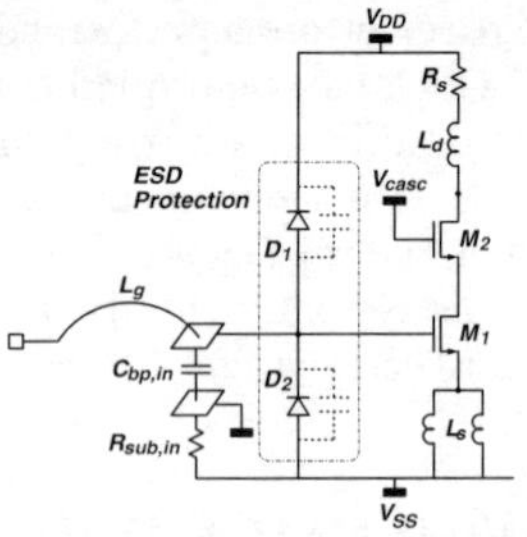

Figure 5: LNA circuit schematic

3.2 PLL

The type II fourth order PLL is shown in Fig. 6[12].

The VCO is built around an on-chip octagonal balanced inductor that is optimized through an in-house inductance simulator-optimizer. The VCO operates at a frequency of 3.14GHz and the quadrature signal is generated from a master-slave frequency divide-by-2. Two DSTC n-latches form a differential dynamic D-flipflop performing the high-speed division. The PLL is locked to a 16.37MHz frequency reference through a divide-by-96 block and a phase frequency detector without dead-zone [6]. The reference frequency spurs are minimized by adding a reference branch in the charge pump core and careful timing of the switch control signals. This way, the charge pump current sources are always on. The current is alternatively flowing in the reference and the output branch of the charge pump. A virtual ground is provided after the charge pump by putting an opamp in the loop filter. This keeps the charge pump switches well in saturation and improves the symmetry between the Up- and the Down- side of the charge pump during locking. For stability reasons, a low frequency zero is inserted in the loop filter. This low-frequency-zero is implemented on-chip without applying any tricks like using a dual-path loop filter [14]. This way, power is saved since we do not need to implement a current summation circuit. This comes at the cost of an increased chip area to implement the loop filter capacitor of almost 2nF on chip.

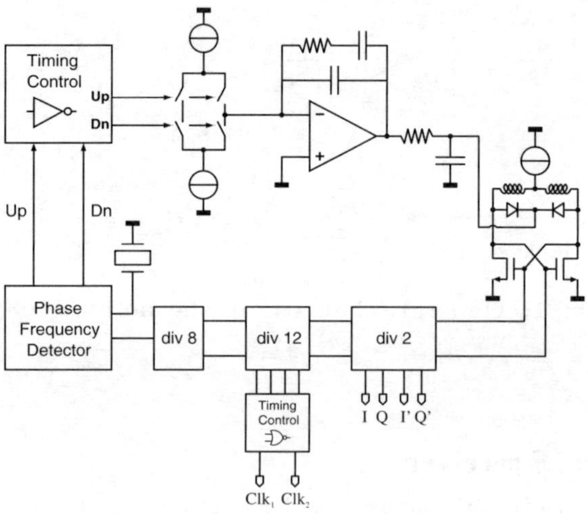

Figure 6: type II PLL schematic

3.3 AD converter

For a given RF down conversion system with an IF frequency BW_c and a bandwidth $2.BW$, different filters can be used to process the IF signal. In the case of a low-IF down conversion stage, the wanted signal is positioned only in the positive frequency band between the frequencies $BW_c - BW$ and $BW_c + BW$. It can be shown [16] that the estimated power consumption of a filter at the low IF frequency is proportional to the width of the passband, integrated over both positive and negative frequencies. Therefore, a quadrature bandpass filter will have the lowest power consumption, as compared to a low pass or regular bandpass implementation. On the other hand, since a low IF receiver as shown in Fig. 3 will have only one VGA stage, a power decrease can be expected when this VGA function is covered by extra dynamic range in the AD converter, in case the extra power necessary for the DR increase in the AD converter doesn't supersede that of the VGA stage. This will of course depend on the system specifications.

3.4 $\Delta\Sigma$ implementation with complex bandpass filter

Figure 7 depicts the equivalent schematic of a first order $\Delta\Sigma$ ADC with a complex bandpass filter. K_R,G_R and K_C,G_C are respectively the real and imaginary feedback coefficients and determine the poles and zeros in the noise transfer function. The indices $''R''$ are used for "real" coefficients, e.g. the coefficients which represent the feedback from an I(Q) node to an I(Q) node, whereas the "complex" $''C''$ indices indicate the coefficients of the feedback from an I(Q) node to a Q(I) node. Note that in the latter case the sign of the feedback coefficient is different for the "I-to-

Q" feedback and the "Q-to-I" feedback. The complex noise transfer function of this modulator has a pole at $-(G_r + K_r).gbw - i(G_c + K_c).gbw$ and a zero at $-K_r.gbw - K_c.gbw$, where gbw is the GBW of the integrator. This can be generalized for higher order loop filters

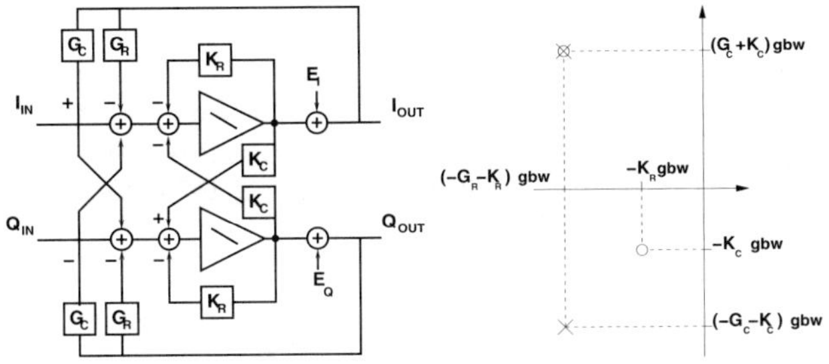

Figure 7: Simplified first order complex ΣΔ (l); poles and zeros plot (r)

3.4.1 ΔΣ architecture

In figure 8 the full ΔΣ architecture is depicted. The loop-filter is a second order complex bandpass filter consisting of two cascaded integrators and two feed-forward OTAs as shown in figure 8. Both blocks of the cascaded structure introduce complex non-conjugate poles at two different frequencies in order to guarantee a broad quantization noise suppression band under varying process conditions. For given specifications, optimal pole positions can be determined. The feed-forward OTAs introduce zeros in the filter frequency characteristics and guarantee the stability of the ΔΣ loop at higher frequencies.

First filterblock

The first filter slice is shown in Fig. 9. The first filter block is implemented as an RC filter and is the most important block in the loop-filter for noise and linearity considerations. The input resistor linearizes the ota transconductance and the degeneration loop gain $gm * Rin$ has to be high enough for the good functionality both in the strength of the virtual ground and the linearization of the transconductance.

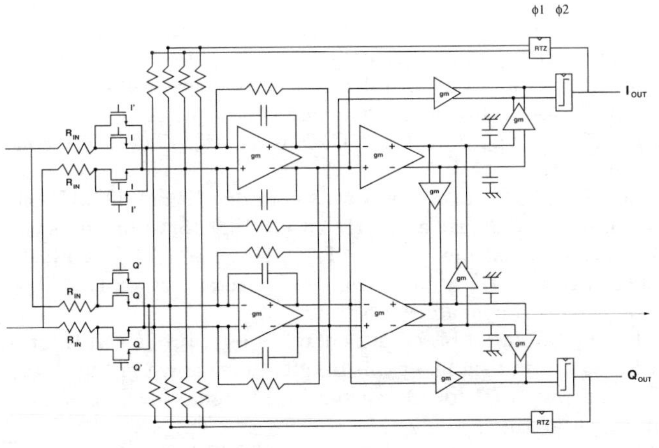

Figure 8: mixer-ΔΣ architecture

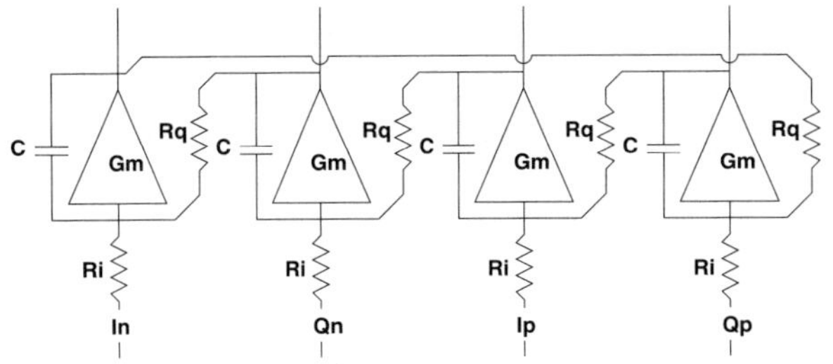

Figure 9: first filter slice

The complex feedback, shifting the integrator characteristic to a complex bandpass filter, is done by the R_Q resistors which are coupled between the outputs of the I(Q) phases and the inputs of the Q(I) phases. The input voltage-to-current conversion is mainly performed by the input degeneration resistor requiring no extra components. The 2kΩ input impedance does not provide a considerable loading on the preceding block. The value of the resistor determines the NF of the entire receiver.

The first OTA is a folded cascode transconductor as shown in Fig. 10. The output common mode voltage is controlled

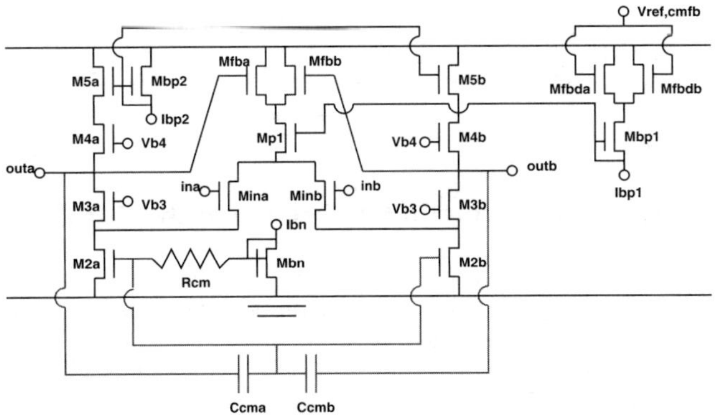

Figure 10: First filter slice ota with CMFB

by both a high frequency feedback loop and a low frequency feedback loop. The first loop is closed by the transistors $M_{fba,b}$ and controls the I_{bp1} biasing. Although reducing the available headroom, this approach is feasible because of the small signal swing level in this receiver AD converter. The GBW of this loop is always smaller than the GBW of the OTA, therefore a second parallel loop with high GBW is needed as well. This loop is closed by the $C_{cma,b}$ capacitors and controls the I_{bn} biasing source. The open loop CMFB transfer function of this loop has a band pass characteristic. The capacitor value is a tradeoff between extra capacitive load on the output and the open loop CMFB gain.

Second filter block

The second filter slice is shown in Fig. 11. The filter is implemented as a gmC filter. The complex feedback is performed by transconductances, which are coupled between I and Q phases. These transconductances have their inputs connected to the outputs of the I(Q) filter stage and their outputs are summed together with the outputs of the Q(I) integrator transconductances.

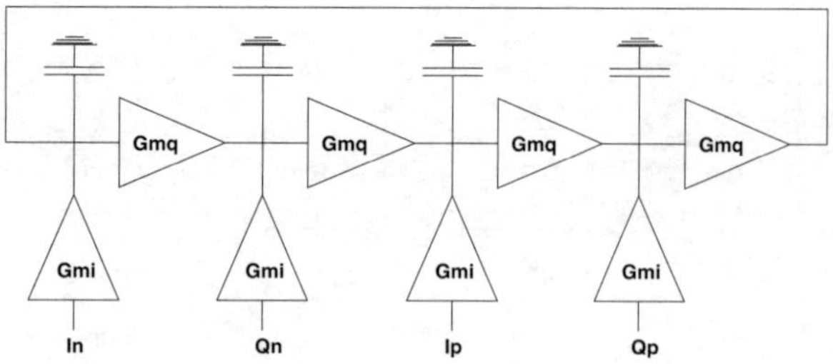

Figure 11: second filter slice

The power requirements for the second filter block are much more relaxed, particularly the noise and linearity specifications are lowered by the gain of the previous block. A simple low power gmC filter implementation with a minor degeneration loop gain is suitable.
The OTAs in the second stage are folded cascode transcon-

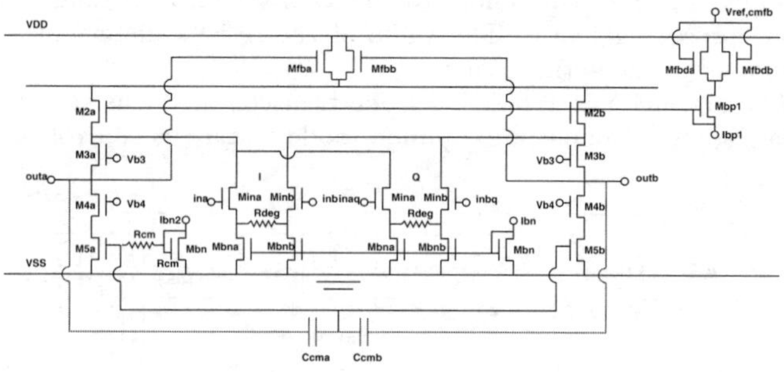

Figure 12: Second filter slice ota with CMFB

ductors, similar to the circuits used in the first stage. This is shown in Fig. 12 where the currents of the degenerated transconductances marked "I" and "Q" are summed.
The CMFB of this stage is implemented in the same way as in the first stage.

Input stage

The input of the ADC is a summing point for both the IF input signal and the DAC return-to-zero [3] feedback pulses. In this continuous time implementation this is done by current summation on that node. Resistors R_{in} placed between the input of the system and the low impedant ADC input provide the necessary V-I conversion of the input signal. The down conversion is performed by the switching transistors in series with the input resistors. The transistors are connected to the virtual ground inputs of the loop filter. The I and Q LO drive signals are square-waves and can be derived from a PLL. The switches have a much lower on-resistance than the value of the input resistor in order to achieve a very linear mixing operation.
The value of this resistor R_{in} determines the NF of the entire receiver $\Delta\Sigma$ architecture. The linearity is determined by the loop gain of the input transconductance degeneration. This implies that for a certain NF, the power drain in the filter ota is proportional to the desired DR.
In receiver architectures, a single-ended input signal is often preferred to a differential one in order to save power. Both single ended and differential signals can be applied to the input of the AD converter. In the latter case, the CMFB has to be good enough to ensure the necessary single-ended-to-differential conversion and to maintain the linearity requirements. At the dummy input node a replica of the previous block (LNA) output has been placed to make the structure loading as symmetrical as possible. Due to this design choice power and area of an extra active balun at RF frequencies are saved.

Comparator

The comparator[15] is shown in Fig. 13. The feed forward blocks of the loopfilter (B1,B2) are indicated on the left. the cascode transistors $M_{cpp,n}$ reduce the comparator kickback effects. The signals at the output of the comparator are sampled in the SR flip flop.

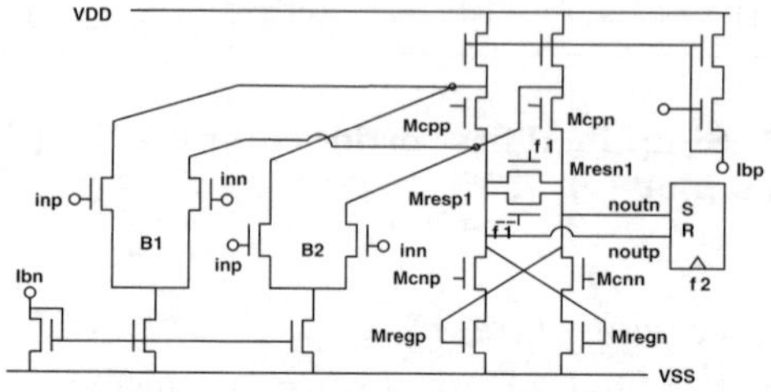

Figure 13: Comparator circuit

4. MEASUREMENTS

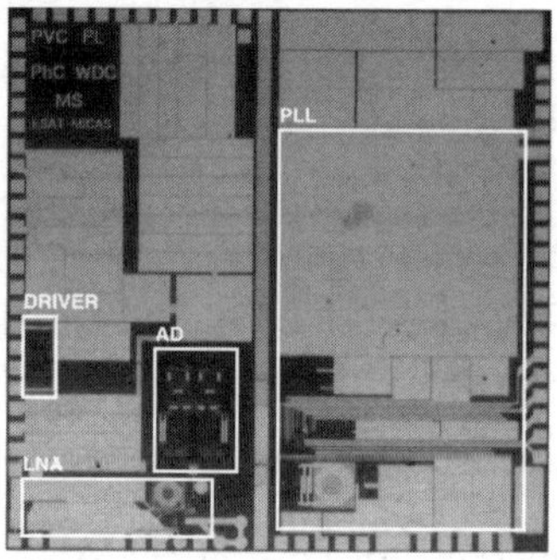

Figure 14: Chip photograph of the receiver

In fig. 14, a photograph of the quadrature modulator receiver with the most important building blocks indicated is shown. It is laid out in a $0.25\mu m$ CMOS process.
The Noise figure vs. frequency of the LNA is shown in Fig. 15(a). The NF at 1.57GHz is 1.5 dB. The LNA consumes 4 mA.
The PLL phase noise is as low as -115 dBc/Hz at 600kHz offset and -138 dBc/Hz at 3MHz offset. Fig. 15(b) presents the PLL phase noise measured at 1.57GHz. The PLL has a locking range of 10% around the center frequency of 1.57GHz. The full PLL consumes 8.5 mA.
The demodulated baseband output spectrum $(I + jQ)$ for a 1.57 GHz input signal at this frequency is shown in Fig. 15(c). The output SNR for increasing input signal levels is shown in fig. Fig. 15(d).
The performance is summarized in Table 1.

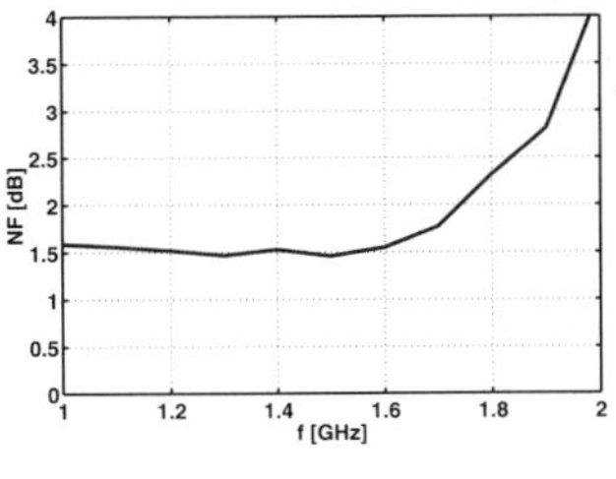

(a) LNA noise figure

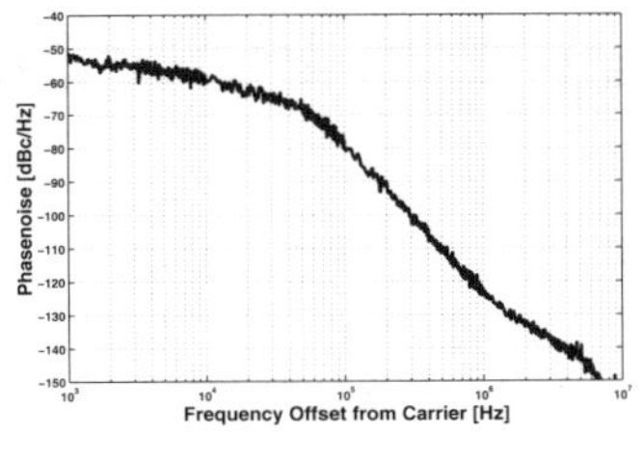

(b) PLL phase noise

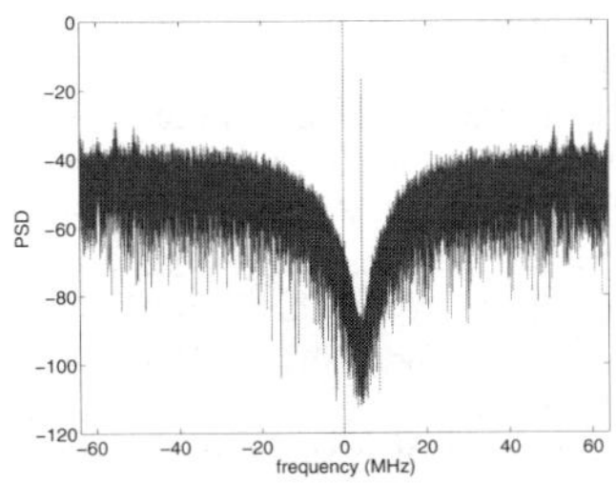

(c) Demodulated baseband output spectrum

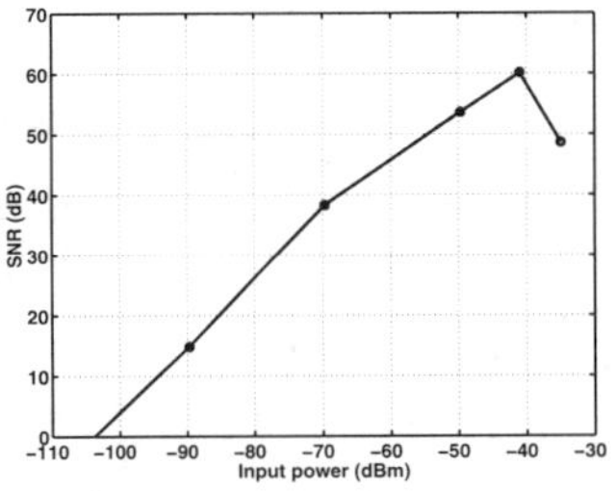

(d) Output SNR vs. input power levels

Figure 15: Measurement results

Signal path	
Input Sensitivity (100 Hz)	-130 dBm
Maximum Signal Level	-68 dBm
Specifications	
NF LNA	1.5 dB
DR	62 dB
IMRR	32 dB
Power dissipation	
LNA	8mW
ADC	14.2mW
PLL/VCO	17mW
LO-buffer	2mW
Implementation	
Die area	16 mm^2
Technology	0.25μm CMOS

Table 1: Summary of measurements results

5. CONCLUSIONS

An overview and comparison of different topologies for wireless architectures has been discussed. The low IF topology, as a good candidate for low power wireless receiver integration has been highlighted. An implementation of a fully integrated GPS receiver has been shown as an example of a wireless low power design in a CMOS technology.

6. ACKNOWLEDGEMENTS

The authors would like to thank Kawasaki Microelectronics Inc. for providing the chip's processing and particularly K. Akeyama and Y. Segawa for their help and support with the layout and design of the circuit.

7. REFERENCES

[1] A. Abidi. Direct-Conversion Radio Transceivers for Digital Communications. *Journal of solid State Circuits*, pages 1399–1410, Dec. 1995.

[2] M. Banu, H. Wang, M. Seidel, M. Tarsia, W. Fischer, J. Glas, A. Dec, and V. Bocuzzi. A bicmos double-low-if receiver for gsm. *Proc. of the 1997 Custom integrated Circuits Conference*, pages 521–524, 1997.

[3] L. J. Breems, E. van der Zwan, E. C. Dijkmans, and H. J. A 1.8mW CMOS $\Sigma\Delta$ modulator with integrated mixer for A/D conversion of IF signals. *Proc. of the 1999 International Solid-State Circuits Conference*, February 1999.

[4] J. Crols and M. Steyaert. *CMOS wireless transceiver design.* Kluwer, 1997.

[5] D. G. Tucker. The history of the homodyne and synchrodyne. *J. British Inst. Radio Engineers*, pages 143–154, 1954.

[6] F. M. Gardner. *Phaselock Techniques.* John Wiley & Sons Ltd., 2 edition, 1979.

[7] http://www.silabs.com/products/aero.asp. Silicon Laboratories Aero Transceiver.

[8] http://www.spirea.com. Spirea AB CMOS RF Products.

[9] http://www.synad.com/Press/mercury5g_release.htm. Synad Mercury 5G.

[10] http://www.valencesemi.com/products.asp. Valence Semiconductor GPS front-end.

[11] P. Leroux, J. Janssens, and M. Steyaert. A 0.8 dB NF ESD-protected 9 mW CMOS LNA. *Proc. of the 2001 International Solid-State Circuits Conference*, February 2001.

[12] B. D. Muer and M. Steyaert. *Fully Integrated CMOS Frequency Synthesizers for Wireless Communications*, pages 287–323. Analog Circuit Design, W. Sansen, J. H. Huijsing, R. J. van de Plassche (eds.) Kluwer Academic Publishers, 2000.

[13] J. Rudell, T. Ou, J. J.and Cho, G. Chien, J. Weldon, et al. A 1.9 GHz Wideband IF Double Conversion CMOS Integration Receiver for Cordless Telephone Applications. *Proc. of the 1997 International Solid-State Circuits Conference*, February 1997.

[14] M. Steyaert, J. Janssens, B. Demuer, M. Borremans, and N. Itoh. A 2V CMOS Cellular Tranceiver Front-End. *Proc. of the 2000 International Solid-State Circuits Conference*, February 2000.

[15] E. van der Zwan and E. Dijkmans. A 0.2-mW CMOS $\Sigma\Delta$ Modulator for Speech Coding with 80 dB Dynamic Range. *IEEE Journal of Solid State Circuits*, 31(12):1873–1880, December 1996.

[16] P. Vancorenland and M. Steyaert. *Continuous time quadrature modulator receivers.* Analog Circuit Design, J. H. Huijsing, A. van Roermund and M. Steyaert (eds.) Kluwer Academic Publishers, 2002.

[17] S. Yantzi, K. Martin, and A. Sedra. Quadrature bandpass delta-sigma modulation for digital radio. *IEEE Journal of Solid State Circuits*, 32(12):1935–1950, December 1997.

TCG-S: Orthogonal Coupling of P*-admissible Representations for General Floorplans *

Jai-Ming Lin[1] and Yao-Wen Chang[2]
[1]Department of Computer and Information Science, National Chiao Tung University, Hsinchu 300, Taiwan
[2]Department of Electrical Engineering & Graduate Institute of Electronics Engineering, National Taiwan University, Taipei 106, Taiwan

Abstract

We extend in this paper the concept of the P-admissible floorplan representation to that of the P*-admissible one. A P*-admissible representation can model the most general floorplans. Each of the currently existing P*-admissible representations, SP, BSG, and TCG, has its strengths as well as weaknesses. We show the equivalence of the two most promising P*-admissible representations, TCG and SP, and integrate TCG with a packing sequence (part of SP) into a new representation, called *TCG-S*. TCG-S combines the advantages of SP and TCG and at the same time eliminates their disadvantages. With the property of SP, faster packing and perturbation schemes are possible. Inherited nice properties from TCG, the geometric relations among modules are transparent to TCG-S (implying faster convergence to a desired solution), placement with position constraints becomes much easier, and incremental update for cost evaluation can be realized. These nice properties make TCG-S a superior representation which exhibits an elegant solution structure to facilitate the search for a desired floorplan/placement. Extensive experiments show that TCG-S results in the best area utilization, wirelength optimization, convergence speed, and stability among existing works and is very flexible in handling placement with special constraints.

Categories and Subject Descriptors

J.6 [COMPUTER-AIDED ENGINEERING]: Computer-aided design (CAD)

General Terms

Algorithms

1 Introduction

As technology advances, the circuit size in modern VLSI design increases dramatically. To handle the increasing design complexity, hierarchical designs and IP modules are widely used to optimize area and timing for design convergence. Further, the need to integrate heterogeneous systems or special modules imposes some placement constraints, e.g., the boundary-module constraint which requires some modules to be placed along the chip boundaries for shorter connections to pads, the preplaced-module constraint which pre-assigns modules to specific positions, etc. These trends make floorplanning/placement much more important than ever, and it is of particular significance to consider the floorplanning/placement with various constraints. To cope with these challenges, it is desired to develop an efficient and effective floorplan representation that can model the geometric relations among regular as well as constrained modules.

Many floorplan representations have been proposed in the literature, e.g., slicing tree [13], normalized Polish expression (NPE) [18], Sequence Pair (SP) [10], Bounded-Sliceline Grid (BSG) [12], O-tree [3], B*-tree [1], Corner Block List (CBL) [4], Transitive Closure Graph (TCG) [8], and Q-sequence [15]. Unlike the traditional classification of the slicing and non-slicing structures, we can alternatively classify them into two categories, *P*-admissible* and *non-P*-admissible* representations. A representation is said to be P-admissible if it satisfies the following four conditions [10]: (1) the solution space is finite, (2) every solution is feasible, (3) packing and cost evaluation can be performed in polynomial time, and (4) the best evaluated packing in the space corresponds to an optimal placement. We extend in this paper the concept of the P-admissible representation to that of the P*-admissible one by adding the fifth condition: (5) the geometric relation between each pair of modules is defined in the representation. With this condition, general floorplans/placements can be modeled. Therefore, a P*-admissible representation contains a complete structure for searching for an optimal floorplan/placement solution. Therefore, it is desirable to develop an effective and flexible P*-admissible representation.

Among the existing popular representations, SP, BSG, and TCG are P*-admissible while slicing tree, NPE, O-tree, B*-tree, CBL, and Q-sequence are not. The slicing tree and NPE are intended for slicing floorplans only. Since an optimal placement could be a non-slicing structure, the two representations are not P-admissible and thus not P*-admissible (i.e., violation of P*-admissibility Condition (4)). An O-tree defines only one-dimensional geometrical relation between *compacted* modules and thus can obtain the relation in the other dimension only after packing (i.e., violation of Condition (5)). (Note that O-tree is undefined for some uncompacted placements which may correspond to the best solutions for wirelength optimization. Therefore, as far as wirelength optimization is concerned, O-tree is not even P-admissible since Condition (4) is violated.) A B*-tree requires a placement to be left and/or bottom compacted. However, the space intended for placing a module may be occupied by previously placed modules during packing, resulting in a mismatch between the original representation and its *compacted* placement. Therefore, it may not be feasible to find a *compacted* placement corresponding to the original B*-tree, and thus it is not P-admissible (i.e., violation of Condition (2)). CBL and Q-sequence can represent only *mosaic* floorplans, in which each region in the floorplan contains exactly one module. CBL and Q-sequence are not P-admissible because it cannot guarantee a feasible solution after a perturbation (i.e., violation of Conditions (2) and (4)). Non-P*-admissible representations intrinsically have a smaller solution space and lower packing cost since their corresponding floorplanning/placement structures are more restricted (e.g., slicing structures [slicing tree, NPE], mosaic floorplans [CBL, Q-sequence], compacted placements [O-tree, B*-tree], etc).[1] However, lack of the definition on the geometric relation between each pair of modules, the guarantee in the feasibility, and/or the optimality of their representations would inevitably complicate floorplan sizing and placement with constraints, lead to longer running times, and/or lower solution quality.

The existing P*-admissible representations, SP, BSG, and TCG, have their own distinct properties as well as common ones. Nevertheless, researchers tend to favor SP over BSG because BSG incurs many redundancies and thus a much larger solution space, implying a longer running time to search for a good solution. Therefore, we shall focus on SP and TCG. Both SP and TCG are considered very flexible representations and construct constraint graphs to evaluate their packing cost. SP consists of two sequences of modules (Γ_+, Γ_-), where Γ_+ (Γ_-) specifies the module ordering from top-left (bottom-left) to bottom-right (top-right). Hence, Γ_- corresponds to the ordering for packing modules to the bottom-left direction and thus can be used to guide module packing.

*This work was partially supported by the National Science Council of Taiwan ROC under Grant No. NSC-89-2215-E-009-117. E-mails: gis87808@cis.nctu.edu.tw; ywchang@cc.ee.ntu.edu.tw.

[1]Generality from the most general to the least general: general floorplan $\succ$ compacted floorplan $\succ$ mosaic floorplan $\succ$ slicing floorplan.

However, like most existing representations (e.g., NPE, BSG, O-tree, B*-tree, CBL, Q-sequence), the geometric relations between modules are *not transparent* to the operations of SP (i.e., the effect of an operation on the change of module relation is not clear before packing), and thus we need to construct constraint graphs from scratch after each perturbation to evaluate the packing cost; this deficiency makes SP harder to converge to a desired solution and to handle placement with constraints (e.g., boundary modules, pre-placed modules, etc).

TCG consists of a horizontal transitive closure graph C_h to define the horizontal geometric relations between modules and a vertical one C_v for vertical geometric relations. Contrast to SP, the geometric relations between modules are transparent to TCG as well as its operations, facilitating the convergence to a desired solution. Further, TCG supports incremental update during operations and keeps the information of boundary modules as well as the shapes and the relative positions of modules in the representation. Nevertheless, like SP, constraint graphs are also needed for TCG to evaluate its packing cost, and unlike SP, we need to perform extra operations to obtain the module packing sequence.

Therefore, an interesting question arises: Is it possible to develop a representation that can combine the advantages of SP and TCG and at the same time eliminate their disadvantages? We answer this question in affirmation by showing the equivalence of TCG and SP, and integrating them into TCG-S = (C_h, C_v, Γ_-). The orthogonal combination leads to a representation with at least the following advantages:

- With the property of SP, faster $O(m \lg m)$-time packing and $O(m)$-time perturbation schemes are possible for a P*-admissible representation, where m is the number of modules. (Note that a linear-time packing scheme is possible for the tree-based representations, but they can represent only more restricted compacted floorplans.)
- Inherited from TCG, the geometric relations among modules are transparent to TCG-S, implying faster convergence to a desired solution.
- Inherited from TCG, placement with position constraints becomes much easier.
- TCG-S can support incremental update for cost evaluation.

These nice properties make TCG-S an effective, efficient, and flexible representation. Extensive experiments based on a set of commonly used MCNC benchmarks show that TCG-S results in the best area utilization, wirelength optimization, convergence speed, and solution stability among existing works.

To show the flexibility of TCG-S, we also consider placement with preplaced and boundary modules. For placement with pre-placed modules, Murata et al. [11] proposed an *adaptation algorithm* to transform an infeasible SP with preplaced modules into a feasible one. However, the process incurs expensive computations. For placement with the boundary-module constraint, Tang and Wong in [16] and Ma et al. in [9] handled this problem using SP and CBL, respectively. However, they cannot guarantee a feasible solution in each perturbation and their final placements. Lai et al. in [6] gave the feasibility conditions for SP with boundary modules and transformed an infeasible solution into a feasible one. However, the method is very complex, and many rules are needed to cope with the constraints .

We present in this paper the methods for handling placement with preplaced and boundary modules. Different from the previous works on boundary modules that cannot guarantee a feasible solution or need to transform an infeasible solution into a feasible one, TCG-S can easily maintain the feasibility during each perturbation. We compared our work with [6] on placement with boundary modules. (Note that there are no common benchmark circuits for this constraint.) Experimental results show that TCG-S results in smaller areas than [6].

The remainder of this paper is organized as follows. Section 2 formulates the floorplan/placement design problem. Section 3 compares SP and TCG. Section 4 presents the procedures to build the TCG-S from a placement and construct the placement from a TCG-S. Section 5 gives the operations to perturb a TCG-S. Section 6 presents our methods to handle placement with boundary and preplaced modules. Experimental results are reported in Section 7. Finally, we give concluding remarks in Section 8.

2 Problem Definition

Let $B = \{b_1, b_2, ..., b_m\}$ be a set of m rectangular modules whose width, height, and area are denoted by W_i, H_i, and A_i, $1 \leq i \leq m$. Let (x_i, y_i) $((x'_i, y'_i))$ denote coordinate of the bottom-left (top-right) corner of module b_i, $1 \leq i \leq m$, on a chip. A placement $\mathcal{P}$ is an assignment of (x_i, y_i) for each b_i, $1 \leq i \leq m$, such that no two modules overlap. The goal of floorplanning/placement is to optimize a predefined cost metric such as a combination of the area (i.e., the minimum bounding rectangle of $\mathcal{P}$) and/or the wirelength (i.e., the summation of half bounding box of interconnections) induced by the assignment of b_i's on the chip.

3 P*-admissible Representations

In this section, we first review the two P*-admissible representations, TCG and SP, then show their equivalence, and compare their properties.

3.1 Review of TCG and SP

TCG describes the geometric relations between modules based on two graphs, namely a *horizontal transitive closure graph* C_h and a *vertical transitive closure graph* C_v, in which a node n_i represents a module b_i and an edge (n_i, n_j) in C_h (C_v) denotes that module b_i is left to (below) module b_j. TCG has the following three *feasibility properties* [8]:

1. C_h and C_v are acyclic.
2. Each pair of nodes must be connected by exactly one edge either in C_h or in C_v.
3. The transitive closure of C_h (C_v) is equal to C_h (C_v) itself.[2]

Figure 1(a) shows a placement with seven modules a, b, c, d, e, f, and g whose widths and heights are (3.5, 1.5), (2, 2.5), (2, 3.5), (3, 2), (1.5, 1.5), (5, 1.5), and (1, 2), respectively. Figure 1(c) shows the TCG $= (C_h, C_v)$ corresponding to the placement of Figure 1(a). The value associated with a node in C_h (C_v) gives the width (height) of the corresponding module, and the edge (n_i, n_j) in C_h (C_v) denotes the horizontal (vertical) relation of b_i and b_j. Since there exists an edge (n_a, n_g) in C_h, module b_a is left to b_g. Similarly, b_a is below b_b since there exists an edge (n_a, n_b) in C_v.

SP uses a pair of sequences (Γ_+, Γ_-) to represent a floorplan/placement, where Γ_+ and Γ_- give two permutations of module names. The geometric relation of modules can be derived from an SP as follows. Module b_a is left (right) to module b_b if a appears before (after) b in both Γ_+ and Γ_-. Module b_a is below (above) module b_b if b appears before (after) a in Γ_+ and a appears before (after) b in Γ_-. Figure 1(b) shows the corresponding SP. Since a is before g in both Γ_+ and Γ_-, module b_a is left to module b_g. Similarly, b_a is below b_b since a is after b in Γ_+ and before b in Γ_-.

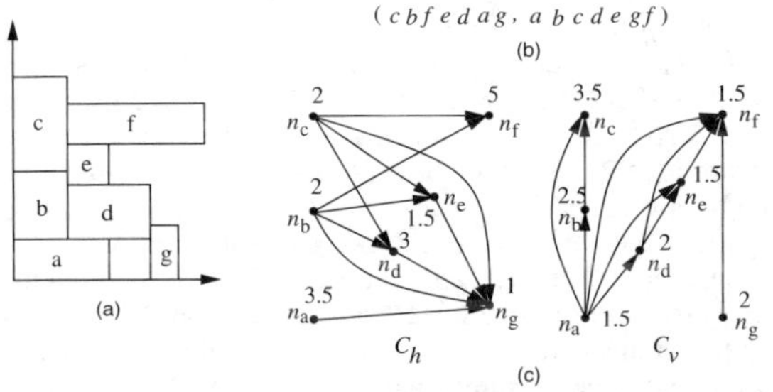

Figure 1: (a) A placement. (b) The corresponding SP of (a). (c) The corresponding TCG of (a).

3.2 Equivalence of SP and TCG

Like the relations between a skewed slicing tree [13] and an NPE [18] for slicing floorplans as well as O-tree and B*-tree for non-slicing floorplans, TCG and SP are equivalent.

We can transform between TCG and SP as follows: Let the *fan-in* (*fan-out*) of a node n_i, denoted by $F_{in}(n_i)$ $(F_{out}(n_i))$, be the nodes n_j's with edges (n_j, n_i) $((n_i, n_j))$. Given a TCG, we can obtain a sequence Γ_+ by repeatedly extracting a node n_i with $F_{in}(n_i) = \emptyset$ in C_h and $F_{out}(n_i) = \emptyset$ in C_v, and then deleting the edges (n_i, n_j)'s

[2] The transitive closure of a directed acyclic graph G is defined as the graph $G' = (V, E')$, where $E' = \{(n_i, n_j)$: there is a path from node n_i to node n_j in $G\}$.

$((n_j, n_i)$'s) from C_h (C_v) until no node is left in C_h (C_v). Similarly, we can transform a TCG into another sequence Γ_- by repeatedly extracting the node n_i with $F_{in}(n_i) = \emptyset$ both in C_v and C_h, and then deleting the edges (n_i, n_j)'s from both C_h and C_v until no node is left in C_h and C_v. Given an SP = (Γ_+, Γ_-), we can obtain a unique TCG = (C_h, C_v) from the two constraint graphs of the SP by removing the source, sink, and associated edges. For example, the SP of Figures 1(b) is equivalent to the TCG of Figures 1(c).

3.3 Comparison between TCG and SP

Although TCG and SP are equivalent, their properties and induced operations are significantly different. (Similar situations are also with skewed slicing trees/NPE's and O-trees/B*-trees.) Both SP and TCG are considered very flexible representations and construct constraint graphs to evaluate their packing cost. Γ_- of an SP corresponds to the ordering for packing modules to the bottom-left direction and thus can be used for guiding module packing. However, like most existing representations, the geometric relations among modules are not transparent to the operations of SP (i.e., the effect of an operation on the change of module relation is not clear before packing), and thus we need to construct constraint graphs from scratch after each perturbation to evaluate the packing cost; this deficiency makes SP harder to converge to a desired solution and to handle placement with constraints (e.g., boundary modules, pre-placed modules, etc).

Contrast to SP, the geometric relations among modules are transparent to TCG as well as its operations, facilitating the convergence to a desired solution. Further, TCG supports incremental update during operations and keeps the information of boundary modules as well as the shapes and the relative positions of modules in the representation. Unlike SP, nevertheless, we need to perform extra operations to obtain the module packing sequence and an additional $O(m^2)$ time to find a special type of edges, called *reduction edges*, in C_h (C_v) for some operations. (We will define the edges later.)

For both SP and TCG, the packing scheme by applying the longest path algorithm is time-consuming since all edges in the constraint graphs are processed, even though they are not on the longest path. As shown in C_h of Figure 1(c), if we add a source with zero weight and connect it to those nodes with zero in-degree, the x coordinate of each module can be obtained by applying the longest path algorithm on the resulting directed acyclic graph. Therefore, we have $x_g = \max\{x'_a, x'_b, x'_c, x'_d, x'_e\}$. However, if we place modules based on the sequence Γ_- and maintain a horizontal and a vertical contours, denoted by R_h and R_v respectively, for the placed modules, the number of nodes need to be considered can be reduced. Let R_h (R_v) be a list of modules b_i's for which there exists no module b_j with $y_j \geq y'_i$ ($x_j \geq x'_i$) and $x'_j \geq x'_i$ ($y'_j \geq y'_i$). For the placement of Figure 1(a), for example, $R_h = < b_c, b_f >$ and $R_v = < b_g, b_d, b_e, b_f, b_c >$. Suppose we have packed the modules b_a, b_b, b_c, b_d, and b_e based on the sequence Γ_-. Then, the resulting horizontal contour $R_h = < b_c, b_e, b_d >$. Keeping R_h, we only need to traverse the contour from b_e, the successor of b_e, to the last module b_d, which have horizontal relations with b_g (since there are edges (n_e, n_g) and (n_d, n_g) in C_h). Thus, we have $x_g = x'_d$. Packing modules in this way, we only need to consider x_e and x_d, and can get rid of the computation for a maximum value, leading to a faster packing scheme. We will show later how to apply a balanced binary tree to implement the contour operation to get a loglinear-time packing scheme.

4 The TCG-S Representation

Combining TCG = (C_h, C_v) and SP = (Γ_+, Γ_-), we develop a representation, called TCG-S = (C_h, C_v, Γ_-), which uses a horizontal and a vertical transitive closure graphs as well as a packing sequence Γ_- to represent a placement. In this section, we first introduce how to construct Γ_-, C_h, and C_v from a placement. Then, we propose an $O(m \lg m)$-time packing scheme for TCG-S, where m is the number of modules.

4.1 From a placement to TCG-S

In this subsection, we first extract Γ_- from a placement, and then construct C_h and C_v according to Γ_-.

For two non-overlapped modules b_i and b_j, b_i is said to be *horizontally (vertically) related* to b_j, denoted by $b_i \vdash b_j$ ($b_i \perp b_j$), if b_i is left to (below) b_j and their projections on the y (x) axis overlap. For two non-overlapped modules b_i and b_j, b_i is said to be *diagonally related* to b_j if b_i is left to b_j and their projections on the x and the y axes do not overlap. To simplify the operations on geometric relations, a diagonal relation for modules b_i and b_j is treated as a horizontal one unless there exists a chain of vertical relations from b_i (b_j), followed by the modules overlapped with the rectangle defined by the two closest corners of b_i and b_j, and finally to b_j (b_i), for which it is considered as $b_i \perp b_j$ ($b_j \perp b_i$).

Given a placement, Γ_- can be extracted based on the procedure described in [10]. For example, the Γ_- for the placement of Figure 2(a) is $< abcdegf >$. After extracting Γ_-, we can construct C_h and C_v based on Γ_-. For each module b_i in Γ_-, we introduce a node n_i with the weight being b_i's width (height) in C_h (C_v). Also, for each module b_i before b_j in Γ_-, we introduce an edge (n_i, n_j) in C_h (C_v) if $b_i \vdash b_j$ ($b_i \perp b_j$). As shown in Figures 2(b), for the first two modules b_a, b_b in Γ_-, we introduce the nodes n_a and n_b in C_h (C_v) and assign the weights as their widths (heights). Also, we construct a directed edge (n_a, n_b) in C_v since module b_a is before b_b and $b_a \perp b_b$. The process repeats for all modules in Γ_-, resulting in the TCG-S shown in Figure 2(b). We have the following theorem.

Theorem 1 *There exists a TCG-S corresponding to a placement.*

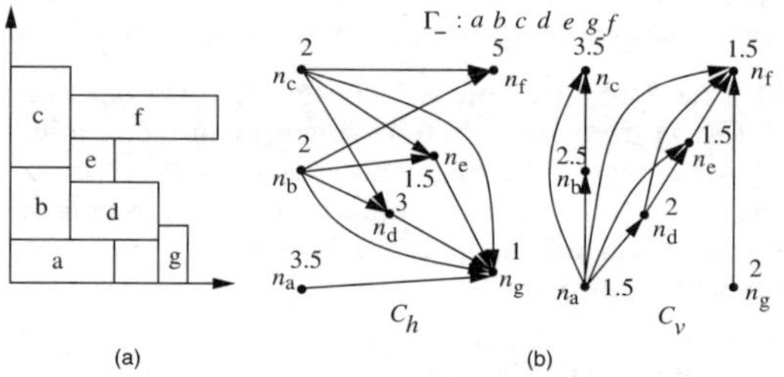

Figure 2: (a) A placement. (b) The corresponding TCG-S of (a).

4.2 From TCG-S to a placement

In this subsection, we propose an $O(m \lg m)$-time packing scheme based on Γ_- as well as a horizontal and a vertical contours R_h and R_v, where m is the number of modules. The basic idea is to process the modules based on the sequence defined in Γ_-, and then pack the current module to a corner formed by two previously placed modules in R_h (R_v) according to the geometric relations defined in C_h and C_v.

We detail the packing scheme as follows. Recall that R_h (R_v) is a list of modules b_i's for which there exists no module b_j with $y_j \geq y'_i$ ($x_j \geq x'_i$) and $x'_j \geq x'_i$ ($y'_j \geq y'_i$). (R_h (R_v) consists of modules along the top (left) boundary of a placement.) We can keep the modules b_i's in R_h (R_v) in a balanced binary search tree (e.g., the red-black tree [2]) T_h (T_v) in the increasing order according to their right (top) boundaries. For easier presentation, we add a boundary *dummy module* b_s (b_t) to R_h (R_v) to denote the left (bottom) boundary module of a placement. We have $b_s \vdash b_i$ and $b_t \perp b_i$, $\forall b_i$. Let $(x'_s, y'_s) = (0, \infty)$ and $(x'_t, y'_t) = (\infty, 0)$. R_h (R_v) consists of b_s (b_t) initially, and so does the corresponding T_h (T_v). To pack a module b_j in Γ_-, we traverse the modules b_k's in T_h (T_v) from its root, and go to the right child if $b_k \vdash b_j$ ($b_k \perp b_j$) and the left child if $b_k \perp b_j$ ($b_k \vdash b_j$). The process is repeated for the newly encountered module until a leaf node is met. Then, b_j is connected to the leaf node, and $x_j = x'_p$ ($y_j = y'_p$), where b_p is the last module with $b_p \vdash b_j$ ($b_p \perp b_j$) in the path. After b_j is inserted into T_h (T_v), every successor b_l with $x'_l \leq x'_j$ ($y'_l \leq y'_j$) in T_h (T_v) is deleted since b_l is no longer in the contour. (Note that the ordering of nodes in T_h (T_v) can be obtained by depth-first search.) This process repeats for all modules in Γ_-. We have $W = x'_v$ ($H = y'_v$) if b_v is the module in the resulting T_h (T_v) with the largest value, where W (H) denotes the width (height) of the placement. We have the following theorems and lemmas.

Theorem 2 *There exists a unique placement corresponding to a TCG-S.*

Lemma 1 *For each module b_i in Γ_-, b_i must be placed adjacent to the right (top) boundary of some module b_j in R_h (R_v) during the packing.*

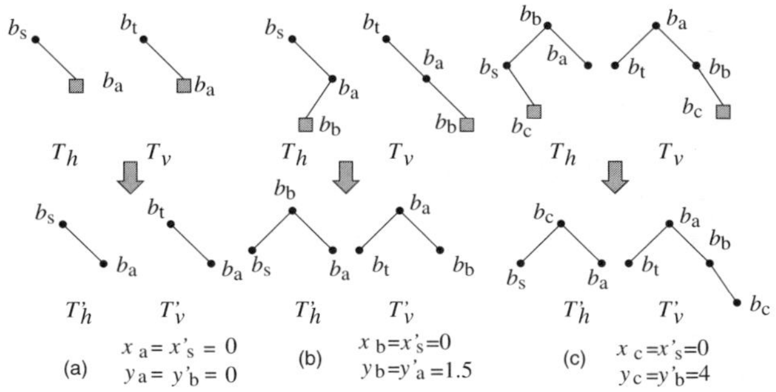

Figure 3: The packing scheme for the TCG-S of Figure 2(b).

Lemma 2 *Given a module b_i in Γ_- to be placed, if $b_j \in R_h(R_v)$, $b_j \perp b_i$ ($b_j \vdash b_i$), and b_j has the largest x'_j (y'_j), any module $b_k \in R_h(R_v)$ with $x'_k > x'_j$ ($y'_k > y'_j$)) cannot have the relation $b_k \vdash b_i$ ($b_k \perp b_i$).*

Theorem 3 *The proposed scheme correctly packs all modules in $O(m \lg m)$ time, where m is the number of modules.*

Figures 3 shows the packing scheme for the TCG-S of Figure 2(b). T_h (T_v) consists of b_s (b_t) initially. To pack the first module b_a in Γ_-, we traverse T_h (T_v) from the root b_s (b_t) and insert it to the right child of b_s (b_t) since $b_s \vdash b_a$ ($b_t \perp b_a$). Therefore, the first module b_a in Γ_- is placed at the bottom-left corner (i.e., $(x_a, y_a) = (0, 0)$) since b_s (b_t) is the last module that is horizontally (vertically) related to b_a and $x'_s = 0$ ($y'_t = 0$). (Note that T'_h (T'_v) in Figures 3(a) denotes a balanced binary search tree after b_a is inserted into T_h (T_v).) Similarly, to pack the second module b_b in Γ_-, we traverse T_h from the root b_s and then its right child since $b_s \vdash b_b$. Then, b_b is inserted to the left child of b_a since $b_s \perp b_a$. Because b_s is the last module with $b_s \vdash b_b$ in the path, $x_b = x'_s = 0$. Similarly, we traverse T_v from the root b_t and then its right child b_a since $b_a \perp b_t$. Then, b_b is inserted to the right child of b_a in T_v since $b_a \perp b_b$. Therefore, $y_b = y'_a = 1.5$ because b_a is the last module with $b_a \perp b_b$ in the path. The resulting balanced binary search trees after performing tree rotations T'_h, T'_v are shown in Figure 3(b) (see [2] for the rotation operations for keeping a tree balanced). As shown in Figure 3(c), after b_c is inserted, b_b in T_h is deleted since b_b is the successor of b_c and $x'_b \leq x'_c$ (i.e., b_b is no longer in the contour). The resulting T'_h and T'_v are shown in Figure 3(c). The process is repeated for all modules in Γ_-.

According to this packing scheme, if the coordinate of a module b_i in Γ_- is changed, we only need to recompute the coordinates of modules after b_i in Γ_- since the coordinates of modules before b_i do not change.

5 Floorplanning Algorithm

We develop a simulated annealing based algorithm [5] by using TCG-S for general floorplan design. Given an initial TCG-S, the algorithm perturbs the TCG-S into a new TCG-S to find a desired solution. To ensure the correctness of the new C_h and C_v, they must satisfy the three feasibility conditions given in Section 3.1. To identify feasible C_h and C_v for perturbation, we describe the concept of *reduction edges* in the following subsection.

5.1 Reduction Edge

An edge (n_i, n_j) is said to be a *reduction edge* if there does not exist another path from n_i to n_j, except the edge (n_i, n_j) itself; otherwise, it is a *closure edge*, for some operations. In Figure 2(b), for example, edges (n_a, n_g), (n_d, n_g), and (n_e, n_g) are reduction edges while (n_b, n_g) and (n_c, n_g) are closure ones. With Γ_-, we can find a *set* of reduction edges in $O(m)$ time (where m is the number of modules), a significant improvement from $O(m^2)$ time using TCG alone [8].

Given an arbitrary node n_i in a transitive closure graph C_h (C_v), we can find all the nodes n_j's that form reduction edges (n_i, n_j)'s using a *Linear Scan* method as follows. First, we extract from Γ_- those nodes n_j's in $F_{out}(n_i)$ of C_h (C_v) and keep their original ordering in Γ_-. Let the resulting sequence be S. The first node n_k in S and n_i must form a reduction edge (n_i, n_k). Then, we continue to traverse S until a node n_l with (n_k, n_l) *not* in C_h (C_v) is encountered. (n_i, n_l) must also be a reduction edge. Starting from n_l, we continue the same process until no node is left in S.

As an example shown in C_h of Figure 2(b), we are to extract all reduction edges emanating from n_c. We first find $S = < n_d, n_e, n_g, n_f >$ by extracting nodes in $F_{out}(n_c)$ based on the sequence in Γ_-. n_c and the first node n_d in S form a reduction edge (n_c, n_d). Traversing S, we have another reduction edge (n_c, n_e) since edge (n_d, n_e) is not in C_h. Starting from n_e, we search the next node n with (n_e, n) not in C_h. We find node n_f, implying that (n_c, n_f) is also a reduction edge. Therefore, we have found all reduction edges emanating from n_c: (n_c, n_d), (n_c, n_e), and (n_c, n_f). (Note that (n_c, n_g) is not a reduction edge because we found (n_e, n_g) in C_h during the processing.)

Theorem 4 *Given a node n_i in C_h (C_v), the Linear Scan method finds all reduction edges emanating from n_i in $O(m)$ time, where m is the number of modules.*

5.2 Solution Perturbation

We extend the four operations *Rotation*, *Swap*, *Reverse*, and *Move* presented in [8] to perturb C_h and C_v. During each perturbation, we must maintain the three feasibility properties for C_h and C_v. Unlike the Rotation operation, Swap, Reverse, and Move may change the configurations of C_h and C_v and thus their properties. Further, we also need to maintain Γ_- to conform to the topological ordering of the new C_h and C_v.

5.2.1 Rotation

To Rotate a module b_i, we exchange the weights of the corresponding node n_i in C_h and C_v. Since the configurations of C_h and C_v do not change, so dose Γ_-. Figure 4(a) shows the resulting TCG-S after rotating the module g shown in Figure 2(b). Notice that the new Γ_- is the same as that in Figure 4(a).

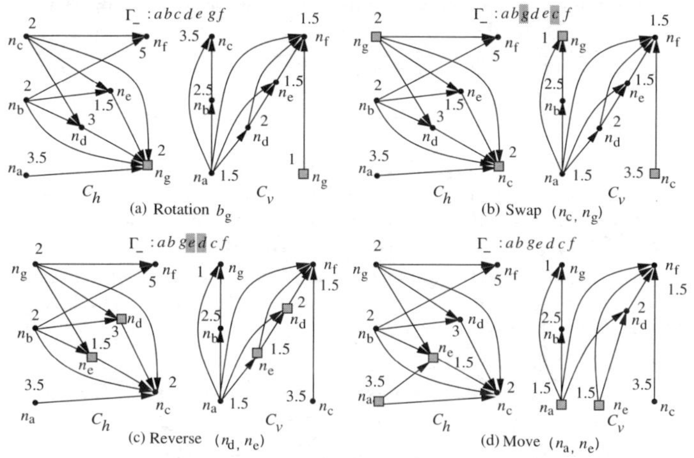

Figure 4: Four types of perturbation. (a) The resulting TCG-S after rotating the module b_g shown in Figure 2(b). (b) The resulting TCG-S after swapping the nodes n_c and n_g shown in (a). (c) The resulting TCG-S after reversing the reduction edge (n_d, n_e) shown in (b). (d) The resulting TCG-S after moving the reduction edge (n_a, n_e) from the C_v of (c) to C_h.

5.2.2 Swap

Swapping n_i and n_j does not change the topologies of C_h and C_v, except that nodes n_i and n_j in both C_h and C_v are exchanged. Therefore, we only need to exchange b_i and b_j in Γ_-. Figure 4(b) shows the resulting TCG-S after swapping the nodes n_c and n_g shown in Figure 4(a). Notice that the modules b_c and b_g in Γ_- in Figure 4(b) are exchanged.

5.2.3 Reverse

Reverse changes the geometric relation between b_i and b_j from $b_i \vdash b_j$ ($b_i \perp b_j$) to $b_j \vdash b_i$ ($b_j \perp b_i$). To reverse a reduction edge (n_i, n_j) in one transitive closure graph, we first delete the edge (n_i, n_j) from the graph, and then add the edge (n_j, n_i) to the graph. To keep C_h and C_v feasible, for each node $n_k \in F_{in}(n_j) \cup \{n_j\}$ and $n_l \in F_{out}(n_i) \cup \{n_i\}$ in the new graph, we have to keep the edge (n_k, n_l) in the new graph.

If the edge does not exist in the graph, we add the edge to the graph and delete the corresponding edge (n_k, n_l) (or (n_l, n_k)) in the other graph. To make Γ_- conform to the topological ordering of the new C_h and C_v, we delete b_i from Γ_- and insert b_i after b_j. For each module b_k between b_i and b_j in Γ_-, we shall check whether the edge (n_i, n_k) exists in the same graph. We do nothing if the edge (n_i, n_k) does not exist in the same graph; otherwise, we delete b_k from Γ_- and insert it after the most recently inserted module.

Figure 4(c) shows the resulting TCG-S after reversing the *reduction edge* (n_d, n_e) of the C_v in Figure 4(b). Since there exists no module between b_d and b_e in Γ_-, we only need to delete b_d from Γ_- and insert it after b_e, and the resulting Γ_- is shown in Figure 4(c).

5.2.4 Move

Move changes the geometric relation between b_i and b_j from $b_i \vdash b_j$ ($b_i \perp b_j$) to $b_i \perp b_j$ ($b_i \vdash b_j$). To move a reduction edge (n_i, n_j) from a transitive closure graph G to the other G', we delete the edge from G and then add it to G'. Similar to Reverse, for each node $n_k \in F_{in}(n_i) \cup \{n_i\}$ and $n_l \in F_{out}(n_j) \cup \{n_j\}$ in G', we must move the edge (n_k, n_l) to G' if the the corresponding edge (n_k, n_l) (or (n_l, n_k)) is in G. Since the operation changes only the edges in C_h or C_v but not the topological ordering among nodes, Γ_- remains unchanged.

Figure 4(d) shows the resulting TCG-S after moving the *reduction edge* (n_a, n_e) from C_v to C_h in Figure 4(c). Notice that the resulting Γ_- is the same as that in Figure 4(c).

For the above operations, Rotation and Move do not change the topological ordering of Γ_- while Swap and Reverse need respective $O(1)$ and $O(m)$ times to maintain the topological ordering of Γ_-, where m is the number of modules. We have the following theorem.

Theorem 5 *TCG-S is closed under the Rotation, Swap, Reverse, and Move operations.*

In particular, it suffices to apply the four perturbations to explore the whole solution space. Extending the similar work by [17] for TCG, we have the following theorem.

Theorem 6 *Given two arbitrary TCG-S's S_1 and S_2, we can obtain S_2 from S_1 by applying a finite number of Rotation, Swap, Reverse, and Move operations, and vice versa.*

6 Placement with Constraints

In this section, we demonstrate the flexibility of TCG-S by extending it to handle placement with boundary and pre-placed modules.

6.1 TCG-S with Boundary Modules

The placement with boundary constraints is to place a set of prespecified modules along the designated boundaries of a chip, which can be formulated as follows:

Definition 1 Boundary Constraint: *Given a boundary module b, it must be placed in one of the four sides: on the left, on the right, at the bottom or at the top in a chip in the final packing.*

TCG-S keeps the following properties that make placement with boundary constraints much easier than other representations.

Theorem 7 *If a module b_i is placed along the left (right) boundary, the in-degree (out-degree) of the node n_i in C_h equals zero. If a module b_i is placed along the bottom (top) boundary, the in-degree (out-degree) of node n_i in C_v equals zero.*

For each perturbation, we can guarantee a feasible placement by checking whether the conditions of boundary modules are satisfied. We discuss the modifications for the four perturbation operations as follows.

6.1.1 Rotation

Since Rotation does not change module location, the operation remains the same as before.

6.1.2 Swap

We can swap two nodes n_a and n_b if

1. b_a and b_b are not boundary modules,
2. b_a and b_b are boundary modules of the same type, or
3. b_a is a boundary module and b_b is not, and n_b satisfies the boundary constraint of b_a.

6.1.3 Reverse

If b_a is a left boundary module or b_b is a right boundary module, then the reduction edge (n_a, n_b) in C_h cannot be reversed. Similarly, we cannot reverse the reduction edge (n_a, n_b) in C_v if b_a is a bottom boundary module or b_b is a top boundary module

6.1.4 Move

If b_a is a top boundary module or b_b is a bottom boundary module, we cannot move the reduction edge (n_a, n_b) from C_h to C_v. Similarly, we cannot move the reduction edge (n_a, n_b) from C_v to C_h if b_a is a right boundary module or b_b is a left boundary module

We have the following theorem.

Theorem 8 *TCG-S is closed under the Rotation, Swap, Reverse, and Move operations with boundary constraints.*

6.2 TCG-S with Pre-placed Modules

The placement with pre-placed modules is to place a set of prespecified modules at the designated locations of a chip, which can be formulated as follows:

Definition 2 Pre-placed Constraint: *Given a module b_i with a fixed coordinate (x_i, y_i) and an orientation, b_i must be placed at the designated location with the same orientation in the final packing.*

Whether a pre-placed module is packed at a correct location is not known until packing. Also, changing the coordinate of a module b_i may affect the packing for other modules after b_i in Γ_-. Therefore, we may need to modify a TCG-S to guarantee a feasible placement with the pre-placed constraint after each perturbation.

Given a TCG-S, modules are packed one by one based on the sequence of Γ_-. A module b_i *interacts with* another module b_j if (1) b_i overlaps b_j, (2) $b_j \vdash b_i$ and their projections on the y axis overlap, or (3) $b_j \perp b_i$ and their projections on the x axis overlap. If b_i interacts with a pre-placed module b_j and b_j was not placed, n_i and n_j are swapped in the TCG-S to make b_j placed at the designated location. If a pre-placed module b_i was placed and the resulting placement of b_i does not interact with itself at the designated location, we swap b_i with the node b_j right after b_i in Γ_-; otherwise, b_i is placed at the designated location if there exists no module behind b_i in Γ_-.

7 Experimental Results

Based on a simulated annealing method [5], we implemented the TCG-S representation in the C++ programming language on a 433 MHz SUN Sparc Ultra-60 workstation with 1 GB memory. The source code is available at http://cc.ee.ntu.edu.tw/~ywchang/research.html. Based on the five commonly used MCNC benchmark circuits, we conducted four experiments: (1) area optimization, (2) wirelength optimization, (3) solution convergence speed and stability, and (4) placement with boundary constraints. In Table 1, Columns 2 and 3 list the respective numbers of modules and nets of the five circuits.

For Experiment (1), the area and runtime comparisons among SP, O-tree, B*-tree, enhanced O-tree, CBL, and TCG are listed in Table 1. As shown in Table 1, TCG-S achieves the best area utilization for the benchmark circuits in very efficient running times. Figure 6 (left) shows the resulting placement for ami49 with area optimization.

For Experiment (2), we estimated the wirelength of a net by half the perimeter of the minimum bounding box enclosing the net. The wirelength of a placement is given by the summation of the wirelengths of all nets. As shown in Table 2, TCG-S achieves better average results in wirelength than O-tree, enhanced O-tree, and TCG in smaller running times. (Note that we did not compare with B*-tree and CBL here since they did not report the results on optimizing wirelength alone.)

In addition to the area and timing optimization, in Experiment(3), we also compared the solution convergence speed and stability among SP, TCG, and TCG-S to eliminate the possible unfairness due to the non-deterministic behavior of simulated annealing, which were neglected in most previous works. (Note that other tools are not available to us for the comparative study.) We randomly ran SP, TCG, and TCG-S on ami49 ten times based on the same initial placement whose area is $102mm^2$. The resulting areas are plotted as functions of the running times (sec). Figures 5(a), (b), and (c) show the resulting curves of SP, TCG, and TCG-S, respectively. To see the detailed convergence rates, we show in

Circuit	# of modules	# of nets	SP Area mm^2	SP Time sec	O-tree Area mm^2	O-tree Time sec	B*-tree Area mm^2	B*-tree Time sec	Enhanced O-tree Area mm^2	Enhanced O-tree Time sec	CBL Area mm^2	CBL Time sec	TCG Area mm^2	TCG Time sec	TCG-S Area mm^2	TCG-S Time sec
apte	9	97	48.12	13	47.1	38	**46.92**	7	**46.92**	11	NA	NA	**46.92**	1	**46.92**	1
xerox	10	203	20.69	15	20.1	118	19.83	25	20.21	38	20.96	30	19.83	18	**19.796**	5
hp	11	83	9.93	5	9.21	57	**8.947**	55	9.16	19	(66.14)	(32)	**8.947**	20	**8.947**	7
ami33	33	123	1.22	676	1.25	1430	1.27	3417	1.24	118	1.20	36	1.20	306	**1.185**	84
ami49	49	408	38.84	1580	37.6	7428	36.80	4752	37.73	406	38.58	65	36.77	434	**36.40**	369

Table 1: Area and runtime comparisons among SP (on Sun Sparc Ultra60), O-tree (Sparc Ultra 60), B*-tree (Sparc Ultra-I), enhanced O-tree (Sparc Ultra60), CBL (Sparc 20), TCG (Sparc Ultra60) and TCG-S (Sparc Ultra60) for area optimization. The best areas are in boldface.

Figures 5 only the potions whose areas are smaller than 47 mm^2. As illustrated in Figure 5(c), TCG-S converges very fast to desired solutions, and the results are very stable ($\leq 37.5\ mm^2$ for all runs). In contrast, the convergence speed of SP is much slower than TCG-S and TCG, and the resulting areas are often larger than 39 mm^2. Further, there is a large variance in its final solutions. Based on the experimental results, we rank the convergence speed from the fastest to the slowest and the solution stability from the most stable to the least stable as follows: TCG-S $\succ$ TCG $\succ$ SP. We note that the stability and convergence speed should be very important metrics to evaluate the quality of a floorplan representation because they reveal the corresponding solution structure for optimization. However, they were often ignored in previous works. (Most previous works focus on the comparison of solution space and packing complexity. Nevertheless, we find that the solution structure induced by a representation plays an even more important role in floorplan optimization.)

Circuit	O-tree Wire mm	O-tree Time sec	enhanced O-tree Wire mm	enhanced O-tree Time sec	TCG Wire mm	TCG Time sec	TCG-S Wire mm	TCG-S Time sec
apte	**317**	47	**317**	15	363	2	363	2
xerox	368	160	372	39	**366**	15	**366**	6
hp	153	90	150	19	**143**	10	**143**	4
ami33	52	2251	52	177	44	52	**43**	89
ami49	636	14112	629	688	604	767	**579**	570

Table 2: Wirelength and runtime comparisons among O-tree, enhanced O-tree, TCG, and TCG-S for wirelength optimization. All ran on Sun Sparc Ultra60.

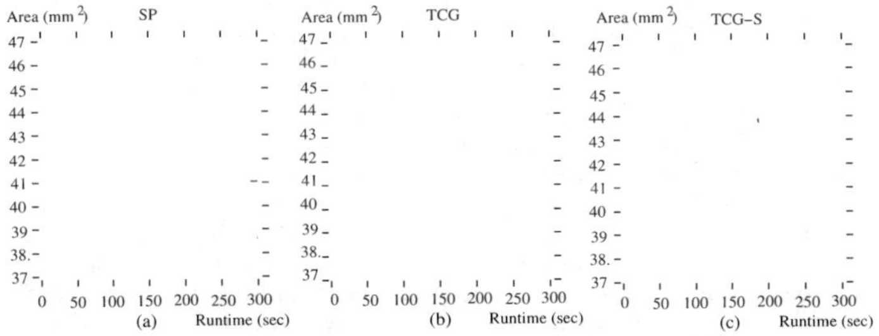

Figure 5: Comparison for stability and convergence among SP, TCG, and TCG-S for ami49.

For the experiments with boundary modules, we compared TCG-S with the SP-based method in [6] using the same data generated by [6]. The second column of Table 3 shows the numbers of the top, bottom, left, and right modules, denoted by $|T|$, $|B|$, $|L|$, and $|R|$, respectively. As shown in Table 3, TCG-S obtains smaller silicon areas than [6]. Figure 6 (right) shows the resulting placement for ami49 with boundary modules.

8 Concluding Remarks

We have presented the TCG-S representation for general floorplans by combining the advantages of two most promising P*-admissible representations TCG and SP, and at the same time eliminating their disadvantages. We also have proposed a loglinear-time packing scheme and a linear-time perturbation scheme for TCG-S. We note that these schemes can also be applied to most existing representations such as SP and BSG. Based on our theoretical study and extensive experiments, we also have shown the superior capability, efficiency, stability, and flexibility of TCG-S for floorplan design.

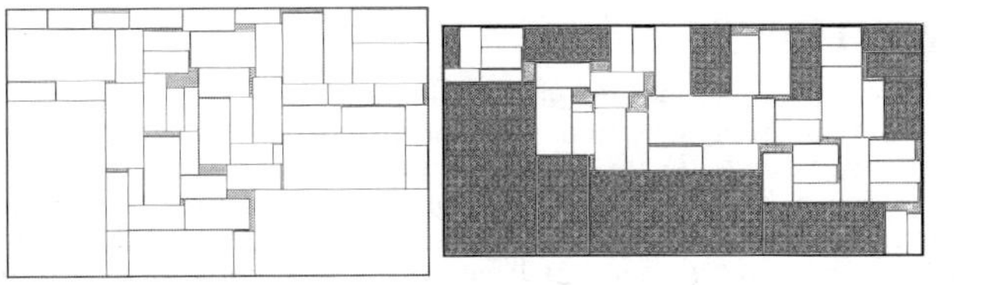

Figure 6: Resulting placements of ami49 (1) left: area optimization (area = $36.398mm^2$); (2) right: placement with boundary modules being heavily shaded ($(|T|, |B|, |L|, |R|) = (3, 3, 2, 3)$, area = $36.765mm^2$). Note that the lightly shaded regions denote dead spaces.

Circuit	$\|T\|, \|B\|, \|L\|, \|R\|$	Lai et al. [6] Area mm^2	Lai et al. [6] Time sec	TCG-S Area mm^2	TCG-S Time sec
apte	1, 1, 1, 1	**46.92**	15	**46.92**	3
xerox	1, 1, 1, 1	20.4	19	**19.977**	33
hp	1, 1, 1, 1	9.24	23	**9.158**	26
ami33	2, 2, 2, 2	1.21	290	**1.190**	238
ami49	3, 3, 2, 3	36.84	601	**36.765**	584

Table 3: Area and runtime comparisons between [6] (on Pentium-II 350) and TCG-S (on Sun Sparc Ultra60) for placement with boundary modules.

References

[1] Y.-C. Chang, Y.-W. Chang, G.-M. Wu, and S.-W. Wu, "B*-trees: A new representation for non-slicing floorplans," *Proc. DAC*, pp. 458–463, 2000.

[2] T. Cormen, C. Leiserson, and R. Rivest, *Introduction to Algorithms*, McGraw-Hill Book Company, 1990.

[3] P.-N. Guo, C.-K. Cheng, and T. Yoshimura, "An O-Tree representation of non-slicing floorplan and its applications," *Proc. DAC*, pp. 268–273, 1999.

[4] X. Hong, G. Huang, T. Cai, J. Gu, S. Dong, C.-K. Cheng, and J. Gu, "Corner Block List: An effective and efficient topological representation of non-slicing floorplan," *Proc. ICCAD*, pp. 8–12, 2000.

[5] S. Kirkpatrick, C. D. Gelatt, and M. P. Vecchi, "Optimization by simulated annealing," *Science*, vol. 220, no. 4598, May 13, 1983, pp.671–680.

[6] J.-B. Lai, M.-S Lin, T.-C Wang, and L.-C. Wang, "Module placement with boundary constraints using the sequence-pair," *ASP-DAC*, pp.515–520, 2001.

[7] E. Lawler, *Combinatorial Optimization: Networks and Matroids*, Holt, Rinehart, and Winston, 1976.

[8] J.-M Lin and Y.-W Chang, "TCG: A transitive closure graph-based representation for non-slicing floorplans," *Proc. DAC*, pp.764–769, 2001.

[9] Y. Ma, S. Dong, X. Hong, Y. Cai, C.-K. Cheng, and J. Gu "VLSI floorplanning with boundary constraints based on corner block list," *Proc. ASP-DAC*, pp. 509–514, 2001.

[10] H. Murata, K. Fujiyoshi, S. Nakatake, and Y. Kajitani, "Rectangle-packing based module placement," *Proc. ICCAD*, pp. 472–479, 1995.

[11] H. Murata, K. Fujiyoshi, and M. Kaneko, "VLSI/PCB placement with obstacles based on sequence pair," *Proc. ISPD*, pp. 26–31, 1997.

[12] S. Nakatake, K. Fujiyoshi, H. Murata, and Y. Kajitani, "Module placement on BSG-Structure and IC layout applications," *Proc. ICCAD*, pp. 484–491, 1996.

[13] R. H. J. M. Otten, "Automatic floorplan design," *Proc. DAC*, pp.261–267, 1982.

[14] Y. Pang, C.-K. Cheng, and T. Yoshimura, "An enhanced perturbing algorithm for floorplan design using the O-tree representation," *Proc. ISPD*, pp. 168-173, 2000.

[15] K. Sakanushi and Y. Kajitani, "The quarter-state sequence (Q-sequence) to represent teh floorplan and applications to layout optimization," *Proc. Asia Pacific Conf. Circuits and Systems*, pp. 829–832, 2000.

[16] X. Tang and D. F. Wong, " FAST-SP: A fast algorithm for block placement based on sequence pair," *Proc ASP-DAC*, pp. 521–526, 2001.

[17] T.-C. Wang, Personal communication, 2001.

[18] D. F. Wong, and C. L. Liu, "A new algorithm for floorplan design," *Proc. DAC*, pp. 101–107, 1986.

Floorplanning with Alignment and Performance Constraints *

Xiaoping Tang†‡ and D.F. Wong†
†University of Texas at Austin, Austin, TX 78712
‡Silicon Perspective, A Cadence Company, Santa Clara, CA 95054
{tang, wong}@cs.utexas.edu

ABSTRACT

In this paper, we present a floorplanning algorithm based on sequence pair representation. Our floorplanner has the following important features: 1) It is explicitly designed for fixed-frame floorplanning, which is different from traditional well-researched min-area floorplanning. Moreover, we also show that it can be adapted to minimize total area. 2) It addresses the problem of handling alignment constraint which arises in bus structure. 3) It deals with performance constraint such as bounded net delay, while many existing floorplanners just minimize total wire length. 4) More importantly, even with all these constraints the algorithm is very fast in that it evaluates the feasibility of a sequence pair and translates to a floorplan in $O(n \log\log n)$ time typically where n is the number of blocks and the number of constrained blocks is $O(n)$, which is significantly faster than the $O(n^3)$ method operating on constraint graph. Our algorithm is based on computing the longest common subsequence of a pair of weighted sequences. Experimental results on MCNC benchmark for block placement show the promise of the method.

Categories and Subject Descriptors

B.7.2 [**Integrated Circuits**]: Design Aids—*Placement and routing*; J.6 [**Computer Applications**]: Computer-Aided Engineering—*Computer-aided design*

General Terms

Algorithms, Design, Performance, Theory

Keywords

longest common subsequence, floorplanning, sequence pair

1. INTRODUCTION

Due to the enormous complexity of VLSI design with continuous scaling-down of technology, a hierarchical approach is needed for the circuit design in order to reduce runtime and improve solution quality. Also, IP (module reuse) based design methodology becomes widely used. This trend makes floorplanning more and more important. Floorplanning is to decide the positions of circuit blocks or IP blocks on a chip subject to various objectives. It is the early stage of design and it determines the overall chip performance. Most floorplanning algorithms use simulated annealing to search for an optimal solution. The implementation of simulated annealing scheme relies on a floorplan representation where a neighbor solution is generated and examined by perturbing the representation (called 'move'). As a result, many researchers explore in this area [2, 5, 6, 10, 11, 12, 13, 14, 17].

Floorplans can be classified into two categories, slicing and non-slicing. For slicing structure, Otten[13] proposed a binary tree representation, and later Wong and Liu[17] presented a normalized Polish expression to represent a slicing floorplan. Slicing structure has advantages, but typical floorplan is non-slicing. Lai and Wong[9] showed that based on 1-D compaction, slicing floorplan can be transformed to non-slicing floorplan without missing any min-area floorplan.

Since mid-1990s, several representations for non-slicing structure are invented, such as BSG[11], sequence pair[10], O-tree[5], B*-tree[2], CBL[6], and Q-sequence[14]. They can be further categorized into three classes. BSG and sequence pair specify the topological relative positions between blocks such as left-of, right-of, below, and above. These two can represent general floorplan. Block placement can be obtained by compacting blocks subject to relative positions. O-tree and B*-tree describe 1-D relative positions (horizontal adjacency) among blocks. The final placement is obtained by compacting blocks to the left and bottom. Although O-tree and B*-tree have smaller solution space, they are dimension dependent, i.e., geometric relation between blocks can not be obtained directly from representation. CBL and Q-sequence are two coding schemes for non-slicing floorplans called Mosaic[6] where a chip is dissected into n rectangular rooms. However, they can not represent some general floorplan unless empty rooms are inserted purposely.

Among these representations, some have larger solution space implying more redundancy, and some have smaller solution space implying less redundancy. All floorplanning algorithms based on the representations use simulated annealing. One observation we made is that redundancy is not a major issue in simulated annealing. Instead, the effectiveness of simulated annealing is more related to the diameter of solution space (diameter is the maximum distance between any two solutions where the distance between two solutions is the minimum number of moves transforming one to the other), how moves (neighbor solution) are generated, the complexity of evaluation of each representation (i.e. transforming a representation to a floorplan), and the cooling schedule in annealing process. Among the evaluations of these representations, some require quadratic runtime, some require linear runtime, and some require runtime between linear and quadratic. Another observation we made is that the difference in complexity could be nullified by implementation overhead and annealing process. Instead, the hidden constant in implementation complexity plays a major role in overall runtime. Recently, Tang et al sped up the original $O(n^2)$-time evaluation algorithm for sequence pair to $O(n \log n)$ in [15], and later further to $O(n \log\log n)$ in [16]. The experimental results in [16] show that it outperforms existing floorplanning algorithms, which implies that sequence pair is a competitive representation. In the paper, we use the sequence pair representation.

Kahng[7] suggested a fixed-frame floorplanning with no dead space. Zero dead space usually requires non-rectangular blocks. However, arbitrary block shape is undesirable in floorplanning stage. Instead the rectilinear polygon with bounded number of edges is more preferable. Of course, post-processing can be applied for floorplanning to entail arbitrary block shape. Actually, most existing algorithms handle rectilinear block by partitioning into a set

*This work was partially supported by the National Science Foundation under grant CCR-9912390.

of rectangular subblocks. Traditional floorplanners minimize the area of bounding box, or use min-area as one of the objectives. A fixed-frame floorplanning algorithm was proposed in [16] to handle range and boundary constraints. Adya and Markov considered both frame-free and fixed-frame floorplanning optimizations in [1]. In the paper, we explicitly design the strategy for fixed-frame floorplanning. To be consistent with traditional floorplanning, we also show that it can minimize area by shrinking the frame.

With bus structures or pipelines, it is better that the blocks are aligned in a row, abutting with each other (alignment constraint). The blocks can be aligned horizontally or vertically. Alignment constraint specifies that several blocks are aligned in a row with a range. It is different from abutment constraint because several abutted blocks may not be aligned. It is also different from rectilinear shape because it has more freedom in positioning. The techniques operating on constraint graph [4] can be applied to alignment constraint, but the complexity is high ($O(n^3)$).

Many existing floorplanners minimize total wire length, where there is no distinguishing between critical and non-critical nets. Even if the total wire length is optimally minimized, there is no guarantee for critical nets. Timing has become a more and more important issue in deep submicron design. Bounded delay for critical nets should be considered in early floorplanning stage, so minimizing total wire length is no longer sufficient. In the paper, we refer to bounded net delay as performance constraint.

With all these constraints (fixed-frame, alignment, and performance) taken into account, it is clear that some sequence pair is not feasible. It can be shown that any placement that satisfies the given constraints is represented by a sequence pair. We first derive necessary conditions of feasible sequence pair satisfying alignment constraint in terms of the composition of the sequence pair. Then only the sequence pairs which satisfy the necessary conditions are evaluated, implying reduced solution space. We also derive a necessary and sufficient condition of feasible sequence pair with respect to all of the three constraints, fixed-frame, alignment, and performance. Moreover, a fast algorithm is designed to evaluate the feasibility of a sequence pair and translate to a floorplan in $O(n \log\log n)$ time typically. The algorithm is based on computing the longest common subsequence of a pair of weighted sequences. Finally, since the feasibility of a sequence pair depends on the actual dimension of blocks, it is impossible to search in solution space including only feasible sequence pairs. We use simulated annealing to search for optimal floorplan satisfying given constraints, where a novel cost function is used to unify the evaluation of feasible and infeasible sequence pairs. The algorithm is very fast. Our experimental results on MCNC benchmark for block placement show the promise of the method.

2. PRELIMINARY

A sequence pair is a pair of sequences of n elements representing a list of n blocks. The two sequences specify the geometric relations (such as left-of, right-of, below, above) between each pair of blocks as follows:

$$(...b_i...b_j...,...b_i...b_j...) \Rightarrow b_i \text{ is to the left of } b_j \quad (1)$$

$$(...b_j...b_i...,...b_i...b_j...) \Rightarrow b_i \text{ is below } b_j \quad (2)$$

The original paper which proposed sequence pair [10] presented an algorithm to translate a sequence pair to a placement by constructing two constraint graphs, G_h and G_v. Both G_h and G_v have $n+2$ vertices representing n blocks plus source node and sink node (representing boundaries). G_h has a directed edge (b_i, b_j) if block b_i is to the left of block b_j. Similarly, if block b_i is below block b_j, G_v has the corresponding directed edge (b_i, b_j). For any pair of blocks (e.g. b_i, b_j), there exists exactly one edge connecting the two nodes either in G_h or in G_v. Both G_h and G_v are vertex weighted, directed, acyclic graphs. The weights in G_h represent the widths of blocks, and the weights in G_v represent the heights of blocks. Given that the coordinates of a block are the coordinates of the lower-left corner of the block, a longest path algorithm can be applied to determine the coordinates of each block and the total width and height of the bounding box. The evaluation of sequence pair takes $\Theta(n^2)$ time.

Tang et al[15] showed that the coordinates of blocks and the total width and height of floorplan can be obtained by computing longest common subsequence in terms of the two sequences. Given a sequence pair (X,Y), the total width of floorplan equals the length of the longest common subsequence of X and Y where weights are blocks' widths. Analogously, the total height of floorplan is determined by dealing with the longest common subsequence of X^R and Y where X^R is the reverse of X and weights are blocks' heights. The coordinates of a block are calculated as follows. Let (X,Y)=(X_1bX_2, Y_1bY_2) and $lcs(X,Y)$ denote the length of the longest common subsequence of X and Y. Then $(X^R,Y) = (X_2^R b X_1^R, Y_1 b Y_2)$. The x-coordinate of block b equals $lcs(X_1,Y_1)$ with blocks' widths as weights. The y-coordinate of block b is $lcs(X_2^R,Y_1)$ with blocks' heights as weights. In addition, all the computations of blocks' x/y coordinates can be integrated into a single longest common subsequence computation for a sequence pair.

3. ALIGNMENT CONSTRAINT

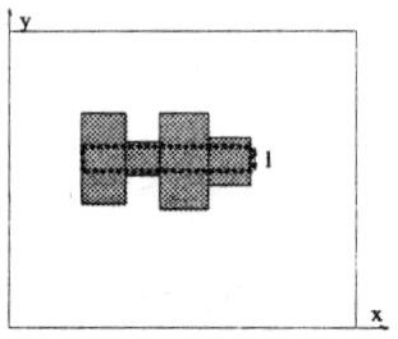

Figure 1: 4 blocks are aligned horizontally. l is the alignment range (e.g. bus width). The relative vertical positions are not fixed as long as the blocks are aligned within the given range l.

Alignment constraint specifies that several blocks are aligned in a range (e.g. bus width), abutting one by one horizontally or vertically. It is useful to facilitate data transfer in bus structure or a pipeline. As an example, Fig 1 illustrates that 4 blocks are aligned horizontally. Aligned blocks are abutted with each other in a row. But alignment is different from abutment in the sense that several abutted blocks may not be aligned . Moreover, it has more freedom in positioning than rectilinear shape because the relative positions are not fixed.

Alignment is classified into H-alignment and V-alignment according to the orientation, horizontal and vertical respectively. H-alignment can be formally defined as follows.

DEFINITION 1. **H-alignment**: *Given a range l and k blocks, b_i, $i = 1,2,...,k$, with dimension $w_i \times h_i$ and coordinates (x_i, y_i) referring to the lower-left corner, $i = 1,2,...,k$ respectively, the k blocks are H-aligned iff $x_i + w_i = x_{i+1}$, $1 \le i \le k-1$ (abutting one by one) and $y_{max} + l \le y_i + h_i$, $1 \le i \le k$, where $y_{max} = max\{y_i | i = 1,2,...,k\}$ (aligning horizontally).*

Similarly, we can define V-alignment.

DEFINITION 2. **V-alignment**: *Given a range l and k blocks, b_i, $i = 1,2,...,k$, with dimension $w_i \times h_i$ and coordinates (x_i, y_i), $i = 1,2,...,k$ respectively, the k blocks are V-aligned iff $y_i + h_i = y_{i+1}$, $1 \le i \le k-1$ (abutting one by one) and $x_{max} + l \le x_i + w_i$, $1 \le i \le k$, where $x_{max} = max\{x_i | i = 1,2,...,k\}$ (aligning vertically).*

In the following, we first discuss H-alignment. Then we show V-alignment can be handled similarly. If k blocks, b_i, $i = 1,2,...,k$, are H-aligned, then in sequence pair representation, it must be like $(X,Y) = (...b_1...b_2......b_i......b_k...,...b_1...b_2......b_i......b_k...)$ maintaining the same order. Moreover, b_i must be **strictly ahead** of b_{i+1}. "Strictly ahead" is defined as follows.

DEFINITION 3. **strictly ahead**: *Given 2 blocks, a and b, and two sequences $(X,Y) = (X_1aX_2bX_3, Y_1aY_2bY_3)$, a is strictly ahead of b in (X,Y) iff $lcs(X_2,Y_2) = 0$.*

Since we are dealing with blocks which only have non-zero dimension, $lcs(X_2,Y_2) = 0$ means no common elements between X_2 and Y_2. For example, in two sequences $(a\ b\ c\ d\ e, a\ d\ c\ b\ e)$, a

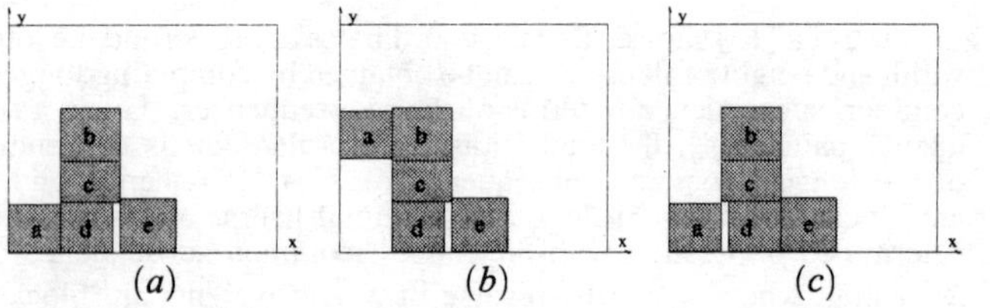

Figure 2: **Given a sequence pair (X,Y)=(a b c d e,a d c b e), (a) Block a abuts with d (a is strictly ahead of d in (X,Y)); (b) a can be moved up to abut with b (a is strictly ahead of b in (X,Y)); (c) d can be moved right to abut with e since d is strictly ahead of e in (X,Y).**

is strictly ahead of b, c or d, but not strictly ahead of e. "Strictly ahead" implies that the two blocks can possibly be abutted with each other, otherwise abutment is impossible. For example, as shown in Fig 2(a), block a abuts with d given that sequence pair be (a b c d e,a d c b e). By moving a up (aligning), a can abut with b (in Fig 2(b)). However, it can not abut with e. d can abut with e (in Fig 2(c)).

Thus we have the following necessary condition for H-alignment constraint in sequence pair.

THEOREM 1. **(adjacency-order)** *If k blocks, b_i, $i = 1,2,...,k$, are H-aligned, then the corresponding sequence pair representation must satisfy that $(X,Y) = (...b_1...b_2......b_i......b_k...,...b_1...b_2......b_i......b_k...)$ and b_i is strictly ahead of b_{i+1} in (X,Y), $1 \le i \le k-1$.*

If a sequence pair satisfies the above condition given in Theorem 1, the constrained blocks can possibly be abutted one by one in x direction and then be aligned in y direction (moving upward). Note that the alignment operation does not change the topological relation specified by sequence pair. Generally, the H-alignment operation is performed as follows:

$$y_{max} \leftarrow max\{y_i | i = 1,2,...,k\}$$
$$y_i \leftarrow max(y_i, y_{max} + l - h_i) \quad \forall i \in \{1,2,...,k\}$$

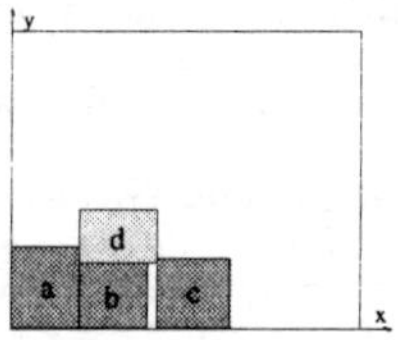

Figure 3: The sequence pair $(a\ d\ b\ c, a\ b\ d\ c)$ satisfies the condition in Theorem 1, but the constrained blocks a,b and c can not be H-aligned.

Note that the condition in Theorem 1 is only necessary, but not sufficient. For example, as shown in Fig 3, the sequence pair $(a\ d\ b\ c, a\ b\ d\ c)$ satisfies the condition, but the constrained blocks a,b and c can not be H-aligned because b and c are not tightly abutted (d is wider than b). We solve the problem by adjusting the x coordinate of the first block in an H-alignment constraint to make them tightly abutted . The adjustment is performed as follows: $x_1 \leftarrow x_k - \sum_{i=1}^{k-1} w_i$. And then we add a "dummy" block at the right as shown in Fig 4(a) and re-evaluate the sequence pair. If the constrained blocks can not be H-aligned, then the dummy block will be pushed out of frame as shown in Fig 4(b). Formally, the dummy block is determined as follows. Given a problem that k blocks $\{b_i | i = 1,2,...,k\}$ need to be H-aligned in a frame $W \times H$, where b_i has dimension $w_i \times h_i$ respectively. Let δ denote the dummy block and x_1 be the x coordinate of block b_1. The width of δ is

$$w_\delta = W - x_1 - \sum_{i=1}^{k} w_i$$

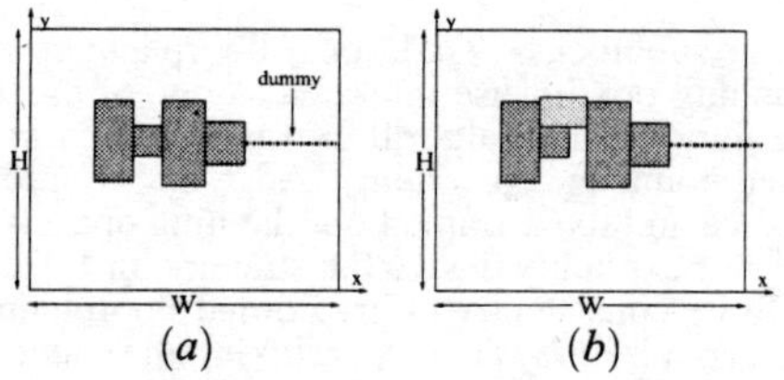

Figure 4: (a) Dummy block is introduced at the right by H-alignment constraint; (b) In violation of H-alignment constraint, the dummy block is pushed out of frame.

and the height of δ is $h_\delta = 0$.

If we handle multiple H-alignment constraints, another condition (called **non-crossing**) must be satisfied. For example, as shown in Fig 5, the two sets of blocks (a_i and b_i, $i = 1,2,3$) can not be H-aligned because a_i and b_i are crossing. However, both a_i and b_i satisfy the adjacency-order condition as stated in Theorem 1. Then we have the following necessary condition (non-crossing).

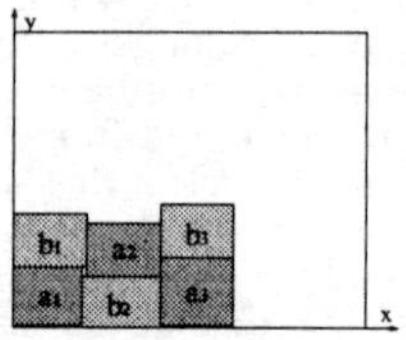

Figure 5: The two sets of blocks (a_i and b_i, $i = 1,2,3$) can not be H-aligned although they satisfy the adjacency-order condition in sequence pair representation. The sequence pair is $(X,Y) = (b_1\ a_1\ a_2\ b_2\ b_3\ a_3, a_1\ b_1\ b_2\ a_2\ a_3\ b_3)$.

THEOREM 2. **(non-crossing)** *Given any two sets of blocks, a_i, $i = 1,...,k$ and b_j, $j = 1,...,k'$, to be H-aligned, the corresponding sequence pair representation must satisfy that no relative position like $(...a_{i_1}...b_{j_1}...b_{j_2}...a_{i_2}...,...b_{j_1}...a_{i_1}...a_{i_2}...b_{j_2}...)$ appears in sequence pair (X,Y).*

Note that the two necessary conditions (Theorem 1 and Theorem 2) guarantee the topological relation of blocks for H-alignment constraints. The feasibility of constrained placement is actual-dimension dependent. However, the two conditions can be utilized to eliminate lots of infeasible sequence pairs.

For V-alignment constraint, we are dealing with the two sequences (X^R,Y) similarly. If the k blocks are V-aligned, the sequence pair must appear like $(X,Y) = (...b_k...b_{k-1}......b_i......b_1...,...b_1...b_2......b_i......b_k...)$ maintaining the reverse order in the first sequence and the obverse order in the second sequence. Analogously we have the following necessary conditions for V-alignment constraint in terms of sequence pair composition.

THEOREM 3. *If k blocks, b_i, $i = 1,2,...,k$, are V-aligned, then the corresponding sequence pair representation must satisfy that $(X^R,Y) = (...b_1...b_2......b_i......b_k...,...b_1...b_2......b_i......b_k...)$ and b_i is strictly ahead of b_{i+1} in (X^R,Y), $1 \le i \le k-1$.*

THEOREM 4. *Given any two sets of blocks, a_i, $i = 1,...,k$ and b_j, $j = 1,...,k'$, to be V-aligned, the corresponding sequence pair representation must satisfy that no relative position like $(...a_{i_1}...b_{j_1}...b_{j_2}...a_{i_2}...,...b_{j_1}...a_{i_1}...a_{i_2}...b_{j_2}...)$ appears in (X^R,Y).*

Note that "strictly ahead" for V-alignment is defined on sequences (X^R,Y). We have the following lemma.

LEMMA 1. *Given two sequences X_i and Y_j with positive weights, $lcs(X_i^R,Y_j) = 0$ implies $lcs(X_i,Y_j) = 0$.*

Let $(X,Y) = (X_0 b_k X_1 b_{k-1} X_2 ... b_i X_{k-i+1} ... b_1 X_k, Y_0 b_1 Y_1 b_2 Y_2 ... b_i Y_i ... b_k Y_k)$. Then $(X^R, Y) = (X_k^R b_1 X_{k-1}^R b_2 X_{k-2}^R ... b_i X_{k-i}^R ... b_k X_0^R, Y_0 b_1 Y_1 b_2 Y_2 ... b_i Y_i ... b_k Y_k)$. Therefore $lcs(X_{k-i}^R, Y_i) = 0, 1 \leq i \leq k-1$, which implies $lcs(X_{k-i}, Y_i) = 0, 1 \leq i \leq k-1$.

4. PERFORMANCE CONSTRAINT

Minimizing total wire length as traditional floorplanners did, can not guarantee bounded delay for critical nets. This becomes more important because timing convergence is a big issue in deep submicron design.

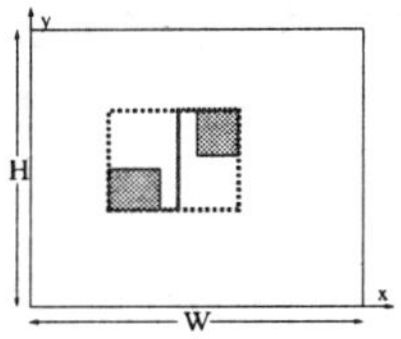

Figure 6: The bounding box of the farthest pin locations in a net.

Given a two-pin net with source at location (x_1, y_1) and sink at location (x_2, y_2), the timing requirement can be described as

$$t_1 + \lambda|x_2 - x_1| + \lambda|y_2 - y_1| \leq t_2$$

where t_1 is the arrival time at source, t_2 denotes the required latest arrival time at sink, and λ represents the constant parameter for converting from Manhattan distance to timing domain. Thus we get

$$|x_2 - x_1| + |y_2 - y_1| \leq (t_2 - t_1)/\lambda$$

Note that we use a linear function in terms of distance to estimate delay. This is reasonable[1]. For floorplanning where pin locations are not known, we assume the pin locations to be as far as possible, i.e., the source-sink distance equal to the half perimeter of the bounding box of the two blocks as shown in Fig 6. Note that this is a pessimistic estimation. An alternative way is to assume pin location at the center of a block or assigning a pin location on the closest boundary of a block, either of which can be handled in the same way.

For multi-pin net, we consider the longest distance between two nodes. In the paper, we use the half perimeter of the bounding box of the blocks belonging to the net to estimate the delay.

Based on the above formula, we can determine two values (how to determine will be discussed later), u and v, such that

$$u + v = (t_2 - t_1)/\lambda$$
$$\text{and} \quad |x_2 - x_1| \leq u$$
$$\text{and} \quad |y_2 - y_1| \leq v$$

Equivalently, the blocks belonging to the net are to be placed within a rectangular box (called range box) with width u and height v. It should be noted that the box is floating, and its shape can be changing with different ratio of u and v. Given a net connecting k blocks, b_i, $i = 1, 2, ..., k$, with coordinate (x_i, y_i) respectively, let $(\bar{x}, \bar{y})$ denote the coordinate of the box referring to the lower-left corner. Thus the coordinate of the box is

$$\bar{x} \leftarrow min\{x_i | i = 1, 2, ..., k\}$$
$$\bar{y} \leftarrow min\{y_i | i = 1, 2, ..., k\}$$

Bounded delay may not be satisfied for some placement obtained by some sequence pair. If the delay of a net is out of bound, then the involved blocks are placed out of range box by the above measurement. In order to solve the problem, we add a dummy block at the right and another dummy block on top of all constrained blocks in a critical net, as shown in Fig 7(a). If one or more of the constrained blocks is placed out of range box, then either or both of the dummy blocks will be pushed out of frame as shown in Fig 7(b). Formally, the dummy blocks are determined as follows. Given a problem that a critical net connects k blocks $\{b_i | i = 1, 2, ..., k\}$ in a frame $W \times H$, a range box is determined at location $(\bar{x}, \bar{y})$ and with width u and height v where $(\bar{x}, \bar{y})$ is computed by above formula. Thus the right dummy block has the dimension:

[1]Although interconnect delay is quadratic in terms of wire length, with appropriate buffer insertions the actual delay is close to linear in terms of source-sink distance.

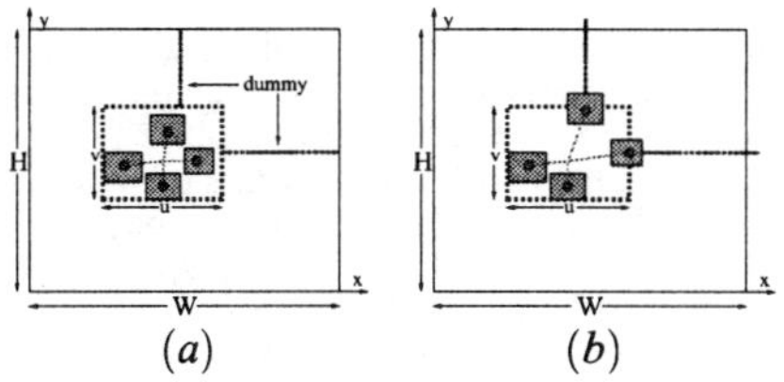

Figure 7: (a) Dummy blocks are introduced at the right or on top by performance constraint; (b) In violation of performance constraint, one or more of the dummy blocks is placed out of frame.

$$\text{width} = W - \bar{x} - u \qquad \text{and} \qquad \text{height} = 0$$

and the dimension of the above dummy block is:

$$\text{width} = 0 \qquad \text{and} \qquad \text{height} = H - \bar{y} - v$$

Now the remaining problem is how to determine u and v since the shape of the range box can change. We adjust the value of u and v dynamically in simulated annealing as follows. At high temperature in annealing process, we let u and v be half-half:

$$u = \frac{t_2 - t_1}{2\lambda} \qquad \text{and} \qquad v = \frac{t_2 - t_1}{2\lambda}$$

At low temperature, letting s be the smallest span in x axis that covers all of the constrained blocks $\{b_i | i = 1, 2, ..., k\}$, i.e. $s = max\{x_i + w_i | i = 1, 2, ..., k\} - min\{x_i | i = 1, 2, ..., k\}$, we adjust u and v:

$$u = min((t_2 - t_1)/\lambda, s) \qquad \text{and} \qquad v = (t_2 - t_1)/\lambda - u$$

This adjustment is certified by the following reasons. At high temperature simulated annealing is characterized as chaotic process where a square range-box is appropriate to use for approximate guidance. On the other hand, at low temperature a specific range box is used to exactly capture the meaning of delay bound.

5. ALGORITHM

5.1 Feasible Sequence Pair

Sequence pair always entails a packing if no constraints are concerned. However, when constraints are introduced, there may not exist corresponding packing for some sequence pairs. *Feasible sequence pair* can be defined as follows.

DEFINITION 4. **Feasible sequence pair**: *If there exists a packing which meets all the constraints and the topological relations imposed by a sequence pair, then the sequence pair is feasible. Otherwise, it is infeasible.*

5.2 Optimality

Although not all sequence pairs are feasible for constraints such as fixed-frame, alignment and performance constraint, we can still show that sequence pair does not miss any placement. Given a placement satisfying the constraints, a sequence pair can be obtained by gridding operation in [10]. This observation is stated in the following theorem.

THEOREM 5. *Any placement that satisfies the given constraints is represented by a sequence pair.*

Thus, an optimal packing with no violation of constraints is included in solution space of feasible sequence pairs. "Optimal packing" means that at least one area-minimal placement is represented by a sequence pair. This is not true in terms of wire length because packing blocks to the left and to the bottom may not represent the optimal wire length. However, packing is still a good method to minimize wire length because all blocks are packed, implying at least near-optimal wire length is represented.

5.3 Evaluation Algorithm

Note that the calculations of x and y coordinates of all the blocks are done independently by evaluating (X,Y) and (X^R,Y) respectively. In the following we mainly describe the algorithm to evaluate (X,Y) in the presence of constraints since the evaluation of (X^R,Y) can be done similarly.

The algorithm is based on the engine of computing longest common subsequence presented in [16]. We still compute longest common subsequence (lcs) in terms of the two sequence (X,Y) to determine the x coordinates of blocks. The alignment operation presented in Section 3 and the computation for performance-constrained blocks described in Section 4 can be embedded into lcs computation. Dummy blocks would not necessarily appear in sequence pair. Instead, a "out-of-frame" variable is introduced to record the intermediate *lcs* imposed by dummy blocks in constraints (alignment/performance). Note that the alignment operations for different alignment constraints are affected by each other. One iteration of lcs computation may not align all blocks imposed by alignment constraints. However, since there is no crossing between alignment constraints, one iteration of lcs computation will align at least one sequence of blocks related to an alignment constraint. Thus at most $k+1$ iterations are needed, where k is the total number of alignment constraints. On the other hand, if alignment constrained blocks are not aligned after $k+1$ iterations, then it indicates that there exists crossing between different alignment constraints. In typical applications, the number of alignment constraints, i.e. k, is bounded. Let $lcs'(X,Y)$ denote the return value of the algorithm, i.e. the length of the longest common subsequence with the presence of dummy blocks introduced by constraints. Note that we are dealing with fixed-frame floorplanning and thus it is enforced that $lcs'(X,Y) \geq W$, the width of the frame.

Similarly $lcs'(X^R,Y)$ is computed with dummy blocks in y direction, and $lcs'(X^R,Y) \geq H$, the height of the frame. If the total number of constrained blocks is $O(n)$ and the number of alignment constraints is bounded, the complexity of the algorithm is $O(n \log\log n)$, as stated in the following theorem.

THEOREM 6. *With the presence of fixed-frame/alignment/performance constraints, a sequence pair can be evaluated in $O(n \log\log n)$ time if the number of constrained blocks is $O(n)$ and the number of alignment constraints is bounded.*

In typical applications, the number of constrained blocks (e.g. the involved blocks in critical nets) is less than n. Note that a block may repeat in several critical nets. In general case, where there are k alignment constraints and m blocks (including repeated) are constrained, the time complexity of the algorithm will be $O(m + kn \log\log n)$. Note that even in the case where k is bounded, the method operating on constraint graph still needs $O(n^3)$ runtime.

5.4 A Necessary and Sufficient Condition

As presented in Section 3 and Section 4, the dimension of dummy blocks is determined by the frame for floorplanning. If there is any violation of given constraints (fixed-frame/alignment/performance), one or more of dummy blocks or real blocks will be pushed out of the frame. On the other hand, if all dummy blocks and real blocks are packed within the frame, then we can derive the corresponding sequence pair and compute a packing satisfying all constraints. Thus, we get the following necessary and sufficient condition.

THEOREM 7. *A sequence pair (X,Y) is feasible with constraints if and only if $lcs'(X,Y) \leq W$ and $lcs'(X^R,Y) \leq H$.*

Since we have enforced $lcs'(X,Y) \geq W$ and $lcs'(X^R,Y) \geq H$ in the evaluation, the following result is directly implied.

COROLLARY 1. *The necessary and sufficient condition of feasible sequence pair (X,Y) is that $lcs'(X,Y) = W$ and $lcs'(X^R,Y) = H$.*

5.5 Unified Cost Function

With all these constraints taken into account, the feasibility of a sequence pair depends on the actual dimension of blocks. Thus it is hard to judge the feasibility without evaluating a sequence pair. In other word, it is impossible to just go through solution space including feasible sequence pairs only. The necessary conditions for alignment constraint presented in Section 3 can be used to reduce the solution space although it can not eliminate all infeasible sequence pairs. An intuitive way to evaluate infeasible sequence pairs is to assign them an infinite cost. However, this will seriously affect the smoothness of stochastic search process. Note that if a sequence pair (X,Y) is infeasible then $lcs'(X,Y) > W$ or $lcs'(X^R,Y) > H$. Thus, we treat an infeasible sequence pair as if it is feasible, and regard the "area" as

$$A = lcs'(X,Y) \cdot lcs'(X^R,Y)$$

Therefore, we get a unified cost function

$$C = \alpha A + \beta W$$

where W is interconnect cost, and α and β are coefficients for balancing between A and W.

The unified cost function interprets the violation of infeasible sequence pair so naturally that we get the following benefits: (i) no need to accumulate the error for each constrained block; (ii) no need to add an additional penalty term into the cost function; and (iii) fixing an infeasible sequence pair (adaption) is not necessary. As a result, it helps handle constraints very fast and very well.

5.6 Perturbation (Move)

We use the following operations to generate a neighbor sequence pair in simulated annealing.

Swap is to swap two blocks in either the first sequence or the second sequence. We maintain the relative ordering of alignment-constrained blocks in swapping. Swap can be done in linear time.

Rotation is to rotate an unconstrained block (e.g. exchange the width and height of the block). Rotation does not change the relative ordering imposed by alignment constraint. This operation can be done in constant time.

Flip is used to change the orientation of an alignment constraint. If a set of blocks is to be aligned horizontally, then flip the blocks to be aligned vertically, vice versa. The operation on sequence pair is to reverse the order of the set of blocks involved in the alignment constraint in the first sequence. In the implementation, we restrict that a block can only be associated with at most one alignment constraint. Thus, after flipping, a sequence pair still meets the ordering condition imposed by all alignment constraints. Suppose that an alignment constraint involves r blocks. Thus the flip operation can be done in $O(r)$ time. Of course, $r \leq n$.

In our implementation of the algorithm, the probability of the three operations is $P(\text{swap}) > P(\text{rotation}) > P(\text{flip})$.

5.7 Solution Space

The solution space we explore in simulated annealing includes only sequence pairs that satisfy the ordering condition imposed by alignment constraints. A set of move operation is said to be "complete" if any sequence pair in solution space can be reached by other sequence pair through a sequence of move operations. Then, the set of move operations proposed above is complete because a sequence pair can be transformed to another sequence pair via a sequence of swap and flip operations. However, rotation is also necessary for better packing.

Given any two different sequence pairs in solution space, one can be changed to the other as follows. First, we use flip operation to make the ordering of alignment-constrained blocks be the same in the two sequences. Then, one sequence can be "sorted" into the other via swap operations. Note that swapping two adjacent blocks does not change the relative ordering of alignment-constrained blocks unless the swap is invalid (i.e. the two adjacent blocks belong to the same alignment constraint). Since one alignment constraint involves at least 2 blocks, there are at most $n/2$ alignment constraints. Thus we need at most $n/2$ flip operations. Sorting requires at most $1+2+...+(n-1) = n(n-1)/2$ adjacent swaps. In total, at most $n/2+n(n-1)/2 = n^2/2$ move operations are needed to transform a sequence pair to another in solution space. Therefore, the diameter of solution space is at most $n^2/2$ for the set of move operations.

5.8 Min-area Optimization

Traditional floorplanners use minimizing area as an objective. Although our algorithm is explicitly designed for fixed-frame floorplanning, it can also be adapted for min-area optimization. An initial frame, W and H under some given aspect ratio, is determined, e.g. $W \cdot H$ equals 120% of the sum of the areas of all the blocks. Then, we decrease W and H when a feasible sequence pair is met in simulated annealing. Three kinds of decreasing schemes are randomly chosen in our experiments: decrease W only, or H only or both.

6. EXPERIMENTAL RESULTS

Table 1: Min-area results of floorplanning with alignment/performance constraints.

circuit	block	constrained blocks		Results		
		align	perf	time (s)	area (mm^2)	dead space
apte	9	4	0	8	47.08	1.1%
xerox-1	10	4	0	9	20.16	4.0%
xerox-2	10	4	2	16	20.93	7.5%
hp-1	11	4	0	10	9.342	5.5%
hp-2	11	4	2	17	9.342	5.5%
ami33-1	33	4	0	31	1.221	5.3%
ami33-2	33	4	3	45	1.226	5.7%
ami49-1	49	5	0	41	38.20	7.2%
ami49-2	49	4	3	241	38.51	7.8%
ami49-3	49	4	6	278	38.72	8.4%

We have implemented the algorithm and tested on problems with various constraints. The experiments were carried out on a SUN Sparc Ultra 10(440Mhz). The technique of simulated annealing is used to search for an optimal placement with a special annealing schedule where a very large number of temperatures is used but only a small number of moves are made within each temperature.

The test problems are derived from MCNC benchmarks for block placement. We impose alignment/performance constraints on a set of blocks (determined by the total number of blocks in the benchmark). Table 1 lists the min-area results. Note that many other floorplanning algorithms give the results for area minimization. Our results are very comparable even though we have handled various alignment/performance constraints. There is no uniform standard for wire length measure because some assume pins at the center of blocks and some let pin locations on the boundary of blocks. Therefore, the results for wire length are omitted. Experimental results show that the algorithm performs very well in dealing with various alignment/performance constraints. The algorithm is very fast – we can easily evaluate a million sequence pairs with constraints in one minute for typical input size of placement problems. The experience in experiments tells us for min-area optimization it is preferable that the initial frame be close to the final one. As illustrations, Fig 8 displays the final packing results for ami49-3 .

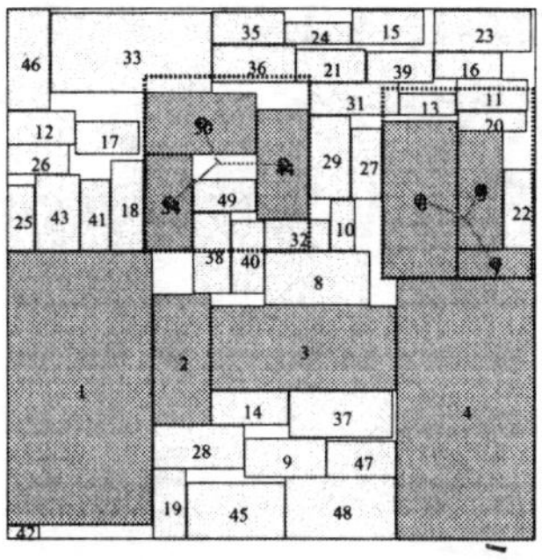

Figure 8: The result packing of ami49-3. Block 1, 2, 3 and 4 are to be aligned, and there are two performance constraints (block 5, 6 and 7 are connected in a critical net, and block 30, 34 and 44 are belonging to another critical net).

7. CONCLUDING REMARKS

In this paper, we have presented an algorithm to handle alignment and performance constraints. The algorithm is based on sequence pair representation and evaluates a sequence pair with constraints in $O(n \log \log n)$ time typically where n is the number of blocks and the number of constrained blocks is $O(n)$. It is designed mainly for fixed-frame floorplanning optimization, and can also be adapted for minimizing area.

We have derived the necessary and sufficient condition of feasible sequence pair and a unified cost function for the evaluation of both feasible and infeasible sequence pairs. We also propose a set of move operations which is complete. Experimental results show that the algorithm performs very well in handling alignment and performance constraints.

Finally, we would like to point out that the algorithm to handle alignment constraint can apply to abutment constraint because abutment is a special case of alignment. The method proposed in the paper has been integrated in a system, FAST-SP, which previously handled constraints such as pre-placed, range, and boundary[16]. After integration, it is capable of handling more constraints (e.g., abutment, alignment, performance, etc.).

8. REFERENCES

[1] S.N. Adya and I.L. Markov. "Fixed-outline floorplanning through better local search", ICCD-01, pp. 328-334, 2001.

[2] Y.C. Chang, Y.W. Chang, G.M. Wu, and S.W. Wu. " B*-trees: a new representation for non-slicing floorplans", DAC-2000, pp. 458-463, 2000.

[3] P. Chen and E.S. Kuh. "Floorplan sizing by linear programming approximation", DAC-00, pp. 468-471, 2000.

[4] K. Fujiyoshi, and H. Murata. "Arbitrary convex and concave rectilinear block packing using sequence pair", ISPD-99, pp. 103-110, 1999.

[5] P.N. Guo, C.K. Cheng, and T. Yoshimura, "An O-tree representation of non-slicing floorplans and its applications", DAC-99, pp. 268-273, 1999.

[6] X. Hong, G. Huang, Y. Cai, J. Gu, S. Dong, C.K. Cheng, and J. Gu. "Corner block list: an effective and efficient topological representation of non-slicing floorplan", ICCAD-00, pp. 8-12, 2000.

[7] A.B. Kahng. "Classical floorplanning harmful", ISPD-00, pp. 207-213, 2000.

[8] M. Kang and W.W.M. Dai. "Arbitrary rectilinear block packing based on sequence pair", ICCAD-98, pp. 259-266, 1998.

[9] M. Lai and D.F. Wong. "Slicing tree is a complete floorplan representation", DATE-01, pp. 228-232, 2001.

[10] H. Murata, K. Fujiyoshi, S. Nakatake, and Y. Kajitani. "VLSI module placement based on rectangle-packing by the sequence pair", *IEEE Transaction on CAD* , vol. 15:12, pp. 1518-1524, 1996.

[11] S. Nakatake, H. Murata, K. Fujiyoshi, and Y. Kajitani. "Module placement on BSG-structure and IC layout applications", ICCAD-96, pp. 484-491, 1996.

[12] T. Ohtsuki, N. Sugiyama, and H. Kawanishi. "An optimization technique for integrated circuit layout design", ICCST Kyoto, Japan, pp. 67-68, 1970.

[13] R.H.J.M. Otten. "Automatic floorplan design", DAC-82, pp. 261-267, 1982.

[14] K. Sakanushi and Y. Kajitani. "The quarter-state sequence (Q-sequence) to represent the floorplan and applications to layout optimization", IEEE APCCAS, pp. 829-832, 2000.

[15] X. Tang, R. Tian, and D.F. Wong. "Fast evaluation of sequence pair in block placement by longest common subsequence computation", DATE-00, pp. 106-111, 2000.

[16] X. Tang and D.F. Wong. "FAST-SP: A fast algorithm for block placement based sequence pair", ASPDAC-01, pp. 521-526, 2001.

[17] D.F. Wong and C.L. Liu. "A new algorithm for floorplan design", DAC-86, pp. 101-107, 1986.

Algorithms for Simultaneous Satisfaction of Multiple Constraints and Objective Optimization in a Placement Flow with Application to Congestion Control

Ke Zhong and Shantanu Dutt
Department of Electrical and Computer Engineering
Univ. of Illinois at Chicago
Chicago, IL 60607-7053
{kzhong,dutt}@ece.uic.edu

ABSTRACT

This paper addresses the problem of tackling multiple constraints simultaneously during a partitioning driven placement (PDP) process, where a larger solution space is available for constraint-satisfying optimization compared to post-placement methods. A general methodology of multi-constraint satisfaction that balances violation correction and primary optimization is presented. A number of techniques are introduced to ensure its convergence and enhance its solution search capability with intermediate relaxation. Application of our approach to congestion control modeled as pin density and external net distribution balance constraints shows it effectively reduces overall congestion by 14.3% and improves chip area by 8.9%, with reasonable running time and only 1.6% increase in wire length. As far as we know, this is the first time an approach to congestion reduction *during* placement optimization produced good congestion improvement with very small wire length increase.

Categories and Subject Descriptors

J.6 [**Computer-Aided Engineering**]:

General Terms

Algorithms, Experimentation

Keywords

Multi-constraint satisfaction, partitioning-driven placement, intermediate relaxation, congestion reduction, connector generation, minimization of objective deterioration

1. INTRODUCTION

Placement is an important phase of physical design automation for VLSI circuits.With the maturity of deep submicron (DSM) technology, very complex circuits and systems with 10 to 100 million plus transistors can now be realized on a single chip. For these circuits, routability becomes an important issue that needs to be tackled during the placement phase, or subsequent routing can become difficult and inefficient, for example, [8, 9, 12]. Also, in the DSM regime, a number of debilitating electrical and thermal effects like signal crosstalk, increased temperatures and hot-spots become prominent, and negatively impact chip reliability [13]. These effects can be significantly alleviated by proper circuit placement. Thus in the DSM regime, not only is placement required to optimize *primary objectives* (e.g., timing, power and/or wire area), but it also needs to address *secondary requirements* (e.g., routability, uniform thermal distribution and crosstalk bounding) which can be most effectively modeled as *constraints* on the placement process.

A number of excellent placement algorithms and tools have been developed that tackle only a single objective such as wire area, timing, and congestion [1, 6, 4, 8, 9, 14, 11, 10, 15]. There has been little work in developing a general theory and algorithms for exploring the design space to optimize primary objective(s) and satisfy a few design constraints simultaneously [5]. Various important design constraints are thus currently tackled in the post-placement phases where a much smaller solution space is available than during placement, and where there is not much control over possible deterioration of primary optimization objective(s).

To develop our multi-constraint satisfaction algorithm, we will use the partitioning driven placement (PDP) approach to tackle these problems, due to a number of beneficial features: a) Being a divide-and-conquer approach with low time-complexity, it can practically handle huge DSM designs. b) Being a top-down approach, it is cognizant of the state of the optimization objective(s) and constraints at each partitioning level, which enables balanced placement decisions. c) It is inherently very flexible—the gain functions can be appropriately formulated to optimize the objective of interest, and to model the problem accurately by taking various DSM effects into account.

The outline of the rest of the paper is as follows. Sec. 2 describes promising approaches to multi-constraint satisfaction with a view to minimizing deterioration of the primary objective. Here we also discuss relevant past research in constraint-satisfaction methods. Constraint-based modeling of congestion control follows in Sec. 3. Experimental results are discussed in Sec. 4, and conclusions are in Sec. 5.

2. MULTI-CONSTRAINT SATISFACTION

Two core issues regarding multiple constraint satisfaction are: a) Reaching feasible constraint-satisfying solutions from infeasible states in a multi-constraint environment where different constraints can be in conflict with one another; b) Minimizing degradation of the primary optimization objective(s) due to constraint satisfaction.

Unless otherwise stated, we will describe our approaches

in the context 2-way recursive PDP for standard-cell circuits (although our approaches are extendible to multi-way partitioning and a number of other design styles). A 2-way recursive PDP flow starts with an input circuit and its assigned rectangular layout area (a single *region*). It then recursively bisects each region (horizontally and/or vertically) and its residing (sub)circuit, until subcircuit sizes in every region become sufficiently small. For a standard-cell circuit, its modules are of equal height, and are normally placed into a number of fixed-height rows. Horizontal partitioning steps thus naturally subdivide regions along row boundaries, and separate them into perfectly aligned groups, each covering a unique set of underlying rows.

To study PDP under multiple constraints, we extensively used the standard cell placement tool SPADE [15]. The SPADE approach features a 'simultaneous level processing' paradigm, facilitating direct, more accurate wire length minimization. It produced very promising results on a set of MCNC benchmark circuits [15].

2.1 Constraint Categorization and Modeling

To develop a general constraint-satisfaction method, constraints must be classified and modeled in a normalized fashion. Most PDP constraints can be categorized into *balance* and *non-balance* types. In this paper we focus on the former type, which requires the ratio of (aggregate) constraint metric values on the two subsets of a partition to lie within a given (generally small) range. To formally define a balance-type constraint, let the constraint metric values in two subsets of a partition be B_0 and B_1, their specified ratio be $r : (1-r)$, tolerances be t_0 and t_1, and their sum be B, respectively. Then the following set of inequalities should be satisfied:

$$B_0 \leq B \cdot (r + t_0)\,, \qquad B_1 \leq B \cdot (1 - r + t_1) \qquad (1)$$

where $0 < r < 1$, and $0 < \{t_0, t_1\} < 1$. Further, if we divide both sides of (1) by B, we get its *normalized form*, with b_0 and b_1 representing B_0/B and B_1/B, respectively:

$$b_0 \leq r + t_0\,, \qquad b_1 \leq 1 - r + t_1 \qquad (2)$$

Two primary balance-type PDP constraints, which have been implicitly included in the standard-cell placement tool, SPADE [15], are *local-size* and *row-size* constraints. These two constraints are for balanced distribution of cells into sub-regions and row-sets, respectively. For local-size constraint, B is the sum of cell sizes within a region, while B_0 and B_1 are total cell sizes of its two subsets. Similarly, for row-size constraint, B is the total cell size of all regions spanning the same underlying set of rows, and B_0 (B_1) is the combined cell sizes in subsets 0 (subsets 1) of those regions.

Balance constraints can be further divided into *monotonic* and *non-monotonic* classes. For monotonic constraints, move of a cell from its source subset s to its target subset t only increases constraint metric value in subset t, and decreases that in subset s. For instance, the local-size and row-size constraints are monotonic. Non-monotonic examples include the *pin-density* constraint for congestion control. Pin density is measured as total number if pins within a region divided by its total cell size. When a cell is moved from subset s to t, it is possible to increase the pin density of s, while decreasing that of t. Therefore, the pin density constraint is non-monotonic. Another categorization of balance constraints is based on the invariant nature of B (e.g., for local-size constraint, sum of cell sizes across both subsets remains unchanged at any cell move, while sum of pin densities is variable).

To address a wide range of constraints in a generic manner, we introduce the *normalized cell constraint weight*, or *c-weight* to model contributions of each cell to the various constraint measures. Each cell u has a vector of c-weights, the ith element $w_i(u)$ of which corresponds to a particular constraint c_i. A c-weight $w_i(u)$ represents how much u's move can change the current distribution of normalized metric values of constraint c_i. Specifically, if cell u is moved from subset s to t, $b_t(c_i)$ (denoting the b_t measure for constraint c_i) changes by $w_i(u)$, while $b_s(c_i)$ changes by $-w_i(u)$.

For local-size (row-size) constraint of a region, c-weights are computed as individual cell sizes relative to total cell size within the region (row-set). They remain constant throughout the cell move process. For any other constraint c_i whose B is not fixed (e.g., pin density constraint), if cell u is to be moved from subset s to t, its $w_i(u)$ is computed using the following set of equations:

$$\Delta(c_i) = b_t(c_i) - b_s(c_i)\,, \qquad \Delta'(c_i) = b'_t(c_i) - b'_s(c_i)\,,$$
$$w_i(u) = (\Delta(c_i) - \Delta'(c_i))/2 \qquad (3)$$

where $b'_s(c_i)$ and $b'_t(c_i)$ are normalized metric values of constraint c_i for subset s and t, respectively, *after* u's move.

Based on Eqn. 3, c-weights of non-monotonic constraints might take negative values, while those of monotonic balance constraints are always non-negative. Signs of the c-weights can be used in the following way to correct constraint violation. For any constraint c_i, we term the subset violating Ineq. (1), if any, as its 'overflowing' subset and the other subset as 'underflowing'. The amount of violation is measured as the difference between the actual constraint metric value and its specified upper bound in the overflowing subset. *The move of cell u can reduce the violation of constraint c_i (if any) if $w_i(u)$ is positive and u is moving out of the overflowing subset, or if $w_i(u)$ is negative and u is moving into the overflowing subset.*

2.2 Constraint-Driven Optimization with Intermediate Relaxation

Most placement algorithms perform some type of search on the discrete solution space of the problem for the required solutions. If additional constraints are introduced to the problem, the new solution space normally becomes a fragmented subset of the original one. There are two basic paradigms for tackling extra constraints imposed on the search process, *post-placement* or *simultaneous-with-placement* (*with-placement* for short). The post-placement paradigm has been extensively used in past work for tackling many types of constraints (e.g., [12, 14, 16]). In this paper, we espouse the use of the with-placement paradigm.

With-placement methods can be of two types. The first is the simpler formulation of disallowing a search path to traverse any illegal solution. This type of method, which has been employed in almost all previous partitioning and PDP algorithms, is prone to get stuck at local optima. The second approach is to allow the search to navigate through illegal solutions, with the goal of reaching an end solution that is legal and represents an optimal or near-optimal solution. It means the constraint(s) are relaxed temporarily (*intermediate relaxation* introduced in [3]). This solution method can potentially yield global optimum or near global-optimum solution, since it provides a mechanism to virtually connect the disjoint components of a fragmented solution space.

Intermediate relaxation allows two different patterns of violating moves: a) One-step violation: if a move produces a newly-violating solution, any move following it must at least generate a next solution with progressively lesser amount of violation. b) Multi-step violation: a sequence of moves are allowed that may produce increasing amount of violation, up to certain upper bounds, in the interest of better optimizing the primary objective, and this is followed by a series of moves to completely correct the violations, once the violation upper bounds are reached.

The crux of navigating through illegal portions of solu-

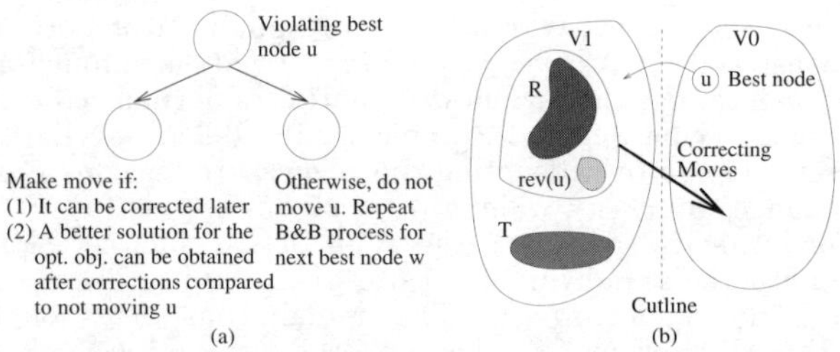

Figure 1: B&B approach to constraint-satisfaction during partitioning: (a) B&B decision process; (b) The set of correcting and violating moves.

tion space "profitably" are effective techniques for estimating whether: *(a)* a legal solution can be finally reached, without backtracking, whenever the search, currently at a legal solution, is inclined to visit an illegal one (based on cost improvement); and *(b)* whether the lowest cost legal solution reachable with relaxation (in either single or multiple moves) will have a better cost than the best legal solution reachable without relaxation. The search would then proceed if both (a) and (b) are satisfied. These correspond to the two core issues noted at the beginning of Sec. 2.

2.3 Tackling a Single General Constraint

In [3], effective techniques have been developed for formally addressing, for the first time, the problem of partitioning and placement with any **single** given constraint. The challenge in such a problem is to satisfy any given constraint without significantly compromising the optimality of the main objective(s), and [3] introduced the concept of *intermediate relaxation* to achieve this goal. Previous works that studied this problem have had limited success [7].

This promising approach of [3] is a non-backtracking branch-and-bound (B&B) search technique with multi-step violation, which follows the paradigm of making violating moves if both conditions (a) and (b) stated above are satisfied; see Fig. 1(a). In the context of 2-way partitioning with the commonly used and important size balance constraint, an average of 14.5% better cutsizes were obtained with as little as 13% time overhead using these techniques compared to standard methods that disallow violating moves.

Referring to Fig. 1(b), let V_0, V_1 be the two subsets into which the current subcircuit is being partitioned (at some level of the PDP hierarchy). Let T be the current set of uncorrected violating moves made into, say, V_1, and let R be the set of required correcting moves, say, from V_1 to V_0 that can correct the violations due to T and will result in the minimum deterioration of the optimization objective. Let u be the current best move to make, say, from V_0 to V_1. If u does not add to the accumulated violations due to T, then we move u. Otherwise, we search for the best set of nodes $rev(u)$ (with respect to the optimization objective), $rev(u) \cap R = \emptyset$, that can be moved to compensate for u's violation. We then check if the following inequality holds:

$$gain(u) + gain(R \cup rev(u) \mid u) > gain(R) \qquad (4)$$

where $gain(y)$ is the estimated improvement in the optimization objective due to the move of cell y, and for a set S of cells, $gain(S) = \sum_{y \in S} gain(y)$ [1]. Further, $gain(y \mid x)$ is the estimated improvement obtained by moving cell y after x has been moved ($gain(y)$ can be affected by x's move), and $gain(S \mid x) = \sum_{y \in S} gain(y \mid x)$. The LHS (left-hand side) of Ineq. 4 is the estimated objective improvement on pursuing the path of moving u followed by $R \cup rev(u)$ to correct the total violation, while the RHS is the estimated objective improvement obtained by only moving cells in R to correct the current violation; note that a legal partition point must be free of violations. Cell u would then be moved if $rev(u)$ exists and Ineq. 4 is satisfied. Otherwise, the B&B process is repeated for the next best free cell in $V = V_0 \cup V_1$.

[1]The $\sum$ function for computing $gain(S)$ and for use in Ineq. 4, should actually be replaced by the appropriate function ψ that defines the quality of the cutset vis-a-vis the optimization objective. For e.g., $\psi = \sum$ for the min-cut & ψ = weighted max for the timing objectives .

2.4 Tackling Multiple Constraints

When we go from single to multiple constraints, the complexity of constraint-satisfying algorithms (with intermediate relaxation) goes up by at least an order of magnitude. This is so, because multiple constraints can be mutually conflicting; i.e., if we correct the violation of one, we could end up worsening that of another. Also, with more constraints, it becomes more difficult to control the compromise in primary objective(s) during violation correction stages. We describe here a number of techniques for dealing with issues of conflicting constraints and with trade-off estimates between gain and constraint correction amount when determining a set of moves to correct multiple constraint violations.

2.4.1 Balancing Constraint Violation Correction and Primary Objective Optimization

The SPADE algorithm relies on global gain-based priority queues to select cell moves, and avoids any row-size and local-size violation [15]. Now, with intermediate relaxation of multiple constraints, additional priority queues are introduced for each constraint c_i at its effective level (for row-size, at row level; for local-size, at regional level, etc.). Cells in each such priority queue are sorted according to some mixed measure based on their gains and contributions to violation reduction, making it easier to find cells that have reasonable gains and can help resolve violations (if any) quickly. For any non-monotonic constraint, cells are sorted into two priority queues (with different signs) in each subset, so that no matter which one is overflowing, the algorithm can quickly find correcting cells to move. Also, when violations are present, there is always a *critically violated* constraint c_m that has the max amount of (normalized) violation. While looking for a cell u to reduce c_m's violation, it is also necessary to check u's impact (as indicated by its c-weights) on other constraints lest some other violations grow too much.

Based on the above consideration of balancing optimization objective improvement and violation correction, we introduce the following *gain-c-weight* measure $GW(u, c_i)$ (each constraint-based priority queue is sorted by decreasing $GW(u, c_i)$ measure). If there are k constraints, then for each constraint c_i,

$$\begin{aligned} GW(u, c_i) &= (g(u) - g_{th}) \cdot (1 + \frac{\alpha \cdot |w_i(u)|}{max_j |w_j(u)|}) \;\; \textbf{if } g(u) \geq g_{th} \\ GW(u, c_i) &= (g(u) - g_{th}) \cdot (1 - \frac{\alpha \cdot |w_i(u)|}{max_j |w_j(u)|}) \;\; \textbf{if } g(u) < g_{th} \end{aligned} \qquad (5)$$

where $g(u)$ is the gain of u, g_{th} is a threshold on cell gains, α is a constant for trade-off between gain and c-weight considerations, and $1 \leq \{i, j\} \leq k$. The threshold g_{th} is designed to penalize cells with gains below it. The max operator in the denominator helps convergence in the violation correction process, since a cell with relatively higher w_i (among its other c-weights) is more likely to be chosen if c_i is the current critically violated constraint.

A cell selection algorithm using the above data structures and gain-c-weight measures is depicted in Fig. 2. When violations occur, the algorithm refers to relevant positive and/or negative weight priority queues for selecting the best correcting move. Note that no move is allowed to produce a higher amount of violation than the current critical amount (to be discussed next).

```
Algorithm Cell_Select() /* multi-constraint cell selection */
  if (current state is feasible) then /* any move allowed */
    Bc = the best cell from gain-based priority queues;
  else /* at least one constraint is already violated */
    Get the type c_m, amount v_a, overflowing subset s_o and
    underflowing subset s_u of the current critical violation;
    Locate relevant positive-weight queue q+ in s_o, and
    negative-weight queue q- in s_u for c_m; /* q- exists for
    non-monotonic constraints only */
    Bc = the cell with maximum gain-c-weight value from q+
    and/or q-;
    if (Bc produces new critical violation >= v_a) then
      Pick the next best cell from q+ (or q-) as the new Bc,
      whose post-move critical overflow < v_a;
  return Bc;
```

Figure 2: Multi-constraint cell selection algorithm.

2.4.2 Mechanisms to Ensure Fast Convergence to Feasible States

With increasing number of constraints, the problem of convergence becomes critical even under one-step violation. Cell selection based on the mixed gain-c-weight measure cannot ensure correctability of violation (not to mention a speedy correction). Hence, the following rules and techniques are introduced to help solve the convergence problem.

⋆ *Decreasing violation:* Any correcting move should produce a new amount of critical violation strictly less than that of the current violation. Such a rule maintains a trend of convergence, but does not guarantee its speed. This is included as part of the cell selection process in Fig. 2.

⋆ *Feasibility estimation:* Whenever a candidate move introduces a fresh violation, the possibility of correcting the new violation within a certain number of steps is estimated. The estimation process tentatively accepts the violating move, and repeatedly looks for the best cell v to resolve the current most critical violation (in the same way as the cell selection algorithm in Fig. 2). These cells are then 'virtually' moved (they are moved, but no cell gains are updated) to correct violations. Only when violations are completely resolved within a certain number of 'virtual' moves, can the initial violating move be actually accepted. This can help us exclude those moves that can lead to a very long (if not impossible) correction process.

⋆ *Learning-based backtracking:* Sometimes existing violations are not corrected after the number of cell moves reaches a user-specified upper bound. Our backtracking mechanism then divides the current sequence of violating moves into sub-sequences, each focusing on resolving critical violations on a certain type of constraint. It first backtracks to the starting point of the current sub-sequence (focusing on, say, constraint c_i) and selects an alternative cell v within c_i's priority queue(s). Cell v is required to produce a low enough violation on c_i so that it can be resolved before the move process hitting the move limit again, assuming the same violation reduction gradient with the failed sub-sequence of moves (this is where learning comes into the picture). Subsequent cells are then picked using the selection algorithm of Sec 3.4.1. If violation still cannot be resolved, the move process backtracks one more level to the starting move of the immediately preceding sub-sequence, and so on. Finally, if backtracking fails for all sub-sequences, the initial violating move is revoked, and replaced by a feasible move (if any).

In summary, the above three mechanisms in combination ensure convergence to feasible PDP results, within a user-specified number of cell moves.

2.4.3 Intermediate Relaxation of Multi-Step Violation under Multiple Constraints

All discussions in Sec. 2.4 so far assume a one-step violation approach for multiple constraints, and convergence ensuring mechanisms require that the amount of critical violation decrease for every rectifying move. Multi-step violation based relaxation, on the other hand, allows temporarily increasing violation, and more thorough exploration of the solution space for better results.

Even under multiple constraints, the basic issues for multi-step relaxation are still centered around violation correctability and minimal primary objective deterioration (which requires good gain estimation methods). Also the same terminologies in Sec. 2.3 apply here: T for the set of violating moves that are not corrected, R for the set of required correcting moves due to T, and $rev(u)$ for the set of best correcting moves for the current violating move.

For the violation correctability problem, the feasibility estimation scheme for one-step violation, with necessary adaptations, provides a good solution. For any candidate move u that *increases* the current critical violation (otherwise the move is allowed unconditionally), as long as the upper bounds on critical violation are not reached, the algorithm proceeds to determine $rev(u)$. To do that, it repeatedly looks for the best free cell v to rectify the current critical violation, and adds v to $rev(u)$. Any such cell v is required to decrease the current critical violation (for fast convergence). If the final number of cells in $rev(u)$ is within a certain limit, u's violation is deemed correctable.

Once $rev(u)$ is decided, the *only* thing we need to do for gain estimation is to check if that Ineq. 4 holds true. If so, move of u is accepted. Hence, gain estimation is *independent* of the number of constraints involved. Note that in general $rev(u)$ can include cells from both subsets simultaneously.

To develop a high-quality gain estimation scheme, it is very important to estimate the aggregate gain of a set of moves accurately. Two special cases are noteworthy: a) some cells are connected to a common net in one subset; b) some cells share a net in both subsets. In the former case, the combined effect of moving those cells should be considered. For the latter, cell gains from different subsets should (partially) cancel each other out.

3. CONSTRAINT-BASED CONGESTION REDUCTION

Some previous work on congestion reduction for placement include [12], [14] and [9]. It is concluded in [12] that inclusion of congestion cost into the primary objective generally does not produce better results (in terms of both wire length and congestion) than a pure wire length objective. Thus, in [12, 14], congestion reduction is performed in a post-processing stage using a number of local improvement methods on wire-length optimized placement. In [9], congestion is estimated after each level of partitioning, and the results are used to influence the quadratic placer at next level. Thus, congestion reduction is also performed in a post-processing manner, albeit level by level.

In our approach, congestion is modeled as a set of constraints on the PDP process so that congestion reduction can be achieved simultaneously with wire length minimization (an *in-processing* approach). The two constraints considered for congestion control are as follows: *(i)* Balance the pin density of each resulting subcircuit of the current partition, as has been explained in Sec. 2.1. This constraint keeps connection density in all regions of the chip evenly distributed, and can thus control overall congestion. *(ii)* Partition a subcircuit so as to distribute its external nets (parallel to the current cutline) among its two subsets in proportion to their dimensions. As shown in Fig. 3, this constraint yields a uniform net distribution across routing channels, and hence a more routable placement. Imposing this constraint is equivalent to performing approximate *global route planning* as we are deciding in a top-down manner through

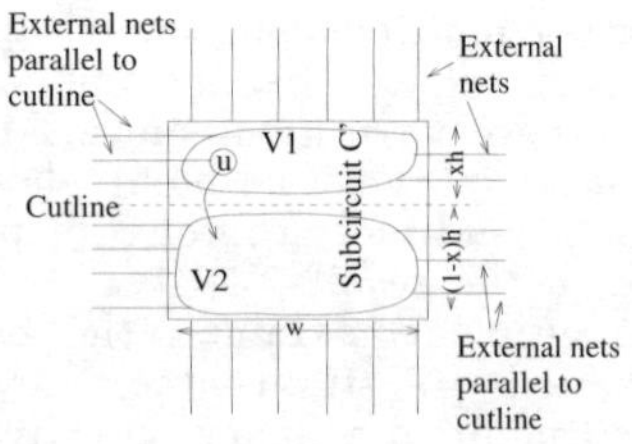

Figure 3: Proportionate distribution of external nets into subcircuit C' when bi-partitioning C' with a size balance ratio of $x : (1-x)$ for low-congestion global route planning.

which channels each net should cross to minimize wiring congestion.

The key to modeling external-net constraint is to keep track of the most likely crossing points of external nets in any region, as the PDP flow proceeds to finer levels of partitioning. To this end, a *connector* is introduced to indicate cross-region connections for any pair of net segments (a *net segment* is the part of a net spanning a particular region) that are likely to be directly connected to each other. Specifically, two net segments of a common net in adjacent containing regions are linked to a connector if the likely route for the net will connect these two segments *directly* through a routing channel. The connector is then positioned at the intersection of a rectilinear route connecting its pair of net segments and the shared boundary of containing regions. For example, in Fig. 4(a), two net segments ns_0 and ns_1 are expected to be connected directly to each other after partitioning, and a connector x is generated at the cutline.

After each level of partitioning, we generate *new* connectors and/or update *existing* ones based on the resultant cell distribution of each net. As seen in Fig. 4(a), a new connector is automatically generated at the cutline whenever a net segment ns_i is locally cut. Existing connectors, on the other hand, have to be re-assigned to new subsets and subsegments at the end of a partitioning process. An existing connector is associated with two net segments, so its new affiliations only depend on the partitioning status of these two segments (in subset 0, subset 1, or cut). Based on this fact, we identify one sub-segment from each associated net segment that are physically (rectilinearly) closest to each other, and re-assign the connector to them. As an example, the only sub-segment of ns_1, ns_{11}, is closest to the right sub-segment ns_{01} of ns_0 in Fig.4(b) as a result of the current level of partitioning (with the vertical dashed cutline), and the existing connector x is re-assigned to them. In some cases there can be two equal-length routes, like in Fig. 4(c) or (d), which lead to two possible new locations. In such cases we will pick a location that gives a better balanced final distribution of external nets. In general, existing connectors are 'inherited' down the PDP hierarchy. Note that in standard-cell placement scenarios, cells have to be abutted horizontally, leading to mis-aligned regions (as shown in Fig. 4(e)). Such *horizontal deviations* may require generation of additional connectors along the dimension *parallel* to the current cutline, if a rectilinear connection between the two closest sub-segments crosses region edges or cutlines other than the one where the existing connector is located.

After connector generation, the next level of partitioning can be started. During the partitioning process, we dynamically determine possible *locations* of each existing connector affiliated to regional edges *perpendicular to* the current cutline, in the same way as we determine its final location after partitioning at this level is completed. If a location on the region boundary of either subset is possible (like in Fig. 4(c)), both are accepted with a 0.5 probability. With locations known, a connector is counted as 1 in its residing subset (those with 0.5 probabilities are counted as 0.5 in both possible subsets). The external-net constraint says that connector counts must be balanced in any region, with the same ratio as local-size balances. This is a balance constraint that can be readily incorporated into our multi-constraint satisfaction framework.

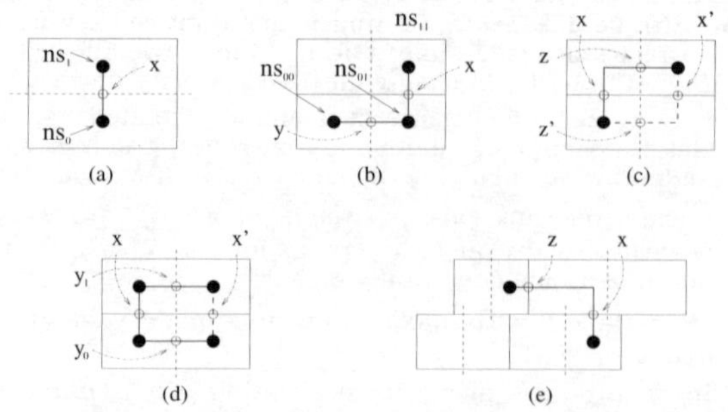

Figure 4: Examples of connector generation for global route planning. Current cutlines are shown as dashed lines, net segments as dark dots, and connectors as circles. Most likely routes are in solid forms, and possible alternative routes in dashed forms. (a) A new connector x is generated for net segments ns_0 and ns_1 after partitioning. (b) x is then 'inherited' after partitioning of next level, connected to new sub-segments ns_{01} and ns_{11}; connector y is newly generated. (c) In this case, x can be re-assigned to two possible locations (x or x'), and an additional connector (z or z') is generated accordingly to complete a rectilinear route. (d) There are again two possible locations for x, besides two new connectors y_0 and y_1. (e) Horizontal deviations may require generation of additional connector(s) z while updating the location of an existing connector x.

4. EXPERIMENTAL RESULTS

To test the efficacy of our proposed multi-constraint-satisfaction framework, we used an implementation of the SPADE algorithm based on the SHRINK-PROP [2] partitioner modified for wire length gains. We then developed the following two algorithms, both of which can be incorporated into SPADE: (1) An algorithm that allows one-step violation, and performs feasibility estimation and learning-based backtracking for fast convergence. (2) The second algorithm is an enhanced version of the first, featuring multi-step violation based intermediate relaxation. We conducted experiments first with SPADE alone (implicitly with local-size and row-size constraints), and then the above two multi-constraint (MC) algorithms with the inclusion of the two congestion constraints described in Sec. 3.

Experiments were run on ten MCNC and five IBM-PLACE standard-cell benchmarks whose characteristics are given in Table 1. All experiments were conducted on a 550MHz Pentium-III Linux workstation. Wire length (WL) results are reported in meters, and run times in seconds. The two congestion-related constraints are denoted as *p-dns* (pin-density) and *n-ext* (external-net). An overall congestion metric *ovfl* (overflow) is included to measure total exceeding of wiring capacity (supply) in all regions. Wiring demand at any edge of a region is estimated as the number of connectors along that edge. To determine wiring capacities, we ran SPADE (with local-size and row-size constraints only), performed connector generation, and then collected the average wiring density (number of net crossings at an edge divided by its length) across all regions. We then derived the maximum allowed wiring density as $(1+\epsilon) \cdot log(P)/K$ times the average density, where ϵ and K are adjustable constants, and P is the total number of pins of a circuit. The $log(P)$ factor stems from our expectation that larger circuits with higher standard deviations in regional wiring density tend to require a higher channel capacity. Wiring

Circuit	*Circuit Statistics*			*PDP w/o p-dns or n-ext*		
	cell	*net*	*row*	*WL*	*time*	*ovfl*
fract	125	147	6	0.021	1	15
prim1	752	902	16	0.74	28	72
struct	1888	1920	21	0.255	68	161
prim2	2907	3029	28	3.05	312	228
biomed	6417	5742	46	1.25	549	402
in2	12142	13419	72	12.01	2565	596
in3	15059	21939	54	36.09	4073	498
avq_s	21854	22124	80	5.40	3017	355
avq_l	25114	25384	86	5.91	3519	461
golem3	99932	144949	192	19.94	24340	1023
ibm01	12028	11507	64	3.95	2730	276
ibm02	19062	18429	64	11.81	6377	632
ibm03	21879	21621	64	10.50	6983	384
ibm04	26332	26163	64	15.14	8081	483
ibm05	29347	28446	64	37.77	12993	626
Total	–	–	–	163.83	75636	6212
Imprv.	–	–	–	–	–	–

Table 1: The left columns show the MCNC and IBM-PLACE benchmark statistics. The right columns are for SPADE results without congestion related constraints.

capacity of a regional edge is then the product of the edge length and maximum wiring density. In the experiments we used $\epsilon = 0.2$, and $K = 6.14$ (equal to logarithm of the pin number of circuit *fract*), which generally give reasonable total overflow without congestion control. All of our reported results here are obtained when the PDP process ends with no region containing more than 16 cells. A direct comparison of our results with those of [9, 12, 14] is difficult, since their congestion measurements are obtained after routing (global and/or detailed) is done, and there is a lack of general methods to determine wiring supply values for various benchmark circuits.

As shown in Table 2, by using of the first MC algorithm with feasibility estimation and learning-based backtracking mechanisms (with $\alpha = 0.5$), we obtained significant improvement on congestion, with very little WL increase. A separate experiment revealed that individually the pin-density and external-net balance constraints produced 5.6% and 11.2% reduction in congestion respectively.

Overall, we achieved about 14.3% reduction in total overflow, and above 20% improvement for some circuits, at the cost of about 1.6% average increase in wire length. We also obtained 10.5% improvement in the sum of maximum horizontal wiring demand per channel (not shown here due to page limit), which translates to at least $10.5 - 1.6 = 8.9\%$ chip area reduction. These results are quite promising. In comparison, previous methods attempting to reduce congestion during placement have generally obtained worse congestion as well as wire length. To the best of our knowledge, we are the first to show good congestion improvement with very small degradation in WL, by performing with-placement congestion control.

The second algorithm introduced intermediate relaxation with multi-step violation, and resulted in more than 50% less wire length increase compared to algorithm 2 (the deterioration is reduced from 1.6% to 0.7% after relaxation).

5. CONCLUSION

This paper has made two major contributions. The first is a general methodology to tackle multiple extra-objective constraints *during* a PDP optimization process. It includes techniques for quickly resolving violation of multiple constraints, and well-balanced consideration of multi-constraint satisfaction and primary objective preservation. The second contribution is a modeling of congestion control using pin-density and external-net balance constraints, and a technique for approximate global route planning (i.e., connector generation). Experiments on congestion reduction using these two constraints demonstrate the efficacy of both our general MC algorithms and constraint-based congestion control method. We achieved 14.3% reduction in congestion, at a wire length cost of only 1.6%, which was further reduced to 0.7% by intermediate relaxation with multi-step violations.

Circuit	*PDP w/ p-dns and n-ext FE+BT*			*PDP w/ p-dns and n-ext FE+BT, relaxation*		
	WL	*time*	*ovfl*	*WL*	*time*	*ovfl*
fract	0.020	2	12	0.019	2	13
prim1	0.76	33	60	0.75	38	62
struct	0.260	86	133	0.250	96	139
prim2	3.12	381	200	3.11	423	198
biomed	1.26	656	356	1.24	755	334
in2	12.26	3099	484	12.11	3422	503
in3	36.71	5110	414	36.50	6042	428
avq_s	5.51	3886	333	5.45	4514	321
avq_l	6.07	4570	391	5.94	5433	403
golem3	20.18	31979	884	20.05	37403	910
ibm01	3.97	3459	232	3.96	3857	236
ibm02	12.06	7998	581	11.87	9214	569
ibm03	10.69	9384	332	10.63	10548	346
ibm04	15.39	10517	413	15.19	12104	397
ibm05	38.11	16795	496	37.94	19813	504
Total	166.41	97955	5321	165.01	113664	5364
Imprv.	−1.6%	−29.5%	14.3%	−0.7%	−50.3%	13.7%

Table 2: The left set of results are with pin-density and external-net constraints, as well as feasibility estimation and learning-based backtracking. After introduction of multi-step violation (with relaxation limit set as 1% beyond original tolerances for each constraint), the right set of results show reduced wire length deterioration.

6. REFERENCES

[1] A. E. Dunlop and B. W. Kernighan. A procedure for placement of standard-cell VLSI circuits. *IEEE Trans. CAD*, pp. 92-98, Jan. 1985.

[2] S. Dutt and W. Deng, "Probability-based Approaches to VLSI Circuit Partitioning," *IEEE Trans. CAD*, pp. 534-549, Mar. 2000.

[3] S. Dutt and H. Theny, "Partitioning Around Roadblocks: Tackling Constraints with Intermediate Relaxations", *Proc. ICCAD*, pp. 349-355, 1997.

[4] H. Eisenmann and F. M. Johannes, "Generic Global Placement and Floorplanning," *Proc. DAC*, pp. 269-274, 1998.

[5] H. Esbensen and E. Kuh, "Exploring the design space for building block placements considering area, aspect ratio, path delay and routing congestion", *Proc. Physical Design Workshop*, 1996, pp. 126-133.

[6] D. J-H. Huang and A.B. Kahng, "Partitioning-based standard-cell global placement with an exact objective", *Proc. ISPD*, pp. 18-25, 1997.

[7] R. Kuznar, F. Brglez and K. Kozminski, "Cost minimization of partitions into multiple devices", *Proc. DAC*, pp. 315-320, 1993.

[8] S. Mayrhofer and U. Lauther, "Congestion-driven placement using a new multi-partitioning heuristic", *Proc. ICCAD*, pp. 332-335, 1990.

[9] P.N. Parakh, R.B. Brown and K.A. Sakallah, "Congestion-driven quadratic placement", *Proc. DAC.*, pp. 275-258, 1998.

[10] W-J. Sun and C. Sechen, "Efficient and Effective Placement for Very Large Circuits", *IEEE Trans. CAD*, pp. 349-359, Mar. 1995.

[11] M. Wang, X. Yang and M. Sarrafzadeh, "Dragon2000: Fast Standard-cell Placement for Large Circuits", *Proc. ICCAD*, pp. 260-263, 2000.

[12] M. Wang, X. Yang and M. Sarrafzadeh, "Congestion Minimization During Placement", *IEEE Trans. CAD*, pp. 1140-1148, Oct. 2000.

[13] W. Wolf, *Modern VLSI Design*, 2nd ed., Prentice-Hall, 1998.

[14] X. Yang, R. Kastner and M. Sarrafzadeh, "Congestion Reduction During Placement Based on Integer Programming", *Proc. ICCAD*, pp. 573-576, 2001.

[15] K. Zhong and S. Dutt, "Effective partition-driven placement with simultaneous level processing and global net views", *Proc. ICCAD*, pp. 254-259, 2000.

[16] H. Zhou and D.F. Wong, "Global Routing with Crosstalk Constraints", *Proc. DAC*, pp. 374-377, 1998.

Coping with Buffer Delay Change Due to Power and Ground Noise

Lauren Hui Chen*, Malgorzata Marek-Sadowska**, Forrest Brewer**
*Avant! Corp., Fremont, CA 94538, USA
** Electrical and Computer Engineering Department, University of California, Santa Barbara, CA 93106, USA
*lauren@avanticorp.com, **{mms, forrest}@ece.ucsb.edu

ABSTRACT

Variation of power and ground levels affect VLSI circuit performance. Trends in device technology and in packaging have necessitated a revision in conventional delay models. In particular, simple scalable models are needed to predict delays in the presence of uncorrelated power and ground noise. In this paper, we analyze the effect of such noise on signal propagation through a buffer and present simple, closed-form formulas to estimate the corresponding change of delay. The model captures both positive (slowdown) and negative (speedup) delay changes. It is consistent with short-channel MOSFET behavior, including carrier velocity saturation effects. An application shows that repeater chains using buffers instead of inherently faster inverters tend to have superior supply level-induced jitter characteristics.

Categories & Subject Descriptors:

J.6 [Computer-Aided Engineering]: Computer-aided design (CAD).

General Terms:

Algorithms.

Keywords:

Power and ground noise, differential mode noise, common mode noise, incremental delay change.

1. INTRODUCTION

This paper describes a new model for the change in buffer delay caused by both power and ground supply level variations and level variations between stages in sequences of repeaters. These delay changes are a large component of the total timing jitter for a signal where the jitter accounts for all noise sources such as substrate noise and coupling noise as well as power level noise. There is a substantial amount of previous work in this area, notably papers: [3][5][10]. However, for several reasons described below, we believe that the problem bears re-examination and an effort made to create a fast, simple model suitable for mass implementation in a modern design flow.

Growth in design sizes and scaling of interconnections have lead to the requirement for insertion of very large numbers of buffer/repeaters in recent designs [1][7]. Because of their preponderance in number, use in heavily loaded nets, and use in clock and timing circuits, buffer delays account for a large percentage of all critical timing nets in a design. In some of these applications, total timing uncertainty (not just worst case delay) is important. At the same time, scaling of power supply levels and improving transconductance of devices have increased the sensitivity of buffers to supply level induced delays. Finally, increases in chip-level design scales and modern packaging strategies such as bump bonding have localized supply variations so that buffers in one set of supply levels are driving buffers in another zone with differing supply levels. Since power loading is logic switching dependent and supply sources are localized, power and ground levels need not be inversely correlated as is typical in wire bonded die.

Under such conditions, power level induced delay changes may either increase or decrease the effective delay of a buffer, and successive stages may or may not accumulate incremental delays. One must consider both power and ground levels at the signal source and at the current buffer to derive an equivalent delay change. This value can be substantially smaller than that predicted by superposing ground-bounce and power level changes [3][11][12]. Second, the delay effects of *common mode* voltage shifts will be shown to be larger in scale to equivalent differential mode changes. (Differential mode voltage shifts are the commonly studied model). Lastly, changes in power distribution and clocking strategies and the potential for future changes, create the need for a timing model which is independent of common assumptions about power level noise sources. We *do* assume that large scale power level changes result from ensemble effect of many devices and occur at a somewhat slower time scale than the typical switching changes in buffers.

In the following, we analyze the effect of P/G noise on buffer delay, and present linear, closed-form formulas for the corresponding incremental changes in delay based on a short-channel transistor model. The expressions simultaneously model both the power supply and ground levels, resulting in positive (slowdown) or negative (speedup) delay changes. They are suitable for estimation of both upper and lower bounds on signal arrival time. Furthermore, they are shown to be largely independent of the buffer load circuit structure, increasing their applicability. These expressions are suitable for inclusion in timing analysis tools and to statistical delay estimators due to their rapid evaluation. Lastly, the expressions make no assumptions about the specific shape of the P/G noise waveform.

The paper is organized as follows: sections 2-3 defines P/G noise, buffer delay nomenclature and illustrate P/G induced buffer delays. Section 4 presents the new model. Section 5 demonstrates the accuracy and fidelity of the model. Applications of the model and concluding remarks are presented in sections 6-7.

2. BUFFER DELAY CONVENTIONS

A *buffer* is a chain of tapered inverters. Here, we consider buffers consisting of one or two inverters.

2.1 Variation of V_{dd} and V_{ss}

We use V_{dd}, V_{ss}, V_i (input), etc. to represent voltages related to ideal power and ground levels, and V_{dd}', V_{ss}', V_i', etc. to represent corresponding values in the presence of power and ground noise. ΔV_{dd} and ΔV_{ss} denote the variation of power and ground levels, respectively.

$$\Delta V_{dd} = V_{dd}' - V_{dd} \text{ (power noise)} \quad (1)$$

$$\Delta V_{ss} = V_{ss}' - V_{ss} = V_{ss}' \ (V_{ss} = 0)\text{(ground/ noise)} \quad (2)$$

In wire-bond packaging styles, a dominant supply level noise source is bond wire inductance in the package. Neglecting I/O current drives, the power and ground noise of the chip due to simultaneous switching typically follow an inverse pattern, and ΔV_{dd} is often symmetric to ΔV_{ss}. However, in modern bump-

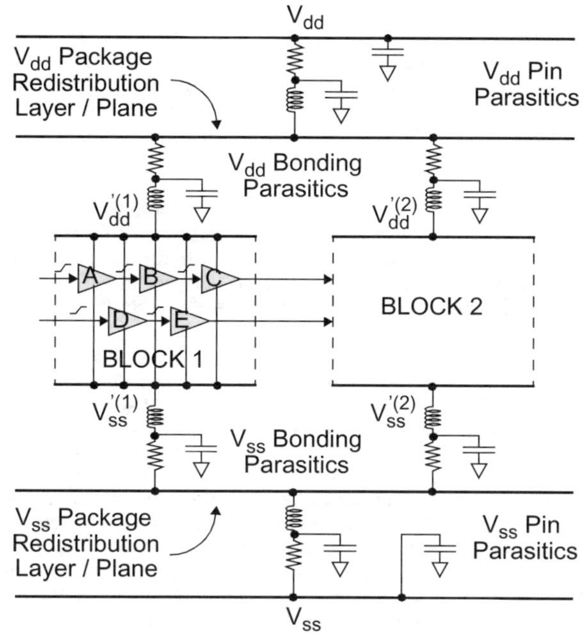

Figure 1. Power distribution: bump-bond packaging

bonded and low-inductance package styles, the package distributes power over the whole area of the chip (figure 1). Every bump connects to a local power/ground network. To save chip metallization area and improve density, chip global power distribution metal is reduced in lieu of thicker package distribution layers. Increasing design scales causes an increase in long wire loading, and in more wires connecting between different power domains. Logic-level-dependent currents flow between such blocks, causing asymmetric power and ground noise within a block. This noise is increased by the inclusion (within a bump block) of long wire repeater buffers which are often added in a post placement timing optimization step.

Figure 1 shows a simple equivalent circuit for bump-bond packaging. Each block is defined by the subcircuit supplied by a pair of bumps (V_{dd}/V_{ss}). Suppose that the cells A to E have transitions. Switching of the buffers A, B and D has a symmetric effect on the power and ground noise ($\Delta V_{dd}^{(1)}$ and $\Delta V_{ss}^{(1)}$), because they drive loads (consisting largely of parasitic interconnect capacitance) within the same block. On the other hand, switching of the C and E buffers has a non-symmetric effect on $\Delta V_{dd}^{(1)}$ and $\Delta V_{ss}^{(1)}$, because they drive loads which are outside of block 1, causing different switching currents to flow through the power and ground ports of block 1. Since wires leaving a block are likely to be physically long, these currents are proportionally large.

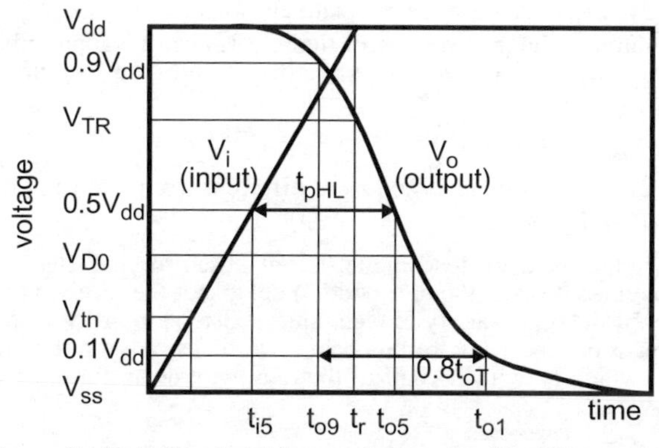

Figure 2. Notation for delay and slope for ideal V_{dd} and V_{ss}

2.2 Incremental buffer delay change

We define a buffer's *ideal* delay as the time interval between its input and output voltage reaching 50% of the power level. Figure 2 illustrates this definition. t_{pHL} is the high-to-low delay when the input of the inverter has a rising transition. Input and output transition times are t_r and t_{oT} respectively. Other time values are: t_{i5}, t_{o5}, t_{o1} and t_{o9} which are times when the input or output voltage reaches 50%, 10%, and 90% of V_{dd}, respectively.

$$t_{pHL} = t_{o5} - t_{i5}\,,\ t_{oT} = (t_{o1} - t_{o9})/0.8 \qquad (3)$$

Figure 3 illustrates the delay and slope with P/G noise. The disturbed output voltage points are defined as follows:

$$V'_{o1} = V'_{ss} + 0.1(V'_{dd} - V'_{ss})\,,\ V'_{o9} = V'_{ss} + 0.9(V'_{dd} - V'_{ss})$$

$$V'_{o5} = V'_{ss} + 0.5(V'_{dd} - V'_{ss})$$

The disturbed high-to-low delay and slope are given by:

$$t'_{pHL} = t'_{o5} - t'_{i5}\,,\ t'_{oT} = (t'_{o1} - t'_{o9})/0.8 \qquad (4)$$

An alternative delay, $t'^{(2)}_{pHL}$, shown in figure 3, measures the delay referenced to the *disturbed* power and ground level V'_{o5}. In this paper, we will use t'_{pHL} in our analysis and results. Results for other definitions of delay change are similar but are omitted due to space limitations.

With a rising transition at the input, the changes of delay and output transition time are defined as follows:

$$\Delta t_{pHL} = t'_{pHL} - t_{pHL}\,,\ \Delta t_{oT} = t'_{oT} - t_{oT} \qquad (5)$$

3. EFFECT OF P/G NOISE ON BUFFER DELAY

Changes of power and ground levels affect signal propagation through an inverter in several ways.

3.1 Differential mode noise

We define *differential mode noise* (DMN) ΔV_{dif} as:

$$\Delta V_{dif} = V'_{dif} - V_{dif} = \Delta V_{dd} - \Delta V_{ss}$$

where

$$V'_{dif} = V'_{dd} - V'_{ss} \text{ and } V_{dif} = V_{dd} - V_{ss}$$

Differential mode noise ΔV_{dif} may be positive or negative, depending on the directions and amplitudes of ΔV_{dd} and ΔV_{ss}. The voltage difference (V_{dif}) between power supply and ground levels determine how fast the buffer charges/discharges its capacitive load.

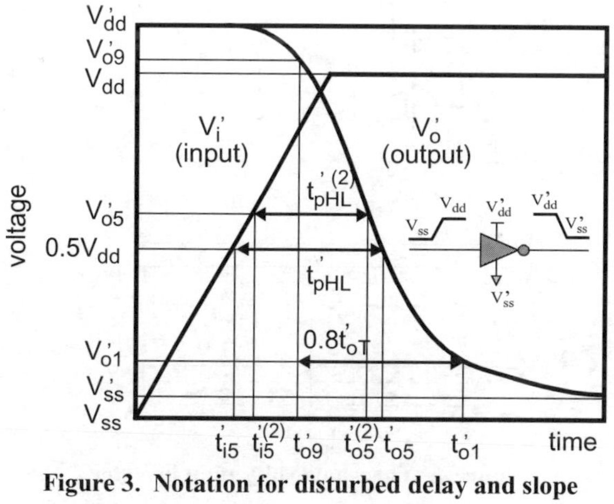

Figure 3. Notation for disturbed delay and slope

Observation1: The buffer delay change is *linearly dependent* on the differential mode noise (DMN) as will be shown in section 4:

$$\Delta t_{pHL}\big|_{DMN} = -k_d \cdot \Delta V_{dif} = -k_d \cdot (\Delta V_{dd} - \Delta V_{ss}) \quad (6)$$

where k_d is a positive constant dependent on the device and technology parameters, input transition times, and the gate load. Similar effects hold for both Δt_{pHL} and Δt_{pLH}.

3.2 Common mode noise

We define the *common mode noise* (CMN) ΔV_{com} as:

$$\Delta V_{com} = \Delta V_{dd} + \Delta V_{ss}$$

CMN modifies the effective switching threshold of the gate. This threshold shift changes the gate delay as illustrated in figure 4. Figure 4(a) shows an rising transition arriving at the buffer. Figure 4(b) illustrates the gate threshold shift and the corresponding delay change.

In figure 4(b), N and P are the original points when the nfet switches from cutoff to saturation region and pfet switches from saturation to cutoff, respectively. The corresponding switching times are indicated t_n and t_p. For noise of limited amplitude, the transistor thresholds (V_{tn} and V_{tp}) do not change significantly, so:

$$V'^{(N)}_{GS}\big|_{CMN} = V_i - V'_{ss} = V_{tn} > 0$$

$$V'^{(P)}_{GS}\big|_{CMN} = V_i - V'_{dd} = V_{tp} < 0$$

This causes a shift of the nfet and pfet switching points from N and P to N' and P'. The corresponding switching time shifts to t'_n and t'_p, respectively. We observe in figure 4(b) that:

$$\Delta V_{ss} > 0 \Rightarrow V^{(shift)}_{tn} > V_{tn} \Rightarrow t'_n > t_n \Rightarrow \text{increasing delay}$$

$$\Delta V_{dd} > 0 \Rightarrow V^{(shift)}_{tp} > V_{tp} \Rightarrow t'_p > t_p \Rightarrow \text{increasing delay}$$

And vise versa when power supply and ground level decrease. Therefore, we make the following observation:

Observation 2: For an input with a rising transition, the dependency between the common mode noise (CMN) and the buffer delay change can be expressed by:

$$\Delta t_{pHL}\big|_{CMN} = k_{cr} \cdot \Delta V_{com} = k_{cr} \cdot (\Delta V_{dd} + \Delta V_{ss}) \quad (7)$$

where k_{cr} is a positive constant determined by the device and technology parameters, input transition time, and the gate load. Similarly, for an input with a falling transition, the dependency between the common mode noise (CMN) and the buffer delay change can be expressed by:

$$\Delta t_{pLH}\big|_{CMN} = -k_{cf} \cdot \Delta V_{com} = -k_{cf} \cdot (\Delta V_{dd} + \Delta V_{ss}) \quad (8)$$

where k_{cf} is a positive constant.

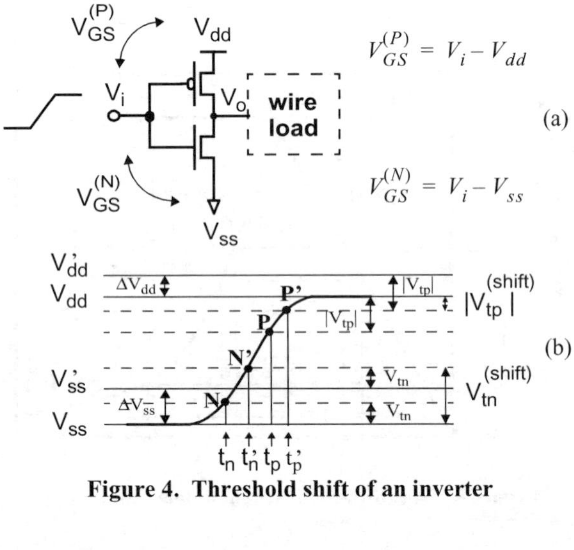

Figure 4. Threshold shift of an inverter

3.3 Loading effects

Both differential mode noise (DMN) and common mode noise (CMN) change buffer delays. Since the delay change can be of either sign, the noise sources need to be modeled together. Figure 5 shows alternative load configurations and the corresponding simulated delay change (both rising and falling transition) in 0.18μm technology. Note: $\Delta delay = 0$ when $\Delta V_{dd} = \Delta V_{ss} = 0$.

In figure 5 (a), the wire load of the inverter is a distributed RC tree network, including vias, extracted from the layout of a real circuit. In figure 5 (b), the wire load is simplified to an RC π-model. In figure 5 (c), the wire load is further simplified to a single resistor plus a single capacitor. In figure 5 (d), an effective loading capacitor is used to replace the inverter's output load. These simplified wire-load models in figures 5 (b)-(d) can be obtained using the methods described in [8]. The delay is measured when the input (V_i) and output (V_o / V_o' / V_o'') voltage reach 50% of the ideal power supply voltage, respectively. The range for the power and ground noise is from -20% (-0.36 volt) to 20% (0.36 volt) of the power supply voltage, which is set to 1.8 volt for the selected technology. The range for the change of delay is from -30ps to 30ps. It is interesting to note that each of the four wire load models displays a linear relationship between the change of power/ground level and the change of inverter delay. Furthermore, the linearity improves when the change of power and ground level is smaller than 20%. In practical designs, the tolerable range for the power and ground levels is less than ±10%. Thus, the CMN and DMN induced delays can be superposed as noted below:

Observation 3: For a rising input transition and any of the wire load models described in figure 5, the incremental change of buffer delay is expressed as:

$$\Delta t_{pHL} = k_{1r} \cdot (\Delta V_{dd} + \Delta V_{ss}) - k_{2r} \cdot (\Delta V_{dd} - \Delta V_{ss}) \quad (9)$$

A similar result applies to a falling input transition:

$$\Delta t_{pLH} = -k_{1f} \cdot (\Delta V_{dd} + \Delta V_{ss}) - k_{2f} \cdot (\Delta V_{dd} - \Delta V_{ss}) \quad (10)$$

where k_{1r}, k_{1f}, k_{2r} and k_{2f} are positive constants dependent only on the input transition time, gate load, and the device technology parameters.

Observation 4: Buffer delay change is more sensitive to common mode noise than to differential mode noise in deep submicron designs, and may be dominated by common mode noise in some instances.

Observation 3 indicates that an appropriate simplified wire load model can be used for delay modeling of the buffer itself. With appropriate techniques [2][8], the distributed RC load of a gate can be simplified into the nearly equivalent π-model shown in figure 5(b). This π-model is further simplified into an effective capacitance load, shown in figure 5(d), by equating the average currents for the two load models. Such a capacitance model is inaccurate when the gate is behaving like a resistor [8]. This inaccuracy occurs primarily in the tail portion of the output waveform. However, for our purposes, the buffer delay is measured at the midpoint of the logic swing. Therefore, the inaccuracy in the gate delay (not the wire delay) caused by the effective capacitance model is relatively small.

The output voltage at node o in figure 5 (a) can be approximated by the voltage at node o' in figure 5 (b), o'' in figure 5 (c) and o''' in figure 5 (d):

$$V_o \approx V'_o \approx V''_o \approx V'''_o$$

However, the voltage at node q is different from that of node o:

$$V_q \neq V_o$$

The simplified wire load model is only used to characterize the interconnect's driving point (node o) delay, not the receiving node (node q) delay. A variety of techniques exist [3] in order to model the interconnect delay. In this paper, we focus only on the buffer delay which is typically half of the total wire delay for optimized long wires.

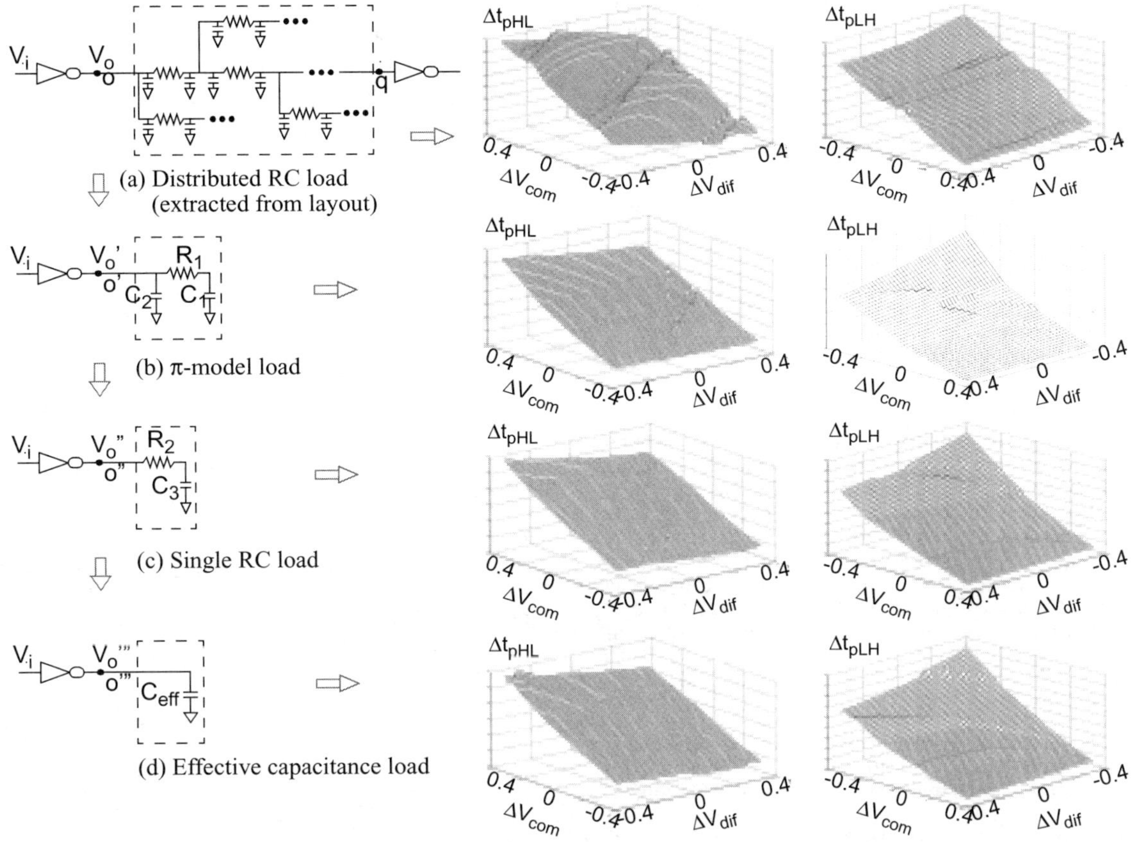

Figure 5. Buffer delay change for different load models

Power and ground noise amplitudes can either be correlated or largely independent depending on the relative magnitude of the power distribution parasitics and the relative number and activity of signals crossing between power distribution blocks. Such noise contributes to the local buffer/inverter delay in a complex way which can either increase or decrease the signal delay. For application in performance estimation, optimization or analysis it is useful to develop simple models which can be quickly evaluated and which can be linked to theoretical device models.

4. THEORETICAL MODEL

In deep submicron circuits, carrier velocity saturation effects predominate. To capture these effects, we use a short channel alpha-power law MOSFET model [9]. This model is based on four parameters: α (velocity saturation index), V_{tn} (threshold voltage), I_{D0} (drain current at $V_{GS} = V_{DS} = V_{dd}$), V_{D0} (drain saturation voltage at $V_{GS} = V_{dd}$). Note, n and p subscripts denote parameters related to nfet and pfet transistors, respectively. We assume that for a given transistor with a given loading capacitance, the above four values remain unchanged given a small disturbance of power supply and ground levels. Below, we present an abbreviated derivation of the model.

Referring to notations given in figures 2 and 3, we have:

$$t_{i5} = t_r/2 \tag{11}$$

$$t_{o5} = (C_L V_{dd})/(2I_{D0}) + (v_T + \alpha)/(1+\alpha) \cdot t_r \tag{12}$$

$$t_{i5}^{'(1)} = t_r/2 \tag{13}$$

$$t_{o5}^{'(1)} = [C_L/(I_{D0} k_{v1})] \cdot (V_{dd}/2 + \Delta V_{dd}) + t_{\alpha}' \tag{14}$$

where

$$t_{\alpha}' = t_r \cdot \left(1 - \left(1 - \frac{\Delta V_{ss}}{V_{dd}} - v_T\right)/(1+\alpha)\right), \quad v_T = V_{tn}/V_{dd}$$

$$k_{v1} = \left(\frac{V_{dd} - \Delta V_{ss} - V_{tn}}{V_{dd} + \Delta V_{dd} - V_{tn}}\right)^{\alpha} \approx 1$$

From equations (11) - (14) and equations (3), (4) and (5), we obtain the incremental change of buffer delay:

$$\begin{aligned}\Delta t_{pHL} &= k_{1n} \cdot \Delta V_{com} - k_{2n} \cdot \Delta V_{dif} \\ &= k_{1n} \cdot (\Delta V_{dd} + \Delta V_{ss}) - k_{2n} \cdot (\Delta V_{dd} - \Delta V_{ss})\end{aligned} \tag{15}$$

The change of delay can also be expressed as:

$$\Delta t_{pHL} = k_{3n} \cdot \Delta V_{dd} + k_{4n} \cdot \Delta V_{ss} \tag{16}$$

where

$$k_{1n} = \frac{t_r}{2V_{dd}(1+\alpha)} + \frac{C_L}{2I_{D0}}, \quad k_{2n} = \frac{t_r}{2V_{dd}(1+\alpha)} - \frac{C_L}{2I_{D0}}$$

$$k_{3n} = k_{1n} - k_{2n}, \quad k_{4n} = k_{1n} + k_{2n} \tag{17}$$

Similar equations are derived for Δt_{pLH} with α, V_{tp}, I_{D0}, V_{D0} from the corresponding pfet, and the polarity of k_{1p} is reversed:

$$\begin{aligned}\Delta t_{pLH} &= -k_{1p} \cdot \Delta V_{com} - k_{2p} \cdot \Delta V_{dif} \\ &= k_{3p} \cdot \Delta V_{dd} + k_{4p} \cdot \Delta V_{ss}\end{aligned} \tag{18}$$

where

$$k_{3p} = -k_{1p} - k_{2p}, \quad k_{4n} = -k_{1p} + k_{2p} \tag{19}$$

Theorem 1: Equations (15) to (19) demonstrate that the incremental change of buffer delay is *linear* with respect to the power and ground variations.

The coefficients k_{1n} and k_{1p} quantify the effect of common mode noise while k_{2n} and k_{2p} characterize the effect of differential mode noise on buffer delay. This theorem shows why the observations in sections 3.1 through 3.3 hold. k_{1n}, k_{2n}, k_{1p} and k_{2p} are equivalent to k_{1r}, k_{2r}, k_{1f} and k_{2f} defined in observation 3.

Finally, with h_3 and h_4 determined by technology parameters, the change of slope can be expressed as:

$$\Delta t_{oT} = h_3 \cdot \Delta V_{dd} + h_4 \cdot \Delta V_{ss} \qquad (20)$$

5. MODEL VALIDATION

We validated our model in both 0.25μm and 0.18μm technologies. As mentioned in section 4.1, the alpha-power law MOSFET model relies on four parameters: α, V_{tn}, I_{D0} (drain saturation current), and V_{DO} (drain saturation voltage). We determined these values for each transistor through HSPICE simulation. We follow the method in [9] to extract α and V_{tn}.

The results in Table 1 show that the model provides accurate estimation, with less than 5% error relative to HSPICE over a ±20% supply variation range. The model is not as accurate in estimating the change of transition time (Δt_{oT}). For delay estimation, this is acceptable because Δt_{oT} has only a second order effect on the delay of the next stage. Note that this modeling technique applies to arbitrary size inverters, loading capacitance, and input transition times. Comparison between the technologies show the trend of increasing sensitivity to supply level noise with scaling.

6. APPLICATION

An important feature of the model is relative lack of dependence on the circuit loading structure. This simplifies inclusion in a design flow as a modification to the existing delay calculation.

6.1 Delay change for clock buffers

To preserve duty cycle, clock buffer chain designs often presume equal input and output transition times. Thus:

$$t_r = \frac{0.9C_L}{0.8I_{D0}} V_{dd} + \frac{C_L V_{D0}}{0.8 I_{D0}} \cdot \ln\left(\frac{10V_{D0}}{eV_{dd}}\right)$$

$$\Delta t_r = \frac{0.9C_L}{0.8I_{D0}}(\Delta V_{dd} - \Delta V_{ss}) + \frac{C_L V_{D0}}{0.8 I_{D0}} \cdot \ln\left(\frac{V_{dd}}{V_{dd} + \Delta V_{dd} - \Delta V_{ss}}\right)$$

Substituting t_r and Δt_r into equation (15), we have:

$$\begin{aligned}\Delta t_{pHL} &= (C_L/I_{D0}) \cdot (f_{1n} \cdot \Delta V_{com} - f_{2n} \cdot \Delta V_{dif}) \\ &= (C_L/I_{D0}) \cdot (f_{3n} \cdot \Delta V_{dd} + f_{4n} \cdot \Delta V_{ss})\end{aligned} \qquad (21)$$

where

$$f_{1n} = (f_{3n} + f_{4n})/2 \ , \ f_{2n} = (f_{4n} - f_{3n})/2 \qquad (22)$$

When Δt_r is ignored, we have:

$$f_{3n} = 1 \ , \ f_{4n} = f_{01}/(1+\alpha)$$

If Δt_r is not negligible, we have:

$$\Delta t_{pHL} = (C_L/I_{D0}) \cdot (f'_{3n} \cdot \Delta V_{dd} + f'_{4n} \cdot \Delta V_{ss}) \qquad (23)$$

where

$$f'_{3n} = f_{3n} + f_{03} \ , \ f'_{4n} = f_{4n} - f_{03}$$

where

$$f_{01} = \frac{0.9}{0.8} + \frac{V_{D0}}{0.8V_{dd}} \cdot \ln\left(\frac{10V_{D0}}{eV_{dd}}\right)$$

$$f_{02} = \frac{0.9}{0.8} - \frac{V_{D0}}{0.8V_{dd}} \ , \ f_{03} = \left(\frac{1}{2} - \frac{1 - v_T}{1+\alpha}\right) \cdot f_{02}$$

For a falling transition, we have:

$$\Delta t_{pLH} = (C_L/I_{D0}) \cdot (-f_{1p} \cdot \Delta V_{com} - f_{2p} \cdot \Delta V_{dif}) \qquad (24)$$

Note that equations (21) ~ (24) are independent of input slope. This result is used below to determine the cumulative jitter in a buffer chain.

6.2 Clock buffer chains

The linear relationship between the P/G noise and delay change can be used to analyze the delivered jitter for a chain of single-inverter-buffers and a chain of double-inverter-buffers, see fig. 6.

Assume the input is a rising transition. For one stage of a single-inverter-buffer, we have from (21):

$$\Delta delay^{(s)} = \Delta t_{pHL}^{(s)} = \frac{C_L}{I_{D0}^{(s)}} \cdot (f_{1n}^{(s)} \cdot \Delta V_{com} - f_{2n}^{(s)} \cdot \Delta V_{dif}) \qquad (25)$$

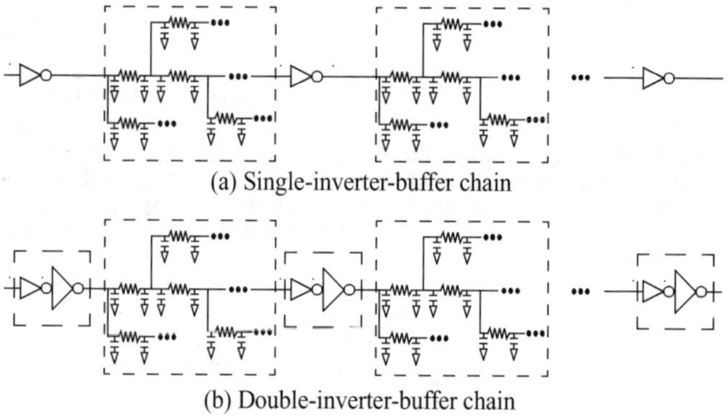

(a) Single-inverter-buffer chain

(b) Double-inverter-buffer chain

Figure 6. Delay change for single-inverter chain v.s. double-inverter chain

Table 1: Model Validation for Delay and Slope

Parameter					0.25μm						0.18μm					
					Simulation		Our Method				Simulation		Our Method			
W_p/W_n (μm)	C_L (ff)	t_r (ps)	ΔV_{dd} (volt)	ΔV_{ss} (volt)	Δt_{pHL} (ps)	Δt_{oT} (ps)	Δt_{pHL} (ps)		Δt_{oT} (ps)		Δt_{pHL} (ps)	Δt_{oT} (ps)	Δt_{pHL} (ps)		Δt_{oT} (ps)	
10/5	100	100	-0.25	-0.250	-21.9	-4.2	-21	3.2%	-4.8	15%	-24.9	-10.7	-24.5	1.8%	-12.9	20%
10/5	100	100	0.00	-0.100	-4.7	0.2	-4.7	0.7%	0.2	1.1%	-6.2	-2.9	-6.0	4.0%	-2.6	12%
10/5	100	100	0.025	0.100	5.4	0.4	5.6	4.5%	0.4	0.7%	7.0	3.0	6.9	0.3%	3.1	6.2%
10/5	100	100	0.100	0.00	3.7	2.0	3.8	1.0%	2.0	1.8%	3.7	2.7	3.8	3.2%	2.9	8.0%
10/5	20	50	-0.500	0.025	-6.0	-1.7	-5.7	5.6%	-1.8	5.1%	-8.1	-6.9	-7.8	3.5%	-6.2	9.9%
10/5	20	50	0.500	0.100	8.6	2.7	8.2	4.7%	2.5	7.0%	9.7	1.6	10.2	5.0%	1.5	5.1%
5/5	100	100	0.250	0.100	14.1	5.4	14.4	1.6%	4.9	8.2%	17.3	3.6	16.5	4.9%	3.1	14%
5/5	100	100	0.500	-0.025	16.5	10.2	17.0	3.4%	10.2	0.8%	19.2	7.7	18.1	5.5%	7.2	6.7%
5/5	100	50	0.025	-0.250	8.3	-4.9	8.7	4.7%	-5.1	5.8%	8.5	1.7	8.9	4.9%	1.5	12%
10/10	100	100	-0.25	0.050	-4.7	-10.1	-4.6	1.1%	-11.8	17%	-4.7	-2.2	-4.5	5.1%	-2.3	3.2%

For one stage of a double-inverter-buffer, we assume a tapered-buffer design. From (21) and (24) we have:

$$\begin{aligned}\Delta delay^{(d)} &= \Delta t_{pHL}^{(d)} + \Delta t_{pLH}^{(d)} \\ &= C_L/I_{D0}^{(d)} \cdot [(f_{1n}^{(d)} - f_{1p}^{(d)}) \cdot \Delta V_{com} - (f_{2n}^{(d)} + f_{2p}^{(d)}) \cdot \Delta V_{dif}]\end{aligned} \quad (26)$$

where the superscripts denote the parameters for single (s) and double (d) inverter buffers.

In deep submicron technologies, the buffer delay change is more sensitive to common mode noise than to differential mode noise. This has been experimentally demonstrated by our simulation results in figure 5, and theoretically shown by equation (15) which indicates $k_{1n} > k_{2n}$. In other words, we have

$$f_{1n}^{(s)} > f_{2n}^{(s)}, \quad f_{1n}^{(d)} > f_{2n}^{(d)}, \quad f_{1p}^{(d)} > f_{2p}^{(d)}$$

When the amplitude of the common mode noise is at least as large as that of the differential mode noise, the delay change of both buffer designs will be dominated by common mode noise. However, when $f_{1n}^{(d)}$ and $f_{1p}^{(d)}$ are comparable, the delay induced by common mode noise from (26) will cancel while the delay change given by (25) will be dominate. Hence, we make the following observation:

Observation 5: The delay of a double-inverter-buffer chain is less sensitive to the power/ground noise variations than that of a single-inverter-buffer chain.

Figure 7 shows simulation results of buffer delay change for the buffer chains shown in figure 6. The power/ground noise of each buffer in the chain is independent of the others and range over $\pm 10\%$ of V_{dd}. Inverter sizes are determined such that both buffer chains in figure 6 have a similar nominal delay (around 280ps). This is done to simplify comparison of the delay changes, because buffer chains with un-correlated delay would be difficult to compare. We choose similar wire loads for each stage. We randomly simulated 20000 combinations of P/G noise induced jitter. The statistics in figure 7 clearly show that the overall delivered jitter (total delay change of the buffer chain) for the double-inverter buffers is smaller. This provides us a new guideline for design: in terms of power/ground noise avoidance, the double-inverter buffer chain is a better choice. Double inverter buffers have slightly larger current requirements than inverters due both to tapering and to the domination of load capacitance by the interconnect, so the effect of additional current is minor. This result is effected by rapidly changing power levels primarily in the slow rise-time (RC dominated) extents of the interconnect. However, such effects should be similar for both styles of repeater.

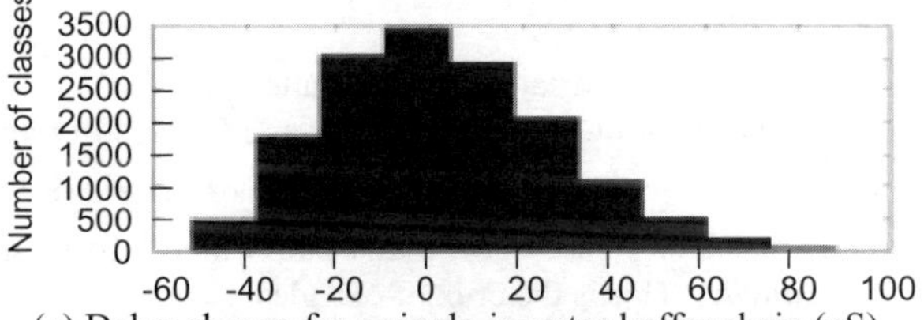

(a) Delay change for a single-inverter-buffer chain (pS)

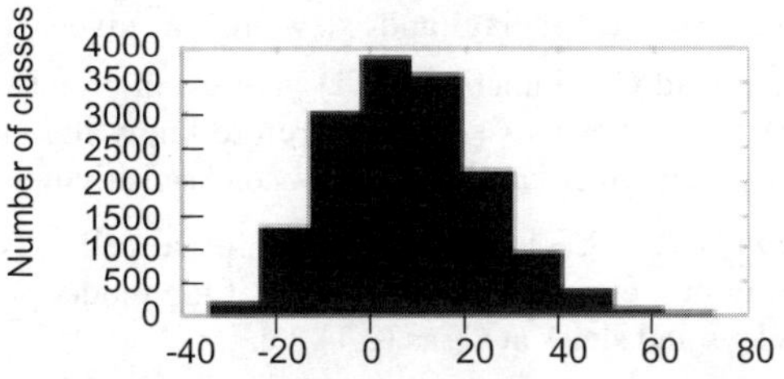

(b) Delay change for a double-inverter-buffer chain (pS)

Figure 7. Comparison of buffer delay change (histogram)

7. CONCLUSIONS

Maintaining signal integrity in deep submicron circuits is a difficult problem. Variations of power and ground levels play an important role because this type of noise significantly degrades circuit performance. Deep submicron circuits have decreased power supply level V_{dd} and decreased velocity saturation index α, leading to increased sensitivity of delay to P/G noise. Thus, despite the reduction of noise from lower inductance packaging, the relative magnitude of the delay changes is still a serious potential problem.

We studied the effects of differential and common mode power/ground noise on buffer delay. Using the α-power law MOSFET model, we derived general formulas to estimate the influence of power and ground noise on delay and slope. As our model does not rely on the circuit structure, it can be incorporated into any existing gate delay calculation techniques. It is simple and accurate. An application in clock buffer chain design shows that repeater chains using buffers instead of inherently faster inverters tend to have superior level - induced delay characteristics.

ACKNOWLEDGEMENT

This work was supported in part by the NSF grant # CCR-0098069 and in part by the California MICRO program through Mindspeed and Synopsys.

REFERENCE

[1] C. J. Alpert, A. Devgan, S. T. Quay, *Buffer insertion for noise and delay optimization*, IEEE Trans. on Computer-Aided Design of Integrated Circuits and Systems, vol.18, no.11, Nov.1999, pp.1633-45.

[2] R. Arunachalam, F. Dartu and L. T. Pillage, CMOS gate delay models for general RLC loading, ICCD, pp.224-229, 1997.

[3] H. H. Chen and J. S. Neely, *Interconnect and circuit modeling techniques for full-chip power supply noise analysis*, IEEE Trans. on Components, Packaging, and Manufacturing Technology-Part B, vol.21, no.3, Aug.1998, pp.209-215.

[4] N. Hedenstierna and K. O. Jeppson, *CMOS circuit speed and buffer optimization*, IEEE Trans. Computer-Aided Design, vol.CAD-6, no.2, pp.270-280, Mar.1987.

[5] P. Heydari and M. Pedram, *Analysis and optimization of ground bounce in digital CMOS circuits*, Proceedings of IEEE International Conference on Computer Design: VLSI in Computers & Processors, 2000, pp.121-126.

[6] A. Kabbani and A. J. Al-Khalili, *Estimation of ground bounce effects on CMOS circuits*, IEEE Trans. on Components and Packaging Technology, vol.22, no.2, June 1999, pp.316-325.

[7] J. Lillis, C. K. Cheng, and T.-T. Y. Lin, *Optimal wire sizing and buffer insertion for low power and a generalized delay model*, IEEE Journal of Solid-State Circuits, vol.31, no.3, March 1996, pp.437-47.

[8] J. Qian, S. Pullela, and L. Pillage, *Modeling the "effective capacitance" for the RC interconnect of CMOS gates*, IEEE Trans. on CAD, vol.13, no.12, Dec.1994.

[9] T. Sakurai and A.R. Newton, *Alpha-power law MOSFET model and its applications to CMOS inverter delay and other formulas*, IEEE Journal of Solid-State Circuits, vol.25, no.2, April 1990, pp.584-594.

[10] R. Saleh, S. Z. Hussain, S. Rochel, and D. Overhauser, *Clock skew verification in the presence of IR-drop in the power distribution network*, IEEE Trans. on Computer-Aided Design of Integrated Circuits and Systems, vol.19, no.6, June 2000, pp.635-644.

[11] S. R. Vemuru, *Effects of simultaneous switching noise on the tapered buffer design*, IEEE Trans. on VLSI Systems, vol.5, no.3, Sep.1997, pp.290-300.

[12] Y. Yang and J.R. Brews, *Design for velocity saturated, short-channel CMOS drivers with simultaneous switching noise and switching time considerations*, IEEE Journal of Solid-State Circuits, vol.31, no.9, September 1996, pp.1357-1360.

Osculating Thevenin Model for Predicting Delay and Slew of Capacitively Characterized Cells

Bernard N. Sheehan
Mentor Graphics Corporation, Wilsonville, Oregon USA

ABSTRACT

To extrapolate from one point to another using a line, one had better get the slope right. In this paper we apply a similar concept to the important problem in Static Timing Analysis (STA) of predicting cell timing for RC loads using capacitive characterization data. Instead of a line we have a Thevenin circuit, and instead of matching slopes we match load sensitivities. We present a table driven, highly accurate cell delay and slew prediction procedure that can improve STA when interconnect effects dominate. It is significantly more accurate (especially slew) than previously published C_{eff} methods.

Categories and Subject Descriptors

B.6.3 [**Logic Design**]: Design Aids – *simulation, verification.*

General Terms

Algorithms, Verification.

Keywords

Static Timing Analysis, Effective Capacitance

1. INTRODUCTION

While static timing analysis (STA) has traditionally treated interconnect as a lumped capacitor, the realities of deep sub-micron design require interconnect to be treated as a distributed RC or RLC network. Various methods for extrapolating cell timing to non-capacitive interconnect have been proposed: [1] suggests characterizing cells for π loads; [2] proposes an empirically derived C_{eff} function; [3] models the cell as a Thevenin circuit and derives a C_{eff} from this; [4] uses a Thevenin model with a piecewise linear ramp; [5] extends the Thevenin approach to networks with coupling; [6] uses a two region driver/load model; recently, [7] computes a C_{eff} from a current source approximation.

This paper presents a new method for calculating Thevenin-C_{eff} pairs based on a simple but effective idea—the recognition, namely, that the Thevenin model should not only match delay and slew at C_{eff}, but should also match delay and slew *sensitivities*. By analogy with fitted curves that match both ordinates and slopes, we call such a model an *osculating* model. In this paper we present a practical, accurate C_{eff} procedure based on the concept of osculating Thevenin models.

2. TIMING MODELS

In order to lend precision to our discussion, we introduce notation for the data normally available in a gate-level timing library. A **timing model** for a cell or gate G consists of a set of tables or equations which give delay and output slew as functions of capacitive load and input slew, i.e.,

$$t^G_{delay} = f_1(C_L, t_{slewIn}), \quad t^G_{slewOut} = f_2(C_L, t_{slewIn}) \qquad (2.1)$$

In practice, delay and slew are computed from the times a cell's input and output waveforms cross certain voltage thresholds-- $\alpha_L V_{DD}$=.2, $\alpha_D V_{DD}$=.5, $\alpha_H V_{DD}$=.8 typically. Slew as we use the term is an interval, not a rate: it is the difference between output crossing times at $\alpha_L V_{DD}$ and $\alpha_H V_{DD}$.

3. THEVENIN TEMPLATE

As in [3], our Thevenin model $M(t_0,t_r,R)$ consists of a resistance R in series with a saturating voltage ramp that begins at time t_0 and rises to V_{DD} after an additional time tr. The response of M when driving a capacitor C is

$$v_C(t) = V_{DD} v(t - t_0, t_r, RC) \qquad (3.1)$$

where $v(t, t_r, RC) =$

$$\begin{cases} \dfrac{t}{t_r} - \dfrac{RC}{t_r}\left(1 - e^{-t/RC}\right) & 0 \le t < t_r \\ 1 - \dfrac{RC}{t_r}\left(e^{-(t-t_r)/RC} - e^{-t/RC}\right) & t_r \le t \end{cases} \qquad (3.2)$$

4. OSCULATION

In a nutshell, here is what we will be attempting. A Thevenin model can have various admixtures of tr and R and still replicate the observed $t^G_{slewOut}$. Among *all* pairs of tr and R which correctly match *slew*, we choose that pair which also replicates *slew sensitivity*. This is the crux of our idea.

To develop this idea we start with the recognition that (2.1) tells us more than just delay and slew for a given t_{slewIn} and capacitive load C_L. Functions (2.1) give us, in effect, *curves* of t_{delay} and $t_{slewOut}$ versus C_L. We therefore know the tangents to these curves; in other words, we know load sensitivities.

Definition 4.1. A circuit model M for a cell is called an **osculating model** at operating point C_0 if the model predicts the same delays and slews at C_0 as (2.1), i.e. if

$$t^M_{delay} = f_1(C_o, t_{slewIn}) \qquad (4.1)$$

$$t^M_{slewOut} = f_2(C_o, t_{slewIn}) \quad (4.2)$$

and, moreover, it replicates one or both of the gate's load sensitivities at C_0:

$$\frac{d}{dC_L} t^M_{delay}\bigg|_{C_L=C_0} = \frac{df_1}{dC_L}(C_o, t_{slewIn}) \quad (4.3)$$

$$\frac{d}{dC_L} t^M_{slewOut}\bigg|_{C_L=C_0} = \frac{df_2}{dC_L}(C_o, t_{slewIn}) \quad (4.4)$$

We expect an osculating model to closely follow cell timing not just at the single load C_0 but also for loads in the *vicinity* of C_0.

5. OSCULATING THEVENIN MODEL

We now elaborate the concept of osculating model into an efficient design procedure for creating an osculating model *based on a Thevenin template.* Our procedure involves constructing special design tables from which we read off the Thevenin model parameters that match delay, slew, and *slew* sensitivity. We do not match *delay* sensitivity, since our Thevenin model $M(t_0,t_r,R)$ has only 3 parameters and therefore cannot satisfy all 4 conditions in definition 4.1.

6. DESIGN TABLES

The specification of suitable design tables is central to making our osculating approach practical. These tables are only computed *once* for a given technology. We use them to *directly* lookup of Thevenin parameters, rather than having to compute them for each cell by (3.2), which would be expensive.

The trick is to collapse *all* design problems (for given $\alpha_L, \alpha_D, \alpha_H$) into *one* set of tables. We do this by scaling time. Specifically, we divide all time quantities by

$$UnitTime \equiv \frac{t^G_{slewOut}}{\alpha_H - \alpha_L} \quad (6.1)$$

We denote the resulting normalized quantity by placing a bar over its symbol. With this scaling, (6.1) takes the form

$$\bar{t}^G_{slewOut} = \alpha_H - \alpha_L \quad (6.2)$$

This, of course, is also the slew of a linear ramp with $\bar{t}_r = 1$, which is the motivation for our choice of *UnitTime*.

Consider now the following argument. Given (3.2) and a set of relative timing thresholds $\alpha_L, \alpha_D, \alpha_H$, the equations

$$\begin{aligned} v(\bar{t}_L, \bar{t}_r, \overline{RC}) &= \alpha_L \\ v(\bar{t}_D, \bar{t}_r, \overline{RC}) &= \alpha_D \\ v(\bar{t}_H, \bar{t}_r, \overline{RC}) &= \alpha_H \end{aligned} \quad (6.3)$$

for threshold crossing times, together with the normalization constraint (6.2),

$$\bar{t}_H - \bar{t}_L = \alpha_H - \alpha_L \quad (6.4)$$

give 4 equations in 5 unknowns ($\alpha_L, \alpha_D, \alpha_H$ are known); such a system can, in theory, be solved for $\bar{t}_L$, $\bar{t}_D$, $\bar{t}_H$, $\overline{RC}$ in terms of $\bar{t}_r$. Since delay, slew, and load sensitivities, in turn, can be computed from $\bar{t}_L$, $\bar{t}_D$, $\bar{t}_H$, $\overline{RC}$ and $\bar{t}_r$, they also can be regarded as functions of $\bar{t}_r$.

Definition 6.1. **Osculating Thevenin design tables** consist of tabulations of the following four functions of $\bar{t}_r$:

$$\begin{aligned} \overline{RC} &= A_1(\bar{t}_r) \\ \bar{t}_D &= A_2(\bar{t}_r) \\ \overline{RC}(\partial \bar{t}_{delay} / \partial \overline{RC}) &= A_3(\bar{t}_r) \\ \overline{RC}(\partial \bar{t}_{slewOut} / \partial \overline{RC}) &= A_4(\bar{t}_r) \end{aligned} \quad (6.5)$$

obtained from (3.2), (6.3), and (6.4).

Normalization moves our design into the space of Thevenin models with $\bar{t}_r$ in [0,1]. At one extreme is $\bar{t}_r$ =1: here R= 0 and the gate's output slew is entirely accounted for by the ramp time of the Thevenin voltage source. At the other extreme is $\bar{t}_r$ =0: here the Thevenin model consists of a *step* voltage and a resistor such that the gate's output slew is fully accounted for by the time constant of R charging C. Between these extremes, A_1 tells how to trade off R and $\bar{t}_r$ so as to hold output slew constant; A_4 gives the *slew* sensitivities of these various admixture of R and $\bar{t}_r$.

It turns out that A_3 (our normalized *delay* sensitivity) is nearly flat for much of $\bar{t}_r$ in [0,1]; it does not, therefore, provide a good basis for selecting $\bar{t}_r$. This is why in practice we match slew sensitivity rather than delay sensitivity.

7. BUILDING THE TABLES

In this section we describe how the design tables can be computed. In the next two sections we explain their use.

To calculate a set of design tables, we begin with the trivial solution of (6.3) and (6.4) at $\bar{t}_r = 1$:

$$\mathbf{x} \equiv \begin{bmatrix} \bar{t}_L \\ \bar{t}_D \\ \bar{t}_H \\ \overline{RC} \end{bmatrix} = \begin{bmatrix} \alpha_L \\ \alpha_D \\ \alpha_H \\ 0 \end{bmatrix} \quad (7.1)$$

If $\bar{t}_r$ is now replace by $\bar{t}_r - d\bar{t}_r$, a new vector $\mathbf{x}$ is found by iteratively solving

$$\mathbf{Mdx} = -\mathbf{F} \quad (7.2)$$

where

$$\mathbf{M} = \begin{bmatrix} \partial v_L/\partial \bar{t} & 0 & 0 & \partial v_L/\partial \overline{RC} \\ 0 & \partial v_D/\partial \bar{t} & 0 & \partial v_D/\partial \overline{RC} \\ 0 & 0 & \partial v_H/\partial \bar{t} & \partial v_H/\partial \overline{RC} \\ -1 & 0 & 1 & 0 \end{bmatrix} \quad (7.3)$$

is the differential of (6.3) and (6.4) and

$$\mathbf{F} = \begin{bmatrix} v_L - \alpha_L \\ v_D - \alpha_D \\ v_H - \alpha_H \\ \bar{t}_H - \bar{t}_L - (\alpha_H - \alpha_L) \end{bmatrix} \tag{7.4}$$

is the residual error. Our notation here is that the subscript on v indicates the crossing time at which a quantity is evaluated—e.g., $\partial v_L / \partial \bar{t}$ is the partial of equations (3.2) with respect to t evaluated at $\bar{t} = \bar{t}_L$.

Once Newton-Raphson iterations (7.2) converge, values of A_1 and A_2 at $\bar{t}_r - d\bar{t}_r$ are immediately available from the 2nd and 4th components of **x**. The delay and slew sensitivities--and hence A_3 and A_4--can be calculated from (3.2) using

$$\frac{\partial \bar{t}_{delay}}{\partial \overline{RC}} = -\frac{\partial v_D / \partial \overline{RC}}{\partial v_D / \partial \bar{t}}, \qquad \frac{\partial \bar{t}_{slewOut}}{\partial \overline{RC}} = \frac{\partial v_L / \partial \overline{RC}}{\partial v_L / \partial \bar{t}} - \frac{\partial v_H / \partial \overline{RC}}{\partial v_H / \partial \bar{t}} \tag{7.5}$$

with the subscript on v again denoting the crossing-time at which a partial of v is evaluated.[1] Continuing in this way we compute the design tables decrementally for all values of $\bar{t}_r$ in [0,1].

8. USING THE TABLES

Given the design tables, it is a simple procedure to construct an osculating Thevenin model $M(t_0, t_r, R)$ for gate targets

$$t^G_{delay} = f_1(C_0, t_{slewIn}), \quad t^G_{slewOut} = f_2(C_0, t_{slewIn}) \tag{8.1}$$

$$S^G_{slewOut} = \frac{df_2(C_o, t_{slewIn})}{dC_L} \tag{8.2}$$

Normally, $S^G_{slewOut}$ is had by numerical differential of (2.1).

Algorithm 8.1. Osculating Thevenin Model Construction

Given: $C_0, t^G_{delay}, t^G_{slewOut}, S^G_{slewOut}$, and design tables.

Find: Osculating Thevenin model $M(t_0, t_r, R)$.

1) Inverse interpolate table A_4 to get $\bar{t}_r$ such that[2]

$$A_4(\bar{t}_r) = \frac{C_0 S^G_{slewOut}}{UnitTime} \tag{8.3}$$

2) $t_r = \bar{t}_r * UnitTime$ (8.4)

3) $R = \dfrac{\overline{RC}}{C_0} * UnitTime$ (8.5)

4) $t_0 = t^G_{delay} - \bar{t}_D * UnitTime$ (8.6)

where $\overline{RC} = A_1(\bar{t}_r)$ and $\bar{t}_D = A_2(\bar{t}_r)$.

Theorem 8.2. The Thevenin model so constructed matches the target delay, slew, and slew sensitivity at C_0.

Proof. Normalization ensures that M and G have the same output slew. Steps (1) and (4) above ensure that M and G have the same slew sensitivities and delays, respectively.

9. EFFECTIVE CAPACITANCE

We now apply the above theory to calculate the effective capacitance of an arbitrary RC load.

Algorithm 9.1. Effective Capacitance Using Osculating Models

Given: A gate timing model like (2.1), an RC load, and an initial estimate C_0 for C_{eff} (e.g. total capacitance to ground).

Find: The effective capacitance C_{eff} of an RC network driven by the gate, and a Thevenin model M for the gate when loaded by the RC network.

1) Calculate the osculating Thevenin model M for the gate at C_0 (Alg. 8.1). Let $t^M_{delay}(C_0)$ be the gate delay (same as M's delay, by construction) when loaded by C_0.
2) Drive the RC network with M to get $t^M_{delay}(RC)$, the delay of the Thevenin model when loaded by the RC network. One option here is to compute the pole/residue representation of the RC network, express the ramp response analytically, then find threshold crossings by Newton-Raphson.
3) Estimate C_{eff} from the following equation
$$t^M_{delay}(RC) - t^M_{delay}(C_0) \approx S^M_{delay} * (C_{eff} - C_0) \tag{9.1}$$
where
$$S^M_{delay} \equiv \partial t^M_{delay} / \partial C = A_3(\bar{t}_r)/C_0 \tag{9.2}$$
4) If $C_{eff} \approx C_0$ and $t^M_{delay}(RC) \approx t^M_{delay}(C_{eff})$, stop, otherwise set $C_0 = C_{eff}$ and go to step 1.

Actually, osculation is so powerful that the above algorithm gives excellent results without iteration (step 4). The iterative version almost always converges in 2 or 3 cycles even with stringent convergence criteria in step 4. Although we do not develop the point here, the same algorithm applies to RLC and coupled networks, with appropriate modifications to step (2).

10. RESULTS

We apply algorithm 9.1 to a set of drivers and RC networks of various lengths and topologies representative of interconnect for a 0.18 μm process. Figure 1 plots (o) gate delays predicted by our algorithm using a timing model like (2.1) against delays obtained by full SPICE-level simulation using transistor models

---

[1] Equations (7.5) make use the identity,

$$F(x, y) = 0 \Rightarrow \frac{dy}{dx} = -\left(\frac{\partial F}{\partial x}\right) \Big/ \left(\frac{\partial F}{\partial y}\right)$$

[2] Since R is constant (but unknown at this point in the calculation), we have

$$\overline{RC}\frac{\partial \bar{t}_{slewOut}}{\partial \overline{RC}} = \overline{C}\frac{\partial \bar{t}_{slewOut}}{\partial \overline{C}} = C\frac{\partial \bar{t}_{slewOut}}{\partial C}$$

an equivalence which underlies (8.3)

for the gates. Also plotted are: (*) delays from algorithm [3], and (+) delays obtained by substituting total load capacitance into (2.1). Figure 2 is a similar set of plots for gate output slew. Both figures are based on threshold levels of $\alpha_L = .2, \alpha_D = .5, \alpha_H = .9$, as preferred by [3][3].

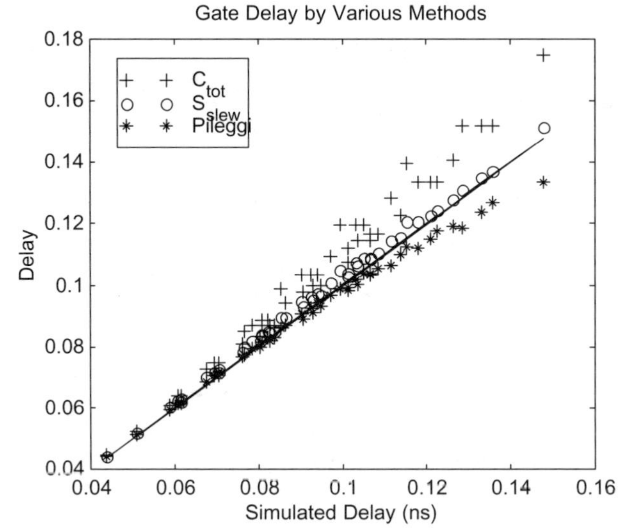

Figure 1. Cell Delay by Various Methods

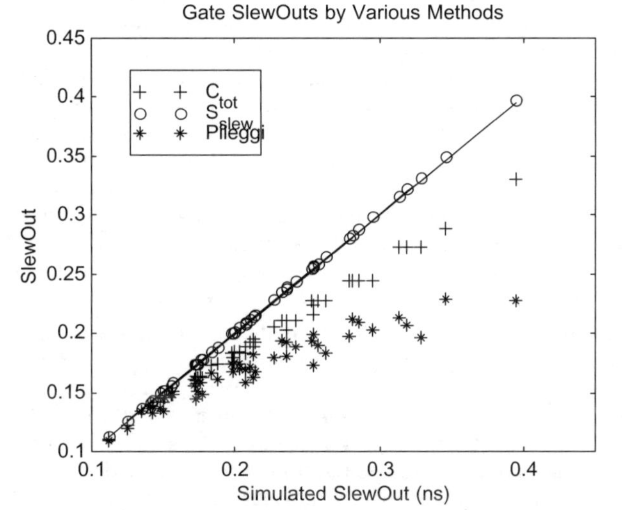

Figure 2. Output Slew by Various Methods

In all cases the osculating algorithm converged to C_{eff} in at most 3 iterations (usually 1 or 2 iterations).

Judging from Figures 1 and 2, we conclude that the osculating Thevenin algorithm is quite robust over a wide range of interconnect scales and topologies. It consistently predicts *slightly* longer delays than simulation--a desirable circumstance when timing long paths. Moreover, it predicts slews as accurately as delays. By contrast, most published C_{eff} algorithms do not even report on the accuracy of their slew predictions. Finally, it does not produce outlying points--a well-tempered consistency that makes it ideal for timing verification.

For the same data-set, algorithm [3] is inferior in all these respects; especially, [3] underestimates slew by as much as 50%. Slew, of course, influences downstream delays and is a key quantity governing noise, so correct slew is important for accurate interconnect analysis.

11. CONCLUSION

The problem we address in this paper is vitally important to Static Timing Analysis of deep sub-micron designs— namely, how to predict delays *and* slews of a cell for *distributed* loads using characterization data for *capacitive* loads. Previous researchers have attacked this problem in various ways, mostly by constructing a Thevenin model that *interpolates* cell timing.

Our contribution is to design a Thevenin model to approximate not only cell timing but also slew sensitivity. In this way we tune the Thevenin model to better mimic cell behavior in a *region* of loads; as confirmed by our tests, the resulting circuit is better at *extrapolating* to non-capacitive loads.

The osculating Thevenin method presented here has many advantages. It is efficient because it uses normalized design tables, the construction and use of which is described in the paper. The method gives surpassingly accurate results for both delay and slew. As an important pragmatic merit, the method does not require special, non-standard cell characterization, unlike many published techniques.

12. ACKNOWLEDGMENTS

I wish to thank Ellis Cohen and Archie Lachner for various discussions during commercial integration of this method.

[3] In applying algorithm [3], we computed the Thevenin resistance anew for each load using the formula $R = \dfrac{t_{90}(C_L) - t_{50}(C_L)}{C_L \ln 5}$ as given in equation (22) of [3].

[1] P. OBrien and T. Savarino, "Modeling the Driving Point Characteristic of Resistive Interconnect for Accurate Delay Estimation," ICCAD 89, pp. 512-15

[2] R. Macys, S. McCormick, "A New Algorithm for Computing the Effective Capacitance'in Deep Submicron Circuits," IEEE Custom Integrated Circuits Conf. 1998, pp. 313-16

[3] F. Dartu, N. Menezes, L. T. Pileggi, "Performance computation for Precharacterized CMOS Gates with RC Loads," ," IEEE Trans. Computer Aided Design, May 1996, pp. 544-53.

[4] R. Arunachalam, F. Dartu, and L. Pileggi, "CMOS Gate Delay Models for General RLC Loading," ICCAD 97.

[5] F. Dartu, L. Pileggi, "Calculating Worst-Case Gate Delays Due to Dominant Capacitance Coupling,", DAC97, pp.46-51.

[6] J. Qian, S. Pullela, L. Pillage, "Modeling the Effective Capacitance'for the RC Interconnect of CMOS Gates," IEEE Trans. Computer Aided Design, Dec. 1994, pp. 1526-35.

[7] B. Sheehan, "Library Compatible Ceff for Gate-Level Timing", DATE 2002, pp. 826-30.

Timed Pattern Generation for Noise-on-Delay Calculation

Seung Hoon Choi
Dept. of ECE, Purdue University
West Lafayette, IN 47907
choi5@ecn.purdue.edu

Florentin Dartu
Strategic CAD Labs, Intel Corporation
Hillsboro, OR 97124
florentin.dartu@intel.com

Kaushik Roy
Dept. of ECE, Purdue University
West Lafayette, IN 47907
kaushik@ecn.purdue.edu

ABSTRACT
Computing the noise on delay effects is required for all circuits from simple ASIC designs to microprocessors. Transistor-level simulation engines make accurate delay calculation affordable only if the number of simulation per stage is very small. We propose a solution that predicts the alignment of aggressor signals with respect to the victim signal to induce the worst-case noise effect on delay. The aggressor alignment can be used to setup a detailed simulation. The worst-case delay in the presence of noise is predicted within 5% error for more than 99% of the cases tested using an average of 1.27 simulations per stage transition.

Categories and Subject Descriptors
B.8.2 [**Hardware**] : Performance Analysis and Design Aids

General Terms : Performance, Verification

1. INTRODUCTION
The practical solution to reducing the interconnect delay in deep sub-micron technologies by increasing the aspect ratio has an unfortunate side effect: the increased coupling capacitance to adjacent wires [1]. Therefore, the electrical signals in a circuit are increasingly affected by their layout neighbors. Crosstalk noise, the noise due to capacitive coupling, can affect circuit performance functionally and/or temporally. Functional failure is possible if the crosstalk-induced noise glitch is propagated and wrongfully evaluated. Delay effects are observed when two neighboring signals undergo switching at the same time. We will call the increase and the decrease of the delay due to crosstalk as *delay push-out* and *pull-in*, respectively. While the analysis of functional failure [2],[3] is very important for high speed circuits, more often the designers will have to deal with the other type of noise-activated problem: noise-on-timing effects. Our experiments with a large synthetic benchmark representative of a high-end microprocessor design environment showed that the coupling could increase the delay by 30% on average, making performance predictions a fortune-telling experience. This problem does not scale well for future process technologies as the timing margins are getting smaller [4]. The noise-on-timing scenario is described in Figure 1 for a gate with n aggressors.

In static timing analysis (STA) [5], the circuit performance is estimated in a pattern independent manner where the delay calculation engine is called for every stage, sometimes multiple times, and therefore it has to be very efficient. To account for crosstalk in STA, the coupling capacitance is multiplied by a factor and modeled as grounded capacitance [6],[7]. The timing analysis is further complicated due to the circular aspects of the problem: aggressor switching affects the victim switching and vice versa [8],[9].

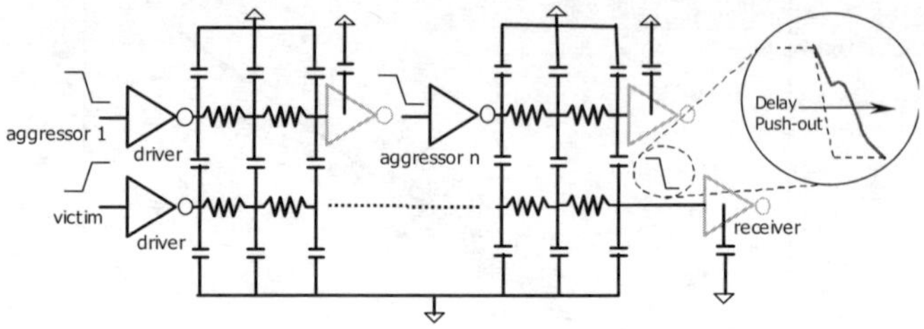

Figure 1 Noise-on-delay setup

The worst-case delay and the corresponding signal alignment analysis based on a linear driver model [10],[11] has been very attractive for its simplicity although it can bring about inaccuracy. A new output resistance for the driver was proposed in [12] to improve the accuracy. However, it requires on-the-fly nonlinear simulations during analysis for gate modeling with cell level analysis. And at least four non-linear simulations are necessary to obtain the resistance value, which increases the run time significantly. In [12] a method is also proposed to align aggressors using the look-up table with given victim slopes, noise pulse widths and heights. Even though the table can be pre-characterized, the noise height and width should be extracted from the noise signal obtained by the iterative simulations explained above. Although [12] makes an important step toward a more accurate analysis, their priorities are sub-optimal considering that tool run time is very crucial.

We find the idea of transistor-level simulation for checking circuit performance very appealing. Many nonlinear effects are difficult to capture inside a simplified model. We propose to make assumptions in determining the signal alignment, and devoting the simulation cycles for the final verification. Our solution predicts the alignments of aggressors to result in the worst-case delay. The solution is the function of signal parameters for the noiseless case. We will show that by properly modeling the driver resistance, no nonlinear simulation is necessary to predict the alignment. Only one nonlinear simulation is necessary to obtain the delay with the predicted alignment. Once the initial delay values are obtained from the first simulation, we can test it to determine whether another run is needed to refine the alignments. The solution is tested with more than 8000 aggressor-victim combinations. The solution is extended to the delay pull-in. We show the results for both inverter drivers and gates that are more complex.

In section 2, delay push-out will be presented in more detail. In section 3, we describe our basic solution and explain the accuracy of the model. Improvements to the basic solution and the modification for pull-in are explained in section 4. Results are shown in section 5 and concluding remarks are in section 6.

This research was funded in part by MARCO GSRC under contract # SA3273JB and Intel Corporation.

2. DELAY PUSH-OUT

The signal whose delay is of interest is called a victim and all the other signals coupled to the victim line are considered aggressors. The delay is measured from the 50% crossing point of the voltage swing from a driver input to the 50% crossing point of a receiver input in Figure 1. This way, both interconnect and gate delay can be considered at the same time. One victim can have more than one aggressor and we assume that one aggressor transition is not affected by the switching of other aggressors.

Figure 2 shows a typical delay variation of an inverter with a single aggressor. The x-axis represents the normalized slack - difference between aggressor and victim signal alignments divided by victim output transition time. Y-axis is the percentage delay increase. The sharp drop in push-out observable at the right side of this plot is an artifact of the delay definition using the 50% threshold. Because it is impractical to get an accurate delay profile like this via comprehensive search in a large circuit, an efficient way of doing it is essential to minimize the overhead.

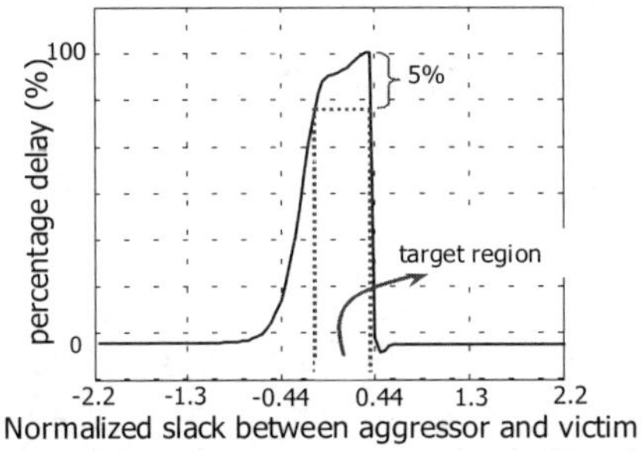

Figure 2 Delay variation vs. aggressor alignment

As it can be seen from the plot, the delay as a function of alignment is very difficult to model due to the large number of factors influencing the behavior. However, if we restrict the goal to finding out the alignment of aggressors to generate the delay within 5% of the worst case, the prediction problem becomes feasible. The solution that we provide will position the aggressor within the target region in Figure 2.

3. ALIGNMENT PREDICTION

3.1 Preliminary experiments

In [10], the victim driver was assumed linear and superposition was applied to obtain the composite noise waveform at the victim. This idea is depicted in Figure 3 for falling transition at driver output and rising transition for an aggressor. The "pushed-out" waveform is obtained by summing the "noiseless" waveform with the noise. The implicit aggressor alignment for worst-case delay requires the noise peak to occur when the "noiseless" transition crosses (VCC/2 – peak_v). Proof to this is provided in [10],[11].

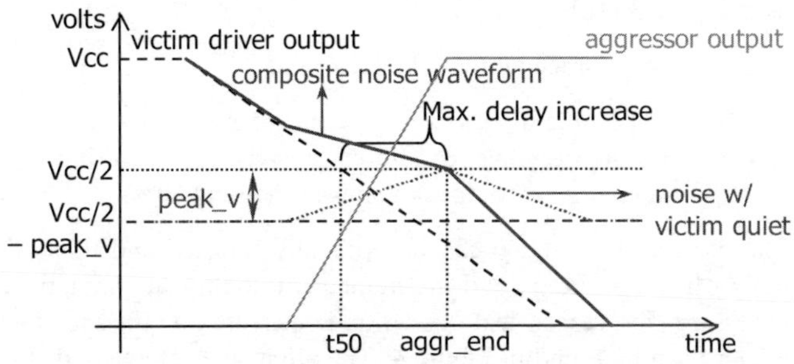

Figure 3 Worst-case delay and corresponding alignments

In order to attain some insights on how the alignment deviates from the linearity assumption in real circuits, inverter test-case circuits were generated and simulated. First, depending on the interconnect lengths, three groups – short, medium and long – are selected. For each group, different combinations for aggressor-victim driver strengths – strong-weak, weak-strong and similar – are considered for two different metal layers. This way, 18 test cases are generated. These circuits are for only one type of signal switching direction. Then another circuit of different switching direction is added just to ensure that the aggressor alignment problem can be approached regardless of the switching direction. Now, the setup for 19 example circuits is completed to cover the possible combination of circuit parameters. There is only one aggressor affecting a victim for this set of experiments.

We decided to use the end-of-transition point as the time origin for the aggressor because for a single pole linear circuit assumption it represents the time when the noise will peak. Our goal is to find the timing relationship between the aggressor origin and the victim output transition for the worst-case alignment.

We first checked the relative timing position between the victim noiseless transition and the 5% max push-out window (target region) defined in Figure 2. The results are plotted in Figure 4 for the 19 examples. The x-axis is the normalized victim transition time such that the start-point and end-point correspond to zero and one, respectively and 0.5 is equivalent to the 50% crossing point, t50. If the linearity assumption depicted in Figure 3 is correct, the max push-out should show up between 0.5 and 1 for noise with its peak smaller than Vcc/2 (as is the case with our circuits). However, the plot strongly implies that non-linearity plays an important part as most worst-case alignment fall outside this range. Some important aspects are to be extracted from the figure in the following subsection.

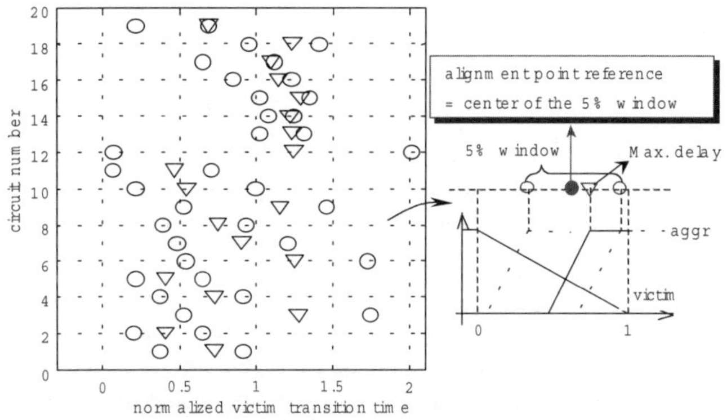

Figure 4 Aggressor alignments examples

3.2 Alignment prediction

Our primary goal is to correlate the intuitive understanding of the physics involved with these experimental results. A single number, *aggrAlign*, defined as follows, will represent each aggressor alignment:

Definition 1: *aggrAlign* is the time when the aggressor transition ends, referenced to the beginning of the victim output transition and normalized to the victim's output transition time. (As an example, if the aggressor ends its transition in the middle of the victim output transition, then *aggrAlign* is 0.5.)

The target being aligning the aggressors to have delay within 5% from the worst case, we will not even try to get the maximum push-out, in place we will try to match the middle point of the 5% window. This way we are obtaining the most robust solution.

The effect of crosstalk, which can be measured as the delay push-out, is dependent on each driver's strength, interconnect parameters, receiver size and the aggressor alignment. The driver strength combined with interconnect parameters and receiver size can be captured by looking at the driver output signal waveform, especially the transition time without coupling. The amount of coupling from an aggressor to a victim can be measured by the noise induced on a victim net when the victim is quiet. The same intuition can be applied when aligning the aggressor with slight variations. Basically, the aggressor alignment points within the normalized victim window, *aggrAlign*, is dependent on:

- $\frac{\text{aggressor driver output transition time}}{\text{victim driver output transition time}}$

As it can be seen from Figure 5, if the aggressor transition time is relatively larger than that of a victim, the *aggrAlign* moves to the right implying larger *aggrAlign*. This is because *aggrAlign* is defined as the aggressor switching end-point.

- peak noise, *peakV*, measured when the victim is quiet

Larger peak noise means the shift of aggressor alignment to the right, as shown in Figure 5. This holds true if we assume the driver is linear (see Figure 3).

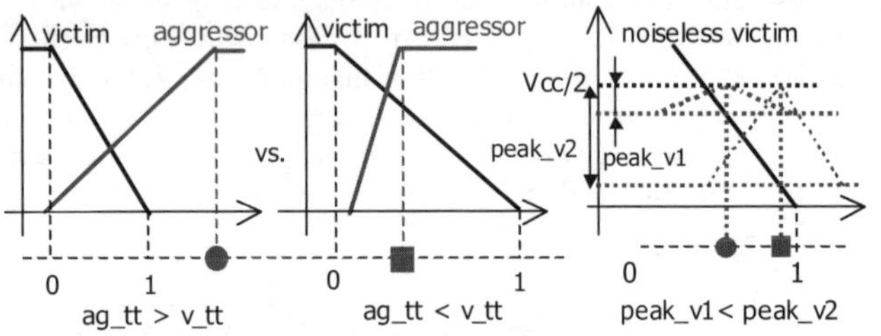

Figure 5 Shift of *aggrAlign*

To include these intuitive effects into a mathematical form we explored various empirical functions that preserve the theoretical relationships. The best model that we found is shown below.

$$aggrAlign = \alpha \bullet \frac{ag\_tt}{v\_tt} \bullet peakV^{\beta} + \gamma \qquad (1)$$

Where *peakV* is the peak of the noise waveform when victim is quiet and *ag_tt* and *v_tt* are the aggressor and victim driver output transition time without coupling, respectively. The fitting coefficients, α, β and γ, are obtained through a standard nonlinear regression [13], from the data of the 19 circuit examples for which the results were plotted in Figure 4. The procedure of applying the model to a circuit can be summarized as follows :

1. *v_tt* and *ag_tt* are measured at the victim and the aggressor receiver input, respectively. Both values can be obtained through nonlinear simulation or from regular characterization.
2. For multiple aggressors, align all the aggressors so that the noise peaks on a victim can also be aligned for max *peakV*. This assumes that the *peakV* is obtained at the time when the aggressor reaches the end of its transition. This holds for low to medium resistive interconnect.
3. Run the non-linear simulation and measure the peak noise, *peakV*, at the receiver input node.
4. Insert *ag_tt*, *v_tt* and *peakV* into (1) and obtain *aggrAlign*.
5. For each aggressor, align the input signal so that the output has its end-point at *aggrAlign*.

4. IMPROVING THE SOLUTION

In this section, several methods to improve the accuracy of the model and its extension to delay pull-in problem are presented.

4.1 Multiple Aggressors

If the results obtained using the procedure presented in Section 3.2 are grouped according to the number of aggressors, 0.80% of single-aggressor examples, 6.50% of two-aggressor examples and 9.56% of four-aggressor examples fail to align the aggressors for a delay within 5% of the worst case. This is because the coefficients were obtained through the regression with single-aggressor test cases. Thus, another set of coefficients for equation (1) can be obtained with multiple-aggressor test cases and used separately for multiple-aggressor cases to improve the accuracy.

4.2 Estimation of Noise Peak

For the procedure explained above, nonlinear simulations are needed for both *peakV* and final delay computation. However, proper modeling of the nonlinear driver with linear resistance will allow us to skip the first simulation. Assuming ramp signaling for aggressor switching and replacing the transistor of a victim driver with a pre-characterized resistance, solving for the noise waveform on a victim driver output reduces to a linear system solution. The solution procedure is omitted here. By using this *peakV* in the prediction procedure, we finally need only one non-linear simulation to get the delay with the worst-case alignment.

4.3 Confidence of the Initial Prediction

Once the nonlinear simulation using the alignment defined by eqn. (1) is over, plenty of data is available and that can be utilized to further refine the results if necessary. The most suitable method that we found to test the quality of the results is to compare the noise peak time from the simulation with the 50% crossing point of the "pushed-out" waveform.

The actual peak noise is calculated by subtracting the noiseless waveform from the "pushed-out" waveform and taking the max value. In order to measure the noise at 50% crossing point, the 50%-point when the "pushed-out" noise waveform crosses VCC/2 is measured (T50 in Figure 6), and then the value is measured from the 'differential' noise at the T50. This procedure is described in Figure 6. For the delay to be the maximum, the peak noise and the noise at 50% should be the same assuming linearity as in Figure 3 [10].

The refinement procedure will look like this:

1. Determine if peak noise - noise at 50% > ε
2. If the previous condition is satisfied, then increase *aggrAlign* by ϕ*(peak voltage – noise at 50%)

The parameters ε and ϕ can be optimally selected and this will trade off the accuracy and the number of additional simulations necessary. The reason that *aggrAlign* is increased by the specified amount can be found in Figure 4. The alignment reference that is the center of the 5% windows is usually located to the left of the worst-case alignment.

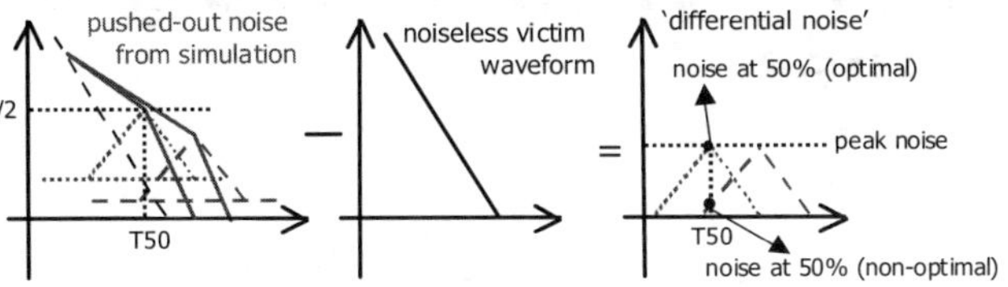

Figure 6 Decision on the reliability of the initial prediction

4.4 Delay Pull-In

When aggressor and victim signals are switching in the same direction, the delay of the victim will decrease. Accordingly, the alignment of an aggressor will be different from the one for push-out. However, the basic procedure used for the push-out problem can be applied with a little modification. From Figure 5 and eqn. (1), it is shown that *aggrAlign* is dependent on *peakV*, so that β in eqn. (1) has positive value. But for the pull-in case, it becomes an opposite story. The larger *peakV* is, the smaller *aggrAlign* should be. This is verified when we get new coefficients fitted from pull-in test cases.

5. RESULTS

In order to check the accuracy of the model, we constructed a synthetic benchmark that is representative for Intel Corporation high-end microprocessor built on 0.18μm technology. The total number of circuits is 8159 and those are composed of inverters. The number of aggressors varies from one to four. The lengths of interconnect varies from 100μm to 2000μm. The transition time range used in experiments was the same as the range used in the microprocessor design. Tape out process parameters (transistor and interconnect) were used.

The accuracy of the model developed in Section 3.2 is shown in the second row of Table 1. The error was measured against the worst-case delay obtained by exhaustive search. Both delays at the input and output of the receiver are presented. The first two columns give the percentage of test cases for which the model was within the desired error range. The fourth and fifth columns show the mean and the standard deviation of the error for the entire benchmark, respectively. Using the basic model, out of 8159 cases, only 6.6% fail to locate the aggressors for delays within 5% of the worst case at the receiver input. Only 2% fail for the delay measured at the receiver output. The improvements introduced in Section 4 were also tested on the same benchmark and the results are also summarized in Table 1. The results for confidence check (section 4.3) reflect a second iteration applied selectively to 27% of the total circuits. The same percentage represents the run time overhead for this method. It can be deduced from the table that the usage of two sets of coefficients for multiple aggressor cases makes a difference. Replacing the victim driver with the resistance does not degrade the accuracy of the model. This is because *peakV* is least sensitive to the variation in the model. If we are allowed to run another simulation with the confidence check, it is possible to align the aggressors to induce the delay within 5% from the worst case for more than 99% of the test cases. The last row of Table 1 represents the results for complex gates benchmarks which consists of 512 different circuits with NAND, NOR and AOI gates considering single-input switching. The same fitting coefficients as the ones for inverter circuits are used to align the aggressor signals.

Table 1 Accuracy of the alignment solution

Delay at receiver input/output	within 3% (%)	within 5% (%)	Mean error (%)	Std. dev. error (%)
Basic model (Sec. 3.2)	84.2/91.1	93.4/98	1.59/1.06	1.98/1.30
Two models (Sec. 4.1)	92.7/96.5	98.7/99.7	1.23/0.71	1.13/0.94
Linear driver (Sec. 4.2)	92.7/96.6	98.5/99.6	1.21/0.67	1.15/0.95
Refining (Sec.4.3, ε=0.025, ϕ=0.2)	96.1/98.7	99.6/99.9	1.02/0.50	0.91/0.74
Complex gates	93.2/99.1	99.0/100	0.84/0.46	1.12/0.62

6. CONCLUSION

We approached the problem of aggressor alignment for the worst-case delay variation due to crosstalk. Our solution is useful when a transistor-level simulation engine is available during STA. Our goal was to use as few simulations as possible to determine the worst-case alignment. Due to the nonlinear simulations, the delay result has all the deep sub-micron electrical effects included.

Our model was obtained from the properties of the aggressor and victim signals and by scrutinizing their effects on timing. A simple analytical formula is used to map the signal parameters for the noiseless case into the aggressor alignment. The coefficients of this model are obtained once per design through a standard nonlinear fitting procedure. The model was also further improved for both runtime and accuracy. By using an estimated value of the peak voltage, we removed one simulation per gate while the accuracy was minimally reduced. We also introduced a convergence criterion for our algorithm. With a 27% runtime overhead, we were able to get 99.6% of the examples within 5% of the worst-case delay.

The alignment problem related to multiple-input-switching has to be considered for more generic cases. We consider this topic to be a natural extension of the work presented in this paper.

REFERENCES

[1] M. Bohr, "Interconnect scaling–the real limiter to high performance ULSI," *International Electronic Device Meeting*, pp.241-244, 1995.

[2] D. Somasekhar, et al, "Dynamic Noise Analysis for Precharge-Evaluate Logic Family," *DAC*, pp.243-246, 2000.

[3] K.L. Shepard, et al, "Harmony: Static Noise Analysis of Deep Submicron Digital Integrated Circuits," *IEEE TCAD*, pp. 1132-1150, Aug 1999.

[4] L. Chen, et al, "A New Gate Delay Model for Simultaneous Switching and Its Applications," *DAC*, pp.289-294, 2001.

[5] R.B. Hitchcock, "Timing verification and timing analysis program," *DAC*, pp.594-604, 1982.

[6] P. Chen, et al, "Miller Factor for Gate-Level Coupling Delay Calculation," *ICCAD*, pp.68-74, 2000.

[7] A.B. Kahng, et al, "On Switch Factor Based Analysis of Coupled RC Interconnects," *DAC*, pp.79-84, 2000.

[8] P. Chen, et al, "Switching Window Computation for Static Timing Analysis in Presence of Crosstalk Noise," *ICCAD*, pp.331-337,2000.

[9] R. Arunachalam, et al, " TACO : Timing Analysis With Coupling", *DAC*, pp.266-269, 2000.

[10] F. Dartu, L.T. Pileggi, "Calculating Worst-Case Gate Delays Due to Dominant Capacitance Coupling," *DAC*, pp.576-580, 1997.

[11] P.D.Gross, et al, "Determination of Worst-Case Aggressor Alignment for Delay Calculation," *ICCAD*, pp.212-219,1998.

[12] S. Sirichotiyakul, et al,"Driver Modeling and Alignment for Worst-Case Delay Noise," *DAC*, pp.720-725, 2001.

[13] N.R. Draper, H. Smith, "Applied Regression Analysis," John Wiley & Sons, Inc., 1998.

VeriCDF: A New Verification Methodology for Charged Device Failures

Jaesik Lee*
Univ of Illinois
Urbana, IL 61801
jlee25@lucent.com

Ki-Wook Kim
Pluris Incorporation
Cupertino, CA 95014
kiwook.kim@pluris.com

Sung-Mo Kang
University of California
Santa Cruz, CA 95064
kang@soe.ucsc.edu

ABSTRACT

A novel tool for full-chip verification is reported for CDM-ESD protection. Until recently, ESD protection has been simulated in device level, leading to the well known limitations on capturing global features such as the power protection circuits and package parasitics. In practice, fatal failures occur due to unexpected discharged paths in multi-power supply chips, which can only be verified by chip-level simulation. Associated with the new concept of macromodelling, hierarchical approach provides effective analysis methodology for mixed-signal chips. The hierarchical approach provides the analysis of chip-level discharging paths and reliability of gate oxide. Simulation results on a CMOS ASIC chip processed in a 0.25-μm technology are in accordance with the measurement data. Scanning electron microscope locates a gate oxide fault as our analysis predicted.

Categories and Subject Descriptors

B.8 [**Hardware**]: Performance and Reliability; C.4 [**Computer Systems Organization**]: Performance of Systems

General Terms

Verification

Keywords

Reliability, Modeling, Simulation

1. INTRODUCTION

An electrical surge of voltage or current that exceeds a device's rating can trigger either a catastrophic or parametric device failure. Such a sudden transfer of a voltage or current via inductance or direct contact is called *electrostatic discharge (ESD)*. ESD is a direct cause of electronic device failure when static charge or discharge attacks sensitive devices during manufacturing processes. The common failure modes caused by ESD are metalization meltdown, thermal secondary breakdown and dielectric breakdown.

As technology scaling continues, supply voltages are lowered, which substantially diminishes the tolerance of deep submicron devices. Thus, the risk of malfunctioning caused by ESD increases as the trends in silicon electronics miniaturization proceed. Estimated average product losses due to ESD damage range from 8% to 33% [1]. If such an ESD failure is investigated and controlled during design stage earlier than manufacturing, then valuable resources and cost can be saved and the chip reliability can be improved proportionally. This is the motivation of ESD failure modeling and simulation.

1.1 Issues on Electrostatic Discharge Models

Three primary models of ESD events have been developed: human body model (HBM), machine model (MM), and charged-device model (CDM). The direct transfer of electrostatic charge from the human body as a charge-storage capacitance can possibly cause device damage. The HBM is characterized by a relatively slow stimulus, and often energy-destructive failure occurs as a result. The CDM is intended to simulate charging and discharging events that occur in production equipment and processes. A chip can be charged from sliding down the feeder in an automated assembler, and then a rapid discharge can occur when it makes contact with low-impedance metal plate [2]. Most distinguishable feature of the CDM is the rapid flow of high current through the device that leads to dielectric destruction. As the dielectric failure mode has gained importance with scaled semiconductor devices and increased automated manufacturing lines, the importance of HBM and MM decreases and the CDM accounts for the majority of ESD failure during chip manufacturing. In this work, we focus on the CDM and its failure mechanism.

Previous works on CDM modeling and simulation have concentrated on the compact modeling of MOS transistor as an ESD protection device and have contributed to the analysis of MOS transistor behavior under CDM ESD stress [3, 4, 5]. These modeling and simulation techniques are limited to small ESD protection circuits because of the complexity of the extracted models. However, the overall CDM performance of a chip depends not only on the robustness of an ESD protection circuit, but also on global components such as the power protection network, inductance and resistance of power lines, and parasitic in the package. Moreover, in multiple power supply ICs, unexpected discharge paths can cause many failures that cannot be identified in the analysis of local ESD protection circuits. Therefore, CDM performance needs to be evaluated at the chip level, and large-scale CDM simulation should be used to predict CDM behavior in a chip with reliable accuracy.

*He is now with Bell Labs, Lucent Technologies, 600 Murray Hill, NJ 07974.

DAC 2002, June 10-14, 2002, New Orleans, Louisiana, USA.

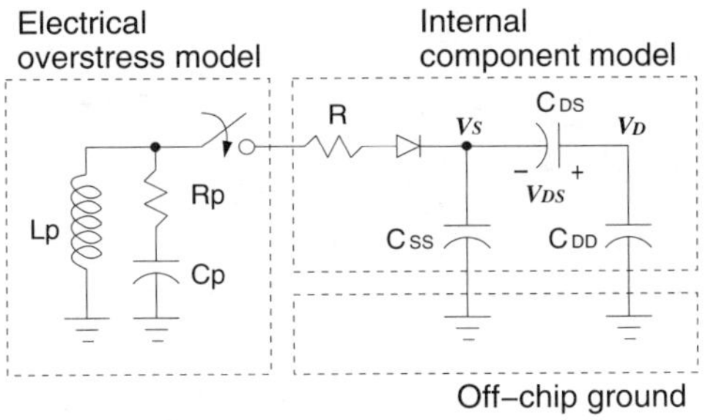

Figure 1: An equivalent circuit model of charged-device ESD event.

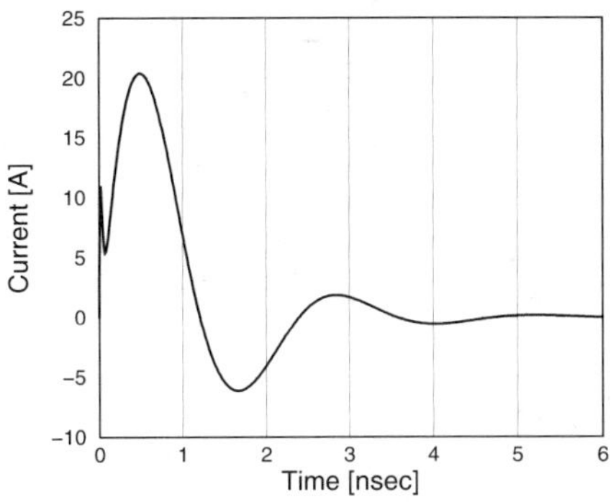

Figure 2: High current transient waveform for 1000 V CDM pulse.

1.2 Proposed Chip-Level CDM Model

In this work, *full-chip* modeling of charged-device failure (CDF) is proposed to overcome the limitations of localized approaches. Chip-level CDF modeling captures the impact of package capacitance and the distribution of power protection circuits on overall CDM performance. Chip-level modeling is accomplished by the divide-and-conquer strategy. For efficient modeling and simulation, the full-chip structure is partitioned into a set of subsystems based on the chip floorplan, the subsystem functionality, and the power bus architecture. A novel *CDM macromodel* is proposed to characterize each subsystem in terms of failure behavior for a given electrostatic overstress. The CDM macromodel incorporates lumped CDM capacitors, and internal resistors and capacitors that can be derived from power-ground network by using a layout extractor and analyzer such as iLEX [6]. Meanwhile, the accuracy of modeling is maintained by considering communication between groups of subsystems. For instance, the gate-oxide reliability of interface buffers used to span different power supply boundaries accounts for interactions between components. Chip-level CDF analysis is performed with a full set of macromodels and the ESD simulator, iETSIM [4].

In order to verify our CDM macromodel and chip-level failure modeling, the CDM tests are carried out for a mixed-signal CMOS ASIC chip processed in 0.25-μm technology. Simulated waveform based on our model is in accordance with the actual measurement data in terms of rise time and peak discharge voltage that determine the CDF. Furthermore, emission microscopy was used to inspect the static *Idd* failure location. As our chip-level simulations predict, a scanning electron microscope (SEM) photograph asserts oxide destructions in the power domain on a high-speed network interface subsystem.

This paper is organized as follows. Section 2 provides necessary details to understand the charging/discharging mechanism of CDM event and corresponding circuit-level CDF model. Section 3 presents the proposed circuit-level CDM macromodel. The hierarchical approach for modeling a chip is also described in this section. Section 4 presents CDM simulation and experimental results. Stress discharging paths are thoroughly investigated in a multi-power IC and the vulnerable devices to CDM stress are presented. Finally, conclusions are drawn in Section 5.

2. CHARGED-DEVICE MODEL (CDM)

The CDM assumes that the charge residing on lead frames and metal parts of a chip are quickly discharged through one pin to off-chip ground and causes failures in junctions, dielectrics, and components that are part of the discharge paths.

The CDM equivalent circuit we propose is illustrated in Fig. 1. The electrostatic charge on the package during the charging event can be modeled by the package capacitances C_{DD} and C_{SS}. C_{DD} (C_{SS}) is a capacitance between on-chip power (ground) metal planes and off-chip ground. We assume that the charge stored on package lead frames are equally distributed to C_{DD} and C_{SS}, and the capacitance between signal lines and off-chip ground is ignored due to little capacitance values. Internal capacitance C_{DS} represents all the lumped capacitance formed between internal power and ground planes.

Chip and tester inductance L_p from 1 to 50-nH as well as the device resistance R of 50 to 300-Ω have significant roles in the CDF. The parameters in the CDM equivalent circuit are largely dependent on the type of package and internal metal configuration. The current waveform of the CDM equivalent circuit after application of 1000 V positive CDM stress is shown in the Fig. 2. The damped oscillation occurs because of an *L-R-C* transition characteristic due to inductance and resistance of the discharging line and the capacitance of the device. The discharging current flowing as the switch turns on is expressed by [7]

$$I_{CDM}(t) = \frac{V_{CDM}}{\sqrt{\frac{L_P}{C_{eq}}}} \exp\left(\frac{-Rt}{2L_P}\right) \sin(\omega_S t) \quad (1)$$

where V_{CDM} is the charging voltage and $C_{eq} = C_P + C_{SS} + \frac{C_{DD} \cdot C_{DS}}{C_{DD}+C_{DS}}$. The series resonant frequency is given by $\omega_S = \frac{1}{\sqrt{L_P C_{eq}}}$. Thus, the peak CDM current is approximately:

$$I_{peak} = \frac{V_{CDM}}{\sqrt{\frac{L_P}{C_{eq}}}} \quad (2)$$

Note that the peak current is a strong function of V_{CDM}, C_{eq}, and L_P, and the damping factor is determined by the values of L_P and R.

2.1 CDM Discharging Mechanism

A CDM behavioral tree as shown in Fig. 3 illustrates a current flow from a package capacitance C_{DV} (representing either C_{DD} or C_{SS}) to off-chip ground. The charge stored in the package capacitance can be discharged through three paths. At the beginning, the charge current flows from a package capacitance C_{DV} into the internal capacitance C_{DS}. After charge is shared by C_{DV} and C_{DS}, if a node voltage V_{ref} is higher than the trigger voltages of internal devices, either internal devices or ESD protection circuits will be activated and discharge the charge to ground. Associated current flows are represented by I_{PC} and I_{Dev}, respectively.

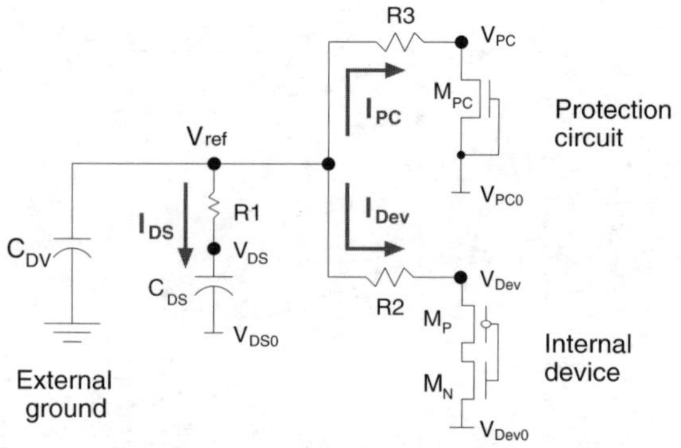

Figure 3: A basic CDM behavioral tree.

To protect internal devices against CDM-ESD stress, protection device M_{PC} should satisfy the following conditions:

$$\tau_{f,pc} < \tau_{f,int} \qquad (3)$$

$$R_3 < R_2 \qquad (4)$$

where τ_f represents the base transit time of an MOS device. Consequently, the optimization of the protection network which involves the design of protection circuit and the distribution of protection network is essential for robust CDM performance.

3. CHIP-LEVEL CDM MODELING

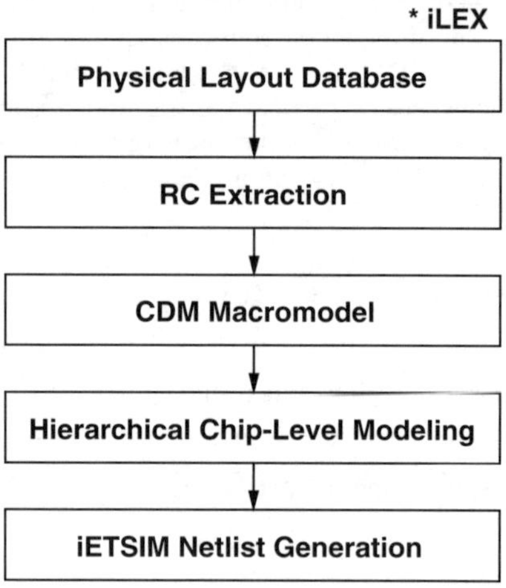

Figure 4: Chip-level CDF model extraction and simulation.

The basic concept of CDM behavior is represented by a CDM tree in a circuit level. It is feasible to apply the CDM behavioral tree for small I/O circuits, but impractical for full-chip simulation with this model. To overcome this inefficiency while maintaining accuracy, we propose a hierarchical method to encompass a full chip. The procedure for hierarchical modeling is outlined in Fig. 4. This procedure involves three major steps in translating a mixed-signal design into a netlist for circuit-level CDF simulation: (1) partitioning a chip into subsystems modeled by a circuit-level CDM macromodel, (2) extracting physical layout parameters for each macromodel, and then (3) simulating full-chip CDM behavior with combined macromodels.

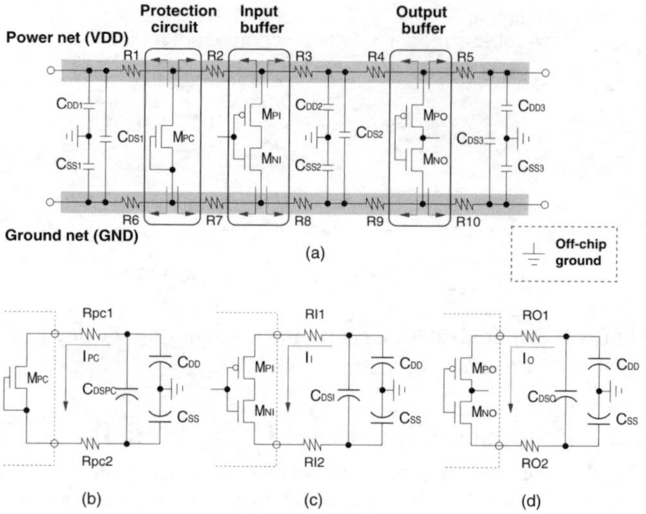

Figure 5: (a) A macromodel for each subsystem in a chip, and equivalent circuits driven by (b) protection circuits, (c) input buffer, and (d) output buffer.

3.1 Hierarchical CDM Modeling

The CDM tree can be built for each internal circuit node, and then CDM behavior is analyzed by using circuit simulator such as HSPICE. However, computational inefficiency of CDM tree-based method prohibits from full-chip simulation to appraise the overall effects of the package capacitance on the CDM performance.

In order to enhance efficiency, we exploit a hierarchical approach to full-chip CDF modeling and simulation. The first step of the simulation process is to partition the full-chip structure into smaller subsystems. Partitioning into subsystems serves two purposes. First, it simplifies the parasitic circuit, namely, the simulation variables within a subsystem. A partition treats each subsystem as a single CDM source and several discharging paths, and the variables are extracted as lumped elements. Second, the flow of CDM charge shifting among subsystems with different power domains can be accurately modeled because partitioning is carried out on the distinct power boundaries.

Partitioning is performed based on the chip floorplan, functions of each circuit block, and power bus architecture. Since the CDM is a charge driven phenomenon and the stress current is discharged through the least-impedance paths along the power-ground networks, major discharge paths such as ESD power protection circuits and I/O buffers in subsystems should be included. In a multi-functioned system, remote blocks on the chip floorplan can contribute to the CDM failures each other and hence should be modeled separately. While each power supply creates a subsystem, a big subsystem can be divided into several different subsystems to simplify the modeling. As a result, hierarchical modeling is carried out based on the following strategies:

- only one subsystem is defined in a specific power domain;
- heterogeneous functional blocks such as analog, RF, digital, and memory should be assigned as different subsystems;
- a block with the same function but remote chip placement can be mapped into different subsystems.

The next section describes the second step of the simulation process, namely the generation and characterization of the CDM macromodel.

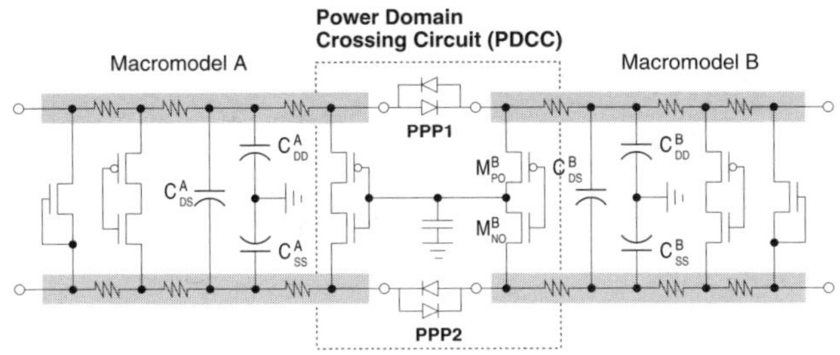

Figure 6: Two macromodels connected with power-to-power protection (PPP) circuits.

3.2 CDM Macromodel for Subsystem

The CDM macromodel captures the behavior of a subsystem with electrical overstress. The CDM macromodel is motivated by an important observation of CDF. The CDM-ESD test involves an ESD waveform of rapid change and high current. Under the high electric fields, most of the CDF occur in the gate oxide and dielectric layers of protection circuits and neighboring I/O buffers [3, 5, 7, 8]. A high electric field can be easily formed across the dielectric at these destructive sites because they are in the discharge path, where the current can be collected from many metal lines.

A macromodel includes a set of simulation variables such as package capacitances, internal capacitance, and the finite resistance of power metal. In addition, the CDM macromodel focuses on ESD protection circuits, input and output buffers, as shown in Fig. 5(a). Other functional devices are simplified into lumped capacitors C_{DS} between the power and ground nets. This simplification is justified by the fact that all the overstress are injected through the power and ground nets. Hence, I/O buffers and protection circuits located in the boundary of the subsystem are attacked and fail ahead of internal components.

Equivalent circuits for protection circuits, input buffer, and output buffer are shown in Fig. 5(b), (c), and (d), respectively. For a protection circuit, R_{PC1} represents the lumped resistance of the power net, R_{PC2} represents the lumped resistance of the ground net, and C_{DSPC} represents the effective internal capacitance driven by protection devices. Those parameters are determined by solving the driving point admittance for power and ground net [9, 10, 11, 12]. Similarly, RC parameters are resolved for I/O buffers. It is important to note that all the three components have the same package capacitance C_{DD} and C_{SS}.

The total quantity of charge stored in a chip is related to the area of the internal conductive parts. The quantity of charge is derived from a simple equation $Q = CV$ and CDM test conditions, as shown in the following equation,

$$Q_{CDM} = \varepsilon_0 A_C V_{CDM} \left(t + \frac{T}{\varepsilon_R} \right)^{-1} \quad (5)$$

where A_C is the total area of the internal conductors, V_{CDM} is the CDM charging voltage, ε_R is relative permitivity of resin used for packaging, and T and t are thickness of resin and air gap, respectively. Accordingly, the package capacitance of a macromodel increases as the conductor area in a macromodel increases. The approximate package capacitance of each macromodel is determined by calculating the area ratio of conductors within a macromodel to the total conductor area within a chip.

Partitioning into subsystems and associated macromodels can improve the computational efficiency. In combining procedure to generate a full-chip model, interactions between subsystems are taken into account to maintain the simulation accuracy. Fig. 6 shows the connection of two adjacent macromodels in different power supplies (namely, different power domains). The diode-based power-to-power protections (PPPs) are inserted between two power domains to create ESD current paths. The box represents a power domain crossing circuit (PDCC), which is the interface circuits between different power supplies.

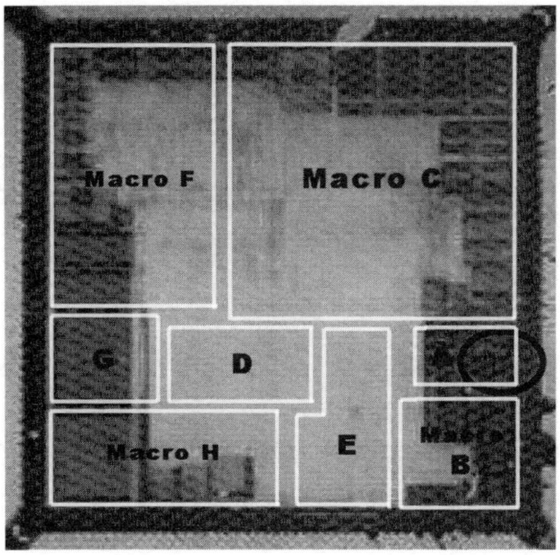

Figure 7: Functional systems of a CMOS ASIC

4. SIMULATION AND EXPERIMENTAL RESULTS

4.1 A Simulation Example

The CDM modeling and simulation methodology described in this work has been applied to several product ICs. As a part of example, a mixed-signal ASIC device processed in a 0.25-μm CMOS technology is presented. The ASIC contains a high-speed I/O interface, data/clock controllers, data converters, embedded static RAM cells, and digital signal processing (DSP) engines. Fig. 7 shows the chip floorplan. In the on-chip power bus architecture with multiple supplies, "power splitting" is employed to distribute different supplies. Each power supply delivers power to all the devices in a functional system. To reduce the complexity of modeling and simulation, a hierarchical approach is employed in such a way that the ASIC is subdivided into seven subsystems (subsystem I $\sim$ VII). Then, each subsystem is converted into a CDM macromodel (Macromodel $A \sim H$). Macromodel A represents a high-speed I/O interface occupying the smallest layout area in this chip. Macromodels C and D denote a DSP, mapped from subsystem III having the largest chip area.

The extracted simulation variables for chip-level CDF analysis are shown in Table 1. The measured total package capacitance is in the range 12.5 to 14.0-pF. The package capacitances for macromodels in the table represent a normalized capacitance ratio of each macromodel when the capacitance of Macromodel A is set to one. All I/O pins are protected by two-stage ESD circuits with a grounded gate nMOSFET (ggNMOS) as the primary protection device. The V_{DD}-to-V_{SS} protection is implemented with an RC-triggered ggNMOS. A symmetric diode string is employed for connecting two different power domains and providing a discharge path.

Table 1: Summary of the variable extraction. * represents the normalized value.

L_P	11.5 nH
C_{pin}	0.64 pF
$C^a_{DD} = C^a_{SS}$	1.0*
$C^b_{DD} = C^b_{SS}$	3.4*
$C^c_{DD} = C^c_{SS}$	10.6*
$C^d_{DD} = C^d_{SS}$	3.2*
$C^e_{DD} = C^e_{SS}$	4.5*
$C^f_{DD} = C^f_{SS}$	9.8*
$C^g_{DD} = C^g_{SS}$	2.5*
$C^h_{DD} = C^h_{SS}$	6.3*
Total package cap. (Meas.)	12.5 ~ 14.0 pF

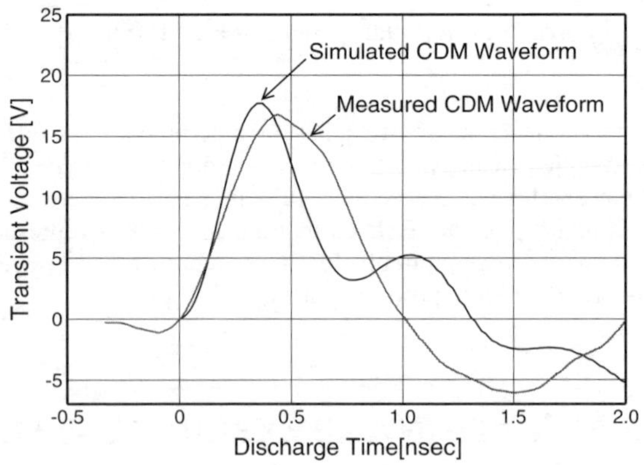

Figure 8: Comparison of measurement and simulation voltages appearing on a grounded pin after application of a 500 V positive CDM stress.

4.2 Simulation Results

Chip-level simulation is performed for analyzing the failure mechanism of a positive CDM event with respect to a stress pin of Macromodel *A*. The simulation can be used to find the maximum tolerable stress in small circuits with a few I/Os by analyzing current waveforms corresponding to all possible stress discharge paths.

Fig. 8 shows the comparison of measurement and simulation for a short rise time CDM pulse. The waveforms are the transient voltages appearing on a stress pin after application of a 500 V CDM pulse. The simulation shows the result of approximately a 400-ps rise time with peak discharge voltage of 18 V, whereas the measurement has a rise time of 450-ps (± 80-ps) and peak voltage of 17 V. The simulation matches with the measurement except for a small deviation of the pulse width. In practice, the simulation variables can be either overestimated or underestimated. The main reason for this discrepancy comes from the inaccuracies and distortions resulting from the relay and the voltage probe as well as from dispersion and frequency dependent cable losses affect the measurement of the transient behavior. In addition, for the simplified modeling, we ignore parasitic inductance effect along the PG conductive busses.

The effects of high electric fields and very short pulses which are presented during CDM-ESD pulses on MOSFETs with thin gate oxides have been extensively studied due to their importance in MOS technology [13]. The oxide hardness to ESD pulses has been explained by the 1/E model, referring to the logarithm of time-to-failure as inversely proportional to the applied electric field (or voltage) [14]. The oxide electric field, E_{ox}, of a 3.5-nm oxide for CDM pulse duration of 0.8-ns is extrapolated to 54.5 MV/cm, corresponding to an oxide breakdown voltage of 19.0 V. Using this result, we predict CDF damage by monitoring voltages across all gate oxides during a CDM stress. Fig. 9 shows the voltages across the gate oxides in the PDCC of Macromodel *A* with respect to a stress pin in Macromodel *A*. Note that the induced gate voltages are approximately proportional to the package capacitances of corresponding macromodels. The peak transient voltage is observed at the gate oxide connected to Macromodel *C* and *D* which has the largest package capacitance, as shown in Table 1. The induced voltage exceeds the dielectric strength of the oxide, most likely resulting in a rupture. This result suggests that the reliability of gate oxide must correlate with the package capacitance of connected macromodels. As a result, the gate oxide connected to a macromodel with larger package capacitance suffers severer damage than the gate oxide with smaller package capacitance for a given CDM stress voltage. Note also that the gate oxide is more susceptible to damage when a pin in a macromodel with the smallest package capacitance is grounded.

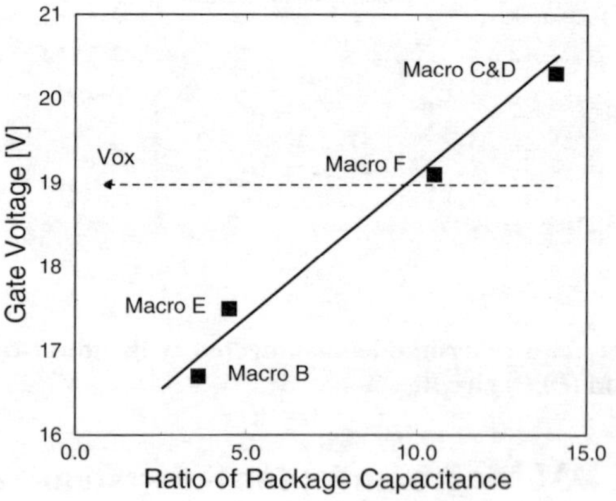

Figure 9: Peak voltages across gate oxides in the PDCC of Macromodel *A* under a 500 V positive CDM stress.

Table 2: Summary of CDM tests.

Device Code	Package Code	Failure Description
CDM1	352-pin PBGA	Functional failures
CDM2	100-pin TQFP	Pass over ± 1000 V
CDM3	272-pin BGA	Oxide destructions
CDM4	217-pin PBGA	Failures by oxide shorts
CDM5	176-pin PBGA	Oxide destructions

4.3 Experimental Results

The proposed method can easily be applied to the CDM failure mechanism analysis (FMA) of any product ICs. The products and the relevant test results are listed in Table 2. The failures occurring below 500 V have been observed in three ICs. The results usually show various failure modes. Some failure modes have been randomly observed throughout a whole chip so that the hierarchical analysis hardly predicts the failure locations and their mechanisms. The most susceptible locations against CDM stress are the gate oxides of input circuits neighboring I/O protection circuits. The next frequently observed failure mode is the gate oxide rupture around the interface circuit located between different power-ground nets.

Figure 10: SEM photograph of gate-oxide damage observed after removing the poly layer

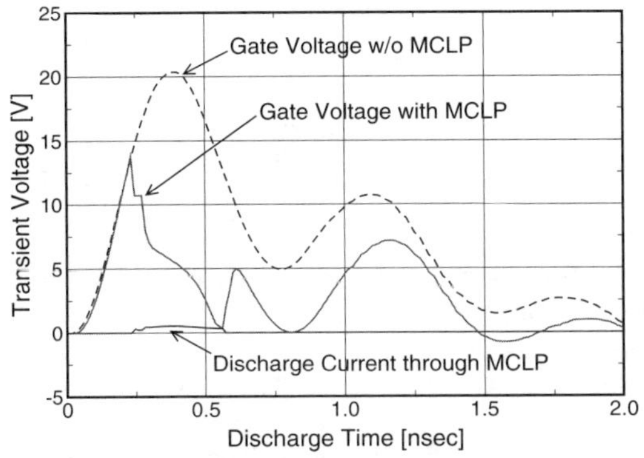

Figure 11: Simulated gate voltage waveforms with and without a clamp device M_{CLP}.

Direct charge CDM tests were performed for various positive and negative voltages. While I/O protection and power protection circuits withstand CDM stress of over 1000 V, the overall CDM robustness of the ASIC is lower than 500 V. Emission microscopy was carried out by inspection of the static *Idd* failure location. Fig. 10 shows SEM views of internal gate oxide destruction observed in the power domain crossing circuit of Macromodel *A* which our chip-level simulations predicted. The gate edge on the source side was damaged and appeared to be melted.

To protect the gate oxide from being damaged, an alternative discharge path, comprised of a ggNMOS M_{CLP} and resistor R_{CDM}, is placed in front of the vulnerable gate oxide. Fig. 11 shows the influence of the protection circuit on the gate oxide protection. Just beyond the turn-on voltage of the protection circuit, the charge transferred from the Macromodels *C* and *D* is discharged through protection device M_{CLP}, and as a result the gate voltage is clamped to a voltage less than the oxide breakdown voltage BV_{OX}. With careful layout design and CDM protection circuits in front of sensitive gate oxides predicted by simulation, the overall CDM performance exceeds 750 V.

5. CONCLUSIONS

Even it may not be well appreciated, estimated ESD failure has been ranked among the top three problems in the state-of-the-art VLSI chip manufacturing. The CDM is the major failure cause of LSIs among various type of ESD failures. One crucial prerequisite of CDM-ESD protection engineering is an efficient and accurate CDM-ESD model in order to identify stress-sensitive components in early design stages. However, it is computationally prohibitive to simulate the CDM behavior of a chip based on such *circuit-level* models.

In this paper, a new methodology was proposed to simulate *full-chip* CDF for a system-on-a-chip with multiple supplies. Hierarchical approach to chip modeling and CDM macromodels yields manageable netlists while generating very accurate prediction to CDF. This hierarchical approach associated with novel macromodel for each subsystem can be easily incorporated into the physical design phase and applied to any complicated chip with multiple functional blocks and power supplies. Chip-level modeling and analysis allow one to investigate the impact of package capacitance and power protection networks on the CDM performance. Thus, this method can help a designer to find vulnerable internal gate oxide in early design stages. We verified our methodology with CDM test results for a CMOS ASIC fabricated on a 0.25-μm technology. As predicted in our simulation results, an oxide destruction was identified in the test chip using a scanning electron microscope.

6. ACKNOWLEDGMENTS

The authors would like to thank Drs. Bendix and Huh at Device Group in LSI Logic for their help in discussion and testing. This work was supported in part by Semiconductor Research Corporation under Grant SRC99-HJ-734

7. REFERENCES

[1] S. Halperin, "Guidelines for Static Control Management," in *Eurostat*, 1990.

[2] Lucent Technologies, "Electrostatic Discharge Control Handbook," February, 1997.

[3] C. Duvvury and A. Amerasekera, "Advanced CMOS Protection Device Trigger Mechanisms During CDM," in *Proc. EOS/ESD Symp.*, pp. 162–174, 1995.

[4] S. Ramaswamy, E. Li, E. Rosenbaum, and S. M. Kang, "Circuit-level Simulation of CDM-ESD and EOS in Submicron MOS Device," in *Proc. EOS/ESD Symp.*, pp. 316–321, 1996.

[5] S. Beebe, "Simulation of Complete CMOS I/O Circuit Response to CDM Stress," in *Proc. EOS/ESD Symp.*, pp. 259–270, 1998.

[6] T. Li and S.M. Kang, "Layout Extraction and Verification Methodology for CMOS I/O Circuits," in *Proc. of the DAC*, pp. 291–296, 1998.

[7] S. Dabral and T. Maloney, "Basic ESD and I/O Design," New York: Wiley, 1998.

[8] M. Ker, C. Wu, H. Chang, and T. Wu, "Whole Chip ESD Protection Scheme for CMOS Mixed-Mode IC's in Deep-Submicron CMOS Technology," in *Proc. of IEEE Custom Integrated Circuit Conference*, pp. 31–34, 1997.

[9] P.J.H. Elias and N.P. van der Meijs, "Extracting Circuit Models for Large RC Interconnections that are Accurate up to a Predefined Signal Frequency," in *Proc. of the DAC*, 1996.

[10] M. Zhao, R.V. Panda, S.S. Sapatnekar, T. Edwards, R. Chaudhary, and D. Blaauw, "Hierarchical Analysis of Power Distribution Networks," in *Proc. of the DAC*, pp. 150–155, 2000.

[11] A. Dharchoudhury, R. Panda, D. Blaauw, R. Vaidyanathan, B. Tutuianu, and D. Bearden, "Design and analysis of power distribution networks in PowerPC microprocessors," in *Proc. of the DAC*, pp. 738-743, 1998.

[12] G. Steele, D. Overhauser, S. Rochel, and Z. Hussain, "Full-chip verification methods for DSM power distribution systems," in *Proc. of the DAC*, pp. 744-749, 1998.

[13] C. Leroux et al., "Analysis of Oxide Breakdown Mechanism Occurring During ESD Pulse," in *Proc. of IEEE Int. Reliability Physics Symp.*, pp. 276–282, 2000.

[14] Y. Fong and C. Hu, "The Effects of High Electric Field Transients on Thin Gate Oxide MOSFETs," in *Proc. of EOS/ESD Symp.*, pp. 252–257, 1987.

A Framework for Evaluating Design Tradeoffs in Packet Processing Architectures

Lothar Thiele, Samarjit Chakraborty, Matthias Gries, Simon Künzli
Computer Engineering and Networks Laboratory
Swiss Federal Institute of Technology (ETH) Zürich
CH-8092 Zürich, Switzerland
{thiele, samarjit, gries, kuenzli}@tik.ee.ethz.ch

ABSTRACT

We present an analytical method to evaluate embedded network packet processor architectures, and to explore their design space. Our approach is in contrast to those based on simulation, which tend to be infeasible when the design space is very large. We illustrate the feasibility of our method using a detailed case study.

Categories and Subject Descriptors

C.0 [**Computer Systems Organization**]: General—*Modeling of computer architecture, System architectures*; B.4.1 [**Hardware**]: Input/output and data communications—*Data communication devices*

General Terms

Performance, Design

1. INTRODUCTION

Architectures for embedded packet processing devices like network processors consist of a heterogeneous combination of different hardware and software components. These include microprocessors, domain specific instruction set processors, reconfigurable hardware modules, IP cores, application specific units, communication primitives, and memory modules. A high-level design space exploration of such packet processors is concerned with resource allocation and partitioning, mapping of different packet processing functions to the different hardware and software units, and determining a scheduling policy at each of these units. These must take into account issues such as cost, power consumption, memory requirements, delays experienced by the packets that are to be processed and their possible throughputs.

There are many issues that are specific to such a design space exploration in the domain of network packet processors–heterogeneity of the processing elements that constitute the target architecture, QoS guarantees that the processed real-time traffic streams (such as voice and video) must meet, different application scenarios where such a processor might be deployed. For example, a network processor to be used in a backbone network can be characterized by very high throughput demands but relatively low processing requirements per packet, whereas access networks have lower throughput demands but high computational requirements. Due to these

DAC 2002, June 10-14, 2002, New Orleans, Louisiana, USA.

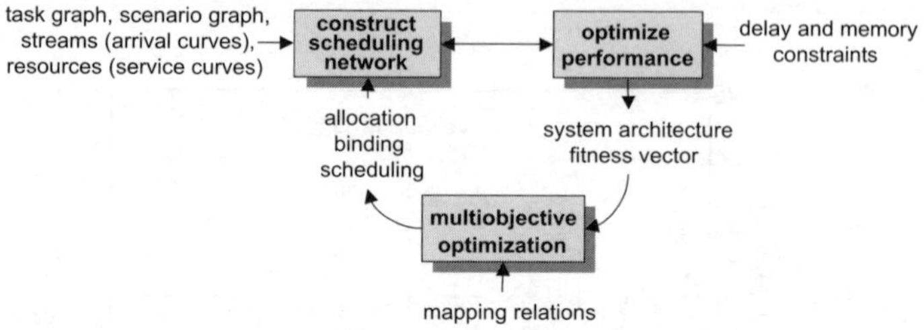

Figure 1: Basic concept for the design space exploration of packet processing systems.

differences with other target domains for system-level design space exploration, several questions arise. These involve appropriate modeling of packet streams, packet processing tasks, and hardware and software resources. Other issues include performance modeling in the case of several (possibly conflicting) usage scenarios, and strategies to efficiently explore large design spaces and to obtain a reasonable compromise between multiple conflicting criteria.

In an attempt to address these issues, we present a framework which can be used to analytically evaluate packet processing architectures and identify the different design tradeoffs involved. When interfaced with a search strategy this gives an efficient means for exploring large design spaces. The search strategy comes up with possible alternatives from the design space, which are evaluated using our framework and the feedback guides further search.

The essential ingredients of our framework are a task and a resource model, and a *real-time calculus* [2, 12] for reasoning about packet streams and their processing. The task model represents different packet processing functions such as header processing, encryption, processing for special packets such as voice and video, etc. The resource model captures the information about different available hardware resources, the possible mappings of packet processing functions to these resources and the costs associated with each of these mappings. Traffic streams are specified using their *arrival curves* [5] and deadlines are associated with the real-time traffic. Given any architecture, we can now analytically determine properties such as delay and throughput experienced by the different traffic streams, taking into account the underlying scheduling disciplines at each of the resources.

The overall scheme used to illustrate the feasibility of our approach for design space exploration is shown in Fig. 1. Here the search strategy is a multiobjective optimizer (such as an evolutionary algorithm [6]) which comes up with possible architectures including allocation, binding, and scheduling information. Using this information, a *scheduling network* is constructed (see Section 3.2) for each usage scenario of the packet processor, which enables the computation of the memory requirements and delay properties of each traffic stream. Then the packet rates for each of the input streams are maximized until one of the delay constraints associated with the streams, or a memory constraint associated with the architecture is violated (see Section 3.3). These values for each usage scenario indicate the optimum performance attainable by the architecture and provide the parameters to guide further exploration by the multiobjective optimizer.

All the previous work on the design space exploration of network processor architectures (such as [4] and [14]) relied on simulation techniques, where different architectures are simulated and evaluated using benchmark workloads. In many cases the search space being explored is very large, and it might be too expensive to evaluate all design alternatives using simulation. The focus of our approach is on a high level of abstraction where the goal is to quickly identify interesting tradeoffs and architectures which can then be further evaluated, by simulation for example, taking finer details into account.

The next section describes our models, following which we describe our method for performance estimation in Section 3. Section 4 briefly introduces the design space exploration scheme and in Section 5 we present our experimental results.

2. MODELING

2.1 Packet Processing

A packet processor operates on interleaved streams of packets which enter the device. All packets belonging to a stream are processed in the same way, i.e. the same set and sequence of tasks are executed for them. In order to know the load on the processing device, it is necessary to have information about the set of tasks associated with a packet stream and the number of packets arriving per time unit. This information can be formalized as follows:

Definition 1. We define a set of streams F and a set of tasks T. To each stream $f \in F$ there is associated a directed acyclic graph $G(f) = (V(f), E(f))$ with task nodes $V(f) \subseteq T$ and edges $E(f)$. The tasks $t \in V(f)$ must be executed for each packet of stream f while respecting the precedence relations in $E(f)$.

In other words, the tasks $V(f)$ must be executed by the packet processor for each packet in stream f. If there is an edge from task t_1 to t_2, then t_1 must be executed before t_2. All tasks can be combined into one conditional task graph where depending on the stream to which a packet belongs, the packet may take different paths through this graph.

As mentioned in the introduction, a packet processing device may be used in a variety of usage scenarios. These scenarios may lead to conflicting design objectives. One of the main ideas in this paper is to exploit of this fact to design architectures which make the best out of possible tradeoffs.

Definition 2. We define a set of scenarios B. To each scenario $b \in B$ there is associated a memory constraint $m(b) \in \mathbb{R}_{\geq 0}$. There is a scenario-stream relation C where $c = (b, f) \in C$ denotes that a stream f is present in scenario b. To each relation c there is associated an end-to-end deadline $d(c) \in \mathbb{R}_{\geq 0}$, a lower arrival curve α_c^l and an upper arrival curve α_c^u.

Here, $m(b)$ denotes the maximum number of packets that can be stored in the processing device in scenario b at any point in time. The end-to-end deadline $d(c)$ with $c = (b, f)$ denotes the maximum allowed time span between the packet entering the processor and the end of the last task execution for a packet of stream f in scenario b. The lower and upper arrival curves are defined as follows.

Definition 3. For any $\Delta \in \mathbb{R}_{\geq 0}$ and any scenario-stream relation $c = (b, f) \in C$, the value $\alpha_c^l(\Delta)$ is smaller than the number of packets arriving in any time interval of length Δ in the stream f in scenario b. In a similar way, the maximum number of packets in an interval of length Δ is always smaller than $\alpha_c^u(\Delta)$.

Therefore, instead of using a stochastic characterization of the input streams, we use deterministic bounds. This approach not only nicely fits into the underlying concept of the SPI model (see [15]) but also follows the usual specification of traffic in the networking context, see the T-SPEC model from IETF [10].

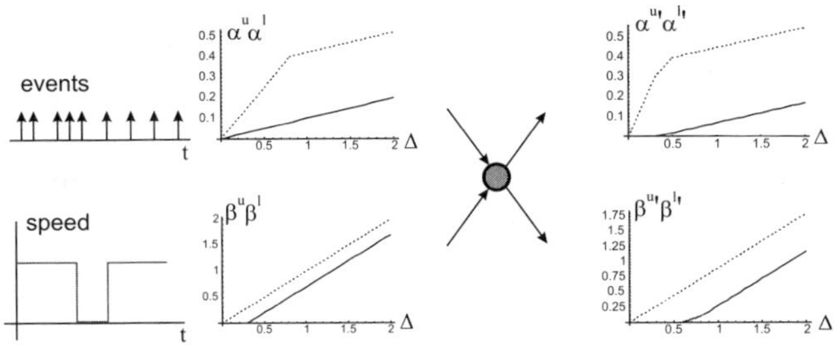

Figure 2: Representation of an arrival curve and a service curve as processed according to Theorem 1.

In Fig. 2, on the upper left corner an example of a packet stream and associated arrival curves is given. Note that in this figure, the arrival curves have been scaled with the computation time for a packet. The steepness of the upper arrival curve until $\Delta = 0.75$ is a measure for the burstiness of packet arrivals.

There are different possibilities to determine the input characterizations of the packet streams:

- The input streams have been characterized in a traffic agreement using T-SPECs.
- Characteristics such as burstiness, burst length, and sustained rates are determined by considering the properties of the generating components, e.g. a sensor and the network connection. In a similar way, properties such as jitter can be easily rephrased in terms of upper and lower arrival curves.
- Measured traffic streams are analyzed with respect to their maximum and minimum number of packets in any time interval.

Methods using the concept of arrival curves to design and analyze communication networks can be found in [5].

In summary, we use a formal means for describing packet rates (using arrival curves), and we gave a model for representing the sequence of tasks to be executed for each packet in a traffic stream, and a model for describing different usage scenarios. In the next section we introduce a representation for packet processing architectures.

2.2 System Architecture

It is widely accepted that packet processing devices will be heterogeneous in nature. In dedicated or application specific instruction set components, simple tasks with high data rate requirements will be executed. Longer and more complicated execution chains will be transferred to software tasks running on (homogeneous) multiprocessors. In this case, usually run-time scheduling methods are used in order to fairly share the available resources among packets belonging to different streams [3]. A sketch of the heterogeneous architecture with different packet paths discussed in this paper is shown in Fig. 3 on the left hand side.

Therefore, our model of a feasible system architecture consists of (1) available resource types including their processing capability and performance described by service curves, (2) cost of implementing a resource on the packet processor and (3) the scheduling policies and associated parameters. The logical structure of a system architecture is shown on the right hand side of Fig. 3. Here we see that the processing components have (logically) associated memory which stores packets waiting for the next task to be executed on them. A corresponding scheduling policy selects a packet

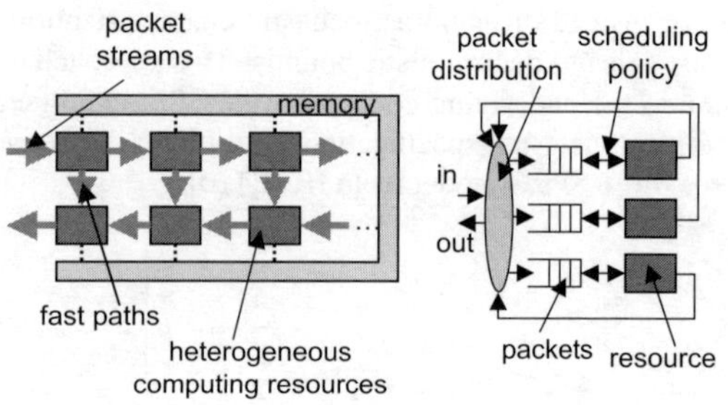

Figure 3: Example of a physical (left) and logical (right) structure of a packet processing architecture.

and starts the execution. The processing of the current packet may be preempted in favor of another task processing a different packet.

Definition 4. We define a set of resource types S. To each type $s \in S$ there is associated a relative implementation cost $cost(s) \in \mathbb{R}_{\geq 0}$ and the number of available instances $inst(s) \in \mathbb{Z}_{\geq 0}$. To each resource type there is associated a finite set of scheduling policies $sched(s)$ which the component supports, a lower service curve β_s^l and an upper service curve β_s^u. The mapping relation $M \subseteq T \times S$ defines possible mappings of tasks to resource types, i.e. if $m = (t,s) \in M$ then task t could be executed on resource type s. If a task t is executed on a resource of type t, then it needs $w(t,s) \in \mathbb{R}_{\geq 0}$ relative computing units for finishing, called the request.

Service curves are defined in a similar way as arrival curves. They describe bounds on the computation capability of the available resource components.

Definition 5. For any $\Delta \in \mathbb{R}_{\geq 0}$ and any resource type $s \in S$, the value $\beta_s^l(\Delta)$ is smaller than the number of available computing units in any time interval of length Δ. In a similar way, the maximum number computing units in any interval of length Δ is always smaller than $\beta_s^u(\Delta)$.

For example, let us suppose that we have a processing resource which is not used by any other task, then it is reasonable to assume that the lower and upper service curves are equal and proportional to Δ: $\beta_s^l(\Delta) = \beta_s^u(\Delta) \propto \Delta$. But if the resource is loaded with some tasks, it is clear that the available computing power within an interval may vary and hence also the time required for executing a task.

An example is given in Fig. 2 on the lower left side. Here, the computing power available is zero in a certain time interval. Therefore, the lower service curve is smaller than the upper one and is zero at the beginning.

Now we can informally describe the design space exploration procedure in the context of packet processors. It selects resources from a given reserve (allocation), associates (or maps) tasks to these resources (binding) and determines a scheduling policy and associated parameters for selecting active packets (scheduling). Although we have chosen a particularly simple cost model, it is not at all clear how to determine the maximum number of stored packets or the maximum end-to-end delays since all packet streams share common resources. For example, the computation time for a task t depends on its request $w(t,s)$, on the available processing power of the resource, i.e. β_s^l and β_s^u, and on the scheduling policy applied. In addition, as the packets may flow from one resource to the next one, there might be intermediate bursts and packet jams.

It is surprising to see, that there exists a *computationally very efficient* possibility to derive *provably correct bounds* on the *end-to-end delays* of packets and on the necessary memory requirements of the processor.

3. ESTIMATING PERFORMANCE AND MEMORY

Our estimation of end-to-end delays and memory requirements is based on the work network calculus [5] from the communication networks area. Recently this approach has been re-formulated in an algebraic setting [2]. In [13], a comparable approach was used to describe the behavior of processing (and not communication) resources. But the resulting formulas for quantities like intermediate arrival and service curves, delays and memory were computationally too expensive to be used in a design space exploration. One of the new results applied in this paper is a simple and efficient piecewise linear approximation. To explain this new methodology, we start with a description of the basic building blocks of a *scheduling network*, describing the flow of packets and resources.

3.1 Basic Building Blocks

The basic idea behind our performance estimation is based on providing a network theoretic view of the system architecture. In particular, packet streams and resource streams flow through a network and thereby respectively change their arrival and service curves. Inputs to a single network node are arrival curves of streams and service curves. Outputs describe the arrival curves of the processed packet streams and the remaining service curves of the (partly) used resources. These resulting arrival and service curves can then serve as inputs to other nodes of the scheduling network.

In order to understand the basic concept, let us describe a very simple example of such a node, namely the preemptive processing of packets of one stream by a single processing resource. Following the previous discussion related to Fig. 3, a packet memory is attached to a processing resource which stores the packets that have to wait for being processed. During the processing of a packet there is no remaining computing power, otherwise it is equal to the original one.

In [11], the following Theorem has been derived which describes the processing of a packet stream in terms of the already defined arrival and service curves.

THEOREM 1. *Given a packet stream described by the arrival curves α^l and α^u and a resource stream described by the service curves β^l and β^u. Then the following expressions bound the remaining service of the resource node and the arrival function of the processed event stream:*

$$\alpha^{l'}(\Delta) = \inf_{0 \leq u \leq \Delta} \left\{ \alpha^l(u) + \beta^l(\Delta - u) \right\} \quad (1)$$

$$\alpha^{u'}(\Delta) = \inf_{0 \leq u \leq \Delta} \{ \sup_{v \geq 0} \left\{ \alpha^u(u+v) - \beta^l(v) \right\} + \beta^u(\Delta - u), \beta^u(\Delta) \} \quad (2)$$

$$\beta^{l'}(\Delta) = \sup_{0 \leq u \leq \Delta} \left\{ \beta^l(u) - \alpha^u(u) \right\} \quad (3)$$

$$\beta^{u'}(\Delta) = \sup_{0 \leq u \leq \Delta} \left\{ \beta^u(u) - \alpha^l(u) \right\} \quad (4)$$

Note that the arrival function as used above describes bounds on the *computing request* and *not* on the *number of packets*. In Fig. 2, an example of the remaining arrival and service curves is given.

Since we deal with packet streams in the system architecture, we need to convert these streams into corresponding computing requests. Given bounds on a packet stream of the form $[\overline{\alpha}^l, \overline{\alpha}^u]$ we can determine bounds on the related computing requests

$$[\alpha^l, \alpha^u] = [w\overline{\alpha}^l, w\overline{\alpha}^u] \quad (5)$$

using the computing request w for each packet. The notation $[\alpha^l, \alpha^u]$ represents the fact that α^l and α^u are the lower and the upper curves of a single stream.

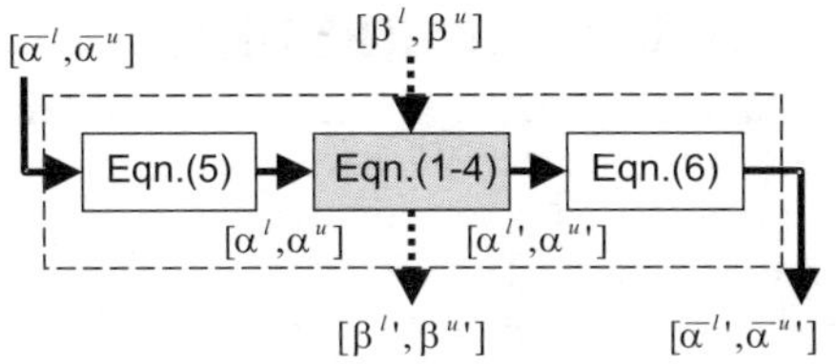

Figure 4: Block diagram depicting the transformation of packet and resource streams by a processing device. The dotted arrows represent the resource flow while the others show the flow of packets and requests.

The conversion of the output stream is slightly more involved, as we usually suppose that a next component can start processing after the whole packet arrived:

$$[\overline{\alpha}^{l'}, \overline{\alpha}^{u'}] = [\lfloor \alpha^{l'}/w \rfloor, \lceil \alpha^{u'}/w \rceil] \tag{6}$$

The whole transformation is depicted in Fig. 4.

Using this building block, we can describe commonly used scheduling approaches such as *fixed priority preemptive scheduling* [3]. In this case, to each input stream $f \in F$ there is assigned a fixed priority $prio(f) \in \mathbb{Z}_{\geq 0}$. At any point in time, a processing device operates on the packet in its memory (see Fig. 3, scheduling policy) which has the highest priority. If there are packets with the same priority, they are served following an FCFS (first come first serve) strategy. This can be modeled directly using our basic building block shown in Fig. 4.

3.2 Scheduling Network Construction

Following the overview of the design space exploration shown in Fig. 1, the scheduling network which enables the estimation of performance measures (such as end-to-end deadlines and memory requirements) is constructed using the specification data in Section 2 and the system architecture provided by the optimization algorithm. In order to simplify the explanation, we restrict ourselves to the use of a fixed priority scheduling policy for all resource types. However, our approach can be generalized to other policies as well, such as GPS [8].

Therefore, the data provided by the multiobjective optimization block in Fig. 1 are as follows.

Definition 6. The allocation of resources can be described by the function $alloc(s) \in \mathbb{Z}_{\geq 0}$ which denotes the number of allocated instances of resource type s. The binding of tasks $t \in T$ to resources in a specific usage scenario $b \in B$ is specified by the function $bind(c,t) \in S$ which maps a task in a specific scenario-stream relation $c = (b, f) \in C$ to a resource type s. The scheduling policy is described by a function $prio(c) \in \mathbb{Z}_{\geq 0}$ which associates a priority to each stream in a scenario-stream relation $c = (b, f) \in C$.

Note that a system architecture is not only described by the type and the number of resource components but also by the scenario dependent mapping of tasks to those components.

Now, we can describe the construction of a scheduling network for a given scenario $b \in B$. The basic idea is that the packet streams pass from one resource to the next one. The order is determined by the precedence relations in $E(f)$, see Def. 1. The resource flows, i.e. the capabilities of the resources, also pass through the network. The order is mainly determined by the priorities associated with packet streams.

The procedure for creating a scheduling network is given as follows: Add source and target nodes for all allocated resources. Add source and target nodes for all packet streams f in scenario b. For all tasks $t \in V(f)$, add a scheduling node as shown in Fig. 4 to the scheduling network and connect its packet stream input/output according to the precedence relations in $E(f)$. Connect the resource stream input/output of the scheduling nodes according to the precedence relations and according to the stream priorities in scenario b.

As a result of applying this procedure we get a scheduling network for each scenario, containing source and target nodes for the different packet streams, and resource flows and scheduling nodes which represent the computations described in Fig. 4. A concrete example to illustrate this is given in Fig. 9.

Given the arrival curves for all source packet nodes, i.e. $[\alpha_c^l, \alpha_c^u]$ with $c = (b, f)$ for stream f in scenario b, and the initial service curves for the allocated resources, i.e. $[alloc(s)\beta_s^l, alloc(s)\beta_s^u]$ for resource type s with $alloc(s)$ allocated resources, we can determine the properties of all internal packet streams and resource flows. It remains to be seen, how we can determine the end-to-end delays of packets and the necessary memory.

3.3 Performance Optimization

In order to estimate the properties of a packet processing system architecture we need quantities like bounds on end-to-end delays of packets to be processed and on memory requirements.

Using well known results from the area of communication networks, see e.g. [5], the bounds derived in Theorem 1 can be used to determine the maximal delay of events and the necessary memory to store packets waiting to be processed.

The following two equations give bounds on both quantities:

$$delay_c \leq \sup_{u \geq 0} \left\{ \inf\{\tau \geq 0 : \alpha_c^u(u) \leq \beta_c^l(u+\tau)\} \right\} \tag{7}$$

$$backlog_c \leq \sup_{u \geq 0} \{\alpha_c^u(u) - \beta_c^l(u)\} \tag{8}$$

In other words, the delay can be bounded by the maximal horizontal distance between the curves α^u and β^l, whereas the backlog is bounded by the maximal vertical distance between them.

In case of the scheduling network constructed above, we have to know which curves to use in (7) and (8). For α_c^u, we need the initial upper arrival curve of an incoming packet stream f. The service curve β_c^l to be used is the *accumulated* curve of all scheduling nodes through which the packets of stream f pass in the scenario b. As described in [2], this quantity can be determined through an iterated convolution.

Finally, we need to determine some measure for the performance of a given architecture under a certain usage scenario. Therefore, we introduce a parameter ψ_b, which is the largest scaling factor of the packet input streams f with $c = (b, f) \in C$ according to $[\psi_b\alpha_c^l, \psi_b\alpha_c^u]$, such that the constraints on end-to-end delays and memory are still satisfied.

3.4 Piecewise Linear Approximation

The only problem that we are left with is the efficient computation of the expensive equations in Theorem 1. Note that this set of equations has to be computed for all scheduling nodes in any scheduling network. In addition, if the design space exploration is to be based on evolutionary multiobjective algorithms, the performance of many system architectures need to be estimated.

To this end, we propose a piecewise linear approximation of all arrival and service curves. The approximation consists of a combination of two line segments and is shown in Fig. 5. As a shorthand notation we can denote curves γ^u and γ^l by the tuples $U(q,r,s)$ and $L(q,r,s)$, respectively.

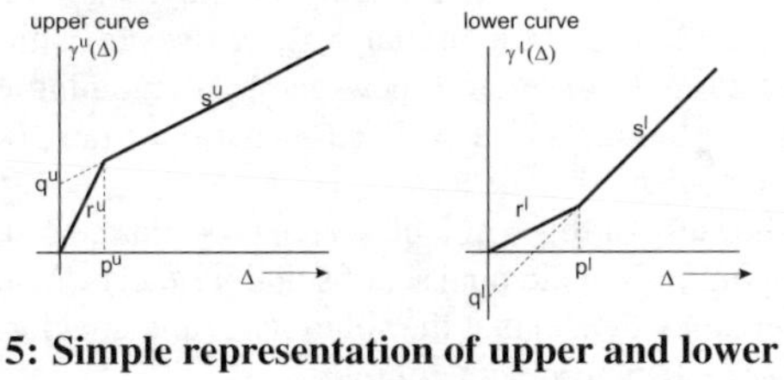

Figure 5: Simple representation of upper and lower curves.

Based on this approach, the equations in Theorem 1 and the equations (7) and (8) can be computed *symbolically*. We omit the explicit formulas because of space restrictions.

Referring to our initial Figure 1, we have now finished the description of the input specification, the construction of the scheduling network, and the performance optimization. It remains to be shown, how a multiobjective optimization algorithm can be used to perform the exploration task and come up with candidate architectures which are then evaluated using our scheme.

4. MULTIOBJECTIVE DESIGN SPACE EXPLORATION

As described in the introduction, in the case of network processor design we are faced with conflicting objectives. Here we show how to obtain a tradeoff between the performance ψ_b in different usage scenarios b and the cost of the system architecture

$$cost = \sum_{s \in S} alloc(s) cost(s) \quad (9)$$

We can formulate this task as a multiobjective optimization problem where with each implementation there is an associated objective vector $v = (v_0, ..., v_{k-1})$ with k elements. The goal is to determine implementations with Pareto-optimal objective vectors. The following definition supposes that we are trying to find vectors with small elements, i.e. we are considering a minimization problem. Note that our design space exploration problem can be formulated in this way by, for example, using $v_0 = cost$ and $v_i = 1/\psi_{b_i}$ for all $b_i \in B$, $i > 0$.

Definition 7. Given a set V of k-dimensional vectors $v \in \mathbb{R}^k$. A vector $v \in V$ dominates a vector $g \in V$ if for all elements $0 \leq i < k$ we have $v_i \leq g_i$ and for at least one element, say k, we have $v_k < g_k$. A vector is called Pareto-optimal if it is not dominated by any other vector in V.

The architectures with Pareto-optimal objective vectors represent the tradeoffs in the network processor design. There are many different approaches to multiobjective optimization, e.g. Tabu search, simulated annealing or evolutionary algorithms [6]. We have chosen evolutionary algorithms, but other options also could have been used.

The evolutionary multiobjective optimizer that we have used, called SPEA2 [6], maintains a population of current solutions and an archive which stores the best solutions found so far. Independent of the optimization principle used, the optimizer generates new solutions (system architectures) based on the already known set. These new solutions are then evaluated for their objective vector. The following methods are used in order to include domain specific knowledge into the search process:

(1) System architectures are represented according to Def. 6. In particular, the representation contains the allocation $alloc(s)$ for all resource types $s \in S$, the binding of tasks to resources *for each scenario*, and the stream priorities *for each scenario*.

(2) New system architectures are determined using *mutation* and *crossover* operations [6]. In case of the mutation, either allocation, or the binding for a scenario, or the priorities are mutated. In case an infeasible system architecture is generated (see Section 3.2) a repair strategy is invoked which attempts to maintain a high diversity in the population. In a similar way, crossover combines two selected solutions to generate a new one by combining either the allocations, the bindings of a selected scenario, or the priorities of a selected scenario.

Clearly, because of this heuristic search procedure, no statements about the optimality of the final set of solutions can be made. There is experimental evidence, that the solutions found are close to optimal even for realistic problem complexities.

5. EXPERIMENTAL RESULTS

Our input traffic consists of a mixture of real-time (RT) and non-real-time (NRT) streams. We have considered two scenarios b_1, b_2 where the first scenario b_1 models a configuration in a network backbone where aggregates of streams are forwarded with the smallest possible computation requirement but at high bandwidths. This scenario therefore contains a single stream class for packet forwarding (NRT_Forward). The second scenario b_2 however resembles an access link from a customer to a service provider. Typically an access network has relatively low bandwidth but high computation requirements. The computations are in particular required to perform encryption, e.g. Triple-DES [1] to allow sensitive information to be transmitted over untrusted networks of a service provider. Another application is voice coding/decoding (e.g. in accordance with ITU G.723.1) to use telephony services over the Internet. Of course, a network processor for an access link could also be used to simply forward traffic from the customer to the service provider. Hence, we define a set of five traffic streams F = {NRT_Forward, RT_Send, RT_Recv, NRT_Encrypt, NRT_Decrypt}, containing two real-time flows based on voice services and three flows with relaxed quality of service requirements where two of them apply Triple-DES. To process these five streams we define 25 tasks, i.e. $|T| = 25$. The task graph and the dependencies are shown in Fig. 6.

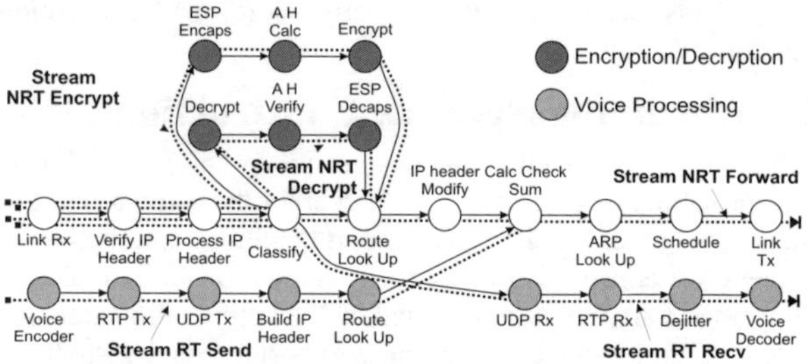

Figure 6: Task graph for a network processor.

The scenario-stream relation C can be represented as in Fig. 7.a). Here, one can also see an example of an end-to-end deadline $d(c)$, a memory constraint $m(b)$, and arrival curves $[\alpha_c^l, \alpha_c^u]$ represented in the piecewise linear form defined in Section 3.4.

Further, we use 8 different resource types s with S = {Classifier, PowerPC, ARM9, MEngine, CheckSum, Cipher, DSP, LookUp}. Each one of these has different computation capabilities that can be represented in form of the mapping relation M, see Def. 4. Part of this specification is represented in Fig. 7.b), including an example for the implementation cost $cost(s)$, number of instances $inst(s)$ and request $w(t,s)$. The initial service curves are simply set to $[\beta_c^l, \beta_c^u] = [L(0,0,1), U(0,0,1)]$, i.e. $\beta_c^l(\Delta) = \beta_c^u(\Delta) = \Delta$, reflecting the fact that the resources are fully available for the processing of the tasks.

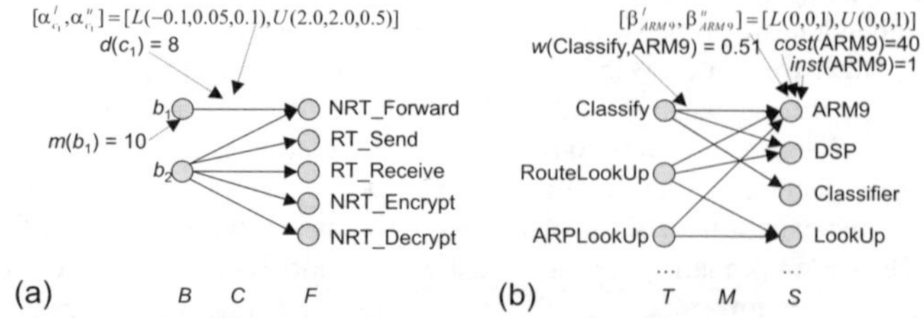

Figure 7: Graphical representation of (a) the scenario-stream relation C and (b) a part of the task-resource relation M.

Fig. 8 (a screen shot from our tool) shows the initial *population* on the left and the final population of a specific design space exploration run on the right. Each dot represents one system architecture (resources, binding and scheduling policy). The axes represent the total cost of the implementations (see Eqn. (9)) and the maximum scaling factor for scenario b_1 of the packet input streams until either the end-to-end deadline or the maximum memory is reached. Note that we are looking at projections of the three-dimensional objective space on two dimensions. Hence, we do not see the "typical" Pareto-tradeoff curves. Nevertheless, no objective vector (or system architecture) dominates another one (see Def. 7). We will not discuss the details of the tradeoff curves, but would like to point out that the different branches that can be seen in the projection are caused by the availability of specialized hardware components which are only useful for some of the input streams.

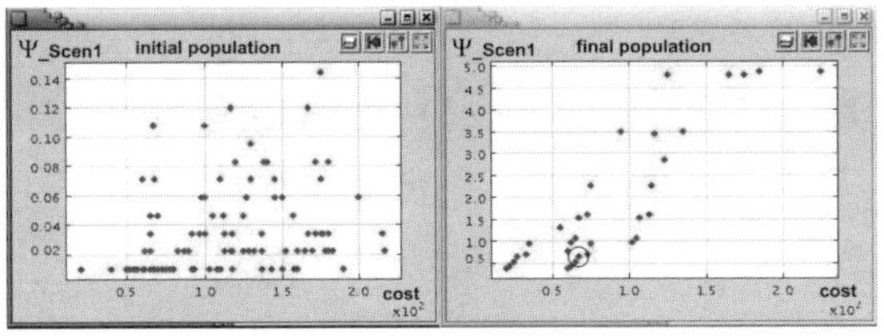

Figure 8: Initial population and final population of a design space exploration run.

Finally, let us look more closely at one of the implementations. It is marked in Fig. 8 and its objective vector consists of $cost = 67$, the maximal scaling for the input streams for scenario 1 (b_1) is $\psi_{b_1} = 0.644$ and that for scenario 2 (b_2) is $\psi_{b_2} = 0.400$. The allocated resource components are CheckSum, Cipher, ARM9 and LookUp. Therefore, we have $alloc(\text{CheckSum}) = alloc(\text{Cipher}) = alloc(\text{ARM9}) = alloc(\text{LookUp}) = 1$ (see Def. 6). In scenario 2 (b_2) all the 5 streams in F are present. The ordering of the priorities as determined by the design space exploration for this scenario is: NRT_Encrypt, NRT_Forward, RT_Send, RT_Receive and NRT_Decrypt. Here the stream NRT_Encrypt has priority 1 (i.e. $prio(\text{NRT\_Encrypt}) = 1$) and NRT_Decrypt has priority 5. Due to space restrictions, not all bindings are given here. The task VerifyIPHeader is mapped to the CheckSum resource, tasks ProcessIPHeader, Classify, VoiceEncoder, RTPTx, UDPTx, BuildIPHeader are executed on the ARM9 resource and tasks RouteLookUp are mapped to the LookUp resource. Based on this information, the system internally builds the scheduling graph and evaluates it according to the methods described in Section 3. Part of the scheduling network (see Section 3.2) is represented in Fig. 9.

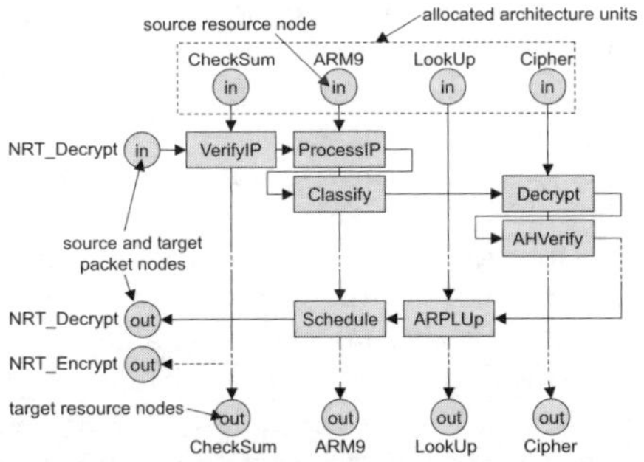

Figure 9: Part of a schedule network constructed for the performance estimation of a given system architecture. The scheduling policy is *fixed priority* and the internal scheduling nodes correspond to the basic blocks shown in Fig. 4.

Our tool performed the above described design space exploration. It uses the evolutionary optimization tool SPEA [6], and an implementation of performance optimization, population and archive handling, mutation, crossover and repair in Java. The graphical output used the Ptolemy-Plot routine from UC Berkeley [9], the graphical input and tool integration was based on the modeling and simulation tool MOSES [7]. The experiments have been run on a Pentium III under LINUX. The population size was 100 system architectures and the optimization was stopped after 300 generations. Each generation, i.e. the mutation, crossover and performance optimization (including scheduling network construction) takes about 2 sec which leads to an overall optimization time of 10 minutes.

6. CONCLUDING REMARKS

We presented several new results concerning the modeling and design space exploration of packet processing devices, such as network processors. Our work involves a careful combination of some known and some new solutions: (1) modeling the flow of packet streams through a heterogeneous system architecture consisting of computation and communication components using a scheduling network, (2) using a real-time calculus to determine end-to-end deadlines and memory requirements and using piecewise linear approximations for fast performance estimation, and finally (3) combining these models and methods in a design exploration system for packet processors.

These results are based on several abstractions. In particular, we neglect effects of caching, separate memories, and shared communication resources in the processor architecture. The extension of the presented work to incorporate these additional constraints is a subject of further investigation.

Acknowledgement

The work presented in this paper has been supported by IBM Research. The authors are grateful to Andreas Herkersdorf of IBM Research Zürich, for discussions that influenced the presented results.

7. REFERENCES

[1] American National Standards Institute. *Triple Data Encryption Algorithm Modes of Operation, ANSI X9.52-1998*, 1998.

[2] J. L. Boudec and P. Thiran. *Network Calculus - A Theory of Deterministic Queuing Systems for the Internet*. LNCS 2050, Springer Verlag, 2001.

[3] G. Buttazzo. *Hard Real-Time Computing Systems - Predictable Scheduling Algorithms and Applications*. Kluwer Academic Publishers, 1997.

[4] P. Crowley, M. Fiuczynski, J.-L. Baer, and B. Bershad. Characterizing processor architectures for programmable network interfaces. In *Proc. International Conference on Supercomputing*, Santa Fe, 2000.

[5] R. Cruz. A calculus for network delay. *IEEE Trans. Information Theory*, 37(1):114–141, 1991.

[6] K. Deb. *Multi-objective optimization using evolutionary algorithms*. John Wiley, Chichester, 2001.

[7] MOSES project. www.tik.ee.ethz.ch/~moses/.

[8] A. Parekh and R. Gallager. A generalized processor sharing approach to flow control in integrated services networks. *IEEE/ACM Trans. Networking*, 2-2:137–150, 1994.

[9] Ptolemy project. ptolemy.eecs.berkeley.edu/.

[10] S. Shenker and J. Wroclawski. General characterization parameters for integrated service network elements. RFC 2215, Internet Engineering Task Force (IETF), Sept. 1997.

[11] L. Thiele, S. Chakraborty, M. Gries, A. Maxiaguine, and J. Greutert. Embedded software in network processors – models and algorithms. In *First Workshop on Embedded Software*, LNCS 2211. Springer-Verlag, 2001.

[12] L. Thiele, S. Chakraborty, and M. Naedele. Real-time calculus for scheduling hard real-time systems. In *Proc. IEEE International Symposium on Circuits and Systems*, volume 4, pages 101–104, 2000. Invited paper.

[13] L. Thiele, S. Chakraborty, and M. Naedele. Real-time calculus for scheduling hard real-time systems. In *Proc. IEEE International Conference on Circuits and Systems*, volume 4, pages 101–104, 2000.

[14] T. Wolf, M. Franklin, and E. Spitznagel. Design tradeoffs for embedded network processors. Technical Report WUCS-00-24, Department of Computer Science, Washington University in St. Louis, 2000.

[15] D. Ziegenbein, K. Richter, R. Ernst, L. Thiele, and J. Teich. SPI – a system model for heterogeneously specified embedded systems. *IEEE Trans. Very Large Scale Integration (VLSI) Systems*, 2002. Accepted for publication.

Energy Estimation and Optimization of Embedded VLIW Processors based on Instruction Clustering

A. Bona[§] M. Sami[‡] D. Sciuto[‡] C. Silvano[♯] V. Zaccaria[‡] R. Zafalon[†]
[§]ALaRI, Lugano, Switzerland [‡] Politecnico di Milano, Milano, Italy
[♯] Università degli Studi di Milano, Milano, Italy [†]STMicroelectronics, Agrate B. (MI), Italy

ABSTRACT

Aim of this paper is to propose a methodology for the definition of an instruction-level energy estimation framework for VLIW (Very Long Instruction Word) processors. The power modeling methodology is the key issue to define an effective energy-aware software optimisation strategy for state-of-the-art ILP (Instruction Level Parallelism) processors. The methodology is based on an energy model for VLIW processors that exploits instruction clustering to achieve an efficient and fine grained energy estimation. The approach aims at reducing the complexity of the characterization problem for VLIW processors from exponential, with respect to the number of parallel operations in the same very long instruction, to quadratic, with respect to the number of instruction clusters. Furthermore, the paper proposes a spatial scheduling algorithm based on a low-power reordering of the parallel operations within the same long instruction. Experimental results have been carried out on the Lx processor, a 4-issue VLIW core jointly designed by HPLabs and STMicroelectronics. The results have shown an average error of 1.9% between the cluster-based estimation model and the reference design, with a standard deviation of 5.8%. For the Lx architecture, the spatial instruction scheduling algorithm provides an average energy saving of 12%.

Categories and Subject Descriptors

B.7.0 [**Design Aids**]: General; B.6.0 [**Logic Design**]: General

General Terms

Experimentation

Keywords

power estimation, vliw architectures

1. INTRODUCTION

The overall goal of this work is to propose a methodology for the definition of an instruction-level energy estimation framework for VLIW processors. This methodology is the key issue to enable an effective low-power hardware and software exploration framework for state-of-the-art ILP processors. The approach presented in this paper aims at extending and integrating into an overall power estimation and optimisation framework our work previously proposed in [1–3], targeting an instruction-level energy model to evaluate the energy consumption associated with a software execution on a pipelined VLIW core. A pipelined VLIW processor executes a set of explicitly parallel *operations* (also called *syllables*) during each clock cycle; this set of operations are statically scheduled to constitute the *Very Long Instruction Word* (also called instruction or *bundle*).

The proposed strategy is based on the combination and interaction of different well known techniques such as clustering and instruction scheduling to obtain an overall hardware/software power optimization for VLIW embedded systems.

First, we aim at increasing the efficiency of the power estimation by reducing the complexity of the VLIW instruction-level model. With this purpose, we apply the instruction clustering concept to the VLIW instruction set, by characterizing only each single operation in isolation, then by clustering operations considering their average energy cost and by characterizing the inter-instruction effects on the clusters.

Second, the power optimization methodology acts at the software level through the definition of a technique to statically optimize the executable applications. The proposed technique consists of a *spatial scheduling* algorithm based on the reordering of parallel operations within the same bundle with respect to the previous bundle. The algorithm is based on the minimization of a cost function that considers inter-instruction effects and power figures carried out from the power characterization phase.

The experimental results validating the characterization and scheduling approaches have been carried out on the Lx architecture, a scalable and customizable processor technology [4] designed for multimedia and signal processing embedded applications. The Lx processor is a statically scheduled VLIW multi-clustered architecture jointly designed by Hewlett-Packard and STMicroelectronics, based on a multiple-issue VLIW core.

The paper is organized as follows. Some relevant background works on instruction-level power analysis and optimization are summarized in Section 2. An overall description of the proposed VLIW power model is given in Section 3, while Section 4 introduces the methodology to cluster operations with the same power behavior. Some experimental results are reported in Section 5, while Section 6 contains some concluding remarks and outlines some future research directions.

2. PREVIOUS WORK

Instruction-level power estimation techniques are based on the functional simulation of the application software on the target ISA (Instruction Set Architecture). During the instruction set simulation (ISS), a black-box instruction-level model correlates the power cost to each individual instruction, by considering both internal state effects, pipeline stalls and cache misses occurred at run time. One of the most significant works on instruction-level power analysis ($ILPA$, for brevity) is based on electrical measurements techniques [5–7] to characterize the *average energy per instruction* and an *average inter-instruction effect* energy. The work outlines that the spatial complexity of an instruction-level energy model that considers y inter-instruction effects (i.e., between instruction i and $i-y$) and x instructions in the ISA is $O(x^{y+1})$. The authors recognized that instruction clustering could be a feasible way to reduce this complexity: grouping the instructions in c clusters, the complexity is reduced to $O(c^{y+1})$.

The instruction-level power model proposed by Russel *et al.* [8] considers the average instruction energy as invariant for all instructions in the ISA. More specifically, this model is based on the observation that, for a given class of processors, the *energy per instruction* shows a very small variance. Sarta *et al.* [9] propose a processor power model by considering possible inter-instruction effects as well as the actual data statistics. Although the developed power model is quite accurate, it lacks general applicability, being developed only for a specific embedded processor. Klass *et al.* [10] propose a power model in which inter-instruction effects are measured by considering only the additional energy consumption observed when a generic instruction is executed after a NOP (the proposed power model is also called the *NOP model*). This model could be an effective solution to the problem of spatial complexity proper of instruction-level power models.

Concerning low-power scheduling algorithms, some approaches have been proposed by Tiwari *et al.* [5–7], but these works target either scalar or DSP processors with packed operations and their extension to VLIW code generation is not straightforward. Other authors [11,12] introduced power optimization methodologies from a software-level perspective, such as pre-processing and restructuring the source code to reduce the power consumption of the executable code. Other techniques, such as spatial and temporal scheduling, have been proposed by Lee *et al.* [13]. Spatial scheduling tries to directly minimize the switching activity on the instruction bus by choosing suitable pairs of instructions through a bipartite matching scheme. Since the temporal scheduling algorithm is an NP-hard problem, it is heuristically solved by reducing it to a bipartite matching scheme on a limited instructions window. Finally, Parikh *et al.* [14] modified list-scheduling by trading-off energy and speed simultaneously.

3. VLIW POWER MODEL

Concerning k-issue ILP processors, where each instruction is composed of k parallel operations, we have that, considering x operations in the ISA, and y inter-instruction effects, the complexity of the instruction-level energy model is $O(x^{k*(y+1)})$. In [3], we introduced the *temporal* and *spatial additive properties*, to deal with such a complexity. The temporal additive property enables the decomposition of the instruction energy into the pipeline stages of the processor. The spatial additive property enables the decomposition of the energy dissipated by the processor in the single energy dissipated by the processor parallel paths followed by the operations.

More in detail, the model considers the energy associated with the n-th instruction $\mathbf{w_n}$ as dependent on its own properties (e.g., class of the instruction, addressing mode and so on) as well as on its *execution context*, i.e., the previous instruction $\mathbf{w_{n-1}}$ and the additional stall/latency cycles introduced during the execution of the instruction. Besides, the inter-instruction effect between $\mathbf{w_n}$ and $\mathbf{w_{n-1}}$ can be decomposed into the sum of the inter-operation effects $(w_n^k|w_{n-1}^k)$ on the k-th parallel path (or *lane*) of the processor [3]:

$$E(\mathcal{W}) \approx \sum_{1\leq n\leq N} \sum_{\forall s\in S} \Big[U_s(\mathbf{0}|\mathbf{0}) + \sum_{\forall k\in K} \nu_s(w_n^k|w_{n-1}^k) + + m_s^n * p_s^n * S_s + l_s^n * q_s^n * M_s \Big] \quad (1)$$

where the term S is the set of pipeline stages of the processor, $U_s(\mathbf{0}|\mathbf{0})$ is the base energy cost that represents the energy consumed by stage s during an execution of a bundle constituted entirely by NOPs ($\mathbf{0} = [\texttt{NOP}\dots\texttt{NOP}]^T$), $\nu_s(w_n^k|w_{n-1}^k)$ is the additional energy contribution due to the change of operation $(w_{n-1}^k \rightarrow w_n^k)$ on the same lane k, m_s^n (l_s^n) is the average number of additional cycles due to a data (instruction) cache miss during the execution of the $\mathbf{w_n}$ in s, p_s^n (q_s^n) is the probability that this event occurs, and S_s (M_s) is the energy consumption per stage of the processor modules that are active due to a data (instruction) cache miss.

The term $\sum_s U_s(\mathbf{0}|\mathbf{0})$ corresponds to the power consumption of the core during NOPs and and we assume it can be substituted by the average base cost $U(\mathbf{0}|\mathbf{0})$ (as confirmed by the experimental results where this approach has been applied). Besides, $\sum_s \sum_k \nu_s(w_n^k|w_{n-1}^k)$ can be substituted by a cost dependent only on the pair of instructions $(\sum_k \nu(w_n^k|w_{n-1}^k))$ that corresponds to the energy consumption of the core while it is executing the same pair of instructions $(\mathbf{w_n}, \mathbf{w_{n-1}})$. For instruction and data cache misses, we assume that the probabilities per stage (p and q) and their penalties (m and l) can be averaged for each instruction of the stream:

$$\sum_s (m_s^n * p_s^n * S_s + l_s^n * q_s^n * M_s) \rightarrow (m * p * S + l * q * M) \quad (2)$$

where $m(l)$ is the average data (instruction) cache miss length, $p(q)$ is the average probability per stage and per instruction that a data (instruction) cache miss can affect one instruction and S (M) is the average energy consumption of the *entire* processor during these events.

Based on these assumptions, the equation (1) can be reduced to:

$$E(\mathcal{W}) \approx \sum_{1 \le n \le N} \Big[U(\mathbf{0}|\mathbf{0}) + \sum_{\forall k \in K} \nu(w_n^k | w_{n-1}^k) + \\ + m * p * S + l * q * M \Big] \quad (3)$$

For a k-issue VLIW core, the complexity of the model is now quadratic with respect to the number of operations into the instruction set ($O(k * |ISA|^2)$), while a black-box model would have a complexity $O(|ISA|^k)$. This implies a significant reduction of the characterization effort, since the number of experiments to be done is reduced exponentially.

In the next section, we show how this model can be further simplified by clustering the operations of the ISA with respect to their average energy consumption to achieve a faster and more effective characterization of the core's power consumption, while preserving the model accuracy.

4. CLUSTERING OF THE OPERATIONS

In this work, we consider two operations in the ISA as "different" if they differ either in terms of functionality (i.e., opcode) or in terms of addressing mode (immediate, register, indirect, etc.). Even without considering data differences (in terms of register names or immediate values), the number of operations pairs to be considered in equation (3) would be too large to be characterized with a transistor-level or even gate-level simulation engine. In the case of the Lx processor, for example, the ISA is composed of about 70 operations, but considering also the possible addressing modes for each operation, we would need to characterize 6126 pairs of operations to completely define our model (see the characterization of the matrix ν in equation (3)).

To give a rough idea of the time required for the characterization phase, a Sun Ultra Sparc 60 workstation at 450MHz with 1GB RAM performs a gate-level power simulation of the Lx core in 25 minutes. Given such simulation time, we would need approximately 108 days to perform the complete characterization phase on a single workstation.

To reduce the number of experiments to be generated during the characterization phase, we propose to apply the well known cluster analysis on the operations of the ISA. The basic idea of the cluster analysis consists of grouping into the same cluster the operations showing a power cost close to each other. The power cost of an operation is defined as the power consumed by the processor when it executes that operation.

Among the various clustering algorithms appeared in literature so far (see [15, 16] for a survey), we have chosen the k-mean clustering algorithm since it requires a lower computational cost.

Given a set of energy values $\Theta = \{e_1 \dots e_t \dots e_\theta\}$, where e_t is the energy consumption of instruction t measured by executing a loop composed of only t instructions, the k-mean clustering algorithm tries to partition Θ into a set of K clusters ($C_1 \dots C_K$) to minimize the mean-square error:

$$\sum_{j=1}^{K} \sum_{i=1}^{n_j} (x_{i,j} - c_j)^2, \quad x_{i,j} \in C_j \quad (4)$$

where n_j is the number of elements of cluster C_j, $x_{i,j}$ is the i-th element of cluster j and c_j is the center of gravity of the j-th cluster. The k-mean clustering algorithm receives as input the number K of classes in which the original population must be partitioned, and it randomly splits the whole set into K subsets. Then, each element is moved into the subset j whose center of gravity is closest. This procedure iterates until the stopping criterion is met.

To provide experimental evidence on the accuracy of the clustering algorithm applied on the instructions of the Lx ISA, Figure 1 shows the mean and the variance of the energy values associated with each instruction in the ISA and the clusters in which they have been grouped. In the experimental section of this paper, we have proved that the maximum variance in a cluster for the Lx architecture can be considered very limited (within 13% for 11 clusters).

Once the instructions have been clustered into a set of $C_1 \dots C_j$ clusters, we compute the matrix ν in equation (3) in the following way:

$$\nu(w_n^k | w_{n-1}^k) = \begin{cases} E_i & \text{if } w_n^k = w_{n-1}^k \wedge w_n^k, w_{n-1}^k \in C_i \\ D_{i,j} & \text{if } w_n^k \neq w_{n-1}^k \wedge w_n^k \in C_i, w_{n-1}^k \in C_j \end{cases} \quad (5)$$

Without considering the switching activity due to data dependency, this decomposition tries to model the fact that when operations are equal ($w_n^k = w_{n-1}^k$), they generate the least switching activity possible (first case). In the second case, when the operations are different, they generate an increased switching activity even if they are in the same cluster (accounted by the matrix $D_{i,j}$). Note that the complexity of matrix ν has been reduced from $O(|ISA^2|)$ to $O(|C^2|)$, which represents the upper bound on complexity given by $D_{i,j}$.

To successfully apply the clustering algorithm, one of the most crucial parameters to choose is the number of clusters. In general, a small number of clusters implies a high variance within them (i.e., poor accuracy of the model), though it also implies a small number of experiments during the regression. On the contrary, a large number of clusters implies good accuracy, but would result in a huge number of experiments.

In our methodology, the number of clusters is automatically determined by a tradeoff between the maximum standard deviation of the elements within the same cluster and the number of experiments that must be done to characterize the model. Section 5.2 reports the application of the tradeoff analysis performed on the Lx processor to determine the number of clusters.

The minimum number of experiments necessary to characterize the energy model is a typical cost function that must be reduced during the design of the experiments. In this work, we assume that the linear regression [17] is used to perform the characterization of the model. In this case, the minimum number of experiments is linearly dependent on the number of parameters of the model. Thus, the characterization cost has the following upper bound:

$$O\left(\frac{K \cdot (K-1)}{2}\right) \quad (6)$$

where K is the number of clusters.

5. LX CASE STUDY

In this section, we describe how the proposed methodology for power estimation and optimization has been applied to

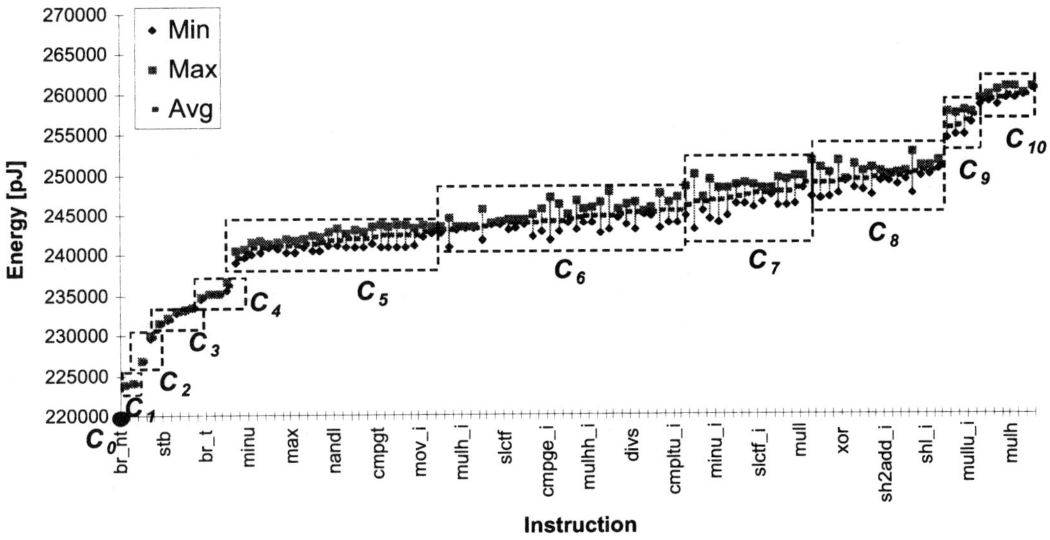

Figure 1: Energy values associated with each instruction and the corresponding clusters (where cluster C_0 corresponds to the NOP operation).

the Lx processor. The Lx architecture is a scalable and customizable processor technology [4] designed for multimedia and signal processing embedded applications. The Lx processor is a statically scheduled VLIW architecture designed by Hewlett-Packard and STMicroelectronics, that supports a multi-cluster organization based on a single PC and a unified I-cache. The basic processor is 4-issue VLIW core featuring four 32-bit integer ALUs, two 16x32 multipliers, one load/store unit and one branch unit. The cluster also includes 64 32-bit GPRs an 8 1-bit Branch Registers. Lx has an in-order 6-stage pipeline and a very simple integer RISC *ISA*. For the first generation, the scalable Lx architecture is planned to span from one to four clusters (i.e., from 4 to 16 instructions issued per cycle).

Lx comes with a complete software tool-chain, where no visible changes are exposed to the programmer when the core is scaled and customized. The tool-chain includes a sophisticated *ILP* compiler technology (derived from the Multiflow compiler [18]) and GNU tools and libraries. The Multiflow compiler includes both traditional high-level optimization algorithms and aggressive code motion technology based on trace scheduling.

A mix of synthesizable RTL and gate-level netlist of the core processor has been used to perform the characterization needed for the power model presented in this work. The experiments have been carried out by using Synopsys VCS 5.2 and a set of PLI routines to elaborate toggle statistics over the whole gate-level netlist. PowerCompiler (by Synopsys) has been used to combine the toggle statistics with the power models of the standard cells library provided by STMicrolectronic, to compute the power figures of the entire core.

5.1 Lx Power Characterization

To characterize the power consumption model expressed by equation (3), we first proceed to group the operations into clusters through the k-mean clustering algorithm. For each operation o, we generated a set of assembly programs composed of repeated cycles of o operations (by varying register names and values) and we measured the energy consumption of the core at the gate-level. Then we applied, to the measured values, the k-means clustering algorithm for several values of K. In order to determine the most suitable number of clusters, we analyzed the minimum number of experiments that would be needed to characterize the energy model with K clusters and we performed a tradeoff analysis with respect to the maximum variance within the clusters. As mentioned above, the minimum number of necessary experiments depends linearly on the number and the size of the parameters involved in the model. In our case, the shape of the curve defining the number of experiments is quadratic, due to the quadratic dependence of the size of the matrix ν with respect to the number of clusters K.

Figure 2 shows the results concerning the maximum variance within the operations clusters and the number of experiments required to characterize each corresponding model. From the reported results, we can note how, for a number of clusters equal to 11, we can obtain a good tradeoff between the maximum variation (that drops to 13%) and the minimum number of experiments (that reaches 78). For this reason, we selected 11 clusters to characterize the Lx power model.

Once selected the number of clusters and, therefore, the number of coefficients in the model (i.e., the size of the ν matrix), we realized a set of experiments in which each possible pairs of clusters have been generated several times by varying register names and immediate values.

These experiments have been performed by keeping the value of p and q (the data and instruction cache miss probabilities per instruction) as lowest as possible to characterize by regression only the terms $U(\mathbf{0}|\mathbf{0})$ and $\nu(w_n^k|w_{n-1}^k)$. The values of M and S (the energy consumed per cycle during an I-cache and D-cache miss) have been characterized by generating a set of experiments with a large number of data and instruction cache misses, and by measuring the power only during these events. Finally the m and l miss penalties have been extrapolated by looking at the behavior of the microarchitecture during these events.

The model has been tested against a set of validation benchmarks that are different from the experimental setup used to perform the characterization. The validation benchmarks include a set of the Mediabench applications [19] (namely, G721 encoder and decoder, EPIC encoder and decoder, MPEG2), a set of finite impulse response filters, discrete cosine transforms and matrix elaboration algorithms. Figure 3 shows the scatter plot of the measured power values for the validation benchmarks and the power values estimated with our model. The power model has shown, on the validation benchmarks, a mean error of 1.9% and a standard deviation on the error of 5.8%. The multiple correlation coefficient, that explains the percentage of the total variation exploited by the model, has been computed as in [17] and is equal to:

$$\sqrt{\frac{SSR}{SSR+SSE}} = 90\% \qquad (7)$$

As a matter of fact, there are only three benchmarks whose error on the prediction is in the neighborhood of 10% probably because of the high switching activity of the data consumed by instructions that is not captured by the model. Note that for high-level/instruction-level power macro models (whose main concern is efficiency), this can be considered an acceptable value in terms of accuracy.

5.2 Low-Power Spatial Scheduling

Once the model has been fully characterized, our basic idea consists of using it as a cost function to statically reorder instructions to minimize the power associated with the executable application.

The basic idea consists of defining an algorithm that considers each basic block of the code generated by the compiler and aims at rescheduling operations within the same bundle (*spatially*). The algorithm starts from bundle $\mathbf{w_{n=1}}$ and attempts to find a suitable reordering of each w_n^k in order to minimize the cost function:

$$\sum_k \nu(w_n^k | w_{n-1}^k), \quad \forall (w_n^k | w_{n-1}^k) \in \text{Basic block}$$

being w_{n-1}^k fixed (see equation 3). This step is iterated on all the adjacent bundles $\mathbf{w_n}$ of the basic block. All the possible permutations of operations within the bundle are evaluated, according to architectural constraints. In actual VLIW processors, the complexity of this approach is acceptable, since the number of k parallel operations ranges from 2 to 8, requiring the evaluation of $k!$ permutations (neglecting architectural constrains that decrease this number). For example, for the 4-issue VLIW Lx processor core, the total number of permutations to be checked is limited to $4! = 24$.

The choice of experimental benchmarks includes the compiled code of two FIR filters ($fir1$ and $fir2$), the fast discrete cosine transform ($fastdct$), the matrix multiply algorithm ($matrix$) and the bubble sort algorithm ($sort$). Figure 4 shows the power and energy savings obtained by apply-

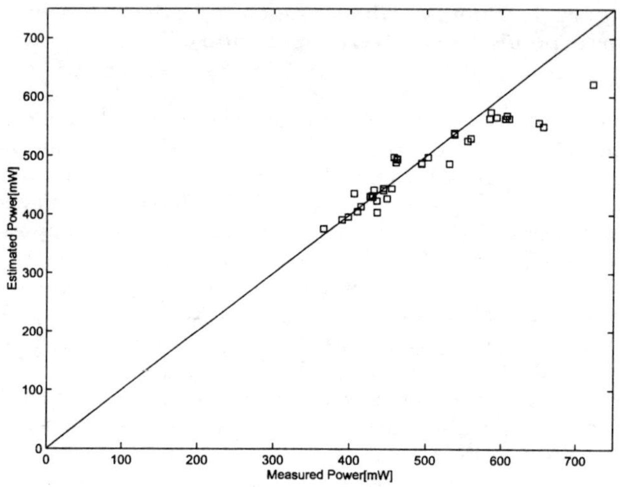

Figure 3: Scatter plot of the measured power values with respect to the estimated power values

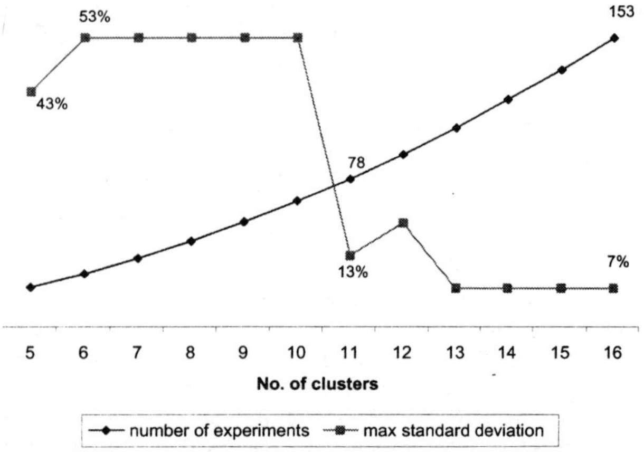

Figure 2: Tradeoff between the number of experiments required to characterize the model for a given number of clusters and the maximum variance within each cluster.

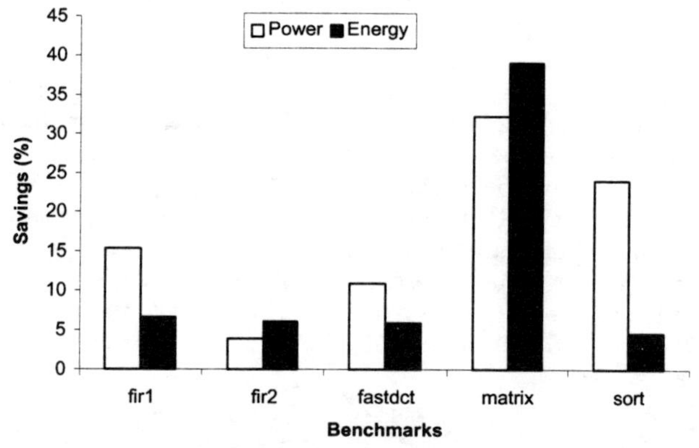

Figure 4: Power and energy savings obtained by our spatial rescheduling algorithm applied to the Lx processor executing the selected set of benchmarks.

ing the rescheduling algorithm to the selected set of benchmarks. The power savings have been computed by simulating, on the reference description of the processor core (gate-level), both the original and the rescheduled code of the benchmarks.

The average power consumption decreases by 17%, while the average energy consumption decreases only of 12%, since the rescheduling algorithm slightly increases the latency of the code. This is due to the fact that operation rescheduling could impact the efficiency of the Lx's instruction compression mechanism, leading to an increment of cache misses. However, in some cases ($fir2$ and $matrix$) the instruction cache misses are reduced since, due to the particular structure of the code, the rescheduling algorithm can lead to a more regular code in terms of instruction cache access patterns.

6. CONCLUDING REMARKS

We have presented an instruction-level methodology to estimate and to optimize the energy consumption in embedded systems based on VLIW architectures. The first goal of the proposed work is the reduction of the complexity of the energy model for VLIW cores, while preserving a good level of accuracy. The second goal of the work is to show how the proposed energy model can be further simplified by automatically clustering the operations in the ISA, based on the average energy behavior of the operations. Moreover, we have also shown how the general power-aware methodology has been successfully applied to the Lx VLIW embedded core. Some results have also been discussed derived from the application of the proposed low-power instruction scheduling algorithm. Future directions of our work target more complex dynamic instruction scheduling algorithms taking advantage of the accurate results derived from the proposed power characterization methodology.

7. REFERENCES

[1] L. Benini, D. Bruni, M. Chinosi, C. Silvano, V. Zaccaria, and R. Zafalon, "A power modeling and estimation framework for VLIW-based embedded systems," in *Proceedings of International Workshop-Power And Timing Modeling, Optimization and Simulation, PATMOS'01*, 26–28 2001.

[2] M. Sami, D. Sciuto, C. Silvano, and V. Zaccaria, "Instruction level power estimation for embedded VLIW cores," in *Proceedings of the 8th International Workshop on Hardware/Software Codesign (CODES-00)*, NY, May 3–5 2000, pp. 34–38, ACM.

[3] M. Sami, D. Sciuto, C. Silvano, and V. Zaccaria, "Power exploration for embedded VLIW architectures," in *Proceedings of the IEEE/ACM International Conference on Computer Aided Design (ICCAD-2000)*, Nov. 5–9 2000, pp. 498–503.

[4] P. Faraboschi, G. Brown, J. Fisher, G. Desoli, and F. Homewood, "Lx: a technology platform for customizable vliw embedded processing," in *Proceedings of the International Symposium on Computer Architecture*, June 2000, pp. 203–213.

[5] T. Lee, V. Tiwari, and S. Malik, "Power analysis and minimization techniques for embedded dsp software," *IEEE Transactions on VLSI Systems*, vol. 5, no. 1, pp. 123–135, 1997.

[6] V. Tiwari, S. Malik, and A. Wolfe, "Power analysis of embedded software: a first step towards software power minimization," *IEEE Transactions on Very Large Scale Integration (VLSI) Systems*, vol. 2, no. 4, pp. 437–445, 1994.

[7] V. Tiwari, S. Malik, A. Wolfe, and M. Lee, "Instruction level power analysis and optimization of software," *J. VLSI Signal Processing*, pp. 1–18, 1996.

[8] J. Russell and M. Jacome, "Software power estimation and optimization for high performance, 32-bit embedded processors," in *International Conference on Computer Design: VLSI in Computers and Processors*, 1998, pp. 328–333.

[9] D. Trifone D. Sarta and G. Ascia, "A data dependent approach to instruction level power estimation," in *Proc. IEEE Alessandro Volta Memorial Workshop on Low Power Design*, Como, Italy, March 1999, pp. 182–190.

[10] B. Klass, D. Thomas, H. Schmit, and D. Nagle, "Modeling inter-instruction energy effects in a digital signal processor," in *Power-Driven Microarchitecture Workshop*, June 1998.

[11] N. Zervas, K. Masselos, and C. Goutis, "Code transformations for embedded multimedia applications: Impact of power and performance," in *Proceedings of the Power-Driven Microarchitecture Workshop In Conjunction With ISCA98*, June 1998.

[12] H. Mehta, R. Owens, M. Irwin, R. Chen, and D. Ghosh, "Techniques for low energy software," in *Proceedings of the International Symposium on Low Power Electronics and Design*, August 1997, pp. 72–75.

[13] C. Lee, J. K. Lee, and T. Hwang, "Compiler optimization on instruction scheduling for low power," in *Proceedings of The 13th International Symposium on System Synthesis*. Sept. 20–22 2000, pp. 55–60, IEEE Computer Society Press.

[14] A. Parikh, M. Kandemir, N. Vijaykrishnan, and M.J. Irwin, "Instruction scheduling based on energy and performance constraints," in *Proceedings of the IEEE Computer Society Workshop on VLSI*. Apr. 27–28 2000, pp. 37–42, IEEE Computer Society Press.

[15] A. K. Jain, M. N. Murty, and P. J. Flynn, "Data clustering: a review.," *ACM Comp. Surveys*, vol. 31, no. 3, pp. 264–323, Sept. 1999.

[16] John H. Gennari, "A survey of clustering methods," Technical Report ICS-TR-89-38, University of California, Irvine, Department of Information and Computer Science, Oct. 1989.

[17] Q. Wu, Q. Qiu, M. Pedram, and C. Ding, "Cycle-accurate macromodels for rt-level power analysis," in *Proc. International Symposium on Low Power Electronics and Design*, 1997, pp. 125–130.

[18] P. Geoffrey Lowney, Stefan M. Freudenberger, Thomas J. Karzes, W. D. Lichtenstein, Robert P. Nix, John S. O'Donnell, and John C. Ruttenberg, "The Multiflow Trace Scheduling compiler," *The Journal of Supercomputing*, vol. 7, no. 1-2, pp. 51–142, 1993.

[19] C. Lee, M. Potkonjak, and W. H. Mangione-Smith, "Mediabench: A tool for evaluating multimedia and communication systems," in *Proceedings of Micro 30*, 1997.

Energy Exploration and Reduction of SDRAM Memory Systems

Yongsoo Joo, Yongseok Choi, Hojun Shim,
Hyung Gyu Lee, Kwanho Kim, Naehyuck Chang*
School of Computer Science & Engineering, Seoul National University, Korea
naehyuck@snu.ac.kr

ABSTRACT

In this paper, we introduce a precise energy characterization of SDRAM main memory systems and explore the amount of energy associated with design parameters, leading to energy reduction techniques that we are able to recommend for practical use.

We build an in-house energy simulator for SDRAM main memory systems based on cycle-accurate energy measurement and state-machine-based characterizations which independently characterize dynamic and static energy. We explore energy behavior of the memory systems by changing design parameters such as processor clock, memory clock and cache configuration. Finally we propose new energy reduction techniques for the address bus and practical mode control schemes for the SDRAM devices. We save 10.8mJ and 12mJ, 40.2% and 14.5% of the total energy, for 24M instructions of an MP3 decoder and a JPEG compressor, using a typical 32-bit, 64MB SDRAM memory system.

Categories and Subject Descriptors

B.8.2 [**Performance and Reliability**]: Performance Analysis and Design Aids; C.4 [**Performance of Systems**]

General Terms

Design, Experimentation, Measurement, Performance

Keywords

low power, memory system, SDRAM

1. INTRODUCTION

Recently, even battery-operated portable systems have become equipped with high-performance memory systems to accommodate high-end applications that have previously been running on desktop computers. Thus, memory systems become dominant power consumers, which has inspired the development of low-power memory buses and memory devices. Off-chip main memory systems are major energy consumers even in hand-held embedded systems.

Previous power reduction work tends to focus on optimization techniques rather than the cost function, *i.e.*, power consumption models. Although these power reduction techniques are innovative, they are not ready for real implementation because they are based on simple energy models and do not consider cooperation with memory buses, bus drivers and memory devices. Simple capacitance models have been regarded as sufficient, and even transition counts are often treated as a power metric. System buses are mostly modeled as if they were composed of large capacitors, and even peripheral devices are often considered as capacitive loads [6]. SRAM-based memory is often used and the models of these systems have constant energy per access [4], while main memory systems composed of DRAMs or SDRAMs are more dominant power consumers than cache memories.

However, simple power consumption models may lead to seriously misguided power reduction practices [1], because it is important to take into account device-level access protocols in power reduction for memory systems. In this paper, we build a cycle-accurate energy simulator for off-chip SDRAM main memory systems and introduce energy reduction methods that are generally applicable to real implementations. In order to build a precise simulator, we perform intensive energy measurement and characterization. This paper presents state-machine-based energy characterization of memory systems, taking simultaneous account of memory buses, bus buffers, memory devices, and interaction between the processor, cache and main memory. We modify a switched-capacitor energy measurement technique [2] to account for leakage current as well. We borrow the energy consumption models of modern bus specifications introduced by chang *et al.*[1].

In this paper, we explore the energy behavior of SDRAM main memory systems considering these factors. Real application traces, which are acquired by special hardware support, drive the simulator. The simulator precisely replays the memory transactions together with the configurable cache memories and clock frequencies of the processor and memory system. Finally, we introduce practical energy reduction schemes.

*Corresponding author

The RIACT at Seoul National University provides research facilities for this study. This work was partly supported by the Brain Korea 21 Project.

DAC 2002, June 18-22, 2002, New Orleans, Louisiana, USA.

2. ENERGY CHARACTERIZATION

2.1 Energy state machine

Finite state machines are widely used to describe the operation of digital systems. In this paper, we introduce an energy state machine to describe the energy consumption behavior of digital systems. A finite state machine M is four-tuple (S, Σ, δ, s_0) where $S = (s_0, ..., s_n)$ is a set of finite states, Σ is a finite input alphabet, $\delta : \Sigma \times S \rightarrow S$ is a state transition function, and s_0 is the initial state. We label each arc that denotes state transition δ with a finite set of state transitions $T = (t_0, ..., t_m)$. Fig.1 illustrates the variation in power supply current i_{dd} of asynchronous and synchronous devices, and justifies the static and dynamic energy association of the energy state machine.

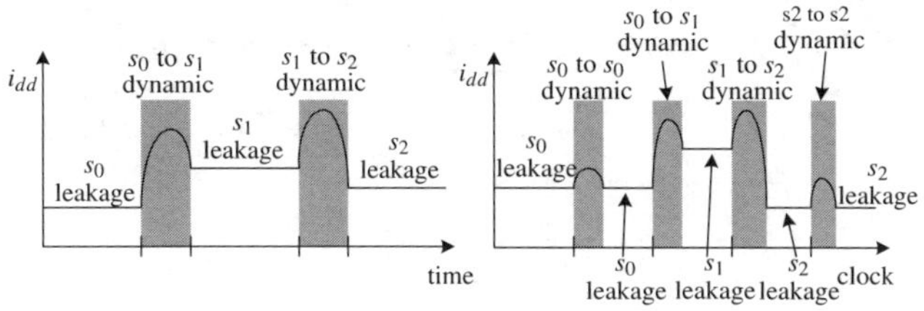

(a) i_{dd} of an asynchronous device (b) i_{dd} of a synchronous device

Figure 1: i_{dd} variation due to state change.

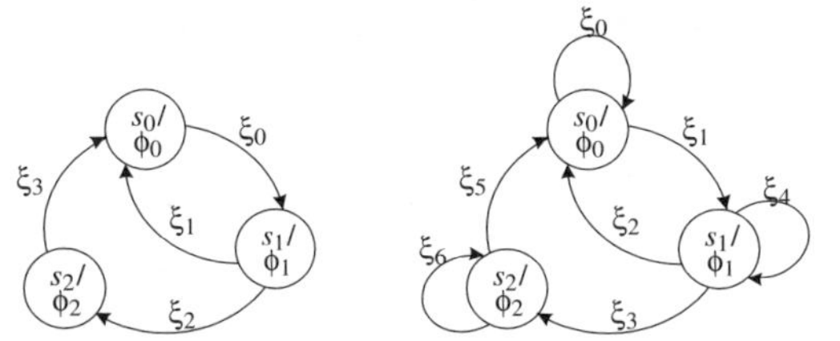

(a) asynchronous energy state machine (b) synchronous energy state machine

Figure 2: The energy state machine.

Asynchronous devices consume dynamic energy when strobe signals are issued. The strobe signal changes the device state, thus leading to variation of the static energy consumption. Synchronous devices consume dynamic energy at each clock edge rather than at the logical state change.

Definition 1. *The energy state machine Π is a three-tuple (M, Φ, Ξ) where M is a finite state machine, $\Phi = (\phi_0, \ldots, \phi_n)$ is the leakage energy associated with the state $S = (s_0, \ldots, s_n)$, and $\Xi = (\xi_0, \ldots, \xi_m)$ is the dynamic energy associated with the transition $T = (t_0, \ldots, t_m)$.*

Fig. 2 shows examples of an asynchronous and a synchronous energy state machine. Each state and transition is associated with a static energy consumption ϕ_i and a dynamic energy consumption ξ_j, respectively.

Remark 1. *Let us consider an asynchronous state machine M_a. If the tenure time of $s_i = t$ such that $s_i \in M_a$, the static energy consumption is $t\phi_i$.*

Remark 2. *Let the clock period of a synchronous state machine M_s be τ. If M_s remains in s_i such that $s_i \in M_s$ for n consecutive clock cycles, the tenure time of state s_i is $n\tau$ and thus the static energy consumption is $n\tau\phi_i$.*

Remark 3. *The dynamic energy consumption Ξ does not vary with the tenure time of states and the clock frequency.*

Remark 4. *Previous current models associate a value of i_{dd} with each state. Let τ be a memory clock period. When a mode change occurs such that $s_i \xrightarrow{t_j} s_k$, the i_{dd} of s_k is determined by $\frac{\xi_j + n\phi_k}{n}$ where n is the tenure time of s_k in number of clock cycles. Because previous current models do not take n into account, the dynamic energy is under-estimated or the static energy is over-estimated.*

2.2 Annotation of the energy state machine

To complete the energy state machine, we have to annotate the energy values Φ and Ξ. To start with, we need to distinguish the energy consumption behavior of target devices before deciding a characterization method. So we further classify dynamic energy consumption into *HDD* (Hamming-distance-Dependent Dynamic) energy, *WDD* (Weight-Dependent Dynamic) energy and *CD* (Common-mode Dynamic) energy. We also classify static energy, Φ, into *WDS* (Weight-Dependent Static) energy and *CL* (Constant Leakage) energy. Switching activity causes dynamic energy consumption by CMOS circuits. This switching activity is largely dependent on the Hamming distance of the data between the current and previous clock cycles in the static CMOS circuits; this is HDD energy. The dynamic CMOS circuits precharge before every evaluation. They draw a large current if the circuit has been discharged in the previous evaluation (usually the previous clock cycle); but they draw only a small current if the circuit has not been discharged. The energy consumption is mainly proportional to the weight[1] of the current data: the WDD energy. We do not need to annotate the energy values of each internal component of a device. So, a significant portion of the dynamic energy is represented in the form of common-mode values: the CD energy. Passive pull-ups or BiCMOS drivers consume more static energy when the signal level is 0. This energy is inversely proportional to the duty ratio of the signal which is the WDS energy. Devices may also consume a constant leakage power supply current which is variable to a state of a device: the CL energy.

While some previous characterizations have ignored WDD or CL energy, most modern memory devices consume significant WDD and CL energy. In addition, previous characterizations have often averaged out instantaneous energy variations, but these variations are essential information in high-level energy reduction. For various reasons, devices are often operated slower than their maximum speed, (*i.e.* we issue strobe signals that are longer than necessary for asynchronous devices, or use slower clocks for synchronous devices). When we do not operate a device at full speed, the static energy consumption increases but dynamic energy does not change. This forces us to use intensive energy characterization to distinguish static energy from dynamic energy, which is difficult to come up with using conventional average power measurement.

Taking these problems into account, we decided to use a real-time cycle-accurate energy measurement technique [2] to characterize the state-machine-based energy characterization for memory devices. We sample the the supply voltage, v_s, at many points to determine the static energy consumption by measuring the slew rate. Because there is static energy consumption in a given state s_j, v_s continuously decreases with a constant slope. The leakage energy consumption is denoted by

$$\phi_j = \frac{C_s}{2} \frac{v_s^2(t+\Delta t) - v_s^2(t)}{\Delta t}, \quad (1)$$

where C_s is the switched capacitor.

3. ENERGY CONSUMPTION OF MEMORY SYSTEM COMPONENTS

3.1 Energy consumption of memory buses

The energy consumption of memory buses is denoted by an energy state machine $\Pi_B = (M_B, \Phi_B, \Xi_B)$. The set of states M_B is (s_0, s_1), which represents driven low and driven high states respectively. The power consumption of LVT and GTL+ bus and bus drivers have already been studied [1]. We compose an energy state machine for an LVT bus (these are commonly used high-performance memory buses for embedded systems) by converting the power values to cycle-accurate energy values. We assume that the LVT drivers have bus-hold functions, which have been devised to avoid excessive WDS energy by eliminating external pull-up resistors. State s_2 and s_3 in Fig. 3 (c) are bus hold states. We control the output enable no later than the input change in order to keep rise and fall time. Thus no state change such as $s_0 \rightarrow s_3 \rightarrow s_1$ is allowed. This guarantees that $\xi_4 = \xi_5 = \xi_6 = \xi_7 = 0$. In addition, $\phi_2 = \phi_3 = 0$.

[1] the number of 1's or 0's according to the logic structures.

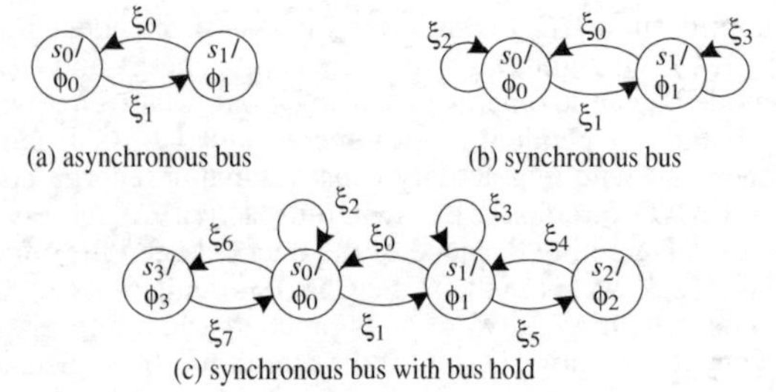

Figure 3: Energy state machine of off-chip memory buses.

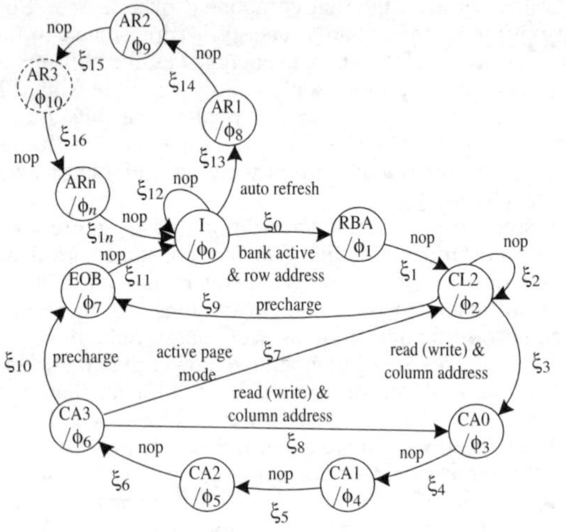

Figure 4: Energy state machine of an SDRAM.

We have constructed a 2-inch bi-directional bus, with a transmission line capacitance of 2.7pF, using a Fairchild 74LVT245, which has values of $\Xi_{aB} = (0.55, 0.55)$ and $\Phi_{aB} = (0.0053t, 0)$ for asynchronous buses, and $\Xi_{sB} = (0.55, 0.55, 0, 0)$ and $\Phi_{sB} = (0.0053\tau, 0)$ for synchronous buses. The units are nJ/bit, t is the tenure time of the asynchronous bus in nS, and τ is the clock period of the synchronous bus in nS.

3.2 Energy consumption of memory devices

In this paper, we focus on SDRAM devices, which are popular in equipment ranging from hand-held devices to desk-top computers. SDRAM has various modes of operation, but we consider two of the most important operations: burst-mode access and $\overline{\text{CAS}}$ before $\overline{\text{RAS}}$ refresh (Fig. 4).

Table 1 shows value of coefficient Ξ for our SDRAM. Let $f_1(\cdot)$ be the number of 1's in $\cdot$. Among Ξ, CD is the dominant component. Table 2 shows the CD energy of K4S280832B-TC1L and MT48LC16M8A2-7E SDRAM devices from two major vendors, Samsung and Micron respectively. They both have a 128Mbit capacity with a 4M address space $(12 \times 10) \times$ 8bit data $\times$ 4 banks.

Table 1: Dynamic energy consumption of an SDRAM, Ξ (nJ/bit).

Ξ	Energy Cost
ξ_0	$\Theta_{ra} + c_{ra} f_1(A_r)$
$\xi_1 = \xi_2 = \xi_7 = \xi_{11} = \xi_{12}$	0
$\xi_3 = \xi_8$	read: $\Theta_{car} + c_{do} f_1(D_0) + c_{car} f_1(A_c)$ write: $\Theta_{caw} + c_{caw} f_1(D_0) + c_{caw} f_1(A_c)$
$\xi_4 = \xi_5 = \xi_6$	read: $\Theta_{car} + c_{do} f_1(D_i)$ write: $\Theta_{car} + c_{di} f_1(D_i)$
$\xi_9 = \xi_{10}$	Θ_{pr}
$\xi_{13} + \ldots + \xi_{1n}$	Θ_{rf}

Table 2: Common Mode Dynamic Energy, Θ (nJ/bit).

Symbol	Description	Samsung	Micron
Θ_{ra}	row active	1.625	1.158
Θ_{car}	column active (read)	1.80	0.929
Θ_{caw}	column active (write)	0.681	0.783
Θ_{pr}	precharge	0.149	0.261
Θ_{rf}	refresh	4.51	5.30

Table 3: Coefficient of Ξ (nJ/bit).

Coefficient	Description	Samsung	Micron
c_{ra}	row address input	0.16	0.05
c_{do}	data output (read)	0.11	0.12
c_{di}	data input (write)	0.07	0.09
c_{car}	column address input to read	0.15	0.19
c_{caw}	column address input to write	0.14	0.16

4. HYBRID ENERGY ESTIMATOR

The energy consumption of main memory systems is widely variable depending on the processor, cache memory, system bus, bus clock frequency and so on. This section introduces a cycle-accurate energy simulator for off-chip main memory systems. Fig. 5 illustrates the organization of the simulator.

4.1 Processor

The processor generates a memory trace for a given application program and data. This is the most complex and time consuming part of system-level energy simulation. We analyze a real target processor connected over a PCI bus interface, and obtain exact memory traces in virtually real time. Our target processor is ARM7TDMI. Once a trace has been captured by the hardware, it is transferred to the cache simulator.

4.2 Cache

The cache module has separated instruction cache and data cache. We can configure index size, associativity and cache block size. The LRU replacement policy and critical-word-first scheme remain at their default setting.

4.3 Memory system

Memory system is composed of SDRAM controller, memory bus and memory device. We built a SDRAM controller with configurable bus clock frequency, burst length, $\overline{\text{CAS}}$ latency, burst refresh, precharge mode, and so on. We can also configure bus drivers, driver I/O capacitance, transmission line capacitance and load capacitance. We can configure capacity and organization of the memory devices.

5. ENERGY REDUCTION PRACTICES

Table 4: Static energy consumption, Φ (nJ/bit).

Φ	samsung	micron
$\phi_1 = \phi_2 = \phi_3 = \phi_4 = \phi_5 = \phi_6$	0.0058τ	0.0044τ
$\phi_0 = \phi_7 = \phi_8 = \phi_9 = \phi_{10} = \phi_n$	0.0019τ	0.0044τ

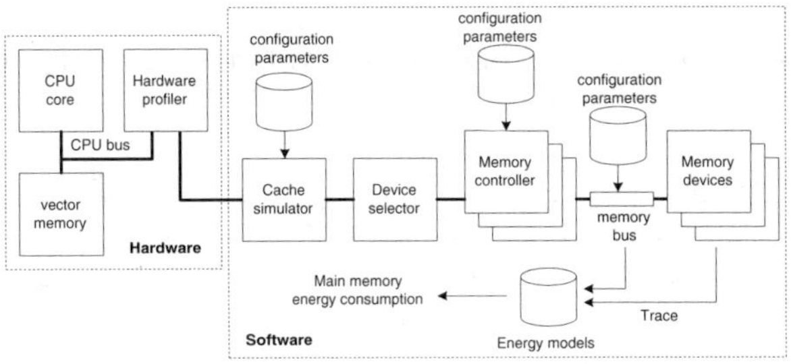

Figure 5: Estimation of main memory energy consumption.

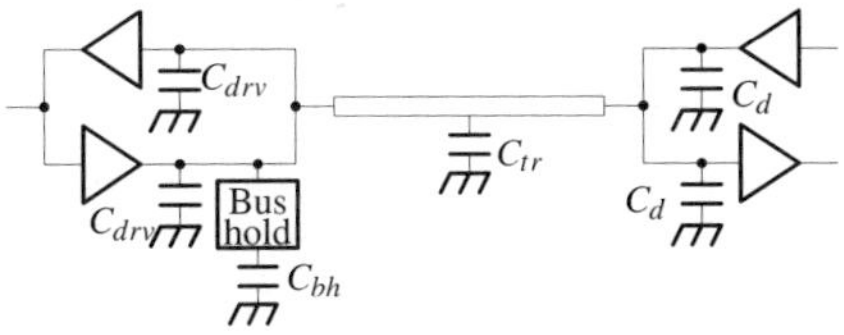

Figure 6: Data bus model.

5.1 Experimental setup

We have set up a typical SDRAM main memory systems for embedded applications. It has 32bit data width and 64MB capacity with four K4S280832B-TC1L devices. The length of the bus is 2 inches.

Fig. 6 shows the data bus model. C_{drv}, C_{bh}, C_{tr} and C_d are the I/O capacitance of the LVT driver, the effective bus-hold logic capacitance, the 2-inch transmission line capacitance and the I/O capacitance of the SDRAM. The energy state machine follows the arrangement shown in Fig. 3 (b). The energy consumption is given by

$$\xi_0 = \xi_1 = \frac{1}{2}(2C_{drv} + C_{bh} + C_{tr} + 2C_d)V_{DD}(V_{OH} - V_{OL}) + E_{drv},$$
$$\xi_2 = \xi_3 = 0, \phi_0 = 0.0053\tau, \phi_1 = 0. \quad (2)$$

E_{drv} is the dynamic energy of the LVT driver, and includes the bias-overlapping current. Typically, $E_{drv} = 0.44$nJ, $C_{drv} = 4$pF, $C_{bh} = 0.5$pF, $C_{tr} = 2.6$pF, $2C_d = 5.25$pF and $V_{DD} = 3.3$V. V_{OH} and V_{OL} are output high and output low voltage of the bus driver, respectively. In order to accommodate many memory chips for large capacity, CBT bus switches are commonly used to reduce the load capacitance, and thus these values are still justified.

Fig. 7 shows the common configuration of SDRAM address buses. In this case too, the state machine also obeys that Fig. 3 (b). Energy consumption is the same to that of data bus except

$$\xi_0 = \xi_1 = \frac{1}{2}(C_{drv} + C_{tr} + 4C_a)V_{DD}(V_{OH} - V_{OL}) + E_{drv} \quad (3)$$

where C_a is the input capacitance of the SDRAM, with a typical value of 3.75pF.

We have chosen two embedded application programs for the reduction practices: an MP3 decoder and a JPEG compressor with a 512×512 gray-scale Lenna image.

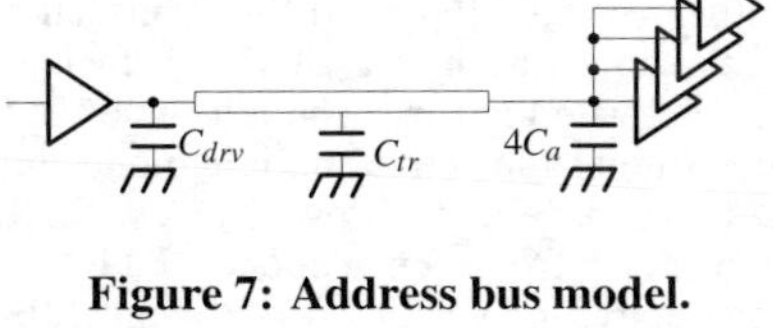

Figure 7: Address bus model.

Table 5: SDRAM memory bus energy consumption by design parameters (CJPEG, mJ/mW, 24M instructions).

Parameter		Address bus HDD	Address bus WDS	Data bus HDD	Data bus WDS
Processor clock (f_P) MHz	100	1.27/4.46	11.79/41.36	4.13/14.48	1.05/3.67
	133	1.27/5.71	9.20/41.32	4.13/18.54	1.05/4.70
	200	1.27/7.97	6.58/41.24	4.13/25.86	1.05/6.55
	266	1.27/9.90	5.29/41.17	4.13/32.13	1.05/8.14
	400	1.27/13.06	4.00/41.04	4.13/42.37	1.05/10.74
f_M =66MHz, 8KB/4word/2way-set-associative cache					
Memory clock (f_M) MHz	33	1.27/7.75	6.72/40.94	4.13/25.16	2.09/12.75
	66	1.27/9.90	5.29/41.17	4.13/32.13	1.05/8.14
	83	1.27/10.50	5.00/41.23	4.13/34.05	0.84/6.90
	100	1.27/10.68	4.92/41.31	4.13/34.63	0.70/5.85
	133	1.27/11.22	4.69/41.35	4.13/36.41	0.52/4.61
f_P =266MHz, 8KB/4word/2way-set-associative cache					
Cache size KB	1	5.78/22.44	10.09/39.18	16.44/63.83	4.48/17.40
	2	4.63/20.55	8.82/39.20	13.70/60.84	3.41/15.13
	4	3.91/19.01	8.05/39.15	12.14/59.05	2.73/13.29
	8	1.27/9.90	5.29/41.17	4.13/32.13	1.05/8.14
	16	0.62/5.59	4.65/42.03	2.17/19.61	0.53/4.79
f_M =66MHz, f_P =266MHz, 4word/2way-set-associative cache					
Cache set	1	1.30/10.04	5.31/41.15	4.11/31.86	1.14/8.86
	2	1.27/9.90	5.29/41.17	4.13/32.13	1.05/8.14
	4	1.36/10.36	5.40/41.13	4.45/33.90	1.09/8.27
	8	1.47/10.87	5.60/41.46	4.81/35.63	1.15/8.52
f_M =66MHz, f_P =266MHz, 8KB/4word cache					
Block size	4	1.27/9.90	5.29/41.17	4.13/32.13	1.05/8.14
	8	0.80/6.68	5.13/42.58	5.15/42.79	1.48/12.28
f_M =66MHz, f_P =266MHz, 8KB/2way-set-associative cache					

5.2 Energy reduction of memory buses

5.2.1 Exploration of memory bus energy

We perform a SDRAM memory system simulation with design parameters and have the energy information of bus and device. Table 5 shows the energy consumption behavior of memory bus according to the system design parameters. The memory bus energy consumption is composed of WDS and HDD energy.

WDS energy in the address bus is directly proportional to the execution time, which is in turn determined by the processor clock frequency and cache miss ratio. A higher memory clock frequency also reduces the WDS energy but not so significantly. The HDD energy of address bus is directly proportional to the number of addresses issued. Therefore, the more memory transaction that are performed, the more HDD energy is consumed. The memory transaction count is closely related to the cache miss ratio, but it is independent of the clock speed of the CPU and the memory system. Generally, the burst length of memory transaction is equal to the cache block size. Thus, a large block size in the cache reduces the number of memory transactions.

A conventional data bus configuration does not waste WDS energy in the idle period because it utilizes the bus-hold logic of the bi-directional bus driver. The WDS energy consumption of the data bus is exactly proportional to the cache miss ratio times the clock period of the memory bus. The HDD energy of the data bus grows as the count of memory transaction increases.

To reduce the energy consumption of the memory bus, we need to provide enough cache memory, or associativity exceeding the knee-point. Higher CPU and memory clock frequency is also helpful in reducing energy consumption of memory bus.

5.2.2 Energy reduction for the address bus

The major portion of energy consumption in the address bus is the WDS energy. Due to the conventional address bus architecture of the SDRAM, the address bus remains driven in most idle time and wastes WDS energy. The shaded areas in Fig. 8 represent the valid range of the address bus.

We propose to minimize static energy consumed during idle time

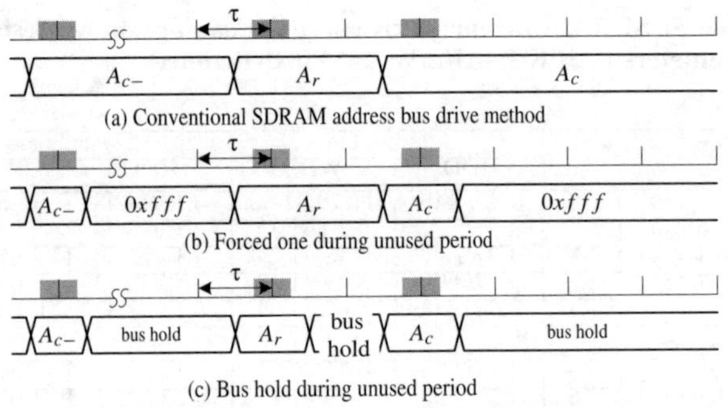

Figure 8: Address bus energy reduction schemes.

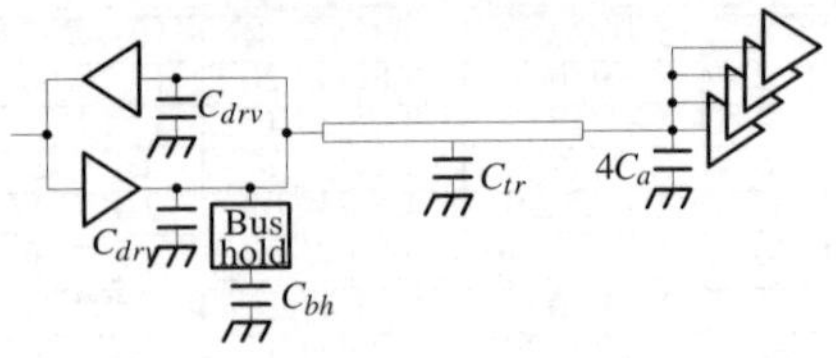

Figure 9: Address bus energy reduction using bus-hold function.

when the address bus is not valid. We suggest using two static energy reduction schemes: *forced 1* and *bus hold* (Fig. 9). *Forced 1* scheme makes the idle period (white areas) high and thus saves WDS energy. The *forced 1* is economic but slightly increases dynamic energy. The *bus hold* scheme is a costly solution because we have to use bi-directional buffers for uni-directional drive in order to utilize the bus-hold function. The *bus hold* scheme induces a minor increase in switching capacitance (0.5pF). Table 6 shows that the address bus energy of MP3 decoder and JPEG compressor is reduced to 8.5% and 24.6% respectively.

5.3 Energy reduction for memory devices

5.3.1 *Exploration of memory device energy*

In practice, it is important to analyze energy behavior by changing conventional design parameters such as processor clock speed, memory clock speed and cache configuration depending on the application program. Table 7 shows how the energy behavior of the SDRAM devices is affected by changing these parameters.

When we increase the processor and memory clock frequencies, the total execution time decreases and thus both CL energy and CD energy decrease. A higher processor clock frequency results in smaller CL energy in idle mode while higher memory clock frequency reduces the CL energy in active mode. The CL energy rapidly decreases as the clock frequencies increase, because the total execution time directly affects CL energy. On the other hand, the CD energy slightly decreases because the number of refresh cycles is reduced. These clock frequencies do not affect the WDD energy. The cache configuration also significantly affects the CL, CD and WDD energy. The energy consumption shows a distinct knee-point as we increase memory size and the associativity. The knee-points are at 4KB/2way and 8KB/direct-mapped for the MP3 and JPEG compressor respectively. A larger block size results in energy loss.

Table 6: Energy reduction for the SDRAM memory buses (mJ, 24M instructions, f_M =66MHz, f_P =266MHz, 8KB/4word/2way-set-associative cache).

App.	Reduction technique	HDD	WDS	Total	%
MP3	conventional	0.32	4.27	4.59	100.0
	forced 1	0.37	0.09	0.46	10.1
	bus hold	0.33	0.06	0.39	8.5
CJPEG	conventional	1.27	5.29	6.56	100.0
	forced 1	1.73	0.44	2.17	33.2
	bus hold	1.34	0.27	1.61	24.6

Table 7: Design parameters for SDRAM memory energy consumption (CJPEG, mJ/mW, 24M instructions).

Parameter		CL	CD	WDD	Total
Processor clock (f_P) MHz	100	19.0/66.6	65.8/230.8	2.3/8.1	87.1/305.5
	133	15.3/68.9	63.5/285.2	2.3/10.4	81.2/364.4
	200	11.7/73.0	61.2/383.1	2.3/14.5	75.1/470.6
	266	9.8/76.5	60.0/467.1	2.3/18.0	72.2/561.6
	400	8.0/82.3	58.9/604.3	2.3/23.7	69.2/710.3
f_M =66MHz, 8KB/4word/2way-set-associative cache					
Memory clock (f_M) MHz	33	14.2/86.8	61.1/372.3	2.3/14.1	77.7/473.2
	66	9.8/76.5	60.0/467.1	2.3/18.0	72.2/561.6
	83	8.9/73.8	59.8/493.3	2.3/19.0	71.1/586.1
	100	8.8/73.7	59.7/501.1	2.3/19.4	70.8/594.2
	133	8.0/70.5	59.6/525.4	2.3/20.4	69.9/616.2
f_P =266MHz, 8KB/4word/2way-set-associative cache					
Cache size KB	1	24.7/96.1	275.8/1070.9	10.8/41.8	311.3/1208.7
	2	20.9/93.0	222.9/990.0	9.0/40.1	252.9/1123.1
	4	18.6/90.6	191.4/930.7	8.0/39.0	218.1/1060.4
	8	9.8/76.5	60.0/467.1	2.3/18.0	72.2/561.6
	16	7.6/69.1	30.0/270.9	1.1/9.8	38.7/349.8
f_M =66MHz, f_P =266MHz, 4word/2way-set-associative cache					
Cache set	1	9.9/77.0	60.6/469.7	2.2/17.2	72.8/563.9
	2	9.8/76.5	60.0/467.1	2.3/18.0	72.2/561.6
	4	10.1/77.2	64.9/494.3	2.5/19.3	77.6/590.8
	8	10.5/78.1	70.8/524.5	2.8/20.8	84.2/623.4
f_M =66MHz, f_P =266MHz, 8KB/4word cache					
Block size	4	9.8/76.5	60.0/467.1	2.3/18.0	72.2/561.6
	8	9.5/78.9	66.2/550.0	2.2/17.8	77.9/646.8
f_M @66MHz, f_P @266MHz, 8KB/2way-set-associative cache					

5.3.2 *SDRAM mode control*

Table 1 shows that HDD energy is not consumed by SDRAM devices. Although WDD energy is consumed, its possible variation is only 10 to 15% of the CD energy. Actual variation are much less. This implies that the switching activity reduction of the SDRAM address bus and data bus would not be effective in energy reduction for SDRAM devices although it is a powerful in energy reduction for memory buses. Although we aggressively eliminated one third of WDD energy with our elaborated encoding scheme, we only reduced the total energy by around 1%. Thus, SDRAM mode control schemes must be introduced to reduce energy consumption significantly. SDRAM mode control schemes may be divided into two categories. The first is to shut down devices that are not being used, as to send them to a low-energy state [3]. The second is to force SDRAM devices to active or idle mode in order to minimize the cycle time [5]. This policy is not driven by the energy model, but reducing the cycle time reduces the total energy consumption as well as run time. The scheme requires correct break-even time for shutting down the devices, and the second scheme requires a correct estimation of the row hit behavior of SDRAM devices.

We have investigated the idle time distributions needed for effective mode control as shown in Fig. 10. At first, we investigate idle time distribution between consecutive memory operations (Fig. 10 (a)). Shutting down the SDRAM is conducted according to the time distribution. Second, we also investigate idle time distribution between consecutive row hits (Fig. 10 (c)). If consecutive burst-mode access refers the same row in the SDRAM device, we need not precharge and reactivate the row. The idle time between consecutive row hits determines the active and idle mode change scheme.

Modern SDRAMs have power down modes. SDRAMs may enter power down mode from active mode or idle mode. We call these transitions active power down *APD* and idle power down

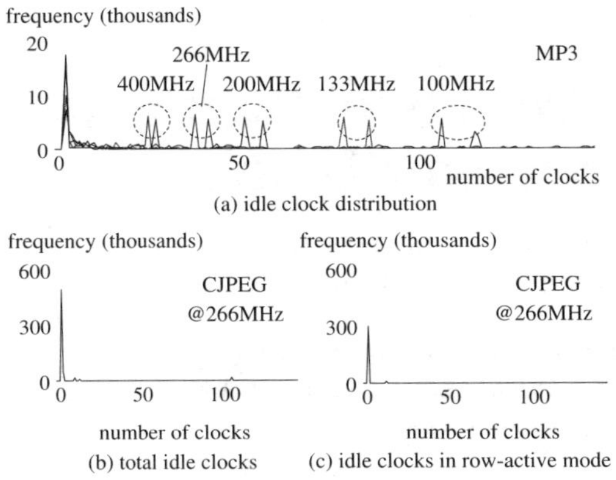

Figure 10: Idle clock distributions (8KB/2way/4word cache).

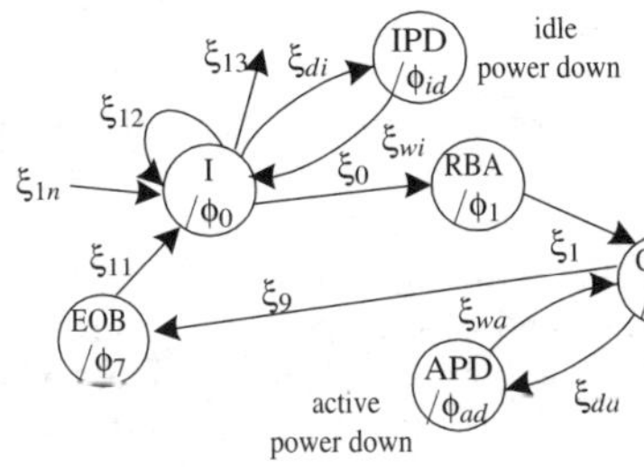

Figure 11: Power down mode in SDRAMs.

IPD, separately. For the Samsung K4S280832B-TC1L, $\phi_{id} = 1.6 \cdot 10^{-4}\tau$ and $\phi_{ad} = 7.5 \cdot 10^{-4}\tau$, while $\xi_{di} = \xi_{da} = \xi_{wi} = 0$ and $\xi_{wa} = 0.54$. It is very important to include dynamic cost Ξ in calculating mode control energy overhead. Otherwise, the mode control energy overhead will be greatly under-estimated or the CL energy will be greatly over-estimated as in [5]. For $I \rightarrow IPD \rightarrow I$, the energy cost is $(0.0038 + 1.6 \cdot 10^{-4}n)\tau$, where n is the length of time spent in *IPD* state, in clock cycles. Since $\xi_{di} = \xi_{wi} = 0$, there is no additional energy overhead in mode control although there is a loss of performance.

The precharge command changes the SDRAM's state from active to idle. The default mode of commercial SDRAM controller is auto precharge that immediately sends the SDRAM devices into the idle mode after every burst-mode transaction. We can optionally configure the SDRAM devices to stay in the row-active mode after burst mode transactions. The time spent in the row-active mode is determined by the refresh period, row miss and the forced precharge command. Delayed precharge sends the SDRAM device from the row-active mode to the idle mode after a predefined time-out value. Commercial SDRAM controllers commonly set up the time-out value to be 256 clock steps. However, Fig. 10 (c) shows that the dominant spectrum is located at 3 clock which is desirable time-out value. Table 8 shows the aggregate energy reduction of a SDRAM memory devices when we run the MP3 decoder and the JPEG compressor.

5.4 Low-power SDRAM main memory system

The energy reduction schemes for memory buses and devices are orthogonal (independent). The aggregate energy reduction amount will be the summation of all the schemes. We applied these reduction schemes to 24M instructions of MP3 decoder and 24M instructions of JPEG compressor with a 512×512 gray-scale Lenna image, using a 266MHz processor clock, a 66MHz memory clock, a 8KB/4word/2way-set-associative cache and a 32-bit 64MB Samsung SDRAM main memory system. We use bus-invert coding for

Table 8: Energy reduction of SDRAM memory device(mJ, 24M instructions, f_M @66MHz, f_P @266MHz, 8KB/4word/2way-set-associative cache).

App.	Reduction	CL	CD	WDD	Total	%
MP3	active page	12.0	14.3	0.3	26.6	123.3
	auto prechg	6.5	14.7	0.4	21.6	100.0
	delayed prechg	7.1	12.9	0.2	20.2	93.7
	auto prechg, power down	1.6	14.7	0.4	16.6	77.0
	delayed prechg, power down	2.3	12.9	0.2	15.4	71.5
CJPEG	active page	21.0	56.5	2.2	79.7	111.8
	auto prechg	9.8	59.2	2.3	71.3	100.0
	delayed prechg	12.6	54.6	2.0	69.2	97.0
	auto prechg, power down	4.9	59.2	2.3	66.4	93.1
	delayed prechg, power down	8.4	54.6	2.0	65.0	91.1

the data bus, *bus hold* for the address bus, delayed precharge and idle-mode power down for the SDRAM devices. The result for MP3 is 40.2% reduction: 26.8mJ to 16.0mJ. For JPEG compressor, we have 14.5% reduction: 83.0mJ to 71.0mJ.

6. CONCLUSION

Memory systems are dominant energy consumers, and thus many energy reduction techniques for memory buses and devices have been proposed. For practical energy reduction, we have to take into account interactions between a processor and cache memories, as well as application programs; and the energy characterization of memory systems must be sufficiently accurate. In this paper, we introduced an energy simulator for memory systems which is accelerated by special hardware support, while maintaining accuracy. We explored the energy behavior of memory systems and achieved low-power consumption for given applications by changing the processor and memory clock frequencies and cache configuration.

Our simulator is based on the precise energy characterization of memory systems including buses, bus drivers and memory devices by a cycle-accurate energy measurement technique. We characterize the energy consumption of each component by an energy state machine whose states and transitions are associated with dynamic and static energy costs. Our approach easily characterizes the energy consumption of complex SDRAMs. The energy simulator enables us to devise practical energy reduction schemes that are justified by actual data and achieve real reductions of the total energy consumption of main memory systems.

7. REFERENCES

[1] N. Chang, K.-H. Kim, J. Cho, and H. Shin. Bus encoding for low-power high-performance memory systems. In *Proceedings of ACM/IEEE Design Automation Conference*, pages 800–805, 2000.

[2] N. Chang, K.-H. Kim, and H. G. Lee. Cycle-accurate energy consumption measurement and analysis: case study of ARM7TDMI. In *Proceedings of International Symposium on Low Power Electronics and Design*, pages 185–190, July 2000.

[3] X. Fan, C. S. Ellis, and A. R. Lebeck. Memory controller policies for dram power management. In *Proceedings of International Symposium on Low Power Electronics and Design*, pages 129–134, April 2001.

[4] P. Hicks, M. Walnock, and R. M. Owens. Analysis of power consumption in memory hierarchies. In *Proceedings of International Symposium on Low Power Electronics and Design*, pages 239–242, 1997.

[5] S. Miura, K. Ayukawa, and T. Watanabe. A dynamic-SDRAM-mode-control scheme for low-power systems with a 32-bit risc cpu. In *Proceeding of International Symposium on Low Power Electronics and Design*, pages 358–363, April 2001.

[6] M. R. Stan and W. P. Burleson. Bus-invert coding for low power I/O. *IEEE Transactions on VLSI*, 3(1):49–58, March 1995.

Coordinated Transformations for High-Level Synthesis of High Performance Microprocessor Blocks

Sumit Gupta[‡] Timothy Kam[§] Michael Kishinevsky[§] Shai Rotem[§]
Nick Savoiu[†] Nikil Dutt[†] Rajesh Gupta[†] Alex Nicolau[†]

[†]Center for Embedded Computer Systems, University of California, Irvine
{sumitg,savoiu,dutt,rgupta,nicolau}@cecs.uci.edu; http://www.cecs.uci.edu/~spark
[§]Strategic CAD Labs, Intel Incorporated, Hillsboro, Oregon
{timothy.kam,michael.kishinevsky,shai.rotem}@intel.com; http://www.intel.com/research/scl

ABSTRACT

High performance microprocessor designs are partially characterized by functional blocks consisting of a large number of operations that are packed into very few cycles (often single-cycle) with little or no resource constraints but tight bounds on the cycle time. Extreme parallelization, conditional and speculative execution of operations is essential to meet the processor performance goals. However, this is a tedious task for which classical high-level synthesis (HLS) formulations are inadequate and thus rarely used. In this paper, we present a new methodology for application of HLS targeted to such microprocessor functional blocks that can potentially speed up the design space exploration for microprocessor designs. Our methodology consists of a coordinated set of source-level and fine-grain parallelizing compiler transformations that targets these behavioral descriptions, specifically loop constructs in them and enables efficient chaining of operations and high-level synthesis of the functional blocks. As a case study in understanding the complexity and challenges in the use of HLS, we walk the reader through the detailed design of an instruction length decoder drawn from the *Pentium*®-family of processors. The chief contribution of this paper is formulation of a domain-specific methodology for application of high-level synthesis techniques to a domain that rarely, if ever, finds use for it.

Categories and Subject Descriptors: B.5.1 [Register-Transfer-Level Implementation] Design Aids C.5.3 [Computer System Implementation] Microcomputers - Microprocessors
General Terms: Design
Keywords: High-level synthesis, microprocessor design

1. INTRODUCTION

The classical high-level synthesis problem is one of transforming a behavioral description of an application through scheduling and binding tasks with constraints on the number of resources, into a multi-cycle schedule of operations. HLS is currently a mature field with a number of university and commercially-available tools. The current application of HLS is heavily concentrated on the design of moderately complex ASIC designs. The target architecture is a multi-cycle design with latencies in the 10s and 100s of cycles. These designs are usually area constrained, which often limits the extent of parallelism in operations the final design can support. Pipelining is generally the preferred means to improve system performance though it does not help latency. Accordingly, support for automatic pipelining (to a limited extent) of behavioral descriptions can be found in most commercial HLS tools.

High performance microprocessor designs are typically considered to lie on the other end of the spectrum where much of the HLS optimizations in scheduling, resource binding and allocation do not find much use. There exist a good number of functional blocks within microprocessors, which are most naturally and succinctly described by a behavioral description. However, the lack of responsiveness to design constraints in HLS formulations leads to little or no use of traditional HLS tools in such high-performance functional blocks. The chief problem is that, one major microprocessor design challenge - especially in the high end - is of identifying maximum parallelism and creating additional parallelization opportunities above and beyond afforded by the algorithmic specification, and then packing all the resulting operations in a safe manner in the smallest number of cycles and in the shortest cycle time. Pure pipelining is of limited value since functional block latencies are critical in the presence of significant control in the behavior.

Our work in this area has been motivated by the advances in parallelizing compiler technology that enable exploitation of extreme amounts of parallelization through a range of code motion techniques. While we have found no single code motion technique (including the ones we have developed specifically for HLS) to be universally useful, we have found that a judicious balance of a number of these techniques driven by well considered heuristics is likely to yield HLS results that compare in quality to the manually designed functional blocks. The challenge then is to identify and isolate a useful set of transformations and couple these with the rest of a high-level synthesis system. This system then provides a working environment for the microprocessor block designer to explore alternative designs and speed up the overall design process. This in essence is the contribution of the Spark synthesis system.

This paper demonstrates the utility of the Spark system through a case study of an instruction length decoder block derived from an advanced microprocessor. The choice of this block is made for a few reasons: (a) it is moderately complex and yet small enough that a detailed walk through the design - in an attempt to understand the challenges in application of HLS to high-performance microprocessor block designs - is possible; (b) the synthesis of this design employs several parallelizing transformations that validate the underlying motivation for building Spark; (c) designs of this nature are most naturally described by a behavioral description rather than a structural model, making them ideal for HLS.

[*]This author was an intern at Intel in the summer of 2001.

DAC 2002, June 10-14, 2002, New Orleans, Louisiana, USA.

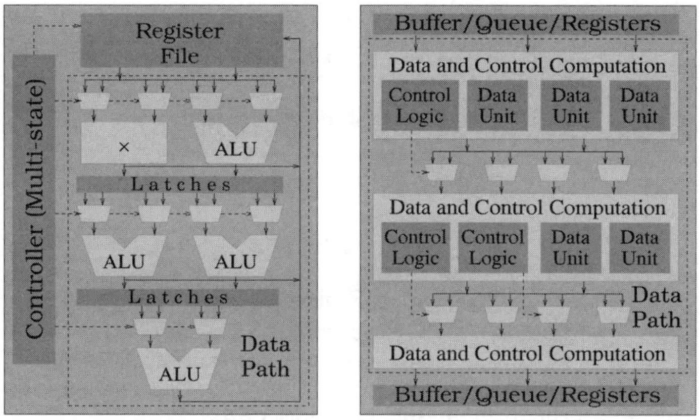

(a) (b)

Figure 1: Generic architectures (a) for ASIC designs and (b) for high performance microprocessor blocks

We can begin our understanding of the microprocessor synthesis domain, by comparing the generic architectures of microprocessor blocks and ASICs. ASICs (as shown in Figure 1(a)) are typically multi-cycle and pipelined, consisting of several functional units, steering logic (multiplexors), a controller (often a finite state machine) and a register file. Intermediate results are usually stored in latches and inter-stage forwarding paths may exist in the data path. On the other hand, as shown in Figure 1(b), microprocessor blocks are often single cycle and have several small computation blocks that operate in tandem and whose results are steered by, and even used to generate, control logic. Inputs and outputs to these type of blocks are stored in memory elements such as buffers and queues.

Keeping these differences in mind, we present Spark's synthesis methodology for microprocessor blocks. The rest of this paper is organized as follows: the next section reviews relevant previous work, followed by an outline of Spark's synthesis strategy and several of its transformations. The Spark system itself is briefly described in Section 4. Sections 5 and 6 describe the instruction length decoder and the steps employed in synthesizing this design. We conclude with a discussion and an outline of future work.

2. PREVIOUS WORK

There has been much work in the field of high-level synthesis (HLS). Early work concentrated on data-flow designs and applied optimizations such as algebraic transformations, retiming and code motions across multiplexors for improved synthesis results [1, 2]. Pipelining has been the primary technique to improve performance [3, 4]. Subsequent work has demonstrated the use of speculative code motions on mixed control-data flow type of designs to reduce schedule lengths [5, 6, 7, 8, 9].

A range of code motion techniques similar to those required for HLS has also been developed previously for software compilers (especially parallelizing compilers) [10, 11]. Although the basic transformations (e.g. dead code elimination, copy propagation) can be used in synthesis as well, other compiler transformations need to be re-instrumented for synthesis. They have to be modified to incorporate the ideas of resource sharing, steering logic and control costs, when used for synthesis. Unlike compilers, in synthesis, mutually exclusive operations can be scheduled in the same clock cycle on the same resource. Also, mapping an operation to a resource can lead to the generation of additional steering logic and associated control logic. So, the cost models in compilers and synthesis tools for the various transformations have to be different. Previous work in HLS for microprocessor designs is limited. Brayton et al. [12] synthesized the 801 processor; a small processor with a simple data path, whereas Gupta et al. synthesized long latency functional units to embed into VLIW processors [13].

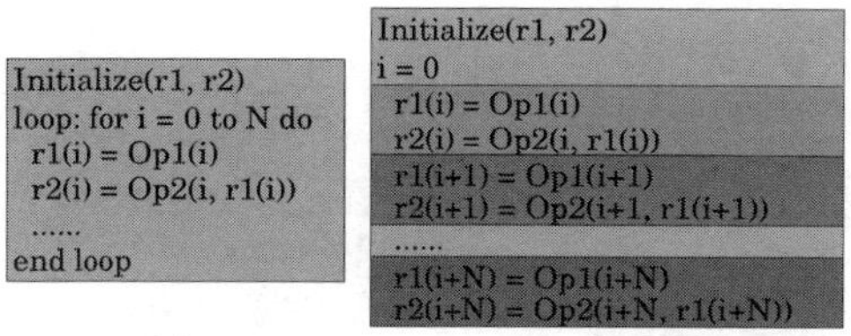

(a) (b)

Figure 2: (a) A sample piece of pseudo-code with a loop and some operations (b) the loop is completely unrolled

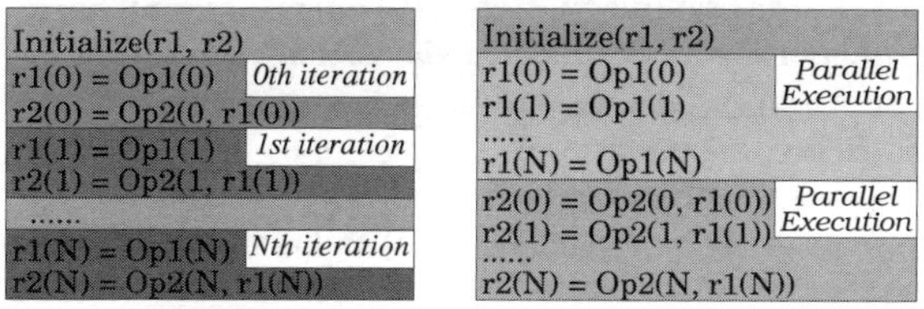

(a) (b)

Figure 3: (a) Constant propagation of loop index variable (b) all Op_1 operations are executed in parallel followed by concurrent execution of all the Op_2 operations

3. SYNTHESIS TRANSFORMATIONS FOR MICROPROCESSOR BLOCKS

Among the parallelizing transformations, the most useful are a number of loop transformations and beyond basic block code motions such as speculation and conditional speculation. In speculative execution, operations are executed before the conditions they depend on, have been evaluated. Conditional execution duplicates operations into the branches of conditional blocks to enhance resource utilization. These transformations have been explored and extended to a set of code motions that include reverse speculation and early condition execution [9, 14] and have been shown to be effective in improving the quality of synthesis results. The effectiveness of these code motions is often limited by the number of resources available to the design; a constraint that is more lax for microprocessor blocks.

The scope for application of code motions can be further increased by loop transformations such as *loop unrolling*. Loop unrolling was developed to enable software compilers to perform optimizations across loop iterations and facilitate global code optimizations [15]. However, loop unrolling can lead to code explosion; so, loops are unrolled one iteration at a time, followed by code compaction by parallelizing transformations, until no further improvements can be obtained. Loops are seldom unrolled fully.

On the other hand, for microprocessor functional blocks, loops are only a programming convenience and latency constraints generally dictate the amount of unrolling a loop has to undergo. For instance, if a design is targeted to, say, three clock cycles, it implies that *all* the operations within *all* the iterations of the loop have to be executed in these three cycles. Hence, when this design is mapped to hardware, it will generate a design in which the loop is, in essence, unrolled within these three cycles. Loops in single cycle designs must, of course, be unrolled completely.

Loop unrolling is demonstrated in Figure 2. Figure 2(a) shows the pseudo-code of a synthetic example, which has a loop and some operations within this loop. Operations Op_1 use the loop index variable i and some other inputs (not shown) to generate a set of results $r1$. These results are used by operations Op_2 to generate the final results $r2$ of the loop. This loop can be unrolled completely, i.e., N times to obtain the pseudo-code shown in Figure 2(b). In this figure, only the ith, $(i+1)$th and $(i+N)$th iterations are shown.

The Op_1 operations still have a dependency with the loop index variable i. However, since the loop is unrolled completely, the

```
1: t1 = a + b;
if (cond)
  2: t2 = t1;
  3: t3 = c + d;
else
  4: t2 = e;
  5: t3 = c - d;
6: f = t2 + t3;
```

(a)

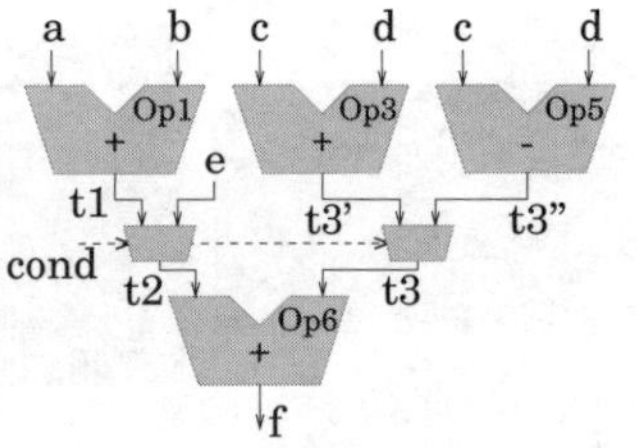

(b)

Figure 4: An example of operation chaining across conditional boundaries; (a) sample "C" code (b) corresponding hardware, with functional units connected via steering logic.

value of the loop index variable is known statically in all the iterations. Hence, the initial value assigned to the loop index variable can be propagated as a constant throughout all the iterations. This is known as *constant propagation* [16] and is shown in Figure 3(a) for the example from Figure 2. In this figure, the constant assignment $i = 0$ has been propagated through all the unrolled iterations and the loop index variable is completely eliminated from the code. This frees up more operations for the application of code parallelizing transformations. In this example, the code motion transformations can execute the Op_1 operations concurrently followed by the concurrent execution of all Op_2 operations (assuming that the resources to do so are available), as shown in Figure 3(b).

The coarse-grain transformations presented above can also be coupled with fine-grain transformations, such as copy propagation, dead code elimination and other standard compiler transformations. Also, another enabling transformation for synthesis is operation chaining, which is discussed in the next section.

3.1 Chaining Operations Across Conditional Boundaries

Operation chaining is an important technique that is supported by most high-level synthesis tools [17]. *Chaining* of operations means that the result of one operation is used immediately by the next operation without storing it in an intermediary latch or register. In the generated hardware, the resources corresponding to the operations have to be connected to each other in combinational blocks. However, in designs that are a mix of control and data operations, such as microprocessor functional blocks, these resources are often connected via steering logic such as multiplexors. In terms of the behavioral description, this implies that operations have to be chained across basic blocks, i.e., across conditional blocks.

Consider the sample fragment of "C" code in Figure 4(a). To achieve a single cycle schedule for this description, all the operations in the description have to be chained together, across the if-then-else conditional block. One possible hardware implementation for this is shown in Figure 4(b). The operations Op_1 to Op_6 correspond to the line numbers in Figure 4(a). As shown in this figure, the inputs to the operation Op_6 are obtained by multiplexing the outputs of the Op_1, Op_3 and Op_5, based on the condition *cond*.

This example demonstrates that operations in several basic blocks may be scheduled in the same clock cycle. Thus, resource utilization in a clock cycle has to keep track of the resource utilization of multiple operations in several basic blocks. Hence, scheduling with operation chaining across conditional boundaries has to use a modified resource utilization and operation scheduling model that looks across the conditional boundaries. Furthermore, chaining an operation with operations that are in the branches of a conditional check requires a more detailed analysis.

3.1.1 Chaining Operations with Operations in the Branches of a Conditional Block

Consider the hierarchical task graph (HTG) [14] representation of a design in Figure 5; we want to schedule operation 4 in the same

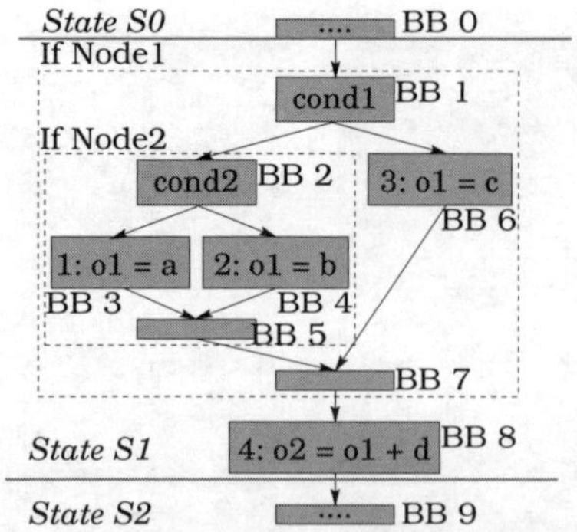

Figure 5: Operation 4 is scheduled in the same cycle as operations 1, 2 and 3. Hence, we have to check that chaining is possible on all *trails* up from basic block BB_8.

cycle as operation 1. The chaining heuristic has to validate that operation 4 can be chained with the other operations in this cycle. It traverses all the paths or *trails* backwards from the basic block that operation 4 is in (BB_8), looking for operations that are scheduled in the same cycle. In this example, there are three trails comprising the basic blocks: $<BB_8, BB_7, BB_5, BB_3, BB_2, BB_1>$,$<BB_8, BB_7, BB_5, BB_4, BB_2, BB_1>$ and $<BB_8, BB_7, BB_6, BB_1>$. These trails have the three operations 1, 2 and 3 respectively, each of which writes to the variable $o1$. The heuristic determines that operation 4 can be executed in the same cycle by using the appropriate value of $o1$ depending on the evaluation of the condition.

Besides checking that there is enough time in the cycle for chaining the operation with other operations in the chaining trails, the chaining heuristic also has to ensure that the correct hardware is generated to implement the schedule, as discussed next.

3.1.2 Creating Wire-Variables to enable Chaining on each Chaining Trail

The Spark synthesis tool initially assumes that each variable in the input behavioral description is mapped to a virtual register. After scheduling, during register binding, a variable life-time analysis pass determines which variables are actually mapped to registers. However, since registers can only be read in the next cycle after being written, therefore, to enable operation chaining, we introduce the notion of a *wire-variable*. Wire-variables are explicitly marked as being wires and are not mapped to registers, and thus, can be read in the same cycle as they are written to.

Consider an operation Op_1 that writes a result, r_1 and another operation Op_2 that reads this result, i.e., a situation that looks like: $r_1 = Op_1(arguments); r_2 = Op_2(r_1)$. To chain operations Op_1 and Op_2, the code has to be modified to: $temp = Op_1(arguments); r_2 = Op_2(temp); r_1 = temp$, where variable $temp$ is marked as being a wire and r_1 is (potentially) mapped to a register[1].

Often, as was the case in the example in Figure 5, a variable may be written by several operations in different basic blocks. When operations are chained across conditional checks, writes to "wire-variables" have to be inserted in all the trails leading back from the chained operation, i.e., in all the branches of the preceding conditional blocks. This is explained by an example in Figure 6(a). In this hierarchical task graph (HTG) representation, variable $o1$ is written to by operations 1 and 2 in basic blocks BB_0 and BB_2 respectively. Operation 3 in basic block BB_5 reads the value of this variable to produce another variable $o2$. Consider that the scheduling algorithm schedules the entire fragment of code in this figure within one clock cycle. Then, to enable the operation chaining, a wire-variable $t1$ is introduced and the copy operations 4 and 5 are

[1]In the RTL VHDL generated after synthesis, r_1 is mapped to a VHDL signal and $temp$ is mapped to a VHDL variable

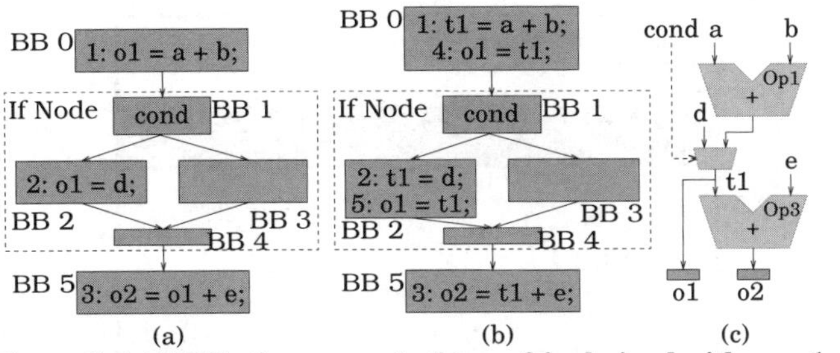

Figure 6: (a) HTG of an example (b) op. 3 is chained with ops. 1 and 2; so, wire-variable $t1$ and copy ops. 4 and 5 are inserted (c) corresponding hardware; $t1$ becomes a wire and $o1$ a register.

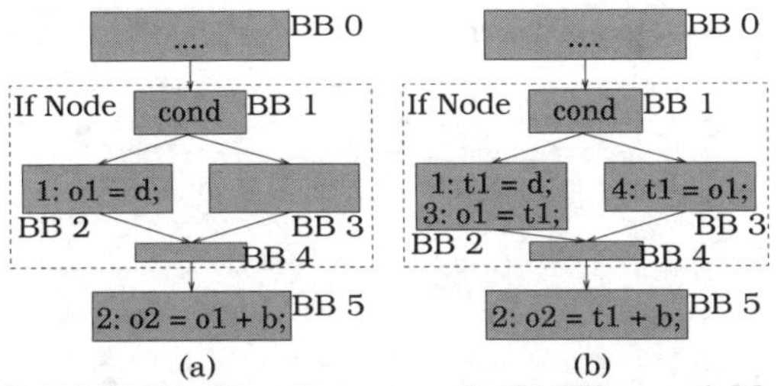

Figure 7: (a) HTG of another example (b) Wire-variable $t1$ and copy operations (3 and 4) are added in all chaining trails.

inserted, as shown in Figure 6(b). In the resulting hardware, shown in Figure 6(c), variable $t1$ becomes a wire and the variables $o1$ and $o2$ are bound to registers. Operation 3 uses the multiplexed result of both the operations that write to wire-variable $t1$.

Similarly consider the fragment of code in the Figure 7(a). In this example, variable $o1$ is written to only in the true branch of a conditional block and is read by operation 2 in basic block BB_5. This code implies that if the condition evaluates to "false", then a value of $o1$ from a previous write (not shown here) will be used by operation 2. In order to chain the operations in this code, a variable copy to wire-variable $t1$ has to be inserted in both branches of the conditional block, as shown in Figure 7(b). So, the operation 2 now reads the variable $t1$ instead. In hardware, variable $t1$ will be mapped to a wire and variable $o1$ to a register. In this way, wire-variables are introduced as and when required and a dead code elimination pass later removes any unnecessary variables and variable copies.

4. THE SPARK SYNTHESIS SYSTEM

The code transformations presented in this paper have been implemented in a high-level synthesis research framework called *Spark* [14, 9]. This synthesis system takes a behavioral description in ANSI-C as input and generates synthesizable register-transfer level VHDL. This enables the system to evaluate the effects of several coarse and fine-grain optimizations on logic synthesis results. Code motion techniques such as Trailblazing [18] are used to enable the parallelizing, speculative code motion transformations. Loop unrolling has been implemented as part of the resource directed loop pipelining (RDLP) technique [11]. These transformations are supported by standard compiler transformations such as constant propagation and dead code elimination.

The rich set of tunable transformations in Spark enable the system to aid in exploration of several alternative designs. Although Spark can apply the various transformations automatically, it also allows the designer to control the various passes and the degree of parallelization through script files. For example, the designer may specify which loops to unroll and by how much. This enables Spark to provide design alternatives that may not be obvious to a designer from the design's behavioral description. In the next few sections, we show how we have used Spark to explore the architecture of a functional block from a modern microprocessor.

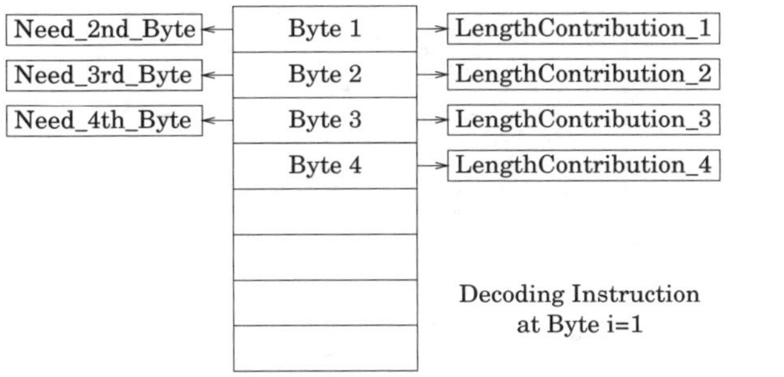

Figure 8: Instruction Length Decoder (ILD) of a modern microprocessor; calculating the length of the first instruction

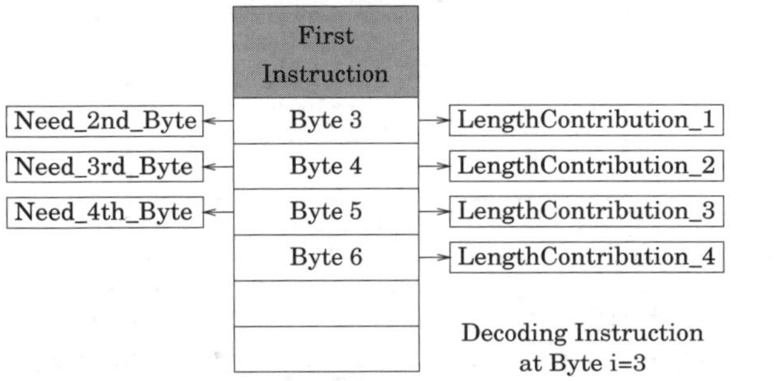

Figure 9: ILD: Calculating the length of the second instruction

5. A CASE STUDY: INSTRUCTION LENGTH DECODER

An important component of the *Pentium*® microprocessor architecture is the *instruction length decoder* [19]. This component determines the starting byte and the length of each instruction from a stream of bytes it receives from the instruction cache. We consider an implementation of this microprocessor architecture in which the instructions can be of variable length ranging from 1 to 11 bytes and the decoder has to look at up to 4 bytes to determine an instruction's length. Instead of processing a stream of bytes, the decoder can alternatively look at a set of bytes in an instruction buffer at every cycle. The instruction length decode then works as shown in Figure 8. The decoder looks at the length contribution of the first byte ($LengthContribution_1$), and checks to see whether it needs to look at the next byte as well (*Need_2nd_Byte*). If it does, then it looks at $LengthContribution_2$ of the second byte, and checks to see if it needs the third byte, and so on. In this way, say, the ILD calculates that the first instruction is two bytes long, then it must determine the length of the next instruction by (potentially) looking at the next 4 bytes as shown in Figure 9. This continues until the length of all the instructions in the buffer are determined.

A representation of this behavior in "C" is shown in Figure 10. In this code, a loop (indexed by *i*) iterates over the entire instruction buffer (of size *n*). If the start of the current instruction $NextStartByte$ is the current byte *i*, then, it marks this as the starting point of an instruction ($Mark[i] = 1$) and calculates the length of the instruction at that byte by calling the function $CalculateLength(i)$. This function is the same as the behavior described above[2]. The final output of this description is a bit vector ($Mark[1..n]$) that contains a 1 at only those bit positions where an instruction starts.

There are several simplifications in this model of the ILD [19]. Since the ILD is decoding a stream of instructions arriving from memory, the behavioral description should have an infinite outer loop, that synthesis should break into chunks of *n* iterations each. Also, consider that an instruction starts at the $(n-1)$th byte. Then the length calculation may need to check bytes from the next set

[2]We assume a zero length contribution from the $n+1$ to $n+3$ bytes

```
ResetArray(Mark);
NextStartByte = 1;
for (i = 1; i <= n; i++) {
  if (i == NextStartByte) {
    Mark[i] = 1;
    len[i] = CalculateLength(i);
    NextStartByte += len[i];
  }
}
int CalculateLength(i) {
  lc1 = LengthContribution_1(i);
  if (Need_2nd_Byte(i)) {
    lc2 = LengthContribution_2(i+1);
    if (Need_3rd_Byte(i+1)) {
      lc3 = LengthContribution_3(i+2);
      if (Need_4th_Byte(i+2)) {
        lc4 = LengthContribution_4(i+3);
        Length = lc1 + lc2 + lc3 + lc4;
      } else Length = lc1 + lc2 + lc3;
    } else Length = lc1 + lc2;
  } else Length = lc1;
  return Length;
}
```

Figure 10: Behavioral "C" code for the ILD

of bytes that fill the buffer. So, the intermediate length calculation information must be saved across buffer decodes and passed to the next cycle. These simplifications are made to keep the discussion focused on the important code transformations used and do not alter the nature and applicability of the transformations presented here.

The processor architectural requirements imply that the whole buffer must be decoded in one cycle. Hence, a designer may choose to compute as much as possible in parallel and then, do the instruction marking after all the information has been calculated. The following section describes how Spark's synthesis methodology achieves this kind of a single cycle architecture for the decoder starting from a natural behavioral description.

6. TRANSFORMATIONS APPLIED BY SPARK TO SYNTHESIZE THE DECODER

In order to achieve a single cycle architecture for the ILD design, the Spark synthesis tool is given an unlimited resource allocation and full freedom to unroll loops. The first transformation Spark applies, tries to speculatively compute all the data and control calculations in the function *CalculateLength*. As shown in Figure 11, the length contributions due to the bytes, i through $i+3$, are calculated speculatively and so are the control variables *need*2 to *need*4, that determine which bytes contribute to the length of the current instruction. The lengths of the instruction for each case of these control variables (*TempLength*1 to *TempLength*3) are also speculatively computed. This results in a behavior where all the data calculation is performed up-front and speculatively, followed

```
int CalculateLength(i) {
  lc1 = LengthContribution_1(i);          [Data Calculation]
  lc2 = LengthContribution_2(i+
  lc3 = LengthContribution_3(i+2);
  lc4 = LengthContribution_4(i+3);
  need2 = Need_2nd_Byte(i);
  need3 = Need_3rd_Byte(i+1);
  need4 = Need_4th_Byte(i+2);

  TempLength1 = lc1 + lc2 + lc3 + lc4;
  TempLength2 = lc1 + lc2 + lc3;
  TempLength3 = lc1 + lc2;
  if (need2) {                            [Control Logic]
    if (need3) {
      if (need4) {
        Length = TempLength1;
      } else Length = TempLength2;
    } else Length = TempLength3;
  } else Length = lc1;
  return Length;
}
```

Figure 11: ILD: All the data operations in the CalculateLength function are *speculatively* executed

```
ResetArray(Mark);
NextStartByte = 1;
for (i = 1; i <= n; i++) {
  Results(i) = DataCalculation(i,i+1,i+2,i+3);
  Length(i) = ControlLogic(Results(i));
  len[i] = Length(i);

  if (i == NextStartByte) {
    Mark[i] = 1;
    NextStartByte += len[i];
  }
}
```

Figure 12: ILD: The instruction length calculation function is *inlined* into the main function

```
ResetArray(Mark);
NextStartByte = 1;
i = 1;

Results(i) = DataCalculation(i,i+1,i+2,i+3);
len[i] = ControlLogic(Results(i));
if (i == NextStartByte) {
  Mark[i] = 1;                              [1st Iteration]
  NextStartByte += len[i];
}

Results(i+1) = DataCalculation(i+1,i+2,i+3,i+4);
len[i+1] = ControlLogic(Results(i+1));
if (i+1 == NextStartByte) {
  Mark[i+1] = 1;                            [2nd Iteration]
  NextStartByte += len[i+1];
}

......... till nth iteration
```

Figure 13: ILD: The loop with index variable *i* is *unrolled* fully; the figure only shows two iterations

by a control logic structure that uses this data and assigns the correct result to the output. This control logic maps to multiplexors in hardware. Hereafter, for brevity of presentation, we will refer to these data and control components in the *CalculateLength* function as *DataCalculation* and *ControlLogic* respectively.

In the next step, the *CalculateLength* function can be *inlined* into the main calling function as shown in Figure 12. *Inlining* refers to replacing a call to a function or a subroutine with the body of the function or subroutine [16]. This transformation allows the optimization of the inlined function with the rest of the code. In practice, Spark performs inlining first, but speculation within the *CalculateLength* has been shown first to simplify explanation.

Next, the loop is fully unrolled as shown by the code in Figure 13. However, the parallelization transformations are still limited due to a dependency that still exists between the operations and the loop index variable i. However, since the loop has been completely unrolled, the constant assignment of $i=1$ can be propagated throughout the code and the loop index variable i can be eliminated. The resulting code is shown in Figure 14. This exposes further opportunities for early and speculative calculation of the lengths of instructions, as shown in Figure 15(a). In this description, the lengths of the instructions are calculated assuming that a new instruction starts at each byte.

This leads to a design, where all the data for all the bytes is calculated concurrently, followed by a control logic unit, which determines the length of the instructions if they were to start at each byte and finally, a ripple control logic unit that determines the actual instruction start bytes. The hardware architecture corresponding to this code is shown in Figure 15(b). This is a maximally parallel architecture that can be targeted for implementation in a single cycle.

In this way, Spark achieves the single cycle architecture that is required by starting with the "C" behavioral description of the ILD shown in Figure 10, and producing the register-transfer level VHDL code corresponding to the architecture shown in Figure 15.

```
ResetArray(Mark);
NextStartByte = 1;
Results(1) = DataCalculation(1,2,3,4);
len[1] = ControlLogic(Results(1));
if (1 == NextStartByte) {
  Mark[1] = 1;
  NextStartByte += len[1];
}
Results(2) = DataCalculation(2,3,4,5);
len[2] = ControlLogic(Results(2));
if (2 == NextStartByte) {
  Mark[2] = 1;
  NextStartByte += len[2];
}
...
Results(n)=DataCalculation(n,...,n+3);
len[n] = ControlLogic(Results(n));
if (n == NextStartByte) {
  Mark[n] = 1;
  NextStartByte += len[n];
}
```

Figure 14: ILD: Constant Propagation of loop index variable *i* after the loop has been completely unrolled

```
ResetArray(Mark);
                                   Data Calculation
Results(1) = DataCalculation(1, 2, 3, 4);
Results(2) = DataCalculation(2, 3, 4,5);
...
Results(n) = DataCalculation(n, ..., n+3)
                                   Control Logic
len[1] = ControlLogic(Results(1));
len[2] = ControlLogic(Results(2));
...
len[n] = ControlLogic(Results(n));
                                   Ripple Control Logic
NextStartByte = 1;
if (1 == NextStartByte) {
  Mark[1] = 1;
  NextStartByte += len[1];
}
...
...
if (n == NextStartByte) {
  Mark[n] = 1;
  NextStartByte += len[n];
}
```

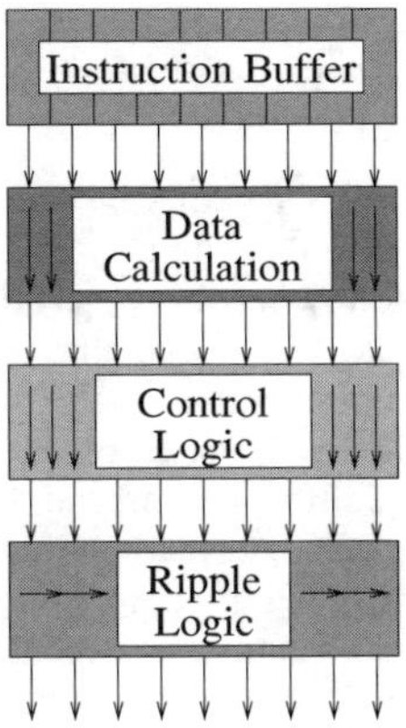

(a) (b)

Figure 15: (a) Speculative calculation of all instruction lengths assuming an instruction starts at each byte (b) the final ILD architecture produced by the Spark system

Although not demonstrated in this section, Spark chains the operations in the decoder together as described in Section 3.1, to achieve the single cycle architecture. We are unable to compare the synthesis results of our design with any hand-design since the model of the ILD used has several simplifications (as discussed earlier).

7. CONCLUSIONS AND FUTURE WORK

In this paper, we have explored a new domain for the application of high-level synthesis; the domain of high performance microprocessor functional blocks. A synthesis methodology has been presented that exploits the liberal resource allocation available to these blocks and achieves the tight cycle bounds by employing a set of aggressive coarse and fine-grain transformations. This methodology systematically extracts parallelism from a behavioral description by employing parallelizing techniques such as loop unrolling and speculative code motions. Operations are packed into a few cycles by chaining operations, often, across conditional boundaries. This synthesis methodology and the associated transformations have been implemented in the Spark high-level synthesis framework. Finally, we have also demonstrated the effectiveness of this methodology by walking through the transformations applied by Spark to synthesize an instruction length decoder inspired by an advanced microprocessor.

While this case study is instructive in understanding the code

```
ResetArray(Mark);
NextStartByte = 1;
while(1) {
  Mark[NextStartByte] = 1;
  len = CalculateLength(NextStartByte);
  NextStartByte += len;
}
```

Figure 16: ILD: A succinct and natural behavioral description

transformations needed for high performance microprocessor blocks, there are several areas of HLS that require further investigation. For instance, the behavioral description we have used as a starting point for our work (Figure 10) may not be the most simple way to describe the design. A more natural and succinct way to describe the ILD's behavior could be as shown in Figure 16. Similar, short behavioral descriptions can be used to describe several such low latency functional blocks in microprocessors. This leads us to future work in developing a new set of source-level transformations that can transform these sort of descriptions into more easily synthesizable behavioral descriptions.

8. REFERENCES

[1] M. Potkonjak and J. Rabaey. Optimizing resource utlization using tranformations. *IEEE Trans. on CAD*, March 1994.

[2] R. Walker and D. Thomas. Behavioral transformation for algorithmic level IC design. *IEEE Trans. on CAD*, Oct. 1989.

[3] N. Park and A. Parker. Sehwa: A software package for synthesis of pipelines from behavioral specifications. *IEEE Trans. on Computer-Aided Design*, March 1988.

[4] P. G. Paulin and J. P. Knight. Force-Directed Scheduling for the Behavioral Synthesis of ASIC's. *IEEE TCAD*, June 1989.

[5] K. Wakabayashi and H. Tanaka. Global scheduling independent of control dependencies based on condition vectors. In *Design Automation Conference*, 1992.

[6] I. Radivojevic and F. Brewer. A new symbolic technique for control-dependent scheduling. *IEEE TCAD*, January 1996.

[7] G. Lakshminarayana, A. Raghunathan, and N.K. Jha. Incorporating speculative execution into scheduling of control-flow intensive behavioral descriptions. *DAC*, 1998.

[8] L.C.V. dos Santos and J.A.G. Jess. A reordering technique for efficient code motion. In *Design Automation Conf.*, 1999.

[9] S. Gupta, N. Savoiu, N.D. Dutt, R.K. Gupta, and A. Nicolau. Conditional speculation and its effects on performance and area for high-level synthesis. In *ISSS*, 2001.

[10] J. Fisher. Trace scheduling: A technique for global microcode compaction. *IEEE Trans. on Comps.*, July 1981.

[11] S. Novack and A. Nicolau. An efficient, global resource-directed approach to exploiting instruction-level parallelism. In *Conference on Parallel Architectures and Compilation Techniques*, 1996.

[12] R.K. Brayton, R. Camposano, G. De Micheli, R.H.J.M. Otten, and J. van Eijndhoven. *The Yorktown Silicon Compiler System*, Silicon Compilation. Addison-Wesley, 1988.

[13] R. L. Gupta, A. Kumar, A. Van Der Werf, and G. N. Busa. Synthesizing a long latency unit within VLIW processor. In *Intl. Conf. on VLSI Design*, 2000.

[14] S. Gupta, N. Savoiu, S. Kim, N.D. Dutt, R.K. Gupta, and A. Nicolau. Speculation techniques for high level synthesis of control intensive designs. In *DAC*, 2001.

[15] S. S. Muchnick. *Advanced Compiler Design and Implementation.* Morgan Kaufmann, 1997.

[16] A. Aho, R. Sethi, and J. Ullman. *Compilers: Principles and Techniques and Tools.* Addison-Wesley, 1986.

[17] D. D. Gajski, N. D. Dutt, A. C-H. Wu, and S. Y-L. Lin. *High-Level Synthesis: Introduction to Chip and System Design.* Kluwer Academic, 1992.

[18] A. Nicolau and S. Novack. Trailblazing: A hierarchical approach to percolation scheduling. In *International Conference on Parallel Processing*, 1993.

[19] Intel Inc., *PentiumPro® Programmer's Reference Manual.*

Forward-Looking Objective Functions: Concept & Applications in High Level Synthesis

Jennifer L. Wong
Univ. of California, Los Angeles
3563F Boelter Hall
Los Angeles, CA 90095
(310) 825-1872
jwong@cs.ucla.edu

Seapahn Megerian
Univ. of California, Los Angeles
3750 Boelter Hall
Los Angeles, CA 90095
(310) 206-3864
seapahn@cs.ucla.edu

Miodrag Potkonjak
Univ. of California, Los Angeles
3532G Boelter Hall
Los Angeles, CA 90095
(310) 825-0790
miodrag@cs.ucla.edu

ABSTRACT

The effectiveness of traditional CAD optimization algorithms is proportional to the accuracy of the targeted objective functions. However, behavioral synthesis tools are not used in isolation; they form a strongly connected design flow where each tool optimizes its own objective function without considering the consequences on the optimization goals of the subsequently applied tools. Therefore, efforts to optimize one aspect of a design often have unforeseen negative impacts on other phases of the design process.

Our objective is to establish a systematic way of developing and validating new types of objective functions that consider the effects on subsequently applied synthesis steps. We demonstrate the generic forward-looking objective function (FLOF) strategy on three main steps in behavioral synthesis: (i) Transformation, (ii) Scheduling, and (iii) Register Assignment. We show how the FLOF can be used in the first two phases to reduce the total number of registers required in the third phase.

Categories and Subject Descriptors

B.6.3 [**Logic Design**]: Design Aids – Optimization.

General Terms

Algorithms, Design.

Keywords

Objective Functions, Behavioral Synthesis, Transformations, Scheduling, Register Assignment

1. INTRODUCTION

The effectiveness of each step in behavioral synthesis is dependent on two main components: (i) algorithmic optimization mechanisms and (ii) targeted objective functions. Tremendous efforts have been dedicated to the development of increasingly sophisticated optimization mechanisms, resulting in intricate and effective schemes that leverage on complex mathematical results. However, the effectiveness of these algorithms is limited by the accuracy of the targeted objective functions. Interestingly, while essentially all CAD tools use some type of objective functions, there has been very little effort to develop systematic theory and practice of designing and evaluating objective functions. Often, the objective functions used are too simplistic. Examples are the use of the number of literals to predict area in logic synthesis and the number of execution units to predict area in behavioral synthesis. Increasing complexity of deep-submicron is bound to make these objective functions more irrelevant. Also, the need for predicting more complex design metrics, such as power and testability, provide additional impetus to study objective functions during synthesis.

One of the key difficulties in developing high quality objective functions is that synthesis tools are mainly applied following the waterfall model and that optimization goals do not take into account the consequences of current choices on optimization in later steps. For example, often slight minimization in the number of execution units has as direct consequence an additional overhead during the variable to register assignment phase. A small saving, if any, during scheduling is compounded with significant additional area due to registers. Note that registers are relatively expensive in comparison with the majority of execution units in terms of both area and power.

Our goal is to develop a systematic approach for developing objective functions that span more than one design step. We want to demonstrate the effectiveness of this approach by utilizing the same optimization mechanisms with both traditional and the new forward-looking objective function in three behavioral synthesis steps: transformations, scheduling, and register assignment.

1.1 Motivational Example

In order to illustrate how an objective function can capture the impacts of optimization choices on future steps, we use a simple example from behavioral synthesis. Figure 1(a) shows the schedule of a computation involving four variables: A, B, C, and D. The results of the computation are the variables X and Y. The circles represent an arbitrary operation and the directed line segments represent the variables and precedence relationships in the computation. If we assume that each operation takes one clock cycle and that the available time is four clock cycles, the solution produced by the ALAP algorithm uses 2 execution units and is optimal. The subsequent application of register assignment, results in 6 registers. The numbers on the right side of each scheduling instance in Figure 1 indicates the size of the variable cut in each control step, i.e. the number of variables that are alive. Since simultaneously alive variables must be stored in separate registers, the maximum variable cut (shown in bold) is a lower bound on the number of required registers.

However, if we considered the impact of scheduling on the lifetimes of the variables, we will obtain the schedule shown in Figure 1b. For example, if the objective function used by scheduling has as component proportional to the sum of lifetimes

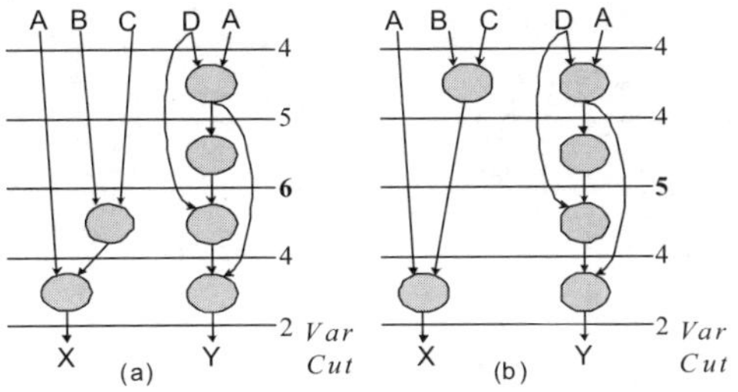

Figure 1. Simple scheduling example.

of variable, we will schedule the operation with B and C inputs in the first clock cycle so that we have only one variable alive after that instead of two as in the ALAP schedule. If we follow this objective function, the resulting schedule (Figure 1b) will also require two execution units. However, the sum of the lifetimes of the variables will be reduced and more importantly the maximal cut and eventually the required number of register will be reduced. The key observation is that although the register assignment is done later using a completely different algorithm (graph coloring), the effectiveness of the step is directly impacted by targeted objective function during scheduling. Therefore, if one wants to optimize across boundaries of individual synthesis tasks, he has to use objective functions that captures the requirements of consecutive synthesis steps.

2. RELATED WORK

Objective functions and their use in optimization has been the focus of intense study for decades [6]. Scheduling is one of the mandatory tasks that must be performed in behavioral synthesis. A number of Scheduling algorithms have been proposed and include [1,5]. Register assignment, typically performed following Scheduling, has also attracted a great deal of attention. Due to its importance, a variety of approaches have also been proposed for Register Assignment [11]. Finally, transformations are considered by many as the behavioral synthesis task with the highest potential to optimize a design. In the last two decades, a number of approaches have been proposed [8,9]. A more comprehensive analysis of high level synthesis, Scheduling, Register Assignment, and Transformations can be found in [3].

3. TECHNICAL PRELIMINARIES

3.1 Global Flow: High-Level Synthesis

Traditionally, the global flow of high-level synthesis has consisted of two major steps: (i) Scheduling and (ii) Resource Allocation. Scheduling refers to the assignment of operations to control steps, while resource allocation refers to the mapping of objects to physical units. One of the main tasks performed in the resource allocation step is *register assignment* which refers to the assignment of variables to physical registers. In order to make the technical discussions in this paper self-contained, we briefly discuss Scheduling, Register Assignment, and Transformations.

3.2 Scheduling

The main goal of Scheduling is to create a plan for the execution of operations subject to a specified set of constraints. The constraints of this optimization process are two fold: (i) the inherent execution order of the operation imposed by the computation, and (ii) physical hardware limits. Having a limited number of adders, multipliers, and other physical functional units that the high-level operations must ultimately be mapped to, restricts the number of operations that can be scheduled in the same control step. Typically, the goal of Scheduling is to schedule a design using a given set of functional units, while reducing the number of control steps required to perform the computation.

The computational model behind our discussion is the synchronous data flow (SDF) model [7]. For the sake of simplicity, we focus our attention on the case where each node consumes two and produces one sample on every execution. The SDF model is well suited for specification of computations in numerous application domains such as DSP, communications, and multimedia. The syntax of a targeted computation is defined as a hierarchical control-data flow graph (CDFG) [10]. A CDFG represents the computation as a flow graph, with nodes, data edges, and control edges. Operation Scheduling is thus the process of partitioning the set of operations in a CDFG into groups such that the operations in the same group are executed concurrently in the same control step while all precedence constraints are satisfied. The three main types of scheduling in high-level synthesis are [5]:

1. *Time-Constrained*: scheduling a CDFG in a given number of control steps while minimizing the required number of functional units.
2. *Resource-Constrained*: scheduling a CDFG using a given set of functional units while minimizing the required number of control steps.
3. *Hybrid*: attempting to schedule a CDFG for a given maximum number of control steps and the number of available functional units.

Unless otherwise specified, we assume that the number of functional units of each type and the maximum allowed number of control steps are specified a-priori (i.e. Hybrid). The simplified problem where only one type of operation is used, which is computationally intractable [4], can formally be stated as:

Problem: ***Scheduling***
Instance : *A CDFG C, a time constraint T, parameter N.*
Question: *Does there exist a Scheduling of C into T control steps such that the number of operations from C in each control step is less than N and all precedence constraints in C are satisfied?*

Two well-known heuristics for Scheduling are the as-soon-as-possible (ASAP) and the as-late-as-possible (ALAP) techniques. In Section 4, we use these heuristics in building the Scheduling FLOF. These heuristics are very efficient in determining the range of valid control steps that each operation can be scheduled in, and approximately the lifetimes of variables. According to the ASAP scheduling heuristic, operations are scheduled in the earliest possible control steps such that no constraints are violated. This is achieved by a simple traversal of the computation CDFG in breadth-first order. Analogously, following the ALAP heuristic, operations are scheduled in the latest possible control steps such that no constraints are violated. Once the process is complete, the schedule is simply "translated" where the first operations in the scheduled CDFG are in the first control step. A non-deterministic choice exists in both ASAP and ALAP scheduling when the number of operations allowed in each control step is fixed (i.e. limited hardware). The heuristics don't specify the order that independent operations should be scheduled. Different implementations address this issue in varying ways.

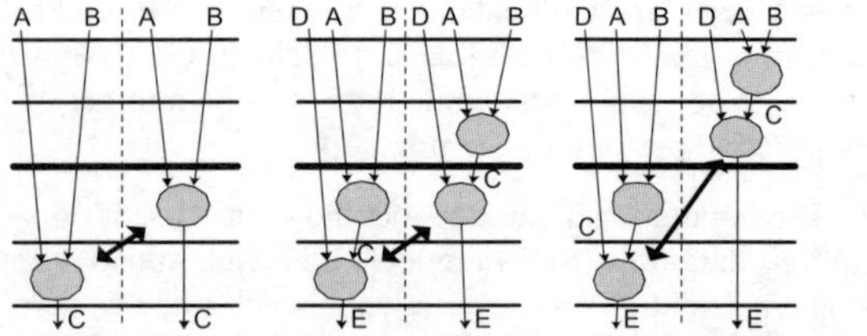

Figure 2. Basic blocks of scheduling and variable-cut trade-offs.

In our subsequent discussions, when we refer to a *cut,* we are referring to *variable cuts.* A ***variable cut*** refers to the variables that are alive in a scheduled CDFG between control steps. We assume that a variable is alive at the *end* of the control step where it's created until the *beginning* of the control step where it's consumed. Similarly, an ***operation cut*** in a scheduled CDFG refers to the operations scheduled in a particular control step. For example, the maximum operation cut refers to the cut in the control step with the largest number of operations of a particular type. Note that the maximum operation cut indicates the minimum number of required functional units (of each type).

3.3 Register Assignment

Register assignment is the process of mapping variables in a scheduled CDFG to physical storage units (registers). Clearly when two variables are simultaneously alive in a scheduled CDFG, they cannot be mapped to the same register. The register assignment problem can easily be modeled as a Graph Coloring problem where nodes in the graph correspond to variables in the CDFG and an edge connects any two nodes (variables) that are alive at the same time. Each color in a graph coloring solution corresponds to a unique register. The NP-complete Graph Coloring decision problem [4] can formally be states as:

Problem: ***Graph Coloring***
Instance : *A graph G(V,E) with vertex set V and edge set E, number of colors k.*
Question: *Does there exist a coloring of the graph G that assigns colors to each vertex* $v \in V$ *such that for any edge* $e \in E$ *the vertices connected by e have different colors, and that no more than k unique colors are used?*

Although graph coloring is a difficult problem, the *maximum variable cut* in a scheduled CDFG can provide a lower bound limit on the number of required colors. This is because all variables in a *cut* are alive by definition and thus must be mapped to separate registers and therefore separate colors. We use this lower bound on the required number of registers to build our Scheduling FLOF.

3.4 Transformations

Transformations can serve as powerful tools in optimizing a design for Scheduling. Some examples of transformations are associativity, distributivity, and retiming.

Retiming is a special transformation action applied to looping CDFGs (repeating) and can be accomplished by doing either of the following: (i) remove one delay element from each input of an operation and add a delay at each of its outputs; (ii) remove one delay element from each output of an operation and introduce a new delay element at each of its inputs.

A delay element is a special type of operation that "delays" the variable at its input until the next iteration of the scheduled CDFG. We discuss examples of retiming and its impact on register minimization in Section 4.2.

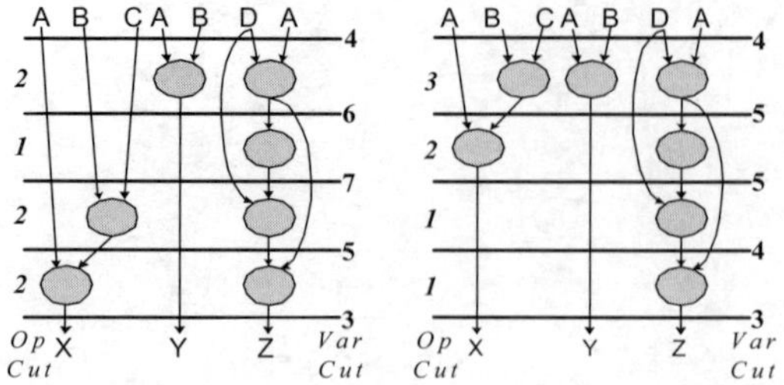

Figure 3. Example of how adding functional units can reduce the variable cut (& potentially the number of registers).

4. FORWARD-LOOKING OBJECTIVE FUNCTIONS

The term *objective function* refers to a function that quantifies the quality of a solution (or a partial solution) in an optimization problem. Objective functions essentially form the backbone of any optimization algorithm and serve as the guide in obtaining desired solutions. In a sorting problem for example, the objective function $OF_{Sort}(\,)$ can be defined as the percentage of items that are correctly sorted in a solution instance. Thus, the objective of a sorting algorithm would be to find a solution such that the objective function $OF_{Sort}(\,)$ is maximized.

The success of an optimization algorithm often depends heavily on carefully designed and tuned objective functions. However, modern layered optimization flows, that are necessary to approach and solve complex problems, often hinder or severely limit the usefulness of objective functions. Layering complex problems into logical steps serves as a tool to abstract away details that can make a problem very difficult to approach otherwise. Although each individual level may work well separately, final results may not be as good as possible due to the interaction between the levels. This is mostly due to the fact that the effects of actions in one layer will impact subsequent layers. For example, in system-on-chip design, optimizing the high-level communication among different modules may have adverse effects that will not be apparent until the floorplanning or place-&-route stages when new constraints (such as physical placement and mapping) take effect.

Here, we propose *forward-looking objective functions* (FLOF) as a generic means to overcome the restrictions imposed by layered optimization flows. To build a successful FLOF, one must carefully analyze the problem specific components in each step that may impact future steps, formulate quantitative expressions, and establish a method for combining the components into the final objective function format. Hence, the outline for designing an FLOF is:

1) Identify problem specific components and their effect on future steps (interaction between layers);
2) Quantify the components;
3) Combine components into a final objective function with appropriate normalization.

We demonstrate this process in detail in the following subsections to design a Scheduling FLOF with the goal of minimizing the required number of registers subject to the traditional scheduling constraints (total number of available functional units and control steps).

4.1 Scheduling

As stated above, the key factor in formulating an objective function in one level that will be successful in achieving good quality results in the overall flow is by the inclusion of properties and insights that can predict the impact of decisions from one optimization level onto future optimization levels. In order to make this clear, here we detail the steps required to build a FLOF for our example problem, namely Scheduling and its impact on Register Assignment in high-level synthesis.

In order to build the Scheduling FLOF, we must first itemize the ways in which specific schedules will potentially impact the register assignment process. As discussed in Section 3, register assignment is a hard optimization problem so we do not expect to be able to predict exact outcomes. Our goal here is to combine careful analysis and problem specific insights to formulate our objective function such that we favor scheduling solutions that are promising in terms of their expected required number of registers.

Consider the three scheduling instances in Figure 2. The shaded circles represent an arbitrary operation that consumes two inputs and produces one output. Suppose that the bold horizontal line is where the maximum variable cut lies. Let us assume that the location of the maximum cut does not change as we explore the different options (i.e. we assume that the maximum cut does not move). We use the term "move" to signify a scheduling option. The term "move" can mean re-scheduling as in iterative improvement-type algorithms or simply the decision process of where to schedule an operation in a constructive algorithm.

The first instance (left) in Figure 2 illustrates the trivial observation that moving a single operation can only impact the maximum cut by at most one variable. The middle and rightmost cases shows that a group of two operations must "move" by at least two control steps in order to increase the effect on the max cut to two variables. The main observations here are two fold: (i) since operations consume more variables than they produce, the variable cut can be reduced by scheduling more operations *before* the cut and (ii) the size of the maximum variable cut can be reduced by the scheduling such that the variable lifetimes are decreased. Note that in Figure 2, moving operations "up" effectively reduces the lifetime of the input variables while potentially increasing the lifetime of the output variables. Since there are more input variables than output variables for each operation, the total lifetime of variables is reduced. In the following subsection we analyze how variable lifetimes can play an important role in determining the required number of registers for scheduling.

Throughout our discussions we assume that the number of available functional units in each control step is fixed. Let us briefly consider how the addition of new functional units can serve to reduce the maximum variable cut using an example. The two diagrams in Figure 3 show the same scheduled CDFG. The schedule on the left requires only 2 functional units while the schedule on the right requires 3. However, the maximum variable cut in the left is 7 compared to 5 on the right. The main observation here is that variables A, B, and C live for relatively long periods of time. The scheduling flexibility for the consumption of these variables can be leveraged to greatly reduce their lifetimes. Although in the process, one extra functional unit is used, the variable cut is reduced by 2. This alone, may potentially translate to a savings of 2 registers in the register assignment step.

As the example in Figure 3 suggests, in certain instances, there is clearly a trade-off in the maximum variable cut and the maximum operation cut. For larger cases (most real designs) the savings in register requirements can be substantial relative to the extra overhead required for the additional functional units. Furthermore, since registers are much more costly in terms of power and area than adders for example, the benefits of trading functional units for register savings can be substantial. However, as stated earlier, for the purpose of brevity and simplicity, we assume the number of functional units is fixed when developing the Scheduling FLOF with register minimization as the goal.

We have identified two main attributes of variables in scheduling that can ultimately impact the minimum number of required registers: variable *lifetimes, and* scheduling *difficulty.* Generally, variables that are alive for shorter time periods are better for minimizing the number of registers since the likelihood that they will overlap with other variables is reduced. We refer to this property as the "*Small-Potato Principle*". It's a well known fact in the agricultural community that relatively smaller potatoes can be packed more densely in a given volume than larger ones. Analogously, it's not unreasonable to expect that shorter lived variables, *in general*, will be easier to schedule and "pack" into fewer registers than variables that live for relatively long periods, potentially overlapping with many other variables. Throughout the rest of this discussion, we will focus on properties and aspects of the scheduling problem that can be leveraged to build a FLOF to minimize the number of required registers.

Due to the nature of the register assignment problem, in addition to the lifetime of a specific variable, we must also consider the interplay among all variables. Although the lifetime of a variable x may be short, it could potentially have adverse effects on the required number of registers if many other variables must also be alive in the same time frame as x. In order to address this, we propose a *weighted lifetime* to characterize individual variable lifetimes. The goal here is to introduce a scaling factor to model the observation that variables that are alive while many others are also alive are more likely to adversely impact the ultimate number of required registers than otherwise.

However, variable lifetimes are not known prior to actually having a schedule in place. Therefore, the objective function we propose here relies on the aforementioned ASAP and ALAP scheduling heuristics to compute the initial approximate variable lifetimes. For a given variable x we define its lifetime:

$$L(x) = \mathrm{ALAP}(x) - \mathrm{ASAP}(x)$$

where ASAP( ) indicates the earliest and ALAP( ) indicates the latest control step at which a variable can be alive. We also define the set Vc(x) for a variable x:

$$V_C(x) = \{V_1, V_2, \ldots, V_T\}$$

where V_i is the cardinality of the variable-cut in control step i, if variable x is alive in control step i and 0 otherwise. Note that here cut cardinalities must again be approximated using the ASAP/ALAP scheduling heuristics.

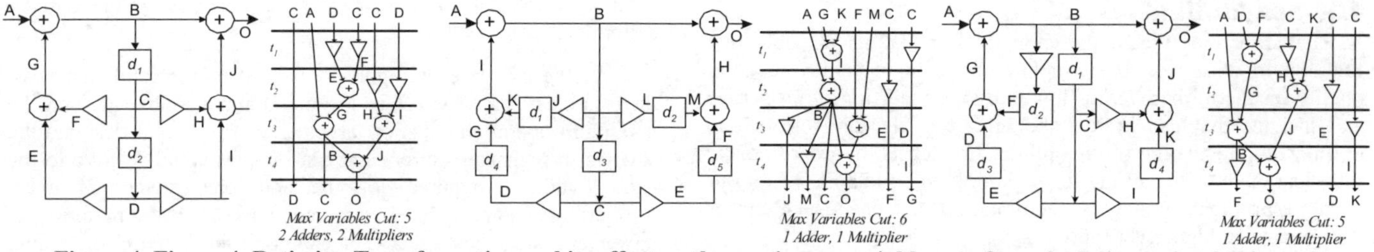

Figure 4. Figure 4. Retiming Transformation and its effect on the maximum variable cut after scheduling a simple bi-quad.

As mentioned above, the cardinality of variable cuts play an important role in determining how many registers are required for a given scheduled computation. In general, if a variable belongs to several large cuts, then it's more likely to remain in large cuts. There are also many other factors such as complex variable dependencies that one can consider. However, ultimately, the size of the largest variable cut will determine the minimum number of required registers for a given schedule. We thus formulate the *weighted lifetime* for the variable x:

$$WL(x) = \mathbf{max}(\ V_C(x)\ ) \cdot L(x)$$

In the definitions above, we compute the longest possible variable lifetimes and worst-case cut cardinalities using the ASAP/ALAP heuristics. Another alternative is to use the mid-point of ASAP/ALAP ranges for each operation as the scheduling heuristic and compute lifetimes and variable cut-cardinalities. As each operation is scheduled during the scheduling algorithm, we replace the heuristic based approximates in our above formulations with the appropriate data (variable lifetimes and variable cut cardinalities) and re-compute the WL for each variable.

To quantify the scheduling *difficulty* of a variable, we first note that each variable has one producer and at least one consumer. The only exceptions are input variables that have no producers and output variables that have no consumers. Generally, the higher the number of the consumers a variable has, the more difficult it's to manipulate the schedule in order to reduce its lifetime.

Suppose the set $C(x)$ denotes the set of operations that consume variable x. We define the set $C_T(x)$:

$$C_T(x) = \{C_{T1}, C_{T2}, \ldots, C_{T|C(x)|}\}$$

where C_{Ti} is the control step where the i-th consumer of variable x is scheduled. If the operation is not scheduled yet, then we use the ALAP heuristic here. We represent the scheduling (or re-scheduling) *difficulty* $D(x)$ of a variable x as:

$$D(x) = \sum_{i=i}^{|C_T(x)|} \left(C_{Ti}(x) - \min\left(C_T(x)\right)\right)$$

The *difficulty* $D(x)$, as defined, increases with the number of consumers of x as well as relative difference between each consumption time and the earliest consumption time of variable x. Now that we have a quantitative expression for each of the identified components of scheduling for register minimization FLOF, we are faced with the difficult task of combining. Combining refers to the process of putting together each component to form the objective function. Since components often represent varying and unrelated aspects of a problem, proper normalization is a critical and often difficult task to tackle.

Since we are interested in reducing *lifetime* (*Small-Potato Principle*) and *difficulty* of all variables while satisfying the given hardware limitations we specify the combined lifetime cWL and combined difficulty cD:

$$cWL = \sum_{i=1}^{n} WL(x_i),\ cD = \sum_{i=1}^{n} D(x_i)$$

where x_i represents the i-th variable in the given computation CDFG, and n is the total number of variables. The combination scheme for cWL and cD are not restricted to the ones specified above. Many other reasonable alternatives exist that can perform just as well, such as median, root-mean-square (L_2 *norm*), and simply taking the largest variable weighted lifetime (L_∞ *norm*).

A popular method that has proven very effective in forming optimization objective functions is the linear combination of components. Following this well established and proven school of thought we formulate the ***Scheduling For Register Minimization*** $\boldsymbol{FLOF_{S\text{-}RM}}$ as: $FLOF_{S\text{-}RM} = \alpha \cdot cWL + \beta \cdot cD$ where α and β are experimentally determined and tuned parameters. We set α equal to the largest weighted lifetime $WL(x_i)$, and β equal to the largest difficulty $D(x_j)$ as computed by the above heuristics. This choice seemed to perform best among other selections.

4.2 Transformations – Register Assignment

We now consider how the insights learned in the Scheduling domain and the corresponding FLOF can be applied to the Transformations domain. Consider the three retimed "bi-quads" in Figure 4. The "bi-quad" computation structure is widely used as a basic building block for designing infinite-impulse-response (IIR) filters. A circle represents an addition operation, a triangle represents a constant multiplication, and a square represents a delay.

Although all three cases in Figure 4 perform the same computation, the use of retiming as the transformation has resulted in three different schedules as shown to the right of each figure. The number of required adders and multipliers as well as the maximum variable cut size are also listed. Although during the Transformations phase of the design process a schedule does not exist, the scheduling FLOF developed in the previous subsection can still guide the optimization process in making choices such that the required number of registers in the future may be minimized. The use of the FLOF above is possible since initially we only rely on the ASAP/ALAP scheduling heuristics that can be computed very efficiently. In the cases presented, the scheduling FLOF can indicate the solution that is the most promising as a candidate for minimizing the required number of registers. The same insights from the scheduling step can be applied to this step using the approximate schedules.

5. Experimental Results

We used a variety of CDFG structures adopted from [2] to evaluate the performance of our proposed scheduling FLOF. The selected set of designs includes the elementary sine function, a linear controller (LinearCntrl13), two eighth-order Avenhaus IIR filters with different structures (cascade8, parallel8), a volterra filter, a differentiator, several transforms (Hilbert, Wavelet, Winogradfft11), a seventh-order IIR filter, and three FIR filters (Dsfir51, Dskais55, fir00). We use 0.25 micron technology.

To obtain the results, we use HYPER which is a resource utilization driven complex heuristic synthesis package [10]. For each case, we synthesized and compared the results with and without the $FLOF_{S\text{-}RM}$ as the objective function in HYPER. The experimental results are in Table 1. The first column contains the design name, followed by the number of nodes, the number of different types of operations in the graph, and the active area used without the default HYPER heuristics. The results of area and percentage improvement in area are shown in the second half of the table. The fifth column presents the active area obtained by using the scheduling $FLOF_{S\text{-}RM}$. The final column shows the percentage improvement.

On average, the approach for Scheduling and Register Assignment using the $FLOF_{S\text{-}RM}$ reduced the area requirement by 11.24% with a median of 9.05%. For the smallest design, the improvement was 5.2%, while for the largest design the improvement was approximately 20%. For Transformations (by using the corresponding FLOF in HYPER), Scheduling, and Register Assignment (results shown in the final two columns) there is an average improvement of 23.37% and a median improvement of 16.15%. Although some of the active area improvement was a result of a reduction in execution units, more that 80% of the improvement is due to register minimization.

6. CONCLUSIONS

The success of an optimization algorithm often depends heavily on carefully designed and tuned objective functions. However, modern layered optimization flows, often hinder the usefulness of objective functions since the effects of actions in one layer on subsequent optimization steps are not considered. We proposed *forward-looking objective functions* as a generic means to overcome the restrictions imposed by layered optimization flows. FLOFs are built by carefully analysis of problem specific components in each step that may impact future steps, formulating quantitative expressions, and establishing a method for combining the components into the final objective function format. We demonstrated the details involved in FLOF design using the Scheduling, Register Assignment, and Transformations examples from High Level synthesis. Experimental results indicate that through the use of the proposed Scheduling FLOF for register minimization, significant area savings can be achieved, due to the fewer required registers in the final design.

7. ACKNOWLEDGMENT

This material is based upon work supported by the National Science Foundation Career award under Grant No. ANI-9734166.

8. REFERENCES

[1] R. Camposano, W. Wolf, *High Level VLSI Synthesis,* Kluwer Academic, Norwell, MA, 1991.

[2] M. R. Corazao, et. al. "Performance Optimization using Template Mapping for Datapath-Intensive High-Level Synthesis", IEEE Trans. on CAD, Vol.15, pp. 877-888, 1996.

[3] G. De Micheli, *Synthesis and optimization of Digital Circuits.* McGraw-Hill, 1994.

[4] M. R. Garey, D. S. Johnson. Computers and Intractability: A Guide To The Theory Of NP-Completeness. W.H. Freeman, 1979.

[5] C.-T. Hwang, J.-H. Lee, Y.-C. Hsu. "A formal approach to the scheduling problem in high level synthesis." IEEE Trans. on CAD, Vol.10, (no.4), pp. 464–475, 1991.

[6] P. K. Jha, et. al. "An empirical study on the effects of physical design in high-level synthesis." Procs of the Seventh International Conference on VLSI Design, p. 11-16, 1994.

[7] E.A. Lee, D.G. Messerschmitt. "Synchronous Dataflow." Procs. of the IEEE, Vol.75, (no.9), pp. 1235-45, 1987.

[8] A. Orailogulu, D. D. Gajski. "Flow graph representation." ACM/IEEE Design Automation Conference (DAC), pp. 503–509, 1986.

[9] M. Potkonjak., J. Rabaey. "Optimizing resource utilization using transformation." IEEE Trans. on CAD, Vol.13, (no.3), pp. 277–292, 1994.

[10] J. Rabaey, et. al. "Fast Prototyping Of Data Path Intensive Architectures." IEEE Design and Test of Computers, Vol.8, pp.40-51, 1991.

[11] C. J. Tseng, D. P. Sieworek. "Automatic synthesis of data path on digital systems." IEEE Trans. on CAD, Vol.5, (no.3), pp. 379–395, 1986.

Table 1. Experimental results: Area savings obtained for Scheduling and Transformations (for minimizing number of registers).

Design Name	# Nodes	# Op. Types	Active Area (mm^2)	Activ Area S/R	% Area Improved S/R	Active Area T/S/R	% Area Improved T/S/R
Volterra (mult)	30	4	2.13	2.02	5.2	1.92	9.9
iit7	33	5	5.46	5.11	6.4	4.88	10.7
cascade8	34	4	2.41	2.25	6.7	2.14	11.0
parallel8	39	4	2.85	2.65	7.1	2.53	11.4
sine	49	6	1.45	1.32	8.9	1.25	13.6
Hilbert	49	5	3.81	3.48	8.6	3.32	12.9
Wavelet	53	5	4.33	3.93	9.2	3.66	15.4
LinearCntrl3	56	3	13.03	11.90	8.7	10.83	16.9
Differentiator	80	5	4.56	3.99	12.5	3.61	20.8
Winogradfft11	104	4	8.85	7.57	14.5	5.72	35.4
Dsfir51	127	4	10.88	9.12	16.2	6.73	38.1
fir1333	130	3	11.96	10.08	15.7	7.15	40.2
Dskais55	137	4	13.12	10.82	17.5	7.52	42.7
fir100	302	3	12.59	10.06	20.1	6.52	48.2

ILP-Based Engineering Change

Farinaz Koushanfar
University of California, Berkeley
EECS Department, Berkeley, CA 94720
farinaz@eecs.berkeley.edu

Jennifer L. Wong, Jessica Feng, and Miodrag Potkonjak
University of California, Los Angeles
CS Department, Los Angeles, CA 90095
{jwong, jessicaf, miodrag}@cs.ucla.edu

ABSTRACT

We have developed a generic integer linear programming(ILP)-based engineering change(EC) methodology. The EC methodology has three components: enabling, fast, and preserving. Enabling EC provides a user with the means to specify the amount of flexibility and how this flexibility should be distributed throughout the solution so that one can guarantee that a specific set of EC demands can be satisfied while preserving the quality of the initially obtained solution. Fast EC conducts changes in a fraction of the time needed to solve the problem while preserving or in some cases improving the quality of the initial solution. Preserving EC maintains either user specified components of the solution or as much as possible of the initial solution while still guaranteeing an optimal solution to the altered problem instance. We applied the generic methodology to Boolean Satisfiability (SAT) problem. The effectiveness of all proposed approaches and algorithms is demonstrated on standard benchmarks.

Categories and Subject Descriptors

B.6 [**Logic Design**]: General; B.6.3 [**Design Aids**]: Software Engineering; B.7 [**Integrated circuits**]: [VLSI (very large scale integration)]

General Terms

Automatic synthesis, optimization

Keywords

Engineering change, satisfiability(SAT), integer linear programming, synthesis

1. INTRODUCTION

Design changes are a standard part of the design process today. EC is crucial to the design process due to the large number of design changes and common requests for new features. By allowing the original design to easily accommodate EC, one facilitates the incorporation of these changes with minimal design time overhead. At the same time, many optimization problems can be formulated as ILPs and LPs. High speed ILP and LP solvers are readily available.

DAC 2002, June 10-14, 2002, New Orleans, Louisiana, USA.

We have developed a new EC approach. The technique is ILP-based and has three components: enabling EC, fast EC, and preserving EC. The essential idea is to formulate EC requirements either as new constraints or as a new part of the objective function, so that ILP solvers can be used to produce new designs with the requested features. We introduce the key ideas behind the three ILP-based EC techniques using the Boolean Satisfiability (SAT) problem.

The essential idea behind enabling EC is to produce a solution such that the addition of new constraints can be handled locally. We illustrate this point using the following SAT example. Our instance of SAT is denoted by F and is given using the standard CNF format. We denote the clauses by f_i, where, for example, $f_1 = (v_1+v_3'+v_5')$. For F, there are at least two different solutions, that we denote by S and E.

$$F = (v_1+v_3'+v_5')(v_2+v_3'+v_5')(v_2+v_4+v_5)(v_3'+v_4')$$
$$S = \{v_1 = 0, v_2 = 1, v_3 = 1, v_4 = 0, v_5 = 0\}$$
$$E = \{v_1 = 1, v_2 = 1, v_3 = 0, v_4 = 1, v_5 = 0\}$$

The key observation is that solution E provides a much better starting point for EC than solution S. Assume that design changes are equivalent to eliminating a single variable. If we eliminate variable v_1, or v_3, then solution S is still correct. However, if we eliminate v_2, then clauses f_2 and f_3 are not satisfied. In addition, if we eliminate v_4, then clause f_4 is not satisfied or if we eliminate v_5, then clause f_1 is not satisfied.

Solution E always has the correct solution, regardless of which variable is being eliminated. If we eliminate one of v_1, v_2, v_4, or v_5, all clauses are still satisfied. An interesting case is when v_3 is being eliminated. In this case, clause f_4 is not anymore satisfied. But, if we change the assignment of variable v_4 from 1 to 0, this clause will again be satisfied without making any other clauses unsatisfied.

Our goal in this paper is to show how to produce solutions such as E, where we can handle changes in specification, by using only local restructuring of the solution. Essentially, we impose additional constraints on an instance so that it has the requested tolerance against specified types of changes.

To introduce fast EC, consider the SAT formula F in CNF format and the satisfying truth assignment, S.

$$F = (v_1+v_2+v_3)(v_1+v_2'+v_3'+v_4)(v_1+v_3+v_6)(v_1+v_4+v_5)$$
$$(v_1'+v_3+v_4)\ (v_2+v_3'+v_5)(v_2+v_6')(v_2'+v_5)(v_3+v_4'+v_5)$$
$$(v_3'+v_5)$$
$$S = \{v_1 = 1, v_2 = 1, v_3 = 0, v_4 = 0, v_5 = 1, v_6 = 0\}$$

If, as an EC, we add the clauses $f_{11}=(v_5'+v_6)$ and $f_{12} = (v_1+v_3'+v_4)$, we must find a new satisfying truth assignment, S', because the original assignment does not satisfy f_{11}. In order to fast solve the new instance we would like to minimize the size of the new formula, F', to be solved.

We construct F' from two types of clauses, the unsatisfied added clauses and clauses that are affected by these unsatisfied clauses. We first identify all variables that are contained in the unsatisfied clauses, we will call these variables, b_i. We examine all clauses that contain variables b_i, and add the clause to F' only if there is no alternate variable not in b_i that satisfies the clause. In our example, we examine f_4, f_5, f_6, f_7, f_8, f_9 and f_{10} as possible clauses to add to F'. We only add f_7 and f_8, for they are not satisfied by any other variables.

$$F' = (v_5'+v_6)(v_2+v_6')(v_2'+v_5)$$

We see that both clauses f_7 and f_8 contain v_2. As a result of adding these clauses to F', we must now consider all of the clauses that contain v_2 as possible clauses to be added to F'. In this case, we look at f_1, f_2 and f_6. All of these clauses are satisfied by variables other that v_2, v_5, v_6, and therefore we do not add any of the clauses to F'.

We can now solve F' over the variables v_2, v_5, v_6. The size of our instance is decreased in size from ten clauses to three (even with the EC). Therefore, the additional overhead for solving the modified instance would be minimal.

Often, a single synthesis step is followed by a number of consecutive synthesis steps. Therefore, if we want to avoid numerous changes to all steps, we have to preserve as much as possible of the initial solution at the higher levels of abstraction.

The goal is to resolve a modified instance in such a way that we minimize the number of changes to the original solution. For example, assume that we have the following SAT instance, F, and satisfying truth assignment, S.

$$F = (v_1+v_2+v_4)(v_1+v_4+v_5')(v_1'+v_3'+v_4)$$
$$(v_2+v_3+v_5)(v_2'+v_4+v_5)(v_3+v_4'+v_5)$$
$$S = \{v_1 = 1, v_2 = 1, v_3 = 0, v_4 = 0, v_5 = 1\}$$

If we add the clauses $(v_2'+v_3+v_4)(v_1+v_2'+v_5')$ the formula F becomes unsatisfied. We now must resolve the instance. There are multiple satisfying assignments for F, for instance consider the following two assignments.

$$S_1 = \{v_1 = 0, v_2 = 1, v_3 = 1, v_4 = 1, v_5 = 0\}$$
$$S_2 = \{v_1 = 1, v_2 = 0, v_3 = 0, v_4 = 0, v_5 = 1\}$$

If we select the S_2, we preserve four out of the five assignments, and therefore make the minimal amount of design changes. However, if we select the S_1, only one of the five variables assignments is preserved, and as a result the design would need to be almost completely redesigned.

2. RELATED WORK

In the last two decades, combinatorial optimization in general, and integer and linear programming in particular, have attracted a great deal of attention. Numerous high quality books appeared including [10, 12]. Quality textbooks on integer programming include [11, 14]. ILP has been widely used in CAD for a great variety of optimization tasks, ranging from physical CAD [8] to behavioral synthesis[2].

It appears that one of the first efforts for EC was performed at Hitachi. Shinsha et al. [13] described an incremental logic synthesis technique for supporting changes in the physical design stage. They demonstrated the effectiveness of the approach on the Hitachi M68XH design. Later, many efforts were made at the logic synthesis level [4, 7].

In a sense, our work is closest to the work by Kirovski et al [5]. They defined the engineering change problem as constraint manipulation where the graph is restructured in such a way that it supports EC. However, there are numerous differences and novelties. First, for the first time we introduce the notion of preserving EC and the quantified notion of enabling EC. More importantly, the technique in [5] is restricted to graph coloring and scheduling, does not guarantee optimality, and does not opt for successive application to new requests. Most importantly, their technique can only handle some specific changes in specification, whereas the new technique is completely general.

Boolean Satisfiability is among the most popular generic NP-complete problems with numerous applications both in CAD and other application domains. Excellent surveys on SAT include [3, 9].

3. PRELIMINARIES

In order to make the paper self-sufficient, we survey relevant background material about ILP methodology and its use, the terminologies that we use throughout the paper, along with the formal definition of engineering change.

There are several techniques how one can solve ILP or LP. For LP, the most widely used is the SIMPLEX approach that has an exponential runtime in the worst case. However, on an overwhelming number of practical instances, SIMPLEX is not just of polynomial complexity, but also very fast. During the last decade, a number of practical and powerful interior point algorithms have been successfully implemented. These algorithms have guaranteed polynomial runtimes.

ILP deals with problems where a function is to be maximized or minimized and the variables are constrained by inequality and equality constraints and/or integral restrictions. The objective function as well as the inequality or equality constraints are linear.

$$max\{cx + hy : Ax + Gy \leq b, x =\in Z_+^n, y \in \Re_+^p\} \quad (1)$$

We define the mixed integer programming problem (MIP) as 1, where Z_+^n is the set of non-negative integral n-dimensional vectors, $\Re_+^p$ is the set of non-negative integral p-dimensional vectors, and $x = \{x_1, \cdots, x_n\}$ and $y = \{y_1, \cdots, y_p\}$ are the variables and unknowns of the problem.

We assume that all numbers are rational. For our purposes, we use a special case of the MIP where all variables are integer values. The integer linear programming (ILP) is defined as:

$$max\{cx : Ax \leq b, x =\in Z_+^n\} \quad (2)$$

More specifically, we restrict the integer values of x to be 0 or 1. This is called 0-1 ILP where x is redefined to be $x \in B^n$ where B^n is a set of n-dimensional binary vector.

For the sake of completeness and due to the fact that most often ILP are usually solved using relaxation to the LP problem, we conclude this section by stating the LP problem.

$$max\{hy : Gy \leq b, y \in \Re_+^p\} \quad (3)$$

A very important and often used special case of MIP is one where all variables are 0-1 variables that means that each variable represents a binary decision. Specifically, it is common to model a variety of problems using the dichotomy formulation for variables with respect to a specific event.

$$x = \begin{cases} 1, & \text{if the event occurs} \\ 0, & \text{otherwise.} \end{cases}$$

We define ILP using the following form.

$$Y = MAX(cx) \quad (4)$$

$$Ax \leq b \quad (5)$$

We define $x = \{x_1, \cdots, x_n\}$ and $p = \{p_1, \cdots, p_n\}$ as 0-1 variables. We denote the original or previous assignments of the instance by P=(p_1,p_2, ..., p_n) and the new assignments, after EC, by X=(x_1,x_2, ..., x_n).

Boolean Satisfiability is a NP-complete problem that is often used in CAD tasks. The problem can be formulated as follows[1].

Satisfiability
Instance: A set U of variables and a collection C of clauses over U.
Question: Is there a satisfying truth assignment for C?

We formulate the SAT problem in the form of an ILP by using the ILP formulation of the set cover problem as an intermediate step. The set cover problem can be formally defined as follows:

Set Cover
Instance: Collection C of subsets of a finite set S, positive $K \leq \mid C \mid$.
Question: Does C contain a cover for S of size K or less, i.e., a subset C' of C with $\mid C' \mid \leq K$ such that every element of S belongs to at least one member of C'?

The ILP formulation for the set cover problem can be specified in the following way.

$$x_i = \begin{cases} 1, & \text{if subset } C_i \text{ is selected} \\ 0, & \text{otherwise.} \end{cases}$$

$$A_{ij} = \begin{cases} 1, & \text{if element } S_j \text{ is covered by subset } C_i \\ 0, & \text{otherwise.} \end{cases}$$

We use the ILP formulation (4) and (5), where we define c as a negative identity vector and b as a positive identity vector. The objective function for the set cover problem is to minimize the number of subsets used to cover the set S. The constraints are that each element in S must be covered by at least one subset.

We now formulate the SAT problem using the set cover ILP formulation. We define each element in the finite set S as a single clause in the SAT formula. The collection of subsets C are the variables in both complemented and uncomplemented forms independently. Each subset C_i contains all clauses as its elements, i.e. elements of S, in which the variable i appears in. Therefore, we have twice as many subsets as there are variables in the instance, one subset to represent the complemented and one to represent the uncomplemented version of the variable. Because no variable can be assigned to both the uncomplemented form and complemented form at the same time, we add a constraint (6), one for each variable, where n is the number of variables in the SAT problem. We define the objective function and the other constraints in the same way as the set cover problem.

$$x_i + x_{i+n} \leq 1 \qquad (6)$$

To illustrate the formulation, consider the following SAT instance, F, with three variables and three clauses.

$$F = (v_1' + v_2)(v_2 + v_3)(v_1 + v_3')$$

In this case, we define $x = \{x_1, \cdots, x_6\}$, where $x_1, \cdots, x_3$ represent uncomplemented versions of the variables, and $x_4, \cdots, x_6$ represent the complemented versions of the variables. We define the elements of S and the subsets C below. The translation to ILP form is now straightforward.

$$S_1=(x_4,\, x_2),\ S_2=(x_2,\, x_3),\ S_3=(x_1,\, x_6)$$
$$C_1=\{S_3\},\ C_2=\{S_1,\, S_2\},\ C_3=\{S_2\}, C_4=\{S_1\},\ C_5=\{\emptyset\},\ C_6=\{S_3\}$$

4. GENERIC ILP-BASED ENGINEERING CHANGE

In the modern design processes, there is often a need to handle small specification alterations and therefore to redesign or update the hardware and/or software implementations(EC). Until now, the EC methodologies and algorithms have been restricted to a particular task. In addition, EC

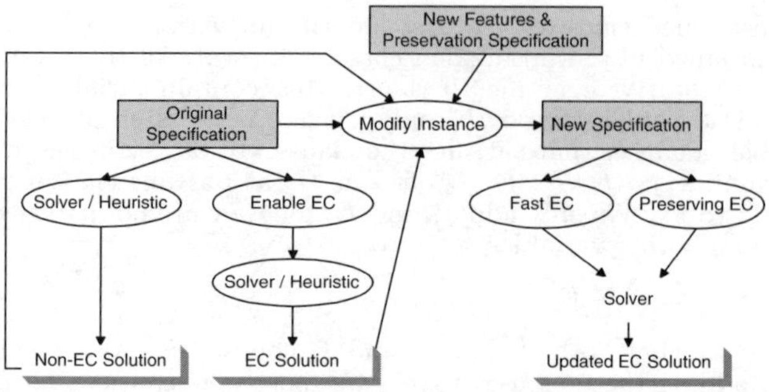

Figure 1: Generic ILP-Based EC Flow.

has been ad-hoc in nature is the sense that a few, if any, guarantees about its optimality could be provided. Our goal is to provide the first systematic, provably optimal, generic EC approach that can be easily applied to a variety of design and compilation tasks. The key enabling step of this effort is the formulation of EC requirements as ILP constraints. An important observation is that even when an optimization/synthesis problem is defined in ILP form, one can use not just an ILP solver, but also an arbitrary algorithm, such as simulated annealing or a heuristic, to solve it.

There are two ways how we can address EC requirements. One is to design for EC, where we embed flexibility into the solution to make it amenable for future changes or improvements. The other aspect is to apply EC after we have a valid and highly optimized solution. The goal here is to minimally alter the initial solution in order to obtain a solution to the altered specification. This task can be performed in at least two different setups: one where the goal is to preserve as much as possible from the original design (so the results of the consequent design steps are preserved maximally) and the other, where the speed of redesign is of prime importance.

Both types of EC require thoughtful considerations. For the first task, one should efficiently predict the parts of the design that would undergo alterations and updates and find mechanisms to make them flexible. For the second, the goal is to figure out how to specify a small and easy to solve instance of ILP (in the case of fast EC), or how to specify that the new solution to the new specification should preserve as much of the original as possible, or conserve the requested parts (in the case of preserving EC).

We now informally introduce the new EC approach. We conduct enabling EC in two different generic ways. The first one is to specify additional constraints that guarantee that the solution will have flexibility as requested by the user. The second approach is that we add a new component, in terms of constraints, to the objective function. Now the objective function has two weighted components, the original part for the quality of solution and the new part for flexibility.

For fast EC, we follow the following procedure. We first isolate constraints that are impacted by the change. Using these constraints and new constraints, we specify a new instance of ILP. Finally, we combine the preserved parts of the initial solution and a partial new solution to create a complete new solution. For preserving EC, we define an ILP where either constraints or the objective function have components that guarantee preservation of either specified components of the previous solution, or as much as possible of the previous solution.

We conclude this section by presenting the generic ILP-based EC flow. The flow for a generic problem with EC can be seen in Figure 1. We begin with the original specification of the problem. We have two options. We can solve the instance with a standard ILP solver or the heuristic iterative improvement-based ILP solver presented in [6], or we can apply enabling EC to the original problem specification. The result of the solver/heuristic on the original instance is denoted as the non-EC solution. Another option is to use an EC enabling procedure to produce the EC solution. Either the non-EC solution or the EC solution can be used in conjunction with the original problem specification and the new features and preservation specification to generate a new problem specification.

The new problem specification is a modified instance of the original problem specification that includes the original solution (EC or non-EC) and, if desired, the preservation specification. The EC or non-EC solutions are the starting solution for solving the new problem specification. We can resolve the new problem in two ways: fast EC or preserving EC. With fast EC, the new problem specification is simplified in such a way that the new ILP is much easier to solve. For preserving EC, we follow the preservation specification. Next, we find the solutions using a standard ILP solver. In this case, we do not use the heuristic ILP solver because the ILP solver will provide an optimal solution in a reasonable amount of time (the new specification is non-trivially smaller than the original instance).

5. ENABLING EC

In this section, we describe how to enable EC for the SAT problem by reformulating the ILP in such a way that we can adapt to changes in the problem.

For example, a variation of the SAT problem would be to remove new variables or clauses. When we remove variables from the problem, we modify the number of subsets, x, and therefore the number of columns, or subsets, in matrix A is changed. On the other hand, if we add clauses, the number of rows, or number of elements in S, in matrix A changes. Note that, in a sense adding clauses is a more general case than removing variables.

If we remove clauses or add variables to the problem we will not need to resolve the problem, because we are loosening the constraints on the problem. If we remove a variable or add clauses, we constrain the problem even more and the original solution is not necessarily adequate. By enabling EC we allow for the removal of variables or the addition of clauses such that their effects can be localized, and a minimum number of clauses and variables that are in direct relationship to them are changed.

One way to enable EC is to make sure that at least two variables per clause are satisfied. We call these clauses k-Satisfied, where k is the number of literals in the clause that evaluate to true. This can be very expensive or impossible in the general case. Therefore, we make a modification to the enabling condition. We say that we want to maximize the number of clauses that are at least 2-satisfiable. For all clauses that are 1-satisfiable there must be other literals in the clause that can switch their assignments to make the clause satisfied.

The ILP formulation for enabling EC is best introduced and explained using a small example. Consider again the SAT formula, F introduced above, along with its satisfying truth assignment. We label the clauses, c_1, c_2, and c_3 respectively. The ILP formulation is the same as presented in Section 3, except that we impose additional constraints in the following way.

For each clause we form a number of constraints. The first constraint is that each clause must be at least 2-Satisfiable (denoted by the first summation), or have at least one variable that can flip its assignment in such a way that it does not make the problem unsatisfiable (the second summation). Note that Z_i variables must evaluate to false in the considered clause in order to be included in the constraint.

$$\sum_i (x_i) + \sum_i (Z_i) \geq 2 \tag{7}$$

Recall that the constraint from (5) guarantees that at least one variable in each clause must be satisfied. A set of constraints is created for each inverse occurrence of each literal of the instance.

In our example, we begin with v_1 in c_1. We say that it is permitted for variable v_1 to be eliminated from c_1 if either v_2 or v_3 can switch its assignment to satisfy the clause. We write equations for each occurrence of the complement of v_1; in our example it is v_1' or x_5 in ILP form. Note that v_1' appears only in c_3.

We define two auxiliary variables, Q_i and Z. Z_i represents the variable that receives support from all variables in clause c_i. Q_i is used to ensure that all variables, Z_i, have non-negative values. Z_{ij} represents the support for variable x_i in clause c_j. Finally, we introduce variable Z_{ijk} which indicates whether variable x_i receives support from clause c_j through variable x_k when x_k flips its value.

The first two constraints specify that either v_1' must satisfy the clause or one of the other literals in the clause must. We enforce this by creating a constraint that specifies that the complements of v_2 and v_3 must be greater than Z_{ijk} in the case that test variables are needed to satisfy the first two constraints. The fifth equation states that either one of the variables or none of them must be able to flip their value. If it is none of the variables, then Q_1 must be 1. Q_1 then is used in the next equation to specify that the value of variable Z_{53} must be zero. If this is the case, then Z_5 must be zero, and none of the variables can flip their value. From (7) we can see then at least two variables in c_1 must evaluate to true in order to satisfy the constraint. If either v_2' or v_4 or both can flip their values then Z_{536} or/and Z_{534} will be 1. As a result then Z_{53} will be greater than zero, that results in equation (7) being satisfied.

$$\begin{aligned}
x_5 + Z_{536} &\leq 1 \\
x_5 + Z_{534} &\leq 1 \\
x_6 &\geq Z_{536} \\
x_4 &\geq Z_{534} \\
Z_{536} + Z_{534} + Q &\geq 1 \\
Z_{536} + Z_{534} + Q - 1 &\geq Z_{53} \\
Z_5 &\leq Z_{53}
\end{aligned}$$

By adding these constraints, we have now enforced that each clause must be either 2-Satisfied or has flexibility in terms of an alternate variable that can flip its assignment to support the unsatisfied clause, in case of an EC.

6. FAST SOLVING EC

While doing fast EC, there are two cases to consider for changing the constraints to the SAT problem. The first case is when variables are added or clauses are deleted. The other is when we add clauses or delete variables. The first case is trivial to handle in the sense that if a variable is added, it can automatically be assigned a DC value. The problem was originally satisfied and therefore the addition of a variable has no effect on the solution. The same is for the deletion of clauses. When clauses are deleted, the idea is to increase the enabling of the problem such that the next EC can be easily and properly handled.

We can increase the EC flexibility of the problem in two ways. First, we try and recover as many DC variables from the initial solution as possible. The second way is to reconstruct the solution in such a way that more clauses are of 2-satisfiability or higher.

If we add more clauses or remove variables, modifications must be made in order to quickly resolve the problem. We have developed the following approach, presented in Figure

2, that performs the minimization of the problem, in terms of the number of clauses and variables, in order to fast solve the instance.

Input: F' a modified SAT Formula.
p the original satisfying truth assignment to F.
Output: A simplified SAT instance with the minimum number of clauses and variables.

Algorithm:
Check the Original Solution(F', p);
If p satisfies F' then quit;
Mark All Unsatisfied Clauses from F';
Add all variables which appear in the marked clauses to list, V;
While (V increases in size) {
Check All Clauses that have variables from V
If the clause is not satisfied by a variable not in V
mark clause;
add any new variables to V; }
Create new ILP, F'', with V and marked Clauses;
Solve F'';
Combine p and new solution p';

Figure 2: Pseudo code for simplifying a SAT instance in order to be fast solved.

The intuition is the following. We solve the instance with all clauses that are not satisfied and the clauses that they could affect. If a clause is solved by other variables that need not be modified, we do not include them into consideration in the new SAT formula F'. If EC was enabled on the instance, then many clauses will not be included due to the fact that many of them will be 2-Satisfied or more. In this case, a minimal number of variables and clauses will have to be resolved.

7. PRESERVING EC

This section presents the ILP-based SAT formulation for preserving EC. We can preserve in two different ways. The first is to preserve the maximum amount possible, and the second is to preserve user specified parts of the solutions. This approach focuses on the quality of a solution rather than the overhead.

We only need to consider the case when variables are removed or clauses are added. In this case, we are removing constraints from the instance and therefore the problem is a simplified version of the original problem that the initial solution will satisfy.

When variables are removed or clauses are added we need to redefine the ILP such that the new solution, n, preserves as much as possible of the original solution, p. The key is that we formulate the ILP in such away that we are maximizing the number of assignments that stay the same as well as finding a satisfying truth assignment for the modified instance. We do this by adding additional constraints to the ILP or by modifying the objective function to give benefit to variable assignments which are preserved. The ILP for preserving EC is defined as follows.

$$x_i = \begin{cases} 1, & \text{if subset } C_i \text{ is selected} \\ 0, & \text{otherwise.} \end{cases}$$

$$N_i = \begin{cases} 1, & \text{if subset } C_i \text{ is selected} \\ 0, & \text{otherwise.} \end{cases}$$

$$A_{ij} = \begin{cases} 1, & \text{if element } S_j \text{ is covered by subset } C_i \\ 0, & \text{otherwise.} \end{cases}$$

The objective function is now defined as follows, where c is an identity vector. In this case, we are maximizing the number of variables which have the same assignment as the original assignment, p. We keep the same constraints (5) and (6), where b is an identity vector. We add the following constraint that evaluates whether or not the new variable assignment for N_i is the same as the assignment as p_i.

$$Y = MAX(cZ)$$

$$Z_i = p_i x_i + p_{n+i} x_{n+i}$$

8. EXPERIMENTAL RESULTS

We use standard DIMACS SAT benchmarks to test our EC methodology. We use CPLEX as our main ILP solver. Both CPLEX and the heuristic ILP solver were ran on a 1 GHz Pentium III computer.

The results for applying EC to DIMACS SAT instances are shown in Table 1. The first three columns indicate the name of the instance, the number of variables and the number of clauses in the instance, respectively. The next column presents the runtime for the original instance. The fifth column represents the runtime values when specified constraints are imposed. The last column shows the runtime when the objective function is augmented with EC requirements. Both of these columns represent the normalized runtime against the original runtime. The average and median for the specified constraints are 0.69 and 2.62 respectively and for objective function EC are 0.75 and 2.34 respectively. As we can see, the overhead is not significant. In all the tests, we used $k = 2$, that means that each clause is directly either 2-Satisfied or is satisfied by the support of other variables in the clause being able to flip their assignment without jeopardizing the satisfiability of other clauses. The second set of results, the five examples shown on the bottom of the table, are solved heuristically using the heuristic ILP solver [6].

For fast solving, we ran ten trials. For each of the trials, we eliminated three variables and added ten clauses. The results are shown in Table 2.

The first three columns represent the DIMACS instance information. The third column indicates the original runtime for solving the instance. The following two columns list the average number of variables and clauses in the fast EC instance. The last column represents the runtime of the fast EC instance. On average, for the smaller examples, we needed 23.25 variables and 99.6 clauses, and on average reduced the runtime to 0.0068 of the original runtime. For the larger examples at the bottom of the table, we initially solved them using our heuristic ILP solver. Once an initial solution was generated, we then used an off-the-shelf solver to obtain a new optimal solution to the original instance.

To evaluate our preserving EC, we randomly added and deleted five variables and randomly added and deleted five clauses, making sure that we did not make the instance non-satisfiable. We compared the percentage of preserved variable assignment for two cases: when new instance were just again evaluated with no consideration for preserving the initial assignment as when preserving EC approach is used. We present the results in Table 3. We begin the table with the DIMACS instance information, name, number of variables and number of clauses in the instance. The fourth column indicates the percentage of the original solution after complete recalculation with no EC goals. The percentage of the original solution preserved when evaluating the instance with preserving EC is shown in the last column. The last two rows present the average and median values. It is clear that preserving EC preserve significantly higher percentage of the initial assignment. As was done for fast solving, the larger instances presented at the bottom of the table were originally ran using the heuristic ILP solver to generate the initial solution. An off-the-shelf solver was used to optimally solve the instance with preserving EC.

In addition to validating the new ILP-based engineering

change approach on SAT benchmarks, we conducted comprehensive experimentation on the graph coloring problem. The results can be found in [6].

9. CONCLUSION

We introduced a generic ILP-based EC methodology. We demonstrated the three components, enabling EC, fast EC, and preserving EC on a CAD related problems, SAT. The EC techniques are applicable to many other CAD and optimization problems. We demonstrated the effectiveness of the methodology on standard DIMACS benchmark instances.

10. ACKNOWLEDGEMENTS

This research was supported by the NSF CARRIER award 9734166 at UCLA. Farinaz Koushanfar is supported by a NSF graduate research fellowship.

11. REFERENCES

[1] M. Garey and D. Johnson. *Computers and Intractability.* W.H. Freeman, 1979.

[2] C.H. Gebotys and M.I. Elmasry. Global optimization approach for architectural synthesis. *IEEE Transactions on Computer-Aided Design of Integrated Circuits and Systems*, 12(9):1266–1278, September 1993.

[3] J. Gu, P.W. Purdom, J. Franco, and B.W. Wah. Algorithms for the satisfiability (sat) problem: a survey. In *Satisfiability Problem: Theory and Applications. DIMACS Workshop*, pages 19–51, 1997.

[4] S.P. Khatri, A. Narayan, S.C. Krishnan, K.L. McMillan, et al. Engineering change in a non-deterministic fsm setting. In *Proceedings of IEEE/ACM Design Automation Conference*, pages 451–456, June 1996.

[5] D. Kirovski and M. Potkonjak. Engineering change: methodology and applications to behavioral and system synthesis. In *Proceedings of IEEE/ACM Design Automation Conference*, pages 604–609, June 1999.

[6] F. Koushanfar, J.L. Wong, J. Feng, and M. Potkonjak. Ilp-based engineering change. In *Technical Report, CS Department, UCLA*, pages 1–11, April 2001.

[7] C. Lin, K. Chen, and M. Marek-Sadowska. Logic synthesis for engineering change. *IEEE Transactions on Computer-Aided Design of Integrated Circuits and Systems*, 18(3):282–292, March 1999.

[8] M. Lin, H. Perng, C. Hwang, and Y. Lin. Channel density reduction by routing over the cells. In *Proceedings of IEEE/ACM International Conference on Computer-Aided Design*, pages 120–125, June 1991.

[9] J.P. Marques-Silva and K.A. Sakallah. Boolean satisfiability in electronic design automation. In *Proceedings of IEEE/ACM Design Automation Conference*, pages 675–680, July 2000.

[10] C.H. Papadimitriou and K. Steiglitz. *Combinatorial optimization: algorithms and complexity.* New Jersey: Prentice Hall, 1982.

[11] H.M. Salkin. *Integer programming.* Massachusetts: Addison-Wesley, 1975.

[12] A. Schrijver. *Theory of linear and integer programming.* New York: Wiley, 1986.

[13] T. Shinsha, T. Kubo, Y. Sakataya, J. Koshishita, and K. Ishihara. Incremental logic synthesis through gate logic structure identification. In *Proceedings of IEEE/ACM Design Automation Conference*, pages 391–397, June 1986.

[14] L.A. Wolsey. *Integer Programming.* New York: Wiley, 1998.

Instance	# Vars	# Clauses	Orig. Runtime	EC (SC) N.R.	EC (OF) N.R.
par8-1-c	64	254	21.14	0.90	2.12
ii8a1	66	186	0.14	0.82	1.37
par8-3-c	75	298	57.9	0.81	1.78
jnh201	100	800	527.83	0.77	2.28
jnh1	100	850	1476.59	0.73	2.40
ii8a2	180	800	3023.94	0.67	3.01
ii8b2	576	4088	20089.8	0.45	4.42
f600	600	2550	18989.8	0.37	3.56
average	-	-	5523.393	0.69	2.62
median	-	-	1002.21	0.75	2.34
par32-5-c	1339	5350	5.2	0.77	1.93
ii16a1	1650	19368	12.6	0.82	2.41
par32-5	3176	10325	22.4	0.94	3.87
g250.15	3750	233965	74.8	0.98	4.37
g250.29	7250	454622	96.9	0.92	3.98
average	-	-	42.38	0.88	3.31
median	-	-	32.39	0.90	3.59

Table 1: Experimental Results for Enabling EC on SAT.

Instance	# Vars	# Clauses	Orig. Runtime	Ave. # Vars/Clauses	New Runtime
par8-1-c	64	254	201.14	11.2/40.8	0.036
ii8a1	66	186	0.14	10.8/27.3	0.005
par8-3-c	75	298	57.9	16.0/38.5	0.007
jnh201	100	800	527.83	21.0/98.9	0.002
jnh1	100	850	1476.59	17.7/67.3	0.001
ii8a2	180	800	3023.94	25.7/165.7	0.002
ii8b2	576	4088	20089.8	56.4/191.4	0.001
f600	600	2550	18989.8	27.2/167.0	0.001
average	-	-	5523.393	23.25/99.61	0.007
median	-	-	1002.21	19.35/83.1	0.002
par32-5-c	1339	5350	5.2	52.6/387.2	261.2
ii16a1	1650	19368	12.6	68.1/401.6	76.2
par32-5	3176	10325	22.4	70.9/476.2	102.9
g250.15	3750	233965	74.8	74.3/639.1	202.7
g250.29	7250	454622	96.9	102.5/876.4	952.1
average	-	-	42.38	73.68/1416.1	319.02
median	-	-	32.39	72.29/757.75	202.7

Table 2: Experimental Results for fast EC on SAT.

Instance	# Vars	# Clauses	% Solution Original	% Solution with EC
par8-1-c	64	254	72.2	98.2
ii8a1	66	186	71.6	99.3
par8-3-c	75	298	68.2	94.7
jnh201	100	800	63.6	93.7
jnh1	100	850	72.8	98.2
ii8a2	180	800	82.7	99.4
ii8b2	576	4088	83.0	99.7
f600	600	2550	76.6	98.6
par32-5-c	1339	5350	62.4	92.8
ii16a1	1650	19368	73.5	99.3
par32-5	3176	10325	71.6	94.1
g250.15	3750	233965	68.7	97.8
g250.29	7250	454622	84.1	94.7
average	-	-	73.19	96.96
median	-	-	72.75	98

Table 3: Experimental Results for preserving EC on SAT.

Conference Author/Panelist Index

G

H

I

J

K

L

M